Handbook of Tables of Functions for Applied Optics

Handbook of Tables of Functions for Applied Optics

Leo Levi
Jerusalem College of Technology
Jerusalem

Published by

CRC PRESS

CRC PRESS, Inc.
18901 Cranwood Parkway · Cleveland, Ohio 44128

International Standard Book Number 0-87819-371-5
Library of Congress Card Number 73-88627

Printed in the U.S.A.

PREFACE

The optical system designer frequently needs numerical data which are not readily available in the standard compilations. Handbook of *Tables of Functions of Applied Optics* was compiled to fill this need.

The majority of the material is not available elsewhere, or available only with considerably lower precision; for much of it, even computation on a large-memory computer, is quite difficult. (Cf. References 20 and 25.) Some available material was also included if it is presently available only in the form of a whole volume devoted to that function. Its presentation in a much more compact format, too, would seem to be an important advantage.

In an introductory section, the use of the tables and the methods used for computing them are explained. The bibliographic notes included there are meant to guide the reader to some of the more important related tables. No effort has been made to provide a full review of these. The interested reader should be able to find a far more complete listing in References 1 and 2.

The "mathematical" tables, representing the bulk of the volume, fall into roughly three categories. The first includes relevant elementary and advanced functions, such as the error, Fresnel, and "sinc" functions, for which especially detailed tables are provided. (It is hoped that these will be of use also to workers in nonoptical fields, such as communication theory.)

A second section provides detailed data on blackbody radiation in radiant, photon, and luminous forms. The last section deals with functions important in the analysis of imaging systems: the spread and transfer functions of a perfect lens. Although such lenses are most uncommon in our imperfect world, their performance is of fundamental importance in that they represent a limit and, at that, a limit which can often be approached quite closely.

As a rule, only mathematical tables, that is, tables not primarily based on experimental data, were included. Deviations from this rule were restricted to the relatively very short Tables 26 to 31. Material for inclusion there was judged on the basis of its probable usefulness to the potential user vs. its consumption of space.

The inclusion of the arithmetic tables requires particular justification. These are of great importance in optical design (see, e.g., A. E. Conrady, *Applied Optics and Optical Design,* p. 11). *Barlow's Tables* do cover this material in sufficient detail and are available in a 250-page volume; I felt, however, that the convenience of having all the required tables in a single volume warrants their inclusion here. (In this connection, only trigonometric, exponential, and logarithmic tables were not included, since these are widely available to sufficient precision.)

I would also like to acknowledge here the help of many colleagues and friends. Dr. Walter G. Driscoll, editor of the *Handbook of Optics* being published by the Optical Society of America, was kind enough to make available the projected table of contents. This enabled me to avoid unnecessary duplication. Indeed, I hope that that Handbook and the present volume will supplement each other well.

I also wish to thank Professor R. H. Austing (University of Maryland) for program modifications and running of Table 23; Mr. Henry Lehman (Information Concepts) for running Tables 17, 18, and 28; and the following for reviewing the outline and making many valuable suggestions: Dr. G. Goertzel (I.B.M. Corp.), Mr. E. Speyer (Pitney-Bowes), and Mr. M. Halberstam (Naval Applied Science Laboratory), who also supported the development of Table 23. Special thanks are due to Professor Harold Shulman (City University of New York) for reviewing the whole manuscript and Professor S. Aranoff (Haifa, Israel), who reviewed part of it. The Office of Naval Research supported, in part, the compilation of Tables 25 to 27. Lastly, it is a pleasant obligation to acknowledge the dedication and moral support which my dear wife provided to make this work possible.

Leo Levi

THE AUTHOR

Leo Levi organized and is head of the Departments of Physics and Electro-Optics, Jerusalem College of Technology.

Prof. Levi received his B.E.E. degree from City College, New York, and his M.Sc. and Ph.D. (Physics) degrees from Polytechnic Institute of Brooklyn.

He was formerly head of the Applied Physics Department, Fairchild Camera and Instrument Corp., Independent Consultant for IBM, NCR, and New York State University, and Adjunct Associate Professor of Physics, City College.

Dr. Levi has written *Applied Optics*, John Wiley & Sons, 1968, and was a contributing author to D. Levine, *Radargrammetry,* McGraw-Hill, 1960, and *Progress in Optics,* Volume 8, North-Holland Publishing Co., 1970.

To a great leader and inspiring teacher

Rav **DR. JOSEPH BREUER**

TABLE OF CONTENTS

Introduction

He who knows how to calculate [astronomical events] and does not do so – of him Scripture says: "The act of God they did not observe and the work of His hands they did not see." (Isaiah 5:12)

Bar Kappara, *Babyl. Talmud, Shabbath* 75a

GENERAL INTRODUCTION

Instructions for the User

In most cases, values are tabulated in the customary manner – ten on a line. In some tables, the second column gives the first few places which apply to all the entries in that row. When an entry has a bar placed over it, the value heading the next row should be applied.

For compactness, tables involving very large or very small values are given in "floating point" notation. This is indicated by a column headed m, which gives the power-of-ten to be applied to all entries in that row.

The column Δ^2 gives the second differences (central, for the last entry in the row, unless otherwise noted). These values can be used in conjunction with Table 1 to correct the result obtained from linear interpolation. Note that if Δ^2 is less than 5, no correction will usually be desirable. (The value of Δ^2 will frequently be low by one unit due to round-off errors; but this is of no practical significance.)

In the description of the tables, the ranges are given in the form

$$x = a\ (b)\ c$$

where x is the argument, a is the lowest and c the highest* value of x covered and b is the argument interval at which the function is tabulated.

Examples

1. Find $828\left(\frac{\pi}{2}\right)$.

We find the answer in Table 5 to be:
1300.619 3586.

2. Find the emittance of a blackbody at 3000°K.

From Table 13 we find
$M(3000) = 4.59246 \times 10^6\ \mathrm{W/m^2}$.

Computation of Tables

The tables were computed on a digital electronic computer which works with an accuracy of at least seven significant places. In some cases, e.g. when very large exponents occurred in the computation, computation at double this precision was used. Unity errors in the last given place must be expected, however.

Tables involving natural constants, such as the blackbody radiation tables and Table 25, are obviously no more accurate than permitted by the accuracy to which these constants are known.

Extensive discussions and bibliographies covering Tables 2 to 9 can be found in References 1 and 2.

TABLE 1. LINEAR INTERPOLATION CORRECTION TERMS

Definition

$$C(\Delta^2, \xi) = \tfrac{1}{2}\xi(1-\xi)\Delta^2. \qquad (1)$$

Range

$\Delta^2 = 0(1)200$

Instructions for the User

Linear interpolation can be used throughout this volume, with obvious exceptions.

The accuracy of linear interpolation will be within half a unit of the last given place as long as the second difference**(Δ^2) is less than 4. When the second difference exceeds this value, a correction term, $C(\Delta^2, \xi)$, should be *subtracted* from the result of the linear interpolation. Table 1 gives the required correction term in the row headed by the second difference (Δ^2) and the column headed by the fractional difference***(ξ) of the argument used in the interpolation. When used with the second differences tabulated in this volume, the correction term will be in units of the last decimal place given in the tables.

When the second difference is negative, the term should be *added* to the result of the linear interpolation. When less precision is required, and the last n places of the tabulated values are dropped, the tabulated value of Δ^2 must be divided by 10^n.

Example

Find the reciprocal of 0.130825.

From Table 2: $f_1 = 1/.1308 = 7.6452599$
$f_2 = 1/.1309 = 7.6394194$
$\Delta^2 = +89;\ \Delta f = f_2 - f_1 = -.0058405$

*Occasionally $c^* = c + b$ is given in place of c, for ease in reading.

**Let $x_1, x_2, x_3 \ldots$ denote the tabulated arguments in the order of increasing magnitude. Then $\Delta^2\,[f(x_n)] = f(x_{n+1}) - 2(x_n) + f(x_{n-1})$.

***$\xi = \delta x/\Delta x$, where δx is the difference between the given and the tabulated argument immediately preceding x, and Δx is the argument difference at which the function is tabulated.

Therefore:

$$1/0.130825 = f_1 + 0.25\Delta f - C(89, 0.25) = 7.6452599 + (.25)(-.0058405) - .000\,00083 = 7.6437990$$

Theory and Applications

Formula (1) is based on Newton's forward interpolation formula:

$$f(x_0 + \xi\Delta x) = f(x_0) + \xi\Delta f + \frac{1}{2!}\xi(\xi - 1)\Delta^2 f + \frac{1}{3!}\xi(\xi - 1)(\xi - 2)\,\Delta^3 f + \cdots$$

It is readily shown that, for $0 < \xi < 1$, the coefficient of the third order term has a maximum value of 0.064 . . . (when $\xi = .4226$ . .).

Thus this term will be less than 0.5 for $\Delta^3 f < 7.8$. Assuming that the fourth difference is negligible, this implies a change of 78 in Δ^2 from one line to the next.

TABLES 2-6. ELEMENTARY FUNCTIONS

Ranges

Table 2: $f = 1/x$,	$x = 0.1(.0001)1$
Table 3: $f = x^2$,	$x = 0.1(.0001)1$
Table 4: $f = \sqrt{x}$,	$x = 0.1(.0001)1(.001)10$
Table 5: $f = n\pi/2$,	$n = 0(1)1000$
Table 6: $f = \dfrac{\sin 2\pi x}{2\pi x}$,	$x = 0(.0001)5(.001)10(.01)100$

Here values of function are given to at least eight decimal or significant places.

Instructions for the User

The use of Tables 2 and 6 is self-evident.

Table 3: The table gives x^2 with absolute precision. Interpolation can be made as precise as desired by using linear interpolation and subtracting from the result, the correction term (in units of the last decimal place given in the table):

$$C = \xi(1 - \xi),$$

where ξ is the fractional part of the argument as defined under Table 1.

Table 4: To find the square root of any number, express it in "floating point" notation, with an *even* power-of-ten:

$$x = x^* \times 10^{2n}, \text{ with } 0.1 \leqslant x^* < 10.$$

The answer is, then:

$$\sqrt{x} = \sqrt{x^*} \times 10^n,$$

with the value of $\sqrt{x^*}$ readily found from Table 4.

Bibliographic Note

The "sinc" function (Table 6), sometimes called the interpolatory function because of its role in the sampling theorem of communication theory, plays a fundamental role there. Tabulations of it, however, do not seem to be generally available. Extra space has therefore been devoted to it here.

Note that Reference 4 (Table 18) tabulates $\frac{\sin x}{x}$, $x = 0(.01)50$.

TABLES 7 AND 8. THE ERROR FUNCTION AND ITS DERIVATIVES

Definition

$$\text{erf}\,(x) = \frac{2}{\sqrt{\pi}}\int_0^x e^{-u^2}\,du \equiv \frac{1}{\sqrt{\pi}}\int_{-x}^{x} e^{-u^2}\,du. \qquad (2)$$

Ranges

$$\text{erf}\,(x), \quad x = 0(.0001)1$$

$$\text{erfc}\,(x) = 1 - \text{erf}\,(x), x = 1(.001)10 \qquad (3)$$

$$\text{erf}'\,(x) = \frac{d}{dx}[\text{erf}\,(x)] = \frac{2}{\sqrt{\pi}}e^{-x^2}, x = 0(.0001)1(.001)10$$

Note that: erf $(x) = 1 -$ erfc (x).

Instructions for the User

The use of these tables for the listed functions is self-evident.

Functions closely related to the error function occur frequently in the literature and can be evaluated by means of Table 7 by using the following relationships:

$$H_1(x) = \frac{1}{\sqrt{\pi}}\int_{-\infty}^{x} e^{-u^2}\,du = \tfrac{1}{2}[\text{erf}\,(x) + 1]; \qquad (4a)$$

$$H_2(x) = \frac{1}{\sqrt{2\pi}}\int_{-x}^{x} e^{-u^2/2}\,du = \sqrt{\frac{2}{\pi}}\int_0^x e^{-u^2/2}\,du = \text{erf}\left(\frac{x}{\sqrt{2}}\right) \qquad (4b)$$

$$H_3(x) = \frac{1}{\sqrt{2\pi}}\int_{-\infty}^{x} e^{-u^2/2}\,du = \tfrac{1}{2}\left[\text{erf}\left(\frac{x}{\sqrt{2}}\right) + 1\right] \qquad (4c)$$

Note that for all these: $\lim_{x\to\infty} H_i(x) = 1$.

Higher order derivatives of the error function can readily be evaluated from Table 8 by means of Equation 8. E.g.

$$\text{erf}''(x) = -2x\,\text{erf}'(x) \qquad (5a)$$

$$\text{erf}'''(x) = (4x^2 - 2)\,\text{erf}'(x) \qquad (5b)$$

$$\text{erf}^{IV}(x) = -4x(2x^2 - 3)\,\text{erf}'\,x. \qquad (5c)$$

Computation of Table

Table 7 was computed in three ranges.

$0 < x < 3$

The series expansion was used – with double precision – in the following form:

$$\text{erf}(x) = \frac{2x}{\sqrt{\pi}} \sum_{n=0} \frac{x^{4n}}{(2n)!}\left[\frac{1}{4n+1} - \frac{x^2}{(2n+1)(4n+3)}\right]. \qquad (6)$$

$3 < x < 5$

In this range, the Taylor series expansion about points $x_n = 3 + n/10$, $n = 0, 1, 2, \ldots 19$ was used. Values for erfc (x_n)* were fed into the computer, and the following formula – again with double precision – was used to find points at closer intervals:

$$\text{erfc}(x + \delta) = \text{erfc}\,x - \sum_{n=1} \frac{\delta^n}{n!}\text{erf}^{(n)}\,x, \qquad (7)$$

where erf′ x is given by Equation 3 and, for $n > 1$,

$$\text{erf}^{(n)}x = -2[x\,\text{erf}^{(n-1)}\,x + (n-2)\,\text{erf}^{(n-2)}\,x] \equiv -(-1)^n H_{n-1}(x)\,\text{erf}'\,x \qquad (8)$$

where $H_n(x)$ is called the Hermite polynomial.

The computed values of erfc x_{n+1} were used to check the values of erfc x_n originally used.

$5 < x < 10$

For values $x > 5$ the asymptotic expansion was used in the form:

$$\text{erfc}(x) = \frac{\text{erf}'x}{2x} \sum_{n=0} \frac{(4n)!}{(2n)!(2x)^{4n}}\left[1 - \frac{(4n+1)(4n+2)}{(2n+1)(2x)^2}\right]. \qquad (9)$$

Range Extension

Values of erf (x) for very small values of x may be found accurately by using the formula:

$$\text{erf}(x) = \frac{2}{\sqrt{\pi}}x(1+\epsilon) \approx 1.128\,37,9167x, \qquad (10)$$

where $|\epsilon| < \frac{x^2}{3}$.

For large values of x the following approximation may be used:

$$\text{erfc}(x) = \frac{\text{erf}'\,x}{2x}\left(1 - \frac{1}{2x^2} + \epsilon\right)$$

where $|\epsilon| < \frac{3}{4x^4}$.

Thus, for $x \geqslant 10$, the following formula may be used with an error of less than one part in 10^4:

$$\text{erfc}(x) \approx \left(1 - \frac{1}{2x^2}\right)e^{-x^2}\sqrt{\pi}x. \qquad (11)$$

Bibliographic Note

More precise values (15 decimal places) of erf x and erf′ x have been published.[3a] The intervals there are the same as ours over the useful range ($0 < x < 5$). In the range ($4 < x < 10$), the intervals are ten times as large as they are in the present volume.

Similarly precise tables of the normal probability function and its derivative:

$$H(x) = \frac{1}{\sqrt{2\pi}}\int_{-x}^{x} e^{-u^2/2}\,du, \quad H'(x) = \frac{1}{\sqrt{2\pi}}e^{-x^2/2}$$

are also available.[3b]

TABLES 9. FRESNEL INTEGRALS

Definitions

$$C(u) = \int_0^u \cos\left(\frac{\pi}{2}v^2\right)dv \qquad (12)$$

$$S(u) = \int_0^u \sin\left(\frac{\pi}{2}v^2\right)dv. \qquad (13)$$

*These were taken from Reference 3a.

The following are equivalent definitions

$$\hat{C}(x) \triangleq C(\sqrt{2x/\pi}) = \frac{1}{\sqrt{2\pi}} \int_0^x \frac{\cos v}{\sqrt{v}}\, dv \tag{12a}$$

$$\hat{S}(x) \triangleq S(\sqrt{2x/\pi}) = \frac{1}{\sqrt{2\pi}} \int_0^x \frac{\sin v}{\sqrt{v}}\, dv. \tag{13a}$$

Ranges

$C(u), S(u)$ $u = 0(.001)10$

Applications

The Fresnel integrals occur in optical diffraction theory. For formulae covering several simple aperture shapes, see Reference 4a.

Computation of Tables

In writing series expansions for the Fresnel integrals, it is convenient to introduce a modified variable:

$$v = \pi u^2 . \tag{14}$$

In the range $(0 < u < 3)$, the tables were computed from the series expansion in the form:

$$C(u) = u \sum_{n=0} \frac{(v/2)^{4n}}{(4n)!} \left[\frac{1}{8n+1} - \frac{(v/2)^2}{(8n+5)(4n+1)(4n+2)}\right] \tag{15}$$

$$S(u) = u \left(\frac{v}{2}\right) \sum_{n=0} \frac{(v/2)^{4n}}{(4n+1)!} \left[\frac{1}{8n+3} - \frac{(v/2)^2}{(8n+7)(4n+2)(4n+3)}\right] . \tag{16}$$

Double precision was used.

For $u > 3$, the following asymptotic expressions were used:

$$C(u) \approx \tfrac{1}{2} + \frac{u}{v^3} \left\{\sin \frac{v}{2} \left[(v^2 - 3) + \frac{105}{v^2}\left(1 - \frac{99}{v^2}\right)\right] - \frac{\cos v/2}{v} \left[(v^2 - 15) + \frac{945}{v^2}\left(1 - \frac{143}{v^2}\right)\right]\right\} \tag{17}$$

$$S(u) \approx \tfrac{1}{2} - \frac{u}{v^3} \left\{\cos \frac{v}{2} \left[(v^2 - 3) + \frac{105}{v^2}\left(1 - \frac{99}{v^2}\right)\right] + \frac{\sin v/2}{v} \left[(v^2 - 15) + \frac{945}{v^2}\left(1 - \frac{143}{v^2}\right)\right]\right\} \tag{18}$$

Range Extension

For $u \geqslant 10$ the following approximation will yield results with an error of less than 10^{-6}:

$$C(u) = \frac{u}{v}\left(\sin \frac{v}{2} - \frac{1}{v} \cos \frac{v}{2}\right) \tag{19}$$

$$S(u) = \tfrac{1}{2} - \frac{u}{v}\left(\cos \frac{v}{2} + \frac{1}{v} \sin \frac{v}{2}\right). \tag{20}$$

Bibliographic Note

Table of $C(u), S(u), u = 0(.001)25$ have been published in Russia.[5]

Tables of $\hat{C}(x)$ and $\hat{S}(x)$ (Eqs. 12a and 13a) have been published for $x = 0(.01)50$.[6] A table supplementary thereto, covering $x = 0(.001)1$, too, is available.[4b]

TABLES 10-14. BLACKBODY RADIATION, RADIANT EMITTANCE[4c]

Definitions

$$\text{Let } \lambda T = y. \tag{21}$$

Table 10. Spectral Emittance, Relative

$$m_\lambda(y) = \frac{M_\lambda(\lambda, T)}{M_{\lambda\max}(T)} = \frac{k_1 (c_2/y)^5}{e^{c_2/y} - 1} \tag{22}$$

Table 11a. Fractional Emittance, Relative

$$m_f(y) = \int_0^\lambda M_\lambda(\lambda, T)\, d\lambda / M(T) = \frac{15}{\pi^4} c_2^{\,4} \int_0^y \frac{dy'}{y'^5 (e^{c_2/y'} - 1)} \tag{23}$$

Table 11b. Its complement:

$$m_{fc}(y) = 1 - m_f(y) \tag{23a}$$

Table 12. Peak Spectral Emittance

$$M_{\lambda max}(T) = \frac{c_1 T^5}{k_1 c_2^{\,5}} \tag{24}$$

Table 13. Total Emittance

$$M(T) = \frac{\pi^4}{15} \frac{c_1}{c_2^4} T^4 \quad (25)$$

Table 14. Peak Wavelength

$$\lambda_p(T) = \frac{c_2/k_2}{T} \quad (26)$$

The significance of the physical constants c_1 and c_2 and the mathematical constants k_1 and k_2 is explained under THEORY.

Ranges

$m_\lambda(y)$,	$y = 100(1)10^4(10^2)10^5(10^3)10^6(10^4)10^7\ \mu.°K$
$m_f(y)$,	$y = 100(1)10^4\ \mu.°K$
$m_{fc}(y)$,	$y = 10^4(10^2)10^5(10^3)10^6(10^4)10^7\ \mu.°K$
$M_{\lambda max}(T)$ and $M(T)$,	$T = 100(10)10^4(100)10^5\ °K$
$\lambda_p(T)$	$T = 100(100)10^5\ °K$

Instructions for the User

To find *peak spectral emittance, total emittance,* or *peak wavelength*, use Tables 12, 13, or 14, respectively.

To find the spectral emittance, $M_\lambda(\lambda, T)$ of a blackbody at any temperature (T) and wavelength (λ), calculate first: $y = \lambda T$. Now, find $m_\lambda(y)$ from Table 10 and $M_{\lambda max}(T)$ from Table 12. Then:

$$M_\lambda(\lambda, T) = m_\lambda(y) M_{\lambda max}(T) \text{W/m}^2 \cdot \mu. \quad (27)$$

To find the *fractional emittance*, $M_f(\lambda_1, \lambda_2; T)$ in the spectral band between λ_1 and λ_2 for a blackbody at temperature T, calculate first

$$y_1 = \lambda_1 T, \quad y_2 = \lambda_2 T$$

and find the corresponding values of $m_f(y_1)$, $m_f(y_2)$ from Table 11. Also find the total emittance $M(T)$ from Table 13. Then the desired fractional emittance is

$$M_f(\lambda_1, \lambda_2; T) = [m_f(y_2) - m_f(y_1)] M(T). \quad (28)$$

When the complement of a function is given, this must be subtracted from unity to yield the function:

$$m_f(y) = 1 - m_{fc}(y). \quad (23b)$$

Example

1. Find the spectral emittance, at $\lambda = 1\mu$, of a blackbody at $T = 3{,}000°$K. Using Equation 21, we find $y = 3{,}000$. Using Equation 27, we find

$$M_\lambda(1\mu, 3{,}000° \text{K}) = (.997141)(3.12619 \times 10^6) = 3.1173 \times 10^6 \frac{\text{W}}{\text{m}^2 \cdot \mu}.$$

2. For the same radiator find the fractional emittance between $\lambda_1 = 1\mu$ and $\lambda_2 = 2\mu$.

Here $y_1 = 3{,}000$ and $y_2 = 6{,}000$.
Hence, from Equation 28:

$$M_f(1\mu, 2\mu; 3{,}000° \text{K}) = [.737785 - .273223]\,[4.59246 \times 10^6]$$

$$= 2.1335 \times 10^6 \text{W/m}^2.$$

Note that the precision is significant only to five places.

Theory

The tables are based on Planck's radiation law:

$$M_\lambda(\lambda, T) = c_1/[\lambda^5(e^{c_2/\lambda T} - 1)], \quad (29)$$

where c_1 and c_2 are the radiation constants, whose presently accepted values are*[7]

$$c_1 = 3.7415 \times 10^8 \text{W}\mu^4/\text{m}^2, \; c_2 = 14{,}387.9\mu°\text{K}.$$

On integrating Equation 29 we find [based on (37) below]

$$M(T) = \int_0^\infty M_\lambda(\lambda, T)\, d\lambda = \frac{\pi^4}{15} \frac{c_1}{c_2^4} T^4 = \sigma T^4. \quad (30)$$

*There is a typographical error in the value of c_1 as given in Reference 7. This can readily be confirmed by evaluating c_1 from the value of the Stefan-Boltzmann constant, there given as in Equation 30a here.

Here

$$\sigma = \frac{\pi^4}{15}\frac{c_1}{c_2^{\,4}} = 5.6697 \times 10^8 \frac{\text{W}}{\text{m}^2}\,{}^\circ\text{K}^{-4} \tag{30a}$$

is the Stefan-Boltzmann constant.

On differentiating Equation 29 and equating the derivative to zero we find that the maximum occurs at

$$\lambda_p = \frac{c_2}{k_2 T} = 2897.8\mu^\circ\text{K}/T \tag{31}$$

where $k_2 = 4.965114232\ldots$ is the positive real root of the equation

$$5 - x = 5e^{-x} \tag{32}$$

When this value of λ is substituted into Equation (29), it is found that the peak spectral emittance using Equation 32 is

$$M_{\lambda\max}(T) = \frac{c_1 k_2^{\,5} T^5}{c_2^{\,5}(e^{k_2} - 1)} = \frac{c_1}{c_2^{\,5}}[k_2^{\,4}(5 - k_2)]\,T^5 = \frac{c_1}{k_1 c_2^{\,5}}\,T^5, \tag{33}$$

$$= 1.2865\cdots \times 10^{-11}\,T^5\,[\text{W/m}^2\cdot\mu(^\circ\text{K})^5]$$

where $k_1 = 0.047166617\ldots.$

When Equation 29 is divided by Equation 33, we obtain Equation 22, which is independent of λ and T individually.

Similarly, on dividing an integral of $M_\lambda(\lambda)$ by Equation 30, we obtain Equation 23 which is also a function only of $y = \lambda T$.

Computation of Tables

The evaluation of Equations 22 to 26 is straightforward, excepting only the integral of Equation 23. For values of $y < 10^4$, this was evaluated from the following series expansion of the integrand. On setting

$$\nu = c_2/y,\ \nu' = c_2/y' \tag{34}$$

Equation 23 becomes

$$m_f(y) = \frac{15}{\pi^4}\int_\nu^\infty \frac{\nu'^3}{e^{\nu'} - 1}\,d\nu' = \frac{15}{\pi^4}\int_\nu^\infty \nu'^3 \sum_{n=1}^{\infty} e^{-n\nu'}\,d\nu'$$

$$= \frac{15}{\pi^4}\sum_{n=1}^{\infty} e^{-n\nu}\left(\frac{\nu^2}{n} + \frac{3\nu^2}{n^2} + \frac{6\nu}{n^3} + \frac{6}{n^4}\right) \tag{35}$$

This expansion converges only slowly for small values of ν. Therefore, for $y \geqslant 10^4$, another expansion was used. This is obtained by expanding the exponential in the second member of Equation 35 and dividing out (Reference 8a)

$$m_{fc}(y) = 1 - m_f(y) = 1 - \frac{15}{\pi^4}\int_\nu^\infty \nu'^3\,d\nu' \Bigg/ \sum_{n=1}^{\infty}\frac{\nu'^n}{n!}$$

$$= \frac{15}{\pi^4}\int_0^\nu \nu'^3\,d\nu' \Bigg| \sum_{n=1}^{\infty}\frac{\nu'^n}{n!}$$

$$= \frac{15}{\pi}\nu^3\left(1/3 - \frac{\nu}{8} + \frac{\nu^2}{60} - \frac{\nu^4}{5040} + \frac{\nu^6}{272{,}160} - \frac{\nu^8}{13{,}305{,}600} + \cdots\right) \tag{36}$$

where we have used the definite integral [cf. Reference 9]

$$\int_0^\infty \frac{\nu^3\,d\nu}{e^\nu - 1} = \frac{\pi^4}{15}. \tag{37}$$

Table 11b was evaluated by means of (36).

Inspection of Equations 22 and 35 shows that we can obtain purely mathematical functions, involving no physical constants, simply by writing

$$\nu = c_2/\lambda T. \tag{34}$$

We have then

$$m_\lambda(\nu) = k_1 \nu^5 (e^\nu - 1)^{-1} \tag{38}$$

$$m_f(\nu) = \frac{15}{\pi^4}\int_0^\nu \frac{u^3}{e^u - 1}\,du. \tag{39}$$

Tables of (38) and (39) would be independent of the specific values of the physical constants c_1 and c_2.

Range Extension

1. Small Values of λT

Values of $\lambda T < 100\mu^\circ$K are not covered by Tables 10 and 11 because they do not lend themselves to the type of tabulation used there and also were deemed less likely to be required. To calculate these, the following approximations will yield precise results.

From Equation 22

$$m_\lambda(y) \approx k_1(c_2/y)^5 e^{-c_2/y},\ y < 100 \tag{22a}$$

$$= (.2908183) \times 10^{20}/y^5 e^{\nu}.$$

From Equation 35

$$m_f(y) \approx \frac{15}{\pi^4} e^{-\nu}(\nu^3 + 3\nu^2 + 6\nu + 6),\ y < 100 \tag{35a}$$

where $\nu = c_2/y = 14387.9\mu°K/y$ and $15/\pi^4 = 0.15398973\cdots$. The error introduced by dropping higher order terms in Equations 22a and 35a is about 10^{-62}!

2. Large Values of λT

For $\lambda T \geqslant 10^7$, the following approximations may be used. From Equation 22

$$m_\lambda(y) = k_1\nu^5[\nu + \frac{1}{2!}\nu^2 + \frac{1}{3!}\nu^3 + \cdots]^{-1}$$

$$\approx k_1\nu^4/(1+\frac{\nu}{2}) = 0.0471666\cdots\nu^4/(1+\frac{\nu}{2}), \tag{40}$$

with an error of less than one part in 10^6.

Again, from Equation 36

$$m_{fc}(y) \approx \frac{15}{\pi^4}\nu^3(\frac{1}{3} - \frac{\nu}{8}) = 0.0513299\nu^3(1 - \frac{3}{8}\nu), \tag{41}$$

with an error of less than one part in 10^7. Here

$$\nu = \frac{c_2}{y} = 14{,}387.9/y.$$

3. New Values of c_2

Clearly, as better values of c_2 become available, the tables need revision. These revisions can be readily made by the user as follows.

Note that Tables 10 and 11 are functions only of

$$\nu = c_2/y.$$

Thus a change in c_2 can be compensated for simply by entering the tables with y' –

$$y' = \frac{c_{20}y}{c_2} \tag{42}$$

in place of y. Here $c_{20} = 14{,}387.9\mu°K$.

If the changes in c_1 and c_2 are very small so that we can write

$$c_1 = c_{10}(1+\epsilon_1) \tag{43}$$

$$c_2 = c_{20}(1+\epsilon_2);\ \epsilon_1, \epsilon_2 << 1,$$

we can approximate

$$y' = y(1-\epsilon_2) = y - \epsilon_2 y. \tag{42a}$$

Tables 12, 13, and 14 are still entered with the actual temperature; but values read from Table 12 must be multiplied by

$$\left(\frac{c_1}{c_{10}}\right)\left(\frac{c_{20}}{c_2}\right)^5 \approx 1 + \epsilon_1 - 5\epsilon_2, \tag{44}$$

those from Table 13 by

$$\left(\frac{c_1}{c_{10}}\right)\left(\frac{c_{20}}{c_2}\right)^4 \approx 1 + \epsilon_1 - 4\epsilon_2, \tag{45}$$

and those from Table 14 by

$$\frac{c_2}{c_{20}} \approx 1 + \epsilon_2. \tag{46}$$

Bibliographic Note

The present tables of spectral emittance and fractional emittance give values at 1°K temperature intervals at $\lambda = 1\mu$ – shorter intervals at longer wavelengths and *vice versa*. They give values at wavelength intervals of 0.2 nm at 5000°K – larger or smaller intervals at other temperatures.

Previously published tables give entries for variations in temperature and wavelength individually.[8,10] However, this convenience is bought at the expense of much larger intervals and much greater size. Furthermore, these tables are based on the value $c_2 = 14380\ \mu°K$, which is not the best presently available value and they can not be readily adapted to compensate for the change in c_2.

Tables using our compact form, too, have been published. Indeed, we have already noted that the functions of Tables 10 and 11 can be expressed in terms of a variable ν which is independent of any physical constant (cf. Equations 38 and 39). Tables of these functions (without k_1) are available.[11] It is hoped, however, that the reader will find the present tables considerably more convenient to use in many applications, more detailed, and more compactly presented.

TABLES 15 AND 16. BLACKBODY RADIATION, PHOTON EMITTANCE

Definitions

Table 15. Fractional Photon Emittance, Relative

$$n_f(y) = \frac{1}{2\zeta(3)} \int_{\nu_1}^{\infty} \frac{\nu^2\, d\nu}{e^{\nu} - 1} \tag{47}$$

Table 16. Total Photon Emittance

$$N_T(T) = 4\pi\zeta(3) \frac{c}{c_2{}^3} T^3 \text{ photons/m}^2\cdot\text{s} \tag{48}$$

where

$$\zeta(3) \equiv \sum_{k=1}^{\infty} k^{-3}$$

and $\nu = c_2/y$.

Ranges

$n_f(y)$, $y = 100(1)10^4(10^2)10^5(10^3)10^6(10^4)10^7$ $\mu.°K$

$N_T(T)$, $T = 100(10)10^4(100)10^5 °K$

Instructions for the User

To find the total number of photons emitted by a blackbody at temperature T, use Table 16.

To find the number of photons emitted by such a blackbody in the spectral range between λ_1 and λ_2, calculate first

$y_1 = \lambda_1 T$ and $y_2 = \lambda_2 T$.

Find the corresponding values of $n_f(y_1)$ and $n_f(y_2)$ from Table 15. Also find the total photon emittance $N_T(T)$ from Table 16. The desired fractional photon emittance is, then,

$$N_f(\lambda_1, \lambda_2; T) = [n_f(y_2) - n_f(y_1)] N_T(T). \tag{49}$$

The *spectral photon emittance* can be found from the spectral radiant emittance (Equation 27), by dividing the latter by the photon energy as given by Equation 50 below.

Example

How long must you wait until a square-meter surface of a blackbody at room temperature (say $T = 290°K$) emits a photon in the visible range of the spectrum (say $\lambda_1 = .4\ \mu$, $\lambda_2 = .7\ \mu$)?

From Table 16: $N_T(290°K) = 3.708 \times 10^{22}$

Now $y_1 = .4 \times 290 = 116$, $y_2 = .7 \times 290 = 203$.

From Table 15: $n_f(y_1) = .8831 \times 10^{-50}$; $n_f(y_2) = .35571 \times 10^{-27}$.

Hence

$$N_f(.4, .7; 290) = (.35571 \times 10^{-27} - .8831 \times 10^{-50}) 3.708 \times 10^{22}$$
$$= 1.319 \times 10^{-5} \text{ photons/m}^2\cdot\text{s}$$

$[1.319 \times 10^{-5}]^{-1} = 75{,}815$ seconds

Thus we must wait 21 hours, on the average.

Theory

The *energy of a photon* at wavelength λ is:

$$E_p = h\nu = hc/\lambda \tag{50}$$

Here Planck's constant $h = 6.6256 \times 10^{-34}$ J.s and the speed of light $c = 2.997925 \times 10^8$ m/s are the accepted values.[7]

Thus, using equations 29 and 50, and noting that

$$c_1 = 2\pi c^2 h, \tag{51}$$

we find for the spectral photon emittance:

$$N_\lambda(T) = \frac{M_\lambda}{E_p} = 2\pi c/\lambda^4 (e^{c_2/\lambda T} - 1). \tag{52}$$

In terms of ν (Equation 34) we have

$$\lambda = c_2/\nu T,\ d\lambda = -\frac{c_2}{T} \cdot \frac{d\nu}{\nu^2}, \tag{53}$$

so that Equation 52 may be written

$$N_\lambda(T) = 2\pi c T^4 \nu^4 / c_2{}^4 (e^{\nu} - 1). \tag{54}$$

By integrating $N_\lambda(T)$ over λ, we obtain the *fractional photon emittance*:

$$N_f(\lambda, T) = \int_0^\lambda N_{\lambda'}(T)\,d\lambda' = \frac{2\pi c}{c_2^{\,4}} T^4 \int_\nu^\infty \frac{\nu^4 c_2}{(e^\nu - 1)T\nu^2}\,d\nu \qquad (55)$$

$$= \frac{2\pi c}{c_2^{\,3}} T^3 \int_\nu^\infty \frac{\nu^2\,d\nu}{e^\nu - 1}.$$

To obtain the *total photon emittance*, we integrate (55) from zero to infinity:

$$N_T(T) = \frac{2\pi c T^3}{c_2^{\,3}} \int_0^\infty \frac{\nu^2\,d\nu}{e^\nu - 1} = 4\pi\zeta(3)cT^3/c_2^{\,3}, \qquad (56)$$

where $\zeta(3) = \sum_{k=1}^{\infty} k^{-3} = 1.2020569\cdots$ [Reference 1, p. 811] is the zeta-function. We may also write

$$N_T(T) = c_p T^3 \qquad (57)$$

where

$$c_p = 4\pi c\zeta(3)/c_2^{\,3} = 1.52042 \times 10^{15} \text{ photons/m}^2.\text{s}.(^\circ\text{K})^3. \qquad (58)$$

Equation 57 is tabulated in Table 16.

The *relative fractional photon emittance* is obtained by dividing the fractional by the total photon emittance:

$$n_f(\lambda, T) = N_f(\lambda, T)/N_T(T) = \frac{1}{2\zeta(3)} \int_\nu^\infty \frac{\nu^2\,d\nu}{e^\nu - 1}. \qquad (59)$$

Equation 59 is tabulated in Table 15.

Computation of Tables

The procedure for evaluating Equation 57 is obvious.

Equation 59 was evaluated by the method used for Table 11. For low values of y, the integrand was expanded in a series of exponentials

$$\int_\nu^\infty \frac{\nu^2\,d\nu}{e^\nu - 1} = \sum_{n=1}^{\infty} \int_\nu^\infty \nu^2 e^{-n\nu} d\nu = \sum_{n=1}^{\infty} e^{-n\nu}\left(\frac{\nu^2}{n} + \frac{2\nu}{n^2} + \frac{2}{n^3}\right). \qquad (60)$$

For large values of y, a power series expansion, analogous to Equation 36, was used:

$$n_c = [2\zeta(3)]^{-1}\nu^2\left[\frac{1}{2} - \frac{\nu}{6} + \frac{\nu^2}{48} - \frac{\nu^4}{4320} + \frac{\nu^6}{241{,}920} - \frac{\nu^8}{12{,}096{,}000} + \cdots\right]. \qquad (61)$$

Note that $n_f = 1 - n_c$.

Range Extension

1. Values for $\lambda T < 100$ are not covered in Table 15. These can be found accurately from

$$n_f(\lambda, T) \approx e^{-\nu}(\nu^2 + 2\nu + 2)/2\zeta(3) \qquad (62)$$

$$= .4159537e^{-\nu}(\nu^2 + 2\nu + 2),\ y < 100.$$

The error introduced by dropping the higher order terms is of the order one part in 10^{59}!

2. Values for $\lambda T \geqslant 10^7$ may be obtained by means of the following approximation:

$$n_c \approx \nu^2\left(1 - \frac{\nu}{3}\right)/4\zeta(3),\ y \geqslant 10^7. \qquad (63)$$

This is accurate to better than one part in 10^7.

3. New Values of c and c_2.

To use the tables when better values for the natural constants become available, the methods given under Range Extension (3) in the preceding section may again be used.

Denoting, again, the fractional correction term to c by ϵ and that to c_2 by ϵ_2, values taken from Table 16 should be multiplied by

$$[1 + \epsilon - 3\epsilon_2]. \qquad (64)$$

The value of y with which Table 15 is entered should be multiplied by

$$[1 - \epsilon_2]. \qquad (64a)$$

Bibliographic Note

Tables of the relative spectral photon emittance (n_λ) and relative fractional photon emittance (n_f) have been tabulated[12] for the ranges

y = 500(10)1000(50)4000(100)10,000(500)20,000

i.e., for 191 values.

These were based on c_2 = 1.436, i.e., on a value 0.2% below the presently accepted value.

More extensive tables of spectral photon emittance have recently been published.[13] But these give neither fractional nor total emittances.

TABLE 17 AND 18. BLACKBODY RADIATION, LUMINANCE

Definitions

Table 17. Spectral Luminance

a. Photopic

$$L_{v\lambda}(T, \lambda) = K_{max} V(\lambda) L_e(T, \lambda) = \frac{1}{\pi} K_{max} V(\lambda) M_e(T, \lambda) \quad (65)$$

b. Scotopic

$$L_{v'\lambda}(T, \lambda) = K'_{max} V'(\lambda) L_e(T, \lambda). \quad (66)$$

Table 18. (Incomplete) Fractional Luminance

a. Photopic

$$L_{vf}^*(T, \lambda) = \int_{\lambda_0}^{\lambda} L_{v\lambda}(\lambda') d\lambda' \quad (67)$$

b. Scotopic

$$L_{v'f}^*(T, \lambda) = \int_{\lambda_0}^{\lambda} L_{v'\lambda}(\lambda') d\lambda'. \quad (68)$$

Here $K_{max} = 680$ lm/W is the maximum value of the photopic luminous efficacy, $K_{max}' = 1746$ is the corresponding scotopic value, and V' denotes scotopic spectral efficiency. In Table 18, λ_0 is the integral multiple of 0.1 μ immediately below λ.

Ranges

All these tables cover:

$T = 500(100)10{,}000(1000)\ 20{,}000°K$

$= 0.36(.01)0.83\ \mu$ (photopic)

$= 0.38(.01)0.78\ \mu$ (scotopic).

Instructions for the User

Because of the wide range in some rows of Tables 17 and 18, luminance values there are expressed in "exponential" or "E"-form, i.e., the power-of-ten by which the number is to be multiplied is given following the letter E. For instance:

$aEn = a \times 10^n$; $1.234E2 = 123.4$; $5.678E-4 = .0005678$.

To find the spectral luminance, at wavelength λ, of a blackbody at temperature T, find the appropriate column in Tables 17. The desired value is then found in the row headed by the value of λ.

In the same manner, Tables 18 give the incomplete* fractional luminance, L_{vf}^*.

The complete fractional luminance, $L_{vf}(n/10)$, for every tenth micron, is given in the row headed "TOTALS." To find the complete fractional luminance $L_{vf}(\lambda)$ for any other wavelength, the incomplete value must be added to the preceding value of $L_{vf}(n/10)$:

$$L_{vf}(\lambda) = L_{vf}^*(\lambda) + L_{vf}(n/10). \quad (69)$$

The fractional luminance $[L_{vf}(\lambda_1, \lambda_2)]$ in the wavelength range λ_1 to λ_2 is found as the difference between the corresponding values of complete fractional luminance:

$$L_{vf}(\lambda_1, \lambda_2) = L_{vf}(\lambda_2) - L_{vf}(\lambda_1). \quad (70)$$

The total blackbody luminance for any temperature is given as the last entry in the column headed by that temperature, in Table 18.

Examples

1. What is the peak spectral luminance of a blackbody at 2800°K and at what wavelength is it observed?

   In Table 17, in column 2800, we find the largest value at .57 μ.

It is 1.558×10^8, cd/m$^2 \cdot \mu$.

2. A blackbody at 1300°K is viewed through a sharp-cut-off filter which passes 50% of the incident radiation in its passband, which extends from .39 μ to .41 μ. What will the observed luminance be?

   The observed luminance will be below .03 cd/m^2, so that the scotopic data in Tables 18 should be used. Thus from the second of Table 18 we find:

$$L = 0.5[L_{v'f}^*(.41) + L_{v'f}(.4) - L_{v'f}^*(.39) - 0]$$

$$= .5[10.05 + 1.70 - .21] \times 10^{-3} = 5.77 \times 10^{-3}\ \text{sec.cd/m}^2.$$

*The reasons for tabulating the fractional luminance in this "incomplete" form are presented under Computation of Tables.

Theory

The photopic spectral luminance differs from the spectral radiance* only by $V(\lambda)$, the spectral luminous efficiency factor. This factor is a measure of the visibility of radiation at wavelength λ, normalized for unity at 0.555 μ. (This is the wavelength of maximum efficiency, i.e., the photopic human visual system is normally most sensitive at that wavelength.)

The values of V, at intervals of 0.005 μ, were standardized by the International Commission on Illumination (C.I.E.) in 1924 and 1931.[14] Interpolations at intervals of 10^{-3} μ have been published.[15-17] (See Reference 16 for an evaluation of many.) The corresponding values of scotopic spectral luminous efficiency factors, $V'(\lambda)$, are normalized at .507 μ and were adopted by the C.I.E. in 1951.[18] Both sets are given in Table 28.

Photopic vision normally occurs for luminances above 3 cd/m^2 and scotopic vision below .03 cd/m^2. In the intermediate, mesopic region, the response changes continuously from scotopic to photopic as the luminance increases and the fovea becomes more exclusively involved.

Computation of Tables

The method for computing Tables 17 is obvious.

Tables 18 were computed by numerical quadrature using Simpson's rule with intervals of 10^{-3} μ. The values of V and V', as given in Table 28, were used.

In Tables 18, the integration is started anew at 0.1 μ-intervals, to avoid the inaccuracies which occur when tabulating small changes in large numbers. It is hoped that the greatly increased accuracy will adequately compensate for the slight inconvenience introduced by this.

Bibliographic Note

Extensive tables of blackbody luminance have been published previously;[10] but these did not cover scotopic vision – nor did they cover fractional luminance, either photopic or scotopic.

NOTE: *A summary of blackbody radiation formulae is given in Table 31.*

TABLES 19 AND 20. POINT SPREAD FUNCTION AND ENCIRCLED ENERGY

Definitions

Point Spread Function:

$$P(r_r) = \left[\frac{2J_1(\pi r_r)}{\pi r_r}\right]^2 \tag{71}$$

Encircled Energy Complement:

$$E_c(r_r) = J_0^2(\pi r_r) + J_1^2(\pi r_r) \tag{72}$$

Ranges

$P(r_r)$, $E_c(r_r)$; $r_r = 0(.001)1(.01)10(.1)100$

Instructions for the User

Table 19 gives the distribution of illumination in the Fraunhofer diffraction pattern of a circular aperture.[4d,19] This is equivalent to the point spread function of a perfect lens with circular aperture – i.e., the illumination distribution near the focal point of a perfect lens when illuminated by a distant monochromatic (or quasi-monochromatic) point source of light on its axis. Thus, this table gives the illumination at a point, a distance r from the center of the pattern.

Table 20 gives the fraction of the total energy beyond radius r, under the same conditions.

The argument with which the tables must be entered – the "reduced radius" is

$$r_r = \frac{2Rr}{\lambda' f} = \frac{r}{F\lambda'} \approx \frac{2r\sin\theta}{\lambda'} = \frac{2Ar}{\lambda_0},\ R << f, \tag{73}$$

where R is the radius of the circular aperture,

λ_0, λ' are, respectively, the wavelengths of the radiation in a vacuum and in the image space,

f is the distance of the image point from the aperture,

$F = f/2R$ is the "effective" F/number,

θ is the apex half-angle of the cone the aperture rim subtends at the image point,

$A = n\sin\theta$ is the numerical aperture, n, the refractive index.

The point spread function is normalized for unity at the origin and the encircled energy for unity at infinity.

The encircled energy proper is found from:

$$E(r_r) = 1 - E_c(r_r) \tag{72a}$$

*The spectral radiance may be obtained by dividing by π the spectral emittance, as given, e.g., by Tables 10 and 12:

$L_v(\lambda) = \frac{1}{\pi} M_e(\lambda) V(\lambda)$.

Computation of Tables

In the range $(0 < r_r < 3)$, the Bessel functions entering Equations 71 and 72 were evaluated by series expansion:

$$J_0(2x) = 1 - \frac{x^2}{(1!)^2} + \frac{x^4}{(2!)^2} - \frac{x^6}{(3!)^2} + \cdots = \sum_{n=0}^{\infty} (-1)^n \frac{x^{2n}}{(n!)^2} \quad (74)$$

$$J_1(2x) = \frac{x}{(0!)(1!)} - \frac{x^3}{(1!)(2!)} + \frac{x^5}{(2!)(3!)} - \cdots$$

$$= x \sum_{n=0}^{\infty} (-1)^n \frac{x^{2n}}{n!(n+1)!} \quad (75)$$

For $(r_r > 3)$, the asymptotic expansions

$$J_0(x) \approx (\pi x)^{-1/2}[(P_0 - Q_0)\sin x + (P_0 + Q_0)\cos x] \quad (76)$$

$$J_1(x) \approx (\pi x)^{-1/2}[(P_1 + Q_1)\sin x - (P_1 - Q_1)\cos x] \quad (77)$$

were used. Here

$$P_0 = 1 - \frac{(1\cdot 3)^2}{2!(8x)^2} + \frac{(1\cdot 3\cdot 5\cdot 7)^2}{4!(8x)^4} - \frac{(1\cdot 3\cdot 5\cdot 7\cdot 9\cdot 11)^2}{6!(8x)^6} + \cdots \quad (78a)$$

$$Q_0 = -\frac{1}{8x}\left[1 - \frac{(1\cdot 3\cdot 5)^2}{3!(8x)^2} + \frac{(1\cdot 3\cdot 5\cdot 7\cdot 9)^2}{5!(8x)^4} - \cdots\right] \quad (78b)$$

$$P_1 = 1 - \frac{(4-1)(4-9)}{2!(8x)^2} + \frac{(4-1)(4-9)(4-25)(4-49)}{4!(8x)^4} - \cdots \quad (78c)$$

$$Q_1 = \frac{1}{8x}\left[(4-1) - \frac{(4-1)(4-9)(4-25)}{3!(8x)^2} + \cdots\right] \quad (78d)$$

The first nine terms in these expansions were used.

Range Extension

For values of $r_r \geqslant 100$ the following approximations will yield values with an error of less than one part in 10^6:

$$J_0(\pi r) \approx \frac{1}{\pi\sqrt{r}}\left[\left(1 + \frac{1}{8\pi r}\right)\sin \pi r + \left(1 - \frac{1}{8\pi r}\right)\cos \pi r\right] \quad (79a)$$

$$J_1(\pi r) \approx \frac{1}{\pi\sqrt{r}}\left[\left(1 + \frac{3}{8\pi r}\right)\sin \pi r - \left(1 - \frac{3}{8\pi r}\right)\cos \pi r\right] \quad (79b)$$

These values may then be substituted into Equations 71 and 72 for $\mathbf{P}(r)$ and $\mathbf{E}_c(r)$.

TABLE 21. LINE SPREAD FUNCTION [4e,]

Definition

The line spread function

$$\mathbf{L}(x_r) = \frac{3}{8\pi x_r^2} H_1(2\pi x_r) \quad (80)$$

where H_1 represents the first order Struve function.

Range

$\mathbf{L}(x_r)$, $x_r = 0(.001)1(.01)10(.1)100$

Instructions for the User

Equation 80 represents the line spread function (**L**) of a perfect lens with circular aperture, radius R, in monochromatic light.

To find the illumination in the image of a line object, formed under the above conditions, at a distance x from the ideal (Gauss) image, evaluate the reduced distance

$$x_r = \frac{2R}{f\lambda'} x = \frac{x}{F\lambda'} \approx \frac{2 \sin\theta}{\lambda'} x = \frac{2A}{\lambda_0} x, \quad R << x, \quad (81)$$

where the symbols are defined above, following Equation 73, and enter Table 21 with this value. The table then gives the illumination at x relative to that at the ideal line image.

Theory

The line spread function can be found as a line integral through the point spread function

$$\mathbf{L}(x) \sim \int \mathbf{P}(x, y)\, dy$$

or as a Fourier transform of the optical transfer function. See Reference 20 for a brief review of the literature concerning this function.

Computation of Table

For $x_r < 2.5$, $\mathbf{L}(x_r)$ was evaluated from the series expansion

$$\mathbf{L}(x_r) = 3\left[\frac{1}{1^2\cdot 3} - \frac{\xi^2}{1^2\cdot 3^2\cdot 5} + \frac{\xi^4}{1^2\cdot 3^2\cdot 5^2\cdot 7} - \cdots\right] \quad (82)$$

$$= -\frac{3}{\xi^2}\sum_{n=1}^{\infty} \frac{(-1)^n}{2n+1}\left[\frac{2^n n!}{(2n)!}\right]^2 \xi^{2n} \quad (83)$$

where $\xi = 2\pi x_r$.

For $x_r > 2.5$, the following asymptotic series expansion was used.

$$L(x_r) \approx \frac{3}{\xi^2}\left\{1 + \frac{1}{\xi^2} - \frac{1^2\cdot 3}{\xi^4} + \frac{1^2\cdot 3^2\cdot 5}{\xi^6} - \cdots\right.$$

$$-\tfrac{1}{2}\sqrt{\frac{\pi}{\xi}}\left[\sin\xi\left(1 + \frac{1\cdot 3}{8\xi} + \frac{1^2\cdot 3\cdot 5}{2!(8\xi)^2} - \frac{1^2\cdot 3^2\cdot 5\cdot 7}{3!(8\xi)^3} - \frac{1^2\cdot 3^2\cdot 5^2\cdot 7\cdot 9}{4!(8\xi)^4} + \cdots\right)\right.$$

$$\left.\left. + \cos\xi\left(1 - \frac{1\cdot 3}{8\xi} + \frac{1^2\cdot 3\cdot 5}{2!(8\xi)^2} + \frac{1^2\cdot 3^2\cdot 5\cdot 7}{3!(8\xi)^3} - \cdots\right)\right]\right\}$$

$$= \frac{3}{\xi^2}\left\{1 + \sum_{n=1} \frac{(-1)^n}{2n-1}\left[\frac{2n!}{2^n n!}\right]^2 \xi^{-2n} - \sqrt{\frac{\pi}{\xi}}(\sin\xi + \cos\xi)\right.$$

$$\left[1 - 4\sum_{n=1}(-1)^n \frac{16n^2 - 1}{(2n)!}\left[\frac{(4n-2)!}{2^{2n}(2n-1)!}\right]^2 (8\xi)^{-2n}\right] \tag{84}$$

$$\left. - \sqrt{\frac{\pi}{\xi}}(\sin\xi - \cos\xi)\sum_{n=0}(-1)^n \cdot \frac{(4n+1)(4n+3)}{(2n+1)!}\left[\frac{(4n)!}{2^{2n}(2n)!}\right]^2 (8\xi)^{-(2n+1)}\right\}.$$

Range Extension

For values of $x_r > 100$ the following approximation may be used:

$$L(x_r) \approx \frac{3}{\xi^2}\left[1 - \sqrt{\frac{\pi}{\xi}}(\sin\xi + \cos\xi)\right]. \tag{85}$$

The error will then be less than 10^{-9}.

TABLE 22. MODULATION TRANSFER FUNCTION (MTF) OF A PERFECT LENS

Definitions

$$T(\nu_r) = \frac{2}{\pi}(\cos^{-1}\nu_r - \nu_r\sqrt{1 - \nu_r^2}),\ \nu_r < 1 \tag{86}$$

$$= 0 \qquad \nu_r > 1.$$

Range

$T(\nu_r)$, $\nu_r = 0(.0001)1$

Instructions for the User

To find the modulation transfer function (mtf) of a perfect lens with circular aperture (radius R) in incoherent quasi-monochromatic light at a spatial frequency of ν cycles/mm, evaluate the reduced frequency

$$\nu_r = \frac{\lambda' f}{2R}\nu = \lambda' F\nu \approx \frac{\lambda'\nu}{2\sin\theta} = \frac{\lambda_0\nu}{2A},\quad R << f, \tag{87}$$

where the symbols have the same meaning as in Equation 73, following which they are defined. Then enter Table 22 with ν_r to find the mtf at ν.

Linear interpolation is accurate throughout, with the possible exception of the last line.

Applications

The mft represents the ratio of image modulation to object modulation for a one-dimensional sinusoidal object. The modulation is here defined as

$$M = (L_{max} - L_{min})/(L_{max} + L_{min})$$

where L_{max} and L_{min} represent the peak and trough object luminance – or image illumination.

Since the spread function of this lens is symmetrical, the mtf is identical with the *optical* transfer function (otf), which represents the Fourier transform of the line spread function.

TABLES 23 AND 24. MTF OF A DEFOCUSED PERFECT LENS[20]

Definitions

Tables 23:

$$T(\nu_r, \Delta) = \frac{4}{\pi}\int_{\nu_r}^{1}\sqrt{1 - u^2}\cos[2\pi\nu_r\Delta(u - \nu_r)]\,du,\ \nu_r < 1$$

$$= 0, \qquad , \nu_r \geqslant 1. \tag{88}$$

Table 24:

$$\int_0^1 T(\nu_r, \Delta)d\nu_r,\quad \int_0^1 T^2(\nu_r, \Delta)d\nu_r$$

Ranges

$T(\nu_r, \Delta), \nu_r = 0(.001)1$

$$\Delta = 0(.1).5(.5)5(1)10(5)50(10)100$$

Instructions for the User

Tables 23 give the mtf of a perfect lens with circular aperture and incoherent, quasi-monochromatic light – in the presence of defocusing. To use the tables, find*

$$\Delta = 2\left(\frac{R}{f}\right)^2 \frac{d}{\lambda'} = \tfrac{1}{2} \frac{d}{\lambda' F^2} \approx \frac{2 \sin^2 \theta}{\lambda'} = \frac{2A^2}{n\lambda_0} d, \quad R << f, \tag{89}$$

where d is the amount of defocusing in linear measure and the other variables have been defined earlier (following Equation 73). Then find ν_r by means of Equation 87 and enter with it the table whose Δ value is nearest the desired value. For more accurate results, use interpolation between the T-values found from the two Tables 23, whose Δ-values straddle the desired one.

Table 23 is based on approximations valid only at low relative aperatures. [Cf. Reference 21.]

Table 24 gives the integrals of the mtf and the $(\mathrm{mtf})^2$ for the above lenses. These provide measures, respectively, for the "correlation quality" and the "relative structural content" of Linfoot.[22] The integral of $(\mathrm{mtf})^2$ has been called "structural resolution"[23] and "equivalent pass band."[24]

Note the following definite integrals:

$$\int_0^1 \frac{2}{\pi} (\cos^{-1} u - u\sqrt{1-u^2})\, du = 4/3\pi$$

$$\int_0^1 \left[\frac{2}{\pi} (\cos^{-1} u - u\sqrt{1-u^2})\right]^2 du = 8(15\pi - 32)/45\pi^2.$$

Computation of Tables

Equation 88 was evaluated by Simpson's rule over the range $(0 < u < .995)$ and by expanding in a Taylor series and integrating term-by-term over the remainder.[20]

In the Simpson's rule portion the differences were 0.0005, and in the expansion portion, terms up to the fifth order – in some places – were retained.

The expansion portion has the form:

$$\Delta T = \frac{4}{\pi}\left[\cos a(1-\nu_r) \sum_{n=0}^{\infty} a^{2n} A_{2n} + \sin a(1-\nu_r) \sum_{n=0}^{\infty} a^{2n+1} A_{2n+1}\right] \tag{90}$$

where $a = 2\pi\nu_r\Delta$, (90a)

$$A_0 = \sqrt{\frac{\epsilon}{2}} \cdot \epsilon \left[\frac{4}{3} - \frac{1}{5}\epsilon - \frac{1}{56}\epsilon^2 - \cdots\right] \tag{90b}$$

$$A_1 = \sqrt{\frac{\epsilon}{2}} \cdot \epsilon^2 \left[\frac{4}{5} - \frac{1}{7}\epsilon - \frac{1}{72}\epsilon^2 - \cdots\right] \tag{90c}$$

$$A_2 = -\sqrt{\frac{\epsilon}{2}} \cdot \epsilon^3 \left[\frac{2}{7} - \frac{1}{18}\epsilon - \frac{1}{176}\epsilon^2 - \cdots\right] \tag{90d}$$

$$A_3 = -\sqrt{\frac{\epsilon}{2}} \cdot \epsilon^4 \left[\frac{1}{27} - \frac{1}{132}\epsilon - \cdots\right] \tag{90e}$$

$$A_4 = \sqrt{\frac{\epsilon}{2}} \cdot \epsilon^5 \left[\frac{1}{132} - \cdots\right] \tag{90f}$$

The error should nowhere reach two units in the last place.

TABLE 25. PERFECT-LENS MTF IN POLYCHROMATIC LIGHT[25]

Definition

$$T[\nu_0, c(\lambda)] = \int c(\lambda) T(\nu_0 \lambda)\, d\lambda \Big/ \int c(\lambda)\, d\lambda \tag{91}$$

where c is the product of the source spectral emittance (M_λ), the system transmittance (τ), and the relative detector sensitivity, all at λ. Thus

*When the defocusing is a substantial distance (d) of the image distance (f), a more accurate value for Δ is obtained by
$\Delta_c = \Delta/(1 + d/f)$
Where d is positive when the defocusing is in the down-light direction. This introduces an asymmetry into the blurring experienced with defocusing in the two directions.[21]

$$c(\lambda) = M_\lambda(\lambda) r(\lambda) S(\lambda)/S(\lambda_r) \tag{92}$$

where $S(\lambda)$ is the detector sensitivity (output/input) at wavelength λ, and λ_r is a reference wavelength.

Range

$T[\nu_0, c(\lambda)]$, ν_0 = 0(50)2800 cycles/mm, with the following radiation sources included:

(1-2) Blackbodies at 3000°K and 5000°K
(3-4) Standard Illuminants (C.I.E., 1931): *A* and *B*[4f]
(5-6) Phosphors (JEDEC): P 11 and P 16[26a]

The following detectors are included:

(1-2) Normal observers: photopic (C.I.E., 1931)[4g,14] and scotopic (C.I.E., 1951)[18]
(3-6) Standard Photosensitive devices*: S10, S11, S12, S20[26b]
(7) Cd Se detector**
(8) Silicon diode**
(9) Vidicon**
(10-12) Photographic emulsions: Blue-sensitive, orthochromatic, and panchromatic.

Instructions for the User

Find the "specific" spatial frequency (ν_0) from the actual spatial frequency (ν) and the relative aperture by means of the following definition:

$$\nu_0 = \frac{f\nu}{2R} = F\nu \approx \frac{\nu}{2 \sin \theta} \left(= \frac{\nu}{2A} \right) \text{for } R << f\ (n \equiv 1), \tag{93}$$

where the variables are defined following Equation 73.

In Table 25 labeled with the radiation source under consideration, find the column headed by the detector under consideration. In that column, and in the row corresponding to ν_0, the mtf will be found.

The source and detector characteristics for which the weighted mtf (Equation 91) was computed, are listed in Table 27. The system transmittance was arbitrarily assumed unity above $\lambda = 0.35\mu$ and zero below this value. See Reference 25 for the rationale for this.

*S10 is representative of some image orthicon responses and S12 of some CdS photo cells.
**Representative specimen. See Reference 25 for particulars.

Computation of Tables

Tables 25 were computed by Simpson's rule with a wavelength interval of $.01\mu$.

TABLE 26. SPECTRAL EFFICIENCY OF SOURCE-DETECTOR COMBINATIONS[25,27]

Definitions

System spectral efficiency:

$$k_s = \int c(\lambda)\, d\lambda \Big/ \int M_\lambda(\lambda)\, d\lambda. \tag{94}$$

Visual system efficiency (photopic):

$$k_v = \int c_v(\lambda)\, d\lambda \Big/ \int M_\lambda(\lambda)\, d\lambda, \tag{95}$$

where $c(\lambda) = \tau(\lambda) M_\lambda(\lambda) s(\lambda)$ is defined in Equation 92 and c_v is the value of c with $s = V$, the normal observer photopic response.

Luminous efficacy (photopic):

$$K_v = K_{0v} k_v \tag{96}$$

where K_{0v} = 680 lm/W relates lumens to watts at the peak of the photopic response ($\lambda = 0.555\mu$).

Luminous and visual spectral efficiency are similarly defined for scotopic vision with $s = V'$ the normal observer scotopic response. Here K_{0v}' = 1746 scotopic lumens per watt is the efficacy at the peak of the scotopic response curve ($\lambda = 0.507\mu$).[18]

Range

Table 26a

K_v, K_v', k_s, k_v, k_v', λ_r are given for all the radiation sources and detectors listed in Table 25.

Table 26b

Efficiencies for additional sources and detectors as given in Reference 27.

Table 26c

Blackbody – S Curve and Visual Response Combinations
$k_s(S, T)$, $\lambda_r(S)$; S = 1(1)25, photopic, scotopic
T = 0(100) 10,000;2854

where S is either the S-number assigned by JEDEC[26b] or the type of visual response, T is the temperature in degrees kelvin.
Data are given for two transmittance characteristics:

1. $\tau(\lambda) = 1$ (unfiltered)
2. $\tau(\lambda) = 1, \lambda \geqslant 0.35\mu$; $= 0, \lambda < 0.35\mu$ (filtered)

Notes: 1. The data for the filtered detectors are given only in cases where filtering reduces $k_s(2854°K)$ by more than 1%.
2. Very low values of $k_s (k_s \lesssim 10^{-30})$ are not given.
3. Low values of k_s are given in "E-format." This has the form $n_1 E \pm n_2$ where n_1 is the mantissa and $\pm n_2$ the exponent of ten associated with the number. E.g. the number (0.123 E-04) is (0.123×10^{-4}).

Instructions for the User

The system spectral efficiency relates the total system response to the response which would result, were an equal total energy radiated at the reference wavelength λ_r and $\tau(\lambda) = 1$. Thus, in a linear system, the system response may be calculated for an equal amount of energy input – all at λ_r – and then multiplied by k_s to get the actual output with that energy:

$$O = k_s E_{in} S(\lambda_r). \tag{97}$$

In a system with nonlinear response the following procedure must be followed. First calculate the input energy (E_0) required to produce the desired response on the assumption that all the energy is radiated at $\lambda = \lambda_r$ and $\tau = 1$. The actual required input energy is then found as

$$E_{in} = E_0 / k_s. \tag{98}$$

When the input is given in luminous quantities (e.g., lumens or candela), the factor

$$k_s' = k_s / K_{0\nu} k_\nu \tag{99}$$

must be used instead of k_s.

The reference wavelengths, λ_r, on which the entries in Table 27 are based, are given there in the last line. Manufacturers' data at that wavelength should be used in calculating the detector response.

Examples

1. A No. 7265 multiplier phototube receives 10^{-8} lm of sunlight. It is operated such that its sensitivity is 0.6×10^6 A/W at $\lambda = 0.42\mu$. What current output will it yield?

This tube type has an S20 response and the radiation approximates a blackbody at 5000°K. Thus we find from Table 26a:

Spectral efficiency: $k_s = 0.2366$
Luminous efficacy: $k_\nu K_{0\nu} = 81$ lm/W
Reference wavelength: $\lambda_r = 0.42\mu$,

so that

$$S(\lambda_r) = 0.6 \times 10^6 \text{ A/W}.$$

From Equations 97 and 99 the current will be:

$$I = \Phi S(\lambda_r) k_s / k_\nu K_{0\nu} = (10^{-8})(0.6 \times 10^6)(0.2366)/(81) = 0.175 \times 10^{-3} \text{ A}$$

2. A P11 crt display is to be recorded on Panatomic-X film. The illumination at the film plane is found to be $E = 2$ lm/m².

How long must the exposure be made to produce unity density?

From manufacturers data we find that, at the reference wavelength (0.4μ), Panatomic-X film requires, for unity density, an exposure of

$$Q_0 = 3.5 \times 10^{-3} \text{ J/m}^2.$$

From Table 26 we find

For P 11: $k_\nu K_{0\nu} = 136.7$ lm/W

For panchromatic emulsions in conjunction with P 11 phosphor: $k_s = 0.688$. Thus, from Equations 98 and 99:

$$Q_{in} = Q_0 K_{0\nu} k_\nu / k_s = (3.5 \times 10^{-3})(136.7)/0.688 = 0.7 \text{ lm}\cdot\text{s/m}^2.$$

The required exposure time is, then,

$$t = Q_{in}/E = 0.7/2 = 0.35 \text{ s}.$$

Theory

The theory underlying the concept of "spectral efficiency" is presented in Reference 25 and is summarized here.

In general terms, with an input spectral characteristic $M_\lambda(\lambda)$, the output will be

$$O = \eta \int \tau(\lambda) M_\lambda(\lambda) S(\lambda)\, d\lambda \qquad (100)$$

where η is a factor relating absorbed flux or irradiation to M_λ.

The total radiated input will be

$$I_T = \eta \int M_\lambda(\lambda)\, d\lambda \qquad (101)$$

on multiplying Equation 100 by

$$I_T/\eta \int M_\lambda(\lambda)\, d\lambda$$

we find, using Equations 92 and 94,

$$O = k_s I_T S(\lambda_r). \qquad (102)$$

This confirms the use of k_s described in Instructions.

If the input is known in luminous quantities, Equation 101 becomes

$$I_{\nu T} = K_{o\,\nu} \int M_\lambda(\lambda) s_\nu(\lambda)\, d\lambda. \qquad (103)$$

On applying this to Equation 100, as above, we find

$$O = (k_s/K_{o\,\nu} k_\nu) I_{\nu T} S(\lambda_r) = k_s' I_{\nu T} S\,(\lambda_r), \qquad (104)$$

confirming Equation 99.

TABLE 27. SPECTRAL DISTRIBUTIONS OF SOURCES AND DETECTORS

Definitions

The spectral sensitivity relative to that at reference wavelength, λ_r:

$$s(\lambda) = S(\lambda)/S(\lambda_r).$$

The spectral emittance, relative to that at λ_r:

$$m(\lambda) = M_{e\lambda}(\lambda)/M_{e\lambda}(\lambda_r).$$

Range

$s(\lambda)$, $m(\lambda)$; $\lambda = \lambda_{min}(.01)\lambda_{max}$, S-number $\neq$ 19.
$\lambda = \lambda_{min}(.05)\lambda_{max}$, S-number = 19.

Table 27a
Sources and detectors listed in Table 25.

Table 27b
All active S-curves

Instructions for the User

Sensitivities for wavelengths below 0.3μ and above 0.89μ are given at the end of the table.

Examples

The relative sensitivity of an S1 detector equals 0.329 at $\lambda = 1\mu$. An S17 detector peaks (s = 1) at 0.49μ.

TABLE 28. SPECTRAL LUMINOUS EFFICIENCY, C.I.E. STANDARD OBSERVER

Definitions

Photopic Spectral Luminous Efficiency:

$$V(\lambda) = L_{\nu\lambda}(\lambda)/L_{e\lambda}(\lambda); \qquad (105)$$

Scotopic Spectral Luminous Efficiency

$$V'(\lambda) = L_{\nu'\lambda}(\lambda)/L_{e\lambda}(\lambda), \qquad (106)$$

where $L_{e\lambda}$ is the spectral radiance and $L_{\nu\lambda}$, $L_{\nu'\lambda}$ the spectral luminances, photopic and scotopic, respectively.

Ranges

V, $\lambda = .36(.001).83\mu$
V', $\lambda = .38(.001).78\mu$

Example

The scotopic efficiency at 0.402μ is 0.01231.

Theory

The spectral luminous efficiency data are an attempt to standardize these values by using the mean for a large number of observers.

The photopic data given in Table 28a are based on those adopted by the C.I.E. in 1924[14] with interpolations as given in Reference 17. The scotopic data are derived from the logarithmic values adopted by the C.I.E. in 1951.[18] Values there were defined to only four significant figures and this limits the accuracy of the tabulated data.

TABLE 29. TRISTIMULUS VALUES, EQUAL-ENERGY SPECTRUM (1931 C.I.E. STANDARD OBSERVER) [Reference 4, pp. 18-23]

Tristimulus colorimetry is based on the selection of three "primary" colors which may be combined in various proportions to match the sensations produced by some other color. This color is then defined in terms of the contribution required from each of the primaries.

The C.I.E. tristimulus system is based on three fictitious* primaries **X**, **Y**, **Z**, roughly corresponding to red, green and blue. The amount of each required to match pure spectral colors for an average observer has been standardized at .005μ intervals[14] and is denoted by $\bar{x}_\lambda$, $\bar{y}_\lambda$, $\bar{z}_\lambda$, respectively. These are tabulated in Table 29.

The x-component of any luminous flux is obtained from

$$X = \int \bar{x}_\lambda \Phi_\lambda \, d\lambda \qquad (107)$$

where Φ_λ is the spectral luminous flux value; the Y- and Z-components are obtained analogously. The chromaticity coordinates are then obtained from

$$x = X/(X + Y + Z), \text{ etc.} \qquad (108)$$

For the spectral colors this reduces to

$$x = \bar{x}_\lambda/(\bar{x}_\lambda + \bar{y}_\lambda + \bar{z}_\lambda), \text{ etc.} \qquad (109)$$

TABLE 30. PHOTOMETRIC UNITS, CONVERSION TABLES

By way of introduction to Table 30, we present here a very cursory review of photometric units. This is meant only to serve as a reminder to those readers who are already somewhat familiar with the subject and not as a substitute for a more rigorous and extensive presentation (cf. Reference 4h).

We may take as our basic quantity *luminous flux*. This is, essentially, visible radiation-power; its unit is the lumen (lm), which is 1/680 of a watt radiated at the wavelength (.555μ) of maximum visibility.

The amount of flux radiated in a given direction is called *intensity*; its unit is the candela (cd)** which corresponds to one lumen per steradian.

The intensity radiated per unit area is called *luminance*; its standard unit is the cd/m^2 (or nit).

The luminous flux incident per unit area is called *illumination* (or *illuminance*); its standard unit is the lux (i.e., lumen/m^2). The luminous flux emitted per unit area is called *emittance* and is measured in the same units.

Many units are in use for the latter quantities, and conversion factors for them are presented in Tables 30.

*These primaries are fictitious in the sense that they include negative spectral luminances and therefore are not physically realizable. Indeed, the X- and Z-primaries are chosen so that their total luminance vanishes. This makes the Y-value (y_λ) a measure of spectral luminance and identical with the photopic spectral luminous efficiency (Table 28).

**Internationally, the candela is defined as the intensity of a projected area of 1/60 cm^2 of a blackbody radiator at the temperature of freezing platinum.[28]

REFERENCES

1. **Abramowitz, M. and Stegun, I. A.,** *Handbook of Mathematical Functions,* National Bureau of Standards, Appl. Math. Ser. No. 55, Washington, 1964.
2. **Fletcher, A., Miller, J. C. P., Rosenhead, L., and Comrie, L. J.,** *An Index of Mathematical Tables,* 2nd ed., Addison-Wesley, Reading, 1962.

3a. *Tables of the Error Function and Its Derivative,* National Bureau of Standards, Appl. Math. Ser. No. 41, Washington, 1954.

3b. *Tables of Normal Probability Functions,* National Bureau of Standards, Appl. Math. Ser. No. 23, Washington, 1953.

4. **Levi, L.,** *Applied Optics,* Vol. 1, Wiley, New York, 1968.
   a. Sect. 2.3.1.4;
   b. Table 14;
   c. Sect. 4.4;
   d. Sect. 2.3.2.2 and Eq. 9.109;
   e. Sect. 3.2 and App. 9.1.3;
   f. Sect. 1.3;
   g. Sect. 1.1.4.;
   h. Table 61;
   i. Sect. 9.5.4.4;
   j. Sect. 9.5.2.
5. *Tablitsy Integralov Frenelya,* Izdat. Akad. Nauk SSSR, Moscow, 1953.
6. **Pearcy, T.,** *Tables of the Fresnel Integral,* University Press, Cambridge, 1956.
7. The new values for the physical constants, *N.B.S. Tech. News Bull.,* 47, 175, 1963.*
8. **Pivovonsky, M. and Nagel, M. R.,** *Tables of Blackbody Radiation Functions,* Macmillan, New York, 1961.
   a. p. XIV.
9. Cf. **Gradshteyn, I. S. and Ryzhik, I. M.,** *Tables of Integrals, Series and Products,* Academic, New York, 1965. Eq. 3.411.2.
10. **Hahn, D., Metzdorf, J., Schley, U., and Verch, J.,** *Seven-Place Tables of the Planck Function for the Visible Spectrum,* Vieweg, Braunschweig, 1964 [Academic, New York].
11. **Czerny, M. and Walther, A.,** *Tables of the Fractional Function for the Planck Radiation Law,* Springer, Berlin, 1961.
12. **Lowan, A. N. and Blanch, G.,** Tables of Planck's radiation and photon functions, *J. Opt. Soc. Am.,* 30, 70, 1940.
13. **Apanasowich, P. A. and Aizenshtadt, S. A.,** *Tables for the Energy and Photon Distribution in Equilibrium Radiation* (transl. from Russian), Macmillan, New York, 1965.
14. Commission International de l'Eclairage, *Recueil des Travaux et Compte Rendu* Sixth Session (Geneva, 1924), Cambridge U. (1926) p. 67; Eighth Session (Cambridge, 1931), Cambridge U. (1932); pp. 25-26.
15. **Judd, D. B.,** Extension of the standard visibility function to intervals of one millimicron by third-difference osculatory interpolation, *Bu. Stds. J. Res.,* 6, 465, 1931.
16. **Mahr, K.,** Korrekturvorschläge zu den Normspektralwert-Funktionen, *Farbe,* 10, 323, 1961.
17. **Wyszecki, G. W. and Stiles, W. S.,** *Color Science,* Wiley, New York, 1967, Table 3.3.
18. International Commission on Illumination, *J. Opt. Soc. Am.,* 41, 734, 1951.
19. **Born, M. and Wolf, E.,** *Principles of Optics,* 4th ed., Pergamon, New York, 1970; Sect. 8.5.2.
20. **Levi, L. and Austing, R. H.,** Tables of the modulation transfer function of a defocused perfect lens, *Appl. Opt.,* 7, 967, 1968.
21. **Stokseth, P. A.,** Properties of a defocused optical system, *J. Opt. Soc. Am.,* 59, 1314, 1969.
22. **Linfoot, E. H.,** Transmission factors in optical design, *J. Opt. Soc. Am.,* 46, 740, 1956.
23. **Linfoot, E. H.,** Image quality and optical resolution, *Opt. Acta,* 4, 12, 1957.
24. **Schade, O. H.,** A new system of measuring and specifying image definition, in *Optical Image Evaluation,* U.S. Nat. Bur. Stds., Circular No. 256, Washington, 1954, 231.
25. **Levi, L.,** Detector response and perfect-lens-MTF in polychromatic light, *Appl. Opt.,* 8, 607, 1969.
26. Electronic Industries Association, Washington.
    a. JEDEC Publication No. 16, 1960: Optical Characteristics of Cathode Ray Tube Screens.
    b. JEDEC Publication No. 50, 1964: Relative Spectral Response Data for Photosensitive Devices.
27. **Eberhardt, E. H.,** Source-detector spectral matching factors, *Appl. Opt.,* 7, 2037, 1968.
28. National Bureau of Standards, Circ. 459, Announcement of Changes in Electric and Photometric Units, 1947.
29. **Papoulis, A.,** *Systems and Transforms with Application in Optics,* McGraw-Hill, New York, 1968.
30. *Tables of Integral Transforms,* Vol. 1, Erdelyi, A., Ed., McGraw-Hill, New York, 1954.

*Note added in proof: For more recent values, cf. *ibid* 55, 71, 1971.

A. Elementary Functions

Table 1—Linear Interpolation, Correction Terms

Table 2—Reciprocals

Table 3—Squares

Table 4—Square Roots

Table 5—Multiples of Half-Pi

Table 6—"Sinc" Function

TABLE 1

Linear Interpolation Correction Term

Δ_2	0.05 / 0.95	0.10 / 0.90	0.15 / 0.85	0.20 / 0.80	0.25 / 0.75	0.30 / 0.70	0.35 / 0.65	0.40 / 0.60	0.45 / 0.55	0.50 / 0.50
1	0.0	0.0	0.1	0.1	0.1	0.1	0.1	0.1	0.1	0.1
2	0.0	0.1	0.1	0.2	0.2	0.2	0.2	0.2	0.2	0.3
3	0.1	0.1	0.2	0.2	0.3	0.3	0.3	0.4	0.4	0.4
4	0.1	0.2	0.3	0.3	0.4	0.4	0.5	0.5	0.5	0.5
5	0.1	0.2	0.3	0.4	0.5	0.5	0.6	0.6	0.6	0.6
6	0.1	0.3	0.4	0.5	0.6	0.6	0.7	0.7	0.7	0.8
7	0.2	0.3	0.4	0.6	0.7	0.7	0.8	0.8	0.9	0.9
8	0.2	0.4	0.5	0.6	0.8	0.8	0.9	1.0	1.0	1.0
9	0.2	0.4	0.6	0.7	0.8	0.9	1.0	1.1	1.1	1.1
10	0.2	0.4	0.6	0.8	0.9	1.0	1.1	1.2	1.2	1.3
11	0.3	0.5	0.7	0.9	1.0	1.2	1.3	1.3	1.4	1.4
12	0.3	0.5	0.8	1.0	1.1	1.3	1.4	1.4	1.5	1.5
13	0.3	0.6	0.8	1.0	1.2	1.4	1.5	1.6	1.6	1.6
14	0.3	0.6	0.9	1.1	1.3	1.5	1.6	1.7	1.7	1.8
15	0.4	0.7	1.0	1.2	1.4	1.6	1.7	1.8	1.9	1.9
16	0.4	0.7	1.0	1.3	1.5	1.7	1.8	1.9	2.0	2.0
17	0.4	0.8	1.1	1.4	1.6	1.8	1.9	2.0	2.1	2.1
18	0.4	0.8	1.1	1.4	1.7	1.9	2.0	2.2	2.2	2.3
19	0.5	0.9	1.2	1.5	1.8	2.0	2.2	2.3	2.4	2.4
20	0.5	0.9	1.3	1.6	1.9	2.1	2.3	2.4	2.5	2.5
21	0.5	0.9	1.3	1.7	2.0	2.2	2.4	2.5	2.6	2.6
22	0.5	1.0	1.4	1.8	2.1	2.3	2.5	2.6	2.7	2.8
23	0.5	1.0	1.5	1.8	2.2	2.4	2.6	2.8	2.8	2.9
24	0.6	1.1	1.5	1.9	2.3	2.5	2.7	2.9	3.0	3.0
25	0.6	1.1	1.6	2.0	2.3	2.6	2.8	3.0	3.1	3.1
26	0.6	1.2	1.7	2.1	2.4	2.7	3.0	3.1	3.2	3.3
27	0.6	1.2	1.7	2.2	2.5	2.8	3.1	3.2	3.3	3.4
28	0.7	1.3	1.8	2.2	2.6	2.9	3.2	3.4	3.5	3.5
29	0.7	1.3	1.8	2.3	2.7	3.0	3.3	3.5	3.6	3.6
30	0.7	1.3	1.9	2.4	2.8	3.1	3.4	3.6	3.7	3.8
31	0.7	1.4	2.0	2.5	2.9	3.3	3.5	3.7	3.8	3.9
32	0.8	1.4	2.0	2.6	3.0	3.4	3.6	3.8	4.0	4.0
33	0.8	1.5	2.1	2.6	3.1	3.5	3.8	4.0	4.1	4.1
34	0.8	1.5	2.2	2.7	3.2	3.6	3.9	4.1	4.2	4.3
35	0.8	1.6	2.2	2.8	3.3	3.7	4.0	4.2	4.3	4.4
36	0.9	1.6	2.3	2.9	3.4	3.8	4.1	4.3	4.5	4.5
37	0.9	1.7	2.4	3.0	3.5	3.9	4.2	4.4	4.6	4.6
38	0.9	1.7	2.4	3.0	3.6	4.0	4.3	4.6	4.7	4.8
39	0.9	1.8	2.5	3.1	3.7	4.1	4.4	4.7	4.8	4.9
40	0.9	1.8	2.5	3.2	3.8	4.2	4.5	4.8	4.9	5.0
41	1.0	1.8	2.6	3.3	3.8	4.3	4.7	4.9	5.1	5.1
42	1.0	1.9	2.7	3.4	3.9	4.4	4.8	5.0	5.2	5.3
43	1.0	1.9	2.7	3.4	4.0	4.5	4.9	5.2	5.3	5.4
44	1.0	2.0	2.8	3.5	4.1	4.6	5.0	5.3	5.4	5.5
45	1.1	2.0	2.9	3.6	4.2	4.7	5.1	5.4	5.6	5.6
46	1.1	2.1	2.9	3.7	4.3	4.8	5.2	5.5	5.7	5.8
47	1.1	2.1	3.0	3.8	4.4	4.9	5.3	5.6	5.8	5.9
48	1.1	2.2	3.1	3.8	4.5	5.0	5.5	5.8	5.9	6.0
49	1.2	2.2	3.1	3.9	4.6	5.1	5.6	5.9	6.1	6.1
50	1.2	2.2	3.2	4.0	4.7	5.2	5.7	6.0	6.2	6.3

TABLE 1

Linear Interpolation Correction Term

Δ_2	0.05 / 0.95	0.10 / 0.90	0.15 / 0.85	0.20 / 0.80	0.25 / 0.75	0.30 / 0.70	0.35 / 0.65	0.40 / 0.60	0.45 / 0.55	0.50 / 0.50
51	1.2	2.3	3.3	4.1	4.8	5.4	5.8	6.1	6.3	6.4
52	1.2	2.3	3.3	4.2	4.9	5.5	5.9	6.2	6.4	6.5
53	1.3	2.4	3.4	4.2	5.0	5.6	6.0	6.4	6.6	6.6
54	1.3	2.4	3.4	4.3	5.1	5.7	6.1	6.5	6.7	6.8
55	1.3	2.5	3.5	4.4	5.2	5.8	6.3	6.6	6.8	6.9
56	1.3	2.5	3.6	4.5	5.3	5.9	6.4	6.7	6.9	7.0
57	1.4	2.6	3.6	4.6	5.3	6.0	6.5	6.8	7.1	7.1
58	1.4	2.6	3.7	4.6	5.4	6.1	6.6	7.0	7.2	7.3
59	1.4	2.7	3.8	4.7	5.5	6.2	6.7	7.1	7.3	7.4
60	1.4	2.7	3.8	4.8	5.6	6.3	6.8	7.2	7.4	7.5
61	1.4	2.7	3.9	4.9	5.7	6.4	6.9	7.3	7.5	7.6
62	1.5	2.8	4.0	5.0	5.8	6.5	7.1	7.4	7.7	7.8
63	1.5	2.8	4.0	5.0	5.9	6.6	7.2	7.6	7.8	7.9
64	1.5	2.9	4.1	5.1	6.0	6.7	7.3	7.7	7.9	8.0
65	1.5	2.9	4.1	5.2	6.1	6.8	7.4	7.8	8.0	8.1
66	1.6	3.0	4.2	5.3	6.2	6.9	7.5	7.9	8.2	8.3
67	1.6	3.0	4.3	5.4	6.3	7.0	7.6	8.0	8.3	8.4
68	1.6	3.1	4.3	5.4	6.4	7.1	7.7	8.2	8.4	8.5
69	1.6	3.1	4.4	5.5	6.5	7.2	7.8	8.3	8.5	8.6
70	1.7	3.1	4.5	5.6	6.6	7.3	8.0	8.4	8.7	8.8
71	1.7	3.2	4.5	5.7	6.7	7.5	8.1	8.5	8.8	8.9
72	1.7	3.2	4.6	5.8	6.8	7.6	8.2	8.6	8.9	9.0
73	1.7	3.3	4.7	5.8	6.8	7.7	8.3	8.8	9.0	9.1
74	1.8	3.3	4.7	5.9	6.9	7.8	8.4	8.9	9.2	9.3
75	1.8	3.4	4.6	6.0	7.0	7.9	8.5	9.0	9.3	9.4
76	1.8	3.4	4.8	6.1	7.1	8.0	8.6	9.1	9.4	9.5
77	1.8	3.5	4.9	6.2	7.2	8.1	8.8	9.2	9.5	9.6
78	1.9	3.5	5.0	6.2	7.3	8.2	8.9	9.4	9.7	9.8
79	1.9	3.6	5.0	6.3	7.4	8.3	9.0	9.5	9.8	9.9
80	1.9	3.6	5.1	6.4	7.5	8.4	9.1	9.6	9.9	10.0
81	1.9	3.6	5.2	6.5	7.6	8.5	9.2	9.7	10.0	10.1
82	1.9	3.7	5.2	6.6	7.7	8.6	9.3	9.8	10.1	10.3
83	2.0	3.7	5.3	6.6	7.8	8.7	9.4	10.0	10.3	10.4
84	2.0	3.8	5.4	6.7	7.9	8.8	9.6	10.1	10.4	10.5
85	2.0	3.8	5.4	6.8	8.0	8.9	9.7	10.2	10.5	10.6
86	2.0	3.9	5.5	6.9	8.1	9.0	9.8	10.3	10.6	10.8
87	2.1	3.9	5.5	7.0	8.2	9.1	9.9	10.4	10.8	10.9
88	2.1	4.0	5.6	7.0	8.3	9.2	10.0	10.6	10.9	11.0
89	2.1	4.0	5.7	7.1	8.3	9.3	10.1	10.7	11.0	11.1
90	2.1	4.0	5.7	7.2	8.4	9.4	10.2	10.8	11.1	11.3
91	2.2	4.1	5.8	7.3	8.5	9.6	10.4	10.9	11.3	11.4
92	2.2	4.1	5.9	7.4	8.6	9.7	10.5	11.0	11.4	11.5
93	2.2	4.2	5.9	7.4	8.7	9.8	10.6	11.2	11.5	11.6
94	2.2	4.2	6.0	7.5	8.8	9.9	10.7	11.3	11.6	11.8
95	2.3	4.3	6.1	7.6	8.9	10.0	10.8	11.4	11.8	11.9
96	2.3	4.3	6.1	7.7	9.0	10.1	10.9	11.5	11.9	12.0
97	2.3	4.4	6.2	7.8	9.1	10.2	11.0	11.6	12.0	12.1
98	2.3	4.4	6.2	7.8	9.2	10.3	11.1	11.8	12.1	12.3
99	2.4	4.5	6.3	7.9	9.3	10.4	11.3	11.9	12.3	12.4
100	2.4	4.5	6.4	8.0	9.4	10.5	11.4	12.0	12.4	12.5

TABLE 1

Linear Interpolation Correction Term

	0.05	0.10	0.15	0.20	0.25	0.30	0.35	0.40	0.45	0.50
Δ_2	0.95	0.90	0.85	0.80	0.75	0.70	0.65	0.60	0.55	0.50
101	2.4	4.5	6.4	8.1	9.5	10.6	11.5	12.1	12.5	12.6
102	2.4	4.6	6.5	8.2	9.6	10.7	11.6	12.2	12.6	12.8
103	2.4	4.6	6.6	8.2	9.7	10.8	11.7	12.4	12.7	12.9
104	2.5	4.7	6.6	8.3	9.8	10.9	11.8	12.5	12.9	13.0
105	2.5	4.7	6.7	8.4	9.8	11.0	11.9	12.6	13.0	13.1
106	2.5	4.8	6.8	8.5	9.9	11.1	12.1	12.7	13.1	13.3
107	2.5	4.8	6.8	8.6	10.0	11.2	12.2	12.8	13.2	13.4
108	2.6	4.9	6.9	8.6	10.1	11.3	12.3	13.0	13.4	13.5
109	2.6	4.9	6.9	8.7	10.2	11.4	12.4	13.1	13.5	13.6
110	2.6	4.9	7.0	8.8	10.3	11.5	12.5	13.2	13.6	13.8
111	2.6	5.0	7.1	8.9	10.4	11.7	12.6	13.3	13.7	13.9
112	2.7	5.0	7.1	9.0	10.5	11.8	12.7	13.4	13.9	14.0
113	2.7	5.1	7.2	9.0	10.6	11.9	12.9	13.6	14.0	14.1
114	2.7	5.1	7.3	9.1	10.7	12.0	13.0	13.7	14.1	14.3
115	2.7	5.2	7.3	9.2	10.8	12.1	13.1	13.8	14.2	14.4
116	2.8	5.2	7.4	9.3	10.9	12.2	13.2	13.9	14.4	14.5
117	2.8	5.3	7.5	9.4	11.0	12.3	13.3	14.0	14.5	14.6
118	2.8	5.3	7.5	9.4	11.1	12.4	13.4	14.2	14.6	14.8
119	2.8	5.4	7.6	9.5	11.2	12.5	13.5	14.3	14.7	14.9
120	2.8	5.4	7.6	9.6	11.3	12.6	13.6	14.4	14.8	15.0
121	2.9	5.4	7.7	9.7	11.3	12.7	13.8	14.5	15.0	15.1
122	2.9	5.5	7.8	9.8	11.4	12.8	13.9	14.6	15.1	15.3
123	2.9	5.5	7.8	9.8	11.5	12.9	14.0	14.8	15.2	15.4
124	2.9	5.6	7.9	9.9	11.6	13.0	14.1	14.9	15.3	15.5
125	3.0	5.6	8.0	10.0	11.7	13.1	14.2	15.0	15.5	15.6
126	3.0	5.7	8.0	10.1	11.8	13.2	14.3	15.1	15.6	15.8
127	3.0	5.7	8.1	10.2	11.9	13.3	14.4	15.2	15.7	15.9
128	3.0	5.8	8.2	10.2	12.0	13.4	14.6	15.4	15.8	16.0
129	3.1	5.8	8.2	10.3	12.1	13.5	14.7	15.5	16.0	16.1
130	3.1	5.8	8.3	10.4	12.2	13.6	14.8	15.6	16.1	16.3
131	3.1	5.9	8.4	10.5	12.3	13.8	14.9	15.7	16.2	16.4
132	3.1	5.9	8.4	10.6	12.4	13.9	15.0	15.8	16.3	16.5
133	3.2	6.0	8.5	10.6	12.5	14.0	15.1	16.0	16.5	16.6
134	3.2	6.0	8.5	10.7	12.6	14.1	15.2	16.1	16.6	16.8
135	3.2	6.1	8.6	10.8	12.7	14.2	15.4	16.2	16.7	16.9
136	3.2	6.1	8.7	10.9	12.8	14.3	15.5	16.3	16.8	17.0
137	3.3	6.2	8.7	11.0	12.8	14.4	15.6	16.4	17.0	17.1
138	3.3	6.2	8.8	11.0	12.9	14.5	15.7	16.6	17.1	17.3
139	3.3	6.3	8.9	11.1	13.0	14.6	15.8	16.7	17.2	17.4
140	3.3	6.3	8.9	11.2	13.1	14.7	15.9	16.8	17.3	17.5
141	3.3	6.3	9.0	11.3	13.2	14.8	16.0	16.9	17.4	17.6
142	3.4	6.4	9.1	11.4	13.3	14.9	16.2	17.0	17.6	17.8
143	3.4	6.4	9.1	11.4	13.4	15.0	16.3	17.2	17.7	17.9
144	3.4	6.5	9.2	11.5	13.5	15.1	16.4	17.3	17.8	18.0
145	3.4	6.5	9.2	11.6	13.6	15.2	16.5	17.4	17.9	18.1
146	3.5	6.6	9.3	11.7	13.7	15.3	16.6	17.5	18.1	18.3
147	3.5	6.6	9.4	11.8	13.8	15.4	16.7	17.6	18.2	18.4
148	3.5	6.7	9.4	11.8	13.9	15.5	16.8	17.8	18.3	18.5
149	3.5	6.7	9.5	11.9	14.0	15.6	16.9	17.9	18.4	18.6
150	3.6	6.7	9.6	12.0	14.1	15.7	17.1	18.0	18.6	18.8

TABLE 1

Linear Interpolation Correction Term

	0.05	0.10	0.15	0.20	0.25	0.30	0.35	0.40	0.45	0.50
Δ_2	0.95	0.90	0.85	0.80	0.75	0.70	0.65	0.60	0.55	0.50
151	3.6	6.8	9.6	12.1	14.2	15.9	17.2	18.1	18.7	18.9
152	3.6	6.8	9.7	12.2	14.3	16.0	17.3	18.2	18.8	19.0
153	3.6	6.9	9.8	12.2	14.3	16.1	17.4	18.4	18.9	19.1
154	3.7	6.9	9.8	12.3	14.4	16.2	17.5	18.5	19.1	19.3
155	3.7	7.0	9.9	12.4	14.5	16.3	17.6	18.6	19.2	19.4
156	3.7	7.0	9.9	12.5	14.6	16.4	17.7	18.7	19.3	19.5
157	3.7	7.1	10.0	12.6	14.7	16.5	17.9	18.8	19.4	19.6
158	3.8	7.1	10.1	12.6	14.8	16.6	18.0	19.0	19.6	19.8
159	3.8	7.2	10.1	12.7	14.9	16.7	18.1	19.1	19.7	19.9
160	3.8	7.2	10.2	12.8	15.0	16.8	18.2	19.2	19.8	20.0
161	3.8	7.2	10.3	12.9	15.1	16.9	18.3	19.3	19.9	20.1
162	3.8	7.3	10.3	13.0	15.2	17.0	18.4	19.4	20.0	20.3
163	3.9	7.3	10.4	13.0	15.3	17.1	18.5	19.6	20.2	20.4
164	3.9	7.4	10.5	13.1	15.4	17.2	18.7	19.7	20.3	20.5
165	3.9	7.4	10.5	13.2	15.5	17.3	18.8	19.8	20.4	20.6
166	3.9	7.5	10.6	13.3	15.6	17.4	18.9	19.9	20.5	20.8
167	4.0	7.5	10.6	13.4	15.7	17.5	19.0	20.0	20.7	20.9
168	4.0	7.6	10.7	13.4	15.8	17.6	19.1	20.2	20.8	21.0
169	4.0	7.6	10.8	13.5	15.8	17.7	19.2	20.3	20.9	21.1
170	4.0	7.6	10.8	13.6	15.9	17.8	19.3	20.4	21.0	21.3
171	4.1	7.7	10.9	13.7	16.0	18.0	19.5	20.5	21.2	21.4
172	4.1	7.7	11.0	13.8	16.1	18.1	19.6	20.6	21.3	21.5
173	4.1	7.8	11.0	13.8	16.2	18.2	19.7	20.8	21.4	21.6
174	4.1	7.8	11.1	13.9	16.3	18.3	19.8	20.9	21.5	21.8
175	4.2	7.9	11.2	14.0	16.4	18.4	19.9	21.0	21.7	21.9
176	4.2	7.9	11.2	14.1	16.5	18.5	20.0	21.1	21.8	22.0
177	4.2	8.0	11.3	14.2	16.6	18.6	20.1	21.2	21.9	22.1
178	4.2	8.0	11.3	14.2	16.7	18.7	20.2	21.4	22.0	22.3
179	4.3	8.1	11.4	14.3	16.8	18.8	20.4	21.5	22.2	22.4
180	4.3	8.1	11.5	14.4	16.9	18.9	20.5	21.6	22.3	22.5
181	4.3	8.1	11.5	14.5	17.0	19.0	20.6	21.7	22.4	22.6
182	4.3	8.2	11.6	14.6	17.1	19.1	20.7	21.8	22.5	22.8
183	4.3	8.2	11.7	14.6	17.2	19.2	20.8	22.0	22.6	22.9
184	4.4	8.3	11.7	14.7	17.3	19.3	20.9	22.1	22.8	23.0
185	4.4	8.3	11.8	14.8	17.3	19.4	21.0	22.2	22.9	23.1
186	4.4	8.4	11.9	14.9	17.4	19.5	21.2	22.3	23.0	23.3
187	4.4	8.4	11.9	15.0	17.5	19.6	21.3	22.4	23.1	23.4
188	4.5	8.5	12.0	15.0	17.6	19.7	21.4	22.6	23.3	23.5
189	4.5	8.5	12.0	15.1	17.7	19.8	21.5	22.7	23.4	23.6
190	4.5	8.5	12.1	15.2	17.8	19.9	21.6	22.8	23.5	23.8
191	4.5	8.6	12.2	15.3	17.9	20.1	21.7	22.9	23.6	23.9
192	4.6	8.6	12.2	15.4	18.0	20.2	21.8	23.0	23.8	24.0
193	4.6	8.7	12.3	15.4	18.1	20.3	22.0	23.2	23.9	24.1
194	4.6	8.7	12.4	15.5	18.2	20.4	22.1	23.3	24.0	24.3
195	4.6	8.8	12.4	15.6	18.3	20.5	22.2	23.4	24.1	24.4
196	4.7	8.8	12.5	15.7	18.4	20.6	22.3	23.5	24.3	24.5
197	4.7	8.9	12.6	15.8	18.5	20.7	22.4	23.6	24.4	24.6
198	4.7	8.9	12.6	15.8	18.6	20.8	22.5	23.8	24.5	24.8
199	4.7	9.0	12.7	15.9	18.7	20.9	22.6	23.9	24.6	24.9
200	4.7	9.0	12.7	16.0	18.8	21.0	22.7	24.0	24.7	25.0

TABLE 2

Reciprocals

X		.0000	.0001	.0002	.0003	.0004	.0005	.0006	.0007	.0008	.0009	Δ_2
0.100		1.0000000	9900100	9800399	9700897	9601594	9502488	9403578	9304866	9206349	9108028	195
0.101	9.	9009901	8911968	8814229	8716683	8619329	8522167	8425197	8328417	8231827	8135427	190
0.102	9.	8039216	7943193	7847358	7751711	7656250	7560976	7465887	7370983	7276265	7181730	184
0.103	9.	7087379	6993210	6899225	6805421	6711799	6618357	6525097	6432015	6339114	6246391	179
0.104	9.	6153846	6061479	5969290	5877277	5785441	5693780	5602294	5510984	5419847	5328885	174
0.105	9.	5238095	5147479	5057034	4966762	4876660	4786730	4696970	4607379	4517958	4428706	169
0.106	9.	4339623	4250707	4161959	4073377	3984962	3896714	3808630	3720712	3632959	3545369	164
0.107	9.	3457944	3370682	3283582	3196645	3109870	3023256	2936803	2850511	2764378	2678406	160
0.108	9.	2592593	2506938	2421442	2336103	2250923	2165899	2081031	1996320	1911765	1827365	155
0.109	9.	1743119	1659028	1575092	1491308	1407678	1324201	1240876	1157703	1074681	0991811	151
0.110	9.	0909091	0826521	0744102	0661831	0579710	0497738	0415913	0334237	0252708	0171326	147
0.111	9.	0090090	0009001	9928057	9847260	9766607	9686099	9605735	9525515	9445438	9365505	143
0.112	8.	9285714	9206066	9126560	9047195	8967972	8888889	8809947	8731145	8652482	8573959	139
0.113	8.	8495575	8417330	8339223	8261253	8183421	8105727	8028169	7950748	7873462	7796312	136
0.114	8.	7719298	7642419	7565674	7489064	7412587	7336245	7260035	7183958	7108014	7032202	132
0.115	8.	6956522	6880973	6805556	6730269	6655113	6580087	6505190	6430423	6355786	6281277	129
0.116	8.	6206897	6132644	6058520	5984523	5910653	5836910	5763293	5689803	5616438	5543199	126
0.117	8.	5470085	5397096	5324232	5251492	5178876	5106383	5034014	4961767	4889643	4817642	122
0.118	8.	4745763	4674005	4602369	4530854	4459459	4388186	4317032	4245998	4175084	4104289	119
0.119	8.	4033613	3963056	3892617	3822297	3752094	3682008	3612040	3542189	3472454	3402836	116
0.120	8.	3333333	3263947	3194676	3125520	3056478	2987552	2918740	2850041	2781457	2712986	113
0.121	8.	2644628	2576383	2508251	2440231	2372323	2304527	2236842	2169269	2101806	2034454	111
0.122	8.	1967213	1900082	1833061	1766149	1699346	1632653	1566069	1499592	1433225	1366965	108
0.123	8.	1300813	1234768	1168831	1103001	1037277	0971660	0906149	0840744	0775444	0710250	105
0.124	8.	0645161	0580177	0515298	0450523	0385852	0321285	0256822	0192462	0128205	0064051	103
0.125	8.	0000000	9936051	9872204	9808460	9744817	9681275	9617834	9554495	9491256	9428117	100
0.126	7.	9365079	9302141	9239303	9176564	9113924	9051383	8988942	8926598	8864353	8802206	98
0.127	7.	8740157	8678206	8616352	8554595	8492936	8431372	8369906	8308536	8247261	8186083	96
0.128	7.	8125000	8064012	8003120	7942323	7881620	7821012	7760498	7700078	7639752	7579519	94
0.129	7.	7519380	7459334	7399381	7339520	7279753	7220077	7160494	7101002	7041602	6982294	91
0.130	7.	6923077	6863951	6804915	6745971	6687117	6628352	6569678	6511094	6452599	6394194	89
0.131	7.	6335878	6277651	6219512	6161462	6103501	6045627	5987842	5930144	5872534	5815011	87
0.132	7.	5757576	5700227	5642965	5585790	5528701	5471698	5414781	5357950	5301205	5244545	85
0.133	7.	5187970	5131480	5075075	5018755	4962519	4906367	4850299	4794316	4738416	4682599	83
0.134	7.	4626866	4571216	4515648	4460164	4404762	4349442	4294205	4239050	4183976	4128984	82
0.135	7.	4074074	4019245	3964497	3909830	3855244	3800738	3746313	3691968	3637703	3583517	80
0.136	7.	3529412	3475386	3421439	3367572	3313783	3260073	3206442	3152890	3099415	3046019	78
0.137	7.	2992701	2939460	2886297	2833212	2780204	2727273	2674419	2621641	2568940	2516316	76
0.138	7.	2463768	2411296	2358900	2306580	2254335	2202166	2150072	2098053	2046109	1994240	75
0.139	7.	1942446	1890726	1839080	1787509	1736011	1684588	1633238	1581961	1530758	1479628	73
0.140	7.	1428571	1377587	1326676	1275837	1225071	1174377	1123755	1073205	1022727	0972321	72
0.141	7.	0921986	0871722	0821530	0771408	0721358	0671378	0621469	0571630	0521862	0472163	70
0.142	7.	0422535	0372977	0323488	0274069	0224719	0175439	0126227	0077085	0028011	9979006	69
0.143	6.	9930070	9881202	9832402	9783671	9735007	9686411	9637883	9589422	9541029	9492703	67
0.144	6.	9444444	9396253	9348128	9300069	9252077	9204152	9156293	9108500	9060773	9013112	66
0.145	6.	8965517	8917988	8870523	8823125	8775791	8728522	8681319	8634180	8587106	8540096	65
0.146	6.	8493151	8446270	8399453	8352700	8306011	8259386	8212824	8166326	8119891	8073519	63
0.147	6.	8027211	7980965	7934783	7888663	7842605	7796610	7750677	7704807	7658999	7613252	62
0.148	6.	7567568	7521945	7476383	7430883	7385445	7340067	7294751	7249496	7204301	7159167	61
0.149	6.	7114094	7069081	7024129	6979236	6934404	6889632	6844920	6800267	6755674	6711141	59

TABLE 2

Reciprocals

X		.0000	.0001	.0002	.0003	.0004	.0005	.0006	.0007	.0008	.0009	Δ_2
0.150	6.	6666667	6622252	6577896	6533599	6489362	6445183	6401062	6357001	6312997	6269052	58
0.151	6.	6225166	6181337	6137566	6093853	6050198	6006601	5963061	5919578	5876153	5832785	57
0.152	6.	5789474	5746220	5703022	5659882	5616798	5573770	5530799	5487885	5445026	5402224	56
0.153	6.	5359477	5316786	5274151	5231572	5189048	5146580	5104167	5061809	5019506	4977258	55
0.154	6.	4935065	4892927	4850843	4808814	4766839	4724919	4683053	4641241	4599483	4557779	54
0.155	6.	4516129	4474533	4432990	4391500	4350064	4308682	4267352	4226076	4184852	4143682	53
0.156	6.	4102564	4061499	4020487	3979527	3938619	3897764	3856960	3816209	3775510	3734863	52
0.157	6.	3694267	3653724	3613232	3572791	3532402	3492063	3451777	3411541	3371356	3331222	51
0.158	6.	3291139	3251107	3211125	3171194	3131313	3091483	3051702	3011972	2972292	2932662	50
0.159	6.	2893082	2853551	2814070	2774639	2735257	2695925	2656642	2617408	2578223	2539087	49
0.160	6.	2500000	2460962	2421973	2383032	2344140	2305296	2266501	2227754	2189055	2150404	48
0.161	6.	2111801	2073246	2034739	1996280	1957869	1919505	1881188	1842919	1804697	1766523	47
0.162	6.	1728395	1690315	1652281	1614294	1576355	1538462	1500615	1462815	1425061	1387354	46
0.163	6.	1349693	1312078	1274510	1236987	1199510	1162080	1124694	1087355	1050061	1012813	46
0.164	6.	0975610	0938452	0901340	0864273	0827251	0790274	0753341	0716454	0679612	0642814	45
0.165	6.	0606061	0569352	0532688	0496068	0459492	0422961	0386473	0350030	0313631	0277275	44
0.166	6.	0240964	0204696	0168472	0132291	0096154	0060060	0024010	9988002	9952038	9916117	43
0.167	5.	9880240	9844405	9808612	9772863	9737156	9701492	9665871	9630292	9594756	9559261	42
0.168	5.	9523809	9488400	9453032	9417706	9382423	9347181	9311981	9276823	9241706	9206631	42
0.169	5.	9171598	9136605	9101655	9066745	9031877	8997050	8962264	8927519	8892815	8858152	41
0.170	5.	8823529	8788948	8754407	8719906	8685446	8651026	8616647	8582308	8548009	8513751	40
0.171	5.	8479532	8445354	8411215	8377116	8343057	8309038	8275058	8241118	8207218	8173357	39
0.172	5.	8139535	8105752	8072009	8038305	8004640	7971014	7937428	7903880	7870370	7836900	39
0.173	5.	7803468	7770075	7736721	7703404	7670127	7636888	7603687	7570524	7537399	7504313	38
0.174	5.	7471264	7438254	7405281	7372347	7339450	7306590	7273769	7240985	7208238	7175529	37
0.175	5.	7142857	7110223	7077626	7045066	7012543	6980057	6947608	6915196	6882821	6850483	37
0.176	5.	6818182	6785917	6753689	6721497	6689342	6657224	6625142	6593096	6561086	6529112	36
0.177	5.	6497175	6465274	6433409	6401579	6369786	6338028	6306306	6274620	6242970	6211355	36
0.178	5.	6179775	6148231	6116723	6085250	6053812	6022409	5991041	5959709	5928412	5897149	35
0.179	5.	5865922	5834729	5803571	5772448	5741360	5710306	5679287	5648303	5617353	5586437	34
0.180	5.	5555556	5524708	5493896	5463117	5432372	5401662	5370986	5340343	5309734	5279160	34
0.181	5.	5248619	5218112	5187638	5157198	5126792	5096419	5066079	5035773	5005501	4975261	33
0.182	5.	4945055	4914882	4884742	4854635	4824561	4794521	4764513	4734537	4704595	4674686	33
0.183	5.	4644809	4614965	4585153	4555374	4525627	4495913	4466231	4436581	4406964	4377379	32
0.184	5.	4347826	4318305	4288816	4259360	4229935	4200542	4171181	4141852	4112554	4083288	32
0.185	5.	4054054	4024851	3995680	3966541	3937433	3908356	3879310	3850296	3821313	3792361	31
0.186	5.	3763441	3734551	3705693	3676865	3648069	3619303	3590568	3561864	3533191	3504548	31
0.187	5.	3475936	3447354	3418803	3390283	3361793	3333333	3304904	3276505	3248136	3219798	30
0.188	5.	3191489	3163211	3134963	3106745	3078556	3050398	3022269	2994171	2966102	2938062	30
0.189	5.	2910053	2882073	2854123	2826202	2798310	2770449	2742616	2714813	2687039	2659294	29
0.190	5.	2631579	2603893	2576236	2548607	2521008	2493438	2465897	2438385	2410901	2383447	29
0.191	5.	2356021	2328624	2301255	2273915	2246604	2219321	2192067	2164841	2137643	2110474	28
0.192	5.	2083333	2056221	2029136	2002080	1975052	1948052	1921080	1894136	1867220	1840332	28
0.193	5.	1813471	1786639	1759834	1733057	1706308	1679587	1652893	1626226	1599587	1572976	27
0.194	5.	1546392	1519835	1493306	1466804	1440329	1413882	1387461	1361068	1334702	1308363	27
0.195	5.	1282051	1255766	1229508	1203277	1177073	1150895	1124744	1098620	1072523	1046452	27
0.196	5.	1020408	0994391	0968400	0942435	0916497	0890585	0864700	0838841	0813008	0787202	26
0.197	5.	0761421	0735667	0709939	0684237	0658561	0632911	0607287	0581689	0556117	0530571	26
0.198	5.	0505051	0479556	0454087	0428643	0403226	0377834	0352467	0327126	0301811	0276521	25
0.199	5.	0251256	0226017	0200803	0175615	0150451	0125313	0100200	0075113	0050050	0025013	25

TABLE 2

Reciprocals

X		.0000	.0001	.0002	.0003	.0004	.0005	.0006	.0007	.0008	.0009	Δ_2
0.200	5.	0000000	9975012	9950050	9925112	9900200	9875312	9850449	9825610	9800797	9776008	25
0.201	4.	9751244	9726504	9701789	9677099	9652433	9627791	9603175	9578582	9554014	9529470	24
0.202	4.	9504950	9480455	9455984	9431537	9407115	9382716	9358342	9333991	9309665	9285362	24
0.203	4.	9261084	9236829	9212598	9188391	9164208	9140049	9115914	9091802	9067713	9043649	24
0.204	4.	9019608	8995590	8971596	8947626	8923679	8899755	8875855	8851978	8828125	8804295	23
0.205	4.	8780488	8756704	8732943	8709206	8685492	8661800	8638132	8614487	8590865	8567266	23
0.206	4.	8543689	8520136	8496605	8473097	8449612	8426150	8402710	8379294	8355899	8332528	23
0.207	4.	8309179	8285852	8262548	8239267	8216008	8192771	8169557	8146365	8123195	8100048	22
0.208	4.	8076923	8053820	8030740	8007681	7984645	7961631	7938638	7915668	7892720	7869794	22
0.209	4.	7846890	7824008	7801147	7778309	7755492	7732697	7709924	7687172	7664442	7641734	22
0.210	4.	7619048	7596383	7573739	7551117	7528517	7505938	7483381	7460845	7438330	7415837	21
0.211	4.	7393365	7370914	7348485	7326077	7303690	7281324	7258979	7236656	7214353	7192072	21
0.212	4.	7169811	7147572	7125353	7103156	7080979	7058823	7036689	7014574	6992481	6970409	21
0.213	4.	6948357	6926326	6904315	6882325	6860356	6838407	6816479	6794572	6772685	6750818	20
0.214	4.	6728972	6707146	6685341	6663556	6641791	6620047	6598322	6576618	6554935	6533271	20
0.215	4.	6511628	6490005	6468401	6446818	6425255	6403712	6382189	6360686	6339203	6317740	20
0.216	4.	6296296	6274873	6253465	6232085	6210721	6189376	6168052	6146747	6125461	6104195	20
0.217	4.	6082949	6061723	6040516	6019328	5998160	5977011	5955882	5934773	5913682	5892611	19
0.218	4.	5871560	5850527	5829514	5808520	5787546	5766590	5745654	5724737	5703839	5682960	19
0.219	4.	5662100	5641260	5620438	5599635	5578851	5558087	5537341	5516613	5495905	5475216	19
0.220	4.	5454545	5433894	5413261	5392646	5372051	5351474	5330916	5310376	5289855	5269353	19
0.221	4.	5248869	5228403	5207957	5187528	5167118	5146727	5126354	5105999	5085663	5065345	18
0.222	4.	5045045	5024764	5004500	4984256	4964029	4943820	4923630	4903458	4883303	4863167	18
0.223	4.	4843049	4822949	4802867	4782803	4762757	4742729	4722719	4702727	4682752	4662796	18
0.224	4.	4642857	4622936	4603033	4583148	4563280	4543430	4523597	4503783	4483986	4464206	18
0.225	4.	4444444	4424700	4404973	4385264	4365572	4345898	4326241	4306602	4286980	4267375	17
0.226	4.	4247788	4228218	4208665	4189129	4169611	4150110	4130627	4111160	4091711	4072279	17
0.227	4.	4052863	4033465	4014084	3994721	3975374	3956044	3936731	3917435	3898156	3878894	17
0.228	4.	3859649	3840421	3821209	3802015	3782837	3763676	3744532	3725404	3706294	3687200	17
0.229	4.	3668122	3649062	3630017	3610990	3591979	3572985	3554007	3535046	3516101	3497173	16
0.230	4.	3478261	3459365	3440487	3421624	3402778	3383948	3365134	3346337	3327556	3308792	16
0.231	4.	3290043	3271311	3252595	3233895	3215212	3196544	3177893	3159258	3140638	3122035	16
0.232	4.	3103448	3084877	3066322	3047783	3029260	3010753	2992261	2973786	2955326	2936883	16
0.233	4.	2918455	2900043	2881647	2863266	2844901	2826552	2808219	2789902	2771600	2753313	16
0.234	4.	2735043	2716788	2698548	2680324	2662116	2643923	2625746	2607584	2589438	2571307	15
0.235	4.	2553191	2535091	2517007	2498938	2480884	2462845	2444822	2426814	2408821	2390844	15
0.236	4.	2372881	2354934	2337003	2319086	2301184	2283298	2265427	2247571	2229730	2211904	15
0.237	4.	2194093	2176297	2158516	2140750	2122999	2105263	2087542	2069836	2052145	2034468	15
0.238	4.	2016807	1999160	1981528	1963911	1946309	1928721	1911148	1893590	1876047	1858518	15
0.239	4.	1841004	1823505	1806020	1788550	1771094	1753653	1736227	1718815	1701418	1684035	15
0.240	4.	1666667	1649313	1631973	1614648	1597338	1580042	1562760	1545492	1528239	1511000	14
0.241	4.	1493776	1476566	1459370	1442188	1425021	1407867	1390728	1373604	1356493	1339396	14
0.242	4.	1322314	1305246	1288192	1271151	1254125	1237113	1220115	1203131	1186161	1169205	14
0.243	4.	1152263	1135335	1118421	1101521	1084634	1067762	1050903	1034058	1017227	1000410	14
0.244	4.	0983607	0966817	0950041	0933279	0916530	0899795	0883074	0866367	0849673	0832993	14
0.245	4.	0816327	0799674	0783034	0766408	0749796	0733198	0716612	0700041	0683483	0666938	13
0.246	4.	0650406	0633889	0617384	0600893	0584416	0567951	0551500	0535063	0518639	0502228	13
0.247	4.	0485830	0469446	0453074	0436717	0420372	0404040	0387722	0371417	0355125	0338846	13
0.248	4.	0322581	0306328	0290089	0273862	0257649	0241449	0225261	0209087	0192926	0176778	13
0.249	4.	0160643	0144520	0128411	0112314	0096231	0080160	0064103	0048058	0032026	0016006	13

TABLE 2

Reciprocals

X		.0000	.0001	.0002	.0003	.0004	.0005	.0006	.0007	.0008	.0009	Δ_2
0.250	4.	0000000	9984006	9968026	9952058	9936102	9920160	9904230	9888313	9872408	9856516	13
0.251	3.	9840637	9824771	9808917	9793076	9777247	9761431	9745628	9729837	9714059	9698293	13
0.252	3.	9682540	9666799	9651071	9635355	9619651	9603960	9588282	9572616	9556962	9541321	12
0.253	3.	9525692	9510075	9494471	9478879	9463299	9447732	9432177	9416634	9401103	9385585	12
0.254	3.	9370079	9354585	9339103	9323633	9308176	9292731	9277298	9261877	9246468	9231071	12
0.255	3.	9215686	9200314	9184953	9169604	9154268	9138943	9123631	9108330	9093041	9077765	12
0.256	3.	9062500	9047247	9032006	9016777	9001560	8986355	8971161	8955980	8940810	8925652	12
0.257	3.	8910506	8895371	8880249	8865138	8850039	8834951	8819876	8804812	8789760	8774719	12
0.258	3.	8759690	8744673	8729667	8714673	8699690	8684720	8669760	8654812	8639876	8624952	12
0.259	3.	8610039	8595137	8580247	8565368	8550501	8535645	8520801	8505968	8491147	8476337	11
0.260	3.	8461538	8446751	8431975	8417211	8402458	8387716	8372985	8358266	8343558	8328862	11
0.261	3.	8314176	8299502	8284839	8270187	8255547	8240918	8226300	8211693	8197097	8182512	11
0.262	3.	8167939	8153377	8138825	8124285	8109756	8095238	8080731	8066235	8051750	8037276	11
0.263	3.	8022814	8008362	7993921	7979491	7965072	7950664	7936267	7921881	7907506	7893141	11
0.264	3.	7878788	7864445	7850114	7835793	7821483	7807183	7792895	7778617	7764350	7750094	11
0.265	3.	7735849	7721614	7707391	7693177	7678975	7664783	7650602	7636432	7622272	7608123	11
0.266	3.	7593985	7579857	7565740	7551633	7537538	7523452	7509377	7495313	7481259	7467216	11
0.267	3.	7453183	7439161	7425150	7411148	7397158	7383178	7369208	7355248	7341299	7327361	10
0.268	3.	7313433	7299515	7285608	7271711	7257824	7243948	7230082	7216226	7202381	7188546	10
0.269	3.	7174721	7160907	7147103	7133309	7119525	7105751	7091988	7078235	7064492	7050759	10
0.270	3.	7037037	7023325	7009622	6995930	6982248	6968577	6954915	6941263	6927622	6913990	10
0.271	3.	6900369	6886758	6873156	6859565	6845984	6832412	6818851	6805300	6791759	6778227	10
0.272	3.	6764706	6751194	6737693	6724201	6710719	6697248	6683786	6670334	6656891	6643459	10
0.273	3.	6630037	6616624	6603221	6589828	6576445	6563071	6549708	6536354	6523009	6509675	10
0.274	3.	6496350	6483035	6469730	6456435	6443149	6429872	6416606	6403349	6390102	6376864	10
0.275	3.	6363636	6350418	6337209	6324010	6310821	6297641	6284470	6271309	6258158	6245016	10
0.276	3.	6231884	6218761	6205648	6192544	6179450	6166365	6153290	6140224	6127168	6114121	9
0.277	3.	6101083	6088055	6075036	6062027	6049027	6036036	6023055	6010083	5997120	5984167	9
0.278	3.	5971223	5958288	5945363	5932447	5919540	5906643	5893754	5880875	5868006	5855145	9
0.279	3.	5842294	5829452	5816619	5803795	5790981	5778175	5765379	5752592	5739814	5727045	9
0.280	3.	5714286	5701535	5688794	5676061	5663338	5650624	5637919	5625223	5612536	5599858	9
0.281	3.	5587189	5574529	5561878	5549236	5536603	5523979	5511364	5498758	5486160	5473572	9
0.282	3.	5460993	5448423	5435861	5423309	5410765	5398230	5385704	5373187	5360679	5348180	9
0.283	3.	5335689	5323207	5310734	5298270	5285815	5273369	5260931	5248502	5236082	5223670	9
0.284	3.	5211268	5198874	5186488	5174112	5161744	5149385	5137034	5124693	5112360	5100035	9
0.285	3.	5087719	5075412	5063114	5050824	5038542	5026270	5014006	5001750	4989503	4977265	9
0.286	3.	4965035	4952814	4940601	4928397	4916201	4904014	4891835	4879665	4867503	4855350	8
0.287	3.	4843206	4831069	4818942	4806822	4794711	4782609	4770515	4758429	4746352	4734283	8
0.288	3.	4722222	4710170	4698126	4686091	4674064	4662045	4650035	4638033	4626039	4614053	8
0.289	3.	4602076	4590107	4578147	4566194	4554250	4542314	4530387	4518467	4506556	4494653	8
0.290	3.	4482759	4470872	4458994	4447124	4435262	4423408	4411562	4399725	4387895	4376074	8
0.291	3.	4364261	4352456	4340659	4328871	4317090	4305317	4293553	4281796	4270048	4258308	8
0.292	3.	4246575	4234851	4223135	4211427	4199726	4188034	4176350	4164674	4153005	4141345	8
0.293	3.	4129693	4118048	4106412	4094783	4083163	4071550	4059945	4048349	4036760	4025179	8
0.294	3.	4013605	4002040	3990483	3978933	3967391	3955857	3944331	3932813	3921303	3909800	8
0.295	3.	3898305	3886818	3875339	3863867	3852403	3840948	3829499	3818059	3806626	3795201	8
0.296	3.	3783784	3772374	3760972	3749578	3738192	3726813	3715442	3704078	3692722	3681374	8
0.297	3.	3670034	3658701	3647375	3636058	3624748	3613445	3602151	3590863	3579584	3568311	8
0.298	3.	3557047	3545790	3534541	3523299	3512064	3500837	3489618	3478406	3467202	3456005	7
0.299	3.	3444816	3433634	3422460	3411293	3400134	3388982	3377837	3366700	3355570	3344448	7

TABLE 2

Reciprocals

X		.0000	.0001	.0002	.0003	.0004	.0005	.0006	.0007	.0008	.0009	Δ_2
0.300	3.	3333333	3322226	3311126	3300033	3288948	3277870	3266800	3255737	3244681	3233632	7
0.301	3.	3222591	3211558	3200531	3189512	3178500	3167496	3156499	3145509	3134526	3123551	7
0.302	3.	3112583	3101622	3090668	3079722	3068783	3057851	3046927	3036009	3025099	3014196	7
0.303	3.	3003300	2992412	2981530	2970656	2959789	2948929	2938076	2927231	2916392	2905561	7
0.304	3.	2894737	2883920	2873110	2862307	2851511	2840722	2829941	2819166	2808399	2797639	7
0.305	3.	2786885	2776139	2765400	2754668	2743942	2733224	2722513	2711809	2701112	2690422	7
0.306	3.	2679739	2669062	2658393	2647731	2637076	2626427	2615786	2605152	2594524	2583904	7
0.307	3.	2573290	2562683	2552083	2541490	2530904	2520325	2509753	2499188	2488629	2478077	7
0.308	3.	2467532	2456994	2446463	2435939	2425422	2414911	2404407	2393910	2383420	2372936	7
0.309	3.	2362460	2351990	2341527	2331070	2320621	2310178	2299742	2289312	2278890	2268474	7
0.310	3.	2258065	2247662	2237266	2226877	2216495	2206119	2195750	2185388	2175032	2164683	7
0.311	3.	2154341	2144005	2133676	2123354	2113038	2102729	2092426	2082130	2071841	2061558	7
0.312	3.	2051282	2041012	2030750	2020493	2010243	2000000	1989763	1979533	1969309	1959092	7
0.313	3.	1948882	1938678	1928480	1918289	1908105	1897927	1887755	1877590	1867431	1857279	6
0.314	3.	1847134	1836995	1826862	1816736	1806616	1796502	1786395	1776295	1766201	1756113	6
0.315	3.	1746032	1735957	1725888	1715826	1705770	1695721	1685678	1675641	1665611	1655587	6
0.316	3.	1645570	1635558	1625553	1615555	1605563	1595577	1585597	1575624	1565657	1555696	6
0.317	3.	1545741	1535793	1525851	1515916	1505986	1496063	1486146	1476235	1466331	1456433	6
0.318	3.	1446541	1436655	1426776	1416902	1407035	1397174	1387320	1377471	1367629	1357792	6
0.319	3.	1347962	1338139	1328321	1318509	1308704	1298905	1289111	1279324	1269543	1259769	6
0.320	3.	1250000	1240237	1230481	1220731	1210986	1201248	1191516	1181790	1172070	1162356	6
0.321	3.	1152648	1142946	1133250	1123561	1113877	1104199	1094527	1084862	1075202	1065548	6
0.322	3.	1055901	1046259	1036623	1026993	1017370	1007752	0998140	0988534	0978934	0969340	6
0.323	3.	0959752	0950170	0940594	0931024	0921459	0911901	0902349	0892802	0883261	0873726	6
0.324	3.	0864198	0854674	0845157	0835646	0826141	0816641	0807147	0797659	0788177	0778701	6
0.325	3.	0769231	0759766	0750307	0740855	0731407	0721966	0712531	0703101	0693677	0684259	6
0.326	3.	0674847	0665440	0656039	0646644	0637255	0627871	0618494	0609122	0599755	0590395	6
0.327	3.	0581040	0571691	0562347	0553009	0543677	0534351	0525031	0515716	0506406	0497103	6
0.328	3.	0487805	0478513	0469226	0459945	0450670	0441400	0432136	0422878	0413625	0404378	6
0.329	3.	0395137	0385901	0376671	0367446	0358227	0349014	0339806	0330604	0321407	0312216	6
0.330	3.	0303030	0293850	0284676	0275507	0266344	0257186	0248034	0238887	0229746	0220610	6
0.331	3.	0211480	0202356	0193237	0184123	0175015	0165913	0156815	0147724	0138638	0129557	5
0.332	3.	0120482	0111412	0102348	0093289	0084236	0075188	0066146	0057109	0048077	0039051	5
0.333	3.	0030030	0021015	0012005	0003000	9994001	9985007	9976019	9967036	9958059	9949087	5
0.334	2.	9940120	9931158	9922202	9913252	9904306	9895366	9886432	9877502	9868578	9859660	5
0.335	2.	9850746	9841838	9832936	9824038	9815146	9806259	9797378	9788502	9779631	9770765	5
0.336	2.	9761905	9753050	9744200	9735355	9726516	9717682	9708853	9700030	9691211	9682398	5
0.337	2.	9673590	9664788	9655990	9647198	9638411	9629630	9620853	9612082	9603316	9594555	5
0.338	2.	9585799	9577048	9568303	9559563	9550827	9542097	9533373	9524653	9515939	9507229	5
0.339	2.	9498525	9489826	9481132	9472443	9463760	9455081	9446407	9437739	9429076	9420418	5
0.340	2.	9411765	9403117	9394474	9385836	9377203	9368576	9359953	9351335	9342723	9334116	5
0.341	2.	9325513	9316916	9308324	9299736	9291154	9282577	9274005	9265437	9256875	9248318	5
0.342	2.	9239766	9231219	9222677	9214140	9205607	9197080	9188558	9180041	9171529	9163021	5
0.343	2.	9154519	9146022	9137529	9129042	9120559	9112081	9103609	9095141	9086678	9078220	5
0.344	2.	9069767	9061319	9052876	9044438	9036005	9027576	9019153	9010734	9002320	8993911	5
0.345	2.	8985507	8977108	8968714	8960324	8951940	8943560	8935185	8926815	8918450	8910090	5
0.346	2.	8901734	8893383	8885038	8876696	8868360	8860029	8851702	8843380	8835063	8826751	5
0.347	2.	8818444	8810141	8801843	8793550	8785262	8776978	8768700	8760426	8752156	8743892	5
0.348	2.	8735632	8727377	8719127	8710881	8702641	8694405	8686173	8677947	8669725	8661508	5
0.349	2.	8653295	8645087	8636884	8628686	8620492	8612303	8604119	8595939	8587764	8579594	5

TABLE 2

Reciprocals

X		.0000	.0001	.0002	.0003	.0004	.0005	.0006	.0007	.0008	.0009	Δ_2
0.350	2.	8571429	8563268	8555111	8546960	8538813	8530670	8522533	8514400	8506271	8498148	5
0.351	2.	8490028	8481914	8473804	8465699	8457598	8449502	8441411	8433324	8425242	8417164	5
0.352	2.	8409091	8401022	8392958	8384899	8376844	8368794	8360749	8352708	8344671	8336639	5
0.353	2.	8328612	8320589	8312571	8304557	8296548	8288543	8280543	8272547	8264556	8256570	5
0.354	2.	8248588	8240610	8232637	8224668	8216704	8208745	8200790	8192839	8184893	8176951	4
0.355	2.	8169014	8161081	8153153	8145229	8137310	8129395	8121485	8113579	8105677	8097780	4
0.356	2.	8089888	8081999	8074116	8066236	8058361	8050491	8042625	8034763	8026906	8019053	4
0.357	2.	8011204	8003360	7995521	7987685	7979854	7972028	7964206	7956388	7948575	7940766	4
0.358	2.	7932961	7925161	7917365	7909573	7901786	7894003	7886224	7878450	7870680	7862914	4
0.359	2.	7855153	7847396	7839644	7831895	7824151	7816412	7808676	7800945	7793218	7785496	4
0.360	2.	7777778	7770064	7762354	7754649	7746948	7739251	7731558	7723870	7716186	7708507	4
0.361	2.	7700831	7693160	7685493	7677830	7670171	7662517	7654867	7647221	7639580	7631942	4
0.362	2.	7624309	7616680	7609056	7601435	7593819	7586207	7578599	7570995	7563396	7555800	4
0.363	2.	7548209	7540622	7533040	7525461	7517887	7510316	7502750	7495188	7487631	7480077	4
0.364	2.	7472527	7464982	7457441	7449904	7442371	7434842	7427318	7419797	7412281	7404768	4
0.365	2.	7397260	7389756	7382256	7374760	7367269	7359781	7352298	7344818	7337343	7329872	4
0.366	2.	7322404	7314941	7307482	7300027	7292576	7285130	7277687	7270248	7262813	7255383	4
0.367	2.	7247956	7240534	7233115	7225701	7218291	7210884	7203482	7196084	7188689	7181299	4
0.368	2.	7173913	7166531	7159153	7151778	7144408	7137042	7129680	7122322	7114967	7107617	4
0.369	2.	7100271	7092929	7085590	7078256	7070926	7063599	7056277	7048959	7041644	7034334	4
0.370	2.	7027027	7019724	7012426	7005131	6997840	6990553	6983270	6975991	6968716	6961445	4
0.371	2.	6954178	6946915	6939655	6932400	6925148	6917900	6910657	6903417	6896181	6888949	4
0.372	2.	6881720	6874496	6867276	6860059	6852846	6845638	6838433	6831231	6824034	6816841	4
0.373	2.	6809651	6802466	6795284	6788106	6780932	6773762	6766595	6759433	6752274	6745119	4
0.374	2.	6737968	6730821	6723677	6716537	6709402	6702270	6695141	6688017	6680896	6673780	4
0.375	2.	6666667	6659557	6652452	6645350	6638253	6631158	6624068	6616982	6609899	6602820	4
0.376	2.	6595745	6588673	6581606	6574542	6567481	6560425	6553372	6546323	6539278	6532237	4
0.377	2.	6525199	6518165	6511135	6504108	6497085	6490066	6483051	6476039	6469031	6462027	4
0.378	2.	6455026	6448030	6441036	6434047	6427061	6420079	6413101	6406126	6399155	6392188	4
0.379	2.	6385224	6378264	6371308	6364355	6357406	6350461	6343519	6336581	6329647	6322716	4
0.380	2.	6315789	6308866	6301946	6295030	6288118	6281209	6274304	6267402	6260504	6253610	4
0.381	2.	6246719	6239832	6232949	6226069	6219192	6212320	6205451	6198585	6191723	6184865	4
0.382	2.	6178010	6171159	6164312	6157468	6150628	6143791	6136958	6130128	6123302	6116479	4
0.383	2.	6109661	6102845	6096033	6089225	6082420	6075619	6068822	6062028	6055237	6048450	4
0.384	2.	6041667	6034887	6028110	6021337	6014568	6007802	6001040	5994281	5987526	5980774	4
0.385	2.	5974026	5967281	5960540	5953802	5947068	5940337	5933610	5926886	5920166	5913449	3
0.386	2.	5906736	5900026	5893319	5886617	5879917	5873221	5866529	5859840	5853154	5846472	3
0.387	2.	5839793	5833118	5826446	5819778	5813113	5806452	5799794	5793139	5786488	5779840	3
0.388	2.	5773196	5766555	5759917	5753284	5746653	5740026	5733402	5726782	5720165	5713551	3
0.389	2.	5706941	5700334	5693731	5687131	5680534	5673941	5667351	5660765	5654182	5647602	3
0.390	2.	5641026	5634453	5627883	5621317	5614754	5608195	5601638	5595086	5588536	5581990	3
0.391	2.	5575448	5568908	5562372	5555839	5549310	5542784	5536261	5529742	5523226	5516713	3
0.392	2.	5510204	5503698	5497195	5490696	5484200	5477707	5471217	5464731	5458248	5451769	3
0.393	2.	5445293	5438820	5432350	5425884	5419420	5412961	5406504	5400051	5393601	5387154	3
0.394	2.	5380711	5374270	5367834	5361400	5354970	5348542	5342119	5335698	5329281	5322867	3
0.395	2.	5316456	5310048	5303644	5297243	5290845	5284450	5278059	5271670	5265285	5258904	3
0.396	2.	5252525	5246150	5239778	5233409	5227043	5220681	5214322	5207966	5201613	5195263	3
0.397	2.	5188917	5182574	5176234	5169897	5163563	5157233	5150905	5144581	5138260	5131943	3
0.398	2.	5125628	5119317	5113008	5106703	5100402	5094103	5087807	5081515	5075226	5068940	3
0.399	2.	5062657	5056377	5050100	5043827	5037556	5031289	5025025	5018764	5012506	5006252	3

TABLE 2

Reciprocals

X		.0000	.0001	.0002	.0003	.0004	.0005	.0006	.0007	.0008	.0009	Δ_2
0.400	2.	5000000	4993752	4987506	4981264	4975025	4968789	4962556	4956326	4950100	4943876	3
0.401	2.	4937656	4931439	4925224	4919013	4912805	4906600	4900398	4894200	4888004	4881811	3
0.402	2.	4875622	4869435	4863252	4857072	4850895	4844720	4838549	4832381	4826216	4820055	3
0.403	2.	4813896	4807740	4801587	4795438	4789291	4783147	4777007	4770869	4764735	4758604	3
0.404	2.	4752475	4746350	4740228	4734108	4727992	4721879	4715769	4709661	4703557	4697456	3
0.405	2.	4691358	4685263	4679171	4673082	4666996	4660912	4654832	4648755	4642681	4636610	3
0.406	2.	4630542	4624477	4618415	4612355	4606299	4600246	4594196	4588148	4582104	4576063	3
0.407	2.	4570025	4563989	4557957	4551927	4545901	4539877	4533857	4527839	4521824	4515813	3
0.408	2.	4509804	4503798	4497795	4491795	4485798	4479804	4473813	4467825	4461839	4455857	3
0.409	2.	4449878	4443901	4437928	4431957	4425989	4420024	4414063	4408103	4402147	4396194	3
0.410	2.	4390244	4384296	4378352	4372410	4366472	4360536	4354603	4348673	4342746	4336822	3
0.411	2.	4330900	4324982	4319066	4313153	4307244	4301337	4295432	4289531	4283633	4277737	3
0.412	2.	4271845	4265955	4260068	4254184	4248303	4242424	4236549	4230676	4224806	4218939	3
0.413	2.	4213075	4207214	4201355	4195500	4189647	4183797	4177950	4172105	4166264	4160425	3
0.414	2.	4154589	4148756	4142926	4137099	4131274	4125452	4119633	4113817	4108004	4102193	3
0.415	2.	4096386	4090581	4084778	4078979	4073182	4067389	4061598	4055809	4050024	4044241	3
0.416	2.	4038462	4032684	4026910	4021139	4015370	4009604	4003841	3998080	3992322	3986568	3
0.417	2.	3980815	3975066	3969319	3963575	3957834	3952096	3946360	3940627	3934897	3929170	3
0.418	2.	3923445	3917723	3912004	3906287	3900574	3894863	3889154	3883449	3877746	3872046	3
0.419	2.	3866348	3860654	3854962	3849273	3843586	3837902	3832221	3826543	3820867	3815194	3
0.420	2.	3809524	3803856	3798191	3792529	3786870	3781213	3775559	3769907	3764259	3758612	3
0.421	2.	3752969	3747328	3741690	3736055	3730422	3724792	3719165	3713540	3707918	3702299	3
0.422	2.	3696682	3691068	3685457	3679848	3674242	3668639	3663038	3657440	3651845	3646252	3
0.423	2.	3640662	3635074	3629490	3623907	3618328	3612751	3607177	3601605	3596036	3590469	3
0.424	2.	3584906	3579344	3573786	3568230	3562677	3557126	3551578	3546032	3540490	3534949	3
0.425	2.	3529412	3523877	3518344	3512814	3507287	3501763	3496241	3490721	3485204	3479690	3
0.426	2.	3474178	3468669	3463163	3457659	3452158	3446659	3441163	3435669	3430178	3424690	3
0.427	2.	3419204	3413720	3408240	3402762	3397286	3391813	3386342	3380874	3375409	3369946	3
0.428	2.	3364486	3359028	3353573	3348120	3342670	3337223	3331778	3326335	3320896	3315458	3
0.429	2.	3310023	3304591	3299161	3293734	3288309	3282887	3277467	3272050	3266636	3261224	3
0.430	2.	3255814	3250407	3245002	3239600	3234201	3228804	3223409	3218017	3212628	3207241	3
0.431	2.	3201856	3196474	3191095	3185718	3180343	3174971	3169601	3164234	3158870	3153508	2
0.432	2.	3148148	3142791	3137436	3132084	3126734	3121387	3116043	3110700	3105360	3100023	2
0.433	2.	3094688	3089356	3084026	3078698	3073373	3068051	3062731	3057413	3052098	3046785	2
0.434	2.	3041475	3036167	3030861	3025558	3020258	3014960	3009664	3004371	2999080	2993792	2
0.435	2.	2988506	2983222	2977941	2972663	2967386	2962113	2956841	2951572	2946306	2941042	2
0.436	2.	2935780	2930521	2925264	2920009	2914757	2909507	2904260	2899015	2893773	2888533	2
0.437	2.	2883295	2878060	2872827	2867597	2862369	2857143	2851920	2846699	2841480	2836264	2
0.438	2.	2831050	2825839	2820630	2815423	2810219	2805017	2799818	2794620	2789426	2784233	2
0.439	2.	2779043	2773856	2768670	2763487	2758307	2753129	2747953	2742779	2737608	2732439	2
0.440	2.	2727273	2722109	2716947	2711787	2706630	2701476	2696323	2691173	2686025	2680880	2
0.441	2.	2675737	2670596	2665458	2660322	2655188	2650057	2644928	2639801	2634676	2629554	2
0.442	2.	2624434	2619317	2614202	2609089	2603978	2598870	2593764	2588660	2583559	2578460	2
0.443	2.	2573363	2568269	2563177	2558087	2553000	2547914	2542831	2537751	2532672	2527596	2
0.444	2.	2522523	2517451	2512382	2507315	2502250	2497188	2492128	2487070	2482014	2476961	2
0.445	2.	2471910	2466861	2461815	2456771	2451729	2446689	2441652	2436617	2431584	2426553	2
0.446	2.	2421525	2416499	2411475	2406453	2401434	2396417	2391402	2386389	2381379	2376371	2
0.447	2.	2371365	2366361	2361360	2356360	2351363	2346369	2341376	2336386	2331398	2326412	2
0.448	2.	2321429	2316447	2311468	2306491	2301516	2296544	2291574	2286606	2281640	2276676	2
0.449	2.	2271715	2266756	2261799	2256844	2251891	2246941	2241993	2237047	2232103	2227162	2

TABLE 2

Reciprocals

X		.0000	.0001	.0002	.0003	.0004	.0005	.0006	.0007	.0008	.0009	Δ_2
0.450	2.	2222222	2217285	2212350	2207417	2202487	2197558	2192632	2187708	2182786	2177866	
0.451	2.	2172949	2168034	2163121	2158210	2153301	2148394	2143490	2138588	2133687	2128790	2
0.452	2.	2123894	2119000	2114109	2109220	2104332	2099448	2094565	2089684	2084806	2079929	2
0.453	2.	2075055	2070183	2065313	2060446	2055580	2050717	2045855	2040996	2036139	2031284	2
0.454	2.	2026432	2021581	2016733	2011886	2007042	2002200	1997360	1992523	1987687	1982853	2
0.455	2.	1978022	1973193	1968366	1963541	1958718	1953897	1949078	1944262	1939447	1934635	2
0.456	2.	1929825	1925016	1920210	1915407	1910605	1905805	1901007	1896212	1891419	1886627	2
0.457	2.	1881838	1877051	1872266	1867483	1862702	1857923	1853147	1848372	1843600	1838829	2
0.458	2.	1834061	1829295	1824531	1819769	1815009	1810251	1805495	1800741	1795990	1791240	2
0.459	2.	1786492	1781747	1777003	1772262	1767523	1762786	1758050	1753317	1748586	1743857	2
0.460	2.	1739130	1734406	1729683	1724962	1720243	1715527	1710812	1706099	1701389	1696680	2
0.461	2.	1691974	1687270	1682567	1677867	1673169	1668472	1663778	1659086	1654396	1649708	2
0.462	2.	1645022	1640338	1635656	1630976	1626298	1621622	1616948	1612276	1607606	1602938	2
0.463	2.	1598272	1593608	1588946	1584287	1579629	1574973	1570319	1565667	1561018	1556370	2
0.464	2.	1551724	1547080	1542439	1537799	1533161	1528525	1523892	1519260	1514630	1510002	2
0.465	2.	1505376	1500753	1496131	1491511	1486893	1482277	1477663	1473051	1468441	1463833	2
0.466	2.	1459227	1454623	1450021	1445421	1440823	1436227	1431633	1427041	1422451	1417862	2
0.467	2.	1413276	1408692	1404110	1399529	1394951	1390374	1385800	1381227	1376657	1372088	2
0.468	2.	1367521	1362957	1358394	1353833	1349274	1344717	1340162	1335609	1331058	1326509	2
0.469	2.	1321962	1317416	1312873	1308332	1303792	1299255	1294719	1290185	1285653	1281124	2
0.470	2.	1276596	1272070	1267546	1263024	1258503	1253985	1249469	1244954	1240442	1235931	2
0.471	2.	1231422	1226916	1222411	1217908	1213407	1208908	1204411	1199915	1195422	1190930	2
0.472	2.	1186441	1181953	1177467	1172983	1168501	1164021	1159543	1155067	1150592	1146120	2
0.473	2.	1141649	1137180	1132713	1128248	1123785	1119324	1114865	1110407	1105952	1101498	2
0.474	2.	1097046	1092596	1088148	1083702	1079258	1074816	1070375	1065936	1061500	1057065	2
0.475	2.	1052632	1048200	1043771	1039344	1034918	1030494	1026072	1021652	1017234	1012818	2
0.476	2.	1008403	1003991	0999580	0995171	0990764	0986359	0981956	0977554	0973154	0968757	2
0.477	2.	0964361	0959966	0955574	0951184	0946795	0942408	0938023	0933640	0929259	0924880	2
0.478	2.	0920502	0916126	0911752	0907380	0903010	0898642	0894275	0889910	0885547	0881186	2
0.479	2.	0876827	0872469	0868114	0863760	0859408	0855057	0850709	0846362	0842018	0837675	2
0.480	2.	0833333	0828994	0824656	0820321	0815987	0811655	0807324	0802996	0798669	0794344	2
0.481	2.	0790021	0785699	0781380	0777062	0772746	0768432	0764120	0759809	0755500	0751193	2
0.482	2.	0746888	0742585	0738283	0733983	0729685	0725389	0721094	0716801	0712510	0708221	2
0.483	2.	0703934	0699648	0695364	0691082	0686802	0682523	0678246	0673971	0669698	0665427	2
0.484	2.	0661157	0656889	0652623	0648358	0644096	0639835	0635576	0631318	0627063	0622809	2
0.485	2.	0618557	0614306	0610058	0605811	0601566	0597322	0593081	0588841	0584603	0580366	2
0.486	2.	0576132	0571899	0567668	0563438	0559211	0554985	0550760	0546538	0542317	0538098	2
0.487	2.	0533881	0529665	0525452	0521239	0517029	0512821	0508614	0504408	0500205	0496003	2
0.488	2.	0491803	0487605	0483408	0479214	0475020	0470829	0466639	0462451	0458265	0454081	2
0.489	2.	0449898	0445717	0441537	0437359	0433183	0429009	0424837	0420666	0416497	0412329	2
0.490	2.	0408163	0403999	0399837	0395676	0391517	0387360	0383204	0379050	0374898	0370748	2
0.491	2.	0366599	0362452	0358306	0354162	0350020	0345880	0341741	0337604	0333469	0329335	2
0.492	2.	0325203	0321073	0316944	0312817	0308692	0304569	0300447	0296326	0292208	0288091	2
0.493	2.	0283976	0279862	0275750	0271640	0267531	0263425	0259319	0255216	0251114	0247014	2
0.494	2.	0242915	0238818	0234723	0230629	0226537	0222447	0218358	0214271	0210186	0206102	2
0.495	2.	0202020	0197940	0193861	0189784	0185709	0181635	0177563	0173492	0169423	0165356	2
0.496	2.	0161290	0157226	0153164	0149103	0145044	0140987	0136931	0132877	0128824	0124774	2
0.497	2.	0120724	0116677	0112631	0108586	0104544	0100503	0096463	0092425	0088389	0084354	2
0.498	2.	0080321	0076290	0072260	0068232	0064205	0060181	0056157	0052136	0048115	0044097	2
0.499	2.	0040080	0036065	0032051	0028039	0024029	0020020	0016013	0012007	0008003	0004001	2

TABLE 2

Reciprocals

X		.0000	.0001	.0002	.0003	.0004	.0005	.0006	.0007	.0008	.0009	Δ_2
0.500	2.	0000000	9996001	9992003	9988007	9984013	9980020	9976029	9972039	9968051	9964065	2
0.501	1.	9960080	9956097	9952115	9948135	9944156	9940179	9936204	9932230	9928258	9924288	2
0.502	1.	9920319	9916351	9912385	9908421	9904459	9900497	9896538	9892580	9888624	9884669	2
0.503	1.	9880716	9876764	9872814	9868865	9864919	9860973	9857029	9853087	9849146	9845207	2
0.504	1.	9841270	9837334	9833399	9829467	9825535	9821606	9817677	9813751	9809826	9805902	2
0.505	1.	9801980	9798060	9794141	9790224	9786308	9782394	9778481	9774570	9770660	9766752	2
0.506	1.	9762846	9758941	9755038	9751136	9747235	9743337	9739439	9735544	9731650	9727757	2
0.507	1.	9723866	9719976	9716088	9712202	9708317	9704433	9700552	9696671	9692792	9688915	2
0.508	1.	9685039	9681165	9677292	9673421	9669551	9665683	9661817	9657952	9654088	9650226	2
0.509	1.	9646365	9642506	9638649	9634793	9630938	9627085	9623234	9619384	9615535	9611689	2
0.510	1.	9607843	9603999	9600157	9596316	9592476	9588639	9584802	9580967	9577134	9573302	2
0.511	1.	9569472	9565643	9561815	9557989	9554165	9550342	9546521	9542701	9538882	9535065	1
0.512	1.	9531250	9527436	9523624	9519813	9516003	9512195	9508389	9504584	9500780	9496978	1
0.513	1.	9493177	9489378	9485581	9481784	9477990	9474197	9470405	9466615	9462826	9459039	1
0.514	1.	9455253	9451469	9447686	9443904	9440124	9436346	9432569	9428793	9425019	9421247	1
0.515	1.	9417476	9413706	9409938	9406171	9402406	9398642	9394880	9391119	9387359	9383601	1
0.516	1.	9379845	9376090	9372336	9368584	9364833	9361084	9357336	9353590	9349845	9346102	1
0.517	1.	9342360	9338619	9334880	9331142	9327406	9323671	9319938	9316206	9312476	9308747	1
0.518	1.	9305019	9301293	9297569	9293845	9290123	9286403	9282684	9278967	9275251	9271536	1
0.519	1.	9267823	9264111	9260401	9256692	9252984	9249278	9245574	9241870	9238169	9234468	1
0.520	1.	9230769	9227072	9223376	9219681	9215988	9212296	9208605	9204916	9201229	9197543	1
0.521	1.	9193858	9190175	9186493	9182812	9179133	9175455	9171779	9168104	9164431	9160759	1
0.522	1.	9157088	9153419	9149751	9146085	9142420	9138756	9135094	9131433	9127773	9124115	1
0.523	1.	9120459	9116804	9113150	9109497	9105846	9102197	9098548	9094902	9091256	9087612	1
0.524	1.	9083969	9080328	9076688	9073050	9069413	9065777	9062143	9058510	9054878	9051248	1
0.525	1.	9047619	9043992	9040366	9036741	9033118	9029496	9025875	9022256	9018638	9015022	1
0.526	1.	9011407	9007793	9004181	9000570	8996960	8993352	8989746	8986140	8982536	8978933	1
0.527	1.	8975332	8971732	8968133	8964536	8960940	8957346	8953753	8950161	8946571	8942982	1
0.528	1.	8939394	8935808	8932223	8928639	8925057	8921476	8917896	8914318	8910741	8907166	1
0.529	1.	8903592	8900019	8896447	8892877	8889309	8885741	8882175	8878611	8875047	8871485	1
0.530	1.	8867925	8864365	8860807	8857251	8853695	8850141	8846589	8843037	8839488	8835939	1
0.531	1.	8832392	8828846	8825301	8821758	8818216	8814675	8811136	8807598	8804062	8800526	1
0.532	1.	8796992	8793460	8789929	8786399	8782870	8779343	8775817	8772292	8768769	8765247	1
0.533	1.	8761726	8758207	8754689	8751172	8747656	8744142	8740630	8737118	8733608	8730099	1
0.534	1.	8726592	8723086	8719581	8716077	8712575	8709074	8705574	8702076	8698579	8695083	1
0.535	1.	8691589	8688096	8684604	8681113	8677624	8674136	8670650	8667164	8663680	8660198	1
0.536	1.	8656716	8653236	8649758	8646280	8642804	8639329	8635855	8632383	8628912	8625442	1
0.537	1.	8621974	8618507	8615041	8611576	8608113	8604651	8601190	8597731	8594273	8590816	1
0.538	1.	8587361	8583906	8580453	8577002	8573551	8570102	8566654	8563208	8559762	8556318	1
0.539	1.	8552876	8549434	8545994	8542555	8539118	8535681	8532246	8528812	8525380	8521948	1
0.540	1.	8518519	8515090	8511662	8508236	8504811	8501388	8497965	8494544	8491124	8487706	1
0.541	1.	8484288	8480872	8477457	8474044	8470632	8467221	8463811	8460402	8456995	8453589	1
0.542	1.	8450184	8446781	8443379	8439978	8436578	8433180	8429782	8426387	8422992	8419598	1
0.543	1.	8416206	8412815	8409426	8406037	8402650	8399264	8395879	8392496	8389114	8385733	1
0.544	1.	8382353	8378974	8375597	8372221	8368846	8365473	8362101	8358730	8355360	8351991	1
0.545	1.	8348624	8345258	8341893	8338529	8335167	8331806	8328446	8325087	8321730	8318373	1
0.546	1.	8315018	8311664	8308312	8304961	8301610	8298262	8294914	8291568	8288222	8284878	1
0.547	1.	8281536	8278194	8274854	8271515	8268177	8264840	8261505	8258170	8254837	8251506	1
0.548	1.	8248175	8244846	8241518	8238191	8234865	8231541	8228217	8224895	8221574	8218255	1
0.549	1.	8214936	8211619	8208303	8204988	8201675	8198362	8195051	8191741	8188432	8185125	1

TABLE 2

Reciprocals

X		.0000	.0001	.0002	.0003	.0004	.0005	.0006	.0007	.0008	.0009	Δ_2
0.550	1.	8181818	8178513	8175209	8171906	8168605	8165304	8162005	8158707	8155410	8152115	1
0.551	1.	8148820	8145527	8142235	8138944	8135655	8132366	8129079	8125793	8122508	8119224	1
0.552	1.	8115942	8112661	8109381	8106102	8102824	8099547	8096272	8092998	8089725	8086453	1
0.553	1.	8083183	8079913	8076645	8073378	8070112	8066847	8063584	8060321	8057060	8053800	1
0.554	1.	8050542	8047284	8044027	8040772	8037518	8034265	8031013	8027763	8024513	8021265	1
0.555	1.	8018018	8014772	8011527	8008284	8005041	8001800	7998560	7995321	7992083	7988847	1
0.556	1.	7985611	7982377	7979144	7975912	7972682	7969452	7966223	7962996	7959770	7956545	1
0.557	1.	7953321	7950099	7946877	7943657	7940438	7937220	7934003	7930787	7927573	7924359	1
0.558	1.	7921147	7917936	7914726	7911517	7908309	7905103	7901898	7898693	7895490	7892288	1
0.559	1.	7889088	7885888	7882690	7879492	7876296	7873101	7869907	7866714	7863523	7860332	1
0.560	1.	7857143	7853955	7850768	7847582	7844397	7841213	7838031	7834849	7831669	7828490	1
0.561	1.	7825312	7822135	7818959	7815785	7812611	7809439	7806268	7803098	7799929	7796761	1
0.562	1.	7793594	7790429	7787264	7784101	7780939	7777778	7774618	7771459	7768301	7765145	1
0.563	1.	7761989	7758835	7755682	7752530	7749379	7746229	7743080	7739933	7736786	7733641	1
0.564	1.	7730496	7727353	7724211	7721070	7717931	7714792	7711654	7708518	7705382	7702248	1
0.565	1.	7699115	7695983	7692852	7689722	7686594	7683466	7680339	7677214	7674090	7670967	1
0.566	1.	7667845	7664724	7661604	7658485	7655367	7652251	7649135	7646021	7642908	7639795	1
0.567	1.	7636684	7633574	7630465	7627358	7624251	7621145	7618041	7614937	7611835	7608734	1
0.568	1.	7605634	7602535	7599437	7596340	7593244	7590149	7587056	7583963	7580872	7577782	1
0.569	1.	7574692	7571604	7568517	7565431	7562346	7559262	7556180	7553098	7550018	7546938	1
0.570	1.	7543860	7540782	7537706	7534631	7531557	7528484	7525412	7522341	7519271	7516202	1
0.571	1.	7513135	7510068	7507003	7503938	7500875	7497813	7494752	7491691	7488632	7485574	1
0.572	1.	7482517	7479462	7476407	7473353	7470300	7467249	7464198	7461149	7458101	7455053	1
0.573	1.	7452007	7448962	7445918	7442875	7439833	7436792	7433752	7430713	7427675	7424638	1
0.574	1.	7421603	7418568	7415535	7412502	7409471	7406440	7403411	7400383	7397356	7394329	1
0.575	1.	7391304	7388280	7385257	7382235	7379214	7376195	7373176	7370158	7367141	7364126	1
0.576	1.	7361111	7358098	7355085	7352074	7349063	7346054	7343045	7340038	7337032	7334027	1
0.577	1.	7331022	7328019	7325017	7322016	7319016	7316017	7313019	7310022	7307027	7304032	1
0.578	1.	7301038	7298045	7295054	7292063	7289073	7286085	7283097	7280111	7277125	7274141	1
0.579	1.	7271157	7268175	7265193	7262213	7259234	7256255	7253278	7250302	7247327	7244352	1
0.580	1.	7241379	7238407	7235436	7232466	7229497	7226529	7223562	7220596	7217631	7214667	1
0.581	1.	7211704	7208742	7205781	7202821	7199862	7196904	7193948	7190992	7188037	7185083	1
0.582	1.	7182131	7179179	7176228	7173278	7170330	7167382	7164435	7161490	7158545	7155601	1
0.583	1.	7152659	7149717	7146776	7143837	7140898	7137961	7135024	7132088	7129154	7126220	1
0.584	1.	7123288	7120356	7117426	7114496	7111567	7108640	7105713	7102788	7099863	7096940	1
0.585	1.	7094017	7091096	7088175	7085255	7082337	7079419	7076503	7073587	7070673	7067759	1
0.586	1.	7064846	7061935	7059024	7056115	7053206	7050298	7047392	7044486	7041581	7038678	1
0.587	1.	7035775	7032873	7029973	7027073	7024174	7021277	7018380	7015484	7012589	7009695	1
0.588	1.	7006803	7003911	7001020	6998130	6995241	6992353	6989466	6986581	6983696	6980812	1
0.589	1.	6977929	6975047	6972166	6969286	6966406	6963528	6960651	6957775	6954900	6952026	1
0.590	1.	6949152	6946280	6943409	6940539	6937669	6934801	6931934	6929067	6926202	6923337	1
0.591	1.	6920474	6917611	6914750	6911889	6909029	6906171	6903313	6900456	6897601	6894746	1
0.592	1.	6891892	6889039	6886187	6883336	6880486	6877637	6874789	6871942	6869096	6866251	1
0.593	1.	6863406	6860563	6857721	6854879	6852039	6849200	6846361	6843524	6840687	6837851	1
0.594	1.	6835017	6832183	6829350	6826519	6823688	6820858	6818029	6815201	6812374	6809548	1
0.595	1.	6806723	6803898	6801075	6798253	6795432	6792611	6789792	6786973	6784156	6781339	1
0.596	1.	6778523	6775709	6772895	6770082	6767270	6764459	6761649	6758840	6756032	6753225	1
0.597	1.	6750419	6747613	6744809	6742006	6739203	6736402	6733601	6730801	6728003	6725205	1
0.598	1.	6722408	6719612	6716817	6714023	6711230	6708438	6705646	6702856	6700067	6697278	1
0.599	1.	6694491	6691704	6688919	6686134	6683350	6680567	6677785	6675004	6672224	6669445	1

TABLE 2

Reciprocals

X		.0000	.0001	.0002	.0003	.0004	.0005	.0006	.0007	.0008	.0009	Δ_2
0.600	1.	6666667	6663889	6661113	6658337	6655563	6652789	6650017	6647245	6644474	6641704	1
0.601	1.	6638935	6636167	6633400	6630634	6627868	6625104	6622340	6619578	6616816	6614055	1
0.602	1.	6611296	6608537	6605779	6603022	6600266	6597510	6594756	6592003	6589250	6586499	1
0.603	1.	6583748	6580998	6578249	6575501	6572754	6570008	6567263	6564519	6561775	6559033	1
0.604	1.	6556291	6553551	6550811	6548072	6545334	6542597	6539861	6537126	6534391	6531658	1
0.605	1.	6528926	6526194	6523463	6520733	6518005	6515277	6512550	6509823	6507098	6504374	1
0.606	1.	6501650	6498928	6496206	6493485	6490765	6488046	6485328	6482611	6479895	6477179	1
0.607	1.	6474465	6471751	6469038	6466326	6463615	6460905	6458196	6455488	6452781	6450074	1
0.608	1.	6447368	6444664	6441960	6439257	6436555	6433854	6431153	6428454	6425756	6423058	1
0.609	1.	6420361	6417665	6414970	6412276	6409583	6406891	6404199	6401509	6398819	6396130	1
0.610	1.	6393443	6390756	6388069	6385384	6382700	6380016	6377334	6374652	6371971	6369291	1
0.611	1.	6366612	6363934	6361257	6358580	6355904	6353230	6350556	6347883	6345211	6342540	1
0.612	1.	6339869	6337200	6334531	6331863	6329197	6326531	6323865	6321201	6318538	6315875	1
0.613	1.	6313214	6310553	6307893	6305234	6302576	6299918	6297262	6294606	6291952	6289298	1
0.614	1.	6286645	6283993	6281342	6278691	6276042	6273393	6270745	6268098	6265452	6262807	1
0.615	1.	6260163	6257519	6254876	6252235	6249594	6246954	6244314	6241676	6239039	6236402	1
0.616	1.	6233766	6231131	6228497	6225864	6223232	6220600	6217970	6215340	6212711	6210083	1
0.617	1.	6207455	6204829	6202203	6199579	6196955	6194332	6191710	6189089	6186468	6183848	1
0.618	1.	6181230	6178612	6175995	6173379	6170763	6168149	6165535	6162922	6160310	6157699	1
0.619	1.	6155089	6152479	6149871	6147263	6144656	6142050	6139445	6136840	6134237	6131634	1
0.620	1.	6129032	6126431	6123831	6121232	6118633	6116035	6113439	6110843	6108247	6105653	1
0.621	1.	6103060	6100467	6097875	6095284	6092694	6090105	6087516	6084928	6082342	6079756	1
0.622	1.	6077170	6074586	6072003	6069420	6066838	6064257	6061677	6059097	6056519	6053941	1
0.623	1.	6051364	6048788	6046213	6043639	6041065	6038492	6035920	6033349	6030779	6028210	1
0.624	1.	6025641	6023073	6020506	6017940	6015375	6012810	6010247	6007684	6005122	6002560	1
0.625	1.	6000000	5997440	5994882	5992324	5989766	5987210	5984655	5982100	5979546	5976993	1
0.626	1.	5974441	5971889	5969339	5966789	5964240	5961692	5959145	5956598	5954052	5951507	1
0.627	1.	5948963	5946420	5943878	5941336	5938795	5936255	5933716	5931177	5928640	5926103	1
0.628	1.	5923567	5921032	5918497	5915964	5913431	5910899	5908368	5905837	5903308	5900779	1
0.629	1.	5898251	5895724	5893198	5890672	5888147	5885623	5883100	5880578	5878057	5875536	1
0.630	1.	5873016	5870497	5867978	5865461	5862944	5860428	5857913	5855399	5852885	5850372	1
0.631	1.	5847861	5845349	5842839	5840329	5837821	5835313	5832806	5830299	5827794	5825289	1
0.632	1.	5822785	5820282	5817779	5815278	5812777	5810277	5807777	5805279	5802781	5800284	1
0.633	1.	5797788	5795293	5792798	5790305	5787812	5785320	5782828	5780338	5777848	5775359	1
0.634	1.	5772871	5770383	5767897	5765411	5762926	5760441	5757958	5755475	5752993	5750512	1
0.635	1.	5748031	5745552	5743073	5740595	5738118	5735641	5733165	5730691	5728216	5725743	1
0.636	1.	5723270	5720799	5718328	5715857	5713388	5710919	5708451	5705984	5703518	5701052	1
0.637	1.	5698587	5696123	5693660	5691197	5688735	5686274	5683814	5681355	5678896	5676438	1
0.638	1.	5673981	5671525	5669069	5666614	5664160	5661707	5659255	5656803	5654352	5651902	1
0.639	1.	5649452	5647004	5644556	5642109	5639662	5637217	5634772	5632328	5629884	5627442	1
0.640	1.	5625000	5622559	5620119	5617679	5615240	5612802	5610365	5607929	5605493	5603058	1
0.641	1.	5600624	5598191	5595758	5593326	5590895	5588465	5586035	5583606	5581178	5578751	1
0.642	1.	5576324	5573898	5571473	5569049	5566625	5564202	5561780	5559359	5556938	5554519	1
0.643	1.	5552099	5549681	5547264	5544847	5542431	5540015	5537601	5535187	5532774	5530362	1
0.644	1.	5527950	5525539	5523129	5520720	5518312	5515904	5513497	5511090	5508685	5506280	1
0.645	1.	5503876	5501473	5499070	5496668	5494267	5491867	5489467	5487068	5484670	5482273	1
0.646	1.	5479876	5477480	5475085	5472691	5470297	5467904	5465512	5463120	5460730	5458340	1
0.647	1.	5455950	5453562	5451174	5448787	5446401	5444015	5441631	5439247	5436863	5434481	1
0.648	1.	5432099	5429718	5427337	5424958	5422579	5420200	5417823	5415446	5413070	5410695	1
0.649	1.	5408320	5405947	5403574	5401201	5398830	5396459	5394089	5391719	5389351	5386983	1

TABLE 2

Reciprocals

X		.0000	.0001	.0002	.0003	.0004	.0005	.0006	.0007	.0008	.0009	Δ_2
0.650	1.	5384615	5382249	5379883	5377518	5375154	5372790	5370427	5368065	5365704	5363343	1
0.651	1.	5360983	5358624	5356265	5353907	5351551	5349194	5346839	5344484	5342129	5339776	1
0.652	1.	5337423	5335071	5332720	5330369	5328020	5325670	5323322	5320974	5318627	5316281	1
0.653	1.	5313936	5311591	5309247	5306903	5304561	5302219	5299878	5297537	5295197	5292858	1
0.654	1.	5290520	5288182	5285845	5283509	5281174	5278839	5276505	5274171	5271839	5269507	1
0.655	1.	5267176	5264845	5262515	5260186	5257858	5255530	5253203	5250877	5248551	5246226	1
0.656	1.	5243902	5241579	5239256	5236934	5234613	5232292	5229973	5227653	5225335	5223017	1
0.657	1.	5220700	5218384	5216068	5213753	5211439	5209125	5206813	5204500	5202189	5199878	1
0.658	1.	5197568	5195259	5192950	5190643	5188335	5186029	5183723	5181418	5179114	5176810	1
0.659	1.	5174507	5172204	5169903	5167602	5165302	5163002	5160703	5158405	5156108	5153811	1
0.660	1.	5151515	5149220	5146925	5144631	5142338	5140045	5137753	5135462	5133172	5130882	1
0.661	1.	5128593	5126305	5124017	5121730	5119444	5117158	5114873	5112589	5110305	5108022	1
0.662	1.	5105740	5103459	5101178	5098898	5096618	5094340	5092062	5089784	5087508	5085232	1
0.663	1.	5082956	5080682	5078408	5076134	5073862	5071590	5069319	5067048	5064778	5062509	1
0.664	1.	5060241	5057973	5055706	5053440	5051174	5048909	5046645	5044381	5042118	5039856	1
0.665	1.	5037594	5035333	5033073	5030813	5028554	5026296	5024038	5021782	5019525	5017270	1
0.666	1.	5015015	5012761	5010507	5008254	5006002	5003751	5001500	4999250	4997001	4994752	1
0.667	1.	4992504	4990256	4988010	4985764	4983518	4981273	4979029	4976786	4974543	4972301	1
0.668	1.	4970060	4967819	4965579	4963340	4961101	4958863	4956626	4954389	4952153	4949918	1
0.669	1.	4947683	4945449	4943216	4940983	4938751	4936520	4934289	4932059	4929830	4927601	1
0.670	1.	4925373	4923146	4920919	4918693	4916468	4914243	4912019	4909796	4907573	4905351	1
0.671	1.	4903130	4900909	4898689	4896470	4894251	4892033	4889815	4887599	4885383	4883167	1
0.672	1.	4880952	4878738	4876525	4874312	4872100	4869888	4867678	4865468	4863258	4861049	1
0.673	1.	4858841	4856633	4854427	4852220	4850015	4847810	4845606	4843402	4841199	4838997	1
0.674	1.	4836795	4834594	4832394	4830194	4827995	4825797	4823599	4821402	4819206	4817010	1
0.675	1.	4814815	4812620	4810427	4808233	4806041	4803849	4801658	4799467	4797277	4795088	1
0.676	1.	4792899	4790711	4788524	4786337	4784151	4781966	4779781	4777597	4775414	4773231	1
0.677	1.	4771049	4768867	4766686	4764506	4762327	4760148	4757969	4755792	4753615	4751438	1
0.678	1.	4749263	4747087	4744913	4742739	4740566	4738394	4736222	4734050	4731880	4729710	1
0.679	1.	4727541	4725372	4723204	4721036	4718870	4716703	4714538	4712373	4710209	4708045	1
0.680	1.	4705882	4703720	4701558	4699397	4697237	4695077	4692918	4690759	4688602	4686444	1
0.681	1.	4684288	4682132	4679976	4677822	4675668	4673514	4671361	4669209	4667058	4664907	1
0.682	1.	4662757	4660607	4658458	4656310	4654162	4652015	4649868	4647722	4645577	4643432	1
0.683	1.	4641288	4639145	4637002	4634860	4632719	4630578	4628438	4626298	4624159	4622021	1
0.684	1.	4619883	4617746	4615609	4613474	4611338	4609204	4607070	4604936	4602804	4600672	1
0.685	1.	4598540	4596409	4594279	4592149	4590020	4587892	4585764	4583637	4581511	4579385	1
0.686	1.	4577259	4575135	4573011	4570887	4568765	4566642	4564521	4562400	4560280	4558160	1
0.687	1.	4556041	4553922	4551804	4549687	4547571	4545455	4543339	4541224	4539110	4536997	1
0.688	1.	4534884	4532771	4530660	4528549	4526438	4524328	4522219	4520110	4518002	4515895	1
0.689	1.	4513788	4511682	4509576	4507471	4505367	4503263	4501160	4499058	4496956	4494854	1
0.690	1.	4492754	4490653	4488554	4486455	4484357	4482259	4480162	4478066	4475970	4473875	1
0.691	1.	4471780	4469686	4467593	4465500	4463408	4461316	4459225	4457135	4455045	4452956	1
0.692	1.	4450867	4448779	4446692	4444605	4442519	4440433	4438348	4436264	4434180	4432097	1
0.693	1.	4430014	4427932	4425851	4423770	4421690	4419611	4417532	4415453	4413376	4411298	1
0.694	1.	4409222	4407146	4405071	4402996	4400922	4398848	4396775	4394703	4392631	4390560	1
0.695	1.	4388489	4386419	4384350	4382281	4380213	4378145	4376078	4374012	4371946	4369881	1
0.696	1.	4367816	4365752	4363689	4361626	4359563	4357502	4355441	4353380	4351320	4349261	1
0.697	1.	4347202	4345144	4343087	4341030	4338973	4336918	4334862	4332808	4330754	4328700	1
0.698	1.	4326648	4324595	4322544	4320493	4318442	4316392	4314343	4312294	4310246	4308199	1
0.699	1.	4306152	4304105	4302059	4300014	4297970	4295926	4293882	4291839	4289797	4287755	1

TABLE 2

Reciprocals

X		.0000	.0001	.0002	.0003	.0004	.0005	.0006	.0007	.0008	.0009	Δ_2
0.700	1.	4285714	4283674	4281634	4279594	4277556	4275517	4273480	4271443	4269406	4267371	1
0.701	1.	4265335	4263301	4261266	4259233	4257200	4255167	4253136	4251104	4249074	4247044	1
0.702	1.	4245014	4242985	4240957	4238929	4236902	4234875	4232849	4230824	4228799	4226775	1
0.703	1.	4224751	4222728	4220705	4218683	4216662	4214641	4212621	4210601	4208582	4206563	1
0.704	1.	4204545	4202528	4200511	4198495	4196479	4194464	4192450	4190436	4188422	4186409	1
0.705	1.	4184397	4182385	4180374	4178364	4176354	4174344	4172336	4170327	4168320	4166312	1
0.706	1.	4164306	4162300	4160295	4158290	4156285	4154282	4152278	4150276	4148274	4146272	1
0.707	1.	4144272	4142271	4140271	4138272	4136274	4134276	4132278	4130281	4128285	4126289	1
0.708	1.	4124294	4122299	4120305	4118311	4116318	4114326	4112334	4110343	4108352	4106362	1
0.709	1.	4104372	4102383	4100395	4098407	4096419	4094433	4092446	4090461	4088476	4086491	1
0.710	1.	4084507	4082524	4080541	4078558	4076577	4074595	4072615	4070635	4068655	4066676	1
0.711	1.	4064698	4062720	4060742	4058766	4056789	4054814	4052839	4050864	4048890	4046917	1
0.712	1.	4044944	4042971	4041000	4039028	4037058	4035088	4033118	4031149	4029181	4027213	1
0.713	1.	4025245	4023279	4021312	4019347	4017382	4015417	4013453	4011489	4009526	4007564	1
0.714	1.	4005602	4003641	4001680	3999720	3997760	3995801	3993843	3991885	3989927	3987970	1
0.715	1.	3986014	3984058	3982103	3980148	3978194	3976240	3974287	3972335	3970383	3968431	1
0.716	1.	3966480	3964530	3962580	3960631	3958682	3956734	3954786	3952839	3950893	3948947	1
0.717	1.	3947001	3945056	3943112	3941168	3939225	3937282	3935340	3933398	3931457	3929517	1
0.718	1.	3927577	3925637	3923698	3921760	3919822	3917884	3915948	3914011	3912076	3910140	1
0.719	1.	3908206	3906272	3904338	3902405	3900473	3898541	3896609	3894678	3892748	3890818	1
0.720	1.	3888889	3886960	3885032	3883104	3881177	3879251	3877324	3875399	3873474	3871549	1
0.721	1.	3869626	3867702	3865779	3863857	3861935	3860014	3858093	3856173	3854253	3852334	1
0.722	1.	3850415	3848497	3846580	3844663	3842746	3840830	3838915	3837000	3835086	3833172	1
0.723	1.	3831259	3829346	3827434	3825522	3823611	3821700	3819790	3817880	3815971	3814063	1
0.724	1.	3812155	3810247	3808340	3806434	3804528	3802622	3800718	3798813	3796909	3795006	1
0.725	1.	3793103	3791201	3789300	3787398	3785498	3783597	3781698	3779799	3777900	3776002	1
0.726	1.	3774105	3772208	3770311	3768415	3766520	3764625	3762731	3760837	3758943	3757050	1
0.727	1.	3755158	3753266	3751375	3749484	3747594	3745704	3743815	3741927	3740038	3738151	1
0.728	1.	3736264	3734377	3732491	3730606	3728720	3726836	3724952	3723068	3721185	3719303	1
0.729	1.	3717421	3715540	3713659	3711778	3709899	3708019	3706140	3704262	3702384	3700507	1
0.730	1.	3698630	3696754	3694878	3693003	3691128	3689254	3687380	3685507	3683634	3681762	1
0.731	1.	3679891	3678019	3676149	3674279	3672409	3670540	3668671	3666803	3664936	3663069	1
0.732	1.	3661202	3659336	3657471	3655606	3653741	3651877	3650014	3648151	3646288	3644426	1
0.733	1.	3642565	3640704	3638843	3636983	3635124	3633265	3631407	3629549	3627691	3625835	1
0.734	1.	3623978	3622122	3620267	3618412	3616558	3614704	3612850	3610998	3609145	3607293	1
0.735	1.	3605442	3603591	3601741	3599891	3598042	3596193	3594345	3592497	3590650	3588803	1
0.736	1.	3586956	3585111	3583265	3581421	3579576	3577733	3575889	3574046	3572204	3570362	1
0.737	1.	3568521	3566680	3564840	3563000	3561161	3559322	3557484	3555646	3553809	3551972	0
0.738	1.	3550135	3548300	3546464	3544630	3542795	3540961	3539128	3537295	3535463	3533631	0
0.739	1.	3531800	3529969	3528139	3526309	3524479	3522650	3520822	3518994	3517167	3515340	0
0.740	1.	3513514	3511688	3509862	3508037	3506213	3504389	3502565	3500743	3498920	3497098	0
0.741	1.	3495277	3493456	3491635	3489815	3487996	3486177	3484358	3482540	3480723	3478905	0
0.742	1.	3477089	3475273	3473457	3471642	3469828	3468013	3466200	3464387	3462574	3460762	0
0.743	1.	3458950	3457139	3455328	3453518	3451708	3449899	3448090	3446282	3444474	3442667	0
0.744	1.	3440860	3439054	3437248	3435443	3433638	3431833	3430030	3428226	3426423	3424621	0
0.745	1.	3422819	3421017	3419216	3417416	3415616	3413816	3412017	3410219	3408420	3406623	0
0.746	1.	3404826	3403029	3401233	3399437	3397642	3395847	3394053	3392259	3390466	3388673	0
0.747	1.	3386881	3385089	3383298	3381507	3379716	3377926	3376137	3374348	3372559	3370771	0
0.748	1.	3368984	3367197	3365410	3363624	3361839	3360053	3358269	3356485	3354701	3352918	0
0.749	1.	3351135	3349353	3347571	3345789	3344009	3342228	3340448	3338669	3336890	3335111	0

TABLE 2

Reciprocals

X		.0000	.0001	.0002	.0003	.0004	.0005	.0006	.0007	.0008	.0009	Δ_2
0.750	1.	3333333	3331556	3329779	3328002	3326226	3324450	3322675	3320900	3319126	3317352	0
0.751	1.	3315579	3313806	3312034	3310262	3308491	3306720	3304949	3303179	3301410	3299641	0
0.752	1.	3297872	3296104	3294337	3292569	3290803	3289037	3287271	3285506	3283741	3281976	0
0.753	1.	3280212	3278449	3276686	3274924	3273162	3271400	3269639	3267878	3266118	3264359	0
0.754	1.	3262599	3260841	3259082	3257325	3255567	3253810	3252054	3250298	3248543	3246788	0
0.755	1.	3245033	3243279	3241525	3239772	3238020	3236267	3234516	3232764	3231013	3229263	0
0.756	1.	3227513	3225764	3224015	3222266	3220518	3218771	3217023	3215277	3213531	3211785	0
0.757	1.	3210040	3208295	3206550	3204807	3203063	3201320	3199578	3197836	3196094	3194353	0
0.758	1.	3192612	3190872	3189132	3187393	3185654	3183916	3182178	3180440	3178703	3176967	0
0.759	1.	3175231	3173495	3171760	3170025	3168291	3166557	3164824	3163091	3161358	3159626	0
0.760	1.	3157895	3156164	3154433	3152703	3150973	3149244	3147515	3145787	3144059	3142331	0
0.761	1.	3140604	3138878	3137152	3135426	3133701	3131976	3130252	3128528	3126805	3125082	0
0.762	1.	3123360	3121638	3119916	3118195	3116474	3114754	3113034	3111315	3109596	3107878	0
0.763	1.	3106160	3104442	3102725	3101009	3099293	3097577	3095862	3094147	3092433	3090719	0
0.764	1.	3089005	3087292	3085580	3083868	3082156	3080445	3078734	3077024	3075314	3073604	0
0.765	1.	3071895	3070187	3068479	3066771	3065064	3063357	3061651	3059945	3058240	3056535	0
0.766	1.	3054830	3053126	3051423	3049719	3048017	3046314	3044613	3042911	3041210	3039510	0
0.767	1.	3037810	3036110	3034411	3032712	3031014	3029316	3027619	3025922	3024225	3022529	0
0.768	1.	3020833	3019138	3017443	3015749	3014055	3012362	3010669	3008976	3007284	3005592	0
0.769	1.	3003901	3002210	3000520	2998830	2997141	2995452	2993763	2992075	2990387	2988700	0
0.770	1.	2987013	2985327	2983641	2981955	2980270	2978585	2976901	2975217	2973534	2971851	0
0.771	1.	2970169	2968487	2966805	2965124	2963443	2961763	2960083	2958403	2956725	2955046	0
0.772	1.	2953368	2951690	2950013	2948336	2946660	2944984	2943308	2941633	2939959	2938284	0
0.773	1.	2936611	2934937	2933264	2931592	2929920	2928248	2926577	2924906	2923236	2921566	0
0.774	1.	2919897	2918228	2916559	2914891	2913223	2911556	2909889	2908223	2906557	2904891	0
0.775	1.	2903226	2901561	2899897	2898233	2896569	2894906	2893244	2891582	2889920	2888259	0
0.776	1.	2886598	2884937	2883277	2881618	2879959	2878300	2876642	2874984	2873326	2871669	0
0.777	1.	2870013	2868357	2866701	2865046	2863391	2861736	2860082	2858429	2856775	2855123	0
0.778	1.	2853470	2851819	2850167	2848516	2846865	2845215	2843565	2841916	2840267	2838619	0
0.779	1.	2836970	2835323	2833676	2832029	2830382	2828736	2827091	2825446	2823801	2822157	0
0.780	1.	2820513	2818869	2817226	2815584	2813942	2812300	2810658	2809018	2807377	2805737	0
0.781	1.	2804097	2802458	2800819	2799181	2797543	2795905	2794268	2792631	2790995	2789359	0
0.782	1.	2787724	2786089	2784454	2782820	2781186	2779553	2777920	2776287	2774655	2773023	0
0.783	1.	2771392	2769761	2768131	2766501	2764871	2763242	2761613	2759985	2758357	2756729	0
0.784	1.	2755102	2753475	2751849	2750223	2748598	2746973	2745348	2743724	2742100	2740476	0
0.785	1.	2738853	2737231	2735609	2733987	2732366	2730745	2729124	2727504	2725884	2724265	0
0.786	1.	2722646	2721028	2719410	2717792	2716175	2714558	2712942	2711326	2709710	2708095	0
0.787	1.	2706480	2704866	2703252	2701638	2700025	2698413	2696800	2695189	2693577	2691966	0
0.788	1.	2690355	2688745	2687135	2685526	2683917	2682308	2680700	2679092	2677485	2675878	0
0.789	1.	2674271	2672665	2671059	2669454	2667849	2666244	2664640	2663037	2661433	2659830	0
0.790	1.	2658228	2656626	2655024	2653423	2651822	2650221	2648621	2647022	2645422	2643823	0
0.791	1.	2642225	2640627	2639029	2637432	2635835	2634239	2632643	2631047	2629452	2627857	0
0.792	1.	2626263	2624669	2623075	2621482	2619889	2618297	2616705	2615113	2613522	2611931	0
0.793	1.	2610340	2608750	2607161	2605572	2603983	2602394	2600806	2599219	2597632	2596045	0
0.794	1.	2594458	2592872	2591287	2589702	2588117	2586532	2584948	2583365	2581782	2580199	0
0.795	1.	2578616	2577034	2575453	2573871	2572291	2570710	2569130	2567551	2565971	2564392	0
0.796	1.	2562814	2561236	2559658	2558081	2556504	2554928	2553352	2551776	2550201	2548626	0
0.797	1.	2547051	2545477	2543904	2542330	2540757	2539185	2537613	2536041	2534470	2532899	0
0.798	1.	2531328	2529758	2528188	2526619	2525050	2523482	2521913	2520346	2518778	2517211	0
0.799	1.	2515645	2514078	2512513	2510947	2509382	2507817	2506253	2504689	2503126	2501563	0

TABLE 2

Reciprocals

X		.0000	.0001	.0002	.0003	.0004	.0005	.0006	.0007	.0008	.0009	Δ_2
0.800	1.	2500000	2498438	2496876	2495314	2493753	2492192	2490632	2489072	2487512	2485953	0
0.801	1.	2484394	2482836	2481278	2479720	2478163	2476606	2475050	2473494	2471938	2470383	0
0.802	1.	2468828	2467273	2465719	2464166	2462612	2461059	2459507	2457954	2456403	2454851	0
0.803	1.	2453300	2451749	2450199	2448649	2447100	2445551	2444002	2442454	2440906	2439358	0
0.804	1.	2437811	2436264	2434718	2433172	2431626	2430081	2428536	2426991	2425447	2423904	0
0.805	1.	2422360	2420817	2419275	2417733	2416191	2414649	2413108	2411568	2410027	2408487	0
0.806	1.	2406948	2405409	2403870	2402332	2400794	2399256	2397719	2396182	2394646	2393109	0
0.807	1.	2391574	2390038	2388503	2386969	2385435	2383901	2382367	2380834	2379302	2377770	0
0.808	1.	2376238	2374706	2373175	2371644	2370114	2368584	2367054	2365525	2363996	2362468	0
0.809	1.	2360939	2359412	2357884	2356357	2354831	2353304	2351779	2350253	2348728	2347203	0
0.810	1.	2345679	2344155	2342631	2341108	2339585	2338063	2336541	2335019	2333498	2331977	0
0.811	1.	2330456	2328936	2327416	2325897	2324378	2322859	2321341	2319823	2318305	2316788	0
0.812	1.	2315271	2313754	2312238	2310723	2309207	2307692	2306178	2304663	2303150	2301636	0
0.813	1.	2300123	2298610	2297098	2295586	2294074	2292563	2291052	2289542	2288031	2286522	0
0.814	1.	2285012	2283503	2281995	2280486	2278978	2277471	2275964	2274457	2272950	2271444	0
0.815	1.	2269939	2268433	2266928	2265424	2263920	2262416	2260912	2259409	2257906	2256404	0
0.816	1.	2254902	2253400	2251899	2250398	2248898	2247397	2245898	2244398	2242899	2241400	0
0.817	1.	2239902	2238404	2236906	2235409	2233912	2232416	2230920	2229424	2227929	2226434	0
0.818	1.	2224939	2223445	2221951	2220457	2218964	2217471	2215978	2214486	2212995	2211503	0
0.819	1.	2210012	2208522	2207031	2205541	2204052	2202563	2201074	2199585	2198097	2196609	0
0.820	1.	2195122	2193635	2192148	2190662	2189176	2187690	2186205	2184720	2183236	2181752	0
0.821	1.	2180268	2178785	2177301	2175819	2174336	2172855	2171373	2169892	2168411	2166930	0
0.822	1.	2165450	2163970	2162491	2161012	2159533	2158055	2156577	2155099	2153622	2152145	0
0.823	1.	2150668	2149192	2147716	2146241	2144766	2143291	2141816	2140342	2138869	2137395	0
0.824	1.	2135922	2134450	2132977	2131506	2130034	2128563	2127092	2125621	2124151	2122682	0
0.825	1.	2121212	2119743	2118274	2116806	2115338	2113870	2112403	2110936	2109470	2108003	0
0.826	1.	2106538	2105072	2103607	2102142	2100678	2099214	2097750	2096286	2094823	2093361	0
0.827	1.	2091898	2090436	2088975	2087514	2086053	2084592	2083132	2081672	2080213	2078753	0
0.828	1.	2077295	2075836	2074378	2072920	2071463	2070006	2068549	2067093	2065637	2064181	0
0.829	1.	2062726	2061271	2059817	2058362	2056909	2055455	2054002	2052549	2051097	2049645	0
0.830	1.	2048193	2046741	2045290	2043840	2042389	2040939	2039490	2038040	2036591	2035143	0
0.831	1.	2033694	2032246	2030799	2029352	2027905	2026458	2025012	2023566	2022121	2020676	0
0.832	1.	2019231	2017786	2016342	2014898	2013455	2012012	2010569	2009127	2007685	2006243	0
0.833	1.	2004802	2003361	2001920	2000480	1999040	1997600	1996161	1994722	1993284	1991846	0
0.834	1.	1990408	1988970	1987533	1986096	1984660	1983223	1981788	1980352	1978917	1977482	0
0.835	1.	1976048	1974614	1973180	1971747	1970314	1968881	1967449	1966016	1964585	1963153	0
0.836	1.	1961722	1960292	1958862	1957432	1956002	1954573	1953144	1951715	1950287	1948859	0
0.837	1.	1947431	1946004	1944577	1943151	1941724	1940298	1938873	1937448	1936023	1934598	0
0.838	1.	1933174	1931750	1930327	1928904	1927481	1926058	1924636	1923214	1921793	1920372	0
0.839	1.	1918951	1917531	1916111	1914691	1913271	1911852	1910434	1909015	1907597	1906179	0
0.840	1.	1904762	1903345	1901928	1900512	1899096	1897680	1896265	1894850	1893435	1892020	0
0.841	1.	1890606	1889193	1887779	1886366	1884954	1883541	1882129	1880718	1879306	1877895	0
0.842	1.	1876485	1875074	1873664	1872255	1870845	1869436	1868028	1866619	1865211	1863804	0
0.843	1.	1862396	1860989	1859583	1858176	1856770	1855365	1853959	1852554	1851150	1849745	0
0.844	1.	1848341	1846938	1845534	1844131	1842729	1841326	1839924	1838523	1837121	1835720	0
0.845	1.	1834320	1832919	1831519	1830119	1828720	1827321	1825922	1824524	1823126	1821728	0
0.846	1.	1820331	1818934	1817537	1816141	1814745	1813349	1811954	1810559	1809164	1807769	0
0.847	1.	1806375	1804982	1803588	1802195	1800802	1799410	1798018	1796626	1795235	1793844	0
0.848	1.	1792453	1791062	1789672	1788282	1786893	1785504	1784115	1782727	1781338	1779951	0
0.849	1.	1778563	1777176	1775789	1774402	1773016	1771630	1770245	1768860	1767475	1766090	0

TABLE 2

Reciprocals

X		.0000	.0001	.0002	.0003	.0004	.0005	.0006	.0007	.0008	.0009	Δ_2
0.850	1.	1764706	1763322	1761938	1760555	1759172	1757790	1756407	1755025	1753644	1752262	0
0.851	1.	1750881	1749501	1748120	1746740	1745361	1743981	1742602	1741223	1739845	1738467	0
0.852	1.	1737089	1735712	1734335	1732958	1731581	1730205	1728829	1727454	1726079	1724704	0
0.853	1.	1723329	1721955	1720581	1719208	1717835	1716462	1715089	1713717	1712345	1710973	0
0.854	1.	1709602	1708231	1706860	1705490	1704120	1702750	1701381	1700012	1698643	1697275	0
0.855	1.	1695906	1694539	1693171	1691804	1690437	1689071	1687705	1686339	1684973	1683608	0
0.856	1.	1682243	1680878	1679514	1678150	1676787	1675423	1674060	1672698	1671335	1669973	0
0.857	1.	1668611	1667250	1665889	1664528	1663168	1661808	1660448	1659088	1657729	1656370	0
0.858	1.	1655012	1653653	1652295	1650938	1649581	1648224	1646867	1645511	1644155	1642799	0
0.859	1.	1641444	1640088	1638734	1637379	1636025	1634671	1633318	1631965	1630612	1629259	0
0.860	1.	1627907	1626555	1625203	1623852	1622501	1621150	1619800	1618450	1617100	1615751	0
0.861	1.	1614402	1613053	1611705	1610356	1609009	1607661	1606314	1604967	1603620	1602274	0
0.862	1.	1600928	1599582	1598237	1596892	1595547	1594203	1592859	1591515	1590172	1588828	0
0.863	1.	1587486	1586143	1584801	1583459	1582117	1580776	1579435	1578094	1576754	1575414	0
0.864	1.	1574074	1572735	1571396	1570057	1568718	1567380	1566042	1564705	1563367	1562030	0
0.865	1.	1560694	1559357	1558021	1556686	1555350	1554015	1552680	1551346	1550012	1548678	0
0.866	1.	1547344	1546011	1544678	1543345	1542013	1540681	1539349	1538018	1536687	1535356	0
0.867	1.	1534025	1532695	1531365	1530036	1528706	1527378	1526049	1524721	1523392	1522065	0
0.868	1.	1520737	1519410	1518083	1516757	1515431	1514105	1512779	1511454	1510129	1508804	0
0.869	1.	1507480	1506156	1504832	1503509	1502185	1500863	1499540	1498218	1496896	1495574	0
0.870	1.	1494253	1492932	1491611	1490291	1488971	1487651	1486331	1485012	1483693	1482375	0
0.871	1.	1481056	1479738	1478421	1477103	1475786	1474469	1473153	1471837	1470521	1469205	0
0.872	1.	1467890	1466575	1465260	1463946	1462632	1461318	1460005	1458691	1457379	1456066	0
0.873	1.	1454754	1453442	1452130	1450819	1449508	1448197	1446886	1445576	1444266	1442957	0
0.874	1.	1441648	1440339	1439030	1437722	1436414	1435106	1433798	1432491	1431184	1429878	0
0.875	1.	1428571	1427265	1425960	1424654	1423349	1422045	1420740	1419436	1418132	1416828	0
0.876	1.	1415525	1414222	1412919	1411617	1410315	1409013	1407712	1406410	1405109	1403809	0
0.877	1.	1402509	1401209	1399909	1398609	1397310	1396011	1394713	1393415	1392117	1390819	0
0.878	1.	1389522	1388225	1386928	1385631	1384335	1383039	1381744	1380448	1379153	1377859	0
0.879	1.	1376564	1375270	1373976	1372683	1371390	1370097	1368804	1367512	1366220	1364928	0
0.880	1.	1363636	1362345	1361054	1359764	1358473	1357183	1355894	1354604	1353315	1352026	0
0.881	1.	1350738	1349450	1348162	1346874	1345587	1344299	1343013	1341726	1340440	1339154	0
0.882	1.	1337868	1336583	1335298	1334013	1332729	1331445	1330161	1328877	1327594	1326311	0
0.883	1.	1325028	1323746	1322464	1321182	1319900	1318619	1317338	1316057	1314777	1313497	0
0.884	1.	1312217	1310938	1309658	1308380	1307101	1305822	1304544	1303267	1301989	1300712	0
0.885	1.	1299435	1298158	1296882	1295606	1294330	1293055	1291780	1290505	1289230	1287956	0
0.886	1.	1286682	1285408	1284134	1282861	1281588	1280316	1279044	1277771	1276500	1275228	0
0.887	1.	1273957	1272686	1271416	1270145	1268875	1267606	1266336	1265067	1263798	1262530	0
0.888	1.	1261261	1259993	1258726	1257458	1256191	1254924	1253657	1252391	1251125	1249859	0
0.889	1.	1248594	1247329	1246064	1244799	1243535	1242271	1241007	1239744	1238481	1237218	0
0.890	1.	1235955	1234693	1233431	1232169	1230907	1229646	1228385	1227125	1225864	1224604	0
0.891	1.	1223345	1222085	1220826	1219567	1218308	1217050	1215792	1214534	1213277	1212019	0
0.892	1.	1210762	1209506	1208249	1206993	1205737	1204482	1203227	1201972	1200717	1199462	0
0.893	1.	1198208	1196954	1195701	1194448	1193195	1191942	1190689	1189437	1188185	1186934	0
0.894	1.	1185682	1184431	1183180	1181930	1180680	1179430	1178180	1176931	1175682	1174433	0
0.895	1.	1173184	1171936	1170688	1169440	1168193	1166946	1165699	1164452	1163206	1161960	0
0.896	1.	1160714	1159469	1158224	1156979	1155734	1154490	1153246	1152002	1150758	1149515	0
0.897	1.	1148272	1147029	1145787	1144545	1143303	1142061	1140820	1139579	1138338	1137098	0
0.898	1.	1135857	1134618	1133378	1132138	1130899	1129661	1128422	1127184	1125946	1124708	0
0.899	1.	1123471	1122233	1120996	1119760	1118523	1117287	1116052	1114816	1113581	1112346	0

TABLE 2

Reciprocals

X		.0000	.0001	.0002	.0003	.0004	.0005	.0006	.0007	.0008	.0009	Δ_2
0.900	1.	1111111	1109877	1108643	1107409	1106175	1104942	1103709	1102476	1101243	1100011	0
0.901	1.	1098779	1097547	1096316	1095085	1093854	1092623	1091393	1090163	1088933	1087704	0
0.902	1.	1086474	1085246	1084017	1082788	1081560	1080332	1079105	1077877	1076650	1075424	0
0.903	1.	1074197	1072971	1071745	1070519	1069294	1068069	1066844	1065619	1064395	1063171	0
0.904	1.	1061947	1060723	1059500	1058277	1057054	1055832	1054610	1053388	1052166	1050945	0
0.905	1.	1049724	1048503	1047282	1046062	1044842	1043622	1042403	1041184	1039965	1038746	0
0.906	1.	1037528	1036309	1035092	1033874	1032657	1031440	1030223	1029006	1027790	1026574	0
0.907	1.	1025358	1024143	1022928	1021713	1020498	1019284	1018070	1016856	1015642	1014429	0
0.908	1.	1013216	1012003	1010791	1009578	1008366	1007155	1005943	1004732	1003521	1002310	0
0.909	1.	1001100	0999890	0998680	0997471	0996261	0995052	0993843	0992635	0991427	0990219	0
0.910	1.	0989011	0987804	0986596	0985389	0984183	0982976	0981770	0980564	0979359	0978153	0
0.911	1.	0976948	0975744	0974539	0973335	0972131	0970927	0969724	0968520	0967317	0966115	0
0.912	1.	0964912	0963710	0962508	0961307	0960105	0958904	0957703	0956503	0955302	0954102	0
0.913	1.	0952903	0951703	0950504	0949305	0948106	0946907	0945709	0944511	0943314	0942116	0
0.914	1.	0940919	0939722	0938525	0937329	0936133	0934937	0933742	0932546	0931351	0930156	0
0.915	1.	0928962	0927767	0926573	0925380	0924186	0922993	0921800	0920607	0919415	0918223	0
0.916	1.	0917031	0915839	0914647	0913456	0912265	0911075	0909884	0908694	0907504	0906315	0
0.917	1.	0905125	0903936	0902747	0901559	0900371	0899183	0897995	0896807	0895620	0894433	0
0.918	1.	0893246	0892060	0890873	0889687	0888502	0887316	0886131	0884946	0883761	0882577	0
0.919	1.	0881393	0880209	0879025	0877842	0876659	0875476	0874293	0873111	0871929	0870747	0
0.920	1.	0869565	0868384	0867203	0866022	0864841	0863661	0862481	0861301	0860122	0858942	0
0.921	1.	0857763	0856585	0855406	0854228	0853050	0851872	0850694	0849517	0848340	0847163	0
0.922	1.	0845987	0844811	0843635	0842459	0841284	0840108	0838933	0837759	0836584	0835410	0
0.923	1.	0834236	0833063	0831889	0830716	0829543	0828370	0827198	0826026	0824854	0823682	0
0.924	1.	0822511	0821340	0820169	0818998	0817828	0816658	0815488	0814318	0813149	0811980	0
0.925	1.	0810811	0809642	0808474	0807306	0806138	0804970	0803803	0802636	0801469	0800302	0
0.926	1.	0799136	0797970	0796804	0795639	0794473	0793308	0792143	0790979	0789814	0788650	0
0.927	1.	0787487	0786323	0785160	0783997	0782834	0781671	0780509	0779347	0778185	0777023	0
0.928	1.	0775862	0774701	0773540	0772380	0771219	0770059	0768899	0767740	0766581	0765421	0
0.929	1.	0764263	0763104	0761946	0760788	0759630	0758472	0757315	0756158	0755001	0753844	0
0.930	1.	0752688	0751532	0750376	0749221	0748065	0746910	0745755	0744601	0743446	0742292	0
0.931	1.	0741139	0739985	0738832	0737679	0736526	0735373	0734221	0733069	0731917	0730765	0
0.932	1.	0729614	0728463	0727312	0726161	0725011	0723861	0722711	0721561	0720412	0719263	0
0.933	1.	0718114	0716965	0715817	0714668	0713520	0712373	0711225	0710078	0708931	0707785	0
0.934	1.	0706638	0705492	0704346	0703200	0702055	0700910	0699765	0698620	0697475	0696331	0
0.935	1.	0695187	0694043	0692900	0691757	0690614	0689471	0688328	0687186	0686044	0684902	0
0.936	1.	0683761	0682619	0681478	0680337	0679197	0678057	0676917	0675777	0674637	0673498	0
0.937	1.	0672359	0671220	0670081	0668943	0667805	0666667	0665529	0664392	0663254	0662117	0
0.938	1.	0660981	0659844	0658708	0657572	0656436	0655301	0654166	0653031	0651896	0650762	0
0.939	1.	0649627	0648493	0647359	0646226	0645093	0643960	0642827	0641694	0640562	0639430	0
0.940	1.	0638298	0637166	0636035	0634904	0633773	0632642	0631512	0630382	0629252	0628122	0
0.941	1.	0626993	0625863	0624734	0623606	0622477	0621349	0620221	0619093	0617966	0616838	0
0.942	1.	0615711	0614584	0613458	0612332	0611205	0610080	0608954	0607829	0606703	0605579	0
0.943	1.	0604454	0603329	0602205	0601081	0599958	0598834	0597711	0596588	0595465	0594343	0
0.944	1.	0593220	0592098	0590976	0589855	0588734	0587612	0586492	0585371	0584251	0583130	0
0.945	1.	0582011	0580891	0579771	0578652	0577533	0576415	0575296	0574178	0573060	0571942	0
0.946	1.	0570825	0569707	0568590	0567473	0566357	0565240	0564124	0563008	0561893	0560777	0
0.947	1.	0559662	0558547	0557432	0556318	0555204	0554090	0552976	0551862	0550749	0549636	0
0.948	1.	0548523	0547411	0546298	0545186	0544074	0542963	0541851	0540740	0539629	0538518	0
0.949	1.	0537408	0536298	0535188	0534078	0532968	0531859	0530750	0529641	0528532	0527424	0

TABLE 2

Reciprocals

X		.0000	.0001	.0002	.0003	.0004	.0005	.0006	.0007	.0008	.0009	Δ_2
0.950	1.	0526316	0525208	0524100	0522993	0521886	0520779	0519672	0518565	0517459	0516353	0
0.951	1.	0515247	0514142	0513036	0511931	0510826	0509721	0508617	0507513	0506409	0505305	0
0.952	1.	0504202	0503098	0501995	0500893	0499790	0498688	0497586	0496484	0495382	0494281	0
0.953	1.	0493179	0492078	0490978	0489877	0488777	0487677	0486577	0485478	0484378	0483279	0
0.954	1.	0482180	0481082	0479983	0478885	0477787	0476689	0475592	0474495	0473398	0472301	0
0.955	1.	0471204	0470108	0469012	0467916	0466820	0465725	0464630	0463535	0462440	0461345	0
0.956	1.	0460251	0459157	0458063	0456970	0455876	0454783	0453690	0452597	0451505	0450413	0
0.957	1.	0449321	0448229	0447137	0446046	0444955	0443864	0442774	0441683	0440593	0439503	0
0.958	1.	0438413	0437324	0436235	0435146	0434057	0432968	0431880	0430792	0429704	0428616	0
0.959	1.	0427529	0426441	0425354	0424268	0423181	0422095	0421009	0419923	0418837	0417752	0
0.960	1.	0416667	0415582	0414497	0413412	0412328	0411244	0410160	0409077	0407993	0406910	0
0.961	1.	0405827	0404745	0403662	0402580	0401498	0400416	0399334	0398253	0397172	0396091	0
0.962	1.	0395010	0393930	0392850	0391770	0390690	0389610	0388531	0387452	0386373	0385294	0
0.963	1.	0384216	0383138	0382060	0380982	0379905	0378827	0377750	0376673	0375597	0374520	0
0.964	1.	0373444	0372368	0371292	0370217	0369141	0368066	0366991	0365917	0364842	0363768	0
0.965	1.	0362694	0361621	0360547	0359474	0358401	0357328	0356255	0355183	0354111	0353039	0
0.966	1.	0351967	0350895	0349824	0348753	0347682	0346611	0345541	0344471	0343401	0342331	0
0.967	1.	0341262	0340192	0339123	0338054	0336986	0335917	0334849	0333781	0332713	0331646	0
0.968	1.	0330579	0329511	0328445	0327378	0326311	0325245	0324179	0323113	0322048	0320983	0
0.969	1.	0319917	0318853	0317788	0316723	0315659	0314595	0313531	0312468	0311404	0310341	0
0.970	1.	0309278	0308216	0307153	0306091	0305029	0303967	0302905	0301844	0300783	0299722	0
0.971	1.	0298661	0297601	0296540	0295480	0294420	0293361	0292301	0291242	0290183	0289124	0
0.972	1.	0288066	0287008	0285949	0284891	0283834	0282776	0281719	0280662	0279605	0278549	0
0.973	1.	0277492	0276436	0275380	0274324	0273269	0272214	0271159	0270104	0269049	0267995	0
0.974	1.	0266940	0265886	0264833	0263779	0262726	0261673	0260620	0259567	0258515	0257462	0
0.975	1.	0256410	0255358	0254307	0253255	0252204	0251153	0250103	0249052	0248002	0246952	0
0.976	1..	0245902	0244852	0243802	0242753	0241704	0240655	0239607	0238558	0237510	0236462	0
0.977	1.	0235415	0234367	0233320	0232273	0231226	0230179	0229133	0228086	0227040	0225994	0
0.978	1.	0224949	0223903	0222858	0221813	0220769	0219724	0218680	0217636	0216592	0215548	0
0.979	1.	0214505	0213461	0212418	0211375	0210333	0209290	0208248	0207206	0206165	0205123	0
0.980	1.	0204082	0203041	0202000	0200959	0199918	0198878	0197838	0196798	0195759	0194719	0
0.981	1.	0193680	0192641	0191602	0190564	0189525	0188487	0187449	0186411	0185374	0184336	0
0.982	1.	0183299	0182262	0181226	0180189	0179153	0178117	0177081	0176046	0175010	0173975	0
0.983	1.	0172940	0171905	0170871	0169836	0168802	0167768	0166734	0165701	0164668	0163635	0
0.984	1.	0162602	0161569	0160536	0159504	0158472	0157440	0156409	0155377	0154346	0153315	0
0.985	1.	0152284	0151254	0150223	0149193	0148163	0147133	0146104	0145075	0144045	0143017	0
0.986	1.	0141988	0140959	0139931	0138903	0137875	0136847	0135820	0134793	0133766	0132739	0
0.987	1.	0131712	0130686	0129660	0128634	0127608	0126582	0125557	0124532	0123507	0122482	0
0.988	1.	0121457	0120433	0119409	0118385	0117361	0116338	0115315	0114291	.0113269	0112246	0
0.989	1.	0111223	0110201	0109179	0108157	0107136	0106114	0105093	0104072	0103051	0102031	0
0.990	1.	0101010	0099990	0098970	0097950	0096931	0095911	0094892	0093873	0092854	0091836	0
0.991	1.	0090817	0089799	0088781	0087764	0086746	0085729	0084712	0083695	0082678	0081661	0
0.992	1.	0080645	0079629	0078613	0077598	0076582	0075567	0074552	0073537	0072522	0071508	0
0.993	1.	0070493	0069479	0068466	0067452	0066438	0065425	0064412	0063399	0062387	0061374	0
0.994	1.	0060362	0059350	0058338	0057327	0056315	0055304	0054293	0053282	0052272	0051261	0
0.995	1.	0050251	0049241	0048232	0047222	0046213	0045203	0044194	0043186	0042177	0041169	0
0.996	1.	0040161	0039153	0038145	0037137	0036130	0035123	0034116	0033109	0032103	0031096	0
0.997	1.	0030090	0029084	0028079	0027073	0026068	0025063	0024058	0023053	0022049	0021044	0
0.998	1.	0020040	0019036	0018032	0017029	0016026	0015023	0014020	0013017	0012014	0011012	0
0.999	1.	0010010	0009008	0008006	0007005	0006004	0005003	0004002	0003001	0002000	0001000	0

TABLE 3

Squares

X		.0000	.0001	.0002	.0003	.0004	.0005	.0006	.0007	.0008	.0009	Δ_2
0.100	0.0	1000000	1002001	1004004	1006009	1008016	1010025	1012036	1014049	1016064	1018081	2
0.101	0.0	1020100	1022121	1024144	1026169	1028196	1030225	1032256	1034289	1036324	1038361	2
0.102	0.0	1040400	1042441	1044484	1046529	1048576	1050625	1052676	1054729	1056784	1058841	2
0.103	0.0	1060900	1062961	1065024	1067089	1069156	1071225	1073296	1075369	1077444	1079521	2
0.104	0.0	1081600	1083681	1085764	1087849	1089936	1092025	1094116	1096209	1098304	1100401	2
0.105	0.0	1102500	1104601	1106704	1108809	1110916	1113025	1115136	1117249	1119364	1121481	2
0.106	0.0	1123600	1125721	1127844	1129969	1132096	1134225	1136356	1138489	1140624	1142761	2
0.107	0.0	1144900	1147041	1149184	1151329	1153476	1155625	1157776	1159929	1162084	1164241	2
0.108	0.0	1166400	1168561	1170724	1172889	1175056	1177225	1179396	1181569	1183744	1185921	2
0.109	0.0	1188100	1190281	1192464	1194649	1196836	1199025	1201216	1203409	1205604	1207801	2
0.110	0.0	1210000	1212201	1214404	1216609	1218816	1221025	1223236	1225449	1227664	1229881	2
0.111	0.0	1232100	1234321	1236544	1238769	1240996	1243225	1245456	1247689	1249924	1252161	2
0.112	0.0	1254400	1256641	1258884	1261129	1263376	1265625	1267876	1270129	1272384	1274641	2
0.113	0.0	1276900	1279161	1281424	1283689	1285956	1288225	1290496	1292769	1295044	1297321	2
0.114	0.0	1299600	1301881	1304164	1306449	1308736	1311025	1313316	1315609	1317904	1320201	2
0.115	0.0	1322500	1324801	1327104	1329409	1331716	1334025	1336336	1338649	1340964	1343281	2
0.116	0.0	1345600	1347921	1350244	1352569	1354896	1357225	1359556	1361889	1364224	1366561	2
0.117	0.0	1368900	1371241	1373584	1375929	1378276	1380625	1382976	1385329	1387684	1390041	2
0.118	0.0	1392400	1394761	1397124	1399489	1401856	1404225	1406596	1408969	1411344	1413721	2
0.119	0.0	1416100	1418481	1420864	1423249	1425636	1428025	1430416	1432809	1435204	1437601	2
0.120	0.0	1440000	1442401	1444804	1447209	1449616	1452025	1454436	1456849	1459264	1461681	2
0.121	0.0	1464100	1466521	1468944	1471369	1473796	1476225	1478656	1481089	1483524	1485961	2
0.122	0.0	1488400	1490841	1493284	1495729	1498176	1500625	1503076	1505529	1507984	1510441	2
0.123	0.0	1512900	1515361	1517824	1520289	1522756	1525225	1527696	1530169	1532644	1535121	2
0.124	0.0	1537600	1540081	1542564	1545049	1547536	1550025	1552516	1555009	1557504	1560001	2
0.125	0.0	1562500	1565001	1567504	1570009	1572516	1575025	1577536	1580049	1582564	1585081	2
0.126	0.0	1587600	1590121	1592644	1595169	1597696	1600225	1602756	1605289	1607824	1610361	2
0.127	0.0	1612900	1615441	1617984	1620529	1623076	1625625	1628176	1630729	1633284	1635841	2
0.128	0.0	1638400	1640961	1643524	1646089	1648656	1651225	1653796	1656369	1658944	1661521	2
0.129	0.0	1664100	1666681	1669264	1671849	1674436	1677025	1679616	1682209	1684804	1687401	2
0.130	0.0	1690000	1692601	1695204	1697809	1700416	1703025	1705636	1708249	1710864	1713481	2
0.131	0.0	1716100	1718721	1721344	1723969	1726596	1729225	1731856	1734489	1737124	1739761	2
0.132	0.0	1742400	1745041	1747684	1750329	1752976	1755625	1758276	1760929	1763584	1766241	2
0.133	0.0	1768900	1771561	1774224	1776889	1779556	1782225	1784896	1787569	1790244	1792921	2
0.134	0.0	1795600	1798281	1800964	1803649	1806336	1809025	1811716	1814409	1817104	1819801	2
0.135	0.0	1822500	1825201	1827904	1830609	1833316	1836025	1838736	1841449	1844164	1846881	2
0.136	0.0	1849600	1852321	1855044	1857769	1860496	1863225	1865956	1868689	1871424	1874161	2
0.137	0.0	1876900	1879641	1882384	1885129	1887876	1890625	1893376	1896129	1898884	1901641	2
0.138	0.0	1904400	1907161	1909924	1912689	1915456	1918225	1920996	1923769	1926544	1929321	2
0.139	0.0	1932100	1934881	1937664	1940449	1943236	1946025	1948816	1951609	1954404	1957201	2
0.140	0.0	1960000	1962801	1965604	1968409	1971216	1974025	1976836	1979649	1982464	1985281	2
0.141	0.0	1988100	1990921	1993744	1996569	1999396	2002225	2005056	2007889	2010724	2013561	2
0.142	0.0	2016400	2019241	2022084	2024929	2027776	2030625	2033476	2036329	2039184	2042041	2
0.143	0.0	2044900	2047761	2050624	2053489	2056356	2059225	2062096	2064969	2067844	2070721	2
0.144	0.0	2073600	2076481	2079364	2082249	2085136	2088025	2090916	2093809	2096704	2099601	2
0.145	0.0	2102500	2105401	2108304	2111209	2114116	2117025	2119936	2122849	2125764	2128681	2
0.146	0.0	2131600	2134521	2137444	2140369	2143296	2146225	2149156	2152089	2155024	2157961	2
0.147	0.0	2160900	2163841	2166784	2169729	2172676	2175625	2178576	2181529	2184484	2187441	2
0.148	0.0	2190400	2193361	2196324	2199289	2202256	2205225	2208196	2211169	2214144	2217121	2
0.149	0.0	2220100	2223081	2226064	2229049	2232036	2235025	2238016	2241009	2244004	2247001	2

TABLE 3

Squares

X		.0000	.0001	.0002	.0003	.0004	.0005	.0006	.0007	.0008	.0009	Δ_2
0.150	0.0	2250000	2253001	2256004	2259009	2262016	2265025	2268036	2271049	2274064	2277081	2
0.151	0.0	2280100	2283121	2286144	2289169	2292196	2295225	2298256	2301289	2304324	2307361	2
0.152	0.0	2310400	2313441	2316484	2319529	2322576	2325625	2328676	2331729	2334784	2337841	2
0.153	0.0	2340900	2343961	2347024	2350089	2353156	2356225	2359296	2362369	2365444	2368521	2
0.154	0.0	2371600	2374681	2377764	2380849	2383936	2387025	2390116	2393209	2396304	2399401	2
0.155	0.0	2402500	2405601	2408704	2411809	2414916	2418025	2421136	2424249	2427364	2430481	2
0.156	0.0	2433600	2436721	2439844	2442969	2446096	2449225	2452356	2455489	2458624	2461761	2
0.157	0.0	2464900	2468041	2471184	2474329	2477476	2480625	2483776	2486929	2490084	2493241	2
0.158	0.0	2496400	2499561	2502724	2505889	2509056	2512225	2515396	2518569	2521744	2524921	2
0.159	0.0	2528100	2531281	2534464	2537649	2540836	2544025	2547216	2550409	2553604	2556801	2
0.160	0.0	2560000	2563201	2566404	2569609	2572816	2576025	2579236	2582449	2585664	2588881	2
0.161	0.0	2592100	2595321	2598544	2601769	2604996	2608225	2611456	2614689	2617924	2621161	2
0.162	0.0	2624400	2627641	2630884	2634129	2637376	2640625	2643876	2647129	2650384	2653641	2
0.163	0.0	2656900	2660161	2663424	2666689	2669956	2673225	2676496	2679769	2683044	2686321	2
0.164	0.0	2689600	2692881	2696164	2699449	2702736	2706025	2709316	2712609	2715904	2719201	2
0.165	0.0	2722500	2725801	2729104	2732409	2735716	2739025	2742336	2745649	2748964	2752281	2
0.166	0.0	2755600	2758921	2762244	2765569	2768896	2772225	2775556	2778889	2782224	2785561	2
0.167	0.0	2788900	2792241	2795584	2798929	2802276	2805625	2808976	2812329	2815684	2819041	2
0.168	0.0	2822400	2825761	2829124	2832489	2835856	2839225	2842596	2845969	2849344	2852721	2
0.169	0.0	2856100	2859481	2862864	2866249	2869636	2873025	2876416	2879809	2883204	2886601	2
0.170	0.0	2890000	2893401	2896804	2900209	2903616	2907025	2910436	2913849	2917264	2920681	2
0.171	0.0	2924100	2927521	2930944	2934369	2937796	2941225	2944656	2948089	2951524	2954961	2
0.172	0.0	2958400	2961841	2965284	2968729	2972176	2975625	2979076	2982529	2985984	2989441	2
0.173	0.0	2992900	2996361	2999824	3003289	3006756	3010225	3013696	3017169	3020644	3024121	2
0.174	0.0	3027600	3031081	3034564	3038049	3041536	3045025	3048516	3052009	3055504	3059001	2
0.175	0.0	3062500	3066001	3069504	3073009	3076516	3080025	3083536	3087049	3090564	3094081	2
0.176	0.0	3097600	3101121	3104644	3108169	3111696	3115225	3118756	3122289	3125824	3129361	2
0.177	0.0	3132900	3136441	3139984	3143529	3147076	3150625	3154176	3157729	3161284	3164841	2
0.178	0.0	3168400	3171961	3175524	3179089	3182656	3186225	3189796	3193369	3196944	3200521	2
0.179	0.0	3204100	3207681	3211264	3214849	3218436	3222025	3225616	3229209	3232804	3236401	2
0.180	0.0	3240000	3243601	3247204	3250809	3254416	3258025	3261636	3265249	3268864	3272481	2
0.181	0.0	3276100	3279721	3283344	3286969	3290596	3294225	3297856	3301489	3305124	3308761	2
0.182	0.0	3312400	3316041	3319684	3323329	3326976	3330625	3334276	3337929	3341584	3345241	2
0.183	0.0	3348900	3352561	3356224	3359889	3363556	3367225	3370896	3374569	3378244	3381921	2
0.184	0.0	3385600	3389281	3392964	3396649	3400336	3404025	3407716	3411409	3415104	3418801	2
0.185	0.0	3422500	3426201	3429904	3433609	3437316	3441025	3444736	3448449	3452164	3455881	2
0.186	0.0	3459600	3463321	3467044	3470769	3474496	3478225	3481956	3485689	3489424	3493161	2
0.187	0.0	3496900	3500641	3504384	3508129	3511876	3515625	3519376	3523129	3526884	3530641	2
0.188	0.0	3534400	3538161	3541924	3545689	3549456	3553225	3556996	3560769	3564544	3568321	2
0.189	0.0	3572100	3575881	3579664	3583449	3587236	3591025	3594816	3598609	3602404	3606201	2
0.190	0.0	3610000	3613801	3617604	3621409	3625216	3629025	3632836	3636649	3640464	3644281	2
0.191	0.0	3648100	3651921	3655744	3659569	3663396	3667225	3671056	3674889	3678724	3682561	2
0.192	0.0	3686400	3690241	3694084	3697929	3701776	3705625	3709476	3713329	3717184	3721041	2
0.193	0.0	3724900	3728761	3732624	3736489	3740356	3744225	3748096	3751969	3755844	3759721	2
0.194	0.0	3763600	3767481	3771364	3775249	3779136	3783025	3786916	3790809	3794704	3798601	2
0.195	0.0	3802500	3806401	3810304	3814209	3818116	3822025	3825936	3829849	3833764	3837681	2
0.196	0.0	3841600	3845521	3849444	3853369	3857296	3861225	3865156	3869089	3873024	3876961	2
0.197	0.0	3880900	3884841	3888784	3892729	3896676	3900625	3904576	3908529	3912484	3916441	2
0.198	0.0	3920400	3924361	3928324	3932289	3936256	3940225	3944196	3948169	3952144	3956121	2
0.199	0.0	3960100	3964081	3968064	3972049	3976036	3980025	3984016	3988009	3992004	3996001	2

TABLE 3

Squares

X		.0000	.0001	.0002	.0003	.0004	.0005	.0006	.0007	.0008	.0009	Δ_2
0.200	0.0	4000000	4004001	4008004	4012009	4016016	4020025	4024036	4028049	4032064	4036081	2
0.201	0.0	4040100	4044121	4048144	4052169	4056196	4060225	4064256	4068289	4072324	4076361	2
0.202	0.0	4080400	4084441	4088484	4092529	4096576	4100625	4104676	4108729	4112784	4116841	2
0.203	0.0	4120900	4124961	4129024	4133089	4137156	4141225	4145296	4149369	4153444	4157521	2
0.204	0.0	4161600	4165681	4169764	4173849	4177936	4182025	4186116	4190209	4194304	4198401	2
0.205	0.0	4202500	4206601	4210704	4214809	4218916	4223025	4227136	4231249	4235364	4239481	2
0.206	0.0	4243600	4247721	4251844	4255969	4260096	4264225	4268356	4272489	4276624	4280761	2
0.207	0.0	4284900	4289041	4293184	4297329	4301476	4305625	4309776	4313929	4318084	4322241	2
0.208	0.0	4326400	4330561	4334724	4338889	4343056	4347225	4351396	4355569	4359744	4363921	2
0.209	0.0	4368100	4372281	4376464	4380649	4384836	4389025	4393216	4397409	4401604	4405801	2
0.210	0.0	4410000	4414201	4418404	4422609	4426816	4431025	4435236	4439449	4443664	4447881	2
0.211	0.0	4452100	4456321	4460544	4464769	4468996	4473225	4477456	4481689	4485924	4490161	2
0.212	0.0	4494400	4498641	4502884	4507129	4511376	4515625	4519876	4524129	4528384	4532641	2
0.213	0.0	4536900	4541161	4545424	4549689	4553956	4558225	4562496	4566769	4571044	4575321	2
0.214	0.0	4579600	4583881	4588164	4592449	4596736	4601025	4605316	4609609	4613904	4618201	2
0.215	0.0	4622500	4626801	4631104	4635409	4639716	4644025	4648336	4652649	4656964	4661281	2
0.216	0.0	4665600	4669921	4674244	4678569	4682896	4687225	4691556	4695889	4700224	4704561	2
0.217	0.0	4708900	4713241	4717584	4721929	4726276	4730625	4734976	4739329	4743684	4748041	2
0.218	0.0	4752400	4756761	4761124	4765489	4769856	4774225	4778596	4782969	4787344	4791721	2
0.219	0.0	4796100	4800481	4804864	4809249	4813636	4818025	4822416	4826809	4831204	4835601	2
0.220	0.0	4840000	4844401	4848804	4853209	4857616	4862025	4866436	4870849	4875264	4879681	2
0.221	0.0	4884100	4888521	4892944	4897369	4901796	4906225	4910656	4915089	4919524	4923961	2
0.222	0.0	4928400	4932841	4937284	4941729	4946176	4950625	4955076	4959529	4963984	4968441	2
0.223	0.0	4972900	4977361	4981824	4986289	4990756	4995225	4999696	5004169	5008644	5013121	2
0.224	0.0	5017600	5022081	5026564	5031049	5035536	5040025	5044516	5049009	5053504	5058001	2
0.225	0.0	5062500	5067001	5071504	5076009	5080516	5085025	5089536	5094049	5098564	5103081	2
0.226	0.0	5107600	5112121	5116644	5121169	5125696	5130225	5134756	5139289	5143824	5148361	2
0.227	0.0	5152900	5157441	5161984	5166529	5171076	5175625	5180176	5184729	5189284	5193841	2
0.228	0.0	5198400	5202961	5207524	5212089	5216656	5221225	5225796	5230369	5234944	5239521	2
0.229	0.0	5244100	5248681	5253264	5257849	5262436	5267025	5271616	5276209	5280804	5285401	2
0.230	0.0	5290000	5294601	5299204	5303809	5308416	5313025	5317636	5322249	5326864	5331481	2
0.231	0.0	5336100	5340721	5345344	5349969	5354596	5359225	5363856	5368489	5373124	5377761	2
0.232	0.0	5382400	5387041	5391684	5396329	5400976	5405625	5410276	5414929	5419584	5424241	2
0.233	0.0	5428900	5433561	5438224	5442889	5447556	5452225	5456896	5461569	5466244	5470921	2
0.234	0.0	5475600	5480281	5484964	5489649	5494336	5499025	5503716	5508409	5513104	5517801	2
0.235	0.0	5522500	5527201	5531904	5536609	5541316	5546025	5550736	5555449	5560164	5564881	2
0.236	0.0	5569600	5574321	5579044	5583769	5588496	5593225	5597956	5602689	5607424	5612161	2
0.237	0.0	5616900	5621641	5626384	5631129	5635876	5640625	5645376	5650129	5654884	5659641	2
0.238	0.0	5664400	5669161	5673924	5678689	5683456	5688225	5692996	5697769	5702544	5707321	2
0.239	0.0	5712100	5716881	5721664	5726449	5731236	5736025	5740816	5745609	5750404	5755201	2
0.240	0.0	5760000	5764801	5769604	5774409	5779216	5784025	5788836	5793649	5798464	5803281	2
0.241	0.0	5808100	5812921	5817744	5822569	5827396	5832225	5837056	5841889	5846724	5851561	2
0.242	0.0	5856400	5861241	5866084	5870929	5875776	5880625	5885476	5890329	5895184	5900041	2
0.243	0.0	5904900	5909761	5914624	5919489	5924356	5929225	5934096	5938969	5943844	5948721	2
0.244	0.0	5953600	5958481	5963364	5968249	5973136	5978025	5982916	5987809	5992704	5997601	2
0.245	0.0	6002500	6007401	6012304	6017209	6022116	6027025	6031936	6036849	6041764	6046681	2
0.246	0.0	6051600	6056521	6061444	6066369	6071296	6076225	6081156	6086089	6091024	6095961	2
0.247	0.0	6100900	6105841	6110784	6115729	6120676	6125625	6130576	6135529	6140484	6145441	2
0.248	0.0	6150400	6155361	6160324	6165289	6170256	6175225	6180196	6185169	6190144	6195121	2
0.249	0.0	6200100	6205081	6210064	6215049	6220036	6225025	6230016	6235009	6240004	6245001	2

TABLE 3

Squares

X		.0000	.0001	.0002	.0003	.0004	.0005	.0006	.0007	.0008	.0009	Δ_2
0.250	0.0	6250000	6255001	6260004	6265009	6270016	6275025	6280036	6285049	6290064	6295081	2
0.251	0.0	6300100	6305121	6310144	6315169	6320196	6325225	6330256	6335289	6340324	6345361	2
0.252	0.0	6350400	6355441	6360484	6365529	6370576	6375625	6380676	6385729	6390784	6395841	2
0.253	0.0	6400900	6405961	6411024	6416089	6421156	6426225	6431296	6436369	6441444	6446521	2
0.254	0.0	6451600	6456681	6461764	6466849	6471936	6477025	6482116	6487209	6492304	6497401	2
0.255	0.0	6502500	6507601	6512704	6517809	6522916	6528025	6533136	6538249	6543364	6548481	2
0.256	0.0	6553600	6558721	6563844	6568969	6574096	6579225	6584356	6589489	6594624	6599761	2
0.257	0.0	6604900	6610041	6615184	6620329	6625476	6630625	6635776	6640929	6646084	6651241	2
0.258	0.0	6656400	6661561	6666724	6671889	6677056	6682225	6687396	6692569	6697744	6702921	2
0.259	0.0	6708100	6713281	6718464	6723649	6728836	6734025	6739216	6744409	6749604	6754801	2
0.260	0.0	6760000	6765201	6770404	6775609	6780816	6786025	6791236	6796449	6801664	6806881	2
0.261	0.0	6812100	6817321	6822544	6827769	6832996	6838225	6843456	6848689	6853924	6859161	2
0.262	0.0	6864400	6869641	6874884	6880129	6885376	6890625	6895876	6901129	6906384	6911641	2
0.263	0.0	6916900	6922161	6927424	6932689	6937956	6943225	6948496	6953769	6959044	6964321	2
0.264	0.0	6969600	6974881	6980164	6985449	6990736	6996025	7001316	7006609	7011904	7017201	2
0.265	0.0	7022500	7027801	7033104	7038409	7043716	7049025	7054336	7059649	7064964	7070281	2
0.266	0.0	7075600	7080921	7086244	7091569	7096896	7102225	7107556	7112889	7118224	7123561	2
0.267	0.0	7128900	7134241	7139584	7144929	7150276	7155625	7160976	7166329	7171684	7177041	2
0.268	0.0	7182400	7187761	7193124	7198489	7203856	7209225	7214596	7219969	7225344	7230721	2
0.269	0.0	7236100	7241481	7246864	7252249	7257636	7263025	7268416	7273809	7279204	7284601	2
0.270	0.0	7290000	7295401	7300804	7306209	7311616	7317025	7322436	7327849	7333264	7338681	2
0.271	0.0	7344100	7349521	7354944	7360369	7365796	7371225	7376656	7382089	7387524	7392961	2
0.272	0.0	7398400	7403841	7409284	7414729	7420176	7425625	7431076	7436529	7441984	7447441	2
0.273	0.0	7452900	7458361	7463824	7469289	7474756	7480225	7485696	7491169	7496644	7502121	2
0.274	0.0	7507600	7513081	7518564	7524049	7529536	7535025	7540516	7546009	7551504	7557001	2
0.275	0.0	7562500	7568001	7573504	7579009	7584516	7590025	7595536	7601049	7606564	7612081	2
0.276	0.0	7617600	7623121	7628644	7634169	7639696	7645225	7650756	7656289	7661824	7667361	2
0.277	0.0	7672900	7678441	7683984	7689529	7695076	7700625	7706176	7711729	7717284	7722841	2
0.278	0.0	7728400	7733961	7739524	7745089	7750656	7756225	7761796	7767369	7772944	7778521	2
0.279	0.0	7784100	7789681	7795264	7800849	7806436	7812025	7817616	7823209	7828804	7834401	2
0.280	0.0	7840000	7845601	7851204	7856809	7862416	7868025	7873636	7879249	7884864	7890481	2
0.281	0.0	7896100	7901721	7907344	7912969	7918596	7924225	7929856	7935489	7941124	7946761	2
0.282	0.0	7952400	7958041	7963684	7969329	7974976	7980625	7986276	7991929	7997584	8003241	2
0.283	0.0	8008900	8014561	8020224	8025889	8031556	8037225	8042896	8048569	8054244	8059921	2
0.284	0.0	8065600	8071281	8076964	8082649	8088336	8094025	8099716	8105409	8111104	8116801	2
0.285	0.0	8122500	8128201	8133904	8139609	8145316	8151025	8156736	8162449	8168164	8173881	2
0.286	0.0	8179600	8185321	8191044	8196769	8202496	8208225	8213956	8219689	8225424	8231161	2
0.287	0.0	8236900	8242641	8248384	8254129	8259876	8265625	8271376	8277129	8282884	8288641	2
0.288	0.0	8294400	8300161	8305924	8311689	8317456	8323225	8328996	8334769	8340544	8346321	2
0.289	0.0	8352100	8357881	8363664	8369449	8375236	8381025	8386816	8392609	8398404	8404201	2
0.290	0.0	8410000	8415801	8421604	8427409	8433216	8439025	8444836	8450649	8456464	8462281	2
0.291	0.0	8468100	8473921	8479744	8485569	8491396	8497225	8503056	8508889	8514724	8520561	2
0.292	0.0	8526400	8532241	8538084	8543929	8549776	8555625	8561476	8567329	8573184	8579041	2
0.293	0.0	8584900	8590761	8596624	8602489	8608356	8614225	8620096	8625969	8631844	8637721	2
0.294	0.0	8643600	8649481	8655364	8661249	8667136	8673025	8678916	8684809	8690704	8696601	2
0.295	0.0	8702500	8708401	8714304	8720209	8726116	8732025	8737936	8743849	8749764	8755681	2
0.296	0.0	8761600	8767521	8773444	8779369	8785296	8791225	8797156	8803089	8809024	8814961	2
0.297	0.0	8820900	8826841	8832784	8838729	8844676	8850625	8856576	8862529	8868484	8874441	2
0.298	0.0	8880400	8886361	8892324	8898289	8904256	8910225	8916196	8922169	8928144	8934121	2
0.299	0.0	8940100	8946081	8952064	8958049	8964036	8970025	8976016	8982009	8988004	8994001	2

TABLE 3

Squares

X		.0000	.0001	.0002	.0003	.0004	.0005	.0006	.0007	.0008	.0009	Δ_2
0.300	0.0	9000000	9006001	9012004	9018009	9024016	9030025	9036036	9042049	9048064	9054081	2
0.301	0.0	9060100	9066121	9072144	9078169	9084196	9090225	9096256	9102289	9108324	9114361	2
0.302	0.0	9120400	9126441	9132484	9138529	9144576	9150625	9156676	9162729	9168784	9174841	2
0.303	0.0	9180900	9186961	9193024	9199089	9205156	9211225	9217296	9223369	9229444	9235521	2
0.304	0.0	9241600	9247681	9253764	9259849	9265936	9272025	9278116	9284209	9290304	9296401	2
0.305	0.0	9302500	9308601	9314704	9320809	9326916	9333025	9339136	9345249	9351364	9357481	2
0.306	0.0	9363600	9369721	9375844	9381969	9388096	9394225	9400356	9406489	9412624	9418761	2
0.307	0.0	9424900	9431041	9437184	9443329	9449476	9455625	9461776	9467929	9474084	9480241	2
0.308	0.0	9486400	9492561	9498724	9504889	9511056	9517225	9523396	9529569	9535744	9541921	2
0.309	0.0	9548100	9554281	9560464	9566649	9572836	9579025	9585216	9591409	9597604	9603801	2
0.310	0.0	9610000	9616201	9622404	9628609	9634816	9641025	9647236	9653449	9659664	9665881	2
0.311	0.0	9672100	9678321	9684544	9690769	9696996	9703225	9709456	9715689	9721924	9728161	2
0.312	0.0	9734400	9740641	9746884	9753129	9759376	9765625	9771876	9778129	9784384	9790641	2
0.313	0.0	9796900	9803161	9809424	9815689	9821956	9828225	9834496	9840769	9847044	9853321	2
0.314	0.0	9859600	9865881	9872164	9878449	9884736	9891025	9897316	9903609	9909904	9916201	2
0.315	0.0	9922500	9928801	9935104	9941409	9947716	9954025	9960336	9966649	9972964	9979281	2
0.316	0.0	9985600	9991921	9998244	0004569	0010896	0017225	0023556	0029889	0036224	0042561	2
0.317	0.1	0048900	0055241	0061584	0067929	0074276	0080625	0086976	0093329	0099684	0106041	2
0.318	0.1	0112400	0118761	0125124	0131489	0137856	0144225	0150596	0156969	0163344	0169721	2
0.319	0.1	0176100	0182481	0188864	0195249	0201636	0208025	0214416	0220809	0227204	0233601	2
0.320	0.1	0240000	0246401	0252804	0259209	0265616	0272025	0278436	0284849	0291264	0297681	2
0.321	0.1	0304100	0310521	0316944	0323369	0329796	0336225	0342656	0349089	0355524	0361961	2
0.322	0.1	0368400	0374841	0381284	0387729	0394176	0400625	0407076	0413529	0419984	0426441	2
0.323	0.1	0432900	0439361	0445824	0452289	0458756	0465225	0471696	0478169	0484644	0491121	2
0.324	0.1	0497600	0504081	0510564	0517049	0523536	0530025	0536516	0543009	0549504	0556001	2
0.325	0.1	0562500	0569001	0575504	0582009	0588516	0595025	0601536	0608049	0614564	0621081	2
0.326	0.1	0627600	0634121	0640644	0647169	0653696	0660225	0666756	0673289	0679824	0686361	2
0.327	0.1	0692900	0699441	0705984	0712529	0719076	0725625	0732176	0738729	0745284	0751841	2
0.328	0.1	0758400	0764961	0771524	0778089	0784656	0791225	0797796	0804369	0810944	0817521	2
0.329	0.1	0824100	0830681	0837264	0843849	0850436	0857025	0863616	0870209	0876804	0883401	2
0.330	0.1	0890000	0896601	0903204	0909809	0916416	0923025	0929636	0936249	0942864	0949481	2
0.331	0.1	0956100	0962721	0969344	0975969	0982596	0989225	0995856	1002489	1009124	1015761	2
0.332	0.1	1022400	1029041	1035684	1042329	1048976	1055625	1062276	1068929	1075584	1082241	2
0.333	0.1	1088900	1095561	1102224	1108889	1115556	1122225	1128896	1135569	1142244	1148921	2
0.334	0.1	1155600	1162281	1168964	1175649	1182336	1189025	1195716	1202409	1209104	1215801	2
0.335	0.1	1222500	1229201	1235904	1242609	1249316	1256025	1262736	1269449	1276164	1282881	2
0.336	0.1	1289600	1296321	1303044	1309769	1316496	1323225	1329956	1336689	1343424	1350161	2
0.337	0.1	1356900	1363641	1370384	1377129	1383876	1390625	1397376	1404129	1410884	1417641	2
0.338	0.1	1424400	1431161	1437924	1444689	1451456	1458225	1464996	1471769	1478544	1485321	2
0.339	0.1	1492100	1498881	1505664	1512449	1519236	1526025	1532816	1539609	1546404	1553201	2
0.340	0.1	1560000	1566801	1573604	1580409	1587216	1594025	1600836	1607649	1614464	1621281	2
0.341	0.1	1628100	1634921	1641744	1648569	1655396	1662225	1669056	1675889	1682724	1689561	2
0.342	0.1	1696400	1703241	1710084	1716929	1723776	1730625	1737476	1744329	1751184	1758041	2
0.343	0.1	1764900	1771761	1778624	1785489	1792356	1799225	1806096	1812969	1819844	1826721	2
0.344	0.1	1833600	1840481	1847364	1854249	1861136	1868025	1874916	1881809	1888704	1895601	2
0.345	0.1	1902500	1909401	1916304	1923209	1930116	1937025	1943936	1950849	1957764	1964681	2
0.346	0.1	1971600	1978521	1985444	1992369	1999296	2006225	2013156	2020089	2027024	2033961	2
0.347	0.1	2040900	2047841	2054784	2061729	2068676	2075625	2082576	2089529	2096484	2103441	2
0.348	0.1	2110400	2117361	2124324	2131289	2138256	2145225	2152196	2159169	2166144	2173121	2
0.349	0.1	2180100	2187081	2194064	2201049	2208036	2215025	2222016	2229009	2236004	2243001	2

TABLE 3

Squares

X		.0000	.0001	.0002	.0003	.0004	.0005	.0006	.0007	.0008	.0009	Δ_2
0.350	0.1	2250000	2257001	2264004	2271009	2278016	2285025	2292036	2299049	2306064	2313081	2
0.351	0.1	2320100	2327121	2334144	2341169	2348196	2355225	2362256	2369289	2376324	2383361	2
0.352	0.1	2390400	2397441	2404484	2411529	2418576	2425625	2432676	2439729	2446784	2453841	2
0.353	0.1	2460900	2467961	2475024	2482089	2489156	2496225	2503296	2510369	2517444	2524521	2
0.354	0.1	2531600	2538681	2545764	2552849	2559936	2567025	2574116	2581209	2588304	2595401	2
0.355	0.1	2602500	2609601	2616704	2623809	2630916	2638025	2645136	2652249	2659364	2666481	2
0.356	0.1	2673600	2680721	2687844	2694969	2702096	2709225	2716356	2723489	2730624	2737761	2
0.357	0.1	2744900	2752041	2759184	2766329	2773476	2780625	2787776	2794929	2802084	2809241	2
0.358	0.1	2816400	2823561	2830724	2837889	2845056	2852225	2859396	2866569	2873744	2880921	2
0.359	0.1	2888100	2895281	2902464	2909649	2916836	2924025	2931216	2938409	2945604	2952801	2
0.360	0.1	2960000	2967201	2974404	2981609	2988816	2996025	3003236	3010449	3017664	3024881	2
0.361	0.1	3032100	3039321	3046544	3053769	3060996	3068225	3075456	3082689	3089924	3097161	2
0.362	0.1	3104400	3111641	3118884	3126129	3133376	3140625	3147876	3155129	3162384	3169641	2
0.363	0.1	3176900	3184161	3191424	3198689	3205956	3213225	3220496	3227769	3235044	3242321	2
0.364	0.1	3249600	3256881	3264164	3271449	3278736	3286025	3293316	3300609	3307904	3315201	2
0.365	0.1	3322500	3329801	3337104	3344409	3351716	3359025	3366336	3373649	3380964	3388281	2
0.366	0.1	3395600	3402921	3410244	3417569	3424896	3432225	3439556	3446889	3454224	3461561	2
0.367	0.1	3468900	3476241	3483584	3490929	3498276	3505625	3512976	3520329	3527684	3535041	2
0.368	0.1	3542400	3549761	3557124	3564489	3571856	3579225	3586596	3593969	3601344	3608721	2
0.369	0.1	3616100	3623481	3630864	3638249	3645636	3653025	3660416	3667809	3675204	3682601	2
0.370	0.1	3690000	3697401	3704804	3712209	3719616	3727025	3734436	3741849	3749264	3756681	2
0.371	0.1	3764100	3771521	3778944	3786369	3793796	3801225	3808656	3816089	3823524	3830961	2
0.372	0.1	3838400	3845841	3853284	3860729	3868176	3875625	3883076	3890529	3897984	3905441	2
0.373	0.1	3912900	3920361	3927824	3935289	3942756	3950225	3957696	3965169	3972644	3980121	2
0.374	0.1	3987600	3995081	4002564	4010049	4017536	4025025	4032516	4040009	4047504	4055001	2
0.375	0.1	4062500	4070001	4077504	4085009	4092516	4100025	4107536	4115049	4122564	4130081	2
0.376	0.1	4137600	4145121	4152644	4160169	4167696	4175225	4182756	4190289	4197824	4205361	2
0.377	0.1	4212900	4220441	4227984	4235529	4243076	4250625	4258176	4265729	4273284	4280841	2
0.378	0.1	4288400	4295961	4303524	4311089	4318656	4326225	4333796	4341369	4348944	4356521	2
0.379	0.1	4364100	4371681	4379264	4386849	4394436	4402025	4409616	4417209	4424804	4432401	2
0.380	0.1	4440000	4447601	4455204	4462809	4470416	4478025	4485636	4493249	4500864	4508481	2
0.381	0.1	4516100	4523721	4531344	4538969	4546596	4554225	4561856	4569489	4577124	4584761	2
0.382	0.1	4592400	4600041	4607684	4615329	4622976	4630625	4638276	4645929	4653584	4661241	2
0.383	0.1	4668900	4676561	4684224	4691889	4699556	4707225	4714896	4722569	4730244	4737921	2
0.384	0.1	4745600	4753281	4760964	4768649	4776336	4784025	4791716	4799409	4807104	4814801	2
0.385	0.1	4822500	4830201	4837904	4845609	4853316	4861025	4868736	4876449	4884164	4891881	2
0.386	0.1	4899600	4907321	4915044	4922769	4930496	4938225	4945956	4953689	4961424	4969161	2
0.387	0.1	4976900	4984641	4992384	5000129	5007876	5015625	5023376	5031129	5038884	5046641	2
0.388	0.1	5054400	5062161	5069924	5077689	5085456	5093225	5100996	5108769	5116544	5124321	2
0.389	0.1	5132100	5139881	5147664	5155449	5163236	5171025	5178816	5186609	5194404	5202201	2
0.390	0.1	5210000	5217801	5225604	5233409	5241216	5249025	5256836	5264649	5272464	5280281	2
0.391	0.1	5288100	5295921	5303744	5311569	5319396	5327225	5335056	5342889	5350724	5358561	2
0.392	0.1	5366400	5374241	5382084	5389929	5397776	5405625	5413476	5421329	5429184	5437041	2
0.393	0.1	5444900	5452761	5460624	5468489	5476356	5484225	5492096	5499969	5507844	5515721	2
0.394	0.1	5523600	5531481	5539364	5547249	5555136	5563025	5570916	5578809	5586704	5594601	2
0.395	0.1	5602500	5610401	5618304	5626209	5634116	5642025	5649936	5657849	5665764	5673681	2
0.396	0.1	5681600	5689521	5697444	5705369	5713296	5721225	5729156	5737089	5745024	5752961	2
0.397	0.1	5760900	5768841	5776784	5784729	5792676	5800625	5808576	5816529	5824484	5832441	2
0.398	0.1	5840400	5848361	5856324	5864289	5872256	5880225	5888196	5896169	5904144	5912121	2
0.399	0.1	5920100	5928081	5936064	5944049	5952036	5960025	5968016	5976009	5984004	5992001	2

TABLE 3

Squares

X		.0000	.0001	.0002	.0003	.0004	.0005	.0006	.0007	.0008	.0009	Δ_2
0.400	0.1	6000000	6008001	6016004	6024009	6032016	6040025	6048036	6056049	6064064	6072081	2
0.401	0.1	6080100	6088121	6096144	6104169	6112196	6120225	6128256	6136289	6144324	6152361	2
0.402	0.1	6160400	6168441	6176484	6184529	6192576	6200625	6208676	6216729	6224784	6232841	2
0.403	0.1	6240900	6248961	6257024	6265089	6273156	6281225	6289296	6297369	6305444	6313521	2
0.404	0.1	6321600	6329681	6337764	6345849	6353936	6362025	6370116	6378209	6386304	6394401	2
0.405	0.1	6402500	6410601	6418704	6426809	6434916	6443025	6451136	6459249	6467364	6475481	2
0.406	0.1	6483600	6491721	6499844	6507969	6516096	6524225	6532356	6540489	6548624	6556761	2
0.407	0.1	6564900	6573041	6581184	6589329	6597476	6605625	6613776	6621929	6630084	6638241	2
0.408	0.1	6646400	6654561	6662724	6670889	6679056	6687225	6695396	6703569	6711744	6719921	2
0.409	0.1	6728100	6736281	6744464	6752649	6760836	6769025	6777216	6785409	6793604	6801801	2
0.410	0.1	6810000	6818201	6826404	6834609	6842816	6851025	6859236	6867449	6875664	6883881	2
0.411	0.1	6892100	6900321	6908544	6916769	6924996	6933225	6941456	6949689	6957924	6966161	2
0.412	0.1	6974400	6982641	6990884	6999129	7007376	7015625	7023876	7032129	7040384	7048641	2
0.413	0.1	7056900	7065161	7073424	7081689	7089956	7098225	7106496	7114769	7123044	7131321	2
0.414	0.1	7139600	7147881	7156164	7164449	7172736	7181025	7189316	7197609	7205904	7214201	2
0.415	0.1	7222500	7230801	7239104	7247409	7255716	7264025	7272336	7280649	7288964	7297281	2
0.416	0.1	7305600	7313921	7322244	7330569	7338896	7347225	7355556	7363889	7372224	7380561	2
0.417	0.1	7388900	7397241	7405584	7413929	7422276	7430625	7438976	7447329	7455684	7464041	2
0.418	0.1	7472400	7480761	7489124	7497489	7505856	7514225	7522596	7530969	7539344	7547721	2
0.419	0.1	7556100	7564481	7572864	7581249	7589636	7598025	7606416	7614809	7623204	7631601	2
0.420	0.1	7640000	7648401	7656804	7665209	7673616	7682025	7690436	7698849	7707264	7715681	2
0.421	0.1	7724100	7732521	7740944	7749369	7757796	7766225	7774656	7783089	7791524	7799961	2
0.422	0.1	7808400	7816841	7825284	7833729	7842176	7850625	7859076	7867529	7875984	7884441	2
0.423	0.1	7892900	7901361	7909824	7918289	7926756	7935225	7943696	7952169	7960644	7969121	2
0.424	0.1	7977600	7986081	7994564	8003049	8011536	8020025	8028516	8037009	8045504	8054001	2
0.425	0.1	8062500	8071001	8079504	8088009	8096516	8105025	8113536	8122049	8130564	8139081	2
0.426	0.1	8147600	8156121	8164644	8173169	8181696	8190225	8198756	8207289	8215824	8224361	2
0.427	0.1	8232900	8241441	8249984	8258529	8267076	8275625	8284176	8292729	8301284	8309841	2
0.428	0.1	8318400	8326961	8335524	8344089	8352656	8361225	8369796	8378369	8386944	8395521	2
0.429	0.1	8404100	8412681	8421264	8429849	8438436	8447025	8455616	8464209	8472804	8481401	2
0.430	0.1	8490000	8498601	8507204	8515809	8524416	8533025	8541636	8550249	8558864	8567481	2
0.431	0.1	8576100	8584721	8593344	8601969	8610596	8619225	8627856	8636489	8645124	8653761	2
0.432	0.1	8662400	8671041	8679684	8688329	8696976	8705625	8714276	8722929	8731584	8740241	2
0.433	0.1	8748900	8757561	8766224	8774889	8783556	8792225	8800896	8809569	8818244	8826921	2
0.434	0.1	8835600	8844281	8852964	8861649	8870336	8879025	8887716	8896409	8905104	8913801	2
0.435	0.1	8922500	8931201	8939904	8948609	8957316	8966025	8974736	8983449	8992164	9000881	2
0.436	0.1	9009600	9018321	9027044	9035769	9044496	9053225	9061956	9070689	9079424	9088161	2
0.437	0.1	9096900	9105641	9114384	9123129	9131876	9140625	9149376	9158129	9166884	9175641	2
0.438	0.1	9184400	9193161	9201924	9210689	9219456	9228225	9236996	9245769	9254544	9263321	2
0.439	0.1	9272100	9280881	9289664	9298449	9307236	9316025	9324816	9333609	9342404	9351201	2
0.440	0.1	9360000	9368801	9377604	9386409	9395216	9404025	9412836	9421649	9430464	9439281	2
0.441	0.1	9448100	9456921	9465744	9474569	9483396	9492225	9501056	9509889	9518724	9527561	2
0.442	0.1	9536400	9545241	9554084	9562929	9571776	9580625	9589476	9598329	9607184	9616041	2
0.443	0.1	9624900	9633761	9642624	9651489	9660356	9669225	9678096	9686969	9695844	9704721	2
0.444	0.1	9713600	9722481	9731364	9740249	9749136	9758025	9766916	9775809	9784704	9793601	2
0.445	0.1	9802500	9811401	9820304	9829209	9838116	9847025	9855936	9864849	9873764	9882681	2
0.446	0.1	9891600	9900521	9909444	9918369	9927296	9936225	9945156	9954089	9963024	9971961	2
0.447	0.1	9980900	9989841	9998784	0007729	0016676	0025625	0034576	0043529	0052484	0061441	2
0.448	0.2	0070400	0079361	0088324	0097289	0106256	0115225	0124196	0133169	0142144	0151121	2
0.449	0.2	0160100	0169081	0178064	0187049	0196036	0205025	0214016	0223009	0232004	0241001	2

TABLE 3

Squares

X		.0000	.0001	.0002	.0003	.0004	.0005	.0006	.0007	.0008	.0009	Δ_2
0.450	0.2	0250000	0259001	0268004	0277009	0286016	0295025	0304036	0313049	0322064	0331081	2
0.451	0.2	0340100	0349121	0358144	0367169	0376196	0385225	0394256	0403289	0412324	0421361	2
0.452	0.2	0430400	0439441	0448484	0457529	0466576	0475625	0484676	0493729	0502784	0511841	2
0.453	0.2	0520900	0529961	0539024	0548089	0557156	0566225	0575296	0584369	0593444	0602521	2
0.454	0.2	0611600	0620681	0629764	0638849	0647936	0657025	0666116	0675209	0684304	0693401	2
0.455	0.2	0702500	0711601	0720704	0729809	0738916	0748025	0757136	0766249	0775364	0784481	2
0.456	0.2	0793600	0802721	0811844	0820969	0830096	0839225	0848356	0857489	0866624	0875761	2
0.457	0.2	0884900	0894041	0903184	0912329	0921476	0930625	0939776	0948929	0958084	0967241	2
0.458	0.2	0976400	0985561	0994724	1003889	1013056	1022225	1031396	1040569	1049744	1058921	2
0.459	0.2	1068100	1077281	1086464	1095649	1104836	1114025	1123216	1132409	1141604	1150801	2
0.460	0.2	1160000	1169201	1178404	1187609	1196816	1206025	1215236	1224449	1233664	1242881	2
0.461	0.2	1252100	1261321	1270544	1279769	1288996	1298225	1307456	1316689	1325924	1335161	2
0.462	0.2	1344400	1353641	1362884	1372129	1381376	1390625	1399876	1409129	1418384	1427641	2
0.463	0.2	1436900	1446161	1455424	1464689	1473956	1483225	1492496	1501769	1511044	1520321	2
0.464	0.2	1529600	1538881	1548164	1557449	1566736	1576025	1585316	1594609	1603904	1613201	2
0.465	0.2	1622500	1631801	1641104	1650409	1659716	1669025	1678336	1687649	1696964	1706281	2
0.466	0.2	1715600	1724921	1734244	1743569	1752896	1762225	1771556	1780889	1790224	1799561	2
0.467	0.2	1808900	1818241	1827584	1836929	1846276	1855625	1864976	1874329	1883684	1893041	2
0.468	0.2	1902400	1911761	1921124	1930489	1939856	1949225	1958596	1967969	1977344	1986721	2
0.469	0.2	1996100	2005481	2014864	2024249	2033636	2043025	2052416	2061809	2071204	2080601	2
0.470	0.2	2090000	2099401	2108804	2118209	2127616	2137025	2146436	2155849	2165264	2174681	2
0.471	0.2	2184100	2193521	2202944	2212369	2221796	2231225	2240656	2250089	2259524	2268961	2
0.472	0.2	2278400	2287841	2297284	2306729	2316176	2325625	2335076	2344529	2353984	2363441	2
0.473	0.2	2372900	2382361	2391824	2401289	2410756	2420225	2429696	2439169	2448644	2458121	2
0.474	0.2	2467600	2477081	2486564	2496049	2505536	2515025	2524516	2534009	2543504	2553001	2
0.475	0.2	2562500	2572001	2581504	2591009	2600516	2610025	2619536	2629049	2638564	2648081	2
0.476	0.2	2657600	2667121	2676644	2686169	2695696	2705225	2714756	2724289	2733824	2743361	2
0.477	0.2	2752900	2762441	2771984	2781529	2791076	2800625	2810176	2819729	2829284	2838841	2
0.478	0.2	2848400	2857961	2867524	2877089	2886656	2896225	2905796	2915369	2924944	2934521	2
0.479	0.2	2944100	2953681	2963264	2972849	2982436	2992025	3001616	3011209	3020804	3030401	2
0.480	0.2	3040000	3049601	3059204	3068809	3078416	3088025	3097636	3107249	3116864	3126481	2
0.481	0.2	3136100	3145721	3155344	3164969	3174596	3184225	3193856	3203489	3213124	3222761	2
0.482	0.2	3232400	3242041	3251684	3261329	3270976	3280625	3290276	3299929	3309584	3319241	2
0.483	0.2	3328900	3338561	3348224	3357889	3367556	3377225	3386896	3396569	3406244	3415921	2
0.484	0.2	3425600	3435281	3444964	3454649	3464336	3474025	3483716	3493409	3503104	3512801	2
0.485	0.2	3522500	3532201	3541904	3551609	3561316	3571025	3580736	3590449	3600164	3609881	2
0.486	0.2	3619600	3629321	3639044	3648769	3658496	3668225	3677956	3687689	3697424	3707161	2
0.487	0.2	3716900	3726641	3736384	3746129	3755876	3765625	3775376	3785129	3794884	3804641	2
0.488	0.2	3814400	3824161	3833924	3843689	3853456	3863225	3872996	3882769	3892544	3902321	2
0.489	0.2	3912100	3921881	3931664	3941449	3951236	3961025	3970816	3980609	3990404	4000201	2
0.490	0.2	4010000	4019801	4029604	4039409	4049216	4059025	4068836	4078649	4088464	4098281	2
0.491	0.2	4108100	4117921	4127744	4137569	4147396	4157225	4167056	4176889	4186724	4196561	2
0.492	0.2	4206400	4216241	4226084	4235929	4245776	4255625	4265476	4275329	4285184	4295041	2
0.493	0.2	4304900	4314761	4324624	4334489	4344356	4354225	4364096	4373969	4383844	4393721	2
0.494	0.2	4403600	4413481	4423364	4433249	4443136	4453025	4462916	4472809	4482704	4492601	2
0.495	0.2	4502500	4512401	4522304	4532209	4542116	4552025	4561936	4571849	4581764	4591681	2
0.496	0.2	4601600	4611521	4621444	4631369	4641296	4651225	4661156	4671089	4681024	4690961	2
0.497	0.2	4700900	4710841	4720784	4730729	4740676	4750625	4760576	4770529	4780484	4790441	2
0.498	0.2	4800400	4810361	4820324	4830289	4840256	4850225	4860196	4870169	4880144	4890121	2
0.499	0.2	4900100	4910081	4920064	4930049	4940036	4950025	4960016	4970009	4980004	4990001	2

TABLE 3

Squares

X		.0000	.0001	.0002	.0003	.0004	.0005	.0006	.0007	.0008	.0009	Δ_2
0.500	0.2	5000000	5010001	5020004	5030009	5040016	5050025	5060036	5070049	5080064	5090081	2
0.501	0.2	5100100	5110121	5120144	5130169	5140196	5150225	5160256	5170289	5180324	5190361	2
0.502	0.2	5200400	5210441	5220484	5230529	5240576	5250625	5260676	5270729	5280784	5290841	2
0.503	0.2	5300900	5310961	5321024	5331089	5341156	5351225	5361296	5371369	5381444	5391521	2
0.504	0.2	5401600	5411681	5421764	5431849	5441936	5452025	5462116	5472209	5482304	5492401	2
0.505	0.2	5502500	5512601	5522704	5532809	5542916	5553025	5563136	5573249	5583364	5593481	2
0.506	0.2	5603600	5613721	5623844	5633969	5644096	5654225	5664356	5674489	5684624	5694761	2
0.507	0.2	5704900	5715041	5725184	5735329	5745476	5755625	5765776	5775929	5786084	5796241	2
0.508	0.2	5806400	5816561	5826724	5836889	5847056	5857225	5867396	5877569	5887744	5897921	2
0.509	0.2	5908100	5918281	5928464	5938649	5948836	5959025	5969216	5979409	5989604	5999801	2
0.510	0.2	6010000	6020201	6030404	6040609	6050816	6061025	6071236	6081449	6091664	6101881	2
0.511	0.2	6112100	6122321	6132544	6142769	6152996	6163225	6173456	6183689	6193924	6204161	2
0.512	0.2	6214400	6224641	6234884	6245129	6255376	6265625	6275876	6286129	6296384	6306641	2
0.513	0.2	6316900	6327161	6337424	6347689	6357956	6368225	6378496	6388769	6399044	6409321	2
0.514	0.2	6419600	6429881	6440164	6450449	6460736	6471025	6481316	6491609	6501904	6512201	2
0.515	0.2	6522500	6532801	6543104	6553409	6563716	6574025	6584336	6594649	6604964	6615281	2
0.516	0.2	6625600	6635921	6646244	6656569	6666896	6677225	6687556	6697889	6708224	6718561	2
0.517	0.2	6728900	6739241	6749584	6759929	6770276	6780625	6790976	6801329	6811684	6822041	2
0.518	0.2	6832400	6842761	6853124	6863489	6873856	6884225	6894596	6904969	6915344	6925721	2
0.519	0.2	6936100	6946481	6956864	6967249	6977636	6988025	6998416	7008809	7019204	7029601	2
0.520	0.2	7040000	7050401	7060804	7071209	7081616	7092025	7102436	7112849	7123264	7133681	2
0.521	0.2	7144100	7154521	7164944	7175369	7185796	7196225	7206656	7217089	7227524	7237961	2
0.522	0.2	7248400	7258841	7269284	7279729	7290176	7300625	7311076	7321529	7331984	7342441	2
0.523	0.2	7352900	7363361	7373824	7384289	7394756	7405225	7415696	7426169	7436644	7447121	2
0.524	0.2	7457600	7468081	7478564	7489049	7499536	7510025	7520516	7531009	7541504	7552001	2
0.525	0.2	7562500	7573001	7583504	7594009	7604516	7615025	7625536	7636049	7646564	7657081	2
0.526	0.2	7667600	7678121	7688644	7699169	7709696	7720225	7730756	7741289	7751824	7762361	2
0.527	0.2	7772900	7783441	7793984	7804529	7815076	7825625	7836176	7846729	7857284	7867841	2
0.528	0.2	7878400	7888961	7899524	7910089	7920656	7931225	7941796	7952369	7962944	7973521	2
0.529	0.2	7984100	7994681	8005264	8015849	8026436	8037025	8047616	8058209	8068804	8079401	2
0.530	0.2	8090000	8100601	8111204	8121809	8132416	8143025	8153636	8164249	8174864	8185481	2
0.531	0.2	8196100	8206721	8217344	8227969	8238596	8249225	8259856	8270489	8281124	8291761	2
0.532	0.2	8302400	8313041	8323684	8334329	8344976	8355625	8366276	8376929	8387584	8398241	2
0.533	0.2	8408900	8419561	8430224	8440889	8451556	8462225	8472896	8483569	8494244	8504921	2
0.534	0.2	8515600	8526281	8536964	8547649	8558336	8569025	8579716	8590409	8601104	8611801	2
0.535	0.2	8622500	8633201	8643904	8654609	8665316	8676025	8686736	8697449	8708164	8718881	2
0.536	0.2	8729600	8740321	8751044	8761769	8772496	8783225	8793956	8804689	8815424	8826161	2
0.537	0.2	8836900	8847641	8858384	8869129	8879876	8890625	8901376	8912129	8922884	8933641	2
0.538	0.2	8944400	8955161	8965924	8976689	8987456	8998225	9008996	9019769	9030544	9041321	2
0.539	0.2	9052100	9062881	9073664	9084449	9095236	9106025	9116816	9127609	9138404	9149201	2
0.540	0.2	9160000	9170801	9181604	9192409	9203216	9214025	9224836	9235649	9246464	9257281	2
0.541	0.2	9268100	9278921	9289744	9300569	9311396	9322225	9333056	9343889	9354724	9365561	2
0.542	0.2	9376400	9387241	9398084	9408929	9419776	9430625	9441476	9452329	9463184	9474041	2
0.543	0.2	9484900	9495761	9506624	9517489	9528356	9539225	9550096	9560969	9571844	9582721	2
0.544	0.2	9593600	9604481	9615364	9626249	9637136	9648025	9658916	9669809	9680704	9691601	2
0.545	0.2	9702500	9713401	9724304	9735209	9746116	9757025	9767936	9778849	9789764	9800681	2
0.546	0.2	9811600	9822521	9833444	9844369	9855296	9866225	9877156	9888089	9899024	9909961	2
0.547	0.2	9920900	9931841	9942784	9953729	9964676	9975625	9986576	9997529	0008484	0019441	2
0.548	0.3	0030400	0041361	0052324	0063289	0074256	0085225	0096196	0107169	0118144	0129121	2
0.549	0.3	0140100	0151081	0162064	0173049	0184036	0195025	0206016	0217009	0228004	0239001	2

TABLE 3

Squares

X		.0000	.0001	.0002	.0003	.0004	.0005	.0006	.0007	.0008	.0009	Δ_2
0.550	0.3	0250000	0261001	0272004	0283009	0294016	0305025	0316036	0327049	0338064	0349081	2
0.551	0.3	0360100	0371121	0382144	0393169	0404196	0415225	0426256	0437289	0448324	0459361	2
0.552	0.3	0470400	0481441	0492484	0503529	0514576	0525625	0536676	0547729	0558784	0569841	2
0.553	0.3	0580900	0591961	0603024	0614089	0625156	0636225	0647296	0658369	0669444	0680521	2
0.554	0.3	0691600	0702681	0713764	0724849	0735936	0747025	0758116	0769209	0780304	0791401	2
0.555	0.3	0802500	0813601	0824704	0835809	0846916	0858025	0869136	0880249	0891364	0902481	2
0.556	0.3	0913600	0924721	0935844	0946969	0958096	0969225	0980356	0991489	1002624	1013761	2
0.557	0.3	1024900	1036041	1047184	1058329	1069476	1080625	1091776	1102929	1114084	1125241	2
0.558	0.3	1136400	1147561	1158724	1169889	1181056	1192225	1203396	1214569	1225744	1236921	2
0.559	0.3	1248100	1259281	1270464	1281649	1292836	1304025	1315216	1326409	1337604	1348801	2
0.560	0.3	1360000	1371201	1382404	1393609	1404816	1416025	1427236	1438449	1449664	1460881	2
0.561	0.3	1472100	1483321	1494544	1505769	1516996	1528225	1539456	1550689	1561924	1573161	2
0.562	0.3	1584400	1595641	1606884	1618129	1629376	1640625	1651876	1663129	1674384	1685641	2
0.563	0.3	1696900	1708161	1719424	1730689	1741956	1753225	1764496	1775769	1787044	1798321	2
0.564	0.3	1809600	1820881	1832164	1843449	1854736	1866025	1877316	1888609	1899904	1911201	2
0.565	0.3	1922500	1933801	1945104	1956409	1967716	1979025	1990336	2001649	2012964	2024281	2
0.566	0.3	2035600	2046921	2058244	2069569	2080896	2092225	2103556	2114889	2126224	2137561	2
0.567	0.3	2148900	2160241	2171584	2182929	2194276	2205625	2216976	2228329	2239684	2251041	2
0.568	0.3	2262400	2273761	2285124	2296489	2307856	2319225	2330596	2341969	2353344	2364721	2
0.569	0.3	2376100	2387481	2398864	2410249	2421636	2433025	2444416	2455809	2467204	2478601	2
0.570	0.3	2490000	2501401	2512804	2524209	2535616	2547025	2558436	2569849	2581264	2592681	2
0.571	0.3	2604100	2615521	2626944	2638369	2649796	2661225	2672656	2684089	2695524	2706961	2
0.572	0.3	2718400	2729841	2741284	2752729	2764176	2775625	2787076	2798529	2809984	2821441	2
0.573	0.3	2832900	2844361	2855824	2867289	2878756	2890225	2901696	2913169	2924644	2936121	2
0.574	0.3	2947600	2959081	2970564	2982049	2993536	3005025	3016516	3028009	3039504	3051001	2
0.575	0.3	3062500	3074001	3085504	3097009	3108516	3120025	3131536	3143049	3154564	3166081	2
0.576	0.3	3177600	3189121	3200644	3212169	3223696	3235225	3246756	3258289	3269824	3281361	2
0.577	0.3	3292900	3304441	3315984	3327529	3339076	3350625	3362176	3373729	3385284	3396841	2
0.578	0.3	3408400	3419961	3431524	3443089	3454656	3466225	3477796	3489369	3500944	3512521	2
0.579	0.3	3524100	3535681	3547264	3558849	3570436	3582025	3593616	3605209	3616804	3628401	2
0.580	0.3	3640000	3651601	3663204	3674809	3686416	3698025	3709636	3721249	3732864	3744481	2
0.581	0.3	3756100	3767721	3779344	3790969	3802596	3814225	3825856	3837489	3849124	3860761	2
0.582	0.3	3872400	3884041	3895684	3907329	3918976	3930625	3942276	3953929	3965584	3977241	2
0.583	0.3	3988900	4000561	4012224	4023889	4035556	4047225	4058896	4070569	4082244	4093921	2
0.584	0.3	4105600	4117281	4128964	4140649	4152336	4164025	4175716	4187409	4199104	4210801	2
0.585	0.3	4222500	4234201	4245904	4257609	4269316	4281025	4292736	4304449	4316164	4327881	2
0.586	0.3	4339600	4351321	4363044	4374769	4386496	4398225	4409956	4421689	4433424	4445161	2
0.587	0.3	4456900	4468641	4480384	4492129	4503876	4515625	4527376	4539129	4550884	4562641	2
0.588	0.3	4574400	4586161	4597924	4609689	4621456	4633225	4644996	4656769	4668544	4680321	2
0.589	0.3	4692100	4703881	4715664	4727449	4739236	4751025	4762816	4774609	4786404	4798201	2
0.590	0.3	4810000	4821801	4833604	4845409	4857216	4869025	4880836	4892649	4904464	4916281	2
0.591	0.3	4928100	4939921	4951744	4963569	4975396	4987225	4999056	5010889	5022724	5034561	2
0.592	0.3	5046400	5058241	5070084	5081929	5093776	5105625	5117476	5129329	5141184	5153041	2
0.593	0.3	5164900	5176761	5188624	5200489	5212356	5224225	5236096	5247969	5259844	5271721	2
0.594	0.3	5283600	5295481	5307364	5319249	5331136	5343025	5354916	5366809	5378704	5390601	2
0.595	0.3	5402500	5414401	5426304	5438209	5450116	5462025	5473936	5485849	5497764	5509681	2
0.596	0.3	5521600	5533521	5545444	5557369	5569296	5581225	5593156	5605089	5617024	5628961	2
0.597	0.3	5640900	5652841	5664784	5676729	5688676	5700625	5712576	5724529	5736484	5748441	2
0.598	0.3	5760400	5772361	5784324	5796289	5808256	5820225	5832196	5844169	5856144	5868121	2
0.599	0.3	5880100	5892081	5904064	5916049	5928036	5940025	5952016	5964009	5976004	5988001	2

TABLE 3

Squares

X		.0000	.0001	.0002	.0003	.0004	.0005	.0006	.0007	.0008	.0009	Δ_2
0.600	0.3	6000000	6012001	6024004	6036009	6048016	6060025	6072036	6084049	6096064	6108081	2
0.601	0.3	6120100	6132121	6144144	6156169	6168196	6180225	6192256	6204289	6216324	6228361	2
0.602	0.3	6240400	6252441	6264484	6276529	6288576	6300625	6312676	6324729	6336784	6348841	2
0.603	0.3	6360900	6372961	6385024	6397089	6409156	6421225	6433296	6445369	6457444	6469521	2
0.604	0.3	6481600	6493681	6505764	6517849	6529936	6542025	6554116	6566209	6578304	6590401	2
0.605	0.3	6602500	6614601	6626704	6638809	6650916	6663025	6675136	6687249	6699364	6711481	2
0.606	0.3	6723600	6735721	6747844	6759969	6772096	6784225	6796356	6808489	6820624	6832761	2
0.607	0.3	6844900	6857041	6869184	6881329	6893476	6905625	6917776	6929929	6942084	6954241	2
0.608	0.3	6966400	6978561	6990724	7002889	7015056	7027225	7039396	7051569	7063744	7075921	2
0.609	0.3	7088100	7100281	7112464	7124649	7136836	7149025	7161216	7173409	7185604	7197801	2
0.610	0.3	7210000	7222201	7234404	7246609	7258816	7271025	7283236	7295449	7307664	7319881	2
0.611	0.3	7332100	7344321	7356544	7368769	7380996	7393225	7405456	7417689	7429924	7442161	2
0.612	0.3	7454400	7466641	7478884	7491129	7503376	7515625	7527876	7540129	7552384	7564641	2
0.613	0.3	7576900	7589161	7601424	7613689	7625956	7638225	7650496	7662769	7675044	7687321	2
0.614	0.3	7699600	7711881	7724164	7736449	7748736	7761025	7773316	7785609	7797904	7810201	2
0.615	0.3	7822500	7834801	7847104	7859409	7871716	7884025	7896336	7908649	7920964	7933281	2
0.616	0.3	7945600	7957921	7970244	7982569	7994896	8007225	8019556	8031889	8044224	8056561	2
0.617	0.3	8068900	8081241	8093584	8105929	8118276	8130625	8142976	8155329	8167684	8180041	2
0.618	0.3	8192400	8204761	8217124	8229489	8241856	8254225	8266596	8278969	8291344	8303721	2
0.619	0.3	8316100	8328481	8340864	8353249	8365636	8378025	8390416	8402809	8415204	8427601	2
0.620	0.3	8440000	8452401	8464804	8477209	8489616	8502025	8514436	8526849	8539264	8551681	2
0.621	0.3	8564100	8576521	8588944	8601369	8613796	8626225	8638656	8651089	8663524	8675961	2
0.622	0.3	8688400	8700841	8713284	8725729	8738176	8750625	8763076	8775529	8787984	8800441	2
0.623	0.3	8812900	8825361	8837824	8850289	8862756	8875225	8887696	8900169	8912644	8925121	2
0.624	0.3	8937600	8950081	8962564	8975049	8987536	9000025	9012516	9025009	9037504	9050001	2
0.625	0.3	9062500	9075001	9087504	9100009	9112516	9125025	9137536	9150049	9162564	9175081	2
0.626	0.3	9187600	9200121	9212644	9225169	9237696	9250225	9262756	9275289	9287824	9300361	2
0.627	0.3	9312900	9325441	9337984	9350529	9363076	9375625	9388176	9400729	9413284	9425841	2
0.628	0.3	9438400	9450961	9463524	9476089	9488656	9501225	9513796	9526369	9538944	9551521	2
0.629	0.3	9564100	9576681	9589264	9601849	9614436	9627025	9639616	9652209	9664804	9677401	2
0.630	0.3	9690000	9702601	9715204	9727809	9740416	9753025	9765636	9778249	9790864	9803481	2
0.631	0.3	9816100	9828721	9841344	9853969	9866596	9879225	9891856	9904489	9917124	9929761	2
0.632	0.3	9942400	9955041	9967684	9980329	9992976	0005625	0018276	0030929	0043584	0056241	2
0.633	0.4	0068900	0081561	0094224	0106889	0119556	0132225	0144896	0157569	0170244	0182921	2
0.634	0.4	0195600	0208281	0220964	0233649	0246336	0259025	0271716	0284409	0297104	0309801	2
0.635	0.4	0322500	0335201	0347904	0360609	0373316	0386025	0398736	0411449	0424164	0436881	2
0.636	0.4	0449600	0462321	0475044	0487769	0500496	0513225	0525956	0538689	0551424	0564161	2
0.637	0.4	0576900	0589641	0602384	0615129	0627876	0640625	0653376	0666129	0678884	0691641	2
0.638	0.4	0704400	0717161	0729924	0742689	0755456	0768225	0780996	0793769	0806544	0819321	2
0.639	0.4	0832100	0844881	0857664	0870449	0883236	0896025	0908816	0921609	0934404	0947201	2
0.640	0.4	0960000	0972801	0985604	0998409	1011216	1024025	1036836	1049649	1062464	1075281	2
0.641	0.4	1088100	1100921	1113744	1126569	1139396	1152225	1165056	1177889	1190724	1203561	2
0.642	0.4	1216400	1229241	1242084	1254929	1267776	1280625	1293476	1306329	1319184	1332041	2
0.643	0.4	1344900	1357761	1370624	1383489	1396356	1409225	1422096	1434969	1447844	1460721	2
0.644	0.4	1473600	1486481	1499364	1512249	1525136	1538025	1550916	1563809	1576704	1589601	2
0.645	0.4	1602500	1615401	1628304	1641209	1654116	1667025	1679936	1692849	1705764	1718681	2
0.646	0.4	1731600	1744521	1757444	1770369	1783296	1796225	1809156	1822089	1835024	1847961	2
0.647	0.4	1860900	1873841	1886784	1899729	1912676	1925625	1938576	1951529	1964484	1977441	2
0.648	0.4	1990400	2003361	2016324	2029289	2042256	2055225	2068196	2081169	2094144	2107121	2
0.649	0.4	2120100	2133081	2146064	2159049	2172036	2185025	2198016	2211009	2224004	2237001	2

TABLE 3

Squares

X		.0000	.0001	.0002	.0003	.0004	.0005	.0006	.0007	.0008	.0009	Δ_2
0.650	0.4	2250000	2263001	2276004	2289009	2302016	2315025	2328036	2341049	2354064	2367081	2
0.651	0.4	2380100	2393121	2406144	2419169	2432196	2445225	2458256	2471289	2484324	2497361	2
0.652	0.4	2510400	2523441	2536484	2549529	2562576	2575625	2588676	2601729	2614784	2627841	2
0.653	0.4	2640900	2653961	2667024	2680089	2693156	2706225	2719296	2732369	2745444	2758521	2
0.654	0.4	2771600	2784681	2797764	2810849	2823936	2837025	2850116	2863209	2876304	2889401	2
0.655	0.4	2902500	2915601	2928704	2941809	2954916	2968025	2981136	2994249	3007364	3020481	2
0.656	0.4	3033600	3046721	3059844	3072969	3086096	3099225	3112356	3125489	3138624	3151761	2
0.657	0.4	3164900	3178041	3191184	3204329	3217476	3230625	3243776	3256929	3270084	3283241	2
0.658	0.4	3296400	3309561	3322724	3335889	3349056	3362225	3375396	3388569	3401744	3414921	2
0.659	0.4	3428100	3441281	3454464	3467649	3480836	3494025	3507216	3520409	3533604	3546801	2
0.660	0.4	3560000	3573201	3586404	3599609	3612816	3626025	3639236	3652449	3665664	3678881	2
0.661	0.4	3692100	3705321	3718544	3731769	3744996	3758225	3771456	3784689	3797924	3811161	2
0.662	0.4	3824400	3837641	3850884	3864129	3877376	3890625	3903876	3917129	3930384	3943641	2
0.663	0.4	3956900	3970161	3983424	3996689	4009956	4023225	4036496	4049769	4063044	4076321	2
0.664	0.4	4089600	4102881	4116164	4129449	4142736	4156025	4169316	4182609	4195904	4209201	2
0.665	0.4	4222500	4235801	4249104	4262409	4275716	4289025	4302336	4315649	4328964	4342281	2
0.666	0.4	4355600	4368921	4382244	4395569	4408896	4422225	4435556	4448889	4462224	4475561	2
0.667	0.4	4488900	4502241	4515584	4528929	4542276	4555625	4568976	4582329	4595684	4609041	2
0.668	0.4	4622400	4635761	4649124	4662489	4675856	4689225	4702596	4715969	4729344	4742721	2
0.669	0.4	4756100	4769481	4782864	4796249	4809636	4823025	4836416	4849809	4863204	4876601	2
0.670	0.4	4890000	4903401	4916804	4930209	4943616	4957025	4970436	4983849	4997264	5010681	2
0.671	0.4	5024100	5037521	5050944	5064369	5077796	5091225	5104656	5118089	5131524	5144961	2
0.672	0.4	5158400	5171841	5185284	5198729	5212176	5225625	5239076	5252529	5265984	5279441	2
0.673	0.4	5292900	5306361	5319824	5333289	5346756	5360225	5373696	5387169	5400644	5414121	2
0.674	0.4	5427600	5441081	5454564	5468049	5481536	5495025	5508516	5522009	5535504	5549001	2
0.675	0.4	5562500	5576001	5589504	5603009	5616516	5630025	5643536	5657049	5670564	5684081	2
0.676	0.4	5697600	5711121	5724644	5738169	5751696	5765225	5778756	5792289	5805824	5819361	2
0.677	0.4	5832900	5846441	5859984	5873529	5887076	5900625	5914176	5927729	5941284	5954841	2
0.678	0.4	5968400	5981961	5995524	6009089	6022656	6036225	6049796	6063369	6076944	6090521	2
0.679	0.4	6104100	6117681	6131264	6144849	6158436	6172025	6185616	6199209	6212804	6226401	2
0.680	0.4	6240000	6253601	6267204	6280809	6294416	6308025	6321636	6335249	6348864	6362481	2
0.681	0.4	6376100	6389721	6403344	6416969	6430596	6444225	6457856	6471489	6485124	6498761	2
0.682	0.4	6512400	6526041	6539684	6553329	6566976	6580625	6594276	6607929	6621584	6635241	2
0.683	0.4	6648900	6662561	6676224	6689889	6703556	6717225	6730896	6744569	6758244	6771921	2
0.684	0.4	6785600	6799281	6812964	6826649	6840336	6854025	6867716	6881409	6895104	6908801	2
0.685	0.4	6922500	6936201	6949904	6963609	6977316	6991025	7004736	7018449	7032164	7045881	2
0.686	0.4	7059600	7073321	7087044	7100769	7114496	7128225	7141956	7155689	7169424	7183161	2
0.687	0.4	7196900	7210641	7224384	7238129	7251876	7265625	7279376	7293129	7306884	7320641	2
0.688	0.4	7334400	7348161	7361924	7375689	7389456	7403225	7416996	7430769	7444544	7458321	2
0.689	0.4	7472100	7485881	7499664	7513449	7527236	7541025	7554816	7568609	7582404	7596201	2
0.690	0.4	7610000	7623801	7637604	7651409	7665216	7679025	7692836	7706649	7720464	7734281	2
0.691	0.4	7748100	7761921	7775744	7789569	7803396	7817225	7831056	7844889	7858724	7872561	2
0.692	0.4	7886400	7900241	7914084	7927929	7941776	7955625	7969476	7983329	7997184	8011041	2
0.693	0.4	8024900	8038761	8052624	8066489	8080356	8094225	8108096	8121969	8135844	8149721	2
0.694	0.4	8163600	8177481	8191364	8205249	8219136	8233025	8246916	8260809	8274704	8288601	2
0.695	0.4	8302500	8316401	8330304	8344209	8358116	8372025	8385936	8399849	8413764	8427681	2
0.696	0.4	8441600	8455521	8469444	8483369	8497296	8511225	8525156	8539089	8553024	8566961	2
0.697	0.4	8580900	8594841	8608784	8622729	8636676	8650625	8664576	8678529	8692484	8706441	2
0.698	0.4	8720400	8734361	8748324	8762289	8776256	8790225	8804196	8818169	8832144	8846121	2
0.699	0.4	8860100	8874081	8888064	8902049	8916036	8930025	8944016	8958009	8972004	8986001	2

TABLE 3

Squares

X		.0000	.0001	.0002	.0003	.0004	.0005	.0006	.0007	.0008	.0009	Δ_2
0.700	0.4	9000000	9014001	9028004	9042009	9056016	9070025	9084036	9098049	9112064	9126081	2
0.701	0.4	9140100	9154121	9168144	9182169	9196196	9210225	9224256	9238289	9252324	9266361	2
0.702	0.4	9280400	9294441	9308484	9322529	9336576	9350625	9364676	9378729	9392784	9406841	2
0.703	0.4	9420900	9434961	9449024	9463089	9477156	9491225	9505296	9519369	9533444	9547521	2
0.704	0.4	9561600	9575681	9589764	9603849	9617936	9632025	9646116	9660209	9674304	9688401	2
0.705	0.4	9702500	9716601	9730704	9744809	9758916	9773025	9787136	9801249	9815364	9829481	2
0.706	0.4	9843600	9857721	9871844	9885969	9900096	9914225	9928356	9942489	9956624	9970761	2
0.707	0.4	9984900	9999041	0013184	0027329	0041476	0055625	0069776	0083929	0098084	0112241	2
0.708	0.5	0126400	0140561	0154724	0168889	0183056	0197225	0211396	0225569	0239744	0253921	2
0.709	0.5	0268100	0282281	0296464	0310649	0324836	0339025	0353216	0367409	0381604	0395801	2
0.710	0.5	0410000	0424201	0438404	0452609	0466816	0481025	0495236	0509449	0523664	0537881	2
0.711	0.5	0552100	0566321	0580544	0594769	0608996	0623225	0637456	0651689	0665924	0680161	2
0.712	0.5	0694400	0708641	0722884	0737129	0751376	0765625	0779876	0794129	0808384	0822641	2
0.713	0.5	0836900	0851161	0865424	0879689	0893956	0908225	0922496	0936769	0951044	0965321	2
0.714	0.5	0979600	0993881	1008164	1022449	1036736	1051025	1065316	1079609	1093904	1108201	2
0.715	0.5	1122500	1136801	1151104	1165409	1179716	1194025	1208336	1222649	1236964	1251281	2
0.716	0.5	1265600	1279921	1294244	1308569	1322896	1337225	1351556	1365889	1380224	1394561	2
0.717	0.5	1408900	1423241	1437584	1451929	1466276	1480625	1494976	1509329	1523684	1538041	2
0.718	0.5	1552400	1566761	1581124	1595489	1609856	1624225	1638596	1652969	1667344	1681721	2
0.719	0.5	1696100	1710481	1724864	1739249	1753636	1768025	1782416	1796809	1811204	1825601	2
0.720	0.5	1840000	1854401	1868804	1883209	1897616	1912025	1926436	1940849	1955264	1969681	2
0.721	0.5	1984100	1998521	2012944	2027369	2041796	2056225	2070656	2085089	2099524	2113961	2
0.722	0.5	2128400	2142841	2157284	2171729	2186176	2200625	2215076	2229529	2243984	2258441	2
0.723	0.5	2272900	2287361	2301824	2316289	2330756	2345225	2359696	2374169	2388644	2403121	2
0.724	0.5	2417600	2432081	2446564	2461049	2475536	2490025	2504516	2519009	2533504	2548001	2
0.725	0.5	2562500	2577001	2591504	2606009	2620516	2635025	2649536	2664049	2678564	2693081	2
0.726	0.5	2707600	2722121	2736644	2751169	2765696	2780225	2794756	2809289	2823824	2838361	2
0.727	0.5	2852900	2867441	2881984	2896529	2911076	2925625	2940176	2954729	2969284	2983841	2
0.728	0.5	2998400	3012961	3027524	3042089	3056656	3071225	3085796	3100369	3114944	3129521	2
0.729	0.5	3144100	3158681	3173264	3187849	3202436	3217025	3231616	3246209	3260804	3275401	2
0.730	0.5	3290000	3304601	3319204	3333809	3348416	3363025	3377636	3392249	3406864	3421481	2
0.731	0.5	3436100	3450721	3465344	3479969	3494596	3509225	3523856	3538489	3553124	3567761	2
0.732	0.5	3582400	3597041	3611684	3626329	3640976	3655625	3670276	3684929	3699584	3714241	2
0.733	0.5	3728900	3743561	3758224	3772889	3787556	3802225	3816896	3831569	3846244	3860921	2
0.734	0.5	3875600	3890281	3904964	3919649	3934336	3949025	3963716	3978409	3993104	4007801	2
0.735	0.5	4022500	4037201	4051904	4066609	4081316	4096025	4110736	4125449	4140164	4154881	2
0.736	0.5	4169600	4184321	4199044	4213769	4228496	4243225	4257956	4272689	4287424	4302161	2
0.737	0.5	4316900	4331641	4346384	4361129	4375876	4390625	4405376	4420129	4434884	4449641	2
0.738	0.5	4464400	4479161	4493924	4508689	4523456	4538225	4552996	4567769	4582544	4597321	2
0.739	0.5	4612100	4626881	4641664	4656449	4671236	4686025	4700816	4715609	4730404	4745201	2
0.740	0.5	4760000	4774801	4789604	4804409	4819216	4834025	4848836	4863649	4878464	4893281	2
0.741	0.5	4908100	4922921	4937744	4952569	4967396	4982225	4997056	5011889	5026724	5041561	2
0.742	0.5	5056400	5071241	5086084	5100929	5115776	5130625	5145476	5160329	5175184	5190041	2
0.743	0.5	5204900	5219761	5234624	5249489	5264356	5279225	5294096	5308969	5323844	5338721	2
0.744	0.5	5353600	5368481	5383364	5398249	5413136	5428025	5442916	5457809	5472704	5487601	2
0.745	0.5	5502500	5517401	5532304	5547209	5562116	5577025	5591936	5606849	5621764	5636681	2
0.746	0.5	5651600	5666521	5681444	5696369	5711296	5726225	5741156	5756089	5771024	5785961	2
0.747	0.5	5800900	5815841	5830784	5845729	5860676	5875625	5890576	5905529	5920484	5935441	2
0.748	0.5	5950400	5965361	5980324	5995289	6010256	6025225	6040196	6055169	6070144	6085121	2
0.749	0.5	6100100	6115081	6130064	6145049	6160036	6175025	6190016	6205009	6220004	6235001	2

TABLE 3

Squares

X		.0000	.0001	.0002	.0003	.0004	.0005	.0006	.0007	.0008	.0009	Δ_2
0.750	0.5	6250000	6265001	6280004	6295009	6310016	6325025	6340036	6355049	6370064	6385081	2
0.751	0.5	6400100	6415121	6430144	6445169	6460196	6475225	6490256	6505289	6520324	6535361	2
0.752	0.5	6550400	6565441	6580484	6595529	6610576	6625625	6640676	6655729	6670784	6685841	2
0.753	0.5	6700900	6715961	6731024	6746089	6761156	6776225	6791296	6806369	6821444	6836521	2
0.754	0.5	6851600	6866681	6881764	6896849	6911936	6927025	6942116	6957209	6972304	6987401	2
0.755	0.5	7002500	7017601	7032704	7047809	7062916	7078025	7093136	7108249	7123364	7138481	2
0.756	0.5	7153600	7168721	7183844	7198969	7214096	7229225	7244356	7259489	7274624	7289761	2
0.757	0.5	7304900	7320041	7335184	7350329	7365476	7380625	7395776	7410929	7426084	7441241	2
0.758	0.5	7456400	7471561	7486724	7501889	7517056	7532225	7547396	7562569	7577744	7592921	2
0.759	0.5	7608100	7623281	7638464	7653649	7668836	7684025	7699216	7714409	7729604	7744801	2
0.760	0.5	7760000	7775201	7790404	7805609	7820816	7836025	7851236	7866449	7881664	7896881	2
0.761	0.5	7912100	7927321	7942544	7957769	7972996	7988225	8003456	8018689	8033924	8049161	2
0.762	0.5	8064400	8079641	8094884	8110129	8125376	8140625	8155876	8171129	8186384	8201641	2
0.763	0.5	8216900	8232161	8247424	8262689	8277956	8293225	8308496	8323769	8339044	8354321	2
0.764	0.5	8369600	8384881	8400164	8415449	8430736	8446025	8461316	8476609	8491904	8507201	2
0.765	0.5	8522500	8537801	8553104	8568409	8583716	8599025	8614336	8629649	8644964	8660281	2
0.766	0.5	8675600	8690921	8706244	8721569	8736896	8752225	8767556	8782889	8798224	8813561	2
0.767	0.5	8828900	8844241	8859584	8874929	8890276	8905625	8920976	8936329	8951684	8967041	2
0.768	0.5	8982400	8997761	9013124	9028489	9043856	9059225	9074596	9089969	9105344	9120721	2
0.769	0.5	9136100	9151481	9166864	9182249	9197636	9213025	9228416	9243809	9259204	9274601	2
0.770	0.5	9290000	9305401	9320804	9336209	9351616	9367025	9382436	9397849	9413264	9428681	2
0.771	0.5	9444100	9459521	9474944	9490369	9505796	9521225	9536656	9552089	9567524	9582961	2
0.772	0.5	9598400	9613841	9629284	9644729	9660176	9675625	9691076	9706529	9721984	9737441	2
0.773	0.5	9752900	9768361	9783824	9799289	9814756	9830225	9845696	9861169	9876644	9892121	2
0.774	0.5	9907600	9923081	9938564	9954049	9969536	9985025	0000516	0016009	0031504	0047001	2
0.775	0.6	0062500	0078001	0093504	0109009	0124516	0140025	0155536	0171049	0186564	0202081	2
0.776	0.6	0217600	0233121	0248644	0264169	0279696	0295225	0310756	0326289	0341824	0357361	2
0.777	0.6	0372900	0388441	0403984	0419529	0435076	0450625	0466176	0481729	0497284	0512841	2
0.778	0.6	0528400	0543961	0559524	0575089	0590656	0606225	0621796	0637369	0652944	0668521	2
0.779	0.6	0684100	0699681	0715264	0730849	0746436	0762025	0777616	0793209	0808804	0824401	2
0.780	0.6	0840000	0855601	0871204	0886809	0902416	0918025	0933636	0949249	0964864	0980481	2
0.781	0.6	0996100	1011721	1027344	1042969	1058596	1074225	1089856	1105489	1121124	1136761	2
0.782	0.6	1152400	1168041	1183684	1199329	1214976	1230625	1246276	1261929	1277584	1293241	2
0.783	0.6	1308900	1324561	1340224	1355889	1371556	1387225	1402896	1418569	1434244	1449921	2
0.784	0.6	1465600	1481281	1496964	1512649	1528336	1544025	1559716	1575409	1591104	1606801	2
0.785	0.6	1622500	1638201	1653904	1669609	1685316	1701025	1716736	1732449	1748164	1763881	2
0.786	0.6	1779600	1795321	1811044	1826769	1842496	1858225	1873956	1889689	1905424	1921161	2
0.787	0.6	1936900	1952641	1968384	1984129	1999876	2015625	2031376	2047129	2062884	2078641	2
0.788	0.6	2094400	2110161	2125924	2141689	2157456	2173225	2188996	2204769	2220544	2236321	2
0.789	0.6	2252100	2267881	2283664	2299449	2315236	2331025	2346816	2362609	2378404	2394201	2
0.790	0.6	2410000	2425801	2441604	2457409	2473216	2489025	2504836	2520649	2536464	2552281	2
0.791	0.6	2568100	2583921	2599744	2615569	2631396	2647225	2663056	2678889	2694724	2710561	2
0.792	0.6	2726400	2742241	2758084	2773929	2789776	2805625	2821476	2837329	2853184	2869041	2
0.793	0.6	2884900	2900761	2916624	2932489	2948356	2964225	2980096	2995969	3011844	3027721	2
0.794	0.6	3043600	3059481	3075364	3091249	3107136	3123025	3138916	3154809	3170704	3186601	2
0.795	0.6	3202500	3218401	3234304	3250209	3266116	3282025	3297936	3313849	3329764	3345681	2
0.796	0.6	3361600	3377521	3393444	3409369	3425296	3441225	3457156	3473089	3489024	3504961	2
0.797	0.6	3520900	3536841	3552784	3568729	3584676	3600625	3616576	3632529	3648484	3664441	2
0.798	0.6	3680400	3696361	3712324	3728289	3744256	3760225	3776196	3792169	3808144	3824121	2
0.799	0.6	3840100	3856081	3872064	3888049	3904036	3920025	3936016	3952009	3968004	3984001	2

TABLE 3

Squares

X		.0000	.0001	.0002	.0003	.0004	.0005	.0006	.0007	.0008	.0009	Δ_2
0.800	0.6	4000000	4016001	4032004	4048009	4064016	4080025	4096036	4112049	4128064	4144081	2
0.801	0.6	4160100	4176121	4192144	4208169	4224196	4240225	4256256	4272289	4288324	4304361	2
0.802	0.6	4320400	4336441	4352484	4368529	4384576	4400625	4416676	4432729	4448784	4464841	2
0.803	0.6	4480900	4496961	4513024	4529089	4545156	4561225	4577296	4593369	4609444	4625521	2
0.804	0.6	4641600	4657681	4673764	4689849	4705936	4722025	4738116	4754209	4770304	4786401	2
0.805	0.6	4802500	4818601	4834704	4850809	4866916	4883025	4899136	4915249	4931364	4947481	2
0.806	0.6	4963600	4979721	4995844	5011969	5028096	5044225	5060356	5076489	5092624	5108761	2
0.807	0.6	5124900	5141041	5157184	5173329	5189476	5205625	5221776	5237929	5254084	5270241	2
0.808	0.6	5286400	5302561	5318724	5334889	5351056	5367225	5383396	5399569	5415744	5431921	2
0.809	0.6	5448100	5464281	5480464	5496649	5512836	5529025	5545216	5561409	5577604	5593801	2
0.810	0.6	5610000	5626201	5642404	5658609	5674816	5691025	5707236	5723449	5739664	5755881	2
0.811	0.6	5772100	5788321	5804544	5820769	5836996	5853225	5869456	5885689	5901924	5918161	2
0.812	0.6	5934400	5950641	5966884	5983129	5999376	6015625	6031876	6048129	6064384	6080641	2
0.813	0.6	6096900	6113161	6129424	6145689	6161956	6178225	6194496	6210769	6227044	6243321	2
0.814	0.6	6259600	6275881	6292164	6308449	6324736	6341025	6357316	6373609	6389904	6406201	2
0.815	0.6	6422500	6438801	6455104	6471409	6487716	6504025	6520336	6536649	6552964	6569281	2
0.816	0.6	6585600	6601921	6618244	6634569	6650896	6667225	6683556	6699889	6716224	6732561	2
0.817	0.6	6748900	6765241	6781584	6797929	6814276	6830625	6846976	6863329	6879684	6896041	2
0.818	0.6	6912400	6928761	6945124	6961489	6977856	6994225	7010596	7026969	7043344	7059721	2
0.819	0.6	7076100	7092481	7108864	7125249	7141636	7158025	7174416	7190809	7207204	7223601	2
0.820	0.6	7240000	7256401	7272804	7289209	7305616	7322025	7338436	7354849	7371264	7387681	2
0.821	0.6	7404100	7420521	7436944	7453369	7469796	7486225	7502656	7519089	7535524	7551961	2
0.822	0.6	7568400	7584841	7601284	7617729	7634176	7650625	7667076	7683529	7699984	7716441	2
0.823	0.6	7732900	7749361	7765824	7782289	7798756	7815225	7831696	7848169	7864644	7881121	2
0.824	0.6	7897600	7914081	7930564	7947049	7963536	7980025	7996516	8013009	8029504	8046001	2
0.825	0.6	8062500	8079001	8095504	8112009	8128516	8145025	8161536	8178049	8194564	8211081	2
0.826	0.6	8227600	8244121	8260644	8277169	8293696	8310225	8326756	8343289	8359824	8376361	2
0.827	0.6	8392900	8409441	8425984	8442529	8459076	8475625	8492176	8508729	8525284	8541841	2
0.828	0.6	8558400	8574961	8591524	8608089	8624656	8641225	8657796	8674369	8690944	8707521	2
0.829	0.6	8724100	8740681	8757264	8773849	8790436	8807025	8823616	8840209	8856804	8873401	2
0.830	0.6	8890000	8906601	8923204	8939809	8956416	8973025	8989636	9006249	9022864	9039481	2
0.831	0.6	9056100	9072721	9089344	9105969	9122596	9139225	9155856	9172489	9189124	9205761	2
0.832	0.6	9222400	9239041	9255684	9272329	9288976	9305625	9322276	9338929	9355584	9372241	2
0.833	0.6	9388900	9405561	9422224	9438889	9455556	9472225	9488896	9505569	9522244	9538921	2
0.834	0.6	9555600	9572281	9588964	9605649	9622336	9639025	9655716	9672409	9689104	9705801	2
0.835	0.6	9722500	9739201	9755904	9772609	9789316	9806025	9822736	9839449	9856164	9872881	2
0.836	0.6	9889600	9906321	9923044	9939769	9956496	9973225	9989956	0006689	0023424	0040161	2
0.837	0.7	0056900	0073641	0090384	0107129	0123876	0140625	0157376	0174129	0190884	0207641	2
0.838	0.7	0224400	0241161	0257924	0274689	0291456	0308225	0324996	0341769	0358544	0375321	2
0.839	0.7	0392100	0408881	0425664	0442449	0459236	0476025	0492816	0509609	0526404	0543201	2
0.840	0.7	0560000	0576801	0593604	0610409	0627216	0644025	0660836	0677649	0694464	0711281	2
0.841	0.7	0728100	0744921	0761744	0778569	0795396	0812225	0829056	0845889	0862724	0879561	2
0.842	0.7	0896400	0913241	0930084	0946929	0963776	0980625	0997476	1014329	1031184	1048041	2
0.843	0.7	1064900	1081761	1098624	1115489	1132356	1149225	1166096	1182969	1199844	1216721	2
0.844	0.7	1233600	1250481	1267364	1284249	1301136	1318025	1334916	1351809	1368704	1385601	2
0.845	0.7	1402500	1419401	1436304	1453209	1470116	1487025	1503936	1520849	1537764	1554681	2
0.846	0.7	1571600	1588521	1605444	1622369	1639296	1656225	1673156	1690089	1707024	1723961	2
0.847	0.7	1740900	1757841	1774784	1791729	1808676	1825625	1842576	1859529	1876484	1893441	2
0.848	0.7	1910400	1927361	1944324	1961289	1978256	1995225	2012196	2029169	2046144	2063121	2
0.849	0.7	2080100	2097081	2114064	2131049	2148036	2165025	2182016	2199009	2216004	2233001	2

TABLE 3

Squares

X		.0000	.0001	.0002	.0003	.0004	.0005	.0006	.0007	.0008	.0009	Δ_2
0.850	0.7	2250000	2267001	2284004	2301009	2318016	2335025	2352036	2369049	2386064	2403081	2
0.851	0.7	2420100	2437121	2454144	2471169	2488196	2505225	2522256	2539289	2556324	2573361	2
0.852	0.7	2590400	2607441	2624484	2641529	2658576	2675625	2692676	2709729	2726784	2743841	2
0.853	0.7	2760900	2777961	2795024	2812089	2829156	2846225	2863296	2880369	2897444	2914521	2
0.854	0.7	2931600	2948681	2965764	2982849	2999936	3017025	3034116	3051209	3068304	3085401	2
0.855	0.7	3102500	3119601	3136704	3153809	3170916	3188025	3205136	3222249	3239364	3256481	2
0.856	0.7	3273600	3290721	3307844	3324969	3342096	3359225	3376356	3393489	3410624	3427761	2
0.857	0.7	3444900	3462041	3479184	3496329	3513476	3530625	3547776	3564929	3582084	3599241	2
0.858	0.7	3616400	3633561	3650724	3667889	3685056	3702225	3719396	3736569	3753744	3770921	2
0.859	0.7	3788100	3805281	3822464	3839649	3856836	3874025	3891216	3908409	3925604	3942801	2
0.860	0.7	3960000	3977201	3994404	4011609	4028816	4046025	4063236	4080449	4097664	4114881	2
0.861	0.7	4132100	4149321	4166544	4183769	4200996	4218225	4235456	4252689	4269924	4287161	2
0.862	0.7	4304400	4321641	4338884	4356129	4373376	4390625	4407876	4425129	4442384	4459641	2
0.863	0.7	4476900	4494161	4511424	4528689	4545956	4563225	4580496	4597769	4615044	4632321	2
0.864	0.7	4649600	4666881	4684164	4701449	4718736	4736025	4753316	4770609	4787904	4805201	2
0.865	0.7	4822500	4839801	4857104	4874409	4891716	4909025	4926336	4943649	4960964	4978281	2
0.866	0.7	4995600	5012921	5030244	5047569	5064896	5082225	5099556	5116889	5134224	5151561	2
0.867	0.7	5168900	5186241	5203584	5220929	5238276	5255625	5272976	5290329	5307684	5325041	2
0.868	0.7	5342400	5359761	5377124	5394489	5411856	5429225	5446596	5463969	5481344	5498721	2
0.869	0.7	5516100	5533481	5550864	5568249	5585636	5603025	5620416	5637809	5655204	5672601	2
0.870	0.7	5690000	5707401	5724804	5742209	5759616	5777025	5794436	5811849	5829264	5846681	2
0.871	0.7	5864100	5881521	5898944	5916369	5933796	5951225	5968656	5986089	6003524	6020961	2
0.872	0.7	6038400	6055841	6073284	6090729	6108176	6125625	6143076	6160529	6177984	6195441	2
0.873	0.7	6212900	6230361	6247824	6265289	6282756	6300225	6317696	6335169	6352644	6370121	2
0.874	0.7	6387600	6405081	6422564	6440049	6457536	6475025	6492516	6510009	6527504	6545001	2
0.875	0.7	6562500	6580001	6597504	6615009	6632516	6650025	6667536	6685049	6702564	6720081	2
0.876	0.7	6737600	6755121	6772644	6790169	6807696	6825225	6842756	6860289	6877824	6895361	2
0.877	0.7	6912900	6930441	6947984	6965529	6983076	7000625	7018176	7035729	7053284	7070841	2
0.878	0.7	7088400	7105961	7123524	7141089	7158656	7176225	7193796	7211369	7228944	7246521	2
0.879	0.7	7264100	7281681	7299264	7316849	7334436	7352025	7369616	7387209	7404804	7422401	2
0.880	0.7	7440000	7457601	7475204	7492809	7510416	7528025	7545636	7563249	7580864	7598481	2
0.881	0.7	7616100	7633721	7651344	7668969	7686596	7704225	7721856	7739489	7757124	7774761	2
0.882	0.7	7792400	7810041	7827684	7845329	7862976	7880625	7898276	7915929	7933584	7951241	2
0.883	0.7	7968900	7986561	8004224	8021889	8039556	8057225	8074896	8092569	8110244	8127921	2
0.884	0.7	8145600	8163281	8180964	8198649	8216336	8234025	8251716	8269409	8287104	8304801	2
0.885	0.7	8322500	8340201	8357904	8375609	8393316	8411025	8428736	8446449	8464164	8481881	2
0.886	0.7	8499600	8517321	8535044	8552769	8570496	8588225	8605956	8623689	8641424	8659161	2
0.887	0.7	8676900	8694641	8712384	8730129	8747876	8765625	8783376	8801129	8818884	8836641	2
0.888	0.7	8854400	8872161	8889924	8907689	8925456	8943225	8960996	8978769	8996544	9014321	2
0.889	0.7	9032100	9049881	9067664	9085449	9103236	9121025	9138816	9156609	9174404	9192201	2
0.890	0.7	9210000	9227801	9245604	9263409	9281216	9299025	9316836	9334649	9352464	9370281	2
0.891	0.7	9388100	9405921	9423744	9441569	9459396	9477225	9495056	9512889	9530724	9548561	2
0.892	0.7	9566400	9584241	9602084	9619929	9637776	9655625	9673476	9691329	9709184	9727041	2
0.893	0.7	9744900	9762761	9780624	9798489	9816356	9834225	9852096	9869969	9887844	9905721	2
0.894	0.7	9923600	9941481	9959364	9977249	9995136	0013025	0030916	0048809	0066704	0084601	2
0.895	0.8	0102500	0120401	0138304	0156209	0174116	0192025	0209936	0227849	0245764	0263681	2
0.896	0.8	0281600	0299521	0317444	0335369	0353296	0371225	0389156	0407089	0425024	0442961	2
0.897	0.8	0460900	0478841	0496784	0514729	0532676	0550625	0568576	0586529	0604484	0622441	2
0.898	0.8	0640400	0658361	0676324	0694289	0712256	0730225	0748196	0766169	0784144	0802121	2
0.899	0.8	0820100	0838081	0856064	0874049	0892036	0910025	0928016	0946009	0964004	0982001	2

TABLE 3

Squares

X		.0000	.0001	.0002	.0003	.0004	.0005	.0006	.0007	.0008	.0009	Δ_2
0.900	0.8	1000000	1018001	1036004	1054009	1072016	1090025	1108036	1126049	1144064	1162081	2
0.901	0.8	1180100	1198121	1216144	1234169	1252196	1270225	1288256	1306289	1324324	1342361	2
0.902	0.8	1360400	1378441	1396484	1414529	1432576	1450625	1468676	1486729	1504784	1522841	2
0.903	0.8	1540900	1558961	1577024	1595089	1613156	1631225	1649296	1667369	1685444	1703521	2
0.904	0.8	1721600	1739681	1757764	1775849	1793936	1812025	1830116	1848209	1866304	1884401	2
0.905	0.8	1902500	1920601	1938704	1956809	1974916	1993025	2011136	2029249	2047364	2065481	2
0.906	0.8	2083600	2101721	2119844	2137969	2156096	2174225	2192356	2210489	2228624	2246761	2
0.907	0.8	2264900	2283041	2301184	2319329	2337476	2355625	2373776	2391929	2410084	2428241	2
0.908	0.8	2446400	2464561	2482724	2500889	2519056	2537225	2555396	2573569	2591744	2609921	2
0.909	0.8	2628100	2646281	2664464	2682649	2700836	2719025	2737216	2755409	2773604	2791801	2
0.910	0.8	2810000	2828201	2846404	2864609	2882816	2901025	2919236	2937449	2955664	2973881	2
0.911	0.8	2992100	3010321	3028544	3046769	3064996	3083225	3101456	3119689	3137924	3156161	2
0.912	0.8	3174400	3192641	3210884	3229129	3247376	3265625	3283876	3302129	3320384	3338641	2
0.913	0.8	3356900	3375161	3393424	3411689	3429956	3448225	3466496	3484769	3503044	3521321	2
0.914	0.8	3539600	3557881	3576164	3594449	3612736	3631025	3649316	3667609	3685904	3704201	2
0.915	0.8	3722500	3740801	3759104	3777409	3795716	3814025	3832336	3850649	3868964	3887281	2
0.916	0.8	3905600	3923921	3942244	3960569	3978896	3997225	4015556	4033889	4052224	4070561	2
0.917	0.8	4088900	4107241	4125584	4143929	4162276	4180625	4198976	4217329	4235684	4254041	2
0.918	0.8	4272400	4290761	4309124	4327489	4345856	4364225	4382596	4400969	4419344	4437721	2
0.919	0.8	4456100	4474481	4492864	4511249	4529636	4548025	4566416	4584809	4603204	4621601	2
0.920	0.8	4640000	4658401	4676804	4695209	4713616	4732025	4750436	4768849	4787264	4805681	2
0.921	0.8	4824100	4842521	4860944	4879369	4897796	4916225	4934656	4953089	4971524	4989961	2
0.922	0.8	5008400	5026841	5045284	5063729	5082176	5100625	5119076	5137529	5155984	5174441	2
0.923	0.8	5192900	5211361	5229824	5248289	5266756	5285225	5303696	5322169	5340644	5359121	2
0.924	0.8	5377600	5396081	5414564	5433049	5451536	5470025	5488516	5507009	5525504	5544001	2
0.925	0.8	5562500	5581001	5599504	5618009	5636516	5655025	5673536	5692049	5710564	5729081	2
0.926	0.8	5747600	5766121	5784644	5803169	5821696	5840225	5858756	5877289	5895824	5914361	2
0.927	0.8	5932900	5951441	5969984	5988529	6007076	6025625	6044176	6062729	6081284	6099841	2
0.928	0.8	6118400	6136961	6155524	6174089	6192656	6211225	6229796	6248369	6266944	6285521	2
0.929	0.8	6304100	6322681	6341264	6359849	6378436	6397025	6415616	6434209	6452804	6471401	2
0.930	0.8	6490000	6508601	6527204	6545809	6564416	6583025	6601636	6620249	6638864	6657481	2
0.931	0.8	6676100	6694721	6713344	6731969	6750596	6769225	6787856	6806489	6825124	6843761	2
0.932	0.8	6862400	6881041	6899684	6918329	6936976	6955625	6974276	6992929	7011584	7030241	2
0.933	0.8	7048900	7067561	7086224	7104889	7123556	7142225	7160896	7179569	7198244	7216921	2
0.934	0.8	7235600	7254281	7272964	7291649	7310336	7329025	7347716	7366409	7385104	7403801	2
0.935	0.8	7422500	7441201	7459904	7478609	7497316	7516025	7534736	7553449	7572164	7590881	2
0.936	0.8	7609600	7628321	7647044	7665769	7684496	7703225	7721956	7740689	7759424	7778161	2
0.937	0.8	7796900	7815641	7834384	7853129	7871876	7890625	7909376	7928129	7946884	7965641	2
0.938	0.8	7984400	8003161	8021924	8040689	8059456	8078225	8096996	8115769	8134544	8153321	2
0.939	0.8	8172100	8190881	8209664	8228449	8247236	8266025	8284816	8303609	8322404	8341201	2
0.940	0.8	8360000	8378801	8397604	8416409	8435216	8454025	8472836	8491649	8510464	8529281	2
0.941	0.8	8548100	8566921	8585744	8604569	8623396	8642225	8661056	8679889	8698724	8717561	2
0.942	0.8	8736400	8755241	8774084	8792929	8811776	8830625	8849476	8868329	8887184	8906041	2
0.943	0.8	8924900	8943761	8962624	8981489	9000356	9019225	9038096	9056969	9075844	9094721	2
0.944	0.8	9113600	9132481	9151364	9170249	9189136	9208025	9226916	9245809	9264704	9283601	2
0.945	0.8	9302500	9321401	9340304	9359209	9378116	9397025	9415936	9434849	9453764	9472681	2
0.946	0.8	9491600	9510521	9529444	9548369	9567296	9586225	9605156	9624089	9643024	9661961	2
0.947	0.8	9680900	9699841	9718784	9737729	9756676	9775625	9794576	9813529	9832484	9851441	2
0.948	0.8	9870400	9889361	9908324	9927289	9946256	9965225	9984196	0003169	0022144	0041121	2
0.949	0.9	0060100	0079081	0098064	0117049	0136036	0155025	0174016	0193009	0212004	0231001	2

TABLE 3

Squares

X		.0000	.0001	.0002	.0003	.0004	.0005	.0006	.0007	.0008	.0009	Δ_2
0.950	0.9	0250000	0269001	0288004	0307009	0326016	0345025	0364036	0383049	0402064	0421081	2
0.951	0.9	0440100	0459121	0478144	0497169	0516196	0535225	0554256	0573289	0592324	0611361	2
0.952	0.9	0630400	0649441	0668484	0687529	0706576	0725625	0744676	0763729	0782784	0801841	2
0.953	0.9	0820900	0839961	0859024	0878089	0897156	0916225	0935296	0954369	0973444	0992521	2
0.954	0.9	1011600	1030681	1049764	1068849	1087936	1107025	1126116	1145209	1164304	1183401	2
0.955	0.9	1202500	1221601	1240704	1259809	1278916	1298025	1317136	1336249	1355364	1374481	2
0.956	0.9	1393600	1412721	1431844	1450969	1470096	1489225	1508356	1527489	1546624	1565761	2
0.957	0.9	1584900	1604041	1623184	1642329	1661476	1680625	1699776	1718929	1738084	1757241	2
0.958	0.9	1776400	1795561	1814724	1833889	1853056	1872225	1891396	1910569	1929744	1948921	2
0.959	0.9	1968100	1987281	2006464	2025649	2044836	2064025	2083216	2102409	2121604	2140801	2
0.960	0.9	2160000	2179201	2198404	2217609	2236816	2256025	2275236	2294449	2313664	2332881	2
0.961	0.9	2352100	2371321	2390544	2409769	2428996	2448225	2467456	2486689	2505924	2525161	2
0.962	0.9	2544400	2563641	2582884	2602129	2621376	2640625	2659876	2679129	2698384	2717641	2
0.963	0.9	2736900	2756161	2775424	2794689	2813956	2833225	2852496	2871769	2891044	2910321	2
0.964	0.9	2929600	2948881	2968164	2987449	3006736	3026025	3045316	3064609	3083904	3103201	2
0.965	0.9	3122500	3141801	3161104	3180409	3199716	3219025	3238336	3257649	3276964	3296281	2
0.966	0.9	3315600	3334921	3354244	3373569	3392896	3412225	3431556	3450889	3470224	3489561	2
0.967	0.9	3508900	3528241	3547584	3566929	3586276	3605625	3624976	3644329	3663684	3683041	2
0.968	0.9	3702400	3721761	3741124	3760489	3779856	3799225	3818596	3837969	3857344	3876721	2
0.969	0.9	3896100	3915481	3934864	3954249	3973636	3993025	4012416	4031809	4051204	4070601	2
0.970	0.9	4090000	4109401	4128804	4148209	4167616	4187025	4206436	4225849	4245264	4264681	2
0.971	0.9	4284100	4303521	4322944	4342369	4361796	4381225	4400656	4420089	4439524	4458961	2
0.972	0.9	4478400	4497841	4517284	4536729	4556176	4575625	4595076	4614529	4633984	4653441	2
0.973	0.9	4672900	4692361	4711824	4731289	4750756	4770225	4789696	4809169	4828644	4848121	2
0.974	0.9	4867600	4887081	4906564	4926049	4945536	4965025	4984516	5004009	5023504	5043001	2
0.975	0.9	5062500	5082001	5101504	5121009	5140516	5160025	5179536	5199049	5218564	5238081	2
0.976	0.9	5257600	5277121	5296644	5316169	5335696	5355225	5374756	5394289	5413824	5433361	2
0.977	0.9	5452900	5472441	5491984	5511529	5531076	5550625	5570176	5589729	5609284	5628841	2
0.978	0.9	5648400	5667961	5687524	5707089	5726656	5746225	5765796	5785369	5804944	5824521	2
0.979	0.9	5844100	5863681	5883264	5902849	5922436	5942025	5961616	5981209	6000804	6020401	2
0.980	0.9	6040000	6059601	6079204	6098809	6118416	6138025	6157636	6177249	6196864	6216481	2
0.981	0.9	6236100	6255721	6275344	6294969	6314596	6334225	6353856	6373489	6393124	6412761	2
0.982	0.9	6432400	6452041	6471684	6491329	6510976	6530625	6550276	6569929	6589584	6609241	2
0.983	0.9	6628900	6648561	6668224	6687889	6707556	6727225	6746896	6766569	6786244	6805921	2
0.984	0.9	6825600	6845281	6864964	6884649	6904336	6924025	6943716	6963409	6983104	7002801	2
0.985	0.9	7022500	7042201	7061904	7081609	7101316	7121025	7140736	7160449	7180164	7199881	2
0.986	0.9	7219600	7239321	7259044	7278769	7298496	7318225	7337956	7357689	7377424	7397161	2
0.987	0.9	7416900	7436641	7456384	7476129	7495876	7515625	7535376	7555129	7574884	7594641	2
0.988	0.9	7614400	7634161	7653924	7673689	7693456	7713225	7732996	7752769	7772544	7792321	2
0.989	0.9	7812100	7831881	7851664	7871449	7891236	7911025	7930816	7950609	7970404	7990201	2
0.990	0.9	8010000	8029801	8049604	8069409	8089216	8109025	8128836	8148649	8168464	8188281	2
0.991	0.9	8208100	8227921	8247744	8267569	8287396	8307225	8327056	8346889	8366724	8386561	2
0.992	0.9	8406400	8426241	8446084	8465929	8485776	8505625	8525476	8545329	8565184	8585041	2
0.993	0.9	8604900	8624761	8644624	8664489	8684356	8704225	8724096	8743969	8763844	8783721	2
0.994	0.9	8803600	8823481	8843364	8863249	8883136	8903025	8922916	8942809	8962704	8982601	2
0.995	0.9	9002500	9022401	9042304	9062209	9082116	9102025	9121936	9141849	9161764	9181681	2
0.996	0.9	9201600	9221521	9241444	9261369	9281296	9301225	9321156	9341089	9361024	9380961	2
0.997	0.9	9400900	9420841	9440784	9460729	9480676	9500625	9520576	9540529	9560484	9580441	2
0.998	0.9	9600400	9620361	9640324	9660289	9680256	9700225	9720196	9740169	9760144	9780121	2
0.999	0.9	9800100	9820081	9840064	9860049	9880036	9900025	9920016	9940009	9960004	9980001	2

TABLE 4

Square Roots

X		.0000	.0001	.0002	.0003	.0004	.0005	.0006	.0007	.0008	.0009	Δ_2
0.100	0.3	1622777	1638584	1654384	1670175	1685959,	1701735	1717503	1733263	1749016	1764760	−7
0.101	0.3	1780497	1796226	1811947	1827661	1843367	1859065	1874755	1890437	1906112	1921779	−7
0.102	0.3	1937439	1953091	1968735	1984371	2000000	2015621	2031235	2046841	2062439	2078030	−7
0.103	0.3	2093613	2109189	2124757	2140317	2155870	2171416	2186954	2202484	2218007	2233523	−6
0.104	0.3	2249031	2264532	2280025	2295511	2310989	2326460	2341923	2357379	2372828	2388269	−6
0.105	0.3	2403703	2419130	2434549	2449961	2465366	2480764	2496154	2511536	2526912	2542280	−6
0.106	0.3	2557641	2572995	2588341	2603681	2619013	2634338	2649655	2664966	2680269	2695565	−6
0.107	0.3	2710854	2726136	2741411	2756679	2771939	2787193	2802439	2817678	2832910	2848135	−6
0.108	0.3	2863353	2878564	2893768	2908965	2924155	2939338	2954514	2969683	2984845	3000000	−6
0.109	0.3	3015148	3030289	3045423	3060551	3075671	3090784	3105891	3120990	3136083	3151169	−6
0.110	0.3	3166248	3181320	3196385	3211444	3226495	3241540	3256578	3271610	3286634	3301652	−6
0.111	0.3	3316662	3331667	3346664	3361655	3376639	3391616	3406586	3421550	3436507	3451457	−6
0.112	0.3	3466401	3481338	3496268	3511192	3526109	3541020	3555923	3570821	3585711	3600595	−6
0.113	0.3	3615473	3630343	3645208	3660065	3674916	3689761	3704599	3719431	3734256	3749074	−6
0.114	0.3	3763886	3778691	3793490	3808283	3823069	3837849	3852622	3867388	3882149	3896903	−5
0.115	0.3	3911650	3926391	3941125	3955854	3970575	3985291	4000000	4014703	4029399	4044089	−5
0.116	0.3	4058773	4073450	4088121	4102786	4117444	4132096	4146742	4161382	4176015	4190642	−5
0.117	0.3	4205263	4219877	4234486	4249088	4263683	4278273	4292856	4307434	4322005	4336569	−5
0.118	0.3	4351128	4365681	4380227	4394767	4409301	4423829	4438351	4452866	4467376	4481879	−5
0.119	0.3	4496377	4510868	4525353	4539832	4554305	4568772	4583233	4597688	4612137	4626579	−5
0.120	0.3	4641016	4655447	4669872	4684290	4698703	4713110	4727511	4741906	4756294	4770677	−5
0.121	0.3	4785054	4799425	4813790	4828150	4842503	4856850	4871192	4885527	4899857	4914180	−5
0.122	0.3	4928498	4942810	4957117	4971417	4985711	5000000	5014283	5028560	5042831	5057096	−5
0.123	0.3	5071356	5085610	5099858	5114100	5128336	5142567	5156792	5171011	5185224	5199432	−5
0.124	0.3	5213634	5227830	5242020	5256205	5270384	5284557	5298725	5312887	5327043	5341194	−5
0.125	0.3	5355339	5369478	5383612	5397740	5411862	5425979	5440090	5454196	5468296	5482390	−5
0.126	0.3	5496479	5510562	5524639	5538711	5552778	5566838	5580894	5594943	5608988	5623026	−5
0.127	0.3	5637059	5651087	5665109	5679125	5693137	5707142	5721142	5735137	5749126	5763109	−4
0.128	0.3	5777088	5791060	5805028	5818989	5832946	5846897	5860842	5874782	5888717	5902646	−4
0.129	0.3	5916570	5930488	5944401	5958309	5972211	5986108	6000000	6013886	6027767	6041643	−4
0.130	0.3	6055513	6069378	6083237	6097091	6110940	6124784	6138622	6152455	6166283	6180105	−4
0.131	0.3	6193922	6207734	6221540	6235342	6249138	6262929	6276714	6290495	6304270	6318040	−4
0.132	0.3	6331804	6345564	6359318	6373067	6386811	6400549	6414283	6428011	6441734	6455452	−4
0.133	0.3	6469165	6482873	6496575	6510272	6523965	6537652	6551334	6565011	6578682	6592349	−4
0.134	0.3	6606010	6619667	6633318	6646964	6660606	6674242	6687873	6701499	6715119	6728735	−4
0.135	0.3	6742346	6755952	6769553	6783148	6796739	6810325	6823905	6837481	6851052	6864617	−4
0.136	0.3	6878178	6891733	6905284	6918830	6932371	6945906	6959437	6972963	6986484	7000000	−4
0.137	0.3	7013511	7027017	7040518	7054015	7067506	7080992	7094474	7107951	7121422	7134889	−4
0.138	0.3	7148351	7161808	7175261	7188708	7202150	7215588	7229021	7242449	7255872	7269290	−4
0.139	0.3	7282704	7296112	7309516	7322915	7336309	7349699	7363083	7376463	7389838	7403208	−4
0.140	0.3	7416574	7429935	7443290	7456642	7469988	7483330	7496666	7509999	7523326	7536649	−4
0.141	0.3	7549967	7563280	7576588	7589892	7603191	7616486	7629775	7643060	7656341	7669616	−4
0.142	0.3	7682887	7696154	7709415	7722672	7735924	7749172	7762415	7775654	7788887	7802116	−4
0.143	0.3	7815341	7828561	7841776	7854986	7868192	7881394	7894591	7907783	7920970	7934153	−4
0.144	0.3	7947332	7960506	7973675	7986840	8000000	8013156	8026307	8039453	8052595	8065733	−4
0.145	0.3	8078865	8091994	8105118	8118237	8131352	8144462	8157568	8170669	8183766	8196858	−3
0.146	0.3	8209946	8223030	8236109	8249183	8262253	8275318	8288379	8301436	8314488	8327536	−3
0.147	0.3	8340579	8353618	8366652	8379682	8392708	8405727	8418745	8431758	8444766	8457769	−3
0.148	0.3	8470768	8483763	8496753	8509739	8522721	8535698	8548670	8561639	8574603	8587563	−3
0.149	0.3	8600518	8613469	8626416	8639358	8652296	8665230	8678159	8691084	8704005	8716921	−3

TABLE 4

Square Roots

X		.0000	.0001	.0002	.0003	.0004	.0005	.0006	.0007	.0008	.0009	Δ,
0.150	0.3	8729833	8742741	8755645	8768544	8781439	8794329	8807216	8820098	8832976	8845849	-3
0.151	0.3	8858718	8871583	8884444	8897301	8910153	8923001	8935845	8948684	8961519	8974350	-3
0.152	0.3	8987177	9000000	9012818	9025633	9038443	9051248	9064050	9076847	9089641	9102430	-3
0.153	0.3	9115214	9127995	9140772	9153544	9166312	9179076	9191836	9204592	9217343	9230090	-3
0.154	0.3	9242834	9255573	9268308	9281039	9293765	9306488	9319206	9331921	9344631	9357337	-3
0.155	0.3	9370039	9382737	9395431	9408121	9420807	9433488	9446166	9458839	9471509	9484174	-3
0.156	0.3	9496835	9509493	9522146	9534795	9547440	9560081	9572718	9585351	9597980	9610605	-3
0.157	0.3	9623225	9635842	9648455	9661064	9673669	9686270	9698866	9711459	9724048	9736633	-3
0.158	0.3	9749214	9761791	9774364	9786932	9799497	9812058	9824615	9837169	9849718	9862263	-3
0.159	0.3	9874804	9887341	9899875	9912404	9924930	9937451	9949969	9962482	9974992	9987498	-3
0.160	0.4	0000000	0012498	0024992	0037482	0049969	0062451	0074930	0087405	0099875	0112342	-3
0.161	0.4	0124805	0137264	0149720	0162171	0174619	0187063	0199502	0211939	0224371	0236799	-3
0.162	0.4	0249224	0261644	0274061	0286474	0298883	0311289	0323690	0336088	0348482	0360872	-3
0.163	0.4	0373258	0385641	0398020	0410395	0422766	0435133	0447497	0459857	0472213	0484565	-3
0.164	0.4	0496913	0509258	0521599	0533936	0546270	0558600	0570926	0583248	0595566	0607881	-3
0.165	0.4	0620192	0632499	0644803	0657103	0669399	0681691	0693980	0706265	0718546	0730824	-3
0.166	0.4	0743098	0755368	0767634	0779897	0792156	0804412	0816663	0828911	0841156	0853396	-3
0.167	0.4	0865633	0877867	0890097	0902323	0914545	0926764	0938979	0951190	0963398	0975602	-3
0.168	0.4	0987803	1000000	1012193	1024383	1036569	1048752	1060930	1073106	1085277	1097445	-3
0.169	0.4	1109610	1121770	1133928	1146081	1158231	1170378	1182521	1194660	1206796	1218928	-3
0.170	0.4	1231056	1243181	1255303	1267421	1279535	1291646	1303753	1315857	1327957	1340053	-3
0.171	0.4	1352146	1364236	1376322	1388404	1400483	1412558	1424630	1436699	1448764	1460825	-3
0.172	0.4	1472883	1484937	1496988	1509035	1521079	1533119	1545156	1557190	1569219	1581246	-2
0.173	0.4	1593269	1605288	1617304	1629317	1641326	1653331	1665333	1677332	1689327	1701319	-2
0.174	0.4	1713307	1725292	1737274	1749251	1761226	1773197	1785165	1797129	1809090	1821047	-2
0.175	0.4	1833001	1844952	1856899	1868843	1880783	1892720	1904654	1916584	1928511	1940434	-2
0.176	0.4	1952354	1964271	1976184	1988094	2000000	2011903	2023803	2035699	2047592	2059482	-2
0.177	0.4	2071368	2083251	2095130	2107007	2118879	2130749	2142615	2154478	2166337	2178193	-2
0.178	0.4	2190046	2201896	2213742	2225585	2237424	2249260	2261093	2272923	2284749	2296572	-2
0.179	0.4	2308392	2320208	2332021	2343831	2355637	2367440	2379240	2391037	2402830	2414620	-2
0.180	0.4	2426407	2438190	2449971	2461747	2473521	2485292	2497059	2508823	2520583	2532341	-2
0.181	0.4	2544095	2555846	2567593	2579338	2591079	2602817	2614551	2626283	2638011	2649736	-2
0.182	0.4	2661458	2673177	2684892	2696604	2708313	2720019	2731721	2743421	2755117	2766810	-2
0.183	0.4	2778499	2790186	2801869	2813549	2825226	2836900	2848571	2860238	2871902	2883563	-2
0.184	0.4	2895221	2906876	2918527	2930176	2941821	2953463	2965102	2976738	2988371	3000000	-2
0.185	0.4	3011626	3023250	3034870	3046486	3058100	3069711	3081318	3092923	3104524	3116122	-2
0.186	0.4	3127717	3139309	3150898	3162484	3174066	3185646	3197222	3208795	3220366	3231933	-2
0.187	0.4	3243497	3255057	3266615	3278170	3289722	3301270	3312816	3324358	3335897	3347434	-2
0.188	0.4	3358967	3370497	3382024	3393548	3405069	3416587	3428101	3439613	3451122	3462628	-2
0.189	0.4	3474130	3485630	3497126	3508620	3520110	3531598	3543082	3554563	3566042	3577517	-2
0.190	0.4	3588989	3600459	3611925	3623388	3634848	3646306	3657760	3669211	3680659	3692105	-2
0.191	0.4	3703547	3714986	3726422	3737855	3749286	3760713	3772137	3783559	3794977	3806392	-2
0.192	0.4	3817805	3829214	3840620	3852024	3863424	3874822	3886217	3897608	3908997	3920382	-2
0.193	0.4	3931765	3943145	3954522	3965896	3977267	3988635	4000000	4011362	4022721	4034078	-2
0.194	0.4	4045431	4056782	4068129	4079474	4090815	4102154	4113490	4124823	4136153	4147480	-2
0.195	0.4	4158804	4170126	4181444	4192760	4204072	4215382	4226689	4237993	4249294	4260592	-2
0.196	0.4	4271887	4283180	4294469	4305756	4317040	4328321	4339599	4350874	4362146	4373415	-2
0.197	0.4	4384682	4395946	4407207	4418465	4429720	4440972	4452222	4463468	4474712	4485953	-2
0.198	0.4	4497191	4508426	4519659	4530888	4542115	4553339	4564560	4575778	4586994	4598206	-2
0.199	0.4	4609416	4620623	4631827	4643029	4654227	4665423	4676616	4687806	4698993	4710178	-2

TABLE 4

Square Roots __

X	.0000	.0001	.0002	.0003	.0004	.0005	.0006	.0007	.0008	.0009	Δ_2
0.200 0.4	4721360	4732538	4743715	4754888	4766059	4777226	4788391	4799554	4810713	4821870	-2
0.201 0.4	4833024	4844175	4855323	4866469	4877611	4888751	4899889	4911023	4922155	4933284	-2
0.202 0.4	4944410	4955534	4966654	4977772	4988887	5000000	5011110	5022217	5033321	5044422	-2
0.203 0.4	5055521	5066617	5077711	5088801	5099889	5110974	5122057	5133136	5144213	5155288	-2
0.204 0.4	5166359	5177428	5188494	5199557	5210618	5221676	5232731	5243784	5254834	5265881	-2
0.205 0.4	5276926	5287967	5299007	5310043	5321077	5332108	5343136	5354162	5365185	5376205	-2
0.206 0.4	5387223	5398238	5409250	5420260	5431267	5442271	5453273	5464272	5475268	5486262	-2
0.207 0.4	5497253	5508241	5519227	5530210	5541190	5552168	5563143	5574115	5585085	5596052	-2
0.208 0.4	5607017	5617979	5628938	5639895	5650849	5661800	5672749	5683695	5694639	5705580	-2
0.209 0.4	5716518	5727453	5738386	5749317	5760245	5771170	5782093	5793013	5803930	5814845	-2
0.210 0.4	5825757	5836667	5847574	5858478	5869380	5880279	5891176	5902070	5912961	5923850	-2
0.211 0.4	5934736	5945620	5956501	5967380	5978256	5989129	6000000	6010868	6021734	6032597	-2
0.212 0.4	6043458	6054316	6065171	6076024	6086874	6097722	6108567	6119410	6130250	6141088	-2
0.213 0.4	6151923	6162755	6173586	6184413	6195238	6206060	6216880	6227697	6238512	6249324	-2
0.214 0.4	6260134	6270941	6281746	6292548	6303348	6314145	6324939	6335731	6346521	6357308	-2
0.215 0.4	6368092	6378874	6389654	6400431	6411206	6421978	6432747	6443514	6454279	6465041	-1
0.216 0.4	6475800	6486557	6497312	6508064	6518813	6529560	6540305	6551047	6561787	6572524	-1
0.217 0.4	6583259	6593991	6604721	6615448	6626173	6636895	6647615	6658333	6669048	6679760	-1
0.218 0.4	6690470	6701178	6711883	6722585	6733286	6743984	6754679	6765372	6776062	6786750	-1
0.219 0.4	6797436	6808119	6818800	6829478	6840154	6850827	6861498	6872167	6882833	6893496	-1
0.220 0.4	6904158	6914816	6925473	6936127	6946778	6957427	6968074	6978719	6989360	7000000	-1
0.221 0.4	7010637	7021272	7031904	7042534	7053161	7063786	7074409	7085029	7095647	7106263	-1
0.222 0.4	7116876	7127487	7138095	7148701	7159304	7169906	7180504	7191101	7201695	7212287	-1
0.223 0.4	7222876	7233463	7244047	7254629	7265209	7275787	7286362	7296934	7307505	7318073	-1
0.224 0.4	7328638	7339201	7349762	7360321	7370877	7381431	7391982	7402532	7413078	7423623	-1
0.225 0.4	7434165	7444705	7455242	7465777	7476310	7486840	7497368	7507894	7518417	7528939	-1
0.226 0.4	7539457	7549974	7560488	7571000	7581509	7592016	7602521	7613023	7623524	7634021	-1
0.227 0.4	7644517	7655010	7665501	7675990	7686476	7696960	7707442	7717921	7728398	7738873	-1
0.228 0.4	7749346	7759816	7770284	7780749	7791213	7801674	7812132	7822589	7833043	7843495	-1
0.229 0.4	7853944	7864392	7874837	7885280	7895720	7906158	7916594	7927028	7937459	7947888	-1
0.230 0.4	7958315	7968740	7979162	7989582	8000000	8010416	8020829	8031240	8041649	8052055	-1
0.231 0.4	8062459	8072861	8083261	8093659	8104054	8114447	8124838	8135226	8145612	8155997	-1
0.232 0.4	8166378	8176758	8187135	8197510	8207883	8218254	8228622	8238988	8249352	8259714	-1
0.233 0.4	8270074	8280431	8290786	8301139	8311489	8321838	8332184	8342528	8352870	8363209	-1
0.234 0.4	8373546	8383882	8394215	8404545	8414874	8425200	8435524	8445846	8456166	8466483	-1
0.235 0.4	8476799	8487112	8497423	8507731	8518038	8528342	8538644	8548944	8559242	8569538	-1
0.236 0.4	8579831	8590122	8600411	8610698	8620983	8631266	8641546	8651824	8662100	8672374	-1
0.237 0.4	8682646	8692915	8703183	8713448	8723711	8733972	8744230	8754487	8764741	8774994	-1
0.238 0.4	8785244	8795492	8805737	8815981	8826222	8836462	8846699	8856934	8867167	8877398	-1
0.239 0.4	8887626	8897853	8908077	8918299	8928519	8938737	8948953	8959167	8969378	8979588	-1
0.240 0.4	8989795	9000000	9010203	9020404	9030603	9040799	9050994	9061186	9071377	9081565	-1
0.241 0.4	9091751	9101935	9112117	9122296	9132474	9142649	9152823	9162994	9173163	9183330	-1
0.242 0.4	9193495	9203658	9213819	9223978	9234134	9244289	9254441	9264592	9274740	9284886	-1
0.243 0.4	9295030	9305172	9315312	9325450	9335586	9345719	9355851	9365980	9376108	9386233	-1
0.244 0.4	9396356	9406477	9416596	9426713	9436828	9446941	9457052	9467161	9477267	9487372	-1
0.245 0.4	9497475	9507575	9517674	9527770	9537864	9547957	9558047	9568135	9578221	9588305	-1
0.246 0.4	9598387	9608467	9618545	9628621	9638695	9648766	9658836	9668904	9678969	9689033	-1
0.247 0.4	9699095	9709154	9719212	9729267	9739320	9749372	9759421	9769469	9779514	9789557	-1
0.248 0.4	9799598	9809638	9819675	9829710	9839743	9849774	9859803	9869831	9879856	9889879	-1
0.249 0.4	9899900	9909919	9919936	9929951	9939964	9949975	9959984	9969991	9979996	9989999	-1

TABLE 4

Square Roots

X		.0000	.0001	.0002	.0003	.0004	.0005	.0006	.0007	.0008	.0009	Δ_2
0.250	0.5	0000000	0009999	0019996	0029991	0039984	0049975	0059964	0069951	0079936	0089919	
0.251	0.5	0099900	0109879	0119856	0129831	0139805	0149776	0159745	0169712	0179677	0189640	−1
0.252	0.5	0199602	0209561	0219518	0229473	0239427	0249378	0259327	0269275	0279220	0289164	−1
0.253	0.5	0299105	0309045	0318982	0328918	0338852	0348783	0358713	0368641	0378567	0388491	−1
0.254	0.5	0398413	0408333	0418251	0428167	0438081	0447993	0457903	0467812	0477718	0487622	−1
0.255	0.5	0497525	0507425	0517324	0527220	0537115	0547008	0556899	0566788	0576674	0586559	−1
0.256	0.5	0596443	0606324	0616203	0626080	0635956	0645829	0655701	0665570	0675438	0685304	−1
0.257	0.5	0695167	0705029	0714889	0724747	0734604	0744458	0754310	0764161	0774009	0783856	−1
0.258	0.5	0793700	0803543	0813384	0823223	0833060	0842895	0852729	0862560	0872389	0882217	−1
0.259	0.5	0892043	0901866	0911688	0921508	0931326	0941143	0950957	0960769	0970580	0980388	−1
0.260	0.5	0990195	1000000	1009803	1019604	1029403	1039201	1048996	1058790	1068581	1078371	−1
0.261	0.5	1088159	1097945	1107729	1117512	1127292	1137071	1146847	1156622	1166395	1176166	−1
0.262	0.5	1185936	1195703	1205468	1215232	1224994	1234754	1244512	1254268	1264022	1273775	−1
0.263	0.5	1283526	1293274	1303021	1312766	1322510	1332251	1341991	1351728	1361464	1371198	−1
0.264	0.5	1380930	1390661	1400389	1410116	1419841	1429563	1439285	1449004	1458721	1468437	−1
0.265	0.5	1478151	1487863	1497573	1507281	1516987	1526692	1536395	1546096	1555795	1565492	−1
0.266	0.5	1575188	1584882	1594573	1604263	1613952	1623638	1633323	1643005	1652686	1662365	−1
0.267	0.5	1672043	1681718	1691392	1701064	1710734	1720402	1730069	1739733	1749396	1759057	−1
0.268	0.5	1768716	1778374	1788030	1797683	1807335	1816986	1826634	1836281	1845926	1855569	−1
0.269	0.5	1865210	1874849	1884487	1894123	1903757	1913389	1923020	1932649	1942276	1951901	−1
0.270	0.5	1961524	1971146	1980766	1990384	2000000	2009614	2019227	2028838	2038447	2048055	−1
0.271	0.5	2057660	2067264	2076866	2086467	2096065	2105662	2115257	2124850	2134442	2144031	−1
0.272	0.5	2153619	2163205	2172790	2182372	2191953	2201533	2211110	2220686	2230259	2239832	−1
0.273	0.5	2249402	2258971	2268537	2278102	2287666	2297227	2306787	2316345	2325902	2335456	−1
0.274	0.5	2345009	2354560	2364110	2373657	2383203	2392748	2402290	2411831	2421370	2430907	−1
0.275	0.5	2440442	2449976	2459508	2469038	2478567	2488094	2497619	2507142	2516664	2526184	−1
0.276	0.5	2535702	2545219	2554733	2564246	2573758	2583267	2592775	2602281	2611786	2621288	−1
0.277	0.5	2630789	2640289	2649786	2659282	2668776	2678269	2687759	2697248	2706736	2716221	−1
0.278	0.5	2725705	2735187	2744668	2754147	2763624	2773099	2782573	2792045	2801515	2810984	−1
0.279	0.5	2820451	2829916	2839379	2848841	2858301	2867760	2877216	2886671	2896125	2905576	−1
0.280	0.5	2915026	2924474	2933921	2943366	2952809	2962251	2971691	2981129	2990565	3000000	−1
0.281	0.5	3009433	3018865	3028294	3037722	3047149	3056574	3065997	3075418	3084838	3094256	−1
0.282	0.5	3103672	3113087	3122500	3131911	3141321	3150729	3160135	3169540	3178943	3188345	−1
0.283	0.5	3197744	3207142	3216539	3225933	3235327	3244718	3254108	3263496	3272882	3282267	−1
0.284	0.5	3291650	3301032	3310412	3319790	3329166	3338541	3347915	3357286	3366656	3376025	−1
0.285	0.5	3385391	3394756	3404120	3413481	3422842	3432200	3441557	3450912	3460266	3469618	−1
0.286	0.5	3478968	3488316	3497663	3507009	3516353	3525695	3535035	3544374	3553711	3563047	−1
0.287	0.5	3572381	3581713	3591044	3600373	3609701	3619026	3628351	3637673	3646994	3656314	−1
0.288	0.5	3665631	3674948	3684262	3693575	3702886	3712196	3721504	3730811	3740115	3749419	−1
0.289	0.5	3758720	3768020	3777319	3786615	3795911	3805204	3814496	3823787	3833075	3842363	−1
0.290	0.5	3851648	3860932	3870214	3879495	3888774	3898052	3907328	3916602	3925875	3935146	−1
0.291	0.5	3944416	3953684	3962950	3972215	3981478	3990740	4000000	4009258	4018515	4027771	−1
0.292	0.5	4037024	4046276	4055527	4064776	4074023	4083269	4092513	4101756	4110997	4120236	−1
0.293	0.5	4129474	4138711	4147945	4157179	4166410	4175640	4184869	4194096	4203321	4212545	−1
0.294	0.5	4221767	4230987	4240206	4249424	4258640	4267854	4277067	4286278	4295488	4304696	−1
0.295	0.5	4313902	4323107	4332311	4341513	4350713	4359912	4369109	4378304	4387499	4396691	−1
0.296	0.5	4405882	4415071	4424259	4433446	4442630	4451814	4460995	4470175	4479354	4488531	−1
0.297	0.5	4497706	4506880	4516053	4525223	4534393	4543561	4552727	4561891	4571055	4580216	−1
0.298	0.5	4589376	4598535	4607692	4616847	4626001	4635153	4644304	4653454	4662601	4671748	−1
0.299	0.5	4680892	4690036	4699177	4708317	4717456	4726593	4735729	4744863	4753995	4763126	−1

TABLE 4

Square Roots

X	.0000	.0001	.0002	.0003	.0004	.0005	.0006	.0007	.0008	.0009	Δ_2
0.300 0.5	4772256	4781384	4790510	4799635	4808758	4817880	4827001	4836119	4845237	4854353	-1
0.301 0.5	4863467	4872580	4881691	4890801	4899909	4909016	4918121	4927225	4936327	4945427	-1
0.302 0.5	4954527	4963624	4972720	4981815	4990908	5000000	5009090	5018179	5027266	5036352	-1
0.303 0.5	5045436	5054518	5063600	5072679	5081757	5090834	5099909	5108983	5118055	5127126	-0
0.304 0.5	5136195	5145263	5154329	5163394	5172457	5181519	5190579	5199638	5208695	5217751	-0
0.305 0.5	5226805	5235858	5244909	5253959	5263008	5272054	5281100	5290144	5299186	5308227	-0
0.306 0.5	5317267	5326305	5335341	5344376	5353410	5362442	5371473	5380502	5389530	5398556	-0
0.307 0.5	5407581	5416604	5425626	5434646	5443665	5452683	5461698	5470713	5479726	5488738	-0
0.308 0.5	5497748	5506756	5515763	5524769	5533773	5542776	5551778	5560777	5569776	5578773	-0
0.309 0.5	5587768	5596762	5605755	5614746	5623736	5632724	5641711	5650696	5659680	5668663	-0
0.310 0.5	5677644	5686623	5695601	5704578	5713553	5722527	5731499	5740470	5749439	5758407	-0
0.311 0.5	5767374	5776339	5785303	5794265	5803226	5812185	5821143	5830099	5839054	5848008	-0
0.312 0.5	5856960	5865911	5874860	5883808	5892754	5901699	5910643	5919585	5928526	5937465	-0
0.313 0.5	5946403	5955339	5964274	5973208	5982140	5991071	6000000	6008928	6017854	6026779	-0
0.314 0.5	6035703	6044625	6053546	6062465	6071383	6080300	6089215	6098128	6107040	6115951	-0
0.315 0.5	6124861	6133769	6142675	6151581	6160484	6169387	6178288	6187187	6196085	6204982	-0
0.316 0.5	6213877	6222771	6231664	6240555	6249444	6258333	6267220	6276105	6284989	6293872	-0
0.317 0.5	6302753	6311633	6320511	6329388	6338264	6347138	6356011	6364883	6373753	6382621	-0
0.318 0.5	6391489	6400355	6409219	6418082	6426944	6435804	6444663	6453521	6462377	6471232	-0
0.319 0.5	6480085	6488937	6497788	6506637	6515485	6524331	6533176	6542020	6550862	6559703	-0
0.320 0.5	6568542	6577381	6586217	6595053	6603887	6612719	6621551	6630381	6639209	6648036	-0
0.321 0.5	6656862	6665686	6674509	6683331	6692151	6700970	6709787	6718604	6727418	6736232	-0
0.322 0.5	6745044	6753854	6762664	6771472	6780278	6789083	6797887	6806690	6815491	6824291	-0
0.323 0.5	6833089	6841886	6850682	6859476	6868269	6877060	6885851	6894639	6903427	6912213	-0
0.324 0.5	6920998	6929781	6938563	6947344	6956123	6964901	6973678	6982453	6991227	7000000	-0
0.325 0.5	7008771	7017541	7026310	7035077	7043843	7052607	7061370	7070132	7078893	7087652	-0
0.326 0.5	7096410	7105166	7113921	7122675	7131427	7140178	7148928	7157677	7166424	7175169	-0
0.327 0.5	7183914	7192657	7201399	7210139	7218878	7227616	7236352	7245087	7253821	7262553	-0
0.328 0.5	7271284	7280014	7288742	7297469	7306195	7314919	7323643	7332364	7341085	7349804	-0
0.329 0.5	7358522	7367238	7375953	7384667	7393379	7402091	7410800	7419509	7428216	7436922	-0
0.330 0.5	7445626	7454330	7463032	7471732	7480431	7489129	7497826	7506521	7515215	7523908	-0
0.331 0.5	7532599	7541289	7549978	7558666	7567352	7576037	7584720	7593402	7602083	7610763	-0
0.332 0.5	7619441	7628118	7636794	7645468	7654141	7662813	7671483	7680153	7688820	7697487	-0
0.333 0.5	7706152	7714816	7723479	7732140	7740800	7749459	7758116	7766772	7775427	7784081	-0
0.334 0.5	7792733	7801384	7810034	7818682	7827329	7835975	7844619	7853263	7861905	7870545	-0
0.335 0.5	7879184	7887823	7896459	7905095	7913729	7922362	7930993	7939624	7948253	7956880	-0
0.336 0.5	7965507	7974132	7982756	7991379	8000000	8008620	8017239	8025856	8034472	8043087	-0
0.337 0.5	8051701	8060313	8068925	8077534	8086143	8094750	8103356	8111961	8120564	8129167	-0
0.338 0.5	8137767	8146367	8154965	8163562	8172158	8180753	8189346	8197938	8206529	8215118	-0
0.339 0.5	8223706	8232293	8240879	8249464	8258047	8266628	8275209	8283788	8292367	8300943	-0
0.340 0.5	8309519	8318093	8326666	8335238	8343809	8352378	8360946	8369513	8378078	8386642	-0
0.341 0.5	8395205	8403767	8412327	8420887	8429445	8438001	8446557	8455111	8463664	8472216	-0
0.342 0.5	8480766	8489315	8497863	8506410	8514955	8523500	8532043	8540584	8549125	8557664	-0
0.343 0.5	8566202	8574739	8583274	8591808	8600341	8608873	8617404	8625933	8634461	8642988	-0
0.344 0.5	8651513	8660037	8668561	8677082	8685603	8694122	8702640	8711157	8719673	8728187	-0
0.345 0.5	8736701	8745213	8753723	8762233	8770741	8779248	8787754	8796258	8804762	8813264	-0
0.346 0.5	8821765	8830264	8838763	8847260	8855756	8864251	8872744	8881236	8889727	8898217	-0
0.347 0.5	8906706	8915193	8923679	8932164	8940648	8949131	8957612	8966092	8974571	8983048	-0
0.348 0.5	8991525	9000000	9008474	9016947	9025418	9033889	9042358	9050826	9059292	9067758	-0
0.349 0.5	9076222	9084685	9093147	9101607	9110067	9118525	9126982	9135438	9143892	9152346	-0

TABLE 4

Square Roots

X		.0000	.0001	.0002	.0003	.0004	.0005	.0006	.0007	.0008	.0009	Δ_2
0.350	0.5	9160798	9169249	9177698	9186147	9194594	9203040	9211485	9219929	9228372	9236813	−0
0.351	0.5	9245253	9253692	9262130	9270566	9279001	9287435	9295868	9304300	9312731	9321160	−0
0.352	0.5	9329588	9338015	9346440	9354865	9363288	9371710	9380131	9388551	9396970	9405387	−0
0.353	0.5	9413803	9422218	9430632	9439044	9447456	9455866	9464275	9472683	9481089	9489495	−0
0.354	0.5	9497899	9506302	9514704	9523105	9531504	9539903	9548300	9556696	9565090	9573484	−0
0.355	0.5	9581876	9590268	9598658	9607047	9615434	9623821	9632206	9640590	9648973	9657355	−0
0.356	0.5	9665736	9674115	9682493	9690870	9699246	9707621	9715994	9724367	9732738	9741108	−0
0.357	0.5	9749477	9757845	9766211	9774577	9782941	9791304	9799666	9808026	9816386	9824744	−0
0.358	0.5	9833101	9841457	9849812	9858166	9866518	9874870	9883220	9891569	9899916	9908263	−0
0.359	0.5	9916609	9924953	9933296	9941638	9949979	9958319	9966657	9974995	9983331	9991666	−0
0.360	0.6	0000000	0008333	0016664	0024995	0033324	0041652	0049979	0058305	0066630	0074953	−0
0.361	0.6	0083276	0091597	0099917	0108236	0116553	0124870	0133186	0141500	0149813	0158125	−0
0.362	0.6	0166436	0174746	0183054	0191362	0199668	0207973	0216277	0224580	0232881	0241182	−0
0.363	0.6	0249481	0257780	0266077	0274373	0282667	0290961	0299254	0307545	0315835	0324125	−0
0.364	0.6	0332413	0340699	0348985	0357270	0365553	0373835	0382117	0390397	0398675	0406953	−0
0.365	0.6	0415230	0423505	0431780	0440053	0448325	0456596	0464866	0473135	0481402	0489669	−0
0.366	0.6	0497934	0506198	0514461	0522723	0530984	0539243	0547502	0555759	0564016	0572271	−0
0.367	0.6	0580525	0588778	0597030	0605280	0613530	0621778	0630026	0638272	0646517	0654761	−0
0.368	0.6	0663004	0671245	0679486	0687725	0695964	0704201	0712437	0720672	0728906	0737139	−0
0.369	0.6	0745370	0753601	0761830	0770058	0778286	0786512	0794737	0802960	0811183	0819405	−0
0.370	0.6	0827625	0835845	0844063	0852280	0860496	0868711	0876925	0885138	0893349	0901560	−0
0.371	0.6	0909769	0917978	0926185	0934391	0942596	0950800	0959003	0967204	0975405	0983604	−0
0.372	0.6	0991803	1000000	1008196	1016391	1024585	1032778	1040970	1049161	1057350	1065539	−0
0.373	0.6	1073726	1081912	1090097	1098281	1106464	1114646	1122827	1131007	1139185	1147363	−0
0.374	0.6	1155539	1163715	1171889	1180062	1188234	1196405	1204575	1212744	1220911	1229078	−0
0.375	0.6	1237244	1245408	1253571	1261734	1269895	1278055	1286214	1294372	1302528	1310684	−0
0.376	0.6	1318839	1326992	1335145	1343296	1351447	1359596	1367744	1375891	1384037	1392182	−0
0.377	0.6	1400326	1408468	1416610	1424751	1432890	1441029	1449166	1457302	1465437	1473572	−0
0.378	0.6	1481705	1489837	1497967	1506097	1514226	1522354	1530480	1538606	1546730	1554854	−0
0.379	0.6	1562976	1571097	1579217	1587336	1595454	1603571	1611687	1619802	1627916	1636028	−0
0.380	0.6	1644140	1652251	1660360	1668468	1676576	1684682	1692787	1700891	1708994	1717096	−0
0.381	0.6	1725197	1733297	1741396	1749494	1757591	1765686	1773781	1781874	1789967	1798058	−0
0.382	0.6	1806149	1814238	1822326	1830413	1838499	1846584	1854668	1862751	1870833	1878914	−0
0.383	0.6	1886994	1895072	1903150	1911227	1919302	1927377	1935450	1943523	1951594	1959664	−0
0.384	0.6	1967734	1975802	1983869	1991935	2000000	2008064	2016127	2024189	2032250	2040309	−0
0.385	0.6	2048368	2056426	2064483	2072538	2080593	2088646	2096699	2104750	2112801	2120850	−0
0.386	0.6	2128898	2136946	2144992	2153037	2161081	2169124	2177166	2185207	2193247	2201286	−0
0.387	0.6	2209324	2217361	2225397	2233432	2241465	2249498	2257530	2265560	2273590	2281618	−0
0.388	0.6	2289646	2297672	2305698	2313722	2321746	2329768	2337789	2345810	2353829	2361847	−0
0.389	0.6	2369865	2377881	2385896	2393910	2401923	2409935	2417946	2425956	2433965	2441973	−0
0.390	0.6	2449980	2457986	2465991	2473995	2481997	2489999	2498000	2506000	2513998	2521996	−0
0.391	0.6	2529993	2537988	2545983	2553977	2561969	2569961	2577951	2585941	2593929	2601917	−0
0.392	0.6	2609903	2617889	2625873	2633857	2641839	2649820	2657801	2665780	2673758	2681736	−0
0.393	0.6	2689712	2697687	2705662	2713635	2721607	2729578	2737549	2745518	2753486	2761453	−0
0.394	0.6	2769419	2777384	2785349	2793312	2801274	2809235	2817195	2825154	2833112	2841069	−0
0.395	0.6	2849025	2856980	2864935	2872888	2880840	2888791	2896741	2904690	2912638	2920585	−0
0.396	0.6	2928531	2936476	2944420	2952363	2960305	2968246	2976186	2984125	2992063	3000000	−0
0.397	0.6	3007936	3015871	3023805	3031738	3039670	3047601	3055531	3063460	3071388	3079315	−0
0.398	0.6	3087241	3095166	3103090	3111013	3118935	3126856	3134776	3142696	3150614	3158531	−0
0.399	0.6	3166447	3174362	3182276	3190189	3198101	3206012	3213922	3221832	3229740	3237647	−0

TABLE 4

Square Roots

X		.0000	.0001	.0002	.0003	.0004	.0005	.0006	.0007	.0008	.0009	Δ,
0.400	0.6	3245553	3253458	3261363	3269266	3277168	3285069	3292970	3300869	3308767	3316664	−0
0.401	0.6	3324561	3332456	3340350	3348244	3356136	3364028	3371918	3379808	3387696	3395583	−0
0.402	0.6	3403470	3411355	3419240	3427124	3435006	3442888	3450768	3458648	3466527	3474404	−0
0.403	0.6	3482281	3490157	3498031	3505905	3513778	3521650	3529521	3537391	3545259	3553127	−0
0.404	0.6	3560994	3568860	3576725	3584589	3592452	3600314	3608176	3616036	3623895	3631753	−0
0.405	0.6	3639610	3647467	3655322	3663176	3671030	3678882	3686733	3694584	3702433	3710282	−0
0.406	0.6	3718129	3725976	3733821	3741666	3749510	3757352	3765194	3773035	3780875	3788714	−0
0.407	0.6	3796552	3804389	3812225	3820060	3827894	3835727	3843559	3851390	3859220	3867049	−0
0.408	0.6	3874878	3882705	3890531	3898357	3906181	3914005	3921827	3929649	3937469	3945289	−0
0.409	0.6	3953108	3960926	3968742	3976558	3984373	3992187	4000000	4007812	4015623	4023433	−0
0.410	0.6	4031242	4039051	4046858	4054664	4062469	4070274	4078077	4085880	4093681	4101482	−0
0.411	0.6	4109282	4117080	4124878	4132675	4140471	4148266	4156060	4163853	4171645	4179436	−0
0.412	0.6	4187226	4195015	4202804	4210591	4218377	4226163	4233947	4241731	4249514	4257295	−0
0.413	0.6	4265076	4272856	4280635	4288413	4296190	4303966	4311741	4319515	4327288	4335060	−0
0.414	0.6	4342832	4350602	4358372	4366140	4373908	4381674	4389440	4397205	4404969	4412732	−0
0.415	0.6	4420494	4428255	4436015	4443774	4451532	4459289	4467046	4474801	4482556	4490309	−0
0.416	0.6	4498062	4505814	4513564	4521314	4529063	4536811	4544558	4552304	4560050	4567794	−0
0.417	0.6	4575537	4583280	4591021	4598762	4606501	4614240	4621978	4629715	4637450	4645185	−0
0.418	0.6	4652919	4660653	4668385	4676116	4683846	4691576	4699304	4707032	4714759	4722484	−0
0.419	0.6	4730209	4737933	4745656	4753378	4761099	4768820	4776539	4784257	4791975	4799691	−0
0.420	0.6	4807407	4815122	4822835	4830548	4838260	4845971	4853681	4861391	4869099	4876806	−0
0.421	0.6	4884513	4892218	4899923	4907627	4915329	4923031	4930732	4938432	4946132	4953830	−0
0.422	0.6	4961527	4969223	4976919	4984614	4992307	5000000	5007692	5015383	5023073	5030762	−0
0.423	0.6	5038450	5046137	5053824	5061509	5069194	5076878	5084560	5092242	5099923	5107603	−0
0.424	0.6	5115282	5122961	5130638	5138314	5145990	5153664	5161338	5169011	5176683	5184354	−0
0.425	0.6	5192024	5199693	5207362	5215029	5222695	5230361	5238026	5245689	5253352	5261014	−0
0.426	0.6	5268675	5276336	5283995	5291653	5299311	5306967	5314623	5322278	5329932	5337585	−0
0.427	0.6	5345237	5352888	5360539	5368188	5375836	5383484	5391131	5398777	5406422	5414066	−0
0.428	0.6	5421709	5429351	5436993	5444633	5452273	5459911	5467549	5475186	5482822	5490457	−0
0.429	0.6	5498092	5505725	5513357	5520989	5528620	5536250	5543878	5551506	5559134	5566760	−0
0.430	0.6	5574385	5582010	5589633	5597256	5604878	5612499	5620119	5627738	5635356	5642974	−0
0.431	0.6	5650590	5658206	5665821	5673434	5681047	5688660	5696271	5703881	5711491	5719099	−0
0.432	0.6	5726707	5734314	5741920	5749525	5757129	5764732	5772335	5779936	5787537	5795137	−0
0.433	0.6	5802735	5810333	5817931	5825527	5833122	5840717	5848311	5855903	5863495	5871086	−0
0.434	0.6	5878676	5886266	5893854	5901441	5909028	5916614	5924199	5931783	5939366	5946948	−0
0.435	0.6	5954530	5962110	5969690	5977269	5984847	5992424	6000000	6007575	6015150	6022723	−0
0.436	0.6	6030296	6037868	6045439	6053009	6060578	6068147	6075714	6083281	6090847	6098411	−0
0.437	0.6	6105976	6113539	6121101	6128662	6136223	6143783	6151342	6158900	6166457	6174013	−0
0.438	0.6	6181568	6189123	6196677	6204229	6211781	6219333	6226883	6234432	6241981	6249528	−0
0.439	0.6	6257075	6264621	6272166	6279710	6287254	6294796	6302338	6309879	6317419	6324958	−0
0.440	0.6	6332496	6340033	6347570	6355105	6362640	6370174	6377707	6385239	6392771	6400301	−0
0.441	0.6	6407831	6415360	6422888	6430415	6437941	6445466	6452991	6460515	6468037	6475559	−0
0.442	0.6	6483081	6490601	6498120	6505639	6513157	6520673	6528189	6535705	6543219	6550732	−0
0.443	0.6	6558245	6565757	6573268	6580778	6588287	6595796	6603303	6610810	6618316	6625821	−0
0.444	0.6	6633325	6640828	6648331	6655832	6663333	6670833	6678332	6685831	6693328	6700825	−0
0.445	0.6	6708320	6715815	6723309	6730802	6738295	6745786	6753277	6760767	6768256	6775744	−0
0.446	0.6	6783231	6790718	6798204	6805688	6813172	6820655	6828138	6835619	6843100	6850580	−0
0.447	0.6	6858059	6865537	6873014	6880490	6887966	6895441	6902915	6910388	6917860	6925332	−0
0.448	0.6	6932802	6940272	6947741	6955209	6962676	6970143	6977608	6985073	6992537	7000000	−0
0.449	0.6	7007462	7014924	7022384	7029844	7037303	7044761	7052218	7059675	7067131	7074585	−0

TABLE 4

Square Roots

X		.0000	.0001	.0002	.0003	.0004	.0005	.0006	.0007	.0008	.0009	Δ_1
0.450	0.6	7082039	7089492	7096945	7104396	7111847	7119297	7126746	7134194	7141641	7149088	−0
0.451	0.6	7156534	7163978	7171422	7178866	7186308	7193750	7201190	7208630	7216069	7223508	−0
0.452	0.6	7230945	7238382	7245818	7253253	7260687	7268120	7275553	7282984	7290415	7297845	−0
0.453	0.6	7305275	7312703	7320131	7327558	7334983	7342409	7349833	7357256	7364679	7372101	−0
0.454	0.6	7379522	7386942	7394362	7401780	7409198	7416615	7424031	7431447	7438861	7446275	−0
0.455	0.6	7453688	7461100	7468511	7475922	7483331	7490740	7498148	7505555	7512962	7520367	−0
0.456	0.6	7527772	7535176	7542579	7549981	7557383	7564784	7572184	7579583	7586981	7594378	−0
0.457	0.6	7601775	7609171	7616566	7623960	7631354	7638746	7646138	7653529	7660919	7668309	−0
0.458	0.6	7675697	7683085	7690472	7697858	7705243	7712628	7720012	7727395	7734777	7742158	−0
0.459	0.6	7749539	7756918	7764297	7771675	7779053	7786429	7793805	7801180	7808554	7815927	−0
0.460	0.6	7823300	7830672	7838042	7845412	7852782	7860150	7867518	7874885	7882251	7889616	−0
0.461	0.6	7896981	7904344	7911707	7919069	7926431	7933791	7941151	7948510	7955868	7963225	−0
0.462	0.6	7970582	7977938	7985293	7992647	8000000	8007353	8014704	8022055	8029405	8036755	−0
0.463	0.6	8044103	8051451	8058798	8066144	8073490	8080834	8088178	8095521	8102863	8110205	−0
0.464	0.6	8117545	8124885	8132224	8139563	8146900	8154237	8161573	8168908	8176242	8183576	−0
0.465	0.6	8190908	8198240	8205572	8212902	8220232	8227560	8234888	8242216	8249542	8256868	−0
0.466	0.6	8264193	8271517	8278840	8286163	8293484	8300805	8308125	8315445	8322763	8330081	−0
0.467	0.6	8337398	8344714	8352030	8359345	8366659	8373972	8381284	8388596	8395906	8403216	−0
0.468	0.6	8410525	8417834	8425142	8432448	8439754	8447060	8454364	8461668	8468971	8476273	−0
0.469	0.6	8483575	8490875	8498175	8505474	8512772	8520070	8527367	8534663	8541958	8549252	−0
0.470	0.6	8556546	8563839	8571131	8578422	8585713	8593003	8600292	8607580	8614867	8622154	−0
0.471	0.6	8629440	8636725	8644009	8651293	8658576	8665858	8673139	8680419	8687699	8694978	−0
0.472	0.6	8702256	8709534	8716810	8724086	8731361	8738635	8745909	8753182	8760454	8767725	−0
0.473	0.6	8774995	8782265	8789534	8796802	8804070	8811336	8818602	8825867	8833131	8840395	−0
0.474	0.6	8847658	8854920	8862181	8869442	8876701	8883960	8891219	8898476	8905733	8912989	−0
0.475	0.6	8920244	8927498	8934752	8942005	8949257	8956508	8963759	8971008	8978257	8985506	−0
0.476	0.6	8992753	9000000	9007246	9014491	9021736	9028979	9036222	9043465	9050706	9057947	−0
0.477	0.6	9065187	9072426	9079664	9086902	9094139	9101375	9108610	9115845	9123079	9130312	−0
0.478	0.6	9137544	9144776	9152006	9159236	9166466	9173694	9180922	9188149	9195376	9202601	−0
0.479	0.6	9209826	9217050	9224273	9231496	9238717	9245939	9253159	9260378	9267597	9274815	−0
0.480	0.6	9282032	9289249	9296465	9303680	9310894	9318107	9325320	9332532	9339743	9346954	−0
0.481	0.6	9354163	9361373	9368581	9375788	9382995	9390201	9397406	9404611	9411815	9419018	−0
0.482	0.6	9426220	9433421	9440622	9447822	9455021	9462220	9469418	9476615	9483811	9491007	−0
0.483	0.6	9498201	9505395	9512589	9519781	9526973	9534164	9541355	9548544	9555733	9562921	−0
0.484	0.6	9570109	9577295	9584481	9591666	9598851	9606034	9613217	9620399	9627581	9634761	−0
0.485	0.6	9641941	9649121	9656299	9663477	9670654	9677830	9685006	9692180	9699354	9706528	−0
0.486	0.6	9713700	9720872	9728043	9735213	9742383	9749552	9756720	9763887	9771054	9778220	−0
0.487	0.6	9785385	9792550	9799713	9806876	9814039	9821200	9828361	9835521	9842680	9849839	−0
0.488	0.6	9856997	9864154	9871310	9878466	9885621	9892775	9899928	9907081	9914233	9921384	−0
0.489	0.6	9928535	9935685	9942834	9949982	9957130	9964277	9971423	9978568	9985713	9992857	−0
0.490	0.7	0000000	0007142	0014284	0021425	0028566	0035705	0042844	0049982	0057120	0064256	−0
0.491	0.7	0071392	0078527	0085662	0092796	0099929	0107061	0114193	0121323	0128454	0135583	−0
0.492	0.7	0142712	0149840	0156967	0164093	0171219	0178344	0185469	0192592	0199715	0206837	−0
0.493	0.7	0213959	0221079	0228199	0235319	0242437	0249555	0256672	0263789	0270904	0278019	−0
0.494	0.7	0285134	0292247	0299360	0306472	0313583	0320694	0327804	0334913	0342022	0349129	−0
0.495	0.7	0356236	0363343	0370448	0377553	0384657	0391761	0398864	0405966	0413067	0420168	−0
0.496	0.7	0427267	0434367	0441465	0448563	0455660	0462756	0469852	0476947	0484041	0491134	−0
0.497	0.7	0498227	0505319	0512410	0519501	0526591	0533680	0540768	0547856	0554943	0562029	−0
0.498	0.7	0569115	0576200	0583284	0590368	0597450	0604532	0611614	0618694	0625774	0632854	−0
0.499	0.7	0639932	0647010	0654087	0661163	0668239	0675314	0682388	0689462	0696535	0703607	−0

TABLE 4

Square Roots

X	.0000	.0001	.0002	.0003	.0004	.0005	.0006	.0007	.0008	.0009	Δ_2
0.500 0.7	0710678	0717749	0724819	0731888	0738957	0746025	0753092	0760158	0767224	0774289	−0
0.501 0.7	0781353	0788417	0795480	0802542	0809604	0816665	0823725	0830784	0837843	0844901	−0
0.502 0.7	0851958	0859015	0866071	0873126	0880181	0887234	0894287	0901340	0908392	0915443	−0
0.503 0.7	0922493	0929543	0936591	0943640	0950687	0957734	0964780	0971825	0978870	0985914	−0
0.504 0.7	0992957	1000000	1007042	1014083	1021124	1028163	1035203	1042241	1049279	1056316	−0
0.505 0.7	1063352	1070388	1077423	1084457	1091490	1098523	1105555	1112587	1119618	1126648	−0
0.506 0.7	1133677	1140706	1147734	1154761	1161787	1168813	1175839	1182863	1189887	1196910	−0
0.507 0.7	1203932	1210954	1217975	1224996	1232015	1239034	1246053	1253070	1260087	1267103	−0
0.508 0.7	1274119	1281134	1288148	1295161	1302174	1309186	1316197	1323208	1330218	1337227	−0
0.509 0.7	1344236	1351244	1358251	1365258	1372264	1379269	1386273	1393277	1400280	1407283	−0
0.510 0.7	1414284	1421285	1428286	1435285	1442284	1449283	1456280	1463277	1470274	1477269	−0
0.511 0.7	1484264	1491258	1498252	1505245	1512237	1519228	1526219	1533209	1540198	1547187	−0
0.512 0.7	1554175	1561163	1568149	1575135	1582121	1589105	1596089	1603073	1610055	1617037	−0
0.513 0.7	1624018	1630999	1637979	1644958	1651936	1658914	1665891	1672868	1679844	1686819	−0
0.514 0.7	1693793	1700767	1707740	1714713	1721684	1728655	1735626	1742595	1749564	1756533	−0
0.515 0.7	1763500	1770467	1777434	1784399	1791364	1798329	1805292	1812255	1819217	1826179	−0
0.516 0.7	1833140	1840100	1847060	1854019	1860977	1867934	1874891	1881847	1888803	1895758	−0
0.517 0.7	1902712	1909666	1916618	1923571	1930522	1937473	1944423	1951372	1958321	1965269	−0
0.518 0.7	1972217	1979164	1986110	1993055	2000000	2006944	2013888	2020830	2027772	2034714	−0
0.519 0.7	2041655	2048595	2055534	2062473	2069411	2076348	2083285	2090221	2097157	2104091	−0
0.520 0.7	2111026	2117959	2124892	2131824	2138755	2145686	2152616	2159545	2166474	2173402	−0
0.521 0.7	2180330	2187256	2194183	2201108	2208033	2214957	2221880	2228803	2235725	2242647	−0
0.522 0.7	2249567	2256488	2263407	2270326	2277244	2284161	2291078	2297994	2304910	2311825	−0
0.523 0.7	2318739	2325652	2332565	2339477	2346389	2353300	2360210	2367120	2374028	2380937	−0
0.524 0.7	2387844	2394751	2401657	2408563	2415468	2422372	2429276	2436179	2443081	2449983	−0
0.525 0.7	2456884	2463784	2470684	2477583	2484481	2491379	2498276	2505172	2512068	2518963	−0
0.526 0.7	2525857	2532751	2539644	2546537	2553429	2560320	2567210	2574100	2580989	2587878	−0
0.527 0.7	2594766	2601653	2608539	2615425	2622311	2629195	2636079	2642962	2649845	2656727	−0
0.528 0.7	2663608	2670489	2677369	2684249	2691127	2698005	2704883	2711760	2718636	2725511	−0
0.529 0.7	2732386	2739260	2746134	2753007	2759879	2766751	2773622	2780492	2787362	2794231	−0
0.530 0.7	2801099	2807967	2814834	2821700	2828566	2835431	2842295	2849159	2856022	2862885	−0
0.531 0.7	2869747	2876608	2883469	2890329	2897188	2904047	2910905	2917762	2924619	2931475	−0
0.532 0.7	2938330	2945185	2952039	2958893	2965745	2972598	2979449	2986300	2993150	3000000	−0
0.533 0.7	3006849	3013697	3020545	3027392	3034239	3041084	3047929	3054774	3061618	3068461	−0
0.534 0.7	3075304	3082146	3088987	3095828	3102668	3109507	3116346	3123184	3130021	3136858	−0
0.535 0.7	3143694	3150530	3157365	3164199	3171033	3177865	3184698	3191530	3198361	3205191	−0
0.536 0.7	3212021	3218850	3225679	3232506	3239334	3246160	3252986	3259812	3266636	3273460	−0
0.537 0.7	3280284	3287107	3293929	3300750	3307571	3314391	3321211	3328030	3334848	3341666	−0
0.538 0.7	3348483	3355300	3362116	3368931	3375745	3382559	3389373	3396185	3402997	3409809	−0
0.539 0.7	3416619	3423430	3430239	3437048	3443856	3450664	3457471	3464277	3471083	3477888	−0
0.540 0.7	3484692	3491496	3498299	3505102	3511904	3518705	3525506	3532306	3539105	3545904	−0
0.541 0.7	3552702	3559500	3566297	3573093	3579889	3586684	3593478	3600272	3607065	3613857	−0
0.542 0.7	3620649	3627441	3634231	3641021	3647811	3654599	3661387	3668175	3674962	3681748	−0
0.543 0.7	3688534	3695319	3702103	3708887	3715670	3722452	3729234	3736016	3742796	3749576	−0
0.544 0.7	3756356	3763134	3769913	3776690	3783467	3790243	3797019	3803794	3810568	3817342	−0
0.545 0.7	3824115	3830888	3837660	3844431	3851202	3857972	3864741	3871510	3878278	3885046	−0
0.546 0.7	3891813	3898579	3905345	3912110	3918874	3925638	3932402	3939164	3945926	3952688	−0
0.547 0.7	3959448	3966208	3972968	3979727	3986485	3993243	4000000	4006756	4013512	4020267	−0
0.548 0.7	4027022	4033776	4040529	4047282	4054034	4060786	4067537	4074287	4081037	4087786	−0
0.549 0.7	4094534	4101282	4108029	4114776	4121522	4128267	4135012	4141756	4148500	4155243	−0

TABLE 4

Square Roots

X		.0000	.0001	.0002	.0003	.0004	.0005	.0006	.0007	.0008	.0009	Δ_2
0.550	0.7	4161985	4168727	4175468	4182208	4188948	4195687	4202426	4209164	4215901	4222638	-0
0.551	0.7	4229374	4236110	4242845	4249579	4256313	4263046	4269779	4276510	4283242	4289972	-0
0.552	0.7	4296702	4303432	4310161	4316889	4323617	4330344	4337070	4343796	4350521	4357246	-0
0.553	0.7	4363970	4370693	4377416	4384138	4390860	4397581	4404301	4411021	4417740	4424458	-0
0.554	0.7	4431176	4437894	4444610	4451326	4458042	4464757	4471471	4478185	4484898	4491610	-0
0.555	0.7	4498322	4505033	4511744	4518454	4525164	4531872	4538581	4545288	4551995	4558702	-0
0.556	0.7	4565408	4572113	4578817	4585521	4592225	4598928	4605630	4612331	4619032	4625733	-0
0.557	0.7	4632433	4639132	4645830	4652528	4659226	4665923	4672619	4679314	4686009	4692704	-0
0.558	0.7	4699398	4706091	4712783	4719475	4726167	4732858	4739548	4746237	4752926	4759615	-0
0.559	0.7	4766303	4772990	4779676	4786362	4793048	4799733	4806417	4813100	4819783	4826466	-0
0.560	0.7	4833148	4839829	4846510	4853190	4859869	4866548	4873226	4879904	4886581	4893257	-0
0.561	0.7	4899933	4906608	4913283	4919957	4926631	4933304	4939976	4946648	4953319	4959989	-0
0.562	0.7	4966659	4973329	4979997	4986665	4993333	5000000	5006666	5013332	5019997	5026662	-0
0.563	0.7	5033326	5039989	5046652	5053314	5059976	5066637	5073297	5079957	5086617	5093275	-0
0.564	0.7	5099933	5106591	5113248	5119904	5126560	5133215	5139870	5146523	5153177	5159830	-0
0.565	0.7	5166482	5173133	5179784	5186435	5193085	5199734	5206383	5213031	5219678	5226325	-0
0.566	0.7	5232971	5239617	5246262	5252907	5259551	5266194	5272837	5279479	5286121	5292762	-0
0.567	0.7	5299402	5306042	5312681	5319320	5325958	5332596	5339233	5345869	5352505	5359140	-0
0.568	0.7	5365775	5372409	5379042	5385675	5392307	5398939	5405570	5412201	5418831	5425460	-0
0.569	0.7	5432089	5438717	5445344	5451971	5458598	5465224	5471849	5478474	5485098	5491721	-0
0.570	0.7	5498344	5504967	5511588	5518210	5524830	5531450	5538070	5544689	5551307	5557925	-0
0.571	0.7	5564542	5571158	5577774	5584390	5591005	5597619	5604233	5610846	5617458	5624070	-0
0.572	0.7	5630682	5637292	5643903	5650512	5657121	5663730	5670338	5676945	5683552	5690158	-0
0.573	0.7	5696763	5703368	5709973	5716577	5723180	5729783	5736385	5742986	5749587	5756188	-0
0.574	0.7	5762788	5769387	5775986	5782584	5789181	5795778	5802375	5808970	5815566	5822160	-0
0.575	0.7	5828754	5835348	5841941	5848533	5855125	5861716	5868307	5874897	5881487	5888075	-0
0.576	0.7	5894664	5901252	5907839	5914425	5921012	5927597	5934182	5940766	5947350	5953933	-0
0.577	0.7	5960516	5967098	5973680	5980261	5986841	5993421	6000000	6006579	6013157	6019734	-0
0.578	0.7	6026311	6032888	6039463	6046039	6052613	6059187	6065761	6072334	6078906	6085478	-0
0.579	0.7	6092049	6098620	6105190	6111760	6118329	6124897	6131465	6138033	6144599	6151165	-0
0.580	0.7	6157731	6164296	6170861	6177424	6183988	6190551	6197113	6203674	6210236	6216796	-0
0.581	0.7	6223356	6229915	6236474	6243032	6249590	6256147	6262704	6269260	6275815	6282370	-0
0.582	0.7	6288924	6295478	6302031	6308584	6315136	6321688	6328239	6334789	6341339	6347888	-0
0.583	0.7	6354437	6360985	6367532	6374079	6380626	6387172	6393717	6400262	6406806	6413350	-0
0.584	0.7	6419893	6426435	6432977	6439519	6446059	6452600	6459139	6465679	6472217	6478755	-0
0.585	0.7	6485293	6491830	6498366	6504902	6511437	6517972	6524506	6531039	6537572	6544105	-0
0.586	0.7	6550637	6557168	6563699	6570229	6576759	6583288	6589817	6596345	6602872	6609399	-0
0.587	0.7	6615925	6622451	6628976	6635501	6642025	6648549	6655072	6661594	6668116	6674637	-0
0.588	0.7	6681158	6687678	6694198	6700717	6707236	6713754	6720271	6726788	6733304	6739820	-0
0.589	0.7	6746335	6752850	6759364	6765878	6772391	6778903	6785415	6791927	6798437	6804948	-0
0.590	0.7	6811457	6817967	6824475	6830983	6837491	6843998	6850504	6857010	6863515	6870020	-0
0.591	0.7	6876524	6883028	6889531	6896034	6902536	6909037	6915538	6922038	6928538	6935038	-0
0.592	0.7	6941536	6948034	6954532	6961029	6967526	6974022	6980517	6987012	6993506	7000000	-0
0.593	0.7	7006493	7012986	7019478	7025970	7032461	7038951	7045441	7051931	7058419	7064908	-0
0.594	0.7	7071395	7077883	7084369	7090855	7097341	7103826	7110311	7116795	7123278	7129761	-0
0.595	0.7	7136243	7142725	7149206	7155687	7162167	7168646	7175125	7181604	7188082	7194559	-0
0.596	0.7	7201036	7207513	7213988	7220464	7226938	7233412	7239886	7246359	7252832	7259304	-0
0.597	0.7	7265775	7272246	7278716	7285186	7291655	7298124	7304592	7311060	7317527	7323994	-0
0.598	0.7	7330460	7336925	7343390	7349854	7356318	7362782	7369244	7375707	7382168	7388630	-0
0.599	0.7	7395090	7401550	7408010	7414469	7420927	7427385	7433843	7440300	7446756	7453212	-0

TABLE 4

Square Roots

X		.0000	.0001	.0002	.0003	.0004	.0005	.0006	.0007	.0008	.0009	Δ_2
0.600	0.7	7459667	7466122	7472576	7479029	7485482	7491935	7498387	7504839	7511289	7517740	−0
0.601	0.7	7524190	7530639	7537088	7543536	7549984	7556431	7562878	7569324	7575769	7582214	−0
0.602	0.7	7588659	7595103	7601546	7607989	7614432	7620873	7627315	7633756	7640196	7646635	−0
0.603	0.7	7653075	7659513	7665951	7672389	7678826	7685262	7691698	7698134	7704569	7711003	−0
0.604	0.7	7717437	7723870	7730303	7736735	7743167	7749598	7756029	7762459	7768888	7775317	−0
0.605	0.7	7781746	7788174	7794601	7801028	7807455	7813880	7820306	7826731	7833155	7839579	−0
0.606	0.7	7846002	7852424	7858847	7865268	7871689	7878110	7884530	7890949	7897368	7903787	−0
0.607	0.7	7910205	7916622	7923039	7929455	7935871	7942286	7948701	7955115	7961529	7967942	−0
0.608	0.7	7974355	7980767	7987178	7993589	8000000	8006410	8012819	8019228	8025637	8032045	−0
0.609	0.7	8038452	8044859	8051265	8057671	8064076	8070481	8076885	8083289	8089692	8096095	−0
0.610	0.7	8102497	8108898	8115299	8121700	8128100	8134499	8140898	8147297	8153695	8160092	−0
0.611	0.7	8166489	8172885	8179281	8185676	8192071	8198465	8204859	8211252	8217645	8224037	−0
0.612	0.7	8230429	8236820	8243211	8249601	8255990	8262379	8268768	8275156	8281543	8287930	−0
0.613	0.7	8294316	8300702	8307088	8313473	8319857	8326241	8332624	8339007	8345389	8351771	−0
0.614	0.7	8358152	8364533	8370913	8377293	8383672	8390050	8396428	8402806	8409183	8415560	−0
0.615	0.7	8421936	8428311	8434686	8441061	8447435	8453808	8460181	8466553	8472925	8479297	−0
0.616	0.7	8485667	8492038	8498408	8504777	8511146	8517514	8523882	8530249	8536616	8542982	−0
0.617	0.7	8549348	8555713	8562077	8568441	8574805	8581168	8587531	8593893	8600254	8606615	−0
0.618	0.7	8612976	8619336	8625695	8632054	8638413	8644771	8651128	8657485	8663842	8670198	−0
0.619	0.7	8676553	8682908	8689262	8695616	8701969	8708322	8714675	8721026	8727378	8733728	−0
0.620	0.7	8740079	8746428	8752778	8759126	8765475	8771822	8778170	8784516	8790862	8797208	−0
0.621	0.7	8803553	8809898	8816242	8822586	8828929	8835271	8841613	8847955	8854296	8860637	−0
0.622	0.7	8866977	8873316	8879655	8885994	8892332	8898669	8905006	8911343	8917679	8924014	−0
0.623	0.7	8930349	8936683	8943017	8949351	8955684	8962016	8968348	8974679	8981010	8987341	−0
0.624	0.7	8993671	9000000	9006329	9012657	9018985	9025312	9031639	9037966	9044291	9050617	−0
0.625	0.7	9056941	9063266	9069590	9075913	9082236	9088558	9094880	9101201	9107522	9113842	−0
0.626	0.7	9120162	9126481	9132800	9139118	9145436	9151753	9158070	9164386	9170702	9177017	−0
0.627	0.7	9183332	9189646	9195959	9202273	9208585	9214898	9221209	9227520	9233831	9240141	−0
0.628	0.7	9246451	9252760	9259069	9265377	9271685	9277992	9284299	9290605	9296910	9303216	−0
0.629	0.7	9309520	9315824	9322128	9328431	9334734	9341036	9347338	9353639	9359939	9366240	−0
0.630	0.7	9372539	9378838	9385137	9391435	9397733	9404030	9410327	9416623	9422919	9429214	0
0.631	0.7	9435508	9441803	9448096	9454389	9460682	9466974	9473266	9479557	9485848	9492138	0
0.632	0.7	9498428	9504717	9511006	9517294	9523581	9529869	9536155	9542441	9548727	9555012	0
0.633	0.7	9561297	9567581	9573865	9580148	9586431	9592713	9598995	9605276	9611557	9617837	0
0.634	0.7	9624117	9630396	9636675	9642953	9649231	9655508	9661785	9668061	9674337	9680612	0
0.635	0.7	9686887	9693162	9699435	9705709	9711981	9718254	9724526	9730797	9737068	9743338	0
0.636	0.7	9749608	9755877	9762146	9768415	9774683	9780950	9787217	9793483	9799749	9806015	0
0.637	0.7	9812280	9818544	9824808	9831072	9837335	9843597	9849859	9856121	9862382	9868642	0
0.638	0.7	9874902	9881162	9887421	9893679	9899937	9906195	9912452	9918709	9924965	9931220	0
0.639	0.7	9937476	9943730	9949984	9956238	9962491	9968744	9974996	9981248	9987499	9993750	0
0.640	0.8	0000000	0006250	0012499	0018748	0024996	0031244	0037491	0043738	0049984	0056230	0
0.641	0.8	0062476	0068720	0074965	0081209	0087452	0093695	0099938	0106180	0112421	0118662	0
0.642	0.8	0124902	0131143	0137382	0143621	0149860	0156098	0162335	0168572	0174809	0181045	0
0.643	0.8	0187281	0193516	0199751	0205985	0212219	0218452	0224684	0230917	0237149	0243380	0
0.644	0.8	0249611	0255841	0262071	0268300	0274529	0280757	0286985	0293213	0299440	0305666	0
0.645	0.8	0311892	0318118	0324343	0330567	0336791	0343015	0349238	0355460	0361682	0367904	0
0.646	0.8	0374125	0380346	0386566	0392786	0399005	0405224	0411442	0417660	0423877	0430094	0
0.647	0.8	0436310	0442526	0448741	0454956	0461171	0467385	0473598	0479811	0486024	0492236	0
0.648	0.8	0498447	0504658	0510869	0517079	0523289	0529498	0535706	0541915	0548122	0554329	0
0.649	0.8	0560536	0566743	0572948	0579154	0585358	0591563	0597767	0603970	0610173	0616376	0

TABLE 4

Square Roots

X		.0000	.0001	.0002	.0003	.0004	.0005	.0006	.0007	.0008	.0009	Δ_2
0.650	0.8	0622577	0628779	0634980	0641181	0647381	0653580	0659779	0665978	0672176	0678374	0
0.651	0.8	0684571	0690768	0696964	0703160	0709355	0715550	0721744	0727938	0734132	0740324	0
0.652	0.8	0746517	0752709	0758900	0765091	0771282	0777472	0783662	0789851	0796040	0802228	0
0.653	0.8	0808415	0814603	0820789	0826976	0833162	0839347	0845532	0851716	0857900	0864083	0
0.654	0.8	0870266	0876449	0882631	0888813	0894994	0901174	0907354	0913534	0919713	0925892	0
0.655	0.8	0932070	0938248	0944425	0950602	0956779	0962954	0969130	0975305	0981479	0987653	0
0.656	0.8	0993827	1000000	1006173	1012345	1018516	1024688	1030858	1037029	1043198	1049368	0
0.657	0.8	1055537	1061705	1067873	1074040	1080207	1086374	1092540	1098705	1104870	1111035	0
0.658	0.8	1117199	1123363	1129526	1135689	1141851	1148013	1154174	1160335	1166496	1172655	0
0.659	0.8	1178815	1184974	1191133	1197291	1203448	1209605	1215762	1221918	1228074	1234229	0
0.660	0.8	1240384	1246538	1252692	1258846	1264999	1271151	1277303	1283455	1289606	1295756	0
0.661	0.8	1301906	1308056	1314205	1320354	1326502	1332650	1338798	1344945	1351091	1357237	0
0.662	0.8	1363382	1369527	1375672	1381816	1387960	1394103	1400246	1406388	1412530	1418671	0
0.663	0.8	1424812	1430952	1437092	1443232	1449371	1455509	1461647	1467785	1473922	1480059	0
0.664	0.8	1486195	1492331	1498466	1504601	1510735	1516869	1523003	1529136	1535268	1541401	0
0.665	0.8	1547532	1553663	1559794	1565924	1572054	1578183	1584312	1590441	1596569	1602696	0
0.666	0.8	1608823	1614950	1621076	1627201	1633327	1639451	1645576	1651699	1657823	1663946	0
0.667	0.8	1670068	1676190	1682311	1688432	1694553	1700673	1706793	1712912	1719031	1725149	0
0.668	0.8	1731267	1737384	1743501	1749618	1755734	1761849	1767964	1774079	1780193	1786307	0
0.669	0.8	1792420	1798533	1804645	1810757	1816869	1822980	1829090	1835200	1841310	1847419	0
0.670	0.8	1853528	1859636	1865744	1871851	1877958	1884064	1890170	1896276	1902381	1908486	0
0.671	0.8	1914590	1920693	1926797	1932899	1939002	1945104	1951205	1957306	1963406	1969507	0
0.672	0.8	1975606	1981705	1987804	1993902	2000000	2006097	2012194	2018291	2024387	2030482	0
0.673	0.8	2036577	2042672	2048766	2054860	2060953	2067046	2073138	2079230	2085321	2091412	0
0.674	0.8	2097503	2103593	2109683	2115772	2121861	2127949	2134037	2140124	2146211	2152298	0
0.675	0.8	2158384	2164469	2170554	2176639	2182723	2188807	2194890	2200973	2207056	2213138	0
0.676	0.8	2219219	2225300	2231381	2237461	2243541	2249620	2255699	2261777	2267855	2273933	0
0.677	0.8	2280010	2286086	2292162	2298238	2304313	2310388	2316463	2322536	2328610	2334683	0
0.678	0.8	2340755	2346827	2352899	2358970	2365041	2371111	2377181	2383251	2389320	2395388	0
0.679	0.8	2401456	2407524	2413591	2419658	2425724	2431790	2437855	2443920	2449985	2456049	0
0.680	0.8	2462113	2468176	2474238	2480301	2486362	2492424	2498485	2504545	2510605	2516665	0
0.681	0.8	2522724	2528783	2534841	2540899	2546956	2553013	2559070	2565126	2571181	2577237	0
0.682	0.8	2583291	2589346	2595399	2601453	2607506	2613558	2619610	2625662	2631713	2637764	0
0.683	0.8	2643814	2649864	2655913	2661962	2668011	2674059	2680106	2686154	2692200	2698247	0
0.684	0.8	2704292	2710338	2716383	2722427	2728471	2734515	2740558	2746601	2752643	2758685	0
0.685	0.8	2764727	2770768	2776808	2782848	2788888	2794927	2800966	2807005	2813042	2819080	0
0.686	0.8	2825117	2831154	2837190	2843225	2849261	2855296	2861330	2867364	2873397	2879430	0
0.687	0.8	2885463	2891495	2897527	2903558	2909589	2915620	2921650	2927679	2933708	2939737	0
0.688	0.8	2945765	2951793	2957821	2963848	2969874	2975900	2981926	2987951	2993976	3000000	0
0.689	0.8	3006024	3012047	3018070	3024093	3030115	3036137	3042158	3048179	3054199	3060219	0
0.690	0.8	3066239	3072258	3078276	3084295	3090312	3096330	3102347	3108363	3114379	3120395	0
0.691	0.8	3126410	3132424	3138439	3144453	3150466	3156479	3162492	3168504	3174515	3180527	0
0.692	0.8	3186537	3192548	3198558	3204567	3210576	3216585	3222593	3228601	3234608	3240615	0
0.693	0.8	3246622	3252628	3258633	3264638	3270643	3276647	3282651	3288655	3294658	3300660	0
0.694	0.8	3306662	3312664	3318665	3324666	3330667	3336667	3342666	3348665	3354664	3360662	0
0.695	0.8	3366660	3372657	3378654	3384651	3390647	3396643	3402638	3408633	3414627	3420621	0
0.696	0.8	3426614	3432608	3438600	3444592	3450584	3456576	3462566	3468557	3474547	3480537	0
0.697	0.8	3486526	3492515	3498503	3504491	3510478	3516465	3522452	3528438	3534424	3540409	0
0.698	0.8	3546394	3552379	3558363	3564346	3570330	3576312	3582295	3588277	3594258	3600239	0
0.699	0.8	3606220	3612200	3618180	3624159	3630138	3636117	3642095	3648072	3654049	3660026	0

TABLE 4

Square Roots

X		.0000	.0001	.0002	.0003	.0004	.0005	.0006	.0007	.0008	.0009	Δ,
0.700	0.8	3666003	3671979	3677954	3683929	3689904	3695878	3701852	3707825	3713798	3719771	0
0.701	0.8	3725743	3731714	3737686	3743656	3749627	3755597	3761566	3767535	3773504	3779472	0
0.702	0.8	3785440	3791408	3797375	3803341	3809307	3815273	3821238	3827203	3833168	3839132	0
0.703	0.8	3845095	3851058	3857021	3862983	3868945	3874907	3880868	3886828	3892789	3898748	0
0.704	0.8	3904708	3910667	3916625	3922583	3928541	3934498	3940455	3946411	3952367	3958323	0
0.705	0.8	3964278	3970233	3976187	3982141	3988094	3994047	4000000	4005952	4011904	4017855	0
0.706	0.8	4023806	4029757	4035707	4041656	4047606	4053554	4059503	4065451	4071398	4077345	0
0.707	0.8	4083292	4089238	4095184	4101130	4107075	4113019	4118963	4124907	4130850	4136793	0
0.708	0.8	4142736	4148678	4154620	4160561	4166502	4172442	4178382	4184322	4190261	4196199	0
0.709	0.8	4202138	4208076	4214013	4219950	4225887	4231823	4237759	4243694	4249629	4255564	0
0.710	0.8	4261498	4267431	4273365	4279298	4285230	4291162	4297094	4303025	4308956	4314886	0
0.711	0.8	4320816	4326745	4332675	4338603	4344532	4350459	4356387	4362314	4368240	4374167	0
0.712	0.8	4380092	4386018	4391943	4397867	4403791	4409715	4415638	4421561	4427484	4433406	0
0.713	0.8	4439327	4445249	4451169	4457090	4463010	4468929	4474848	4480767	4486685	4492603	0
0.714	0.8	4498521	4504438	4510354	4516271	4522186	4528102	4534017	4539931	4545846	4551759	0
0.715	0.8	4557673	4563586	4569498	4575410	4581322	4587233	4593144	4599054	4604964	4610874	0
0.716	0.8	4616783	4622692	4628600	4634508	4640416	4646323	4652230	4658136	4664042	4669947	0
0.717	0.8	4675853	4681757	4687661	4693565	4699469	4705372	4711274	4717177	4723078	4728980	0
0.718	0.8	4734881	4740781	4746681	4752581	4758480	4764379	4770278	4776176	4782074	4787971	0
0.719	0.8	4793868	4799764	4805660	4811556	4817451	4823346	4829240	4835134	4841028	4846921	0
0.720	0.8	4852814	4858706	4864598	4870490	4876381	4882271	4888162	4894052	4899941	4905830	0
0.721	0.8	4911719	4917607	4923495	4929382	4935269	4941156	4947042	4952928	4958814	4964699	0
0.722	0.8	4970583	4976467	4982351	4988234	4994117	5000000	5005882	5011764	5017645	5023526	0
0.723	0.8	5029407	5035287	5041166	5047046	5052925	5058803	5064681	5070559	5076436	5082313	0
0.724	0.8	5088190	5094066	5099941	5105816	5111691	5117566	5123440	5129313	5135187	5141059	0
0.725	0.8	5146932	5152804	5158675	5164547	5170417	5176288	5182158	5188027	5193896	5199765	0
0.726	0.8	5205634	5211502	5217369	5223236	5229103	5234969	5240835	5246701	5252566	5258431	0
0.727	0.8	5264295	5270159	5276022	5281885	5287748	5293611	5299472	5305334	5311195	5317056	0
0.728	0.8	5322916	5328776	5334635	5340494	5346353	5352211	5358069	5363927	5369784	5375641	0
0.729	0.8	5381497	5387353	5393208	5399063	5404918	5410772	5416626	5422479	5428333	5434185	0
0.730	0.8	5440037	5445889	5451741	5457592	5463442	5469293	5475143	5480992	5486841	5492690	0
0.731	0.8	5498538	5504386	5510233	5516080	5521927	5527773	5533619	5539465	5545310	5551154	0
0.732	0.8	5556998	5562842	5568686	5574529	5580372	5586214	5592056	5597897	5603738	5609579	0
0.733	0.8	5615419	5621259	5627098	5632938	5638776	5644614	5650452	5656290	5662127	5667964	0
0.734	0.8	5673800	5679636	5685471	5691306	5697141	5702975	5708809	5714643	5720476	5726309	0
0.735	0.8	5732141	5737973	5743804	5749636	5755466	5761297	5767127	5772956	5778785	5784614	0
0.736	0.8	5790442	5796270	5802098	5807925	5813752	5819578	5825404	5831230	5837055	5842880	0
0.737	0.8	5848704	5854528	5860352	5866175	5871998	5877820	5883642	5889464	5895285	5901106	0
0.738	0.8	5906926	5912746	5918566	5924385	5930204	5936023	5941841	5947658	5953476	5959293	0
0.739	0.8	5965109	5970925	5976741	5982556	5988371	5994186	6000000	6005814	6011627	6017440	0
0.740	0.8	6023253	6029065	6034877	6040688	6046499	6052310	6058120	6063930	6069739	6075548	0
0.741	0.8	6081357	6087165	6092973	6098780	6104588	6110394	6116201	6122006	6127812	6133617	0
0.742	0.8	6139422	6145226	6151030	6156834	6162637	6168440	6174242	6180044	6185846	6191647	0
0.743	0.8	6197448	6203248	6209048	6214848	6220647	6226446	6232245	6238043	6243840	6249638	0
0.744	0.8	6255435	6261231	6267027	6272823	6278618	6284413	6290208	6296002	6301796	6307589	0
0.745	0.8	6313382	6319175	6324967	6330759	6336551	6342342	6348133	6353923	6359713	6365502	0
0.746	0.8	6371292	6377080	6382869	6388657	6394444	6400231	6406018	6411805	6417591	6423376	0
0.747	0.8	6429162	6434947	6440731	6446515	6452299	6458082	6463865	6469648	6475430	6481212	0
0.748	0.8	6486993	6492774	6498555	6504335	6510115	6515894	6521674	6527452	6533231	6539009	0
0.749	0.8	6544786	6550563	6556340	6562116	6567892	6573668	6579443	6585218	6590993	6596767	0

TABLE 4

Square Roots

X		.0000	.0001	.0002	.0003	.0004	.0005	.0006	.0007	.0008	.0009	Δ_2
0.750	0.8	6602540	6608314	6614087	6619859	6625631	6631403	6637174	6642945	6648716	6654486	0
0.751	0.8	6660256	6666026	6671795	6677563	6683332	6689100	6694867	6700634	6706401	6712167	0
0.752	0.8	6717933	6723699	6729464	6735229	6740994	6746758	6752522	6758285	6764048	6769810	0
0.753	0.8	6775573	6781334	6787096	6792857	6798617	6804378	6810138	6815897	6821656	6827415	0
0.754	0.8	6833173	6838931	6844689	6850446	6856203	6861959	6867715	6873471	6879226	6884981	0
0.755	0.8	6890736	6896490	6902244	6907997	6913750	6919503	6925255	6931007	6936759	6942510	0
0.756	0.8	6948260	6954011	6959761	6965510	6971260	6977008	6982757	6988505	6994253	7000000	0
0.757	0.8	7005747	7011493	7017240	7022985	7028731	7034476	7040221	7045965	7051709	7057452	0
0.758	0.8	7063195	7068938	7074681	7080423	7086164	7091905	7097646	7103387	7109127	7114867	0
0.759	0.8	7120606	7126345	7132084	7137822	7143560	7149297	7155034	7160771	7166507	7172243	0
0.760	0.8	7177979	7183714	7189449	7195183	7200917	7206651	7212384	7218117	7223850	7229582	0
0.761	0.8	7235314	7241045	7246776	7252507	7258237	7263967	7269697	7275426	7281155	7286883	0
0.762	0.8	7292611	7298339	7304066	7309793	7315520	7321246	7326972	7332697	7338422	7344147	0
0.763	0.8	7349871	7355595	7361319	7367042	7372765	7378487	7384209	7389931	7395652	7401373	0
0.764	0.8	7407093	7412814	7418533	7424253	7429972	7435691	7441409	7447127	7452844	7458562	0
0.765	0.8	7464278	7469995	7475711	7481427	7487142	7492857	7498571	7504286	7509999	7515713	0
0.766	0.8	7521426	7527139	7532851	7538563	7544274	7549986	7555697	7561407	7567117	7572827	0
0.767	0.8	7578536	7584245	7589954	7595662	7601370	7607077	7612784	7618491	7624198	7629904	0
0.768	0.8	7635609	7641314	7647019	7652724	7658428	7664132	7669835	7675538	7681241	7686943	0
0.769	0.8	7692645	7698347	7704048	7709749	7715449	7721149	7726849	7732548	7738247	7743946	0
0.770	0.8	7749644	7755342	7761039	7766736	7772433	7778129	7783825	7789521	7795216	7800911	0
0.771	0.8	7806606	7812300	7817994	7823687	7829380	7835073	7840765	7846457	7852148	7857840	0
0.772	0.8	7863531	7869221	7874911	7880601	7886290	7891979	7897668	7903356	7909044	7914731	0
0.773	0.8	7920419	7926105	7931792	7937478	7943163	7948849	7954534	7960218	7965902	7971586	0
0.774	0.8	7977270	7982953	7988636	7994318	8000000	8005682	8011363	8017044	8022724	8028404	0
0.775	0.8	8034084	8039764	8045443	8051121	8056800	8062478	8068155	8073833	8079510	8085186	0
0.776	0.8	8090862	8096538	8102213	8107888	8113563	8119237	8124911	8130585	8136258	8141931	0
0.777	0.8	8147603	8153276	8158947	8164619	8170290	8175960	8181631	8187301	8192970	8198639	0
0.778	0.8	8204308	8209977	8215645	8221313	8226980	8232647	8238314	8243980	8249646	8255311	0
0.779	0.8	8260977	8266641	8272306	8277970	8283634	8289297	8294960	8300623	8306285	8311947	0
0.780	0.8	8317609	8323270	8328931	8334591	8340251	8345911	8351570	8357229	8362888	8368546	0
0.781	0.8	8374204	8379862	8385519	8391176	8396832	8402489	8408144	8413800	8419455	8425110	0
0.782	0.8	8430764	8436418	8442071	8447725	8453377	8459030	8464682	8470334	8475985	8481636	0
0.783	0.8	8487287	8492938	8498588	8504237	8509886	8515535	8521184	8526832	8532480	8538127	0
0.784	0.8	8543774	8549421	8555068	8560714	8566359	8572005	8577650	8583294	8588938	8594582	0
0.785	0.8	8600226	8605869	8611512	8617154	8622796	8628438	8634079	8639720	8645361	8651001	0
0.786	0.8	8656641	8662281	8667920	8673559	8679197	8684835	8690473	8696110	8701747	8707384	0
0.787	0.8	8713020	8718656	8724292	8729927	8735562	8741197	8746831	8752465	8758098	8763731	0
0.788	0.8	8769364	8774996	8780629	8786260	8791891	8797522	8803153	8808783	8814413	8820043	0
0.789	0.8	8825672	8831301	8836929	8842557	8848185	8853813	8859440	8865066	8870693	8876318	0
0.790	0.8	8881944	8887569	8893194	8898819	8904443	8910067	8915690	8921314	8926936	8932559	0
0.791	0.8	8938181	8943802	8949424	8955045	8960665	8966286	8971906	8977525	8983144	8988763	0
0.792	0.8	8994382	9000000	9005618	9011235	9016852	9022469	9028085	9033701	9039317	9044932	0
0.793	0.8	9050547	9056162	9061776	9067390	9073004	9078617	9084230	9089842	9095454	9101066	0
0.794	0.8	9106678	9112289	9117899	9123510	9129120	9134729	9140339	9145948	9151556	9157165	0
0.795	0.8	9162772	9168380	9173987	9179594	9185201	9190807	9196412	9202018	9207623	9213228	0
0.796	0.8	9218832	9224436	9230040	9235643	9241246	9246849	9252451	9258053	9263654	9269256	0
0.797	0.8	9274856	9280457	9286057	9291657	9297256	9302855	9308454	9314053	9319651	9325248	0
0.798	0.8	9330846	9336443	9342039	9347636	9353232	9358827	9364422	9370017	9375612	9381206	0
0.799	0.8	9386800	9392393	9397987	9403579	9409172	9414764	9420356	9425947	9431538	9437129	0

TABLE 4

Square Roots

X		.0000	.0001	.0002	.0003	.0004	.0005	.0006	.0007	.0008	.0009	Δ_2
0.800	0.8	9442719	9448309	9453899	9459488	9465077	9470666	9476254	9481842	9487429	9493016	0
0.801	0.8	9498603	9504190	9509776	9515362	9520947	9526532	9532117	9537702	9543286	9548869	0
0.802	0.8	9554453	9560036	9565618	9571201	9576783	9582364	9587946	9593526	9599107	9604687	0
0.803	0.8	9610267	9615847	9621426	9627005	9632583	9638162	9643739	9649317	9654894	9660471	0
0.804	0.8	9666047	9671623	9677199	9682774	9688349	9693924	9699498	9705072	9710646	9716219	0
0.805	0.8	9721792	9727365	9732937	9738509	9744081	9749652	9755223	9760793	9766363	9771933	0
0.806	0.8	9777503	9783072	9788641	9794209	9799777	9805345	9810912	9816480	9822046	9827613	0
0.807	0.8	9833179	9838744	9844310	9849875	9855439	9861004	9866568	9872131	9877695	9883258	0
0.808	0.8	9888820	9894382	9899944	9905506	9911067	9916628	9922189	9927749	9933309	9938868	0
0.809	0.8	9944427	9949986	9955545	9961103	9966660	9972218	9977775	9983332	9988888	9994444	0
0.810	0.9	0000000	0005555	0011110	0016665	0022219	0027773	0033327	0038880	0044433	0049986	0
0.811	0.9	0055538	0061090	0066642	0072193	0077744	0083295	0088845	0094395	0099945	0105494	0
0.812	0.9	0111043	0116591	0122139	0127687	0133235	0138782	0144329	0149875	0155421	0160967	0
0.813	0.9	0166513	0172058	0177603	0183147	0188691	0194235	0199778	0205321	0210864	0216406	0
0.814	0.9	0221949	0227490	0233032	0238573	0244113	0249654	0255194	0260733	0266273	0271812	0
0.815	0.9	0277350	0282889	0288427	0293964	0299502	0305039	0310575	0316112	0321647	0327183	0
0.816	0.9	0332718	0338253	0343788	0349322	0354856	0360390	0365923	0371456	0376988	0382520	0
0.817	0.9	0388052	0393584	0399115	0404646	0410176	0415707	0421236	0426766	0432295	0437824	0
0.818	0.9	0443352	0448881	0454408	0459936	0465463	0470990	0476516	0482042	0487568	0493094	0
0.819	0.9	0498619	0504144	0509668	0515192	0520716	0526239	0531762	0537285	0542808	0548330	0
0.820	0.9	0553851	0559373	0564894	0570415	0575935	0581455	0586975	0592494	0598013	0603532	0
0.821	0.9	0609050	0614568	0620086	0625603	0631120	0636637	0642154	0647670	0653185	0658701	0
0.822	0.9	0664216	0669730	0675245	0680759	0686272	0691786	0697299	0702811	0708324	0713836	0
0.823	0.9	0719347	0724859	0730370	0735880	0741391	0746901	0752410	0757920	0763429	0768937	0
0.824	0.9	0774446	0779954	0785461	0790969	0796476	0801982	0807489	0812995	0818500	0824006	0
0.825	0.9	0829511	0835015	0840520	0846024	0851527	0857031	0862534	0868036	0873538	0879040	0
0.826	0.9	0884542	0890043	0895544	0901045	0906545	0912045	0917545	0923044	0928543	0934042	0
0.827	0.9	0939540	0945038	0950536	0956033	0961530	0967027	0972523	0978019	0983515	0989010	0
0.828	0.9	0994505	1000000	1005494	1010988	1016482	1021975	1027468	1032961	1038453	1043945	0
0.829	0.9	1049437	1054928	1060420	1065910	1071401	1076891	1082380	1087870	1093359	1098847	0
0.830	0.9	1104336	1109824	1115312	1120799	1126286	1131773	1137259	1142745	1148231	1153716	0
0.831	0.9	1159201	1164686	1170171	1175655	1181138	1186622	1192105	1197588	1203070	1208552	0
0.832	0.9	1214034	1219515	1224997	1230477	1235958	1241438	1246918	1252397	1257876	1263355	0
0.833	0.9	1268834	1274312	1279790	1285267	1290744	1296221	1301698	1307174	1312650	1318125	0
0.834	0.9	1323600	1329075	1334550	1340024	1345498	1350972	1356445	1361918	1367390	1372862	0
0.835	0.9	1378334	1383806	1389277	1394748	1400219	1405689	1411159	1416629	1422098	1427567	0
0.836	0.9	1433036	1438504	1443972	1449440	1454907	1460374	1465841	1471307	1476773	1482239	0
0.837	0.9	1487704	1493169	1498634	1504098	1509562	1515026	1520490	1525953	1531415	1536878	0
0.838	0.9	1542340	1547802	1553263	1558724	1564185	1569646	1575106	1580566	1586025	1591484	0
0.839	0.9	1596943	1602402	1607860	1613318	1618775	1624233	1629690	1635146	1640602	1646058	0
0.840	0.9	1651514	1656969	1662424	1667879	1673333	1678787	1684241	1689694	1695147	1700600	0
0.841	0.9	1706052	1711504	1716956	1722407	1727858	1733309	1738760	1744210	1749659	1755109	0
0.842	0.9	1760558	1766007	1771455	1776903	1782351	1787799	1793246	1798693	1804139	1809586	0
0.843	0.9	1815031	1820477	1825922	1831367	1836812	1842256	1847700	1853144	1858587	1864030	0
0.844	0.9	1869473	1874915	1880357	1885799	1891240	1896681	1902122	1907562	1913002	1918442	0
0.845	0.9	1923882	1929321	1934759	1940198	1945636	1951074	1956511	1961949	1967386	1972822	0
0.846	0.9	1978258	1983694	1989130	1994565	2000000	2005435	2010869	2016303	2021737	2027170	0
0.847	0.9	2032603	2038036	2043468	2048900	2054332	2059763	2065194	2070625	2076056	2081486	0
0.848	0.9	2086915	2092345	2097774	2103203	2108632	2114060	2119488	2124915	2130342	2135769	0
0.849	0.9	2141196	2146622	2152048	2157474	2162899	2168324	2173749	2179173	2184597	2190021	0

TABLE 4

Square Roots

X		.0000	.0001	.0002	.0003	.0004	.0005	.0006	.0007	.0008	.0009	Δ_2
0.850	0.9	2195445	2200868	2206290	2211713	2217135	2222557	2227978	2233400	2238820	2244241	0
0.851	0.9	2249661	2255081	2260501	2265920	2271339	2276758	2282176	2287594	2293012	2298429	0
0.852	0.9	2303846	2309263	2314679	2320095	2325511	2330927	2336342	2341757	2347171	2352585	0
0.853	0.9	2357999	2363413	2368826	2374239	2379651	2385064	2390476	2395887	2401299	2406710	0
0.854	0.9	2412120	2417531	2422941	2428351	2433760	2439169	2444578	2449986	2455395	2460802	0
0.855	0.9	2466210	2471617	2477024	2482431	2487837	2493243	2498649	2504054	2509459	2514864	0
0.856	0.9	2520268	2525672	2531076	2536479	2541882	2547285	2552688	2558090	2563492	2568893	0
0.857	0.9	2574294	2579695	2585096	2590496	2595896	2601296	2606695	2612094	2617493	2622891	0
0.858	0.9	2628289	2633687	2639085	2644482	2649879	2655275	2660671	2666067	2671463	2676858	0
0.859	0.9	2682253	2687647	2693042	2698436	2703829	2709223	2714616	2720009	2725401	2730793	0
0.860	0.9	2736185	2741576	2746968	2752358	2757749	2763139	2768529	2773919	2779308	2784697	0
0.861	0.9	2790086	2795474	2800862	2806250	2811637	2817024	2822411	2827798	2833184	2838570	0
0.862	0.9	2843955	2849340	2854725	2860110	2865494	2870878	2876262	2881645	2887028	2892411	0
0.863	0.9	2897793	2903175	2908557	2913939	2919320	2924701	2930081	2935461	2940841	2946221	0
0.864	0.9	2951600	2956979	2962358	2967736	2973114	2978492	2983870	2989247	2994623	3000000	0
0.865	0.9	3005376	3010752	3016128	3021503	3026878	3032252	3037627	3043001	3048375	3053748	0
0.866	0.9	3059121	3064494	3069866	3075238	3080610	3085982	3091353	3096724	3102094	3107465	0
0.867	0.9	3112835	3118204	3123574	3128943	3134312	3139680	3145048	3150416	3155783	3161151	0
0.868	0.9	3166518	3171884	3177250	3182616	3187982	3193347	3198712	3204077	3209442	3214806	0
0.869	0.9	3220169	3225533	3230896	3236259	3241622	3246984	3252346	3257707	3263069	3268430	0
0.870	0.9	3273791	3279151	3284511	3289871	3295230	3300589	3305948	3311307	3316665	3322023	0
0.871	0.9	3327381	3332738	3338095	3343452	3348808	3354164	3359520	3364876	3370231	3375586	0
0.872	0.9	3380940	3386294	3391648	3397002	3402355	3407708	3413061	3418414	3423766	3429118	0
0.873	0.9	3434469	3439820	3445171	3450522	3455872	3461222	3466572	3471921	3477270	3482619	0
0.874	0.9	3487967	3493315	3498663	3504011	3509358	3514705	3520051	3525398	3530744	3536089	0
0.875	0.9	3541435	3546780	3552125	3557469	3562813	3568157	3573501	3578844	3584187	3589529	0
0.876	0.9	3594872	3600214	3605555	3610897	3616238	3621579	3626919	3632259	3637599	3642939	0
0.877	0.9	3648278	3653617	3658956	3664294	3669632	3674970	3680307	3685645	3690981	3696318	0
0.878	0.9	3701654	3706990	3712326	3717661	3722996	3728331	3733665	3738999	3744333	3749667	0
0.879	0.9	3755000	3760333	3765665	3770998	3776330	3781661	3786993	3792324	3797655	3802985	0
0.880	0.9	3808315	3813645	3818975	3824304	3829633	3834961	3840290	3845618	3850946	3856273	0
0.881	0.9	3861600	3866927	3872254	3877580	3882906	3888231	3893557	3898882	3904206	3909531	0
0.882	0.9	3914855	3920179	3925502	3930826	3936149	3941471	3946793	3952115	3957437	3962759	0
0.883	0.9	3968080	3973400	3978721	3984041	3989361	3994681	4000000	4005319	4010638	4015956	0
0.884	0.9	4021274	4026592	4031909	4037227	4042544	4047860	4053176	4058492	4063808	4069123	0
0.885	0.9	4074439	4079753	4085068	4090382	4095696	4101010	4106323	4111636	4116949	4122261	0
0.886	0.9	4127573	4132885	4138196	4143507	4148818	4154129	4159439	4164749	4170059	4175368	0
0.887	0.9	4180677	4185986	4191295	4196603	4201911	4207218	4212526	4217833	4223139	4228446	0
0.888	0.9	4233752	4239058	4244363	4249668	4254973	4260278	4265582	4270886	4276190	4281493	0
0.889	0.9	4286797	4292099	4297402	4302704	4308006	4313308	4318609	4323910	4329211	4334511	0
0.890	0.9	4339811	4345111	4350411	4355710	4361009	4366308	4371606	4376904	4382202	4387499	0
0.891	0.9	4392796	4398093	4403390	4408686	4413982	4419278	4424573	4429868	4435163	4440457	0
0.892	0.9	4445752	4451046	4456339	4461632	4466925	4472218	4477511	4482803	4488094	4493386	0
0.893	0.9	4498677	4503968	4509259	4514549	4519839	4525129	4530418	4535708	4540996	4546285	0
0.894	0.9	4551573	4556861	4562149	4567436	4572723	4578010	4583297	4588583	4593869	4599154	0
0.895	0.9	4604440	4609725	4615009	4620294	4625578	4630862	4636145	4641429	4646711	4651994	0
0.896	0.9	4657276	4662559	4667840	4673122	4678403	4683684	4688964	4694245	4699525	4704804	0
0.897	0.9	4710084	4715363	4720642	4725920	4731199	4736477	4741754	4747032	4752309	4757585	0
0.898	0.9	4762862	4768138	4773414	4778690	4783965	4789240	4794515	4799789	4805063	4810337	0
0.899	0.9	4815610	4820884	4826157	4831429	4836702	4841974	4847246	4852517	4857788	4863059	0

TABLE 4

Square Roots

X		.0000	.0001	.0002	.0003	.0004	.0005	.0006	.0007	.0008	.0009	Δ_2
0.900	0.9	4868330	4873600	4878870	4884140	4889409	4894678	4899947	4905216	4910484	4915752	0
0.901	0.9	4921020	4926287	4931554	4936821	4942088	4947354	4952620	4957885	4963151	4968416	0
0.902	0.9	4973681	4978945	4984209	4989473	4994737	5000000	5005263	5010526	5015788	5021050	0
0.903	0.9	5026312	5031574	5036835	5042096	5047357	5052617	5057877	5063137	5068396	5073656	0
0.904	0.9	5078915	5084173	5089432	5094690	5099947	5105205	5110462	5115719	5120976	5126232	0
0.905	0.9	5131488	5136744	5141999	5147254	5152509	5157764	5163018	5168272	5173526	5178779	0
0.906	0.9	5184032	5189285	5194538	5199790	5205042	5210294	5215545	5220796	5226047	5231297	0
0.907	0.9	5236548	5241797	5247047	5252297	5257546	5262794	5268043	5273291	5278539	5283787	0
0.908	0.9	5289034	5294281	5299528	5304774	5310020	5315266	5320512	5325757	5331002	5336247	0
0.909	0.9	5341491	5346736	5351980	5357223	5362466	5367709	5372952	5378195	5383437	5388679	0
0.910	0.9	5393920	5399161	5404402	5409643	5414883	5420124	5425363	5430603	5435842	5441081	0
0.911	0.9	5446320	5451558	5456796	5462034	5467272	5472509	5477746	5482983	5488219	5493455	0
0.912	0.9	5498691	5503927	5509162	5514397	5519631	5524866	5530100	5535334	5540567	5545800	0
0.913	0.9	5551033	5556266	5561498	5566731	5571962	5577194	5582425	5587656	5592887	5598117	0
0.914	0.9	5603347	5608577	5613806	5619036	5624265	5629493	5634722	5639950	5645178	5650405	0
0.915	0.9	5655632	5660859	5666086	5671312	5676538	5681764	5686990	5692215	5697440	5702664	0
0.916	0.9	5707889	5713113	5718337	5723560	5728784	5734006	5739229	5744452	5749674	5754895	0
0.917	0.9	5760117	5765338	5770559	5775780	5781000	5786220	5791440	5796660	5801879	5807098	0
0.918	0.9	5812317	5817535	5822753	5827971	5833188	5838406	5843623	5848839	5854056	5859272	0
0.919	0.9	5864488	5869703	5874918	5880133	5885348	5890563	5895777	5900991	5906204	5911417	0
0.920	0.9	5916630	5921843	5927056	5932268	5937480	5942691	5947903	5953114	5958324	5963535	0
0.921	0.9	5968745	5973955	5979164	5984374	5989583	5994791	6000000	6005208	6010416	6015624	0
0.922	0.9	6020831	6026038	6031245	6036451	6041658	6046863	6052069	6057275	6062480	6067684	0
0.923	0.9	6072889	6078093	6083297	6088501	6093704	6098907	6104110	6109313	6114515	6119717	0
0.924	0.9	6124919	6130120	6135321	6140522	6145723	6150923	6156123	6161323	6166522	6171721	0
0.925	0.9	6176920	6182119	6187317	6192515	6197713	6202911	6208108	6213305	6218501	6223698	0
0.926	0.9	6228894	6234090	6239285	6244480	6249675	6254870	6260064	6265258	6270452	6275646	0
0.927	0.9	6280839	6286032	6291225	6296417	6301610	6306801	6311993	6317184	6322375	6327566	0
0.928	0.9	6332757	6337947	6343137	6348326	6353516	6358705	6363894	6369082	6374270	6379458	0
0.929	0.9	6384646	6389833	6395021	6400207	6405394	6410580	6415766	6420952	6426137	6431323	0
0.930	0.9	6436508	6441692	6446877	6452061	6457244	6462428	6467611	6472794	6477977	6483159	0
0.931	0.9	6488341	6493523	6498705	6503886	6509067	6514248	6519428	6524608	6529788	6534968	0
0.932	0.9	6540147	6545326	6550505	6555683	6560862	6566040	6571217	6576395	6581572	6586749	0
0.933	0.9	6591925	6597101	6602277	6607453	6612629	6617804	6622979	6628153	6633328	6638502	0
0.934	0.9	6643675	6648849	6654022	6659195	6664368	6669540	6674712	6679884	6685056	6690227	0
0.935	0.9	6695398	6700569	6705739	6710909	6716079	6721249	6726418	6731587	6736756	6741925	0
0.936	0.9	6747093	6752261	6757429	6762596	6767763	6772930	6778097	6783263	6788429	6793595	0
0.937	0.9	6798760	6803925	6809090	6814255	6819419	6824584	6829747	6834911	6840074	6845237	0
0.938	0.9	6850400	6855563	6860725	6865887	6871048	6876210	6881371	6886532	6891692	6896852	0
0.939	0.9	6902012	6907172	6912332	6917491	6922650	6927808	6932966	6938125	6943282	6948440	0
0.940	0.9	6953597	6958754	6963911	6969067	6974223	6979379	6984535	6989690	6994845	7000000	0
0.941	0.9	7005154	7010309	7015463	7020616	7025770	7030923	7036076	7041228	7046381	7051533	0
0.942	0.9	7056684	7061836	7066987	7072138	7077289	7082439	7087589	7092739	7097889	7103038	0
0.943	0.9	7108187	7113336	7118484	7123633	7128780	7133928	7139076	7144223	7149369	7154516	0
0.944	0.9	7159662	7164808	7169954	7175100	7180245	7185390	7190534	7195679	7200823	7205967	0
0.945	0.9	7211110	7216254	7221397	7226540	7231682	7236824	7241966	7247108	7252249	7257390	0
0.946	0.9	7262531	7267672	7272812	7277952	7283092	7288232	7293371	7298510	7303648	7308787	0
0.947	0.9	7313925	7319063	7324200	7329338	7334475	7339612	7344748	7349884	7355020	7360156	0
0.948	0.9	7365291	7370427	7375562	7380696	7385831	7390965	7396098	7401232	7406365	7411498	0
0.949	0.9	7416631	7421763	7426896	7432028	7437159	7442291	7447422	7452553	7457683	7462813	0

TABLE 4

Square Roots

X		.0000	.0001	.0002	.0003	.0004	.0005	.0006	.0007	.0008	.0009	Δ_2
0.950	0.9	7467943	7473073	7478203	7483332	7488461	7493590	7498718	7503846	7508974	7514101	0
0.951	0.9	7519229	7524356	7529483	7534609	7539735	7544861	7549987	7555113	7560238	7565363	0
0.952	0.9	7570487	7575612	7580736	7585860	7590983	7596106	7601229	7606352	7611475	7616597	0
0.953	0.9	7621719	7626841	7631962	7637083	7642204	7647325	7652445	7657565	7662685	7667804	0
0.954	0.9	7672924	7678043	7683161	7688280	7693398	7698516	7703633	7708751	7713868	7718985	0
0.955	0.9	7724101	7729218	7734334	7739450	7744565	7749680	7754795	7759910	7765024	7770139	0
0.956	0.9	7775252	7780366	7785479	7790593	7795705	7800818	7805930	7811042	7816154	7821266	0
0.957	0.9	7826377	7831488	7836598	7841709	7846819	7851929	7857039	7862148	7867257	7872366	0
0.958	0.9	7877474	7882583	7887691	7892798	7897906	7903013	7908120	7913227	7918333	7923439	0
0.959	0.9	7928545	7933651	7938756	7943861	7948966	7954071	7959175	7964279	7969383	7974486	0
0.960	0.9	7979590	7984693	7989795	7994898	8000000	8005102	8010204	8015305	8020406	8025507	0
0.961	0.9	8030607	8035708	8040808	8045908	8051007	8056106	8061205	8066304	8071403	8076501	0
0.962	0.9	8081599	8086696	8091794	8096891	8101988	8107084	8112181	8117277	8122373	8127468	0
0.963	0.9	8132563	8137658	8142753	8147848	8152942	8158036	8163129	8168223	8173316	8178409	0
0.964	0.9	8183502	8188594	8193686	8198778	8203870	8208961	8214052	8219143	8224233	8229323	0
0.965	0.9	8234413	8239503	8244593	8249682	8254771	8259860	8264948	8270036	8275124	8280212	0
0.966	0.9	8285299	8290386	8295473	8300559	8305646	8310732	8315818	8320903	8325988	8331073	0
0.967	0.9	8336158	8341243	8346327	8351411	8356494	8361578	8366661	8371744	8376826	8381909	0
0.968	0.9	8386991	8392073	8397154	8402236	8407317	8412398	8417478	8422558	8427638	8432718	0
0.969	0.9	8437798	8442877	8447956	8453034	8458113	8463191	8468269	8473347	8478424	8483501	0
0.970	0.9	8488578	8493655	8498731	8503807	8508883	8513958	8519034	8524109	8529183	8534258	0
0.971	0.9	8539332	8544406	8549480	8554553	8559627	8564700	8569772	8574845	8579917	8584989	0
0.972	0.9	8590060	8595132	8600203	8605274	8610344	8615415	8620485	8625554	8630624	8635693	0
0.973	0.9	8640762	8645831	8650900	8655968	8661036	8666104	8671171	8676238	8681305	8686372	0
0.974	0.9	8691438	8696504	8701570	8706636	8711701	8716767	8721831	8726896	8731960	8737024	0
0.975	0.9	8742088	8747152	8752215	8757278	8762341	8767404	8772466	8777528	8782590	8787651	0
0.976	0.9	8792712	8797773	8802834	8807894	8812955	8818014	8823074	8828134	8833193	8838252	0
0.977	0.9	8843310	8848369	8853427	8858485	8863542	8868600	8873657	8878714	8883770	8888826	0
0.978	0.9	8893882	8898938	8903994	8909049	8914104	8919159	8924213	8929268	8934322	8939375	0
0.979	0.9	8944429	8949482	8954535	8959588	8964640	8969692	8974744	8979796	8984847	8989898	0
0.980	0.9	8994949	9000000	9005050	9010100	9015150	9020200	9025249	9030298	9035347	9040396	0
0.981	0.9	9045444	9050492	9055540	9060588	9065635	9070682	9075729	9080775	9085821	9090867	0
0.982	0.9	9095913	9100959	9106004	9111049	9116094	9121138	9126182	9131226	9136270	9141313	0
0.983	0.9	9146356	9151399	9156442	9161484	9166527	9171568	9176610	9181651	9186693	9191734	0
0.984	0.9	9196774	9201814	9206855	9211894	9216934	9221973	9227012	9232051	9237090	9242128	0
0.985	0.9	9247166	9252204	9257241	9262279	9267316	9272353	9277389	9282425	9287461	9292497	0
0.986	0.9	9297533	9302568	9307603	9312638	9317672	9322706	9327740	9332774	9337807	9342841	0
0.987	0.9	9347874	9352906	9357939	9362971	9368003	9373035	9378066	9383097	9388128	9393159	0
0.988	0.9	9398189	9403219	9408249	9413279	9418308	9423337	9428366	9433395	9438423	9443451	0
0.989	0.9	9448479	9453507	9458534	9463561	9468588	9473615	9478641	9483667	9488693	9493718	0
0.990	0.9	9498744	9503769	9508794	9513818	9518842	9523866	9528890	9533914	9538937	9543960	0
0.991	0.9	9548983	9554005	9559028	9564050	9569071	9574093	9579114	9584135	9589156	9594176	0
0.992	0.9	9599197	9604217	9609236	9614256	9619275	9624294	9629313	9634331	9639350	9644368	0
0.993	0.9	9649385	9654403	9659420	9664437	9669454	9674470	9679486	9684502	9689518	9694533	0
0.994	0.9	9699549	9704564	9709578	9714593	9719607	9724621	9729634	9734648	9739661	9744674	0
0.995	0.9	9749687	9754699	9759711	9764723	9769735	9774746	9779757	9784768	9789779	9794789	0
0.996	0.9	9799800	9804809	9809819	9814828	9819838	9824847	9829855	9834864	9839872	9844880	0
0.997	0.9	9849887	9854895	9859902	9864909	9869915	9874922	9879928	9884934	9889939	9894945	0
0.998	0.9	9899950	9904955	9909959	9914964	9919968	9924972	9929975	9934979	9939982	9944985	0
0.999	0.9	9949987	9954990	9959992	9964994	9969995	9974997	9979998	9984999	9989999	9995000	0

TABLE 4

Square Roots

X	.0000	.0001	.0002	.0003	.0004	.0005	.0006	.0007	.0008	.0009	Δ_2
1.00 1.	0000000	0004999	0009995	0014989	0019980	0024969	0029955	0034939	0039920	0044899	-1
1.01 1.	0049876	0054850	0059821	0064790	0069757	0074721	0079683	0084642	0089599	0094553	-1
1.02 1.	0099505	0104454	0109402	0114346	0119289	0124228	0129166	0134101	0139033	0143964	-1
1.03 1.	0148892	0153817	0158740	0163661	0168579	0173495	0178409	0183320	0188228	0193135	-1
1.04 1.	0198039	0202941	0207840	0212737	0217632	0222524	0227414	0232302	0237187	0242070	-1
1.05 1.	0246951	0251829	0256705	0261579	0266450	0271319	0276186	0281051	0285913	0290773	-1
1.06 1.	0295630	0300485	0305338	0310189	0315038	0319884	0324728	0329569	0334409	0339246	-1
1.07 1.	0344080	0348913	0353743	0358571	0363397	0368221	0373042	0377861	0382678	0387492	-1
1.08 1.	0392305	0397115	0401923	0406729	0411532	0416333	0421132	0425929	0430724	0435516	-1
1.09 1.	0440307	0445095	0449880	0454664	0459445	0464225	0469002	0473777	0478550	0483320	-1
1.10 1.	0488088	0492855	0497619	0502381	0507140	0511898	0516653	0521407	0526158	0530907	-1
1.11 1.	0535654	0540398	0545141	0549882	0554620	0559356	0564090	0568822	0573552	0578280	-1
1.12 1.	0583005	0587729	0592450	0597169	0601887	0606602	0611315	0616026	0620734	0625441	-1
1.13 1.	0630146	0634848	0639549	0644247	0648944	0653638	0658330	0663020	0667708	0672394	-1
1.14 1.	0677078	0681760	0686440	0691118	0695794	0700467	0705139	0709809	0714476	0719142	-1
1.15 1.	0723805	0728467	0733126	0737784	0742439	0747093	0751744	0756393	0761041	0765686	-1
1.16 1.	0770330	0774971	0779610	0784248	0788883	0793517	0798148	0802777	0807405	0812030	-1
1.17 1.	0816654	0821275	0825895	0830512	0835128	0839742	0844353	0848963	0853571	0858177	-1
1.18 1.	0862780	0867382	0871982	0876580	0881176	0885771	0890363	0894953	0899541	0904128	-1
1.19 1.	0908712	0913295	0917875	0922454	0927031	0931606	0936178	0940750	0945319	0949886	-1
1.20 1.	0954451	0959015	0963576	0968136	0972693	0977249	0981803	0986355	0990905	0995454	-1
1.21 1.	1000000	1004545	1009087	1013628	1018167	1022704	1027239	1031772	1036304	1040833	-1
1.22 1.	1045361	1049887	1054411	1058933	1063453	1067972	1072488	1077003	1081516	1086027	-1
1.23 1.	1090536	1095044	1099550	1104053	1108555	1113055	1117554	1122050	1126545	1131038	-1
1.24 1.	1135529	1140018	1144505	1148991	1153475	1157957	1162437	1166915	1171392	1175867	-1
1.25 1.	1180340	1184811	1189281	1193748	1198214	1202678	1207141	1211601	1216060	1220517	-1
1.26 1.	1224972	1229426	1233877	1238327	1242775	1247222	1251667	1256109	1260551	1264990	-1
1.27 1.	1269428	1273864	1278298	1282730	1287161	1291590	1296017	1300442	1304866	1309288	-1
1.28 1.	1313708	1318127	1322544	1326959	1331372	1335784	1340194	1344602	1349009	1353414	-1
1.29 1.	1357817	1362218	1366618	1371016	1375412	1379807	1384200	1388591	1392980	1397368	-1
1.30 1.	1401754	1406139	1410521	1414903	1419282	1423660	1428036	1432410	1436783	1441154	-1
1.31 1.	1445523	1449891	1454257	1458621	1462984	1467345	1471704	1476062	1480418	1484773	-1
1.32 1.	1489125	1493476	1497826	1502174	1506520	1510864	1515207	1519549	1523888	1528226	-1
1.33 1.	1532563	1536897	1541230	1545562	1549892	1554220	1558547	1562872	1567195	1571517	-1
1.34 1.	1575837	1580155	1584472	1588788	1593101	1597413	1601724	1606033	1610340	1614646	-1
1.35 1.	1618950	1623253	1627553	1631853	1636151	1640447	1644741	1649034	1653326	1657616	-1
1.36 1.	1661904	1666190	1670476	1674759	1679041	1683321	1687600	1691878	1696153	1700427	-1
1.37 1.	1704700	1708971	1713240	1717508	1721775	1726039	1730303	1734564	1738824	1743083	-1
1.38 1.	1747340	1751596	1755850	1760102	1764353	1768602	1772850	1777096	1781341	1785584	-1
1.39 1.	1789826	1794066	1798305	1802542	1806778	1811012	1815244	1819475	1823705	1827933	-1
1.40 1.	1832160	1836385	1840608	1844830	1849051	1853270	1857487	1861703	1865918	1870131	-0
1.41 1.	1874342	1878552	1882761	1886968	1891173	1895377	1899580	1903781	1907981	1912179	-0
1.42 1.	1916375	1920570	1924764	1928956	1933147	1937336	1941524	1945711	1949895	1954079	-0
1.43 1.	1958261	1962441	1966620	1970798	1974974	1979149	1983322	1987493	1991664	1995833	-0
1.44 1.	2000000	2004166	2008330	2012493	2016655	2020815	2024974	2029131	2033287	2037442	-0
1.45 1.	2041595	2045746	2049896	2054045	2058192	2062338	2066482	2070625	2074767	2078907	-0
1.46 1.	2083046	2087183	2091319	2095454	2099587	2103718	2107849	2111978	2116105	2120231	-0
1.47 1.	2124356	2128479	2132601	2136721	2140840	2144958	2149074	2153189	2157302	2161414	-0
1.48 1.	2165525	2169634	2173742	2177849	2181954	2186058	2190160	2194261	2198361	2202459	-0
1.49 1.	2206556	2210651	2214745	2218838	2222929	2227019	2231108	2235195	2239281	2243366	-0

TABLE 4

Square Roots

X		.0000	.0001	.0002	.0003	.0004	.0005	.0006	.0007	.0008	.0009	Δ_2
1.50	1.	2247449	2251531	2255611	2259690	2263768	2267844	2271919	2275993	2280065	2284136	−0
1.51	1.	2288206	2292274	2296341	2300406	2304471	2308534	2312595	2316655	2320714	2324772	−0
1.52	1.	2328828	2332883	2336936	2340989	2345039	2349089	2353137	2357184	2361230	2365274	−0
1.53	1.	2369317	2373358	2377399	2381438	2385475	2389512	2393547	2397580	2401613	2405644	−0
1.54	1.	2409674	2413702	2417729	2421755	2425780	2429803	2433825	2437845	2441865	2445883	−0
1.55	1.	2449900	2453915	2457929	2461942	2465954	2469964	2473973	2477981	2481987	2485992	−0
1.56	1.	2489996	2493999	2498000	2502000	2505999	2509996	2513992	2517987	2521981	2525973	−0
1.57	1.	2529964	2533954	2537942	2541930	2545916	2549900	2553884	2557866	2561847	2565827	−0
1.58	1.	2569805	2573782	2577758	2581733	2585706	2589678	2593649	2597619	2601587	2605554	−0
1.59	1.	2609520	2613485	2617448	2621410	2625371	2629331	2633289	2637247	2641202	2645157	−0
1.60	1.	2649111	2653063	2657014	2660964	2664912	2668859	2672806	2676750	2680694	2684636	−0
1.61	1.	2688578	2692517	2696456	2700394	2704330	2708265	2712199	2716131	2720063	2723993	−0
1.62	1.	2727922	2731850	2735776	2739702	2743626	2747549	2751470	2755391	2759310	2763228	−0
1.63	1.	2767145	2771061	2774976	2778889	2782801	2786712	2790622	2794530	2798437	2802344	−0
1.64	1.	2806248	2810152	2814055	2817956	2821856	2825755	2829653	2833550	2837445	2841339	−0
1.65	1.	2845233	2849124	2853015	2856905	2860793	2864680	2868566	2872451	2876335	2880217	−0
1.66	1.	2884099	2887979	2891858	2895736	2899612	2903488	2907362	2911235	2915107	2918978	−0
1.67	1.	2922848	2926717	2930584	2934450	2938315	2942179	2946042	2949903	2953764	2957623	−0
1.68	1.	2961481	2965338	2969194	2973049	2976903	2980755	2984606	2988456	2992305	2996153	−0
1.69	1.	3000000	3003846	3007690	3011533	3015376	3019217	3023056	3026895	3030733	3034569	−0
1.70	1.	3038405	3042239	3046072	3049904	3053735	3057565	3061393	3065221	3069047	3072873	−0
1.71	1.	3076697	3080520	3084342	3088163	3091982	3095801	3099618	3103435	3107250	3111064	−0
1.72	1.	3114877	3118689	3122500	3126309	3130118	3133926	3137732	3141537	3145341	3149144	−0
1.73	1.	3152946	3156747	3160547	3164346	3168143	3171940	3175735	3179530	3183323	3187115	−0
1.74	1.	3190906	3194696	3198485	3202273	3206059	3209845	3213629	3217413	3221195	3224976	−0
1.75	1.	3228757	3232536	3236314	3240091	3243866	3247641	3251415	3255188	3258959	3262730	−0
1.76	1.	3266499	3270268	3274035	3277801	3281566	3285330	3289093	3292855	3296616	3300376	−0
1.77	1.	3304135	3307892	3311649	3315405	3319159	3322913	3326665	3330416	3334167	3337916	−0
1.78	1.	3341664	3345411	3349157	3352902	3356646	3360389	3364131	3367872	3371612	3375350	−0
1.79	1.	3379088	3382825	3386560	3390295	3394029	3397761	3401492	3405223	3408952	3412681	−0
1.80	1.	3416408	3420134	3423859	3427584	3431307	3435029	3438750	3442470	3446189	3449907	−0
1.81	1.	3453624	3457340	3461055	3464769	3468482	3472194	3475904	3479614	3483323	3487031	−0
1.82	1.	3490738	3494443	3498148	3501852	3505554	3509256	3512957	3516656	3520355	3524053	−0
1.83	1.	3527749	3531445	3535139	3538833	3542526	3546217	3549908	3553597	3557286	3560973	−0
1.84	1.	3564660	3568345	3572030	3575714	3579396	3583078	3586758	3590438	3594116	3597794	−0
1.85	1.	3601470	3605146	3608821	3612494	3616167	3619838	3623509	3627179	3630847	3634515	−0
1.86	1.	3638182	3641847	3645512	3649176	3652839	3656500	3660161	3663821	3667480	3671137	−0
1.87	1.	3674794	3678450	3682105	3685759	3689412	3693064	3696715	3700365	3704014	3707662	−0
1.88	1.	3711309	3714955	3718600	3722245	3725888	3729530	3733171	3736812	3740451	3744090	−0
1.89	1.	3747727	3751364	3754999	3758634	3762267	3765900	3769532	3773162	3776792	3780421	−0
1.90	1.	3784049	3787676	3791302	3794927	3798551	3802174	3805796	3809417	3813037	3816657	−0
1.91	1.	3820275	3823892	3827509	3831124	3834739	3838352	3841965	3845577	3849188	3852798	−0
1.92	1.	3856406	3860014	3863621	3867228	3870833	3874437	3878040	3881643	3885244	3888844	−0
1.93	1.	3892444	3896043	3899640	3903237	3906833	3910428	3914022	3917615	3921207	3924798	−0
1.94	1.	3928388	3931978	3935566	3939153	3942740	3946326	3949910	3953494	3957077	3960659	−0
1.95	1.	3964240	3967820	3971399	3974978	3978555	3982131	3985707	3989282	3992855	3996428	−0
1.96	1.	4000000	4003571	4007141	4010710	4014278	4017846	4021412	4024978	4028542	4032106	−0
1.97	1.	4035669	4039231	4042792	4046352	4049911	4053469	4057027	4060583	4064139	4067693	−0
1.98	1.	4071247	4074800	4078352	4081903	4085453	4089003	4092551	4096099	4099645	4103191	−0
1.99	1.	4106736	4110280	4113823	4117365	4120906	4124447	4127986	4131525	4135063	4138600	−0

TABLE 4

Square Roots

X		.0000	.0001	.0002	.0003	.0004	.0005	.0006	.0007	.0008	.0009	Δ_2
2.00	1.	4142136	4145671	4149205	4152738	4156271	4159802	4163333	4166863	4170392	4173920	−0
2.01	1.	4177447	4180973	4184499	4188023	4191547	4195070	4198591	4202113	4205633	4209152	−0
2.02	1.	4212670	4216188	4219705	4223220	4226735	4230249	4233763	4237275	4240786	4244297	−0
2.03	1.	4247807	4251316	4254824	4258331	4261837	4265343	4268847	4272351	4275854	4279356	−0
2.04	1.	4282857	4286357	4289857	4293355	4296853	4300350	4303846	4307341	4310835	4314328	−0
2.05	1.	4317821	4321313	4324804	4328294	4331783	4335271	4338759	4342245	4345731	4349216	−0
2.06	1.	4352700	4356183	4359666	4363147	4366628	4370108	4373587	4377065	4380542	4384019	−0
2.07	1.	4387495	4390969	4394443	4397916	4401389	4404860	4408331	4411801	4415270	4418738	−0
2.08	1.	4422205	4425672	4429137	4432602	4436066	4439529	4442991	4446453	4449913	4453373	−0
2.09	1.	4456832	4460290	4463748	4467204	4470660	4474115	4477569	4481022	4484474	4487926	−0
2.10	1.	4491377	4494827	4498276	4501724	4505171	4508618	4512064	4515509	4518953	4522396	−0
2.11	1.	4525839	4529281	4532722	4536162	4539601	4543040	4546477	4549914	4553350	4556785	−0
2.12	1.	4560220	4563653	4567086	4570518	4573949	4577380	4580809	4584238	4587666	4591093	−0
2.13	1.	4594520	4597945	4601370	4604794	4608217	4611639	4615061	4618481	4621901	4625321	−0
2.14	1.	4628739	4632156	4635573	4638989	4642404	4645819	4649232	4652645	4656057	4659468	−0
2.15	1.	4662878	4666288	4669697	4673105	4676512	4679918	4683324	4686729	4690133	4693536	−0
2.16	1.	4696938	4700340	4703741	4707141	4710540	4713939	4717337	4720734	4724130	4727525	−0
2.17	1.	4730920	4734314	4737707	4741099	4744490	4747881	4751271	4754660	4758049	4761436	−0
2.18	1.	4764823	4768209	4771594	4774979	4778363	4781745	4785128	4788509	4791890	4795269	−0
2.19	1.	4798649	4802027	4805404	4808781	4812157	4815532	4818907	4822281	4825653	4829026	−0
2.20	1.	4832397	4835768	4839137	4842506	4845875	4849242	4852609	4855975	4859340	4862705	−0
2.21	1.	4866069	4869432	4872794	4876155	4879516	4882876	4886235	4889594	4892951	4896308	−0
2.22	1.	4899664	4903020	4906374	4909728	4913081	4916434	4919785	4923136	4926486	4929836	−0
2.23	1.	4933184	4936532	4939880	4943226	4946571	4949916	4953260	4956604	4959946	4963288	−0
2.24	1.	4966630	4969970	4973310	4976648	4979987	4983324	4986661	4989997	4993332	4996666	−0
2.25	1.	5000000	5003333	5006665	5009997	5013327	5016657	5019987	5023315	5026643	5029970	−0
2.26	1.	5033296	5036622	5039947	5043271	5046594	5049917	5053239	5056560	5059880	5063200	−0
2.27	1.	5066519	5069837	5073155	5076472	5079788	5083103	5086418	5089732	5093045	5096357	−0
2.28	1.	5099669	5102980	5106290	5109600	5112908	5116216	5119524	5122830	5126136	5129441	−0
2.29	1.	5132746	5136050	5139353	5142655	5145957	5149257	5152557	5155857	5159156	5162454	−0
2.30	1.	5165751	5169047	5172343	5175638	5178933	5182226	5185519	5188812	5192103	5195394	−0
2.31	1.	5198684	5201974	5205262	5208550	5211837	5215124	5218410	5221695	5224979	5228263	−0
2.32	1.	5231546	5234829	5238110	5241391	5244671	5247951	5251229	5254508	5257785	5261061	−0
2.33	1.	5264338	5267613	5270887	5274161	5277434	5280707	5283979	5287250	5290520	5293790	−0
2.34	1.	5297059	5300327	5303594	5306861	5310127	5313393	5316658	5319922	5323185	5326448	−0
2.35	1.	5329710	5332971	5336232	5339492	5342751	5346009	5349267	5352524	5355781	5359036	−0
2.36	1.	5362291	5365546	5368799	5372052	5375305	5378556	5381807	5385058	5388307	5391556	−0
2.37	1.	5394804	5398052	5401299	5404545	5407790	5411035	5414279	5417522	5420765	5424007	−0
2.38	1.	5427249	5430489	5433729	5436969	5440207	5443445	5446682	5449919	5453155	5456390	−0
2.39	1.	5459625	5462859	5466092	5469324	5472556	5475788	5479018	5482248	5485477	5488706	−0
2.40	1.	5491933	5495161	5498387	5501613	5504838	5508062	5511286	5514509	5517732	5520954	−0
2.41	1.	5524175	5527395	5530615	5533834	5537052	5540270	5543487	5546704	5549920	5553135	−0
2.42	1.	5556349	5559563	5562776	5565988	5569200	5572411	5575622	5578832	5582041	5585249	−0
2.43	1.	5588457	5591664	5594871	5598077	5601282	5604486	5607690	5610894	5614096	5617298	−0
2.44	1.	5620499	5623700	5626900	5630099	5633298	5636496	5639693	5642890	5646086	5649281	−0
2.45	1.	5652476	5655670	5658863	5662056	5665248	5668440	5671630	5674821	5678010	5681199	−0
2.46	1.	5684387	5687575	5690762	5693948	5697133	5700318	5703503	5706686	5709869	5713052	−0
2.47	1.	5716234	5719415	5722595	5725775	5728954	5732133	5735311	5738488	5741664	5744840	−0
2.48	1.	5748016	5751190	5754364	5757538	5760711	5763883	5767054	5770225	5773395	5776565	−0
2.49	1.	5779734	5782902	5786070	5789237	5792403	5795569	5798734	5801899	5805062	5808226	−0

TABLE 4

Square Roots

X	.0000	.0001	.0002	.0003	.0004	.0005	.0006	.0007	.0008	.0009	Δ_2
2.50 1.	5811388	5814550	5817712	5820872	5824032	5827192	5830351	5833509	5836666	5839823	−0
2.51 1.	5842979	5846135	5849290	5852445	5855598	5858751	5861904	5865056	5868207	5871358	−0
2.52 1.	5874508	5877657	5880806	5883954	5887102	5890249	5893395	5896540	5899685	5902830	−0
2.53 1.	5905974	5909117	5912259	5915401	5918543	5921683	5924823	5927963	5931102	5934240	−0
2.54 1.	5937377	5940514	5943651	5946786	5949922	5953056	5956190	5959323	5962456	5965588	−0
2.55 1.	5968719	5971850	5974980	5978110	5981239	5984367	5987495	5990622	5993749	5996875	−0
2.56 1.	6000000	6003125	6006249	6009372	6012495	6015617	6018739	6021860	6024980	6028100	−0
2.57 1.	6031220	6034338	6037456	6040573	6043690	6046806	6049922	6053037	6056151	6059265	−0
2.58 1.	6062378	6065491	6068603	6071714	6074825	6077935	6081045	6084154	6087262	6090370	−0
2.59 1.	6093477	6096583	6099689	6102795	6105900	6109004	6112107	6115210	6118313	6121414	−0
2.60 1.	6124515	6127616	6130716	6133815	6136914	6140012	6143110	6146207	6149303	6152399	−0
2.61 1.	6155494	6158589	6161683	6164776	6167869	6170962	6174053	6177144	6180235	6183325	−0
2.62 1.	6186414	6189503	6192591	6195678	6198765	6201852	6204937	6208023	6211107	6214191	−0
2.63 1.	6217275	6220358	6223440	6226521	6229603	6232683	6235763	6238842	6241921	6244999	−0
2.64 1.	6248077	6251154	6254230	6257306	6260381	6263456	6266530	6269604	6272676	6275749	−0
2.65 1.	6278821	6281892	6284962	6288032	6291102	6294171	6297239	6300307	6303374	6306440	−0
2.66 1.	6309506	6312572	6315637	6318701	6321765	6324828	6327890	6330952	6334014	6337074	−0
2.67 1.	6340135	6343194	6346253	6349312	6352370	6355427	6358484	6361540	6364596	6367651	−0
2.68 1.	6370706	6373759	6376813	6379866	6382918	6385970	6389021	6392071	6395121	6398171	−0
2.69 1.	6401219	6404268	6407315	6410363	6413409	6416455	6419501	6422545	6425590	6428633	−0
2.70 1.	6431677	6434719	6437761	6440803	6443844	6446884	6449924	6452963	6456002	6459040	−0
2.71 1.	6462078	6465115	6468151	6471187	6474222	6477257	6480291	6483325	6486358	6489391	−0
2.72 1.	6492422	6495454	6498485	6501515	6504545	6507574	6510603	6513631	6516658	6519685	−0
2.73 1.	6522712	6525737	6528763	6531788	6534812	6537835	6540858	6543881	6546903	6549924	−0
2.74 1.	6552945	6555966	6558985	6562005	6565023	6568042	6571059	6574076	6577093	6580109	−0
2.75 1.	6583124	6586139	6589153	6592167	6595180	6598193	6601205	6604216	6607227	6610238	−0
2.76 1.	6613248	6616257	6619266	6622274	6625282	6628289	6631296	6634302	6637307	6640312	−0
2.77 1.	6643317	6646321	6649324	6652327	6655329	6658331	6661332	6664333	6667333	6670333	−0
2.78 1.	6673332	6676330	6679328	6682326	6685323	6688319	6691315	6694310	6697305	6700299	−0
2.79 1.	6703293	6706286	6709279	6712271	6715262	6718253	6721244	6724234	6727223	6730212	−0
2.80 1.	6733200	6736188	6739176	6742162	6745149	6748134	6751119	6754104	6757088	6760072	−0
2.81 1.	6763055	6766037	6769019	6772000	6774981	6777962	6780942	6783921	6786900	6789878	−0
2.82 1.	6792856	6795833	6798809	6801786	6804761	6807736	6810711	6813685	6816658	6819631	−0
2.83 1.	6822604	6825576	6828547	6831518	6834488	6837458	6840428	6843396	6846365	6849332	−0
2.84 1.	6852299	6855266	6858232	6861198	6864163	6867128	6870092	6873055	6876018	6878981	−0
2.85 1.	6881943	6884904	6887865	6890826	6893786	6896745	6899704	6902662	6905620	6908578	−0
2.86 1.	6911535	6914491	6917447	6920402	6923357	6926311	6929265	6932218	6935170	6938123	−0
2.87 1.	6941074	6944025	6946976	6949926	6952876	6955825	6958774	6961722	6964669	6967616	−0
2.88 1.	6970563	6973509	6976454	6979399	6982344	6985288	6988231	6991174	6994117	6997059	−0
2.89 1.	7000000	7002941	7005881	7008821	7011761	7014700	7017638	7020576	7023513	7026450	−0
2.90 1.	7029386	7032322	7035258	7038192	7041127	7044061	7046994	7049927	7052859	7055791	−0
2.91 1.	7058722	7061653	7064583	7067513	7070442	7073371	7076299	7079227	7082154	7085081	−0
2.92 1.	7088007	7090933	7093858	7096783	7099708	7102631	7105555	7108477	7111400	7114321	0
2.93 1.	7117243	7120164	7123084	7126004	7128923	7131842	7134760	7137678	7140595	7143512	0
2.94 1.	7146428	7149344	7152259	7155174	7158088	7161002	7163916	7166828	7169741	7172653	0
2.95 1.	7175564	7178475	7181385	7184295	7187205	7190113	7193022	7195930	7198837	7201744	0
2.96 1.	7204650	7207556	7210462	7213367	7216271	7219175	7222079	7224982	7227884	7230786	0
2.97 1.	7233688	7236589	7239489	7242390	7245289	7248188	7251087	7253985	7256883	7259780	0
2.98 1.	7262676	7265573	7268468	7271364	7274258	7277153	7280046	7282940	7285832	7288725	0
2.99 1.	7291616	7294508	7297399	7300289	7303179	7306068	7308957	7311846	7314734	7317621	0

TABLE 4

Square Roots

X		.0000	.0001	.0002	.0003	.0004	.0005	.0006	.0007	.0008	.0009	Δ_2
3.00	1.	7320508	7323395	7326281	7329166	7332051	7334936	7337820	7340704	7343587	7346469	0
3.01	1.	7349352	7352233	7355115	7357995	7360876	7363755	7366635	7369514	7372392	7375270	0
3.02	1.	7378147	7381024	7383901	7386777	7389652	7392527	7395402	7398276	7401149	7404023	0
3.03	1.	7406895	7409767	7412639	7415510	7418381	7421251	7424121	7426990	7429859	7432728	0
3.04	1.	7435596	7438463	7441330	7444197	7447063	7449928	7452793	7455658	7458522	7461386	0
3.05	1.	7464249	7467112	7469974	7472836	7475697	7478558	7481419	7484279	7487138	7489997	0
3.06	1.	7492856	7495714	7498571	7501428	7504285	7507141	7509997	7512852	7515707	7518562	0
3.07	1.	7521415	7524269	7527122	7529974	7532826	7535678	7538529	7541380	7544230	7547079	0
3.08	1.	7549929	7552778	7555626	7558474	7561321	7564168	7567014	7569861	7572706	7575551	0
3.09	1.	7578396	7581240	7584084	7586927	7589770	7592612	7595454	7598295	7601136	7603977	0
3.10	1.	7606817	7609656	7612495	7615334	7618172	7621010	7623847	7626684	7629521	7632357	0
3.11	1.	7635192	7638027	7640862	7643696	7646529	7649363	7652195	7655028	7657859	7660691	0
3.12	1.	7663522	7666352	7669182	7672012	7674841	7677670	7680498	7683325	7686153	7688980	0
3.13	1.	7691806	7694632	7697457	7700282	7703107	7705931	7708755	7711578	7714401	7717223	0
3.14	1.	7720045	7722867	7725688	7728508	7731328	7734148	7736967	7739786	7742604	7745422	0
3.15	1.	7748239	7751056	7753873	7756689	7759504	7762320	7765134	7767949	7770762	7773576	0
3.16	1.	7776389	7779201	7782013	7784825	7787636	7790447	7793257	7796067	7798876	7801685	0
3.17	1.	7804494	7807302	7810109	7812917	7815723	7818530	7821335	7824141	7826946	7829750	0
3.18	1.	7832554	7835358	7838161	7840964	7843766	7846568	7849370	7852171	7854971	7857771	0
3.19	1.	7860571	7863370	7866169	7868968	7871765	7874563	7877360	7880157	7882953	7885748	0
3.20	1.	7888544	7891339	7894133	7896927	7899721	7902514	7905306	7908099	7910891	7913682	0
3.21	1.	7916473	7919263	7922053	7924843	7927632	7930421	7933209	7935997	7938785	7941572	0
3.22	1.	7944358	7947145	7949930	7952716	7955500	7958285	7961069	7963853	7966636	7969418	0
3.23	1.	7972201	7974983	7977764	7980545	7983326	7986106	7988885	7991665	7994444	7997222	0
3.24	1.	8000000	8002778	8005555	8008331	8011108	8013884	8016659	8019434	8022209	8024983	0
3.25	1.	8027756	8030530	8033303	8036075	8038847	8041619	8044390	8047160	8049931	8052701	0
3.26	1.	8055470	8058239	8061008	8063776	8066544	8069311	8072078	8074844	8077610	8080376	0
3.27	1.	8083141	8085906	8088670	8091434	8094198	8096961	8099724	8102486	8105248	8108009	0
3.28	1.	8110770	8113531	8116291	8119051	8121810	8124569	8127327	8130085	8132843	8135600	0
3.29	1.	8138357	8141114	8143869	8146625	8149380	8152135	8154889	8157643	8160396	8163150	0
3.30	1.	8165902	8168654	8171406	8174157	8176908	8179659	8182409	8185159	8187908	8190657	0
3.31	1.	8193405	8196153	8198901	8201648	8204395	8207141	8209887	8212633	8215378	8218123	0
3.32	1.	8220867	8223611	8226354	8229098	8231840	8234582	8237324	8240066	8242807	8245547	0
3.33	1.	8248288	8251027	8253767	8256506	8259244	8261982	8264720	8267457	8270194	8272931	0
3.34	1.	8275667	8278403	8281138	8283873	8286607	8289341	8292075	8294808	8297541	8300273	0
3.35	1.	8303005	8305737	3308468	8311199	8313929	8316659	8319389	8322118	8324846	8327575	0
3.36	1.	8330303	8333030	8335757	8338484	8341210	8343936	8346662	8349387	8352112	8354836	0
3.37	1.	8357560	8360283	8363006	8365729	8368451	8371173	8373894	8376616	8379336	8382056	0
3.38	1.	8384776	8387496	8390215	8392933	8395652	8398369	8401087	8403804	8406521	8409237	0
3.39	1.	8411953	8414668	8417383	8420098	8422812	8425526	8428239	8430952	8433665	8436377	0
3.40	1.	8439089	8441800	8444511	8447222	8449932	8452642	8455352	8458061	8460769	8463477	0
3.41	1.	8466185	8468893	8471600	8474306	8477013	8479719	8482424	8485129	8487834	8490538	0
3.42	1.	8493242	8495945	8498649	8501351	8504054	8506755	8509457	8512158	8514859	8517559	0
3.43	1.	8520259	8522959	8525658	8528357	8531055	8533753	8536451	8539148	8541845	8544541	0
3.44	1.	8547237	8549933	8552628	8555323	8558017	8560711	8563405	8566098	8568791	8571483	0
3.45	1.	8574176	8576867	8579559	8582250	8584940	8587630	8590320	8593009	8595698	8598387	0
3.46	1.	8601075	8603763	8606450	8609138	8611824	8614510	8617196	8619882	8622567	8625252	0
3.47	1.	8627936	8630620	8633303	8635987	8638669	8641352	8644034	8646716	8649397	8652078	0
3.48	1.	8654758	8657438	8660118	8662797	8665476	8668155	8670833	8673511	8676188	8678865	0
3.49	1.	8681542	8684218	8686894	8689569	8692244	8694919	8697593	8700267	8702941	8705614	0

TABLE 4

Square Roots

X		.0000	.0001	.0002	.0003	.0004	.0005	.0006	.0007	.0008	.0009	Δ_2
3.50	1.	8708287	8710959	8713631	8716303	8718974	8721645	8724316	8726986	8729656	8732325	0
3.51	1.	8734994	8737663	8740331	8742999	8745666	8748333	8751000	8753666	8756332	8758998	0
3.52	1.	8761663	8764328	8766992	8769656	8772320	8774983	8777646	8780309	8782971	8785633	0
3.53	1.	8788294	8790955	8793616	8796276	8798936	8801596	8804255	8806914	8809572	8812230	0
3.54	1.	8814888	8817545	8820202	8822858	8825515	8828170	8830826	8833481	8836135	8838790	0
3.55	1.	8841444	8844097	8846750	8849403	8852056	8854708	8857359	8860011	8862661	8865312	0
3.56	1.	8867962	8870612	8873261	8875911	8878559	8881208	8883855	8886503	8889150	8891797	0
3.57	1.	8894444	8897090	8899735	8902381	8905026	8907670	8910315	8912959	8915602	8918245	0
3.58	1.	8920888	8923530	8926172	8928814	8931455	8934096	8936737	8939377	8942017	8944656	0
3.59	1.	8947295	8949934	8952572	8955210	8957848	8960485	8963122	8965759	8968395	8971030	0
3.60	1.	8973666	8976301	8978936	8981570	8984204	8986837	8989471	8992104	8994736	8997368	0
3.61	1.	9000000	9002631	9005262	9007893	9010523	9013153	9015783	9018412	9021041	9023669	0
3.62	1.	9026298	9028925	9031553	9034180	9036806	9039433	9042059	9044684	9047309	9049934	0
3.63	1.	9052559	9055183	9057807	9060430	9063053	9065676	9068298	9070920	9073542	9076163	0
3.64	1.	9078784	9081405	9084025	9086645	9089264	9091883	9094502	9097120	9099738	9102356	0
3.65	1.	9104973	9107590	9110207	9112823	9115439	9118054	9120669	9123284	9125899	9128513	0
3.66	1.	9131126	9133740	9136353	9138965	9141578	9144190	9146801	9149413	9152023	9154634	0
3.67	1.	9157244	9159854	9162463	9165072	9167681	9170289	9172897	9175505	9178113	9180719	0
3.68	1.	9183326	9185932	9188538	9191144	9193749	9196354	9198958	9201562	9204166	9206770	0
3.69	1.	9209373	9211975	9214578	9217180	9219781	9222383	9224984	9227584	9230185	9232785	0
3.70	1.	9235384	9237983	9240582	9243181	9245779	9248376	9250974	9253571	9256168	9258764	0
3.71	1.	9261360	9263956	9266551	9269146	9271741	9274335	9276929	9279523	9282116	9284709	0
3.72	1.	9287301	9289894	9292486	9295077	9297668	9300259	9302849	9305440	9308029	9310619	0
3.73	1.	9313208	9315797	9318385	9320973	9323561	9326148	9328735	9331322	9333908	9336494	0
3.74	1.	9339080	9341665	9344250	9346834	9349419	9352002	9354586	9357169	9359752	9362335	0
3.75	1.	9364917	9367499	9370080	9372661	9375242	9377822	9380402	9382982	9385562	9388141	0
3.76	1.	9390719	9393298	9395876	9398454	9401031	9403608	9406185	9408761	9411337	9413913	0
3.77	1.	9416488	9419063	9421637	9424212	9426786	9429359	9431932	9434505	9437078	9439650	0
3.78	1.	9442222	9444794	9447365	9449936	9452506	9455076	9457646	9460216	9462785	9465354	0
3.79	1.	9467922	9470490	9473058	9475626	9478193	9480760	9483326	9485892	9488458	9491024	0
3.80	1.	9493589	9496153	9498718	9501282	9503846	9506409	9508972	9511535	9514097	9516660	0
3.81	1.	9519221	9521783	9524344	9526904	9529465	9532025	9534585	9537144	9539703	9542262	0
3.82	1.	9544820	9547378	9549936	9552493	9555050	9557607	9560164	9562720	9565275	9567831	0
3.83	1.	9570386	9572940	9575495	9578049	9580603	9583156	9585709	9588262	9590814	9593366	0
3.84	1.	9595918	9598469	9601020	9603571	9606121	9608672	9611221	9613771	9616320	9618868	0
3.85	1.	9621417	9623965	9626513	9629060	9631607	9634154	9636700	9639246	9641792	9644338	0
3.86	1.	9646883	9649427	9651972	9654516	9657060	9659603	9662146	9664689	9667232	9669774	0
3.87	1.	9672316	9674857	9677398	9679939	9682480	9685020	9687559	9690099	9692638	9695177	0
3.88	1.	9697716	9700254	9702792	9705329	9707866	9710403	9712940	9715476	9718012	9720548	0
3.89	1.	9723083	9725618	9728152	9730687	9733221	9735754	9738288	9740821	9743353	9745886	0
3.90	1.	9748418	9750949	9753481	9756012	9758542	9761073	9763603	9766133	9768662	9771191	0
3.91	1.	9773720	9776248	9778776	9781304	9783832	9786359	9788886	9791412	9793938	9796464	0
3.92	1.	9798990	9801515	9804040	9806565	9809089	9811613	9814136	9816660	9819183	9821705	0
3.93	1.	9824228	9826750	9829271	9831793	9834314	9836834	9839355	9841875	9844395	9846914	0
3.94	1.	9849433	9851952	9854470	9856989	9859506	9862024	9864541	9867058	9869575	9872091	0
3.95	1.	9874607	9877122	9879638	9882153	9884667	9887182	9889696	9892209	9894723	9897236	0
3.96	1.	9899749	9902261	9904773	9907285	9909797	9912308	9914819	9917329	9919839	9922349	0
3.97	1.	9924859	9927368	9929877	9932386	9934894	9937402	9939910	9942417	9944924	9947431	0
3.98	1.	9949937	9952443	9954949	9957455	9959960	9962465	9964969	9967474	9969977	9972481	0
3.99	1.	9974984	9977487	9979990	9982492	9984994	9987496	9989997	9992499	9994999	9997500	0

TABLE 4

Square Roots

X	.0000	.0001	.0002	.0003	.0004	.0005	.0006	.0007	.0008	.0009	Δ_2
4.00 2.	0000000	0002500	0004999	0007499	0009998	0012496	0014994	0017492	0019990	0022487	0
4.01 2.	0024984	0027481	0029978	0032474	0034969	0037465	0039960	0042455	0044949	0047444	0
4.02 2.	0049938	0052431	0054925	0057418	0059910	0062403	0064895	0067386	0069878	0072369	0
4.03 2.	0074860	0077350	0079841	0082331	0084820	0087309	0089798	0092287	0094775	0097263	0
4.04 2.	0099751	0102239	0104726	0107213	0109699	0112185	0114671	0117157	0119642	0122127	0
4.05 2.	0124612	0127096	0129580	0132064	0134547	0137031	0139513	0141996	0144478	0146960	0
4.06 2.	0149442	0151923	0154404	0156885	0159365	0161845	0164325	0166804	0169284	0171762	0
4.07 2.	0174241	0176719	0179197	0181675	0184152	0186629	0189106	0191582	0194059	0196534	0
4.08 2.	0199010	0201485	0203960	0206435	0208909	0211383	0213857	0216330	0218803	0221276	0
4.09 2.	0223748	0226221	0228692	0231164	0233635	0236106	0238577	0241047	0243517	0245987	0
4.10 2.	0248457	0250926	0253395	0255863	0258332	0260800	0263267	0265735	0268202	0270668	0
4.11 2.	0273135	0275601	0278067	0280533	0282998	0285463	0287927	0290392	0292856	0295320	0
4.12 2.	0297783	0300246	0302709	0305172	0307634	0310096	0312558	0315019	0317480	0319941	0
4.13 2.	0322401	0324862	0327322	0329781	0332240	0334699	0337158	0339617	0342075	0344532	0
4.14 2.	0346990	0349447	0351904	0354361	0356817	0359273	0361729	0364184	0366639	0369094	0
4.15 2.	0371549	0374003	0376457	0378911	0381364	0383817	0386270	0388722	0391175	0393626	0
4.16 2.	0396078	0398529	0400980	0403431	0405882	0408332	0410781	0413231	0415680	0418129	0
4.17 2.	0420578	0423026	0425474	0427922	0430370	0432817	0435264	0437710	0440157	0442603	0
4.18 2.	0445048	0447494	0449939	0452384	0454828	0457273	0459717	0462160	0464604	0467047	0
4.19 2.	0469489	0471932	0474374	0476816	0479258	0481699	0484140	0486581	0489021	0491462	0
4.20 2.	0493902	0496341	0498780	0501219	0503658	0506097	0508535	0510973	0513410	0515848	0
4.21 2.	0518285	0520721	0523158	0525594	0528030	0530465	0532900	0535335	0537770	0540204	0
4.22 2.	0542639	0545072	0547506	0549939	0552372	0554805	0557237	0559669	0562101	0564533	0
4.23 2.	0566964	0569395	0571825	0574256	0576686	0579116	0581545	0583974	0586403	0588832	0
4.24 2.	0591260	0593688	0596116	0598544	0600971	0603398	0605824	0608251	0610677	0613103	0
4.25 2.	0615528	0617953	0620378	0622803	0625227	0627651	0630075	0632499	0634922	0637345	0
4.26 2.	0639767	0642190	0644612	0647034	0649455	0651876	0654297	0656718	0659138	0661558	0
4.27 2.	0663978	0666398	0668817	0671236	0673655	0676073	0678491	0680909	0683327	0685744	0
4.28 2.	0688161	0690578	0692994	0695410	0697826	0700242	0702657	0705072	0707487	0709901	0
4.29 2.	0712315	0714729	0717143	0719556	0721969	0724382	0726794	0729206	0731618	0734030	0
4.30 2.	0736441	0738852	0741263	0743674	0746084	0748494	0750904	0753313	0755722	0758131	0
4.31 2.	0760539	0762948	0765356	0767763	0770171	0772578	0774985	0777392	0779798	0782204	0
4.32 2.	0784610	0787015	0789420	0791825	0794230	0796634	0799038	0801442	0803846	0806249	0
4.33 2.	0808652	0811055	0813457	0815859	0818261	0820663	0823064	0825465	0827866	0830266	0
4.34 2.	0832667	0835067	0837466	0839866	0842265	0844664	0847062	0849460	0851858	0854256	0
4.35 2.	0856654	0859051	0861448	0863844	0866241	0868637	0871033	0873428	0875823	0878218	0
4.36 2.	0880613	0883007	0885402	0887795	0890189	0892582	0894975	0897368	0899761	0902153	0
4.37 2.	0904545	0906937	0909328	0911719	0914110	0916501	0918891	0921281	0923671	0926060	0
4.38 2.	0928450	0930838	0933227	0935616	0938004	0940392	0942779	0945166	0947554	0949940	0
4.39 2.	0952327	0954713	0957099	0959485	0961870	0964255	0966640	0969025	0971409	0973793	0
4.40 2.	0976177	0978560	0980944	0983327	0985709	0988092	0990474	0992856	0995238	0997619	0
4.41 2.	1000000	1002381	1004761	1007142	1009522	1011901	1014281	1016660	1019039	1021418	0
4.42 2.	1023796	1026174	1028552	1030930	1033307	1035684	1038061	1040437	1042813	1045189	0
4.43 2.	1047565	1049941	1052316	1054691	1057065	1059440	1061814	1064188	1066561	1068934	0
4.44 2.	1071308	1073680	1076053	1078425	1080797	1083169	1085540	1087911	1090282	1092653	0
4.45 2.	1095023	1097393	1099763	1102133	1104502	1106871	1109240	1111608	1113976	1116344	0
4.46 2.	1118712	1121080	1123447	1125814	1128180	1130547	1132913	1135279	1137644	1140009	0
4.47 2.	1142375	1144739	1147104	1149468	1151832	1154196	1156559	1158922	1161285	1163648	0
4.48 2.	1166010	1168373	1170735	1173096	1175457	1177819	1180179	1182540	1184900	1187260	0
4.49 2.	1189620	1191980	1194339	1196698	1199057	1201415	1203773	1206131	1208489	1210846	0

TABLE 4

Square Roots

X	.0000	.0001	.0002	.0003	.0004	.0005	.0006	.0007	.0008	.0009	Δ,
4.50 2.	1213203	1215560	1217917	1220273	1222629	1224985	1227341	1229696	1232051	1234406	0
4.51 2.	1236761	1239115	1241469	1243823	1246176	1248529	1250882	1253235	1255587	1257940	0
4.52 2.	1260292	1262643	1264995	1267346	1269697	1272047	1274398	1276748	1279098	1281447	0
4.53 2.	1283797	1286146	1288495	1290843	1293191	1295539	1297887	1300235	1302582	1304929	0
4.54 2.	1307276	1309622	1311968	1314314	1316660	1319006	1321351	1323696	1326040	1328385	0
4.55 2.	1330729	1333073	1335417	1337760	1340103	1342446	1344789	1347131	1349473	1351815	0
4.56 2.	1354156	1356498	1358839	1361180	1363520	1365861	1368201	1370540	1372880	1375219	0
4.57 2.	1377558	1379897	1382236	1384574	1386912	1389250	1391587	1393924	1396261	1398598	0
4.58 2.	1400935	1403271	1405607	1407942	1410278	1412613	1414948	1417283	1419617	1421951	0
4.59 2.	1424285	1426619	1428952	1431286	1433618	1435951	1438284	1440616	1442948	1445279	0
4.60 2.	1447611	1449942	1452273	1454603	1456934	1459264	1461594	1463923	1466253	1468582	0
4.61 2.	1470911	1473239	1475568	1477896	1480223	1482551	1484878	1487205	1489532	1491859	0
4.62 2.	1494185	1496511	1498837	1501163	1503488	1505813	1508138	1510463	1512787	1515111	0
4.63 2.	1517435	1519758	1522082	1524405	1526728	1529050	1531372	1533695	1536016	1538338	0
4.64 2.	1540659	1542980	1545301	1547622	1549942	1552262	1554582	1556901	1559221	1561540	0
4.65 2.	1563859	1566177	1568496	1570814	1573131	1575449	1577766	1580083	1582400	1584717	0
4.66 2.	1587033	1589349	1591665	1593981	1596296	1598611	1600926	1603240	1605555	1607869	0
4.67 2.	1610183	1612496	1614810	1617123	1619436	1621748	1624061	1626373	1628685	1630996	0
4.68 2.	1633308	1635619	1637930	1640240	1642551	1644861	1647171	1649480	1651790	1654099	0
4.69 2.	1656408	1658716	1661025	1663333	1665641	1667949	1670256	1672563	1674870	1677177	0
4.70 2.	1679483	1681790	1684096	1686401	1688707	1691012	1693317	1695622	1697926	1700230	0
4.71 2.	1702534	1704838	1707142	1709445	1711748	1714051	1716353	1718656	1720958	1723259	0
4.72 2.	1725561	1727862	1730163	1732464	1734765	1737065	1739365	1741665	1743965	1746264	0
4.73 2.	1748563	1750862	1753161	1755459	1757757	1760055	1762353	1764650	1766947	1769244	0
4.74 2.	1771541	1773838	1776134	1778430	1780725	1783021	1785316	1787611	1789906	1792200	0
4.75 2.	1794495	1796789	1799083	1801376	1803669	1805962	1808255	1810548	1812840	1815132	0
4.76 2.	1817424	1819716	1822007	1824298	1826589	1828880	1831170	1833461	1835750	1838040	0
4.77 2.	1840330	1842619	1844908	1847197	1849485	1851773	1854061	1856349	1858637	1860924	0
4.78 2.	1863211	1865498	1867785	1870071	1872357	1874643	1876928	1879214	1881499	1883784	0
4.79 2.	1886069	1888353	1890637	1892921	1895205	1897488	1899772	1902055	1904337	1906620	0
4.80 2.	1908902	1911184	1913466	1915748	1918029	1920310	1922591	1924872	1927152	1929432	0
4.81 2.	1931712	1933992	1936271	1938551	1940830	1943108	1945387	1947665	1949943	1952221	0
4.82 2.	1954498	1956776	1959053	1961330	1963606	1965883	1968159	1970435	1972710	1974986	0
4.83 2.	1977261	1979536	1981811	1984085	1986359	1988633	1990907	1993181	1995454	1997727	0
4.84 2.	2000000	2002273	2004545	2006817	2009089	2011361	2013632	2015903	2018174	2020445	0
4.85 2.	2022716	2024986	2027256	2029526	2031795	2034065	2036334	2038602	2040871	2043140	0
4.86 2.	2045408	2047676	2049943	2052211	2054478	2056745	2059012	2061278	2063545	2065811	0
4.87 2.	2068076	2070342	2072607	2074873	2077137	2079402	2081667	2083931	2086195	2088459	0
4.88 2.	2090722	2092985	2095248	2097511	2099774	2102036	2104298	2106560	2108822	2111083	0
4.89 2.	2113344	2115605	2117866	2120127	2122387	2124647	2126907	2129166	2131426	2133685	0
4.90 2.	2135944	2138202	2140461	2142719	2144977	2147235	2149492	2151749	2154006	2156263	0
4.91 2.	2158520	2160776	2163032	2165288	2167544	2169799	2172054	2174309	2176564	2178819	0
4.92 2.	2181073	2183327	2185581	2187834	2190088	2192341	2194594	2196847	2199099	2201351	0
4.93 2.	2203603	2205855	2208107	2210358	2212609	2214860	2217111	2219361	2221611	2223861	0
4.94 2.	2226111	2228360	2230610	2232859	2235107	2237356	2239604	2241852	2244100	2246348	0
4.95 2.	2248595	2250843	2253090	2255336	2257583	2259829	2262075	2264321	2266567	2268812	0
4.96 2.	2271057	2273302	2275547	2277792	2280036	2282280	2284524	2286767	2289011	2291254	0
4.97 2.	2293497	2295740	2297982	2300224	2302466	2304708	2306950	2309191	2311432	2313673	0
4.98 2.	2315914	2318154	2320394	2322634	2324874	2327114	2329353	2331592	2333831	2336069	0
4.99 2.	2338308	2340546	2342784	2345022	2347259	2349497	2351734	2353971	2356207	2358444	0

TABLE 4

Square Roots

X	.0000	.0001	.0002	.0003	.0004	.0005	.0006	.0007	.0008	.0009	Δ_2
5.00 2.	2360680	2362916	2365151	2367387	2369622	2371857	2374092	2376327	2378561	2380795	0
5.01 2.	2383029	2385263	2387496	2389730	2391963	2394196	2396428	2398661	2400893	2403125	0
5.02 2.	2405356	2407588	2409819	2412050	2414281	2416512	2418742	2420972	2423202	2425432	0
5.03 2.	2427661	2429891	2432120	2434349	2436577	2438806	2441034	2443262	2445490	2447717	0
5.04 2.	2449944	2452171	2454398	2456625	2458851	2461077	2463303	2465529	2467755	2469980	0
5.05 2.	2472205	2474430	2476655	2478879	2481103	2483327	2485551	2487774	2489998	2492221	0
5.06 2.	2494444	2496666	2498889	2501111	2503333	2505555	2507776	2509998	2512219	2514440	0
5.07 2.	2516660	2518881	2521101	2523321	2525541	2527761	2529980	2532199	2534418	2536637	0
5.08 2.	2538855	2541074	2543292	2545510	2547727	2549945	2552162	2554379	2556595	2558812	0
5.09 2.	2561028	2563244	2565460	2567676	2569891	2572107	2574322	2576536	2578751	2580965	0
5.10 2.	2583180	2585393	2587607	2589821	2592034	2594247	2596460	2598673	2600885	2603097	0
5.11 2.	2605309	2607521	2609732	2611944	2614155	2616366	2618576	2620787	2622997	2625207	0
5.12 2.	2627417	2629627	2631836	2634045	2636254	2638463	2640671	2642880	2645088	2647296	0
5.13 2.	2649503	2651711	2653918	2656125	2658332	2660538	2662745	2664951	2667157	2669363	0
5.14 2.	2671568	2673773	2675978	2678183	2680388	2682592	2684797	2687001	2689204	2691408	0
5.15 2.	2693611	2695815	2698018	2700220	2702423	2704625	2706827	2709029	2711231	2713432	0
5.16 2.	2715633	2717834	2720035	2722236	2724436	2726636	2728836	2731036	2733236	2735435	0
5.17 2.	2737634	2739833	2742032	2744230	2746428	2748626	2750824	2753022	2755219	2757416	0
5.18 2.	2759613	2761810	2764007	2766203	2768399	2770595	2772791	2774986	2777182	2779377	0
5.19 2.	2781571	2783766	2785961	2788155	2790349	2792543	2794736	2796930	2799123	2801316	0
5.20 2.	2803508	2805701	2807893	2810085	2812277	2814469	2816661	2818852	2821043	2823234	0
5.21 2.	2825424	2827615	2829805	2831995	2834185	2836374	2838564	2840753	2842942	2845131	0
5.22 2.	2847319	2849508	2851696	2853884	2856071	2858259	2860446	2862633	2864820	2867007	0
5.23 2.	2869193	2871379	2873566	2875751	2877937	2880122	2882308	2884493	2886677	2888862	0
5.24 2.	2891046	2893230	2895414	2897598	2899782	2901965	2904148	2906331	2908514	2910696	0
5.25 2.	2912878	2915061	2917242	2919424	2921606	2923787	2925968	2928149	2930329	2932510	0
5.26 2.	2934690	2936870	2939050	2941229	2943409	2945588	2947767	2949946	2952124	2954302	0
5.27 2.	2956481	2958658	2960836	2963014	2965191	2967368	2969545	2971722	2973898	2976074	0
5.28 2.	2978251	2980426	2982602	2984778	2986953	2989128	2991303	2993477	2995652	2997826	0
5.29 2.	3000000	3002174	3004347	3006521	3008694	3010867	3013040	3015212	3017385	3019557	0
5.30 2.	3021729	3023901	3026072	3028244	3030415	3032586	3034756	3036927	3039097	3041267	0
5.31 2.	3043437	3045607	3047776	3049946	3052115	3054284	3056452	3058621	3060789	3062957	0
5.32 2.	3065125	3067293	3069460	3071628	3073795	3075962	3078128	3080295	3082461	3084627	0
5.33 2.	3086793	3088958	3091124	3093289	3095454	3097619	3099784	3101948	3104112	3106276	0
5.34 2.	3108440	3110604	3112767	3114930	3117093	3119256	3121419	3123581	3125743	3127905	0
5.35 2.	3130067	3132229	3134390	3136551	3138712	3140873	3143033	3145194	3147354	3149514	0
5.36 2.	3151674	3153833	3155993	3158152	3160311	3162470	3164628	3166787	3168945	3171103	0
5.37 2.	3173260	3175418	3177575	3179733	3181889	3184046	3186203	3188359	3190515	3192671	0
5.38 2.	3194827	3196983	3199138	3201293	3203448	3205603	3207757	3209912	3212066	3214220	0
5.39 2.	3216374	3218527	3220680	3222834	3224987	3227139	3229292	3231444	3233596	3235748	0
5.40 2.	3237900	3240052	3242203	3244354	3246505	3248656	3250806	3252957	3255107	3257257	0
5.41 2.	3259407	3261556	3263706	3265855	3268004	3270153	3272301	3274449	3276598	3278746	0
5.42 2.	3280893	3283041	3285188	3287336	3289483	3291629	3293776	3295922	3298069	3300215	0
5.43 2.	3302360	3304506	3306651	3308797	3310942	3313086	3315231	3317375	3319520	3321664	0
5.44 2.	3323808	3325951	3328095	3330238	3332381	3334524	3336666	3338809	3340951	3343093	0
5.45 2.	3345235	3347377	3349518	3351659	3353801	3355941	3358082	3360223	3362363	3364503	0
5.46 2.	3366643	3368783	3370922	3373061	3375201	3377339	3379478	3381617	3383755	3385893	0
5.47 2.	3388031	3390169	3392306	3394444	3396581	3398718	3400855	3402991	3405128	3407264	0
5.48 2.	3409400	3411536	3413671	3415807	3417942	3420077	3422212	3424346	3426481	3428615	0
5.49 2.	3430749	3432883	3435016	3437150	3439283	3441416	3443549	3445682	3447814	3449947	0

TABLE 4

Square Roots

X	.0000	.0001	.0002	.0003	.0004	.0005	.0006	.0007	.0008	.0009	Δ_2
5.50 2.	3452079	3454211	3456342	3458474	3460605	3462736	3464867	3466998	3469129	3471259	0
5.51 2.	3473389	3475519	3477649	3479779	3481908	3484037	3486166	3488295	3490424	3492552	0
5.52 2.	3494680	3496808	3498936	3501064	3503191	3505319	3507446	3509572	3511699	3513826	0
5.53 2.	3515952	3518078	3520204	3522330	3524455	3526581	3528706	3530831	3532956	3535080	0
5.54 2.	3537205	3539329	3541453	3543577	3545700	3547824	3549947	3552070	3554193	3556315	0
5.55 2.	3558438	3560560	3562682	3564804	3566926	3569047	3571169	3573290	3575411	3577532	0
5.56 2.	3579652	3581773	3583893	3586013	3588133	3590252	3592372	3594491	3596610	3598729	0
5.57 2.	3600847	3602966	3605084	3607202	3609320	3611438	3613555	3615673	3617790	3619907	0
5.58 2.	3622024	3624140	3626257	3628373	3630489	3632605	3634720	3636836	3638951	3641066	0
5.59 2.	3643181	3645296	3647410	3649524	3651638	3653752	3655866	3657980	3660093	3662206	0
5.60 2.	3664319	3666432	3668545	3670657	3672769	3674881	3676993	3679105	3681216	3683327	0
5.61 2.	3685439	3687549	3689660	3691771	3693881	3695991	3698101	3700211	3702321	3704430	0
5.62 2.	3706539	3708648	3710757	3712866	3714974	3717082	3719191	3721298	3723406	3725514	0
5.63 2.	3727621	3729728	3731835	3733942	3736049	3738155	3740261	3742367	3744473	3746579	0
5.64 2.	3748684	3750789	3752895	3754999	3757104	3759209	3761313	3763417	3765521	3767625	0
5.65 2.	3769729	3771832	3773935	3776038	3778141	3780244	3782346	3784449	3786551	3788653	0
5.66 2.	3790754	3792856	3794957	3797059	3799160	3801260	3803361	3805462	3807562	3809662	0
5.67 2.	3811762	3813861	3815961	3818060	3820160	3822258	3824357	3826456	3828554	3830652	0
5.68 2.	3832751	3834848	3836946	3839044	3841141	3843238	3845335	3847432	3849528	3851625	0
5.69 2.	3853721	3855817	3857913	3860008	3862104	3864199	3866294	3868389	3870484	3872578	0
5.70 2.	3874673	3876767	3878861	3880955	3883048	3885142	3887235	3889328	3891421	3893514	0
5.71 2.	3895606	3897699	3899791	3901883	3903975	3906066	3908158	3910249	3912340	3914431	0
5.72 2.	3916521	3918612	3920702	3922792	3924882	3926972	3929062	3931151	3933240	3935330	0
5.73 2.	3937418	3939507	3941596	3943684	3945772	3947860	3949948	3952035	3954123	3956210	0
5.74 2.	3958297	3960384	3962471	3964557	3966643	3968730	3970816	3972901	3974987	3977072	0
5.75 2.	3979158	3981243	3983328	3985412	3987497	3989581	3991665	3993749	3995833	3997917	0
5.76 2.	4000000	4002083	4004166	4006249	4008332	4010414	4012497	4014579	4016661	4018743	0
5.77 2.	4020824	4022906	4024987	4027068	4029149	4031230	4033310	4035391	4037471	4039551	0
5.78 2.	4041631	4043710	4045790	4047869	4049948	4052027	4054106	4056184	4058263	4060341	0
5.79 2.	4062419	4064497	4066574	4068652	4070729	4072806	4074883	4076960	4079036	4081113	0
5.80 2.	4083189	4085265	4087341	4089417	4091492	4093568	4095643	4097718	4099793	4101867	0
5.81 2.	4103942	4106016	4108090	4110164	4112238	4114311	4116384	4118458	4120531	4122603	0
5.82 2.	4124676	4126749	4128821	4130893	4132965	4135037	4137108	4139180	4141251	4143322	0
5.83 2.	4145393	4147464	4149534	4151604	4153675	4155745	4157814	4159884	4161954	4164023	0
5.84 2.	4166092	4168161	4170230	4172298	4174367	4176435	4178503	4180571	4182638	4184706	0
5.85 2.	4186773	4188840	4190907	4192974	4195041	4197107	4199174	4201240	4203306	4205371	0
5.86 2.	4207437	4209502	4211567	4213632	4215697	4217762	4219827	4221891	4223955	4226019	0
5.87 2.	4228083	4230146	4232210	4234273	4236336	4238399	4240462	4242525	4244587	4246649	0
5.88 2.	4248711	4250773	4252835	4254896	4256958	4259019	4261080	4263141	4265201	4267262	0
5.89 2.	4269322	4271382	4273442	4275502	4277562	4279621	4281680	4283739	4285798	4287857	0
5.90 2.	4289916	4291974	4294032	4296090	4298148	4300206	4302263	4304321	4306378	4308435	0
5.91 2.	4310492	4312548	4314605	4316661	4318717	4320773	4322829	4324884	4326940	4328995	0
5.92 2.	4331050	4333105	4335160	4337214	4339269	4341323	4343377	4345431	4347484	4349538	0
5.93 2.	4351591	4353644	4355697	4357750	4359803	4361855	4363908	4365960	4368012	4370064	0
5.94 2.	4372115	4374167	4376218	4378269	4380320	4382371	4384421	4386472	4388522	4390572	0
5.95 2.	4392622	4394672	4396721	4398770	4400820	4402869	4404918	4406966	4409015	4411063	0
5.96 2.	4413111	4415159	4417207	4419255	4421302	4423349	4425397	4427444	4429490	4431537	0
5.97 2.	4433583	4435630	4437676	4439722	4441768	4443813	4445859	4447904	4449949	4451994	0
5.98 2.	4454039	4456083	4458127	4460172	4462216	4464260	4466303	4468347	4470390	4472433	0
5.99 2.	4474476	4476519	4478562	4480605	4482647	4484689	4486731	4488773	4490815	4492856	0

TABLE 4

Square Roots

X	.0000	.0001	.0002	.0003	.0004	.0005	.0006	.0007	.0008	.0009	Δ_2
6.00 2.	4494897	4496939	4498980	4501020	4503061	4505101	4507142	4509182	4511222	4513262	0
6.01 2.	4515301	4517341	4519380	4521419	4523458	4525497	4527536	4529574	4531612	4533650	0
6.02 2.	4535688	4537726	4539764	4541801	4543838	4545875	4547912	4549949	4551986	4554022	0
6.03 2.	4556058	4558094	4560130	4562166	4564202	4566237	4568272	4570307	4572342	4574377	0
6.04 2.	4576411	4578446	4580480	4582514	4584548	4586582	4588615	4590649	4592682	4594715	0
6.05 2.	4596748	4598780	4600813	4602845	4604878	4606910	4608941	4610973	4613005	4615036	0
6.06 2.	4617067	4619098	4621129	4623160	4625190	4627221	4629251	4631281	4633311	4635340	0
6.07 2.	4637370	4639399	4641429	4643458	4645486	4647515	4649544	4651572	4653600	4655628	0
6.08 2.	4657656	4659684	4661711	4663739	4665766	4667793	4669820	4671846	4673873	4675899	0
6.09 2.	4677925	4679951	4681977	4684003	4686028	4688054	4690079	4692104	4694129	4696154	0
6.10 2.	4698178	4700202	4702227	4704251	4706274	4708298	4710322	4712345	4714368	4716391	0
6.11 2.	4718414	4720437	4722459	4724482	4726504	4728526	4730548	4732570	4734591	4736613	0
6.12 2.	4738634	4740655	4742676	4744696	4746717	4748737	4750758	4752778	4754797	4756817	0
6.13 2.	4758837	4760856	4762875	4764895	4766913	4768932	4770951	4772969	4774987	4777005	0
6.14 2.	4779023	4781041	4783059	4785076	4787093	4789110	4791127	4793144	4795161	4797177	0
6.15 2.	4799194	4801210	4803226	4805241	4807257	4809272	4811288	4813303	4815318	4817333	0
6.16 2.	4819347	4821362	4823376	4825390	4827404	4829418	4831432	4833445	4835459	4837472	0
6.17 2.	4839485	4841498	4843510	4845523	4847535	4849547	4851559	4853571	4855583	4857594	0
6.18 2.	4859606	4861617	4863628	4865639	4867650	4869660	4871671	4873681	4875691	4877701	0
6.19 2.	4879711	4881720	4883730	4885739	4887748	4889757	4891766	4893774	4895783	4897791	0
6.20 2.	4899799	4901807	4903815	4905823	4907830	4909837	4911845	4913852	4915858	4917865	0
6.21 2.	4919872	4921878	4923884	4925890	4927896	4929902	4931907	4933913	4935918	4937923	0
6.22 2.	4939928	4941933	4943937	4945942	4947946	4949950	4951954	4953958	4955961	4957965	0
6.23 2.	4959968	4961971	4963974	4965977	4967979	4969982	4971984	4973986	4975988	4977990	0
6.24 2.	4979992	4981994	4983995	4985996	4987997	4989998	4991999	4993999	4996000	4998000	0
6.25 2.	5000000	5002000	5004000	5005999	5007999	5009998	5011997	5013996	5015995	5017994	0
6.26 2.	5019992	5021990	5023988	5025986	5027984	5029982	5031980	5033977	5035974	5037971	0
6.27 2.	5039968	5041965	5043961	5045958	5047954	5049950	5051946	5053942	5055937	5057933	0
6.28 2.	5059928	5061923	5063918	5065913	5067908	5069902	5071897	5073891	5075885	5077879	0
6.29 2.	5079872	5081866	5083859	5085853	5087846	5089839	5091831	5093824	5095816	5097809	0
6.30 2.	5099801	5101793	5103785	5105776	5107768	5109759	5111750	5113741	5115732	5117723	0
6.31 2.	5119713	5121704	5123694	5125684	5127674	5129664	5131653	5133643	5135632	5137621	0
6.32 2.	5139610	5141599	5143588	5145576	5147564	5149553	5151541	5153529	5155516	5157504	0
6.33 2.	5159491	5161478	5163466	5165452	5167439	5169426	5171412	5173399	5175385	5177371	0
6.34 2.	5179357	5181342	5183328	5185313	5187298	5189283	5191268	5193253	5195238	5197222	0
6.35 2.	5199206	5201190	5203174	5205158	5207142	5209125	5211109	5213092	5215075	5217058	0
6.36 2.	5219040	5221023	5223005	5224988	5226970	5228952	5230933	5232915	5234896	5236878	0
6.37 2.	5238859	5240840	5242821	5244801	5246782	5248762	5250743	5252723	5254702	5256682	0
6.38 2.	5258662	5260641	5262621	5264600	5266579	5268558	5270536	5272515	5274493	5276471	0
6.39 2.	5278449	5280427	5282405	5284382	5286360	5288337	5290314	5292291	5294268	5296245	0
6.40 2.	5298221	5300198	5302174	5304150	5306126	5308101	5310077	5312052	5314028	5316003	0
6.41 2.	5317978	5319953	5321927	5323902	5325876	5327850	5329824	5331798	5333772	5335745	0
6.42 2.	5337719	5339692	5341665	5343638	5345611	5347584	5349556	5351529	5353501	5355473	0
6.43 2.	5357445	5359416	5361388	5363359	5365331	5367302	5369273	5371244	5373214	5375185	0
6.44 2.	5377155	5379125	5381095	5383065	5385035	5387005	5388974	5390943	5392912	5394881	0
6.45 2.	5396850	5398819	5400787	5402756	5404724	5406692	5408660	5410628	5412595	5414563	0
6.46 2.	5416530	5418497	5420464	5422431	5424398	5426364	5428331	5430297	5432263	5434229	0
6.47 2.	5436195	5438160	5440126	5442091	5444056	5446021	5447986	5449951	5451915	5453880	0
6.48 2.	5455844	5457808	5459772	5461736	5463700	5465663	5467626	5469590	5471553	5473516	0
6.49 2.	5475478	5477441	5479403	5481366	5483328	5485290	5487252	5489213	5491175	5493136	0

TABLE 4

Square Roots

X	.0000	.0001	.0002	.0003	.0004	.0005	.0006	.0007	.0008	.0009	Δ_2
6.50 2.	5495098	5497059	5499020	5500980	5502941	5504901	5506862	5508822	5510782	5512742	
6.51 2.	5514702	5516661	5518621	5520580	5522539	5524498	5526457	5528416	5530374	5532332	0
6.52 2.	5534291	5536249	5538207	5540164	5542122	5544080	5546037	5547994	5549951	5551908	0
6.53 2.	5553865	5555821	5557778	5559734	5561690	5563646	5565602	5567558	5569513	5571468	0
6.54 2.	5573424	5575379	5577334	5579288	5581243	5583198	5585152	5587106	5589060	5591014	0
6.55 2.	5592968	5594921	5596875	5598828	5600781	5602734	5604687	5606640	5608592	5610545	0
6.56 2.	5612497	5614449	5616401	5618353	5620304	5622256	5624207	5626158	5628109	5630060	0
6.57 2.	5632011	5633962	5635912	5637863	5639813	5641763	5643713	5645662	5647612	5649561	0
6.58 2.	5651511	5653460	5655409	5657358	5659306	5661255	5663203	5665151	5667100	5669047	0
6.59 2.	5670995	5672943	5674890	5676838	5678785	5680732	5682679	5684626	5686572	5688519	0
6.60 2.	5690465	5692411	5694357	5696303	5698249	5700195	5702140	5704085	5706030	5707975	0
6.61 2.	5709920	5711865	5713810	5715754	5717698	5719642	5721586	5723530	5725474	5727417	0
6.62 2.	5729361	5731304	5733247	5735190	5737133	5739075	5741018	5742960	5744902	5746844	0
6.63 2.	5748786	5750728	5752670	5754611	5756553	5758494	5760435	5762376	5764316	5766257	0
6.64 2.	5768197	5770138	5772078	5774018	5775958	5777897	5779837	5781776	5783716	5785655	0
6.65 2.	5787594	5789533	5791471	5793410	5795348	5797287	5799225	5801163	5803101	5805038	0
6.66 2.	5806976	5808913	5810850	5812787	5814724	5816661	5818598	5820534	5822471	5824407	0
6.67 2.	5826343	5828279	5830215	5832151	5834086	5836021	5837957	5839892	5841826	5843761	0
6.68 2.	5845696	5847630	5849565	5851499	5853433	5855367	5857301	5859234	5861168	5863101	0
6.69 2.	5865034	5866967	5868900	5870833	5872766	5874698	5876630	5878563	5880495	5882426	0
6.70 2.	5884358	5886290	5888221	5890153	5892084	5894015	5895946	5897876	5899807	5901737	0
6.71 2.	5903668	5905598	5907528	5909458	5911387	5913317	5915246	5917176	5919105	5921034	0
6.72 2.	5922963	5924891	5926820	5928748	5930677	5932605	5934533	5936461	5938388	5940316	0
6.73 2.	5942243	5944171	5946098	5948025	5949952	5951879	5953805	5955731	5957658	5959584	0
6.74 2.	5961510	5963436	5965362	5967287	5969213	5971138	5973063	5974988	5976913	5978838	0
6.75 2.	5980762	5982687	5984611	5986535	5988459	5990383	5992307	5994230	5996154	5998077	0
6.76 2.	6000000	6001923	6003846	6005769	6007691	6009614	6011536	6013458	6015380	6017302	0
6.77 2.	6019224	6021145	6023067	6024988	6026909	6028830	6030751	6032672	6034592	6036513	0
6.78 2.	6038433	6040353	6042273	6044193	6046113	6048032	6049952	6051871	6053790	6055709	0
6.79 2.	6057628	6059547	6061466	6063384	6065303	6067221	6069139	6071057	6072975	6074892	0
6.80 2.	6076810	6078727	6080644	6082561	6084478	6086395	6088312	6090228	6092144	6094061	0
6.81 2.	6095977	6097893	6099808	6101724	6103640	6105555	6107470	6109385	6111300	6113215	0
6.82 2.	6115130	6117044	6118959	6120873	6122787	6124701	6126615	6128528	6130442	6132355	0
6.83 2.	6134269	6136182	6138095	6140008	6141920	6143833	6145745	6147658	6149570	6151482	0
6.84 2.	6153394	6155305	6157217	6159128	6161040	6162951	6164862	6166773	6168684	6170594	0
6.85 2.	6172505	6174415	6176325	6178235	6180145	6182055	6183965	6185874	6187783	6189693	0
6.86 2.	6191602	6193511	6195419	6197328	6199237	6201145	6203053	6204961	6206869	6208777	0
6.87 2.	6210685	6212592	6214500	6216407	6218314	6220221	6222128	6224035	6225941	6227848	0
6.88 2.	6229754	6231660	6233566	6235472	6237378	6239283	6241189	6243094	6244999	6246905	0
6.89 2.	6248809	6250714	6252619	6254523	6256428	6258332	6260236	6262140	6264044	6265948	0
6.90 2.	6267851	6269754	6271658	6273561	6275464	6277367	6279269	6281172	6283074	6284977	0
6.91 2.	6286879	6288781	6290683	6292584	6294486	6296388	6298289	6300190	6302091	6303992	0
6.92 2.	6305893	6307793	6309694	6311594	6313495	6315395	6317295	6319195	6321094	6322994	0
6.93 2.	6324893	6326792	6328692	6330591	6332489	6334388	6336287	6338185	6340083	6341982	0
6.94 2.	6343880	6345778	6347675	6349573	6351471	6353368	6355265	6357162	6359059	6360956	0
6.95 2.	6362853	6364749	6366646	6368542	6370438	6372334	6374230	6376126	6378021	6379917	0
6.96 2.	6381812	6383707	6385602	6387497	6389392	6391286	6393181	6395075	6396969	6398864	0
6.97 2.	6400758	6402651	6404545	6406439	6408332	6410225	6412118	6414011	6415904	6417797	0
6.98 2.	6419690	6421582	6423474	6425367	6427259	6429151	6431042	6432934	6434826	6436717	0
6.99 2.	6438608	6440499	6442390	6444281	6446172	6448062	6449953	6451843	6453733	6455623	0

TABLE 4

Square Roots

X		.0000	.0001	.0002	.0003	.0004	.0005	.0006	.0007	.0008	.0009	Δ_2
7.00	2.	6457513	6459403	6461292	6463182	6465071	6466961	6468850	6470739	6472627	6474516	0
7.01	2.	6476405	6478293	6480181	6482069	6483957	6485845	6487733	6489621	6491508	6493395	0
7.02	2.	6495283	6497170	6499057	6500943	6502830	6504717	6506603	6508489	6510375	6512261	0
7.03	2.	6514147	6516033	6517918	6519804	6521689	6523574	6525459	6527344	6529229	6531114	0
7.04	2.	6532998	6534883	6536767	6538651	6540535	6542419	6544303	6546186	6548070	6549953	0
7.05	2.	6551836	6553719	6555602	6557485	6559367	6561250	6563132	6565015	6566897	6568779	0
7.06	2.	6570660	6572542	6574424	6576305	6578187	6580068	6581949	6583830	6585710	6587591	0
7.07	2.	6589472	6591352	6593232	6595112	6596992	6598872	6600752	6602631	6604511	6606390	0
7.08	2.	6608269	6610148	6612027	6613906	6615785	6617663	6619542	6621420	6623298	6625176	0
7.09	2.	6627054	6628932	6630809	6632687	6634564	6636441	6638318	6640195	6642072	6643949	0
7.10	2.	6645825	6647702	6649578	6651454	6653330	6655206	6657082	6658957	6660833	6662708	0
7.11	2.	6664583	6666458	6668333	6670208	6672083	6673957	6675832	6677706	6679580	6681454	0
7.12	2.	6683328	6685202	6687075	6688949	6690822	6692696	6694569	6696442	6698315	6700187	0
7.13	2.	6702060	6703932	6705805	6707677	6709549	6711421	6713293	6715164	6717036	6718907	0
7.14	2.	6720778	6722650	6724521	6726391	6728262	6730133	6732003	6733874	6735744	6737614	0
7.15	2.	6739484	6741354	6743223	6745093	6746962	6748832	6750701	6752570	6754439	6756308	0
7.16	2.	6758176	6760045	6761913	6763781	6765650	6767518	6769385	6771253	6773121	6774988	0
7.17	2.	6776856	6778723	6780590	6782457	6784324	6786190	6788057	6789923	6791790	6793656	0
7.18	2.	6795522	6797388	6799254	6801119	6802985	6804850	6806716	6808581	6810446	6812311	0
7.19	2.	6814175	6816040	6817904	6819769	6821633	6823497	6825361	6827225	6829089	6830952	0
7.20	2.	6832816	6834679	5836542	6838405	6840268	6842131	6843994	6845856	6847719	6849581	0
7.21	2.	6851443	6853305	6855167	6857029	6858891	6860752	6862613	6864475	6866336	6868197	0
7.22	2.	6870058	6871918	6873779	6875639	6877500	6879360	6881220	6883080	6884940	6886800	0
7.23	2.	6888659	6890519	6892378	6894237	6896096	6897955	6899814	6901673	6903531	6905390	0
7.24	2.	6907248	6909106	6910964	6912822	6914680	6916538	6918395	6920253	6922110	6923967	0
7.25	2.	6925824	6927681	6929538	6931394	6933251	6935107	6936963	6938820	6940676	6942531	0
7.26	2.	6944387	6946243	6948098	6949954	6951809	6953664	6955519	6957374	6959228	6961083	0
7.27	2.	6962937	6964792	6966646	6968500	6970354	6972208	6974062	6975915	6977769	6979622	0
7.28	2.	6981475	6983328	6985181	6987034	6988887	6990739	6992592	6994444	6996296	6998148	0
7.29	2.	7000000	7001852	7003703	7005555	7007406	7009258	7011109	7012960	7014811	7016661	0
7.30	2.	7018512	7020363	7022213	7024063	7025913	7027764	7029613	7031463	7033313	7035162	0
7.31	2.	7037012	7038861	7040710	7042559	7044408	7046257	7048105	7049954	7051802	7053650	0
7.32	2.	7055499	7057346	7059194	7061042	7062890	7064737	7066585	7068432	7070279	7072126	0
7.33	2.	7073973	7075819	7077666	7079512	7081359	7083205	7085051	7086897	7088743	7090589	0
7.34	2.	7092434	7094280	7096125	7097970	7099815	7101660	7103505	7105350	7107195	7109039	0
7.35	2.	7110883	7112728	7114572	7116416	7118260	7120103	7121947	7123790	7125634	7127477	0
7.36	2.	7129320	7131163	7133006	7134848	7136691	7138533	7140376	7142218	7144060	7145902	0
7.37	2.	7147744	7149586	7151427	7153269	7155110	7156951	7158792	7160633	7162474	7164315	0
7.38	2.	7166155	7167996	7169836	7171676	7173516	7175356	7177196	7179036	7180876	7182715	0
7.39	2.	7184554	7186394	7188233	7190072	7191911	7193749	7195588	7197426	7199265	7201103	0
7.40	2.	7202941	7204779	7206617	7208455	7210292	7212130	7213967	7215804	7217641	7219478	0
7.41	2.	7221315	7223152	7224988	7226825	7228661	7230498	7232334	7234170	7236006	7237841	0
7.42	2.	7239677	7241512	7243348	7245183	7247018	7248853	7250688	7252523	7254357	7256192	0
7.43	2.	7258026	7259861	7261695	7263529	7265363	7267196	7269030	7270864	7272697	7274530	0
7.44	2.	7276363	7278196	7280029	7281862	7283695	7285527	7287360	7289192	7291024	7292856	0
7.45	2.	7294688	7296520	7298352	7300183	7302015	7303846	7305677	7307508	7309339	7311170	0
7.46	2.	7313001	7314831	7316662	7318492	7320322	7322152	7323982	7325812	7327642	7329471	0
7.47	2.	7331301	7333130	7334959	7336788	7338617	7340446	7342275	7344104	7345932	7347760	0
7.48	2.	7349589	7351417	7353245	7355073	7356900	7358728	7360556	7362383	7364210	7366037	0
7.49	2.	7367864	7369691	7371518	7373345	7375171	7376998	7378824	7380650	7382476	7384302	0

TABLE 4

Square Roots

X		.0000	.0001	.0002	.0003	.0004	.0005	.0006	.0007	.0008	.0009	Δ_2
7.50	2.	7386128	7387954	7389779	7391604	7393430	7395255	7397080	7398905	7400730	7402555	0
7.51	2.	7404379	7406204	7408028	7409852	7411676	7413500	7415324	7417148	7418972	7420795	0
7.52	2.	7422618	7424442	7426265	7428088	7429911	7431733	7433556	7435379	7437201	7439023	0
7.53	2.	7440845	7442667	7444489	7446311	7448133	7449954	7451776	7453597	7455418	7457239	0
7.54	2.	7459060	7460881	7462702	7464522	7466343	7468163	7469984	7471804	7473624	7475444	0
7.55	2.	7477263	7479083	7480902	7482722	7484541	7486360	7488179	7489998	7491817	7493636	0
7.56	2.	7495454	7497273	7499091	7500909	7502727	7504545	7506363	7508181	7509998	7511816	0
7.57	2.	7513633	7515450	7517267	7519084	7520901	7522718	7524534	7526351	7528167	7529984	0
7.58	2.	7531800	7533616	7535432	7537247	7539063	7540879	7542694	7544509	7546325	7548140	0
7.59	2.	7549955	7551769	7553584	7555399	7557213	7559028	7560842	7562656	7564470	7566284	0
7.60	2.	7568097	7569911	7571725	7573538	7575351	7577164	7578977	7580790	7582603	7584416	0
7.61	2.	7586228	7588041	7589853	7591665	7593477	7595289	7597101	7598913	7600725	7602536	0
7.62	2.	7604347	7606159	7607970	7609781	7611592	7613403	7615213	7617024	7618834	7620644	0
7.63	2.	7622455	7624265	7626075	7627884	7629694	7631504	7633313	7635123	7636932	7638741	0
7.64	2.	7640550	7642359	7644168	7645976	7647785	7649593	7651401	7653210	7655018	7656826	0
7.65	2.	7658633	7660441	7662249	7664056	7665863	7667671	7669478	7671285	7673092	7674898	0
7.66	2.	7676705	7678512	7680318	7682124	7683930	7685736	7687542	7689348	7691154	7692959	0
7.67	2.	7694765	7696570	7698375	7700180	7701985	7703790	7705595	7707400	7709204	7711009	0
7.68	2.	7712813	7714617	7716421	7718225	7720029	7721833	7723636	7725440	7727243	7729046	0
7.69	2.	7730849	7732652	7734455	7736258	7738060	7739863	7741665	7743468	7745270	7747072	0
7.70	2.	7748874	7750676	7752477	7754279	7756080	7757882	7759683	7761484	7763285	7765086	0
7.71	2.	7766887	7768687	7770488	7772288	7774089	7775889	7777689	7779489	7781289	7783088	0
7.72	2.	7784888	7786687	7788487	7790286	7792085	7793884	7795683	7797482	7799281	7801079	0
7.73	2.	7802878	7804676	7806474	7808272	7810070	7811868	7813666	7815463	7817261	7819058	0
7.74	2.	7820855	7822653	7824450	7826247	7828043	7829840	7831637	7833433	7835229	7837026	0
7.75	2.	7838822	7840618	7842414	7844209	7846005	7847801	7849596	7851391	7853187	7854982	0
7.76	2.	7856776	7858571	7860366	7862161	7863955	7865750	7867544	7869338	7871132	7872926	0
7.77	2.	7874720	7876513	7878307	7880100	7881894	7883687	7885480	7887273	7889066	7890859	0
7.78	2.	7892651	7894444	7896236	7898029	7899821	7901613	7903405	7905197	7906988	7908780	0
7.79	2.	7910571	7912363	7914154	7915945	7917736	7919527	7921318	7923109	7924899	7926690	0
7.80	2.	7928480	7930270	7932060	7933850	7935640	7937430	7939220	7941009	7942799	7944588	0
7.81	2.	7946377	7948166	7949955	7951744	7953533	7955321	7957110	7958898	7960687	7962475	0
7.82	2.	7964263	7966051	7967839	7969626	7971414	7973201	7974989	7976776	7978563	7980350	0
7.83	2.	7982137	7983924	7985711	7987497	7989284	7991070	7992856	7994642	7996428	7998214	0
7.84	2.	8000000	8001786	8003571	8005357	8007142	8008927	8010712	8012497	8014282	8016067	0
7.85	2.	8017851	8019636	8021420	8023205	8024989	8026773	8028557	8030341	8032124	8033908	0
7.86	2.	8035692	8037475	8039258	8041041	8042824	8044607	8046390	8048173	8049955	8051738	0
7.87	2.	8053520	8055303	8057085	8058867	8060649	8062430	8064212	8065994	8067775	8069556	0
7.88	2.	8071338	8073119	8074900	8076681	8078461	8080242	8082023	8083803	8085583	8087364	0
7.89	2.	8089144	8090924	8092704	8094483	8096263	8098043	8099822	8101601	8103381	8105160	0
7.90	2.	8106939	8108718	8110496	8112275	8114053	8115832	8117610	8119388	8121166	8122944	0
7.91	2.	8124722	8126500	8128278	8130055	8131832	8133610	8135387	8137164	8138941	8140718	0
7.92	2.	8142494	8144271	8146048	8147824	8149600	8151376	8153153	8154928	8156704	8158480	0
7.93	2.	8160256	8162031	8163806	8165582	8167357	8169132	8170907	8172682	8174457	8176231	0
7.94	2.	8178006	8179780	8181554	8183328	8185102	8186876	8188650	8190424	8192197	8193971	0
7.95	2.	8195744	8197518	8199291	8201064	8202837	8204610	8206382	8208155	8209927	8211700	0
7.96	2.	8213472	8215244	8217016	8218788	8220560	8222332	8224103	8225875	8227646	8229417	0
7.97	2.	8231188	8232959	8234730	8236501	8238272	8240042	8241813	8243583	8245354	8247124	0
7.98	2.	8248894	8250664	8252433	8254203	8255973	8257742	8259512	8261281	8263050	8264819	0
7.99	2.	8266588	8268357	8270126	8271894	8273663	8275431	8277199	8278967	8280735	8282503	0

TABLE 4

Square Roots

X		.0000	.0001	.0002	.0003	.0004	.0005	.0006	.0007	.0008	.0009	Δ,
8.00	2.	8284271	8286039	8287807	8289574	8291341	8293109	8294876	8296643	8298410	8300177	0
8.01	2.	8301943	8303710	8305476	8307243	8309009	8310775	8312541	8314307	8316073	8317839	0
8.02	2.	8319604	8321370	8323135	8324901	8326666	8328431	8330196	8331961	8333725	8335490	0
8.03	2.	8337255	8339019	8340783	8342547	8344312	8346076	8347839	8349603	8351367	8353130	0
8.04	2.	8354894	8356657	8358420	8360183	8361946	8363709	8365472	8367235	8368997	8370760	0
8.05	2.	8372522	8374284	8376046	8377808	8379570	8381332	8383094	8384855	8386617	8388378	0
8.06	2.	8390139	8391900	8393661	8395422	8397183	8398944	8400704	8402465	8404225	8405985	0
8.07	2.	8407745	8409505	8411265	8413025	8414785	8416544	8418304	8420063	8421823	8423582	0
8.08	2.	8425341	8427100	8428859	8430617	8432376	8434134	8435893	8437651	8439409	8441167	0
8.09	2.	8442925	8444683	8446441	8448199	8449956	8451713	8453471	8455228	8456985	8458742	0
8.10	2.	8460499	8462256	8464012	8465769	8467525	8469282	8471038	8472794	8474550	8476306	0
8.11	2.	8478062	8479817	8481573	8483328	8485084	8486839	8488594	8490349	8492104	8493859	0
8.12	2.	8495614	8497368	8499123	8500877	8502631	8504386	8506140	8507894	8509647	8511401	0
8.13	2.	8513155	8514908	8516662	8518415	8520168	8521921	8523674	8525427	8527180	8528933	0
8.14	2.	8530685	8532438	8534190	8535942	8537694	8539446	8541198	8542950	8544702	8546453	0
8.15	2.	8548205	8549956	8551707	8553459	8555210	8556961	8558711	8560462	8562213	8563963	0
8.16	2.	8565714	8567464	8569214	8570964	8572714	8574464	8576214	8577963	8579713	8581463	0
8.17	2.	8583212	8584961	8586710	8588459	8590208	8591957	8593706	8595454	8597203	8598951	0
8.18	2.	8600699	8602447	8604195	8605943	8607691	8609439	8611187	8612934	8614682	8616429	0
8.19	2.	8618176	8619923	8621670	8623417	8625164	8626910	8628657	8630403	8632150	8633896	0
8.20	2.	8635642	8637388	8639134	8640880	8642626	8644371	8646117	8647862	8649607	8651352	0
8.21	2.	8653098	8654842	8656587	8658332	8660077	8661821	8663566	8665310	8667054	8668798	0
8.22	2.	8670542	8672286	8674030	8675774	8677517	8679261	8681004	8682747	8684491	8686234	0
8.23	2.	8687977	8689719	8691462	8693205	8694947	8696690	8698432	8700174	8701916	8703658	0
8.24	2.	8705400	8707142	8708884	8710625	8712367	8714108	8715849	8717590	8719331	8721072	0
8.25	2.	8722813	8724554	8726295	8728035	8729775	8731516	8733256	8734996	8736736	8738476	0
8.26	2.	8740216	8741955	8743695	8745434	8747174	8748913	8750652	8752391	8754130	8755869	0
8.27	2.	8757608	8759346	8761085	8762823	8764561	8766300	8768038	8769776	8771514	8773251	0
8.28	2.	8774989	8776727	8778464	8780202	8781939	8783676	8785413	8787150	8788887	8790623	0
8.29	2.	8792360	8794097	8795833	8797569	8799305	8801042	8802778	8804514	8806249	8807985	0
8.30	2.	8809721	8811456	8813191	8814927	8816662	8818397	8820132	8821867	8823601	8825336	0
8.31	2.	8827071	8828805	8830539	8832274	8834008	8835742	8837476	8839209	8840943	8842677	0
8.32	2.	8844410	8846144	8847877	8849610	8851343	8853076	8854809	8856542	8858274	8860007	0
8.33	2.	8861739	8863472	8865204	8866936	8868668	8870400	8872132	8873864	8875595	8877327	0
8.34	2.	8879058	8880789	8882521	8884252	8885983	8887714	8889444	8891175	8892906	8894636	0
8.35	2.	8896367	8898097	8899827	8901557	8903287	8905017	8906747	8908476	8910206	8911935	0
8.36	2.	8913665	8915394	8917123	8918852	8920581	8922310	8924038	8925767	8927496	8929224	0
8.37	2.	8930952	8932680	8934409	8936137	8937864	8939592	8941320	8943048	8944775	8946502	0
8.38	2.	8948230	8949957	8951684	8953411	8955138	8956864	8958591	8960318	8962044	8963770	0
8.39	2.	8965497	8967223	8968949	8970675	8972401	8974126	8975852	8977578	8979303	8981028	0
8.40	2.	8982753	8984479	8986204	8987928	8989653	8991378	8993103	8994827	8996551	8998276	0
8.41	2.	9000000	9001724	9003448	9005172	9006896	9008619	9010343	9012066	9013790	9015513	0
8.42	2.	9017236	9018959	9020682	9022405	9024128	9025851	9027573	9029296	9031018	9032740	0
8.43	2.	9034462	9036184	9037906	9039628	9041350	9043071	9044793	9046514	9048236	9049957	0
8.44	2.	9051678	9053399	9055120	9056841	9058562	9060282	9062003	9063723	9065443	9067164	0
8.45	2.	9068884	9070604	9072324	9074043	9075763	9077483	9079202	9080922	9082641	9084360	0
8.46	2.	9086079	9087798	9089517	9091236	9092954	9094673	9096391	9098110	9099828	9101546	0
8.47	2.	9103264	9104982	9106700	9108418	9110136	9111853	9113571	9115288	9117005	9118722	0
8.48	2.	9120440	9122157	9123873	9125590	9127307	9129023	9130740	9132456	9134172	9135889	0
8.49	2.	9137605	9139320	9141036	9142752	9144468	9146183	9147899	9149614	9151329	9153044	0

TABLE 4

Square Roots

X		.0000	.0001	.0002	.0003	.0004	.0005	.0006	.0007	.0008	.0009	Δ_2
8.50	2.	9154759	9156474	9158189	9159904	9161619	9163333	9165048	9166762	9168476	9170190	0
8.51	2.	9171904	9173618	9175332	9177046	9178759	9180473	9182186	9183900	9185613	9187326	0
8.52	2.	9189039	9190752	9192465	9194177	9195890	9197603	9199315	9201027	9202740	9204452	0
8.53	2.	9206164	9207876	9209587	9211299	9213011	9214722	9216434	9218145	9219856	9221567	0
8.54	2.	9223278	9224989	9226700	9228411	9230121	9231832	9233542	9235253	9236963	9238673	0
8.55	2.	9240383	9242093	9243803	9245512	9247222	9248932	9250641	9252350	9254059	9255769	0
8.56	2.	9257478	9259187	9260895	9262604	9264313	9266021	9267730	9269438	9271146	9272854	0
8.57	2.	9274562	9276270	9277978	9279686	9281393	9283101	9284808	9286516	9288223	9289930	0
8.58	2.	9291637	9293344	9295051	9296757	9298464	9300171	9301877	9303583	9305290	9306996	0
8.59	2.	9308702	9310408	9312114	9313819	9315525	9317230	9318936	9320641	9322346	9324052	0
8.60	2.	9325757	9327461	9329166	9330871	9332576	9334280	9335985	9337689	9339393	9341097	0
8.61	2.	9342801	9344505	9346209	9347913	9349617	9351320	9353024	9354727	9356430	9358133	0
8.62	2.	9359837	9361539	9363242	9364945	9366648	9368350	9370053	9371755	9373457	9375160	0
8.63	2.	9376862	9378564	9380265	9381967	9383669	9385370	9387072	9388773	9390475	9392176	0
8.64	2.	9393877	9395578	9397279	9398980	9400680	9402381	9404081	9405782	9407482	9409182	0
8.65	2.	9410882	9412582	9414282	9415982	9417682	9419381	9421081	9422780	9424480	9426179	0
8.66	2.	9427878	9429577	9431276	9432975	9434673	9436372	9438071	9439769	9441467	9443166	0
8.67	2.	9444864	9446562	9448260	9449958	9451655	9453353	9455050	9456748	9458445	9460143	0
8.68	2.	9461840	9463537	9465234	9466931	9468627	9470324	9472021	9473717	9475413	9477110	0
8.69	2.	9478806	9480502	9482198	9483894	9485590	9487285	9488981	9490676	9492372	9494067	0
8.70	2.	9495762	9497457	9499153	9500847	9502542	9504237	9505932	9507626	9509321	9511015	0
8.71	2.	9512709	9514403	9516097	9517791	9519485	9521179	9522872	9524566	9526259	9527953	0
8.72	2.	9529646	9531339	9533032	9534725	9536418	9538111	9539804	9541496	9543189	9544881	0
8.73	2.	9546573	9548266	9549958	9551650	9553342	9555033	9556725	9558417	9560108	9561800	0
8.74	2.	9563491	9565182	9566873	9568564	9570255	9571946	9573637	9575327	9577018	9578708	0
8.75	2.	9580399	9582089	9583779	9585469	9587159	9588849	9590539	9592229	9593918	9595608	0
8.76	2.	9597297	9598986	9600676	9602365	9604054	9605743	9607431	9609120	9610809	9612497	0
8.77	2.	9614186	9615874	9617562	9619250	9620939	9622626	9624314	9626002	9627690	9629377	0
8.78	2.	9631065	9632752	9634439	9636127	9637814	9639501	9641187	9642874	9644561	9646248	0
8.79	2.	9647934	9649621	9651307	9652993	9654679	9656365	9658051	9659737	9661423	9663108	0
8.80	2.	9664794	9666479	9668165	9669850	9671535	9673220	9674905	9676590	9678275	9679960	0
8.81	2.	9681644	9683329	9685013	9686697	9688382	9690066	9691750	9693434	9695117	9696801	0
8.82	2.	9698485	9700168	9701852	9703535	9705218	9706902	9708585	9710268	9711950	9713633	0
8.83	2.	9715316	9716998	9718681	9720363	9722046	9723728	9725410	9727092	9728774	9730456	0
8.84	2.	9732137	9733819	9735501	9737182	9738863	9740545	9742226	9743907	9745588	9747269	0
8.85	2.	9748950	9750630	9752311	9753991	9755672	9757352	9759032	9760712	9762392	9764072	0
8.86	2.	9765752	9767432	9769111	9770791	9772470	9774150	9775829	9777508	9779187	9780866	0
8.87	2.	9782545	9784224	9785903	9787581	9789260	9790938	9792617	9794295	9795973	9797651	0
8.88	2.	9799329	9801007	9802684	9804362	9806040	9807717	9809394	9811072	9812749	9814426	0
8.89	2.	9816103	9817780	9819457	9821133	9822810	9824487	9826163	9827839	9829516	9831192	0
8.90	2.	9832868	9834544	9836220	9837895	9839571	9841247	9842922	9844597	9846273	9847948	0
8.91	2.	9849623	9851298	9852973	9854648	9856323	9857997	9859672	9861346	9863021	9864695	0
8.92	2.	9866369	9868043	9869717	9871391	9873065	9874738	9876412	9878086	9879759	9881432	0
8.93	2.	9883106	9884779	9886452	9888125	9889798	9891470	9893143	9894816	9896488	9898160	0
8.94	2.	9899833	9901505	9903177	9904849	9906521	9908193	9909865	9911536	9913208	9914879	0
8.95	2.	9916551	9918222	9919893	9921564	9923235	9924906	9926577	9928247	9929918	9931589	0
8.96	2.	9933259	9934929	9936600	9938270	9939940	9941610	9943280	9944949	9946619	9948289	0
8.97	2.	9949958	9951628	9953297	9954966	9956635	9958304	9959973	9961642	9963311	9964980	0
8.98	2.	9966648	9968317	9969985	9971653	9973321	9974990	9976658	9978325	9979993	9981661	0
8.99	2.	9983329	9984996	9986664	9988331	9989998	9991665	9993333	9995000	9996666	9998333	0

TABLE 4

Square Roots

X	.0000	.0001	.0002	.0003	.0004	.0005	.0006	.0007	.0008	.0009	Δ_2
9.00 3.	0000000	0001667	0003333	0005000	0006666	0008332	0009998	0011664	0013330	0014996	0
9.01 3.	0016662	0018328	0019993	0021659	0023324	0024990	0026655	0028320	0029985	0031650	0
9.02 3.	0033315	0034980	0036644	0038309	0039973	0041638	0043302	0044966	0046630	0048294	0
9.03 3.	0049958	0051622	0053286	0054950	0056613	0058277	0059940	0061603	0063267	0064930	0
9.04 3.	0066593	0068256	0069919	0071581	0073244	0074906	0076569	0078231	0079894	0081556	0
9.05 3.	0083218	0084880	0086542	0088204	0089865	0091527	0093189	0094850	0096511	0098173	0
9.06 3.	0099834	0101495	0103156	0104817	0106478	0108138	0109799	0111460	0113120	0114780	0
9.07 3.	0116441	0118101	0119761	0121421	0123081	0124741	0126400	0128060	0129720	0131379	0
9.08 3.	0133038	0134698	0136357	0138016	0139675	0141334	0142993	0144651	0146310	0147968	0
9.09 3.	0149627	0151285	0152943	0154602	0156260	0157918	0159576	0161233	0162891	0164549	0
9.10 3.	0166206	0167864	0169521	0171178	0172835	0174493	0176150	0177806	0179463	0181120	0
9.11 3.	0182777	0184433	0186090	0187746	0189402	0191058	0192714	0194370	0196026	0197682	0
9.12 3.	0199338	0200993	0202649	0204304	0205960	0207615	0209270	0210925	0212580	0214235	0
9.13 3.	0215890	0217545	0219199	0220854	0222508	0224163	0225817	0227471	0229125	0230779	0
9.14 3.	0232433	0234087	0235740	0237394	0239048	0240701	0242354	0244008	0245661	0247314	0
9.15 3.	0248967	0250620	0252273	0253925	0255578	0257231	0258883	0260535	0262188	0263840	0
9.16 3.	0265492	0267144	0268796	0270448	0272099	0273751	0275403	0277054	0278705	0280357	0
9.17 3.	0282008	0283659	0285310	0286961	0288612	0290262	0291913	0293564	0295214	0296865	0
9.18 3.	0298515	0300165	0301815	0303465	0305115	0306765	0308415	0310064	0311714	0313363	0
9.19 3.	0315013	0316662	0318311	0319960	0321609	0323258	0324907	0326556	0328205	0329853	0
9.20 3.	0331502	0333150	0334798	0336447	0338095	0339743	0341391	0343039	0344687	0346334	0
9.21 3.	0347982	0349629	0351277	0352924	0354571	0356218	0357866	0359513	0361159	0362806	0
9.22 3.	0364453	0366100	0367746	0369392	0371039	0372685	0374331	0375977	0377623	0379269	0
9.23 3.	0380915	0382561	0384206	0385852	0387497	0389143	0390788	0392433	0394078	0395723	0
9.24 3.	0397368	0399013	0400658	0402303	0403947	0405592	0407236	0408880	0410524	0412169	0
9.25 3.	0413813	0415457	0417100	0418744	0420388	0422031	0423675	0425318	0426962	0428605	0
9.26 3.	0430248	0431891	0433534	0435177	0436820	0438463	0440105	0441748	0443390	0445032	0
9.27 3.	0446675	0448317	0449959	0451601	0453243	0454885	0456526	0458168	0459810	0461451	0
9.28 3.	0463092	0464734	0466375	0468016	0469657	0471298	0472939	0474580	0476220	0477861	0
9.29 3.	0479501	0481142	0482782	0484422	0486062	0487702	0489342	0490982	0492622	0494262	0
9.30 3.	0495901	0497541	0499180	0500820	0502459	0504098	0505737	0507376	0509015	0510654	0
9.31 3.	0512293	0513931	0515570	0517208	0518847	0520485	0522123	0523761	0525399	0527037	0
9.32 3.	0528675	0530313	0531950	0533588	0535226	0536863	0538500	0540138	0541775	0543412	0
9.33 3.	0545049	0546686	0548322	0549959	0551596	0553232	0554869	0556505	0558141	0559777	0
9.34 3.	0561414	0563050	0564686	0566321	0567957	0569593	0571228	0572864	0574499	0576134	0
9.35 3.	0577770	0579405	0581040	0582675	0584310	0585944	0587579	0589214	0590848	0592483	0
9.36 3.	0594117	0595751	0597386	0599020	0600654	0602287	0603921	0605555	0607189	0608822	0
9.37 3.	0610456	0612089	0613722	0615356	0616989	0618622	0620255	0621888	0623520	0625153	0
9.38 3.	0626786	0628418	0630051	0631683	0633315	0634947	0636579	0638211	0639843	0641475	0
9.39 3.	0643107	0644739	0646370	0648002	0649633	0651264	0652895	0654527	0656158	0657789	0
9.40 3.	0659419	0661050	0662681	0664311	0665942	0667572	0669203	0670833	0672463	0674093	0
9.41 3.	0675723	0677353	0678983	0680613	0682242	0683872	0685501	0687131	0688760	0690389	0
9.42 3.	0692019	0693648	0695277	0696905	0698534	0700163	0701791	0703420	0705048	0706677	0
9.43 3.	0708305	0709933	0711561	0713189	0714817	0716445	0718073	0719701	0721328	0722956	0
9.44 3.	0724583	0726210	0727838	0729465	0731092	0732719	0734346	0735972	0737599	0739226	0
9.45 3.	0740852	0742479	0744105	0745731	0747358	0748984	0750610	0752236	0753862	0755487	0
9.46 3.	0757113	0758739	0760364	0761990	0763615	0765240	0766865	0768490	0770115	0771740	0
9.47 3.	0773365	0774990	0776614	0778239	0779864	0781488	0783112	0784736	0786361	0787985	0
9.48 3.	0789609	0791233	0792856	0794480	0796104	0797727	0799351	0800974	0802597	0804220	0
9.49 3.	0805844	0807467	0809090	0810712	0812335	0813958	0815580	0817203	0818825	0820448	0

TABLE 4

Square Roots

X		.0000	.0001	.0002	.0003	.0004	.0005	.0006	.0007	.0008	.0009	Δ_2
9.50	3.	0822070	0823692	0825314	0826936	0828558	0830180	0831802	0833423	0835045	0836666	0
9.51	3.	0838288	0839909	0841530	0843152	0844773	0846394	0848015	0849635	0851256	0852877	0
9.52	3.	0854497	0856118	0857738	0859358	0860979	0862599	0864219	0865839	0867459	0869078	0
9.53	3.	0870698	0872318	0873937	0875557	0877176	0878795	0880414	0882034	0883653	0885272	0
9.54	3.	0886890	0888509	0890128	0891746	0893365	0894983	0896602	0898220	0899838	0901456	0
9.55	3.	0903074	0904692	0906310	0907928	0909545	0911163	0912781	0914398	0916015	0917633	0
9.56	3.	0919250	0920867	0922484	0924101	0925717	0927334	0928951	0930567	0932184	0933800	0
9.57	3.	0935417	0937033	0938649	0940265	0941881	0943497	0945113	0946728	0948344	0949960	0
9.58	3.	0951575	0953190	0954806	0956421	0958036	0959651	0961266	0962881	0964496	0966110	0
9.59	3.	0967725	0969340	0970954	0972569	0974183	0975797	0977411	0979025	0980639	0982253	0
9.60	3.	0983867	0985480	0987094	0988708	0990321	0991934	0993548	0995161	0996774	0998387	0
9.61	3.	1000000	1001613	1003226	1004838	1006451	1008063	1009676	1011288	1012901	1014513	0
9.62	3.	1016125	1017737	1019349	1020961	1022572	1024184	1025796	1027407	1029019	1030630	0
9.63	3.	1032241	1033852	1035464	1037075	1038686	1040296	1041907	1043518	1045128	1046739	0
9.64	3.	1048349	1049960	1051570	1053180	1054790	1056400	1058010	1059620	1061230	1062840	0
9.65	3.	1064449	1066059	1067668	1069277	1070887	1072496	1074105	1075714	1077323	1078932	0
9.66	3.	1080541	1082149	1083758	1085366	1086975	1088583	1090191	1091800	1093408	1095016	0
9.67	3.	1096624	1098231	1099839	1101447	1103055	1104662	1106269	1107877	1109484	1111091	0
9.68	3.	1112698	1114305	1115912	1117519	1119126	1120733	1122339	1123946	1125552	1127159	0
9.69	3.	1128765	1130371	1131977	1133583	1135189	1136795	1138401	1140006	1141612	1143218	0
9.70	3.	1144823	1146428	1148034	1149639	1151244	1152849	1154454	1156059	1157664	1159268	0
9.71	3.	1160873	1162477	1164082	1165686	1167291	1168895	1170499	1172103	1173707	1175311	0
9.72	3.	1176915	1178518	1180122	1181725	1183329	1184932	1186536	1188139	1189742	1191345	0
9.73	3.	1192948	1194551	1196154	1197756	1199359	1200962	1202564	1204166	1205769	1207371	0
9.74	3.	1208973	1210575	1212177	1213779	1215381	1216983	1218584	1220186	1221787	1223389	0
9.75	3.	1224990	1226591	1228192	1229793	1231394	1232995	1234596	1236197	1237798	1239398	0
9.76	3.	1240999	1242599	1244199	1245800	1247400	1249000	1250600	1252200	1253800	1255400	0
9.77	3.	1256999	1258599	1260198	1261798	1263397	1264996	1266596	1268195	1269794	1271393	0
9.78	3.	1272992	1274590	1276189	1277788	1279386	1280985	1282583	1284181	1285780	1287378	0
9.79	3.	1288976	1290574	1292172	1293769	1295367	1296965	1298562	1300160	1301757	1303354	0
9.80	3.	1304952	1306549	1308146	1309743	1311340	1312937	1314533	1316130	1317727	1319323	0
9.81	3.	1320920	1322516	1324112	1325708	1327304	1328900	1330496	1332092	1333688	1335284	0
9.82	3.	1336879	1338475	1340070	1341666	1343261	1344856	1346451	1348046	1349641	1351236	0
9.83	3.	1352831	1354426	1356020	1357615	1359209	1360804	1362398	1363992	1365586	1367180	0
9.84	3.	1368774	1370368	1371962	1373556	1375149	1376743	1378336	1379930	1381523	1383116	0
9.85	3.	1384710	1386303	1387896	1389489	1391082	1392674	1394267	1395860	1397452	1399045	0
9.86	3.	1400637	1402229	1403821	1405414	1407006	1408598	1410189	1411781	1413373	1414965	0
9.87	3.	1416556	1418148	1419739	1421330	1422922	1424513	1426104	1427695	1429286	1430877	0
9.88	3.	1432467	1434058	1435649	1437239	1438829	1440420	1442010	1443600	1445190	1446780	0
9.89	3.	1448370	1449960	1451550	1453140	1454729	1456319	1457908	1459498	1461087	1462676	0
9.90	3.	1464265	1465854	1467443	1469032	1470621	1472210	1473799	1475387	1476976	1478564	0
9.91	3.	1480152	1481741	1483329	1484917	1486505	1488093	1489681	1491269	1492856	1494444	0
9.92	3.	1496031	1497619	1499206	1500794	1502381	1503968	1505555	1507142	1508729	1510316	0
9.93	3.	1511903	1513489	1515076	1516662	1518249	1519835	1521421	1523007	1524594	1526180	0
9.94	3.	1527766	1529351	1530937	1532523	1534109	1535694	1537280	1538865	1540450	1542035	0
9.95	3.	1543621	1545206	1546791	1548376	1549960	1551545	1553130	1554714	1556299	1557883	0
9.96	3.	1559468	1561052	1562636	1564220	1565804	1567388	1568972	1570556	1572140	1573723	0
9.97	3.	1575307	1576890	1578474	1580057	1581640	1583223	1584806	1586389	1587972	1589555	0
9.98	3.	1591138	1592721	1594303	1595886	1597468	1599051	1600633	1602215	1603797	1605379	0
9.99	3.	1606961	1608543	1610125	1611707	1613288	1614870	1616451	1618033	1619614	1621195	0

TABLE 5

Multiples of Half-Pi

N		0	1	2	3	4	5	6	7	8	9
0	0+	0.0	1.5707963	3.1415927	4.7123890	6.2831853	7.8539816	9.4247780	10.9955743	12.5663706	14.1371669
10	0+	15.7079633	17.2787596	18.8495559	20.4203522	21.9911486	23.5619449	25.1327412	26.7035376	28.2743339	29.8451302
20	0+	31.4159265	32.9867229	34.5575192	36.1283155	37.6991118	39.2699082	40.8407045	42.4115008	43.9822972	45.5530935
30	0+	47.1238898	48.6946861	50.2654825	51.8362788	53.4070751	54.9778714	56.5486678	58.1194641	59.6902604	61.2610567
40	0+	62.8318531	64.4026494	65.9734457	67.5442421	69.1150384	70.6858347	72.2566310	73.8274274	75.3982237	76.9690200
50	0+	78.5398163	80.1106127	81.6814090	83.2522053	84.8230016	86.3937980	87.9645943	89.5353906	91.1061870	92.6769833
60	0+	94.2477796	95.8185759	97.3893723	98.9601686	0.5309649	2.1017612	3.6725576	5.2433539	6.8141502	8.3849465
70	100+	9.9557429	11.5265392	13.0973355	14.6681319	16.2389282	17.8097245	19.3805208	20.9513172	22.5221135	24.0929098
80	100+	25.6637061	27.2345025	28.8052988	30.3760951	31.9468915	33.5176878	35.0884841	36.6592804	38.2300768	39.8008731
90	100+	41.3716694	42.9424657	44.5132621	46.0840584	47.6548547	49.2256510	50.7964474	52.3672437	53.9380400	55.5088364
100	100+	57.0796327	58.6504290	60.2212253	61.7920217	63.3628180	64.9336143	66.5044106	68.0752070	69.6460033	71.2167996
110	100+	72.7875959	74.3583923	75.9291886	77.4999849	79.0707813	80.6415776	82.2123739	83.7831702	85.3539666	86.9247629
120	100+	88.4955592	90.0663555	91.6371519	93.2079482	94.7787445	96.3495408	97.9203372	99.4911335	1.0619298	2.6327262
130	200+	4.2035225	5.7743188	7.3451151	8.9159115	10.4867078	12.0575041	13.6283004	15.1990968	16.7698931	18.3406894
140	200+	19.9114858	21.4822821	23.0530784	24.6238747	26.1946711	27.7654674	29.3362637	30.9070600	32.4778564	34.0486527
150	200+	35.6194490	37.1902453	38.7610417	40.3318380	41.9026343	43.4734307	45.0442270	46.6150233	48.1858196	49.7566160
160	200+	51.3274123	52.8982086	54.4690049	56.0398013	57.6105976	59.1813939	60.7521902	62.3229866	63.8937829	65.4645792
170	200+	67.0353756	68.6061719	70.1769682	71.7477645	73.3185609	74.8893572	76.4601535	78.0309498	79.6017462	81.1725425
180	200+	82.7433388	84.3141351	85.8849315	87.4557278	89.0265241	90.5973205	92.1681168	93.7389131	95.3097094	96.8805058
190	200+	98.4513021	0.0220984	1.5928947	3.1636911	4.7344874	6.3052837	7.8760801	9.4468764	11.0176727	12.5884690
200	300+	14.1592654	15.7300617	17.3008580	18.8716543	20.4424507	22.0132470	23.5840433	25.1548396	26.7256360	28.2964323
210	300+	29.8672286	31.4380250	33.0088213	34.5796176	36.1504139	37.7212103	39.2920066	40.8628029	42.4335992	44.0043956
220	300+	45.5751919	47.1459882	48.7167845	50.2875809	51.8583772	53.4291735	54.9999699	56.5707662	58.1415625	59.7123588
230	300+	61.2831552	62.8539515	64.4247478	65.9955441	67.5663405	69.1371368	70.7079331	72.2787295	73.8495258	75.4203221
240	300+	76.9911184	78.5619148	80.1327111	81.7035074	83.2743037	84.8451001	86.4158964	87.9866927	89.5574890	91.1282854
250	300+	92.6990817	94.2698780	95.8406744	97.4114707	98.9822670	0.5530633	2.1238597	3.6946560	5.2654523	6.8362486
260	400+	8.4070450	9.9778413	11.5486376	13.1194339	14.6902303	16.2610266	17.8318229	19.4026193	20.9734156	22.5442119
270	400+	24.1150082	25.6858046	27.2566009	28.8273972	30.3981935	31.9689899	33.5397862	35.1105825	36.6813788	38.2521752
280	400+	39.8229715	41.3937678	42.9645642	44.5353605	46.1061568	47.6769531	49.2477495	50.8185458	52.3893421	53.9601384
290	400+	55.5309348	57.1017311	58.6725274	60.2433238	61.8141201	63.3849164	64.9557127	66.5265091	68.0973054	69.6681017
300	400+	71.2388980	72.8096944	74.3804907	75.9512870	77.5220833	79.0928797	80.6636760	82.2344723	83.8052687	85.3760650
310	400+	86.9468613	88.5176576	90.0884540	91.6592503	93.2300466	94.8008429	96.3716393	97.9424356	99.5132319	1.0840282
320	500+	2.6548246	4.2256209	5.7964172	7.3672136	8.9380099	10.5088062	12.0796025	13.6503989	15.2211952	16.7919915
330	500+	18.3627878	19.9335842	21.5043805	23.0751768	24.6459731	26.2167695	27.7875658	29.3583621	30.9291585	32.4999548
340	500+	34.0707511	35.6415474	37.2123438	38.7831401	40.3539364	41.9247327	43.4955291	45.0663254	46.6371217	48.2079181
350	500+	49.7787144	51.3495107	52.9203070	54.4911034	56.0618997	57.6326960	59.2034923	60.7742887	62.3450850	63.9158813
360	500+	65.4866776	67.0574740	68.6282703	70.1990666	71.7698630	73.3406593	74.9114556	76.4822519	78.0530483	79.6238446
370	500+	81.1946409	82.7654372	84.3362336	85.9070299	87.4778262	89.0486225	90.6194189	92.1902152	93.7610115	95.3318079
380	500+	96.9026042	98.4734005	0.0441968	1.6149932	3.1857895	4.7565858	6.3273821	7.8981785	9.4689748	11.0397711
390	600+	12.6105675	14.1813638	15.7521601	17.3229564	18.8937528	20.4645491	22.0353454	23.6061417	25.1769381	26.7477344
400	600+	28.3185307	29.8893270	31.4601234	33.0309197	34.6017160	36.1725124	37.7433087	39.3141050	40.8849013	42.4556977
410	600+	44.0264940	45.5972903	47.1680866	48.7388830	50.3096793	51.8804756	53.4512719	55.0220683	56.5928646	58.1636609
420	600+	59.7344573	61.3052536	62.8760499	64.4468462	66.0176426	67.5884389	69.1592352	70.7300315	72.3008279	73.8716242
430	600+	75.4424205	77.0132168	78.5840132	80.1548095	81.7256058	83.2964022	84.8671985	86.4379948	88.0087911	89.5795875
440	600+	91.1503838	92.7211801	94.2919764	95.8627728	97.4335691	99.0043654	0.5751618	2.1459581	3.7167544	5.2875507
450	700+	6.8583471	8.4291434	9.9999397	11.5707360	13.1415324	14.7123287	16.2831250	17.8539213	19.4247177	20.9955140
460	700+	22.5663103	24.1371067	25.7079030	27.2786993	28.8494956	30.4202920	31.9910883	33.5618846	35.1326809	36.7034773
470	700+	38.2742736	39.8450699	41.4158662	42.9866626	44.5574589	46.1282552	47.6990516	49.2698479	50.8406442	52.4114405
480	700+	53.9822369	55.5530332	57.1238295	58.6946258	60.2654222	61.8362185	63.4070148	64.9778111	66.5486075	68.1194038
490	700+	69.6902001	71.2609965	72.8317928	74.4025891	75.9733854	77.5441818	79.1149781	80.6857744	82.2565707	83.8273671

$\frac{1}{2}\pi = 1.57079\ 63267\ 94897\ldots$

TABLE 5

Multiples of Half-Pi

N		0	1	2	3	4	5	6	7	8	9
500	700+	85.3981634	86.9689597	88.5397561	90.1105524	91.6813487	93.2521450	94.8229414	96.3937377	97.9645340	99.5353303
510	700+	1.1061267	2.6769230	4.2477193	5.8185156	7.3893120	8.9601083	10.5309046	12.1017010	13.6724973	15.2432936
520	800+	16.8140899	18.3848863	19.9556826	21.5264789	23.0972752	24.6680716	26.2388679	27.8096642	29.3804605	30.9512569
530	800+	32.5220532	34.0928495	35.6636459	37.2344422	38.8052385	40.3760348	41.9468312	43.5176275	45.0884238	46.6592201
540	800+	48.2300165	49.8008128	51.3716091	52.9424054	54.5132018	56.0839981	57.6547944	59.2255908	60.7963871	62.3671834
550	800+	63.9379797	65.5087761	67.0795724	68.6503687	70.2211650	71.7919614	73.3627577	74.9335540	76.5043504	78.0751467
560	800+	79.6459430	81.2167393	82.7875357	84.3583320	85.9291283	87.4999246	89.0707210	90.6415173	92.2123136	93.7831099
570	800+	95.3539063	96.9247026	98.4954989	0.0662953	1.6370916	3.2078879	4.7786842	6.3494806	7.9202769	9.4910732
580	900+	11.0618695	12.6326659	14.2034622	15.7742585	17.3450548	18.9158512	20.4866475	22.0574438	23.6282402	25.1990365
590	900+	26.7698328	28.3406291	29.9114255	31.4822218	33.0530181	34.6238144	36.1946108	37.7654071	39.3362034	40.9069998
600	900+	42.4777961	44.0485924	45.6193887	47.1901851	48.7609814	50.3317777	51.9025740	53.4733704	55.0441667	56.6149630
610	900+	58.1857593	59.7565557	61.3273520	62.8981483	64.4689447	66.0397410	67.6105373	69.1813336	70.7521300	72.3229263
620	900+	73.8937226	75.4645189	77.0353153	78.6061116	80.1769079	81.7477042	83.3185006	84.8892969	86.4600932	88.0308896
630	900+	89.6016859	91.1724822	92.7432785	94.3140749	95.8848712	97.4556675	99.0264638	0.5972602	2.1680565	3.7388528
640	1000+	5.3096491	6.8804455	8.4512418	10.0220381	11.5928345	13.1636308	14.7344271	16.3052234	17.8760198	19.4468161
650	1000+	21.0176124	22.5884087	24.1592051	25.7300014	27.3007977	28.8715941	30.4423904	32.0131867	33.5839830	35.1547794
660	1000+	36.7255757	38.2963720	39.8671683	41.4379647	43.0087610	44.5795573	46.1503536	47.7211500	49.2919463	50.8627426
670	1000+	52.4335390	54.0043353	55.5751316	57.1459279	58.7167243	60.2875206	61.8583169	63.4291132	64.9999096	66.5707059
680	1000+	68.1415022	69.7122985	71.2830949	72.8538912	74.4246875	75.9954839	77.5662802	79.1370765	80.7078728	82.2786692
690	1000+	83.8494655	85.4202618	86.9910581	88.5618545	90.1326508	91.7034471	93.2742434	94.8450398	96.4158361	97.9866324
700	1000+	99.5574288	1.1282251	2.6990214	4.2698177	5.8406141	7.4114104	8.9822067	10.5530030	12.1237994	13.6945957
710	1100+	15.2653920	16.8361884	18.4069847	19.9777810	21.5485773	23.1193737	24.6901700	26.2609663	27.8317626	29.4025590
720	1100+	30.9733553	32.5441516	34.1149479	35.6857443	37.2565406	38.8273369	40.3981333	41.9689296	43.5397259	45.1105222
730	1100+	46.6813186	48.2521149	49.8229112	51.3937075	52.9645039	54.5353002	56.1060965	57.6768928	59.2476892	60.8184855
740	1100+	62.3892818	63.9600782	65.5308745	67.1016708	68.6724671	70.2432635	71.8140598	73.3848561	74.9556524	76.5264488
750	1100+	78.0972451	79.6680414	81.2388377	82.8096341	84.3804304	85.9512267	87.5220231	89.0928194	90.6636157	92.2344120
760	1100+	93.8052084	95.3760047	96.9468010	98.5175973	0.0883937	1.6591900	3.2299863	4.8007827	6.3715790	7.9423753
770	1200+	9.5131716	11.0839680	12.6547643	14.2255606	15.7963569	17.3671533	18.9379496	20.5087459	22.0795422	23.6503386
780	1200+	25.2211349	26.7919312	28.3627276	29.9335239	31.5043202	33.0751165	34.6459129	36.2167092	37.7875055	39.3583018
790	1200+	40.9290982	42.4998945	44.0706908	45.6414871	47.2122835	48.7830798	50.3538761	51.9246725	53.4954688	55.0662651
800	1200+	56.6370614	58.2078578	59.7786541	61.3494504	62.9202467	64.4910431	66.0618394	67.6326357	69.2034321	70.7742284
810	1200+	72.3450247	73.9158210	75.4866174	77.0574137	78.6282100	80.1990063	81.7698027	83.3405990	84.9113953	86.4821916
820	1200+	88.0529880	89.6237843	91.1945806	92.7653770	94.3361733	95.9069696	97.4777659	99.0485623	0.6193586	2.1901549
830	1300+	3.7609512	5.3317476	6.9025439	8.4733402	10.0441365	11.6149329	13.1857292	14.7565255	16.3273219	17.8981182
840	1300+	19.4689145	21.0397108	22.6105072	24.1813035	25.7520998	27.3228961	28.8936925	30.4644888	32.0352851	33.6060814
850	1300+	35.1768778	36.7476741	38.3184704	39.8892668	41.4600631	43.0308594	44.6016557	46.1724521	47.7432484	49.3140447
860	1300+	50.8848410	52.4556374	54.0264337	55.5972300	57.1680264	58.7388227	60.3096190	61.8804153	63.4512117	65.0220080
870	1300+	66.5928043	68.1636006	69.7343970	71.3051933	72.8759896	74.4467859	76.0175823	77.5883786	79.1591749	80.7299713
880	1300+	82.3007676	83.8715639	85.4423602	87.0131566	88.5839529	90.1547492	91.7255455	93.2963419	94.8671382	96.4379345
890	1300+	98.0087308	99.5795272	1.1503235	2.7211198	4.2919162	5.8627125	7.4335088	9.0043051	10.5751015	12.1458978
900	1400+	13.7166941	15.2874904	16.8582868	18.4290831	19.9998794	21.5706757	23.1414721	24.7122684	26.2830647	27.8538611
910	1400+	29.4246574	30.9954537	32.5662500	34.1370464	35.7078427	37.2786390	38.8494353	40.4202317	41.9910280	43.5618243
920	1400+	45.1326207	46.7034170	48.2742133	49.8450096	51.4158060	52.9866023	54.5573986	56.1281949	57.6989913	59.2697876
930	1400+	60.8405839	62.4113802	63.9821766	65.5529729	67.1237692	68.6945656	70.2653619	71.8361582	73.4069545	74.9777509
940	1400+	76.5485472	78.1193435	79.6901398	81.2609362	82.8317325	84.4025288	85.9733251	87.5441215	89.1149178	90.6857141
950	1400+	92.2565105	93.8273068	95.3981031	96.9688994	98.5396958	0.1104921	1.6812884	3.2520847	4.8228811	6.3936774
960	1500+	7.9644737	9.5352700	11.1060664	12.6768627	14.2476590	15.8184554	17.3892517	18.9600480	20.5308443	22.1016407
970	1500+	23.6724370	25.2432333	26.8140296	28.3848260	29.9556223	31.5264186	33.0972150	34.6680113	36.2388076	37.8096039
980	1500+	39.3804003	40.9511966	42.5219929	44.0927892	45.6635856	47.2343819	48.8051782	50.3759745	51.9467709	53.5175672
990	1500+	55.0883635	56.6591599	58.2299562	59.8007525	61.3715488	62.9423452	64.5131415	66.0839378	67.6547341	69.2255305

$\frac{1}{2}\pi = 1.57079\ 63267\ 94897\ \ldots$

$\frac{\sin 2\pi x}{2\pi x}$

TABLE 6

Sinc Function

x		.0	.0001	.0002	.0003	.0004	.0005	.0006	.0007	.0008	.0009	Δ_2
0.0	1.	00000000	99999993	99999974	99999941	99999895	99999836	99999763	99999678	99999579	99999467	-13
0.001	0.	99999342	99999204	99999053	99998888	99998710	99998520	99998316	99998098	99997868	99997625	-13
0.002	0.	99997368	99997098	99996815	99996519	99996210	99995888	99995552	99995203	99994842	99994467	-13
0.003	0.	99994078	99993677	99993262	99992835	99992394	99991940	99991473	99990993.	99990499	99989993	-13
0.004	0.	99989473	99988940	99988394	99987835	99987262	99986877	99986078	99985466	99984841	99984203	-13
0.005	0.	99983551	99982887	99982209	99981519	99980815	99980097	99979367	99978624	99977867	99977098	-13
0.006	0.	99976315	99975519	99974709	99973887	99973052	99972203	99971341	99970466	99969578	99968677	-13
0.007	0.	99967762	99966835	99965894	99964940	99963973	99962993	99962000	99960993	99959974	99958941	-13
0.008	0.	99957895	99956836	99955764	99954678	99953580	99952468	99951343	99950205	99949054	99947890	-13
0.009	0.	99946713	99945522	99944318	99943102	99941872	99940628	99939372	99938103	99936820	99935524	-13
0.010	0.	99934216	99932894	99931558	99930210	99928849	99927474	99926086	99924686	99923272	99921844	-13
0.011	0.	99920404	99918951	99917484	99916005	99914512	99913006	99911487	99909954	99908409	99906850	-13
0.012	0.	99905279	99903694	99902096	99900485	99898861	99897223	99895573	99893909	99892232	99890543	-13
0.013	0.	99888840	99837123	99885394	99883652	99881896	99880127	99878346	99876551	99874743	99872921	-13
0.014	0.	99871087	99869240	99867379	99865505	99363618	99861718	99859805	99857879	99855940	99853987	-13
0.015	0.	99852022	99850043	99848051	99846046	99844028	99841997	99839952	99837895	99835824	99833741	-13
0.016	0.	99831644	99829534	99827411	99825275	99823125	99820963	99818787	99816599	99814397	99812182	-13
0.017	0.	99809954	99807713	99805459	99803191	99800911	99798617	99796311	99793991	99791658	99789312	-13
0.018	0.	99786953	99784581	99782195	99779797	99777385	99774961	99772523	99770072	99767608	99765131	-13
0.019	0.	99762641	99760137	99757621	99755092	99752549	99749993	99747424	99744843	99742248	99739639	-13
0.020	0.	99737018	99734384	99731737	99729076	99726403	99723716	99721016	99718303	99715577	99712838	-13
0.021	0.	99710086	99707321	99704543	99701751	99698947	99696129	99693298	99690455	99687598	99684728	-13
0.022	0.	99681845	99678949	99676040	99673117	99670182	99667234	99664272	99661297	99658310	99655309	-13
0.023	0.	99652295	99649268	99646228	99643175	99640109	99637030	99633938	99630832	99627714	99624582	-13
0.024	0.	99621438	99618280	99615110	99611926	99608729	99605519	99602296	99599060	99595811	99592549	-13
0.025	0.	99589274	99585985	99582684	99579369	99576042	99572701	99569348	99565981	99562602	99559209	-13
0.026	0.	99555803	99552384	99548952	99545507	99542049	99538578	99535094	99531597	99528087	99524563	-13
0.027	0.	99521027	99517478	99513915	99510340	99506751	99503150	99499535	99495908	99492267	99488613	-13
0.028	0.	99484946	99481267	99477574	99473868	99470149	99466417	99462672	99458914	99455143	99451359	-13
0.029	0.	99447562	99443752	99439929	99436093	99432243	99428381	99424506	99420618	99416716	99412802	-13
0.030	0.	99408875	99404935	99400981	99397015	99393035	99389043	99385038	99381019	99376988	99372943	-13
0.031	0.	99368886	99364815	99360732	99356635	99352526	99348403	99344268	99340119	99335958	99331783	-13
0.032	0.	99327596	99323395	99319182	99314955	99310715	99306463	99302197	99297919	99293627	99289323	-13
0.033	0.	99285005	99280675	99276331	99271975	99267606	99263223	99258828	99254419	99249998	99245564	-13
0.034	0.	99241116	99236656	99232183	99227696	99223197	99218685	99214160	99209621	99205070	99200506	-13
0.035	0.	99195929	99191339	99186736	99182120	99177491	99172849	99168194	99163526	99158846	99154152	-13
0.036	0.	99149445	99144725	99139993	99135247	99130488	99125717	99120932	99116135	99111325	99106501	-13
0.037	0.	99101665	99096816	99091954	99087079	99082191	99077290	99072376	99067449	99062509	99057556	-13
0.038	0.	99052591	99047612	99042620	99037616	99032598	99027568	99022525	99017469	99012400	99007317	-13
0.039	0.	99002222	98997115	98991994	98986860	98981713	98976554	98971381	98966196	98960998	98955786	-13
0.040	0.	98950562	98945325	98940075	98934812	98929536	98924248	98918946	98913632	98908304	98902964	-13
0.041	0.	98897611	98892244	98886865	98881474	98876069	98870651	98865220	98859777	98854321	98848851	-13
0.042	0.	98843369	98837874	98832366	98826846	98821312	98815765	98810206	98804634	98799048	98793450	-13
0.043	0.	98787839	98782216	98776579	98770929	98765267	98759592	98753903	98748202	98742489	98736762	-13
0.044	0.	98731022	98725270	98719504	98713726	98707935	98702131	98696315	98690485	98684643	98678787	-13
0.045	0.	98672919	98667038	98661144	98655238	98649318	98643386	98637441	98631483	98625512	98619528	-13
0.046	0.	98613532	98607522	98601500	98595465	98589417	98583356	98577283	98571197	98565098	98558986	-13
0.047	0.	98552861	98546723	98540573	98534410	98528234	98522045	98515843	98509629	98503401	98497161	-13
0.048	0.	98490908	98484643	98478364	98472073	98465769	98459452	98453122	98446780	98440425	98434057	-13
0.049	0.	98427676	98421282	98414876	98408456	98402025	98395580	98389122	98382652	98376169	98369673	-13

TABLE 6

Sinc Function

x		.0	.0001	.0002	.0003	.0004	.0005	.0006	.0007	.0008	.0009	Δ_2
0.050	0.	98363164	98356643	98350109	98343562	98337002	98330430	98323844	98317246	98310636	98304012	-13
0.051	0.	98297376	98290727	98284065	98277390	98270703	98264003	98257290	98250565	98243826	98237075	-13
0.052	0.	98230311	98223535	98216746	98209944	98203129	98196301	98189461	98182608	98175743	98168864	-13
0.053	0.	98161973	98155069	98148153	98141223	98134281	98127327	98120359	98113379	98106386	98099381	-13
0.054	0.	98092362	98085331	98078288	98071231	98064162	98057080	98049986	98042879	98035759	98028626	-13
0.055	0.	98021481	98014323	98007152	97999969	97992773	97985564	97978343	97971108	97963862	97956602	-13
0.056	0.	97949330	97942045	97934748	97927438	97920115	97912779	97905431	97898070	97890697	97883311	-13
0.057	0.	97875912	97868500	97861076	97853640	97846190	97838728	97831253	97823766	97816266	97808753	-13
0.058	0.	97801228	97793690	97786140	97778576	97771001	97763412	97755811	97748197	97740571	97732932	-13
0.059	0.	97725280	97717616	97709939	97702250	97694548	97686833	97679106	97671366	97663613	97655848	-13
0.060	0.	97648070	97640280	97632477	97624662	97616833	97608993	97601139	97593273	97585395	97577504	-13
0.061	0.	97569600	97561684	97553755	97545814	97537860	97529893	97521914	97513922	97505918	97497901	-13
0.062	0.	97489872	97481829	97473775	97465708	97457628	97449536	97441431	97433314	97425184	97417041	-13
0.063	0.	97408886	97400719	97392539	97384346	97376141	97367923	97359693	97351450	97343195	97334927	-13
0.064	0.	97326646	97318353	97310048	97301730	97293400	97285057	97276701	97268333	97259952	97251559	-13
0.065	0.	97243154	97234736	97226305	97217862	97209406	97200938	97192458	97183965	97175459	97166941	-12
0.066	0.	97158410	97149867	97141312	97132744	97124163	97115570	97106965	97098347	97089716	97081074	-12
0.067	0.	97072418	97063750	97055070	97046377	97037672	97028955	97020224	97011482	97002727	96993959	-12
0.068	0.	96985179	96976387	96967582	96958765	96949935	96941093	96932239	96923372	96914492	96905600	-12
0.069	0.	96896696	96887779	96878850	96869908	96860955	96851988	96843009	96834018	96825014	96815998	-12
0.070	0.	96806970	96797929	96788876	96779810	96770732	96761642	96752539	96743423	96734296	96725156	-12
0.071	0.	96716003	96706839	96697661	96688472	96679270	96670056	96660829	96651590	96642338	96633075	-12
0.072	0.	96623799	96614510	96605209	96595896	96586570	96577232	96567882	96558520	96549145	96539757	-12
0.073	0.	96530358	96520946	96511521	96502085	96492636	96483174	96473701	96464215	96454716	96445206	-12
0.074	0.	96435683	96426147	96416600	96407040	96397468	96387883	96378286	96368677	96359056	96349422	-12
0.075	0.	96339776	96330118	96320447	96310764	96301069	96291362	96281642	96271910	96262166	96252409	-12
0.076	0.	96242640	96232859	96223066	96213260	96203442	96193612	96183769	96173915	96164048	96154169	-12
0.077	0.	96144277	96134373	96124457	96114529	96104589	96094636	96084671	96074694	96064705	96054703	-12
0.078	0.	96044689	96034663	96024625	96014574	96004512	95994437	95984349	95974250	95964139	95954015	-12
0.079	0.	95943879	95933731	95923570	95913398	95903213	95893016	95882807	95872585	95862352	95852106	-12
0.080	0.	95841848	95831578	95821296	95811002	95800695	95790376	95780045	95769702	95759347	95748980	-12
0.081	0.	95738600	95728208	95717804	95707388	95696960	95686520	95676068	95665603	95655126	95644638	-12
0.082	0.	95634137	95623624	95613098	95602561	95592011	95581450	95570876	95560290	95549693	95539082	-12
0.083	0.	95528460	95517826	95507180	95496521	95485851	95475168	95464473	95453767	95443048	95432317	-12
0.084	0.	95421574	95410819	95400051	95389272	95378481	95367677	95356862	95346034	95335195	95324343	-12
0.085	0.	95313479	95302604	95291716	95280816	95269904	95258980	95248044	95237096	95226136	95215164	-12
0.086	0.	95204180	95193184	95182176	95171155	95160123	95149079	95138023	95126954	95115874	95104782	-12
0.087	0.	95093678	95082561	95071433	95060293	95049141	95037976	95026800	95015612	95004412	94993200	-12
0.088	0.	94981975	94970739	94959491	94948231	94936959	94925675	94914379	94903071	94891751	94880419	-12
0.089	0.	94869076	94857720	94846352	94834973	94823581	94812178	94800762	94789335	94777896	94766444	-12
0.090	0.	94754981	94743506	94732019	94720520	94709009	94697487	94685952	94674406	94662847	94651277	-12
0.091	0.	94639695	94628100	94616494	94604877	94593247	94581605	94569952	94558286	94546609	94534920	-12
0.092	0.	94523219	94511506	94499781	94488044	94476296	94464536	94452763	94440979	94429183	94417376	-12
0.093	0.	94405556	94393725	94381881	94370026	94358159	94346281	94334390	94322488	94310573	94298647	-12
0.094	0.	94286710	94274760	94262799	94250825	94238840	94226843	94214835	94202814	94190782	94178738	-12
0.095	0.	94166682	94154615	94142535	94130444	94118341	94106226	94094100	94081962	94069812	94057650	-12
0.096	0.	94045476	94033291	94021094	94008885	93996665	93984433	93972189	93959933	93947665	93935386	-12
0.097	0.	93923095	93910793	93898478	93886152	93873814	93861465	93849104	93836731	93824346	93811950	-12
0.098	0.	93799542	93787122	93774690	93762247	93749792	93737326	93724848	93712358	93699856	93687343	-12
0.099	0.	93674818	93662282	93649734	93637174	93624602	93612019	93599424	93586818	93574200	93561570	-12

$$\frac{\sin 2\pi x}{2\pi x}$$

TABLE 6

Sinc Function

x		.0	.0001	.0002	.0003	.0004	.0005	.0006	.0007	.0008	.0009	Δ_2
0.100	0.	93548928	93536275	93523611	93510934	93498246	93485547	93472836	93460113	93447378	93434632	−12
0.101	0.	93421875	93409105	93396325	93383532	93370728	93357912	93345085	93332246	93319396	93306534	−12
0.102	0.	93293660	93280775	93267878	93254970	93242050	93229119	93216176	93203221	93190255	93177277	−12
0.103	0.	93164288	93151287	93138275	93125251	93112216	93099169	93086111	93073041	93059959	93046866	−12
0.104	0.	93033761	93020645	93007518	92994379	92981228	92968066	92954892	92941707	92928511	92915303	−11
0.105	0.	92902083	92888852	92875609	92862355	92849090	92835813	92822525	92809225	92795913	92782590	−11
0.106	0.	92769256	92755910	92742553	92729185	92715805	92702413	92689010	92675596	92662170	92648733	−11
0.107	0.	92635284	92621824	92608352	92594869	92581375	92567869	92554352	92540823	92527283	92513732	−11
0.108	0.	92500169	92486595	92473010	92459413	92445804	92432185	92418554	92404911	92391257	92377592	−11
0.109	0.	92363916	92350228	92336529	92322818	92309096	92295363	92281618	92267862	92254095	92240316	−11
0.110	0.	92226526	92212725	92198912	92185088	92171253	92157406	92143549	92129679	92115799	92101907	−11
0.111	0.	92088004	92074090	92060164	92046227	92032279	92018319	92004348	91990366	91976373	91962368	−11
0.112	0.	91948352	91934325	91920287	91906237	91892176	91878104	91864021	91849926	91835820	91821703	−11
0.113	0.	91807575	91793435	91779284	91765122	91750949	91736765	91722569	91708362	91694144	91679915	−11
0.114	0.	91665674	91651423	91637160	91622886	91608600	91594304	91579996	91565678	91551348	91537006	−11
0.115	0.	91522654	91508291	91493916	91479530	91465133	91450725	91436306	91421876	91407434	91392982	−11
0.116	0.	91378518	91364043	91349557	91335060	91320552	91306032	91291502	91276960	91262408	91247844	−11
0.117	0.	91233269	91218683	91204086	91189478	91174859	91160228	91145587	91130934	91116271	91101596	−11
0.118	0.	91086911	91072214	91057506	91042787	91028058	91013317	90998565	90983802	90969028	90954243	−11
0.119	0.	90939446	90924639	90909821	90894992	90880152	90865301	90850439	90835565	90820681	90805786	−11
0.120	0.	90790880	90775963	90761034	90746095	90731145	90716184	90701212	90686229	90671235	90656230	−11
0.121	0.	90641214	90626187	90611149	90596101	90581041	90565970	90550889	90535796	90520693	90505578	−11
0.122	0.	90490453	90475317	90460169	90445011	90429842	90414662	90399472	90384270	90369057	90353834	−11
0.123	0.	90338600	90323354	90308098	90292831	90277553	90262265	90246965	90231655	90216333	90201001	−11
0.124	0.	90185658	90170304	90154939	90139564	90124177	90108780	90093372	90077953	90062524	90047083	−11
0.125	0.	90031632	90016169	90000696	89985213	89969718	89954213	89938697	89923170	89907632	89892083	−11
0.126	0.	89876524	89860954	89845373	89829781	89814179	89798566	89782942	89767307	89751662	89736006	−11
0.127	0.	89720339	89704661	89688973	89673274	89657564	89641843	89626112	89610370	89594617	89578854	−11
0.128	0.	89563080	89547295	89531499	89515693	89499876	89484048	89468210	89452361	89436501	89420631	−11
0.129	0.	89404750	89388858	89372956	89357043	89341120	89325185	89309240	89293285	89277319	89261342	−11
0.130	0.	89245354	89229356	89213347	89197328	89181298	89165257	89149206	89133144	89117072	89100989	−11
0.131	0.	89084895	89068791	89052676	89036551	89020415	89004269	88988111	88971944	88955766	88939577	−11
0.132	0.	88923377	88907168	88890947	88874716	88858475	88842223	88825960	88809687	88793403	88777109	−11
0.133	0.	88760804	88744489	88728163	88711827	88695480	88679123	88662755	88646377	88629988	88613589	−10
0.134	0.	88597180	88580759	88564329	88547888	88531436	88514974	88498502	88482019	88465525	88449021	−10
0.135	0.	88432507	88415982	88399447	88382902	88366346	88349779	88333202	88316615	88300017	88283409	−10
0.136	0.	88266791	88250162	88233523	88216873	88200213	88183543	88166862	88150170	88133469	88116757	−10
0.137	0.	88100035	88083302	88066559	88049806	88033042	88016268	87999483	87982689	87965884	87949068	−10
0.138	0.	87932242	87915406	87898560	87881703	87864836	87847959	87831071	87814173	87797265	87780347	−10
0.139	0.	87763418	87746479	87729529	87712570	87695600	87678620	87661629	87644629	87627618	87610597	−10
0.140	0.	87593565	87576524	87559472	87542410	87525337	87508255	87491162	87474059	87456945	87439822	−10
0.141	0.	87422688	87405544	87388390	87371226	87354052	87336867	87319672	87302467	87285252	87268027	−10
0.142	0.	87250791	87233545	87216290	87199024	87181747	87164461	87147165	87129858	87112541	87095215	−10
0.143	0.	87077878	87060530	87043173	87025806	87008429	86991041	86973643	86956236	86938818	86921390	−10
0.144	0.	86903952	86886504	86869046	86851577	86834099	86816611	86799112	86781604	86764085	86746557	−10
0.145	0.	86729018	86711469	86693911	86676342	86658763	86641174	86623575	86605967	86588348	86570719	−10
0.146	0.	86553080	86535431	86517772	86500103	86482425	86464736	86447037	86429328	86411609	86393881	−10
0.147	0.	86376142	86358393	86340635	86322866	86305088	86287299	86269501	86251693	86233874	86216046	−10
0.148	0.	86198208	86180360	86162502	86144634	86126757	86108869	86090971	86073064	86055147	86037219	−10
0.149	0.	86019282	86001335	85983379	85965412	85947435	85929449	85911453	85893447	85875431	85857405	−10

TABLE 6

Sinc Function

x		.0	.0001	.0002	.0003	.0004	.0005	.0006	.0007	.0008	.0009	Δ_2
0.150	0.	85839369	85821324	85803268	85785203	85767128	85749043	85730949	85712845	85694730	85676606	−10
0.151	0.	85658473	85640329	85622176	85604012	85585839	85567657	85549464	85531262	85513050	85494828	−10
0.152	0.	85476597	85458355	85440104	85421844	85403573	85385293	85367003	85348703	85330394	85312075	−10
0.153	0.	85293746	85275407	85257059	85238701	85220334	85201956	85183569	85165173	85146766	85128350	−10
0.154	0.	85109924	85091489	85073044	85054589	85036125	85017651	84999167	84980674	84962171	84943659	−10
0.155	0.	84925137	84906605	84888063	84869512	84850952	84832382	84813802	84795212	84776613	84758005	−10
0.156	0.	84739387	84720759	84702122	84683475	84664818	84646152	84627477	84608791	84590097	84571393	−10
0.157	0.	84552679	84533955	84515223	84496480	84477728	84458967	84440196	84421416	84402626	84383826	−9
0.158	0.	84365017	84346199	84327371	84308534	84289687	84270831	84251965	84233090	84214205	84195311	−9
0.159	0.	84176407	84157494	84138571	84119639	84100698	84081747	84062787	84043817	84024838	84005850	−9
0.160	0.	83986852	83967845	83948828	83929802	83910766	83891721	83872667	83853603	83834530	83815448	−9
0.161	0.	83796356	83777255	83758145	83739025	83719896	83700757	83681609	83662452	83643286	83624110	−9
0.162	0.	83604925	83585730	83566527	83547314	83528091	83508860	83489619	83470368	83451109	83431840	−9
0.163	0.	83412562	83393275	83373978	83354672	83335357	83316033	83296699	83277356	83258004	83238643	−9
0.164	0.	83219272	83199892	83180503	83161105	83141698	83122281	83102855	83083420	83063976	83044522	−9
0.165	0.	83025060	83005588	82986107	82966617	82947117	82927609	82908091	82888564	82869028	82849483	−9
0.166	0.	82829929	82810366	82790793	82771212	82751621	82732021	82712412	82692794	82673167	82653530	−9
0.167	0.	82633885	82614231	82594567	82574894	82555213	82535522	82515822	82496113	82476395	82456668	−9
0.168	0.	82436932	82417187	82397433	82377670	82357897	82338116	82318326	82298527	82278718	82258901	−9
0.169	0.	82239075	82219239	82199395	82179542	82159680	82139808	82119928	82100039	82080141	82060234	−9
0.170	0.	82040318	82020392	82000459	81980516	81960564	81940603	81920633	81900655	81880667	81860670	−9
0.171	0.	81840665	81820651	81800628	81780595	81760554	81740505	81720446	81700378	81680302	81660216	−9
0.172	0.	81640122	81620019	81599907	81579786	81559657	81539518	81519371	81499215	81479050	81458876	−9
0.173	0.	81438693	81418502	81398301	81378092	81357874	81337648	81317412	81297168	81276915	81256653	−9
0.174	0.	81236383	81216103	81195815	81175518	81155213	81134899	81114575	81094244	81073903	81053554	−9
0.175	0.	81033196	81012829	80992454	80972069	80951677	80931275	80910865	80890446	80870018	80849582	−9
0.176	0.	80829137	80808683	80788221	80767750	80747270	80726782	80706285	80685779	80665265	80644742	−9
0.177	0.	80624211	80603671	80583122	80562565	80541999	80521424	80500841	80480249	80459649	80439040	−9
0.178	0.	80418422	80397796	80377161	80356518	80335866	80315206	80294537	80273860	80253174	80232479	−9
0.179	0.	80211776	80191064	80170344	80149616	80128878	80108133	80087378	80066616	80045845	80025065	−8
0.180	0.	80004277	79983480	79962675	79941861	79921039	79900209	79879370	79858522	79837666	79816802	−8
0.181	0.	79795929	79775048	79754158	79733260	79712354	79691439	79670516	79649584	79628644	79607695	−8
0.182	0.	79586738	79565773	79544800	79523817	79502827	79481828	79460821	79439806	79418782	79397750	−8
0.183	0.	79376709	79355660	79334603	79313538	79292464	79271381	79250291	79229192	79208085	79186970	−8
0.184	0.	79165846	79144714	79123574	79102425	79081268	79060103	79038930	79017748	78996559	78975361	−8
0.185	0.	78954154	78932940	78911717	78890486	78869246	78847999	78826743	78805479	78784207	78762927	−8
0.186	0.	78741638	78720342	78699037	78677724	78656402	78635073	78613735	78592390	78571036	78549674	−8
0.187	0.	78528303	78506925	78485539	78464144	78442741	78421330	78399911	78378484	78357049	78335606	−8
0.188	0.	78314154	78292695	78271227	78249752	78228268	78206776	78185276	78163768	78142252	78120728	−8
0.189	0.	78099196	78077656	78056108	78034552	78012987	77991415	77969835	77948246	77926650	77905046	−8
0.190	0.	77883434	77861813	77840185	77818549	77796904	77775252	77753592	77731924	77710248	77688564	−8
0.191	0.	77666872	77645172	77623464	77601748	77580024	77558293	77536553	77514805	77493050	77471287	−8
0.192	0.	77449515	77427736	77405949	77384154	77362352	77340541	77318723	77296896	77275062	77253220	−8
0.193	0.	77231370	77209512	77187647	77165773	77143892	77122003	77100106	77078201	77056289	77034368	−8
0.194	0.	77012440	76990504	76968560	76946609	76924650	76902683	76880708	76858725	76836735	76814737	−8
0.195	0.	76792731	76770717	76748696	76726667	76704630	76682586	76660533	76638474	76616406	76594331	−8
0.196	0.	76572247	76550157	76528058	76505952	76483838	76461717	76439588	76417451	76395307	76373155	−8
0.197	0.	76350995	76328828	76306653	76284470	76262280	76240082	76217876	76195663	76173442	76151214	−8
0.198	0.	76128978	76106735	76084484	76062225	76039959	76017685	75995404	75973115	75950818	75928514	−8
0.199	0.	75906202	75883883	75861557	75839223	75816881	75794532	75772175	75749811	75727439	75705060	−7

$\frac{\sin 2\pi x}{2\pi x}$

TABLE 6

Sinc Function

x		.0	.0001	.0002	.0003	.0004	.0005	.0006	.0007	.0008	.0009	Δ_2
0.200	0.	75682673	75660279	75637877	75615468	75593051	75570627	75548195	75525756	75503310	75480856	−7
0.201	0.	75458394	75435926	75413449	75390966	75368475	75345976	75323470	75300957	75278436	75255908	−7
0.202	0.	75233372	75210830	75188279	75165722	75143157	75120584	75098004	75075417	75052823	75030221	−7
0.203	0.	75007612	74984995	74962372	74939740	74917102	74894456	74871803	74849143	74826475	74803800	−7
0.204	0.	74781118	74758429	74735732	74713028	74690316	74667598	74644872	74622139	74599398	74576651	−7
0.205	0.	74553896	74531134	74508365	74485588	74462805	74440014	74417216	74394410	74371598	74348778	−7
0.206	0.	74325951	74303117	74280276	74257427	74234572	74211709	74188839	74165962	74143078	74120187	−7
0.207	0.	74097288	74074383	74051470	74028551	74005624	73982690	73959749	73936800	73913845	73890883	−7
0.208	0.	73867913	73844937	73821953	73798963	73775965	73752960	73729948	73706930	73683904	73660871	−7
0.209	0.	73637831	73614784	73591730	73568669	73545601	73522526	73499444	73476355	73453259	73430156	−7
0.210	0.	73407047	73383930	73360806	73337675	73314537	73291393	73268241	73245083	73221917	73198745	−7
0.211	0.	73175566	73152379	73129186	73105986	73082779	73059565	73036345	73013117	72989883	72966641	−7
0.212	0.	72943393	72920138	72896876	72873607	72850332	72827049	72803760	72780464	72757161	72733851	−7
0.213	0.	72710534	72687211	72663881	72640544	72617200	72593849	72570492	72547128	72523757	72500379	−7
0.214	0.	72476995	72453604	72430206	72406801	72383390	72359971	72336547	72313115	72289677	72266232	−7
0.215	0.	72242780	72219321	72195856	72172384	72148906	72125421	72101929	72078430	72054925	72031413	−7
0.216	0.	72007895	71984370	71960838	71937299	71913754	71890203	71866644	71843079	71819508	71795930	−7
0.217	0.	71772345	71748753	71725155	71701551	71677940	71654322	71630698	71607067	71583430	71559786	−7
0.218	0.	71536135	71512478	71488815	71465145	71441468	71417785	71394095	71370399	71346697	71322988	−6
0.219	0.	71299272	71275550	71251821	71228086	71204345	71180597	71156842	71133081	71109314	71085540	−6
0.220	0.	71061760	71037973	71014180	70990380	70966574	70942762	70918943	70895118	70871287	70847449	−6
0.221	0.	70823604	70799754	70775897	70752033	70728163	70704287	70680405	70656516	70632621	70608719	−6
0.222	0.	70584811	70560897	70536976	70513050	70489116	70465177	70441231	70417279	70393321	70369356	−6
0.223	0.	70345385	70321408	70297425	70273435	70249439	70225437	70201429	70177414	70153393	70129366	−6
0.224	0.	70105333	70081293	70057247	70033195	70009137	69985073	69961002	69936926	69912843	69888754	−6
0.225	0.	69864659	69840557	69816450	69792336	69768216	69744090	69719958	69695820	69671675	69647525	−6
0.226	0.	69623368	69599206	69575037	69550862	69526681	69502494	69478301	69454101	69429896	69405685	−6
0.227	0.	69381467	69357244	69333014	69308779	69284537	69260290	69236036	69211776	69187511	69163239	−6
0.228	0.	69138961	69114678	69090388	69066092	69041791	69017483	68993170	68968850	58944525	68920193	−6
0.229	0.	68895856	68871512	68847163	68822808	68798447	68774080	68749707	68725328	68700943	68676553	−6
0.230	0.	68652156	68627754	68603345	68578931	68554511	68530085	68505653	68481216	68456772	68432323	−6
0.231	0.	68407868	68383407	68358940	68334467	68309989	68285505	68261015	68236519	68212017	68187510	−6
0.232	0.	68162996	68138477	58113953	68089422	68064886	68040344	68015796	67991242	67966683	67942118	−6
0.233	0.	67917548	67892971	57868389	67843801	67819208	67794608	67770003	67745393	67720776	67696154	−6
0.234	0.	67671527	67646893	67622254	67597610	67572959	67548304	67523642	67498975	67474302	67449624	−6
0.235	0.	67424939	67400250	67375555	67350854	67326147	67301435	67276718	67251994	67227266	67202531	−6
0.236	0.	67177791	67153046	67128295	67103539	67078776	67054009	67029236	67004457	66979673	66954883	−6
0.237	0.	66930088	66905288	66880481	66855670	66830853	66806030	66781202	66756369	66731530	66706685	−5
0.238	0.	66681835	66656980	66632119	66607253	66582381	66557504	66532622	66507734	66482841	66457942	−5
0.239	0.	66433038	66408129	66383214	66358294	66333369	66308438	66283501	56258560	66233613	66208661	−5
0.240	0.	66183703	66158740	66133772	66108798	66083819	66058835	66033846	66008851	65983851	65958845	−5
0.241	0.	65933835	65908819	65883798	65858771	65833740	65808703	65783660	65758613	65733560	65708503	−5
0.242	0.	65683439	65658371	65633298	65608219	65583135	65558046	65532952	65507852	65482747	65457638	−5
0.243	0.	65432523	65407402	65382277	65357147	65332011	65306870	65281724	65256573	65231417	65206256	−5
0.244	0.	65181090	65155918	65130742	65105560	65080374	65055182	65029985	65004783	64979576	64954364	−5
0.245	0.	64929147	64903925	54878698	64853465	64828228	64802986	64777738	64752486	64727229	64701967	−5
0.246	0.	64676699	64651427	64626150	64600867	64575580	64550288	64524991	64499689	64474382	64449070	−5
0.247	0.	64423753	64398431	64373104	64347773	64322436	64297094	64271748	64246397	64221040	64195679	−5
0.248	0.	64170313	64144942	64119567	64094186	64068801	64043410	64018015	63992615	63967210	63941801	−5
0.249	0.	63916386	63890967	63865543	63840114	63814680	63789242	63763798	63738350	63712897	63687440	−5

TABLE 6

Sinc Function

x		.0	.0001	.0002	.0003	.0004	.0005	.0006	.0007	.0008	.0009	Δ_2
0.250	0.	63661977	63636510	63611038	63585561	63560080	63534594	63509103	63483607	63458107	.63432602	−5
0.251	0.	63407092	63381578	63356059	63330535	63305006	63279473	63253935	63228393	63202845	.63177293	−5
0.252	0.	63151737	63126176	63100610	63075039	63049464	63023885	62998300	62972711	62947118	62921520	−5
0.253	0.	62895917	62870309	62844698	62819081	62793460	62767834	62742204	62716569	62690930	62665286	−5
0.254	0.	62639638	62613985	62588328	62562666	62536999	62511328	62485653	62459973	62434288	62408599	−4
0.255	0.	62382906	62357208	62331505	62305799	62280087	62254372	62228651	62202927	62177198	62151464	−4
0.256	0.	62125726	62099984	62074237	62048486	62022730	61996970	61971206	61945437	61919664	61893887	−4
0.257	0.	61868105	61842319	61816528	61790733	61764934	61739131	61713323	61687510	61661694	61635873	−4
0.258	0.	61610048	61584218	61558385	61532547	61506704	61480858	61455007	61429152	61403292	61377428	−4
0.259	0.	61351561	61325688	61299812	61273931	61248047	61222157	61196264	61170367	61144465	61118559	−4
0.260	0.	61092649	61066735	61040816	61014894	60988967	60963036	60937101	60911161	60885218	60859270	−4
0.261	0.	60833319	60807363	60781403	60755439	60729471	60703498	60677522	60651542	60625557	60599568	−4
0.262	0.	60573576	60547579	60521578	60495573	60469564	60443551	60417534	60391513	60365488	60339459	−4
0.263	0.	60313425	60287388	60261347	60235302	60209253	60183200	60157142	60131081	60105016	60078947	−4
0.264	0.	60052874	60026797	60000716	59974631	59948543	59922450	59896353	59870253	59844148	59818040	−4
0.265	0.	59791927	59765811	59739691	59713567	59687439	59661308	59635172	59609033	59582889	59556742	−4
0.266	0.	59530591	59504436	59478278	59452115	59425949	59399779	59373605	59347427	59321245	59295060	−4
0.267	0.	59268871	59242678	59216481	59190281	59164077	59137869	59111657	59085442	59059223	59033000	−4
0.268	0.	59006773	58980543	58954309	58928071	58901829	58875584	58849335	58823083	58796826	58770566	−4
0.269	0.	58744303	58718035	58691765	58665490	58639212	58612930	58586644	58560355	58534063	58507766	−4
0.270	0.	58481466	58455163	58428855	58402545	58376230	58349912	58323591	58297266	58270937	58244605	−4
0.271	0.	58218269	58191930	58165587	58139241	58112891	58086537	58060180	58033820	58007456	57981089	−3
0.272	0.	57954718	57928343	57901965	57875584	57849199	57822811	57796419	57770024	57743625	57717223	−3
0.273	0.	57690817	57664408	57637996	57611580	57585161	57558738	57532312	57505882	57479450	57453013	−3
0.274	0.	57426574	57400131	57373684	57347235	57320782	57294325	57267866	57241403	57214936	57188466	−3
0.275	0.	57161993	57135517	57109037	57082554	57056068	57029579	57003086	56976590	56950090	56923588	−3
0.276	0.	56897082	56870572	56844060	56817544	56791026	56764503	56737978	56711450	56684918	56658383	−3
0.277	0.	56631845	56605303	56578759	56552211	56525660	56499106	56472549	56445988	56419425	56392858	−3
0.278	0.	56366288	56339715	56313139	56286560	56259977	56233392	56206803	56180211	56153617	56127019	−3
0.279	0.	56100418	56073814	56047207	56020596	55993983	55967367	55940748	55914125	55887500	55860871	−3
0.280	0.	55834240	55807605	55780968	55754327	55727684	55701037	55674388	55647735	55621080	55594421	−3
0.281	0.	55567760	55541096	55514428	55487758	55461085	55434409	55407730	55381048	55354363	55327675	−3
0.282	0.	55300984	55274290	55247594	55220894	55194192	55167487	55140779	55114068	55087354	55060638	−3
0.283	0.	55033918	55007196	54980471	54953743	54927012	54900278	54873542	54846802	54820060	54793315	−3
0.284	0.	54766568	54739817	54713064	54686308	54659549	54632788	54606024	54579257	54552487	54525714	−3
0.285	0.	54498939	54472161	54445381	54418597	54391811	54365022	54338231	54311437	54284640	54257840	−3
0.286	0.	54231038	54204233	54177426	54150616	54123803	54096987	54070169	54043348	54016525	53989699	−3
0.287	0.	53962870	53936039	53909205	53882369	53855530	53828688	53801844	53774998	53748148	53721296	−3
0.288	0.	53694442	53667585	53640726	53613864	53586999	53560132	53533262	53506390	53479516	53452639	−3
0.289	0.	53425759	53398877	53371992	53345105	53318216	53291324	53264429	53237532	53210633	53183731	−2
0.290	0.	53156827	53129920	53103011	53076100	53049186	53022269	52995351	52968429	52941506	52914580	−2
0.291	0.	52887652	52860721	52833788	52806853	52779915	52752975	52726033	52699088	52672141	52645191	−2
0.292	0.	52618240	52591286	52564329	52537371	52510410	52483447	52456481	52429513	52402543	52375571	−2
0.293	0.	52348596	52321619	52294640	52267659	52240676	52213690	52186702	52159712	52132719	52105724	−2
0.294	0.	52078728	52051729	52024727	51997724	51970718	51943711	51916701	51889689	51862675	51835658	−2
0.295	0.	51808640	51781619	51754596	51727571	51700544	51673515	51646484	51619451	51592415	51565378	−2
0.296	0.	51538338	51511296	51484253	51457207	51430159	51403109	51376057	51349003	51321947	51294889	−2
0.297	0.	51267829	51240767	51213703	51186637	51159569	51132498	51105426	51078352	51051276	51024198	−2
0.298	0.	50997118	50970036	50942952	50915867	50888779	50861689	50834597	50807504	50780408	50753311	−2
0.299	0.	50726212	50699110	50672007	50644902	50617796	50590687	50563576	50536464	50509349	50482233	−2

$$\frac{\sin 2\pi x}{2\pi x}$$

TABLE 6

Sinc Function

x		.0	.0001	.0002	.0003	.0004	.0005	.0006	.0007	.0008	.0009	Δ_2
0.300	0.	50455115	50427995	50400874	50373750	50346625	50319498	50292369	50265238	50238105	50210971	−2
0.301	0.	50183835	50156697	50129557	50102416	50075273	50048128	50020981	49993832	49966682	49939530	−2
0.302	0.	49912376	49885221	49858064	49830905	49803745	49776582	49749418	49722253	49695085	49667916	−2
0.303	0.	49640746	49613574	49586400	49559224	49532047	49504868	49477687	49450505	49423321	49396136	−2
0.304	0.	49368949	49341760	49314570	49287378	49260185	49232990	49205794	49178596	49151396	49124195	−2
0.305	0.	49096992	49069788	49042582	49015374	48988165	48960955	48933743	48906530	48879315	48852098	0
0.306	0.	48824880	48797661	48770440	48743218	48715994	48688768	48661542	48634313	48607084	48579853	0
0.307	0.	48552620	48525386	48498151	48470914	48443676	48416436	48389195	48361953	48334709	48307464	0
0.308	0.	48280217	48252970	48225720	48198470	48171218	48143964	48116710	48089454	48062196	48034938	0
0.309	0.	48007678	47980417	47953154	47925890	47898625	47871359	47844091	47816822	47789552	47762280	0
0.310	0.	47735008	47707734	47680458	47653182	47625904	47598625	47571345	47544064	47516781	47489497	0
0.311	0.	47462212	47434926	47407639	47380350	47353061	47325770	47298478	47271185	47243890	47216595	0
0.312	0.	47189298	47162000	47134702	47107402	47080100	47052798	47025495	46998190	46970885	46943578	0
0.313	0.	46916271	46888962	46861652	46834341	46807029	46779716	46752402	46725087	46697771	46670454	0
0.314	0.	46643136	46615817	46588497	46561176	46533854	46506530	46479206	46451881	46424555	46397228	0
0.315	0.	46369900	46342571	46315241	46287910	46260579	46233246	46205912	46178578	46151242	46123906	0
0.316	0.	46096569	46069230	46041891	46014551	45987211	45959869	45932526	45905183	45877839	45850493	0
0.317	0.	45823148	45795801	45768453	45741105	45713755	45686405	45659054	45631703	45604350	45576997	0
0.318	0.	45549643	45522288	45494932	45467576	45440219	45412861	45385502	45358143	45330783	45303422	0
0.319	0.	45276060	45248698	45221335	45193971	45166607	45139242	45111876	45084509	45057142	45029774	0
0.320	0.	45002406	44975036	44947667	44920296	44892925	44865553	44838181	44810808	44783434	44756060	0
0.321	0.	44728685	44701310	44673934	44646557	44619180	44591802	44564423	44537044	44509665	44482285	0
0.322	0.	44454904	44427523	44400141	44372759	44345376	44317993	44290609	44263225	44235840	44208455	0
0.323	0.	44181069	44153683	44126296	44098909	44071521	44044133	44016744	43989355	43961966	43934576	0
0.324	0.	43907185	43879795	43852403	43825012	43797620	43770227	43742835	43715441	43688048	43660654	0
0.325	0.	43633259	43605865	43578469	43551074	43523678	43496282	43468886	43441489	43414092	43386694	0
0.326	0.	43359296	43331898	43304500	43277101	43249702	43222303	43194903	43167503	43140103	43112703	0
0.327	0.	43085302	43057902	43030500	43003099	42975698	42948296	42920894	42893491	42866089	42838686	0
0.328	0.	42811284	42783880	42756477	42729074	42701670	42674266	42646863	42619458	42592054	42564650	0
0.329	0.	42537245	42509841	42482436	42455031	42427626	42400221	42372816	42345410	42318005	42290599	0
0.330	0.	42263194	42235788	42208382	42180977	42153571	42126165	42098759	42071353	42043947	42016541	0
0.331	0.	41989135	41961729	41934323	41906917	41879510	41852104	41824698	41797292	41769886	41742480	0
0.332	0.	41715074	41687668	41660262	41632856	41605451	41578045	41550639	41523234	41495828	41468423	0
0.333	0.	41441017	41413612	41386207	41358802	41331397	41303992	41276588	41249183	41221779	41194375	0
0.334	0.	41166971	41139567	41112163	41084760	41057356	41029953	41002550	40975147	40947744	40920342	0
0.335	0.	40892940	40865538	40838136	40810734	40783333	40755932	40728531	40701131	40673730	40646330	0
0.336	0.	40618930	40591531	40564132	40536733	40509334	40481936	40454538	40427140	40399742	40372345	0
0.337	0.	40344948	40317552	40290156	40262760	40235365	40207969	40180575	40153180	40125786	40098393	0
0.338	0.	40071000	40043607	40016214	39988822	39961431	39934039	39906649	39879258	39851868	39824479	0
0.339	0.	39797090	39769701	39742313	39714925	39687538	39660151	39632765	39605379	39577994	39550609	0
0.340	0.	39523224	39495841	39468457	39441074	39413692	39386310	39358929	39331548	39304168	39276789	1
0.341	0.	39249410	39222031	39194653	39167276	39139899	39112523	39085148	39057773	39030398	39003024	1
0.342	0.	38975651	38948279	38920907	38893536	38866165	38838795	38811426	38784057	38756689	38729322	1
0.343	0.	38701955	38674589	38647224	38619859	38592495	38565132	38537769	38510408	38483046	38455686	1
0.344	0.	38428326	38400968	38373609	38346252	38318895	38291539	38264184	38236830	38209476	38182124	1
0.345	0.	38154772	38127420	38100070	38072720	38045371	38018023	37990676	37963330	37935984	37908640	1
0.346	0.	37881296	37853953	37826611	37799269	37771929	37744590	37717251	37689913	37662576	37635240	1
0.347	0.	37607905	37580571	37553238	37525906	37498574	37471244	37443914	37416586	37389258	37361931	1
0.348	0.	37334605	37307281	37279957	37252634	37225312	37197991	37170672	37143353	37116035	37088718	1
0.349	0.	37061402	37034087	37006774	36979461	36952149	36924836	36897529	36870220	36842913	36815606	1

TABLE 6

Sinc Function

x		.0	.0001	.0002	.0003	.0004	.0005	.0006	.0007	.0008	.0009	Δ_2
0.350	0.	36788301	36760997	36733694	36706392	36679091	36651791	36624492	36597194	36569898	36542602	1
0.351	0.	36515308	36488015	36460723	36433432	36406142	36378854	36351566	36324280	36296995	36269711	1
0.352	0.	36242428	36215147	36187867	36160588	36133310	36106033	36078758	36051483	36024210	35996939	1
0.353	0.	35969668	35942399	35915131	35887864	35860599	35833335	35806072	35778810	35751550	35724291	1
0.354	0.	35697033	35669777	35642521	35615268	35588015	35560764	35533514	35506266	35479019	35451773	1
0.355	0.	35424528	35397285	35370044	35342803	35315565	35288327	35261091	35233856	35206623	35179391	1
0.356	0.	35152160	35124931	35097704	35070477	35043252	35016029	34988807	34961587	34934368	34907150	1
0.357	0.	34879934	34852720	34825506	34798295	34771085	34743876	34716669	34689463	34662259	34635057	2
0.358	0.	34607856	34580656	34553458	34526262	34499067	34471874	34444682	34417492	34390303	34363116	2
0.359	0.	34335930	34308747	34281564	34254384	34227205	34200027	34172851	34145677	34118505	34091334	2
0.360	0.	34064164	34036997	34009831	33982666	33955504	33928342	33901183	33874025	33846869	33819715	2
0.361	0.	33792563	33765412	33738262	33711115	33683969	33656825	33629683	33602542	33575403	33548266	2
0.362	0.	33521131	33493997	33466866	33439736	33412607	33385481	33358356	33331233	33304112	33276993	2
0.363	0.	33249875	33222760	33195646	33168534	33141424	33114315	33087209	33060104	33033001	33005900	2
0.364	0.	32978801	32951704	32924608	32897515	32870423	32843334	32816246	32789160	32762076	32734994	2
0.365	0.	32707914	32680835	32653759	32626685	32599612	32572542	32545473	32518407	32491342	32464279	2
0.366	0.	32437219	32410160	32383103	32356049	32328996	32301945	32274897	32247850	32220805	32193763	2
0.367	0.	32166722	32139683	32112647	32085612	32058580	32031550	32004521	31977495	31950471	31923449	2
0.368	0.	31896429	31869411	31842395	31815381	31788370	31761360	31734353	31707348	31680345	31653344	2
0.369	0.	31626345	31599348	31572354	31545361	31518371	31491383	31464397	31437414	31410432	31383453	2
0.370	0.	31356476	31329501	31302528	31275558	31248589	31221623	31194660	31167698	31140739	31113782	2
0.371	0.	31086827	31059874	31032924	31005976	30979030	30952087	30925146	30898207	30871270	30844336	2
0.372	0.	30817404	30790474	30763547	30736622	30709699	30682779	30655860	30628945	30602031	30575120	2
0.373	0.	30548212	30521306	30494402	30467500	30440601	30413704	30386810	30359918	30333029	30306141	2
0.374	0.	30279257	30252374	30225495	30198617	30171742	30144870	30118000	30091132	30064267	30037404	2
0.375	0.	30010544	29983686	29956831	29929978	29903128	29876280	29849435	29822592	29795752	29768914	3
0.376	0.	29742079	29715246	29688416	29661588	29634763	29607941	29581121	29554303	29527489	29500676	3
0.377	0.	29473867	29447060	29420255	29393453	29366654	29339857	29313063	29286272	29259483	29232697	3
0.378	0.	29205913	29179132	29152354	29125579	29098806	29072035	29045268	29018503	28991740	28964981	3
0.379	0.	28938224	28911470	28884718	28857969	28831223	28804480	28777739	28751001	28724266	28697534	3
0.380	0.	28670804	28644077	28617353	28590632	28563913	28537197	28510484	28483773	28457066	28430361	3
0.381	0.	28403659	28376960	28350264	28323570	28296879	28270191	28243506	28216824	28190145	28163468	3
0.382	0.	28136794	28110124	28083456	28056790	28030128	28003469	27976812	27950159	27923508	27896860	3
0.383	0.	27870215	27843573	27816934	27790298	27763665	27737035	27710407	27683783	27657161	27630543	3
0.384	0.	27603927	27577315	27550705	27524098	27497495	27470894	27444296	27417702	27391110	27364521	3
0.385	0.	27337936	27311353	27284773	27258197	27231623	27205053	27178485	27151921	27125359	27098801	3
0.386	0.	27072246	27045694	27019144	26992598	26966055	26939516	26912979	26886445	26859915	26833387	3
0.387	0.	26806863	26780342	26753824	26727309	26700797	26674288	26647783	26621280	26594781	26568285	3
0.388	0.	26541792	26515303	26488816	26462333	26435853	26409376	26382902	26356432	26329965	26303501	3
0.389	0.	26277040	26250582	26224128	26197677	26171229	26144784	26118343	26091905	26065470	26039038	3
0.390	0.	26012610	25986185	25959763	25933345	25906930	25880518	25854109	25827704	25801302	25774904	3
0.391	0.	25748508	25722117	25695728	25669343	25642961	25616582	25590207	25563836	25537467	25511102	3
0.392	0.	25484741	25458382	25432027	25405676	25379328	25352983	25326642	25300304	25273970	25247639	3
0.393	0.	25221311	25194987	25168667	25142349	25116036	25089726	25063419	25037115	25010816	24984519	3
0.394	0.	24958226	24931937	24905651	24879369	24853090	24826814	24800543	24774274	24748009	24721748	4
0.395	0.	24695490	24669236	24642986	24616739	24590495	24564255	24538019	24511786	24485557	24459331	4
0.396	0.	24433109	24406891	24380676	24354465	24328257	24302053	24275853	24249656	24223463	24197273	4
0.397	0.	24171087	24144905	24118727	24092552	24066381	24040213	24014049	23987889	23961732	23935580	4
0.398	0.	23909431	23883285	23857143	23831005	23804871	23778741	23752614	23726491	23700371	23674256	4
0.399	0.	23648144	23622036	23595931	23569831	23543734	23517641	23491551	23465466	23439384	23413306	4

$$\frac{\sin 2\pi x}{2\pi x}$$

TABLE 6

Sinc Function

x		.0	.0001	.0002	.0003	.0004	.0005	.0006	.0007	.0008	.0009	Δ_2
0.400	0.	23387232	23361162	23335095	23309033	23282974	23256919	23230867	23204820	23178776	23152737	4
0.401	0.	23126701	23100669	23074641	23048616	23022596	22996579	22970567	22944558	22918553	22892552	4
0.402	0.	22866555	22840562	22814572	22788587	22762606	22736628	22710655	22684685	22658719	22632757	4
0.403	0.	22606800	22580846	22554896	22528950	22503008	22477070	22451136	22425206	22399280	22373358	4
0.404	0.	22347440	22321526	22295616	22269710	22243808	22217910	22192016	22166126	22140240	22114359	4
0.405	0.	22088481	22062607	22036738	22010872	21985011	21959153	21933300	21907451	21881606	21855765	4
0.406	0.	21829928	21804095	21778266	21752442	21726621	21700805	21674993	21649185	21623381	21597581	4
0.407	0.	21571786	21545994	21520207	21494424	21468645	21442870	21417099	21391333	21365571	21339813	4
0.408	0.	21314059	21288310	21262564	21236823	21211086	21185353	21159625	21133901	21108181	21082465	4
0.409	0.	21056754	21031046	21005343	20979645	20953950	20928260	20902574	20876893	20851216	20825543	4
0.410	0.	20799874	20774210	20748550	20722894	20697242	20671595	20645953	20620314	20594680	20569050	4
0.411	0.	20543425	20517804	20492188	20466575	20440967	20415364	20389765	20364170	20338580	20312994	4
0.412	0.	20287412	20261835	20236262	20210694	20185130	20159571	20134016	20108465	20082919	20057377	4
0.413	0.	20031840	20006307	19980779	19955255	19929736	19904221	19878710	19853205	19827703	19802206	4
0.414	0.	19776714	19751226	19725742	19700263	19674789	19649319	19623854	19598393	19572937	19547485	5
0.415	0.	19522038	19496595	19471157	19445724	19420295	19394871	19369451	19344036	19318625	19293219	5
0.416	0.	19267818	19242421	19217029	19191641	19166258	19140880	19115506	19090137	19064773	19039413	5
0.417	0.	19014058	18988708	18963362	18938021	18912684	18887352	18862025	18836703	18811385	18786072	5
0.418	0.	18760764	18735460	18710161	18684867	18659577	18634292	18609012	18583737	18558466	18533201	5
0.419	0.	18507939	18482683	18457431	18432185	18406942	18381705	18356473	18331245	18306022	18280804	5
0.420	0.	18255590	18230382	18205178	18179979	18154785	18129595	18104411	18079231	18054056	18028886	5
0.421	0.	18003721	17978560	17953405	17928254	17903108	17877967	17852831	17827700	17802574	17777452	5
0.422	0.	17752336	17727224	17702118	17677016	17651919	17626827	17601740	17576657	17551580	17526508	5
0.423	0.	17501441	17476378	17451321	17426268	17401220	17376178	17351140	17326107	17301080	17276057	5
0.424	0.	17251039	17226027	17201019	17176016	17151018	17126025	17101038	17076055	17051077	17026105	5
0.425	0.	17001137	16976174	16951217	16926264	16901317	16876374	16851437	16826505	16801578	16776655	5
0.426	0.	16751738	16726826	16701919	16677018	16652121	16627229	16602343	16577461	16552585	16527714	5
0.427	0.	16502848	16477987	16453131	16428281	16403435	16378595	16353760	16328930	16304105	16279285	5
0.428	0.	16254471	16229661	16204857	16180058	16155265	16130476	16105693	16080914	16056142	16031374	5
0.429	0.	16006611	15981854	15957102	15932355	15907613	15882877	15858146	15833420	15808700	15783984	5
0.430	0.	15759274	15734569	15709870	15685176	15660487	15635803	15611125	15586451	15561784	15537121	5
0.431	0.	15512464	15487812	15463166	15438525	15413889	15389258	15364633	15340013	15315399	15290789	5
0.432	0.	15266186	15241587	15216994	15192406	15167824	15143247	15118676	15094110	15069549	15044993	5
0.433	0.	15020443	14995899	14971360	14946826	14922298	14897775	14873257	14848745	14824239	14799738	5
0.434	0.	14775242	14750752	14726267	14701788	14677314	14652846	14628383	14603925	14579474	14555027	5
0.435	0.	14530586	14506151	14481721	14457296	14432878	14408464	14384056	14359654	14335257	14310866	6
0.436	0.	14286480	14262100	14237725	14213356	14188993	14164635	14140282	14115936	14091594	14067259	6
0.437	0.	14042929	14018604	13994285	13969972	13945664	13921362	13897066	13872775	13848490	13824210	6
0.438	0.	13799936	13775668	13751405	13727148	13702897	13678651	13654411	13630176	13605948	13581725	6
0.439	0.	13557507	13533295	13509089	13484889	13460694	13436505	13412322	13388144	13363973	13339806	6
0.440	0.	13315646	13291491	13267342	13243199	13219062	13194930	13170804	13146683	13122569	13098460	6
0.441	0.	13074357	13050260	13026169	13002083	12978003	12953929	12929861	12905798	12881741	12857690	6
0.442	0.	12833645	12809606	12785573	12761545	12737523	12713507	12689497	12665493	12641494	12617501	6
0.443	0.	12593515	12569534	12545559	12521589	12497626	12473669	12449717	12425771	12401832	12377898	6
0.444	0.	12353970	12330047	12306131	12282221	12258316	12234418	12210525	12186639	12162758	12138883	6
0.445	0.	12115014	12091152	12067295	12043444	12019599	11995760	11971926	11948099	11924278	11900463	6
0.446	0.	11876654	11852851	11829053	11805262	11781477	11757698	11733924	11710157	11686396	11662641	6
0.447	0.	11638892	11615149	11591411	11567680	11543955	11520236	11496524	11472817	11449116	11425421	6
0.448	0.	11401733	11378050	11354373	11330703	11307039	11283380	11259728	11236082	11212442	11188808	6
0.449	0.	11165181	11141559	11117944	11094334	11070731	11047134	11023543	10999958	10976379	10952807	6

TABLE 6

Sinc Function

x		.0	.0001	.0002	.0003	.0004	.0005	.0006	.0007	.0008	.0009	Δ_2
0.450	0.	10929240	10905680	10882126	10858578	10835037	10811501	10787972	10764448	10740931	10717421	6
0.451	0.	10693916	10670418	10646925	10623439	10599960	10576486	10553019	10529557	10506103	10482654	6
0.452	0.	10459212	10435775	10412345	10388922	10365504	10342093	10318688	10295290	10271897	10248511	6
0.453	0.	10225131	10201758	10178390	10155030	10131675	10108327	10084984	10061649	10038319	10014996	6
0.454	0.	09991679	09968369	09945065	09921767	09898476	09875190	09851912	09828639	09805373	09782113	6
0.455	0.	09758860	09735613	09712373	09689138	09665910	09642689	09619474	09596265	09573063	09549867	6
0.456	0.	09526678	09503495	09480318	09457148	09433984	09410826	09387675	09364531	09341393	09318261	6
0.457	0.	09295136	09272017	09248905	09225799	09202700	09179607	09156520	09133440	09110367	09087300	6
0.458	0.	09064239	09041185	09018138	08995096	08972062	08949034	08926012	08902997	08879989	08856987	7
0.459	0.	08833991	08811002	08788020	08765044	08742075	08719112	08696156	08673206	08650263	08627327	7
0.460	0.	08604397	08581473	08558556	08535646	08512743	08489846	08466955	08444071	08421194	08398323	7
0.461	0.	08375459	08352602	08329751	08306907	08284069	08261238	08238414	08215596	08192785	08169981	7
0.462	0.	08147183	08124392	08101607	08078829	08056058	08033294	08010536	07987785	07965040	07942303	7
0.463	0.	07919572	07896847	07874129	07851418	07828714	07806017	07783326	07760642	07737964	07715293	7
0.464	0.	07692629	07669972	07647322	07624678	07602041	07579410	07556787	07534170	07511560	07488957	7
0.465	0.	07466360	07443771	07421188	07398611	07376042	07353479	07330923	07308374	07285832	07263297	7
0.466	0.	07240768	07218246	07195731	07173223	07150722	07128227	07105739	07083259	07060784	07038317	7
0.467	0.	07015857	06993403	06970957	06948517	06926084	06903658	06881239	06858826	06836421	06814022	7
0.468	0.	06791630	06769246	06746868	06724497	06702132	06679775	06657425	06635081	06612745	06590415	7
0.469	0.	06568093	06545777	06523468	06501166	06478871	06456583	06434302	06412028	06389761	06367501	7
0.470	0.	06345247	06323001	06300762	06278529	06256304	06234086	06211874	06189670	06167472	06145282	7
0.471	0.	06123098	06100922	06078752	06056590	06034435	06012286	05990145	05968010	05945883	05923763	7
0.472	0.	05901649	05879543	05857444	05835352	05813267	05791189	05769118	05747054	05724997	05702947	7
0.473	0.	05680904	05658869	05636840	05614819	05592804	05570797	05548797	05526804	05504818	05482839	7
0.474	0.	05460867	05438902	05416945	05394994	05373051	05351115	05329186	05307264	05285349	05263441	7
0.475	0.	05241541	05219647	05197761	05175882	05154010	05132146	05110288	05088438	05066594	05044758	7
0.476	0.	05022930	05001108	04979294	04957486	04935686	04913893	04892108	04870329	04848558	04826794	7
0.477	0.	04805037	04783288	04761545	04739810	04718082	04696362	04674648	04652942	04631243	04609552	7
0.478	0.	04587867	04566190	04544520	04522858	04501202	04479554	04457914	04436280	04414654	04393035	7
0.479	0.	04371423	04349819	04328222	04306632	04285050	04263475	04241907	04220347	04198793	04177248	7
0.480	0.	04155709	04134178	04112654	04091138	04069628	04048127	04026632	04005145	03983665	03962193	7
0.481	0.	03940728	03919270	03897820	03876377	03854942	03833514	03812093	03790679	03769273	03747875	7
0.482	0.	03726484	03705100	03683724	03662355	03640993	03619639	03598292	03576953	03555621	03534297	7
0.483	0.	03512980	03491670	03470368	03449073	03427786	03406506	03385234	03363969	03342712	03321462	7
0.484	0.	03300220	03278985	03257757	03236537	03215325	03194120	03172922	03151732	03130549	03109374	8
0.485	0.	03088207	03067047	03045894	03024749	03003612	02982482	02961359	02940244	02919137	02898037	8
0.486	0.	02876945	02855860	02834783	02813713	02792651	02771597	02750550	02729510	02708478	02687454	8
0.487	0.	02666437	02645428	02624426	02603432	02582446	02561467	02540496	02519532	02498576	02477628	8
0.488	0.	02456687	02435754	02414828	02393910	02373000	02352097	02331202	02310315	02289435	02268563	8
0.489	0.	02247698	02226841	02205992	02185150	02164317	02143490	02122672	02101861	02081057	02060262	8
0.490	0.	02039474	02018694	01997921	01977156	01956399	01935649	01914907	01894173	01873447	01852728	8
0.491	0.	01832017	01811314	01790618	01769930	01749250	01728578	01707913	01687256	01666607	01645965	8
0.492	0.	01625332	01604706	01584087	01563477	01542874	01522279	01501692	01481112	01460541	01439977	8
0.493	0.	01419421	01398872	01378332	01357799	01337274	01316756	01296247	01275745	01255252	01234765	8
0.494	0.	01214287	01193817	01173354	01152899	01132452	01112013	01091582	01071158	01050743	01030335	8
0.495	0.	01009935	00989543	00969158	00948782	00928413	00908053	00887700	00867355	00847017	00826688	8
0.496	0.	00806367	00786053	00765747	00745450	00725160	00704878	00684604	00664337	00644079	00623829	8
0.497	0.	00603586	00583351	00563125	00542906	00522695	00502492	00482297	00462110	00441930	00421759	8
0.498	0.	00401596	00381440	00361293	00341153	00321022	00300898	00280783	00260675	00240575	00220483	8
0.499	0.	00200399	00180324	00160256	00140196	00120144	00100100	00080064	00060036	00040016	00020004	-8

$$\frac{\sin 2\pi x}{2\pi x}$$

$$\frac{\sin 2\pi x}{2\pi x}$$

TABLE 6

Sinc Function

x		.0	.0001	.0002	.0003	.0004	.0005	.0006	.0007	.0008	.0009	Δ_2
0.500	-0.	00000000	00019996	00039984	00059964	00079936	00099900	00119856	00139804	00159744	00179676	8
0.501	-0.	00199599	00219515	00239423	00259323	00279215	00299098	00318974	00338841	00358701	00378552	8
0.502	-0.	00398396	00418231	00438059	00457878	00477689	00497492	00517287	00537074	00556853	00576623	8
0.503	-0.	00596386	00616141	00635887	00655626	00675356	00695078	00714792	00734498	00754196	00773886	8
0.504	-0.	00793567	00813241	00832906	00852563	00872212	00891853	00911486	00931111	00950727	00970336	8
0.505	-0.	00989936	01009528	01029112	01048688	01068256	01087815	01107366	01126910	01146445	01165971	8
0.506	-0.	01185490	01205000	01224503	01243997	01263482	01282960	01302430	01321891	01341344	01360789	8
0.507	-0.	01380226	01399654	01419074	01438486	01457890	01477286	01496673	01516052	01535423	01554786	8
0.508	-0.	01574140	01593486	01612824	01632154	01651475	01670789	01690093	01709390	01728679	01747959	8
0.509	-0.	01767231	01786494	01805750	01824997	01844236	01863466	01882688	01901902	01921108	01940305	8
0.510	-0.	01959494	01978675	01997848	02017012	02036168	02055315	02074455	02093586	02112708	02131822	8
0.511	-0.	02150928	02170026	02189115	02208196	02227269	02246333	02265389	02284437	02303476	02322507	8
0.512	-0.	02341530	02360544	02379550	02398548	02417537	02436518	02455490	02474454	02493410	02512357	8
0.513	-0.	02531296	02550227	02569149	02588063	02606968	02625865	02644754	02663634	02682506	02701369	8
0.514	-0.	02720224	02739071	02757909	02776739	02795560	02814373	02833178	02851974	02870761	02889541	8
0.515	-0.	02908311	02927074	02945828	02964573	02983310	03002039	03020759	03039471	03058174	03076869	8
0.516	-0.	03095555	03114233	03132902	03151563	03170215	03188859	03207495	03226122	03244740	03263351	8
0.517	-0.	03281952	03300545	03319130	03337706	03356273	03374832	03393383	03411925	03430459	03448984	9
0.518	-0.	03467500	03486008	03504508	03522999	03541481	03559955	03578421	03596877	03615326	03633766	9
0.519	-0.	03652197	03670620	03689034	03707439	03725836	03744225	03762605	03780976	03799339	03817693	9
0.520	-0.	03836039	03854376	03872705	03891025	03909336	03927639	03945933	03964219	03982496	04000765	9
0.521	-0.	04019025	04037276	04055519	04073753	04091978	04110195	04128403	04146603	04164794	04182977	9
0.522	-0.	04201151	04219316	04237472	04255620	04273760	04291890	04310013	04328126	04346231	04364327	9
0.523	-0.	04382415	04400493	04418564	04436625	04454678	04472722	04490758	04508785	04526803	04544813	9
0.524	-0.	04562814	04580806	04598790	04616765	04634731	04652689	04670638	04688578	04706509	04724432	9
0.525	-0.	04742346	04760252	04778149	04796037	04813916	04831787	04849649	04887502	04885346	04903182	9
0.526	-0.	04921009	04938828	04956637	04974438	04992230	05010014	05027789	05045555	05063312	05081060	9
0.527	-0.	05098800	05116531	05134254	05151967	05169672	05187368	05205055	05222734	05240404	05258065	9
0.528	-0.	05275717	05293360	05310995	05328621	05346238	05363847	05381446	05399037	05416619	05434192	9
0.529	-0.	05451757	05469312	05486859	05504397	05521927	05539447	05556959	05574462	05591956	05609441	9
0.530	-0.	05626917	05644385	05661844	05679294	05696735	05714167	05731591	05749006	05766411	05783809	9
0.531	-0.	05801197	05818576	05835947	05853308	05870661	05888005	05905340	05922666	05939984	05957293	9
0.532	-0.	05974592	05991883	06009165	06026438	06043702	06060958	06078204	06095442	06112671	06129891	9
0.533	-0.	06147102	06164304	06181497	06198681	06215857	06233023	06250181	06267330	06284470	06301601	9
0.534	-0.	06318723	06335836	06352940	06370036	06387122	06404199	06421268	06438328	06455379	06472420	9
0.535	-0.	06489453	06506477	06523492	06540498	06557496	06574484	06591463	06608434	06625395	06642348	9
0.536	-0.	06659291	06676226	06693151	06710068	06726976	06743874	06760764	06777645	06794517	06811380	9
0.537	-0.	06828234	06845079	06861915	06878742	06895560	06912369	06929169	06945960	06962742	06979515	9
0.538	-0.	06996280	07013035	07029781	07046518	07063246	07079965	07096676	07113377	07130069	07146752	9
0.539	-0.	07163426	07180091	07196747	07213394	07230032	07246661	07263282	07279892	07296494	07313087	9
0.540	-0.	07329671	07346246	07362812	07379369	07395916	07412455	07428985	07445505	07462017	07478520	9
0.541	-0.	07495013	07511497	07527973	07544439	07560896	07577344	07593784	07610214	07626635	07643046	9
0.542	-0.	07659449	07675843	07692228	07708603	07724970	07741327	07757676	07774015	07790345	07806666	9
0.543	-0.	07822978	07839281	07855575	07871859	07888135	07904402	07920659	07936907	07953146	07969376	9
0.544	-0.	07985597	08001809	08018012	08034206	08050390	08066565	08082732	08098889	08115037	08131176	9
0.545	-0.	08147305	08163426	08179537	08195640	08211733	08227817	08243892	08259957	08276014	08292061	9
0.546	-0.	08308100	08324129	08340149	08356160	08372161	08388154	08404137	08420111	08436076	08452032	9
0.547	-0.	08467979	08483916	08499845	08515764	08531674	08547575	08563466	08579349	08595222	08611086	9
0.548	-0.	08626941	08642787	08658623	08674450	08690268	08706077	08721877	08737668	08753449	08769221	9
0.549	-0.	08784984	08800737	08816482	08832217	08847943	08863660	08879368	08895066	08910755	08926435	9

TABLE 6

Sinc Function

x		.0	.0001	.0002	.0003	.0004	.0005	.0006	.0007	.0008	.0009	Δ_2
0.550	-0.	08942106	08957767	08973420	08989063	09004696	09020321	09035936	09051542	09067139	09082727	9
0.551	-0.	09098305	09113874	09129434	09144985	09160526	09176058	09191581	09207095	09222599	09238094	9
0.552	-0.	09253580	09269057	09284524	09299982	09315431	09330870	09346301	09361721	09377133	09392536	9
0.553	-0.	09407929	09423313	09438687	09454052	09469408	09484755	09500093	09515421	09530740	09546049	9
0.554	-0.	09561349	09576640	09591922	09607194	09622457	09637711	09652956	09668191	09683417	09698633	9
0.555	-0.	09713840	09729038	09744227	09759406	09774576	09789737	09804888	09820030	09835163	09850286	9
0.556	-0.	09865400	09880505	09895600	09910686	09925763	09940830	09955888	09970937	09985976	10001006	9
0.557	-0.	10016027	10031038	10046040	10061033	10076016	10090990	10105954	10120910	10135855	10150792	9
0.558	-0.	10165719	10180637	10195545	10210444	10225334	10240214	10255085	10269947	10284799	10299642	9
0.559	-0.	10314475	10329299	10344114	10358919	10373715	10388501	10403279	10418046	10432805	10447554	9
0.560	-0.	10462293	10477023	10491744	10506456	10521158	10535850	10550533	10565207	10579872	10594527	9
0.561	-0.	10609172	10623808	10638435	10653053	10667660	10682259	10696848	10711428	10725998	10740559	9
0.562	-0.	10755110	10769652	10784185	10798708	10813222	10827726	10842221	10856707	10871183	10885649	9
0.563	-0.	10900106	10914554	10928992	10943421	10957840	10972250	10986651	11001042	11015424	11029796	9
0.564	-0.	11044158	11058512	11072856	11087190	11101515	11115830	11130136	11144433	11158720	11172997	9
0.565	-0.	11187265	11201524	11215773	11230013	11244243	11258464	11272675	11286877	11301070	11315253	9
0.566	-0.	11329426	11343590	11357744	11371889	11386025	11400151	11414267	11428374	11442472	11456560	9
0.567	-0.	11470638	11484708	11498767	11512817	11526858	11540889	11554910	11568923	11582925	11596918	10
0.568	-0.	11610902	11624876	11638840	11652795	11666741	11680677	11694604	11708521	11722428	11736326	10
0.569	-0.	11750214	11764093	11777963	11791823	11805673	11819514	11833345	11847167	11860979	11874782	10
0.570	-0.	11888575	11902359	11916133	11929898	11943653	11957398	11971134	11984861	11998578	12012285	10
0.571	-0.	12025983	12039671	12053350	12067019	12080679	12094329	12107969	12121600	12135222	12148834	10
0.572	-0.	12162436	12176029	12189612	12203186	12216750	12230305	12243850	12257385	12270911	12284427	10
0.573	-0.	12297934	12311431	12324919	12338397	12351865	12365324	12378773	12392213	12405643	12419064	10
0.574	-0.	12432475	12445876	12459268	12472650	12486023	12499386	12512740	12526084	12539418	12552743	10
0.575	-0.	12566058	12579363	12592659	12605946	12619222	12632490	12645747	12658995	12672234	12685462	10
0.576	-0.	12698682	12711891	12725091	12738282	12751462	12764634	12777795	12790947	12804090	12817222	10
0.577	-0.	12830345	12843459	12856563	12869657	12882742	12895817	12908882	12921938	12934984	12948021	10
0.578	-0.	12961048	12974065	12987073	13000071	13013060	13026039	13039008	13051967	13064917	13077858	10
0.579	-0.	13090788	13103709	13116621	13129523	13142415	13155297	13168170	13181033	13193887	13206731	10
0.580	-0.	13219565	13232390	13245205	13258010	13270806	13283592	13296369	13309135	13321893	13334640	10
0.581	-0.	13347378	13360106	13372825	13385534	13398233	13410922	13423602	13436273	13448933	13461584	10
0.582	-0.	13474225	13486857	13499479	13512091	13524694	13537287	13549870	13562444	13575008	13587562	10
0.583	-0.	13600107	13612642	13625167	13637683	13650188	13662685	13675171	13687648	13700115	13712573	10
0.584	-0.	13725021	13737459	13749888	13762306	13774716	13787115	13799505	13811885	13824255	13836616	10
0.585	-0.	13848967	13861308	13873640	13885962	13898274	13910577	13922870	13935153	13947427	13959690	10
0.586	-0.	13971944	13984189	13996424	14008649	14020864	14033070	14045266	14057452	14069628	14081795	10
0.587	-0.	14093952	14106100	14118237	14130365	14142484	14154592	14166691	14178780	14190860	14202929	10
0.588	-0.	14214990	14227040	14239080	14251111	14263133	14275144	14287146	14299138	14311120	14323093	10
0.589	-0.	14335056	14347009	14358952	14370886	14382810	14394724	14406629	14418524	14430409	14442284	10
0.590	-0.	14454150	14466006	14477852	14489688	14501515	14513332	14525139	14536937	14548725	14560503	10
0.591	-0.	14572271	14584030	14595779	14607518	14619247	14630967	14642677	14654377	14666067	14677748	10
0.592	-0.	14689419	14701080	14712732	14724374	14736006	14747628	14759240	14770843	14782436	14794020	10
0.593	-0.	14805593	14817157	14828711	14840255	14851790	14863315	14874830	14886335	14897831	14909316	10
0.594	-0.	14920792	14932259	14943715	14955162	14966599	14978026	14989444	15000852	15012250	15023638	10
0.595	-0.	15035016	15046385	15057744	15069093	15080433	15091763	15103083	15114393	15125693	15136984	10
0.596	-0.	15148265	15159536	15170797	15182049	15193291	15204523	15215745	15226957	15238160	15249353	10
0.597	-0.	15260536	15271710	15282874	15294027	15305172	15316306	15327431	15338545	15349650	15360746	10
0.598	-0.	15371831	15382907	15393973	15405029	15416075	15427112	15438139	15449156	15460163	15471161	10
0.599	-0.	15482149	15493127	15504095	15515053	15526002	15536941	15547870	15558789	15569698	15580598	10

$\frac{\sin 2\pi x}{2\pi x}$

TABLE 6

Sinc Function

x		.0	.0001	.0002	.0003	.0004	.0005	.0006	.0007	.0008	.0009	Δ_2
0.600	-0.	15591488	15602368	15613239	15624099	15634950	15645791	15656622	15667444	15678255	15689057	10
0.601	-0.	15699849	15710631	15721404	15732167	15742920	15753663	15764396	15775120	15785833	15796537	10
0.602	-0.	15807231	15817916	15828590	15839255	15849910	15860555	15871191	15881817	15892432	15903038	10
0.603	-0.	15913635	15924221	15934798	15945365	15955922	15966469	15977006	15987534	15998052	16008560	10
0.604	-0.	16019058	16029547	16040025	16050494	16060953	16071403	16081842	16092272	16102692	16113102	10
0.605	-0.	16123502	16133892	16144273	16154644	16165005	16175356	16185698	16196029	16206351	16216663	10
0.606	-0.	16226966	16237258	16247541	16257814	16268077	16278330	16288573	16298807	16309031	16319245	10
0.607	-0.	16329449	16339643	16349828	16360002	16370167	16380323	16390468	16400603	16410729	16420845	10
0.608	-0.	16430951	16441047	16451134	16461211	16471277	16481334	16491382	16501419	16511447	16521465	10
0.609	-0.	16531473	16541471	16551459	16561438	16571406	16581365	16591315	16601254	16611183	16621103	10
0.610	-0.	16631013	16640913	16650803	16660684	16670554	16680415	16690266	16700107	16709939	16719760	10
0.611	-0.	16729572	16739374	16749166	16758948	16768721	16778483	16788236	16797979	16807712	16817436	10
0.612	-0.	16827149	16836853	16846547	16856231	16865906	16875570	16885225	16894870	16904505	16914130	10
0.613	-0.	16923745	16933351	16942947	16952533	16962109	16971675	16981232	16990779	17000315	17009842	10
0.614	-0.	17019360	17028867	17038365	17047853	17057331	17066799	17076257	17085706	17095144	17104573	10
0.615	-0.	17113992	17123401	17132801	17142191	17151570	17160940	17170301	17179651	17188991	17198322	10
0.616	-0.	17207643	17216954	17226255	17235547	17244828	17254100	17263362	17272614	17281857	17291089	10
0.617	-0.	17300312	17309525	17318728	17327921	17337105	17346278	17355442	17364596	17373740	17382875	10
0.618	-0.	17391999	17401114	17410219	17419314	17428399	17437475	17446540	17455596	17464642	17473678	10
0.619	-0.	17482705	17491721	17500728	17509725	17518712	17527689	17536657	17545614	17554562	17563500	10
0.620	-0.	17572428	17581347	17590255	17599154	17608043	17616922	17625791	17634651	17643501	17652340	10
0.621	-0.	17661171	17669991	17678801	17687602	17696393	17705174	17713945	17722706	17731458	17740199	10
0.622	-0.	17748931	17757653	17766366	17775068	17783761	17792444	17801117	17809780	17818433	17827077	10
0.623	-0.	17835711	17844335	17852949	17861553	17870148	17878732	17887307	17895872	17904428	17912973	10
0.624	-0.	17921509	17930035	17938551	17947057	17955554	17964040	17972517	17980984	17989441	17997889	10
0.625	-0.	18006326	18014754	18023172	18031580	18039979	18048367	18056746	18065115	18073474	18081823	10
0.626	-0.	18090163	18098493	18106813	18115123	18123423	18131714	18139994	18148265	18156526	18164778	10
0.627	-0.	18173019	13181251	18189473	18197685	18205887	18214080	18222262	18230435	18238598	18246752	10
0.628	-0.	18254895	13263029	18271153	18279267	18287371	18295466	18303550	18311625	18319691	18327746	10
0.629	-0.	18335791	13343827	18351853	18359869	18367876	18375872	18383859	18391836	18399803	18407760	10
0.630	-0.	18415708	18423646	18431574	18439492	18447400	18455299	18463188	18471067	18478936	18486796	10
0.631	-0.	18494645	18502485	18510315	18518136	18525946	18533747	18541538	18549319	18557091	18564852	10
0.632	-0.	18572604	18580346	18588078	18595801	18603514	18611217	18618910	18626593	18634267	18641930	10
0.633	-0.	18649584	18657229	18664863	18672488	18680103	18687708	18695303	18702889	18710465	18718031	10
0.634	-0.	18725587	18733133	18740670	18748197	18755714	18763221	18770719	18778207	18785685	18793153	10
0.635	-0.	18800612	18808060	18815499	18822929	18830348	18837758	18845158	18852548	18859928	18867299	10
0.636	-0.	18874660	18882011	18889352	18896684	18904005	18911317	18918620	18925912	18933195	18940468	10
0.637	-0.	18947731	18954985	18962228	18969462	18976686	18983901	18991106	18998300	19005486	19012661	10
0.638	-0.	19019827	19026983	19034129	19041265	19048392	19055509	19062616	19069713	19076801	19083879	10
0.639	-0.	19090947	19098005	19105054	19112093	19119122	19126141	19133151	19140151	19147141	19154122	10
0.640	-0.	19161092	19168053	19175005	19181946	19188878	19195800	19202712	19209615	19216507	19223390	10
0.641	-0.	19230264	19237127	19243981	19250825	19257660	19264484	19271299	19278104	19284900	19291686	10
0.642	-0.	19298462	19305228	19311984	19318731	19325468	19332196	19338913	19345621	19352319	19359008	10
0.643	-0.	19365687	19372356	19379015	19385665	19392304	19398935	19405555	19412166	19418767	19425358	10
0.644	-0.	19431940	19438511	19445074	19451626	19458169	19464702	19471225	19477739	19484242	19490737	10
0.645	-0.	19497221	19503696	19510161	19516616	19523062	19529498	19535924	19542341	19548747	19555145	10
0.646	-0.	19561532	19567910	19574278	19580636	19586985	19593324	19599653	19605973	19612282	19618583	10
0.647	-0.	19624873	19631154	19637425	19643686	19649938	19656180	19662412	19668635	19674848	19681051	10
0.648	-0.	19687245	19693429	19699603	19705768	19711923	19718068	19724203	19730329	19736445	19742552	10
0.649	-0.	19748649	19754736	19760813	19766881	19772939	19778988	19785027	19791056	19797075	19803085	10

TABLE 6

Sinc Function

x		.0	.0001	.0002	.0003	.0004	.0005	.0006	.0007	.0008	.0009	Δ_2
0.650	-0.	19809085	19815076	19821056	19827028	19832989	19838941	19844883	19850816	19856738	19862652	10
0.651	-0.	19868555	19874449	19880333	19886208	19892073	19897928	19903773	19909609	19915436	19921252	10
0.652	-0.	19927059	19932857	19938644	19944422	19950191	19955950	19961699	19967438	19973168	19978888	10
0.653	-0.	19984599	19990300	19995991	20001673	20007345	20013007	20018660	20024303	20029937	20035561	10
0.654	-0.	20041175	20046780	20052375	20057960	20063536	20069102	20074658	20080205	20085742	20091270	10
0.655	-0.	20096788	20102296	20107795	20113284	20118764	20124234	20129694	20135145	20140586	20146018	10
0.656	-0.	20151439	20156852	20162254	20167648	20173031	20178405	20183769	20189124	20194469	20199804	10
0.657	-0.	20205130	20210447	20215753	20221050	20226338	20231616	20236884	20242143	20247392	20252631	10
0.658	-0.	20257861	20263082	20268292	20273494	20278685	20283867	20289040	20294203	20299356	20304500	10
0.659	-0.	20309634	20314758	20319873	20324979	20330075	20335161	20340238	20345305	20350362	20355410	10
0.660	-0.	20360449	20365478	20370497	20375507	20380507	20385498	20390479	20395450	20400412	20405365	10
0.661	-0.	20410308	20415241	20420165	20425079	20429984	20434879	20439764	20444640	20449507	20454364	10
0.662	-0.	20459211	20464049	20468877	20473696	20478505	20483305	20488095	20492876	20497647	20502409	10
0.663	-0.	20507161	20511903	20516636	20521360	20526074	20530778	20535473	20540158	20544834	20549501	10
0.664	-0.	20554158	20558805	20563443	20568071	20572690	20577299	20581899	20586489	20591070	20595641	10
0.665	-0.	20600203	20604755	20609298	20613831	20618355	20622869	20627374	20631869	20636355	20640831	9
0.666	-0.	20645298	20649755	20654203	20658641	20663070	20667489	20671899	20676299	20680690	20685072	9
0.667	-0.	20689443	20693806	20698159	20702502	20706836	20711161	20715476	20719782	20724078	20728364	9
0.668	-0.	20732642	20736909	20741168	20745416	20749656	20753886	20758106	20762317	20766519	20770711	9
0.669	-0.	20774893	20779066	20783230	20787384	20791529	20795665	20799791	20803907	20808014	20812112	9
0.670	-0.	20816200	20820279	20824348	20828408	20832458	20836499	20840531	20844553	20848566	20852569	9
0.671	-0.	20856563	20860547	20864522	20868488	20872444	20876391	20880328	20884256	20888175	20892084	9
0.672	-0.	20895984	20899874	20903755	20907626	20911488	20915341	20919184	20923018	20926843	20930658	9
0.673	-0.	20934463	20938260	20942047	20945824	20949592	20953351	20957100	20960840	20964571	20968292	9
0.674	-0.	20972004	20975706	20979399	20983083	20986757	20990422	20994078	20997724	21001361	21004988	9
0.675	-0.	21008606	21012215	21015814	21019404	21022985	21026556	21030118	21033671	21037214	21040748	9
0.676	-0.	21044272	21047787	21051293	21054790	21058277	21061755	21065223	21068682	21072132	21075572	9
0.677	-0.	21079003	21082425	21085838	21089241	21092634	21096019	21099394	21102760	21106116	21109463	9
0.678	-0.	21112801	21116130	21119449	21122759	21126059	21129350	21132632	21135905	21139168	21142422	9
0.679	-0.	21145667	21148902	21152129	21155345	21158553	21161751	21164940	21168120	21171290	21174451	9
0.680	-0.	21177603	21180745	21183878	21187002	21190117	21193222	21196318	21199405	21202482	21205551	9
0.681	-0.	21208610	21211659	21214700	21217731	21220753	21223765	21226769	21229763	21232748	21235723	9
0.682	-0.	21238690	21241647	21244595	21247533	21250463	21253383	21256293	21259195	21262087	21264971	9
0.683	-0.	21267844	21270709	21273565	21276411	21279248	21282075	21284894	21287703	21290503	21293294	9
0.684	-0.	21296076	21298848	21301611	21304365	21307110	21309845	21312572	21315289	21317997	21320695	9
0.685	-0.	21323385	21326065	21328736	21331398	21334051	21336694	21339328	21341953	21344569	21347176	9
0.686	-0.	21349774	21352362	21354941	21357511	21360072	21362624	21365166	21367699	21370223	21372738	9
0.687	-0.	21375244	21377741	21380228	21382706	21385176	21387636	21390086	21392528	21394960	21397384	9
0.688	-0.	21399798	21402203	21404599	21406986	21409363	21411732	21414091	21416441	21418782	21421114	9
0.689	-0.	21423437	21425751	21428055	21430351	21432637	21434914	21437182	21439441	21441691	21443931	9
0.690	-0.	21446163	21448385	21450599	21452803	21454998	21457184	21459361	21461529	21463687	21465837	9
0.691	-0.	21467978	21470109	21472231	21474345	21476449	21478544	21480630	21482707	21484774	21486833	9
0.692	-0.	21488883	21490923	21492955	21494977	21496991	21498995	21500990	21502977	21504954	21506922	9
0.693	-0.	21508881	21510831	21512772	21514703	21516626	21518540	21520445	21522340	21524227	21526105	9
0.694	-0.	21527973	21529833	21531683	21533525	21535357	21537180	21538995	21540800	21542596	21544384	9
0.695	-0.	21546162	21547931	21549691	21551443	21553185	21554918	21556642	21558357	21560064	21561761	9
0.696	-0.	21563449	21565128	21566798	21568460	21570112	21571755	21573389	21575015	21576631	21578238	9
0.697	-0.	21579836	21581426	21583006	21584578	21586140	21587693	21589238	21590773	21592300	21593818	9
0.698	-0.	21595326	21596826	21598317	21599798	21601271	21602735	21604190	21605636	21607073	21608501	9
0.699	-0.	21609920	21611331	21612732	21614124	21615508	21616882	21618248	21619604	21620952	21622291	9

$\frac{\sin 2\pi x}{2\pi x}$

$\frac{\sin 2\pi x}{2\pi x}$

TABLE 6

Sinc Function

x		.0	.0001	.0002	.0003	.0004	.0005	.0006	.0007	.0008	.0009	Δ_2
0.700	-0.	21623621	21624942	21626254	21627557	21628851	21630137	21631413	21632681	21633939	21635189	9
0.701	-0.	21636430	21637662	21638885	21640099	21641304	21642501	21643688	21644867	21646036	21647197	9
0.702	-0.	21648349	21649492	21650627	21651752	21652869	21653976	21655075	21656165	21657246	21658318	9
0.703	-0.	21659382	21660436	21661482	21662518	21663546	21664566	21665576	21666577	21667570	21668554	9
0.704	-0.	21669529	21670495	21671452	21672400	21673340	21674271	21675193	21676106	21677010	21677906	9
0.705	-0.	21678792	21679670	21680539	21681400	21682251	21683094	21683928	21684753	21685569	21686377	9
0.706	-0.	21687176	21687965	21688747	21689519	21690283	21691037	21691783	21692521	21693249	21693969	9
0.707	-0.	21694680	21695382	21696076	21696760	21697436	21698103	21698762	21699411	21700052	21700684	9
0.708	-0.	21701308	21701923	21702528	21703126	21703714	21704294	21704865	21705427	21705981	21706526	9
0.709	-0.	21707062	21707589	21708108	21708618	21709119	21709611	21710095	21710570	21711037	21711494	9
0.710	-0.	21711943	21712384	21712815	21713238	21713652	21714058	21714455	21714843	21715222	21715593	9
0.711	-0.	21715955	21716309	21716654	21716990	21717317	21717636	21717946	21718248	21718540	21718825	9
0.712	-0.	21719100	21719367	21719625	21719875	21720116	21720348	21720572	21720787	21720993	21721191	9
0.713	-0.	21721380	21721560	21721732	21721895	21722050	21722196	21722333	21722462	21722582	21722694	9
0.714	-0.	21722797	21722891	21722977	21723054	21723123	21723162	21723234	21723277	21723311	21723336	9
0.715	-0.	21723353	21723362	21723362	21723353	21723336	21723310	21723275	21723232	21723181	21723121	9
0.716	-0.	21723052	21722975	21722889	21722795	21722692	21722580	21722460	21722332	21722195	21722049	9
0.717	-0.	21721895	21721732	21721561	21721382	21721193	21720997	21720791	21720578	21720355	21720124	9
0.718	-0.	21719885	21719637	21719381	21719116	21718843	21718561	21718271	21717972	21717664	21717349	8
0.719	-0.	21717024	21716692	21716350	21716001	21715642	21715276	21714901	21714517	21714125	21713724	8
0.720	-0.	21713315	21712898	21712472	21712038	21711595	21711144	21710684	21710216	21709739	21709254	8
0.721	-0.	21708761	21708259	21707749	21707230	21706703	21706167	21705623	21705071	21704510	21703941	8
0.722	-0.	21703363	21702777	21702182	21701580	21700968	21700349	21699721	21699084	21698439	21697786	8
0.723	-0.	21697124	21696454	21695776	21695089	21694394	21693691	21692979	21692259	21691530	21690793	8
0.724	-0.	21690048	21689294	21688532	21687762	21686983	21686196	21685400	21684597	21683785	21682964	8
0.725	-0.	21682135	21681298	21680453	21679599	21678737	21677867	21676988	21676101	21675206	21674302	8
0.726	-0.	21673390	21672470	21671541	21670604	21669659	21668706	21667744	21666774	21665796	21664809	8
0.727	-0.	21663814	21662811	21661800	21660780	21659752	21658716	21657672	21656619	21655558	21654488	8
0.728	-0.	21653411	21652325	21651231	21650129	21649018	21647900	21646773	21645637	21644494	21643342	8
0.729	-0.	21642182	21641014	21639838	21638653	21637461	21636260	21635050	21633833	21632607	21631373	8
0.730	-0.	21630131	21628881	21627623	21626356	21625081	21623798	21622507	21621208	21619900	21618584	8
0.731	-0.	21617261	21615928	21614588	21613240	21611883	21610519	21609146	21607765	21606375	21604978	8
0.732	-0.	21603573	21602159	21600737	21599307	21597869	21596423	21594969	21593506	21592036	21590557	8
0.733	-0.	21589070	21587575	21586072	21584561	21583042	21581515	21579979	21578436	21576884	21575324	8
0.734	-0.	21573756	21572181	21570597	21569004	21567404	21565796	21564180	21562555	21560923	21559282	8
0.735	-0.	21557634	21555977	21554312	21552640	21550959	21549270	21547573	21545868	21544155	21542434	8
0.736	-0.	21540705	21538968	21537223	21535469	21533708	21531939	21530162	21528377	21526583	21524782	8
0.737	-0.	21522973	21521155	21519330	21517497	21515656	21513806	21511949	21510084	21508211	21506329	8
0.738	-0.	21504440	21502543	21500638	21498725	21496804	21494875	21492937	21490993	21489040	21487079	8
0.739	-0.	21485110	21483133	21481148	21479156	21477155	21475146	21473130	21471106	21469073	21467033	8
0.740	-0.	21464985	21462929	21460865	21458793	21456713	21454625	21452529	21450426	21448314	21446195	8
0.741	-0.	21444068	21441933	21439790	21437639	21435480	21433313	21431139	21428956	21426766	21424568	8
0.742	-0.	21422362	21420148	21417926	21415696	21413459	21411214	21408960	21406699	21404431	21402154	8
0.743	-0.	21399869	21397577	21395277	21392969	21390653	21388329	21385998	21383659	21381312	21378957	8
0.744	-0.	21376594	21374223	21371845	21369459	21367065	21364663	21362254	21359837	21357412	21354979	8
0.745	-0.	21352538	21350090	21347634	21345170	21342698	21340219	21337731	21335236	21332734	21330223	8
0.746	-0.	21327705	21325179	21322645	21320104	21317555	21314998	21312433	21309861	21307281	21304693	8
0.747	-0.	21302098	21299494	21296884	21294265	21291639	21289004	21286363	21283713	21281056	21278391	8
0.748	-0.	21275719	21273039	21270351	21267655	21264952	21262241	21259523	21256796	21254062	21251321	8
0.749	-0.	21248572	21245815	21243050	21240278	21237498	21234711	21231916	21229113	21226303	21223485	8

TABLE 6

Sinc Function

x		.0	.0001	.0002	.0003	.0004	.0005	.0006	.0007	.0008	.0009	Δ_2
0.750	-0.	21220659	21217826	21214985	21212137	21209280	21206417	21203545	21200667	21197780	21194886	
0.751	-0.	21191984	21189075	21186158	21183234	21180302	21177362	21174415	21171460	21168498	21165528	8
0.752	-0.	21162550	21159565	21156572	21153572	21150565	21147549	21144526	21141496	21138458	21135413	8
0.753	-0.	21132360	21129299	21126231	21123156	21120073	21116982	21113884	21110778	21107665	21104545	8
0.754	-0.	21101416	21098281	21095138	21091987	21088829	21085663	21082490	21079310	21076122	21072926	8
0.755	-0.	21069723	21066513	21063295	21060069	21056837	21053596	21050348	21047093	21043831	21040560	7
0.756	-0.	21037283	21033998	21030706	21027406	21024098	21020784	21017462	21014132	21010795	21007451	7
0.757	-0.	21004099	21000740	20997373	20993999	20990618	20987229	20983833	20980430	20977019	20973600	7
0.758	-0.	20970175	20966742	20963301	20959853	20956398	20952936	20949466	20945989	20942504	20939012	7
0.759	-0.	20935513	20932006	20928492	20924971	20921442	20917906	20914363	20910612	20907255	20903689	7
0.760	-0.	20900117	20896537	20892950	20889355	20885753	20882144	20878528	20874904	20871273	20867635	7
0.761	-0.	20863990	20860337	20856677	20853010	20849335	20845653	20841964	20838268	20834564	20830853	7
0.762	-0.	20827135	20823410	20819677	20815937	20812190	20808436	20804674	20800905	20797129	20793346	7
0.763	-0.	20789556	20785758	20781953	20778141	20774322	20770495	20766662	20762821	20758973	20755117	7
0.764	-0.	20751255	20747385	20743509	20739625	20735733	20731835	20727930	20724017	20720097	20716170	7
0.765	-0.	20712236	20708295	20704347	20700391	20696429	20692459	20688482	20684498	20680507	20676508	7
0.766	-0.	20672503	20668490	20664471	20660444	20656410	20652369	20648321	20644266	20640204	20636135	7
0.767	-0.	20632058	20627975	20623884	20619786	20615682	20611570	20607451	20603325	20599192	20595052	7
0.768	-0.	20590905	20586751	20582590	20578422	20574246	20570064	20565875	20561679	20557475	20553265	7
0.769	-0.	20549047	20544823	20540592	20536353	20532108	20527855	20523596	20519329	20515056	20510776	7
0.770	-0.	20506488	20502194	20497892	20493584	20489269	20484947	20480617	20476281	20471938	20467588	7
0.771	-0.	20463231	20458867	20454496	20450118	20445733	20441341	20436943	20432537	20428125	20423705	7
0.772	-0.	20419279	20414845	20410405	20405958	20401504	20397043	20392575	20388101	20383619	20379131	7
0.773	-0.	20374635	20370133	20365624	20361108	20356585	20352055	20347519	20342975	20338425	20333868	7
0.774	-0.	20329304	20324733	20320155	20315571	20310979	20306381	20301776	20297164	20292546	20287920	7
0.775	-0.	20283288	20278649	20274003	20269350	20264691	20260024	20255351	20250671	20245985	20241291	7
0.776	-0.	20236591	20231884	20227170	20222449	20217722	20212968	20208247	20203500	20198745	20193984	7
0.777	-0.	20189216	20184442	20179660	20174872	20170077	20165276	20160467	20155652	20150831	20146002	7
0.778	-0.	20141167	20136325	20131477	20126622	20121760	20116891	20112016	20107134	20102245	20097350	7
0.779	-0.	20092447	20087539	20082623	20077701	20072773	20067837	20062895	20057946	20052991	20048029	7
0.780	-0.	20043060	20038085	20033103	20028115	20023120	20018118	20013109	20008094	20003073	19998045	7
0.781	-0.	19993010	19987968	19982920	19977866	19972804	19967736	19962662	19957581	19952493	19947399	7
0.782	-0.	19942299	19937191	19932077	19926957	19921830	19916697	19911557	19906410	19901257	19896097	7
0.783	-0.	19890931	19885758	19880579	19875393	19870201	19865002	19859796	19854584	19849366	19844141	7
0.784	-0.	19838910	19833672	19828427	19823177	19817919	19812655	19807385	19802108	19796825	19791535	6
0.785	-0.	19786239	19780936	19775627	19770312	19764990	19759661	19754326	19748985	19743637	19738283	6
0.786	-0.	19732922	19727555	19722182	19716802	19711415	19706023	19700624	19695218	19689806	19684388	6
0.787	-0.	19678963	19673532	19668094	19662650	19657200	19651743	19646280	19640811	19635335	19629853	6
0.788	-0.	19624365	19618870	19613369	19607861	19602347	19596827	19591300	19585768	19580228	19574683	6
0.789	-0.	19569131	19563573	19558008	19552438	19546860	19541277	19535687	19530091	19524489	19518880	6
0.790	-0.	19513266	19507644	19502017	19496383	19490743	19485097	19479445	19473786	19468121	19462450	6
0.791	-0.	19456772	19451088	19445398	19439702	19434000	19428291	19422576	19416855	19411127	19405394	6
0.792	-0.	19399654	19393908	19388156	19382397	19376633	19370862	19365085	19359302	19353512	19347717	6
0.793	-0.	19341915	19336107	19330293	19324473	19318647	19312814	19306975	19301131	19295280	19289422	6
0.794	-0.	19283559	19277690	19271814	19265932	19260045	19254151	19248251	19242345	19236432	19230514	6
0.795	-0.	19224590	19218659	19212722	19206780	19200831	19194876	19188915	19182948	19176975	19170995	6
0.796	-0.	19165010	19159019	19153021	19147018	19141008	19134993	19128971	19122944	19116910	19110870	6
0.797	-0.	19104825	19098773	19092715	19086651	19080582	19074506	19068424	19062336	19056242	19050142	6
0.798	-0.	19044037	19037925	19031807	19025683	19019554	19013418	19007276	19001129	18994975	18988815	6
0.799	-0.	18982650	18976478	18970301	18964118	18957928	18951733	18945532	18939325	18933112	18926893	6

$$\frac{\sin 2\pi x}{2\pi x}$$

TABLE 6

Sinc Function

x		.0	.0001	.0002	.0003	.0004	.0005	.0006	.0007	.0008	.0009	Δ_2
0.800	-0.	18920668	18914437	18908201	18901958	18895710	18889455	18883195	18876929	18870657	18864379	6
0.801	-0.	18858095	18851806	18845510	18839209	18832901	18826588	18820269	18813944	18807614	18801277	6
0.802	-0.	18794935	18788587	18782232	18775873	18769507	18763135	18756758	18750375	18743986	18737591	6
0.803	-0.	18731191	18724784	18718372	18711954	18705530	18699101	18692665	18686224	18679777	18673325	6
0.804	-0.	18666866	18660402	18653932	18647457	18640975	18634488	18627995	18621496	18614992	18608482	6
0.805	-0.	18601966	18595444	18588917	18582384	18575845	18569301	18562751	18556195	18549633	18543066	6
0.806	-0.	18536493	18529914	18523330	18516740	18510144	18503543	18496936	18490323	18483705	18477081	6
0.807	-0.	18470452	18463816	18457175	18450529	18443877	18437219	18430555	18423886	18417212	18410531	6
0.808	-0.	18403845	18397154	18390457	18383754	18377045	18370332	18363612	18356887	18350156	18343420	6
0.809	-0.	18336678	18329931	18323178	18316419	18309655	18302885	18296110	18289329	18282543	18275751	6
0.810	-0.	18268954	18262151	18255342	18248528	18241709	18234884	18228053	18221217	18214375	18207528	6
0.811	-0.	18200676	18193818	18186954	18180085	18173211	18166331	18159445	18152554	18145658	18138756	5
0.812	-0.	18131849	18124936	18118018	18111094	18104165	18097230	18090290	18083345	18076394	18069437	5
0.813	-0.	18062476	18055509	18048536	18041558	18034575	18027586	18020592	18013592	18006587	17999577	5
0.814	-0.	17992561	17985540	17978514	17971482	17964444	17957402	17950354	17943301	17936242	17929178	5
0.815	-0.	17922109	17915034	17907954	17900869	17893778	17886682	17879581	17872474	17865362	17858245	5
0.816	-0.	17851122	17843994	17836861	17829723	17822579	17815430	17808276	17801116	17793951	17786781	5
0.817	-0.	17779606	17772425	17765239	17758048	17750851	17743650	17736443	17729231	17722013	17714791	5
0.818	-0.	17707563	17700330	17693091	17685848	17678599	17671345	17664086	17656822	17649552	17642278	5
0.819	-0.	17634998	17627713	17620422	17613127	17605826	17598521	17591210	17583894	17576572	17569246	5
0.820	-0.	17561914	17554578	17547236	17539889	17532537	17525180	17517817	17510450	17503077	17495699	5
0.821	-0.	17488317	17480929	17473536	17466137	17458734	17451326	17443913	17436494	17429071	17421642	5
0.822	-0.	17414208	17406769	17399326	17391877	17384423	17376964	17369500	17362031	17354557	17347078	5
0.823	-0.	17339593	17332104	17324610	17317111	17309606	17302097	17294583	17287064	17279539	17272010	5
0.824	-0.	17264476	17256937	17249392	17241843	17234289	17226730	17219166	17211597	17204023	17196444	5
0.825	-0.	17188860	17181271	17173677	17166078	17158475	17150866	17143252	17135634	17128010	17120382	5
0.826	-0.	17112749	17105111	17097468	17089820	17082167	17074509	17066847	17059179	17051507	17043830	5
0.827	-0.	17036147	17028460	17020769	17013072	17005370	16997664	16989953	16982237	16974516	16966790	5
0.828	-0.	16959059	16951324	16943583	16935838	16928088	16920334	16912574	16904810	16897041	16889267	5
0.829	-0.	16881488	16873705	16865916	16858123	16850326	16842523	16834716	16826903	16819087	16811265	5
0.830	-0.	16803439	16795607	16787771	16779931	16772086	16764235	16756381	16748521	16740657	16732788	5
0.831	-0.	16724914	16717036	16709153	16701265	16693372	16685475	16677573	16669667	16661755	16653840	5
0.832	-0.	16645919	16637994	16630064	16622129	16614190	16606246	16598298	16590344	16582387	16574424	5
0.833	-0.	16566457	16558485	16550509	16542528	16534543	16526552	16518558	16510558	16502554	16494546	5
0.834	-0.	16486533	16478515	16470493	16462466	16454434	16446398	16438358	16430312	16422263	16414208	5
0.835	-0.	16406150	16398086	16390013	16381946	16373869	16365787	16357701	16349611	16341515	16333416	5
0.836	-0.	16325312	16317203	16309090	16300972	16292850	16284724	16276593	16268457	16260317	16252172	4
0.837	-0.	16244023	16235870	16227712	16219550	16211383	16203212	16195036	16186856	16178671	16170482	4
0.838	-0.	16162289	16154091	16145888	16137682	16129471	16121255	16113035	16104811	16096582	16088349	4
0.839	-0.	16080111	16071869	16063623	16055372	16047117	16038858	16030594	16022326	16014054	16005777	4
0.840	-0.	15997496	15989210	15980920	15972626	15964327	15956025	15947717	15939406	15931090	15922770	4
0.841	-0.	15914446	15906117	15897784	15889446	15881105	15872759	15864409	15856054	15847696	15839333	4
0.842	-0.	15830965	15822594	15814218	15805838	15797454	15789065	15780672	15772275	15763874	15755469	4
0.843	-0.	15747059	15738645	15730227	15721804	15713378	15704947	15696512	15688073	15679630	15671182	4
0.844	-0.	15662730	15654274	15645814	15637350	15628882	15620409	15611932	15603451	15594966	15586477	4
0.845	-0.	15577984	15569486	15560985	15552479	15543969	15535455	15526937	15518415	15509888	15501358	4
0.846	-0.	15492823	15484284	15475742	15467195	15458644	15450089	15441530	15432967	15424399	15415828	4
0.847	-0.	15407253	15398673	15390090	15381502	15372911	15364315	15355716	15347112	15338504	15329893	4
0.848	-0.	15321277	15312657	15304033	15295405	15286774	15278138	15269498	15260854	15252207	15243555	4
0.849	-0.	15234899	15226239	15217576	15208908	15200237	15191561	15182881	15174198	15165511	15156819	4

TABLE 6

Sinc Function

x		.0	.0001	.0002	.0003	.0004	.0005	.0006	.0007	.0008	.0009	Δ_2
0.850	-0.	15148124	15139425	15130722	15122015	15113304	15104589	15095870	15087147	15078420	15069690	4
0.851	-0.	15060955	15052217	15043475	15034729	15025979	15017225	15008467	14999706	14990940	14982171	4
0.852	-0.	14973398	14964621	14955840	14947055	14938266	14929474	14920678	14911878	14903074	14894266	4
0.853	-0.	14885455	14876639	14867820	14858997	14850171	14841340	14832506	14823668	14814826	14805980	4
0.854	-0.	14797131	14788278	14779421	14770560	14761695	14752827	14743955	14735080	14726200	14717317	4
0.855	-0.	14708430	14699539	14690645	14681747	14672845	14663939	14655030	14646117	14637201	14628280	4
0.856	-0.	14619356	14610429	14601497	14592562	14583624	14574681	14565735	14556785	14547832	14538875	4
0.857	-0.	14529914	14520950	14511982	14503010	14494035	14485056	14476074	14467088	14458098	14449105	4
0.858	-0.	14440108	14431107	14422103	14413095	14404084	14395069	14386050	14377028	14368003	14358973	4
0.859	-0.	14349941	14340904	14331864	14322821	14313774	14304723	14295669	14286612	14277551	14268486	4
0.860	-0.	14259418	14250346	14241271	14232192	14223110	14214024	14204935	14195842	14186745	14177646	3
0.861	-0.	14168542	14159436	14150326	14141212	14132095	14122974	14113850	14104723	14095592	14086457	3
0.862	-0.	14077320	14068178	14059034	14049885	14040734	14031579	14022421	14013259	14004094	13994925	3
0.863	-0.	13985753	13976577	13967399	13958217	13949031	13939842	13930650	13921454	13912255	13903052	3
0.864	-0.	13893847	13884638	13875425	13866209	13856990	13847768	13838542	13829313	13820080	13810844	3
0.865	-0.	13801605	13792363	13783117	13773868	13764616	13755360	13746101	13736839	13727573	13718304	3
0.866	-0.	13709032	13699757	13690478	13681197	13671911	13662623	13653331	13644037	13634739	13625437	3
0.867	-0.	13616133	13606825	13597514	13588200	13578882	13569561	13560237	13550910	13541580	13532247	3
0.868	-0.	13522910	13513570	13504227	13494881	13485531	13476179	13466823	13457464	13448102	13438737	3
0.869	-0.	13429369	13419997	13410622	13401245	13391864	13382480	13373093	13363702	13354309	13344912	3
0.870	-0.	13335513	13326110	13316704	13307295	13297883	13288468	13279050	13269629	13260204	13250777	3
0.871	-0.	13241346	13231913	13222476	13213037	13203594	13194148	13184699	13175248	13165793	13156335	3
0.872	-0.	13146874	13137410	13127943	13118473	13109000	13099524	13090045	13080563	13071079	13061591	3
0.873	-0.	13052100	13042606	13033109	13023609	13014106	13004601	12995092	12985580	12976066	12966548	3
0.874	-0.	12957027	12947504	12937978	12928448	12918916	12909381	12899843	12890302	12880758	12871211	3
0.875	-0.	12861662	12852109	12842554	12832995	12823434	12813870	12804303	12794733	12785160	12775585	3
0.876	-0.	12766006	12756425	12746841	12737254	12727664	12718072	12708476	12698878	12689277	12679673	3
0.877	-0.	12670066	12660456	12650844	12641229	12631611	12621990	12612366	12602740	12593111	12583479	3
0.878	-0.	12573844	12564207	12554567	12544924	12535278	12525629	12515978	12506324	12496667	12487008	3
0.879	-0.	12477346	12467681	12458013	12448343	12438670	12428994	12419315	12409634	12399950	12390264	3
0.880	-0.	12380575	12370883	12361188	12351491	12341791	12332088	12322383	12312675	12302964	12293251	3
0.881	-0.	12283535	12273816	12264095	12254371	12244645	12234916	12225184	12215449	12205712	12195973	3
0.882	-0.	12186231	12176486	12166739	12156989	12147236	12137481	12127723	12117963	12108200	12098434	3
0.883	-0.	12088666	12078396	12069123	12059347	12049569	12039788	12030005	12020219	12010431	12000640	3
0.884	-0.	11990846	11981050	11971252	11961451	11951647	11941841	11932033	11922222	11912409	11902593	2
0.885	-0.	11892774	11882953	11873130	11863304	11853476	11843645	11833812	11823976	11814138	11804298	2
0.886	-0.	11794455	11784609	11774761	11764911	11755058	11745203	11735346	11725486	11715624	11705759	2
0.887	-0.	11695892	11686022	11676150	11666276	11656399	11646520	11636639	11626755	11616869	11606980	2
0.888	-0.	11597089	11587196	11577301	11567403	11557503	11547600	11537695	11527788	11517878	11507966	2
0.889	-0.	11498052	11488136	11478217	11468296	11458372	11448447	11438519	11428589	11418656	11408721	2
0.890	-0.	11398784	11388845	11378903	11368959	11359013	11349065	11339114	11329161	11319206	11309249	2
0.891	-0.	11299289	11289328	11279364	11269397	11259429	11249458	11239485	11229510	11219533	11209554	2
0.892	-0.	11199572	11189588	11179602	11169614	11159624	11149631	11139637	11129640	11119641	11109640	2
0.893	-0.	11099636	11089631	11079624	11069614	11059602	11049588	11039572	11029554	11019534	11009511	2
0.894	-0.	10999487	10989460	10979431	10969401	10959368	10949333	10939296	10929257	10919216	10909172	2
0.895	-0.	10899127	10889080	10879030	10868979	10858925	10848870	10838812	10828753	10818691	10808627	2
0.896	-0.	10798562	10788494	10778424	10768352	10753279	10748203	10738125	10728046	10717964	10707880	2
0.897	-0.	10697795	10687707	10677617	10667526	10657432	10647337	10637239	10627140	10617039	10606935	2
0.898	-0.	10596830	10586723	10576614	10566503	10556390	10546275	10536158	10526040	10515919	10505797	2
0.899	-0.	10495672	10485546	10475418	10465288	10455156	10445022	10434887	10424749	10414610	10404468	2

$$\frac{\sin 2\pi x}{2\pi x}$$

$$\frac{\sin 2\pi x}{2\pi x}$$

TABLE 6

Sinc Function

x		.0	.0001	.0002	.0003	.0004	.0005	.0006	.0007	.0008	.0009	Δ_2
0.900	-0.	10394325	10384180	10374034	10363885	10353735	10343582	10333428	10323272	10313115	10302955	2
0.901	-0.	10292794	10282630	10272465	10262299	10252130	10241960	10231787	10221614	10211438	10201260	2
0.902	-0.	10191081	10180900	10170717	10160533	10150346	10140158	10129968	10119777	10109584	10099389	2
0.903	-0.	10089192	10078993	10068793	10058591	10048388	10038182	10027975	10017767	10007556	09997344	2
0.904	-0.	09987130	09976915	09966698	09956479	09946258	09936036	09925812	09915587	09905360	09895131	2
0.905	-0.	09884900	09874668	09864435	09854199	09843962	09833724	09823483	09813241	09802998	09792753	2
0.906	-0.	09782506	09772258	09762008	09751757	09741504	09731249	09720993	09710735	09700476	09690215	2
0.907	-0.	09679952	09669688	09659423	09649155	09638887	09628617	09618345	09608071	09597797	09587520	2
0.908	-0.	09577242	09566963	09556682	09546400	09536116	09525830	09515543	09505255	09494965	09484674	0
0.909	-0.	09474381	09464087	09453791	09443493	09433195	09422895	09412593	09402290	09391985	09381679	0
0.910	-0.	09371372	09361063	09350753	09340441	09330128	09319813	09309497	09299180	09288861	09278541	0
0.911	-0.	09268219	09257896	09247572	09237246	09226919	09216591	09206261	09195929	09185597	09175263	0
0.912	-0.	09164927	09154591	09144253	09133913	09123573	09113231	09102887	09092543	09082197	09071849	0
0.913	-0.	09061501	09051151	09040799	09030447	09020093	09009738	08999381	08989023	08978664	08968304	0
0.914	-0.	08957943	08947580	08937216	08926850	08916484	08906116	08895747	08885376	08875005	08864632	0
0.915	-0.	08854258	08843882	08833506	08823128	08812749	08802369	08791988	08781605	08771221	08760836	0
0.916	-0.	08750450	08740063	08729674	08719284	08708893	08698501	08688108	08677714	08667318	08656922	0
0.917	-0.	08646524	08636125	08625725	08615323	08604921	08594517	08584113	08573707	08563300	08552892	0
0.918	-0.	08542483	08532073	08521661	08511249	08500835	08490421	08480005	08469588	08459170	08448751	0
0.919	-0.	08438331	08427910	08417488	08407065	08396641	08386216	08375789	08365362	08354934	08344504	0
0.920	-0.	08334074	08323642	08313210	08302776	08292342	08281906	08271470	08261032	08250594	08240154	0
0.921	-0.	08229714	08219272	08208830	08198386	08187942	08177497	08167050	08156603	08146155	08135706	0
0.922	-0.	08125256	08114805	08104353	08093900	08083446	08072991	08062535	08052079	08041621	08031163	0
0.923	-0.	08020704	08010243	07999782	07989320	07978857	07968393	07957929	07947463	07936997	07926529	0
0.924	-0.	07916061	07905592	07895122	07884652	07874180	07863708	07853234	07842760	07832286	07821810	0
0.925	-0.	07811333	07800856	07790378	07779899	07769419	07758938	07748457	07737975	07727492	07717008	0
0.926	-0.	07706523	07696038	07685552	07675065	07664577	07654089	07643600	07633110	07622619	07612128	0
0.927	-0.	07601635	07591143	07580649	07570155	07559660	07549164	07538667	07528170	07517672	07507173	0
0.928	-0.	07496674	07486174	07475673	07465172	07454670	07444167	07433663	07423159	07412654	07402149	0
0.929	-0.	07391643	07381136	07370629	07360121	07349612	07339102	07328592	07318082	07307571	07297059	0
0.930	-0.	07286546	07276033	07265519	07255005	07244490	07233974	07223458	07212942	07202424	07191906	0
0.931	-0.	07181388	07170869	07160349	07149829	07139308	07128787	07118265	07107742	07097219	07086696	0
0.932	-0.	07076172	07065647	07055122	07044597	07034070	07023544	07013016	07002489	06991960	06981432	0
0.933	-0.	06970902	06960373	06949843	06939312	06928781	06918249	06907717	06897184	06886651	06876118	0
0.934	-0.	06865584	06855049	06844514	06833979	06823443	06812907	06802370	06791833	06781296	06770758	0
0.935	-0.	06760219	06749680	06739141	06728602	06718062	06707521	06696980	06686439	06675898	06665356	0
0.936	-0.	06654813	06644271	06633728	06623184	06612640	06602096	06591552	06581007	06570462	06559916	0
0.937	-0.	06549370	06538824	06528277	06517730	06507183	06496636	06486088	06475540	06464991	06454442	0
0.938	-0.	06443893	06433344	06422794	06412244	06401694	06391144	06380593	06370042	06359491	06348939	0
0.939	-0.	06338387	06327835	06317283	06306730	06296177	06285624	06275071	06264517	06253964	06243410	0
0.940	-0.	06232856	06222301	06211747	06201192	06190637	06180081	06169526	06158970	06148415	06137859	0
0.941	-0.	06127302	06116746	06106190	06095633	06085076	06074519	06063962	06053404	06042847	06032289	0
0.942	-0.	06021732	06011174	06000616	05990058	05979499	05968941	05958382	05947824	05937265	05926706	0
0.943	-0.	05916147	05905588	05895029	05884470	05873911	05863351	05852792	05842232	05831673	05821113	0
0.944	-0.	05810553	05799994	05789434	05778874	05768314	05757754	05747194	05736634	05726074	05715514	0
0.945	-0.	05704954	05694394	05683834	05673274	05662713	05652153	05641593	05631033	05620473	05609913	0
0.946	-0.	05599353	05588793	05578232	05567672	05557112	05546553	05535993	05525433	05514873	05504313	0
0.947	-0.	05493754	05483194	05472634	05462075	05451515	05440956	05430397	05419838	05409278	05398719	0
0.948	-0.	05388161	05377602	05367043	05356484	05345926	05335368	05324809	05314251	05303693	05293135	0
0.949	-0.	05282578	05272020	05261463	05250905	05240348	05229791	05219234	05208678	05198121	05187565	0

TABLE 6

Sinc Function

x		.0	.0001	.0002	.0003	.0004	.0005	.0006	.0007	.0008	.0009	Δ_2
0.950	-0.	05177009	05166453	05155897	05145341	05134786	05124231	05113676	05103121	05092566	05082012	0
0.951	-0.	05071458	05060904	05050350	05039796	05029243	05018690	05008137	04997584	04987032	04976480	0
0.952	-0.	04965928	04955377	04944825	04934274	04923723	04913173	04902623	04892073	04881523	04870974	0
0.953	-0.	04860424	04849876	04839327	04828779	04818231	04807683	04797136	04786589	04776042	04765496	0
0.954	-0.	04754950	04744405	04733859	04723314	04712770	04702225	04691681	04681138	04670595	04660052	0
0.955	-0.	04649509	04638967	04628425	04617884	04607343	04596803	04586262	04575723	04565183	04554644	0
0.956	-0.	04544106	04533567	04523030	04512492	04501955	04491419	04480883	04470347	04459812	04449277	0
0.957	-0.	04438743	04428209	04417676	04407143	04396610	04386078	04375547	04365016	04354485	04343955	0
0.958	-0.	04333425	04322896	04312368	04301839	04291312	04280785	04270258	04259732	04249206	04238681	-1
0.959	-0.	04228156	04217632	04207109	04196586	04186063	04175541	04165020	04154499	04143979	04133459	-1
0.960	-0.	04122940	04112421	04101903	04091386	04080869	04070353	04059837	04049322	04038807	04028293	-1
0.961	-0.	04017780	04007267	03996755	03986244	03975733	03965222	03954713	03944204	03933695	03923187	-1
0.962	-0.	03912680	03902174	03891668	03881163	03870658	03860154	03849651	03839148	03828646	03818145	-1
0.963	-0.	03807644	03797145	03786645	03776147	03765649	03755152	03744655	03734159	03723664	03713170	-1
0.964	-0.	03702676	03692183	03681691	03671200	03660709	03650219	03639730	03629241	03618753	03608266	-1
0.965	-0.	03597780	03587294	03576809	03566325	03555842	03545360	03534878	03524397	03513917	03503437	-1
0.966	-0.	03492959	03482481	03472004	03461527	03451052	03440577	03430103	03419630	03409158	03398687	-1
0.967	-0.	03388216	03377747	03367278	03356810	03346342	03335876	03325410	03314946	03304482	03294019	-1
0.968	-0.	03283557	03273096	03262635	03252176	03241717	03231259	03220803	03210347	03199891	03189437	-1
0.969	-0.	03178984	03168532	03158080	03147630	03137180	03126731	03116283	03105836	03095390	03084945	-1
0.970	-0.	03074501	03064058	03053616	03043175	03032734	03022295	03011857	03001419	02990983	02980547	-1
0.971	-0.	02970113	02959679	02949246	02938815	02928384	02917955	02907526	02897098	02886672	02876246	-1
0.972	-0.	02865822	02855398	02844975	02834554	02824133	02813714	02803295	02792878	02782461	02772046	-1
0.973	-0.	02761632	02751219	02740806	02730395	02719985	02709576	02699168	02688761	02678355	02667951	-1
0.974	-0.	02657547	02647145	02636743	02626343	02615943	02605545	02595148	02584752	02574357	02563964	-1
0.975	-0.	02553571	02543180	02532789	02522400	02512012	02501625	02491239	02480855	02470471	02460089	-1
0.976	-0.	02449707	02439327	02428949	02418571	02408194	02397819	02387445	02377072	02366700	02356329	-1
0.977	-0.	02345960	02335592	02325225	02314859	02304494	02294131	02283768	02273408	02263048	02252689	-1
0.978	-0.	02242332	02231976	02221621	02211267	02200915	02190564	02180214	02169865	02159518	02149172	-1
0.979	-0.	02138827	02128484	02118141	02107800	02097461	02087122	02076785	02066449	02056115	02045781	-1
0.980	-0.	02035449	02025119	02014789	02004461	01994135	01983809	01973485	01963162	01952841	01942521	-1
0.981	-0.	01932202	01921885	01911569	01901254	01890940	01880628	01870318	01860008	01849701	01839394	-1
0.982	-0.	01829089	01818785	01808483	01798181	01787882	01777584	01767287	01756991	01746697	01736404	-1
0.983	-0.	01726113	01715823	01705535	01695248	01684962	01674678	01664395	01654114	01643834	01633556	-1
0.984	-0.	01623279	01613003	01602729	01592457	01582185	01571916	01561647	01551381	01541115	01530852	-1
0.985	-0.	01520589	01510328	01500069	01489811	01479555	01469300	01459046	01448794	01438544	01428295	-2
0.986	-0.	01418048	01407802	01397558	01387315	01377074	01366834	01356596	01346359	01336124	01325890	-2
0.987	-0.	01315658	01305428	01295199	01284972	01274746	01264522	01254299	01244078	01233859	01223641	-2
0.988	-0.	01213424	01203210	01192997	01182785	01172575	01162367	01152160	01141955	01131751	01121550	-2
0.989	-0.	01111349	01101151	01090954	01080758	01070565	01060372	01050182	01039993	01029806	01019620	-2
0.990	-0.	01009437	00999254	00989074	00978895	00968718	00958542	00948368	00938196	00928026	00917857	-2
0.991	-0.	00907690	00897524	00887360	00877198	00867038	00856879	00846723	00836567	00826414	00816262	-2
0.992	-0.	00806112	00795964	00785817	00775672	00765529	00755388	00745248	00735110	00724974	00714840	-2
0.993	-0.	00704707	00694576	00684447	00674320	00664195	00654071	00643949	00633829	00623710	00613594	-2
0.994	-0.	00603479	00593366	00583255	00573145	00563037	00552932	00542828	00532725	00522625	00512527	-2
0.995	-0.	00502430	00492335	00482242	00472151	00462061	00451974	00441888	00431804	00421722	00411642	-2
0.996	-0.	00401564	00391488	00381413	00371341	00361270	00351201	00341134	00331069	00321006	00310944	-2
0.997	-0.	00300885	00290827	00280772	00270718	00260666	00250616	00240568	00230522	00220478	00210436	-2
0.998	-0.	00200396	00190357	00180321	00170286	00160254	00150223	00140194	00130168	00120143	00110120	-2
0.999	-0.	00100099	00090081	00080064	00070049	00060036	00050025	00040016	00030009	00020004	.00010001	2

$$\frac{\sin 2\pi x}{2\pi x}$$

$\frac{\sin 2\pi x}{2\pi x}$

TABLE 6

Sinc Function

x		.0	.0001	.0002	.0003	.0004	.0005	.0006	.0007	.0008	.0009	Δ_2
1.000	0.	00000000	00009999	00019996	00029991	00039984	00049975	00059964	00069951	00079936	00089919	-2
1.001	0.	00099899	00109878	00119855	00129830	00139802	00149773	00159742	00169708	00179673	00189635	-2
1.002	0.	00199596	00209554	00219510	00229464	00239416	00249366	00259314	00269260	00279204	00289145	-2
1.003	0.	00299085	00309022	00318958	00328891	00338822	00348751	00358678	00368603	00378525	00388446	-2
1.004	0.	00398364	00408281	00418195	00428107	00438017	00447924	00457830	00467733	00477635	00487534	-2
1.005	0.	00497431	00507325	00517218	00527108	00536997	00546883	00556767	00566648	00576528	00586405	-2
1.006	0.	00596280	00606153	00616024	00625892	00635759	00645623	00655485	00665344	00675202	00685057	-2
1.007	0.	00694910	00704761	00714609	00724456	00734300	00744141	00753981	00763818	00773653	00783486	-2
1.008	0.	00793317	00803145	00812971	00822795	00832616	00842435	00852252	00862067	00871879	00881689	-2
1.009	0.	00891497	00901302	00911106	00920906	00930705	00940501	00950295	00960087	00969876	00979663	-2
1.010	0.	00989448	00999230	01009010	01018788	01028563	01038336	01048107	01057875	01067641	01077404	-2
1.011	0.	01087166	01096924	01106681	01116435	01126187	01135936	01145683	01155428	01165170	01174910	-2
1.012	0.	01184648	01194383	01204115	01213846	01223574	01233299	01243022	01252743	01262461	01272177	-2
1.013	0.	01281890	01291601	01301310	01311016	01320720	01330421	01340120	01349816	01359510	01369202	-2
1.014	0.	01378891	01388577	01398261	01407943	01417622	01427299	01436973	01446645	01456314	01465981	-2
1.015	0.	01475646	01485308	01494967	01504624	01514278	01523930	01533579	01543226	01552871	01562513	-3
1.016	0.	01572152	01581789	01591423	01601055	01610684	01620311	01629935	01639557	01649176	01658792	-3
1.017	0.	01668406	01678018	01687627	01697233	01706837	01716438	01726037	01735633	01745226	01754817	-3
1.018	0.	01764406	01773992	01783575	01793156	01802734	01812309	01821882	01831452	01841020	01850585	-3
1.019	0.	01860147	01869707	01879264	01888819	01898371	01907920	01917467	01927011	01936553	01946092	-3
1.020	0.	01955628	01965161	01974692	01984221	01993746	02003269	02012789	02022307	02031822	02041334	-3
1.021	0.	02050844	02060351	02069855	02079357	02088856	02098352	02107846	02117337	02126825	02136310	-3
1.022	0.	02145793	02155273	02164751	02174225	02183697	02193167	02202633	02212097	02221558	02231016	-3
1.023	0.	02240472	02249925	02259375	02268822	02278267	02287709	02297148	02306585	02316018	02325449	-3
1.024	0.	02334877	02344303	02353725	02363145	02372562	02381977	02391388	02400797	02410203	02419606	-3
1.025	0.	02429007	02438404	02447799	02457191	02466580	02475967	02485350	02494731	02504109	02513484	-3
1.026	0.	02522857	02532226	02541593	02550957	02560318	02569676	02579031	02588384	02597733	02607080	-3
1.027	0.	02616424	02625765	02635104	02644439	02653772	02663101	02672428	02681752	02691073	02700391	-3
1.028	0.	02709707	02719019	02728329	02737635	02746939	02756240	02765538	02774833	02784125	02793415	-3
1.029	0.	02802701	02811984	02821265	02830543	02839817	02849089	02858358	02867624	02876887	02886147	-3
1.030	0.	02895404	02904658	02913910	02923158	02932403	02941646	02950885	02960122	02969355	02978586	-3
1.031	0.	02987813	02997038	03006260	03015478	03024694	03033907	03043116	03052323	03061527	03070728	-3
1.032	0.	03079925	03089120	03098312	03107501	03116687	03125869	03135049	03144226	03153399	03162570	-3
1.033	0.	03171738	03180902	03190064	03199223	03208378	03217531	03226680	03235827	03244970	03254110	-3
1.034	0.	03263248	03272382	03281513	03290641	03299766	03308888	03318007	03327123	03336235	03345345	-3
1.035	0.	03354452	03363555	03372656	03381753	03390847	03399938	03409026	03418111	03427193	03436272	-3
1.036	0.	03445348	03454420	03463489	03472556	03481619	03490679	03499736	03508790	03517840	03526888	-3
1.037	0.	03535932	03544973	03554011	03563046	03572078	03581107	03590132	03599155	03608174	03617190	-3
1.038	0.	03626203	03635212	03644219	03653222	03662222	03671219	03680213	03689204	03698191	03707176	-3
1.039	0.	03716157	03725134	03734109	03743081	03752049	03761014	03769976	03778934	03787890	03796842	-3
1.040	0.	03805791	03814737	03823679	03832618	03841554	03850487	03859417	03868343	03877266	03886186	-3
1.041	0.	03895103	03904016	03912926	03921833	03930737	03939637	03948534	03957428	03966318	03975206	-3
1.042	0.	03984090	03992970	04001848	04010722	04019593	04028460	04037325	04046186	04055043	04063898	-3
1.043	0.	04072749	04081597	04090441	04099282	04108120	04116955	04125786	04134614	04143438	04152260	-3
1.044	0.	04161078	04169892	04178703	04187511	04196316	04205117	04213915	04222710	04231501	04240289	-3
1.045	0.	04249073	04257854	04266632	04275406	04284177	04292945	04301709	04310470	04319228	04327982	-3
1.046	0.	04336733	04345480	04354224	04362965	04371702	04380436	04389166	04397893	04406617	04415337	-3
1.047	0.	04424054	04432767	04441477	04450184	04458887	04467587	04476283	04484976	04493665	04502351	-3
1.048	0.	04511034	04519713	04528389	04537061	04545730	04554395	04563057	04571716	04580371	04589022	-3
1.049	0.	04597670	04606315	04614956	04623594	04632228	04640859	04649486	04658110	04666730	04675347	-3

TABLE 6

Sinc Function

x		.0	.0001	.0002	.0003	.0004	.0005	.0006	.0007	.0008	.0009	Δ_2
1.050	0.	04683960	04692570	04701176	04709779	04718379	04726974	04735567	04744156	04752741	04761323	−4
1.051	0.	04769901	04778476	04787047	04795615	04804179	04812740	04821297	04829851	04838401	04846948	−4
1.052	0.	04855491	04864030	04872566	04881099	04889627	04898153	04906675	04915193	04923707	04932219	−4
1.053	0.	04940726	04949230	04957730	04966227	04974721	04983210	04991696	05000179	05008658	05017133	−4
1.054	0.	05025605	05034073	05042538	05050999	05059456	05067910	05076360	05084807	05093250	05101689	−4
1.055	0.	05110125	05118557	05126985	05135410	05143831	05152249	05160663	05169073	05177480	05185883	−4
1.056	0.	05194283	05202678	05211071	05219459	05227844	05236225	05244603	05252977	05261347	05269714	−4
1.057	0.	05278077	05286436	05294791	05303143	05311492	05319836	05328177	05336514	05344848	05353178	−4
1.058	0.	05361504	05369826	05378145	05386460	05394772	05403079	05411383	05419684	05427980	05436273	−4
1.059	0.	05444562	05452848	05461130	05469408	05477682	05485952	05494219	05502482	05510742	05518997	−4
1.060	0.	05527249	05535497	05543742	05551983	05560219	05568453	05576682	05584908	05593130	05601348	−4
1.061	0.	05609562	05617773	05625980	05634183	05642382	05650578	05658770	05666958	05675142	05683322	−4
1.062	0.	05691499	05699672	05707841	05716006	05724168	05732326	05740480	05748630	05756776	05764919	−4
1.063	0.	05773057	05781192	05789323	05797450	05805574	05813694	05821809	05829921	05838030	05846134	−4
1.064	0.	05854234	05862331	05870424	05878513	05886598	05894679	05902757	05910830	05918900	05926966	−4
1.065	0.	05935028	05943086	05951141	05959191	05967238	05975281	05983319	05991354	05999386	06007413	−4
1.066	0.	06015436	06023456	06031471	06039483	06047491	06055495	06063495	06071491	06079484	06087472	−4
1.067	0.	06095456	06103437	06111414	06119386	06127355	06135320	06143281	06151238	06159192	06167141	−4
1.068	0.	06175086	06183028	06190965	06198899	06206829	06214754	06222676	06230594	06238508	06246418	−4
1.069	0.	06254324	06262226	06270124	06278018	06285908	06293794	06301677	06309555	06317429	06325300	−4
1.070	0.	06333166	06341029	06348887	06356742	06364592	06372439	06380281	06388120	06395955	06403785	−4
1.071	0.	06411612	06419434	06427253	06435068	06442878	06450685	06458488	06466286	06474081	06481872	−4
1.072	0.	06489658	06497441	06505219	06512994	06520764	06528531	06536293	06544052	06551806	06559557	−4
1.073	0.	06567303	06575045	06582784	06590518	06598248	06605974	06613696	06621414	06629128	06636838	−4
1.074	0.	06644544	06652246	06659944	06667638	06675327	06683013	06690694	06698372	06706045	06713715	−4
1.075	0.	06721380	06729041	06736698	06744351	06752000	06759645	06767285	06774922	06782555	06790183	−4
1.076	0.	06797807	06805428	06813044	06820656	06828264	06835867	06843467	06851063	06858654	06866242	−4
1.077	0.	06873825	06881404	06888979	06896550	06904117	06911679	06919238	06926792	06934342	06941888	−4
1.078	0.	06949430	06956968	06964502	06972031	06979556	06987078	06994595	07002108	07009616	07017121	−4
1.079	0.	07024621	07032118	07039610	07047098	07054581	07062061	07069536	07077008	07084475	07091937	−4
1.080	0.	07099396	07106851	07114301	07121747	07129189	07136627	07144060	07151490	07158915	07166336	−4
1.081	0.	07173753	07181165	07188574	07195978	07203378	07210773	07218165	07225552	07232935	07240314	−4
1.082	0.	07247689	07255059	07262425	07269787	07277145	07284498	07291848	07299193	07306534	07313870	−4
1.083	0.	07321202	07328530	07335854	07343174	07350489	07357800	07365107	07372410	07379708	07387002	−4
1.084	0.	07394292	07401577	07408858	07416135	07423408	07430677	07437941	07445201	07452456	07459708	−4
1.085	0.	07466955	07474197	07481436	07488670	07495900	07503126	07510347	07517564	07524777	07531985	−4
1.086	0.	07539189	07546389	07553585	07560776	07567963	07575145	07582324	07589498	07596667	07603833	−4
1.087	0.	07610994	07618150	07625303	07632451	07639594	07646734	07653869	07661000	07668126	07675248	−4
1.088	0.	07682366	07689479	07696588	07703693	07710793	07717889	07724981	07732068	07739151	07746229	−4
1.089	0.	07753304	07760374	07767439	07774500	07781557	07788609	07795657	07802701	07809740	07816775	−4
1.090	0.	07823806	07830832	07837854	07844871	07851884	07858893	07865897	07872897	07879892	07886883	−4
1.091	0.	07893870	07900852	07907830	07914804	07921773	07928737	07935698	07942654	07949605	07956552	−4
1.092	0.	07963495	07970433	07977367	07984296	07991221	07998141	08005058	08011969	08018876	08025779	−4
1.093	0.	08032678	08039572	08046461	08053346	08060227	08067103	08073975	08080842	08087705	08094563	−4
1.094	0.	08101417	08108267	08115112	08121953	08128789	08135621	08142448	08149271	08156089	08162903	−4
1.095	0.	08169712	08176517	08183318	08190113	08196905	08203692	08210475	08217253	08224026	08230795	−4
1.096	0.	08237560	08244320	08251076	08257827	08264574	08271316	08278053	08284787	08291515	08298240	−4
1.097	0.	08304959	08311674	08318385	08325091	08331793	08338490	08345183	08351871	08358554	08365234	−5
1.098	0.	08371908	08378578	08385244	08391905	08398561	08405213	08411861	08418503	08425142	08431776	−5
1.099	0.	08438405	08445030	08451650	08458266	08464877	08471483	08478085	08484683	08491276	08497864	−5

TABLE 6

Sinc Function

x		.0	.0001	.0002	.0003	.0004	.0005	.0006	.0007	.0008	.0009	Δ_2
1.100	0.	08504448	08511027	08517602	08524172	08530738	08537299	08543855	08550407	08556955	08563498	−5
1.101	0.	08570036	08576569	08583098	08589623	08596143	08602658	08609169	08615675	08622177	08628674	−5
1.102	0.	08635166	08641654	08648138	08654616	08661090	08667560	08674025	08680485	08686941	08693392	−5
1.103	0.	08699838	08706280	08712718	08719150	08725578	08732002	08738421	08744835	08751245	08757650	−5
1.104	0.	08764050	08770446	08776837	08783223	08789605	08795983	08802355	08808723	08815087	08821446	−5
1.105	0.	08827800	08834149	08840494	08846834	08853170	08859501	08865827	08872149	08878466	08884778	−5
1.106	0.	08891086	08897389	08903688	08909981	08916270	08922555	08928835	08935110	08941380	08947646	−5
1.107	0.	08953907	08960164	08966416	08972663	08978905	08985143	08991376	08997605	09003828	09010048	−5
1.108	0.	09016262	09022472	09028677	09034877	09041073	09047264	09053450	09059632	09065809	09071981	−5
1.109	0.	09078149	09084311	09090470	09096623	09102772	09108916	09115055	09121190	09127320	09133445	−5
1.110	0.	09139566	09145681	09151793	09157899	09164001	09170098	09176190	09182277	09188360	09194438	−5
1.111	0.	09200512	09206580	09212644	09218703	09224758	09230808	09236853	09242893	09248928	09254959	−5
1.112	0.	09260985	09267006	09273023	09279035	09285042	09291044	09297042	09303035	09309023	09315006	−5
1.113	0.	09320985	09326959	09332928	09338892	09344851	09350806	09356756	09362702	09368642	09374578	−5
1.114	0.	09380509	09386435	09392357	09398273	09404185	09410092	09415995	09421892	09427785	09433673	−5
1.115	0.	09439556	09445435	09451308	09457177	09463041	09468901	09474755	09480605	09486450	09492290	−5
1.116	0.	09498126	09503956	09509782	09515603	09521419	09527230	09533037	09538839	09544636	09550428	−5
1.117	0.	09556215	09561998	09567776	09573549	09579317	09585080	09590838	09596592	09602341	09608085	−5
1.118	0.	09613824	09619559	09625288	09631013	09636733	09642448	09648158	09653864	09659564	09665260	−5
1.119	0.	09670951	09676637	09682318	09687995	09693666	09699333	09704995	09710652	09716304	09721952	−5
1.120	0.	09727594	09733232	09738865	09744493	09750116	09755734	09761348	09766956	09772560	09778159	−5
1.121	0.	09783753	09789342	09794926	09800506	09806080	09811650	09817215	09822775	09828330	09833880	−5
1.122	0.	09839425	09844966	09850501	09856032	09861558	09867079	09872595	09878106	09883613	09889114	−5
1.123	0.	09894611	09900102	09905589	09911071	09916548	09922020	09927487	09932950	09938407	09943860	−5
1.124	0.	09949307	09954750	09960188	09965621	09971049	09976472	09981891	09987304	09992712	09998116	−5
1.125	0.	10003515	10008908	10014297	10019681	10025060	10030434	10035803	10041168	10046527	10051881	−5
1.126	0.	10057231	10062576	10067915	10073250	10078580	10083905	10089225	10094540	10099850	10105155	−5
1.127	0.	10110455	10115751	10121041	10126326	10131607	10136882	10142153	10147419	10152680	10157935	−5
1.128	0.	10163186	10168432	10173673	10178909	10184140	10189367	10194588	10199804	10205015	10210222	−5
1.129	0.	10215423	10220620	10225811	10230998	10236179	10241356	10246528	10251694	10256856	10262013	−5
1.130	0.	10267165	10272311	10277453	10282590	10287722	10292849	10297971	10303088	10308200	10313308	−5
1.131	0.	10318410	10323507	10328599	10333686	10338768	10343846	10348918	10353985	10359047	10364105	−5
1.132	0.	10369157	10374204	10379247	10384284	10389317	10394344	10399366	10404384	10409396	10414404	−5
1.133	0.	10419406	10424403	10429396	10434383	10439366	10444343	10449316	10454283	10459245	10464203	−5
1.134	0.	10469155	10474103	10479045	10483982	10488915	10493842	10498765	10503682	10508594	10513502	−5
1.135	0.	10518404	10523301	10528194	10533081	10537963	10542840	10547712	10552580	10557442	10562299	−5
1.136	0.	10567151	10571998	10576840	10581677	10586509	10591336	10596158	10600975	10605787	10610594	−5
1.137	0.	10615396	10620192	10624984	10629771	10634552	10639329	10644101	10648867	10653629	10658385	−5
1.138	0.	10663137	10667883	10672624	10677361	10682092	10686818	10691539	10696255	10700966	10705672	−5
1.139	0.	10710373	10715069	10719760	10724446	10729126	10733802	10738473	10743138	10747799	10752454	−5
1.140	0.	10757105	10761750	10766390	10771025	10775655	10780280	10784900	10789515	10794125	10798730	−5
1.141	0.	10803330	10807924	10812514	10817098	10821678	10826252	10830821	10835385	10839945	10844499	−5
1.142	0.	10849048	10853591	10858130	10862664	10867193	10871716	10876235	10880748	10885256	10889760	−5
1.143	0.	10894258	10898751	10903239	10907722	10912199	10916672	10921140	10925602	10930059	10934512	−5
1.144	0.	10938959	10943401	10947838	10952270	10956697	10961119	10965535	10969947	10974353	10978755	−5
1.145	0.	10983151	10987542	10991928	10996309	11000685	11005055	11009421	11013781	11018137	11022487	−5
1.146	0.	11026832	11031172	11035507	11039837	11044162	11048481	11052796	11057105	11061409	11065708	−5
1.147	0.	11070002	11074291	11078575	11082854	11087127	11091396	11095659	11099917	11104170	11108418	−5
1.148	0.	11112661	11116899	11121131	11125359	11129581	11133798	11138010	11142217	11146419	11150615	−5
1.149	0.	11154807	11158993	11163174	11167351	11171522	11175687	11179848	11184004	11188154	11192299	−5

TABLE 6

Sinc Function

x		.0	.0001	.0002	.0003	.0004	.0005	.0006	.0007	.0008	.0009	Δ_2
1.150	0.	11196439	11200574	11204704	11208829	11212949	11217063	11221172	11225276	11229376	11233469	−5
1.151	0.	11237558	11241642	11245720	11249793	11253861	11257924	11261982	11266035	11270082	11274125	−5
1.152	0.	11278162	11282194	11286221	11290243	11294259	11298271	11302277	11306278	11310274	11314265	−5
1.153	0.	11318251	11322231	11326207	11330177	11334142	11338102	11342056	11346006	11349950	11353890	−5
1.154	0.	11357824	11361752	11365676	11369595	11373508	11377416	11381319	11385217	11389110	11392997	−5
1.155	0.	11396880	11400757	11404629	11408496	11412358	11416214	11420065	11423912	11427753	11431588	−5
1.156	0.	11435419	11439244	11443065	11446880	11450690	11454494	11458294	11462088	11465878	11469662	−5
1.157	0.	11473440	11477214	11480983	11484746	11488504	11492257	11496005	11499747	11503484	11507217	−5
1.158	0.	11510944	11514665	11518382	11522093	11525800	11529501	11533197	11536887	11540573	11544253	−5
1.159	0.	11547928	11551598	11555263	11558922	11562577	11566226	11569870	11573508	11577142	11580770	−5
1.160	0.	11584393	11588011	11591624	11595232	11598834	11602431	11606023	11609610	11613191	11616768	−5
1.161	0.	11620339	11623905	11627465	11631021	11634571	11638116	11641656	11645191	11648721	11652245	−5
1.162	0.	11655764	11659278	11662787	11666290	11669788	11673281	11676769	11680252	11683729	11687202	−5
1.163	0.	11690669	11694130	11697587	11701038	11704485	11707926	11711361	11714792	11718217	11721637	−5
1.164	0.	11725052	11728462	11731866	11735265	11738659	11742048	11745432	11748810	11752183	11755551	−5
1.165	0.	11758914	11762272	11765624	11768971	11772313	11775649	11778981	11782307	11785628	11788944	−5
1.166	0.	11792254	11795559	11798859	11802154	11805444	11808728	11812007	11815281	11818550	11821814	−5
1.167	0.	11825072	11828325	11831573	11834815	11838053	11841285	11844512	11847733	11850950	11854161	−5
1.168	0.	11857367	11860568	11863763	11866954	11870139	11873318	11876493	11879662	11882827	11885985	−5
1.169	0.	11889139	11892288	11895431	11898569	11901701	11904829	11907951	11911068	11914180	11917287	−5
1.170	0.	11920388	11923484	11926575	11929661	11932741	11935816	11938886	11941951	11945010	11948064	−5
1.171	0.	11951113	11954157	11957196	11960229	11963257	11966280	11969297	11972309	11975316	11978318	−5
1.172	0.	11981315	11984306	11987292	11990273	11993249	11996219	11999184	12002144	12005099	12008048	−5
1.173	0.	12010992	12013931	12016865	12019793	12022716	12025634	12028547	12031454	12034357	12037254	−5
1.174	0.	12040145	12043032	12045913	12048789	12051660	12054525	12057385	12060240	12063090	12065935	−5
1.175	0.	12068774	12071608	12074437	12077260	12080078	12082891	12085699	12088502	12091299	12094091	−5
1.176	0.	12096878	12099659	12102435	12105206	12107972	12110733	12113488	12116238	12118983	12121722	−5
1.177	0.	12124457	12127186	12129909	12132628	12135341	12138049	12140752	12143449	12146142	12148829	−5
1.178	0.	12151510	12154187	12156858	12159524	12162185	12164840	12167491	12170136	12172775	12175410	−5
1.179	0.	12178039	12180663	12183282	12185895	12188503	12191106	12193704	12196296	12198884	12201466	−5
1.180	0.	12204042	12206614	12209180	12211741	12214296	12216847	12219392	12221932	12224467	12226996	−5
1.181	0.	12229520	12232039	12234553	12237061	12239564	12242062	12244555	12247042	12249524	12252001	−5
1.182	0.	12254472	12256939	12259400	12261856	12264306	12266752	12269192	12271626	12274056	12276480	−5
1.183	0.	12278899	12281313	12283721	12286125	12288523	12290916	12293303	12295685	12298062	12300434	−5
1.184	0.	12302800	12305162	12307518	12309868	12312214	12314554	12316889	12319218	12321543	12323862	−5
1.185	0.	12326176	12328485	12330788	12333086	12335379	12337667	12339949	12342226	12344498	12346765	−5
1.186	0.	12349026	12351282	12353533	12355778	12358019	12360254	12362484	12364708	12366927	12369141	−5
1.187	0.	12371350	12373554	12375752	12377945	12380133	12382315	12384493	12386665	12388831	12390993	−5
1.188	0.	12393149	12395300	12397446	12399586	12401721	12403851	12405976	12408095	12410210	12412319	−5
1.189	0.	12414422	12416521	12418614	12420702	12422784	12424862	12426934	12429001	12431063	12433119	−5
1.190	0.	12435170	12437216	12439257	12441292	12443322	12445347	12447367	12449381	12451390	12453394	−5
1.191	0.	12455393	12457386	12459374	12461357	12463334	12465307	12467274	12469236	12471192	12473144	−5
1.192	0.	12475090	12477031	12478966	12480897	12482822	12484741	12486656	12488565	12490469	12492368	−5
1.193	0.	12494262	12496150	12498033	12499911	12501784	12503651	12505513	12507370	12509222	12511068	−5
1.194	0.	12512909	12514745	12516575	12518401	12520221	12522036	12523845	12525650	12527449	12529243	−5
1.195	0.	12531031	12532815	12534593	12536366	12538133	12539896	12541653	12543405	12545152	12546893	−5
1.196	0.	12548629	12550360	12552086	12553806	12555521	12557231	12558936	12560636	12562330	12564019	−5
1.197	0.	12565703	12567381	12569054	12570722	12572385	12574043	12575695	12577342	12578984	12580620	−5
1.198	0.	12582252	12583878	12585499	12587114	12588725	12590330	12591930	12593525	12595114	12596698	−5
1.199	0.	12598277	12599851	12601419	12602983	12604541	12606093	12607641	12609183	12610720	12612252	−5

$$\frac{\sin 2\pi x}{2\pi x}$$

TABLE 6

Sinc Function

x		.0	.0001	.0002	.0003	.0004	.0005	.0006	.0007	.0008	.0009	Δ_2
1.200	0.	12613779	12615300	12616816	12618327	12619833	12621333	12622829	12624319	12625803	12627283	−5
1.201	0.	12628757	12630226	12631690	12633149	12634602	12636050	12637493	12638931	12640363	12641790	−5
1.202	0.	12643212	12644629	12646041	12647447	12648848	12650244	12651635	12653020	12654400	12655775	−5
1.203	0.	12657145	12658509	12659869	12661223	12662572	12663915	12665254	12666587	12667915	12669237	−5
1.204	0.	12670555	12671867	12673174	12674476	12675773	12677064	12678350	12679631	12680907	12682178	−5
1.205	0.	12683443	12684703	12685958	12687208	12688452	12689691	12690925	12692154	12693378	12694596	−5
1.206	0.	12695809	12697017	12698220	12699417	12700610	12701797	12702979	12704155	12705327	12706493	−5
1.207	0.	12707654	12708810	12709961	12711106	12712246	12713381	12714511	12715636	12716755	12717869	−5
1.208	0.	12718978	12720082	12721181	12722274	12723362	12724445	12725523	12726596	12727663	12728725	−5
1.209	0.	12729782	12730834	12731881	12732922	12733958	12734989	12736015	12737035	12738051	12739061	−5
1.210	0.	12740066	12741066	12742060	12743050	12744034	12745013	12745987	12746955	12747919	12748877	−5
1.211	0.	12749830	12750778	12751721	12752658	12753590	12754518	12755440	12756356	12757268	12758174	−5
1.212	0.	12759075	12759971	12760862	12761748	12762628	12763503	12764374	12765238	12766098	12766953	−5
1.213	0.	12767802	12768646	12769485	12770319	12771148	12771971	12772789	12773602	12774410	12775213	−5
1.214	0.	12776011	12776803	12777590	12778372	12779149	12779921	12780687	12781449	12782205	12782956	−5
1.215	0.	12783702	12784442	12785178	12785908	12786633	12787353	12788068	12788778	12789483	12790182	−5
1.216	0.	12790876	12791565	12792249	12792928	12793601	12794269	12794933	12795591	12796244	12796891	−5
1.217	0.	12797534	12798171	12798804	12799431	12800053	12800669	12801281	12801888	12802489	12803085	−5
1.218	0.	12803676	12804262	12804843	12805418	12805989	12806554	12807114	12807669	12808219	12808764	−5
1.219	0.	12809303	12809838	12810367	12810891	12811410	12811924	12812432	12812936	12813434	12813928	−5
1.220	0.	12814416	12814899	12815377	12815849	12816317	12816779	12817237	12817689	12818136	12818578	−5
1.221	0.	12819014	12819446	12819873	12820294	12820710	12821121	12821527	12821928	12822324	12822714	−5
1.222	0.	12823100	12823480	12823855	12824226	12824591	12824950	12825305	12825655	12825999	12826339	−5
1.223	0.	12826673	12827002	12827326	12827645	12827959	12828267	12828571	12828869	12829163	12829451	−5
1.224	0.	12829734	12830012	12830285	12830553	12830815	12831073	12831325	12831573	12831815	12832052	−5
1.225	0.	12832284	12832511	12832733	12832950	12833161	12833368	12833569	12833766	12833957	12834143	−5
1.226	0.	12834324	12834500	12834671	12834837	12834997	12835153	12835303	12835449	12835589	12835724	−5
1.227	0.	12835854	12835979	12836099	12836214	12836324	12836428	12836528	12836623	12836712	12836796	−5
1.228	0.	12836876	12836950	12837019	12837083	12837142	12837196	12837244	12837288	12837327	12837360	−5
1.229	0.	12837389	12837412	12837431	12837444	12837452	12837455	12837453	12837446	12837434	12837417	−5
1.230	0.	12837395	12837368	12837335	12837298	12837256	12837208	12837156	12837098	12837035	12836968	−5
1.231	0.	12836895	12836817	12836734	12836646	12836553	12836455	12836352	12836244	12836130	12836012	−5
1.232	0.	12835889	12835761	12835627	12835489	12835345	12835197	12835043	12834885	12834721	12834552	−5
1.233	0.	12834378	12834200	12834016	12833827	12833633	12833434	12833230	12833021	12832807	12832588	−5
1.234	0.	12832364	12832135	12831901	12831662	12831417	12831168	12830914	12830655	12830390	12830121	−5
1.235	0.	12829847	12829567	12829283	12828994	12828699	12828400	12828095	12827786	12827471	12827152	−5
1.236	0.	12826827	12826498	12826163	12825824	12825479	12825130	12824775	12824416	12824051	12823682	−5
1.237	0.	12823307	12822928	12822543	12822153	12821759	12821359	12820955	12820545	12820131	12819711	−5
1.238	0.	12819287	12818857	12818423	12817983	12817539	12817089	12816635	12816175	12815711	12815241	−5
1.239	0.	12814767	12814287	12813803	12813314	12812819	12812320	12811816	12811307	12810792	12810273	−5
1.240	0.	12809749	12809220	12808686	12808147	12807603	12807053	12806499	12805941	12805377	12804808	−5
1.241	0.	12804234	12803655	12803071	12802482	12801889	12801290	12800687	12800078	12799464	12798846	−5
1.242	0.	12798223	12797594	12796961	12796322	12795679	12795031	12794378	12793720	12793057	12792389	−5
1.243	0.	12791716	12791038	12790355	12789668	12788975	12788277	12787575	12786867	12786155	12785438	−5
1.244	0.	12784715	12783988	12783256	12782519	12781777	12781030	12780278	12779521	12778760	12777993	−5
1.245	0.	12777222	12776445	12775664	12774878	12774086	12773290	12772489	12771683	12770872	12770057	−5
1.246	0.	12769236	12768410	12767580	12766744	12765904	12765059	12764209	12763354	12762494	12761629	−5
1.247	0.	12760759	12759885	12759005	12758121	12757232	12756337	12755438	12754534	12753625	12752712	−5
1.248	0.	12751793	12750869	12749941	12749008	12748070	12747127	12746179	12745226	12744268	12743305	−5
1.249	0.	12742338	12741366	12740388	12739406	12738419	12737428	12736431	12735429	12734423	12733412	−5

TABLE 6

Sinc Function

x		.0	.0001	.0002	.0003	.0004	.0005	.0006	.0007	.0008	.0009	Δ_2
1.250	0.	12732395	12731374	12730349	12729318	12728282	12727242	12726196	12725146	12724091	12723031	−5
1.251	0.	12721967	12720897	12719823	12718743	12717659	12716570	12715476	12714378	12713274	12712166	−5
1.252	0.	12711052	12709934	12708812	12707684	12706551	12705414	12704272	12703125	12701973	12700816	−5
1.253	0.	12699654	12698488	12697317	12696141	12694960	12693774	12692584	12691388	12690188	12688983	−5
1.254	0.	12687774	12686559	12685340	12684115	12682886	12681652	12680414	12679170	12677922	12676669	−5
1.255	0.	12675411	12674148	12672881	12671609	12670332	12669050	12667763	12666472	12665175	12663874	−5
1.256	0.	12662568	12661258	12659942	12658622	12657297	12655967	12654633	12653293	12651949	12650600	−5
1.257	0.	12649247	12647888	12646525	12645157	12643784	12642406	12641024	12639637	12638245	12636848	−5
1.258	0.	12635447	12634041	12632630	12631214	12629794	12628368	12626938	12625504	12624064	12622620	−5
1.259	0.	12621171	12619717	12618259	12616795	12615327	12613855	12612377	12610895	12609408	12607916	−5
1.260	0.	12606420	12604918	12603412	12601902	12600386	12598866	12597341	12595812	12594277	12592738	−5
1.261	0.	12591194	12589646	12588093	12586535	12584972	12583405	12581832	12580256	12578674	12577088	−5
1.262	0.	12575497	12573901	12572301	12570695	12569086	12567471	12565852	12564228	12562599	12560966	−5
1.263	0.	12559328	12557685	12556037	12554385	12552729	12551067	12549401	12547730	12546054	12544374	−5
1.264	0.	12542689	12540999	12539305	12537606	12535902	12534194	12532481	12530763	12529041	12527313	−5
1.265	0.	12525582	12523845	12522104	12520358	12518608	12516853	12515093	12513329	12511559	12509786	−5
1.266	0.	12508007	12506224	12504437	12502644	12500847	12499045	12497239	12495428	12493612	12491792	−5
1.267	0.	12489967	12488138	12486304	12484465	12482621	12480773	12478920	12477063	12475201	12473334	−5
1.268	0.	12471463	12469587	12467707	12465821	12463932	12462037	12460138	12458235	12456326	12454414	−5
1.269	0.	12452496	12450574	12448647	12446716	12444780	12442839	12440894	12438944	12436990	12435031	−5
1.270	0.	12433068	12431099	12429127	12427149	12425167	12423181	12421190	12419194	12417194	12415189	−5
1.271	0.	12413179	12411165	12409147	12407123	12405096	12403063	12401026	12398985	12396939	12394888	−5
1.272	0.	12392833	12390773	12388708	12386640	12384566	12382488	12380405	12378318	12376226	12374130	−5
1.273	0.	12372029	12369924	12367814	12365699	12363580	12361456	12359328	12357196	12355058	12352917	−5
1.274	0.	12350770	12348619	12346464	12344304	12342139	12339970	12337797	12335619	12333436	12331249	−5
1.275	0.	12329057	12326861	12324660	12322455	12320246	12318031	12315812	12313589	12311361	12309129	−5
1.276	0.	12306892	12304651	12302405	12300155	12297900	12295641	12293377	12291108	12288836	12286558	−4
1.277	0.	12284276	12281990	12279699	12277404	12275104	12272800	12270491	12268178	12265860	12263538	−4
1.278	0.	12261211	12258880	12256545	12254204	12251860	12249511	12247157	12244799	12242437	12240070	−4
1.279	0.	12237699	12235323	12232943	12230558	12228169	12225775	12223377	12220974	12218567	12216156	−4
1.280	0.	12213740	12211320	12208895	12206466	12204032	12201594	12199151	12196704	12194253	12191797	−4
1.281	0.	12189337	12186372	12184403	12181930	12179452	12176969	12174482	12171991	12169496	12166996	−4
1.282	0.	12164491	12161982	12159469	12156951	12154429	12151903	12149372	12146836	12144297	12141753	−4
1.283	0.	12139204	12136651	12134094	12131532	12128966	12126396	12123821	12121242	12118658	12116070	−4
1.284	0.	12113478	12110881	12108280	12105674	12103064	12100450	12097831	12095209	12092581	12089949	−4
1.285	0.	12087313	12084673	12082028	12079379	12076725	12074068	12071405	12068739	12066068	12063393	−4
1.286	0.	12060713	12058029	12055341	12052648	12049951	12047250	12044544	12041834	12039120	12036401	−4
1.287	0.	12033678	12030951	12028219	12025483	12022743	12019998	12017249	12014496	12011739	12008977	−4
1.288	0.	12006211	12003440	12000665	11997886	11995103	11992315	11989523	11986727	11983926	11981121	−4
1.289	0.	11978312	11975499	11972681	11969859	11967032	11964202	11961367	11958528	11955684	11952836	−4
1.290	0.	11949984	11947128	11944267	11941403	11938533	11935660	11932782	11929900	11927014	11924124	−4
1.291	0.	11921229	11918330	11915427	11912519	11909608	11906692	11903771	11900847	11897918	11894985	−4
1.292	0.	11892048	11889107	11886161	11883211	11880257	11877298	11874336	11871369	11868398	11865422	−4
1.293	0.	11862443	11859459	11856471	11853479	11850483	11847482	11844477	11841468	11838455	11835437	−4
1.294	0.	11832416	11829390	11826360	11823325	11820287	11817244	11814197	11811146	11808091	11805032	−4
1.295	0.	11801968	11798900	11795828	11792752	11789672	11786587	11783498	11780406	11777309	11774207	−4
1.296	0.	11771102	11767992	11764879	11761761	11758639	11755512	11752382	11749248	11746109	11742966	−4
1.297	0.	11739819	11736668	11733513	11730353	11727190	11724022	11720850	11717674	11714494	11711310	−4
1.298	0.	11708121	11704929	11701732	11698531	11695326	11692117	11688904	11685687	11682465	11679240	−4
1.299	0.	11676010	11672776	11669539	11666297	11663051	11659800	11656546	11653288	11650025	11646759	−4

$\dfrac{\sin 2\pi x}{2\pi x}$

TABLE 6

Sinc Function

x		.0	.0001	.0002	.0003	.0004	.0005	.0006	.0007	.0008	.0009	Δ_2
1.300	0.	11643488	11640213	11636935	11633652	11630365	11627073	11623778	11620479	11617176	11613868	−4
1.301	0.	11610557	11607241	11603921	11600598	11597270	11593938	11590602	11587262	11583918	11580570	−4
1.302	0.	11577218	11573862	11570501	11567137	11563769	11560396	11557020	11553639	11550255	11546866	−4
1.303	0.	11543474	11540077	11536676	11533271	11529863	11526450	11523033	11519612	11516187	11512758	−4
1.304	0.	11509326	11505889	11502448	11499003	11495554	11492101	11488644	11485183	11481718	11478249	−4
1.305	0.	11474776	11471299	11467818	11464333	11460844	11457351	11453854	11450353	11446848	11443339	−4
1.306	0.	11439826	11436310	11432789	11429264	11425735	11422202	11418666	11415125	11411580	11408032	−4
1.307	0.	11404479	11400923	11397362	11393798	11390229	11386657	11383081	11379501	11375916	11372328	−4
1.308	0.	11368736	11365140	11361540	11357936	11354329	11350717	11347101	11343482	11339858	11336231	−4
1.309	0.	11332599	11328964	11325325	11321682	11318035	11314384	11310729	11307070	11303408	11299741	−4
1.310	0.	11296071	11292396	11288718	11285036	11281350	11277660	11273966	11270268	11266567	11262861	−4
1.311	0.	11259152	11255439	11251721	11248001	11244276	11240547	11236814	11233078	11229338	11225593	−4
1.312	0.	11221845	11218093	11214338	11210578	11206815	11203047	11199276	11195501	11191722	11187939	−4
1.313	0.	11184153	11180363	11176568	11172770	11168968	11165163	11161353	11157540	11153723	11149901	−4
1.314	0.	11146077	11142248	11138416	11134579	11130739	11126895	11123048	11119196	11115341	11111482	−4
1.315	0.	11107619	11103752	11099881	11096007	11092129	11088247	11084361	11080472	11076579	11072682	−4
1.316	0.	11068781	11064876	11060968	11057056	11053140	11049220	11045297	11041370	11037439	11033504	−4
1.317	0.	11029566	11025623	11021677	11017728	11013774	11009817	11005856	11001891	10997923	10993950	−4
1.318	0.	10989975	10985995	10982011	10978024	10974033	10970039	10966040	10962038	10958033	10954023	−4
1.319	0.	10950010	10945993	10941972	10937948	10933920	10929888	10925853	10921814	10917771	10913724	−4
1.320	0.	10909674	10905620	10901563	10897501	10893436	10889368	10885295	10881219	10877139	10873056	−4
1.321	0.	10868969	10864878	10860784	10856686	10852584	10848478	10844369	10840257	10836140	10832020	−4
1.322	0.	10827896	10823769	10819638	10815503	10811365	10807223	10803078	10798928	10794776	10790619	−4
1.323	0.	10786459	10782295	10778128	10773957	10769782	10765604	10761422	10757237	10753048	10748855	−4
1.324	0.	10744659	10740459	10736255	10732048	10727837	10723623	10719405	10715184	10710959	10706730	−4
1.325	0.	10702498	10698262	10694022	10689779	10685533	10681282	10677029	10672771	10668510	10664246	−4
1.326	0.	10659978	10655706	10651431	10647152	10642870	10638584	10634295	10630002	10625705	10621405	−4
1.327	0.	10617102	10612795	10608484	10604170	10599852	10595531	10591206	10586877	10582546	10578210	−4
1.328	0.	10573871	10569529	10565183	10560833	10556480	10552124	10547764	10543400	10539033	10534663	−4
1.329	0.	10530289	10525911	10521530	10517146	10512758	10508366	10503971	10499573	10495171	10490765	−3
1.330	0.	10486356	10481944	10477528	10473109	10468686	10464260	10459830	10455397	10450960	10446520	−3
1.331	0.	10442076	10437629	10433179	10428725	10424268	10419807	10415342	10410875	10406404	10401929	−3
1.332	0.	10397451	10392969	10388485	10383996	10379504	10375009	10370511	10366009	10361503	10356994	−3
1.333	0.	10352482	10347967	10343448	10338925	10334399	10329870	10325337	10320801	10316262	10311719	−3
1.334	0.	10307173	10302623	10298070	10293514	10288954	10284391	10279824	10275254	10270681	10266104	−3
1.335	0.	10261524	10256941	10252354	10247764	10243171	10238574	10233974	10229370	10224763	10220153	−3
1.336	0.	10215539	10210922	10206302	10201679	10197052	10192421	10187788	10183151	10178511	10173867	−3
1.337	0.	10169220	10164570	10159917	10155260	10150600	10145936	10141269	10136599	10131926	10127249	−3
1.338	0.	10122569	10117886	10113200	10108510	10103817	10099120	10094420	10089717	10085011	10080302	−3
1.339	0.	10075589	10070873	10066153	10061431	10056705	10051976	10047243	10042507	10037768	10033026	−3
1.340	0.	10028281	10023532	10018780	10014025	10009266	10004505	09999740	09994972	09990200	09985426	−3
1.341	0.	09980648	09975867	09971082	09966295	09961504	09956710	09951913	09947113	09942309	09937502	−3
1.342	0.	09932692	09927879	09923062	09918243	09913420	09908594	09903765	09898932	09894097	09889258	−3
1.343	0.	09884416	09879571	09874722	09869871	09865016	09860158	09855297	09850433	09845566	09840695	−3
1.344	0.	09835822	09830945	09826065	09821182	09816295	09811406	09806513	09801618	09796719	09791817	−3
1.345	0.	09786912	09782003	09777092	09772177	09767260	09762339	09757415	09752488	09747558	09742625	−3
1.346	0.	09737688	09732749	09727806	09722860	09717912	09712960	09708005	09703047	09698085	09693121	−3
1.347	0.	09688154	09683183	09678210	09673233	09668253	09663271	09658285	09653296	09648304	09643309	−3
1.348	0.	09638311	09633309	09628305	09623298	09618287	09613274	09608258	09603238	09598215	09593190	−3
1.349	0.	09588161	09583129	09578095	09573057	09568016	09562972	09557925	09552875	09547823	09542767	−3

TABLE 6

Sinc Function

x		.0	.0001	.0002	.0003	.0004	.0005	.0006	.0007	.0008	.0009	Δ_2
1.350	0.0	95377077	95326457	95275807	95225127	95174417	95123677	95072907	95022107	94971277	94920417	−30
1.351	0.0	94869527	94818607	94767657	94716677	94665668	94614628	94563559	94512460	94461332	94410173	−30
1.352	0.0	94358985	94307767	94256520	94205243	94153936	94102600	94051234	93999839	93948414	93896960	−29
1.353	0.0	93845476	93793962	93742420	93690848	93639246	93587616	93535956	93484266	93432548	93380800	−29
1.354	0.0	93329023	93277216	93225381	93173516	93121623	93069700	93017748	92965767	92913757	92861718	−29
1.355	0.0	92809650	92757553	92705428	92653273	92601089	92548877	92496635	92444365	92392066	92339739	−29
1.356	0.0	92287383	92234997	92182584	92130141	92077670	92025171	91972642	91920085	91867500	91814886	−29
1.357	0.0	91762244	91709573	91656874	91604146	91551390	91498606	91445793	91392952	91340082	91287185	−28
1.358	0.0	91234259	91181305	91128322	91075312	91022273	90969206	90916111	90862988	90809837	90756658	−28
1.359	0.0	90703451	90650216	90596953	90543663	90490344	90436997	90383623	90330220	90276790	90223332	−28
1.360	0.0	90169847	90116333	90062792	90009223	89955627	89902003	89848351	89794672	89740965	89687231	−28
1.361	0.0	89633469	89579679	89525863	89472018	89418147	89364248	89310321	89256367	89202386	89148378	−27
1.362	0.0	89094342	89040280	88986190	88932072	88877928	88823756	88769558	88715332	88661079	88606799	−27
1.363	0.0	88552493	88498159	88443798	88389410	88334996	88280554	88226086	88171590	88117068	88062520	−27
1.364	0.0	88007944	87953342	87898713	87844057	87789374	87734665	87679930	87625167	87570379	87515563	−27
1.365	0.0	87460721	87405853	87350958	87296037	87241089	87186115	87131115	87076088	87021035	86965955	−26
1.366	0.0	86910850	86855718	86800560	86745375	86690165	86634928	86579665	86524377	86469062	86413721	−26
1.367	0.0	86358354	86302961	86247542	86192097	86136626	86081130	86025607	85970059	85914485	85858885	−26
1.368	0.0	85803259	85747607	85691930	85636227	85580499	85524745	85468965	85413160	85357329	85301472	−26
1.369	0.0	85245590	85189683	85133750	85077791	85021808	84965798	84909764	84853704	84797619	84741508	−25
1.370	0.0	84685373	84629212	84573025	84516814	84460578	84404316	84348029	84291717	84235380	84179018	−25
1.371	0.0	84122631	84066219	84009783	83953321	83896834	83840322	83783786	83727225	83670639	83614028	−25
1.372	0.0	83557392	83500732	83444047	83387337	83330602	83273843	83217060	83160252	83103419	83046561	−25
1.373	0.0	82989680	82932773	82875843	82818888	82761908	82704904	82647876	82590823	82533746	82476645	−24
1.374	0.0	82419520	82362370	82305196	82247998	82190776	82133530	82076260	82018965	81961647	81904304	−24
1.375	0.0	81846938	81789547	81732133	81674695	81617233	81559747	81502237	81444703	81387145	81329564	−24
1.376	0.0	81271959	81214331	81156678	81099002	81041303	80983579	80925832	80868062	80810268	80752451	−24
1.377	0.0	80694610	80636745	80578857	80520946	80463012	80405054	80347072	80289068	80231040	80172989	−23
1.378	0.0	80114915	80056817	79998696	79940553	79882386	79824196	79765983	79707746	79649487	79591205	−23
1.379	0.0	79532900	79474572	79416221	79357847	79299450	79241031	79182588	79124123	79065635	79007124	−23
1.380	0.0	78948591	78890035	78831456	78772855	78714231	78655584	78596915	78538224	78479510	78420773	−22
1.381	0.0	78362014	78303233	78244429	78185602	78126754	78067883	78008990	77950074	77891136	77832177	−22
1.382	0.0	77773194	77714190	77655164	77596115	77537044	77477952	77418837	77359700	77300542	77241361	−22
1.383	0.0	77182158	77122934	77063687	77004419	76945129	76885817	76826483	76767128	76707751	76648352	−22
1.384	0.0	76588931	76529489	76470025	76410540	76351033	76291504	76231954	76172383	76112790	76053175	−21
1.385	0.0	75993540	75933882	75874204	75814504	75754783	75695040	75635276	75575491	75515685	75455858	−21
1.386	0.0	75396009	75336139	75276249	75216337	75156404	75096450	75036475	74976479	74916462	74856425	−21
1.387	0.0	74796366	74736286	74676186	74616065	74555923	74495760	74435577	74375372	74315148	74254902	−21
1.388	0.0	74194636	74134349	74074042	74013714	73953366	73892997	73832607	73772197	73711767	73651317	−20
1.389	0.0	73590846	73530354	73469842	73409311	73348758	73288186	73227593	73166980	73106347	73045694	−20
1.390	0.0	72985021	72924327	72863614	72802880	72742127	72681354	72620560	72559747	72498914	72438061	−20
1.391	0.0	72377188	72316295	72255382	72194450	72133498	72072526	72011535	71950524	71889493	71828443	−20
1.392	0.0	71767373	71706283	71645174	71584046	71522898	71461730	71400543	71339337	71278111	71216866	−19
1.393	0.0	71155602	71094319	71033016	70971694	70910352	70848992	70787612	70726213	70664795	70603358	−19
1.394	0.0	70541902	70480427	70418933	70357420	70295888	70234337	70172767	70111178	70049571	69987944	−19
1.395	0.0	69926299	69864635	69802952	69741251	69679531	69617792	69556035	69494259	69432464	69370651	−19
1.396	0.0	69308819	69246969	69185101	69123214	69061308	68999384	68937442	68875481	68813502	68751505	−18
1.397	0.0	68689490	68627456	68565404	68503334	68441245	68379139	68317015	68254872	68192711	68130533	−18
1.398	0.0	68068336	68006121	67943889	67881638	67819370	67757083	67694779	67632457	67570118	67507760	−18
1.399	0.0	67445385	67382992	67320581	67258153	67195707	67133244	67070763	67008264	66945748	66883214	−17

$\frac{\sin 2\pi x}{2\pi x}$

TABLE 6

Sinc Function

x		.0	.0001	.0002	.0003	.0004	.0005	.0006	.0007	.0008	.0009	Δ_2
1.400	0.0	66820663	66758095	66695509	66632905	66570285	66507646	66444991	66382318	66319629	66256921	−17
1.401	0.0	66194197	66131456	66068697	66005921	65943128	65880318	65817491	65754647	65691786	65628908	−17
1.402	0.0	65566013	65503102	65440173	65377227	65314265	65251286	65188290	65125277	65062248	64999201	−17
1.403	0.0	64936139	64873059	64809963	64746850	64683721	64620575	64557413	64494234	64431039	64367827	−16
1.404	0.0	64304599	64241355	64178094	64114817	64051523	63988214	63924888	63861546	63798187	63734813	−16
1.405	0.0	63671422	63608015	63544592	63481154	63417699	63354228	63290741	63227238	63163719	63100184	−16
1.406	0.0	63036634	62973067	62909485	62845887	62782273	62718644	62654998	62591338	62527661	62463969	−16
1.407	0.0	62400261	62336538	62272799	62209044	62145274	62081489	62017688	61953872	61890040	61826193	−15
1.408	0.0	61762330	61698453	61634560	61570651	61506728	61442789	61378835	61314866	61250882	61186883	−15
1.409	0.0	61122869	61058839	60994795	60930736	60866661	60802572	60738468	60674349	60610215	60546066	−15
1.410	0.0	60481903	60417725	60353532	60289324	60225101	60160864	60096612	60032346	59968065	59903770	−15
1.411	0.0	59839459	59775135	59710796	59646442	59582074	59517692	59453295	59388884	59324459	59260019	−14
1.412	0.0	59195565	59131097	59066615	59002118	58937608	58873083	58808544	58743991	58679424	58614843	−14
1.413	0.0	58550247	58485638	58421015	58356378	58291728	58227063	58162384	58097692	58032986	57968266	−14
1.414	0.0	57903532	57838785	57774024	57709249	57644461	57579659	57514844	57450015	57385172	57320316	−13
1.415	0.0	57255447	57190564	57125668	57060758	56995835	56930899	56865949	56800986	56736010	56671021	−13
1.416	0.0	56606018	56541003	56475974	56410932	56345877	56280809	56215728	56150633	56085526	56020406	−13
1.417	0.0	55955273	55890127	55824969	55759797	55694613	55629415	55564205	55498983	55433747	55368499	−13
1.418	0.0	55303238	55237965	55172679	55107380	55042069	54976746	54911410	54846061	54780700	54715327	−12
1.419	0.0	54649941	54584543	54519132	54453709	54388274	54322827	54257367	54191896	54126412	54060916	−12
1.420	0.0	53995407	53929887	53864355	53798810	53733254	53667686	53602105	53536513	53470909	53405293	−12
1.421	0.0	53339665	53274025	53208374	53142711	53077036	53011349	52945651	52879941	52814219	52748486	−12
1.422	0.0	52682741	52616984	52551217	52485437	52419646	52353844	52288030	52222205	52156369	52090521	−11
1.423	0.0	52024662	51958791	51892909	51827017	51761112	51695197	51629271	51563333	51497385	51431425	−11
1.424	0.0	51365454	51299472	51233480	51167476	51101461	51035436	50969399	50903352	50837294	50771225	−11
1.425	0.0	50705145	50639055	50572954	50506842	50440720	50374587	50308443	50242289	50176124	50109948	−11
1.426	0.0	50043762	49977566	49911359	49845142	49778915	49712677	49646428	49580170	49513901	49447622	−10
1.427	0.0	49381332	49315033	49248723	49182403	49116073	49049733	48983383	48917022	48850652	48784272	−10
1.428	0.0	48717882	48651481	48585071	48518652	48452222	48385782	48319333	48252874	48186405	48119926	−10
1.429	0.0	48053438	47986940	47920432	47853915	47787388	47720852	47654306	47587750	47521185	47454611	−9
1.430	0.0	47388027	47321434	47254832	47188220	47121598	47054968	46988328	46921679	46855021	46788354	−9
1.431	0.0	46721677	46654992	46588297	46521593	46454880	46388158	46321428	46254688	46187939	46121181	−9
1.432	0.0	46054415	45987639	45920855	45854062	45787260	45720450	45653631	45586803	45519966	45453121	−9
1.433	0.0	45386267	45319404	45252533	45185654	45118766	45051869	44984964	44918051	44851129	44784199	−8
1.434	0.0	44717260	44650313	44583358	44516395	44449423	44382443	44315455	44248459	44181454	44114442	−8
1.435	0.0	44047422	43980393	43913356	43846312	43779259	43712199	43645130	43578054	43510970	43443878	−8
1.436	0.0	43376778	43309671	43242555	43175432	43108302	43041163	42974017	42906864	42839702	42772534	−8
1.437	0.0	42705357	42638173	42570982	42503783	42436577	42369363	42302142	42234914	42167678	42100435	−7
1.438	0.0	42033185	41965928	41898663	41831391	41764112	41696826	41629533	41562232	41494925	41427611	−7
1.439	0.0	41360289	41292961	41225625	41158283	41090934	41023578	40956215	40888846	40821469	40754086	−7
1.440	0.0	40686696	40619299	40551896	40484486	40417070	40349646	40282217	40214780	40147338	40079888	−6
1.441	0.0	40012433	39944970	39877502	39810027	39742546	39675058	39607564	39540064	39472557	39405045	−6
1.442	0.0	39337526	39270001	39202470	39134932	39067389	38999840	38932284	38864723	38797155	38729582	−6
1.443	0.0	38662003	38594417	38526826	38459229	38391627	38324018	38256404	38188784	38121158	38053527	−6
1.444	0.0	37985890	37918247	37850599	37782945	37715286	37647621	37579950	37512274	37444593	37376906	−5
1.445	0.0	37309214	37241517	37173814	37106106	37038392	36970674	36902950	36835221	36767486	36699747	−5
1.446	0.0	36632003	36564253	36496498	36428739	36360974	36293204	36225429	36157650	36089865	36022076	−5
1.447	0.0	35954282	35886483	35818679	35750870	35683057	35615239	35547416	35479589	35411757	35343920	−5
1.448	0.0	35276079	35208233	35140383	35072528	35004668	34936805	34868936	34801064	34733187	34665306	−4
1.449	0.0	34597420	34529530	34461636	34393738	34325835	34257928	34190017	34122102	34054183	33986260	−4

TABLE 6

Sinc Function

x		.0	.0001	.0002	.0003	.0004	.0005	.0006	.0007	.0008	.0009	Δ_2
1.450	0.0	33918333	33850401	33782466	33714527	33646583	33578636	33510685	33442730	33374772	33306809	−4
1.451	0.0	33238843	33170873	33102899	33034922	32966940	32898956	32830967	32762975	32694980	32626981	−4
1.452	0.0	32558978	32490972	32422962	32354949	32286933	32218913	32150890	32082863	32014834	31946801	−3
1.453	0.0	31878764	31810725	31742682	31674636	31606587	31538535	31470480	31402422	31334360	31266296	−3
1.454	0.0	31198229	31130158	31062085	30994009	30925930	30857848	30789764	30721676	30653586	30585493	−3
1.455	0.0	30517398	30449299	30381198	30313095	30244989	30176880	30108768	30040654	29972538	29904419	−2
1.456	0.0	29836298	29768174	29700048	29631920	29563789	29495656	29427520	29359382	29291242	29223100	−2
1.457	0.0	29154956	29086810	29018661	28950510	28882357	28814203	28746046	28677887	28609726	28541563	−2
1.458	0.0	28473399	28405232	28337064	28268893	28200721	28132547	28064372	27996194	27928015	27859834	−2
1.459	0.0	27791652	27723468	27655282	27587095	27518906	27450716	27382524	27314331	27246137	27177941	0
1.460	0.0	27109743	27041544	26973344	26905143	26836940	26768736	26700531	26632324	26564117	26495908	0
1.461	0.0	26427698	26359487	26291275	26223062	26154848	26086633	26018417	25950200	25881982	25813763	0
1.462	0.0	25745543	25677323	25609102	25540880	25472657	25404433	25336209	25267984	25199759	25131532	0
1.463	0.0	25063306	24995078	24926850	24858622	24790393	24722164	24653934	24585704	24517473	24449242	0
1.464	0.0	24381011	24312780	24244548	24176316	24108083	24039851	23971618	23903385	23835152	23766919	0
1.465	0.0	23698686	23630453	23562220	23493987	23425754	23357520	23289288	23221055	23152822	23084589	0
1.466	0.0	23016357	22948125	22879893	22811662	22743430	22675199	22606969	22538738	22470509	22402279	0
1.467	0.0	22334050	22265822	22197594	22129367	22061140	21992913	21924688	21856463	21788238	21720015	1
1.468	0.0	21651792	21583570	21515348	21447128	21378908	21310689	21242471	21174254	21106038	21037823	1
1.469	0.0	20969608	20901395	20833183	20764972	20696762	20628553	20560345	20492138	20423933	20355728	1
1.470	0.0	20287525	20219324	20151123	20082924	20014726	19946530	19878335	19810142	19741949	19673759	1
1.471	0.0	19605570	19537382	19469196	19401012	19332829	19264648	19196468	19128290	19060114	18991940	2
1.472	0.0	18923767	18855596	18787427	18719260	18651095	18582931	18514770	18446610	18378453	18310297	2
1.473	0.0	18242143	18173992	18105843	18037695	17969550	17901407	17833266	17765127	17696991	17628857	2
1.474	0.0	17560725	17492596	17424468	17356343	17288221	17220101	17151983	17083868	17015756	16947645	3
1.475	0.0	16879538	16811433	16743330	16675231	16607134	16539039	16470947	16402858	16334772	16266688	3
1.476	0.0	16198608	16130530	16062455	15994383	15926314	15858247	15790184	15722124	15654066	15586012	3
1.477	0.0	15517961	15449913	15381867	15313826	15245787	15177751	15109719	15041690	14973664	14905642	3
1.478	0.0	14837622	14769607	14701594	14633585	14565579	14497577	14429578	14361583	14293591	14225603	4
1.479	0.0	14157619	14089638	14021660	13953687	13885717	13817750	13749788	13681829	13613874	13545923	4
1.480	0.0	13477975	13410032	13342092	13274157	13206225	13138297	13070373	13002453	12934538	12866626	4
1.481	0.0	12798718	12730815	12662915	12595020	12527129	12459242	12391360	12323482	12255608	12187738	4
1.482	0.0	12119873	12052012	11984155	11916303	11848456	11780613	11712774	11644940	11577110	11509285	5
1.483	0.0	11441465	11373649	11305838	11238031	11170230	11102433	11034640	10966853	10899070	10831293	5
1.484	0.0	10763520	10695751	10627988	10560230	10492477	10424729	10356985	10289247	10221514	10153786	5
1.485	0.0	10086063	10018345	09950632	09882925	09815223	09747526	09679834	09612147	09544466	09476790	5
1.486	0.0	09409120	09341455	09273795	09206141	09138492	09070849	09003211	08935579	08867953	08800332	6
1.487	0.0	08732716	08665107	08597503	08529904	08462312	08394725	08327144	08259568	08191999	08124435	6
1.488	0.0	08056877	07989325	07921780	07854240	07786705	07719177	07651655	07584140	07516630	07449126	6
1.489	0.0	07381628	07314137	07246651	07179172	07111699	07044233	06976772	06909318	06841871	06774429	6
1.490	0.0	06706994	06639566	06572144	06504728	06437319	06369916	06302520	06235130	06167747	06100371	7
1.491	0.0	06033001	05965638	05898281	05830931	05763588	05696252	05628922	05561600	05494284	05426975	7
1.492	0.0	05359673	05292377	05225089	05157808	05090533	05023266	04956006	04888752	04821506	04754267	7
1.493	0.0	04687035	04619810	04552593	04485382	04418179	04350983	04283794	04216613	04149439	04082272	7
1.494	0.0	04015113	03947961	03880817	03813680	03746550	03679428	03612314	03545207	03478107	03411016	8
1.495	0.0	03343931	03276855	03209786	03142725	03075672	03008626	02941588	02874558	02807536	02740522	8
1.496	0.0	02673515	02606517	02539526	02472543	02405569	02338602	02271643	02204693	02137750	02070816	8
1.497	0.0	02003889	01936971	01870061	01803159	01736266	01669380	01602503	01535635	01468774	01401922	8
1.498	0.0	01335078	01268243	01201416	01134598	01067788	01000986	00934193	00867409	00800633	00733866	9
1.499	0.0	00667107	00600357	00533616	00466883	00400159	00333444	00266738	00200040	00133351	00066671	−9

$$\frac{\sin 2\pi x}{2\pi x}$$

TABLE 6

Sinc Function

x	.0	.0001	.0002	.0003	.0004	.0005	.0006	.0007	.0008	.0009	Δ_2
1.500 -0.0	00000000	00066662	00133316	00199960	00266595	00333222	00399839	00466447	00533047	00599637	9
1.501 -0.0	00666218	00732790	00799353	00865907	00932451	00998986	01065512	01132029	01198536	01265034	9
1.502 -0.0	01331523	01398002	01464472	01530933	01597384	01663825	01730257	01796680	01863093	01929496	10
1.503 -0.0	01995890	02062274	02128648	02195013	02261368	02327714	02394050	02460376	02526692	02592998	10
1.504 -0.0	02659294	02725581	02791858	02858125	02924382	02990628	03056865	03123092	03189309	03255516	10
1.505 -0.0	03321713	03387899	03454076	03520242	03586398	03652544	03718680	03784805	03850921	03917026	10
1.506 -0.0	03983120	04049204	04115278	04181342	04247395	04313437	04379469	04445491	04511502	04577503	11
1.507 -0.0	04643493	04709472	04775441	04841399	04907346	04973283	05039209	05105125	05171029	05236923	11
1.508 -0.0	05302806	05368678	05434540	05500390	05566230	05632058	05697876	05763683	05829478	05895263	11
1.509 -0.0	05961037	06026799	06092551	06158291	06224020	06289738	06355445	06421141	06486825	06552498	11
1.510 -0.0	06618160	06683811	06749450	06815078	06880694	06946299	07011893	07077475	07143046	07208605	11
1.511 -0.0	07274153	07339689	07405213	07470726	07536228	07601717	07667195	07732662	07798116	07863559	12
1.512 -0.0	07928990	07994410	08059817	08125213	08190597	08255969	08321329	08386677	08452013	08517338	12
1.513 -0.0	08582650	08647950	08713238	08778514	08843778	08909030	08974270	09039498	09104713	09169916	12
1.514 -0.0	09235107	09300286	09365452	09430606	09495748	09560878	09625995	09691099	09756192	09821271	12
1.515 -0.0	09886339	09951394	10016436	10081466	10146483	10211488	10276480	10341459	10406426	10471380	13
1.516 -0.0	10536321	10601250	10666166	10731069	10795959	10860837	10925701	10990553	11055392	11120218	13
1.517 -0.0	11185031	11249831	11314618	11379392	11444153	11508902	11573636	11638358	11703067	11767763	13
1.518 -0.0	11832445	11897114	11961770	12026413	12091042	12155659	12220262	12284851	12349427	12413990	13
1.519 -0.0	12478540	12543076	12607598	12672107	12736603	12801085	12865554	12930009	12994450	13058878	14
1.520 -0.0	13123292	13187692	13252079	13316452	13380812	13445157	13509489	13573807	13638112	13702402	14
1.521 -0.0	13766679	13830941	13895190	13959425	14023646	14087853	14152046	14216224	14280389	14344540	14
1.522 -0.0	14408677	14472799	14536908	14601002	14665082	14729148	14793199	14857237	14921260	14985269	14
1.523 -0.0	15049263	15113243	15177209	15241161	15305098	15369020	15432928	15496822	15560701	15624565	15
1.524 -0.0	15683415	15752251	15816072	15879878	15943670	16007446	16071209	16134956	16198689	16262407	15
1.525 -0.0	16326110	16389799	16453473	16517131	16580775	16644404	16708019	16771618	16835202	16898771	15
1.526 -0.0	16962326	17025865	17089389	17152898	17216392	17279871	17343335	17406783	17470217	17533635	15
1.527 -0.0	17597038	17660426	17723798	17787156	17850497	17913824	17977135	18040431	18103711	18166976	15
1.528 -0.0	18230226	18293460	18356678	18419881	18483069	18546241	18609397	18672538	18735663	18798772	16
1.529 -0.0	18861866	18924944	18988006	19051053	19114084	19177099	19240098	19303081	19366049	19429000	16
1.530 -0.0	19491936	19554856	19617760	19680648	19743520	19806376	19869216	19932040	19994847	20057639	16
1.531 -0.0	20120414	20183174	20245917	20308644	20371355	20434050	20496728	20559390	20622036	20684665	16
1.532 -0.0	20747278	20809875	20872456	20935020	20997567	21060098	21122613	21185111	21247593	21310058	17
1.533 -0.0	21372506	21434938	21497353	21559752	21622134	21684499	21746848	21809180	21871495	21933794	17
1.534 -0.0	21996075	22058340	22120588	22182819	22245034	22307231	22369412	22431575	22493722	22555852	17
1.535 -0.0	22617964	22680060	22742139	22804200	22866245	22928272	22990282	23052275	23114251	23176210	17
1.536 -0.0	23238151	23300075	23361982	23423872	23485744	23547600	23609437	23671258	23733061	23794846	17
1.537 -0.0	23856614	23918365	23980098	24041814	24103512	24165193	24226856	24288501	24350129	24411739	18
1.538 -0.0	24473332	24534907	24596464	24658004	24719525	24781029	24842516	24903984	24965435	25026868	18
1.539 -0.0	25088283	25149680	25211059	25272420	25333763	25395088	25456396	25517685	25578956	25640210	18
1.540 -0.0	25701445	25762662	25823861.	25885041	25946204	26007348	26068475	26129583	26190672	26251744	18
1.541 -0.0	26312797	26373832	26434849.	26495847	26556827	26617788	26678731	26739656	26800562	26861449	18
1.542 -0.0	26922318	26983169	27044001.	27104815	27165610	27226386	27287144	27347883	27408603	27469305	19
1.543 -0.0	27529988	27590652	27651297.	27711924	27772532	27833121	27893691	27954243	28014775	28075289	19
1.544 -0.0	28135784	28196259	28256716.	28317154	28377573	28437972	28498353	28558715	28619057	28679381	19
1.545 -0.0	28739685	28799970	28860236.	28920483	28980711	29040919	29101108	29161278	29221429	29281560	19
1.546 -0.0	29341672	29401764	29461837.	29521891	29581925	29641940	29701936	29761912	29821868	29881805	20
1.547 -0.0	29941722	30001620	30061498.	30121357	30181196	30241015	30300815	30360595	30420355	30480096	20
1.548 -0.0	30539817	30599518	30659199.	30718860	30778502	30838123	30897725	30957307	31016869	31076411	20
1.549 -0.0	31135934	31195436	31254918.	31314380	31373822	31433244	31492646	31552028	31611390	31670732	20

TABLE 6

Sinc Function

x		.0	.0001	.0002	.0003	.0004	.0005	.0006	.0007	.0008	.0009	Δ_2
1.550	-0.0	31730053	31789354	31848635	31907896	31967137	32026357	32085557	32144737	32203897	32263036	20
1.551	-0.0	32322154	32381253	32440331	32499388	32558426	32617442	32676438	32735414	32794369	32853304	21
1.552	-0.0	32912218	32971111	33029984	33088836	33147668	33206479	33265269	33324038	33382787	33441515	21
1.553	-0.0	33500223	33558909	33617575	33676220	33734844	33793447	33852029	33910591	33969131	34027651	21
1.554	-0.0	34086149	34144627	34203083	34261519	34319933	34378327	34436699	34495050	34553380	34611689	21
1.555	-0.0	34669977	34728244	34786489	34844713	34902916	34961098	35019258	35077398	35135515	35193612	21
1.556	-0.0	35251687	35309741	35367773	35425784	35483773	35541741	35599688	35657613	35715516	35773398	22
1.557	-0.0	35831259	35889098	35946915	36004710	36062484	36120237	36177967	36235676	36293364	36351029	22
1.558	-0.0	36408673	36466295	36523895	36581473	36639030	36696565	36754078	36811568	36869038	36926485	22
1.559	-0.0	36983910	37041313	37098694	37156053	37213391	37270706	37327999	37385270	37442519	37499746	22
1.560	-0.0	37556950	37614133	37671293	37728431	37785547	37842641	37899712	37956761	38013788	38070792	22
1.561	-0.0	38127775	38184734	38241672	38298587	38355480	38412350	38469198	38526023	38582826	38639606	23
1.562	-0.0	38696364	38753099	38809812	38866502	38923169	38979814	39036437	39093036	39149613	39206167	23
1.563	-0.0	39262699	39319208	39375694	39432157	39488597	39545015	39601410	39657782	39714131	39770457	23
1.564	-0.0	39826761	39883041	39939299	39995533	40051745	40107933	40164099	40220241	40276361	40332457	23
1.565	-0.0	40388530	40444580	40500608	40556611	40612592	40668550	40724484	40780395	40836283	40892148	23
1.566	-0.0	40947989	41003807	41059602	41115373	41171121	41226846	41282547	41338225	41393880	41449511	23
1.567	-0.0	41505118	41560702	41616263	41671800	41727313	41782803	41838270	41893713	41949132	42004527	24
1.568	-0.0	42059899	42115247	42170572	42225873	42281150	42336403	42391633	42446839	42502021	42557179	24
1.569	-0.0	42612314	42667424	42722511	42777574	42832613	42887628	42942619	42997586	43052529	43107448	24
1.570	-0.0	43162343	43217214	43272061	43326884	43381683	43436458	43491209	43545935	43600638	43655316	24
1.571	-0.0	43709970	43764599	43819205	43873786	43928343	43982876	44037385	44091869	44146328	44200764	24
1.572	-0.0	44255175	44309562	44363924	44418262	44472575	44526864	44581128	44635368	44689584	44743775	25
1.573	-0.0	44797941	44852083	44906200	44960292	45014360	45068404	45122422	45176416	45230385	45284330	25
1.574	-0.0	45338250	45392145	45446015	45499861	45553681	45607477	45661248	45714995	45768716	45822412	25
1.575	-0.0	45876084	45929731	45983352	46036949	46090521	46144067	46197589	46251086	46304557	46358004	25
1.576	-0.0	46411425	46464822	46518193	46571539	46624860	46678156	46731427	46784672	46837892	46891087	25
1.577	-0.0	46944257	46997401	47050521	47103614	47156683	47209726	47262744	47315736	47368703	47421645	25
1.578	-0.0	47474561	47527452	47580317	47633157	47685971	47738760	47791523	47844261	47896973	47949660	26
1.579	-0.0	48002321	48054956	48107566	48160150	48212708	48265241	48317748	48370229	48422684	48475114	26
1.580	-0.0	48527518	48579896	48632249	48684575	48736876	48789151	48841400	48893623	48945820	48997991	26
1.581	-0.0	49050137	49102256	49154349	49206417	49258458	49310474	49362463	49414426	49466363	49518274	26
1.582	-0.0	49570159	49622018	49673851	49725657	49777438	49829192	49880920	49932622	49984297	50035946	26
1.583	-0.0	50087569	50139166	50190736	50242280	50293798	50345289	50396754	50448193	50499605	50550990	26
1.584	-0.0	50602350	50653683	50704989	50756269	50807522	50858749	50909949	50961123	51012270	51063390	27
1.585	-0.0	51114484	51165552	51216592	51267606	51318593	51369554	51420488	51471395	51522276	51573129	27
1.586	-0.0	51623956	51674756	51725530	51776276	51826996	51877689	51928355	51978994	52029606	52080191	27
1.537	-0.0	52130750	52181281	52231785	52282263	52332713	52383137	52433533	52483902	52534245	52584560	27
1.588	-0.0	52634848	52685109	52735343	52785549	52835729	52885881	52936006	52986104	53036175	53086219	27
1.589	-0.0	53136235	53186224	53236186	53286120	53336027	53385907	53435759	53485584	53535382	53585152	27
1.590	-0.0	53634895	53684610	53734298	53783959	53833592	53883197	53932775	53982326	54031849	54081344	28
1.591	-0.0	54130812	54180252	54229665	54279050	54328407	54377737	54427039	54476314	54525560	54574779	28
1.592	-0.0	54623971	54673134	54722270	54771378	54820458	54869510	54918535	54967532	55016501	55065442	28
1.593	-0.0	55114355	55163240	55212097	55260927	55309728	55358502	55407247	55455965	55504654	55553316	28
1.594	-0.0	55601949	55650555	55699132	55747681	55796202	55844695	55893160	55941597	55990006	56038386	28
1.595	-0.0	56086739	56135063	56183358	56231626	56279865	56328077	56376259	56424414	56472540	56520638	28
1.596	-0.0	56568708	56616749	56664762	56712746	56760702	56808630	56856529	56904400	56952242	57000056	28
1.597	-0.0	57047841	57095598	57143326	57191026	57238697	57286340	57333954	57381540	57429096	57476625	29
1.598	-0.0	57524124	57571595	57619038	57666451	57713836	57761192	57808520	57855819	57903089	57950330	29
1.599	-0.0	57997542	58044726	58091881	58139007	58186104	58233172	58280212	58327222	58374204	58421156	29

$$\frac{\sin 2\pi x}{2\pi x}$$

TABLE 6

Sinc Function

x	.0	0001	.0002	.0003	.0004	.0005	.0006	.0007	.0008	.0009	Δ_2
1.600 -0.0	58468080	58514975	58561841	58608678	58655486	58702265	58749015	58795735	58842427	58889090	29
1.601 -0.0	58935724	58982328	59028904	59075450	59121967	59168455	59214914	59261344	59307744	59354116	29
1.602 -0.0	59400458	59446771	59493054	59539309	59585534	59631730	59677896	59724033	59770141	59816220	29
1.603 -0.0	59862269	59908289	59954279	60000240	60046171	60092074	60137946	60183789	60229603	60275387	29
1.604 -0.0	60321142	60366867	60412563	60458229	60503866	60549473	60595050	60640598	60686116	60731605	30
1.605 -0.0	60777064	60822493	60867893	60913263	60958603	61003913	61049194	61094445	61139666	61184858	30
1.606 -0.0	61230020	61275152	61320254	61365326	61410369	61455381	61500364	61545317	61590240	61635133	30
1.607 -0.0	61679996	61724829	61769633	61814406	61859149	61903863	61948546	61993200	62037823	62082416	30
1.608 -0.0	62126979	62171513	62216016	62260489	62304932	62349344	62393727	62438079	62482402	62526694	30
1.609 -0.0	62570956	62615188	62659389	62703561	62747702	62791812	62835893	62879943	62923963	62967953	30
1.610 -0.0	63011912	63055841	63099740	63143608	63187446	63231254	63275031	63318778	63362494	63406180	30
1.611 -0.0	63449835	63493460	63537055	63580619	63624152	63667655	63711127	63754569	63797980	63841361	31
1.612 -0.0	63884711	63928031	63971320	64014578	64057806	64101003	64144169	64187305	64230410	64273484	31
1.613 -0.0	64316528	64359541	64402523	64445474	64488395	64531285	64574144	64616972	64659769	64702536	31
1.614 -0.0	64745272	64787977	64830651	64873294	64915906	64958487	65001038	65043557	65086046	65128504	31
1.615 -0.0	65170930	65213326	65255691	65298024	65340327	65382598	65424839	65467049	65509227	65551374	31
1.616 -0.0	65593491	65635576	65677630	65719653	65761645	65803605	65845535	65887433	65929300	65971136	31
1.617 -0.0	66012940	66054714	66096456	66138167	66179847	66221495	66263112	66304698	66346252	66387776	31
1.618 -0.0	66429267	66470728	66512157	66553555	66594921	66636256	66677559	66718832	66760072	66801281	31
1.619 -0.0	66842459	66883605	66924720	66965803	67006855	67047875	67088864	67129821	67170747	67211641	32
1.620 -0.0	67252504	67293334	67334134	67374901	67415637	67456342	67497014	67537656	67578265	67618843	32
1.621 -0.0	67659389	67699903	67740385	67780836	67821255	67861643	67901998	67942322	67982614	68022874	32
1.622 -0.0	68063103	68103299	68143464	68183597	68223698	68263767	68303804	68343809	68383783	68423724	32
1.623 -0.0	68463634	68503511	68543357	68583171	68622952	68662702	68702420	68742106	68781759	68821381	32
1.624 -0.0	68860970	68900528	68940053	68979547	69019008	69058437	69097834	69137199	69176532	69215833	32
1.625 -0.0	69255101	69294338	69333542	69372714	69411853	69450961	69490036	69529079	69568090	69607069	32
1.626 -0.0	69646015	69684929	69723811	69762660	69801477	69840262	69879014	69917734	69956422	69995077	32
1.627 -0.0	70033700	70072291	70110849	70149375	70187868	70226329	70264757	70303153	70341517	70379848	32
1.628 -0.0	70418146	70456412	70494646	70532847	70571015	70609151	70647254	70685325	70723363	70761369	33
1.629 -0.0	70799342	70837282	70875190	70913065	70950908	70988717	71026495	71064239	71101951	71139630	33
1.630 -0.0	71177276	71214890	71252471	71290019	71327535	71365017	71402467	71439885	71477269	71514620	33
1.631 -0.0	71551939	71589225	71626478	71663699	71700886	71738041	71775162	71812251	71849307	71886330	33
1.632 -0.0	71923320	71960277	71997201	72034093	72070951	72107776	72144569	72181328	72218054	72254748	33
1.633 -0.0	72291408	72328036	72364630	72401191	72437719	72474214	72510677	72547105	72583501	72619864	33
1.634 -0.0	72656194	72692490	72728754	72764984	72801181	72837345	72873476	72909573	72945637	72981669	33
1.635 -0.0	73017666	73053631	73089563	73125461	73161326	73197157	73232956	73268721	73304453	73340151	33
1.636 -0.0	73375816	73411448	73447047	73482612	73518144	73553642	73589107	73624539	73659937	73695302	33
1.637 -0.0	73730634	73765932	73801196	73836428	73871625	73906790	73941921	73977018	74012082	74047112	33
1.638 -0.0	74082109	74117072	74152002	74186898	74221761	74256590	74291386	74326148	74360876	74395571	34
1.639 -0.0	74430232	74464860	74499454	74534014	74568541	74603034	74637494	74671919	74706312	74740670	34
1.640 -0.0	74774995	74809286	74843543	74877767	74911956	74946113	74980235	75014323	75048378	75082399	34
1.641 -0.0	75116387	75150340	75184260	75218146	75251998	75285816	75319600	75353351	75387067	75420750	34
1.642 -0.0	75454399	75488014	75521595	75555142	75588656	75622135	75655581	75688992	75722370	75755713	34
1.643 -0.0	75789023	75822299	75855540	75888748	75921922	75955062	75988167	76021239	76054277	76087280	34
1.644 -0.0	76120250	76153185	76186087	76218954	76251787	76284586	76317352	76350082	76382779	76415442	34
1.645 -0.0	76448071	76480665	76513225	76545751	76578243	76610701	76643125	76675514	76707869	76740190	34
1.646 -0.0	76772477	76804729	76836947	76869132	76901281	76933397	76965478	76997525	77029538	77061516	34
1.647 -0.0	77093460	77125370	77157245	77189086	77220893	77252665	77284404	77316107	77347777	77379411	34
1.648 -0.0	77411012	77442578	77474110	77505607	77537070	77568499	77599893	77631253	77662578	77693868	34
1.649 -0.0	77725125	77756347	77737534	77818687	77849805	77880889	77911938	77942953	77973933	78004879	35

TABLE 6

Sinc Function

x	.0	.0001	.0002	.0003	.0004	.0005	.0006	.0007	.0008	.0009	Δ_2
1.650 -0.0	78035790	78066667	78097509	78128316	78159089	78189828	78220531	78251201	78281835	78312435	35
1.651 -0.0	78343000	78373531	78404027	78434488	78464915	78495307	78525665	78555987	78586276	78616529	35
1.652 -0.0	78646748	78676932	78707081	78737195	78767275	78797320	78827331	78857306	78887247	78917153	35
1.653 -0.0	78947024	78976861	79006663	79036430	79066162	79095859	79125522	79155149	79184742	79214300	35
1.654 -0.0	79243823	79273312	79302765	79332184	79361567	79390916	79420230	79449509	79478753	79507963	35
1.655 -0.0	79537137	79566276	79595381	79624450	79653485	79682485	79711449	79740379	79769274	79798134	35
1.656 -0.0	79826958	79855748	79884503	79913223	79941907	79970557	79999172	80027752	80056296	80084806	35
1.657 -0.0	80113280	80141720	80170124	80198494	80226828	80255127	80283391	80311620	80339814	80367973	35
1.658 -0.0	80396096	80424185	80452238	80480256	80508239	80536187	80564100	80591978	80619820	80647627	35
1.659 -0.0	80675399	80703136	80730838	80758504	80786135	80813731	80841292	80868817	80896308	80923763	35
1.660 -0.0	80951182	80978567	81005916	81033230	81060509	81087752	81114960	81142133	81169271	81196373	35
1.661 -0.0	81223440	81250471	81277468	81304429	81331354	81358244	81385099	81411919	81438703	81465452	35
1.662 -0.0	81492165	81518843	81545486	81572093	81598665	81625201	81651702	81678168	81704598	81730993	35
1.663 -0.0	81757352	81783676	81809964	81836217	81862434	81888616	81914763	81940874	81966949	81992990	35
1.664 -0.0	82018994	82044963	82070897	82096795	82122657	82148484	82174276	82200032	82225752	82251437	36
1.665 -0.0	82277086	82302700	82328278	82353821	82379328	82404799	82430235	82455635	82481000	82506329	36
1.666 -0.0	82531622	82556880	82582102	82607289	82632439	82657555	82682634	82707678	82732687	82757659	36
1.667 -0.0	82782596	82807498	82832363	82857193	82881988	82906746	82931469	82956156	82980808	83005424	36
1.668 -0.0	83030004	83054548	83079056	83103529	83127966	83152368	83176733	83201063	83225357	83249616	36
1.669 -0.0	83273838	83298025	83322176	83346291	83370371	83394415	83418422	83442394	83466331	83490231	36
1.670 -0.0	83514096	83537924	83561717	83585474	83609196	83632881	83656531	83680144	83703722	83727264	36
1.671 -0.0	83750770	83774241	83797675	83821073	83844436	83867763	83891054	83914308	83937527	83960710	36
1.672 -0.0	83983858	84006969	84030044	84053084	84076087	84099054	84121986	84144882	84167741	84190565	36
1.673 -0.0	84213353	84236104	84258820	84281500	84304144	84326752	84349323	84371859	84394359	84416823	36
1.674 -0.0	84439251	84461643	84483999	84506318	84528602	84550850	84573061	84595237	84617377	84639480	36
1.675 -0.0	84661548	84683579	84705575	84727534	84749457	84771345	84793196	84815011	84836790	84858532	36
1.676 -0.0	84880239	84901910	84923545	84945143	84966705	84988232	85009722	85031176	85052594	85073975	36
1.677 -0.0	85095321	85116630	85137904	85159141	85180342	85201507	85222636	85243728	85264785	85285805	36
1.678 -0.0	85306789	85327737	85348648	85369524	85390363	85411166	85431933	85452664	85473359	85494017	36
1.679 -0.0	85514639	85535225	85555775	85576288	85596765	85617206	85637611	85657980	85678312	85698608	36
1.680 -0.0	85718868	85739091	85759279	85779430	85799545	85819623	85839665	85859671	85879641	85899575	36
1.681 -0.0	85919472	85939333	85959157	85978945	85998697	86018413	86038093	86057736	86077342	86096913	36
1.682 -0.0	86116447	86135945	86155406	86174832	86194220	86213573	86232889	86252169	86271413	86290620	36
1.683 -0.0	86309791	86328925	86348023	86367085	86386110	86405100	86424052	86442969	86461848	86480692	36
1.684 -0.0	86499499	86518270	86537005	86555703	86574364	86592989	86611578	86630131	86648647	86667127	36
1.685 -0.0	86685570	86703977	86722347	86740681	86758979	86777240	86795465	86813653	86831805	86849921	36
1.686 -0.0	86868000	86886042	86904048	86922018	86939951	86957848	86975709	86993533	87011320	87029071	36
1.687 -0.0	87046786	87064464	87082105	87099711	87117279	87134812	87152307	87169767	87187189	87204576	36
1.688 -0.0	87221925	87239239	87256516	87273756	87290960	87308127	87325258	87342353	87359410	87376432	37
1.689 -0.0	87393417	87410365	87427277	87444152	87460991	87477793	87494559	87511288	87527981	87544637	37
1.690 -0.0	87561257	87577840	87594386	87610896	87627370	87643807	87660207	87676571	87692898	87709189	37
1.691 -0.0	87725443	87741661	87757842	87773987	87790095	87806166	87822201	87838199	87854161	87870086	37
1.692 -0.0	87885975	87901827	87917642	87933421	87949164	87964869	87980539	87996171	88011767	88027327	37
1.693 -0.0	88042849	88058336	88073785	88089198	88104575	88119915	88135218	88150484	88165714	88180908	37
1.694 -0.0	88196065	88211185	88226269	88241316	88256326	88271300	88286237	88301138	88316002	88330829	37
1.695 -0.0	88345620	88360374	88375091	88389772	88404416	88419024	88433595	88448129	88462627	88477088	37
1.696 -0.0	88491512	88505900	88520251	88534566	88548844	88563085	88577290	88591458	88605589	88619684	37
1.697 -0.0	88633742	88647763	88661748	88675696	88689608	88703482	88717321	88731122	88744887	88758615	37
1.698 -0.0	88772307	88785962	88799580	88813161	88826706	88840215	88853686	88867121	88880519	88893881	37
1.699 -0.0	88907206	88920494	88933746	88946961	88960139	88973281	88986386	88999454	89012485	89025480	37

TABLE 6

Sinc Function

x		.0	.0001	.0002	.0003	.0004	.0005	.0006	.0007	.0008	.0009	Δ_2
1.700	-0.0	89038439	89051360	89064245	89077093	89089905	89102680	89115418	89128120	89140784	89153413	37
1.701	-0.0	89166004	89178559	89191077	89203558	89216003	89228411	89240783	89253117	89265415	89277677	37
1.702	-0.0	89289901	89302089	89314241	89326355	89338433	89350475	89362479	89374447	89386378	89398273	37
1.703	-0.0	89410130	89421952	89433736	89445484	89457195	89468869	89480507	89492108	89503672	89515200	37
1.704	-0.0	89526691	89538145	89549562	89560943	89572287	89583595	89594866	89606100	89617297	89628458	37
1.705	-0.0	89639582	89650669	89661720	89672734	89683711	89694651	89705555	89716422	89727253	89738047	37
1.706	-0.0	89748804	89759524	89770208	89780855	89791465	89802039	89812576	89823076	89833540	89843967	37
1.707	-0.0	89854357	89864710	89875027	89885307	89895551	89905758	89915928	89926061	89936158	89946218	37
1.708	-0.0	89956241	89966228	89976178	89986091	89995968	90005808	90015611	90025377	90035107	90044800	37
1.709	-0.0	90054457	90064077	90073660	90083206	90092716	90102189	90111626	90121025	90130388	90139715	37
1.710	-0.0	90149005	90158258	90167474	90176654	90185797	90194903	90203973	90213006	90222002	90230962	37
1.711	-0.0	90239885	90248771	90257621	90266434	90275210	90283950	90292653	90301320	90309949	90318542	37
1.712	-0.0	90327099	90335619	90344102	90352548	90360958	90369331	90377667	90385967	90394230	90402457	37
1.713	-0.0	90410647	90418800	90426917	90434997	90443040	90451047	90459017	90466950	90474847	90482707	37
1.714	-0.0	90490530	90498317	90506067	90513781	90521458	90529098	90536702	90544269	90551799	90559293	37
1.715	-0.0	90566750	90574171	90581555	90588902	90596213	90603487	90610724	90617925	90625089	90632217	37
1.716	-0.0	90639308	90646362	90653380	90660361	90667306	90674214	90681085	90687920	90694718	90701480	37
1.717	-0.0	90708205	90714894	90721545	90728161	90734739	90741281	90747787	90754256	90760688	90767084	37
1.718	-0.0	90773443	90779766	90786052	90792301	90798514	90804690	90810830	90816933	90823000	90829030	37
1.719	-0.0	90835024	90840981	90846901	90852785	90858632	90864443	90870217	90875955	90881656	90887320	37
1.720	-0.0	90892949	90898540	90904095	90909614	90915095	90920541	90925950	90931322	90936658	90941957	37
1.721	-0.0	90947220	90952446	90957636	90962789	90967906	90972986	90978030	90983037	90988008	90992942	36
1.722	-0.0	90997840	91002701	91007526	91012314	91017066	91021782	91026460	91031103	91035709	91040278	36
1.723	-0.0	91044811	91049308	91053768	91058191	91062578	91066929	91071243	91075521	91079762	91083967	36
1.724	-0.0	91088135	91092267	91096363	91100422	91104445	91108431	91112381	91116294	91120171	91124011	36
1.725	-0.0	91127815	91131583	91135314	91139009	91142668	91146290	91149875	91153425	91156938	91160414	36
1.726	-0.0	91163854	91167258	91170625	91173956	91177251	91180509	91183731	91186916	91190065	91193178	36
1.727	-0.0	91196254	91199294	91202298	91205265	91208196	91211091	91213949	91216771	91219556	91222306	36
1.728	-0.0	91225018	91227695	91230335	91232939	91235507	91238038	91240533	91242992	91245414	91247800	36
1.729	-0.0	91250150	91252464	91254741	91256982	91259186	91261355	91263487	91265583	91267642	91269665	36
1.730	-0.0	91271653	91273603	91275518	91277396	91279238	91281044	91282813	91284546	91286243	91287904	36
1.731	-0.0	91289529	91291117	91292669	91294185	91295665	91297108	91298516	91299887	91301222	91302520	36
1.732	-0.0	91303783	91305009	91306199	91307353	91308471	91309552	91310598	91311607	91312580	91313517	36
1.733	-0.0	91314418	91315282	91316111	91316903	91317659	91318379	91319063	91319711	91320322	91320898	36
1.734	-0.0	91321437	91321941	91322408	91322839	91323234	91323593	91323916	91324202	91324453	91324668	36
1.735	-0.0	91324846	91324988	91325095	91325165	91325199	91325197	91325159	91325085	91324975	91324829	36
1.736	-0.0	91324647	91324429	91324175	91323885	91323559	91323197	91322798	91322364	91321894	91321388	36
1.737	-0.0	91320846	91320268	91319653	91319003	91318317	91317595	91316837	91316043	91315213	91314347	36
1.738	-0.0	91313445	91312508	91311534	91310524	91309479	91308397	91307280	91306126	91304937	91303712	36
1.739	-0.0	91302451	91301154	91299821	91298452	91297047	91295607	91294130	91292618	91291070	91289486	36
1.740	-0.0	91287866	91286210	91284519	91282792	91281028	91279229	91277394	91275524	91273617	91271675	36
1.741	-0.0	91269697	91267683	91265633	91263547	91261426	91259269	91257076	91254847	91252583	91250283	36
1.742	-0.0	91247947	91245575	91243168	91240725	91238246	91235731	91233181	91230594	91227973	91225315	36
1.743	-0.0	91222622	91219893	91217128	91214328	91211492	91208620	91205713	91202770	91199791	91196777	36
1.744	-0.0	91193727	91190641	91187520	91184363	91181170	91177942	91174678	91171378	91168043	91164673	36
1.745	-0.0	91161266	91157824	91154347	91150834	91147285	91143701	91140081	91136426	91132735	91129008	36
1.746	-0.0	91125246	91121449	91117616	91113747	91109843	91105903	91101928	91097917	91093871	91089790	35
1.747	-0.0	91085672	91081520	91077332	91073108	91068849	91064554	91060224	91055859	91051458	91047021	35
1.748	-0.0	91042550	91038042	91033500	91028922	91024308	91019659	91014975	91010256	91005500	91000710	35
1.749	-0.0	90995884	90991023	90986127	90981195	90976227	90971225	90966187	90961114	90956005	90950861	35

TABLE 6

Sinc Function

x	.0	.0001	.0002	.0003	.0004	.0005	.0006	.0007	.0008	.0009	Δ_2
1.750 -0.0	90945682	90940467	90935217	90929932	90924612	90919256	90913865	90908439	90902977	90897480	35
1.751 -0.0	90891948	90886381	90880778	90875141	90869468	90863759	90858016	90852237	90846423	90840574	35
1.752 -0.0	90834690	90828771	90822816	90816826	90810801	90804741	90798646	90792515	90786350	90780149	35
1.753 -0.0	90773913	90767642	90761336	90754995	90748618	90742207	90735761	90729279	90722762	90716211	35
1.754 -0.0	90709624	90703002	90696345	90689653	90682926	90676164	90669367	90662535	90655668	90648766	35
1.755 -0.0	90641829	90634857	90627850	90620808	90613731	90606619	90599472	90592290	90585073	90577822	35
1.756 -0.0	90570535	90563213	90555857	90548465	90541039	90533578	90526082	90518551	90510985	90503384	35
1.757 -0.0	90495748	90488078	90480373	90472632	90464857	90457048	90449203	90441324	90433409	90425460	35
1.758 -0.0	90417476	90409458	90401404	90393316	90385193	90377036	90368843	90360616	90352354	90344057	35
1.759 -0.0	90335726	90327360	90318959	90310524	90302053	90293548	90285009	90276435	90267826	90259182	35
1.760 -0.0	90250504	90241791	90233044	90224262	90215445	90206594	90197708	90188787	90179832	90170842	35
1.761 -0.0	90161818	90152759	90143666	90134538	90125375	90116178	90106947	90097681	90088380	90079045	35
1.762 -0.0	90069675	90060271	90050833	90041360	90031852	90022310	90012734	90003123	89993477	89983798	34
1.763 -0.0	89974083	89964335	89954552	89944734	89934883	89924996	89915076	89905121	89895131	89885108	34
1.764 -0.0	89875050	89864957	89854831	89844670	89834474	89824245	89813981	89803683	89793350	89782983	34
1.765 -0.0	89772582	89762147	89751677	89741174	89730636	89720063	89709457	89698816	89688141	89677432	34
1.766 -0.0	89666689	89655911	89645100	89634254	89623374	89612460	89601512	89590529	89579513	89568462	34
1.767 -0.0	89557377	89546258	89535106	89523919	89512697	89501442	89490153	89478830	89467473	89456081	34
1.768 -0.0	89444656	89433196	89421703	89410176	89398614	89387019	89375390	89363726	89352029	89340298	34
1.769 -0.0	89328533	89316734	89304901	89293034	89281133	89269198	89257230	89245227	89233191	89221120	34
1.770 -0.0	89209016	89196878	89184707	89172501	89160261	89147988	89135681	89123340	89110966	89098557	34
1.771 -0.0	89086115	89073639	89061129	89048586	89036009	89023398	89010753	88998075	88985363	88972617	34
1.772 -0.0	88959837	88947024	88934177	88921297	88908383	88895435	88882454	88869439	88856390	88843308	34
1.773 -0.0	88830192	88817043	88803860	88790643	88777393	88764110	88750792	88737442	88724057	88710640	34
1.774 -0.0	88697188	88683704	88670185	88656634	88643048	88629430	88615778	88602092	88588373	88574621	33
1.775 -0.0	88560835	88547015	88533163	88519277	88505357	88491405	88477418	88463399	88449346	88435260	33
1.776 -0.0	88421140	88406988	88392801	88378582	88364329	88350043	88335724	88321372	88306986	88292567	33
1.777 -0.0	88278114	88263629	88249110	88234558	88219973	88205355	88190704	88176019	88161301	88146550	33
1.778 -0.0	88131766	88116949	88102099	88087216	88072299	88057349	88042367	88027351	88012302	87997220	33
1.779 -0.0	87982105	87966958	87951777	87936563	87921316	87906036	87890723	87875377	87859998	87844586	33
1.780 -0.0	87829141	87813664	87798153	87782609	87767033	87751423	87735781	87720106	87704398	87688657	33
1.781 -0.0	87672883	87657077	87641238	87625365	87609460	87593523	87577552	87561549	87545512	87529443	33
1.782 -0.0	87513342	87497207	87481040	87464840	87448608	87432343	87416045	87399714	87383351	87366955	33
1.783 -0.0	87350526	87334065	87317571	87301045	87284485	87267894	87251269	87234613	87217923	87201201	33
1.784 -0.0	87184447	87167659	87150840	87133988	87117103	87100186	87083236	87066254	87049240	87032192	32
1.785 -0.0	87015113	86998001	86980857	86963680	86946471	86929225	86911955	86894649	86877310	86859939	32
1.786 -0.0	86842536	86825100	86807632	86790132	86772599	86755034	86737437	86719807	86702146	86684452	32
1.787 -0.0	86666725	86648967	86631176	86613353	86595498	86577611	86559691	86541740	86523756	86505740	32
1.788 -0.0	86487692	86469612	86451499	86433355	86415179	86396970	86378729	86360457	86342152	86323815	32
1.789 -0.0	86305446	86287045	86268613	86250148	86231651	86213122	86194561	86175969	86157344	86138687	32
1.790 -0.0	86119999	86101278	86082526	86063742	86044926	86026078	86007198	85988286	85969343	85950368	32
1.791 -0.0	85931361	85912322	85893251	85874148	85855014	85835848	85816650	85797421	85778160	85758867	32
1.792 -0.0	85739542	85720186	85700798	85681378	85661927	85642444	85622930	85603383	85583805	85564196	32
1.793 -0.0	85544555	85524882	85505178	85485443	85465675	85445876	85426046	85406184	85386291	85366366	31
1.794 -0.0	85346410	85326422	85306403	85286352	85266270	85246157	85226012	85205835	85185628	85165389	31
1.795 -0.0	85145118	85124816	85104483	85084119	85063723	85043296	85022837	85002348	84981827	84961274	31
1.796 -0.0	84940691	84920076	84899430	84878753	84858045	84837305	84816534	84795732	84774899	84754035	31
1.797 -0.0	84733140	84712213	84691256	84670267	84649247	84628196	84607114	84586001	84564857	84543682	31
1.798 -0.0	84522476	84501239	84479971	84458672	84437342	84415981	84394589	84373166	84351712	84330228	31
1.799 -0.0	84308712	84287165	84265588	84243980	84222341	84200671	84178970	84157238	84135476	84113683	31

TABLE 6

Sinc Function

x	.0	.0001	.0002	.0003	.0004	.0005	.0006	.0007	.0008	.0009	Δ_2
1.800 -0.0	84091859	84070004	84048118	84026202	84004255	83982277	83960269	83938230	83916160	83894060	31
1.801 -0.0	83871928	83849767	83827574	83805351	83783097	83760813	83738498	83716153	83693777	83671370	31
1.802 -0.0	83648933	83626465	83603967	83581438	83558879	83536289	83513669	83491019	83468337	83445626	30
1.803 -0.0	83422884	83400112	83377309	83354476	83331612	83308719	83285794	83262840	83239855	83216840	30
1.804 -0.0	83193794	83170719	83147613	83124476	83101310	83078113	83054886	83031629	83008341	82985024	30
1.805 -0.0	82961676	82938298	82914890	82891452	82867983	82844485	82820956	82797398	82773809	82750190	30
1.806 -0.0	82726541	82702862	82679154	82655415	82631646	82607847	82584018	82560159	82536270	82512352	30
1.807 -0.0	82488403	82464424	82440416	82416377	82392309	82368211	82344083	82319925	82295738	82271520	30
1.808 -0.0	82247273	82222996	82198689	82174353	82149986	82125590	82101165	82076709	82052224	82027709	30
1.809 -0.0	82003165	81978590	81953987	81929353	81904690	81879997	81855275	81830523	81805742	81780931	30
1.810 -0.0	81756090	81731220	81706321	81681392	81656433	81631445	81606428	81581381	81556304	81531199	29
1.811 -0.0	81506063	81480899	81455705	81430481	81405228	81379946	81354635	81329294	81303924	81278525	29
1.812 -0.0	81253096	81227636	81202151	81176635	81151089	81125514	81099910	81074277	81048614	81022923	29
1.813 -0.0	80997202	80971452	80945673	80919865	80894028	80868162	80842266	80816342	80790388	80764406	29
1.814 -0.0	80738394	80712354	80686284	80660186	80634058	80607902	80581717	80555502	80529259	80502987	29
1.815 -0.0	80476686	80450356	80423998	80397610	80371194	80344748	80318274	80291772	80265240	80238680	29
1.816 -0.0	80212091	80185473	80158826	80132151	80105447	80078715	80051953	80025163	79998345	79971498	29
1.817 -0.0	79944622	79917717	79890784	79863823	79836833	79809814	79782767	79755691	79728587	79701454	28
1.818 -0.0	79674293	79647103	79619885	79592638	79565363	79538060	79510728	79483368	79455979	79428562	28
1.819 -0.0	79401117	79373643	79346142	79318611	79291053	79263466	79235851	79208208	79180536	79152837	28
1.820 -0.0	79125109	79097353	79069568	79041756	79013915	78986047	78958150	78930225	78902272	78874291	28
1.821 -0.0	78846282	78818244	78790179	78762086	78733965	78705815	78677638	78649433	78621200	78592939	28
1.822 -0.0	78564650	78536333	78507988	78479615	78451215	78422786	78394330	78365846	78337334	78308794	28
1.823 -0.0	78280227	78251631	78223008	78194358	78165679	78136973	78108239	78079478	78050688	78021871	28
1.824 -0.0	77993027	77964155	77935255	77906328	77877373	77848390	77819380	77790342	77761277	77732185	28
1.825 -0.0	77703064	77673917	77644742	77615539	77586309	77557052	77527767	77498455	77469115	77439748	27
1.826 -0.0	77410354	77380932	77351483	77322007	77292503	77262972	77233414	77203829	77174216	77144576	27
1.827 -0.0	77114909	77085215	77055493	77025745	76995969	76966166	76936336	76906479	76876595	76846683	27
1.828 -0.0	76816745	76786780	76756787	76726768	76696721	76666648	76636547	76606420	76576266	76546084	27
1.829 -0.0	76515876	76485641	76455379	76425090	76394774	76364432	76334062	76303666	76273243	76242793	27
1.830 -0.0	76212317	76181814	76151283	76120727	76090143	76059533	76028896	75998233	75967542	75936826	27
1.831 -0.0	75906082	75875312	75844515	75813692	75782842	75751966	75721063	75690134	75659178	75628196	26
1.832 -0.0	75597187	75566151	75535090	75504001	75472887	75441746	75410578	75379385	75348165	75316918	26
1.833 -0.0	75285645	75254346	75223021	75191669	75160291	75128887	75097457	75066000	75034518	75003009	26
1.834 -0.0	74971473	74939912	74908325	74876711	74845071	74813406	74781714	74749996	74718252	74686482	26
1.835 -0.0	74654686	74622864	74591016	74559142	74527242	74495316	74463364	74431386	74399382	74367353	26
1.836 -0.0	74335297	74303216	74271109	74238976	74206817	74174633	74142423	74110187	74077925	74045637	26
1.837 -0.0	74013324	73980985	73948620	73916230	73883814	73851372	73818905	73786412	73753894	73721350	26
1.838 -0.0	73688780	73656185	73623565	73590918	73558247	73525550	73492827	73460079	73427306	73394507	25
1.839 -0.0	73361682	73328833	73295957	73263057	73230131	73197180	73164204	73131202	73098175	73065123	25
1.840 -0.0	73032045	72998942	72965814	72932661	72899483	72866279	72833050	72799797	72766518	72733214	25
1.841 -0.0	72699884	72666530	72633151	72599746	72566317	72532863	72499383	72465879	72432349	72398795	25
1.842 -0.0	72365216	72331612	72297983	72264329	72230650	72196946	72163217	72129464	72095686	72061883	25
1.843 -0.0	72028055	71994203	71960325	71926423	71892497	71858545	71824569	71790568	71756543	71722493	25
1.844 -0.0	71688418	71654319	71620195	71586047	71551874	71517676	71483454	71449207	71414936	71380641	24
1.845 -0.0	71346321	71311976	71277607	71243214	71208796	71174354	71139888	71105397	71070882	71036343	24
1.846 -0.0	71001779	70967191	70932579	70897942	70863281	70828596	70793887	70759154	70724396	70689614	24
1.847 -0.0	70654809	70619979	70585124	70550246	70515344	70480418	70445467	70410493	70375495	70340472	24
1.848 -0.0	70305426	70270356	70235261	70200143	70165001	70129835	70094645	70059431	70024194	69988932	24
1.849 -0.0	69953647	69918338	69883005	69847649	69812268	69776864	69741436	69705985	69670510	69635011	24

TABLE 6

Sinc Function

x	.0	.0001	.0002	.0003	.0004	.0005	.0006	.0007	.0008	.0009	Δ_2
1.850 −0.0	69599488	69563942	69528373	69492779	69457162	69421522	69385858	69350170	69314459	69278724	24
1.851 −0.0	69242966	69207185	69171380	69135551	69099699	69063824	69027925	68992003	68956058	68920089	23
1.852 −0.0	68884097	68848082	68812043	68775981	68739896	68703787	68667656	68631501	68595323	68559122	23
1.853 −0.0	68522897	68486650	68450379	68414085	68377768	68341429	68305066	68268680	68232270	68195838	23
1.854 −0.0	68159383	68122905	68086404	68049880	68013334	67976764	67940171	67903556	67866917	67830256	23
1.855 −0.0	67793572	67756865	67720135	67683383	67646608	67609810	67572989	67536146	67499280	67462391	23
1.856 −0.0	67425480	67388546	67351589	67314610	67277608	67240584	67203537	67166467	67129375	67092261	23
1.857 −0.0	67055124	67017964	66980782	66943578	66906351	66869102	66831830	66794537	66757220	66719882	22
1.858 −0.0	66682521	66645137	66607732	66570304	66532854	66495382	66457887	66420370	66382831	66345270	22
1.859 −0.0	66307687	66270082	66232454	66194805	66157133	66119439	66081724	66043986	66006226	65968444	22
1.860 −0.0	65930640	65892815	65854967	65817097	65779206	65741292	65703357	65665400	65627421	65589420	22
1.861 −0.0	65551398	65513353	65475287	65437199	65399089	65360958	65322805	65284630	65246433	65208215	22
1.862 −0.0	65169976	65131714	65093431	65055127	65016800	64978453	64940083	64901693	64863280	64824847	21
1.863 −0.0	64786392	64747915	64709417	64670897	64632356	64593794	64555211	64516606	64477979	64439332	21
1.864 −0.0	64400663	64361973	64323261	64284529	64245775	64207000	64168203	64129386	64090547	64051688	21
1.865 −0.0	64012807	63973905	63934982	63896038	63857073	63818086	63779079	63740051	63701002	63661932	21
1.866 −0.0	63622841	63583729	63544596	63505442	63466267	63427072	63387855	63348618	63309360	63270081	21
1.867 −0.0	63230782	63191462	63152120	63112759	63073376	63033973	62994549	62955105	62915640	62876154	21
1.868 −0.0	62836648	62797121	62757574	62718006	62678417	62638808	62599179	62559529	62519858	62480168	20
1.869 −0.0	62440456	62400725	62360973	62321200	62281408	62241594	62201761	62161907	62122033	62082139	20
1.870 −0.0	62042225	62002290	61962335	61922360	61882365	61842350	61802314	61762258	61722183	61682087	20
1.871 −0.0	61641971	61601835	61561679	61521503	61481307	61441091	61400855	61360599	61320324	61280028	20
1.872 −0.0	61239713	61199377	61159022	61118647	61078252	61037837	60997403	60956948	60916474	60875981	20
1.873 −0.0	60835467	60794934	60754381	60713809	60673217	60632605	60591974	60551323	60510653	60469963	20
1.874 −0.0	60429253	60388524	60347776	60307008	60266220	60225413	60184587	60143741	60102876	60061992	19
1.875 −0.0	60021088	59980165	59939222	59898260	59857279	59816279	59775260	59734221	59693163	59652086	19
1.876 −0.0	59610989	59569874	59528739	59487586	59446413	59405221	59364010	59322780	59281531	59240263	19
1.877 −0.0	59198976	59157670	59116345	59075001	59033638	58992256	58950856	58909436	58867998	58826541	19
1.878 −0.0	58785065	58743570	58702057	58660525	58618974	58577404	58535816	58494209	58452583	58410939	19
1.879 −0.0	58369276	58327594	58285894	58244175	58202438	58160682	58118908	58077115	58035303	57993474	18
1.880 −0.0	57951625	57909759	57867874	57825970	57784048	57742108	57700150	57658173	57616178	57574164	18
1.881 −0.0	57532133	57490083	57448014	57405928	57363824	57321701	57279560	57237401	57195224	57153029	18
1.882 −0.0	57110815	57068584	57026335	56984067	56941782	56899479	56857157	56814818	56772461	56730085	18
1.883 −0.0	56687692	56645281	56602853	56560406	56517942	56475459	56432959	56390442	56347906	56305353	18
1.884 −0.0	56262782	56220193	56177587	56134963	56092321	56049662	56006985	55964291	55921579	55878849	18
1.885 −0.0	55836102	55793337	55750555	55707756	55664939	55622104	55579253	55536383	55493497	55450593	17
1.886 −0.0	55407672	55364733	55321777	55278804	55235813	55192806	55149781	55106739	55063679	55020603	17
1.887 −0.0	54977509	54934398	54891270	54848125	54804963	54761784	54718588	54675374	54632144	54588897	17
1.888 −0.0	54545633	54502352	54459054	54415739	54372407	54329058	54285692	54242310	54198911	54155495	17
1.889 −0.0	54112062	54068612	54025146	53981663	53938163	53894646	53851113	53807563	53763997	53720414	17
1.890 −0.0	53676814	53633198	53589565	53545916	53502250	53458568	53414869	53371154	53327422	53283674	16
1.891 −0.0	53239909	53196128	53152331	53108517	53064687	53020841	52976978	52933099	52889204	52845293	16
1.892 −0.0	52801365	52757421	52713461	52669485.	52625493	52581484	52537460	52493419	52449363	52405290	16
1.893 −0.0	52361201	52317096	52272976	52228839	52184686	52140517	52096333	52052132	52007916	51963684	16
1.894 −0.0	51919436	51875172	51830892	51786597	51742286	51697959	51653616	51609258	51564883	51520494	16
1.895 −0.0	51476088	51431667	51387231	51342778	51298310	51253827	51209328	51164814	51120284	51075738	15
1.896 −0.0	51031177	50986601	50942009	50897402	50852779	50808141	50763488	50718819	50674135	50629436	15
1.897 −0.0	50584722	50539992	50495247	50450487	50405711	50360921	50316115	50271294	50226458	50181607	15
1.898 −0.0	50136741	50091859	50046963	50002052	49957125	49912184	49867228	49822256	49777270	49732269	15
1.899 −0.0	49687253	49642222	49597176	49552116	49507040	49461950	49416845	49371726	49326591	49281442	15

$$\frac{\sin 2\pi x}{2\pi x}$$

TABLE 6

Sinc Function

x	.0	.0001	.0002	.0003	.0004	.0005	.0006	.0007	.0008	.0009	Δ_2
1.900 -0.0	49236278	49191100	49145906	49100698	49055476	49010239	48964987	48919721	48874440	48829145	15
1.901 -0.0	48783835	48738511	48693172	48647818	48602451	48557069	48511672	48466261	48420836	48375396	14
1.902 -0.0	48329943	48284474	48238992	48193495	48147984	48102459	48056920	48011366	47965798	47920216	14
1.903 -0.0	47874620	47829010	47783386	47737748	47692095	47646429	47600749	47555054	47509346	47463623	14
1.904 -0.0	47417887	47372137	47326373	47280595	47234803	47188998	47143178	47097345	47051498	47005637	14
1.905 -0.0	46959763	46913874	46867973	46822057	46776128	46730185	46684228	46638258	46592274	46546277	14
1.906 -0.0	46500266	46454242	46408204	46362152	46316087	46270009	46223917	46177812	46131694	46085562	13
1.907 -0.0	46039416	45993258	45947086	45900901	45854702	45808490	45762265	45716027	45669776	45623511	13
1.908 -0.0	45577233	45530942	45484638	45438321	45391991	45345648	45299291	45252922	45206540	45160144	13
1.909 -0.0	45113736	45067315	45020881	44974433	44927974	44881501	44835015	44788516	44742005	44695481	13
1.910 -0.0	44648944	44602394	44555832	44509257	44462669	44416069	44369455	44322830	44276191	44229540	13
1.911 -0.0	44182877	44136201	44089512	44042811	43996097	43949371	43902632	43855881	43809118	43762342	12
1.912 -0.0	43715554	43668753	43621940	43575115	43528277	43481427	43434565	43387691	43340804	43293905	12
1.913 -0.0	43246994	43200071	43153135	43106188	43059228	43012257	42965273	42918277	42871269	42824250	12
1.914 -0.0	42777218	42730174	42683118	42636050	42588971	42541879	42494776	42447661	42400534	42353395	12
1.915 -0.0	42306244	42259082	42211907	42164722	42117524	42070315	42023093	41975861	41928616	41881360	12
1.916 -0.0	41834093	41786814	41739523	41692221	41644907	41597582	41550245	41502897	41455537	41408166	11
1.917 -0.0	41360784	41313390	41265984	41218568	41171140	41123701	41076250	41028788	40981315	40933831	11
1.918 -0.0	40886336	40838829	40791311	40743782	40696242	40648691	40601129	40553555	40505971	40458376	11
1.919 -0.0	40410769	40363152	40315523	40267884	40220234	40172572	40124900	40077217	40029524	39981819	11
1.920 -0.0	39934103	39886377	39838640	39790892	39743134	39695365	39647585	39599794	39551993	39504181	11
1.921 -0.0	39456358	39408525	39360681	39312827	39264962	39217087	39169201	39121305	39073398	39025481	10
1.922 -0.0	38977553	38929615	38881667	38833708	38785739	38737760	38689770	38641770	38593760	38545739	10
1.923 -0.0	38497708	38449668	38401616	38353555	38305484	38257402	38209311	38161209	38113097	38064975	10
1.924 -0.0	38016843	37968701	37920550	37872388	37824216	37776034	37727843	37679641	37631430	37583209	10
1.925 -0.0	37534978	37486737	37438486	37390226	37341956	37293676	37245386	37197087	37148778	37100460	10
1.926 -0.0	37052131	37003794	36955446	36907089	36858723	36810347	36761961	36713566	36665162	36616748	9
1.927 -0.0	36568324	36519892	36471449	36422998	36374537	36326067	36277587	36229098	36180600	36132093	9
1.928 -0.0	36083576	36035050	35986515	35937971	35889418	35840855	35792284	35743703	35695114	35646515	9
1.929 -0.0	35597907	35549290	35500664	35452029	35403386	35354733	35306071	35257401	35208722	35160033	9
1.930 -0.0	35111336	35062630	35013916	34965192	34916460	34867719	34818969	34770211	34721444	34672668	9
1.931 -0.0	34623884	34575091	34526290	34477480	34428661	34379834	34330998	34282154	34233301	34184440	8
1.932 -0.0	34135570	34086692	34037806	33988911	33940008	33891097	33842177	33793249	33744312	33695368	8
1.933 -0.0	33646415	33597454	33548485	33499507	33450522	33401528	33352526	33303516	33254498	33205472	8
1.934 -0.0	33156438	33107396	33058346	33009287	32960221	32911147	32862065	32812976	32763878	32714772	8
1.935 -0.0	32665659	32616538	32567409	32518272	32469127	32419975	32370815	32321647	32272472	32223289	8
1.936 -0.0	32174098	32124900	32075694	32026480	31977259	31928031	31878794	31829551	31780300	31731041	7
1.937 -0.0	31681775	31632502	31583221	31533933	31484637	31435334	31386024	31336706	31287382	31238049	7
1.938 -0.0	31188710	31139364	31090010	31040649	30991281	30941906	30892523	30843134	30793737	30744334	7
1.939 -0.0	30694923	30645506	30596081	30546649	30497211	30447765	30398313	30348853	30299387	30249914	7
1.940 -0.0	30200434	30150947	30101454	30051953	30002446	29952933	29903412	29853885	29804351	29754810	7
1.941 -0.0	29705263	29655709	29606149	29556582	29507008	29457428	29407841	29358248	29308648	29259042	6
1.942 -0.0	29209430	29159811	29110185	29060553	29010915	28961271	28911620	28861963	28812300	28762630	6
1.943 -0.0	28712954	28663272	28613584	28563889	28514189	28464482	28414769	28365050	28315325	28265594	6
1.944 -0.0	28215856	28166113	28116364	28066609	28016848	27967080	27917307	27867529	27817744	27767953	6
1.945 -0.0	27718157	27668354	27618546	27568732	27518912	27469087	27419256	27369419	27319576	27269728	6
1.946 -0.0	27219874	27170015	27120150	27070279	27020403	26970521	26920634	26870741	26820843	26770939	5
1.947 -0.0	26721030	26671116	26621196	26571270	26521340	26471403	26421462	26371515	26321563	26271606	5
1.948 -0.0	26221644	26171676	26121703	26071725	26021742	25971753	25921760	25871761	25821758	25771749	5
1.949 -0.0	25721735	25671716	25621692	25571664	25521630	25471591	25421548	25371499	25321446	25271387	5

TABLE 6

Sinc Function

x	.0	.0001	.0002	.0003	.0004	.0005	.0006	.0007	.0008	.0009	Δ_2
1.950 -0.0	25221324	25171256	25121184	25071106	25021024	24970937	24920845	24870748	24820647	24770542	5
1.951 -0.0	24720431	24670316	24620196	24570072	24519943	24469810	24419672	24369530	24319383	24269232	4
1.952 -0.0	24219076	24168916	24118751	24068582	24018409	23968231	23918049	23867863	23817672	23767477	4
1.953 -0.0	23717278	23667075	23616868	23566656	23516440	23466220	23415996	23365768	23315535	23265299	4
1.954 -0.0	23215059	23164814	23114566	23064313	23014057	22963797	22913532	22863264	22812992	22762716	4
1.955 -0.0	22712437	22662153	22611866	22561575	22511280	22460981	22410679	22360373	22310063	22259749	4
1.956 -0.0	22209432	22159112	22108787	22058460	22008128	21957793	21907455	21857113	21806767	21756418	3
1.957 -0.0	21706066	21655710	21605351	21554988	21504622	21454253	21403880	21353504	21303125	21252743	3
1.958 -0.0	21202357	21151968	21101576	21051181	21000782	20950380	20899976	20849568	20799157	20748743	3
1.959 -0.0	20698326	20647906	20597483	20547057	20496628	20446196	20395761	20345323	20294882	20244439	3
1.960 -0.0	20193992	20143543	20093091	20042636	19992179	19941718	19891255	19840790	19790321	19739850	3
1.961 -0.0	19689376	19638900	19588421	19537939	19487455	19436969	19386479	19335988	19285493	19234997	2
1.962 -0.0	19184498	19133996	19083492	19032986	18982477	18931966	18881453	18830937	18780419	18729899	2
1.963 -0.0	18679377	18628852	18578325	18527796	18477265	18426731	18376196	18325658	18275118	18224577	2
1.964 -0.0	18174033	18123487	18072939	18022389	17971837	17921283	17870728	17820170	17769611	17719049	2
1.965 -0.0	17668486	17617921	17567354	17516786	17466215	17415643	17365069	17314493	17263916	17213337	2
1.966 -0.0	17162757	17112174	17061591	17011005	16960418	16909830	16859239	16808648	16758055	16707460	0
1.967 -0.0	16656864	16606267	16555668	16505068	16454466	16403863	16353259	16302653	16252046	16201438	0
1.968 -0.0	16150829	16100218	16049606	15998993	15948379	15897763	15847147	15796529	15745910	15695291	0
1.969 -0.0	15644670	15594048	15543425	15492801	15442176	15391550	15340923	15290296	15239667	15189038	0
1.970 -0.0	15138407	15087776	15037144	14986511	14935878	14885244	14834609	14783973	14733336	14682699	0
1.971 -0.0	14632061	14581423	14530784	14480144	14429504	14378863	14328222	14277580	14226938	14176295	0
1.972 -0.0	14125652	14075008	14024364	13973719	13923074	13872429	13821783	13771137	13720491	13669845	0
1.973 -0.0	13619198	13568551	13517904	13467256	13416608	13365961	13315313	13264665	13214016	13163368	0
1.974 -0.0	13112720	13062071	13011423	12960775	12910126	12859478	12808829	12758181	12707533	12656885	0
1.975 -0.0	12606237	12555589	12504942	12454294	12403647	12353000	12302354	12251707	12201061	12150415	0
1.976 -0.0	12099770	12049125	11998480	11947835	11897192	11846548	11795905	11745262	11694620	11643978	-1
1.977 -0.0	11593337	11542697	11492057	11441417	11390779	11340140	11289503	11238866	11188230	11137594	-1
1.978 -0.0	11086959	11036325	10985692	10935060	10884428	10833797	10783167	10732538	10681910	10631282	-1
1.979 -0.0	10580656	10530030	10479406	10428782	10378160	10327538	10276918	10226298	10175680	10125062	-1
1.980 -0.0	10074446	10023831	09973217	09922605	09871993	09821383	09770774	09720166	09669559	09618954	-1
1.981 -0.0	09568350	09517748	09467146	09416546	09365948	09315351	09264755	09214161	09163568	09112977	-1
1.982 -0.0	09062387	09011799	08961212	08910627	08860044	08809462	08758882	08708303	08657726	08607151	-2
1.983 -0.0	08556577	08506005	08455435	08404867	08354300	08303735	08253172	08202611	08152052	08101495	-2
1.984 -0.0	08050939	08000385	07949834	07899284	07848736	07798191	07747647	07697105	07646566	07596028	-2
1.985 -0.0	07545493	07494960	07444428	07393899	07343372	07292848	07242325	07191805	07141287	07090771	-2
1.986 -0.0	07040258	06989747	06939238	06888732	06838227	06787726	06737226	06686730	06636235	06585743	-2
1.987 -0.0	06535254	06484767	06434282	06383800	06333321	06282844	06232370	06181898	06131430	06080963	-3
1.988 -0.0	06030500	05980039	05929581	05879125	05828673	05778223	05727775	05677331	05626890	05576451	-3
1.989 -0.0	05526015	05475582	05425152	05374725	05324301	05273880	05223462	05173047	05122635	05072226	-3
1.990 -0.0	05021820	04971417	04921017	04870620	04820227	04769837	04719449	04669065	04618685	04568307	-3
1.991 -0.0	04517933	04467562	04417194	04366830	04316469	04266111	04215757	04165406	04115058	04064714	-3
1.992 -0.0	04014373	03964036	03913702	03863372	03813046	03762722	03712403	03662087	03611775	03561466	-4
1.993 -0.0	03511161	03460859	03410562	03360268	03309977	03259691	03209408	03159129	03108853	03058582	-4
1.994 -0.0	03008314	02958051	02907791	02857535	02807283	02757035	02706790	02656550	02606314	02556082	-4
1.995 -0.0	02505853	02455629	02405409	02355193	02304981	02254773	02204570	02154370	02104175	02053984	-4
1.996 -0.0	02003797	01953614	01903436	01853262	01803092	01752927	01702765	01652609	01602456	01552308	-4
1.997 -0.0	01502164	01452025	01401890	01351760	01301634	01251513	01201396	01151284	01101176	01051073	-5
1.998 -0.0	01000975	00950881	00900792	00850707	00800627	00750552	00700481	00650416	00600355	00550298	-5
1.999 -0.0	00500247	00450200	00400158	00350121	00300089	00250062	00200040	00150022	00100010	00050002	5

$$\frac{\sin 2\pi x}{2\pi x}$$

$\frac{\sin 2\pi x}{2\pi x}$

TABLE 6

Sinc Function

x		.0	.0001	.0002	.0003	.0004	.0005	.0006	.0007	.0008	.0009	Δ_2
2.000	0.0	00000000	00049997	00099990	00149977	00199960	00249937	00299909	00349876	00399838	00449795	-5
2.001	0.0	00499747	00549693	00599635	00649571	00699501	00749427	00799347	00849262	00899172	00949076	-5
2.002	0.0	00998975	01048868	01098756	01148639	01198516	01248388	01298254	01348115	01397971	01447820	-6
2.003	0.0	01497665	01547503	01597336	01647164	01696986	01746802	01796613	01846417	01896217	01946010	-6
2.004	0.0	01995798	02045580	02095356	02145126	02194891	02244650	02294403	02344150	02393891	02443626	-6
2.005	0.0	02493355	02543079	02592796	02642508	02692213	02741912	02791606	02841293	02890974	02940649	-6
2.006	0.0	02990318	03039981	03089638	03139289	03188933	03238571	03288203	03337829	03387448	03437061	-6
2.007	0.0	03486668	03536269	03585863	03635451	03685032	03734607	03784176	03833738	03883294	03932843	-6
2.008	0.0	03982386	04031923	04081452	04130976	04180492	04230002	04279506	04329003	04378493	04427977	-7
2.009	0.0	04477454	04526924	04576387	04625844	04675294	04724737	04774174	04823604	04873026	04922442	-7
2.010	0.0	04971852	05021254	05070649	05120038	05169419	05218794	05268161	05317522	05366876	05416222	-7
2.011	0.0	05465562	05514894	05564219	05613538	05662849	05712153	05761450	05810740	05860022	05909297	-7
2.012	0.0	05958565	06007826	06057080	06106326	06155565	06204796	06254021	06303238	06352447	06401649	-7
2.013	0.0	06450844	06500031	06549211	06598384	06647548	06696706	06745856	06794998	06844133	06893260	-8
2.014	0.0	06942379	06991491	07040596	07089692	07138781	07187863	07236936	07286002	07335060	07384110	-8
2.015	0.0	07433153	07482188	07531215	07580234	07629245	07678248	07727244	07776231	07825211	07874183	-8
2.016	0.0	07923146	07972102	08021050	08069989	08118921	08167845	08216760	08265668	08314567	08363458	-8
2.017	0.0	08412341	08461216	08510083	08558941	08607791	08656633	08705467	08754293	08803110	08851919	-8
2.018	0.0	08900719	08949511	08998295	09047071	09095838	09144596	09193347	09242088	09290821	09339546	-8
2.019	0.0	09388262	09436970	09485669	09534360	09583042	09631715	09680380	09729036	09777683	09826322	-9
2.020	0.0	09874952	09923574	09972186	10020790	10069385	10117972	10166549	10215118	10263678	10312229	-9
2.021	0.0	10360771	10409304	10457829	10506344	10554850	10603348	10651836	10700316	10748786	10797248	-9
2.022	0.0	10845700	10894144	10942578	10991003	11039419	11087826	11136223	11184612	11232991	11281361	-9
2.023	0.0	11329722	11378074	11426416	11474749	11523073	11571387	11619692	11667988	11716274	11764551	-9
2.024	0.0	11812819	11861077	11909325	11957565	12005794	12054014	12102225	12150426	12198618	12246800	-10
2.025	0.0	12294972	12343135	12391288	12439431	12487565	12535689	12583804	12631909	12680004	12728089	-10
2.026	0.0	12776164	12824230	12872286	12920332	12968368	13016394	13064411	13112417	13160414	13208401	-10
2.027	0.0	13256378	13304344	13352301	13400248	13448185	13496112	13544028	13591935	13639831	13687718	-10
2.028	0.0	13735594	13783460	13831316	13879162	13926998	13974823	14022638	14070443	14118238	14166022	-10
2.029	0.0	14213796	14261560	14309314	14357057	14404789	14452512	14500224	14547925	14595616	14643297	-10
2.030	0.0	14690967	14738626	14786275	14833914	14881542	14929159	14976766	15024362	15071948	15119523	-11
2.031	0.0	15167087	15214641	15262184	15309716	15357238	15404749	15452249	15499738	15547217	15594684	-11
2.032	0.0	15642141	15689587	15737022	15784446	15831860	15879262	15926654	15974034	16021404	16068762	-11
2.033	0.0	16116110	16163447	16210772	16258087	16305390	16352682	16399964	16447234	16494493	16541741	-11
2.034	0.0	16588977	16636203	16683417	16730620	16777812	16824992	16872161	16919319	16966466	17013601	-11
2.035	0.0	17060725	17107837	17154939	17202028	17249107	17296174	17343229	17390273	17437306	17484327	-12
2.036	0.0	17531336	17578334	17625320	17672295	17719258	17766210	17813150	17860078	17906995	17953900	-12
2.037	0.0	18000793	18047675	18094545	18141403	18188249	18235084	18281907	18328718	18375517	18422304	-12
2.038	0.0	18469080	18515843	18562595	18609335	18656063	18702778	18749482	18796174	18842854	18889522	-12
2.039	0.0	18936178	18982822	19029453	19076073	19122681	19169276	19215859	19262431	19308990	19355536	-12
2.040	0.0	19402071	19448593	19495104	19541601	19588087	19634560	19681021	19727470	19773906	19820330	-12
2.041	0.0	19866742	19913141	19959528	20005902	20052264	20098614	20144951	20191275	20237587	20283887	-13
2.042	0.0	20330174	20376448	20422710	20468959	20515196	20561420	20607631	20653830	20700016	20746189	-13
2.043	0.0	20792350	20838498	20884633	20930755	20976865	21022962	21069046	21115117	21161175	21207221	-13
2.044	0.0	21253253	21299273	21345280	21391274	21437255	21483223	21529177	21575119	21621048	21666964	-13
2.045	0.0	21712867	21758757	21804634	21850498	21896348	21942186	21988010	22033821	22079619	22125404	-13
2.046	0.0	22171175	22216934	22262679	22308410	22354129	22399834	22445526	22491205	22536870	22582522	-13
2.047	0.0	22628161	22673786	22719397	22764996	22810581	22856152	22901710	22947254	22992785	23038303	-14
2.048	0.0	23083807	23129297	23174774	23220237	23265686	23311122	23356545	23401953	23447348	23492730	-14
2.049	0.0	23538097	23583451	23628791	23674118	23719430	23764729	23810014	23855285	23900543	23945786	-14

TABLE 6

Sinc Function

x		.0	.0001	.0002	.0003	.0004	.0005	.0006	.0007	.0008	.0009	Δ_2
2.050	0.0	23991016	24036231	24081433	24126621	24171795	24216955	24262101	24307234	24352352	24397456	−14
2.051	0.0	24442546	24487622	24532684	24577732	24622766	24667785	24712791	24757782	24802760	24847723	−14
2.052	0.0	24892672	24937606	24982527	25027433	25072325	25117203	25162066	25206915	25251750	25296570	−14
2.053	0.0	25341376	25386168	25430945	25475708	25520457	25565191	25609911	25654616	25699307	25743983	−14
2.054	0.0	25788644	25833292	25877924	25922542	25967146	26011735	26056309	26100869	26145414	26189944	−15
2.055	0.0	26234460	26278961	26323447	26367918	26412375	26456817	26501245	26545657	26590055	26634438	−15
2.056	0.0	26678806	26723159	26767497	26811821	26856130	26900423	26944702	26988966	27033215	27077448	−15
2.057	0.0	27121667	27165871	27210060	27254234	27298393	27342536	27386665	27430779	27474877	27518960	−15
2.058	0.0	27563028	27607081	27651119	27695142	27739149	27783141	27827118	27871080	27915026	27958957	−15
2.059	0.0	28002873	28046773	28090659	28134528	28178383	28222222	28266045	28309854	28353646	28397424	−15
2.060	0.0	28441186	28484932	28528663	28572378	28616078	28659762	28703431	28747084	28790722	28834344	−16
2.061	0.0	28877951	28921541	28965115	29008676	29052220	29095748	29139260	29182757	29226238	29269703	−16
2.062	0.0	29313152	29356586	29400004	29443406	29486792	29530162	29573517	29616855	29660178	29703485	−16
2.063	0.0	29746776	29790051	29833310	29876553	29919780	29962991	30006186	30049365	30092527	30135674	−16
2.064	0.0	30178805	30221920	30265018	30308101	30351167	30394217	30437251	30480269	30523271	30566256	−16
2.065	0.0	30609225	30652178	30695115	30738035	30780939	30823827	30866698	30909554	30952392	30995215	−16
2.066	0.0	31038021	31080810	31123584	31166340	31209081	31251805	31294512	31337203	31379877	31422535	−17
2.067	0.0	31465177	31507802	31550410	31593002	31635577	31678135	31720677	31763202	31805711	31848203	−17
2.068	0.0	31890678	31933136	31975578	32018003	32060412	32102803	32145178	32187536	32229878	32272202	−17
2.069	0.0	32314510	32356800	32399074	32441331	32483571	32525795	32568001	32610190	32652363	32694518	−17
2.070	0.0	32736656	32778778	32820882	32862970	32905040	32947094	32989130	33031149	33073151	33115136	−17
2.071	0.0	33157104	33199055	33240988	33282905	33324804	33366686	33408551	33450398	33492229	33534042	−17
2.072	0.0	33575837	33617616	33659377	33701121	33742847	33784557	33826248	33867923	33909580	33951220	−17
2.073	0.0	33992842	34034447	34076034	34117604	34159156	34200691	34242209	34283708	34325191	34366656	−18
2.074	0.0	34408103	34449532	34490945	34532339	34573716	34615075	34656417	34697740	34739047	34780335	−18
2.075	0.0	34821606	34862859	34904094	34945312	34986512	35027694	35068858	35110004	35151133	35192244	−18
2.076	0.0	35233336	35274412	35315469	35356508	35397529	35438533	35479518	35520486	35561435	35602367	−18
2.077	0.0	35643280	35684176	35725053	35765913	35806754	35847578	35888383	35929170	35969939	36010690	−18
2.078	0.0	36051423	36092138	36132834	36173513	36214173	36254815	36295439	36336044	36376631	36417200	−18
2.079	0.0	36457751	36498283	36538797	36579293	36619771	36660230	36700670	36741093	36781497	36821882	−18
2.080	0.0	36862249	36902598	36942928	36983240	37023533	37063808	37104064	37144302	37184522	37224722	−19
2.081	0.0	37264904	37305068	37345213	37385339	37425447	37465536	37505607	37545659	37585692	37625706	−19
2.082	0.0	37665702	37705679	37745638	37785577	37825498	37865400	37905284	37945148	37984994	38024821	−19
2.083	0.0	38064629	38104418	38144189	38183940	38223673	38263386	38303081	38342757	38382414	38422052	−19
2.084	0.0	38461671	38501271	38540852	38580414	38619957	38659481	38698986	38738471	38777938	38817386	−19
2.085	0.0	38856814	38896224	38935614	38974985	39014337	39053670	39092983	39132278	39171553	39210809	−19
2.086	0.0	39250045	39289263	39328461	39367640	39406799	39445940	39485061	39524162	39563245	39602307	−19
2.087	0.0	39641351	39680375	39719380	39758365	39797331	39836278	39875205	39914112	39953000	39991869	−19
2.088	0.0	40030718	40069547	40108357	40147147	40185918	40224670	40263401	40302113	40340806	40379479	−20
2.089	0.0	40418132	40456765	40495379	40533973	40572548	40611103	40649638	40688153	40726649	40765124	−20
2.090	0.0	40803580	40842017	40880433	40918830	40957207	40995563	41033901	41072218	41110515	41148793	−20
2.091	0.0	41187050	41225288	41263506	41301703	41339881	41378039	41416177	41454295	41492393	41530470	−20
2.092	0.0	41568528	41606566	41644584	41682581	41720559	41758516	41796454	41834371	41872268	41910145	−20
2.093	0.0	41948002	41985838	42023654	42061451	42099227	42136982	42174718	42212433	42250128	42287803	−20
2.094	0.0	42325457	42363091	42400705	42438298	42475871	42513424	42550957	42588469	42625960	42663431	−20
2.095	0.0	42700882	42738312	42775722	42813112	42850481	42887829	42925157	42962465	42999752	43037018	−21
2.096	0.0	43074264	43111489	43148694	43185878	43223042	43260185	43297307	43334409	43371490	43408550	−21
2.097	0.0	43445590	43482609	43519607	43556585	43593542	43630478	43667394	43704288	43741162	43778015	−21
2.098	0.0	43814648	43851659	43888450	43925220	43961969	43998697	44035405	44072091	44108756	44145401	−21
2.099	0.0	44182025	44218628	44255209	44291770	44328310	44364829	44401327	44437804	44474260	44510695	−21

$$\frac{\sin 2\pi x}{2\pi x}$$

$$\frac{\sin 2\pi x}{2\pi x}$$

TABLE 6

Sinc Function

x		.0	.0001	.0002	.0003	.0004	.0005	.0006	.0007	.0008	.0009	Δ_2
2.100	0.0	44547109	44583502	44619873	44656224	44692554	44728862	44765149	44801416	44837661	44873885	-21
2.101	0.0	44910087	44946269	44982429	45018569	45054687	45090783	45126859	45162913	45198946	45234958	-21
2.102	0.0	45270948	45306918	45342865	45378792	45414697	45450581	45486444	45522285	45558105	45593903	-21
2.103	0.0	45629680	45665435	45701170	45736882	45772574	45808243	45843892	45879519	45915124	45950708	-22
2.104	0.0	45986270	46021811	46057330	46092828	46128304	46163758	46199191	46234602	46269992	46305360	-22
2.105	0.0	46340707	46376031	46411334	46446616	46481876	46517114	46552330	46587525	46622697	46657849	-22
2.106	0.0	46692978	46728086	46763171	46798235	46833278	46868298	46903297	46938273	46973228	47008161	-22
2.107	0.0	47043073	47077962	47112829	47147675	47182498	47217300	47252080	47286837	47321573	47356287	-22
2.108	0.0	47390979	47425648	47460296	47494922	47529526	47564107	47598667	47633205	47667720	47702213	-22
2.109	0.0	47736685	47771134	47805561	47839966	47874349	47908709	47943048	47977364	48011658	48045930	-22
2.110	0.0	48080180	48114407	48148612	48182795	48216956	48251094	48285210	48319304	48353376	48387425	-22
2.111	0.0	48421452	48455456	48489438	48523398	48557336	48591251	48625143	48659014	48692862	48726687	-22
2.112	0.0	48760490	48794270	48828029	48861764	48895477	48929168	48962836	48996482	49030105	49063705	-23
2.113	0.0	49097283	49130839	49164372	49197882	49231369	49264835	49298277	49331697	49365094	49398469	-23
2.114	0.0	49431821	49465150	49498456	49531740	49565001	49598240	49631455	49664648	49697819	49730966	-23
2.115	0.0	49764091	49797193	49830272	49863328	49896362	49929373	49962361	49995326	50028268	50061187	-23
2.116	0.0	50094084	50126957	50159808	50192636	50225440	50258222	50290981	50323717	50356431	50389121	-23
2.117	0.0	50421788	50454432	50487053	50519651	50552226	50584778	50617307	50649813	50682296	50714756	-23
2.118	0.0	50747193	50779607	50811997	50844365	50876709	50909030	50941328	50973603	51005855	51038083	-23
2.119	0.0	51070288	51102471	51134630	51166765	51198878	51230967	51263033	51295076	51327095	51359091	-23
2.120	0.0	51391064	51423014	51454940	51486843	51518722	51550579	51582411	51614221	51646007	51677770	-23
2.121	0.0	51709509	51741225	51772918	51804587	51836232	51867855	51899453	51931029	51962581	51994109	-23
2.122	0.0	52025614	52057095	52088553	52119987	52151398	52182785	52214149	52245489	52276805	52308098	-24
2.123	0.0	52339368	52370613	52401835	52433034	52464209	52495360	52526487	52557591	52588672	52619728	-24
2.124	0.0	52650761	52681770	52712755	52743717	52774655	52805569	52836459	52867326	52898169	52928988	-24
2.125	0.0	52959783	52990555	53021302	53052026	53082726	53113402	53144055	53174683	53205288	53235869	-24
2.126	0.0	53266425	53296958	53327467	53357952	53388413	53418851	53449264	53479653	53510018	53540360	-24
2.127	0.0	53570677	53600970	53631240	53661485	53691706	53721904	53752077	53782226	53812351	53842452	-24
2.128	0.0	53872529	53902582	53932611	53962615	53992596	54022552	54052484	54082392	54112276	54142136	-24
2.129	0.0	54171972	54201783	54231570	54261333	54291072	54320787	54350477	54380143	54409785	54439402	-24
2.130	0.0	54468996	54498564	54528109	54557630	54587126	54616597	54646045	54675468	54704867	54734241	-24
2.131	0.0	54763591	54792917	54822218	54851495	54880748	54909976	54939179	54968359	54997513	55026644	-24
2.132	0.0	55055750	55084831	55113888	55142921	55171929	55200912	55229871	55258805	55287715	55316601	-25
2.133	0.0	55345462	55374298	55403110	55431897	55460659	55489397	55518111	55546800	55575464	55604103	-25
2.134	0.0	55632718	55661308	55689874	55718415	55746931	55775423	55803890	55832332	55860749	55889142	-25
2.135	0.0	55917510	55945854	55974172	56002466	56030735	56058980	56087199	56115394	56143564	56171709	-25
2.136	0.0	56199829	56227925	56255996	56284042	56312063	56340059	56368030	56395977	56423898	56451795	-25
2.137	0.0	56479667	56507513	56535335	56563132	56590905	56618652	56646374	56674071	56701744	56729391	-25
2.138	0.0	56757013	56784611	56812183	56839730	56867253	56894750	56922222	56949670	56977092	57004489	-25
2.139	0.0	57031861	57059208	57086530	57113827	57141099	57168345	57195567	57222763	57249934	57277081	-25
2.140	0.0	57304202	57331297	57358368	57385414	57412434	57439429	57466399	57493344	57520263	57547157	-25
2.141	0.0	57574026	57600870	57627689	57654482	57681250	57707993	57734710	57761403	57788069	57814711	-25
2.142	0.0	57841327	57867918	57894484	57921024	57947539	57974029	58000493	58026932	58053346	58079734	-25
2.143	0.0	58106097	58132434	58158746	58185032	58211294	58237529	58263739	58289924	58316084	58342217	-25
2.144	0.0	58368326	58394409	58420466	58446498	58472505	58498486	58524441	58550371	58576275	58602154	-26
2.145	0.0	58628007	58653835	58679637	58705414	58731165	58756890	58782590	58808265	58833913	58859536	-26
2.146	0.0	58885134	58910705	58936251	58961772	58987267	59012736	59038179	59063597	59088989	59114356	-26
2.147	0.0	59139697	59165012	59190301	59215565	59240802	59266015	59291201	59316362	59341496	59366606	-26
2.148	0.0	59391689	59416747	59441778	59466784	59491764	59516719	59541647	59566550	59591427	59616278	-26
2.149	0.0	59641103	59665903	59690676	59715424	59740145	59764841	59789511	59814155	59838773	59863366	-26

TABLE 6

Sinc Function

x		.0	.0001	.0002	.0003	.0004	.0005	.0006	.0007	.0008	.0009	Δ_2
2.150	0.0	59887932	59912472	59936987	59961475	59985938	60010375	60034785	60059170	60083529	60107861	-26
2.151	0.0	60132168	60156449	60180704	60204932	60229135	60253312	60277452	60301587	60325685	60349758	-26
2.152	0.0	60373804	60397825	60421819	60445787	60469729	60493645	60517535	60541399	60565237	60589048	-26
2.153	0.0	60612834	60636593	60660326	60684033	60707714	60731369	60754997	60778600	60802176	60825726	-26
2.154	0.0	60849250	60872747	60896219	60919664	60943083	60966475	60989842	61013182	61036496	61059783	-26
2.155	0.0	61083045	61106280	61129489	61152671	61175828	61198958	61222061	61245139	61268190	61291215	-26
2.156	0.0	61314213	61337165	61360131	61383050	61405943	61428810	61451650	61474464	61497251	61520012	-26
2.157	0.0	61542747	61565455	61588137	61610793	61633422	61656025	61678601	61701151	61723674	61746171	-26
2.158	0.0	61768641	61791085	61813503	61835894	61858258	61880596	61902908	61925193	61947451	61969683	-27
2.159	0.0	61991888	62014067	62036220	62058346	62080445	62102518	62124564	62146583	62168576	62190543	-27
2.160	0.0	62212483	62234396	62256283	62278143	62299976	62321783	62343564	62365317	62387044	62408744	-27
2.161	0.0	62430418	62452065	62473686	62495279	62516846	62538387	62559901	62581388	62602848	62624281	-27
2.162	0.0	62645688	62667069	62688422	62709749	62731049	62752322	62773569	62794789	62815982	62837148	-27
2.163	0.0	62858288	62879400	62900486	62921546	62942578	62963584	62984563	63005515	63026440	63047338	-27
2.164	0.0	63068210	63089055	63109873	63130664	63151428	63172165	63192876	63213560	63234217	63254846	-27
2.165	0.0	63275450	63296026	63316575	63337098	63357593	63378062	63398503	63418918	63439306	63459667	-27
2.166	0.0	63480001	63500308	63520588	63540841	63561068	63581267	63601439	63621585	63641703	63661794	-27
2.167	0.0	63681859	63701896	63721907	63741890	63761846	63781776	63801678	63821554	63841402	63861223	-27
2.168	0.0	63881018	63900785	63920525	63940238	63959924	63979583	63999215	64018820	64038398	64057948	-27
2.169	0.0	64077472	64096968	64116438	64135880	64155295	64174683	64194044	64213378	64232685	64251964	-27
2.170	0.0	64271216	64290442	64309640	64328811	64347955	64367071	64386161	64405223	64424258	64443266	-27
2.171	0.0	64462247	64481200	64500126	64519025	64537897	64556742	64575560	64594350	64613113	64631849	-27
2.172	0.0	64650557	64669238	64687892	64706519	64725119	64743691	64762236	64780754	64799244	64817707	-27
2.173	0.0	64836143	64854552	64872933	64891287	64909614	64927913	64946185	64964430	64982647	65000837	-27
2.174	0.0	65019000	65037135	65055243	65073324	65091378	65109404	65127402	65145373	65163317	65181234	-27
2.175	0.0	65199123	65216985	65234819	65252626	65270406	65288158	65305883	65323580	65341250	65358893	-27
2.176	0.0	65376508	65394096	65411656	65429189	65446694	65464172	65481622	65499045	65516441	65533809	-27
2.177	0.0	65551150	65568463	65585749	65603007	65620238	65637441	65654617	65671765	65688886	65705979	-28
2.178	0.0	65723045	65740083	65757094	65774077	65791033	65807961	65824862	65841735	65858580	65875398	-28
2.179	0.0	65892189	65908952	65925687	65942395	65959075	65975728	65992353	66008950	66025520	66042062	-28
2.180	0.0	66058577	66075064	66091524	66107956	66124360	66140737	66157086	66173407	66189701	66205968	-28
2.181	0.0	66222206	66238417	66254601	66270756	66286885	66302985	66319058	66335103	66351120	66367110	-28
2.182	0.0	66383072	66399007	66414914	66430793	66446644	66462468	66478264	66494033	66509773	66525486	-28
2.183	0.0	66541172	66556829	66572459	66588061	66603636	66619183	66634702	66650193	66665656	66681092	-28
2.184	0.0	66696500	66711881	66727233	66742558	66757855	66773125	66788366	66803580	66818766	66833924	-28
2.185	0.0	66849055	66864158	66879233	66894280	66909299	66924291	66939255	66954191	66969099	66983980	-28
2.186	0.0	66998832	67013657	67028454	67043223	67057965	67072678	67087364	67102022	67116652	67131254	-28
2.187	0.0	67145829	67160375	67174894	67189385	67203848	67218283	67232691	67247070	67261422	67275745	-28
2.188	0.0	67290041	67304309	67318549	67332762	67346946	67361103	67375231	67389332	67403405	67417450	-28
2.189	0.0	67431467	67445456	67459417	67473350	67487256	67501133	67514983	67528805	67542598	67556364	-28
2.190	0.0	67570102	67583812	67597494	67611148	67624774	67638373	67651943	67665485	67679000	67692486	-28
2.191	0.0	67705945	67719375	67732778	67746152	67759499	67772818	67786109	67799371	67812606	67825813	-28
2.192	0.0	67838992	67852142	67865265	67878360	67891427	67904466	67917477	67930460	67943415	67956341	-28
2.193	0.0	67969240	67982111	67994954	68007769	68020556	68033314	68046045	68058748	68071423	68084069	-28
2.194	0.0	68096688	68109279	68121841	68134376	68146882	68159361	68171811	68184234	68196628	68208994	-28
2.195	0.0	68221333	68233643	68245925	68258179	68270405	68282603	68294773	68306915	68319028	68331114	-28
2.196	0.0	68343172	63355201	68367203	68379176	68391121	68403038	68414928	68426789	68438621	68450426	-28
2.197	0.0	68462203	68473952	68485672	68497365	68509029	68520665	68532273	68543853	68555405	68566929	-28
2.198	0.0	68578424	68589892	68601331	68612743	68624126	68635481	68646808	68658106	68669377	68680620	-28
2.199	0.0	68691834	68703020	68714178	68725308	68736410	68747484	68758529	68769547	68780536	68791497	-28

$$\frac{\sin 2\pi x}{2\pi x}$$

TABLE 6

Sinc Function

x		.0	.0001	.0002	.0003	.0004	.0005	.0006	.0007	.0008	.0009	Δ_2
2.200	0.0	68802430	68813335	68824211	68835060	68845880	68856672	68867436	68878172	68888880	68899559	-28
2.201	0.0	68910210	68920833	68931428	68941995	68952534	68963044	68973526	68983981	68994406	69004804	-28
2.202	0.0	69015174	69025515	69035828	69046113	69056370	69066598	69076799	69086971	69097115	69107231	-28
2.203	0.0	69117318	69127378	69137409	69147412	69157387	69167333	69177251	69187142	69197004	69206837	-28
2.204	0.0	69216643	69226420	69236169	69245890	69255583	69265247	69274883	69284491	69294071	69303623	-28
2.205	0.0	69313146	69322641	69332108	69341547	69350957	69360339	69369693	69379019	69388316	69397586	-28
2.206	0.0	69406827	69416039	69425224	69434380	69443508	69452608	69461680	69470723	69479738	69488725	-28
2.207	0.0	69497683	69506614	69515516	69524390	69533235	69542053	69550842	69559602	69568335	69577039	-28
2.208	0.0	69585715	69594363	69602983	69611574	69620137	69628672	69637179	69645657	69654107	69662529	-28
2.209	0.0	69670922	69679287	69687624	69695933	69704213	69712465	69720689	69728885	69737052	69745191	-28
2.210	0.0	69753302	69761385	69769439	69777465	69785463	69793432	69801373	69809286	69817171	69825027	-28
2.211	0.0	69832855	69840655	69848427	69856170	69863885	69871572	69879230	69886860	69894462	69902036	-28
2.212	0.0	69909581	69917098	69924587	69932047	69939479	69946883	69954259	69961606	69968925	69976216	-28
2.213	0.0	69983479	69990713	69997919	70005096	70012246	70019367	70026460	70033524	70040560	70047568	-28
2.214	0.0	70054548	70061499	70068422	70075317	70082184	70089022	70095832	70102613	70109367	70116092	-28
2.215	0.0	70122789	70129457	70136097	70142709	70149293	70155848	70162375	70168874	70175345	70181787	-28
2.216	0.0	70188201	70194586	70200944	70207273	70213573	70219846	70226090	70232306	70238494	70244653	-28
2.217	0.0	70250784	70256887	70262961	70269007	70275025	70281015	70286976	70292909	70298814	70304691	-28
2.218	0.0	70310539	70316359	70322150	70327914	70333649	70339356	70345034	70350684	70356306	70361900	-28
2.219	0.0	70367465	70373002	70378511	70383992	70389444	70394868	70400264	70405631	70410970	70416281	-28
2.220	0.0	70421564	70426818	70432044	70437242	70442411	70447553	70452666	70457750	70462807	70467835	-28
2.221	0.0	70472835	70477806	70482750	70487665	70492551	70497410	70502240	70507042	70511816	70516561	-28
2.222	0.0	70521278	70525967	70530628	70535260	70539865	70544440	70548988	70553507	70557998	70562461	-28
2.223	0.0	70566896	70571302	70575680	70580030	70584352	70588645	70592910	70597147	70601355	70605536	-28
2.224	0.0	70609688	70613811	70617907	70621974	70626013	70630024	70634007	70637961	70641887	70645785	-28
2.225	0.0	70649655	70653496	70657309	70661094	70664851	70668579	70672279	70675951	70679595	70683211	-28
2.226	0.0	70686798	70690357	70693888	70697390	70700865	70704311	70707729	70711119	70714480	70717813	-28
2.227	0.0	70721118	70724395	70727644	70730864	70734057	70737221	70740356	70743464	70746543	70749595	-28
2.228	0.0	70752618	70755612	70758579	70761517	70764427	70767309	70770163	70772989	70775786	70778555	-28
2.229	0.0	70781296	70784009	70786694	70789350	70791979	70794579	70797151	70799694	70802210	70804697	-28
2.230	0.0	70807156	70809587	70811990	70814365	70816712	70819030	70821320	70823582	70825816	70828022	-28
2.231	0.0	70830199	70832349	70834470	70836563	70838628	70840665	70842673	70844654	70846606	70848530	-28
2.232	0.0	70850426	70852294	70854134	70855946	70857729	70859485	70861212	70862911	70864582	70866225	-28
2.233	0.0	70867840	70869426	70870985	70872515	70874017	70875492	70876938	70878355	70879745	70881107	-28
2.234	0.0	70882441	70883746	70885024	70886273	70887494	70888687	70889852	70890989	70892098	70893179	-28
2.235	0.0	70894232	70895256	70896253	70897221	70898162	70899074	70899958	70900814	70901643	70902443	-28
2.236	0.0	70903215	70903959	70904674	70905362	70906022	70906654	70907257	70907833	70908381	70908900	-28
2.237	0.0	70909392	70909855	70910291	70910698	70911077	70911429	70911752	70912047	70912314	70912554	-28
2.238	0.0	70912765	70912948	70913103	70913231	70913330	70913401	70913444	70913459	70913447	70913406	-28
2.239	0.0	70913337	70913240	70913116	70912963	70912782	70912573	70912337	70912072	70911779	70911459	-28
2.240	0.0	70911110	70910734	70910329	70909897	70909437	70908948	70908432	70907888	70907316	70906715	-28
2.241	0.0	70906087	70905431	70904748	70904036	70903296	70902528	70901733	70900909	70900058	70899178	-28
2.242	0.0	70898271	70897336	70896373	70895382	70894363	70893316	70892241	70891139	70890008	70888850	-28
2.243	0.0	70887664	70886450	70885208	70883938	70882640	70881315	70879961	70878580	70877171	70875734	-28
2.244	0.0	70874269	70872776	70871256	70869707	70868131	70866527	70864895	70863235	70861548	70859832	-28
2.245	0.0	70858089	70856318	70854519	70852692	70850838	70848956	70847046	70845108	70843142	70841149	-28
2.246	0.0	70839127	70837078	70835002	70832897	70830765	70828605	70826417	70824201	70821958	70819686	-28
2.247	0.0	70817387	70815061	70812706	70810324	70807914	70805477	70803011	70800518	70797997	70795449	-28
2.248	0.0	70792872	70790268	70787637	70784977	70782290	70779575	70776833	70774062	70771264	70768439	-28
2.249	0.0	70765585	70762704	70759796	70756859	70753895	70750904	70747884	70744837	70741763	70738660	-28

TABLE 6

Sinc Function

x		.0	.0001	.0002	.0003	.0004	.0005	.0006	.0007	.0008	.0009	Δ_2
2.250	0.0	70735530	70732373	70729187	70725974	70722734	70719466	70716170	70712847	70709496	70706117	−28
2.251	0.0	70702711	70699277	70695815	70692326	70688809	70685265	70681693	70678094	70674467	70670812	−28
2.252	0.0	70667130	70663420	70659683	70655918	70652126	70648306	70644458	70640583	70636680	70632750	−28
2.253	0.0	70628793	70624807	70620795	70616754	70612686	70608591	70604468	70600318	70596140	70591935	−27
2.254	0.0	70587702	70583442	70579154	70574839	70570496	70566126	70561728	70557303	70552850	70548370	−27
2.255	0.0	70543862	70539327	70534765	70530175	70525558	70520913	70516241	70511541	70506814	70502060	−27
2.256	0.0	70497278	70492469	70487632	70482768	70477877	70472958	70468012	70463038	70458037	70453009	−27
2.257	0.0	70447953	70442870	70437759	70432622	70427457	70422264	70417044	70411797	70406523	70401221	−27
2.258	0.0	70395892	70390535	70385152	70379740	70374302	70368836	70363343	70357823	70352276	70346701	−27
2.259	0.0	70341099	70335469	70329813	70324129	70318418	70312679	70306913	70301121	70295300	70289453	−27
2.260	0.0	70283578	70277677	70271747	70265791	70259808	70253797	70247759	70241694	70235602	70229482	−27
2.261	0.0	70223336	70217162	70210961	70204733	70198477	70192195	70185885	70179548	70173184	70166793	−27
2.262	0.0	70160375	70153930	70147457	70140957	70134431	70127877	70121296	70114688	70108053	70101391	−27
2.263	0.0	70094701	70087985	70081241	70074471	70067673	70060849	70053997	70047118	70040212	70033280	−27
2.264	0.0	70026320	70019333	70012319	70005278	69998210	69991115	69983993	69976844	69969668	69962465	−27
2.265	0.0	69955235	69947978	69940694	69933384	69926046	69918681	69911289	69903871	69896425	69888952	−27
2.266	0.0	69881453	69873927	69866373	69858793	69851186	69843552	69835891	69828203	69820488	69812747	−27
2.267	0.0	69804978	69797183	69789361	69781512	69773636	69765733	69757803	69749847	69741863	69733853	−27
2.268	0.0	69725816	69717753	69709662	69701545	69693400	69685229	69677032	69668807	69660556	69652278	−27
2.269	0.0	69643973	69635641	69627283	69618898	69610486	69602047	69593582	69585090	69576571	69568025	−27
2.270	0.0	69559453	69550854	69542229	69533576	69524897	69516192	69507459	69498700	69489915	69481102	−27
2.271	0.0	69472263	69463397	69454505	69445586	69436641	69427668	69418670	69409644	69400592	69391514	−27
2.272	0.0	69382408	69373277	69364118	69354933	69345722	69336484	69327219	69317928	69308610	69299266	−27
2.273	0.0	69289895	69280497	69271074	69261623	69252146	69242643	69233113	69223556	69213974	69204364	−26
2.274	0.0	69194728	69185066	69175377	69165662	69155920	69146152	69136358	69126537	69116689	69106815	−26
2.275	0.0	69096915	69086988	69077035	69067056	69057050	69047018	69036959	69026874	69016763	69006625	−26
2.276	0.0	68996461	68986271	68976054	68965811	68955541	68945246	68934924	68924575	68914201	68903800	−26
2.277	0.0	68893373	68882919	68872439	68861933	68851401	68840842	68830258	68819647	68809009	68798346	−26
2.278	0.0	68787656	68776940	68766198	68755430	68744635	68733814	68722967	68712094	68701195	68690270	−26
2.279	0.0	68679318	68668340	68657336	68646306	68635250	68624168	68613059	68601925	68590764	68579578	−26
2.280	0.0	68568365	68557126	68545861	68534570	68523253	68511909	68500540	68489145	68477724	68466276	−26
2.281	0.0	68454803	68443304	68431778	68420227	68408649	68397046	68385417	68373761	68362080	68350373	−26
2.282	0.0	68338639	68326880	68315095	68303284	68291447	68279584	68267695	68255781	68243840	68231873	−26
2.283	0.0	68219881	68207863	68195818	68183748	68171652	68159531	68147383	68135210	68123010	68110785	−26
2.284	0.0	68098534	68086258	68073955	68061627	68049273	68036893	68024487	68012056	67999599	67987116	−26
2.285	0.0	67974607	67962072	67949512	67936926	67924315	67911677	67899014	67886326	67873611	67860871	−26
2.286	0.0	67848105	67835314	67822497	67809654	67796786	67783892	67770972	67758027	67745056	67732059	−26
2.287	0.0	67719037	67705990	67692916	67679817	67666693	67653543	67640367	67627166	67613940	67600687	−26
2.288	0.0	67587410	67574106	67560778	67547423	67534044	67520638	67507208	67493751	67480270	67466763	−25
2.289	0.0	67453230	67439672	67426088	67412479	67398845	67385165	67371500	67357790	67344054	67330292	−25
2.290	0.0	67316506	67302694	67288856	67274993	67261105	67247192	67233253	67219289	67205299	67191284	−25
2.291	0.0	67177244	67163179	67149088	67134972	67120831	67106665	67092473	67078256	67064014	67049746	−25
2.292	0.0	67035454	67021136	67006793	66992424	66978031	66963612	66949168	66934699	66920205	56905686	−25
2.293	0.0	66891141	66876572	66861977	66847357	66832712	66818042	66803347	66788627	66773881	66759111	−25
2.294	0.0	66744315	66729495	66714649	66699778	66684883	66669962	66655016	66640046	66625050	66610029	−25
2.295	0.0	66594983	66579913	66564817	66549696	66534551	66519380	66504185	66488964	66473719	66458449	−25
2.296	0.0	66443154	66427834	66412489	66397119	66381724	66366305	66350860	66335391	66319897	66304378	−25
2.297	0.0	66288834	66273266	66257672	66242054	66226411	66210743	66195051	66179334	66163592	66147825	−25
2.298	0.0	66132033	66116217	66100376	66084510	66068620	66052705	66036765	66020800	66004811	65988797	−25
2.299	0.0	65972759	65956696	65940608	65924496	65908359	65892197	65876011	65859800	65843564	65827304	−25

$\frac{\sin 2\pi x}{2\pi x}$

TABLE 6

Sinc Function

x		.0	.0001	.0002	.0003	.0004	.0005	.0006	.0007	.0008	.0009	Δ_2
2.300	0.0	65811020	65794711	65778377	65762019	65745636	65729229	65712797	65696340	65679860	65663354	-24
2.301	0.0	65646824	65630270	65613691	65597088	65580460	65563808	65547132	65530431	65513705	65496955	-24
2.302	0.0	65480181	65463382	65446559	65429712	65412840	65395944	65379024	65362079	65345110	65328116	-24
2.303	0.0	65311099	65294056	65276990	65259899	65242785	65225645	65208482	65191294	65174082	65156846	-24
2.304	0.0	65139586	65122301	65104992	65087659	65070302	65052920	65035515	65018085	65000631	64983153	-24
2.305	0.0	64965651	64948125	64930574	64913000	64895401	64877778	64860131	64842460	64824766	64807046	-24
2.306	0.0	64789303	64771536	64753745	64735930	64718091	64700228	64682341	64664429	64646494	64628535	-24
2.307	0.0	64610552	64592545	64574514	64556459	64538381	64520278	64502151	64484001	64465827	64447628	-24
2.308	0.0	64429406	64411160	64392891	64374597	64356279	64337938	64319573	64301184	64282771	64264335	-24
2.309	0.0	64245875	64227391	64208883	64190352	64171796	64153217	64134615	64115988	64097338	64078664	-24
2.310	0.0	64059967	64041246	64022501	64003733	63984941	63966125	63947286	63928423	63909536	63890626	-24
2.311	0.0	63871692	63852735	63833754	63814750	63795722	63776670	63757595	63738496	63719374	63700229	-24
2.312	0.0	63681060	63661867	63642651	63623412	63604149	63584862	63565553	63546219	63526863	63507483	-23
2.313	0.0	63488079	63468652	63449202	63429729	63410232	63390712	63371168	63351601	63332011	63312397	-23
2.314	0.0	63292760	63273100	63253417	63233710	63213980	63194227	63174450	63154651	63134828	63114982	-23
2.315	0.0	63095112	63075220	63055304	63035365.	63015403	62995418	62975410	62955378	62935324	62915246	-23
2.316	0.0	62895145	62875022	62854875	62834705	62814511	62794295	62774056	62753794	62733509	62713200	-23
2.317	0.0	62692869	62672515	62652138	62631737	62611314	62590868	62570399	62549907	62529392	62508854	-23
2.318	0.0	62488293	62467710	62447103	62426474	62405821	62385146	62364448	62343728	62322984	62302217	-23
2.319	0.0	62281428	62260616	62239781	62218924	62198043	62177140	62156214	62135266	62114294	62093300	-23
2.320	0.0	62072284	62051244	62030182	62009097	61987990	61966860	61945707	61924532	61903334	61882113	-23
2.321	0.0	61860870	61839604	61818316	61797005	61775671	61754315	61732937	61711536	61690112	61668666	-23
2.322	0.0	61647197	61625706	61604192	61582656	61561098	61539517	61517913	61496288	61474639	61452969	-22
2.323	0.0	61431276	61409560	61387822	61366062	61344280	61322475	61300648	61278798	61256926	61235032	-22
2.324	0.0	61213116	61191177	61169216	61147233	61125227	61103200	61081150	61059078	61036983	61014867	-22
2.325	0.0	60992728	60970567	60948384	60926179	60903951	60881702	60859430	60837137	60814821	60792483	-22
2.326	0.0	60770123	60747741	60725337	60702911	60680462	60657992	60635500	60612986	60590450	60567891	-22
2.327	0.0	60545311	60522709	60500085	60477439	60454771	60432081	60409369	60386636	60363880	60341103	-22
2.328	0.0	60318303	60295482	60272639	60249774	60226888	60203979	60181049	60158097	60135123	60112127	-22
2.329	0.0	60089110	60066071	60043010	60019928	59996823	59973697	59950550	59927380	59904189	59880977	-22
2.330	0.0	59857742	59834486	59811209	59787910	59764589	59741247	59717883	59694497	59671090	59647661	-22
2.331	0.0	59624211	59600740	59577246	59553732	59530195	59506636	59483059	59459458	59435836	59412192	-21
2.332	0.0	59388527	59364841	59341133	59317404	59293654	59269882	59246089	59222274	59198438	59174581	-21
2.333	0.0	59150702	59126802	59102881	59078939	59054975	59030990	59006984	58982956	58958908	58934838	-21
2.334	0.0	58910746	58886634	58862501	58838346	58814170	58789973	58765755	58741516	58717256	58692974	-21
2.335	0.0	58668672	58644348	58620003	58595638	58571251	58546843	58522414	58497964	58473494	58449002	-21
2.336	0.0	58424489	58399955	58375400	58350825	58326228	58301611	58276972	58252313	58227633	58202932	-21
2.337	0.0	58178210	58153467	58128703	58103919	58079114	58054287	58029441	58004573	57979684	57954775	-21
2.338	0.0	57929845	57904895	57879923	57854931	57829918	57804885	57779831	57754756	57729660	57704544	-21
2.339	0.0	57679407	57654250	57629072	57603873	57578654	57553414	57528154	57502873	57477572	57452250	-21
2.340	0.0	57426907	57401544	57376161	57350757	57325332	57299888	57274422	57248936	57223430	57197904	-20
2.341	0.0	57172357	57146789	57121202	57095594	57069965	57044316	57018647	56992958	56967248	56941518	-20
2.342	0.0	56915768	56889997	56864206	56838395	56812564	56786712	56760840	56734948	56709036	56683104	-20
2.343	0.0	56657151	56631179	56605186	56579173	56553140	56527087	56501014	56474920	56448807	56422674	-20
2.344	0.0	56396520	56370347	56344153	56317940	56291706	56265452	56239179	56212886	56186572	56160239	-20
2.345	0.0	56133886	56107512	56081119	56054706	56028274	56001821	55975348	55948856	55922344	55895812	-20
2.346	0.0	55869260	55842688	55816097	55789486	55762855	55736204	55709534	55682844	55656134	55629404	-20
2.347	0.0	55602655	55575886	55549098	55522289	55495462	55468614	55441747	55414860	55387954	55361028	-20
2.348	0.0	55334083	55307118	55280134	55253130	55226106	55199063	55172001	55144919	55117817	55090696	-19
2.349	0.0	55063556	55036396	55009217	54982019	54954801	54927564	54900307	54873031	54845735	54818421	-19

TABLE 6

Sinc Function

x		.0	.0001	.0002	.0003	.0004	.0005	.0006	.0007	.0008	.0009	Δ_2
2.350	0.0	54791087	54763733	54736361	54708969	54681558	54654127	54626678	54599209	54571721	54544213	-19
2.351	0.0	54516687	54489141	54461576	54433992	54406389	54378767	54351125	54323465	54295785	54268087	-19
2.352	0.0	54240369	54212632	54184876	54157101	54129308	54101495	54073663	54045812	54017942	53990053	-19
2.353	0.0	53962146	53934219	53906273	53878309	53850326	53822323	53794302	53766262	53738203	53710126	-19
2.354	0.0	53682029	53653914	53625780	53597627	53569455	53541265	53513056	53484828	53456581	53428316	-19
2.355	0.0	53400032	53371730	53343408	53315068	53286710	53258333	53229937	53201522	53173089	53144638	-19
2.356	0.0	53116168	53087679	53059172	53030646	53002101	52973539	52944957	52916358	52887739	52859103	-18
2.357	0.0	52830448	52801774	52773082	52744372	52715643	52686896	52658130	52629347	52600544	52571724	-18
2.358	0.0	52542885	52514028	52485153	52456259	52427347	52398417	52369469	52340502	52311517	52282514	-18
2.359	0.0	52253493	52224454	52195396	52166321	52137227	52108115	52078985	52049837	52020671	51991487	-18
2.360	0.0	51962284	51933064	51903826	51874569	51845295	51816003	51786693	51757364	51728018	51698654	-18
2.361	0.0	51669272	51639872	51610454	51581018	51551565	51522093	51492604	51463097	51433572	51404029	-18
2.362	0.0	51374468	51344890	51315294	51285680	51256049	51226399	51196732	51167047	51137345	51107625	-18
2.363	0.0	51077887	51048132	51018359	50988568	50958760	50928934	50899091	50869230	50839351	50809455	-18
2.364	0.0	50779541	50749610	50719661	50689695	50659712	50629711	50599692	50569656	50539603	50509532	-17
2.365	0.0	50479444	50449338	50419215	50389075	50358917	50328742	50298550	50268340	50238114	50207869	-17
2.366	0.0	50177608	50147329	50117033	50086720	50056390	50026042	49995678	49965296	49934897	49904480	-17
2.367	0.0	49874047	49843597	49813129	49782645	49752143	49721624	49691088	49660535	49629965	49599378	-17
2.368	0.0	49568775	49538154	49507516	49476861	49446189	49415501	49384795	49354072	49323333	49292577	-17
2.369	0.0	49261804	49231014	49200207	49169383	49138543	49107686	49076812	49045921	49015013	48984089	-17
2.370	0.0	48953148	48922190	48891216	48860225	48829217	48798192	48767151	48736093	48705019	48673928	-17
2.371	0.0	48642821	48611696	48580556	48549399	48518225	48487034	48455828	48424604	48393365	48362108	-16
2.372	0.0	48330836	48299546	48268241	48236919	48205580	48174226	48142854	48111467	48080063	48048643	-16
2.373	0.0	48017206	47985753	47954284	47922799	47891297	47859779	47828245	47796695	47765128	47733545	-16
2.374	0.0	47701946	47670331	47638700	47607052	47575389	47543709	47512013	47480301	47448573	47416829	-16
2.375	0.0	47385069	47353293	47321501	47289693	47257869	47226029	47194173	47162301	47130413	47098509	-16
2.376	0.0	47066589	47034654	47002702	46970735	46938751	46906752	46874737	46842707	46810660	46778598	-16
2.377	0.0	46746520	46714426	46682316	46650191	46618050	46585893	46553721	46521533	46489329	46457110	-16
2.378	0.0	46424875	46392624	46360358	46328076	46295779	46263466	46231137	46198793	46166434	46134059	-16
2.379	0.0	46101668	46069262	46036841	46004404	45971952	45939484	45907001	45874502	45841988	45809459	-15
2.380	0.0	45776914	45744354	45711779	45679188	45646582	45613961	45581325	45548673	45516006	45483324	-15
2.381	0.0	45450626	45417914	45385186	45352443	45319685	45286912	45254123	45221320	45188501	45155668	-15
2.382	0.0	45122819	45089955	45057076	45024182	44991274	44958350	44925411	44892457	44859488	44826505	-15
2.383	0.0	44793506	44760492	44727464	44694421	44661362	44628239	44595201	44562099	44528981	44495849	-15
2.384	0.0	44462702	44429540	44396363	44363172	44329966	44296745	44263509	44230259	44196994	44163714	-15
2.385	0.0	44130420	44097111	44063788	44030450	43997097	43963730	43930348	43896952	43863541	43830116	-15
2.386	0.0	43796676	43763222	43729753	43696269	43662772	43629260	43595733	43562192	43528637	43495067	-14
2.387	0.0	43461483	43427884	43394272	43360645	43327003	43293348	43259678	43225993	43192295	43158582	-14
2.388	0.0	43124855	43091114	43057359	43023590	42989806	42956008	42922196	42888371	42854530	42820676	-14
2.389	0.0	42786808	42752926	42719029	42685119	42651195	42617256	42583304	42549338	42515357	42481363	-14
2.390	0.0	42447355	42413333	42379297	42345247	42311184	42277106	42243015	42208910	42174791	42140658	-14
2.391	0.0	42106511	42072351	42038177	42003989	41969787	41935572	41901343	41867100	41832844	41798574	-14
2.392	0.0	41764291	41729993	41695682	41661358	41627020	41592669	41558303	41523925	41489533	41455127	-14
2.393	0.0	41420708	41386275	41351829	41317370	41282897	41248410	41213910	41179397	41144871	41110331	-13
2.394	0.0	41075778	41041211	41006631	40972038	40937431	40902812	40868179	40833533	40798873	40764200	-13
2.395	0.0	40729515	40694815	40660103	40625378	40590639	40555888	40521123	40486345	40451554	40416750	-13
2.396	0.0	40381933	40347103	40312260	40277404	40242535	40207653	40172758	40137850	40102929	40067995	-13
2.397	0.0	40033048	39998089	39963116	39928131	39893133	39858122	39823098	39788061	39753012	39717950	-13
2.398	0.0	39682875	39647787	39612687	39577574	39542448	39507309	39472158	39436994	39401818	39366629	-13
2.399	0.0	39331427	39296213	39260986	39225747	39190495	39155230	39119953	39084664	39049362	39014047	-12

$$\frac{\sin 2\pi x}{2\pi x}$$

TABLE 6

Sinc Function

x		.0	.0001	.0002	.0003	.0004	.0005	.0006	.0007	.0008	.0009	Δ_2
2.400	0.0	38978720	38943381	38908029	38872665	38837288	38801899	38766498	38731084	38695658	38660220	-12
2.401	0.0	38624769	38589306	38553831	38518343	38482843	38447331	38411807	38376271	38340722	38305161	-12
2.402	0.0	38269588	38234003	38198406	38162796	38127175	38091541	38055896	38020238	37984568	37948887	-12
2.403	0.0	37913193	37877487	37841769	37806040	37770298	37734544	37698779	37663001	37627212	37591411	-12
2.404	0.0	37555598	37519773	37483936	37448088	37412227	37376355	37340471	37304576	37268668	37232749	-12
2.405	0.0	37196818	37160876	37124922	37088956	37052978	37016989	36980988	36944976	36908952	36872916	-12
2.406	0.0	36836869	36800810	36764740	36728659	36692565	36656461	36620344	36584217	36548078	36511927	-11
2.407	0.0	36475765	36439592	36403407	36367211	36331004	36294785	36258555	36222314	36186061	36149797	-11
2.408	0.0	36113522	36077236	36040938	36004629	35968309	35931978	35895635	35859282	35822917	35786541	-11
2.409	0.0	35750154	35713756	35677347	35640927	35604496	35568054	35531600	35495136	35458661	35422175	-11
2.410	0.0	35385678	35349170	35312650	35276121	35239580	35203028	35166465	35129892	35093308	35056713	-11
2.411	0.0	35020107	34983490	34946863	34910224	34873575	34836915	34800245	34763564	34726873	34690170	-11
2.412	0.0	34653457	34616733	34579999	34543254	34506498	34469732	34432956	34396168	34359371	34322562	-11
2.413	0.0	34285744	34248914	34212075	34175225	34138364	34101493	34064612	34027720	33990818	33953905	-10
2.414	0.0	33916982	33880049	33843105	33806151	33769187	33732213	33695228	33658233	33621228	33584213	-10
2.415	0.0	33547187	33510152	33473106	33436050	33398984	33361907	33324821	33287725	33250618	33213502	-10
2.416	0.0	33176375	33139238	33102092	33064935	33027769	32990592	32953405	32916209	32879003	32841786	-10
2.417	0.0	32804560	32767324	32730078	32692823	32655557	32618282	32580997	32543702	32506397	32469083	-10
2.418	0.0	32431758	32394425	32357081	32319728	32282365	32244992	32207610	32170218	32132817	32095406	-10
2.419	0.0	32057985	32020555	31983115	31945666	31908207	31870739	31833261	31795774	31758277	31720771	-9
2.420	0.0	31683256	31645731	31608196	31570653	31533100	31495537	31457965	31420384	31382794	31345194	-9
2.421	0.0	31307585	31269967	31232340	31194703	31157057	31119402	31081738	31044065	31006382	30968691	-9
2.422	0.0	30930990	30893280	30855561	30817833	30780096	30742350	30704595	30666831	30629058	30591276	-9
2.423	0.0	30553485	30515685	30477876	30440058	30402231	30364396	30326551	30288698	30250836	30212965	-9
2.424	0.0	30175085	30137197	30099299	30061393	30023479	29985555	29947623	29909682	29871732	29833774	-9
2.425	0.0	29795807	29757832	29719847	29681855	29643853	29605843	29567825	29529798	29491762	29453718	-9
2.426	0.0	29415666	29377605	29339535	29301457	29263371	29225276	29187173	29149062	29110942	29072814	-8
2.427	0.0	29034677	28996532	28958379	28920217	28882048	28843870	28805683	28767489	28729286	28691075	-8
2.428	0.0	28652856	28614629	28576394	28538150	28499899	28461639	28423371	28385095	28346811	28308519	-8
2.429	0.0	28270219	28231911	28193595	28155271	28116939	28078599	28040252	28001896	27963532	27925161	-8
2.430	0.0	27886781	27848394	27809999	27771596	27733186	27694767	27656341	27617907	27579465	27541016	-8
2.431	0.0	27502559	27464094	27425621	27387141	27348653	27310158	27271655	27233144	27194626	27156100	-8
2.432	0.0	27117567	27079026	27040477	27001921	26963358	26924787	26886209	26847623	26809030	26770429	-7
2.433	0.0	26731821	26693205	26654583	26615953	26577315	26538670	26500018	26461359	26422692	26384018	-7
2.434	0.0	26345337	26306649	26267953	26229251	26190541	26151824	26113099	26074368	26035630	25996884	-7
2.435	0.0	25958131	25919372	25880605	25841831	25803050	25764263	25725468	25686666	25647857	25609042	-7
2.436	0.0	25570219	25531390	25492553	25453710	25414860	25376003	25337139	25298269	25259391	25220507	-7
2.437	0.0	25181616	25142718	25103814	25064903	25025985	24987060	24948129	24909191	24870247	24831296	-7
2.438	0.0	24792338	24753374	24714403	24675425	24636441	24597451	24558454	24519450	24480440	24441424	-6
2.439	0.0	24402401	24363372	24324336	24285294	24246245	24207190	24168129	24129061	24089987	24050907	-6
2.440	0.0	24011821	23972728	23933629	23894523	23855412	23816294	23777170	23738040	23698904	23659761	-6
2.441	0.0	23620613	23581458	23542297	23503130	23463957	23424778	23385593	23346402	23307205	23268002	-6
2.442	0.0	23228793	23189578	23150357	23111130	23071897	23032659	22993414	22954164	22914907	22875645	-6
2.443	0.0	22836377	22797104	22757824	22718539	22679248	22639951	22600649	22561340	22522026	22482707	-6
2.444	0.0	22443382	22404051	22364714	22325372	22286025	22246671	22207312	22167948	22128578	22089203	-6
2.445	0.0	22049822	22010435	21971044	21931646	21892243	21852835	21813422	21774003	21734578	21695149	-5
2.446	0.0	21655714	21616273	21576828	21537377	21497920	21458459	21418992	21379520	21340043	21300561	-5
2.447	0.0	21261073	21221580	21182082	21142579	21103071	21063558	21024040	20984516	20944988	20905454	-5
2.448	0.0	20865916	20826372	20786824	20747270	20707712	20668148	20628580	20589007	20549429	20509846	-5
2.449	0.0	20470258	20430665	20391068	20351465	20311858	20272246	20232629	20193008	20153382	20113751	-5

TABLE 6

Sinc Function

x		.0	.0001	.0002	.0003	.0004	.0005	.0006	.0007	.0008	.0009	Δ_2
2.450	0.0	20074115	20034475	19994830	19955180	19915526	19875867	19836203	19796535	19756863	19717185	−5
2.451	0.0	19677504	19637817	19598126	19558431	19518731	19479027	19439318	19399605	19359887	19320165	−4
2.452	0.0	19280439	19240708	19200973	19161233	19121490	19081742	19041989	19002233	18962472	18922706	−4
2.453	0.0	18882937	18843163	18803386	18763604	18723818	18684027	18644233	18604434	18564632	18524825	−4
2.454	0.0	18485014	18445199	18405380	18365557	18325730	18285900	18246065	18206226	18166383	18126536	−4
2.455	0.0	18086686	18046831	18006973	17967111	17927245	17887375	17847501	17807623	17767742	17727857	−4
2.456	0.0	17687968	17648076	17608179	17568279	17528376	17488468	17448557	17408643	17368724	17328803	−4
2.457	0.0	17288877	17248948	17209015	17169079	17129140	17089196	17049250	17009300	16969346	16929389	−3
2.458	0.0	16889428	16849465	16809497	16769527	16729552	16689575	16649594	16609610	16569623	16529632	−3
2.459	0.0	16489638	16449641	16409641	16369637	16329630	16289620	16249607	16209590	16169571	16129548	−3
2.460	0.0	16089522	16049493	16009461	15969426	15929388	15889347	15849303	15809256	15769206	15729153	−3
2.461	0.0	15689097	15649038	15608976	15568911	15528843	15488773	15448699	15408623	15368544	15328462	−3
2.462	0.0	15288377	15248290	15208199	15168106	15128011	15087912	15047811	15007707	14967601	14927491	−3
2.463	0.0	14887380	14847265	14807148	14767029	14726906	14686782	14646654	14606525	14566392	14526257	−2
2.464	0.0	14486120	14445980	14405838	14365694	14325547	14285397	14245245	14205091	14164935	14124776	−2
2.465	0.0	14084615	14044451	14004285	13964117	13923947	13883775	13843600	13803423	13763244	13723062	−2
2.466	0.0	13682879	13642693	13602506	13562316	13522124	13481930	13441734	13401536	13361335	13321133	−2
2.467	0.0	13280929	13240723	13200515	13160305	13120093	13079879	13039663	12999445	12959225	12919004	−2
2.468	0.0	12878781	12838555	12798329	12758100	12717869	12677637	12637403	12597167	12556930	12516691	−2
2.469	0.0	12476450	12436207	12395963	12355717	12315470	12275221	12234970	12194718	12154464	12114209	−2
2.470	0.0	12073952	12033694	11993434	11953173	11912910	11872646	11832381	11792114	11751845	11711575	0
2.471	0.0	11671304	11631032	11590758	11550483	11510206	11469929	11429650	11389370	11349088	11308805	0
2.472	0.0	11268521	11228236	11187950	11147663	11107374	11067084	11026794	10986502	10946209	10905915	0
2.473	0.0	10865620	10825323	10785026	10744728	10704429	10664129	10623828	10583526	10543223	10502919	0
2.474	0.0	10462615	10422309	10382003	10341696	10301387	10261079	10220769	10180458	10140147	10099835	0
2.475	0.0	10059523	10019209	09978895	09938580	09898265	09857949	09817632	09777315	09736997	09696678	0
2.476	0.0	09656359	09616039	09575719	09535398	09495077	09454756	09414433	09374111	09333788	09293464	0
2.477	0.0	09253140	09212816	09172491	09132166	09091840	09051515	09011189	08970862	08930535	08890209	0
2.478	0.0	08849881	08809554	08769226	08728898	08688570	08648242	08607914	08567585	08527256	08486928	0
2.479	0.0	08446599	08406270	08365941	08325612	08285282	08244953	08204624	08164295	08123966	08083637	0
2.480	0.0	08043308	08002979	07962650	07922321	07881993	07841664	07801336	07761008	07720680	07680352	0
2.481	0.0	07640025	07599698	07559371	07519044	07478717	07438391	07398065	07357740	07317415	07277090	0
2.482	0.0	07236765	07196441	07156118	07115794	07075472	07035149	06994827	06954506	06914185	06873865	1
2.483	0.0	06833545	06793226	06752907	06712589	06672271	06631954	06591638	06551323	06511008	06470693	1
2.484	0.0	06430380	06390067	06349755	06309443	06269132	06228823	06188513	06148205	06107898	06067591	1
2.485	0.0	06027285	05986980	05946676	05906373	05866070	05825769	05785469	05745169	05704871	05664573	1
2.486	0.0	05624277	05583981	05543687	05503394	05463101	05422810	05382520	05342231	05301943	05261656	1
2.487	0.0	05221371	05181086	05140803	05100521	05060240	05019961	04979683	04939406	04899130	04858856	1
2.488	0.0	04818583	04778311	04738040	04697771	04657504	04617238	04576973	04536709	04496447	04456187	1
2.489	0.0	04415928	04375670	04335414	04295160	04254907	04214655	04174406	04134157	04093911	04053666	2
2.490	0.0	04013422	03973181	03932940	03892702	03852465	03812230	03771997	03731766	03691536	03651308	2
2.491	0.0	03611082	03570857	03530635	03490414	03450195	03409978	03369763	03329549	03289338	03249129	2
2.492	0.0	03208921	03168716	03128512	03088311	03048111	03007914	02967718	02927525	02887333	02847144	2
2.493	0.0	02806957	02766772	02726589	02686408	02646229	02606053	02565879	02525707	02485537	02445369	2
2.494	0.0	02405204	02365041	02324880	02284722	02244566	02204412	02164260	02124111	02083965	02043820	2
2.495	0.0	02003678	01963539	01923402	01883267	01843135	01803006	01762878	01722754	01682632	01642512	3
2.496	0.0	01602395	01562281	01522169	01482060	01441954	01401850	01361749	01321650	01281554	01241461	3
2.497	0.0	01201371	01161283	01121198	01081116	01041036	01000960	00960886	00920815	00880747	00840682	3
2.498	0.0	00800619	00760560	00720503	00680450	00640399	00600351	00560307	00520265	00480226	00440190	3
2.499	0.0	00400157	00360128	00320101	00280078	00240057	00200040	00160025	00120014	00080006	00040002	−3

$$\frac{\sin 2\pi x}{2\pi x}$$

$\frac{\sin 2\pi x}{2\pi x}$

TABLE 6

Sinc Function

x	.0	.0001	.0002	.0003	.0004	.0005	.0006	.0007	.0008	.0009	Δ_2
2.500 −0.0	00000000	00039998	00079994	00119986	00159974	00199960	00239942	00279921	00319896	00359869	3
2.501 −0.0	00399837	00439803	00479765	00519724	00559679	00599631	00639580	00679525	00719467	00759405	3
2.502 −0.0	00799339	00839271	00879198	00919122	00959043	00998960	01038873	01078783	01118689	01158592	4
2.503 −0.0	01198491	01238386	01278278	01318165	01358050	01397930	01437807	01477680	01517549	01557414	4
2.504 −0.0	01597276	01637134	01676983	01716838	01756684	01796526	01836365	01876200	01916030	01955857	4
2.505 −0.0	01995680	02035498	02075313	02115124	02154931	02194734	02234532	02274327	02314118	02353904	4
2.506 −0.0	02393687	02433465	02473239	02513009	02552775	02592537	02632294	02672047	02711796	02751541	4
2.507 −0.0	02791282	02831018	02870750	02910478	02950201	02989920	03029635	03069345	03109051	03148752	4
2.508 −0.0	03188450	03228142	03267831	03307514	03347194	03386869	03426539	03466205	03505866	03545523	5
2.509 −0.0	03585175	03624823	03664466	03704104	03743738	03783367	03822992	03862611	03902227	03941837	5
2.510 −0.0	03981443	04021044	04060640	04100232	04139818	04179400	04218978	04258550	04298117	04337680	5
2.511 −0.0	04377238	04416791	04456339	04495882	04535420	04574953	04614482	04654005	04693523	04733037	5
2.512 −0.0	04772545	04812048	04851547	04891040	04930528	04970011	05009489	05048962	05088430	05127892	5
2.513 −0.0	05167349	05206802	05246249	05285690	05325127	05364558	05403984	05443405	05482821	05522231	5
2.514 −0.0	05561636	05601035	05640429	05679818	05719202	05758580	05797953	05837320	05876682	05916038	5
2.515 −0.0	05955389	05994734	06034074	06073409	06112738	06152061	06191379	06230691	06269998	06309299	6
2.516 −0.0	06348594	06387884	06427168	06466447	06505720	06544987	06584248	06623504	06662754	06701998	6
2.517 −0.0	06741236	06780469	06819696	06858917	06898132	06937342	06976545	07015743	07054935	07094121	6
2.518 −0.0	07133301	07172475	07211643	07250805	07289961	07329112	07368256	07407394	07446526	07485652	6
2.519 −0.0	07524772	07563886	07602994	07642096	07681192	07720281	07759365	07798442	07837513	07876578	6
2.520 −0.0	07915636	07954689	07993735	08032775	08071808	08110836	08149857	08188872	08227880	08266882	6
2.521 −0.0	08305878	08344867	08383850	08422827	08461797	08500761	08539718	08578669	08617613	08656551	6
2.522 −0.0	08695482	08734407	08773325	08812237	08851142	08890041	08928933	08967818	09006697	09045569	7
2.523 −0.0	09084434	09123293	09162145	09200991	09239829	09278661	09317487	09356305	09395117	09433922	7
2.524 −0.0	09472720	09511511	09550296	09589073	09627844	09666608	09705365	09744115	09782858	09821595	7
2.525 −0.0	09860324	09899046	09937762	09976470	10015172	10053866	10092553	10131234	10169907	10208573	7
2.526 −0.0	10247232	10285884	10324529	10363167	10401797	10440421	10479037	10517646	10556248	10594843	7
2.527 −0.0	10633430	10672010	10710583	10749148	10787707	10826258	10864801	10903338	10941867	10980388	7
2.528 −0.0	11018902	11057409	11095908	11134400	11172885	11211362	11249832	11288294	11326748	11365195	8
2.529 −0.0	11403635	11442067	11480492	11518908	11557318	11595720	11634114	11672500	11710879	11749250	8
2.530 −0.0	11787614	11825969	11864318	11902658	11940991	11979316	12017633	12055942	12094244	12132538	8
2.531 −0.0	12170824	12209102	12247372	12285635	12323889	12362136	12400375	12438606	12476829	12515044	8
2.532 −0.0	12553251	12591450	12629641	12667824	12705999	12744166	12782325	12820476	12858619	12896754	8
2.533 −0.0	12934880	12972999	13011109	13049212	13087306	13125392	13163469	13201539	13239600	13277653	8
2.534 −0.0	13315698	13353735	13391763	13429783	13467795	13505798	13543794	13581780	13619759	13657729	8
2.535 −0.0	13695690	13733644	13771588	13809525	13847453	13885372	13923283	13961186	13999080	14036965	9
2.536 −0.0	14074842	14112711	14150571	14188422	14226265	14264099	14301924	14339741	14377549	14415349	9
2.537 −0.0	14453140	14490922	14528696	14566461	14604217	14641964	14679703	14717432	14755153	14792866	9
2.538 −0.0	14830569	14868264	14905949	14943626	14981294	15018954	15056604	15094245	15131878	15169501	9
2.539 −0.0	15207116	15244721	15282318	15319905	15357484	15395054	15432614	15470166	15507708	15545241	9
2.540 −0.0	15582766	15620281	15657787	15695284	15732771	15770250	15807719	15845180	15882631	15920072	9
2.541 −0.0	15957505	15994928	16032342	16069747	16107143	16144529	16181906	16219273	16256632	16293980	9
2.542 −0.0	16331320	16363650	16405971	16443282	16480584	16517876	16555159	16592433	16629697	16666951	9
2.543 −0.0	16704196	16741432	16778658	16815874	16853081	16890278	16927466	16964644	17001812	17038971	10
2.544 −0.0	17076120	17113260	17150389	17187510	17224620	17261721	17298812	17335893	17372964	17410026	10
2.545 −0.0	17447078	17484120	17521152	17558175	17595188	17632190	17669183	17706166	17743139	17780103	10
2.546 −0.0	17817056	17853999	17890933	17927856	17964770	18001673	18038567	18075450	18112324	18149187	10
2.547 −0.0	18186040	18222884	18259717	18296540	18333353	18370156	18406948	18443731	18480503	18517265	10
2.548 −0.0	18554017	18590759	18627491	18664212	18700923	18737624	18774314	18810995	18847664	18884324	10
2.549 −0.0	18920973	18957612	18994241	19030859	19067467	19104064	19140651	19177228	19213794	19250350	10

TABLE 6

Sinc Function

x		.0	.0001	.0002	.0003	.0004	.0005	.0006	.0007	.0008	.0009	Δ_2
2.550	-0.0	19286895	19323430	19359954	19396468	19432971	19469464	19505946	19542417	19578878	19615329	11
2.551	-0.0	19651769	19638198	19724616	19761024	19797422	19833808	19870184	19906549	19942904	19979248	11
2.552	-0.0	20015581	20051903	20088215	20124515	20160805	20197085	20233353	20269610	20305857	20342093	11
2.553	-0.0	20378318	20414532	20450735	20486928	20523109	20559279	20595439	20631587	20667725	20703851	11
2.554	-0.0	20739967	20776072	20812165	20848247	20884319	20920379	20956428	20992467	21028494	21064510	11
2.555	-0.0	21100514	21136508	21172491	21208462	21244422	21280371	21316309	21352235	21388150	21424054	11
2.556	-0.0	21459947	21495829	21531699	21567558	21603405	21639241	21675066	21710880	21746682	21782472	11
2.557	-0.0	21818252	21854020	21889776	21925521	21961255	21996977	22032688	22068387	22104074	22139751	12
2.558	-0.0	22175415	22211068	22246710	22282340	22317958	22353565	22389160	22424744	22460316	22495876	12
2.559	-0.0	22531425	22566961	22602487	22638000	22673502	22708992	22744471	22779937	22815392	22850835	12
2.560	-0.0	22886266	22921686	22957094	22992490	23027874	23063246	23098606	23133954	23169291	23204616	12
2.561	-0.0	23239928	23275229	23310518	23345795	23381060	23416312	23451553	23486782	23521999	23557204	12
2.562	-0.0	23592397	23627577	23662746	23697903	23733047	23768179	23803300	23838408	23873504	23908588	12
2.563	-0.0	23943659	23978719	24013766	24048801	24083824	24118834	24153832	24188818	24223792	24258754	12
2.564	-0.0	24293703	24328640	24363564	24398476	24433376	24468263	24503138	24538001	24572851	24607689	12
2.565	-0.0	24642515	24677328	24712128	24746916	24781692	24816455	24851205	24885943	24920669	24955382	13
2.566	-0.0	24990082	25024770	25059445	25094108	25128758	25163395	25198020	25232632	25267232	25301819	13
2.567	-0.0	25336393	25370954	25405503	25440039	25474562	25509073	25543571	25578056	25612528	25646987	13
2.568	-0.0	25681434	25715868	25750289	25784697	25819092	25853474	25887844	25922200	25956544	25990875	13
2.569	-0.0	26025193	26059498	26093790	26128069	26162335	26196588	26230828	26265055	26299268	26333469	13
2.570	-0.0	26367657	26401832	26435994	26470142	26504278	26538400	26572509	26606605	26640688	26674758	13
2.571	-0.0	26708815	26742858	26776888	26810905	26844909	26878899	26912877	26946841	26980791	27014729	13
2.572	-0.0	27048653	27082564	27116461	27150345	27184216	27218073	27251917	27285748	27319565	27353369	13
2.573	-0.0	27387159	27420936	27454700	27488450	27522186	27555910	27589619	27623315	27656998	27690667	14
2.574	-0.0	27724322	27757964	27791592	27825207	27858808	27892396	27925970	27959530	27993077	28026610	14
2.575	-0.0	28060129	28093635	28127127	28160605	28194069	28227520	28260957	28294380	28327790	28361186	14
2.576	-0.0	28394568	28427936	28461290	28494631	28527958	28561270	28594569	28627854	28661126	28694383	14
2.577	-0.0	28727626	28760856	28794072	28827273	28860461	28893635	28926794	28959940	28993072	29026189	14
2.578	-0.0	29059293	29092383	29125458	29158520	29191567	29224601	29257620	29290625	29323616	29356593	14
2.579	-0.0	29389556	29422504	29455439	29488359	29521265	29554157	29587034	29619898	29652747	29685582	14
2.580	-0.0	29718403	29751209	29784001	29816779	29849542	29882291	29915026	29947747	29980453	30013144	14
2.581	-0.0	30045822	30078485	30111133	30143767	30176387	30208992	30241583	30274160	30306721	30339269	14
2.582	-0.0	30371802	30404320	30436824	30469313	30501788	30534248	30566694	30599125	30631542	30663943	15
2.583	-0.0	30696331	30728703	30761061	30793405	30825733	30858047	30890347	30922631	30954901	30987157	15
2.584	-0.0	31019397	31051623	31083834	31116030	31148212	31180378	31212530	31244667	31276789	31308897	15
2.585	-0.0	31340989	31373067	31405130	31437178	31469211	31501229	31533232	31565221	31597194	31629153	15
2.586	-0.0	31661096	31693025	31724938	31756837	31788720	31820589	31852442	31884281	31916104	31947913	15
2.587	-0.0	31979706	32011434	32043247	32074995	32106728	32138446	32170149	32201836	32233509	32265166	15
2.588	-0.0	32296808	32328434	32360046	32391642	32423224	32454789	32486340	32517875	32549396	32580900	15
2.589	-0.0	32612390	32643864	32675323	32706767	32738195	32769608	32801005	32832387	32863754	32895105	15
2.590	-0.0	32926441	32957762	32989067	33020357	33051631	33082890	33114133	33145361	33176573	33207770	16
2.591	-0.0	33238951	33270117	33301267	33332402	33363521	33394624	33425712	33456785	33487841	33518882	16
2.592	-0.0	33549908	33580918	33611912	33642890	33673853	33704801	33735732	33766648	33797548	33828432	16
2.593	-0.0	33859301	33890154	33920991	33951812	33982618	34013408	34044181	34074940	34105682	34136408	16
2.594	-0.0	34167119	34197314	34228493	34259156	34289803	34320434	34351050	34381649	34412233	34442800	16
2.595	-0.0	34473352	34503887	34534407	34564911	34595399	34625870	34656326	34686766	34717189	34747597	16
2.596	-0.0	34777988	34808364	34838723	34869066	34899393	34929704	34959999	34990278	35020541	35050787	16
2.597	-0.0	35081017	35111232	35141430	35171611	35201777	35231926	35262059	35292176	35322277	35352361	16
2.598	-0.0	35382429	35412481	35442516	35472536	35502538	35532525	35562495	35592449	35622387	35652308	16
2.599	-0.0	35682212	35712101	35741973	35771828	35801668	35831490	35861297	35891086	35920860	35950617	16

$$\frac{\sin 2\pi x}{2\pi x}$$

TABLE 6

Sinc Function

x	.0	.0001	.0002	.0003	.0004	.0005	.0006	.0007	.0008	.0009	Δ_2
2.600 -0.0	35980357	36010081	36039788	36069479	36099154	36128812	36158453	36188078	36217686	36247278	17
2.601 -0.0	36276853	36306411	36335953	36365478	36394987	36424479	36453954	36483413	36512855	36542280	17
2.602 -0.0	36571689	36601080	36630456	36659814	36689156	36718481	36747789	36777081	36806356	36835614	17
2.603 -0.0	36864855	36894079	36923287	36952477	36981651	37010808	37039949	37069072	37098179	37127268	17
2.604 -0.0	37156341	37185397	37214436	37243458	37272463	37301451	37330422	37359376	37388314	37417234	17
2.605 -0.0	37446137	37475023	37503893	37532745	37561580	37590398	37619199	37647983	37676750	37705500	17
2.606 -0.0	37734233	37762949	37791647	37820329	37848993	37877640	37906271	37934883	37963479	37992058	17
2.607 -0.0	38020619	38049163	38077690	38106200	38134692	38163168	38191626	38220066	38248490	38276896	17
2.608 -0.0	38305285	38333656	38362011	38390348	38418667	38446970	38475255	38503522	38531772	38560005	17
2.609 -0.0	38588221	38616419	38644600	38672763	38700909	38729037	38757148	38785242	38813318	38841376	17
2.610 -0.0	38869417	38897441	38925447	38953435	38981406	39009360	39037296	39065214	39093115	39120999	18
2.611 -0.0	39148864	39176712	39204543	39232356	39260151	39287929	39315689	39343431	39371156	39398863	18
2.612 -0.0	39426552	39454224	39481878	39509514	39537133	39564734	39592317	39619882	39647430	39674960	18
2.613 -0.0	39702472	39729966	39757443	39784902	39812343	39839766	39867171	39894559	39921928	39949280	18
2.614 -0.0	39976614	40003930	40031228	40058508	40085771	40113015	40140242	40167450	40194641	40221814	18
2.615 -0.0	40248968	40276105	40303224	40330325	40357408	40384472	40411519	40438548	40465559	40492552	18
2.616 -0.0	40519526	40546483	40573422	40600342	40627244	40654129	40680995	40707843	40734673	40761485	18
2.617 -0.0	40788278	40815054	40841811	40868551	40895272	40921975	40948659	40975326	41001974	41028604	18
2.618 -0.0	41055216	41081809	41108385	41134942	41161480	41188001	41214503	41240987	41267453	41293900	18
2.619 -0.0	41320329	41346740	41373132	41399506	41425861	41452199	41478518	41504818	41531100	41557364	18
2.620 -0.0	41583609	41609836	41636044	41662234	41688406	41714559	41740694	41766810	41792907	41818987	18
2.621 -0.0	41845047	41871089	41897113	41923118	41949105	41975073	42001022	42026953	42052866	42078760	19
2.622 -0.0	42104635	42130491	42156329	42182149	42207950	42233732	42259495	42285240	42310966	42336674	19
2.623 -0.0	42362363	42388033	42413684	42439317	42464931	42490527	42516103	42541661	42567200	42592721	19
2.624 -0.0	42618223	42643705	42669170	42694615	42720041	42745449	42770838	42796208	42821559	42846892	19
2.625 -0.0	42872206	42897500	42922776	42948033	42973271	42998491	43023691	43048872	43074035	43099179	19
2.626 -0.0	43124303	43149409	43174496	43199564	43224612	43249642	43274653	43299645	43324618	43349572	19
2.627 -0.0	43374507	43399423	43424320	43449198	43474057	43498896	43523717	43548519	43573301	43598065	19
2.628 -0.0	43622809	43647534	43672240	43696927	43721595	43746244	43770874	43795484	43820075	43844647	19
2.629 -0.0	43869200	43893734	43918249	43942744	43967220	43991677	44016115	44040533	44064932	44089312	19
2.630 -0.0	44113673	44138015	44162337	44186640	44210923	44235188	44259433	44283658	44307865	44332052	19
2.631 -0.0	44356219	44380368	44404497	44428606	44452696	44476767	44500819	44524851	44548864	44572857	19
2.632 -0.0	44596831	44620785	44644720	44668636	44692532	44716408	44740265	44764103	44787921	44811720	19
2.633 -0.0	44835499	44859259	44882999	44906720	44930421	44954103	44977765	45001407	45025030	45048634	20
2.634 -0.0	45072217	45095782	45119326	45142851	45166357	45189843	45213309	45236755	45260182	45283589	20
2.635 -0.0	45306977	45330345	45353693	45377022	45400331	45423620	45446890	45470140	45493370	45516580	20
2.636 -0.0	45539771	45562942	45586093	45609224	45632336	45655428	45678500	45701552	45724585	45747598	20
2.637 -0.0	45770591	45793564	45816517	45839451	45862364	45885258	45908132	45930986	45953820	45976635	20
2.638 -0.0	45999429	46022204	46044959	46067693	46090408	46113103	46135778	46158434	46181069	46203684	20
2.639 -0.0	46226279	46248855	46271410	46293945	46316461	46338956	46361432	46383887	46406322	46428738	20
2.640 -0.0	46451133	46473508	46495864	46518199	46540514	46562809	46585084	46607339	46629574	46651789	20
2.641 -0.0	46673984	46696158	46718313	46740447	46762561	46784655	46806729	46828783	46850817	46872830	20
2.642 -0.0	46894823	46916796	46938749	46960682	46982595	47004487	47026359	47048211	47070043	47091854	20
2.643 -0.0	47113645	47135416	47157167	47178898	47200608	47222298	47243967	47265617	47287246	47308854	20
2.644 -0.0	47330443	47352011	47373559	47395086	47416593	47438080	47459546	47480992	47502418	47523823	20
2.645 -0.0	47545208	47566573	47587917	47609241	47630544	47651827	47673090	47694332	47715553	47736755	20
2.646 -0.0	47757935	47779096	47800235	47821355	47842454	47863532	47884590	47905627	47926644	47947641	21
2.647 -0.0	47968617	47989572	48010507	48031421	48052315	48073188	48094041	48114873	48135685	48156476	21
2.648 -0.0	48177246	48197996	48218725	48239434	48260122	48280789	48301436	48322062	48342668	48363253	21
2.649 -0.0	48383817	48404360	48424883	48445386	48465867	48486328	48506768	48527188	48547587	48567965	21

TABLE 6

Sinc Function

x		.0	.0001	.0002	.0003	.0004	.0005	.0006	.0007	.0008	.0009	Δ_2
2.650	-0.0	48588322	48608659	48628975	48649270	48669545	48689798	48710031	48730244	48750435	48770606	21
2.651	-0.0	48790756	48810885	48830993	48851081	48871148	48891194	48911219	48931223	48951207	48971169	21
2.652	-0.0	48991111	49011032	49030932	49050812	49070670	49090508	49110324	49130120	49149895	49169649	21
2.653	-0.0	49189382	49209095	49228786	49248456	49268106	49287734	49307342	49326929	49346494	49366039	21
2.654	-0.0	49385563	49405066	49424548	49444006	49463448	49482867	49502265	49521642	49540998	49560333	21
2.655	-0.0	49579647	49598939	49618211	49637462	49656692	49675900	49695088	49714254	49733400	49752524	21
2.656	-0.0	49771628	49790710	49809771	49828811	49847830	49866828	49885805	49904760	49923695	49942608	21
2.657	-0.0	49961500	49980371	49999221	50018050	50036857	50055644	50074409	50093153	50111876	50130577	21
2.658	-0.0	50149258	50167917	50186555	50205172	50223768	50242342	50260895	50279427	50297938	50316427	21
2.659	-0.0	50334896	50353342	50371768	50390172	50408556	50426917	50445258	50463577	50481875	50500152	21
2.660	-0.0	50518407	50536641	50554854	50573045	50591215	50609364	50627491	50645597	50663682	50681745	21
2.661	-0.0	50699787	50717808	50735807	50753785	50771741	50789676	50807590	50825482	50843353	50861202	21
2.662	-0.0	50879030	50896837	50914622	50932386	50950128	50967849	50985548	51003226	51020882	51038517	21
2.663	-0.0	51056131	51073723	51091293	51108842	51126370	51143876	51161360	51178823	51196265	51213685	22
2.664	-0.0	51231083	51248460	51265816	51283150	51300462	51317753	51335022	51352270	51369496	51386700	22
2.665	-0.0	51403883	51421044	51438184	51455302	51472399	51489474	51506527	51523559	51540569	51557558	22
2.666	-0.0	51574524	51591470	51608393	51625295	51642176	51659034	51675871	51692687	51709480	51726252	22
2.667	-0.0	51743003	51759731	51776438	51793124	51809787	51826429	51843049	51859648	51876224	51892779	22
2.668	-0.0	51909313	51925824	51942314	51958782	51975228	51991653	52008056	52024437	52040796	52057134	22
2.669	-0.0	52073449	52089743	52106015	52122266	52138494	52154701	52170886	52187049	52203191	52219310	22
2.670	-0.0	52235408	52251484	52267538	52283570	52299581	52315569	52331536	52347481	52363404	52379305	22
2.671	-0.0	52395184	52411042	52426877	52442691	52458483	52474252	52490000	52505726	52521431	52537113	22
2.672	-0.0	52552773	52568412	52584028	52599623	52615195	52630746	52646275	52661782	52677266	52692729	22
2.673	-0.0	52708170	52723589	52738986	52754361	52769714	52785045	52800355	52815642	52830907	52846150	22
2.674	-0.0	52861371	52876570	52891747	52906902	52922035	52937146	52952235	52967302	52982347	52997370	22
2.675	-0.0	53012371	53027350	53042307	53057241	53072154	53087045	53101913	53116759	53131584	53146386	22
2.676	-0.0	53161166	53175924	53190660	53205374	53220066	53234736	53249383	53264009	53278612	53293193	22
2.677	-0.0	53307752	53322289	53336804	53351297	53365767	53380216	53394642	53409046	53423428	53437788	22
2.678	-0.0	53452125	53466441	53480734	53495005	53509254	53523481	53537685	53551867	53566027	53580165	22
2.679	-0.0	53594281	53608375	53622446	53636495	53650522	53664526	53678509	53692469	53706407	53720322	22
2.680	-0.0	53734216	53748087	53761936	53775762	53789567	53803349	53817109	53830846	53844562	53858255	22
2.681	-0.0	53871925	53885574	53899200	53912804	53926385	53939945	53953482	53966996	53980489	53993959	22
2.682	-0.0	54007406	54020832	54034235	54047616	54060974	54074310	54087624	54100915	54114184	54127431	22
2.683	-0.0	54140655	54153857	54167037	54180194	54193329	54206441	54219531	54232599	54245644	54258667	22
2.684	-0.0	54271668	54284646	54297602	54310535	54323446	54336335	54349201	54362045	54374866	54387665	22
2.685	-0.0	54400441	54413195	54425927	54438636	54451323	54463987	54476629	54489249	54501846	54514420	22
2.686	-0.0	54526972	54539502	54552009	54564494	54576956	54589396	54601813	54614208	54626580	54638930	22
2.687	-0.0	54651257	54663562	54675844	54688104	54700341	54712556	54724748	54736918	54749065	54761190	23
2.688	-0.0	54773292	54785372	54797429	54809464	54821476	54833466	54845433	54857377	54869299	54881199	23
2.689	-0.0	54893076	54904930	54916762	54928571	54940358	54952122	54963863	54975582	54987279	54998952	23
2.690	-0.0	55010604	55022232	55033838	55045422	55056983	55068521	55080037	55091530	55103000	55114448	23
2.691	-0.0	55125873	55137276	55148656	55160013	55171348	55182660	55193950	55205217	55216461	55227683	23
2.692	-0.0	55238882	55250058	55261212	55272343	55283451	55294537	55305600	55316641	55327659	55338654	23
2.693	-0.0	55349626	55360576	55371503	55382408	55393290	55404149	55414985	55425799	55436590	55447359	23
2.694	-0.0	55458105	55468828	55479528	55490206	55500861	55511493	55522102	55532689	55543253	55553795	23
2.695	-0.0	55564314	55574810	55585283	55595734	55606161	55616566	55626949	55637309	55647645	55657960	23
2.696	-0.0	55668251	55678520	55688766	55698989	55709190	55719367	55729522	55739655	55749764	55759851	23
2.697	-0.0	55769915	55779956	55789974	55799970	55809943	55819893	55829820	55839725	55849607	55859466	23
2.698	-0.0	55869302	55879115	55888906	55898674	55908419	55918141	55927841	55937518	55947171	55956803	23
2.699	-0.0	55966411	55975996	55985559	55995099	56004616	56014110	56023582	56033030	56042456	56051859	23

$$\frac{\sin 2\pi x}{2\pi x}$$

TABLE 6

Sinc Function

x		.0	.0001	.0002	.0003	.0004	.0005	.0006	.0007	.0008	.0009	Δ_2
2.700	-0.0	56061239	56070596	56079931	56089243	56098531	56107797	56117041	56126261	56135458	56144633	23
2.701	-0.0	56153785	56162914	56172020	56181103	56190164	56199201	56208216	56217208	56226177	56235123	23
2.702	-0.0	56244046	56252946	56261824	56270679	56279510	56288319	56297105	56305869	56314609	56323326	23
2.703	-0.0	56332021	56340692	56349341	56357967	56366570	56375150	56383707	56392242	56400753	56409242	23
2.704	-0.0	56417707	56426150	56434570	56442967	56451341	56459692	56468020	56476326	56484608	56492868	23
2.705	-0.0	56501104	56509318	56517509	56525677	56533821	56541943	56550043	56558119	56566172	56574202	23
2.706	-0.0	56582210	56590194	56598156	56606094	56614010	56621903	56629772	56637619	56645443	56653244	23
2.707	-0.0	56661022	56668777	56676509	56684219	56691905	56699568	56707209	56714826	56722420	56729992	23
2.708	-0.0	56737541	56745066	56752569	56760048	56767505	56774939	56782350	56789738	56797102	56804444	23
2.709	-0.0	56811763	56819059	56826332	56833582	56840809	56848013	56855195	56862353	56869488	56876600	23
2.710	-0.0	56883689	56890755	56897799	56904819	56911816	56918791	56925742	56932670	56939576	56946458	23
2.711	-0.0	56953317	56960154	56966967	56973757	56980525	56987269	56993991	57000689	57007365	57014017	23
2.712	-0.0	57020646	57027253	57033836	57040397	57046934	57053449	57059940	57066408	57072854	57079276	23
2.713	-0.0	57085676	57092052	57098405	57104736	57111043	57117328	57123589	57129827	57136043	57142235	23
2.714	-0.0	57148404	57154550	57160674	57166774	57172851	57178905	57184937	57190945	57196930	57202892	23
2.715	-0.0	57208831	57214747	57220640	57226510	57232357	57238181	57243982	57249760	57255515	57261247	23
2.716	-0.0	57266956	57272642	57278305	57283944	57289561	57295155	57300726	57306273	57311798	57317300	23
2.717	-0.0	57322778	57328234	57333666	57339076	57344462	57349825	57355166	57360483	57365777	57371049	23
2.718	-0.0	57376297	57381522	57386724	57391903	57397059	57402193	57407303	57412389	57417453	57422494	23
2.719	-0.0	57427512	57432507	57437479	57442427	57447353	57452256	57457135	57461992	57466825	57471636	23
2.720	-0.0	57476423	57481188	57485929	57490647	57495343	57500015	57504664	57509290	57513893	57518473	23
2.721	-0.0	57523030	57527564	57532075	57536563	57541028	57545470	57549889	57554284	57558657	57563007	23
2.722	-0.0	57567333	57571637	57575917	57580175	57584409	57588620	57592809	57596974	57601116	57605235	23
2.723	-0.0	57609331	57613405	57617455	57621482	57625486	57629466	57633424	57637359	57641271	57645160	23
2.724	-0.0	57649025	57652868	57656688	57660484	57664258	57668008	57671736	57675440	57679122	57682780	23
2.725	-0.0	57686415	57690028	57693617	57697183	57700726	57704246	57707743	57711217	57714668	57718096	23
2.726	-0.0	57721501	57724883	57728242	57731578	57734890	57738180	57741447	57744691	57747911	57751109	23
2.727	-0.0	57754283	57757435	57760563	57763669	57766751	57769811	57772847	57775861	57778851	57781818	23
2.728	-0.0	57784762	57787684	57790582	57793457	57796309	57799138	57801944	57804728	57807488	57810225	23
2.729	-0.0	57812939	57815630	57818298	57820943	57823565	57826163	57828739	57831292	57833822	57836329	23
2.730	-0.0	57838813	57841274	57843711	57846126	57848518	57850887	57853232	57855555	57857855	57860132	23
2.731	-0.0	57862385	57864616	57866824	57869008	57871170	57873309	57875425	57877517	57879587	57881634	23
2.732	-0.0	57883657	57885658	57887636	57889590	57891522	57893431	57895316	57897179	57899019	57900836	23
2.733	-0.0	57902629	57904400	57906148	57907873	57909574	57911253	57912909	57914542	57916152	57917738	23
2.734	-0.0	57919302	57920843	57922361	57923856	57925328	57926777	57928203	57929606	57930986	57932343	23
2.735	-0.0	57933677	57934989	57936277	57937542	57938784	57940004	57941200	57942373	57943524	57944651	23
2.736	-0.0	57945756	57946837	57947896	57948932	57949944	57950934	57951901	57952845	57953766	57954664	23
2.737	-0.0	57955539	57956391	57957220	57958026	57958809	57959570	57960307	57961021	57961713	57962381	23
2.738	-0.0	57963027	57963650	57964250	57964826	57965380	57965911	57966419	57966905	57967367	57967806	23
2.739	-0.0	57968223	57968616	57968987	57969334	57969659	57969961	57970240	57970496	57970729	57970939	23
2.740	-0.0	57971127	57971291	57971433	57971551	57971647	57971720	57971770	57971797	57971801	57971783	23
2.741	-0.0	57971741	57971676	57971589	57971479	57971346	57971190	57971011	57970809	57970585	57970337	23
2.742	-0.0	57970067	57969774	57969458	57969119	57968757	57968372	57967965	57967535	57967081	57966605	23
2.743	-0.0	57966106	57965585	57965040	57964473	57963882	57963269	57962633	57961975	57961293	57960588	23
2.744	-0.0	57959861	57959111	57958338	57957542	57956724	57955882	57955018	57954131	57953221	57952289	23
2.745	-0.0	57951333	57950355	57949354	57948330	57947283	57946214	57945122	57944006	57942869	57941708	23
2.746	-0.0	57940524	57939318	57938089	57936837	57935563	57934265	57932945	57931602	57930237	57928848	23
2.747	-0.0	57927437	57926003	57924546	57923067	57921565	57920040	57918492	57916921	57915328	57913712	23
2.748	-0.0	57912073	57910412	57908727	57907020	57905291	57903538	57901763	57899965	57898144	57896301	23
2.749	-0.0	57894435	57892546	57890635	57888700	57886743	57884764	57882761	57880736	57878688	57876618	23

TABLE 6

Sinc Function

x		.0	.0001	.0002	.0003	.0004	.0005	.0006	.0007	.0008	.0009	Δ_2
2.750	-0.0	57874525	57872409	57870270	57868109	57865925	57863718	57861489	57859237	57856963	57854665	23
2.751	-0.0	57852345	57850002	57847637	57845249	57842838	57840405	57837949	57835470	57832969	57830445	23
2.752	-0.0	57827899	57825329	57822737	57820123	57817486	57814826	57812144	57809439	57806711	57803961	23
2.753	-0.0	57801188	57798392	57795574	57792733	57789870	57786984	57784075	57781144	57778190	57775214	23
2.754	-0.0	57772215	57769194	57766149	57763083	57759993	57756881	57753747	57750590	57747410	57744208	23
2.755	-0.0	57740984	57737736	57734466	57731174	57727859	57724522	57721162	57717779	57714374	57710946	23
2.756	-0.0	57707496	57704023	57700528	57697010	57693470	57689907	57686322	57682714	57679084	57675431	23
2.757	-0.0	57671755	57668058	57664337	57660594	57656829	57653041	57649231	57645398	57641543	57637665	22
2.758	-0.0	57633765	57629842	57625897	57621929	57617939	57613927	57609892	57605834	57601754	57597652	22
2.759	-0.0	57593527	57589380	57585210	57581018	57576804	57572567	57568308	57564026	57559722	57555395	22
2.760	-0.0	57551046	57546675	57542281	57537865	57533426	57528965	57524482	57519976	57515448	57510897	22
2.761	-0.0	57506324	57501729	57497112	57492472	57487809	57483124	57478417	57473688	57468936	57464162	22
2.762	-0.0	57459366	57454547	57449706	57444842	57439957	57435049	57430118	57425165	57420190	57415193	22
2.763	-0.0	57410173	57405131	57400067	57394981	57389872	57384741	57379587	57374412	57369214	57363994	22
2.764	-0.0	57358751	57353486	57348199	57342890	57337559	57332205	57326829	57321430	57316010	57310567	22
2.765	-0.0	57305102	57299615	57294106	57288574	57283020	57277444	57271846	57266225	57260583	57254918	22
2.766	-0.0	57249231	57243522	57237790	57232037	57226261	57220463	57214643	57208800	57202936	57197049	22
2.767	-0.0	57191140	57185210	57179256	57173281	57167284	57161264	57155223	57149159	57143073	57136965	22
2.768	-0.0	57130835	57124683	57118509	57112312	57106094	57099853	57093590	57087306	57080999	57074670	22
2.769	-0.0	57068319	57061946	57055550	57049133	57042694	57036233	57029749	57023244	57016716	57010167	22
2.770	-0.0	57003595	56997002	56990386	56983748	56977089	56970407	56963704	56956978	56950230	56943461	22
2.771	-0.0	56936669	56929855	56923020	56916162	56909283	56902381	56895458	56888512	56881545	56874555	22
2.772	-0.0	56867544	56860511	56853455	56846378	56839279	56832158	56825015	56817851	56810664	56803455	22
2.773	-0.0	56796225	56788972	56781698	56774402	56767083	56759743	56752382	56744998	56737592	56730165	22
2.774	-0.0	56722715	56715244	56707751	56700236	56692699	56685141	56677560	56669958	56662334	56654688	22
2.775	-0.0	56647020	56639331	56631620	56623886	56616132	56608355	56600556	56592736	56584894	56577030	22
2.776	-0.0	56569145	56561237	56553308	56545357	56537385	56529390	56521374	56513336	56505277	56497195	22
2.777	-0.0	56489092	56480968	56472821	56464653	56456463	56448252	56440018	56431763	56423487	56415188	22
2.778	-0.0	56406868	56398527	56390164	56381779	56373372	56364944	56356494	56348022	56339529	56331014	22
2.779	-0.0	56322478	56313920	56305340	56296739	56288116	56279471	56270805	56262118	56253408	56244677	22
2.780	-0.0	56235925	56227151	56218355	56209538	56200700	56191839	56182958	56174054	56165129	56156183	22
2.781	-0.0	56147215	56138226	56129215	56120182	56111128	56102053	56092956	56083838	56074698	56065536	21
2.782	-0.0	56056353	56047149	56037923	56028676	56019407	56010117	56000805	55991472	55982118	55972742	21
2.783	-0.0	55963345	55953926	55944486	55935024	55925541	55916037	55906511	55896964	55887395	55877805	21
2.784	-0.0	55868194	55858562	55848908	55839232	55829535	55819817	55810078	55800317	55790535	55780732	21
2.785	-0.0	55770907	55761061	55751194	55741305	55731395	55721464	55711512	55701538	55691543	55681527	21
2.786	-0.0	55671489	55661430	55651350	55641249	55631127	55620983	55610818	55600631	55590424	55580195	21
2.787	-0.0	55569946	55559674	55549382	55539069	55528734	55518378	55508001	55497603	55487184	55476743	21
2.788	-0.0	55466282	55455799	55445295	55434770	55424224	55413656	55403068	55392458	55381828	55371176	21
2.789	-0.0	55360503	55349809	55339094	55328358	55317601	55306823	55296023	55285203	55274362	55263499	21
2.790	-0.0	55252616	55241711	55230786	55219839	55208872	55197883	55186873	55175843	55164791	55153719	21
2.791	-0.0	55142625	55131511	55120375	55109219	55098041	55086843	55075624	55064383	55053122	55041840	21
2.792	-0.0	55030537	55019213	55007863	54996503	54985116	54973709	54962280	54950831	54939361	54927870	21
2.793	-0.0	54916358	54904825	54893271	54881697	54870102	54858486	54846849	54835191	54823512	54811813	21
2.794	-0.0	54800093	54788352	54776590	54764808	54753004	54741180	54729335	54717470	54705583	54693676	21
2.795	-0.0	54681748	54669800	54657831	54645841	54633830	54621798	54609746	54597673	54585580	54573466	21
2.796	-0.0	54561331	54549175	54536999	54524802	54512585	54500346	54488088	54475808	54463508	54451187	21
2.797	-0.0	54438846	54426484	54414101	54401698	54389275	54376830	54364365	54351880	54339374	54326847	21
2.798	-0.0	54314300	54301733	54289144	54276536	54263906	54251257	54238586	54225896	54213184	54200452	20
2.799	-0.0	54187700	54174927	54162134	54149321	54136486	54123632	54110757	54097861	54084945	54072009	20

TABLE 6

Sinc Function

x	.0	.0001	.0002	.0003	.0004	.0005	.0006	.0007	.0008	.0009	Δ_2
2.800 -0.0	54059052	54046075	54033077	54020059	54007021	53993962	53980883	53967783	53954663	53941523	20
2.801 -0.0	53928362	53915181	53901980	53888758	53875516	53862254	53848971	53835668	53822345	53809002	20
2.802 -0.0	53795638	53782254	53768849	53755424	53741980	53728514	53715029	53701523	53687997	53674451	20
2.803 -0.0	53660885	53647298	53633691	53620064	53606417	53592750	53579062	53565354	53551627	53537878	20
2.804 -0.0	53524110	53510322	53496513	53482685	53468836	53454967	53441078	53427169	53413240	53399290	20
2.805 -0.0	53385321	53371331	53357322	53343292	53329243	53315173	53301083	53286973	53272844	53258694	20
2.806 -0.0	53244524	53230334	53216124	53201894	53187644	53173374	53159085	53144775	53130445	53116095	20
2.807 -0.0	53101726	53087336	53072927	53058497	53044048	53029578	53015089	53000580	52986051	52971502	20
2.808 -0.0	52956934	52942345	52927737	52913108	52898460	52883792	52869104	52854397	52839669	52824922	20
2.809 -0.0	52810155	52795368	52780561	52765735	52750889	52736023	52721137	52706231	52691306	52676361	20
2.810 -0.0	52661396	52646412	52631408	52616384	52601340	52586277	52571194	52556091	52540969	52525827	20
2.811 -0.0	52510665	52495484	52480283	52465063	52449822	52434563	52419283	52403984	52388666	52373327	20
2.812 -0.0	52357969	52342592	52327195	52311779	52296343	52280887	52265412	52249917	52234403	52218869	20
2.813 -0.0	52203316	52187743	52172151	52156539	52140908	52125257	52109587	52093897	52078188	52062460	19
2.814 -0.0	52046712	52030945	52015158	51999352	51983526	51967681	51951817	51935933	51920030	51904107	19
2.815 -0.0	51888165	51872204	51856223	51840224	51824204	51808166	51792108	51776031	51759934	51743819	19
2.816 -0.0	51727684	51711529	51695356	51679163	51662951	51646719	51630469	51614199	51597910	51581602	19
2.817 -0.0	51565274	51548928	51532562	51516177	51499773	51483349	51466907	51450445	51433964	51417464	19
2.818 -0.0	51400945	51384407	51367850	51351273	51334678	51318063	51301430	51284777	51268105	51251414	19
2.819 -0.0	51234704	51217976	51201228	51184461	51167675	51150870	51134046	51117203	51100341	51083460	19
2.820 -0.0	51066560	51049641	51032703	51015746	50998770	50981776	50964762	50947730	50930678	50913608	19
2.821 -0.0	50896519	50879411	50862284	50845138	50827973	50810790	50793587	50776366	50759126	50741867	19
2.822 -0.0	50724590	50707293	50689978	50672644	50655291	50637919	50620529	50603120	50585692	50568246	19
2.823 -0.0	50550780	50533296	50515794	50498272	50480732	50463173	50445596	50428000	50410385	50392752	19
2.824 -0.0	50375100	50357429	50339739	50322031	50304305	50286560	50268796	50251013	50233213	50215393	19
2.825 -0.0	50197555	50179698	50161823	50143929	50126017	50108086	50090137	50072169	50054183	50036178	18
2.826 -0.0	50018155	50000113	49982053	49963974	49945877	49927762	49909628	49891476	49873305	49855116	18
2.827 -0.0	49836908	49818682	49800438	49782175	49763894	49745595	49727277	49708941	49690586	49672214	18
2.828 -0.0	49653822	49635413	49616985	49598540	49580075	49561593	49543092	49524573	49506036	49487480	18
2.829 -0.0	49468907	49450315	49431705	49413076	49394430	49375765	49357082	49338381	49319662	49300925	18
2.830 -0.0	49282170	49263396	49244604	49225794	49206967	49188121	49169256	49150374	49131474	49112556	18
2.831 -0.0	49093619	49074665	49055692	49036702	49017694	48998667	48979623	48960560	48941480	48922381	18
2.832 -0.0	48903265	48884130	48864978	48845808	48826620	48807414	48788190	48768948	48749688	48730410	18
2.833 -0.0	48711115	48691801	48672470	48653121	48633754	48614369	48594967	48575546	48556108	48536652	18
2.834 -0.0	48517178	48497686	48478177	48458650	48439105	48419542	48399962	48380364	48360748	48341114	18
2.835 -0.0	48321463	48301794	48282108	48262403	48242682	48222942	48203185	48183410	48163617	48143807	18
2.836 -0.0	48123980	48104134	48084271	48064391	48044493	48024577	48004644	47984693	47964725	47944739	18
2.837 -0.0	47924736	47904715	47884677	47864621	47844548	47824457	47804349	47784223	47764080	47743919	17
2.838 -0.0	47723741	47703546	47683333	47663103	47642855	47622590	47602308	47582008	47561691	47541357	17
2.839 -0.0	47521005	47500635	47480250	47459846	47439425	47418987	47398531	47378058	47357568	47337061	17
2.840 -0.0	47316536	47295994	47275435	47254859	47234266	47213655	47193027	47172382	47151720	47131040	17
2.841 -0.0	47110344	47089630	47068899	47048152	47027387	47006604	46985805	46964989	46944156	46923305	17
2.842 -0.0	46902438	46881553	46860651	46839733	46818797	46797845	46776875	46755888	46734885	46713864	17
2.843 -0.0	46692826	46671772	46650701	46629612	46608507	46587385	46566245	46545089	46523917	46502727	17
2.844 -0.0	46481520	46460297	46439056	46417799	46396525	46375234	46353926	46332602	46311261	46289903	17
2.845 -0.0	46268528	46247136	46225728	46204303	46182861	46161403	46139927	46118435	46096927	46075401	17
2.846 -0.0	46053859	46032301	46010725	45989133	45967525	45945899	45924258	45902599	45880924	45859232	17
2.847 -0.0	45837524	45815799	45794058	45772300	45750526	45728735	45706927	45685103	45663262	45641405	16
2.848 -0.0	45619532	45597642	45575736	45553813	45531873	45509917	45487945	45465957	45443952	45421930	16
2.849 -0.0	45399892	45377838	45355768	45333681	45311578	45289456	45267322	45245170	45223001	45200816	16

TABLE 6

Sinc Function

x	.0	.0001	.0002	.0003	.0004	.0005	.0006	.0007	.0008	.0009	Δ_2
2.850 -0.0	45178615	45156398	45134164	45111914	45089648	45067366	45045067	45022752	45000421	44978074	16
2.851 -0.0	44955711	44933331	44910935	44888523	44866095	44843651	44821190	44798714	44776221	44753713	16
2.852 -0.0	44731188	44708647	44686090	44663517	44640928	44618323	44595702	44573065	44550412	44527742	16
2.853 -0.0	44505057	44482356	44459639	44436906	44414157	44391392	44368611	44345815	44323002	44300173	16
2.854 -0.0	44277329	44254469	44231592	44208700	44185792	44162869	44139929	44116974	44094002	44071015	16
2.855 -0.0	44048013	44024994	44001960	43978910	43955844	43932762	43909665	43886552	43863423	43840279	16
2.856 -0.0	43817119	43793943	43770751	43747544	43724321	43701083	43677829	43654559	43631274	43607973	16
2.857 -0.0	43584657	43561325	43537977	43514614	43491236	43467841	43444432	43421007	43397566	43374110	15
2.858 -0.0	43350638	43327151	43303648	43280130	43256597	43233048	43209483	43185904	43162308	43138698	15
2.859 -0.0	43115072	43091431	43067774	43044102	43020415	42996712	42972994	42949261	42925512	42901748	15
2.860 -0.0	42877969	42854175	42830365	42806540	42782700	42758844	42734974	42711088	42687187	42663271	15
2.861 -0.0	42639340	42615393	42591432	42567455	42543463	42519456	42495434	42471397	42447344	42423277	15
2.862 -0.0	42399194	42375097	42350984	42326857	42302714	42278557	42254384	42230196	42205994	42181776	15
2.863 -0.0	42157544	42133296	42109034	42084756	42060464	42036157	42011835	41987498	41963146	41938779	15
2.864 -0.0	41914398	41890002	41865590	41841164	41816723	41792268	41767797	41743312	41718812	41694297	15
2.865 -0.0	41669768	41645224	41620665	41596091	41571502	41546899	41522282	41497649	41473002	41448340	15
2.866 -0.0	41423664	41398973	41374267	41349547	41324812	41300063	41275299	41250520	41225727	41200919	15
2.867 -0.0	41176097	41151260	41126409	41101543	41076663	41051768	41026859	41001935	40976997	40952044	14
2.868 -0.0	40927077	40902096	40877100	40852090	40827065	40802027	40776973	40751906	40726824	40701727	14
2.869 -0.0	40676617	40651492	40626352	40601199	40576031	40550849	40525653	40500442	40475217	40449978	14
2.870 -0.0	40424725	40399457	40374176	40348880	40323570	40298246	40272908	40247555	40222189	40196808	14
2.871 -0.0	40171413	40146005	40120582	40095145	40069694	40044229	40018749	39993256	39967749	39942228	14
2.872 -0.0	39916693	39891144	39865581	39840004	39814413	39788808	39763189	39737556	39711909	39686249	14
2.873 -0.0	39660574	39634886	39609184	39583468	39557738	39531994	39506237	39480466	39454680	39428882	14
2.874 -0.0	39403069	39377243	39351402	39325549	39299681	39273800	39247905	39221996	39196073	39170137	14
2.875 -0.0	39144188	39118224	39092247	39066257	39040252	39014235	38988203	38962158	38936099	38910027	14
2.876 -0.0	38883942	38857842	38831730	38805603	38779463	38753310	38727143	38700963	38674770	38648562	13
2.877 -0.0	38622342	38596108	38569860	38543600	38517325	38491038	38464737	38438423	38412095	38385754	13
2.878 -0.0	38359400	38333032	38306651	38280257	38253850	38227429	38200995	38174548	38148087	38121613	13
2.879 -0.0	38095126	38068626	38042113	38015587	37989047	37962494	37935928	37909349	37882757	37856152	13
2.880 -0.0	37829533	37802902	37776257	37749600	37722929	37696245	37669548	37642839	37616116	37589380	13
2.881 -0.0	37562631	37535870	37509095	37482307	37455507	37428693	37401867	37375028	37348176	37321311	13
2.882 -0.0	37294433	37267542	37240638	37213722	37186792	37159850	37132895	37105928	37078947	37051954	13
2.883 -0.0	37024948	36997929	36970898	36943853	36916797	36889727	36862645	36835550	36808442	36781322	13
2.884 -0.0	36754189	36727043	36699885	36672714	36645531	36618335	36591127	36563906	36536672	36509426	13
2.885 -0.0	36482167	36454896	36427612	36400316	36373007	36345686	36318353	36291007	36263648	36236277	12
2.886 -0.0	36208894	36181499	36154090	36126670	36099237	36071792	36044335	36016865	35989383	35961888	12
2.887 -0.0	35934382	35906863	35879331	35851788	35824232	35796664	35769084	35741491	35713887	35686270	12
2.888 -0.0	35658641	35631000	35603346	35575681	35548003	35520314	35492612	35464898	35437172	35409434	12
2.889 -0.0	35381684	35353922	35326147	35298361	35270563	35242753	35214931	35187096	35159250	35131392	12
2.890 -0.0	35103522	35075640	35047746	35019840	34991923	34963993	34936052	34908098	34880133	34852156	12
2.891 -0.0	34824167	34796167	34768154	34740130	34712094	34684046	34655987	34627916	34599833	34571738	12
2.892 -0.0	34543632	34515514	34487384	34459242	34431089	34402925	34374748	34346560	34318361	34290149	12
2.893 -0.0	34261927	34233692	34205446	34177189	34148920	34120639	34092347	34064044	34035729	34007402	11
2.894 -0.0	33979064	33950715	33922354	33893981	33865598	33837203	33808796	33780378	33751949	33723508	11
2.895 -0.0	33695056	33666593	33638118	33609632	33581135	33552626	33524106	33495575	33467033	33438479	11
2.896 -0.0	33409914	33381338	33352751	33324153	33295543	33266922	33238290	33209647	33180993	33152327	11
2.897 -0.0	33123651	33094963	33066265	33037555	33008834	32980102	32951360	32922606	32893841	32865065	11
2.898 -0.0	32836278	32807480	32778671	32749851	32721021	32692179	32663327	32634463	32605589	32576703	11
2.899 -0.0	32547807	32518900	32489983	32461054	32432114	32403164	32374203	32345231	32316249	32287255	11

$$\frac{\sin 2\pi x}{2\pi x}$$

$\frac{\sin 2\pi x}{2\pi x}$

TABLE 6

Sinc Function

x		.0	.0001	.0002	.0003	.0004	.0005	.0006	.0007	.0008	.0009	Δ_2
2.900	-0.0	32258251	32229236	32200211	32171174	32142127	32113070	32084001	32054922	32025833	31996732	11
2.901	-0.0	31967622	31938500	31909368	31880225	31851072	31821908	31792734	31763549	31734353	31705147	11
2.902	-0.0	31675931	31646704	31617466	31588218	31558960	31529691	31500412	31471122	31441822	31412511	10
2.903	-0.0	31383191	31353859	31324518	31295166	31265804	31236431	31207048	31177655	31148251	31118838	10
2.904	-0.0	31089414	31059979	31030535	31001080	30971615	30942140	30912655	30883159	30853654	30824138	10
2.905	-0.0	30794612	30765076	30735530	30705974	30676407	30646831	30617244	30587648	30558041	30528425	10
2.906	-0.0	30498798	30469161	30439515	30409858	30380192	30350515	30320829	30291132	30261426	30231710	10
2.907	-0.0	30201984	30172248	30142502	30112746	30082981	30053206	30023420	29993626	29963821	29934006	10
2.908	-0.0	29904182	29874348	29844504	29814651	29784787	29754914	29725032	29695140	29665238	29635326	10
2.909	-0.0	29605405	29575474	29545533	29515583	29485623	29455654	29425675	29395687	29365689	29335681	10
2.910	-0.0	29305664	29275638	29245602	29215556	29185501	29155437	29125363	29095280	29065187	29035085	9
2.911	-0.0	29004973	28974852	28944722	28914583	28884434	28854275	28824108	28793931	28763744	28733549	9
2.912	-0.0	28703344	28673130	28642907	28612674	28582433	28552182	28521922	28491652	28461374	28431086	9
2.913	-0.0	28400789	28370484	28340168	28309844	28279511	28249169	28218817	28188457	28158088	28127709	9
2.914	-0.0	28097321	28066925	28036519	28006105	27975681	27945249	27914808	27884357	27853898	27823430	9
2.915	-0.0	27792953	27762467	27731972	27701469	27670956	27640435	27609905	27579366	27548818	27518261	9
2.916	-0.0	27487696	27457122	27426539	27395948	27365348	27334739	27304121	27273495	27242860	27212216	9
2.917	-0.0	27181564	27150903	27120234	27089555	27058869	27028173	26997470	26966757	26936036	26905307	9
2.918	-0.0	26874569	26843822	26813067	26782304	26751532	26720752	26689963	26659166	26628360	26597546	8
2.919	-0.0	26566724	26535893	26505054	26474206	26443350	26412486	26381613	26350733	26319844	26288946	8
2.920	-0.0	26258041	26227127	26196205	26165274	26134336	26103389	26072434	26041471	26010500	25979520	8
2.921	-0.0	25948533	25917537	25886533	25855521	25824501	25793473	25762437	25731393	25700341	25669281	8
2.922	-0.0	25638213	25607137	25576052	25544960	25513860	25482752	25451636	25420512	25389380	25358241	8
2.923	-0.0	25327093	25295938	25264774	25233603	25202424	25171238	25140043	25108841	25077631	25046413	8
2.924	-0.0	25015187	24983954	24952712	24921464	24890207	24858943	24827671	24796391	24765104	24733809	8
2.925	-0.0	24702507	24671197	24639879	24608554	24577221	24545880	24514532	24483177	24451814	24420443	8
2.926	-0.0	24389065	24357680	24326287	24294886	24263479	24232063	24200641	24169210	24137773	24106328	7
2.927	-0.0	24074876	24043416	24011949	23980475	23948993	23917504	23886008	23854504	23822994	23791476	7
2.928	-0.0	23759950	23728418	23696878	23665331	23633777	23602216	23570648	23539072	23507489	23475900	7
2.929	-0.0	23444303	23412699	23381087	23349469	23317844	23286212	23254572	23222926	23191273	23159612	7
2.930	-0.0	23127945	23096271	23064589	23032901	23001206	22969504	22937795	22906079	22874357	22842627	7
2.931	-0.0	22810891	22779147	22747397	22715640	22683877	22652106	22620329	22588545	22556754	22524956	7
2.932	-0.0	22493152	22461341	22429523	22397699	22365868	22334030	22302186	22270335	22238478	22206613	7
2.933	-0.0	22174743	22142865	22110981	22079091	22047194	22015290	21983380	21951464	21919541	21887611	6
2.934	-0.0	21855675	21823733	21791784	21759828	21727867	21695899	21663924	21631943	21599956	21567962	6
2.935	-0.0	21535962	21503956	21471944	21439925	21407900	21375868	21343831	21311787	21279737	21247680	6
2.936	-0.0	21215618	21183549	21151474	21119393	21087306	21055212	21023113	20991007	20958896	20926778	6
2.937	-0.0	20894654	20862524	20830388	20798246	20766098	20733944	20701784	20669618	20637446	20605268	6
2.938	-0.0	20573084	20540894	20508698	20476497	20444289	20412076	20379856	20347631	20315400	20283163	6
2.939	-0.0	20250921	20218672	20186418	20154158	20121892	20089621	20057344	20025061	19992772	19960478	6
2.940	-0.0	19928178	19895872	19863561	19831244	19798921	19766593	19734259	19701919	19669574	19637224	6
2.941	-0.0	19604868	19572506	19540139	19507766	19475388	19443004	19410615	19378220	19345820	19313414	5
2.942	-0.0	19281003	19248587	19216165	19183738	19151306	19118868	19086425	19053976	19021522	18989063	5
2.943	-0.0	18956599	18924129	18891654	18859174	18826688	18794197	18761702	18729200	18696694	18664183	5
2.944	-0.0	18631666	18599144	18566617	18534085	18501548	18469006	18436459	18403906	18371349	18338786	5
2.945	-0.0	18306219	18273646	18241069	18208486	18175899	18143306	18110709	18078107	18045499	18012887	5
2.946	-0.0	17980270	17947648	17915021	17382390	17849753	17817112	17784466	17751815	17719159	17686498	5
2.947	-0.0	17653833	17621163	17588488	17555809	17523124	17490435	17457742	17425043	17392340	17359633	5
2.948	-0.0	17326921	17294204	17261482	17228756	17196026	17163290	17130551	17097806	17065057	17032304	5
2.949	-0.0	16999546	16966784	16934017	16901246	16868470	16835690	16802905	16770116	16737323	16704525	4

TABLE 6

Sinc Function

x		.0	.0001	.0002	.0003	.0004	.0005	.0006	.0007	.0008	.0009	Δ_2
2.950	-0.0	16671723	16638916	16606105	16573290	16540471	16507647	16474819	16441986	16409150	16376309	4
2.951	-0.0	16343464	16310614	16277761	16244903	16212041	16179175	16146304	16113430	16080551	16047669	4
2.952	-0.0	16014782	15981891	15948996	15916097	15883194	15850287	15817375	15784460	15751541	15718618	4
2.953	-0.0	15685691	15652760	15619824	15586885	15553943	15520996	15488045	15455090	15422132	15389169	4
2.954	-0.0	15356203	15323233	15290259	15257282	15224300	15191315	15158326	15125333	15092337	15059337	4
2.955	-0.0	15026333	14993325	14960314	14927299	14894280	14861258	14828232	14795203	14762170	14729133	4
2.956	-0.0	14696093	14663049	14630001	14596950	14563896	14530838	14497776	14464711	14431643	14398571	3
2.957	-0.0	14365496	14332417	14299335	14266249	14233160	14200068	14166972	14133873	14100770	14067665	3
2.958	-0.0	14034555	14001443	13968327	13935208	13902086	13868961	13835832	13802700	13769565	13736427	3
2.959	-0.0	13703285	13670140	13636992	13603841	13570687	13537530	13504370	13471206	13438040	13404870	3
2.960	-0.0	13371698	13338522	13305343	13272162	13238977	13205789	13172598	13139405	13106208	13073009	3
2.961	-0.0	13039806	13006601	12973393	12940182	12906968	12873751	12840531	12807309	12774084	12740856	3
2.962	-0.0	12707625	12674391	12641155	12607916	12574674	12541429	12508182	12474932	12441679	12408424	3
2.963	-0.0	12375166	12341905	12308642	12275376	12242107	12208836	12175563	12142287	12109008	12075727	3
2.964	-0.0	12042443	12009156	11975868	11942576	11909283	11875986	11842688	11809387	11776083	11742777	2
2.965	-0.0	11709469	11676158	11642845	11609530	11576212	11542692	11509570	11476245	11442918	11409589	2
2.966	-0.0	11376257	11342924	11309588	11276250	11242909	11209567	11176222	11142875	11109526	11076175	2
2.967	-0.0	11042822	11009466	10976109	10942749	10909387	10876024	10842658	10809290	10775920	10742549	2
2.968	-0.0	10709175	10675799	10642421	10609041	10575660	10542276	10508891	10475503	10442114	10408723	2
2.969	-0.0	10375330	10341935	10308538	10275140	10241740	10208337	10174934	10141528	10108121	10074711	2
2.970	-0.0	10041300	10007888	09974474	09941058	09907640	09874221	09840800	09807377	09773953	09740527	2
2.971	-0.0	09707100	09673671	09640240	09606808	09573374	09539939	09506502	09473064	09439624	09406183	0
2.972	-0.0	09372741	09339296	09305851	09272404	09238956	09205506	09172055	09138602	09105148	09071693	0
2.973	-0.0	09038237	09004779	08971320	08937859	08904397	08870934	08837470	08804005	08770538	08737070	0
2.974	-0.0	08703601	08670130	08636659	08603186	08569713	08536238	08502762	08469284	08435806	08402327	0
2.975	-0.0	08368847	08335365	08301883	08268399	08234915	08201429	08167943	08134455	08100967	08067477	0
2.976	-0.0	08033987	08000496	07967004	07933510	07900017	07866522	07833026	07799530	07766032	07732534	0
2.977	-0.0	07699035	07665536	07632035	07598534	07565032	07531529	07498026	07464521	07431016	07397511	0
2.978	-0.0	07364005	07330498	07296990	07263482	07229973	07196464	07162954	07129443	07095932	07062421	0
2.979	-0.0	07028908	06995396	06961882	06928369	06894854	06861340	06827825	06794309	06760793	06727276	0
2.980	-0.0	06693760	06660242	06626725	06593207	06559683	06526170	06492651	06459131	06425612	06392092	0
2.981	-0.0	06358572	06325051	06291530	06258009	06224488	06190967	06157445	06123924	06090402	06056880	0
2.982	-0.0	06023357	05989835	05956312	05922790	05889267	05855745	05822222	05788699	05755176	05721653	0
2.983	-0.0	05688130	05654607	05621084	05587561	05554039	05520516	05486993	05453470	05419948	05386425	0
2.984	-0.0	05352903	05319381	05285859	05252337	05218815	05185294	05151773	05118251	05084731	05051210	0
2.985	-0.0	05017690	04984169	04950650	04917130	04883611	04850092	04816573	04783055	04749537	04716020	0
2.986	-0.0	04682502	04648986	04615469	04581953	04548438	04514923	04481408	04447894	04414381	04380868	-1
2.987	-0.0	04347355	04313843	04280331	04246820	04213310	04179800	04146291	04112782	04079274	04045767	-1
2.988	-0.0	04012260	03978754	03945249	03911744	03878240	03844737	03811234	03777732	03744231	03710731	-1
2.989	-0.0	03677231	03643733	03610235	03576737	03543241	03509746	03476251	03442757	03409264	03375773	-1
2.990	-0.0	03342281	03308791	03275302	03241814	03208327	03174840	03141355	03107871	03074387	03040905	-1
2.991	-0.0	03007424	02973943	02940464	02906986	02873509	02840033	02806559	02773085	02739612	02706141	-1
2.992	-0.0	02672671	02639202	02605734	02572268	02538802	02505338	02471875	02438414	02404953	02371494	-1
2.993	-0.0	02338037	02304580	02271125	02237671	02204219	02170768	02137318	02103870	02070423	02036977	-1
2.994	-0.0	02003533	01970091	01936650	01903210	01869772	01836335	01802900	01769466	01736034	01702604	-2
2.995	-0.0	01669174	01635747	01602321	01568897	01535474	01502053	01468634	01435216	01401800	01368386	-2
2.996	-0.0	01334973	01301562	01268153	01234745	01201339	01167935	01134533	01101132	01067734	01034337	-2
2.997	-0.0	01000942	00967548	00934157	00900768	00867380	00833994	00800610	00767228	00733848	00700470	-2
2.998	-0.0	00667094	00633720	00600347	00566977	00533609	00500243	00466879	00433516	00400156	00366798	-2
2.999	-0.0	00333442	00300088	00266737	00233387	00200040	00166694	00133351	00100010	00066671	00033334	2

$$\frac{\sin 2\pi x}{2\pi x}$$

$$\frac{\sin 2\pi x}{2\pi x}$$

TABLE 6

Sinc Function

x		.0	.0001	.0002	.0003	.0004	.0005	.0006	.0007	.0008	.0009	Δ_2
3.000	0.0	00000000	00033332	00066662	00099990	00133315	00166639	00199960	00233278	00266594	00299908	−2
3.001	0.0	00333220	00366529	00399836	00433141	00466443	00499743	00533040	00566335	00599627	00632917	−2
3.002	0.0	00666205	00699490	00732773	00766053	00799330	00832605	00865878	00899148	00932415	00965680	−3
3.003	0.0	00998942	01032201	01065458	01098713	01131964	01165213	01198460	01231703	01264944	01298182	−3
3.004	0.0	01331418	01364650	01397880	01431108	01464332	01497554	01530773	01563989	01597202	01630412	−3
3.005	0.0	01663620	01696824	01730026	01763225	01796421	01829614	01862804	01895991	01929176	01962357	−3
3.006	0.0	01995535	02028711	02061883	02095052	02128218	02161381	02194542	02227699	02260853	02294003	−3
3.007	0.0	02327151	02360296	02393437	02426576	02459711	02492843	02525972	02559097	02592220	02625339	−3
3.008	0.0	02658455	02691567	02724677	02757783	02790886	02823985	02857082	02890174	02923264	02956350	−3
3.009	0.0	02989433	03022513	03055589	03088661	03121731	03154796	03187859	03220918	03253973	03287025	−3
3.010	0.0	03320074	03353119	03386160	03419198	03452232	03485263	03518290	03551314	03584334	03617351	−4
3.011	0.0	03650363	03683373	03716378	03749380	03782378	03815373	03848364	03881351	03914334	03947314	−4
3.012	0.0	03980290	04013262	04046231	04079195	04112156	04145113	04178066	04211016	04243961	04276903	−4
3.013	0.0	04309840	04342774	04375704	04408630	04441552	04474471	04507385	04540295	04573201	04606104	−4
3.014	0.0	04639002	04671896	04704787	04737673	04770555	04803433	04836307	04869177	04902043	04934905	−4
3.015	0.0	04967762	05000616	05033465	05066310	05099151	05131988	05164820	05197649	05230473	05263293	−4
3.016	0.0	05296108	05328920	05361727	05394530	05427328	05460122	05492912	05525698	05558479	05591256	−4
3.017	0.0	05624028	05656796	05689559	05722319	05755073	05787824	05820570	05853311	05886048	05918780	−4
3.018	0.0	05951508	05984232	06016950	06049665	06082374	06115080	06147780	06180476	06213168	06245854	−5
3.019	0.0	06278537	06311214	06343887	06376555	06409219	06441877	06474531	06507181	06539825	06572465	−5
3.020	0.0	06605101	06637731	06670357	06702977	06735593	06768205	06800811	06833412	06866009	06898601	−5
3.021	0.0	06931188	06963770	06996347	07028919	07061486	07094049	07126606	07159158	07191706	07224248	−5
3.022	0.0	07256786	07289318	07321845	07354368	07386885	07419397	07451904	07484406	07516903	07549395	−5
3.023	0.0	07581882	07614363	07646839	07679311	07711777	07744237	07776693	07809144	07841589	07874029	−5
3.024	0.0	07906463	07938893	07971317	08003736	08036149	08068558	08100960	08133358	08165750	08198137	−5
3.025	0.0	08230518	08262895	08295265	08327630	08359990	08392345	08424694	08457037	08489375	08521708	−5
3.026	0.0	08554035	08586356	08618672	08650983	08683287	08715587	08747880	08780169	08812451	08844728	−6
3.027	0.0	08876999	08909265	08941525	08973780	09006028	09038271	09070509	09102740	09134966	09167186	−6
3.028	0.0	09199401	09231609	09263812	09296009	09328200	09360386	09392566	09424739	09456907	09489070	−6
3.029	0.0	09521226	09553376	09585521	09617659	09649792	09681919	09714039	09746154	09778263	09810366	−6
3.030	0.0	09842463	09874554	09906639	09938717	09970790	10002857	10034918	10066972	10099021	10131063	−6
3.031	0.0	10163099	10195130	10227154	10259172	10291183	10323189	10355188	10387181	10419168	10451149	−6
3.032	0.0	10483124	10515092	10547054	10579009	10610959	10642902	10674839	10706769	10738694	10770611	−6
3.033	0.0	10802523	10834428	10866327	10898219	10930105	10961984	10993858	11025724	11057584	11089438	−6
3.034	0.0	11121285	11153126	11184960	11216788	11248609	11280424	11312232	11344034	11375829	11407617	−7
3.035	0.0	11439399	11471174	11502942	11534704	11566460	11598208	11629950	11661686	11693414	11725136	−7
3.036	0.0	11756851	11788560	11820261	11851956	11883644	11915326	11947000	11978668	12010329	12041983	−7
3.037	0.0	12073631	12105271	12136905	12168531	12200151	12231764	12263370	12294969	12326561	12358147	−7
3.038	0.0	12389725	12421296	12452861	12484418	12515968	12547512	12579048	12610577	12642099	12673614	−7
3.039	0.0	12705122	12736623	12768117	12799604	12831083	12862556	12894021	12925479	12956930	12988374	−7
3.040	0.0	13019811	13051240	13082662	13114077	13145485	13176885	13208279	13239665	13271043	13302414	−7
3.041	0.0	13333778	13365135	13396484	13427826	13459161	13490488	13521808	13553121	13584426	13615723	−7
3.042	0.0	13647014	13678296	13709572	13740839	13772100	13803353	13834598	13865836	13897066	13928289	−8
3.043	0.0	13959504	13990712	14021912	14053104	14084289	14115467	14146636	14177798	14208953	14240099	−8
3.044	0.0	14271238	14302370	14333494	14364610	14395718	14426818	14457911	14488996	14520074	14551143	−8
3.045	0.0	14582205	14613259	14644305	14675343	14706374	14737396	14768411	14799418	14830417	14861408	−8
3.046	0.0	14892391	14923367	14954334	14985294	15016245	15047189	15078124	15109052	15139972	15170883	−8
3.047	0.0	15201787	15232682	15263570	15294449	15325321	15356184	15387039	15417887	15448726	15479556	−8
3.048	0.0	15510379	15541194	15572000	15602799	15633589	15664371	15695144	15725910	15756667	15787416	−8
3.049	0.0	15818157	15848890	15879614	15910330	15941038	15971737	16002428	16033111	16063785	16094451	−8

TABLE 6

Sinc Function

x		.0	.0001	.0002	.0003	.0004	.0005	.0006	.0007	.0008	.0009	Δ_2
3.050	0.0	16125109	16155758	16186399	16217032	16247656	16278271	16308879	16339477	16370068	16400650	−8
3.051	0.0	16431223	16461788	16492344	16522892	16553432	16583962	16614485	16644998	16675504	16706000	−9
3.052	0.0	16736488	16766968	16797438	16827900	16858354	16888799	16919235	16949662	16980081	17010491	−9
3.053	0.0	17040893	17071286	17101669	17132045	17162411	17192769	17223118	17253458	17283789	17314112	−9
3.054	0.0	17344426	17374730	17405026	17435314	17465592	17495861	17526122	17556374	17586616	17616850	−9
3.055	0.0	17647075	17677291	17707498	17737696	17767885	17798065	17828236	17858398	17888551	17918695	−9
3.056	0.0	17948830	17978956	18009073	18039181	18069279	18099369	18129449	18159520	18189583	18219636	−9
3.057	0.0	18249679	18279714	18309740	18339756	18369763	18399761	18429749	18459729	18489699	18519660	−9
3.058	0.0	18549612	18579554	18609487	18639411	18669325	18699230	18729126	18759013	18788890	18818757	−9
3.059	0.0	18848616	18878465	18908304	18938134	18967955	18997766	19027568	19057360	19087143	19116917	−9
3.060	0.0	19146680	19176435	19206180	19235915	19265641	19295357	19325064	19354761	19384449	19414127	−10
3.061	0.0	19443795	19473454	19503103	19532742	19562372	19591992	19621603	19651204	19680795	19710376	−10
3.062	0.0	19739948	19769510	19799062	19828605	19858137	19887660	19917174	19946677	19976171	20005654	−10
3.063	0.0	20035128	20064593	20094047	20123491	20152926	20182351	20211765	20241170	20270565	20299950	−10
3.064	0.0	20329326	20358691	20388046	20417391	20446727	20476052	20505367	20534673	20563968	20593253	−10
3.065	0.0	20622529	20651794	20681049	20710294	20739529	20768754	20797969	20827173	20856368	20885552	−10
3.066	0.0	20914726	20943890	20973044	21002188	21031322	21060445	21089558	21118661	21147754	21176836	−10
3.067	0.0	21205908	21234970	21264022	21293063	21322094	21351115	21380125	21409125	21438115	21467094	−10
3.068	0.0	21496063	21525022	21553970	21582908	21611835	21640752	21669659	21698555	21727441	21756316	−10
3.069	0.0	21785181	21814035	21842879	21871712	21900535	21929347	21958149	21986940	22015721	22044491	−11
3.070	0.0	22073250	22101999	22130738	22159465	22188182	22216889	22245585	22274270	22302944	22331608	−11
3.071	0.0	22360261	22388904	22417535	22446156	22474767	22503366	22531955	22560533	22589101	22617657	−11
3.072	0.0	22646203	22674738	22703262	22731775	22760278	22788769	22817250	22845720	22874179	22902627	−11
3.073	0.0	22931064	22959491	22987906	23016311	23044704	23073087	23101459	23129820	23158169	23186508	−11
3.074	0.0	23214836	23243153	23271458	23299753	23328037	23356309	23384571	23412821	23441061	23469289	−11
3.075	0.0	23497506	23525713	23553908	23582091	23610264	23638426	23666576	23694715	23722843	23750960	−11
3.076	0.0	23779066	23807160	23835244	23863315	23891376	23919426	23947464	23975491	24003506	24031511	−11
3.077	0.0	24059504	24087486	24115456	24143415	24171363	24199299	24227224	24255138	24283040	24310931	−11
3.078	0.0	24338810	24366678	24394535	24422380	24450213	24478036	24505846	24533646	24561433	24589210	−12
3.079	0.0	24616974	24644728	24672469	24700199	24727918	24755625	24783321	24811004	24838677	24866337	−12
3.080	0.0	24893987	24921624	24949250	24976864	25004467	25032057	25059637	25087204	25114760	25142304	−12
3.081	0.0	25169836	25197357	25224866	25252363	25279849	25307322	25334784	25362234	25389673	25417099	−12
3.082	0.0	25444514	25471917	25499308	25526687	25554054	25581410	25608754	25636085	25663405	25690713	−12
3.083	0.0	25718009	25745293	25772565	25799826	25827074	25854310	25881535	25908747	25935947	25963136	−12
3.084	0.0	25990312	26017476	26044629	26071769	26098897	26126013	26153117	26180209	26207289	26234357	−12
3.085	0.0	26261412	26288456	26315487	26342507	26369514	26396509	26423492	26450462	26477421	26504367	−12
3.086	0.0	26531301	26558223	26585132	26612030	26638915	26665788	26692648	26719496	26746332	26773156	−12
3.087	0.0	26799967	26826767	26853553	26880328	26907090	26933839	26960577	26987302	27014014	27040715	−12
3.088	0.0	27067402	27094078	27120741	27147391	27174029	27200655	27227268	27253869	27280457	27307033	−12
3.089	0.0	27333596	27360146	27386685	27413210	27439723	27466224	27492712	27519187	27545650	27572101	−13
3.090	0.0	27598538.	27624963	27651376	27677776	27704163	27730537	27756899	27783248	27809585	27835909	−13
3.091	0.0	27862220.	27888519	27914804	27941077	27967338	27993585	28019820	28046042	28072251	28098448	−13
3.092	0.0	28124632.	28150803	28176961	28203106	28229239	28255358	28281465	28307559	28333640	28359708	−13
3.093	0.0	28385764.	28411806	28437836	28463852	28489856	28515847	28541825	28567790	28593742	28619681	−13
3.094	0.0	28645607.	28671520	28697420	28723307	28749181	28775042	28800890	28826725	28852547	28878355	−13
3.095	0.0	28904151.	28929934	28955703	28981460	29007203	29032934	29058651	29084355	29110046	29135723	−13
3.096	0.0	29161388.	29187039	29212678	29238303	29263915	29289513	29315099	29340671	29366230	29391775	−13
3.097	0.0	29417308.	29442827	29468333	29493826	29519305	29544771	29570224	29595663	29621089	29646502	−13
3.098	0.0	29671901.	29697288	29722660	29748020	29773366	29798698	29824017	29849323	29874615	29899894	−13
3.099	0.0	29925160.	29950412	29975650	30000876	30026087	30051285	30076470	30101641	30126799	30151943	−14

$$\frac{\sin 2\pi x}{2\pi x}$$

TABLE 6

Sinc Function

x		.0	.0001	.0002	.0003	.0004	.0005	.0006	.0007	.0008	.0009	Δ_2
3.100	0.0	30177074	30202191	30227294	30252384	30277461	30302524	30327573	30352609	30377631	30402639	−14
3.101	0.0	30427634	30452615	30477583	30502537	30527477	30552404	30577317	30602216	30627102	30651974	−14
3.102	0.0	30676832	30701677	30726508	30751325	30776128	30800918	30825693	30850455	30875204	30899938	−14
3.103	0.0	30924659	30949366	30974059	30998738	31023404	31048055	31072693	31097317	31121927	31146523	−14
3.104	0.0	31171106	31195674	31220229	31244769	31269296	31293809	31318308	31342792	31367263	31391720	−14
3.105	0.0	31416163	31440592	31465007	31489409	31513796	31538169	31562528	31586873	31611204	31635521	−14
3.106	0.0	31659823	31684112	31708387	31732648	31756894	31781127	31805345	31829549	31853739	31877915	−14
3.107	0.0	31902077	31926225	31950358	31974478	31998583	32022674	32046751	32070813	32094862	32118896	−14
3.108	0.0	32142916	32166922	32190913	32214890	32238853	32262802	32286737	32310657	32334563	32358454	−14
3.109	0.0	32382331	32406194	32430043	32453877	32477697	32501503	32525294	32549071	32572833	32596581	−14
3.110	0.0	32620315	32644034	32667739	32691429	32715105	32738767	32762414	32786047	32809665	32833269	−14
3.111	0.0	32856858	32880433	32903993	32927539	32951070	32974587	32998089	33021576	33045049	33068508	−15
3.112	0.0	33091952	33115381	33138796	33162197	33185582	33208953	33232310	33255652	33278979	33302291	−15
3.113	0.0	33325589	33348873	33372141	33395395	33418634	33441859	33465069	33488264	33511444	33534610	−15
3.114	0.0	33557761	33580898	33604019	33627126	33650218	33673295	33696358	33719405	33742438	33765456	−15
3.115	0.0	33788460	33811448	33834422	33857381	33880325	33903254	33926168	33949068	33971952	33994822	−15
3.116	0.0	34017677	34040517	34063342	34086152	34108947	34131727	34154492	34177243	34199978	34222699	−15
3.117	0.0	34245404	34268095	34290770	34313431	34336076	34358707	34381322	34403923	34426508	34449079	−15
3.118	0.0	34471634	34494174	34516699	34539210	34561705	34584185	34606650	34629099	34651534	34673954	−15
3.119	0.0	34696358	34718748	34741122	34763481	34785825	34808153	34830467	34852765	34875048	34897316	−15
3.120	0.0	34919569	34941807	34964029	34986236	35008428	35030605	35052766	35074912	35097043	35119159	−15
3.121	0.0	35141259	35163344	35185414	35207468	35229507	35251531	35273539	35295532	35317510	35339473	−15
3.122	0.0	35361420	35383351	35405268	35427169	35449054	35470924	35492779	35514619	35536442	35558251	−15
3.123	0.0	35580044	35601822	35623584	35645331	35667062	35688778	35710478	35732163	35753832	35775486	−16
3.124	0.0	35797124	35818747	35840354	35861946	35883522	35905083	35926628	35948157	35969671	35991170	−16
3.125	0.0	36012653	36034120	36055571	36077007	36098428	36119833	36141222	36162595	36183953	36205295	−16
3.126	0.0	36226622	36247933	36269228	36290508	36311771	36333020	36354252	36375469	36396670	36417855	−16
3.127	0.0	36439025	36460179	36481317	36502439	36523546	36544636	36565711	36586771	36607814	36628842	−16
3.128	0.0	36649854	36670850	36691830	36712794	36733743	36754675	36775592	36796493	36817379	36838248	−16
3.129	0.0	36859101	36879939	36900760	36921566	36942356	36963130	36983888	37004630	37025356	37046066	−16
3.130	0.0	37066761	37087439	37108101	37128748	37149378	37169992	37190591	37211173	37231740	37252290	−16
3.131	0.0	37272824	37293343	37313845	37334331	37354801	37375256	37395694	37416116	37436522	37456912	−16
3.132	0.0	37477286	37497643	37517985	37538310	37558620	37578913	37599190	37619451	37639696	37659925	−16
3.133	0.0	37680137	37700333	37720514	37740678	37760826	37780957	37801073	37821172	37841255	37861322	−16
3.134	0.0	37881372	37901407	37921425	37941427	37961412	37981381	38001334	38021271	38041192	38061096	−16
3.135	0.0	38080984	38100856	38120711	38140550	38160373	38180179	38199969	38219743	38239500	38259241	−16
3.136	0.0	38278965	38298674	38318366	38338041	38357700	38377343	38396969	38416579	38436172	38455749	−16
3.137	0.0	38475310	38494854	38514382	38533893	38553383	38572866	38592328	38611774	38631203	38650615	−16
3.138	0.0	38670011	38689390	38708753	38728100	38747430	38766743	38786040	38805320	38824584	38843831	−17
3.139	0.0	38863062	38882276	38901473	38920654	38939819	38958966	38978097	38997212	39016310	39035391	−17
3.140	0.0	39054456	39073504	39092535	39111550	39130548	39149530	39168494	39187442	39206374	39225289	−17
3.141	0.0	39244187	39263068	39281933	39300781	39319612	39338426	39357224	39376005	39394770	39413517	−17
3.142	0.0	39432248	39450962	39469659	39488340	39507004	39525651	39544281	39562894	39581491	39600071	−17
3.143	0.0	39618633	39637180	39655709	39674221	39692717	39711196	39729658	39748103	39766531	39784942	−17
3.144	0.0	39803337	39821714	39840075	39858419	39876746	39895056	39913349	39931625	39949884	39968126	−17
3.145	0.0	39986352	40004560	40022752	40040926	40059084	40077224	40095348	40113454	40131544	40149617	−17
3.146	0.0	40167672	40185711	40203732	40221737	40239725	40257695	40275649	40293585	40311505	40329407	−17
3.147	0.0	40347292	40365160	40383012	40400846	40418663	40436463	40454246	40472011	40489760	40507491	−17
3.148	0.0	40525206	40542903	40560583	40578246	40595892	40613521	40631132	40648727	40666304	40683864	−17
3.149	0.0	40701407	40718933	40736441	40753933	40771407	40788864	40806303	40823726	40841131	40858519	−17

TABLE 6

Sinc Function

x		.0	.0001	.0002	.0003	.0004	.0005	.0006	.0007	.0008	.0009	Δ_2
3.150	0.0	40875890	40893244	40910580	40927899	40945201	40962485	40979753	40997003	41014236	41031451	-17
3.151	0.0	41048649	41065830	41082994	41100140	41117269	41134380	41151475	41168552	41185611	41202654	-17
3.152	0.0	41219679	41236686	41253676	41270649	41287605	41304543	41321464	41338367	41355253	41372122	-17
3.153	0.0	41388973	41405807	41422623	41439422	41456203	41472967	41489714	41506443	41523155	41539849	-17
3.154	0.0	41556526	41573186	41589828	41606452	41623059	41639648	41656220	41672775	41689312	41705831	-18
3.155	0.0	41722333	41738818	41755285	41771734	41788166	41804580	41820977	41837356	41853718	41870062	-18
3.156	0.0	41886389	41902698	41918989	41935263	41951519	41967758	41983979	42000183	42016368	42032537	-18
3.157	0.0	42048687	42064820	42080936	42097033	42113113	42129176	42145221	42161248	42177257	42193249	-18
3.158	0.0	42209223	42225180	42241119	42257040	42272943	42288829	42304697	42320547	42336380	42352195	-18
3.159	0.0	42367992	42383772	42399533	42415277	42431004	42446712	42462403	42478076	42493731	42509369	-18
3.160	0.0	42524988	42540590	42556174	42571741	42587289	42602820	42618333	42633828	42649306	42664765	-18
3.161	0.0	42680207	42695631	42711037	42726425	42741795	42757148	42772483	42787799	42803098	42818379	-18
3.162	0.0	42833643	42848888	42864116	42879325	42894517	42909691	42924847	42939985	42955105	42970207	-18
3.163	0.0	42985291	43000358	43015406	43030436	43045449	43060444	43075420	43090379	43105320	43120243	-18
3.164	0.0	43135147	43150034	43164903	43179754	43194587	43209402	43224199	43238978	43253739	43268482	-18
3.165	0.0	43283206	43297913	43312602	43327273	43341926	43356561	43371177	43385776	43400357	43414919	-18
3.166	0.0	43429464	43443990	43458499	43472989	43487461	43501915	43516351	43530769	43545169	43559551	-18
3.167	0.0	43573915	43588260	43602588	43616897	43631188	43645461	43659716	43673953	43688172	43702373	-18
3.168	0.0	43716555	43730719	43744865	43758993	43773103	43787195	43801268	43815323	43829360	43843379	-18
3.169	0.0	43857380	43871362	43885326	43899273	43913200	43927110	43941001	43954875	43968730	43982566	-18
3.170	0.0	43996385	44010185	44023967	44037731	44051476	44065204	44078913	44092603	44106276	44119930	-18
3.171	0.0	44133566	44147183	44160783	44174364	44187927	44201471	44214997	44228505	44241995	44255466	-18
3.172	0.0	44268919	44282353	44295769	44309167	44322547	44335908	44349251	44362576	44375882	44389170	-18
3.173	0.0	44402439	44415690	44428923	44442137	44455333	44468511	44481670	44494811	44507933	44521037	-18
3.174	0.0	44534123	44547190	44560239	44573269	44586281	44599275	44612250	44625207	44638145	44651065	-18
3.175	0.0	44663966	44676849	44689714	44702560	44715387	44728196	44740987	44753759	44766513	44779248	-18
3.176	0.0	44791965	44804663	44817343	44830004	44842647	44855272	44867877	44880465	44893034	44905584	-19
3.177	0.0	44918116	44930629	44943124	44955600	44968057	44980497	44992917	45005319	45017703	45030068	-19
3.178	0.0	45042414	45054742	45067051	45079342	45091614	45103867	45116102	45128319	45140517	45152696	-19
3.179	0.0	45164857	45176999	45189122	45201227	45213313	45225381	45237430	45249460	45261472	45273465	-19
3.180	0.0	45285440	45297396	45309333	45321251	45333151	45345033	45356895	45368739	45380565	45392372	-19
3.181	0.0	45404160	45415929	45427680	45439412	45451125	45462820	45474496	45486153	45497792	45509412	-19
3.182	0.0	45521013	45532596	45544160	45555705	45567231	45578739	45590228	45601698	45613150	45624583	-19
3.183	0.0	45635997	45647392	45658769	45670127	45681466	45692786	45704088	45715371	45726635	45737880	-19
3.184	0.0	45749107	45760315	45771504	45782674	45793826	45804959	45816073	45827168	45838244	45849302	-19
3.185	0.0	45860341	45871361	45882362	45893344	45904308	45915253	45926179	45937086	45947974	45958844	-19
3.186	0.0	45969695	45980527	45991340	46002134	46012909	46023666	46034403	46045122	46055822	46066504	-19
3.187	0.0	46077166	46087809	46098434	46109040	46119626	46130194	46140743	46151274	46161785	46172277	-19
3.188	0.0	46182751	46193206	46203641	46214058	46224456	46234835	46245196	46255537	46265859	46276163	-19
3.189	0.0	46286447	46296713	46306960	46317187	46327396	46337586	46347757	46357909	46368043	46378157	-19
3.190	0.0	46388252	46398328	46408386	46418424.	46428443	46438444	46448425	46458388	46468332	46478256	-19
3.191	0.0	46488162	46498049	46507916	46517765.	46527595	46537406	46547197	46556970	46566724	46576459	-19
3.192	0.0	46586175	46595872	46605549	46615208	46624848	46634469	46644071	46653653	46663217	46672762	-19
3.193	0.0	46682287	46691794	46701282	46710750	46720200	46729631	46739042	46748435	46757808	46767162	-19
3.194	0.0	46776498	46785814	46795111	46804389	46813649	46822889	46832110	46841312	46850494	46859658	-19
3.195	0.0	46868803	46877929	46887035	46896123	46905191	46914240	46923271	46932232	46941274	46950247	-19
3.196	0.0	46959201	46968135	46977051	46985948	46994825	47003683	47012523	47021343	47030144	47038926	-19
3.197	0.0	47047688	47056432	47065157	47073862	47082548	47091215	47099863	47108492	47117102	47125693	-19
3.198	0.0	47134264	47142816	47151350	47159864	47168359	47176834	47185291	47193728	47202147	47210546	-19
3.199	0.0	47218926	47227286	47235628	47243950	47252254	47260538	47268803	47277048	47285275	47293482	-19

TABLE 6

Sinc Function

x		.0	.0001	.0002	.0003	.0004	.0005	.0006	.0007	.0008	.0009	Δ_2
3.200	0.0	47301671	47309840	47317989	47326120	47334231	47342324	47350397	47358451	47366485	47374501	-19
3.201	0.0	47382497	47390474	47398432	47406371	47414290	47422190	47430071	47437933	47445775	47453599	-19
3.202	0.0	47461403	47469188	47476954	47484700	47492427	47500135	47507824	47515493	47523144	47530775	-19
3.203	0.0	47538387	47545979	47553552	47561106	47568641	47576157	47583653	47591130	47598588	47606027	-19
3.204	0.0	47613446	47620846	47628227	47635588	47642931	47650254	47657557	47664842	47672107	47679353	-19
3.205	0.0	47686579	47693787	47700975	47708144	47715293	47722423	47729534	47736626	47743698	47750751	-19
3.206	0.0	47757785	47764800	47771795	47778771	47785727	47792665	47799583	47806481	47813361	47820221	-19
3.207	0.0	47827062	47833883	47840685	47847468	47854232	47860976	47867701	47874407	47881093	47887760	-19
3.208	0.0	47894408	47901036	47907645	47914235	47920805	47927356	47933888	47940400	47946893	47953367	-19
3.209	0.0	47959821	47966256	47972672	47979068	47985446	47991803	47998142	48004461	48010760	48017041	-19
3.210	0.0	48023302	48029543	48035765	48041968	48048152	48054316	48060461	48066586	48072693	48078779	-19
3.211	0.0	48084847	48090895	48096924	48102933	48108923	48114894	48120845	48126777	48132689	48138582	-19
3.212	0.0	48144456	48150311	48156146	48161961	48167758	48173534	48179292	48185030	48190749	48196448	-19
3.213	0.0	48202128	48207789	48213430	48219052	48224654	48230238	48235801	48241346	48246870	48252376	-19
3.214	0.0	48257862	48263329	48268776	48274204	48279613	48285002	48290372	48295722	48301053	48306365	-19
3.215	0.0	48311657	48316930	48322183	48327417	48332632	48337827	48343002	48348159	48353296	48358413	-19
3.216	0.0	48363511	48368590	48373649	48378689	48383710	48388711	48393693	48398655	48403598	48408521	-19
3.217	0.0	48413425	48418310	48423175	48428020	48432847	48437654	48442441	48447209	48451958	48456687	-19
3.218	0.0	48461397	48466087	48470758	48475410	48480042	48484654	48489248	48493822	48498376	48502911	-19
3.219	0.0	48507426	48511922	48516399	48520856	48525294	48529713	48534111	48538491	48542851	48547192	-19
3.220	0.0	48551513	48555815	48560097	48564360	48568603	48572827	48577032	48581217	48585383	48589529	-19
3.221	0.0	48593656	48597763	48601851	48605920	48609969	48613998	48618009	48621999	48625971	48629923	-19
3.222	0.0	48633855	48637768	48641661	48645536	48649390	48653225	48657041	48660837	48664614	48668372	-19
3.223	0.0	48672110	48675828	48679527	48683207	48686867	48690508	48694129	48697731	48701313	48704876	-19
3.224	0.0	48708420	48711944	48715448	48718934	48722399	48725845	48729272	48732680	48736068	48739436	-19
3.225	0.0	48742785	48746115	48749425	48752715	48755987	48759238	48762471	48765683	48768877	48772051	-19
3.226	0.0	48775205	48778340	48781456	48784552	48787629	48790686	48793724	48796742	48799741	48802720	-19
3.227	0.0	48805680	48808621	48811542	48814444	48817326	48820189	48823032	48825856	48828660	48831445	-19
3.228	0.0	48834211	48836957	48839683	48842390	48845078	48847746	48850395	48853024	48855634	48858225	-19
3.229	0.0	48860796	48863347	48865879	48868392	48870885	48873359	48875813	48878248	48880664	48883060	-19
3.230	0.0	48885436	48887793	48890131	48892449	48894748	48897027	48899287	48901527	48903748	48905950	-19
3.231	0.0	48908132	48910295	48912438	48914562	48916666	48918751	48920816	48922862	48924889	48926896	-19
3.232	0.0	48928884	48930852	48932801	48934730	48936640	48938530	48940401	48942253	48944085	48945898	-19
3.233	0.0	48947691	48949465	48951220	48952955	48954670	48956366	48958043	48959700	48961338	48962957	-19
3.234	0.0	48964556	48966135	48967695	48969236	48970757	48972259	48973741	48975204	48976648	48978072	-19
3.235	0.0	48979477	48980862	48982228	48983575	48984902	48986209	48987497	48988766	48990016	48991246	-19
3.236	0.0	48992456	48993647	48994819	48995971	48997104	48998218	48999312	49000386	49001441	49002477	-19
3.237	0.0	49003494	49004491	49005468	49006426	49007365	49008285	49009185	49010065	49010926	49011768	-19
3.238	0.0	49012590	49013393	49014177	49014941	49015686	49016411	49017117	49017804	49018471	49019119	-19
3.239	0.0	49019747	49020356	49020946	49021516	49022067	49022599	49023111	49023603	49024077	49024531	-19
3.240	0.0	49024965	49025380	49025776	49026153	49026510	49026847	49027166	49027464	49027744	49028004	-19
3.241	0.0	49028245	49028466	49028668	49028851	49029014	49029158	49029283	49029388	49029474	49029541	-19
3.242	0.0	49029588	49029616	49029624	49029613	49029583	49029533	49029464	49029376	49029263	49029141	-19
3.243	0.0	49028995	49028829	49028644	49028439	49028216	49027973	49027710	49027428	49027127	49026807	-19
3.244	0.0	49026467	49026108	49025729	49025332	49024915	49024478	49024022	49023547	49023053	49022539	-19
3.245	0.0	49022006	49021454	49020882	49020291	49019681	49019051	49018402	49017734	49017046	49016339	-19
3.246	0.0	49015613	49014868	49014103	49013319	49012515	49011693	49010851	49009989	49009109	49008209	-19
3.247	0.0	49007290	49006351	49005393	49004416	49003420	49002404	49001370	49000315	48999242	48998149	-19
3.248	0.0	48997037	48995906	48994755	48993586	48992396	48991188	48989960	48988714	48987447	48986162	-19
3.249	0.0	48984857	48983533	48982190	48980828	48979446	48978045	48976625	48975186	48973727	48972249	-19

TABLE 6

Sinc Function

x		.0	.0001	.0002	.0003	.0004	.0005	.0006	.0007	.0008	.0009	Δ_2
3.250	0.0	48970752	48969235	48967700	48966145	48964571	48962977	48961365	48959733	48958082	48956412	−19
3.251	0.0	48954722	48953013	48951285	48949538	48947772	48945986	48944182	48942358	48940514	48938652	−19
3.252	0.0	48936770	48934869	48932949	48931010	48929052	48927074	48925077	48923061	48921026	48918972	−19
3.253	0.0	48916898	48914805	48912693	48910562	48908412	48906243	48904054	48901846	48899619	48897373	−19
3.254	0.0	48895108	48892823	48890520	48888197	48885855	48883494	48881113	48878714	48876295	48873858	−19
3.255	0.0	48871401	48868925	48866430	48863915	48861382	48858829	48856258	48853667	48851057	48848428	−19
3.256	0.0	48845780	48843113	48840426	48837721	48834996	48832252	48829489	48826707	48823906	48821086	−19
3.257	0.0	48818247	48815388	48812511	48809614	48806699	48803764	48800810	48797837	48794845	48791834	−19
3.258	0.0	48788804	48785755	48782686	48779599	48776493	48773367	48770223	48767059	48763876	48760675	−19
3.259	0.0	48757454	48754214	48750955	48747677	48744380	48741064	48737729	48734375	48731002	48727610	−19
3.260	0.0	48724199	48720768	48717319	48713851	48710364	48706857	48703332	48699788	48696224	48692642	−19
3.261	0.0	48689041	48685420	48681781	48678123	48674445	48670749	48667034	48663300	48659546	48655774	−19
3.262	0.0	48651983	48648173	48644343	48640495	48636628	48632742	48628837	48624913	48620970	48617008	−19
3.263	0.0	48613028	48609028	48605009	48600971	48596915	48592839	48588745	48584631	48580499	48576348	−19
3.264	0.0	48572178	48567989	48563781	48559554	48555308	48551043	48546759	48542457	48538135	48533795	−19
3.265	0.0	48529436	48525058	48520661	48516245	48511810	48507356	48502884	48498392	48493882	48489353	−19
3.266	0.0	48484805	48480238	48475652	48471048	48466424	48461782	48457121	48452441	48447742	48443024	−19
3.267	0.0	48438288	48433532	48428758	48423965	48419153	48414323	48409473	48404605	48399718	48394812	−19
3.268	0.0	48389887	48384944	48379981	48375000	48370001	48364982	48359944	48354888	48349813	48344719	−19
3.269	0.0	48339607	48334475	48329325	48324156	48318969	48313762	48308537	48303293	48298031	48292749	−19
3.270	0.0	48287449	48282130	48276793	48271436	48266061	48260668	48255255	48249824	48244374	48238905	−19
3.271	0.0	48233418	48227912	48222387	48216843	48211281	48205700	48200101	48194483	48188846	48183190	−19
3.272	0.0	48177516	48171823	48166111	48160381	48154632	48148864	48143078	48137273	48131450	48125608	−19
3.273	0.0	48119747	48113867	48107969	48102052	48096117	48090163	48084190	48078199	48072189	48066161	−19
3.274	0.0	48060114	48054048	48047964	48041861	48035739	48029599	48023441	48017263	48011068	48004853	−19
3.275	0.0	47998620	47992369	47986099	47979810	47973503	47967177	47960833	47954470	47948089	47941689	−19
3.276	0.0	47935270	47928833	47922378	47915904	47909411	47902900	47896370	47889822	47883256	47876671	−19
3.277	0.0	47870067	47863445	47856804	47850145	47843468	47836772	47830057	47823324	47816573	47809803	−18
3.278	0.0	47803014	47796207	47789382	47782538	47775676	47768795	47761896	47754979	47748043	47741089	−18
3.279	0.0	47734116	47727125	47720115	47713087	47706040	47698976	47691892	47684791	47677671	47670532	−18
3.280	0.0	47663375	47656200	47649007	47641795	47634564	47627316	47620049	47612763	47605460	47598138	−18
3.281	0.0	47590797	47583438	47576061	47568666	47561252	47553820	47546370	47538901	47531414	47523909	−18
3.282	0.0	47516385	47508843	47501283	47493704	47486107	47478492	47470859	47463207	47455537	47447849	−18
3.283	0.0	47440143	47432418	47424675	47416914	47409134	47401337	47393521	47385686	47377834	47369963	−18
3.284	0.0	47362074	47354167	47346242	47338299	47330337	47322357	47314359	47306343	47298308	47290256	−18
3.285	0.0	47282185	47274096	47265989	47257863	47249720	47241558	47233378	47225180	47216964	47208730	−18
3.286	0.0	47200477	47192207	47183918	47175612	47167287	47158944	47150583	47142203	47133806	47125391	−18
3.287	0.0	47116957	47108506	47100036	47091548	47083042	47074518	47065976	47057416	47048838	47040242	−18
3.288	0.0	47031628	47022996	47014346	47005677	46996991	46988287	46979564	46970824	46962066	46953289	−18
3.289	0.0	46944495	46935682	46926852	46918004	46909137	46900253	46891351	46882430	46873492	46864536	−18
3.290	0.0	46855562	46846570	46837559	46828531	46819486	46810422	46801340	46792240	46783122	46773987	−18
3.291	0.0	46764833	46755662	46746473	46737266	46728041	46718798	46709537	46700258	46690961	46681647	−18
3.292	0.0	46672315	46662964	46653596	46644211	46634807	46625385	46615946	46606489	46597014	46587521	−18
3.293	0.0	46578010	46568482	46558935	46549371	46539789	46530190	46520572	46510937	46501284	46491613	−18
3.294	0.0	46481925	46472218	46462494	46452752	46442993	46433215	46423420	46413608	46403777	46393929	−18
3.295	0.0	46384063	46374179	46364278	46354359	46344422	46334467	46324495	46314505	46304498	46294473	−18
3.296	0.0	46284430	46274369	46264291	46254195	46244082	46233951	46223802	46213636	46203452	46193250	−18
3.297	0.0	46183031	46172794	46162539	46152267	46141978	46131670	46121345	46111003	46100643	46090265	−18
3.298	0.0	46079870	46069458	46059027	46048580	46038114	46027631	46017131	46006613	45996077	45985524	−18
3.299	0.0	45974954	45964366	45953760	45943137	45932497	45921839	45911163	45900470	45889760	45879032	−17

$\frac{\sin 2\pi x}{2\pi x}$

TABLE 6

Sinc Function

x		.0	.0001	.0002	.0003	.0004	.0005	.0006	.0007	.0008	.0009	Δ_2
3.300	0.0	45868287	45857524	45846743	45835946	45825131	45814298	45803448	45792581	45781696	45770793	−17
3.301	0.0	45759874	45748937	45737982	45727010	45716021	45705014	45693990	45682949	45671890	45660814	−17
3.302	0.0	45649720	45638610	45627481	45616336	45605173	45593993	45582795	45571581	45560348	45549099	−17
3.303	0.0	45537832	45526548	45515247	45503928	45492592	45481239	45469869	45458481	45447076	45435654	−17
3.304	0.0	45424215	45412758	45401284	45389793	45378285	45366759	45355216	45343656	45332079	45320485	−17
3.305	0.0	45308873	45297244	45285598	45273935	45262255	45250557	45238843	45227111	45215362	45203596	−17
3.306	0.0	45191813	45180013	45168195	45156361	45144509	45132640	45120754	45108852	45096932	45084994	−17
3.307	0.0	45073040	45061069	45049081	45037075	45025053	45013013	45000957	44988883	44976793	44964685	−17
3.308	0.0	44952560	44940419	44928260	44916084	44903892	44891682	44879455	44867212	44854951	44842674	−17
3.309	0.0	44830379	44818068	44805739	44793394	44781032	44768652	44756256	44743843	44731413	44718966	−17
3.310	0.0	44706503	44694022	44681524	44669010	44656479	44643930	44631365	44618783	44606185	44593569	−17
3.311	0.0	44580936	44568287	44555621	44542938	44530238	44517522	44504788	44492038	44479271	44466487	−17
3.312	0.0	44453687	44440869	44428035	44415184	44402317	44389432	44376531	44363613	44350679	44337727	−17
3.313	0.0	44324759	44311774	44298773	44285755	44272720	44259668	44246600	44233515	44220414	44207295	−17
3.314	0.0	44194160	44181009	44167840	44154656	44141454	44128236	44115001	44101750	44088482	44075197	−17
3.315	0.0	44061896	44048578	44035244	44021893	44008525	43995141	43981741	43968323	43954890	43941439	−17
3.316	0.0	43927972	43914489	43900989	43887473	43873940	43860390	43846825	43833242	43819643	43806028	−16
3.317	0.0	43792396	43778748	43765083	43751402	43737704	43723990	43710259	43696512	43682749	43668969	−16
3.318	0.0	43655173	43641360	43627531	43613686	43599824	43585946	43572051	43558141	43544213	43530270	−16
3.319	0.0	43516310	43502333	43488341	43474332	43460307	43446265	43432207	43418133	43404042	43389936	−16
3.320	0.0	43375813	43361673	43347518	43333346	43319158	43304954	43290733	43276496	43262243	43247974	−16
3.321	0.0	43233688	43219387	43205069	43190735	43176384	43162018	43147635	43133237	43118822	43104391	−16
3.322	0.0	43089943	43075480	43061000	43046505	43031993	43017465	43002921	42988361	42973785	42959193	−16
3.323	0.0	42944584	42929960	42915319	42900663	42885990	42871301	42856597	42841876	42827139	42812386	−16
3.324	0.0	42797618	42782833	42768032	42753215	42738382	42723534	42708669	42693788	42678892	42663979	−16
3.325	0.0	42649050	42634106	42619145	42604169	42589177	42574169	42559145	42544105	42529049	42513977	−16
3.326	0.0	42498889	42483786	42468666	42453531	42438380	42423213	42408030	42392832	42377618	42362387	−16
3.327	0.0	42347141	42331879	42316602	42301308	42285999	42270674	42255334	42239977	42224605	42209217	−16
3.328	0.0	42193813	42178394	42162958	42147508	42132041	42116559	42101061	42085547	42070018	42054472	−16
3.329	0.0	42038912	42023335	42007743	41992136	41976512	41960873	41945219	41929549	41913863	41898161	−16
3.330	0.0	41882444	41866712	41850964	41835200	41819421	41803626	41787815	41771989	41756148	41740291	−16
3.331	0.0	41724418	41708530	41692626	41676707	41660773	41644822	41628857	41612876	41596879	41580867	−15
3.332	0.0	41564840	41548797	41532738	41516665	41500576	41484471	41468351	41452215	41436065	41419898	−15
3.333	0.0	41403717	41387520	41371307	41355080	41338837	41322578	41306305	41290016	41273711	41257391	−15
3.334	0.0	41241056	41224706	41208341	41191960	41175564	41159152	41142725	41126283	41109826	41093354	−15
3.335	0.0	41076866	41060363	41043845	41027312	41010763	40994199	40977620	40961026	40944417	40927793	−15
3.336	0.0	40911153	40894498	40877828	40861143	40844443	40827728	40810997	40794252	40777491	40760715	−15
3.337	0.0	40743924	40727119	40710298	40693462	40676611	40659744	40642863	40625967	40609056	40592129	−15
3.338	0.0	40575188	40558232	40541261	40524274	40507273	40490257	40473226	40456180	40439119	40422043	−15
3.339	0.0	40404952	40387846	40370725	40353589	40336439	40319273	40302093	40284897	40267687	40250462	−15
3.340	0.0	40233222	40215968	40198698	40181414	40164114	40146800	40129472	40112128	40094769	40077396	−15
3.341	0.0	40060008	40042605	40025188	40007755	39990308	39972846	39955370	39937879	39920373	39902852	−15
3.342	0.0	39885316	39867766	39850202	39832622	39815028	39797419	39779796	39762157	39744505	39726837	−15
3.343	0.0	39709155	39691458	39673747	39656021	39638281	39620526	39602756	39584972	39567173	39549360	−15
3.344	0.0	39531532	39513690	39495833	39477961	39460075	39442175	39424260	39406330	39388386	39370428	−14
3.345	0.0	39352455	39334468	39316466	39298449	39280419	39262374	39244314	39226240	39208152	39190049	−14
3.346	0.0	39171932	39153800	39135654	39117494	39099319	39081130	39062927	39044709	39026477	39008231	−14
3.347	0.0	38989971	38971696	38953406	38935103	38916785	38898453	38880107	38861746	38843371	38824982	−14
3.348	0.0	38806579	38788162	38769730	38751284	38732824	38714350	38695861	38677359	38658842	38640311	−14
3.349	0.0	38621766	38603207	38584633	38566046	38547444	38528828	38510199	38491555	38472897	38454225	−14

TABLE 6

Sinc Function

x		.0	.0001	.0002	.0003	.0004	.0005	.0006	.0007	.0008	.0009	Δ_2
3.350	0.0	38435538	38416838	38398124	38379396	38360653	38341897	38323127	38304342	38285544	38266732	−14
3.351	0.0	38247905	38229065	38210211	38191342	38172460	38153564	38134654	38115730	38096792	38077840	−14
3.352	0.0	38058875	38039895	38020902	38001894	37982873	37963838	37944789	37925726	37906649	37887559	−14
3.353	0.0	37868455	37849337	37830205	37811059	37791900	37772726	37753539	37734339	37715124	37695896	−14
3.354	0.0	37676654	37657398	37638129	37618845	37599549	37580238	37560914	37541576	37522224	37502859	−14
3.355	0.0	37483480	37464088	37444681	37425262	37405828	37386381	37366921	37347446	37327959	37308457	−14
3.356	0.0	37288942	37269414	37249872	37230316	37210747	37191165	37171568	37151959	37132336	37112699	−13
3.357	0.0	37093049	37073385	37053708	37034018	37014314	36994596	36974865	36955121	36935364	36915592	−13
3.358	0.0	36895808	36876010	36856199	36836374	36816536	36796685	36776820	36756942	36737051	36717146	−13
3.359	0.0	36697228	36677297	36657353	36637395	36617424	36597439	36577442	36557431	36537407	36517369	−13
3.360	0.0	36497319	36477255	36457178	36437088	36416984	36396868	36376738	36356595	36336439	36316270	−13
3.361	0.0	36296088	36275892	36255684	36235462	36215227	36194979	36174718	36154444	36134157	36113857	−13
3.362	0.0	36093544	36073218	36052878	36032526	36012161	35991782	35971391	35950987	35930569	35910139	−13
3.363	0.0	35889696	35869240	35848771	35828289	35807794	35787286	35766765	35746231	35725685	35705125	−13
3.364	0.0	35684553	35663968	35643370	35622759	35602135	35581498	35560849	35540187	35519512	35498824	−13
3.365	0.0	35478123	35457410	35436684	35415945	35395193	35374429	35353652	35332862	35312059	35291244	−13
3.366	0.0	35270416	35249575	35228722	35207856	35186978	35166086	35145182	35124266	35103337	35082395	−13
3.367	0.0	35061440	35040473	35019494	34998502	34977497	34956480	34935450	34914407	34893352	34872285	−13
3.368	0.0	34851205	34830112	34809007	34787890	34766760	34745618	34724463	34703295	34682116	34660924	−12
3.369	0.0	34639719	34618502	34597272	34576031	34554776	34533510	34512231	34490939	34469636	34448320	−12
3.370	0.0	34426991	34405651	34384297	34362932	34341555	34320165	34298762	34277348	34255921	34234482	−12
3.371	0.0	34213031	34191568	34170092	34148604	34127104	34105592	34084067	34062530	34040982	34019421	−12
3.372	0.0	33997848	33976262	33954665	33933055	33911434	33889800	33868154	33846496	33824826	33803144	−12
3.373	0.0	33781450	33759744	33738025	33716295	33694553	33672799	33651032	33629254	33607464	33585661	−12
3.374	0.0	33563847	33542021	33520183	33498333	33476471	33454597	33432711	33410813	33388904	33366982	−12
3.375	0.0	33345049	33323104	33301146	33279177	33257197	33235204	33213200	33191183	33169155	33147116	−12
3.376	0.0	33125064	33103001	33080925	33058839	33036740	33014630	32992508	32970374	32948228	32926071	−12
3.377	0.0	32903902	32881722	32859529	32837325	32815110	32792883	32770644	32748394	32726132	32703858	−12
3.378	0.0	32681573	32659276	32636967	32614648	32592316	32569973	32547618	32525252	32502875	32480486	−12
3.379	0.0	32458085	32435673	32413249	32390814	32368368	32345910	32323440	32300959	32278467	32255964	−11
3.380	0.0	32233448	32210922	32188384	32165835	32143274	32120702	32098119	32075524	32052919	32030301	−11
3.381	0.0	32007673	31985033	31962382	31939719	31917046	31894361	31871664	31848957	31826238	31803508	−11
3.382	0.0	31780767	31758015	31735251	31712477	31689691	31666894	31644086	31621266	31598436	31575594	−11
3.383	0.0	31552742	31529878	31507003	31484117	31461220	31438312	31415393	31392462	31369521	31346569	−11
3.384	0.0	31323605	31300631	31277646	31254650	31231642	31208624	31185595	31162555	31139503	31116441	−11
3.385	0.0	31093368	31070285	31047190	31024084	31000968	30977840	30954702	30931553	30908393	30885222	−11
3.386	0.0	30862040	30838848	30815645	30792431	30769206	30745970	30722724	30699467	30676199	30652920	−11
3.387	0.0	30629631	30606331	30583020	30559698	30536366	30513024	30489670	30466306	30442931	30419546	−11
3.388	0.0	30396150	30372743	30349326	30325898	30302459	30279010	30255551	30232080	30208600	30185108	−11
3.389	0.0	30161607	30138094	30114571	30091038	30067494	30043940	30020375	29996800	29973214	29949618	−10
3.390	0.0	29926011	29902394	29878767	29855129	29831481	29807323	29784154	29760474	29736785	29713085	−10
3.391	0.0	29689374	29665654	29641923	29618181	29594430	29570668	29546896	29523113	29499321	29475518	−10
3.392	0.0	29451705	29427882	29404048	29380204	29356350	29332486	29308612	29284727	29260833	29236928	−10
3.393	0.0	29213013	29189038	29165153	29141208	29117253	29093287	29069312	29045326	29021331	28997325	−10
3.394	0.0	28973309	28949284	28925248	28901202	28877146	28853081	28829005	28804920	28780824	28756718	−10
3.395	0.0	28732603	28708478	28684342	28660197	28636042	28611877	28587702	28563518	28539323	28515119	−10
3.396	0.0	28490905	28466681	28442447	28418203	28393950	28369686	28345413	28321131	28296838	28272536	−10
3.397	0.0	28248224	28223902	28199571	28175230	28150879	28126519	28102148	28077769	28053379	28028980	−10
3.398	0.0	28004571	27980153	27955725	27931288	27906840	27882384	27857917	27833442	27808956	27784461	−10
3.399	0.0	27759957	27735443	27710919	27686386	27661844	27637292	27612731	27588160	27563580	27538990	−9

$$\frac{\sin 2\pi x}{2\pi x}$$

TABLE 6

Sinc Function

x		.0	.0001	.0002	.0003	.0004	.0005	.0006	.0007	.0008	.0009	Δ_2
3.400	0.0	27514391	27489782	27465164	27440537	27415900	27391254	27366598	27341934	27317259	27292576	-9
3.401	0.0	27267883	27243181	27218469	27193749	27169019	27144279	27119531	27094773	27070006	27045230	-9
3.402	0.0	27020444	26995649	26970845	26946032	26921210	26896379	26871538	26846688	26821829	26796961	-9
3.403	0.0	26772084	26747198	26722303	26697398	26672485	26647562	26622630	26597690	26572740	26547781	-9
3.404	0.0	26522814	26497837	26472851	26447856	26422853	26397840	26372818	26347788	26322748	26297700	-9
3.405	0.0	26272643	26247576	26222501	26197417	26172324	26147223	26122112	26096993	26071865	26046727	-9
3.406	0.0	26021582	25996427	25971264	25946091	25920910	25895721	25870522	25845315	25820099	25794874	-9
3.407	0.0	25769641	25744399	25719148	25693889	25668621	25643344	25618059	25592765	25567462	25542151	-9
3.408	0.0	25516831	25491503	25466166	25440820	25415466	25390104	25364733	25339353	25313965	25288568	-9
3.409	0.0	25263163	25237749	25212327	25186896	25161457	25136010	25110554	25085090	25059617	25034136	-8
3.410	0.0	25008646	24983148	24957642	24932127	24906604	24881073	24855533	24829985	24804429	24778864	-8
3.411	0.0	24753292	24727710	24702121	24676523	24650917	24625303	24599681	24574050	24548412	24522765	-8
3.412	0.0	24497110	24471446	24445775	24420095	24394408	24368712	24343008	24317296	24291576	24265848	-8
3.413	0.0	24240111	24214367	24188614	24162854	24137086	24111309	24085525	24059732	24033932	24008123	-8
3.414	0.0	23982307	23956482	23930650	23904810	23878961	23853105	23827241	23801369	23775490	23749602	-8
3.415	0.0	23723706	23697803	23671892	23645973	23620046	23594111	23568169	23542219	23516261	23490295	-8
3.416	0.0	23464321	23438340	23412351	23386354	23360350	23334338	23308318	23282291	23256255	23230213	-8
3.417	0.0	23204162	23178104	23152038	23125965	23099884	23073796	23047699	23021596	22995484	22969366	-8
3.418	0.0	22943239	22917105	22890964	22864815	22838659	22812495	22786324	22760145	22733958	22707765	-7
3.419	0.0	22681564	22655355	22629139	22602916	22576685	22550447	22524201	22497948	22471688	22445421	-7
3.420	0.0	22419146	22392864	22366574	22340277	22313973	22287662	22261343	22235017	22208684	22182344	-7
3.421	0.0	22155997	22129642	22103280	22076911	22050534	22024151	21997760	21971363	21944958	21918546	-7
3.422	0.0	21892127	21865700	21839267	21812827	21786379	21759925	21733463	21706995	21680519	21654037	-7
3.423	0.0	21627547	21601050	21574547	21548036	21521519	21494994	21468463	21441925	21415379	21388827	-7
3.424	0.0	21362268	21335702	21309130	21282550	21255963	21229370	21202770	21176163	21149549	21122929	-7
3.425	0.0	21096301	21069667	21043026	21016379	20989724	20963063	20936395	20909721	20883040	20856352	-7
3.426	0.0	20829657	20802956	20776248	20749533	20722812	20696084	20669350	20642609	20615861	20589107	-7
3.427	0.0	20562346	20535579	20508805	20482025	20455238	20428445	20401645	20374838	20348025	20321206	-6
3.428	0.0	20294380	20267548	20240709	20213864	20187012	20160155	20133290	20106419	20079542	20052659	-6
3.429	0.0	20025769	19998873	19971971	19945062	19918147	19891225	19864298	19837364	19810424	19783477	-6
3.430	0.0	19756524	19729566	19702600	19675629	19648652	19621668	19594678	19567682	19540680	19513671	-6
3.431	0.0	19486657	19459636	19432610	19405577	19378538	19351493	19324442	19297385	19270322	19243253	-6
3.432	0.0	19216178	19189097	19162009	19134916	19107817	19080712	19053601	19026484	18999361	18972232	-6
3.433	0.0	18945098	18917957	18890811	18863658	18836500	18809336	18782166	18754990	18727808	18700621	-6
3.434	0.0	18673428	18646229	18619024	18591813	18564597	18537375	18510147	18482914	18455675	18428430	-6
3.435	0.0	18401179	18373923	18346661	18319393	18292120	18264841	18237557	18210266	18182971	18155669	-6
3.436	0.0	18128363	18101050	18073732	18046409	18019079	17991745	17964405	17937059	17909708	17882351	-5
3.437	0.0	17854989	17827622	17800249	17772870	17745487	17718097	17690703	17663303	17635897	17608487	-5
3.438	0.0	17581070	17553649	17526222	17498790	17471353	17443910	17416462	17389009	17361550	17334086	-5
3.439	0.0	17306617	17279143	17251663	17224178	17196689	17169193	17141693	17114188	17086677	17059161	-5
3.440	0.0	17031640	17004114	16976583	16949047	16921505	16893959	16866407	16838851	16811289	16783723	-5
3.441	0.0	16756151	16728574	16700993	16673406	16645814	16618218	16590616	16563010	16535398	16507782	-5
3.442	0.0	16480160	16452534	16424903	16397267	16369626	16341981	16314330	16286675	16259015	16231350	-5
3.443	0.0	16203680	16176005	16148326	16120642	16092953	16065259	16037561	16009858	15982150	15954438	-5
3.444	0.0	15926720	15898998	15871272	15843541	15815805	15788064	15760319	15732570	15704815	15677056	-5
3.445	0.0	15649293	15621525	15593752	15565975	15538194	15510407	15482617	15454822	15427022	15399218	-4
3.446	0.0	15371409	15343596	15315778	15287957	15260130	15232299	15204464	15176625	15148781	15120932	-4
3.447	0.0	15093080	15065223	15037361	15009496	14981626	14953751	14925873	14897990	14870103	14842212	-4
3.448	0.0	14814316	14786416	14758512	14730604	14702692	14674775	14646854	14618929	14591000	14563067	-4
3.449	0.0	14535129	14507188	14479242	14451293	14423339	14395381	14367419	14339453	14311483	14283509	-4

TABLE 6

Sinc Function

x		.0	.0001	.0002	.0003	.0004	.0005	.0006	.0007	.0008	.0009	Δ_2
3.450	0.0	14255531	14227549	14199563	14171573	14143579	14115581	14087579	14059573	14031563	14003550	−4
3.451	0.0	13975532	13947511	13919485	13891456	13863423	13835386	13807345	13779300	13751252	13723200	−4
3.452	0.0	13695144	13667084	13639020	13610953	13582882	13554807	13526728	13498646	13470560	13442471	−4
3.453	0.0	13414377	13386280	13358180	13330075	13301967	13273856	13245741	13217622	13189499	13161373	−4
3.454	0.0	13133244	13105111	13076974	13048834	13020690	12992543	12964393	12936238	12908081	12879920	−3
3.455	0.0	12851755	12823587	12795416	12767241	12739062	12710881	12682696	12654507	12626315	12598120	−3
3.456	0.0	12569922	12541720	12513515	12485306	12457095	12428880	12400661	12372440	12344215	12315987	−3
3.457	0.0	12287756	12259521	12231283	12203042	12174798	12146551	12118301	12090047	12061790	12033530	−3
3.458	0.0	12005268	11977001	11948732	11920460	11892185	11863906	11835625	11807341	11779053	11750763	−3
3.459	0.0	11722469	11694173	11665873	11637571	11609265	11580957	11552646	11524331	11496014	11467694	−3
3.460	0.0	11439371	11411046	11382717	11354385	11326051	11297714	11269374	11241031	11212685	11184337	−3
3.461	0.0	11155986	11127632	11099275	11070916	11042553	11014189	10985821	10957451	10929078	10900702	−3
3.462	0.0	10872324	10843943	10815559	10787173	10758784	10730392	10701998	10673602	10645202	10616801	−3
3.463	0.0	10588396	10559989	10531580	10503168	10474754	10446337	10417917	10389495	10361071	10332644	−2
3.464	0.0	10304215	10275783	10247349	10218913	10190474	10162032	10133589	10105143	10076694	10048244	−2
3.465	0.0	10019791	09991336	09962878	09934418	09905956	09877491	09849025	09820556	09792084	09763611	−2
3.466	0.0	09735135	09706658	09678178	09649695	09621211	09592725	09564236	09535745	09507252	09478757	−2
3.467	0.0	09450260	09421761	09393260	09364756	09336251	09307743	09279234	09250723	09222209	09193694	−2
3.468	0.0	09165176	09136657	09108135	09079612	09051087	09022559	08994030	08965499	08936966	08908431	−2
3.469	0.0	08879895	08851356	08822816	08794273	08765729	08737183	08708636	08680086	08651535	08622982	−2
3.470	0.0	08594427	08565871	08537312	08508752	08480191	08451627	08423062	08394495	08365927	08337357	−2
3.471	0.0	08308785	08280212	08251637	08223060	08194482	08165902	08137321	08108738	08080153	08051567	−2
3.472	0.0	08022980	07994390	07965800	07937208	07908614	07880019	07851423	07822825	07794225	07765624	0
3.473	0.0	07737022	07708418	07679813	07651207	07622599	07593990	07565379	07536767	07508154	07479539	0
3.474	0.0	07450924	07422307	07393688	07365068	07336447	07307825	07279202	07250577	07221951	07193324	0
3.475	0.0	07164696	07136066	07107436	07078804	07050171	07021537	06992902	06964266	06935628	06906990	0
3.476	0.0	06878350	06849710	06821068	06792425	06763781	06735137	06706491	06677844	06649196	06620547	0
3.477	0.0	06591898	06563247	06534595	06505943	06477289	06448635	06419980	06391323	06362666	06334008	0
3.478	0.0	06305350	06276690	06248030	06219368	06190705	06162043	06133380	06104715	06076050	06047384	0
3.479	0.0	06018717	05990050	05961382	05932713	05904044	05875373	05846703	05818031	05789359	05760686	0
3.480	0.0	05732013	05703338	05674664	05645989	05617313	05588636	05559959	05531282	05502604	05473925	0
3.481	0.0	05445246	05416567	05387886	05359206	05330525	05301843	05273161	05244479	05215796	05187113	0
3.482	0.0	05158429	05129746	05101061	05072376	05043691	05015006	04986320	04957634	04928948	04900261	0
3.483	0.0	04871574	04842887	04814199	04785511	04756823	04728135	04699447	04670758	04642069	04613380	0
3.484	0.0	04584691	04556001	04527312	04498622	04469932	04441243	04412552	04383862	04355172	04326482	0
3.485	0.0	04297791	04269101	04240411	04211720	04183030	04154339	04125649	04096958	04068268	04039577	0
3.486	0.0	04010887	03982197	03953507	03924816	03896126	03867436	03838747	03810057	03781367	03752678	0
3.487	0.0	03723989	03695300	03666611	03637922	03609234	03580546	03551858	03523170	03494482	03465795	0
3.488	0.0	03437108	03408422	03379735	03351049	03322363	03293678	03264993	03236308	03207624	03178940	0
3.489	0.0	03150256	03121573	03092890	03064208	03035526	03006845	02978164	02949483	02920803	02892124	0
3.490	0.0	02863445	02834766	02806088	02777411	02748734	02720057	02691381	02662706	02634032	02605358	1
3.491	0.0	02576684	02548011	02519339	02490668	02461997	02433327	02404657	02375988	02347320	02318653	1
3.492	0.0	02289986	02261320	02232655	02203991	02175327	02146664	02118002	02089341	02060680	02032021	1
3.493	0.0	02003362	01974704	01946047	01917391	01888735	01860081	01831428	01802775	01774123	01745473	1
3.494	0.0	01716823	01688174	01659526	01630879	01602233	01573589	01544945	01516302	01487660	01459019	1
3.495	0.0	01430380	01401741	01373104	01344467	01315832	01287198	01258565	01229933	01201302	01172673	1
3.496	0.0	01144044	01115417	01086791	01058166	01029543	01000920	00972299	00943679	00915061	00886443	1
3.497	0.0	00857827	00829213	00800599	00771987	00743376	00714767	00686159	00657552	00628947	00600343	1
3.498	0.0	00571740	00543139	00514539	00485941	00457344	00428749	00400155	00371562	00342971	00314382	1
3.499	0.0	00285794	00257208	00228623	00200039	00171458	00142877	00114299	00085722	00057146	00028572	−2

$$\frac{\sin 2\pi x}{2\pi x}$$

TABLE 6

Sinc Function

x		.0	.0001	.0002	.0003	.0004	.0005	.0006	.0007	.0008	.0009	Δ_2
3.500	-0.0	00000000	00028571	00057140	00085707	00114273	00142837	00171399	00199959	00228518	00257075	2
3.501	-0.0	00285631	00314184	00342736	00371287	00399835	00428381	00456926	00485469	00514010	00542550	2
3.502	-0.0	00571087	00599623	00628157	00656688	00685218	00713747	00742273	00770797	00799319	00827840	2
3.503	-0.0	00856358	00884875	00913389	00941902	00970412	00998920	01027427	01055931	01084434	01112934	2
3.504	-0.0	01141432	01169929	01198423	01226915	01255405	01283892	01312378	01340861	01369343	01397822	2
3.505	-0.0	01426299	01454774	01483246	01511717	01540185	01568651	01597114	01625576	01654035	01682492	2
3.506	-0.0	01710947	01739399	01767849	01796297	01824742	01853185	01881626	01910064	01938500	01966933	2
3.507	-0.0	01995365	02023793	02052220	02080643	02109065	02137484	02165900	02194314	02222726	02251135	2
3.508	-0.0	02279542	02307946	02336347	02364746	02393142	02421536	02449927	02478316	02506702	02535086	3
3.509	-0.0	02563467	02591845	02620220	02648593	02676964	02705331	02733696	02762058	02790418	02818775	3
3.510	-0.0	02847129	02875480	02903829	02932174	02960517	02988858	03017195	03045530	03073862	03102191	3
3.511	-0.0	03130517	03158840	03187161	03215478	03243793	03272105	03300414	03328720	03357023	03385323	3
3.512	-0.0	03413620	03441914	03470205	03498494	03526779	03555061	03583341	03611617	03639890	03668160	3
3.513	-0.0	03696427	03724691	03752952	03781210	03809465	03837717	03865965	03894211	03922453	03950692	3
3.514	-0.0	03978928	04007161	04035390	04063616	04091840	04120060	04148276	04176490	04204700	04232907	3
3.515	-0.0	04261110	04289311	04317508	04345702	04373892	04402079	04430263	04458443	04486620	04514794	3
3.516	-0.0	04542964	04571131	04599295	04627455	04655612	04683765	04711915	04740061	04768204	04796343	3
3.517	-0.0	04824479	04852611	04880740	04908865	04936987	04965105	04993220	05021331	05049439	05077543	4
3.518	-0.0	05105643	05133740	05161833	05189922	05218008	05246090	05274168	05302243	05330314	05358382	4
3.519	-0.0	05386446	05414505	05442562	05470614	05498663	05526708	05554749	05582787	05610820	05638850	4
3.520	-0.0	05666876	05694898	05722917	05750931	05778942	05806948	05834951	05862950	05890945	05918936	4
3.521	-0.0	05946924	05974907	06002886	06030862	06058833	06086800	06114764	06142723	06170679	06198630	4
3.522	-0.0	06226577	06254521	06282460	06310395	06338326	06366253	06394176	06422095	06450010	06477920	4
3.523	-0.0	06505827	06533729	06561627	06589521	06617411	06645296	06673178	06701055	06728928	06756797	4
3.524	-0.0	06784661	06812521	06840377	06868229	06896076	06923919	06951758	06979592	07007422	07035248	4
3.525	-0.0	07063069	07090886	07118699	07146507	07174311	07202110	07229905	07257696	07285482	07313263	4
3.526	-0.0	07341040	07368813	07396581	07424345	07452104	07479859	07507609	07535355	07563096	07590833	5
3.527	-0.0	07618565	07646292	07674015	07701733	07729447	07757156	07784860	07812560	07840255	07867945	5
3.528	-0.0	07895631	07923312	07950988	07978660	08006326	08033989	08061646	08089299	08116947	08144590	5
3.529	-0.0	08172228	08199862	08227490	08255114	08282733	08310348	08337957	08365552	08393162	08420756	5
3.530	-0.0	08448346	08475931	08503512	08531087	08558657	08586222	08613783	08641338	08668889	08696434	5
3.531	-0.0	08723975	08751510	08779041	08806566	08834087	08861602	08889112	08916617	08944118	08971613	5
3.532	-0.0	08999103	09026588	09054067	09081542	09109011	09136476	09163935	09191389	09218838	09246281	5
3.533	-0.0	09273720	09301153	09328581	09356004	09383421	09410833	09438240	09465642	09493038	09520430	5
3.534	-0.0	09547815	09575196	09602571	09629941	09657305	09684664	09712018	09739366	09766709	09794047	5
3.535	-0.0	09821379	09848706	09876027	09903343	09930653	09957958	09985258	10012552	10039840	10067123	6
3.536	-0.0	10094401	10121672	10148939	10176200	10203455	10230705	10257949	10285187	10312420	10339647	6
3.537	-0.0	10366869	10394085	10421296	10448500	10475599	10502893	10530081	10557263	10584439	10611610	6
3.538	-0.0	10638775	10665934	10693087	10720235	10747377	10774513	10801643	10828768	10855886	10882999	6
3.539	-0.0	10910106	10937208	10964303	10991393	11018476	11045554	11072626	11099692	11126752	11153806	6
3.540	-0.0	11180854	11207897	11234933	11261963	11288983	11316006	11343019	11370025	11397026	11424020	6
3.541	-0.0	11451008	11477991	11504967	11531937	11558901	11585859	11612811	11639757	11666697	11693630	6
3.542	-0.0	11720558	11747479	11774394	11801303	11828206	11855102	11881993	11908877	11935755	11962627	6
3.543	-0.0	11989492	12016351	12043204	12070051	12096892	12123726	12150554	12177375	12204190	12230999	6
3.544	-0.0	12257802	12284598	12311388	12338171	12364948	12391719	12418483	12445241	12471993	12498738	6
3.545	-0.0	12525476	12552208	12578934	12605653	12632366	12659072	12685772	12712465	12739152	12765832	7
3.546	-0.0	12792505	12819173	12845833	12872487	12899134	12925775	12952409	12979037	13005657	13032272	7
3.547	-0.0	13058879	13085480	13112074	13138662	13165243	13191817	13218385	13244945	13271499	13298047	7
3.548	-0.0	13324587	13351121	13377648	13404169	13430682	13457189	13483689	13510182	13536668	13563147	7
3.549	-0.0	13589620	13616086	13642544	13668996	13695442	13721880	13748311	13774735	13801153	13827563	7

TABLE 6

Sinc Function

x		.0	.0001	.0002	.0003	.0004	.0005	.0006	.0007	.0008	.0009	Δ_2
3.550	-0.0	13853967	13880363	13906753	13933136	13959511	13985880	14012242	14038596	14064944	14091284	7
3.551	-0.0	14117618	14143945	14170264	14196576	14222881	14249180	14275471	14301755	14328031	14354301	7
3.552	-0.0	14380564	14406819	14433067	14459308	14485542	14511769	14537988	14564200	14590405	14616603	7
3.553	-0.0	14642794	14668977	14695153	14721322	14747483	14773637	14799784	14825924	14852056	14878181	7
3.554	-0.0	14904298	14930408	14956511	14982607	15008695	15034775	15060849	15086914	15112973	15139024	7
3.555	-0.0	15165067	15191103	15217132	15243153	15269167	15295173	15321172	15347163	15373147	15399123	8
3.556	-0.0	15425091	15451052	15477006	15502952	15528890	15554821	15580744	15606660	15632567	15658468	8
3.557	-0.0	15684360	15710245	15736123	15761992	15787854	15813709	15839555	15865394	15891225	15917049	8
3.558	-0.0	15942865	15968673	15994473	16020265	16046050	16071827	16097596	16123357	16149111	16174857	8
3.559	-0.0	16200594	16226324	16252047	16277761	16303467	16329166	16354856	16380539	16406214	16431881	8
3.560	-0.0	16457540	16483191	16508834	16534469	16560096	16585716	16611327	16636930	16662525	16688113	8
3.561	-0.0	16713692	16739263	16764826	16790381	16815928	16841467	16866998	16892520	16918035	16943542	8
3.562	-0.0	16969040	16994530	17020012	17045486	17070952	17096410	17121859	17147301	17172734	17198159	8
3.563	-0.0	17223575	17248984	17274384	17299776	17325159	17350535	17375902	17401261	17426612	17451954	8
3.564	-0.0	17477288	17502613	17527931	17553240	17578540	17603833	17629117	17654392	17679659	17704918	8
3.565	-0.0	17730168	17755410	17780644	17805869	17831085	17856294	17881493	17906684	17931867	17957041	8
3.566	-0.0	17982207	18007364	18032513	18057653	18082785	18107908	18133023	18158129	18183226	18208315	9
3.567	-0.0	18233395	18258467	18283530	18308584	18333630	18358667	18383695	18408715	18433726	18458729	9
3.568	-0.0	18483723	18508708	18533684	18558652	18583611	18608561	18633502	18658435	18683359	18708274	9
3.569	-0.0	18733180	18758078	16782967	18807847	18832718	18857580	18882433	18907278	18932114	18956941	9
3.570	-0.0	18981759	19006568	19031368	19056160	19080942	19105716	19130480	19155236	19179982	19204720	9
3.571	-0.0	19229449	19254169	19278860	19303581	19328274	19352958	19377633	19402299	19426955	19451603	9
3.572	-0.0	19476242	19500871	19525492	19550103	19574705	19599298	19623882	19648457	19673023	19697580	9
3.573	-0.0	19722127	19746666	19771195	19795715	19820226	19844727	19869220	19893703	19918177	19942642	9
3.574	-0.0	19967097	19991543	20015980	20040408	20064827	20089236	20113636	20138026	20162407	20186779	9
3.575	-0.0	20211142	20235495	20259839	20284174	20308499	20332814	20357121	20381418	20405705	20429984	9
3.576	-0.0	20454252	20478512	20502762	20527002	20551233	20575455	20599667	20623869	20648062	20672246	10
3.577	-0.0	20696420	20720584	20744739	20768885	20793021	20817147	20841264	20865371	20889468	20913556	10
3.578	-0.0	20937635	20961704	20985763	21009812	21033852	21057883	21081903	21105914	21129915	21153907	10
3.579	-0.0	21177889	21201861	21225824	21249776	21273719	21297653	21321576	21345490	21369394	21393288	10
3.580	-0.0	21417173	21441047	21464912	21488767	21512613	21536448	21560274	21584090	21607896	21631692	10
3.581	-0.0	21655478	21679254	21703021	21726777	21750524	21774260	21797987	21821704	21845411	21869108	10
3.582	-0.0	21892795	21916472	21940139	21963796	21987443	22011081	22034708	22058325	22081932	22105529	10
3.583	-0.0	22129116	22152693	22176260	22199816	22223363	22246900	22270426	22293943	22317449	22340945	10
3.584	-0.0	22364431	22387907	22411373	22434829	22458274	22481709	22505135	22528549	22551954	22575349	10
3.585	-0.0	22598733	22622107	22645471	22668824	22692168	22715501	22738824	22762136	22785438	22808730	10
3.586	-0.0	22832012	22855283	22878544	22901795	22925035	22948265	22971485	22994694	23017893	23041082	10
3.587	-0.0	23064260	23087427	23110585	23133732	23156868	23179994	23203110	23226215	23249310	23272394	10
3.588	-0.0	23295468	23318531	23341584	23364626	23387658	23410679	23433690	23456690	23479680	23502659	11
3.589	-0.0	23525628	23548586	23571533	23594470	23617396	23640312	23663217	23686111	23708995	23731868	11
3.590	-0.0	23754731	23777583	23800424	23823254	23846074	23868883	23891682	23914470	23937247	23960013	11
3.591	-0.0	23982769	24005514	24028248	24050971	24073684	24096386	24119077	24141757	24164426	24187085	11
3.592	-0.0	24209733	24232370	24254996	24277612	24300216	24322810	24345393	24367965	24390526	24413076	11
3.593	-0.0	24435616	24458144	24480662	24503168	24525664	24548149	24570623	24593085	24615537	24637978	11
3.594	-0.0	24660408	24682827	24705235	24727632	24750018	24772393	24794757	24817110	24839452	24861783	11
3.595	-0.0	24884102	24906411	24928709	24950995	24973271	24995535	25017788	25040030	25062261	25084481	11
3.596	-0.0	25106690	25128888	25151074	25173249	25195413	25217566	25239708	25261839	25283958	25306066	11
3.597	-0.0	25328163	25350249	25372323	25394386	25416438	25438479	25460508	25482527	25504533	25526529	11
3.598	-0.0	25548513	25570486	25592448	25614398	25636337	25658265	25680181	25702086	25723980	25745862	11
3.599	-0.0	25767733	25789592	25811440	25833277	25855102	25876916	25898718	25920509	25942289	25964057	11

$$\frac{\sin 2\pi x}{2\pi x}$$

TABLE 6

Sinc Function

x		.0	.0001	.0002	.0003	.0004	.0005	.0006	.0007	.0008	.0009	Δ_2
3.600	0.0	25985813	26007559	26029292	26051014	26072725	26094424	26116112	26137788	26159453	26181106	12
3.601	-0.0	26202747	26224377	26245996	26267603	26289198	26310782	26332354	26353915	26375464	26397001	12
3.602	-0.0	26418527	26440041	26461543	26483034	26504514	26525981	26547437	26568881	26590314	26611735	12
3.603	-0.0	26633144	26654541	26675927	26697301	26718663	26740014	26761353	26782680	26803995	26825299	12
3.604	-0.0	26846590	26867870	26889139	26910395	26931640	26952873	26974094	26995303	27016500	27037686	12
3.605	-0.0	27058859	27080021	27101171	27122309	27143435	27164549	27185652	27206742	27227821	27248887	12
3.606	-0.0	27269942	27290985	27312016	27333035	27354042	27375037	27396020	27416991	27437950	27458897	12
3.607	-0.0	27479832	27500755	27521666	27542565	27563452	27584327	27605190	27626041	27646879	27667706	12
3.608	-0.0	27688521	27709323	27730114	27750892	27771658	27792412	27813154	27833884	27854602	27875308	12
3.609	-0.0	27896001	27916682	27937352	27958008	27978653	27999286	28019906	28040514	28061110	28081694	12
3.610	-0.0	28102266	28122825	28143372	28163907	28184429	28204940	28225438	28245923	28266397	28286858	12
3.611	-0.0	28307307	28327743	28348167	28368579	28388979	28409366	28429741	28450104	28470454	28490792	12
3.612	-0.0	28511117	28531430	28551731	28572019	28592295	28612558	28632809	28653048	28673274	28693488	12
3.613	-0.0	28713689	28733878	28754055	28774218	28794370	28814509	28834635	28854749	28874851	28894940	13
3.614	-0.0	28915016	28935080	28955132	28975170	28995197	29015210	29035212	29055200	29075176	29095140	13
3.615	-0.0	29115091	29135029	29154954	29174868	29194768	29214656	29234531	29254393	29274243	29294081	13
3.616	-0.0	29313905	29333717	29353516	29373303	29393077	29412838	29432586	29452322	29472045	29491755	13
3.617	-0.0	29511453	29531138	29550810	29570469	29590116	29609749	29629370	29648979	29668574	29688157	13
3.618	-0.0	29707727	29727284	29746828	29766359	29785878	29805384	29824876	29844356	29863824	29883278	13
3.619	-0.0	29902719	29922148	29941564	29960966	29980356	29999733	30019097	30038448	30057787	30077112	13
3.620	-0.0	30096424	30115724	30135010	30154284	30173544	30192792	30212026	30231248	30250456	30269652	13
3.621	-0.0	30288834	30308004	30327160	30346304	30365434	30384552	30403656	30422747	30441825	30460891	13
3.622	-0.0	30479943	30498982	30518008	30537020	30556020	30575007	30593980	30612940	30631888	30650822	13
3.623	-0.0	30669743	30688650	30707545	30726426	30745295	30764150	30782992	30801820	30820636	30839438	13
3.624	-0.0	30858227	30877003	30895766	30914515	30933252	30951974	30970684	30989381	31008064	31026734	13
3.625	-0.0	31045390	31064034	31082664	31101280	31119884	31138474	31157051	31175614	31194164	31212701	13
3.626	-0.0	31231225	31249735	31268231	31286715	31305185	31323641	31342085	31360515	31378931	31397334	13
3.627	-0.0	31415724	31434100	31452463	31470812	31489148	31507471	31525780	31544075	31562358	31580626	13
3.628	-0.0	31598881	31617123	31635351	31653566	31671767	31689955	31708129	31726290	31744437	31762571	14
3.629	-0.0	31780691	31798798	31816891	31834970	31853036	31871088	31889127	31907152	31925164	31943162	14
3.630	-0.0	31961146	31979117	31997074	32015018	32032947	32050864	32068766	32086655	32104531	32122392	14
3.631	-0.0	32140240	32158075	32175895	32193702	32211496	32229275	32247041	32264793	32282532	32300257	14
3.632	-0.0	32317968	32335665	32353348	32371018	32388674	32406317	32423945	32441560	32459161	32476748	14
3.633	-0.0	32494321	32511881	32529427	32546959	32564477	32581981	32599472	32616949	32634411	32651861	14
3.634	-0.0	32669296	32686717	32704124	32721518	32738898	32756264	32773616	32790954	32808278	32825588	14
3.635	-0.0	32842884	32860167	32877435	32894690	32911930	32929157	32946370	32963569	32980754	32997924	14
3.636	-0.0	33015081	33032224	33049353	33066468	33083569	33100656	33117729	33134788	33151833	33168864	14
3.637	-0.0	33185881	33202883	33219872	33236847	33253808	33270754	33287687	33304605	33321510	33338400	14
3.638	-0.0	33355276	33372138	33388986	33405820	33422640	33439446	33456237	33473015	33489778	33506527	14
3.639	-0.0	33523262	33539983	33556690	33573382	33590061	33606725	33623375	33640011	33656632	33673240	14
3.640	-0.0	33689833	33706412	33722977	33739527	33756063	33772586	33789093	33805587	33822066	33838531	14
3.641	-0.0	33854982	33871419	33887841	33904249	33920643	33937022	33953387	33969738	33986075	34002397	14
3.642	-0.0	34018705	34034998	34051278	34067543	34083793	34100029	34116251	34132459	34148652	34164831	14
3.643	-0.0	34180995	34197145	34213281	34229402	34245509	34261601	34277679	34293743	34309792	34325827	14
3.644	-0.0	34341847	34357853	34373844	34389821	34405784	34421732	34437666	34453585	34469490	34485380	14
3.645	-0.0	34501255	34517117	34532963	34548796	34564613	34580417	34596205	34611979	34627739	34643484	15
3.646	-0.0	34659215	34674931	34690632	34706319	34721991	34737649	34753292	34768921	34784535	34800135	15
3.647	-0.0	34815719	34831290	34846845	34862386	34877913	34893425	34908922	34924404	34939872	34955326	15
3.648	-0.0	34970764	34986188	35001598	35016992	35032372	35047738	35063088	35078424	35093745	35109052	15
3.649	-0.0	35124344	35139621	35154883	35170131	35185364	35200583	35215786	35230975	35246149	35261308	15

TABLE 6

Sinc Function

x		.0	.0001	.0002	.0003	.0004	.0005	.0006	.0007	.0008	.0009	Δ_2
3.650	-0.0	35276453	35291583	35306698	35321798	35336884	35351955	35367011	35382052	35397078	35412090	15
3.651	-0.0	35427087	35442069	35457036	35471988	35486926	35501849	35516756	35531649	35546528	35561391	15
3.652	-0.0	35576240	35591073	35605892	35620696	35635485	35650259	35665018	35679763	35694492	35709207	15
3.653	-0.0	35723907	35738592	35753261	35767916	35782556	35797182	35811792	35826387	35840967	35855533	15
3.654	-0.0	35870083	35884619	35899139	35913645	35928135	35942611	35957071	35971517	35985948	36000363	15
3.655	-0.0	36014764	36029149	36043520	36057876	36072216	36086542	36100852	36115148	36129428	36143694	15
3.656	-0.0	36157944	36172179	36186399	36200605	36214795	36228970	36243130	36257275	36271404	36285519	15
3.657	-0.0	36299619	36313703	36327773	36341827	36355866	36369890	36383899	36397893	36411871	36425835	15
3.658	-0.0	36439783	36453717	36467635	36481538	36495425	36509298	36523155	36536998	36550825	36564637	15
3.659	-0.0	36578433	36592215	36605981	36619732	36633468	36647189	36660894	36674584	36688259	36701919	15
3.660	-0.0	36715564	36729193	36742807	36756406	36769989	36783558	36797111	36810648	36824171	36837678	15
3.661	-0.0	36851170	36864647	36878108	36891554	36904985	36918400	36931801	36945185	36958555	36971909	15
3.662	-0.0	36985248	36998572	37011880	37025173	37038450	37051712	37064959	37078191	37091407	37104608	15
3.663	-0.0	37117793	37130963	37144118	37157257	37170381	37183489	37196582	37209660	37222722	37235769	15
3.664	-0.0	37248801	37261817	37274818	37287803	37300772	37313727	37326666	37339589	37352497	37365390	15
3.665	-0.0	37378267	37391129	37403975	37416806	37429621	37442421	37455205	37467974	37480727	37493465	15
3.666	-0.0	37506187	37518894	37531585	37544261	37556922	37569566	37582196	37594809	37607408	37619990	16
3.667	-0.0	37632557	37645109	37657645	37670166	37682671	37695160	37707634	37720092	37732535	37744962	16
3.668	-0.0	37757373	37769769	37782150	37794515	37806864	37819197	37831515	37843818	37856105	37868376	16
3.669	-0.0	37880631	37892871	37905096	37917304	37929497	37941675	37953836	37965983	37978113	37990228	16
3.670	-0.0	38002327	38014410	38026478	38038530	38050567	38062588	38074593	38086582	38098556	38110514	16
3.671	-0.0	38122456	38134383	38146294	38158189	38170069	38181933	38193781	38205613	38217430	38229231	16
3.672	-0.0	38241016	38252785	38264539	38276277	38287999	38299706	38311396	38323071	38334730	38346374	16
3.673	-0.0	38358001	38369613	38381209	38392790	38404354	38415903	38427436	38438953	38450454	38461940	16
3.674	-0.0	38473409	38484863	38496301	38507724	38519130	38530521	38541895	38553254	38564597	38575925	16
3.675	-0.0	38587236	38598532	38609811	38621075	38632323	38643555	38654772	38665972	38677157	38688325	16
3.676	-0.0	38699478	38710615	38721736	38732841	38743930	38755003	38766061	38777102	38788128	38799138	16
3.677	-0.0	38810131	38821109	38832071	38843017	38853947	38864861	38875760	38886642	38897508	38908359	16
3.678	-0.0	38919193	38930011	38940814	38951600	38962371	38973126	38983864	38994587	39005294	39015984	16
3.679	-0.0	39026659	39037318	39047961	39058587	39069198	39079793	39090372	39100934	39111481	39122012	16
3.680	-0.0	39132527	39143025	39153508	39163975	39174425	39184860	39195278	39205681	39216067	39226438	16
3.681	-0.0	39236792	39247130	39257453	39267759	39278049	39288323	39298581	39308823	39319049	39329259	16
3.682	-0.0	39339452	39349630	39359792	39369937	39380066	39390180	39400277	39410358	39420423	39430472	16
3.683	-0.0	39440504	39450521	39460521	39470506	39480474	39490426	39500362	39510282	39520186	39530073	16
3.684	-0.0	39539945	39549800	39559639	39569462	39579269	39589060	39598834	39608593	39618335	39628061	16
3.685	-0.0	39637771	39647464	39657142	39666803	39676448	39686077	39695690	39705287	39714867	39724431	16
3.686	-0.0	39733979	39743511	39753027	39762526	39772009	39781476	39790927	39800361	39809779	39819181	16
3.687	-0.0	39828567	39837937	39847290	39856627	39865948	39875253	39884541	39893813	39903069	39912309	16
3.688	-0.0	39921532	39930739	39939930	39949105	39958263	39967405	39976531	39985640	39994733	40003810	16
3.689	-0.0	40012871	40021915	40030943	40039955	40048950	40057930	40066893	40075839	40084769	40093683	16
3.690	-0.0	40102581	40111462	40120327	40129176	40138008	40146824	40155624	40164408	40173175	40181925	16
3.691	-0.0	40190660	40199378	40208079	40216765	40225434	40234086	40242723	40251343	40259946	40268534	16
3.692	-0.0	40277104	40285659	40294197	40302719	40311224	40319713	40328186	40336642	40345082	40353506	16
3.693	-0.0	40361913	40370304	40378678	40387036	40395377	40403703	40412011	40420304	40428580	40436839	16
3.694	-0.0	40445082	40453309	40461519	40469713	40477891	40486052	40494196	40502324	40510436	40518531	16
3.695	-0.0	40526610	40534673	40542719	40550748	40558762	40566758	40574738	40582702	40590650	40598580	16
3.696	-0.0	40606495	40614393	40622274	40630139	40637988	40645820	40653636	40661435	40669218	40676984	16
3.697	-0.0	40684734	40692467	40700184	40707884	40715568	40723235	40730886	40738520	40746138	40753739	16
3.698	-0.0	40761324	40768893	40776444	40783980	40791499	40799001	40806487	40813956	40821409	40828845	17
3.699	-0.0	40836265	40843668	40851055	40858425	40865778	40873115	40880436	40887740	40895028	40902298	17

TABLE 6

Sinc Function

x		.0	.0001	.0002	.0003	.0004	.0005	.0006	.0007	.0008	.0009	Δ_2
3.700	-0.0	40909553	40916791	40924012	40931217	40938405	40945577	40952732	40959871	40966993	40974098	17
3.701	-0.0	40981187	40988259	40995315	41002354	41009377	41016383	41023373	41030345	41037302	41044242	17
3.702	-0.0	41051165	41058071	41064962	41071835	41078692	41085532	41092356	41099163	41105954	41112727	17
3.703	-0.0	41119485	41126226	41132950	41139657	41146348	41153022	41159680	41166321	41172946	41179554	17
3.704	-0.0	41186145	41192720	41199278	41205819	41212344	41218852	41225344	41231819	41238277	41244719	17
3.705	-0.0	41251144	41257552	41263944	41270319	41276677	41283019	41289345	41295653	41301945	41308220	17
3.706	-0.0	41314479	41320721	41326946	41333155	41339347	41345523	41351681	41357823	41363949	41370058	17
3.707	-0.0	41376150	41382225	41388284	41394326	41400352	41406360	41412352	41418328	41424287	41430229	17
3.708	-0.0	41436154	41442063	41447955	41453830	41459689	41465531	41471356	41477165	41482957	41488732	17
3.709	-0.0	41494491	41500233	41505958	41511666	41517358	41523033	41528692	41534334	41539959	41545567	17
3.710	-0.0	41551158	41556733	41562292	41567833	41573358	41578866	41584357	41589832	41595290	41600731	17
3.711	-0.0	41606156	41611563	41616954	41622329	41627686	41633027	41638351	41643659	41648950	41654224	17
3.712	-0.0	41659481	41664721	41669945	41675152	41680343	41685516	41690673	41695813	41700937	41706043	17
3.713	-0.0	41711133	41716207	41721263	41726303	41731326	41736332	41741321	41746294	41751250	41756189	17
3.714	-0.0	41761112	41766017	41770906	41775779	41780634	41785473	41790295	41795100	41799888	41804660	17
3.715	-0.0	41809415	41814153	41818875	41823579	41828267	41832938	41837592	41842230	41846851	41851455	17
3.716	-0.0	41856042	41860613	41865166	41869703	41874223	41878727	41883214	41887683	41892136	41896573	17
3.717	-0.0	41900992	41905395	41909781	41914150	41918503	41922838	41927157	41931459	41935744	41940013	17
3.718	-0.0	41944264	41948499	41952717	41956919	41961103	41965271	41969422	41973556	41977674	41981774	17
3.719	-0.0	41985858	41989925	41993975	41998008	42002025	42006025	42010008	42013974	42017923	42021856	17
3.720	-0.0	42025772	42029671	42033553	42037419	42041267	42045099	42048914	42052712	42056493	42060258	17
3.721	-0.0	42064006	42067737	42071451	42075148	42078829	42082493	42086139	42089770	42093383	42096979	17
3.722	-0.0	42100559	42104122	42107668	42111197	42114710	42118205	42121684	42125146	42128591	42132019	17
3.723	-0.0	42135431	42138826	42142204	42145565	42148909	42152236	42155547	42158841	42162118	42165378	17
3.724	-0.0	42168621	42171848	42175057	42178250	42181426	42184586	42187728	42190854	42193962	42197054	17
3.725	-0.0	42200129	42203188	42206229	42209254	42212261	42215252	42218227	42221184	42224124	42227048	17
3.726	-0.0	42229955	42232845	42235718	42238575	42241414	42244237	42247043	42249832	42252604	42255359	17
3.727	-0.0	42258098	42260820	42263525	42266213	42268884	42271538	42274176	42276797	42279401	42281988	17
3.728	-0.0	42284558	42287111	42289648	42292168	42294671	42297157	42299626	42302078	42304514	42306933	17
3.729	-0.0	42309335	42311720	42314088	42316440	42318774	42321092	42323393	42325677	42327945	42330195	17
3.730	-0.0	42332429	42334645	42336845	42339029	42341195	42343344	42345477	42347593	42349692	42351774	17
3.731	-0.0	42353839	42355888	42357919	42359934	42361932	42363913	42365878	42367825	42369756	42371670	17
3.732	-0.0	42373567	42375447	42377310	42379157	42380987	42382800	42384596	42386375	42388137	42389883	17
3.733	-0.0	42391611	42393323	42395018	42396697	42398358	42400003	42401631	42403241	42404836	42406413	17
3.734	-0.0	42407973	42409517	42411044	42412554	42414047	42415523	42416983	42418426	42419851	42421260	17
3.735	-0.0	42422653	42424028	42425387	42426728	42428053	42429362	42430653	42431927	42433185	42434426	17
3.736	-0.0	42435650	42436857	42438047	42439221	42440378	42441518	42442641	42443747	42444837	42445909	17
3.737	-0.0	42446965	42448004	42449027	42450032	42451021	42451992	42452947	42453886	42454807	42455712	17
3.738	-0.0	42456599	42457470	42458324	42459162	42459982	42460786	42461573	42462343	42463096	42463833	17
3.739	-0.0	42464552	42465255	42465941	42466611	42467263	42467899	42468518	42469120	42469705	42470274	17
3.740	-0.0	42470825	42471360	42471878	42472380	42472864	42473332	42473783	42474217	42474634	42475035	17
3.741	-0.0	42475419	42475786	42476136	42476469	42476786	42477086	42477369	42477635	42477885	42478117	17
3.742	-0.0	42478333	42478533	42478715	42478880	42479029	42479161	42479277	42479375	42479457	42479522	17
3.743	-0.0	42479570	42479601	42479616	42479614	42479595	42479559	42479507	42479437	42479351	42479249	17
3.744	-0.0	42479129	42478993	42478840	42478670	42478483	42478280	42478060	42477823	42477569	42477299	17
3.745	-0.0	42477012	42476708	42476387	42476050	42475696	42475325	42474937	42474533	42474112	42473674	17
3.746	-0.0	42473220	42472748	42472260	42471755	42471234	42470695	42470140	42469569	42468980	42468375	17
3.747	-0.0	42467753	42467114	42466459	42465786	42465098	42464392	42463670	42462930	42462175	42461402	17
3.748	-0.0	42460613	42459807	42458984	42458145	42457289	42456416	42455526	42454620	42453697	42452757	17
3.749	-0.0	42451801	42450828	42449838	42448831	42447808	42446768	42445712	42444638	42443548	42442442	17

TABLE 6

Sinc Function

x		.0	.0001	.0002	.0003	.0004	.0005	.0006	.0007	.0008	.0009	Δ_2
3.750	-0.0	42441318	42440178	42439021	42437848	42436658	42435451	42434227	42432987	42431730	42430456	17
3.751	-0.0	42429166	42427859	42426535	42425195	42423838	42422464	42421074	42419667	42418243	42416803	17
3.752	-0.0	42415346	42413872	42412382	42410874	42409351	42407810	42406253	42404680	42403089	42401482	17
3.753	-0.0	42399859	42398218	42396561	42394888	42393198	42391491	42389767	42388027	42386270	42384497	17
3.754	-0.0	42382707	42380900	42379076	42377236	42375380	42373506	42371617	42369710	42367787	42365847	17
3.755	-0.0	42363891	42361918	42359928	42357922	42355899	42353860	42351803	42349731	42347641	42345535	17
3.756	-0.0	42343413	42341274	42339118	42336946	42334757	42332551	42330329	42328091	42325835	42323563	17
3.757	-0.0	42321275	42318970	42316648	42314310	42311955	42309584	42307196	42304791	42302370	42299933	17
3.758	-0.0	42297478	42295008	42292520	42290016	42287496	42284959	42282405	42279835	42277248	42274645	17
3.759	-0.0	42272025	42269389	42266736	42264066	42261380	42258678	42255959	42253223	42250471	42247702	17
3.760	-0.0	42244917	42242115	42239297	42236462	42233611	42230743	42227858	42224957	42222040	42219106	17
3.761	-0.0	42216156	42213189	42210205	42207205	42204189	42201156	42198107	42195041	42191958	42188859	16
3.762	-0.0	42185744	42182612	42179463	42176299	42173117	42169919	42166705	42163474	42160227	42156963	16
3.763	-0.0	42153683	42150386	42147073	42143744	42140397	42137035	42133656	42130261	42126849	42123420	16
3.764	-0.0	42119976	42116514	42113037	42109543	42106032	42102505	42098962	42095402	42091825	42088233	16
3.765	-0.0	42084624	42080998	42077356	42073698	42070023	42066332	42062624	42058900	42055160	42051403	16
3.766	-0.0	42047629	42043840	42040034	42036211	42032372	42028517	42024646	42020758	42016853	42012933	16
3.767	-0.0	42008995	42005042	42001072	41997086	41993083	41989064	41985029	41980977	41976909	41972825	16
3.768	-0.0	41968724	41964607	41960473	41956323	41952157	41947975	41943776	41939561	41935329	41931081	16
3.769	-0.0	41926817	41922537	41918240	41913927	41909597	41905252	41900890	41896511	41892117	41887706	16
3.770	-0.0	41883278	41878835	41874375	41869899	41865406	41860897	41856372	41851831	41847273	41842699	16
3.771	-0.0	41838109	41833503	41828880	41824241	41819586	41814914	41810226	41805522	41800802	41796066	16
3.772	-0.0	41791313	41786544	41781758	41776957	41772139	41767305	41762455	41757589	41752706	41747807	16
3.773	-0.0	41742892	41737960	41733013	41728049	41723069	41718073	41713061	41708032	41702987	41697926	16
3.774	-0.0	41692849	41687756	41682646	41677520	41672378	41667220	41662046	41656856	41651649	41646426	16
3.775	-0.0	41641187	41635932	41630661	41625373	41620070	41614750	41609414	41604062	41598694	41593310	16
3.776	-0.0	41587909	41582493	41577060	41571611	41566146	41560665	41555168	41549655	41544125	41538580	16
3.777	-0.0	41533018	41527440	41521847	41516237	41510611	41504969	41499310	41493636	41487946	41482239	16
3.778	-0.0	41476517	41470778	41465024	41459253	41453466	41447663	41441844	41436010	41430159	41424292	16
3.779	-0.0	41418408	41412509	41406594	41400663	41394716	41388753	41382773	41376778	41370767	41364739	16
3.780	-0.0	41358696	41352637	41346562	41340470	41334363	41328240	41322100	41315945	41309774	41303586	16
3.781	-0.0	41297383	41291164	41284929	41278678	41272410	41266127	41259828	41253513	41247182	41240835	16
3.782	-0.0	41234473	41228094	41221699	41215288	41208862	41202419	41195961	41189486	41182996	41176490	16
3.783	-0.0	41169968	41163430	41156876	41150306	41143720	41137119	41130501	41123868	41117218	41110553	16
3.784	-0.0	41103872	41097175	41090463	41083734	41076989	41070229	41063453	41056661	41049853	41043029	16
3.785	-0.0	41036189	41029334	41022463	41015575	41008673	41001754	40994819	40987869	40980902	40973920	16
3.786	-0.0	40966923	40959909	40952880	40945834	40938773	40931696	40924604	40917495	40910371	40903231	16
3.787	-0.0	40896076	40888904	40881717	40874514	40867295	40860061	40852810	40845544	40838263	40830965	16
3.788	-0.0	40823652	40816323	40808978	40801618	40794242	40786850	40779442	40772019	40764580	40757125	16
3.789	-0.0	40749655	40742169	40734667	40727150	40719617	40712068	40704504	40696923	40689328	40681716	16
3.790	-0.0	40674089	40666446	40658788	40651114	40643424	40635719	40627998	40620261	40612509	40604741	16
3.791	-0.0	40596958	40589159	40581344	40573514	40565668	40557806	40549929	40542036	40534128	40526204	16
3.792	-0.0	40518265	40510310	40502339	40494353	40486351	40478334	40470301	40462253	40454189	40446109	16
3.793	-0.0	40438014	40429903	40421777	40413636	40405478	40397306	40389118	40380914	40372695	40364460	16
3.794	-0.0	40356210	40347944	40339663	40331366	40323054	40314726	40306383	40298024	40289650	40281261	15
3.795	-0.0	40272856	40264435	40255999	40247548	40239081	40230599	40222101	40213588	40205059	40196515	15
3.796	-0.0	40187956	40179381	40170791	40162185	40153564	40144928	40136276	40127609	40118926	40110228	15
3.797	-0.0	40101515	40092786	40084042	40075283	40066508	40057718	40048912	40040091	40031255	40022404	15
3.798	-0.0	40013537	40004654	39995757	39986844	39977916	39968972	39960014	39951040	39942050	39933045	15
3.799	-0.0	39924025	39914990	39905940	39896874	39887793	39878696	39869585	39860458	39851316	39842158	15

TABLE 6

Sinc Function

x	.0	.0001	.0002	.0003	.0004	.0005	.0006	.0007	.0008	.0009	Δ_2
3.800 -0.0	39832986	39823798	39814595	39805376	39796143	39786894	39777630	39768351	39759056	39749747	15
3.801 -0.0	39740422	39731082	39721726	39712356	39702970	39693570	39684154	39674722	39665276	39655815	15
3.802 -0.0	39646338	39636846	39627339	39617817	39608280	39598728	39589160	39579577	39569980	39560367	15
3.803 -0.0	39550739	39541096	39531438	39521764	39512076	39502373	39492654	39482920	39473172	39463408	15
3.804 -0.0	39453629	39443835	39434026	39424202	39414363	39404509	39394640	39384756	39374857	39364942	15
3.805 -0.0	39355013	39345069	39335110	39325135	39315146	39305142	39295123	39285088	39275039	39264975	15
3.806 -0.0	39254896	39244802	39234693	39224569	39214430	39204276	39194107	39183923	39173724	39163510	15
3.807 -0.0	39153282	39143038	39132780	39122506	39112218	39101915	39091597	39081264	39070916	39060553	15
3.808 -0.0	39050176	39039783	39029376	39018954	39008517	38998065	38987598	38977117	38966620	38956109	15
3.809 -0.0	38945583	38935042	38924486	38913916	38903330	38892730	38882115	38871485	38860841	38850182	15
3.810 -0.0	38839508	38828819	38818115	38807397	38796664	38785916	38775153	38764376	38753583	38742777	15
3.811 -0.0	38731955	38721119	38710268	38699402	38688522	38677626	38666717	38655792	38644853	38633899	15
3.812 -0.0	38622930	38611947	38600949	38589937	38578909	38567867	38556811	38545740	38534654	38523553	15
3.813 -0.0	38512438	38501309	38490164	38479005	38467832	38456644	38445441	38434224	38422992	38411745	15
3.814 -0.0	38400484	38389209	38377919	38366614	38355295	38343961	38332612	38321249	38309872	38298480	15
3.815 -0.0	38287073	38275652	38264217	38252767	38241302	38229823	38218330	38206822	38195299	38183762	14
3.816 -0.0	38172211	38160645	38149065	38137470	38125861	38114237	38102599	38090946	38079279	38067598	14
3.817 -0.0	38055902	38044192	38032467	38020728	38008975	37997207	37985424	37973628	37961817	37949992	14
3.818 -0.0	37938152	37926298	37914429	37902547	37890650	37878738	37866812	37854872	37842918	37830949	14
3.819 -0.0	37818966	37806969	37794957	37782931	37770891	37758837	37746768	37734685	37722588	37710476	14
3.820 -0.0	37698350	37686210	37674056	37661887	37649705	37637508	37625296	37613071	37600831	37588578	14
3.821 -0.0	37576310	37564027	37551731	37539420	37527096	37514757	37502404	37490036	37477655	37465259	14
3.822 -0.0	37452850	37440426	37427988	37415536	37403070	37390589	37378095	37365586	37353064	37340527	14
3.823 -0.0	37327976	37315411	37302832	37290239	37277632	37265011	37252376	37239727	37227063	37214386	14
3.824 -0.0	37201695	37188989	37176270	37163537	37150789	37138028	37125253	37112463	37099660	37086842	14
3.825 -0.0	37074011	37061166	37048307	37035433	37022546	37009645	36996730	36983801	36970858	36957902	14
3.826 -0.0	36944931	36931946	36918948	36905935	36892909	36879869	36866815	36853747	36840665	36827570	14
3.827 -0.0	36814460	36801337	36788200	36775049	36761884	36748705	36735513	36722306	36709086	36695852	14
3.828 -0.0	36682604	36669343	36656068	36642779	36629476	36616159	36602829	36589485	36576127	36562755	14
3.829 -0.0	36549370	36535971	36522558	36509132	36495691	36482237	36468770	36455288	36441793	36428285	14
3.830 -0.0	36414762	36401226	36387677	36374113	36360536	36346946	36333341	36319723	36306092	36292447	14
3.831 -0.0	36278788	36265115	36251429	36237730	36224017	36210290	36196550	36182796	36169028	36155247	14
3.832 -0.0	36141452	36127644	36113823	36099987	36086139	36072276	36058401	36044511	36030608	36016692	13
3.833 -0.0	36002762	35988819	35974862	35960892	35946908	35932911	35918901	35904876	35890839	35876788	13
3.834 -0.0	35862724	35848646	35834555	35820450	35806332	35792200	35778056	35763897	35749726	35735541	13
3.835 -0.0	35721342	35707131	35692906	35678667	35664416	35650150	35635872	35621580	35607275	35592957	13
3.836 -0.0	35578625	35564280	35549922	35535550	35521166	35506767	35492356	35477931	35463494	35449042	13
3.837 -0.0	35434578	35420100	35405610	35391106	35376588	35362058	35347514	35332957	35318387	35303804	13
3.838 -0.0	35289207	35274598	35259975	35245339	35230690	35216028	35201353	35186664	35171962	35157248	13
3.839 -0.0	35142520	35127779	35113025	35098258	35083477	35068684	35053878	35039058	35024226	35009380	13
3.840 -0.0	34994522	34979650	34964765	34949868	34934957	34920033	34905096	34890146	34875184	34860208	13
3.841 -0.0	34845219	34830218	34815203	34800175	34785135	34770081	34755015	34739935	34724843	34709738	13
3.842 -0.0	34694619	34679488	34664344	34649187	34634018	34618835	34603639	34588431	34573210	34557975	13
3.843 -0.0	34542729	34527469	34512196	34496911	34481612	34466301	34450977	34435640	34420291	34404928	13
3.844 -0.0	34389553	34374165	34358765	34343351	34327925	34312486	34297035	34281570	34266093	34250603	13
3.845 -0.0	34235101	34219585	34204057	34188517	34172963	34157397	34141818	34126227	34110623	34095006	13
3.846 -0.0	34079377	34063735	34048080	34032413	34016733	34001041	33985336	33969618	33953888	33938145	13
3.847 -0.0	33922389	33906621	33890841	33875047	33859242	33843423	33827593	33811749	33795893	33780025	13
3.848 -0.0	33764144	33748251	33732345	33716427	33700496	33684552	33668597	33652628	33636648	33620654	12
3.849 -0.0	33604649	33588631	33572600	33556557	33540502	33524434	33508354	33492262	33476157	33460040	12

TABLE 6

Sinc Function

x		.0	.0001	.0002	.0003	.0004	.0005	.0006	.0007	.0008	.0009	Δ_2
3.850	-0.0	33443910	33427768	33411614	33395447	33379208	33363077	33346873	33330657	33314428	33298188	12
3.851	-0.0	33281935	33265669	33249392	33233102	33216800	33200486	33184159	33167820	33151469	33135106	12
3.852	-0.0	33118730	33102342	33085942	33069530	33053105	33036669	33020220	33003759	32987286	32970800	12
3.853	-0.0	32954303	32937793	32921271	32904737	32888191	32871633	32855063	32838480	32821886	32805279	12
3.854	-0.0	32788660	32772029	32755387	32738732	32722065	32705386	32688694	32671991	32655276	32638549	12
3.855	-0.0	32621810	32605058	32588295	32571520	32554733	32537933	32521122	32504299	32487464	32470617	12
3.856	-0.0	32453758	32436887	32420004	32403109	32386203	32369284	32352354	32335411	32318457	32301491	12
3.857	-0.0	32284513	32267523	32250521	32233507	32216482	32199444	32182395	32165334	32148262	32131177	12
3.858	-0.0	32114081	32096973	32079853	32062721	32045577	32028422	32011255	31994076	31976886	31959684	12
3.859	-0.0	31942470	31925244	31908007	31890758	31873497	31856224	31838940	31821644	31804337	31787018	12
3.860	-0.0	31769687	31752344	31734990	31717625	31700247	31682858	31665458	31648046	31630622	31613187	12
3.861	-0.0	31595740	31578281	31560811	31543330	31525836	31508332	31490816	31473288	31455749	31438198	12
3.862	-0.0	31420636	31403062	31385477	31367880	31350272	31332652	31315021	31297378	31279724	31262059	11
3.863	-0.0	31244382	31226694	31208994	31191283	31173560	31155826	31138081	31120325	31102557	31084777	11
3.864	-0.0	31066986	31049184	31031371	31013546	30995710	30977863	30960004	30942134	30924253	30906360	11
3.865	-0.0	30888457	30870541	30852615	30834678	30816729	30798769	30780797	30762815	30744821	30726816	11
3.866	-0.0	30708800	30690773	30672734	30654684	30636624	30618551	30600468	30582374	30564269	30546152	11
3.867	-0.0	30528024	30509885	30491735	30473574	30455402	30437219	30419025	30400819	30382603	30364376	11
3.868	-0.0	30346137	30327888	30309627	30291355	30273073	30254779	30236475	30218159	30199832	30181495	11
3.869	-0.0	30163146	30144787	30126416	30108035	30089643	30071239	30052825	30034400	30015964	29997517	11
3.870	-0.0	29979060	29960591	29942111	29923621	29905120	29886608	29868085	29849551	29831006	29812451	11
3.871	-0.0	29793885	29775308	29756720	29738121	29719512	29700892	29682261	29663619	29644967	29626304	11
3.872	-0.0	29607630	29588945	29570250	29551544	29532827	29514099	29495361	29476613	29457853	29439083	11
3.873	-0.0	29420302	29401511	29382709	29363896	29345073	29326239	29307394	29288539	29269673	29250797	11
3.874	-0.0	29231910	29213013	29194105	29175186	29156257	29137318	29118367	29099407	29080436	29061454	11
3.875	-0.0	29042462	29023459	29004446	28985422	28966388	28947344	28928289	28909224	28890148	28871062	10
3.876	-0.0	28851965	28832858	28813740	28794613	28775474	28756326	28737167	28717998	28698818	28679628	10
3.877	-0.0	28660428	28641217	28621996	28602765	28583523	28564271	28545009	28525737	28506454	28487161	10
3.878	-0.0	28467858	28448544	28429221	28409887	28390543	28371188	28351824	28332449	28313064	28293669	10
3.879	-0.0	28274264	28254848	28235423	28215987	28196541	28177085	28157619	28138143	28118656	28099160	10
3.880	-0.0	28079654	28060137	28040610	28021074	28001527	27981970	27962403	27942826	27923239	27903642	10
3.881	-0.0	27884035	27864418	27844792	27825155	27805508	27785851	27766184	27746507	27726821	27707124	10
3.882	-0.0	27687417	27667701	27647975	27628238	27608492	27588736	27568970	27549195	27529409	27509613	10
3.883	-0.0	27489808	27469993	27450168	27430333	27410489	27390634	27370770	27350896	27331012	27311119	10
3.884	-0.0	27291215	27271302	27251380	27231447	27211505	27191553	27171591	27151620	27131639	27111648	10
3.885	-0.0	27091648	27071638	27051618	27031589	27011550	26991501	26971443	26951375	26931298	26911211	10
3.886	-0.0	26891114	26871008	26850892	26830766	26810632	26790487	26770333	26750169	26729996	26709814	10
3.887	-0.0	26689622	26669420	26649209	26628988	26608758	26588519	26568270	26548012	26527744	26507466	9
3.888	-0.0	26487180	26466884	26446578	26426263	26405939	26385605	26365262	26344910	26324548	26304177	9
3.889	-0.0	26283796	26263407	26243008	26222599	26202182	26181755	26161318	26140873	26120418	26099954	9
3.890	-0.0	26079480	26058998	26038506	26018005	25997495	25976975	25956447	25935909	25915362	25894805	9
3.891	-0.0	25874240	25853666	25833082	25812489	25791887	25771276	25750656	25730026	25709388	25688741	9
3.892	-0.0	25668084	25647418	25626744	25606060	25585367	25564665	25543954	25523234	25502506	25481768	9
3.893	-0.0	25461021	25440265	25419500	25398726	25377943	25357152	25336351	25315541	25294723	25273895	9
3.894	-0.0	25253059	25232214	25211360	25190496	25169625	25148744	25127854	25106956	25086048	25065132	9
3.895	-0.0	25044207	25023273	25002331	24981380	24960419	24939450	24918473	24897486	24876491	24855487	9
3.896	-0.0	24834474	24813453	24792423	24771384	24750336	24729280	24708215	24687142	24666059	24644968	9
3.897	-0.0	24623869	24602761	24581644	24560518	24539384	24518242	24497090	24475931	24454762	24433585	9
3.898	-0.0	24412400	24391205	24370003	24348792	24327572	24306344	24285107	24263862	24242608	24221346	8
3.899	-0.0	24200075	24178796	24157509	24136213	24114908	24093595	24072274	24050944	24029606	24008260	8

$\frac{\sin 2\pi x}{2\pi x}$

TABLE 6

Sinc Function

x	.0	.0001	.0002	.0003	.0004	.0005	.0006	.0007	.0008	.0009	Δ_2
3.900 -0.0	23986905	23965541	23944170	23922790	23901402	23880005	23858600	23837187	23815765	23794335	8
3.901 -0.0	23772897	23751450	23729995	23708532	23687061	23665581	23644094	23622598	23601093	23579581	8
3.902 -0.0	23558060	23536531	23514994	23493449	23471896	23450334	23428764	23407186	23385600	23364006	8
3.903 -0.0	23342404	23320794	23299175	23277548	23255914	23234271	23212620	23190962	23169295	23147620	8
3.904 -0.0	23125937	23104246	23082547	23060840	23039125	23017402	22995671	22973932	22952185	22930431	8
3.905 -0.0	22908668	22886897	22865119	22843332	22821538	22799735	22777925	22756107	22734281	22712448	8
3.906 -0.0	22690606	22668757	22646899	22625034	22603161	22581281	22559392	22537496	22515592	22493680	8
3.907 -0.0	22471760	22449833	22427898	22405955	22384004	22362046	22340080	22318107	22296125	22274136	8
3.908 -0.0	22252139	22230135	22208123	22186104	22164076	22142041	22119999	22097949	22075891	22053826	8
3.909 -0.0	22031753	22009672	21987584	21965489	21943386	21921275	21899157	21877031	21854898	21832758	7
3.910 -0.0	21810609	21788454	21766291	21744120	21721942	21699757	21677564	21655364	21633156	21610941	7
3.911 -0.0	21588718	21566488	21544251	21522007	21499755	21477495	21455229	21432954	21410673	21388384	7
3.912 -0.0	21366089	21343785	21321475	21299157	21276832	21254500	21232160	21209813	21187459	21165098	7
3.913 -0.0	21142729	21120354	21097971	21075581	21053183	21030779	21008367	20985949	20963523	20941090	7
3.914 -0.0	20918650	20896202	20873748	20851287	20828818	20806343	20783860	20761370	20738873	20716370	7
3.915 -0.0	20693859	20671341	20648816	20626284	20603746	20581200	20558647	20536087	20513520	20490947	7
3.916 -0.0	20468366	20445779	20423184	20400583	20377975	20355359	20332737	20310109	20287473	20264830	7
3.917 -0.0	20242181	20219524	20196861	20174191	20151515	20128831	20106141	20083444	20060740	20038029	7
3.918 -0.0	20015312	19992588	19969857	19947119	19924375	19901624	19878866	19856102	19833331	19810553	7
3.919 -0.0	19787769	19764978	19742180	19719376	19696565	19673747	19650923	19628093	19605255	19582411	7
3.920 -0.0	19559561	19536704	19513840	19490970	19468094	19445210	19422321	19399425	19376522	19353613	6
3.921 -0.0	19330697	19307775	19284847	19261912	19238970	19216023	19193068	19170108	19147141	19124167	6
3.922 -0.0	19101188	19078201	19055209	19032210	19009205	18986193	18963175	18940151	18917121	18894084	6
3.923 -0.0	18871041	18847992	18824936	18801874	18778806	18755732	18732651	18709564	18686471	18663372	6
3.924 -0.0	18640267	18617155	18594037	18570913	18547783	18524647	18501505	18478356	18455202	18432041	6
3.925 -0.0	18408874	18385702	18362523	18339338	18316147	18292949	18269746	18246537	18223322	18200101	6
3.926 -0.0	18176873	18153640	18130401	18107156	18083905	18060648	18037385	18014116	17990841	17967560	6
3.927 -0.0	17944273	17920981	17897682	17874378	17851067	17827751	17804429	17781102	17757768	17734429	6
3.928 -0.0	17711083	17687732	17664375	17641013	17617644	17594270	17570890	17547504	17524113	17500716	6
3.929 -0.0	17477313	17453904	17430490	17407070	17383644	17360213	17336776	17313333	17289885	17266431	6
3.930 -0.0	17242972	17219507	17196036	17172559	17149078	17125590	17102097	17078598	17055094	17031584	6
3.931 -0.0	17008069	16984548	16961022	16937490	16913953	16890410	16866862	16843309	16819750	16796185	5
3.932 -0.0	16772615	16749040	16725459	16701872	16678281	16654684	16631081	16607474	16583861	16560242	5
3.933 -0.0	16536618	16512989	16489355	16465715	16442070	16418420	16394764	16371103	16347437	16323766	5
3.934 -0.0	16300089	16276407	16252720	16229028	16205330	16181628	16157920	16134207	16110489	16086765	5
3.935 -0.0	16063037	16039303	16015564	15991821	15968072	15944317	15920558	15896794	15873025	15849250	5
3.936 -0.0	15825471	15801686	15777897	15754102	15730303	15706498	15682689	15658874	15635055	15611231	5
3.937 -0.0	15587401	15563567	15539728	15515883	15492034	15468180	15444321	15420458	15396589	15372715	5
3.938 -0.0	15348837	15324954	15301066	15277173	15253275	15229373	15205465	15181553	15157636	15133715	5
3.939 -0.0	15109788	15085857	15061921	15037981	15014035	14990085	14966130	14942171	14918207	14894238	5
3.940 -0.0	14870265	14846286	14822304	14798316	14774324	14750328	14726326	14702320	14678310	14654295	5
3.941 -0.0	14630275	14606251	14582223	14558189	14534152	14510109	14486063	14462011	14437956	14413895	4
3.942 -0.0	14389831	14365761	14341688	14317610	14293527	14269440	14245349	14221253	14197153	14173049	4
3.943 -0.0	14148940	14124827	14100709	14076587	14052461	14028330	14004195	13980056	13955913	13931765	4
3.944 -0.0	13907613	13883456	13859296	13835131	13810962	13786789	13762611	13738429	13714244	13690053	4
3.945 -0.0	13665859	13641661	13617458	13593251	13569040	13544825	13520606	13496383	13472156	13447924	4
3.946 -0.0	13423689	13399449	13375205	13350958	13326706	13302450	13278190	13253927	13229659	13205387	4
3.947 -0.0	13181111	13156831	13132548	13108260	13083968	13059673	13035373	13011070	12986763	12962451	4
3.948 -0.0	12938136	12913817	12889494	12865168	12840837	12816503	12792165	12767823	12743477	12719127	4
3.949 -0.0	12694774	12670417	12646056	12621691	12597322	12572950	12548574	12524195	12499811	12475424	4

TABLE 6

Sinc Function

x	.0	.0001	.0002	.0003	.0004	.0005	.0006	.0007	.0008	.0009	Δ_2
3.950 -0.0	12451033	12426639	12402241	12377839	12353434	12329025	12304612	12280196	12255776	12231352	4
3.951 -0.0	12206925	12182494	12158060	12133622	12109181	12084736	12060288	12035836	12011380	11986921	4
3.952 -0.0	11962459	11937993	11913523	11889050	11864574	11840094	11815611	11791124	11766634	11742140	3
3.953 -0.0	11717643	11693143	11668639	11644132	11619622	11595108	11570591	11546071	11521547	11497020	3
3.954 -0.0	11472490	11447956	11423419	11398879	11374336	11349789	11325239	11300686	11276130	11251570	3
3.955 -0.0	11227007	11202441	11177872	11153300	11128724	11104146	11079564	11054979	11030391	11005800	3
3.956 -0.0	10981206	10956608	10932008	10907405	10882798	10858188	10833576	10808960	10784341	10759720	3
3.957 -0.0	10735095	10710467	10685837	10661203	10636566	10611927	10587284	10562639	10537990	10513339	3
3.958 -0.0	10488685	10464028	10439368	10414705	10390039	10365371	10340699	10316025	10291348	10266668	3
3.959 -0.0	10241985	10217300	10192612	10167921	10143227	10118530	10093831	10069129	10044424	10019716	3
3.960 -0.0	09995006	09970293	09945578	09920859	09896139	09871415	09846689	09821960	09797228	09772494	3
3.961 -0.0	09747757	09723018	09698276	09673532	09648784	09624035	09599283	09574528	09549771	09525011	3
3.962 -0.0	09500248	09475484	09450716	09425947	09401174	09376400	09351622	09326843	09302061	09277276	2
3.963 -0.0	09252490	09227700	09202909	09178115	09153318	09128519	09103718	09078915	09054109	09029301	2
3.964 -0.0	09004490	08979678	08954863	08930045	08905226	08880404	08855580	08830753	08805925	08781094	2
3.965 -0.0	08756261	08731426	08706588	08681749	08656907	08632063	08607217	08582369	08557518	08532666	2
3.966 -0.0	08507811	08482955	08458096	08433235	08408372	08383507	08358640	08333771	08308899	08284026	2
3.967 -0.0	08259151	08234274	08209394	08184513	08159630	08134745	08109858	08084969	08060078	08035185	2
3.968 -0.0	08010290	07985393	07960495	07935594	07910692	07885787	07860881	07835973	07811063	07786152	2
3.969 -0.0	07761238	07736323	07711406	07686487	07661566	07636644	07611720	07586794	07561866	07536937	2
3.970 -0.0	07512006	07487073	07462138	07437202	07412264	07387325	07362383	07337441	07312496	07287550	2
3.971 -0.0	07262602	07237653	07212702	07187749	07162795	07137839	07112882	07087923	07062963	07038001	2
3.972 -0.0	07013038	06988073	06963106	06938138	06913169	06888198	06863226	06838252	06813277	06788300	0
3.973 -0.0	06763322	06738342	06713361	06688379	06663395	06638410	06613424	06588436	06563447	06538456	0
3.974 -0.0	06513465	06488472	06463477	06438482	06413485	06388486	06363487	06338486	06313484	06288481	0
3.975 -0.0	06263476	06238471	06213464	06188456	06163446	06138436	06113424	06088412	06063398	06038383	0
3.976 -0.0	06013366	05988349	05963331	05938311	05913291	05888269	05863246	05838223	05813198	05788172	0
3.977 -0.0	05763145	05738117	05713088	05688058	05663027	05637996	05612963	05587929	05562894	05537859	0
3.978 -0.0	05512822	05487784	05462746	05437707	05412667	05387625	05362583	05337541	05312497	05287452	0
3.979 -0.0	05262407	05237361	05212314	05187266	05162218	05137168	05112118	05087067	05062016	05036964	0
3.980 -0.0	05011910	04986857	04961802	04936747	04911691	04886635	04861577	04836519	04811461	04786402	0
3.981 -0.0	04761342	04736281	04711220	04686159	04661096	04636034	04610970	04585906	04560842	04535777	0
3.982 -0.0	04510711	04485645	04460578	04435511	04410444	04385376	04360307	04335238	04310168	04285099	0
3.983 -0.0	04260028	04234957	04209886	04184815	04159743	04134670	04109597	04084524	04059451	04034377	0
3.984 -0.0	04009303	03984229	03959154	03934079	03909003	03883928	03858852	03833775	03808699	03783622	0
3.985 -0.0	03758545	03733468	03708391	03683313	03658235	03633157	03608079	03583001	03557923	03532844	0
3.986 -0.0	03507765	03482686	03457607	03432528	03407449	03382370	03357290	03332211	03307132	03282052	0
3.987 -0.0	03256972	03231893	03206813	03181734	03156654	03131574	03106495	03081415	03056336	03031256	0
3.988 -0.0	03006177	02981098	02956018	02930939	02905860	02880781	02855702	02830624	02805545	02780467	0
3.989 -0.0	02755388	02730310	02705232	02680155	02655077	02630000	02604923	02579846	02554769	02529693	0
3.990 -0.0	02504617	02479541	02454466	02429390	02404315	02379241	02354166	02329092	02304019	02278945	0
3.991 -0.0	02253872	02228800	02203727	02178656	02153584	02128513	02103442	02078372	02053302	02028233	0
3.992 -0.0	02003164	01978096	01953028	01927960	01902893	01877827	01852761	01827696	01802631	01777567	−1
3.993 -0.0	01752503	01727440	01702377	01677315	01652253	01627192	01602132	01577073	01552014	01526955	−1
3.994 -0.0	01501898	01476841	01451784	01426728	01401673	01376619	01351566	01326513	01301461	01276409	−1
3.995 -0.0	01251359	01226309	01201260	01176211	01151164	01126117	01101071	01076026	01050982	01025938	−1
3.996 -0.0	01000896	00975854	00950813	00925773	00900734	00875696	00850658	00825622	00800587	00775552	−1
3.997 -0.0	00750518	00725486	00700454	00675424	00650394	00625365	00600337	00575311	00550285	00525261	−1
3.998 -0.0	00500237	00475214	00450193	00425173	00400153	00375135	00350118	00325102	00300087	00275073	−1
3.999 -0.0	00250061	00225049	00200039	00175030	00150022	00125015	00100010	00075006	00050002	00025001	1

$$\frac{\sin 2\pi x}{2\pi x}$$

$$\frac{\sin 2\pi x}{2\pi x}$$

TABLE 6

Sinc Function

x		.0	.0001	.0002	.0003	.0004	.0005	.0006	.0007	.0008	.0009	Δ_2
4.000	0.0	00000000	00024999	00049997	00074994	00099990	00124984	00149977	00174969	00199959	00224948	0
4.001	0.0	00249936	00274922	00299907	00324891	00349873	00374854	00399833	00424811	00449788	00474763	0
4.002	0.0	00499737	00524709	00549680	00574650	00599617	00624584	00649549	00674512	00699474	00724435	−2
4.003	0.0	00749394	00774351	00799307	00824261	00849214	00874165	00899114	00924062	00949008	00973953	−2
4.004	0.0	00998896	01023837	01048777	01073715	01098651	01123586	01148519	01173450	01198380	01223308	−2
4.005	0.0	01248234	01273159	01298081	01323002	01347921	01372839	01397754	01422668	01447580	01472490	−2
4.006	0.0	01497399	01522305	01547210	01572113	01597014	01621913	01646810	01671705	01696599	01721490	−2
4.007	0.0	01746380	01771267	01796153	01821037	01845919	01870798	01895676	01920552	01945426	01970298	−2
4.008	0.0	01995168	02020035	02044901	02069765	02094626	02119486	02144344	02169199	02194052	02218903	−2
4.009	0.0	02243753	02268600	02293444	02318287	02343128	02367966	02392802	02417636	02442468	02467298	−2
4.010	0.0	02492125	02516950	02541773	02566594	02591412	02616229	02641043	02665854	02690664	02715471	−2
4.011	0.0	02740275	02765078	02789878	02814676	02839471	02864264	02889055	02913843	02938629	02963413	−2
4.012	0.0	02988194	03012973	03037749	03062523	03087294	03112063	03136830	03161594	03186355	03211114	−2
4.013	0.0	03235871	03260625	03285376	03310125	03334872	03359616	03384357	03409096	03433832	03458566	−3
4.014	0.0	03483297	03508025	03532751	03557474	03582194	03606912	03631627	03656340	03681050	03705757	−3
4.015	0.0	03730462	03755163	03779862	03804559	03829252	03853943	03878631	03903317	03927999	03952679	−3
4.016	0.0	03977356	04002031	04026702	04051371	04076036	04100699	04125359	04150017	04174671	04199323	−3
4.017	0.0	04223971	04248617	04273260	04297900	04322537	04347171	04371802	04396430	04421055	04445677	−3
4.018	0.0	04470297	04494913	04519526	04544136	04568744	04593348	04617949	04642547	04667142	04691734	−3
4.019	0.0	04716323	04740909	04765491	04790071	04814648	04839221	04863791	04888358	04912922	04937483	−3
4.020	0.0	04962041	04986595	05011146	05035694	05060239	05084781	05109319	05133854	05158386	05182915	−3
4.021	0.0	05207440	05231963	05256481	05280997	05305509	05330018	05354524	05379026	05403525	05428020	−3
4.022	0.0	05452513	05477001	05501487	05525969	05550448	05574923	05599395	05623863	05648328	05672790	−3
4.023	0.0	05697248	05721702	05746154	05770601	05795045	05819486	05843923	05868357	05892787	05917213	−4
4.024	0.0	05941636	05966056	05990472	06014884	06039293	06063698	06088099	06112497	06136892	06161282	−4
4.025	0.0	06185669	06210052	06234432	06258808	06283180	06307549	06331914	06356275	06380633	06404986	−4
4.026	0.0	06429337	06453683	06478025	06502364	06526699	06551030	06575358	06599681	06624001	06648317	−4
4.027	0.0	06672629	06696937	06721242	06745542	06769839	06794132	06818421	06842706	06866987	06891264	−4
4.028	0.0	06915537	06939807	06964072	06988334	07012591	07036845	07061094	07085340	07109581	07133819	−4
4.029	0.0	07158052	07182282	07206507	07230729	07254946	07279159	07303369	07327574	07351775	07375972	−4
4.030	0.0	07400164	07424353	07448538	07472718	07496894	07521066	07545234	07569398	07593558	07617713	−4
4.031	0.0	07641864	07666011	07690154	07714292	07738427	07762557	07786682	07810804	07834921	07859034	−4
4.032	0.0	07883143	07907247	07931347	07955442	07979534	08003621	08027703	08051782	08075855	08099925	−4
4.033	0.0	08123990	08148051	08172107	08196159	08220206	08244249	08268288	08292322	08316352	08340377	−4
4.034	0.0	08364398	08388414	08412425	08436433	08460435	08484433	08508427	08532416	08556400	08580380	−5
4.035	0.0	08604356	08628326	08652293	08676254	08700211	08724163	08748111	08772054	08795993	08819926	−5
4.036	0.0	08843855	08867780	08891699	08915614	08939525	08963430	08987331	09011227	09035119	09059005	−5
4.037	0.0	09082887	09106764	09130637	09154504	09178367	09202225	09226078	09249926	09273770	09297609	−5
4.038	0.0	09321442	09345271	09369095	09392915	09416729	09440538	09464343	09488142	09511937	09535727	−5
4.039	0.0	09559511	09583291	09607066	09630836	09654601	09678361	09702116	09725866	09749611	09773351	−5
4.040	0.0	09797085	09820815	09844540	09868260	09891974	09915684	09939388	09963088	09986782	10010471	−5
4.041	0.0	10034155	10057834	10081508	10105176	10128840	10152498	10176151	10199799	10223442	10247079	−5
4.042	0.0	10270711	10294338	10317960	10341577	10365188	10388794	10412395	10435990	10459581	10483166	−5
4.043	0.0	10506745	10530320	10553889	10577452	10601010	10624563	10648111	10671653	10695190	10718722	−5
4.044	0.0	10742248	10765768	10789284	10812793	10836298	10859797	10883290	10906778	10930261	10953738	−5
4.045	0.0	10977210	11000676	11024137	11047592	11071041	11094485	11117924	11141357	11164784	11188206	−6
4.046	0.0	11211622	11235033	11258438	11281838	11305232	11328620	11352003	11375380	11398751	11422117	−6
4.047	0.0	11445477	11468831	11492180	11515523	11538860	11562192	11585518	11608838	11632152	11655461	−6
4.048	0.0	11678764	11702061	11725352	11748638	11771918	11795192	11818460	11841722	11864979	11888230	−6
4.049	0.0	11911475	11934714	11957947	11981174	12004396	12027611	12050821	12074025	12097223	12120415	−6

TABLE 6

Sinc Function

x		.0	.0001	.0002	.0003	.0004	.0005	.0006	.0007	.0008	.0009	Δ_2
4.050	0.0	12143601	12166781	12189955	12213123	12236285	12259441	12282592	12305736	12328874	12352006	−6
4.051	0.0	12375132	12398253	12421367	12444475	12467577	12490673	12513763	12536847	12559924	12582996	−6
4.052	0.0	12606062	12629121	12652174	12675222	12698263	12721298	12744326	12767349	12790365	12813375	−6
4.053	0.0	12836379	12859377	12882369	12905354	12928333	12951306	12974273	12997233	13020187	13043135	−6
4.054	0.0	13066077	13089012	13111941	13134864	13157780	13180690	13203594	13226491	13249382	13272267	−6
4.055	0.0	13295145	13318017	13340883	13363742	13386595	13409441	13432281	13455114	13477941	13500762	−6
4.056	0.0	13523576	13546384	13569185	13591980	13614768	13637550	13660325	13683094	13705856	13728611	−7
4.057	0.0	13751361	13774103	13796839	13819569	13842291	13865008	13887717	13910420	13933117	13955807	−7
4.058	0.0	13978490	14001167	14023836	14046500	14069156	14091806	14114450	14137086	14159716	14182339	−7
4.059	0.0	14204956	14227565	14250169	14272765	14295354	14317937	14340513	14363082	14385645	14408200	−7
4.060	0.0	14430749	14453291	14475827	14498355	14520877	14543391	14565899	14588400	14610894	14633382	−7
4.061	0.0	14655862	14678336	14700802	14723262	14745715	14768161	14790600	14813031	14835457	14857875	−7
4.062	0.0	14880286	14902690	14925087	14947477	14969860	14992236	15014605	15036967	15059322	15081670	−7
4.063	0.0	15104011	15126345	15148672	15170992	15193304	15215610	15237908	15260200	15282484	15304761	−7
4.064	0.0	15327031	15349294	15371549	15393798	15416039	15438273	15460500	15482720	15504932	15527138	−7
4.065	0.0	15549336	15571527	15593710	15615887	15638056	15660218	15682372	15704519	15726659	15748792	−7
4.066	0.0	15770918	15793036	15815146	15837250	15859346	15881435	15903516	15925590	15947657	15969716	−7
4.067	0.0	15991768	16013812	16035850	16057879	16079901	16101916	16123924	16145924	16167916	16189901	−7
4.068	0.0	16211879	16233849	16255811	16277766	16299714	16321654	16343586	16365511	16387429	16409339	−8
4.069	0.0	16431241	16453136	16475023	16496903	16518775	16540639	16562496	16584345	16606187	16628021	−8
4.070	0.0	16649847	16671666	16693477	16715281	16737076	16758864	16780645	16802417	16824182	16845940	−8
4.071	0.0	16867689	16889431	16911165	16932891	16954610	16976321	16998024	17019719	17041407	17063086	−8
4.072	0.0	17084758	17106422	17128078	17149727	17171367	17193000	17214625	17236242	17257851	17279453	−8
4.073	0.0	17301046	17322632	17344209	17365779	17387341	17408895	17430441	17451979	17473509	17495031	−8
4.074	0.0	17516545	17538051	17559550	17581040	17602522	17623996	17645463	17666921	17688371	17709813	−8
4.075	0.0	17731247	17752673	17774091	17795501	17816903	17838297	17859682	17881060	17902429	17923791	−8
4.076	0.0	17945144	17966489	17987826	18009155	18030475	18051788	18073092	18094388	18115676	18136956	−8
4.077	0.0	18158227	18179491	18200746	18221993	18243231	18264462	18285684	18306898	18328103	18349301	−8
4.078	0.0	18370490	18391671	18412843	18434007	18455163	18476311	18497450	18518581	18539703	18560817	−8
4.079	0.0	18581923	18603020	18624110	18645190	18666262	18687326	18708382	18729429	18750467	18771498	−8
4.080	0.0	18792519	18813533	18834537	18855534	18876522	18897501	18918472	18939434	18960388	18981334	−9
4.081	0.0	19002271	19023199	19044119	19065030	19085933	19106827	19127712	19148589	19169458	19190318	−9
4.082	0.0	19211169	19232012	19252846	19273671	19294488	19315296	19336096	19356886	19377669	19398442	−9
4.083	0.0	19419207	19439963	19460711	19481449	19502179	19522901	19543613	19564317	19585013	19605699	−9
4.084	0.0	19626377	19647045	19667706	19688357	19709000	19729633	19750258	19770875	19791482	19812080	−9
4.085	0.0	19832670	19853251	19873823	19894386	19914941	19935486	19956023	19976550	19997069	20017579	−9
4.086	0.0	20038080	20058572	20079055	20099529	20119995	20140451	20160898	20181337	20201766	20222187	−9
4.087	0.0	20242598	20263001	20283394	20303779	20324154	20344521	20364878	20385227	20405566	20425897	−9
4.088	0.0	20446218	20466530	20486833	20507127	20527412	20547688	20567955	20588212	20608461	20628700	−9
4.089	0.0	20648931	20669152	20689364	20709567	20729760	20749945	20770120	20790286	20810443	20830591	−9
4.090	0.0	20850729	20870859	20890979	20911090	20931191	20951284	20971367	20991441	21011505	21031560	−9
4.091	0.0	21051606	21071643	21091671	21111689	21131698	21151697	21171687	21191668	21211640	21231602	−9
4.092	0.0	21251555	21271498	21291432	21311357	21331272	21351178	21371074	21390961	21410839	21430707	−9
4.093	0.0	21450566	21470415	21490255	21510086	21529907	21549718	21569521	21589313	21609096	21628870	−9
4.094	0.0	21648634	21668388	21688134	21707869	21727595	21747312	21767018	21786716	21806404	21826082	−10
4.095	0.0	21845750	21865409	21885059	21904699	21924329	21943950	21963561	21983162	22002754	22022336	−10
4.096	0.0	22041909	22061471	22081024	22100568	22120102	22139626	22159140	22178645	22198140	22217625	−10
4.097	0.0	22237101	22256567	22276023	22295469	22314906	22334333	22353750	22373157	22392555	22411942	−10
4.098	0.0	22431320	22450688	22470047	22489395	22508734	22528063	22547382	22566691	22585991	22605280	−10
4.099	0.0	22624560	22643829	22663089	22682339	22701579	22720810	22740030	22759240	22778441	22797631	−10

$$\frac{\sin 2\pi x}{2\pi x}$$

TABLE 6

Sinc Function

x		.0	.0001	.0002	.0003	.0004	.0005	.0006	.0007	.0008	.0009	Δ_2
4.100	0.0	22816812	22835982	22855143	22874294	22893435	22912565	22931686	22950797	22969898	22988989	−10
4.101	0.0	23008070	23027140	23046201	23065252	23084293	23103323	23122344	23141355	23160355	23179346	−10
4.102	0.0	23198326	23217296	23236257	23255207	23274147	23293077	23311996	23330906	23349806	23368695	−10
4.103	0.0	23387574	23406443	23425302	23444151	23462990	23481818	23500636	23519444	23538242	23557030	−10
4.104	0.0	23575807	23594574	23613331	23632078	23650814	23669541	23688256	23706962	23725658	23744343	−10
4.105	0.0	23763018	23781682	23800336	23818980	23837614	23856237	23874850	23893453	23912045	23930627	−10
4.106	0.0	23949199	23967760	23986311	24004852	24023382	24041902	24060411	24078910	24097399	24115877	−10
4.107	0.0	24134345	24152802	24171249	24189685	24208111	24226527	24244932	24263327	24281711	24300084	−10
4.108	0.0	24318448	24336800	24355143	24373474	24391795	24410106	24428406	24446696	24464975	24483243	−11
4.109	0.0	24501501	24519748	24537985	24556212	24574427	24592632	24610827	24629011	24647184	24665346	−11
4.110	0.0	24683498	24701640	24719771	24737891	24756000	24774099	24792187	24810265	24828332	24846388	−11
4.111	0.0	24864433	24882468	24900492	24918505	24936508	24954500	24972481	24990451	25008411	25026360	−11
4.112	0.0	25044298	25062226	25080142	25098048	25115944	25133828	25151701	25169564	25187416	25205257	−11
4.113	0.0	25223088	25240907	25258716	25276514	25294301	25312077	25329842	25347596	25365340	25383073	−11
4.114	0.0	25400795	25418505	25436205	25453895	25471573	25489240	25506896	25524542	25542176	25559800	−11
4.115	0.0	25577412	25595014	25612605	25630185	25647753	25665311	25682858	25700394	25717919	25735432	−11
4.116	0.0	25752935	25770427	25787908	25805377	25822836	25840284	25857720	25875146	25892560	25909964	−11
4.117	0.0	25927356	25944737	25962107	25979467	25996815	26014151	26031477	26048792	26066095	26083388	−11
4.118	0.0	26100669	26117939	26135198	26152446	26169682	26186908	26204122	26221325	26238517	26255698	−11
4.119	0.0	26272868	26290026	26307173	26324309	26341434	26358547	26375649	26392740	26409820	26426888	−11
4.120	0.0	26443946	26460991	26478026	26495050	26512062	26529062	26546052	26563030	26579997	26596953	−11
4.121	0.0	26613897	26630830	26647751	26664662	26681561	26698448	26715324	26732189	26749043	26765885	−11
4.122	0.0	26782715	26799535	26816342	26833139	26849924	26866698	26883460	26900211	26916950	26933678	−11
4.123	0.0	26950395	26967100	26983793	27000476	27017146	27033805	27050453	27067089	27083714	27100327	−11
4.124	0.0	27116929	27133519	27150098	27166665	27183221	27199765	27216298	27232819	27249328	27265826	−12
4.125	0.0	27282313	27298787	27315251	27331702	27348142	27364571	27380988	27397393	27413787	27430169	−12
4.126	0.0	27446539	27462898	27479245	27495581	27511904	27528217	27544517	27560806	27577083	27593349	−12
4.127	0.0	27609603	27625845	27642075	27658294	27674501	27690697	27706880	27723052	27739212	27755361	−12
4.128	0.0	27771498	27787623	27803736	27819837	27835927	27852005	27868071	27884125	27900168	27916199	−12
4.129	0.0	27932218	27948225	27964221	27980204	27996176	28012136	28028084	28044020	28059945	28075857	−12
4.130	0.0	28091758	28107647	28123524	28139389	28155242	28171084	28186913	28202731	28218536	28234330	−12
4.131	0.0	28250112	28265882	28281640	28297386	28313120	28328843	28344553	28360251	28375938	28391612	−12
4.132	0.0	28407274	28422925	28438563	28454190	28469805	28485407	28500998	28516576	28532143	28547697	−12
4.133	0.0	28563240	28578770	28594289	28609795	28625289	28640772	28656242	28671700	28687146	28702580	−12
4.134	0.0	28718002	28733412	28748810	28764195	28779569	28794930	28810280	28825617	28840942	28856255	−12
4.135	0.0	28871556	28886845	28902121	28917386	28932638	28947878	28963106	28978322	28993526	29008717	−12
4.136	0.0	29023896	29039063	29054218	29069361	29084491	29099610	29114716	29129810	29144891	29159960	−12
4.137	0.0	29175018	29190062	29205095	29220115	29235123	29250119	29265103	29280074	29295033	29309980	−12
4.138	0.0	29324914	29339836	29354746	29369643	29384529	29399401	29414262	29429110	29443946	29458770	−12
4.139	0.0	29473581	29488380	29503166	29517940	29532702	29547451	29562188	29576913	29591625	29606325	−12
4.140	0.0	29621012	29635687	29650350	29665000	29679638	29694263	29708876	29723477	29738065	29752641	−12
4.141	0.0	29767204	29781754	29796293	29810818	29825332	29839833	29854321	29868797	29883260	29897711	−12
4.142	0.0	29912150	29926575	29940989	29955390	29969778	29984154	29998517	30012868	30027206	30041532	−13
4.143	0.0	30055845	30070145	30084433	30098709	30112972	30127222	30141460	30155685	30169897	30184097	−13
4.144	0.0	30198284	30212459	30226621	30240771	30254908	30269032	30283143	30297242	30311329	30325402	−13
4.145	0.0	30339463	30353512	30367548	30381571	30395581	30409579	30423564	30437536	30451496	30465443	−13
4.146	0.0	30479377	30493298	30507207	30521103	30534987	30548858	30562715	30576561	30590393	30604213	−13
4.147	0.0	30618020	30631814	30645596	30659364	30673120	30686863	30700594	30714312	30728016	30741708	−13
4.148	0.0	30755388	30769054	30782708	30796349	30809977	30823592	30837194	30850784	30864360	30877924	−13
4.149	0.0	30891475	30905013	30918539	30932051	30945551	30959038	30972511	30985972	30999420	31012856	−13

TABLE 6

Sinc Function

x		.0	.0001	.0002	.0003	.0004	.0005	.0006	.0007	.0008	.0009	Δ_2
4.150	0.0	31026278	31039687	31053084	31066468	31079838	31093196	31106541	31119873	31133192	31146498	-13
4.151	0.0	31159791	31173071	31186339	31199593	31212834	31226063	31239278	31252481	31265670	31278847	-13
4.152	0.0	31292010	31305161	31318298	31331423	31344535	31357633	31370719	31383791	31396851	31409897	-13
4.153	0.0	31422931	31435951	31448959	31461953	31474934	31487902	31500858	31513800	31526729	31539645	-13
4.154	0.0	31552548	31565438	31578314	31591178	31604029	31616866	31629691	31642502	31655300	31668085	-13
4.155	0.0	31680857	31693616	31706362	31719094	31731814	31744520	31757213	31769893	31782560	31795214	-13
4.156	0.0	31807854	31820482	31833096	31845697	31858285	31870860	31883421	31895969	31908505	31921026	-13
4.157	0.0	31933535	31946031	31958513	31970982	31983438	31995880	32008310	32020726	32033129	32045519	-13
4.158	0.0	32057895	32070258	32082608	32094945	32107268	32119578	32131875	32144159	32156429	32168686	-13
4.159	0.0	32180930	32193160	32205377	32217581	32229772	32241949	32254113	32266263	32278401	32290525	-13
4.160	0.0	32302635	32314733	32326817	32338887	32350944	32362988	32375019	32387036	32399040	32411030	-13
4.161	0.0	32423007	32434971	32446921	32458858	32470782	32482692	32494589	32506472	32518342	32530199	-13
4.162	0.0	32542042	32553872	32565688	32577491	32589280	32601056	32612819	32624568	32636304	32648026	-13
4.163	0.0	32659735	32671430	32683112	32694780	32706435	32718077	32729705	32741319	32752921	32764508	-13
4.164	0.0	32776082	32787643	32799190	32810723	32822243	32833750	32845243	32856723	32868189	32879641	-14
4.165	0.0	32891080	32902505	32913917	32925316	32936700	32948072	32959429	32970773	32982104	32993421	-14
4.166	0.0	33004725	33016014	33027291	33038553	33049802	33061038	33072260	33083468	33094663	33105844	-14
4.167	0.0	33117012	33128166	33139306	33150433	33161546	33172645	33183731	33194803	33205862	33216907	-14
4.168	0.0	33227938	33238956	33249960	33260950	33271927	33282890	33293839	33304775	33315697	33326605	-14
4.169	0.0	33337500	33348381	33359248	33370102	33380941	33391768	33402580	33413379	33424164	33434935	-14
4.170	0.0	33445693	33456437	33467167	33477884	33488586	33499275	33509951	33520612	33531260	33541894	-14
4.171	0.0	33552514	33563121	33573714	33584293	33594858	33605409	33615947	33626471	33636981	33647478	-14
4.172	0.0	33657960	33668429	33678884	33689325	33699753	33710166	33720566	33730952	33741324	33751683	-14
4.173	0.0	33762027	33772358	33782675	33792978	33803267	33813542	33823804	33834052	33844285	33854505	-14
4.174	0.0	33864712	33874904	33385082	33895247	33905397	33915534	33925657	33935766	33945862	33955943	-14
4.175	0.0	33966010	33976064	33986103	33996129	34006141	34016139	34026123	34036093	34046049	34055991	-14
4.176	0.0	34065920	34075834	34085735	34095621	34105494	34115353	34125197	34135028	34144845	34154648	-14
4.177	0.0	34164437	34174212	34183973	34193720	34203453	34213172	34222878	34232569	34242246	34251909	-14
4.178	0.0	34261559	34271194	34280815	34290422	34300016	34309595	34319160	34328712	34338249	34347772	-14
4.179	0.0	34357281	34366777	34376258	34385725	34395178	34404617	34414042	34423453	34432850	34442233	-14
4.180	0.0	34451602	34460957	34470298	34479625	34488938	34498236	34507521	34516791	34526048	34535290	-14
4.181	0.0	34544519	34553733	34562933	34572119	34581291	34590449	34599593	34608722	34617838	34626939	-14
4.182	0.0	34636027	34645100	34654159	34663204	34672235	34681252	34690255	34699243	34708218	34717178	-14
4.183	0.0	34726124	34735056	34743974	34752878	34761768	34770643	34779504	34788351	34797184	34806003	-14
4.184	0.0	34814808	34823599	34832375	34841137	34849885	34858619	34867339	34876044	34884735	34893412	-14
4.185	0.0	34902075	34910724	34919359	34927979	34936585	34945177	34953755	34962318	34970867	34979403	-14
4.186	0.0	34987923	34996430	35004922	35013401	35021865	35030314	35038750	35047171	35055578	35063971	-14
4.187	0.0	35072350	35080714	35089064	35097400	35105721	35114029	35122322	35130600	35138865	35147115	-14
4.188	0.0	35155351	35163573	35171780	35179973	35188152	35196317	35204467	35212603	35220725	35228832	-14
4.189	0.0	35236925	35245004	35253069	35261119	35269155	35277177	35285184	35293177	35301156	35309120	-14
4.190	0.0	35317070	35325006	35332927	35340834	35348727	35356606	35364470	35372319	35380155	35387976	-14
4.191	0.0	35395783	35403575	35411353	35419117	35426866	35434601	35442322	35450028	35457720	35465397	-14
4.192	0.0	35473061	35480709	35488344	35495964	35503569	35511161	35518738	35526300	35533848	35541382	-14
4.193	0.0	35548901	35556406	35563897	35571373	35578835	35586282	35593715	35601134	35608538	35615928	-14
4.194	0.0	35623303	35630664	35638011	35645343	35652660	35659964	35667252	35674527	35681787	35689032	-14
4.195	0.0	35696263	35703480	35710682	35717870	35725043	35732202	35739347	35746477	35753592	35760693	-14
4.196	0.0	35767780	35774852	35781910	35788953	35795982	35802996	35809996	35816982	35823952	35830909	-14
4.197	0.0	35837851	35844778	35851691	35858590	35865474	35872343	35879198	35886039	35892865	35899677	-14
4.198	0.0	35906474	35913256	35920024	35926778	35933517	35940242	35946952	35953647	35960328	35966995	-15
4.199	0.0	35973647	35980284	35986907	35993516	36000110	36006689	36013254	36019804	36026340	36032861	-15

$$\frac{\sin 2\pi x}{2\pi x}$$

$$\frac{\sin 2\pi x}{2\pi x}$$

TABLE 6

Sinc Function

x		.0	.0001	.0002	.0003	.0004	.0005	.0006	.0007	.0008	.0009	Δ_2
4.200	0.0	36039368	36045860	36052338	36058801	36065250	36071684	36078103	36084508	36090898	36097274	-15
4.201	0.0	36103636	36109982	36116314	36122632	36128935	36135224	36141497	36147757	36154002	36160232	-15
4.202	0.0	36166447	36172649	36178835	36185007	36191164	36197307	36203435	36209549	36215648	36221732	-15
4.203	0.0	36227802	36233857	36239898	36245924	36251935	36257932	36263915	36269882	36275835	36281774	-15
4.204	0.0	36287698	36293607	36299501	36305382	36311247	36317098	36322934	36328755	36334562	36340355	-15
4.205	0.0	36346132	36351896	36357644	36363378	36369097	36374802	36380492	36386167	36391827	36397474	-15
4.206	0.0	36403105	36408722	36414324	36419911	36425484	36431042	36436586	36442115	36447629	36453129	-15
4.207	0.0	36458614	36464084	36469539	36474980	36480407	36485818	36491215	36496598	36501965	36507318	-15
4.208	0.0	36512657	36517980	36523289	36528584	36533863	36539128	36544379	36549614	36554835	36560042	-15
4.209	0.0	36565233	36570410	36575572	36580720	36585853	36590971	36596074	36601163	36606237	36611297	-15
4.210	0.0	36616342	36621372	36626387	36631388	36636373	36641345	36646301	36651243	36656170	36661083	-15
4.211	0.0	36665980	36670863	36675732	36680585	36685424	36690248	36695058	36699853	36704633	36709398	-15
4.212	0.0	36714148	36718884	36723605	36728312	36733004	36737681	36742343	36746990	36751623	36756241	-15
4.213	0.0	36760845	36765433	36770007	36774566	36779111	36783640	36788155	36792655	36797141	36801612	-15
4.214	0.0	36806068	36810509	36814935	36819347	36823744	36828126	36832494	36836847	36841185	36845508	-15
4.215	0.0	36849817	36854110	36858389	36862654	36866903	36871138	36875358	36879563	36883754	36887929	-15
4.216	0.0	36892090	36896236	36900368	36904485	36908587	36912674	36916746	36920804	36924846	36928875	-15
4.217	0.0	36932888	36936886	36940870	36944839	36948793	36952733	36956657	36960567	36964463	36968343	-15
4.218	0.0	36972208	36976059	36979895	36983716	36987523	36991315	36995091	36998854	37002601	37006333	-15
4.219	0.0	37010051	37013754	37017442	37021116	37024774	37028418	37032047	37035661	37039261	37042845	-15
4.220	0.0	37046415	37049970	37053510	37057036	37060546	37064042	37067523	37070990	37074441	37077878	-15
4.221	0.0	37081300	37084707	37088099	37091476	37094839	37098187	37101520	37104838	37108142	37111430	-15
4.222	0.0	37114704	37117963	37121207	37124437	37127651	37130851	37134036	37137206	37140362	37143502	-15
4.223	0.0	37146628	37149739	37152835	37155916	37158983	37162034	37165071	37168093	37171100	37174093	-15
4.224	0.0	37177070	37180033	37182981	37185914	37188833	37191736	37194625	37197499	37200358	37203202	-15
4.225	0.0	37206031	37208846	37211645	37214430	37217200	37219956	37222696	37225422	37228133	37230829	-15
4.226	0.0	37233510	37236176	37238828	37241464	37244086	37246693	37249285	37251863	37254425	37256973	-15
4.227	0.0	37259506	37262024	37264527	37267015	37269489	37271948	37274392	37276821	37279235	37281634	-15
4.228	0.0	37284019	37286389	37288744	37291084	37293409	37295719	37298015	37300296	37302562	37304813	-15
4.229	0.0	37307049	37309270	37311477	37313669	37315846	37318008	37320155	37322287	37324405	37326508	-15
4.230	0.0	37328595	37330669	37332727	37334770	37336799	37338812	37340811	37342795	37344765	37346719	-15
4.231	0.0	37348659	37350583	37352493	37354388	37356268	37358134	37359984	37361820	37363641	37365447	-15
4.232	0.0	37367238	37369014	37370776	37372523	37374254	37375971	37377674	37379361	37381033	37382691	-15
4.233	0.0	37384334	37385962	37387575	37389173	37390757	37392326	37393879	37395418	37396942	37398452	-15
4.234	0.0	37399946	37401426	37402891	37404341	37405776	37407196	37408602	37409992	37411368	37412729	-15
4.235	0.0	37414075	37415406	37416723	37418024	37419311	37420583	37421840	37423083	37424310	37425523	-15
4.236	0.0	37426720	37427903	37429072	37430225	37431363	37432487	37433596	37434690	37435769	37436833	-15
4.237	0.0	37437883	37438917	37439937	37440942	37441932	37442908	37443868	37444814	37445745	37446661	-15
4.238	0.0	37447562	37448448	37449320	37450177	37451019	37451846	37452658	37453455	37454238	37455006	-15
4.239	0.0	37455759	37456497	37457220	37457929	37458622	37459301	37459965	37460615	37461249	37461869	-15
4.240	0.0	37462473	37463063	37463638	37464199	37464744	37465275	37465791	37466292	37466778	37467250	-15
4.241	0.0	37467706	37468148	37468575	37468987	37469385	37469767	37470135	37470488	37470826	37471149	-15
4.242	0.0	37471458	37471751	37472030	37472294	37472544	37472778	37472998	37473203	37473393	37473568	-15
4.243	0.0	37473728	37473874	37474005	37474121	37474222	37474309	37474380	37474437	37474479	37474507	-15
4.244	0.0	37474519	37474517	37474500	37474468	37474421	37474360	37474283	37474192	37474086	37473966	-15
4.245	0.0	37473830	37473680	37473515	37473335	37473141	37472931	37472707	37472468	37472215	37471946	-15
4.246	0.0	37471663	37471365	37471052	37470724	37470382	37470025	37469653	37469266	37468865	37468448	-15
4.247	0.0	37468017	37467572	37467111	37466636	37466146	37465641	37465121	37464587	37464037	37463473	-15
4.248	0.0	37462895	37462301	37461693	37461070	37460432	37459780	37459113	37458431	37457734	37457022	-15
4.249	0.0	37456296	37455555	37454799	37454029	37453243	37452443	37451628	37450799	37449955	37449096	-15

TABLE 6

Sinc Function

x		.0	.0001	.0002	.0003	.0004	.0005	.0006	.0007	.0008	.0009	Δ_2
4.250	0.0	37448222	37447333	37446430	37445512	37444579	37443632	37442670	37441693	37440701	37439695	−15
4.251	0.0	37438674	37437538	37436587	37435522	37434442	37433347	37432237	37431113	37429974	37428821	−15
4.252	0.0	37427652	37426469	37425271	37424059	37422831	37421589	37420333	37419061	37417775	37416474	−15
4.253	0.0	37415159	37413828	37412483	37411124	37409749	37408360	37406956	37405538	37404105	37402657	−15
4.254	0.0	37401194	37399717	37398225	37396718	37395197	37393661	37392110	37390545	37388965	37387370	−15
4.255	0.0	37385760	37384136	37382497	37380844	37379175	37377493	37375795	37374083	37372356	37370614	−15
4.256	0.0	37368858	37367087	37365301	37363501	37361686	37359857	37358012	37356153	37354280	37352391	−15
4.257	0.0	37350489	37348571	37346639	37344692	37342730	37340754	37338763	37336758	37334738	37332703	−15
4.258	0.0	37330654	37328590	37326511	37324418	37322310	37320187	37318050	37315898	37313732	37311550	−15
4.259	0.0	37309355	37307144	37304919	37302680	37300426	37298157	37295873	37293575	37291262	37288935	−15
4.260	0.0	37286593	37284237	37281866	37279480	37277079	37274665	37272235	37269791	37267332	37264859	−15
4.261	0.0	37262371	37259868	37257351	37254819	37252273	37249712	37247137	37244547	37241942	37239323	−15
4.262	0.0	37236689	37234041	37231378	37228700	37226008	37223301	37220580	37217844	37215094	37212329	−15
4.263	0.0	37209549	37206755	37203947	37201124	37198286	37195434	37192567	37189685	37186789	37183879	−15
4.264	0.0	37180954	37178014	37175060	37172092	37169109	37166111	37163099	37160072	37157031	37153975	−15
4.265	0.0	37150905	37147820	37144720	37141606	37138478	37135335	37132178	37129006	37125819	37122618	−15
4.266	0.0	37119403	37116173	37112928	37109669	37106396	37103108	37099806	37096489	37093157	37089811	−14
4.267	0.0	37086451	37083076	37079687	37076283	37072864	37069432	37065984	37062523	37059046	37055556	−14
4.268	0.0	37052051	37048531	37044997	37041448	37037885	37034308	37030716	37027110	37023489	37019854	−14
4.269	0.0	37016204	37012540	37008861	37005168	37001461	36997739	36994003	36990252	36986487	36982707	−14
4.270	0.0	36978913	36975105	36971282	36967445	36963593	36959727	36955846	36951951	36948042	36944118	−14
4.271	0.0	36940180	36936228	36932261	36928279	36924284	36920274	36916249	36912210	36908157	36904089	−14
4.272	0.0	36900007	36895911	36891800	36887675	36883536	36879382	36875214	36871031	36866834	36862623	−14
4.273	0.0	36858397	36854157	36849903	36845634	36841351	36837054	36832742	36828416	36824075	36819721	−14
4.274	0.0	36815351	36810968	36806570	36802158	36797732	36793291	36788836	36784367	36779883	36775385	−14
4.275	0.0	36770873	36766346	36761805	36757250	36752681	36748097	36743499	36738887	36734260	36729619	−14
4.276	0.0	36724964	36720294	36715611	36710913	36706200	36701474	36696733	36691978	36687208	36682425	−14
4.277	0.0	36677627	36672815	36667988	36663147	36658293	36653423	36648540	36643642	36638731	36633804	−14
4.278	0.0	36628864	36623910	36618941	36613958	36608960	36603949	36598923	36593883	36588829	36583761	−14
4.279	0.0	36578679	36573582	36568471	36563346	36558207	36553053	36547885	36542703	36537507	36532297	−14
4.280	0.0	36527073	36521834	36516581	36511314	36506033	36500738	36495429	36490105	36484767	36479415	−14
4.281	0.0	36474049	36468569	36463275	36457866	36452444	36447007	36441556	36436091	36430612	36425119	−14
4.282	0.0	36419611	36414090	36408554	36403004	36397440	36391862	36386270	36380664	36375044	36369409	−14
4.283	0.0	36363761	36358098	36352422	36346731	36341026	36335307	36329574	36323827	36318066	36312291	−14
4.284	0.0	36306502	36300698	36294881	36289049	36283204	36277344	36271471	36265583	36259681	36253766	−14
4.285	0.0	36247836	36241892	36235934	36229962	36223977	36217977	36211963	36205935	36199893	36193837	−14
4.286	0.0	36187767	36181683	36175585	36169473	36163347	36157207	36151053	36144885	36138703	36132508	−14
4.287	0.0	36126298	36120074	36113836	36107584	36101319	36095039	36088745	36082438	36076116	36069781	−14
4.288	0.0	36063431	36057068	36050691	36044299	36037894	36031475	36025042	36018595	36012134	36005659	−14
4.289	0.0	35999171	35992668	35986152	35979621	35973077	35966519	35959947	35953361	35946761	35940147	−14
4.290	0.0	35933519	35926878	35920222	35913553	35906870	35900173	35893462	35886737	35879999	35873247	−14
4.291	0.0	35866480	35859700	35852906	35846098	35839277	35832441	35825592	35818729	35811852	35804961	−14
4.292	0.0	35798057	35791138	35784206	35777260	35770301	35763327	35756340	35749338	35742324	35735295	−14
4.293	0.0	35728252	35721196	35714126	35707042	35699945	35692833	35685708	35678569	35671417	35664250	−14
4.294	0.0	35657070	35649876	35642669	35635447	35628212	35620964	35613701	35606425	35599135	35591831	−14
4.295	0.0	35584514	35577183	35569838	35562479	35555107	35547721	35540322	35532909	35525482	35518041	−14
4.296	0.0	35510587	35503119	35495637	35488142	35480633	35473110	35465574	35458024	35450461	35442883	−14
4.297	0.0	35435293	35427688	35420070	35412438	35404793	35397134	35389461	35381775	35374075	35366362	−14
4.298	0.0	35358635	35350894	35343140	35335372	35327591	35319796	35311987	35304165	35296329	35288480	−14
4.299	0.0	35280617	35272741	35264851	35256947	35249030	35241099	35233155	35225198	35217226	35209242	−14

$$\frac{\sin 2\pi x}{2\pi x}$$

TABLE 6

Sinc Function

x		.0	.0001	.0002	.0003	.0004	.0005	.0006	.0007	.0008	.0009	Δ_2
4.300	0.0	35201243	35193231	35185206	35177167	35169115	35161049	35152969	35144877	35136770	35128650	-13
4.301	0.0	35120517	35112370	35104210	35096036	35087848	35079648	35071433	35063206	35054965	35046710	-13
4.302	0.0	35038442	35030160	35021865	35013557	35005235	34996900	34988551	34980189	34971813	34963424	-13
4.303	0.0	34955022	34946606	34938177	34929734	34921278	34912809	34904326	34895830	34887321	34878798	-13
4.304	0.0	34870261	34861712	34853149	34844572	34835983	34827379	34818763	34810133	34801490	34792834	-13
4.305	0.0	34784164	34775481	34766784	34758074	34749351	34740615	34731865	34723102	34714326	34705536	-13
4.306	0.0	34696733	34687917	34679088	34670245	34661389	34652520	34643637	34634741	34625832	34616910	-13
4.307	0.0	34607974	34599025	34590063	34581088	34572099	34563097	34554082	34545054	34536012	34526958	-13
4.308	0.0	34517890	34508809	34499714	34490607	34481486	34472352	34463205	34454045	34444872	34435685	-13
4.309	0.0	34426485	34417272	34408046	34398807	34389555	34380289	34371010	34361718	34352414	34343095	-13
4.310	0.0	34333764	34324420	34315062	34305692	34296308	34286911	34277501	34268079	34258642	34249193	-13
4.311	0.0	34239731	34230256	34220767	34211266	34201751	34192224	34182683	34173129	34163563	34153983	-13
4.312	0.0	34144390	34134784	34125165	34115533	34105889	34096231	34086560	34076876	34067179	34057469	-13
4.313	0.0	34047746	34038010	34028261	34018499	34008724	33998936	33989135	33979321	33969495	33959655	-13
4.314	0.0	33949802	33939937	33930058	33920167	33910262	33900345	33890415	33880471	33870515	33860546	-13
4.315	0.0	33850564	33840570	33830562	33820541	33810508	33800461	33790402	33780330	33770245	33760147	-13
4.316	0.0	33750036	33739913	33729776	33719627	33709465	33699290	33689102	33678902	33668688	33658462	-13
4.317	0.0	33648223	33637971	33627706	33617429	33607138	33596835	33586519	33576191	33565849	33555495	-13
4.318	0.0	33545128	33534749	33524356	33513951	33503533	33493102	33482659	33472203	33461734	33451252	-13
4.319	0.0	33440758	33430250	33419731	33409198	33398653	33388095	33377524	33366941	33356345	33345737	-13
4.320	0.0	33335115	33324481	33313835	33303175	33292503	33281819	33271122	33260412	33249689	33238954	-13
4.321	0.0	33228206	33217446	33206673	33195887	33185089	33174278	33163455	33152619	33141770	33130909	-13
4.322	0.0	33120035	33109149	33098250	33087338	33076414	33065478	33054529	33043567	33032593	33021606	-13
4.323	0.0	33010607	32999595	32988571	32977534	32966485	32955423	32944348	32933262	32922162	32911050	-12
4.324	0.0	32899926	32888789	32877640	32866479	32855304	32844118	32832919	32821707	32810483	32799247	-12
4.325	0.0	32787998	32776737	32765463	32754177	32742879	32731568	32720245	32708909	32697561	32686201	-12
4.326	0.0	32674828	32663443	32652045	32640635	32629213	32617779	32606332	32594872	32583401	32571917	-12
4.327	0.0	32560420	32548912	32537391	32525858	32514312	32502754	32491184	32479602	32468007	32456400	-12
4.328	0.0	32444780	32433149	32421505	32409849	32398181	32386500	32374807	32363102	32351385	32339655	-12
4.329	0.0	32327913	32316159	32304393	32292615	32280824	32269021	32257206	32245379	32233540	32221688	-12
4.330	0.0	32209824	32197949	32186060	32174160	32162248	32150323	32138387	32126438	32114477	32102504	-12
4.331	0.0	32090519	32078521	32066512	32054490	32042457	32030411	32018353	32006283	31994201	31982107	-12
4.332	0.0	31970001	31957883	31945753	31933611	31921456	31909290	31897112	31884921	31872719	31860504	-12
4.333	0.0	31848278	31836039	31823789	31811526	31799252	31786965	31774667	31762357	31750034	31737700	-12
4.334	0.0	31725354	31712995	31700625	31688243	31675849	31663443	31651025	31638595	31626153	31613699	-12
4.335	0.0	31601234	31588756	31576267	31563766	31551252	31538727	31526190	31513642	31501081	31488508	-12
4.336	0.0	31475924	31463328	31450720	31438100	31425468	31412824	31400169	31387502	31374823	31362132	-12
4.337	0.0	31349430	31336715	31323989	31311251	31298501	31285740	31272967	31260182	31247385	31234576	-12
4.338	0.0	31221756	31208924	31196081	31183225	31170358	31157479	31144589	31131686	31118772	31105847	-12
4.339	0.0	31092909	31079960	31067000	31054027	31041043	31028048	31015040	31002021	30988991	30975949	-12
4.340	0.0	30962895	30949829	30936752	30923663	30910563	30897451	30884328	30871193	30858046	30844888	-12
4.341	0.0	30831718	30818536	30805344	30792139	30778923	30765695	30752456	30739206	30725943	30712670	-12
4.342	0.0	30699385	30686088	30672780	30659460	30646129	30632786	30619432	30606066	30592689	30579300	-11
4.343	0.0	30565900	30552489	30539066	30525632	30512186	30498729	30485260	30471780	30458289	30444786	-11
4.344	0.0	30431271	30417746	30404209	30390660	30377101	30363529	30349947	30336353	30322748	30309131	-11
4.345	0.0	30295503	30281864	30268213	30254552	30240878	30227194	30213498	30199791	30186073	30172343	-11
4.346	0.0	30158602	30144850	30131086	30117311	30103525	30089728	30075920	30062100	30048269	30034427	-11
4.347	0.0	30020573	30006708	29992833	29978946	29965047	29951138	29937217	29923285	29909342	29895388	-11
4.348	0.0	29881423	29867446	29853459	29839460	29825450	29811429	29797397	29783354	29769299	29755234	-11
4.349	0.0	29741157	29727070	29712971	29698861	29684740	29670608	29656465	29642311	29628146	29613970	-11

TABLE 6

Sinc Function

x		.0	.0001	.0002	.0003	.0004	.0005	.0006	.0007	.0008	.0009	Δ_2
4.350	0.0	29599782	29585584	29571375	29557155	29542923	29528681	29514428	29500163	29485888	29471602	-11
4.351	0.0	29457304	29442996	29428677	29414347	29400005	29385653	29371290	29356916	29342531	29328136	-11
4.352	0.0	29313729	29299311	29284883	29270443	29255993	29241532	29227059	29212577	29198083	29183578	-11
4.353	0.0	29169062	29154536	29139999	29125451	29110892	29096322	29081741	29067150	29052548	29037935	-11
4.354	0.0	29023311	29008677	28994031	28979375	28964708	28950031	28935342	28920643	28905933	28891213	-11
4.355	0.0	28876481	28861739	28846986	28832223	28817449	28802664	28787868	28773062	28758245	28743417	-11
4.356	0.0	28728579	28713730	28698871	28684000	28669119	28654228	28639326	28624413	28609490	28594556	-11
4.357	0.0	28579611	28564656	28549690	28534714	28519727	28504729	28489721	28474702	28459673	28444633	-11
4.358	0.0	28429583	28414522	28399451	28384369	28369277	28354174	28339060	28323937	28308802	28293657	-10
4.359	0.0	28278502	28263336	28248160	28232973	28217776	28202569	28187351	28172122	28156883	28141634	-10
4.360	0.0	28126374	28111104	28095824	28080533	28065231	28049920	28034598	28019265	28003922	27988569	-10
4.361	0.0	27973206	27957832	27942448	27927054	27911649	27896234	27880808	27865373	27849927	27834470	-10
4.362	0.0	27819004	27803527	27788040	27772543	27757035	27741517	27725989	27710451	27694902	27679343	-10
4.363	0.0	27663774	27648195	27632606	27617006	27601396	27585776	27570146	27554506	27538856	27523195	-10
4.364	0.0	27507524	27491843	27476152	27460451	27444740	27429018	27413287	27397545	27381794	27366032	-10
4.365	0.0	27350260	27334478	27318686	27302884	27287072	27271250	27255417	27239575	27223723	27207861	-10
4.366	0.0	27191988	27176106	27160214	27144311	27128399	27112477	27096544	27080602	27064650	27048688	-10
4.367	0.0	27032716	27016734	27000742	26984740	26968728	26952706	26936675	26920633	26904582	26888520	-10
4.368	0.0	26872449	26856368	26840277	26824176	26808066	26791945	26775815	26759675	26743525	26727365	-10
4.369	0.0	26711195	26695016	26678827	26662628	26646419	26630200	26613972	26597734	26581486	26565229	-10
4.370	0.0	26548961	26532684	26516397	26500101	26483794	26467479	26451153	26434817	26418472	26402118	-10
4.371	0.0	26385753	26369379	26352995	26336602	26320199	26303786	26287364	26270932	26254491	26238039	-10
4.372	0.0	26221579	26205108	26188628	26172139	26155640	26139131	26122613	26106085	26089547	26073001	-10
4.373	0.0	26056444	26039878	26023303	26006718	25990123	25973519	25956906	25940283	25923650	25907008	-9
4.374	0.0	25890357	25873696	25857025	25840346	25823656	25806958	25790250	25773532	25756805	25740069	-9
4.375	0.0	25723323	25706568	25689804	25673030	25656247	25639454	25622652	25605841	25589020	25572190	-9
4.376	0.0	25555351	25538502	25521644	25504777	25487901	25471015	25454120	25437215	25420302	25403379	-9
4.377	0.0	25386447	25369505	25352555	25335595	25318626	25301647	25284660	25267663	25250657	25233642	-9
4.378	0.0	25216618	25199584	25182542	25165490	25148429	25131359	25114279	25097191	25080093	25062987	-9
4.379	0.0	25045871	25028746	25011612	24994469	24977317	24960156	24942985	24925806	24908618	24891420	-9
4.380	0.0	24874214	24856998	24839774	24822540	24805297	24788046	24770785	24753516	24736237	24718949	-9
4.381	0.0	24701653	24684347	24667033	24649710	24632377	24615036	24597686	24580327	24562959	24545582	-9
4.382	0.0	24528196	24510801	24493398	24475985	24458564	24441134	24423695	24406247	24388790	24371324	-9
4.383	0.0	24353850	24336367	24318875	24301374	24283864	24266346	24248819	24231283	24213738	24196185	-9
4.384	0.0	24178622	24161051	24143472	24125883	24108286	24090680	24073066	24055442	24037810	24020170	-9
4.385	0.0	24002520	23984862	23967196	23949520	23931836	23914144	23896443	23878733	23861014	23843287	-9
4.386	0.0	23825551	23807807	23790054	23772293	23754523	23736744	23718957	23701161	23683357	23665544	-9
4.387	0.0	23647723	23629893	23612054	23594208	23576352	23558488	23540616	23522735	23504846	23486948	-8
4.388	0.0	23469042	23451127	23433204	23415272	23397332	23379384	23361427	23343462	23325488	23307506	-8
4.389	0.0	23289516	23271517	23253510	23235494	23217470	23199438	23181398	23163349	23145292	23127226	-8
4.390	0.0	23109152	23091070	23072980	23054881	23036774	23018659	23000535	22982404	22964264	22946116	-8
4.391	0.0	22927959	22909794	22891621	22873440	22855251	22837053	22818848	22800634	22782412	22764182	-8
4.392	0.0	22745943	22727697	22709442	22691179	22672908	22654629	22636342	22618047	22599744	22581432	-8
4.393	0.0	22563113	22544785	22526449	22508106	22489754	22471394	22453026	22434650	22416266	22397875	-8
4.394	0.0	22379475	22361067	22342651	22324227	22305795	22287355	22268907	22250452	22231988	22213516	-8
4.395	0.0	22195037	22176549	22158054	22139551	22121040	22102521	22083994	22065459	22046916	22028365	-8
4.396	0.0	22009807	21991241	21972667	21954085	21935495	21916898	21898292	21879679	21861058	21842429	-8
4.397	0.0	21823793	21805149	21786496	21767837	21749169	21730494	21711811	21693120	21674422	21655716	-8
4.398	0.0	21637002	21618280	21599551	21580814	21562070	21543317	21524557	21505790	21487015	21468232	-8
4.399	0.0	21449442	21430644	21411838	21393025	21374204	21355376	21336540	21317696	21298845	21279986	-8

TABLE 6

Sinc Function

x		.0	.0001	.0002	.0003	.0004	.0005	.0006	.0007	.0008	.0009	Δ_2
4.400	0.0	21261120	21242246	21223365	21204476	21185580	21166676	21147765	21128846	21109920	21090986	−7
4.401	0.0	21072045	21053096	21034140	21015177	20996206	20977227	20958241	20939248	20920248	20901240	−7
4.402	0.0	20882224	20863201	20844171	20825134	20806089	20787036	20767977	20748910	20729836	20710754	−7
4.403	0.0	20691665	20672569	20653466	20634355	20615237	20596112	20576979	20557839	20538692	20519538	−7
4.404	0.0	20500376	20481208	20462032	20442848	20423658	20404460	20385256	20366044	20346825	20327598	−7
4.405	0.0	20308365	20289124	20269877	20250622	20231360	20212091	20192815	20173532	20154241	20134944	−7
4.406	0.0	20115639	20096328	20077009	20057684	20038351	20019011	19999664	19980311	19960950	19941582	−7
4.407	0.0	19922207	19902825	19883437	19864041	19844638	19825229	19805812	19786389	19766958	19747521	−7
4.408	0.0	19728076	19708625	19689167	19669702	19650230	19630752	19611266	19591774	19572274	19552768	−7
4.409	0.0	19533255	19513735	19494209	19474675	19455135	19435588	19416034	19396474	19376906	19357332	−7
4.410	0.0	19337751	19318164	19298569	19278968	19259360	19239746	19220125	19200497	19180862	19161221	−7
4.411	0.0	19141573	19121918	19102257	19082589	19062914	19043233	19023545	19003851	18984150	18964442	−7
4.412	0.0	18944728	18925007	18905279	18885545	18865805	18846058	18826304	18806544	18786777	18767004	−7
4.413	0.0	18747224	18727438	18707645	18687846	18668040	18648228	18628409	18608584	18588752	18568914	−6
4.414	0.0	18549070	18529219	18509362	18489498	18469628	18449752	18429869	18409979	18390084	18370182	−6
4.415	0.0	18350274	18330359	18310438	18290511	18270577	18250637	18230691	18210738	18190779	18170814	−6
4.416	0.0	18150843	18130865	18110881	18090891	18070895	18050892	18030883	18010868	17990847	17970820	−6
4.417	0.0	17950786	17930746	17910700	17890648	17870590	17850525	17830455	17810378	17790295	17770206	−6
4.418	0.0	17750111	17730010	17709903	17689790	17669670	17649545	17629413	17609276	17589132	17568983	−6
4.419	0.0	17548827	17528665	17508498	17488324	17468144	17447959	17427767	17407569	17387366	17367156	−6
4.420	0.0	17346941	17326719	17306492	17286259	17266020	17245775	17225524	17205267	17185004	17164736	−6
4.421	0.0	17144461	17124181	17103895	17083603	17063305	17043002	17022692	17002377	16982056	16961730	−6
4.422	0.0	16941397	16921059	16900715	16880365	16860009	16839648	16819281	16798908	16778530	16758146	−6
4.423	0.0	16737756	16717360	16696959	16676552	16656139	16635721	16615297	16594868	16574433	16553992	−6
4.424	0.0	16533546	16513094	16492636	16472173	16451704	16431230	16410750	16390265	16369774	16349278	−6
4.425	0.0	16328776	16308268	16287755	16267237	16246713	16226183	16205648	16185108	16164562	16144010	−5
4.426	0.0	16123454	16102891	16082324	16061750	16041172	16020588	15999999	15979404	15958804	15938199	−5
4.427	0.0	15917588	15896972	15876350	15855723	15835091	15814454	15793811	15773163	15752509	15731851	−5
4.428	0.0	15711187	15690518	15669843	15649163	15628478	15607788	15587093	15566392	15545686	15524975	−5
4.429	0.0	15504259	15483537	15462811	15442079	15421342	15400600	15379853	15359100	15338343	15317580	−5
4.430	0.0	15296812	15276040	15255262	15234479	15213691	15192897	15172099	15151296	15130488	15109674	−5
4.431	0.0	15088856	15068032	15047204	15026371	15005532	14984689	14963841	14942987	14922129	14901266	−5
4.432	0.0	14880398	14859524	14838646	14817764	14796876	14775983	14755085	14734183	14713275	14692363	−5
4.433	0.0	14671446	14650524	14629597	14608666	14587729	14566788	14545842	14524891	14503935	14482975	−5
4.434	0.0	14462010	14441040	14420065	14399085	14378101	14357112	14336119	14315120	14294117	14273109	−5
4.435	0.0	14252097	14231080	14210058	14189032	14168000	14146965	14125924	14104879	14083829	14062775	−5
4.436	0.0	14041716	14020653	13999585	13978512	13957435	13936353	13915267	13894176	13873081	13851981	−5
4.437	0.0	13830876	13809767	13788654	13767536	13746414	13725287	13704155	13683020	13661879	13640735	−4
4.438	0.0	13619585	13598432	13577274	13556112	13534945	13513774	13492598	13471418	13450234	13429045	−4
4.439	0.0	13407852	13386655	13365453	13344247	13323037	13301822	13280604	13259380	13238153	13216921	−4
4.440	0.0	13195685	13174445	13153200	13131952	13110699	13089442	13068180	13046915	13025645	13004371	−4
4.441	0.0	12983093	12961811	12940524	12919234	12897939	12876640	12855337	12834030	12812719	12791403	−4
4.442	0.0	12770084	12748760	12727433	12706101	12684765	12663426	12642082	12620734	12599382	12578026	−4
4.443	0.0	12556667	12535303	12513935	12492563	12471187	12449808	12428424	12407036	12385645	12364249	−4
4.444	0.0	12342850	12321447	12300039	12278628	12257213	12235794	12214372	12192945	12171515	12150080	−4
4.445	0.0	12128642	12107200	12085754	12064305	12042852	12021394	11999933	11978469	11957000	11935528	−4
4.446	0.0	11914052	11892572	11871089	11849602	11828111	11806616	11785118	11763616	11742111	11720601	−4
4.447	0.0	11699088	11677572	11656051	11634528	11613000	11591469	11569934	11548396	11526854	11505309	−4
4.448	0.0	11483759	11462207	11440651	11419091	11397528	11375961	11354390	11332817	11311239	11289658	−3
4.449	0.0	11268074	11246486	11224895	11203300	11181702	11160100	11138495	11116887	11095275	11073660	−3

TABLE 6

Sinc Function

x		.0	.0001	.0002	.0003	.0004	.0005	.0006	.0007	.0008	.0009	Δ_2
4.450	0.0	11052041	11030419	11008793	10987164	10965532	10943897	10922258	10900615	10878970	10857321	−3
4.451	0.0	10835669	10814013	10792354	10770692	10749027	10727358	10705686	10684011	10662332	10640651	−3
4.452	0.0	10618966	10597278	10575586	10553892	10532194	10510493	10488789	10467082	10445372	10423658	−3
4.453	0.0	10401941	10380221	10358499	10336772	10315043	10293311	10271576	10249837	10228096	10206351	−3
4.454	0.0	10184604	10162853	10141099	10119343	10097583	10075820	10054054	10032286	10010514	09988739	−3
4.455	0.0	09966962	09945181	09923397	09901611	09879821	09858029	09836234	09814436	09792635	09770831	−3
4.456	0.0	09749024	09727214	09705401	09683586	09661768	09639947	09618123	09596296	09574466	09552634	−3
4.457	0.0	09530799	09508961	09487120	09465277	09443431	09421582	09399730	09377876	09356018	09334158	−3
4.458	0.0	09312296	09290431	09268563	09246692	09224819	09202943	09181064	09159183	09137299	09115412	−3
4.459	0.0	09093523	09071632	09049737	09027840	09005941	08984039	08962134	08940227	08918317	08896405	−3
4.460	0.0	08874490	08852573	08830653	08808730	08786805	08764878	08742948	08721016	08699081	08677144	−2
4.461	0.0	08655204	08633262	08611318	08589371	08567422	08545470	08523516	08501559	08479600	08457639	−2
4.462	0.0	08435676	08413710	08391741	08369771	08347798	08325823	08303845	08281865	08259883	08237899	−2
4.463	0.0	08215912	08193923	08171932	08149939	08127943	08105945	08083945	08061943	08039938	08017932	−2
4.464	0.0	07995923	07973912	07951899	07929883	07907866	07885846	07863825	07841801	07819775	07797747	−2
4.465	0.0	07775717	07753684	07731650	07709614	07687575	07665535	07643492	07621448	07599401	07577353	−2
4.466	0.0	07555302	07533250	07511195	07489139	07467080	07445020	07422957	07400893	07378826	07356758	−2
4.467	0.0	07334688	07312616	07290542	07268466	07246389	07224309	07202228	07180144	07158059	07135972	−2
4.468	0.0	07113883	07091793	07069700	07047606	07025510	07003412	06981313	06959211	06937108	06915003	−2
4.469	0.0	06892897	06870788	06848678	06826566	06804453	06782338	06760221	06738102	06715982	06693860	−2
4.470	0.0	06671737	06649611	06627485	06605356	06583226	06561094	06538961	06516826	06494690	06472552	−2
4.471	0.0	06450412	06428271	06406128	06383984	06361838	06339691	06317542	06295392	06273240	06251087	0
4.472	0.0	06228932	06206776	06184618	06162459	06140299	06118137	06095973	06073808	06051642	06029474	0
4.473	0.0	06007305	05985135	05962963	05940790	05918615	05896440	05874263	05852084	05829904	05807723	0
4.474	0.0	05785541	05763357	05741172	05718986	05696798	05674609	05652419	05630228	05608035	05585842	0
4.475	0.0	05563647	05541450	05519253	05497054	05474855	05452654	05430452	05408249	05386044	05363839	0
4.476	0.0	05341632	05319424	05297215	05275006	05252794	05230582	05208369	05186155	05163940	05141723	0
4.477	0.0	05119506	05097287	05075068	05052848	05030626	05008404	04986180	04963956	04941730	04919504	0
4.478	0.0	04897277	04875049	04852820	04830589	04808358	04786127	04763894	04741660	04719426	04697190	0
4.479	0.0	04674954	04652717	04630479	04608240	04586000	04563760	04541518	04519276	04497034	04474790	0
4.480	0.0	04452545	04430300	04408054	04385808	04363560	04341312	04319063	04296814	04274563	04252312	0
4.481	0.0	04230061	04207808	04185555	04163302	04141047	04118792	04096537	04074280	04052024	04029766	0
4.482	0.0	04007508	03985250	03962990	03940731	03918470	03896209	03873948	03851686	03829423	03807160	0
4.483	0.0	03784897	03762633	03740368	03718103	03695833	03673572	03651305	03629038	03606771	03584503	0
4.484	0.0	03562235	03539967	03517698	03495428	03473159	03450889	03428618	03406347	03384076	03361804	0
4.485	0.0	03339532	03317260	03294988	03272715	03250442	03228168	03205895	03183621	03161346	03139072	0
4.486	0.0	03116797	03094522	03072247	03049972	03027696	03005420	02983144	02960868	02938591	02916315	0
4.487	0.0	02894038	02871761	02849484	02827207	02804930	02782652	02760375	02738097	02715820	02693542	0
4.488	0.0	02671264	02648986	02626708	02604430	02582152	02559874	02537596	02515318	02493040	02470762	0
4.489	0.0	02448484	02426206	02403928	02381650	02359372	02337094	02314817	02292539	02270261	02247984	0
4.490	0.0	02225706	02203429	02181152	02158875	02136598	02114321	02092045	02069768	02047492	02025216	0
4.491	0.0	02002940	01980664	01958389	01936114	01913839	01891564	01869289	01847015	01824741	01802467	0
4.492	0.0	01780194	01757921	01735648	01713375	01691103	01668831	01646560	01624288	01602018	01579747	0
4.493	0.0	01557477	01535207	01512938	01490669	01468400	01446132	01423864	01401597	01379330	01357063	0
4.494	0.0	01334797	01312532	01290267	01268002	01245738	01223474	01201211	01178949	01156687	01134425	1
4.495	0.0	01112164	01089904	01067644	01045384	01023126	01000868	00978610	00956353	00934097	00911841	1
4.496	0.0	00889586	00867332	00845078	00822825	00800572	00778320	00756069	00733819	00711569	00689320	1
4.497	0.0	00667072	00644824	00622578	00600331	00578086	00555842	00533598	00511355	00489112	00466871	1
4.498	0.0	00444630	00422391	00400152	00377913	00355676	00333440	00311204	00288969	00266735	00244502	1
4.499	0.0	00222270	00200039	00177809	00155579	00133351	00111123	00088897	00066671	00044446	00022223	−1

$\frac{\sin 2\pi x}{2\pi x}$

TABLE 6

Sinc Function

x		.0	.0001	.0002	.0003	.0004	.0005	.0006	.0007	.0008	.0009	Δ_2
4.500	-0.0	00000000	00022222	00044442	00066662	00088881	00111099	00133315	00155531	00177745	00199959	0
4.501	-0.0	00222171	00244383	00266593	00288802	00311010	00333217	00355423	00377628	00399832	00422034	0
4.502	-0.0	00444235	00466435	00488634	00510832	00533029	00555224	00577418	00599611	00621803	00643994	0
4.503	-0.0	00666183	00688371	00710558	00732743	00754928	00777111	00799292	00821473	00843652	00865830	0
4.504	-0.0	00888006	00910181	00932355	00954527	00976698	00998868	01021036	01043203	01065369	01087533	0
4.505	-0.0	01109695	01131857	01154016	01176175	01198332	01220487	01242641	01264794	01286945	01309094	0
4.506	-0.0	01331243	01353389	01375534	01397678	01419820	01441960	01464099	01486236	01508372	01530506	2
4.507	-0.0	01552639	01574770	01596899	01619027	01641153	01663278	01685401	01707522	01729641	01751759	2
4.508	-0.0	01773876	01795990	01818103	01840214	01862324	01884432	01906538	01928642	01950744	01972845	2
4.509	-0.0	01994944	02017042	02039137	02061231	02083323	02105413	02127501	02149588	02171672	02193755	2
4.510	-0.0	02215836	02237915	02259993	02282068	02304142	02326213	02348283	02370351	02392417	02414481	2
4.511	-0.0	02436543	02458603	02480661	02502717	02524772	02546824	02568874	02590922	02612969	02635013	2
4.512	-0.0	02657055	02679096	02701134	02723170	02745204	02767236	02789266	02811294	02833320	02855344	2
4.513	-0.0	02877365	02899385	02921402	02943417	02965431	02987441	03009450	03031457	03053461	03075464	2
4.514	-0.0	03097464	03119462	03141458	03163451	03185442	03207431	03229418	03251403	03273385	03295365	2
4.515	-0.0	03317343	03339318	03361292	03383262	03405231	03427197	03449161	03471123	03493082	03515039	2
4.516	-0.0	03536994	03558946	03580896	03602843	03624788	03646731	03668671	03690609	03712544	03734477	2
4.517	-0.0	03756407	03778335	03800261	03822184	03844105	03866023	03887938	03909852	03931762	03953670	3
4.518	-0.0	03975576	03997479	04019379	04041277	04063173	04085065	04106956	04128843	04150728	04172611	3
4.519	-0.0	04194490	04216367	04238242	04260114	04281983	04303850	04325714	04347575	04369433	04391289	3
4.520	-0.0	04413142	04434993	04456841	04478686	04500528	04522367	04544204	04566038	04587869	04609698	3
4.521	-0.0	04631524	04653346	04675167	04696984	04718798	04740610	04762419	04784225	04806028	04827828	3
4.522	-0.0	04849625	04871420	04893211	04915000	04936786	04958569	04980349	05002126	05023900	05045671	3
4.523	-0.0	05067439	05089205	05110967	05132726	05154482	05176236	05197986	05219733	05241478	05263219	3
4.524	-0.0	05284957	05306692	05328424	05350153	05371879	05393602	05415322	05437038	05458752	05480462	3
4.525	-0.0	05502170	05523874	05545575	05567273	05588968	05610659	05632348	05654033	05675715	05697394	3
4.526	-0.0	05719070	05740742	05762411	05784077	05805740	05827399	05849056	05870709	05892358	05914005	3
4.527	-0.0	05935648	05957288	05978924	06000557	06022187	06043814	06065437	06087057	06108673	06130286	3
4.528	-0.0	06151896	06173502	06195105	06216705	06238301	06259894	06281483	06303069	06324651	06346230	3
4.529	-0.0	06367806	06389378	06410947	06432512	06454073	06475631	06497186	06518737	06540285	06561829	4
4.530	-0.0	06583369	06604906	06626440	06647969	06669496	06691018	06712537	06734053	06755565	06777073	4
4.531	-0.0	06798577	06820078	06841576	06863070	06884560	06906046	06927529	06949008	06970483	06991955	4
4.532	-0.0	07013422	07034887	07056347	07077804	07099257	07120706	07142152	07163593	07185031	07206465	4
4.533	-0.0	07227896	07249322	07270745	07292164	07313579	07334991	07356398	07377802	07399201	07420597	4
4.534	-0.0	07441989	07463377	07484762	07506142	07527518	07548891	07570260	07591624	07612985	07634342	4
4.535	-0.0	07655695	07677044	07698388	07719729	07741066	07762399	07783728	07805053	07826374	07847691	4
4.536	-0.0	07869004	07890312	07911617	07932918	07954214	07975507	07996795	08018080	08039360	08060636	4
4.537	-0.0	08081908	08103176	08124439	08145699	08166954	08188206	08209453	08230696	08251935	08273169	4
4.538	-0.0	08294399	08315626	08336847	08358065	08379279	08400488	08421693	08442893	08464090	08485282	4
4.539	-0.0	08506470	08527653	08548833	08570008	08591178	08612345	08633507	08654664	08675818	08696967	4
4.540	-0.0	08718111	08739251	08760387	08781519	08802646	08823768	08844887	08866000	08887110	08908215	4
4.541	-0.0	08929315	08950411	08971503	08992590	09013673	09034751	09055824	09076893	09097958	09119018	5
4.542	-0.0	09140074	09161125	09182171	09203213	09224251	09245283	09266312	09287335	09308354	09329369	5
4.543	-0.0	09350379	09371384	09392385	09413381	09434372	09455359	09476341	09497318	09518291	09539259	5
4.544	-0.0	09560222	09581181	09602135	09623084	09644029	09664968	09685903	09706834	09727759	09748680	5
4.545	-0.0	09769596	09790507	09811414	09832315	09853212	09874104	09894991	09915874	09936751	09957624	5
4.546	-0.0	09978492	09999355	10020213	10041066	10061915	10082758	10103597	10124431	10145259	10166083	5
4.547	-0.0	10186902	10207716	10228525	10249329	10270129	10290923	10311712	10332496	10353275	10374050	5
4.548	-0.0	10394819	10415583	10436342	10457096	10477845	10498589	10519328	10540062	10560791	10581515	5
4.549	-0.0	10602234	10622947	10643656	10664359	10685057	10705751	10726439	10747121	10767799	10788472	5

TABLE 6

Sinc Function

x		.0	.0001	.0002	.0003	.0004	.0005	.0006	.0007	.0008	.0009	Δ_2
4.550	-0.0	10809139	10829801	10850458	10871110	10891757	10912398	10933034	10953665	10974291	10994911	5
4.551	-0.0	11015527	11036137	11056741	11077341	11097935	11118524	11139108	11159686	11180259	11200826	5
4.552	-0.0	11221389	11241946	11262498	11283044	11303585	11324120	11344651	11365176	11385695	11406209	5
4.553	-0.0	11426718	11447221	11467719	11488211	11508698	11529180	11549656	11570126	11590592	11611051	5
4.554	-0.0	11631505	11651954	11672397	11692835	11713267	11733694	11754115	11774531	11794941	11815345	6
4.555	-0.0	11835744	11856137	11876525	11896907	11917284	11937655	11958021	11978380	11998735	12019083	6
4.556	-0.0	12039426	12059763	12080095	12100421	12120741	12141056	12161365	12181668	12201965	12222257	6
4.557	-0.0	12242543	12262824	12283098	12303367	12323630	12343888	12364139	12384385	12404625	12424860	6
4.558	-0.0	12445088	12465311	12485528	12505739	12525944	12546144	12566337	12586525	12606707	12626883	6
4.559	-0.0	12647053	12667217	12687376	12707528	12727675	12747816	12767950	12788079	12808202	12828319	6
4.560	-0.0	12848430	12868535	12888635	12908728	12928815	12948896	12968971	12989040	13009104	13029161	6
4.561	-0.0	13049212	13069257	13089296	13109329	13129356	13149377	13169392	13189401	13209404	13229400	6
4.562	-0.0	13249391	13269375	13289353	13309326	13329292	13349251	13369205	13389153	13409094	13429030	6
4.563	-0.0	13448959	13468882	13488798	13508709	13528613	13548511	13568403	13588289	13608168	13628041	6
4.564	-0.0	13647908	13667769	13687623	13707472	13727313	13747149	13766978	13786801	13806618	13826428	6
4.565	-0.0	13846232	13866030	13885821	13905606	13925385	13945157	13964923	13984682	14004436	14024182	6
4.566	-0.0	14043923	14063657	14083384	14103105	14122820	14142528	14162230	14181925	14201614	14221296	6
4.567	-0.0	14240972	14260642	14280305	14299961	14319611	14339254	14358891	14378522	14398145	14417763	7
4.568	-0.0	14437373	14456978	14476575	14496166	14515751	14535329	14554900	14574465	14594023	14613574	7
4.569	-0.0	14633119	14652657	14672189	14691714	14711232	14730743	14750248	14769747	14789238	14808723	7
4.570	-0.0	14828201	14847673	14867137	14886595	14906047	14925491	14944929	14964360	14983784	15003202	7
4.571	-0.0	15022613	15042017	15061414	15080804	15100188	15119565	15138935	15158298	15177654	15197004	7
4.572	-0.0	15216346	15235682	15255011	15274333	15293648	15312956	15332258	15351552	15370840	15390121	7
4.573	-0.0	15409395	15428661	15447921	15467174	15486420	15505659	15524891	15544117	15563335	15582546	7
4.574	-0.0	15601750	15620947	15640138	15659321	15678497	15697666	15716828	15735983	15755131	15774272	7
4.575	-0.0	15793406	15812533	15831652	15850765	15869871	15888969	15908060	15927145	15946222	15965292	7
4.576	-0.0	15984355	16003410	16022459	16041500	16060534	16079561	16098581	16117594	16136600	16155598	7
4.577	-0.0	16174589	16193573	16212549	16231519	16250481	16269436	16288384	16307324	16326257	16345183	7
4.578	-0.0	16364102	16383013	16401917	16420814	16439703	16458585	16477460	16496328	16515188	16534040	7
4.579	-0.0	16552886	16571724	16590555	16609378	16628194	16647002	16665804	16684597	16703384	16722163	7
4.580	-0.0	16740934	16759698	16778455	16797204	16815946	16834680	16853407	16872126	16890838	16909543	7
4.581	-0.0	16928240	16946929	16965611	16984285	17002952	17021612	17040263	17058908	17077544	17096174	8
4.582	-0.0	17114795	17133409	17152016	17170615	17189206	17207790	17226366	17244934	17263495	17282048	8
4.583	-0.0	17300594	17319132	17337662	17356185	17374700	17393207	17411707	17430199	17448683	17467160	8
4.584	-0.0	17485629	17504090	17522543	17540989	17559427	17577857	17596280	17614695	17633102	17651501	8
4.585	-0.0	17669893	17688276	17706652	17725020	17743381	17761733	17780078	17798415	17816744	17835065	8
4.586	-0.0	17853379	17871684	17889982	17908272	17926554	17944828	17963094	17981353	17999603	18017845	8
4.587	-0.0	18036080	18054307	18072526	18090737	18108939	18127134	18145322	18163501	18181672	18199835	8
4.588	-0.0	18217990	18236137	18254276	18272408	18290531	18308646	18326753	18344852	18362944	18381027	8
4.589	-0.0	18399102	18417169	18435228	18453278	18471321	18489356	18507383	18525401	18543412	18561414	8
4.590	-0.0	18579408	18597394	18615372	18633342	18651304	18669257	18687203	18705140	18723069	18740990	8
4.591	-0.0	18758903	18776807	18794704	18812592	18830472	18848343	18866207	18884062	18901909	18919748	8
4.592	-0.0	18937579	18955401	18973215	18991021	19008818	19026608	19044389	19062161	19079926	19097682	8
4.593	-0.0	19115429	19133169	19150900	19168623	19186337	19204043	19221741	19239430	19257111	19274784	8
4.594	-0.0	19292448	19310104	19327752	19345391	19363021	19380644	19398257	19415863	19433460	19451048	8
4.595	-0.0	19468629	19486200	19503763	19521318	19538864	19556402	19573931	19591452	19608965	19626468	9
4.596	-0.0	19643964	19661451	19678929	19696399	19713860	19731312	19748757	19766192	19783619	19801037	9
4.597	-0.0	19818447	19835849	19853241	19870625	19888001	19905368	19922726	19940076	19957417	19974749	9
4.598	-0.0	19992073	20009388	20026694	20043992	20061281	20078562	20095833	20113097	20130351	20147597	9
4.599	-0.0	20164834	20182062	20199282	20216492	20233695	20250888	20268073	20285248	20302416	20319574	9

$\dfrac{\sin 2\pi x}{2\pi x}$

TABLE 6

Sinc Function

x	.0	.0001	.0002	.0003	.0004	.0005	.0006	.0007	.0008	.0009	Δ_2
4.600 -0.0	20336724	20353864	20370996	20388120	20405234	20422340	20439437	20456525	20473604	20490674	9
4.601 -0.0	20507736	20524789	20541833	20558868	20575894	20592911	20609920	20626919	20643910	20660892	9
4.602 -0.0	20677865	20694829	20711784	20728730	20745667	20762596	20779515	20796426	20813327	20830220	9
4.603 -0.0	20847103	20863978	20880844	20897700	20914548	20931387	20948217	20965037	20981849	20998652	9
4.604 -0.0	21015446	21032230	21049006	21065773	21082530	21099279	21116018	21132749	21149470	21166182	9
4.605 -0.0	21182885	21199579	21216264	21232940	21249607	21266265	21282913	21299553	21316183	21332804	9
4.606 -0.0	21349416	21366019	21382613	21399197	21415773	21432339	21448896	21465444	21481983	21498512	9
4.607 -0.0	21515032	21531543	21548045	21564538	21581021	21597495	21613960	21630416	21646862	21663299	9
4.608 -0.0	21679727	21696146	21712555	21728955	21745346	21761727	21778099	21794462	21810816	21827160	9
4.609 -0.0	21843495	21859820	21876137	21892444	21908741	21925029	21941308	21957577	21973838	21990088	9
4.610 -0.0	22006329	22022561	22038784	22054997	22071201	22087395	22103580	22119755	22135921	22152078	9
4.611 -0.0	22168225	22184362	22200491	22216609	22232719	22248818	22264909	22280989	22297061	22313123	9
4.612 -0.0	22329175	22345218	22361251	22377275	22393289	22409294	22425289	22441274	22457251	22473217	10
4.613 -0.0	22489174	22505121	22521059	22536987	22552906	22568815	22584714	22600604	22616484	22632355	10
4.614 -0.0	22648216	22664067	22679909	22695741	22711564	22727376	22743179	22758973	22774757	22790531	10
4.615 -0.0	22806295	22822050	22837795	22853530	22869256	22884972	22900678	22916375	22932061	22947738	10
4.616 -0.0	22963406	22979063	22994711	23010349	23025977	23041596	23057205	23072804	23088393	23103972	10
4.617 -0.0	23119542	23135102	23150652	23166192	23181722	23197243	23212753	23228254	23243745	23259227	10
4.618 -0.0	23274698	23290159	23305611	23321053	23336485	23351907	23367319	23382721	23398113	23413496	10
4.619 -0.0	23428868	23444231	23459583	23474926	23490259	23505582	23520895	23536198	23551491	23566774	10
4.620 -0.0	23582047	23597310	23612563	23627806	23643039	23658262	23673476	23688679	23703872	23719055	10
4.621 -0.0	23734228	23749391	23764544	23779688	23794821	23809943	23825056	23840159	23855252	23870335	10
4.622 -0.0	23885407	23900470	23915522	23930565	23945597	23960619	23975631	23990633	24005625	24020607	10
4.623 -0.0	24035578	24050540	24065491	24080432	24095363	24110284	24125194	24140095	24154985	24169865	10
4.624 -0.0	24184735	24199595	24214445	24229284	24244113	24258932	24273741	24288539	24303328	24318106	10
4.625 -0.0	24332873	24347631	24362378	24377115	24391842	24406559	24421265	24435961	24450647	24465322	10
4.626 -0.0	24479987	24494642	24509286	24523921	24538544	24553158	24567761	24582354	24596937	24611509	10
4.627 -0.0	24626071	24640622	24655164	24669694	24684215	24698725	24713225	24727714	24742193	24756662	10
4.628 -0.0	24771120	24785567	24800005	24814432	24828848	24843254	24857650	24872035	24886410	24900774	10
4.629 -0.0	24915128	24929471	24943804	24958127	24972439	24986740	25001032	25015312	25029582	25043842	10
4.630 -0.0	25058091	25072329	25086557	25100775	25114982	25129178	25143364	25157540	25171705	25185859	11
4.631 -0.0	25200003	25214136	25228259	25242371	25256472	25270563	25284643	25298713	25312772	25326821	11
4.632 -0.0	25340859	25354886	25368903	25382909	25396904	25410889	25424864	25438827	25452780	25466722	11
4.633 -0.0	25480654	25494575	25508485	25522385	25536274	25550152	25564020	25577876	25591723	25605558	11
4.634 -0.0	25619383	25633197	25647000	25660793	25674575	25688346	25702107	25715856	25729595	25743323	11
4.635 -0.0	25757041	25770748	25784444	25798129	25811803	25825467	25839120	25852762	25866393	25880013	11
4.636 -0.0	25893623	25907222	25920810	25934387	25947953	25961509	25975054	25988587	26002110	26015623	11
4.637 -0.0	26029124	26042614	26056094	26069563	26083021	26096468	26109904	26123329	26136743	26150147	11
4.638 -0.0	26163539	26176921	26190291	26203651	26217000	26230338	26243665	26256981	26270286	26283580	11
4.639 -0.0	26296864	26310136	26323397	26336648	26349887	26363116	26376333	26389539	26402735	26415919	11
4.640 -0.0	26429093	26442255	26455407	26468547	26481677	26494795	26507903	26520999	26534084	26547159	11
4.641 -0.0	26560222	26573274	26586315	26599346	26612365	26625373	26638369	26651355	26664330	26677294	11
4.642 -0.0	26690246	26703188	26716118	26729037	26741946	26754843	26767729	26780603	26793467	26806320	11
4.643 -0.0	26819161	26831991	26844811	26857618	26870415	26883201	26895976	26908739	26921491	26934232	11
4.644 -0.0	26946962	26959680	26972388	26985084	26997769	27010443	27023106	27035757	27048397	27061026	11
4.645 -0.0	27073644	27086250	27098846	27111430	27124003	27136564	27149114	27161653	27174181	27186698	11
4.646 -0.0	27199203	27211697	27224180	27236651	27249111	27261560	27273997	27286424	27298838	27311242	11
4.647 -0.0	27323634	27336015	27348385	27360743	27373090	27385426	27397750	27410063	27422365	27434655	11
4.648 -0.0	27446934	27459201	27471457	27483702	27495936	27508158	27520368	27532567	27544755	27556932	11
4.649 -0.0	27569097	27581250	27593393	27605523	27617643	27629751	27641847	27653932	27666006	27678068	11

TABLE 6

Sinc Function

x	.0	.0001	.0002	.0003	.0004	.0005	.0006	.0007	.0008	.0009	Δ_2
4.650 -0.0	27690119	27702158	27714186	27726203	27738208	27750201	27762183	27774154	27786113	27798060	11
4.651 -0.0	27809996	27821921	27833834	27845736	27857626	27869504	27881372	27893227	27905071	27916904	12
4.652 -0.0	27928725	27940534	27952332	27964118	27975893	27987656	27999408	28011148	28022877	28034594	12
4.653 -0.0	28046299	28057993	28069676	28081346	28093005	28104653	28116289	28127913	28139526	28151127	12
4.654 -0.0	28162717	28174295	28185861	28197416	28208959	28220490	28232010	28243518	28255014	28266499	12
4.655 -0.0	28277972	28289434	28300884	28312322	28323749	28335163	28346567	28357958	28369338	28380706	12
4.656 -0.0	28392063	28403407	28414740	28426062	28437371	28448669	28459955	28471230	28482493	28493744	12
4.657 -0.0	28504983	28516211	28527426	28538630	28549823	28561003	28572172	28583329	28594475	28605608	12
4.658 -0.0	28616730	28627840	28638938	28650024	28661099	28672162	28683213	28694252	28705280	28716295	12
4.659 -0.0	28727299	28738291	28749271	28760240	28771196	28782141	28793074	28803995	28814904	28825802	12
4.660 -0.0	28836687	28847561	28858423	28869273	28880111	28890937	28901752	28912554	28923345	28934123	12
4.661 -0.0	28944890	28955645	28966388	28977120	28987839	28998546	29009242	29019925	29030597	29041257	12
4.662 -0.0	29051904	29062540	29073164	29083776	29094376	29104964	29115541	29126105	29136657	29147198	12
4.663 -0.0	29157726	29168242	29178747	29189239	29199720	29210188	29220645	29231089	29241522	29251943	12
4.664 -0.0	29262351	29272748	29283132	29293505	29303866	29314214	29324551	29334875	29345188	29355488	12
4.665 -0.0	29365777	29376053	29386318	29396570	29406810	29417038	29427255	29437459	29447651	29457831	12
4.666 -0.0	29467999	29478155	29488298	29498430	29508550	29518657	29528753	29538836	29548908	29558967	12
4.667 -0.0	29569014	29579049	29589072	29599083	29609081	29619068	29629042	29639005	29648955	29658893	12
4.668 -0.0	29668819	29678733	29688634	29698524	29708401	29718266	29728119	29737960	29747789	29757605	12
4.669 -0.0	29767410	29777202	29786982	29796750	29806506	29816249	29825980	29835700	29845406	29855101	12
4.670 -0.0	29864784	29874454	29884112	29893758	29903392	29913013	29922622	29932219	29941804	29951377	12
4.671 -0.0	29960937	29970485	29980021	29989545	29999056	30008555	30018042	30027517	30036979	30046429	12
4.672 -0.0	30055867	30065292	30074706	30084107	30093495	30102872	30112236	30121588	30130927	30140255	12
4.673 -0.0	30149570	30158872	30168163	30177441	30186707	30195960	30205201	30214430	30223646	30232851	12
4.674 -0.0	30242042	30251222	30260389	30269544	30278686	30287816	30296934	30306040	30315133	30324213	12
4.675 -0.0	30333282	30342338	30351381	30360413	30369432	30378438	30387432	30396414	30405383	30414340	12
4.676 -0.0	30423285	30432217	30441137	30450045	30458940	30467822	30476692	30485550	30494396	30503228	12
4.677 -0.0	30512049	30520857	30529653	30538436	30547207	30555965	30564711	30573445	30582166	30590874	12
4.678 -0.0	30599571	30608254	30616926	30625585	30634231	30642865	30651486	30660095	30668692	30677276	12
4.679 -0.0	30685847	30694406	30702953	30711487	30720008	30728518	30737014	30745498	30753970	30762429	13
4.680 -0.0	30770876	30779310	30787731	30796140	30804537	30812921	30821293	30829651	30837998	30846332	13
4.681 -0.0	30854653	30862962	30871258	30879542	30887813	30896072	30904318	30912552	30920773	30928981	13
4.682 -0.0	30937177	30945361	30953531	30961690	30969835	30977968	30986089	30994197	31002292	31010375	13
4.683 -0.0	31018445	31026503	31034547	31042580	31050600	31058607	31066601	31074583	31082553	31090509	13
4.684 -0.0	31098454	31106385	31114304	31122210	31130104	31137985	31145853	31153709	31161552	31169383	13
4.685 -0.0	31177201	31185006	31192798	31200578	31208346	31216100	31223842	31231572	31239288	31246992	13
4.686 -0.0	31254684	31262362	31270028	31277682	31285322	31292950	31300565	31308168	31315758	31323335	13
4.687 -0.0	31330900	31338452	31345991	31353517	31361031	31368532	31376021	31383496	31390959	31398409	13
4.688 -0.0	31405847	31413272	31420684	31428083	31435470	31442844	31450205	31457553	31464889	31472212	13
4.689 -0.0	31479522	31486820	31494105	31501377	31508636	31515883	31523116	31530337	31537546	31544741	13
4.690 -0.0	31551924	31559094	31566251	31573396	31580527	31587646	31594753	31601846	31608927	31615994	13
4.691 -0.0	31623049	31630092	31637121	31644138	31651142	31658133	31665111	31672077	31679029	31685969	13
4.692 -0.0	31692896	31699811	31706712	31713601	31720477	31727340	31734190	31741027	31747852	31754664	13
4.693 -0.0	31761463	31768249	31775022	31781782	31788530	31795265	31801987	31808696	31815392	31822075	13
4.694 -0.0	31828746	31835404	31842049	31848681	31855300	31861906	31868499	31875080	31881648	31888202	13
4.695 -0.0	31894744	31901274	31907790	31914293	31920784	31927261	31933726	31940178	31946617	31953043	13
4.696 -0.0	31959456	31965856	31972244	31978618	31984980	31991328	31997664	32003987	32010297	32016594	13
4.697 -0.0	32022878	32029150	32035408	32041654	32047886	32054106	32060312	32066506	32072687	32078855	13
4.698 -0.0	32085010	32091152	32097281	32103397	32109501	32115591	32121668	32127733	32133784	32139823	13
4.699 -0.0	32145849	32151861	32157861	32163848	32169822	32175783	32181731	32187666	32193588	32199497	13

TABLE 6

Sinc Function

x		.0	.0001	.0002	.0003	.0004	.0005	.0006	.0007	.0008	.0009	Δ_2
4.700	-0.0	32205393	32211276	32217146	32223003	32228847	32234679	32240497	32246302	32252095	32257874	13
4.701	-0.0	32263640	32269394	32275134	32280861	32286576	32292277	32297966	32303641	32309304	32314953	13
4.702	-0.0	32320590	32326213	32331824	32337421	32343005	32348577	32354135	32359681	32365213	32370733	13
4.703	-0.0	32376239	32381732	32387213	32392680	32398134	32403576	32409004	32414419	32419822	32425211	13
4.704	-0.0	32430587	32435950	32441300	32446637	32451961	32457272	32462570	32467855	32473127	32478386	13
4.705	-0.0	32483632	32488864	32494084	32499291	32504484	32509665	32514832	32519987	32525128	32530256	13
4.706	-0.0	32535372	32540474	32545563	32550639	32555702	32560752	32565789	32570813	32575823	32580821	13
4.707	-0.0	32585806	32590777	32595736	32600681	32605613	32610532	32615438	32620331	32625211	32630078	13
4.708	-0.0	32634932	32639773	32644600	32649415	32654216	32659004	32663780	32668542	32673291	32678027	13
4.709	-0.0	32682749	32687459	32692156	32696839	32701510	32706167	32710811	32715442	32720060	32724665	13
4.710	-0.0	32729256	32733835	32738401	32742953	32747492	32752018	32756531	32761031	32765518	32769991	13
4.711	-0.0	32774452	32778899	32783334	32787755	32792163	32796558	32800939	32805308	32809663	32814006	13
4.712	-0.0	32818335	32822651	32826954	32831243	32835520	32839783	32844034	32848271	32852495	32856706	13
4.713	-0.0	32860903	32865088	32869259	32873418	32877563	32881695	32885814	32889919	32894012	32898091	13
4.714	-0.0	32902157	32906210	32910250	32914277	32918290	32922291	32926278	32930252	32934213	32938160	13
4.715	-0.0	32942095	32946016	32949924	32953819	32957701	32961570	32965425	32969267	32973096	32976912	13
4.716	-0.0	32980715	32984505	32988281	32992044	32995794	32999531	33003255	33006965	33010662	33014347	13
4.717	-0.0	33018017	33021675	33025320	33028951	33032569	33036174	33039766	33043344	33046910	33050462	13
4.718	-0.0	33054001	33057526	33061039	33064538	33068024	33071497	33074957	33078404	33081837	33085257	13
4.719	-0.0	33088664	33092058	33095438	33098806	33102160	33105500	33108828	33112143	33115444	33118732	13
4.720	-0.0	33122007	33125268	33128517	33131752	33134974	33138183	33141378	33144560	33147729	33150885	13
4.721	-0.0	33154028	33157157	33160273	33163376	33166466	33169543	33172606	33175656	33178693	33181717	13
4.722	-0.0	33184727	33187724	33190708	33193679	33196636	33199580	33202511	33205429	33208334	33211225	13
4.723	-0.0	33214103	33216968	33219820	33222658	33225483	33228295	33231094	33233879	33236651	33239410	13
4.724	-0.0	33242156	33244889	33247608	33250314	33253007	33255686	33258352	33261005	33263645	33266272	13
4.725	-0.0	33268885	33271485	33274072	33276645	33279206	33281753	33284287	33286807	33289315	33291809	13
4.726	-0.0	33294289	33296757	33299211	33301653	33304080	33306495	33308896	33311284	33313659	33316021	13
4.727	-0.0	33318369	33320704	33323026	33325335	33327630	33329912	33332181	33334436	33336679	33338908	13
4.728	-0.0	33341123	33343326	33345515	33347691	33349854	33352004	33354140	33356263	33358373	33360469	13
4.729	-0.0	33362552	33364622	33366679	33368722	33370753	33372769	33374773	33376764	33378741	33380705	13
4.730	-0.0	33382655	33384593	33386517	33388427	33390325	33392209	33394080	33395938	33397783	33399614	13
4.731	-0.0	33401432	33403237	33405028	33406806	33408571	33410323	33412061	33413787	33415498	33417197	13
4.732	-0.0	33418882	33420555	33422213	33423859	33425491	33427110	33428716	33430309	33431888	33433454	13
4.733	-0.0	33435006	33436546	33438072	33439585	33441085	33442571	33444044	33445504	33446951	33448384	13
4.734	-0.0	33449804	33451211	33452604	33453985	33455352	33456705	33458046	33459373	33460687	33461988	13
4.735	-0.0	33463275	33464549	33465810	33467058	33468292	33469513	33470721	33471916	33473097	33474265	13
4.736	-0.0	33475420	33476561	33477689	33478804	33479906	33480995	33482070	33483132	33484180	33485216	13
4.737	-0.0	33486238	33487247	33488242	33489225	33490194	33491150	33492092	33493022	33493938	33494840	13
4.738	-0.0	33495730	33496606	33497469	33498319	33499155	33499979	33500788	33501585	33502369	33503139	13
4.739	-0.0	33503896	33504639	33505370	33506087	33506791	33507481	33508159	33508823	33509474	33510111	13
4.740	-0.0	33510736	33511347	33511945	33512529	33513101	33513659	33514203	33514735	33515253	33515758	13
4.741	-0.0	33516250	33516729	33517194	33517646	33518085	33518510	33518923	33519322	33519708	33520080	13
4.742	-0.0	33520439	33520785	33521118	33521438	33521744	33522037	33522317	33522584	33522837	33523077	13
4.743	-0.0	33523304	33523517	33523718	33523905	33524079	33524239	33524387	33524521	33524642	33524749	13
4.744	-0.0	33524844	33524925	33524993	33525048	33525089	33525117	33525132	33525134	33525123	33525098	13
4.745	-0.0	33525060	33525009	33524944	33524867	33524776	33524672	33524554	33524424	33524280	33524123	13
4.746	-0.0	33523953	33523769	33523573	33523363	33523140	33522903	33522654	33522391	33522115	33521825	13
4.747	-0.0	33521523	33521207	33520878	33520536	33520181	33519812	33519430	33519035	33518627	33518206	13
4.748	-0.0	33517771	33517323	33516862	33516388	33515900	33515400	33514886	33514358	33513818	33513265	13
4.749	-0.0	33512698	33512118	33511525	33510918	33510299	33509666	33509020	33508361	33507688	33507003	13

TABLE 6

Sinc Function

x	.0	.0001	.0002	.0003	.0004	.0005	.0006	.0007	.0008	.0009	Δ_2
4.750 -0.0	33506304	33505592	33504867	33504128	33503377	33502612	33501834	33501043	33500238	33499421	13
4.751 -0.0	33498590	33497746	33496889	33496019	33495135	33494239	33493329	33492406	33491469	33490520	13
4.752 -0.0	33489557	33488582	33487593	33486591	33485575	33484547	33483505	33482450	33481382	33480301	13
4.753 -0.0	33479207	33478099	33476979	33475845	33474698	33473537	33472364	33471178	33469978	33468765	13
4.754 -0.0	33467539	33466300	33465047	33463782	33462503	33461212	33459907	33458588	33457257	33455913	13
4.755 -0.0	33454555	33453184	33451801	33450404	33448993	33447570	33446134	33444684	33443221	33441745	13
4.756 -0.0	33440256	33438754	33437239	33435710	33434169	33432614	33431046	33429465	33427871	33426264	13
4.757 -0.0	33424644	33423010	33421364	33419704	33418031	33416345	33414646	33412934	33411208	33409470	13
4.758 -0.0	33407718	33405954	33404176	33402385	33400581	33398764	33396933	33395090	33393234	33391364	13
4.759 -0.0	33389481	33387586	33385677	33383755	33381820	33379872	33377910	33375936	33373948	33371948	13
4.760 -0.0	33369934	33367908	33365868	33363815	33361749	33359670	33357578	33355472	33353354	33351223	13
4.761 -0.0	33349078	33346921	33344750	33342566	33340370	33338160	33335937	33333701	33331452	33329190	13
4.762 -0.0	33326915	33324627	33322325	33320011	33317684	33315343	33312990	33310623	33308243	33305851	13
4.763 -0.0	33303445	33301026	33298595	33296150	33293692	33291221	33288737	33286240	33283730	33281207	13
4.764 -0.0	33278671	33276122	33273560	33270984	33268396	33265795	33263181	33260553	33257913	33255260	13
4.765 -0.0	33252593	33249914	33247222	33244516	33241798	33239067	33236322	33233565	33230794	33228011	13
4.766 -0.0	33225215	33222405	33219583	33216747	33213899	33211038	33208163	33205276	33202375	33199462	13
4.767 -0.0	33196536	33193596	33190644	33187679	33184701	33181709	33178705	33175688	33172658	33169615	13
4.768 -0.0	33166559	33163490	33160408	33157313	33154205	33151084	33147950	33144803	33141643	33138471	13
4.769 -0.0	33135285	33132086	33128875	33125650	33122413	33119163	33115899	33112623	33109334	33106032	13
4.770 -0.0	33102717	33099389	33096048	33092694	33089327	33085948	33082555	33079150	33075731	33072300	13
4.771 -0.0	33068856	33065398	33061928	33058445	33054949	33051441	33047919	33044384	33040837	33037277	13
4.772 -0.0	33033703	33030117	33026518	33022906	33019281	33015644	33011993	33008330	33004653	33000964	13
4.773 -0.0	32997262	32993547	32989819	32986078	32982325	32978556	32974779	32970987	32967182	32963364	13
4.774 -0.0	32959533	32955690	32951833	32947964	32944082	32940187	32936279	32932359	32928425	32924479	13
4.775 -0.0	32920520	32916548	32912563	32908565	32904555	32900532	32896496	32892447	32888385	32884310	13
4.776 -0.0	32880223	32876123	32872010	32867884	32863746	32859594	32855430	32851253	32847063	32842861	13
4.777 -0.0	32838645	32834417	32830176	32825923	32821656	32817377	32813085	32808780	32804463	32800132	13
4.778 -0.0	32795789	32791433	32787065	32782683	32778289	32773882	32769463	32765030	32760585	32756127	13
4.779 -0.0	32751656	32747173	32742677	32738168	32733646	32729112	32724565	32720005	32715433	32710847	13
4.780 -0.0	32706249	32701639	32697015	32692379	32687730	32683069	32678394	32673707	32669008	32664295	13
4.781 -0.0	32659570	32654833	32650082	32645319	32640543	32635755	32630953	32626139	32621313	32616474	13
4.782 -0.0	32611622	32606757	32601880	32596990	32592087	32587172	32582244	32577304	32572350	32567385	13
4.783 -0.0	32562406	32557415	32552411	32547395	32542366	32537324	32532269	32527202	32522123	32517031	13
4.784 -0.0	32511926	32506808	32501678	32496535	32491380	32486212	32481032	32475838	32470633	32465414	13
4.785 -0.0	32460183	32454940	32449683	32444415	32439133	32433839	32428533	32423214	32417882	32412538	13
4.786 -0.0	32407181	32401812	32396430	32391035	32385628	32380209	32374776	32369332	32363874	32358405	13
4.787 -0.0	32352922	32347427	32341920	32336400	32330867	32325322	32319764	32314194	32308612	32303016	13
4.788 -0.0	32297409	32291789	32286156	32280511	32274853	32269183	32263500	32257805	32252097	32246377	12
4.789 -0.0	32240644	32234899	32229141	32223371	32217588	32211793	32205985	32200165	32194333	32188488	12
4.790 -0.0	32182630	32176760	32170878	32164983	32159075	32153156	32147223	32141279	32135322	32129352	12
4.791 -0.0	32123370	32117376	32111369	32105350	32099318	32093274	32087217	32081148	32075067	32068973	12
4.792 -0.0	32062867	32056749	32050618	32044474	32038319	32032150	32025970	32019777	32013572	32007354	12
4.793 -0.0	32001124	31994882	31988627	31982360	31976080	31969788	31963484	31957167	31950838	31944497	12
4.794 -0.0	31938143	31931777	31925399	31919008	31912605	31906190	31899762	31893322	31886870	31880405	12
4.795 -0.0	31873928	31867439	31860938	31854424	31847898	31841359	31834808	31828245	31821670	31815082	12
4.796 -0.0	31808482	31801870	31795246	31788609	31781960	31775298	31768625	31761939	31755241	31748530	12
4.797 -0.0	31741808	31735073	31728326	31721566	31714795	31708011	31701215	31694407	31687586	31680753	12
4.798 -0.0	31673908	31667051	31660182	31653300	31646406	31639500	31632582	31625652	31618709	31611754	12
4.799 -0.0	31604787	31597808	31590816	31583813	31576797	31569769	31562729	31555677	31548612	31541536	12

TABLE 6

Sinc Function

x	.0	.0001	.0002	.0003	.0004	.0005	.0006	.0007	.0008	.0009	Δ_2
4.800 -0.0	31534447	31527346	31520233	31513108	31505971	31498821	31491659	31484486	31477300	31470102	12
4.801 -0.0	31462892	31455669	31448435	31441189	31433930	31426659	31419376	31412082	31404775	31397455	12
4.802 -0.0	31390124	31382781	31375426	31368058	31360679	31353287	31345883	31338468	31331040	31323600	12
4.803 -0.0	31316148	31308684	31301208	31293720	31286220	31278708	31271184	31263648	31256099	31248539	12
4.804 -0.0	31240967	31233383	31225786	31218178	31210558	31202925	31195281	31187625	31179956	31172276	12
4.805 -0.0	31164584	31156880	31149163	31141435	31133695	31125943	31118179	31110402	31102614	31094814	12
4.806 -0.0	31087002	31079179	31071343	31063495	31055635	31047763	31039880	31031984	31024077	31016158	12
4.807 -0.0	31008226	31000283	30992328	30984361	30976382	30968391	30960389	30952374	30944348	30936309	12
4.808 -0.0	30928259	30920197	30912123	30904037	30895939	30887830	30879709	30871575	30863430	30855273	12
4.809 -0.0	30847104	30838924	30830731	30822527	30814311	30806083	30797843	30789591	30781328	30773053	12
4.810 -0.0	30764766	30756467	30748156	30739834	30731500	30723154	30714796	30706427	30698045	30689652	12
4.811 -0.0	30681247	30672831	30664402	30655962	30647510	30639047	30630571	30622084	30613585	30605075	12
4.812 -0.0	30596552	30588018	30579473	30570915	30562346	30553765	30545172	30536568	30527952	30519324	12
4.813 -0.0	30510685	30502034	30493371	30484697	30476011	30467313	30458604	30449882	30441150	30432405	12
4.814 -0.0	30423649	30414882	30406102	30397311	30388509	30379694	30370868	30362031	30353182	30344321	12
4.815 -0.0	30335449	30326565	30317669	30308762	30299843	30290913	30281971	30273017	30264052	30255076	12
4.816 -0.0	30246087	30237088	30228076	30219053	30210019	30200973	30191915	30182846	30173765	30164673	12
4.817 -0.0	30155569	30146454	30137327	30128189	30119039	30109878	30100705	30091521	30082325	30073118	11
4.818 -0.0	30063899	30054668	30045427	30036173	30026909	30017632	30008345	29999046	29989735	29980413	11
4.819 -0.0	29971079	29961735	29952378	29943010	29933631	29924240	29914838	29905425	29896000	29886564	11
4.820 -0.0	29877116	29867657	29858186	29848704	29839211	29829706	29820190	29810662	29801124	29791573	11
4.821 -0.0	29782012	29772439	29762855	29753259	29743652	29734034	29724404	29714763	29705110	29695447	11
4.822 -0.0	29685772	29676085	29666388	29656679	29646959	29637227	29627484	29617730	29607965	29598188	11
4.823 -0.0	29588400	29578601	29568790	29558968	29549135	29539291	29529435	29519569	29509691	29499801	11
4.824 -0.0	29489901	29479989	29470066	29460132	29450186	29440230	29430262	29420283	29410292	29400291	11
4.825 -0.0	29390278	29380254	29370219	29360173	29350116	29340047	29329968	29319877	29309775	29299661	11
4.826 -0.0	29289537	29279402	29269255	29259097	29248928	29238748	29228557	29218355	29208142	29197917	11
4.827 -0.0	29187682	29177435	29167177	29156908	29146628	29136337	29126035	29115722	29105398	29095063	11
4.828 -0.0	29084716	29074359	29063990	29053611	29043220	29032819	29022406	29011983	29001548	28991102	11
4.829 -0.0	28980646	28970178	28959699	28949209	28938709	28928197	28917674	28907141	28896596	28886041	11
4.830 -0.0	28875474	28864897	28854308	28843709	28833098	28822477	28811845	28801202	28790547	28779882	11
4.831 -0.0	28769206	28758520	28747822	28737113	28726393	28715663	28704922	28694169	28683406	28672632	11
4.832 -0.0	28661847	28651051	28640245	28629427	28618599	28607760	28596910	28586049	28575177	28564295	11
4.833 -0.0	28553401	28542497	28531582	28520656	28509720	28498772	28487814	28476845	28465865	28454874	11
4.834 -0.0	28443873	28432861	28421838	28410804	28399760	28388705	28377639	28366562	28355475	28344376	11
4.835 -0.0	28333267	28322148	28311018	28299876	28288725	28277562	28266389	28255205	28244011	28232805	11
4.836 -0.0	28221589	28210363	28199125	28187877	28176619	28165350	28154070	28142779	28131478	28120166	11
4.837 -0.0	28108844	28097510	28086167	28074812	28063447	28052072	28040686	28029289	28017881	28006463	11
4.838 -0.0	27995035	27983596	27972146	27960686	27949215	27937733	27926241	27914739	27903226	27891702	11
4.839 -0.0	27880168	27868623	27857068	27845503	27833926	27822340	27810742	27799135	27787516	27775888	10
4.840 -0.0	27764249	27752599	27740939	27729268	27717587	27705895	27694193	27682481	27670758	27659025	10
4.841 -0.0	27647281	27635527	27623762	27611987	27600202	27588406	27576599	27564783	27552956	27541118	10
4.842 -0.0	27529270	27517412	27505544	27493665	27481775	27469876	27457966	27446045	27434115	27422173	10
4.843 -0.0	27410222	27398260	27386288	27374306	27362313	27350310	27338297	27326273	27314240	27302195	10
4.844 -0.0	27290141	27278076	27266001	27253916	27241821	27229715	27217599	27205473	27193336	27181189	10
4.845 -0.0	27169032	27156865	27144688	27132500	27120302	27108094	27095876	27083648	27071409	27059160	10
4.846 -0.0	27046901	27034632	27022353	27010063	26997764	26985454	26973134	26960804	26948464	26936114	10
4.847 -0.0	26923753	26911383	26899002	26886611	26874210	26861799	26849378	26836947	26824506	26812054	10
4.848 -0.0	26799593	26787122	26774640	26762148	26749647	26737135	26724613	26712082	26699540	26686988	10
4.849 -0.0	26674426	26661855	26649273	26636681	26624079	26611467	26598845	26586214	26573572	26560920	10

TABLE 6

Sinc Function

x		.0	.0001	.0002	.0003	.0004	.0005	.0006	.0007	.0008	.0009	Δ_2
4.850	-0.0	26548258	26535587	26522905	26510214	26497512	26484801	26472079	26459348	26446607	26433856	10
4.851	-0.0	26421095	26408324	26395543	26382752	26369952	26357141	26344321	26331491	26318650	26305800	10
4.852	-0.0	26292941	26280071	26267191	26254302	26241403	26228494	26215575	26202646	26189708	26176760	10
4.853	-0.0	26163801	26150834	26137856	26124869	26111871	26098864	26085848	26072821	26059785	26046739	10
4.854	-0.0	26033683	26020617	26007542	25994457	25981362	25968258	25955144	25942020	25928886	25915743	10
4.855	-0.0	25902590	25889428	25876255	25863073	25849882	25836681	25823470	25810249	25797019	25783779	10
4.856	-0.0	25770529	25757270	25744001	25730723	25717435	25704138	25690830	25677514	25664187	25650851	10
4.857	-0.0	25637506	25624151	25610786	25597412	25584028	25570634	25557232	25543819	25530397	25516966	10
4.858	-0.0	25503525	25490074	25476614	25463145	25449666	25436177	25422679	25409172	25395655	25382128	9
4.859	-0.0	25368592	25355047	25341492	25327928	25314354	25300771	25287179	25273577	25259965	25246344	9
4.860	-0.0	25232714	25219075	25205426	25191767	25178099	25164422	25150736	25137040	25123335	25109620	9
4.861	-0.0	25095896	25082163	25068420	25054668	25040907	25027136	25013356	24999567	24985769	24971961	9
4.862	-0.0	24958144	24944317	24930482	24916637	24902782	24888919	24875046	24861164	24847273	24833372	9
4.863	-0.0	24819463	24805544	24791616	24777678	24763732	24749776	24735811	24721837	24707854	24693861	9
4.864	-0.0	24679859	24665848	24651828	24637799	24623761	24609713	24595657	24581591	24567516	24553432	9
4.865	-0.0	24539339	24525237	24511126	24497005	24482876	24468737	24454589	24440433	24426267	24412092	9
4.866	-0.0	24397908	24383715	24369513	24355302	24341082	24326853	24312615	24298368	24284111	24269846	9
4.867	-0.0	24255572	24241289	24226997	24212696	24198386	24184067	24169739	24155402	24141056	24126701	9
4.868	-0.0	24112337	24097965	24083583	24069193	24054793	24040385	24025968	24011541	23997106	23982662	9
4.869	-0.0	23968210	23953748	23939278	23924798	23910310	23895813	23881307	23866792	23852269	23837736	9
4.870	-0.0	23823195	23808645	23794086	23779519	23764942	23750357	23735763	23721161	23706549	23691929	9
4.871	-0.0	23677300	23662662	23648016	23633360	23618697	23604024	23589342	23574652	23559954	23545246	9
4.872	-0.0	23530530	23515805	23501072	23486329	23471578	23456819	23442051	23427274	23412488	23397694	9
4.873	-0.0	23382891	23368080	23353260	23338432	23323594	23308748	23293894	23279031	23264160	23249279	9
4.874	-0.0	23234391	23219493	23204588	23189673	23174750	23159819	23144879	23129930	23114973	23100008	9
4.875	-0.0	23085034	23070051	23055060	23040061	23025053	23010036	22995011	22979978	22964936	22949886	8
4.876	-0.0	22934827	22919760	22904684	22889600	22874508	22859407	22844297	22829180	22814054	22798919	8
4.877	-0.0	22783776	22768625	22753466	22738298	22723121	22707937	22692744	22677542	22662333	22647115	8
4.878	-0.0	22631889	22616654	22601411	22586160	22570900	22555633	22540357	22525072	22509780	22494479	8
4.879	-0.0	22479170	22463852	22448527	22433193	22417851	22402501	22387142	22371775	22356401	22341017	8
4.880	-0.0	22325626	22310227	22294819	22279403	22263979	22248547	22233107	22217658	22202202	22186737	8
4.881	-0.0	22171264	22155783	22140294	22124797	22109292	22093779	22078257	22062728	22047190	22031644	8
4.882	-0.0	22016091	22000529	21984959	21969381	21953795	21938201	21922599	21906989	21891372	21875746	8
4.883	-0.0	21860112	21844470	21828820	21813162	21797496	21781822	21766140	21750450	21734753	21719047	8
4.884	-0.0	21703334	21687612	21671883	21656145	21640400	21624647	21608886	21593117	21577340	21561556	8
4.885	-0.0	21545763	21529963	21514154	21498338	21482514	21466683	21450843	21434996	21419140	21403277	8
4.886	-0.0	21387407	21371528	21355642	21339747	21323845	21307936	21292018	21276093	21260160	21244219	8
4.887	-0.0	21228271	21212315	21196351	21180379	21164400	21148413	21132418	21116416	21100405	21084388	8
4.888	-0.0	21068362	21052329	21036288	21020240	21004184	20988120	20972049	20955970	20939883	20923789	8
4.889	-0.0	20907688	20891578	20875461	20859337	20843205	20827065	20810918	20794763	20778601	20762431	8
4.890	-0.0	20746253	20730068	20713876	20697676	20681468	20665253	20649031	20632801	20616564	20600319	7
4.891	-0.0	20584066	20567806	20551539	20535264	20518982	20502693	20486395	20470091	20453779	20437460	7
4.892	-0.0	20421133	20404799	20388457	20372109	20355752	20339389	20323018	20306639	20290254	20273861	7
4.893	-0.0	20257460	20241053	20224638	20208215	20191786	20175349	20158905	20142453	20125994	20109528	7
4.894	-0.0	20093055	20076575	20060087	20043592	20027089	20010580	19994063	19977539	19961008	19944469	7
4.895	-0.0	19927924	19911371	19894811	19878244	19861670	19845088	19828500	19811904	19795301	19778691	7
4.896	-0.0	19762074	19745449	19728818	19712179	19695534	19678881	19662221	19645554	19628880	19612199	7
4.897	-0.0	19595511	19578816	19562114	19545404	19528688	19511965	19495234	19478497	19461753	19445001	7
4.898	-0.0	19428243	19411478	19394705	19377926	19361140	19344346	19327546	19310739	19293925	19277104	7
4.899	-0.0	19260276	19243441	19226600	19209751	19192896	19176033	19159164	19142288	19125405	19108515	7

TABLE 6

Sinc Function

x		.0	.0001	.0002	.0003	.0004	.0005	.0006	.0007	.0008	.0009	Δ_2
4.900	-0.0	19091618	19074714	19057804	19040887	19023963	19007032	18990094	18973149	18956198	18939240	7
4.901	-0.0	18922275	18905303	18888325	18871340	18854348	18837349	18820344	18803331	18786312	18769287	7
4.902	-0.0	18752254	18735215	18718169	18701117	18684058	18666992	18649919	18632840	18615754	18598662	7
4.903	-0.0	18581563	18564457	18547345	18530226	18513100	18495968	18478829	18461683	18444531	18427373	7
4.904	-0.0	18410207	18393036	18375857	18358672	18341481	18324283	18307079	18289867	18272650	18255426	7
4.905	-0.0	18238195	18220958	18203715	18186465	18169208	18151945	18134676	18117400	18100117	18082829	6
4.906	-0.0	18065533	18048232	18030924	18013609	17996288	17978961	17961627	17944287	17926941	17909588	6
4.907	-0.0	17892229	17874863	17857491	17840113	17822729	17805338	17787941	17770537	17753127	17735711	6
4.908	-0.0	17718289	17700860	17683425	17665984	17648536	17631083	17613623	17596156	17578684	17561205	6
4.909	-0.0	17543720	17526229	17508732	17491228	17473718	17456202	17438680	17421152	17403618	17386077	6
4.910	-0.0	17368530	17350977	17333418	17315853	17298282	17280704	17263121	17245531	17227936	17210334	6
4.911	-0.0	17192726	17175112	17157492	17139866	17122234	17104596	17086952	17069301	17051645	17033983	6
4.912	-0.0	17016315	16998641	16980960	16963274	16945582	16927884	16910180	16892470	16874754	16857032	6
4.913	-0.0	16839304	16821570	16803830	16786085	16768333	16750576	16732812	16715043	16697268	16679487	6
4.914	-0.0	16661700	16643908	16626109	16608305	16590494	16572678	16554857	16537029	16519195	16501356	6
4.915	-0.0	16483511	16465660	16447804	16429941	16412073	16394199	16376320	16358435	16340543	16322647	6
4.916	-0.0	16304744	16286836	16268922	16251002	16233077	16215146	16197209	16179267	16161319	16143365	6
4.917	-0.0	16125406	16107441	16089471	16071495	16053513	16035525	16017532	15999534	15981530	15963520	6
4.918	-0.0	15945505	15927484	15909457	15891425	15873388	15855345	15837296	15819242	15801183	15783118	6
4.919	-0.0	15765047	15746971	15728889	15710802	15692710	15674612	15656508	15638400	15620285	15602166	5
4.920	-0.0	15584040	15565910	15547774	15529632	15511486	15493334	15475176	15457013	15438845	15420671	5
4.921	-0.0	15402492	15384308	15366118	15347923	15329723	15311517	15293306	15275090	15256869	15238642	5
4.922	-0.0	15220410	15202173	15183930	15165682	15147429	15129171	15110907	15092638	15074364	15056085	5
4.923	-0.0	15037801	15019511	15001216	14982916	14964611	14946301	14927986	14909665	14891339	14873008	5
4.924	-0.0	14854672	14836331	14817985	14799634	14781277	14762916	14744549	14726177	14707801	14689419	5
4.925	-0.0	14671032	14652640	14634243	14615841	14597434	14579022	14560605	14542183	14523756	14505324	5
4.926	-0.0	14486887	14468445	14449998	14431546	14413089	14394628	14376161	14357689	14339213	14320731	5
4.927	-0.0	14302245	14283754	14265258	14246757	14228251	14209740	14191224	14172704	14154179	14135648	5
4.928	-0.0	14117113	14098574	14080029	14061480	14042925	14024366	14005803	13987234	13968661	13950083	5
4.929	-0.0	13931500	13912912	13894320	13875723	13857121	13838515	13819903	13801287	13782667	13764042	5
4.930	-0.0	13745412	13726777	13708138	13689494	13670845	13652192	13633534	13614871	13596204	13577533	5
4.931	-0.0	13558856	13540175	13521490	13502800	13484105	13465406	13446702	13427994	13409281	13390563	5
4.932	-0.0	13371841	13353115	13334384	13315649	13296909	13278164	13259415	13240662	13221904	13203141	4
4.933	-0.0	13184375	13165603	13146828	13128048	13109263	13090474	13071681	13052883	13034081	13015274	4
4.934	-0.0	12996463	12977648	12958829	12940005	12921176	12902344	12883507	12864665	12845820	12826970	4
4.935	-0.0	12808115	12789257	12770394	12751527	12732656	12713780	12694900	12676016	12657128	12638235	4
4.936	-0.0	12619338	12600437	12581532	12562623	12543709	12524791	12505869	12486943	12468013	12449078	4
4.937	-0.0	12430139	12411197	12392250	12373299	12354344	12335384	12316421	12297453	12278482	12259506	4
4.938	-0.0	12240527	12221543	12202555	12183563	12164567	12145567	12126563	12107555	12088543	12069527	4
4.939	-0.0	12050507	12031483	12012455	11993424	11974388	11955348	11936304	11917256	11898205	11879149	4
4.940	-0.0	11860090	11841026	11821959	11802888	11783812	11764733	11745651	11726564	11707473	11688379	4
4.941	-0.0	11669281	11650178	11631073	11611963	11592849	11573732	11554611	11535486	11516357	11497225	4
4.942	-0.0	11478088	11458948	11439804	11420657	11401506	11382351	11363192	11344030	11324864	11305694	4
4.943	-0.0	11286520	11267343	11248162	11228978	11209790	11190598	11171402	11152203	11133001	11113794	4
4.944	-0.0	11094584	11075371	11056154	11036933	11017709	10998481	10979249	10960014	10940776	10921534	4
4.945	-0.0	10902288	10883039	10863786	10844530	10825270	10806007	10786741	10767470	10748197	10728920	3
4.946	-0.0	10709639	10690355	10671068	10651777	10632483	10613185	10593884	10574579	10555272	10535960	3
4.947	-0.0	10516646	10497328	10478006	10458681	10439353	10420022	10400687	10381349	10362008	10342663	3
4.948	-0.0	10323315	10303964	10284609	10265251	10245890	10226526	10207158	10187787	10168413	10149035	3
4.949	-0.0	10129655	10110271	10090884	10071494	10052100	10032704	10013304	09993901	09974495	09955086	3

TABLE 6

Sinc Function

x		.0	.0001	.0002	.0003	.0004	.0005	.0006	.0007	.0008	.0009	Δ_2
4.950	-0.0	09935673	09916258	09896839	09877417	09857992	09838584	09819133	09799699	09780262	09760821	3
4.951	-0.0	09741378	09721931	09702482	09683029	09663573	09644115	09624653	09605188	09585721	09566250	3
4.952	-0.0	09546776	09527300	09507820	09488337	09468852	09449363	09429872	09410377	09390880	09371380	3
4.953	-0.0	09351877	09332370	09312862	09293350	09273835	09254317	09234797	09215273	09195747	09176218	3
4.954	-0.0	09156686	09137152	09117614	09098074	09078531	09058985	09039436	09019885	09000330	08980773	3
4.955	-0.0	08961214	08941651	08922086	08902518	08882947	08863374	08843798	08824219	08804637	08785053	3
4.956	-0.0	08765466	08745876	08726284	08706689	08687092	08667492	08647889	08628283	08608675	08589065	3
4.957	-0.0	08569451	08549836	08530217	08510596	08490973	08471346	08451718	08432087	08412453	08392817	3
4.958	-0.0	08373178	08353536	08333893	08314246	08294597	08274946	08255292	08235636	08215977	08196316	2
4.959	-0.0	08176653	08156987	08137318	08117647	08097974	08078298	08058620	08038940	08019257	07999572	2
4.960	-0.0	07979884	07960194	07940502	07920807	07901110	07881411	07861709	07842005	07822299	07802591	2
4.961	-0.0	07782880	07763167	07743451	07723734	07704014	07684292	07664567	07644841	07625112	07605381	2
4.962	-0.0	07585648	07565912	07546175	07526435	07506693	07488949	07467203	07447454	07427703	07407951	2
4.963	-0.0	07388196	07368439	07348680	07328919	07309155	07289390	07269622	07249853	07230081	07210308	2
4.964	-0.0	07190532	07170754	07150974	07131192	07111409	07091623	07071835	07052045	07032253	07012459	2
4.985	-0.0	06992664	06972866	06953066	06933265	06913461	06893655	06873848	06854039	06834227	06814414	2
4.966	-0.0	06794599	06774782	06754963	06735143	06715320	06695496	06675670	06655841	06636012	06616180	2
4.967	-0.0	06596346	06576511	06556674	06536835	06516994	06497152	06477307	06457461	06437614	06417764	2
4.968	-0.0	06397913	06378060	06358205	06338349	06318491	06298631	06278769	06258906	06239041	06219175	2
4.969	-0.0	06199307	06179437	06159565	06139692	06119818	06099941	06080063	06060184	06040303	06020420	2
4.970	-0.0	06000536	05980650	05960762	05940873	05920983	05901091	05881197	05861302	05841406	05821508	2
4.971	-0.0	05801608	05781707	05761804	05741900	05721995	05702088	05682179	05662269	05642358	05622445	0
4.972	-0.0	05602531	05582616	05562699	05542780	05522861	05502939	05483017	05463093	05443168	05423241	0
4.973	-0.0	05403313	05383384	05363454	05343522	05323588	05303654	05283718	05263781	05243843	05223903	0
4.974	-0.0	05203962	05184020	05164077	05144132	05124186	05104239	05084291	05064342	05044391	05024439	0
4.975	-0.0	05004486	04984532	04964577	04944620	04924662	04904703	04884744	04864782	04844820	04824857	0
4.976	-0.0	04804893	04784927	04764960	04744993	04725024	04705054	04685083	04665111	04645138	04625164	0
4.977	-0.0	04605189	04585213	04565236	04545258	04525279	04505299	04485318	04465336	04445354	04425370	0
4.978	-0.0	04405385	04385399	04365412	04345425	04325436	04305447	04285457	04265465	04245473	04225480	0
4.979	-0.0	04205487	04185492	04165496	04145500	04125503	04105505	04085506	04065506	04045506	04025505	0
4.980	-0.0	04005503	03985500	03965496	03945492	03925487	03905481	03885475	03865467	03845459	03825450	0
4.981	-0.0	03805441	03785431	03765420	03745409	03725396	03705383	03685370	03665356	03645341	03625325	0
4.982	-0.0	03605309	03585293	03565275	03545257	03525239	03505220	03485200	03465180	03445159	03425138	0
4.983	-0.0	03405116	03385093	03365070	03345047	03325023	03304998	03284973	03264948	03244922	03224895	0
4.984	-0.0	03204868	03184841	03164813	03144785	03124756	03104727	03084697	03064667	03044636	03024606	0
4.985	-0.0	03004574	02984543	02964511	02944478	02924446	02904413	02884379	02864345	02844311	02824277	0
4.986	-0.0	02804242	02784207	02764172	02744136	02724101	02704064	02684028	02663991	02643954	02623917	0
4.987	-0.0	02603880	02583842	02563804	02543766	02523728	02503690	02483651	02463612	02443573	02423534	0
4.988	-0.0	02403495	02383456	02363416	02343376	02323337	02303297	02283257	02263217	02243176	02223136	0
4.989	-0.0	02203096	02183055	02163015	02142974	02122934	02102893	02082852	02062812	02042771	02022730	0
4.990	-0.0	02002690	01982649	01962608	01942568	01922527	01902487	01882446	01862406	01842365	01822325	0
4.991	-0.0	01802285	01782245	01762205	01742165	01722125	01702086	01682046	01662007	01641967	01621928	0
4.992	-0.0	01601889	01581851	01561812	01541774	01521735	01501697	01481660	01461622	01441585	01421548	0
4.993	-0.0	01401511	01381474	01361438	01341402	01321366	01301330	01281295	01261260	01241226	01221191	0
4.994	-0.0	01201157	01181123	01161090	01141057	01121024	01100992	01080960	01060929	01040898	01020867	0
4.995	-0.0	01000836	00980806	00960777	00940748	00920719	00900691	00880663	00860636	00840609	00820582	0
4.996	-0.0	00800556	00780531	00760506	00740481	00720457	00700434	00680411	00660389	00640367	00620345	-1
4.997	-0.0	00600325	00580304	00560285	00540266	00520247	00500230	00480212	00460196	00440180	00420164	-1
4.998	-0.0	00400150	00380135	00360122	00340109	00320097	00300086	00280075	00260065	00240055	00220047	-1
4.999	-0.0	00200039	00180031	00160025	00140019	00120014	00100010	00080006	00060004	00040002	00020000	1

$$\frac{\sin 2\pi x}{2\pi x}$$

TABLE 6

Sinc Function

x		.0	.001	.002	.003	.004	.005	.006	.007	.008	.009	Δ_2
5.00	0.0	00000000	00199959	00399830	00599605	00799276	00998837	01198278	01397592	01596771	01795808	-142
5.01	0.0	01994695	02193423	02391986	02590375	02788582	02986601	03184422	03382040	03579444	03776629	-219
5.02	0.0	03973586	04170308	04366787	04563016	04758986	04954690	05150121	05345271	05540132	05734697	-295
5.03	0.0	05928959	06122909	06316540	06509845	06702817	06895447	07087728	07279654	07471216	07662407	-370
5.04	0.0	07853219	08043646	08233680	08423314	08612540	08801351	08989739	09177699	09365221	09552300	-443
5.05	0.0	09738927	09925096	10110800	10296031	10480783	10665047	10848818	11032088	11214850	11397097	-514
5.06	0.0	11578823	11760019	11940680	12120798	12300366	12479378	12657827	12835706	13013008	13189726	-583
5.07	0.0	13365854	13541385	13716312	13890629	14064329	14237405	14409852	14581661	14752827	14923344	-649
5.08	0.0	15093204	15262402	15430931	15598785	15765956	15932440	16098229	16263318	16427700	16591369	-712
5.09	0.0	16754319	16916543	17078036	17238792	17398804	17558066	17716573	17874319	18031297	18187501	-772
5.10	0.0	18342927	18497568	18651418	18804471	18956723	19108166	19258796	19408607	19557593	19705748	-829
5.11	0.0	19853068	19999547	20145179	20289959	20433881	20576941	20719132	20860450	21000890	21140446	-883
5.12	0.0	21279112	21416885	21553759	21689728	21824788	21958935	22092162	22224465	22355839	22486280	-933
5.13	0.0	22615782	22744341	22871952	22998611	23124313	23249053	23372826	23495629	23617457	23738305	-979
5.14	0.0	23858170	23977045	24094929	24211815	24327700	24442580	24556451	24669308	24781148	24891966	-1021
5.15	0.0	25001758	25110521	25218251	25324943	25430594	25535201	25638759	25741265	25842716	25943107	-1058
5.16	0.0	26042435	26140696	26237888	26334007	26429048	26523010	26615889	26707681	26798384	26887993	-1092
5.17	0.0	26976507	27063921	27150234	27235442	27319541	27402530	27484405	27565164	27644803	27723321	-1121
5.18	0.0	27800714	27876980	27952116	28026120	28098989	28170720	28241313	28310763	28379069	28446229	-1145
5.19	0.0	28512240	28577100	28640807	28703359	28764754	28824990	28884066	28941978	28998726	29054307	-1165
5.20	0.0	29108720	29161964	29214035	29264934	29314658	29363206	29410576	29456767	29501778	29545607	-1181
5.21	0.0	29588253	29629715	29669991	29709081	29746983	29783696	29819220	29853553	29886695	29918644	-1191
5.22	0.0	29949401	29978963	30007331	30034503	30060480	30085260	30108843	30131228	30152416	30172406	-1197
5.23	0.0	30191197	30208789	30225182	30240376	30254370	30267165	30278760	30289156	30298352	30306349	-1198
5.24	0.0	30313146	30318745	30323145	30326346	30328348	30329153	30328761	30327172	30324386	30320404	-1195
5.25	0.0	30315227	30308856	30301290	30292532	30282581	30271439	30259106	30245583	30230872	30214973	-1187
5.26	0.0	30197887	30179616	30160161	30139523	30117703	30094702	30070523	30045165	30018632	29990923	-1174
5.27	0.0	29962042	29931988	29900765	29868373	29834815	29800091	29764205	29727157	29688950	29649586	-1156
5.28	0.0	29609067	29567394	29524569	29480596	29435476	29389210	29341803	29293255	29243569	29192748	-1134
5.29	0.0	29140794	29087709	29033496	28978157	28921696	28864115	28805416	28745602	28684676	28622641	-1108
5.30	0.0	28559499	28495254	28429909	28363466	28295929	28227300	28157583	28086780	28014896	27941933	-1078
5.31	0.0	27867895	27792785	27716606	27639361	27561055	27481691	27401271	27319800	27237282	27153719	-1043
5.32	0.0	27069116	26983477	26896804	26809103	26720376	26630628	26539862	26448083	26355295	26261501	-1004
5.33	0.0	26166705	26070913	25974127	25876353	25777593	25677853	25577138	25475450	25372795	25269177	-962
5.34	0.0	25164600	25059069	24952588	24845163	24736797	24627495	24517262	24406103	24294022	24181023	-916
5.35	0.0	24067113	23952295	23836575	23719957	23602446	23484048	23364767	23244608	23123577	23001678	-867
5.36	0.0	22878916	22755298	22630827	22505510	22379350	22252355	22124529	21995877	21866404	21736117	-814
5.37	0.0	21605021	21473120	21340421	21206929	21072650	20937589	20801751	20665144	20527771	20389639	-758
5.38	0.0	20250754	20111121	19970746	19829635	19687794	19545228	19401944	19257947	19113243	18967839	-700
5.39	0.0	18821740	18674952	18527482	18379335	18230518	18081036	17930895	17780103	17628665	17476587	-639
5.40	0.0	17323876	17170537	17016577	16862003	16706820	16551036	16394655	16237686	16080133	15922004	-575
5.41	0.0	15763306	15604043	15444224	15283854	15122940	14961488	14799506	14636999	14473974	14310438	-510
5.42	0.0	14146398	13981860	13816831	13651317	13485326	13318863	13151936	12984552	12816716	12648437	-443
5.43	0.0	12479720	12310573	12141002	11971014	11800617	11629816	11458619	11287033	11115064	10942720	-374
5.44	0.0	10770008	10596934	10423505	10249728	10075611	09901160	09726382	09551284	09375874	09200158	-305
5.45	0.0	09024144	08847837	08671247	08494378	08317240	08139838	07962179	07784272	07606122	07427738	-234
5.46	0.0	07249125	07070292	06891246	06711993	06532540	06352896	06173066	05993059	05812881	05632539	-163
5.47	0.0	05452041	05271393	05090604	04909680	04728628	04547455	04366170	04184778	04003287	03821704	-91
5.48	0.0	03640037	03458293	03276478	03094600	02912666	02730684	02548661	02366603	02184518	02002413	-19
5.49	0.0	01820295	01638172	01456051	01273938	01091842	00909768	00727725	00545720	00363759	00181850	52

TABLE 6

Sinc Function

x		.0	.001	.002	.003	.004	.005	.006	.007	.008	.009	Δ_2
5.50	-0.0	00000000	00181784	00363495	00545125	00726668	00908116	01089462	01270700	01451821	01632820	-122
5.51	-0.0	01813688	01994419	02175006	02355442	02535719	02715830	02895769	03075529	03255102	03434481	-193
5.52	-0.0	03613660	03792631	03971388	04149924	04328230	04506302	04684131	04861711	05039035	05216096	-262
5.53	-0.0	05392887	05569401	05745631	05921571	06097214	06272552	06447580	06622289	06796675	06970729	-330
5.54	-0.0	07144445	07317816	07490836	07663498	07835795	08007721	08179269	08350432	08521203	08691577	-397
5.55	-0.0	08861546	09031105	09200245	09368962	09537248	09705097	09872503	10039458	10205958	10371994	-462
5.56	-0.0	10537562	10702653	10867264	11031386	11195013	11358140	11520760	11682867	11844454	12005516	-525
5.57	-0.0	12166046	12326039	12485487	12644386	12802728	12960508	13117720	13274358	13430415	13585887	-585
5.58	-0.0	13740767	13895049	14048727	14201795	14354248	14506080	14657285	14807857	14957791	15107081	-643
5.59	-0.0	15255721	15403706	15551030	15697688	15843673	15988981	16133606	16277542	16420784	16563327	-698
5.60	-0.0	16705166	16846294	16986707	17126400	17265366	17403602	17541101	17677859	17813870	17949130	-751
5.61	-0.0	18083633	18217374	18350348	18482551	18613977	18744622	18874480	19003547	19131818	19259288	-800
5.62	-0.0	19385953	19511807	19636847	19761067	19884462	20007029	20128763	20249659	20369712	20488919	-846
5.63	-0.0	20607275	20724776	20841417	20957193	21072102	21186137	21299297	21411575	21522968	21633472	-888
5.64	-0.0	21743084	21851798	21959611	22066519	22172518	22277604	22381774	22485024	22587349	22688747	-927
5.65	-0.0	22789213	22888744	22987337	23084987	23181691	23277447	23372249	23466096	23558983	23650908	-962
5.66	-0.0	23741866	23831855	23920872	24008913	24095976	24182056	24267152	24351260	24434376	24516500	-993
5.67	-0.0	24597626	24677753	24756878	24834997	24912109	24988210	25063298	25137371	25210425	25282458	-1020
5.68	-0.0	25353468	25423452	25492408	25560334	25627227	25693084	25757905	25821686	25884425	25946121	-1043
5.69	-0.0	26006770	26066372	26124924	26182425	26238871	26294263	26348596	26401871	26454085	26505237	-1062
5.70	-0.0	26555324	26604345	26652300	26699185	26745000	26789744	26833414	26876010	26917530	26957973	-1076
5.71	-0.0	26997338	27035623	27072828	27108951	27143992	27177949	27210821	27242608	27273308	27302921	-1086
5.72	-0.0	27331446	27358882	27385229	27410486	27434652	27457726	27479709	27500599	27520397	27539101	-1092
5.73	-0.0	27556712	27573229	27588652	27602980	27616214	27628354	27639398	27649348	27658203	27665963	-1094
5.74	-0.0	27672628	27678199	27682676	27686058	27688346	27689540	27689642	27688650	27686565	27683389	-1091
5.75	-0.0	27679121	27673761	27667312	27659772	27651144	27641427	27630622	27618730	27605753	27591690	-1084
5.76	-0.0	27576543	27560313	27543000	27524607	27505133	27484581	27462950	27440244	27416462	27391606	-1073
5.77	-0.0	27365677	27338678	27310608	27281471	27251266	27219997	27187664	27154269	27119813	27084299	-1058
5.78	-0.0	27047729	27010103	26971424	26931694	26890915	26849088	26806217	26762301	26717345	26671350	-1038
5.79	-0.0	26624317	26576251	26527151	26477022	26425865	26373682	26320476	26266250	26211006	26154746	-1015
5.80	-0.0	26097473	26039190	25979899	25919604	25858305	25796008	25732713	25668425	25603146	25536878	-987
5.81	-0.0	25469625	25401391	25332177	25261986	25190823	25118690	25045591	24971528	24896505	24820524	-956
5.82	-0.0	24743591	24665707	24586876	24507102	24426388	24344737	24262153	24178641	24094202	24008842	-921
5.83	-0.0	23922563	23835369	23747264	23658253	23568338	23477523	23385813	23293212	23199722	23105349	-883
5.84	-0.0	23010096	22913968	22816968	22719101	22620370	22520780	22420336	22319041	22216899	22113916	-841
5.85	-0.0	22010095	21905440	21799957	21693649	21586521	21478578	21369824	21260264	21149902	21038742	-796
5.86	-0.0	20926790	20814051	20700528	20586227	20471152	20355309	20238701	20121335	20003214	19884344	-748
5.87	-0.0	19764729	19644375	19523287	19401469	19278928	19155666	19031691	18907007	18781618	18655531	-698
5.88	-0.0	18528751	18401282	18273131	18144301	18014800	17884631	17753800	17622313	17490176	17357393	-644
5.89	-0.0	17223969	17089911	16955225	16819914	16683986	16547445	16410297	16272548	16134204	15995269	-589
5.90	-0.0	15855751	15715653	15574983	15433746	15291947	15149593	15006689	14863241	14719255	14574737	-531
5.91	-0.0	14429693	14284127	14138048	13991459	13844368	13696781	13548702	13400139	13251097	13101582	-472
5.92	-0.0	12951601	12801159	12650263	12498918	12347131	12194908	12042255	11889179	11735684	11581778	-410
5.93	-0.0	11427467	11272757	11117654	10962164	10806294	10650051	10493439	10336466	10179138	10021461	-348
5.94	-0.0	09863441	09705086	09546401	09387392	09228066	09068430	08908489	08748251	08587721	08426906	-284
5.95	-0.0	08265812	08104446	07942815	07780925	07618781	07456392	07293763	07130900	06967811	06804501	-219
5.96	-0.0	06640977	06477247	06313315	06149189	05984876	05820381	05655712	05490875	05325876	05160723	-154
5.97	-0.0	04995421	04829977	04664398	04498690	04332861	04166915	04000861	03834704	03668452	03502110	-88
5.98	-0.0	03335686	03169186	03002616	02835984	02669295	02502557	02335776	02168958	02002110	01835239	-22
5.99	-0.0	01668351	01501453	01334551	01167653	01000764	00833891	00667041	00500221	00333436	00166693	42

$$\frac{\sin 2\pi x}{2\pi x}$$

$\frac{\sin 2\pi x}{2\pi x}$

TABLE 6

Sinc Function

x		.0	.001	.002	.003	.004	.005	.006	.007	.008	.009	Δ_2
6.00	0.0	00000000	00166638	00333213	00499721	00666152	00832503	00998764	01164931	01330997	01496955	-107
6.01	0.0	01662799	01828522	01994117	02159579	02324901	02490075	02655097	02819959	02984655	03149178	-172
6.02	0.0	03313522	03477681	03641648	03805417	03968982	04132335	04295471	04458383	04621066	04783512	-235
6.03	0.0	04945715	05107669	05269368	05430806	05591975	05752871	05913486	06073814	06233850	06393586	-298
6.04	0.0	06553017	06712137	06870939	07029418	07187566	07345379	07502849	07659971	07816739	07973146	-359
6.05	0.0	08129187	08284856	08440146	08595051	08749567	08903685	09057402	09210710	09363604	09516078	-419
6.06	0.0	09668126	09819742	09970921	10121656	10271942	10421772	10571143	10720046	10868478	11016431	-477
6.07	0.0	11163901	11310882	11457367	11603353	11748832	11893800	12038250	12182178	12325577	12468443	-533
6.08	0.0	12610770	12752552	12893784	13034460	13174576	13314126	13453105	13591506	13729326	13866559	-586
6.09	0.0	14003199	14139242	14274682	14409514	14543733	14677334	14810311	14942661	15074377	15205455	-637
6.10	0.0	15335890	15465677	15594810	15723286	15851100	15978245	16104719	16230515	16355629	16480057	-686
6.11	0.0	16603794	16726834	16849175	16970810	17091735	17211946	17331439	17450208	17568250	17685560	-731
6.12	0.0	17802133	17917966	18033053	18147391	18260976	18373802	18485867	18597165	18707693	18817446	-773
6.13	0.0	18926421	19034613	19142019	19248634	19354454	19459476	19563695	19667109	19769712	19871502	-813
6.14	0.0	19972474	20072625	20171951	20270448	20368114	20464943	20560933	20656081	20750382	20843833	-849
6.15	0.0	20936431	21028173	21119055	21209074	21298226	21386509	21473919	21560452	21646107	21730880	-881
6.16	0.0	21814767	21897766	21979873	22061086	22141403	22220819	22299332	22376940	22453639	22529427	-910
6.17	0.0	22604301	22678259	22751298	22823415	22894607	22964873	23034210	23102615	23170086	23236621	-935
6.18	0.0	23302217	23366871	23430583	23493349	23555168	23616036	23675953	23734916	23792923	23849973	-957
6.19	0.0	23906062	23961190	24015354	24068552	24120784	24172046	24222338	24271657	24320003	24367373	-975
6.20	0.0	24413765	24459180	24503614	24547066	24589536	24631021	24671521	24711034	24749559	24787094	-988
6.21	0.0	24823639	24859192	24893753	24927320	24959892	24991468	25022048	25051631	25080215	25107800	-998
6.22	0.0	25134384	25159969	25184552	25208133	25230711	25252286	25272858	25292425	25310988	25328545	-1004
6.23	0.0	25345098	25360644	25375185	25388719	25401247	25412768	25423282	25432790	25441290	25448784	-1006
6.24	0.0	25455270	25460750	25465223	25468690	25471150	25472604	25473052	25472494	25470931	25468363	-1004
6.25	0.0	25464791	25460215	25454635	25448052	25440467	25431880	25422292	25411704	25400116	25387529	-998
6.26	0.0	25373944	25359361	25343783	25327209	25309640	25291079	25271524	25250979	25229443	25206919	-988
6.27	0.0	25183407	25158908	25133423	25106955	25079505	25051073	25021661	24991271	24959905	24927563	-974
6.28	0.0	24894247	24859959	24824702	24788475	24751281	24713123	24674001	24633917	24592874	24550873	-957
6.29	0.0	24507917	24464007	24419145	24373333	24326574	24278870	24230222	24180634	24130107	24078643	-935
6.30	0.0	24026245	23972916	23918657	23863471	23807361	23750329	23692378	23633509	23573727	23513033	-911
6.31	0.0	23451430	23388921	23325509	23261196	23195985	23129879	23062881	22994994	22926221	22856565	-882
6.32	0.0	22786028	22714615	22642327	22569169	22495142	22420252	22344500	22267890	22190425	22112109	-850
6.33	0.0	22032945	21952936	21872086	21790398	21707875	21624522	21540342	21455338	21369514	21282873	-815
6.34	0.0	21195420	21107158	21018090	20928221	20837554	20746093	20653842	20560805	20466986	20372388	-777
6.35	0.0	20277016	20180874	20083965	19986294	19887865	19788682	19688749	19588071	19486650	19384493	-736
6.36	0.0	19281602	19177983	19073639	18968576	18862796	18756306	18649108	18541208	18432610	18323318	-692
6.37	0.0	18213338	18102673	17991328	17879308	17766618	17653261	17539243	17424569	17309243	17193270	-646
6.38	0.0	17076654	16959402	16841516	16723002	16603866	16484112	16363744	16242768	16121189	15999011	-597
6.39	0.0	15876241	15752882	15628940	15504419	15379326	15253665	15127441	15000659	14873325	14745444	-546
6.40	0.0	14617020	14488060	14358568	14228549	14098010	13966955	13835390	13703319	13570749	13437685	-493
6.41	0.0	13304132	13170095	13035580	12900592	12765138	12629222	12492849	12356027	12218758	12081051	-439
6.42	0.0	11942909	11804340	11665347	11525937	11386116	11245888	11105261	10964239	10822827	10681033	-382
6.43	0.0	10538861	10396318	10253408	10110138	09966514	09822541	09678225	09533571	09388587	09243276	-325
6.44	0.0	09097646	08951702	08805451	08658896	08512046	08364905	08217480	08069776	07921799	07773556	-266
6.45	0.0	07625051	07476292	07327284	07178033	07028544	06878825	06728880	06578716	06428339	06277755	-206
6.46	0.0	06126970	05975989	05824820	05673467	05521937	05370236	05218370	05066345	04914167	04761842	-146
6.47	0.0	04609376	04456775	04304046	04151194	03998225	03845146	03691962	03538680	03385305	03231844	-85
6.48	0.0	03078303	02924688	02771005	02617259	02463458	02309607	02155713	02001780	01847817	01693827	-24
6.49	0.0	01539818	01385796	01231767	01077737	00923711	00769696	00615699	00461724	00307779	00153869	35

TABLE 6

Sinc Function

x		.0	.001	.002	.003	.004	.005	.006	.007	.008	.009	Δ_2
6.50	-0.0	00000000	00153821	00307590	00461298	00614941	00768513	00922007	01075418	01228739	01381964	-95
6.51	-0.0	01535088	01688104	01841006	01993789	02146446	02298972	02451360	02603605	02755700	02907639	-154
6.52	-0.0	03059418	03211029	03362466	03513725	03664798	03815681	03966367	04116850	04267124	04417184	-213
6.53	-0.0	04567023	04716637	04866018	05015162	05164062	05312712	05461108	05609242	05757110	05904705	-271
6.54	-0.0	06052022	06199055	06345799	06492247	06638394	06784234	06929762	07074972	07219859	07364416	-328
6.55	-0.0	07508638	07652520	07796056	07939241	08082068	08224533	08366630	08508353	08649697	08790656	-384
6.56	-0.0	08931226	09071400	09211173	09350541	09489496	09628035	09766152	09903840	10041097	10177914	-437
6.57	-0.0	10314289	10450215	10585687	10720700	10855249	10989328	11122933	11256058	11388698	11520849	-489
6.58	-0.0	11652504	11783660	11914311	12044451	12174077	12303183	12431764	12559815	12687331	12814308	-539
6.59	-0.0	12940741	13066625	13191954	13316725	13440932	13564571	13687638	13810126	13932033	14053352	-586
6.60	-0.0	14174080	14294212	14413743	14532670	14650986	14768688	14885772	15002233	15118066	15233268	-631
6.61	-0.0	15347833	15461758	15575039	15687670	15799648	15910968	16021626	16131619	16240942	16349591	-673
6.62	-0.0	16457561	16564850	16671452	16777363	16882581	16987100	17090918	17194029	17296431	17398119	-713
6.63	-0.0	17499089	17599338	17698863	17797659	17895722	17993050	18089638	18185483	18280581	18374929	-749
6.64	-0.0	18468523	18561360	18653436	18744749	18835294	18925068	19014068	19102291	19189733	19276392	-783
6.65	-0.0	19362264	19447345	19531634	19615126	19697818	19779709	19860794	19941070	20020536	20099187	-813
6.66	-0.0	20177021	20254036	20330228	20405594	20480133	20553841	20626715	20698753	20769953	20840311	-840
6.67	-0.0	20909826	20978495	21046314	21113283	21179399	21244658	21309059	21372600	21435279	21497092	-864
6.68	-0.0	21558039	21618116	21677322	21735654	21793111	21849691	21905391	21960210	22014146	22067197	-884
6.69	-0.0	22119361	22170636	22221021	22270513	22319112	22366815	22413621	22459528	22504535	22548641	-901
6.70	-0.0	22591843	22634140	22675532	22716016	22755591	22794256	22832010	22868852	22904779	22939792	-914
6.71	-0.0	22973889	23007069	23039332	23070675	23101098	23130600	23159180	23186838	23213572	23239382	-923
6.72	-0.0	23264267	23288226	23311259	23333364	23354543	23374793	23394114	23412506	23429969	23446502	-929
6.73	-0.0	23462104	23476775	23490516	23503325	23515202	23526148	23536162	23545244	23553394	23560612	-931
6.74	-0.0	23566897	23572251	23576672	23580162	23582719	23584345	23585040	23584804	23583636	23581538	-930
6.75	-0.0	23578510	23574552	23569665	23563849	23557104	23549431	23540832	23531305	23520853	23509475	-924
6.76	-0.0	23497173	23483946	23469797	23454726	23438733	23421819	23403986	23385235	23365566	23344981	-915
6.77	-0.0	23323480	23301065	23277737	23253496	23228345	23202285	23175317	23147441	23118660	23088976	-903
6.78	-0.0	23058388	23026899	22994511	22961225	22927042	22891964	22855993	22819130	22781378	22742737	-887
6.79	-0.0	22703210	22662799	22621505	22579330	22536276	22492345	22447540	22401861	22355312	22307894	-868
6.80	-0.0	22259610	22210461	22160449	22109578	22057849	22005265	21951827	21897538	21842402	21786419	-845
6.81	-0.0	21729592	21671925	21613419	21554077	21493902	21432896	21371062	21308402	21244920	21180618	-819
6.82	-0.0	21115498	21049564	20982819	20915265	20846905	20777743	20707780	20637021	20565467	20493123	-790
6.83	-0.0	20419991	20346075	20271377	20195900	20119649	20042626	19964834	19886277	19806958	19726880	-758
6.84	-0.0	19646047	19564462	19482129	19399051	19315231	19230674	19145382	19059359	18972609	18885136	-723
6.85	-0.0	18796942	18708033	18618410	18528079	18437044	18345306	18252872	18159744	18065926	17971423	-685
6.86	-0.0	17876238	17780375	17683838	17586631	17488758	17390224	17291031	17191185	17090690	16989549	-644
6.87	-0.0	16887767	16785348	16682297	16578616	16474312	16369388	16263848	16157696	16050938	15943577	-602
6.88	-0.0	15835619	15727066	15617924	15508198	15397891	15287008	15175555	15063534	14950952	14837812	-556
6.89	-0.0	14724119	14609878	14495093	14379770	14263912	14147525	14030614	13913182	13795235	13676778	-509
6.90	-0.0	13557816	13438352	13318393	13197943	13077007	12955590	12833696	12711332	12588500	12465208	-460
6.91	-0.0	12341459	12217259	12092613	11967525	11842001	11716046	11589665	11462863	11335645	11208016	-410
6.92	-0.0	11079982	10951548	10822719	10693499	10563895	10433911	10303553	10172825	10041734	09910285	-358
6.93	-0.0	09778482	09646331	09513838	09381007	09247844	09114355	08980544	08846418	08711980	08577238	-304
6.94	-0.0	08442196	08306860	08171235	08035326	07899140	07762680	07625954	07488966	07351722	07214227	-250
6.95	-0.0	07076487	06938507	06800293	06661850	06523184	06384301	06245206	06105904	05966401	05826702	-194
6.96	-0.0	05686814	05546741	05406490	05266066	05125474	04984720	04843810	04702749	04561543	04420197	-139
6.97	-0.0	04278718	04137110	03995379	03853532	03711573	03569508	03427343	03285084	03142735	03000303	-82
6.98	-0.0	02857794	02715213	02572565	02429857	02287094	02144281	02001425	01858530	01715603	01572649	-26
6.99	-0.0	01429674	01286683	01143683	01000678	00857675	00714679	00571695	00428730	00285788	00142877	30

$$\frac{\sin 2\pi x}{2\pi x}$$

$$\frac{\sin 2\pi x}{2\pi x}$$

TABLE 6

Sinc Function

x		.0	.001	.002	.003	.004	.005	.006	.007	.008	.009	Δ_2
7.00	0.0	00000000	00142836	00285625	00428362	00571042	00713658	00856206	00998679	01141072	01283379	-85
7.01	0.0	01425595	01567714	01709731	01851640	01993435	02135111	02276662	02418084	02559369	02700513	-140
7.02	0.0	02841510	02982356	03123043	03263567	03403922	03544102	03684103	03823919	03963544	04102972	-195
7.03	0.0	04242199	04381220	04520027	04658617	04796983	04935121	05073024	05210689	05348108	05485277	-249
7.04	0.0	05622191	05758844	05895231	06031346	06167185	06302741	06438011	06572988	06707667	06842043	-302
7.05	0.0	06976111	07109865	07243301	07376414	07509197	07641646	07773756	07905522	08036939	08168000	-353
7.06	0.0	08298703	08429041	08559009	08688602	08817816	08946645	09075085	09203130	09330776	09458017	-403
7.07	0.0	09584849	09711266	09837265	09962839	10087985	10212697	10336971	10460802	10584184	10707115	-452
7.08	0.0	10829587	10951598	11073142	11194215	11314811	11434927	11554558	11673698	11792345	11910492	-498
7.09	0.0	12028136	12145272	12261895	12378002	12493587	12608647	12723176	12837171	12950627	13063540	-542
7.10	0.0	13175905	13287719	13398977	13509674	13619807	13729372	13838364	13946778	14054612	14161861	-584
7.11	0.0	14268520	14374586	14480056	14584923	14689186	14792839	14895880	14998303	15100106	15201284	-624
7.12	0.0	15301834	15401751	15501032	15599674	15697672	15795023	15891723	15987769	16083157	16177883	-661
7.13	0.0	16271944	16365336	16458056	16550101	16641466	16732149	16822146	16911454	17000069	17087989	-695
7.14	0.0	17175209	17261727	17347539	17432642	17517034	17600710	17683669	17765906	17847419	17928204	-726
7.15	0.0	18008259	18087581	18166167	18244014	18321119	18397479	18473092	18547954	18622063	18695417	-755
7.16	0.0	18768012	18839846	18910916	18981220	19050755	19119518	19187508	19254721	19321156	19386809	-780
7.17	0.0	19451679	19515763	19579059	19641564	19703277	19764194	19824315	19883636	19942155	19999872	-802
7.18	0.0	20056782	20112886	20168179	20222662	20276330	20329184	20381220	20432437	20482834	20532408	-822
7.19	0.0	20581158	20629081	20676178	20722445	20767881	20812484	20856254	20899189	20941286	20982545	-837
7.20	0.0	21022965	21062543	21101279	21139171	21176219	21212420	21247774	21282279	21315935	21348740	-850
7.21	0.0	21380693	21411794	21442040	21471432	21499968	21527648	21554470	21580433	21605538	21629783	-859
7.22	0.0	21653168	21675691	21697353	21718152	21738088	21757160	21775368	21792712	21809191	21824804	-864
7.23	0.0	21839552	21853433	21866448	21878596	21889877	21900291	21909838	21918517	21926329	21933273	-867
7.24	0.0	21939349	21944558	21948899	21952372	21954978	21956716	21957588	21957592	21956730	21955001	-865
7.25	0.0	21952406	21948945	21944619	21939428	21933372	21926452	21918669	21910022	21900513	21890142	-861
7.26	0.0	21878910	21866817	21853865	21840054	21825384	21809857	21793473	21776233	21758139	21739190	-853
7.27	0.0	21719389	21698736	21677232	21654878	21631676	21607626	21582730	21556989	21530404	21502976	-842
7.28	0.0	21474708	21445599	21415652	21384867	21353247	21320793	21287506	21253388	21218440	21182665	-827
7.29	0.0	21146063	21108636	21070387	21031316	20991426	20950718	20909194	20866857	20823707	20779747	-809
7.30	0.0	20734979	20689405	20643026	20595846	20547865	20499086	20449512	20399144	20347984	20296036	-788
7.31	0.0	20243300	20189780	20135477	20080395	20024535	19967900	19910492	19852313	19793368	19733657	-764
7.32	0.0	19673183	19611949	19549958	19487212	19423714	19359467	19294473	19228735	19162256	19095039	-737
7.33	0.0	19027086	18958401	18888986	18818845	18747980	18676394	18604090	18531072	18457342	18382904	-707
7.34	0.0	18307761	18231915	18155370	18078130	18000196	17921574	17842266	17762275	17681604	17600258	-675
7.35	0.0	17518239	17435550	17352196	17268180	17183505	17098175	17012193	16925563	16838288	16750372	-640
7.36	0.0	16661819	16572633	16482816	16392374	16301308	16209625	16117326	16024416	15930898	15836777	-602
7.37	0.0	15742057	15646741	15550833	15454337	15357258	15259599	15161363	15062556	14963181	14863243	-563
7.38	0.0	14762745	14661691	14560086	14457934	14355238	14252004	14148236	14043937	13939111	13833764	-521
7.39	0.0	13727900	13621522	13514635	13407244	13299352	13190965	13082087	12972721	12862873	12752547	-477
7.40	0.0	12641747	12530478	12418745	12306552	12193903	12080803	11967257	11853269	11738845	11623987	-431
7.41	0.0	11508702	11392994	11276867	11160326	11043377	10926023	10808269	10690120	10571582	10452658	-384
7.42	0.0	10333353	10213673	10093621	09973204	09852425	09731291	09609804	09487971	09365796	09243285	-336
7.43	0.0	09120441	08997271	08873779	08749969	08625848	08501419	08376688	08251660	08126340	08000733	-286
7.44	0.0	07874844	07748678	07622240	07495535	07368569	07241345	07113870	06986148	06858185	06729985	-235
7.45	0.0	06601555	06472898	06344020	06214926	06085622	05956112	05826402	05696496	05566401	05436120	-184
7.46	0.0	05305660	05175026	05044222	04913254	04782128	04650847	04519419	04387847	04256137	04124294	-132
7.47	0.0	03992324	03860232	03728023	03595701	03463274	03330745	03198120	03065404	02932603	02799722	-79
7.48	0.0	02666765	02533739	02400648	02267499	02134295	02001043	01867747	01734413	01601046	01467652	-27
7.49	0.0	01334235	01200802	01067356	00933904	00800451	00667002	00533562	00400136	00266731	00133350	25

TABLE 6

Sinc Function

x		.0	.001	.002	.003	.004	.005	.006	.007	.008	.009	Δ_2
7.50	-0.0	00000000	00133315	00266589	00399816	00532993	00666113	00799171	00932162	01065081	01197923	-76
7.51	-0.0	01330682	01463353	01595931	01728411	01860787	01993054	02125208	02257243	02389153	02520934	-128
7.52	-0.0	02652580	02784087	02915449	03046661	03177717	03308614	03439345	03569905	03700290	03830494	-180
7.53	-0.0	03960513	04090341	04219972	04349403	04478627	04607641	04736438	04865015	04993365	05121484	-230
7.54	-0.0	05249367	05377008	05504404	05631549	05758437	05885065	06011427	06137517	06263333	06388868	-280
7.55	-0.0	06514117	06639076	06763740	06888104	07012163	07135912	07259347	07382463	07505254	07627717	-328
7.56	-0.0	07749847	07871638	07993087	08114187	08234936	08355327	08475357	08595021	08714313	08833230	-375
7.57	-0.0	08951767	09069920	09187683	09305052	09422023	09538592	09654753	09770502	09885835	10000747	-420
7.58	-0.0	10115235	10229292	10342916	10456102	10568845	10681141	10792986	10904376	11015306	11125771	-463
7.59	-0.0	11235769	11345293	11454342	11562909	11670991	11778584	11885684	11992287	12098388	12203983	-504
7.60	-0.0	12309070	12413642	12517697	12621231	12724239	12826718	12928663	13030071	13130939	13231261	-544
7.61	-0.0	13331035	13430257	13528922	13627027	13724569	13821543	13917947	14013775	14109025	14203694	-581
7.62	-0.0	14297776	14391270	14484171	14576476	14668182	14759284	14849780	14939666	15028938	15117594	-616
7.63	-0.0	15205630	15293043	15379829	15465986	15551509	15636396	15720644	15804249	15887208	15969518	-648
7.64	-0.0	16051177	16132180	16212526	16292211	16371231	16449585	16527269	16604280	16680616	16756273	-677
7.65	-0.0	16831249	16905541	16979146	17052062	17124286	17195815	17266646	17336778	17406206	17474930	-704
7.66	-0.0	17542946	17610251	17676844	17742722	17807882	17872322	17936040	17999033	18061300	18122837	-728
7.67	-0.0	18183643	18243715	18303051	18361650	18419508	18476624	18532996	18588622	18643500	18697627	-749
7.68	-0.0	18751002	18803623	18855489	18906596	18956944	19006530	19055354	19103412	19150704	19197227	-767
7.69	-0.0	19242981	19287963	19332172	19375607	19418265	19460146	19501248	19541569	19581109	19619865	-782
7.70	-0.0	19657837	19695023	19731422	19767033	19801854	19835884	19869123	19901568	19933220	19964077	-794
7.71	-0.0	19994137	20023401	20051866	20079533	20106400	20132466	20157731	20182194	20205853	20228709	-803
7.72	-0.0	20250761	20272007	20292448	20312082	20330910	20348930	20366142	20382546	20398141	20412927	-808
7.73	-0.0	20426903	20440069	20452425	20463971	20474706	20484629	20493742	20502043	20509533	20516212	-811
7.74	-0.0	20522078	20527134	20531377	20534809	20537430	20539239	20540238	20540425	20539801	20538366	-810
7.75	-0.0	20536122	20533067	20529202	20524529	20519046	20512754	20505655	20497748	20489034	20479513	-806
7.76	-0.0	20469186	20458055	20446118	20433377	20419834	20405487	20390339	20374390	20357641	20340092	-798
7.77	-0.0	20321745	20302601	20282660	20261923	20240393	20218068	20194952	20171044	20146346	20120859	-788
7.78	-0.0	20094585	20067524	20039678	20011048	19981636	19951442	19920469	19888717	19856188	19822884	-774
7.79	-0.0	19788806	19753955	19718334	19681944	19644786	19606862	19568174	19528723	19488511	19447541	-758
7.80	-0.0	19405814	19363331	19320094	19276107	19231369	19185884	19139653	19092679	19044963	18996507	-739
7.81	-0.0	18947314	18897386	18846724	18795332	18743211	18690363	18636791	18582497	18527483	18471752	-716
7.82	-0.0	18415307	18358148	18300280	18241704	18182423	18122440	18061756	18000375	17938300	17875532	-691
7.83	-0.0	17812074	17747930	17683101	17617591	17551402	17484537	17416999	17348791	17279916	17210376	-664
7.84	-0.0	17140174	17069313	16997797	16925629	16852810	16779345	16705236	16630487	16555100	16479079	-633
7.85	-0.0	16402427	16325147	16247242	16168716	16089572	16009812	15929441	15848462	15766877	15684691	-601
7.86	-0.0	15601907	15518528	15434558	15350000	15264857	15179133	15092833	15005958	14918513	14830501	-566
7.87	-0.0	14741926	14652792	14563102	14472860	14382070	14290735	14198859	14106446	14013500	13920024	-529
7.88	-0.0	13826022	13731499	13636457	13540901	13444835	13348263	13251188	13153615	13055547	12956989	-489
7.89	-0.0	12857944	12758417	12658411	12557932	12456981	12355565	12253687	12151351	12048561	11945321	-449
7.90	-0.0	11841637	11737510	11632947	11527952	11422527	11316679	11210411	11103727	10996631	10889129	-406
7.91	-0.0	10781224	10672921	10564224	10455137	10345665	10235813	10125584	10014983	09904015	09792684	-362
7.92	-0.0	09680995	09568951	09456559	09343821	09230743	09117329	09003584	08889512	08775118	08660406	-317
7.93	-0.0	08545382	08430049	08314413	08198477	08082247	07965728	07848923	07731838	07614477	07496845	-270
7.94	-0.0	07378947	07260788	07142371	07023703	06904787	06785628	06666232	06546602	06426744	06306663	-223
7.95	-0.0	06186363	06065848	05945125	05824198	05703071	05581749	05460238	05338541	05216664	05094612	-174
7.96	-0.0	04972390	04850002	04727453	04604749	04481894	04358892	04235749	04112470	03989060	03865523	-125
7.97	-0.0	03741865	03618090	03494203	03370209	03246113	03121921	02997636	02873264	02748810	02624278	-76
7.98	-0.0	02499675	02375003	02250270	02125478	02000634	01875742	01750808	01625836	01500830	01375797	-27
7.99	-0.0	01250741	01125667	01000580	00875484	00750385	00625288	00500197	00375118	00250056	00125015	21

$$\frac{\sin 2\pi x}{2\pi x}$$

$\frac{\sin 2\pi x}{2\pi x}$

TABLE 6

Sinc Function

x		.0	.001	.002	.003	.004	.005	.006	.007	.008	.009	Δ_2
8.00	0.0	00000000	00124984	00249931	00374837	00499698	00624507	00749260	00873953	00998580	01123137	-70
8.01	0.0	01247618	01372019	01496335	01620560	01744691	01868722	01992648	02116464	02240166	02363749	-118
8.02	0.0	02487207	02610537	02733733	02856790	02979704	03102470	03225082	03347537	03469829	03591953	-166
8.03	0.0	03713906	03835681	03957275	04078682	04199898	04320918	04441737	04562351	04682755	04802944	-214
8.04	0.0	04922914	05042659	05162175	05281459	05400503	05519306	05637860	05756163	05874209	05991994	-260
8.05	0.0	06109513	06226762	06343736	06460430	06576841	06692963	06808792	06924323	07039552	07154475	-305
8.06	0.0	07269087	07383384	07497360	07611013	07724337	07837328	07949982	08062293	08174259	08285874	-349
8.07	0.0	08397135	08508036	08618575	08728745	08838544	08947967	09057009	09165667	09273936	09381813	-392
8.08	0.0	09489292	09596370	09703043	09809306	09915156	10020588	10125599	10230184	10334340	10438061	-433
8.09	0.0	10541345	10644188	10746584	10848532	10950026	11051062	11151638	11251748	11351389	11450558	-472
8.10	0.0	11549250	11647462	11745191	11842431	11939180	12035433	12131188	12226441	12321187	12415423	-509
8.11	0.0	12509147	12602353	12695039	12787201	12878835	12969939	13060508	13150539	13240029	13328974	-544
8.12	0.0	13417371	13505217	13592508	13679241	13765413	13851020	13936060	14020528	14104422	14187739	-576
8.13	0.0	14270475	14352627	14434193	14515169	14595552	14675339	14754527	14833114	14911095	14988469	-607
8.14	0.0	15065232	15141382	15216915	15291829	15366121	15439788	15512828	15585237	15657014	15728154	-635
8.15	0.0	15798657	15868518	15937736	16006308	16074232	16141504	16208122	16274084	16339388	16404031	-660
8.16	0.0	16468010	16531324	16593969	16655945	16717247	16777875	16837825	16897097	16955686	17013592	-683
8.17	0.0	17070813	17127345	17183188	17238338	17292795	17346556	17399619	17451982	17503643	17554601	-703
8.18	0.0	17604853	17654398	17703234	17751360	17798773	17845472	17891455	17936720	17981267	18025092	-720
8.19	0.0	18068196	18110576	18152230	18193158	18233358	18272828	18311567	18349574	18386848	18423386	-734
8.20	0.0	18459189	18494253	18528580	18562167	18595012	18627116	18658477	18689093	18718964	18748090	-745
8.21	0.0	18776467	18804097	18830978	18857109	18882490	18907118	18930995	18954118	18976488	18998103	-754
8.22	0.0	19018962	19039066	19058414	19077005	19094838	19111913	19128229	19143786	19158584	19172622	-759
8.23	0.0	19185900	19198417	19210174	19221169	19231403	19240875	19249586	19257534	19264721	19271145	-761
8.24	0.0	19276807	19281706	19285844	19289219	19291832	19293682	19294771	19295098	19294663	19293466	-761
8.25	0.0	19291508	19288789	19285310	19281070	19276070	19270310	19263791	19256513	19248477	19239683	-757
8.26	0.0	19230132	19219823	19208759	19196939	19184364	19171035	19156953	19142118	19126530	19110192	-750
8.27	0.0	19093103	19075264	19056677	19037342	19017260	18996433	18974860	18952544	18929485	18905685	-741
8.28	0.0	18881144	18855863	18829845	18803089	18775598	18747372	18718413	18688722	18658300	18627150	-728
8.29	0.0	18595271	18562666	18529337	18495284	18460509	18425013	18388800	18351868	18314222	18275861	-713
8.30	0.0	18236789	18197006	18156514	18115315	18073411	18030804	17987495	17943487	17898781	17853379	-695
8.31	0.0	17807283	17760496	17713019	17664854	17616003	17566468	17516253	17465357	17413785	17361538	-674
8.32	0.0	17308618	17255027	17200768	17145843	17090255	17034005	16977097	16919532	16861312	16802442	-651
8.33	0.0	16742922	16682756	16621945	16560493	16498402	16435675	16372314	16308321	16243701	16178455	-625
8.34	0.0	16112585	16046096	15978989	15911268	15842934	15773992	15704443	15634292	15563539	15492190	-596
8.35	0.0	15420246	15347711	15274587	15200877	15126586	15051715	14976267	14900247	14823657	14746500	-566
8.36	0.0	14668779	14590498	14511659	14432267	14352324	14271833	14190799	14109223	14027110	13944463	-533
8.37	0.0	13861286	13777581	13693352	13608603	13523337	13437557	13351267	13264471	13177173	13089375	-498
8.38	0.0	13001081	12912295	12823020	12733261	12643020	12552302	12461110	12369448	12277319	12184728	-462
8.39	0.0	12091678	11998173	11904216	11809812	11714964	11619677	11523953	11427798	11331214	11234206	-423
8.40	0.0	11136777	11038932	10940675	10842009	10742939	10643468	10543601	10443341	10342693	10241660	-383
8.41	0.0	10140248	10038459	09936298	09833769	09730876	09627624	09524016	09420057	09315751	09211102	-342
8.42	0.0	09106114	09000791	08895139	08789160	08682859	08576241	08469310	08362070	08254525	08146680	-299
8.43	0.0	08038538	07930106	07821385	07712382	07603101	07493545	07383719	07273628	07163276	07052667	-256
8.44	0.0	06941806	06830697	06719345	06607754	06495929	06383874	06271593	06159091	06046373	05933443	-211
8.45	0.0	05820306	05706965	05593426	05479693	05365771	05251663	05137376	05022912	04908278	04793477	-166
8.46	0.0	04678514	04563393	04448119	04332697	04217131	04101426	03985587	03869618	03753523	03637307	-120
8.47	0.0	03520975	03404532	03287982	03171330	03054580	02937737	02820805	02703790	02586696	02469527	-73
8.48	0.0	02352288	02234984	02117620	02000199	01882728	01765210	01647649	01530052	01412421	01294763	-27
8.49	0.0	01177081	01059381	00941666	00823942	00706214	00588485	00470760	00353045	00235343	00117660	18

TABLE 6

Sinc Function

x		.0	.001	.002	.003	.004	.005	.006	.007	.008	.009	Δ_2
8.50	-0.0	.00000000	00117632	00235233	00352796	00470317	00587793	00705217	00822587	00939896	01057140	-64
8.51	-0.0	01174315	01291416	01408439	01525379	01642231	01758990	01875653	01992215	02108670	02225015	-110
8.52	-0.0	02341245	02457355	02573340	02689197	02804921	02920507	03035950	03151246	03266391	03381380	-155
8.53	-0.0	03496209	03610873	03725367	03839687	03953829	04067789	04181560	04295141	04408525	04521708	-200
8.54	-0.0	04634687	04747456	04860011	04972348	05084463	05196350	05308007	05419427	05530608	05641544	-243
8.55	-0.0	05752232	05862667	05972844	06082760	06192410	06301790	06410896	06519723	06628267	06736525	-286
8.56	-0.0	06844491	06952162	07059533	07166600	07273360	07379807	07485939	07591750	07697236	07802395	-327
8.57	-0.0	07907220	08011710	08115858	08219662	08323117	08426219	08528965	08631351	08733371	08835023	-368
8.58	-0.0	08936303	09037206	09137729	09237868	09337619	09436978	09535942	09634506	09732666	09830420	-406
8.59	-0.0	09927763	10024691	10121201	10217289	10312952	10408185	10502985	10597348	10691271	10784751	-443
8.60	-0.0	10877782	10970363	11062489	11154157	11245364	11336105	11426378	11516179	11605505	11694351	-478
8.61	-0.0	11782715	11870594	11957984	12044881	12131283	12217185	12302586	12387481	12471868	12555742	-511
8.62	-0.0	12639102	12721943	12804263	12886058	12967326	13048063	13128266	13207932	13287059	13365642	-542
8.63	-0.0	13443680	13521169	13598107	13674490	13750315	13825580	13900282	13974418	14047985	14120981	-571
8.64	-0.0	14193402	14265246	14336510	14407193	14477290	14546799	14615718	14684044	14751775	14818907	-597
8.65	-0.0	14885440	14951369	15016693	15081409	15145515	15209008	15271886	15334146	15395787	15456806	-621
8.66	-0.0	15517201	15576970	15636109	15694618	15752494	15809734	15866338	15922302	15977624	16032303	-642
8.67	-0.0	16086337	16139723	16192460	16244545	16295977	16346754	16396874	16446336	16495136	16543274	-661
8.68	-0.0	16590749	16637557	16683698	16729169	16773970	16818098	16861553	16904331	16946433	16987856	-678
8.69	-0.0	17028599	17068660	17108038	17146732	17184740	17222062	17258694	17294637	17329889	17364449	-691
8.70	-0.0	17398316	17431488	17463964	17495743	17526825	17557207	17586889	17615871	17644150	17671727	-702
8.71	-0.0	17698599	17724767	17750229	17774984	17799032	17822373	17845004	17866925	17888137	17908637	-710
8.72	-0.0	17928426	17947502	17965866	17983516	18000452	18016674	18032181	18046973	18061049	18074408	-715
8.73	-0.0	18087051	18098978	18110187	18120679	18130453	18139509	18147847	18155466	18162368	18168550	-718
8.74	-0.0	18174015	18178760	18182787	18186095	18188685	18190556	18191708	18192142	18191858	18190856	-717
8.75	-0.0	18189136	18186699	18183544	18179672	18175083	18169778	18163757	18157021	18149569	18141402	-714
8.76	-0.0	18132521	18122927	18112619	18101599	18089866	18077422	18064268	18050403	18035829	18020547	-708
8.77	-0.0	18004556	17987859	17970455	17952346	17933532	17914015	17893795	17872873	17851251	17828929	-699
8.78	-0.0	17805908	17782189	17757774	17732664	17706859	17680362	17653172	17625292	17596722	17567464	-687
8.79	-0.0	17537520	17506890	17475575	17443579	17410901	17377543	17343506	17308793	17273404	17237342	-673
8.80	-0.0	17200607	17163202	17125128	17086386	17046979	17006908	16966175	16924781	16882728	16840019	-656
8.81	-0.0	16796654	16752637	16707968	16662649	16616683	16570072	16522817	16474921	16426385	16377212	-636
8.82	-0.0	16327403	16276962	16225889	16174187	16121859	16068906	16015330	15961135	15906322	15850893	-614
8.83	-0.0	15794852	15738199	15680938	15623071	15564601	15505529	15445859	15385592	15324732	15263280	-590
8.84	-0.0	15201240	15138614	15075405	15011614	14947246	14882302	14816785	14750699	14684045	14616826	-564
8.85	-0.0	14549046	14480706	14411811	14342362	14272362	14201815	14130724	14059091	13986918	13914210	-535
8.86	-0.0	13840970	13767199	13692902	13618080	13542739	13466879	13390505	13313620	13236227	13158328	-504
8.87	-0.0	13079928	13001029	12921635	12841748	12761373	12680511	12599168	12517346	12435048	12352277	-471
8.88	-0.0	12269038	12185333	12101166	12016540	11931459	11845926	11759945	11673519	11586651	11499346	-437
8.89	-0.0	11411606	11323436	11234838	11145817	11056376	10966519	10876249	10785570	10694486	10603000	-401
8.90	-0.0	10511116	10418837	10326168	10233113	10139674	10045856	09951663	09857097	09762164	09666867	-363
8.91	-0.0	09571210	09475197	09378831	09282116	09185057	09087656	08989919	08891850	08793451	08694727	-324
8.92	-0.0	08595681	08496319	08396644	08296660	08196370	08095780	07994892	07893711	07792242	07690487	-284
8.93	-0.0	07588452	07486140	07383556	07280703	07177586	07074208	06970574	06866689	06762555	06658178	-243
8.94	-0.0	06553562	06448710	06343627	06238317	06132785	06027033	05921068	05814893	05708512	05601929	-201
8.95	-0.0	05495149	05388176	05281014	05173667	05066141	04958438	04850564	04742522	04634317	04525954	-158
8.96	-0.0	04417436	04308768	04199954	04090998	03981905	03872680	03763326	03653848	03544250	03434536	-114
8.97	-0.0	03324712	03214780	03104746	02994615	02884389	02774074	02663675	02553195	02442638	02332010	-71
8.98	-0.0	02221314	02110556	01999739	01888867	01777946	01666979	01555971	01444926	01333849	01222744	-27
8.99	-0.0	01111615	01000468	00889305	00778132	00666953	00555773	00444595	00333425	00222266	00111123	16

$\frac{\sin 2\pi x}{2\pi x}$

TABLE 6

Sinc Function

x		.0	.001	.002	.003	.004	.005	.006	.007	.008	.009	Δ_2
9.00	0.0	00000000	00111098	00222167	00333203	00444200	00555156	00666065	00776923	00887726	00998469	−59
9.01	0.0	01109148	01219759	01330297	01440758	01551137	01661431	01771635	01881745	01991756	02101663	−102
9.02	0.0	02211464	02321153	02430726	02540178	02649506	02758706	02867772	02976701	03085488	03194129	−145
9.03	0.0	03302620	03410957	03519135	03627151	03734999	03842676	03950177	04057499	04164636	04271586	−187
9.04	0.0	04378343	04484904	04591265	04697420	04803367	04909100	05014617	05119912	05224982	05329822	−228
9.05	0.0	05434429	05538798	05642926	05746807	05850439	05953817	06056937	06159796	06262388	06364711	−269
9.06	0.0	06466760	06568531	06670020	06771223	06872137	06972758	07073081	07173102	07272819	07372226	−308
9.07	0.0	07471321	07570098	07668555	07766688	07864493	07961965	08059102	08155899	08252353	08348460	−346
9.08	0.0	08444216	08539617	08634661	08729343	08823659	08917607	09011181	09104380	09197198	09289633	−382
9.09	0.0	09381681	09473339	09564602	09655468	09745932	09835992	09925644	10014884	10103710	10192117	−417
9.10	0.0	10280102	10367662	10454794	10541494	10627758	10713585	10798969	10883908	10968400	11052439	−451
9.11	0.0	11136024	11219151	11301817	11384018	11465753	11547016	11627806	11708119	11787953	11867304	−482
9.12	0.0	11946168	12024544	12102428	12179818	12256709	12333100	12408988	12484368	12559240	12633599	−511
9.13	0.0	12707444	12780770	12853576	12925859	12997616	13068843	13139540	13209702	13279327	13348413	−538
9.14	0.0	13416957	13484957	13552409	13619311	13685662	13751457	13816695	13881374	13945491	14009043	−563
9.15	0.0	14072028	14134444	14196288	14257558	14318253	14378368	14437903	14496855	14555222	14613002	−586
9.16	0.0	14670192	14726791	14782796	14838205	14893017	14947228	15000838	15053844	15106244	15158036	−607
9.17	0.0	15209219	15259790	15309748	15359091	15407816	15455923	15503409	15550273	15596512	15642126	−625
9.18	0.0	15687113	15731471	15775198	15818292	15860753	15902579	15943767	15984318	16024228	16063498	−640
9.19	0.0	16102124	16140107	16177444	16214135	16250178	16285571	16320314	16354405	16387843	16420626	−653
9.20	0.0	16452755	16484227	16515042	16545197	16574694	16603529	16631703	16659214	16686062	16712245	−663
9.21	0.0	16737763	16762615	16786799	16810316	16833164	16855342	16876850	16897688	16917854	16937347	−671
9.22	0.0	16956168	16974316	16991789	17008588	17024712	17040161	17054933	17069029	17082448	17095190	−676
9.23	0.0	17107254	17118641	17129349	17139379	17148730	17157402	17165395	17172709	17179343	17185297	−679
9.24	0.0	17190572	17195167	17199083	17202318	17204874	17206751	17207947	17208464	17208302	17207460	−678
9.25	0.0	17205940	17203740	17200862	17197306	17193071	17188159	17182569	17176302	17169359	17161739	−675
9.26	0.0	17153444	17144473	17134827	17124507	17113513	17101847	17089507	17076496	17062813	17048460	−670
9.27	0.0	17033437	17017745	17001384	16984356	16966661	16948300	16929274	16909584	16889231	16868215	−661
9.28	0.0	16846538	16824201	16801204	16777549	16753237	16728269	16702646	16676369	16649439	16621858	−650
9.29	0.0	16593627	16564747	16535220	16505045	16474226	16442764	16410658	16377912	16344527	16310504	−637
9.30	0.0	16275844	16240549	16204620	16168060	16130869	16093049	16054603	16015531	15975835	15935517	−621
9.31	0.0	15894578	15853021	15810847	15768058	15724656	15680643	15636019	15590789	15544952	15498512	−603
9.32	0.0	15451470	15403828	15355588	15306752	15257323	15207302	15156692	15105494	15053710	15001344	−582
9.33	0.0	14948397	14894870	14840768	14786091	14730842	14675024	14618638	14561687	14504174	14446101	−559
9.34	0.0	14387469	14328283	14268543	14208253	14147415	14086031	14024105	13961638	13898634	13835094	−534
9.35	0.0	13771022	13706420	13641290	13575637	13509461	13442766	13375555	13307830	13239594	13170851	−507
9.36	0.0	13101602	13031850	12961599	12890852	12819611	12747879	12675659	12602954	12529767	12456101	−478
9.37	0.0	12381959	12307345	12232260	12156709	12080693	12004218	11927284	11849896	11772057	11693770	−447
9.38	0.0	11615038	11535864	11456252	11376204	11295725	11214816	11133483	11051727	10969552	10886962	−415
9.39	0.0	10803959	10720548	10636732	10552513	10467896	10382883	10297479	10211686	10125509	10038950	−380
9.40	0.0	09952014	09864703	09777021	09688972	09600559	09511786	09422657	09333174	09243342	09153164	−345
9.41	0.0	09062644	08971786	08880593	08789068	08697216	08605041	08512545	08419733	08326608	08233174	−308
9.42	0.0	08139435	08045395	07951057	07856425	07761503	07666295	07570805	07475036	07378992	07282677	−270
9.43	0.0	07186095	07089250	06992146	06894786	06797175	06699316	06601214	06502872	06404293	06305483	−231
9.44	0.0	06206445	06107183	06007701	05908003	05808092	05707974	05607651	05507129	05406410	05305499	−191
9.45	0.0	05204400	05103117	05001654	04900015	04798204	04696226	04594083	04491781	04389323	04286713	−151
9.46	0.0	04183956	04081056	03978016	03874841	03771534	03668101	03564545	03460870	03357080	03253179	−110
9.47	0.0	03149172	03045063	02940856	02836554	02732163	02627685	02523126	02418490	02313780	02209001	−68
9.48	0.0	02104157	01999251	01894289	01789275	01684212	01579104	01473957	01368773	01263557	01158314	−27
9.49	0.0	01053048	00947761	00842460	00737148	00631828	00526506	00421186	00315871	00210565	00105274	14

TABLE 6

Sinc Function

x	.0	.001	.002	.003	.004	.005	.006	.007	.008	.009	Δ_2
9.50 −0.0	00000000	00105251	00210476	00315671	00420831	00525952	00631031	00736062	00841042	00945967	−54
9.51 −0.0	01050833	01155635	01260369	01365032	01469619	01574125	01678548	01782882	01887125	01991270	−95
9.52 −0.0	02095316	02199256	02303088	02406808	02510410	02613892	02717248	02820476	02923571	03026529	−136
9.53 −0.0	03129345	03232017	03334540	03436909	03539121	03641172	03743058	03844775	03946318	04047685	−176
9.54 −0.0	04148871	04249871	04350683	04451302	04551724	04651945	04751962	04851770	04951365	05050745	−215
9.55 −0.0	05149904	05248839	05347546	05446022	05544262	05642262	05740019	05837530	05934789	06031794	−254
9.56 −0.0	06128540	06225024	06321242	06417191	06512866	06608264	06703382	06798215	06892759	06987012	−291
9.57 −0.0	07080970	07174628	07267983	07361032	07453771	07546197	07638305	07730092	07821555	07912691	−327
9.58 −0.0	08003495	08093964	08184094	08273883	08363327	08452421	08541164	08629550	08717578	08805243	−361
9.59 −0.0	08892543	08979473	09066030	09152212	09238014	09323434	09408468	09493113	09577365	09661222	−395
9.60 −0.0	09744680	09827736	09910387	09992629	10074460	10155876	10236874	10317451	10397604	10477330	−426
9.61 −0.0	10556626	10635489	10713915	10791903	10869448	10946547	11023199	11099399	11175146	11250436	−456
9.62 −0.0	11325266	11399633	11473535	11546968	11619931	11692420	11764432	11835964	11907015	11977581	−484
9.63 −0.0	12047659	12117248	12186343	12254943	12323046	12390647	12457745	12524338	12590423	12655997	−510
9.64 −0.0	12721057	12785602	12849629	12913135	12976119	13038577	13100508	13161908	13222777	13283110	−534
9.65 −0.0	13342907	13402165	13460881	13519054	13576682	13633761	13690290	13746268	13801691	13856557	−555
9.66 −0.0	13910866	13964614	14017799	14070421	14122476	14173963	14224879	14275224	14324994	14374189	−575
9.67 −0.0	14422807	14470845	14518301	14565175	14611464	14657167	14702282	14746807	14790741	14834082	−592
9.68 −0.0	14876828	14918979	14960531	15001485	15041838	15081589	15120736	15159278	15197214	15234542	−607
9.69 −0.0	15271261	15307370	15342867	15377751	15412021	15445676	15478713	15511133	15542934	15574115	−619
9.70 −0.0	15604675	15634612	15663926	15692616	15720680	15748118	15774929	15801111	15826665	15851588	−629
9.71 −0.0	15875880	15899541	15922569	15944964	15966725	15987851	16008342	16028197	16047415	16065995	−636
9.72 −0.0	16083937	16101241	16117906	16133931	16149316	16164060	16178163	16191624	16204444	16216621	−641
9.73 −0.0	16228156	16239048	16249296	16258901	16267862	16276180	16283852	16290881	16297265	16303005	−644
9.74 −0.0	16308099	16312549	16316354	16319515	16322030	16323900	16325126	16325707	16325644	16324936	−643
9.75 −0.0	16323584	16321588	16318948	16315664	16311737	16307167	16301954	16296098	16289601	16282462	−641
9.76 −0.0	16274681	16266260	16257198	16247497	16237156	16226176	16214558	16202302	16189409	16175880	−635
9.77 −0.0	16161715	16146915	16131481	16115413	16098712	16081379	16063415	16044820	16025596	16005743	−628
9.78 −0.0	15985263	15964156	15942422	15920064	15897082	15873477	15849251	15824404	15798937	15772851	−617
9.79 −0.0	15746149	15718830	15690897	15662349	15633190	15603419	15573038	15542049	15510452	15478250	−605
9.80 −0.0	15445443	15412034	15378023	15343411	15308201	15272394	15235992	15198995	15161406	15123226	−590
9.81 −0.0	15084457	15045100	15005158	14964631	14923522	14881832	14839564	14796718	14753296	14709302	−573
9.82 −0.0	14664735	14619599	14573894	14527624	14480790	14433394	14385437	14336923	14287852	14238227	−553
9.83 −0.0	14188051	14137324	14086050	14034230	13981867	13928963	13875519	13821539	13767024	13711976	−531
9.84 −0.0	13656399	13600293	13543663	13486509	13428834	13370641	13311932	13252710	13192976	13132734	−508
9.85 −0.0	13071985	13010733	12948980	12886728	12823980	12760738	12697006	12632785	12568079	12502890	−482
9.86 −0.0	12437220	12371073	12304451	12237357	12169793	12101762	12033268	11964312	11894898	11825029	−454
9.87 −0.0	11754707	11683936	11612717	11541054	11468951	11396409	11323432	11250023	11176185	11101920	−425
9.88 −0.0	11027232	10952124	10876599	10800660	10724310	10647552	10570389	10492825	10414862	10336504	−394
9.89 −0.0	10257753	10178614	10099088	10019181	09938893	09858230	09777194	09695788	09614016	09531881	−362
9.90 −0.0	09449387	09366536	09283332	09199778	09115878	09031635	08947053	08862135	08776883	08691303	−328
9.91 −0.0	08605397	08519168	08432621	08345758	08258583	08171100	08083312	07995223	07906835	07818154	−293
9.92 −0.0	07729181	07639922	07550379	07460556	07370456	07280084	07189442	07098535	07007367	06915939	−257
9.93 −0.0	06824258	06732325	06640145	06547722	06455058	06362159	06269027	06175667	06082081	05988274	−220
9.94 −0.0	05894250	05800012	05705563	05610909	05516052	05420997	05325747	05230305	05134677	05038864	−182
9.95 −0.0	04942873	04846705	04750365	04653858	04557186	04460353	04363364	04266222	04168931	04071495	−144
9.96 −0.0	03973918	03876204	03778356	03680379	03582276	03484052	03385709	03287253	03188687	03090015	−105
9.97 −0.0	02991240	02892367	02793400	02694342	02595198	02495972	02396667	02297287	02197836	02098318	−66
9.98 −0.0	01998738	01899098	01799404	01699659	01599866	01500030	01400155	01300245	01200304	01100335	−26
9.99 −0.0	01000342	00900331	00800303	00700265	00600218	00500168	00400118	00300072	00200035	00100009	12

$$\frac{\sin 2\pi x}{2\pi x}$$

TABLE 6

Sinc Function

x		.0	.01	.02	.03	.04	.05	.06	.07	.08	.09
10.0	0.0	00000000	00998344	01990759	02973346	03942253	04893690	05823941	06729382	07606496	08451881
10.1	0.0	09262270	10034538	10765717	11453007	12093786	12685621	13226276	13713721	14146139	14521936
10.2	0.0	14839740	15098413	15297052	15434991	15511805	15527312	15481568	15374874	15207770	14981030
10.3	0.0	14695665	14352912	13954234	13501311	12996031	12440488	11836968	11187942	10496055	09764117
10.4	0.0	08995089	08192073	07358299	06497112	05611958	04706371	03783960	02848392	01903378	00952662
10.5	-0.0	00000000	00950849	01896141	02832162	03755240	04661761	05548186	06411058	07247021	08052831
10.6	-0.0	08825371	09561657	10258856	10914295	11525469	12090052	12605906	13071091	13483867	13842706
10.7	-0.0	14146294	14393539	14583570	14715746	14789654	14805111	14762164	14661092	14502400	14286821
10.8	-0.0	14015310	13689040	13309399	12877982	12396583	11867194	11291988	10673317	10013700	09315811
10.9	-0.0	08582470	07816635	07021381	06199897	05355470	04491469	03611334	02718565	01816703	00909320
11.0	0.0	00000000	00907668	01810109	02703777	03585165	04450822	05297364	06121489	06919989	07689764
11.1	0.0	08427831	09131339	09797577	10423986	11008168	11547897	12041126	12485993	12880832	13224175
11.2	0.0	13514763	13751543	13933678	14060548	14131752	14147106	14106651	14010644	13859563	13654101
11.3	0.0	13395163	13083866	12721528	12309668	11849997	11344410	10794982	10203954	09573731	08906864
11.4	0.0	08206046	07474100	06713965	05928686	05121402	04295335	03453772	02600058	01737579	00869749
11.5	-0.0	00000000	00868238	01731545	02586528	03429829	04258146	05068239	05856947	06621198	07358023
11.6	-0.0	08064563	08738086	09375994	09975835	10535309	11052279	11524782	11951032	12329426	12658556
11.7	-0.0	12937209	13164372	13339238	13461207	13529888	13545102	13506878	13415460	13271296	13075046
11.8	-0.0	12827572	12529934	12183392	11789395	11349575	10865743	10339881	09774133	09170796	08532311
11.9	-0.0	07861254	07160326	06432339	05680208	04906938	04115614	03309383	02491450	01665059	00833480
12.0	0.0	00000000	00832092	01659518	02479024	03287394	04081459	04858113	05614323	06347142	07053721
12.1	0.0	07731316	08377306	08989196	09564630	10101400	10597453	11050902	11460028	11823292	12139337
12.2	0.0	12406996	12625291	12793443	12910872	12977197	12992240	12956027	12868782	12730934	12543108
12.3	0.0	12306126	12021001	11688937	11311317	10889705	10425834	09921601	09379059	08800408	08187989
12.4	0.0	07544268	06871836	06173388	05451720	04709714	03950328	03176583	02391553	01598350	00800114
12.5	-0.0	00000000	00798835	01593243	02380101	03156318	03918851	04664717	05391001	06094871	06773589
12.6	-0.0	07424518	08045137	08633047	09185983	09701819	10178581	10614452	11007777	11357074	11661034
12.7	-0.0	11918531	12128623	12290556	12403767	12467888	12482741	12448345	12364915	12232854	12052760
12.8	-0.0	11825418	11551797	11233050	10870502	10465651	10020160	09535847	09014682	08458778	07870378
12.9	-0.0	07251855	06605692	05934480	05240903	04527731	03797806	03054030	02299357	01536780	00769317
13.0	0.0	00000000	00768134	01532059	02288769	03035293	03768704	04486129	05184765	05861887	06514857
13.1	0.0	07141140	07738305	08304044	08836174	09332648	09791563	10211167	10589866	10926229	11218993
13.2	0.0	11467072	11669553	11825709	11934993	11997046	12011694	11978951	11899017	11772279	11599308
13.3	0.0	11380853	11117846	10811389	10462756	10073386	09644873	09178966	08677559	08142680	07576488
13.4	0.0	06981263	06359395	05713374	05045784	04359289	03656623	02940581	02214006	01479778	00740802
13.5	-0.0	00000000	00739706	01475400	02204188	02923207	03629637	04320711	04993727	05646059	06275164
13.6	-0.0	06878598	07454018	07999196	08512029	08990542	09432898	09837406	10202527	10526879	10809242
13.7	-0.0	11048565	11243968	11394743	11500361	11560472	11574905	11543669	11466954	11345129	11178738
13.8	-0.0	10968503	10715317	10420239	10084493	09709463	09296683	08847835	08364741	07849356	07303757
13.9	-0.0	06730139	06130804	05508152	04864672	04202930	03525561	02835260	02134765	01426853	00714326
14.0	0.0	00000000	00713306	01422782	02125635	02819104	03500468	04167058	04816267	05445560	06052483
14.1	0.0	06634676	07189878	07715939	08210825	08672630	09099580	09490040	09842522	10155691	10428367
14.2	0.0	10659531	10848332	10994084	11096273	11154557	11168768	11138912	11065169	10947890	10787600
14.3	0.0	10584989	10340917	10056403	09732627	09370918	08972756	08539763	08073692	07576429	07049978
14.4	0.0	06496453	05918077	05317162	04696111	04057399	03403570	02737222	02060999	01377583	00689677
14.5	-0.0	00000000	00688727	01373788	02052489	02722161	03380177	04023959	04650987	05258812	05845064
14.6	-0.0	06407461	06943818	07452056	07930209	08376434	08789014	09166369	09507058	09809789	10073419
14.7	-0.0	10296962	10479592	10620643	10719617	10776180	10790166	10761578	10690586	10577529	10422907
14.8	-0.0	10227388	09991798	09717119	09404487	09055186	08670643	08252422	07802217	07321845	06813242
14.9	-0.0	06278452	05719617	05138973	04538840	03921609	03289738	02645737	01992162	01331602	00666673

TABLE 6

Sinc Function

x		.0	.01	.02	.03	.04	.05	.06	.07	.08	.09
15.0	0.0	00000000	00665784	01328056	01984209	02631664	03267879	03890361	04496674	05084448	05651391
15.1	0.0	06195293	06714042	07205625	07668140	08099801	08498947	08864048	09193707	09486673	09741838
15.2	0.0	09958246	10135095	10271739	10367693	10422630	10436390	10408970	10340534	10231405	10082067
15.3	0.0	09893160	09665482	09399980	09097752	08760037	08388212	07983788	07548403	07083814	06591889
15.4	0.0	06074606	05534035	04972340	04391761	03794614	03183274	02560170	01927774	01288592	00645153
15.5	-0.0	00000000	00644321	01285271	01920326	02546990	03162803	03765350	04352272	04921276	05470140
15.6	-0.0	05996726	06498986	06974972	07422838	07840856	08227416	08581032	08900354	09184164	09431391
15.7	-0.0	09641105	09812527	09945030	10038141	10091543	10105076	10078736	10012680	09907216	09762812
15.8	-0.0	09580085	09359805	09102889	08810394	08483520	08123600	07732093	07310584	06860772	06384467
15.9	-0.0	05883580	05360118	04816173	04253916	03675586	03083485	02479964	01867418	01248273	00624979
16.0	0.00	00000000	06241987	12451563	18604281	24675951	30642730	36481222	42168562	47682512	53001543
16.1	0.00	58104924	62972799	67586263	71927440	75979549	79726968	83155299	86251416	89003522	91401188
16.2	0.00	93435399	95098580	96384631	97288946	97808428	97941503	97688122	97049760	96029405	94631552
16.3	0.00	92862175	90728709	88240011	85406332	82239267	78751715	74957819	70872914	66513465	61896997
16.4	0.00	57042030	51967997	46695176	41244601	35637982	29897618	24046309	18107263	12104007	06060292
16.5	-0.00	00000000	06052951	12074700	18041538	23930003	29716968	35379736	40896125	46244559	51404149
16.6	-0.00	56354776	61077170	65552982	69764859	73696509	77332765	80659642	83664391	86335550	88662986
16.7	-0.00	90637932	92253021	93502315	94381326	94887029	95017876	94773799	94156207	93167981	91813459
16.8	-0.00	90098420	88030056	85616943	82869008	79797484	76414869	72734870	68772354	64543279	60064641
16.9	-0.00	55354395	50431392	45315295	40026509	34586093	29015683	23337397	17573755	11747588	05881943
17.0	0.00	00000000	05875027	11719979	17511839	23227831	28845503	34342815	39698230	44890795	49900224
17.1	0.00	54706976	59292331	63638467	67728523	71546669	75078165	78309419	81228037	83822874	86084074
17.2	0.00	88003108	89572805	90787382	91642460	92135085	92263735	92028324	91430202	90472148	89158356
17.3	0.00	87494420	85487304	83145322	80478096	77496518	74212711	70639972	66792723	62686453	58337653
17.4	0.00	53763752	48983046	44014626	38878301	33594520	28184288	22669086	17070786	11411558	05713792
17.5	-0.00	00000000	05707265	11385504	17012357	22565693	28023694	33364944	38568514	43614038	48481798
17.6	-0.00	53152800	57608847	61832608	65807692	69518703	72951305	76092278	78929564	81452318	83650946
17.7	-0.00	85517144	87043929	88225661	89058070	89538268	89664757	89437436	88857602	87927937	86652500
17.8	-0.00	85036711	83087324	80812401	78221279	75324531	72133924	68662369	64923873	60933476	56707199
17.9	-0.00	52261971	47615568	42786539	37794132	32658218	27399210	22037987	16595805	11094218	05554987
18.0	0.00	00000000	05548818	11069591	16540578	21940258	27247414	32441219	37501317	42407897	47141782
18.1	0.00	51684491	56018321	60126410	63992808	67602531	70941627	73997226	76757589	79212155	81351580
18.2	0.00	83167772	84653925	85804540	86615446	87083820	87208188	86988438	86425812	85522906	84283651
18.3	0.00	82713304	80818418	78606822	76087583	73270972	70168422	66792479	63156756	59275874	55165404
18.4	0.00	50841809	46322370	41625124	36768789	31772691	26656684	21441075	16146542	10794050	05404771
18.5	-0.00	00000000	05398931	10770736	16094259	21348557	26512982	31567264	36491588	41266673	45873848
18.6	-0.00	50295123	54513261	58511845	62275341	65789158	69039707	72014450	74701950	77091915	79175240
18.7	-0.00	80944035	82391661	83512752	84303235	84760345	84882636	84669982	84123580	83245938	82040872
18.8	-0.00	80513482	78670135	76518437	74067201	71326413	68307190	65021735	61483286	57706068	53705230
18.9	-0.00	49496787	45097558	40525094	35797612	30933919	25953341	20875646	15720961	10509696	05262465
19.0	0.00	00000000	05256929	10487594	15671394	20787933	25817104	30739162	35534808	40185261	44672333
19.1	0.00	48978497	53086959	56981724	60647653	64070528	67237104	70135158	72753542	75082220	77112310
19.2	0.00	78836118	80247162	81340204	82111263	82557634	82677893	82471904	81940819	81087070	79914359
19.3	0.00	78427640	76633104	74538146	72151340	69482401	66542147	63342454	59896211	56217263	52320360
19.4	0.00	48221097	43935849	39481709	34876417	30138293	25286161	20339273	15317238	10239940	05127461
19.5	-0.00	00000000	05122205	10218957	15270180	20256000	25156820	29953396	34626918	39159080	43532151
19.6	-0.00	47729045	51733391	55529590	59102884	62439405	65526236	68351456	70904189	73174643	75154151
19.7	-0.00	76835201	78211465	79277825	80030390	80466508	80584781	80385064	79868467	79037347	77895300
19.8	-0.00	76447144	74698901	72657769	70332093	67731332	64866022	61747730	58389009	54803348	51005118
19.9	-0.00	47009512	42832488	38490702	34001445	29382569	24652422	19829772	14933732	09983686	04999210

$$\frac{\sin 2\pi x}{2\pi x}$$

$$\frac{\sin 2\pi x}{2\pi x}$$

TABLE 6

Sinc Function

x		.0	.01	.02	.03	.04	.05	.06	.07	.08	.09
20.0	0.00	00000000	04994214	09963738	14888998	19750611	24529467	29206801	33764265	38184003	42448722
20.1	0.00	46541755	50447130	54149630	57634854	60889271	63900275	66656232	69146524	71361595	73292979
20.2	0.00	74933339	76276496	77317444	78052377	78478699	78595034	78401228	77898352	77088694	75975751
20.3	0.00	74564210	72859933	70869930	68602331	66066353	63272262	60231332	56955798	53458810	49754379
20.4	0.00	45857318	41783186	37548227	33169300	28663817	24049673	19345173	14568961	09739943	04877219
20.5	-0.00	00000000	04872463	09720957	14526382	19269827	23932643	28496519	32943548	37256306	41417913
20.6	-0.00	45412101	49223279	52836593	56237984	59414240	62353053	65043060	67473894	69636218	71521761
20.7	-0.00	73123355	74434958	75451675	76169782	76586734	76701177	76512951	76023090	75233817	74148532
20.8	-0.00	72771801	71109334	69167963	66955612	64481268	61754942	58787628	55591260	52178667	48563513
20.9	-0.00	44760253	40784066	36650802	32376913	27979390	23475696	18883695	14221584	09507819	04761039
21.0	0.00	00000000	04756507	09489726	14181009	18811894	23364172	27819963	32161784	36372618	40435981
21.1	0.00	44335985	48057404	51585727	54907222	58008984	60878985	63506126	65880274	67992303	69834131
21.2	0.00	71398748	72680244	73673832	74375864	74783845	74896444	74713494	74235994	73466105	72407139
21.3	0.00	71063543	69440884	67545825	65386095	62970461	60308690	57411513	54290576	50958398	47428321
21.4	0.00	43714453	39831613	35795275	31621502	27326885	22928477	18443721	13890388	09286501	04650266
21.5	-0.00	00000000	04645942	09269240	13851678	18375220	22822080	27174788	31416263	35529879	39499529
21.6	-0.00	43309689	46945478	50392718	53637984	56668665	59473004	62040149	64360194	66424215	68224308
21.7	-0.00	69753616	71006356	71977841	72664500	73063886	73174686	72996731	72530987	71779555	70745662
21.8	-0.00	69433645	67848933	65998028	63888474	61528829	58928629	56098349	53049365	49795901	46344988
21.9	-0.00	42716406	38922630	34978777	30900538	26704121	22406188	18023782	13574266	09075252	04544530
22.0	0.00	00000000	04540401	09058766	13537296	17958360	22304572	26558859	30704521	34725307	38605470
22.1	0.00	42329832	45883844	49253642	52426101	55388885	58130498	60640326	62908678	64926825	66687032
22.2	0.00	68182588	69407833	70358178	71030121	71421262	71530312	71357092	70902541	70168704	69158725
22.3	0.00	67876837	66328339	64519578	62457922	60151729	57610315	54843914	51863639	48681437	45310040
22.4	0.00	41762914	38054209	34198697	30211716	26109110	21907164	17622540	13272213	08873400	04443496
22.5	-0.00	00000000	04439548	08857639	13236867	17559993	21810014	25970231	30024315	33956368	37750989
22.6	-0.00	41393331	44869164	48164923	51267769	54165632	56847264	59302278	61521191	63495458	65217507
22.7	-0.00	66680769	67879700	68809803	69467646	69850874	69958217	69789493	69345612	68628565	67641421
22.8	-0.00	66388310	64874408	63105915	61090031	58834922	56349695	53644353	50729760	47617594	44320305
22.9	-0.00	40851060	37223694	33452652	29552935	25540036	21429883	17238774	12983310	08680332	04346856
23.0	0.00	00000000	04343078	08665249	12949484	17178917	21336912	25407130	29373593	33220745	36933514
23.1	0.00	40497372	43898390	47123294	50159516	52995243	55619462	58022005	60193586	62125840	63811351
23.2	0.00	65243684	66417405	67328110	67972432	68348058	68453739	68289289	67855590	67154584	66189265
23.3	0.00	64963668	63482850	61752872	59780771	57574534	55143064	52496144	49644399	46599254	43372885
23.4	0.00	39978175	36428656	32738462	28922270	24995240	20972956	16871366	12706716	08495487	04254330
23.5	-0.00	00000000	04250711	08481039	12674315	16814029	20883899	24867930	28750479	32516318	36150692
23.6	-0.00	39639376	42968733	46125764	49098164	51874362	54443574	56795842	58922070	60814062	62464552
23.7	-0.00	63867234	65016785	65908883	66540227	66908546	67012608	66852225	66428254	65742587	64798150
23.8	-0.00	63598885	62149737	60456632	58526454	56367015	53987025	51396057	48604508	45623558	42465123
23.9	-0.00	39141811	35666869	32054130	28317960	24473201	20535107	16519292	12441661	08318350	04165661
24.0	0.00	00000000	04162191	08304498	12410596	16464320	20449722	24351140	28153253	31841146	35400367
24.1	0.00	38816983	42077635	45169592	48080796	50799914	53316378	55620432	57703161	59556534	61173429
24.2	0.00	62547664	63674018	64548254	65167131	65528419	65630904	65474397	65059728	64388744	63464305
24.3	0.00	62290266	60871462	59213692	57323691	55209106	52878461	50341129	47607288	44687882	41594579
24.4	0.00	38339725	34936290	31397821	27738387	23972521	20115167	16181613	12187439	08148449	04080613
24.5	-0.00	00000000	04077283	08135156	12157628	16128861	20033231	23855392	27580333	31193441	34680554
24.6	-0.00	38028020	41222746	44252257	47104734	49769071	52234910	54492686	56533660	58349959	59934598
24.7	-0.00	61281516	62385592	63242666	63849559	64204077	64305028	64152216	63746451	63089537	62184267
24.8	-0.00	61034414	59644709	58020829	56169368	54097811	51814509	49328637	46650165	43789813	40759011
24.9	-0.00	37569851	34235039	30767849	27182061	23491917	19712057	15857462	11943397	07985350	03998968

TABLE 6

Sinc Function

x		.0	.01	.02	.03	.04	.05	.06	.07	.08	.09
25.0	0.00	00000000	03995770	07972583	11914767	15806799	19633366	23379426	27030267	30571562	33989431
25.1	0.00	37270489	40401903	43371439	46167513	48779233	51196443	53409763	55410624	57191302	58744948
25.2	0.00	60065613	61148274	61988847	62584209	62932206	63031661	62882378	62485144	61841721	60954843
25.3	0.00	59828200	58466426	56875078	55060616	53030372	50792526	48356069	45730769	42927130	39956352
25.4	0.00	36830287	33561386	30162659	26647613	23030205	19324787	15546043	11708937	07828651	03920526
25.5	-0.00	00000000	03917453	07816381	11681419	15497347	19249152	22922082	26501713	29973995	33325316
25.6	-0.00	36542550	39613112	42525002	45266859	47828000	50198461	52369042	54331336	56077764	57601605
25.7	-0.00	58897022	59959081	60783776	61368037	61709746	61807745	61661835	61272782	60642309	59773090
25.8	-0.00	58668739	57333795	55773702	53994789	52004243	49810079	47421110	44846912	42097780	39184696
25.9	-0.00	36119277	32913733	29580817	26133775	22586292	18952440	15246620	11483505	07677984	03845103
26.0	0.00	00000000	03842146	07666181	11457035	15199779	18879686	22482288	25993433	29399340	32686655
26.1	0.00	35842501	38854530	41710971	44400674	46913157	49238644	51368105	53293290	55006760	56501918
26.2	0.00	57773033	58815261	59624665	60198231	60533875	60630455	60487771	60106570	59488536	58636287
26.3	0.00	57553363	56244213	54714171	52969442	51017070	48864916	46521620	43996572	41299869	38442281
26.4	0.00	35435200	32290603	29020999	25639379	22159169	18594171	14958513	11266590	07533007	03772526
26.5	-0.00	00000000	03769680	07521645	11241109	14913423	18524136	22059052	25504283	28846305	32072013
26.6	-0.00	35168770	38124457	40927519	43567015	46032654	48314842	50404712	52294166	53975899	55443433
26.7	-0.00	56691141	57714264	58508934	59072188	59401977	59497175	59357581	58983922	58377846	57541918
26.8	-0.00	56479607	55195272	53694145	51982311	50066678	47954955	45655618	43177879	40531643	37727474
26.9	-0.00	34776553	31690629	28481976	25163342	21747900	18249196	14681092	11057717	07393404	03702639
27.0	0.00	00000000	03699897	07382459	11033171	14637657	18181731	21651457	25033202	28313692	31480060
27.1	0.00	34519900	37421313	40172956	42764084	45184595	47425066	49476790	51331814	52982965	54423878
27.2	0.00	55649024	56653729	57434192	57987499	58311633	58405484	58268851	57902442	57307871	56487650
27.3	0.00	55445182	54184740	52711456	51031299	49151047	47078265	44821269	42389098	39791474	37038766
27.4	0.00	34141945	31112544	27962611	24704659	21351619	17916788	14413775	10856448	07258881	03635293
27.5	-0.00	00000000	03632651	07248330	10832787	14371904	17851754	21258651	24579209	27800391	30909563
27.6	-0.00	33894539	36743636	39445712	41990214	44367218	46567470	48582416	50404243	52025902	53441143
27.7	-0.00	54644529	55631468	56398222	56941925	57260594	57353133	57219340	56859906	56276412	55471320
27.8	-0.00	54447966	53210544	51764090	50114459	48268306	46233053	44016867	41628619	39077854	36374750
27.9	-0.00	33530082	30555171	27461848	24262398	20969521	17596273	14156017	10662375	07129165	03570354
28.0	0.00	00000000	03567805	07118988	10639551	14115629	17533541	20879844	24141389	27305370	30359375
28.1	0.00	33291434	36090067	38744330	41243854	43578888	45740339	47719802	49509599	51102803	52493268
28.2	0.00	53675655	54645444	55398962	55933390	56246773	56338033	56206966	55854248	55281426	54490915
28.3	0.00	53485988	52270761	50850176	49229982	47416712	45417656	43240829	40894946	38389378	35734124
28.4	0.00	32939764	30017417	26978705	23835694	20600859	17287024	13907317	10475119	07004004	03507694
28.5	-0.00	00000000	03505234	06994181	10453089	13868334	17226474	20514300	23718894	26827669	29828431
28.6	-0.00	32709416	35459342	38067455	40523563	42818084	44942078	46887286	48646160	50211889	51578433
28.7	-0.00	52740539	53693764	54434496	54959958	55268228	55358241	55229794	54883545	54321012	53544563
28.8	-0.00	52557412	51363597	49967973	48376184	46594647	44630521	42491681	40186685	37724742	35115673
28.9	-0.00	32369871	29498265	26512268	23423740	20244935	16988457	13667205	10294326	06883162	03447196
29.0	0.00	00000000	03444820	06873675	10273049	13629554	16929977	20161336	23310932	26366396	29315738
29.1	0.00	32147398	34850285	37413824	39827999	42083388	44171202	46083321	47812321	49351507	50694938
29.2	0.00	51837447	52774665	53503036	54019829	54323149	54411946	54286017	53946006	53393399	52630522
29.3	0.00	51660528	50487384	49115859	47551497	45800601	43870206	41768049	39502540	37082728	34518264
29.4	0.00	31819363	28996764	26061685	23025783	19901101	16700028	13435243	10119668	06766419	03388749
29.5	-0.00	00000000	03386453	06757251	10099107	13398857	16643513	19820312	22916767	25920716	28820373
29.6	-0.00	31604368	34261796	36782261	39155910	41373479	43426325	45306461	47006586	48520114	49841200
29.7	-0.00	50964763	51886502	52602918	53111321	53409848	53497460	53373954	53039959	52496935	51747163
29.8	-0.00	50793740	49640565	48292320	46754455	45033165	43135361	41068651	38841299	36462201	33940843
29.9	-0.00	31287267	28512030	25626163	22641122	19568752	16421229	13211023	09950838	06653570	03332251

$$\frac{\sin 2\pi x}{2\pi x}$$

$$\frac{\sin 2\pi x}{2\pi x}$$

TABLE 6

Sinc Function

x		.0	.01	.02	.03	.04	.05	.06	.07	.08	.09
30.0	0.00	00000000	03330031	06644705	09930957	13175841	16366583	19490633	22535710	25489853	28341470
30.1	0.00	31079378	33692852	36171665	38506127	40687124	42706154	44555359	46227557	47716268	49015742
30.2	0.00	50120975	51027738	51732585	52232868	52526748	52613204	52492031	52163845	51630076	50892967
30.3	0.00	49955560	48821684	47495943	45983693	44291023	42424729	40392290	38201831	35862099	33382421
30.4	0.00	30772674	28043237	25204957	22269103	19247320	16151587	12994164	09787549	06544424	03277606
30.5	-0.00	00000000	03275458	06535847	09768314	12960126	16098718	19171742	22167118	25073080	27878223
30.6	-0.00	30571545	33142496	35581011	37877558	40023169	42009479	43828755	45473929	46938624	48217179
30.7	-0.00	49304673	50196938	50890583	51383000	51672377	51757705	51638780	51316204	50791381	50066514
30.8	-0.00	49144593	48029381	46725405	45237931	43572945	41737132	39737846	37583078	35281430	32842078
30.9	-0.00	30274734	27589610	24797373	21909111	18936277	15890657	12784310	09629533	06438800	03224725
31.0	0.00	00000000	03222645	06430498	09610913	12751361	15839479	18863117	21810389	24669716	27429876
31.1	0.00	30080041	32609829	35009337	37269181	39380537	41335170	43125470	44744479	46185920	47444220
31.2	0.00	48514534	49392758	50075551	50560345	50845354	50929582	50812824	50495670	49979499	49266474
31.3	0.00	48359535	47262384	45979469	44515972	42877780	41071468	39104270	36984049	34719266	32318948
31.4	0.00	29792652	27150424	24402762	21560572	18635128	15638023	12581127	09476537	06336532	03173522
31.5	-0.00	00000000	03171508	06328491	09458504	12549215	15588457	18564272	21464960	24279126	26995721
31.6	-0.00	29604091	32094014	34455742	36680038	38758215	40682165	42444398	44038061	45456975	46695653
31.7	-0.00	47749320	48613938	49286214	49763618	50044388	50127541	50012874	49700963	49193163	48491600
31.8	-0.00	47599165	46519498	45256976	43816695	42204448	40426704	38490581	36403816	34174735	31812223
31.9	-0.00	29325683	26725003	24020513	21222950	18343407	15393296	12384301	09328327	06237462	03123920
32.0	0.00	00000000	03121969	06229670	09310853	12353379	15345267	18274748	21130302	23900710	26575096
32.1	0.00	29142968	31594263	33919382	36109231	38155256	40049472	41784503	43353603	44750683	45970340
32.2	0.00	47007871	47859298	48521375	48991610	49268265	49350370	49237721	48930883	48431187	47740724
32.3	0.00	46862336	45799605	44556837	43139047	41551937	39801871	37895856	35841508	33647022	31321142
32.4	0.00	28873126	26312707	23650055	20895738	18060679	15156112	12193538	09184682	06141442	03075845
32.5	-0.00	00000000	03073953	06133888	09167741	12163560	15109549	17994116	20805919	23533910	26167377
32.6	-0.00	28695990	31109837	33399465	35555918	37570769	39436157	41144814	42690095	44066003	45267214
32.7	-0.00	46289097	47127728	47779912	48243189	48515848	48596929	48486229	48184302	47692456	47012747
32.8	-0.00	46147971	45101653	43878031	42482041	40919294	39196059	37319230	35296307	33135358	30844992
32.9	-0.00	28434325	25912939	23290850	20578463	17786534	14926125	12008563	09045394	06048333	03029228
33.0	0.00	00000000	03027392	06041007	09028962	11979487	14880963	17721973	20491345	23178198	25771980
33.1	0.00	28262516	30640042	32895246	35019306	37003920	38841344	40524416	42046590	43401958	44585274
33.2	0.00	45591972	46418187	47060768	47517291	47786067	47866148	47757332	47460162	46975923	46306638
33.3	0.00	45455059	44424654	43219597	41844746	40305628	38608412	36759889	34767444	32639022	30383102
33.4	0.00	28008661	25525137	22942393	20270679	17520587	14703014	11829117	08910267	05958006	02984002
33.5	-0.00	00000000	02982221	05950896	08894322	11800902	14659190	17457939	20186142	22833079	25388355
33.6	-0.00	27841943	30184225	32406025	34498650	36453921	38264206	39922449	41422198	42757630	43923575
33.7	-0.00	44915533	45729694	46362951	46812914	47077916	47157020	47050026	46757465	46280601	45621426
33.8	-0.00	44782647	43767679	42580632	41226290	39710095	38038125	36217068	34254196	32157336	29934842
33.9	-0.00	27595554	25148771	22604209	19971965	17262476	14486475	11654954	08779118	05870337	02940106
34.0	0.00	00000000	02938377	05863434	08763639	11627563	14443930	17201657	19889897	22498086	25015982
34.1	0.00	27433703	29741770	31931142	33993249	35920033	37703969	39338104	40816078	42132153	43281230
34.2	0.00	44258873	45061326	45685526	46129114	46390446	46468597	46363365	46075273	45605563	44956197
34.3	0.00	44129838	43129852	41960285	40625849	39131905	37484441	35690044	33755880	31689661	29499616
34.4	0.00	27194456	24783343	22275851	19681928	17011859	14276221	11485846	08651773	05785210	02897484
34.5	-0.00	00000000	02895805	05778506	08636740	11459243	14234901	16952790	19602221	22172782	24654375
34.6	-0.00	27037263	29312100	31469976	33502443	35401556	37159900	38770618	40227442	41524711	42657401
34.7	-0.00	43621137	44412215	45027613	45465004	45722765	45799984	45696458	45412700	44949934	44310088
34.8	-0.00	43495789	42510349	41357754	40042647	38570311	36946644	35178139	33271856	31235394	29076864
34.9	-0.00	26804851	24428382	21956895	19400194	16768415	14071983	11321575	08528070	05702517	02856079

TABLE 6

Sinc Function

x		.0	.01	.02	.03	.04	.05	.06	.07	.08	.09
35.0	0.00	00000000	02854448	05696003	08513463	11295726	14031835	16711022	19322748	21856750	24303073
35.1	0.00	26652116	28894668	31021941	33025608	34897835	36631310	38219273	39655542	40934536	42051300
35.2	0.00	43001519	43781539	44388379	44819744	45074032	45150338	45048465	44768914	44312889	43682289
35.3	0.00	42879701	41908390	40772281	39475952	38024608	36424060	34680710	32801516	30793967	28666058
35.4	0.00	26426251	24083446	21646945	19126412	16531840	13873507	11161936	08407855	05622154	02815842
35.5	−0.00	00000000	02814256	05615823	08393657	11136811	13834482	16476052	19051133	21549600	23961642
35.6	−0.00	26277789	28488958	30586484	32562156	34408247	36117547	37683388	39099675	40360902	41462181
35.7	−0.00	42399256	43168524	43767041	44192544	44443449	44518865	44418593	44143125	43693648	43072031
35.8	−0.00	42280823	41323240	40203154	38925074	37494130	35916054	34197153	32344288	30364843	28266698
35.9	−0.00	26058197	23748116	21345623	18860250	16301848	13680551	11006737	08290982	05544025	02776722
36.0	0.00	00000000	02775180	05537869	08277175	10982304	13642603	16247599	18787047	21250964	23629671
36.1	0.00	25913831	28094483	30163083	32111531	33932206	35617996	37162324	38559176	39803123	40889341
36.2	0.00	41813631	42572438	43162858	43582655	43830267	43904812	43806091	43534590	43091475	42478589
36.3	0.00	41698442	40754206	39649697	38389359	36978251	35422023	33726895	31899632	29947514	27878312
36.4	0.00	25700255	23421995	21052575	18601394	16078167	13492889	10855794	08177314	05468038	02738674
36.5	−0.00	00000000	02737174	05462049	08163882	10832026	13455973	16025394	18530183	20960492	23306773
36.6	−0.00	25559816	27710784	29751244	31673208	33469157	35132075	36655473	38033417	39260550	40332113
36.7	−0.00	41243963	41992590	42575128	42989371	43233774	43307467	43210252	42942605	42505675	41901277
36.8	−0.00	41131887	40200631	39111271	37868189	36476374	34941399	33269395	31467036	29541501	27500455
36.9	−0.00	25352013	23104710	20767464	18349547	15860542	13310306	10708935	08066720	05394106	02701655
37.0	0.00	00C00000	02700195	05388278	08053649	10685806	13274381	15809186	18280248	20677853	22992581
37.1	0.00	25215345	27337424	29350500	31246690	33018576	34659234	36162261	37521802	38732571	39789869
37.2	0.00	40689609	41428325	42003190	42412022	42653299	42726159	42630405	42366504	41935588	41339447
37.3	0.00	40580522	39661893	38587272	37360980	35987939	34473642	32824141	31046016	29146350	27132704
37.4	0.00	25013082	22795906	20489973	18104429	15648729	13132599	10565997	07959077	05322146	02665623
37.5	−0.00	00000000	02664202	05316472	07946353	10543480	13097625	15598733	18036965	20402735	22686747
37.6	−0.00	24880034	26973991	28960408	30831507	32579966	34198952	35682146	37023770	38218604	39262012
37.7	−0.00	40149959	40879024	41446413	41849976	42088205	42160250	42065913	41805655	41380591	40792484
37.8	−0.00	40043742	39137404	38077128	36867179	35512411	34018244	32390647	30636113	28761630	26774658
37.9	−0.00	24683095	22495247	20219799	17865773	15442499	12959574	10426824	07854270	05252081	02630540
38.0	0.00	00000000	02629156	05246555	07841878	10404896	12925514	15393810	17800073	20134842	22388943
38.1	0.00	24553525	26620094	28580550	30427212	32152856	33750735	35214613	36538784	37718098	38747977
38.2	0.00	39624436	40344098	40904205	41302631	41537889	41609135	41516175	41259461	40840092	40259806
38.3	0.00	39520978	38626605	37580297	36386261	35049286	33574721	31968454	30236894	28386935	26425939
38.4	0.00	24361700	22202417	19956658	17633328	15241634	12791049	10291270	07752187	05183837	02596368
38.5	−0.00	00000000	02595020	05178454	07740115	10269908	12757868	15194202	17569323	19873893	22098855
38.6	−0.00	24235474	26275364	28210527	30033383	31736799	33314115	34759173	36066341	37230532	38247228
38.7	−0.00	39112492	39822991	40376000	40769419	41001778	41072243	40980621	40727356	40313531	39740860
38.8	−0.00	39011687	38128968	37096264	35917728	34598085	33142614	31557126	29847945	28021876	26086186
38.9	−0.00	24048568	21917112	19700277	17406853	15045928	12626850	10159195	07652723	05117343	02563073
39.0	0.00	00000000	02561759	05112097	07640959	10138377	12594515	14999704	17344479	19619621	21816189
39.1	0.00	23925557	25939447	27849963	29649619	31331372	32888647	34315363	35605959	36755410	37759256
39.2	0.00	38613609	39315174	39861263	40249798	40479329	40549030	40458708	40208800	39800375	39235123
39.3	0.00	38515355	37643990	36624542	35461109	34158354	32721488	31156248	29468875	27666088	25755059
39.4	0.00	23743383	21639047	19450400	17186122	14855183	12466814	10030468	07555780	05052534	02530621
39.5	−0.00	00000000	02529340	05047420	07544311	10010173	12435293	14810122	17125317	19371773	21540663
39.6	−0.00	23623467	25612012	27498500	29275539	30936173	32473910	33882744	35157182	36292263	37283579
39.7	−0.00	38127291	38820146	39359484	39743257	39970027	40038979	39949921	39703284	39300119	38742095
39.8	−0.00	38031494	37171194	36164665	35015953	33729659	32310929	30765427	29099313	27319222	25432233
39.9	−0.00	23445847	21367949	19206783	16970919	14669214	12310784	09904961	07461262	04989346	02498980

$$\frac{\sin 2\pi x}{2\pi x}$$

TABLE 6

Sinc Function

x		.0	.01	.02	.03	.04	.05	.06	.07	.08	.09
40.0	0.00	00000000	02497731	04984359	07450078	09885171	12280045	14625273	16911624	19130109	21272009
40.1	0.00	23328910	25292740	27155797	28910780	30550820	32069503	33460897	34719577	35840642	36819737
40.2	0.00	37653071	38337428	38870182	39249306	39473382	39541601	39453772	39210320	38812282	38261305
40.3	0.00	37559639	36710127	35716195	34581835	33311592	31910546	30384289	28738905	26980945	25117400
40.4	0.00	23155675	21103559	18969193	16761039	14487844	12158611	09782557	07369079	04927718	02468121
40.5	−0.00	00000000	02466902	04922854	07358170	09763252	12128627	14444981	16703199	18894401	21009974
40.6	−0.00	23041608	24981329	26821530	28554999	30174949	31675044	33049425	34292732	35400122	36367295
40.7	−0.00	37190503	37866568	38392896	38767483	38988927	39056428	38969796	38729448	38336408	37792302
40.8	−0.00	37099349	36260359	35278711	34158349	32903762	31519964	30012480	28387316	26650943	24810266
40.9	−0.00	22872599	20845633	18737409	16556286	14310904	12010154	09663141	07279146	04867595	02438015
41.0	0.00	00000000	02436826	04862848	07268502	09644304	11980897	14269080	16499849	18664430	20754316
41.1	0.00	22761296	24677494	26495393	28207868	29808214	31290171	32647950	33876255	34970301	35925837
41.2	0.00	36739162	37407134	37927189	38297346	38516219	38583017	38497549	38260227	37872062	37334657
41.3	0.00	36650205	35821478	34851815	33745110	32505797	31138828	29649659	28044225	26328916	24510553
41.4	0.00	22596360	20593935	18511221	16356476	14138234	11865279	09546605	07191382	04808921	02408634
41.5	−0.00	00000000	02407473	04804288	07180993	09528220	11836723	14097411	16301390	18439990	20504805
41.6	−0.00	22487723	24380961	26177092	27869075	29450286	30914539	32256112	33469772	34550791	35494968
41.7	−0.00	36298644	36958714	37472644	37838476	38054836	38120944	38036611	37802241	37418830	36887963
41.8	−0.00	36211805	35393093	34435126	33341750	32117343	30766799	29295507	27709329	26014579	24217994
41.9	−0.00	22326713	20348242	18290429	16161431	13969681	11723857	09432847	07105709	04751645	02379953
42.0	0.00	00000000	02378820	04747121	07095566	09414896	11695977	13929825	16107649	18220884	20261222
42.1	0.00	22220648	24091470	25866348	27538324	29100852	30547818	31873568	33072929	34141228	35074312
42.2	0.00	35868565	36520919	37028866	37390471	37604377	37669809	37586580	37355088	36976318	36451832
42.3	0.00	35783770	34974834	34028284	32947919	31738064	30403555	28949715	27382337	25707658	23932338
42.4	0.00	22063427	20108343	18074842	15970983	13805099	11585767	09321768	07022054	04695717	02351947
42.5	−0.00	00000000	02350840	04691299	07012147	09304237	11558539	13766175	15918459	18006923	20023358
42.6	−0.00	21959842	23808772	25562894	27215332	28759613	30189696	31499991	32685385	33741260	34663510
42.7	−0.00	35448559	36093373	36595476	36952951	37164457	37229226	37147074	36918391	36544149	36025893
42.8	−0.00	35365735	34566345	33630943	32563283	31367638	30048787	28611990	27062972	25407895	23653341
42.9	−0.00	21806277	19874035	17864277	15784971	13644351	11450892	09213274	06940345	04641090	02324592
43.0	0.00	00000000	02323511	04636774	06930668	09196149	11424293	13606327	15733661	17797929	19791015
43.1	0.00	21705088	23532633	25266479	26899829	28426285	29839873	31135070	32306820	33350555	34262219
43.2	0.00	35038274	35675723	36172113	36525551	36734710	36798831	36717727	36491786	36121967	35609794
43.3	0.00	34957355	34167288	33242774	32187524	31005760	29702204	28282055	26750971	25115043	23380774
43.4	0.00	21555053	19645124	17658563	15603242	13487303	11319121	09107277	06860516	04587719	02297867
43.5	−0.00	00000000	02296810	04583503	06851060	09090543	11293130	13450147	15553105	17593731	19564002
43.6	−0.00	21456176	23262825	24976858	26591556	28100594	29498065	30778507	31936922	32968795	33870113
43.7	−0.00	34637379	35267627	35758434	36107926	36314789	36378273	36298192	36074928	35709427	35203197
43.8	−0.00	34558298	33777339	32863464	31820338	30652136	29363524	27959642	26446082	24828864	23114418
43.9	−0.00	21309551	19421426	17457532	15425650	13333828	11190349	09003691	06782502	04535562	02271748
44.0	0.00	00000000	02270716	04531441	06773260	08987335	11164945	13297513	15376646	17394165	19342137
44.1	0.00	21212909	22999134	24693802	26290270	27782282	29163999	30430019	31575400	32595676	33486880
44.2	0.00	34245553	34868762	35354109	35699742	35904360	35967219	35888135	35667486	35306204	34805780
44.3	0.00	34168250	33396191	32492712	31461435	30306487	29032481	27644498	26148064	24549134	22854061
44.4	0.00	21069578	19202766	17261026	15252055	13183808	11064473	08902435	06706243	04484578	02246217
44.5	−0.00	00000000	02245208	04480549	06697207	08886445	11039637	13148304	15204146	17199075	19125248
44.6	−0.00	20975096	22741354	24417090	25995734	27471100	28837414	30089333	31221970	32230908	33112223
44.7	−0.00	33862493	34478819	34958826	35300684	35503104	35565350	35487240	35269144	34911986	34417236
44.8	−0.00	33786908	33023549	32130232	31110538	29968547	28708819	27336378	25856688	24275636	22599505
44.9	−0.00	20834951	18988974	17068895	15082323	13037126	10941398	08803431	06631679	04434727	02221254

TABLE 6

Sinc Function

x		.0	.01	.02	.03	.04	.05	.06	.07	.08	.09
45.0	0.00	00000000	02220267	04430787	06622843	08787794	10917110	13002406	15035473	17008314	18913170
45.1	0.00	20742556	22489288	24146511	25707724	27166812	28518063	29756192	30876365	31874214	32745856
45.2	0.00	33487908	34097500	34572285	34910449	35110718	35172363	35095203	34879602	34526473	34037270
45.3	0.00	33413984	32659131	31775750	30767381	29638060	28392294	27035051	25571735	24008166	22350557
45.4	0.00	20605491	18779891	16880995	14916328	12893671	10821030	08706605	06558756	04385973	02196839
45.5	-0.00	00000000	02195874	04382119	06550113	08691310	10797274	12859711	14870502	16821737	18705743
45.6	-0.00	20515116	22242749	23881862	25426027	26869192	28205707	29430347	30538327	31525328	32387508
45.7	-0.00	33121520	33724524	34194198	34528747	34726910	34787966	34711732	34498571	34149382	33665603
45.8	-0.00	33049202	32302668	31429004	30431713	29314782	28082673	26740295	25292993	23746525	22107034
45.9	-0.00	20381030	18575361	16697186	14753947	12753340	10703282	08611885	06487418	04338278	02172955
46.0	0.00	00000000	02172011	04334508	06478962	08596921	10680040	12720113	14709112	16639210	18502817
46.1	0.00	20292609	22001557	23622952	25150436	26578022	27900120	29111560	30207611	31183997	32036918
46.2	0.00	32763062	33359619	33824291	34155302	34351403	34411880	34336551	34125775	33780439	33301965
46.3	0.00	32692299	31953903	31089745	30103289	28998481	27779731	26451896	25020263	23490525	21868760
46.4	0.00	20161407	18375239	16517337	14595063	12616030	10588069	08519205	06417616	04291610	02149585
46.5	-0.00	00000000	02148661	04287920	06409341	08504561	10565324	12583514	14551187	16460601	18304246
46.6	-0.00	20074877	21765539	23369596	24880755	26293094	27601083	28799606	29883981	30849978	31693837
46.7	-0.00	32412280	33002526	33462301	33789848	33983930	34043838	33969394	33760949	33419383	32946099
46.8	-0.00	32343023	31612588	30757731	29781879	28688933	27483256	26169652	24753352	23239986	21635568
46.9	-0.00	19946467	18179382	16341321	14439565	12481645	10475310	08428498	06349300	04245935	02126712
47.0	0.00	00000000	02125808	04242323	06341200	08414163	10453046	12449818	14396618	16285786	18109892
47.1	0.00	19861768	21534532	23121616	24616796	26014211	27308389	28494267	29567212	30523039	31358026
47.2	0.00	32068929	32652997	33107978	33432132	33624235	33683586	33610006	33403842	33065963	32597758
47.3	0.00	32001130	31278487	30432734	29467260	28385924	27193042	25893368	24492075	22994735	21407297
47.4	0.00	19736061	17987657	16169017	14287345	12350093	10364928	08339702	06282423	04201222	02104321
47.5	-0.00	00000000	02103435	04197686	06274492	08325668	10343130	12318932	14245297	16114645	17919622
47.6	-0.00	19653136	21308376	22878844	24358379	25741182	27021837	28195334	29257088	30202957	31029256
47.7	-0.00	31732777	32310794	32761080	33081910	33272075	33330878	33258142	33054209	32719940	32256706
47.8	-0.00	31666390	30951375	30114533	29159218	28089248	26908893	25622857	24236257	22754606	21183792
47.9	-0.00	19530048	17799934	16000309	14138301	12221285	10256847	08252757	06216940	04157441	02082397
48.0	0.00	00000000	02081529	04153978	06209174	08239014	10235501	12190770	14097125	15947063	17733309
48.1	0.00	19448842	21086921	22641117	24105331	25473825	26741237	27902609	28953402	29889518	30707309
48.2	0.00	31403599	31975689	32421375	32738951	32927215	32985480	32913570	32711821	32381084	31922716
48.3	0.00	31338581	30631034	29802918	28857550	27798710	26630621	25357939	23985727	22519441	20964906
48.4	0.00	19328291	17616088	15835084	13992335	12095137	10150997	08167607	06152808	04114563	02060924
48.5	-0.00	00000000	02060075	04111171	06145201	08154146	10130089	12065248	13952003	15782931	17550830
48.6	-0.00	19248751	2C870022	22408280	23857487	25211964	26466404	27615899	28655956	29582518	30391974
48.7	-0.00	31081180	31647464	32088644	32403029	32589431	32647168	32576064	32376452	32049174	31595572
48.8	-0.00	31017489	30317255	29497685	28562060	27514120	26358046	25098443	23740323	22289087	20750497
48.9	-0.00	19130660	17436001	15673238	13849352	11971566	10047310	08084196	06089986	04072561	02039890
49.0	0.00	00000000	02039058	04069238	06082534	08071008	10026826	11942283	13809839	15622143	17372068
49.1	0.00	19052735	20657540	22180182	23614688	24955432	26197163	27335021	28364560	29281760	30083050
49.2	0.00	30765314	31325909	31762672	32073930	32258507	32315724	32245410	32047891	31724000	31275066
49.3	0.00	30702910	30009841	29198641	28272560	27235299	26090994	24844204	23499891	22063397	20540429
49.4	0.00	18937030	17259559	15514666	13709261	11850494	09945719	08002472	06028434	04031407	02019281
49.5	-0.00	00000000	02018465	04028151	06021131	07989549	09925647	11821800	13670542	15464598	17196911
49.6	-0.00	18860671	20449341	21956682	23376780	24704068	25933344	27059799	28079030	28987057	29780343
49.7	-0.00	30455804	31010822	31443257	31751450	31934235	31990943	31921400	31725931	31405358	30960996
49.8	-0.00	30394648	29708597	28905600	27988870	26962071	25829299	24595064	23264279	21842232	20334572
49.9	-0.00	18747280	17086653	15359271	13571977	11731847	09846163	07922383	05968113	03991077	01999084

$$\frac{\sin 2\pi x}{2\pi x}$$

TABLE 6

Sinc Function

x		.0	.01	.02	.03	.04	.05	.06	.07	.08	.09
50.0	0.00	00000000	01998285	03987886	05960956	07909717	09826490	11703724	13534028	15310199	17025251
50.1	0.00	18672441	20245296	21737641	23143619	24457717	25674786	26790064	27799191	28698226	29483667
50.2	0.00	30152459	30702011	31130201	31435389	31616419	31672626	31603837	31410376	31093053	30653171
50.3	0.00	30092514	29413342	28618382	27710817	26694272	25572801	24350872	23033345	21625458	20132800
50.4	0.00	18561295	16917176	15206957	13437414	11615552	09748579	07843881	05908988	03951546	01979287
50.5	-0.00	00000000	01978504	03948417	05901972	07831465	09729294	11587983	13400213	15158853	16856984
50.6	-0.00	18487930	20045283	21522927	22915062	24216231	25421333	26525654	27524875	28415094	29192843
50.7	-0.00	29855098	30399290	30823319	31125559	31304865	31360580	31292531	31101036	30786899	30351407
50.8	-0.00	29796328	29123898	28336816	27438233	26431739	25321348	24111481	22806951	21412943	19934993
50.9	-0.00	18378964	16751028	15057635	13305494	11501539	09652911	07766920	05851023	03912790	01959879
51.0	0.00	00000000	01959110	03909722	05844143	07754746	09634002	11474509	13269019	15010470	16692011
51.1	0.00	18307031	19849184	21312413	22690976	23979466	25172836	26266412	27255919	28137495	28907701
51.2	0.00	29563544	30102480	30522427	30821776	30999393	31054623	30987298	30797729	30486714	30055527
51.3	0.00	29505915	28840094	28060736	27170960	26174321	25074791	23876751	22584964	21204565	19741035
51.4	0.00	18200181	16588112	14911217	13176138	11389744	09559102	07691455	05794184	03874787	01940847
51.5	-0.00	00000000	01940093	03871779	05787437	07679516	09540559	11363235	13140368	14864963	16530235
51.6	-0.00	18129637	19656884	21105977	22471230	23747287	24929149	26012188	26992169	27865267	28628076
51.7	-0.00	29277630	29811409	30227353	30523866	30699824	30754578	30687961	30500282	30192327	29765360
51.8	-0.00	29221109	28561769	27789984	26908844	25921868	24832990	23646547	22367257	21000203	19550815
51.9	-0.00	18024842	16428334	14767619	13049274	11280101	09467100	07617441	05738438	03837515	01922181
52.0	0.00	00000000	01921442	03834564	05731821	07605731	09448911	11254100	13014188	14722250	16371565
52.1	0.00	17955648	19468275	20903503	22255699	23519561	24690135	25762838	26733475	27598256	28353808
52.2	0.00	28997193	29525914	29937930	30231660	30405989	30460276	30394353	30208525	29903571	29480742
52.3	0.00	28941749	28288764	27524407	26651737	25674238	24595808	23420739	22153706	20799743	19364226
52.4	0.00	17852849	16271605	14626761	12924829	11172548	09376851	07544839	05683755	03800953	01903872
52.5	-0.00	00000000	01903146	03798059	05677263	07533351	09359007	11147040	12890409	14582252	16215912
52.6	-0.00	17784967	19283250	20704876	22044264	23296161	24455661	25518223	26479692	27336313	28084745
52.7	-0.00	28722077	29245835	29653997	29944995	30117726	30171553	30106309	29922297	29620286	29201515
52.8	-0.00	28667679	28020929	27263858	26399496	25431295	24363113	23199204	21944195	20603074	19181164
52.9	-0.00	17684107	16117838	14488564	12802735	11067027	09288306	07473607	05630104	03765082	01885907
53.0	0.00	00000000	01885196	03762241	05623734	07462335	09270798	11041998	12768962	14444890	16063191
53.1	0.00	17617501	19101709	20509988	21836808	23076965	24225598	25278210	26230683	27079296	27820741
53.2	0.00	28452133	28971020	29375399	29663716	29834877	29888252	29823674	29641442	29342318	28927528
53.3	0.00	28398752	27758117	27008195	26151986	25192906	24134780	22981820	21738610	20410089	19001532
53.4	0.00	17518526	15966951	14352954	12682927	10963481	09201419	07403708	05577457	03729881	01868278
53.5	-0.00	00000000	01867580	03727093	05571205	07392646	09184236	10938918	12649781	14310093	15913320
53.6	-0.00	17453158	18923555	20318735	21633220	22861855	23999824	25042669	25986313	26827067	27561655
53.7	-0.00	28187215	28701322	29101986	29387671	29557292	29610222	29546296	29365810	29069519	28658635
53.8	-0.00	28134823	27500190	26757283	25909073	24958946	23910688	22768472	21536841	20220686	18825233
53.9	-0.00	17356016	15818862	14219859	12565340	10861854	09116141	07335105	05525785	03695332	01850976
54.0	0.00	00000000	01850291	03692596	05519649	07324246	09099275	10837744	12532805	14177788	15766220
54.1	0.00	17291854	18748693	20131015	21433394	22650719	23778219	24811478	25746454	26579494	27307349
54.2	0.00	27927186	28436598	28833617	29116717	29284824	29337317	29274030	29095257	28801745	28394695
54.3	0.00	27875754	27247012	26510990	25670631	24729290	23690718	22559049	21338783	20034766	18652175
54.4	0.00	17196494	15673494	14089210	12449913	10762094	09032430	07267761	05475062	03661418	01833992
54.5	-0.00	00000000	01833319	03658731	05469038	07257100	09015872	10738424	12417973	14047907	15621814
54.6	-0.00	17133503	18577033	19946733	21237225	22443446	23560669	24584516	25510982	26336448	27057693
54.7	-0.00	27671910	28176713	28570152	28850714	29017334	29069396	29006736	28829644	28538859	28135572
54.8	-0.00	27621413	26998454	26269190	25436538	24503823	23474759	22353444	21144334	19852233	18482270
54.9	-0.00	17039878	15530775	13960939	12336588	10664150	08950242	07201642	05425261	03628120	01817316

TABLE 6

Sinc Function

x		.0	.01	.02	.03	.04	.05	.06	.07	.08	.09
55.0	0.00	00000000	01816655	03625482	05419346	07191175	08933984	10640909	12305226	13920385	15480030
55.1	0.00	16978027	18408488	19765794	21044615	22239933	23347063	24361668	25279779	26097807	26812561
55.2	0.00	27421258	27921536	28311458	28589527	28754686	28806325	28744279	28568836	28280729	27881135
55.3	0.00	27371672	26754389	26031760	25206676	24282429	23262702	22151552	20953397	19672997	18315432
55.4	0.00	16886088	15390630	13834984	12225307	10567973	08869537	07136716	05376359	03595422	01800941
55.5	−0.00	00000000	01800292	03592832	05370550	07126436	08853570	10545148	12194508	13795156	15340796
55.6	−0.00	16825347	18242974	19588108	20855467	22040078	23137296	24142825	25052729	25863452	26571830
55.7	−0.00	27175107	27670938	28057407	28333027	28496750	28547972	28486529	28312706	28027227	27631260
55.8	−0.00	27126406	26514697	25798584	24980931	24065001	23054441	21953274	20765878	19496968	18151580
55.9	−0.00	16735050	15252993	13711280	12116016	10473515	08790274	07072949	05328330	03563309	01784858
56.0	0.00	00000000	01784221	03560765	05322624	07062852	08774591	10451096	12085764	13672161	15204044
56.1	0.00	16675388	18080410	19413588	20669688	21843782	22931265	23927878	24829720	25633268	26335384
56.2	0.00	26933336	27424799	27807875	28081088	28243401	28294212	28233361	28061127	27778229	27385823
56.3	0.00	26885497	26279262	25569549	24759194	23851431	22849876	21758515	20581685	19324061	17990633
56.4	0.00	16586689	15117795	13589769	12008662	10380730	08712415	07010313	05281151	03531764	01769060
56.5	−0.00	00000000	01768434	03529265	05275546	07000393	08697008	10358706	11978943	13551339	15069709
56.6	−0.00	16528079	17920717	19242150	20487191	21650952	22728871	23716725	24610648	25407145	26103109
56.7	−0.00	26695828	27183001	27562742	27833590	27994517	28044924	27984652	27813979	27533616	27144708
56.8	−0.00	26648828	26047971	25344544	24541358	23641619	22648910	21567181	20400732	19154194	17832515
56.9	−0.00	16440936	14984973	13470393	11903193	10289575	08635923	06948775	05234801	03500773	01753539
57.0	0.00	00000000	01752924	03498317	05229294	06939030	08620786	10267936	11873993	13432635	14937727
57.1	0.00	16383350	17763820	19073714	20307887	21461497	22530018	23509266	24395407	25184977	25874895
57.2	0.00	26462473	26945429	27321893	27590417	27749980	27799990	27740288	27571147	27293274	26907802
57.3	0.00	26416291	25820716	25123464	24327322	23435466	22451448	21379183	20222932	18987288	17677153
57.4	0.00	16297723	14854465	13353096	11799561	10200007	08560763	06888309	05189257	03470321	01738289
57.5	−0.00	00000000	01737684	03467907	05183845	06878732	08545887	10178743	11770867	13315991	14808037
57.6	−0.00	16241133	17609647	18908201	20131695	21275328	22334615	23305405	24183898	24966661	25650637
57.7	−0.00	26233162	26711973	27085217	27351457	27509679	27559297	27500154	27332518	27057091	26674995
57.8	−0.00	26187776	25597392	24906209	24116988	23232877	22257399	21194433	20048205	18823265	17524474
57.9	−0.00	16156982	14726210	13237824	11697718	10111985	08486899	06828886	05144499	03440394	01723301
58.0	0.00	00000000	01722707	03438022	05139180	06819474	08472279	10091085	11669516	13201357	14680579
58.1	0.00	16101365	17458127	18745536	19958534	21092362	22142572	23105049	23976025	24752097	25430233
58.2	0.00	26007791	26482528	26852606	27116599	27273504	27322737	27264141	27097985	26824961	26446183
58.3	0.00	25963181	25377898	24692678	23910259	23033761	22066676	21012850	19876471	18662051	17374410
58.4	0.00	16018652	14600151	13124526	11597617	10025469	08414300	06770480	05100507	03410979	01708569
58.5	−0.00	00000000	01707985	03408647	05095278	06761227	08399929	10004925	11569896	13088678	14555297
58.6	−0.00	15963981	17309193	18585646	19788327	20912516	21953803	22908108	23771696	24541189	25213584
58.7	−0.00	25786260	26256992	26623956	26885741	27041350	27090203	27032146	26867442	26596780	26221262
58.8	−0.00	25742406	25162136	24482778	23707044	22838029	21879193	20834351	19707654	18503576	17226894
58.9	−0.00	15882670	14476232	13013150	11499216	09940421	08342932	06713064	05057260	03382062	01694087
59.0	0.00	00000000	01693513	03379770	05052120	06703968	08328803	09920224	11471962	12977908	14432135
59.1	0.00	15828922	17162778	18428460	19620998	20735710	21768225	22714497	23570820	24333846	25000595
59.2	0.00	25568471	26035264	26399168	26658781	26813114	26861594	26804065	26640789	26372448	26000135
59.3	0.00	25525353	24950012	24276416	23507254	22645595	21694870	20658860	19541681	18347770	17081862
59.4	0.00	15748978	14354399	12903648	11402470	09856804	08272764	06656614	05014741	03353632	01679849
59.5	−0.00	00000000	01679284	03351378	05009686	06647670	08258872	09836945	11375672	12868996	14311039
59.6	−0.00	15696129	17018819	18273911	19456475	20561870	21585759	22524131	23373310	24129976	24791175
59.7	−0.00	25354329	25817250	26178143	26435620	26588699	26636811	26579800	26417929	26151869	25782706
59.8	−0.00	25311931	24741435	24073504	23310804	22456377	21513626	20486300	19378480	18194565	16939252
59.9	−0.00	15617517	14234599	12795974	11307338	09774582	08203767	06601105	04972930	03325676	01665848

$$\frac{\sin 2\pi x}{2\pi x}$$

TABLE 6

Sinc Function

x		.0	.01	.02	.03	.04	.05	.06	.07	.08	.09
60.0	0.00	00000000	01665293	03323459	04967960	06592309	08190105	09755052	11280985	12761897	14191959
60.1	0.00	15565545	16877255	18121932	19294688	20390920	21406326	22336929	23179083	23929495	24585234
60.2	0.00	25143745	25602856	25960789	26216165	26368009	26415758	26359258	26198765	25934949	25568883
60.3	0.00	25102047	24536316	23873955	23117610	22270295	21335386	20316599	19217983	18043898	16799003
60.4	0.00	15488233	14116783	12690083	11213781	09693720	08135911	06546514	04931811	03298182	01652078
60.5	-0.00	00000000	01651532	03296002	04926923	06537863	08122474	09674512	11187862	12656566	14074845
60.6	-0.00	15437117	16738027	17972461	19135570	20222789	21229852	22152813	22988057	23732317	24382686
60.7	-0.00	24936630	25391994	25747014	26000323	26150953	26198345	26142345	25983209	25721598	25358578
60.8	-0.00	24895616	24334571	23677688	22927592	22087272	21160075	20149686	19060122	17895706	16661057
60.9	-0.00	15361072	14000900	12585929	11121759	09614185	08069169	06492819	04891367	03271139	01638534
61.0	0.00	00000000	01637997	03268994	04886558	06484309	08055951	09595290	11096263	12552960	13959647
61.1	0.00	15310790	16601077	17825435	18979055	20057408	21056264	21971707	22800154	23538362	24183449
61.2	0.00	24732900	25184577	25536732	25788006	25937441	25984481	25928973	25771170	25511728	25151705
61.3	0.00	24692552	24136115	23484621	22740672	21907232	20987621	19985494	18904833	17749928	16525359
61.4	0.00	15235982	13886905	12483471	11031235	09535944	08003512	06439998	04851580	03244535	01625211
61.5	-0.00	00000000	01624682	03242426	04846849	06431626	07990509	09517356	11006152	12451036	13846320
61.6	-0.00	15186514	16466349	17680794	18825079	19894710	20885491	21793539	22615298	23347552	23987441
61.7	-0.00	24532471	24980521	25329856	25579128	25727387	25774080	25719056	25562564	25305256	24948179
61.8	-0.00	24492774	23940871	23294678	22556775	21730104	20817955	19823956	18752054	17606505	16391853
61.9	-0.00	15112912	13774751	12382668	10942173	09458967	07938916	06388029	04812435	03218361	01612102
62.0	0.00	00000000	01611582	03216286	04807781	06379791	07926121	09440677	10917493	12350754	13734818
62.1	0.00	15064240	16333791	17538483	18673581	19734630	20717466	21618237	22433415	23159810	23794585
62.2	0.00	24335265	24779746	25126305	25373607	25520708	25567059	25512510	25357308	25102099	24747921
62.3	0.00	24296203	23748760	23107782	22375829	21555817	20651011	19665008	18601725	17465382	16260487
62.4	0.00	14991815	13664394	12283479	10854538	09383223	07875353	06336892	04773917	03192606	01599203
62.5	-0.00	00000000	01598692	03190564	04769337	06328786	07862763	09365224	10830251	12252074	13625097
62.6	-0.00	14943918	16203351	17398444	18524503	19577106	20552123	21445733	22254434	22975063	23604805
62.7	-0.00	24141203	24582172	24926000	25171363	25317323	25363338	25309255	25155322	24902178	24550852
62.8	-0.00	24102762	23559708	22923862	22197762	21384303	20486723	19508589	18453787	17326504	16131210
62.9	-0.00	14872644	13555791	12185868	10768295	09308682	07812801	06286567	04736011	03167260	01586509
63.0	0.00	00000000	01586006	03165250	04731503	06278589	07800410	09290968	10744392	12154959	13517116
63.1	0.00	14825504	16074977	17260624	18377786	19422077	20389399	21275960	22078287	22793241	23418029
63.2	0.00	23950213	24387723	24728863	24972317	25117155	25162837	25109214	24956529	24705416	24356897
63.3	0.00	23912377	23373641	22742846	22022507	21215498	20325028	19354639	18308184	17189816	16003972
63.4	0.00	14755351	13448901	12089795	10683411	09235316	07751234	06237035	04698702	03142313	01574015
63.5	-0.00	00000000	01573519	03140334	04694265	06229182	07739037	09217880	10659883	12059371	13410832
63.6	-0.00	14708951	15948621	17124969	18233374	19269483	20229231	21108854	21904907	22614274	23234185
63.7	-0.00	23762221	24196327	24534820	24776394	24920127	24965481	24912310	24760853	24511739	24165982
63.8	-0.00	23724976	23190491	22564666	21849998	21049336	20165866	19203099	18164860	17055269	15878726
63.9	-0.00	14639895	13343684	11995225	10599856	09163097	07690631	06188278	04661976	03117756	01561716
64.0	0.00	00000000	01561228	03115808	04657608	06180547	07678623	09145932	10576694	11965274	13306207
64.1	0.00	14594217	15824236	16991431	18091215	19119269	20071559	20944352	21734228	22438096	23053205
64.2	0.00	23577157	24007911	24343798	24583522	24726166	24771197	24718470	24568221	24321075	23978037
64.3	0.00	23540489	23010189	22389257	21680171	20885757	20009177	19053914	18023763	16922811	15755425
64.4	0.00	14526231	13240100	11902123	10517597	09091999	07630967	06140277	04625820	03093580	01549608
64.5	0.00	00000000	01549127	03091662	04621519	06132666	07619145	09075099	10494793	11872635	13203202
64.6	-0.00	14481258	15701777	16859959	17951255	18971379	19916327	20782395	21566188	22264641	22875023
64.7	-0.00	23394953	23822407	24155728	24393629	24535200	24579914	24527623	24378564	24133355	23792992
64.8	-0.00	23358850	22832668	22216553	21512963	20724701	19854904	18907029	17884841	16792395	15634024
64.9	-0.00	14414319	13138112	11810456	10436605	09021996	07572222	06093015	04590220	03069776	01537686

TABLE 6

Sinc Function

x		.0	.01	.02	.03	.04	.05	.06	.07	.08	.09
65.0	0.00	00000000	01537213	03067887	04585985	06085520	07560581	09005355	10414151	11781420	13101780
65.1	0.00	14370035	15581198	16730506	17813444	18825759	19763477	20622922	21400727	22093848	22699574
65.2	0.00	23215544	23639748	23970541	24206647	24347162	24391562	24339701	24191812	23948510	23610782
65.3	0.00	23179992	22657866	22046494	21348315	20566110	19702992	18762392	17748044	16663973	15514479
65.4	0.00	14304117	13037683	11720189	10356851	08953063	07514375	06046475	04555165	03046335	01525946
65.5	-0.00	00000000	01525480	03044476	04550994	06039094	07502911	08936675	10334738	11691595	13001903
65.6	-0.00	14260507	15462457	16603026	17677733	18682357	19612956	20465879	21237786	21925654	22526796
65.7	-0.00	23038865	23459869	23788173	24022510	24161985	24206075	24154636	24007900	23766475	23431342
65.8	-0.00	23003852	22485720	21879018	21186167	20409928	19553387	18619950	17613323	16537501	15396749
65.9	-0.00	14195589	12938778	11631292	10278307	08885175	07457404	06000641	04520640	03023250	01514384
66.C	0.00	00000000	01513925	03021418	04516532	05993371	07446114	08869035	10256528	11603129	12903538
66.1	0.00	14152637	15345512	16477474	17544074	18541124	19464710	20311210	21077307	21760003	22356628
66.2	0.00	22864856	23282706	23608558	23841153	23979603	24023388	23972365	23826763	23587186	23254608
66.3	0.00	22830369	22316170	21714068	21026465	20256099	19406037	18479655	17480633	16412934	15280792
66.4	0.00	14088694	12841362	11543734	10200945	08818309	07401292	05955496	04486635	03000512	01502996
66.5	-0.00	00000000	01502544	02998708	04482589	05948336	07390170	08802410	10179492	11515993	12806650
66.6	-0.00	14046386	15230323	16353806	17412421	18402009	19318688	20158860	20919235	21596835	22189012
66.7	-0.00	22693455	23108199	23431635	23662514	23799953	23843437	23792823	23648339	23410583	23080521
66.8	-0.00	22659483	22149158	21551586	20869152	20104573	19260891	18341458	17349927	16290230	15166569
66.9	-0.00	13983397	12745402	11457483	10124739	08752441	07346017	05911025	04453138	02978113	01491778
67.0	0.00	00000000	01491333	02976336	04449152	05903971	07335061	08736779	10103605	11430155	12711206
67.1	0.00	13941718	15116850	16231981	17282729	18264967	19174840	20008779	20763516	21436097	22023891
67.2	0.00	22524605	22936289	23257345	23486533	23622975	23666163	23615951	23472567	23236604	22909020
67.3	0.00	22491136	21984627	21391518	20714175	19955296	19117900	18205313	17221161	16169346	15054040
67.4	0.00	13879663	12650865	11372512	10049663	08687551	07291561	05867214	04420137	02956047	01480726
67.5	0.00	00000000	01480288	02954296	04416209	05860264	07280767	08672120	10028841	11345587	12617175
67.6	-0.00	13838599	15005055	16111957	17154955	18129951	19033120	19860917	20610099	21277733	21861209
67.7	-0.00	22358249	22766917	23085628	23313149	23448610	23491505	23441689	23299389	23065192	22740050
67.8	-0.00	22325272	21822522	21233810	20561483	19808220	18977016	18071175	17094292	16050244	14943170
67.9	-0.00	13777456	12557721	11288792	09975692	08623615	07237908	05824047	04387621	02934305	01469837
68.0	0.00	00000000	01469405	02932579	04383752	05817199	07227271	08608411	09955175	11262262	12524524
68.1	0.00	13736994	14894902	15993696	17029056	17996917	18893478	19715223	20458932	21121692	21700913
68.2	0.00	22194332	22600029	22916428	23142307	23276801	23319406	23269981	23128747	22896291	22573554
68.3	0.00	22161837	21662791	21078410	20411026	19663296	18838194	17938998	16969279	15932883	14833920
68.4	0.00	13676744	12465938	11206296	09902803	08560614	07185038	05781511	04355581	02912880	01459107
68.5	0.00	00000000	01458681	02911180	04351767	05774763	07174556	08545630	09882584	11180151	12433224
68.6	-0.00	13636870	14786355	15877158	16904992	17865820	18755871	19571652	20309966	20967923	21542950
68.7	-0.00	22032801	22435569	22749690	22973950	23107490	23149810	23100769	22960587	22729845	22409478
68.8	-0.00	22000777	21505381	20925269	20262755	19520477	18701388	17808741	16846081	15817226	14726256
68.9	-0.00	13577493	12375487	11124997	09830970	08498527	07132934	05739592	04324005	02891766	01448532
69.0	0.00	00000000	01448112	02890090	04320247	05732941	07122604	08483759	09811044	11099230	12343245
69.1	0.00	13538195	14679378	15762306	16782722	17736620	18620254	19430157	20163155	20816377	21387270
69.2	0.00	21873605	22273486	22585361	22808025	22940625	22982663	22934000	22794855	22565801	22247770
69.3	0.00	21842041	21350241	20774336	20116622	19379718	18566554	17680362	16724659	15703237	14620144
69.4	0.00	13479673	12286340	11044869	09760173	08437333	07081581	05698276	04292884	02870956	01438109
69.5	0.00	00000000	01437696	02869304	04289179	05691721	07071399	08422778	09740532	11019471	12254560
69.6	-0.00	13440938	14573937	15649103	16662209	17609275	18486583	19290692	20018450	20667006	21233825
69.7	-0.00	21716692	22113728	22423389	22644480	22776152	22817913	22769623	22631498	22404109	22088379
69.8	-0.00	21685580	21197325	20625565	19972582	19240974	18433651	17553821	16604975	15590878	14515550
69.9	-0.00	13383252	12198467	10965887	09690387	08377015	07030962	05657551	04262207	02850443	01427836

$$\frac{\sin 2\pi x}{2\pi x}$$

TABLE 6

Sinc Function

x		.0	.01	.02	.03	.04	.05	.06	.07	.08	.09
70.0	0.00	00000000	01427428	02848815	04258555	05651089	07020925	08362667	09671026	10940850	12167140
70.1	0.00	13345068	14470001	15537515	16543414	17483746	18354819	19153216	19875807	20519763	21082565
70.2	0.00	21562015	21956245	22263724	22483263	22614022	22655508	22607584	22470465	22244717	21931256
70.3	0.00	21531344	21046583	20478910	19830590	19104203	18302637	17429078	16486992	15480116	14412442
70.4	0.00	13288200	12111843	10888026	09621593	08317553	06981062	05617403	04231966	02830222	01417708
70.5	−0.00	00000000	01417306	02828617	04228366	05611033	06971167	08303407	09602505	10863344	12080958
70.6	−0.00	13250556	14367537	15427507	16426300	17359993	18224919	19017685	19735183	20374604	20933445
70.7	−0.00	21409526	21800990	22106317	22324326	22454183	22495398	22447836	22311708	22087577	21776352
70.8	−0.00	21379286	20897970	20334326	19690603	18969362	18173473	17306095	16370673	15370916	14310788
70.9	−0.00	13194489	12026440	10811263	09553768	08258929	06931865	05577822	04202151	02810285	01407722
71.0	0.00	00000000	01407326	02808702	04198601	05571541	06922109	08244982	09534949	10786927	11995989
71.1	0.00	13157374	14266514	15319046	16310834	17237980	18096845	18884059	19596535	20231483	20786420
71.2	0.00	21259178	21647914	21951119	22167620	22296587	22337536	22290329	22155179	21932642	21623622
71.3	0.00	21229361	20751441	20191769	19552578	18836412	18046118	17184836	16255984	15263247	14210559
71.4	0.00	13102091	11942233	10735575	09486893	08201126	06883357	05538794	04172753	02790627	01397877
71.5	−0.00	00000000	01397486	02789066	04169252	05532601	06873736	08187373	09468336	10711578	11912206
71.6	−0.00	13065493	14166901	15212099	16196979	17117670	17970559	18752297	19459821	20090360	20641446
71.7	−0.00	21110927	21496974	21798086	22013099	22141189	22181874	22135018	22000830	21779865	21473018
71.8	−0.00	21081524	20606952	20051197	19416475	18705312	17920536	17065265	16142891	15157075	14111723
71.9	−0.00	13010978	11859197	10660940	09420948	08144126	06835522	05500309	04143763	02771243	01388168
72.0	0.00	00000000	01387782	02769703	04140311	05494201	06826035	08130564	09402647	10637275	11829586
72.1	0.00	12974886	14068670	15106636	16084703	16999028	17846023	18622362	19325002	19951191	20498480
72.2	0.00	20964729	21348123	21647171	21860717	21987941	22028366	21981855	21848618	21629202	21324498
72.3	0.00	20935732	20464462	19912569	19282254	18576025	17796690	16947345	16031361	15052370	14014253
72.4	0.00	12921123	11777307	10587335	09355913	08087913	06788348	05462355	04115174	02752125	01378593
72.5	−0.00	00000000	01378213	02750607	04111769	05456331	06778991	08074537	09337864	10563995	11748103
72.6	−0.00	12885527	13971792	15002624	15973972	16882020	17723201	18494215	19192038	19813938	20357480
72.7	−0.00	20820543	21201320	21498332	21710430	21836801	21876968	21830798	21698496	21480609	21178019
72.8	−0.00	20791943	20323929	19775844	19149875	18448512	17674544	16831045	15921361	14949102	13918120
72.9	−0.00	12832500	11696541	10514739	09291770	08032471	06741821	05424921	04086976	02733270	01369149
73.0	0.00	00000000	01368774	02731773	04083618	05418979	06732592	08019278	09273967	10491718	11667736
73.1	0.00	12797391	13876238	14900035	15864756	16766611	17602058	18367819	19060891	19678559	20218407
73.2	0.00	20678326	21056522	21351526	21562196	21687724	21727637	21681803	21550424	21334044	21033538
73.3	0.00	20650115	20185312	19640984	19019302	18322738	17554063	16716329	15812861	14847241	13823297
73.4	0.00	12745086	11616876	10443132	09228500	07977784	06695927	05387997	04059162	02714671	01359834
73.5	−0.00	00000000	01359464	02713194	04055850	05382136	06686823	07964769	09210939	10420424	11588461
73.6	−0.00	12710452	13781983	14798839	15757023	16652769	17482560	18243139	18931524	19545019	20081222
73.7	−0.00	20538039	20913688	21206711	21415972	21540668	21580331	21534827	21404359	21189465	20891015
73.8	−0.00	20510209	20048574	19507951	18890497	18198668	17435214	16603167	15705829	14746759	13729758
73.9	−0.00	12658854	11538288	10372494	09166087	07923836	06650653	05351572	04031724	02696324	01350645
74.0	0.00	00000000	01350280	02694867	04028456	05345789	06641672	07910997	09148762	10350092	11510256
74.1	0.00	12624687	13689000	14699009	15650743	16540463	17364673	18120141	18803902	19413278	19945885
74.2	0.00	20399642	20772780	21063847	21271717	21395594	21435009	21389831	21260261	21046832	20750410
74.3	0.00	20372187	19913676	19376709	18763425	18076266	17317963	16491527	15600237	14647628	13637475
74.4	0.00	12573781	11460756	10302806	09104511	07870613	06605988	05315636	04004655	02678223	01341579
74.5	−0.00	00000000	01341219	02676785	04001431	05309931	06597127	07857946	09087418	10280702	11433099
74.6	−0.00	12540071	13597263	14600517	15545888	16429661	17248366	17998790	18677988	19283302	19812361
74.7	−0.00	20263098	20633757	20922895	21129394	21252460	21291631	21246775	21118090	20906107	20611686
74.8	−0.00	20236009	19780581	19247220	18638052	17955500	17202278	16381377	15496055	14549820	13546425
74.9	−0.00	12489844	11384259	10234047	09043758	07818100	06561919	05280179	03977946	02660363	01332634

TABLE 6

Sinc Function

x		.0	.01	.02	.03	.04	.05	.06	.07	.08	.09
75.0	0.00	00000000	01332279	02658945	03974765	05274550	06553176	07805601	09026892	10212237	11356969
75.1	0.00	12456582	13506747	14503335	15442428	16320334	17133607	17879053	18553750	19155054	19680612
75.2	0.00	20128370	20496583	20783817	20988962	21111229	21150159	21105619	20977808	20767252	20474804
75.3	0.00	20101640	19649253	19119450	18514342	17836337	17088129	16272690	15393255	14453311	13456583
75.4	0.00	12407020	11308776	10166200	08983810	07766283	06518434	05245193	03951592	02642740	01323807
75.5	-0.00	00000000	01323457	02641341	03948453	05239638	06509806	07753949	08967167	10144678	11281847
75.6	-0.00	12374197	13417429	14407439	15340336	16212453	17020364	17760899	18431154	19028501	19550604
75.7	-0.00	19995422	20361220	20646576	20850384	20971863	21010554	20966326	20839377	20630228	20339728
75.8	-0.00	19969043	19519658	18993366	18392264	17718745	16975485	16165435	15291810	14358073	13367924
75.9	-0.00	12325287	11234288	10099246	08924652	07715149	06475521	05210667	03925584	02625349	01315097
76.0	0.00	00000000	01314751	02623968	03922486	05205185	06467006	07702977	08908227	10078007	11207712
76.1	0.00	12292895	13329284	14312803	15239585	16105988	16908609	17644297	18310167	18903610	19422303
76.2	0.00	19864219	20227634	20511135	20713624	20834324	20872779	20828860	20702761	20495002	20206423
76.3	0.00	19838184	19391760	18868933	18271786	17602694	16864316	16059585	15191693	14264082	13280427
76.4	0.00	12244624	11160775	10033169	08866267	07664684	06433170	05176592	03899917	02608186	01306500
76.5	-0.00	00000000	01306159	02606822	03896859	05171182	06424766	07652670	08850056	10012207	11134545
76.6	-0.00	12212654	13242289	14219402	15140149	16000912	16798311	17529215	18190758	18780347	19295674
76.7	-0.00	19734726	20095789	20377460	20578647	20698578	20736800	20693185	20567925	20361536	20074853
76.8	-0.00	19709029	19265528	18746121	18152875	17488152	16754594	15955112	15092879	14171313	13194067
76.9	-0.00	12165010	11088218	09967951	08808641	07614874	06391369	05142961	03874583	02591245	01298016
77.0	0.00	00000000	01297678	02589899	03871565	05137620	06383074	07603016	08792640	09947260	11062328
77.1	0.00	12133454	13156423	14127212	15042002	15897199	16689443	17415625	18072896	18658681	19170686
77.2	0.00	19606910	19965652	20245516	20445417	20564589	20602582	20559266	20434834	20229797	19944986
77.3	0.00	19581545	19140929	18624896	18035502	17375092	16646290	15851990	14995342	14079744	13108823
77.4	0.00	12086425	11016598	09903575	08751760	07565708	06350107	05109763	03849576	02574523	01289640
77.5	-0.00	00000000	01289307	02573194	03846596	05104491	06341919	07554002	08735965	09883150	10991040
77.6	-0.00	12055274	13071663	14036209	14945119	15794821	16581977	17303498	17956552	18538581	19047306
77.7	-0.00	19480739	19837189	20115269	20313902	20432324	20470089	20427069	20303454	20099752	19816789
77.8	-0.00	19455700	19017931	18505230	17919638	17263484	16539377	15750192	14899057	13989350	13024673
77.9	-0.00	12008848	10945897	09840025	08695609	07517172	06309376	05076991	03824889	02558015	01281372
78.0	0.00	00000000	01281044	02556704	03821948	05071787	06301292	07505616	08680015	09819862	10920666
78.1	0.00	11978096	12987989	13946372	14849477	15693754	16475887	17192805	17841696	18420018	18925505
78.2	0.00	19356182	19710369	19986688	20184067	20301749	20339290	20296561	20173752	19971368	19690228
78.3	0.00	19331462	18896504	18387091	17805252	17153301	16433829	15649693	14804002	13900109	12941597
78.4	0.00	11932261	10876098	09777286	08640173	07469256	06269163	05044637	03800518	02541718	01273210
78.5	-0.00	00000000	01272885	02540423	03797614	05039499	06261182	07457847	08624778	09757378	10851188
78.6	-0.00	11901899	12905378	13857677	14755050	15593971	16371145	17083519	17728300	18302961	18805251
78.7	-0.00	19233208	19585160	19859740	20055882	20172833	20210151	20167710	20045697	19844614	19565275
78.8	-0.00	19208800	18776618	18270451	17692318	17044516	16329620	15550468	14710151	13812000	12859574
78.9	-0.00	11856645	10807183	09715342	08585440	07421946	06229459	05012693	03776455	02525627	01265150
79.0	0.00	00000000	01264830	02524349	03773588	05007620	06221579	07410681	08570239	09695685	10782587
79.1	0.00	11826666	12823812	13770103	14661817	15495450	16267726	16975614	17616337	18187383	18686516
79.2	0.00	19111786	19461532	19734394	19929315	20045544	20082643	20040485	19919258	19719459	19441897
79.3	0.00	19087685	18658243	18155282	17580807	16937101	16226724	15452494	14617483	13725001	12778584
79.4	0.00	11781981	10739137	09654178	08531396	07375232	06190256	04981151	03752694	02509739	01257192
79.5	-0.00	00000000	01256876	02508476	03749863	04976141	06182474	07364108	08516385	09634767	10714849
79.6	-0.00	11752378	12743271	13683629	14569755	15398166	16165606	16869064	17505779	18073255	18569271
79.7	-0.00	18991888	19339455	19610621	19804334	19919850	19956733	19914855	19794404	19595873	19320065
79.8	-0.00	18968088	18541351	18041556	17470693	16831032	16125116	15355746	14525975	13639091	12698608
79.9	-0.0C	11708251	10671941	09593779	08478028	07329102	06151542	04950003	03729231	02494049	01249334

$$\frac{\sin 2\pi x}{2\pi x}$$

TABLE 6

Sinc Function

x		.0	.01	.02	.03	.04	.05	.06	.07	.08	.09
80.0	0.00	00000000	01249022	02492802	03726435	04945056	06143858	07318117	08463205	09574610	10647956
80.1	0.00	11679017	12663735	13598235	14478842	15302095	16064760	16763843	17396600	17960551	18453488
80.2	0.00	18873485	19218900	19488391	19680912	19795724	19832392	19790791	19671105	19473826	19199751
80.3	0.00	18849981	18425915	17929245	17361949	16726284	16024773	15260203	14435605	13554249	12619627
80.4	0.00	11635439	10605582	09534131	08425324	07283546	06113310	04919242	03706060	02478554	01241573
80.5	−0.00	00000000	01241265	02477323	03703298	04914356	06105721	07272696	08410684	09515200	10581894
80.6	−0.00	11606567	12585185	13513899	14389056	15207216	15965165	16659926	17288774	17849244	18339140
80.7	−0.00	18756548	19099839	19367675	19559019	19673134	19709590	19668262	19549333	19353289	19080926
80.8	−0.00	18733335	18311907	17818324	17254551	16622831	15925671	15165841	14346353	13470457	12541622
80.9	−0.00	11563526	10540042	09475220	08373271	07238552	06075551	04888862	03683174	02463251	01233908
81.0	0.00	00000000	01233603	02462035	03680447	04884036	06068055	07227836	08358811	09456522	10516646
81.1	0.00	11535010	12507604	13430604	14300377	15113506	15866797	16557290	17182277	17739307	18226201
81.2	0.00	18641052	18982243	19248445	19438626	19552054	19588301	19547242	19429059	19234236	18963562
81.3	0.00	18618124	18199302	17708768	17148474	16520649	15827788	15072639	14258198	13387694	12464575
81.4	0.00	11492497	10475308	09417033	08321857	07194111	06038254	04858854	03660570	02448135	01226337
81.5	−0.00	00000000	01226036	02446934	03657876	04854087	06030850	07183527	08307574	09398563	10452198
81.6	−0.00	11464329	12430974	13348328	14212785	15020945	15769633	16455910	17077083	17630717	18114644
81.7	−0.00	18526970	18866087	19130674	19319706	19432455	19468495	19427701	19310255	19116639	18847634
81.8	−0.00	18504321	18088073	17600550	17043693	16419717	15731100	14980576	14171120	13305942	12388470
81.9	−0.00	11422336	10411364	09359555	08271070	07150213	06001413	04829212	03638241	02433204	01218859
82.0	0.00	00000000	01218561	02432017	03635580	04824503	05994099	07139757	08256961	09341311	10388535
82.1	0.00	11394510	12355277	13267055	14126258	14929510	15673652	16355765	16973170	17523448	18004444
82.2	0.00	18414276	18751344	19014336	19202233	19314310	19350145	19309614	19192896	19000471	18733114
82.3	0.00	18391901	17978195	17493647	16940185	16320010	15635586	14889630	14085099	13225183	12313288
82.4	0.00	11353025	10348196	09302776	08220900	07106846	05965019	04799930	03616183	02418453	01211471
82.5	−0.00	00000000	01211177	02417281	03613554	04795278	05957793	07096517	08206961	09284752	10325643
82.6	−0.00	11325536	12280496	13186765	14040779	14839181	15578833	16256831	16870514	17417477	17895577
82.7	−0.00	18302944	18637988	18899404	19086179	19197593	19233226	19192954	19076955	18885706	18619978
82.8	−0.00	18280839	17869644	17388034	16837926	16221507	15541226	14799782	14000116	13145398	12239013
82.9	−0.00	11284551	10285790	09246681	08171335	07064003	05929064	04771001	03594391	02403881	01204172
83.0	0.00	00000000	01203882	02402723	03591794	04766405	05921924	07053798	08157563	09228873	10263507
83.1	0.00	11257392	12206615	13107442	13956329	14749939	15485154	16159086	16769092	17312779	17788018
83.2	0.00	18192950	18525994	18785853	18971520	19082279	19117711	19077695	18962406	18772319	18508200
83.3	0.00	18171110	17762396	17283689	16736894	16124186	15447997	14711011	13916152	13066569	12165629
83.4	0.00	11216898	10224132	09191259	08122364	07021673	05893539	04742419	03572860	02389483	01196960
83.5	−0.00	00000000	01196674	02388339	03570294	04737877	05886485	07011590	08108757	09173663	10202115
83.6	−0.00	11190063	12133618	13029067	13872888	14661764	15392595	16062511	16668882	17209333	17681745
83.7	−0.00	18084271	18415338	18673659	18858230	18968341	19003575	18963812	18849225	18660285	18397756
83.8	−0.00	18062690	17656428	17180589	16637068	16028025	15355880	14623300	13833189	12988681	12093119
83.9	−0.00	11150051	10163209	09136496	08073976	06979848	05858437	04714176	03551585	02375256	01189835
84.0	0.00	00000000	01189551	02374126	03549049	04709689	05851467	06969884	08060530	09119110	10141454
84.1	0.00	11123535	12061488	12951623	13790439	14574636	15301135	15967082	16569863	17107115	17576734
84.2	0.00	17976882	18305997	18562796	18746285	18855756	18890794	18851280	18737387	18549581	18288622
84.3	0.00	17955557	17551717	17078712	16538425	15933005	15264855	14536628	13751210	12911716	12021469
84.4	0.00	11083996	10103007	09082383	08026161	06938518	05823752	04686269	03530563	02361198	01182793
84.5	−0.00	00000000	01182513	02360081	03528057	04681834	05816864	06928671	08012874	09065202	10081509
84.6	−0.00	11057793	11990211	12875095	13708964	14488539	15210756	15872781	16472014	17006105	17472963
84.7	−0.00	17870761	18197946	18453243	18635661	18744499	18779344	18740076	18626868	18440183	18180776
84.8	−0.00	17849687	17448240	16978036	16440945	15839105	15174903	14450977	13670197	12835657	11950663
84.9	−0.00	11018719	10043515	09028907	07978910	06897674	05789474	04658689	03509787	02347306	01175835

TABLE 6

Sinc Function

x		.0	.01	.02	.03	.04	.05	.06	.07	.08	.09
85.0	0.00	00000000	01175558	02346201	03507311	04654307	05782667	06887943	07965779	09011927	10022269
85.1	0.00	10992824	11919772	12799466	13628446	14403452	15121439	15779587	16375313	16906281	17370410
85.2	0.00	17765886	18091163	18344974	18526336	18634548	18669202	18630177	18517645	18332067	18074194
85.3	0.00	17745058	17345976	16878539	16344608	15746305	15086005	14366330	13590132	12760489	11880686
85.4	0.00	10954207	09984719	08976057	07932211	06857308	05755598	04631433	03489255	02333576	01168958
85.5	−0.00	00000000	01168685	02332484	03486807	04627101	05748870	06847691	07919233	08959275	09963720
85.6	−0.00	10928613	11850155	12724720	13548868	14319359	15033164	15687481	16279741	16807621	17269054
85.7	−0.00	17662234	17985626	18237969	18418285	18525879	18560343	18521559	18409696	18225212	17968854
85.8	−0.00	17641649	17244904	16780203	16249393	15654586	14998143	14282668	13511000	12686197	11811524
85.9	−0.00	10890446	09926607	08923822	07886056	06817412	05722115	04604493	03468962	02320005	01162161
86.0	0.00	00000000	01161891	02318926	03466542	04600212	05715466	06807906	07873229	08907235	09905852
86.1	0.00	10865148	11781347	12650843	13470215	14236242	14945915	15596444	16185278	16710107	17168874
86.2	0.00	17559785	17881313	18132205	18311488	18418470	18452747	18414200	18302997	18119596	17864735
86.3	0.00	17539438	17145003	16683005	16155281	15563929	14911297	14199976	13432785	12612764	11743162
86.4	0.00	10827422	09869168	08872191	07840435	06777978	05689021	04577865	03448903	02306592	01155442
86.5	−0.00	00000000	01155175	02305525	03446511	04573634	05682447	06768582	07827755	08855796	09848653
86.6	−0.00	10802417	11713333	12577818	13392469	14154085	14859672	15506458	16091905	16613717	17069849
86.7	−0.00	17458517	17778203	18027660	18205922	18312300	18346391	18308078	18197529	18015196	17761816
86.8	−0.00	17438404	17046253	16586927	16062253	15474316	14825452	14118235	13355469	12540177	11675587
86.9	−0.00	10765124	09812390	08821155	07795339	06738997	05656306	04551544	03429075	02293332	01148801
87.0	0.00	00000000	01148537	02292278	03426711	04547360	05649808	06729709	07782805	08804947	09792110
87.1	0.00	10740405	11646100	12505631	13315616	14072870	14774418	15417504	15999603	16518433	16971961
87.2	0.00	17358411	17676275	17924315	18101566	18207346	18241254	18203173	18093269	17911993	17660075
87.3	0.00	17338528	16948634	16491949	15970290	15385730	14740590	14037430	13279039	12468420	11608786
87.4	0.00	10703539	09756262	08770702	07750758	06700462	05623966	04525523	03409473	02280224	01142236
87.5	−0.00	00000000	01141975	02279182	03407136	04521387	05617542	06691279	07738367	08754679	09736212
87.6	−0.00	10679101	11579635	12434268	13239639	13992582	14690137	15329565	15908354	16424236	16875188
87.7	−0.00	17259446	17575510	17822147	17998399	18103589	18137315	18099463	17990197	17809965	17559494
87.8	−0.00	17239789	16852127	16398053	15879374	15298152	14656694	13957545	13203478	12397480	11542744
87.9	−0.00	10642654	09700772	08720823	07706685	06662365	05591993	04499798	03390095	02267266	01135745
88.0	0.00	00000000	01135487	02266235	03387784	04495709	05585643	06653287	07694434	08704982	09680949
88.1	0.00	10618494	11513923	12363715	13164525	13913205	14606813	15242623	15818140	16331107	16779513
88.2	0.00	17161604	17475887	17721137	17896402	18001007	18034554	17996928	17888293	17709093	17460052
88.3	0.00	17142168	16756712	16305219	15789487	15211565	14573747	13878564	13128772	12327343	11477450
88.4	0.00	10582458	09645909	08671509	07663110	06624699	05560382	04474364	03370935	02254453	01129328
88.5	−0.00	00000000	01129073	02253435	03368650	04470321	05554103	06615723	07650997	08655845	09626310
88.6	−0.00	10558570	11448954	12293958	13090258	13834724	14524428	15156662	15728943	16239028	16684916
88.7	−0.00	17064864	17377387	17621266	17795555	17899582	17932951	17895548	17787536	17609357	17361730
88.8	−0.00	17045647	16662372	16213431	15700612	15125953	14491734	13800472	13054907	12257995	11412890
88.9	−0.00	10522939	09591664	08622748	07620025	06587457	05529127	04449216	03351991	02241785	01122983
89.0	0.00	00000000	01122730	02240778	03349732	04445218	05522918	06578581	07608047	08607261	09572285
89.1	0.00	10499319	11384713	12224984	13016825	13757123	14442967	15071665	15640747	16147981	16591381
89.2	0.00	16969209	17279991	17522514	17695838	17799293	17832487	17795304	17687908	17510738	17264509
89.3	0.00	16950207	16569088	16122671	15612733	15041299	14410638	13723253	12981869	12189422	11349052
89.4	0.00	10464086	09538025	08574534	07577421	06550631	05498220	04424349	03333258	02229258	01116708
89.5	−0.00	00000000	01116459	02228262	03331025	04420396	05492081	06541854	07565578	08559218	09518862
89.6	−0.00	10440729	11321189	12156779	12944211	13680387	14362415	14987616	15553534	16057950	16498888
89.7	−0.00	16874620	17183681	17424863	17597232	17700121	17733141	17696177	17589391	17413218	17168370
89.8	−0.00	16855829	16476843	16032921	15525831	14957587	14330446	13646894	12909643	12121613	11285925
89.9	−0.00	10405887	09484983	08526855	07535292	06514214	05467658	04399758	03314734	02216871	01110504

$\frac{\sin 2\pi x}{2\pi x}$

TABLE 6

Sinc Function

x		.0	.01	.02	.03	.04	.05	.06	.07	.08	.09
90.0	0.00	00000000	01110257	02215886	03312525	04395849	05461586	06505534	07523579	08511709	09466032
90.1	0.00	10382789	11258371	12089332	12872402	13604503	14282757	14904499	15467288	15968918	16407420
90.2	0.00	16781080	17088438	17328294	17499718	17602049	17634897	17598148	17491964	17316778	17073297
90.3	0.00	16762497	16385619	15944165	15439891	14874802	14251140	13571380	12838216	12054554	11223496
90.4	0.00	10348333	09432528	08479703	07493628	06478200	05437433	04375439	03296415	02204620	01104368
90.5	-0.00	00000000	01104123	02203646	03294230	04371573	05431428	06469616	07482045	08464725	09413786
90.6	-0.00	10325489	11196245	12022628	12801386	13529456	14203977	14822299	15381994	15880867	16316961
90.7	-0.00	16688572	16994245	17232790	17403280	17505057	17537735	17501200	17395611	17221400	16979271
90.8	-0.00	16670192	16295400	15856386	15354898	14792929	14172708	13496697	12767576	11988232	11161754
90.9	-0.00	10291411	09380649	08433071	07452423	06442582	05407541	04351388	03278296	02192504	01098299
91.0	0.00	00000000	01098058	02191541	03276136	04347564	05401602	06434092	07440966	08418256	09362113
91.1	0.00	10268818	11134802	11956657	12731149	13455233	14126062	14741001	15297635	15793781	16227495
91.2	0.00	16597077	16901085	17138333	17307899	17409128	17441638	17405313	17300313	17127067	16886274
91.3	0.00	16578899	16206168	15769568	15270835	14711951	14095135	13422832	12697708	11922637	11100687
91.4	0.00	10235113	09329338	08386948	07411668	06407354	05377975	04327599	03260376	02180521	01092297
91.5	-0.00	00000000	01092058	02179568	03258239	04323817	05372101	06398956	07400336	08372295	09311004
91.6	-0.00	10212765	11074029	11891405	12661679	13381819	14048997	14660589	15214197	15707646	16139004
91.7	-0.00	16506581	16808941	17044905	17213557	17314245	17346588	17310472	17206054	17033762	16794291
91.8	-0.00	16488600	16117909	15683696	15187688	14631856	14018405	13349770	12628601	11857755	11040285
91.9	-0.00	10179426	09278586	08341327	07371356	06372508	05348731	04304070	03242651	02168667	01086360
92.0	0.00	00000000	01086123	02167725	03240537	04300329	05342920	06364202	07360148	08326833	09260450
92.1	0.00	10157321	11013916	11826862	12592962	13309202	13972768	14581051	15131663	15622445	16051472
92.2	0.00	16417066	16717796	16952491	17120238	17220391	17252568	17216658	17112817	16941469	16703305
92.3	0.00	16399279	16030606	15598754	15105441	14552628	13942507	13277500	12560243	11793576	10980537
92.4	0.00	10124343	09228383	08296200	07331481	06338040	05319803	04280794	03225118	02156942	01080487
92.5	-0.00	00000000	01080253	02156010	03223026	04277094	05314055	06329823	07320393	08281862	09210442
92.6	-0.00	10102476	10954452	11763016	12524988	13237370	13897361	14502370	15050020	15538163	15964885
92.7	-0.00	16328516	16627634	16861073	17027926	17127549	17159563	17123856	17020584	16850169	16613299
92.8	-0.00	16310921	15944243	15514727	15024081	14474253	13867426	13206008	12492620	11730088	10921432
92.9	-0.00	10069852	09178720	08251558	07292035	06303943	05291187	04257769	03207773	02145343	01074677
93.0	0.00	00000000	01074446	02144421	03205704	04254108	05285501	06295814	07281066	08237374	09160971
93.1	0.00	10048220	10895627	11699856	12457743	13166308	13822765	14424534	14969254	15454786	15879228
93.2	0.00	16240917	16538440	16770636	16936604	17035702	17067554	17032049	16929341	16759849	16524257
93.3	0.00	16223510	15858807	15431601	14943592	14396718	13793150	13135282	12425721	11667279	10862960
93.4	0.00	10015945	09129588	08207394	07253011	06270210	05262877	04234991	03190613	02133869	01068929
93.5	-0.00	00000000	01068701	02132956	03188566	04231369	05257251	06262168	07242159	08193362	09112029
93.6	-0.00	09994544	10837430	11637370	12391217	13096005	13748965	14347530	14889350	15372299	15794484
93.7	-0.00	16154252	16450197	16681164	16846256	16944835	16976527	16941221	16839070	16670492	16436166
93.8	-0.00	16137030	15774280	15349360	14863960	14320009	13719665	13065309	12359536	11605140	10805110
93.9	-0.00	09962612	09080980	08163701	07214402	06236837	05234868	04212455	03173637	02122516	01063243
94.0	0.00	00000000	01063017	02121613	03171611	04208871	05229302	06228880	07203665	08149817	09063608
94.1	0.00	09941438	10779851	11575548	12325397	13026449	13675948	14271343	14810294	15290688	15710641
94.2	0.00	16068508	16362891	16592642	16756867	16854933	16886466	16851357	16749757	16582082	16349008
94.3	0.00	16051468	15690650	15267992	14785173	14244113	13646959	12996078	12294051	11543659	10747874
94.4	0.00	09909844	09032887	08120470	07176202	06203816	05207155	04190157	03156839	02111283	01057617
94.5	-0.00	00000000	01057393	02110390	03154836	04186611	05201648	06195944	07165579	08106733	09015698
94.6	-0.00	09888893	10722881	11514379	12260273	12957628	13603703	14195961	14732073	15209939	15627682
94.7	-0.00	15983669	16276507	16505054	16668422	16765979	16797355	16762441	16661386	16494606	16262770
94.8	-0.00	15966809	15607903	15187481	14707217	14169018	13575019	12927577	12229257	11482826	10691240
94.9	-0.00	09857632	08985300	08077695	07138405	06171144	05179735	04168094	03140219	02100169	01052050

TABLE 6

Sinc Function

x		.0	.01	.02	.03	.04	.05	.06	.07	.08	.09
95.0	0.00	00000000	01051828	02099285	03138237	04164586	05174285	06163354	07127893	08064102	08968291
95.1	0.00	09836901	10666510	11453854	12195833	12889530	13532218	14121371	14654675	15130038	15545596
95.2	0.00	15899721	16191030	16418386	16580905	16677960	16709180	16674458	16573943	16408047	16177437
95.3	0.00	15883037	15526023	15107816	14630079	14094710	13503834	12859794	12165142	11422631	10635201
95.4	0.00	09805967	08938212	08035368	07101004	06138814	05152602	04146263	03123773	02089171	01046541
95.5	-0.00	00000000	01046322	02088296	03121811	04142791	05147209	06131105	07090602	08021917	08921381
95.6	-0.00	09785453	10610729	11393961	12132067	12822145	13461480	14047560	14578085	15050972	15464367
95.7	-0.00	15816651	16106446	16332623	16494303	16590859	16621926	16587394	16487413	16322392	16092995
95.8	-0.00	15800140	15444998	15028981	14553745	14021177	13433391	12792718	12101696	11363064	10579745
95.9	-0.00	09754841	08891615	07993482	07063992	06106821	05125751	04124659	03107498	02078288	01041090
96.0	0.00	00000000	01040873	02077422	03105557	04121223	05120415	06099192	07053698	07980171	08874959
96.1	0.00	09734540	10555528	11334692	12068965	12755460	13391477	13974518	14502292	14972728	15383982
96.2	0.00	15734443	16022742	16247752	16408600	16504664	16535579	16501235	16401782	16237627	16009430
96.3	0.00	15718104	15364814	14950965	14478204	13948408	13363680	12726338	12038908	11304115	10524866
96.4	0.00	09704246	08845502	07952031	07027365	06075160	05099179	04103279	03091392	02067517	01035695
96.5	-0.00	00000000	01035480	02066660	03089471	04099878	05093898	06067610	07017177	07938857	08829018
96.6	-0.00	09684154	10500898	11276036	12006516	12689465	13322199	13902231	14427282	14895294	15304429
96.7	-0.00	15653086	15939903	16163758	16323784	16419360	16450123	16415966	16317036	16153737	15926728
96.8	-0.00	15636916	15285458	14873755	14403443	13876390	13294688	12660643	11976769	11245774	10470552
96.9	-0.00	09654172	08799864	07911007	06991115	06043825	05072881	04082119	03075452	02056857	01030356
97.0	0.00	00000000	01030143	02056009	03073551	04078754	05067654	06036353	06981032	07897969	08783550
97.1	0.00	09634287	10446831	11217984	11944709	12624150	13253634	13830688	14353045	14818656	15225694
97.2	0.00	15572566	15857916	16080629	16239839	16334933	16365547	16331574	16233161	16070710	15844876
97.3	0.00	15556562	15206918	14797338	14329450	13805112	13226405	12595624	11915268	11188032	10416796
97.4	0.00	09604613	08754695	07870404	06955238	06012812	05046853	04061176	03059676	02046307	01025071
97.5	-0.00	00000000	01024861	02045468	03057794	04057845	05041679	06005416	06945258	07857499	08738547
97.6	-0.00	09584931	10393318	11160526	11883536	12559503	13185771	13759877	14279568	14742803	15147766
97.7	-0.00	15492871	15776768	15998350	16156754	16251370	16281836	16248045	16150144	15988533	15763861
97.8	-0.00	15477029	15129181	14721703	14256214	13734563	13158820	12531268	11854395	11130880	10363590
97.9	-0.00	09555560	08709987	07830216	06919726	05982116	05021091	04040448	03044061	02035865	01019841
98.0	0.00	00000000	01019633	02035034	03042198	04037151	05015970	05974795	06909848	07817443	08694004
98.1	0.00	09536078	10340351	11103654	11822986	12495516	13118599	13689788	14206839	14667722	15070631
98.2	0.00	15413986	15696446	15916908	16074515	16168657	16198976	16165366	16067972	15907191	15683671
98.3	0.00	15398306	15052235	14646837	14183722	13664731	13091922	12467567	11794141	11074309	10310924
98.4	0.00	09507005	08665733	07790437	06884576	05951731	04995590	04019929	03028604	02025528	01014664
98.5	-0.00	00000000	01014458	02024706	03026760	04016666	04990521	05944485	06874797	07777793	08649912
98.6	-0.00	09487721	10287920	11047359	11763050	12432177	13052109	13620410	14134847	14593403	14994277
98.7	-0.00	15335901	15616938	15836292	15993108	16086782	16116956	16083524	15986631	15826673	15604292
98.8	-0.00	15320379	14976068	14572728	14111964	13595605	13025701	12404511	11734496	11018311	10258790
98.9	-0.00	09458941	08621927	07751059	06849781	05921654	04970347	03999619	03013303	02015296	01009538
99.0	0.00	00000000	01009335	02014482	03011478	03996388	04965329	05914480	06840101	07738542	08606265
99.1	0.00	09439852	10236018	10991632	11703718	12369477	12986289	13551731	14063582	14519832	14918694
99.2	0.00	15258603	15538232	15756488	15912522	16005732	16035763	16002507	15906110	15746965	15525712
99.3	0.00	15243237	14900667	14499365	14040928	13527176	12960146	12342089	11675451	10962875	10207182
99.4	0.00	09411361	08578562	07712078	06815335	05891879	04945358	03979512	02998156	02005167	01004465
99.5	-0.00	00000000	01004263	02004361	02996349	03976314	04940390	05884777	06805753	07699687	08563057
99.6	-0.00	09392463	10184638	10936464	11644982	12307406	12921129	13483741	13993031	14447000	14843868
99.7	-0.00	15182081	15460315	15677484	15832744	15925495	15955383	15922302	15826397	15668057	15447920
99.8	-0.00	15166868	14826022	14426738	13970604	13459431	12895248	12280292	11616998	10907995	10156090
99.9	-0.00	09364257	08535630	07673487	06781235	05862402	04920619	03959606	02983161	01995139	00999442

B. Error Function and Fresnel Integrals

Table 7—Error Function

Table 8—Error Function Derivative

Table 9—Fresnel Integrals

TABLE 7A

Error Function

X	.0	.0001	.0002	.0003	.0004	.0005	.0006	.0007	.0008	.0009	m	Δ_2
0.0	.0	.0112838	.0225676	.0338514	.0451352	.0564190	.0677027	.0789865	.0902703	.1015541	-2	0
0.001	.1128379	.1241217	.1354054	.1466892	.1579730	.1692567	.1805405	.1918243	.2031080	.2143918	-2	0
0.002	.2256755	.2369593	.2482430	.2595268	.2708105	.2820942	.2933779	.3046616	.3159453	.3272290	-2	0
0.003	.3385127	.3497964	.3610801	.3723638	.3836474	.3949311	.4062147	.4174984	.4287820	.4400656	-2	0
0.004	.4513493	.4626329	.4739165	.4852001	.4964836	.5077672	.5190508	.5303343	.5416178	.5529014	-2	0
0.005	.5641849	.5754684	.5867519	.5980354	.6093188	.6206023	.6318857	.6431692	.6544526	.6657360	-2	0
0.006	.6770194	.6883028	.6995861	.7108695	.7221528	.7334361	.7447194	.7560027	.7672860	.7785693	-2	0
0.007	.7898525	.8011357	.8124190	.8237022	.8349853	.8462685	.8575517	.8688348	.8801179	.8914010	-2	0
0.008	.0902684	.0913967	.0925250	.0936533	.0947816	.0959099	.0970382	.0981665	.0992948	.1004231	-1	0
0.009	.1015514	.1026797	.1038080	.1049362	.1060645	.1071928	.1083211	.1094493	.1105776	.1117059	-1	0
0.010	.1128342	.1139624	.1150907	.1162189	.1173472	.1184755	.1196037	.1207320	.1218602	.1229885	-1	0
0.011	.1241167	.1252449	.1263732	.1275014	.1286297	.1297579	.1308861	.1320143	.1331426	.1342708	-1	0
0.012	.1353990	.1365272	.1376554	.1387836	.1399118	.1410401	.1421683	.1432965	.1444246	.1455528	-1	0
0.013	.1466810	.1478092	.1489374	.1500656	.1511938	.1523219	.1534501	.1545783	.1557064	.1568346	-1	0
0.014	.1579628	.1590909	.1602191	.1613472	.1624754	.1636035	.1647317	.1658598	.1669879	.1681161	-1	0
0.015	.1692442	.1703723	.1715004	.1726285	.1737567	.1748848	.1760129	.1771410	.1782691	.1793972	-1	0
0.016	.1805253	.1816534	.1827814	.1839095	.1850376	.1861657	.1872937	.1884218	.1895499	.1906779	-1	0
0.017	.1918060	.1929340	.1940621	.1951901	.1963182	.1974462	.1985742	.1997023	.2008303	.2019583	-1	0
0.018	.2030863	.2042143	.2053423	.2064703	.2075983	.2087263	.2098543	.2109823	.2121103	.2132383	-1	0
0.019	.2143662	.2154942	.2166222	.2177501	.2188781	.2200061	.2211340	.2222619	.2233899	.2245178	-1	0
0.020	.2256457	.2267737	.2279016	.2290295	.2301574	.2312853	.2324132	.2335411	.2346690	.2357969	-1	0
0.021	.2369248	.2380527	.2391806	.2403084	.2414363	.2425641	.2436920	.2448199	.2459477	.2470755	-1	0
0.022	.2482034	.2493312	.2504590	.2515868	.2527147	.2538425	.2549703	.2560981	.2572259	.2583537	-1	0
0.023	.2594815	.2606092	.2617370	.2628648	.2639925	.2651203	.2662481	.2673758	.2685035	.2696313	-1	0
0.024	.2707590	.2718867	.2730145	.2741422	.2752699	.2763976	.2775253	.2786530	.2797807	.2809084	-1	0
0.025	.2820360	.2831637	.2842914	.2854190	.2865467	.2876743	.2888020	.2899296	.2910572	.2921849	-1	0
0.026	.2933125	.2944401	.2955677	.2966953	.2978229	.2989505	.3000781	.3012057	.3023332	.3034608	-1	0
0.027	.3045884	.3057159	.3068435	.3079710	.3090985	.3102261	.3113536	.3124811	.3136086	.3147361	-1	0
0.028	.3158636	.3169911	.3181186	.3192461	.3203735	.3215010	.3226285	.3237559	.3248834	.3260108	-1	0
0.029	.3271382	.3282657	.3293931	.3305205	.3316479	.3327753	.3339027	.3350301	.3361575	.3372849	-1	0
0.030	.3384122	.3395396	.3406669	.3417943	.3429216	.3440490	.3451763	.3453036	.3474309	.3485582	-1	0
0.031	.3496855	.3508128	.3519401	.3530674	.3541946	.3553219	.3564492	.3575764	.3587037	.3598309	-1	0
0.032	.3609581	.3620853	.3632126	.3643398	.3654670	.3665942	.3677213	.3688485	.3699757	.3711028	-1	0
0.033	.3722300	.3733571	.3744843	.3756114	.3767385	.3778657	.3789928	.3801199	.3812470	.3823741	-1	0
0.034	.3835011	.3846282	.3857553	.3868823	.3880094	.3891364	.3902634	.3913905	.3925175	.3936445	-1	0
0.035	.3947715	.3958985	.3970255	.3981525	.3992794	.4004064	.4015333	.4026603	.4037872	.4049142	-1	0
0.036	.4060411	.4071680	.4082949	.4094218	.4105487	.4116756	.4128024	.4139293	.4150562	.4161830	-1	0
0.037	.4173099	.4184367	.4195635	.4206903	.4218171	.4229439	.4240707	.4251975	.4263243	.4274510	-1	0
0.038	.4285778	.4297045	.4308313	.4319580	.4330847	.4342114	.4353381	.4364648	.4375915	.4387182	-1	0
0.039	.4398449	.4409715	.4420982	.4432248	.4443514	.4454781	.4466047	.4477313	.4488579	.4499845	-1	0
0.040	.4511111	.4522376	.4533642	.4544907	.4556173	.4567438	.4578703	.4589969	.4601234	.4612499	-1	0
0.041	.4623764	.4635028	.4646293	.4657558	.4668822	.4680087	.4691351	.4702615	.4713879	.4725143	-1	0
0.042	.4736407	.4747671	.4758935	.4770199	.4781462	.4792726	.4803989	.4815252	.4826516	.4837779	-1	0
0.043	.4849042	.4860305	.4871567	.4882830	.4894093	.4905355	.4916618	.4927880	.4939142	.4950404	-1	0
0.044	.4961666	.4972928	.4984190	.4995452	.5006713	.5017975	.5029236	.5040498	.5051759	.5063020	-1	0
0.045	.5074281	.5085542	.5096803	.5108063	.5119324	.5130584	.5141845	.5153105	.5164365	.5175625	-1	0
0.046	.5186885	.5198145	.5209405	.5220665	.5231924	.5243184	.5254443	.5265702	.5276962	.5288221	-1	0
0.047	.5299480	.5310738	.5321997	.5333256	.5344514	.5355773	.5367031	.5378289	.5389547	.5400805	-1	0
0.048	.5412063	.5423321	.5434579	.5445836	.5457094	.5468351	.5479608	.5490865	.5502122	.5513379	-1	0
0.049	.5524636	.5535893	.5547149	.5558406	.5569662	.5580918	.5592174	.5603430	.5614686	.5625942	-1	0

TABLE 7A

Error Function

X	.0	.0001	.0002	.0003	.0004	.0005	.0006	.0007	.0008	.0009	m	Δ_2
0.050	.5637198	.5648453	.5659709	.5670964	.5682219	.5693474	.5704729	.5715984	.5727239	.5738494	-1	0
0.051	.5749748	.5761003	.5772257	.5783511	.5794765	.5806019	.5817273	.5828527	.5839780	.5851034	-1	0
0.052	.5862287	.5873541	.5884794	.5896047	.5907300	.5918552	.5929805	.5941058	.5952310	.5963562	-1	0
0.053	.5974815	.5986067	.5997319	.6008571	.6019822	.6031074	.6042325	.6053577	.6064828	.6076079	-1	0
0.054	.6087330	.6098581	.6109832	.6121082	.6132333	.6143583	.6154833	.6166084	.6177334	.6188584	-1	0
0.055	.6199833	.6211083	.6222332	.6233582	.6244831	.6256080	.6267329	.6278578	.6289827	.6301076	-1	0
0.056	.6312324	.6323573	.6334821	.6346069	.6357317	.6368565	.6379813	.6391060	.6402308	.6413555	-1	0
0.057	.6424802	.6436050	.6447297	.6458543	.6469790	.6481037	.6492283	.6503530	.6514776	.6526022	-1	0
0.058	.6537268	.6548514	.6559759	.6571005	.6582250	.6593496	.6604741	.6615986	.6627231	.6638476	-1	0
0.059	.6649720	.6660965	.6672209	.6683453	.6694698	.6705942	.6717185	.6728429	.6739673	.6750916	-1	0
0.060	.6762159	.6773403	.6784646	.6795889	.6807131	.6818374	.6829616	.6840859	.6852101	.6863343	-1	0
0.061	.6874585	.6885827	.6897069	.6908310	.6919552	.6930793	.6942034	.6953275	.6964516	.6975755	-1	0
0.062	.6986997	.6998237	.7009478	.7020718	.7031958	.7043198	.7054438	.7065677	.7076917	.7088156	-1	0
0.063	.7099395	.7110634	.7121873	.7133112	.7144350	.7155589	.7166827	.7178065	.7189303	.7200541	-1	0
0.064	.7211779	.7223016	.7234254	.7245491	.7256728	.7267965	.7279202	.7290439	.7301676	.7312912	-1	0
0.065	.7324148	.7335384	.7346620	.7357856	.7369092	.7380328	.7391563	.7402798	.7414033	.7425268	-1	0
0.066	.7436503	.7447738	.7458972	.7470207	.7481441	.7492675	.7503909	.7515143	.7526376	.7537610	-1	0
0.067	.7548843	.7560076	.7571309	.7582542	.7593775	.7605008	.7616240	.7627472	.7638704	.7649936	-1	0
0.068	.7661168	.7672400	.7683631	.7694863	.7706094	.7717325	.7728556	.7739787	.7751017	.7762248	-1	0
0.069	.7773478	.7784708	.7795938	.7807168	.7818397	.7829627	.7840856	.7852085	.7863314	.7874543	-1	0
0.070	.7885772	.7897001	.7908229	.7919457	.7930685	.7941913	.7953141	.7964369	.7975596	.7986823	-1	0
0.071	.7998050	.8009277	.8020504	.8031731	.8042957	.8054184	.8065410	.8076636	.8087862	.8099087	-1	0
0.072	.8110313	.8121538	.8132764	.8143989	.8155213	.8166438	.8177663	.8188887	.8200111	.8211335	-1	0
0.073	.8222559	.8233783	.8245007	.8256230	.8267453	.8278676	.8289899	.8301122	.8312345	.8323567	-1	0
0.074	.8334789	.8346011	.8357233	.8368455	.8379677	.8390898	.8402119	.8413340	.8424561	.8435782	-1	0
0.075	.8447003	.8458223	.8469443	.8480663	.8491883	.8503103	.8514323	.8525542	.8536761	.8547980	-1	0
0.076	.8559199	.8570418	.8581636	.8592855	.8604073	.8615291	.8626509	.8637727	.8648944	.8660162	-1	0
0.077	.8671379	.8682596	.8693813	.8705029	.8716246	.8727462	.8738678	.8749894	.8761110	.8772325	-1	0
0.078	.8783541	.8794756	.8805971	.8817186	.8828401	.8839615	.8850830	.8862044	.8873258	.8884472	-1	0
0.079	.8895686	.8906899	.8918112	.8929326	.8940539	.8951751	.8962964	.8974176	.8985389	.8996601	-1	0
0.080	.9007813	.9019024	.9030236	.9041447	.9052658	.9063869	.9075080	.9086291	.9097501	.9108712	-1	0
0.081	.9119922	.9131132	.9142341	.9153551	.9164760	.9175969	.9187178	.9198387	.9209596	.9220804	-1	0
0.082	.9232013	.9243221	.9254429	.9265636	.9276844	.9288051	.9299258	.9310465	.9321672	.9332879	-1	0
0.083	.9344085	.9355291	.9366497	.9377703	.9388909	.9400114	.9411320	.9422525	.9433730	.9444934	-1	0
0.084	.9456139	.9467343	.9478547	.9489751	.9500955	.9512159	.9523362	.9534566	.9545769	.9556971	-1	0
0.085	.9568174	.9579377	.9590579	.9601781	.9612983	.9624184	.9635386	.9646587	.9657788	.9668989	-1	0
0.086	.9680190	.9691391	.9702591	.9713791	.9724991	.9736191	.9747390	.9758590	.9769789	.9780988	-1	0
0.087	.9792187	.9803385	.9814584	.9825782	.9836980	.9848178	.9859376	.9870573	.9881770	.9892967	-1	0
0.088	.0990416	.0991536	.0992656	.0993775	.0994895	.0996015	.0997134	.0998254	.0999373	.1000493	0	0
0.089	.1001612	.1002732	.1003851	.1004971	.1006090	.1007209	.1008329	.1009448	.1010567	.1011687	0	0
0.090	.1012806	.1013925	.1015044	.1016164	.1017283	.1018402	.1019521	.1020640	.1021760	.1022879	0	0
0.091	.1023998	.1025117	.1026236	.1027355	.1028474	.1029593	.1030712	.1031831	.1032950	.1034069	0	0
0.092	.1035187	.1036306	.1037425	.1038544	.1039663	.1040781	.1041900	.1043019	.1044138	.1045256	0	0
0.093	.1046375	.1047494	.1048612	.1049731	.1050850	.1051968	.1053087	.1054205	.1055324	.1056442	0	0
0.094	.1057561	.1058679	.1059797	.1060916	.1062034	.1063153	.1064271	.1065389	.1066508	.1067626	0	0
0.095	.1068744	.1069862	.1070981	.1072099	.1073217	.1074335	.1075453	.1076571	.1077689	.1078807	0	0
0.096	.1079925	.1081043	.1082161	.1083279	.1084397	.1085515	.1086633	.1087751	.1088869	.1089987	0	0
0.097	.1091105	.1092222	.1093340	.1094458	.1095576	.1096693	.1097811	.1098929	.1100046	.1101164	0	0
0.098	.1102282	.1103399	.1104517	.1105634	.1106752	.1107869	.1108987	.1110104	.1111222	.1112339	0	0
0.099	.1113457	.1114574	.1115691	.1116809	.1117926	.1119043	.1120160	.1121278	.1122395	.1123512	0	0

TABLE 7A

Error Function

X	.0	.0001	.0002	.0003	.0004	.0005	.0006	.0007	.0008	.0009	m	Δ_2
0.100	.1124629	.1125746	.1126863	.1127981	.1129098	.1130215	.1131332	.1132449	.1133566	.1134683	0	0
0.101	.1135800	.1136916	.1138033	.1139150	.1140267	.1141384	.1142501	.1143617	.1144734	.1145851	0	0
0.102	.1146968	.1148084	.1149201	.1150318	.1151434	.1152551	.1153667	.1154784	.1155901	.1157017	0	0
0.103	.1158134	.1159250	.1160366	.1161483	.1162599	.1163716	.1164832	.1165948	.1167065	.1168181	0	0
0.104	.1169297	.1170413	.1171530	.1172646	.1173762	.1174878	.1175994	.1177110	.1178226	.1179342	0	0
0.105	.1180458	.1181574	.1182690	.1183806	.1184922	.1186038	.1187154	.1188270	.1189386	.1190501	0	0
0.106	.1191617	.1192733	.1193849	.1194964	.1196080	.1197196	.1198311	.1199427	.1200543	.1201658	0	0
0.107	.1202774	.1203889	.1205005	.1206120	.1207236	.1208351	.1209467	.1210582	.1211697	.1212813	0	0
0.108	.1213928	.1215043	.1216158	.1217274	.1218389	.1219504	.1220619	.1221734	.1222850	.1223965	0	0
0.109	.1225080	.1226195	.1227310	.1228425	.1229540	.1230655	.1231770	.1232884	.1233999	.1235114	0	0
0.110	.1236229	.1237344	.1238459	.1239573	.1240688	.1241803	.1242917	.1244032	.1245147	.1246261	0	0
0.111	.1247376	.1248490	.1249605	.1250719	.1251834	.1252948	.1254063	.1255177	.1256292	.1257406	0	0
0.112	.1258520	.1259634	.1260749	.1261863	.1262977	.1264091	.1265206	.1266320	.1267434	.1268548	0	0
0.113	.1269662	.1270776	.1271890	.1273004	.1274118	.1275232	.1276346	.1277460	.1278574	.1279688	0	0
0.114	.1280801	.1281915	.1283029	.1284143	.1285256	.1286370	.1287484	.1288597	.1289711	.1290825	0	0
0.115	.1291938	.1293052	.1294165	.1295279	.1296392	.1297506	.1298619	.1299733	.1300846	.1301959	0	0
0.116	.1303073	.1304186	.1305299	.1306412	.1307525	.1308639	.1309752	.1310865	.1311978	.1313091	0	0
0.117	.1314204	.1315317	.1316430	.1317543	.1318656	.1319769	.1320882	.1321995	.1323108	.1324220	0	0
0.118	.1325333	.1326446	.1327559	.1328671	.1329784	.1330897	.1332009	.1333122	.1334235	.1335347	0	0
0.119	.1336460	.1337572	.1338685	.1339797	.1340910	.1342022	.1343134	.1344247	.1345359	.1346471	0	0
0.120	.1347584	.1348696	.1349808	.1350920	.1352032	.1353144	.1354257	.1355369	.1356481	.1357593	0	0
0.121	.1358705	.1359817	.1360929	.1362040	.1363152	.1364264	.1365376	.1366488	.1367600	.1368711	0	0
0.122	.1369823	.1370935	.1372046	.1373158	.1374270	.1375381	.1376493	.1377604	.1378716	.1379827	0	0
0.123	.1380939	.1382050	.1383162	.1384273	.1385384	.1386496	.1387607	.1388718	.1389829	.1390941	0	0
0.124	.1392052	.1393163	.1394274	.1395385	.1396496	.1397607	.1398718	.1399829	.1400940	.1402051	0	0
0.125	.1403162	.1404273	.1405384	.1406495	.1407605	.1408716	.1409827	.1410938	.1412048	.1413159	0	0
0.126	.1414270	.1415380	.1416491	.1417601	.1418712	.1419822	.1420933	.1422043	.1423153	.1424264	0	0
0.127	.1425374	.1426484	.1427595	.1428705	.1429815	.1430925	.1432036	.1433146	.1434256	.1435366	0	0
0.128	.1436476	.1437586	.1438696	.1439806	.1440916	.1442026	.1443136	.1444246	.1445355	.1446465	0	0
0.129	.1447575	.1448685	.1449794	.1450904	.1452014	.1453123	.1454233	.1455343	.1456452	.1457562	0	0
0.130	.1458671	.1459781	.1460890	.1461999	.1463109	.1464218	.1465327	.1466437	.1467546	.1468655	0	0
0.131	.1469764	.1470874	.1471983	.1473092	.1474201	.1475310	.1476419	.1477528	.1478637	.1479746	0	0
0.132	.1480855	.1481964	.1483072	.1484181	.1485290	.1486399	.1487508	.1488616	.1489725	.1490834	0	0
0.133	.1491942	.1493051	.1494159	.1495268	.1496376	.1497485	.1498593	.1499702	.1500810	.1501918	0	0
0.134	.1503027	.1504135	.1505243	.1506351	.1507460	.1508568	.1509676	.1510784	.1511892	.1513000	0	0
0.135	.1514108	.1515216	.1516324	.1517432	.1518540	.1519648	.1520756	.1521863	.1522971	.1524079	0	0
0.136	.1525187	.1526294	.1527402	.1528510	.1529617	.1530725	.1531832	.1532940	.1534047	.1535155	0	0
0.137	.1536262	.1537370	.1538477	.1539584	.1540691	.1541799	.1542906	.1544013	.1545120	.1546227	0	0
0.138	.1547335	.1548442	.1549549	.1550656	.1551763	.1552870	.1553977	.1555084	.1556190	.1557297	0	0
0.139	.1558404	.1559511	.1560618	.1561724	.1562831	.1563938	.1565044	.1566151	.1567257	.1568364	0	0
0.140	.1569470	.1570577	.1571683	.1572790	.1573896	.1575002	.1576109	.1577215	.1578321	.1579427	0	0
0.141	.1580534	.1581640	.1582746	.1583852	.1584958	.1586064	.1587170	.1588276	.1589382	.1590488	0	0
0.142	.1591594	.1592700	.1593805	.1594911	.1596017	.1597123	.1598228	.1599334	.1600439	.1601545	0	0
0.143	.1602651	.1603756	.1604862	.1605967	.1607073	.1608178	.1609283	.1610389	.1611494	.1612599	0	0
0.144	.1613704	.1614810	.1615915	.1617020	.1618125	.1619230	.1620335	.1621440	.1622545	.1623650	0	0
0.145	.1624755	.1625860	.1626965	.1628070	.1629174	.1630279	.1631384	.1632489	.1633593	.1634698	0	0
0.146	.1635802	.1636907	.1638012	.1639116	.1640221	.1641325	.1642429	.1643534	.1644638	.1645742	0	0
0.147	.1646847	.1647951	.1649055	.1650159	.1651263	.1652368	.1653472	.1654576	.1655680	.1656784	0	0
0.148	.1657888	.1658992	.1660095	.1661199	.1662303	.1663407	.1664511	.1665614	.1666718	.1667822	0	0
0.149	.1668925	.1670029	.1671132	.1672236	.1673339	.1674443	.1675546	.1676650	.1677753	.1678856	0	0

TABLE 7A

Error Function

X	.0	.0001	.0002	.0003	.0004	.0005	.0006	.0007	.0008	.0009	m	Δ_2
0.150	.1679960	.1581063	.1682166	.1683269	.1684373	.1685476	.1686579	.1687682	.1688785	.1689889	0	0
0.151	.1690991	.1692094	.1693197	.1694299	.1695402	.1696505	.1697608	.1698711	.1699813	.1700916	0	0
0.152	.1702019	.1703121	.1704224	.1705326	.1706429	.1707531	.1708634	.1709736	.1710838	.1711941	0	0
0.153	.1713043	.1714145	.1715247	.1716350	.1717452	.1718554	.1719656	.1720758	.1721860	.1722962	0	0
0.154	.1724064	.1725166	.1726268	.1727370	.1728471	.1729573	.1730675	.1731777	.1732878	.1733980	0	0
0.155	.1735082	.1736183	.1737285	.1738386	.1739488	.1740589	.1741691	.1742792	.1743893	.1744995	0	0
0.156	.1746096	.1747197	.1748298	.1749399	.1750501	.1751602	.1752703	.1753804	.1754905	.1756006	0	0
0.157	.1757107	.1758207	.1759308	.1760409	.1761510	.1762611	.1763711	.1764812	.1765913	.1767013	0	0
0.158	.1768114	.1769214	.1770315	.1771415	.1772516	.1773616	.1774717	.1775817	.1776917	.1778018	0	0
0.159	.1779118	.1780218	.1781318	.1782418	.1783518	.1784618	.1785718	.1786818	.1787918	.1789018	0	0
0.160	.1790118	.1791218	.1792318	.1793418	.1794517	.1795617	.1796717	.1797816	.1798916	.1800015	0	0
0.161	.1801115	.1802214	.1803314	.1804413	.1805513	.1806612	.1807711	.1808811	.1809910	.1811009	0	0
0.162	.1812108	.1813207	.1814306	.1815406	.1816505	.1817604	.1818703	.1819801	.1820900	.1821999	0	0
0.163	.1823098	.1824197	.1825295	.1826394	.1827493	.1828592	.1829690	.1830789	.1831887	.1832986	0	0
0.164	.1834084	.1835183	.1836281	.1837379	.1838478	.1839576	.1840674	.1841772	.1842870	.1843969	0	0
0.165	.1845067	.1846165	.1847263	.1848361	.1849459	.1850557	.1851654	.1852752	.1853350	.1854948	0	0
0.166	.1856046	.1857143	.1858241	.1859339	.1860436	.1861534	.1862631	.1863729	.1864826	.1865923	0	0
0.167	.1867021	.1868118	.1869215	.1870313	.1871410	.1872507	.1873604	.1874701	.1875798	.1876895	0	0
0.168	.1877992	.1879089	.1880186	.1381283	.1882380	.1883477	.1884574	.1885670	.1886767	.1887864	0	0
0.169	.1888960	.1890057	.1891154	.1892250	.1893347	.1894443	.1895539	.1896636	.1897732	.1898828	0	0
0.170	.1899925	.1901021	.1902117	.1903213	.1904309	.1905405	.1906501	.1907597	.1908693	.1909789	0	0
0.171	.1910885	.1911981	.1913077	.1914173	.1915268	.1916364	.1917460	.1918555	.1919651	.1920746	0	0
0.172	.1921842	.1922937	.1924033	.1925128	.1926223	.1927319	.1928414	.1929509	.1930605	.1931700	0	0
0.173	.1932795	.1933890	.1934985	.1936080	.1937175	.1938270	.1939365	.1940460	.1941554	.1942649	0	0
0.174	.1943744	.1944839	.1945933	.1947028	.1948123	.1949217	.1950312	.1951406	.1952501	.1953595	0	0
0.175	.1954689	.1955784	.1956878	.1957972	.1959066	.1960161	.1961255	.1962349	.1963443	.1964537	0	0
0.176	.1965631	.1966725	.1967819	.1968913	.1970006	.1971100	.1972194	.1973288	.1974381	.1975475	0	0
0.177	.1976569	.1977662	.1978756	.1979849	.1980943	.1982036	.1983129	.1984223	.1985316	.1986409	0	0
0.178	.1987502	.1988596	.1989689	.1990782	.1991875	.1992968	.1994061	.1995154	.1996247	.1997340	0	0
0.179	.1998432	.1999525	.2000618	.2001711	.2002803	.2003896	.2004988	.2006081	.2007174	.2008266	0	0
0.180	.2009358	.2010451	.2011543	.2012635	.2013728	.2014820	.2015912	.2017004	.2018096	.2019188	0	0
0.181	.2020280	.2021372	.2022464	.2023556	.2024648	.2025740	.2026832	.2027924	.2029015	.2030107	0	0
0.182	.2031199	.2032290	.2033382	.2034473	.2035565	.2036656	.2037748	.2038839	.2039930	.2041022	0	0
0.183	.2042113	.2043204	.2044295	.2045386	.2046477	.2047568	.2048659	.2049750	.2050841	.2051932	0	0
0.184	.2053023	.2054114	.2055204	.2056295	.2057386	.2058477	.2059567	.2060658	.2061748	.2062839	0	0
0.185	.2063929	.2065019	.2066110	.2067200	.2068290	.2069381	.2070471	.2071561	.2072651	.2073741	0	0
0.186	.2074831	.2075921	.2077011	.2078101	.2079191	.2080281	.2081371	.2082460	.2083550	.2084640	0	0
0.187	.2085729	.2086819	.2087908	.2088998	.2090087	.2091177	.2092266	.2093355	.2094445	.2095534	0	0
0.188	.2096623	.2097712	.2098802	.2099891	.2100980	.2102069	.2103158	.2104247	.2105335	.2106424	0	0
0.189	.2107513	.2108602	.2109691	.2110779	.2111868	.2112957	.2114045	.2115134	.2116222	.2117311	0	0
0.190	.2118399	.2119487	.2120576	.2121664	.2122752	.2123840	.2124928	.2126017	.2127105	.2128193	0	0
0.191	.2129281	.2130368	.2131456	.2132544	.2133632	.2134720	.2135808	.2136895	.2137983	.2139070	0	0
0.192	.2140158	.2141246	.2142333	.2143420	.2144508	.2145595	.2146683	.2147770	.2148857	.2149944	0	0
0.193	.2151031	.2152118	.2153206	.2154293	.2155380	.2156466	.2157553	.2158640	.2159727	.2160814	0	0
0.194	.2161900	.2162987	.2164074	.2165160	.2166247	.2167333	.2168420	.2169506	.2170593	.2171679	0	0
0.195	.2172765	.2173852	.2174938	.2176024	.2177110	.2178196	.2179282	.2180368	.2181454	.2182540	0	0
0.196	.2183626	.2184712	.2185798	.2186883	.2187969	.2189055	.2190140	.2191226	.2192311	.2193397	0	0
0.197	.2194482	.2195568	.2196653	.2197739	.2198824	.2199909	.2200994	.2202079	.2203164	.2204250	0	0
0.198	.2205335	.2206420	.2207504	.2208589	.2209674	.2210759	.2211844	.2212928	.2214013	.2215098	0	0
0.199	.2216182	.2217267	.2218351	.2219436	.2220520	.2221605	.2222689	.2223773	.2224858	.2225942	0	0

TABLE 7A

Error Function

X	.0	.0001	.0002	.0003	.0004	.0005	.0006	.0007	.0008	.0009	m	Δ_2
0.200	.2227026	.2228110	.2229194	.2230278	.2231362	.2232445	.2233530	.2234613	.2235698	.2236781	0	0
0.201	.2237865	.2238948	.2240032	.2241116	.2242199	.2243283	.2244366	.2245449	.2246533	.2247616	0	0
0.202	.2248700	.2249783	.2250866	.2251949	.2253032	.2254115	.2255198	.2256281	.2257364	.2258447	0	0
0.203	.2259530	.2260613	.2261696	.2262778	.2263861	.2264944	.2266026	.2267109	.2268191	.2269273	0	0
0.204	.2270356	.2271438	.2272521	.2273603	.2274686	.2275767	.2276850	.2277932	.2279014	.2280096	0	0
0.205	.2281178	.2282259	.2283341	.2284423	.2285505	.2286587	.2287669	.2288750	.2289832	.2290913	0	0
0.206	.2291995	.2293077	.2294158	.2295239	.2296321	.2297402	.2298483	.2299564	.2300646	.2301726	0	0
0.207	.2302808	.2303889	.2304969	.2306051	.2307131	.2308213	.2309293	.2310374	.2311454	.2312535	0	0
0.208	.2313616	.2314696	.2315777	.2316858	.2317938	.2319018	.2320099	.2321179	.2322260	.2323340	0	0
0.209	.2324420	.2325500	.2326580	.2327660	.2328740	.2329820	.2330900	.2331980	.2333059	.2334139	0	0
0.210	.2335219	.2336299	.2337378	.2338458	.2339537	.2340617	.2341696	.2342775	.2343855	.2344934	0	0
0.211	.2346014	.2347093	.2348172	.2349251	.2350330	.2351409	.2352488	.2353567	.2354646	.2355725	0	0
0.212	.2356804	.2357883	.2358961	.2360040	.2361119	.2362197	.2363276	.2364354	.2365432	.2366511	0	0
0.213	.2367589	.2368668	.2369746	.2370824	.2371902	.2372981	.2374058	.2375137	.2376215	.2377293	0	0
0.214	.2378370	.2379448	.2380526	.2381604	.2382681	.2383759	.2384837	.2385914	.2386992	.2388070	0	0
0.215	.2389147	.2390224	.2391301	.2392379	.2393456	.2394533	.2395610	.2396687	.2397764	.2398841	0	0
0.216	.2399918	.2400995	.2402073	.2403149	.2404226	.2405303	.2406380	.2407456	.2408533	.2409609	0	0
0.217	.2410685	.2411762	.2412838	.2413915	.2414991	.2416068	.2417144	.2418220	.2419296	.2420372	0	0
0.218	.2421448	.2422524	.2423600	.2424676	.2425752	.2426828	.2427903	.2428979	.2430055	.2431130	0	0
0.219	.2432206	.2433282	.2434357	.2435432	.2436507	.2437583	.2438658	.2439733	.2440808	.2441884	0	0
0.220	.2442959	.2444034	.2445109	.2446184	.2447259	.2448334	.2449408	.2450483	.2451558	.2452633	0	0
0.221	.2453707	.2454782	.2455856	.2456931	.2458005	.2459080	.2460154	.2461228	.2462302	.2463377	0	0
0.222	.2464451	.2465525	.2466599	.2467673	.2468747	.2469820	.2470894	.2471969	.2473042	.2474115	0	0
0.223	.2475190	.2476263	.2477337	.2478410	.2479483	.2480557	.2481630	.2482704	.2483777	.2484850	0	0
0.224	.2485923	.2486997	.2488070	.2489142	.2490216	.2491289	.2492362	.2493435	.2494507	.2495580	0	0
0.225	.2496653	.2497725	.2498798	.2499871	.2500943	.2502015	.2503088	.2504160	.2505233	.2506305	0	0
0.226	.2507377	.2508449	.2509521	.2510594	.2511665	.2512738	.2513809	.2514881	.2515953	.2517025	0	0
0.227	.2518097	.2519168	.2520240	.2521312	.2522383	.2523454	.2524526	.2525597	.2526668	.2527740	0	0
0.228	.2528811	.2529882	.2530953	.2532024	.2533095	.2534167	.2535238	.2536308	.2537379	.2538450	0	0
0.229	.2539521	.2540591	.2541662	.2542733	.2543803	.2544874	.2545944	.2547015	.2548085	.2549155	0	0
0.230	.2550226	.2551296	.2552366	.2553436	.2554506	.2555577	.2556646	.2557716	.2558786	.2559856	0	0
0.231	.2560925	.2561995	.2563065	.2564135	.2565205	.2566274	.2567343	.2568412	.2569482	.2570552	0	0
0.232	.2571620	.2572690	.2573759	.2574828	.2575897	.2576966	.2578035	.2579104	.2580173	.2581242	0	0
0.233	.2582310	.2583379	.2584448	.2585517	.2586585	.2587654	.2588722	.2589791	.2590859	.2591928	0	0
0.234	.2592996	.2594064	.2595132	.2596200	.2597268	.2598336	.2599404	.2600472	.2601540	.2602608	0	0
0.235	.2603676	.2604743	.2605811	.2606879	.2607946	.2609014	.2610081	.2611149	.2612216	.2613283	0	0
0.236	.2614351	.2615418	.2616485	.2617552	.2618619	.2619686	.2620754	.2621821	.2622887	.2623954	0	0
0.237	.2625021	.2626088	.2627154	.2628221	.2629287	.2630354	.2631420	.2632487	.2633553	.2634619	0	0
0.238	.2635686	.2636752	.2637818	.2638884	.2639951	.2641016	.2642082	.2643148	.2644214	.2645280	0	0
0.239	.2646345	.2647411	.2648477	.2649543	.2650608	.2651674	.2652739	.2653804	.2654870	.2655935	0	0
0.240	.2657000	.2658066	.2659131	.2660196	.2661261	.2662326	.2663391	.2664456	.2665520	.2666585	0	0
0.241	.2667650	.2668715	.2669779	.2670844	.2671908	.2672973	.2674037	.2675102	.2676166	.2677230	0	0
0.242	.2678294	.2679359	.2680423	.2681487	.2682551	.2683614	.2684678	.2685742	.2686806	.2687870	0	0
0.243	.2688934	.2689998	.2691061	.2692125	.2693188	.2694252	.2695315	.2696378	.2697442	.2698505	0	0
0.244	.2699568	.2700631	.2701694	.2702757	.2703820	.2704883	.2705946	.2707009	.2708071	.2709134	0	0
0.245	.2710197	.2711260	.2712322	.2713385	.2714447	.2715510	.2716572	.2717634	.2718697	.2719759	0	0
0.246	.2720821	.2721883	.2722945	.2724007	.2725069	.2726130	.2727193	.2728254	.2729316	.2730378	0	0
0.247	.2731439	.2732501	.2733563	.2734624	.2735685	.2736747	.2737808	.2738869	.2739931	.2740992	0	0
0.248	.2742053	.2743114	.2744175	.2745236	.2746297	.2747357	.2748418	.2749479	.2750540	.2751600	0	0
0.249	.2752661	.2753721	.2754782	.2755842	.2756903	.2757963	.2759023	.2760083	.2761143	.2762204	0	0

TABLE 7A

Error Function

X	.0	.0001	.0002	.0003	.0004	.0005	.0006	.0007	.0008	.0009	m	Δ_2
0.250	.2763264	.2764323	.2765384	.2766443	.2767503	.2768563	.2769623	.2770683	.2771742	.2772802	0	0
0.251	.2773861	.2774920	.2775980	.2777039	.2778099	.2779158	.2780217	.2781276	.2782335	.2783394	0	0
0.252	.2784453	.2785512	.2786571	.2787630	.2788689	.2789747	.2790806	.2791865	.2792923	.2793981	0	0
0.253	.2795040	.2796099	.2797157	.2798215	.2799273	.2800332	.2801390	.2802448	.2803506	.2804564	0	0
0.254	.2805622	.2806680	.2807737	.2808795	.2809852	.2810910	.2811968	.2813025	.2814083	.2815140	0	0
0.255	.2816198	.2817255	.2818312	.2819369	.2820427	.2821484	.2822541	.2823598	.2824655	.2825711	0	0
0.256	.2826768	.2827825	.2828882	.2829939	.2830995	.2832052	.2833108	.2834165	.2835221	.2836277	0	0
0.257	.2837334	.2838390	.2839446	.2840502	.2841558	.2842614	.2843670	.2844726	.2845782	.2846838	0	0
0.258	.2847894	.2848949	.2850005	.2851061	.2852116	.2853171	.2854227	.2855282	.2856337	.2857393	0	0
0.259	.2858448	.2859503	.2860558	.2861613	.2862668	.2863723	.2864778	.2865833	.2866887	.2867942	0	0
0.260	.2868997	.2870051	.2871106	.2872161	.2873215	.2874269	.2875324	.2876378	.2877432	.2878487	0	0
0.261	.2879540	.2880595	.2881649	.2882702	.2883756	.2884810	.2885864	.2886918	.2887971	.2889025	0	0
0.262	.2890078	.2891132	.2892185	.2893239	.2894292	.2895345	.2896398	.2897452	.2898505	.2899558	0	0
0.263	.2900611	.2901664	.2902716	.2903770	.2904822	.2905875	.2906927	.2907980	.2909033	.2910085	0	0
0.264	.2911138	.2912190	.2913243	.2914295	.2915347	.2916399	.2917451	.2918503	.2919555	.2920607	0	0
0.265	.2921659	.2922711	.2923763	.2924814	.2925866	.2926918	.2927969	.2929021	.2930072	.2931123	0	0
0.266	.2932175	.2933226	.2934278	.2935328	.2936380	.2937431	.2938482	.2939532	.2940583	.2941634	0	0
0.267	.2942685	.2943736	.2944787	.2945837	.2946888	.2947938	.2948989	.2950039	.2951089	.2952139	0	0
0.268	.2953190	.2954240	.2955290	.2956340	.2957390	.2958440	.2959490	.2960539	.2961589	.2962639	0	0
0.269	.2963688	.2964738	.2965788	.2966837	.2967886	.2968936	.2969985	.2971035	.2972084	.2973133	0	0
C.270	.2974182	.2975231	.2976280	.2977329	.2978377	.2979426	.2980475	.2981524	.2982572	.2983621	0	0
0.271	.2984670	.2985718	.2986766	.2987815	.2988863	.2989911	.2990959	.2992008	.2993056	.2994103	0	0
0.272	.2995151	.2996199	.2997247	.2998295	.2999343	.3000391	.3001438	.3002486	.3003533	.3004580	0	0
0.273	.3005628	.3006675	.3007722	.3008770	.3009816	.3010864	.3011911	.3012958	.3014005	.3015051	0	0
0.274	.3016098	.3017145	.3018191	.3019238	.3020285	.3021331	.3022378	.3023424	.3024471	.3025517	0	0
0.275	.3026563	.3027609	.3028655	.3029701	.3030747	.3031793	.3032839	.3033885	.3034931	.3035976	0	0
0.276	.3037022	.3038068	.3039113	.3040159	.3041204	.3042250	.3043295	.3044340	.3045385	.3046430	0	0
0.277	.3047475	.3048520	.3049566	.3050610	.3051655	.3052700	.3053744	.3054789	.3055834	.3056878	0	0
0.278	.3057923	.3058967	.3060012	.3061056	.3062100	.3063145	.3064188	.3065233	.3066277	.3067321	0	0
0.279	.3068364	.3069409	.3070452	.3071496	.3072540	.3073583	.3074627	.3075670	.3076714	.3077757	0	0
0.280	.3078800	.3079844	.3080887	.3081930	.3082973	.3084016	.3085059	.3086102	.3087145	.3088188	0	0
0.281	.3089231	.3090273	.3091316	.3092358	.3093401	.3094443	.3095486	.3096528	.3097570	.3098612	0	0
0.282	.3099654	.3100697	.3101739	.3102781	.3103822	.3104864	.3105906	.3106948	.3107989	.3109031	0	0
0.283	.3110073	.3111115	.3112156	.3113197	.3114238	.3115280	.3116321	.3117362	.3118403	.3119444	0	0
0.284	.3120485	.3121526	.3122567	.3123608	.3124648	.3125689	.3126730	.3127770	.3128811	.3129851	0	0
0.285	.3130892	.3131932	.3132972	.3134012	.3135052	.3136092	.3137133	.3138173	.3139213	.3140252	0	0
0.286	.3141292	.3142332	.3143371	.3144411	.3145451	.3146490	.3147529	.3148569	.3149608	.3150647	0	0
0.287	.3151687	.3152726	.3153765	.3154804	.3155843	.3156882	.3157921	.3158959	.3159998	.3161037	0	0
0.288	.3162075	.3163114	.3164153	.3165191	.3166229	.3167267	.3168306	.3169344	.3170382	.3171420	0	0
0.289	.3172458	.3173496	.3174534	.3175572	.3176609	.3177647	.3178685	.3179722	.3180760	.3181797	0	0
0.290	.3182834	.3183872	.3184909	.3185946	.3186983	.3188021	.3189058	.3190095	.3191131	.3192168	0	0
0.291	.3193205	.3194242	.3195279	.3196315	.3197352	.3198388	.3199425	.3200461	.3201497	.3202534	0	0
0.292	.3203570	.3204606	.3205642	.3206678	.3207714	.3208750	.3209786	.3210822	.3211857	.3212893	0	0
0.293	.3213928	.3214964	.3216000	.3217035	.3218070	.3219106	.3220140	.3221176	.3222211	.3223246	0	0
0.294	.3224281	.3225316	.3226351	.3227385	.3228420	.3229455	.3230489	.3231524	.3232558	.3233593	0	0
0.295	.3234627	.3235661	.3236696	.3237730	.3238764	.3239798	.3240832	.3241866	.3242900	.3243934	0	0
0.296	.3244967	.3246001	.3247035	.3248069	.3249102	.3250135	.3251169	.3252202	.3253235	.3254269	0	0
0.297	.3255302	.3256335	.3257368	.3258401	.3259434	.3260466	.3261499	.3262532	.3263565	.3264597	0	0
0.298	.3265630	.3266662	.3267695	.3268727	.3269759	.3270791	.3271824	.3272856	.3273888	.3274919	0	0
0.299	.3275952	.3276983	.3278015	.3279047	.3280079	.3281110	.3282142	.3283173	.3284205	.3285236	0	0

TABLE 7A

Error Function

X	.0	.0001	.0002	.0003	.0004	.0005	.0006	.0007	.0008	.0009	m	Δ_2
0.300	.3286268	.3287299	.3288330	.3289361	.3290392	.3291423	.3292454	.3293484	.3294516	.3295546	0	0
0.301	.3296577	.3297607	.3298638	.3299668	.3300699	.3301729	.3302760	.3303789	.3304820	.3305850	0	0
0.302	.3306880	.3307910	.3308940	.3309970	.3311000	.3312029	.3313059	.3314089	.3315119	.3316148	0	0
0.303	.3317177	.3318207	.3319236	.3320265	.3321294	.3322324	.3323352	.3324382	.3325410	.3326439	0	0
0.304	.3327468	.3328497	.3329526	.3330554	.3331583	.3332611	.3333640	.3334668	.3335696	.3336725	0	0
0.305	.3337753	.3338781	.3339809	.3340837	.3341865	.3342893	.3343920	.3344948	.3345976	.3347003	0	0
0.306	.3348031	.3349059	.3350086	.3351113	.3352141	.3353168	.3354195	.3355222	.3356249	.3357276	0	0
0.307	.3358303	.3359330	.3360357	.3361384	.3362410	.3363436	.3364463	.3365490	.3366516	.3367543	0	0
0.308	.3368569	.3369595	.3370621	.3371647	.3372673	.3373699	.3374725	.3375751	.3376777	.3377802	0	0
0.309	.3378828	.3379854	.3380879	.3381905	.3382930	.3383955	.3384981	.3386006	.3387031	.3388056	0	0
0.310	.3389081	.3390106	.3391131	.3392156	.3393180	.3394206	.3395230	.3396255	.3397279	.3398303	0	0
0.311	.3399328	.3400352	.3401377	.3402401	.3403425	.3404449	.3405473	.3406497	.3407521	.3408545	0	0
0.312	.3409568	.3410592	.3411615	.3412639	.3413662	.3414686	.3415709	.3416733	.3417756	.3418779	0	0
0.313	.3419802	.3420825	.3421848	.3422871	.3423894	.3424917	.3425940	.3426962	.3427985	.3429008	0	0
0.314	.3430030	.3431052	.3432075	.3433097	.3434119	.3435141	.3436163	.3437185	.3438207	.3439229	0	0
0.315	.3440251	.3441272	.3442295	.3443316	.3444337	.3445359	.3446380	.3447402	.3448423	.3449444	0	0
0.316	.3450466	.3451487	.3452508	.3453529	.3454550	.3455570	.3456591	.3457612	.3458633	.3459653	0	0
0.317	.3460674	.3461694	.3462715	.3463735	.3464755	.3465775	.3466796	.3467816	.3468836	.3469856	0	0
0.318	.3470876	.3471895	.3472915	.3473935	.3474954	.3475974	.3476993	.3478013	.3479033	.3480052	0	0
0.319	.3481071	.3482090	.3483109	.3484128	.3485147	.3486166	.3487185	.3488204	.3489223	.3490241	0	0
0.320	.3491260	.3492278	.3493297	.3494315	.3495333	.3496352	.3497370	.3498388	.3499406	.3500424	0	0
0.321	.3501442	.3502460	.3503478	.3504495	.3505513	.3506531	.3507548	.3508565	.3509583	.3510600	0	0
0.322	.3511618	.3512635	.3513652	.3514669	.3515686	.3516703	.3517720	.3518737	.3519753	.3520770	0	0
0.323	.3521787	.3522803	.3523820	.3524836	.3525853	.3526869	.3527885	.3528901	.3529918	.3530933	0	0
0.324	.3531950	.3532965	.3533981	.3534997	.3536013	.3537028	.3538044	.3539059	.3540075	.3541090	0	0
0.325	.3542106	.3543121	.3544136	.3545151	.3546166	.3547181	.3548196	.3549211	.3550225	.3551240	0	0
0.326	.3552255	.3553270	.3554284	.3555298	.3556313	.3557327	.3558341	.3559356	.3560370	.3561383	0	0
0.327	.3562398	.3563412	.3564426	.3565439	.3566453	.3567467	.3568480	.3569494	.3570507	.3571520	0	0
0.328	.3572534	.3573547	.3574560	.3575574	.3576586	.3577600	.3578612	.3579625	.3580638	.3581651	0	0
0.329	.3582664	.3583676	.3584688	.3585701	.3586713	.3587726	.3588738	.3589750	.3590762	.3591774	0	0
0.330	.3592786	.3593798	.3594810	.3595822	.3596833	.3597845	.3598857	.3599868	.3600880	.3601891	0	0
0.331	.3602902	.3603914	.3604925	.3605936	.3606947	.3607958	.3608969	.3609980	.3610991	.3612002	0	0
0.332	.3613012	.3614023	.3615033	.3616043	.3617054	.3618064	.3619074	.3620085	.3621095	.3622105	0	0
0.333	.3623115	.3624125	.3625134	.3626145	.3627154	.3628163	.3629173	.3630183	.3631192	.3632202	0	0
0.334	.3633211	.3634220	.3635229	.3636239	.3637247	.3638256	.3639265	.3640274	.3641283	.3642291	0	0
0.335	.3643300	.3644308	.3645317	.3646325	.3647334	.3648342	.3649350	.3650358	.3651367	.3652375	0	0
0.336	.3653383	.3654391	.3655398	.3656406	.3657414	.3658422	.3659429	.3660436	.3661444	.3662451	0	0
0.337	.3663458	.3664466	.3665473	.3666480	.3667487	.3668494	.3669500	.3670508	.3671514	.3672521	0	0
0.338	.3673527	.3674534	.3675541	.3676547	.3677553	.3678560	.3679566	.3680572	.3681577	.3682584	0	0
0.339	.3683590	.3684595	.3685601	.3686607	.3687612	.3688618	.3689623	.3690629	.3691635	.3692639	0	0
0.340	.3693645	.3694650	.3695655	.3696660	.3697665	.3698670	.3699675	.3700680	.3701684	.3702689	0	0
0.341	.3703694	.3704698	.3705702	.3706707	.3707711	.3708715	.3709719	.3710724	.3711727	.3712732	0	0
0.342	.3713735	.3714739	.3715743	.3716747	.3717750	.3718753	.3719757	.3720760	.3721763	.3722767	0	0
0.343	.3723770	.3724773	.3725776	.3726779	.3727782	.3728785	.3729787	.3730791	.3731793	.3732796	0	0
0.344	.3733798	.3734800	.3735803	.3736805	.3737807	.3738809	.3739811	.3740813	.3741815	.3742817	0	0
0.345	.3743819	.3744821	.3745822	.3746824	.3747826	.3748827	.3749828	.3750830	.3751831	.3752832	0	0
0.346	.3753833	.3754834	.3755835	.3756836	.3757837	.3758838	.3759838	.3760839	.3761839	.3762840	0	0
0.347	.3763840	.3764841	.3765841	.3766841	.3767841	.3768842	.3769841	.3770841	.3771841	.3772841	0	0
0.348	.3773841	.3774840	.3775840	.3776839	.3777839	.3778838	.3779837	.3780836	.3781836	.3782835	0	0
0.349	.3783834	.3784833	.3785832	.3786830	.3787829	.3788828	.3789827	.3790825	.3791823	.3792822	0	0

TABLE 7A

Error Function

X	.0	.0001	.0002	.0003	.0004	.0005	.0006	.0007	.0008	.0009	m	Δ_2
0.350	.3793820	.3794819	.3795817	.3796815	.3797813	.3798811	.3799809	.3800806	.3801804	.3802802	0	0
0.351	.3803800	.3804797	.3805795	.3806792	.3807789	.3808787	.3809784	.3810781	.3811778	.3812775	0	0
0.352	.3813772	.3814769	.3815765	.3816762	.3817759	.3818756	.3819752	.3820748	.3821745	.3822741	0	0
0.353	.3823738	.3824733	.3825729	.3826725	.3827721	.3828717	.3829713	.3830709	.3831704	.3832700	0	0
0.354	.3833696	.3834691	.3835686	.3836682	.3837677	.3838672	.3839667	.3840662	.3841657	.3842652	0	0
0.355	.3843647	.3844641	.3845636	.3846631	.3847625	.3848619	.3849614	.3850608	.3851603	.3852597	0	0
0.356	.3853591	.3854585	.3855579	.3856573	.3857567	.3858560	.3859554	.3860548	.3861541	.3862535	0	0
0.357	.3863528	.3864521	.3865514	.3866508	.3867501	.3868494	.3869487	.3870480	.3871472	.3872465	0	0
0.358	.3873458	.3874451	.3875443	.3876436	.3877428	.3878421	.3879412	.3880405	.3881397	.3882389	0	0
0.359	.3883381	.3884373	.3885365	.3886356	.3887348	.3888340	.3889331	.3890322	.3891314	.3892305	0	0
0.360	.3893297	.3894288	.3895279	.3896270	.3897261	.3898252	.3899243	.3900234	.3901224	.3902215	0	0
0.361	.3903205	.3904196	.3905186	.3906177	.3907167	.3908157	.3909147	.3910137	.3911127	.3912117	0	0
0.362	.3913107	.3914096	.3915086	.3916076	.3917065	.3918055	.3919044	.3920034	.3921023	.3922012	0	0
0.363	.3923001	.3923990	.3924979	.3925968	.3926957	.3927945	.3928934	.3929923	.3930911	.3931900	0	0
0.364	.3932889	.3933877	.3934865	.3935853	.3936841	.3937829	.3938817	.3939805	.3940793	.3941780	0	0
0.365	.3942768	.3943756	.3944743	.3945731	.3946718	.3947706	.3948693	.3949680	.3950667	.3951654	0	0
0.366	.3952641	.3953628	.3954614	.3955601	.3956588	.3957574	.3958561	.3959548	.3960534	.3961520	0	0
0.367	.3962507	.3963493	.3964479	.3965465	.3966451	.3967437	.3968422	.3969408	.3970394	.3971379	0	0
0.368	.3972365	.3973350	.3974336	.3975321	.3976306	.3977291	.3978276	.3979262	.3980246	.3981231	0	0
0.369	.3982216	.3983201	.3984185	.3985170	.3986154	.3987139	.3988123	.3989107	.3990091	.3991076	0	0
0.370	.3992060	.3993043	.3994027	.3995011	.3995995	.3996978	.3997962	.3998946	.3999929	.4000913	0	0
0.371	.4001896	.4002879	.4003862	.4004846	.4005828	.4006811	.4007794	.4008777	.4009760	.4010742	0	0
0.372	.4011725	.4012707	.4013690	.4014673	.4015655	.4016637	.4017619	.4018601	.4019583	.4020565	0	0
0.373	.4021547	.4022529	.4023511	.4024492	.4025474	.4026455	.4027436	.4028418	.4029399	.4030380	0	0
0.374	.4031361	.4032342	.4033324	.4034305	.4035285	.4036266	.4037247	.4038227	.4039208	.4040188	0	0
0.375	.4041169	.4042149	.4043129	.4044110	.4045089	.4046069	.4047050	.4048029	.4049009	.4049989	0	0
0.376	.4050969	.4051948	.4052927	.4053907	.4054887	.4055866	.4056845	.4057824	.4058803	.4059782	0	0
0.377	.4060761	.4061740	.4062719	.4063697	.4064676	.4065655	.4066633	.4067612	.4068590	.4069568	0	0
0.378	.4070546	.4071524	.4072502	.4073480	.4074458	.4075436	.4076414	.4077392	.4078369	.4079347	0	0
0.379	.4080324	.4081301	.4082279	.4083256	.4084233	.4085210	.4086187	.4087164	.4088141	.4089118	0	0
0.380	.4090094	.4091071	.4092047	.4093024	.4094000	.4094977	.4095953	.4096929	.4097905	.4098881	0	0
0.381	.4099857	.4100833	.4101809	.4102784	.4103760	.4104736	.4105711	.4106687	.4107662	.4108638	0	0
0.382	.4109613	.4110588	.4111563	.4112538	.4113513	.4114488	.4115462	.4116437	.4117411	.4118386	0	0
0.383	.4119360	.4120335	.4121310	.4122283	.4123257	.4124232	.4125206	.4126180	.4127154	.4128127	0	0
0.384	.4129101	.4130075	.4131048	.4132022	.4132995	.4133968	.4134942	.4135915	.4136888	.4137861	0	0
0.385	.4138834	.4139807	.4140780	.4141753	.4142725	.4143698	.4144670	.4145643	.4146615	.4147587	0	0
0.386	.4148560	.4149532	.4150504	.4151476	.4152448	.4153420	.4154391	.4155363	.4156335	.4157306	0	0
0.387	.4158278	.4159249	.4160221	.4161192	.4162163	.4163134	.4164105	.4165076	.4166047	.4167017	0	0
0.388	.4167988	.4168959	.4169930	.4170900	.4171870	.4172841	.4173811	.4174781	.4175751	.4176722	0	0
0.389	.4177691	.4178661	.4179631	.4180601	.4181570	.4182540	.4183509	.4184479	.4185448	.4186417	0	0
0.390	.4187387	.4188356	.4189325	.4190294	.4191263	.4192231	.4193200	.4194169	.4195138	.4196106	0	0
0.391	.4197075	.4198043	.4199011	.4199979	.4200948	.4201916	.4202884	.4203852	.4204819	.4205787	0	0
0.392	.4206755	.4207723	.4208690	.4209657	.4210625	.4211592	.4212559	.4213527	.4214494	.4215461	0	0
0.393	.4216428	.4217395	.4218361	.4219328	.4220294	.4221261	.4222228	.4223194	.4224160	.4225127	0	0
0.394	.4226093	.4227059	.4228025	.4228991	.4229957	.4230922	.4231888	.4232854	.4233819	.4234785	0	0
0.395	.4235750	.4236715	.4237681	.4238646	.4239611	.4240576	.4241541	.4242506	.4243471	.4244435	0	0
0.396	.4245400	.4246365	.4247329	.4248294	.4249258	.4250222	.4251186	.4252151	.4253114	.4254079	0	0
0.397	.4255043	.4256006	.4256970	.4257933	.4258897	.4259861	.4260824	.4261788	.4262751	.4263714	0	0
0.398	.4264677	.4265640	.4266603	.4267566	.4268529	.4269491	.4270454	.4271417	.4272379	.4273342	0	0
0.399	.4274304	.4275267	.4276229	.4277191	.4278153	.4279115	.4280077	.4281038	.4282000	.4282961	0	0

TABLE 7A

Error Function

X	.0	.0001	.0002	.0003	.0004	.0005	.0006	.0007	.0008	.0009	m	Δ_2
0.400	.4283924	.4284885	.4285846	.4286808	.4287769	.4288730	.4289691	.4290652	.4291613	.4292574	0	0
0.401	.4293535	.4294496	.4295456	.4296417	.4297377	.4298338	.4299298	.4300258	.4301218	.4302179	0	0
0.402	.4303139	.4304098	.4305059	.4306018	.4306978	.4307938	.4308897	.4309857	.4310816	.4311776	0	0
0.403	.4312735	.4313694	.4314653	.4315612	.4316571	.4317530	.4318489	.4319447	.4320406	.4321365	0	0
0.404	.4322323	.4323282	.4324240	.4325199	.4326156	.4327115	.4328073	.4329031	.4329988	.4330946	0	0
0.405	.4331904	.4332861	.4333819	.4334776	.4335734	.4336691	.4337649	.4338606	.4339563	.4340520	0	0
0.406	.4341477	.4342434	.4343390	.4344347	.4345304	.4346260	.4347217	.4348173	.4349130	.4350086	0	0
0.407	.4351042	.4351998	.4352954	.4353910	.4354866	.4355822	.4356777	.4357733	.4358689	.4359644	0	0
0.408	.4360600	.4361554	.4362510	.4363465	.4364420	.4365375	.4366330	.4367285	.4368240	.4369195	0	0
0.409	.4370149	.4371104	.4372058	.4373012	.4373966	.4374921	.4375875	.4376829	.4377783	.4378737	0	0
0.410	.4379691	.4380645	.4381598	.4382551	.4383505	.4384459	.4385412	.4386365	.4387318	.4388272	0	0
0.411	.4389225	.4390178	.4391130	.4392083	.4393036	.4393989	.4394941	.4395894	.4396846	.4397798	0	0
0.412	.4398751	.4399703	.4400655	.4401607	.4402559	.4403511	.4404463	.4405414	.4406366	.4407318	0	0
0.413	.4408269	.4409220	.4410172	.4411123	.4412074	.4413025	.4413976	.4414927	.4415878	.4416829	0	0
0.414	.4417779	.4418730	.4419680	.4420631	.4421581	.4422532	.4423482	.4424432	.4425382	.4426332	0	0
0.415	.4427282	.4428232	.4429181	.4430131	.4431081	.4432030	.4432979	.4433929	.4434878	.4435827	0	0
0.416	.4436777	.4437726	.4438674	.4439623	.4440572	.4441521	.4442469	.4443418	.4444367	.4445315	0	0
0.417	.4446263	.4447212	.4448159	.4449108	.4450055	.4451004	.4451951	.4452899	.4453847	.4454795	0	0
0.418	.4455742	.4456689	.4457637	.4458584	.4459531	.4460478	.4461426	.4462373	.4463319	.4464266	0	0
0.419	.4465213	.4466159	.4467106	.4468052	.4468999	.4469945	.4470891	.4471838	.4472784	.4473730	0	0
0.420	.4474676	.4475622	.4476568	.4477513	.4478459	.4479405	.4480350	.4481295	.4482241	.4483186	0	0
0.421	.4484131	.4485076	.4486021	.4486966	.4487911	.4488856	.4489800	.4490744	.4491689	.4492634	0	0
0.422	.4493578	.4494522	.4495466	.4496410	.4497355	.4498299	.4499242	.4500186	.4501130	.4502074	0	0
0.423	.4503017	.4503961	.4504904	.4505847	.4506791	.4507734	.4508677	.4509619	.4510562	.4511505	0	0
0.424	.4512448	.4513391	.4514334	.4515276	.4516218	.4517161	.4518103	.4519045	.4519987	.4520929	0	0
0.425	.4521871	.4522813	.4523755	.4524696	.4525638	.4526580	.4527521	.4528463	.4529404	.4530345	0	0
0.426	.4531286	.4532228	.4533169	.4534109	.4535050	.4535991	.4536932	.4537872	.4538813	.4539753	0	0
0.427	.4540694	.4541634	.4542574	.4543514	.4544454	.4545394	.4546334	.4547274	.4548213	.4549153	0	0
0.428	.4550093	.4551032	.4551972	.4552911	.4553850	.4554789	.4555728	.4556667	.4557606	.4558545	0	0
0.429	.4559484	.4560422	.4561361	.4562299	.4563238	.4564176	.4565114	.4566053	.4566991	.4567929	0	0
0.430	.4568866	.4569805	.4570742	.4571680	.4572617	.4573555	.4574493	.4575430	.4576367	.4577305	0	0
0.431	.4578242	.4579179	.4580116	.4581053	.4581989	.4582926	.4583862	.4584799	.4585736	.4586672	0	0
0.432	.4587609	.4588544	.4589481	.4590417	.4591353	.4592289	.4593225	.4594160	.4595096	.4596032	0	0
0.433	.4596967	.4597902	.4598838	.4599773	.4600708	.4601644	.4602578	.4603513	.4604448	.4605383	0	0
0.434	.4606318	.4607252	.4608187	.4609122	.4610056	.4610990	.4611924	.4612858	.4613792	.4614726	0	0
0.435	.4615660	.4616594	.4617528	.4618462	.4619395	.4620329	.4621262	.4622195	.4623129	.4624062	0	0
0.436	.4624995	.4625928	.4626861	.4627793	.4628726	.4629659	.4630591	.4631524	.4632456	.4633389	0	0
0.437	.4634321	.4635253	.4636185	.4637117	.4638049	.4638981	.4639913	.4640844	.4641776	.4642708	0	0
0.438	.4643639	.4644570	.4645502	.4646433	.4647364	.4648295	.4649226	.4650157	.4651088	.4652019	0	0
0.439	.4652949	.4653879	.4654810	.4655740	.4656671	.4657601	.4658531	.4659461	.4660391	.4661321	0	0
0.440	.4662251	.4663181	.4664110	.4665040	.4665969	.4666899	.4667828	.4668757	.4669687	.4670615	0	0
0.441	.4671544	.4672474	.4673402	.4674331	.4675260	.4676188	.4677117	.4678045	.4678974	.4679902	0	0
0.442	.4680830	.4681758	.4682686	.4683614	.4684542	.4685469	.4686397	.4687325	.4688252	.4689180	0	0
0.443	.4690107	.4691035	.4691961	.4692889	.4693816	.4694743	.4695669	.4696596	.4697523	.4698449	0	0
0.444	.4699376	.4700302	.4701229	.4702155	.4703081	.4704008	.4704934	.4705859	.4706785	.4707711	0	0
0.445	.4708637	.4709563	.4710488	.4711413	.4712339	.4713264	.4714189	.4715114	.4716039	.4716964	0	0
0.446	.4717889	.4718814	.4719739	.4720663	.4721588	.4722512	.4723437	.4724361	.4725286	.4726210	0	0
0.447	.4727134	.4728057	.4728981	.4729905	.4730829	.4731753	.4732676	.4733600	.4734523	.4735447	0	0
0.448	.4736370	.4737293	.4738216	.4739139	.4740062	.4740984	.4741907	.4742830	.4743752	.4744675	0	0
0.449	.4745597	.4746520	.4747442	.4748364	.4749286	.4750208	.4751130	.4752052	.4752974	.4753895	0	0

TABLE 7A

Error Function

X	.0	.0001	.0002	.0003	.0004	.0005	.0006	.0007	.0008	.0009	m	Δ_2
0.450	.4754817	.4755738	.4756660	.4757581	.4758502	.4759424	.4760345	.4761266	.4762186	.4763107	0	0
0.451	.4764028	.4764948	.4765869	.4766790	.4767710	.4768630	.4769551	.4770471	.4771391	.4772311	0	0
0.452	.4773231	.4774151	.4775071	.4775990	.4776910	.4777829	.4778749	.4779668	.4780587	.4781507	0	0
0.453	.4782426	.4783344	.4784263	.4785182	.4786101	.4787019	.4787938	.4788857	.4789775	.4790694	0	0
0.454	.4791611	.4792530	.4793448	.4794366	.4795284	.4796202	.4797119	.4798037	.4798955	.4799872	0	0
0.455	.4800789	.4801707	.4802624	.4803541	.4804459	.4805375	.4806292	.4807209	.4808126	.4809043	0	0
0.456	.4809959	.4810876	.4811792	.4812708	.4813625	.4814541	.4815457	.4816373	.4817289	.4818205	0	0
0.457	.4819120	.4820036	.4820952	.4821867	.4822782	.4823698	.4824613	.4825528	.4826443	.4827358	0	0
0.458	.4828273	.4829188	.4830103	.4831017	.4831932	.4832847	.4833761	.4834675	.4835590	.4836503	0	0
0.459	.4837418	.4838331	.4839246	.4840159	.4841073	.4841987	.4842900	.4843814	.4844727	.4845641	0	0
0.460	.4846554	.4847467	.4848380	.4849293	.4850206	.4851118	.4852031	.4852944	.4853857	.4854769	0	0
0.461	.4855681	.4856594	.4857506	.4858418	.4859330	.4860242	.4861154	.4862065	.4862977	.4863889	0	0
0.462	.4864801	.4865712	.4866623	.4867535	.4868446	.4869357	.4870268	.4871179	.4872090	.4873000	0	0
0.463	.4873911	.4874822	.4875733	.4876643	.4877553	.4878464	.4879374	.4880284	.4881194	.4882104	0	0
0.464	.4883013	.4883924	.4884833	.4885743	.4886652	.4887562	.4888471	.4889380	.4890289	.4891198	0	0
0.465	.4892108	.4893016	.4893925	.4894834	.4895743	.4896652	.4897560	.4898468	.4899377	.4900285	0	0
0.466	.4901193	.4902101	.4903009	.4903917	.4904825	.4905733	.4906640	.4907548	.4908456	.4909363	0	0
0.467	.4910270	.4911177	.4912084	.4912992	.4913899	.4914805	.4915712	.4916619	.4917526	.4918432	0	0
0.468	.4919339	.4920245	.4921151	.4922057	.4922963	.4923870	.4924775	.4925681	.4926587	.4927493	0	0
0.469	.4928399	.4929304	.4930210	.4931115	.4932020	.4932926	.4933830	.4934736	.4935641	.4936545	0	0
0.470	.4937450	.4938355	.4939259	.4940164	.4941068	.4941973	.4942877	.4943781	.4944685	.4945589	0	0
0.471	.4946494	.4947397	.4948301	.4949204	.4950108	.4951012	.4951915	.4952818	.4953722	.4954625	0	0
0.472	.4955528	.4956431	.4957334	.4958236	.4959139	.4960042	.4960945	.4961847	.4962749	.4963652	0	0
0.473	.4964554	.4965456	.4966358	.4967260	.4968162	.4969063	.4969965	.4970867	.4971768	.4972670	0	0
0.474	.4973571	.4974473	.4975374	.4976275	.4977176	.4978077	.4978977	.4979879	.4980779	.4981680	0	0
0.475	.4982580	.4983481	.4984381	.4985281	.4986181	.4987081	.4987981	.4988881	.4989781	.4990681	0	0
0.476	.4991581	.4992480	.4993380	.4994279	.4995179	.4996077	.4996977	.4997876	.4998775	.4999673	0	0
0.477	.5000572	.5001471	.5002370	.5003268	.5004167	.5005065	.5005963	.5006862	.5007760	.5008658	0	0
0.478	.5009556	.5010453	.5011351	.5012249	.5013146	.5014044	.5014942	.5015839	.5016736	.5017633	0	0
0.479	.5018530	.5019428	.5020324	.5021221	.5022118	.5023015	.5023911	.5024807	.5025704	.5026600	0	0
0.480	.5027496	.5028393	.5029289	.5030184	.5031080	.5031976	.5032872	.5033768	.5034663	.5035558	0	0
0.481	.5036454	.5037349	.5038244	.5039139	.5040035	.5040929	.5041824	.5042719	.5043614	.5044508	0	0
0.482	.5045403	.5046297	.5047191	.5048085	.5048980	.5049874	.5050768	.5051662	.5052556	.5053449	0	0
0.483	.5054343	.5055236	.5056130	.5057023	.5057917	.5058810	.5059703	.5060596	.5061489	.5062382	0	0
0.484	.5063275	.5064167	.5065060	.5065953	.5066845	.5067737	.5068629	.5069522	.5070413	.5071306	0	0
0.485	.5072197	.5073090	.5073981	.5074873	.5075764	.5076656	.5077547	.5078439	.5079330	.5080221	0	0
0.486	.5081112	.5082003	.5082893	.5083784	.5084675	.5085565	.5086456	.5087346	.5088237	.5089127	0	0
0.487	.5090017	.5090908	.5091798	.5092688	.5093578	.5094467	.5095357	.5096246	.5097136	.5098025	0	0
0.488	.5098915	.5099804	.5100693	.5101582	.5102471	.5103360	.5104249	.5105137	.5106026	.5106914	0	0
0.489	.5107803	.5108691	.5109580	.5110468	.5111356	.5112244	.5113131	.5114020	.5114907	.5115795	0	0
0.490	.5116682	.5117570	.5118457	.5119345	.5120232	.5121119	.5122006	.5122893	.5123780	.5124667	0	0
0.491	.5125553	.5126440	.5127326	.5128213	.5129099	.5129985	.5130872	.5131758	.5132644	.5133529	0	0
0.492	.5134416	.5135301	.5136187	.5137072	.5137958	.5138843	.5139729	.5140614	.5141499	.5142384	0	0
0.493	.5143269	.5144154	.5145038	.5145923	.5146808	.5147693	.5148577	.5149461	.5150346	.5151230	0	0
0.494	.5152113	.5152997	.5153881	.5154765	.5155649	.5156533	.5157416	.5158300	.5159183	.5160066	0	0
0.495	.5160950	.5161833	.5162716	.5163599	.5164481	.5165364	.5166247	.5167130	.5168012	.5168895	0	0
0.496	.5169777	.5170659	.5171542	.5172424	.5173305	.5174187	.5175069	.5175951	.5176833	.5177714	0	0
0.497	.5178596	.5179477	.5180358	.5181239	.5182120	.5183002	.5183882	.5184763	.5185644	.5186524	0	0
0.498	.5187405	.5188286	.5189166	.5190046	.5190927	.5191807	.5192687	.5193567	.5194447	.5195327	0	0
0.499	.5196206	.5197086	.5197966	.5198845	.5199724	.5200604	.5201483	.5202362	.5203241	.5204120	0	0

TABLE 7A

Error Function

X	.0	.0001	.0002	.0003	.0004	.0005	.0006	.0007	.0008	.0009	m	Δ_2
0.500	.5204998	.5205877	.5206756	.5207635	.5208513	.5209391	.5210270	.5211148	.5212026	.5212904	0	0
0.501	.5213782	.5214660	.5215538	.5216415	.5217293	.5218170	.5219048	.5219925	.5220802	.5221679	0	0
0.502	.5222557	.5223433	.5224310	.5225187	.5226064	.5226941	.5227817	.5228693	.5229570	.5230446	0	0
0.503	.5231323	.5232198	.5233074	.5233951	.5234826	.5235702	.5236577	.5237453	.5238329	.5239204	0	0
0.504	.5240079	.5240955	.5241830	.5242705	.5243580	.5244455	.5245329	.5246204	.5247079	.5247953	0	0
0.505	.5248827	.5249702	.5250576	.5251451	.5252324	.5253198	.5254073	.5254946	.5255820	.5256693	0	0
0.506	.5257567	.5258440	.5259314	.5260187	.5261060	.5261933	.5262806	.5263679	.5264552	.5265425	0	0
0.507	.5266297	.5267170	.5268043	.5268915	.5269787	.5270659	.5271531	.5272403	.5273275	.5274147	0	0
0.508	.5275019	.5275891	.5276762	.5277634	.5278505	.5279377	.5280248	.5281119	.5281990	.5282861	0	0
0.509	.5283732	.5284603	.5285473	.5286344	.5287215	.5288085	.5288956	.5289826	.5290696	.5291566	0	0
0.510	.5292436	.5293306	.5294176	.5295045	.5295915	.5296785	.5297654	.5298523	.5299393	.5300262	0	0
0.511	.5301131	.5302000	.5302869	.5303738	.5304607	.5305475	.5306344	.5307212	.5308081	.5308949	0	0
0.512	.5309817	.5310685	.5311553	.5312421	.5313289	.5314157	.5315025	.5315892	.5316760	.5317627	0	0
0.513	.5318494	.5319362	.5320229	.5321096	.5321963	.5322830	.5323697	.5324563	.5325430	.5326297	0	0
0.514	.5327163	.5328029	.5328895	.5329762	.5330628	.5331494	.5332360	.5333226	.5334091	.5334957	0	0
0.515	.5335822	.5336688	.5337553	.5338418	.5339284	.5340149	.5341014	.5341879	.5342743	.5343608	0	0
0.516	.5344473	.5345337	.5346202	.5347067	.5347931	.5348795	.5349659	.5350523	.5351387	.5352251	0	0
0.517	.5353115	.5353978	.5354842	.5355706	.5356569	.5357432	.5358295	.5359159	.5360022	.5360885	0	0
0.518	.5361747	.5362610	.5363473	.5364335	.5365198	.5366060	.5366923	.5367785	.5368647	.5369509	0	0
0.519	.5370371	.5371233	.5372095	.5372956	.5373818	.5374680	.5375541	.5376403	.5377264	.5378125	0	0
0.520	.5378986	.5379847	.5380708	.5381569	.5382429	.5383290	.5384151	.5385011	.5385872	.5386732	0	0
0.521	.5387592	.5388452	.5389312	.5390172	.5391032	.5391892	.5392751	.5393611	.5394470	.5395330	0	0
0.522	.5396189	.5397048	.5397907	.5398766	.5399625	.5400484	.5401343	.5402201	.5403060	.5403919	0	0
0.523	.5404777	.5405635	.5406494	.5407351	.5408210	.5409067	.5409925	.5410783	.5411640	.5412498	0	0
0.524	.5413356	.5414213	.5415071	.5415928	.5416785	.5417642	.5418499	.5419356	.5420213	.5421069	0	0
0.525	.5421926	.5422782	.5423639	.5424495	.5425351	.5426207	.5427064	.5427920	.5428776	.5429631	0	0
0.526	.5430487	.5431342	.5432198	.5433053	.5433909	.5434764	.5435619	.5436474	.5437329	.5438184	0	0
0.527	.5439039	.5439894	.5440748	.5441602	.5442457	.5443311	.5444165	.5445020	.5445874	.5446728	0	0
0.528	.5447582	.5448436	.5449289	.5450143	.5450996	.5451850	.5452703	.5453556	.5454410	.5455263	0	0
0.529	.5456116	.5456969	.5457821	.5458674	.5459527	.5460379	.5461232	.5462084	.5462936	.5463789	0	0
0.530	.5464641	.5465493	.5466344	.5467196	.5468048	.5468900	.5469751	.5470603	.5471454	.5472305	0	0
0.531	.5473157	.5474008	.5474859	.5475709	.5476561	.5477411	.5478262	.5479112	.5479963	.5480813	0	0
0.532	.5481663	.5482514	.5483364	.5484214	.5485064	.5485914	.5486763	.5487613	.5488462	.5489312	0	0
0.533	.5490161	.5491011	.5491860	.5492709	.5493558	.5494407	.5495256	.5496104	.5496953	.5497802	0	0
0.534	.5498650	.5499498	.5500347	.5501195	.5502043	.5502891	.5503739	.5504587	.5505435	.5506282	0	0
0.535	.5507130	.5507977	.5508825	.5509672	.5510519	.5511366	.5512213	.5513060	.5513907	.5514754	0	0
0.536	.5515600	.5516447	.5517293	.5518140	.5518986	.5519832	.5520679	.5521525	.5522370	.5523216	0	0
0.537	.5524062	.5524908	.5525753	.5526599	.5527444	.5528290	.5529135	.5529979	.5530825	.5531670	0	0
0.538	.5532514	.5533359	.5534204	.5535048	.5535893	.5536737	.5537581	.5538426	.5539270	.5540114	0	0
0.539	.5540958	.5541801	.5542645	.5543489	.5544333	.5545176	.5546020	.5546863	.5547706	.5548549	0	0
0.540	.5549392	.5550235	.5551078	.5551921	.5552763	.5553606	.5554448	.5555291	.5556133	.5556975	0	0
0.541	.5557817	.5558659	.5559502	.5560343	.5561185	.5562027	.5562868	.5563710	.5564551	.5565392	0	0
0.542	.5566233	.5567074	.5567915	.5568756	.5569597	.5570438	.5571279	.5572119	.5572960	.5573800	0	0
0.543	.5574641	.5575480	.5576321	.5577161	.5578001	.5578840	.5579680	.5580520	.5581359	.5582199	0	0
0.544	.5583038	.5583878	.5584717	.5585556	.5586395	.5587234	.5588073	.5588911	.5589750	.5590588	0	0
0.545	.5591427	.5592265	.5593104	.5593942	.5594780	.5595618	.5596456	.5597294	.5598131	.5598969	0	0
0.546	.5599806	.5600644	.5601481	.5602319	.5603156	.5603993	.5604830	.5605667	.5606503	.5607340	0	0
0.547	.5608177	.5609013	.5609850	.5610687	.5611523	.5612359	.5613195	.5614031	.5614867	.5615703	0	0
0.548	.5616539	.5617374	.5618209	.5619045	.5619880	.5620716	.5621551	.5622386	.5623221	.5624056	0	0
0.549	.5624890	.5625725	.5626560	.5627394	.5628229	.5629063	.5629897	.5630732	.5631565	.5632399	0	0

TABLE 7A

Error Function

X	.0	.0001	.0002	.0003	.0004	.0005	.0006	.0007	.0008	.0009	m	Δ_2
0.550	.5633233	.5634067	.5634901	.5635734	.5636568	.5637401	.5638235	.5639068	.5639901	.5640734		
0.551	.5641567	.5642400	.5643233	.5644066	.5644898	.5645730	.5646563	.5647395	.5648227	.5649059	0	0
0.552	.5649891	.5650724	.5651556	.5652387	.5653219	.5654051	.5654882	.5655714	.5656545	.5657376	0	0
0.553	.5658207	.5659038	.5659869	.5660700	.5661530	.5662361	.5663192	.5664023	.5664853	.5665683	0	0
0.554	.5666513	.5667344	.5668173	.5669004	.5669833	.5670663	.5671493	.5672323	.5673152	.5673981	0	0
0.555	.5674810	.5675640	.5676469	.5677298	.5678127	.5678955	.5679784	.5680613	.5681441	.5682270	0	0
0.556	.5683098	.5683926	.5684755	.5685583	.5686411	.5687239	.5688066	.5688894	.5689722	.5690550	0	0
0.557	.5691377	.5692204	.5693032	.5693859	.5694686	.5695513	.5696340	.5697166	.5697993	.5698820	0	0
0.558	.5699646	.5700473	.5701299	.5702125	.5702952	.5703778	.5704603	.5705429	.5706255	.5707081	0	0
0.559	.5707906	.5708732	.5709558	.5710382	.5711208	.5712033	.5712858	.5713683	.5714508	.5715333	0	0
0.560	.5716158	.5716982	.5717806	.5718631	.5719455	.5720279	.5721104	.5721927	.5722752	.5723575		
0.561	.5724399	.5725223	.5726047	.5726870	.5727693	.5728517	.5729340	.5730163	.5730986	.5731809	0	0
0.562	.5732632	.5733454	.5734277	.5735099	.5735922	.5736744	.5737566	.5738389	.5739211	.5740033	0	0
0.563	.5740855	.5741677	.5742498	.5743320	.5744141	.5744963	.5745784	.5746605	.5747427	.5748248	0	0
0.564	.5749069	.5749890	.5750710	.5751531	.5752352	.5753172	.5753993	.5754813	.5755633	.5756453	0	0
0.565	.5757273	.5758094	.5758913	.5759733	.5760553	.5761372	.5762192	.5763011	.5763831	.5764650	0	0
0.566	.5765469	.5766288	.5767107	.5767925	.5768744	.5769563	.5770382	.5771200	.5772018	.5772837	0	0
0.567	.5773655	.5774473	.5775291	.5776109	.5776927	.5777745	.5778562	.5779380	.5780197	.5781015	0	0
0.568	.5781832	.5782649	.5783466	.5784283	.5785100	.5785917	.5786734	.5787550	.5788367	.5789183	0	0
0.569	.5789999	.5790816	.5791632	.5792448	.5793264	.5794080	.5794895	.5795711	.5796527	.5797342	0	0
0.570	.5798158	.5798973	.5799788	.5800604	.5801418	.5802233	.5803048	.5803863	.5804678	.5805492		
0.571	.5806307	.5807121	.5807936	.5808750	.5809564	.5810378	.5811192	.5812005	.5812819	.5813633	0	0
0.572	.5814446	.5815260	.5816073	.5816886	.5817700	.5818513	.5819326	.5820139	.5820951	.5821764	0	0
0.573	.5822577	.5823389	.5824202	.5825014	.5825827	.5826638	.5827451	.5828263	.5829074	.5829886	0	0
0.574	.5830698	.5831510	.5832321	.5833132	.5833944	.5834755	.5835566	.5836377	.5837188	.5837999	0	0
0.575	.5838810	.5839620	.5840431	.5841241	.5842052	.5842862	.5843672	.5844482	.5845292	.5846102	0	0
0.576	.5846912	.5847722	.5848532	.5849341	.5850151	.5850960	.5851769	.5852578	.5853388	.5854197	0	0
0.577	.5855005	.5855814	.5856623	.5857431	.5858240	.5859048	.5859857	.5860665	.5861473	.5862281	0	0
0.578	.5863089	.5863897	.5864705	.5865512	.5866320	.5867128	.5867935	.5868742	.5869550	.5870357	0	0
0.579	.5871164	.5871971	.5872777	.5873584	.5874391	.5875197	.5876004	.5876810	.5877616	.5878423	0	0
0.580	.5879229	.5880035	.5880840	.5881646	.5882452	.5883258	.5884063	.5884869	.5885674	.5886479		
0.581	.5887284	.5888090	.5888894	.5889699	.5890504	.5891309	.5892113	.5892918	.5893722	.5894527	0	0
0.582	.5895331	.5896135	.5896939	.5897743	.5898547	.5899351	.5900154	.5900958	.5901761	.5902565	0	0
0.583	.5903368	.5904171	.5904974	.5905777	.5906580	.5907383	.5908186	.5908988	.5909791	.5910593	0	0
0.584	.5911396	.5912198	.5913000	.5913802	.5914604	.5915406	.5916207	.5917009	.5917811	.5918612	0	0
0.585	.5919414	.5920215	.5921016	.5921817	.5922619	.5923420	.5924220	.5925021	.5925822	.5926622	0	0
0.586	.5927423	.5928223	.5929024	.5929824	.5930624	.5931424	.5932224	.5933024	.5933823	.5934623	0	0
0.587	.5935422	.5936222	.5937021	.5937821	.5938619	.5939419	.5940217	.5941017	.5941815	.5942614	0	0
0.588	.5943413	.5944211	.5945010	.5945808	.5946606	.5947404	.5948202	.5949000	.5949798	.5950596	0	0
0.589	.5951393	.5952191	.5952988	.5953786	.5954583	.5955380	.5956177	.5956974	.5957771	.5958568	0	0
0.590	.5959365	.5960161	.5960958	.5961754	.5962551	.5963347	.5964143	.5964939	.5965735	.5966531		
0.591	.5967327	.5968122	.5968918	.5969713	.5970509	.5971304	.5972099	.5972894	.5973690	.5974485	0	0
0.592	.5975279	.5976074	.5976869	.5977663	.5978457	.5979252	.5980046	.5980840	.5981634	.5982428	0	0
0.593	.5983222	.5984016	.5984810	.5985603	.5986397	.5987191	.5987984	.5988777	.5989570	.5990363	0	0
0.594	.5991156	.5991949	.5992742	.5993534	.5994327	.5995119	.5995912	.5996704	.5997496	.5998288	0	0
0.595	.5999081	.5999872	.6000664	.6001456	.6002247	.6003039	.6003830	.6004622	.6005413	.6006204	0	0
0.596	.6006995	.6007786	.6008577	.6009368	.6010159	.6010949	.6011740	.6012530	.6013321	.6014110	0	0
0.597	.6014901	.6015691	.6016481	.6017271	.6018060	.6018850	.6019639	.6020429	.6021218	.6022007	0	0
0.598	.6022797	.6023586	.6024375	.6025164	.6025953	.6026741	.6027530	.6028318	.6C29107	.6029895	0	0
0.599	.6030684	.6031471	.6032259	.6033047	.6033835	.6034623	.6035411	.6036199	.6036986	.6037773	0	0

TABLE 7A

Error Function

X	.0	.0001	.0002	.0003	.0004	.0005	.0006	.0007	.0008	.0009	m	Δ_2
0.600	.6038561	.6039348	.6040135	.6040922	.6041709	.6042495	.6043282	.6044069	.6044856	.6045642	0	0
0.601	.6046428	.6047215	.6048000	.6048787	.6049573	.6050358	.6051145	.6051930	.6052716	.6053501	0	0
0.602	.6054286	.6055072	.6055857	.6056642	.6057428	.6058212	.6058997	.6059782	.6060566	.6061351	0	0
0.603	.6062135	.6062919	.6063704	.6064488	.6065272	.6066056	.6066840	.6067624	.6068408	.6069191	0	0
0.604	.6069975	.6070758	.6071541	.6072325	.6073108	.6073891	.6074674	.6075457	.6076239	.6077022	0	0
0.605	.6077805	.6078587	.6079369	.6080152	.6080934	.6081716	.6082498	.6083280	.6084062	.6084843	0	0
0.606	.6085625	.6086407	.6087188	.6087970	.6088750	.6089532	.6090313	.6091093	.6091874	.6092655	0	0
0.607	.6093436	.6094217	.6094997	.6095777	.6096557	.6097338	.6098118	.6098898	.6099678	.6100458	0	0
0.608	.6101238	.6102017	.6102797	.6103576	.6104355	.6105134	.6105914	.6106693	.6107472	.6108251	0	0
0.609	.6109030	.6109808	.6110587	.6111365	.6112143	.6112922	.6113700	.6114478	.6115256	.6116034	0	0
0.610	.6116812	.6117589	.6118367	.6119145	.6119922	.6120700	.6121477	.6122254	.6123031	.6123808	0	0
0.611	.6124585	.6125362	.6126139	.6126915	.6127691	.6128468	.6129244	.6130021	.6130797	.6131573	0	0
0.612	.6132348	.6133124	.6133900	.6134676	.6135451	.6136227	.6137002	.6137777	.6138552	.6139327	0	0
0.613	.6140102	.6140878	.6141652	.6142427	.6143202	.6143976	.6144750	.6145524	.6146299	.6147073	0	0
0.614	.6147847	.6148621	.6149395	.6150168	.6150942	.6151716	.6152489	.6153262	.6154036	.6154809	0	0
0.615	.6155582	.6156355	.6157128	.6157901	.6158673	.6159446	.6160218	.6160991	.6161763	.6162536	0	0
0.616	.6163307	.6164079	.6164851	.6165623	.6166395	.6167167	.6167938	.6168709	.6169481	.6170253	0	0
0.617	.6171023	.6171795	.6172565	.6173337	.6174107	.6174878	.6175649	.6176419	.6177189	.6177959	0	0
0.618	.6178730	.6179500	.6180270	.6181040	.6181810	.6182579	.6183349	.6184119	.6184888	.6185657	0	0
0.619	.6186427	.6187196	.6187965	.6188734	.6189503	.6190272	.6191040	.6191809	.6192577	.6193346	0	0
0.620	.6194114	.6194882	.6195651	.6196418	.6197187	.6197954	.6198722	.6199490	.6200258	.6201025	0	0
0.621	.6201792	.6202559	.6203327	.6204094	.6204861	.6205627	.6206394	.6207161	.6207927	.6208694	0	0
0.622	.6209460	.6210227	.6210993	.6211759	.6212525	.6213291	.6214057	.6214823	.6215588	.6216354	0	0
0.623	.6217120	.6217885	.6218650	.6219415	.6220180	.6220945	.6221710	.6222475	.6223240	.6224004	0	0
0.624	.6224769	.6225533	.6226298	.6227062	.6227826	.6228590	.6229354	.6230118	.6230881	.6231645	0	0
0.625	.6232408	.6233172	.6233935	.6234698	.6235462	.6236225	.6236988	.6237751	.6238514	.6239276	0	0
0.626	.6240039	.6240801	.6241564	.6242326	.6243088	.6243850	.6244612	.6245374	.6246136	.6246898	0	0
0.627	.6247659	.6248421	.6249182	.6249944	.6250705	.6251466	.6252227	.6252988	.6253749	.6254510	0	0
0.628	.6255271	.6256031	.6256791	.6257552	.6258312	.6259072	.6259832	.6260593	.6261352	.6262112	0	0
0.629	.6262872	.6263632	.6264392	.6265151	.6265910	.6266669	.6267428	.6268188	.6268947	.6269705	0	0
0.630	.6270464	.6271223	.6271982	.6272740	.6273499	.6274257	.6275015	.6275773	.6276531	.6277289	0	0
0.631	.6278046	.6278805	.6279562	.6280320	.6281077	.6281834	.6282591	.6283349	.6284106	.6284863	0	0
0.632	.6285620	.6286376	.6287133	.6287889	.6288646	.6289402	.6290159	.6290915	.6291671	.6292427	0	0
0.633	.6293183	.6293939	.6294695	.6295450	.6296206	.6296961	.6297716	.6298472	.6299226	.6299981	0	0
0.634	.6300737	.6301491	.6302246	.6303001	.6303756	.6304510	.6305264	.6306018	.6306773	.6307527	0	0
0.635	.6308281	.6309035	.6309788	.6310542	.6311296	.6312049	.6312802	.6313556	.6314309	.6315062	0	0
0.636	.6315815	.6316568	.6317321	.6318074	.6318827	.6319579	.6320332	.6321084	.6321836	.6322588	0	0
0.637	.6323341	.6324092	.6324844	.6325596	.6326348	.6327099	.6327851	.6328602	.6329353	.6330104	0	0
0.638	.6330856	.6331607	.6332358	.6333109	.6333860	.6334610	.6335360	.6336111	.6336861	.6337612	0	0
0.639	.6338362	.6339112	.6339862	.6340612	.6341361	.6342111	.6342860	.6343610	.6344360	.6345109	0	0
0.640	.6345858	.6346607	.6347356	.6348105	.6348854	.6349602	.6350351	.6351100	.6351848	.6352596	0	0
0.641	.6353345	.6354093	.6354841	.6355589	.6356336	.6357085	.6357832	.6358579	.6359327	.6360074	0	0
0.642	.6360822	.6361569	.6362316	.6363063	.6363810	.6364557	.6365303	.6366050	.6366796	.6367543	0	0
0.643	.6368289	.6369035	.6369781	.6370528	.6371273	.6372020	.6372765	.6373511	.6374257	.6375002	0	0
0.644	.6375747	.6376492	.6377237	.6377982	.6378728	.6379473	.6380217	.6380962	.6381707	.6382451	0	0
0.645	.6383196	.6383940	.6384684	.6385428	.6386172	.6386916	.6387660	.6388404	.6389147	.6389891	0	0
0.646	.6390634	.6391377	.6392121	.6392864	.6393607	.6394350	.6395093	.6395836	.6396579	.6397321	0	0
0.647	.6398063	.6398806	.6399548	.6400290	.6401032	.6401774	.6402516	.6403258	.6404000	.6404741	0	0
0.648	.6405483	.6406224	.6406966	.6407707	.6408448	.6409189	.6409930	.6410671	.6411412	.6412152	0	0
0.649	.6412893	.6413633	.6414374	.6415114	.6415854	.6416594	.6417334	.6418074	.6418813	.6419553	0	0

TABLE 7A

Error Function

X	.0	.0001	.0002	.0003	.0004	.0005	.0006	.0007	.0008	.0009	m	Δ_2
0.650	.6420293	.6421033	.6421772	.6422511	.6423250	.6423990	.6424729	.6425468	.6426206	.6426945	0	0
0.651	.6427684	.6428422	.6429161	.6429899	.6430637	.6431375	.6432114	.6432852	.6433589	.6434327	0	0
0.652	.6435065	.6435802	.6436540	.6437277	.6438015	.6438752	.6439489	.6440226	.6440963	.6441699	0	0
0.653	.6442436	.6443173	.6443909	.6444646	.6445382	.6446118	.6446854	.6447591	.6448326	.6449062	0	0
0.654	.6449798	.6450533	.6451269	.6452004	.6452740	.6453475	.6454210	.6454945	.6455680	.6456415	0	0
0.655	.6457150	.6457885	.6458619	.6459354	.6460088	.6460823	.6461557	.6462291	.6463025	.6463759	0	0
0.656	.6464493	.6465226	.6465960	.6466693	.6467427	.6468160	.6468893	.6469626	.6470360	.6471093	0	0
0.657	.6471826	.6472558	.6473291	.6474023	.6474756	.6475489	.6476220	.6476953	.6477685	.6478417	0	0
0.658	.6479149	.6479881	.6480612	.6481344	.6482075	.6482807	.6483538	.6484269	.6485001	.6485732	0	0
0.659	.6486462	.6487193	.6487924	.6488655	.6489385	.6490116	.6490846	.6491576	.6492307	.6493037	0	0
0.660	.6493766	.6494496	.6495226	.6495956	.6496686	.6497415	.6498144	.6498874	.6499603	.6500332	0	0
0.661	.6501061	.6501790	.6502519	.6503248	.6503976	.6504704	.6505433	.6506161	.6506889	.6507618	0	0
0.662	.6508346	.6509074	.6509802	.6510530	.6511257	.6511984	.6512712	.6513439	.6514167	.6514894	0	0
0.663	.6515621	.6516348	.6517075	.6517801	.6518528	.6519255	.6519981	.6520708	.6521434	.6522160	0	0
0.664	.6522886	.6523612	.6524338	.6525064	.6525790	.6526515	.6527241	.6527966	.6528692	.6529417	0	0
0.665	.6530142	.6530867	.6531592	.6532317	.6533042	.6533766	.6534491	.6535215	.6535940	.6536664	0	0
0.666	.6537389	.6538113	.6538836	.6539561	.6540284	.6541008	.6541731	.6542455	.6543179	.6543902	0	0
0.667	.6544625	.6545348	.6546071	.6546794	.6547517	.6548240	.6548962	.6549685	.6550407	.6551130	0	0
0.668	.6551852	.6552574	.6553296	.6554018	.6554740	.6555462	.6556183	.6556905	.6557627	.6558348	0	0
0.669	.6559069	.6559790	.6560512	.6561232	.6561953	.6562674	.6563395	.6564115	.6564836	.6565557	0	0
0.670	.6566277	.6566997	.6567717	.6568437	.6569157	.6569877	.6570597	.6571316	.6572036	.6572756	0	0
0.671	.6573474	.6574194	.6574913	.6575632	.6576351	.6577070	.6577789	.6578507	.6579226	.6579944	0	0
0.672	.6580663	.6581382	.6582099	.6582817	.6583536	.6584253	.6584972	.6585689	.6586407	.6587124	0	0
0.673	.6587842	.6588559	.6589276	.6589993	.6590710	.6591427	.6592144	.6592861	.6593578	.6594294	0	0
0.674	.6595011	.6595727	.6596443	.6597160	.6597875	.6598591	.6599308	.6600023	.6600739	.6601455	0	0
0.675	.6602170	.6602885	.6603600	.6604316	.6605031	.6605746	.6606461	.6607175	.6607891	.6608605	0	0
0.676	.6609319	.6610034	.6610748	.6611463	.6612177	.6612891	.6613605	.6614318	.6615033	.6615746	0	0
0.677	.6616459	.6617173	.6617886	.6618600	.6619313	.6620026	.6620739	.6621452	.6622165	.6622877	0	0
0.678	.6623590	.6624302	.6625015	.6625727	.6626440	.6627151	.6627864	.6628575	.6629288	.6629999	0	0
0.679	.6630711	.6631422	.6632134	.6632845	.6633556	.6634268	.6634979	.6635689	.6636400	.6637111	0	0
0.680	.6637822	.6638532	.6639243	.6639953	.6640663	.6641374	.6642084	.6642793	.6643503	.6644213	0	0
0.681	.6644923	.6645633	.6646342	.6647052	.6647761	.6648470	.6649179	.6649888	.6650597	.6651306	0	0
0.682	.6652015	.6652724	.6653432	.6654140	.6654849	.6655557	.6656265	.6656973	.6657681	.6658389	0	0
0.683	.6659097	.6659805	.6660512	.6661220	.6661927	.6662634	.6663342	.6664048	.6664755	.6665462	0	0
0.684	.6666169	.6666876	.6667582	.6668289	.6668996	.6669702	.6670408	.6671114	.6671820	.6672526	0	0
0.685	.6673232	.6673937	.6674643	.6675349	.6676054	.6676760	.6677465	.6678170	.6678875	.6679580	0	0
0.686	.6680285	.6680990	.6681694	.6682399	.6683103	.6683808	.6684512	.6685216	.6685920	.6686624	0	0
0.687	.6687328	.6688032	.6688736	.6689439	.6690143	.6690846	.6691549	.6692253	.6692956	.6693659	0	0
0.688	.6694362	.6695065	.6695768	.6696470	.6697173	.6697875	.6698577	.6699280	.6699982	.6700684	0	0
0.689	.6701386	.6702088	.6702790	.6703491	.6704193	.6704894	.6705596	.6706297	.6706998	.6707699	0	0
0.690	.6708400	.6709101	.6709802	.6710503	.6711203	.6711904	.6712604	.6713305	.6714005	.6714705	0	0
0.691	.6715405	.6716105	.6716805	.6717504	.6718204	.6718904	.6719603	.6720303	.6721002	.6721701	0	0
0.692	.6722400	.6723099	.6723798	.6724496	.6725195	.6725894	.6726592	.6727291	.6727989	.6728687	0	0
0.693	.6729385	.6730083	.6730781	.6731479	.6732177	.6732875	.6733572	.6734269	.6734967	.6735664	0	0
0.694	.6736361	.6737058	.6737755	.6738452	.6739148	.6739845	.6740541	.6741238	.6741934	.6742631	0	0
0.695	.6743327	.6744023	.6744719	.6745415	.6746110	.6746807	.6747502	.6748197	.6748893	.6749588	0	0
0.696	.6750283	.6750978	.6751673	.6752368	.6753063	.6753758	.6754453	.6755147	.6755841	.6756536	0	0
0.697	.6757230	.6757924	.6758618	.6759312	.6760006	.6760700	.6761393	.6762087	.6762780	.6763474	0	0
0.698	.6764167	.6764860	.6765553	.6766246	.6766939	.6767632	.6768324	.6769017	.6769710	.6770402	0	0
0.699	.6771094	.6771786	.6772478	.6773170	.6773862	.6774554	.6775246	.6775938	.6776629	.6777321	0	0

TABLE 7A

Error Function

X	.0	.0001	.0002	.0003	.0004	.0005	.0006	.0007	.0008	.0009	m	Δ_2
0.700	.6778012	.6778703	.6779394	.6780085	.6780776	.6781467	.6782157	.6782848	.6783538	.6784229	0	0
0.701	.6784920	.6785610	.6786300	.6786990	.6787680	.6788370	.6789060	.6789749	.6790439	.6791129	0	0
0.702	.6791818	.6792507	.6793196	.6793885	.6794574	.6795263	.6795952	.6796641	.6797329	.6798018	0	0
0.703	.6798706	.6799394	.6800083	.6800771	.6801459	.6802147	.6802835	.6803523	.6804210	.6804898	0	0
0.704	.6805585	.6806272	.6806960	.6807647	.6808334	.6809021	.6809708	.6810395	.6811081	.6811768	0	0
0.705	.6812454	.6813141	.6813827	.6814513	.6815199	.6815885	.6816571	.6817257	.6817943	.6818628	0	0
0.706	.6819314	.6819999	.6820685	.6821370	.6822055	.6822740	.6823425	.6824110	.6824794	.6825479	0	0
0.707	.6826164	.6826848	.6827533	.6828217	.6828901	.6829585	.6830269	.6830953	.6831636	.6832320	0	0
0.708	.6833004	.6833687	.6834370	.6835054	.6835737	.6836420	.6837103	.6837786	.6838469	.6839152	0	0
0.709	.6839834	.6840517	.6841199	.6841882	.6842564	.6843246	.6843928	.6844610	.6845292	.6845974	0	0
0.710	.6846655	.6847337	.6848018	.6848699	.6849381	.6850062	.6850743	.6851424	.6852105	.6852786	0	0
0.711	.6853466	.6854147	.6854827	.6855507	.6856188	.6856868	.6857548	.6858228	.6858909	.6859588	0	0
0.712	.6860268	.6860948	.6861627	.6862307	.6862985	.6863665	.6864344	.6865023	.6865702	.6866381	0	0
0.713	.6867059	.6867738	.6868417	.6869095	.6869773	.6870452	.6871130	.6871808	.6872486	.6873164	0	0
0.714	.6873842	.6874520	.6875197	.6875874	.6876552	.6877229	.6877906	.6878583	.6879261	.6879937	0	0
0.715	.6880614	.6881291	.6881967	.6882644	.6883320	.6883997	.6884673	.6885349	.6886025	.6886701	0	0
0.716	.6887377	.6888052	.6888728	.6889403	.6890079	.6890755	.6891430	.6892105	.6892780	.6893455	0	0
0.717	.6894130	.6894805	.6895480	.6896154	.6896828	.6897503	.6898177	.6898851	.6899526	.6900199	0	0
0.718	.6900873	.6901547	.6902221	.6902894	.6903568	.6904241	.6904914	.6905588	.6906261	.6906934	0	0
0.719	.6907607	.6908280	.6908953	.6909625	.6910298	.6910970	.6911643	.6912315	.6912987	.6913659	0	0
0.720	.6914331	.6915003	.6915675	.6916347	.6917018	.6917689	.6918361	.6919032	.6919703	.6920374	0	0
0.721	.6921045	.6921716	.6922387	.6923057	.6923729	.6924399	.6925069	.6925740	.6926410	.6927080	0	0
0.722	.6927750	.6928420	.6929090	.6929759	.6930429	.6931099	.6931768	.6932437	.6933107	.6933776	0	0
0.723	.6934445	.6935114	.6935783	.6936452	.6937120	.6937789	.6938457	.6939126	.6939794	.6940462	0	0
0.724	.6941130	.6941798	.6942466	.6943134	.6943802	.6944469	.6945137	.6945804	.6946471	.6947139	0	0
0.725	.6947806	.6948473	.6949140	.6949807	.6950474	.6951140	.6951807	.6952473	.6953139	.6953806	0	0
0.726	.6954472	.6955138	.6955804	.6956470	.6957136	.6957802	.6958467	.6959133	.6959798	.6960463	0	0
0.727	.6961128	.6961793	.6962458	.6963123	.6963788	.6964453	.6965117	.6965782	.6966446	.6967111	0	0
0.728	.6967775	.6968439	.6969103	.6969767	.6970431	.6971095	.6971758	.6972422	.6973085	.6973749	0	0
0.729	.6974412	.6975075	.6975738	.6976401	.6977064	.6977727	.6978390	.6979052	.6979715	.6980377	0	0
0.730	.6981039	.6981701	.6982363	.6983026	.6983687	.6984349	.6985011	.6985673	.6986334	.6986995	0	0
0.731	.6987657	.6988318	.6988979	.6989640	.6990301	.6990962	.6991622	.6992283	.6992944	.6993604	0	0
0.732	.6994265	.6994925	.6995585	.6996245	.6996905	.6997565	.6998225	.6998885	.6999544	.7000204	0	0
0.733	.7000863	.7001522	.7002181	.7002841	.7003500	.7004158	.7004817	.7005476	.7006135	.7006793	0	0
0.734	.7007452	.7008110	.7008768	.7009426	.7010084	.7010742	.7011400	.7012058	.7012715	.7013373	0	0
0.735	.7014031	.7014688	.7015345	.7016003	.7016659	.7017316	.7017973	.7018630	.7019287	.7019944	0	0
0.736	.7020600	.7021256	.7021912	.7022569	.7023225	.7023881	.7024537	.7025192	.7025849	.7026504	0	0
0.737	.7027159	.7027815	.7028470	.7029126	.7029781	.7030436	.7031091	.7031745	.7032400	.7033055	0	0
0.738	.7033709	.7034364	.7035018	.7035673	.7036327	.7036981	.7037635	.7038289	.7038943	.7039596	0	0
0.739	.7040250	.7040904	.7041557	.7042210	.7042863	.7043517	.7044169	.7044823	.7045475	.7046128	0	0
0.740	.7046781	.7047433	.7048085	.7048738	.7049390	.7050042	.7050694	.7051346	.7051998	.7052650	0	0
0.741	.7053301	.7053953	.7054604	.7055256	.7055907	.7056558	.7057210	.7057860	.7058511	.7059162	0	0
0.742	.7059813	.7060463	.7061114	.7061765	.7062415	.7063065	.7063715	.7064365	.7065015	.7065665	0	0
0.743	.7066314	.7066964	.7067614	.7068263	.7068912	.7069561	.7070211	.7070860	.7071509	.7072158	0	0
0.744	.7072806	.7073455	.7074104	.7074752	.7075400	.7076049	.7076697	.7077345	.7077993	.7078641	0	0
0.745	.7079289	.7079937	.7080584	.7081231	.7081879	.7082527	.7083174	.7083821	.7084468	.7085115	0	0
0.746	.7085761	.7086408	.7087055	.7087702	.7088348	.7088994	.7089640	.7090287	.7090933	.7091579	0	0
0.747	.7092225	.7092870	.7093516	.7094162	.7094807	.7095453	.7096098	.7096743	.7097389	.7098033	0	0
0.748	.7098678	.7099323	.7099968	.7100613	.7101257	.7101901	.7102545	.7103190	.7103834	.7104478	0	0
0.749	.7105122	.7105766	.7106410	.7107053	.7107697	.7108340	.7108983	.7109627	.7110270	.7110913	0	0

TABLE 7A

Error Function

X	.0	.0001	.0002	.0003	.0004	.0005	.0006	.0007	.0008	.0009	m	Δ_2
0.750	.7111556	.7112199	.7112842	.7113484	.7114127	.7114769	.7115412	.7116054	.7116696	.7117339	0	0
0.751	.7117981	.7118623	.7119264	.7119906	.7120548	.7121189	.7121831	.7122472	.7123113	.7123755	0	0
0.752	.7124395	.7125036	.7125677	.7126318	.7126958	.7127599	.7128240	.7128880	.7129520	.7130160	0	0
0.753	.7130800	.7131441	.7132080	.7132720	.7133360	.7133999	.7134639	.7135279	.7135918	.7136557	0	0
0.754	.7137196	.7137835	.7138474	.7139113	.7139752	.7140390	.7141029	.7141668	.7142305	.7142944	0	0
0.755	.7143582	.7144220	.7144858	.7145496	.7146134	.7146772	.7147409	.7148046	.7148684	.7149321	0	0
0.756	.7149959	.7150595	.7151232	.7151870	.7152506	.7153143	.7153779	.7154416	.7155052	.7155689	0	0
0.757	.7156325	.7156961	.7157597	.7158233	.7158869	.7159505	.7160140	.7160776	.7161412	.7162047	0	0
0.758	.7162682	.7163317	.7163953	.7164587	.7165222	.7165857	.7166492	.7167127	.7167761	.7168395	0	0
0.759	.7169030	.7169664	.7170298	.7170932	.7171566	.7172199	.7172833	.7173467	.7174100	.7174734	0	0
0.760	.7175367	.7176000	.7176633	.7177266	.7177899	.7178532	.7179165	.7179798	.7180430	.7181063	0	0
0.761	.7181695	.7182328	.7182960	.7183592	.7184224	.7184856	.7185488	.7186120	.7186751	.7187383	0	0
0.762	.7188014	.7188645	.7189276	.7189907	.7190539	.7191169	.7191800	.7192431	.7193062	.7193692	0	0
0.763	.7194323	.7194953	.7195584	.7196214	.7196844	.7197474	.7198104	.7198733	.7199363	.7199993	0	0
0.764	.7200622	.7201251	.7201881	.7202510	.7203139	.7203768	.7204397	.7205026	.7205654	.7206283	0	0
0.765	.7206911	.7207540	.7208169	.7208797	.7209425	.7210053	.7210681	.7211308	.7211937	.7212564	0	0
0.766	.7213192	.7213819	.7214447	.7215074	.7215701	.7216328	.7216955	.7217582	.7218209	.7218835	0	0
0.767	.7219462	.7220089	.7220715	.7221342	.7221968	.7222594	.7223220	.7223846	.7224472	.7225097	0	0
0.768	.7225723	.7226349	.7226974	.7227600	.7228225	.7228850	.7229475	.7230100	.7230725	.7231349	0	0
0.769	.7231974	.7232599	.7233223	.7233848	.7234472	.7235096	.7235720	.7236344	.7236968	.7237592	0	0
0.770	.7238216	.7238839	.7239463	.7240086	.7240710	.7241333	.7241956	.7242579	.7243202	.7243825	0	0
0.771	.7244448	.7245070	.7245693	.7246315	.7246938	.7247560	.7248182	.7248805	.7249426	.7250049	0	0
0.772	.7250670	.7251292	.7251914	.7252535	.7253156	.7253778	.7254399	.7255020	.7255641	.7256262	0	0
0.773	.7256883	.7257504	.7258124	.7258745	.7259365	.7259986	.7260606	.7261226	.7261846	.7262467	0	0
0.774	.7263086	.7263706	.7264326	.7264946	.7265565	.7266184	.7266803	.7267423	.7268042	.7268661	0	0
0.775	.7269280	.7269899	.7270517	.7271136	.7271755	.7272373	.7272992	.7273610	.7274228	.7274846	0	0
0.776	.7275464	.7276081	.7276700	.7277317	.7277935	.7278552	.7279170	.7279787	.7280404	.7281021	0	0
0.777	.7281638	.7282255	.7282872	.7283489	.7284105	.7284722	.7285338	.7285955	.7286571	.7287187	0	0
0.778	.7287803	.7288419	.7289035	.7289650	.7290266	.7290882	.7291498	.7292113	.7292728	.7293344	0	0
0.779	.7293958	.7294573	.7295188	.7295803	.7296418	.7297032	.7297647	.7298262	.7298875	.7299490	0	0
0.780	.7300104	.7300718	.7301332	.7301946	.7302560	.7303174	.7303787	.7304400	.7305014	.7305627	0	0
0.781	.7306240	.7306853	.7307466	.7308079	.7308692	.7309304	.7309917	.7310530	.7311142	.7311754	0	0
0.782	.7312366	.7312979	.7313591	.7314203	.7314814	.7315426	.7316038	.7316650	.7317261	.7317872	0	0
0.783	.7318484	.7319095	.7319705	.7320317	.7320927	.7321538	.7322149	.7322760	.7323370	.7323980	0	0
0.784	.7324591	.7325201	.7325811	.7326421	.7327031	.7327641	.7328251	.7328860	.7329470	.7330079	0	0
0.785	.7330689	.7331298	.7331907	.7332516	.7333125	.7333734	.7334343	.7334952	.7335560	.7336168	0	0
0.786	.7336777	.7337385	.7337993	.7338601	.7339209	.7339817	.7340425	.7341033	.7341641	.7342248	0	0
0.787	.7342855	.7343463	.7344070	.7344677	.7345284	.7345892	.7346498	.7347105	.7347712	.7348318	0	0
0.788	.7348925	.7349531	.7350137	.7350743	.7351350	.7351956	.7352561	.7353168	.7353773	.7354379	0	0
0.789	.7354984	.7355590	.7356195	.7356800	.7357405	.7358010	.7358615	.7359220	.7359825	.7360430	0	0
0.790	.7361034	.7361639	.7362243	.7362847	.7363452	.7364056	.7364660	.7365264	.7365867	.7366471	0	0
0.791	.7367074	.7367678	.7368281	.7368885	.7369488	.7370091	.7370694	.7371297	.7371900	.7372503	0	0
0.792	.7373105	.7373708	.7374310	.7374913	.7375515	.7376118	.7376720	.7377322	.7377923	.7378525	0	0
0.793	.7379127	.7379729	.7380330	.7380931	.7381533	.7382134	.7382735	.7383336	.7383937	.7384538	0	0
0.794	.7385139	.7385739	.7386340	.7386940	.7387541	.7388141	.7388741	.7389342	.7389941	.7390541	0	0
0.795	.7391141	.7391741	.7392340	.7392940	.7393539	.7394139	.7394738	.7395337	.7395936	.7396535	0	0
0.796	.7397134	.7397733	.7398331	.7398930	.7399528	.7400126	.7400725	.7401323	.7401921	.7402519	0	0
0.797	.7403117	.7403715	.7404312	.7404910	.7405508	.7406105	.7406702	.7407299	.7407897	.7408494	0	0
0.798	.7409090	.7409688	.7410284	.7410881	.7411478	.7412074	.7412670	.7413266	.7413863	.7414459	0	0
0.799	.7415054	.7415650	.7416247	.7416842	.7417437	.7418033	.7418629	.7419224	.7419819	.7420414	0	0

TABLE 7A

Error Function

X	.0	.0001	.0002	.0003	.0004	.0005	.0006	.0007	.0008	.0009	m	Δ_2
0.800	.7421010	.7421604	.7422199	.7422794	.7423388	.7423983	.7424577	.7425172	.7425766	.7426360	0	0
0.801	.7426955	.7427548	.7428142	.7428736	.7429330	.7429923	.7430517	.7431110	.7431704	.7432297	0	0
0.802	.7432890	.7433483	.7434076	.7434669	.7435262	.7435854	.7436447	.7437039	.7437631	.7438224	0	0
0.803	.7438816	.7439408	.7440000	.7440592	.7441184	.7441776	.7442367	.7442959	.7443550	.7444142	0	0
0.804	.7444733	.7445324	.7445915	.7446506	.7447097	.7447687	.7448278	.7448869	.7449459	.7450050	0	0
0.805	.7450640	.7451230	.7451820	.7452410	.7453000	.7453589	.7454180	.7454769	.7455359	.7455948	0	0
0.806	.7456537	.7457126	.7457716	.7458305	.7458894	.7459483	.7460071	.7460660	.7461249	.7461837	0	0
0.807	.7462425	.7463014	.7463602	.7464190	.7464778	.7465366	.7465954	.7466542	.7467129	.7467716	0	0
0.808	.7468304	.7468891	.7469478	.7470065	.7470652	.7471240	.7471827	.7472413	.7473000	.7473586	0	0
0.809	.7474173	.7474759	.7475346	.7475932	.7476518	.7477104	.7477690	.7478276	.7478861	.7479447	0	0
0.810	.7480032	.7480618	.7481203	.7481788	.7482374	.7482959	.7483544	.7484128	.7484713	.7485298	0	0
0.811	.7485883	.7486467	.7487051	.7487636	.7488220	.7488804	.7489388	.7489972	.7490556	.7491140	0	0
0.812	.7491723	.7492307	.7492890	.7493473	.7494057	.7494640	.7495223	.7495806	.7496389	.7496971	0	0
0.813	.7497554	.7498137	.7498720	.7499302	.7499884	.7500466	.7501048	.7501631	.7502213	.7502794	0	0
0.814	.7503376	.7503958	.7504539	.7505121	.7505702	.7506284	.7506865	.7507446	.7508026	.7508608	0	0
0.815	.7509188	.7509769	.7510350	.7510930	.7511510	.7512091	.7512671	.7513251	.7513831	.7514411	0	0
0.816	.7514991	.7515571	.7516150	.7516730	.7517309	.7517889	.7518468	.7519047	.7519626	.7520205	0	0
0.817	.7520784	.7521363	.7521942	.7522520	.7523099	.7523677	.7524256	.7524834	.7525412	.7525990	0	0
0.818	.7526568	.7527146	.7527723	.7528301	.7528879	.7529457	.7530034	.7530611	.7531188	.7531765	0	0
0.819	.7532342	.7532919	.7533496	.7534073	.7534649	.7535226	.7535802	.7536379	.7536955	.7537531	0	0
0.820	.7538107	.7538683	.7539259	.7539835	.7540411	.7540986	.7541562	.7542137	.7542712	.7543287	0	0
0.821	.7543862	.7544438	.7545013	.7545587	.7546162	.7546737	.7547311	.7547886	.7548460	.7549034	0	0
0.822	.7549609	.7550183	.7550757	.7551331	.7551904	.7552478	.7553052	.7553625	.7554199	.7554772	0	0
0.823	.7555345	.7555918	.7556491	.7557064	.7557637	.7558210	.7558783	.7559355	.7559928	.7560500	0	0
0.824	.7561072	.7561644	.7562217	.7562789	.7563360	.7563933	.7564504	.7565076	.7565647	.7566219	0	0
0.825	.7566790	.7567362	.7567933	.7568504	.7569075	.7569646	.7570216	.7570787	.7571357	.7571928	0	0
0.826	.7572498	.7573069	.7573639	.7574209	.7574779	.7575349	.7575919	.7576488	.7577058	.7577628	0	0
0.827	.7578197	.7578766	.7579336	.7579905	.7580474	.7581043	.7581612	.7582181	.7582750	.7583318	0	0
0.828	.7583886	.7584455	.7585024	.7585592	.7586160	.7586728	.7587296	.7587864	.7588431	.7588999	0	0
0.829	.7589567	.7590134	.7590702	.7591269	.7591836	.7592403	.7592970	.7593537	.7594104	.7594671	0	0
0.830	.7595237	.7595804	.7596370	.7596937	.7597503	.7598069	.7598635	.7599201	.7599767	.7600332	0	0
0.831	.7600899	.7601464	.7602029	.7602595	.7603160	.7603726	.7604291	.7604856	.7605421	.7605985	0	0
0.832	.7606550	.7607115	.7607679	.7608244	.7608808	.7609373	.7609937	.7610501	.7611065	.7611629	0	0
0.833	.7612193	.7612756	.7613320	.7613884	.7614447	.7615010	.7615573	.7616137	.7616700	.7617263	0	0
0.834	.7617826	.7618389	.7618951	.7619514	.7620077	.7620639	.7621201	.7621763	.7622325	.7622887	0	0
0.835	.7623450	.7624011	.7624573	.7625135	.7625696	.7626258	.7626819	.7627380	.7627941	.7628503	0	0
0.836	.7629064	.7629625	.7630185	.7630746	.7631307	.7631867	.7632428	.7632988	.7633548	.7634109	0	0
0.837	.7634668	.7635229	.7635788	.7636348	.7636908	.7637467	.7638027	.7638586	.7639146	.7639705	0	0
0.838	.7640264	.7640823	.7641382	.7641941	.7642500	.7643058	.7643617	.7644175	.7644734	.7645292	0	0
0.839	.7645850	.7646408	.7646966	.7647524	.7648082	.7648640	.7649197	.7649755	.7650312	.7650869	0	0
0.840	.7651427	.7651984	.7652541	.7653098	.7653655	.7654212	.7654768	.7655325	.7655882	.7656438	0	0
0.841	.7656994	.7657551	.7658107	.7658663	.7659218	.7659774	.7660330	.7660886	.7661442	.7661997	0	0
0.842	.7662552	.7663108	.7663663	.7664218	.7664773	.7665328	.7665883	.7666437	.7666992	.7667546	0	0
0.843	.7668101	.7668656	.7669210	.7669764	.7670318	.7670872	.7671426	.7671980	.7672533	.7673087	0	0
0.844	.7673640	.7674194	.7674747	.7675300	.7675853	.7676407	.7676960	.7677512	.7678065	.7678618	0	0
0.845	.7679170	.7679723	.7680275	.7680827	.7681380	.7681932	.7682484	.7683036	.7683588	.7684139	0	0
0.846	.7684691	.7685242	.7685794	.7686346	.7686896	.7687448	.7687999	.7688550	.7689101	.7689651	0	0
0.847	.7690202	.7690753	.7691303	.7691854	.7692404	.7692955	.7693505	.7694055	.7694604	.7695155	0	0
0.848	.7695704	.7696254	.7696804	.7697353	.7697902	.7698452	.7699001	.7699550	.7700099	.7700648	0	0
0.849	.7701197	.7701746	.7702295	.7702843	.7703391	.7703940	.7704488	.7705036	.7705584	.7706133	0	0

TABLE 7A

Error Function

X	.0	.0001	.0002	.0003	.0004	.0005	.0006	.0007	.0008	.0009	m	Δ_2
0.850	.7706680	.7707228	.7707776	.7708324	.7708871	.7709419	.7709966	.7710513	.7711060	.7711607	0	0
0.851	.7712154	.7712701	.7713248	.7713795	.7714341	.7714888	.7715434	.7715980	.7716527	.7717073	0	0
0.852	.7717619	.7718165	.7718711	.7719257	.7719802	.7720348	.7720894	.7721439	.7721984	.7722529	0	0
0.853	.7723075	.7723619	.7724165	.7724710	.7725254	.7725798	.7726343	.7726887	.7727432	.7727976	0	0
0.854	.7728521	.7729065	.7729609	.7730153	.7730696	.7731240	.7731784	.7732328	.7732871	.7733414	0	0
0.855	.7733957	.7734501	.7735044	.7735587	.7736130	.7736672	.7737215	.7737758	.7738300	.7738842	0	0
0.856	.7739385	.7739927	.7740470	.7741011	.7741553	.7742095	.7742637	.7743179	.7743720	.7744262	0	0
0.857	.7744803	.7745345	.7745886	.7746427	.7746968	.7747509	.7748049	.7748591	.7749131	.7749672	0	0
0.858	.7750212	.7750753	.7751293	.7751833	.7752373	.7752913	.7753453	.7753993	.7754533	.7755072	0	0
0.859	.7755612	.7756152	.7756691	.7757230	.7757769	.7758308	.7758847	.7759386	.7759925	.7760464	0	0
0.860	.7761002	.7761541	.7762079	.7762617	.7763156	.7763694	.7764232	.7764770	.7765308	.7765846	0	0
0.861	.7766383	.7766921	.7767459	.7767996	.7768533	.7769071	.7769608	.7770145	.7770682	.7771218	0	0
0.862	.7771755	.7772292	.7772829	.7773365	.7773902	.7774438	.7774974	.7775511	.7776046	.7776582	0	0
0.863	.7777118	.7777654	.7778190	.7778725	.7779261	.7779796	.7780331	.7780867	.7781402	.7781937	0	0
0.864	.7782472	.7783006	.7783541	.7784076	.7784610	.7785145	.7785679	.7786213	.7786748	.7787282	0	0
0.865	.7787816	.7788349	.7788883	.7789417	.7789951	.7790484	.7791018	.7791551	.7792084	.7792618	0	0
0.866	.7793151	.7793683	.7794216	.7794749	.7795282	.7795815	.7796347	.7796879	.7797412	.7797944	0	0
0.867	.7798476	.7799008	.7799540	.7800072	.7800604	.7801136	.7801667	.7802199	.7802730	.7803262	0	0
0.868	.7803793	.7804324	.7804855	.7805386	.7805917	.7806448	.7806978	.7807509	.7808039	.7808570	0	0
0.869	.7809100	.7809631	.7810161	.7810690	.7811220	.7811750	.7812280	.7812810	.7813339	.7813869	0	0
0.870	.7814398	.7814928	.7815456	.7815986	.7816515	.7817044	.7817572	.7818101	.7818630	.7819158	0	0
0.871	.7819687	.7820215	.7820744	.7821272	.7821800	.7822328	.7822856	.7823384	.7823911	.7824439	0	0
0.872	.7824967	.7825494	.7826021	.7826549	.7827076	.7827603	.7828130	.7828657	.7829184	.7829710	0	0
0.873	.7830237	.7830763	.7831290	.7831816	.7832342	.7832869	.7833395	.7833921	.7834447	.7834973	0	0
0.874	.7835498	.7836024	.7836549	.7837075	.7837600	.7838125	.7838650	.7839175	.7839701	.7840226	0	0
0.875	.7840750	.7841275	.7841800	.7842324	.7842848	.7843373	.7843897	.7844421	.7844945	.7845469	0	0
0.876	.7845993	.7846517	.7847041	.7847564	.7848088	.7848611	.7849134	.7849658	.7850181	.7850704	0	0
0.877	.7851227	.7851750	.7852272	.7852795	.7853318	.7853840	.7854363	.7854885	.7855407	.7855929	0	0
0.878	.7856451	.7856973	.7857495	.7858017	.7858539	.7859060	.7859582	.7860103	.7860624	.7861146	0	0
0.879	.7861667	.7862188	.7862709	.7863230	.7863750	.7864271	.7864792	.7865312	.7865832	.7866353	0	0
0.880	.7866873	.7867393	.7867913	.7868433	.7868953	.7869473	.7869992	.7870512	.7871031	.7871551	0	0
0.881	.7872070	.7872589	.7873108	.7873628	.7874146	.7874665	.7875184	.7875702	.7876221	.7876740	0	0
0.882	.7877258	.7877776	.7878294	.7878813	.7879331	.7879848	.7880366	.7880884	.7881402	.7881919	0	0
0.883	.7882437	.7882954	.7883471	.7883989	.7884505	.7885023	.7885540	.7886056	.7886573	.7887090	0	0
0.884	.7887606	.7888123	.7888639	.7889155	.7889671	.7890188	.7890704	.7891219	.7891735	.7892251	0	0
0.885	.7892767	.7893282	.7893798	.7894313	.7894828	.7895343	.7895858	.7896373	.7896888	.7897403	0	0
0.886	.7897918	.7898433	.7898947	.7899461	.7899976	.7900490	.7901005	.7901518	.7902033	.7902547	0	0
0.887	.7903060	.7903574	.7904087	.7904601	.7905114	.7905628	.7906141	.7906654	.7907168	.7907680	0	0
0.888	.7908193	.7908706	.7909219	.7909731	.7910244	.7910756	.7911268	.7911781	.7912293	.7912805	0	0
0.889	.7913317	.7913829	.7914341	.7914852	.7915365	.7915876	.7916387	.7916899	.7917410	.7917921	0	0
0.890	.7918432	.7918943	.7919454	.7919965	.7920476	.7920986	.7921497	.7922007	.7922518	.7923028	0	0
0.891	.7923538	.7924048	.7924558	.7925068	.7925578	.7926087	.7926597	.7927107	.7927616	.7928125	0	0
0.892	.7928635	.7929144	.7929653	.7930162	.7930671	.7931179	.7931688	.7932197	.7932705	.7933214	0	0
0.893	.7933722	.7934231	.7934739	.7935247	.7935755	.7936262	.7936770	.7937278	.7937786	.7938293	0	0
0.894	.7938801	.7939308	.7939816	.7940323	.7940830	.7941337	.7941844	.7942351	.7942857	.7943364	0	0
0.895	.7943870	.7944377	.7944883	.7945389	.7945896	.7946402	.7946908	.7947413	.7947919	.7948425	0	0
0.896	.7948931	.7949436	.7949942	.7950447	.7950953	.7951458	.7951962	.7952468	.7952973	.7953478	0	0
0.897	.7953982	.7954487	.7954991	.7955496	.7956000	.7956504	.7957008	.7957513	.7958016	.7958521	0	0
0.898	.7959024	.7959528	.7960032	.7960535	.7961039	.7961542	.7962046	.7962549	.7963052	.7963555	0	0
0.899	.7964057	.7964560	.7965063	.7965566	.7966068	.7966571	.7967073	.7967576	.7968078	.7968580	0	0

TABLE 7A

Error Function

X	.0	.0001	.0002	.0003	.0004	.0005	.0006	.0007	.0008	.0009	m	Δ_2
0.900	.7969082	.7969584	.7970086	.7970588	.7971089	.7971591	.7972092	.7972593	.7973095	.7973596	0	0
0.901	.7974097	.7974598	.7975099	.7975600	.7976100	.7976601	.7977102	.7977602	.7978103	.7978603	0	0
0.902	.7979103	.7979603	.7980103	.7980603	.7981103	.7981603	.7982103	.7982602	.7983102	.7983601	0	0
0.903	.7984100	.7984599	.7985098	.7985598	.7986097	.7986596	.7987094	.7987593	.7988092	.7988590	0	0
0.904	.7989088	.7989587	.7990085	.7990583	.7991081	.7991579	.7992077	.7992575	.7993072	.7993570	0	0
0.905	.7994068	.7994565	.7995062	.7995560	.7996057	.7996554	.7997050	.7997547	.7998044	.7998541	0	0
0.906	.7999038	.7999534	.8000031	.8000527	.8001023	.8001519	.8002015	.8002511	.8003007	.8003503	0	0
0.907	.8003998	.8004494	.8004990	.8005485	.8005980	.8006476	.8006971	.8007466	.8007961	.8008456	0	0
0.908	.8008951	.8009446	.8009940	.8010435	.8010929	.8011423	.8011917	.8012412	.8012906	.8013400	0	0
0.909	.8013894	.8014387	.8014882	.8015375	.8015869	.8016362	.8016855	.8017349	.8017842	.8018335	0	0
0.910	.8018828	.8019321	.8019814	.8020306	.8020799	.8021291	.8021784	.8022277	.8022769	.8023261	0	0
0.911	.8023753	.8024245	.8024737	.8025229	.8025721	.8026212	.8026704	.8027195	.8027687	.8028178	0	0
0.912	.8028669	.8029160	.8029652	.8030142	.8030633	.8031124	.8031615	.8032106	.8032596	.8033086	0	0
0.913	.8033577	.8034067	.8034557	.8035047	.8035537	.8036027	.8036516	.8037006	.8037496	.8037986	0	0
0.914	.8038475	.8038964	.8039454	.8039942	.8040432	.8040920	.8041409	.8041899	.8042387	.8042876	0	0
0.915	.8043364	.8043852	.8044341	.8044829	.8045318	.8045806	.8046293	.8046781	.8047269	.8047757	0	0
0.916	.8048245	.8048732	.8049220	.8049707	.8050194	.8050681	.8051168	.8051655	.8052142	.8052629	0	0
0.917	.8053116	.8053603	.8054090	.8054576	.8055062	.8055549	.8056034	.8056521	.8057007	.8057493	0	0
0.918	.8057979	.8058465	.8058950	.8059435	.8059921	.8060406	.8060892	.8061377	.8061862	.8062347	0	0
0.919	.8062832	.8063317	.8063802	.8064287	.8064771	.8065256	.8065740	.8066224	.8066708	.8067193	0	0
0.920	.8067677	.8068161	.8068645	.8069128	.8069612	.8070096	.8070579	.8071063	.8071547	.8072029	0	0
0.921	.8072513	.8072996	.8073479	.8073962	.8074445	.8074927	.8075410	.8075892	.8076375	.8076857	0	0
0.922	.8077340	.8077822	.8078304	.8078786	.8079268	.8079750	.8080232	.8080713	.8081195	.8081676	0	0
0.923	.8082157	.8082639	.8083120	.8083602	.8084083	.8084563	.8085044	.8085525	.8086005	.8086486	0	0
0.924	.8086967	.8087447	.8087928	.8088408	.8088888	.8089368	.8089848	.8090328	.8090808	.8091288	0	0
0.925	.8091767	.8092247	.8092726	.8093205	.8093685	.8094164	.8094643	.8095122	.8095601	.8096080	0	0
0.926	.8096558	.8097037	.8097516	.8097994	.8098472	.8098951	.8099429	.8099907	.8100385	.8100863	0	0
0.927	.8101341	.8101819	.8102297	.8102774	.8103251	.8103729	.8104206	.8104683	.8105161	.8105637	0	0
0.928	.8106114	.8106592	.8107068	.8107545	.8108022	.8108498	.8108975	.8109451	.8109927	.8110403	0	0
0.929	.8110879	.8111355	.8111831	.8112307	.8112783	.8113258	.8113734	.8114209	.8114685	.8115160	0	0
0.930	.8115636	.8116111	.8116586	.8117060	.8117535	.8118010	.8118485	.8118959	.8119434	.8119908	0	0
0.931	.8120382	.8120857	.8121331	.8121805	.8122279	.8122753	.8123226	.8123700	.8124174	.8124647	0	0
0.932	.8125120	.8125594	.8126068	.8126540	.8127013	.8127487	.8127959	.8128432	.8128905	.8129377	0	0
0.933	.8129850	.8130323	.8130795	.8131267	.8131739	.8132212	.8132684	.8133156	.8133627	.8134099	0	0
0.934	.8134571	.8135042	.8135514	.8135985	.8136457	.8136928	.8137399	.8137870	.8138341	.8138812	0	0
0.935	.8139282	.8139753	.8140224	.8140694	.8141165	.8141635	.8142105	.8142576	.8143046	.8143516	0	0
0.936	.8143986	.8144456	.8144925	.8145395	.8145865	.8146334	.8146803	.8147272	.8147742	.8148211	0	0
0.937	.8148680	.8149149	.8149618	.8150086	.8150555	.8151024	.8151492	.8151960	.8152429	.8152897	0	0
0.938	.8153365	.8153833	.8154302	.8154770	.8155237	.8155705	.8156173	.8156640	.8157107	.8157575	0	0
0.939	.8158042	.8158509	.8158976	.8159443	.8159910	.8160377	.8160844	.8161311	.8161777	.8162243	0	0
0.940	.8162710	.8163176	.8163642	.8164108	.8164575	.8165041	.8165506	.8165972	.8166438	.8166903	0	0
0.941	.8167369	.8167834	.8168300	.8168765	.8169230	.8169695	.8170160	.8170625	.8171090	.8171555	0	0
0.942	.8172019	.8172484	.8172948	.8173413	.8173877	.8174341	.8174806	.8175269	.8175733	.8176197	0	0
0.943	.8176661	.8177125	.8177588	.8178052	.8178515	.8178979	.8179442	.8179905	.8180368	.8180831	0	0
0.944	.8181294	.8181757	.8182219	.8182682	.8183144	.8183607	.8184069	.8184532	.8184994	.8185456	0	0
0.945	.8185918	.8186380	.8186842	.8187304	.8187765	.8188227	.8188688	.8189149	.8189611	.8190072	0	0
0.946	.8190534	.8190994	.8191456	.8191916	.8192377	.8192838	.8193299	.8193759	.8194219	.8194680	0	0
0.947	.8195140	.8195601	.8196060	.8196520	.8196980	.8197440	.8197900	.8198360	.8198819	.8199279	0	0
0.948	.8199738	.8200197	.8200657	.8201116	.8201575	.8202034	.8202493	.8202952	.8203410	.8203869	0	0
0.949	.8204327	.8204786	.8205244	.8205702	.8206161	.8206618	.8207077	.8207535	.8207992	.8208450	0	0

TABLE 7A

Error Function

X	.0	.0001	.0002	.0003	.0004	.0005	.0006	.0007	.0008	.0009	m	Δ_2
0.950	.8208908	.8209366	.8209823	.8210281	.8210738	.8211195	.8211652	.8212109	.8212566	.8213023	0	0
0.951	.8213480	.8213936	.8214393	.8214849	.8215306	.8215762	.8216218	.8216675	.8217131	.8217587	0	0
0.952	.8218043	.8218499	.8218954	.8219410	.8219866	.8220321	.8220776	.8221232	.8221687	.8222142	0	0
0.953	.8222597	.8223052	.8223507	.8223962	.8224416	.8224871	.8225326	.8225780	.8226234	.8226689	0	0
0.954	.8227143	.8227597	.8228051	.8228505	.8228959	.8229412	.8229867	.8230320	.8230773	.8231227	0	0
0.955	.8231680	.8232133	.8232586	.8233039	.8233492	.8233945	.8234398	.8234851	.8235304	.8235756	0	0
0.956	.8236209	.8236661	.8237113	.8237566	.8238018	.8238470	.8238922	.8239374	.8239825	.8240277	0	0
0.957	.8240728	.8241180	.8241631	.8242083	.8242534	.8242985	.8243436	.8243887	.8244338	.8244789	0	0
0.958	.8245240	.8245690	.8246141	.8246591	.8247042	.8247492	.8247942	.8248392	.8248842	.8249292	0	0
0.959	.8249742	.8250192	.8250642	.8251091	.8251541	.8251990	.8252439	.8252889	.8253338	.8253787	0	0
0.960	.8254236	.8254685	.8255134	.8255582	.8256031	.8256480	.8256928	.8257377	.8257825	.8258274	0	0
0.961	.8258721	.8259169	.8259618	.8260065	.8260513	.8260961	.8261409	.8261856	.8262303	.8262751	0	0
0.962	.8263198	.8263645	.8264093	.8264539	.8264986	.8265433	.8265880	.8266327	.8266773	.8267220	0	0
0.963	.8267666	.8268113	.8268559	.8269005	.8269451	.8269897	.8270343	.8270789	.8271235	.8271680	0	0
0.964	.8272126	.8272571	.8273017	.8273462	.8273907	.8274353	.8274797	.8275242	.8275687	.8276132	0	0
0.965	.8276576	.8277021	.8277466	.8277910	.8278354	.8278799	.8279243	.8279687	.8280131	.8280575	0	0
0.966	.8281019	.8281463	.8281906	.8282350	.8282794	.8283237	.8283680	.8284124	.8284566	.8285010	0	0
0.967	.8285453	.8285896	.8286338	.8286781	.8287224	.8287666	.8288109	.8288552	.8288994	.8289436	0	0
0.968	.8289878	.8290320	.8290762	.8291204	.8291646	.8292087	.8292529	.8292971	.8293412	.8293853	0	0
0.969	.8294294	.8294736	.8295177	.8295618	.8296059	.8296500	.8296940	.8297381	.8297822	.8298262	0	0
0.970	.8298703	.8299143	.8299583	.8300024	.8300464	.8300903	.8301343	.8301783	.8302223	.8302662	0	0
0.971	.8303102	.8303542	.8303981	.8304420	.8304859	.8305299	.8305738	.8306177	.8306615	.8307055	0	0
0.972	.8307493	.8307932	.8308370	.8308809	.8309247	.8309686	.8310124	.8310562	.8311000	.8311438	0	0
0.973	.8311875	.8312314	.8312751	.8313189	.8313626	.8314064	.8314501	.8314939	.8315375	.8315812	0	0
0.974	.8316250	.8316687	.8317123	.8317560	.8317997	.8318433	.8318870	.8319306	.8319743	.8320179	0	0
0.975	.8320615	.8321051	.8321487	.8321923	.8322359	.8322794	.8323230	.8323666	.8324101	.8324537	0	0
0.976	.8324972	.8325407	.8325842	.8326277	.8326712	.8327147	.8327582	.8328017	.8328452	.8328886	0	0
0.977	.8329321	.8329754	.8330189	.8330624	.8331057	.8331491	.8331925	.8332359	.8332793	.8333226	0	0
0.978	.8333660	.8334094	.8334527	.8334960	.8335394	.8335827	.8336260	.8336693	.8337126	.8337559	0	0
0.979	.8337992	.8338425	.8338857	.8339289	.8339722	.8340154	.8340586	.8341019	.8341451	.8341883	0	0
0.980	.8342315	.8342746	.8343178	.8343610	.8344042	.8344473	.8344904	.8345336	.8345767	.8346198	0	0
0.981	.8346629	.8347060	.8347491	.8347922	.8348353	.8348783	.8349214	.8349645	.8350075	.8350505	0	0
0.982	.8350936	.8351365	.8351796	.8352225	.8352655	.8353085	.8353515	.8353944	.8354374	.8354803	0	0
0.983	.8355233	.8355662	.8356091	.8356521	.8356950	.8357379	.8357807	.8358237	.8358665	.8359094	0	0
0.984	.8359522	.8359951	.8360379	.8360807	.8361235	.8361664	.8362092	.8362520	.8362948	.8363375	0	0
0.985	.8363803	.8364230	.8364658	.8365086	.8365513	.8365940	.8366367	.8366795	.8367221	.8367648	0	0
0.986	.8368075	.8368502	.8368928	.8369355	.8369782	.8370208	.8370634	.8371061	.8371487	.8371913	0	0
0.987	.8372339	.8372765	.8373191	.8373617	.8374043	.8374468	.8374894	.8375319	.8375744	.8376169	0	0
0.988	.8376595	.8377020	.8377445	.8377870	.8378295	.8378719	.8379144	.8379568	.8379993	.8380417	0	0
0.989	.8380842	.8381266	.8381690	.8382114	.8382538	.8382962	.8383386	.8383810	.8384233	.8384657	0	0
0.990	.8385081	.8385504	.8385927	.8386350	.8386773	.8387197	.8387620	.8388042	.8388465	.8388888	0	0
0.991	.8389311	.8389733	.8390156	.8390578	.8391001	.8391423	.8391845	.8392267	.8392689	.8393111	0	0
0.992	.8393533	.8393955	.8394376	.8394797	.8395219	.8395640	.8396062	.8396483	.8396904	.8397325	0	0
0.993	.8397746	.8398167	.8398588	.8399009	.8399429	.8399850	.8400270	.8400691	.8401111	.8401532	0	0
0.994	.8401951	.8402371	.8402792	.8403211	.8403631	.8404051	.8404471	.8404890	.8405309	.8405729	0	0
0.995	.8406149	.8406568	.8406987	.8407406	.8407825	.8408244	.8408663	.8409081	.8409500	.8409919	0	0
0.996	.8410337	.8410755	.8411174	.8411592	.8412010	.8412428	.8412846	.8413264	.8413681	.8414099	0	0
0.997	.8414517	.8414934	.8415352	.8415769	.8416187	.8416604	.8417021	.8417438	.8417855	.8418272	0	0
0.998	.8418689	.8419105	.8419522	.8419939	.8420355	.8420772	.8421188	.8421604	.8422020	.8422437	0	0
0.999	.8422852	.8423268	.8423684	.8424100	.8424516	.8424931	.8425347	.8425762	.8426178	.8426592	0	0

TABLE 7B

Error Function Complement

X	.0	.001	.002	.003	.004	.005	.006	.007	.008	.009	m	Δ_2
1.00	.157299	.156884	.156471	.156058	.155645	.155234	.154823	.154414	.154005	.153597	0	1
1.01	.153189	.152783	.152377	.151973	.151569	.151165	.150763	.150362	.149961	.149561	0	1
1.02	.149162	.148764	.148366	.147970	.147574	.147179	.146785	.146391	.145999	.145607	0	1
1.03	.145216	.144826	.144436	.144048	.143660	.143273	.142887	.142501	.142117	.141733	0	1
1.04	.141350	.140968	.140586	.140206	.139826	.139447	.139069	.138691	.138315	.137939	0	1
1.05	.137564	.137190	.136816	.136443	.136072	.135700	.135330	.134960	.134592	.134224	0	1
1.06	.133856	.133490	.133124	.132759	.132395	.132032	.131669	.131307	.130946	.130586	0	1
1.07	.130227	.129868	.129510	.129153	.128796	.128441	.128086	.127732	.127378	.127026	0	1
1.08	.126674	.126323	.125972	.125623	.125274	.124926	.124579	.124232	.123886	.123541	0	1
1.09	.123197	.122853	.122510	.122168	.121827	.121487	.121147	.120808	.120469	.120132	0	1
1.10	.119795	.119459	.119123	.118789	.118455	.118122	.117789	.117458	.117127	.116796	0	1
1.11	.116467	.116138	.115810	.115483	.115156	.114830	.114505	.114181	.113857	.113534	0	1
1.12	.113212	.112891	.112570	.112250	.111930	.111612	.111294	.110977	.110660	.110344	0	1
1.13	.110029	.109715	.109401	.109088	.108776	.108465	.108154	.107844	.107534	.107226	0	1
1.14	.106918	.106610	.106304	.105998	.105693	.105388	.105084	.104781	.104479	.104177	0	1
1.15	.103876	.103576	.103276	.102977	.102679	.102381	.102084	.101788	.101493	.101198	0	1
1.16	.100904	.100610	.100317	.100025	.099734	.099443	.099153	.098864	.098575	.098287	0	1
1.17	.979996	.977129	.974269	.971415	.968568	.965727	.962894	.960067	.957247	.954433	-1	8
1.18	.951626	.948825	.946031	.943244	.940463	.937689	.934922	.932161	.929407	.926658	-1	7
1.19	.923917	.921182	.918454	.915732	.913016	.910308	.907606	.904910	.902220	.899537	-1	6
1.20	.896860	.894190	.891526	.888869	.886218	.883573	.880935	.878303	.875677	.873058	-1	6
1.21	.870445	.867839	.865238	.862644	.860056	.857474	.854900	.852331	.849768	.847211	-1	7
1.22	.844661	.842117	.839579	.837048	.834522	.832003	.829490	.826983	.824482	.821988	-1	7
1.23	.819499	.817017	.814540	.812070	.809606	.807148	.804696	.802249	.799810	.797376	-1	7
1.24	.794948	.792526	.790110	.787701	.785297	.782899	.780507	.778121	.775741	.773367	-1	7
1.25	.770999	.768636	.766280	.763930	.761585	.759247	.756914	.754587	.752265	.749950	-1	5
1.26	.747640	.745337	.743039	.740747	.738461	.736180	.733905	.731636	.729373	.727115	-1	6
1.27	.724864	.722618	.720377	.718143	.715913	.713690	.711472	.709260	.707054	.704853	-1	5
1.28	.702658	.700469	.698285	.696106	.693934	.691766	.689604	.687449	.685298	.683153	-1	6
1.29	.681014	.678880	.676751	.674628	.672510	.670399	.668292	.666191	.664095	.662005	-1	6
1.30	.659920	.657841	.655767	.653699	.651635	.649578	.647525	.645478	.643436	.641400	-1	5
1.31	.639368	.637343	.635323	.633307	.631297	.629293	.627293	.625299	.623311	.621327	-1	5
1.32	.619349	.617375	.615407	.613445	.611487	.609535	.607587	.605645	.603708	.601777	-1	6
1.33	.599850	.597928	.596012	.594100	.592194	.590293	.588397	.586506	.584620	.582739	-1	5
1.34	.580863	.578992	.577126	.575265	.573409	.571558	.569712	.567871	.566035	.564204	-1	5
1.35	.562378	.560557	.558741	.556929	.555123	.553321	.551524	.549732	.547945	.546163	-1	5
1.36	.544386	.542613	.540846	.539083	.537324	.535571	.533823	.532079	.530340	.528606	-1	5
1.37	.526876	.525151	.523431	.521716	.520005	.518299	.516598	.514902	.513210	.511522	-1	5
1.38	.509840	.508162	.506488	.504820	.503156	.501496	.499841	.498191	.496545	.494904	-1	5
1.39	.493267	.491635	.490007	.488384	.486766	.485152	.483542	.481937	.480337	.478741	-1	5
1.40	.477149	.475562	.473979	.472401	.470827	.469257	.467692	.466132	.464575	.463023	-1	4
1.41	.461476	.459933	.458394	.456859	.455329	.453803	.452282	.450764	.449251	.447743	-1	4
1.42	.446238	.444738	.443242	.441751	.440263	.438780	.437301	.435826	.434356	.432890	-1	4
1.43	.431428	.429970	.428516	.427066	.425621	.424179	.422742	.421309	.419880	.418455	-1	4
1.44	.417034	.415618	.414205	.412796	.411392	.409992	.408595	.407203	.405814	.404430	-1	4
1.45	.403050	.401673	.400301	.398933	.397568	.396208	.394852	.393499	.392150	.390806	-1	4
1.46	.389465	.388128	.386795	.385466	.384141	.382820	.381502	.380189	.378879	.377573	-1	4
1.47	.376271	.374973	.373678	.372388	.371101	.369818	.368539	.367263	.365992	.364724	-1	4
1.48	.363459	.362199	.360942	.359689	.358440	.357194	.355952	.354714	.353480	.352249	-1	4
1.49	.351021	.349798	.348578	.347362	.346149	.344940	.343734	.342533	.341334	.340140	-1	4

TABLE 7B

Error Function Complement

X	.0	.001	.002	.003	.004	.005	.006	.007	.008	.009	m	Δ_2
1.50	.338949	.337761	.336577	.335397	.334220	.333046	.331877	.330710	.329548	.328388	−1	4
1.51	.327233	.326080	.324931	.323786	.322644	.321506	.320371	.319239	.318111	.316986	−1	3
1.52	.315865	.314747	.313633	.312522	.311414	.310309	.309208	.308111	.307017	.305926	−1	3
1.53	.304838	.303754	.302673	.301595	.300521	.299450	.298382	.297317	.296256	.295198	−1	3
1.54	.294143	.293092	.292043	.290998	.289957	.288918	.287882	.286850	.285821	.284795	−1	3
1.55	.283773	.282753	.281737	.280724	.279714	.278707	.277703	.276702	.275705	.274710	−1	3
1.56	.273719	.272731	.271745	.270763	.269784	.268808	.267835	.266865	.265898	.264935	−1	3
1.57	.263974	.263016	.262061	.261109	.260161	.259215	.258272	.257332	.256395	.255461	−1	3
1.58	.254530	.253602	.252677	.251754	.250835	.249919	.249005	.248094	.247187	.246282	−1	3
1.59	.245380	.244481	.243584	.242691	.241800	.240913	.240028	.239146	.238266	.237390	−1	3
1.60	.236516	.235645	.234777	.233912	.233049	.232189	.231332	.230478	.229627	.228778	−1	3
1.61	.227932	.227088	.226248	.225410	.224574	.223742	.222912	.222085	.221260	.220438	−1	3
1.62	.219619	.218803	.217989	.217177	.216369	.215563	.214759	.213958	.213160	.212365	−1	3
1.63	.211572	.210781	.209993	.209208	.208425	.207645	.206867	.206092	.205320	.204550	−1	3
1.64	.203782	.203017	.202255	.201495	.200737	.199982	.199230	.198480	.197732	.196987	−1	2
1.65	.196244	.195504	.194766	.194031	.193298	.192567	.191839	.191113	.190390	.189669	−1	2
1.66	.188951	.188235	.187521	.186810	.186100	.185394	.184689	.183988	.183288	.182591	−1	2
1.67	.181896	.181203	.180513	.179825	.179139	.178455	.177774	.177095	.176419	.175744	−1	2
1.68	.175072	.174402	.173735	.173069	.172406	.171745	.171087	.170430	.169776	.169124	−1	2
1.69	.168474	.167827	.167181	.166538	.165897	.165258	.164621	.163986	.163354	.162724	−1	2
1.70	.162095	.161469	.160845	.160224	.159604	.158986	.158371	.157758	.157146	.156537	−1	2
1.71	.155930	.155325	.154722	.154121	.153522	.152925	.152330	.151738	.151147	.150558	−1	2
1.72	.149972	.149387	.148804	.148224	.147645	.147069	.146494	.145921	.145350	.144782	−1	2
1.73	.144215	.143650	.143087	.142526	.141967	.141410	.140855	.140302	.139751	.139201	−1	2
1.74	.138654	.138108	.137565	.137023	.136483	.135945	.135409	.134875	.134342	.133812	−1	2
1.75	.133283	.132756	.132231	.131708	.131187	.130668	.130150	.129634	.129120	.128608	−1	2
1.76	.128097	.127589	.127082	.126577	.126073	.125572	.125072	.124574	.124078	.123583	−1	2
1.77	.123091	.122600	.122110	.121623	.121137	.120653	.120170	.119690	.119211	.118734	−1	2
1.78	.118258	.117784	.117312	.116841	.116373	.115905	.115440	.114976	.114514	.114053	−1	2
1.79	.113594	.113137	.112682	.112228	.111775	.111325	.110875	.110428	.109982	.109538	−1	2
1.80	.109095	.108654	.108214	.107776	.107340	.106905	.106472	.106040	.105610	.105182	−1	1
1.81	.104755	.104329	.103905	.103483	.103062	.102642	.102225	.101808	.101394	.100980	−1	1
1.82	.100568	.100158	.099749	.099342	.098936	.098532	.098129	.097727	.097327	.096929	−1	1
1.83	.965320	.961364	.957423	.953496	.949583	.945684	.941800	.937931	.934075	.930233	−2	14
1.84	.926406	.922592	.918793	.915008	.911236	.907479	.903735	.900005	.896289	.892586	−2	13
1.85	.888898	.885222	.881561	.877913	.874278	.870657	.867049	.863455	.859874	.856306	−2	13
1.86	.852752	.849211	.845682	.842167	.838665	.835177	.831701	.828237	.824787	.821350	−2	12
1.87	.817926	.814514	.811115	.807729	.804356	.800995	.797646	.794311	.790987	.787676	−2	12
1.88	.784378	.781092	.777819	.774557	.771308	.768071	.764846	.761634	.758433	.755245	−2	12
1.89	.752069	.748904	.745752	.742611	.739483	.736366	.733261	.730167	.727086	.724016	−2	12
1.90	.720958	.717911	.714876	.711852	.708840	.705839	.702850	.699872	.696906	.693950	−2	11
1.91	.691007	.688074	.685152	.682242	.679342	.676454	.673577	.670711	.667856	.665011	−2	11
1.92	.662178	.659355	.656543	.653743	.650953	.648173	.645405	.642646	.639899	.637162	−2	10
1.93	.634435	.631720	.629014	.626319	.623634	.620960	.618296	.615643	.612999	.610366	−2	10
1.94	.607743	.605130	.602528	.599935	.597353	.594780	.592218	.589665	.587123	.584590	−2	10
1.95	.582067	.579554	.577051	.574557	.572073	.569599	.567135	.564680	.562235	.559799	−2	10
1.96	.557373	.554956	.552549	.550151	.547763	.545384	.543014	.540654	.538302	.535961	−2	10
1.97	.533628	.531304	.528990	.526685	.524388	.522101	.519823	.517554	.515294	.513043	−2	9
1.98	.510800	.508567	.506342	.504126	.501919	.499721	.497531	.495350	.493178	.491014	−2	9
1.99	.488859	.486712	.484574	.482445	.480324	.478211	.476107	.474011	.471924	.469844	−2	8

TABLE 7B

Error Function Complement

X	.0	.001	.002	.003	.004	.005	.006	.007	.008	.009	m	Δ_2
2.00	.467774	.465711	.463657	.461611	.459573	.457543	.455521	.453508	.451502	.449505	-2	8
2.01	.447515	.445534	.443560	.441595	.439637	.437687	.435745	.433811	.431884	.429966	-2	8
2.02	.428055	.426152	.424256	.422369	.420488	.418616	.416750	.414893	.413043	.411201	-2	7
2.03	.409365	.407538	.405718	.403904	.402099	.400301	.398510	.396727	.394950	.393181	-2	7
2.04	.391419	.389665	.387917	.386177	.384443	.382717	.380998	.379286	.377580	.375882	-2	7
2.05	.374191	.372506	.370829	.369158	.367494	.365837	.364187	.362544	.360907	.359277	-2	7
2.06	.357654	.356038	.354428	.352824	.351228	.349638	.348054	.346477	.344907	.343343	-2	6
2.07	.341785	.340234	.338689	.337151	.335619	.334093	.332574	.331061	.329554	.328054	-2	6
2.08	.326559	.325071	.323589	.322114	.320644	.319181	.317723	.316272	.314827	.313387	-2	6
2.09	.311954	.310527	.309106	.307690	.306281	.304877	.303479	.302088	.300702	.299321	-2	6
2.10	.297947	.296578	.295215	.293858	.292507	.291161	.289820	.288486	.287157	.285834	-2	6
2.11	.284516	.283203	.281897	.280596	.279300	.278010	.276725	.275445	.274171	.272903	-2	5
2.12	.271640	.270382	.269129	.267882	.266640	.265403	.264172	.262945	.261724	.260509	-2	5
2.13	.259298	.258092	.256892	.255697	.254506	.253321	.252141	.250966	.249796	.248631	-2	5
2.14	.247471	.246316	.245166	.244020	.242880	.241744	.240614	.239488	.238367	.237251	-2	5
2.15	.236139	.235033	.233931	.232834	.231741	.230654	.229571	.228492	.227419	.226349	-2	5
2.16	.225285	.224225	.223170	.222119	.221073	.220031	.218994	.217961	.216933	.215909	-2	4
2.17	.214889	.213874	.212864	.211857	.210856	.209858	.208865	.207876	.206892	.205911	-2	4
2.18	.204935	.203963	.202996	.202033	.201073	.200118	.199168	.198221	.197278	.196340	-2	4
2.19	.195406	.194476	.193549	.192627	.191709	.190795	.189885	.188979	.188077	.187179	-2	4
2.20	.186285	.185395	.184508	.183626	.182747	.181872	.181002	.180135	.179271	.178412	-2	4
2.21	.177556	.176704	.175856	.175012	.174171	.173335	.172501	.171672	.170846	.170024	-2	4
2.22	.169205	.168390	.167579	.166771	.165967	.165167	.164370	.163576	.162786	.162000	-2	4
2.23	.161217	.160437	.159661	.158889	.158120	.157354	.156592	.155833	.155078	.154326	-2	3
2.24	.153577	.152832	.152090	.151351	.150615	.149883	.149155	.148429	.147707	.146988	-2	3
2.25	.146272	.145559	.144850	.144143	.143440	.142740	.142044	.141350	.140660	.139972	-2	3
2.26	.139288	.138607	.137929	.137254	.136582	.135913	.135247	.134584	.133924	.133267	-2	3
2.27	.132613	.131962	.131314	.130669	.130026	.129387	.128751	.128117	.127487	.126859	-2	3
2.28	.126234	.125612	.124993	.124376	.123763	.123152	.122544	.121939	.121336	.120736	-2	3
2.29	.120139	.119545	.118954	.118365	.117779	.117195	.116614	.116036	.115461	.114888	-2	3
2.30	.114318	.113750	.113185	.112623	.112063	.111506	.110951	.110399	.109849	.109302	-2	3
2.31	.108758	.108216	.107676	.107139	.106605	.106073	.105543	.105016	.104491	.103969	-2	2
2.32	.103449	.102931	.102416	.101903	.101393	.100885	.100380	.099876	.099375	.098877	-2	2
2.33	.983806	.978867	.973950	.969056	.964185	.959337	.954511	.949707	.944926	.940168	-3	22
2.34	.935431	.930717	.926025	.921355	.916706	.912079	.907474	.902890	.898328	.893788	-3	21
2.35	.889268	.884770	.880293	.875837	.871401	.866987	.862593	.858220	.853868	.849536	-3	20
2.36	.845224	.840933	.836662	.832411	.828180	.823969	.819778	.815607	.811456	.807324	-3	19
2.37	.803211	.799118	.795045	.790990	.786955	.782939	.778942	.774964	.771005	.767065	-3	18
2.38	.763143	.759240	.755355	.751489	.747642	.743812	.740001	.736208	.732433	.728676	-3	18
2.39	.724937	.721216	.717513	.713827	.710159	.706508	.702875	.699259	.695660	.692079	-3	17
2.40	.688515	.684968	.681438	.677924	.674428	.670948	.667485	.664039	.660609	.657196	-3	16
2.41	.653799	.650419	.647054	.643706	.640374	.637059	.633759	.630475	.627206	.623954	-3	15
2.42	.620717	.617496	.614291	.611100	.607926	.604767	.601623	.598494	.595380	.592282	-3	14
2.43	.589198	.586130	.583076	.580037	.577013	.574004	.571009	.568029	.565063	.562111	-3	14
2.44	.559175	.556252	.553344	.550449	.547569	.544703	.541851	.539013	.536188	.533378	-3	14
2.45	.530581	.527798	.525028	.522272	.519529	.516800	.514085	.511382	.508693	.506017	-3	13
2.46	.503354	.500704	.498068	.495444	.492833	.490235	.487650	.485077	.482517	.479970	-3	13
2.47	.477435	.474913	.472403	.469905	.467420	.464947	.462486	.460038	.457602	.455177	-3	12
2.48	.452765	.450364	.447976	.445599	.443234	.440881	.438539	.436209	.433891	.431584	-3	11
2.49	.429288	.427004	.424732	.422470	.420220	.417981	.415753	.413537	.411331	.409136	-3	11

TABLE 7B

Error Function Complement

X	.0	.001	.002	.003	.004	.005	.006	.007	.008	.009	m	Δ₂
2.50	.406952	.404780	.402618	.400466	.398326	.396196	.394077	.391968	.389871	.387783	−3	10
2.51	.385706	.383639	.381583	.379537	.377501	.375476	.373460	.371455	.369460	.367475	−3	10
2.52	.365500	.363534	.361579	.359634	.357698	.355772	.353855	.351949	.350052	.348164	−3	10
2.53	.346286	.344418	.342559	.340709	.338869	.337038	.335216	.333404	.331600	.329806	−3	9
2.54	.328021	.326245	.324478	.322720	.320971	.319231	.317499	.315776	.314062	.312357	−3	9
2.55	.310661	.308973	.307293	.305623	.303960	.302307	.300661	.299024	.297396	.295775	−3	8
2.56	.294163	.292560	.290964	.289376	.287797	.286226	.284663	.283107	.281560	.280021	−3	8
2.57	.278489	.276966	.275450	.273942	.272442	.270949	.269464	.267987	.266517	.265055	−3	8
2.58	.263601	.262153	.260714	.259282	.257857	.256439	.255029	.253626	.252230	.250842	−3	7
2.59	.249461	.248087	.246720	.245360	.244007	.242661	.241322	.239990	.238664	.237346	−3	7
2.60	.236035	.234730	.233432	.232141	.230857	.229579	.228308	.227043	.225785	.224534	−3	6
2.61	.223289	.222050	.220819	.219593	.218374	.217161	.215955	.214754	.213560	.212373	−3	6
2.62	.211191	.210016	.208847	.207684	.206527	.205376	.204231	.203092	.201959	.200832	−3	6
2.63	.199711	.198596	.197487	.196383	.195286	.194194	.193107	.192027	.190952	.189883	−3	6
2.64	.188820	.187762	.186709	.185663	.184621	.183585	.182555	.181530	.180511	.179497	−3	5
2.65	.178488	.177485	.176486	.175494	.174506	.173524	.172547	.171575	.170608	.169646	−3	5
2.66	.168690	.167738	.166792	.165850	.164914	.163982	.163056	.162134	.161218	.160306	−3	5
2.67	.159399	.158497	.157600	.156707	.155819	.154936	.154058	.153184	.152315	.151451	−3	5
2.68	.150591	.149736	.148886	.148040	.147198	.146361	.145529	.144701	.143877	.143058	−3	4
2.69	.142244	.141433	.140627	.139825	.139028	.138235	.137446	.136661	.135881	.135105	−3	4
2.70	.134333	.133565	.132801	.132042	.131286	.130535	.129787	.129044	.128305	.127569	−3	4
2.71	.126838	.126111	.125387	.124668	.123952	.123240	.122532	.121828	.121128	.120431	−3	4
2.72	.119739	.119050	.118364	.117683	.117005	.116331	.115661	.114994	.114331	.113671	−3	4
2.73	.113015	.112363	.111714	.111069	.110427	.109788	.109154	.108522	.107894	.107270	−3	3
2.74	.106649	.106031	.105417	.104806	.104198	.103594	.102993	.102395	.101801	.101210	−3	3
2.75	.100622	.100037	.099456	.098878	.098303	.097731	.097162	.096596	.096034	.095474	−3	3
2.76	.949178	.943645	.938142	.932670	.927228	.921815	.916433	.911080	.905757	.900463	−4	29
2.77	.895199	.889963	.884757	.879579	.874430	.869309	.864217	.859153	.854117	.849109	−4	28
2.78	.844128	.839176	.834250	.829352	.824482	.819638	.814821	.810031	.805268	.800531	−4	27
2.79	.795820	.791136	.786477	.781845	.777238	.772657	.768102	.763572	.759067	.754588	−4	25
2.80	.750133	.745704	.741299	.736918	.732563	.728231	.723924	.719641	.715382	.711146	−4	23
2.81	.706935	.702747	.698582	.694441	.690323	.686228	.682157	.678108	.674082	.670078	−4	23
2.82	.666097	.662138	.658202	.654288	.650395	.646525	.642677	.638850	.635045	.631261	−4	21
2.83	.627499	.623757	.620037	.616338	.612660	.609003	.605366	.601750	.598155	.594579	−4	20
2.84	.591024	.587489	.583974	.580480	.577004	.573549	.570113	.566697	.563300	.559922	−4	19
2.85	.556564	.553224	.549904	.546603	.543320	.540056	.536811	.533584	.530375	.527185	−4	18
2.86	.524013	.520859	.517723	.514605	.511504	.508422	.505357	.502309	.499279	.496267	−4	17
2.87	.493271	.490293	.487332	.484387	.481460	.478549	.475655	.472778	.469917	.467073	−4	16
2.88	.464245	.461433	.458637	.455857	.453094	.450346	.447614	.444898	.442197	.439512	−4	15
2.89	.436843	.434188	.431549	.428926	.426317	.423724	.421145	.418582	.416033	.413499	−4	15
2.90	.410980	.408475	.405984	.403508	.401047	.398600	.396166	.393747	.391342	.388951	−4	14
2.91	.386574	.384210	.381861	.379525	.377202	.374893	.372598	.370315	.368046	.365791	−4	13
2.92	.363548	.361318	.359102	.356898	.354708	.352529	.350364	.348211	.346071	.343944	−4	13
2.93	.341828	.339726	.337635	.335557	.333490	.331437	.329394	.327364	.325346	.323340	−4	12
2.94	.321345	.319362	.317391	.315431	.313483	.311546	.309620	.307706	.305804	.303912	−4	11
2.95	.302031	.300162	.298303	.296456	.294619	.292793	.290978	.289174	.287380	.285597	−4	11
2.96	.283824	.282062	.280310	.278569	.276837	.275117	.273406	.271705	.270015	.268334	−4	10
2.97	.266663	.265003	.263352	.261711	.260079	.258458	.256846	.255243	.253650	.252067	−4	10
2.98	.250493	.248928	.247372	.245826	.244289	.242761	.241243	.239733	.238232	.236740	−4	9
2.99	.235257	.233783	.232318	.230861	.229414	.227974	.226544	.225122	.223708	.222303	−4	9

TABLE 7B

Error Function Complement

X	.0	.001	.002	.003	.004	.005	.006	.007	.008	.009	m	Δ_2
3.00	.220905	.219517	.218137	.216765	.215401	.214046	.212698	.211359	.210028	.208705	−4	8
3.01	.207390	.206082	.204783	.203491	.202207	.200931	.199662	.198401	.197148	.195902	−4	8
3.02	.194664	.193433	.192210	.190994	.189785	.188583	.187389	.186202	.185022	.183850	−4	7
3.03	.182684	.181525	.180374	.179229	.178092	.176961	.175837	.174720	.173609	.172506	−4	7
3.04	.171409	.170318	.169235	.168157	.167087	.166023	.164965	.163914	.162869	.161830	−4	6
3.05	.160798	.159772	.158753	.157739	.156732	.155731	.154736	.153747	.152764	.151787	−4	6
3.06	.150816	.149851	.148891	.147938	.146991	.146049	.145113	.144183	.143258	.142339	−4	6
3.07	.141426	.140518	.139616	.138719	.137828	.136943	.136062	.135188	.134318	.133454	−4	5
3.08	.132595	.131742	.130893	.130050	.129212	.128380	.127552	.126729	.125912	.125100	−4	5
3.09	.124292	.123490	.122692	.121899	.121112	.120329	.119551	.118778	.118009	.117246	−4	5
3.10	.116487	.115732	.114983	.114238	.113497	.112761	.112030	.111304	.110581	.109864	−4	4
3.11	.109150	.108441	.107737	.107037	.106341	.105650	.104962	.104280	.103601	.102927	−4	4
3.12	.102256	.101590	.100928	.100271	.099617	.098967	.098322	.09768C	.097043	.096409	−4	4
3.13	.957795	.951538	.945320	.939141	.933001	.926899	.920835	.914809	.908821	.902870	−5	37
3.14	.896957	.891080	.885240	.879437	.873670	.867940	.862245	.856586	.850962	.845374	−5	34
3.15	.839821	.834303	.828819	.823370	.817955	.812574	.807227	.801914	.796634	.791388	−5	32
3.16	.786174	.780993	.775845	.770730	.765646	.760595	.755575	.750588	.745631	.740706	−5	31
3.17	.735813	.730950	.726117	.721316	.716545	.711804	.707093	.702411	.697760	.693138	−5	29
3.18	.688545	.683981	.679446	.674941	.670463	.666014	.661593	.657201	.652836	.648499	−5	28
3.19	.644190	.639908	.635653	.631425	.627225	.623051	.618903	.614782	.610687	.606619	−5	26
3.20	.602576	.598559	.594568	.590602	.586662	.582746	.578856	.574991	.571150	.567334	−5	25
3.21	.563542	.559775	.556031	.552312	.548616	.544945	.541296	.537671	.534070	.530491	−5	23
3.22	.526935	.523403	.519893	.516405	.512940	.509497	.506076	.502677	.499301	.495945	−5	22
3.23	.492612	.489300	.486009	.482740	.479491	.476263	.473057	.469871	.466705	.463560	−5	20
3.24	.460435	.457331	.454246	.451182	.448137	.445112	.442107	.439121	.436154	.433206	−5	19
3.25	.430278	.427368	.424478	.421606	.418753	.415918	.413102	.410304	.407524	.404762	−5	18
3.26	.402018	.399292	.396584	.393893	.391220	.388564	.385926	.383304	.380700	.378113	−5	17
3.27	.375542	.372989	.370452	.367931	.365427	.362939	.360468	.358013	.355573	.353150	−5	16
3.28	.350742	.348351	.345975	.343614	.341269	.338939	.336625	.334325	.332041	.329772	−5	15
3.29	.327517	.325278	.323053	.320842	.318646	.316465	.314298	.312145	.310006	.307882	−5	14
3.30	.305771	.303674	.301591	.299522	.297466	.295424	.293395	.291380	.289378	.287389	−5	13
3.31	.285414	.283451	.281501	.279564	.277640	.275729	.273830	.271944	.270071	.268209	−5	12
3.32	.266360	.264524	.262699	.260886	.259086	.257297	.255521	.253756	.252003	.250261	−5	12
3.33	.248531	.246812	.245105	.243409	.241725	.240052	.238389	.236738	.235098	.233469	−5	11
3.34	.231850	.230243	.228646	.227059	.225484	.223919	.222364	.220819	.219285	.217761	−5	10
3.35	.216248	.214744	.213251	.211767	.210293	.208829	.207375	.205931	.204497	.203072	−5	10
3.36	.201656	.200250	.198853	.197466	.196088	.194720	.193360	.192010	.190668	.189336	−5	9
3.37	.188013	.186698	.185393	.184096	.182808	.181528	.180257	.178995	.177741	.176496	−5	8
3.38	.175259	.174030	.172810	.171597	.170393	.169198	.168010	.166830	.165658	.164494	−5	8
3.39	.163338	.162190	.161050	.159917	.158792	.157674	.156564	.155462	.154367	.153279	−5	7
3.40	.152199	.151126	.150061	.149003	.147951	.146907	.145870	.144841	.143818	.142802	−5	7
3.41	.141793	.140790	.139795	.138806	.137824	.136849	.135881	.134919	.133963	.133014	−5	6
3.42	.132072	.131136	.130206	.129283	.128366	.127455	.126551	.125652	.124760	.123874	−5	6
3.43	.122994	.122120	.121252	.120390	.119533	.118683	.117839	.117000	.116167	.115339	−5	6
3.44	.114518	.113702	.112891	.112086	.111287	.110493	.109705	.108922	.108144	.107372	−5	5
3.45	.106605	.105843	.105087	.104336	.103590	.102849	.102113	.101382	.100657	.099936	−5	5
3.46	.992201	.985093	.978034	.971024	.964062	.957148	.950282	.943463	.936691	.929967	−6	47
3.47	.923288	.916656	.910070	.903529	.897034	.890583	.884178	.877816	.871499	.865226	−6	44
3.48	.858996	.852809	.846665	.840564	.834505	.828488	.822513	.816580	.810687	.804836	−6	41
3.49	.799025	.793255	.787525	.781835	.776184	.770573	.765001	.759468	.753973	.748517	−6	38

TABLE 7B

Error Function Complement

X	.0	001	.002	.003	.004	.005	.006	.007	.008	.009	m	Δ_2
3.50	.743098	.737718	.732375	.727069	.721800	.716568	.711373	.706214	.701091	.696004	-6	36
3.51	.690952	.685936	.680955	.676009	.671097	.666220	.661377	.656567	.651792	.647050	-6	32
3.52	.642341	.637666	.633023	.628412	.623834	.619289	.614775	.610293	.605842	.601423	-6	31
3.53	.597035	.592677	.588351	.584054	.579788	.575552	.571346	.567170	.563023	.558905	-6	29
3.54	.554816	.550757	.546725	.542723	.538748	.534802	.530883	.526992	.523129	.519293	-6	28
3.55	.515484	.511702	.507947	.504218	.500516	.496840	.493190	.489566	.485967	.482394	-6	25
3.56	.478847	.475325	.471827	.468355	.464907	.461483	.458084	.454709	.451358	.448031	-6	24
3.57	.444728	.441448	.438191	.434958	.431747	.428560	.425395	.422253	.419133	.416035	-6	22
3.58	.412960	.409906	.406875	.403864	.400876	.397908	.394962	.392037	.389133	.386250	-6	21
3.59	.383387	.380545	.377723	.374921	.372139	.369378	.366636	.363913	.361211	.358527	-6	19
3.60	.355863	.353218	.350592	.347985	.345396	.342826	.340275	.337742	.335227	.332730	-6	18
3.61	.330251	.327790	.325347	.322921	.320513	.318122	.315748	.313391	.311052	.308729	-6	17
3.62	.306423	.304134	.301861	.299604	.297364	.295140	.292932	.290740	.288564	.286404	-6	16
3.63	.284259	.282130	.280016	.277917	.275834	.273766	.271713	.269674	.267651	.265642	-6	15
3.64	.263647	.261667	.259702	.257751	.255813	.253890	.251981	.250086	.248205	.246337	-6	14
3.65	.244483	.242642	.240815	.239001	.237200	.235412	.233637	.231876	.230127	.228391	-6	13
3.66	.226667	.224956	.223258	.221572	.219898	.218236	.216587	.214950	.213324	.211711	-6	12
3.67	.210109	.208519	.206941	.205374	.203819	.202275	.200742	.199221	.197710	.196211	-6	11
3.68	.194723	.193246	.191779	.190324	.188878	.187444	.186020	.184607	.183204	.181811	-6	10
3.69	.180429	.179056	.177694	.176342	.175000	.173667	.172345	.171032	.169729	.168435	-6	10
3.70	.167151	.165877	.164611	.163356	.162109	.160872	.159643	.158424	.157214	.156013	-6	9
3.71	.154821	.153637	.152462	.151296	.150139	.148990	.147850	.146718	.145594	.144479	-6	8
3.72	.143372	.142273	.141183	.140100	.139026	.137959	.136900	.135850	.134807	.133772	-6	8
3.73	.132744	.131724	.130712	.129707	.128710	.127720	.126738	.125762	.124794	.123834	-6	7
3.74	.122880	.121934	.120994	.120062	.119137	.118218	.117306	.116401	.115503	.114612	-6	7
3.75	.113727	.112849	.111978	.111112	.110254	.109402	.108556	.107716	.106883	.106056	-6	6
3.76	.105236	.104421	.103613	.102810	.102014	.101223	.100439	.099660	.098887	.098120	-6	6
3.77	.973592	.966037	.958538	.951096	.943711	.936380	.929105	.921884	.914718	.907606	-7	53
3.78	.900547	.893542	.886589	.879689	.872840	.866044	.859298	.852603	.845959	.839365	-7	50
3.79	.832821	.826327	.819881	.813484	.807135	.800835	.794582	.788376	.782217	.776105	-7	46
3.80	.770039	.764019	.758045	.752116	.746232	.740392	.734597	.728845	.723138	.717473	-7	43
3.81	.711852	.706273	.700737	.695242	.689790	.684378	.679008	.673679	.668390	.663142	-7	40
3.82	.657933	.652764	.647635	.642544	.637493	.632479	.627504	.622567	.617668	.612806	-7	37
3.83	.607981	.603193	.598441	.593726	.589046	.584403	.579795	.575222	.570684	.566180	-7	34
3.84	.561712	.557277	.552876	.548509	.544176	.539875	.535608	.531373	.527171	.523001	-7	32
3.85	.518863	.514757	.510682	.506638	.502626	.498644	.494693	.490772	.486881	.483021	-7	30
3.86	.479190	.475388	.471616	.467872	.464158	.460472	.456814	.453185	.449584	.446010	-7	28
3.87	.442464	.438945	.435453	.431989	.428551	.425139	.421754	.418395	.415062	.411755	-7	26
3.88	.408473	.405217	.401986	.398780	.395598	.392442	.389309	.386201	.383117	.380057	-7	24
3.89	.377021	.374008	.371018	.368052	.365109	.362188	.359290	.356415	.353562	.350731	-7	22
3.90	.347922	.345135	.342370	.339626	.336904	.334202	.331522	.328862	.326224	.323605	-7	20
3.91	.321007	.318430	.315872	.313335	.310817	.308319	.305840	.303381	.300940	.298519	-7	19
3.92	.296117	.293734	.291369	.289022	.286694	.284384	.282093	.279819	.277563	.275324	-7	17
3.93	.273103	.270900	.268714	.266544	.264392	.262257	.260139	.258037	.255951	.253882	-7	16
3.94	.251829	.249793	.247772	.245767	.243778	.241804	.239846	.237904	.235977	.234064	-7	15
3.95	.232167	.230285	.228418	.226565	.224727	.222903	.221094	.219299	.217518	.215752	-7	14
3.96	.213999	.212260	.210534	.208823	.207125	.205440	.203768	.202110	.200465	.198833	-7	13
3.97	.197214	.195607	.194014	.192432	.190864	.189308	.187764	.186232	.184713	.183205	-7	12
3.98	.181710	.180226	.178754	.177294	.175845	.174408	.172983	.171568	.170165	.168773	-7	11
3.99	.167392	.166022	.164663	.163315	.161977	.160650	.159334	.158028	.156733	.155447	-7	10

TABLE 7B

Error Function Complement

X	.0	.001	.002	.003	.004	.005	.006	.007	.008	.009	m	Δ_2
4.00	.154173	.152908	.151653	.150408	.149174	.147949	.146734	.145528	.144332	.143146	−7	9
4.01	.141969	.140802	.139644	.138495	.137356	.136225	.135104	.133991	.132887	.131793	−7	9
4.02	.130707	.129629	.128561	.127501	.126449	.125406	.124371	.123344	.122326	.121316	−7	8
4.03	.120314	.119320	.118334	.117356	.116386	.115423	.114469	.113522	.112582	.111650	−7	8
4.04	.110726	.109809	.108900	.107997	.107103	.106215	.105334	.104461	.103594	.102735	−7	7
4.05	.101882	.101037	.100198	.099366	.098541	.097722	.096910	.096104	.095305	.094513	−7	6
4.06	.937269	.929471	.921737	.914065	.906455	.898907	.891420	.883993	.876627	.869321	−8	59
4.07	.862073	.854885	.847754	.840682	.833667	.826709	.819807	.812961	.806171	.799436	−8	54
4.08	.792756	.786130	.779558	.773040	.766574	.760161	.753800	.747491	.741233	.735027	−8	51
4.09	.728871	.722764	.716708	.710701	.704743	.698834	.692973	.687160	.681394	.675675	−8	47
4.10	.670003	.664377	.658797	.653263	.647774	.642330	.636930	.631574	.626263	.620995	−8	44
4.11	.615769	.610587	.605447	.600349	.595293	.590279	.585305	.580373	.575480	.570628	−8	40
4.12	.565816	.561043	.556309	.551614	.546958	.542340	.537760	.533218	.528712	.524244	−8	36
4.13	.519813	.515418	.511060	.506737	.502449	.498197	.493981	.489798	.485651	.481537	−8	34
4.14	.477458	.473412	.469399	.465419	.461473	.457559	.453677	.449827	.446009	.442223	−8	31
4.15	.438468	.434744	.431050	.427388	.423755	.420153	.416580	.413037	.409524	.406039	−8	29
4.16	.402583	.399156	.395758	.392387	.389045	.385730	.382442	.379182	.375949	.372743	−8	27
4.17	.369564	.366410	.363284	.360183	.357107	.354058	.351034	.348034	.345060	.342111	−8	25
4.18	.339186	.336286	.333409	.330557	.327728	.324923	.322141	.319383	.316647	.313935	−8	23
4.19	.311245	.308577	.305932	.303309	.300707	.298128	.295570	.293033	.290517	.288023	−8	21
4.20	.285549	.283096	.280664	.278252	.275860	.273489	.271137	.268804	.266492	.264198	−8	19
4.21	.261924	.259669	.257433	.255216	.253017	.250837	.248675	.246531	.244405	.242297	−8	18
4.22	.240207	.238134	.236079	.234041	.232020	.230016	.228029	.226059	.224105	.222168	−8	16
4.23	.220247	.218343	.216454	.214581	.212724	.210883	.209057	.207247	.205452	.203672	−8	15
4.24	.201907	.200157	.198422	.196701	.194995	.193303	.191626	.189963	.188314	.186679	−8	14
4.25	.185057	.183450	.181856	.180275	.178708	.177155	.175614	.174086	.172572	.171070	−8	13
4.26	.169581	.168105	.166641	.165189	.163750	.162323	.160908	.159506	.158115	.156736	−8	12
4.27	.155369	.154013	.152669	.151336	.150015	.148705	.147406	.146118	.144841	.143575	−8	11
4.28	.142319	.141075	.139841	.138618	.137405	.136202	.135009	.133827	.132655	.131493	−8	10
4.29	.130341	.129198	.128066	.126943	.125830	.124726	.123632	.122547	.121471	.120404	−8	9
4.30	.119347	.118299	.117260	.116229	.115208	.114195	.113191	.112195	.111208	.110229	−8	8
4.31	.109259	.108298	.107344	.106399	.105461	.104532	.103611	.102698	.101792	.100895	−8	8
4.32	.100005	.099122	.098248	.097381	.096521	.095669	.094824	.093986	.093156	.092332	−8	7
4.33	.915161	.907070	.899048	.891096	.883212	.875396	.867647	.859966	.852351	.844801	−9	65
4.34	.837317	.829898	.822542	.815250	.808022	.800856	.793751	.786709	.779727	.772806	−9	59
4.35	.765944	.759142	.752400	.745715	.739088	.732519	.726007	.719551	.713152	.706808	−9	54
4.36	.700519	.694284	.688104	.681977	.675904	.669883	.663915	.657999	.652134	.646320	−9	51
4.37	.640556	.634843	.629180	.623565	.618000	.612483	.607015	.601593	.596220	.590893	−9	47
4.38	.585612	.580378	.575189	.570045	.564946	.559892	.554882	.549916	.544993	.540113	−9	43
4.39	.535276	.530481	.525728	.521017	.516346	.511717	.507128	.502579	.498071	.493601	−9	39
4.40	.489171	.484780	.480427	.476112	.471835	.467596	.463393	.459228	.455099	.451006	−9	36
4.41	.446950	.442929	.438943	.434992	.431076	.427195	.423347	.419534	.415754	.412007	−9	33
4.42	.408293	.404612	.400963	.397346	.393762	.390208	.386686	.383196	.379735	.376306	−9	30
4.43	.372907	.369537	.366198	.362887	.359606	.356354	.353131	.349936	.346770	.343631	−9	28
4.44	.340521	.337437	.334381	.331352	.328350	.325374	.322425	.319502	.316604	.313733	−9	25
4.45	.310886	.308065	.305269	.302498	.299751	.297029	.294331	.291657	.289006	.286379	−9	23
4.46	.283776	.281195	.278637	.276103	.273590	.271100	.268632	.266186	.263762	.261360	−9	21
4.47	.258978	.256618	.254279	.251961	.249663	.247386	.245130	.242893	.240676	.238479	−9	20
4.48	.236302	.234144	.232005	.229885	.227785	.225703	.223639	.221594	.219568	.217559	−9	18
4.49	.215569	.213596	.211641	.209703	.207783	.205879	.203993	.202124	.200271	.198435	−9	16

TABLE 7B

Error Function Complement

X	.0	.001	.002	.003	.004	.005	.006	.007	.008	.009	m	Δ_2
4.50	.196616	.194813	.193026	.191255	.189500	.187760	.186037	.184328	.182635	.180957	-9	15
4.51	.179295	.177647	.176014	.174396	.172792	.171202	.169627	.168066	.166520	.164986	-9	14
4.52	.163467	.161962	.160470	.158991	.157526	.156074	.154635	.153209	.151796	.150396	-9	13
4.53	.149008	.147633	.146270	.144919	.143581	.142255	.140941	.139638	.138347	.137069	-9	12
4.54	.135801	.134545	.133300	.132067	.130845	.129634	.128434	.127244	.126066	.124898	-9	11
4.55	.123740	.122594	.121457	.120331	.119215	.118109	.117014	.115928	.114852	.113785	-9	10
4.56	.112729	.111682	.110644	.109616	.108598	.107588	.106588	.105597	.104615	.103642	-9	9
4.57	.102677	.101722	.100775	.099836	.098907	.097985	.097072	.096168	.095272	.094383	-9	8
4.58	.935034	.926314	.917673	.909111	.900627	.892221	.883891	.875637	.867459	.859355	-10	74
4.59	.851326	.843370	.835486	.827675	.819935	.812266	.804667	.797137	.789677	.782284	-10	68
4.60	.774960	.767702	.760511	.753386	.746326	.739331	.732400	.725533	.718728	.711986	-10	62
4.61	.705306	.698687	.692129	.685631	.679193	.672814	.666493	.660231	.654026	.647878	-10	56
4.62	.641787	.635752	.629772	.623847	.617977	.612161	.606398	.600689	.595032	.589427	-10	51
4.63	.583874	.578372	.572921	.567520	.562169	.556867	.551614	.546410	.541253	.536144	-10	47
4.64	.531083	.526068	.521100	.516177	.511300	.506468	.501681	.496938	.492239	.487583	-10	43
4.65	.482970	.478400	.473873	.469387	.464943	.460540	.456178	.451856	.447575	.443333	-10	39
4.66	.439130	.434967	.430842	.426755	.422706	.418695	.414721	.410784	.406884	.403020	-10	36
4.67	.399192	.395399	.391641	.387919	.384231	.380578	.376958	.373372	.369820	.366301	-10	33
4.68	.362814	.359360	.355938	.352548	.349190	.345863	.342567	.339301	.336066	.332862	-10	30
4.69	.329687	.326542	.323426	.320340	.317282	.314253	.311252	.308279	.305334	.302416	-10	27
4.70	.299526	.296663	.293826	.291017	.288233	.285475	.282744	.280038	.277357	.274701	-10	25
4.71	.272071	.269465	.266883	.264326	.261792	.259283	.256797	.254334	.251894	.249478	-10	23
4.72	.247084	.244712	.242363	.240036	.237731	.235447	.233185	.230944	.228725	.226526	-10	21
4.73	.224348	.222190	.220053	.217936	.215838	.213761	.211703	.209665	.207645	.205645	-10	19
4.74	.203664	.201701	.199757	.197831	.195924	.194034	.192162	.190308	.188472	.186652	-10	17
4.75	.184850	.183066	.181297	.179546	.177811	.176093	.174391	.172705	.171035	.169381	-10	16
4.76	.167742	.166119	.164512	.162919	.161342	.159780	.158232	.156699	.155181	.153677	-10	14
4.77	.152187	.150712	.149251	.147803	.146369	.144949	.143542	.142149	.140769	.139402	-10	13
4.78	.138048	.136707	.135379	.134063	.132760	.131469	.130191	.128925	.127670	.126428	-10	12
4.79	.125198	.123979	.122772	.121577	.120392	.119220	.118058	.116907	.115768	.114639	-10	11
4.80	.113521	.112414	.111318	.110231	.109156	.108090	.107035	.105990	.104955	.103929	-10	10
4.81	.102914	.101908	.100912	.099925	.098948	.097980	.097022	.096073	.095133	.094201	-10	9
4.82	.932791	.923657	.914611	.905652	.896778	.887990	.879286	.870666	.862128	.853673	-11	81
4.83	.845299	.837005	.828791	.820656	.812599	.804620	.796718	.788892	.781141	.773464	-11	73
4.84	.765862	.758333	.750876	.743491	.736178	.728935	.721762	.714658	.707622	.700655	-11	67
4.85	.693754	.686920	.680153	.673450	.666813	.660239	.653729	.647282	.640897	.634574	-11	61
4.86	.628312	.622111	.615970	.609888	.603865	.597900	.591993	.586143	.580350	.574613	-11	55
4.87	.568932	.563306	.557734	.552216	.546752	.541341	.535982	.530675	.525420	.520216	-11	51
4.88	.515062	.509959	.504905	.499900	.494944	.490035	.485175	.480362	.475596	.470876	-11	46
4.89	.466202	.461573	.456990	.452451	.447957	.443506	.439098	.434734	.430412	.426132	-11	42
4.90	.421894	.417697	.413541	.409426	.405351	.401315	.397319	.393362	.389444	.385564	-11	38
4.91	.381722	.377917	.374150	.370419	.366725	.363067	.359445	.355858	.352306	.348789	-11	34
4.92	.345307	.341859	.338444	.335063	.331715	.328400	.325117	.321866	.318648	.315461	-11	31
4.93	.312305	.309180	.306086	.303022	.299988	.296984	.294010	.291064	.288148	.285260	-11	28
4.94	.282401	.279570	.276767	.273991	.271243	.268521	.265827	.263159	.260517	.257901	-11	26
4.95	.255311	.252746	.250207	.247693	.245203	.242738	.240298	.237881	.235489	.233120	-11	23
4.96	.230774	.228451	.226152	.223875	.221620	.219388	.217178	.214990	.212823	.210678	-11	21
4.97	.208554	.206451	.204369	.202307	.200266	.198245	.196244	.194263	.192302	.190360	-11	19
4.98	.188437	.186533	.184648	.182782	.180934	.179105	.177294	.175500	.173725	.171967	-11	17
4.99	.170227	.168503	.166797	.165108	.163436	.161780	.160141	.158518	.156912	.155321	-11	16

TABLE 7B

Error Function Complement

X	.0	.001	.002	.003	.004	.005	.006	.007	.008	.009	m	Δ_2
5.00	.153746	.152187	.150643	.149114	.147601	.146103	.144620	.143152	.141698	.140259	-11	14
5.01	.138834	.137423	.136026	.134643	.133274	.131919	.130577	.129249	.127934	.126632	-11	13
5.02	.125343	.124067	.122804	.121553	.120315	.119089	.117875	.116674	.115484	.114307	-11	12
5.03	.113141	.111987	.110845	.109713	.108594	.107485	.106388	.105301	.104226	.103161	-11	11
5.04	.102107	.101063	.100030	.099008	.097995	.096993	.096001	.095019	.094046	.093084	-11	10
5.05	.921309	.911876	.902538	.893293	.884141	.875082	.866113	.857235	.848446	.839745	-12	88
5.06	.831132	.822606	.814166	.805810	.797539	.789351	.781246	.773222	.765280	.757417	-12	79
5.07	.749634	.741929	.734302	.726752	.719278	.711880	.704556	.697307	.690130	.683026	-12	72
5.08	.675994	.669033	.662142	.655321	.648569	.641886	.635270	.628721	.622238	.615821	-12	65
5.09	.609469	.603181	.596957	.590796	.584697	.578660	.572684	.566769	.560914	.555119	-12	59
5.10	.549382	.543703	.538082	.532518	.527011	.521560	.516163	.510822	.505535	.500302	-12	52
5.11	.495122	.489994	.484919	.479896	.474923	.470001	.465129	.460307	.455534	.450809	-12	48
5.12	.446133	.441504	.436923	.432388	.427899	.423456	.419059	.414706	.410398	.406133	-12	43
5.13	.401912	.397735	.393600	.389507	.385455	.381446	.377477	.373549	.369661	.365813	-12	39
5.14	.362004	.358234	.354502	.350809	.347154	.343536	.339955	.336410	.332902	.329430	-12	36
5.15	.325994	.322592	.319226	.315894	.312596	.309332	.306102	.302904	.299740	.296608	-12	32
5.16	.293508	.290440	.287403	.284398	.281424	.278480	.275566	.272682	.269828	.267003	-12	29
5.17	.264208	.261441	.258702	.255992	.253310	.250655	.248027	.245427	.242853	.240306	-12	26
5.18	.237786	.235291	.232822	.230378	.227959	.225566	.223197	.220853	.218532	.216236	-12	24
5.19	.213964	.211715	.209489	.207286	.205106	.202948	.200813	.198700	.196608	.194539	-12	21
5.20	.192490	.190463	.188457	.186472	.184507	.182563	.180638	.178734	.176849	.174984	-12	19
5.21	.173138	.171312	.169504	.167715	.165944	.164192	.162459	.160743	.159045	.157364	-12	17
5.22	.155701	.154055	.152427	.150815	.149220	.147642	.146080	.144534	.143004	.141490	-12	16
5.23	.139992	.138510	.137043	.135591	.134155	.132733	.131326	.129934	.128556	.127193	-12	14
5.24	.125844	.124509	.123188	.121880	.120587	.119306	.118040	.116786	.115545	.114318	-12	13
5.25	.113103	.111901	.110711	.109534	.108370	.107217	.106076	.104948	.103831	.102726	-12	12
5.26	.101632	.100550	.099479	.098420	.097371	.096333	.095307	.094291	.093286	.092291	-12	10
5.27	.913066	.903326	.893688	.884152	.874716	.865378	.856139	.846997	.837950	.828998	-13	94
5.28	.820141	.811376	.802704	.794122	.785631	.777229	.768916	.760690	.752550	.744496	-13	85
5.29	.736527	.728642	.720839	.713119	.705481	.697922	.690443	.683043	.675721	.668476	-13	76
5.30	.661308	.654215	.647197	.640253	.633392	.626584	.619857	.613201	.606616	.600100	-13	70
5.31	.593654	.587275	.580963	.574719	.568540	.562427	.556378	.550393	.544472	.538613	-13	62
5.32	.532816	.527081	.521406	.515791	.510236	.504740	.499301	.493921	.488597	.483330	-13	56
5.33	.478119	.472963	.467862	.462814	.457821	.452880	.447992	.443155	.438370	.433636	-13	50
5.34	.428952	.424318	.419733	.415197	.410709	.406269	.401876	.397529	.393229	.388975	-13	45
5.35	.384766	.380602	.376482	.372406	.368373	.364383	.360436	.356531	.352667	.348845	-13	41
5.36	.345063	.341322	.337621	.333959	.330336	.326752	.323206	.319698	.316227	.312793	-13	37
5.37	.309397	.306036	.302711	.299422	.296168	.292949	.289764	.286613	.283496	.280412	-13	33
5.38	.277362	.274344	.271358	.268404	.265482	.262591	.259731	.256902	.254103	.251334	-13	30
5.39	.248595	.245885	.243204	.240552	.237929	.235333	.232765	.230225	.227713	.225227	-13	27
5.40	.222768	.220335	.217929	.215548	.213193	.210863	.208558	.206278	.204023	.201792	-13	24
5.41	.199585	.197401	.195241	.193105	.190991	.188900	.186832	.184786	.182762	.180759	-13	22
5.42	.178779	.176820	.174881	.172964	.171068	.169192	.167336	.165500	.163684	.161888	-13	19
5.43	.160111	.158353	.156614	.154894	.153192	.151509	.149845	.148198	.146569	.144957	-13	18
5.44	.143363	.141787	.140227	.138684	.137158	.135648	.134155	.132678	.131217	.129772	-13	16
5.45	.128343	.126929	.125530	.124146	.122778	.121424	.120085	.118761	.117451	.116155	-13	14
5.46	.114873	.113605	.112351	.111110	.109883	.108670	.107469	.106282	.105107	.103946	-13	13
5.47	.102797	.101660	.100536	.099424	.098324	.097236	.096160	.095096	.094043	.093002	-13	11
5.48	.919719	.909532	.899456	.889491	.879633	.869884	.860240	.850702	.841268	.831937	-14	102
5.49	.822708	.813579	.804550	.795621	.786789	.778052	.769412	.760866	.752413	.744053	-14	92

TABLE 7B

Error Function Complement

X	.0	.001	.002	.003	.004	.005	.006	.007	.008	.009	m	Δ_2
5.50	.735785	.727606	.719518	.711517	.703604	.695778	.688038	.680383	.672810	.665322	-14	82
5.51	.657915	.650590	.643345	.636179	.629091	.622082	.615149	.608292	.601511	.594804	-14	74
5.52	.588171	.581611	.575122	.568705	.562359	.556082	.549873	.543734	.537662	.531656	-14	66
5.53	.525717	.519843	.514033	.508288	.502606	.496986	.491428	.485932	.480495	.475119	-14	59
5.54	.469802	.464543	.459343	.454200	.449113	.444083	.439108	.434188	.429322	.424510	-14	53
5.55	.419751	.415045	.410390	.405787	.401235	.396733	.392281	.387878	.383524	.379217	-14	48
5.56	.374959	.370747	.366582	.362463	.358390	.354362	.350379	.346439	.342543	.338690	-14	43
5.57	.334880	.331112	.327386	.323702	.320058	.316454	.312890	.309366	.305881	.302435	-14	38
5.58	.299027	.295656	.292324	.289028	.285768	.282545	.279358	.276206	.273089	.270007	-14	34
5.59	.266959	.263945	.260965	.258017	.255103	.252220	.249370	.246552	.243765	.241009	-14	31
5.60	.238284	.235589	.232924	.230288	.227683	.225106	.222558	.220038	.217546	.215083	-14	27
5.61	.212646	.210237	.207855	.205499	.203170	.200866	.198589	.196337	.194110	.191907	-14	25
5.62	.189730	.187577	.185448	.183342	.181260	.179202	.177166	.175154	.173164	.171196	-14	22
5.63	.169250	.167326	.165423	.163542	.161682	.159843	.158024	.156226	.154447	.152689	-14	20
5.64	.150951	.149232	.147532	.145851	.144190	.142546	.140922	.139315	.137727	.136156	-14	18
5.65	.134604	.133068	.131550	.130049	.128564	.127097	.125646	.124211	.122792	.121390	-14	16
5.66	.120003	.118632	.117276	.115935	.114610	.113299	.112004	.110722	.109456	.108203	-14	14
5.67	.106965	.105741	.104530	.103333	.102150	.100980	.099823	.098679	.097548	.096430	-14	13
5.68	.953249	.942320	.931513	.920829	.910266	.899822	.889496	.879287	.869193	.859213	-15	113
5.69	.849347	.839592	.829948	.820412	.810985	.801664	.792449	.783338	.774331	.765426	-15	101
5.70	.756621	.747916	.739310	.730802	.722390	.714073	.705851	.697722	.689686	.681740	-15	90
5.71	.673885	.666119	.658441	.650851	.643346	.635927	.628593	.621341	.614172	.607085	-15	80
5.72	.600078	.593151	.586303	.579532	.572839	.566222	.559680	.553213	.546819	.540498	-15	72
5.73	.534249	.528072	.521964	.515927	.509958	.504057	.498224	.492457	.486756	.481120	-15	64
5.74	.475548	.470040	.464595	.459212	.453890	.448629	.443429	.438287	.433205	.428180	-15	57
5.75	.423213	.418303	.413449	.408651	.403907	.399218	.394582	.389999	.385469	.380991	-15	52
5.76	.376564	.372187	.367862	.363585	.359357	.355178	.351047	.346963	.342926	.338935	-15	46
5.77	.334990	.331091	.327236	.323425	.319658	.315934	.312254	.308615	.305018	.301463	-15	41
5.78	.297948	.294474	.291039	.287645	.284289	.280972	.277692	.274451	.271247	.268080	-15	36
5.79	.264949	.261855	.258796	.255772	.252783	.249828	.246908	.244021	.241168	.238347	-15	32
5.80	.235559	.232803	.230079	.227386	.224724	.222093	.219493	.216922	.214381	.211870	-15	29
5.81	.209387	.206934	.204508	.202111	.199741	.197398	.195083	.192795	.190533	.188297	-15	26
5.82	.186087	.183903	.181743	.179609	.177500	.175415	.173354	.171317	.169304	.167314	-15	23
5.83	.165347	.163402	.161481	.159582	.157704	.155849	.154015	.152202	.150410	.148639	-15	21
5.84	.146889	.145159	.143449	.141759	.140089	.138438	.136806	.135193	.133599	.132023	-15	18
5.85	.130466	.128927	.127406	.125902	.124416	.122947	.121496	.120061	.118643	.117242	-15	16
5.86	.115856	.114487	.113134	.111797	.110475	.109169	.107878	.106602	.105341	.104094	-15	15
5.87	.102862	.101645	.100442	.099252	.098077	.096916	.095767	.094633	.093511	.092403	-15	13
5.88	.913079	.902254	.891555	.880982	.870532	.860205	.849998	.839911	.829942	.820089	-16	115
5.89	.810352	.800729	.791219	.781820	.772531	.763351	.754279	.745313	.736452	.727695	-16	102
5.90	.719041	.710488	.702036	.693682	.685427	.677269	.669207	.661239	.653365	.645583	-16	92
5.91	.637893	.630293	.622782	.615360	.608025	.600776	.593612	.586533	.579537	.572623	-16	81
5.92	.565791	.559039	.552366	.545773	.539256	.532817	.526453	.520165	.513950	.507809	-16	72
5.93	.501740	.495743	.489816	.483959	.478171	.472452	.466800	.461215	.455696	.450242	-16	64
5.94	.444852	.439526	.434263	.429062	.423922	.418843	.413825	.408865	.403965	.399122	-16	58
5.95	.394336	.389607	.384934	.380317	.375754	.371245	.366789	.362386	.358035	.353736	-16	51
5.96	.349488	.345290	.341142	.337043	.332992	.328990	.325035	.321127	.317265	.313450	-16	45
5.97	.309679	.305953	.302272	.298634	.295039	.291487	.287978	.284509	.281082	.277696	-16	40
5.98	.274351	.271044	.267778	.264550	.261360	.258209	.255095	.252018	.248977	.245973	-16	36
5.99	.243004	.240071	.237173	.234309	.231480	.228684	.225922	.223192	.220495	.217830	-16	32

TABLE 7B

Error Function Complement

X	.0	.001	.002	.003	.004	.005	.006	.007	.008	.009	m	Δ_2
6.00	.215197	.212596	.210025	.207485	.204975	.202496	.200046	.197625	.195233	.192869	-16	29
6.01	.190534	.188227	.185948	.183695	.181470	.179271	.177098	.174952	.172831	.170735	-16	25
6.02	.168665	.166619	.164598	.162601	.160628	.158678	.156752	.154849	.152969	.151111	-16	22
6.03	.149276	.147463	.145671	.143901	.142152	.140424	.138716	.137030	.135363	.133717	-16	20
6.04	.132090	.130483	.128895	.127326	.125776	.124245	.122732	.121237	.119760	.118301	-16	18
6.05	.116859	.115435	.114028	.112638	.111265	.109908	.108567	.107243	.105934	.104642	-16	16
6.06	.103364	.102103	.100856	.099625	.098408	.097206	.096018	.094845	.093686	.092541	-16	14
6.07	.914100	.902924	.891883	.880975	.870199	.859553	.849036	.838645	.828380	.818240	-17	123
6.08	.808221	.798324	.788546	.778887	.769344	.759917	.750604	.741404	.732315	.723336	-17	109
6.09	.714465	.705702	.697045	.688493	.680044	.671698	.663453	.655308	.647261	.639313	-17	96
6.10	.631460	.623703	.616040	.608469	.600991	.593603	.586305	.579096	.571974	.564938	-17	86
6.11	.557988	.551123	.544340	.537641	.531022	.524484	.518026	.511646	.505343	.499118	-17	76
6.12	.492968	.486893	.480891	.474963	.469107	.463322	.457608	.451963	.446387	.440879	-17	67
6.13	.435438	.430063	.424754	.419509	.414329	.409211	.404156	.399163	.394231	.389358	-17	59
6.14	.384546	.379792	.375096	.370457	.365875	.361349	.356878	.352462	.348100	.343791	-17	53
6.15	.339535	.335330	.331178	.327075	.323024	.319021	.315068	.311163	.307306	.303496	-17	47
6.16	.299733	.296016	.292344	.288717	.285135	.281596	.278101	.274649	.271239	.267871	-17	41
6.17	.264544	.261258	.258013	.254807	.251640	.248513	.245423	.242372	.239358	.236381	-17	37
6.18	.233441	.230537	.227668	.224835	.222037	.219273	.216543	.213846	.211183	.208552	-17	32
6.19	.205954	.203388	.200853	.198350	.195877	.193435	.191023	.188640	.186287	.183963	-17	29
6.20	.181667	.179400	.177161	.174949	.172765	.170608	.168477	.166372	.164294	.162241	-17	25
6.21	.160213	.158211	.156233	.154279	.152350	.150445	.148563	.146704	.144868	.143055	-17	22
6.22	.141265	.139496	.137750	.136025	.134321	.132638	.130977	.129335	.127714	.126114	-17	20
6.23	.124533	.122971	.121429	.119906	.118402	.116916	.115449	.114000	.112570	.111156	-17	18
6.24	.109761	.108382	.107021	.105677	.104349	.103038	.101743	.100464	.099201	.097954	-17	15
6.25	.967220	.955055	.943040	.931176	.919459	.907888	.896459	.885174	.874029	.863023	-18	136
6.26	.852153	.841419	.830818	.820349	.810010	.799800	.789717	.779760	.769927	.760216	-18	121
6.27	.750627	.741156	.731804	.722569	.713448	.704441	.695547	.686764	.678090	.669524	-18	107
6.28	.661065	.652712	.644463	.636317	.628273	.620329	.612485	.604739	.597089	.589535	-18	94
6.29	.582075	.574709	.567434	.560251	.553158	.546153	.539236	.532405	.525660	.518999	-18	84
6.30	.512422	.505927	.499514	.493180	.486926	.480751	.474653	.468631	.462685	.456813	-18	74
6.31	.451015	.445289	.439636	.434053	.428541	.423097	.417722	.412414	.407173	.401998	-18	65
6.32	.396888	.391842	.386859	.381939	.377081	.372284	.367547	.362870	.358251	.353691	-18	57
6.33	.349188	.344742	.340351	.336016	.331735	.327509	.323335	.319214	.315145	.311127	-18	51
6.34	.307160	.303243	.299375	.295556	.291785	.288062	.284385	.280755	.277171	.273632	-18	45
6.35	.270137	.266687	.263280	.259916	.256595	.253316	.250078	.246881	.243724	.240607	-18	39
6.36	.237530	.234491	.231491	.228529	.225604	.222717	.219865	.217050	.214271	.211527	-18	35
6.37	.208817	.206142	.203500	.200892	.198317	.195775	.193265	.190787	.188340	.185924	-18	31
6.38	.183539	.181184	.178859	.176563	.174296	.172059	.169849	.167668	.165514	.163388	-18	27
6.39	.161289	.159216	.157170	.155149	.153155	.151185	.149241	.147322	.145426	.143555	-18	24
6.40	.141708	.139884	.138084	.136306	.134551	.132818	.131107	.129418	.127751	.126105	-18	21
6.41	.124480	.122875	.121291	.119727	.118183	.116659	.115154	.113669	.112202	.110754	-18	19
6.42	.109324	.107913	.106520	.105144	.103786	.102446	.101122	.099816	.098526	.097252	-18	16
6.43	.959953	.947542	.935289	.923194	.911253	.899464	.887827	.876338	.864996	.853799	-19	144
6.44	.842746	.831833	.821061	.810427	.799929	.789565	.779334	.769233	.759263	.749420	-19	127
6.45	.739704	.730111	.720641	.711294	.702065	.692956	.683964	.675086	.666322	.657672	-19	111
6.46	.649132	.640701	.632379	.624164	.616054	.608048	.600145	.592344	.584643	.577041	-19	96
6.47	.569537	.562129	.554816	.547598	.540473	.533438	.526495	.519641	.512875	.506196	-19	86
6.48	.499603	.493095	.486671	.480330	.474070	.467891	.461792	.455771	.449827	.443961	-19	76
6.49	.438170	.432454	.426811	.421241	.415744	.410317	.404960	.399672	.394452	.389300	-19	67

TABLE 7B

Error Function Complement

X	.0	.001	.002	.003	.004	.005	.006	.007	.008	.009	m	Δ_2
6.50	.384215	.379195	.374240	.369349	.364521	.359756	.355052	.350409	.345826	.341302	-19	59
6.51	.336837	.332429	.328079	.323785	.319546	.315363	.311233	.307157	.303134	.299163	-19	51
6.52	.295243	.291374	.287555	.283786	.280065	.276393	.272768	.269191	.265659	.262174	-19	45
6.53	.258734	.255338	.251987	.248679	.245414	.242191	.239010	.235870	.232772	.229713	-19	40
6.54	.226694	.223715	.220774	.217872	.215007	.212179	.209388	.206634	.203915	.201232	-19	35
6.55	.198583	.195969	.193390	.190843	.188330	.185850	.183401	.180985	.178600	.176247	-19	31
6.56	.173924	.171631	.169368	.167135	.164931	.162755	.160608	.158489	.156398	.154333	-19	27
6.57	.152296	.150286	.148301	.146343	.144410	.142502	.140620	.138761	.136928	.135118	-19	24
6.58	.133332	.131569	.129829	.128112	.126417	.124745	.123094	.121466	.119858	.118271	-19	21
6.59	.116706	.115160	.113635	.112130	.110645	.109179	.107732	.106304	.104895	.103505	-19	18
6.60	.102132	.100778	.099442	.098123	.096821	.095536	.094268	.093017	.091783	.090564	-19	16
6.61	.893614	.881747	.870036	.858479	.847073	.835817	.824709	.813748	.802930	.792255	-20	139
6.62	.781719	.771323	.761063	.750938	.740947	.731087	.721356	.711754	.702279	.692927	-20	124
6.63	.683700	.674593	.665607	.656739	.647988	.639353	.630831	.622421	.614122	.605933	-20	108
6.64	.597853	.589878	.582008	.574243	.566580	.559019	.551556	.544193	.536927	.529756	-20	95
6.65	.522681	.515699	.508809	.502010	.495302	.488681	.482149	.475702	.469342	.463064	-20	84
6.66	.456871	.450759	.444728	.438777	.432905	.427110	.421392	.415750	.410182	.404689	-20	72
6.67	.399268	.393919	.388641	.383432	.378293	.373222	.368218	.363281	.358409	.353602	-20	64
6.68	.348859	.344178	.339560	.335003	.330506	.326069	.321691	.317371	.313109	.308903	-20	56
6.69	.304753	.300659	.296618	.292632	.288698	.284817	.280987	.277209	.273480	.269801	-20	49
6.70	.266171	.262590	.259056	.255569	.252129	.248734	.245385	.242080	.238820	.235602	-20	43
6.71	.232428	.229296	.226206	.223157	.220148	.217180	.214251	.211362	.208511	.205698	-20	38
6.72	.202922	.200184	.197482	.194816	.192186	.189591	.187031	.184505	.182012	.179553	-20	33
6.73	.177127	.174733	.172372	.170041	.167742	.165474	.163236	.161028	.158850	.156701	-20	29
6.74	.154580	.152488	.150424	.148388	.146379	.144396	.142441	.140511	.138608	.136730	-20	25
6.75	.134877	.133049	.131245	.129466	.127710	.125978	.124270	.122584	.120921	.119280	-20	22
6.76	.117661	.116064	.114489	.112935	.111401	.109888	.108396	.106923	.105470	.104037	-20	19
6.77	.102623	.101228	.099852	.098495	.097155	.095834	.094530	.093244	.091976	.090724	-20	17
6.78	.894893	.882712	.870694	.858839	.847144	.835606	.824223	.812995	.801917	.790989	-21	147
6.79	.780208	.769573	.759081	.748730	.738520	.728447	.718510	.708707	.699037	.689497	-21	128
6.80	.680086	.670802	.661644	.652609	.643696	.634904	.626231	.617675	.609235	.600908	-21	113
6.81	.592695	.584592	.576599	.568715	.560937	.553264	.545695	.538229	.530864	.523598	-21	98
6.82	.516431	.509361	.502387	.495508	.488721	.482027	.475423	.468909	.462483	.456144	-21	86
6.83	.449892	.443724	.437640	.431638	.425718	.419878	.414118	.408436	.402830	.397301	-21	75
6.84	.391848	.386468	.381162	.375927	.370763	.365670	.360646	.355691	.350803	.345981	-21	66
6.85	.341225	.336534	.331906	.327342	.322839	.318398	.314017	.309696	.305434	.301230	-21	57
6.86	.297083	.292993	.288958	.284979	.281054	.277182	.273363	.269596	.265881	.262216	-21	50
6.87	.258601	.255035	.251518	.248050	.244628	.241253	.237925	.234641	.231403	.228209	-21	44
6.88	.225059	.221951	.218886	.215863	.212881	.209940	.207040	.204179	.201357	.198573	-21	38
6.89	.195828	.193120	.190450	.187816	.185218	.182655	.180128	.177635	.175177	.172752	-21	33
6.90	.170360	.168002	.165675	.163380	.161117	.158885	.156684	.154512	.152371	.150259	-21	29
6.91	.148175	.146121	.144094	.142096	.140125	.138181	.136263	.134372	.132507	.130668	-21	25
6.92	.128854	.127065	.125300	.123560	.121843	.120151	.118481	.116835	.115211	.113609	-21	22
6.93	.112030	.110472	.108936	.107420	.105926	.104452	.102999	.101566	.100152	.098758	-21	19
6.94	.973829	.960269	.946896	.933707	.920701	.907873	.895223	.882747	.870443	.858309	-22	168
6.95	.846343	.834541	.822903	.811425	.800106	.788943	.777935	.767078	.756371	.745813	-22	146
6.96	.735400	.725131	.715005	.705018	.695169	.685457	.675879	.666433	.657118	.647933	-22	127
6.97	.638874	.629941	.621131	.612443	.603876	.595427	.587095	.578879	.570777	.562787	-22	111
6.98	.554908	.547137	.539475	.531919	.524468	.517120	.509874	.502728	.495682	.488733	-22	96
6.99	.481881	.475124	.468461	.461891	.455411	.449022	.442721	.436508	.430382	.424340	-22	84

TABLE 7B

Error Function Complement

X	.0	.001	.002	.003	.004	.005	.006	.007	.008	.009	m	Δ_2
7.00	.418382	.412508	.406715	.401002	.395369	.389814	.384337	.378936	.373610	.368358	-22	73
7.01	.363179	.358072	.353037	.348071	.343175	.338347	.333586	.328892	.324262	.319698	-22	63
7.02	.315197	.310759	.306383	.302067	.297812	.293617	.289479	.285400	.281377	.277411	-22	55
7.03	.273500	.269644	.265841	.262092	.258395	.254749	.251155	.247611	.244116	.240670	-22	48
7.04	.237272	.233922	.230619	.227361	.224150	.220983	.217861	.214782	.211747	.208753	-22	42
7.05	.205802	.202892	.200023	.197194	.194405	.191655	.188943	.186269	.183633	.181034	-22	37
7.06	.178471	.175944	.173452	.170996	.168574	.166186	.163831	.161509	.159221	.156964	-22	32
7.07	.154738	.152544	.150381	.148249	.146146	.144073	.142029	.140013	.138026	.136067	-22	27
7.08	.134135	.132231	.130353	.128502	.126677	.124878	.123103	.121354	.119629	.117929	-22	24
7.09	.116253	.114600	.112970	.111364	.109780	.108218	.106679	.105161	.103664	.102189	-22	21
7.10	.100734	.099300	.097886	.096492	.095118	.093763	.092427	.091110	.089812	.088532	-22	18
7.11	.872696	.860255	.847989	.835897	.823975	.812221	.800634	.789211	.777949	.766846	-23	157
7.12	.755900	.745109	.734470	.723982	.713642	.703449	.693399	.683492	.673726	.664097	-23	136
7.13	.654605	.645247	.636021	.626927	.617961	.609122	.600408	.591818	.583350	.575002	-23	118
7.14	.566772	.558658	.550660	.542775	.535002	.527339	.519785	.512339	.504997	.497761	-23	103
7.15	.490626	.483593	.476660	.469826	.463088	.456447	.449899	.443445	.437082	.430810	-23	89
7.16	.424627	.418532	.412524	.406600	.400762	.395006	.389332	.383739	.378226	.372791	-23	77
7.17	.367433	.362152	.356946	.351814	.346755	.341768	.336852	.332007	.327230	.322521	-23	67
7.18	.317880	.313305	.308795	.304349	.299967	.295647	.291389	.287191	.283054	.278975	-23	58
7.19	.274955	.270992	.267086	.263236	.259440	.255699	.252011	.248376	.244793	.241261	-23	50
7.20	.237779	.234348	.230965	.227631	.224344	.221105	.217912	.214764	.211662	.208604	-23	44
7.21	.205589	.202618	.199690	.196803	.193958	.191153	.188389	.185664	.182978	.180331	-23	38
7.22	.177722	.175150	.172615	.170116	.167654	.165226	.162834	.160475	.158151	.155860	-23	33
7.23	.153601	.151376	.149182	.147019	.144888	.142787	.140717	.138676	.136665	.134682	-23	28
7.24	.132728	.130802	.128904	.127033	.125189	.123372	.121580	.119815	.118074	.116359	-23	24
7.25	.114669	.113003	.111361	.109742	.108147	.106575	.105025	.103498	.101993	.100509	-23	21
7.26	.990471	.976060	.961857	.947859	.934063	.920466	.907065	.893857	.880840	.868010	-24	185
7.27	.855366	.842904	.830622	.818518	.806589	.794831	.783244	.771824	.760568	.749476	-24	160
7.28	.738544	.727769	.717151	.706686	.696373	.686208	.676191	.666318	.656589	.647000	-24	139
7.29	.637550	.628236	.619058	.610012	.601098	.592312	.583654	.575121	.566712	.558425	-24	120
7.30	.550257	.542208	.534276	.526459	.518755	.511163	.503681	.496307	.489041	.481880	-24	104
7.31	.474823	.467868	.461014	.454260	.447604	.441044	.434580	.428209	.421932	.415745	-24	90
7.32	.409648	.403640	.397719	.391885	.386135	.380468	.374885	.369382	.363959	.358616	-24	78
7.33	.353350	.348160	.343047	.338007	.333041	.328148	.323325	.318573	.313890	.309275	-24	67
7.34	.304728	.300247	.295831	.291479	.287191	.282965	.278802	.274698	.270655	.266671	-24	58
7.35	.262744	.258875	.255063	.251306	.247604	.243956	.240361	.236819	.233329	.229889	-24	50
7.36	.226500	.223161	.219870	.216627	.213432	.210283	.207180	.204123	.201111	.198142	-24	43
7.37	.195217	.192335	.189495	.186696	.183939	.181222	.178544	.175906	.173307	.170745	-24	37
7.38	.168221	.165734	.163284	.160869	.158490	.156146	.153836	.151560	.149317	.147107	-24	32
7.39	.144930	.142784	.140670	.138587	.136535	.134513	.132520	.130557	.128623	.126716	-24	28
7.40	.124838	.122988	.121165	.119368	.117598	.115854	.114136	.112443	.110774	.109131	-24	24
7.41	.107511	.105915	.104343	.102794	.101268	.099764	.098282	.096822	.095384	.093967	-24	21
7.42	.925704	.911946	.898391	.885035	.871877	.858912	.846139	.833554	.821154	.808937	-25	180
7.43	.796901	.785042	.773358	.761846	.750504	.739330	.728321	.717473	.706787	.696257	-25	156
7.44	.685884	.675664	.665594	.655674	.645900	.636270	.626783	.617436	.608227	.599154	-25	134
7.45	.590216	.581410	.572734	.564186	.555765	.547468	.539294	.531242	.523308	.515492	-25	115
7.46	.507792	.500205	.492731	.485368	.478114	.470967	.463926	.456989	.450156	.443423	-25	100
7.47	.436791	.430257	.423819	.417478	.411230	.405075	.399011	.393037	.387152	.381355	-25	86
7.48	.375643	.370016	.364473	.359012	.353632	.348333	.343111	.337968	.332901	.327909	-25	74
7.49	.322992	.318147	.313375	.308673	.304042	.299479	.294985	.290557	.286195	.281898	-25	64

TABLE 7B

Error Function Complement

X	.0	.001	.002	.003	.004	.005	.006	.007	.008	.009	m	Δ_2
7.50	.277665	.273495	.269387	.265340	.261354	.257427	.253558	.249747	.245993	.242295	-25	55
7.51	.238652	.235063	.231528	.228045	.224615	.221235	.217906	.214627	.211396	.208214	-25	47
7.52	.205079	.201991	.198950	.195953	.193002	.190094	.187230	.184408	.181629	.178892	-25	41
7.53	.176195	.173538	.170922	.168344	.165805	.163304	.160840	.158413	.156023	.153668	-25	35
7.54	.151349	.149064	.146813	.144596	.142413	.140262	.138143	.136056	.134000	.131975	-25	30
7.55	.129980	.128015	.126080	.124174	.122296	.120447	.118625	.116830	.115063	.113322	-25	26
7.56	.111607	.109917	.108254	.106615	.105000	.103410	.101844	.100302	.098782	.097286	-25	23
7.57	.958114	.943594	.929292	.915205	.901330	.887663	.874202	.860943	.847884	.835022	-26	194
7.58	.822352	.809873	.797582	.785477	.773553	.761809	.750241	.738848	.727626	.716573	-26	167
7.59	.705687	.694965	.684405	.674003	.663758	.653668	.643730	.633941	.624300	.614806	-26	143
7.60	.605453	.596242	.587170	.578235	.569435	.560767	.552230	.543823	.535542	.527386	-26	124
7.61	.519353	.511442	.503650	.495976	.488418	.480974	.473643	.466422	.459311	.452307	-26	106
7.62	.445409	.438615	.431924	.425335	.418845	.412453	.406158	.399958	.393853	.387839	-26	91
7.63	.381917	.376084	.370340	.364682	.359111	.353624	.348220	.342897	.337656	.332494	-26	78
7.64	.327411	.322404	.317473	.312617	.307835	.303125	.298487	.293919	.289421	.284991	-26	67
7.65	.280628	.276331	.272099	.267932	.263828	.259787	.255807	.251887	.248027	.244225	-26	58
7.66	.240482	.236795	.233164	.229589	.226068	.222600	.219185	.215822	.212511	.209249	-26	50
7.67	.206038	.202875	.199761	.196693	.193673	.190699	.187769	.184885	.182044	.179247	-26	43
7.68	.176493	.173780	.171109	.168478	.165888	.163337	.160825	.158351	.155915	.153516	-26	37
7.69	.151154	.148828	.146537	.144281	.142060	.139873	.137719	.135598	.133509	.131453	-26	31
7.70	.129427	.127433	.125469	.123535	.121631	.119756	.117909	.116091	.114301	.112538	-26	27
7.71	.110802	.109092	.107409	.105751	.104119	.102512	.100929	.099371	.097837	.096326	-26	23
7.72	.948378	.933727	.919301	.905097	.891109	.877337	.863775	.850421	.837273	.824326	-27	199
7.73	.811577	.799024	.786664	.774493	.762509	.750709	.739090	.727649	.716385	.705293	-27	171
7.74	.694371	.683618	.673029	.662603	.652338	.642230	.632278	.622478	.612829	.603329	-27	146
7.75	.593975	.584764	.575695	.566766	.557974	.549318	.540794	.532402	.524139	.516004	-27	125
7.76	.507993	.500106	.492340	.484694	.477166	.469754	.462456	.455270	.448196	.441230	-27	108
7.77	.434372	.427619	.420970	.414425	.407980	.401634	.395387	.389236	.383179	.377217	-27	92
7.78	.371346	.365566	.359875	.354272	.348756	.343325	.337978	.332713	.327530	.322427	-27	79
7.79	.317402	.312456	.307586	.302791	.298070	.293423	.288847	.284342	.279907	.275540	-27	68
7.80	.271241	.267008	.262841	.258739	.254700	.250724	.246809	.242955	.239160	.235425	-27	58
7.81	.231747	.228126	.224562	.221052	.217597	.214196	.210847	.207551	.204305	.201110	-27	49
7.82	.197964	.194868	.191819	.188818	.185863	.182954	.180090	.177271	.174495	.171763	-27	42
7.83	.169073	.166425	.163818	.161251	.158725	.156237	.153789	.151378	.149005	.146669	-27	36
7.84	.144369	.142105	.139876	.137682	.135522	.133396	.131303	.129242	.127213	.125216	-27	31
7.85	.123250	.121315	.119410	.117535	.115688	.113871	.112082	.110321	.108587	.106880	-27	27
7.86	.105200	.103546	.101918	.100316	.098738	.097185	.095656	.094151	.092670	.091211	-27	23
7.87	.897756	.883625	.869714	.856021	.842541	.829272	.816211	.803353	.790697	.778238	-28	195
7.88	.765974	.753902	.742019	.730322	.718807	.707473	.696316	.685333	.674523	.663881	-28	167
7.89	.653407	.643096	.632947	.622956	.613123	.603443	.593915	.584536	.575304	.566216	-28	143
7.90	.557272	.548467	.539801	.531270	.522873	.514608	.506472	.498464	.490582	.482823	-28	122
7.91	.475186	.467669	.460271	.452988	.445819	.438763	.431818	.424982	.418254	.411631	-28	104
7.92	.405112	.398696	.392380	.386164	.380045	.374023	.368095	.362261	.356518	.350866	-28	89
7.93	.345302	.339827	.334437	.329132	.323911	.318772	.313713	.308735	.303835	.299012	-28	76
7.94	.294265	.289592	.284994	.280468	.276013	.271628	.267313	.263065	.258885	.254770	-28	65
7.95	.250721	.246735	.242812	.238951	.235151	.231411	.227730	.224107	.220541	.217032	-28	55
7.96	.213578	.210178	.206833	.203540	.200299	.197109	.193970	.190880	.187840	.184847	-28	47
7.97	.181902	.179003	.176150	.173342	.170578	.167859	.165182	.162548	.159955	.157404	-28	40
7.98	.154892	.152421	.149989	.147595	.145239	.142920	.140639	.138393	.136183	.134008	-28	34
7.99	.131867	.129761	.127688	.125647	.123639	.121663	.119718	.117804	.115921	.114067	-28	29

TABLE 7B

Error Function Complement

X	.0	.001	.002	.003	.004	.005	.006	.007	.008	.009	m	Δ_2
8.00	.112243	.110448	.108681	.106942	.105231	.103547	.101890	.100259	.098654	.097074	-28	25
8.01	.955200	.939903	.924849	.910035	.895456	.881109	.866990	.853095	.839422	.825966	-29	215
8.02	.812724	.799693	.786869	.774249	.761831	.749610	.737583	.725748	.714102	.702641	-29	182
8.03	.691362	.680264	.669342	.658594	.648017	.637609	.627367	.617289	.607371	.597611	-29	155
8.04	.588006	.578555	.569255	.560103	.551098	.542236	.533515	.524934	.516489	.508180	-29	133
8.05	.500003	.491956	.484039	.476247	.468581	.461036	.453613	.446308	.439119	.432046	-29	113
8.06	.425085	.418237	.411497	.404865	.398340	.391919	.385600	.379383	.373265	.367245	-29	96
8.07	.361322	.355493	.349758	.344114	.338561	.333097	.327720	.322430	.317224	.312102	-29	82
8.08	.307062	.302102	.297222	.292421	.287696	.283047	.278473	.273972	.269543	.265186	-29	70
8.09	.260898	.256679	.252528	.248443	.244425	.240470	.236579	.232751	.228984	.225278	-29	59
8.10	.221631	.218043	.214512	.211038	.207620	.204257	.200948	.197692	.194489	.191337	-29	51
8.11	.188236	.185185	.182183	.179229	.176323	.173463	.170650	.167881	.165158	.162478	-29	43
8.12	.159841	.157247	.154695	.152184	.149713	.147282	.144891	.142537	.140222	.137944	-29	37
8.13	.135703	.133498	.131329	.129194	.127094	.125028	.122995	.120995	.119028	.117092	-29	31
8.14	.115187	.113313	.111470	.109656	.107871	.106115	.104388	.102689	.101017	.099372	-29	27
8.15	.977535	.961613	.945948	.930537	.915375	.900459	.885783	.871346	.857141	.843166	-30	228
8.16	.829419	.815893	.802587	.789495	.776616	.763946	.751481	.739217	.727152	.715283	-30	193
8.17	.703606	.692118	.680817	.669699	.658761	.648000	.637414	.626999	.616754	.606675	-30	164
8.18	.596759	.587004	.577407	.567967	.558679	.549542	.540554	.531712	.523013	.514455	-30	139
8.19	.506037	.497755	.489608	.481593	.473708	.465952	.458322	.450815	.443431	.436167	-30	118
8.20	.429021	.421991	.415076	.408273	.401581	.394998	.388522	.382151	.375884	.369719	-30	100
8.21	.363655	.357689	.351820	.346047	.340368	.334782	.329287	.323881	.318563	.313332	-30	85
8.22	.308186	.303125	.298145	.293247	.288429	.283690	.279027	.274441	.269930	.265492	-30	72
8.23	.261127	.256833	.252609	.248454	.244367	.240347	.236392	.232502	.228676	.224912	-30	62
8.24	.221209	.217567	.213985	.210461	.206995	.203586	.200232	.196933	.193688	.190496	-30	52
8.25	.187357	.184268	.181231	.178243	.175304	.172413	.169569	.166772	.164021	.161315	-30	44
8.26	.158653	.156035	.153460	.150926	.148435	.145984	.143574	.141203	.138870	.136576	-30	38
8.27	.134320	.132101	.129918	.127771	.125659	.123582	.121539	.119529	.117553	.115609	-30	32
8.28	.113697	.111816	.109966	.108147	.106357	.104597	.102866	.101163	.099488	.097841	-30	27
8.29	.962207	.946272	.930599	.915183	.900021	.885109	.870441	.856016	.841827	.827872	-31	230
8.30	.814147	.800648	.787371	.774313	.761470	.748838	.736414	.724195	.712177	.700358	-31	196
8.31	.688733	.677300	.666055	.654996	.644119	.633422	.622900	.612553	.602376	.592366	-31	165
8.32	.582522	.572841	.563319	.553955	.544745	.535687	.526779	.518018	.509401	.500928	-31	139
8.33	.492593	.484397	.476336	.468408	.460611	.452944	.445403	.437986	.430692	.423519	-31	119
8.34	.416465	.409527	.402704	.395993	.389394	.382904	.376522	.370245	.364072	.358001	-31	101
8.35	.352031	.346160	.340386	.334707	.329123	.323631	.318230	.312919	.307696	.302559	-31	85
8.36	.297507	.292540	.287654	.282850	.278125	.273479	.268910	.264416	.259997	.255652	-31	72
8.37	.251379	.247176	.243044	.238980	.234983	.231053	.227188	.223387	.219650	.215974	-31	61
8.38	.212360	.208806	.205310	.201873	.198493	.195170	.191901	.188687	.185526	.182418	-31	52
8.39	.179362	.176357	.173401	.170495	.167637	.164827	.162063	.159346	.156673	.154045	-31	44
8.40	.151461	.148921	.146422	.143965	.141549	.139173	.136837	.134540	.132281	.130060	-31	37
8.41	.127876	.125728	.123616	.121539	.119497	.117489	.115515	.113573	.111664	.109787	-31	31
8.42	.107941	.106126	.104341	.102586	.100861	.099164	.097496	.095855	.094242	.092656	-31	27
8.43	.910961	.895625	.880546	.865719	.851140	.836804	.822709	.808849	.795221	.781822	-32	224
8.44	.768646	.755691	.742953	.730428	.718113	.706004	.694098	.682391	.670881	.659563	-32	190
8.45	.648436	.637494	.626736	.616158	.605757	.595531	.585477	.575590	.565870	.556313	-32	160
8.46	.546916	.537677	.528593	.519661	.510879	.502245	.493756	.485408	.477202	.469133	-32	135
8.47	.461199	.453399	.445730	.438190	.430776	.423487	.416320	.409274	.402347	.395536	-32	114
8.48	.388839	.382255	.375782	.369418	.363160	.357008	.350960	.345013	.339167	.333418	-32	97
8.49	.327767	.322211	.316748	.311377	.306097	.300906	.295802	.290784	.285851	.281001	-32	82

TABLE 7B

Error Function Complement

X	.0	.001	.002	.003	.004	.005	.006	.007	.008	.009	m	Δ_2
8.50	.276232	.271544	.266935	.262404	.257949	.253569	.249263	.245030	.240868	.236777	-32	69
8.51	.232754	.228799	.224911	.221089	.217331	.213637	.210005	.206434	.202924	.199473	-32	58
8.52	.196080	.192745	.189466	.186242	.183073	.179957	.176894	.173883	.170923	.168013	-32	49
8.53	.165152	.162339	.159575	.156856	.154184	.151557	.148975	.146436	.143940	.141487	-32	42
8.54	.139075	.136703	.134372	.132081	.129828	.127614	.125437	.123297	.121193	.119125	-32	35
8.55	.117091	.115093	.113128	.111197	.109298	.107431	.105597	.103793	.102020	.100277	-32	30
8.56	.985636	.968793	.952236	.935959	.919960	.904232	.888771	.873573	.858632	.843946	-33	250
8.57	.829510	.815318	.801368	.787655	.774175	.760924	.747899	.735095	.722509	.710137	-33	210
8.58	.697975	.686021	.674269	.662718	.651363	.640202	.629231	.618446	.607845	.597425	-33	177
8.59	.587182	.577113	.567216	.557488	.547925	.538526	.529286	.520204	.511277	.502502	-33	150
8.60	.493877	.485399	.477065	.468873	.460822	.452907	.445128	.437481	.429965	.422578	-33	126
8.61	.415316	.408178	.401162	.394266	.387488	.380825	.374277	.367840	.361513	.355294	-33	106
8.62	.349182	.343174	.337269	.331465	.325760	.320152	.314640	.309223	.303899	.298665	-33	89
8.63	.293521	.288465	.283495	.278611	.273810	.269092	.264454	.259896	.255415	.251012	-33	75
8.64	.246683	.242429	.238248	.234139	.230100	.226130	.222228	.218393	.214624	.210919	-33	64
8.65	.207279	.203700	.200183	.196726	.193329	.189990	.186708	.183482	.180312	.177196	-33	53
8.66	.174134	.171124	.168166	.165259	.162402	.159593	.156833	.154121	.151455	.148835	-33	45
8.67	.146260	.143729	.141242	.138797	.136395	.134034	.131713	.129432	.127191	.124988	-33	38
8.68	.122823	.120696	.118605	.116550	.114530	.112545	.110594	.108677	.106793	.104941	-33	32
8.69	.103122	.101333	.099576	.097848	.096151	.094483	.092843	.091232	.089649	.088093	-33	27
8.70	.865632	.850603	.835833	.821318	.807054	.793035	.779259	.765720	.752416	.739340	-34	225
8.71	.726491	.713863	.701454	.689259	.677275	.665497	.653923	.642550	.631372	.620388	-34	190
8.72	.609594	.598986	.588562	.578318	.568252	.558359	.548638	.539084	.529696	.520471	-34	159
8.73	.511405	.502496	.493741	.485138	.476684	.468376	.460212	.452189	.444306	.436559	-34	134
8.74	.428946	.421465	.414114	.406890	.399792	.392816	.385961	.379226	.372607	.366103	-34	113
8.75	.359711	.353431	.347260	.341195	.335236	.329380	.323626	.317972	.312416	.306957	-34	95
8.76	.301592	.296320	.291140	.286050	.281048	.276134	.271305	.266559	.261896	.257315	-34	80
8.77	.252812	.248388	.244042	.239770	.235573	.231449	.227397	.223415	.219502	.215658	-34	67
8.78	.211880	.208169	.204522	.200938	.197417	.193957	.190557	.187217	.183934	.180709	-34	56
8.79	.177540	.174427	.171367	.168361	.165408	.162505	.159654	.156852	.154099	.151394	-34	47
8.80	.148736	.146125	.143559	.141038	.138561	.136127	.133736	.131386	.129078	.126809	-34	40
8.81	.124581	.122391	.120239	.118126	.116049	.114008	.112003	.110033	.108098	.106196	-34	33
8.82	.104327	.102492	.100688	.098916	.097175	.095464	.093783	.092132	.090509	.088915	-34	28
8.83	.873493	.858106	.842988	.828135	.813542	.799204	.785118	.771278	.757681	.744322	-35	235
8.84	.731197	.718302	.705633	.693186	.680957	.668943	.657140	.645543	.634150	.622956	-35	196
8.85	.611959	.601155	.590541	.580113	.569867	.559802	.549913	.540198	.530654	.521277	-35	165
8.86	.512065	.503014	.494123	.485387	.476806	.468375	.460092	.451955	.443960	.436107	-35	138
8.87	.428391	.420811	.413364	.406049	.398862	.391801	.384865	.378051	.371356	.364780	-35	116
8.88	.358319	.351972	.345737	.339611	.333593	.327682	.321874	.316169	.310564	.305058	-35	97
8.89	.299649	.294335	.289115	.283987	.278950	.274001	.269139	.264364	.259672	.255063	-35	81
8.90	.250536	.246088	.241719	.237427	.233210	.229069	.225000	.221003	.217076	.213219	-35	68
8.91	.209431	.205709	.202052	.198461	.194932	.191467	.188062	.184717	.181432	.178205	-35	57
8.92	.175035	.171920	.168861	.165856	.162905	.160005	.157157	.154359	.151610	.148911	-35	48
8.93	.146259	.143654	.141095	.138581	.136112	.133687	.131304	.128964	.126665	.124407	-35	40
8.94	.122189	.120010	.117870	.115768	.113703	.111675	.109683	.107726	.105803	.103915	-35	33
8.95	.102061	.100239	.098449	.096692	.094965	.093269	.091603	.089967	.088360	.086781	-35	28
8.96	.852309	.837078	.822119	.807424	.792991	.778814	.764889	.751213	.737778	.724583	-36	234
8.97	.711622	.698892	.686388	.674106	.662043	.650195	.638557	.627126	.615899	.604871	-36	196
8.98	.594040	.583402	.572952	.562689	.552609	.542708	.532984	.523432	.514051	.504837	-36	164
8.99	.495787	.486899	.478169	.469594	.461172	.452901	.444776	.436797	.428960	.421263	-36	137

TABLE 7B

Error Function Complement

X	.0	.001	.002	.003	.004	.005	.006	.007	.008	.009	m	Δ_2
9.00	.413703	.406278	.398985	.391823	.384788	.377879	.371094	.364429	.357883	.351454	−36	115
9.01	.345140	.338939	.332849	.326867	.320992	.315222	.309556	.303990	.298524	.293156	−36	96
9.02	.287883	.282705	.277620	.272625	.267720	.262902	.258171	.253524	.248961	.244479	−36	80
9.03	.240077	.235754	.231509	.227339	.223244	.219223	.215273	.211394	.207585	.203844	−36	67
9.04	.200170	.196562	.193018	.189538	.186120	.182764	.179468	.176230	.173051	.169929	−36	56
9.05	.166863	.163852	.160895	.157991	.155139	.152338	.149588	.146887	.144234	.141629	−36	47
9.06	.139071	.136558	.134091	.131668	.129289	.126952	.124658	.122404	.120192	.118018	−36	39
9.07	.115884	.113789	.111731	.109710	.107725	.105776	.103862	.101983	.100137	.098324	−36	33
9.08	.965444	.947967	.930803	.913947	.897396	.881143	.865182	.849509	.834118	.819004	−37	273
9.09	.804162	.789588	.775276	.761223	.747423	.733871	.720563	.707495	.694663	.682063	−37	226
9.10	.669690	.657539	.645608	.633893	.622388	.611092	.599998	.589105	.578409	.567907	−37	189
9.11	.557593	.547465	.537521	.527756	.518167	.508752	.499507	.490429	.481515	.472762	−37	158
9.12	.464167	.455727	.447441	.439304	.431314	.423468	.415764	.408200	.400773	.393480	−37	132
9.13	.386318	.379287	.372383	.365603	.358946	.352410	.345992	.339691	.333503	.327428	−37	110
9.14	.321462	.315605	.309854	.304206	.298662	.293217	.287872	.282623	.277470	.272409	−37	92
9.15	.267441	.262563	.257773	.253070	.248452	.243919	.239467	.235096	.230805	.226591	−37	76
9.16	.222454	.218392	.214404	.210488	.206643	.202868	.199162	.195522	.191950	.188442	−37	64
9.17	.184997	.181616	.178295	.175035	.171835	.168692	.165607	.162578	.159604	.156684	−37	53
9.18	.153817	.151002	.148239	.145526	.142862	.140246	.137679	.135158	.132682	.130252	−37	44
9.19	.127867	.125524	.123225	.120967	.118750	.116574	.114437	.112339	.110280	.108258	−37	37
9.20	.106273	.104324	.102411	.100532	.098688	.096878	.095100	.093355	.091642	.089960	−37	31
9.21	.883087	.866875	.850960	.835335	.819995	.804936	.790151	.775636	.761387	.747397	−38	256
9.22	.733664	.720181	.706945	.693950	.681193	.668670	.656375	.644305	.632455	.620823	−38	213
9.23	.609403	.598192	.587186	.576381	.565774	.555361	.545139	.535104	.525252	.515581	−38	176
9.24	.506087	.496767	.487617	.478635	.469818	.461162	.452664	.444323	.436134	.428095	−38	147
9.25	.420203	.412456	.404852	.397387	.390058	.382864	.375802	.368869	.362064	.355383	−38	122
9.26	.348825	.342387	.336068	.329864	.323775	.317797	.311928	.306168	.300513	.294963	−38	102
9.27	.289514	.284165	.278915	.273761	.268702	.263735	.258860	.254075	.249377	.244766	−38	85
9.28	.240240	.235797	.231435	.227154	.222952	.218827	.214778	.210803	.206901	.203072	−38	71
9.29	.199312	.195622	.192000	.188445	.184955	.181529	.178167	.174866	.171626	.168446	−38	59
9.30	.165324	.162260	.159253	.156301	.153403	.150559	.147767	.145027	.142337	.139696	−38	49
9.31	.137105	.134561	.132064	.129614	.127208	.124847	.122530	.120255	.118022	.115831	−38	40
9.32	.113680	.111568	.109496	.107462	.105466	.103506	.101583	.099695	.097842	.096023	−38	34
9.33	.942381	.924861	.907664	.890785	.874218	.857958	.841999	.826334	.810960	.795870	−39	278
9.34	.781059	.766522	.752255	.738251	.724507	.711017	.697777	.684782	.672028	.659510	−39	232
9.35	.647224	.635166	.623331	.611715	.600314	.589125	.578143	.567365	.556787	.546405	−39	193
9.36	.536215	.526215	.516399	.506766	.497312	.488033	.478926	.469988	.461216	.452607	−39	160
9.37	.444158	.435865	.427727	.419739	.411900	.404207	.396656	.389246	.381973	.374836	−39	132
9.38	.367831	.360957	.354210	.347588	.341090	.334713	.328454	.322311	.316283	.310367	−39	110
9.39	.304561	.298863	.293271	.287783	.282397	.277111	.271924	.266833	.261838	.256935	−39	91
9.40	.252123	.247401	.242767	.238220	.233757	.229377	.225079	.220861	.216721	.212659	−39	76
9.41	.208673	.204760	.200921	.197153	.193456	.189828	.186267	.182772	.179343	.175978	−39	63
9.42	.172676	.169435	.166255	.163134	.160071	.157066	.154117	.151223	.148382	.145595	−39	52
9.43	.142860	.140176	.137543	.134958	.132422	.129933	.127490	.125094	.122742	.120434	−39	43
9.44	.118169	.115947	.113766	.111626	.109526	.107465	.105443	.103459	.101512	.099601	−39	36
9.45	.977264	.958866	.940812	.923096	.905712	.888654	.871916	.855491	.839373	.823558	−40	297
9.46	.808039	.792811	.777868	.763206	.748818	.734700	.720847	.707254	.693916	.680827	−40	246
9.47	.667985	.655383	.643018	.630885	.618979	.607297	.595834	.584587	.573551	.562721	−40	203
9.48	.552096	.541669	.531439	.521401	.511551	.501887	.492404	.483099	.473969	.465011	−40	168
9.49	.456221	.447597	.439135	.430831	.422684	.414690	.406847	.399151	.391600	.384191	−40	140

TABLE 7B

Error Function Complement

X	.0	.001	.002	.003	.004	.005	.006	.007	.008	.009	m	Δ_2
9.50	.376921	.369789	.362790	.355923	.349186	.342575	.336089	.329725	.323481	.317354	-40	116
9.51	.311343	.305445	.299659	.293981	.288410	.282944	.277582	.272320	.267158	.262093	-40	95
9.52	.257123	.252248	.247464	.242770	.238165	.233647	.229214	.224865	.220598	.216411	-40	79
9.53	.212304	.208274	.204320	.200440	.196634	.192900	.189237	.185642	.182116	.178656	-40	65
9.54	.175262	.171931	.168664	.165458	.162313	.159228	.156201	.153231	.150317	.147458	-40	54
9.55	.144654	.141902	.139203	.136554	.133956	.131407	.128906	.126453	.124046	.121684	-40	44
9.56	.119367	.117095	.114865	.112677	.110531	.108426	.106360	.104334	.102346	.100395	-40	37
9.57	.984820	.966049	.947634	.929568	.911845	.894457	.877400	.860666	.844250	.828145	-41	306
9.58	.812346	.796846	.781641	.766724	.752090	.737735	.723652	.709836	.696283	.682987	-41	253
9.59	.669944	.657148	.644596	.632282	.620202	.608351	.596726	.585322	.574135	.563160	-41	209
9.60	.552394	.541833	.531473	.521310	.511340	.501559	.491965	.482554	.473321	.464264	-41	172
9.61	.455380	.446665	.438115	.429729	.421502	.413432	.405515	.397750	.390132	.382659	-41	143
9.62	.375329	.368138	.361085	.354166	.347379	.340721	.334190	.327784	.321500	.315335	-41	117
9.63	.309289	.303357	.297539	.291832	.286234	.280742	.275355	.270072	.264889	.259804	-41	97
9.64	.254818	.249926	.245127	.240421	.235804	.231275	.226833	.222476	.218202	.214010	-41	80
9.65	.209898	.205865	.201908	.198027	.194221	.190487	.186825	.183232	.179709	.176253	-41	66
9.66	.172863	.169537	.166276	.163077	.159939	.156861	.153842	.150881	.147977	.145128	-41	55
9.67	.142333	.139593	.136905	.134268	.131682	.129145	.126657	.124217	.121823	.119475	-41	45
9.68	.117173	.114914	.112699	.110526	.108395	.106305	.104255	.102244	.100272	.098338	-41	37
9.69	.964407	.945799	.927549	.909649	.892091	.874872	.857983	.841419	.825172	.809238	-42	306
9.70	.793610	.778282	.763249	.748504	.734044	.719860	.705950	.692307	.678926	.665803	-42	252
9.71	.652932	.640308	.627928	.615785	.603876	.592196	.580742	.569507	.558489	.547682	-42	208
9.72	.537084	.526690	.516496	.506498	.496693	.487076	.477645	.468396	.459324	.450428	-42	171
9.73	.441703	.433146	.424754	.416523	.408452	.400536	.392773	.385159	.377692	.370369	-42	141
9.74	.363188	.356145	.349238	.342464	.335821	.329306	.322917	.316651	.310506	.304480	-42	116
9.75	.298570	.292774	.287091	.281516	.276050	.270689	.265432	.260277	.255221	.250262	-42	96
9.76	.245400	.240632	.235955	.231370	.226873	.222462	.218137	.213896	.209737	.205658	-42	79
9.77	.201659	.197736	.193890	.190118	.186419	.182791	.179234	.175746	.172325	.168970	-42	65
9.78	.165681	.162455	.159292	.156190	.153148	.150165	.147239	.144371	.141558	.138800	-42	53
9.79	.136095	.133442	.130841	.128291	.125790	.123337	.120932	.118574	.116261	.113993	-42	44
9.80	.111770	.109589	.107451	.105354	.103298	.101282	.099305	.097367	.095466	.093602	-42	36
9.81	.917742	.899820	.882246	.865014	.848116	.831547	.815300	.799368	.783747	.768429	-43	298
9.82	.753409	.738681	.724240	.710080	.696194	.682580	.669230	.656140	.643305	.630719	-43	245
9.83	.618379	.606278	.594413	.582780	.571373	.560188	.549221	.538468	.527924	.517585	-43	201
9.84	.507448	.497508	.487762	.478207	.468837	.459651	.450643	.441811	.433151	.424660	-43	166
9.85	.416334	.408171	.400168	.392320	.384626	.377081	.369684	.362432	.355321	.348349	-43	136
9.86	.341513	.334810	.328238	.321795	.315477	.309283	.303210	.297256	.291418	.285694	-43	112
9.87	.280082	.274579	.269184	.263895	.258709	.253624	.248639	.243752	.238960	.234261	-43	92
9.88	.229655	.225139	.220711	.216370	.212114	.207941	.203849	.199838	.195906	.192050	-43	76
9.89	.188270	.184564	.180930	.177368	.173876	.170452	.167095	.163803	.160577	.157413	-43	62
9.90	.154312	.151271	.148290	.145368	.142503	.139694	.136940	.134240	.131593	.128998	-43	51
9.91	.126454	.123960	.121514	.119117	.116767	.114463	.112204	.109990	.107819	.105691	-43	42
9.92	.103604	.1C1559	.099553	.097588	.095660	.093771	.091919	.090103	.088323	.086577	-43	34
9.93	.848666	.831894	.815452	.799333	.783531	.768040	.752854	.737967	.723372	.709065	-44	281
9.94	.695040	.681290	.667811	.654598	.641645	.628946	.616498	.604295	.592332	.580606	-44	231
9.95	.569110	.557840	.546793	.535963	.525347	.514940	.504739	.494738	.484934	.475324	-44	190
9.96	.465903	.456669	.447616	.438742	.430043	.421516	.413156	.404962	.396930	.389056	-44	156
9.97	.381337	.373771	.366355	.359085	.351958	.344972	.338124	.331411	.324831	.318381	-44	128
9.98	.312059	.305861	.299786	.293831	.287994	.282272	.276663	.271165	.265776	.260494	-44	104
9.99	.255316	.250240	.245265	.240388	.235608	.230922	.226329	.221827	.217414	.213089	-44	86

TABLE 8

Error Function Derivative

X	.0	.0001	.0002	.0003	.0004	.0005	.0006	.0007	.0008	.0009	Δ_2
0.0	1.128379	1.128379	1.128379	1.128379	1.128379	1.128379	1.128378	1.128378	1.128378	1.128378	0
0.001	1.128378	1.128377	1.128377	1.128377	1.128376	1.128376	1.128376	1.128375	1.128375	1.128375	0
0.002	1.128374	1.128374	1.128373	1.128373	1.128372	1.128371	1.128371	1.128370	1.128370	1.128369	0
0.003	1.128368	1.128367	1.128367	1.128366	1.128366	1.128365	1.128364	1.128364	1.128363	1.128362	0
0.004	1.128361	1.128360	1.128359	1.128358	1.128357	1.128356	1.128355	1.128354	1.128353	1.128351	1
0.005	1.128350	1.128349	1.128348	1.128347	1.128345	1.128345	1.128344	1.128342	1.128341	1.128340	0
0.006	1.128338	1.128337	1.128335	1.128334	1.128332	1.128331	1.128329	1.128328	1.128326	1.128325	1
0.007	1.128324	1.128322	1.128320	1.128319	1.128317	1.128315	1.128313	1.128312	1.128310	1.128308	0
0.008	1.128306	1.128304	1.128303	1.128301	1.128299	1.128297	1.128295	1.128293	1.128291	1.128289	0
0.009	1.128287	1.128285	1.128284	1.128281	1.128279	1.128277	1.128275	1.128272	1.128270	1.128268	0
0.010	1.128266	1.128263	1.128262	1.128259	1.128257	1.128254	1.128252	1.128249	1.128247	1.128244	1
0.011	1.128242	1.128240	1.128237	1.128235	1.128232	1.128229	1.128227	1.128224	1.128222	1.128219	0
0.012	1.128216	1.128214	1.128211	1.128208	1.128205	1.128202	1.128200	1.128197	1.128194	1.128191	0
0.013	1.128188	1.128185	1.128182	1.128180	1.128176	1.128173	1.128170	1.128167	1.128164	1.128160	1
0.014	1.128158	1.128155	1.128151	1.128148	1.128144	1.128141	1.128139	1.128135	1.128132	1.128128	1
0.015	1.128125	1.128121	1.128118	1.128115	1.128111	1.128108	1.128104	1.128100	1.128097	1.128094	0
0.016	1.128090	1.128086	1.128082	1.128078	1.128076	1.128072	1.128068	1.128064	1.128060	1.128057	0
0.017	1.128053	1.128049	1.128045	1.128041	1.128037	1.128034	1.128029	1.128025	1.128021	1.128017	0
0.018	1.128014	1.128009	1.128005	1.128001	1.127996	1.127993	1.127989	1.127984	1.127980	1.127975	1
0.019	1.127972	1.127967	1.127963	1.127958	1.127954	1.127950	1.127945	1.127941	1.127936	1.127932	1
0.020	1.127928	1.127923	1.127918	1.127913	1.127910	1.127905	1.127900	1.127895	1.127891	1.127886	0
0.021	1.127881	1.127876	1.127872	1.127867	1.127862	1.127857	1.127852	1.127848	1.127843	1.127837	1
0.022	1.127832	1.127828	1.127823	1.127818	1.127812	1.127808	1.127803	1.127797	1.127792	1.127787	1
0.023	1.127782	1.127777	1.127771	1.127767	1.127761	1.127756	1.127750	1.127745	1.127740	1.127734	0
0.024	1.127728	1.127724	1.127718	1.127712	1.127707	1.127702	1.127696	1.127690	1.127685	1.127679	1
0.025	1.127674	1.127668	1.127663	1.127657	1.127651	1.127645	1.127640	1.127634	1.127627	1.127622	0
0.026	1.127616	1.127610	1.127604	1.127599	1.127592	1.127586	1.127581	1.127575	1.127568	1.127563	0
0.027	1.127557	1.127550	1.127544	1.127538	1.127532	1.127525	1.127520	1.127513	1.127507	1.127501	1
0.028	1.127495	1.127488	1.127481	1.127476	1.127469	1.127462	1.127456	1.127450	1.127443	1.127437	0
0.029	1.127430	1.127423	1.127417	1.127410	1.127404	1.127398	1.127391	1.127384	1.127378	1.127371	0
0.030	1.127363	1.127357	1.127350	1.127343	1.127337	1.127330	1.127322	1.127316	1.127309	1.127302	0
0.031	1.127295	1.127288	1.127281	1.127274	1.127267	1.127259	1.127253	1.127245	1.127238	1.127231	1
0.032	1.127224	1.127216	1.127210	1.127202	1.127194	1.127188	1.127180	1.127172	1.127166	1.127158	0
0.033	1.127151	1.127143	1.127135	1.127129	1.127121	1.127113	1.127106	1.127098	1.127090	1.127083	0
0.034	1.127075	1.127068	1.127060	1.127052	1.127045	1.127036	1.127028	1.127021	1.127013	1.127006	0
0.035	1.126997	1.126989	1.126982	1.126973	1.126966	1.126958	1.126949	1.126942	1.126933	1.126925	0
0.036	1.126917	1.126909	1.126901	1.126893	1.126884	1.126877	1.126868	1.126860	1.126852	1.126843	0
0.037	1.126835	1.126826	1.126819	1.126810	1.126801	1.126793	1.126784	1.126776	1.126767	1.126759	1
0.038	1.126751	1.126742	1.126734	1.126724	1.126716	1.126707	1.126698	1.126690	1.126681	1.126673	0
0.039	1.126664	1.126655	1.126646	1.126637	1.126628	1.126619	1.126611	1.126601	1.126593	1.126584	0
0.040	1.126575	1.126566	1.126556	1.126548	1.126538	1.126530	1.126520	1.126511	1.126502	1.126493	1
0.041	1.126484	1.126474	1.126465	1.126455	1.126447	1.126437	1.126428	1.126418	1.126409	1.126399	1
0.042	1.126390	1.126381	1.126371	1.126362	1.126352	1.126343	1.126333	1.126323	1.126313	1.126304	0
0.043	1.126294	1.126285	1.126275	1.126266	1.126255	1.126245	1.126236	1.126225	1.126216	1.126206	0
0.044	1.126196	1.126186	1.126177	1.126166	1.126157	1.126146	1.126137	1.126126	1.126116	1.126106	0
0.045	1.126096	1.126086	1.126076	1.126065	1.126056	1.126045	1.126035	1.126024	1.126015	1.126004	0
0.046	1.125994	1.125983	1.125973	1.125962	1.125952	1.125941	1.125931	1.125920	1.125910	1.125899	0
0.047	1.125889	1.125878	1.125868	1.125857	1.125846	1.125835	1.125825	1.125814	1.125803	1.125793	0
0.048	1.125782	1.125772	1.125760	1.125750	1.125738	1.125728	1.125716	1.125706	1.125694	1.125684	0
0.049	1.125672	1.125662	1.125650	1.125640	1.125628	1.125617	1.125607	1.125595	1.125584	1.125572	1

TABLE 8

Error Function Derivative

X	.0	.0001	.0002	.0003	.0004	.0005	.0006	.0007	.0008	.0009	Δ_2
0.050	1.125562	1.125550	1.125539	1.125527	1.125516	1.125504	1.125493	1.125482	1.125470	1.125459	0
0.051	1.125447	1.125436	1.125424	1.125413	1.125401	1.125390	1.125379	1.125366	1.125355	1.125343	0
0.052	1.125332	1.125320	1.125308	1.125297	1.125285	1.125273	1.125261	1.125249	1.125237	1.125225	1
0.053	1.125214	1.125201	1.125190	1.125177	1.125166	1.125154	1.125141	1.125130	1.125117	1.125105	1
0.054	1.125093	1.125081	1.125069	1.125056	1.125045	1.125032	1.125020	1.125008	1.124995	1.124983	0
0.055	1.124970	1.124958	1.124946	1.124933	1.124921	1.124908	1.124896	1.124884	1.124871	1.124858	1
0.056	1.124846	1.124833	1.124821	1.124807	1.124795	1.124783	1.124769	1.124757	1.124744	1.124731	1
0.057	1.124719	1.124705	1.124693	1.124680	1.124667	1.124654	1.124641	1.124628	1.124616	1.124602	0
0.058	1.124589	1.124577	1.124563	1.124550	1.124537	1.124524	1.124511	1.124497	1.124484	1.124471	0
0.059	1.124457	1.124444	1.124431	1.124417	1.124404	1.124391	1.124377	1.124364	1.124351	1.124337	0
0.060	1.124324	1.124310	1.124297	1.124283	1.124269	1.124256	1.124243	1.124228	1.124215	1.124202	0
0.061	1.124187	1.124174	1.124161	1.124146	1.124133	1.124119	1.124105	1.124091	1.124078	1.124063	0
0.062	1.124049	1.124036	1.124022	1.124007	1.123994	1.123980	1.123965	1.123951	1.123938	1.123923	0
0.063	1.123909	1.123895	1.123880	1.123866	1.123852	1.123837	1.123824	1.123810	1.123795	1.123780	0
0.064	1.123766	1.123752	1.123737	1.123723	1.123709	1.123694	1.123679	1.123665	1.123651	1.123636	0
0.065	1.123621	1.123607	1.123592	1.123577	1.123563	1.123549	1.123533	1.123519	1.123504	1.123489	0
0.066	1.123474	1.123459	1.123445	1.123429	1.123415	1.123400	1.123384	1.123370	1.123355	1.123340	0
0.067	1.123324	1.123310	1.123295	1.123280	1.123264	1.123249	1.123234	1.123219	1.123203	1.123188	0
0.068	1.123173	1.123158	1.123142	1.123127	1.123112	1.123096	1.123081	1.123066	1.123050	1.123034	0
0.069	1.123019	1.123004	1.122988	1.122972	1.122957	1.122941	1.122926	1.122910	1.122894	1.122879	0
0.070	1.122863	1.122848	1.122831	1.122816	1.122800	1.122785	1.122768	1.122752	1.122737	1.122721	0
0.071	1.122705	1.122689	1.122673	1.122657	1.122641	1.122624	1.122609	1.122593	1.122577	1.122561	0
0.072	1.122544	1.122528	1.122512	1.122496	1.122479	1.122463	1.122447	1.122431	1.122415	1.122398	0
0.073	1.122381	1.122365	1.122349	1.122333	1.122315	1.122299	1.122283	1.122267	1.122250	1.122233	0
0.074	1.122216	1.122200	1.122183	1.122167	1.122150	1.122133	1.122116	1.122100	1.122083	1.122066	0
0.075	1.122049	1.122032	1.122016	1.121999	1.121982	1.121965	1.121948	1.121931	1.121914	1.121897	0
0.076	1.121880	1.121863	1.121846	1.121829	1.121812	1.121795	1.121778	1.121760	1.121743	1.121726	0
0.077	1.121708	1.121691	1.121674	1.121656	1.121639	1.121622	1.121604	1.121587	1.121570	1.121552	1
0.078	1.121534	1.121517	1.121499	1.121482	1.121465	1.121447	1.121429	1.121411	1.121394	1.121376	1
0.079	1.121359	1.121341	1.121323	1.121305	1.121287	1.121269	1.121252	1.121234	1.121216	1.121198	1
0.080	1.121181	1.121162	1.121144	1.121126	1.121108	1.121090	1.121072	1.121054	1.121036	1.121017	0
0.081	1.120999	1.120981	1.120963	1.120945	1.120927	1.120909	1.120891	1.120872	1.120853	1.120835	0
0.082	1.120817	1.120798	1.120780	1.120762	1.120743	1.120725	1.120707	1.120687	1.120669	1.120650	1
0.083	1.120632	1.120613	1.120595	1.120576	1.120558	1.120539	1.120520	1.120502	1.120482	1.120463	1
0.084	1.120445	1.120426	1.120407	1.120388	1.120369	1.120351	1.120332	1.120313	1.120294	1.120275	0
0.085	1.120255	1.120236	1.120217	1.120198	1.120179	1.120160	1.120141	1.120122	1.120103	1.120083	1
0.086	1.120064	1.120045	1.120026	1.120007	1.119987	1.119967	1.119948	1.119928	1.119909	1.119889	1
0.087	1.119870	1.119851	1.119831	1.119812	1.119792	1.119773	1.119753	1.119733	1.119714	1.119694	1
0.088	1.119675	1.119655	1.119635	1.119615	1.119596	1.119576	1.119555	1.119535	1.119516	1.119496	0
0.089	1.119476	1.119456	1.119436	1.119416	1.119396	1.119376	1.119356	1.119336	1.119316	1.119296	0
0.090	1.119276	1.119256	1.119235	1.119215	1.119195	1.119175	1.119155	1.119134	1.119114	1.119094	0
0.091	1.119073	1.119053	1.119033	1.119012	1.118992	1.118971	1.118951	1.118930	1.118910	1.118889	1
0.092	1.118869	1.118848	1.118827	1.118807	1.118786	1.118765	1.118745	1.118724	1.118703	1.118682	1
0.093	1.118662	1.118641	1.118620	1.118599	1.118578	1.118557	1.118536	1.118515	1.118494	1.118473	0
0.094	1.118452	1.118431	1.118410	1.118389	1.118368	1.118347	1.118325	1.118304	1.118283	1.118262	0
0.095	1.118241	1.118219	1.118198	1.118177	1.118155	1.118134	1.118113	1.118092	1.118071	1.118049	1
0.096	1.118028	1.118006	1.117985	1.117963	1.117941	1.117920	1.117898	1.117876	1.117855	1.117833	0
0.097	1.117811	1.117790	1.117768	1.117746	1.117724	1.117702	1.117682	1.117660	1.117638	1.117616	0
0.098	1.117594	1.117572	1.117550	1.117528	1.117506	1.117484	1.117462	1.117439	1.117417	1.117395	0
0.099	1.117373	1.117352	1.117330	1.117307	1.117285	1.117263	1.117240	1.117218	1.117196	1.117173	1

TABLE 8

Error Function Derivative

X	.0	.0001	.0002	.0003	.0004	.0005	.0006	.0007	.0008	.0009	Δ_2
0.100	1.117151	1.117128	1.117106	1.117084	1.117062	1.117039	1.117017	1.116994	1.116972	1.116949	0
0.101	1.116926	1.116904	1.116881	1.116858	1.116837	1.116814	1.116791	1.116768	1.116745	1.116723	0
0.102	1.116700	1.116677	1.116654	1.116632	1.116609	1.116586	1.116563	1.116540	1.116517	1.116494	0
0.103	1.116471	1.116448	1.116425	1.116402	1.116379	1.116356	1.116333	1.116309	1.116286	1.116263	0
0.104	1.116240	1.116217	1.116194	1.116170	1.116147	1.116123	1.116100	1.116076	1.116054	1.116030	1
0.105	1.116007	1.115983	1.115960	1.115936	1.115912	1.115890	1.115866	1.115842	1.115818	1.115795	0
0.106	1.115771	1.115747	1.115724	1.115700	1.115676	1.115653	1.115629	1.115605	1.115582	1.115558	0
0.107	1.115534	1.115510	1.115486	1.115462	1.115438	1.115414	1.115390	1.115366	1.115342	1.115317	1
0.108	1.115294	1.115270	1.115246	1.115221	1.115197	1.115173	1.115149	1.115125	1.115100	1.115076	0
0.109	1.115051	1.115027	1.115003	1.114979	1.114954	1.114930	1.114905	1.114882	1.114857	1.114832	0
0.110	1.114807	1.114783	1.114758	1.114734	1.114709	1.114685	1.114660	1.114635	1.114611	1.114586	0
0.111	1.114561	1.114536	1.114511	1.114487	1.114462	1.114437	1.114412	1.114388	1.114363	1.114338	0
0.112	1.114313	1.114287	1.114263	1.114238	1.114213	1.114187	1.114162	1.114138	1.114112	1.114087	0
0.113	1.114062	1.114037	1.114012	1.113986	1.113961	1.113935	1.113911	1.113885	1.113860	1.113834	1
0.114	1.113810	1.113784	1.113758	1.113733	1.113708	1.113682	1.113656	1.113631	1.113605	1.113580	0
0.115	1.113554	1.113528	1.113503	1.113477	1.113452	1.113426	1.113400	1.113375	1.113349	1.113323	0
0.116	1.113297	1.113271	1.113245	1.113219	1.113194	1.113168	1.113142	1.113115	1.113090	1.113064	0
0.117	1.113037	1.113011	1.112986	1.112959	1.112933	1.112907	1.112881	1.112855	1.112828	1.112803	0
0.118	1.112776	1.112750	1.112723	1.112698	1.112671	1.112644	1.112618	1.112592	1.112565	1.112538	1
0.119	1.112513	1.112486	1.112459	1.112432	1.112406	1.112380	1.112353	1.112327	1.112300	1.112273	0
0.120	1.112247	1.112220	1.112193	1.112166	1.112140	1.112113	1.112086	1.112059	1.112032	1.112005	0
0.121	1.111979	1.111952	1.111924	1.111897	1.111871	1.111844	1.111816	1.111790	1.111762	1.111735	0
0.122	1.111709	1.111681	1.111654	1.111627	1.111600	1.111572	1.111546	1.111518	1.111490	1.111464	0
0.123	1.111436	1.111408	1.111382	1.111354	1.111326	1.111299	1.111272	1.111244	1.111217	1.111189	0
0.124	1.111161	1.111134	1.111106	1.111078	1.111051	1.111023	1.110995	1.110968	1.110940	1.110912	0
0.125	1.110885	1.110857	1.110829	1.110802	1.110773	1.110745	1.110718	1.110690	1.110662	1.110634	0
0.126	1.110606	1.110578	1.110550	1.110522	1.110494	1.110466	1.110437	1.110410	1.110381	1.110353	0
0.127	1.110325	1.110297	1.110269	1.110240	1.110212	1.110184	1.110155	1.110127	1.110099	1.110070	0
0.128	1.110042	1.110014	1.109985	1.109957	1.109928	1.109900	1.109871	1.109842	1.109814	1.109785	0
0.129	1.109756	1.109728	1.109699	1.109671	1.109642	1.109613	1.109585	1.109556	1.109527	1.109498	0
0.130	1.109469	1.109441	1.109411	1.109383	1.109354	1.109324	1.109296	1.109267	1.109238	1.109209	0
0.131	1.109179	1.109151	1.109121	1.109093	1.109063	1.109035	1.109005	1.108975	1.108947	1.108917	0
0.132	1.108888	1.108859	1.108829	1.108800	1.108770	1.108742	1.108712	1.108683	1.108653	1.108624	0
0.133	1.108594	1.108564	1.108535	1.108506	1.108477	1.108446	1.108417	1.108387	1.108357	1.108328	0
0.134	1.108298	1.108269	1.108239	1.108209	1.108179	1.108150	1.108120	1.108089	1.108060	1.108030	0
0.135	1.108000	1.107970	1.107941	1.107910	1.107881	1.107850	1.107821	1.107790	1.107760	1.107730	1
0.136	1.107700	1.107670	1.107639	1.107610	1.107579	1.107549	1.107519	1.107489	1.107458	1.107428	1
0.137	1.107398	1.107368	1.107337	1.107306	1.107276	1.107245	1.107215	1.107184	1.107154	1.107123	0
0.138	1.107093	1.107062	1.107032	1.107001	1.106971	1.106940	1.106910	1.106878	1.106848	1.106817	0
0.139	1.106787	1.106756	1.106725	1.106694	1.106664	1.106632	1.106602	1.106570	1.106540	1.106509	0
0.140	1.106478	1.106447	1.106416	1.106384	1.106354	1.106322	1.106292	1.106260	1.106229	1.106198	0
0.141	1.106167	1.106135	1.106105	1.106073	1.106042	1.106010	1.105979	1.105948	1.105917	1.105886	0
0.142	1.105854	1.105823	1.105791	1.105760	1.105728	1.105697	1.105665	1.105634	1.105602	1.105571	0
0.143	1.105538	1.105507	1.105475	1.105444	1.105412	1.105380	1.105349	1.105317	1.105285	1.105253	0
0.144	1.105222	1.105189	1.105158	1.105126	1.105094	1.105062	1.105030	1.104999	1.104966	1.104934	1
0.145	1.104902	1.104870	1.104838	1.104806	1.104774	1.104742	1.104710	1.104677	1.104645	1.104613	0
0.146	1.104581	1.104548	1.104516	1.104484	1.104451	1.104419	1.104386	1.104355	1.104322	1.104290	0
0.147	1.104257	1.104224	1.104192	1.104159	1.104127	1.104095	1.104062	1.104030	1.103996	1.103964	0
0.148	1.103931	1.103899	1.103866	1.103833	1.103801	1.103767	1.103735	1.103703	1.103669	1.103637	0
0.149	1.103603	1.103571	1.103538	1.103505	1.103472	1.103439	1.103406	1.103373	1.103340	1.103307	0

TABLE 8

Error Function Derivative

X	.0	.0001	.0002	.0003	.0004	.0005	.0006	.0007	.0008	.0009	Δ_2
0.150	1.103273	1.103241	1.103208	1.103174	1.103141	1.103107	1.103075	1.103042	1.103008	1.102975	0
0.151	1.102942	1.102908	1.102875	1.102841	1.102808	1.102775	1.102741	1.102708	1.102674	1.102641	0
0.152	1.102608	1.102574	1.102540	1.102507	1.102473	1.102440	1.102406	1.102372	1.102339	1.102305	0
0.153	1.102271	1.102238	1.102203	1.102170	1.102137	1.102102	1.102069	1.102035	1.102001	1.101967	0
0.154	1.101933	1.101899	1.101865	1.101831	1.101797	1.101763	1.101728	1.101695	1.101661	1.101626	0
0.155	1.101592	1.101559	1.101524	1.101490	1.101456	1.101421	1.101387	1.101353	1.101318	1.101284	0
0.156	1.101250	1.101215	1.101181	1.101147	1.101112	1.101078	1.101044	1.101008	1.100974	1.100940	0
0.157	1.100905	1.100871	1.100836	1.100801	1.100767	1.100732	1.100698	1.100663	1.100628	1.100594	0
0.158	1.100558	1.100524	1.100489	1.100454	1.100419	1.100385	1.100349	1.100314	1.100280	1.100245	0
0.159	1.100209	1.100175	1.100140	1.100104	1.100070	1.100035	1.099999	1.099964	1.099929	1.099894	1
0.160	1.099859	1.099824	1.099789	1.099753	1.099718	1.099683	1.099648	1.099612	1.099577	1.099542	0
0.161	1.099505	1.099470	1.099435	1.099400	1.099364	1.099328	1.099293	1.099257	1.099222	1.099186	1
0.162	1.099151	1.099115	1.099079	1.099044	1.099009	1.098972	1.098937	1.098901	1.098866	1.098829	1
0.163	1.098794	1.098758	1.098722	1.098686	1.098650	1.098614	1.098578	1.098542	1.098506	1.098471	0
0.164	1.098434	1.098398	1.098362	1.098326	1.098290	1.098254	1.098218	1.098182	1.098145	1.098109	0
0.165	1.098073	1.098037	1.098001	1.097964	1.097928	1.097892	1.097856	1.097818	1.097782	1.097746	0
0.166	1.097710	1.097673	1.097636	1.097600	1.097564	1.097528	1.097490	1.097454	1.097418	1.097381	1
0.167	1.097344	1.097307	1.097271	1.097234	1.097198	1.097160	1.097124	1.097087	1.097051	1.097013	0
0.168	1.096976	1.096940	1.096903	1.096866	1.096829	1.096792	1.096755	1.096718	1.096681	1.096644	0
0.169	1.096607	1.096570	1.096533	1.096496	1.096458	1.096421	1.096384	1.096347	1.096310	1.096272	0
0.170	1.096235	1.096198	1.096161	1.096124	1.096086	1.096048	1.096011	1.095974	1.095937	1.095899	1
0.171	1.095861	1.095824	1.095786	1.095749	1.095712	1.095674	1.095636	1.095598	1.095561	1.095523	1
0.172	1.095486	1.095448	1.095410	1.095372	1.095335	1.095297	1.095260	1.095222	1.095183	1.095145	1
0.173	1.095108	1.095070	1.095032	1.094994	1.094956	1.094918	1.094880	1.094842	1.094804	1.094766	0
0.174	1.094728	1.094689	1.094651	1.094613	1.094575	1.094537	1.094499	1.094460	1.094422	1.094384	0
0.175	1.094346	1.094308	1.094269	1.094231	1.094193	1.094154	1.094115	1.094077	1.094039	1.094000	1
0.176	1.093962	1.093924	1.093884	1.093846	1.093807	1.093769	1.093730	1.093692	1.093653	1.093615	0
0.177	1.093575	1.093537	1.093498	1.093459	1.093421	1.093382	1.093343	1.093305	1.093266	1.093226	0
0.178	1.093187	1.093148	1.093110	1.093071	1.093032	1.092993	1.092954	1.092915	1.092875	1.092836	0
0.179	1.092797	1.092758	1.092719	1.092680	1.092641	1.092602	1.092563	1.092523	1.092484	1.092444	0
0.180	1.092405	1.092366	1.092326	1.092287	1.092248	1.092208	1.092169	1.092130	1.092090	1.092051	0
0.181	1.092010	1.091971	1.091931	1.091892	1.091852	1.091813	1.091773	1.091734	1.091694	1.091654	1
0.182	1.091615	1.091575	1.091535	1.091496	1.091455	1.091415	1.091375	1.091336	1.091296	1.091256	0
0.183	1.091216	1.091176	1.091136	1.091096	1.091056	1.091016	1.090976	1.090936	1.090896	1.090856	0
0.184	1.090816	1.090775	1.090735	1.090695	1.090655	1.090614	1.090574	1.090534	1.090494	1.090454	0
0.185	1.090413	1.090373	1.090333	1.090292	1.090252	1.090211	1.090171	1.090131	1.090090	1.090050	0
0.186	1.090009	1.089969	1.089928	1.089887	1.089847	1.089806	1.089766	1.089725	1.089684	1.089643	0
0.187	1.089602	1.089561	1.089520	1.089480	1.089439	1.089398	1.089357	1.089316	1.089275	1.089234	0
0.188	1.089193	1.089153	1.089112	1.089071	1.089030	1.088989	1.088947	1.088906	1.088865	1.088824	0
0.189	1.088783	1.088742	1.088701	1.088659	1.088618	1.088577	1.088536	1.088494	1.088453	1.088412	0
0.190	1.088370	1.088329	1.088288	1.088246	1.088205	1.088163	1.088122	1.088080	1.088039	1.087997	1
0.191	1.087956	1.087914	1.087873	1.087831	1.087790	1.087748	1.087707	1.087665	1.087623	1.087582	0
0.192	1.087540	1.087498	1.087456	1.087414	1.087372	1.087331	1.087289	1.087247	1.087205	1.087163	0
0.193	1.087121	1.087079	1.087037	1.086995	1.086953	1.086911	1.086868	1.086826	1.086784	1.086742	0
0.194	1.086700	1.086658	1.086616	1.086574	1.086532	1.086489	1.086447	1.086405	1.086362	1.086320	0
0.195	1.086278	1.086235	1.086193	1.086150	1.086108	1.086065	1.086023	1.085980	1.085938	1.085896	0
0.196	1.085853	1.085811	1.085768	1.085725	1.085683	1.085640	1.085597	1.085554	1.085512	1.085469	0
0.197	1.085426	1.085383	1.085340	1.085298	1.085255	1.085212	1.085170	1.085127	1.085084	1.085041	0
0.198	1.084997	1.084954	1.084911	1.084868	1.084826	1.084783	1.084740	1.084697	1.084653	1.084610	0
0.199	1.084567	1.084524	1.084480	1.084437	1.084394	1.084351	1.084308	1.084265	1.084221	1.084178	0

$$\frac{2}{\sqrt{\pi}}\exp(-x^2)$$

$$\frac{2}{\sqrt{\pi}}\exp(-x^2)$$

TABLE 8

Error Function Derivative

X	.0	.0001	.0002	.0003	.0004	.0005	.0006	.0007	.0008	.0009	Δ_2
0.200	1.084134	1.084091	1.084047	1.084004	1.083961	1.083918	1.083874	1.083830	1.083787	1.083743	0
0.201	1.083699	1.083656	1.083612	1.083569	1.083526	1.083482	1.083438	1.083394	1.083350	1.083306	1
0.202	1.083263	1.083220	1.083176	1.083132	1.083088	1.083044	1.083000	1.082956	1.082912	1.082869	0
0.203	1.082825	1.082781	1.082736	1.082692	1.082648	1.082604	1.082561	1.082516	1.082472	1.082428	0
0.204	1.082384	1.082339	1.082295	1.082252	1.082207	1.082163	1.082118	1.082074	1.082030	1.081985	1
0.205	1.081942	1.081897	1.081852	1.081808	1.081763	1.081719	1.081675	1.081630	1.081586	1.081541	0
0.206	1.081496	1.081452	1.081408	1.081363	1.081318	1.081273	1.081229	1.081184	1.081140	1.081095	0
0.207	1.081050	1.081005	1.080960	1.080915	1.080871	1.080826	1.080781	1.080736	1.080691	1.080647	0
0.208	1.080602	1.080557	1.080511	1.080466	1.080421	1.080377	1.080332	1.080286	1.080241	1.080196	0
0.209	1.080151	1.080106	1.080061	1.080015	1.079970	1.079925	1.079880	1.079834	1.079789	1.079743	1
0.210	1.079699	1.079653	1.079608	1.079562	1.079517	1.079472	1.079426	1.079381	1.079335	1.079289	0
0.211	1.079244	1.079199	1.079153	1.079107	1.079062	1.079016	1.078970	1.078925	1.078879	1.078834	0
0.212	1.078788	1.078742	1.078696	1.078650	1.078605	1.078559	1.078513	1.078467	1.078421	1.078375	0
0.213	1.078329	1.078283	1.078238	1.078192	1.078145	1.078099	1.078053	1.078007	1.077961	1.077915	0
0.214	1.077868	1.077823	1.077777	1.077730	1.077684	1.077638	1.077592	1.077545	1.077499	1.077453	1
0.215	1.077407	1.077360	1.077313	1.077268	1.077221	1.077174	1.077128	1.077082	1.077035	1.076989	0
0.216	1.076942	1.076896	1.076849	1.076802	1.076756	1.076710	1.076663	1.076616	1.076570	1.076523	0
0.217	1.076476	1.076429	1.076383	1.076336	1.076289	1.076242	1.076196	1.076149	1.076101	1.076055	0
0.218	1.076008	1.075961	1.075914	1.075867	1.075820	1.075773	1.075727	1.075679	1.075632	1.075584	1
0.219	1.075538	1.075491	1.075443	1.075397	1.075349	1.075302	1.075254	1.075208	1.075160	1.075113	0
0.220	1.075066	1.075019	1.074971	1.074924	1.074877	1.074829	1.074781	1.074734	1.074687	1.074639	0
0.221	1.074592	1.074544	1.074496	1.074450	1.074402	1.074354	1.074306	1.074259	1.074211	1.074163	0
0.222	1.074116	1.074068	1.074020	1.073973	1.073925	1.073877	1.073830	1.073781	1.073733	1.073686	0
0.223	1.073638	1.073590	1.073542	1.073494	1.073446	1.073399	1.073350	1.073302	1.073255	1.073206	1
0.224	1.073158	1.073110	1.073062	1.073013	1.072966	1.072918	1.072869	1.072821	1.072773	1.072724	1
0.225	1.072677	1.072628	1.072579	1.072532	1.072483	1.072434	1.072387	1.072338	1.072289	1.072241	1
0.226	1.072193	1.072145	1.072096	1.072047	1.071999	1.071950	1.071901	1.071853	1.071804	1.071755	0
0.227	1.071707	1.071658	1.071609	1.071561	1.071512	1.071464	1.071415	1.071366	1.071318	1.071268	1
0.228	1.071219	1.071171	1.071122	1.071073	1.071024	1.070975	1.070926	1.070877	1.070828	1.070779	1
0.229	1.070730	1.070681	1.070632	1.070583	1.070534	1.070485	1.070436	1.070386	1.070337	1.070288	1
0.230	1.070239	1.070189	1.070140	1.070091	1.070042	1.069992	1.069943	1.069894	1.069844	1.069795	0
0.231	1.069745	1.069696	1.069647	1.069597	1.069548	1.069498	1.069448	1.069399	1.069349	1.069300	0
0.232	1.069250	1.069201	1.069151	1.069101	1.069052	1.069002	1.068953	1.068903	1.068852	1.068803	0
0.233	1.068753	1.068704	1.068654	1.068604	1.068554	1.068504	1.068454	1.068404	1.068355	1.068304	1
0.234	1.068254	1.068204	1.068154	1.068104	1.068054	1.068004	1.067954	1.067904	1.067854	1.067803	1
0.235	1.067754	1.067703	1.067653	1.067603	1.067553	1.067502	1.067451	1.067402	1.067351	1.067301	0
0.236	1.067250	1.067201	1.067150	1.067100	1.067049	1.066998	1.066948	1.066897	1.066847	1.066796	0
0.237	1.066746	1.066695	1.066645	1.066594	1.066544	1.066493	1.066442	1.066392	1.066340	1.066290	0
0.238	1.066239	1.066189	1.066137	1.066087	1.066036	1.065986	1.065934	1.065884	1.065833	1.065782	1
0.239	1.065731	1.065680	1.065629	1.065578	1.065527	1.065475	1.065425	1.065373	1.065323	1.065271	1
0.240	1.065221	1.065169	1.065118	1.065067	1.065016	1.064964	1.064914	1.064862	1.064811	1.064759	0
0.241	1.064708	1.064657	1.064606	1.064554	1.064503	1.064451	1.064400	1.064348	1.064297	1.064245	0
0.242	1.064194	1.064142	1.064091	1.064039	1.063988	1.063936	1.063885	1.063833	1.063782	1.063730	0
0.243	1.063678	1.063626	1.063575	1.063523	1.063471	1.063419	1.063368	1.063315	1.063264	1.063212	0
0.244	1.063160	1.063108	1.063056	1.063004	1.062952	1.062901	1.062848	1.062797	1.062744	1.062693	0
0.245	1.062640	1.062589	1.062536	1.062484	1.062432	1.062380	1.062327	1.062276	1.062223	1.062171	0
0.246	1.062119	1.062066	1.062015	1.061962	1.061910	1.061857	1.061805	1.061752	1.061700	1.061647	0
0.247	1.061595	1.061543	1.061490	1.061438	1.061385	1.061333	1.061280	1.061228	1.061175	1.061123	0
0.248	1.061069	1.061017	1.060965	1.060912	1.060860	1.060806	1.060754	1.060701	1.060648	1.060596	0
0.249	1.060543	1.060490	1.060437	1.060384	1.060331	1.060279	1.060225	1.060173	1.060120	1.060066	1

TABLE 8

Error Function Derivative

X	.0	.0001	.0002	.0003	.0004	.0005	.0006	.0007	.0008	.0009	Δ_2
0.250	1.060014	1.059960	1.059908	1.059855	1.059801	1.059749	1.059695	1.059642	1.059589	1.059536	0
0.251	1.059483	1.059429	1.059377	1.059323	1.059270	1.059216	1.059163	1.059110	1.059056	1.059003	0
0.252	1.058949	1.058897	1.058844	1.058790	1.058736	1.058682	1.058629	1.058576	1.058522	1.058469	0
0.253	1.058415	1.058362	1.058309	1.058254	1.058201	1.058147	1.058094	1.058040	1.057986	1.057933	0
0.254	1.057878	1.057825	1.057772	1.057717	1.057664	1.057610	1.057556	1.057502	1.057448	1.057394	1
0.255	1.057341	1.057286	1.057233	1.057178	1.057125	1.057071	1.057016	1.056963	1.056909	1.056854	0
0.256	1.056800	1.056746	1.056692	1.056638	1.056583	1.056530	1.056476	1.056421	1.056367	1.056313	0
0.257	1.056258	1.056204	1.056149	1.056095	1.056041	1.055986	1.055932	1.055878	1.055823	1.055769	0
0.258	1.055715	1.055660	1.055606	1.055551	1.055496	1.055442	1.055387	1.055333	1.055278	1.055223	0
0.259	1.055169	1.055114	1.055059	1.055005	1.054950	1.054895	1.054840	1.054786	1.054731	1.054676	1
0.260	1.054622	1.054566	1.054512	1.054457	1.054401	1.054347	1.054292	1.054237	1.054182	1.054127	1
0.261	1.054072	1.054017	1.053962	1.053906	1.053852	1.053797	1.053741	1.053686	1.053631	1.053576	0
0.262	1.053521	1.053466	1.053411	1.053355	1.053300	1.053245	1.053189	1.053134	1.053079	1.053023	0
0.263	1.052968	1.052913	1.052857	1.052802	1.052746	1.052691	1.052635	1.052580	1.052525	1.052468	1
0.264	1.052413	1.052358	1.052302	1.052246	1.052191	1.052135	1.052079	1.052024	1.051968	1.051912	0
0.265	1.051857	1.051801	1.051745	1.051689	1.051634	1.051578	1.051521	1.051466	1.051410	1.051354	0
0.266	1.051298	1.051242	1.051187	1.051130	1.051074	1.051019	1.050962	1.050906	1.050850	1.050794	1
0.267	1.050738	1.050682	1.050626	1.050570	1.050513	1.050457	1.050401	1.050344	1.050288	1.050232	0
0.268	1.050176	1.050119	1.050063	1.050007	1.049951	1.049894	1.049838	1.049782	1.049726	1.049668	1
0.269	1.049612	1.049556	1.049500	1.049442	1.049386	1.049330	1.049273	1.049216	1.049160	1.049104	0
0.270	1.049047	1.048990	1.048933	1.048877	1.048820	1.048763	1.048706	1.048650	1.048593	1.048536	0
0.271	1.048479	1.048423	1.049366	1.048308	1.048252	1.048195	1.048138	1.048081	1.048024	1.047967	0
0.272	1.047910	1.047853	1.047796	1.047739	1.047682	1.047625	1.047567	1.047510	1.047454	1.047397	0
0.273	1.047339	1.047282	1.047225	1.047168	1.047111	1.047053	1.046996	1.046938	1.046881	1.046824	0
0.274	1.046766	1.046709	1.046652	1.046595	1.046536	1.046479	1.046422	1.046365	1.046307	1.046249	0
0.275	1.046192	1.046134	1.046077	1.046020	1.045961	1.045904	1.045846	1.045789	1.045731	1.045673	0
0.276	1.045615	1.045558	1.045500	1.045443	1.045384	1.045327	1.045269	1.045211	1.045154	1.045095	0
0.277	1.045037	1.044980	1.044922	1.044864	1.044806	1.044747	1.044690	1.044632	1.044574	1.044516	0
0.278	1.044457	1.044399	1.044341	1.044283	1.044226	1.044168	1.044109	1.044051	1.043993	1.043934	1
0.279	1.043876	1.043818	1.043759	1.043701	1.043643	1.043585	1.043527	1.043468	1.043409	1.043351	0
0.280	1.043293	1.043234	1.043176	1.043118	1.043059	1.043000	1.042942	1.042883	1.042825	1.042767	0
0.281	1.042707	1.042649	1.042590	1.042532	1.042473	1.042415	1.042356	1.042297	1.042238	1.042179	1
0.282	1.042121	1.042062	1.042003	1.041945	1.041885	1.041826	1.041768	1.041709	1.041650	1.041591	1
0.283	1.041533	1.041473	1.041414	1.041355	1.041296	1.041237	1.041178	1.041119	1.041059	1.041000	0
0.284	1.040941	1.040882	1.040823	1.040764	1.040705	1.040646	1.040586	1.040527	1.040468	1.040409	0
0.285	1.040350	1.040290	1.040231	1.040172	1.040112	1.040052	1.039993	1.039934	1.039875	1.039815	1
0.286	1.039756	1.039696	1.039637	1.039577	1.039517	1.039458	1.039398	1.039339	1.039279	1.039220	0
0.287	1.039160	1.039101	1.039041	1.038981	1.038921	1.038861	1.038802	1.038742	1.038682	1.038623	0
0.288	1.038563	1.038503	1.038443	1.038383	1.038323	1.038263	1.038203	1.038143	1.C38084	1.038024	0
0.289	1.037964	1.037904	1.037844	1.037784	1.037724	1.037663	1.037603	1.037543	1.037483	1.037423	0
0.290	1.037363	1.037303	1.037243	1.037182	1.037122	1.037062	1.037002	1.036942	1.036880	1.036820	0
0.291	1.036760	1.036700	1.036639	1.036579	1.036519	1.036458	1.036398	1.036338	1.036277	1.036217	0
0.292	1.036156	1.036096	1.036035	1.035975	1.035913	1.035853	1.035792	1.035732	1.035671	1.035611	0
0.293	1.035550	1.035489	1.035429	1.035368	1.035307	1.035247	1.035186	1.035125	1.035064	1.035004	0
0.294	1.034943	1.034882	1.034821	1.034760	1.034698	1.034637	1.034577	1.034516	1.034455	1.034394	0
0.295	1.034333	1.034272	1.034211	1.034150	1.034089	1.034028	1.033966	1.033905	1.033844	1.033783	0
0.296	1.033722	1.033661	1.033600	1.033538	1.033477	1.033416	1.033355	1.033293	1.033232	1.033171	0
0.297	1.033109	1.033048	1.032987	1.032925	1.032864	1.032802	1.032741	1.032680	1.032618	1.032557	0
0.298	1.032495	1.032433	1.032372	1.032310	1.032248	1.032187	1.032125	1.032063	1.032001	1.031940	0
0.299	1.031878	1.031816	1.031754	1.031693	1.031631	1.031569	1.031507	1.031446	1.031384	1.031322	0

$$\frac{2}{\sqrt{\pi}}\exp(-x^2)$$

TABLE 8

Error Function Derivative

X	.0	.0001	.0002	.0003	.0004	.0005	.0006	.0007	.0008	.0009	Δ_2
0.300	1.031260	1.031199	1.031137	1.031075	1.031013	1.030951	1.030889	1.030827	1.030765	1.030703	0
0.301	1.030641	1.030579	1.030517	1.030455	1.030393	1.030331	1.030269	1.030206	1.030144	1.030082	0
0.302	1.030020	1.029958	1.029895	1.029833	1.029771	1.029708	1.029646	1.029584	1.029521	1.029459	0
0.303	1.029397	1.029334	1.029272	1.029209	1.029147	1.029084	1.029022	1.028959	1.028897	1.028834	1
0.304	1.028772	1.028709	1.028646	1.028584	1.028522	1.028459	1.028397	1.028334	1.028271	1.028209	0
0.305	1.028146	1.028083	1.028020	1.027957	1.027895	1.027832	1.027769	1.027706	1.027643	1.027580	0
0.306	1.027517	1.027454	1.027391	1.027328	1.027266	1.027203	1.027140	1.027077	1.027014	1.026951	0
0.307	1.026888	1.026825	1.026762	1.026699	1.026635	1.026572	1.026509	1.026446	1.026383	1.026320	1
0.308	1.026257	1.026194	1.026130	1.026067	1.026004	1.025940	1.025877	1.025814	1.025750	1.025687	0
0.309	1.025623	1.025560	1.025496	1.025434	1.025370	1.025307	1.025243	1.025180	1.025116	1.025052	1
0.310	1.024989	1.024925	1.024861	1.024798	1.024734	1.024671	1.024608	1.024544	1.024480	1.024416	1
0.311	1.024353	1.024289	1.024225	1.024161	1.024097	1.024034	1.023970	1.023906	1.023842	1.023779	0
0.312	1.023715	1.023651	1.023587	1.023523	1.023458	1.023395	1.023331	1.023267	1.023203	1.023139	0
0.313	1.023075	1.023011	1.022947	1.022882	1.022819	1.022755	1.022691	1.022626	1.022562	1.022498	0
0.314	1.022433	1.022369	1.022305	1.022241	1.022177	1.022113	1.022048	1.021984	1.021919	1.021855	0
0.315	1.021791	1.021727	1.021662	1.021598	1.021533	1.021469	1.021404	1.021339	1.021276	1.021211	0
0.316	1.021146	1.021082	1.021017	1.020952	1.020888	1.020823	1.020759	1.020694	1.020630	1.020565	0
0.317	1.020500	1.020435	1.020370	1.020306	1.020241	1.020176	1.020112	1.020047	1.019982	1.019917	0
0.318	1.019853	1.019788	1.019723	1.019657	1.019592	1.019527	1.019463	1.019398	1.019333	1.019268	0
0.319	1.019203	1.019137	1.019073	1.019008	1.018943	1.018877	1.018812	1.018747	1.018682	1.018617	1
0.320	1.018552	1.018487	1.018421	1.018356	1.018291	1.018226	1.018160	1.018095	1.018030	1.017964	1
0.321	1.017900	1.017834	1.017768	1.017703	1.017637	1.017572	1.017507	1.017441	1.017376	1.017310	0
0.322	1.017244	1.017179	1.017114	1.017048	1.016983	1.016917	1.016851	1.016786	1.016720	1.016655	0
0.323	1.016589	1.016523	1.016458	1.016392	1.016326	1.016260	1.016194	1.016129	1.016063	1.015997	0
0.324	1.015931	1.015865	1.015800	1.015734	1.015668	1.015602	1.015536	1.015471	1.015404	1.015338	0
0.325	1.015272	1.015206	1.015141	1.015074	1.015008	1.014942	1.014876	1.014810	1.014744	1.014678	0
0.326	1.014611	1.014545	1.014479	1.014413	1.014347	1.014280	1.014215	1.014148	1.014082	1.014015	1
0.327	1.013949	1.013883	1.013817	1.013750	1.013683	1.013618	1.013551	1.013485	1.013418	1.013351	1
0.328	1.013286	1.013219	1.013152	1.013085	1.013020	1.012953	1.012886	1.012819	1.012753	1.012687	0
0.329	1.012620	1.012553	1.012486	1.012420	1.012353	1.012286	1.012219	1.012153	1.012086	1.012019	0
0.330	1.011952	1.011886	1.011819	1.011752	1.011685	1.011619	1.011552	1.011485	1.011417	1.011351	0
0.331	1.011284	1.011217	1.011150	1.011083	1.011016	1.010949	1.010882	1.010815	1.010748	1.010681	0
0.332	1.010613	1.010547	1.010480	1.010412	1.010345	1.010278	1.010211	1.010143	1.010077	1.010009	1
0.333	1.009942	1.009874	1.009808	1.009740	1.009673	1.009605	1.009538	1.009471	1.009403	1.009336	1
0.334	1.009269	1.009201	1.009133	1.009067	1.008999	1.008931	1.008863	1.008796	1.008729	1.008661	0
0.335	1.008594	1.008526	1.008458	1.008390	1.008323	1.008255	1.008187	1.008121	1.008053	1.007985	0
0.336	1.007917	1.007849	1.007781	1.007713	1.007646	1.007578	1.007510	1.007442	1.007375	1.007307	0
0.337	1.007238	1.007171	1.007103	1.007035	1.006968	1.006899	1.006831	1.006763	1.006695	1.006627	0
0.338	1.006559	1.006491	1.006423	1.006355	1.006287	1.006219	1.006150	1.006083	1.006014	1.005946	0
0.339	1.005877	1.005810	1.005741	1.005673	1.005605	1.005537	1.005468	1.005401	1.005332	1.005263	1
0.340	1.005196	1.005127	1.005058	1.004990	1.004922	1.004853	1.004785	1.004716	1.004648	1.004580	0
0.341	1.004511	1.004442	1.004374	1.004305	1.004237	1.004169	1.004100	1.004031	1.003963	1.003894	0
0.342	1.003825	1.003757	1.003688	1.003619	1.003551	1.003482	1.003413	1.003344	1.003275	1.003206	0
0.343	1.003138	1.003069	1.003000	1.002932	1.002862	1.002793	1.002725	1.002656	1.002586	1.002518	0
0.344	1.002449	1.002379	1.002311	1.002242	1.002172	1.002104	1.002035	1.001966	1.001897	1.001827	1
0.345	1.001759	1.001689	1.001620	1.001551	1.001482	1.001412	1.001344	1.001274	1.001204	1.001136	0
0.346	1.001066	1.000997	1.000928	1.000858	1.000789	1.000720	1.000650	1.000581	1.000512	1.000443	0
0.347	1.000373	1.000303	1.000234	1.000164	1.000095	1.000026	0.999956	0.999887	0.999817	0.999748	0
0.348	0.999678	0.999609	0.999539	0.999469	0.999400	0.999330	0.999261	0.999191	0.999121	0.999051	0
0.349	0.998982	0.998912	0.998842	0.998772	0.998703	0.998633	0.998563	0.998493	0.998423	0.998354	0

TABLE 8

Error Function Derivative

X	.0	.0001	.0002	.0003	.0004	.0005	.0006	.0007	.0008	.0009	Δ_2
0.350	0.998284	0.998214	0.998144	0.998074	0.998004	0.997934	0.997864	0.997794	0.997724	0.997654	0
0.351	0.997584	0.997514	0.997444	0.997374	0.997304	0.997234	0.997164	0.997094	0.997023	0.996953	0
0.352	0.996883	0.996813	0.996743	0.996672	0.996602	0.996532	0.996462	0.996391	0.996321	0.996251	0
0.353	0.996181	0.996110	0.996040	0.995969	0.995899	0.995829	0.995758	0.995688	0.995617	0.995547	0
0.354	0.995476	0.995406	0.995335	0.995265	0.995194	0.995124	0.995053	0.994983	0.994912	0.994842	0
0.355	0.994771	0.994700	0.994630	0.994559	0.994488	0.994418	0.994347	0.994276	0.994205	0.994135	0
0.356	0.994064	0.993993	0.993922	0.993851	0.993781	0.993710	0.993639	0.993568	0.993497	0.993426	0
0.357	0.993355	0.993284	0.993213	0.993143	0.993072	0.993001	0.992930	0.992859	0.992787	0.992716	0
0.358	0.992645	0.992574	0.992503	0.992432	0.992361	0.992290	0.992219	0.992148	0.992076	0.992005	0
0.359	0.991934	0.991863	0.991791	0.991720	0.991649	0.991578	0.991506	0.991435	0.991364	0.991292	0
0.360	0.991221	0.991150	0.991078	0.991007	0.990935	0.990864	0.990793	0.990721	0.990650	0.990578	0
0.361	0.990507	0.990435	0.990363	0.990292	0.990220	0.990149	0.990077	0.990006	0.989934	0.989862	0
0.362	0.989791	0.989719	0.989647	0.989576	0.989504	0.989432	0.989360	0.989289	0.989217	0.989145	0
0.363	0.989073	0.989002	0.988930	0.988858	0.988786	0.988714	0.988642	0.988570	0.988498	0.988427	0
0.364	0.988355	0.988283	0.988211	0.988139	0.988067	0.987995	0.987923	0.987851	0.987778	0.987706	0
0.365	0.987634	0.987562	0.987490	0.987418	0.987346	0.987274	0.987201	0.987129	0.987057	0.986985	0
0.366	0.986913	0.986840	0.986768	0.986696	0.986624	0.986551	0.986479	0.986407	0.986334	0.986262	0
0.367	0.986189	0.986117	0.986045	0.985972	0.985900	0.985827	0.985755	0.985682	0.985610	0.985537	0
0.368	0.985465	0.985392	0.985320	0.985247	0.985175	0.985102	0.985029	0.984957	0.984884	0.984812	0
0.369	0.984739	0.984666	0.984594	0.984521	0.984448	0.984375	0.984303	0.984230	0.984157	0.984084	0
0.370	0.984011	0.983939	0.983866	0.983793	0.983720	0.983647	0.983574	0.983501	0.983428	0.983355	0
0.371	0.983283	0.983210	0.983137	0.983064	0.982991	0.982918	0.982845	0.982771	0.982698	0.982625	0
0.372	0.982552	0.982479	0.982406	0.982333	0.982260	0.982187	0.982113	0.982040	0.981967	0.981894	0
0.373	0.981820	0.981747	0.981674	0.981601	0.981527	0.981454	0.981381	0.981307	0.981234	0.981161	0
0.374	0.981087	0.981014	0.980941	0.980867	0.980794	0.980720	0.980647	0.980573	0.980500	0.980426	0
0.375	0.980353	0.980279	0.980206	0.980132	0.980059	0.979985	0.979911	0.979838	0.979764	0.979690	0
0.376	0.979617	0.979543	0.979469	0.979396	0.979322	0.979248	0.979175	0.979101	0.979027	0.978953	0
0.377	0.978879	0.978806	0.978732	0.978658	0.978584	0.978510	0.978436	0.978362	0.978289	0.978215	0
0.378	0.978141	0.978067	0.977993	0.977919	0.977845	0.977771	0.977697	0.977623	0.977549	0.977475	0
0.379	0.977400	0.977326	0.977252	0.977178	0.977104	0.977030	0.976956	0.976882	0.976807	0.976733	0
0.380	0.976659	0.976585	0.976510	0.976436	0.976362	0.976288	0.976213	0.976139	0.976065	0.975990	0
0.381	0.975916	0.975842	0.975767	0.975693	0.975618	0.975544	0.975470	0.975395	0.975321	0.975246	0
0.382	0.975172	0.975097	0.975023	0.974948	0.974873	0.974799	0.974724	0.974650	0.974575	0.974501	0
0.383	0.974426	0.974351	0.974277	0.974202	0.974127	0.974053	0.973978	0.973903	0.973828	0.973754	0
0.384	0.973679	0.973604	0.973529	0.973454	0.973380	0.973305	0.973230	0.973155	0.973080	0.973005	0
0.385	0.972930	0.972855	0.972780	0.972706	0.972631	0.972556	0.972481	0.972406	0.972331	0.972256	0
0.386	0.972180	0.972105	0.972030	0.971955	0.971880	0.971805	0.971730	0.971655	0.971580	0.971504	0
0.387	0.971429	0.971354	0.971279	0.971204	0.971128	0.971053	0.970978	0.970903	0.970827	0.970752	0
0.388	0.970677	0.970601	0.970526	0.970451	0.970375	0.970300	0.970225	0.970149	0.970074	0.969998	0
0.389	0.969923	0.969847	0.969772	0.969696	0.969621	0.969545	0.969470	0.969394	0.969319	0.969243	0
0.390	0.969168	0.969092	0.969016	0.968941	0.968865	0.968789	0.968714	0.968638	0.968562	0.968487	0
0.391	0.968411	0.968335	0.968259	0.968184	0.968108	0.968032	0.967956	0.967880	0.967805	0.967729	0
0.392	0.967653	0.967577	0.967501	0.967425	0.967349	0.967273	0.967198	0.967122	0.967046	0.966970	0
0.393	0.966894	0.966818	0.966742	0.966666	0.966590	0.966513	0.966437	0.966361	0.966285	0.966209	0
0.394	0.966133	0.966057	0.965981	0.965905	0.965828	0.965752	0.965676	0.965600	0.965523	0.965447	0
0.395	0.965371	0.965295	0.965218	0.965142	0.965066	0.964990	0.964913	0.964837	0.964760	0.964684	0
0.396	0.964608	0.964531	0.964455	0.964378	0.964302	0.964226	0.964149	0.964073	0.963996	0.963920	0
0.397	0.963843	0.963767	0.963690	0.963613	0.963537	0.963460	0.963384	0.963307	0.963230	0.963154	0
0.398	0.963077	0.963000	0.962924	0.962847	0.962770	0.962694	0.962617	0.962540	0.962463	0.962387	0
0.399	0.962310	0.962233	0.962156	0.962079	0.962003	0.961926	0.961849	0.961772	0.961695	0.961618	0

$$\frac{2}{\sqrt{\pi}}\exp(-x^2)$$

$$\frac{2}{\sqrt{\pi}}\exp(-x^2)$$

TABLE 8

Error Function Derivative

X	.0	.0001	.0002	.0003	.0004	.0005	.0006	.0007	.0008	.0009	Δ_2
0.400	0.961541	0.961464	0.961387	0.961310	0.961233	0.961156	0.961079	0.961002	0.960925	0.960848	0
0.401	0.960771	0.960694	0.960617	0.960540	0.960463	0.960386	0.960309	0.960232	0.960155	0.960077	0
0.402	0.960000	0.959923	0.959846	0.959769	0.959691	0.959614	0.959537	0.959460	0.959382	0.959305	0
0.403	0.959228	0.959150	0.959073	0.958996	0.958918	0.958841	0.958764	0.958686	0.958609	0.958531	0
0.404	0.958454	0.958376	0.958299	0.958222	0.958144	0.958067	0.957989	0.957911	0.957834	0.957756	0
0.405	0.957679	0.957601	0.957524	0.957446	0.957368	0.957291	0.957213	0.957135	0.957058	0.956980	0
0.406	0.956903	0.956825	0.956747	0.956669	0.956592	0.956514	0.956436	0.956358	0.956280	0.956203	0
0.407	0.956125	0.956047	0.955969	0.955891	0.955813	0.955736	0.955658	0.955580	0.955502	0.955424	0
0.408	0.955346	0.955268	0.955190	0.955112	0.955034	0.954956	0.954878	0.954800	0.954722	0.954644	0
0.409	0.954566	0.954488	0.954410	0.954331	0.954253	0.954175	0.954097	0.954019	0.953941	0.953862	0
0.410	0.953784	0.953706	0.953628	0.953550	0.953471	0.953393	0.953315	0.953236	0.953158	0.953080	0
0.411	0.953001	0.952923	0.952845	0.952766	0.952688	0.952610	0.952531	0.952453	0.952374	0.952296	0
0.412	0.952218	0.952139	0.952061	0.951982	0.951904	0.951825	0.951747	0.951668	0.951589	0.951511	0
0.413	0.951432	0.951354	0.951275	0.951196	0.951118	0.951039	0.950961	0.950882	0.950803	0.950724	0
0.414	0.950646	0.950567	0.950488	0.950410	0.950331	0.950252	0.950173	0.950094	0.950016	0.949937	0
0.415	0.949858	0.949779	0.949700	0.949621	0.949543	0.949464	0.949385	0.949306	0.949227	0.949148	0
0.416	0.949069	0.948990	0.948911	0.948832	0.948753	0.948674	0.948595	0.948516	0.948437	0.948358	0
0.417	0.948279	0.948200	0.948121	0.948041	0.947962	0.947883	0.947804	0.947725	0.947646	0.947566	0
0.418	0.947487	0.947408	0.947329	0.947250	0.947170	0.947091	0.947012	0.946932	0.946853	0.946774	0
0.419	0.946695	0.946615	0.946536	0.946456	0.946377	0.946298	0.946218	0.946139	0.946059	0.945980	0
0.420	0.945901	0.945821	0.945742	0.945662	0.945583	0.945503	0.945424	0.945344	0.945265	0.945185	0
0.421	0.945105	0.945026	0.944946	0.944867	0.944787	0.944707	0.944628	0.944548	0.944468	0.944389	0
0.422	0.944309	0.944229	0.944150	0.944070	0.943990	0.943910	0.943831	0.943751	0.943671	0.943591	0
0.423	0.943511	0.943432	0.943352	0.943272	0.943192	0.943112	0.943032	0.942952	0.942872	0.942793	0
0.424	0.942713	0.942633	0.942553	0.942473	0.942393	0.942313	0.942233	0.942153	0.942073	0.941993	0
0.425	0.941913	0.941833	0.941752	0.941672	0.941592	0.941512	0.941432	0.941352	0.941272	0.941192	0
0.426	0.941111	0.941031	0.940951	0.940871	0.940791	0.940710	0.940630	0.940550	0.940470	0.940389	0
0.427	0.940309	0.940229	0.940148	0.940068	0.939988	0.939907	0.939827	0.939747	0.939666	0.939586	0
0.428	0.939505	0.939425	0.939344	0.939264	0.939184	0.939103	0.939023	0.938942	0.938862	0.938781	0
0.429	0.938700	0.938620	0.938539	0.938459	0.938378	0.938298	0.938217	0.938136	0.938056	0.937975	0
0.430	0.937895	0.937814	0.937733	0.937652	0.937572	0.937491	0.937410	0.937330	0.937249	0.937168	0
0.431	0.937087	0.937007	0.936926	0.936845	0.936764	0.936683	0.936602	0.936522	0.936441	0.936360	0
0.432	0.936279	0.936198	0.936117	0.936036	0.935955	0.935874	0.935793	0.935712	0.935631	0.935550	0
0.433	0.935469	0.935388	0.935307	0.935226	0.935145	0.935064	0.934983	0.934902	0.934821	0.934740	0
0.434	0.934659	0.934578	0.934496	0.934415	0.934334	0.934253	0.934172	0.934091	0.934009	0.933928	0
0.435	0.933847	0.933766	0.933684	0.933603	0.933522	0.933441	0.933359	0.933278	0.933197	0.933115	0
0.436	0.933034	0.932952	0.932871	0.932790	0.932708	0.932627	0.932545	0.932464	0.932383	0.932301	0
0.437	0.932220	0.932138	0.932057	0.931975	0.931894	0.931812	0.931731	0.931649	0.931567	0.931486	0
0.438	0.931404	0.931323	0.931241	0.931159	0.931078	0.930996	0.930915	0.930833	0.930751	0.930670	0
0.439	0.930588	0.930506	0.930424	0.930343	0.930261	0.930179	0.930097	0.930016	0.929934	0.929852	0
0.440	0.929770	0.929688	0.929607	0.929525	0.929443	0.929361	0.929279	0.929197	0.929115	0.929033	0
0.441	0.928951	0.928870	0.928788	0.928706	0.928624	0.928542	0.928460	0.928378	0.928296	0.928214	0
0.442	0.928132	0.928050	0.927967	0.927885	0.927803	0.927721	0.927639	0.927557	0.927475	0.927393	0
0.443	0.927311	0.927228	0.927146	0.927064	0.926982	0.926900	0.926817	0.926735	0.926653	0.926571	0
0.444	0.926488	0.926406	0.926324	0.926242	0.926159	0.926077	0.925995	0.925912	0.925830	0.925747	0
0.445	0.925665	0.925583	0.925500	0.925418	0.925335	0.925253	0.925171	0.925088	0.925006	0.924923	0
0.446	0.924841	0.924758	0.924676	0.924593	0.924511	0.924428	0.924345	0.924263	0.924180	0.924098	0
0.447	0.924015	0.923933	0.923850	0.923767	0.923685	0.923602	0.923519	0.923437	0.923354	0.923271	0
0.448	0.923189	0.923106	0.923023	0.922940	0.922858	0.922775	0.922692	0.922609	0.922526	0.922444	0
0.449	0.922361	0.922278	0.922195	0.922112	0.922029	0.921947	0.921864	0.921781	0.921698	0.921615	0

TABLE 8

Error Function Derivative

X	.0	.0001	.0002	.0003	.0004	.0005	.0006	.0007	.0008	.0009	Δ_2
0.450	0.921532	0.921449	0.921366	0.921283	0.921200	0.921117	0.921034	0.920951	0.920868	0.920785	0
0.451	0.920702	0.920619	0.920536	0.920453	0.920370	0.920287	0.920204	0.920120	0.920037	0.919954	0
0.452	0.919871	0.919788	0.919705	0.919622	0.919538	0.919455	0.919372	0.919289	0.919205	0.919122	0
0.453	0.919039	0.918956	0.918872	0.918789	0.918706	0.918622	0.918539	0.918456	0.918372	0.918289	0
0.454	0.918206	0.918122	0.918039	0.917956	0.917872	0.917789	0.917705	0.917622	0.917538	0.917455	0
0.455	0.917371	0.917288	0.917204	0.917121	0.917037	0.916954	0.916870	0.916787	0.916703	0.916620	0
0.456	0.916536	0.916453	0.916369	0.916285	0.916202	0.916118	0.916034	0.915951	0.915867	0.915783	0
0.457	0.915700	0.915616	0.915532	0.915449	0.915365	0.915281	0.915197	0.915114	0.915030	0.914946	0
0.458	0.914862	0.914778	0.914695	0.914611	0.914527	0.914443	0.914359	0.914275	0.914191	0.914108	0
0.459	0.914024	0.913940	0.913856	0.913772	0.913688	0.913604	0.913520	0.913436	0.913352	0.913268	0
0.460	0.913184	0.913100	0.913016	0.912932	0.912848	0.912764	0.912680	0.912596	0.912512	0.912428	0
0.461	0.912343	0.912259	0.912175	0.912091	0.912007	0.911923	0.911839	0.911754	0.911670	0.911586	0
0.462	0.911502	0.911417	0.911333	0.911249	0.911165	0.911080	0.910996	0.910912	0.910828	0.910743	0
0.463	0.910659	0.910575	0.910490	0.910406	0.910322	0.910237	0.910153	0.910068	0.909984	0.909900	0
0.464	0.909815	0.909731	0.909646	0.909562	0.909477	0.909393	0.909308	0.909224	0.909139	0.909055	0
0.465	0.908970	0.908886	0.908801	0.908717	0.908632	0.908548	0.908463	0.908378	0.908294	0.908209	0
0.466	0.908125	0.908040	0.907955	0.907871	0.907786	0.907701	0.907616	0.907532	0.907447	0.907362	0
0.467	0.907278	0.907193	0.907108	0.907023	0.906939	0.906854	0.906769	0.906684	0.906599	0.906515	0
0.468	0.906430	0.906345	0.906260	0.906175	0.906090	0.906005	0.905920	0.905836	0.905751	0.905666	0
0.469	0.905581	0.905496	0.905411	0.905326	0.905241	0.905156	0.905071	0.904986	0.904901	0.904816	0
0.470	0.904731	0.904646	0.904561	0.904476	0.904391	0.904306	0.904220	0.904135	0.904050	0.903965	0
0.471	0.903880	0.903795	0.903710	0.903624	0.903539	0.903454	0.903369	0.903284	0.903198	0.903113	0
0.472	0.903028	0.902943	0.902857	0.902772	0.902687	0.902602	0.902516	0.902431	0.902346	0.902260	0
0.473	0.902175	0.902090	0.902004	0.901919	0.901834	0.901748	0.901663	0.901577	0.901492	0.901406	0
0.474	0.901321	0.901236	0.901150	0.901065	0.900979	0.900894	0.900808	0.900723	0.900637	0.900552	0
0.475	0.900466	0.900380	0.900295	0.900209	0.900124	0.900038	0.899953	0.899867	0.899781	0.899696	0
0.476	0.899610	0.899525	0.899439	0.899353	0.899267	0.899182	0.899096	0.899010	0.898925	0.898839	0
0.477	0.898753	0.898667	0.898582	0.898496	0.898410	0.898324	0.898239	0.898153	0.898067	0.897981	0
0.478	0.897895	0.897809	0.897724	0.897638	0.897552	0.897466	0.897380	0.897294	0.897208	0.897122	0
0.479	0.897036	0.896950	0.896865	0.896779	0.896693	0.896607	0.896521	0.896435	0.896349	0.896263	0
0.480	0.896177	0.896091	0.896004	0.895918	0.895832	0.895746	0.895660	0.895574	0.895488	0.895402	0
0.481	0.895316	0.895230	0.895144	0.895057	0.894971	0.894885	0.894799	0.894713	0.894626	0.894540	0
0.482	0.894454	0.894368	0.894282	0.894195	0.894109	0.894023	0.893936	0.893850	0.893764	0.893678	0
0.483	0.893591	0.893505	0.893419	0.893332	0.893246	0.893160	0.893073	0.892987	0.892900	0.892814	0
0.484	0.892728	0.892641	0.892555	0.892468	0.892382	0.892295	0.892209	0.892122	0.892036	0.891949	0
0.485	0.891863	0.891776	0.891690	0.891603	0.891517	0.891430	0.891344	0.891257	0.891171	0.891084	0
0.486	0.890997	0.890911	0.890824	0.890738	0.890651	0.890564	0.890478	0.890391	0.890304	0.890218	0
0.487	0.890131	0.890044	0.889957	0.889871	0.889784	0.889697	0.889610	0.889524	0.889437	0.889350	0
0.488	0.889263	0.889177	0.889090	0.889003	0.888916	0.888829	0.888743	0.888656	0.888569	0.888482	0
0.489	0.888395	0.888308	0.888221	0.888134	0.888047	0.887960	0.887874	0.887787	0.887700	0.887613	0
0.490	0.887526	0.887439	0.887352	0.887265	0.887178	0.887091	0.887004	0.886917	0.886830	0.886743	0
0.491	0.886655	0.886568	0.886481	0.886394	0.886307	0.886220	0.886133	0.886046	0.885959	0.885871	0
0.492	0.885784	0.885697	0.885610	0.885523	0.885436	0.885348	0.885261	0.885174	0.885087	0.885000	0
0.493	0.884912	0.884825	0.884738	0.884650	0.884563	0.884476	0.884389	0.884301	0.884214	0.884127	0
0.494	0.884039	0.883952	0.883865	0.883777	0.883690	0.883602	0.883515	0.883428	0.883340	0.883253	0
0.495	0.883165	0.883078	0.882990	0.882903	0.882816	0.882728	0.882641	0.882553	0.882466	0.882378	0
0.496	0.882291	0.882203	0.882116	0.882028	0.881940	0.881853	0.881765	0.881678	0.881590	0.881503	0
0.497	0.881415	0.881327	0.881240	0.881152	0.881064	0.880977	0.880889	0.880801	0.880714	0.880626	0
0.498	0.880538	0.880451	0.880363	0.880275	0.880187	0.880100	0.880012	0.879924	0.879836	0.879749	0
0.499	0.879661	0.879573	0.879485	0.879397	0.879310	0.879222	0.879134	0.879046	0.878958	0.878870	0

$$\frac{2}{\sqrt{\pi}} \exp(-x^2)$$

$\frac{2}{\sqrt{\pi}} \exp(-x^2)$

TABLE 8

Error Function Derivative

X	.0	.0001	.0002	.0003	.0004	.0005	.0006	.0007	.0008	.0009	Δ_2
0.500	0.878783	0.878695	0.878607	0.878519	0.878431	0.878343	0.878255	0.878167	0.878079	0.877991	0
0.501	0.877903	0.877815	0.877727	0.877639	0.877551	0.877463	0.877375	0.877287	0.877199	0.877111	0
0.502	0.877023	0.876935	0.876847	0.876759	0.876671	0.876583	0.876495	0.876407	0.876319	0.876230	0
0.503	0.876142	0.876054	0.875966	0.875878	0.875790	0.875701	0.875613	0.875525	0.875437	0.875349	0
0.504	0.875260	0.875172	0.875084	0.874996	0.874907	0.874819	0.874731	0.874643	0.874554	0.874466	0
0.505	0.874378	0.874289	0.874201	0.874113	0.874024	0.873936	0.873848	0.873759	0.873671	0.873583	0
0.506	0.873494	0.873406	0.873317	0.873229	0.873141	0.873052	0.872964	0.872875	0.872787	0.872698	0
0.507	0.872610	0.872521	0.872433	0.872344	0.872256	0.872167	0.872079	0.871990	0.871902	0.871813	0
0.508	0.871725	0.871636	0.871547	0.871459	0.871370	0.871282	0.871193	0.871104	0.871016	0.870927	0
0.509	0.870838	0.870750	0.870661	0.870572	0.870484	0.870395	0.870306	0.870218	0.870129	0.870040	0
0.510	0.869952	0.869863	0.869774	0.869685	0.869596	0.869508	0.869419	0.869330	0.869241	0.869153	0
0.511	0.869064	0.868975	0.868886	0.868797	0.868708	0.868620	0.868531	0.868442	0.868353	0.868264	0
0.512	0.868175	0.868086	0.867997	0.867908	0.867819	0.867731	0.867642	0.867553	0.867464	0.867375	0
0.513	0.867286	0.867197	0.867108	0.867019	0.866930	0.866841	0.866752	0.866663	0.866574	0.866485	0
0.514	0.866395	0.866306	0.866217	0.866128	0.866039	0.865950	0.865861	0.865772	0.865683	0.865594	0
0.515	0.865504	0.865415	0.865326	0.865237	0.865148	0.865059	0.864969	0.864880	0.864791	0.864702	0
0.516	0.864613	0.864523	0.864434	0.864345	0.864256	0.864166	0.864077	0.863988	0.863898	0.863809	0
0.517	0.863720	0.863631	0.863541	0.863452	0.863363	0.863273	0.863184	0.863095	0.863005	0.862916	0
0.518	0.862826	0.862737	0.862648	0.862558	0.862469	0.862379	0.862290	0.862200	0.862111	0.862022	0
0.519	0.861932	0.861843	0.861753	0.861664	0.861574	0.861485	0.861395	0.861306	0.861216	0.861127	0
0.520	0.861037	0.860947	0.860858	0.860768	0.860679	0.860589	0.860500	0.860410	0.860320	0.860231	0
0.521	0.860141	0.860052	0.859962	0.859872	0.859783	0.859693	0.859603	0.859514	0.859424	0.859334	0
0.522	0.859244	0.859155	0.859065	0.858975	0.858886	0.858796	0.858706	0.858616	0.858527	0.858437	0
0.523	0.858347	0.858257	0.858167	0.858078	0.857988	0.857898	0.857808	0.857718	0.857629	0.857539	0
0.524	0.857449	0.857359	0.857269	0.857179	0.857089	0.856999	0.856910	0.856820	0.856730	0.856640	0
0.525	0.856550	0.856460	0.856370	0.856280	0.856190	0.856100	0.856010	0.855920	0.855830	0.855740	0
0.526	0.855650	0.855560	0.855470	0.855380	0.855290	0.855200	0.855110	0.855020	0.854930	0.854840	0
0.527	0.854750	0.854659	0.854569	0.854479	0.854389	0.854299	0.854209	0.854119	0.854029	0.853938	0
0.528	0.853848	0.853758	0.853668	0.853578	0.853488	0.853397	0.853307	0.853217	0.853127	0.853036	0
0.529	0.852946	0.852856	0.852766	0.852675	0.852585	0.852495	0.852405	0.852314	0.852224	0.852134	0
0.530	0.852043	0.851953	0.851863	0.851772	0.851682	0.851592	0.851501	0.851411	0.851321	0.851230	0
0.531	0.851140	0.851049	0.850959	0.850869	0.850778	0.850688	0.850597	0.850507	0.850416	0.850326	0
0.532	0.850236	0.850145	0.850055	0.849964	0.849874	0.849783	0.849693	0.849602	0.849512	0.849421	0
0.533	0.849331	0.849240	0.849149	0.849059	0.848968	0.848878	0.848787	0.848697	0.848606	0.848515	0
0.534	0.848425	0.848334	0.848244	0.848153	0.848062	0.847972	0.847881	0.847790	0.847700	0.847609	0
0.535	0.847518	0.847428	0.847337	0.847246	0.847156	0.847065	0.846974	0.846883	0.846793	0.846702	0
0.536	0.846611	0.846520	0.846430	0.846339	0.846248	0.846157	0.846066	0.845976	0.845885	0.845794	0
0.537	0.845703	0.845612	0.845522	0.845431	0.845340	0.845249	0.845158	0.845067	0.844976	0.844885	0
0.538	0.844795	0.844704	0.844613	0.844522	0.844431	0.844340	0.844249	0.844158	0.844067	0.843976	0
0.539	0.843885	0.843794	0.843703	0.843612	0.843521	0.843430	0.843339	0.843248	0.843157	0.843066	0
0.540	0.842975	0.842884	0.842793	0.842702	0.842611	0.842520	0.842429	0.842338	0.842247	0.842156	0
0.541	0.842064	0.841973	0.841882	0.841791	0.841700	0.841609	0.841518	0.841426	0.841335	0.841244	0
0.542	0.841153	0.841062	0.840971	0.840879	0.840788	0.840697	0.840606	0.840514	0.840423	0.840332	0
0.543	0.840241	0.840150	0.840058	0.839967	0.839876	0.839784	0.839693	0.839602	0.839511	0.839419	0
0.544	0.839328	0.839237	0.839145	0.839054	0.838963	0.838871	0.838780	0.838689	0.838597	0.838506	0
0.545	0.838414	0.838323	0.838232	0.838140	0.838049	0.837957	0.837866	0.837775	0.837683	0.837592	0
0.546	0.837500	0.837409	0.837317	0.837226	0.837134	0.837043	0.836951	0.836860	0.836768	0.836677	0
0.547	0.8365 5	0.836494	0.836402	0.836311	0.836219	0.836128	0.836036	0.835944	0.835853	0.835761	0
0.548	0.835670	0.835578	0.835487	0.835395	0.835303	0.835212	0.835120	0.835028	0.834937	0.834845	0
0.549	0.834754	0.834662	0.834570	0.834478	0.834387	0.834295	0.834203	0.834112	0.834020	0.833928	0

TABLE 8

Error Function Derivative

X	.0	.0001	.0002	.0003	.0004	.0005	.0006	.0007	.0008	.0009	Δ_2
0.550	0.833837	0.833745	0.833653	0.833561	0.833470	0.833378	0.833286	0.833194	0.833103	0.833011	0
0.551	0.832919	0.832827	0.832735	0.832644	0.832552	0.832460	0.832368	0.832276	0.832185	0.832093	0
0.552	0.832001	0.831909	0.831817	0.831725	0.831633	0.831542	0.831450	0.831358	0.831266	0.831174	0
0.553	0.831082	0.830990	0.830898	0.830806	0.830714	0.830622	0.830530	0.830438	0.830346	0.830254	0
0.554	0.830163	0.830071	0.829979	0.829887	0.829795	0.829702	0.829611	0.829518	0.829426	0.829334	0
0.555	0.829242	0.829150	0.829058	0.828966	0.828874	0.828782	0.828690	0.828598	0.828506	0.828414	0
0.556	0.828322	0.828229	0.828137	0.828045	0.827953	0.827861	0.827769	0.827677	0.827585	0.827492	0
0.557	0.827400	0.827308	0.827216	0.827124	0.827031	0.826939	0.826847	0.826755	0.826663	0.826570	0
0.558	0.826478	0.826386	0.826294	0.826201	0.826109	0.826017	0.825925	0.825832	0.825740	0.825648	0
0.559	0.825556	0.825463	0.825371	0.825279	0.825186	0.825094	0.825002	0.824909	0.824817	0.824725	0
0.560	0.824632	0.824540	0.824447	0.824355	0.824263	0.824170	0.824078	0.823986	0.823893	0.823801	0
0.561	0.823708	0.823616	0.823523	0.823431	0.823339	0.823246	0.823154	0.823061	0.822969	0.822876	0
0.562	0.822784	0.822691	0.822599	0.822506	0.822414	0.822321	0.822229	0.822136	0.822044	0.821951	0
0.563	0.821859	0.821766	0.821674	0.821581	0.821488	0.821396	0.821303	0.821211	0.821118	0.821026	0
0.564	0.820933	0.820840	0.820748	0.820655	0.820563	0.820470	0.820377	0.820285	0.820192	0.820099	0
0.565	0.820007	0.819914	0.819821	0.819729	0.819636	0.819543	0.819451	0.819358	0.819265	0.819173	0
0.566	0.819080	0.818987	0.818894	0.818802	0.818709	0.818616	0.818523	0.818431	0.818338	0.818245	0
0.567	0.818152	0.818060	0.817967	0.817874	0.817781	0.817688	0.817595	0.817503	0.817410	0.817317	0
0.568	0.817224	0.817131	0.817039	0.816946	0.816853	0.816760	0.816667	0.816574	0.816481	0.816388	0
0.569	0.816296	0.816203	0.816110	0.816017	0.815924	0.815831	0.815738	0.815645	0.815552	0.815459	0
0.570	0.815366	0.815273	0.815180	0.815087	0.814994	0.814901	0.814808	0.814716	0.814623	0.814530	0
0.571	0.814436	0.814344	0.814250	0.814157	0.814064	0.813971	0.813878	0.813785	0.813692	0.813599	0
0.572	0.813506	0.813413	0.813320	0.813227	0.813134	0.813041	0.812948	0.812855	0.812761	0.812668	0
0.573	0.812575	0.812482	0.812389	0.812296	0.812203	0.812110	0.812016	0.811923	0.811830	0.811737	0
0.574	0.811644	0.811551	0.811457	0.811364	0.811271	0.811178	0.811085	0.810991	0.810898	0.810805	0
0.575	0.810712	0.810618	0.810525	0.810432	0.810339	0.810245	0.810152	0.810059	0.809966	0.809872	0
0.576	0.809779	0.809686	0.809592	0.809499	0.809406	0.809313	0.809219	0.809126	0.809033	0.808939	0
0.577	0.808846	0.808753	0.808659	0.808566	0.808473	0.808379	0.808286	0.808192	0.808099	0.808006	0
0.578	0.807912	0.807819	0.807725	0.807632	0.807539	0.807445	0.807352	0.807258	0.807165	0.807072	0
0.579	0.806978	0.806885	0.806791	0.806698	0.806604	0.806511	0.806417	0.806324	0.806230	0.806137	0
0.580	0.806043	0.805950	0.805856	0.805763	0.805669	0.805576	0.805482	0.805389	0.805295	0.805202	0
0.581	0.805108	0.805014	0.804921	0.804827	0.804734	0.804640	0.804547	0.804453	0.804359	0.804266	0
0.582	0.804172	0.804079	0.803985	0.803891	0.803798	0.803704	0.803611	0.803517	0.803423	0.803330	0
0.583	0.803236	0.803142	0.803049	0.802955	0.802861	0.802768	0.802674	0.802580	0.802486	0.802393	0
0.584	0.802299	0.802205	0.802112	0.802018	0.801924	0.801830	0.801737	0.801643	0.801549	0.801455	0
0.585	0.801362	0.801268	0.801174	0.801080	0.800987	0.800893	0.800799	0.800705	0.800611	0.800518	0
0.586	0.800424	0.800330	0.800236	0.800142	0.800049	0.799955	0.799861	0.799767	0.799673	0.799579	0
0.587	0.799486	0.799392	0.799298	0.799204	0.799110	0.799016	0.798922	0.798828	0.798735	0.798641	0
0.588	0.798547	0.798453	0.798359	0.798265	0.798171	0.798077	0.797983	0.797889	0.797795	0.797701	0
0.589	0.797607	0.797513	0.797419	0.797325	0.797231	0.797138	0.797044	0.796950	0.796856	0.796762	0
0.590	0.796668	0.796574	0.796480	0.796386	0.796291	0.796197	0.796103	0.796009	0.795915	0.795821	0
0.591	0.795727	0.795633	0.795539	0.795445	0.795351	0.795257	0.795163	0.795069	0.794975	0.794881	0
0.592	0.794786	0.794692	0.794598	0.794504	0.794410	0.794316	0.794222	0.794128	0.794033	0.793939	0
0.593	0.793845	0.793751	0.793657	0.793563	0.793469	0.793374	0.793280	0.793186	0.793092	0.792998	0
0.594	0.792903	0.792809	0.792715	0.792621	0.792527	0.792432	0.792338	0.792244	0.792150	0.792055	0
0.595	0.791961	0.791867	0.791773	0.791678	0.791584	0.791490	0.791396	0.791301	0.791207	0.791113	0
0.596	0.791019	0.790924	0.790830	0.790736	0.790641	0.790547	0.790453	0.790358	0.790264	0.790170	0
0.597	0.790075	0.789981	0.789887	0.789792	0.789698	0.789604	0.789509	0.789415	0.789321	0.789226	0
0.598	0.789132	0.789038	0.788943	0.788849	0.788754	0.788660	0.788566	0.788471	0.788377	0.788282	0
0.599	0.788188	0.788093	0.787999	0.787905	0.787810	0.787716	0.787621	0.787527	0.787432	0.787338	0

$\frac{2}{\sqrt{\pi}}\exp(-x^2)$

TABLE 8

Error Function Derivative

X	.0	.0001	.0002	.0003	.0004	.0005	.0006	.0007	.0008	.0009	Δ_2
0.600	0.787243	0.787149	0.787054	0.786960	0.786865	0.786771	0.786677	0.786582	0.786488	0.786393	0
0.601	0.786299	0.786204	0.786109	0.786015	0.785920	0.785826	0.785731	0.785637	0.785542	0.785448	0
0.602	0.785353	0.785259	0.785164	0.785069	0.784975	0.784880	0.784786	0.784691	0.784597	0.784502	0
0.603	0.784407	0.784313	0.784218	0.784124	0.784029	0.783934	0.783840	0.783745	0.783650	0.783556	0
0.604	0.783461	0.783367	0.783272	0.783177	0.783083	0.782988	0.782893	0.782799	0.782704	0.782609	0
0.605	0.782515	0.782420	0.782325	0.782230	0.782136	0.782041	0.781946	0.781852	0.781757	0.781662	0
0.606	0.781567	0.781473	0.781378	0.781283	0.781189	0.781094	0.780999	0.780904	0.780810	0.780715	0
0.607	0.780620	0.780525	0.780430	0.780336	0.780241	0.780146	0.780051	0.779957	0.779862	0.779767	0
0.608	0.779672	0.779577	0.779482	0.779388	0.779293	0.779198	0.779103	0.779008	0.778914	0.778819	0
0.609	0.778724	0.778629	0.778534	0.778439	0.778344	0.778250	0.778155	0.778060	0.777965	0.777870	0
0.610	0.777775	0.777680	0.777585	0.777490	0.777396	0.777301	0.777206	0.777111	0.777016	0.776921	0
0.611	0.776826	0.776731	0.776636	0.776541	0.776446	0.776351	0.776256	0.776161	0.776067	0.775972	0
0.612	0.775877	0.775782	0.775687	0.775592	0.775497	0.775402	0.775307	0.775212	0.775117	0.775022	0
0.613	0.774927	0.774832	0.774737	0.774642	0.774547	0.774452	0.774357	0.774262	0.774167	0.774072	0
0.614	0.773976	0.773881	0.773786	0.773691	0.773596	0.773501	0.773406	0.773311	0.773216	0.773121	0
0.615	0.773026	0.772931	0.772836	0.772741	0.772645	0.772550	0.772455	0.772360	0.772265	0.772170	0
0.616	0.772075	0.771980	0.771885	0.771789	0.771694	0.771599	0.771504	0.771409	0.771314	0.771219	0
0.617	0.771123	0.771028	0.770933	0.770838	0.770743	0.770648	0.770552	0.770457	0.770362	0.770267	0
0.618	0.770172	0.770077	0.769981	0.769886	0.769791	0.769696	0.769600	0.769505	0.769410	0.769315	0
0.619	0.769220	0.769124	0.769029	0.768934	0.768839	0.768743	0.768648	0.768553	0.768458	0.768362	0
0.620	0.768267	0.768172	0.768077	0.767981	0.767886	0.767791	0.767695	0.767600	0.767505	0.767410	0
0.621	0.767314	0.767219	0.767124	0.767028	0.766933	0.766838	0.766742	0.766647	0.766552	0.766456	0
0.622	0.766361	0.766266	0.766170	0.766075	0.765980	0.765884	0.765789	0.765694	0.765598	0.765503	0
0.623	0.765408	0.765312	0.765217	0.765121	0.765026	0.764931	0.764835	0.764740	0.764645	0.764549	0
0.624	0.764454	0.764358	0.764263	0.764167	0.764072	0.763977	0.763881	0.763786	0.763690	0.763595	0
0.625	0.763499	0.763404	0.763309	0.763213	0.763118	0.763022	0.762927	0.762831	0.762736	0.762640	0
0.626	0.762545	0.762450	0.762354	0.762259	0.762163	0.762068	0.761972	0.761877	0.761781	0.761686	0
0.627	0.761590	0.761495	0.761399	0.761304	0.761208	0.761113	0.761017	0.760921	0.760826	0.760730	0
0.628	0.760635	0.760539	0.760444	0.760348	0.760253	0.760157	0.760062	0.759966	0.759871	0.759775	0
0.629	0.759679	0.759584	0.759488	0.759393	0.759297	0.759202	0.759106	0.759010	0.758915	0.758819	0
0.630	0.758724	0.758628	0.758532	0.758437	0.758341	0.758246	0.758150	0.758054	0.757959	0.757863	0
0.631	0.757767	0.757672	0.757576	0.757481	0.757385	0.757289	0.757194	0.757098	0.757002	0.756907	0
0.632	0.756811	0.756715	0.756620	0.756524	0.756428	0.756333	0.756237	0.756141	0.756046	0.755950	0
0.633	0.755854	0.755758	0.755663	0.755567	0.755471	0.755376	0.755280	0.755184	0.755089	0.754993	0
0.634	0.754897	0.754801	0.754706	0.754610	0.754514	0.754418	0.754323	0.754227	0.754131	0.754036	0
0.635	0.753940	0.753844	0.753748	0.753653	0.753557	0.753461	0.753365	0.753269	0.753174	0.753078	0
0.636	0.752982	0.752886	0.752791	0.752695	0.752599	0.752503	0.752407	0.752312	0.752216	0.752120	0
0.637	0.752024	0.751928	0.751833	0.751737	0.751641	0.751545	0.751449	0.751353	0.751258	0.751162	0
0.638	0.751066	0.750970	0.750874	0.750778	0.750683	0.750587	0.750491	0.750395	0.750299	0.750203	0
0.639	0.750107	0.750012	0.749916	0.749820	0.749724	0.749628	0.749532	0.749436	0.749340	0.749245	0
0.640	0.749149	0.749053	0.748957	0.748861	0.748765	0.748669	0.748573	0.748477	0.748381	0.748286	0
0.641	0.748190	0.748094	0.747998	0.747902	0.747806	0.747710	0.747614	0.747518	0.747422	0.747326	0
0.642	0.747230	0.747134	0.747038	0.746942	0.746847	0.746751	0.746655	0.746559	0.746463	0.746367	0
0.643	0.746271	0.746175	0.746079	0.745983	0.745887	0.745791	0.745695	0.745599	0.745503	0.745407	0
0.644	0.745311	0.745215	0.745119	0.745023	0.744927	0.744831	0.744735	0.744639	0.744543	0.744447	0
0.645	0.744351	0.744255	0.744159	0.744063	0.743967	0.743871	0.743775	0.743679	0.743583	0.743487	0
0.646	0.743391	0.743294	0.743198	0.743102	0.743006	0.742910	0.742814	0.742718	0.742622	0.742526	0
0.647	0.742430	0.742334	0.742238	0.742142	0.742046	0.741950	0.741853	0.741757	0.741661	0.741565	0
0.648	0.741469	0.741373	0.741277	0.741181	0.741085	0.740989	0.740892	0.740796	0.740700	0.740604	0
0.649	0.740508	0.740412	0.740316	0.740220	0.740124	0.740027	0.739931	0.739835	0.739739	0.739643	0

TABLE 8

Error Function Derivative

X	.0	.0001	.0002	.0003	.0004	.0005	.0006	.0007	.0008	.0009	Δ_2
0.650	0.739547	0.739451	0.739354	0.739258	0.739162	0.739066	0.738970	0.738874	0.738778	0.738681	0
0.651	0.738585	0.738489	0.738393	0.738297	0.738201	0.738104	0.738008	0.737912	0.737816	0.737720	0
0.652	0.737623	0.737527	0.737431	0.737335	0.737239	0.737143	0.737046	0.736950	0.736854	0.736758	0
0.653	0.736661	0.736565	0.736469	0.736373	0.736277	0.736180	0.736084	0.735988	0.735892	0.735796	0
0.654	0.735699	0.735603	0.735507	0.735411	0.735314	0.735218	0.735122	0.735026	0.734929	0.734833	0
0.655	0.734737	0.734641	0.734544	0.734448	0.734352	0.734256	0.734159	0.734063	0.733967	0.733871	0
0.656	0.733774	0.733678	0.733582	0.733485	0.733389	0.733293	0.733197	0.733100	0.733004	0.732908	0
0.657	0.732811	0.732715	0.732619	0.732523	0.732426	0.732330	0.732234	0.732137	0.732041	0.731945	0
0.658	0.731848	0.731752	0.731656	0.731560	0.731463	0.731367	0.731271	0.731174	0.731078	0.730982	0
0.659	0.730885	0.730789	0.730693	0.730596	0.730500	0.730404	0.730307	0.730211	0.730115	0.730018	0
0.660	0.729922	0.729825	0.729729	0.729633	0.729536	0.729440	0.729344	0.729247	0.729151	0.729055	0
0.661	0.728958	0.728862	0.728766	0.728669	0.728573	0.728476	0.728380	0.728284	0.728187	0.728091	0
0.662	0.727995	0.727898	0.727802	0.727705	0.727609	0.727513	0.727416	0.727320	0.727223	0.727127	0
0.663	0.727031	0.726934	0.726838	0.726741	0.726645	0.726548	0.726452	0.726356	0.726259	0.726163	0
0.664	0.726066	0.725970	0.725874	0.725777	0.725681	0.725584	0.725488	0.725391	0.725295	0.725199	0
0.665	0.725102	0.725006	0.724909	0.724813	0.724716	0.724620	0.724523	0.724427	0.724331	0.724234	0
0.666	0.724138	0.724041	0.723945	0.723848	0.723752	0.723655	0.723559	0.723462	0.723366	0.723269	0
0.667	0.723173	0.723077	0.722980	0.722884	0.722787	0.722691	0.722594	0.722498	0.722401	0.722305	0
0.668	0.722208	0.722112	0.722015	0.721919	0.721822	0.721726	0.721629	0.721533	0.721436	0.721340	0
0.669	0.721243	0.721147	0.721050	0.720954	0.720857	0.720761	0.720664	0.720568	0.720471	0.720375	0
0.670	0.720278	0.720182	0.720085	0.719989	0.719892	0.719796	0.719699	0.719603	0.719506	0.719409	0
0.671	0.719313	0.719216	0.719120	0.719023	0.718927	0.718830	0.718734	0.718637	0.718541	0.718444	0
0.672	0.718348	0.718251	0.718154	0.718058	0.717961	0.717865	0.717768	0.717672	0.717575	0.717479	0
0.673	0.717382	0.717285	0.717189	0.717092	0.716996	0.716899	0.716803	0.716706	0.716609	0.716513	0
0.674	0.716416	0.716320	0.716223	0.716127	0.716030	0.715933	0.715837	0.715740	0.715644	0.715547	0
0.675	0.715451	0.715354	0.715257	0.715161	0.715064	0.714968	0.714871	0.714774	0.714678	0.714581	0
0.676	0.714485	0.714388	0.714291	0.714195	0.714098	0.714002	0.713905	0.713808	0.713712	0.713615	0
0.677	0.713519	0.713422	0.713325	0.713229	0.713132	0.713036	0.712939	0.712842	0.712746	0.712649	0
0.678	0.712552	0.712456	0.712359	0.712263	0.712166	0.712069	0.711973	0.711876	0.711779	0.711683	0
0.679	0.711586	0.711489	0.711393	0.711296	0.711200	0.711103	0.711006	0.710910	0.710813	0.710716	0
0.680	0.710620	0.710523	0.710426	0.710330	0.710233	0.710137	0.710040	0.709943	0.709847	0.709750	0
0.681	0.709653	0.709557	0.709460	0.709363	0.709267	0.709170	0.709073	0.708977	0.708880	0.708783	0
0.682	0.708687	0.708590	0.708493	0.708397	0.708300	0.708203	0.708107	0.708010	0.707913	0.707817	0
0.683	0.707720	0.707623	0.707527	0.707430	0.707333	0.707237	0.707140	0.707043	0.706947	0.706850	0
0.684	0.706753	0.706656	0.706560	0.706463	0.706366	0.706270	0.706173	0.706076	0.705980	0.705883	0
0.685	0.705786	0.705690	0.705593	0.705496	0.705400	0.705303	0.705206	0.705109	0.705013	0.704916	0
0.686	0.704819	0.704723	0.704626	0.704529	0.704432	0.704336	0.704239	0.704142	0.704046	0.703949	0
0.687	0.703852	0.703756	0.703659	0.703562	0.703465	0.703369	0.703272	0.703175	0.703079	0.702982	0
0.688	0.702885	0.702788	0.702692	0.702595	0.702498	0.702402	0.702305	0.702208	0.702111	0.702015	0
0.689	0.701918	0.701821	0.701724	0.701628	0.701531	0.701434	0.701338	0.701241	0.701144	0.701047	0
0.690	0.700951	0.700854	0.700757	0.700660	0.700564	0.700467	0.700370	0.700274	0.700177	0.700080	0
0.691	0.699983	0.699887	0.699790	0.699693	0.699596	0.699500	0.699403	0.699306	0.699209	0.699113	0
0.692	0.699016	0.698919	0.698822	0.698726	0.698629	0.698532	0.698435	0.698339	0.698242	0.698145	0
0.693	0.698048	0.697952	0.697855	0.697758	0.697661	0.697565	0.697468	0.697371	0.697274	0.697178	0
0.694	0.697081	0.696984	0.696887	0.696791	0.696694	0.696597	0.696500	0.696404	0.696307	0.696210	0
0.695	0.696113	0.696017	0.695920	0.695823	0.695726	0.695630	0.695533	0.695436	0.695339	0.695242	0
0.696	0.695146	0.695049	0.694952	0.694855	0.694759	0.694662	0.694565	0.694468	0.694372	0.694275	0
0.697	0.694178	0.694081	0.693985	0.693888	0.693791	0.693694	0.693597	0.693501	0.693404	0.693307	0
0.698	0.693210	0.693114	0.693017	0.692920	0.692823	0.692726	0.692630	0.692533	0.692436	0.692339	0
0.699	0.692243	0.692146	0.692049	0.691952	0.691855	0.691759	0.691662	0.691565	0.691468	0.691372	0

$$\frac{2}{\sqrt{\pi}} \exp(-x^2)$$

$$\frac{2}{\sqrt{\pi}}\exp(-x^2)$$

TABLE 8

Error Function Derivative

X	.0	.0001	.0002	.0003	.0004	.0005	.0006	.0007	.0008	.0009	Δ_2
0.700	0.691275	0.691178	0.691081	0.690984	0.690888	0.690791	0.690694	0.690597	0.690501	0.690404	0
0.701	0.690307	0.690210	0.690113	0.690017	0.689920	0.689823	0.689726	0.689630	0.689533	0.689436	0
0.702	0.689339	0.689242	0.689146	0.689049	0.688952	0.688855	0.688758	0.688662	0.688565	0.688468	0
0.703	0.688371	0.688275	0.688178	0.688081	0.687984	0.687887	0.687791	0.687694	0.687597	0.687500	0
0.704	0.687404	0.687307	0.687210	0.687113	0.687016	0.686920	0.686823	0.686726	0.686629	0.686532	0
0.705	0.686436	0.686339	0.686242	0.686145	0.686049	0.685952	0.685855	0.685758	0.685661	0.685565	0
0.706	0.685468	0.685371	0.685274	0.685177	0.685081	0.684984	0.684887	0.684790	0.684693	0.684597	0
0.707	0.684500	0.684403	0.684306	0.684210	0.684113	0.684016	0.683919	0.683822	0.683726	0.683629	0
0.708	0.683532	0.683435	0.683338	0.683242	0.683145	0.683048	0.682951	0.682854	0.682758	0.682661	0
0.709	0.682564	0.682467	0.682371	0.682274	0.682177	0.682080	0.681983	0.681887	0.681790	0.681693	0
0.710	0.681596	0.681499	0.681403	0.681306	0.681209	0.681112	0.681016	0.680919	0.680822	0.680725	0
0.711	0.680628	0.680532	0.680435	0.680338	0.680241	0.680144	0.680048	0.679951	0.679854	0.679757	0
0.712	0.679661	0.679564	0.679467	0.679370	0.679273	0.679177	0.679080	0.678983	0.678886	0.678789	0
0.713	0.678693	0.678596	0.678499	0.678402	0.678306	0.678209	0.678112	0.678015	0.677918	0.677822	0
0.714	0.677725	0.677628	0.677531	0.677435	0.677338	0.677241	0.677144	0.677047	0.676951	0.676854	0
0.715	0.676757	0.676660	0.676564	0.676467	0.676370	0.676273	0.676176	0.676080	0.675983	0.675886	0
0.716	0.675789	0.675693	0.675596	0.675499	0.675402	0.675306	0.675209	0.675112	0.675015	0.674918	0
0.717	0.674822	0.674725	0.674628	0.674531	0.674435	0.674338	0.674241	0.674144	0.674048	0.673951	0
0.718	0.673854	0.673757	0.673660	0.673564	0.673467	0.673370	0.673273	0.673177	0.673080	0.672983	0
0.719	0.672886	0.672790	0.672693	0.672596	0.672499	0.672403	0.672306	0.672209	0.672112	0.672016	0
0.720	0.671919	0.671822	0.671725	0.671629	0.671532	0.671435	0.671338	0.671242	0.671145	0.671048	0
0.721	0.670951	0.670855	0.670758	0.670661	0.670564	0.670467	0.670371	0.670274	0.670177	0.670080	0
0.722	0.669984	0.669887	0.669790	0.669694	0.669597	0.669500	0.669403	0.669307	0.669210	0.669113	0
0.723	0.669016	0.668920	0.668823	0.668726	0.668629	0.668533	0.668436	0.668339	0.668242	0.668146	0
0.724	0.668049	0.667952	0.667856	0.667759	0.667662	0.667565	0.667469	0.667372	0.667275	0.667178	0
0.725	0.667082	0.666985	0.666888	0.666791	0.666695	0.666598	0.666501	0.666405	0.666308	0.666211	0
0.726	0.666114	0.666018	0.665921	0.665824	0.665728	0.665631	0.665534	0.665437	0.665341	0.665244	0
0.727	0.665147	0.665051	0.664954	0.664857	0.664760	0.664664	0.664567	0.664470	0.664374	0.664277	0
0.728	0.664180	0.664083	0.663987	0.663890	0.663793	0.663697	0.663600	0.663503	0.663407	0.663310	0
0.729	0.663213	0.663116	0.663020	0.662923	0.662826	0.662730	0.662633	0.662536	0.662440	0.662343	0
0.730	0.662246	0.662150	0.662053	0.661956	0.661860	0.661763	0.661666	0.661569	0.661473	0.661376	0
0.731	0.661279	0.661183	0.661086	0.660989	0.660893	0.660796	0.660699	0.660603	0.660506	0.660409	0
0.732	0.660313	0.660216	0.660119	0.660023	0.659926	0.659829	0.659733	0.659636	0.659539	0.659443	0
0.733	0.659346	0.659249	0.659153	0.659056	0.658959	0.658863	0.658766	0.658669	0.658573	0.658476	0
0.734	0.658379	0.658283	0.658186	0.658090	0.657993	0.657896	0.657800	0.657703	0.657606	0.657510	0
0.735	0.657413	0.657316	0.657220	0.657123	0.657026	0.656930	0.656833	0.656737	0.656640	0.656543	0
0.736	0.656447	0.656350	0.656253	0.656157	0.656060	0.655964	0.655867	0.655770	0.655674	0.655577	0
0.737	0.655480	0.655384	0.655287	0.655191	0.655094	0.654997	0.654901	0.654804	0.654708	0.654611	0
0.738	0.654514	0.654418	0.654321	0.654225	0.654128	0.654031	0.653935	0.653838	0.653742	0.653645	0
0.739	0.653548	0.653452	0.653355	0.653259	0.653162	0.653065	0.652969	0.652872	0.652776	0.652679	0
0.740	0.652582	0.652486	0.652389	0.652293	0.652196	0.652100	0.652003	0.651906	0.651810	0.651713	0
0.741	0.651617	0.651520	0.651424	0.651327	0.651230	0.651134	0.651037	0.650941	0.650844	0.650748	0
0.742	0.650651	0.650554	0.650458	0.650361	0.650265	0.650168	0.650072	0.649975	0.649879	0.649782	0
0.743	0.649686	0.649589	0.649492	0.649396	0.649299	0.649203	0.649106	0.649010	0.648913	0.648817	0
0.744	0.648720	0.648624	0.648527	0.648431	0.648334	0.648238	0.648141	0.648045	0.647948	0.647851	0
0.745	0.647755	0.647658	0.647562	0.647465	0.647369	0.647272	0.647176	0.647079	0.646983	0.646886	0
0.746	0.646790	0.646693	0.646597	0.646500	0.646404	0.646307	0.646211	0.646114	0.646018	0.645921	0
0.747	0.645825	0.645728	0.645632	0.645535	0.645439	0.645343	0.645246	0.645150	0.645053	0.644957	0
0.748	0.644860	0.644764	0.644667	0.644571	0.644474	0.644378	0.644281	0.644185	0.644088	0.643992	0
0.749	0.643896	0.643799	0.643703	0.643606	0.643510	0.643413	0.643317	0.643220	0.643124	0.643027	0

TABLE 8

Error Function Derivative

X	.0	.0001	.0002	.0003	.0004	.0005	.0006	.0007	.0008	.0009	Δ_2
0.750	0.642931	0.642835	0.642738	0.642642	0.642545	0.642449	0.642352	0.642256	0.642160	0.642063	0
0.751	0.641967	0.641870	0.641774	0.641677	0.641581	0.641485	0.641388	0.641292	0.641195	0.641099	0
0.752	0.641003	0.640906	0.640810	0.640713	0.640617	0.640521	0.640424	0.640328	0.640231	0.640135	0
0.753	0.640039	0.639942	0.639846	0.639749	0.639653	0.639557	0.639460	0.639364	0.639268	0.639171	0
0.754	0.639075	0.638978	0.638882	0.638786	0.638689	0.638593	0.638497	0.638400	0.638304	0.638207	0
0.755	0.638111	0.638015	0.637918	0.637822	0.637726	0.637629	0.637533	0.637437	0.637340	0.637244	0
0.756	0.637148	0.637051	0.636955	0.636859	0.636762	0.636666	0.636570	0.636473	0.636377	0.636281	0
0.757	0.636184	0.636088	0.635992	0.635895	0.635799	0.635703	0.635607	0.635510	0.635414	0.635318	0
0.758	0.635221	0.635125	0.635029	0.634932	0.634836	0.634740	0.634644	0.634547	0.634451	0.634355	0
0.759	0.634258	0.634162	0.634066	0.633970	0.633873	0.633777	0.633681	0.633584	0.633488	0.633392	0
0.760	0.633296	0.633199	0.633103	0.633007	0.632911	0.632814	0.632718	0.632622	0.632526	0.632429	0
0.761	0.632333	0.632237	0.632141	0.632044	0.631948	0.631852	0.631756	0.631660	0.631563	0.631467	0
0.762	0.631371	0.631275	0.631178	0.631082	0.630986	0.630890	0.630794	0.630697	0.630601	0.630505	0
0.763	0.630409	0.630313	0.630216	0.630120	0.630024	0.629928	0.629832	0.629735	0.629639	0.629543	0
0.764	0.629447	0.629351	0.629255	0.629158	0.629062	0.628966	0.628870	0.628774	0.628677	0.628581	0
0.765	0.628485	0.628389	0.628293	0.628197	0.628101	0.628004	0.627908	0.627812	0.627716	0.627620	0
0.766	0.627524	0.627428	0.627331	0.627235	0.627139	0.627043	0.626947	0.626851	0.626755	0.626659	0
0.767	0.626562	0.626466	0.626370	0.626274	0.626178	0.626082	0.625986	0.625890	0.625794	0.625697	0
0.768	0.625601	0.625505	0.625409	0.625313	0.625217	0.625121	0.625025	0.624929	0.624833	0.624737	0
0.769	0.624641	0.624545	0.624448	0.624352	0.624256	0.624160	0.624064	0.623968	0.623872	0.623776	0
0.770	0.623680	0.623584	0.623488	0.623392	0.623296	0.623200	0.623104	0.623008	0.622912	0.622816	0
0.771	0.622720	0.622624	0.622528	0.622432	0.622336	0.622240	0.622144	0.622048	0.621952	0.621856	0
0.772	0.621760	0.621664	0.621568	0.621472	0.621376	0.621280	0.621184	0.621088	0.620992	0.620896	0
0.773	0.620800	0.620704	0.620608	0.620512	0.620416	0.620320	0.620224	0.620128	0.620032	0.619936	0
0.774	0.619840	0.619744	0.619648	0.619552	0.619456	0.619360	0.619264	0.619168	0.619072	0.618977	0
0.775	0.618881	0.618785	0.618689	0.618593	0.618497	0.618401	0.618305	0.618209	0.618113	0.618017	0
0.776	0.617922	0.617826	0.617730	0.617634	0.617538	0.617442	0.617346	0.617250	0.617154	0.617059	0
0.777	0.616963	0.616867	0.616771	0.616675	0.616579	0.616483	0.616387	0.616292	0.616196	0.616100	0
0.778	0.616004	0.615908	0.615812	0.615716	0.615621	0.615525	0.615429	0.615333	0.615237	0.615141	0
0.779	0.615046	0.614950	0.614854	0.614758	0.614662	0.614567	0.614471	0.614375	0.614279	0.614183	0
0.780	0.614088	0.613992	0.613896	0.613800	0.613704	0.613609	0.613513	0.613417	0.613321	0.613225	0
0.781	0.613130	0.613034	0.612938	0.612842	0.612747	0.612651	0.612555	0.612459	0.612364	0.612268	0
0.782	0.612172	0.612076	0.611981	0.611885	0.611789	0.611693	0.611598	0.611502	0.611406	0.611311	0
0.783	0.611215	0.611119	0.611023	0.610928	0.610832	0.610736	0.610641	0.610545	0.610449	0.610353	0
0.784	0.610258	0.610162	0.610066	0.609971	0.609875	0.609779	0.609684	0.609588	0.609492	0.609397	0
0.785	0.609301	0.609205	0.609110	0.609014	0.608918	0.608823	0.608727	0.608631	0.608536	0.608440	0
0.786	0.608345	0.608249	0.608153	0.608058	0.607962	0.607866	0.607771	0.607675	0.607580	0.607484	0
0.787	0.607388	0.607293	0.607197	0.607102	0.607006	0.606910	0.606815	0.606719	0.606624	0.606528	0
0.788	0.606432	0.606337	0.606241	0.606146	0.606050	0.605955	0.605859	0.605764	0.605668	0.605572	0
0.789	0.605477	0.605381	0.605286	0.605190	0.605095	0.604999	0.604904	0.604808	0.604713	0.604617	0
0.790	0.604522	0.604426	0.604331	0.604235	0.604140	0.604044	0.603949	0.603853	0.603758	0.603662	0
0.791	0.603567	0.603471	0.603376	0.603280	0.603185	0.603089	0.602994	0.602898	0.602803	0.602707	0
0.792	0.602612	0.602517	0.602421	0.602326	0.602230	0.602135	0.602039	0.601944	0.601848	0.601753	0
0.793	0.601658	0.601562	0.601467	0.601371	0.601276	0.601180	0.601085	0.600990	0.600894	0.600799	0
0.794	0.600703	0.600608	0.600513	0.600417	0.600322	0.600227	0.600131	0.600036	0.599940	0.599845	0
0.795	0.599750	0.599654	0.599559	0.599464	0.599368	0.599273	0.599178	0.599082	0.598987	0.598892	0
0.796	0.598796	0.598701	0.598606	0.598510	0.598415	0.598320	0.598224	0.598129	0.598034	0.597938	0
0.797	0.597843	0.597748	0.597653	0.597557	0.597462	0.597367	0.597271	0.597176	0.597081	0.596986	0
0.798	0.596890	0.596795	0.596700	0.596605	0.596509	0.596414	0.596319	0.596224	0.596128	0.596033	0
0.799	0.595938	0.595843	0.595747	0.595652	0.595557	0.595462	0.595367	0.595271	0.595176	0.595081	0

$$\frac{2}{\sqrt{\pi}}\exp(-x^2)$$

TABLE 8

Error Function Derivative

X	.0	.0001	.0002	.0003	.0004	.0005	.0006	.0007	.0008	.0009	Δ_2
0.800	0.594986	0.594891	0.594795	0.594700	0.594605	0.594510	0.594415	0.594319	0.594224	0.594129	0
0.801	0.594034	0.593939	0.593844	0.593748	0.593653	0.593558	0.593463	0.593368	0.593273	0.593178	0
0.802	0.593082	0.592987	0.592892	0.592797	0.592702	0.592607	0.592512	0.592417	0.592322	0.592226	0
0.803	0.592131	0.592036	0.591941	0.591846	0.591751	0.591656	0.591561	0.591466	0.591371	0.591276	0
0.804	0.591181	0.591085	0.590990	0.590895	0.590800	0.590705	0.590610	0.590515	0.590420	0.590325	0
0.805	0.590230	0.590135	0.590040	0.589945	0.589850	0.589755	0.589660	0.589565	0.589470	0.589375	0
0.806	0.589280	0.589185	0.589090	0.588995	0.588900	0.588805	0.588710	0.588615	0.588520	0.588425	0
0.807	0.588330	0.588235	0.588140	0.588045	0.587950	0.587856	0.587761	0.587666	0.587571	0.587476	0
0.808	0.587381	0.587286	0.587191	0.587096	0.587001	0.586906	0.586811	0.586717	0.586622	0.586527	0
0.809	0.586432	0.586337	0.586242	0.586147	0.586052	0.585957	0.585863	0.585768	0.585673	0.585578	0
0.810	0.585483	0.585388	0.585293	0.585199	0.585104	0.585009	0.584914	0.584819	0.584725	0.584630	0
0.811	0.584535	0.584440	0.584345	0.584250	0.584156	0.584061	0.583966	0.583871	0.583777	0.583682	0
0.812	0.583587	0.583492	0.583397	0.583303	0.583208	0.583113	0.583018	0.582924	0.582829	0.582734	0
0.813	0.582639	0.582545	0.582450	0.582355	0.582260	0.582166	0.582071	0.581976	0.581882	0.581787	0
0.814	0.581692	0.581598	0.581503	0.581408	0.581313	0.581219	0.581124	0.581029	0.580935	0.580840	0
0.815	0.580745	0.580651	0.580556	0.580461	0.580367	0.580272	0.580177	0.580083	0.579988	0.579894	0
0.816	0.579799	0.579704	0.579610	0.579515	0.579421	0.579326	0.579231	0.579137	0.579042	0.578948	0
0.817	0.578853	0.578758	0.578664	0.578569	0.578475	0.578380	0.578286	0.578191	0.578096	0.578002	0
0.818	0.577907	0.577813	0.577718	0.577624	0.577529	0.577435	0.577340	0.577246	0.577151	0.577057	0
0.819	0.576962	0.576868	0.576773	0.576679	0.576584	0.576490	0.576395	0.576301	0.576206	0.576112	0
0.820	0.576017	0.575923	0.575828	0.575734	0.575639	0.575545	0.575450	0.575356	0.575262	0.575167	0
0.821	0.575073	0.574978	0.574884	0.574789	0.574695	0.574601	0.574506	0.574412	0.574317	0.574223	0
0.822	0.574129	0.574034	0.573940	0.573846	0.573751	0.573657	0.573562	0.573468	0.573374	0.573279	0
0.823	0.573185	0.573091	0.572996	0.572902	0.572808	0.572713	0.572619	0.572525	0.572430	0.572336	0
0.824	0.572242	0.572147	0.572053	0.571959	0.571865	0.571770	0.571676	0.571582	0.571487	0.571393	0
0.825	0.571299	0.571205	0.571110	0.571016	0.570922	0.570828	0.570733	0.570639	0.570545	0.570451	0
0.826	0.570356	0.570262	0.570168	0.570074	0.569980	0.569885	0.569791	0.569697	0.569603	0.569509	0
0.827	0.569414	0.569320	0.569226	0.569132	0.569038	0.568944	0.568849	0.568755	0.568661	0.568567	0
0.828	0.568473	0.568379	0.568285	0.568190	0.568096	0.568002	0.567908	0.567814	0.567720	0.567626	0
0.829	0.567532	0.567438	0.567343	0.567249	0.567155	0.567061	0.566967	0.566873	0.566779	0.566685	0
0.830	0.566591	0.566497	0.566403	0.566309	0.566215	0.566121	0.566027	0.565933	0.565839	0.565745	0
0.831	0.565651	0.565557	0.565463	0.565369	0.565275	0.565181	0.565087	0.564993	0.564899	0.564805	0
0.832	0.564711	0.564617	0.564523	0.564429	0.564335	0.564241	0.564147	0.564053	0.563959	0.563865	0
0.833	0.563771	0.563677	0.563583	0.563489	0.563396	0.563302	0.563208	0.563114	0.563020	0.562926	0
0.834	0.562832	0.562738	0.562644	0.562551	0.562457	0.562363	0.562269	0.562175	0.562081	0.561987	0
0.835	0.561894	0.561800	0.561706	0.561612	0.561518	0.561424	0.561331	0.561237	0.561143	0.561049	0
0.836	0.560955	0.560862	0.560768	0.560674	0.560580	0.560487	0.560393	0.560299	0.560205	0.560111	0
0.837	0.560018	0.559924	0.559830	0.559737	0.559643	0.559549	0.559455	0.559362	0.559268	0.559174	0
0.838	0.559080	0.558987	0.558893	0.558799	0.558706	0.558612	0.558518	0.558425	0.558331	0.558237	0
0.839	0.558144	0.558050	0.557956	0.557863	0.557769	0.557675	0.557582	0.557488	0.557395	0.557301	0
0.840	0.557207	0.557114	0.557020	0.556927	0.556833	0.556739	0.556646	0.556552	0.556459	0.556365	0
0.841	0.556271	0.556178	0.556084	0.555991	0.555897	0.555804	0.555710	0.555617	0.555523	0.555430	0
0.842	0.555336	0.555243	0.555149	0.555056	0.554962	0.554869	0.554775	0.554682	0.554588	0.554495	0
0.843	0.554401	0.554308	0.554214	0.554121	0.554027	0.553934	0.553840	0.553747	0.553653	0.553560	0
0.844	0.553467	0.553373	0.553280	0.553186	0.553093	0.553000	0.552906	0.552813	0.552719	0.552626	0
0.845	0.552533	0.552439	0.552346	0.552253	0.552159	0.552066	0.551972	0.551879	0.551786	0.551692	0
0.846	0.551599	0.551506	0.551412	0.551319	0.551226	0.551133	0.551039	0.550946	0.550853	0.550759	0
0.847	0.550666	0.550573	0.550479	0.550386	0.550293	0.550200	0.550106	0.550013	0.549920	0.549827	0
0.848	0.549733	0.549640	0.549547	0.549454	0.549361	0.549267	0.549174	0.549081	0.548988	0.548895	0
0.849	0.548801	0.548708	0.548615	0.548522	0.548429	0.548335	0.548242	0.548149	0.548056	0.547963	0

TABLE 8

Error Function Derivative

X	.0	.0001	.0002	.0003	.0004	.0005	.0006	.0007	.0008	.0009	Δ_2
0.850	0.547870	0.547777	0.547683	0.547590	0.547497	0.547404	0.547311	0.547218	0.547125	0.547032	0
0.851	0.546939	0.546845	0.546752	0.546659	0.546566	0.546473	0.546380	0.546287	0.546194	0.546101	0
0.852	0.546008	0.545915	0.545822	0.545729	0.545636	0.545543	0.545450	0.545357	0.545264	0.545171	0
0.853	0.545078	0.544985	0.544892	0.544799	0.544706	0.544613	0.544520	0.544427	0.544334	0.544241	0
0.854	0.544148	0.544055	0.543962	0.543869	0.543776	0.543683	0.543591	0.543498	0.543405	0.543312	0
0.855	0.543219	0.543126	0.543033	0.542940	0.542847	0.542755	0.542662	0.542569	0.542476	0.542383	0
0.856	0.542290	0.542197	0.542105	0.542012	0.541919	0.541826	0.541733	0.541641	0.541548	0.541455	0
0.857	0.541362	0.541269	0.541177	0.541084	0.540991	0.540898	0.540806	0.540713	0.540620	0.540527	0
0.858	0.540434	0.540342	0.540249	0.540156	0.540064	0.539971	0.539878	0.539785	0.539693	0.539600	0
0.859	0.539507	0.539415	0.539322	0.539229	0.539137	0.539044	0.538951	0.538859	0.538766	0.538673	0
0.860	0.538581	0.538488	0.538396	0.538303	0.538210	0.538118	0.538025	0.537932	0.537840	0.537747	0
0.861	0.537655	0.537562	0.537470	0.537377	0.537284	0.537192	0.537099	0.537007	0.536914	0.536822	0
0.862	0.536729	0.536637	0.536544	0.536452	0.536359	0.536267	0.536174	0.536081	0.535989	0.535896	0
0.863	0.535804	0.535712	0.535619	0.535527	0.535434	0.535342	0.535249	0.535157	0.535064	0.534972	0
0.864	0.534880	0.534787	0.534695	0.534602	0.534510	0.534417	0.534325	0.534233	0.534140	0.534048	0
0.865	0.533955	0.533863	0.533771	0.533678	0.533586	0.533494	0.533401	0.533309	0.533217	0.533124	0
0.866	0.533032	0.532940	0.532847	0.532755	0.532663	0.532570	0.532478	0.532386	0.532294	0.532201	0
0.867	0.532109	0.532017	0.531925	0.531832	0.531740	0.531648	0.531556	0.531463	0.531371	0.531279	0
0.868	0.531187	0.531094	0.531002	0.530910	0.530818	0.530726	0.530633	0.530541	0.530449	0.530357	0
0.869	0.530265	0.530173	0.530080	0.529988	0.529896	0.529804	0.529712	0.529620	0.529528	0.529436	0
0.870	0.529343	0.529251	0.529159	0.529067	0.528975	0.528883	0.528791	0.528699	0.528607	0.528515	0
0.871	0.528423	0.528331	0.528239	0.528147	0.528054	0.527962	0.527870	0.527778	0.527686	0.527594	0
0.872	0.527502	0.527410	0.527318	0.527226	0.527134	0.527043	0.526951	0.526859	0.526767	0.526675	0
0.873	0.526583	0.526491	0.526399	0.526307	0.526215	0.526123	0.526031	0.525939	0.525847	0.525755	0
0.874	0.525664	0.525572	0.525480	0.525388	0.525296	0.525204	0.525112	0.525021	0.524929	0.524837	0
0.875	0.524745	0.524653	0.524561	0.524470	0.524378	0.524286	0.524194	0.524102	0.524011	0.523919	0
0.876	0.523827	0.523735	0.523643	0.523552	0.523460	0.523368	0.523276	0.523185	0.523093	0.523001	0
0.877	0.522910	0.522818	0.522726	0.522634	0.522543	0.522451	0.522359	0.522268	0.522176	0.522084	0
0.878	0.521993	0.521901	0.521809	0.521718	0.521626	0.521534	0.521443	0.521351	0.521259	0.521168	0
C.879	0.521076	0.520985	0.520893	0.520801	0.520710	0.520618	0.520527	0.520435	0.520344	0.520252	0
0.880	0.520160	0.520069	0.519977	0.519886	0.519794	0.519703	0.519611	0.519520	0.519428	0.519337	0
0.881	0.519245	0.519154	0.519062	0.518971	0.518879	0.518788	0.518696	0.518605	0.518514	0.518422	0
0.882	0.518331	0.518239	0.518148	0.518056	0.517965	0.517874	0.517782	0.517691	0.517599	0.517508	0
0.883	0.517417	0.517325	0.517234	0.517143	0.517051	0.516960	0.516869	0.516777	0.516686	0.516595	0
0.884	0.516503	0.516412	0.516321	0.516229	0.516138	0.516047	0.515955	0.515864	0.515773	0.515682	0
0.885	0.515590	0.515499	0.515408	0.515317	0.515225	0.515134	0.515043	0.514952	0.514860	0.514769	0
0.886	0.514678	0.514587	0.514496	0.514404	0.514313	0.514222	0.514131	0.514040	0.513949	0.513857	0
0.887	0.513766	0.513675	0.513584	0.513493	0.513402	0.513311	0.513220	0.513128	0.513037	0.512946	0
0.888	0.512855	0.512764	0.512673	0.512582	0.512491	0.512400	0.512309	0.512218	0.512127	0.512036	0
0.889	0.511945	0.511854	0.511763	0.511672	0.511581	0.511490	0.511399	0.511308	0.511217	0.511126	0
0.890	0.511035	0.510944	0.510853	0.510762	0.510671	0.510580	0.510489	0.510398	0.510307	0.510216	0
0.891	0.510125	0.510034	0.509944	0.509853	0.509762	0.509671	0.509580	0.509489	0.509398	0.509307	0
0.892	0.509217	0.509126	0.509035	0.508944	0.508853	0.508762	0.508672	0.508581	0.508490	0.508399	0
0.893	0.508308	0.508218	0.508127	0.508036	0.507945	0.507855	0.507764	0.507673	0.507582	0.507492	0
0.894	0.507401	0.507310	0.507219	0.507129	0.507038	0.506947	0.506857	0.506766	0.506675	0.506585	0
0.895	0.506494	0.506403	0.506313	0.506222	0.506131	0.506041	0.505950	0.505859	0.505769	0.505678	0
0.896	0.505588	0.505497	0.505406	0.505316	0.505225	0.505135	0.505044	0.504954	0.504863	0.504772	0
0.897	0.504682	0.504591	0.504501	0.504410	0.504320	0.504229	0.504139	0.504048	0.503958	0.503867	0
0.898	0.503777	0.503686	0.503596	0.503505	0.503415	0.503325	0.503234	0.503144	0.503053	0.502963	0
0.899	0.502872	0.502782	0.502692	0.502601	0.502511	0.502420	0.502330	0.502240	0.502149	0.502059	0

$$\frac{2}{\sqrt{\pi}} \exp(-x^2)$$

$\frac{2}{\sqrt{\pi}} \exp(-x^2)$

TABLE 8

Error Function Derivative

X	.0	.0001	.0002	.0003	.0004	.0005	.0006	.0007	.0008	.0009	Δ_2
0.900	0.501969	0.501878	0.501788	0.501697	0.501607	0.501517	0.501427	0.501336	0.501246	0.501156	0
0.901	0.501065	0.500975	0.500885	0.500794	0.500704	0.500614	0.500524	0.500433	0.500343	0.500253	0
0.902	0.500163	0.500072	0.499982	0.499892	0.499802	0.499712	0.499621	0.499531	0.499441	0.499351	0
0.903	0.499261	0.499171	0.499080	0.498990	0.498900	0.498810	0.498720	0.498630	0.498540	0.498450	0
0.904	0.498359	0.498269	0.498179	0.498089	0.497999	0.497909	0.497819	0.497729	0.497639	0.497549	0
0.905	0.497459	0.497369	0.497279	0.497189	0.497099	0.497009	0.496919	0.496829	0.496739	0.496649	0
0.906	0.496559	0.496469	0.496379	0.496289	0.496199	0.496109	0.496019	0.495929	0.495839	0.495749	0
0.907	0.495659	0.495569	0.495479	0.495389	0.495300	0.495210	0.495120	0.495030	0.494940	0.494850	0
0.908	0.494760	0.494671	0.494581	0.494491	0.494401	0.494311	0.494221	0.494132	0.494042	0.493952	0
0.909	0.493862	0.493772	0.493683	0.493593	0.493503	0.493413	0.493324	0.493234	0.493144	0.493054	0
0.910	0.492965	0.492875	0.492785	0.492696	0.492606	0.492516	0.492426	0.492337	0.492247	0.492157	0
0.911	0.492068	0.491978	0.491888	0.491799	0.491709	0.491620	0.491530	0.491440	0.491351	0.491261	0
0.912	0.491172	0.491082	0.490992	0.490903	0.490813	0.490724	0.490634	0.490545	0.490455	0.490366	0
0.913	0.490276	0.490186	0.490097	0.490007	0.489918	0.489828	0.489739	0.489649	0.489560	0.489471	0
0.914	0.489381	0.489292	0.489202	0.489113	0.489023	0.488934	0.488844	0.488755	0.488666	0.488576	0
0.915	0.488487	0.488397	0.488308	0.488219	0.488129	0.488040	0.487951	0.487861	0.487772	0.487683	0
0.916	0.487593	0.487504	0.487415	0.487325	0.487236	0.487147	0.487057	0.486968	0.486879	0.486790	0
0.917	0.486700	0.486611	0.486522	0.486432	0.486343	0.486254	0.486165	0.486076	0.485986	0.485897	0
0.918	0.485808	0.485719	0.485630	0.485540	0.485451	0.485362	0.485273	0.485184	0.485095	0.485005	0
0.919	0.484916	0.484827	0.484738	0.484649	0.484560	0.484471	0.484382	0.484293	0.484204	0.484115	0
0.920	0.484025	0.483936	0.483847	0.483758	0.483669	0.483580	0.483491	0.483402	0.483313	0.483224	0
0.921	0.483135	0.483046	0.482957	0.482868	0.482779	0.482690	0.482601	0.482512	0.482423	0.482334	0
0.922	0.482246	0.482157	0.482068	0.481979	0.481890	0.481801	0.481712	0.481623	0.481534	0.481445	0
0.923	0.481357	0.481268	0.481179	0.481090	0.481001	0.480912	0.480824	0.480735	0.480646	0.480557	0
0.924	0.480468	0.480380	0.480291	0.480202	0.480113	0.480025	0.479936	0.479847	0.479758	0.479670	0
0.925	0.479581	0.479492	0.479403	0.479315	0.479226	0.479137	0.479049	0.478960	0.478871	0.478783	0
0.926	0.478694	0.478605	0.478517	0.478428	0.478339	0.478251	0.478162	0.478074	0.477985	0.477896	0
0.927	0.477808	0.477719	0.477631	0.477542	0.477453	0.477365	0.477276	0.477188	0.477099	0.477011	0
0.928	0.476922	0.476834	0.476745	0.476657	0.476568	0.476480	0.476391	0.476303	0.476214	0.476126	0
0.929	0.476037	0.475949	0.475861	0.475772	0.475684	0.475595	0.475507	0.475418	0.475330	0.475242	0
0.930	0.475153	0.475065	0.474977	0.474888	0.474800	0.474711	0.474623	0.474535	0.474446	0.474358	0
0.931	0.474270	0.474182	0.474093	0.474005	0.473917	0.473828	0.473740	0.473652	0.473564	0.473475	0
0.932	0.473387	0.473299	0.473211	0.473122	0.473034	0.472946	0.472858	0.472770	0.472681	0.472593	0
0.933	0.472505	0.472417	0.472329	0.472241	0.472152	0.472064	0.471976	0.471888	0.471800	0.471712	0
0.934	0.471624	0.471536	0.471448	0.471359	0.471271	0.471183	0.471095	0.471007	0.470919	0.470831	0
0.935	0.470743	0.470655	0.470567	0.470479	0.470391	0.470303	0.470215	0.470127	0.470039	0.469951	0
0.936	0.469863	0.469775	0.469687	0.469599	0.469511	0.469423	0.469335	0.469248	0.469160	0.469072	0
0.937	0.468984	0.468896	0.468808	0.468720	0.468632	0.468545	0.468457	0.468369	0.468281	0.468193	0
0.938	0.468105	0.468018	0.467930	0.467842	0.467754	0.467666	0.467579	0.467491	0.467403	0.467315	0
0.939	0.467228	0.467140	0.467052	0.466964	0.466877	0.466789	0.466701	0.466614	0.466526	0.466438	0
0.940	0.466350	0.466263	0.466175	0.466087	0.466000	0.465912	0.465825	0.465737	0.465649	0.465562	0
0.941	0.465474	0.465387	0.465299	0.465211	0.465124	0.465036	0.464949	0.464861	0.464774	0.464686	0
0.942	0.464598	0.464511	0.464423	0.464336	0.464248	0.464161	0.464073	0.463986	0.463898	0.463811	0
0.943	0.463723	0.463636	0.463549	0.463461	0.463374	0.463286	0.463199	0.463111	0.463024	0.462937	0
0.944	0.462849	0.462762	0.462674	0.462587	0.462500	0.462412	0.462325	0.462238	0.462150	0.462063	0
0.945	0.461976	0.461888	0.461801	0.461714	0.461627	0.461539	0.461452	0.461365	0.461277	0.461190	0
0.946	0.461103	0.461016	0.460929	0.460841	0.460754	0.460667	0.460580	0.460492	0.460405	0.460318	0
0.947	0.460231	0.460144	0.460057	0.459969	0.459882	0.459795	0.459708	0.459621	0.459534	0.459447	0
0.948	0.459360	0.459273	0.459185	0.459098	0.459011	0.458924	0.458837	0.458750	0.458663	0.458576	0
0.949	0.458489	0.458402	0.458315	0.458228	0.458141	0.458054	0.457967	0.457880	0.457793	0.457706	0

TABLE 8

Error Function Derivative

X	.0	.0001	.0002	.0003	.0004	.0005	.0006	.0007	.0008	.0009	Δ_2
0.950	0.457619	0.457532	0.457445	0.457358	0.457272	0.457185	0.457098	0.457011	0.456924	0.456837	0
0.951	0.456750	0.456663	0.456576	0.456490	0.456403	0.456316	0.456229	0.456142	0.456055	0.455969	0
0.952	0.455882	0.455795	0.455708	0.455621	0.455535	0.455448	0.455361	0.455274	0.455188	0.455101	0
0.953	0.455014	0.454927	0.454841	0.454754	0.454667	0.454581	0.454494	0.454407	0.454321	0.454234	0
0.954	0.454147	0.454061	0.453974	0.453887	0.453801	0.453714	0.453627	0.453541	0.453454	0.453368	0
0.955	0.453281	0.453194	0.453108	0.453021	0.452935	0.452848	0.452762	0.452675	0.452589	0.452502	0
0.956	0.452416	0.452329	0.452243	0.452156	0.452070	0.451983	0.451897	0.451810	0.451724	0.451638	0
0.957	0.451551	0.451465	0.451378	0.451292	0.451205	0.451119	0.451033	0.450946	0.450860	0.450774	0
0.958	0.450687	0.450601	0.450514	0.450428	0.450342	0.450256	0.450169	0.450083	0.449997	0.449910	0
0.959	0.449824	0.449738	0.449651	0.449565	0.449479	0.449393	0.449306	0.449220	0.449134	0.449048	0
0.960	0.448962	0.448875	0.448789	0.448703	0.448617	0.448531	0.448445	0.448358	0.448272	0.448186	0
0.961	0.448100	0.448014	0.447928	0.447842	0.447756	0.447670	0.447583	0.447497	0.447411	0.447325	0
0.962	0.447239	0.447153	0.447067	0.446981	0.446895	0.446809	0.446723	0.446637	0.446551	0.446465	0
0.963	0.446379	0.446293	0.446207	0.446121	0.446035	0.445949	0.445863	0.445777	0.445692	0.445606	0
0.964	0.445520	0.445434	0.445348	0.445262	0.445176	0.445090	0.445004	0.444919	0.444833	0.444747	0
0.965	0.444661	0.444575	0.444489	0.444404	0.444318	0.444232	0.444146	0.444061	0.443975	0.443889	0
0.966	0.443803	0.443718	0.443632	0.443546	0.443460	0.443375	0.443289	0.443203	0.443118	0.443032	0
0.967	0.442946	0.442861	0.442775	0.442689	0.442604	0.442518	0.442432	0.442347	0.442261	0.442176	0
0.968	0.442090	0.442004	0.441919	0.441833	0.441748	0.441662	0.441577	0.441491	0.441406	0.441320	0
0.969	0.441234	0.441149	0.441063	0.440978	0.440893	0.440807	0.440722	0.440636	0.440551	0.440465	0
0.970	0.440380	0.440294	0.440209	0.440123	0.440038	0.439953	0.439867	0.439782	0.439697	0.439611	0
0.971	0.439526	0.439440	0.439355	0.439270	0.439184	0.439099	0.439014	0.438928	0.438843	0.438758	0
0.972	0.438673	0.438587	0.438502	0.438417	0.438332	0.438246	0.438161	0.438076	0.437991	0.437905	0
0.973	0.437820	0.437735	0.437650	0.437565	0.437479	0.437394	0.437309	0.437224	0.437139	0.437054	0
0.974	0.436969	0.436884	0.436798	0.436713	0.436628	0.436543	0.436458	0.436373	0.436288	0.436203	0
0.975	0.436118	0.436033	0.435948	0.435863	0.435778	0.435693	0.435608	0.435523	0.435438	0.435353	0
0.976	0.435268	0.435183	0.435098	0.435013	0.434928	0.434843	0.434758	0.434673	0.434588	0.434503	0
0.977	0.434419	0.434334	0.434249	0.434164	0.434079	0.433994	0.433909	0.433825	0.433740	0.433655	0
0.978	0.433570	0.433485	0.433401	0.433316	0.433231	0.433146	0.433061	0.432977	0.432892	0.432807	0
0.979	0.432722	0.432638	0.432553	0.432468	0.432384	0.432299	0.432214	0.432130	0.432045	0.431960	0
0.980	0.431876	0.431791	0.431706	0.431622	0.431537	0.431452	0.431368	0.431283	0.431199	0.431114	0
0.981	0.431029	0.430945	0.430860	0.430776	0.430691	0.430607	0.430522	0.430438	0.430353	0.430269	0
0.982	0.430184	0.430100	0.430015	0.429931	0.429846	0.429762	0.429677	0.429593	0.429509	0.429424	0
0.983	0.429340	0.429255	0.429171	0.429087	0.429002	0.428918	0.428833	0.428749	0.428665	0.428580	0
0.984	0.428496	0.428412	0.428327	0.428243	0.428159	0.428074	0.427990	0.427906	0.427822	0.427737	0
0.985	0.427653	0.427569	0.427485	0.427400	0.427316	0.427232	0.427148	0.427064	0.426979	0.426895	0
0.986	0.426811	0.426727	0.426643	0.426559	0.426474	0.426390	0.426306	0.426222	0.426138	0.426054	0
0.987	0.425970	0.425886	0.425802	0.425718	0.425633	0.425549	0.425465	0.425381	0.425297	0.425213	0
0.988	0.425129	0.425045	0.424961	0.424877	0.424793	0.424709	0.424625	0.424541	0.424458	0.424374	0
0.989	0.424290	0.424206	0.424122	0.424038	0.423954	0.423870	0.423786	0.423702	0.423619	0.423535	0
0.990	0.423451	0.423367	0.423283	0.423199	0.423116	0.423032	0.422948	0.422864	0.422780	0.422697	0
0.991	0.422613	0.422529	0.422445	0.422362	0.422278	0.422194	0.422110	0.422027	0.421943	0.421859	0
0.992	0.421776	0.421692	0.421608	0.421525	0.421441	0.421357	0.421274	0.421190	0.421106	0.421023	0
0.993	0.420939	0.420856	0.420772	0.420688	0.420605	0.420521	0.420438	0.420354	0.420271	0.420187	0
0.994	0.420104	0.420020	0.419937	0.419853	0.419770	0.419686	0.419603	0.419519	0.419436	0.419352	0
0.995	0.419269	0.419185	0.419102	0.419019	0.418935	0.418852	0.418768	0.418685	0.418602	0.418518	0
0.996	0.418435	0.418352	0.418268	0.418185	0.418102	0.418018	0.417935	0.417852	0.417768	0.417685	0
0.997	0.417602	0.417519	0.417435	0.417352	0.417269	0.417186	0.417102	0.417019	0.416936	0.416853	0
0.998	0.416770	0.416686	0.416603	0.416520	0.416437	0.416354	0.416271	0.416187	0.416104	0.416021	0
0.999	0.415938	0.415855	0.415772	0.415689	0.415606	0.415523	0.415440	0.415357	0.415274	0.415191	0

$$\frac{2}{\sqrt{\pi}}\exp(-x^2)$$

$\frac{2}{\sqrt{\pi}}\exp(-x^2)$

TABLE 8

Error Function Derivative

X	.0	.001	.002	.003	.004	.005	.006	.007	.008	.009	m	Δ_2
1.00	.4151075	.4142777	.4134487	.4126205	.4117933	.4109668	.4101412	.4093164	.4084924	.4076694	0	1
1.01	.4068471	.4060257	.4052051	.4043854	.4035665	.4027486	.4019314	.4011151	.4002997	.3994851	0	1
1.02	.3986714	.3978585	.3970465	.3962354	.3954251	.3946157	.3938072	.3929995	.3921928	.3913869	0	1
1.03	.3905818	.3897777	.3889744	.3881720	.3873705	.3865698	.3857700	.3849711	.3841732	.3833761	0	1
1.04	.3825799	.3817846	.3809901	.3801966	.3794039	.3786122	.3778213	.3770314	.3762423	.3754542	0	1
1.05	.3746669	.3738806	.3730952	.3723106	.3715270	.3707442	.3699624	.3691815	.3684015	.3676224	0	1
1.06	.3668443	.3660670	.3652907	.3645152	.3637407	.3629672	.3621945	.3614228	.3606520	.3598821	0	1
1.07	.3591131	.3583451	.3575780	.3568118	.3560466	.3552822	.3545188	.3537564	.3529949	.3522342	0	1
1.08	.3514746	.3507159	.3499581	.3492013	.3484454	.3476904	.3469364	.3461833	.3454312	.3446800	0	1
1.09	.3439298	.3431805	.3424321	.3416847	.3409383	.3401928	.3394482	.3387046	.3379620	.3372203	0	1
1.10	.3364795	.3357398	.3350010	.3342631	.3335262	.3327903	.3320553	.3313212	.3305882	.3298561	0	1
1.11	.3291249	.3283948	.3276656	.3269373	.3262100	.3254837	.3247584	.3240340	.3233106	.3225881	0	1
1.12	.3218667	.3211462	.3204266	.3197081	.3189905	.3182739	.3175583	.3168436	.3161300	.3154173	0	1
1.13	.3147056	.3139948	.3132851	.3125762	.3118684	.3111616	.3104557	.3097509	.3090470	.3083441	0	1
1.14	.3076422	.3069413	.3062413	.3055423	.3048444	.3041474	.3034514	.3027564	.3020623	.3013693	0	1
1.15	.3006772	.2999862	.2992961	.2986071	.2979189	.2972319	.2965457	.2958606	.2951765	.2944934	0	1
1.16	.2938112	.2931300	.2924499	.2917708	.2910926	.2904154	.2897393	.2890641	.2883899	.2877167	0	1
1.17	.2870445	.2863734	.2857032	.2850340	.2843658	.2836986	.2830324	.2823672	.2817030	.2810398	0	1
1.18	.2803776	.2797164	.2790563	.2783971	.2777389	.2770817	.2764255	.2757704	.2751161	.2744630	0	1
1.19	.2738108	.2731596	.2725095	.2718603	.2712122	.2705650	.2699189	.2692737	.2686296	.2679864	0	1
1.20	.2673443	.2667032	.2660631	.2654240	.2647858	.2641488	.2635127	.2628776	.2622435	.2616104	0	1
1.21	.2609783	.2603472	.2597172	.2590882	.2584601	.2578331	.2572070	.2565820	.2559580	.2553350	0	1
1.22	.2547130	.2540920	.2534720	.2528530	.2522351	.2516181	.2510021	.2503872	.2497732	.2491603	0	1
1.23	.2485483	.2479374	.2473275	.2467186	.2461107	.2455038	.2448979	.2442930	.2436891	.2430863	0	1
1.24	.2424844	.2418835	.2412837	.2406849	.2400870	.2394902	.2388943	.2382995	.2377057	.2371129	0	1
1.25	.2365211	.2359303	.2353405	.2347517	.2341639	.2335771	.2329913	.2324066	.2318228	.2312400	0	1
1.26	.2306583	.2300776	.2294978	.2289190	.2283413	.2277645	.2271888	.2266141	.2260403	.2254676	0	1
1.27	.2248958	.2243251	.2237554	.2231867	.2226189	.2220522	.2214864	.2209217	.2203580	.2197952	0	1
1.28	.2192335	.2186728	.2181130	.2175543	.2169966	.2164398	.2158840	.2153293	.2147755	.2142227	0	1
1.29	.2136710	.2131202	.2125704	.2120216	.2114739	.2109271	.2103813	.2098364	.2092927	.2087498	0	1
1.30	.2082080	.2076671	.2071273	.2065884	.2060505	.2055137	.2049778	.2044428	.2039089	.2033760	0	1
1.31	.2028440	.2023131	.2017831	.2012541	.2007261	.2001991	.1996731	.1991480	.1986240	.1981009	0	1
1.32	.1975788	.1970577	.1965375	.1960183	.1955002	.1949830	.1944668	.1939515	.1934373	.1929240	0	1
1.33	.1924117	.1919004	.1913900	.1908807	.1903722	.1898648	.1893584	.1888529	.1883484	.1878449	0	1
1.34	.1873423	.1868407	.1863401	.1858404	.1853417	.1848440	.1843473	.1838515	.1833567	.1828628	0	1
1.35	.1823699	.1818780	.1813871	.1808971	.1804081	.1799200	.1794329	.1789468	.1784616	.1779774	0	1
1.36	.1774941	.1770118	.1765304	.1760501	.1755706	.1750922	.1746146	.1741381	.1736624	.1731878	0	1
1.37	.1727141	.1722413	.1717695	.1712986	.1708287	.1703597	.1698917	.1694247	.1689585	.1684933	0	1
1.38	.1680291	.1675658	.1671035	.1666421	.1661816	.1657221	.1652635	.1648059	.1643492	.1638934	0	1
1.39	.1634386	.1629847	.1625317	.1620798	.1616287	.1611785	.1607293	.1602809	.1598336	.1593872	0	1
1.40	.1589417	.1584971	.1580535	.1576107	.1571689	.1567281	.1562881	.1558491	.1554110	.1549739	0	1
1.41	.1545376	.1541023	.1536678	.1532343	.1528018	.1523701	.1519393	.1515095	.1510805	.1506526	0	1
1.42	.1502255	.1497993	.1493740	.1489496	.1485262	.1481037	.1476820	.1472613	.1468414	.1464225	0	1
1.43	.1460045	.1455874	.1451712	.1447558	.1443414	.1439279	.1435152	.1431035	.1426927	.1422828	0	1
1.44	.1418737	.1414655	.1410583	.1406519	.1402465	.1398419	.1394382	.1390353	.1386334	.1382324	0	1
1.45	.1378322	.1374330	.1370346	.1366371	.1362404	.1358447	.1354498	.1350558	.1346627	.1342705	0	1
1.46	.1338791	.1334887	.1330990	.1327103	.1323224	.1319354	.1315492	.1311640	.1307796	.1303961	0	1
1.47	.1300133	.1296316	.1292506	.1288705	.1284913	.1281130	.1277354	.1273588	.1269830	.1266081	0	1
1.48	.1262340	.1258608	.1254884	.1251169	.1247462	.1243764	.1240074	.1236393	.1232720	.1229056	0	1
1.49	.1225400	.1221752	.1218113	.1214483	.1210860	.1207246	.1203641	.1200044	.1196455	.1199875	0	1

TABLE 8

Error Function Derivative

X	.0	.001	.002	.003	.004	.005	.006	.007	.008	.009	m	Δ_2
1.50	.1189303	.1185739	.1182184	.1178637	.1175097	.1171567	.1168045	.1164531	.1161025	.1157528		
1.51	.1154038	.1150557	.1147084	.1143619	.1140163	.1136714	.1133274	.1129842	.1126419	.1123003	0	1
1.52	.1119595	.1116195	.1112804	.1109421	.1106045	.1102678	.1099319	.1095968	.1092625	.1089290	0	1
1.53	.1085963	.1082644	.1079333	.1076030	.1072735	.1069447	.1066168	.1062897	.1059633	.1056378	0	1
1.54	.1053131	.1049891	.1046659	.1043435	.1040218	.1037011	.1033810	.1030617	.1027433	.1024256	0	1
1.55	.1021087	.1017925	.1014771	.1011625	.1008487	.1005356	.1002234	.0999119	.0996011	.0992911	0	1
1.56	.9898195	.9867350	.9836583	.9805892	.9775277	.9744738	.9714274	.9683887	.9653576	.9623340	0	1
1.57	.9593180	.9563095	.9533085	.9503151	.9473291	.9443507	.9413797	.9384162	.9354602	.9325116	-1	1
1.58	.9295704	.9266367	.9237104	.9207914	.9178799	.9149758	.9120790	.9091895	.9063075	.9034327	-1	1
1.59	.9005652	.8977051	.8948522	.8920066	.8891683	.8863373	.8835135	.8806970	.8778876	.8750855	-1	1
1.60	.8722906	.8695028	.8667223	.8639489	.8611826	.8584235	.8556715	.8529267	.8501889	.8474582	-1	1
1.61	.8447347	.8420181	.8393087	.8366063	.8339109	.8312225	.8285412	.8258669	.8231995	.8205391	-1	1
1.62	.8178857	.8152392	.8125997	.8099670	.8073413	.8047225	.8021107	.7995057	.7969075	.7943162	-1	1
1.63	.7917317	.7891541	.7865832	.7840192	.7814620	.7789116	.7763680	.7738311	.7713009	.7687775	-1	1
1.64	.7662608	.7637508	.7612475	.7587509	.7562610	.7537777	.7513012	.7488312	.7463679	.7439111	-1	1
1.65	.7414610	.7390175	.7365806	.7341502	.7317263	.7293090	.7268983	.7244940	.7220963	.7197051	-1	1
1.66	.7173204	.7149421	.7125703	.7102050	.7078460	.7054935	.7031474	.7008078	.6984744	.6961475	-1	1
1.67	.6938269	.6915128	.6892049	.6869034	.6846081	.6823192	.6800365	.6777602	.6754901	.6732263	-1	1
1.68	.6709688	.6687174	.6664723	.6642334	.6620007	.6597741	.6575538	.6553396	.6531315	.6509296	-1	1
1.69	.6487339	.6465442	.6443607	.6421832	.6400118	.6378464	.6356872	.6335340	.6313868	.6292456	-1	1
1.70	.6271104	.6249812	.6228580	.6207408	.6186295	.6165242	.6144248	.6123313	.6102438	.6081622	-1	1
1.71	.6060864	.6040165	.6019525	.5998943	.5978420	.5957956	.5937549	.5917200	.5896909	.5876676	-1	1
1.72	.5856501	.5836384	.5816324	.5796321	.5776376	.5756487	.5736656	.5716881	.5697163	.5677502	-1	1
1.73	.5657898	.5638350	.5618858	.5599422	.5580043	.5560719	.5541451	.5522239	.5503082	.5483981	-1	1
1.74	.5464936	.5445945	.5427010	.5408130	.5389305	.5370534	.5351818	.5333157	.5314550	.5295998	-1	1
1.75	.5277500	.5259055	.5240665	.5222328	.5204046	.5185817	.5167642	.5149519	.5131450	.5113435	-1	1
1.76	.5095472	.5077563	.5059706	.5041902	.5024151	.5006452	.4988805	.4971210	.4953668	.4936178	-1	1
1.77	.4918740	.4901353	.4884018	.4866735	.4849504	.4832323	.4815194	.4798116	.4781089	.4764113	-1	0
1.78	.4747187	.4730313	.4713489	.4696715	.4679992	.4663318	.4646696	.4630123	.4613600	.4597126	-1	0
1.79	.4580702	.4564328	.4548004	.4531728	.4515502	.4499325	.4483197	.4467118	.4451087	.4435105	-1	0
1.80	.4419172	.4403287	.4387451	.4371662	.4355922	.4340230	.4324586	.4308989	.4293441	.4277939	-1	0
1.81	.4262485	.4247078	.4231719	.4216407	.4201142	.4185923	.4170752	.4155627	.4140549	.4125517	-1	0
1.82	.4110532	.4095592	.4080699	.4065852	.4051051	.4036295	.4021586	.4006922	.3992304	.3977730	-1	0
1.83	.3963202	.3948720	.3934282	.3919889	.3905541	.3891238	.3876979	.3862765	.3848596	.3834471	-1	0
1.84	.3820390	.3806353	.3792359	.3778410	.3764505	.3750643	.3736825	.3723050	.3709319	.3695631	-1	0
1.85	.3681986	.3668385	.3654826	.3641309	.3627836	.3614405	.3601017	.3587672	.3574368	.3561106	-1	0
1.86	.3547887	.3534710	.3521575	.3508481	.3495430	.3482420	.3469451	.3456523	.3443637	.3430793	-1	0
1.87	.3417989	.3405226	.3392504	.3379823	.3367183	.3354583	.3342023	.3329504	.3317025	.3304586	-1	0
1.88	.3292188	.3279829	.3267511	.3255231	.3242992	.3230792	.3218632	.3206511	.3194429	.3182386	-1	0
1.89	.3170383	.3158418	.3146492	.3134605	.3122757	.3110948	.3099176	.3087443	.3075749	.3064092	-1	0
1.90	.3052474	.3040893	.3029351	.3017846	.3006379	.2994949	.2983558	.2972203	.2960885	.2949605	-1	0
1.91	.2938362	.2927156	.2915987	.2904854	.2893759	.2882700	.2871677	.2860692	.2849742	.2838828	-1	0
1.92	.2827951	.2817109	.2806304	.2795535	.2784801	.2774103	.2763440	.2752813	.2742221	.2731665	-1	0
1.93	.2721144	.2710658	.2700207	.2689791	.2679409	.2669063	.2658750	.2648473	.2638230	.2628022	-1	0
1.94	.2617847	.2607707	.2597601	.2587529	.2577491	.2567486	.2557516	.2547579	.2537675	.2527805	-1	0
1.95	.2517968	.2508165	.2498394	.2488657	.2478953	.2469282	.2459643	.2450038	.2440464	.2430924	-1	0
1.96	.2421415	.2411940	.2402496	.2393085	.2383706	.2374359	.2365043	.2355760	.2346508	.2337288	-1	0
1.97	.2328100	.2318943	.2309817	.2300723	.2291660	.2282628	.2273627	.2264657	.2255718	.2246810	-1	0
1.98	.2237932	.2229085	.2220269	.2211483	.2202727	.2194002	.2185307	.2176642	.2168007	.2159402	-1	0
1.99	.2150826	.2142282	.2133766	.2125279	.2116823	.2108396	.2099998	.2091629	.2083290	.2074980	-1	0

$$\frac{2}{\sqrt{\pi}}\exp(-x^2)$$

$\frac{2}{\sqrt{\pi}}\exp(-x^2)$

TABLE 8

Error Function Derivative

X	.0	.001	.002	.003	.004	.005	.006	.007	.008	.009	m	Δ_2
2.00	.2066698	.2058446	.2050222	.2042028	.2033862	.2025724	.2017615	.2009535	.2001483	.1993459	-1	0
2.01	.1985464	.1977496	.1969556	.1961645	.1953761	.1945905	.1938077	.1930277	.1922504	.1914758	-1	0
2.02	.1907040	.1899349	.1891686	.1884049	.1876440	.1868858	.1861302	.1853774	.1846272	.1838797	-1	0
2.03	.1831348	.1823926	.1816530	.1809161	.1801818	.1794502	.1787211	.1779947	.1772708	.1765496	-1	0
2.04	.1758308	.1751148	.1744012	.1736903	.1729818	.1722759	.1715726	.1708718	.1701735	.1694777	-1	0
2.05	.1687844	.1680937	.1674054	.1667196	.1660363	.1653554	.1646771	.1640012	.1633277	.1626566	-1	0
2.06	.1619880	.1613218	.1606581	.1599967	.1593378	.1586812	.1580271	.1573753	.1567259	.1560789	-1	0
2.07	.1554342	.1547919	.1541519	.1535143	.1528789	.1522459	.1516153	.1509870	.1503609	.1497371	-1	0
2.08	.1491157	.1484965	.1478796	.1472650	.1466526	.1460425	.1454346	.1448290	.1442255	.1436244	-1	0
2.09	.1430254	.1424287	.1418341	.1412418	.1406516	.1400637	.1394779	.1388943	.1383129	.1377336	-1	0
2.10	.1371565	.1365815	.1360086	.1354379	.1348693	.1343029	.1337385	.1331763	.1326161	.1320580	-1	0
2.11	.1315020	.1309481	.1303963	.1298466	.1292989	.1287532	.1282096	.1276680	.1271285	.1265910	-1	0
2.12	.1260555	.1255220	.1249906	.1244611	.1239337	.1234082	.1228847	.1223632	.1218436	.1213260	-1	0
2.13	.1208104	.1202967	.1197850	.1192752	.1187673	.1182614	.1177574	.1172553	.1167551	.1162568	-1	0
2.14	.1157604	.1152658	.1147732	.1142825	.1137936	.1133066	.1128215	.1123381	.1118567	.1113771	-1	0
2.15	.1108993	.1104233	.1099492	.1094769	.1090063	.1085377	.1080708	.1076057	.1071423	.1066808	-1	0
2.16	.1062211	.1057631	.1053069	.1048524	.1043996	.1039487	.1034995	.1030520	.1026062	.1021622	-1	0
2.17	.1017199	.1012792	.1008403	.1004031	.0999676	.0995338	.0991017	.0986712	.0982425	.0978153	-1	0
2.18	.9738991	.9696611	.9654397	.9612347	.9570462	.9528740	.9487181	.9445784	.9404549	.9363475	-2	0
2.19	.9322562	.9281809	.9241216	.9200782	.9160507	.9120389	.9080430	.9040626	.9000980	.8961490	-2	0
2.20	.8922155	.8882974	.8843949	.8805077	.8766358	.8727793	.8689379	.8651117	.8613007	.8575047	-2	0
2.21	.8537238	.8499578	.8462067	.8424705	.8387492	.8350425	.8313506	.8276734	.8240108	.8203628	-2	0
2.22	.8167292	.8131102	.8095056	.8059153	.8023394	.7987777	.7952302	.7916970	.7881778	.7846727	-2	0
2.23	.7811816	.7777045	.7742414	.7707921	.7673566	.7639350	.7605270	.7571328	.7537522	.7503852	-2	0
2.24	.7470317	.7436917	.7403653	.7370521	.7337524	.7304660	.7271928	.7239329	.7206861	.7174525	-2	0
2.25	.7142318	.7110243	.7078298	.7046482	.7014795	.6983237	.6951806	.6920503	.6889328	.6858279	-2	0
2.26	.6827356	.6796559	.6765888	.6735341	.6704919	.6674621	.6644447	.6614396	.6584467	.6554661	-2	0
2.27	.6524977	.6495414	.6465973	.6436651	.6407450	.6378369	.6349407	.6320564	.6291839	.6263233	-2	0
2.28	.6234744	.6206372	.6178117	.6149978	.6121955	.6094047	.6066255	.6038578	.6011014	.5983564	-2	0
2.29	.5956228	.5929005	.5901895	.5874897	.5848010	.5821235	.5794571	.5768018	.5741574	.5715241	-2	0
2.30	.5689017	.5662902	.5636895	.5610997	.5585207	.5559524	.5533948	.5508478	.5483115	.5457858	-2	0
2.31	.5432706	.5407660	.5382718	.5357881	.5333147	.5308517	.5283990	.5259566	.5235244	.5211024	-2	0
2.32	.5186906	.5162890	.5138974	.5115159	.5091444	.5067829	.5044313	.5020896	.4997578	.4974359	-2	0
2.33	.4951237	.4928213	.4905286	.4882457	.4859723	.4837086	.4814544	.4792098	.4769748	.4747492	-2	0
2.34	.4725330	.4703262	.4681289	.4659408	.4637620	.4615926	.4594323	.4572812	.4551393	.4530066	-2	0
2.35	.4508829	.4487683	.4466627	.4445661	.4424784	.4403996	.4383298	.4362688	.4342167	.4321733	-2	0
2.36	.4301386	.4281127	.4260955	.4240870	.4220871	.4200957	.4181129	.4161387	.4141729	.4122157	-2	0
2.37	.4102668	.4083263	.4063942	.4044704	.4025549	.4006478	.3987488	.3968580	.3949755	.3931010	-2	0
2.38	.3912347	.3893765	.3875263	.3856841	.3838499	.3820237	.3802054	.3783950	.3765925	.3747978	-2	0
2.39	.3730109	.3712318	.3694604	.3676968	.3659408	.3641925	.3624519	.3607188	.3589932	.3572753	-2	0
2.40	.3555648	.3538619	.3521664	.3504782	.3487976	.3471242	.3454582	.3437995	.3421481	.3405039	-2	0
2.41	.3388669	.3372372	.3356146	.3339992	.3323909	.3307896	.3291954	.3276083	.3260281	.3244549	-2	0
2.42	.3228887	.3213294	.3197769	.3182313	.3166926	.3151607	.3136355	.3121172	.3106055	.3091006	-2	0
2.43	.3076023	.3061106	.3046256	.3031473	.3016754	.3002101	.2987514	.2972991	.2958533	.2944139	-2	0
2.44	.2929810	.2915544	.2901342	.2887204	.2873128	.2859116	.2845166	.2831278	.2817453	.2803690	-2	0
2.45	.2789989	.2776348	.2762769	.2749251	.2735793	.2722396	.2709059	.2695782	.2682565	.2669407	-2	0
2.46	.2656308	.2643269	.2630288	.2617366	.2604502	.2591696	.2578948	.2566257	.2553624	.2541047	-2	0
2.47	.2528528	.2516066	.2503659	.2491309	.2479015	.2466777	.2454594	.2442467	.2430394	.2418376	-2	0
2.48	.2406414	.2394505	.2382650	.2370850	.2359103	.2347410	.2335770	.2324182	.2312648	.2301167	-2	0
2.49	.2289738	.2278361	.2267036	.2255763	.2244542	.2233372	.2222252	.2211185	.2200167	.2189201	-2	0

TABLE 8

Error Function Derivative

X	.0	.001	.002	.003	.004	.005	.006	.007	.008	.009	m	Δ_2
2.50	.2178284	.2167417	.2156601	.2145834	.2135117	.2124449	.2113830	.2103260	.2092738	.2082265	-2	0
2.51	.2071841	.2061464	.2051135	.2040854	.2030621	.2020434	.2010295	.2000203	.1990157	.1980157	-2	0
2.52	.1970205	.1960298	.1950437	.1940622	.1930852	.1921127	.1911449	.1901814	.1892225	.1882680	-2	0
2.53	.1873180	.1863723	.1854311	.1844943	.1835618	.1826337	.1817099	.1807904	.1798753	.1789643	-2	0
2.54	.1780577	.1771553	.1762571	.1753631	.1744733	.1735876	.1727061	.1718288	.1709555	.1700864	-2	0
2.55	.1692213	.1683604	.1675034	.1666505	.1658015	.1649566	.1641157	.1632787	.1624457	.1616166	-2	0
2.56	.1607913	.1599700	.1591526	.1583390	.1575293	.1567234	.1559213	.1551230	.1543285	.1535378	-2	0
2.57	.1527507	.1519675	.1511879	.1504120	.1496398	.1488714	.1481065	.1473452	.1465877	.1458336	-2	0
2.58	.1450832	.1443363	.1435931	.1428533	.1421171	.1413844	.1406552	.1399295	.1392072	.1384884	-2	0
2.59	.1377730	.1370611	.1363525	.1356474	.1349456	.1342471	.1335521	.1328604	.1321719	.1314868	-2	0
2.60	.1308050	.1301264	.1294512	.1287791	.1281103	.1274447	.1267823	.1261231	.1254671	.1248142	-2	0
2.61	.1241645	.1235179	.1228745	.1222342	.1215969	.1209627	.1203316	.1197036	.1190786	.1184566	-2	0
2.62	.1178376	.1172217	.1166086	.1159986	.1153916	.1147875	.1141863	.1135881	.1129927	.1124003	-2	0
2.63	.1118107	.1112241	.1106402	.1100592	.1094810	.1089057	.1083332	.1077634	.1071965	.1066323	-2	0
2.64	.1060709	.1055122	.1049562	.1044030	.1038525	.1033046	.1027595	.1022170	.1016772	.1011401	-2	0
2.65	.1006055	.1000736	.0995443	.0990177	.0984936	.0979721	.0974531	.0969367	.0964229	.0959116	-2	0
2.66	.9540278	.9489648	.9439269	.9389139	.9339256	.9289619	.9240228	.9191081	.9142177	.9093515	-3	0
2.67	.9045095	.8996913	.8948971	.8901266	.8853799	.8806566	.8759568	.8712803	.8666271	.8619969	-3	0
2.68	.8573899	.8528057	.8482444	.8437057	.8391897	.8346962	.8302250	.8257762	.8213496	.8169450	-3	0
2.69	.8125625	.8082018	.8038629	.7995458	.7952502	.7909761	.7867234	.7824920	.7782819	.7740928	-3	0
2.70	.7699247	.7657776	.7616512	.7575456	.7534606	.7493961	.7453521	.7413284	.7373250	.7333417	-3	0
2.71	.7293785	.7254352	.7215118	.7176082	.7137243	.7098600	.7060152	.7021898	.6983837	.6945969	-3	0
2.72	.6908293	.6870807	.6833511	.6796404	.6759484	.6722752	.6686206	.6649845	.6613669	.6577677	-3	0
2.73	.6541867	.6506239	.6470792	.6435526	.6400439	.6365530	.6330799	.6296246	.6261868	.6227666	-3	0
2.74	.6193637	.6159783	.6126102	.6092592	.6059253	.6026085	.5993087	.5960256	.5927595	.5895100	-3	0
2.75	.5862772	.5830610	.5798612	.5766778	.5735108	.5703600	.5672254	.5641069	.5610044	.5579178	-3	0
2.76	.5548472	.5517923	.5487531	.5457297	.5427217	.5397293	.5367523	.5337906	.5308443	.5279132	-3	0
2.77	.5249971	.5220962	.5192102	.5163391	.5134829	.5106415	.5078148	.5050027	.5022052	.4994221	-3	0
2.78	.4966536	.4938993	.4911594	.4884337	.4857221	.4830247	.4803412	.4776717	.4750161	.4723743	-3	0
2.79	.4697463	.4671319	.4645312	.4619440	.4593703	.4568101	.4542632	.4517296	.4492092	.4467020	-3	0
2.80	.4442079	.4417269	.4392588	.4368036	.4343613	.4319318	.4295150	.4271109	.4247194	.4223405	-3	0
2.81	.4199740	.4176199	.4152783	.4129488	.4106317	.4083267	.4060339	.4037532	.4014844	.3992276	-3	0
2.82	.3969827	.3947496	.3925284	.3903188	.3881208	.3859345	.3837597	.3815965	.3794447	.3773042	-3	0
2.83	.3751751	.3730572	.3709505	.3688551	.3667706	.3646973	.3626350	.3605835	.3585430	.3565134	-3	0
2.84	.3544945	.3524863	.3504888	.3485019	.3465256	.3445598	.3426045	.3406596	.3387250	.3368008	-3	0
2.85	.3348868	.3329831	.3310895	.3292060	.3273326	.3254691	.3236157	.3217722	.3199385	.3181146	-3	0
2.86	.3163005	.3144961	.3127014	.3109163	.3091407	.3073748	.3056182	.3038712	.3021334	.3004050	-3	0
2.87	.2986860	.2969761	.2952754	.2935840	.2919016	.2902282	.2885640	.2869086	.2852622	.2836246	-3	0
2.88	.2819960	.2803760	.2787649	.2771624	.2755686	.2739834	.2724068	.2708387	.2692792	.2677280	-3	0
2.89	.2661853	.2646509	.2631249	.2616071	.2600975	.2585962	.2571030	.2556179	.2541409	.2526718	-3	0
2.90	.2512109	.2497578	.2483127	.2468754	.2454459	.2440243	.2426103	.2412041	.2398056	.2384147	-3	0
2.91	.2370314	.2356557	.2342874	.2329267	.2315733	.2302274	.2288889	.2275577	.2262337	.2249171	-3	0
2.92	.2236076	.2223053	.2210101	.2197221	.2184411	.2171672	.2159003	.2146403	.2133873	.2121411	-3	0
2.93	.2109018	.2096694	.2084436	.2072247	.2060125	.2048069	.2036080	.2024157	.2012300	.2000509	-3	0
2.94	.1988782	.1977121	.1965523	.1953990	.1942521	.1931115	.1919772	.1908492	.1897275	.1886119	-3	0
2.95	.1875026	.1863994	.1853023	.1842113	.1831264	.1820475	.1809746	.1799076	.1788467	.1777915	-3	0
2.96	.1767423	.1756989	.1746613	.1736295	.1726034	.1715831	.1705685	.1695595	.1685561	.1675584	-3	0
2.97	.1665662	.1655796	.1645984	.1636228	.1626526	.1616879	.1607285	.1597745	.1588259	.1578826	-3	0
2.98	.1569446	.1560118	.1550843	.1541619	.1532448	.1523328	.1514260	.1505241	.1496274	.1487358	-3	0
2.99	.1478491	.1469675	.1460909	.1452191	.1443523	.1434903	.1426333	.1417810	.1409336	.1400909	-3	0

$$\frac{2}{\sqrt{\pi}} \exp(-x^2)$$

 $\frac{2}{\sqrt{\pi}} \exp(-x^2)$

TABLE 8

Error Function Derivative

X	.0	.001	.002	.003	.004	.005	.006	.007	.008	.009	m	Δ_2
3.00	.1392530	.1384199	.1375914	.1367677	.1359485	.1351341	.1343243	.1335190	.1327183	.1319221	-3	0
3.01	.1311305	.1303433	.1295606	.1287823	.1280085	.1272390	.1264740	.1257132	.1249568	.1242048	-3	0
3.02	.1234570	.1227134	.1219741	.1212389	.1205081	.1197813	.1190587	.1183402	.1176258	.1169155	-3	0
3.03	.1162093	.1155071	.1148089	.1141146	.1134244	.1127381	.1120558	.1113773	.1107028	.1100320	-3	0
3.04	.1093652	.1087021	.1080429	.1073875	.1067358	.1060879	.1054437	.1048031	.1041663	.1035331	-3	0
3.05	.1029036	.1022777	.1016554	.1010367	.1004215	.0998099	.0992019	.0985973	.0979962	.0973986	-3	0
3.06	.9680443	.9621370	.9562638	.9504246	.9446192	.9388473	.9331089	.9274036	.9217314	.9160920	-4	0
3.07	.9104854	.9049112	.8993694	.8938597	.8883820	.8829361	.8775219	.8721390	.8667875	.8614671	-4	0
3.08	.8561776	.8509189	.8456908	.8404932	.8353258	.8301886	.8250812	.8200037	.8149558	.8099373	-4	0
3.09	.8049482	.7999881	.7950571	.7901548	.7852812	.7804361	.7756193	.7708308	.7660702	.7613376	-4	0
3.10	.7566326	.7519553	.7473053	.7426826	.7380871	.7335185	.7289768	.7244617	.7199731	.7155109	-4	0
3.11	.7110749	.7066651	.7022812	.6979231	.6935906	.6892836	.6850021	.6807457	.6765144	.6723082	-4	0
3.12	.6681267	.6639699	.6598377	.6557298	.6516463	.6475868	.6435513	.6395398	.6355519	.6315877	-4	0
3.13	.6276470	.6237295	.6198353	.6159642	.6121160	.6082907	.6044880	.6007079	.5969503	.5932150	-4	0
3.14	.5895019	.5858108	.5821417	.5784944	.5748688	.5712648	.5676823	.5641211	.5605810	.5570621	-4	0
3.15	.5535643	.5500872	.5466309	.5431952	.5397801	.5363854	.5330109	.5296566	.5263224	.5230080	-4	0
3.16	.5197136	.5164388	.5131837	.5099480	.5067318	.5035348	.5003570	.4971982	.4940584	.4909375	-4	0
3.17	.4878353	.4847517	.4816867	.4786400	.4756117	.4726017	.4696097	.4666357	.4636796	.4607413	-4	0
3.18	.4578208	.4549178	.4520324	.4491643	.4463136	.4434801	.4406636	.4378642	.4350817	.4323160	-4	0
3.19	.4295670	.4268347	.4241189	.4214195	.4187365	.4160697	.4134191	.4107845	.4081659	.4055632	-4	0
3.20	.4029763	.4004051	.3978495	.3953094	.3927848	.3902755	.3877814	.3853025	.3828387	.3803899	-4	0
3.21	.3779560	.3755369	.3731326	.3707429	.3683678	.3660071	.3636609	.3613290	.3590113	.3567078	-4	0
3.22	.3544183	.3521428	.3498812	.3476335	.3453996	.3431792	.3409725	.3387793	.3365995	.3344331	-4	0
3.23	.3322799	.3301400	.3280132	.3258994	.3237986	.3217107	.3196357	.3175734	.3155237	.3134866	-4	0
3.24	.3114622	.3094501	.3074504	.3054630	.3034879	.3015249	.2995741	.2976353	.2957084	.2937934	-4	0
3.25	.2918902	.2899988	.2881191	.2862509	.2843943	.2825492	.2807155	.2788932	.2770821	.2752822	-4	0
3.26	.2734935	.2717158	.2699492	.2681935	.2664487	.2647147	.2629915	.2612790	.2595771	.2578858	-4	0
3.27	.2562050	.2545346	.2528746	.2512249	.2495856	.2479563	.2463373	.2447283	.2431294	.2415404	-4	0
3.28	.2399613	.2383921	.2368326	.2352830	.2337429	.2322125	.2306916	.2291802	.2276783	.2261858	-4	0
3.29	.2247026	.2232287	.2217640	.2203085	.2188621	.2174248	.2159964	.2145770	.2131665	.2117649	-4	0
3.30	.2103721	.2089880	.2076126	.2062458	.2048876	.2035380	.2021968	.2008641	.1995398	.1982238	-4	0
3.31	.1969161	.1956166	.1943253	.1930422	.1917671	.1905001	.1892411	.1879900	.1867468	.1855115	-4	0
3.32	.1842839	.1830642	.1818521	.1806477	.1794509	.1782617	.1770800	.1759058	.1747391	.1735797	-4	0
3.33	.1724277	.1712829	.1701455	.1690152	.1678922	.1667762	.1656673	.1645655	.1634707	.1623828	-4	0
3.34	.1613019	.1602278	.1591606	.1581001	.1570464	.1559995	.1549592	.1539255	.1528984	.1518779	-4	0
3.35	.1508638	.1498563	.1488551	.1478604	.1468720	.1458899	.1449142	.1439446	.1429813	.1420241	-4	0
3.36	.1410730	.1401281	.1391891	.1382563	.1373293	.1364083	.1354932	.1345840	.1336806	.1327831	-4	0
3.37	.1318913	.1310052	.1301247	.1292500	.1283809	.1275174	.1266594	.1258070	.1249600	.1241185	-4	0
3.38	.1232824	.1224517	.1216264	.1208063	.1199916	.1191821	.1183779	.1175788	.1167849	.1159961	-4	0
3.39	.1152124	.1144338	.1136602	.1128917	.1121281	.1113694	.1106157	.1098668	.1091228	.1083836	-4	0
3.40	.1076492	.1069195	.1061946	.1054744	.1047589	.1040480	.1033418	.1026401	.1019430	.1012504	-4	0
3.41	.1005623	.0998787	.0991995	.0985248	.0978545	.0971885	.0965269	.0958695	.0952165	.0945678	-4	0
3.42	.9392324	.9328290	.9264675	.9201475	.9138687	.9076310	.9014341	.8952777	.8891616	.8830854	-5	0
3.43	.8770491	.8710523	.8650947	.8591762	.8532964	.8474552	.8416523	.8358874	.8301604	.8244710	-5	0
3.44	.8188189	.8132039	.8076259	.8020845	.7965795	.7911108	.7856779	.7802809	.7749193	.7695931	-5	0
3.45	.7643020	.7590457	.7538240	.7486368	.7434838	.7383648	.7332795	.7282278	.7232095	.7182243	-5	0
3.46	.7132721	.7083526	.7034656	.6986109	.6937883	.6889977	.6842388	.6795114	.6748152	.6701502	-5	0
3.47	.6655162	.6609128	.6563400	.6517975	.6472852	.6428028	.6383501	.6339271	.6295334	.6251690	-5	0
3.48	.6208335	.6165269	.6122489	.6079994	.6037782	.5995851	.5954199	.5912825	.5871727	.5830902	-5	0
3.49	.5790350	.5750068	.5710055	.5670310	.5630829	.5591612	.5552658	.5513964	.5475528	.5437350	-5	0

TABLE 8

Error Function Derivative

X	.0	.001	.002	.003	.004	.005	.006	.007	.008	.009	m	Δ_2
3.50	.5399426	.5361757	.5324340	.5287173	.5250255	.5213585	.5177161	.5140980	.5105042	.5069346		
3.51	.5033888	.4998669	.4963686	.4928939	.4894425	.4860142	.4826090	.4792268	.4758673	.4725303	-5	0
3.52	.4692159	.4659237	.4626538	.4594058	.4561797	.4529755	.4497927	.4466315	.4434916	.4403729	-5	0
3.53	.4372753	.4341986	.4311426	.4281073	.4250926	.4220982	.4191241	.4161701	.4132360	.4103219	-5	0
3.54	.4074275	.4045526	.4016973	.3988613	.3960446	.3932470	.3904684	.3877085	.3849675	.3822451	-5	0
3.55	.3795411	.3768556	.3741882	.3715390	.3689079	.3662946	.3636991	.3611213	.3585610	.3560182	-5	0
3.56	.3534927	.3509845	.3484933	.3460191	.3435618	.3411212	.3386973	.3362900	.3338991	.3315246	-5	0
3.57	.3291662	.3268240	.3244978	.3221875	.3198931	.3176143	.3153512	.3131035	.3108712	.3086543	-5	0
3.58	.3064525	.3042659	.3020942	.2999374	.2977955	.2956682	.2935556	.2914574	.2893737	.2873043	-5	0
3.59	.2852491	.2832081	.2811811	.2791680	.2771689	.2751834	.2732117	.2712535	.2693089	.2673776	-5	0
3.60	.2654597	.2635549	.2616634	.2597849	.2579193	.2560667	.2542268	.2523997	.2505852	.2487832		
3.61	.2469937	.2452166	.2434518	.2416992	.2399587	.2382303	.2365139	.2348093	.2331166	.2314357	-5	0
3.62	.2297663	.2281086	.2264624	.2248276	.2232041	.2215920	.2199910	.2184012	.2168224	.2152547	-5	0
3.63	.2136978	.2121518	.2106165	.2090919	.2075779	.2060745	.2045816	.2030991	.2016269	.2001650	-5	0
3.64	.1987132	.1972716	.1958402	.1944187	.1930071	.1916053	.1902134	.1888313	.1874588	.1860958	-5	0
3.65	.1847425	.1833986	.1820641	.1807390	.1794231	.1781165	.1768190	.1755307	.1742514	.1729810	-5	0
3.66	.1717196	.1704670	.1692232	.1679882	.1667619	.1655442	.1643350	.1631343	.1619421	.1607583	-5	0
3.67	.1595828	.1584156	.1572566	.1561058	.1549631	.1538284	.1527018	.1515831	.1504723	.1493693	-5	0
3.68	.1482741	.1471867	.1461070	.1450348	.1439703	.1429132	.1418637	.1408216	.1397868	.1387594	-5	0
3.69	.1377393	.1367264	.1357207	.1347221	.1337305	.1327460	.1317685	.1307979	.1298343	.1288774	-5	0
3.70	.1279274	.1269841	.1260475	.1251175	.1241942	.1232775	.1223673	.1214635	.1205662	.1196752	-5	0
3.71	.1187906	.1179124	.1170403	.1161745	.1153149	.1144614	.1136140	.1127726	.1119373	.1111079	-5	0
3.72	.1102844	.1094669	.1086551	.1078492	.1070490	.1062546	.1054658	.1046827	.1039052	.1031333	-5	0
3.73	.1023668	.1016059	.1008505	.1001004	.0993558	.0986164	.0978824	.0971537	.0964301	.0957118	-5	0
3.74	.9499868	.9429064	.9358770	.9288981	.9219694	.9150906	.9082612	.9014811	.8947497	.8880669	-6	0
3.75	.8814322	.8748453	.8683059	.8618136	.8553682	.8489693	.8426166	.8363097	.8300484	.8238323	-6	0
3.76	.8176612	.8115346	.8054523	.7994140	.7934195	.7874683	.7815601	.7756947	.7698718	.7640911	-6	0
3.77	.7583523	.7526551	.7469991	.7413843	.7358101	.7302763	.7247828	.7193291	.7139151	.7085404	-6	0
3.78	.7032047	.6979078	.6926494	.6874293	.6822472	.6771027	.6719958	.6669260	.6618931	.6568969	-6	0
3.79	.6519371	.6470134	.6421257	.6372736	.6324569	.6276753	.6229287	.6182167	.6135391	.6088957	-6	0
3.80	.6042863	.5997105	.5951682	.5906591	.5861830	.5817397	.5773289	.5729503	.5686039	.5642893		
3.81	.5600063	.5557547	.5515343	.5473449	.5431861	.5390579	.5349600	.5308922	.5268543	.5228460	-6	0
3.82	.5188672	.5149177	.5109972	.5071055	.5032424	.4994078	.4956014	.4918231	.4880725	.4843496	-6	0
3.83	.4806542	.4769859	.4733447	.4697304	.4661428	.4625816	.4590467	.4555379	.4520551	.4485979	-6	0
3.84	.4451663	.4417601	.4383790	.4350231	.4316918	.4283853	.4251032	.4218454	.4186118	.4154021	-6	0
3.85	.4122162	.4090539	.4059151	.4027995	.3997071	.3966376	.3935909	.3905668	.3875651	.3845859	-6	0
3.86	.3816286	.3786935	.3757800	.3728883	.3700181	.3671693	.3643416	.3615351	.3587494	.3559844	-6	0
3.87	.3532401	.3505162	.3478127	.3451293	.3424659	.3398224	.3371986	.3345944	.3320096	.3294442	-6	0
3.88	.3268979	.3243707	.3218623	.3193728	.3169018	.3144493	.3120152	.3095993	.3072015	.3048217	-6	0
3.89	.3024597	.3001153	.2977886	.2954794	.2931874	.2909126	.2886549	.2864142	.2841903	.2819831	-6	0
3.90	.2797924	.2776182	.2754604	.2733188	.2711933	.2690839	.2669902	.2649123	.2628501	.2608034	-6	0
3.91	.2587721	.2567562	.2547554	.2527698	.2507991	.2488432	.2469021	.2449757	.2430638	.2411664	-6	0
3.92	.2392833	.2374144	.2355596	.2337189	.2318920	.2300791	.2282798	.2264941	.2247220	.2229633	-6	0
3.93	.2212179	.2194856	.2177666	.2160606	.2143675	.2126873	.2110198	.2093650	.2077227	.2060929	-6	0
3.94	.2044755	.2028703	.2012774	.1996965	.1981277	.1965709	.1950258	.1934925	.1919709	.1904609	-6	0
3.95	.1889624	.1874753	.1859995	.1845350	.1830816	.1816393	.1802080	.1787876	.1773782	.1759794	-6	0
3.96	.1745913	.1732138	.1718469	.1704904	.1691443	.1678084	.1664828	.1651673	.1638619	.1625665	-6	0
3.97	.1612809	.1600053	.1587394	.1574832	.1562366	.1549997	.1537721	.1525540	.1513453	.1501458	-6	0
3.98	.1489555	.1477744	.1466023	.1454393	.1442852	.1431400	.1420035	.1408758	.1397568	.1386464	-6	0
3.99	.1375446	.1364512	.1353662	.1342896	.1332213	.1321613	.1311094	.1300656	.1290299	.1280021	-6	0

$$\frac{2}{\sqrt{\pi}}\exp(-x^2)$$

$$\frac{2}{\sqrt{\pi}}\exp(-x^2)$$

TABLE 8

Error Function Derivative

X	.0	.001	.002	.003	.004	.005	.006	.007	.008	.009	m	Δ_2
4.00	.1269823	.1259704	.1249663	.1239699	.1229813	.1220002	.1210268	.1200609	.1191024	.1181514	-6	0
4.01	.1172077	.1162714	.1153423	.1144204	.1135055	.1125979	.1116972	.1108035	.1099168	.1090369	-6	0
4.02	.1081639	.1072977	.1064381	.1055853	.1047390	.1038994	.1030662	.1022395	.1014193	.1006055	-6	0
4.03	.9979799	.9899675	.9820175	.9741294	.9663027	.9585370	.9508318	.9431866	.9356011	.9280746	-7	0
4.04	.9206069	.9131975	.9058458	.8985516	.8913143	.8841336	.8770089	.8699399	.8629262	.8559673	-7	0
4.05	.8490628	.8422123	.8354154	.8286718	.8219809	.8153424	.8087559	.8022210	.7957373	.7893044	-7	0
4.06	.7829220	.7765897	.7703070	.7640736	.7578892	.7517532	.7456655	.7396256	.7336332	.7276878	-7	0
4.07	.7217892	.7159370	.7101308	.7043703	.6986551	.6929848	.6873593	.6817780	.6762407	.6707470	-7	0
4.08	.6652967	.6598893	.6545246	.6492021	.6439217	.6386829	.6334855	.6283292	.6232135	.6181383	-7	0
4.09	.6131031	.6081077	.6031519	.5982352	.5933575	.5885184	.5837175	.5789546	.5742295	.5695418	-7	0
4.10	.5648912	.5602775	.5557003	.5511594	.5466546	.5421855	.5377518	.5333533	.5289898	.5246609	-7	0
4.11	.5203664	.5161059	.5118794	.5076865	.5035269	.4994004	.4953067	.4912456	.4872168	.4832200	-7	0
4.12	.4792551	.4753218	.4714199	.4675490	.4637089	.4598995	.4561206	.4523717	.4486527	.4449635	-7	0
4.13	.4413036	.4376730	.4340714	.4304986	.4269543	.4234384	.4199506	.4164907	.4130585	.4096537	-7	0
4.14	.4062762	.4029257	.3996020	.3963050	.3930344	.3897900	.3865716	.3833790	.3802120	.3770705	-7	0
4.15	.3739541	.3708628	.3677962	.3647543	.3617369	.3587437	.3557745	.3528292	.3499076	.3470095	-7	0
4.16	.3441347	.3412830	.3384543	.3356484	.3328650	.3301041	.3273655	.3246489	.3219541	.3192812	-7	0
4.17	.3166298	.3139997	.3113909	.3088031	.3062363	.3036902	.3011646	.2986595	.2961746	.2937098	-7	0
4.18	.2912649	.2888398	.2864343	.2840483	.2816815	.2793339	.2770054	.2746958	.2724048	.2701324	-7	0
4.19	.2678784	.2656426	.2634251	.2612255	.2590438	.2568797	.2547333	.2526042	.2504925	.2483979	-7	0
4.20	.2463204	.2442597	.2422158	.2401885	.2381777	.2361833	.2342051	.2322429	.2302969	.2283666	-7	0
4.21	.2264520	.2245530	.2226696	.2208015	.2189486	.2171109	.2152881	.2134802	.2116870	.2099085	-7	0
4.22	.2081446	.2063950	.2046598	.2029387	.2012317	.1995386	.1978595	.1961940	.1945422	.1929039	-7	0
4.23	.1912790	.1896674	.1880690	.1864837	.1849114	.1833521	.1818054	.1802715	.1787502	.1772413	-7	0
4.24	.1757448	.1742606	.1727886	.1713287	.1698808	.1684448	.1670206	.1656080	.1642072	.1628178	-7	0
4.25	.1614399	.1600733	.1587180	.1573738	.1560407	.1547186	.1534074	.1521069	.1508173	.1495382	-7	0
4.26	.1482697	.1470117	.1457640	.1445267	.1432995	.1420825	.1408756	.1396786	.1384915	.1373142	-7	0
4.27	.1361467	.1349888	.1338405	.1327018	.1315724	.1304523	.1293416	.1282400	.1271476	.1260643	-7	0
4.28	.1249899	.1239244	.1228678	.1218199	.1207808	.1197502	.1187282	.1177147	.1167096	.1157129	-7	0
4.29	.1147244	.1137442	.1127721	.1118081	.1108521	.1099041	.1089639	.1080316	.1071070	.1061902	-7	0
4.30	.1052810	.1043794	.1034852	.1025985	.1017193	.1008474	.0999827	.0991252	.0982749	.0974317	-7	0
4.31	.9659560	.9576643	.9494418	.9412881	.9332025	.9251845	.9172336	.9093493	.9015308	.8937778	-8	0
4.32	.8860897	.8784660	.8709062	.8634096	.8559759	.8486045	.8412949	.8340465	.8268591	.8197318	-8	0
4.33	.8126644	.8056563	.7987071	.7918161	.7849831	.7782075	.7714888	.7648267	.7582205	.7516699	-8	0
4.34	.7451743	.7387335	.7323468	.7260140	.7197344	.7135077	.7073334	.7012113	.6951407	.6891212	-8	0
4.35	.6831526	.6772342	.6713659	.6655470	.6597772	.6540561	.6483834	.6427586	.6371813	.6316511	-8	0
4.36	.6261677	.6207306	.6153395	.6099941	.6046938	.5994384	.5942275	.5890607	.5839376	.5788580	-8	0
4.37	.5738214	.5688275	.5638759	.5589663	.5540984	.5492717	.5444860	.5397409	.5350361	.5303712	-8	0
4.38	.5257460	.5211601	.5166131	.5121047	.5076348	.5032027	.4988084	.4944516	.4901317	.4858487	-8	0
4.39	.4816021	.4773917	.4732171	.4690781	.4649743	.4609056	.4568715	.4528719	.4489063	.4449746	-8	0
4.40	.4410765	.4372116	.4333797	.4295805	.4258138	.4220793	.4183767	.4147058	.4110662	.4074577	-8	0
4.41	.4038802	.4003332	.3968166	.3933301	.3898734	.3864464	.3830487	.3796802	.3763404	.3730294	-8	0
4.42	.3697467	.3664922	.3632656	.3600667	.3568952	.3537509	.3506337	.3475433	.3444794	.3414418	-8	0
4.43	.3384303	.3354447	.3324848	.3295504	.3266411	.3237570	.3208976	.3180629	.3152526	.3124664	-8	0
4.44	.3097044	.3069661	.3042514	.3015600	.2988920	.2962469	.2936246	.2910250	.2884478	.2858928	-8	0
4.45	.2833600	.2808490	.2783597	.2758919	.2734454	.2710201	.2686158	.2662323	.2638693	.2615269	-8	0
4.46	.2592047	.2569026	.2546205	.2523581	.2501153	.2478920	.2456879	.2435030	.2413369	.2391897	-8	0
4.47	.2370611	.2349510	.2328593	.2307856	.2287300	.2266922	.2246721	.2226696	.2206845	.2187167	-8	0
4.48	.2167659	.2148322	.2129152	.2110150	.2091312	.2072639	.2054129	.2035780	.2017590	.1999559	-8	0
4.49	.1981686	.1963968	.1946405	.1928995	.1911736	.1894629	.1877670	.1860861	.1844197	.1827679	-8	0

TABLE 8

Error Function Derivative

X	.0	.001	.002	.003	.004	.005	.006	.007	.008	.009	m	Δ_2
4.50	.1811306	.1795075	.1778986	.1763039	.1747230	.1731560	.1716027	.1700630	.1685368	.1670240	-8	0
4.51	.1655243	.1640378	.1625644	.1611038	.1596561	.1582211	.1567986	.1553886	.1539910	.1526057	-8	0
4.52	.1512324	.1498713	.1485221	.1471848	.1458592	.1445453	.1432430	.1419520	.1406724	.1394041	-8	0
4.53	.1381469	.1369008	.1356657	.1344415	.1332280	.1320252	.1308330	.1296514	.1284801	.1273192	-8	0
4.54	.1261685	.1250279	.1238974	.1227769	.1216663	.1205655	.1194744	.1183929	.1173210	.1162586	-8	0
4.55	.1152056	.1141618	.1131274	.1121020	.1110857	.1100784	.1090801	.1080905	.1071097	.1061377	-8	0
4.56	.1051742	.1042193	.1032728	.1023347	.1014050	.1004834	.0995701	.0986649	.0977677	.0968784	-8	0
4.57	.9599712	.9512361	.9425786	.9339980	.9254937	.9170650	.9087112	.9004317	.8922259	.8840930	-9	0
4.58	.8760326	.8680439	.8601264	.8522793	.8445022	.8367943	.8291551	.8215841	.8140805	.8066439	-9	0
4.59	.7992736	.7919691	.7847297	.7775549	.7704443	.7633971	.7564129	.7494910	.7426310	.7358323	-9	0
4.60	.7290944	.7224168	.7157989	.7092403	.7027403	.6962984	.6899142	.6835873	.6773170	.6711028	-9	0
4.61	.6649443	.6588411	.6527925	.6467982	.6408576	.6349704	.6291359	.6233538	.6176237	.6119449	-9	0
4.62	.6063172	.6007400	.5952130	.5897356	.5843074	.5789281	.5735971	.5683141	.5630786	.5578903	-9	0
4.63	.5527486	.5476533	.5426037	.5375997	.532[illegible]408	.5277266	.5228567	.5180306	.5132481	.5085087	-9	0
4.64	.5038121	.4991578	.4945456	.4899750	.4854456	.4809572	.4765093	.4721016	.4677338	.4634054	-9	0
4.65	.4591162	.4548658	.4506537	.4464799	.4423438	.4382451	.4341835	.4301587	.4261704	.4222182	-9	0
4.66	.4183018	.4144210	.4105753	.4067644	.4029882	.3992462	.3955382	.3918638	.3882228	.3846148	-9	0
4.67	.3810396	.3774969	.3739864	.3705077	.3670607	.3636451	.3602605	.3569067	.3535833	.3502903	-9	0
4.68	.3470272	.3437939	.3405899	.3374152	.3342694	.3311523	.3280635	.3250030	.3219703	.3189653	-9	0
4.69	.3159877	.3130373	.3101138	.3072170	.3043466	.3015025	.2986843	.2958919	.2931250	.2903835	-9	0
4.70	.2876669	.2849752	.2823082	.2796655	.2770470	.2744525	.2718818	.2693346	.2668107	.2643099	-9	0
4.71	.2618321	.2593769	.2569443	.2545339	.2521457	.2497794	.2474349	.2451118	.2428100	.2405294	-9	0
4.72	.2382697	.2360308	.2338125	.2316145	.2294367	.2272789	.2251411	.2230228	.2209241	.2188447	-9	0
4.73	.2167844	.2147431	.2127205	.2107166	.2087312	.2067640	.2048150	.2028840	.2009707	.1990751	-9	0
4.74	.1971970	.1953362	.1934926	.1916660	.1898562	.1880632	.1862867	.1845267	.1827829	.1810552	-9	0
4.75	.1793435	.1776477	.1759675	.1743028	.1726536	.1710196	.1694008	.1677969	.1662078	.1646336	-9	0
4.76	.1630738	.1615286	.1599976	.1584809	.1569782	.1554895	.1540146	.1525533	.1511056	.1496714	-9	0
4.77	.1482505	.1468427	.1454481	.1440664	.1426975	.1413414	.1399978	.1386668	.1373482	.1360418	-9	0
4.78	.1347476	.1334654	.1321951	.1309367	.1296900	.1284549	.1272314	.1260192	.1248183	.1236286	-9	0
4.79	.1224501	.1212825	.1201258	.1189799	.1178446	.1167200	.1156059	.1145022	.1134088	.1123256	-9	0
4.80	.1112526	.1101896	.1091365	.1080932	.1070598	.1060359	.1050217	.1040170	.1030216	.1020356	-9	0
4.81	.1010588	.1000912	.0991327	.0981831	.0972424	.0963106	.0953875	.0944730	.0935671	.0926697	-9	0
4.82	.9178082	.9090021	.9002787	.8916373	.8830770	.8745972	.8661970	.8578759	.8496329	.8414675	-10	0
4.83	.8333789	.8253664	.8174293	.8095669	.8017785	.7940635	.7864211	.7788507	.7713517	.7639233	-10	0
4.84	.7565650	.7492760	.7420558	.7349036	.7278190	.7208012	.7138497	.7069638	.7001429	.6933864	-10	0
4.85	.6866937	.6800643	.6734976	.6669929	.6605497	.6541675	.6478456	.6415836	.6353807	.6292366	-10	0
4.86	.6231507	.6171224	.6111512	.6052366	.5993780	.5935749	.5878268	.5821332	.5764937	.5709076	-10	0
4.87	.5653746	.5598940	.5544654	.5490884	.5437624	.5384870	.5332618	.5280861	.5229597	.5178820	-10	0
4.88	.5128525	.5078710	.5029367	.4980494	.4932087	.4884140	.4836649	.4789611	.4743021	.4696874	-10	0
4.89	.4651167	.4605896	.4561056	.4516644	.4472655	.4429086	.4385932	.4343190	.4300857	.4258927	-10	0
4.90	.4217398	.4176264	.4135525	.4095174	.4055209	.4015626	.3976421	.3937591	.3899133	.3861042	-10	0
4.91	.3823316	.3785951	.3748944	.3712291	.3675989	.3640035	.3604425	.3569156	.3534225	.3499629	-10	0
4.92	.3465366	.3431430	.3397821	.3364533	.3331565	.3298913	.3266575	.3234548	.3202828	.3171413	-10	0
4.93	.3140299	.3109485	.3078967	.3048742	.3018808	.2989162	.2959802	.2930723	.2901925	.2873403	-10	0
4.94	.2845157	.2817182	.2789477	.2762039	.2734865	.2707953	.2681301	.2654906	.2628765	.2602876	-10	0
4.95	.2577237	.2551847	.2526700	.2501796	.2477134	.2452709	.2428520	.2404565	.2380842	.2357348	-10	0
4.96	.2334081	.2311039	.2288220	.2265621	.2243242	.2221079	.2199131	.2177395	.2155870	.2134553	-10	0
4.97	.2113442	.2092537	.2071834	.2051331	.2031028	.2010922	.1991010	.1971292	.1951765	.1932428	-10	0
4.98	.1913278	.1894315	.1875535	.1856939	.1838522	.1820285	.1802225	.1784341	.1766630	.1749092	-10	0
4.99	.1731725	.1714527	.1697496	.1680630	.1663930	.1647391	.1631014	.1614797	.1598737	.1582834	-10	0

$$\frac{2}{\sqrt{\pi}}\exp(-x^2)$$

TABLE 8

Error Function Derivative

X	.0	.001	.002	.003	.004	.005	.006	.007	.008	.009	m	Δ_2
5.00	.1567087	.1551492	.1536050	.1520758	.1505616	.1490622	.1475773	.1461070	.1446511	.1432093	-10	0
5.01	.1417816	.1403680	.1389681	.1375819	.1362092	.1348500	.1335041	.1321713	.1308516	.1295449	-10	0
5.02	.1282508	.1269695	.1257008	.1244444	.1232004	.1219686	.1207488	.1195410	.1183450	.1171608	-10	0
5.03	.1159882	.1148270	.1136773	.1125389	.1114117	.1102955	.1091903	.1080959	.1070123	.1059394	-10	0
5.04	.1048770	.1038250	.1027834	.1017520	.1007308	.0997197	.0987185	.0977271	.0967456	.0957736	-10	0
5.05	.9481128	.9385841	.9291494	.9198076	.9105579	.9013994	.8923313	.8833526	.8744626	.8656603	-11	0
5.06	.8569448	.8483154	.8397712	.8313115	.8229352	.8146418	.8064303	.7983000	.7902501	.7822798	-11	0
5.07	.7743884	.7665750	.7588389	.7511794	.7435957	.7360871	.7286529	.7212923	.7140047	.7067893	-11	0
5.08	.6996453	.6925722	.6855692	.6786357	.6717710	.6649744	.6582452	.6515828	.6449866	.6384558	-11	0
5.09	.6319900	.6255883	.6192502	.6129752	.6067625	.6006116	.5945219	.5884927	.5825235	.5766137	-11	0
5.10	.5707627	.5649699	.5592349	.5535569	.5479354	.5423700	.5368600	.5314050	.5260044	.5206575	-11	0
5.11	.5153640	.5101233	.5049349	.4997982	.4947128	.4896782	.4846939	.4797593	.4748740	.4700375	-11	0
5.12	.4652494	.4605091	.4558161	.4511701	.4465706	.4420171	.4375091	.4330462	.4286280	.4242541	-11	0
5.13	.4199239	.4156371	.4113932	.4071919	.4030326	.3989151	.3948388	.3908033	.3868084	.3828535	-11	0
5.14	.3789383	.3750624	.3712254	.3674269	.3636666	.3599440	.3562588	.3526106	.3489991	.3454239	-11	0
5.15	.3418847	.3383810	.3349125	.3314790	.3280800	.3247152	.3213843	.3180869	.3148226	.3115913	-11	0
5.16	.3083925	.3052260	.3020914	.2989883	.2959166	.2928758	.2898656	.2868859	.2839361	.2810162	-11	0
5.17	.2781258	.2752645	.2724321	.2696283	.2668529	.2641054	.2613858	.2586936	.2560287	.2533907	-11	0
5.18	.2507793	.2481945	.2456357	.2431028	.2405956	.2381137	.2356570	.2332252	.2308180	.2284352	-11	0
5.19	.2260765	.2237417	.2214307	.2191430	.2168785	.2146370	.2124183	.2102221	.2080481	.2058963	-11	0
5.20	.2037662	.2016578	.1995709	.1975051	.1954603	.1934363	.1914329	.1894498	.1874869	.1855441	-11	0
5.21	.1836209	.1817173	.1798331	.1779681	.1761221	.1742948	.1724862	.1706960	.1689241	.1671702	-11	0
5.22	.1654342	.1637158	.1620151	.1603316	.1586654	.1570161	.1553836	.1537679	.1521686	.1505857	-11	0
5.23	.1490189	.1474681	.1459333	.1444140	.1429103	.1414220	.1399489	.1384909	.1370478	.1356194	-11	0
5.24	.1342056	.1328064	.1314214	.1300507	.1286940	.1273512	.1260221	.1247067	.1234047	.1221161	-11	0
5.25	.1208407	.1195784	.1183290	.1170925	.1158686	.1146573	.1134585	.1122720	.1110976	.1099353	-11	0
5.26	.1087850	.1076465	.1065196	.1054044	.1043006	.1032082	.1021270	.1010569	.0999979	.0989497	-11	0
5.27	.9791243	.9688576	.9586965	.9486402	.9386875	.9288373	.9190887	.9094405	.8998919	.8904417	-12	0
5.28	.8810890	.8718327	.8626720	.8536059	.8446333	.8357534	.8269652	.8182678	.8096601	.8011415	-12	0
5.29	.7927109	.7843674	.7761102	.7679384	.7598512	.7518476	.7439268	.7360879	.7283303	.7206529	-12	0
5.30	.7130550	.7055358	.6980945	.6907304	.6834425	.6762303	.6690927	.6620291	.6550389	.6481211	-12	0
5.31	.6412751	.6345002	.6277956	.6211606	.6145944	.6080965	.6016661	.5953025	.5890049	.5827729	-12	0
5.32	.5766057	.5705025	.5644628	.5584860	.5525714	.5467182	.5409260	.5351942	.5295219	.5239088	-12	0
5.33	.5183541	.5128573	.5074177	.5020348	.4967081	.4914369	.4862207	.4810588	.4759508	.4708961	-12	0
5.34	.4658942	.4609445	.4560464	.4511995	.4464032	.4416570	.4369604	.4323128	.4277139	.4231631	-12	0
5.35	.4186597	.4142036	.4097940	.4054306	.4011127	.3968402	.3926123	.3884287	.3842889	.3801925	-12	0
5.36	.3761389	.3721279	.3681589	.3642315	.3603452	.3564997	.3526946	.3489294	.3452036	.3415170	-12	0
5.37	.3378691	.3342595	.3306878	.3271535	.3236564	.3201961	.3167721	.3133841	.3100317	.3067145	-12	0
5.38	.3034323	.3001845	.2969710	.2937913	.2906450	.2875318	.2844514	.2814035	.2783877	.2754036	-12	0
5.39	.2724509	.2695294	.2666387	.2637784	.2609484	.2581481	.2553774	.2526360	.2499234	.2472395	-12	0
5.40	.2445839	.2419564	.2393566	.2367843	.2342391	.2317209	.2292292	.2267639	.2243247	.2219113	-12	0
5.41	.2195233	.2171606	.2148230	.2125101	.2102216	.2079574	.2057172	.2035007	.2013077	.1991379	-12	0
5.42	.1969911	.1948670	.1927655	.1906863	.1886291	.1865937	.1845799	.1825875	.1806162	.1786659	-12	0
5.43	.1767362	.1748271	.1729382	.1710694	.1692204	.1673912	.1655813	.1637907	.1620191	.1602664	-12	0
5.44	.1585323	.1568167	.1551193	.1534400	.1517785	.1501348	.1485085	.1468996	.1453078	.1437330	-12	0
5.45	.1421750	.1406335	.1391085	.1375998	.1361071	.1346304	.1331695	.1317241	.1302941	.1288794	-12	0
5.46	.1274799	.1260952	.1247254	.1233702	.1220294	.1207030	.1193908	.1180927	.1168084	.1155378	-12	0
5.47	.1142808	.1130372	.1118070	.1105900	.1093860	.1081948	.1070164	.1058507	.1046974	.1035565	-12	0
5.48	.1024278	.1013113	.1002066	.0991139	.0980328	.0969633	.0959054	.0948588	.0938234	.0927991	-12	0
5.49	.9178589	.9078351	.8979188	.8881092	.8784049	.8688050	.8593082	.8499136	.8406200	.8314263	-13	0

TABLE 8

Error Function Derivative

X	.0	.001	.002	.003	.004	.005	.006	.007	.008	.009	m	Δ_2
5.50	.8223315	.8133347	.8044346	.7956303	.7869208	.7783052	.7697822	.7613512	.7530109	.7447605	-13	0
5.51	.7365990	.7285255	.7205391	.7126388	.7048237	.6970929	.6894455	.6818807	.6743975	.6669951	-13	0
5.52	.6596726	.6524292	.6452641	.6381764	.6311652	.6242298	.6173695	.6105832	.6038705	.5972302	-13	0
5.53	.5906618	.5841646	.5777375	.5713801	.5650915	.5588710	.5527179	.5466314	.5406108	.5346555	-13	0
5.54	.5287648	.5229379	.5171741	.5114729	.5058336	.5002553	.4947377	.4892799	.4838813	.4785414	-13	0
5.55	.4732594	.4680348	.4628670	.4577553	.4526991	.4476979	.4427510	.4378580	.4330181	.4282309	-13	0
5.56	.4234958	.4188122	.4141796	.4095974	.4050651	.4005821	.3961479	.3917621	.3874240	.3831332	-13	0
5.57	.3788891	.3746914	.3705394	.3664326	.3623707	.3583530	.3543792	.3504488	.3465613	.3427162	-13	0
5.58	.3389131	.3351515	.3314310	.3277512	.3241115	.3205116	.3169512	.3134295	.3099465	.3065015	-13	0
5.59	.3030942	.2997242	.2963910	.2930943	.2898338	.2866089	.2834194	.2802647	.2771447	.2740588	-13	0
5.60	.2710067	.2679881	.2650026	.2620498	.2591294	.2562410	.2533844	.2505590	.2477647	.2450011	-13	0
5.61	.2422678	.2395645	.2368909	.2342467	.2316315	.2290450	.2264870	.2239571	.2214550	.2189805	-13	0
5.62	.2165331	.2141127	.2117189	.2093515	.2070101	.2046945	.2024044	.2001395	.1978995	.1956843	-13	0
5.63	.1934934	.1913267	.1891839	.1870647	.1849689	.1828961	.1808463	.1788191	.1768142	.1748315	-13	0
5.64	.1728706	.1709315	.1690137	.1671171	.1652414	.1633865	.1615521	.1597379	.1579438	.1561696	-13	0
5.65	.1544150	.1526797	.1509637	.1492667	.1475884	.1459287	.1442875	.1426643	.1410592	.1394718	-13	0
5.66	.1379020	.1363496	.1348144	.1332963	.1317950	.1303103	.1288421	.1273901	.1259543	.1245345	-13	0
5.67	.1231303	.1217418	.1203687	.1190108	.1176680	.1163402	.1150271	.1137285	.1124445	.1111746	-13	0
5.68	.1099190	.1086773	.1074494	.1062351	.1050344	.1038470	.1026728	.1015117	.1003636	.0992282	-13	0
5.69	.9810553	.9699532	.9589748	.9481188	.9373838	.9267685	.9162716	.9058917	.8956277	.8854781	-14	0
5.70	.8754419	.8655177	.8557042	.8460004	.8364049	.8269166	.8175342	.8082567	.7990830	.7900117	-14	0
5.71	.7810419	.7721723	.7634020	.7547298	.7461545	.7376753	.7292910	.7210004	.7128028	.7046969	-14	0
5.72	.6966818	.6887565	.6809199	.6731713	.6655094	.6579334	.6504424	.6430354	.6357114	.6284696	-14	0
5.73	.6213092	.6142290	.6072283	.6003062	.5934618	.5866942	.5800027	.5733864	.5668443	.5603759	-14	0
5.74	.5539801	.5476562	.5414035	.5352210	.5291081	.5230640	.5170878	.5111790	.5053366	.4995600	-14	0
5.75	.4938485	.4882013	.4826177	.4770970	.4716385	.4662415	.4609054	.4556294	.4504129	.4452553	-14	0
5.76	.4401558	.4351138	.4301288	.4252000	.4203269	.4155088	.4107451	.4060352	.4013785	.3967744	-14	0
5.77	.3922223	.3877217	.3832719	.3788725	.3745228	.3702223	.3659705	.3617668	.3576106	.3535015	-14	0
5.78	.3494389	.3454223	.3414512	.3375251	.3336434	.3298057	.3260115	.3222603	.3185517	.3148851	-14	0
5.79	.3112600	.3076761	.3041329	.3006299	.2971665	.2937425	.2903574	.2870108	.2837021	.2804310	-14	0
5.80	.2771971	.2739999	.2708390	.2677141	.2646247	.2615705	.2585509	.2555658	.2526145	.2496969	-14	0
5.81	.2468125	.2439609	.2411417	.2383546	.2355993	.2328754	.2301825	.2275203	.2248885	.2222866	-14	0
5.82	.2197145	.2171715	.2146577	.2121725	.2097157	.2072869	.2048858	.2025121	.2001655	.1978458	-14	0
5.83	.1955525	.1932853	.1910442	.1888286	.1866384	.1844731	.1823326	.1802166	.1781248	.1760570	-14	0
5.84	.1740128	.1719919	.1699942	.1680194	.1660672	.1641374	.1622296	.1603437	.1584794	.1566365	-14	0
5.85	.1548147	.1530137	.1512334	.1494735	.1477339	.1460142	.1443142	.1426337	.1409725	.1393303	-14	0
5.86	.1377071	.1361024	.1345162	.1329482	.1313982	.1298661	.1283515	.1268544	.1253744	.1239115	-14	0
5.87	.1224654	.1210359	.1196229	.1182262	.1168455	.1154807	.1141316	.1127981	.1114799	.1101769	-14	0
5.88	.1088889	.1076158	.1063573	.1051134	.1038837	.1026683	.1014669	.1002793	.0991055	.0979452	-14	0
5.89	.9679824	.9566455	.9454395	.9343629	.9234142	.9125920	.9018948	.8913212	.8808699	.8705392	-15	0
5.90	.8603281	.8502351	.8402587	.8303978	.8206509	.8110169	.8014943	.7920820	.7827786	.7735829	-15	0
5.91	.7644938	.7555099	.7466301	.7378532	.7291780	.7206033	.7121282	.7037512	.6954714	.6872876	-15	0
5.92	.6791988	.6712039	.6633016	.6554912	.6477714	.6401412	.6325997	.6251458	.6177784	.6104967	-15	0
5.93	.6032996	.5961861	.5891553	.5822063	.5753381	.5685498	.5618404	.5552092	.5486550	.5421772	-15	0
5.94	.5357748	.5294469	.5231927	.5170113	.5109020	.5048639	.4988961	.4929979	.4871684	.4814069	-15	0
5.95	.4757126	.4700847	.4645224	.4590251	.4535919	.4482221	.4429150	.4376698	.4324859	.4273626	-15	0
5.96	.4222991	.4172947	.4123489	.4074609	.4026299	.3978555	.3931369	.3884735	.3838646	.3793097	-15	0
5.97	.3748080	.3703590	.3659621	.3616167	.3573222	.3530779	.3488834	.3447381	.3406413	.3365924	-15	0
5.98	.3325911	.3286367	.3247286	.3208664	.3170494	.3132773	.3095495	.3058653	.3022244	.2986262	-15	0
5.99	.2950704	.2915562	.2880833	.2846513	.2812595	.2779076	.2745951	.2713216	.2680866	.2648895	-15	0

$$\frac{2}{\sqrt{\pi}}\exp(-x^2)$$

TABLE 8

Error Function Derivative

X	.0	.001	.002	.003	.004	.005	.006	.007	.008	.009	m	Δ_2
6.00	.2617301	.2586079	.2555223	.2524731	.2494598	.2464820	.2435392	.2406310	.2377571	.2349171	-15	0
6.01	.2321106	.2293370	.2265962	.2238877	.2212112	.2185662	.2159523	.2133694	.2108169	.2082945	-15	0
6.02	.2058018	.2033386	.2009045	.1984991	.1961222	.1937733	.1914521	.1891584	.1868918	.1846520	-15	0
6.03	.1824386	.1802514	.1780901	.1759544	.1738439	.1717584	.1696975	.1676611	.1656488	.1636603	-15	0
6.04	.1616953	.1597536	.1578349	.1559390	.1540655	.1522142	.1503848	.1485772	.1467910	.1450260	-15	0
6.05	.1432819	.1415585	.1398555	.1381727	.1365100	.1348670	.1332434	.1316392	.1300540	.1284876	-15	0
6.06	.1269399	.1254106	.1238993	.1224062	.1209307	.1194728	.1180322	.1166088	.1152023	.1138125	-15	0
6.07	.1124393	.1110824	.1097417	.1084170	.1071080	.1058146	.1045367	.1032739	.1020262	.1007934	-15	0
6.08	.9957528	.9837167	.9718242	.9600736	.9484632	.9369913	.9256563	.9144565	.9033906	.8924567	-16	0
6.09	.8816534	.8709791	.8604323	.8500116	.8397154	.8295422	.8194906	.8095592	.7997466	.7900513	-16	0
6.10	.7804721	.7710074	.7616560	.7524164	.7432876	.7342679	.7253563	.7165514	.7078519	.6992567	-16	0
6.11	.6907645	.6823741	.6740841	.6658936	.6578013	.6498061	.6419067	.6341021	.6263911	.6187727	-16	0
6.12	.6112457	.6038090	.5964617	.5892025	.5820305	.5749447	.5679440	.5610275	.5541941	.5474428	-16	0
6.13	.5407727	.5341827	.5276721	.5212396	.5148847	.5086061	.5024032	.4962749	.4902203	.4842387	-16	0
6.14	.4783291	.4724907	.4667225	.4610239	.4553940	.4498319	.4443368	.4389080	.4335447	.4282461	-16	0
6.15	.4230113	.4178397	.4127305	.4076830	.4026964	.3977700	.3929031	.3880949	.3833448	.3786522	-16	0
6.16	.3740161	.3694361	.3649115	.3604416	.3560257	.3516632	.3473535	.3430959	.3388898	.3347346	-16	0
6.17	.3306296	.3265744	.3225683	.3186107	.3147010	.3108386	.3070230	.3032538	.2995301	.2958516	-16	0
6.18	.2922177	.2886278	.2850814	.2815781	.2781173	.2746984	.2713211	.2679847	.2646888	.2614329	-16	0
6.19	.2582166	.2550393	.2519007	.2488002	.2457373	.2427116	.2397227	.2367702	.2338535	.2309723	-16	0
6.20	.2281262	.2253147	.2225373	.2197938	.2170837	.2144065	.2117620	.2091496	.2065691	.2040200	-16	0
6.21	.2015019	.1990145	.1965575	.1941304	.1917328	.1893645	.1870251	.1847142	.1824315	.1801767	-16	0
6.22	.1779493	.1757492	.1735759	.1714291	.1693085	.1672139	.1651448	.1631010	.1610822	.1590881	-16	0
6.23	.1571183	.1551726	.1532506	.1513522	.1494771	.1476248	.1457952	.1439880	.1422029	.1404396	-16	0
6.24	.1386980	.1369777	.1352784	.1335999	.1319420	.1303045	.1286870	.1270893	.1255112	.1239524	-16	0
6.25	.1224128	.1208920	.1193899	.1179062	.1164408	.1149933	.1135635	.1121514	.1107566	.1093789	-16	0
6.26	.1080181	.1066740	.1053465	.1040353	.1027401	.1014609	.1001974	.0989495	.0977169	.0964995	-16	0
6.27	.9529706	.9410940	.9293636	.9177776	.9063342	.8950316	.8838683	.8728424	.8619524	.8511965	-17	0
6.28	.8405732	.8300808	.8197178	.8094824	.7993733	.7893889	.7795275	.7697879	.7601684	.7506677	-17	0
6.29	.7412842	.7320165	.7228633	.7138230	.7048944	.6960762	.6873668	.6787651	.6702696	.6618792	-17	0
6.30	.6535925	.6454082	.6373252	.6293421	.6214578	.6136711	.6059806	.5983854	.5908842	.5834758	-17	0
6.31	.5761592	.5689332	.5617967	.5547486	.5477878	.5409133	.5341240	.5274189	.5207968	.5142570	-17	0
6.32	.5077981	.5014195	.4951199	.4888986	.4827544	.4766865	.4706939	.4647757	.4589311	.4531590	-17	0
6.33	.4474586	.4418290	.4362695	.4307789	.4253567	.4200018	.4147136	.4094911	.4043335	.3992401	-17	0
6.34	.3942101	.3892427	.3843371	.3794926	.3747084	.3699837	.3653179	.3607103	.3561600	.3516664	-17	0
6.35	.3472288	.3428466	.3385189	.3342453	.3300248	.3258571	.3217413	.3176770	.3136632	.3096997	-17	0
6.36	.3057855	.3019203	.2981033	.2943339	.2906117	.2869359	.2833061	.2797216	.2761820	.2726865	-17	0
6.37	.2692348	.2658263	.2624604	.2591365	.2558542	.2526131	.2494125	.2462519	.2431309	.2400490	-17	0
6.38	.2370057	.2340004	.2310328	.2281024	.2252088	.2223514	.2195298	.2167436	.2139923	.2112755	-17	0
6.39	.2085928	.2059437	.2033279	.2007449	.1981943	.1956757	.1931888	.1907331	.1883082	.1859137	-17	0
6.40	.1835494	.1812147	.1789095	.1766331	.1743854	.1721660	.1699744	.1678104	.1656737	.1635638	-17	0
6.41	.1614804	.1594233	.1573921	.1553864	.1534060	.1514505	.1495197	.1476131	.1457306	.1438718	-17	0
6.42	.1420365	.1402242	.1384349	.1366680	.1349235	.1332009	.1315001	.1298208	.1281626	.1265254	-17	0
6.43	.1249088	.1233127	.1217366	.1201805	.1186441	.1171270	.1156291	.1141502	.1126899	.1112481	-17	0
6.44	.1098245	.1084189	.1070312	.1056609	.1043079	.1029721	.1016533	.1003511	.0990654	.0977959	-17	0
6.45	.9654257	.9530507	.9408326	.9287691	.9168585	.9050989	.8934882	.8820248	.8707067	.8595321	-18	0
6.46	.8484992	.8376063	.8268515	.8162332	.8057497	.7953992	.7851802	.7750908	.7651296	.7552948	-18	0
6.47	.7455850	.7359986	.7265339	.7171896	.7079639	.6988556	.6898631	.6809849	.6722196	.6635659	-18	0
6.48	.6550222	.6465873	.6382596	.6300380	.6219211	.6139074	.6059958	.5981851	.5904737	.5828606	-18	0
6.49	.5753446	.5679243	.5605986	.5533663	.5462262	.5391771	.5322180	.5253476	.5185649	.5118687	-18	0

TABLE 8

Error Function Derivative

X	.0	.001	.002	.003	.004	.005	.006	.007	.008	.009	m	Δ_2
6.50	.5052580	.4987316	.4922886	.4859278	.4796483	.4734490	.4673288	.4612868	.4553221	.4494336	-18	0
6.51	.4436203	.4378814	.4322159	.4266227	.4211012	.4156502	.4102690	.4049567	.3997123	.3945351	-18	0
6.52	.3894241	.3843786	.3793977	.3744807	.3696265	.3648346	.3601041	.3554342	.3508242	.3462732	-18	0
6.53	.3417806	.3373457	.3329676	.3286456	.3243791	.3201674	.3160098	.3119054	.3078538	.3038542	-18	0
6.54	.2999060	.2960085	.2921610	.2883630	.2846138	.2809128	.2772593	.2736528	.2700927	.2665783	-18	0
6.55	.2631092	.2596847	.2563042	.2529672	.2496732	.2464216	.2432119	.2400435	.2369159	.2338285	-18	0
6.56	.2307810	.2277727	.2248031	.2218719	.2189784	.2161222	.2133029	.2105199	.2077728	.2050611	-18	0
6.57	.2023844	.1997423	.1971343	.1945599	.1920187	.1895104	,1870345	.1845905	.1821781	.1797969	-18	0
6.58	.1774465	.1751264	.1728363	.1705759	.1683446	.1661422	.1639683	.1618225	.1597044	.1576138	-18	0
6.59	.1555503	.1535134	.1515029	.1495185	.1475597	.1456263	.1437180	.1418344	.1399751	.1381400	-18	0
6.60	.1363287	.1345409	.1327762	.1310344	.1293152	.1276183	.1259434	.1242903	.1226586	.1210481	-18	0
6.61	.1194585	.1178895	.1163409	.1148124	.1133038	.1118148	.1103451	.1088945	.1074628	.1060497	-18	0
6.62	.1046550	.1032783	.1019197	.1005787	.0992551	.0979487	.0966594	.0953867	.0941308	.0928911	-18	0
6.63	.9166762	.9046004	.8926818	.8809186	.8693087	.8578500	.8465407	.8353788	.8243624	.8134897	-19	0
6.64	.8027588	.7921678	.7817150	.7713986	.7612168	.7511680	.7412502	.7314620	.7218016	.7122673	-19	0
6.65	.7028576	.6935707	.6844053	.6753595	.6664321	.6576213	.6489257	.6403438	.6318741	.6235152	-19	0
6.66	.6152657	.6071241	.5990890	.5911591	.5833330	.5756094	.5679868	.5604641	.5530400	.5457130	-19	0
6.67	.5384821	.5313459	.5243033	.5173529	.5104936	.5037243	.4970438	.4904509	.4839445	.4775234	-19	0
6.68	.4711866	.4649330	.4587614	.4526708	.4466602	.4407286	.4348748	.4290980	.4233969	.4177709	-19	0
6.69	.4122187	.4067396	.4013325	.3959965	.3907306	.3855339	.3804057	.3753449	.3703506	.3654221	-19	0
6.70	.3605585	.3557589	.3510225	.3463484	.3417359	.3371841	.3326923	.3282598	.3238856	.3195690	-19	0
6.71	.3153093	.3111058	.3069578	.3028644	.2988250	.2948389	.2909054	.2870238	.2831934	.2794136	-19	0
6.72	.2756837	.2720030	.2683709	.2647868	.2612501	.2577600	.2543162	.2509177	.2475643	.2442551	-19	0
6.73	.2409897	.2377675	.2345878	.2314503	.2283543	.2252992	.2222845	.2193097	.2163744	.2134778	-19	0
6.74	.2106197	.2077994	.2050164	.2022703	.1995606	.1968868	.1942485	.1916451	.1890762	.1865414	-19	0
6.75	.1840402	.1815721	.1791369	.1767339	.1743628	.1720232	.1697146	.1674367	.1651890	.1629711	-19	0
6.76	.1607828	.1586235	.1564928	.1543905	.1523161	.1502693	.1482497	.1462570	.1442907	.1423506	-19	0
6.77	.1404363	.1385475	.1366838	.1348449	.1330305	.1312402	.1294737	.1277308	.1260111	.1243143	-19	0
6.78	.1226401	.1209882	.1193583	.1177502	.1161634	.1145979	.1130531	.1115291	.1100253	.1085415	-19	0
6.79	.1070776	.1056333	.1042081	.1028020	.1014147	.1000459	.0986953	.0973629	.0960482	.0947511	-19	0
6.80	.9347128	.9220859	.9096277	.8973359	.8852086	.8732434	.8614382	.8497909	.8382994	.8269616	-20	0
6.81	.8157756	.8047393	.7938507	.7831078	.7725087	.7620516	.7517346	.7415558	.7315133	.7216053	-20	0
6.82	.7118301	.7021860	.6926711	.6832838	.6740224	.6648852	.6558705	.6469767	.6382023	.6295456	-20	0
6.83	.6210051	.6125792	.6042665	.5960653	.5879744	.5799920	.5721169	.5643476	.5566826	.5491207	-20	0
6.84	.5416604	.5343004	.5270394	.5198760	.5128089	.5058369	.4989587	.4921730	.4854786	.4788744	-20	0
6.85	.4723590	.4659314	.4595903	.4533345	.4471630	.4410747	.4350684	.4291431	.4232975	.4175308	-20	0
6.86	.4118418	.4062295	.4006929	.3952309	.3898427	.3845271	.3792832	.3741102	.3690069	.3639725	-20	0
6.87	.3590060	.3541067	.3492735	.3445056	.3398021	.3351621	.3305848	.3260694	.3216150	.3172209	-20	0
6.88	.3128861	.3086100	.3043917	.3002304	.2961255	.2920761	.2880815	.2841409	.2802537	.2764191	-20	0
6.89	.2726365	.2689050	.2652242	.2615931	.2580113	.2544780	.2509925	.2475544	.2441628	.2408172	-20	0
6.90	.2375170	.2342616	.2310503	.2278825	.2247578	.2216754	.2186349	.2156357	.2126771	.2097588	-20	0
6.91	.2068801	.2040405	.2012394	.1984764	.1957510	.1930625	.1904107	.1877949	.1852146	.1826695	-20	0
6.92	.1801589	.1776825	.1752398	.1728303	.1704536	.1681093	.1657968	.1635159	.1612660	.1590468	-20	0
6.93	.1568577	.1546985	.1525688	.1504679	.1483958	.1463519	.1443359	.1423473	.1403859	.1384512	-20	0
6.94	.1365429	.1346607	.1328041	.1309729	.1291666	.1273850	.1256278	.1238945	.1221849	.1204986	-20	0
6.95	.1188354	.1171949	.1155768	.1139808	.1124067	.1108540	.1093226	.1078121	.1063223	.1048529	-20	0
6.96	.1034035	.1019740	.1005641	.0991734	.0978018	.0964490	.0951146	.0937986	.0925006	.0912204	-20	0
6.97	.8995768	.8871228	.8748395	.8627245	.8507756	.8389906	.8273671	.8159029	.8045961	.7934444	-21	0
6.98	.7824456	.7715978	.7608988	.7503467	.7399395	.7296752	.7195517	.7095674	.6997201	.6900082	-21	0
6.99	.6804296	.6709827	.6616657	.6524767	.6434140	.6344759	.6256608	.6169669	.6083926	.5999362	-21	0

TABLE 8

Error Function Derivative

X	.0	.001	.002	.003	.004	.005	.006	.007	.008	.009	m	Δ_2
7.00	.5915962	.5833710	.5752590	.5672587	.5593685	.5515869	.5439125	.5363438	.5288794	.5215178	-21	0
7.01	.5142577	.5070975	.5000362	.4930721	.4862041	.4794307	.4727508	.4661630	.4596661	.4532589	-21	0
7.02	.4469400	.4407084	.4345627	.4285020	.4225249	.4166304	.4108172	.4050844	.3994308	.3938552	-21	0
7.03	.3883567	.3829343	.3775868	.3723132	.3671125	.3619838	.3569260	.3519381	.3470193	.3421685	-21	0
7.04	.3373849	.3326675	.3280153	.3234276	.3189034	.3144419	.3100422	.3057034	.3014247	.2972054	-21	0
7.05	.2930444	.2889413	.2848949	.2809047	.2769698	.2730895	.2692629	.2654895	.2617685	.2580990	-21	0
7.06	.2544805	.2509122	.2473935	.2439237	.2405020	.2371278	.2338005	.2305195	.2272840	.2240935	-21	0
7.07	.2209473	.2178449	.2147856	.2117689	.2087941	.2058606	.2029680	.2001156	.1973029	.1945294	-21	0
7.08	.1917945	.1890976	.1864383	.1838160	.1812302	.1786805	.1761663	.1736871	.1712424	.1688319	-21	0
7.09	.1664549	.1641110	.1617999	.1595209	.1572738	.1550580	.1528731	.1507187	.1485944	.1464997	-21	0
7.10	.1444342	.1423976	.1403894	.1384093	.1364568	.1345316	.1326333	.1307615	.1289159	.1270961	-21	0
7.11	.1253017	.1235324	.1217878	.1200677	.1183715	.1166992	.1150501	.1134242	.1118211	.1102404	-21	0
7.12	.1086818	.1071450	.1056297	.1041358	.1026626	.1012102	.0997781	.0983660	.0969737	.0956010	-21	0
7.13	.9424751	.9291299	.9159718	.9029981	.8902066	.8775944	.8651592	.8528985	.8408098	.8288909	-22	0
7.14	.8171393	.8055526	.7941287	.7828652	.7717599	.7608107	.7500153	.7393715	.7288774	.7185308	-22	0
7.15	.7083296	.6982719	.6883556	.6785787	.6689394	.6594357	.6500658	.6408277	.6317195	.6227396	-22	0
7.16	.6138862	.6051574	.5965515	.5880668	.5797016	.5714543	.5633232	.5553067	.5474031	.5396110	-22	0
7.17	.5319287	.5243548	.5168876	.5095258	.5022678	.4951122	.4880576	.4811026	.4742457	.4674856	-22	0
7.18	.4608209	.4542503	.4477726	.4413863	.4350902	.4288831	.4227638	.4167308	.4107832	.4049196	-22	0
7.19	.3991389	.3934399	.3878216	.3822827	.3768222	.3714389	.3661318	.3608999	.3557419	.3506570	-22	0
7.20	.3456441	.3407021	.3358302	.3310272	.3262923	.3216244	.3170227	.3124862	.3080141	.3036053	-22	0
7.21	.2992590	.2949744	.2907505	.2865865	.2824816	.2784349	.2744457	.2705131	.2666363	.2628145	-22	0
7.22	.2590470	.2553329	.2516717	.2480624	.2445045	.2409970	.2375394	.2341309	.2307709	.2274587	-22	0
7.23	.2241935	.2209747	.2178018	.2146739	.2115906	.2085512	.2055550	.2026014	.1996899	.1968198	-22	0
7.24	.1939905	.1912016	.1884524	.1857423	.1830709	.1804375	.1778416	.1752828	.1727604	.1702739	-22	0
7.25	.1678229	.1654069	.1630253	.1606777	.1583636	.1560825	.1538339	.1516174	.1494326	.1472790	-22	0
7.26	.1451560	.1430635	.1410007	.1389675	.1369634	.1349878	.1330405	.1311210	.1292289	.1273639	-22	0
7.27	.1255255	.1237135	.1219273	.1201667	.1184313	.1167207	.1150346	.1133727	.1117344	.1101198	-22	0
7.28	.1085281	.1069593	.1054129	.1038887	.1023864	.1009055	.0994458	.0980071	.0965890	.0951913	-22	0
7.29	.9381357	.9245560	.9111710	.8979781	.8849743	.8721572	.8595239	.8470719	.8347987	.8227016	-23	0
7.30	.8107783	.7990261	.7874427	.7760257	.7647727	.7536813	.7427493	.7319745	.7213544	.7108871	-23	0
7.31	.7005702	.6904017	.6803794	.6705013	.6607652	.6511692	.6417114	.6323895	.6232020	.6141466	-23	0
7.32	.6052216	.5964251	.5877553	.5792103	.5707884	.5624879	.5543069	.5462439	.5382971	.5304648	-23	0
7.33	.5227454	.5151374	.5076390	.5002488	.4929652	.4857867	.4787117	.4717389	.4648666	.4580936	-23	0
7.34	.4514184	.4448395	.4383557	.4319655	.4256675	.4194606	.4133434	.4073145	.4013728	.3955169	-23	0
7.35	.3897457	.3840580	.3784525	.3729281	.3674836	.3621178	.3568297	.3516181	.3464819	.3414201	-23	0
7.36	.3364315	.3315152	.3266701	.3218951	.3171892	.3125516	.3079811	.3034769	.2990380	.2946634	-23	0
7.37	.2903522	.2861035	.2819164	.2777900	.2737235	.2697160	.2657666	.2618746	.2580389	.2542590	-23	0
7.38	.2505340	.2468630	.2432453	.2396802	.2361668	.2327045	.2292925	.2259301	.2226165	.2193511	-23	0
7.39	.2161331	.2129620	.2098369	.2067572	.2037224	.2007318	.1977846	.1948803	.1920183	.1891979	-23	0
7.40	.1864186	.1836797	.1809807	.1783211	.1757001	.1731173	.1705722	.1680641	.1655926	.1631571	-23	0
7.41	.1607571	.1583921	.1560615	.1537650	.1515020	.1492719	.1470744	.1449089	.1427751	.1406723	-23	0
7.42	.1386003	.1365585	.1345466	.1325640	.1306103	.1286852	.1267883	.1249190	.1230770	.1212620	-23	0
7.43	.1194735	.1177111	.1159745	.1142633	.1125771	.1109156	.1092784	.1076651	.1060754	.1045090	-23	0
7.44	.1029655	.1014447	.0999460	.0984694	.0970143	.0955806	.0941678	.0927758	.0914041	.0900525	-23	0
7.45	.8872082	.8740860	.8611560	.8484157	.8358622	.8234927	.8113047	.7992955	.7874625	.7758031	-24	0
7.46	.7643148	.7529951	.7418416	.7308518	.7200234	.7093540	.6988413	.6884831	.6782770	.6682208	-24	0
7.47	.6583125	.6485497	.6389305	.6294527	.6201142	.6109131	.6018473	.5929148	.5841137	.5754421	-24	0
7.48	.5668982	.5584800	.5501856	.5420134	.5339615	.5260282	.5182116	.5105103	.5029223	.4954461	-24	0
7.49	.4880802	.4808227	.4736723	.4666272	.4596860	.4528472	.4461092	.4394705	.4329298	.4264857	-24	0

TABLE 8

Error Function Derivative

X	.0	.001	.002	.003	.004	.005	.006	.007	.008	.009	m	Δ_2
7.50	.4201365	.4138811	.4077179	.4016458	.3956633	.3897691	.3839620	.3782406	.3726038	.3670502	-24	0
7.51	.3615787	.3561880	.3508770	.3456445	.3404893	.3354104	.3304065	.3254766	.3206197	.3158346	-24	0
7.52	.3111203	.3064758	.3019000	.2973918	.2929505	.2885749	.2842641	.2800171	.2758330	.2717109	-24	0
7.53	.2676499	.2636490	.2597074	.2558242	.2519986	.2482297	.2445167	.2408587	.2372550	.2337047	-24	0
7.54	.2302071	.2267615	.2233669	.2200227	.2167281	.2134824	.2102849	.2071350	.2040318	.2009746	-24	0
7.55	.1979629	.1949959	.1920730	.1891935	.1863569	.1835624	.1808094	.1780974	.1754257	.1727937	-24	0
7.56	.1702009	.1676466	.1651304	.1626516	.1602097	.1578041	.1554344	.1530999	.1508002	.1485347	-24	0
7.57	.1463029	.1441045	.1419387	.1398053	.1377036	.1356332	.1335937	.1315846	.1296055	.1276559	-24	0
7.58	.1257354	.1238435	.1219798	.1201439	.1183355	.1165540	.1147991	.1130704	.1113675	.1096900	-24	0
7.59	.1080376	.1064099	.1048065	.1032270	.1016712	.1001385	.0986288	.0971417	.0956768	.0942338	-24	0
7.60	.9281235	.9141218	.9003295	.8867435	.8733609	.8601784	.8471932	.8344024	.8218031	.8093924	-25	0
7.61	.7971675	.7851257	.7732642	.7615805	.7500717	.7387354	.7275690	.7165699	.7057357	.6950639	-25	0
7.62	.6845521	.6741980	.6639991	.6539532	.6440580	.6343113	.6247108	.6152543	.6059399	.5967652	-25	0
7.63	.5877283	.5788271	.5700595	.5614237	.5529175	.5445391	.5362867	.5281582	.5201519	.5122659	-25	0
7.64	.5044985	.4968478	.4893122	.4818900	.4745793	.4673787	.4602863	.4533007	.4464203	.4396433	-25	0
7.65	.4329684	.4263940	.4199185	.4135406	.4072587	.4010715	.3949775	.3889753	.3830635	.3772408	-25	0
7.66	.3715059	.3658574	.3602942	.3548147	.3494179	.3441026	.3388674	.3337111	.3286327	.3236309	-25	0
7.67	.3187046	.3138527	.3090740	.3043675	.2997320	.2951666	.2906700	.2862415	.2818798	.2775840	-25	0
7.68	.2733532	.2691863	.2650824	.2610406	.2570598	.2531393	.2492780	.2454752	.2417299	.2380412	-25	0
7.69	.2344084	.2308305	.2273068	.2238365	.2204186	.2170526	.2137375	.2104726	.2072572	.2040905	-25	0
7.70	.2009718	.1979004	.1948755	.1918964	.1889625	.1860731	.1832275	.1804251	.1776652	.1749471	-25	0
7.71	.1722703	.1696341	.1670379	.1644811	.1619631	.1594834	.1570413	.1546363	.1522678	.1499352	-25	0
7.72	.1476382	.1453760	.1431482	.1409543	.1387937	.1366659	.1345705	.1325070	.1304749	.1284736	-25	0
7.73	.1265028	.1245620	.1226507	.1207685	.1189150	.1170896	.1152921	.1135219	.1117787	.1100619	-25	0
7.74	.1083714	.1067067	.1050673	.1034528	.1018630	.1002974	.0987557	.0972374	.0957423	.0942701	-25	0
7.75	.9282025	.9139253	.8998660	.8860212	.8723876	.8589622	.8457416	.8327228	.8199029	.8072786	-26	0
7.76	.7948472	.7826056	.7705510	.7586806	.7469916	.7354811	.7241466	.7129853	.7019947	.6911721	-26	0
7.77	.6805150	.6700209	.6596873	.6495118	.6394919	.6296253	.6199098	.6103429	.6009225	.5916463	-26	0
7.78	.5825121	.5735177	.5646612	.5559403	.5473530	.5388972	.5305710	.5223725	.5142996	.5063504	-26	0
7.79	.4985231	.4908158	.4832267	.4757540	.4683959	.4611507	.4540166	.4469921	.4400753	.4332647	-26	0
7.80	.4265587	.4199556	.4134538	.4070520	.4007484	.3945417	.3884304	.3824129	.3764879	.3706539	-26	0
7.81	.3649096	.3592537	.3536847	.3482013	.3428023	.3374863	.3322520	.3270983	.3220239	.3170276	-26	0
7.82	.3121082	.3072644	.3024953	.2977996	.2931762	.2886240	.2841419	.2797289	.2753838	.2711057	-26	0
7.83	.2668935	.2627463	.2586629	.2546425	.2506841	.2467868	.2429495	.2391715	.2354517	.2317893	-26	0
7.84	.2281834	.2246332	.2211378	.2176962	.2143079	.2109719	.2076873	.2044535	.2012697	.1981350	-26	0
7.85	.1950488	.1920103	.1890187	.1860733	.1831735	.1803185	.1775076	.1747403	.1720157	.1693333	-26	0
7.86	.1666923	.1640922	.1615324	.1590122	.1565310	.1540881	.1516832	.1493154	.1469843	.1446893	-26	0
7.87	.1424299	.1402054	.1380154	.1358594	.1337368	.1316471	.1295897	.1275643	.1255703	.1236072	-26	0
7.88	.1216745	.1197718	.1178987	.1160546	.1142391	.1124518	.1106922	.1089600	.1072546	.1055757	-26	0
7.89	.1039230	.1022958	.1006939	.0991170	.0975645	.0960361	.0945315	.0930504	.0915921	.0901566	-26	0
7.90	.8874347	.8735226	.8598268	.8463441	.8330711	.8200046	.8071415	.7944785	.7820127	.7697409	-27	0
7.91	.7576602	.7457676	.7340602	.7225351	.7111896	.7000207	.6890260	.6782026	.6675478	.6570591	-27	0
7.92	.6467339	.6365697	.6265640	.6167144	.6070184	.5974735	.5880777	.5788283	.5697234	.5607606	-27	0
7.93	.5519376	.5432523	.5347027	.5262866	.5180019	.5098466	.5018187	.4939163	.4861372	.4784798	-27	0
7.94	.4709420	.4635221	.4562181	.4490283	.4419510	.4349843	.4281266	.4213762	.4147313	.4081905	-27	0
7.95	.4017520	.3954142	.3891757	.3830348	.3769901	.3710400	.3651831	.3594180	.3537431	.3481572	-27	0
7.96	.3426587	.3372464	.3319190	.3266751	.3215133	.3164325	.3114313	.3065087	.3016631	.2968937	-27	0
7.97	.2921990	.2875779	.2830294	.2785524	.2741455	.2698078	.2655383	.2613357	.2571992	.2531276	-27	0
7.98	.2491201	.2451754	.2412927	.2374711	.2337095	.2300071	.2263628	.2227758	.2192453	.2157702	-27	0
7.99	.2123498	.2089832	.2056696	.2024081	.1991979	.1960382	.1929284	.1898674	.1868546	.1838893	-27	0

$$\frac{2}{\sqrt{\pi}} \exp(-x^2)$$

TABLE 8

Error Function Derivative

X	.0	.001	.002	.003	.004	.005	.006	.007	.008	.009	m	Δ_2
8.00	.1809707	.1780980	.1752706	.1724877	.1697487	.1670528	.1643994	.1617879	.1592174	.1566876	-27	0
8.01	.1541976	.1517469	.1493348	.1469608	.1446242	.1423245	.1400611	.1378334	.1356409	.1334829	-27	0
8.02	.1313591	.1292688	.1272115	.1251866	.1231938	.1212325	.1193021	.1174022	.1155324	.1136921	-27	0
8.03	.1118809	.1100983	.1083440	.1066173	.1049179	.1032455	.1015995	.0999796	.0983852	.0968161	-27	0
8.04	.9527191	.9375209	.9225634	.9078426	.8933551	.8790969	.8650645	.8512545	.8376632	.8242872	-28	0
8.05	.8111233	.7981680	.7854180	.7728702	.7605213	.7483682	.7364078	.7246372	.7130532	.7016531	-28	0
8.06	.6904338	.6793925	.6685265	.6578330	.6473092	.6369525	.6267602	.6167299	.6068587	.5971445	-28	0
8.07	.5875845	.5781764	.5689179	.5598064	.5508398	.5420157	.5333320	.5247863	.5163764	.5081004	-28	0
8.08	.4999560	.4919412	.4840538	.4762920	.4686537	.4611369	.4537399	.4464605	.4392971	.4322478	-28	0
8.09	.4253107	.4184842	.4117664	.4051555	.3986501	.3922483	.3859485	.3797492	.3736487	.3676454	-28	0
8.10	.3617380	.3559247	.3502041	.3445747	.3390352	.3335841	.3282200	.3229414	.3177472	.3126358	-28	0
8.11	.3076061	.3026567	.2977863	.2929937	.2882776	.2836369	.2790704	.2745768	.2701550	.2658039	-28	0
8.12	.2615224	.2573093	.2531636	.2490842	.2450700	.2411200	.2372333	.2334087	.2296453	.2259421	-28	0
8.13	.2222983	.2187127	.2151845	.2117129	.2082968	.2049354	.2016279	.1983734	.1951710	.1920199	-28	0
8.14	.1889193	.1858684	.1828664	.1799126	.1770061	.1741462	.1713321	.1685632	.1658388	.1631580	-28	0
8.15	.1605202	.1579248	.1553710	.1528583	.1503858	.1479530	.1455594	.1432041	.1408867	.1386065	-28	0
8.16	.1363629	.1341554	.1319833	.1298462	.1277434	.1256744	.1236387	.1216357	.1196650	.1177259	-28	0
8.17	.1158180	.1139408	.1120938	.1102765	.1084884	.1067292	.1049983	.1032952	.1016195	.0999709	-28	0
8.18	.9834877	.9675278	.9518250	.9363751	.9211742	.9062182	.8915033	.8770256	.8627812	.8487666	-29	0
8.19	.8349778	.8214115	.8080639	.7949317	.7820113	.7692993	.7567925	.7444876	.7323812	.7204703	-29	0
8.20	.7087516	.6972222	.6858789	.6747189	.6637391	.6529367	.6423088	.6318526	.6215654	.6114445	-29	0
8.21	.6014872	.5916908	.5820528	.5725707	.5632419	.5540640	.5450345	.5361512	.5274115	.5188133	-29	0
8.22	.5103543	.5020322	.4938447	.4857898	.4778654	.4700693	.4623995	.4548538	.4474305	.4401274	-29	0
8.23	.4329426	.4258742	.4189205	.4120795	.4053493	.3987283	.3922146	.3858066	.3795025	.3733007	-29	0
8.24	.3671994	.3611972	.3552924	.3494834	.3437687	.3381468	.3326162	.3271753	.3218228	.3165572	-29	0
8.25	.3113772	.3062813	.3012683	.2963367	.2914852	.2867125	.2820175	.2773988	.2728552	.2683854	-29	0
8.26	.2639884	.2596629	.2554077	.2512218	.2471039	.2430531	.2390683	.2351483	.2312920	.2274986	-29	0
8.27	.2237669	.2200961	.2164850	.2129327	.2094383	.2060008	.2026194	.1992930	.1960208	.1928021	-29	0
8.28	.1896358	.1865211	.1834571	.1804432	.1774784	.1745620	.1716932	.1688712	.1660952	.1633645	-29	0
8.29	.1606784	.1580362	.1554371	.1528804	.1503655	.1478916	.1454582	.1430646	.1407100	.1383939	-29	0
8.30	.1361157	.1338747	.1316703	.1295019	.1273690	.1252711	.1232074	.1211774	.1191807	.1172166	-29	0
8.31	.1152847	.1133844	.1115152	.1096766	.1078681	.1060892	.1043394	.1026183	.1009253	.0992602	-29	0
8.32	.9762223	.9601114	.9442645	.9286773	.9133455	.8982651	.8834320	.8688420	.8544912	.8403759	-30	0
8.33	.8264920	.8128359	.7994038	.7861922	.7731973	.7604157	.7478439	.7354785	.7233161	.7113534	-30	0
8.34	.6995870	.6880140	.6766312	.6654352	.6544232	.6435922	.6329392	.6224612	.6121554	.6020191	-30	0
8.35	.5920495	.5822438	.5725994	.5631135	.5537837	.5446075	.5355821	.5267053	.5179746	.5093876	-30	0
8.36	.5009419	.4926354	.4844655	.4764302	.4685273	.4607545	.4531097	.4455909	.4381959	.4309229	-30	0
8.37	.4237697	.4167345	.4098151	.4030100	.3963169	.3897343	.3832603	.3768930	.3706307	.3644718	-30	0
8.38	.3584145	.3524572	.3465982	.3408359	.3351688	.3295952	.3241137	.3187226	.3134207	.3082063	-30	0
8.39	.3030780	.2980345	.2930743	.2881961	.2833986	.2786803	.2740400	.2694765	.2649884	.2605746	-30	0
8.40	.2562338	.2519647	.2477664	.2436374	.2395768	.2355834	.2316561	.2277939	.2239956	.2202601	-30	0
8.41	.2165865	.2129738	.2094209	.2059269	.2024907	.1991115	.1957883	.1925202	.1893063	.1861456	-30	0
8.42	.1830373	.1799806	.1769746	.1740184	.1711113	.1682524	.1654409	.1626761	.1599572	.1572834	-30	0
8.43	.1546540	.1520682	.1495254	.1470248	.1445657	.1421475	.1397694	.1374309	.1351312	.1328697	-30	0
8.44	.1306458	.1284589	.1263083	.1241935	.1221139	.1200688	.1180578	.1160802	.1141354	.1122231	-30	0
8.45	.1103426	.1084934	.1066749	.1048867	.1031283	.1013992	.0996988	.0980268	.0963826	.0947658	-30	0
8.46	.9317601	.9161264	.9007533	.8856363	.8707713	.8561541	.8417805	.8276467	.8137485	.8000821	-31	0
8.47	.7866437	.7734294	.7604355	.7476586	.7350947	.7227406	.7105927	.6986475	.6869018	.6753522	-31	0
8.48	.6639955	.6528285	.6418479	.6310508	.6204340	.6099948	.5997299	.5896366	.5797120	.5699533	-31	0
8.49	.5603577	.5509226	.5416453	.5325232	.5235536	.5147340	.5060621	.4975352	.4891511	.4809072	-31	0

TABLE 8

Error Function Derivative

X	.0	.001	.002	.003	.004	.005	.006	.007	.008	.009	m	Δ_2
8.50	.4728013	.4648312	.4569945	.4492890	.4417126	.4342630	.4269382	.4197361	.4126548	.4056920	-31	0
8.51	.3988460	.3921147	.3854962	.3789887	.3725902	.3662992	.3601135	.3540316	.3480518	.3421723	-31	0
8.52	.3363914	.3307075	.3251190	.3196243	.3142219	.3089101	.3036875	.2985526	.2935040	.2885402	-31	0
8.53	.2836597	.2788612	.2741434	.2695048	.2649442	.2604603	.2560517	.2517172	.2474556	.2432657	-31	0
8.54	.2391462	.2350961	.2311140	.2271990	.2233498	.2195655	.2158447	.2121866	.2085901	.2050542	-31	0
8.55	.2015778	.1981599	.1947996	.1914959	.1882478	.1850545	.1819149	.1788283	.1757937	.1728103	-31	0
8.56	.1698771	.1669934	.1641583	.1613711	.1586307	.1559367	.1532881	.1506842	.1481242	.1456074	-31	0
8.57	.1431331	.1407006	.1383091	.1359580	.1336466	.1313742	.1291403	.1269440	.1247849	.1226622	-31	0
8.58	.1205754	.1185238	.1165070	.1145242	.1125749	.1106586	.1087747	.1069227	.1051020	.1033121	-31	0
8.59	.1015524	.0998225	.0981219	.0964502	.0948066	.0931909	.0916026	.0900411	.0885060	.0869970	-31	0
8.60	.8551359	.8405525	.8262162	.8121227	.7982681	.7846482	.7712592	.7580971	.7451581	.7324386	-32	0
8.61	.7199346	.7076428	.6955594	.6836810	.6720041	.6605253	.6492413	.6381488	.6272445	.6165254	-32	0
8.62	.6059881	.5956299	.5854474	.5754380	.5655985	.5559261	.5464181	.5370716	.5278839	.5188524	-32	0
8.63	.5099744	.5012472	.4926684	.4842355	.4759460	.4677975	.4597875	.4519138	.4441740	.4365659	-32	0
8.64	.4290873	.4217359	.4145097	.4074065	.4004242	.3935608	.3868142	.3801825	.3736638	.3672561	-32	0
8.65	.3609576	.3547664	.3486806	.3426986	.3368186	.3310387	.3253574	.3197730	.3142838	.3088882	-32	0
8.66	.3035846	.2983715	.2932473	.2882106	.2832597	.2783934	.2736102	.2689086	.2642872	.2597448	-32	0
8.67	.2552798	.2508912	.2465775	.2423375	.2381699	.2340735	.2300472	.2260896	.2221997	.2183762	-32	0
8.68	.2146181	.2109243	.2072936	.2037250	.2002175	.1967700	.1933814	.1900508	.1867772	.1835596	-32	0
8.69	.1803970	.1772887	.1742335	.1712306	.1682792	.1653783	.1625270	.1597247	.1569703	.1542631	-32	0
8.70	.1516023	.1489871	.1464167	.1438903	.1414073	.1389669	.1365682	.1342108	.1318937	.1296164	-32	0
8.71	.1273782	.1251783	.1230162	.1208912	.1188027	.1167501	.1147326	.1127498	.1108010	.1088858	-32	0
8.72	.1070033	.1051533	.1033350	.1015480	.0997916	.0980654	.0963690	.0947016	.0930629	.0914525	-32	0
8.73	.8986967	.8831407	.8678523	.8528268	.8380599	.8235469	.8092837	.7952659	.7814894	.7679499	-33	0
8.74	.7546436	.7415663	.7287142	.7160833	.7036700	.6914706	.6794812	.6676984	.6561186	.6447383	-33	0
8.75	.6335542	.6225628	.6117609	.6011452	.5907125	.5804598	.5703838	.5604817	.5507503	.5411868	-33	0
8.76	.5317883	.5225520	.5134751	.5045548	.4957886	.4871736	.4787074	.4703873	.4622110	.4541759	-33	0
8.77	.4462795	.4385196	.4308937	.4233996	.4160351	.4087977	.4016855	.3946962	.3878278	.3810782	-33	0
8.78	.3744452	.3679270	.3615215	.3552268	.3490410	.3429623	.3369887	.3311186	.3253500	.3196813	-33	0
8.79	.3141107	.3086366	.3032573	.2979711	.2927765	.2876719	.2826557	.2777264	.2728825	.2681226	-33	0
8.80	.2634452	.2588489	.2543322	.2498939	.2455325	.2412468	.2370354	.2328970	.2288305	.2248345	-33	0
8.81	.2209078	.2170493	.2132577	.2095320	.2058709	.2022734	.1987384	.1952648	.1918514	.1884974	-33	0
8.82	.1852017	.1819632	.1787810	.1756541	.1725815	.1695623	.1665956	.1636805	.1608161	.1580015	-33	0
8.83	.1552358	.1525183	.1498480	.1472242	.1446460	.1421127	.1396235	.1371776	.1347743	.1324128	-33	0
8.84	.1300924	.1278125	.1255723	.1233710	.1212081	.1190829	.1169948	.1149430	.1129270	.1109461	-33	0
8.85	.1089997	.1070873	.1052082	.1033619	.1015478	.0997653	.0980139	.0962931	.0946022	.0929409	-33	0
8.86	.9130865	.8970482	.8812899	.8658067	.8505937	.8356465	.8209602	.8065304	.7923527	.7784227	-34	0
8.87	.7647360	.7512884	.7380759	.7250943	.7123395	.6998078	.6874951	.6753976	.6635118	.6518338	-34	0
8.88	.6403601	.6290870	.6180112	.6071292	.5964376	.5859331	.5756125	.5654725	.5555100	.5457219	-34	0
8.89	.5361053	.5266570	.5173743	.5082542	.4992937	.4904904	.4818412	.4733436	.4649950	.4567927	-34	0
8.90	.4487342	.4408169	.4330385	.4253964	.4178885	.4105122	.4032653	.3961455	.3891507	.3822786	-34	0
8.91	.3755271	.3688941	.3623775	.3559754	.3496857	.3435063	.3374355	.3314714	.3256120	.3198556	-34	0
8.92	.3142003	.3086444	.3031861	.2978237	.2925556	.2873800	.2822955	.2773005	.2723932	.2675722	-34	0
8.93	.2628361	.2581832	.2536123	.2491217	.2447101	.2403763	.2361187	.2319360	.2278270	.2237903	-34	0
8.94	.2198247	.2159290	.2121018	.2083421	.2046486	.2010202	.1974558	.1939541	.1905141	.1871349	-34	0
8.95	.1838151	.1805539	.1773502	.1742030	.1711113	.1680742	.1650906	.1621597	.1592805	.1564521	-34	0
8.96	.1536735	.1509441	.1482628	.1456289	.1430414	.1404997	.1380029	.1355501	.1331407	.1307738	-34	0
8.97	.1284488	.1261649	.1239213	.1217173	.1195524	.1174257	.1153365	.1132844	.1112685	.1092883	-34	0
8.98	.1073431	.1054323	.1035554	.1017116	.0999005	.0981214	.0963738	.0946571	.0929709	.0913145	-34	0
8.99	.8968743	.8808917	.8651922	.8497708	.8346227	.8197429	.8051268	.7907697	.7766671	.7628144	-35	0

$$\frac{2}{\sqrt{\pi}} \exp(-x^2)$$

$\frac{2}{\sqrt{\pi}}\exp(-x^2)$

TABLE 8

Error Function Derivative

X	.0	.001	.002	.003	.004	.005	.006	.007	.008	.009	m	Δ_2
9.00	.7492073	.7358415	.7227127	.7098167	.6971494	.6847068	.6724849	.6604799	.6486879	.6371052	-35	0
9.01	.6257280	.6145527	.6035758	.5927939	.5822033	.5718008	.5615831	.5515468	.5416888	.5320060	-35	0
9.02	.5224952	.5131533	.5039775	.4949648	.4861123	.4774172	.4688767	.4604880	.4522485	.4441555	-35	0
9.03	.4362065	.4283988	.4207302	.4131979	.4057997	.3985332	.3913959	.3843858	.3775004	.3707377	-35	0
9.04	.3640953	.3575712	.3511634	.3448697	.3386881	.3326167	.3266534	.3207964	.3150438	.3093937	-35	0
9.05	.3038444	.2983940	.2930408	.2877830	.2826190	.2775471	.2725657	.2676731	.2628679	.2581484	-35	0
9.06	.2535132	.2489606	.2444893	.2400979	.2357848	.2315488	.2273884	.2233024	.2192892	.2153478	-35	0
9.07	.2114769	.2076750	.2039412	.2002741	.1966724	.1931352	.1896613	.1862494	.1828985	.1796076	-35	0
9.08	.1763756	.1732013	.1700839	.1670222	.1640153	.1610622	.1581619	.1553137	.1525163	.1497691	-35	0
9.09	.1470710	.1444213	.1418190	.1392633	.1367534	.1342885	.1318678	.1294904	.1271556	.1248627	-35	
9.10	.1226109	.1203994	.1182276	.1160947	.1140001	.1119431	.1099229	.1079390	.1059907	.1040773	-35	0
9.11	.1021984	.1003531	.0985409	.0967612	.0950136	.0932972	.0916117	.0899565	.0883310	.0867347	-35	0
9.12	.8516715	.8362769	.8211591	.8063129	.7917335	.7774162	.7633563	.7495491	.7359902	.7226751	-36	0
9.13	.7095996	.6967592	.6841497	.6717672	.6596074	.6476665	.6359404	.6244254	.6131176	.6020135	-36	0
9.14	.5911092	.5804013	.5698863	.5595606	.5494209	.5394639	.5296862	.5200847	.5106563	.5013978	-36	0
9.15	.4923062	.4833784	.4746116	.4660029	.4575494	.4492483	.4410970	.4330927	.4252328	.4175147	-36	0
9.16	.4099359	.4024938	.3951861	.3880103	.3809640	.3740449	.3672507	.3605793	.3540283	.3475956	-36	0
9.17	.3412791	.3350768	.3289865	.3230063	.3171341	.3113681	.3057063	.3001468	.2946879	.2893276	-36	0
9.18	.2840644	.2788962	.2738216	.2688388	.2639461	.2591419	.2544246	.2497928	.2452447	.2407790	-36	0
9.19	.2363942	.2320887	.2278612	.2237102	.2196345	.2156326	.2117031	.2078448	.2040564	.2003368	-36	
9.20	.1966845	.1930984	.1895773	.1861200	.1827254	.1793924	.1761199	.1729066	.1697517	.1666540	-36	0
9.21	.1636125	.1606262	.1576940	.1548151	.1519884	.1492131	.1464882	.1438127	.1411858	.1386066	-36	0
9.22	.1360742	.1335879	.1311467	.1287499	.1263966	.1240861	.1218176	.1195903	.1174035	.1152564	-36	0
9.23	.1131485	.1110788	.1090468	.1070517	.1050929	.1031697	.1012816	.0994278	.0976077	.0958208	-36	0
9.24	.9406639	.9234391	.9065280	.8899248	.8736238	.8576198	.8419073	.8264810	.8113357	.7964664	-37	0
9.25	.7818680	.7675356	.7534645	.7396499	.7260870	.7127715	.6996987	.6868644	.6742641	.6618937	-37	0
9.26	.6497489	.6378256	.6261199	.6146278	.6033455	.5922690	.5813947	.5707189	.5602381	.5499486	-37	0
9.27	.5398471	.5299300	.5201941	.5106360	.5012525	.4920405	.4829968	.4741183	.4654022	.4568453	-37	0
9.28	.4484449	.4401981	.4321021	.4241541	.4163515	.4086916	.4011719	.3937896	.3865425	.3794280	-37	0
9.29	.3724437	.3655872	.3588563	.3522485	.3457617	.3393937	.3331423	.3270054	.3209810	.3150669	-37	0
9.30	.3092611	.3035617	.2979667	.2924743	.2870826	.2817897	.2765937	.2714931	.2664860	.2615708	-37	0
9.31	.2567456	.2520090	.2473593	.2427949	.2383142	.2339157	.2295980	.2253595	.2211988	.2171145	-37	0
9.32	.2131052	.2091694	.2053061	.2015136	.1977908	.1941364	.1905491	.1870277	.1835710	.1801779	-37	0
9.33	.1768472	.1735776	.1703682	.1672177	.1641253	.1610896	.1581098	.1551848	.1523136	.1494952	-37	0
9.34	.1467288	.1440132	.1413475	.1387310	.1361626	.1336415	.1311668	.1287377	.1263533	.1240128	-37	0
9.35	.1217154	.1194603	.1172469	.1150742	.1129414	.1108481	.1087933	.1067764	.1047966	.1028534	-37	0
9.36	.1009460	.0990738	.0972360	.0954323	.0936617	.0919238	.0902181	.0885437	.0869003	.0852872	-37	0
9.37	.8370393	.8214984	.8062445	.7912722	.7765764	.7621520	.7479941	.7340976	.7204579	.7070702	-38	0
9.38	.6939300	.6810325	.6683734	.6559483	.6437529	.6317830	.6200344	.6085031	.5971850	.5860763	-38	0
9.39	.5751731	.5644715	.5539680	.5436589	.5335405	.5236094	.5138621	.5042953	.4949056	.4856897	-38	0
9.40	.4766445	.4677668	.4590536	.4505017	.4421083	.4338704	.4257852	.4178498	.4100614	.4024174	-38	0
9.41	.3949152	.3875520	.3803253	.3732327	.3662716	.3594396	.3527343	.3461534	.3396946	.3333557	-38	0
9.42	.3271344	.3210285	.3150360	.3091547	.3033827	.2977178	.2921581	.2867016	.2813465	.2760909	-38	0
9.43	.2709329	.2658706	.2609025	.2560267	.2512416	.2465453	.2419364	.2374132	.2329740	.2286174	-38	0
9.44	.2243418	.2201458	.2160277	.2119864	.2080202	.2041277	.2003077	.1965588	.1928797	.1892691	-38	0
9.45	.1857257	.1822483	.1788356	.1754864	.1721997	.1689742	.1658087	.1627023	.1596537	.1566619	-38	0
9.46	.1537259	.1508446	.1480169	.1452421	.1425189	.1398466	.1372240	.1346504	.1321248	.1296463	-38	0
9.47	.1272140	.1248271	.1224848	.1201862	.1179304	.1157168	.1135445	.1114128	.1093208	.1072679	-38	0
9.48	.1052534	.1032765	.1013365	.0994328	.0975646	.0957313	.0939323	.0921670	.0904346	.0887346	-38	0
9.49	.8706640	.8542938	.8382297	.8224660	.8069971	.7918176	.7769221	.7623053	.7479619	.7338870	-39	0

TABLE 8

Error Function Derivative

X	.0	.001	.002	.003	.004	.005	.006	.007	.008	.009	m	Δ_2
9.50	.7200755	.7065225	.6932232	.6801729	.6673670	.6548008	.6424700	.6303701	.6184968	.6068460	-39	0
9.51	.5954135	.5841951	.5731871	.5623853	.5517859	.5413852	.5311795	.5211652	.5113387	.5016963	-39	0
9.52	.4922349	.4829509	.4738411	.4649022	.4561310	.4475244	.4390793	.4307927	.4226617	.4146833	-39	0
9.53	.4068547	.3991731	.3916357	.3842399	.3769830	.3698624	.3628756	.3560200	.3492934	.3426930	-39	0
9.54	.3362167	.3298622	.3236271	.3175093	.3115065	.3056165	.2998373	.2941668	.2886030	.2831438	-39	0
9.55	.2777874	.2725317	.2673750	.2623153	.2573507	.2524797	.2477004	.2430111	.2384101	.2338957	-39	0
9.56	.2294663	.2251203	.2208562	.2166725	.2125676	.2085400	.2045884	.2007112	.1969071	.1931747	-39	0
9.57	.1895127	.1859197	.1823945	.1789358	.1755423	.1722128	.1689461	.1657411	.1625965	.1595113	-39	0
9.58	.1564844	.1535145	.1506007	.1477419	.1449371	.1421853	.1394854	.1368365	.1342376	.1316879	-39	0
9.59	.1291863	.1267321	.1243241	.1219617	.1196439	.1173699	.1151389	.1129502	.1108028	.1086960	-39	0
9.60	.1066290	.1046012	.1026117	.1006598	.0987449	.0968662	.0950231	.0932148	.0914408	.0897003	-39	0
9.61	.8799290	.8631774	.8467430	.8306198	.8148021	.7992839	.7840597	.7691240	.7544712	.7400962	-40	0
9.62	.7259936	.7121583	.6985852	.6852695	.6722062	.6593907	.6468182	.6344841	.6223840	.6105134	-40	0
9.63	.5988680	.5874436	.5762359	.5652410	.5544547	.5438733	.5334926	.5233091	.5133188	.5035183	-40	0
9.64	.4939040	.4844723	.4752197	.4661429	.4572385	.4485034	.4399342	.4315279	.4232814	.4151917	-40	0
9.65	.4072556	.3994706	.3918335	.3843418	.3769925	.3697829	.3627105	.3557727	.3489670	.3422906	-40	0
9.66	.3357414	.3293168	.3230145	.3168321	.3107675	.3048184	.2989825	.2932578	.2876421	.2821335	-40	0
9.67	.2767297	.2714289	.2662291	.2611284	.2561249	.2512168	.2464022	.2416794	.2370467	.2325023	-40	0
9.68	.2280446	.2236719	.2193826	.2151751	.2110479	.2069994	.2030283	.1991328	.1953118	.1915637	-40	0
9.69	.1878871	.1842807	.1807432	.1772732	.1738695	.1705308	.1672559	.1640435	.1608926	.1578018	-40	0
9.70	.1547701	.1517964	.1488795	.1460183	.1432118	.1404591	.1377589	.1351104	.1325125	.1299644	-40	0
9.71	.1274649	.1250133	.1226086	.1202499	.1179363	.1156670	.1134412	.1112580	.1091166	.1070161	-40	0
9.72	.1049560	.1029353	.1009532	.0990091	.0971023	.0952320	.0933976	.0915982	.0898334	.0881024	-40	0
9.73	.8640463	.8473936	.8310602	.8150401	.7993271	.7839155	.7687995	.7539735	.7394320	.7251694	-41	0
9.74	.7111805	.6974601	.6840030	.6708042	.6578588	.6451619	.6327088	.6204948	.6085154	.5967661	-41	0
9.75	.5852425	.5739402	.5628551	.5519830	.5413198	.5308616	.5206043	.5105442	.5006775	.4910005	-41	0
9.76	.4815096	.4722012	.4630719	.4541181	.4453365	.4367239	.4282770	.4199927	.4118677	.4038992	-41	0
9.77	.3960840	.3884192	.3809021	.3735296	.3662991	.3592079	.3522532	.3454325	.3387432	.3321828	-41	0
9.78	.3257487	.3194386	.3132502	.3071810	.3012288	.2953914	.2896665	.2840520	.2785457	.2731457	-41	0
9.79	.2678497	.2626560	.2575625	.2525672	.2476683	.2428639	.2381523	.2335315	.2290000	.2245560	-41	0
9.80	.2201978	.2159237	.2117322	.2076216	.2035904	.1996371	.1957601	.1919581	.1882296	.1845730	-41	0
9.81	.1809872	.1774706	.1740221	.1706402	.1673237	.1640713	.1608818	.1577541	.1546867	.1516788	-41	0
9.82	.1487290	.1458364	.1429996	.1402178	.1374899	.1348147	.1321913	.1296187	.1270959	.1246220	-41	0
9.83	.1221960	.1198170	.1174840	.1151962	.1129528	.1107528	.1085955	.1064800	.1044055	.1023712	-41	0
9.84	.1003763	.0984201	.0965018	.0946208	.0927762	.0909674	.0891936	.0874544	.0857488	.0840763	-41	0
9.85	.8243634	.8082815	.7925118	.7770482	.7618848	.7470158	.7324356	.7181385	.7041190	.6903719	-42	0
9.86	.6768917	.6636735	.6507121	.6380026	.6255401	.6133197	.6013369	.5895870	.5780656	.5667682	-42	0
9.87	.5556904	.5448281	.5341771	.5237332	.5134925	.5034510	.4936049	.4839504	.4744837	.4652014	-42	0
9.88	.4560997	.4471751	.4384244	.4298440	.4214307	.4131812	.4050925	.3971612	.3893845	.3817593	-42	0
9.89	.3742827	.3669518	.3597637	.3527157	.3458051	.3390293	.3323855	.3258712	.3194841	.3132214	-42	0
9.90	.3070809	.3010602	.2951570	.2893689	.2836937	.2781293	.2726735	.2673243	.2620794	.2569368	-42	0
9.91	.2518947	.2469511	.2421040	.2373515	.2326919	.2281233	.2236440	.2192521	.2149461	.2107242	-42	0
9.92	.2065849	.2025264	.1985473	.1946459	.1908209	.1870706	.1833937	.1797887	.1762542	.1727889	-42	0
9.93	.1693913	.1660601	.1627943	.1595923	.1564530	.1533750	.1503574	.1473988	.1444982	.1416544	-42	0
9.94	.1388662	.1361327	.1334527	.1308252	.1282492	.1257237	.1232476	.1208200	.1184401	.1161067	-42	0
9.95	.1138192	.1115764	.1093777	.1072221	.1051087	.1030368	.1010055	.0990141	.0970618	.0951477	-42	0
9.96	.9327120	.9143153	.8962796	.8785980	.8612635	.8442693	.8276087	.8112753	.7952627	.7795646	-43	0
9.97	.7641748	.7490873	.7342962	.7197958	.7055802	.6916441	.6779818	.6645881	.6514577	.6385854	-43	0
9.98	.6259663	.6135952	.6014675	.5895783	.5779229	.5664968	.5552955	.5443146	.5335498	.5229968	-43	0
9.99	.5126516	.5025100	.4925680	.4828217	.4732674	.4639011	.4547194	.4457185	.4368948	.4282450	-43	0

$$\frac{2}{\sqrt{\pi}}\exp(-x^2)$$

TABLE 9A

Fresnel Sine Integral

U	.000	.001	.002	.003	.004	.005	.006	.007	.008	.009	Δ_2
0.01	.000001	.000001	.000001	.000001	.000001	.000002	.000002	.000003	.000003	.000004	0
0.02	.000004	.000005	.000006	.000006	.000007	.000008	.000009	.000010	.000011	.000013	0
0.03	.000014	.000016	.000017	.000019	.000021	.000022	.000024	.000027	.000029	.000031	0
0.04	.000034	.000036	.000039	.000042	.000045	.000048	.000051	.000054	.000058	.000062	0
0.05	.000065	.000069	.000074	.000078	.000082	.000087	.000092	.000097	.000102	.000108	0
0.06	.000113	.000119	.000125	.000131	.000137	.000144	.000151	.000157	.000165	.000172	0
0.07	.000180	.000187	.000195	.000204	.000212	.000221	.000230	.000239	.000248	.000258	0
0.08	.000268	.000278	.000289	.000299	.000310	.000322	.000333	.000345	.000357	.000369	0
0.09	.000382	.000395	.000408	.000421	.000435	.000449	.000463	.000478	.000493	.000508	0
0.10	.000524	.000539	.000556	.000572	.000589	.000606	.000624	.000641	.000660	.000678	0
0.11	.000697	.000716	.000736	.000755	.000776	.000796	.000817	.000839	.000860	.000882	0
0.12	.000905	.000928	.000951	.000974	.000998	.001023	.001047	.001072	.001098	.001124	0
0.13	.001150	.001177	.001204	.001232	.001260	.001288	.001317	.001346	.001376	.001406	0
0.14	.001437	.001468	.001499	.001531	.001563	.001596	.001629	.001663	.001697	.001732	0
0.15	.001767	.001803	.001839	.001875	.001912	.001950	.001988	.002026	.002065	.002104	0
0.16	.002144	.002185	.002226	.002267	.002309	.002352	.002395	.002438	.002482	.002527	0
0.17	.002572	.002618	.002664	.002711	.002758	.002806	.002854	.002903	.002952	.003002	0
0.18	.003053	.003104	.003156	.003208	.003261	.003315	.003369	.003423	.003478	.003534	0
0.19	.003591	.003648	.003705	.003763	.003822	.003881	.003941	.004002	.004063	.004125	0
0.20	.004188	.004251	.004314	.004379	.004444	.004509	.004576	.004643	.004710	.004778	0
0.21	.004847	.004917	.004987	.005058	.005130	.005202	.005275	.005348	.005422	.005497	0
0.22	.005573	.005649	.005726	.005804	.005882	.005961	.006041	.006122	.006203	.006285	0
0.23	.006367	.006451	.006535	.006620	.006705	.006792	.006879	.006966	.007055	.007144	0
0.24	.007234	.007325	.007416	.007508	.007601	.007695	.007790	.007885	.007981	.008078	0
0.25	.008176	.008274	.008373	.008473	.008574	.008676	.008778	.008881	.008985	.009090	0
0.26	.009195	.009302	.009409	.009517	.009626	.009736	.009846	.009957	.010070	.010183	0
0.27	.010296	.010411	.010527	.010643	.010760	.010878	.010997	.011117	.011238	.011359	0
0.28	.011482	.011605	.011729	.011854	.011980	.012107	.012234	.012363	.012493	.012623	0
0.29	.012754	.012886	.013019	.013153	.013288	.013424	.013561	.013698	.013837	.013977	0
0.30	.014117	.014258	.014401	.014544	.014688	.014833	.014979	.015126	.015274	.015423	0
0.31	.015573	.015724	.015876	.016029	.016182	.016337	.016493	.016650	.016807	.016966	0
0.32	.017126	.017286	.017448	.017611	.017774	.017939	.018105	.018271	.018439	.018608	1
0.33	.018777	.018948	.019120	.019293	.019466	.019641	.019817	.019994	.020172	.020351	1
0.34	.020531	.020712	.020894	.021078	.021262	.021447	.021634	.021821	.022010	.022199	1
0.35	.022390	.022582	.022775	.022969	.023164	.023360	.023557	.023755	.023955	.024155	1
0.36	.024357	.024560	.024763	.024968	.025174	.025382	.025590	.025799	.026010	.026222	1
0.37	.026434	.026648	.026863	.027080	.027297	.027516	.027735	.027956	.028178	.028401	1
0.38	.028626	.028851	.029078	.029305	.029534	.029765	.029996	.030228	.030462	.030697	1
0.39	.030933	.031170	.031409	.031648	.031889	.032131	.032374	.032619	.032864	.033111	1
0.40	.033359	.033609	.033859	.034111	.034364	.034618	.034874	.035130	.035388	.035647	1
0.41	.035908	.036169	.036432	.036696	.036962	.037228	.037496	.037765	.038036	.038307	1
0.42	.038580	.038854	.039130	.039407	.039685	.039964	.040244	.040526	.040809	.041094	1
0.43	.041380	.041667	.041955	.042245	.042535	.042828	.043121	.043416	.043712	.044010	1
0.44	.044308	.044609	.044910	.045213	.045517	.045822	.046129	.046437	.046746	.047057	1
0.45	.047369	.047683	.047997	.048314	.048631	.048950	.049270	.049592	.049914	.050239	1
0.46	.050564	.050891	.051220	.051549	.051880	.052213	.052547	.052882	.053219	.053556	1
0.47	.053896	.054237	.054579	.054922	.055267	.055614	.055961	.056310	.056661	.057013	1
0.48	.057366	.057721	.058077	.058435	.058794	.059154	.059516	.059879	.060244	.060610	1
0.49	.060978	.061347	.061717	.062089	.062462	.062837	.063213	.063591	.063970	.064350	1

TABLE 9A

Fresnel Sine Integral

U	.000	.001	.002	.003	.004	.005	.006	.007	.008	.009	Δ_2
0.50	.064732	.065116	.065501	.065887	.066275	.066664	.067055	.067447	.067841	.068236	1
0.51	.068632	.069030	.069430	.069831	.070233	.070637	.071042	.071449	.071858	.072268	1
0.52	.072679	.073092	.073506	.073922	.074339	.074758	.075178	.075600	.076023	.076448	1
0.53	.076874	.077302	.077732	.078162	.078595	.079029	.079464	.079901	.080339	.080779	1
0.54	.081221	.081664	.082108	.082554	.083002	.083451	.083901	.084353	.084807	.085262	1
0.55	.085719	.086177	.086637	.087098	.087561	.088026	.088492	.088959	.089428	.089899	1
0.56	.090371	.090844	.091320	.091797	.092275	.092755	.093236	.093719	.094204	.094690	1
0.57	.095178	.095667	.096158	.096650	.097144	.097640	.098137	.098635	.099136	.099637	1
0.58	.100141	.100646	.101152	.101660	.102170	.102681	.103194	.103709	.104225	.104742	1
0.59	.105261	.105782	.106304	.106828	.107354	.107881	.108410	.108940	.109472	.110005	1
0.60	.110540	.111077	.111615	.112155	.112696	.113239	.113784	.114330	.114878	.115427	1
0.61	.115978	.116531	.117085	.117641	.118198	.118757	.119318	.119880	.120444	.121009	1
0.62	.121576	.122145	.122715	.123287	.123860	.124435	.125012	.125590	.126170	.126751	1
0.63	.127334	.127919	.128505	.129093	.129682	.130273	.130866	.131460	.132056	.132654	1
0.64	.133253	.133854	.134456	.135060	.135665	.136273	.136881	.137492	.138104	.138717	1
0.65	.139332	.139949	.140568	.141188	.141809	.142433	.143057	.143684	.144312	.144942	1
0.66	.145573	.146206	.146840	.147476	.148114	.148753	.149394	.150037	.150681	.151327	1
0.67	.151974	.152623	.153273	.153926	.154579	.155235	.155892	.156550	.157210	.157872	1
0.68	.158535	.159200	.159867	.160535	.161205	.161876	.162549	.163223	.163900	.164577	1
0.69	.165256	.165937	.166620	.167304	.167989	.168677	.169365	.170056	.170748	.171441	1
0.70	.172136	.172833	.173531	.174231	.174933	.175636	.176340	.177046	.177754	.178463	1
0.71	.179174	.179887	.180601	.181316	.182033	.182752	.183472	.184194	.184918	.185642	1
0.72	.186369	.187097	.187827	.188558	.189290	.190025	.190760	.191498	.192236	.192977	1
0.73	.193719	.194462	.195207	.195954	.196702	.197451	.198202	.198955	.199709	.200465	1
0.74	.201222	.201981	.202741	.203503	.204266	.205031	.205797	.206565	.207334	.208105	1
0.75	.208877	.209651	.210426	.211203	.211981	.212761	.213542	.214325	.215109	.215894	1
0.76	.216682	.217470	.218260	.219052	.219845	.220639	.221435	.222232	.223031	.223831	1
0.77	.224633	.225436	.226241	.227047	.227854	.228663	.229473	.230285	.231098	.231913	1
0.78	.232729	.233546	.234365	.235185	.236007	.236830	.237654	.238480	.239307	.240136	1
0.79	.240966	.241797	.242630	.243464	.244300	.245137	.245975	.246814	.247655	.248498	1
0.80	.249341	.250186	.251033	.251880	.252729	.253580	.254431	.255284	.256139	.256994	1
0.81	.257851	.258710	.259569	.260430	.261292	.262156	.263020	.263886	.264754	.265622	1
0.82	.266492	.267363	.268236	.269109	.269984	.270860	.271738	.272616	.273496	.274377	1
0.83	.275260	.276143	.277028	.277914	.278801	.279690	.280579	.281470	.282362	.283255	1
0.84	.284150	.285045	.285942	.286840	.287739	.288639	.289541	.290443	.291347	.292252	1
0.85	.293158	.294065	.294973	.295882	.296792	.297704	.298617	.299530	.300445	.301361	1
0.86	.302278	.303196	.304115	.305035	.305957	.306879	.307802	.308727	.309652	.310578	1
0.87	.311506	.312434	.313364	.314294	.315226	.316158	.317092	.318026	.318962	.319898	0
0.88	.320836	.321774	.322713	.323653	.324595	.325537	.326480	.327424	.328369	.329314	0
0.89	.330261	.331209	.332157	.333107	.334057	.335008	.335960	.336913	.337866	.338821	0
0.90	.339776	.340733	.341690	.342647	.343606	.344566	.345526	.346487	.347449	.348411	0
0.91	.349375	.350339	.351304	.352269	.353236	.354203	.355171	.356139	.357109	.358079	0
0.92	.359049	.360021	.360993	.361965	.362939	.363913	.364888	.365863	.366839	.367816	0
0.93	.368793	.369771	.370749	.371728	.372708	.373688	.374669	.375651	.376633	.377615	0
0.94	.378598	.379582	.380566	.381550	.382536	.383521	.384507	.385494	.386481	.387469	0
0.95	.388457	.389445	.390434	.391424	.392414	.393404	.394395	.395386	.396377	.397369	0
0.96	.398361	.399354	.400347	.401340	.402334	.403328	.404322	.405317	.406312	.407307	0
0.97	.408303	.409298	.410294	.411291	.412287	.413284	.414281	.415279	.416276	.417274	0
0.98	.418272	.419270	.420269	.421267	.422266	.423265	.424264	.425263	.426262	.427261	0
0.99	.428261	.429260	.430260	.431260	.432260	.433259	.434259	.435259	.436259	.437259	0

TABLE 9A

Fresnel Sine Integral

U	.000	.001	.002	.003	.004	.005	.006	.007	.008	.009	Δ_2
1.00	.438259	.439259	.440259	.441259	.442259	.443259	.444259	.445259	.446258	.447258	-0
1.01	.448257	.449257	.450256	.451255	.452255	.453254	.454252	.455251	.456249	.457248	-0
1.02	.458246	.459244	.460241	.461239	.462236	.463233	.464230	.465226	.466222	.467218	-0
1.03	.468214	.469209	.470204	.471199	.472193	.473187	.474180	.475174	.476166	.477159	-0
1.04	.478151	.479142	.480134	.481124	.482114	.483104	.484094	.485082	.486071	.487059	-0
1.05	.488046	.489033	.490019	.491005	.491990	.492975	.493958	.494942	.495925	.496907	-0
1.06	.497888	.498869	.499849	.500829	.501808	.502786	.503764	.504740	.505717	.506692	-0
1.07	.507666	.508640	.509613	.510586	.511557	.512528	.513498	.514467	.515435	.516402	-0
1.08	.517369	.518334	.519299	.520263	.521226	.522188	.523149	.524109	.525068	.526026	-0
1.09	.526983	.527939	.528894	.529849	.530802	.531754	.532705	.533655	.534604	.535551	-1
1.10	.536498	.537443	.538388	.539331	.540273	.541214	.542154	.543092	.544030	.544966	-1
1.11	.545901	.546834	.547767	.548698	.549628	.550556	.551484	.552409	.553334	.554257	-1
1.12	.555179	.556100	.557019	.557937	.558853	.559768	.560681	.561593	.562504	.563413	-1
1.13	.564321	.565227	.566132	.567035	.567936	.568836	.569735	.570632	.571527	.572421	-1
1.14	.573313	.574203	.575092	.575979	.576865	.577749	.578631	.579511	.580390	.581267	-1
1.15	.582142	.583016	.583888	.584758	.585626	.586492	.587357	.588219	.589080	.589939	-1
1.16	.590797	.591652	.592505	.593357	.594206	.595054	.595900	.596743	.597585	.598425	-1
1.17	.599263	.600098	.600932	.601764	.602593	.603421	.604246	.605070	.605891	.606710	-2
1.18	.607527	.608342	.609155	.609966	.610774	.611580	.612384	.613186	.613985	.614783	-2
1.19	.615578	.616371	.617161	.617949	.618735	.619519	.620300	.621079	.621855	.622629	-2
1.20	.623401	.624170	.624937	.625702	.626464	.627223	.627980	.628735	.629487	.630237	-2
1.21	.630984	.631728	.632470	.633210	.633947	.634681	.635413	.636142	.636868	.637592	-2
1.22	.638313	.639032	.639748	.640461	.641171	.641879	.642584	.643287	.643986	.644683	-2
1.23	.645377	.646068	.646757	.647442	.648125	.648805	.649482	.650156	.650828	.651496	-2
1.24	.652162	.652825	.653484	.654141	.654795	.655446	.656094	.656739	.657381	.658020	-3
1.25	.658656	.659288	.659918	.660545	.661169	.661789	.662407	.663021	.663632	.664241	-3
1.26	.664846	.665447	.666046	.666642	.667234	.667823	.668409	.668992	.669571	.670147	-3
1.27	.670720	.671290	.671856	.672419	.672979	.673535	.674088	.674638	.675185	.675728	-3
1.28	.676267	.676803	.677336	.677866	.678392	.678914	.679433	.679949	.680461	.680970	-3
1.29	.681475	.681977	.682475	.682970	.683461	.683949	.684433	.684913	.685390	.685864	-3
1.30	.686333	.686799	.687262	.687721	.688176	.688628	.689076	.689520	.689960	.690397	-3
1.31	.690830	.691260	.691686	.692108	.692526	.692940	.693351	.693758	.694161	.694561	-3
1.32	.694956	.695348	.695736	.696120	.696500	.696877	.697249	.697618	.697983	.698344	-3
1.33	.698701	.699054	.699403	.699749	.700090	.700427	.700761	.701090	.701416	.701737	-3
1.34	.702055	.702369	.702678	.702984	.703285	.703583	.703876	.704166	.704451	.704732	-4
1.35	.705010	.705283	.705552	.705817	.706078	.706335	.706587	.706836	.707080	.707321	-4
1.36	.707557	.707789	.708016	.708240	.708460	.708675	.708886	.709093	.709296	.709494	-4
1.37	.709688	.709878	.710064	.710246	.710423	.710596	.710765	.710930	.711090	.711246	-4
1.38	.711398	.711545	.711688	.711827	.711962	.712092	.712218	.712340	.712457	.712570	-4
1.39	.712678	.712783	.712883	.712978	.713069	.713156	.713239	.713317	.713391	.713460	-4
1.40	.713525	.713586	.713642	.713694	.713741	.713784	.713823	.713857	.713887	.713912	-4
1.41	.713933	.713949	.713961	.713969	.713972	.713971	.713965	.713955	.713940	.713921	-4
1.42	.713898	.713870	.713837	.713800	.713759	.713713	.713663	.713608	.713549	.713485	-4
1.43	.713417	.713344	.713267	.713185	.713099	.713008	.712913	.712814	.712709	.712601	-4
1.44	.712488	.712370	.712248	.712121	.711990	.711855	.711715	.711570	.711421	.711268	-4
1.45	.711109	.710947	.710780	.710608	.710432	.710252	.710067	.709877	.709683	.709485	-4
1.46	.709282	.709074	.708862	.708646	.708425	.708199	.707969	.707735	.707496	.707253	-4
1.47	.707005	.706753	.706496	.706235	.705969	.705699	.705424	.705145	.704862	.704574	-4
1.48	.704281	.703984	.703683	.703377	.703067	.702753	.702433	.702110	.701782	.701450	-4
1.49	.701113	.700772	.700427	.700077	.699722	.699364	.699001	.698633	.698262	.697885	-4

TABLE 9A

Fresnel Sine Integral

U	.000	.001	.002	.003	.004	.005	.006	.007	.008	.009	Δ_2
1.50	.697505	.697120	.696731	.696337	.695939	.695537	.695131	.694720	.694305	.693885	-4
1.51	.693461	.693033	.692601	.692164	.691723	.691278	.690829	.690375	.689917	.689455	-4
1.52	.688989	.688518	.688043	.687564	.687081	.686594	.686102	.685606	.685107	.684603	-4
1.53	.684094	.683582	.683066	.682545	.682020	.681492	.680959	.680422	.679881	.679336	-4
1.54	.678787	.678234	.677676	.677115	.676550	.675981	.675408	.674830	.674249	.673664	-3
1.55	.673075	.672482	.671886	.671285	.670680	.670072	.669459	.668843	.668223	.667599	-3
1.56	.666971	.666340	.665704	.665065	.664422	.663776	.663125	.662471	.661813	.661152	-3
1.57	.660487	.659818	.659145	.658469	.657789	.657106	.656419	.655728	.655034	.654336	-3
1.58	.653635	.652930	.652221	.651510	.650794	.650075	.649353	.648627	.647898	.647166	-3
1.59	.646430	.645690	.644948	.644202	.643452	.642700	.641944	.641185	.640422	.639657	-3
1.60	.638888	.638116	.637340	.636562	.635780	.634995	.634208	.633417	.632623	.631825	-3
1.61	.631025	.630222	.629416	.628607	.627795	.626980	.626162	.625341	.624517	.623690	-2
1.62	.622861	.622028	.621193	.620355	.619514	.618671	.617825	.616976	.616124	.615270	-2
1.63	.614413	.613553	.612691	.611826	.610959	.610089	.609217	.608342	.607465	.606585	-2
1.64	.605703	.604818	.603931	.603041	.602150	.601256	.600359	.599460	.598559	.597656	-2
1.65	.596751	.595843	.594934	.594022	.593108	.592192	.591273	.590353	.589431	.588507	-1
1.66	.587580	.586652	.585722	.584790	.583856	.582920	.581983	.581043	.580102	.579159	-1
1.67	.578215	.577268	.576320	.575371	.574419	.573466	.572512	.571556	.570598	.569639	-1
1.68	.568678	.567716	.566753	.565788	.564822	.563854	.562885	.561915	.560943	.559971	-1
1.69	.558997	.558021	.557045	.556068	.555089	.554109	.553129	.552147	.551164	.550181	-0
1.70	.549196	.548210	.547224	.546237	.545248	.544259	.543270	.542279	.541288	.540296	-0
1.71	.539304	.538310	.537317	.536322	.535327	.534332	.533336	.532339	.531342	.530345	-0
1.72	.529347	.528349	.527351	.526352	.525353	.524354	.523355	.522355	.521356	.520356	-0
1.73	.519356	.518356	.517356	.516356	.515356	.514356	.513356	.512357	.511357	.510358	0
1.74	.509358	.508359	.507361	.506362	.505364	.504367	.503369	.502373	.501376	.500380	0
1.75	.499385	.498390	.497395	.496402	.495409	.494416	.493424	.492433	.491443	.490454	0
1.76	.489465	.488477	.487490	.486504	.485519	.484535	.483552	.482569	.481588	.480608	1
1.77	.479629	.478652	.477675	.476700	.475726	.474753	.473782	.472811	.471843	.470875	1
1.78	.469909	.468945	.467982	.467020	.466061	.465102	.464145	.463190	.462237	.461285	1
1.79	.460335	.459387	.458441	.457496	.456554	.455613	.454674	.453737	.452802	.451870	2
1.80	.450939	.450010	.449083	.448159	.447237	.446317	.445399	.444483	.443570	.442659	2
1.81	.441750	.440844	.439940	.439039	.438140	.437244	.436350	.435459	.434570	.433684	2
1.82	.432801	.431920	.431042	.430167	.429295	.428425	.427558	.426694	.425833	.424975	2
1.83	.424120	.423268	.422419	.421573	.420731	.419891	.419054	.418221	.417390	.416563	3
1.84	.415740	.414919	.414102	.413288	.412478	.411671	.410867	.410067	.409271	.408478	3
1.85	.407688	.406902	.406120	.405341	.404566	.403795	.403027	.402263	.401503	.400747	3
1.86	.399994	.399246	.398501	.397760	.397023	.396291	.395562	.394837	.394116	.393399	4
1.87	.392687	.391978	.391274	.390574	.389878	.389186	.388499	.387816	.387137	.386463	4
1.88	.385792	.385127	.384466	.383809	.383156	.382509	.381865	.381226	.380592	.379963	4
1.89	.379338	.378717	.378102	.377491	.376884	.376283	.375686	.375094	.374507	.373925	4
1.90	.373347	.372775	.372207	.371644	.371087	.370534	.369986	.369443	.368905	.368373	5
1.91	.367845	.367323	.366805	.366293	.365786	.365284	.364788	.364296	.363810	.363329	5
1.92	.362854	.362383	.361919	.361459	.361005	.360556	.360113	.359675	.359242	.358815	5
1.93	.358393	.357977	.357567	.357162	.356763	.356369	.355980	.355598	.355221	.354849	5
1.94	.354484	.354124	.353769	.353421	.353078	.352741	.352409	.352084	.351764	.351450	5
1.95	.351142	.350839	.350543	.350252	.349967	.349689	.349416	.349149	.348887	.348632	5
1.96	.348383	.348140	.347902	.347671	.347446	.347227	.347013	.346806	.346605	.346410	6
1.97	.346221	.346038	.345861	.345690	.345526	.345367	.345215	.345068	.344928	.344794	6
1.98	.344666	.344545	.344429	.344320	.344217	.344120	.344030	.343945	.343867	.343795	6
1.99	.343729	.343670	.343616	.343569	.343529	.343494	.343466	.343444	.343428	.343419	6

TABLE 9A

Fresnel Sine Integral

U	.000	.001	.002	.003	.004	.005	.006	.007	.008	.009	Δ_2
2.00	.343416	.343419	.343428	.343444	.343466	.343494	.343529	.343570	.343617	.343670	6
2.01	.343730	.343796	.343869	.343947	.344032	.344124	.344221	.344325	.344436	.344552	6
2.02	.344675	.344804	.344939	.345081	.345229	.345383	.345544	.345711	.345884	.346063	6
2.03	.346249	.346441	.346639	.346843	.347054	.347271	.347494	.347723	.347959	.348201	6
2.04	.348449	.348703	.348963	.349230	.349503	.349782	.350067	.350358	.350656	.350959	6
2.05	.351269	.351585	.351906	.352235	.352569	.352909	.353255	.353607	.353966	.354330	6
2.06	.354700	.355077	.355459	.355847	.356242	.356642	.357048	.357460	.357878	.358302	5
2.07	.358732	.359168	.359609	.360057	.360510	.360969	.361433	.361904	.362380	.362862	5
2.08	.363350	.363843	.364342	.364847	.365357	.365873	.366395	.366922	.367455	.367993	5
2.09	.368537	.369086	.369641	.370201	.370767	.371338	.371914	.372496	.373083	.373676	5
2.10	.374273	.374876	.375485	.376098	.376717	.377341	.377970	.378604	.379243	.379888	5
2.11	.380537	.381192	.381851	.382516	.383185	.383859	.384539	.385223	.385911	.386605	4
2.12	.387304	.388007	.388715	.389427	.390145	.390867	.391593	.392324	.393060	.393800	4
2.13	.394545	.395294	.396047	.396805	.397568	.398334	.399105	.399880	.400660	.401443	4
2.14	.402231	.403023	.403819	.404619	.405423	.406231	.407042	.407858	.408678	.409502	3
2.15	.410329	.411160	.411995	.412834	.413676	.414522	.415371	.416224	.417081	.417941	3
2.16	.418804	.419671	.420542	.421415	.422292	.423172	.424055	.424942	.425832	.426724	3
2.17	.427620	.428519	.429420	.430325	.431233	.432143	.433056	.433972	.434891	.435812	2
2.18	.436736	.437663	.438592	.439524	.440458	.441395	.442333	.443275	.444218	.445164	2
2.19	.446112	.447062	.448015	.448969	.449926	.450884	.451845	.452807	.453771	.454737	1
2.20	.455705	.456674	.457645	.458618	.459592	.460568	.461545	.462524	.463504	.464486	1
2.21	.465469	.466453	.467438	.468425	.469412	.470401	.471391	.472381	.473373	.474365	0
2.22	.475358	.476353	.477347	.478343	.479339	.480336	.481333	.482331	.483329	.484327	0
2.23	.485326	.486326	.487325	.488325	.489325	.490325	.491325	.492325	.493324	.494324	−0
2.24	.495324	.496324	.497323	.498322	.499320	.500319	.501316	.502314	.503311	.504307	−0
2.25	.505302	.506297	.507291	.508284	.509277	.510268	.511259	.512249	.513237	.514225	−1
2.26	.515211	.516196	.517180	.518163	.519144	.520124	.521102	.522079	.523055	.524028	−1
2.27	.525000	.525971	.526940	.527907	.528872	.529835	.530796	.531755	.532712	.533667	−2
2.28	.534620	.535571	.536519	.537466	.538410	.539351	.540290	.541226	.542160	.543092	−2
2.29	.544020	.544947	.545870	.546790	.547708	.548623	.549535	.550443	.551349	.552252	−3
2.30	.553152	.554048	.554941	.555831	.556718	.557601	.558481	.559357	.560230	.561099	−3
2.31	.561965	.562827	.563685	.564540	.565391	.566238	.567081	.567920	.568755	.569586	−4
2.32	.570413	.571236	.572054	.572869	.573679	.574485	.575287	.576084	.576877	.577665	−4
2.33	.578449	.579228	.580003	.580773	.581538	.582298	.583054	.583805	.584551	.585292	−4
2.34	.586028	.586760	.587486	.588207	.588923	.589634	.590339	.591039	.591735	.592424	−5
2.35	.593109	.593788	.594461	.595129	.595792	.596449	.597100	.597746	.598386	.599020	−5
2.36	.599649	.600272	.600889	.601500	.602105	.602704	.603298	.603885	.604467	.605042	−6
2.37	.605611	.606174	.606731	.607282	.607826	.608364	.608896	.609421	.609941	.610453	−6
2.38	.610960	.611459	.611953	.612440	.612920	.613394	.613861	.614321	.614775	.615222	−6
2.39	.615662	.616096	.616523	.616943	.617356	.617762	.618162	.618554	.618940	.619318	−6
2.40	.619690	.620055	.620412	.620763	.621106	.621442	.621772	.622094	.622409	.622716	−7
2.41	.623017	.623310	.623596	.623875	.624146	.624411	.624667	.624917	.625159	.625394	−7
2.42	.625621	.625841	.626054	.626259	.626456	.626646	.626829	.627004	.627172	.627332	−7
2.43	.627485	.627630	.627767	.627897	.628020	.628134	.628241	.628341	.628433	.628517	−7
2.44	.628594	.628663	.628724	.628778	.628824	.628862	.628893	.628916	.628931	.628939	−7
2.45	.628939	.628931	.628915	.628892	.628861	.628823	.628776	.628722	.628661	.628591	−7
2.46	.628514	.628429	.628337	.628237	.628129	.628013	.627890	.627759	.627620	.627474	−7
2.47	.627320	.627158	.626989	.626812	.626627	.626435	.626235	.626028	.625813	.625590	−7
2.48	.625360	.625122	.624877	.624624	.624363	.624095	.623820	.623536	.623246	.622948	−7
2.49	.622643	.622330	.622009	.621682	.621347	.621004	.620654	.620297	.619933	.619561	−7

TABLE 9A

Fresnel Sine Integral

U	.000	.001	.002	.003	.004	.005	.006	.007	.008	.009	Δ_2
2.50	.619182	.618795	.618402	.618001	.617593	.617178	.616756	.616326	.615890	.615446	-7
2.51	.614996	.614538	.614073	.613602	.613123	.612638	.612145	.611646	.611140	.610627	-6
2.52	.610108	.609581	.609048	.608508	.607962	.607409	.606849	.606283	.605710	.605131	-6
2.53	.604545	.603953	.603354	.602750	.602138	.601521	.600897	.600267	.599631	.598989	-6
2.54	.598341	.597686	.597026	.596359	.595687	.595009	.594325	.593635	.592939	.592238	-5
2.55	.591531	.590818	.590100	.589376	.588647	.587912	.587172	.586426	.585675	.584919	-5
2.56	.584157	.583391	.582619	.581842	.581060	.580273	.579481	.578685	.577883	.577077	-4
2.57	.576265	.575450	.574629	.573804	.572974	.572140	.571302	.570459	.569612	.568760	-4
2.58	.567904	.567044	.566180	.565312	.564440	.563564	.562684	.561800	.560913	.560022	-3
2.59	.559127	.558228	.557326	.556421	.555512	.554599	.553684	.552765	.551843	.550918	-3
2.60	.549989	.549058	.548124	.547187	.546247	.545304	.544359	.543411	.542460	.541507	-2
2.61	.540551	.539594	.538633	.537671	.536706	.535739	.534770	.533799	.532827	.531852	-1
2.62	.530875	.529897	.528917	.527936	.526952	.525968	.524982	.523995	.523006	.522016	-1
2.63	.521025	.520033	.519040	.518046	.517051	.516056	.515059	.514062	.513065	.512067	-0
2.64	.511068	.510069	.509069	.508070	.507070	.506070	.505070	.504070	.503070	.502070	0
2.65	.501071	.500072	.499073	.498074	.497077	.496079	.495082	.494086	.493091	.492097	0
2.66	.491103	.490111	.489120	.488129	.487140	.486153	.485166	.484181	.483197	.482215	1
2.67	.481235	.480256	.479279	.478304	.477331	.476360	.475390	.474423	.473458	.472496	2
2.68	.471535	.470577	.469622	.468669	.467718	.466770	.465825	.464883	.463943	.463007	2
2.69	.462073	.461143	.460215	.459291	.458370	.457453	.456539	.455628	.454721	.453817	3
2.70	.452917	.452021	.451129	.450241	.449356	.448475	.447599	.446726	.445858	.444994	4
2.71	.444135	.443279	.442429	.441582	.440741	.439904	.439071	.438244	.437421	.436603	4
2.72	.435790	.434982	.434179	.433381	.432589	.431801	.431019	.430242	.429471	.428705	5
2.73	.427945	.427191	.426442	.425698	.424961	.424229	.423503	.422784	.422070	.421362	6
2.74	.420660	.419965	.419275	.418592	.417916	.417245	.416581	.415924	.415273	.414629	6
2.75	.413991	.413360	.412736	.412118	.411507	.410903	.410307	.409717	.409134	.408558	7
2.76	.407989	.407427	.406873	.406326	.405786	.405253	.404728	.404210	.403700	.403197	7
2.77	.402702	.402214	.401734	.401262	.400797	.400340	.399891	.399449	.399016	.398590	7
2.78	.398172	.397763	.397361	.396967	.396581	.396203	.395834	.395472	.395119	.394774	8
2.79	.394437	.394109	.393789	.393477	.393173	.392878	.392591	.392313	.392043	.391781	8
2.80	.391528	.391284	.391048	.390821	.390602	.390392	.390190	.389998	.389813	.389638	8
2.81	.389471	.389313	.389163	.389023	.388891	.388768	.388653	.388548	.388451	.388363	8
2.82	.388284	.388214	.388152	.388100	.388056	.388021	.387995	.387978	.387970	.387970	8
2.83	.387980	.387998	.388026	.388062	.388107	.388161	.388224	.388296	.388376	.388466	8
2.84	.388564	.388672	.388788	.388913	.389047	.389189	.389341	.389501	.389671	.389849	8
2.85	.390036	.390231	.390436	.390649	.390871	.391101	.391341	.391589	.391845	.392111	8
2.86	.392385	.392668	.392959	.393259	.393567	.393884	.394210	.394544	.394886	.395237	8
2.87	.395597	.395965	.396341	.396725	.397118	.397519	.397929	.398346	.398772	.399206	8
2.88	.399648	.400098	.400556	.401023	.401497	.401979	.402469	.402967	.403473	.403987	7
2.89	.404508	.405037	.405574	.406119	.406671	.407231	.407798	.408372	.408954	.409544	7
2.90	.410141	.410745	.411356	.411974	.412600	.413233	.413873	.414519	.415173	.415834	6
2.91	.416501	.417175	.417856	.418544	.419238	.419939	.420646	.421360	.422080	.422806	6
2.92	.423539	.424278	.425023	.425774	.426531	.427294	.428063	.428838	.429619	.430405	5
2.93	.431197	.431995	.432798	.433607	.434421	.435240	.436065	.436894	.437729	.438569	4
2.94	.439414	.440264	.441118	.441978	.442842	.443710	.444584	.445461	.446344	.447230	4
2.95	.448121	.449015	.449914	.450817	.451724	.452635	.453550	.454468	.455390	.456316	3
2.96	.457245	.458177	.459113	.460052	.460994	.461939	.462887	.463838	.464792	.465749	2
2.97	.466708	.467670	.468635	.469602	.470571	.471542	.472516	.473492	.474469	.475449	1
2.98	.476431	.477414	.478399	.479385	.480373	.481363	.482353	.483345	.484339	.485333	0
2.99	.486328	.487324	.488321	.489318	.490316	.491315	.492314	.493313	.494313	.495313	0

TABLE 9A

Fresnel Sine Integral

U	.000	.001	.002	.003	.004	.005	.006	.007	.008	.009	Δ_2
3.00	.496313	.497312	.498312	.499312	.500311	.501310	.502309	.503307	.504304	.505302	−1
3.01	.506297	.507291	.508286	.509279	.510271	.511261	.512251	.513239	.514225	.515210	−2
3.02	.516193	.517174	.518154	.519131	.520107	.521079	.522051	.523020	.523986	.524950	−3
3.03	.525911	.526869	.527825	.528778	.529728	.530675	.531619	.532560	.533497	.534431	−4
3.04	.535362	.536289	.537213	.538132	.539048	.539960	.540868	.541773	.542672	.543568	−5
3.05	.544459	.545345	.546228	.547106	.547980	.548848	.549712	.550572	.551425	.552274	−5
3.06	.553119	.553957	.554791	.555618	.556441	.557258	.558069	.558876	.559675	.560470	−6
3.07	.561259	.562041	.562817	.563587	.564352	.565109	.565861	.566606	.567344	.568077	−7
3.08	.568802	.569520	.570233	.570937	.571636	.572327	.573011	.573689	.574358	.575021	−7
3.09	.575676	.576324	.576965	.577597	.578223	.578841	.579452	.580054	.580649	.581236	−8
3.10	.581815	.582386	.582949	.583504	.584051	.584589	.585119	.585642	.586155	.586661	−8
3.11	.587158	.587646	.588126	.588597	.589059	.589513	.589958	.590395	.590822	.591241	−8
3.12	.591651	.592051	.592443	.592826	.593199	.593563	.593919	.594265	.594602	.594930	−9
3.13	.595248	.595557	.595856	.596146	.596427	.596699	.596961	.597213	.597456	.597689	−9
3.14	.597912	.598126	.598331	.598525	.598710	.598885	.599051	.599207	.599353	.599489	−9
3.15	.599615	.599732	.599838	.599935	.600022	.600099	.600166	.600224	.600271	.600309	−9
3.16	.600336	.600354	.600362	.600360	.600347	.600325	.600293	.600251	.600200	.600138	−9
3.17	.600066	.599984	.599893	.599791	.599680	.599559	.599427	.599286	.599135	.598974	−9
3.18	.598803	.598623	.598433	.598233	.598022	.597803	.597573	.597334	.597085	.596826	−8
3.19	.596558	.596280	.595993	.595696	.595389	.595073	.594748	.594412	.594067	.593713	−8
3.20	.593350	.592977	.592595	.592204	.591803	.591393	.590974	.590546	.590109	.589663	−9
3.21	.589208	.588744	.588271	.587789	.587298	.586799	.586290	.585773	.585248	.584714	−8
3.22	.584171	.583620	.583061	.582492	.581916	.581332	.580739	.580138	.579530	.578912	−7
3.23	.578288	.577655	.577014	.576366	.575710	.575047	.574375	.573697	.573010	.572316	−6
3.24	.571615	.570908	.570193	.569470	.568741	.568006	.567262	.566511	.565755	.564992	−5
3.25	.564222	.563446	.562663	.561875	.561079	.560278	.559471	.558657	.557838	.557012	−2
3.26	.556182	.555346	.554504	.553657	.552804	.551946	.551081	.550212	.549339	.548461	−4
3.27	.547576	.546688	.545796	.544898	.543996	.543090	.542180	.541265	.540345	.539422	−3
3.28	.538494	.537564	.536629	.535692	.534751	.533806	.532857	.531905	.530951	.529992	−1
3.29	.529032	.528069	.527103	.526134	.525162	.524189	.523212	.522234	.521255	.520271	0
3.30	.519287	.518302	.517314	.516326	.515334	.514343	.513349	.512354	.511360	.510362	1
3.31	.509366	.508367	.507368	.506371	.505370	.504371	.503370	.502370	.501372	.500371	2
3.32	.499371	.498372	.497373	.496375	.495378	.494382	.493386	.492391	.491398	.490405	3
3.33	.489413	.488424	.487436	.486450	.485466	.484483	.483502	.482523	.481547	.480574	3
3.34	.479603	.478634	.477667	.476704	.475742	.474785	.473830	.472879	.471931	.470985	5
3.35	.470045	.469107	.468173	.467243	.466316	.465395	.464476	.463561	.462653	.461747	5
3.36	.460846	.459950	.459059	.458173	.457291	.456415	.455544	.454677	.453817	.452962	5
3.37	.452112	.451268	.450429	.449597	.448770	.447949	.447135	.446327	.445525	.444729	7
3.38	.443939	.443158	.442381	.441612	.440850	.440095	.439346	.438604	.437871	.437144	8
3.39	.436424	.435712	.435007	.434311	.433621	.432940	.432266	.431600	.430942	.430292	8
3.40	.429651	.429017	.428393	.427776	.427167	.426568	.425977	.425395	.424821	.424256	11
3.41	.423699	.423153	.422615	.422086	.421566	.421056	.420554	.420062	.419579	.419105	10
3.42	.418642	.418188	.417743	.417309	.416883	.416468	.416062	.415666	.415280	.414904	11
3.43	.414538	.414182	.413836	.413501	.413175	.412860	.412554	.412259	.411975	.411701	10
3.44	.411438	.411184	.410942	.410709	.410488	.410277	.410076	.409886	.409707	.409538	11
3.45	.409380	.409233	.409097	.408971	.408856	.408752	.408658	.408576	.408504	.408443	11
3.46	.408393	.408354	.408325	.408308	.408301	.408306	.408321	.408347	.408384	.408432	11
3.47	.408491	.408560	.408641	.408732	.408834	.408947	.409071	.409206	.409351	.409508	11
3.48	.409675	.409853	.410042	.410241	.410451	.410672	.410903	.411146	.411398	.411662	10
3.49	.411936	.412220	.412515	.412820	.413136	.413462	.413798	.414145	.414502	.414870	10

TABLE 9A

Fresnel Sine Integral

U	.000	.001	.002	.003	.004	.005	.006	.007	.008	.009	Δ_2
3.50	.415248	.415635	.416033	.416441	.416859	.417287	.417725	.418173	.418630	.419097	9
3.51	.419574	.420061	.420557	.421063	.421579	.422103	.422637	.423181	.423733	.424295	10
3.52	.424866	.425446	.426035	.426633	.427240	.427855	.428479	.429113	.429753	.430403	8
3.53	.431062	.431728	.432403	.433086	.433777	.434475	.435182	.435898	.436620	.437349	8
3.54	.438087	.438832	.439584	.440344	.441111	.441885	.442666	.443454	.444248	.445050	6
3.55	.445858	.446672	.447493	.448320	.449154	.449993	.450840	.451691	.452548	.453411	4
3.56	.454280	.455154	.456033	.456917	.457808	.458703	.459603	.460508	.461417	.462332	3
3.57	.463250	.464173	.465100	.466032	.466968	.467906	.468850	.469796	.470746	.471701	0
3.58	.472658	.473618	.474581	.475547	.476517	.477488	.478463	.479440	.480420	.481401	0
3.59	.482383	.483369	.484357	.485345	.486336	.487328	.488321	.489317	.490312	.491310	−1
3.60	.492307	.493305	.494305	.495304	.496303	.497303	.498304	.499305	.500303	.501303	0
3.61	.502302	.503301	.504298	.505295	.506291	.507286	.508279	.509272	.510263	.511252	−2
3.62	.512240	.513226	.514210	.515190	.516169	.517146	.518121	.519094	.520061	.521029	−5
3.63	.521990	.522951	.523907	.524861	.525810	.526756	.527700	.528639	.529572	.530504	−5
3.64	.531431	.532352	.533270	.534182	.535091	.535994	.536892	.537786	.538673	.539556	−6
3.65	.540433	.541304	.542171	.543030	.543885	.544733	.545575	.546411	.547240	.548064	−7
3.66	.548881	.549690	.550493	.551289	.552077	.552859	.553634	.554402	.555161	.555914	−8
3.67	.556658	.557394	.558123	.558844	.559558	.560263	.560960	.561648	.562328	.563000	−9
3.68	.563663	.564316	.564962	.565598	.566226	.566844	.567454	.568054	.568644	.569226	−9
3.69	.569798	.570360	.570913	.571456	.571989	.572512	.573026	.573530	.574023	.574506	−9
3.70	.574979	.575442	.575895	.576337	.576769	.577189	.577600	.578000	.578389	.578768	−10
3.71	.579135	.579491	.579837	.580172	.580496	.580808	.581110	.581401	.581680	.581948	−11
3.72	.582205	.582450	.582685	.582907	.583119	.583318	.583507	.583684	.583849	.584003	−11
3.73	.584145	.584276	.584395	.584502	.584598	.584682	.584754	.584815	.584863	.584901	−10
3.74	.584926	.584940	.584942	.584932	.584910	.584877	.584831	.584775	.584706	.584625	−10
3.75	.584533	.584429	.584314	.584187	.584048	.583897	.583735	.583561	.583375	.583178	−10
3.76	.582969	.582750	.582518	.582275	.582020	.581754	.581476	.581188	.580888	.580576	−10
3.77	.580254	.579920	.579575	.579219	.578851	.578474	.578085	.577684	.577274	.576852	−10
3.78	.576420	.575977	.575523	.575059	.574585	.574100	.573604	.573098	.572582	.572056	−9
3.79	.571519	.570973	.570417	.569852	.569275	.568690	.568095	.567490	.566876	.566252	−7
3.80	.565620	.564978	.564327	.563667	.562998	.562320	.561634	.560938	.560234	.559522	−8
3.81	.558801	.558073	.557335	.556590	.555837	.555077	.554308	.553532	.552749	.551957	−5
3.82	.551158	.550353	.549540	.548722	.547895	.547062	.546222	.545376	.544524	.543665	−5
3.83	.542800	.541930	.541053	.540171	.539283	.538389	.537490	.536586	.535678	.534764	−3
3.84	.533845	.532921	.531993	.531062	.530124	.529184	.528239	.527290	.526337	.525380	−1
3.85	.524420	.523458	.522492	.521523	.520550	.519576	.518598	.517618	.516636	.515651	0
3.86	.514663	.513676	.512684	.511694	.510700	.509706	.508709	.507712	.506715	.505716	1
3.87	.504717	.503718	.502718	.501718	.500718	.499718	.498718	.497719	.496721	.495723	2
3.88	.494727	.493731	.492736	.491744	.490751	.489762	.488772	.487786	.486802	.485820	4
3.89	.484839	.483862	.482888	.481917	.480947	.479981	.479019	.478060	.477105	.476153	4
3.90	.475203	.474261	.473320	.472385	.471453	.470527	.469604	.468687	.467775	.466866	8
3.91	.465964	.465069	.464177	.463292	.462412	.461538	.460671	.459809	.458954	.458105	7
3.92	.457263	.456428	.455599	.454778	.453963	.453155	.452356	.451562	.450778	.449999	10
3.93	.449230	.448468	.447715	.446970	.446232	.445505	.444784	.444072	.443370	.442676	9
3.94	.441990	.441314	.440648	.439990	.439342	.438704	.438075	.437456	.436847	.436246	12
3.95	.435657	.435077	.434508	.433949	.433400	.432862	.432334	.431816	.431310	.430813	12
3.96	.430329	.429855	.429392	.428940	.428499	.428069	.427651	.427243	.426848	.426464	12
3.97	.426091	.425730	.425381	.425043	.424717	.424403	.424101	.423811	.423533	.423266	13
3.98	.423012	.422770	.422540	.422323	.422117	.421924	.421743	.421574	.421418	.421274	12
3.99	.421143	.421024	.420918	.420823	.420742	.420673	.420616	.420572	.420541	.420522	13

TABLE 9A

Fresnel Sine Integral

U	.000	.001	.002	.003	.004	.005	.006	.007	.008	.009	Δ_2
4.00	.420516	.420522	.420541	.420572	.420616	.420673	.420742	.420823	.420918	.421024	12
4.01	.421143	.421275	.421419	.421576	.421745	.421926	.422121	.422327	.422545	.422776	12
4.02	.423019	.423275	.423543	.423822	.424114	.424418	.424734	.425062	.425402	.425753	12
4.03	.426117	.426492	.426879	.427278	.427688	.428110	.428543	.428988	.429443	.429911	10
4.04	.430389	.430879	.431379	.431891	.432413	.432946	.433490	.434045	.434609	.435185	10
4.05	.435771	.436366	.436973	.437588	.438215	.438851	.439497	.440153	.440817	.441492	9
4.06	.442177	.442870	.443573	.444284	.445004	.445733	.446471	.447218	.447972	.448736	7
4.07	.449507	.450286	.451073	.451868	.452671	.453482	.454300	.455125	.455957	.456797	6
4.08	.457643	.458496	.459355	.460221	.461095	.461972	.462858	.463748	.464644	.465547	3
4.09	.466454	.467366	.468284	.469207	.470136	.471067	.472005	.472947	.473892	.474842	3
4.10	.475796	.476754	.477716	.478680	.479648	.480619	.481593	.482570	.483549	.484533	0
4.11	.485517	.486504	.487493	.488483	.489475	.490469	.491465	.492460	.493458	.494456	0
4.12	.495456	.496454	.497454	.498455	.499455	.500454	.501453	.502453	.503451	.504449	−1
4.13	.505445	.506440	.507435	.508426	.509418	.510406	.511394	.512379	.513361	.514342	−4
4.14	.515319	.516294	.517265	.518234	.519199	.520160	.521119	.522073	.523023	.523969	−4
4.15	.524912	.525849	.526781	.527708	.528632	.529550	.530463	.531370	.532271	.533167	−7
4.16	.534057	.534941	.535819	.536690	.537555	.538413	.539265	.540110	.540947	.541777	−7
4.17	.542600	.543415	.544223	.545022	.545815	.546598	.547374	.548142	.548901	.549651	−9
4.18	.550393	.551125	.551849	.552563	.553269	.553964	.554651	.555328	.555994	.556652	−10
4.19	.557300	.557936	.558563	.559179	.559786	.560381	.560966	.561541	.562104	.562657	−10
4.20	.563198	.563728	.564246	.564755	.565251	.565736	.566209	.566671	.567120	.567559	−11
4.21	.567984	.568398	.568800	.569189	.569567	.569932	.570285	.570625	.570953	.571269	−12
4.22	.571572	.571862	.572139	.572404	.572656	.572895	.573121	.573334	.573534	.573721	−12
4.23	.573895	.574056	.574204	.574338	.574460	.574568	.574663	.574745	.574813	.574868	−12
4.24	.574910	.574939	.574954	.574956	.574944	.574920	.574882	.574830	.574765	.574687	−12
4.25	.574596	.574491	.574373	.574242	.574098	.573940	.573770	.573585	.573389	.573178	−12
4.26	.572955	.572719	.572470	.572208	.571933	.571645	.571344	.571030	.570704	.570365	−11
4.27	.570014	.569650	.569273	.568884	.568483	.568070	.567644	.567206	.566757	.566295	−11
4.28	.565821	.565336	.564838	.564330	.563810	.563278	.562735	.562181	.561615	.561038	−9
4.29	.560451	.559853	.559245	.558624	.557993	.557353	.556702	.556040	.555370	.554688	−8
4.30	.553998	.553297	.552587	.551867	.551138	.550400	.549652	.548895	.548131	.547357	−5
4.31	.546575	.545785	.544986	.544179	.543364	.542542	.541711	.540873	.540029	.539176	−5
4.32	.538317	.537452	.536578	.535699	.534813	.533922	.533024	.532120	.531210	.530295	−5
4.33	.529374	.528449	.527518	.526582	.525641	.524697	.523747	.522793	.521836	.520874	−1
4.34	.519909	.518941	.517969	.516995	.516017	.515037	.514053	.513066	.512080	.511089	0
4.35	.510097	.509104	.508109	.507113	.506115	.505118	.504118	.503119	.502119	.501119	2
4.36	.500119	.499119	.498120	.497121	.496124	.495126	.494130	.493136	.492143	.491151	3
4.37	.490161	.489174	.488189	.487207	.486226	.485250	.484276	.483304	.482336	.481373	3
4.38	.480412	.479457	.478505	.477557	.476613	.475675	.474740	.473811	.472887	.471969	7
4.39	.471055	.470149	.469247	.468352	.467463	.466581	.465705	.464835	.463973	.463117	8
4.40	.462269	.461429	.460596	.459771	.458953	.458144	.457343	.456550	.455766	.454990	10
4.41	.454223	.453465	.452716	.451977	.451246	.450527	.449815	.449114	.448424	.447742	10
4.42	.447072	.446412	.445762	.445123	.444494	.443877	.443270	.442674	.442090	.441517	12
4.43	.440955	.440406	.439867	.439341	.438826	.438324	.437833	.437355	.436889	.436435	13
4.44	.435994	.435565	.435150	.434747	.434356	.433979	.433614	.433262	.432924	.432599	13
4.45	.432287	.431988	.431703	.431431	.431173	.430928	.430697	.430479	.430275	.430085	14
4.46	.429909	.429746	.429598	.429463	.429342	.429235	.429142	.429063	.428998	.428947	14
4.47	.428909	.428886	.428878	.428883	.428902	.428935	.428982	.429043	.429119	.429208	14
4.48	.429311	.429429	.429560	.429705	.429864	.430037	.430224	.430425	.430640	.430868	13
4.49	.431110	.431365	.431635	.431918	.432214	.432524	.432847	.433183	.433533	.433896	13

TABLE 9A

Fresnel Sine Integral

U	.000	.001	.002	.003	.004	.005	.006	.007	.008	.009	Δ_2
5.00	.499191	.500190	.501190	.502188	.503187	.504185	.505180	.506176	.507169	.508160	-3
5.01	.509148	.510134	.511119	.512098	.513077	.514050	.515021	.515988	.516949	.517908	-5
5.02	.518862	.519809	.520753	.521691	.522624	.523549	.524470	.525385	.526292	.527193	-7
5.03	.528087	.528973	.529854	.530724	.531589	.532444	.533292	.534131	.534961	.535783	-9
5.04	.536595	.537397	.538191	.538974	.539748	.540511	.541264	.542007	.542739	.543460	-10
5.05	.544170	.544868	.545556	.546231	.546895	.547546	.548187	.548814	.549429	.550032	-12
5.06	.550622	.551198	.551762	.552312	.552849	.553373	.553883	.554379	.554862	.555330	-13
5.07	.555785	.556224	.556650	.557061	.557458	.557840	.558207	.558559	.558896	.559219	-14
5.08	.559526	.559818	.560094	.560355	.560601	.560831	.561046	.561245	.561428	.561596	-15
5.09	.561747	.561883	.562003	.562108	.562196	.562268	.562325	.562365	.562389	.562398	-15
5.10	.562390	.562366	.562327	.562271	.562199	.562112	.562008	.561888	.561753	.561601	-15
5.11	.561434	.561251	.561052	.560838	.560607	.560362	.560100	.559824	.559531	.559224	-14
5.12	.558901	.558563	.558210	.557842	.557459	.557061	.556649	.556221	.555780	.555323	-12
5.13	.554853	.554369	.553870	.553358	.552832	.552292	.551738	.551171	.550592	.549999	-11
5.14	.549392	.548774	.548143	.547499	.546843	.546175	.545495	.544802	.544099	.543384	-10
5.15	.542657	.541921	.541172	.540414	.539645	.538866	.538077	.537277	.536468	.535650	-8
5.16	.534822	.533987	.533141	.532287	.531425	.530555	.529677	.528791	.527898	.526998	-6
5.17	.526091	.525177	.524256	.523331	.522397	.521459	.520515	.519565	.518612	.517653	-3
5.18	.516689	.515721	.514750	.513775	.512796	.511814	.510829	.509841	.508851	.507858	0
5.19	.506865	.505870	.504871	.503874	.502874	.501875	.500875	.499875	.498875	.497876	1
5.20	.496877	.495880	.494883	.493890	.492896	.491906	.490918	.489931	.488949	.487969	4
5.21	.486993	.486021	.485052	.484088	.483128	.482173	.481222	.480276	.479337	.478403	6
5.22	.477475	.476553	.475638	.474730	.473828	.472934	.472047	.471167	.470297	.469434	9
5.23	.468579	.467733	.466896	.466068	.465249	.464441	.463642	.462852	.462074	.461305	11
5.24	.460546	.459799	.459063	.458339	.457625	.456923	.456233	.455554	.454889	.454236	12
5.25	.453595	.452967	.452352	.451750	.451161	.450586	.450024	.449476	.448942	.448422	15
5.26	.447916	.447425	.446947	.446485	.446037	.445605	.445187	.444783	.444396	.444023	16
5.27	.443666	.443325	.442999	.442689	.442394	.442116	.441853	.441607	.441377	.441162	17
5.28	.440964	.440783	.440617	.440468	.440336	.440220	.440120	.440037	.439970	.439920	17
5.29	.439887	.439870	.439870	.439887	.439920	.439970	.440036	.440119	.440219	.440335	16
5.30	.440468	.440616	.440782	.440964	.441162	.441377	.441608	.441855	.442118	.442397	15
5.31	.442693	.443004	.443331	.443673	.444032	.444405	.444794	.445199	.445619	.446054	14
5.32	.446504	.446968	.447448	.447942	.448450	.448973	.449510	.450062	.450626	.451205	13
5.33	.451798	.452403	.453022	.453654	.454299	.454957	.455627	.456310	.457004	.457711	11
5.34	.458429	.459158	.459900	.460652	.461416	.462189	.462974	.463769	.464574	.465389	10
5.35	.466214	.467047	.467891	.468743	.469604	.470472	.471350	.472236	.473128	.474029	6
5.36	.474937	.475851	.476772	.477700	.478634	.479573	.480518	.481470	.482425	.483386	4
5.37	.484352	.485321	.486295	.487271	.488252	.489236	.490223	.491212	.492203	.493198	1
5.38	.494192	.495189	.496187	.497185	.498186	.499185	.500185	.501185	.502184	.503183	-1
5.39	.504180	.505176	.506171	.507162	.508152	.509139	.510125	.511107	.512085	.513060	-5
5.40	.514031	.514996	.515959	.516915	.517868	.518815	.519756	.520691	.521620	.522543	-7
5.41	.523459	.524368	.525270	.526164	.527051	.527930	.528801	.529663	.530516	.531360	-9
5.42	.532194	.533020	.533835	.534641	.535437	.536222	.536996	.537761	.538512	.539234	-11
5.43	.539984	.540701	.541407	.542100	.542782	.543450	.544106	.544749	.545379	.545995	-13
5.44	.546598	.547187	.547762	.548323	.548870	.549402	.549920	.550423	.550911	.551385	-15
5.45	.551843	.552286	.552714	.553125	.553522	.553902	.554267	.554615	.554947	.555264	-16
5.46	.555563	.555847	.556114	.556364	.556598	.556815	.557015	.557198	.557365	.557514	-16
5.47	.557647	.557762	.557861	.557942	.558006	.558053	.558083	.558095	.558090	.558069	-16
5.48	.558029	.557973	.557900	.557809	.557701	.557576	.557434	.557275	.557099	.556906	-16
5.49	.556696	.556470	.556226	.555966	.555689	.555396	.555087	.554760	.554418	.554059	-14

TABLE 9A

Fresnel Sine Integral

U	.000	.001	.002	.003	.004	.005	.006	.007	.008	.009	Δ_2
4.50	.434273	.434661	.435064	.435479	.435907	.436347	.436801	.437267	.437745	.438236	12
4.51	.438739	.439254	.439781	.440320	.440871	.441434	.442009	.442595	.443193	.443802	10
4.52	.444423	.445053	.445696	.446348	.447012	.447686	.448371	.449066	.449771	.450486	10
4.53	.451212	.451947	.452691	.453445	.454207	.454979	.455760	.456551	.457350	.458156	8
4.54	.458972	.459796	.460628	.461467	.462314	.463169	.464031	.464901	.465776	.466659	6
4.55	.467548	.468444	.469346	.470254	.471169	.472088	.473013	.473945	.474881	.475822	4
4.56	.476768	.477717	.478672	.479629	.480593	.481559	.482529	.483504	.484480	.485460	1
4.57	.486442	.487426	.488414	.489403	.490395	.491388	.492382	.493379	.494376	.495375	0
4.58	.496374	.497374	.498374	.499373	.500374	.501373	.502372	.503371	.504368	.505364	0
4.59	.506359	.507352	.508344	.509333	.510321	.511306	.512289	.513270	.514246	.515220	−4
4.60	.516190	.517157	.518121	.519079	.520034	.520983	.521930	.522871	.523807	.524738	−5
4.61	.525663	.526583	.527497	.528404	.529306	.530202	.531092	.531974	.532849	.533717	−7
4.62	.534579	.535431	.536278	.537115	.537946	.538767	.539581	.540386	.541181	.541969	−9
4.63	.542747	.543515	.544275	.545025	.545765	.546496	.547216	.547927	.548626	.549317	−11
4.64	.549995	.550663	.551321	.551966	.552602	.553225	.553837	.554438	.555026	.555603	−11
4.65	.556169	.556721	.557262	.557790	.558306	.558808	.559299	.559777	.560242	.560693	−12
4.66	.561132	.561557	.561969	.562368	.562753	.563125	.563483	.563828	.564158	.564475	−13
4.67	.564777	.565066	.565341	.565601	.565847	.566079	.566297	.566501	.566689	.566864	−14
4.68	.567024	.567169	.567300	.567417	.567518	.567605	.567678	.567735	.567778	.567807	−13
4.69	.567820	.567819	.567803	.567772	.567727	.567667	.567592	.567502	.567398	.567279	−13
4.70	.567146	.566997	.566835	.566657	.566466	.566259	.566038	.565803	.565554	.565290	−13
4.71	.565012	.564720	.564413	.564093	.563759	.563411	.563048	.562672	.562283	.561880	−11
4.72	.561463	.561032	.560589	.560132	.559662	.559180	.558684	.558174	.557653	.557119	−10
4.73	.556572	.556014	.555442	.554859	.554263	.553656	.553037	.552406	.551764	.551111	−9
4.74	.550446	.549771	.549084	.548387	.547678	.546961	.546232	.545493	.544744	.543985	−8
4.75	.543217	.542439	.541652	.540857	.540051	.539237	.538415	.537584	.536745	.535898	−7
4.76	.535043	.534181	.533310	.532434	.531549	.530658	.529760	.528855	.527944	.527027	−4
4.77	.526105	.525176	.524242	.523303	.522359	.521411	.520457	.519498	.518536	.517570	−3
4.78	.516599	.515627	.514649	.513670	.512686	.511702	.510713	.509723	.508731	.507737	0
4.79	.506741	.505745	.504747	.503749	.502748	.501750	.500749	.499749	.498750	.497751	1
4.80	.496752	.495755	.494758	.493764	.492771	.491780	.490790	.489802	.488818	.487836	5
4.81	.486857	.485882	.484909	.483942	.482976	.482016	.481059	.480108	.479161	.478219	5
4.82	.477281	.476350	.475424	.474505	.473591	.472683	.471783	.470887	.470000	.469119	9
4.83	.468246	.467381	.466522	.465672	.464830	.463997	.463171	.462355	.461547	.460748	9
4.84	.459959	.459180	.458409	.457648	.456898	.456158	.455427	.454707	.453998	.453300	12
4.85	.452612	.451936	.451271	.450618	.449977	.449347	.448729	.448123	.447529	.446947	13
4.86	.446378	.445822	.445279	.444748	.444230	.443726	.443234	.442756	.442292	.441842	14
4.87	.441404	.440981	.440571	.440176	.439795	.439428	.439075	.438737	.438413	.438103	15
4.88	.437809	.437529	.437264	.437013	.436777	.436557	.436351	.436160	.435985	.435824	15
4.89	.435679	.435549	.435434	.435335	.435250	.435181	.435128	.435090	.435067	.435059	15
4.90	.435067	.435091	.435130	.435184	.435253	.435338	.435438	.435554	.435685	.435831	15
4.91	.435992	.436159	.436361	.436567	.436789	.437026	.437278	.437545	.437826	.438123	14
4.92	.438434	.438760	.439100	.439455	.439825	.440208	.440606	.441018	.441444	.441884	14
4.93	.442339	.442806	.443287	.443782	.444290	.444812	.445346	.445894	.446455	.447028	12
4.94	.447615	.448213	.448824	.449447	.450082	.450729	.451388	.452059	.452740	.453434	11
4.95	.454139	.454853	.455580	.456316	.457064	.457820	.458588	.459366	.460153	.460949	9
4.96	.461755	.462570	.463394	.464226	.465068	.465917	.466774	.467640	.468513	.469395	5
4.97	.470283	.471177	.472080	.472989	.473904	.474825	.475752	.476685	.477623	.478567	3
4.98	.479517	.480469	.481428	.482390	.483357	.484327	.485302	.486280	.487260	.488244	2
4.99	.489231	.490219	.491211	.492203	.493199	.494194	.495192	.496191	.497190	.498190	0

TABLE 9A

Fresnel Sine Integral

U	.000	.001	.002	.003	.004	.005	.006	.007	.008	.009	Δ_2
5.50	.553684	.553294	.552887	.552465	.552027	.551575	.551106	.550621	.550123	.549609	−12
5.51	.549081	.548538	.547980	.547408	.546822	.546222	.545608	.544981	.544340	.543686	−12
5.52	.543019	.542339	.541647	.540942	.540225	.539496	.538755	.538002	.537239	.536464	−9
5.53	.535677	.534882	.534075	.533258	.532432	.531596	.530750	.529895	.529031	.528158	−7
5.54	.527277	.526388	.525491	.524587	.523675	.522758	.521831	.520899	.519961	.519018	−5
5.55	.518069	.517114	.516154	.515189	.514220	.513248	.512270	.511289	.510305	.509319	−1
5.56	.508328	.507337	.506343	.505347	.504350	.503352	.502353	.501352	.500352	.499352	1
5.57	.498353	.497354	.496356	.495360	.494365	.493372	.492381	.491392	.490407	.489424	4
5.58	.488445	.487471	.486499	.485532	.484570	.483613	.482661	.481713	.480773	.479838	7
5.59	.478910	.477989	.477074	.476167	.475267	.474375	.473491	.472615	.471748	.470890	10
5.60	.470040	.469201	.468371	.467551	.466741	.465943	.465153	.464375	.463610	.462854	13
5.61	.462111	.461379	.460660	.459953	.459259	.458577	.457909	.457253	.456612	.455983	14
5.62	.455368	.454768	.454182	.453610	.453053	.452511	.451983	.451471	.450973	.450491	16
5.63	.450026	.449575	.449141	.448722	.448320	.447933	.447564	.447210	.446874	.446553	17
5.64	.446250	.445964	.445695	.445443	.445208	.444991	.444790	.444607	.444442	.444294	18
5.65	.444164	.444051	.443956	.443879	.443819	.443777	.443753	.443747	.443758	.443788	18
5.66	.443835	.443899	.443982	.444082	.444200	.444335	.444489	.444659	.444847	.445053	17
5.67	.445276	.445517	.445774	.446049	.446341	.446649	.446975	.447318	.447677	.448053	16
5.68	.448446	.448854	.449279	.449720	.450177	.450650	.451139	.451643	.452162	.452697	15
5.69	.453246	.453810	.454390	.454983	.455591	.456212	.456849	.457499	.458161	.458838	12
5.70	.459527	.460230	.460945	.461671	.462411	.463162	.463926	.464700	.465486	.466283	9
5.71	.467090	.467907	.468736	.469573	.470422	.471278	.472145	.473021	.473903	.474796	7
5.72	.475696	.476603	.477518	.478440	.479369	.480304	.481246	.482194	.483146	.484105	4
5.73	.485069	.486036	.487009	.487985	.488966	.489949	.490935	.491926	.492916	.493911	1
5.74	.494907	.495904	.496903	.497902	.498902	.499901	.500901	.501901	.502900	.503897	−1
5.75	.504894	.505887	.506880	.507869	.508857	.509840	.510821	.511798	.512771	.513739	−5
5.76	.514702	.515661	.516615	.517562	.518504	.519440	.520369	.521292	.522206	.523115	−8
5.77	.524015	.524907	.525791	.526665	.527532	.528388	.529236	.530074	.530900	.531718	−11
5.78	.532525	.533320	.534105	.534878	.535640	.536389	.537127	.537852	.538564	.539264	−11
5.79	.539951	.540624	.541284	.541929	.542562	.543179	.543782	.544370	.544944	.545503	−15
5.80	.546046	.546574	.547087	.547584	.548065	.548529	.548978	.549410	.549826	.550225	−16
5.81	.550608	.550973	.551321	.551652	.551966	.552263	.552542	.552803	.553047	.553273	−17
5.82	.553481	.553671	.553843	.553997	.554133	.554251	.554351	.554432	.554495	.554540	−17
5.83	.554567	.554575	.554565	.554537	.554490	.554425	.554342	.554241	.554121	.553983	−16
5.84	.553827	.553653	.553461	.553251	.553023	.552777	.552513	.552232	.551933	.551617	−16
5.85	.551283	.550932	.550564	.550179	.549776	.549358	.548922	.548470	.548002	.547517	−14
5.86	.547017	.546500	.545968	.545421	.544858	.544280	.543687	.543079	.542457	.541820	−12
5.87	.541170	.540505	.539827	.539136	.538430	.537713	.536982	.536239	.535484	.534716	−9
5.88	.533938	.533148	.532346	.531534	.530711	.529878	.529034	.528180	.527318	.526445	−8
5.89	.525564	.524675	.523777	.522872	.521958	.521037	.520109	.519174	.518234	.517286	−4
5.90	.516332	.515374	.514410	.513442	.512469	.511493	.510512	.509527	.508541	.507549	1
5.91	.506557	.505563	.504566	.503569	.502570	.501571	.500571	.499570	.498571	.497571	2
5.92	.496574	.495578	.494582	.493591	.492599	.491612	.490627	.489645	.488668	.487694	5
5.93	.486725	.485762	.484802	.483849	.482901	.481960	.481023	.480095	.479174	.478260	9
5.94	.477352	.476455	.475564	.474683	.473811	.472948	.472094	.471250	.470418	.469594	11
5.95	.468782	.467981	.467191	.466413	.465646	.464892	.464151	.463421	.462705	.462002	14
5.96	.461313	.460637	.459975	.459328	.458694	.458075	.457471	.456882	.456309	.455750	17
5.97	.455208	.454681	.454170	.453676	.453198	.452736	.452292	.451864	.451453	.451059	18
5.98	.450683	.450324	.449983	.449659	.449354	.449066	.448796	.448545	.448312	.448097	18
5.99	.447900	.447722	.447563	.447422	.447300	.447196	.447112	.447046	.446999	.446970	19

TABLE 9A

Fresnel Sine Integral

U	.000	.001	.002	.003	.004	.005	.006	.007	.008	.009	Δ_2
6.00	.446961	.446970	.446998	.447046	.447111	.447196	.447300	.447422	.447563	.447722	19
6.01	.447901	.448097	.448313	.448546	.448798	.449068	.449357	.449663	.449987	.450330	18
6.02	.450690	.451067	.451462	.451874	.452303	.452750	.453213	.453693	.454189	.454702	16
6.03	.455231	.455776	.456337	.456913	.457505	.458111	.458733	.459371	.460021	.460687	13
6.04	.461367	.462060	.462766	.463486	.464219	.464965	.465724	.466495	.467277	.468072	11
6.05	.468877	.469693	.470521	.471358	.472207	.473065	.473934	.474811	.475696	.476592	7
6.06	.477496	.478406	.479325	.480251	.481184	.482123	.483071	.484023	.484980	.485944	3
6.07	.486912	.487884	.488862	.489842	.490827	.491814	.492804	.493798	.494793	.495790	0
6.08	.496788	.497786	.498786	.499786	.500786	.501785	.502784	.503782	.504777	.505772	−2
6.09	.506763	.507751	.508738	.509720	.510699	.511674	.512644	.513610	.514571	.515526	−6
6.10	.516475	.517418	.518354	.519284	.520206	.521120	.522028	.522927	.523815	.524697	−10
6.11	.525569	.526431	.527283	.528125	.528957	.529777	.530587	.531386	.532172	.532947	−12
6.12	.533709	.534459	.535195	.535918	.536629	.537325	.538008	.538676	.539330	.539970	−15
6.13	.540594	.541202	.541796	.542374	.542937	.543483	.544013	.544527	.545024	.545504	−16
6.14	.545967	.546413	.546842	.547253	.547646	.548022	.548379	.548719	.549040	.549344	−18
6.15	.549628	.549894	.550141	.550370	.550579	.550770	.550942	.551094	.551228	.551342	−18
6.16	.551437	.551513	.551569	.551606	.551624	.551622	.551602	.551561	.551501	.551422	−18
6.17	.551324	.551207	.551070	.550914	.550739	.550545	.550331	.550099	.549848	.549578	−17
6.18	.549290	.548983	.548658	.548315	.547953	.547573	.547175	.546759	.546327	.545876	−15
6.19	.545408	.544924	.544421	.543903	.543368	.542817	.542249	.541666	.541067	.540452	−14
6.20	.539822	.539177	.538517	.537844	.537155	.536453	.535736	.535007	.534264	.533508	−10
6.21	.532740	.531960	.531167	.530362	.529547	.528720	.527882	.527033	.526175	.525306	−8
6.22	.524428	.523541	.522645	.521741	.520828	.519909	.518980	.518045	.517103	.516154	−5
6.23	.515200	.514241	.513275	.512305	.511330	.510352	.509369	.508383	.507394	.506402	−1
6.24	.505407	.504412	.503413	.502415	.501415	.500416	.499415	.498416	.497417	.496419	3
6.25	.495423	.494429	.493437	.492449	.491462	.490480	.489501	.488525	.487556	.486591	6
6.26	.485632	.484678	.483730	.482789	.481854	.480927	.480006	.479094	.478191	.477295	10
6.27	.476409	.475532	.474665	.473808	.472961	.472124	.471299	.470484	.469682	.468891	13
6.28	.468113	.467347	.466594	.465855	.465127	.464415	.463717	.463032	.462362	.461707	15
6.29	.461066	.460442	.459832	.459239	.458662	.458101	.457556	.457028	.456517	.456023	18
6.30	.455546	.455087	.454646	.454222	.453816	.453429	.453060	.452709	.452377	.452064	19
6.31	.451769	.451494	.451238	.451001	.450783	.450585	.450406	.450246	.450107	.449987	20
6.32	.449886	.449806	.449745	.449704	.449683	.449682	.449701	.449740	.449798	.449876	20
6.33	.449975	.450092	.450230	.450387	.450564	.450760	.450976	.451212	.451466	.451740	19
6.34	.452033	.452344	.452675	.453025	.453393	.453779	.454184	.454607	.455048	.455507	18
6.35	.455983	.456477	.456988	.457516	.458061	.458623	.459201	.459795	.460405	.461031	15
6.36	.461673	.462329	.463001	.463686	.464387	.465102	.465830	.466573	.467327	.468096	11
6.37	.468877	.469670	.470476	.471292	.472121	.472960	.473811	.474672	.475542	.476422	8
6.38	.477311	.478209	.479117	.480032	.480955	.481886	.482824	.483769	.484720	.485677	6
6.39	.486640	.487607	.488581	.489557	.490539	.491524	.492512	.493504	.494497	.495493	1
6.40	.496490	.497488	.498489	.499488	.500488	.501487	.502486	.503485	.504480	.505474	−2
6.41	.506467	.507456	.508441	.509423	.510402	.511375	.512345	.513309	.514267	.515219	−7
6.42	.516165	.517105	.518037	.518961	.519879	.520787	.521687	.522578	.523459	.524332	−11
6.43	.525194	.526045	.526887	.527716	.528535	.529341	.530135	.530917	.531686	.532442	−13
6.44	.533185	.533914	.534629	.535329	.536016	.536686	.537343	.537983	.538608	.539217	−15
6.45	.539811	.540386	.540946	.541489	.542015	.542523	.543014	.543487	.543942	.544379	−18
6.46	.544797	.545197	.545578	.545940	.546284	.546608	.546913	.547199	.547465	.547711	−19
6.47	.547938	.548145	.548332	.548499	.548646	.548772	.548879	.548965	.549031	.549077	−19
6.48	.549102	.549107	.549092	.549056	.549000	.548923	.548827	.548710	.548572	.548415	−19
6.49	.548237	.548040	.547823	.547585	.547328	.547052	.546755	.546439	.546104	.545750	−18

TABLE 9A

Fresnel Sine Integral

U	.000	.001	.002	.003	.004	.005	.006	.007	.008	.009	Δ_2
6.50	.545377	.544985	.544574	.544145	.543697	.543232	.542748	.542246	.541727	.541190	−15
6.51	.540637	.540066	.539479	.538876	.538255	.537620	.536968	.536301	.535619	.534922	−13
6.52	.534211	.533485	.532746	.531993	.531226	.530447	.529655	.528849	.528033	.527205	−9
6.53	.526365	.525515	.524654	.523783	.522902	.522011	.521111	.520202	.519285	.518360	−6
6.54	.517426	.516488	.515540	.514588	.513628	.512664	.511693	.510719	.509740	.508757	−1
6.55	.507770	.506781	.505788	.504794	.503797	.502799	.501800	.500799	.499800	.498800	1
6.56	.497800	.496803	.495806	.494812	.493819	.492830	.491844	.490861	.489883	.488908	6
6.57	.487939	.486975	.486017	.485065	.484119	.483181	.482249	.481325	.480410	.479503	10
6.58	.478604	.477716	.476836	.475968	.475108	.474261	.473424	.472598	.471785	.470984	13
6.59	.470195	.469420	.468657	.467909	.467173	.466452	.465746	.465054	.464378	.463716	16
6.60	.463070	.462441	.461828	.461232	.460652	.460089	.459543	.459015	.458505	.458012	19
6.61	.457538	.457082	.456645	.456227	.455827	.455447	.455086	.454744	.454422	.454120	20
6.62	.453837	.453575	.453333	.453111	.452909	.452728	.452567	.452427	.452307	.452209	21
6.63	.452131	.452073	.452037	.452021	.452026	.452052	.452099	.452167	.452255	.452364	21
6.64	.452494	.452645	.452816	.453008	.453220	.453452	.453705	.453978	.454271	.454583	20
6.65	.454916	.455268	.455640	.456031	.456441	.456870	.457318	.457785	.458269	.458773	17
6.66	.459294	.459832	.460389	.460962	.461553	.462160	.462784	.463424	.464080	.464752	14
6.67	.465438	.466140	.466857	.467588	.468334	.469093	.469866	.470652	.471450	.472262	11
6.68	.473086	.473920	.474767	.475624	.476492	.477370	.478258	.479156	.480062	.480978	7
6.69	.481902	.482832	.483772	.484717	.485669	.486628	.487593	.488562	.489537	.490516	4
6.70	.491500	.492486	.493477	.494469	.495464	.496461	.497460	.498459	.499458	.500459	0
6.71	.501458	.502457	.503454	.504449	.505444	.506435	.507423	.508408	.509389	.510366	−6
6.72	.511338	.512304	.513265	.514219	.515168	.516109	.517043	.517970	.518888	.519798	−9
6.73	.520699	.521590	.522471	.523343	.524204	.525053	.525892	.526718	.527533	.528336	−12
6.74	.529124	.529901	.530663	.531412	.532146	.532866	.533572	.534262	.534936	.535595	−16
6.75	.536238	.536864	.537474	.538066	.538641	.539199	.539739	.540261	.540765	.541251	−18
6.76	.541717	.542165	.542594	.543003	.543393	.543764	.544114	.544445	.544755	.545046	−19
6.77	.545316	.545565	.545793	.546001	.546188	.546354	.546499	.546623	.546726	.546807	−20
6.78	.546868	.546907	.546924	.546921	.546896	.546850	.546782	.546693	.546583	.546452	−20
6.79	.546300	.546127	.545933	.545718	.545481	.545225	.544948	.544650	.544332	.543994	−19
6.80	.543636	.543258	.542861	.542443	.542007	.541551	.541077	.540583	.540072	.539541	−16
6.81	.538993	.538428	.537844	.537244	.536626	.535992	.535341	.534674	.533992	.533294	−14
6.82	.532580	.531853	.531110	.530353	.529582	.528799	.528001	.527191	.526369	.525534	−10
6.83	.524688	.523831	.522962	.522084	.521195	.520297	.519390	.518473	.517549	.516616	−6
6.84	.515676	.514729	.513775	.512816	.511849	.510878	.509902	.508921	.507937	.506948	−1
6.85	.505956	.504964	.503967	.502970	.501971	.500971	.499971	.498971	.497972	.496973	3
6.86	.495976	.494983	.493990	.493001	.492016	.491034	.490056	.489083	.488116	.487154	7
6.87	.486198	.485249	.484306	.483372	.482445	.481527	.480617	.479716	.478825	.477944	11
6.88	.477074	.476215	.475366	.474529	.473704	.472892	.472092	.471306	.470533	.469774	14
6.89	.469029	.468299	.467584	.466884	.465200	.465532	.464880	.464244	.463626	.463024	18
6.90	.462440	.461874	.461326	.460797	.460285	.459793	.459319	.458864	.458430	.458014	21
6.91	.457619	.457243	.456888	.456554	.456239	.455946	.455673	.455421	.455190	.454980	22
6.92	.454792	.454626	.454480	.454356	.454254	.454174	.454115	.454078	.454062	.454069	22
6.93	.454097	.454147	.454219	.454312	.454427	.454564	.454722	.454901	.455103	.455325	21
6.94	.455568	.455833	.456119	.456425	.456752	.457100	.457468	.457356	.458264	.458693	19
6.95	.459140	.459607	.460094	.460599	.461123	.461665	.462226	.462805	.463401	.464015	17
6.96	.464646	.465293	.465957	.466638	.467334	.468045	.468773	.469515	.470271	.471042	13
6.97	.471826	.472623	.473434	.474257	.475093	.475940	.476798	.477669	.478549	.479440	8
6.98	.480340	.481250	.482169	.483095	.484031	.484972	.485923	.486879	.487841	.488810	5
6.99	.489784	.490761	.491744	.492730	.493721	.494713	.495708	.496706	.497704	.498703	0

TABLE 9A

Fresnel Sine Integral

U	.000	.001	.002	.003	.004	.005	.006	.007	.008	.009	Δ_2
7.00	.499704	.500703	.501702	.502701	.503698	.504693	.505686	.506676	.507661	.508645	-5
7.01	.509622	.510595	.511564	.512526	.513482	.514431	.515374	.516309	.517235	.518154	-9
7.02	.519063	.519962	.520853	.521732	.522601	.523459	.524306	.525140	.525962	.526771	-13
7.03	.527568	.528349	.529119	.529872	.530612	.531336	.532045	.532739	.533416	.534077	-16
7.04	.534721	.535348	.535957	.536549	.537123	.537679	.538216	.538734	.539233	.539713	-19
7.05	.540173	.540614	.541035	.541435	.541815	.542174	.542513	.542831	.543127	.543403	-21
7.06	.543657	.543889	.544100	.544289	.544456	.544602	.544725	.544827	.544906	.544963	-21
7.07	.544998	.545010	.545001	.544969	.544915	.544839	.544740	.544620	.544478	.544313	-21
7.08	.544127	.543919	.543689	.543438	.543165	.542871	.542555	.542219	.541861	.541483	-19
7.09	.541085	.540666	.540227	.539768	.539290	.538792	.538275	.537738	.537184	.536610	-16
7.10	.536018	.535409	.534783	.534139	.533478	.532800	.532107	.531397	.530672	.529931	-13
7.11	.529176	.528406	.527622	.526825	.526014	.525191	.524354	.523506	.522647	.521776	-9
7.12	.520894	.520002	.519099	.518169	.517268	.516340	.515402	.514457	.513506	.512548	-5
7.13	.511584	.510614	.509639	.508659	.507675	.506687	.505695	.504701	.503705	.502707	0
7.14	.501708	.500709	.499708	.498709	.497710	.496713	.495716	.494723	.493733	.492745	5
7.15	.491760	.490782	.489807	.488838	.487874	.486918	.485967	.485023	.484088	.483161	9
7.16	.482241	.481333	.480432	.479543	.478662	.477794	.476937	.476091	.475258	.474437	14
7.17	.473630	.472837	.472057	.471292	.470540	.469805	.469084	.468379	.467690	.467018	17
7.18	.466363	.465725	.465104	.464502	.463917	.463352	.462804	.462276	.461768	.461278	21
7.19	.460808	.460359	.459930	.459521	.459133	.458767	.458421	.458096	.457793	.457512	22
7.20	.457252	.457014	.456798	.456605	.456433	.456285	.456158	.456054	.455972	.455913	23
7.21	.455877	.455863	.455872	.455904	.455958	.456035	.456134	.456257	.456401	.456568	22
7.22	.456757	.456968	.457202	.457457	.457735	.458034	.458354	.458697	.459060	.459444	20
7.23	.459850	.460275	.460722	.461188	.461675	.462181	.462707	.463252	.463816	.464398	17
7.24	.464999	.465618	.466254	.466908	.467580	.468267	.468971	.469692	.470427	.471177	15
7.25	.471944	.472724	.473518	.474325	.475147	.475980	.476827	.477685	.478554	.479435	10
7.26	.480326	.481227	.482138	.483058	.483987	.484924	.485868	.486821	.487780	.488745	5
7.27	.489716	.490691	.491672	.492656	.493645	.494637	.495632	.496629	.497626	.498626	0
7.28	.499626	.500626	.501626	.502623	.503621	.504615	.505608	.506597	.507583	.508565	-4
7.29	.509542	.510513	.511480	.512439	.513394	.514339	.515278	.516209	.517130	.518043	-9
7.30	.518945	.519838	.520720	.521591	.522452	.523299	.524135	.524958	.525767	.526563	-13
7.31	.527345	.528112	.528864	.529600	.530322	.531027	.531716	.532387	.533042	.533679	-17
7.32	.534298	.534898	.535481	.536044	.536588	.537113	.537618	.538103	.538568	.539012	-20
7.33	.539436	.539839	.540220	.540580	.540919	.541235	.541530	.541803	.542053	.542281	-22
7.34	.542487	.542669	.542830	.542967	.543081	.543173	.543241	.543287	.543309	.543309	-23
7.35	.543285	.543238	.543168	.543075	.542960	.542821	.542659	.542475	.542268	.542038	-22
7.36	.541786	.541512	.541215	.540897	.540556	.540194	.539811	.539406	.538980	.538533	-19
7.37	.538066	.537579	.537071	.536544	.535996	.535431	.534845	.534241	.533619	.532979	-16
7.38	.532322	.531647	.530955	.530247	.529522	.528781	.528025	.527254	.526469	.525669	-12
7.39	.524856	.524029	.523190	.522338	.521474	.520600	.519713	.518816	.517910	.516993	-8
7.40	.516067	.515134	.514191	.513242	.512285	.511323	.510353	.509378	.508399	.507415	-2
7.41	.506427	.505436	.504441	.503446	.502447	.501449	.500449	.499449	.498449	.497450	2
7.42	.496454	.495459	.494466	.493478	.492492	.491511	.490534	.489563	.488598	.487638	8
7.43	.486686	.485741	.484804	.483876	.482956	.482046	.481145	.480255	.479376	.478508	13
7.44	.477652	.476809	.475978	.475161	.474357	.473568	.472793	.472032	.471288	.470559	18
7.45	.469846	.469151	.468471	.467810	.467166	.466540	.465933	.465344	.464775	.464224	20
7.46	.463694	.463184	.462694	.462225	.461776	.461349	.460942	.460557	.460195	.459854	22
7.47	.459534	.459238	.458964	.458713	.458484	.458278	.458096	.457936	.457800	.457687	24
7.48	.457597	.457531	.457489	.457469	.457474	.457501	.457553	.457628	.457726	.457847	23
7.49	.457992	.458160	.458352	.458566	.458803	.459063	.459346	.459652	.459979	.460329	21

TABLE 9A

Fresnel Sine Integral

U	.000	.001	.002	.003	.004	.005	.006	.007	.008	.009	Δ_2
7.50	.460701	.461094	.461509	.461945	.462403	.462881	.463380	.463900	.464438	.464997	19
7.51	.465576	.466173	.466789	.467423	.468076	.468746	.469434	.470139	.470859	.471597	16
7.52	.472350	.473118	.473902	.474699	.475511	.476336	.477175	.478026	.478889	.479765	10
7.53	.480651	.481547	.482455	.483372	.484298	.485232	.486175	.487126	.488083	.489048	5
7.54	.490018	.490993	.491974	.492958	.493947	.494939	.495934	.496931	.497929	.498929	0
7.55	.499928	.500927	.501927	.502925	.503921	.504914	.505906	.506894	.507877	.508856	−6
7.56	.509830	.510797	.511760	.512715	.513663	.514602	.515534	.516457	.517371	.518274	−11
7.57	.519167	.520048	.520920	.521778	.522624	.523457	.524277	.525083	.525874	.526652	−15
7.58	.527413	.528159	.528889	.529602	.530299	.530978	.531640	.532283	.532908	.533514	−18
7.59	.534101	.534669	.535217	.535744	.536252	.536739	.537204	.537649	.538072	.538473	−21
7.60	.538853	.539209	.539544	.539856	.540145	.540412	.540655	.540875	.541071	.541245	−23
7.61	.541394	.541520	.541622	.541700	.541755	.541785	.541792	.541774	.541733	.541668	−23
7.62	.541579	.541466	.541329	.541169	.540985	.540778	.540547	.540293	.540016	.539716	−21
7.63	.539394	.539048	.538681	.538291	.537879	.537445	.536990	.536513	.536016	.535498	−18
7.64	.534960	.534402	.533823	.533226	.532609	.531974	.531320	.530648	.529959	.529252	−15
7.65	.528528	.527789	.527033	.526262	.525475	.524675	.523860	.523030	.522189	.521334	−10
7.66	.520466	.519588	.518698	.517798	.516887	.515966	.515037	.514098	.513152	.512198	−6
7.67	.511237	.510271	.509298	.508320	.507337	.506350	.505358	.504364	.503369	.502371	0
7.68	.501371	.500372	.499371	.498372	.497374	.496378	.495383	.494391	.493402	.492418	5
7.69	.491438	.490464	.489495	.488533	.487576	.486629	.485688	.484755	.483833	.482919	10
7.70	.482016	.481123	.480241	.479372	.478514	.477670	.476838	.476020	.475216	.474427	16
7.71	.473653	.472895	.472151	.471426	.470717	.470025	.469350	.468694	.468057	.467437	20
7.72	.466838	.466258	.465697	.465158	.464639	.464140	.463663	.463207	.462774	.462361	23
7.73	.461972	.461604	.461260	.460938	.460640	.460365	.460112	.459884	.459680	.459499	24
7.74	.459342	.459210	.459101	.459017	.458957	.458921	.458910	.458923	.458960	.459022	24
7.75	.459108	.459217	.459352	.459510	.459692	.459899	.460129	.460382	.460659	.460960	23
7.76	.461284	.461630	.462000	.462391	.462806	.463242	.463700	.464180	.464680	.465202	20
7.77	.465745	.466308	.466891	.467493	.468115	.468756	.469415	.470093	.470788	.471501	16
7.78	.472231	.472976	.473739	.474516	.475310	.476117	.476938	.477774	.478623	.479484	12
7.79	.480358	.481242	.482139	.483046	.483963	.484889	.485825	.486768	.487719	.488678	6
7.80	.489643	.490614	.491592	.492573	.493560	.494550	.495543	.496539	.497536	.498536	0
7.81	.499536	.500535	.501535	.502533	.503530	.504524	.505517	.506505	.507489	.508469	−5
7.82	.509443	.510411	.511374	.512328	.513276	.514215	.515146	.516068	.516979	.517881	−11
7.83	.518771	.519650	.520517	.521371	.522212	.523039	.523853	.524652	.525436	.526204	−15
7.84	.526956	.527692	.528411	.529113	.529797	.530462	.531110	.531739	.532347	.532937	−20
7.85	.533506	.534054	.534582	.535088	.535574	.536037	.536479	.536898	.537295	.537669	−23
7.86	.538020	.538348	.538652	.538933	.539190	.539423	.539631	.539816	.539976	.540112	−24
7.87	.540224	.540310	.540372	.540410	.540422	.540410	.540373	.540312	.540225	.540114	−24
7.88	.539979	.539819	.539635	.539427	.539194	.538937	.538656	.538352	.538024	.537673	−22
7.89	.537299	.536902	.536482	.536040	.535576	.535090	.534583	.534053	.533504	.532934	−18
7.90	.532343	.531733	.531103	.530455	.529787	.529101	.528398	.527676	.526939	.526184	−14
7.91	.525413	.524627	.523825	.523010	.522179	.521336	.520479	.519609	.518728	.517835	−9
7.92	.516932	.516019	.515094	.514161	.513219	.512270	.511312	.510347	.509377	.508400	−2
7.93	.507419	.506433	.505444	.504451	.503455	.502458	.501458	.500459	.499459	.498459	2
7.94	.497460	.496464	.495469	.494479	.493490	.492507	.491528	.490553	.489587	.488626	8
7.95	.487672	.486726	.485789	.484861	.483941	.483033	.482135	.481247	.480373	.479510	15
7.96	.478661	.477825	.477003	.476195	.475402	.474625	.473864	.473119	.472391	.471680	19
7.97	.470988	.470314	.469658	.469021	.468404	.467807	.467231	.466674	.466139	.465625	22
7.98	.465133	.464663	.464215	.463790	.463387	.463008	.462651	.462319	.462010	.461725	24
7.99	.461464	.461228	.461016	.460829	.460666	.460528	.460415	.460327	.460265	.460227	25

TABLE 9A

Fresnel Sine Integral

U	.000	.001	.002	.003	.004	.005	.006	.007	.008	.009	Δ_2
8.00	.460214	.460227	.460264	.460327	.460415	.460528	.460666	.460828	.461016	.461228	24
8.01	.461464	.461725	.462010	.462319	.462652	.463009	.463389	.463792	.464218	.464667	22
8.02	.465139	.465632	.466147	.466683	.467241	.467819	.468418	.469037	.469675	.470332	19
8.03	.471009	.471703	.472416	.473146	.473893	.474656	.475436	.476232	.477041	.477867	13
8.04	.478706	.479558	.480423	.481301	.482190	.483091	.484002	.484924	.485855	.486795	7
8.05	.487743	.488698	.489662	.490631	.491607	.492587	.493573	.494562	.495555	.496551	1
8.06	.497548	.498547	.499547	.500546	.501546	.502545	.503541	.504536	.505526	.506514	-5
8.07	.507498	.508476	.509448	.510414	.511374	.512326	.513271	.514207	.515133	.516049	-11
8.08	.516956	.517850	.518734	.519604	.520463	.521308	.522139	.522956	.523758	.524544	-15
8.09	.525315	.526068	.526806	.527525	.528227	.528910	.529575	.530221	.530846	.531453	-21
8.10	.532038	.532603	.533146	.533668	.534168	.534645	.535101	.535533	.535943	.536329	-23
8.11	.536691	.537029	.537344	.537634	.537900	.538140	.538357	.538548	.538714	.538855	-25
8.12	.538970	.539061	.539125	.539165	.539178	.539167	.539129	.539067	.538979	.538865	-24
8.13	.538726	.538562	.538373	.538158	.537919	.537655	.537367	.537054	.536717	.536356	-23
8.14	.535971	.535563	.535131	.534677	.534200	.533701	.533179	.532636	.532072	.531487	-19
8.15	.530880	.530255	.529609	.528944	.528260	.527558	.526837	.526099	.525344	.524573	-15
8.16	.523784	.522982	.522163	.521331	.520484	.519625	.518752	.517866	.516970	.516062	-8
8.17	.515144	.514217	.513279	.512334	.511379	.510419	.509450	.508475	.507496	.506512	-3
8.18	.505523	.504531	.503535	.502539	.501539	.500540	.499540	.498540	.497542	.496545	4
8.19	.495550	.494559	.493572	.492589	.491610	.490639	.489672	.488713	.487762	.486818	10
8.20	.485883	.484959	.484044	.483140	.482246	.481366	.480497	.479641	.478800	.477972	16
8.21	.477160	.476363	.475581	.474816	.474067	.473337	.472623	.471928	.471252	.470595	20
8.22	.469957	.469340	.468744	.468168	.467614	.467081	.466570	.466082	.465617	.465174	24
8.23	.464755	.464359	.463988	.463640	.463317	.463018	.462745	.462496	.462272	.462074	26
8.24	.461901	.461753	.461631	.461535	.461465	.461421	.461402	.461410	.461443	.461502	26
8.25	.461587	.461698	.461835	.461997	.462185	.462398	.462636	.462900	.463188	.463502	24
8.26	.463839	.464201	.464587	.464997	.465430	.465887	.466367	.466869	.467393	.467940	21
8.27	.468508	.469097	.469707	.470338	.470988	.471658	.472347	.473055	.473780	.474523	17
8.28	.475284	.476061	.476854	.477663	.478487	.479325	.480177	.481043	.481921	.482812	11
8.29	.483714	.484626	.485550	.486482	.487425	.488374	.489333	.490299	.491270	.492248	5
8.30	.493231	.494217	.495209	.496202	.497199	.498197	.499197	.500198	.501196	.502196	-3
8.31	.503193	.504188	.505180	.506167	.507152	.508130	.509104	.510071	.511031	.511983	-9
8.32	.512928	.513862	.514788	.515703	.516607	.517500	.518381	.519250	.520104	.520945	-14
8.33	.521772	.522583	.523380	.524159	.524923	.525669	.526397	.527108	.527799	.528471	-19
8.34	.529124	.529756	.530368	.530959	.531529	.532077	.532604	.533107	.533588	.534046	-24
8.35	.534480	.534890	.535276	.535638	.535976	.536288	.536576	.536838	.537075	.537286	-25
8.36	.537472	.537631	.537765	.537872	.537954	.538009	.538038	.538041	.538017	.537967	-26
8.37	.537891	.537789	.537661	.537506	.537326	.537120	.536888	.536630	.536348	.536040	-24
8.38	.535707	.535350	.534968	.534562	.534131	.533678	.533201	.532701	.532178	.531633	-21
8.39	.531066	.530478	.529869	.529239	.528588	.527918	.527228	.526520	.525793	.525048	-16
8.40	.524286	.523508	.522712	.521902	.521076	.520235	.519380	.518512	.517632	.516739	-9
8.41	.515834	.514919	.513992	.513057	.512112	.511160	.510199	.509231	.508258	.507278	-4
8.42	.506294	.505306	.504313	.503318	.502320	.501322	.500322	.499322	.498323	.497324	4
8.43	.496328	.495335	.494345	.493360	.492377	.491403	.490433	.489469	.488515	.487568	10
8.44	.486629	.485701	.484783	.483875	.482979	.482095	.481223	.480365	.479521	.478692	16
8.45	.477876	.477078	.476296	.475530	.474781	.474051	.473339	.472645	.471972	.471318	21
8.46	.470684	.470072	.469480	.468910	.468362	.467837	.467334	.466854	.466398	.465966	24
8.47	.465558	.465174	.464815	.464481	.464172	.463889	.463632	.463400	.463194	.463014	27
8.48	.462860	.462733	.462633	.462559	.462511	.462490	.462496	.462528	.462587	.462673	26
8.49	.462785	.462924	.463089	.463280	.463498	.463741	.464010	.464305	.464625	.464970	25

TABLE 9A

Fresnel Sine Integral

U	.000	.001	.002	.003	.004	.005	.006	.007	.008	.009	Δ_2
8.50	.465341	.465735	.466155	.466598	.467066	.467556	.468070	.468607	.469165	.469746	21
8.51	.470348	.470971	.471616	.472280	.472964	.473667	.474390	.475130	.475888	.476663	16
8.52	.477456	.478263	.479087	.479925	.480778	.481644	.482524	.483416	.484319	.485234	10
8.53	.486159	.487094	.488038	.488991	.489952	.490919	.491893	.492874	.493857	.494847	2
8.54	.495840	.496835	.497833	.498832	.499832	.500831	.501831	.502829	.503824	.504817	-3
8.55	.505806	.506790	.507771	.508744	.509712	.510672	.511625	.512569	.513504	.514430	-11
8.56	.515345	.516247	.517139	.518018	.518884	.519736	.520574	.521397	.522204	.522995	-16
8.57	.523769	.524526	.525265	.525985	.526687	.527369	.528031	.528674	.529294	.529894	-22
8.58	.530472	.531027	.531561	.532070	.532557	.533020	.533459	.533873	.534263	.534628	-24
8.59	.534967	.535281	.535569	.535832	.536068	.536278	.536461	.536618	.536748	.536851	-26
8.60	.536928	.536977	.536999	.536995	.536963	.536904	.536819	.536706	.536567	.536401	-26
8.61	.536208	.535989	.535743	.535472	.535174	.534851	.534502	.534128	.533729	.533305	-23
8.62	.532857	.532385	.531889	.531370	.530828	.530264	.529677	.529068	.528439	.527788	-19
8.63	.527117	.526427	.525717	.524988	.524240	.523476	.522693	.521894	.521080	.520249	-12
8.64	.519404	.518545	.517672	.516786	.515888	.514979	.514058	.513127	.512187	.511237	-7
8.65	.510280	.509315	.508342	.507365	.506382	.505395	.504403	.503407	.502411	.501412	0
8.66	.500411	.499412	.498412	.497415	.496418	.495425	.494435	.493450	.492469	.491494	7
8.67	.490526	.489565	.488612	.487667	.486731	.485806	.484891	.483986	.483095	.482215	14
8.68	.481349	.480498	.479661	.478839	.478033	.477244	.476471	.475715	.474979	.474261	19
8.69	.473561	.472883	.472224	.471586	.470969	.470374	.469801	.469251	.468724	.468220	24
8.70	.467740	.467284	.466853	.466446	.466065	.465709	.465379	.465074	.464796	.464544	27
8.71	.464319	.464121	.463949	.463804	.463687	.463597	.463534	.463498	.463490	.463509	27
8.72	.463556	.463629	.463731	.463859	.464014	.464197	.464406	.464643	.464905	.465194	26
8.73	.465510	.465851	.466217	.466609	.467027	.467468	.467935	.468426	.468940	.469477	22
8.74	.470038	.470621	.471226	.471853	.472501	.473169	.473858	.474567	.475294	.476041	17
8.75	.476805	.477586	.478385	.479200	.480030	.480875	.481735	.482609	.483495	.484394	11
8.76	.485304	.486225	.487158	.488099	.489050	.490008	.490974	.491947	.492925	.493911	3
8.77	.494899	.495891	.496887	.497884	.498884	.499883	.500884	.501883	.502880	.503876	-3
8.78	.504868	.505855	.506840	.507818	.508791	.509755	.510714	.511664	.512605	.513536	-10
8.79	.514457	.515366	.516264	.517149	.518021	.518879	.519723	.520552	.521364	.522160	-16
8.80	.522939	.523701	.524444	.525169	.525874	.526559	.527225	.527869	.528491	.529092	-22
8.81	.529671	.530227	.530760	.531268	.531753	.532213	.532649	.533060	.533445	.533804	-25
8.82	.534138	.534445	.534726	.534980	.535207	.535407	.535580	.535726	.535843	.535934	-27
8.83	.535997	.536032	.536039	.536018	.535970	.535894	.535791	.535660	.535501	.535315	-26
8.84	.535102	.534862	.534595	.534302	.533981	.533635	.533263	.532865	.532442	.531993	-24
8.85	.531520	.531023	.530502	.529958	.529389	.528799	.528186	.527552	.526896	.526220	-18
8.86	.525523	.524807	.524071	.523317	.522545	.521756	.520949	.520126	.519289	.518435	-12
8.87	.517569	.516688	.515794	.514889	.513972	.513044	.512107	.511159	.510204	.509240	-6
8.88	.508270	.507293	.506310	.505323	.504331	.503337	.502340	.501341	.500341	.499341	2
8.89	.498341	.497344	.496348	.495355	.494366	.493382	.492403	.491429	.490463	.489504	1.0
8.90	.488553	.487612	.486680	.485759	.484849	.483952	.483066	.482194	.481336	.480493	16
8.91	.479665	.478853	.478058	.477281	.476521	.475779	.475057	.474354	.473672	.473010	22
8.92	.472370	.471751	.471155	.470581	.470030	.469503	.469000	.468522	.468068	.467640	26
8.93	.467236	.466859	.466508	.466184	.465886	.465615	.465371	.465154	.464965	.464804	28
8.94	.464670	.464564	.464486	.464436	.464415	.464421	.464455	.464518	.464608	.464727	28
8.95	.464873	.465047	.465249	.465478	.465735	.466018	.466329	.466666	.467029	.467419	25
8.96	.467834	.468274	.468740	.469230	.469745	.470284	.470846	.471432	.472039	.472670	21
8.97	.473321	.473994	.474688	.475401	.476134	.476885	.477655	.478444	.479248	.480070	15
8.98	.480907	.481758	.482626	.483505	.484399	.485304	.486221	.487149	.488087	.489035	8
8.99	.489992	.490956	.491927	.492904	.493888	.494875	.495867	.496863	.497860	.498859	0

TABLE 9A

Fresnel Sine Integral

U	.000	.001	.002	.003	.004	.005	.006	.007	.008	.009	Δ_2
9.00	.499860	.500859	.501858	.502856	.503851	.504842	.505831	.506814	.507791	.508763	−7
9.01	.509726	.510682	.511630	.512567	.513496	.514412	.515318	.516211	.517091	.517957	−14
9.02	.518809	.519645	.520465	.521269	.522056	.522825	.523575	.524307	.525019	.525710	−20
9.03	.526382	.527031	.527659	.528264	.528846	.529405	.529941	.530453	.530939	.531401	−25
9.04	.531838	.532248	.532633	.532991	.533322	.533627	.533904	.534155	.534377	.534571	−28
9.05	.534738	.534876	.534987	.535069	.535122	.535148	.535145	.535113	.535053	.534964	−28
9.06	.534847	.534703	.534530	.534329	.534100	.533843	.533560	.533248	.532910	.532546	−25
9.07	.532154	.531737	.531294	.530826	.530333	.529815	.529273	.528707	.528118	.527506	−21
9.08	.526872	.526217	.525539	.524842	.524124	.523386	.522629	.521854	.521062	.520252	−14
9.09	.519426	.518585	.517727	.516857	.515971	.515074	.514163	.513241	.512309	.511367	−7
9.10	.510415	.509456	.508488	.507514	.506534	.505548	.504558	.503564	.502568	.501569	0
9.11	.500569	.499570	.498570	.497572	.496576	.495583	.494593	.493608	.492629	.491655	8
9.12	.490688	.489730	.488779	.487839	.486908	.485988	.485080	.484184	.483301	.482432	16
9.13	.481578	.480739	.479915	.479109	.478320	.477548	.476795	.476061	.475348	.474654	21
9.14	.473981	.473330	.472702	.472095	.471512	.470953	.470417	.469905	.469420	.468958	26
9.15	.468523	.468114	.467731	.467375	.467046	.466744	.466470	.466223	.466005	.465815	28
9.16	.465652	.465519	.465414	.465338	.465290	.465271	.465281	.465320	.465388	.465484	29
9.17	.465609	.465762	.465944	.466154	.466392	.466658	.466951	.467273	.467621	.467996	26
9.18	.468398	.468825	.469280	.469759	.470263	.470792	.471346	.471923	.472523	.473147	22
9.19	.473793	.474460	.475149	.475858	.476588	.477336	.478104	.478890	.479693	.480514	15
9.20	.481351	.482202	.483069	.483950	.484844	.485750	.486669	.487599	.488538	.489488	8
9.21	.490445	.491410	.492384	.493362	.494347	.495336	.496330	.497326	.498324	.499324	0
9.22	.500324	.501323	.502322	.503317	.504311	.505301	.506286	.507266	.508240	.509206	−8
9.23	.510166	.511115	.512056	.512985	.513905	.514813	.515707	.516589	.517457	.518310	−15
9.24	.519147	.519968	.520772	.521558	.522327	.523076	.523806	.524516	.525204	.525872	−21
9.25	.526517	.527140	.527741	.528317	.528870	.529398	.529901	.530379	.530831	.531258	−26
9.26	.531657	.532029	.532375	.532693	.532984	.533246	.533480	.533686	.533864	.534012	−29
9.27	.534132	.534223	.534285	.534317	.534321	.534295	.534240	.534157	.534044	.533902	−28
9.28	.533731	.533532	.533305	.533049	.532765	.532453	.532114	.531747	.531353	.530933	−25
9.29	.530486	.530014	.529515	.528992	.528444	.527873	.527277	.526658	.526016	.525352	−20
9.30	.524667	.523961	.523234	.522489	.521723	.520939	.520137	.519318	.518483	.517631	−13
9.31	.516765	.515885	.514991	.514085	.513166	.512237	.511297	.510347	.509389	.508422	−5
9.32	.507448	.506469	.505484	.504494	.503500	.502504	.501505	.500506	.499506	.498506	2
9.33	.497508	.496513	.495519	.494531	.493547	.492570	.491597	.490632	.489676	.488729	10
9.34	.487792	.486866	.485950	.485047	.484156	.483280	.482418	.481571	.480740	.479926	18
9.35	.479129	.478351	.477590	.476851	.476130	.475430	.474752	.474095	.473461	.472850	24
9.36	.472262	.471699	.471160	.470646	.470158	.469695	.469259	.468849	.468467	.468111	28
9.37	.467783	.467484	.467212	.466969	.466755	.466570	.466413	.466286	.466188	.466119	29
9.38	.466080	.466070	.466089	.466138	.466217	.466325	.466462	.466628	.466823	.467048	29
9.39	.467300	.467581	.467891	.468228	.468593	.468985	.469404	.469850	.470322	.470820	25
9.40	.471343	.471891	.472464	.473061	.473681	.474324	.474989	.475676	.476384	.477113	19
9.41	.477862	.478629	.479417	.480221	.481043	.481880	.482734	.483603	.484487	.485383	13
9.42	.486292	.487213	.488146	.489088	.490041	.491001	.491970	.492946	.493927	.494914	3
9.43	.495905	.496900	.497897	.498896	.499896	.500895	.501895	.502892	.503886	.504878	−5
9.44	.505865	.506846	.507821	.508789	.509749	.510701	.511643	.512576	.513496	.514405	−12
9.45	.515301	.516183	.517051	.517904	.518741	.519561	.520364	.521149	.521915	.522662	−19
9.46	.523389	.524094	.524779	.525441	.526081	.526698	.527292	.527861	.528405	.528924	−26
9.47	.529418	.529885	.530327	.530740	.531127	.531486	.531818	.532121	.532395	.532641	−29
9.48	.532858	.533045	.533204	.533333	.533432	.533501	.533541	.533551	.533531	.533482	−29
9.49	.533402	.533293	.533155	.532987	.532789	.532563	.532307	.532023	.531710	.531369	−27

TABLE 9A

Fresnel Sine Integral

U	.000	.001	.002	.003	.004	.005	.006	.007	.008	.009	Δ_2
9.50	.531000	.530604	.530180	.529730	.529253	.528750	.528222	.527668	.527090	.526487	-22
9.51	.525862	.525214	.524542	.523849	.523134	.522400	.521644	.520870	.520077	.519267	-16
9.52	.518438	.517595	.516735	.515861	.514971	.514069	.513155	.512228	.511291	.510344	-8
9.53	.509387	.508423	.507450	.506472	.505487	.504498	.503505	.502508	.501511	.500510	1
9.54	.499510	.498511	.497513	.496518	.495525	.494537	.493553	.492576	.491606	.490643	9
9.55	.489688	.488744	.487809	.486886	.485973	.485075	.484189	.483317	.482462	.481622	17
9.56	.480798	.479992	.479204	.478435	.477685	.476956	.476247	.475560	.474896	.474253	24
9.57	.473634	.473040	.472469	.471924	.471404	.470910	.470442	.470001	.469588	.469202	28
9.58	.468843	.468513	.468212	.467940	.467696	.467482	.467297	.467142	.467017	.466922	30
9.59	.466857	.466822	.466817	.466842	.466897	.466982	.467098	.467243	.467418	.467622	29
9.60	.467857	.468120	.468412	.468733	.469082	.469459	.469864	.470297	.470757	.471243	26
9.61	.471756	.472293	.472857	.473444	.474056	.474692	.475351	.476032	.476735	.477459	20
9.62	.478204	.478968	.479752	.480553	.481374	.482210	.483063	.483931	.484814	.485710	13
9.63	.486620	.487541	.488474	.489417	.490371	.491332	.492302	.493279	.494261	.495249	4
9.64	.496242	.497236	.498235	.499233	.500234	.501233	.502232	.503229	.504221	.505211	-5
9.65	.506195	.507173	.508146	.509109	.510065	.511011	.511948	.512873	.513786	.514687	-12
9.66	.515574	.516446	.517303	.518143	.518968	.519774	.520563	.521333	.522082	.522811	-21
9.67	.523519	.524205	.524868	.525509	.526126	.526719	.527287	.527831	.528347	.528839	-27
9.68	.529303	.529740	.530150	.530531	.530884	.531209	.531505	.531771	.532008	.532216	-30
9.69	.532393	.532541	.532658	.532745	.532802	.532828	.532824	.532789	.532724	.532629	-30
9.70	.532503	.532347	.532161	.531945	.531700	.531425	.531121	.530788	.530427	.530036	-27
9.71	.529619	.529174	.528701	.528203	.527677	.527127	.526550	.525949	.525325	.524676	-22
9.72	.524004	.523311	.522595	.521859	.521102	.520326	.519531	.518717	.517887	.517039	-13
9.73	.516176	.515298	.514405	.513500	.512581	.511652	.510711	.509759	.508800	.507832	-6
9.74	.506857	.505875	.504888	.503897	.502902	.501905	.500906	.499905	.498906	.497907	3
9.75	.496910	.495917	.494927	.493943	.492964	.491992	.491027	.490071	.489124	.488188	12
9.76	.487263	.486350	.485450	.484564	.483692	.482836	.481995	.481172	.480367	.479580	20
9.77	.478812	.478066	.477339	.476634	.475950	.475291	.474653	.474040	.473452	.472888	27
9.78	.472351	.471839	.471354	.470897	.470466	.470064	.469690	.469344	.469028	.468741	30
9.79	.468484	.468256	.468059	.467892	.467755	.467649	.467573	.467528	.467514	.467530	31
9.80	.467578	.467656	.467764	.467904	.468073	.468273	.468503	.468763	.469053	.469372	28
9.81	.469720	.470096	.470502	.470934	.471395	.471883	.472397	.472938	.473504	.474096	24
9.82	.474712	.475352	.476015	.476701	.477410	.478139	.478890	.479661	.480450	.481259	17
9.83	.482086	.482928	.483788	.484662	.485552	.486455	.487370	.488299	.489237	.490187	8
9.84	.491144	.492111	.493085	.494065	.495052	.496042	.497037	.498035	.499034	.500034	0
9.85	.501034	.502032	.503029	.504022	.505012	.505997	.506976	.507948	.508913	.509868	-9
9.86	.510815	.511750	.512675	.513586	.514486	.515370	.516241	.517096	.517933	.518754	-18
9.87	.519557	.520340	.521104	.521847	.522570	.523271	.523949	.524604	.525235	.525843	-24
9.88	.526424	.526981	.527511	.528015	.528492	.528941	.529363	.529756	.530120	.530455	-29
9.89	.530761	.531037	.531283	.531499	.531684	.531839	.531963	.532056	.532118	.532149	-30
9.90	.532149	.532118	.532056	.531962	.531838	.531683	.531498	.531281	.531035	.530759	-29
9.91	.530452	.530117	.529752	.529358	.528936	.528486	.528008	.527503	.526972	.526414	-24
9.92	.525831	.525223	.524590	.523934	.523254	.522552	.521828	.521082	.520317	.519531	-16
9.93	.518727	.517905	.517065	.516209	.515336	.514450	.513549	.512635	.511709	.510771	-9
9.94	.509823	.508866	.507900	.506927	.505946	.504961	.503970	.502975	.501978	.500979	0
9.95	.499979	.498979	.497981	.496985	.495991	.495001	.494017	.493038	.492067	.491102	10
9.96	.490148	.489203	.488268	.487346	.486435	.485538	.484654	.483787	.482935	.482100	19
9.97	.481282	.480484	.479704	.478944	.478205	.477488	.476793	.476120	.475472	.474847	25
9.98	.474246	.473672	.473124	.472601	.472106	.471639	.471199	.470788	.470405	.470052	30
9.99	.469728	.469434	.469170	.468937	.468734	.468562	.468421	.468311	.468233	.468186	31

S(u)

TABLE 9B

Fresnel Cosine Integral

U	.000	.001	.002	.003	.004	.005	.006	.007	.008	.009	Δ_2
0.01	.010000	.011000	.012000	.013000	.014000	.015000	.016000	.017000	.018000	.019000	0
0.02	.020000	.021000	.022000	.023000	.024000	.025000	.026000	.027000	.028000	.029000	0
0.03	.030000	.031000	.032000	.033000	.034000	.035000	.036000	.037000	.038000	.039000	0
0.04	.040000	.041000	.042000	.043000	.044000	.045000	.046000	.047000	.048000	.049000	0
0.05	.050000	.051000	.052000	.053000	.054000	.055000	.056000	.057000	.058000	.059000	0
0.06	.060000	.061000	.062000	.063000	.064000	.065000	.066000	.067000	.068000	.069000	0
0.07	.070000	.071000	.072000	.072999	.073999	.074999	.075999	.076999	.077999	.078999	0
0.08	.079999	.080999	.081999	.082999	.083999	.084999	.085999	.086999	.087999	.088999	0
0.09	.089999	.090998	.091998	.092998	.093998	.094998	.095998	.096998	.097998	.098998	0
0.10	.099998	.100997	.101997	.102997	.103997	.104997	.105997	.106997	.107996	.108996	0
0.11	.109996	.110996	.111996	.112995	.113995	.114995	.115995	.116995	.117994	.118994	0
0.12	.119994	.120994	.121993	.122993	.123993	.124992	.125992	.126992	.127992	.128991	0
0.13	.129991	.130990	.131990	.132990	.133989	.134989	.135989	.136988	.137988	.138987	0
0.14	.139987	.140986	.141986	.142985	.143985	.144984	.145984	.146983	.147982	.148982	0
0.15	.149981	.150981	.151980	.152979	.153979	.154978	.155977	.156976	.157976	.158975	0
0.16	.159974	.160973	.161972	.162972	.163971	.164970	.165969	.166968	.167967	.168966	0
0.17	.169965	.170964	.171963	.172962	.173961	.174960	.175958	.176957	.177956	.178955	0
0.18	.179953	.180952	.181951	.182949	.183948	.184947	.185945	.186944	.187942	.188941	0
0.19	.189939	.190937	.191936	.192934	.193932	.194930	.195929	.196927	.197925	.198923	0
0.20	.199921	.200919	.201917	.202915	.203913	.204911	.205908	.206906	.207904	.208902	0
0.21	.209899	.210897	.211894	.212892	.213889	.214887	.215884	.216881	.217879	.218876	0
0.22	.219873	.220870	.221867	.222864	.223861	.224858	.225855	.226851	.227848	.228845	0
0.23	.229841	.230838	.231834	.232831	.233827	.234823	.235819	.236816	.237812	.238808	0
0.24	.239804	.240799	.241795	.242791	.243787	.244782	.245778	.246773	.247769	.248764	0
0.25	.249759	.250754	.251749	.252744	.253739	.254734	.255729	.256724	.257718	.258713	0
0.26	.259707	.260701	.261696	.262690	.263684	.264678	.265672	.266665	.267659	.268653	0
0.27	.269646	.270640	.271633	.272626	.273619	.274612	.275605	.276598	.277591	.278583	0
0.28	.279576	.280568	.281560	.282552	.283544	.284536	.285528	.286520	.287512	.288503	0
0.29	.289494	.290486	.291477	.292468	.293458	.294449	.295440	.296430	.297421	.298411	0
0.30	.299401	.300391	.301381	.302370	.303360	.304349	.305339	.306328	.307317	.308306	0
0.31	.309294	.310283	.311271	.312260	.313248	.314236	.315223	.316211	.317199	.318186	0
0.32	.319173	.320160	.321147	.322134	.323120	.324106	.325093	.326079	.327065	.328050	0
0.33	.329036	.330021	.331006	.331991	.332976	.333960	.334945	.335929	.336913	.337897	0
0.34	.338881	.339864	.340847	.341830	.342813	.343796	.344778	.345761	.346743	.347725	0
0.35	.348706	.349688	.350669	.351650	.352631	.353611	.354592	.355572	.356552	.357531	0
0.36	.358511	.359490	.360469	.361448	.362426	.363405	.364383	.365361	.366338	.367316	0
0.37	.368293	.369270	.370246	.371222	.372199	.373174	.374150	.375125	.376100	.377075	0
0.38	.378050	.379024	.379998	.380972	.381945	.382918	.383891	.384864	.385836	.386808	0
0.39	.387780	.388751	.389722	.390693	.391664	.392634	.393604	.394574	.395543	.396512	0
0.40	.397481	.398449	.399417	.400385	.401353	.402320	.403287	.404253	.405219	.406185	0
0.41	.407151	.408116	.409081	.410045	.411009	.411973	.412936	.413900	.414862	.415825	0
0.42	.416787	.417748	.418710	.419671	.420631	.421591	.422551	.423511	.424470	.425429	0
0.43	.426387	.427345	.428302	.429259	.430216	.431173	.432128	.433084	.434039	.434994	0
0.44	.435948	.436902	.437856	.438809	.439761	.440714	.441665	.442617	.443568	.444518	0
0.45	.445468	.446418	.447367	.448316	.449264	.450212	.451159	.452106	.453053	.453998	0
0.46	.454944	.455889	.456834	.457778	.458721	.459664	.460607	.461549	.462491	.463432	0
0.47	.464373	.465313	.466252	.467192	.468130	.469068	.470006	.470943	.471879	.472815	-0
0.48	.473751	.474686	.475620	.476554	.477488	.478420	.479353	.480284	.481215	.482146	-0
0.49	.483076	.484005	.484934	.485863	.486790	.487717	.488644	.489570	.490495	.491420	-0

TABLE 9B

Fresnel Cosine Integral

U	.000	.001	.002	.003	.004	.005	.006	.007	.008	.009	Δ_2
0.50	.492344	.493268	.494191	.495113	.496035	.496956	.497877	.498796	.499716	.500634	−0
0.51	.501552	.502470	.503386	.504303	.505218	.506133	.507047	.507960	.508873	.509785	−0
0.52	.510697	.511608	.512518	.513427	.514336	.515244	.516151	.517058	.517964	.518869	−0
0.53	.519774	.520678	.521581	.522483	.523385	.524286	.525186	.526086	.526985	.527883	−0
0.54	.528780	.529677	.530572	.531467	.532362	.533255	.534148	.535040	.535931	.536821	−0
0.55	.537711	.538600	.539488	.540375	.541262	.542147	.543032	.543916	.544799	.545682	−0
0.56	.546563	.547444	.548324	.549203	.550081	.550958	.551835	.552710	.553585	.554459	−0
0.57	.555332	.556204	.557075	.557946	.558815	.559684	.560551	.561418	.562284	.563149	−0
0.58	.564013	.564876	.565738	.566600	.567460	.568320	.569178	.570036	.570892	.571748	−0
0.59	.572602	.573456	.574309	.575161	.576011	.576861	.577710	.578558	.579405	.580251	−0
0.60	.581095	.581939	.582782	.583624	.584465	.585304	.586143	.586981	.587817	.588653	−0
0.61	.589487	.590321	.591153	.591985	.592815	.593644	.594472	.595299	.596125	.596950	−0
0.62	.597774	.598596	.599418	.600238	.601057	.601876	.602693	.603509	.604323	.605137	−0
0.63	.605949	.606761	.607571	.608380	.609187	.609994	.610800	.611604	.612407	.613209	−0
0.64	.614009	.614809	.615607	.616404	.617200	.617995	.618788	.619580	.620371	.621160	−0
0.65	.621949	.622736	.623522	.624306	.625090	.625872	.626652	.627432	.628210	.628987	−0
0.66	.629763	.630537	.631310	.632081	.632852	.633621	.634388	.635154	.635919	.636683	−0
0.67	.637445	.638206	.638965	.639724	.640480	.641236	.641989	.642742	.643493	.644243	−0
0.68	.644991	.645738	.646484	.647228	.647970	.648711	.649451	.650189	.650926	.651662	−0
0.69	.652396	.653128	.653859	.654588	.655316	.656043	.656768	.657491	.658213	.658933	−1
0.70	.659652	.660370	.661086	.661800	.662513	.663224	.663933	.664641	.665348	.666053	−1
0.71	.666756	.667458	.668158	.668857	.669554	.670249	.670943	.671635	.672325	.673014	−1
0.72	.673701	.674387	.675071	.675753	.676433	.677112	.677790	.678465	.679139	.679811	−1
0.73	.680482	.681151	.681818	.682483	.683147	.683809	.684469	.685127	.685784	.686439	−1
0.74	.687092	.687743	.688393	.689041	.689687	.690331	.690974	.691615	.692254	.692891	−1
0.75	.693526	.694159	.694791	.695421	.696049	.696675	.697299	.697922	.698542	.699161	−1
0.76	.699778	.700393	.701006	.701617	.702226	.702834	.703439	.704043	.704644	.705244	−1
0.77	.705842	.706437	.707031	.707623	.708213	.708801	.709387	.709971	.710553	.711133	−1
0.78	.711711	.712287	.712861	.713433	.714003	.714571	.715137	.715701	.716263	.716823	−1
0.79	.717381	.717937	.718490	.719042	.719591	.720139	.720684	.721227	.721768	.722307	−1
0.80	.722844	.723379	.723912	.724442	.724970	.725497	.726021	.726543	.727062	.727580	−1
0.81	.728095	.728609	.729120	.729628	.730135	.730639	.731142	.731642	.732139	.732635	−1
0.82	.733128	.733619	.734108	.734595	.735079	.735561	.736041	.736518	.736993	.737466	−1
0.83	.737937	.738405	.738871	.739335	.739796	.740255	.740712	.741166	.741618	.742068	−1
0.84	.742515	.742960	.743403	.743843	.744281	.744716	.745149	.745580	.746008	.746434	−1
0.85	.746858	.747279	.747697	.748113	.748527	.748938	.749347	.749754	.750157	.750559	−1
0.86	.750958	.751354	.751748	.752140	.752529	.752915	.753299	.753681	.754060	.754436	−2
0.87	.754810	.755182	.755550	.755917	.756280	.756642	.757000	.757356	.757710	.758061	−2
0.88	.758409	.758755	.759098	.759438	.759776	.760111	.760444	.760774	.761102	.761426	−2
0.89	.761748	.762068	.762385	.762699	.763010	.763319	.763625	.763929	.764230	.764528	−2
0.90	.764823	.765116	.765406	.765693	.765978	.766259	.766538	.766815	.767088	.767359	−2
0.91	.767627	.767893	.768155	.768415	.768672	.768927	.769178	.769427	.769673	.769916	−2
0.92	.770156	.770394	.770629	.770860	.771090	.771316	.771539	.771760	.771978	.772192	−2
0.93	.772405	.772614	.772820	.773024	.773224	.773422	.773617	.773809	.773998	.774184	−2
0.94	.774367	.774548	.774725	.774900	.775071	.775240	.775406	.775569	.775728	.775885	−2
0.95	.776039	.776191	.776339	.776484	.776626	.776765	.776902	.777035	.777165	.777292	−2
0.96	.777417	.777538	.777656	.777772	.777884	.777993	.778100	.778203	.778303	.778401	−2
0.97	.778495	.778586	.778674	.778759	.778841	.778920	.778996	.779069	.779139	.779206	−2
0.98	.779269	.779330	.779388	.779442	.779494	.779542	.779587	.779629	.779668	.779704	−2
0.99	.779737	.779767	.779793	.779817	.779837	.779854	.779868	.779879	.779887	.779892	−2

TABLE 9B

Fresnel Cosine Integral

U	.000	.001	.002	.003	.004	.005	.006	.007	.008	.009	Δ_2
1.00	.779893	.779892	.779887	.779879	.779868	.779854	.779837	.779816	.779793	.779766	-2
1.01	.779736	.779703	.779666	.779627	.779584	.779538	.779489	.779437	.779382	.779323	-2
1.02	.779261	.779196	.779128	.779056	.778982	.778904	.778823	.778739	.778651	.778561	-2
1.03	.778467	.778370	.778269	.778166	.778059	.777949	.777835	.777719	.777599	.777476	-2
1.04	.777350	.777221	.777088	.776952	.776813	.776670	.776525	.776376	.776224	.776068	-2
1.05	.775910	.775748	.775582	.775414	.775242	.775067	.774889	.774707	.774523	.774335	-2
1.06	.774143	.773949	.773751	.773550	.773346	.773138	.772927	.772713	.772495	.772275	-2
1.07	.772051	.771823	.771593	.771359	.771122	.770881	.770638	.770391	.770141	.769887	-2
1.08	.769630	.769370	.769107	.768840	.768570	.768297	.768021	.767741	.767458	.767172	-2
1.09	.766882	.766589	.766293	.765994	.765691	.765385	.765076	.764764	.764448	.764129	-2
1.10	.763807	.763481	.763152	.762820	.762485	.762146	.761804	.761459	.761111	.760759	-2
1.11	.760404	.760046	.759685	.759320	.758952	.758581	.758206	.757829	.757448	.757063	-2
1.12	.756676	.756285	.755891	.755494	.755094	.754690	.754283	.753873	.753460	.753043	-2
1.13	.752624	.752201	.751774	.751345	.750912	.750477	.750038	.749595	.749150	.748701	-2
1.14	.748249	.747794	.747336	.746875	.746410	.745942	.745471	.744997	.744520	.744040	-2
1.15	.743556	.743069	.742579	.742086	.741590	.741091	.740588	.740083	.739574	.739062	-2
1.16	.738547	.738029	.737507	.736983	.736455	.735925	.735391	.734854	.734315	.733772	-2
1.17	.733226	.732676	.732124	.731569	.731011	.730449	.729885	.729317	.728747	.728173	-2
1.18	.727597	.727017	.726435	.725849	.725260	.724669	.724074	.723476	.722876	.722272	-2
1.19	.721666	.721056	.720444	.719828	.719210	.718588	.717964	.717337	.716707	.716074	-2
1.20	.715438	.714799	.714157	.713512	.712865	.712214	.711561	.710905	.710246	.709584	-2
1.21	.708919	.708252	.707581	.706908	.706232	.705553	.704872	.704187	.703500	.702810	-2
1.22	.702118	.701422	.700724	.700023	.699319	.698613	.697904	.697192	.696477	.695760	-2
1.23	.695040	.694317	.693592	.692864	.692133	.691400	.690664	.689926	.689185	.688441	-2
1.24	.687695	.686946	.686194	.685440	.684684	.683924	.683163	.682399	.681632	.680863	-2
1.25	.680091	.679316	.678540	.677761	.676979	.676195	.675408	.674619	.673828	.673034	-1
1.26	.672238	.671439	.670638	.669835	.669029	.668221	.667411	.666598	.665783	.664966	-1
1.27	.664146	.663324	.662500	.661673	.660845	.660014	.659181	.658345	.657508	.656668	-1
1.28	.655826	.654982	.654136	.653288	.652437	.651585	.650730	.649873	.649014	.648153	-1
1.29	.647290	.646425	.645558	.644689	.643818	.642945	.642070	.641193	.640314	.639433	-1
1.30	.638550	.637666	.636779	.635891	.635000	.634108	.633214	.632318	.631420	.630521	-1
1.31	.629619	.628716	.627812	.626905	.625997	.625087	.624175	.623261	.622346	.621429	-1
1.32	.620511	.619591	.618669	.617746	.616821	.615895	.614967	.614037	.613106	.612174	-0
1.33	.611239	.610304	.609367	.608428	.607488	.606547	.605604	.604660	.603715	.602768	-0
1.34	.601819	.600870	.599919	.598967	.598013	.597059	.596103	.595146	.594187	.593228	-0
1.35	.592267	.591305	.590342	.589377	.588412	.587446	.586478	.585509	.584540	.583569	-0
1.36	.582597	.581625	.580651	.579676	.578701	.577724	.576747	.575768	.574789	.573809	-0
1.37	.572828	.571846	.570864	.569880	.568896	.567911	.566926	.565939	.564952	.563964	-0
1.38	.562976	.561987	.560997	.560007	.559016	.558024	.557032	.556040	.555047	.554053	0
1.39	.553059	.552065	.551070	.550074	.549078	.548082	.547085	.546089	.545091	.544094	0
1.40	.543096	.542098	.541099	.540101	.539102	.538103	.537103	.536104	.535104	.534105	0
1.41	.533105	.532105	.531105	.530105	.529105	.528105	.527105	.526105	.525105	.524106	0
1.42	.523106	.522106	.521107	.520107	.519108	.518109	.517111	.516112	.515114	.514116	0
1.43	.513118	.512121	.511124	.510127	.509131	.508135	.507140	.506145	.505150	.504156	1
1.44	.503162	.502169	.501177	.500185	.499193	.498203	.497212	.496223	.495234	.494246	1
1.45	.493259	.492272	.491286	.490301	.489316	.488333	.487350	.486368	.485387	.484407	1
1.46	.483428	.482450	.481472	.480496	.479521	.478547	.477573	.476601	.475630	.474660	1
1.47	.473691	.472724	.471757	.470792	.469828	.468865	.467904	.466943	.465984	.465027	1
1.48	.464070	.463116	.462162	.461210	.460259	.459310	.458362	.457416	.456471	.455528	2
1.49	.454587	.453646	.452708	.451771	.450836	.449903	.448971	.448041	.447113	.446186	2

TABLE 9B

Fresnel Cosine Integral

U	.000	.001	.002	.003	.004	.005	.006	.007	.008	.009	Δ_2
1.50	.445261	.444338	.443417	.442498	.441580	.440665	.439751	.438839	.437930	.437022	2
1.51	.436116	.435212	.434311	.433411	.432514	.431618	.430725	.429834	.428945	.428058	2
1.52	.427173	.426291	.425411	.424533	.423657	.422784	.421913	.421045	.420179	.419315	2
1.53	.418454	.417595	.416739	.415885	.415034	.414185	.413339	.412495	.411654	.410816	3
1.54	.409980	.409147	.408316	.407489	.406664	.405842	.405022	.404206	.403392	.402581	3
1.55	.401773	.400967	.400165	.399365	.398569	.397775	.396985	.396197	.395413	.394631	3
1.56	.393853	.393078	.392305	.391536	.390770	.390007	.389248	.388491	.387738	.386988	3
1.57	.386242	.385498	.384758	.384022	.383288	.382558	.381832	.381109	.380389	.379672	3
1.58	.378960	.378250	.377544	.376842	.376143	.375448	.374757	.374069	.373384	.372704	4
1.59	.372026	.371353	.370683	.370017	.369355	.368697	.368042	.367391	.366744	.366101	4
1.60	.365462	.364826	.364195	.363567	.362943	.362323	.361708	.361096	.360488	.359884	4
1.61	.359284	.358688	.358097	.357509	.356926	.356346	.355771	.355200	.354633	.354071	4
1.62	.353512	.352958	.352408	.351862	.351321	.350783	.350251	.349722	.349198	.348678	4
1.63	.348163	.347652	.347145	.346643	.346145	.345652	.345163	.344679	.344199	.343724	5
1.64	.343253	.342787	.342325	.341868	.341416	.340968	.340524	.340086	.339652	.339223	5
1.65	.338798	.338378	.337963	.337553	.337147	.336746	.336350	.335958	.335572	.335190	5
1.66	.334813	.334441	.334074	.333711	.333354	.333001	.332653	.332311	.331973	.331640	5
1.67	.331312	.330989	.330670	.330357	.330049	.329746	.329448	.329155	.328867	.328584	5
1.68	.328306	.328033	.327766	.327503	.327245	.326993	.326746	.326503	.326266	.326035	5
1.69	.325808	.325586	.325370	.325159	.324953	.324752	.324557	.324366	.324181	.324001	5
1.70	.323827	.323658	.323494	.323335	.323181	.323033	.322890	.322753	.322620	.322494	5
1.71	.322372	.322256	.322145	.322039	.321939	.321844	.321755	.321670	.321592	.321518	5
1.72	.321450	.321388	.321330	.321279	.321232	.321191	.321156	.321126	.321101	.321081	5
1.73	.321068	.321059	.321056	.321059	.321067	.321080	.321099	.321123	.321153	.321188	5
1.74	.321228	.321274	.321326	.321383	.321445	.321513	.321587	.321666	.321750	.321840	5
1.75	.321935	.322036	.322142	.322254	.322371	.322494	.322622	.322755	.322894	.323039	5
1.76	.323189	.323344	.323505	.323671	.323843	.324021	.324203	.324391	.324585	.324784	5
1.77	.324989	.325199	.325414	.325635	.325861	.326093	.326330	.326573	.326820	.327074	5
1.78	.327333	.327597	.327866	.328141	.328421	.328707	.328998	.329295	.329596	.329903	5
1.79	.330216	.330534	.330857	.331185	.331519	.331858	.332203	.332552	.332907	.333267	5
1.80	.333633	.334004	.334380	.334761	.335147	.335539	.335936	.336338	.336745	.337158	5
1.81	.337576	.337998	.338426	.338859	.339298	.339741	.340189	.340643	.341101	.341565	5
1.82	.342034	.342508	.342986	.343470	.343959	.344453	.344952	.345456	.345964	.346478	5
1.83	.346996	.347520	.348048	.348582	.349120	.349663	.350210	.350763	.351320	.351883	5
1.84	.352450	.353021	.353598	.354179	.354765	.355355	.355951	.356550	.357155	.357764	5
1.85	.358378	.358996	.359619	.360246	.360878	.361515	.362155	.362801	.363451	.364105	4
1.86	.364764	.365427	.366094	.366766	.367442	.368122	.368807	.369496	.370189	.370886	4
1.87	.371588	.372294	.373003	.373717	.374436	.375158	.375884	.376614	.377349	.378087	4
1.88	.378829	.379576	.380326	.381080	.381838	.382599	.383365	.384134	.384908	.385684	4
1.89	.386465	.387249	.388037	.388829	.389624	.390423	.391226	.392032	.392841	.393654	3
1.90	.394471	.395290	.396114	.396940	.397770	.398604	.399440	.400280	.401123	.401970	3
1.91	.402819	.403672	.404527	.405386	.406248	.407113	.407981	.408852	.409726	.410603	3
1.92	.411482	.412365	.413250	.414138	.415029	.415923	.416819	.417718	.418620	.419524	3
1.93	.420431	.421340	.422252	.423166	.424083	.425002	.425924	.426848	.427774	.428703	2
1.94	.429633	.430566	.431501	.432439	.433378	.434319	.435263	.436208	.437156	.438105	2
1.95	.439057	.440010	.440965	.441922	.442880	.443841	.444803	.445766	.446732	.447698	2
1.96	.448667	.449637	.450608	.451581	.452555	.453531	.454508	.455486	.456466	.457447	1
1.97	.458429	.459412	.460396	.461381	.462368	.463355	.464343	.465333	.466323	.467314	1
1.98	.468306	.469298	.470291	.471286	.472280	.473275	.474271	.475268	.476265	.477262	0
1.99	.478260	.479258	.480257	.481256	.482255	.483254	.484254	.485254	.486253	.487253	0

TABLE 9B

Fresnel Cosine Integral

U	.000	.001	.002	.003	.004	.005	.006	.007	.008	.009	Δ_2
2.00	.488253	.489253	.490253	.491253	.492253	.493253	.494252	.495251	.496250	.497249	0
2.01	.498247	.499245	.500242	.501239	.502235	.503231	.504226	.505221	.506215	.507208	−0
2.02	.508200	.509192	.510183	.511173	.512162	.513150	.514137	.515123	.516108	.517091	−0
2.03	.518074	.519055	.520036	.521015	.521992	.522968	.523943	.524916	.525888	.526859	−1
2.04	.527827	.528794	.529760	.530724	.531686	.532646	.533605	.534561	.535516	.536469	−1
2.05	.537420	.538368	.539315	.540260	.541202	.542143	.543081	.544017	.544950	.545882	−1
2.06	.546811	.547737	.548661	.549583	.550502	.551418	.552332	.553243	.554151	.555057	−2
2.07	.555960	.556860	.557757	.558652	.559543	.560432	.561317	.562199	.563079	.563955	−2
2.08	.564828	.565698	.566564	.567428	.568287	.569144	.569997	.570847	.571693	.572536	−3
2.09	.573375	.574211	.575043	.575871	.576696	.577517	.578334	.579147	.579957	.580762	−3
2.10	.581564	.582362	.583156	.583945	.584731	.585512	.586290	.587063	.587832	.588596	−3
2.11	.589357	.590113	.590865	.591612	.592355	.593093	.593827	.594557	.595282	.596002	−4
2.12	.596718	.597429	.598135	.598836	.599533	.600225	.600912	.601594	.602272	.602944	−4
2.13	.603612	.604274	.604931	.605584	.606231	.606873	.607510	.608142	.608769	.609390	−4
2.14	.610006	.610617	.611222	.611822	.612417	.613006	.613590	.614168	.614741	.615308	−5
2.15	.615870	.616426	.616976	.617521	.618060	.618593	.619121	.619643	.620159	.620669	−5
2.16	.621173	.621672	.622164	.622651	.623132	.623607	.624075	.624538	.624995	.625445	−5
2.17	.625890	.626328	.626761	.627187	.627607	.628020	.628428	.628829	.629224	.629613	−5
2.18	.629995	.630371	.630741	.631104	.631461	.631812	.632156	.632493	.632824	.633149	−6
2.19	.633467	.633779	.634084	.634382	.634674	.634959	.635238	.635510	.635775	.636034	−6
2.20	.636286	.636531	.636770	.637002	.637227	.637446	.637657	.637862	.638060	.638251	−6
2.21	.638436	.638613	.638784	.638948	.639105	.639256	.639399	.639535	.639665	.639787	−6
2.22	.639903	.640012	.640114	.640209	.640297	.640377	.640451	.640519	.640579	.640632	−6
2.23	.640678	.640717	.640749	.640774	.640792	.640803	.640807	.640804	.640794	.640777	−6
2.24	.640753	.640721	.640683	.640638	.640586	.640526	.640460	.640387	.640306	.640219	−6
2.25	.640124	.640023	.639914	.639799	.639676	.639546	.639410	.639266	.639115	.638957	−6
2.26	.638793	.638621	.638442	.638257	.638064	.637864	.637658	.637444	.637224	.636996	−6
2.27	.636762	.636520	.636272	.636017	.635755	.635486	.635210	.634927	.634638	.634342	−6
2.28	.634038	.633728	.633411	.633088	.632757	.632420	.632076	.631726	.631368	.631004	−6
2.29	.630633	.630256	.629872	.629481	.629084	.628680	.628269	.627852	.627429	.626998	−6
2.30	.626562	.626119	.625669	.625213	.624750	.624281	.623806	.623324	.622836	.622342	−5
2.31	.621841	.621334	.620821	.620302	.619776	.619245	.618707	.618163	.617613	.617057	−5
2.32	.616494	.615926	.615352	.614772	.614186	.613594	.612996	.612392	.611783	.611168	−5
2.33	.610547	.609920	.609287	.608649	.608006	.607356	.606701	.606041	.605375	.604704	−4
2.34	.604027	.603345	.602657	.601964	.601266	.600563	.599854	.599140	.598421	.597697	−4
2.35	.596968	.596234	.595495	.594751	.594002	.593248	.592489	.591726	.590957	.590184	−4
2.36	.589406	.588624	.587837	.587046	.586250	.585449	.584644	.583835	.583021	.582203	−3
2.37	.581381	.580555	.579724	.578889	.578050	.577208	.576361	.575510	.574655	.573797	−3
2.38	.572934	.572068	.571199	.570325	.569448	.568567	.567683	.566795	.565904	.565010	−2
2.39	.564112	.563211	.562306	.561399	.560488	.559575	.558658	.557738	.556815	.555890	−2
2.40	.554961	.554030	.553096	.552160	.551221	.550279	.549335	.548388	.547439	.546487	−1
2.41	.545534	.544578	.543619	.542659	.541697	.540732	.539766	.538797	.537827	.536855	−1
2.42	.535881	.534906	.533928	.532950	.531969	.530988	.530005	.529020	.528034	.527047	−0
2.43	.526059	.525069	.524079	.523087	.522095	.521101	.520107	.519112	.518116	.517120	−0
2.44	.516123	.515125	.514127	.513129	.512130	.511130	.510131	.509131	.508131	.507131	0
2.45	.506131	.505131	.504131	.503132	.502132	.501133	.500134	.499136	.498137	.497140	1
2.46	.496143	.495146	.494151	.493156	.492162	.491168	.490176	.489185	.488194	.487205	1
2.47	.486217	.485230	.484244	.483260	.482277	.481296	.480316	.479338	.478361	.477387	2
2.48	.476413	.475442	.474473	.473505	.472540	.471576	.470615	.469656	.468699	.467745	2
2.49	.466792	.465843	.464895	.463950	.463008	.462069	.461132	.460198	.459267	.458338	3

TABLE 9B

Fresnel Cosine Integral

U	.000	.001	.002	.003	.004	.005	.006	.007	.008	.009	Δ_2
2.50	.457413	.456491	.455571	.454655	.453742	.452832	.451926	.451023	.450123	.449227	4
2.51	.448334	.447445	.446560	.445678	.444800	.443925	.443055	.442189	.441326	.440468	4
2.52	.439613	.438763	.437917	.437075	.436238	.435405	.434576	.433752	.432932	.432117	5
2.53	.431306	.430500	.429699	.428903	.428112	.427325	.426543	.425767	.424995	.424229	5
2.54	.423467	.422711	.421960	.421214	.420474	.419739	.419010	.418286	.417568	.416855	6
2.55	.416148	.415446	.414750	.414060	.413376	.412698	.412026	.411359	.410699	.410045	6
2.56	.409396	.408754	.408119	.407489	.406866	.406249	.405638	.405034	.404436	.403844	7
2.57	.403260	.402681	.402110	.401545	.400986	.400435	.399890	.399352	.398821	.398297	7
2.58	.397779	.397269	.396765	.396269	.395780	.395297	.394822	.394354	.393893	.393440	7
2.59	.392994	.392555	.392123	.391699	.391282	.390872	.390470	.390076	.389689	.389309	8
2.60	.388937	.388573	.388216	.387867	.387526	.387192	.386866	.386548	.386238	.385935	8
2.61	.385640	.385353	.385074	.384803	.384540	.384284	.384037	.383798	.383566	.383343	8
2.62	.383127	.382920	.382721	.382529	.382346	.382171	.382004	.381846	.381695	.381553	8
2.63	.381418	.381292	.381174	.381065	.380963	.380870	.380785	.380708	.380640	.380580	8
2.64	.380528	.380484	.380449	.380422	.380403	.380393	.380391	.380397	.380412	.380435	8
2.65	.380466	.380505	.380553	.380609	.380674	.380746	.380828	.380917	.381015	.381121	8
2.66	.381235	.381358	.381489	.381628	.381775	.381931	.382095	.382267	.382448	.382637	8
2.67	.382833	.383039	.383252	.383474	.383703	.383941	.384188	.384442	.384704	.384975	8
2.68	.385253	.385540	.385835	.386137	.386448	.386767	.387094	.387428	.387771	.388122	8
2.69	.388480	.388846	.389221	.389603	.389992	.390390	.390795	.391209	.391629	.392058	8
2.70	.392494	.392938	.393389	.393848	.394314	.394788	.395270	.395758	.396255	.396758	7
2.71	.397269	.397787	.398313	.398846	.399386	.399933	.400487	.401048	.401616	.402192	7
2.72	.402774	.403363	.403959	.404562	.405172	.405788	.406412	.407041	.407678	.408321	7
2.73	.408971	.409627	.410290	.410959	.411634	.412316	.413003	.413698	.414398	.415104	6
2.74	.415817	.416535	.417260	.417990	.418726	.419468	.420216	.420970	.421729	.422494	6
2.75	.423264	.424040	.424821	.425607	.426399	.427196	.427998	.428806	.429618	.430436	5
2.76	.431258	.432086	.432918	.433755	.434597	.435443	.436294	.437150	.438010	.438874	4
2.77	.439743	.440616	.441493	.442374	.443260	.444149	.445043	.445940	.446841	.447746	4
2.78	.448655	.449567	.450482	.451402	.452324	.453250	.454179	.455112	.456047	.456986	3
2.79	.457927	.458872	.459819	.460769	.461722	.462678	.463636	.464596	.465559	.466524	2
2.80	.467492	.468461	.469433	.470407	.471383	.472360	.473340	.474321	.475304	.476288	2
2.81	.477274	.478262	.479251	.480241	.481232	.482224	.483218	.484212	.485207	.486204	1
2.82	.487200	.488198	.489196	.490195	.491194	.492193	.493193	.494193	.495193	.496193	0
2.83	.497193	.498192	.499192	.500191	.501190	.502189	.503187	.504184	.505181	.506177	−0
2.84	.507172	.508166	.509160	.510152	.511143	.512132	.513121	.514108	.515094	.516078	−1
2.85	.517060	.518041	.519019	.519997	.520972	.521945	.522916	.523884	.524851	.525815	−1
2.86	.526777	.527736	.528692	.529646	.530598	.531546	.532492	.533434	.534374	.535310	−2
2.87	.536243	.537173	.538100	.539023	.539942	.540858	.541771	.542679	.543584	.544485	−3
2.88	.545382	.546275	.547164	.548049	.548929	.549805	.550677	.551544	.552406	.553264	−4
2.89	.554118	.554966	.555810	.556649	.557482	.558311	.559135	.559953	.560766	.561574	−4
2.90	.562376	.563173	.563965	.564751	.565531	.566305	.567074	.567836	.568593	.569344	−5
2.91	.570089	.570827	.571559	.572286	.573005	.573719	.574426	.575126	.575820	.576507	−6
2.92	.577188	.577862	.578529	.579189	.579842	.580488	.581127	.581759	.582384	.583002	−6
2.93	.583613	.584216	.584811	.585400	.585981	.586554	.587120	.587678	.588228	.588771	−7
2.94	.589306	.589833	.590352	.590864	.591367	.591862	.592350	.592829	.593300	.593763	−7
2.95	.594217	.594664	.595102	.595531	.595953	.596366	.596770	.597166	.597553	.597932	−8
2.96	.598302	.598663	.599016	.599360	.599695	.600022	.600340	.600649	.600948	.601240	−8
2.97	.601522	.601795	.602059	.602314	.602560	.602797	.603025	.603244	.603454	.603654	−8
2.98	.603846	.604028	.604201	.604364	.604519	.604664	.604800	.604927	.605044	.605152	−8
2.99	.605250	.605340	.605420	.605490	.605551	.605603	.605645	.605678	.605702	.605716	−8

TABLE 9B

Fresnel Cosine Integral

U	.000	.001	.002	.003	.004	.005	.006	.007	.008	.009	Δ_2
3.00	.605720	.605716	.605702	.605678	.605645	.605603	.605551	.605489	.605419	.605339	−9
3.01	.605249	.605150	.605042	.604924	.604797	.604660	.604514	.604359	.604194	.604020	−8
3.02	.603837	.603645	.603442	.603231	.603011	.602781	.602542	.602294	.602036	.601770	−8
3.03	.601494	.601209	.600915	.600612	.600300	.599979	.599648	.599309	.598961	.598604	−8
3.04	.598238	.597863	.597479	.597087	.596685	.596275	.595857	.595429	.594993	.594548	−8
3.05	.594095	.593634	.593163	.592685	.592198	.591703	.591199	.590687	.590167	.589638	−7
3.06	.589101	.588557	.588004	.587444	.586875	.586299	.585714	.585122	.584522	.583915	−6
3.07	.583299	.582677	.582046	.581409	.580764	.580111	.579451	.578784	.578110	.577429	−5
3.08	.576740	.576045	.575343	.574634	.573918	.573196	.572466	.571730	.570988	.570239	−5
3.09	.569484	.568723	.567954	.567181	.566400	.565615	.564822	.564024	.563220	.562410	−4
3.10	.561595	.560774	.559947	.559116	.558278	.557436	.556588	.555735	.554877	.554014	−3
3.11	.553146	.552274	.551396	.550514	.549627	.548737	.547841	.546941	.546037	.545129	−3
3.12	.544216	.543301	.542380	.541457	.540529	.539598	.538663	.537724	.536783	.535838	−1
3.13	.534891	.533940	.532986	.532029	.531069	.530107	.529142	.528174	.527204	.526231	0
3.14	.525256	.524280	.523301	.522320	.521337	.520353	.519367	.518378	.517390	.516398	0
3.15	.515406	.514413	.513419	.512424	.511427	.510431	.509433	.508434	.507436	.506436	0
3.16	.505436	.504437	.503437	.502437	.501437	.500438	.499438	.498438	.497440	.496441	2
3.17	.495445	.494449	.493452	.492458	.491464	.490472	.489480	.488490	.487502	.486515	2
3.18	.485529	.484546	.483564	.482584	.481606	.480631	.479657	.478686	.477718	.476752	4
3.19	.475788	.474828	.473871	.472916	.471964	.471016	.470071	.469127	.468189	.467253	5
3.20	.466321	.465395	.464469	.463549	.462634	.461722	.460814	.459909	.459010	.458116	3
3.21	.457225	.456341	.455459	.454583	.453711	.452846	.451983	.451128	.450278	.449432	6
3.22	.448592	.447757	.446929	.446105	.445287	.444477	.443672	.442871	.442079	.441291	7
3.23	.440510	.439736	.438968	.438207	.437452	.436705	.435962	.435228	.434501	.433780	8
3.24	.433066	.432360	.431662	.430970	.430285	.429610	.428940	.428278	.427624	.426978	8
3.25	.426340	.425709	.425086	.424473	.423866	.423268	.422678	.422096	.421523	.420957	10
3.26	.420402	.419854	.419314	.418784	.418262	.417749	.417244	.416749	.416263	.415785	9
3.27	.415316	.414857	.414407	.413966	.413534	.413112	.412699	.412295	.411901	.411516	10
3.28	.411140	.410775	.410418	.410072	.409736	.409409	.409091	.408784	.408487	.408198	10
3.29	.407921	.407653	.407395	.407147	.406909	.406682	.406464	.406256	.406059	.405872	10
3.30	.405695	.405528	.405371	.405225	.405089	.404963	.404848	.404742	.404648	.404563	11
3.31	.404489	.404425	.404372	.404329	.404297	.404274	.404263	.404261	.404270	.404290	10
3.32	.404320	.404360	.404411	.404472	.404544	.404626	.404718	.404821	.404934	.405058	10
3.33	.405192	.405336	.405491	.405656	.405831	.406017	.406213	.406419	.406636	.406862	10
3.34	.407099	.407346	.407603	.407871	.408148	.408436	.408733	.409041	.409358	.409686	9
3.35	.410023	.410371	.410728	.411095	.411472	.411858	.412254	.412661	.413076	.413501	9
3.36	.413936	.414380	.414833	.415296	.415769	.416250	.416741	.417241	.417750	.418268	10
3.37	.418795	.419331	.419876	.420430	.420993	.421565	.422145	.422734	.423330	.423937	8
3.38	.424551	.425172	.425804	.426443	.427090	.427744	.428409	.429079	.429758	.430445	8
3.39	.431140	.431842	.432551	.433268	.433992	.434724	.435463	.436210	.436963	.437723	7
3.40	.438490	.439264	.440044	.440832	.441626	.442425	.443233	.444045	.444864	.445691	3
3.41	.446522	.447359	.448201	.449049	.449905	.450763	.451629	.452499	.453376	.454257	3
3.42	.455141	.456032	.456928	.457828	.458733	.459643	.460558	.461475	.462397	.463325	1
3.43	.464256	.465190	.466129	.467070	.468015	.468964	.469917	.470873	.471831	.472793	3
3.44	.473757	.474726	.475695	.476668	.477643	.478620	.479599	.480582	.481566	.482552	1
3.45	.483540	.484529	.485518	.486511	.487504	.488498	.489495	.490492	.491488	.492486	0
3.46	.493484	.494484	.495483	.496483	.497483	.498482	.499483	.500483	.501481	.502481	−2
3.47	.503480	.504477	.505474	.506469	.507465	.508458	.509451	.510442	.511430	.512419	−3
3.48	.513404	.514387	.515371	.516350	.517328	.518303	.519276	.520247	.521213	.522179	−4
3.49	.523141	.524099	.525055	.526007	.526955	.527900	.528842	.529780	.530714	.531645	−5

TABLE 9B

Fresnel Cosine Integral

U	.000	.001	.002	.003	.004	.005	.006	.007	.008	.009	Δ_2
3.50	.532572	.533493	.534410	.535323	.536231	.537136	.538035	.538929	.539818	.540702	−4
3.51	.541580	.542454	.543322	.544184	.545042	.545893	.546738	.547578	.548411	.549238	−4
3.52	.550059	.550874	.551682	.552483	.553279	.554066	.554848	.555623	.556389	.557150	−7
3.53	.557903	.558648	.559386	.560117	.560840	.561554	.562262	.562962	.563653	.564336	−7
3.54	.565012	.565679	.566338	.566987	.567630	.568263	.568888	.569503	.570110	.570708	−9
3.55	.571298	.571877	.572449	.573010	.573563	.574105	.574640	.575164	.575678	.576184	−9
3.56	.576679	.577165	.577641	.578106	.578563	.579009	.579445	.579871	.580286	.580692	−9
3.57	.581087	.581472	.581847	.582211	.582565	.582907	.583240	.583562	.583873	.584174	−10
3.58	.584464	.584743	.585011	.585268	.585515	.585750	.585975	.586189	.586391	.586583	−10
3.59	.586763	.586932	.587091	.587238	.587373	.587498	.587612	.587714	.587805	.587885	−10
3.60	.587953	.588010	.588056	.588091	.588114	.588126	.588126	.588116	.588094	.588060	−11
3.61	.588016	.587960	.587892	.587813	.587723	.587622	.587510	.587386	.587251	.587104	−9
3.62	.586946	.586778	.586598	.586406	.586204	.585991	.585766	.585530	.585284	.585025	−9
3.63	.584757	.584477	.584187	.583885	.583573	.583250	.582915	.582570	.582215	.581849	−9
3.64	.581471	.581084	.580686	.580279	.579859	.579430	.578991	.578541	.578081	.577611	−8
3.65	.577131	.576641	.576140	.575631	.575110	.574581	.574041	.573492	.572935	.572366	−8
3.66	.571788	.571202	.570606	.570000	.569386	.568763	.568130	.567488	.566839	.566180	−7
3.67	.565513	.564837	.564153	.563460	.562758	.562050	.561332	.560607	.559874	.559132	−6
3.68	.558383	.557627	.556863	.556092	.555313	.554528	.553734	.552934	.552128	.551314	−5
3.69	.550493	.549667	.548834	.547994	.547148	.546296	.545438	.544573	.543703	.542829	−6
3.70	.541947	.541061	.540169	.539272	.538370	.537463	.536552	.535634	.534713	.533787	−2
3.71	.532858	.531925	.530985	.530044	.529097	.528148	.527195	.526237	.525277	.524313	−1
3.72	.523346	.522378	.521405	.520431	.519453	.518474	.517491	.516507	.515521	.514533	0
3.73	.513543	.512551	.511559	.510565	.509568	.508573	.507575	.506576	.505578	.504579	0
3.74	.503579	.502579	.501578	.500579	.499579	.498581	.497581	.496583	.495584	.494587	3
3.75	.493592	.492597	.491603	.490612	.489622	.488634	.487646	.486662	.485679	.484699	3
3.76	.483721	.482747	.481773	.480804	.479837	.478873	.477911	.476955	.476000	.475050	4
3.77	.474103	.473161	.472222	.471288	.470357	.469432	.468510	.467593	.466682	.465776	5
3.78	.464873	.463977	.463085	.462200	.461320	.460446	.459576	.458713	.457858	.457007	7
3.79	.456162	.455325	.454494	.453670	.452852	.452042	.451238	.450441	.449653	.448870	9
3.80	.448096	.447330	.446571	.445820	.445077	.444341	.443615	.442896	.442185	.441483	9
3.81	.440790	.440105	.439429	.438762	.438104	.437456	.436815	.436185	.435564	.434952	10
3.82	.434349	.433757	.433174	.432602	.432038	.431485	.430942	.430408	.429886	.429373	10
3.83	.428871	.428379	.427898	.427427	.426967	.426518	.426079	.425652	.425236	.424830	12
3.84	.424435	.424052	.423679	.423318	.422968	.422630	.422303	.421987	.421683	.421390	12
3.85	.421109	.420840	.420582	.420336	.420102	.419879	.419669	.419470	.419283	.419108	12
3.86	.418945	.418794	.418654	.418528	.418412	.418310	.418219	.418140	.418073	.418019	12
3.87	.417976	.417946	.417928	.417922	.417928	.417947	.417977	.418020	.418075	.418142	12
3.88	.418221	.418313	.418416	.418532	.418660	.418799	.418952	.419116	.419291	.419480	12
3.89	.419680	.419892	.420115	.420351	.420599	.420858	.421130	.421413	.421707	.422014	12
3.90	.422332	.422662	.423003	.423356	.423720	.424095	.424483	.424881	.425290	.425711	10
3.91	.426143	.426585	.427039	.427503	.427979	.428465	.428962	.429470	.429988	.430517	10
3.92	.431056	.431605	.432165	.432734	.433314	.433904	.434504	.435114	.435733	.436363	8
3.93	.437001	.437649	.438306	.438973	.439649	.440333	.441027	.441730	.442441	.443161	8
3.94	.443891	.444628	.445372	.446126	.446888	.447657	.448434	.449220	.450012	.450813	5
3.95	.451621	.452435	.453258	.454086	.454922	.455765	.456615	.457471	.458332	.459202	4
3.96	.460075	.460955	.461842	.462734	.463631	.464534	.465443	.466357	.467274	.468197	4
3.97	.469126	.470058	.470995	.471937	.472882	.473831	.474784	.475742	.476702	.477667	1
3.98	.478633	.479603	.480576	.481553	.482531	.483512	.484496	.485482	.486469	.487459	1
3.99	.488451	.489443	.490438	.491433	.492430	.493427	.494426	.495425	.496424	.497425	0

TABLE 9B

Fresnel Cosine Integral

U	.000	.001	.002	.003	.004	.005	.006	.007	.008	.009	Δ_2
4.00	.498426	.499425	.500425	.501423	.502423	.503421	.504419	.505415	.506411	.507405	−1
4.01	.508397	.509389	.510378	.511366	.512351	.513334	.514316	.515295	.516271	.517243	−3
4.02	.518214	.519180	.520144	.521104	.522060	.523012	.523962	.524907	.525847	.526783	−4
4.03	.527715	.528642	.529564	.530481	.531393	.532299	.533201	.534096	.534986	.535871	−6
4.04	.536749	.537621	.538487	.539346	.540198	.541045	.541884	.542717	.543541	.544359	−7
4.05	.545170	.545972	.546768	.547554	.548334	.549106	.549869	.550625	.551371	.552109	−8
4.06	.552840	.553559	.554272	.554974	.555667	.556352	.557027	.557693	.558349	.558995	−9
4.07	.559632	.560258	.560874	.561480	.562077	.562663	.563238	.563803	.564357	.564901	−10
4.08	.565434	.565956	.566466	.566966	.567455	.567932	.568399	.568853	.569296	.569729	−11
4.09	.570148	.570557	.570953	.571338	.571711	.572072	.572421	.572758	.573082	.573395	−12
4.10	.573695	.573983	.574259	.574521	.574772	.575010	.575236	.575449	.575649	.575837	−12
4.11	.576012	.576175	.576324	.576461	.576585	.576696	.576795	.576880	.576952	.577012	−11
4.12	.577059	.577092	.577113	.577121	.577116	.577098	.577067	.577023	.576966	.576896	−12
4.13	.576814	.576718	.576609	.576488	.576353	.576206	.576046	.575873	.575688	.575489	−11
4.14	.575278	.575054	.574818	.574569	.574307	.574033	.573746	.573447	.573135	.572811	−11
4.15	.572474	.572126	.571765	.571393	.571008	.570611	.570202	.569781	.569349	.568904	−10
4.16	.568448	.567980	.567501	.567011	.566509	.565996	.565472	.564936	.564390	.563832	−10
4.17	.563264	.562686	.562096	.561496	.560885	.560265	.559633	.558992	.558341	.557680	−9
4.18	.557009	.556329	.555638	.554939	.554230	.553512	.552784	.552048	.551303	.550549	−7
4.19	.549786	.549016	.548237	.547450	.546654	.545851	.545040	.544221	.543395	.542561	−5
4.20	.541720	.540873	.540017	.539157	.538288	.537414	.536532	.535645	.534752	.533853	−4
4.21	.532949	.532039	.531123	.530203	.529277	.528346	.527410	.526470	.525525	.524576	−2
4.22	.523623	.522665	.521705	.520741	.519773	.518803	.517827	.516850	.515871	.514888	0
4.23	.513903	.512916	.511928	.510937	.509944	.508950	.507954	.506956	.505961	.504962	0
4.24	.503962	.502963	.501963	.500963	.499963	.498964	.497964	.496966	.495968	.494971	1
4.25	.493974	.492981	.491987	.490996	.490006	.489019	.488034	.487050	.486070	.485092	3
4.26	.484118	.483147	.482178	.481214	.480252	.479295	.478341	.477391	.476446	.475504	7
4.27	.474568	.473637	.472711	.471789	.470873	.469963	.469058	.468158	.467266	.466379	7
4.28	.465497	.464623	.463755	.462895	.462041	.461194	.460355	.459522	.458697	.457880	8
4.29	.457071	.456270	.455476	.454692	.453915	.453148	.452389	.451638	.450898	.450165	11
4.30	.449443	.448730	.448026	.447331	.446647	.445973	.445308	.444653	.444010	.443375	12
4.31	.442753	.442140	.441539	.440948	.440368	.439800	.439242	.438695	.438161	.437637	13
4.32	.437126	.436626	.436137	.435662	.435197	.434746	.434305	.433877	.433462	.433059	13
4.33	.432668	.432290	.431925	.431572	.431232	.430905	.430591	.430290	.430002	.429726	13
4.34	.429464	.429216	.428980	.428758	.428549	.428354	.428171	.428002	.427847	.427705	14
4.35	.427577	.427462	.427361	.427274	.427200	.427140	.427093	.427060	.427041	.427036	14
4.36	.427044	.427066	.427101	.427151	.427214	.427290	.427381	.427485	.427602	.427733	13
4.37	.427878	.428037	.428209	.428394	.428593	.428805	.429031	.429270	.429523	.429788	13
4.38	.430067	.430359	.430665	.430983	.431315	.431659	.432016	.432386	.432769	.433165	12
4.39	.433573	.433994	.434427	.434873	.435331	.435801	.436283	.436778	.437284	.437802	11
4.40	.438332	.438873	.439427	.439991	.440568	.441154	.441753	.442363	.442983	.443614	10
4.41	.444256	.444908	.445572	.446245	.446928	.447621	.448325	.449038	.449760	.450493	9
4.42	.451235	.451985	.452745	.453514	.454292	.455078	.455873	.456677	.457488	.458308	7
4.43	.459135	.459970	.460814	.461663	.462521	.463386	.464257	.465136	.466020	.466911	6
4.44	.467809	.468713	.469622	.470536	.471458	.472383	.473315	.474251	.475192	.476137	4
4.45	.477088	.478041	.479000	.479962	.480929	.481898	.482871	.483848	.484826	.485808	2
4.46	.486792	.487778	.488767	.489757	.490750	.491744	.492741	.493738	.494735	.495735	−1
4.47	.496734	.497733	.498733	.499732	.500733	.501732	.502732	.503730	.504727	.505723	−1
4.48	.506718	.507711	.508702	.509691	.510679	.511662	.512645	.513626	.514602	.515576	−5
4.49	.516546	.517513	.518476	.519435	.520390	.521340	.522287	.523229	.524166	.525098	−5

TABLE 9B

Fresnel Cosine Integral

U	.000	.001	.002	.003	.004	.005	.006	.007	.008	.009	Δ_2
4.50	.526025	.526945	.527861	.528771	.529674	.530572	.531463	.532349	.533227	.534099	-7
4.51	.534962	.535819	.536669	.537511	.538345	.539171	.539989	.540800	.541601	.542395	-9
4.52	.543180	.543955	.544722	.545479	.546227	.546965	.547695	.548414	.549123	.549821	-9
4.53	.550510	.551189	.551857	.552513	.553160	.553795	.554419	.555033	.555634	.556225	-10
4.54	.556804	.557371	.557926	.558469	.559001	.559520	.560027	.560522	.561003	.561473	-12
4.55	.561930	.562374	.562805	.563223	.563629	.564021	.564400	.564766	.565118	.565457	-12
4.56	.565782	.566094	.566392	.566676	.566947	.567204	.567447	.567676	.567891	.568092	-13
4.57	.568279	.568452	.568610	.568755	.568885	.569001	.569102	.569190	.569263	.569321	-13
4.58	.569366	.569395	.569411	.569412	.569399	.569371	.569329	.569273	.569202	.569116	-13
4.59	.569017	.568903	.568775	.568632	.568475	.568305	.568119	.567920	.567706	.567479	-13
4.60	.567237	.566981	.566712	.566428	.566131	.565820	.565495	.565156	.564805	.564439	-12
4.61	.564060	.563668	.563262	.562843	.562411	.561966	.561508	.561037	.560554	.560058	-11
4.62	.559548	.559028	.558494	.557949	.557390	.556821	.556239	.555645	.555041	.554424	-10
4.63	.553796	.553157	.552507	.551846	.551174	.550491	.549797	.549093	.548379	.547654	-7
4.64	.546920	.546176	.545422	.544660	.543886	.543105	.542314	.541514	.540706	.539889	-7
4.65	.539064	.538231	.537389	.536541	.535683	.534820	.533948	.533068	.532184	.531292	-5
4.66	.530393	.529488	.528577	.527660	.526737	.525809	.524875	.523936	.522992	.522044	-4
4.67	.521091	.520134	.519172	.518207	.517238	.516266	.515289	.514309	.513328	.512343	0
4.68	.511355	.510366	.509375	.508383	.507387	.506392	.505394	.504395	.503396	.502396	1
4.69	.501396	.500396	.499397	.498397	.497398	.496401	.495402	.494406	.493412	.492419	2
4.70	.491428	.490439	.489452	.488468	.487487	.486509	.485534	.484561	.483593	.482629	4
4.71	.481669	.480712	.479760	.478813	.477870	.476934	.476001	.475074	.474154	.473238	7
4.72	.472329	.471426	.470530	.469641	.468758	.467883	.467014	.466152	.465300	.464454	9
4.73	.463616	.462788	.461966	.461155	.460351	.459557	.458771	.457995	.457228	.456471	11
4.74	.455724	.454987	.454260	.453543	.452837	.452142	.451457	.450783	.450120	.449468	12
4.75	.448828	.448200	.447582	.446978	.446384	.445803	.445235	.444678	.444134	.443603	13
4.76	.443084	.442578	.442085	.441606	.441139	.440685	.440246	.439819	.439406	.439007	14
4.77	.438621	.438250	.437892	.437549	.437219	.436904	.436603	.436316	.436044	.435787	14
4.78	.435543	.435315	.435101	.434902	.434717	.434547	.434392	.434252	.434127	.434017	15
4.79	.433922	.433842	.433776	.433726	.433691	.433671	.433666	.433676	.433701	.433741	15
4.80	.433797	.433867	.433952	.434052	.434168	.434298	.434443	.434604	.434779	.434969	14
4.81	.435173	.435393	.435627	.435876	.436140	.436418	.436710	.437017	.437338	.437674	14
4.82	.438024	.438388	.438766	.439158	.439564	.439984	.440417	.440865	.441325	.441800	12
4.83	.442287	.442787	.443301	.443827	.444367	.444919	.445483	.446061	.446650	.447252	12
4.84	.447866	.448491	.449129	.449778	.450438	.451110	.451793	.452489	.453193	.453909	10
4.85	.454636	.455372	.456120	.456877	.457643	.458420	.459206	.460002	.460806	.461620	8
4.86	.462443	.463273	.464113	.464960	.465816	.466679	.467551	.468429	.469315	.470207	7
4.87	.471107	.472013	.472925	.473844	.474768	.475698	.476634	.477575	.478521	.479472	3
4.88	.480427	.481387	.482351	.483319	.484291	.485266	.486244	.487227	.488210	.489198	2
4.89	.490188	.491179	.492172	.493167	.494164	.495161	.496159	.497159	.498159	.499159	0
4.90	.500160	.501158	.502158	.503157	.504155	.505150	.506145	.507139	.508130	.509120	-2
4.91	.510107	.511091	.512072	.513050	.514026	.514997	.515965	.516929	.517888	.518843	-5
4.92	.519793	.520738	.521679	.522613	.523543	.524466	.525384	.526295	.527199	.528097	-6
4.93	.528989	.529872	.530749	.531618	.532479	.533332	.534177	.535014	.535842	.536661	-9
4.94	.537473	.538273	.539065	.539846	.540619	.541381	.542133	.542876	.543607	.544327	-10
4.95	.545038	.545736	.546424	.547100	.547765	.548417	.549059	.549688	.550305	.550909	-11
4.96	.551501	.552081	.552647	.553201	.553742	.554269	.554783	.555284	.555771	.556245	-14
4.97	.556705	.557150	.557582	.558000	.558403	.558792	.559167	.559527	.559872	.560203	-14
4.98	.560519	.560820	.561106	.561377	.561633	.561873	.562099	.562309	.562504	.562683	-14
4.99	.562848	.562996	.563129	.563247	.563349	.563435	.563505	.563560	.563600	.563623	-14

TABLE 9B

Fresnel Cosine Integral

U	.000	.001	.002	.003	.004	.005	.006	.007	.008	.009	Δ_2
5.00	.563631	.563623	.563600	.563560	.563506	.563435	.563349	.563247	.563129	.562996	-14
5.01	.562847	.562683	.562503	.562308	.562097	.561871	.561629	.561373	.561101	.560814	-14
5.02	.560512	.560195	.559863	.559516	.559154	.558778	.558387	.557982	.557562	.557128	-13
5.03	.556680	.556218	.555741	.555252	.554748	.554231	.553700	.553155	.552598	.552027	-11
5.04	.551444	.550848	.550239	.549618	.548984	.548339	.547681	.547011	.546330	.545636	-9
5.05	.544933	.544218	.543491	.542754	.542006	.541248	.540479	.539700	.538912	.538113	-8
5.06	.537306	.536489	.535663	.534829	.533985	.533133	.532273	.531405	.530529	.529645	-6
5.07	.528754	.527856	.526951	.526040	.525122	.524198	.523267	.522331	.521390	.520443	-3
5.08	.519491	.518535	.517574	.516609	.515640	.514667	.513690	.512709	.511727	.510740	0
5.09	.509753	.508763	.507769	.506775	.505779	.504782	.503783	.502783	.501784	.500784	2
5.10	.499784	.498784	.497785	.496787	.495790	.494794	.493799	.492806	.491816	.490827	3
5.11	.489840	.488858	.487877	.486901	.485927	.484959	.483993	.483032	.482075	.481124	6
5.12	.480177	.479236	.478300	.477371	.476447	.475530	.474619	.473715	.472818	.471927	9
5.13	.471046	.470171	.469304	.468446	.467595	.466754	.465920	.465096	.464282	.463477	10
5.14	.462681	.461896	.461120	.460355	.459600	.458857	.458124	.457401	.456691	.455991	12
5.15	.455303	.454628	.453964	.453313	.452673	.452047	.451433	.450832	.450243	.449669	13
5.16	.449107	.448559	.448024	.447504	.446998	.446505	.446026	.445562	.445112	.444677	15
5.17	.444257	.443851	.443460	.443085	.442724	.442379	.442048	.441733	.441434	.441150	16
5.18	.440882	.440630	.440393	.440173	.439968	.439779	.439606	.439449	.439308	.439184	16
5.19	.439075	.438983	.438907	.438848	.438805	.438778	.438767	.438772	.438794	.438833	16
5.20	.438887	.438958	.439046	.439149	.439269	.439404	.439557	.439725	.439909	.440110	16
5.21	.440326	.440558	.440807	.441070	.441350	.441646	.441957	.442284	.442626	.442983	15
5.22	.443356	.443744	.444147	.444565	.444998	.445445	.445907	.446384	.446875	.447380	13
5.23	.447900	.448433	.448980	.449541	.450115	.450703	.451304	.451918	.452544	.453184	12
5.24	.453837	.454501	.455179	.455867	.456568	.457280	.458004	.458740	.459486	.460242	11
5.25	.461011	.461788	.462577	.463375	.464184	.465001	.465829	.466666	.467511	.468365	8
5.26	.469227	.470098	.470977	.471863	.472758	.473659	.474567	.475483	.476404	.477332	6
5.27	.478266	.479206	.480152	.481102	.482058	.483018	.483984	.484953	.485926	.486903	2
5.28	.487883	.488866	.489853	.490841	.491833	.492825	.493821	.494818	.495815	.496814	0
5.29	.497814	.498813	.499813	.500812	.501813	.502811	.503809	.504806	.505801	.506794	-1
5.30	.507785	.508773	.509759	.510742	.511723	.512699	.513672	.514642	.515606	.516567	-6
5.31	.517522	.518472	.519418	.520356	.521291	.522217	.523139	.524054	.524961	.525862	-8
5.32	.526756	.527640	.528518	.529387	.530248	.531101	.531945	.532779	.533604	.534419	-9
5.33	.535225	.536021	.536806	.537581	.538346	.539099	.539841	.540572	.541291	.541999	-11
5.34	.542695	.543377	.544049	.544707	.545353	.545986	.546606	.547213	.547806	.548385	-13
5.35	.548951	.549503	.550041	.550564	.551074	.551568	.552048	.552513	.552963	.553398	-15
5.36	.553818	.554222	.554611	.554985	.555342	.555684	.556010	.556321	.556614	.556892	-16
5.37	.557154	.557399	.557628	.557841	.558037	.558216	.558378	.558524	.558654	.558766	-16
5.38	.558862	.558941	.559003	.559048	.559076	.559087	.559081	.559059	.559019	.558963	-16
5.39	.558890	.558800	.558693	.558569	.558428	.558271	.558097	.557906	.557699	.557475	-15
5.40	.557234	.556977	.556704	.556415	.556109	.555787	.555449	.555095	.554726	.554340	-14
5.41	.553939	.553523	.553091	.552644	.552181	.551704	.551212	.550705	.550184	.549647	-13
5.42	.549098	.548533	.547955	.547363	.546757	.546137	.545505	.544859	.544201	.543529	-10
5.43	.542845	.542149	.541440	.540720	.539988	.539245	.538490	.537723	.536947	.536159	-9
5.44	.535361	.534553	.533735	.532907	.532070	.531224	.530369	.529504	.528632	.527750	-5
5.45	.526861	.525964	.525060	.524150	.523231	.522307	.521375	.520438	.519495	.518546	-3
5.46	.517593	.516634	.515670	.514702	.513730	.512754	.511774	.510791	.509805	.508816	0
5.47	.507825	.506832	.505836	.504840	.503841	.502843	.501843	.500843	.499843	.498844	1
5.48	.497844	.496846	.495848	.494853	.493859	.492867	.491876	.490889	.489905	.488924	4
5.49	.487945	.486972	.486002	.485037	.484075	.483120	.482169	.481223	.480284	.479350	7

TABLE 9B

Fresnel Cosine Integral

U	.000	.001	.002	.003	.004	.005	.006	.007	.008	.009	Δ_2
5.50	.478422	.477502	.476588	.475683	.474783	.473892	.473009	.472133	.471267	.470408	11
5.51	.469560	.468721	.467891	.467071	.466260	.465461	.464671	.463892	.463124	.462368	12
5.52	.461623	.460890	.460168	.459459	.458762	.458078	.457406	.456746	.456101	.455469	14
5.53	.454850	.454245	.453654	.453077	.452515	.451967	.451433	.450915	.450411	.449922	16
5.54	.449448	.448991	.448548	.448121	.447710	.447316	.446936	.446573	.446227	.445897	17
5.55	.445584	.445286	.445006	.444743	.444496	.444266	.444054	.443858	.443680	.443518	17
5.56	.443374	.443248	.443138	.443046	.442972	.442915	.442875	.442853	.442848	.442861	18
5.57	.442892	.442939	.443005	.443087	.443188	.443305	.443440	.443592	.443762	.443949	17
5.58	.444153	.444374	.444612	.444867	.445140	.445428	.445734	.446057	.446396	.446751	16
5.59	.447123	.447510	.447915	.448335	.448771	.449222	.449690	.450173	.450670	.451184	14
5.60	.451713	.452256	.452814	.453386	.453973	.454573	.455189	.455818	.456459	.457116	12
5.61	.457785	.458466	.459161	.459868	.460587	.461319	.462063	.462819	.463585	.464363	11
5.62	.465153	.465952	.466762	.467582	.468413	.469253	.470103	.470961	.471829	.472705	7
5.63	.473589	.474482	.475382	.476291	.477206	.478128	.479057	.479993	.480934	.481883	4
5.64	.482835	.483793	.484757	.485724	.486696	.487672	.488652	.489635	.490621	.491610	2
5.65	.492602	.493595	.494591	.495588	.496586	.497585	.498584	.499586	.500585	.501584	0
5.66	.502584	.503581	.504578	.505573	.506566	.507556	.508545	.509530	.510512	.511491	−5
5.67	.512465	.513436	.514402	.515363	.516320	.517270	.518216	.519156	.520088	.521016	−8
5.68	.521936	.522848	.523754	.524650	.525540	.526421	.527294	.528158	.529013	.529858	−9
5.69	.530694	.531519	.532334	.533138	.533933	.534716	.535488	.536249	.536997	.537734	−13
5.70	.538458	.539170	.539869	.540555	.541228	.541888	.542534	.543167	.543785	.544389	−14
5.71	.544979	.545554	.546115	.546660	.547190	.547705	.548205	.548689	.549157	.549610	−16
5.72	.550046	.550466	.550870	.551256	.551627	.551980	.552318	.552637	.552940	.553225	−16
5.73	.553494	.553744	.553978	.554193	.554392	.554572	.554735	.554880	.555007	.555116	−16
5.74	.555207	.555281	.555336	.555373	.555392	.555394	.555377	.555342	.555289	.555218	−16
5.75	.555129	.555022	.554897	.554754	.554593	.554415	.554219	.554004	.553773	.553524	−16
5.76	.553257	.552974	.552672	.552354	.552018	.551665	.551296	.550909	.550507	.550087	−15
5.77	.549651	.549199	.548731	.548247	.547747	.547231	.546700	.546153	.545593	.545016	−13
5.78	.544425	.543820	.543199	.542566	.541917	.541256	.540580	.539891	.539190	.538475	−11
5.79	.537748	.537009	.536257	.535494	.534718	.533932	.533134	.532326	.531507	.530677	−8
5.80	.529838	.528989	.528131	.527263	.526386	.525501	.524607	.523705	.522796	.521879	−6
5.81	.520955	.520024	.519086	.518143	.517194	.516239	.515278	.514313	.513343	.512369	−3
5.82	.511390	.510410	.509424	.508437	.507445	.506453	.505458	.504460	.503463	.502464	0
5.83	.501463	.500464	.499464	.498465	.497466	.496468	.495471	.494476	.493484	.492493	4
5.84	.491506	.490521	.489540	.488562	.487589	.486620	.485655	.484695	.483741	.482792	7
5.85	.481849	.480913	.479983	.479061	.478145	.477237	.476337	.475444	.474561	.473686	9
5.86	.472820	.471964	.471117	.470281	.469454	.468638	.467833	.467038	.466256	.465484	13
5.87	.464725	.463978	.463243	.462521	.461811	.461115	.460432	.459763	.459108	.458466	15
5.88	.457840	.457228	.456629	.456047	.455479	.454926	.454389	.453867	.453362	.452873	17
5.89	.452399	.451943	.451502	.451079	.450672	.450282	.449910	.449554	.449216	.448896	18
5.90	.448592	.448307	.448040	.447791	.447559	.447346	.447150	.446973	.446815	.446674	19
5.91	.446552	.446449	.446364	.446297	.446249	.446220	.446209	.446217	.446243	.446288	18
5.92	.446352	.446434	.446534	.446653	.446791	.446947	.447121	.447314	.447525	.447754	18
5.93	.448001	.448266	.448549	.448849	.449168	.449504	.449858	.450229	.450617	.451023	17
5.94	.451446	.451884	.452341	.452813	.453302	.453806	.454328	.454865	.455417	.455985	15
5.95	.456569	.457167	.457781	.458408	.459051	.459708	.460378	.461063	.461760	.462472	13
5.96	.463196	.463932	.464682	.465444	.466218	.467003	.467800	.468608	.469427	.470258	9
5.97	.471098	.471947	.472808	.473676	.474555	.475442	.476337	.477241	.478153	.479073	6
5.98	.479999	.480932	.481872	.482818	.483770	.484728	.485691	.486659	.487631	.488608	4
5.99	.489589	.490572	.491561	.492550	.493542	.494537	.495534	.496532	.497530	.498530	0

TABLE 9B

Fresnel Cosine Integral

U	.000	.001	.002	.003	.004	.005	.006	.007	.008	.009	Δ_2
6.00	.499530	.500530	.501530	.502528	.503526	.504522	.505517	.506510	.507499	.508487	-4
6.01	.509471	.510450	.511427	.512399	.513367	.514329	.515287	.516239	.517185	.518125	-6
6.02	.519058	.519984	.520903	.521814	.522717	.523611	.524498	.525375	.526244	.527102	-9
6.03	.527951	.528789	.529617	.530434	.531240	.532035	.532818	.533590	.534348	.535095	-12
6.04	.535829	.536550	.537257	.537951	.538631	.539297	.539949	.540586	.541208	.541816	-15
6.05	.542408	.542985	.543546	.544091	.544621	.545134	.545631	.546111	.546574	.547021	-16
6.06	.547450	.547862	.548257	.548634	.548993	.549334	.549658	.549964	.550251	.550520	-18
6.07	.550770	.551002	.551216	.551410	.551586	.551743	.551882	.552001	.552101	.552183	-18
6.08	.552245	.552288	.552313	.552318	.552304	.552270	.552218	.552147	.552056	.551947	-18
6.09	.551819	.551671	.551505	.551320	.551116	.550893	.550652	.550392	.550114	.549818	-17
6.10	.549503	.549170	.548819	.548450	.548064	.547660	.547238	.546798	.546343	.545869	-14
6.11	.545379	.544873	.544350	.543811	.543255	.542684	.542097	.541494	.540877	.540244	-13
6.12	.539597	.538935	.538259	.537569	.536864	.536147	.535416	.534673	.533916	.533147	-9
6.13	.532366	.531574	.530769	.529953	.529126	.528289	.527440	.526582	.525715	.524837	-8
6.14	.523951	.523056	.522152	.521241	.520322	.519396	.518461	.517521	.516574	.515620	-4
6.15	.514661	.513698	.512729	.511756	.510777	.509796	.508811	.507822	.506832	.505838	-1
6.16	.504842	.503845	.502847	.501848	.500848	.499848	.498848	.497848	.496851	.495853	3
6.17	.494859	.493866	.492876	.491888	.490903	.489923	.488946	.487972	.487005	.486042	6
6.18	.485084	.484132	.483187	.482248	.481315	.480391	.479473	.478563	.477662	.476769	10
6.19	.475884	.475011	.474145	.473291	.472445	.471612	.470788	.469976	.469175	.468387	12
6.20	.467610	.466846	.466093	.465356	.464630	.463919	.463220	.462536	.461867	.461212	15
6.21	.460572	.459948	.459338	.458744	.458166	.457604	.457058	.456529	.456016	.455520	18
6.22	.455041	.454580	.454136	.453709	.453300	.452910	.452537	.452182	.451846	.451528	19
6.23	.451229	.450949	.450688	.450445	.450221	.450017	.449832	.449666	.449520	.449392	20
6.24	.449285	.449197	.449128	.449079	.449050	.449040	.449050	.449079	.449128	.449197	20
6.25	.449285	.449393	.449520	.449667	.449833	.450018	.450223	.450447	.450690	.450952	19
6.26	.451233	.451532	.451851	.452188	.452544	.452917	.453310	.453720	.454147	.454593	17
6.27	.455056	.455537	.456035	.456549	.457081	.457629	.458194	.458775	.459371	.459983	15
6.28	.460611	.461254	.461913	.462585	.463273	.463974	.464689	.465418	.466160	.466916	13
6.29	.467685	.468465	.469258	.470063	.470879	.471707	.472546	.473395	.474255	.475124	9
6.30	.476003	.476891	.477788	.478694	.479608	.480529	.481459	.482396	.483339	.484288	6
6.31	.485245	.486205	.487172	.488143	.489120	.490099	.491084	.492071	.493061	.494054	2
6.32	.495049	.496045	.497044	.498043	.499043	.500042	.501042	.502042	.503040	.504037	-2
6.33	.505033	.506025	.507016	.508003	.508988	.509967	.510944	.511917	.512884	.513845	-5
6.34	.514801	.515750	.516694	.517631	.518561	.519483	.520398	.521304	.522201	.523090	-9
6.35	.523969	.524839	.525698	.526547	.527386	.528213	.529029	.529834	.530626	.531406	-12
6.36	.532174	.532928	.533669	.534396	.535110	.535809	.536494	.537165	.537820	.538460	-15
6.37	.539085	.539694	.540287	.540863	.541423	.541966	.542493	.543003	.543495	.543969	-17
6.38	.544426	.544864	.545285	.545688	.546072	.546437	.546784	.547112	.547421	.547710	-18
6.39	.547981	.548232	.548463	.548675	.548868	.549040	.549193	.549326	.549439	.549532	-19
6.40	.549604	.549657	.549690	.549703	.549695	.549667	.549620	.549552	.549464	.549356	-18
6.41	.549228	.549080	.548912	.548725	.548517	.548290	.548044	.547778	.547492	.547188	-18
6.42	.546864	.546521	.546159	.545779	.545380	.544963	.544527	.544073	.543602	.543112	-15
6.43	.542605	.542081	.541539	.540981	.540406	.539815	.539208	.538584	.537945	.537290	-14
6.44	.536620	.535936	.535237	.534524	.533796	.533055	.532300	.531532	.530752	.529958	-10
6.45	.529152	.528336	.527507	.526667	.525816	.524956	.524084	.523203	.522312	.521412	-7
6.46	.520505	.519589	.518664	.517733	.516793	.515847	.514895	.513936	.512973	.512004	-3
6.47	.511029	.510051	.509069	.508083	.507094	.506102	.505107	.504110	.503114	.502114	1
6.48	.501114	.500115	.499114	.498116	.497117	.496121	.495125	.494131	.493141	.492153	5
6.49	.491168	.490189	.489213	.488242	.487275	.486314	.485359	.484410	.483468	.482533	8

TABLE 9B

Fresnel Cosine Integral

U	.000	.001	.002	.003	.004	.005	.006	.007	.008	.009	Δ_2
6.50	.481605	.480685	.479773	.478870	.477976	.477091	.476216	.475350	.474496	.473652	12
6.51	.472820	.471999	.471189	.470393	.469608	.468836	.468077	.467332	.466601	.465884	15
6.52	.465180	.464493	.463819	.463162	.462519	.461894	.461283	.460689	.460112	.459551	18
6.53	.459008	.458482	.457974	.457483	.457010	.456556	.456120	.455702	.455304	.454924	19
6.54	.454563	.454221	.453899	.453597	.453314	.453051	.452807	.452584	.452381	.452198	20
6.55	.452035	.451893	.451771	.451669	.451588	.451527	.451488	.451468	.451469	.451491	21
6.56	.451534	.451597	.451680	.451784	.451909	.452054	.452219	.452405	.452611	.452837	20
6.57	.453083	.453349	.453635	.453940	.454266	.454610	.454974	.455358	.455760	.456181	18
6.58	.456621	.457078	.457555	.458049	.458562	.459091	.459639	.460204	.460785	.461384	16
6.59	.461998	.462629	.463276	.463938	.464616	.465309	.466017	.466739	.467475	.468226	13
6.60	.468990	.469767	.470557	.471359	.472174	.473000	.473839	.474688	.475547	.476418	9
6.61	.477299	.478188	.479088	.479995	.480913	.481837	.482770	.483710	.484657	.485610	5
6.62	.486570	.487534	.488505	.489479	.490459	.491442	.492430	.493420	.494412	.495407	1
6.63	.496404	.497402	.498402	.499401	.500401	.501400	.502400	.503398	.504393	.505388	-3
6.64	.506380	.507368	.508353	.509334	.510312	.511283	.512252	.513214	.514170	.515120	-7
6.65	.516063	.516998	.517927	.518847	.519760	.520662	.521557	.522442	.523316	.524181	-11
6.66	.525034	.525876	.526708	.527526	.528333	.529128	.529909	.530678	.531432	.532173	-14
6.67	.532899	.533611	.534308	.534990	.535657	.536307	.536942	.537561	.538162	.538747	-17
6.68	.539315	.539865	.540398	.540912	.541409	.541887	.542347	.542788	.543210	.543613	-19
6.69	.543997	.544361	.544705	.545030	.545334	.545619	.545883	.546127	.546351	.546554	-20
6.70	.546736	.546898	.547039	.547159	.547257	.547335	.547392	.547428	.547443	.547437	-20
6.71	.547410	.547361	.547292	.547202	.547090	.546958	.546805	.546631	.546436	.546221	-19
6.72	.545985	.545728	.545451	.545155	.544837	.544500	.544143	.543767	.543371	.542955	-18
6.73	.542521	.542068	.541595	.541105	.540595	.540069	.539523	.538961	.538381	.537784	-14
6.74	.537170	.536539	.535892	.535230	.534551	.533858	.533148	.532424	.531686	.530934	-12
6.75	.530167	.529388	.528595	.527790	.526972	.526142	.525300	.524447	.523583	.522709	-9
6.76	.521825	.520932	.520028	.519116	.518195	.517267	.516330	.515386	.514436	.513478	-4
6.77	.512515	.511547	.510573	.509595	.508613	.507627	.506637	.505644	.504651	.503654	0
6.78	.502655	.501656	.500656	.499657	.498657	.497659	.496660	.495664	.494670	.493678	4
6.79	.492690	.491705	.490724	.489748	.488776	.487810	.486849	.485894	.484946	.484005	8
6.80	.483072	.482146	.481229	.480320	.479420	.478530	.477650	.476779	.475921	.475073	13
6.81	.474236	.473412	.472599	.471800	.471013	.470241	.469481	.468736	.468005	.467289	15
6.82	.466588	.465903	.465233	.464579	.463942	.463322	.462718	.462131	.461562	.461011	18
6.83	.460478	.459963	.459467	.458989	.458531	.458091	.457671	.457271	.456890	.456529	21
6.84	.456189	.455869	.455569	.455290	.455031	.454793	.454576	.454380	.454206	.454052	22
6.85	.453920	.453809	.453720	.453652	.453606	.453581	.453577	.453596	.453635	.453696	21
6.86	.453779	.453883	.454009	.454156	.454324	.454513	.454724	.454956	.455208	.455482	20
6.87	.455776	.456090	.456425	.456781	.457156	.457551	.457966	.458401	.458855	.459328	18
6.88	.459820	.460330	.460860	.461407	.461972	.462555	.463156	.463774	.464408	.465059	16
6.89	.465727	.466410	.467109	.467823	.468553	.469296	.470055	.470827	.471613	.472412	12
6.90	.473224	.474048	.474884	.475732	.476592	.477462	.478343	.479234	.480134	.481044	7
6.91	.481962	.482889	.483824	.484765	.485715	.486671	.487634	.488601	.489574	.490552	3
6.92	.491534	.492519	.493509	.494501	.495496	.496492	.497491	.498490	.499489	.500490	-1
6.93	.501490	.502488	.503486	.504481	.505475	.506465	.507452	.508437	.509416	.510391	-5
6.94	.511361	.512325	.513284	.514235	.515181	.516118	.517048	.517970	.518882	.519786	-9
6.95	.520680	.521564	.522438	.523301	.524153	.524992	.525821	.526637	.527439	.528229	-13
6.96	.529004	.529766	.530513	.531246	.531964	.532666	.533353	.534024	.534678	.535315	-16
6.97	.535936	.536539	.537124	.537692	.538242	.538772	.539285	.539778	.540252	.540707	-19
6.98	.541142	.541558	.541953	.542328	.542683	.543016	.543330	.543622	.543893	.544143	-21
6.99	.544372	.544579	.544765	.544929	.545072	.545192	.545291	.545368	.545423	.545456	-21

C(u)

TABLE 9B

Fresnel Cosine Integral

U	.000	.001	.002	.003	.004	.005	.006	.007	.008	.009	Δ_2
7.00	.545467	.545456	.545423	.545368	.545291	.545193	.545072	.544929	.544765	.544579	-20
7.01	.544372	.544143	.543893	.543621	.543328	.543015	.542680	.542324	.541949	.541552	-18
7.02	.541136	.540700	.540244	.539768	.539274	.538760	.538227	.537676	.537107	.536519	-16
7.03	.535913	.535291	.534651	.533995	.533322	.532632	.531927	.531206	.530471	.529720	-13
7.04	.528955	.528176	.527383	.526577	.525758	.524927	.524083	.523228	.522362	.521484	-9
7.05	.520597	.519700	.518792	.517876	.516951	.516018	.515077	.514129	.513174	.512212	-4
7.06	.511244	.510273	.509295	.508314	.507327	.506338	.505346	.504351	.503354	.502355	0
7.07	.501355	.500356	.499356	.498357	.497358	.496362	.495366	.494373	.493383	.492397	5
7.08	.491414	.490437	.489463	.488496	.487533	.486578	.485628	.484687	.483753	.482827	9
7.09	.481910	.481003	.480104	.479216	.478338	.477471	.476615	.475770	.474939	.474119	15
7.10	.473312	.472520	.471741	.470976	.470225	.469490	.468770	.468065	.467377	.466704	17
7.11	.466049	.465410	.464789	.464186	.463601	.463034	.462485	.461955	.461445	.460953	20
7.12	.460481	.460029	.459597	.459186	.458795	.458425	.458075	.457747	.457440	.457154	22
7.13	.456890	.456647	.456426	.456228	.456051	.455896	.455763	.455653	.455565	.455499	22
7.14	.455456	.455435	.455437	.455460	.455507	.455576	.455667	.455781	.455916	.456074	22
7.15	.456255	.456457	.456681	.456927	.457195	.457484	.457795	.458127	.458480	.458854	21
7.16	.459249	.459664	.460100	.460556	.461032	.461527	.462042	.462576	.463129	.463701	17
7.17	.464291	.464898	.465524	.466167	.466827	.467504	.468197	.468907	.469632	.470372	15
7.18	.471128	.471898	.472683	.473480	.474292	.475116	.475953	.476803	.477663	.478536	10
7.19	.479419	.480312	.481215	.482128	.483050	.483979	.484918	.485864	.486816	.487777	5
7.20	.488743	.489713	.490690	.491671	.492657	.493645	.494637	.495632	.496628	.497627	0
7.21	.498625	.499625	.500625	.501624	.502623	.503619	.504615	.505608	.506597	.507583	-4
7.22	.508565	.509542	.510515	.511481	.512442	.513396	.514344	.515284	.516215	.517138	-9
7.23	.518053	.518957	.519852	.520736	.521610	.522472	.523323	.524162	.524987	.525800	-13
7.24	.526600	.527385	.528156	.528913	.529655	.530380	.531091	.531785	.532462	.533122	-16
7.25	.533766	.534391	.534999	.535588	.536159	.536710	.537244	.537757	.538251	.538725	-20
7.26	.539178	.539611	.540024	.540415	.540785	.541134	.541462	.541768	.542052	.542315	-21
7.27	.542555	.542773	.542968	.543141	.543292	.543420	.543525	.543608	.543667	.543704	-22
7.28	.543718	.543709	.543678	.543623	.543546	.543446	.543322	.543177	.543009	.542818	-22
7.29	.542604	.542369	.542111	.541832	.541530	.541206	.540861	.540495	.540107	.539698	-20
7.30	.539269	.538819	.538348	.537858	.537347	.536818	.536268	.535699	.535112	.534506	-17
7.31	.533882	.533241	.532582	.531907	.531213	.530505	.529779	.529038	.528282	.527511	-13
7.32	.526725	.525927	.525113	.524287	.523448	.522597	.521734	.520860	.519975	.519079	-9
7.33	.518172	.517257	.516332	.515400	.514459	.513511	.512555	.511593	.510625	.509650	-3
7.34	.508673	.507690	.506702	.505712	.504719	.503724	.502726	.501727	.500727	.499727	1
7.35	.498726	.497728	.496731	.495735	.494742	.493752	.492765	.491781	.490803	.489830	6
7.36	.488862	.487900	.486945	.485998	.485058	.484126	.483202	.482287	.481383	.480488	12
7.37	.479603	.478730	.477869	.477020	.476182	.475359	.474547	.473750	.472967	.472199	16
7.38	.471446	.470709	.469987	.469281	.468592	.467920	.467265	.466628	.466009	.465409	19
7.39	.464827	.464264	.463721	.463197	.462693	.462209	.461746	.461303	.460882	.460481	22
7.40	.460102	.459745	.459410	.459096	.458805	.458536	.458289	.458065	.457864	.457686	23
7.41	.457530	.457398	.457289	.457203	.457140	.457100	.457084	.457091	.457121	.457175	23
7.42	.457251	.457351	.457475	.457621	.457790	.457982	.458198	.458435	.458696	.458979	22
7.43	.459284	.459611	.459961	.460331	.460724	.461138	.461573	.462030	.462506	.463004	20
7.44	.463521	.464058	.464615	.465191	.465786	.466399	.467031	.467681	.468348	.469033	15
7.45	.469735	.470453	.471187	.471936	.472702	.473482	.474276	.475085	.475907	.476742	12
7.46	.477589	.478449	.479321	.480203	.481097	.482001	.482914	.483838	.484769	.485710	7
7.47	.486658	.487612	.488575	.489541	.490516	.491494	.492477	.493465	.494455	.495449	1
7.48	.496445	.497442	.498442	.499441	.500441	.501440	.502439	.503437	.504432	.505425	-4
7.49	.506415	.507400	.508382	.509358	.510330	.511295	.512254	.513207	.514151	.515088	-9

TABLE 9B

Fresnel Cosine Integral

U	.000	.001	.002	.003	.004	.005	.006	.007	.008	.009	Δ_2
7.50	.516017	.516936	.517846	.518745	.519635	.520513	.521380	.522235	.523076	.523906	-12
7.51	.524721	.525523	.526311	.527083	.527841	.528583	.529309	.530019	.530711	.531387	-17
7.52	.532045	.532685	.533307	.533909	.534493	.535058	.535603	.536128	.536632	.537117	-20
7.53	.537580	.538022	.538443	.538842	.539220	.539575	.539908	.540219	.540507	.540773	-23
7.54	.541016	.541235	.541431	.541604	.541754	.541880	.541983	.542062	.542117	.542149	-23
7.55	.542157	.542141	.542102	.542039	.541952	.541841	.541707	.541550	.541369	.541165	-22
7.56	.540937	.540687	.540414	.540118	.539799	.539458	.539095	.538709	.538302	.537873	-20
7.57	.537423	.536952	.536459	.535947	.535414	.534861	.534288	.533696	.533085	.532455	-16
7.58	.531807	.531141	.530458	.529757	.529039	.528306	.527556	.526790	.526010	.525215	-13
7.59	.524406	.523583	.522746	.521897	.521035	.520162	.519277	.518380	.517475	.516559	-8
7.60	.515633	.514699	.513757	.512807	.511849	.510886	.509916	.508940	.507960	.506974	-1
7.61	.505985	.504994	.503999	.503003	.502004	.501005	.500004	.499004	.498006	.497007	3
7.62	.496011	.495018	.494027	.493040	.492057	.491079	.490106	.489139	.488178	.487224	8
7.63	.486278	.485340	.484410	.483489	.482578	.481677	.480786	.479907	.479040	.478184	14
7.64	.477341	.476512	.475696	.474894	.474107	.473335	.472578	.471837	.471113	.470405	18
7.65	.469714	.469042	.468387	.467751	.467133	.466534	.465955	.465395	.464856	.464337	21
7.66	.463838	.463361	.462905	.462471	.462058	.461667	.461299	.460953	.460630	.460330	23
7.67	.460053	.459799	.459568	.459361	.459178	.459018	.458882	.458771	.458683	.458619	24
7.68	.458579	.458563	.458572	.458605	.458661	.458742	.458847	.458976	.459129	.459305	24
7.69	.459505	.459729	.459977	.460247	.460541	.460858	.461198	.461561	.461945	.462352	22
7.70	.462781	.463232	.463704	.464197	.464712	.465246	.465802	.466377	.466971	.467586	18
7.71	.468219	.468870	.469540	.470227	.470932	.471654	.472392	.473147	.473917	.474703	14
7.72	.475503	.476318	.477147	.477988	.478842	.479709	.480588	.481478	.482379	.483290	9
7.73	.484212	.485141	.486081	.487027	.487982	.488942	.489911	.490885	.491863	.492847	3
7.74	.493835	.494825	.495820	.496816	.497814	.498813	.499813	.500814	.501812	.502811	-2
7.75	.503808	.504801	.505792	.506779	.507763	.508741	.509714	.510682	.511642	.512596	-8
7.76	.513542	.514480	.515409	.516329	.517239	.518139	.519028	.519906	.520770	.521624	-13
7.77	.522464	.523290	.524103	.524900	.525684	.526451	.527203	.527939	.528657	.529359	-17
7.78	.530043	.530708	.531356	.531985	.532594	.533184	.533754	.534303	.534832	.535340	-21
7.79	.535827	.536292	.536736	.537158	.537557	.537933	.538287	.538618	.538925	.539210	-23
7.80	.539470	.539707	.539920	.540109	.540274	.540414	.540530	.540622	.540690	.540733	-23
7.81	.540751	.540745	.540715	.540659	.540580	.540476	.540347	.540195	.540017	.539816	-23
7.82	.539591	.539342	.539069	.538773	.538454	.538111	.537745	.537356	.536946	.536512	-20
7.83	.536056	.535579	.535080	.534561	.534020	.533459	.532877	.532276	.531655	.531015	-18
7.84	.530357	.529680	.528985	.528273	.527544	.526798	.526036	.525257	.524464	.523656	-12
7.85	.522833	.521998	.521148	.520287	.519412	.518526	.517629	.516721	.515803	.514875	-7
7.86	.513939	.512994	.512042	.511082	.510115	.509143	.508165	.507182	.506196	.505204	-1
7.87	.504210	.503214	.502216	.501217	.500217	.499218	.498218	.497220	.496224	.495230	4
7.88	.494240	.493253	.492270	.491292	.490319	.489354	.488393	.487440	.486496	.485559	10
7.89	.484632	.483714	.482806	.481910	.481024	.480150	.479289	.478439	.477604	.476783	15
7.90	.475975	.475183	.474406	.473646	.472901	.472173	.471463	.470769	.470095	.469439	20
7.91	.468801	.468184	.467586	.467008	.466450	.465913	.465398	.464903	.464431	.463981	23
7.92	.463553	.463148	.462765	.462406	.462070	.461757	.461469	.461204	.460963	.460746	25
7.93	.460554	.460387	.460244	.460125	.460032	.459963	.459919	.459901	.459907	.459938	25
7.94	.459994	.460074	.460180	.460310	.460466	.460645	.460850	.461078	.461331	.461608	24
7.95	.461909	.462233	.462581	.462952	.463347	.463764	.464204	.464667	.465150	.465657	20
7.96	.466184	.466732	.467301	.467890	.468500	.469129	.469778	.470445	.471130	.471835	16
7.97	.472556	.473294	.474049	.474820	.475607	.476409	.477226	.478057	.478902	.479760	12
7.98	.480631	.481513	.482407	.483311	.484228	.485152	.486087	.487030	.487981	.488939	7
7.99	.489905	.490876	.491854	.492836	.493823	.494813	.495807	.496804	.497801	.498801	0

TABLE 9B

Fresnel Cosine Integral

U	.000	.001	.002	.003	.004	.005	.006	.007	.008	.009	Δ_2
8.00	.499801	.500800	.501800	.502798	.503794	.504787	.505778	.506765	.507747	.508724	-6
8.01	.509695	.510660	.511619	.512569	.513512	.514446	.515371	.516287	.517191	.518085	-11
8.02	.518968	.519837	.520694	.521538	.522368	.523184	.523985	.524771	.525540	.526294	-15
8.03	.527031	.527750	.528452	.529135	.529800	.530445	.531071	.531678	.532263	.532829	-21
8.04	.533374	.533897	.534398	.534877	.535334	.535768	.536180	.536568	.536933	.537274	-23
8.05	.537591	.537884	.538154	.538398	.538618	.538813	.538984	.539129	.539249	.539344	-24
8.06	.539414	.539459	.539478	.539472	.539440	.539384	.539302	.539194	.539062	.538904	-24
8.07	.538722	.538515	.538283	.538026	.537745	.537440	.537110	.536757	.536380	.535980	-22
8.08	.535557	.535111	.534642	.534151	.533638	.533103	.532547	.531970	.531373	.530755	-18
8.09	.530118	.529462	.528786	.528092	.527380	.526650	.525903	.525138	.524359	.523563	-12
8.10	.522752	.521927	.521087	.520235	.519369	.518491	.517600	.516698	.515786	.514863	-8
8.11	.513931	.512990	.512041	.511084	.510120	.509150	.508173	.507191	.506205	.505215	-2
8.12	.504222	.503226	.502228	.501230	.500229	.499230	.498230	.497231	.496236	.495242	5
8.13	.494253	.493267	.492284	.491308	.490337	.489373	.488415	.487464	.486524	.485591	10
8.14	.484667	.483755	.482853	.481962	.481083	.480218	.479364	.478524	.477698	.476887	16
8.15	.476091	.475312	.474548	.473802	.473072	.472361	.471667	.470992	.470337	.469701	20
8.16	.469084	.468489	.467914	.467360	.466827	.466318	.465829	.465363	.464921	.464501	24
8.17	.464105	.463733	.463384	.463060	.462760	.462485	.462234	.462009	.461808	.461633	25
8.18	.461483	.461358	.461259	.461186	.461138	.461116	.461120	.461150	.461205	.461286	26
8.19	.461392	.461524	.461681	.461864	.462072	.462305	.462563	.462846	.463154	.463486	24
8.20	.463842	.464222	.464626	.465052	.465503	.465976	.466472	.466990	.467529	.468091	20
8.21	.468673	.469276	.469900	.470543	.471207	.471888	.472589	.473309	.474045	.474799	17
8.22	.475570	.476357	.477159	.477977	.478809	.479655	.480515	.481388	.482273	.483170	10
8.23	.484078	.484996	.485925	.486862	.487809	.488763	.489724	.490693	.491667	.492648	4
8.24	.493633	.494622	.495615	.496609	.497607	.498606	.499606	.500607	.501605	.502604	-2
8.25	.503601	.504594	.505585	.506571	.507554	.508530	.509502	.510467	.511424	.512374	-9
8.26	.513315	.514246	.515169	.516080	.516982	.517871	.518749	.519614	.520465	.521303	-15
8.27	.522126	.522934	.523726	.524502	.525262	.526004	.526729	.527436	.528124	.528793	-19
8.28	.529442	.530071	.530680	.531268	.531835	.532380	.532904	.533404	.533882	.534338	-23
8.29	.534769	.535177	.535562	.535922	.536257	.536568	.536854	.537116	.537351	.537562	-25
8.30	.537746	.537905	.538039	.538146	.538228	.538283	.538312	.538316	.538293	.538244	-25
8.31	.538169	.538068	.537941	.537789	.537610	.537406	.537176	.536921	.536641	.536336	-24
8.32	.536006	.535652	.535273	.534870	.534443	.533993	.533520	.533023	.532505	.531964	-20
8.33	.531401	.530817	.530211	.529586	.528939	.528274	.527588	.526884	.526163	.525422	-15
8.34	.524665	.523891	.523100	.522294	.521472	.520636	.519785	.518921	.518045	.517155	-9
8.35	.516254	.515342	.514420	.513488	.512547	.511597	.510639	.509673	.508703	.507725	-4
8.36	.506742	.505755	.504763	.503770	.502773	.501775	.500775	.499775	.498776	.497776	3
8.37	.496779	.495785	.494793	.493806	.492821	.491844	.490871	.489904	.488945	.487993	8
8:38	.487051	.486117	.485193	.484280	.483377	.482486	.481606	.480740	.479888	.479050	15
8.39	.478226	.477417	.476624	.475848	.475088	.474346	.473622	.472916	.472230	.471562	20
8.40	.470914	.470287	.469680	.469095	.468532	.467990	.467471	.466974	.466501	.466051	24
8.41	.465624	.465222	.464844	.464491	.464162	.463859	.463581	.463328	.463102	.462901	26
8.42	.462725	.462577	.462454	.462357	.462287	.462243	.462226	.462235	.462271	.462333	26
8.43	.462422	.462537	.462678	.462845	.463038	.463257	.463503	.463774	.464069	.464391	24
8.44	.464737	.465108	.465504	.465923	.466367	.466834	.467325	.467839	.468374	.468933	21
8.45	.469514	.470114	.470737	.471380	.472044	.472726	.473428	.474149	.474888	.475644	17
8.46	.476418	.477207	.478014	.478835	.479672	.480522	.481387	.482265	.483154	.484056	10
8.47	.484970	.485893	.486826	.487768	.488720	.489678	.490645	.491618	.492596	.493580	3
8.48	.494568	.495560	.496555	.497551	.498551	.499550	.500551	.501551	.502548	.503545	-4
8.49	.504539	.505528	.506515	.507496	.508473	.509442	.510406	.511361	.512308	.513247	-10

TABLE 9B

Fresnel Cosine Integral

U	.000	.001	.002	.003	.004	.005	.006	.007	.008	.009	Δ_2
8.50	.514177	.515095	.516003	.516898	.517783	.518654	.519512	.520356	.521185	.521999	-16
8.51	.522797	.523578	.524344	.525091	.525820	.526531	.527223	.527895	.528547	.529179	-21
8.52	.529790	.530378	.530946	.531491	.532014	.532513	.532989	.533441	.533870	.534273	-24
8.53	.534653	.535007	.535336	.535640	.535918	.536170	.536397	.536597	.536770	.536918	-26
8.54	.537038	.537132	.537199	.537240	.537253	.537240	.537200	.537133	.537039	.536919	-26
8.55	.536772	.536599	.536399	.536173	.535920	.535642	.535338	.535009	.534654	.534275	-24
8.56	.533870	.533442	.532989	.532512	.532012	.531489	.530942	.530374	.529784	.529172	-20
8.57	.528539	.527885	.527212	.526518	.525806	.525075	.524326	.523558	.522775	.521974	-14
8.58	.521158	.520327	.519481	.518621	.517747	.516861	.515962	.515052	.514131	.513200	-8
8.59	.512260	.511311	.510353	.509389	.508416	.507439	.506456	.505468	.504477	.503482	0
8.60	.502485.	.501487	.500486	.499487	.498487	.497489	.496493	.495499	.494509	.493523	5
8.61	.492541	.491566	.490596	.489634	.488679	.487733	.486796	.485868	.484952	.484045	13
8.62	.483151	.482270	.481401	.480547	.479706	.478881	.478071	.477278	.476501	.475742	19
8.63	.475001	.474278	.473574	.472889	.472225	.471581	.470958	.470356	.469777	.469219	24
8.64	.468684	.468173	.467685	.467221	.466781	.466366	.465975	.465610	.465270	.464956	26
8.65	.464668	.464406	.464170	.463960	.463778	.463622	.463493	.463391	.463316	.463268	27
8.66	.463248	.463254	.463288	.463349	.463437	.463552	.463695	.463864	.464060	.464282	26
8.67	.464531	.464806	.465108	.465435	.465788	.466165	.466569	.466997	.467449	.467926	23
8.68	.468427	.468951	.469498	.470067	.470659	.471272	.471907	.472563	.473239	.473935	19
8.69	.474650	.475384	.476137	.476906	.477694	.478497	.479317	.480153	.481001	.481866	12
8.70	.482743	.483633	.484535	.485448	.486373	.487307	.488252	.489204	.490164	.491133	5
8.71	.492106	.493086	.494071	.495060	.496054	.497049	.498047	.499047	.500047	.501046	0
8.72	.502046	.503043	.504038	.505029	.506017	.507000	.507978	.508950	.509914	.510872	-9
8.73	.511821	.512761	.513691	.514611	.515520	.516416	.517302	.518173	.519030	.519874	-15
8.74	.520702	.521514	.522311	.523089	.523851	.524594	.525320	.526026	.526711	.527377	-21
8.75	.528022	.528646	.529248	.529828	.530385	.530919	.531430	.531917	.532379	.532818	-24
8.76	.533231	.533618	.533981	.534318	.534628	.534913	.535171	.535402	.535606	.535783	-27
8.77	.535934	.536056	.536152	.536220	.536260	.536273	.536258	.536216	.536146	.536049	-26
8.78	.535924	.535772	.535593	.535387	.535153	.534894	.534607	.534294	.533955	.533591	-25
8.79	.533200	.532785	.532344	.531879	.531389	.530876	.530339	.529779	.529197	.528592	-21
8.80	.527965	.527318	.526649	.525960	.525251	.524523	.523776	.523011	.522229	.521430	-15
8.81	.520613	.519783	.518936	.518076	.517200	.516314	.515413	.514501	.513579	.512645	-9
8.82	.511702	.510751	.509791	.508825	.507850	.506871	.505885	.504896	.503903	.502907	0
8.83	.501909	.500910	.499909	.498910	.497911	.496915	.495919	.494928	.493941	.492958	7
8.84	.491982	.491012	.490047	.489092	.488144	.487207	.486278	.485361	.484455	.483561	13
8.85	.482679	.481812	.480958	.480120	.479297	.478490	.477700	.476926	.476172	.475435	20
8.86	.474717	.474020	.473343	.472686	.472050	.471437	.470845	.470276	.469731	.469208	25
8.87	.468710	.468236	.467787	.467363	.466965	.466592	.466245	.465924	.465630	.465363	27
8.88	465123	.464910	.464724	.464565	.464435	.464332	.464256	.464209	.464190	.464198	28
8.89	.464234	.464299	.464391	.464511	.464658	.464833	.465036	.465266	.465523	.465807	26
8.90	.466118	.466454	.466818	.467207	.467622	.468062	.468527	.469017	.469530	.470068	23
8.91	.470629	.471213	.471819	.472447	.473097	.473768	.474460	.475172	.475902	.476652	17
8.92	.477419	.478205	.479007	.479826	.480661	.481511	.482375	.483254	.484144	.485048	11
8.93	.485964	.486889	.487826	.488772	.489726	.490688	.491658	.492635	.493616	.494604	3
8.94	.495595	.496589	.497586	.498584	.499584	.500584	.501584	.502582	.503577	.504571	-4
8.95	.505560	.506544	.507524	.508497	.509464	.510422	.511373	.512315	.513246	.514168	-11
8.96	.515077	.515974	.516859	.517730	.518588	.519430	.520257	.521068	.521862	.522638	-18
8.97	.523397	.524137	.524857	.525558	.526238	.526897	.527535	.528152	.528745	.529316	-23
8.98	.529863	.530386	.530886	.531360	.531809	.532233	.532632	.533004	.533350	.533669	-26
8.99	.533962	.534227	.534465	.534675	.534858	.535013	.535140	.535239	.535309	.535352	-28

TABLE 9B

Fresnel Cosine Integral

U	.000	.001	.002	.003	.004	.005	.006	.007	.008	.009	Δ_2
9.00	.535366	.535352	.535310	.535239	.535140	.535013	.534859	.534676	.534465	.534227	-26
9.01	.533962	.533669	.533350	.533004	.532631	.532232	.531807	.531357	.530882	.530382	-23
9.02	.529857	.529310	.528738	.528143	.527526	.526887	.526226	.525544	.524841	.524119	-18
9.03	.523377	.522618	.521839	.521043	.520231	.519402	.518557	.517697	.516824	.515937	-11
9.04	.515037	.514125	.513202	.512269	.511325	.510373	.509412	.508443	.507469	.506488	-3
9.05	.505503	.504513	.503518	.502523	.501524	.500525	.499524	.498524	.497527	.496530	4
9.06	.495536	.494548	.493562	.492584	.491610	.490643	.489685	.488734	.487793	.486862	12
9.07	.485941	.485033	.484136	.483252	.482382	.481527	.480687	.479862	.479054	.478263	19
9.08	.477490	.476735	.475999	.475283	.474587	.473912	.473258	.472626	.472016	.471429	24
9.09	.470866	.470327	.469812	.469321	.468855	.468416	.468001	.467613	.467251	.466917	27
9.10	.466609	.466328	.466076	.465851	.465653	.465484	.465344	.465231	.465147	.465092	28
9.11	.465065	.465067	.465097	.465157	.465244	.465360	.465505	.465677	.465878	.466107	28
9.12	.466364	.466648	.466960	.467298	.467664	.468056	.468474	.468918	.469388	.469883	24
9.13	.470402	.470945	.471513	.472104	.472718	.473354	.474012	.474692	.475392	.476112	19
9.14	.476853	.477612	.478389	.479184	.479997	.480824	.481669	.482529	.483402	.484290	11
9.15	.485191	.486103	.487027	.487961	.488906	.489859	.490821	.491790	.492765	.493747	4
9.16	.494735	.495725	.496720	.497716	.498716	.499714	.500715	.501714	.502712	.503707	-4
9.17	.504699	.505687	.506670	.507648	.508619	.509583	.510539	.511486	.512423	.513350	-10
9.18	.514266	.515169	.516061	.516938	.517802	.518650	.519483	.520300	.521099	.521881	-18
9.19	.522645	.523389	.524115	.524819	.525504	.526166	.526807	.527425	.528021	.528593	-24
9.20	.529141	.529664	.530163	.530637	.531084	.531506	.531902	.532271	.532612	.532926	-27
9.21	.533213	.533472	.533703	.533905	.534079	.534225	.534342	.534430	.534489	.534520	-29
9.22	.534521	.534494	.534437	.534352	.534237	.534094	.533923	.533722	.533494	.533237	-27
9.23	.532953	.532641	.532301	.531935	.531541	.531121	.530675	.530203	.529706	.529183	-24
9.24	.528636	.528066	.527471	.526854	.526213	.525552	.524868	.524163	.523438	.522694	-18
9.25	.521930	.521148	.520348	.519532	.518698	.517849	.516985	.516106	.515214	.514309	-9
9.26	.513393	.512466	.511527	.510580	.509623	.508658	.507686	.506706	.505723	.504734	-2
9.27	.503741	.502745	.501747	.500748	.499748	.498748	.497750	.496753	.495760	.494770	6
9.28	.493784	.492804	.491830	.490864	.489905	.488955	.488014	.487084	.486165	.485257	13
9.29	.484363	.483482	.482614	.481762	.480926	.480106	.479303	.478517	.477750	.477002	21
9.30	.476274	.475566	.474879	.474214	.473571	.472950	.472352	.471778	.471229	.470704	26
9.31	.470204	.469729	.469281	.468859	.468464	.468096	.467755	.467441	.467156	.466899	29
9.32	.466670	.466470	.466298	.466156	.466042	.465958	.465903	.465877	.465881	.465913	29
9.33	.465975	.466067	.466187	.466337	.466515	.466722	.466958	.467222	.467514	.467835	27
9.34	.468183	.468558	.468960	.469389	.469844	.470325	.470832	.471364	.471920	.472501	23
9.35	.473106	.473733	.474383	.475055	.475749	.476463	.477198	.477952	.478725	.479517	17
9.36	.480326	.481152	.481995	.482852	.483725	.484611	.485511	.486423	.487347	.488282	9
9.37	.489227	.490181	.491143	.492113	.493090	.494072	.495060	.496053	.497047	.498045	1
9.38	.499044	.500043	.501043	.502041	.503039	.504032	.505023	.506009	.506990	.507965	-7
9.39	.508933	.509892	.510843	.511784	.512715	.513634	.514543	.515438	.516319	.517187	-14
9.40	.518039	.518875	.519695	.520498	.521282	.522048	.522794	.523521	.524227	.524912	-22
9.41	.525575	.526215	.526833	.527427	.527997	.528542	.529062	.529557	.530026	.530469	-26
9.42	.530885	.531273	.531635	.531968	.532274	.532551	.532800	.533020	.533210	.533372	-29
9.43	.533505	.533608	.533681	.533725	.533739	.533724	.533679	.533605	.533501	.533367	-28
9.44	.533204	.533012	.532791	.532541	.532263	.531956	.531621	.531258	.530868	.530451	-26
9.45	.530006	.529536	.529039	.528518	.527970	.527399	.526803	.526183	.525541	.524875	-21
9.46	.524189	.523481	.522752	.522003	.521235	.520448	.519643	.518821	.517982	.517127	-13
9.47	.516257	.515374	.514476	.513566	.512643	.511711	.510767	.509813	.508852	.507883	-5
9.48	.506906	.505925	.504937	.503946	.502950	.501954	.500954	.499954	.498954	.497955	3
9.49	.496958	.495964	.494973	.493988	.493008	.492034	.491067	.490108	.489159	.488219	11

TABLE 9B

Fresnel Cosine Integral

U	.000	.001	.002	.003	.004	.005	.006	.007	.008	.009	Δ_2
9.50	.487289	.486371	.485465	.484573	.483693	.482829	.481980	.481147	.480332	.479533	20
9.51	.478754	.477993	.477252	.476532	.475832	.475154	.474498	.473865	.473256	.472670	25
9.52	.472109	.471573	.471063	.470578	.470120	.469689	.469284	.468908	.468559	.468239	29
9.53	.467946	.467683	.467449	.467244	.467068	.466922	.466806	.466719	.466662	.466636	30
9.54	.466639	.466672	.466735	.466828	.466951	.467103	.467285	.467497	.467737	.468007	29
9.55	.468305	.468632	.468987	.469370	.469781	.470218	.470683	.471174	.471690	.472233	25
9.56	.472800	.473392	.474007	.474646	.475308	.475992	.476698	.477425	.478172	.478939	18
9.57	.479724	.480528	.481350	.482187	.483041	.483911	.484795	.485692	.486602	.487525	10
9.58	.488459	.489403	.490356	.491318	.492288	.493265	.494248	.495236	.496228	.497223	2
9.59	.498221	.499220	.500220	.501219	.502218	.503213	.504207	.505197	.506181	.507160	−7
9.60	.508133	.509097	.510054	.511000	.511938	.512863	.513778	.514680	.515567	.516441	−15
9.61	.517300	.518143	.518969	.519778	.520569	.521340	.522093	.522825	.523536	.524226	−22
9.62	.524894	.525538	.526160	.526757	.527330	.527877	.528400	.528896	.529366	.529808	−27
9.63	.530224	.530611	.530971	.531302	.531605	.531879	.532123	.532338	.532523	.532679	−30
9.64	.532804	.532899	.532965	.533000	.533004	.532979	.532923	.532836	.532720	.532574	−29
9.65	.532397	.532192	.531956	.531691	.531397	.531074	.530722	.530343	.529935	.529500	−26
9.66	.529038	.528549	.528034	.527493	.526926	.526336	.525720	.525081	.524420	.523735	−20
9.67	.523030	.522302	.521555	.520787	.520000	.519195	.518372	.517532	.516677	.515805	−12
9.68	.514919	.514020	.513107	.512184	.511248	.510303	.509347	.508383	.507412	.506433	−5
9.69	.505449	.504460	.503467	.502471	.501473	.500474	.499473	.498473	.497476	.496480	4
9.70	.495488	.494500	.493518	.492542	.491572	.490611	.489658	.488715	.487783	.486862	13
9.71	.485954	.485059	.484177	.483311	.482460	.481626	.480808	.480009	.479228	.478467	21
9.72	.477725	.477006	.476306	.475630	.474976	.474347	.473740	.473158	.472602	.472071	27
9.73	.471566	.471088	.470637	.470213	.469817	.469450	.469111	.468801	.468520	.468269	30
9.74	.468048	.467856	.467695	.467564	.467463	.467393	.467353	.467344	.467365	.467417	31
9.75	.467500	.467613	.467757	.467931	.468135	.468368	.468632	.468925	.469247	.469598	28
9.76	.469977	.470385	.470820	.471283	.471773	.472289	.472831	.473399	.473992	.474609	23
9.77	.475251	.475915	.476602	.477311	.478042	.478792	.479563	.480354	.481161	.481988	16
9.78	.482831	.483690	.484565	.485454	.486357	.487271	.488199	.489138	.490086	.491045	7
9.79	.492011	.492985	.493965	.494951	.495942	.496935	.497933	.498932	.499931	.500932	−1
9.80	.501930	.502926	.503921	.504910	.505896	.506875	.507849	.508815	.509771	.510720	−11
9.81	.511657	.512583	.513498	.514399	.515287	.516159	.517017	.517858	.518682	.519489	−18
9.82	.520277	.521045	.521793	.522520	.523226	.523910	.524571	.525208	.525821	.526410	−25
9.83	.526974	.527511	.528022	.528507	.528964	.529393	.529795	.530168	.530511	.530827	−30
9.84	.531112	.531367	.531593	.531789	.531954	.532088	.532192	.532265	.532307	.532318	−30
9.85	.532298	.532248	.532166	.532054	.531910	.531737	.531533	.531298	.531034	.530740	−28
9.86	.530417	.530064	.529683	.529273	.528835	.528369	.527876	.527357	.526812	.526240	−23
9.87	.525643	.525022	.524377	.523708	.523017	.522303	.521569	.520813	.520038	.519243	−17
9.88	.518430	.517600	.516752	.515889	.515010	.514117	.513209	.512290	.511359	.510417	−9
9.89	.509464	.508503	.507534	.506558	.505575	.504588	.503595	.502599	.501602	.500602	1
9.90	.499601	.498602	.497604	.496608	.495616	.494629	.493646	.492669	.491700	.490739	10
9.91	.489787	.488845	.487914	.486996	.486089	.485196	.484317	.483454	.482607	.481777	19
9.92	.480965	.480172	.479397	.478643	.477909	.477198	.476508	.475841	.475198	.474579	26
9.93	.473985	.473416	.472873	.472357	.471867	.471405	.470971	.470565	.470188	.469840	30
9.94	.469522	.469233	.468974	.468746	.468548	.468381	.468244	.468139	.468065	.468022	31
9.95	.468010	.468029	.468080	.468162	.468275	.468419	.468594	.468800	.469036	.469302	30
9.96	.469599	.469924	.470280	.470665	.471078	.471520	.471989	.472486	.473010	.473560	25
9.97	.474137	.474738	.475364	.476014	.476688	.477384	.478103	.478844	.479604	.480385	19
9.98	.481186	.482003	.482840	.483692	.484561	.485445	.486343	.487255	.488178	.489114	10
9.99	.490061	.491016	.491981	.492953	.493932	.494917	.495907	.496902	.497898	.498897	0

C. Blackbody Radiation Tables - Radiant

Table 10—Relative Spectral Emittance

Table 11—Relative Fractional Emittance

Table 12—Peak Spectral Emittance

Table 13—Total Emittance

Table 14—Peak Wavelength

TABLE 10

Blackbody: Relative Spectral Emittance (Radiant)

$\lambda T(\mu.°K)$	0	1	2	3	4	5	6	7	8	9	m	Δ_2
100	.000010	.000038	.000145	.000541	.001977	.007036	.024444	.082929	.274934	.891284	-48	0
110	.000028	.000088	.000267	.000796	.002328	.006677	.018802	.051997	.141285	.377368	-43	0
120	.000099	.000256	.000651	.001632	.004024	.009781	.023433	.055353	.128972	.296501	-39	0
130	.000673	.001507	.003334	.007286	.015735	.033584	.070867	.147879	.305226	.623297	-36	0
140	.001260	.002520	.004989	.009785	.019004	.036567	.069715	.131720	.246684	.458010	-33	0
150	.000843	.001539	.002788	.005009	.008929	.015795	.027731	.048330	.083623	.143667	-30	0
160	.002451	.004153	.006991	.011690	.019421	.032062	.052602	.085775	.139032	.224031	-28	0
170	.003589	.005717	.009057	.014269	.022360	.034852	.054042	.083370	.127970	.195460	-26	0
180	.002971	.004494	.006767	.010141	.015130	.022471	.033229	.048924	.071726	.104716	-24	0
190	.015225	.022047	.031798	.045682	.065375	.093201	.132374	.187319	.264109	.371047	-23	0
200	.051945	.072469	.100757	.139615	.192819	.265428	.364205	.498157	.679248	.923319	-22	0
210	.012513	.016907	.022776	.030594	.040978	.054732	.072898	.096827	.128263	.169452	-20	0
220	.022328	.029344	.038466	.050296	.065601	.085353	.110782	.143444	.185298	.238806	-19	0
230	.030706	.039392	.050422	.064399	.082070	.104365	.132435	.167703	.211922	.267253	-18	0
240	.033635	.042247	.052959	.066259	.082740	.103124	.128290	.159301	.197448	.244289	-17	0
250	.030170	.037195	.045777	.056241	.068980	.084462	.103249	.126007	.153532	.186771	-16	0
260	.022685	.027509	.033308	.040268	.048609	.058590	.070516	.084747	.101702	.121876	-15	0
270	.145846	.174287	.207987	.247866	.294993	.350614	.416173	.493347	.584078	.690616	-15	0
280	.081556	.096190	.113311	.133316	.156664	.183881	.215571	.252428	.295244	.344929	-14	0
290	.040252	.046920	.054631	.063541	.073823	.085677	.099329	.115035	.133086	.153811	-13	0
300	.177581	.204818	.235997	.271653	.312389	.358886	.411908	.472314	.541069	.619255	-13	0
310	.070809	.080892	.092327	.105285	.119955	.136549	.155303	.176481	.200376	.227314	-12	0
320	.257657	.291808	.330215	.373373	.421831	.476198	.537147	.605423	.681847	.767328	-12	0
330	.086287	.096957	.108866	.122146	.136946	.153428	.171769	.192165	.214832	.240003	-11	0
340	.267938	.298918	.333252	.371279	.413368	.459922	.511382	.568229	.630986	.700226	-11	0
350	.077657	.086069	.095333	.105529	.116744	.129072	.142615	.157485	.173803	.191698	-10	0
360	.211313	.232801	.256326	.282068	.310221	.340993	.374609	.411313	.451368	.495055	-10	0
370	.054268	.059457	.065108	.071259	.077950	.085226	.093133	.101723	.111049	.121169	-9	0
380	.132147	.144049	.156946	.170916	.186040	.202407	.220110	.239249	.259933	.282274	-9	0
390	.306397	.332430	.360514	.390796	.423435	.458598	.496466	.537228	.581089	.628263	-9	0
400	.067898	.073349	.079204	.085491	.092240	.099481	.107247	.115574	.124497	.134057	-8	0
410	.144295	.155253	.166981	.179525	.192938	.207276	.222596	.238959	.256431	.275080	-8	0
420	.294979	.316202	.338832	.362953	.388653	.416028	.445177	.476203	.509218	.544337	-8	0
430	.058168	.062138	.066357	.070838	.075598	.080651	.086014	.091705	.097740	.104140	-7	0
440	.110925	.118114	.125730	.133796	.142336	.151376	.160941	.171059	.181759	.193072	-7	0
450	.205028	.217662	.231007	.245100	.259979	.275683	.292253	.309733	.328167	.347602	-7	0
460	.368087	.389674	.412415	.436367	.461586	.488133	.516072	.545467	.576386	.608902	-7	0
470	.064309	.067902	.071678	.075644	.079810	.084185	.088778	.093599	.098657	.103963	-6	0
480	.109529	.115365	.121483	.127896	.134615	.141655	.149028	.156749	.164833	.173294	-6	0
490	.182149	.191413	.201103	.211238	.221834	.232911	.244489	.256586	.269224	.282425	-6	0

TABLE 10

Blackbody: Relative Spectral Emittance (Radiant)

$\lambda T(\mu \cdot {}^\circ K)$	0	1	2	3	4	5	6	7	8	9	m	Δ_2
500	.296210	.310603	.325627	.341308	.357669	.374738	.392541	.411107	.430464	.450643	-6	851
510	.471674	.493588	.516419	.540201	.564969	.590757	.617605	.645549	.674630	.704889	-6	1219
520	.073637	.076911	.080315	.083855	.087535	.091360	.095335	.099465	.103755	.108211	-5	171
530	.112839	.117643	.122631	.127809	.133181	.138756	.144540	.150539	.156761	.163213	-5	237
540	.169902	.176835	.184022	.191469	.199185	.207178	.215458	.224032	.232911	.242103	-5	323
550	.251619	.261467	.271659	.282205	.293114	.304399	.316070	.328140	.340618	.353519	-5	434
560	.366853	.380635	.394875	.409589	.424789	.440489	.456704	.473449	.490737	.508584	-5	574
570	.527007	.546020	.565640	.585885	.606770	.628314	.650534	.673448	.697076	.721436	-5	751
580	.074655	.077243	.079911	.082660	.085492	.088410	.091415	.094511	.097699	.100981	-4	97
590	.104361	.107840	.111422	.115108	.118901	.122804	.126819	.130950	.135200	.139570	-4	124
600	.144064	.148685	.153436	.158321	.163341	.168501	.173804	.179253	.184852	.190604	-4	156
610	.196512	.202580	.208813	.215213	.221784	.228530	.235456	.242565	.249861	.257349	-4	195
620	.265032	.272916	.281004	.289300	.297810	.306538	.315488	.324666	.334077	.343724	-4	242
630	.353613	.363750	.374139	.384786	.395696	.406874	.418327	.430058	.442076	.454384	-4	296
640	.466989	.479897	.493113	.506645	.520498	.534679	.549194	.564049	.579251	.594808	-4	360
650	.610725	.627010	.643670	.660711	.678141	.695968	.714199	.732841	.751902	.771389	-4	434
660	.791312	.811677	.832492	.853767	.875508	.897726	.920427	.943621	.967317	.991523	-4	519
670	.101625	.104150	.106729	.109363	.112053	.114799	.117603	.120465	.123387	.126370	-3	61
680	.129414	.132520	.135691	.138926	.142227	.145595	.149031	.152535	.156111	.159757	-3	72
690	.163476	.167269	.171137	.175080	.179101	.183200	.187380	.191640	.195982	.200408	-3	84
700	.204918	.209515	.214199	.218971	.223834	.228788	.233835	.238976	.244213	.249547	-3	98
710	.254979	.260511	.266145	.271881	.277722	.283668	.289722	.295885	.302158	.308543	-3	113
720	.315041	.321655	.328386	.335235	.342204	.349295	.356509	.363849	.371315	.378910	-3	130
730	.386635	.394491	.402482	.410609	.418872	.427275	.435819	.444506	.453337	.462315	-3	148
740	.471442	.480719	.490148	.499732	.509472	.519370	.529428	.539649	.550034	.560585	-3	168
750	.571304	.582194	.593257	.604494	.615908	.627501	.639275	.651232	.663374	.675704	-3	189
760	.688224	.700937	.713843	.726946	.740248	.753752	.767458	.781371	.795492	.809824	-3	213
770	.824368	.839128	.854106	.869304	.884725	.900371	.916245	.932349	.948685	.965257	-3	237
780	.098207	.099912	.101641	.103395	.105173	.106977	.108806	.110661	.112542	.114448	-2	26
790	.116381	.118341	.120328	.122342	.124384	.126453	.128551	.130676	.132831	.135014	-2	29
800	.137227	.139469	.141741	.144043	.146375	.148739	.151133	.153558	.156015	.158504	-2	32
810	.161025	.163578	.166165	.168784	.171437	.174123	.176844	.179599	.182388	.185213	-2	35
820	.188073	.190968	.193899	.196867	.199871	.202912	.205991	.209106	.212260	.215452	-2	38
830	.218683	.221952	.225261	.228609	.231997	.235425	.238894	.242404	.245955	.249547	-2	41
840	.253182	.256859	.260578	.264340	.268146	.271996	.275889	.279827	.283809	.287837	-2	45
850	.291910	.296029	.300194	.304406	.308665	.312971	.317324	.321725	.326175	.330674	-2	49
860	.335221	.339819	.344465	.349163	.353910	.358709	.363559	.368461	.373415	.378421	-2	52
870	.383480	.388592	.393758	.398978	.404252	.409582	.414966	.420405	.425901	.431453	-2	56
880	.437061	.442727	.448450	.454231	.460070	.465968	.471925	.477941	.484017	.490154	-2	60
890	.496351	.502609	.508928	.515309	.521752	.528258	.534827	.541459	.548155	.554916	-2	64
900	.561741	.568631	.575586	.582607	.589694	.596848	.604069	.611358	.618714	.626138	-2	68
910	.633631	.641194	.648825	.656527	.664299	.672141	.680055	.688040	.696097	.704226	-2	72
920	.712428	.720704	.729053	.737475	.745973	.754545	.763192	.771915	.780714	.789589	-2	76
930	.798541	.807571	.816678	.825863	.835127	.844470	.853892	.863393	.872975	.882638	-2	80
940	.892381	.902206	.912113	.922102	.932174	.942329	.952567	.962889	.973295	.983786	-2	85
950	.099436	.100502	.101577	.102661	.103753	.104853	.105963	.107081	.108209	.109345	-1	8
960	.110490	.111644	.112807	.113979	.115160	.116350	.117549	.118758	.119976	.121203	-1	9
970	.122440	.123685	.124941	.126205	.127480	.128763	.130057	.131360	.132672	.133995	-1	9
980	.135327	.136668	.138020	.139382	.140753	.142134	.143525	.144927	.146338	.147760	-1	10
990	.149191	.150633	.152085	.153547	.155019	.156502	.157995	.159499	.161013	.162537	-1	10

TABLE 10

Blackbody: Relative Spectral Emittance (Radiant)

$\lambda T(\mu.°K)$	0	1	2	3	4	5	6	7	8	9	m	Δ_2
1000	.164073	.165618	.167174	.168741	.170319	.171907	.173506	.175115	.176736	.178367	-1	10
1010	.180010	.181663	.183327	.185003	.186689	.188386	.190095	.191814	.193545	.195288	-1	11
1020	.197041	.198806	.200582	.202369	.204168	.205978	.207800	.209633	.211478	.213335	-1	11
1030	.215203	.217082	.218974	.220877	.222792	.224719	.226657	.228608	.230570	.232544	-1	12
1040	.234531	.236529	.238539	.240562	.242596	.244643	.246702	.248773	.250856	.252951	-1	12
1050	.255059	.257180	.259312	.261457	.263615	.265784	.267967	.270162	.272369	.274589	-1	12
1060	.276822	.279067	.281325	.283596	.285879	.288176	.290485	.292807	.295141	.297489	-1	12
1070	.299849	.302223	.304609	.307009	.309422	.311847	.314286	.316738	.319203	.321681	-1	13
1080	.324172	.326677	.329195	.331726	.334270	.336828	.339399	.341984	.344582	.347193	-1	13
1090	.349818	.352456	.355108	.357774	.360453	.363145	.365852	.368572	.371305	.374052	-1	13
1100	.376813	.379588	.382377	.385179	.387995	.390825	.393669	.396526	.399398	.402283	-1	13
1110	.405183	.408096	.411023	.413965	.416920	.419890	.422873	.425871	.428883	.431909	-1	14
1120	.434949	.438003	.441071	.444154	.447251	.450362	.453487	.456627	.459781	.462949	-1	14
1130	.466132	.469329	.472540	.475766	.479006	.482260	.485529	.488813	.492111	.495423	-1	14
1140	.498750	.502091	.505447	.508818	.512203	.515603	.519017	.522446	.525889	.529348	-1	14
1150	.532820	.536308	.539810	.543327	.546858	.550405	.553966	.557541	.561132	.564737	-1	14
1160	.568357	.571992	.575642	.579306	.582985	.586679	.590388	.594112	.597850	.601604	-1	14
1170	.605372	.609155	.612953	.616766	.620594	.624437	.628295	.632168	.636055	.639958	-1	14
1180	.643875	.647808	.651755	.655718	.659695	.663688	.667695	.671717	.675755	.679807	-1	15
1190	.683875	.687957	.692055	.696167	.700295	.704437	.708595	.712767	.716955	.721158	-1	14
1200	.725376	.729608	.733856	.738119	.742397	.746690	.750999	.755322	.759660	.764013	-1	15
1210	.768382	.772765	.777164	.781577	.786006	.790449	.794908	.799382	.803871	.808375	-1	15
1220	.812894	.817428	.821977	.826541	.831120	.835715	.840324	.844948	.849588	.854242	-1	15
1230	.858911	.863596	.868295	.873010	.877739	.882484	.887243	.892018	.896807	.901612	-1	15
1240	.906431	.911265	.916115	.920979	.925859	.930753	.935662	.940586	.945525	.950479	-1	14
1250	.095545	.096043	.096543	.097044	.097547	.098052	.098557	.099065	.099573	.100084	-0	1
1260	.100595	.101109	.101623	.102140	.102657	.103176	.103697	.104219	.104743	.105268	-0	1
1270	.105794	.106322	.106852	.107382	.107915	.108449	.108984	.109521	.110059	.110599	-0	1
1280	.111140	.111682	.112226	.112772	.113319	.113867	.114417	.114968	.115521	.116075	-0	1
1290	.116631	.117188	.117747	.118307	.118868	.119431	.119995	.120561	.121128	.121696	-0	1
1300	.122266	.122838	.123411	.123985	.124560	.125138	.125716	.126296	.126877	.127460	-0	1
1310	.128044	.128630	.129217	.129805	.130395	.130986	.131578	.132172	.132767	.133364	-0	1
1320	.133962	.134562	.135163	.135765	.136368	.136973	.137580	.138188	.138797	.139407	-0	1
1330	.140019	.140632	.141247	.141863	.142480	.143099	.143719	.144340	.144963	.145587	-0	1
1340	.146212	.146839	.147467	.148096	.148727	.149359	.149992	.150627	.151263	.151900	-0	1
1350	.152539	153179	.153820	.154463	.155107	.155752	.156398	.157046	.157695	.158346	-0	1
1360	.158997	.159650	.160304	.160960	.161617	.162275	.162934	.163595	.164257	.164920	-0	1
1370	.165584	.166250	.166917	.167585	.168254	.168925	.169597	.170270	.170945	.171620	-0	1
1380	.172297	.172975	.173655	.174335	.175017	.175700	.176384	.177070	.177756	.178444	-0	1
1390	.179133	.179823	.180515	.181207	.181901	.182596	.183292	.183990	.184688	.185388	-0	1
1400	.186089	.186791	.187494	.188199	.188904	.189611	.190319	.191028	.191738	.192449	-0	1
1410	.193162	.193875	.194590	.195306	.196023	.196741	.197460	.198180	.198902	.199624	-0	1
1420	.200348	.201072	.201798	.202525	.203253	.203982	.204712	.205444	.206176	.206909	-0	1
1430	.207644	.208379	.209116	.209854	.210592	.211332	.212073	.212815	.213558	.214302	-0	1
1440	.215047	.215793	.216540	.217288	.218037	.218787	.219538	.220290	.221044	.221798	-0	1
1450	.222553	.223309	.224066	.224824	.225583	.226343	.227104	.227866	.228629	.229393	-0	0
1460	.230158	.230924	.231691	.232459	.233227	.233997	.234768	.235539	.236312	.237085	-0	0
1470	.237859	.238635	.239411	.240188	.240966	.241745	.242525	.243305	.244087	.244869	-0	0
1480	.245653	.246437	.247222	.248008	.248795	.249582	.250371	.251160	.251951	.252742	-0	0
1490	.253534	.254327	.255120	.255915	.256710	.257506	.258303	.259101	.259900	.260699	-0	0

TABLE 10

Blackbody: Relative Spectral Emittance (Radiant)

$\lambda T(\mu.°K)$	0	1	2	3	4	5	6	7	8	9	m	Δ_2
1500	.261499	.262301	.263102	.263905	.264708	.265513	.266318	.267123	.267930	.268737	-0	0
1510	.269545	.270354	.271164	.271974	.272785	.273597	.274410	.275223	.276037	.276852	-0	0
1520	.277668	.278484	.279301	.280119	.280937	.281756	.282576	.283397	.284218	.285040	-0	0
1530	.285862	.286686	.287510	.288334	.289160	.289986	.290812	.291640	.292467	.293296	-0	0
1540	.294125	.294955	.295786	.296617	.297449	.298281	.299114	.299948	.300782	.301617	-0	0
1550	.302453	.303289	.304126	.304963	.305801	.306639	.307478	.308318	.309158	.309999	-0	0
1560	.310840	.311682	.312525	.313368	.314211	.315056	.315900	.316745	.317591	.318437	-0	0
1570	.319284	.320132	.320979	.321828	.322677	.323526	.324376	.325226	.326077	.326929	-0	0
1580	.327780	.328633	.329485	.330339	.331192	.332047	.332901	.333756	.334612	.335468	-0	0
1590	.336325	.337181	.338039	.338897	.339755	.340613	.341472	.342332	.343192	.344052	-0	0
1600	.344913	.345774	.346635	.347497	.348360	.349222	.350085	.350949	.351812	.352677	-0	0
1610	.353541	.354406	.355271	.356137	.357003	.357869	.358736	.359603	.360470	.361337	-0	0
1620	.362205	.363074	.363942	.364811	.365680	.366550	.367420	.368290	.369160	.370031	-0	0
1630	.370902	.371773	.372644	.373516	.374388	.375261	.376133	.377006	.377879	.378752	-0	0
1640	.379626	.380500	.381374	.382248	.383123	.383998	.384873	.385748	.386623	.387499	-0	0
1650	.388375	.389251	.390127	.391003	.391880	.392757	.393634	.394511	.395388	.396266	-0	0
1660	.397143	.398021	.398899	.399777	.400656	.401534	.402413	.403291	.404170	.405049	-0	0
1670	.405929	.406808	.407687	.408567	.409446	.410326	.411206	.412086	.412966	.413846	-0	0
1680	.414726	.415607	.416487	.417367	.418248	.419129	.420009	.420890	.421771	.422652	-0	0
1690	.423533	.424414	.425295	.426176	.427057	.427938	.428819	.429700	.430582	.431463	-0	-0
1700	.432344	.433226	.434107	.434988	.435869	.436751	.437632	.438513	.439395	.440276	-0	0
1710	.441157	.442038	.442920	.443801	.444682	.445563	.446444	.447325	.448206	.449087	-0	-0
1720	.449968	.450849	.451729	.452610	.453491	.454371	.455252	.456132	.457012	.457893	-0	-0
1730	.458773	.459653	.460533	.461413	.462292	.463172	.464051	.464931	.465810	.466689	-0	-0
1740	.467568	.468447	.469326	.470205	.471083	.471962	.472840	.473718	.474596	.475474	-0	-0
1750	.476351	.477229	.478106	.478983	.479860	.480737	.481613	.482490	.483366	.484242	-0	-0
1760	.485118	.485994	.486869	.487744	.488620	.489494	.490369	.491243	.492118	.492992	-0	-0
1770	.493865	.494739	.495612	.496485	.497358	.498231	.499103	.499975	.500847	.501719	-0	-0
1780	.502590	.503461	.504332	.505203	.506073	.506943	.507813	.508682	.509551	.510420	-0	-0
1790	.511289	.512157	.513025	.513893	.514761	.515628	.516494	.517361	.518227	.519093	-0	-0
1800	.519959	.520824	.521689	.522554	.523418	.524282	.525146	.526009	.526872	.527734	-0	-0
1810	.528597	.529459	.530320	.531181	.532042	.532903	.533763	.534622	.535482	.536341	-0	-0
1820	.537199	.538058	.538916	.539773	.540630	.541487	.542343	.543199	.544055	.544910	-0	-0
1830	.545765	.546619	.547473	.548326	.549179	.550032	.550884	.551736	.552587	.553438	-0	-0
1840	.554289	.555139	.555989	.556838	.557687	.558535	.559383	.560230	.561077	.561924	-0	-0
1850	.562770	.563615	.564461	.565305	.566149	.566993	.567836	.568679	.569521	.570363	-0	-0
1860	.571205	.572045	.572886	.573726	.574565	.575404	.576242	.577080	.577917	.578754	-0	-0
1870	.579591	.580426	.581262	.582097	.582931	.583765	.584598	.585431	.586263	.587094	-0	-0
1880	.587926	.588756	.589586	.590416	.591245	.592073	.592901	.593728	.594555	.595381	-0	-0
1890	.596207	.597032	.597856	.598680	.599504	.600326	.601149	.601970	.602791	.603612	-0	-0
1900	.604432	.605251	.606070	.606888	.607706	.608523	.609339	.610155	.610970	.611785	-0	-0
1910	.612599	.613412	.614225	.615037	.615849	.616659	.617470	.618279	.619089	.619897	-0	-0
1920	.620705	.621512	.622319	.623125	.623930	.624735	.625539	.626342	.627145	.627947	-0	-0
1930	.628748	.629549	.630349	.631149	.631948	.632746	.633543	.634340	.635137	.635932	-0	-0
1940	.636727	.637521	.638315	.639108	.639900	.640692	.641482	.642273	.643062	.643851	-0	-0
1950	.644639	.645426	.646213	.646999	.647785	.648569	.649353	.650137	.650919	.651701	-0	-0
1960	.652482	.653263	.654042	.654822	.655600	.656377	.657154	.657931	.658706	.659481	-0	-0
1970	.660255	.661028	.661801	.662573	.663344	.664114	.664884	.665653	.666421	.667189	-0	-0
1980	.667955	.668721	.669486	.670251	.671015	.671778	.672540	.673302	.674062	.674822	-0	-0
1990	.675582	.676340	.677098	.677855	.678611	.679366	.680121	.680875	.681628	.682381	-0	-0

TABLE 10

Blackbody: Relative Spectral Emittance (Radiant)

λT(μ.°K)	0	1	2	3	4	5	6	7	8	9	m	Δ_2
2000	.683132	.683883	.684633	.685382	.686131	.686879	.687626	.688372	.689117	.689862	-0	-0
2010	.690606	.691349	.692091	.692832	.693573	.694313	.695052	.695790	.696528	.697264	-0	-0
2020	.698000	.698735	.699470	.700203	.700936	.701668	.702399	.703129	.703858	.704587	-0	-0
2030	.705315	.706042	.706768	.707493	.708218	.708941	.709664	.710386	.711108	.711828	-0	-0
2040	.712548	.713267	.713984	.714702	.715418	.716133	.716848	.717562	.718275	.718987	-0	-0
2050	.719698	.720408	.721118	.721827	.722535	.723242	.723948	.724653	.725358	.726061	-0	-0
2060	.726764	.727466	.728167	.728868	.729567	.730265	.730963	.731660	.732356	.733051	-0	-0
2070	.733745	.734439	.735131	.735823	.736514	.737203	.737893	.738581	.739268	.739954	-0	-0
2080	.740640	.741325	.742009	.742692	.743374	.744055	.744735	.745415	.746093	.746771	-0	-0
2090	.747448	.748124	.748799	.749473	.750146	.750818	.751490	.752160	.752830	.753499	-0	-0
2100	.754167	.754834	.755500	.756165	.756830	.757493	.758156	.758817	.759478	.760138	-0	-0
2110	.760797	.761455	.762112	.762769	.763424	.764078	.764732	.765385	.766036	.766687	-0	-0
2120	.767337	.767986	.768634	.769282	.769928	.770573	.771218	.771861	.772504	.773146	-0	-0
2130	.773787	.774427	.775066	.775704	.776341	.776977	.777612	.778247	.778880	.779513	-0	-0
2140	.780145	.780775	.781405	.782034	.782662	.783289	.783915	.784540	.785164	.785788	-0	-0
2150	.786410	.787032	.787652	.788272	.788891	.789508	.790125	.790741	.791356	.791970	-0	-0
2160	.792583	.793195	.793807	.794417	.795026	.795635	.796242	.796849	.797454	.798059	-0	-0
2170	.798663	.799266	.799867	.800468	.801068	.801667	.802266	.802863	.803459	.804054	-0	-0
2180	.804649	.805242	.805834	.806426	.807017	.807606	.808195	.808783	.809369	.809955	-0	-0
2190	.810540	.811124	.811707	.812289	.812870	.813450	.814030	.814608	.815185	.815761	-0	-0
2200	.816337	.816911	.817485	.818057	.818629	.819200	.819769	.820338	.820906	.821473	-0	-0
2210	.822039	.822603	.823167	.823730	.824293	.824854	.825414	.825973	.826531	.827089	-0	-0
2220	.827645	.828200	.828755	.829308	.829861	.830412	.830963	.831512	.832061	.832609	-0	-0
2230	.833156	.833701	.834246	.834790	.835333	.835875	.836416	.836956	.837495	.838033	-0	-0
2240	.838571	.839107	.839642	.840176	.840710	.841242	.841773	.842304	.842833	.843362	-0	-0
2250	.843889	.844416	.844942	.845466	.845990	.846513	.847035	.847555	.848075	.848594	-0	-0
2260	.849112	.849629	.850145	.850660	.851174	.851687	.852200	.852711	.853221	.853730	-0	-0
2270	.854239	.854746	.855252	.855758	.856262	.856766	.857268	.857770	.858271	.858770	-0	-0
2280	.859269	.859767	.860263	.860759	.861254	.861748	.862241	.862733	.863224	.863714	-0	-0
2290	.864203	.864691	.865178	.865664	.866150	.866634	.867117	.867600	.868081	.868561	-0	-0
2300	.869041	.869519	.869997	.870473	.870949	.871424	.871897	.872370	.872842	.873313	-0	-1
2310	.873782	.874251	.874719	.875186	.875652	.876117	.876581	.877044	.877507	.877968	-0	-0
2320	.878428	.878887	.879346	.879803	.880259	.880715	.881169	.881623	.882075	.882527	-0	-1
2330	.882978	.883427	.883876	.884324	.884771	.885217	.885662	.886106	.886549	.886991	-0	-0
2340	.887432	.887872	.888311	.888749	.889187	.889623	.890058	.890493	.890926	.891359	-0	-0
2350	.891790	.892221	.892650	.893079	.893507	.893934	.894360	.894784	.895208	.895631	-0	-0
2360	.896053	.896475	.896895	.897314	.897732	.898149	.898566	.898981	.899396	.899809	-0	-0
2370	.900222	.900633	.901044	.901454	.901862	.902270	.902677	.903083	.903488	.903892	-0	-0
2380	.904295	.904697	.905099	.905499	.905898	.906297	.906694	.907091	.907486	.907881	-0	-0
2390	.908275	.908667	.909059	.909450	.909840	.910229	.910617	.911004	.911390	.911775	-0	-0
2400	.912160	.912543	.912926	.913307	.913688	.914067	.914446	.914824	.915201	.915577	-0	-0
2410	.915951	.916326	.916699	.917071	.917442	.917812	.918182	.918550	.918918	.919284	-0	-0
2420	.919650	.920015	.920379	.920741	.921103	.921464	.921825	.922184	.922542	.922899	-0	-0
2430	.923256	.923611	.923966	.924319	.924672	.925024	.925375	.925725	.926074	.926422	-0	-0
2440	.926769	.927116	.927461	.927805	.928149	.928492	.928833	.929174	.929514	.929853	-0	-0
2450	.930191	.930528	.930864	.931200	.931534	.931868	.932200	.932532	.932863	.933192	-0	-0
2460	.933521	.933849	.934177	.934503	.934828	.935153	.935476	.935799	.936120	.936441	-0	-0
2470	.936761	.937080	.937398	.937715	.938032	.938347	.938662	.938975	.939288	.939600	-0	-0
2480	.939911	.940221	.940530	.940838	.941145	.941452	.941757	.942062	.942366	.942669	-0	-0
2490	.942971	.943272	.943572	.943871	.944170	.944467	.944764	.945059	.945354	.945648	-0	-0

TABLE 10

Blackbody: Relative Spectral Emittance (Radiant)

$\lambda T(\mu.^{\circ}K)$	0	1	2	3	4	5	6	7	8	9	m	Δ_2
2500	.945941	.946234	.946525	.946815	.947105	.947394	.947681	.947968	.948254	.948539	-0	-0
2510	.948824	.949107	.949390	.949671	.949952	.950232	.950511	.950789	.951066	.951343	-0	-0
2520	.951618	.951893	.952167	.952440	.952712	.952983	.953253	.953523	.953791	.954059	-0	-0
2530	.954326	.954592	.954857	.955121	.955384	.955647	.955909	.956169	.956429	.956688	-0	-0
2540	.956947	.957204	.957461	.957716	.957971	.958225	.958478	.958730	.958982	.959232	-0	-0
2550	.959482	.959731	.959978	.960226	.960472	.960717	.960962	.961206	.961448	.961690	-0	-0
2560	.961932	.962172	.962411	.962650	.962888	.963125	.963361	.963596	.963831	.964064	-0	-0
2570	.964297	.964529	.964760	.964990	.965220	.965448	.965676	.965903	.966129	.966354	-0	-0
2580	.966579	.966802	.967025	.967247	.967468	.967688	.967908	.968126	.968344	.968561	-0	-0
2590	.968777	.968992	.969207	.969421	.969633	.969845	.970057	.970267	.970477	.970685	-0	-0
2600	.970893	.971100	.971307	.971512	.971717	.971921	.972124	.972326	.972528	.972728	-0	-0
2610	.972928	.973127	.973325	.973523	.973719	.973915	.974110	.974304	.974498	.974690	-0	-0
2620	.974882	.975073	.975263	.975452	.975641	.975829	.976016	.976202	.976387	.976572	-0	-0
2630	.976756	.976939	.977121	.977302	.977483	.977663	.977842	.978020	.978197	.978374	-0	-0
2640	.978550	.978725	.978899	.979073	.979246	.979418	.979589	.979759	.979929	.980098	-0	-0
2650	.980266	.980433	.980600	.980765	.980930	.981095	.981258	.981421	.981583	.981744	-0	-0
2660	.981904	.982063	.982222	.982380	.982538	.982694	.982850	.983005	.983159	.983312	-0	-0
2670	.983465	.983617	.983768	.983919	.984068	.984217	.984365	.984512	.984659	.984805	-0	-0
2680	.984950	.985094	.985238	.985381	.985523	.985664	.985805	.985945	.986084	.986222	-0	-0
2690	.986360	.986496	.986633	.986768	.986902	.987036	.987169	.987302	.987434	.987564	-0	-0
2700	.987695	.987824	.987953	.988081	.988208	.988334	.988460	.988585	.988709	.988833	-0	-0
2710	.988956	.989078	.989199	.989320	.989440	.989559	.989677	.989795	.989912	.990029	-0	-0
2720	.990144	.990259	.990373	.990487	.990599	.990711	.990822	.990933	.991043	.991152	-0	-0
2730	.991260	.991368	.991475	.991581	.991687	.991792	.991896	.991999	.992102	.992204	-0	-0
2740	.992305	.992406	.992506	.992605	.992703	.992801	.992898	.992994	.993090	.993185	-0	-0
2750	.993279	.993373	.993466	.993558	.993650	.993740	.993831	.993920	.994009	.994097	-0	-0
2760	.994184	.994271	.994357	.994442	.994527	.994611	.994694	.994776	.994858	.994939	-0	-0
2770	.995020	.995100	.995179	.995257	.995335	.995412	.995489	.995564	.995639	.995714	-0	-0
2780	.995788	.995861	.995933	.996005	.996076	.996146	.996216	.996285	.996353	.996421	-0	-0
2790	.996488	.996555	.996620	.996685	.996750	.996814	.996877	.996939	.997001	.997062	-0	-0
2800	.997122	.997182	.997241	.997300	.997358	.997415	.997471	.997527	.997583	.997637	-0	-0
2810	.997691	.997744	.997797	.997849	.997900	.997951	.998001	.998051	.998099	.998148	-0	-0
2820	.998195	.998242	.998288	.998334	.998379	.998423	.998467	.998510	.998552	.998594	-0	-0
2830	.998635	.998676	.998716	.998755	.998794	.998832	.998869	.998906	.998942	.998977	-0	-0
2840	.999012	.999047	.999080	.999113	.999146	.999178	.999209	.999239	.999269	.999299	-0	-0
2850	.999327	.999355	.999383	.999410	.999436	.999462	.999487	.999511	.999535	.999558	-0	-0
2860	.999581	.999603	.999624	.999645	.999665	.999685	.999704	.999722	.999740	.999757	-0	-0
2870	.999774	.999790	.999805	.999820	.999834	.999848	.999861	.999873	.999885	.999896	-0	-0
2880	.999907	.999917	.999926	.999935	.999943	.999951	.999958	.999965	.999971	.999976	-0	-0
2890	.999981	.999985	.999989	.999992	.999994	.999996	.999998	.999998	.999999	.999998	-0	-0
2900	.999997	.999996	.999993	.999991	.999988	.999984	.999979	.999974	.999969	.999963	-0	-0
2910	.999956	.999949	.999941	.999933	.999924	.999914	.999904	.999894	.999882	.999871	-0	-0
2920	.999858	.999846	.999832	.999818	.999804	.999789	.999773	.999757	.999740	.999723	-0	-0
2930	.999705	.999687	.999668	.999648	.999628	.999608	.999587	.999565	.999543	.999520	-0	-0
2940	.999497	.999473	.999449	.999424	.999399	.999373	.999346	.999319	.999292	.999263	-0	-0
2950	.999235	.999206	.999176	.999146	.999115	.999084	.999052	.999020	.998987	.998953	-0	-0
2960	.998919	.998885	.998850	.998814	.998778	.998742	.998705	.998667	.998629	.998591	-0	-0
2970	.998551	.998512	.998472	.998431	.998390	.998348	.998306	.998263	.998220	.998176	-0	-0
2980	.998132	.998087	.998042	.997996	.997950	.997903	.997856	.997808	.997760	.997711	-0	-0
2990	.997661	.997612	.997561	.997511	.997459	.997407	.997355	.997302	.997249	.997195	-0	-0

TABLE 10

Blackbody: Relative Spectral Emittance (Radiant)

λT(μ.°K)	0	1	2	3	4	5	6	7	8	9	m	Δ_2
3000	.997141	.997086	.997031	.996975	.996919	.996862	.996805	.996747	.996689	.996630	-0	-0
3010	.996571	.996511	.996451	.996391	.996329	.996268	.996206	.996143	.996080	.996017	-0	-0
3020	.995953	.995888	.995823	.995758	.995692	.995625	.995559	.995491	.995423	.995355	-0	-0
3030	.995286	.995217	.995147	.995077	.995007	.994936	.994864	.994792	.994719	.994647	-0	-0
3040	.994573	.994499	.994425	.994350	.994275	.994199	.994123	.994046	.993969	.993892	-0	-0
3050	.993814	.993735	.993656	.993577	.993497	.993417	.993336	.993255	.993173	.993091	-0	-0
3060	.993009	.992926	.992842	.992758	.992674	.992589	.992504	.992418	.992332	.992246	-0	-0
3070	.992159	.992071	.991984	.991895	.991807	.991717	.991628	.991538	.991447	.991356	-0	-0
3080	.991265	.991173	.991081	.990989	.990895	.990802	.990708	.990614	.990519	.990424	-0	-0
3090	.990328	.990232	.990135	.990039	.989941	.989844	.989745	.989647	.989548	.989448	-0	-0
3100	.989348	.989248	.989147	.989046	.988945	.988843	.988741	.988638	.988535	.988431	-0	-0
3110	.988327	.988223	.988118	.988013	.987907	.987801	.987694	.987587	.987480	.987373	-0	-0
3120	.987264	.987156	.987047	.986938	.986828	.986718	.986608	.986497	.986385	.986274	-0	-0
3130	.986162	.986049	.985936	.985823	.985709	.985595	.985481	.985366	.985251	.985135	-0	-0
3140	.985019	.984902	.984786	.984668	.984551	.984433	.984315	.984196	.984077	.983957	-0	-0
3150	.983837	.983717	.983596	.983475	.983354	.983232	.983110	.982987	.982864	.982741	-0	-0
3160	.982617	.982493	.982369	.982244	.982119	.981993	.981867	.981741	.981614	.981487	-0	-0
3170	.981360	.981232	.981104	.980975	.980846	.980717	.980588	.980458	.980327	.980197	-0	-0
3180	.980065	.979934	.979802	.979670	.979537	.979405	.979271	.979138	.979004	.978869	-0	-0
3190	.978735	.978600	.978464	.978329	.978193	.978056	.977919	.977782	.977645	.977507	-0	-0
3200	.977368	.977230	.977091	.976952	.976812	.976672	.976532	.976391	.976250	.976109	-0	-0
3210	.975967	.975825	.975683	.975540	.975397	.975254	.975110	.974966	.974822	.974677	-0	-0
3220	.974532	.974387	.974241	.974095	.973948	.973802	.973654	.973507	.973359	.973211	-0	-0
3230	.973063	.972914	.972765	.972616	.972466	.972316	.972166	.972015	.971864	.971713	-0	-0
3240	.971561	.971409	.971257	.971104	.970951	.970798	.970644	.970490	.970336	.970181	-0	-0
3250	.970027	.969871	.969716	.969560	.969404	.969247	.969091	.968934	.968776	.968618	-0	-0
3260	.968461	.968302	.968144	.967985	.967825	.967666	.967506	.967346	.967185	.967024	-0	-0
3270	.966863	.966702	.966540	.966378	.966216	.966053	.965890	.965727	.965564	.965400	-0	-0
3280	.965236	.965071	.964907	.964742	.964576	.964411	.964245	.964079	.963912	.963745	-0	-0
3290	.963578	.963411	.963243	.963075	.962907	.962739	.962570	.962401	.962231	.962062	-0	-0
3300	.961892	.961721	.961551	.961380	.961209	.961038	.960866	.960694	.960522	.960349	-0	-0
3310	.960176	.960003	.959830	.959656	.959482	.959308	.959134	.958959	.958784	.958609	-0	-0
3320	.958433	.958257	.958081	.957905	.957728	.957551	.957374	.957196	.957019	.956841	-0	-0
3330	.956662	.956484	.956305	.956126	.955946	.955767	.955587	.955407	.955226	.955046	-0	-0
3340	.954865	.954683	.954502	.954320	.954138	.953956	.953773	.953590	.953407	.953224	-0	-0
3350	.953041	.952857	.952673	.952488	.952304	.952119	.951934	.951748	.951563	.951377	-0	-0
3360	.951191	.951004	.950818	.950631	.950444	.950256	.950069	.949881	.949693	.949504	-0	-0
3370	.949316	.949127	.948938	.948748	.948559	.948369	.948179	.947989	.947798	.947607	-0	-0
3380	.947416	.947225	.947033	.946841	.946649	.946457	.946265	.946072	.945879	.945686	-0	-0
3390	.945492	.945299	.945105	.944911	.944716	.944522	.944327	.944132	.943936	.943741	-0	-0
3400	.943545	.943349	.943153	.942956	.942760	.942563	.942366	.942168	.941971	.941773	-0	-0
3410	.941575	.941376	.941178	.940979	.940780	.940581	.940382	.940182	.939982	.939782	-0	-0
3420	.939582	.939381	.939181	.938980	.938778	.938577	.938375	.938174	.937972	.937769	-0	-0
3430	.937567	.937364	.937161	.936958	.936755	.936551	.936348	.936144	.935940	.935735	-0	-0
3440	.935531	.935326	.935121	.934916	.934710	.934505	.934299	.934093	.933886	.933680	-0	-0
3450	.933473	.933266	.933059	.932852	.932645	.932437	.932229	.932021	.931813	.931604	-0	-0
3460	.931396	.931187	.930978	.930768	.930559	.930349	.930139	.929929	.929719	.929508	-0	-0
3470	.929298	.929087	.928876	.928665	.928453	.928242	.928030	.927818	.927606	.927393	-0	-0
3480	.927181	.926968	.926755	.926542	.926328	.926115	.925901	.925687	.925473	.925259	-0	-0
3490	.925044	.924830	.924615	.924400	.924185	.923969	.923754	.923538	.923322	.923106	-0	-0

TABLE 10

Blackbody: Relative Spectral Emittance (Radiant)

$\lambda T(\mu.°K)$	0	1	2	3	4	5	6	7	8	9	m	Δ_2
3500	.922889	.922673	.922456	.922240	.922022	.921805	.921588	.921370	.921153	.920935	-0	-0
3510	.920717	.920498	.920280	.920061	.919842	.919623	.919404	.919185	.918966	.918746	-0	-0
3520	.918526	.918306	.918086	.917866	.917645	.917424	.917204	.916982	.916761	.916540	-0	-0
3530	.916318	.916097	.915875	.915653	.915431	.915208	.914986	.914763	.914540	.914317	-0	-0
3540	.914094	.913871	.913647	.913424	.913200	.912976	.912752	.912527	.912303	.912078	-0	-0
3550	.911853	.911629	.911403	.911178	.910953	.910727	.910501	.910276	.910050	.909823	-0	-0
3560	.909597	.909371	.909144	.908917	.908690	.908463	.908236	.908008	.907781	.907553	-0	-0
3570	.907325	.907097	.906869	.906641	.906412	.906184	.905955	.905726	.905497	.905268	-0	-0
3580	.905038	.904809	.904579	.904349	.904120	.903889	.903659	.903429	.903198	.902968	-0	-0
3590	.902737	.902506	.902275	.902044	.901812	.901581	.901349	.901118	.900886	.900654	-0	-0
3600	.900421	.900189	.899957	.899724	.899491	.899259	.899026	.898792	.898559	.898326	-0	-0
3610	.898092	.897859	.897625	.897391	.897157	.896923	.896688	.896454	.896219	.895985	-0	-0
3620	.895750	.895515	.895280	.895044	.894809	.894574	.894338	.894102	.893866	.893630	-0	-0
3630	.893394	.893158	.892922	.892685	.892448	.892212	.891975	.891738	.891501	.891264	-0	-0
3640	.891026	.890789	.890551	.890313	.890076	.889838	.889599	.889361	.889123	.888885	-0	-0
3650	.888646	.888407	.888169	.887930	.887691	.887451	.887212	.886973	.886733	.886494	-0	-0
3660	.886254	.886014	.885774	.885534	.885294	.885054	.884813	.884573	.884332	.884091	-0	-0
3670	.883851	.883610	.883369	.883127	.882886	.882645	.882403	.882162	.881920	.881678	-0	-0
3680	.881436	.881194	.880952	.880710	.880467	.880225	.879982	.879740	.879497	.879254	-0	-0
3690	.879011	.878768	.878525	.878282	.878038	.877795	.877551	.877307	.877064	.876820	-0	-0
3700	.876576	.876332	.876087	.875843	.875599	.875354	.875110	.874865	.874620	.874375	-0	-0
3710	.874130	.873885	.873640	.873395	.873150	.872904	.872659	.872413	.872167	.871921	-0	-0
3720	.871675	.871429	.871183	.870937	.870691	.870444	.870198	.869951	.869705	.869458	-0	-0
3730	.869211	.868964	.868717	.868470	.868223	.867976	.867728	.867481	.867233	.866986	-0	-0
3740	.866738	.866490	.866242	.865994	.865746	.865498	.865250	.865002	.864753	.864505	-0	-0
3750	.864256	.864008	.863759	.863510	.863261	.863012	.862763	.862514	.862265	.862015	-0	-0
3760	.861766	.861517	.861267	.861018	.860768	.860518	.860268	.860018	.859768	.859518	-0	-0
3770	.859268	.859018	.858768	.858517	.858267	.858016	.857766	.857515	.857264	.857013	-0	-0
3780	.856762	.856511	.856260	.856009	.855758	.855507	.855255	.855004	.854753	.854501	-0	-0
3790	.854249	.853998	.853746	.853494	.853242	.852990	.852738	.852486	.852234	.851982	-0	-0
3800	.851729	.851477	.851224	.850972	.850719	.850467	.850214	.849961	.849708	.849455	-0	-0
3810	.849203	.848949	.848696	.848443	.848190	.847937	.847683	.847430	.847176	.846923	-0	-0
3820	.846669	.846416	.846162	.845908	.845654	.845400	.845146	.844892	.844638	.844384	-0	-0
3830	.844130	.843876	.843621	.843367	.843113	.842858	.842604	.842349	.842094	.841840	-0	-0
3840	.841585	.841330	.841075	.840820	.840565	.840310	.840055	.839800	.839545	.839289	-0	-0
3850	.839034	.838779	.838523	.838268	.838012	.837757	.837501	.837245	.836990	.836734	-0	-0
3860	.836478	.836222	.835966	.835710	.835454	.835198	.834942	.834686	.834429	.834173	-0	-0
3870	.833917	.833660	.833404	.833148	.832891	.832635	.832378	.832121	.831865	.831608	-0	0
3880	.831351	.831094	.830837	.830580	.830323	.830066	.829809	.829552	.829295	.829038	-0	-0
3890	.828781	.828523	.828266	.828009	.827751	.827494	.827236	.826979	.826721	.826464	-0	-0
3900	.826206	.825949	.825691	.825433	.825175	.824917	.824660	.824402	.824144	.823886	-0	-0
3910	.823628	.823370	.823112	.822853	.822595	.822337	.822079	.821821	.821562	.821304	-0	-0
3920	.821045	.820787	.820529	.820270	.820012	.819753	.819494	.819236	.818977	.818719	-0	-0
3930	.818460	.818201	.817942	.817683	.817425	.817166	.816907	.816648	.816389	.816130	-0	-0
3940	.815871	.815612	.815353	.815094	.814834	.814575	.814316	.814057	.813798	.813538	-0	-0
3950	.813279	.813020	.812760	.812501	.812241	.811982	.811722	.811463	.811203	.810944	-0	-0
3960	.810684	.810425	.810165	.809905	.809646	.809386	.809126	.808867	.808607	.808347	-0	-0
3970	.808087	.807827	.807567	.807308	.807048	.806788	.806528	.806268	.806008	.805748	-0	-0
3980	.805488	.805228	.804968	.804707	.804447	.804187	.803927	.803667	.803407	.803146	-0	-0
3990	.802886	.802626	.802366	.802105	.801845	.801585	.801324	.801064	.800804	.800543	-0	-0

TABLE 10

Blackbody: Relative Spectral Emittance (Radiant)

λT(μ.°K)	0	1	2	3	4	5	6	7	8	9	m	Δ_2
4000	.800283	.800022	.799762	.799501	.799241	.798980	.798720	.798459	.798199	.797938	-0	-0
4010	.797678	.797417	.797157	.796896	.796635	.796375	.796114	.795853	.795593	.795332	-0	-0
4020	.795071	.794811	.794550	.794289	.794028	.793768	.793507	.793246	.792985	.792724	-0	-0
4030	.792464	.792203	.791942	.791681	.791420	.791159	.790898	.790638	.790377	.790116	-0	-0
4040	.789855	.789594	.789333	.789072	.788811	.788550	.788289	.788028	.787767	.787506	-0	-0
4050	.787245	.786984	.786723	.786462	.786201	.785940	.785679	.785418	.785157	.784896	-0	-0
4060	.784635	.784374	.784113	.783852	.783591	.783330	.783069	.782808	.782547	.782286	-0	0
4070	.782025	.781764	.781503	.781241	.780980	.780719	.780458	.780197	.779936	.779675	-0	0
4080	.779414	.779153	.778892	.778631	.778370	.778108	.777847	.777586	.777325	.777064	-0	-0
4090	.776803	.776542	.776281	.776020	.775759	.775498	.775237	.774975	.774714	.774453	-0	-0
4100	.774192	.773931	.773670	.773409	.773148	.772887	.772626	.772365	.772104	.771843	-0	-0
4110	.771582	.771321	.771060	.770799	.770538	.770277	.770016	.769755	.769494	.769233	-0	0
4120	.768972	.768711	.768450	.768189	.767928	.767667	.767406	.767145	.766885	.766624	-0	-0
4130	.766363	.766102	.765841	.765580	.765319	.765058	.764798	.764537	.764276	.764015	-0	0
4140	.763754	.763494	.763233	.762972	.762711	.762450	.762190	.761929	.761668	.761408	-0	0
4150	.761147	.760886	.760626	.760365	.760104	.759844	.759583	.759322	.759062	.758801	-0	0
4160	.758541	.758280	.758020	.757759	.757499	.757238	.756978	.756717	.756457	.756196	-0	0
4170	.755936	.755675	.755415	.755155	.754894	.754634	.754373	.754113	.753853	.753593	-0	-0
4180	.753332	.753072	.752812	.752552	.752291	.752031	.751771	.751511	.751251	.750991	-0	-0
4190	.750730	.750470	.750210	.749950	.749690	.749430	.749170	.748910	.748650	.748390	-0	0
4200	.748130	.747870	.747611	.747351	.747091	.746831	.746571	.746311	.746052	.745792	-0	0
4210	.745532	.745273	.745013	.744753	.744494	.744234	.743974	.743715	.743455	.743196	-0	0
4220	.742936	.742677	.742417	.742158	.741898	.741639	.741380	.741120	.740861	.740602	-0	0
4230	.740342	.740083	.739824	.739565	.739306	.739046	.738787	.738528	.738269	.738010	-0	0
4240	.737751	.737492	.737233	.736974	.736715	.736456	.736197	.735938	.735680	.735421	-0	0
4250	.735162	.734903	.734644	.734386	.734127	.733868	.733610	.733351	.733093	.732834	-0	0
4260	.732576	.732317	.732059	.731800	.731542	.731283	.731025	.730767	.730509	.730250	-0	0
4270	.729992	.729734	.729476	.729218	.728959	.728701	.728443	.728185	.727927	.727669	-0	0
4280	.727411	.727153	.726896	.726638	.726380	.726122	.725864	.725607	.725349	.725091	-0	0
4290	.724834	.724576	.724318	.724061	.723803	.723546	.723289	.723031	.722774	.722516	-0	0
4300	.722259	.722002	.721745	.721487	.721230	.720973	.720716	.720459	.720202	.719945	-0	0
4310	.719688	.719431	.719174	.718917	.718660	.718403	.718147	.717890	.717633	.717377	-0	0
4320	.717120	.716863	.716607	.716350	.716094	.715837	.715581	.715324	.715068	.714812	-0	0
4330	.714555	.714299	.714043	.713787	.713531	.713275	.713019	.712762	.712506	.712251	-0	0
4340	.711995	.711739	.711483	.711227	.710971	.710716	.710460	.710204	.709949	.709693	-0	0
4350	.709438	.709182	.708927	.708671	.708416	.708160	.707905	.707650	.707395	.707139	-0	0
4360	.706884	.706629	.706374	.706119	.705864	.705609	.705354	.705099	.704844	.704590	-0	0
4370	.704335	.704080	.703825	.703571	.703316	.703062	.702807	.702553	.702298	.702044	-0	0
4380	.701790	.701535	.701281	.701027	.700773	.700518	.700264	.700010	.699756	.699502	-0	0
4390	.699248	.698994	.698741	.698487	.698233	.697979	.697726	.697472	.697218	.696965	-0	0
4400	.696711	.696458	.696205	.695951	.695698	.695445	.695191	.694938	.694685	.694432	-0	0
4410	.694179	.693926	.693673	.693420	.693167	.692914	.692661	.692409	.692156	.691903	-0	0
4420	.691651	.691398	.691145	.690893	.690641	.690388	.690136	.689883	.689631	.689379	-0	0
4430	.689127	.688875	.688623	.688371	.688119	.687867	.687615	.687363	.687111	.686859	-0	0
4440	.686608	.686356	.686105	.685853	.685601	.685350	.685099	.684847	.684596	.684345	-0	0
4450	.684093	.683842	.683591	.683340	.683089	.682838	.682587	.682336	.682085	.681835	-0	0
4460	.681584	.681333	.681082	.680832	.680581	.680331	.680080	.679830	.679580	.679329	-0	0
4470	.679079	.678829	.678579	.678329	.678079	.677829	.677579	.677329	.677079	.676829	-0	0
4480	.676579	.676330	.676080	.675830	.675581	.675331	.675082	.674832	.674583	.674334	-0	0
4490	.674085	.673835	.673586	.673337	.673088	.672839	.672590	.672341	.672092	.671844	-0	0

TABLE 10

Blackbody: Relative Spectral Emittance (Radiant)

λT(μ.°K)	0	1	2	3	4	5	6	7	8	9	m	Δ_2
4500	.671595	.671346	.671098	.670849	.670601	.670352	.670104	.669855	.669607	.669359	-0	0
4510	.669110	.668862	.668614	.668366	.668118	.667870	.667622	.667375	.667127	.666879	-0	0
4520	.666631	.666384	.666136	.665889	.665641	.665394	.665146	.664899	.664652	.664405	-0	0
4530	.664157	.663910	.663663	.663416	.663169	.662923	.662676	.662429	.662182	.661936	-0	0
4540	.661689	.661443	.661196	.660950	.660703	.660457	.660211	.659964	.659718	.659472	-0	0
4550	.659226	.658980	.658734	.658488	.658242	.657997	.657751	.657505	.657260	.657014	-0	0
4560	.656769	.656523	.656278	.656033	.655787	.655542	.655297	.655052	.654807	.654562	-0	0
4570	.654317	.654072	.653827	.653582	.653338	.653093	.652848	.652604	.652359	.652115	-0	0
4580	.651871	.651626	.651382	.651138	.650894	.650650	.650406	.650162	.649918	.649674	-0	0
4590	.649430	.649187	.648943	.648699	.648456	.648212	.647969	.647725	.647482	.647239	-0	0
4600	.646996	.646753	.646509	.646266	.646023	.645781	.645538	.645295	.645052	.644810	-0	0
4610	.644567	.644324	.644082	.643839	.643597	.643355	.643113	.642870	.642628	.642386	-0	0
4620	.642144	.641902	.641660	.641418	.641177	.640935	.640693	.640452	.640210	.639969	-0	0
4630	.639727	.639486	.639245	.639003	.638762	.638521	.638280	.638039	.637798	.637557	-0	0
4640	.637316	.637076	.636835	.636594	.636354	.636113	.635873	.635632	.635392	.635152	-0	0
4650	.634912	.634672	.634431	.634191	.633951	.633712	.633472	.633232	.632992	.632753	-0	0
4660	.632513	.632274	.632034	.631795	.631555	.631316	.631077	.630838	.630599	.630360	-0	0
4670	.630121	.629882	.629643	.629404	.629165	.628927	.628688	.628450	.628211	.627973	-0	0
4680	.627734	.627496	.627258	.627020	.626782	.626544	.626306	.626068	.625830	.625592	-0	0
4690	.625354	.625117	.624879	.624642	.624404	.624167	.623930	.623692	.623455	.623218	-0	0
4700	.622981	.622744	.622507	.622270	.622033	.621796	.621560	.621323	.621087	.620850	-0	0
4710	.620614	.620377	.620141	.619905	.619668	.619432	.619196	.618960	.618724	.618489	-0	0
4720	.618253	.618017	.617781	.617546	.617310	.617075	.616839	.616604	.616369	.616134	-0	0
4730	.615898	.615663	.615428	.615193	.614958	.614724	.614489	.614254	.614020	.613785	-0	0
4740	.613550	.613316	.613082	.612847	.612613	.612379	.612145	.611911	.611677	.611443	-0	0
4750	.611209	.610975	.610742	.610508	.610274	.610041	.609807	.609574	.609341	.609108	-0	0
4760	.608874	.608641	.608408	.608175	.607942	.607709	.607477	.607244	.607011	.606779	-0	0
4770	.606546	.606314	.606081	.605849	.605617	.605385	.605152	.604920	.604688	.604456	-0	0
4780	.604225	.603993	.603761	.603529	.603298	.603066	.602835	.602604	.602372	.602141	-0	0
4790	.601910	.601679	.601448	.601217	.600986	.600755	.600524	.600293	.600063	.599832	-0	0
4800	.599602	.599371	.599141	.598911	.598680	.598450	.598220	.597990	.597760	.597530	-0	0
4810	.597300	.597070	.596841	.596611	.596382	.596152	.595923	.595693	.595464	.595235	-0	0
4820	.595006	.594777	.594547	.594319	.594090	.593861	.593632	.593403	.593175	.592946	-0	0
4830	.592718	.592489	.592261	.592033	.591805	.591576	.591348	.591120	.590892	.590665	-0	0
4840	.590437	.590209	.589981	.589754	.589526	.589299	.589071	.588844	.588617	.588390	-0	0
4850	.588163	.587936	.587709	.587482	.587255	.587028	.586801	.586575	.586348	.586122	-0	0
4860	.585895	.585669	.585443	.585217	.584990	.584764	.584538	.584312	.584087	.583861	-0	0
4870	.583635	.583409	.583184	.582958	.582733	.582508	.582282	.582057	.581832	.581607	-0	0
4880	.581382	.581157	.580932	.580707	.580482	.580258	.580033	.579808	.579584	.579360	-0	0
4890	.579135	.578911	.578687	.578463	.578239	.578015	.577791	.577567	.577343	.577119	-0	0
4900	.576896	.576672	.576449	.576225	.576002	.575779	.575555	.575332	.575109	.574886	-0	0
4910	.574663	.574441	.574218	.573995	.573772	.573550	.573327	.573105	.572882	.572660	-0	0
4920	.572438	.572216	.571994	.571772	.571550	.571328	.571106	.570884	.570663	.570441	-0	0
4930	.570220	.569998	.569777	.569555	.569334	.569113	.568892	.568671	.568450	.568229	-0	0
4940	.568008	.567787	.567567	.567346	.567126	.566905	.566685	.566465	.566244	.566024	-0	0
4950	.565804	.565584	.565364	.565144	.564924	.564705	.564485	.564265	.564046	.563826	-0	0
4960	.563607	.563388	.563168	.562949	.562730	.562511	.562292	.562073	.561854	.561636	-0	0
4970	.561417	.561198	.560980	.560761	.560543	.560325	.560106	.559888	.559670	.559452	-0	0
4980	.559234	.559016	.558798	.558580	.558363	.558145	.557928	.557710	.557493	.557275	-0	0
4990	.557058	.556841	.556624	.556407	.556190	.555973	.555756	.555539	.555323	.555106	-0	0

$M_{e\lambda}(\lambda T)/M_{emax}(T)$

TABLE 10

Blackbody: Relative Spectral Emittance (Radiant)

λT(μ.°K)	0	1	2	3	4	5	6	7	8	9	m	Δ_2
5000	.554890	.554673	.554457	.554240	.554024	.553808	.553592	.553376	.553160	.552944	-0	0
5010	.552728	.552512	.552297	.552081	.551866	.551650	.551435	.551219	.551004	.550789	-0	0
5020	.550574	.550359	.550144	.549929	.549714	.549500	.549285	.549070	.548856	.548641	-0	0
5030	.548427	.548213	.547998	.547784	.547570	.547356	.547142	.546928	.546714	.546501	-0	0
5040	.546287	.546074	.545860	.545647	.545433	.545220	.545007	.544794	.544580	.544367	-0	0
5050	.544154	.543942	.543729	.543516	.543303	.543091	.542878	.542666	.542454	.542241	-0	0
5060	.542029	.541817	.541605	.541393	.541181	.540969	.540757	.540546	.540334	.540122	-0	0
5070	.539911	.539700	.539488	.539277	.539066	.538855	.538643	.538433	.538222	.538011	-0	0
5080	.537800	.537589	.537379	.537168	.536958	.536747	.536537	.536327	.536116	.535906	-0	0
5090	.535696	.535486	.535276	.535067	.534857	.534647	.534438	.534228	.534019	.533809	-0	0
5100	.533600	.533391	.533181	.532972	.532763	.532554	.532345	.532137	.531928	.531719	-0	0
5110	.531511	.531302	.531094	.530885	.530677	.530469	.530261	.530053	.529845	.529637	-0	0
5120	.529429	.529221	.529013	.528806	.528598	.528391	.528183	.527976	.527768	.527561	-0	0
5130	.527354	.527147	.526940	.526733	.526526	.526320	.526113	.525906	.525700	.525493	-0	0
5140	.525287	.525080	.524874	.524668	.524462	.524256	.524050	.523844	.523638	.523432	-0	0
5150	.523227	.523021	.522815	.522610	.522405	.522199	.521994	.521789	.521584	.521379	-0	0
5160	.521174	.520969	.520764	.520559	.520355	.520150	.519946	.519741	.519537	.519332	-0	0
5170	.519128	.518924	.518720	.518516	.518312	.518108	.517904	.517701	.517497	.517293	-0	0
5180	.517090	.516887	.516683	.516480	.516277	.516074	.515870	.515667	.515465	.515262	-0	0
5190	.515059	.514856	.514654	.514451	.514249	.514046	.513844	.513642	.513439	.513237	-0	0
5200	.513035	.512833	.512631	.512430	.512228	.512026	.511824	.511623	.511421	.511220	-0	0
5210	.511019	.510818	.510616	.510415	.510214	.510013	.509812	.509612	.509411	.509210	-0	0
5220	.509010	.508809	.508609	.508408	.508208	.508008	.507808	.507608	.507407	.507208	-0	0
5230	.507008	.506808	.506608	.506409	.506209	.506009	.505810	.505611	.505411	.505212	-0	0
5240	.505013	.504814	.504615	.504416	.504217	.504018	.503820	.503621	.503423	.503224	-0	0
5250	.503026	.502827	.502629	.502431	.502233	.502035	.501837	.501639	.501441	.501243	-0	0
5260	.501046	.500848	.500650	.500453	.500256	.500058	.499861	.499664	.499467	.499270	-0	0
5270	.499073	.498876	.498679	.498482	.498286	.498089	.497893	.497696	.497500	.497303	-0	0
5280	.497107	.496911	.496715	.496519	.496323	.496127	.495931	.495736	.495540	.495344	-0	0
5290	.495149	.494953	.494758	.494563	.494368	.494172	.493977	.493782	.493587	.493393	-0	0
5300	.493198	.493003	.492808	.492614	.492419	.492225	.492031	.491836	.491642	.491448	-0	0
5310	.491254	.491060	.490866	.490672	.490478	.490285	.490091	.489898	.489704	.489511	-0	0
5320	.489317	.489124	.488931	.488738	.488545	.488352	.488159	.487966	.487773	.487581	-0	0
5330	.487388	.487195	.487003	.486811	.486618	.486426	.486234	.486042	.485850	.485658	-0	0
5340	.485466	.485274	.485082	.484891	.484699	.484508	.484316	.484125	.483933	.483742	-0	0
5350	.483551	.483360	.483169	.482978	.482787	.482596	.482405	.482215	.482024	.481834	-0	0
5360	.481643	.481453	.481263	.481072	.480882	.480692	.480502	.480312	.480122	.479932	-0	0
5370	.479743	.479553	.479363	.479174	.478985	.478795	.478606	.478417	.478227	.478038	-0	0
5380	.477849	.477660	.477472	.477283	.477094	.476905	.476717	.476528	.476340	.476152	-0	0
5390	.475963	.475775	.475587	.475399	.475211	.475023	.474835	.474647	.474459	.474272	-0	0
5400	.474084	.473897	.473709	.473522	.473335	.473147	.472960	.472773	.472586	.472399	-0	0
5410	.472212	.472026	.471839	.471652	.471466	.471279	.471093	.470906	.470720	.470534	-0	0
5420	.470348	.470162	.469976	.469790	.469604	.469418	.469232	.469047	.468861	.468676	-0	0
5430	.468490	.468305	.468120	.467934	.467749	.467564	.467379	.467194	.467009	.466825	-0	0
5440	.466640	.466455	.466271	.466086	.465902	.465717	.465533	.465349	.465165	.464981	-0	0
5450	.464797	.464613	.464429	.464245	.464061	.463878	.463694	.463511	.463327	.463144	-0	0
5460	.462961	.462777	.462594	.462411	.462228	.462045	.461862	.461679	.461497	.461314	-0	0
5470	.461131	.460949	.460767	.460584	.460402	.460220	.460037	.459855	.459673	.459491	-0	0
5480	.459310	.459128	.458946	.458764	.458583	.458401	.458220	.458038	.457857	.457676	-0	0
5490	.457495	.457314	.457133	.456952	.456771	.456590	.456409	.456228	.456048	.455867	-0	0

TABLE 10

Blackbody: Relative Spectral Emittance (Radiant)

λT(μ.°K)	0	1	2	3	4	5	6	7	8	9	m	Δ_2
5500	.455687	.455506	.455326	.455146	.454966	.454786	.454606	.454426	.454246	.454066	-0	0
5510	.453886	.453706	.453527	.453347	.453168	.452988	.452809	.452630	.452451	.452271	-0	0
5520	.452092	.451913	.451734	.451556	.451377	.451198	.451019	.450841	.450662	.450484	-0	0
5530	.450306	.450127	.449949	.449771	.449593	.449415	.449237	.449059	.448881	.448704	-0	0
5540	.448526	.448348	.448171	.447993	.447816	.447639	.447461	.447284	.447107	.446930	-0	0
5550	.446753	.446576	.446400	.446223	.446046	.445870	.445693	.445517	.445340	.445164	-0	0
5560	.444988	.444811	.444635	.444459	.444283	.444107	.443931	.443756	.443580	.443404	-0	0
5570	.443229	.443053	.442878	.442703	.442527	.442352	.442177	.442002	.441827	.441652	-0	0
5580	.441477	.441302	.441127	.440953	.440778	.440604	.440429	.440255	.440081	.439906	-0	0
5590	.439732	.439558	.439384	.439210	.439036	.438862	.438689	.438515	.438341	.438168	-0	0
5600	.437994	.437821	.437647	.437474	.437301	.437128	.436955	.436782	.436609	.436436	-0	0
5610	.436263	.436090	.435918	.435745	.435573	.435400	.435228	.435056	.434883	.434711	-0	0
5620	.434539	.434367	.434195	.434023	.433851	.433680	.433508	.433336	.433165	.432993	-0	0
5630	.432822	.432650	.432479	.432308	.432137	.431966	.431795	.431624	.431453	.431282	-0	0
5640	.431111	.430941	.430770	.430600	.430429	.430259	.430088	.429918	.429748	.429578	-0	0
5650	.429408	.429238	.429068	.428898	.428728	.428559	.428389	.428219	.428050	.427880	-0	0
5660	.427711	.427542	.427373	.427203	.427034	.426865	.426696	.426527	.426359	.426190	-0	0
5670	.426021	.425853	.425684	.425516	.425347	.425179	.425011	.424842	.424674	.424506	-0	0
5680	.424338	.424170	.424002	.423834	.423667	.423499	.423331	.423164	.422996	.422829	-0	0
5690	.422662	.422494	.422327	.422160	.421993	.421826	.421659	.421492	.421326	.421159	-0	0
5700	.420992	.420826	.420659	.420493	.420326	.420160	.419994	.419827	.419661	.419495	-0	0
5710	.419329	.419163	.418998	.418832	.418666	.418500	.418335	.418169	.418004	.417839	-0	0
5720	.417673	.417508	.417343	.417178	.417013	.416848	.416683	.416518	.416353	.416188	-0	0
5730	.416024	.415859	.415695	.415530	.415366	.415202	.415037	.414873	.414709	.414545	-0	0
5740	.414381	.414217	.414053	.413890	.413726	.413562	.413399	.413235	.413072	.412908	-0	0
5750	.412745	.412582	.412419	.412256	.412093	.411930	.411767	.411604	.411441	.411278	-0	0
5760	.411116	.410953	.410791	.410628	.410466	.410303	.410141	.409979	.409817	.409655	-0	0
5770	.409493	.409331	.409169	.409007	.408846	.408684	.408522	.408361	.408200	.408038	-0	0
5780	.407877	.407716	.407554	.407393	.407232	.407071	.406910	.406749	.406589	.406428	-0	0
5790	.406267	.406107	.405946	.405786	.405625	.405465	.405305	.405145	.404984	.404824	-0	0
5800	.404664	.404504	.404345	.404185	.404025	.403865	.403706	.403546	.403387	.403227	-0	0
5810	.403068	.402909	.402750	.402590	.402431	.402272	.402113	.401954	.401796	.401637	-0	0
5820	.401478	.401320	.401161	.401002	.400844	.400686	.400527	.400369	.400211	.400053	-0	0
5830	.399895	.399737	.399579	.399421	.399263	.399106	.398948	.398790	.398633	.398475	-0	0
5840	.398318	.398161	.398003	.397846	.397689	.397532	.397375	.397218	.397061	.396904	-0	0
5850	.396748	.396591	.396434	.396278	.396121	.395965	.395809	.395652	.395496	.395340	-0	0
5860	.395184	.395028	.394872	.394716	.394560	.394404	.394249	.394093	.393937	.393782	-0	0
5870	.393626	.393471	.393316	.393161	.393005	.392850	.392695	.392540	.392385	.392230	-0	0
5880	.392075	.391921	.391766	.391611	.391457	.391302	.391148	.390994	.390839	.390685	-0	0
5890	.390531	.390377	.390223	.390069	.389915	.389761	.389607	.389454	.389300	.389146	-0	0
5900	.388993	.388839	.388686	.388533	.388379	.388226	.388073	.387920	.387767	.387614	-0	0
5910	.387461	.387308	.387155	.387003	.386850	.386697	.386545	.386392	.386240	.386088	-0	0
5920	.385935	.385783	.385631	.385479	.385327	.385175	.385023	.384871	.384720	.384568	-0	0
5930	.384416	.384265	.384113	.383962	.383810	.383659	.383508	.383357	.383206	.383055	-0	0
5940	.382904	.382753	.382602	.382451	.382300	.382149	.381999	.381848	.381698	.381547	-0	0
5950	.381397	.381247	.381096	.380946	.380796	.380646	.380496	.380346	.380196	.380046	-0	0
5960	.379897	.379747	.379597	.379448	.379298	.379149	.379000	.378850	.378701	.378552	-0	0
5970	.378403	.378254	.378105	.377956	.377807	.377658	.377509	.377361	.377212	.377063	-0	0
5980	.376915	.376767	.376618	.376470	.376322	.376173	.376025	.375877	.375729	.375581	-0	0
5990	.375433	.375286	.375138	.374990	.374842	.374695	.374547	.374400	.374253	.374105	-0	0

TABLE 10

Blackbody: Relative Spectral Emittance (Radiant)

$\lambda T(\mu.°K)$	0	1	2	3	4	5	6	7	8	9	m	Δ_2
6000	.373958	.373811	.373664	.373517	.373370	.373223	.373076	.372929	.372782	.372635	-0	0
6010	.372489	.372342	.372196	.372049	.371903	.371756	.371610	.371464	.371318	.371172	-0	0
6020	.371026	.370880	.370734	.370588	.370442	.370296	.370151	.370005	.369860	.369714	-0	0
6030	.369569	.369423	.369278	.369133	.368988	.368843	.368698	.368553	.368408	.368263	-0	0
6040	.368118	.367973	.367828	.367684	.367539	.367395	.367250	.367106	.366962	.366817	-0	0
6050	.366673	.366529	.366385	.366241	.366097	.365953	.365809	.365665	.365522	.365378	-0	0
6060	.365234	.365091	.364947	.364804	.364661	.364517	.364374	.364231	.364088	.363945	-0	0
6070	.363802	.363659	.363516	.363373	.363230	.363088	.362945	.362803	.362660	.362518	-0	0
6080	.362375	.362233	.362091	.361948	.361806	.361664	.361522	.361380	.361238	.361096	-0	0
6090	.360955	.360813	.360671	.360530	.360388	.360247	.360105	.359964	.359822	.359681	-0	0
6100	.359540	.359399	.359258	.359117	.358976	.358835	.358694	.358553	.358413	.358272	-0	0
6110	.358131	.357991	.357850	.357710	.357569	.357429	.357289	.357149	.357009	.356869	-0	0
6120	.356729	.356589	.356449	.356309	.356169	.356029	.355890	.355750	.355611	.355471	-0	0
6130	.355332	.355192	.355053	.354914	.354775	.354635	.354496	.354357	.354218	.354080	-0	0
6140	.353941	.353802	.353663	.353525	.353386	.353248	.353109	.352971	.352832	.352694	-0	0
6150	.352556	.352418	.352279	.352141	.352003	.351865	.351727	.351590	.351452	.351314	-0	0
6160	.351177	.351039	.350901	.350764	.350626	.350489	.350352	.350215	.350077	.349940	-0	0
6170	.349803	.349666	.349529	.349392	.349255	.349119	.348982	.348845	.348709	.348572	-0	0
6180	.348436	.348299	.348163	.348026	.347890	.347754	.347618	.347482	.347346	.347210	-0	0
6190	.347074	.346938	.346802	.346666	.346531	.346395	.346260	.346124	.345989	.345853	-0	0
6200	.345718	.345583	.345447	.345312	.345177	.345042	.344907	.344772	.344637	.344502	-0	0
6210	.344368	.344233	.344098	.343964	.343829	.343695	.343560	.343426	.343292	.343157	-0	0
6220	.343023	.342889	.342755	.342621	.342487	.342353	.342219	.342085	.341952	.341818	-0	0
6230	.341684	.341551	.341417	.341284	.341150	.341017	.340884	.340750	.340617	.340484	-0	0
6240	.340351	.340218	.340085	.339952	.339819	.339687	.339554	.339421	.339289	.339156	-0	0
6250	.339024	.338891	.338759	.338627	.338494	.338362	.338230	.338098	.337966	.337834	-0	0
6260	.337702	.337570	.337438	.337306	.337175	.337043	.336911	.336780	.336648	.336517	-0	0
6270	.336386	.336254	.336123	.335992	.335861	.335730	.335599	.335468	.335337	.335206	-0	0
6280	.335075	.334944	.334814	.334683	.334552	.334422	.334291	.334161	.334031	.333900	-0	0
6290	.333770	.333640	.333510	.333380	.333250	.333120	.332990	.332860	.332730	.332600	-0	0
6300	.332471	.332341	.332211	.332082	.331952	.331823	.331694	.331564	.331435	.331306	-0	0
6310	.331177	.331048	.330919	.330790	.330661	.330532	.330403	.330274	.330146	.330017	-0	0
6320	.329888	.329760	.329631	.329503	.329374	.329246	.329118	.328990	.328862	.328733	-0	0
6330	.328605	.328477	.328349	.328222	.328094	.327966	.327838	.327711	.327583	.327455	-0	0
6340	.327328	.327200	.327073	.326946	.326818	.326691	.326564	.326437	.326310	.326183	-0	0
6350	.326056	.325929	.325802	.325675	.325549	.325422	.325295	.325169	.325042	.324916	-0	0
6360	.324789	.324663	.324537	.324410	.324284	.324158	.324032	.323906	.323780	.323654	-0	0
6370	.323528	.323402	.323277	.323151	.323025	.322900	.322774	.322649	.322523	.322398	-0	0
6380	.322273	.322147	.322022	.321897	.321772	.321647	.321522	.321397	.321272	.321147	-0	0
6390	.321022	.320897	.320773	.320648	.320523	.320399	.320274	.320150	.320026	.319901	-0	0
6400	.319777	.319653	.319529	.319405	.319281	.319157	.319033	.318909	.318785	.318661	-0	0
6410	.318537	.318414	.318290	.318167	.318043	.317920	.317796	.317673	.317549	.317426	-0	0
6420	.317303	.317180	.317057	.316934	.316811	.316688	.316565	.316442	.316319	.316197	-0	0
6430	.316074	.315951	.315829	.315706	.315584	.315461	.315339	.315217	.315094	.314972	-0	0
6440	.314850	.314728	.314606	.314484	.314362	.314240	.314118	.313996	.313875	.313753	-0	0
6450	.313631	.313510	.313388	.313267	.313145	.313024	.312903	.312781	.312660	.312539	-0	0
6460	.312418	.312297	.312176	.312055	.311934	.311813	.311692	.311572	.311451	.311330	-0	0
6470	.311210	.311089	.310969	.310848	.310728	.310608	.310487	.310367	.310247	.310127	-0	0
6480	.310007	.309887	.309767	.309647	.309527	.309407	.309288	.309168	.309048	.308929	-0	0
6490	.308809	.308689	.308570	.308451	.308331	.308212	.308093	.307974	.307854	.307735	-0	0

TABLE 10

Blackbody: Relative Spectral Emittance (Radiant)

$\lambda T (\mu.°K)$	0	1	2	3	4	5	6	7	8	9	m	Δ_2
6500	.307616	.307497	.307378	.307259	.307141	.307022	.306903	.306784	.306666	.306547	-0	0
6510	.306429	.306310	.306192	.306073	.305955	.305837	.305719	.305600	.305482	.305364	-0	0
6520	.305246	.305128	.305010	.304892	.304775	.304657	.304539	.304421	.304304	.304186	-0	0
6530	.304069	.303951	.303834	.303717	.303599	.303482	.303365	.303248	.303130	.303013	-0	0
6540	.302896	.302779	.302663	.302546	.302429	.302312	.302195	.302079	.301962	.301846	-0	0
6550	.301729	.301613	.301496	.301380	.301264	.301147	.301031	.300915	.300799	.300683	-0	0
6560	.300567	.300451	.300335	.300219	.300103	.299987	.299872	.299756	.299640	.299525	-0	0
6570	.299409	.299294	.299179	.299063	.298948	.298833	.298717	.298602	.298487	.298372	-0	0
6580	.298257	.298142	.298027	.297912	.297798	.297683	.297568	.297453	.297339	.297224	-0	0
6590	.297110	.296995	.296881	.296766	.296652	.296538	.296424	.296309	.296195	.296081	-0	0
6600	.295967	.295853	.295739	.295625	.295512	.295398	.295284	.295170	.295057	.294943	-0	0
6610	.294830	.294716	.294603	.294489	.294376	.294263	.294150	.294036	.293923	.293810	-0	0
6620	.293697	.293584	.293471	.293358	.293245	.293133	.293020	.292907	.292794	.292682	-0	0
6630	.292569	.292457	.292344	.292232	.292120	.292007	.291895	.291783	.291671	.291558	-0	0
6640	.291446	.291334	.291222	.291110	.290999	.290887	.290775	.290663	.290551	.290440	-0	0
6650	.290328	.290217	.290105	.289994	.289882	.289771	.289660	.289548	.289437	.289326	-0	0
6660	.289215	.289104	.288993	.288882	.288771	.288660	.288549	.288438	.288328	.288217	-0	0
6670	.288106	.287996	.287885	.287775	.287664	.287554	.287444	.287333	.287223	.287113	-0	0
6680	.287003	.286893	.286783	.286673	.286563	.286453	.286343	.286233	.286123	.286013	-0	0
6690	.285904	.285794	.285684	.285575	.285465	.285356	.285247	.285137	.285028	.284919	-0	0
6700	.284809	.284700	.284591	.284482	.284373	.284264	.284155	.284046	.283937	.283829	-0	0
6710	.283720	.283611	.283503	.283394	.283285	.283177	.283068	.282960	.282852	.282743	-0	0
6720	.282635	.282527	.282419	.282311	.282202	.282094	.281986	.281878	.281771	.281663	-0	0
6730	.281555	.281447	.281339	.281232	.281124	.281017	.280909	.280802	.280694	.280587	-0	0
6740	.280479	.280372	.280265	.280158	.280050	.279943	.279836	.279729	.279622	.279515	-0	0
6750	.279409	.279302	.279195	.279088	.278981	.278875	.278768	.278662	.278555	.278449	-0	0
6760	.278342	.278236	.278130	.278023	.277917	.277811	.277705	.277599	.277493	.277387	-0	0
6770	.277281	.277175	.277069	.276963	.276857	.276751	.276646	.276540	.276434	.276329	-0	0
6780	.276223	.276118	.276013	.275907	.275802	.275697	.275591	.275486	.275381	.275276	-0	0
6790	.275171	.275066	.274961	.274856	.274751	.274646	.274542	.274437	.274332	.274227	-0	0
6800	.274123	.274018	.273914	.273809	.273705	.273601	.273496	.273392	.273288	.273184	-0	0
6810	.273079	.272975	.272871	.272767	.272663	.272559	.272455	.272352	.272248	.272144	-0	0
6820	.272040	.271937	.271833	.271730	.271626	.271523	.271419	.271316	.271212	.271109	-0	0
6830	.271006	.270903	.270799	.270696	.270593	.270490	.270387	.270284	.270181	.270079	-0	0
6840	.269976	.269873	.269770	.269668	.269565	.269462	.269360	.269257	.269155	.269052	-0	0
6850	.268950	.268848	.268745	.268643	.268541	.268439	.268337	.268235	.268133	.268031	-0	0
6860	.267929	.267827	.267725	.267623	.267522	.267420	.267318	.267217	.267115	.267014	-0	0
6870	.266912	.266811	.266709	.266608	.266507	.266405	.266304	.266203	.266102	.266001	-0	0
6880	.265900	.265799	.265698	.265597	.265496	.265395	.265294	.265193	.265093	.264992	-0	0
6890	.264892	.264791	.264690	.264590	.264489	.264389	.264289	.264188	.264088	.263988	-0	0
6900	.263888	.263788	.263688	.263587	.263487	.263388	.263288	.263188	.263088	.262988	-0	0
6910	.262888	.262789	.262689	.262589	.262490	.262390	.262291	.262191	.262092	.261993	-0	0
6920	.261893	.261794	.261695	.261596	.261496	.261397	.261298	.261199	.261100	.261001	-0	0
6930	.260902	.260804	.260705	.260606	.260507	.260409	.260310	.260211	.260113	.260014	-0	0
6940	.259916	.259817	.259719	.259621	.259522	.259424	.259326	.259228	.259130	.259032	-0	0
6950	.258934	.258836	.258738	.258640	.258542	.258444	.258346	.258248	.258151	.258053	-0	0
6960	.257955	.257858	.257760	.257663	.257565	.257468	.257371	.257273	.257176	.257079	-0	0
6970	.256982	.256884	.256787	.256690	.256593	.256496	.256399	.256302	.256206	.256109	-0	0
6980	.256012	.255915	.255819	.255722	.255625	.255529	.255432	.255336	.255239	.255143	-0	0
6990	.255046	.254950	.254854	.254758	.254661	.254565	.254469	.254373	.254277	.254181	-0	0

$M_{e\lambda}(\lambda T)/M_{emax}(T)$

TABLE 10

Blackbody: Relative Spectral Emittance (Radiant)

$\lambda T(\mu\cdot{}^\circ K)$	0	1	2	3	4	5	6	7	8	9	m	Δ_2
7000	.254085	.253989	.253893	.253798	.253702	.253606	.253510	.253415	.253319	.253223	-0	0
7010	.253128	.253032	.252937	.252842	.252746	.252651	.252556	.252460	.252365	.252270	-0	0
7020	.252175	.252080	.251985	.251890	.251795	.251700	.251605	.251510	.251415	.251321	-0	0
7030	.251226	.251131	.251037	.250942	.250847	.250753	.250658	.250564	.250470	.250375	-0	0
7040	.250281	.250187	.250092	.249998	.249904	.249810	.249716	.249622	.249528	.249434	-0	0
7050	.249340	.249246	.249152	.249059	.248965	.248871	.248778	.248684	.248590	.248497	-0	0
7060	.248403	.248310	.248216	.248123	.248030	.247936	.247843	.247750	.247657	.247564	-0	0
7070	.247471	.247378	.247285	.247192	.247099	.247006	.246913	.246820	.246727	.246635	-0	0
7080	.246542	.246449	.246357	.246264	.246171	.246079	.245987	.245894	.245802	.245709	-0	0
7090	.245617	.245525	.245433	.245340	.245248	.245156	.245064	.244972	.244880	.244788	-0	0
7100	.244696	.244604	.244513	.244421	.244329	.244237	.244146	.244054	.243963	.243871	-0	0
7110	.243780	.243688	.243597	.243505	.243414	.243323	.243231	.243140	.243049	.242958	-0	0
7120	.242867	.242776	.242685	.242594	.242503	.242412	.242321	.242230	.242139	.242048	-0	0
7130	.241958	.241867	.241776	.241686	.241595	.241505	.241414	.241324	.241233	.241143	-0	0
7140	.241053	.240962	.240872	.240782	.240692	.240602	.240512	.240422	.240331	.240241	-0	0
7150	.240152	.240062	.239972	.239882	.239792	.239702	.239613	.239523	.239433	.239344	-0	0
7160	.239254	.239165	.239075	.238986	.238896	.238807	.238718	.238628	.238539	.238450	-0	0
7170	.238361	.238272	.238183	.238094	.238005	.237916	.237827	.237738	.237649	.237560	-0	0
7180	.237471	.237383	.237294	.237205	.237117	.237028	.236939	.236851	.236762	.236674	-0	0
7190	.236586	.236497	.236409	.236321	.236232	.236144	.236056	.235968	.235880	.235792	-0	0
7200	.235704	.235616	.235528	.235440	.235352	.235264	.235176	.235088	.235001	.234913	-0	0
7210	.234825	.234738	.234650	.234563	.234475	.234388	.234300	.234213	.234126	.234038	-0	0
7220	.233951	.233864	.233777	.233689	.233602	.233515	.233428	.233341	.233254	.233167	-0	0
7230	.233080	.232994	.232907	.232820	.232733	.232646	.232560	.232473	.232387	.232300	-0	0
7240	.232213	.232127	.232041	.231954	.231868	.231781	.231695	.231609	.231523	.231436	-0	0
7250	.231350	.231264	.231178	.231092	.231006	.230920	.230834	.230748	.230662	.230577	-0	0
7260	.230491	.230405	.230319	.230234	.230148	.230062	.229977	.229891	.229806	.229720	-0	0
7270	.229635	.229550	.229464	.229379	.229294	.229208	.229123	.229038	.228953	.228868	-0	0
7280	.228783	.228698	.228613	.228528	.228443	.228358	.228273	.228189	.228104	.228019	-0	0
7290	.227934	.227850	.227765	.227681	.227596	.227511	.227427	.227343	.227258	.227174	-0	0
7300	.227090	.227005	.226921	.226837	.226753	.226668	.226584	.226500	.226416	.226332	-0	0
7310	.226248	.226164	.226080	.225997	.225913	.225829	.225745	.225662	.225578	.225494	-0	0
7320	.225411	.225327	.225244	.225160	.225077	.224993	.224910	.224826	.224743	.224660	-0	0
7330	.224577	.224493	.224410	.224327	.224244	.224161	.224078	.223995	.223912	.223829	-0	0
7340	.223746	.223663	.223581	.223498	.223415	.223332	.223250	.223167	.223084	.223002	-0	0
7350	.222919	.222837	.222754	.222672	.222590	.222507	.222425	.222343	.222260	.222178	-0	0
7360	.222096	.222014	.221932	.221850	.221768	.221686	.221604	.221522	.221440	.221358	-0	0
7370	.221276	.221194	.221113	.221031	.220949	.220867	.220786	.220704	.220623	.220541	-0	0
7380	.220460	.220378	.220297	.220216	.220134	.220053	.219972	.219890	.219809	.219728	-0	0
7390	.219647	.219566	.219485	.219404	.219323	.219242	.219161	.219080	.218999	.218918	-0	0
7400	.218838	.218757	.218676	.218595	.218515	.218434	.218354	.218273	.218193	.218112	-0	0
7410	.218032	.217951	.217871	.217790	.217710	.217630	.217550	.217470	.217389	.217309	-0	0
7420	.217229	.217149	.217069	.216989	.216909	.216829	.216749	.216669	.216590	.216510	-0	0
7430	.216430	.216350	.216271	.216191	.216111	.216032	.215952	.215873	.215793	.215714	-0	0
7440	.215634	.215555	.215476	.215396	.215317	.215238	.215159	.215080	.215000	.214921	-0	0
7450	.214842	.214763	.214684	.214605	.214526	.214447	.214369	.214290	.214211	.214132	-0	0
7460	.214053	.213975	.213896	.213817	.213739	.213660	.213582	.213503	.213425	.213346	-0	0
7470	.213268	.213190	.213111	.213033	.212955	.212876	.212798	.212720	.212642	.212564	-0	0
7480	.212486	.212408	.212330	.212252	.212174	.212096	.212018	.211940	.211862	.211785	-0	0
7490	.211707	.211629	.211552	.211474	.211396	.211319	.211241	.211164	.211086	.211009	-0	0

TABLE 10

Blackbody: Relative Spectral Emittance (Radiant)

$\lambda T(\mu .^\circ K)$	0	1	2	3	4	5	6	7	8	9	m	Δ_2
7500	.210932	.210854	.210777	.210700	.210622	.210545	.210468	.210391	.210314	.210236	-0	0
7510	.210159	.210082	.210005	.209928	.209851	.209775	.209698	.209621	.209544	.209467	-0	0
7520	.209390	.209314	.209237	.209160	.209084	.209007	.208931	.208854	.208778	.208701	-0	0
7530	.208625	.208548	.208472	.208396	.208320	.208243	.208167	.208091	.208015	.207939	-0	0
7540	.207863	.207786	.207710	.207634	.207559	.207483	.207407	.207331	.207255	.207179	-0	0
7550	.207103	.207028	.206952	.206876	.206801	.206725	.206650	.206574	.206499	.206423	-0	0
7560	.206348	.206272	.206197	.206121	.206046	.205971	.205896	.205820	.205745	.205670	-0	0
7570	.205595	.205520	.205445	.205370	.205295	.205220	.205145	.205070	.204995	.204920	-0	0
7580	.204845	.204771	.204696	.204621	.204547	.204472	.204397	.204323	.204248	.204174	-0	0
7590	.204099	.204025	.203950	.203876	.203802	.203727	.203653	.203579	.203505	.203430	-0	0
7600	.203356	.203282	.203208	.203134	.203060	.202986	.202912	.202838	.202764	.202690	-0	0
7610	.202616	.202542	.202469	.202395	.202321	.202247	.202174	.202100	.202027	.201953	-0	0
7620	.201879	.201806	.201732	.201659	.201586	.201512	.201439	.201366	.201292	.201219	-0	0
7630	.201146	.201073	.200999	.200926	.200853	.200780	.200707	.200634	.200561	.200488	-0	0
7640	.200415	.200342	.200270	.200197	.200124	.200051	.199978	.199906	.199833	.199760	-0	0
7650	.199688	.199615	.199543	.199470	.199398	.199325	.199253	.199180	.199108	.199036	-0	0
7660	.198963	.198891	.198819	.198747	.198675	.198602	.198530	.198458	.198386	.198314	-0	0
7670	.198242	.198170	.198098	.198026	.197955	.197883	.197811	.197739	.197667	.197596	-0	0
7680	.197524	.197452	.197381	.197309	.197238	.197166	.197095	.197023	.196952	.196880	-0	0
7690	.196809	.196737	.196666	.196595	.196524	.196452	.196381	.196310	.196239	.196168	-0	0
7700	.196097	.196026	.195955	.195884	.195813	.195742	.195671	.195600	.195529	.195458	-0	0
7710	.195388	.195317	.195246	.195175	.195105	.195034	.194963	.194893	.194822	.194752	-0	0
7720	.194681	.194611	.194541	.194470	.194400	.194329	.194259	.194189	.194119	.194048	-0	0
7730	.193978	.193908	.193838	.193768	.193698	.193628	.193558	.193488	.193418	.193348	-0	0
7740	.193278	.193208	.193138	.193069	.192999	.192929	.192859	.192790	.192720	.192650	-0	0
7750	.192581	.192511	.192442	.192372	.192303	.192233	.192164	.192095	.192025	.191956	-0	0
7760	.191887	.191817	.191748	.191679	.191610	.191541	.191472	.191402	.191333	.191264	-0	0
7770	.191195	.191126	.191057	.190989	.190920	.190851	.190782	.190713	.190644	.190576	-0	0
7780	.190507	.190438	.190370	.190301	.190232	.190164	.190095	.190027	.189958	.189890	-0	0
7790	.189821	.189753	.189685	.189616	.189548	.189480	.189412	.189343	.189275	.189207	-0	0
7800	.189139	.189071	.189003	.188935	.188867	.188799	.188731	.188663	.188595	.188527	-0	0
7810	.188459	.188391	.188324	.188256	.188188	.188120	.188053	.187985	.187917	.187850	-0	0
7820	.187782	.187715	.187647	.187580	.187512	.187445	.187378	.187310	.187243	.187176	-0	0
7830	.187108	.187041	.186974	.186907	.186840	.186772	.186705	.186638	.186571	.186504	-0	0
7840	.186437	.186370	.186303	.186236	.186170	.186103	.186036	.185969	.185902	.185836	-0	0
7850	.185769	.185702	.185636	.185569	.185502	.185436	.185369	.185303	.185236	.185170	-0	0
7860	.185103	.185037	.184971	.184904	.184838	.184772	.184705	.184639	.184573	.184507	-0	0
7870	.184441	.184375	.184308	.184242	.184176	.184110	.184044	.183978	.183913	.183847	-0	0
7880	.183781	.183715	.183649	.183583	.183518	.183452	.183386	.183320	.183255	.183189	-0	0
7890	.183124	.183058	.182993	.182927	.182862	.182796	.182731	.182665	.182600	.182535	-0	0
7900	.182469	.182404	.182339	.182273	.182208	.182143	.182078	.182013	.181948	.181883	-0	0
7910	.181818	.181753	.181688	.181623	.181558	.181493	.181428	.181363	.181298	.181234	-0	0
7920	.181169	.181104	.181039	.180975	.180910	.180845	.180781	.180716	.180652	.180587	-0	0
7930	.180523	.180458	.180394	.180329	.180265	.180201	.180136	.180072	.180008	.179943	-0	0
7940	.179879	.179815	.179751	.179687	.179623	.179558	.179494	.179430	.179366	.179302	-0	0
7950	.179238	.179174	.179111	.179047	.178983	.178919	.178855	.178791	.178728	.178664	-0	0
7960	.178600	.178537	.178473	.178409	.178346	.178282	.178219	.178155	.178092	.178028	-0	0
7970	.177965	.177902	.177838	.177775	.177712	.177648	.177585	.177522	.177459	.177395	-0	0
7980	.177332	.177269	.177206	.177143	.177080	.177017	.176954	.176891	.176828	.176765	-0	0
7990	.176702	.176639	.176577	.176514	.176451	.176388	.176325	.176263	.176200	.176137	-0	0

$M_{e\lambda}(\lambda T)/M_{emax}(T)$

TABLE 10

Blackbody: Relative Spectral Emittance (Radiant)

λT(μ.°K)	0	1	2	3	4	5	6	7	8	9	m	Δ_2
8000	.176075	.176012	.175950	.175887	.175825	.175762	.175700	.175637	.175575	.175512	-0	0
8010	.175450	.175388	.175325	.175263	.175201	.175139	.175076	.175014	.174952	.174890	-0	0
8020	.174828	.174766	.174704	.174642	.174580	.174518	.174456	.174394	.174332	.174270	-0	0
8030	.174208	.174146	.174085	.174023	.173961	.173899	.173838	.173776	.173714	.173653	-0	0
8040	.173591	.173530	.173468	.173407	.173345	.173284	.173222	.173161	.173100	.173038	-0	0
8050	.172977	.172916	.172854	.172793	.172732	.172671	.172609	.172548	.172487	.172426	-0	0
8060	.172365	.172304	.172243	.172182	.172121	.172060	.171999	.171938	.171877	.171817	-0	0
8070	.171756	.171695	.171634	.171574	.171513	.171452	.171391	.171331	.171270	.171210	-0	0
8080	.171149	.171089	.171028	.170968	.170907	.170847	.170786	.170726	.170665	.170605	-0	0
8090	.170545	.170485	.170424	.170364	.170304	.170244	.170184	.170123	.170063	.170003	-0	0
8100	.169943	.169883	.169823	.169763	.169703	.169643	.169583	.169524	.169464	.169404	-0	0
8110	.169344	.169284	.169224	.169165	.169105	.169045	.168986	.168926	.168866	.168807	-0	0
8120	.168747	.168688	.168628	.168569	.168509	.168450	.168391	.168331	.168272	.168212	-0	0
8130	.168153	.168094	.168035	.167975	.167916	.167857	.167798	.167739	.167680	.167621	-0	0
8140	.167561	.167502	.167443	.167384	.167325	.167267	.167208	.167149	.167090	.167031	-0	0
8150	.166972	.166913	.166855	.166796	.166737	.166678	.166620	.166561	.166503	.166444	-0	0
8160	.166385	.166327	.166268	.166210	.166151	.166093	.166034	.165976	.165918	.165859	-0	0
8170	.165801	.165743	.165684	.165626	.165568	.165510	.165452	.165393	.165335	.165277	-0	0
8180	.165219	.165161	.165103	.165045	.164987	.164929	.164871	.164813	.164755	.164697	-0	0
8190	.164640	.164582	.164524	.164466	.164408	.164351	.164293	.164235	.164178	.164120	-0	0
8200	.164063	.164005	.163947	.163890	.163832	.163775	.163717	.163660	.163603	.163545	-0	0
8210	.163488	.163430	.163373	.163316	.163259	.163201	.163144	.163087	.163030	.162973	-0	0
8220	.162915	.162858	.162801	.162744	.162687	.162630	.162573	.162516	.162459	.162402	-0	0
8230	.162346	.162289	.162232	.162175	.162118	.162061	.162005	.161948	.161891	.161835	-0	0
8240	.161778	.161721	.161665	.161608	.161552	.161495	.161439	.161382	.161326	.161269	-0	0
8250	.161213	.161156	.161100	.161044	.160987	.160931	.160875	.160818	.160762	.160706	-0	0
8260	.160650	.160594	.160538	.160481	.160425	.160369	.160313	.160257	.160201	.160145	-0	0
8270	.160089	.160033	.159977	.159922	.159866	.159810	.159754	.159698	.159643	.159587	-0	0
8280	.159531	.159475	.159420	.159364	.159308	.159253	.159197	.159142	.159086	.159031	-0	0
8290	.158975	.158920	.158864	.158809	.158753	.158698	.158643	.158587	.158532	.158477	-0	0
8300	.158422	.158366	.158311	.158256	.158201	.158146	.158090	.158035	.157980	.157925	-0	0
8310	.157870	.157815	.157760	.157705	.157650	.157595	.157540	.157486	.157431	.157376	-0	0
8320	.157321	.157266	.157212	.157157	.157102	.157047	.156993	.156938	.156884	.156829	-0	0
8330	.156774	.156720	.156665	.156611	.156556	.156502	.156447	.156393	.156339	.156284	-0	0
8340	.156230	.156175	.156121	.156067	.156013	.155958	.155904	.155850	.155796	.155742	-0	0
8350	.155687	.155633	.155579	.155525	.155471	.155417	.155363	.155309	.155255	.155201	-0	0
8360	.155147	.155094	.155040	.154986	.154932	.154878	.154824	.154771	.154717	.154663	-0	0
8370	.154610	.154556	.154502	.154449	.154395	.154342	.154288	.154234	.154181	.154127	-0	0
8380	.154074	.154021	.153967	.153914	.153860	.153807	.153754	.153700	.153647	.153594	-0	0
8390	.153541	.153487	.153434	.153381	.153328	.153275	.153222	.153169	.153115	.153062	-0	0
8400	.153009	.152956	.152903	.152850	.152798	.152745	.152692	.152639	.152586	.152533	-0	0
8410	.152480	.152428	.152375	.152322	.152269	.152217	.152164	.152111	.152059	.152006	-0	0
8420	.151954	.151901	.151848	.151796	.151743	.151691	.151638	.151586	.151534	.151481	-0	0
8430	.151429	.151376	.151324	.151272	.151220	.151167	.151115	.151063	.151011	.150958	-0	0
8440	.150906	.150854	.150802	.150750	.150698	.150646	.150594	.150542	.150490	.150438	-0	0
8450	.150386	.150334	.150282	.150230	.150178	.150127	.150075	.150023	.149971	.149919	-0	0
8460	.149868	.149816	.149764	.149713	.149661	.149609	.149558	.149506	.149455	.149403	-0	0
8470	.149352	.149300	.149249	.149197	.149146	.149094	.149043	.148992	.148940	.148889	-0	0
8480	.148838	.148786	.148735	.148684	.148633	.148581	.148530	.148479	.148428	.148377	-0	0
8490	.148326	.148275	.148224	.148172	.148121	.148070	.148019	.147969	.147918	.147867	-0	0

TABLE 10

Blackbody: Relative Spectral Emittance (Radiant)

$\lambda T(\mu.°K)$	0	1	2	3	4	5	6	7	8	9	m	Δ_2
8500	.147816	.147765	.147714	.147663	.147612	.147562	.147511	.147460	.147409	.147359	-0	0
8510	.147308	.147257	.147207	.147156	.147106	.147055	.147004	.146954	.146903	.146853	-0	0
8520	.146802	.146752	.146702	.146651	.146601	.146550	.146500	.146450	.146399	.146349	-0	0
8530	.146299	.146249	.146198	.146148	.146098	.146048	.145998	.145948	.145898	.145847	-0	0
8540	.145797	.145747	.145697	.145647	.145597	.145547	.145497	.145448	.145398	.145348	-0	0
8550	.145298	.145248	.145198	.145148	.145099	.145049	.144999	.144950	.144900	.144850	-0	0
8560	.144801	.144751	.144701	.144652	.144602	.144553	.144503	.144454	.144404	.144355	-0	0
8570	.144305	.144256	.144206	.144157	.144108	.144058	.144009	.143960	.143910	.143861	-0	0
8580	.143812	.143763	.143713	.143664	.143615	.143566	.143517	.143468	.143418	.143369	-0	0
8590	.143320	.143271	.143222	.143173	.143124	.143075	.143027	.142978	.142929	.142880	-0	0
8600	.142831	.142782	.142733	.142685	.142636	.142587	.142538	.142490	.142441	.142392	-0	0
8610	.142344	.142295	.142246	.142198	.142149	.142101	.142052	.142004	.141955	.141907	-0	0
8620	.141858	.141810	.141761	.141713	.141665	.141616	.141568	.141520	.141471	.141423	-0	0
8630	.141375	.141327	.141278	.141230	.141182	.141134	.141086	.141038	.140990	.140942	-0	0
8640	.140893	.140845	.140797	.140749	.140701	.140653	.140606	.140558	.140510	.140462	-0	0
8650	.140414	.140366	.140318	.140271	.140223	.140175	.140127	.140079	.140032	.139984	-0	0
8660	.139936	.139889	.139841	.139794	.139746	.139698	.139651	.139603	.139556	.139508	-0	0
8670	.139461	.139413	.139366	.139319	.139271	.139224	.139176	.139129	.139082	.139034	-0	0
8680	.138987	.138940	.138893	.138845	.138798	.138751	.138704	.138657	.138610	.138563	-0	0
8690	.138515	.138468	.138421	.138374	.138327	.138280	.138233	.138186	.138139	.138093	-0	0
8700	.138046	.137999	.137952	.137905	.137858	.137811	.137765	.137718	.137671	.137624	-0	0
8710	.137578	.137531	.137484	.137438	.137391	.137344	.137298	.137251	.137205	.137158	-0	0
8720	.137112	.137065	.137019	.136972	.136926	.136879	.136833	.136787	.136740	.136694	-0	0
8730	.136648	.136601	.136555	.136509	.136462	.136416	.136370	.136324	.136278	.136231	-0	0
8740	.136185	.136139	.136093	.136047	.136001	.135955	.135909	.135863	.135817	.135771	-0	0
8750	.135725	.135679	.135633	.135587	.135541	.135495	.135450	.135404	.135358	.135312	-0	0
8760	.135266	.135221	.135175	.135129	.135083	.135038	.134992	.134947	.134901	.134855	-0	0
8770	.134810	.134764	.134719	.134673	.134628	.134582	.134537	.134491	.134446	.134400	-0	0
8780	.134355	.134309	.134264	.134219	.134173	.134128	.134083	.134038	.133992	.133947	-0	0
8790	.133902	.133857	.133811	.133766	.133721	.133676	.133631	.133586	.133541	.133496	-0	0
8800	.133451	.133406	.133361	.133316	.133271	.133226	.133181	.133136	.133091	.133046	-0	0
8810	.133001	.132957	.132912	.132867	.132822	.132777	.132733	.132688	.132643	.132598	-0	0
8820	.132554	.132509	.132465	.132420	.132375	.132331	.132286	.132242	.132197	.132153	-0	0
8830	.132108	.132064	.132019	.131975	.131930	.131886	.131842	.131797	.131753	.131708	-0	0
8840	.131664	.131620	.131576	.131531	.131487	.131443	.131399	.131354	.131310	.131266	-0	0
8850	.131222	.131178	.131134	.131090	.131046	.131002	.130958	.130914	.130870	.130826	-0	0
8860	.130782	.130738	.130694	.130650	.130606	.130562	.130518	.130474	.130431	.130387	-0	0
8870	.130343	.130299	.130256	.130212	.130168	.130124	.130081	.130037	.129993	.129950	-0	0
8880	.129906	.129863	.129819	.129775	.129732	.129688	.129645	.129601	.129558	.129515	-0	0
8890	.129471	.129428	.129384	.129341	.129298	.129254	.129211	.129168	.129124	.129081	-0	0
8900	.129038	.128995	.128951	.128908	.128865	.128822	.128779	.128735	.128692	.128649	-0	0
8910	.128606	.128563	.128520	.128477	.128434	.128391	.128348	.128305	.128262	.128219	-0	0
8920	.128176	.128133	.128091	.128048	.128005	.127962	.127919	.127876	.127834	.127791	-0	0
8930	.127748	.127705	.127663	.127620	.127577	.127535	.127492	.127449	.127407	.127364	-0	0
8940	.127322	.127279	.127237	.127194	.127152	.127109	.127067	.127024	.126982	.126939	-0	0
8950	.126897	.126855	.126812	.126770	.126728	.126685	.126643	.126601	.126558	.126516	-0	0
8960	.126474	.126432	.126390	.126347	.126305	.126263	.126221	.126179	.126137	.126095	-0	0
8970	.126053	.126011	.125969	.125927	.125885	.125843	.125801	.125759	.125717	.125675	-0	0
8980	.125633	.125591	.125549	.125507	.125466	.125424	.125382	.125340	.125298	.125257	-0	0
8990	.125215	.125173	.125132	.125090	.125048	.125007	.124965	.124923	.124882	.124840	-0	0

TABLE 10

Blackbody: Relative Spectral Emittance (Radiant)

λT(μ.°K)	0	1	2	3	4	5	6	7	8	9	m	Δ_2
9000	.124799	.124757	.124716	.124674	.124633	.124591	.124550	.124508	.124467	.124425	-0	0
9010	.124384	.124343	.124301	.124260	.124219	.124177	.124136	.124095	.124053	.124012	-0	0
9020	.123971	.123930	.123889	.123847	.123806	.123765	.123724	.123683	.123642	.123601	-0	0
9030	.123560	.123519	.123478	.123437	.123396	.123355	.123314	.123273	.123232	.123191	-0	0
9040	.123150	.123109	.123068	.123027	.122987	.122946	.122905	.122864	.122823	.122783	-0	0
9050	.122742	.122701	.122660	.122620	.122579	.122538	.122498	.122457	.122417	.122376	-0	0
9060	.122335	.122295	.122254	.122214	.122173	.122133	.122092	.122052	.122011	.121971	-0	0
9070	.121931	.121890	.121850	.121809	.121769	.121729	.121688	.121648	.121608	.121568	-0	0
9080	.121527	.121487	.121447	.121407	.121367	.121326	.121286	.121246	.121206	.121166	-0	0
9090	.121126	.121086	.121046	.121006	.120966	.120926	.120886	.120846	.120806	.120766	-0	0
9100	.120726	.120686	.120646	.120606	.120566	.120526	.120486	.120447	.120407	.120367	-0	0
9110	.120327	.120288	.120248	.120208	.120168	.120129	.120089	.120049	.120010	.119970	-0	0
9120	.119930	.119891	.119851	.119812	.119772	.119733	.119693	.119654	.119614	.119575	-0	0
9130	.119535	.119496	.119456	.119417	.119377	.119338	.119299	.119259	.119220	.119181	-0	0
9140	.119141	.119102	.119063	.119024	.118984	.118945	.118906	.118867	.118828	.118788	-0	0
9150	.118749	.118710	.118671	.118632	.118593	.118554	.118515	.118476	.118437	.118398	-0	0
9160	.118359	.118320	.118281	.118242	.118203	.118164	.118125	.118086	.118047	.118008	-0	0
9170	.117970	.117931	.117892	.117853	.117814	.117776	.117737	.117698	.117659	.117621	-0	0
9180	.117582	.117543	.117505	.117466	.117427	.117389	.117350	.117312	.117273	.117235	-0	0
9190	.117196	.117158	.117119	.117081	.117042	.117004	.116965	.116927	.116888	.116850	-0	0
9200	.116812	.116773	.116735	.116697	.116658	.116620	.116582	.116543	.116505	.116467	-0	0
9210	.116429	.116390	.116352	.116314	.116276	.116238	.116200	.116161	.116123	.116085	-0	0
9220	.116047	.116009	.115971	.115933	.115895	.115857	.115819	.115781	.115743	.115705	-0	0
9230	.115667	.115629	.115591	.115554	.115516	.115478	.115440	.115402	.115364	.115327	-0	0
9240	.115289	.115251	.115213	.115176	.115138	.115100	.115062	.115025	.114987	.114949	-0	0
9250	.114912	.114874	.114837	.114799	.114761	.114724	.114686	.114649	.114611	.114574	-0	0
9260	.114536	.114499	.114461	.114424	.114387	.114349	.114312	.114274	.114237	.114200	-0	0
9270	.114162	.114125	.114088	.114050	.114013	.113976	.113939	.113901	.113864	.113827	-0	0
9280	.113790	.113753	.113716	.113678	.113641	.113604	.113567	.113530	.113493	.113456	-0	0
9290	.113419	.113382	.113345	.113308	.113271	.113234	.113197	.113160	.113123	.113086	-0	0
9300	.113049	.113012	.112975	.112939	.112902	.112865	.112828	.112791	.112755	.112718	-0	0
9310	.112681	.112644	.112608	.112571	.112534	.112498	.112461	.112424	.112388	.112351	-0	0
9320	.112314	.112278	.112241	.112205	.112168	.112132	.112095	.112059	.112022	.111986	-0	0
9330	.111949	.111913	.111876	.111840	.111803	.111767	.111731	.111694	.111658	.111622	-0	0
9340	.111585	.111549	.111513	.111476	.111440	.111404	.111368	.111331	.111295	.111259	-0	0
9350	.111223	.111187	.111151	.111114	.111078	.111042	.111006	.110970	.110934	.110898	-0	0
9360	.110862	.110826	.110790	.110754	.110718	.110682	.110646	.110610	.110574	.110538	-0	0
9370	.110502	.110466	.110431	.110395	.110359	.110323	.110287	.110251	.110216	.110180	-0	0
9380	.110144	.110108	.110073	.110037	.110001	.109966	.109930	.109894	.109859	.109823	-0	0
9390	.109787	.109752	.109716	.109681	.109645	.109609	.109574	.109538	.109503	.109467	-0	0
9400	.109432	.109397	.109361	.109326	.109290	.109255	.109219	.109184	.109149	.109113	-0	0
9410	.109078	.109043	.109007	.108972	.108937	.108901	.108866	.108831	.108796	.108761	-0	0
9420	.108725	.108690	.108655	.108620	.108585	.108550	.108514	.108479	.108444	.108409	-0	0
9430	.108374	.108339	.108304	.108269	.108234	.108199	.108164	.108129	.108094	.108059	-0	0
9440	.108024	.107989	.107954	.107920	.107885	.107850	.107815	.107780	.107745	.107711	-0	0
9450	.107676	.107641	.107606	.107571	.107537	.107502	.107467	.107433	.107398	.107363	-0	0
9460	.107329	.107294	.107259	.107225	.107190	.107156	.107121	.107086	.107052	.107017	-0	0
9470	.106983	.106948	.106914	.106879	.106845	.106810	.106776	.106742	.106707	.106673	-0	0
9480	.106638	.106604	.106570	.106535	.106501	.106467	.106432	.106398	.106364	.106329	-0	0
9490	.106295	.106261	.106227	.106193	.106158	.106124	.106090	.106056	.106022	.105988	-0	0

TABLE 10

Blackbody: Relative Spectral Emittance (Radiant)

λT(μ.°K)	0	1	2	3	4	5	6	7	8	9	m	Δ_2
9500	.105953	.105919	.105885	.105851	.105817	.105783	.105749	.105715	.105681	.105647	-0	0
9510	.105613	.105579	.105545	.105511	.105477	.105443	.105409	.105375	.105342	.105308	-0	0
9520	.105274	.105240	.105206	.105172	.105139	.105105	.105071	.105037	.105004	.104970	-0	0
9530	.104936	.104902	.104869	.104835	.104801	.104768	.104734	.104700	.104667	.104633	-0	0
9540	.104600	.104566	.104532	.104499	.104465	.104432	.104398	.104365	.104331	.104298	-0	0
9550	.104264	.104231	.104197	.104164	.104131	.104097	.104064	.104030	.103997	.103964	-0	0
9560	.103930	.103897	.103864	.103831	.103797	.103764	.103731	.103697	.103664	.103631	-0	0
9570	.103598	.103565	.103531	.103498	.103465	.103432	.103399	.103366	.103333	.103300	-0	0
9580	.103266	.103233	.103200	.103167	.103134	.103101	.103068	.103035	.103002	.102969	-0	0
9590	.102936	.102903	.102871	.102838	.102805	.102772	.102739	.102706	.102673	.102640	-0	0
9600	.102608	.102575	.102542	.102509	.102476	.102444	.102411	.102378	.102346	.102313	-0	0
9610	.102280	.102247	.102215	.102182	.102149	.102117	.102084	.102052	.102019	.101986	-0	0
9620	.101954	.101921	.101889	.101856	.101824	.101791	.101759	.101726	.101694	.101661	-0	0
9630	.101629	.101596	.101564	.101532	.101499	.101467	.101435	.101402	.101370	.101337	-0	0
9640	.101305	.101273	.101241	.101208	.101176	.101144	.101112	.101079	.101047	.101015	-0	0
9650	.100983	.100951	.100918	.100886	.100854	.100822	.100790	.100758	.100726	.100694	-0	0
9660	.100661	.100629	.100597	.100565	.100533	.100501	.100469	.100437	.100405	.100373	-0	0
9670	.100341	.100309	.100278	.100246	.100214	.100182	.100150	.100118	.100086	.100054	-0	0
9680	1.00023	.999908	.999590	.999273	.998955	.998637	.998320	.998002	.997685	.997368	-1	0
9690	.997051	.996734	.996417	.996101	.995784	.995468	.995151	.994835	.994519	.994203	-1	0
9700	.993888	.993572	.993256	.992941	.992626	.992310	.991995	.991680	.991366	.991051	-1	0
9710	.990736	.990422	.990108	.989793	.989479	.989165	.988851	.988538	.988224	.987911	-1	0
9720	.987597	.987284	.986971	.986658	.986345	.986032	.985720	.985407	.985095	.984782	-1	0
9730	.984470	.984158	.983846	.983534	.983223	.982911	.982600	.982288	.981977	.981666	-1	0
9740	.981355	.981044	.980733	.980423	.980112	.979802	.979492	.979182	.978872	.978562	-1	0
9750	.978252	.977942	.977633	.977323	.977014	.976705	.976396	.976087	.975778	.975469	-1	0
9760	.975161	.974852	.974544	.974235	.973927	.973619	.973311	.973004	.972696	.972388	-1	0
9770	.972081	.971774	.971467	.971159	.970853	.970546	.970239	.969932	.969626	.969320	-1	0
9780	.969013	.968707	.968401	.968095	.967790	.967484	.967178	.966873	.966568	.966262	-1	0
9790	.965957	.965652	.965347	.965043	.964738	.964434	.964129	.963825	.963521	.963217	-1	0
9800	.962913	.962609	.962305	.962002	.961698	.961395	.961092	.960789	.960486	.960183	-1	0
9810	.959880	.959578	.959275	.958973	.958670	.958368	.958066	.957764	.957462	.957161	-1	0
9820	.956859	.956558	.956256	.955955	.955654	.955353	.955052	.954751	.954450	.954150	-1	0
9830	.953849	.953549	.953249	.952949	.952649	.952349	.952049	.951750	.951450	.951151	-1	0
9840	.950851	.950552	.950253	.949954	.949655	.949356	.949058	.948759	.948461	.948163	-1	0
9850	.947865	.947566	.947269	.946971	.946673	.946375	.946078	.945781	.945483	.945186	-1	0
9860	.944889	.944592	.944295	.943999	.943702	.943406	.943109	.942813	.942517	.942221	-1	0
9870	.941925	.941629	.941334	.941038	.940743	.940447	.940152	.939857	.939562	.939267	-1	0
9880	.938972	.938678	.938383	.938089	.937794	.937500	.937206	.936912	.936618	.936324	-1	0
9890	.936031	.935737	.935444	.935150	.934857	.934564	.934271	.933978	.933685	.933393	-1	0
9900	.933100	.932808	.932515	.932223	.931931	.931639	.931347	.931055	.930764	.930472	-1	0
9910	.930181	.929890	.929598	.929307	.929016	.928725	.928435	.928144	.927853	.927563	-1	0
9920	.927273	.926982	.926692	.926402	.926112	.925823	.925533	.925243	.924954	.924665	-1	0
9930	.924375	.924086	.923797	.923508	.923220	.922931	.922642	.922354	.922065	.921777	-1	0
9940	.921489	.921201	.920913	.920625	.920338	.920050	.919763	.919475	.919188	.918901	-1	0
9950	.918614	.918327	.918040	.917753	.917467	.917180	.916894	.916608	.916321	.916035	-1	0
9960	.915749	.915463	.915178	.914892	.914607	.914321	.914036	.913751	.913466	.913181	-1	0
9970	.912896	.912611	.912326	.912042	.911757	.911473	.911189	.910905	.910621	.910337	-1	0
9980	.910053	.909769	.909486	.909202	.908919	.908635	.908352	.908069	.907786	.907503	-1	0
9990	.907221	.906938	.906656	.906373	.906091	.905809	.905527	.905245	.904963	.904681	-1	0

$M_{e\lambda}(\lambda T)/M_{emax}(T)$

TABLE 10

Blackbody: Relative Spectral Emittance (Radiant)

λT(m.°K)	0	1	2	3	4	5	6	7	8	9	m
10	.904399	.876763	.850146	.824504	.799797	.775986	.753033	.730902	.709559	.688972	-1
11	.669110	.649943	.631443	.613583	.596338	.579682	.563593	.548048	.533025	.518504	-1
12	.504466	.490891	.477763	.465063	.452775	.440885	.429376	.418234	.407447	.396999	-1
13	.386880	.377077	.367578	.358372	.349448	.340798	.332409	.324274	.316384	.308729	-1
14	.301301	.294093	.287097	.280305	.273711	.267307	.261087	.255046	.249176	.243473	-1
15	.237930	.232543	.227305	.222213	.217262	.212446	.207761	.203204	.198771	.194456	-1
16	.190257	.186169	.182190	.178316	.174543	.170869	.167290	.163804	.160407	.157098	-1
17	.153872	.150729	.147664	.144677	.141764	.138924	.136154	.133452	.130816	.128245	-1
18	.125736	.123288	.120899	.118567	.116291	.114068	.111899	.109780	.107711	.105690	-1
19	.103716	.101788	.099903	.098063	.096264	.094506	.092787	.091108	.089466	.087860	-1
20	.862907	.847558	.832546	.817864	.803502	.789453	.775708	.762260	.749100	.736222	-2
21	.723618	.711282	.699206	.687384	.675810	.664477	.653380	.642513	.631870	.621445	-2
22	.611234	.601231	.591430	.581828	.572419	.563199	.554163	.545307	.536626	.528117	-2
23	.519775	.511596	.503578	.495715	.488004	.480443	.473026	.465752	.458617	.451617	-2
24	.444750	.438012	.431401	.424914	.418549	.412301	.406169	.400151	.394243	.388443	-2
25	.382750	.377160	.371671	.366282	.360989	.355792	.350687	.345674	.340749	.335912	-2
26	.331160	.326491	.321905	.317398	.312970	.308618	.304341	.300139	.296008	.291948	-2
27	.287957	.284034	.280177	.276386	.272658	.268993	.265388	.261844	.258359	.254931	-2
28	.251560	.248245	.244983	.241775	.238619	.235515	.232460	.229455	.226498	.223589	-2
29	.220725	.217908	.215135	.212406	.209720	.207077	.204475	.201913	.199391	.196909	-2
30	.194465	.192059	.189689	.187356	.185059	.182797	.180569	.178375	.176214	.174086	-2
31	.171990	.169925	.167891	.165887	.163913	.161968	.160052	.158164	.156304	.154471	-2
32	.152664	.150884	.149130	.147401	.145698	.144018	.142363	.140731	.139123	.137538	-2
33	.135975	.134434	.132915	.131417	.129940	.128484	.127048	.125632	.124236	.122859	-2
34	.121501	.120162	.118841	.117539	.116254	.114986	.113736	.112502	.111286	.110085	-2
35	.108901	.107733	.106580	.105443	.104320	.103213	.102120	.101042	.099977	.098927	-2
36	.978907	.968677	.958580	.948614	.938777	.929067	.919482	.910021	.900680	.891459	-3
37	.882356	.873368	.864494	.855733	.847082	.838540	.830105	.821776	.813551	.805428	-3
38	.797407	.789485	.781661	.773934	.766301	.758763	.751317	.743962	.736696	.729519	-3
39	.722429	.715425	.708505	.701669	.694915	.688242	.681649	.675134	.668697	.662336	-3
40	.656051	.649840	.643703	.637637	.631643	.625719	.619864	.614077	.608358	.602705	-3
41	.597118	.591595	.586135	.580739	.575404	.570131	.564917	.559763	.554668	.549630	-3
42	.544650	.539725	.534856	.530042	.525282	.520575	.515921	.511318	.506767	.502266	-3
43	.497815	.493413	.489060	.484754	.480496	.476284	.472118	.467998	.463923	.459891	-3
44	.455904	.451959	.448057	.444197	.440379	.436601	.432864	.429166	.425508	.421889	-3
45	.418308	.414766	.411260	.407791	.404359	.400963	.397603	.394277	.390986	.387730	-3
46	.384507	.381318	.378161	.375037	.371946	.368886	.365857	.362860	.359893	.356956	-3
47	.354049	.351172	.348324	.345504	.342713	.339951	.337215	.334508	.331827	.329174	-3
48	.326546	.323945	.321370	.318820	.316295	.313796	.311321	.308870	.306443	.304040	-3
49	.301661	.299305	.296972	.294661	.292373	.290107	.287863	.285640	.283439	.281259	-3

TABLE 10

Blackbody: Relative Spectral Emittance (Radiant)

λT(m.°K)	0	1	2	3	4	5	6	7	8	9	m
50	.279100	.276962	.274844	.272746	.270669	.268611	.266572	.264553	.262553	.260571	-3
51	.258609	.256664	.254738	.252830	.250940	.249068	.247213	.245375	.243554	.241750	-3
52	.239962	.238192	.236437	.234699	.232976	.231269	.229578	.227902	.226242	.224596	-3
53	.222966	.221350	.219749	.218162	.216590	.215032	.213488	.211957	.210440	.208937	-3
54	.207447	.205971	.204507	.203057	.201619	.200194	.198781	.197381	.195994	.194618	-3
55	.193255	.191903	.190563	.189235	.187919	.186613	.185320	.184037	.182765	.181505	-3
56	.180255	.179016	.177787	.176569	.175362	.174165	.172978	.171801	.170634	.169477	-3
57	.168329	.167192	.166064	.164945	.163836	.162736	.161646	.160564	.159492	.158428	-3
58	.157374	.156328	.155290	.154262	.153242	.152230	.151226	.150231	.149244	.148265	-3
59	.147294	.146331	.145376	.144429	.143489	.142557	.141633	.140716	.139806	.138904	-3
60	.138009	.137121	.136240	.135367	.134500	.133641	.132788	.131942	.131102	.130270	-3
61	.129443	.128624	.127811	.127004	.126204	.125410	.124622	.123840	.123065	.122296	-3
62	.121532	.120775	.120023	.119277	.118537	.117803	.117074	.116351	.115634	.114922	-3
63	.114216	.113515	.112819	.112129	.111444	.110764	.110089	.109420	.108755	.108096	-3
64	.107442	.106792	.106148	.105508	.104873	.104243	.103617	.102997	.102381	.101769	-3
65	.101162	.100560	.099962	.099368	.098779	.098194	.097614	.097038	.096466	.095898	-3
66	.953346	.947752	.942199	.936686	.931214	.925782	.920389	.915035	.909720	.904444	-4
67	.899206	.894006	.888843	.883717	.878629	.873577	.868561	.863581	.858636	.853728	-4
68	.848854	.844014	.839210	.834439	.829702	.824999	.820329	.815692	.811087	.806516	-4
69	.801976	.797468	.792992	.788547	.784133	.779750	.775398	.771076	.766784	.762521	-4
70	.758289	.754085	.749911	.745766	.741649	.737561	.733500	.729468	.725463	.721486	-4
71	.717535	.713612	.709716	.705846	.702003	.698185	.694394	.690628	.686888	.683173	-4
72	.679483	.675818	.672177	.668561	.664970	.661402	.657858	.654338	.650842	.647369	-4
73	.643919	.640492	.637087	.633706	.630346	.627009	.623694	.620401	.617129	.613879	-4
74	.610651	.607443	.604257	.601091	.597946	.594822	.591718	.588634	.585571	.582527	-4
75	.579503	.576499	.573514	.570548	.567601	.564674	.561765	.558875	.556004	.553151	-4
76	.550316	.547499	.544700	.541920	.539156	.536411	.533683	.530972	.528279	.525602	-4
77	.522943	.520300	.517674	.515064	.512471	.509894	.507333	.504789	.502260	.499747	-4
78	.497250	.494768	.492302	.489851	.487416	.484995	.482590	.480199	.477823	.475462	-4
79	.473116	.470784	.468466	.466162	.463873	.461597	.459336	.457088	.454855	.452634	-4
80	.450428	.448234	.446054	.443888	.441734	.439593	.437466	.435351	.433249	.431160	-4
81	.429083	.427019	.424967	.422927	.420900	.418885	.416882	.414890	.412911	.410944	-4
82	.408988	.407043	.405111	.403190	.401280	.399381	.397494	.395618	.393752	.391898	-4
83	.390055	.388222	.386401	.384590	.382789	.380999	.379220	.377451	.375692	.373943	-4
84	.372205	.370476	.368758	.367050	.365351	.363663	.361984	.360314	.358655	.357005	-4
85	.355364	.353733	.352111	.350498	.348895	.347301	.345715	.344139	.342572	.341014	-4
86	.339465	.337924	.336392	.334869	.333354	.331848	.330351	.328862	.327381	.325909	-4
87	.324445	.322989	.321541	.320101	.318670	.317246	.315830	.314423	.313023	.311630	-4
88	.310246	.308869	.307500	.306138	.304784	.303438	.302099	.300767	.299443	.298125	-4
89	.296815	.295513	.294217	.292929	.291647	.290373	.289105	.287845	.286591	.285344	-4
90	.284104	.282870	.281644	.280423	.279210	.278003	.276803	.275609	.274421	.273240	-4
91	.272065	.270897	.269735	.268579	.267429	.266285	.265147	.264016	.262891	.261771	-4
92	.260658	.259550	.258448	.257352	.256262	.255178	.254099	.253026	.251959	.250897	-4
93	.249841	.248791	.247746	.246706	.245672	.244644	.243620	.242602	.241590	.240583	-4
94	.239580	.238584	.237592	.236605	.235624	.234647	.233676	.232710	.231749	.230792	-4
95	.229841	.228894	.227953	.227016	.226084	.225156	.224234	.223316	.222403	.221495	-4
96	.220591	.219692	.218797	.217907	.217022	.216141	.215264	.214392	.213524	.212661	-4
97	.211802	.210947	.210097	.209251	.208409	.207571	.206738	.205909	.205084	.204263	-4
98	.203446	.202633	.201824	.201020	.200219	.199422	.198630	.197841	.197056	.196275	-4
99	.195498	.194725	.193955	.193190	.192428	.191670	.190915	.190164	.189417	.188674	-4

$M_{e\lambda}(\lambda T)/M_{emax}(T)$

TABLE 10

Blackbody: Relative Spectral Emittance (Radiant)

λT(m.°K)	0	1	2	3	4	5	6	7	8	9	m
100	.187934	.180733	.173873	.167336	.161103	.155157	.149483	.144066	.138893	.133949	-4
110	.129223	.124704	.120381	.116243	.112282	.108488	.104853	.101369	.098028	.094824	-4
120	.917493	.887985	.859654	.832444	.806302	.781177	.757024	.733797	.711454	.689955	-5
130	.669261	.649338	.630150	.611666	.593855	.576687	.560135	.544173	.528775	.513918	-5
140	.499579	.485737	.472370	.459461	.446989	.434937	.423289	.412029	.401140	.390608	-5
150	.380420	.370562	.361021	.351785	.342843	.334183	.325795	.317668	.309793	.302160	-5
160	.294761	.287588	.280631	.273883	.267337	.260985	.254821	.248837	.243028	.237388	-5
170	.231910	.226590	.221421	.216399	.211518	.206774	.202162	.197679	.193318	.189078	-5
180	.184953	.180940	.177035	.173235	.169536	.165935	.162430	.159017	.155693	.152455	-5
190	.149301	.146228	.143234	.140316	.137472	.134699	.131996	.129361	.126791	.124284	-5
200	.121840	.119455	.117128	.114857	.112641	.110479	.108368	.106307	.104295	.102330	-5
210	.100412	.098538	.096707	.094919	.093172	.091465	.089797	.088167	.086574	.085016	-5
220	.834938	.820051	.805494	.791259	.777337	.763720	.750399	.737368	.724618	.712143	-6
230	.699935	.687987	.676294	.664848	.653642	.642672	.631931	.621414	.611114	.601027	-6
240	.591147	.581470	.571989	.562701	.553601	.544684	.535946	.527382	.518989	.510762	-6
250	.502697	.494791	.487040	.479440	.471987	.464679	.457512	.450482	.443587	.436823	-6
260	.430188	.423678	.417291	.411024	.404874	.398839	.392915	.387101	.381395	.375793	-6
270	.370293	.364894	.359593	.354388	.349276	.344257	.339327	.334485	.329729	.325058	-6
280	.320469	.315960	.311531	.307179	.302903	.298700	.294571	.290512	.286524	.282603	-6
290	.278749	.274961	.271237	.267576	.263976	.260436	.256956	.253534	.250168	.246858	-6
300	.243603	.240401	.237252	.234154	.231106	.228108	.225158	.222256	.219400	.216591	-6
310	.213826	.211105	.208427	.205791	.203197	.200644	.198130	.195656	.193221	.190823	-6
320	.188462	.186138	.183849	.181595	.179376	.177191	.175039	.172919	.170832	.168775	-6
330	.166750	.164755	.162790	.160854	.158946	.157067	.155215	.153391	.151594	.149822	-6
340	.148077	.146356	.144661	.142990	.141344	.139720	.138121	.136543	.134989	.133456	-6
350	.131945	.130456	.128987	.127539	.126112	.124704	.123316	.121947	.120597	.119265	-6
360	.117952	.116657	.115380	.114120	.112877	.111651	.110442	.109249	.108072	.106911	-6
370	.105766	.104636	.103520	.102420	.101334	.100263	.099206	.098162	.097133	.096116	-6
380	.951134	.941234	.931464	.921819	.912299	.902902	.893625	.884467	.875427	.866501	-7
390	.857689	.848988	.840398	.831916	.823541	.815270	.807104	.799040	.791075	.783210	-7
400	.775442	.767771	.760194	.752710	.745319	.738017	.730805	.723681	.716643	.709691	-7
410	.702823	.696037	.689334	.682711	.676167	.669701	.663313	.657001	.650763	.644599	-7
420	.638508	.632489	.626541	.620662	.614853	.609111	.603435	.597826	.592282	.586802	-7
430	.581385	.576031	.570738	.565505	.560333	.555219	.550164	.545166	.540225	.535340	-7
440	.530509	.525733	.521011	.516342	.511724	.507159	.502644	.498179	.493764	.489397	-7
450	.485079	.480808	.476584	.472406	.468274	.464188	.460145	.456147	.452191	.448279	-7
460	.444409	.440580	.436793	.433046	.429340	.425673	.422045	.418455	.414904	.411390	-7
470	.407913	.404473	.401069	.397701	.394369	.391071	.387807	.384577	.381382	.378219	-7
480	.375088	.371990	.368924	.365890	.362887	.359914	.356972	.354060	.351177	.348324	-7
490	.345500	.342704	.339936	.337197	.334485	.331800	.329142	.326510	.323905	.321326	-7

TABLE 10

Blackbody: Relative Spectral Emittance (Radiant)

λT(m.°K)	0	1	2	3	4	5	6	7	8	9	m
500	.318772	.316244	.313740	.311262	.308808	.306377	.303971	.301589	.299229	.296893	-7
510	.294580	.292289	.290020	.287773	.285548	.283344	.281162	.279000	.276859	.274739	-7
520	.272639	.270559	.268499	.266458	.264437	.262435	.260452	.258487	.256541	.254614	-7
530	.252704	.250812	.248938	.247081	.245242	.243420	.241614	.239826	.238053	.236298	-7
540	.234558	.232834	.231126	.229434	.227757	.226096	.224449	.222818	.221201	.219599	-7
550	.218012	.216439	.214880	.213335	.211803	.210286	.208782	.207292	.205814	.204350	-7
560	.202899	.201461	.200036	.198623	.197222	.195834	.194458	.193094	.191742	.190402	-7
570	.189074	.187757	.186452	.185157	.183875	.182603	.181342	.180092	.178853	.177624	-7
580	.176406	.175199	.174001	.172814	.171637	.170470	.169313	.168166	.167028	.165900	-7
590	.164782	.163673	.162573	.161483	.160401	.159329	.158265	.157211	.156165	.155128	-7
600	.154100	.153079	.152068	.151065	.150070	.149083	.148104	.147134	.146171	.145216	-7
610	.144269	.143329	.142398	.141474	.140557	.139648	.138746	.137851	.136963	.136083	-7
620	.135210	.134343	.133484	.132632	.131786	.130947	.130115	.129289	.128470	.127657	-7
630	.126851	.126051	.125257	.124470	.123689	.122913	.122144	.121381	.120624	.119873	-7
640	.119128	.118388	.117654	.116926	.116204	.115487	.114775	.114069	.113369	.112674	-7
650	.111984	.111299	.110620	.109946	.109276	.108613	.107954	.107300	.106651	.106006	-7
660	.105367	.104733	.104103	.103478	.102858	.102242	.101631	.101025	.100423	.099825	-7
670	.992323	.986436	.980594	.974794	.969037	.963323	.957651	.952020	.946430	.940882	-8
680	.935374	.929907	.924479	.919091	.913742	.908432	.903160	.897927	.892732	.887574	-8
690	.882453	.877369	.872322	.867311	.862335	.857396	.852492	.847623	.842788	.837988	-8
700	.833222	.828490	.823791	.819126	.814494	.809894	.805326	.800792	.796288	.791817	-8
710	.787377	.782967	.778589	.774242	.769924	.765637	.761379	.757151	.752952	.748783	-8
720	.744642	.740529	.736445	.732389	.728362	.724361	.720388	.716443	.712524	.708632	-8
730	.704767	.700928	.697115	.693328	.689567	.685831	.682120	.678434	.674773	.671137	-8
740	.667526	.663938	.660375	.656836	.653320	.649828	.646359	.642913	.639490	.636090	-8
750	.632712	.629357	.626024	.622713	.619424	.616157	.612911	.609687	.606483	.603301	-8
760	.600139	.596998	.593878	.590778	.587698	.584639	.581599	.578579	.575579	.572597	-8
770	.569636	.566693	.563769	.560864	.557978	.555110	.552261	.549430	.546617	.543822	-8
780	.541046	.538286	.535544	.532820	.530113	.527423	.524750	.522094	.519455	.516832	-8
790	.514227	.511637	.509064	.506507	.503966	.501441	.498931	.496437	.493959	.491497	-8
800	.489049	.486617	.484200	.481798	.479411	.477038	.474681	.472337	.470008	.467694	-8
810	.465394	.463108	.460836	.458578	.456333	.454103	.451886	.449682	.447492	.445316	-8
820	.443152	.441002	.438864	.436740	.434628	.432529	.430443	.428370	.426308	.424259	-8
830	.422223	.420199	.418186	.416186	.414198	.412222	.410257	.408304	.406363	.404433	-8
840	.402515	.400608	.398712	.396828	.394954	.393092	.391241	.389400	.387571	.385752	-8
850	.383944	.382146	.380359	.378582	.376815	.375059	.373313	.371578	.369852	.368136	-8
860	.366431	.364735	.363049	.361373	.359706	.358049	.356401	.354764	.353135	.351516	-8
870	.349905	.348304	.346713	.345130	.343557	.341992	.340436	.338890	.337351	.335822	-8
880	.334301	.332789	.331285	.329790	.328303	.326825	.325355	.323893	.322440	.320994	-8
890	.319557	.318128	.316706	.315293	.313888	.312490	.311100	.309718	.308343	.306976	-8
900	.305617	.304265	.302921	.301584	.300254	.298932	.297617	.296309	.295009	.293715	-8
910	.292429	.291150	.289877	.288612	.287353	.286102	.284857	.283619	.282387	.281163	-8
920	.279944	.278733	.277528	.276330	.275138	.273952	.272773	.271600	.270434	.269273	-8
930	.268119	.266971	.265830	.264694	.263564	.262441	.261323	.260212	.259106	.258006	-8
940	.256912	.255824	.254741	.253664	.252593	.251528	.250468	.249414	.248365	.247322	-8
950	.246284	.245252	.244225	.243203	.242187	.241176	.240171	.239170	.238175	.237185	-8
960	.236200	.235220	.234246	.233276	.232311	.231352	.230397	.229447	.228502	.227562	-8
970	.226627	.225697	.224771	.223850	.222934	.222023	.221116	.220213	.219316	.218423	-8
980	.217534	.216650	.215771	.214895	.214025	.213159	.212297	.211439	.210586	.209737	-8
990	.208893	.208052	.207216	.206384	.205557	.204733	.203913	.203098	.202287	.201479	-8

$M_{e\lambda}(\lambda T)/M_{e\max}(T)$

TABLE 10

Blackbody: Relative Spectral Emittance (Radiant)

λT(m.°K)	0	1	2	3	4	5	6	7	8	9	m
1000	.200676	.192860	.185420	.178336	.171586	.165153	.159019	.153167	.147582	.142249	-8
1100	.137154	.132286	.127632	.123181	.118922	.114845	.110942	.107203	.103620	.100186	-8
1200	.968930	.937341	.907029	.877932	.849992	.823156	.797369	.772585	.748757	.725840	-9
1300	.703793	.682577	.662154	.642489	.623550	.605303	.587718	.570767	.554423	.538660	-9
1400	.523453	.508779	.494616	.480942	.467737	.454983	.442660	.430752	.419241	.408113	-9
1500	.397351	.386941	.376871	.367125	.357693	.348562	.339720	.331157	.322862	.314824	-9
1600	.307036	.299487	.292169	.285072	.278190	.271515	.265038	.258753	.252653	.246733	-9
1700	.240984	.235402	.229981	.224715	.219599	.214627	.209796	.205099	.200534	.196094	-9
1800	.191777	.187578	.183493	.179519	.175652	.171888	.168225	.164659	.161187	.157806	-9
1900	.154513	.151305	.148180	.145136	.142169	.139278	.136460	.133712	.131034	.128422	-9
2000	.125875	.123391	.120968	.118604	.116297	.114047	.111850	.109706	.107614	.105571	-9
2100	.103576	.101628	.099725	.097867	.096052	.094279	.092547	.090854	.089200	.087583	-9
2200	.860027	.844579	.829477	.814710	.800269	.786148	.772337	.758827	.745613	.732685	-10
2300	.720034	.707657	.695544	.683688	.672085	.660727	.649608	.638721	.628063	.617624	-10
2400	.607402	.597391	.587585	.577979	.568569	.559350	.550317	.541465	.532790	.524289	-10
2500	.515957	.507790	.499783	.491933	.484237	.476691	.469292	.462035	.454918	.447938	-10
2600	.441091	.434375	.427786	.421321	.414977	.408752	.402645	.396651	.390767	.384993	-10
2700	.379324	.373760	.368298	.362935	.357669	.352498	.347421	.342433	.337537	.332726	-10
2800	.328001	.323360	.318800	.314321	.309920	.305596	.301347	.297172	.293068	.289036	-10
2900	.285071	.281175	.277346	.273581	.269881	.266242	.262664	.259146	.255688	.252286	-10
3000	.248941	.245652	.242416	.239233	.236103	.233024	.229994	.227014	.224082	.221197	-10
3100	.218358	.215565	.212815	.210111	.207449	.204828	.202249	.199710	.197211	.194751	-10
3200	.192330	.189946	.187599	.185288	.183012	.180771	.178564	.176392	.174251	.172143	-10
3300	.170068	.168022	.166008	.164024	.162070	.160145	.158247	.156378	.154537	.152722	-10
3400	.150934	.149173	.147437	.145726	.144039	.142377	.140740	.139125	.137533	.135965	-10
3500	.134418	.132894	.131391	.129909	.128448	.127008	.125587	.124187	.122806	.121444	-10
3600	.120101	.118776	.117470	.116181	.114911	.113657	.112421	.111201	.109998	.108811	-10
3700	.107640	.106484	.105344	.104220	.103110	.102015	.100935	.099869	.098816	.097778	-10
3800	.967536	.957422	.947440	.937591	.927867	.918267	.908792	.899440	.890206	.881095	-11
3900	.872094	.863212	.854441	.845781	.837231	.828790	.820452	.812222	.804093	.796065	-11
4000	.788137	.780311	.772579	.764942	.757399	.749949	.742593	.735323	.728142	.721051	-11
4100	.714046	.707125	.700287	.693534	.686858	.680265	.673749	.667314	.660954	.654669	-11
4200	.648460	.642324	.636259	.630266	.624344	.618489	.612707	.606988	.601337	.595753	-11
4300	.590232	.584776	.579382	.574051	.568782	.563571	.558419	.553328	.548294	.543318	-11
4400	.538396	.533533	.528723	.523966	.519264	.514614	.510015	.505469	.500973	.496526	-11
4500	.492130	.487781	.483480	.479226	.475020	.470859	.466744	.462675	.458649	.454665	-11
4600	.450726	.446830	.442976	.439161	.435390	.431658	.427966	.424314	.420700	.417124	-11
4700	.413588	.410087	.406625	.403198	.399808	.396453	.393132	.389848	.386598	.383381	-11
4800	.380197	.377045	.373927	.370841	.367787	.364765	.361773	.358812	.355880	.352979	-11
4900	.350107	.347266	.344452	.341667	.338908	.336179	.333478	.330803	.328155	.325534	-11

TABLE 10

Blackbody: Relative Spectral Emittance (Radiant)

λT(m.°K)	0	1	2	3	4	5	6	7	8	9	m
5000	.322938	.320367	.317823	.315305	.312811	.310341	.307896	.305476	.303078	.300704	-11
5100	.298353	.296026	.293720	.291437	.289177	.286937	.284721	.282526	.280350	.278196	-11
5200	.276064	.273950	.271859	.269786	.267733	.265699	.263686	.261690	.259714	.257757	-11
5300	.255817	.253896	.251994	.250108	.248242	.246390	.244558	.242741	.240942	.239160	-11
5400	.237394	.235644	.233911	.232192	.230491	.228805	.227133	.225477	.223836	.222211	-11
5500	.220600	.219003	.217420	.215852	.214299	.212759	.211233	.209721	.208222	.206736	-11
5600	.205265	.203805	.202359	.200926	.199505	.198096	.196701	.195317	.193945	.192587	-11
5700	.191238	.189903	.188579	.187266	.185966	.184675	.183396	.182129	.180872	.179626	-11
5800	.178391	.177166	.175953	.174749	.173555	.172372	.171199	.170035	.168882	.167738	-11
5900	.166604	.165479	.164365	.163260	.162164	.161076	.159998	.158929	.157868	.156818	-11
6000	.155775	.154741	.153716	.152700	.151690	.150690	.149698	.148715	.147739	.146772	-11
6100	.145811	.144860	.143916	.142979	.142050	.141128	.140215	.139308	.138409	.137517	-11
6200	.136632	.135755	.134884	.134020	.133163	.132313	.131470	.130634	.129804	.128981	-11
6300	.128164	.127353	.126550	.125752	.124961	.124176	.123397	.122624	.121857	.121097	-11
6400	.120341	.119592	.118849	.118112	.117380	.116654	.115934	.115219	.114510	.113806	-11
6500	.113107	.112414	.111726	.111043	.110366	.109694	.109026	.108365	.107707	.107056	-11
6600	.106408	.105766	.105128	.104496	.103868	.103244	.102626	.102012	.101403	.100798	-11
6700	.100198	.099602	.099010	.098423	.097841	.097262	.096688	.096119	.095553	.094992	-11
6800	.944339	.938806	.933311	.927861	.922447	.917077	.911743	.906446	.901190	.895970	-12
6900	.890786	.885642	.880539	.875464	.870436	.865434	.860473	.855550	.850659	.845801	-12
7000	.840981	.836193	.831436	.826716	.822033	.817381	.812758	.808172	.803615	.799093	-12
7100	.794601	.790143	.785714	.781319	.776952	.772612	.768306	.764033	.759787	.755568	-12
7200	.751380	.747219	.743090	.738992	.734914	.730873	.726851	.722865	.718903	.714966	-12
7300	.711058	.707175	.703321	.699490	.695687	.691908	.688157	.684429	.680729	.677055	-12
7400	.673404	.669775	.666173	.662592	.659038	.655510	.652003	.648521	.645060	.641620	-12
7500	.638209	.634814	.631448	.628102	.624776	.621473	.618190	.614930	.611695	.608477	-12
7600	.605283	.602108	.598955	.595820	.592707	.589617	.586545	.583494	.580461	.577449	-12
7700	.574454	.571480	.568523	.565587	.562672	.559773	.556895	.554037	.551194	.548367	-12
7800	.545565	.542773	.540005	.537253	.534515	.531796	.529097	.526416	.523750	.521097	-12
7900	.518464	.515848	.513251	.510668	.508098	.505546	.503011	.500494	.497990	.495504	-12
8000	.493030	.490573	.488129	.485706	.483295	.480897	.478515	.476150	.473796	.471458	-12
8100	.469136	.466826	.464532	.462249	.459982	.457729	.455492	.453267	.451056	.448856	-12
8200	.446671	.444500	.442340	.440195	.438064	.435943	.433836	.431740	.429661	.427593	-12
8300	.425534	.423489	.421458	.419440	.417431	.415436	.413450	.411481	.409521	.407571	-12
8400	.405633	.403708	.401795	.399891	.398000	.396121	.394251	.392393	.390547	.388709	-12
8500	.386884	.385070	.383264	.381470	.379688	.377913	.376154	.374399	.372658	.370925	-12
8600	.369204	.367493	.365790	.364098	.362417	.360743	.359079	.357426	.355784	.354149	-12
8700	.352524	.350906	.349302	.347704	.346116	.344536	.342965	.341404	.339853	.338308	-12
8800	.336773	.335248	.333729	.332220	.330720	.329229	.327744	.326269	.324803	.323345	-12
8900	.321894	.320452	.319019	.317591	.316172	.314762	.313361	.311965	.310578	.309199	-12
9000	.307826	.306464	.305105	.303757	.302417	.301082	.299755	.298436	.297123	.295817	-12
9100	.294520	.293230	.291945	.290668	.289399	.288137	.286880	.285630	.284388	.283154	-12
9200	.281924	.280701	.279486	.278278	.277075	.275878	.274689	.273507	.272329	.271161	-12
9300	.269994	.268837	.267687	.266540	.265401	.264268	.263140	.262018	.260902	.259793	-12
9400	.258691	.257593	.256500	.255415	.254335	.253260	.252190	.251127	.250070	.249017	-12
9500	.247972	.246931	.245893	.244864	.243838	.242820	.241804	.240796	.239793	.238794	-12
9600	.237801	.236813	.235829	.234853	.233879	.232912	.231950	.230991	.230037	.229089	-12
9700	.228146	.227209	.226274	.225347	.224423	.223505	.222589	.221681	.220775	.219874	-12
9800	.218979	.218086	.217201	.216318	.215440	.214565	.213697	.212832	.211974	.211116	-12
9900	.210265	.209418	.208575	.207736	.206901	.206071	.205246	.204423	.203605	.202791	-12

$M_{e\lambda}(\lambda T)/M_{emax}(T)$

TABLE 11A

Blackbody: Relative Fractional Emittance (Radiant)

λT(μ.°K)	0	1	2	3	4	5	6	7	8	9	m
100	.000002	.000006	.000024	.000093	.000345	.001251	.004429	.015313	.051731	.170859	-51
110	.000055	.000175	.000541	.001642	.004886	.014267	.040883	.115040	.318018	.864053	-47
120	.000231	.000607	.001569	.003994	.010016	.024742	.060240	.144598	.342311	.799466	-43
130	.000184	.000419	.000942	.002090	.004583	.009930	.021270	.045048	.094362	.195539	-39
140	.000401	.000814	.001635	.003252	.006405	.012499	.024165	.046294	.087901	.165451	-36
150	.003088	.005714	.010488	.019094	.034491	.061822	.109970	.194165	.340320	.592227	-34
160	.001023	.001756	.002993	.005069	.008526	.014251	.023670	.039072	.064105	.104553	-31
170	.001695	.002733	.004381	.006984	.011073	.017463	.027394	.042751	.066378	.102549	-29
180	.015765	.024119	.036725	.055658	.083964	.126092	.188516	.280610	.415897	.613795	-28
190	.009021	.013203	.019247	.027946	.040416	.058227	.083569	.119491	.170226	.241623	-26
200	.034174	.048165	.067648	.094688	.132091	.183658	.254523	.351597	.484155	.664609	-25
210	.009095	.012409	.016879	.022892	.030957	.041744	.056129	.075262	.100638	.134207	-23
220	.017849	.023677	.031325	.041338	.054413	.071445	.093577	.122267	.159370	.207240	-22
230	.026886	.034800	.044940	.057905	.074445	.095500	.122245	.156146	.199029	.253161	-21
240	.032136	.040709	.051467	.064939	.081777	.102783	.128939	.161446	.201773	.251710	-20
250	.031344	.038960	.048342	.059877	.074037	.091389	.112618	.138546	.170163	.208655	-19
260	.025544	.031222	.038102	.046426	.056481	.068610	.083218	.100786	.121884	.147185	-18
270	.177482	.213711	.256975	.308568	.370011	.443086	.529880	.632834	.754798	.899098	-18
280	.106961	.127084	.150802	.178725	.211557	.250116	.295345	.348338	.410354	.482847	-17
290	.056749	.066621	.078121	.091504	.107061	.125126	.146080	.170360	.198463	.230960	-16
300	.268497	.311813	.361747	.419253	.485414	.561459	.648779	.748952	.863759	.995216	-16
310	.114560	.131747	.151373	.173763	.199284	.228347	.261416	.299011	.341714	.390177	-15
320	.044513	.050740	.057788	.065762	.074773	.084950	.096434	.109383	.123971	.140395	-14
330	.158871	.179639	.202966	.229149	.258514	.291425	.328284	.369534	.415665	.467220	-14
340	.052479	.058905	.066070	.074057	.082951	.092850	.103860	.116099	.129693	.144783	-13
350	.161525	.180085	.200649	.223420	.248618	.276486	.307286	.341308	.378867	.420304	-13
360	.046599	.051635	.057180	.063284	.069999	.077382	.085496	.094407	.104188	.114918	-12
370	.126684	.139579	.153702	.169164	.186082	.204584	.224808	.246902	.271028	.297359	-12
380	.326083	.357401	.391531	.428708	.469184	.513232	.561143	.613232	.669838	.731322	-12
390	.079808	.087052	.094910	.103429	.112663	.122665	.133496	.145218	.157900	.171614	-11
400	.186439	.202457	.219757	.238434	.258590	.280333	.303777	.329047	.356273	.385595	-11
410	.417163	.451134	.487679	.526976	.569218	.614606	.663359	.715705	.771889	.832169	-11
420	.089682	.096614	.104043	.112002	.120527	.129653	.139420	.149871	.161047	.172996	-10
430	.185767	.199412	.213986	.229546	.246155	.263877	.282780	.302937	.324424	.347322	-10
440	.371715	.397693	.425350	.454787	.486107	.519420	.554845	.592501	.632520	.675035	-10
450	.072019	.076813	.081902	.087303	.093032	.099107	.105549	.112378	.119613	.127278	-9
460	.135396	.143991	.153089	.162717	.172902	.183675	.195066	.207107	.219832	.233277	-9
470	.247478	.262475	.278307	.295018	.312652	.331254	.350874	.371561	.393369	.416353	-9
480	.440570	.466081	.492947	.521235	.551013	.582351	.615325	.650011	.686489	.724845	-9
490	.076516	.080754	.085206	.089884	.094796	.099955	.105370	.111054	.117019	.123277	-8

TABLE 11A

Blackbody: Relative Fractional Emittance (Radiant)

$\lambda T(\mu.^\circ K)$	0	1	2	3	4	5	6	7	8	9	m	Δ_2
500	.129841	.136724	.143941	.151507	.159436	.167744	.176448	.185565	.195111	.205107	-8	467
510	.215569	.226519	.237976	.249963	.262500	.275610	.289318	.303647	.318623	.334273	-8	700
520	.350622	.367700	.385536	.404160	.423603	.443897	.465076	.487174	.510228	.534273	-8	1031
530	.559350	.585496	.612753	.641164	.670771	.701620	.733758	.767233	.802094	.838393	-8	1491
540	.087618	.091552	.095645	.099905	.104337	.108947	.113741	.118727	.123911	.129299	-7	212
550	.134900	.140721	.146769	.153052	.159579	.166358	.173397	.180705	.188291	.196166	-7	298
560	.204338	.212818	.221616	.230742	.240208	.250024	.260203	.270755	.281693	.293030	-7	411
570	.304778	.316951	.329563	.342626	.356157	.370168	.384676	.399696	.415245	.431337	-7	561
580	.447991	.465223	.483052	.501495	.520571	.540300	.560701	.581793	.603599	.626139	-7	756
590	.649434	.673508	.698383	.724082	.750630	.778051	.806370	.835614	.865808	.896980	-7	1006
600	.092916	.096237	.099665	.103201	.106851	.110615	.114499	.118504	.122635	.126894	-6	132
610	.131286	.135814	.140481	.145292	.150249	.155358	.160622	.166045	.171632	.177386	-6	172
620	.183312	.189415	.195699	.202170	.208830	.215687	.222743	.230006	.237479	.245169	-6	222
630	.253080	.261219	.269590	.278200	.287055	.296160	.305522	.315147	.325041	.335212	-6	283
640	.345665	.356407	.367446	.378788	.390441	.402413	.414709	.427339	.440310	.453630	-6	357
650	.467307	.481349	.495765	.510563	.525753	.541342	.557341	.573757	.590602	.607884	-6	447
660	.625613	.643799	.662452	.681583	.701202	.721320	.741947	.763095	.784775	.806998	-6	555
670	.082978	.085312	.087705	.090156	.092668	.095242	.097879	.100580	.103346	.106180	-5	68
680	.109082	.112053	.115096	.118212	.121402	.124667	.128010	.131431	.134933	.138516	-5	84
690	.142183	.145936	.149775	.153703	.157721	.161832	.166036	.170336	.174734	.179231	-5	101
700	.183829	.188531	.193338	.198253	.203277	.208412	.213660	.219024	.224506	.230108	-5	122
710	.235832	.241681	.247656	.253760	.259995	.266364	.272870	.279513	.286298	.293227	-5	146
720	.300302	.307525	.314900	.322429	.330115	.337960	.345968	.354141	.362481	.370993	-5	174
730	.379678	.388540	.397582	.406807	.416218	.425818	.435610	.445597	.455784	.466172	-5	205
740	.476766	.487569	.498583	.509814	.521264	.532936	.544835	.556965	.569327	.581928	-5	241
750	.594770	.607856	.621192	.634781	.648627	.662734	.677106	.691748	.706662	.721855	-5	282
760	.737329	.753090	.769141	.785487	.802133	.819083	.836342	.853914	.871804	.890017	-5	328
770	.090856	.092743	.094664	.096619	.098609	.100635	.102696	.104793	.106927	.109099	-4	38
780	.111308	.113556	.115842	.118169	.120535	.122942	.125390	.127880	.130412	.132988	-4	44
790	.135606	.138270	.140977	.143731	.146530	.149376	.152269	.155210	.158199	.161238	-4	50
800	.164327	.167466	.170657	.173899	.177194	.180542	.183944	.187401	.190913	.194482	-4	57
810	.198107	.201790	.205531	.209331	.213191	.217112	.221093	.225137	.229244	.233415	-4	65
820	.237650	.241951	.246317	.250750	.255252	.259821	.264461	.269170	.273951	.278803	-4	73
830	.283729	.288728	.293802	.298951	.304177	.309480	.314862	.320322	.325863	.331485	-4	82
840	.337188	.342975	.348846	.354801	.360842	.366971	.373187	.379492	.385886	.392372	-4	92
850	.398949	.405620	.412384	.419244	.426200	.433252	.440403	.447654	.455004	.462457	-4	103
860	.470012	.477670	.485434	.493304	.501280	.509365	.517560	.525865	.534282	.542812	-4	114
870	.551456	.560216	.569092	.578086	.587199	.596432	.605787	.615265	.624867	.634594	-4	127
880	.644448	.654429	.664540	.674782	.685155	.695661	.706302	.717079	.727993	.739046	-4	140
890	.750238	.761572	.773048	.784669	.796435	.808348	.820409	.832620	.844982	.857497	-4	154
900	.870166	.882991	.895973	.909113	.922414	.935876	.949501	.963290	.977246	.991370	-4	169
910	.100566	.102013	.103476	.104957	.106456	.107972	.109506	.111058	.112629	.114218	-3	19
920	.115825	.117451	.119096	.120759	.122442	.124145	.125867	.127609	.129370	.131152	-3	20
930	.132954	.134776	.136619	.138482	.140367	.142272	.144199	.146147	.148117	.150109	-3	22
940	.152123	.154159	.156218	.158299	.160403	.162529	.164679	.166852	.169049	.171269	-3	24
950	.173514	.175782	.178075	.180392	.182734	.185101	.187493	.189910	.192352	.194821	-3	26
960	.197315	.199835	.202381	.204954	.207554	.210181	.212835	.215516	.218224	.220960	-3	28
970	.223725	.226517	.229338	.232187	.235066	.237973	.240909	.243875	.246871	.249896	-3	30
980	.252952	.256038	.259154	.262302	.265480	.268690	.271931	.275203	.278508	.281844	-3	32
990	.285214	.288615	.292050	.295517	.299018	.302553	.306121	.309723	.313359	.317030	-3	35

$\int_0^{\lambda T} M_{e\lambda}\, d\lambda / M_e(T)$

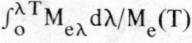

TABLE 11A

Blackbody: Relative Fractional Emittance (Radiant)

$\lambda T(\mu.°K)$	0	1	2	3	4	5	6	7	8	9	m	Δ_2
1000	.320736	.324476	.328252	.332063	.335910	.339793	.343712	.347667	.351659	.355688	-3	37
1010	.359754	.363857	.367998	.372177	.376394	.380650	.384944	.389277	.393649	.398061	-3	40
1020	.402512	.407003	.411534	.416106	.420718	.425372	.430066	.434802	.439580	.444400	-3	42
1030	.449262	.454166	.459114	.464104	.469138	.474215	.479336	.484502	.489711	.494966	-3	45
1040	.500265	.505609	.510999	.516435	.521917	.527445	.533019	.538641	.544309	.550025	-3	48
1050	.555789	.561601	.567461	.573369	.579326	.585333	.591388	.597494	.603649	.609855	-3	50
1060	.616111	.622418	.628776	.635185	.641646	.648159	.654725	.661342	.668013	.674737	-3	53
1070	.681514	.688345	.695230	.702169	.709163	.716211	.723315	.730475	.737690	.744961	-3	56
1080	.752289	.759673	.767114	.774613	.782169	.789783	.797455	.805186	.812975	.820824	-3	59
1090	.828732	.836700	.844728	.852816	.860964	.869174	.877445	.885778	.894172	.902628	-3	63
1100	.911148	.919729	.928374	.937083	.945855	.954691	.963592	.972557	.981587	.990683	-3	66
1110	.099984	.100907	.101836	.102772	.103715	.104665	.105621	.106584	.107553	.108530	-2	7
1120	.109514	.110504	.111501	.112506	.113517	.114535	.115561	.116594	.117633	.118680	-2	7
1130	.119734	.120796	.121864	.122940	.124023	.125114	.126212	.127317	.128430	.129551	-2	8
1140	.130679	.131814	.132957	.134108	.135267	.136433	.137607	.138788	.139978	.141175	-2	8
1150	.142380	.143593	.144814	.146043	.147280	.148524	.149777	.151039	.152308	.153585	-2	8
1160	.154871	.156164	.157467	.158777	.160096	.161423	.162758	.164102	.165454	.166815	-2	9
1170	.168185	.169563	.170949	.172344	.173748	.175161	.176582	.178012	.179451	.180899	-2	9
1180	.182355	.183821	.185295	.186779	.188271	.189773	.191283	.192803	.194332	.195870	-2	9
1190	.197417	.198973	.200539	.202114	.203698	.205292	.206895	.208508	.210130	.211762	-2	10
1200	.213403	.215054	.216714	.218384	.220064	.221753	.223452	.225161	.226880	.228609	-2	10
1210	.230348	.232096	.233855	.235623	.237402	.239190	.240989	.242798	.244617	.246446	-2	10
1220	.248285	.250135	.251995	.253865	.255746	.257637	.259539	.261451	.263374	.265307	-2	11
1230	.267250	.269205	.271170	.273145	.275132	.277129	.279136	.281155	.283185	.285225	-2	11
1240	.287276	.289339	.291412	.293496	.295592	.297698	.299816	.301944	.304084	.306236	-2	11
1250	.308398	.310572	.312757	.314953	.317161	.319380	.321611	.323853	.326106	.328372	-2	12
1260	.330649	.332937	.335237	.337549	.339872	.342208	.344555	.346914	.349285	.351667	-2	12
1270	.354062	.356469	.358887	.361318	.363761	.366215	.368682	.371161	.373653	.376156	-2	12
1280	.378672	.381200	.383740	.386293	.388858	.391436	.394026	.396628	.399243	.401871	-2	13
1290	.404511	.407164	.409830	.412508	.415199	.417902	.420619	.423348	.426090	.428845	-2	13
1300	.431613	.434394	.437188	.439995	.442815	.445648	.448494	.451353	.454225	.457111	-2	13
1310	.460010	.462922	.465847	.468786	.471738	.474704	.477683	.480675	.483681	.486701	-2	14
1320	.489734	.492780	.495840	.498914	.502002	.505103	.508218	.511347	.514489	.517646	-2	14
1330	.520816	.524000	.527198	.530410	.533636	.536876	.540131	.543399	.546681	.549978	-2	14
1340	.553288	.556613	.559952	.563306	.566673	.570055	.573452	.576862	.580287	.583727	-2	14
1350	.587181	.590650	.594133	.597630	.601143	.604670	.608211	.611767	.615338	.618924	-2	15
1360	.622524	.626140	.629770	.633415	.637075	.640749	.644439	.648144	.651863	.655598	-2	15
1370	.659348	.663113	.666893	.670688	.674498	.678324	.682165	.686021	.689892	.693779	-2	15
1380	.697681	.701598	.705531	.709479	.713442	.717422	.721416	.725426	.729452	.733493	-2	16
1390	.737550	.741623	.745711	.749815	.753935	.758070	.762222	.766389	.770572	.774770	-2	16
1400	.778985	.783216	.787462	.791725	.796003	.800298	.804608	.808935	.813277	.817636	-2	16
1410	.822011	.826403	.830810	.835234	.839673	.844130	.848602	.853091	.857596	.862118	-2	16
1420	.866656	.871210	.875781	.880368	.884972	.889592	.894229	.898883	.903553	.908239	-2	17
1430	.912943	.917663	.922400	.927153	.931923	.936710	.941514	.946335	.951172	.956027	-2	17
1440	.096090	.096579	.097069	.097561	.098055	.098551	.099048	.099547	.100048	.100550	-1	2
1450	.101054	.101560	.102068	.102577	.103088	.103601	.104115	.104632	.105150	.105669	-1	2
1460	.106191	.106714	.107239	.107765	.108294	.108824	.109355	.109889	.110424	.110962	-1	2
1470	.111500	.112041	.112583	.113128	.113673	.114221	.114771	.115322	.115875	.116429	-1	2
1480	.116986	.117544	.118104	.118666	.119230	.119795	.120363	.120932	.121502	.122075	-1	2
1490	.122649	.123226	.123804	.124383	.124965	.125548	.126134	.126721	.127310	.127900	-1	2

TABLE 11A

Blackbody: Relative Fractional Emittance (Radiant)

$\lambda T(\mu.°K)$	0	1	2	3	4	5	6	7	8	9	m	Δ_2
1500	.128493	.129087	.129683	.130281	.130881	.131482	.132086	.132691	.133298	.133907	-1	2
1510	.134518	.135130	.135745	.136361	.136979	.137599	.138220	.138844	.139470	.140097	-1	2
1520	.140726	.141357	.141990	.142624	.143261	.143899	.144540	.145182	.145826	.146472	-1	2
1530	.147119	.147769	.148420	.149074	.149729	.150386	.151045	.151706	.152369	.153033	-1	2
1540	.153700	.154368	.155038	.155710	.156384	.157060	.157738	.158418	.159099	.159783	-1	2
1550	.160468	.161155	.161845	.162536	.163229	.163923	.164620	.165319	.166019	.166722	-1	2
1560	.167426	.168133	.168841	.169551	.170263	.170977	.171693	.172410	.173130	.173852	-1	2
1570	.174575	.175301	.176028	.176757	.177489	.178222	.178957	.179694	.180433	.181174	-1	2
1580	.181917	.182661	.183408	.184157	.184907	.185660	.186414	.187170	.187929	.188689	-1	2
1590	.189451	.190215	.190981	.191749	.192519	.193291	.194065	.194841	.195619	.196399	-1	2
1600	.197180	.197964	.198749	.199537	.200327	.201118	.201911	.202707	.203504	.204303	-1	2
1610	.205105	.205908	.206713	.207520	.208329	.209140	.209953	.210768	.211585	.212404	-1	2
1620	.213225	.214048	.214873	.215700	.216529	.217359	.218192	.219027	.219863	.220702	-1	2
1630	.221543	.222385	.223230	.224077	.224925	.225776	.226628	.227483	.228339	.229197	-1	2
1640	.230058	.230920	.231785	.232651	.233519	.234390	.235262	.236136	.237013	.237891	-1	2
1650	.238771	.239654	.240538	.241424	.242312	.243203	.244095	.244989	.245885	.246783	-1	2
1660	.247684	.248586	.249490	.250396	.251304	.252214	.253126	.254041	.254957	.255875	-1	2
1670	.256795	.257717	.258641	.259567	.260495	.261425	.262357	.263292	.264228	.265166	-1	2
1680	.266106	.267048	.267992	.268938	.269886	.270836	.271788	.272742	.273698	.274656	-1	2
1690	.275616	.276578	.277542	.278509	.279477	.280447	.281419	.282393	.283369	.284347	-1	2
1700	.285327	.286309	.287293	.288279	.289267	.290257	.291249	.292243	.293239	.294237	-1	2
1710	.295237	.296239	.297243	.298249	.299258	.300268	.301280	.302294	.303310	.304328	-1	2
1720	.305348	.306370	.307394	.308420	.309448	.310478	.311510	.312544	.313580	.314618	-1	2
1730	.315658	.316700	.317744	.318790	.319838	.320888	.321940	.322994	.324050	.325108	-1	2
1740	.326168	.327230	.328294	.329360	.330428	.331498	.332570	.333644	.334720	.335798	-1	2
1750	.336877	.337959	.339043	.340129	.341217	.342307	.343399	.344493	.345588	.346686	-1	2
1760	.347786	.348888	.349992	.351097	.352205	.353315	.354427	.355540	.356656	.357774	-1	2
1770	.358893	.360015	.361139	.362264	.363392	.364521	.365653	.366786	.367922	.369059	-1	2
1780	.370199	.371340	.372484	.373629	.374776	.375926	.377077	.378230	.379386	.380543	-1	2
1790	.381702	.382863	.384026	.385191	.386358	.387527	.388698	.389871	.391046	.392223	-1	2
1800	.393402	.394583	.395766	.396951	.398137	.399326	.400517	.401709	.402904	.404100	-1	2
1810	.405299	.406499	.407702	.408906	.410112	.411321	.412531	.413743	.414957	.416173	-1	2
1820	.417391	.418611	.419833	.421057	.422283	.423510	.424740	.425972	.427205	.428441	-1	2
1830	.429678	.430917	.432159	.433402	.434647	.435894	.437143	.438394	.439647	.440902	-1	2
1840	.442159	.443418	.444678	.445941	.447205	.448472	.449740	.451011	.452283	.453557	-1	2
1850	.454833	.456111	.457391	.458673	.459956	.461242	.462529	.463819	.465110	.466403	-1	2
1860	.467699	.468996	.470295	.471596	.472899	.474203	.475510	.476818	.478129	.479441	-1	2
1870	.480755	.482071	.483389	.484709	.486031	.487355	.488680	.490008	.491337	.492669	-1	2
1880	.494002	.495337	.496674	.498012	.499353	.500696	.502040	.503386	.504735	.506085	-1	2
1890	.507437	.508790	.510146	.511504	.512863	.514224	.515587	.516952	.518319	.519688	-1	2
1900	.521059	.522431	.523806	.525182	.526560	.527940	.529321	.530705	.532090	.533478	-1	2
1910	.534867	.536258	.537651	.539045	.540442	.541840	.543240	.544642	.546046	.547452	-1	2
1920	.548860	.550269	.551680	.553093	.554508	.555925	.557343	.558764	.560186	.561610	-1	2
1930	.563036	.564463	.565893	.567324	.568757	.570192	.571629	.573067	.574507	.575950	-1	2
1940	.577393	.578839	.580287	.581736	.583187	.584640	.586095	.587551	.589010	.590470	-1	2
1950	.591932	.593395	.594861	.596328	.597797	.599268	.600740	.602215	.603691	.605169	-1	2
1960	.606648	.608130	.609613	.611098	.612585	.614073	.615564	.617056	.618549	.620045	-1	2
1970	.621542	.623041	.624542	.626045	.627549	.629055	.630563	.632073	.633584	.635097	-1	2
1980	.636612	.638129	.639647	.641167	.642689	.644212	.645737	.647264	.648793	.650323	-1	2
1990	.651855	.653389	.654925	.656462	.658001	.659542	.661084	.662628	.664174	.665722	-1	2

$\int_o^{\lambda T} M_{e\lambda} d\lambda / M_e(T)$

TABLE 11A

Blackbody: Relative Fractional Emittance (Radiant)

λT(μ.°K)	0	1	2	3	4	5	6	7	8	9	m	Δ_2
2000	.667271	.668822	.670375	.671929	.673485	.675043	.676602	.678164	.679726	.681291	-1	2
2010	.682857	.684425	.685995	.687566	.689139	.690714	.692290	.693868	.695448	.697029	-1	2
2020	.698612	.700197	.701783	.703371	.704961	.706552	.708145	.709740	.711336	.712934	-1	2
2030	.714534	.716135	.717738	.719343	.720949	.722557	.724166	.725777	.727390	.729004	-1	2
2040	.730620	.732238	.733857	.735478	.737101	.738725	.740351	.741978	.743607	.745238	-1	2
2050	.746870	.748504	.750140	.751777	.753416	.755056	.756698	.758342	.759987	.761633	-1	2
2060	.763282	.764932	.766583	.768236	.769891	.771547	.773205	.774865	.776526	.778188	-1	2
2070	.779852	.781518	.783185	.784854	.786525	.788197	.789870	.791546	.793222	.794901	-1	2
2080	.796580	.798262	.799945	.801629	.803315	.805003	.806692	.808383	.810075	.811769	-1	2
2090	.813464	.815161	.816859	.818559	.820260	.821963	.823668	.825374	.827081	.828790	-1	2
2100	.830501	.832213	.833927	.835642	.837358	.839076	.840796	.842517	.844240	.845964	-1	2
2110	.847689	.849416	.851145	.852875	.854607	.856340	.858074	.859810	.861548	.863287	-1	1
2120	.865027	.866769	.868513	.870257	.872004	.873752	.875501	.877252	.879004	.880757	-1	1
2130	.882512	.884269	.886027	.887786	.889547	.891310	.893074	.894839	.896605	.898374	-1	1
2140	.900143	.901914	.903686	.905460	.907236	.909012	.910790	.912570	.914351	.916133	-1	1
2150	.917917	.919702	.921489	.923277	.925066	.926857	.928649	.930443	.932238	.934034	-1	1
2160	.935832	.937631	.939431	.941233	.943037	.944842	.946648	.948455	.950264	.952074	-1	1
2170	.953886	.955699	.957513	.959329	.961146	.962964	.964784	.966605	.968428	.970251	-1	1
2180	.972077	.973903	.975731	.977560	.979391	.981223	.983056	.984890	.986726	.988564	-1	1
2190	.099040	.099224	.099408	.099593	.099777	.099961	.100146	.100331	.100516	.100701	-0	0
2200	.100886	.101071	.101257	.101442	.101628	.101814	.102000	.102186	.102372	.102558	-0	0
2210	.102745	.102931	.103118	.103305	.103492	.103679	.103866	.104054	.104241	.104429	-0	0
2220	.104617	.104804	.104992	.105181	.105369	.105557	.105746	.105934	.106123	.106312	-0	0
2230	.106501	.106690	.106879	.107069	.107258	.107448	.107637	.107827	.108017	.108207	-0	0
2240	.108398	.108588	.108778	.108969	.109160	.109351	.109541	.109733	.109924	.110115	-0	0
2250	.110306	.110498	.110690	.110881	.111073	.111265	.111458	.111650	.111842	.112035	-0	0
2260	.112227	.112420	.112613	.112806	.112999	.113192	.113385	.113579	.113772	.113966	-0	0
2270	.114160	.114354	.114548	.114742	.114936	.115131	.115325	.115520	.115714	.115909	-0	0
2280	.116104	.116299	.116494	.116689	.116885	.117080	.117276	.117472	.117667	.117863	-0	0
2290	.118059	.118256	.118452	.118648	.118845	.119041	.119238	.119435	.119632	.119829	-0	0
2300	.120026	.120223	.120421	.120618	.120816	.121013	.121211	.121409	.121607	.121805	-0	0
2310	.122003	.122202	.122400	.122599	.122797	.122996	.123195	.123394	.123593	.123792	-0	0
2320	.123991	.124191	.124390	.124590	.124789	.124989	.125189	.125389	.125589	.125789	-0	0
2330	.125990	.126190	.126391	.126591	.126792	.126993	.127194	.127395	.127596	.127797	-0	0
2340	.127998	.128200	.128401	.128603	.128805	.129007	.129208	.129410	.129613	.129815	-0	0
2350	.130017	.130220	.130422	.130625	.130827	.131030	.131233	.131436	.131639	.131842	-0	0
2360	.132046	.132249	.132452	.132656	.132860	.133063	.133267	.133471	.133675	.133879	-0	0
2370	.134084	.134288	.134492	.134697	.134901	.135106	.135311	.135516	.135721	.135926	-0	0
2380	.136131	.136336	.136541	.136747	.136952	.137158	.137364	.137570	.137775	.137981	-0	0
2390	.138 87	.138394	.138600	.138806	.139013	.139219	.139426	.139632	.139839	.140046	-0	0
2400	.140253	.140460	.140667	.140874	.141081	.141289	.141496	.141704	.141911	.142119	-0	0
2410	.142327	.142535	.142743	.142951	.143159	.143367	.143576	.143784	.143992	.144201	-0	0
2420	.144410	.144618	.144827	.145036	.145245	.145454	.145663	.145872	.146082	.146291	-0	0
2430	.146501	.146710	.146920	.147129	.147339	.147549	.147759	.147969	.148179	.148389	-0	0
2440	.148600	.148810	.149020	.149231	.149441	.149652	.149863	.150074	.150284	.150495	-0	0
2450	.150706	.150918	.151129	.151340	.151551	.151763	.151974	.152186	.152397	.152609	-0	0
2460	.152821	.153033	.153245	.153457	.153669	.153881	.154093	.154306	.154518	.154730	-0	0
2470	.154943	.155155	.155368	.155581	.155794	.156007	.156220	.156433	.156646	.156859	-0	0
2480	.157072	.157285	.157499	.157712	.157926	.158139	.158353	.158567	.158781	.158994	-0	0
2490	.159208	.159422	.159636	.159851	.160065	.160279	.160493	.160708	.160922	.161137	-0	0

TABLE 11A

Blackbody: Relative Fractional Emittance (Radiant)

$\lambda T(\mu.^\circ K)$	0	1	2	3	4	5	6	7	8	9	m	Δ_2
2500	.161352	.161566	.161781	.161996	.162211	.162426	.162641	.162856	.163071	.163286	-0	0
2510	.163501	.163717	.163932	.164147	.164363	.164579	.164794	.165010	.165226	.165442	-0	0
2520	.165657	.165873	.166089	.166306	.166522	.166738	.166954	.167171	.167387	.167603	-0	0
2530	.167820	.168036	.168253	.168470	.168687	.168903	.169120	.169337	.169554	.169771	-0	0
2540	.169988	.170206	.170423	.170640	.170857	.171075	.171292	.171510	.171727	.171945	-0	0
2550	.172163	.172380	.172598	.172816	.173034	.173252	.173470	.173688	.173906	.174124	-0	0
2560	.174343	.174561	.174779	.174998	.175216	.175435	.175653	.175872	.176091	.176309	-0	0
2570	.176528	.176747	.176966	.177185	.177404	.177623	.177842	.178061	.178280	.178500	-0	0
2580	.178719	.178938	.179158	.179377	.179597	.179816	.180036	.180255	.180475	.180695	-0	0
2590	.180915	.181135	.181354	.181574	.181794	.182014	.182235	.182455	.182675	.182895	-0	0
2600	.183115	.183336	.183556	.183777	.183997	.184218	.184438	.184659	.184879	.185100	-0	0
2610	.185321	.185542	.185762	.185983	.186204	.186425	.186646	.186867	.187088	.187310	-0	0
2620	.187531	.187752	.187973	.188195	.188416	.188637	.188859	.189080	.189302	.189523	-0	0
2630	.189745	.189967	.190188	.190410	.190632	.190854	.191076	.191297	.191519	.191741	-0	0
2640	.191963	.192186	.192408	.192630	.192852	.193074	.193296	.193519	.193741	.193963	-0	0
2650	.194186	.194408	.194631	.194853	.195076	.195299	.195521	.195744	.195967	.196189	-0	0
2660	.196412	.196635	.196858	.197081	.197304	.197527	.197750	.197973	.198196	.198419	-0	0
2670	.198642	.198865	.199088	.199312	.199535	.199758	.199982	.200205	.200428	.200652	-0	0
2680	.200875	.201099	.201322	.201546	.201770	.201993	.202217	.202441	.202664	.202888	-0	0
2690	.203112	.203336	.203560	.203783	.204007	.204231	.204455	.204679	.204903	.205127	-0	0
2700	.205352	.205576	.205800	.206024	.206248	.206473	.206697	.206921	.207145	.207370	-0	0
2710	.207594	.207819	.208043	.208268	.208492	.208717	.208941	.209166	.209390	.209615	-0	0
2720	.209840	.210064	.210289	.210514	.210739	.210963	.211188	.211413	.211638	.211863	-0	0
2730	.212088	.212313	.212538	.212763	.212988	.213213	.213438	.213663	.213888	.214113	-0	0
2740	.214338	.214563	.214789	.215014	.215239	.215464	.215690	.215915	.216140	.216366	-0	0
2750	.216591	.216816	.217042	.217267	.217493	.217718	.217944	.218169	.218395	.218620	-0	0
2760	.218846	.219072	.219297	.219523	.219748	.219974	.220200	.220426	.220651	.220877	-0	0
2770	.221103	.221329	.221554	.221780	.222006	.222232	.222458	.222684	.222910	.223136	-0	0
2780	.223362	.223588	.223814	.224040	.224266	.224492	.224718	.224944	.225170	.225396	-0	0
2790	.225622	.225848	.226074	.226300	.226527	.226753	.226979	.227205	.227431	.227658	-0	0
2800	.227884	.228110	.228336	.228563	.228789	.229015	.229242	.229468	.229694	.229921	-0	0
2810	.230147	.230373	.230600	.230826	.231053	.231279	.231506	.231732	.231959	.232185	-0	0
2820	.232412	.232638	.232865	.233091	.233318	.233544	.233771	.233997	.234224	.234451	-0	0
2830	.234677	.234904	.235130	.235357	.235584	.235810	.236037	.236264	.236490	.236717	-0	0
2840	.236944	.237170	.237397	.237624	.237850	.238077	.238304	.238531	.238757	.238984	-0	0
2850	.239211	.239438	.239664	.239891	.240118	.240345	.240572	.240798	.241025	.241252	-0	0
2860	.241479	.241706	.241932	.242159	.242386	.242613	.242840	.243067	.243293	.243520	-0	0
2870	.243747	.243974	.244201	.244428	.244655	.244882	.245108	.245335	.245562	.245789	-0	0
2880	.246016	.246243	.246470	.246697	.246924	.247150	.247377	.247604	.247831	.248058	-0	0
2890	.248285	.248512	.248739	.248966	.249193	.249420	.249646	.249873	.250100	.250327	-0	0
2900	.250554	.250781	.251008	.251235	.251462	.251689	.251916	.252142	.252369	.252596	-0	0
2910	.252823	.253050	.253277	.253504	.253731	.253958	.254185	.254411	.254638	.254865	-0	0
2920	.255092	.255319	.255546	.255773	.256000	.256227	.256453	.256680	.256907	.257134	-0	0
2930	.257361	.257588	.257814	.258041	.258268	.258495	.258722	.258949	.259175	.259402	-0	0
2940	.259629	.259856	.260083	.260309	.260536	.260763	.260990	.261216	.261443	.261670	-0	0
2950	.261897	.262123	.262350	.262577	.262804	.263030	.263257	.263484	.263710	.263937	-0	0
2960	.264164	.264390	.264617	.264844	.265070	.265297	.265524	.265750	.265977	.266203	-0	0
2970	.266430	.266657	.266883	.267110	.267336	.267563	.267789	.268016	.268242	.268469	-0	0
2980	.268695	.268922	.269148	.269375	.269601	.269828	.270054	.270281	.270507	.270733	-0	0
2990	.270960	.271186	.271413	.271639	.271865	.272092	.272318	.272544	.272771	.272997	-0	0

$\int_0^{\lambda T} M_{e\lambda}\, d\lambda / M_e(T)$

TABLE 11A

Blackbody: Relative Fractional Emittance (Radiant)

λT(μ.°K)	0	1	2	3	4	5	6	7	8	9	m	Δ_2
3000	.273223	.273449	.273676	.273902	.274128	.274354	.274580	.274807	.275033	.275259	-0	0
3010	.275485	.275711	.275937	.276163	.276389	.276616	.276842	.277068	.277294	.277520	-0	0
3020	.277746	.277972	.278198	.278424	.278650	.278876	.279101	.279327	.279553	.279779	-0	0
3030	.280005	.280231	.280457	.280682	.280908	.281134	.281360	.281585	.281811	.282037	-0	0
3040	.282263	.282488	.282714	.282940	.283165	.283391	.283616	.283842	.284067	.284293	-0	0
3050	.284519	.284744	.284970	.285195	.285420	.285646	.285871	.286097	.286322	.286547	-0	0
3060	.286773	.286998	.287223	.287449	.287674	.287899	.288124	.288350	.288575	.288800	-0	0
3070	.289025	.289250	.289475	.289700	.289925	.290150	.290375	.290600	.290825	.291050	-0	0
3080	.291275	.291500	.291725	.291950	.292175	.292400	.292625	.292849	.293074	.293299	-0	0
3090	.293524	.293748	.293973	.294198	.294422	.294647	.294872	.295096	.295321	.295545	-0	0
3100	.295770	.295994	.296219	.296443	.296668	.296892	.297116	.297341	.297565	.297789	-0	0
3110	.298014	.298238	.298462	.298686	.298910	.299135	.299359	.299583	.299807	.300031	-0	0
3120	.300255	.300479	.300703	.300927	.301151	.301375	.301599	.301823	.302046	.302270	-0	0
3130	.302494	.302718	.302941	.303165	.303389	.303613	.303836	.304060	.304283	.304507	-0	0
3140	.304730	.304954	.305177	.305401	.305624	.305848	.306071	.306294	.306518	.306741	-0	0
3150	.306964	.307187	.307411	.307634	.307857	.308080	.308303	.308526	.308749	.308972	-0	0
3160	.309195	.309418	.309641	.309864	.310087	.310310	.310533	.310755	.310978	.311201	-0	0
3170	.311424	.311646	.311869	.312092	.312314	.312537	.312759	.312982	.313204	.313427	-0	0
3180	.313649	.313871	.314094	.314316	.314538	.314761	.314983	.315205	.315427	.315649	-0	0
3190	.315871	.316093	.316315	.316537	.316759	.316981	.317203	.317425	.317647	.317869	-0	0
3200	.318091	.318312	.318534	.318756	.318978	.319199	.319421	.319642	.319864	.320085	-0	0
3210	.320307	.320528	.320750	.320971	.321192	.321414	.321635	.321856	.322078	.322299	-0	0
3220	.322520	.322741	.322962	.323183	.323404	.323625	.323846	.324067	.324288	.324509	-0	0
3230	.324730	.324950	.325171	.325392	.325613	.325833	.326054	.326274	.326495	.326715	-0	0
3240	.326936	.327156	.327377	.327597	.327817	.328038	.328258	.328478	.328698	.328919	-0	0
3250	.329139	.329359	.329579	.329799	.330019	.330239	.330459	.330679	.330898	.331118	-0	0
3260	.331338	.331558	.331778	.331997	.332217	.332436	.332656	.332875	.333095	.333314	-0	0
3270	.333534	.333753	.333973	.334192	.334411	.334630	.334850	.335069	.335288	.335507	-0	0
3280	.335726	.335945	.336164	.336383	.336602	.336821	.337039	.337258	.337477	.337696	-0	0
3290	.337914	.338133	.338352	.338570	.338789	.339007	.339226	.339444	.339662	.339881	-0	0
3300	.340099	.340317	.340535	.340754	.340972	.341190	.341408	.341626	.341844	.342062	-0	0
3310	.342280	.342497	.342715	.342933	.343151	.343369	.343586	.343804	.344021	.344239	-0	0
3320	.344456	.344674	.344891	.345109	.345326	.345543	.345761	.345978	.346195	.346412	-0	0
3330	.346629	.346846	.347063	.347280	.347497	.347714	.347931	.348148	.348365	.348581	-0	0
3340	.348798	.349015	.349231	.349448	.349664	.349881	.350097	.350314	.350530	.350746	-0	0
3350	.350963	.351179	.351395	.351611	.351827	.352043	.352259	.352475	.352691	.352907	-0	0
3360	.353123	.353339	.353555	.353770	.353986	.354202	.354417	.354633	.354848	.355064	-0	0
3370	.355279	.355495	.355710	.355925	.356141	.356356	.356571	.356786	.357001	.357216	-0	0
3380	.357431	.357646	.357861	.358076	.358291	.358506	.358720	.358935	.359150	.359364	-0	0
3390	.359579	.359794	.360008	.360222	.360437	.360651	.360865	.361080	.361294	.361508	-0	0
3400	.361722	.361936	.362150	.362364	.362578	.362792	.363006	.363220	.363434	.363647	-0	0
3410	.363861	.364075	.364288	.364502	.364715	.364929	.365142	.365356	.365569	.365782	-0	0
3420	.365995	.366209	.366422	.366635	.366848	.367061	.367274	.367487	.367700	.367912	-0	0
3430	.368125	.368338	.368550	.368763	.368976	.369188	.369401	.369613	.369826	.370038	-0	0
3440	.370250	.370463	.370675	.370887	.371099	.371311	.371523	.371735	.371947	.372159	-0	0
3450	.372371	.372583	.372794	.373006	.373218	.373429	.373641	.373852	.374064	.374275	-0	0
3460	.374487	.374698	.374909	.375120	.375332	.375543	.375754	.375965	.376176	.376387	-0	0
3470	.376598	.376809	.377019	.377230	.377441	.377651	.377862	.378073	.378283	.378494	-0	0
3480	.378704	.378914	.379125	.379335	.379545	.379755	.379965	.380176	.380386	.380596	-0	0
3490	.380805	.381015	.381225	.381435	.381645	.381854	.382064	.382274	.382483	.382693	-0	0

TABLE 11A

Blackbody: Relative Fractional Emittance (Radiant)

$\lambda T(\mu.{}^\circ K)$	0	1	2	3	4	5	6	7	8	9	m	Δ_2
3500	.382902	.383111	.383321	.383530	.383739	.383949	.384158	.384367	.384576	.384785	-0	0
3510	.384994	.385203	.385412	.385620	.385829	.386038	.386246	.386455	.386664	.386872	-0	0
3520	.387081	.387289	.387497	.387706	.387914	.388122	.388330	.388538	.388746	.388954	-0	0
3530	.389162	.389370	.389578	.389786	.389994	.390201	.390409	.390617	.390824	.391032	-0	0
3540	.391239	.391446	.391654	.391861	.392068	.392275	.392483	.392690	.392897	.393104	-0	0
3550	.393311	.393518	.393724	.393931	.394138	.394345	.394551	.394758	.394964	.395171	-0	0
3560	.395377	.395584	.395790	.395996	.396202	.396409	.396615	.396821	.397027	.397233	-0	0
3570	.397439	.397645	.397850	.398056	.398262	.398467	.398673	.398879	.399084	.399290	-0	0
3580	.399495	.399700	.399906	.400111	.400316	.400521	.400726	.400931	.401136	.401341	-0	0
3590	.401546	.401751	.401956	.402160	.402365	.402569	.402774	.402979	.403183	.403387	-0	0
3600	.403592	.403796	.404000	.404204	.404409	.404613	.404817	.405021	.405225	.405428	-0	0
3610	.405632	.405836	.406040	.406243	.406447	.406651	.406854	.407058	.407261	.407464	-0	0
3620	.407668	.407871	.408074	.408277	.408480	.408683	.408886	.409089	.409292	.409495	-0	0
3630	.409697	.409900	.410103	.410305	.410508	.410710	.410913	.411115	.411317	.411520	-0	0
3640	.411722	.411924	.412126	.412328	.412530	.412732	.412934	.413136	.413338	.413539	-0	0
3650	.413741	.413943	.414144	.414346	.414547	.414749	.414950	.415151	.415353	.415554	-0	0
3660	.415755	.415956	.416157	.416358	.416559	.416760	.416961	.417161	.417362	.417563	-0	0
3670	.417763	.417964	.418164	.418365	.418565	.418765	.418966	.419166	.419366	.419566	-0	0
3680	.419766	.419966	.420166	.420366	.420566	.420765	.420965	.421165	.421364	.421564	-0	0
3690	.421763	.421963	.422162	.422362	.422561	.422760	.422959	.423158	.423357	.423556	-0	0
3700	.423755	.423954	.424153	.424352	.424550	.424749	.424948	.425146	.425345	.425543	-0	0
3710	.425742	.425940	.426138	.426336	.426534	.426733	.426931	.427129	.427327	.427524	-0	0
3720	.427722	.427920	.428118	.428315	.428513	.428711	.428908	.429105	.429303	.429500	-0	0
3730	.429697	.429895	.430092	.430289	.430486	.430683	.430880	.431077	.431273	.431470	-0	0
3740	.431667	.431864	.432060	.432257	.432453	.432650	.432846	.433042	.433239	.433435	-0	0
3750	.433631	.433827	.434023	.434219	.434415	.434611	.434807	.435002	.435198	.435394	-0	0
3760	.435589	.435785	.435980	.436176	.436371	.436566	.436761	.436957	.437152	.437347	-0	0
3770	.437542	.437737	.437932	.438126	.438321	.438516	.438711	.438905	.439100	.439294	-0	0
3780	.439489	.439683	.439877	.440072	.440266	.440460	.440654	.440848	.441042	.441236	-0	0
3790	.441430	.441624	.441818	.442011	.442205	.442399	.442592	.442785	.442979	.443172	-0	0
3800	.443366	.443559	.443752	.443945	.444138	.444331	.444524	.444717	.444910	.445103	-0	0
3810	.445295	.445488	.445681	.445873	.446066	.446258	.446451	.446643	.446835	.447027	-0	0
3820	.447219	.447412	.447604	.447796	.447987	.448179	.448371	.448563	.448755	.448946	-0	0
3830	.449138	.449329	.449521	.449712	.449903	.450095	.450286	.450477	.450668	.450859	-0	0
3840	.451050	.451241	.451432	.451623	.451814	.452004	.452195	.452386	.452576	.452767	-0	0
3850	.452957	.453147	.453338	.453528	.453718	.453908	.454098	.454288	.454478	.454668	-0	0
3860	.454858	.455048	.455238	.455427	.455617	.455806	.455996	.456185	.456375	.456564	-0	0
3870	.456753	.456942	.457132	.457321	.457510	.457699	.457888	.458076	.458265	.458454	-0	0
3880	.458643	.458831	.459020	.459208	.459397	.459585	.459773	.459962	.460150	.460338	-0	0
3890	.460526	.460714	.460902	.461090	.461278	.461466	.461653	.461841	.462029	.462216	-0	0
3900	.462404	.462591	.462779	.462966	.463153	.463341	.463528	.463715	.463902	.464089	-0	0
3910	.464276	.464463	.464649	.464836	.465023	.465209	.465396	.465583	.465769	.465955	-0	0
3920	.466142	.466328	.466514	.466700	.466886	.467073	.467258	.467444	.467630	.467816	-0	0
3930	.468002	.468188	.468373	.468559	.468744	.468930	.469115	.469300	.469486	.469671	-0	0
3940	.469856	.470041	.470226	.470411	.470596	.470781	.470966	.471151	.471335	.471520	-0	0
3950	.471704	.471889	.472073	.472258	.472442	.472626	.472811	.472995	.473179	.473363	-0	0
3960	.473547	.473731	.473915	.474099	.474282	.474466	.474650	.474833	.475017	.475200	-0	0
3970	.475384	.475567	.475750	.475933	.476117	.476300	.476483	.476666	.476849	.477031	-0	0
3980	.477214	.477397	.477580	.477762	.477945	.478127	.478310	.478492	.478675	.478857	-0	0
3990	.479039	.479221	.479403	.479585	.479767	.479949	.480131	.480313	.480495	.480676	-0	0

$\int_o^{\lambda T} M_{e\lambda}\, d\lambda / M_e(T)$

TABLE 11A

Blackbody: Relative Fractional Emittance (Radiant)

λT(μ.°K)	0	1	2	3	4	5	6	7	8	9	m	Δ_2
4000	.480858	.481040	.481221	.481403	.481584	.481765	.481947	.482128	.482309	.482490	-0	0
4010	.482671	.482852	.483033	.483214	.483395	.483575	.483756	.483937	.484117	.484298	-0	0
4020	.484478	.484658	.484839	.485019	.485199	.485379	.485559	.485740	.485919	.486099	-0	0
4030	.486279	.486459	.486639	.486818	.486998	.487178	.487357	.487537	.487716	.487895	-0	0
4040	.488074	.488254	.488433	.488612	.488791	.488970	.489149	.489328	.489506	.489685	-0	0
4050	.489864	.490042	.490221	.490399	.490578	.490756	.490935	.491113	.491291	.491469	-0	0
4060	.491647	.491825	.492003	.492181	.492359	.492537	.492714	.492892	.493070	.493247	-0	0
4070	.493425	.493602	.493779	.493957	.494134	.494311	.494488	.494665	.494842	.495019	-0	0
4080	.495196	.495373	.495550	.495727	.495903	.496080	.496256	.496433	.496609	.496786	-0	0
4090	.496962	.497138	.497314	.497490	.497666	.497842	.498018	.498194	.498370	.498546	-0	0
4100	.498722	.498897	.499073	.499248	.499424	.499599	.499775	.499950	.500125	.500300	-0	0
4110	.500475	.500650	.500825	.501000	.501175	.501350	.501525	.501699	.501874	.502049	-0	0
4120	.502223	.502398	.502572	.502746	.502921	.503095	.503269	.503443	.503617	.503791	-0	0
4130	.503965	.504139	.504313	.504487	.504660	.504834	.505008	.505181	.505354	.505528	-0	0
4140	.505701	.505874	.506048	.506221	.506394	.506567	.506740	.506913	.507086	.507259	-0	0
4150	.507431	.507604	.507777	.507949	.508122	.508294	.508467	.508639	.508811	.508983	-0	0
4160	.509155	.509328	.509500	.509672	.509843	.510015	.510187	.510359	.510531	.510702	-0	0
4170	.510874	.511045	.511217	.511388	.511559	.511731	.511902	.512073	.512244	.512415	-0	0
4180	.512586	.512757	.512928	.513099	.513269	.513440	.513611	.513781	.513952	.514122	-0	0
4190	.514293	.514463	.514633	.514803	.514974	.515144	.515314	.515484	.515653	.515823	-0	0
4200	.515993	.516163	.516333	.516502	.516672	.516841	.517011	.517180	.517349	.517519	-0	0
4210	.517688	.517857	.518026	.518195	.518364	.518533	.518702	.518871	.519039	.519208	-0	0
4220	.519377	.519545	.519714	.519882	.520050	.520219	.520387	.520555	.520723	.520891	-0	0
4230	.521059	.521227	.521395	.521563	.521731	.521899	.522066	.522234	.522402	.522569	-0	0
4240	.522736	.522904	.523071	.523238	.523406	.523573	.523740	.523907	.524074	.524241	-0	0
4250	.524408	.524574	.524741	.524908	.525074	.525241	.525407	.525574	.525740	.525907	-0	0
4260	.526073	.526239	.526405	.526571	.526737	.526903	.527069	.527235	.527401	.527567	-0	0
4270	.527732	.527898	.528063	.528229	.528394	.528560	.528725	.528890	.529055	.529221	-0	0
4280	.529386	.529551	.529716	.529881	.530045	.530210	.530375	.530540	.530704	.530869	-0	0
4290	.531033	.531198	.531362	.531527	.531691	.531855	.532019	.532183	.532347	.532511	-0	0
4300	.532675	.532839	.533003	.533167	.533330	.533494	.533658	.533821	.533984	.534148	-0	0
4310	.534311	.534474	.534638	.534801	.534964	.535127	.535290	.535453	.535616	.535779	-0	0
4320	.535941	.536104	.536267	.536429	.536592	.536754	.536917	.537079	.537241	.537404	-0	0
4330	.537566	.537728	.537890	.538052	.538214	.538376	.538537	.538699	.538861	.539023	-0	0
4340	.539184	.539346	.539507	.539669	.539830	.539991	.540153	.540314	.540475	.540636	-0	0
4350	.540797	.540958	.541119	.541280	.541440	.541601	.541762	.541922	.542083	.542243	-0	0
4360	.542404	.542564	.542724	.542885	.543045	.543205	.543365	.543525	.543685	.543845	-0	0
4370	.544005	.544165	.544324	.544484	.544644	.544803	.544963	.545122	.545282	.545441	-0	0
4380	.545600	.545759	.545919	.546078	.546237	.546396	.546555	.546714	.546872	.547031	-0	0
4390	.547190	.547348	.547507	.547666	.547824	.547982	.548141	.548299	.548457	.548616	-0	0
4400	.548774	.548932	.549090	.549248	.549406	.549563	.549721	.549879	.550037	.550194	-0	0
4410	.550352	.550509	.550667	.550824	.550981	.551139	.551296	.551453	.551610	.551767	-0	0
4420	.551924	.552081	.552238	.552395	.552551	.552708	.552865	.553021	.553178	.553334	-0	0
4430	.553491	.553647	.553803	.553959	.554116	.554272	.554428	.554584	.554740	.554896	-0	0
4440	.555051	.555207	.555363	.555519	.555674	.555830	.555985	.556141	.556296	.556451	-0	0
4450	.556607	.556762	.556917	.557072	.557227	.557382	.557537	.557692	.557847	.558001	-0	0
4460	.558156	.558311	.558465	.558620	.558774	.558929	.559083	.559237	.559392	.559546	-0	0
4470	.559700	.559854	.560008	.560162	.560316	.560470	.560623	.560777	.560931	.561084	-0	0
4480	.561238	.561391	.561545	.561698	.561852	.562005	.562158	.562311	.562464	.562617	-0	0
4490	.562770	.562923	.563076	.563229	.563382	.563534	.563687	.563840	.563992	.564145	-0	0

TABLE 11A

Blackbody: Relative Fractional Emittance (Radiant)

λT(μ.°K)	0	1	2	3	4	5	6	7	8	9	m	Δ_2
4500	.564297	.564449	.564602	.564754	.564906	.565058	.565210	.565362	.565514	.565666	-0	0
4510	.565818	.565970	.566122	.566273	.566425	.566577	.566728	.566880	.567031	.567182	-0	0
4520	.567334	.567485	.567636	.567787	.567938	.568089	.568240	.568391	.568542	.568693	-0	0
4530	.568844	.568994	.569145	.569295	.569446	.569596	.569747	.569897	.570047	.570198	-0	0
4540	.570348	.570498	.570648	.570798	.570948	.571098	.571248	.571397	.571547	.571697	-0	0
4550	.571847	.571996	.572146	.572295	.572444	.572594	.572743	.572892	.573041	.573191	-0	0
4560	.573340	.573489	.573638	.573786	.573935	.574084	.574233	.574381	.574530	.574679	-0	0
4570	.574827	.574976	.575124	.575272	.575421	.575569	.575717	.575865	.576013	.576161	-0	0
4580	.576309	.576457	.576605	.576753	.576900	.577048	.577196	.577343	.577491	.577638	-0	0
4590	.577785	.577933	.578080	.578227	.578374	.578522	.578669	.578816	.578963	.579110	-0	0
4600	.579256	.579403	.579550	.579697	.579843	.579990	.580136	.580283	.580429	.580575	-0	0
4610	.580722	.580868	.581014	.581160	.581306	.581452	.581598	.581744	.581890	.582036	-0	0
4620	.582182	.582327	.582473	.582618	.582764	.582909	.583055	.583200	.583346	.583491	-0	0
4630	.583636	.583781	.583926	.584071	.584216	.584361	.584506	.584651	.584795	.584940	-0	0
4640	.585085	.585229	.585374	.585518	.585663	.585807	.585952	.586096	.586240	.586384	-0	0
4650	.586528	.586672	.586816	.586960	.587104	.587248	.587392	.587535	.587679	.587823	-0	0
4660	.587966	.588110	.588253	.588397	.588540	.588683	.588826	.588970	.589113	.589256	-0	0
4670	.589399	.589542	.589685	.589828	.589970	.590113	.590256	.590398	.590541	.590683	-0	0
4680	.590826	.590968	.591111	.591253	.591395	.591537	.591680	.591822	.591964	.592106	-0	0
4690	.592248	.592390	.592531	.592673	.592815	.592956	.593098	.593240	.593381	.593523	-0	0
4700	.593664	.593805	.593947	.594088	.594229	.594370	.594511	.594652	.594793	.594934	-0	0
4710	.595075	.595216	.595356	.595497	.595638	.595778	.595919	.596059	.596200	.596340	-0	0
4720	.596480	.596621	.596761	.596901	.597041	.597181	.597321	.597461	.597601	.597741	-0	0
4730	.597881	.598020	.598160	.598300	.598439	.598579	.598718	.598858	.598997	.599136	-0	0
4740	.599276	.599415	.599554	.599693	.599832	.599971	.600110	.600249	.600388	.600526	-0	0
4750	.600665	.600804	.600942	.601081	.601219	.601358	.601496	.601635	.601773	.601911	-0	0
4760	.602049	.602188	.602326	.602464	.602602	.602740	.602877	.603015	.603153	.603291	-0	0
4770	.603428	.603566	.603704	.603841	.603978	.604116	.604253	.604391	.604528	.604665	-0	0
4780	.604802	.604939	.605076	.605213	.605350	.605487	.605624	.605761	.605897	.606034	-0	0
4790	.606171	.606307	.606444	.606580	.606716	.606853	.606989	.607125	.607262	.607398	-0	0
4800	.607534	.607670	.607806	.607942	.608078	.608213	.608349	.608485	.608621	.608756	-0	0
4810	.608892	.609027	.609163	.609298	.609433	.609569	.609704	.609839	.609974	.610109	-0	0
4820	.610244	.610379	.610514	.610649	.610784	.610919	.611054	.611188	.611323	.611457	-0	0
4830	.611592	.611726	.611861	.611995	.612130	.612264	.612398	.612532	.612666	.612800	-0	0
4840	.612934	.613068	.613202	.613336	.613470	.613604	.613737	.613871	.614005	.614138	-0	0
4850	.614272	.614405	.614538	.614672	.614805	.614938	.615071	.615205	.615338	.615471	-0	0
4860	.615604	.615737	.615869	.616002	.616135	.616268	.616400	.616533	.616666	.616798	-0	0
4870	.616931	.617063	.617195	.617328	.617460	.617592	.617724	.617856	.617988	.618120	-0	0
4880	.618252	.618384	.618516	.618648	.618780	.618911	.619043	.619175	.619306	.619438	-0	0
4890	.619569	.619700	.619832	.619963	.620094	.620225	.620357	.620488	.620619	.620750	-0	0
4900	.620881	.621011	.621142	.621273	.621404	.621534	.621665	.621796	.621926	.622057	-0	0
4910	.622187	.622317	.622448	.622578	.622708	.622838	.622969	.623099	.623229	.623359	-0	0
4920	.623489	.623618	.623748	.623878	.624008	.624137	.624267	.624397	.624526	.624656	-0	0
4930	.624785	.624914	.625044	.625173	.625302	.625431	.625560	.625689	.625818	.625947	-0	0
4940	.626076	.626205	.626334	.626463	.626592	.626720	.626849	.626977	.627106	.627234	-0	0
4950	.627363	.627491	.627619	.627748	.627876	.628004	.628132	.628260	.628388	.628516	-0	0
4960	.628644	.628772	.628900	.629028	.629155	.629283	.629411	.629538	.629666	.629793	-0	0
4970	.629921	.630048	.630175	.630303	.630430	.630557	.630684	.630811	.630938	.631065	-0	0
4980	.631192	.631319	.631446	.631572	.631699	.631826	.631953	.632079	.632206	.632332	-0	0
4990	.632459	.632585	.632711	.632838	.632964	.633090	.633216	.633342	.633468	.633594	-0	0

$\int_o^{\lambda T} M_{e\lambda} d\lambda / M_e(T)$

TABLE 11A

Blackbody: Relative Fractional Emittance (Radiant)

λT(μ.°K)	0	1	2	3	4	5	6	7	8	9	m	Δ_2
5000	.633720	.633846	.633972	.634098	.634223	.634349	.634475	.634600	.634726	.634851	-0	0
5010	.634977	.635102	.635228	.635353	.635478	.635603	.635728	.635854	.635979	.636104	-0	0
5020	.636229	.636353	.636478	.636603	.636728	.636853	.636977	.637102	.637226	.637351	-0	0
5030	.637475	.637600	.637724	.637849	.637973	.638097	.638221	.638345	.638469	.638593	-0	0
5040	.638717	.638841	.638965	.639089	.639213	.639337	.639460	.639584	.639708	.639831	-0	0
5050	.639955	.640078	.640201	.640325	.640448	.640571	.640695	.640818	.640941	.641064	-0	0
5060	.641187	.641310	.641433	.641556	.641679	.641801	.641924	.642047	.642169	.642292	-0	0
5070	.642415	.642537	.642659	.642782	.642904	.643026	.643149	.643271	.643393	.643515	-0	0
5080	.643637	.643759	.643881	.644003	.644125	.644247	.644369	.644490	.644612	.644734	-0	0
5090	.644855	.644977	.645098	.645220	.645341	.645462	.645584	.645705	.645826	.645947	-0	0
5100	.646068	.646189	.646310	.646431	.646552	.646673	.646794	.646915	.647036	.647156	-0	0
5110	.647277	.647397	.647518	.647638	.647759	.647879	.648000	.648120	.648240	.648360	-0	0
5120	.648481	.648601	.648721	.648841	.648961	.649081	.649200	.649320	.649440	.649560	-0	0
5130	.649680	.649799	.649919	.650038	.650158	.650277	.650397	.650516	.650635	.650755	-0	0
5140	.650874	.650993	.651112	.651231	.651350	.651469	.651588	.651707	.651826	.651945	-0	0
5150	.652063	.652182	.652301	.652419	.652538	.652656	.652775	.652893	.653012	.653130	-0	0
5160	.653248	.653367	.653485	.653603	.653721	.653839	.653957	.654075	.654193	.654311	-0	0
5170	.654429	.654546	.654664	.654782	.654899	.655017	.655135	.655252	.655370	.655487	-0	0
5180	.655604	.655722	.655839	.655956	.656073	.656190	.656307	.656425	.656542	.656658	-0	0
5190	.656775	.656892	.657009	.657126	.657242	.657359	.657476	.657592	.657709	.657825	-0	0
5200	.657942	.658058	.658175	.658291	.658407	.658523	.658639	.658756	.658872	.658988	-0	0
5210	.659104	.659220	.659335	.659451	.659567	.659683	.659799	.659914	.660030	.660145	-0	0
5220	.660261	.660376	.660492	.660607	.660723	.660838	.660953	.661068	.661183	.661299	-0	0
5230	.661414	.661529	.661644	.661759	.661873	.661988	.662103	.662218	.662333	.662447	-0	0
5240	.662562	.662676	.662791	.662905	.663020	.663134	.663249	.663363	.663477	.663591	-0	0
5250	.663706	.663820	.663934	.664048	.664162	.664276	.664390	.664503	.664617	.664731	-0	0
5260	.664845	.664958	.665072	.665186	.665299	.665413	.665526	.665639	.665753	.665866	-0	0
5270	.665979	.666093	.666206	.666319	.666432	.666545	.666658	.666771	.666884	.666997	-0	0
5280	.667110	.667222	.667335	.667448	.667561	.667673	.667786	.667898	.668011	.668123	-0	0
5290	.668235	.668348	.668460	.668572	.668685	.668797	.668909	.669021	.669133	.669245	-0	0
5300	.669357	.669469	.669581	.669692	.669804	.669916	.670027	.670139	.670251	.670362	-0	0
5310	.670474	.670585	.670697	.670808	.670919	.671031	.671142	.671253	.671364	.671475	-0	0
5320	.671586	.671697	.671808	.671919	.672030	.672141	.672252	.672362	.672473	.672584	-0	0
5330	.672694	.672805	.672915	.673026	.673136	.673247	.673357	.673467	.673578	.673688	-0	0
5340	.673798	.673908	.674018	.674128	.674238	.674348	.674458	.674568	.674678	.674788	-0	0
5350	.674898	.675007	.675117	.675227	.675336	.675446	.675555	.675665	.675774	.675883	-0	0
5360	.675993	.676102	.676211	.676320	.676429	.676539	.676648	.676757	.676866	.676975	-0	0
5370	.677083	.677192	.677301	.677410	.677519	.677627	.677736	.677844	.677953	.678061	-0	0
5380	.678170	.678278	.678387	.678495	.678603	.678711	.678820	.678928	.679036	.679144	-0	0
5390	.679252	.679360	.679468	.679576	.679684	.679791	.679899	.680007	.680115	.680222	-0	0
5400	.680330	.680437	.680545	.680652	.680760	.680867	.680975	.681082	.681189	.681296	-0	0
5410	.681404	.681511	.681618	.681725	.681832	.681939	.682046	.682153	.682259	.682366	-0	0
5420	.682473	.682580	.682686	.682793	.682899	.683006	.683113	.683219	.683325	.683432	-0	0
5430	.683538	.683644	.683751	.683857	.683963	.684069	.684175	.684281	.684387	.684493	-0	0
5440	.684599	.684705	.684811	.684917	.685022	.685128	.685234	.685339	.685445	.685550	-0	0
5450	.685656	.685761	.685867	.685972	.686077	.686183	.686288	.686393	.686498	.686603	-0	0
5460	.686708	.686813	.686918	.687023	.687128	.687233	.687338	.687443	.687548	.687652	-0	0
5470	.687757	.687862	.687966	.688071	.688175	.688280	.688384	.688488	.688593	.688697	-0	0
5480	.688801	.688905	.689010	.689114	.689218	.689322	.689426	.689530	.689634	.689738	-0	0
5490	.689841	.689945	.690049	.690153	.690256	.690360	.690463	.690567	.690671	.690774	-0	0

TABLE 11A

Blackbody: Relative Fractional Emittance (Radiant)

λT(μ.°K)	0	1	2	3	4	5	6	7	8	9	m	Δ_2
5500	.690877	.690981	.691084	.691187	.691291	.691394	.691497	.691600	.691703	.691806	-0	0
5510	.691909	.692012	.692115	.692218	.692321	.692424	.692527	.692629	.692732	.692835	-0	0
5520	.692937	.693040	.693142	.693245	.693347	.693450	.693552	.693654	.693757	.693859	-0	0
5530	.693961	.694063	.694165	.694267	.694370	.694472	.694573	.694675	.694777	.694879	-0	0
5540	.694981	.695083	.695184	.695286	.695388	.695489	.695591	.695692	.695794	.695895	-0	0
5550	.695997	.696098	.696199	.696301	.696402	.696503	.696604	.696705	.696806	.696907	-0	0
5560	.697008	.697109	.697210	.697311	.697412	.697513	.697614	.697714	.697815	.697916	-0	0
5570	.698016	.698117	.698217	.698318	.698418	.698519	.698619	.698719	.698819	.698920	-0	0
5580	.699020	.699120	.699220	.699320	.699420	.699520	.699620	.699720	.699820	.699920	-0	0
5590	.700020	.700119	.700219	.700319	.700418	.700518	.700618	.700717	.700817	.700916	-0	0
5600	.701016	.701115	.701214	.701314	.701413	.701512	.701611	.701710	.701809	.701908	-0	0
5610	.702007	.702106	.702205	.702304	.702403	.702502	.702601	.702699	.702798	.702897	-0	0
5620	.702995	.703094	.703193	.703291	.703389	.703488	.703586	.703685	.703783	.703881	-0	0
5630	.703979	.704078	.704176	.704274	.704372	.704470	.704568	.704666	.704764	.704862	-0	0
5640	.704960	.705057	.705155	.705253	.705351	.705448	.705546	.705643	.705741	.705839	-0	0
5650	.705936	.706033	.706131	.706228	.706325	.706423	.706520	.706617	.706714	.706811	-0	0
5660	.706908	.707005	.707102	.707199	.707296	.707393	.707490	.707587	.707684	.707780	-0	0
5670	.707877	.707974	.708070	.708167	.708263	.708360	.708456	.708553	.708649	.708746	-0	0
5680	.708842	.708938	.709034	.709131	.709227	.709323	.709419	.709515	.709611	.709707	-0	0
5690	.709803	.709899	.709995	.710090	.710186	.710282	.710378	.710473	.710569	.710664	-0	0
5700	.710760	.710856	.710951	.711046	.711142	.711237	.711332	.711428	.711523	.711618	-0	0
5710	.711713	.711809	.711904	.711999	.712094	.712189	.712284	.712379	.712473	.712568	-0	0
5720	.712663	.712758	.712853	.712947	.713042	.713136	.713231	.713326	.713420	.713514	-0	0
5730	.713609	.713703	.713798	.713892	.713986	.714080	.714175	.714269	.714363	.714457	-0	0
5740	.714551	.714645	.714739	.714833	.714927	.715021	.715115	.715208	.715302	.715396	-0	0
5750	.715490	.715583	.715677	.715770	.715864	.715957	.716051	.716144	.716238	.716331	-0	0
5760	.716424	.716518	.716611	.716704	.716797	.716890	.716983	.717076	.717169	.717262	-0	0
5770	.717355	.717448	.717541	.717634	.717727	.717819	.717912	.718005	.718097	.718190	-0	0
5780	.718283	.718375	.718468	.718560	.718653	.718745	.718837	.718930	.719022	.719114	-0	0
5790	.719206	.719298	.719391	.719483	.719575	.719667	.719759	.719851	.719943	.720035	-0	0
5800	.720126	.720218	.720310	.720402	.720493	.720585	.720677	.720768	.720860	.720951	-0	0
5810	.721043	.721134	.721226	.721317	.721408	.721500	.721591	.721682	.721773	.721864	-0	0
5820	.721956	.722047	.722138	.722229	.722320	.722411	.722502	.722592	.722683	.722774	-0	0
5830	.722865	.722956	.723046	.723137	.723228	.723318	.723409	.723499	.723590	.723680	-0	0
5840	.723770	.723861	.723951	.724041	.724132	.724222	.724312	.724402	.724492	.724582	-0	0
5850	.724673	.724762	.724852	.724942	.725032	.725122	.725212	.725302	.725392	.725481	-0	0
5860	.725571	.725661	.725750	.725840	.725929	.726019	.726108	.726198	.726287	.726377	-0	0
5870	.726466	.726555	.726645	.726734	.726823	.726912	.727001	.727090	.727179	.727268	-0	0
5880	.727357	.727446	.727535	.727624	.727713	.727802	.727891	.727979	.728068	.728157	-0	0
5890	.728245	.728334	.728422	.728511	.728599	.728688	.728776	.728865	.728953	.729041	-0	0
5900	.729130	.729218	.729306	.729394	.729483	.729571	.729659	.729747	.729835	.729923	-0	0
5910	.730011	.730099	.730186	.730274	.730362	.730450	.730538	.730625	.730713	.730801	-0	0
5920	.730888	.730976	.731063	.731151	.731238	.731326	.731413	.731500	.731588	.731675	-0	0
5930	.731762	.731849	.731937	.732024	.732111	.732198	.732285	.732372	.732459	.732546	-0	0
5940	.732633	.732720	.732806	.732893	.732980	.733067	.733153	.733240	.733327	.733413	-0	0
5950	.733500	.733586	.733673	.733759	.733846	.733932	.734019	.734105	.734191	.734277	-0	0
5960	.734364	.734450	.734536	.734622	.734708	.734794	.734880	.734966	.735052	.735138	-0	0
5970	.735224	.735310	.735396	.735481	.735567	.735653	.735739	.735824	.735910	.735995	-0	0
5980	.736081	.736166	.736252	.736337	.736423	.736508	.736593	.736679	.736764	.736849	-0	0
5990	.736934	.737020	.737105	.737190	.737275	.737360	.737445	.737530	.737615	.737700	-0	0

$\int_0^{\lambda T} M_{e\lambda}\, d\lambda / M_e(T)$

TABLE 11A

Blackbody: Relative Fractional Emittance (Radiant)

λT(μ.°K)	0	1	2	3	4	5	6	7	8	9	m	Δ_2
6000	.737785	.737870	.737954	.738039	.738124	.738209	.738293	.738378	.738463	.738547	-0	0
6010	.738632	.738716	.738801	.738885	.738969	.739054	.739138	.739222	.739307	.739391	-0	0
6020	.739475	.739559	.739643	.739728	.739812	.739896	.739980	.740064	.740148	.740232	-0	0
6030	.740315	.740399	.740483	.740567	.740651	.740734	.740818	.740902	.740985	.741069	-0	0
6040	.741152	.741236	.741319	.741403	.741486	.741570	.741653	.741736	.741820	.741903	-0	0
6050	.741986	.742069	.742152	.742236	.742319	.742402	.742485	.742568	.742651	.742734	-0	0
6060	.742816	.742899	.742982	.743065	.743148	.743230	.743313	.743396	.743478	.743561	-0	0
6070	.743644	.743726	.743809	.743891	.743974	.744056	.744138	.744221	.744303	.744385	-0	0
6080	.744468	.744550	.744632	.744714	.744796	.744878	.744960	.745042	.745124	.745206	-0	0
6090	.745288	.745370	.745452	.745534	.745616	.745697	.745779	.745861	.745942	.746024	-0	0
6100	.746106	.746187	.746269	.746350	.746432	.746513	.746595	.746676	.746757	.746839	-0	0
6110	.746920	.747001	.747082	.747164	.747245	.747326	.747407	.747488	.747569	.747650	-0	0
6120	.747731	.747812	.747893	.747974	.748054	.748135	.748216	.748297	.748377	.748458	-0	0
6130	.748539	.748619	.748700	.748781	.748861	.748942	.749022	.749102	.749183	.749263	-0	0
6140	.749344	.749424	.749504	.749584	.749665	.749745	.749825	.749905	.749985	.750065	-0	0
6150	.750145	.750225	.750305	.750385	.750465	.750545	.750625	.750704	.750784	.750864	-0	0
6160	.750943	.751023	.751103	.751182	.751262	.751342	.751421	.751501	.751580	.751659	-0	0
6170	.751739	.751818	.751898	.751977	.752056	.752135	.752215	.752294	.752373	.752452	-0	0
6180	.752531	.752610	.752689	.752768	.752847	.752926	.753005	.753084	.753163	.753241	-0	0
6190	.753320	.753399	.753478	.753556	.753635	.753714	.753792	.753871	.753949	.754028	-0	0
6200	.754106	.754185	.754263	.754341	.754420	.754498	.754576	.754655	.754733	.754811	-0	0
6210	.754889	.754967	.755045	.755123	.755201	.755279	.755357	.755435	.755513	.755591	-0	0
6220	.755669	.755747	.755825	.755902	.755980	.756058	.756135	.756213	.756291	.756368	-0	0
6230	.756446	.756523	.756601	.756678	.756756	.756833	.756910	.756988	.757065	.757142	-0	0
6240	.757220	.757297	.757374	.757451	.757528	.757605	.757682	.757759	.757837	.757913	-0	0
6250	.757990	.758067	.758144	.758221	.758298	.758375	.758451	.758528	.758605	.758682	-0	0
6260	.758758	.758835	.758911	.758988	.759064	.759141	.759217	.759294	.759370	.759447	-0	0
6270	.759523	.759599	.759676	.759752	.759828	.759904	.759980	.760057	.760133	.760209	-0	0
6280	.760285	.760361	.760437	.760513	.760589	.760665	.760740	.760816	.760892	.760968	-0	0
6290	.761044	.761119	.761195	.761271	.761346	.761422	.761498	.761573	.761649	.761724	-0	0
6300	.761800	.761875	.761950	.762026	.762101	.762176	.762252	.762327	.762402	.762477	-0	0
6310	.762553	.762628	.762703	.762778	.762853	.762928	.763003	.763078	.763153	.763228	-0	0
6320	.763303	.763377	.763452	.763527	.763602	.763676	.763751	.763826	.763900	.763975	-0	0
6330	.764050	.764124	.764199	.764273	.764348	.764422	.764497	.764571	.764645	.764720	-0	0
6340	.764794	.764868	.764942	.765017	.765091	.765165	.765239	.765313	.765387	.765461	-0	0
6350	.765535	.765609	.765683	.765757	.765831	.765905	.765979	.766052	.766126	.766200	-0	0
6360	.766274	.766347	.766421	.766495	.766568	.766642	.766715	.766789	.766862	.766936	-0	0
6370	.767009	.767083	.767156	.767229	.767303	.767376	.767449	.767522	.767596	.767669	-0	0
6380	.767742	.767815	.767888	.767961	.768034	.768107	.768180	.768253	.768326	.768399	-0	0
6390	.768472	.768545	.768617	.768690	.768763	.768836	.768908	.768981	.769054	.769126	-0	0
6400	.769199	.769271	.769344	.769416	.769489	.769561	.769634	.769706	.769778	.769851	-0	0
6410	.769923	.769995	.770067	.770140	.770212	.770284	.770356	.770428	.770500	.770572	-0	0
6420	.770644	.770716	.770788	.770860	.770932	.771004	.771076	.771148	.771219	.771291	-0	0
6430	.771363	.771435	.771506	.771578	.771650	.771721	.771793	.771864	.771936	.772007	-0	0
6440	.772079	.772150	.772222	.772293	.772364	.772436	.772507	.772578	.772649	.772721	-0	0
6450	.772792	.772863	.772934	.773005	.773076	.773147	.773218	.773289	.773360	.773431	-0	0
6460	.773502	.773573	.773644	.773715	.773785	.773856	.773927	.773998	.774068	.774139	-0	0
6470	.774210	.774280	.774351	.774421	.774492	.774562	.774633	.774703	.774774	.774844	-0	0
6480	.774914	.774985	.775055	.775125	.775196	.775266	.775336	.775406	.775476	.775546	-0	0
6490	.775617	.775687	.775757	.775827	.775897	.775967	.776037	.776106	.776176	.776246	-0	0

TABLE 11A

Blackbody: Relative Fractional Emittance (Radiant)

λT(μ.°K)	0	1	2	3	4	5	6	7	8	9	m	Δ_2
6500	.776316	.776386	.776455	.776525	.776595	.776665	.776734	.776804	.776873	.776943	-0	0
6510	.777013	.777082	.777152	.777221	.777291	.777360	.777429	.777499	.777568	.777637	-0	0
6520	.777707	.777776	.777845	.777914	.777983	.778053	.778122	.778191	.778260	.778329	-0	0
6530	.778398	.778467	.778536	.778605	.778674	.778743	.778811	.778880	.778949	.779018	-0	0
6540	.779087	.779155	.779224	.779293	.779361	.779430	.779498	.779567	.779636	.779704	-0	0
6550	.779773	.779841	.779909	.779978	.780046	.780115	.780183	.780251	.780319	.780388	-0	0
6560	.780456	.780524	.780592	.780660	.780728	.780797	.780865	.780933	.781001	.781069	-0	0
6570	.781137	.781205	.781272	.781340	.781408	.781476	.781544	.781612	.781679	.781747	-0	0
6580	.781815	.781882	.781950	.782018	.782085	.782153	.782220	.782288	.782355	.782423	-0	0
6590	.782490	.782558	.782625	.782692	.782760	.782827	.782894	.782961	.783029	.783096	-0	0
6600	.783163	.783230	.783297	.783364	.783431	.783499	.783566	.783633	.783699	.783766	-0	0
6610	.783833	.783900	.783967	.784034	.784101	.784168	.784234	.784301	.784368	.784434	-0	0
6620	.784501	.784568	.784634	.784701	.784767	.784834	.784900	.784967	.785033	.785100	-0	0
6630	.785166	.785233	.785299	.785365	.785432	.785498	.785564	.785630	.785697	.785763	-0	0
6640	.785829	.785895	.785961	.786027	.786093	.786159	.786225	.786291	.786357	.786423	-0	0
6650	.786489	.786555	.786621	.786686	.786752	.786818	.786884	.786949	.787015	.787081	-0	0
6660	.787146	.787212	.787278	.787343	.787409	.787474	.787540	.787605	.787671	.787736	-0	0
6670	.787801	.787867	.787932	.787997	.788063	.788128	.788193	.788258	.788324	.788389	-0	0
6680	.788454	.788519	.788584	.788649	.788714	.788779	.788844	.788909	.788974	.789039	-0	0
6690	.789104	.789169	.789234	.789298	.789363	.789428	.789493	.789557	.789622	.789687	-0	0
6700	.789751	.789816	.789881	.789945	.790010	.790074	.790139	.790203	.790268	.790332	-0	0
6710	.790396	.790461	.790525	.790590	.790654	.790718	.790782	.790847	.790911	.790975	-0	0
6720	.791039	.791103	.791167	.791231	.791295	.791359	.791423	.791487	.791551	.791615	-0	0
6730	.791679	.791743	.791807	.791871	.791935	.791998	.792062	.792126	.792189	.792253	-0	0
6740	.792317	.792380	.792444	.792508	.792571	.792635	.792698	.792762	.792825	.792889	-0	0
6750	.792952	.793015	.793079	.793142	.793205	.793269	.793332	.793395	.793458	.793522	-0	0
6760	.793585	.793648	.793711	.793774	.793837	.793900	.793963	.794026	.794089	.794152	-0	0
6770	.794215	.794278	.794341	.794404	.794467	.794530	.794592	.794655	.794718	.794781	-0	0
6780	.794843	.794906	.794969	.795031	.795094	.795156	.795219	.795281	.795344	.795406	-0	0
6790	.795469	.795531	.795594	.795656	.795718	.795781	.795843	.795905	.795968	.796030	-0	0
6800	.796092	.796154	.796216	.796279	.796341	.796403	.796465	.796527	.796589	.796651	-0	0
6810	.796713	.796775	.796837	.796899	.796961	.797022	.797084	.797146	.797208	.797270	-0	0
6820	.797331	.797393	.797455	.797516	.797578	.797640	.797701	.797763	.797824	.797886	-0	0
6830	.797947	.798009	.798070	.798132	.798193	.798255	.798316	.798377	.798439	.798500	-0	0
6840	.798561	.798622	.798684	.798745	.798806	.798867	.798928	.798989	.799051	.799112	-0	0
6850	.799173	.799234	.799295	.799356	.799417	.799478	.799538	.799599	.799660	.799721	-0	0
6860	.799782	.799843	.799903	.799964	.800025	.800086	.800146	.800207	.800267	.800328	-0	0
6870	.800389	.800449	.800510	.800570	.800631	.800691	.800752	.800812	.800872	.800933	-0	0
6880	.800993	.801053	.801114	.801174	.801234	.801295	.801355	.801415	.801475	.801535	-0	0
6890	.801595	.801655	.801716	.801776	.801836	.801896	.801956	.802016	.802075	.802135	-0	0
6900	.802195	.802255	.802315	.802375	.802435	.802494	.802554	.802614	.802674	.802733	-0	0
6910	.802793	.802853	.802912	.802972	.803031	.803091	.803150	.803210	.803269	.803329	-0	0
6920	.803388	.803448	.803507	.803567	.803626	.803685	.803745	.803804	.803863	.803922	-0	0
6930	.803981	.804041	.804100	.804159	.804218	.804277	.804336	.804395	.804454	.804513	-0	0
6940	.804572	.804631	.804690	.804749	.804808	.804867	.804926	.804985	.805044	.805102	-0	0
6950	.805161	.805220	.805279	.805337	.805396	.805455	.805513	.805572	.805630	.805689	-0	0
6960	.805748	.805806	.805865	.805923	.805981	.806040	.806098	.806157	.806215	.806273	-0	0
6970	.806332	.806390	.806448	.806507	.806565	.806623	.806681	.806739	.806798	.806856	-0	0
6980	.806914	.806972	.807030	.807088	.807146	.807204	.807262	.807320	.807378	.807436	-0	0
6990	.807494	.807551	.807609	.807667	.807725	.807783	.807840	.807898	.807956	.808014	-0	0

$\int_0^{\lambda T} M_{e\lambda}\, d\lambda / M_e(T)$

$\int_o^{\lambda T} M_{e\lambda} d\lambda / M_e(T)$

TABLE 11A

Blackbody: Relative Fractional Emittance (Radiant)

λT(μ.°K)	0	1	2	3	4	5	6	7	8	9	m	Δ_2
7000	.808071	.808129	.808186	.808244	.808302	.808359	.808417	.808474	.808532	.808589	-0	0
7010	.808647	.808704	.808762	.808819	.808876	.808934	.808991	.809048	.809106	.809163	-0	0
7020	.809220	.809277	.809334	.809392	.809449	.809506	.809563	.809620	.809677	.809734	-0	0
7030	.809791	.809848	.809905	.809962	.810019	.810076	.810133	.810190	.810246	.810303	-0	0
7040	.810360	.810417	.810474	.810530	.810587	.810644	.810700	.810757	.810814	.810870	-0	0
7050	.810927	.810984	.811040	.811097	.811153	.811210	.811266	.811323	.811379	.811435	-0	0
7060	.811492	.811548	.811604	.811661	.811717	.811773	.811830	.811886	.811942	.811998	-0	0
7070	.812054	.812110	.812167	.812223	.812279	.812335	.812391	.812447	.812503	.812559	-0	0
7080	.812615	.812671	.812727	.812783	.812838	.812894	.812950	.813006	.813062	.813117	-0	0
7090	.813173	.813229	.813285	.813340	.813396	.813452	.813507	.813563	.813618	.813674	-0	0
7100	.813729	.813785	.813840	.813896	.813951	.814007	.814062	.814118	.814173	.814228	-0	0
7110	.814284	.814339	.814394	.814450	.814505	.814560	.814615	.814670	.814726	.814781	-0	0
7120	.814836	.814891	.814946	.815001	.815056	.815111	.815166	.815221	.815276	.815331	-0	0
7130	.815386	.815441	.815496	.815551	.815605	.815660	.815715	.815770	.815824	.815879	-0	0
7140	.815934	.815989	.816043	.816098	.816153	.816207	.816262	.816316	.816371	.816425	-0	0
7150	.816480	.816534	.816589	.816643	.816698	.816752	.816806	.816861	.816915	.816969	-0	0
7160	.817024	.817078	.817132	.817187	.817241	.817295	.817349	.817403	.817457	.817512	-0	0
7170	.817566	.817620	.817674	.817728	.817782	.817836	.817890	.817944	.817998	.818052	-0	0
7180	.818106	.818159	.818213	.818267	.818321	.818375	.818428	.818482	.818536	.818590	-0	0
7190	.818643	.818697	.818751	.818804	.818858	.818912	.818965	.819019	.819072	.819126	-0	0
7200	.819179	.819233	.819286	.819340	.819393	.819446	.819500	.819553	.819606	.819660	-0	0
7210	.819713	.819766	.819820	.819873	.819926	.819979	.820032	.820086	.820139	.820192	-0	0
7220	.820245	.820298	.820351	.820404	.820457	.820510	.820563	.820616	.820669	.820722	-0	0
7230	.820775	.820828	.820881	.820933	.820986	.821039	.821092	.821145	.821197	.821250	-0	0
7240	.821303	.821355	.821408	.821461	.821513	.821566	.821619	.821671	.821724	.821776	-0	0
7250	.821829	.821881	.821934	.821986	.822038	.822091	.822143	.822196	.822248	.822300	-0	0
7260	.822353	.822405	.822457	.822509	.822562	.822614	.822666	.822718	.822770	.822823	-0	0
7270	.822875	.822927	.822979	.823031	.823083	.823135	.823187	.823239	.823291	.823343	-0	0
7280	.823395	.823447	.823499	.823550	.823602	.823654	.823706	.823758	.823809	.823861	-0	0
7290	.823913	.823965	.824016	.824068	.824120	.824171	.824223	.824275	.824326	.824378	-0	0
7300	.824429	.824481	.824532	.824584	.824635	.824687	.824738	.824789	.824841	.824892	-0	0
7310	.824944	.824995	.825046	.825097	.825149	.825200	.825251	.825302	.825354	.825405	-0	0
7320	.825456	.825507	.825558	.825609	.825660	.825711	.825763	.825814	.825865	.825916	-0	0
7330	.825967	.826017	.826068	.826119	.826170	.826221	.826272	.826323	.826374	.826424	-0	0
7340	.826475	.826526	.826577	.826627	.826678	.826729	.826779	.826830	.826881	.826931	-0	0
7350	.826982	.827033	.827083	.827134	.827184	.827235	.827285	.827336	.827386	.827436	-0	0
7360	.827487	.827537	.827588	.827638	.827688	.827739	.827789	.827839	.827889	.827940	-0	0
7370	.827990	.828040	.828090	.828140	.828191	.828241	.828291	.828341	.828391	.828441	-0	0
7380	.828491	.828541	.828591	.828641	.828691	.828741	.828791	.828841	.828891	.828941	-0	0
7390	.828990	.829040	.829090	.829140	.829190	.829239	.829289	.829339	.829389	.829438	-0	0
7400	.829488	.829538	.829587	.829637	.829686	.829736	.829786	.829835	.829885	.829934	-0	0
7410	.829984	.830033	.830082	.830132	.830181	.830231	.830280	.830329	.830379	.830428	-0	0
7420	.830477	.830527	.830576	.830625	.830674	.830724	.830773	.830822	.830871	.830920	-0	0
7430	.830969	.831018	.831068	.831117	.831166	.831215	.831264	.831313	.831362	.831411	-0	0
7440	.831460	.831509	.831557	.831606	.831655	.831704	.831753	.831802	.831850	.831899	-0	0
7450	.831948	.831997	.832045	.832094	.832143	.832192	.832240	.832289	.832337	.832386	-0	0
7460	.832435	.832483	.832532	.832580	.832629	.832677	.832726	.832774	.832823	.832871	-0	0
7470	.832919	.832968	.833016	.833065	.833113	.833161	.833209	.833258	.833306	.833354	-0	0
7480	.833402	.833451	.833499	.833547	.833595	.833643	.833691	.833740	.833788	.833836	-0	0
7490	.833884	.833932	.833980	.834028	.834076	.834124	.834172	.834220	.834268	.834315	-0	0

TABLE 11A

Blackbody: Relative Fractional Emittance (Radiant)

$\lambda T(\mu.^\circ K)$	0	1	2	3	4	5	6	7	8	9	m	Δ_2
7500	.834363	.834411	.834459	.834507	.834555	.834602	.834650	.834698	.834746	.834793	-0	0
7510	.834841	.834889	.834936	.834984	.835032	.835079	.835127	.835174	.835222	.835269	-0	0
7520	.835317	.835365	.835412	.835459	.835507	.835554	.835602	.835649	.835697	.835744	-0	0
7530	.835791	.835839	.835886	.835933	.835981	.836028	.836075	.836122	.836169	.836217	-0	0
7540	.836264	.836311	.836358	.836405	.836452	.836499	.836547	.836594	.836641	.836688	-0	0
7550	.836735	.836782	.836829	.836876	.836922	.836969	.837016	.837063	.837110	.837157	-0	0
7560	.837204	.837251	.837297	.837344	.837391	.837438	.837484	.837531	.837578	.837624	-0	0
7570	.837671	.837718	.837764	.837811	.837858	.837904	.837951	.837997	.838044	.838090	-0	0
7580	.838137	.838183	.838230	.838276	.838323	.838369	.838415	.838462	.838508	.838554	-0	0
7590	.838601	.838647	.838693	.838740	.838786	.838832	.838878	.838925	.838971	.839017	-0	0
7600	.839063	.839109	.839155	.839201	.839247	.839294	.839340	.839386	.839432	.839478	-0	0
7610	.839524	.839570	.839616	.839661	.839707	.839753	.839799	.839845	.839891	.839937	-0	0
7620	.839983	.840028	.840074	.840120	.840166	.840211	.840257	.840303	.840348	.840394	-0	0
7630	.840440	.840485	.840531	.840577	.840622	.840668	.840713	.840759	.840804	.840850	-0	0
7640	.840895	.840941	.840986	.841032	.841077	.841123	.841168	.841213	.841259	.841304	-0	0
7650	.841349	.841395	.841440	.841485	.841530	.841576	.841621	.841666	.841711	.841756	-0	0
7660	.841802	.841847	.841892	.841937	.841982	.842027	.842072	.842117	.842162	.842207	-0	0
7670	.842252	.842297	.842342	.842387	.842432	.842477	.842522	.842567	.842612	.842657	-0	0
7680	.842701	.842746	.842791	.842836	.842880	.842925	.842970	.843015	.843059	.843104	-0	0
7690	.843149	.843193	.843238	.843283	.843327	.843372	.843416	.843461	.843505	.843550	-0	0
7700	.843595	.843639	.843683	.843728	.843772	.843817	.843861	.843906	.843950	.843994	-0	0
7710	.844039	.844083	.844127	.844172	.844216	.844260	.844304	.844349	.844393	.844437	-0	0
7720	.844481	.844525	.844570	.844614	.844658	.844702	.844746	.844790	.844834	.844878	-0	0
7730	.844922	.844966	.845010	.845054	.845098	.845142	.845186	.845230	.845274	.845318	-0	0
7740	.845362	.845405	.845449	.845493	.845537	.845581	.845624	.845668	.845712	.845756	-0	0
7750	.845799	.845843	.845887	.845930	.845974	.846018	.846061	.846105	.846148	.846192	-0	0
7760	.846236	.846279	.846323	.846366	.846410	.846453	.846497	.846540	.846583	.846627	-0	0
7770	.846670	.846714	.846757	.846800	.846844	.846887	.846930	.846973	.847017	.847060	-0	0
7780	.847103	.847146	.847190	.847233	.847276	.847319	.847362	.847405	.847449	.847492	-0	0
7790	.847535	.847578	.847621	.847664	.847707	.847750	.847793	.847836	.847879	.847922	-0	0
7800	.847965	.848008	.848051	.848093	.848136	.848179	.848222	.848265	.848308	.848350	-0	0
7810	.848393	.848436	.848479	.848521	.848564	.848607	.848649	.848692	.848735	.848777	-0	0
7820	.848820	.848863	.848905	.848948	.848990	.849033	.849075	.849118	.849160	.849203	-0	0
7830	.849245	.849288	.849330	.849373	.849415	.849457	.849500	.849542	.849584	.849627	-0	0
7840	.849669	.849711	.849754	.849796	.849838	.849880	.849923	.849965	.850007	.850049	-0	0
7850	.850091	.850134	.850176	.850218	.850260	.850302	.850344	.850386	.850428	.850470	-0	0
7860	.850512	.850554	.850596	.850638	.850680	.850722	.850764	.850806	.850848	.850890	-0	0
7870	.850931	.850973	.851015	.851057	.851099	.851141	.851182	.851224	.851266	.851308	-0	0
7880	.851349	.851391	.851433	.851474	.851516	.851558	.851599	.851641	.851682	.851724	-0	0
7890	.851766	.851807	.851849	.851890	.851932	.851973	.852015	.852056	.852097	.852139	-0	0
7900	.852180	.852222	.852263	.852304	.852346	.852387	.852428	.852470	.852511	.852552	-0	0
7910	.852594	.852635	.852676	.852717	.852759	.852800	.852841	.852882	.852923	.852964	-0	0
7920	.853005	.853047	.853088	.853129	.853170	.853211	.853252	.853293	.853334	.853375	-0	0
7930	.853416	.853457	.853498	.853539	.853580	.853620	.853661	.853702	.853743	.853784	-0	0
7940	.853825	.853866	.853906	.853947	.853988	.854029	.854069	.854110	.854151	.854191	-0	0
7950	.854232	.854273	.854313	.854354	.854395	.854435	.854476	.854516	.854557	.854598	-0	0
7960	.854638	.854679	.854719	.854760	.854800	.854841	.854881	.854921	.854962	.855002	-0	0
7970	.855043	.855083	.855123	.855164	.855204	.855244	.855285	.855325	.855365	.855406	-0	0
7980	.855446	.855486	.855526	.855566	.855607	.855647	.855687	.855727	.855767	.855807	-0	0
7990	.855847	.855888	.855928	.855968	.856008	.856048	.856088	.856128	.856168	.856208	-0	0

$\int_o^{\lambda T} M_{e\lambda} d\lambda / M_e(T)$

TABLE 11A

Blackbody: Relative Fractional Emittance (Radiant)

λT(μ.°K)	0	1	2	3	4	5	6	7	8	9	m	Δ_2
8000	.856248	.856288	.856328	.856368	.856407	.856447	.856487	.856527	.856567	.856607	-0	0
8010	.856647	.856686	.856726	.856766	.856806	.856845	.856885	.856925	.856965	.857004	-0	0
8020	.857044	.857084	.857123	.857163	.857203	.857242	.857282	.857321	.857361	.857400	-0	0
8030	.857440	.857479	.857519	.857558	.857598	.857637	.857677	.857716	.857756	.857795	-0	0
8040	.857835	.857874	.857913	.857953	.857992	.858031	.858071	.858110	.858149	.858189	-0	0
8050	.858228	.858267	.858306	.858345	.858385	.858424	.858463	.858502	.858541	.858580	-0	0
8060	.858620	.858659	.858698	.858737	.858776	.858815	.858854	.858893	.858932	.858971	-0	0
8070	.859010	.859049	.859088	.859127	.859166	.859205	.859244	.859283	.859321	.859360	-0	0
8080	.859399	.859438	.859477	.859516	.859554	.859593	.859632	.859671	.859709	.859748	-0	0
8090	.859787	.859825	.859864	.859903	.859941	.859980	.860019	.860057	.860096	.860134	-0	0
8100	.860173	.860212	.860250	.860289	.860327	.860366	.860404	.860443	.860481	.860520	-0	0
8110	.860558	.860596	.860635	.860673	.860712	.860750	.860788	.860827	.860865	.860903	-0	0
8120	.860942	.860980	.861018	.861056	.861095	.861133	.861171	.861209	.861247	.861286	-0	0
8130	.861324	.861362	.861400	.861438	.861476	.861514	.861552	.861591	.861629	.861667	-0	0
8140	.861705	.861743	.861781	.861819	.861857	.861895	.861933	.861971	.862008	.862046	-0	0
8150	.862084	.862122	.862160	.862198	.862236	.862274	.862311	.862349	.862387	.862425	-0	0
8160	.862462	.862500	.862538	.862576	.862613	.862651	.862689	.862726	.862764	.862802	-0	0
8170	.862839	.862877	.862915	.862952	.862990	.863027	.863065	.863102	.863140	.863177	-0	0
8180	.863215	.863252	.863290	.863327	.863365	.863402	.863440	.863477	.863514	.863552	-0	0
8190	.863589	.863627	.863664	.863701	.863738	.863776	.863813	.863850	.863888	.863925	-0	0
8200	.863962	.863999	.864037	.864074	.864111	.864148	.864185	.864222	.864259	.864297	-0	0
8210	.864334	.864371	.864408	.864445	.864482	.864519	.864556	.864593	.864630	.864667	-0	0
8220	.864704	.864741	.864778	.864815	.864852	.864889	.864926	.864963	.864999	.865036	-0	0
8230	.865073	.865110	.865147	.865184	.865220	.865257	.865294	.865331	.865367	.865404	-0	0
8240	.865441	.865478	.865514	.865551	.865588	.865624	.865661	.865697	.865734	.865771	-0	0
8250	.865807	.865844	.865880	.865917	.865954	.865990	.866027	.866063	.866100	.866136	-0	0
8260	.866172	.866209	.866245	.866282	.866318	.866355	.866391	.866427	.866464	.866500	-0	0
8270	.866536	.866573	.866609	.866645	.866682	.866718	.866754	.866790	.866827	.866863	-0	0
8280	.866899	.866935	.866971	.867008	.867044	.867080	.867116	.867152	.867188	.867224	-0	0
8290	.867260	.867296	.867332	.867369	.867405	.867441	.867477	.867513	.867549	.867585	-0	0
8300	.867620	.867656	.867692	.867728	.867764	.867800	.867836	.867872	.867908	.867943	-0	0
8310	.867979	.868015	.868051	.868087	.868123	.868158	.868194	.868230	.868265	.868301	-0	0
8320	.868337	.868373	.868408	.868444	.868480	.868515	.868551	.868587	.868622	.868658	-0	0
8330	.868693	.868729	.868764	.868800	.868835	.868871	.868906	.868942	.868977	.869013	-0	0
8340	.869048	.869084	.869119	.869155	.869190	.869225	.869261	.869296	.869332	.869367	-0	0
8350	.869402	.869438	.869473	.869508	.869544	.869579	.869614	.869649	.869685	.869720	-0	0
8360	.869755	.869790	.869825	.869861	.869896	.869931	.869966	.870001	.870036	.870071	-0	0
8370	.870106	.870141	.870177	.870212	.870247	.870282	.870317	.870352	.870387	.870422	-0	0
8380	.870457	.870492	.870527	.870561	.870596	.870631	.870666	.870701	.870736	.870771	-0	0
8390	.870806	.870840	.870875	.870910	.870945	.870980	.871014	.871049	.871084	.871119	-0	0
8400	.871153	.871188	.871223	.871258	.871292	.871327	.871362	.871396	.871431	.871465	-0	0
8410	.871500	.871535	.871569	.871604	.871638	.871673	.871707	.871742	.871776	.871811	-0	0
8420	.871845	.871880	.871914	.871949	.871983	.872018	.872052	.872087	.872121	.872155	-0	0
8430	.872190	.872224	.872258	.872293	.872327	.872361	.872396	.872430	.872464	.872498	-0	0
8440	.872533	.872567	.872601	.872635	.872670	.872704	.872738	.872772	.872806	.872840	-0	0
8450	.872875	.872909	.872943	.872977	.873011	.873045	.873079	.873113	.873147	.873181	-0	0
8460	.873215	.873249	.873283	.873317	.873351	.873385	.873419	.873453	.873487	.873521	-0	0
8470	.873555	.873589	.873622	.873656	.873690	.873724	.873758	.873792	.873825	.873859	-0	0
8480	.873893	.873927	.873960	.873994	.874028	.874062	.874095	.874129	.874163	.874196	-0	0
8490	.874230	.874264	.874297	.874331	.874365	.874398	.874432	.874465	.874499	.874533	-0	0

TABLE 11A

Blackbody: Relative Fractional Emittance (Radiant)

$\lambda T (\mu \cdot {}^\circ K)$	0	1	2	3	4	5	6	7	8	9	m	Δ_2
8500	.874566	.874600	.874633	.874667	.874700	.874734	.874767	.874801	.874834	.874868	-0	0
8510	.874901	.874934	.874968	.875001	.875035	.875068	.875101	.875135	.875168	.875201	-0	0
8520	.875235	.875268	.875301	.875334	.875368	.875401	.875434	.875468	.875501	.875534	-0	0
8530	.875567	.875600	.875634	.875667	.875700	.875733	.875766	.875799	.875832	.875865	-0	0
8540	.875899	.875932	.875965	.875998	.876031	.876064	.876097	.876130	.876163	.876196	-0	0
8550	.876229	.876262	.876295	.876328	.876361	.876394	.876426	.876459	.876492	.876525	-0	0
8560	.876558	.876591	.876624	.876656	.876689	.876722	.876755	.876788	.876820	.876853	-0	0
8570	.876886	.876919	.876951	.876984	.877017	.877050	.877082	.877115	.877148	.877180	-0	0
8580	.877213	.877245	.877278	.877311	.877343	.877376	.877408	.877441	.877474	.877506	-0	0
8590	.877539	.877571	.877604	.877636	.877669	.877701	.877734	.877766	.877798	.877831	-0	0
8600	.877863	.877896	.877928	.877960	.877993	.878025	.878058	.878090	.878122	.878155	-0	0
8610	.878187	.878219	.878251	.878284	.878316	.878348	.878380	.878413	.878445	.878477	-0	0
8620	.878509	.878541	.878574	.878606	.878638	.878670	.878702	.878734	.878766	.878799	-0	0
8630	.878831	.878863	.878895	.878927	.878959	.878991	.879023	.879055	.879087	.879119	-0	0
8640	.879151	.879183	.879215	.879247	.879279	.879311	.879343	.879374	.879406	.879438	-0	0
8650	.879470	.879502	.879534	.879566	.879597	.879629	.879661	.879693	.879725	.879756	-0	0
8660	.879788	.879820	.879852	.879883	.879915	.879947	.879978	.880010	.880042	.880073	-0	0
8670	.880105	.880137	.880168	.880200	.880232	.880263	.880295	.880326	.880358	.880389	-0	0
8680	.880421	.880453	.880484	.880516	.880547	.880579	.880610	.880642	.880673	.880704	-0	0
8690	.880736	.880767	.880799	.880830	.880862	.880893	.880924	.880956	.880987	.881018	-0	0
8700	.881050	.881081	.881112	.881144	.881175	.881206	.881237	.881269	.881300	.881331	-0	0
8710	.881362	.881394	.881425	.881456	.881487	.881518	.881549	.881581	.881612	.881643	-0	0
8720	.881674	.881705	.881736	.881767	.881798	.881829	.881861	.881892	.881923	.881954	-0	0
8730	.881985	.882016	.882047	.882078	.882109	.882140	.882170	.882201	.882232	.882263	-0	0
8740	.882294	.882325	.882356	.882387	.882418	.882449	.882479	.882510	.882541	.882572	-0	0
8750	.882603	.882633	.882664	.882695	.882726	.882757	.882787	.882818	.882849	.882879	-0	0
8760	.882910	.882941	.882971	.883002	.883033	.883063	.883094	.883125	.883155	.883186	-0	0
8770	.883217	.883247	.883278	.883308	.883339	.883369	.883400	.883430	.883461	.883491	-0	0
8780	.883522	.883552	.883583	.883613	.883644	.883674	.883705	.883735	.883765	.883796	-0	0
8790	.883826	.883857	.883887	.883917	.883948	.883978	.884008	.884039	.884069	.884099	-0	0
8800	.884130	.884160	.884190	.884220	.884251	.884281	.884311	.884341	.884372	.884402	-0	0
8810	.884432	.884462	.884492	.884522	.884553	.884583	.884613	.884643	.884673	.884703	-0	0
8820	.884733	.884763	.884793	.884823	.884853	.884883	.884914	.884944	.884974	.885004	-0	0
8830	.885033	.885063	.885093	.885123	.885153	.885183	.885213	.885243	.885273	.885303	-0	0
8840	.885333	.885363	.885392	.885422	.885452	.885482	.885512	.885542	.885571	.885601	-0	0
8850	.885631	.885661	.885691	.885720	.885750	.885780	.885809	.885839	.885869	.885899	-0	0
8860	.885928	.885958	.885988	.886017	.886047	.886077	.886106	.886136	.886165	.886195	-0	0
8870	.886225	.886254	.886284	.886313	.886343	.886372	.886402	.886431	.886461	.886490	-0	0
8880	.886520	.886549	.886579	.886608	.886638	.886667	.886697	.886726	.886755	.886785	-0	0
8890	.886814	.886843	.886873	.886902	.886932	.886961	.886990	.887019	.887049	.887078	-0	0
8900	.887107	.887137	.887166	.887195	.887224	.887254	.887283	.887312	.887341	.887371	-0	0
8910	.887400	.887429	.887458	.887487	.887516	.887545	.887575	.887604	.887633	.887662	-0	0
8920	.887691	.887720	.887749	.887778	.887807	.887836	.887865	.887894	.887923	.887952	-0	0
8930	.887981	.888010	.888039	.888068	.888097	.888126	.888155	.888184	.888213	.888242	-0	0
8940	.888271	.888300	.888329	.888357	.888386	.888415	.888444	.888473	.888502	.888530	-0	0
8950	.888559	.888588	.888617	.888646	.888674	.888703	.888732	.888761	.888789	.888818	-0	0
8960	.888847	.888875	.888904	.888933	.888961	.888990	.889019	.889047	.889076	.889105	-0	0
8970	.889133	.889162	.889190	.889219	.889248	.889276	.889305	.889333	.889362	.889390	-0	0
8980	.889419	.889447	.889476	.889504	.889533	.889561	.889590	.889618	.889646	.889675	-0	0
8990	.889703	.889732	.889760	.889789	.889817	.889845	.889874	.889902	.889930	.889959	-0	0

$\int_o^{\lambda T} M_{e\lambda} d\lambda / M_e(T)$

TABLE 11A

Blackbody: Relative Fractional Emittance (Radiant)

λT(μ.°K)	0	1	2	3	4	5	6	7	8	9	m	Δ_2
9000	.889987	.890015	.890044	.890072	.890100	.890128	.890157	.890185	.890213	.890241	-0	0
9010	.890270	.890298	.890326	.890354	.890383	.890411	.890439	.890467	.890495	.890523	-0	0
9020	.890552	.890580	.890608	.890636	.890664	.890692	.890720	.890748	.890776	.890804	-0	0
9030	.890832	.890860	.890888	.890916	.890944	.890972	.891000	.891028	.891056	.891084	-0	0
9040	.891112	.891140	.891168	.891196	.891224	.891252	.891280	.891308	.891335	.891363	-0	0
9050	.891391	.891419	.891447	.891475	.891503	.891530	.891558	.891586	.891614	.891642	-0	0
9060	.891669	.891697	.891725	.891753	.891780	.891808	.891836	.891863	.891891	.891919	-0	0
9070	.891946	.891974	.892002	.892029	.892057	.892085	.892112	.892140	.892167	.892195	-0	0
9080	.892223	.892250	.892278	.892305	.892333	.892360	.892388	.892415	.892443	.892470	-0	0
9090	.892498	.892525	.892553	.892580	.892608	.892635	.892663	.892690	.892718	.892745	-0	0
9100	.892772	.892800	.892827	.892854	.892882	.892909	.892937	.892964	.892991	.893019	-0	0
9110	.893046	.893073	.893100	.893128	.893155	.893182	.893209	.893237	.893264	.893291	-0	0
9120	.893318	.893346	.893373	.893400	.893427	.893454	.893482	.893509	.893536	.893563	-0	0
9130	.893590	.893617	.893644	.893671	.893699	.893726	.893753	.893780	.893807	.893834	-0	0
9140	.893861	.893888	.893915	.893942	.893969	.893996	.894023	.894050	.894077	.894104	-0	0
9150	.894131	.894158	.894185	.894212	.894239	.894265	.894292	.894319	.894346	.894373	-0	0
9160	.894400	.894427	.894454	.894480	.894507	.894534	.894561	.894588	.894614	.894641	-0	0
9170	.894668	.894695	.894721	.894748	.894775	.894802	.894828	.894855	.894882	.894909	-0	0
9180	.894935	.894962	.894989	.895015	.895042	.895068	.895095	.895122	.895148	.895175	-0	0
9190	.895202	.895228	.895255	.895281	.895308	.895334	.895361	.895388	.895414	.895441	-0	0
9200	.895467	.895494	.895520	.895547	.895573	.895599	.895626	.895652	.895679	.895705	-0	0
9210	.895732	.895758	.895785	.895811	.895837	.895864	.895890	.895916	.895943	.895969	-0	0
9220	.895995	.896022	.896048	.896074	.896101	.896127	.896153	.896180	.896206	.896232	-0	0
9230	.896258	.896285	.896311	.896337	.896363	.896389	.896416	.896442	.896468	.896494	-0	0
9240	.896520	.896547	.896573	.896599	.896625	.896651	.896677	.896703	.896729	.896755	-0	0
9250	.896782	.896808	.896834	.896860	.896886	.896912	.896938	.896964	.896990	.897016	-0	0
9260	.897042	.897068	.897094	.897120	.897146	.897172	.897198	.897224	.897250	.897275	-0	0
9270	.897301	.897327	.897353	.897379	.897405	.897431	.897457	.897482	.897508	.897534	-0	0
9280	.897560	.897586	.897612	.897637	.897663	.897689	.897715	.897741	.897766	.897792	-0	0
9290	.897818	.897844	.897869	.897895	.897921	.897946	.897972	.897998	.898023	.898049	-0	0
9300	.898075	.898100	.898126	.898152	.898177	.898203	.898228	.898254	.898280	.898305	-0	0
9310	.898331	.898356	.898382	.898407	.898433	.898459	.898484	.898510	.898535	.898561	-0	0
9320	.898586	.898612	.898637	.898663	.898688	.898713	.898739	.898764	.898790	.898815	-0	0
9330	.898841	.898866	.898891	.898917	.898942	.898967	.898993	.899018	.899043	.899069	-0	0
9340	.899094	.899119	.899145	.899170	.899195	.899221	.899246	.899271	.899296	.899322	-0	0
9350	.899347	.899372	.899397	.899423	.899448	.899473	.899498	.899523	.899549	.899574	-0	0
9360	.899599	.899624	.899649	.899674	.899699	.899725	.899750	.899775	.899800	.899825	-0	0
9370	.899850	.899875	.899900	.899925	.899950	.899975	.900000	.900025	.900050	.900075	-0	0
9380	.900100	.900125	.900150	.900175	.900200	.900225	.900250	.900275	.900300	.900325	-0	0
9390	.900350	.900375	.900400	.900425	.900450	.900474	.900499	.900524	.900549	.900574	-0	0
9400	.900599	.900623	.900648	.900673	.900698	.900723	.900747	.900772	.900797	.900822	-0	0
9410	.900847	.900871	.900896	.900921	.900946	.900970	.900995	.901020	.901044	.901069	-0	0
9420	.901094	.901118	.901143	.901168	.901192	.901217	.901242	.901266	.901291	.901315	-0	0
9430	.901340	.901365	.901389	.901414	.901438	.901463	.901487	.901512	.901536	.901561	-0	0
9440	.901585	.901610	.901635	.901659	.901683	.901708	.901732	.901757	.901781	.901806	-0	0
9450	.901830	.901855	.901879	.901903	.901928	.901952	.901977	.902001	.902025	.902050	-0	0
9460	.902074	.902099	.902123	.902147	.902172	.902196	.902220	.902244	.902269	.902293	-0	0
9470	.902317	.902342	.902366	.902390	.902414	.902439	.902463	.902487	.902511	.902535	-0	0
9480	.902560	.902584	.902608	.902632	.902656	.902681	.902705	.902729	.902753	.902777	-0	0
9490	.902801	.902825	.902850	.902874	.902898	.902922	.902946	.902970	.902994	.903018	-0	0

TABLE 11A

Blackbody: Relative Fractional Emittance (Radiant)

$\lambda T(\mu.°K)$	0	1	2	3	4	5	6	7	8	9	m	Δ_2
9500	.903042	.903066	.903090	.903114	.903138	.903162	.903186	.903210	.903234	.903258	−0	0
9510	.903282	.903306	.903330	.903354	.903378	.903402	.903426	.903450	.903474	.903497	−0	0
9520	.903521	.903545	.903569	.903593	.903617	.903641	.903665	.903688	.903712	.903736	−0	0
9530	.903760	.903784	.903807	.903831	.903855	.903879	.903903	.903926	.903950	.903974	−0	0
9540	.903998	.904021	.904045	.904069	.904092	.904116	.904140	.904164	.904187	.904211	−0	0
9550	.904235	.904258	.904282	.904306	.904329	.904353	.904376	.904400	.904424	.904447	−0	0
9560	.904471	.904494	.904518	.904541	.904565	.904589	.904612	.904636	.904659	.904683	−0	0
9570	.904706	.904730	.904753	.904777	.904800	.904824	.904847	.904871	.904894	.904918	−0	0
9580	.904941	.904964	.904988	.905011	.905035	.905058	.905081	.905105	.905128	.905152	−0	0
9590	.905175	.905198	.905222	.905245	.905268	.905292	.905315	.905338	.905362	.905385	−0	0
9600	.905408	.905431	.905455	.905478	.905501	.905524	.905548	.905571	.905594	.905617	−0	0
9610	.905641	.905664	.905687	.905710	.905733	.905757	.905780	.905803	.905826	.905849	−0	0
9620	.905872	.905895	.905919	.905942	.905965	.905988	.906011	.906034	.906057	.906080	−0	0
9630	.906103	.906126	.906149	.906172	.906195	.906218	.906241	.906264	.906287	.906310	−0	0
9640	.906333	.906356	.906379	.906402	.906425	.906448	.906471	.906494	.906517	.906540	−0	0
9650	.906563	.906586	.906609	.906632	.906655	.906677	.906700	.906723	.906746	.906769	−0	0
9660	.906792	.906815	.906837	.906860	.906883	.906906	.906929	.906951	.906974	.906997	−0	0
9670	.907020	.907043	.907065	.907088	.907111	.907134	.907156	.907179	.907202	.907224	−0	0
9680	.907247	.907270	.907293	.907315	.907338	.907361	.907383	.907406	.907428	.907451	−0	0
9690	.907474	.907496	.907519	.907542	.907564	.907587	.907609	.907632	.907655	.907677	−0	0
9700	.907700	.907722	.907745	.907767	.907790	.907812	.907835	.907857	.907880	.907902	−0	0
9710	.907925	.907947	.907970	.907992	.908015	.908037	.908060	.908082	.908104	.908127	−0	0
9720	.908149	.908172	.908194	.908216	.908239	.908261	.908284	.908306	.908328	.908351	−0	0
9730	.908373	.908395	.908418	.908440	.908462	.908485	.908507	.908529	.908551	.908574	−0	0
9740	.908596	.908618	.908641	.908663	.908685	.908707	.908730	.908752	.908774	.908796	−0	0
9750	.908818	.908841	.908863	.908885	.908907	.908929	.908951	.908974	.908996	.909018	−0	0
9760	.909040	.909062	.909084	.909106	.909128	.909151	.909173	.909195	.909217	.909239	−0	0
9770	.909261	.909283	.909305	.909327	.909349	.909371	.909393	.909415	.909437	.909459	−0	0
9780	.909481	.909503	.909525	.909547	.909569	.909591	.909613	.909635	.909657	.909679	−0	0
9790	.909701	.909723	.909745	.909766	.909788	.909810	.909832	.909854	.909876	.909898	−0	0
9800	.909920	.909941	.909963	.909985	.910007	.910029	.910051	.910072	.910094	.910116	−0	0
9810	.910138	.910159	.910181	.910203	.910225	.910247	.910268	.910290	.910312	.910333	−0	0
9820	.910355	.910377	.910399	.910420	.910442	.910464	.910485	.910507	.910529	.910550	−0	0
9830	.910572	.910594	.910615	.910637	.910658	.910680	.910702	.910723	.910745	.910766	−0	0
9840	.910788	.910810	.910831	.910853	.910874	.910896	.910917	.910939	.910960	.910982	−0	0
9850	.911003	.911025	.911046	.911068	.911089	.911111	.911132	.911154	.911175	.911197	−0	0
9860	.911218	.911240	.911261	.911283	.911304	.911325	.911347	.911368	.911390	.911411	−0	0
9870	.911432	.911454	.911475	.911496	.911518	.911539	.911560	.911582	.911603	.911624	−0	0
9880	.911646	.911667	.911688	.911710	.911731	.911752	.911773	.911795	.911816	.911837	−0	0
9890	.911858	.911880	.911901	.911922	.911943	.911965	.911986	.912007	.912028	.912049	−0	0
9900	.912070	.912092	.912113	.912134	.912155	.912176	.912197	.912219	.912240	.912261	−0	0
9910	.912282	.912303	.912324	.912345	.912366	.912387	.912408	.912429	.912451	.912472	−0	0
9920	.912493	.912514	.912535	.912556	.912577	.912598	.912619	.912640	.912661	.912682	−0	0
9930	.912703	.912724	.912745	.912766	.912787	.912808	.912828	.912849	.912870	.912891	−0	0
9940	.912912	.912933	.912954	.912975	.912996	.913017	.913037	.913058	.913079	.913100	−0	0
9950	.913121	.913142	.913163	.913183	.913204	.913225	.913246	.913267	.913287	.913308	−0	0
9960	.913329	.913350	.913371	.913391	.913412	.913433	.913454	.913474	.913495	.913516	−0	0
9970	.913536	.913557	.913578	.913599	.913619	.913640	.913661	.913681	.913702	.913723	−0	0
9980	.913743	.913764	.913785	.913805	.913826	.913846	.913867	.913888	.913908	.913929	−0	0
9990	.913950	.913970	.913991	.914011	.914032	.914052	.914073	.914093	.914114	.914135	−0	0

$\int_0^{\lambda T} M_{e\lambda} d\lambda / M_e(T)$

TABLE 11B

Blackbody: Relative Fractional Emittance (Radiant) Complement

λT(m.°K)	0	1	2	3	4	5	6	7	8	9	m
10	.858449	.838242	.818651	.799653	.781226	.763349	.746003	.729168	.712827	.696961	-1
11	.681554	.666590	.652053	.637928	.624202	.610861	.597890	.585279	.573015	.561085	-1
12	.549480	.538188	.527198	.516502	.506090	.495951	.486078	.476462	.467095	.457968	-1
13	.449075	.440408	.431960	.423725	.415694	.407864	.400226	.392776	.385508	.378416	-1
14	.371495	.364740	.358147	.351710	.345424	.339286	.333292	.327436	.321716	.316127	-1
15	.310665	.305328	.300111	.295011	.290025	.285150	.280383	.275720	.271160	.266698	-1
16	.262334	.258063	.253884	.249794	.245791	.241872	.238036	.234279	.230601	.226999	-1
17	.223471	.220015	.216630	.213313	.210064	.206879	.203758	.200700	.197701	.194762	-1
18	.191881	.189056	.186285	.183568	.180904	.178290	.175727	.173212	.170744	.168323	-1
19	.165947	.163616	.161328	.159082	.156877	.154713	.152588	.150501	.148453	.146441	-1
20	.144465	.142525	.140618	.138746	.136907	.135099	.133324	.131579	.129864	.128179	-1
21	.126523	.124895	.123294	.121721	.120175	.118654	.117159	.115689	.114243	.112821	-1
22	.111422	.110047	.108694	.107363	.106053	.104765	.103497	.102250	.101022	.099814	-1
23	.986253	.974552	.963034	.951697	.940536	.929549	.918731	.908080	.897593	.887266	-2
24	.877096	.867081	.857217	.847502	.837933	.828506	.819220	.810072	.801060	.792180	-2
25	.783430	.774809	.766313	.757941	.749689	.741557	.733542	.725641	.717854	.710177	-2
26	.702608	.695147	.687791	.680537	.673386	.666333	.659379	.652521	.645758	.639087	-2
27	.632508	.626018	.619617	.613302	.607073	.600928	.594865	.588884	.582982	.577158	-2
28	.571412	.565741	.560145	.554623	.549172	.543793	.538484	.533243	.528070	.522964	-2
29	.517923	.512946	.508033	.503183	.498393	.493665	.488995	.484385	.479832	.475335	-2
30	.470895	.466510	.462179	.457901	.453676	.449502	.445380	.441307	.437284	.433310	-2
31	.429383	.425504	.421672	.417885	.414143	.410446	.406792	.403182	.399614	.396088	-2
32	.392604	.389160	.385756	.382392	.379066	.375779	.372530	.369318	.366143	.363004	-2
33	.359901	.356833	.353800	.350801	.347836	.344904	.342005	.339138	.336303	.333500	-2
34	.330727	.327985	.325274	.322592	.319939	.317316	.314721	.312154	.309615	.307104	-2
35	.304619	.302161	.299730	.297324	.294944	.292590	.290260	.287955	.285674	.283418	-2
36	.281185	.278975	.276789	.274625	.272483	.270364	.268267	.266191	.264137	.262104	-2
37	.260091	.258099	.256128	.254176	.252244	.250331	.248438	.246564	.244709	.242872	-2
38	.241053	.239253	.237470	.235706	.233958	.232228	.230515	.228818	.227138	.225475	-2
39	.223827	.222196	.220581	.218981	.217396	.215827	.214273	.212733	.211209	.209698	-2
40	.208203	.206721	.205253	.203800	.202360	.200933	.199520	.198120	.196733	.195359	-2
41	.193998	.192649	.191313	.189989	.188677	.187378	.186090	.184814	.183549	.182297	-2
42	.181055	.179825	.178606	.177397	.176200	.175013	.173837	.172672	.171517	.170372	-2
43	.169237	.168113	.166998	.165893	.164798	.163713	.162637	.161570	.160513	.159465	-2
44	.158426	.157396	.156374	.155362	.154359	.153364	.152377	.151399	.150429	.149468	-2
45	.148515	.147570	.146632	.145703	.144782	.143868	.142962	.142064	.141173	.140289	-2
46	.139413	.138544	.137682	.136828	.135980	.135140	.134306	.133479	.132659	.131846	-2
47	.131039	.130239	.129446	.128659	.127878	.127103	.126335	.125573	.124817	.124067	-2
48	.123323	.122585	.121853	.121126	.120406	.119691	.118982	.118278	.117580	.116887	-2
49	.116200	.115518	.114842	.114171	.113505	.112844	.112188	.111537	.110892	.110251	-2

TABLE 11B

Blackbody: Relative Fractional Emittance (Radiant) Complement

λT(m.°K)	0	1	2	3	4	5	6	7	8	9	m
50	.109615	.108984	.108358	.107737	.107120	.106509	.105901	.105299	.104701	.104107	-2
51	.103518	.102934	.102353	.101777	.101206	.100639	.100076	.099517	.098962	.098411	-2
52	.978649	.973224	.967839	.962493	.957187	.951920	.946692	.941501	.936349	.931234	-3
53	.926156	.921115	.916110	.911142	.906209	.901312	.896450	.891624	.886831	.882073	-3
54	.877349	.872658	.868001	.863377	.858786	.854227	.849701	.845206	.840743	.836311	-3
55	.831910	.827541	.823201	.818892	.814613	.810364	.806144	.801953	.797792	.793659	-3
56	.789555	.785478	.781430	.777410	.773417	.769451	.765513	.761601	.757716	.753857	-3
57	.750025	.746218	.742437	.738682	.734951	.731246	.727566	.723910	.720279	.716672	-3
58	.713089	.709530	.705994	.702482	.698993	.695528	.692085	.688665	.685267	.681891	-3
59	.678538	.675207	.671897	.668609	.665342	.662097	.658873	.655669	.652487	.649325	-3
60	.646183	.643061	.639960	.636878	.633816	.630774	.627751	.624748	.621764	.618798	-3
61	.615851	.612924	.610014	.607123	.604250	.601396	.598559	.595740	.592939	.590155	-3
62	.587388	.584639	.581907	.579192	.576494	.573813	.571148	.568499	.565867	.563252	-3
63	.560652	.558068	.555500	.552948	.550412	.547890	.545385	.542894	.540419	.537959	-3
64	.535513	.533083	.530667	.528265	.525878	.523506	.521148	.518803	.516473	.514157	-3
65	.511855	.509566	.507291	.505029	.502781	.500547	.498325	.496117	.493921	.491739	-3
66	.489569	.487412	.485268	.483136	.481017	.478910	.476815	.474733	.472663	.470604	-3
67	.468558	.466523	.464501	.452490	.460490	.458502	.456526	.454560	.452606	.450664	-3
68	.448732	.446811	.444902	.443003	.441115	.439237	.437371	.435514	.433669	.431833	-3
69	.430008	.428194	.426389	.424595	.422811	.421036	.419272	.417517	.415773	.414037	-3
70	.412312	.410596	.408890	.407193	.405505	.403827	.402158	.400498	.398847	.397206	-3
71	.395573	.393949	.392334	.390728	.389131	.387543	.385963	.384391	.382828	.381274	-3
72	.379728	.378190	.376661	.375140	.373627	.372122	.370625	.369136	.367655	.366183	-3
73	.364718	.363260	.361811	.360369	.358935	.357508	.356089	.354678	.353274	.351877	-3
74	.350488	.349106	.347731	.346363	.345003	.343650	.342304	.340964	.339632	.338307	-3
75	.336988	.335677	.334372	.333074	.331783	.330498	.329220	.327949	.326684	.325425	-3
76	.324174	.322928	.321689	.320456	.319229	.318009	.316795	.315587	.314385	.313190	-3
77	.312000	.310816	.309639	.308467	.307301	.306141	.304987	.303839	.302696	.301559	-3
78	.300428	.299303	.298183	.297069	.295960	.294856	.293759	.292666	.291579	.290498	-3
79	.289422	.288351	.287285	.286225	.285170	.284120	.283075	.282035	.281000	.279971	-3
80	.278946	.277926	.276912	.275902	.274897	.273897	.272902	.271912	.270927	.269946	-3
81	.268970	.267999	.267032	.266070	.265113	.264160	.263212	.262268	.261329	.260394	-3
82	.259464	.258538	.257616	.256699	.255787	.254878	.253974	.253074	.252179	.251287	-3
83	.250400	.249517	.248638	.247764	.246893	.246026	.245164	.244305	.243451	.242600	-3
84	.241754	.240911	.240072	.239238	.238407	.237580	.236756	.235937	.235121	.234309	-3
85	.233501	.232696	.231895	.231098	.230305	.229515	.228729	.227946	.227167	.226391	-3
86	.225619	.224851	.224086	.223324	.222566	.221811	.221060	.220312	.219567	.218826	-3
87	.218088	.217354	.216623	.215895	.215170	.214448	.213730	.213015	.212303	.211594	-3
88	.210889	.210186	.209487	.208791	.208098	.207408	.206721	.206037	.205356	.204678	-3
89	.204003	.203331	.202662	.201995	.201332	.200672	.200014	.199360	.198708	.198059	-3
90	.197413	.196770	.196129	.195492	.194857	.194225	.193595	.192968	.192344	.191723	-3
91	.191104	.190488	.189875	.189264	.188656	.188050	.187448	.186847	.186249	.185654	-3
92	.185061	.184471	.183883	.183298	.182715	.182135	.181557	.180982	.180409	.179839	-3
93	.179270	.178705	.178141	.177580	.177022	.176465	.175911	.175360	.174810	.174263	-3
94	.173719	.173176	.172636	.172098	.171562	.171029	.170497	.169968	.169441	.168916	-3
95	.168394	.167873	.167355	.166839	.166325	.165813	.165303	.164795	.164289	.163786	-3
96	.163284	.162785	.162287	.161792	.161298	.160807	.160317	.159830	.159344	.158861	-3
97	.158379	.157900	.157422	.156946	.156472	.156000	.155530	.155062	.154596	.154131	-3
98	.153669	.153208	.152749	.152292	.151837	.151383	.150932	.150482	.150034	.149588	-3
99	.149143	.148700	.148260	.147820	.147383	.146947	.146513	.146081	.145650	.145221	-3

$1-\int_o^{\lambda T} M_{e\lambda} d\lambda / M_e(T)$

TABLE 11B

Blackbody: Relative Fractional Emittance (Radiant) Complement

λT(m.°K)	0	1	2	3	4	5	6	7	8	9	m
100	.144794	.140612	.136589	.132718	.128992	.125405	.121949	.118619	.115409	.112314	-3
110	.109328	.106448	.103667	.100983	.098391	.095886	.093466	.091127	.088865	.086677	-3
120	.845602	.825120	.805294	.786098	.767508	.749499	.732048	.715136	.698740	.682842	-4
130	.667422	.652463	.637948	.623860	.610184	.596904	.584007	.571479	.559307	.547478	-4
140	.535980	.524802	.513933	.503361	.493078	.483072	.473336	.463859	.454634	.445652	-4
150	.436905	.428385	.420085	.411999	.404118	.396437	.388950	.381650	.374531	.367589	-4
160	.360817	.354210	.347764	.341473	.335333	.329339	.323487	.317773	.312193	.306742	-4
170	.301418	.296216	.291134	.286167	.281312	.276566	.271927	.267391	.262955	.258617	-4
180	.254373	.250222	.246161	.242187	.238298	.234492	.230767	.227120	.223550	.220054	-4
190	.216630	.213277	.209993	.206776	.203625	.200537	.197511	.194546	.191640	.188792	-4
200	.185999	.183262	.180578	.177946	.175365	.172833	.170350	.167915	.165526	.163181	-4
210	.160881	.158624	.156409	.154235	.152101	.150006	.147950	.145931	.143948	.142001	-4
220	.140090	.138212	.136368	.134556	.132777	.131028	.129310	.127623	.125964	.124334	-4
230	.122732	.121157	.119609	.118088	.116592	.115121	.113675	.112253	.110855	.109480	-4
240	.108127	.106797	.105488	.104201	.102934	.101688	.100462	.099256	.098069	.096900	-4
250	.957504	.946187	.935048	.924083	.913289	.902662	.892199	.881898	.871754	.861766	-5
260	.851929	.842241	.832700	.823303	.814046	.804928	.795945	.787095	.778376	.769786	-5
270	.761321	.752980	.744760	.736660	.728677	.720808	.713053	.705408	.697872	.690443	-5
280	.683120	.675899	.668780	.661760	.654839	.648013	.641282	.634644	.628098	.621641	-5
290	.615272	.608990	.602793	.596680	.590649	.584699	.578829	.573038	.567323	.561684	-5
300	.556119	.550628	.545209	.539861	.534582	.529372	.524229	.519153	.514142	.509196	-5
310	.504313	.499492	.494732	.490032	.485392	.480810	.476286	.471818	.467406	.463049	-5
320	.458746	.454496	.450298	.446152	.442057	.438011	.434015	.430067	.426167	.422314	-5
330	.418508	.414747	.411030	.407359	.403730	.400145	.396602	.393101	.389640	.386221	-5
340	.382841	.379500	.376199	.372935	.369709	.366520	.363368	.360252	.357171	.354126	-5
350	.351115	.348137	.345194	.342283	.339406	.336560	.333746	.330964	.328212	.325490	-5
360	.322799	.320137	.317505	.314901	.312325	.309778	.307258	.304766	.302300	.299861	-5
370	.297448	.295061	.292699	.290363	.288051	.285764	.283501	.281262	.279046	.276853	-5
380	.274684	.272537	.270412	.268309	.266228	.264169	.262131	.260113	.258117	.256140	-5
390	.254184	.252248	.250331	.248434	.246556	.244696	.242856	.241033	.239229	.237443	-5
400	.235675	.233924	.232190	.230474	.228774	.227091	.225425	.223775	.222141	.220522	-5
410	.218920	.217333	.215761	.214204	.212663	.211136	.209623	.208125	.206642	.205172	-5
420	.203716	.202274	.200846	.199431	.198029	.196640	.195265	.193902	.192551	.191214	-5
430	.189888	.188575	.187274	.185985	.184708	.183442	.182188	.180945	.179714	.178493	-5
440	.177284	.176086	.174898	.173721	.172555	.171399	.170253	.169118	.167992	.166877	-5
450	.165771	.164675	.163589	.162512	.161445	.160387	.159339	.158299	.157268	.156247	-5
460	.155234	.154230	.153234	.152248	.151269	.150299	.149337	.148384	.147438	.146501	-5
470	.145571	.144649	.143736	.142829	.141931	.141039	.140156	.139279	.138410	.137549	-5
480	.136694	.135846	.135006	.134172	.133345	.132525	.131712	.130905	.130105	.129311	-5
490	.128524	.127743	.126969	.126201	.125439	.124683	.123933	.123189	.122451	.121719	-5

TABLE 11B

Blackbody: Relative Fractional Emittance (Radiant) Complement

λT(m.°K)	0	1	2	3	4	5	6	7	8	9	m
500	.120993	.120272	.119557	.118848	.118145	.117447	.116754	.116067	.115386	.114709	-5
510	.114038	.113372	.112712	.112056	.111406	.110760	.110120	.109484	.108854	.108228	-5
520	.107607	.106991	.106379	.105772	.105170	.104572	.103979	.103390	.102806	.102226	-5
530	.101650	.101079	.100512	.099949	.099390	.098836	.098286	.097739	.097197	.096659	-5
540	.961249	.955946	.950682	.945457	.940270	.935121	.930009	.924934	.919897	.914895	-6
550	.909930	.905001	.900108	.895249	.890426	.885637	.880883	.876162	.871475	.866821	-6
560	.862201	.857613	.853058	.848535	.844044	.839584	.835156	.830759	.826393	.822057	-6
570	.817752	.813477	.809231	.805015	.800828	.796670	.792541	.788440	.784368	.780323	-6
580	.776307	.772317	.768356	.764421	.760513	.756631	.752776	.748947	.745144	.741367	-6
590	.737615	.733889	.730187	.726511	.722859	.719231	.715628	.712049	.708493	.704961	-6
600	.701453	.697968	.694506	.691066	.687650	.684256	.680884	.677534	.674207	.670901	-6
610	.667616	.664353	.661112	.657891	.654691	.651512	.648353	.645215	.642097	.638999	-6
620	.635922	.632863	.629825	.626805	.623805	.620825	.617863	.614920	.611995	.609089	-6
630	.606202	.603332	.600481	.597648	.594832	.592034	.589254	.586491	.583745	.581017	-6
640	.578305	.575610	.572932	.570271	.567626	.564997	.562385	.559788	.557208	.554643	-6
650	.552094	.549561	.547043	.544541	.542054	.539582	.537125	.534682	.532255	.529842	-6
660	.527444	.525060	.522691	.520336	.517995	.515668	.513355	.511055	.508770	.506498	-6
670	.504240	.501995	.499763	.497544	.495339	.493146	.490967	.488800	.486646	.484505	-6
680	.482376	.480260	.478156	.476065	.473985	.471918	.469862	.467819	.465787	.463767	-6
690	.461759	.459763	.457778	.455804	.453841	.451890	.449950	.448021	.446104	.444197	-6
700	.442300	.440415	.438541	.436677	.434823	.432980	.431148	.429325	.427513	.425712	-6
710	.423920	.422138	.420366	.418605	.416853	.415111	.413378	.411655	.409942	.408238	-6
720	.406544	.404859	.403183	.401516	.399859	.398211	.396572	.394942	.393320	.391708	-6
730	.390104	.388510	.386923	.385346	.383777	.382216	.380664	.379121	.377585	.376058	-6
740	.374540	.373029	.371526	.370032	.368546	.367067	.365596	.364134	.362679	.361231	-6
750	.359792	.358360	.356936	.355519	.354110	.352708	.351313	.349926	.348546	.347174	-6
760	.345808	.344450	.343099	.341755	.340418	.339088	.337765	.336448	.335139	.333836	-6
770	.332540	.331251	.329969	.328693	.327423	.326160	.324904	.323654	.322410	.321173	-6
780	.319942	.318718	.317499	.316287	.315081	.313882	.312688	.311500	.310318	.309143	-6
790	.307973	.306809	.305651	.304499	.303352	.302212	.301077	.299947	.298824	.297705	-6
800	.296593	.295486	.294385	.293289	.292198	.291113	.290033	.288959	.287889	.286826	-6
810	.285767	.284714	.283665	.282622	.281584	.280551	.279523	.278500	.277482	.276469	-6
820	.275461	.274458	.273460	.272467	.271478	.270494	.269515	.268541	.267571	.266606	-6
830	.265646	.264690	.263739	.262792	.261850	.260912	.259979	.259050	.258126	.257206	-6
840	.256291	.255379	.254473	.253570	.252672	.251778	.250888	.250002	.249120	.248243	-6
850	.247370	.246501	.245636	.244774	.243917	.243064	.242215	.241370	.240529	.239692	-6
860	.238858	.238029	.237203	.236381	.235563	.234749	.233938	.233131	.232328	.231529	-6
870	.230733	.229941	.229152	.228367	.227586	.226808	.226034	.225263	.224496	.223732	-6
880	.222972	.222215	.221461	.220711	.219965	.219221	.218481	.217745	.217011	.216281	-6
890	.215555	.214831	.214111	.213394	.212680	.211969	.211262	.210558	.209856	.209158	-6
900	.208463	.207771	.207082	.206396	.205714	.205034	.204357	.203683	.203012	.202344	-6
910	.201679	.201017	.200358	.199702	.199048	.198398	.197750	.197105	.196463	.195823	-6
920	.195187	.194553	.193922	.193293	.192668	.192045	.191424	.190807	.190192	.189579	-6
930	.188970	.188363	.187758	.187156	.186557	.185960	.185366	.184774	.184185	.183598	-6
940	.183014	.182432	.181853	.181276	.180702	.180130	.179560	.178993	.178428	.177866	-6
950	.177306	.176748	.176193	.175640	.175089	.174541	.173995	.173451	.172909	.172370	-6
960	.171833	.171298	.170765	.170235	.169707	.169181	.168657	.168135	.167615	.167098	-6
970	.166583	.166070	.165558	.165049	.164543	.164038	.163535	.163034	.162536	.162039	-6
980	.161544	.161052	.160561	.160072	.159586	.159101	.158618	.158138	.157659	.157182	-6
990	.156707	.156234	.155763	.155294	.154826	.154361	.153897	.153435	.152975	.152517	-6

TABLE 11B

Blackbody: Relative Fractional Emittance (Radiant) Complement

λT(m.°K)	0	1	2	3	4	5	6	7	8	9	m
1000	.152061	.147597	.143306	.139179	.135210	.131390	.127712	.124171	.120759	.117471	-6
1100	.114302	.111245	.108297	.105451	.102705	.100053	.097492	.095017	.092625	.090313	-6
1200	.880775	.859150	.838226	.817977	.798375	.779394	.761010	.743199	.725941	.709212	-7
1300	.692993	.677266	.662010	.647210	.632847	.618906	.605372	.592229	.579464	.567063	-7
1400	.555014	.543304	.531920	.520853	.510090	.499622	.489439	.479530	.469887	.460501	-7
1500	.451363	.442465	.433800	.425359	.417137	.409124	.401316	.393705	.386285	.379050	-7
1600	.371995	.365114	.358402	.351853	.345463	.339227	.333140	.327197	.321395	.315730	-7
1700	.310197	.304792	.299512	.294354	.289313	.284387	.279571	.274864	.270262	.265763	-7
1800	.261362	.257058	.252849	.248730	.244701	.240758	.236899	.233123	.229426	.225807	-7
1900	.222263	.218794	.215396	.212069	.208809	.205616	.202488	.199422	.196419	.193475	-7
2000	.190590	.187762	.184990	.182272	.179607	.176994	.174431	.171917	.169452	.167033	-7
2100	.164660	.162332	.160048	.157806	.155606	.153447	.151327	.149246	.147203	.145198	-7
2200	.143228	.141294	.139395	.137530	.135698	.133898	.132130	.130392	.128686	.127008	-7
2300	.125360	.123741	.122149	.120584	.119046	.117534	.116047	.114585	.113148	.111735	-7
2400	.110345	.108978	.107634	.106311	.105010	.103731	.102472	.101233	.100014	.098815	-7
2500	.976349	.964734	.953303	.942052	.930977	.920074	.909342	.898776	.888372	.878129	-8
2600	.868043	.858110	.848329	.838695	.829207	.819862	.810656	.801588	.792654	.783853	-8
2700	.775181	.766637	.758218	.749922	.741747	.733690	.725749	.717922	.710207	.702603	-8
2800	.695107	.687717	.680431	.673248	.666166	.659183	.652297	.645506	.638810	.632205	-8
2900	.625692	.619268	.612931	.606680	.600514	.594432	.588431	.582511	.576670	.570907	-8
3000	.565220	.559609	.554071	.548607	.543214	.537892	.532638	.527453	.522336	.517284	-8
3100	.512297	.507374	.502513	.497715	.492978	.488300	.483682	.479121	.474618	.470171	-8
3200	.465779	.461442	.457159	.452928	.448750	.444623	.440546	.436519	.432540	.428610	-8
3300	.424728	.420892	.417102	.413358	.409658	.406002	.402390	.398821	.395293	.391807	-8
3400	.388362	.384957	.381592	.378266	.374978	.371729	.368517	.365341	.362203	.359100	-8
3500	.356032	.352999	.350001	.347036	.344105	.341207	.338341	.335507	.332705	.329934	-8
3600	.327193	.324483	.321803	.319152	.316530	.313937	.311372	.308835	.306325	.303842	-8
3700	.301387	.298957	.296554	.294176	.291824	.289497	.287194	.284916	.282662	.280431	-8
3800	.278224	.276040	.273879	.271741	.269624	.267530	.265457	.263405	.261375	.259365	-8
3900	.257376	.255407	.253458	.251529	.249620	.247730	.245859	.244006	.242172	.240357	-8
4000	.238560	.236780	.235018	.233274	.231547	.229836	.228143	.226466	.224806	.223162	-8
4100	.221534	.219921	.218324	.216743	.215177	.213626	.212090	.210568	.209061	.207568	-8
4200	.206090	.204625	.203175	.201738	.200315	.198904	.197508	.196124	.194753	.193395	-8
4300	.192049	.190716	.189395	.188087	.186790	.185505	.184232	.182971	.181721	.180483	-8
4400	.179255	.178039	.176834	.175640	.174456	.173283	.172121	.170968	.169827	.168695	-8
4500	.167573	.166461	.165359	.164267	.163185	.162111	.161048	.159993	.158948	.157912	-8
4600	.156884	.155866	.154857	.153856	.152863	.151880	.150904	.149937	.148979	.148028	-8
4700	.147086	.146151	.145225	.144306	.143395	.142491	.141596	.140707	.139826	.138953	-8
4800	.138086	.137227	.136375	.135530	.134692	.133861	.133037	.132219	.131408	.130604	-8
4900	.129806	.129015	.128230	.127452	.126680	.125914	.125154	.124400	.123653	.122911	-8

TABLE 11B

Blackbody: Relative Fractional Emittance (Radiant) Complement

λT(m.°K)	0	1	2	3	4	5	6	7	8	9	m
5000	.122175	.121446	.120722	.120003	.119291	.118584	.117882	.117186	.116496	.115811	-8
5100	.115131	.114457	.113788	.113124	.112465	.111811	.111163	.110519	.109881	.109247	-8
5200	.108618	.107994	.107375	.106760	.106150	.105545	.104945	.104348	.103757	.103170	-8
5300	.102587	.102009	.101435	.100865	.100300	.099739	.099182	.098629	.098080	.097535	-8
5400	.969946	.964579	.959251	.953963	.948714	.943503	.938330	.933195	.928097	.923036	-9
5500	.918012	.913025	.908073	.903158	.898277	.893432	.888622	.883846	.879104	.874396	-9
5600	.869721	.865080	.860472	.855897	.851353	.846842	.842363	.837915	.833499	.829114	-9
5700	.824759	.820435	.816140	.811876	.807642	.803437	.799261	.795113	.790995	.786905	-9
5800	.782843	.778809	.774803	.770824	.766872	.762947	.759049	.755178	.751332	.747513	-9
5900	.743720	.739952	.736210	.732493	.728801	.725133	.721491	.717872	.714278	.710708	-9
6000	.707161	.703638	.700138	.696662	.693209	.689778	.686370	.682984	.679621	.676279	-9
6100	.672960	.669662	.666385	.663130	.659897	.656684	.653492	.650320	.647170	.644039	-9
6200	.640929	.637838	.634768	.631717	.628685	.625673	.622681	.619707	.616752	.613816	-9
6300	.610899	.608000	.605119	.602256	.599412	.596585	.593776	.590985	.588211	.585455	-9
6400	.582716	.579993	.577288	.574600	.571928	.569273	.566634	.564011	.561405	.558814	-9
6500	.556240	.553681	.551138	.548611	.546099	.543602	.541120	.538654	.536203	.533766	-9
6600	.531344	.528937	.526544	.524166	.521802	.519452	.517116	.514794	.512487	.510192	-9
6700	.507912	.505645	.503392	.501152	.498925	.496712	.494511	.492323	.490149	.487987	-9
6800	.485838	.483701	.481577	.479466	.477366	.475279	.473204	.471142	.469091	.467052	-9
6900	.465024	.463009	.461005	.459013	.457032	.455063	.453104	.451157	.449222	.447297	-9
7000	.445383	.443480	.441588	.439707	.437836	.435976	.434127	.432288	.430459	.428641	-9
7100	.426833	.425035	.423247	.421469	.419701	.417943	.416195	.414456	.412727	.411008	-9
7200	.409298	.407598	.405907	.404225	.402553	.400890	.399236	.397592	.395956	.394329	-9
7300	.392711	.391102	.389502	.387910	.386327	.384753	.383187	.381630	.380081	.378540	-9
7400	.377008	.375484	.373969	.372461	.370962	.369470	.367987	.366511	.365044	.363584	-9
7500	.362132	.360687	.359251	.357822	.356400	.354986	.353580	.352181	.350789	.349405	-9
7600	.348028	.346658	.345295	.343939	.342591	.341250	.339915	.338588	.337267	.335953	-9
7700	.334646	.333346	.332053	.330766	.329486	.328213	.326946	.325685	.324431	.323184	-9
7800	.321943	.320708	.319479	.318257	.317041	.315831	.314628	.313430	.312239	.311053	-9
7900	.309874	.308700	.307533	.306371	.305215	.304065	.302921	.301782	.300649	.299522	-9
8000	.298401	.297285	.296174	.295069	.293970	.292876	.291788	.290704	.289627	.288554	-9
8100	.287487	.286425	.285368	.284317	.283271	.282229	.281193	.280162	.279136	.278115	-9
8200	.277099	.276088	.275082	.274081	.273084	.272093	.271106	.270124	.269146	.268174	-9
8300	.267206	.266243	.265284	.264330	.263380	.262436	.261495	.260559	.259628	.258701	-9
8400	.257778	.256860	.255946	.255036	.254131	.253230	.252333	.251441	.250553	.249668	-9
8500	.248788	.247913	.247041	.246173	.245310	.244450	.243595	.242743	.241895	.241052	-9
8600	.240212	.239376	.238544	.237716	.236892	.236071	.235255	.234442	.233633	.232827	-9
8700	.232025	.231227	.230433	.229642	.228855	.228071	.227291	.226515	.225742	.224973	-9
8800	.224207	.223444	.222685	.221930	.221177	.220429	.219683	.218941	.218203	.217467	-9
8900	.216735	.216006	.215281	.214559	.213840	.213124	.212411	.211702	.210995	.210292	-9
9000	.209592	.208895	.208201	.207510	.206823	.206138	.205456	.204778	.204102	.203429	-9
9100	.202759	.202093	.201429	.200768	.200110	.199454	.198802	.198152	.197506	.196862	-9
9200	.196221	.195582	.194947	.194314	.193684	.193056	.192432	.191810	.191190	.190574	-9
9300	.189960	.189349	.188740	.188134	.187530	.186929	.186331	.185735	.185142	.184551	-9
9400	.183963	.183377	.182794	.182213	.181635	.181059	.180485	.179914	.179346	.178779	-9
9500	.178215	.177654	.177095	.176538	.175984	.175431	.174882	.174334	.173789	.173246	-9
9600	.172705	.172167	.171630	.171096	.170564	.170035	.169507	.168982	.168459	.167938	-9
9700	.167419	.166903	.166388	.165876	.165366	.164857	.164351	.163847	.163345	.162845	-9
9800	.162347	.161852	.161358	.160866	.160376	.159888	.159402	.158918	.158436	.157956	-9
9900	.157478	.157002	.156528	.156055	.155585	.155116	.154650	.154185	.153722	.153261	-9

$1-\int_o^{\lambda T} M_{e\lambda}\, d\lambda / M_e(T)$

TABLE 12

Blackbody: Peak Spectral Radiant Emittance ($W/m^2.\mu$)

T(°K)	0	10	20	30	40	50	60	70	80	90	m
100	0.12865	0.20719	0.32012	0.47767	0.69191	0.97694	1.34899	1.82665	2.43093	3.18550	0
200	0.41168	0.52542	0.66301	0.82804	1.02439	1.25635	1.52854	1.84599	2.21411	2.63876	1
300	0.31262	0.36831	0.43168	0.50348	0.58453	0.67569	0.77790	0.89211	1.01936	1.16073	2
400	1.31738	1.49049	1.68134	1.89126	2.12165	2.37395	2.64971	2.95052	3.27805	3.63404	2
500	4.02031	4.43875	4.89132	5.38008	5.90716	6.47475	7.08516	7.74077	8.44403	9.19750	2
600	1.00038	1.08657	1.17860	1.27677	1.38137	1.49271	1.61113	1.73694	1.87049	2.01213	3
700	2.16222	2.32114	2.48927	2.66701	2.85475	3.05292	3.26195	3.48228	3.71435	3.95863	3
800	4.21560	4.48575	4.76957	5.06758	5.38030	5.70827	6.05204	6.41218	6.78927	7.18389	3
900	0.75967	0.80282	0.84791	0.89500	0.94417	0.99547	1.04898	1.10476	1.16289	1.22345	4
1000	1.28650	1.35212	1.42040	1.49141	1.56522	1.64194	1.72163	1.80438	1.89029	1.97944	4
1100	2.07192	2.16783	2.26725	2.37029	2.47705	2.58761	2.70209	2.82058	2.94320	3.07004	4
1200	3.20122	3.33685	3.47703	3.62189	3.77154	3.92609	4.08566	4.25039	4.42038	4.59577	4
1300	4.77668	4.96325	5.15560	5.35387	5.55819	5.76871	5.98555	6.20887	6.43880	6.67550	4
1400	6.91911	7.16977	7.42765	7.69290	7.96567	8.24612	8.53442	8.83073	9.13521	9.44803	4
1500	0.97694	1.00994	1.04383	1.07862	1.11433	1.15098	1.18859	1.22718	1.26676	1.30736	5
1600	1.34899	1.39168	1.43544	1.48029	1.52626	1.57336	1.62162	1.67106	1.72169	1.77355	5
1700	1.82665	1.88101	1.93665	1.99361	2.05190	2.11154	2.17257	2.23499	2.29885	2.36415	5
1800	2.43093	2.49921	2.56902	2.64037	2.71331	2.78784	2.86401	2.94183	3.02134	3.10255	5
1900	3.18550	3.27022	3.35673	3.44506	3.53524	3.62729	3.72126	3.81716	3.91504	4.01490	5
2000	4.11680	4.22075	4.32680	4.43496	4.54528	4.65778	4.77250	4.88947	5.00872	5.13028	5
2100	5.25420	5.38049	5.50921	5.64037	5.77402	5.91020	6.04893	6.19025	6.33421	6.48083	5
2200	6.63015	6.78221	6.93705	7.09470	7.25521	7.41861	7.58494	7.75424	7.92655	8.10191	5
2300	0.82804	0.84619	0.86467	0.88347	0.90259	0.92204	0.94182	0.96195	0.98241	1.00323	6
2400	1.02439	1.04591	1.06779	1.09004	1.11265	1.13564	1.15900	1.18275	1.20689	1.23142	6
2500	1.25635	1.28168	1.30741	1.33356	1.36012	1.38711	1.41452	1.44237	1.47065	1.49937	6
2600	1.52854	1.55816	1.58824	1.61878	1.64979	1.68128	1.71324	1.74569	1.77862	1.81205	6
2700	1.84599	1.88043	1.91538	1.95085	1.98684	2.02336	2.06042	2.09802	2.13616	2.17486	6
2800	2.21411	2.25393	2.29433	2.33530	2.37685	2.41899	2.46173	2.50507	2.54901	2.59358	6
2900	2.63876	2.68457	2.73101	2.77810	2.82583	2.87422	2.92327	2.97298	3.02337	3.07444	6
3000	3.12619	3.17865	3.23180	3.28566	3.34024	3.39554	3.45157	3.50834	3.56585	3.62412	6
3100	3.68314	3.74293	3.80349	3.86484	3.92697	3.98990	4.05364	4.11819	4.18355	4.24975	6
3200	4.31678	4.38465	4.45337	4.52296	4.59341	4.66473	4.73694	4.81004	4.88404	4.95894	6
3300	5.03477	5.11152	5.18920	5.26782	5.34739	5.42792	5.50942	5.59190	5.67536	5.75981	6
3400	5.84527	5.93173	6.01922	6.10774	6.19729	6.28789	6.37955	6.47228	6.56608	6.66096	6
3500	6.75694	6.85402	6.95221	7.05153	7.15198	7.25357	7.35631	7.46021	7.56528	7.67153	6
3600	7.77897	7.88762	7.99747	8.10854	8.22085	8.33440	8.44919	8.56525	8.68258	8.80119	6
3700	0.89211	0.90423	0.91648	0.92887	0.94139	0.95404	0.96683	0.97975	0.99282	1.00602	7
3800	1.01936	1.03284	1.04647	1.06024	1.07415	1.08821	1.10242	1.11677	1.13128	1.14593	7
3900	1.16073	1.17569	1.19080	1.20607	1.22149	1.23707	1.25281	1.26871	1.28477	1.30099	7
4000	1.31738	1.33393	1.35064	1.36752	1.38458	1.40180	1.41919	1.43675	1.45449	1.47240	7
4100	1.49049	1.50876	1.52720	1.54582	1.56463	1.58362	1.60279	1.62215	1.64169	1.66142	7
4200	1.68134	1.70145	1.72176	1.74225	1.76295	1.78383	1.80492	1.82620	1.84769	1.86937	7
4300	1.89126	1.91336	1.93566	1.95816	1.98088	2.00381	2.02695	2.05030	2.07386	2.09765	7
4400	2.12165	2.14587	2.17031	2.19497	2.21986	2.24497	2.27030	2.29587	2.32167	2.34769	7
4500	2.37395	2.40045	2.42718	2.45415	2.48136	2.50880	2.53649	2.56443	2.59261	2.62104	7
4600	2.64971	2.67864	2.70782	2.73725	2.76694	2.79688	2.82709	2.85755	2.88828	2.91927	7
4700	2.95052	2.98205	3.01384	3.04590	3.07823	3.11084	3.14372	3.17689	3.21033	3.24405	7
4800	3.27805	3.31234	3.34692	3.38178	3.41693	3.45238	3.48812	3.52415	3.56048	3.59711	7
4900	3.63404	3.67128	3.70882	3.74666	3.78481	3.82328	3.86205	3.90114	3.94055	3.98027	7

TABLE 12

Blackbody: Peak Spectral Radiant Emittance (W/m^2.μ)

T(°K)	0	10	20	30	40	50	60	70	80	90	m
5000	4.02031	4.06068	4.10136	4.14238	4.18372	4.22539	4.26739	4.30972	4.35240	4.39540	7
5100	4.43875	4.48244	4.52647	4.57085	4.61557	4.66064	4.70607	4.75185	4.79798	4.84447	7
5200	4.89132	4.93854	4.98611	5.03406	5.08237	5.13105	5.18010	5.22953	5.27934	5.32952	7
5300	5.38008	5.43103	5.48236	5.53408	5.58619	5.63870	5.69159	5.74488	5.79857	5.85266	7
5400	5.90716	5.96206	6.01736	6.07308	6.12921	6.18575	6.24271	6.30008	6.35788	6.41611	7
5500	6.47475	6.53383	6.59333	6.65327	6.71365	6.77446	6.83571	6.89741	6.95954	7.02213	7
5600	7.08516	7.14865	7.21259	7.27699	7.34185	7.40716	7.47295	7.53920	7.60591	7.67310	7
5700	7.74077	7.80891	7.87753	7.94663	8.01621	8.08628	8.15684	8.22790	8.29944	8.37149	7
5800	8.44403	8.51707	8.59062	8.66468	8.73925	8.81433	8.88992	8.96603	9.04266	9.11982	7
5900	9.19750	9.27571	9.35445	9.43373	9.51354	9.59389	9.67478	9.75622	9.83820	9.92074	7
6000	1.00038	1.00875	1.01717	1.02564	1.03418	1.04277	1.05141	1.06012	1.06888	1.07770	8
6100	1.08657	1.09551	1.10450	1.11356	1.12267	1.13184	1.14107	1.15036	1.15972	1.16913	8
6200	1.17860	1.18814	1.19774	1.20740	1.21712	1.22690	1.23675	1.24666	1.25663	1.26667	8
6300	1.27677	1.28693	1.29716	1.30746	1.31782	1.32825	1.33874	1.34930	1.35992	1.37061	8
6400	1.38137	1.39219	1.40309	1.41405	1.42508	1.43618	1.44735	1.45858	1.46989	1.48127	8
6500	1.49271	1.50423	1.51582	1.52748	1.53921	1.55102	1.56289	1.57484	1.58686	1.59896	8
6600	1.61113	1.62337	1.63569	1.64808	1.66054	1.67308	1.68570	1.69840	1.71117	1.72401	8
6700	1.73694	1.74994	1.76302	1.77617	1.78941	1.80272	1.81611	1.82959	1.84314	1.85677	8
6800	1.87049	1.88428	1.89815	1.91211	1.92615	1.94027	1.95448	1.96876	1.98313	1.99759	8
6900	2.01213	2.02675	2.04146	2.05625	2.07113	2.08609	2.10114	2.11628	2.13151	2.14682	8
7000	2.16222	2.17771	2.19329	2.20895	2.22471	2.24055	2.25649	2.27252	2.28863	2.30484	8
7100	2.32114	2.33753	2.35402	2.37060	2.38727	2.40403	2.42089	2.43784	2.45489	2.47203	8
7200	2.48927	2.50661	2.52404	2.54157	2.55919	2.57691	2.59473	2.61265	2.63067	2.64879	8
7300	2.66701	2.68532	2.70374	2.72226	2.74088	2.75960	2.77843	2.79735	2.81638	2.83552	8
7400	2.85475	2.87409	2.89354	2.91309	2.93275	2.95251	2.97238	2.99235	3.01244	3.03263	8
7500	3.05292	3.07333	3.09385	3.11447	3.13521	3.15605	3.17701	3.19808	3.21926	3.24055	8
7600	3.26195	3.28347	3.30510	3.32684	3.34870	3.37068	3.39276	3.41497	3.43729	3.45972	8
7700	3.48228	3.50495	3.52774	3.55064	3.57367	3.59682	3.62008	3.64347	3.66697	3.69060	8
7800	3.71435	3.73822	3.76221	3.78633	3.81057	3.83494	3.85942	3.88404	3.90878	3.93364	8
7900	3.95863	3.98375	4.00900	4.03437	4.05987	4.08550	4.11126	4.13715	4.16317	4.18932	8
8000	4.21560	4.24202	4.26856	4.29524	4.32205	4.34900	4.37608	4.40329	4.43064	4.45813	8
8100	4.48575	4.51351	4.54140	4.56943	4.59761	4.62592	4.65437	4.68296	4.71169	4.74056	8
8200	4.76957	4.79872	4.82802	4.85746	4.88704	4.91677	4.94664	4.97665	5.00681	5.03712	8
8300	5.06758	5.09818	5.12893	5.15982	5.19087	5.22206	5.25341	5.28490	5.31655	5.34835	8
8400	5.38030	5.41240	5.44465	5.47706	5.50962	5.54234	5.57521	5.60824	5.64143	5.67477	8
8500	5.70827	5.74193	5.77574	5.80972	5.84385	5.87815	5.91260	5.94722	5.98200	6.01694	8
8600	6.05204	6.08731	6.12274	6.15834	6.19410	6.23003	6.26613	6.30239	6.33882	6.37542	8
8700	6.41218	6.44912	6.48623	6.52350	6.56095	6.59857	6.63637	6.67433	6.71247	6.75078	8
8800	6.78927	6.82793	6.86677	6.90579	6.94498	6.98435	7.02390	7.06363	7.10354	7.14362	8
8900	7.18389	7.22434	7.26497	7.30579	7.34679	7.38797	7.42933	7.47088	7.51262	7.55454	8
9000	7.59665	7.63895	7.68144	7.72411	7.76697	7.81003	7.85327	7.89671	7.94034	7.98416	8
9100	8.02817	8.07238	8.11678	8.16138	8.20617	8.25116	8.29635	8.34174	8.38732	8.43310	8
9200	8.47908	8.52527	8.57165	8.61823	8.66502	8.71201	8.75920	8.80660	8.85421	8.90201	8
9300	8.95003	8.99825	9.04668	9.09532	9.14417	9.19322	9.24249	9.29197	9.34166	9.39156	8
9400	9.44167	9.49200	9.54255	9.59330	9.64428	9.69547	9.74688	9.79850	9.85035	9.90241	8
9500	0.99547	1.00072	1.00599	1.01129	1.01660	1.02194	1.02730	1.03269	1.03810	1.04352	9
9600	1.04898	1.05445	1.05995	1.06547	1.07101	1.07658	1.08217	1.08778	1.09342	1.09908	9
9700	1.10476	1.11047	1.11620	1.12195	1.12773	1.13353	1.13935	1.14520	1.15108	1.15697	9
9800	1.16289	1.16884	1.17481	1.18080	1.18682	1.19286	1.19893	1.20502	1.21114	1.21728	9
9900	1.22345	1.22964	1.23586	1.24210	1.24837	1.25466	1.26097	1.26732	1.27369	1.28008	9

TABLE 12

Blackbody: Peak Spectral Radiant Emittance ($W/m^2 \cdot \mu$)

T(°K)	0	10	20	30	40	50	60	70	80	90	m
10000	1.28650	1.35212	1.42040	1.49141	1.56522	1.64194	1.72163	1.80438	1.89029	1.97944	9
11000	2.07192	2.16783	2.26725	2.37029	2.47705	2.58761	2.70209	2.82058	2.94320	3.07004	9
12000	3.20122	3.33685	3.47703	3.62189	3.77154	3.92609	4.08566	4.25039	4.42038	4.59577	9
13000	4.77668	4.96325	5.15560	5.35387	5.55819	5.76871	5.98555	6.20887	6.43880	6.67550	9
14000	6.91911	7.16977	7.42765	7.69290	7.96567	8.24612	8.53442	8.83073	9.13521	9.44803	9
15000	0.97694	1.00994	1.04382	1.07862	1.11433	1.15098	1.18859	1.22718	1.26676	1.30736	10
16000	1.34899	1.39168	1.43544	1.48029	1.52626	1.57336	1.62162	1.67106	1.72169	1.77355	10
17000	1.82665	1.88101	1.93665	1.99361	2.05190	2.11154	2.17257	2.23499	2.29885	2.36415	10
18000	2.43093	2.49921	2.56902	2.64037	2.71331	2.78784	2.86401	2.94183	3.02134	3.10255	10
19000	3.18550	3.27022	3.35673	3.44506	3.53524	3.62729	3.72126	3.81716	3.91504	4.01490	10
20000	4.11680	4.22075	4.32680	4.43496	4.54528	4.65778	4.77250	4.88947	5.00872	5.13028	10
21000	5.25420	5.38049	5.50921	5.64037	5.77402	5.91020	6.04893	6.19025	6.33421	6.48083	10
22000	6.63015	6.78221	6.93705	7.09470	7.25521	7.41861	7.58494	7.75424	7.92655	8.10191	10
23000	0.82804	0.84619	0.86467	0.88346	0.90259	0.92204	0.94182	0.96195	0.98241	1.00323	11
24000	1.02439	1.04591	1.06779	1.09004	1.11265	1.13564	1.15900	1.18275	1.20689	1.23142	11
25000	1.25635	1.28168	1.30741	1.33356	1.36012	1.38711	1.41452	1.44237	1.47065	1.49937	11
26000	1.52854	1.55816	1.58824	1.61878	1.64979	1.68128	1.71324	1.74569	1.77862	1.81205	11
27000	1.84599	1.88043	1.91538	1.95085	1.98684	2.02336	2.06042	2.09801	2.13616	2.17486	11
28000	2.21411	2.25393	2.29433	2.33530	2.37685	2.41899	2.46173	2.50507	2.54901	2.59358	11
29000	2.63876	2.68457	2.73101	2.77810	2.82583	2.87422	2.92327	2.97298	3.02337	3.07444	11
30000	3.12619	3.17865	3.23180	3.28566	3.34024	3.39554	3.45157	3.50834	3.56585	3.62412	11
31000	3.68314	3.74293	3.80349	3.86484	3.92697	3.98990	4.05364	4.11819	4.18355	4.24975	11
32000	4.31678	4.38465	4.45337	4.52296	4.59341	4.66473	4.73694	4.81004	4.88404	4.95894	11
33000	5.03477	5.11152	5.18920	5.26782	5.34739	5.42792	5.50942	5.59190	5.67536	5.75981	11
34000	5.84527	5.93173	6.01922	6.10774	6.19729	6.28789	6.37955	6.47228	6.56608	6.66096	11
35000	6.75694	6.85402	6.95221	7.05153	7.15198	7.25357	7.35631	7.46021	7.56528	7.67153	11
36000	7.77897	7.88762	7.99747	8.10854	8.22085	8.33439	8.44919	8.56525	8.68258	8.80119	11
37000	0.89211	0.90423	0.91648	0.92887	0.94139	0.95404	0.96683	0.97975	0.99282	1.00602	12
38000	1.01936	1.03284	1.04647	1.06024	1.07415	1.08821	1.10242	1.11677	1.13128	1.14593	12
39000	1.16073	1.17569	1.19080	1.20607	1.22149	1.23707	1.25281	1.26871	1.28477	1.30099	12
40000	1.31738	1.33393	1.35064	1.36752	1.38458	1.40180	1.41919	1.43675	1.45449	1.47240	12
41000	1.49049	1.50876	1.52720	1.54582	1.56463	1.58362	1.60279	1.62215	1.64169	1.66142	12
42000	1.68134	1.70145	1.72176	1.74225	1.76295	1.78383	1.80492	1.82620	1.84769	1.86937	12
43000	1.89126	1.91336	1.93566	1.95816	1.98088	2.00381	2.02695	2.05030	2.07386	2.09765	12
44000	2.12165	2.14587	2.17031	2.19497	2.21985	2.24497	2.27030	2.29587	2.32167	2.34769	12
45000	2.37395	2.40045	2.42718	2.45415	2.48136	2.50880	2.53649	2.56443	2.59261	2.62104	12
46000	2.64971	2.67864	2.70782	2.73725	2.76694	2.79688	2.82709	2.85755	2.88828	2.91927	12
47000	2.95052	2.98205	3.01384	3.04590	3.07823	3.11084	3.14372	3.17689	3.21033	3.24405	12
48000	3.27805	3.31234	3.34692	3.38178	3.41693	3.45238	3.48812	3.52415	3.56048	3.59711	12
49000	3.63404	3.67128	3.70882	3.74666	3.78481	3.82328	3.86205	3.90114	3.94055	3.98027	12

TABLE 12

Blackbody: Peak Spectral Radiant Emittance ($W/m^2.\mu$)

T(°K)	0	10	20	30	40	50	60	70	80	90	m
50000	4.02031	4.06068	4.10136	4.14238	4.18372	4.22539	4.26739	4.30972	4.35239	4.39540	12
51000	4.43875	4.48244	4.52647	4.57085	4.61557	4.66064	4.70607	4.75185	4.79798	4.84447	12
52000	4.89132	4.93854	4.98611	5.03406	5.08237	5.13105	5.18010	5.22953	5.27934	5.32952	12
53000	5.38008	5.43103	5.48236	5.53408	5.58619	5.63870	5.69159	5.74488	5.79857	5.85266	12
54000	5.90716	5.96206	6.01736	6.07308	6.12921	6.18575	6.24271	6.30008	6.35788	6.41610	12
55000	6.47475	6.53383	6.59333	6.65327	6.71365	6.77446	6.83571	6.89740	6.95954	7.02213	12
56000	7.08516	7.14865	7.21259	7.27699	7.34185	7.40716	7.47295	7.53920	7.60591	7.67310	12
57000	7.74077	7.80891	7.87753	7.94663	8.01621	8.08628	8.15684	8.22790	8.29944	8.37149	12
58000	8.44403	8.51707	8.59062	8.66468	8.73925	8.81432	8.88992	8.96603	9.04266	9.11982	12
59000	9.19750	9.27571	9.35445	9.43372	9.51354	9.59389	9.67478	9.75622	9.83820	9.92073	12
60000	1.00038	1.00875	1.01717	1.02564	1.03418	1.04277	1.05141	1.06012	1.06888	1.07770	13
61000	1.08657	1.09551	1.10450	1.11356	1.12267	1.13184	1.14107	1.15036	1.15972	1.16913	13
62000	1.17860	1.18814	1.19774	1.20740	1.21712	1.22690	1.23675	1.24666	1.25663	1.26667	13
63000	1.27677	1.28693	1.29716	1.30746	1.31782	1.32825	1.33874	1.34929	1.35992	1.37061	13
64000	1.38137	1.39219	1.40309	1.41405	1.42508	1.43618	1.44735	1.45858	1.46989	1.48127	13
65000	1.49271	1.50423	1.51582	1.52748	1.53921	1.55102	1.56289	1.57484	1.58686	1.59896	13
66000	1.61113	1.62337	1.63569	1.64808	1.66054	1.67308	1.68570	1.69840	1.71117	1.72401	13
67000	1.73694	1.74994	1.76302	1.77617	1.78941	1.80272	1.81611	1.82959	1.84314	1.85677	13
68000	1.87049	1.88428	1.89815	1.91211	1.92615	1.94027	1.95448	1.96876	1.98313	1.99759	13
69000	2.01213	2.02675	2.04146	2.05625	2.07113	2.08609	2.10114	2.11628	2.13151	2.14682	13
70000	2.16222	2.17771	2.19329	2.20895	2.22471	2.24055	2.25649	2.27252	2.28863	2.30484	13
71000	2.32114	2.33753	2.35402	2.37060	2.38727	2.40403	2.42089	2.43784	2.45489	2.47203	13
72000	2.48927	2.50661	2.52404	2.54157	2.55919	2.57691	2.59473	2.61265	2.63067	2.64879	13
73000	2.66701	2.68532	2.70374	2.72226	2.74088	2.75960	2.77843	2.79735	2.81638	2.83551	13
74000	2.85475	2.87409	2.89354	2.91309	2.93275	2.95251	2.97238	2.99235	3.01244	3.03263	13
75000	3.05292	3.07333	3.09385	3.11447	3.13521	3.15605	3.17701	3.19808	3.21926	3.24055	13
76000	3.26195	3.28347	3.30510	3.32684	3.34870	3.37068	3.39276	3.41497	3.43729	3.45972	13
77000	3.48228	3.50495	3.52774	3.55064	3.57367	3.59682	3.62008	3.64347	3.66697	3.69060	13
78000	3.71435	3.73822	3.76221	3.78633	3.81057	3.83494	3.85942	3.88404	3.90878	3.93364	13
79000	3.95863	3.98375	4.00900	4.03437	4.05987	4.08550	4.11126	4.13715	4.16317	4.18932	13
80000	4.21560	4.24202	4.26856	4.29524	4.32205	4.34900	4.37608	4.40329	4.43064	4.45813	13
81000	4.48575	4.51351	4.54140	4.56943	4.59761	4.62592	4.65437	4.68296	4.71169	4.74056	13
82000	4.76957	4.79872	4.82802	4.85746	4.88704	4.91677	4.94664	4.97665	5.00681	5.03712	13
83000	5.06758	5.09818	5.12893	5.15982	5.19087	5.22206	5.25341	5.28490	5.31655	5.34835	13
84000	5.38030	5.41240	5.44465	5.47706	5.50962	5.54234	5.57521	5.60824	5.64143	5.67477	13
85000	5.70827	5.74193	5.77574	5.80972	5.84385	5.87815	5.91260	5.94722	5.98200	6.01694	13
86000	6.05204	6.08731	6.12274	6.15834	6.19410	6.23003	6.26613	6.30239	6.33882	6.37542	13
87000	6.41218	6.44912	6.48623	6.52350	6.56095	6.59857	6.63637	6.67433	6.71247	6.75078	13
88000	6.78927	6.82793	6.86677	6.90579	6.94498	6.98435	7.02390	7.06363	7.10354	7.14362	13
89000	7.18389	7.22434	7.26497	7.30579	7.34679	7.38797	7.42933	7.47088	7.51262	7.55454	13
90000	7.59665	7.63895	7.68144	7.72411	7.76697	7.81003	7.85327	7.89671	7.94034	7.98416	13
91000	8.02817	8.07238	8.11678	8.16138	8.20617	8.25116	8.29635	8.34174	8.38732	8.43310	13
92000	8.47908	8.52526	8.57165	8.61823	8.66502	8.71201	8.75920	8.80660	8.85421	8.90201	13
93000	8.95003	8.99825	9.04668	9.09532	9.14417	9.19322	9.24249	9.29197	9.34166	9.39156	13
94000	9.44167	9.49200	9.54255	9.59330	9.64428	9.69547	9.74688	9.79850	9.85034	9.90241	13
95000	0.99547	1.00072	1.00599	1.01129	1.01660	1.02194	1.02730	1.03269	1.03810	1.04352	14
96000	1.04898	1.05445	1.05995	1.06547	1.07101	1.07658	1.08217	1.08778	1.09342	1.09908	14
97000	1.10476	1.11047	1.11620	1.12195	1.12773	1.13353	1.13935	1.14520	1.15108	1.15697	14
98000	1.16289	1.16884	1.17481	1.18080	1.18682	1.19286	1.19893	1.20502	1.21114	1.21728	14
99000	1.22345	1.22964	1.23586	1.24210	1.24837	1.25466	1.26097	1.26732	1.27369	1.28008	14

TABLE 13

Blackbody: Total Radiant Emittance (W/m^2)

T(°K)	0	10	20	30	40	50	60	70	80	90	m
100	0.56697	0.83010	1.17567	1.61932	2.17807	2.87029	3.71569	4.73539	5.95182	7.38881	1
200	0.90715	1.10265	1.32816	1.58661	1.88107	2.21473	2.59092	3.01311	3.48492	4.01007	2
300	0.45925	0.52361	0.59451	0.67238	0.75766	0.85081	0.95229	1.06259	1.18221	1.31165	3
400	1.45144	1.60212	1.76424	1.93836	2.12506	2.32493	2.53858	2.76663	3.00971	3.26847	3
500	3.54356	3.83567	4.14547	4.47367	4.82098	5.18813	5.57586	5.98494	6.41611	6.87018	3
600	0.73479	0.78502	0.83777	0.89315	0.95122	1.01208	1.07581	1.14251	1.21226	1.28516	4
700	1.36129	1.44077	1.52367	1.61010	1.70015	1.79393	1.89154	1.99307	2.09864	2.20835	4
800	2.32231	2.44062	2.56339	2.69074	2.82278	2.95962	3.10137	3.24816	3.40009	3.55730	4
900	3.71989	3.88799	4.06173	4.24123	4.42661	4.61801	4.81554	5.01934	5.22955	5.44629	4
1000	5.66970	5.89991	6.13707	6.38130	6.63275	6.89156	7.15787	7.43182	7.71356	8.00324	4
1100	0.83010	0.86070	0.89214	0.92443	0.95759	0.99163	1.02658	1.06244	1.09923	1.13697	5
1200	1.17567	1.21535	1.25603	1.29772	1.34044	1.38420	1.42903	1.47494	1.52195	1.57007	5
1300	1.61932	1.66973	1.72130	1.77405	1.82801	1.88319	1.93962	1.99730	2.05625	2.11650	5
1400	2.17807	2.24097	2.30523	2.37085	2.43787	2.50629	2.57615	2.64746	2.72024	2.79451	5
1500	2.87029	2.94760	3.02646	3.10689	3.18891	3.27255	3.35783	3.44476	3.53336	3.62367	5
1600	3.71569	3.80946	3.90499	4.00231	4.10143	4.20238	4.30519	4.40987	4.51645	4.62495	5
1700	4.73539	4.84780	4.96220	5.07861	5.19705	5.31756	5.44015	5.56485	5.69167	5.82066	5
1800	5.95182	6.08519	6.22079	6.35864	6.49877	6.64121	6.78597	6.93308	7.08258	7.23448	5
1900	7.38881	7.54560	7.70486	7.86664	8.03095	8.19782	8.36728	8.53935	8.71407	8.89145	5
2000	0.90715	0.92543	0.94399	0.96282	0.98193	1.00133	1.02101	1.04098	1.06124	1.08180	6
2100	1.10265	1.12380	1.14526	1.16702	1.18909	1.21147	1.23417	1.25718	1.28052	1.30418	6
2200	1.32816	1.35247	1.37712	1.40210	1.42742	1.45308	1.47909	1.50544	1.53214	1.55920	6
2300	1.58661	1.61439	1.64253	1.67103	1.69990	1.72915	1.75877	1.78877	1.81915	1.84991	6
2400	1.88107	1.91262	1.94456	1.97690	2.00965	2.04279	2.07635	2.11032	2.14470	2.17950	6
2500	2.21473	2.25038	2.28645	2.32296	2.35991	2.39729	2.43512	2.47339	2.51211	2.55129	6
2600	2.59092	2.63101	2.67156	2.71258	2.75407	2.79604	2.83848	2.88141	2.92482	2.96872	6
2700	3.01311	3.05800	3.10339	3.14928	3.19567	3.24258	3.29000	3.33794	3.38641	3.43540	6
2800	3.48491	3.53497	3.58556	3.63669	3.68836	3.74058	3.79336	3.84669	3.90059	3.95504	6
2900	4.01007	4.06567	4.12184	4.17860	4.23594	4.29386	4.35238	4.41150	4.47121	4.53153	6
3000	4.59246	4.65400	4.71615	4.77893	4.84233	4.90636	4.97102	5.03632	5.10226	5.16885	6
3100	5.23609	5.30398	5.37252	5.44174	5.51161	5.58216	5.65338	5.72528	5.79787	5.87114	6
3200	5.94511	6.01977	6.09514	6.17121	6.24799	6.32548	6.40369	6.48263	6.56229	6.64268	6
3300	6.72382	6.80569	6.88831	6.97167	7.05579	7.14068	7.22632	7.31273	7.39992	7.48788	6
3400	7.57662	7.66615	7.75648	7.84759	7.93951	8.03224	8.12577	8.22012	8.31528	8.41127	6
3500	8.50809	8.60575	8.70424	8.80357	8.90375	9.00479	9.10668	9.20943	9.31305	9.41755	6
3600	0.95229	0.96292	0.97363	0.98443	0.99533	1.00631	1.01738	1.02855	1.03980	1.05115	7
3700	1.06259	1.07413	1.08575	1.09748	1.10929	1.12121	1.13321	1.14532	1.15752	1.16981	7
3800	1.18221	1.19470	1.20730	1.21999	1.23278	1.24567	1.25866	1.27176	1.28495	1.29825	7
3900	1.31165	1.32516	1.33876	1.35248	1.36630	1.38022	1.39425	1.40839	1.42263	1.43698	7
4000	1.45144	1.46601	1.48069	1.49548	1.51038	1.52539	1.54051	1.55574	1.57109	1.58655	7
4100	1.60212	1.61781	1.63361	1.64953	1.66556	1.68172	1.69798	1.71437	1.73087	1.74750	7
4200	1.76424	1.78110	1.79808	1.81519	1.83241	1.84976	1.86723	1.88483	1.90255	1.92039	7
4300	1.93836	1.95645	1.97467	1.99302	2.01150	2.03010	2.04883	2.06769	2.08668	2.10580	7
4400	2.12506	2.14444	2.16396	2.18361	2.20339	2.22331	2.24336	2.26355	2.28387	2.30433	7
4500	2.32493	2.34567	2.36654	2.38755	2.40870	2.43000	2.45143	2.47300	2.49472	2.51658	7
4600	2.53858	2.56073	2.58302	2.60546	2.62804	2.65077	2.67364	2.69667	2.71984	2.74316	7
4700	2.76663	2.79025	2.81403	2.83795	2.86202	2.88625	2.91064	2.93517	2.95986	2.98471	7
4800	3.00971	3.03487	3.06019	3.08566	3.11130	3.13709	3.16304	3.18916	3.21543	3.24187	7
4900	3.26847	3.29523	3.32216	3.34925	3.37651	3.40393	3.43152	3.45928	3.48720	3.51530	7

TABLE 13

Blackbody: Total Radiant Emittance (W/m^2)

T(°K)	0	10	20	30	40	50	60	70	80	90	m
5000	3.54356	3.57200	3.60060	3.62938	3.65832	3.68745	3.71674	3.74621	3.77585	3.80567	7
5100	3.83567	3.86584	3.89619	3.92672	3.95742	3.98831	4.01938	4.05063	4.08206	4.11367	7
5200	4.14547	4.17745	4.20961	4.24196	4.27450	4.30722	4.34013	4.37323	4.40652	4.44000	7
5300	4.47367	4.50752	4.54158	4.57582	4.61026	4.64489	4.67971	4.71473	4.74995	4.78537	7
5400	4.82098	4.85679	4.89280	4.92901	4.96542	5.00203	5.03884	5.07586	5.11308	5.15050	7
5500	5.18813	5.22596	5.26401	5.30225	5.34071	5.37938	5.41825	5.45734	5.49663	5.53614	7
5600	5.57586	5.61580	5.65595	5.69631	5.73689	5.77768	5.81870	5.85993	5.90138	5.94305	7
5700	5.98494	6.02705	6.06938	6.11193	6.15471	6.19771	6.24094	6.28439	6.32807	6.37198	7
5800	6.41611	6.46048	6.50507	6.54989	6.59495	6.64024	6.68576	6.73151	6.77750	6.82372	7
5900	6.87018	6.91688	6.96381	7.01098	7.05839	7.10604	7.15394	7.20207	7.25045	7.29907	7
6000	7.34793	7.39704	7.44639	7.49600	7.54584	7.59594	7.64629	7.69688	7.74773	7.79883	7
6100	7.85018	7.90178	7.95364	8.00575	8.05812	8.11074	8.16362	8.21676	8.27016	8.32382	7
6200	8.37774	8.43192	8.48636	8.54107	8.59604	8.65128	8.70678	8.76254	8.81858	8.87488	7
6300	8.93146	8.98830	9.04541	9.10280	9.16046	9.21839	9.27659	9.33507	9.39383	9.45287	7
6400	0.95122	0.95718	0.96316	0.96918	0.97522	0.98129	0.98739	0.99352	0.99968	1.00586	8
6500	1.01208	1.01832	1.02459	1.03089	1.03722	1.04358	1.04997	1.05638	1.06283	1.06931	8
6600	1.07581	1.08235	1.08891	1.09550	1.10213	1.10878	1.11547	1.12218	1.12893	1.13570	8
6700	1.14251	1.14934	1.15621	1.16311	1.17004	1.17700	1.18399	1.19101	1.19806	1.20514	8
6800	1.21226	1.21941	1.22658	1.23379	1.24104	1.24831	1.25562	1.26295	1.27032	1.27772	8
6900	1.28516	1.29262	1.30012	1.30765	1.31522	1.32282	1.33045	1.33811	1.34580	1.35353	8
7000	1.36129	1.36909	1.37692	1.38478	1.39268	1.40061	1.40857	1.41657	1.42460	1.43267	8
7100	1.44077	1.44890	1.45707	1.46527	1.47351	1.48178	1.49009	1.49843	1.50681	1.51522	8
7200	1.52367	1.53215	1.54067	1.54922	1.55781	1.56643	1.57509	1.58379	1.59252	1.60129	8
7300	1.61009	1.61894	1.62781	1.63673	1.64568	1.65466	1.66369	1.67275	1.68184	1.69098	8
7400	1.70015	1.70936	1.71860	1.72789	1.73721	1.74657	1.75596	1.76540	1.77487	1.78438	8
7500	1.79393	1.80352	1.81314	1.82280	1.83251	1.84225	1.85203	1.86185	1.87170	1.88160	8
7600	1.89154	1.90151	1.91152	1.92158	1.93167	1.94181	1.95198	1.96219	1.97245	1.98274	8
7700	1.99307	2.00345	2.01386	2.02431	2.03481	2.04535	2.05592	2.06654	2.07720	2.08790	8
7800	2.09864	2.10943	2.12025	2.13112	2.14202	2.15297	2.16396	2.17500	2.18607	2.19719	8
7900	2.20835	2.21956	2.23080	2.24209	2.25342	2.26479	2.27621	2.28767	2.29917	2.31072	8
8000	2.32231	2.33394	2.34562	2.35734	2.36910	2.38091	2.39277	2.40466	2.41660	2.42859	8
8100	2.44062	2.45269	2.46481	2.47698	2.48919	2.50144	2.51374	2.52609	2.53848	2.55091	8
8200	2.56339	2.57592	2.58849	2.60111	2.61378	2.62649	2.63925	2.65205	2.66490	2.67780	8
8300	2.69074	2.70374	2.71677	2.72986	2.74299	2.75617	2.76940	2.78267	2.79599	2.80936	8
8400	2.82278	2.83625	2.84976	2.86332	2.87693	2.89059	2.90430	2.91806	2.93186	2.94572	8
8500	2.95962	2.97357	2.98757	3.00162	3.01572	3.02987	3.04407	3.05832	3.07262	3.08697	8
8600	3.10137	3.11582	3.13032	3.14487	3.15948	3.17413	3.18883	3.20359	3.21839	3.23325	8
8700	3.24816	3.26312	3.27813	3.29319	3.30831	3.32347	3.33869	3.35396	3.36929	3.38466	8
8800	3.40009	3.41557	3.43111	3.44670	3.46234	3.47803	3.49378	3.50958	3.52543	3.54134	8
8900	3.55730	3.57331	3.58938	3.60550	3.62168	3.63791	3.65420	3.67054	3.68693	3.70338	8
9000	3.71989	3.73645	3.75307	3.76974	3.78646	3.80325	3.82008	3.83698	3.85393	3.87093	8
9100	3.88799	3.90511	3.92229	3.93952	3.95681	3.97415	3.99155	4.00901	4.02653	4.04410	8
9200	4.06173	4.07942	4.09717	4.11497	4.13283	4.15075	4.16873	4.18677	4.20486	4.22302	8
9300	4.24123	4.25950	4.27783	4.29622	4.31467	4.33318	4.35175	4.37037	4.38906	4.40781	8
9400	4.42661	4.44548	4.46441	4.48339	4.50244	4.52155	4.54072	4.55995	4.57924	4.59859	8
9500	4.61801	4.63748	4.65702	4.67662	4.69628	4.71600	4.73578	4.75563	4.77554	4.79551	8
9600	4.81554	4.83564	4.85579	4.87602	4.89630	4.91665	4.93706	4.95754	4.97808	4.99868	8
9700	5.01934	5.04007	5.06087	5.08173	5.10265	5.12364	5.14469	5.16581	5.18699	5.20824	8
9800	5.22955	5.25093	5.27237	5.29388	5.31545	5.33710	5.35880	5.38057	5.40241	5.42432	8
9900	5.44629	5.46833	5.49043	5.51261	5.53485	5.55715	5.57953	5.60197	5.62448	5.64705	8

TABLE 13

Blackbody: Total Radiant Emittance (W/m²)

T(°K)	0	10	20	30	40	50	60	70	80	90	m
10000	5.66970	5.89991	6.13707	6.38130	6.63275	6.89156	7.15787	7.43182	7.71356	8.00324	8
11000	0.83010	0.86070	0.89214	0.92443	0.95759	0.99163	1.02658	1.06244	1.09923	1.13697	9
12000	1.17567	1.21535	1.25603	1.29772	1.34044	1.38420	1.42903	1.47494	1.52195	1.57007	9
13000	1.61932	1.66973	1.72130	1.77405	1.82801	1.88319	1.93962	1.99730	2.05625	2.11650	9
14000	2.17807	2.24097	2.30523	2.37085	2.43787	2.50629	2.57615	2.64746	2.72024	2.79451	9
15000	2.87029	2.94760	3.02646	3.10689	3.18891	3.27255	3.35783	3.44476	3.53336	3.62367	9
16000	3.71569	3.80946	3.90499	4.00231	4.10143	4.20238	4.30519	4.40987	4.51645	4.62495	9
17000	4.73539	4.84780	4.96220	5.07861	5.19705	5.31756	5.44015	5.56485	5.69167	5.82066	9
18000	5.95182	6.08519	6.22079	6.35864	6.49877	6.64121	6.78597	6.93308	7.08258	7.23448	9
19000	7.38881	7.54560	7.70486	7.86664	8.03095	8.19782	8.36728	8.53935	8.71407	8.89144	9
20000	0.90715	0.92543	0.94399	0.96282	0.98193	1.00133	1.02101	1.04098	1.06124	1.08180	10
21000	1.10265	1.12380	1.14526	1.16702	1.18909	1.21147	1.23417	1.25718	1.28052	1.30418	10
22000	1.32816	1.35247	1.37712	1.40210	1.42742	1.45308	1.47909	1.50544	1.53214	1.55920	10
23000	1.58661	1.61439	1.64252	1.67103	1.69990	1.72915	1.75877	1.78877	1.81915	1.84991	10
24000	1.88107	1.91262	1.94456	1.97690	2.00964	2.04279	2.07635	2.11032	2.14470	2.17950	10
25000	2.21473	2.25038	2.28645	2.32296	2.35991	2.39729	2.43512	2.47339	2.51211	2.55129	10
26000	2.59092	2.63101	2.67156	2.71258	2.75407	2.79604	2.83848	2.88141	2.92482	2.96872	10
27000	3.01311	3.05800	3.10339	3.14928	3.19567	3.24258	3.29000	3.33794	3.38641	3.43540	10
28000	3.48491	3.53497	3.58556	3.63669	3.68836	3.74058	3.79336	3.84669	3.90059	3.95504	10
29000	4.01007	4.06567	4.12184	4.17860	4.23594	4.29386	4.35238	4.41150	4.47121	4.53153	10
30000	4.59246	4.65400	4.71615	4.77893	4.84233	4.90636	4.97102	5.03632	5.10226	5.16885	10
31000	5.23609	5.30398	5.37252	5.44174	5.51161	5.58216	5.65338	5.72528	5.79787	5.87114	10
32000	5.94511	6.01977	6.09514	6.17121	6.24799	6.32548	6.40369	6.48263	6.56229	6.64268	10
33000	6.72382	6.80569	6.88831	6.97167	7.05579	7.14068	7.22632	7.31273	7.39992	7.48788	10
34000	7.57662	7.66615	7.75648	7.84759	7.93951	8.03224	8.12577	8.22012	8.31528	8.41127	10
35000	8.50809	8.60575	8.70424	8.80357	8.90375	9.00479	9.10668	9.20943	9.31305	9.41755	10
36000	0.95229	0.96292	0.97363	0.98443	0.99533	1.00631	1.01738	1.02855	1.03980	1.05115	11
37000	1.06259	1.07413	1.08575	1.09748	1.10929	1.12121	1.13321	1.14532	1.15752	1.16981	11
38000	1.18221	1.19470	1.20730	1.21999	1.23278	1.24567	1.25866	1.27176	1.28495	1.29825	11
39000	1.31165	1.32516	1.33876	1.35248	1.36630	1.38022	1.39425	1.40839	1.42263	1.43698	11
40000	1.45144	1.46601	1.48069	1.49548	1.51038	1.52539	1.54051	1.55574	1.57109	1.58655	11
41000	1.60212	1.61781	1.63361	1.64953	1.66556	1.68172	1.69798	1.71437	1.73087	1.74750	11
42000	1.76424	1.78110	1.79808	1.81519	1.83241	1.84976	1.86723	1.88483	1.90255	1.92039	11
43000	1.93836	1.95645	1.97467	1.99302	2.01149	2.03010	2.04883	2.06769	2.08668	2.10580	11
44000	2.12506	2.14444	2.16396	2.18361	2.20339	2.22331	2.24336	2.26355	2.28387	2.30433	11
45000	2.32493	2.34567	2.36654	2.38755	2.40870	2.43000	2.45143	2.47300	2.49472	2.51658	11
46000	2.53858	2.56073	2.58302	2.60546	2.62804	2.65077	2.67364	2.69667	2.71984	2.74316	11
47000	2.76663	2.79025	2.81403	2.83795	2.86202	2.88625	2.91064	2.93517	2.95986	2.98471	11
48000	3.00971	3.03487	3.06019	3.08566	3.11130	3.13709	3.16304	3.18916	3.21543	3.24187	11
49000	3.26847	3.29523	3.32216	3.34925	3.37651	3.40393	3.43152	3.45928	3.48720	3.51530	11

TABLE 13

Blackbody: Total Radiant Emittance (W/m^2)

T(°K)	0	10	20	30	40	50	60	70	80	90	m
50000	3.54356	3.57200	3.60060	3.62938	3.65832	3.68744	3.71674	3.74621	3.77585	3.80567	11
51000	3.83567	3.86584	3.89619	3.92672	3.95742	3.98831	4.01938	4.05063	4.08206	4.11367	11
52000	4.14547	4.17745	4.20961	4.24196	4.27450	4.30722	4.34013	4.37323	4.40652	4.44000	11
53000	4.47367	4.50752	4.54158	4.57582	4.61026	4.64489	4.67971	4.71473	4.74995	4.78537	11
54000	4.82098	4.85679	4.89280	4.92901	4.96542	5.00203	5.03884	5.07586	5.11308	5.15050	11
55000	5.18813	5.22596	5.26401	5.30225	5.34071	5.37938	5.41825	5.45734	5.49663	5.53614	11
56000	5.57586	5.61580	5.65595	5.69631	5.73689	5.77768	5.81870	5.85993	5.90138	5.94305	11
57000	5.98494	6.02705	6.06938	6.11193	6.15471	6.19771	6.24094	6.28439	6.32807	6.37198	11
58000	6.41611	6.46048	6.50507	6.54989	6.59495	6.64024	6.68576	6.73151	6.77750	6.82372	11
59000	6.87018	6.91688	6.96381	7.01098	7.05839	7.10604	7.15394	7.20207	7.25045	7.29907	11
60000	7.34793	7.39704	7.44639	7.49599	7.54584	7.59594	7.64629	7.69688	7.74773	7.79883	11
61000	7.85018	7.90178	7.95364	8.00575	8.05812	8.11074	8.16362	8.21676	8.27016	8.32382	11
62000	8.37774	8.43192	8.48636	8.54107	8.59604	8.65127	8.70678	8.76254	8.81858	8.87488	11
63000	8.93146	8.98830	9.04541	9.10280	9.16046	9.21839	9.27659	9.33507	9.39383	9.45287	11
64000	0.95122	0.95718	0.96316	0.96918	0.97522	0.98129	0.98739	0.99352	0.99968	1.00586	12
65000	1.01208	1.01832	1.02459	1.03089	1.03722	1.04358	1.04997	1.05638	1.06283	1.06931	12
66000	1.07581	1.08235	1.08891	1.09550	1.10213	1.10878	1.11547	1.12218	1.12893	1.13570	12
67000	1.14251	1.14934	1.15621	1.16311	1.17004	1.17700	1.18399	1.19101	1.19806	1.20514	12
68000	1.21226	1.21941	1.22658	1.23379	1.24104	1.24831	1.25561	1.26295	1.27032	1.27772	12
69000	1.28516	1.29262	1.30012	1.30765	1.31522	1.32282	1.33045	1.33811	1.34580	1.35353	12
70000	1.36129	1.36909	1.37692	1.38478	1.39268	1.40061	1.40857	1.41657	1.42460	1.43267	12
71000	1.44077	1.44890	1.45707	1.46527	1.47351	1.48178	1.49009	1.49843	1.50681	1.51522	12
72000	1.52367	1.53215	1.54067	1.54922	1.55781	1.56643	1.57509	1.58379	1.59252	1.60129	12
73000	1.61009	1.61894	1.62781	1.63673	1.64568	1.65466	1.66369	1.67275	1.68184	1.69098	12
74000	1.70015	1.70936	1.71860	1.72789	1.73721	1.74657	1.75596	1.76540	1.77487	1.78438	12
75000	1.79393	1.80352	1.81314	1.82280	1.83251	1.84225	1.85203	1.86185	1.87170	1.88160	12
76000	1.89154	1.90151	1.91152	1.92158	1.93167	1.94181	1.95198	1.96219	1.97245	1.98274	12
77000	1.99307	2.00345	2.01386	2.02431	2.03481	2.04535	2.05592	2.06654	2.07720	2.08790	12
78000	2.09864	2.10943	2.12025	2.13112	2.14202	2.15297	2.16396	2.17500	2.18607	2.19719	12
79000	2.20835	2.21956	2.23080	2.24209	2.25342	2.26479	2.27621	2.28767	2.29917	2.31072	12
80000	2.32231	2.33394	2.34562	2.35734	2.36910	2.38091	2.39277	2.40466	2.41660	2.42859	12
81000	2.44062	2.45269	2.46481	2.47698	2.48919	2.50144	2.51374	2.52609	2.53848	2.55091	12
82000	2.56339	2.57592	2.58849	2.60111	2.61378	2.62649	2.63925	2.65205	2.66490	2.67780	12
83000	2.69074	2.70373	2.71677	2.72986	2.74299	2.75617	2.76940	2.78267	2.79599	2.80936	12
84000	2.82278	2.83625	2.84976	2.86332	2.87693	2.89059	2.90430	2.91806	2.93186	2.94572	12
85000	2.95962	2.97357	2.98757	3.00162	3.01572	3.02987	3.04407	3.05832	3.07262	3.08697	12
86000	3.10137	3.11582	3.13032	3.14487	3.15948	3.17413	3.18883	3.20359	3.21839	3.23325	12
87000	3.24816	3.26312	3.27813	3.29319	3.30831	3.32347	3.33869	3.35396	3.36929	3.38466	12
88000	3.40009	3.41557	3.43111	3.44669	3.46234	3.47803	3.49377	3.50957	3.52543	3.54134	12
89000	3.55730	3.57331	3.58938	3.60550	3.62168	3.63791	3.65420	3.67054	3.68693	3.70338	12
90000	3.71989	3.73645	3.75307	3.76974	3.78646	3.80325	3.82008	3.83698	3.85393	3.87093	12
91000	3.88799	3.90511	3.92229	3.93952	3.95681	3.97415	3.99155	4.00901	4.02653	4.04410	12
92000	4.06173	4.07942	4.09717	4.11497	4.13283	4.15075	4.16873	4.18677	4.20486	4.22302	12
93000	4.24123	4.25950	4.27783	4.29622	4.31467	4.33318	4.35174	4.37037	4.38906	4.40781	12
94000	4.42661	4.44548	4.46441	4.48339	4.50244	4.52155	4.54072	4.55995	4.57924	4.59859	12
95000	4.61801	4.63748	4.65702	4.67662	4.69627	4.71600	4.73578	4.75563	4.77554	4.79551	12
96000	4.81554	4.83564	4.85579	4.87602	4.89630	4.91665	4.93706	4.95754	4.97808	4.99868	12
97000	5.01934	5.04007	5.06087	5.08173	5.10265	5.12364	5.14469	5.16581	5.18699	5.20824	12
98000	5.22955	5.25093	5.27237	5.29388	5.31545	5.33710	5.35880	5.38057	5.40241	5.42432	12
99000	5.44629	5.46833	5.49043	5.51261	5.53485	5.55715	5.57953	5.60197	5.62448	5.64705	12

TABLE 14

Blackbody Radiation: Peak Wavelength (μ)

T(°K)	0	100	200	300	400	500	600	700	800	900
0			14.48900	9.65933	7.24450	5.79560	4.82967	4.13971	3.62225	3.21978
1000	2.89780	2.63436	2.41483	2.22908	2.06986	1.93187	1.81112	1.70459	1.60989	1.52516
2000	1.44890	1.37990	1.31718	1.25991	1.20742	1.15912	1.11454	1.07326	1.03493	0.99924
3000	0.96593	0.93477	0.90556	0.87812	0.85229	0.82794	0.80494	0.78319	0.76258	0.74303
4000	0.72445	0.70678	0.68995	0.67391	0.65859	0.64396	0.62996	0.61655	0.60371	0.59139
5000	0.57956	0.56820	0.55727	0.54675	0.53663	0.52687	0.51746	0.50839	0.49962	0.49115
6000	0.48297	0.47505	0.46739	0.45997	0.45278	0.44582	0.43906	0.43251	0.42615	0.41997
7000	0.41397	0.40814	0.40247	0.39696	0.39159	0.38637	0.38129	0.37634	0.37151	0.36681
8000	0.36222	0.35775	0.35339	0.34913	0.34498	0.34092	0.33695	0.33308	0.32930	0.32560
9000	0.32198	0.31844	0.31498	0.31159	0.30828	0.30503	0.30185	0.29874	0.29569	0.29271
10000	0.28978	0.28691	0.28410	0.28134	0.27863	0.27598	0.27338	0.27082	0.26831	0.26585
11000	0.26344	0.26106	0.25873	0.25644	0.25419	0.25198	0.24981	0.24768	0.24558	0.24351
12000	0.24148	0.23949	0.23752	0.23559	0.23369	0.23182	0.22998	0.22817	0.22639	0.22464
13000	0.22291	0.22121	0.21953	0.21788	0.21625	0.21465	0.21307	0.21152	0.20999	0.20847
14000	0.20699	0.20552	0.20407	0.20264	0.20124	0.19985	0.19848	0.19713	0.19580	0.19448
15000	0.19319	0.19191	0.19064	0.18940	0.18817	0.18695	0.18576	0.18457	0.18341	0.18225
16000	0.18111	0.17999	0.17888	0.17778	0.17670	0.17562	0.17457	0.17352	0.17249	0.17147
17000	0.17046	0.16946	0.16848	0.16750	0.16654	0.16559	0.16465	0.16372	0.16280	0.16189
18000	0.16099	0.16010	0.15922	0.15835	0.15749	0.15664	0.15580	0.15496	0.15414	0.15332
19000	0.15252	0.15172	0.15093	0.15015	0.14937	0.14861	0.14785	0.14710	0.14635	0.14562
20000	0.14489	0.14417	0.14346	0.14275	0.14205	0.14136	0.14067	0.13999	0.13932	0.13865
21000	0.13799	0.13734	0.13669	0.13605	0.13541	0.13478	0.13416	0.13354	0.13293	0.13232
22000	0.13172	0.13112	0.13053	0.12995	0.12937	0.12879	0.12822	0.12766	0.12710	0.12654
23000	0.12599	0.12545	0.12491	0.12437	0.12384	0.12331	0.12279	0.12227	0.12176	0.12125
24000	0.12074	0.12024	0.11974	0.11925	0.11876	0.11828	0.11780	0.11732	0.11685	0.11638
25000	0.11591	0.11545	0.11499	0.11454	0.11409	0.11364	0.11320	0.11275	0.11232	0.11188
26000	0.11145	0.11103	0.11060	0.11018	0.10977	0.10935	0.10894	0.10853	0.10813	0.10772
27000	0.10733	0.10693	0.10654	0.10615	0.10576	0.10537	0.10499	0.10461	0.10424	0.10386
28000	0.10349	0.10312	0.10276	0.10240	0.10204	0.10168	0.10132	0.10097	0.10062	0.10027
29000	0.09992	0.09958	0.09924	0.09890	0.09856	0.09823	0.09790	0.09757	0.09724	0.09692
30000	0.09659	0.09627	0.09595	0.09564	0.09532	0.09501	0.09470	0.09439	0.09408	0.09378
31000	0.09348	0.09318	0.09288	0.09258	0.09229	0.09199	0.09170	0.09141	0.09113	0.09084
32000	0.09056	0.09027	0.08999	0.08972	0.08944	0.08916	0.08889	0.08862	0.08835	0.08808
33000	0.08781	0.08755	0.08728	0.08702	0.08676	0.08650	0.08624	0.08599	0.08573	0.08548
34000	0.08523	0.08498	0.08473	0.08448	0.08424	0.08399	0.08375	0.08351	0.08327	0.08303
35000	0.08279	0.08256	0.08232	0.08209	0.08186	0.08163	0.08140	0.08117	0.08094	0.08072
36000	0.08049	0.08027	0.08005	0.07983	0.07961	0.07939	0.07917	0.07896	0.07874	0.07853
37000	0.07832	0.07811	0.07790	0.07769	0.07748	0.07727	0.07707	0.07686	0.07666	0.07646
38000	0.07626	0.07606	0.07586	0.07566	0.07546	0.07527	0.07507	0.07488	0.07469	0.07449
39000	0.07430	0.07411	0.07392	0.07374	0.07355	0.07336	0.07318	0.07299	0.07281	0.07263
40000	0.07244	0.07226	0.07208	0.07191	0.07173	0.07155	0.07137	0.07120	0.07102	0.07085
41000	0.07068	0.07051	0.07033	0.07016	0.07000	0.06983	0.06966	0.06949	0.06933	0.06916
42000	0.06900	0.06883	0.06867	0.06851	0.06834	0.06818	0.06802	0.06786	0.06771	0.06755
43000	0.06739	0.06723	0.06708	0.06692	0.06677	0.06662	0.06646	0.06631	0.06616	0.06601
44000	0.06586	0.06571	0.06556	0.06541	0.06527	0.06512	0.06497	0.06483	0.06468	0.06454
45000	0.06440	0.06425	0.06411	0.06397	0.06383	0.06369	0.06355	0.06341	0.06327	0.06313
46000	0.06300	0.06286	0.06272	0.06259	0.06245	0.06232	0.06218	0.06205	0.06192	0.06179
47000	0.06166	0.06152	0.06139	0.06126	0.06114	0.06101	0.06088	0.06075	0.06062	0.06050
48000	0.06037	0.06025	0.06012	0.06000	0.05987	0.05975	0.05963	0.05950	0.05938	0.05926
49000	0.05914	0.05902	0.05890	0.05878	0.05866	0.05854	0.05842	0.05831	0.05819	0.05807

TABLE 14

Blackbody Radiation: Peak Wavelength (μ)

T(°K)	0	100	200	300	400	500	600	700	800	900
50000	0.05796	0.05784	0.05773	0.05761	0.05750	0.05738	0.05727	0.05716	0.05704	0.05693
51000	0.05682	0.05671	0.05660	0.05649	0.05638	0.05627	0.05616	0.05605	0.05594	0.05583
52000	0.05573	0.05562	0.05551	0.05541	0.05530	0.05520	0.05509	0.05499	0.05488	0.05478
53000	0.05468	0.05457	0.05447	0.05437	0.05427	0.05416	0.05406	0.05396	0.05386	0.05376
54000	0.05366	0.05356	0.05346	0.05337	0.05327	0.05317	0.05307	0.05298	0.05288	0.05278
55000	0.05269	0.05259	0.05250	0.05240	0.05231	0.05221	0.05212	0.05203	0.05193	0.05184
56000	0.05175	0.05165	0.05156	0.05147	0.05138	0.05129	0.05120	0.05111	0.05102	0.05093
57000	0.05084	0.05075	0.05066	0.05057	0.05048	0.05040	0.05031	0.05022	0.05013	0.05005
58000	0.04996	0.04988	0.04979	0.04970	0.04962	0.04954	0.04945	0.04937	0.04928	0.04920
59000	0.04912	0.04903	0.04895	0.04887	0.04878	0.04870	0.04862	0.04854	0.04846	0.04838
60000	0.04830	0.04822	0.04814	0.04806	0.04798	0.04790	0.04782	0.04774	0.04766	0.04758
61000	0.04750	0.04743	0.04735	0.04727	0.04720	0.04712	0.04704	0.04697	0.04689	0.04681
62000	0.04674	0.04666	0.04659	0.04651	0.04644	0.04636	0.04629	0.04622	0.04614	0.04607
63000	0.04600	0.04592	0.04585	0.04578	0.04571	0.04563	0.04556	0.04549	0.04542	0.04535
64000	0.04528	0.04521	0.04514	0.04507	0.04500	0.04493	0.04486	0.04479	0.04472	0.04465
65000	0.04458	0.04451	0.04444	0.04438	0.04431	0.04424	0.04417	0.04411	0.04404	0.04397
66000	0.04391	0.04384	0.04377	0.04371	0.04364	0.04358	0.04351	0.04345	0.04338	0.04332
67000	0.04325	0.04319	0.04312	0.04306	0.04299	0.04293	0.04287	0.04280	0.04274	0.04268
68000	0.04261	0.04255	0.04249	0.04243	0.04237	0.04230	0.04224	0.04218	0.04212	0.04206
69000	0.04200	0.04194	0.04188	0.04182	0.04176	0.04169	0.04164	0.04158	0.04152	0.04146
70000	0.04140	0.04134	0.04128	0.04122	0.04116	0.04110	0.04105	0.04099	0.04093	0.04087
71000	0.04081	0.04076	0.04070	0.04064	0.04059	0.04053	0.04047	0.04042	0.04036	0.04030
72000	0.04025	0.04019	0.04014	0.04008	0.04002	0.03997	0.03991	0.03986	0.03980	0.03975
73000	0.03970	0.03964	0.03959	0.03953	0.03948	0.03943	0.03937	0.03932	0.03927	0.03921
74000	0.03916	0.03911	0.03905	0.03900	0.03895	0.03890	0.03884	0.03879	0.03874	0.03869
75000	0.03864	0.03859	0.03853	0.03848	0.03843	0.03838	0.03833	0.03828	0.03823	0.03818
76000	0.03813	0.03808	0.03803	0.03798	0.03793	0.03788	0.03783	0.03778	0.03773	0.03768
77000	0.03763	0.03758	0.03754	0.03749	0.03744	0.03739	0.03734	0.03729	0.03725	0.03720
78000	0.03715	0.03710	0.03706	0.03701	0.03696	0.03691	0.03687	0.03682	0.03677	0.03673
79000	0.03668	0.03663	0.03659	0.03654	0.03650	0.03645	0.03640	0.03636	0.03631	0.03627
80000	0.03622	0.03618	0.03613	0.03609	0.03604	0.03600	0.03595	0.03591	0.03586	0.03582
81000	0.03578	0.03573	0.03569	0.03564	0.03560	0.03556	0.03551	0.03547	0.03543	0.03538
82000	0.03534	0.03530	0.03525	0.03521	0.03517	0.03512	0.03508	0.03504	0.03500	0.03496
83000	0.03491	0.03487	0.03483	0.03479	0.03475	0.03470	0.03466	0.03462	0.03458	0.03454
84000	0.03450	0.03446	0.03442	0.03437	0.03433	0.03429	0.03425	0.03421	0.03417	0.03413
85000	0.03409	0.03405	0.03401	0.03397	0.03393	0.03389	0.03385	0.03381	0.03377	0.03373
86000	0.03370	0.03366	0.03362	0.03358	0.03354	0.03350	0.03346	0.03342	0.03338	0.03335
87000	0.03331	0.03327	0.03323	0.03319	0.03316	0.03312	0.03308	0.03304	0.03300	0.03297
88000	0.03293	0.03289	0.03285	0.03282	0.03278	0.03274	0.03271	0.03267	0.03263	0.03260
89000	0.03256	0.03252	0.03249	0.03245	0.03241	0.03238	0.03234	0.03231	0.03227	0.03223
90000	0.03220	0.03216	0.03213	0.03209	0.03206	0.03202	0.03198	0.03195	0.03191	0.03188
91000	0.03184	0.03181	0.03177	0.03174	0.03170	0.03167	0.03164	0.03160	0.03157	0.03153
92000	0.03150	0.03146	0.03143	0.03140	0.03136	0.03133	0.03129	0.03126	0.03123	0.03119
93000	0.03116	0.03113	0.03109	0.03106	0.03103	0.03099	0.03096	0.03093	0.03089	0.03086
94000	0.03083	0.03079	0.03076	0.03073	0.03070	0.03066	0.03063	0.03060	0.03057	0.03054
95000	0.03050	0.03047	0.03044	0.03041	0.03038	0.03034	0.03031	0.03028	0.03025	0.03022
96000	0.03019	0.03015	0.03012	0.03009	0.03006	0.03003	0.03000	0.02997	0.02994	0.02991
97000	0.02987	0.02984	0.02981	0.02978	0.02975	0.02972	0.02969	0.02966	0.02963	0.02960
98000	0.02957	0.02954	0.02951	0.02948	0.02945	0.02942	0.02939	0.02936	0.02933	0.02930
99000	0.02927	0.02924	0.02921	0.02918	0.02915	0.02912	0.02909	0.02907	0.02904	0.02901

D. Blackbody Radiation Tables - Photon

Table 15—Relative Fractional Emittance

Table 16—Total Emittance

TABLE 15A

Blackbody: Relative Fractional Photon Emittance

λT(μ.°K)	0	1	2	3	4	5	6	7	8	9	m
100	.000003	.000012	.000046	.000178	.000668	.002447	.008748	.030531	.104097	.346971	-53
110	.000011	.000036	.000113	.000346	.001037	.003055	.008831	.025062	.069869	.191429	-48
120	.000052	.000137	.000356	.000914	.002311	.005755	.014124	.034169	.081521	.191867	-44
130	.000446	.001022	.002313	.005170	.011420	.024929	.053789	.114751	.242107	.505299	-41
140	.001043	.002133	.004314	.008642	.017142	.033679	.065557	.126443	.241701	.457979	-38
150	.000860	.001603	.002961	.005426	.009864	.017794	.031855	.056599	.099828	.174809	-35
160	.003039	.005248	.009001	.015333	.025949	.043633	.072906	.121062	.199802	.327785	-33
170	.005346	.008668	.013976	.022408	.035730	.056666	.089395	.140292	.219044	.340282	-31
180	.005260	.008092	.012388	.018876	.028629	.043224	.064967	.097218	.144849	.214895	-29
190	.031747	.046708	.068440	.099881	.145192	.210239	.303263	.435803	.623948	.890058	-28
200	.012651	.017918	.025289	.035571	.049862	.069663	.097006	.134645	.186291	.256936	-26
210	.035327	.048424	.066177	.090169	.122499	.165940	.224147	.301921	.405554	.543271	-25
220	.072579	.096706	.128514	.170344	.225213	.297005	.390711	.512721	.671205	.876579	-24
230	.011421	.014846	.019253	.024913	.032164	.041435	.053260	.068314	.087436	.111676	-22
240	.014234	.018106	.022983	.029117	.036815	.046458	.058514	.073559	.092299	.115598	-21
250	.144513	.180335	.224633	.279320	.346715	.429629	.531464	.656331	.809186	.996001	-21
260	.122395	.150166	.183945	.224968	.274714	.334946	.407763	.495667	.601628	.729169	-20
270	.088247	.106646	.128700	.155096	.186646	.224307	.269201	.322647	.386191	.461643	-19
280	.055112	.065709	.078245	.093055	.110530	.131126	.155370	.183875	.217350	.256616	-18
290	.030262	.035646	.041940	.049290	.057862	.067851	.079476	.092991	.108689	.126901	-17
300	.148008	.172446	.200711	.233371	.271070	.314544	.364628	.422272	.488552	.564692	-17
310	.065208	.075227	.086705	.099841	.114862	.132023	.151611	.173951	.199405	.228385	-16
320	.261350	.298816	.341362	.389638	.444370	.506372	.576551	.655925	.745626	.846918	-16
330	.096121	.109008	.123526	.139870	.158257	.178926	.202142	.228202	.257433	.290197	-15
340	.326894	.367969	.413912	.465265	.522624	.586650	.658070	.737683	.826373	.925110	-15
350	.103496	.115709	.129280	.144350	.161073	.179621	.200178	.222949	.248158	.276048	-14
360	.306886	.340964	.378601	.420143	.465971	.516496	.572170	.633483	.700969	.775207	-14
370	.085683	.094652	.104503	.115316	.127179	.140188	.154445	.170063	.187161	.205872	-13
380	.226337	.248710	.273155	.299853	.328996	.360792	.395468	.433265	.474446	.519292	-13
390	.056811	.062122	.067897	.074175	.080997	.088405	.096447	.105173	.114637	.124898	-12
400	.136017	.148061	.161101	.175215	.190484	.206995	.224844	.244129	.264958	.287446	-12
410	.311715	.337897	.366129	.396563	.429356	.464677	.502707	.543639	.587677	.635040	-12
420	.068596	.074068	.079947	.086260	.093038	.100311	.108114	.116482	.125452	.135065	-11
430	.145363	.156391	.168197	.180832	.194349	.208806	.224261	.240780	.258429	.277279	-11
440	.297406	.318890	.341815	.366269	.392347	.420148	.449777	.481343	.514964	.550763	-11
450	.058887	.062942	.067256	.071843	.076721	.081905	.087414	.093266	.099481	.106079	-10
460	.113082	.120513	.128396	.136756	.145619	.155013	.164968	.175514	.186682	.198508	-10
470	.211026	.224273	.238288	.253112	.268787	.285359	.302873	.321380	.340930	.361578	-10
480	.383378	.406391	.430678	.456302	.483333	.511838	.541893	.573574	.606960	.642137	-10
490	.067919	.071821	.075930	.080255	.084806	.089595	.094633	.099931	.105502	.111359	-9

TABLE 15A

Blackbody: Relative Fractional Photon Emittance

λT(μ.°K)	0	1	2	3	4	5	6	7	8	9	m
500	.117514	.123982	.130777	.137915	.145410	.153279	.161539	.170208	.179304	.188846	-9
510	.198854	.209349	.220351	.231884	.243970	.256634	.269901	.283795	.298346	.313579	-9
520	.329525	.346214	.363676	.381945	.401054	.421037	.441932	.463775	.486606	.510465	-9
530	.535393	.561433	.588632	.617034	.646689	.677645	.709955	.743672	.778850	.815548	-9
540	.085382	.089374	.093536	.097874	.102396	.107108	.112019	.117134	.122462	.128011	-8
550	.133790	.139805	.146067	.152585	.159367	.166424	.173764	.181400	.189341	.197597	-8
560	.206182	.215105	.224379	.234017	.244031	.254434	.265240	.276463	.288117	.300217	-8
570	.312778	.325816	.339347	.353388	.367956	.383068	.398744	.415000	.431858	.449335	-8
580	.467454	.486235	.505698	.525867	.546765	.568414	.590839	.614064	.638115	.663018	-8
590	.688800	.715488	.743112	.771698	.801279	.831884	.863545	.896294	.930164	.965190	-8
600	.100141	.103885	.107755	.111756	.115891	.120163	.124578	.129138	.133850	.138716	-7
610	.143741	.148931	.154289	.159820	.165531	.171424	.177507	.183784	.190260	.196942	-7
620	.203835	.210944	.218277	.225838	.233635	.241674	.249961	.258503	.267307	.276380	-7
630	.285730	.295364	.305289	.315513	.326044	.336890	.348060	.359561	.371403	.383595	-7
640	.396145	.409062	.422357	.436038	.450116	.464601	.479503	.494832	.510600	.526817	-7
650	.543494	.560643	.578276	.596404	.615039	.634195	.653883	.674117	.694910	.716275	-7
660	.738227	.760778	.783943	.807738	.832176	.857274	.883046	.909508	.936676	.964568	-7
670	.099320	.102259	.105275	.108370	.111546	.114805	.118149	.121579	.125098	.128708	-6
680	.132410	.136206	.140100	.144092	.148185	.152381	.156683	.161092	.165612	.170244	-6
690	.174991	.179856	.184840	.189946	.195178	.200538	.206027	.211650	.217409	.223306	-6
700	.229345	.235529	.241860	.248341	.254976	.261768	.268720	.275835	.283116	.290568	-6
710	.298192	.305993	.313974	.322140	.330492	.339036	.347774	.356711	.365851	.375197	-6
720	.384753	.394524	.404514	.414727	.425166	.435837	.446744	.457891	.469283	.480924	-6
730	.492819	.504973	.517390	.530076	.543035	.556273	.569794	.583604	.597708	.612111	-6
740	.626819	.641837	.657171	.672826	.688808	.705123	.721777	.738776	.756125	.773831	-6
750	.791901	.810341	.829156	.848353	.867940	.887923	.908308	.929102	.950313	.971947	-6
760	.099401	.101651	.103946	.106286	.108672	.111105	.113585	.116114	.118692	.121320	-5
770	.123998	.126728	.129511	.132346	.135236	.138181	.141182	.144239	.147354	.150528	-5
780	.153761	.157055	.160410	.163827	.167308	.170853	.174463	.178140	.181884	.185696	-5
790	.189578	.193530	.197554	.201650	.205820	.210065	.214386	.218783	.223260	.227815	-5
800	.232451	.237169	.241970	.246855	.251825	.256882	.262027	.267261	.272586	.278002	-5
810	.283512	.289115	.294815	.300612	.306507	.312502	.318598	.324797	.331101	.337509	-5
820	.344025	.350650	.357384	.364230	.371189	.378263	.385452	.392760	.400187	.407735	-5
830	.415405	.423200	.431120	.439169	.447346	.455654	.464095	.472671	.481383	.490232	-5
840	.499222	.508353	.517627	.527047	.536614	.546330	.556197	.566217	.576392	.586723	-5
850	.597213	.607864	.618678	.629657	.640802	.652116	.663602	.675260	.687094	.699105	-5
860	.711296	.723669	.736225	.748968	.761899	.775022	.788337	.801847	.815556	.829464	-5
870	.843575	.857890	.872413	.887145	.902090	.917249	.932626	.948222	.964040	.980083	-5
880	.099635	.101285	.102959	.104655	.106376	.108121	.109890	.111683	.113502	.115345	-4
890	.117214	.119109	.121030	.122977	.124951	.126951	.128979	.131034	.133117	.135228	-4
900	.137368	.139536	.141733	.143960	.146216	.148502	.150818	.153165	.155542	.157951	-4
910	.160392	.162864	.165369	.167906	.170476	.173079	.175716	.178387	.181092	.183831	-4
920	.186606	.189416	.192262	.195143	.198061	.201016	.204008	.207037	.210105	.213210	-4
930	.216355	.219538	.222761	.226023	.229326	.232669	.236053	.239479	.242946	.246456	-4
940	.250008	.253603	.257241	.260924	.264650	.268421	.272237	.276099	.280007	.283961	-4
950	.287961	.292009	.296105	.300248	.304441	.308682	.312972	.317313	.321704	.326145	-4
960	.330638	.335183	.339780	.344429	.349132	.353888	.358698	.363563	.368482	.373458	-4
970	.378489	.383577	.388722	.393924	.399185	.404503	.409881	.415319	.420816	.426374	-4
980	.431993	.437674	.443416	.449222	.455090	.461022	.467019	.473080	.479206	.485398	-4
990	.491657	.497983	.504376	.510837	.517367	.523966	.530635	.537374	.544184	.551065	-4

TABLE 15A

Blackbody: Relative Fractional Photon Emittance

λT(μ.°K)	0	1	2	3	4	5	6	7	8	9	m
1000	.558018	.565044	.572144	.579316	.586564	.593886	.601284	.608758	.616309	.623937	−4
1010	.631643	.639427	.647291	.655235	.663259	.671364	.679550	.687819	.696171	.704607	−4
1020	.713126	.721731	.730421	.739197	.748060	.757010	.766048	.775175	.784392	.793698	−4
1030	.803095	.812584	.822165	.831838	.841605	.851466	.861422	.871473	.881620	.891864	−4
1040	.902206	.912646	.923185	.933824	.944563	.955403	.966345	.977390	.988537	.999789	−4
1050	.101115	.102261	.103418	.104585	.105763	.106952	.108152	.109363	.110585	.111819	−3
1060	.113063	.114319	.115586	.116864	.118154	.119456	.120769	.122094	.123431	.124780	−3
1070	.126141	.127513	.128898	.130296	.131705	.133127	.134561	.136008	.137468	.138940	−3
1080	.140425	.141923	.143434	.144958	.146495	.148045	.149609	.151186	.152776	.154380	−3
1090	.155997	.157629	.159274	.160933	.162606	.164293	.165994	.167709	.169438	.171182	−3
1100	.172941	.174714	.176502	.178304	.180122	.181954	.183801	.185664	.187541	.189434	−3
1110	.191342	.193266	.195205	.197160	.199131	.201117	.203120	.205138	.207173	.209224	−3
1120	.211291	.213374	.215474	.217591	.219724	.221874	.224041	.226225	.228425	.230643	−3
1130	.232879	.235131	.237401	.239689	.241994	.244317	.246658	.249016	.251393	.253788	−3
1140	.256201	.258632	.261082	.263550	.266037	.268543	.271067	.273610	.276173	.278754	−3
1150	.281355	.283975	.286614	.289273	.291952	.294650	.297368	.300106	.302864	.305643	−3
1160	.308441	.311260	.314099	.316959	.319840	.322741	.325663	.328606	.331570	.334556	−3
1170	.337562	.340590	.343640	.346711	.349804	.352919	.356055	.359214	.362395	.365598	−3
1180	.368823	.372071	.375342	.378635	.381951	.385290	.388652	.392037	.395445	.398877	−3
1190	.402332	.405811	.409313	.412839	.416389	.419963	.423562	.427184	.430831	.434502	−3
1200	.438198	.441919	.445664	.449434	.453230	.457050	.460896	.464767	.468663	.472585	−3
1210	.476533	.480507	.484507	.488532	.492584	.496662	.500767	.504898	.509056	.513240	−3
1220	.517452	.521690	.525955	.530248	.534568	.538916	.543291	.547693	.552124	.556582	−3
1230	.561069	.565584	.570127	.574698	.579298	.583926	.588584	.593270	.597985	.602729	−3
1240	.607503	.612306	.617138	.622000	.626892	.631813	.636765	.641747	.646758	.651800	−3
1250	.656873	.661976	.667110	.672275	.677470	.682697	.687955	.693244	.698564	.703916	−3
1260	.709300	.714716	.720163	.725643	.731154	.736698	.742275	.747884	.753525	.759200	−3
1270	.764907	.770647	.776421	.782227	.788068	.793941	.799849	.805790	.811765	.817774	−3
1280	.823817	.829894	.836006	.842153	.848334	.854550	.860800	.867086	.873407	.879763	−3
1290	.886155	.892582	.899045	.905544	.912079	.918649	.925256	.931899	.938579	.945295	−3
1300	.095205	.095884	.096566	.097253	.097943	.098637	.099334	.100036	.100741	.101450	−2
1310	.102162	.102879	.103599	.104323	.105051	.105783	.106519	.107258	.108002	.108749	−2
1320	.109501	.110256	.111015	.111778	.112545	.113317	.114092	.114871	.115654	.116441	−2
1330	.117233	.118028	.118827	.119631	.120439	.121250	.122066	.122886	.123711	.124539	−2
1340	.125371	.126208	.127049	.127894	.128744	.129597	.130455	.131318	.132184	.133055	−2
1350	.133930	.134809	.135693	.136581	.137474	.138371	.139272	.140178	.141088	.142002	−2
1360	.142921	.143844	.144772	.145705	.146642	.147583	.148529	.149479	.150434	.151394	−2
1370	.152358	.153327	.154300	.155278	.156260	.157247	.158239	.159236	.160237	.161243	−2
1380	.162253	.163269	.164289	.165313	.166343	.167377	.168416	.169460	.170509	.171562	−2
1390	.172621	.173684	.174752	.175825	.176902	.177985	.179073	.180165	.181263	.182365	−2
1400	.183472	.184585	.185702	.186824	.187952	.189084	.190222	.191364	.192512	.193664	−2
1410	.194822	.195985	.197153	.198326	.199504	.200687	.201876	.203070	.204268	.205473	−2
1420	.206682	.207896	.209116	.210341	.211572	.212807	.214048	.215295	.216546	.217803	−2
1430	.219065	.220333	.221606	.222884	.224168	.225457	.226752	.228052	.229358	.230669	−2
1440	.231985	.233307	.234634	.235967	.237306	.238650	.240000	.241355	.242715	.244082	−2
1450	.245454	.246831	.248214	.249603	.250998	.252398	.253804	.255215	.256632	.258055	−2
1460	.259484	.260918	.262358	.263804	.265256	.266713	.268177	.269646	.271121	.272601	−2
1470	.274088	.275581	.277079	.278583	.280093	.281609	.283131	.284659	.286193	.287733	−2
1480	.289279	.290831	.292388	.293952	.295522	.297098	.298680	.300268	.301862	.303462	−2
1490	.305068	.306681	.308299	.309924	.311555	.313192	.314835	.316484	.318139	.319801	−2

TABLE 15A

Blackbody: Relative Fractional Photon Emittance

λT(μ.°K)	0	1	2	3	4	5	6	7	8	9	m
1500	.321469	.323143	.324823	.326510	.328203	.329902	.331608	.333319	.335037	.336762	-2
1510	.338493	.340230	.341973	.343723	.345479	.347242	.349011	.350786	.352568	.354356	-2
1520	.356151	.357952	.359760	.361574	.363395	.365222	.367056	.368896	.370743	.372596	-2
1530	.374456	.376323	.378196	.380075	.381962	.383855	.385754	.387660	.389573	.391493	-2
1540	.393419	.395352	.397292	.399238	.401191	.403151	.405118	.407091	.409071	.411058	-2
1550	.413052	.415052	.417060	.419074	.421095	.423123	.425158	.427199	.429248	.431303	-2
1560	.433365	.435435	.437511	.439594	.441684	.443781	.445885	.447996	.450114	.452239	-2
1570	.454371	.456510	.458656	.460809	.462969	.465136	.467310	.469492	.471680	.473876	-2
1580	.476078	.478288	.480505	.482729	.484961	.487199	.489445	.491697	.493958	.496225	-2
1590	.498499	.500781	.503070	.505366	.507670	.509980	.512298	.514624	.516956	.519296	-2
1600	.521644	.523998	.526360	.528730	.531106	.533490	.535882	.538281	.540687	.543101	-2
1610	.545522	.547950	.550386	.552830	.555281	.557739	.560205	.562678	.565159	.567648	-2
1620	.570144	.572647	.575158	.577677	.580203	.582736	.585278	.587827	.590383	.592947	-2
1630	.595519	.598098	.600685	.603280	.605882	.608492	.611109	.613735	.616368	.619008	-2
1640	.621657	.624313	.626977	.629648	.632328	.635015	.637710	.640412	.643123	.645841	-2
1650	.648567	.651301	.654042	.656792	.659549	.662314	.665088	.667868	.670657	.673454	-2
1660	.676258	.679071	.681891	.684719	.687556	.690400	.693252	.696112	.698980	.701856	-2
1670	.704740	.707631	.710531	.713439	.716355	.719279	.722211	.725151	.728099	.731055	-2
1680	.734019	.736991	.739972	.742960	.745956	.748961	.751974	.754994	.758023	.761060	-2
1690	.764105	.767159	.770220	.773290	.776368	.779453	.782548	.785650	.788760	.791879	-2
1700	.795006	.798141	.801285	.804437	.807596	.810765	.813941	.817126	.820319	.823520	-2
1710	.826730	.829947	.833174	.836408	.839651	.842902	.846161	.849429	.852705	.855990	-2
1720	.859283	.862584	.865894	.869212	.872538	.875873	.879216	.882568	.885928	.889296	-2
1730	.892673	.896058	.899452	.902854	.906265	.909684	.913112	.916548	.919993	.923446	-2
1740	.926907	.930377	.933856	.937343	.940839	.944343	.947856	.951377	.954907	.958445	-2
1750	.961992	.965548	.969112	.972685	.976266	.979856	.983454	.987061	.990677	.994301	-2
1760	.099793	.100158	.100523	.100889	.101255	.101623	.101991	.102361	.102731	.103102	-1
1770	.103474	.103847	.104221	.104595	.104970	.105347	.105724	.106102	.106481	.106861	-1
1780	.107241	.107623	.108005	.108389	.108773	.109158	.109544	.109931	.110318	.110707	-1
1790	.111096	.111487	.111878	.112270	.112663	.113057	.113452	.113847	.114244	.114641	-1
1800	.115039	.115438	.115838	.116239	.116641	.117044	.117447	.117852	.118257	.118664	-1
1810	.119071	.119479	.119888	.120297	.120708	.121120	.121532	.121946	.122360	.122775	-1
1820	.123191	.123608	.124026	.124445	.124864	.125285	.125706	.126129	.126552	.126976	-1
1830	.127401	.127827	.128254	.128681	.129110	.129540	.129970	.130401	.130834	.131267	-1
1840	.131701	.132136	.132572	.133008	.133446	.133884	.134324	.134764	.135206	.135648	-1
1850	.136091	.136535	.136980	.137425	.137872	.138320	.138768	.139218	.139668	.140119	-1
1860	.140571	.141024	.141478	.141933	.142389	.142846	.143303	.143762	.144221	.144682	-1
1870	.145143	.145605	.146068	.146532	.146997	.147463	.147930	.148397	.148866	.149335	-1
1880	.149806	.150277	.150749	.151223	.151697	.152172	.152648	.153124	.153602	.154081	-1
1890	.154560	.155041	.155522	.156005	.156488	.156972	.157457	.157943	.158430	.158918	-1
1900	.159407	.159896	.160387	.160879	.161371	.161864	.162359	.162854	.163350	.163847	-1
1910	.164345	.164844	.165344	.165845	.166346	.166849	.167352	.167857	.168362	.168869	-1
1920	.169376	.169884	.170393	.170903	.171414	.171926	.172439	.172952	.173467	.173983	-1
1930	.174499	.175017	.175535	.176054	.176574	.177095	.177617	.178140	.178664	.179189	-1
1940	.179715	.180242	.180769	.181298	.181827	.182358	.182889	.183421	.183954	.184489	-1
1950	.185024	.185560	.186097	.186634	.187173	.187713	.188253	.188795	.189337	.189881	-1
1960	.190425	.190970	.191517	.192064	.192612	.193161	.193711	.194262	.194813	.195366	-1
1970	.195920	.196474	.197030	.197586	.198144	.198702	.199261	.199821	.200382	.200944	-1
1980	.201507	.202071	.202636	.203202	.203768	.204336	.204905	.205474	.206044	.206616	-1
1990	.207188	.207761	.208335	.208910	.209486	.210063	.210641	.211220	.211800	.212380	-1

$N(\lambda,T)/N_T(T)$

TABLE 15A

Blackbody: Relative Fractional Photon Emittance

λT(μ.°K)	0	1	2	3	4	5	6	7	8	9	m
2000	.212962	.213544	.214128	.214712	.215297	.215884	.216471	.217059	.217648	.218238	-1
2010	.218829	.219421	.220013	.220607	.221202	.221797	.222394	.222991	.223589	.224189	-1
2020	.224789	.225390	.225992	.226595	.227199	.227804	.228409	.229016	.229624	.230232	-1
2030	.230842	.231452	.232064	.232676	.233289	.233903	.234518	.235134	.235751	.236369	-1
2040	.236988	.237607	.238228	.238850	.239472	.240096	.240720	.241345	.241971	.242599	-1
2050	.243227	.243856	.244486	.245116	.245748	.246381	.247014	.247649	.248284	.248921	-1
2060	.249558	.250196	.250836	.251476	.252117	.252759	.253402	.254045	.254690	.255336	-1
2070	.255982	.256630	.257278	.257928	.258578	.259229	.259881	.260534	.261188	.261843	-1
2080	.262499	.263156	.263813	.264472	.265131	.265792	.266453	.267115	.267779	.268443	-1
2090	.269108	.269774	.270441	.271108	.271777	.272447	.273117	.273789	.274461	.275134	-1
2100	.275809	.276484	.277160	.277837	.278515	.279194	.279873	.280554	.281236	.281918	-1
2110	.282601	.283286	.283971	.284657	.285344	.286032	.286721	.287411	.288102	.288793	-1
2120	.289486	.290179	.290874	.291569	.292265	.292962	.293660	.294359	.295059	.295760	-1
2130	.296462	.297164	.297868	.298572	.299277	.299984	.300691	.301399	.302108	.302818	-1
2140	.303528	.304240	.304953	.305666	.306380	.307096	.307812	.308529	.309247	.309966	-1
2150	.310686	.311407	.312128	.312851	.313574	.314299	.315024	.315750	.316477	.317205	-1
2160	.317934	.318664	.319394	.320126	.320858	.321592	.322326	.323061	.323797	.324534	-1
2170	.325272	.326011	.326750	.327491	.328232	.328975	.329718	.330462	.331207	.331953	-1
2180	.332700	.333448	.334196	.334946	.335696	.336448	.337200	.337953	.338707	.339462	-1
2190	.340217	.340974	.341732	.342490	.343249	.344009	.344771	.345533	.346295	.347059	-1
2200	.347824	.348589	.349356	.350123	.350891	.351660	.352430	.353201	.353973	.354745	-1
2210	.355519	.356293	.357068	.357845	.358622	.359400	.360178	.360958	.361738	.362520	-1
2220	.363302	.364085	.364869	.365654	.366440	.367227	.368014	.368803	.369592	.370382	-1
2230	.371173	.371965	.372758	.373552	.374346	.375142	.375938	.376735	.377533	.378332	-1
2240	.379132	.379933	.380734	.381536	.382340	.383144	.383949	.384755	.385561	.386369	-1
2250	.387177	.387987	.388797	.389608	.390420	.391233	.392046	.392861	.393676	.394492	-1
2260	.395309	.396127	.396946	.397766	.398586	.399408	.400230	.401053	.401877	.402702	-1
2270	.403527	.404354	.405181	.406009	.406838	.407668	.408499	.409331	.410163	.410997	-1
2280	.411831	.412666	.413502	.414338	.415176	.416014	.416854	.417694	.418535	.419377	-1
2290	.420219	.421063	.421907	.422752	.423598	.424445	.425293	.426141	.426991	.427841	-1
2300	.428692	.429544	.430397	.431250	.432105	.432960	.433816	.434673	.435531	.436390	-1
2310	.437249	.438109	.438970	.439832	.440695	.441559	.442423	.443288	.444155	.445022	-1
2320	.445889	.446758	.447627	.448498	.449369	.450241	.451113	.451987	.452861	.453736	-1
2330	.454612	.455489	.456367	.457245	.458125	.459005	.459886	.460768	.461650	.462534	-1
2340	.463418	.464303	.465189	.466075	.466963	.467851	.468740	.469630	.470521	.471413	-1
2350	.472305	.473198	.474092	.474987	.475883	.476779	.477676	.478574	.479473	.480373	-1
2360	.481273	.482175	.483077	.483980	.484883	.485788	.486693	.487599	.488506	.489414	-1
2370	.490322	.491232	.492142	.493053	.493964	.494877	.495790	.496704	.497619	.498535	-1
2380	.499451	.500368	.501286	.502205	.503125	.504045	.504966	.505888	.506811	.507735	-1
2390	.508659	.509584	.510510	.511437	.512365	.513293	.514222	.515152	.516083	.517014	-1
2400	.517946	.518879	.519813	.520747	.521683	.522619	.523556	.524493	.525432	.526371	-1
2410	.527311	.528252	.529193	.530136	.531079	.532023	.532967	.533913	.534859	.535806	-1
2420	.536754	.537702	.538651	.539601	.540552	.541504	.542456	.543409	.544363	.545318	-1
2430	.546273	.547229	.548186	.549144	.550102	.551061	.552021	.552982	.553943	.554905	-1
2440	.555868	.556832	.557797	.558762	.559728	.560695	.561662	.562630	.563599	.564569	-1
2450	.565539	.566511	.567483	.568455	.569429	.570403	.571378	.572354	.573330	.574307	-1
2460	.575285	.576264	.577243	.578224	.579204	.580186	.581168	.582152	.583135	.584120	-1
2470	.585105	.586091	.587078	.588066	.589054	.590043	.591033	.592023	.593014	.594006	-1
2480	.594999	.595992	.596986	.597981	.598977	.599973	.600970	.601968	.602966	.603965	-1
2490	.604965	.605966	.606967	.607969	.608972	.609976	.610980	.611985	.612990	.613997	-1

TABLE 15A

Blackbody: Relative Fractional Photon Emittance

$\lambda T(\mu.^\circ K)$	0	1	2	3	4	5	6	7	8	9	m
2500	.615004	.616012	.617020	.618029	.619039	.620050	.621061	.622074	.623086	.624100	-1
2510	.625114	.626129	.627145	.628161	.629178	.630196	.631214	.632233	.633253	.634274	-1
2520	.635295	.636317	.637339	.638363	.639387	.640412	.641437	.642463	.643490	.644517	-1
2530	.645546	.646575	.647604	.648634	.649665	.650697	.651730	.652763	.653796	.654831	-1
2540	.655866	.656902	.657938	.658975	.660013	.661052	.662091	.663131	.664172	.665213	-1
2550	.666255	.667297	.668341	.669385	.670429	.671475	.672521	.673568	.674615	.675663	-1
2560	.676712	.677761	.678811	.679862	.680913	.681965	.683018	.684072	.685126	.686180	-1
2570	.687236	.688292	.689349	.690406	.691464	.692523	.693582	.694642	.695703	.696764	-1
2580	.697826	.698889	.699952	.701016	.702081	.703146	.704212	.705279	.706346	.707414	-1
2590	.708483	.709552	.710622	.711692	.712764	.713835	.714908	.715981	.717055	.718129	-1
2600	.719204	.720280	.721356	.722433	.723511	.724589	.725668	.726748	.727828	.728909	-1
2610	.729990	.731072	.732155	.733238	.734322	.735407	.736492	.737578	.738665	.739752	-1
2620	.740840	.741928	.743017	.744107	.745197	.746288	.747379	.748472	.749564	.750658	-1
2630	.751752	.752847	.753942	.755038	.756134	.757231	.758329	.759428	.760527	.761626	-1
2640	.762726	.763827	.764929	.766031	.767133	.768237	.769341	.770445	.771550	.772656	-1
2650	.773762	.774869	.775977	.777085	.778194	.779303	.780413	.781524	.782635	.783747	-1
2660	.784859	.785972	.787085	.788200	.789314	.790430	.791546	.792662	.793779	.794897	-1
2670	.796015	.797134	.798254	.799374	.800495	.801616	.802738	.803860	.804983	.806107	-1
2680	.807231	.808356	.809481	.810607	.811734	.812861	.813989	.815117	.816246	.817375	-1
2690	.818506	.819636	.820767	.821899	.823031	.824164	.825298	.826432	.827567	.828702	-1
2700	.829838	.830974	.832111	.833248	.834386	.835525	.836664	.837804	.838944	.840085	-1
2710	.841227	.842369	.843511	.844654	.845798	.846942	.848087	.849232	.850378	.851525	-1
2720	.852672	.853819	.854968	.856116	.857266	.858415	.859566	.860717	.861868	.863020	-1
2730	.864173	.865326	.866479	.867634	.868788	.869944	.871099	.872256	.873413	.874570	-1
2740	.875728	.876887	.878046	.879205	.880366	.881526	.882688	.883849	.885012	.886175	-1
2750	.887338	.888502	.889666	.890831	.891997	.893163	.894329	.895496	.896664	.897832	-1
2760	.899001	.900170	.901340	.902510	.903681	.904852	.906024	.907196	.908369	.909543	-1
2770	.910716	.911891	.913066	.914241	.915417	.916594	.917771	.918948	.920126	.921305	-1
2780	.922484	.923663	.924843	.926024	.927205	.928387	.929569	.930751	.931935	.933118	-1
2790	.934302	.935487	.936672	.937858	.939044	.940231	.941418	.942605	.943793	.944982	-1
2800	.946171	.947361	.948551	.949742	.950933	.952124	.953316	.954509	.955702	.956896	-1
2810	.958090	.959284	.960479	.961675	.962871	.964068	.965264	.966462	.967660	.968858	-1
2820	.970057	.971257	.972457	.973657	.974858	.976059	.977261	.978463	.979666	.980869	-1
2830	.982073	.983277	.984482	.985687	.986893	.988099	.989305	.990512	.991720	.992928	-1
2840	.099414	.099534	.099655	.099776	.099897	.100019	.100140	.100261	.100382	.100503	-0
2850	.100625	.100746	.100867	.100989	.101110	.101232	.101353	.101475	.101597	.101718	-0
2860	.101840	.101962	.102084	.102206	.102328	.102450	.102572	.102694	.102816	.102938	-0
2870	.103060	.103183	.103305	.103427	.103550	.103672	.103795	.103917	.104040	.104162	-0
2880	.104285	.104408	.104530	.104653	.104776	.104899	.105022	.105145	.105268	.105391	-0
2890	.105514	.105637	.105760	.105883	.106007	.106130	.106253	.106377	.106500	.106624	-0
2900	.106747	.106871	.106994	.107118	.107242	.107365	.107489	.107613	.107737	.107861	-0
2910	.107985	.108109	.108233	.108357	.108481	.108605	.108729	.108853	.108978	.109102	-0
2920	.109226	.109351	.109475	.109600	.109724	.109849	.109973	.110098	.110223	.110347	-0
2930	.110472	.110597	.110722	.110847	.110972	.111097	.111222	.111347	.111472	.111597	-0
2940	.111722	.111847	.111972	.112098	.112223	.112348	.112474	.112599	.112725	.112850	-0
2950	.112976	.113101	.113227	.113353	.113478	.113604	.113730	.113856	.113982	.114108	-0
2960	.114233	.114359	.114485	.114612	.114738	.114864	.114990	.115116	.115242	.115369	-0
2970	.115495	.115621	.115748	.115874	.116001	.116127	.116254	.116380	.116507	.116634	-0
2980	.116760	.116887	.117014	.117141	.117267	.117394	.117521	.117648	.117775	.117902	-0
2990	.118029	.118156	.118283	.118411	.118538	.118665	.118792	.118920	.119047	.119174	-0

TABLE 15A

Blackbody: Relative Fractional Photon Emittance

λT(μ.°K)	0	1	2	3	4	5	6	7	8	9	m
3000	.119302	.119429	.119557	.119684	.119812	.119939	.120067	.120195	.120322	.120450	-0
3010	.120578	.120706	.120834	.120961	.121089	.121217	.121345	.121473	.121601	.121729	-0
3020	.121858	.121986	.122114	.122242	.122370	.122499	.122627	.122755	.122884	.123012	-0
3030	.123141	.123269	.123398	.123526	.123655	.123783	.123912	.124041	.124169	.124298	-0
3040	.124427	.124556	.124685	.124814	.124942	.125071	.125200	.125329	.125458	.125588	-0
3050	.125717	.125846	.125975	.126104	.126233	.126363	.126492	.126621	.126751	.126880	-0
3060	.127009	.127139	.127268	.127398	.127528	.127657	.127787	.127916	.128046	.128176	-0
3070	.128306	.128435	.128565	.128695	.128825	.128955	.129085	.129215	.129345	.129475	-0
3080	.129605	.129735	.129865	.129995	.130125	.130255	.130386	.130516	.130646	.130776	-0
3090	.130907	.131037	.131168	.131298	.131428	.131559	.131689	.131820	.131951	.132081	-0
3100	.132212	.132343	.132473	.132604	.132735	.132866	.132996	.133127	.133258	.133389	-0
3110	.133520	.133651	.133782	.133913	.134044	.134175	.134306	.134437	.134568	.134699	-0
3120	.134831	.134962	.135093	.135224	.135356	.135487	.135619	.135750	.135881	.136013	-0
3130	.136144	.136276	.136407	.136539	.136670	.136802	.136934	.137065	.137197	.137329	-0
3140	.137461	.137592	.137724	.137856	.137988	.138120	.138252	.138384	.138516	.138647	-0
3150	.138779	.138912	.139044	.139176	.139308	.139440	.139572	.139704	.139837	.139969	-0
3160	.140101	.140233	.140366	.140498	.140630	.140763	.140895	.141028	.141160	.141293	-0
3170	.141425	.141558	.141690	.141823	.141955	.142088	.142221	.142353	.142486	.142619	-0
3180	.142752	.142884	.143017	.143150	.143283	.143416	.143549	.143681	.143814	.143947	-0
3190	.144080	.144213	.144346	.144480	.144613	.144746	.144879	.145012	.145145	.145278	-0
3200	.145412	.145545	.145678	.145811	.145945	.146078	.146211	.146345	.146478	.146612	-0
3210	.146745	.146879	.147012	.147146	.147279	.147413	.147546	.147680	.147814	.147947	-0
3220	.148081	.148215	.148348	.148482	.148616	.148750	.148883	.149017	.149151	.149285	-0
3230	.149419	.149553	.149687	.149820	.149954	.150088	.150222	.150356	.150491	.150625	-0
3240	.150759	.150893	.151027	.151161	.151295	.151429	.151564	.151698	.151832	.151966	-0
3250	.152101	.152235	.152369	.152504	.152638	.152772	.152907	.153041	.153176	.153310	-0
3260	.153445	.153579	.153714	.153848	.153983	.154117	.154252	.154387	.154521	.154656	-0
3270	.154791	.154925	.155060	.155195	.155330	.155464	.155599	.155734	.155869	.156004	-0
3280	.156138	.156273	.156408	.156543	.156678	.156813	.156948	.157083	.157218	.157353	-0
3290	.157488	.157623	.157758	.157893	.158028	.158164	.158299	.158434	.158569	.158704	-0
3300	.158839	.158975	.159110	.159245	.159380	.159516	.159651	.159786	.159922	.160057	-0
3310	.160193	.160328	.160463	.160599	.160734	.160870	.161005	.161141	.161276	.161412	-0
3320	.161547	.161683	.161818	.161954	.162090	.162225	.162361	.162497	.162632	.162768	-0
3330	.162904	.163039	.163175	.163311	.163447	.163582	.163718	.163854	.163990	.164126	-0
3340	.164262	.164397	.164533	.164669	.164805	.164941	.165077	.165213	.165349	.165485	-0
3350	.165621	.165757	.165893	.166029	.166165	.166301	.166437	.166573	.166709	.166846	-0
3360	.166982	.167118	.167254	.167390	.167526	.167663	.167799	.167935	.168071	.168208	-0
3370	.168344	.168480	.168617	.168753	.168889	.169026	.169162	.169298	.169435	.169571	-0
3380	.169707	.169844	.169980	.170117	.170253	.170390	.170526	.170663	.170799	.170936	-0
3390	.171072	.171209	.171345	.171482	.171619	.171755	.171892	.172028	.172165	.172302	-0
3400	.172438	.172575	.172712	.172848	.172985	.173122	.173259	.173395	.173532	.173669	-0
3410	.173806	.173942	.174079	.174216	.174353	.174490	.174626	.174763	.174900	.175037	-0
3420	.175174	.175311	.175448	.175585	.175722	.175859	.175995	.176132	.176269	.176406	-0
3430	.176543	.176680	.176817	.176954	.177091	.177228	.177366	.177503	.177640	.177777	-0
3440	.177914	.178051	.178188	.178325	.178462	.178599	.178737	.178874	.179011	.179148	-0
3450	.179285	.179423	.179560	.179697	.179834	.179971	.180109	.180246	.180383	.180520	-0
3460	.180658	.180795	.180932	.181070	.181207	.181344	.181482	.181619	.181756	.181894	-0
3470	.182031	.182168	.182306	.182443	.182581	.182718	.182855	.182993	.183130	.183268	-0
3480	.183405	.183543	.183680	.183818	.183955	.184093	.184230	.184368	.184505	.184643	-0
3490	.184780	.184918	.185055	.185193	.185330	.185468	.185605	.185743	.185881	.186018	-0

TABLE 15A

Blackbody: Relative Fractional Photon Emittance

$\lambda T (\mu . ^\circ K)$	0	1	2	3	4	5	6	7	8	9	m
3500	.186156	.186293	.186431	.186569	.186706	.186844	.186982	.187119	.187257	.187395	-0
3510	.187532	.187670	.187808	.187945	.188083	.188221	.188358	.188496	.188634	.188772	-0
3520	.188909	.189047	.189185	.189322	.189460	.189598	.189736	.189874	.190011	.190149	-0
3530	.190287	.190425	.190563	.190700	.190838	.190976	.191114	.191252	.191389	.191527	-0
3540	.191665	.191803	.191941	.192079	.192217	.192354	.192492	.192630	.192768	.192906	-0
3550	.193044	.193182	.193320	.193458	.193596	.193733	.193871	.194009	.194147	.194285	-0
3560	.194423	.194561	.194699	.194837	.194975	.195113	.195251	.195389	.195527	.195665	-0
3570	.195803	.195941	.196079	.196217	.196355	.196493	.196631	.196769	.196907	.197045	-0
3580	.197183	.197321	.197459	.197597	.197735	.197873	.198011	.198149	.198287	.198425	-0
3590	.198563	.198701	.198840	.198978	.199116	.199254	.199392	.199530	.199668	.199806	-0
3600	.199944	.200082	.200220	.200358	.200497	.200635	.200773	.200911	.201049	.201187	-0
3610	.201325	.201463	.201601	.201740	.201878	.202016	.202154	.202292	.202430	.202568	-0
3620	.202706	.202845	.202983	.203121	.203259	.203397	.203535	.203673	.203812	.203950	-0
3630	.204088	.204226	.204364	.204502	.204641	.204779	.204917	.205055	.205193	.205331	-0
3640	.205470	.205608	.205746	.205884	.206022	.206160	.206299	.206437	.206575	.206713	-0
3650	.206851	.206989	.207128	.207266	.207404	.207542	.207680	.207819	.207957	.208095	-0
3660	.208233	.208371	.208509	.208648	.208786	.208924	.209062	.209200	.209339	.209477	-0
3670	.209615	.209753	.209891	.210029	.210168	.210306	.210444	.210582	.210720	.210859	-0
3680	.210997	.211135	.211273	.211411	.211550	.211688	.211826	.211964	.212102	.212240	-0
3690	.212379	.212517	.212655	.212793	.212931	.213069	.213208	.213346	.213484	.213622	-0
3700	.213760	.213898	.214037	.214175	.214313	.214451	.214589	.214727	.214866	.215004	-0
3710	.215142	.215280	.215418	.215556	.215695	.215833	.215971	.216109	.216247	.216385	-0
3720	.216523	.216662	.216800	.216938	.217076	.217214	.217352	.217490	.217628	.217767	-0
3730	.217905	.218043	.218181	.218319	.218457	.218595	.218733	.218871	.219010	.219148	-0
3740	.219286	.219424	.219562	.219700	.219838	.219976	.220114	.220252	.220390	.220528	-0
3750	.220667	.220805	.220943	.221081	.221219	.221357	.221495	.221633	.221771	.221909	-0
3760	.222047	.222185	.222323	.222461	.222599	.222737	.222875	.223013	.223151	.223289	-0
3770	.223427	.223565	.223703	.223841	.223979	.224117	.224255	.224393	.224531	.224669	-0
3780	.224807	.224945	.225083	.225221	.225359	.225497	.225635	.225773	.225911	.226049	-0
3790	.226187	.226324	.226462	.226600	.226738	.226876	.227014	.227152	.227290	.227428	-0
3800	.227566	.227703	.227841	.227979	.228117	.228255	.228393	.228531	.228668	.228806	-0
3810	.228944	.229082	.229220	.229358	.229495	.229633	.229771	.229909	.230047	.230184	-0
3820	.230322	.230460	.230598	.230736	.230873	.231011	.231149	.231287	.231424	.231562	-0
3830	.231700	.231837	.231975	.232113	.232251	.232388	.232526	.232664	.232801	.232939	-0
3840	.233077	.233214	.233352	.233490	.233627	.233765	.233903	.234040	.234178	.234316	-0
3850	.234453	.234591	.234728	.234866	.235004	.235141	.235279	.235416	.235554	.235691	-0
3860	.235829	.235967	.236104	.236242	.236379	.236517	.236654	.236792	.236929	.237067	-0
3870	.237204	.237342	.237479	.237617	.237754	.237892	.238029	.238166	.238304	.238441	-0
3880	.238579	.238716	.238853	.238991	.239128	.239266	.239403	.239540	.239678	.239815	-0
3890	.239952	.240090	.240227	.240364	.240502	.240639	.240776	.240914	.241051	.241188	-0
3900	.241326	.241463	.241600	.241737	.241875	.242012	.242149	.242286	.242423	.242561	-0
3910	.242698	.242835	.242972	.243109	.243247	.243384	.243521	.243658	.243795	.243932	-0
3920	.244069	.244206	.244344	.244481	.244618	.244755	.244892	.245029	.245166	.245303	-0
3930	.245440	.245577	.245714	.245851	.245988	.246125	.246262	.246399	.246536	.246673	-0
3940	.246810	.246947	.247084	.247221	.247358	.247494	.247631	.247768	.247905	.248042	-0
3950	.248179	.248316	.248452	.248589	.248726	.248863	.249000	.249137	.249273	.249410	-0
3960	.249547	.249684	.249820	.249957	.250094	.250231	.250367	.250504	.250641	.250777	-0
3970	.250914	.251051	.251187	.251324	.251461	.251597	.251734	.251870	.252007	.252144	-0
3980	.252280	.252417	.252553	.252690	.252826	.252963	.253099	.253236	.253372	.253509	-0
3990	.253645	.253782	.253918	.254055	.254191	.254328	.254464	.254600	.254737	.254873	-0

TABLE 15A

Blackbody: Relative Fractional Photon Emittance

λT(μ.°K)	0	1	2	3	4	5	6	7	8	9	m
4000	.255010	.255146	.255282	.255419	.255555	.255691	.255828	.255964	.256100	.256237	-0
4010	.256373	.256509	.256645	.256782	.256918	.257054	.257190	.257326	.257463	.257599	-0
4020	.257735	.257871	.258007	.258143	.258280	.258416	.258552	.258688	.258824	.258960	-0
4030	.259096	.259232	.259368	.259504	.259640	.259776	.259912	.260048	.260184	.260320	-0
4040	.260456	.260592	.260728	.260864	.261000	.261136	.261271	.261407	.261543	.261679	-0
4050	.261815	.261951	.262086	.262222	.262358	.262494	.262630	.262765	.262901	.263037	-0
4060	.263172	.263308	.263444	.263580	.263715	.263851	.263987	.264122	.264258	.264393	-0
4070	.264529	.264665	.264800	.264936	.265071	.265207	.265342	.265478	.265613	.265749	-0
4080	.265884	.266020	.266155	.266291	.266426	.266561	.266697	.266832	.266968	.267103	-0
4090	.267238	.267374	.267509	.267644	.267780	.267915	.268050	.268185	.268321	.268456	-0
4100	.268591	.268726	.268862	.268997	.269132	.269267	.269402	.269537	.269673	.269808	-0
4110	.269943	.270078	.270213	.270348	.270483	.270618	.270753	.270888	.271023	.271158	-0
4120	.271293	.271428	.271563	.271698	.271833	.271968	.272103	.272238	.272372	.272507	-0
4130	.272642	.272777	.272912	.273047	.273181	.273316	.273451	.273586	.273720	.273855	-0
4140	.273990	.274124	.274259	.274394	.274528	.274663	.274798	.274932	.275067	.275202	-0
4150	.275336	.275471	.275605	.275740	.275874	.276009	.276143	.276278	.276412	.276547	-0
4160	.276681	.276816	.276950	.277084	.277219	.277353	.277487	.277622	.277756	.277890	-0
4170	.278025	.278159	.278293	.278427	.278562	.278696	.278830	.278964	.279099	.279233	-0
4180	.279367	.279501	.279635	.279769	.279903	.280037	.280171	.280306	.280440	.280574	-0
4190	.280708	.280842	.280976	.281110	.281243	.281377	.281511	.281645	.281779	.281913	-0
4200	.282047	.282181	.282315	.282448	.282582	.282716	.282850	.282984	.283117	.283251	-0
4210	.283385	.283518	.283652	.283786	.283919	.284053	.284187	.284320	.284454	.284588	-0
4220	.284721	.284855	.284988	.285122	.285255	.285389	.285522	.285656	.285789	.285923	-0
4230	.286056	.286189	.286323	.286456	.286589	.286723	.286856	.286989	.287123	.287256	-0
4240	.287389	.287523	.287656	.287789	.287922	.288055	.288189	.288322	.288455	.288588	-0
4250	.288721	.288854	.288987	.289120	.289253	.289386	.289519	.289652	.289785	.289918	-0
4260	.290051	.290184	.290317	.290450	.290583	.290716	.290849	.290982	.291114	.291247	-0
4270	.291380	.291513	.291646	.291778	.291911	.292044	.292176	.292309	.292442	.292574	-0
4280	.292707	.292840	.292972	.293105	.293238	.293370	.293503	.293635	.293768	.293900	-0
4290	.294033	.294165	.294298	.294430	.294562	.294695	.294827	.294959	.295092	.295224	-0
4300	.295356	.295489	.295621	.295753	.295886	.296018	.296150	.296282	.296414	.296547	-0
4310	.296679	.296811	.296943	.297075	.297207	.297339	.297471	.297603	.297735	.297867	-0
4320	.297999	.298131	.298263	.298395	.298527	.298659	.298791	.298923	.299055	.299186	-0
4330	.299318	.299450	.299582	.299714	.299845	.299977	.300109	.300240	.300372	.300504	-0
4340	.300635	.300767	.300899	.301030	.301162	.301293	.301425	.301556	.301688	.301819	-0
4350	.301951	.302082	.302214	.302345	.302477	.302608	.302739	.302871	.303002	.303133	-0
4360	.303265	.303396	.303527	.303659	.303790	.303921	.304052	.304183	.304315	.304446	-0
4370	.304577	.304708	.304839	.304970	.305101	.305232	.305363	.305494	.305625	.305756	-0
4380	.305887	.306018	.306149	.306280	.306411	.306542	.306673	.306803	.306934	.307065	-0
4390	.307196	.307327	.307457	.307588	.307719	.307849	.307980	.308111	.308241	.308372	-0
4400	.308503	.308633	.308764	.308894	.309025	.309155	.309286	.309416	.309547	.309677	-0
4410	.309808	.309938	.310068	.310199	.310329	.310460	.310590	.310720	.310850	.310981	-0
4420	.311111	.311241	.311371	.311502	.311632	.311762	.311892	.312022	.312152	.312282	-0
4430	.312412	.312542	.312672	.312802	.312932	.313062	.313192	.313322	.313452	.313582	-0
4440	.313712	.313842	.313972	.314102	.314231	.314361	.314491	.314621	.314750	.314880	-0
4450	.315010	.315139	.315269	.315399	.315528	.315658	.315788	.315917	.316047	.316176	-0
4460	.316306	.316435	.316565	.316694	.316824	.316953	.317082	.317212	.317341	.317470	-0
4470	.317600	.317729	.317858	.317988	.318117	.318246	.318375	.318504	.318634	.318763	-0
4480	.318892	.319021	.319150	.319279	.319408	.319537	.319666	.319795	.319924	.320053	-0
4490	.320182	.320311	.320440	.320569	.320698	.320827	.320956	.321084	.321213	.321342	-0

TABLE 15A

Blackbody: Relative Fractional Photon Emittance

$\lambda T(\mu.^\circ K)$	0	1	2	3	4	5	6	7	8	9	m
4500	.321471	.321599	.321728	.321857	.321986	.322114	.322243	.322371	.322500	.322629	-0
4510	.322757	.322886	.323014	.323143	.323271	.323400	.323528	.323657	.323785	.323913	-0
4520	.324042	.324170	.324298	.324427	.324555	.324683	.324812	.324940	.325068	.325196	-0
4530	.325324	.325453	.325581	.325709	.325837	.325965	.326093	.326221	.326349	.326477	-0
4540	.326605	.326733	.326861	.326989	.327117	.327245	.327373	.327501	.327628	.327756	-0
4550	.327884	.328012	.328139	.328267	.328395	.328523	.328650	.328778	.328906	.329033	-0
4560	.329161	.329288	.329416	.329543	.329671	.329798	.329926	.330053	.330181	.330308	-0
4570	.330436	.330563	.330690	.330818	.330945	.331072	.331200	.331327	.331454	.331581	-0
4580	.331708	.331836	.331963	.332090	.332217	.332344	.332471	.332598	.332725	.332852	-0
4590	.332979	.333106	.333233	.333360	.333487	.333614	.333741	.333868	.333995	.334121	-0
4600	.334248	.334375	.334502	.334628	.334755	.334882	.335009	.335135	.335262	.335388	-0
4610	.335515	.335642	.335768	.335895	.336021	.336148	.336274	.336401	.336527	.336654	-0
4620	.336780	.336906	.337033	.337159	.337285	.337412	.337538	.337664	.337790	.337917	-0
4630	.338043	.338169	.338295	.338421	.338547	.338673	.338799	.338926	.339052	.339178	-0
4640	.339304	.339430	.339555	.339681	.339807	.339933	.340059	.340185	.340311	.340437	-0
4650	.340562	.340688	.340814	.340940	.341065	.341191	.341317	.341442	.341568	.341693	-0
4660	.341819	.341945	.342070	.342196	.342321	.342447	.342572	.342697	.342823	.342948	-0
4670	.343074	.343199	.343324	.343450	.343575	.343700	.343825	.343951	.344076	.344201	-0
4680	.344326	.344451	.344576	.344702	.344827	.344952	.345077	.345202	.345327	.345452	-0
4690	.345577	.345702	.345826	.345951	.346076	.346201	.346326	.346451	.346576	.346700	-0
4700	.346825	.346950	.347074	.347199	.347324	.347448	.347573	.347698	.347822	.347947	-0
4710	.348071	.348196	.348320	.348445	.348569	.348694	.348818	.348943	.349067	.349191	-0
4720	.349316	.349440	.349564	.349688	.349813	.349937	.350061	.350185	.350309	.350434	-0
4730	.350558	.350682	.350806	.350930	.351054	.351178	.351302	.351426	.351550	.351674	-0
4740	.351798	.351921	.352045	.352169	.352293	.352417	.352541	.352664	.352788	.352912	-0
4750	.353035	.353159	.353283	.353406	.353530	.353654	.353777	.353901	.354024	.354148	-0
4760	.354271	.354395	.354518	.354642	.354765	.354888	.355012	.355135	.355258	.355382	-0
4770	.355505	.355628	.355751	.355874	.355998	.356121	.356244	.356367	.356490	.356613	-0
4780	.356736	.356859	.356982	.357105	.357228	.357351	.357474	.357597	.357720	.357843	-0
4790	.357966	.358088	.358211	.358334	.358457	.358579	.358702	.358825	.358947	.359070	-0
4800	.359193	.359315	.359438	.359560	.359683	.359806	.359928	.360050	.360173	.360295	-0
4810	.360418	.360540	.360663	.360785	.360907	.361029	.361152	.361274	.361396	.361518	-0
4820	.361641	.361763	.361885	.362007	.362129	.362251	.362373	.362495	.362617	.362739	-0
4830	.362861	.362983	.363105	.363227	.363349	.363471	.363593	.363714	.363836	.363958	-0
4840	.364080	.364201	.364323	.364445	.364567	.364688	.364810	.364931	.365053	.365175	-0
4850	.365296	.365418	.365539	.365661	.365782	.365903	.366025	.366146	.366268	.366389	-0
4860	.366510	.366631	.366753	.366874	.366995	.367116	.367238	.367359	.367480	.367601	-0
4870	.367722	.367843	.367964	.368085	.368206	.368327	.368448	.368569	.368690	.368811	-0
4880	.368932	.369053	.369174	.369294	.369415	.369536	.369657	.369777	.369898	.370019	-0
4890	.370139	.370260	.370381	.370501	.370622	.370742	.370863	.370983	.371104	.371224	-0
4900	.371345	.371465	.371586	.371706	.371826	.371947	.372067	.372187	.372307	.372428	-0
4910	.372548	.372668	.372788	.372908	.373029	.373149	.373269	.373389	.373509	.373629	-0
4920	.373749	.373869	.373989	.374109	.374229	.374348	.374468	.374588	.374708	.374828	-0
4930	.374947	.375067	.375187	.375307	.375426	.375546	.375666	.375785	.375905	.376024	-0
4940	.376144	.376263	.376383	.376502	.376622	.376741	.376861	.376980	.377100	.377219	-0
4950	.377338	.377457	.377577	.377696	.377815	.377934	.378054	.378173	.378292	.378411	-0
4960	.378530	.378649	.378768	.378887	.379006	.379125	.379244	.379363	.379482	.379601	-0
4970	.379720	.379839	.379958	.380076	.380195	.380314	.380433	.380552	.380670	.380789	-0
4980	.380908	.381026	.381145	.381263	.381382	.381500	.381619	.381737	.381856	.381974	-0
4990	.382093	.382211	.382330	.382448	.382566	.382685	.382803	.382921	.383039	.383158	-0

$N(\lambda,T)/N_T(T)$

TABLE 15A

Blackbody: Relative Fractional Photon Emittance

λT(μ.°K)	0	1	2	3	4	5	6	7	8	9	m
5000	.383276	.383394	.383512	.383630	.383748	.383867	.383985	.384103	.384221	.384339	-0
5010	.384457	.384575	.384693	.384810	.384928	.385046	.385164	.385282	.385400	.385517	-0
5020	.385635	.385753	.385871	.385988	.386106	.386224	.386341	.386459	.386576	.386694	-0
5030	.386812	.386929	.387047	.387164	.387281	.387399	.387516	.387634	.387751	.387868	-0
5040	.387986	.388103	.388220	.388337	.388455	.388572	.388689	.388806	.388923	.389040	-0
5050	.389157	.389274	.389391	.389508	.389625	.389742	.389859	.389976	.390093	.390210	-0
5060	.390327	.390444	.390561	.390677	.390794	.390911	.391028	.391144	.391261	.391378	-0
5070	.391494	.391611	.391727	.391844	.391960	.392077	.392193	.392310	.392426	.392543	-0
5080	.392659	.392776	.392892	.393008	.393125	.393241	.393357	.393473	.393590	.393706	-0
5090	.393822	.393938	.394054	.394170	.394286	.394402	.394518	.394634	.394750	.394866	-0
5100	.394982	.395098	.395214	.395330	.395446	.395562	.395678	.395793	.395909	.396025	-0
5110	.396141	.396256	.396372	.396488	.396603	.396719	.396834	.396950	.397065	.397181	-0
5120	.397296	.397412	.397527	.397643	.397758	.397874	.397989	.398104	.398220	.398335	-0
5130	.398450	.398565	.398680	.398796	.398911	.399026	.399141	.399256	.399371	.399486	-0
5140	.399601	.399716	.399831	.399946	.400061	.400176	.400291	.400406	.400521	.400636	-0
5150	.400750	.400865	.400980	.401095	.401209	.401324	.401439	.401553	.401668	.401783	-0
5160	.401897	.402012	.402126	.402241	.402355	.402470	.402584	.402699	.402813	.402927	-0
5170	.403042	.403156	.403270	.403385	.403499	.403613	.403727	.403842	.403956	.404070	-0
5180	.404184	.404298	.404412	.404526	.404640	.404754	.404868	.404982	.405096	.405210	-0
5190	.405324	.405438	.405552	.405665	.405779	.405893	.406007	.406120	.406234	.406348	-0
5200	.406462	.406575	.406689	.406802	.406916	.407029	.407143	.407256	.407370	.407483	-0
5210	.407597	.407710	.407824	.407937	.408050	.408164	.408277	.408390	.408503	.408617	-0
5220	.408730	.408843	.408956	.409069	.409182	.409296	.409409	.409522	.409635	.409748	-0
5230	.409861	.409974	.410087	.410199	.410312	.410425	.410538	.410651	.410764	.410876	-0
5240	.410989	.411102	.411215	.411327	.411440	.411553	.411665	.411778	.411890	.412003	-0
5250	.412115	.412228	.412340	.412453	.412565	.412678	.412790	.412902	.413015	.413127	-0
5260	.413239	.413351	.413464	.413576	.413688	.413800	.413912	.414025	.414137	.414249	-0
5270	.414361	.414473	.414585	.414697	.414809	.414921	.415033	.415145	.415256	.415368	-0
5280	.415480	.415592	.415704	.415815	.415927	.416039	.416151	.416262	.416374	.416486	-0
5290	.416597	.416709	.416820	.416932	.417043	.417155	.417266	.417378	.417489	.417600	-0
5300	.417712	.417823	.417934	.418046	.418157	.418268	.418380	.418491	.418602	.418713	-0
5310	.418824	.418935	.419046	.419158	.419269	.419380	.419491	.419602	.419712	.419823	-0
5320	.419934	.420045	.420156	.420267	.420378	.420489	.420599	.420710	.420821	.420931	-0
5330	.421042	.421153	.421263	.421374	.421485	.421595	.421706	.421816	.421927	.422037	-0
5340	.422148	.422258	.422369	.422479	.422589	.422700	.422810	.422920	.423030	.423141	-0
5350	.423251	.423361	.423471	.423581	.423692	.423802	.423912	.424022	.424132	.424242	-0
5360	.424352	.424462	.424572	.424682	.424792	.424901	.425011	.425121	.425231	.425341	-0
5370	.425450	.425560	.425670	.425780	.425889	.425999	.426109	.426218	.426328	.426437	-0
5380	.426547	.426656	.426766	.426875	.426985	.427094	.427204	.427313	.427422	.427532	-0
5390	.427641	.427750	.427859	.427969	.428078	.428187	.428296	.428405	.428514	.428624	-0
5400	.428733	.428842	.428951	.429060	.429169	.429278	.429387	.429495	.429604	.429713	-0
5410	.429822	.429931	.430040	.430148	.430257	.430366	.430475	.430583	.430692	.430801	-0
5420	.430909	.431018	.431126	.431235	.431343	.431452	.431560	.431669	.431777	.431886	-0
5430	.431994	.432102	.432211	.432319	.432427	.432536	.432644	.432752	.432860	.432968	-0
5440	.433077	.433185	.433293	.433401	.433509	.433617	.433725	.433833	.433941	.434049	-0
5450	.434157	.434265	.434373	.434481	.434588	.434696	.434804	.434912	.435020	.435127	-0
5460	.435235	.435343	.435450	.435558	.435665	.435773	.435881	.435988	.436096	.436203	-0
5470	.436311	.436418	.436525	.436633	.436740	.436848	.436955	.437062	.437170	.437277	-0
5480	.437384	.437491	.437598	.437706	.437813	.437920	.438027	.438134	.438241	.438348	-0
5490	.438455	.438562	.438669	.438776	.438883	.438990	.439097	.439204	.439310	.439417	-0

TABLE 15A

Blackbody: Relative Fractional Photon Emittance

λT(μ.°K)	0	1	2	3	4	5	6	7	8	9	m
5500	.439524	.439631	.439737	.439844	.439951	.440058	.440164	.440271	.440377	.440484	-0
5510	.440591	.440697	.440804	.440910	.441016	.441123	.441229	.441336	.441442	.441548	-0
5520	.441655	.441761	.441867	.441974	.442080	.442186	.442292	.442398	.442505	.442611	-0
5530	.442717	.442823	.442929	.443035	.443141	.443247	.443353	.443459	.443565	.443671	-0
5540	.443776	.443882	.443988	.444094	.444200	.444305	.444411	.444517	.444623	.444728	-0
5550	.444834	.444939	.445045	.445151	.445256	.445362	.445467	.445573	.445678	.445784	-0
5560	.445889	.445994	.446100	.446205	.446310	.446416	.446521	.446626	.446731	.446837	-0
5570	.446942	.447047	.447152	.447257	.447362	.447467	.447572	.447678	.447783	.447887	-0
5580	.447992	.448097	.448202	.448307	.448412	.448517	.448622	.448726	.448831	.448936	-0
5590	.449041	.449145	.449250	.449355	.449459	.449564	.449669	.449773	.449878	.449982	-0
5600	.450087	.450191	.450296	.450400	.450505	.450609	.450713	.450818	.450922	.451026	-0
5610	.451131	.451235	.451339	.451443	.451547	.451652	.451756	.451860	.451964	.452068	-0
5620	.452172	.452276	.452380	.452484	.452588	.452692	.452796	.452900	.453004	.453107	-0
5630	.453211	.453315	.453419	.453523	.453626	.453730	.453834	.453937	.454041	.454145	-0
5640	.454248	.454352	.454455	.454559	.454662	.454766	.454869	.454973	.455076	.455180	-0
5650	.455283	.455386	.455490	.455593	.455696	.455799	.455903	.456006	.456109	.456212	-0
5660	.456315	.456418	.456522	.456625	.456728	.456831	.456934	.457037	.457140	.457243	-0
5670	.457345	.457448	.457551	.457654	.457757	.457860	.457963	.458065	.458168	.458271	-0
5680	.458373	.458476	.458579	.458681	.458784	.458886	.458989	.459092	.459194	.459297	-0
5690	.459399	.459501	.459604	.459706	.459809	.459911	.460013	.460116	.460218	.460320	-0
5700	.460422	.460525	.460627	.460729	.460831	.460933	.461035	.461137	.461240	.461342	-0
5710	.461444	.461546	.461648	.461749	.461851	.461953	.462055	.462157	.462259	.462361	-0
5720	.462462	.462564	.462666	.462768	.462869	.462971	.463073	.463174	.463276	.463378	-0
5730	.463479	.463581	.463682	.463784	.463885	.463987	.464088	.464189	.464291	.464392	-0
5740	.464494	.464595	.464696	.464797	.464899	.465000	.465101	.465202	.465303	.465405	-0
5750	.465506	.465607	.465708	.465809	.465910	.466011	.466112	.466213	.466314	.466415	-0
5760	.466516	.466616	.466717	.466818	.466919	.467020	.467120	.467221	.467322	.467423	-0
5770	.467523	.467624	.467725	.467825	.467926	.468026	.468127	.468227	.468328	.468428	-0
5780	.468529	.468629	.468730	.468830	.468930	.469031	.469131	.469231	.469331	.469432	-0
5790	.469532	.469632	.469732	.469832	.469933	.470033	.470133	.470233	.470333	.470433	-0
5800	.470533	.470633	.470733	.470833	.470933	.471033	.471132	.471232	.471332	.471432	-0
5810	.471532	.471631	.471731	.471831	.471930	.472030	.472130	.472229	.472329	.472429	-0
5820	.472528	.472628	.472727	.472827	.472926	.473026	.473125	.473224	.473324	.473423	-0
5830	.473522	.473622	.473721	.473820	.473920	.474019	.474118	.474217	.474316	.474415	-0
5840	.474515	.474614	.474713	.474812	.474911	.475010	.475109	.475208	.475307	.475405	-0
5850	.475504	.475603	.475702	.475801	.475900	.475998	.476097	.476196	.476295	.476393	-0
5860	.476492	.476591	.476689	.476788	.476886	.476985	.477084	.477182	.477281	.477379	-0
5870	.477477	.477576	.477674	.477773	.477871	.477969	.478068	.478166	.478264	.478362	-0
5880	.478461	.478559	.478657	.478755	.478853	.478951	.479050	.479148	.479246	.479344	-0
5890	.479442	.479540	.479638	.479736	.479833	.479931	.480029	.480127	.480225	.480323	-0
5900	.480420	.480518	.480616	.480714	.480811	.480909	.481007	.481104	.481202	.481300	-0
5910	.481397	.481495	.481592	.481690	.481787	.481885	.481982	.482079	.482177	.482274	-0
5920	.482372	.482469	.482566	.482663	.482761	.482858	.482955	.483052	.483149	.483247	-0
5930	.483344	.483441	.483538	.483635	.483732	.483829	.483926	.484023	.484120	.484217	-0
5940	.484314	.484411	.484507	.484604	.484701	.484798	.484895	.484991	.485088	.485185	-0
5950	.485282	.485378	.485475	.485572	.485668	.485765	.485861	.485958	.486054	.486151	-0
5960	.486247	.486344	.486440	.486536	.486633	.486729	.486826	.486922	.487018	.487114	-0
5970	.487211	.487307	.487403	.487499	.487595	.487692	.487788	.487884	.487980	.488076	-0
5980	.488172	.488268	.488364	.488460	.488556	.488652	.488748	.488844	.488939	.489035	-0
5990	.489131	.489227	.489323	.489418	.489514	.489610	.489706	.489801	.489897	.489992	-0

TABLE 15A

Blackbody: Relative Fractional Photon Emittance

λT(μ.°K)	0	1	2	3	4	5	6	7	8	9	m
6000	.490088	.490184	.490279	.490375	.490470	.490566	.490661	.490757	.490852	.490947	-0
6010	.491043	.491138	.491233	.491329	.491424	.491519	.491615	.491710	.491805	.491900	-0
6020	.491995	.492091	.492186	.492281	.492376	.492471	.492566	.492661	.492756	.492851	-0
6030	.492946	.493041	.493136	.493231	.493325	.493420	.493515	.493610	.493705	.493799	-0
6040	.493894	.493989	.494083	.494178	.494273	.494367	.494462	.494557	.494651	.494746	-0
6050	.494840	.494935	.495029	.495124	.495218	.495312	.495407	.495501	.495596	.495690	-0
6060	.495784	.495878	.495973	.496067	.496161	.496255	.496350	.496444	.496538	.496632	-0
6070	.496726	.496820	.496914	.497008	.497102	.497196	.497290	.497384	.497478	.497572	-0
6080	.497666	.497760	.497853	.497947	.498041	.498135	.498228	.498322	.498416	.498510	-0
6090	.498603	.498697	.498790	.498884	.498978	.499071	.499165	.499258	.499352	.499445	-0
6100	.499539	.499632	.499725	.499819	.499912	.500005	.500099	.500192	.500285	.500379	-0
6110	.500472	.500565	.500658	.500751	.500845	.500938	.501031	.501124	.501217	.501310	-0
6120	.501403	.501496	.501589	.501682	.501775	.501868	.501961	.502053	.502146	.502239	-0
6130	.502332	.502425	.502517	.502610	.502703	.502796	.502888	.502981	.503074	.503166	-0
6140	.503259	.503351	.503444	.503536	.503629	.503721	.503814	.503906	.503999	.504091	-0
6150	.504184	.504276	.504368	.504461	.504553	.504645	.504737	.504830	.504922	.505014	-0
6160	.505106	.505198	.505290	.505382	.505475	.505567	.505659	.505751	.505843	.505935	-0
6170	.506027	.506119	.506210	.506302	.506394	.506486	.506578	.506670	.506761	.506853	-0
6180	.506945	.507037	.507128	.507220	.507312	.507403	.507495	.507587	.507678	.507770	-0
6190	.507861	.507953	.508044	.508136	.508227	.508319	.508410	.508501	.508593	.508684	-0
6200	.508775	.508867	.508958	.509049	.509141	.509232	.509323	.509414	.509505	.509596	-0
6210	.509688	.509779	.509870	.509961	.510052	.510143	.510234	.510325	.510416	.510507	-0
6220	.510598	.510688	.510779	.510870	.510961	.511052	.511142	.511233	.511324	.511415	-0
6230	.511505	.511596	.511687	.511777	.511868	.511959	.512049	.512140	.512230	.512321	-0
6240	.512411	.512502	.512592	.512683	.512773	.512863	.512954	.513044	.513134	.513225	-0
6250	.513315	.513405	.513495	.513586	.513676	.513766	.513856	.513946	.514036	.514126	-0
6260	.514217	.514307	.514397	.514487	.514577	.514667	.514756	.514846	.514936	.515026	-0
6270	.515116	.515206	.515296	.515386	.515475	.515565	.515655	.515744	.515834	.515924	-0
6280	.516014	.516103	.516193	.516282	.516372	.516461	.516551	.516641	.516730	.516819	-0
6290	.516909	.516998	.517088	.517177	.517267	.517356	.517445	.517534	.517624	.517713	-0
6300	.517802	.517891	.517981	.518070	.518159	.518248	.518337	.518426	.518515	.518604	-0
6310	.518694	.518783	.518872	.518961	.519049	.519138	.519227	.519316	.519405	.519494	-0
6320	.519583	.519672	.519760	.519849	.519938	.520027	.520115	.520204	.520293	.520381	-0
6330	.520470	.520558	.520647	.520736	.520824	.520913	.521001	.521090	.521178	.521267	-0
6340	.521355	.521443	.521532	.521620	.521708	.521797	.521885	.521973	.522062	.522150	-0
6350	.522238	.522326	.522414	.522503	.522591	.522679	.522767	.522855	.522943	.523031	-0
6360	.523119	.523207	.523295	.523383	.523471	.523559	.523647	.523735	.523822	.523910	-0
6370	.523998	.524086	.524174	.524261	.524349	.524437	.524524	.524612	.524700	.524787	-0
6380	.524875	.524963	.525050	.525138	.525225	.525313	.525400	.525488	.525575	.525662	-0
6390	.525750	.525837	.525925	.526012	.526099	.526187	.526274	.526361	.526448	.526535	-0
6400	.526623	.526710	.526797	.526884	.526971	.527058	.527145	.527232	.527320	.527407	-0
6410	.527494	.527581	.527667	.527754	.527841	.527928	.528015	.528102	.528189	.528276	-0
6420	.528362	.528449	.528536	.528623	.528709	.528796	.528883	.528969	.529056	.529143	-0
6430	.529229	.529316	.529402	.529489	.529575	.529662	.529748	.529835	.529921	.530008	-0
6440	.530094	.530180	.530267	.530353	.530439	.530526	.530612	.530698	.530784	.530871	-0
6450	.530957	.531043	.531129	.531215	.531301	.531387	.531473	.531559	.531645	.531732	-0
6460	.531818	.531903	.531989	.532075	.532161	.532247	.532333	.532419	.532505	.532591	-0
6470	.532676	.532762	.532848	.532934	.533019	.533105	.533191	.533276	.533362	.533447	-0
6480	.533533	.533619	.533704	.533790	.533875	.533961	.534046	.534132	.534217	.534302	-0
6490	.534388	.534473	.534559	.534644	.534729	.534814	.534900	.534985	.535070	.535155	-0

TABLE 15A

Blackbody: Relative Fractional Photon Emittance

λT(μ.°K)	0	1	2	3	4	5	6	7	8	9	m
6500	.535241	.535326	.535411	.535496	.535581	.535666	.535751	.535836	.535921	.536006	−0
6510	.536091	.536176	.536261	.536346	.536431	.536516	.536601	.536686	.536771	.536855	−0
6520	.536940	.537025	.537110	.537195	.537279	.537364	.537449	.537533	.537618	.537703	−0
6530	.537787	.537872	.537956	.538041	.538125	.538210	.538294	.538379	.538463	.538548	−0
6540	.538632	.538716	.538801	.538885	.538969	.539054	.539138	.539222	.539306	.539391	−0
6550	.539475	.539559	.539643	.539727	.539812	.539896	.539980	.540064	.540148	.540232	−0
6560	.540316	.540400	.540484	.540568	.540652	.540736	.540819	.540903	.540987	.541071	−0
6570	.541155	.541239	.541322	.541406	.541490	.541574	.541657	.541741	.541825	.541908	−0
6580	.541992	.542075	.542159	.542243	.542326	.542410	.542493	.542577	.542660	.542744	−0
6590	.542827	.542910	.542994	.543077	.543160	.543244	.543327	.543410	.543494	.543577	−0
6600	.543660	.543743	.543826	.543910	.543993	.544076	.544159	.544242	.544325	.544408	−0
6610	.544491	.544574	.544657	.544740	.544823	.544906	.544989	.545072	.545155	.545238	−0
6620	.545320	.545403	.545486	.545569	.545652	.545734	.545817	.545900	.545983	.546065	−0
6630	.546148	.546230	.546313	.546396	.546478	.546561	.546643	.546726	.546808	.546891	−0
6640	.546973	.547056	.547138	.547220	.547303	.547385	.547467	.547550	.547632	.547714	−0
6650	.547797	.547879	.547961	.548043	.548125	.548208	.548290	.548372	.548454	.548536	−0
6660	.548618	.548700	.548782	.548864	.548946	.549028	.549110	.549192	.549274	.549356	−0
6670	.549438	.549520	.549601	.549683	.549765	.549847	.549929	.550010	.550092	.550174	−0
6680	.550255	.550337	.550419	.550500	.550582	.550664	.550745	.550827	.550908	.550990	−0
6690	.551071	.551153	.551234	.551316	.551397	.551478	.551560	.551641	.551722	.551804	−0
6700	.551885	.551966	.552048	.552129	.552210	.552291	.552372	.552454	.552535	.552616	−0
6710	.552697	.552778	.552859	.552940	.553021	.553102	.553183	.553264	.553345	.553426	−0
6720	.553507	.553588	.553669	.553750	.553831	.553911	.553992	.554073	.554154	.554235	−0
6730	.554315	.554396	.554477	.554557	.554638	.554719	.554799	.554880	.554960	.555041	−0
6740	.555122	.555202	.555283	.555363	.555444	.555524	.555604	.555685	.555765	.555846	−0
6750	.555926	.556006	.556087	.556167	.556247	.556327	.556408	.556488	.556568	.556648	−0
6760	.556728	.556809	.556889	.556969	.557049	.557129	.557209	.557289	.557369	.557449	−0
6770	.557529	.557609	.557689	.557769	.557849	.557929	.558009	.558088	.558168	.558248	−0
6780	.558328	.558408	.558487	.558567	.558647	.558727	.558806	.558886	.558966	.559045	−0
6790	.559125	.559204	.559284	.559363	.559443	.559522	.559602	.559681	.559761	.559840	−0
6800	.559920	.559999	.560079	.560158	.560237	.560317	.560396	.560475	.560554	.560634	−0
6810	.560713	.560792	.560871	.560951	.561030	.561109	.561188	.561267	.561346	.561425	−0
6820	.561504	.561583	.561662	.561741	.561820	.561899	.561978	.562057	.562136	.562215	−0
6830	.562294	.562373	.562451	.562530	.562609	.562688	.562766	.562845	.562924	.563003	−0
6840	.563081	.563160	.563239	.563317	.563396	.563474	.563553	.563631	.563710	.563789	−0
6850	.563867	.563945	.564024	.564102	.564181	.564259	.564338	.564416	.564494	.564573	−0
6860	.564651	.564729	.564808	.564886	.564964	.565042	.565120	.565199	.565277	.565355	−0
6870	.565433	.565511	.565589	.565667	.565745	.565823	.565901	.565979	.566057	.566135	−0
6880	.566213	.566291	.566369	.566447	.566525	.566603	.566681	.566758	.566836	.566914	−0
6890	.566992	.567069	.567147	.567225	.567303	.567380	.567458	.567535	.567613	.567691	−0
6900	.567768	.567846	.567923	.568001	.568078	.568156	.568233	.568311	.568388	.568466	−0
6910	.568543	.568620	.568698	.568775	.568852	.568930	.569007	.569084	.569162	.569239	−0
6920	.569316	.569393	.569470	.569548	.569625	.569702	.569779	.569856	.569933	.570010	−0
6930	.570087	.570164	.570241	.570318	.570395	.570472	.570549	.570626	.570703	.570780	−0
6940	.570857	.570933	.571010	.571087	.571164	.571240	.571317	.571394	.571471	.571547	−0
6950	.571624	.571701	.571777	.571854	.571931	.572007	.572084	.572160	.572237	.572313	−0
6960	.572390	.572466	.572543	.572619	.572696	.572772	.572848	.572925	.573001	.573077	−0
6970	.573154	.573230	.573306	.573383	.573459	.573535	.573611	.573687	.573764	.573840	−0
6980	.573916	.573992	.574068	.574144	.574220	.574296	.574372	.574448	.574524	.574600	−0
6990	.574676	.574752	.574828	.574904	.574980	.575056	.575132	.575207	.575283	.575359	−0

TABLE 15A

Blackbody: Relative Fractional Photon Emittance

λT(μ.°K)	0	1	2	3	4	5	6	7	8	9	m
7000	.575435	.575511	.575586	.575662	.575738	.575813	.575889	.575965	.576040	.576116	-0
7010	.576192	.576267	.576343	.576418	.576494	.576569	.576645	.576720	.576796	.576871	-0
7020	.576947	.577022	.577097	.577173	.577248	.577323	.577399	.577474	.577549	.577625	-0
7030	.577700	.577775	.577850	.577926	.578001	.578076	.578151	.578226	.578301	.578376	-0
7040	.578451	.578526	.578601	.578676	.578751	.578826	.578901	.578976	.579051	.579126	-0
7050	.579201	.579276	.579351	.579426	.579501	.579575	.579650	.579725	.579800	.579874	-0
7060	.579949	.580024	.580099	.580173	.580248	.580322	.580397	.580472	.580546	.580621	-0
7070	.580695	.580770	.580844	.580919	.580993	.581068	.581142	.581217	.581291	.581365	-0
7080	.581440	.581514	.581588	.581663	.581737	.581811	.581886	.581960	.582034	.582108	-0
7090	.582183	.582257	.582331	.582405	.582479	.582553	.582627	.582701	.582775	.582850	-0
7100	.582924	.582998	.583072	.583145	.583219	.583293	.583367	.583441	.583515	.583589	-0
7110	.583663	.583737	.583810	.583884	.583958	.584032	.584105	.584179	.584253	.584327	-0
7120	.584400	.584474	.584548	.584621	.584695	.584768	.584842	.584916	.584989	.585063	-0
7130	.585136	.585210	.585283	.585356	.585430	.585503	.585577	.585650	.585723	.585797	-0
7140	.585870	.585943	.586017	.586090	.586163	.586237	.586310	.586383	.586456	.586529	-0
7150	.586602	.586676	.586749	.586822	.586895	.586968	.587041	.587114	.587187	.587260	-0
7160	.587333	.587406	.587479	.587552	.587625	.587698	.587771	.587844	.587916	.587989	-0
7170	.588062	.588135	.588208	.588280	.588353	.588426	.588499	.588571	.588644	.588717	-0
7180	.588789	.588862	.588934	.589007	.589080	.589152	.589225	.589297	.589370	.589442	-0
7190	.589515	.589587	.589660	.589732	.589805	.589877	.589949	.590022	.590094	.590166	-0
7200	.590239	.590311	.590383	.590455	.590528	.590600	.590672	.590744	.590816	.590889	-0
7210	.590961	.591033	.591105	.591177	.591249	.591321	.591393	.591465	.591537	.591609	-0
7220	.591681	.591753	.591825	.591897	.591969	.592041	.592113	.592184	.592256	.592328	-0
7230	.592400	.592472	.592543	.592615	.592687	.592759	.592830	.592902	.592974	.593045	-0
7240	.593117	.593189	.593260	.593332	.593403	.593475	.593546	.593618	.593689	.593761	-0
7250	.593832	.593904	.593975	.594047	.594118	.594189	.594261	.594332	.594403	.594475	-0
7260	.594546	.594617	.594689	.594760	.594831	.594902	.594973	.595045	.595116	.595187	-0
7270	.595258	.595329	.595400	.595471	.595542	.595613	.595684	.595755	.595826	.595897	-0
7280	.595968	.596039	.596110	.596181	.596252	.596323	.596394	.596465	.596535	.596606	-0
7290	.596677	.596748	.596819	.596889	.596960	.597031	.597101	.597172	.597243	.597313	-0
7300	.597384	.597455	.597525	.597596	.597666	.597737	.597808	.597878	.597949	.598019	-0
7310	.598090	.598160	.598230	.598301	.598371	.598442	.598512	.598582	.598653	.598723	-0
7320	.598793	.598864	.598934	.599004	.599074	.599144	.599215	.599285	.599355	.599425	-0
7330	.599495	.599565	.599636	.599706	.599776	.599846	.599916	.599986	.600056	.600126	-0
7340	.600196	.600266	.600336	.600406	.600475	.600545	.600615	.600685	.600755	.600825	-0
7350	.600895	.600964	.601034	.601104	.601174	.601243	.601313	.601383	.601452	.601522	-0
7360	.601592	.601661	.601731	.601801	.601870	.601940	.602009	.602079	.602148	.602218	-0
7370	.602287	.602357	.602426	.602496	.602565	.602634	.602704	.602773	.602843	.602912	-0
7380	.602981	.603051	.603120	.603189	.603258	.603328	.603397	.603466	.603535	.603604	-0
7390	.603674	.603743	.603812	.603881	.603950	.604019	.604088	.604157	.604226	.604295	-0
7400	.604364	.604433	.604502	.604571	.604640	.604709	.604778	.604847	.604916	.604984	-0
7410	.605053	.605122	.605191	.605260	.605328	.605397	.605466	.605535	.605603	.605672	-0
7420	.605741	.605809	.605878	.605947	.606015	.606084	.606152	.606221	.606290	.606358	-0
7430	.606427	.606495	.606564	.606632	.606700	.606769	.606837	.606906	.606974	.607042	-0
7440	.607111	.607179	.607247	.607316	.607384	.607452	.607521	.607589	.607657	.607725	-0
7450	.607793	.607862	.607930	.607998	.608066	.608134	.608202	.608270	.608338	.608407	-0
7460	.608475	.608543	.608611	.608679	.608747	.608814	.608882	.608950	.609018	.609086	-0
7470	.609154	.609222	.609290	.609358	.609425	.609493	.609561	.609629	.609696	.609764	-0
7480	.609832	.609900	.609967	.610035	.610103	.610170	.610238	.610305	.610373	.610441	-0
7490	.610508	.610576	.610643	.610711	.610778	.610846	.610913	.610981	.611048	.611116	-0

TABLE 15A

Blackbody: Relative Fractional Photon Emittance

λT(μ.°K)	0	1	2	3	4	5	6	7	8	9	m
7500	.611183	.611250	.611318	.611385	.611452	.611520	.611587	.611654	.611722	.611789	-0
7510	.611856	.611923	.611991	.612058	.612125	.612192	.612259	.612326	.612393	.612461	-0
7520	.612528	.612595	.612662	.612729	.612796	.612863	.612930	.612997	.613064	.613131	-0
7530	.613198	.613265	.613331	.613398	.613465	.613532	.613599	.613666	.613733	.613799	-0
7540	.613866	.613933	.614000	.614066	.614133	.614200	.614266	.614333	.614400	.614466	-0
7550	.614533	.614600	.614666	.614733	.614799	.614866	.614932	.614999	.615065	.615132	-0
7560	.615198	.615265	.615331	.615398	.615464	.615530	.615597	.615663	.615730	.615796	-0
7570	.615862	.615928	.615995	.616061	.616127	.616193	.616260	.616326	.616392	.616458	-0
7580	.616524	.616591	.616657	.616723	.616789	.616855	.616921	.616987	.617053	.617119	-0
7590	.617185	.617251	.617317	.617383	.617449	.617515	.617581	.617647	.617713	.617778	-0
7600	.617844	.617910	.617976	.618042	.618107	.618173	.618239	.618305	.618370	.618436	-0
7610	.618502	.618568	.618633	.618699	.618765	.618830	.618896	.618961	.619027	.619092	-0
7620	.619158	.619224	.619289	.619355	.619420	.619485	.619551	.619616	.619682	.619747	-0
7630	.619813	.619878	.619943	.620009	.620074	.620139	.620205	.620270	.620335	.620400	-0
7640	.620466	.620531	.620596	.620661	.620726	.620792	.620857	.620922	.620987	.621052	-0
7650	.621117	.621182	.621247	.621312	.621377	.621442	.621507	.621572	.621637	.621702	-0
7660	.621767	.621832	.621897	.621962	.622027	.622092	.622156	.622221	.622286	.622351	-0
7670	.622416	.622480	.622545	.622610	.622675	.622739	.622804	.622869	.622933	.622998	-0
7680	.623063	.623127	.623192	.623256	.623321	.623386	.623450	.623515	.623579	.623644	-0
7690	.623708	.623773	.623837	.623902	.623966	.624030	.624095	.624159	.624223	.624288	-0
7700	.624352	.624416	.624481	.624545	.624609	.624674	.624738	.624802	.624866	.624930	-0
7710	.624995	.625059	.625123	.625187	.625251	.625315	.625379	.625444	.625508	.625572	-0
7720	.625636	.625700	.625764	.625828	.625892	.625956	.626020	.626083	.626147	.626211	-0
7730	.626275	.626339	.626403	.626467	.626531	.626594	.626658	.626722	.626786	.626849	-0
7740	.626913	.626977	.627041	.627104	.627168	.627232	.627295	.627359	.627423	.627486	-0
7750	.627550	.627613	.627677	.627740	.627804	.627867	.627931	.627994	.628058	.628121	-0
7760	.628185	.628248	.628312	.628375	.628438	.628502	.628565	.628629	.628692	.628755	-0
7770	.628818	.628882	.628945	.629008	.629071	.629135	.629198	.629261	.629324	.629387	-0
7780	.629451	.629514	.629577	.629640	.629703	.629766	.629829	.629892	.629955	.630018	-0
7790	.630081	.630144	.630207	.630270	.630333	.630396	.630459	.630522	.630585	.630648	-0
7800	.630710	.630773	.630836	.630899	.630962	.631025	.631087	.631150	.631213	.631275	-0
7810	.631338	.631401	.631464	.631526	.631589	.631652	.631714	.631777	.631839	.631902	-0
7820	.631965	.632027	.632090	.632152	.632215	.632277	.632340	.632402	.632464	.632527	-0
7830	.632589	.632652	.632714	.632777	.632839	.632901	.632964	.633026	.633088	.633150	-0
7840	.633213	.633275	.633337	.633400	.633462	.633524	.633586	.633648	.633710	.633773	-0
7850	.633835	.633897	.633959	.634021	.634083	.634145	.634207	.634269	.634331	.634393	-0
7860	.634455	.634517	.634579	.634641	.634703	.634765	.634827	.634889	.634951	.635012	-0
7870	.635074	.635136	.635198	.635260	.635322	.635383	.635445	.635507	.635569	.635630	-0
7880	.635692	.635754	.635815	.635877	.635939	.636000	.636062	.636123	.636185	.636247	-0
7890	.636308	.636370	.636431	.636493	.636554	.636616	.636677	.636739	.636800	.636862	-0
7900	.636923	.636984	.637046	.637107	.637169	.637230	.637291	.637353	.637414	.637475	-0
7910	.637536	.637598	.637659	.637720	.637781	.637843	.637904	.637965	.638026	.638087	-0
7920	.638148	.638209	.638271	.638332	.638393	.638454	.638515	.638576	.638637	.638698	-0
7930	.638759	.638820	.638881	.638942	.639003	.639064	.639125	.639185	.639246	.639307	-0
7940	.639368	.639429	.639490	.639551	.639611	.639672	.639733	.639794	.639854	.639915	-0
7950	.639976	.640036	.640097	.640158	.640218	.640279	.640340	.640400	.640461	.640522	-0
7960	.640582	.640643	.640703	.640764	.640824	.640885	.640945	.641006	.641066	.641127	-0
7970	.641187	.641247	.641308	.641368	.641429	.641489	.641549	.641610	.641670	.641730	-0
7980	.641791	.641851	.641911	.641971	.642032	.642092	.642152	.642212	.642272	.642333	-0
7990	.642393	.642453	.642513	.642573	.642633	.642693	.642753	.642813	.642873	.642933	-0

TABLE 15A

Blackbody: Relative Fractional Photon Emittance

λT(μ.°K)	0	1	2	3	4	5	6	7	8	9	m
8000	.642994	.643054	.643113	.643173	.643233	.643293	.643353	.643413	.643473	.643533	-0
8010	.643593	.643653	.643713	.643772	.643832	.643892	.643952	.644012	.644071	.644131	-0
8020	.644191	.644251	.644310	.644370	.644430	.644489	.644549	.644609	.644668	.644728	-0
8030	.644788	.644847	.644907	.644966	.645026	.645085	.645145	.645204	.645264	.645323	-0
8040	.645383	.645442	.645502	.645561	.645620	.645680	.645739	.645799	.645858	.645917	-0
8050	.645977	.646036	.646095	.646155	.646214	.646273	.646332	.646392	.646451	.646510	-0
8060	.646569	.646628	.646688	.646747	.646806	.646865	.646924	.646983	.647042	.647101	-0
8070	.647160	.647219	.647278	.647337	.647396	.647455	.647514	.647573	.647632	.647691	-0
8080	.647750	.647809	.647868	.647927	.647986	.648045	.648103	.648162	.648221	.648280	-0
8090	.648339	.648397	.648456	.648515	.648574	.648632	.648691	.648750	.648808	.648867	-0
8100	.648926	.648984	.649043	.649102	.649160	.649219	.649277	.649336	.649394	.649453	-0
8110	.649511	.649570	.649628	.649687	.649745	.649804	.649862	.649921	.649979	.650037	-0
8120	.650096	.650154	.650213	.650271	.650329	.650387	.650446	.650504	.650562	.650621	-0
8130	.650679	.650737	.650795	.650853	.650912	.650970	.651028	.651086	.651144	.651202	-0
8140	.651261	.651319	.651377	.651435	.651493	.651551	.651609	.651667	.651725	.651783	-0
8150	.651841	.651899	.651957	.652015	.652073	.652131	.652189	.652246	.652304	.652362	-0
8160	.652420	.652478	.652536	.652593	.652651	.652709	.652767	.652825	.652882	.652940	-0
8170	.652998	.653055	.653113	.653171	.653228	.653286	.653344	.653401	.653459	.653517	-0
8180	.653574	.653632	.653689	.653747	.653804	.653862	.653919	.653977	.654034	.654092	-0
8190	.654149	.654207	.654264	.654321	.654379	.654436	.654494	.654551	.654608	.654666	-0
8200	.654723	.654780	.654838	.654895	.654952	.655009	.655067	.655124	.655181	.655238	-0
8210	.655295	.655353	.655410	.655467	.655524	.655581	.655638	.655695	.655752	.655810	-0
8220	.655867	.655924	.655981	.656038	.656095	.656152	.656209	.656266	.656323	.656380	-0
8230	.656436	.656493	.656550	.656607	.656664	.656721	.656778	.656835	.656891	.656948	-0
8240	.657005	.657062	.657119	.657175	.657232	.657289	.657345	.657402	.657459	.657516	-0
8250	.657572	.657629	.657685	.657742	.657799	.657855	.657912	.657968	.658025	.658082	-0
8260	.658138	.658195	.658251	.658308	.658364	.658421	.658477	.658534	.658590	.658646	-0
8270	.658703	.658759	.658816	.658872	.658928	.658985	.659041	.659097	.659154	.659210	-0
8280	.659266	.659322	.659379	.659435	.659491	.659547	.659604	.659660	.659716	.659772	-0
8290	.659828	.659884	.659940	.659997	.660053	.660109	.660165	.660221	.660277	.660333	-0
8300	.660389	.660445	.660501	.660557	.660613	.660669	.660725	.660781	.660837	.660893	-0
8310	.660949	.661004	.661060	.661116	.661172	.661228	.661284	.661339	.661395	.661451	-0
8320	.661507	.661563	.661618	.661674	.661730	.661785	.661841	.661897	.661952	.662008	-0
8330	.662064	.662119	.662175	.662231	.662286	.662342	.662397	.662453	.662508	.662564	-0
8340	.662619	.662675	.662730	.662786	.662841	.662897	.662952	.663008	.663063	.663119	-0
8350	.663174	.663229	.663285	.663340	.663395	.663451	.663506	.663561	.663617	.663672	-0
8360	.663727	.663782	.663838	.663893	.663948	.664003	.664058	.664114	.664169	.664224	-0
8370	.664279	.664334	.664389	.664444	.664499	.664554	.664610	.664665	.664720	.664775	-0
8380	.664830	.664885	.664940	.664995	.665050	.665105	.665159	.665214	.665269	.665324	-0
8390	.665379	.665434	.665489	.665544	.665598	.665653	.665708	.665763	.665818	.665872	-0
8400	.665927	.665982	.666037	.666091	.666146	.666201	.666256	.666310	.666365	.666419	-0
8410	.666474	.666529	.666583	.666638	.666693	.666747	.666802	.666856	.666911	.666965	-0
8420	.667020	.667074	.667129	.667183	.667238	.667292	.667347	.667401	.667455	.667510	-0
8430	.667564	.667619	.667673	.667727	.667782	.667836	.667890	.667945	.667999	.668053	-0
8440	.668107	.668162	.668216	.668270	.668324	.668379	.668433	.668487	.668541	.668595	-0
8450	.668649	.668704	.668758	.668812	.668866	.668920	.668974	.669028	.669082	.669136	-0
8460	.669190	.669244	.669298	.669352	.669406	.669460	.669514	.669568	.669622	.669676	-0
8470	.669730	.669784	.669837	.669891	.669945	.669999	.670053	.670107	.670160	.670214	-0
8480	.670268	.670322	.670375	.670429	.670483	.670537	.670590	.670644	.670698	.670751	-0
8490	.670805	.670859	.670912	.670966	.671020	.671073	.671127	.671180	.671234	.671287	-0

TABLE 15A

Blackbody: Relative Fractional Photon Emittance

λT(μ.°K)	0	1	2	3	4	5	6	7	8	9	m
8500	.671341	.671394	.671448	.671501	.671555	.671608	.671662	.671715	.671769	.671822	-0
8510	.671876	.671929	.671982	.672036	.672089	.672142	.672196	.672249	.672302	.672356	-0
8520	.672409	.672462	.672515	.672569	.672622	.672675	.672728	.672782	.672835	.672888	-0
8530	.672941	.672994	.673047	.673101	.673154	.673207	.673260	.673313	.673366	.673419	-0
8540	.673472	.673525	.673578	.673631	.673684	.673737	.673790	.673843	.673896	.673949	-0
8550	.674002	.674055	.674108	.674161	.674214	.674266	.674319	.674372	.674425	.674478	-0
8560	.674531	.674583	.674636	.674689	.674742	.674795	.674847	.674900	.674953	.675005	-0
8570	.675058	.675111	.675163	.675216	.675269	.675321	.675374	.675427	.675479	.675532	-0
8580	.675584	.675637	.675689	.675742	.675794	.675847	.675900	.675952	.676004	.676057	-0
8590	.676109	.676162	.676214	.676267	.676319	.676371	.676424	.676476	.676529	.676581	-0
8600	.676633	.676686	.676738	.676790	.676842	.676895	.676947	.676999	.677052	.677104	-0
8610	.677156	.677208	.677260	.677313	.677365	.677417	.677469	.677521	.677573	.677625	-0
8620	.677677	.677730	.677782	.677834	.677886	.677938	.677990	.678042	.678094	.678146	-0
8630	.678198	.678250	.678302	.678354	.678406	.678458	.678509	.678561	.678613	.678665	-0
8640	.678717	.678769	.678821	.678873	.678924	.678976	.679028	.679080	.679131	.679183	-0
8650	.679235	.679287	.679338	.679390	.679442	.679494	.679545	.679597	.679649	.679700	-0
8660	.679752	.679803	.679855	.679907	.679958	.680010	.680061	.680113	.680164	.680216	-0
8670	.680268	.680319	.680370	.680422	.680473	.680525	.680576	.680628	.680679	.680731	-0
8680	.680782	.680833	.680885	.680936	.680988	.681039	.681090	.681141	.681193	.681244	-0
8690	.681295	.681347	.681398	.681449	.681500	.681552	.681603	.681654	.681705	.681756	-0
8700	.681808	.681859	.681910	.681961	.682012	.682063	.682114	.682165	.682217	.682268	-0
8710	.682319	.682370	.682421	.682472	.682523	.682574	.682625	.682676	.682727	.682778	-0
8720	.682829	.682879	.682930	.682981	.683032	.683083	.683134	.683185	.683236	.683286	-0
8730	.683337	.683388	.683439	.683490	.683541	.683591	.683642	.683693	.683744	.683794	-0
8740	.683845	.683896	.683946	.683997	.684048	.684098	.684149	.684200	.684250	.684301	-0
8750	.684351	.684402	.684453	.684503	.684554	.684604	.684655	.684705	.684756	.684806	-0
8760	.684857	.684907	.684958	.685008	.685059	.685109	.685159	.685210	.685260	.685311	-0
8770	.685361	.685411	.685462	.685512	.685562	.685613	.685663	.685713	.685764	.685814	-0
8780	.685864	.685914	.685965	.686015	.686065	.686115	.686165	.686216	.686266	.686316	-0
8790	.686366	.686416	.686466	.686516	.686567	.686617	.686667	.686717	.686767	.686817	-0
8800	.686867	.686917	.686967	.687017	.687067	.687117	.687167	.687217	.687267	.687317	-0
8810	.687367	.687417	.687466	.687516	.687566	.687616	.687666	.687716	.687766	.687815	-0
8820	.687865	.687915	.687965	.688015	.688064	.688114	.688164	.688214	.688263	.688313	-0
8830	.688363	.688412	.688462	.688512	.688561	.688611	.688661	.688710	.688760	.688810	-0
8840	.688859	.688909	.688958	.689008	.689057	.689107	.689156	.689206	.689255	.689305	-0
8850	.689354	.689404	.689453	.689503	.689552	.689602	.689651	.689700	.689750	.689799	-0
8860	.689849	.689898	.689947	.689997	.690046	.690095	.690145	.690194	.690243	.690292	-0
8870	.690342	.690391	.690440	.690489	.690539	.690588	.690637	.690686	.690735	.690785	-0
8880	.690834	.690883	.690932	.690981	.691030	.691079	.691128	.691177	.691226	.691276	-0
8890	.691325	.691374	.691423	.691472	.691521	.691570	.691619	.691668	.691716	.691765	-0
8900	.691814	.691863	.691912	.691961	.692010	.692059	.692108	.692157	.692205	.692254	-0
8910	.692303	.692352	.692401	.692449	.692498	.692547	.692596	.692644	.692693	.692742	-0
8920	.692791	.692839	.692888	.692937	.692985	.693034	.693083	.693131	.693180	.693229	-0
8930	.693277	.693326	.693374	.693423	.693471	.693520	.693569	.693617	.693666	.693714	-0
8940	.693763	.693811	.693860	.693908	.693956	.694005	.694053	.694102	.694150	.694199	-0
8950	.694247	.694295	.694344	.694392	.694440	.694489	.694537	.694585	.694634	.694682	-0
8960	.694730	.694779	.694827	.694875	.694923	.694972	.695020	.695068	.695116	.695164	-0
8970	.695212	.695261	.695309	.695357	.695405	.695453	.695501	.695549	.695597	.695646	-0
8980	.695694	.695742	.695790	.695838	.695886	.695934	.695982	.696030	.696078	.696126	-0
8990	.696174	.696222	.696270	.696318	.696365	.696413	.696461	.696509	.696557	.696605	-0

TABLE 15A

Blackbody: Relative Fractional Photon Emittance

λT(μ.°K)	0	1	2	3	4	5	6	7	8	9	m
9000	.696653	.696701	.696748	.696796	.696844	.696892	.696940	.696987	.697035	.697083	-0
9010	.697131	.697178	.697226	.697274	.697322	.697369	.697417	.697465	.697512	.697560	-0
9020	.697608	.697655	.697703	.697750	.697798	.697846	.697893	.697941	.697988	.698036	-0
9030	.698083	.698131	.698178	.698226	.698273	.698321	.698368	.698416	.698463	.698511	-0
9040	.698558	.698606	.698653	.698700	.698748	.698795	.698843	.698890	.698937	.698985	-0
9050	.699032	.699079	.699127	.699174	.699221	.699268	.699316	.699363	.699410	.699457	-0
9060	.699505	.699552	.699599	.699646	.699693	.699741	.699788	.699835	.699882	.699929	-0
9070	.699976	.700023	.700070	.700118	.700165	.700212	.700259	.700306	.700353	.700400	-0
9080	.700447	.700494	.700541	.700588	.700635	.700682	.700729	.700776	.700823	.700870	-0
9090	.700916	.700963	.701010	.701057	.701104	.701151	.701198	.701245	.701291	.701338	-0
9100	.701385	.701432	.701479	.701525	.701572	.701619	.701666	.701712	.701759	.701806	-0
9110	.701852	.701899	.701946	.701993	.702039	.702086	.702132	.702179	.702226	.702272	-0
9120	.702319	.702366	.702412	.702459	.702505	.702552	.702598	.702645	.702691	.702738	-0
9130	.702784	.702831	.702877	.702924	.702970	.703017	.703063	.703110	.703156	.703202	-0
9140	.703249	.703295	.703342	.703388	.703434	.703481	.703527	.703573	.703620	.703666	-0
9150	.703712	.703758	.703805	.703851	.703897	.703943	.703990	.704036	.704082	.704128	-0
9160	.704175	.704221	.704267	.704313	.704359	.704405	.704451	.704498	.704544	.704590	-0
9170	.704636	.704682	.704728	.704774	.704820	.704866	.704912	.704958	.705004	.705050	-0
9180	.705096	.705142	.705188	.705234	.705280	.705326	.705372	.705418	.705464	.705510	-0
9190	.705556	.705601	.705647	.705693	.705739	.705785	.705831	.705876	.705922	.705968	-0
9200	.706014	.706060	.706105	.706151	.706197	.706243	.706288	.706334	.706380	.706426	-0
9210	.706471	.706517	.706563	.706608	.706654	.706699	.706745	.706791	.706836	.706882	-0
9220	.706928	.706973	.707019	.707064	.707110	.707155	.707201	.707246	.707292	.707337	-0
9230	.707383	.707428	.707474	.707519	.707565	.707610	.707656	.707701	.707746	.707792	-0
9240	.707837	.707882	.707928	.707973	.708019	.708064	.708109	.708155	.708200	.708245	-0
9250	.708290	.708336	.708381	.708426	.708471	.708517	.708562	.708607	.708652	.708698	-0
9260	.708743	.708788	.708833	.708878	.708923	.708969	.709014	.709059	.709104	.709149	-0
9270	.709194	.709239	.709284	.709329	.709374	.709419	.709464	.709509	.709554	.709599	-0
9280	.709644	.709689	.709734	.709779	.709824	.709869	.709914	.709959	.710004	.710049	-0
9290	.710094	.710139	.710184	.710228	.710273	.710318	.710363	.710408	.710453	.710497	-0
9300	.710542	.710587	.710632	.710676	.710721	.710766	.710811	.710855	.710900	.710945	-0
9310	.710990	.711034	.711079	.711124	.711168	.711213	.711258	.711302	.711347	.711391	-0
9320	.711436	.711481	.711525	.711570	.711614	.711659	.711703	.711748	.711792	.711837	-0
9330	.711881	.711926	.711970	.712015	.712059	.712104	.712148	.712193	.712237	.712281	-0
9340	.712326	.712370	.712415	.712459	.712503	.712548	.712592	.712636	.712681	.712725	-0
9350	.712769	.712814	.712858	.712902	.712947	.712991	.713035	.713079	.713124	.713168	-0
9360	.713212	.713256	.713300	.713345	.713389	.713433	.713477	.713521	.713565	.713609	-0
9370	.713654	.713698	.713742	.713786	.713830	.713874	.713918	.713962	.714006	.714050	-0
9380	.714094	.714138	.714182	.714226	.714270	.714314	.714358	.714402	.714446	.714490	-0
9390	.714534	.714578	.714622	.714665	.714709	.714753	.714797	.714841	.714885	.714929	-0
9400	.714972	.715016	.715060	.715104	.715148	.715191	.715235	.715279	.715323	.715366	-0
9410	.715410	.715454	.715498	.715541	.715585	.715629	.715672	.715716	.715760	.715803	-0
9420	.715847	.715891	.715934	.715978	.716021	.716065	.716109	.716152	.716196	.716239	-0
9430	.716283	.716326	.716370	.716413	.716457	.716500	.716544	.716587	.716631	.716674	-0
9440	.716718	.716761	.716805	.716848	.716891	.716935	.716978	.717022	.717065	.717108	-0
9450	.717152	.717195	.717238	.717282	.717325	.717368	.717412	.717455	.717498	.717541	-0
9460	.717585	.717628	.717671	.717714	.717758	.717801	.717844	.717887	.717930	.717974	-0
9470	.718017	.718060	.718103	.718146	.718189	.718232	.718276	.718319	.718362	.718405	-0
9480	.718448	.718491	.718534	.718577	.718620	.718663	.718706	.718749	.718792	.718835	-0
9490	.718878	.718921	.718964	.719007	.719050	.719093	.719136	.719179	.719222	.719264	-0

TABLE 15A

Blackbody: Relative Fractional Photon Emittance

λT(μ.°K)	0	1	2	3	4	5	6	7	8	9	m
9500	.719307	.719350	.719393	.719436	.719479	.719522	.719564	.719607	.719650	.719693	-0
9510	.719736	.719778	.719821	.719864	.719907	.719949	.719992	.720035	.720078	.720120	-0
9520	.720163	.720206	.720248	.720291	.720334	.720376	.720419	.720462	.720504	.720547	-0
9530	.720590	.720632	.720675	.720717	.720760	.720802	.720845	.720888	.720930	.720973	-0
9540	.721015	.721058	.721100	.721143	.721185	.721228	.721270	.721312	.721355	.721397	-0
9550	.721440	.721482	.721525	.721567	.721609	.721652	.721694	.721736	.721779	.721821	-0
9560	.721863	.721906	.721948	.721990	.722033	.722075	.722117	.722160	.722202	.722244	-0
9570	.722286	.722329	.722371	.722413	.722455	.722497	.722540	.722582	.722624	.722666	-0
9580	.722708	.722750	.722792	.722835	.722877	.722919	.722961	.723003	.723045	.723087	-0
9590	.723129	.723171	.723213	.723255	.723297	.723339	.723381	.723423	.723465	.723507	-0
9600	.723549	.723591	.723633	.723675	.723717	.723759	.723801	.723843	.723885	.723927	-0
9610	.723968	.724010	.724052	.724094	.724136	.724178	.724220	.724261	.724303	.724345	-0
9620	.724387	.724428	.724470	.724512	.724554	.724596	.724637	.724679	.724721	.724762	-0
9630	.724804	.724846	.724887	.724929	.724971	.725012	.725054	.725096	.725137	.725179	-0
9640	.725221	.725262	.725304	.725345	.725387	.725428	.725470	.725512	.725553	.725595	-0
9650	.725636	.725678	.725719	.725761	.725802	.725844	.725885	.725927	.725968	.726009	-0
9660	.726051	.726092	.726134	.726175	.726216	.726258	.726299	.726341	.726382	.726423	-0
9670	.726465	.726506	.726547	.726589	.726630	.726671	.726712	.726754	.726795	.726836	-0
9680	.726878	.726919	.726960	.727001	.727042	.727084	.727125	.727166	.727207	.727248	-0
9690	.727290	.727331	.727372	.727413	.727454	.727495	.727536	.727577	.727619	.727660	-0
9700	.727701	.727742	.727783	.727824	.727865	.727906	.727947	.727988	.728029	.728070	-0
9710	.728111	.728152	.728193	.728234	.728275	.728316	.728357	.728398	.728439	.728479	-0
9720	.728520	.728561	.728602	.728643	.728684	.728725	.728766	.728806	.728847	.728888	-0
9730	.728929	.728970	.729010	.729051	.729092	.729133	.729174	.729214	.729255	.729296	-0
9740	.729337	.729377	.729418	.729459	.729499	.729540	.729581	.729621	.729662	.729703	-0
9750	.729743	.729784	.729825	.729865	.729906	.729946	.729987	.730027	.730068	.730109	-0
9760	.730149	.730190	.730230	.730271	.730311	.730352	.730392	.730433	.730473	.730514	-0
9770	.730554	.730595	.730635	.730676	.730716	.730756	.730797	.730837	.730878	.730918	-0
9780	.730958	.730999	.731039	.731079	.731120	.731160	.731200	.731241	.731281	.731321	-0
9790	.731362	.731402	.731442	.731482	.731523	.731563	.731603	.731643	.731684	.731724	-0
9800	.731764	.731804	.731844	.731885	.731925	.731965	.732005	.732045	.732085	.732126	-0
9810	.732166	.732206	.732246	.732286	.732326	.732366	.732406	.732446	.732486	.732526	-0
9820	.732566	.732606	.732646	.732686	.732726	.732766	.732806	.732846	.732886	.732926	-0
9830	.732966	.733006	.733046	.733086	.733126	.733166	.733206	.733246	.733285	.733325	-0
9840	.733365	.733405	.733445	.733485	.733525	.733564	.733604	.733644	.733684	.733724	-0
9850	.733763	.733803	.733843	.733883	.733922	.733962	.734002	.734042	.734081	.734121	-0
9860	.734161	.734200	.734240	.734280	.734319	.734359	.734399	.734438	.734478	.734518	-0
9870	.734557	.734597	.734636	.734676	.734716	.734755	.734795	.734834	.734874	.734913	-0
9880	.734953	.734992	.735032	.735071	.735111	.735150	.735190	.735229	.735269	.735308	-0
9890	.735348	.735387	.735426	.735466	.735505	.735545	.735584	.735623	.735663	.735702	-0
9900	.735742	.735781	.735820	.735860	.735899	.735938	.735978	.736017	.736056	.736095	-0
9910	.736135	.736174	.736213	.736252	.736292	.736331	.736370	.736409	.736449	.736488	-0
9920	.736527	.736566	.736605	.736644	.736684	.736723	.736762	.736801	.736840	.736879	-0
9930	.736918	.736957	.736997	.737036	.737075	.737114	.737153	.737192	.737231	.737270	-0
9940	.737309	.737348	.737387	.737426	.737465	.737504	.737543	.737582	.737621	.737660	-0
9950	.737699	.737738	.737777	.737816	.737854	.737893	.737932	.737971	.738010	.738049	-0
9960	.738088	.738127	.738165	.738204	.738243	.738282	.738321	.738360	.738398	.738437	-0
9970	.738476	.738515	.738553	.738592	.738631	.738670	.738708	.738747	.738786	.738825	-0
9980	.738863	.738902	.738941	.738979	.739018	.739057	.739095	.739134	.739172	.739211	-0
9990	.739250	.739288	.739327	.739366	.739404	.739443	.739481	.739520	.739558	.739597	-0

TABLE 15B

Blackbody: Relative Fractional Photon Emittance Complement

λT(m.°K)	0	1	2	3	4	5	6	7	8	9	m
10	.260364	.256552	.252819	.249163	.245583	.242076	.238640	.235274	.231976	.228744	-0
11	.225577	.222473	.219430	.216447	.213522	.210654	.207842	.205083	.202378	.199724	-0
12	.197121	.194566	.192059	.189600	.187185	.184816	.182489	.180206	.177964	.175762	-0
13	.173600	.171477	.169391	.167342	.165330	.163352	.161409	.159500	.157624	.155780	-0
14	.153968	.152186	.150434	.148712	.147019	.145354	.143716	.142106	.140522	.138964	-0
15	.137431	.135923	.134439	.132979	.131542	.130128	.128736	.127366	.126018	.124690	-0
16	.123383	.122097	.120830	.119582	.118353	.117143	.115951	.114777	.113620	.112480	-0
17	.111358	.110252	.109162	.108088	.107029	.105986	.104958	.103944	.102945	.101960	-0
18	.100989	.100032	.099088	.098157	.097239	.096334	.095441	.094560	.093692	.092835	-0
19	.919897	.911558	.903332	.895215	.887206	.879303	.871504	.863807	.856211	.848714	-1
20	.841314	.834009	.826798	.819680	.812652	.805714	.798863	.792098	.785419	.778822	-1
21	.772308	.765875	.759521	.753245	.747046	.740922	.734873	.728898	.722994	.717161	-1
22	.711398	.705703	.700077	.694516	.689022	.683591	.678225	.672920	.667678	.662496	-1
23	.657373	.652310	.647304	.642355	.637463	.632626	.627843	.623114	.618438	.613814	-1
24	.609241	.604719	.600247	.595824	.591450	.587123	.582843	.578609	.574422	.570279	-1
25	.566180	.562126	.558114	.554146	.550219	.546333	.542489	.538684	.534919	.531194	-1
26	.527506	.523857	.520246	.516671	.513134	.509632	.506165	.502734	.499337	.495975	-1
27	.492646	.489350	.486088	.482857	.479659	.476492	.473356	.470251	.467176	.464131	-1
28	.461116	.458130	.455173	.452244	.449343	.446470	.443624	.440805	.438013	.435247	-1
29	.432507	.429793	.427105	.424441	.421802	.419187	.416597	.414031	.411488	.408968	-1
30	.406471	.403997	.401546	.399116	.396709	.394323	.391958	.389615	.387293	.384991	-1
31	.382709	.380448	.378207	.375985	.373783	.371600	.369436	.367291	.365164	.363056	-1
32	.360965	.358893	.356839	.354802	.352782	.350780	.348794	.346825	.344873	.342937	-1
33	.341018	.339114	.337226	.335354	.333498	.331657	.329831	.328019	.326223	.324442	-1
34	.322675	.320922	.319183	.317459	.315748	.314052	.312368	.310699	.309042	.307399	-1
35	.305769	.304151	.302547	.300955	.299376	.297809	.296254	.294711	.293180	.291661	-1
36	.290154	.288659	.287175	.285702	.284241	.282791	.281352	.279923	.278506	.277099	-1
37	.275703	.274318	.272943	.271578	.270223	.268879	.267544	.266219	.264904	.263599	-1
38	.262303	.261017	.259741	.258473	.257215	.255966	.254726	.253495	.252273	.251060	-1
39	.249855	.248659	.247472	.246293	.245122	.243960	.242806	.241660	.240522	.239393	-1
40	.238271	.237157	.236050	.234952	.233861	.232778	.231702	.230633	.229572	.228519	-1
41	.227472	.226433	.225400	.224375	.223357	.222346	.221341	.220343	.219352	.218368	-1
42	.217390	.216419	.215454	.214496	.213544	.212598	.211659	.210725	.209798	.208877	-1
43	.207963	.207054	.206151	.205254	.204362	.203477	.202597	.201723	.200854	.199992	-1
44	.199134	.198283	.197436	.196595	.195760	.194929	.194104	.193284	.192470	.191660	-1
45	.190856	.190056	.189262	.188473	.187688	.186908	.186134	.185364	.184598	.183838	-1
46	.183082	.182331	.181584	.180842	.180105	.179372	.178643	.177919	.177200	.176484	-1
47	.175773	.175066	.174364	.173666	.172972	.172282	.171596	.170914	.170236	.169563	-1
48	.168893	.168227	.167565	.166907	.166253	.165603	.164956	.164314	.163675	.163040	-1
49	.162408	.161780	.161156	.160535	.159918	.159305	.158695	.158088	.157485	.156886	-1

TABLE 15B

Blackbody: Relative Fractional Photon Emittance Complement

λT(m.°K)	0	1	2	3	4	5	6	7	8	9	m
50	.156289	.155697	.155107	.154521	.153938	.153359	.152783	.152210	.151640	.151073	-1
51	.150510	.149949	.149392	.148838	.148287	.147739	.147194	.146652	.146113	.145577	-1
52	.145044	.144514	.143987	.143463	.142941	.142423	.141907	.141394	.140883	.140376	-1
53	.139871	.139369	.138870	.138373	.137879	.137388	.136899	.136413	.135929	.135448	-1
54	.134970	.134494	.134020	.133549	.133081	.132615	.132151	.131690	.131231	.130775	-1
55	.130321	.129869	.129420	.128973	.128528	.128086	.127646	.127208	.126772	.126339	-1
56	.125908	.125479	.125052	.124628	.124205	.123785	.123367	.122951	.122537	.122125	-1
57	.121715	.121308	.120902	.120498	.120097	.119697	.119299	.118904	.118510	.118118	-1
58	.117729	.117341	.116955	.116571	.116188	.115808	.115430	.115053	.114678	.114305	-1
59	.113934	.113565	.113197	.112832	.112468	.112105	.111745	.111386	.111029	.110674	-1
60	.110320	.109968	.109618	.109270	.108923	.108577	.108234	.107892	.107551	.107213	-1
61	.106876	.106540	.106206	.105874	.105543	.105213	.104885	.104559	.104234	.103911	-1
62	.103589	.103269	.102950	.102633	.102317	.102003	.101690	.101379	.101068	.100760	-1
63	.100453	.100147	.099842	.099539	.099238	.098937	.098638	.098341	.098044	.097749	-1
64	.974559	.971636	.968727	.965830	.962946	.960076	.957218	.954372	.951540	.948720	-2
65	.945913	.943118	.940335	.937565	.934806	.932060	.929326	.926604	.923894	.921196	-2
66	.918510	.915835	.913172	.910521	.907881	.905253	.902636	.900030	.897436	.894853	-2
67	.892280	.889719	.887169	.884630	.882102	.879585	.877078	.874582	.872097	.869622	-2
68	.867158	.864704	.862261	.859828	.857405	.854992	.852590	.850198	.847815	.845443	-2
69	.843081	.840728	.838386	.836053	.833730	.831416	.829112	.826818	.824533	.822258	-2
70	.819992	.817736	.815488	.813250	.811021	.808802	.806591	.804389	.802197	.800013	-2
71	.797839	.795673	.793516	.791367	.789228	.787097	.784974	.782861	.780755	.778659	-2
72	.776570	.774490	.772419	.770355	.768300	.766253	.764215	.762184	.760162	.758147	-2
73	.756141	.754142	.752152	.750169	.748194	.746227	.744267	.742316	.740372	.738435	-2
74	.736507	.734585	.732671	.730765	.728866	.726975	.725091	.723214	.721344	.719482	-2
75	.717627	.715779	.713938	.712104	.710278	.708458	.706645	.704839	.703041	.701249	-2
76	.699463	.697685	.695914	.694149	.692391	.690639	.688894	.687156	.685424	.683699	-2
77	.681981	.680269	.678563	.676863	.675171	.673484	.671804	.670130	.668462	.666800	-2
78	.665145	.663496	.661853	.660216	.658585	.656960	.655341	.653728	.652121	.650520	-2
79	.648925	.647335	.645752	.644174	.642602	.641036	.639476	.637921	.636372	.634828	-2
80	.633290	.631758	.630231	.628710	.627194	.625684	.624179	.622680	.621186	.619697	-2
81	.618214	.616736	.615264	.613796	.612334	.610877	.609425	.607979	.606537	.605101	-2
82	.603670	.602243	.600822	.599406	.597995	.596589	.595188	.593792	.592400	.591014	-2
83	.589632	.588255	.586884	.585516	.584154	.582796	.581443	.580095	.578752	.577413	-2
84	.576079	.574749	.573424	.572104	.570788	.569476	.568169	.566867	.565569	.564276	-2
85	.562987	.561702	.560422	.559146	.557875	.556608	.555345	.554086	.552832	.551582	-2
86	.550336	.549094	.547857	.546624	.545395	.544170	.542949	.541732	.540520	.539311	-2
87	.538107	.536906	.535710	.534517	.533329	.532144	.530964	.529787	.528614	.527446	-2
88	.526281	.525119	.523962	.522809	.521659	.520513	.519371	.518233	.517098	.515967	-2
89	.514840	.513716	.512596	.511480	.510368	.509259	.508153	.507052	.505954	.504859	-2
90	.503768	.502680	.501596	.500516	.499439	.498365	.497295	.496229	.495165	.494106	-2
91	.493049	.491996	.490947	.489900	.488857	.487818	.486781	.485748	.484719	.483692	-2
92	.482669	.481649	.480632	.479619	.478608	.477601	.476597	.475597	.474599	.473604	-2
93	.472613	.471625	.470640	.469657	.468678	.467702	.466729	.465760	.464793	.463829	-2
94	.462868	.461910	.460955	.460003	.459054	.458108	.457165	.456225	.455287	.454353	-2
95	.453421	.452492	.451566	.450643	.449723	.448806	.447891	.446979	.446070	.445164	-2
96	.444260	.443360	.442462	.441566	.440674	.439784	.438897	.438012	.437130	.436251	-2
97	.435375	.434501	.433629	.432761	.431895	.431031	.430170	.429312	.428456	.427603	-2
98	.426753	.425904	.425059	.424216	.423375	.422537	.421702	.420869	.420038	.419210	-2
99	.418384	.417561	.416740	.415921	.415105	.414292	.413480	.412672	.411865	.411061	-2

$1-N(\lambda,T)/N_T(T)$

TABLE 15B

Blackbody: Relative Fractional Photon Emittance Complement

λT(m.°K)	0	1	2	3	4	5	6	7	8	9	m
100	.410259	.402369	.394704	.387255	.380016	.372978	.366133	.359475	.352997	.346692	-2
110	.340555	.334580	.328760	.323090	.317566	.312183	.306935	.301819	.296829	.291962	-2
120	.287213	.282580	.278058	.273643	.269333	.265123	.261012	.256995	.253071	.249236	-2
130	.245487	.241822	.238238	.234734	.231306	.227953	.224673	.221462	.218320	.215244	-2
140	.212233	.209285	.206397	.203569	.200799	.198084	.195425	.192819	.190264	.187760	-2
150	.185305	.182898	.180537	.178222	.175951	.173723	.171537	.169392	.167287	.165222	-2
160	.163194	.161203	.159249	.157329	.155445	.153594	.151775	.149989	.148235	.146510	-2
170	.144816	.143151	.141515	.139906	.138325	.136770	.135242	.133738	.132260	.130806	-2
180	.129376	.127970	.126586	.125224	.123885	.122566	.121269	.119992	.118735	.117498	-2
190	.116280	.115081	.113900	.112738	.111593	.110465	.109355	.108261	.107184	.106122	-2
200	.105076	.104046	.103030	.102030	.101044	.100072	.099114	.098170	.097239	.096321	-2
210	.954168	.945248	.936452	.927779	.919226	.910790	.902470	.894264	.886168	.878183	-3
220	.870305	.862532	.854863	.847296	.839828	.832459	.825187	.818010	.810925	.803933	-3
230	.797030	.790216	.783489	.776848	.770290	.763815	.757422	.751108	.744873	.738716	-3
240	.732634	.726627	.720694	.714833	.709044	.703324	.697673	.692091	.686574	.681124	-3
250	.675738	.670416	.665157	.659959	.654821	.649744	.644725	.639765	.634861	.630013	-3
260	.625221	.620484	.615799	.611168	.606589	.602061	.597584	.593156	.588777	.584447	-3
270	.580164	.575928	.571739	.567595	.563495	.559441	.555429	.551461	.547535	.543651	-3
280	.539808	.536005	.532243	.528520	.524836	.521191	.517583	.514012	.510479	.506981	-3
290	.503520	.500094	.496703	.493346	.490023	.486733	.483477	.480253	.477061	.473901	-3
300	.470772	.467674	.464607	.461569	.458562	.455583	.452634	.449713	.446820	.443955	-3
310	.441118	.438307	.435524	.432767	.430036	.427331	.424651	.421996	.419366	.416761	-3
320	.414180	.411622	.409089	.406578	.404091	.401627	.399185	.396765	.394367	.391991	-3
330	.389636	.387302	.384989	.382697	.380425	.378174	.375942	.373730	.371538	.369365	-3
340	.367211	.365075	.362958	.360860	.358779	.356717	.354672	.352645	.350635	.348643	-3
350	.346667	.344708	.342766	.340840	.338930	.337036	.335158	.333295	.331448	.329617	-3
360	.327800	.325999	.324212	.322440	.320683	.318939	.317210	.315495	.313794	.312107	-3
370	.310433	.308773	.307125	.305491	.303871	.302262	.300667	.299084	.297514	.295956	-3
380	.294410	.292876	.291355	.289845	.288347	.286860	.285385	.283921	.282468	.281027	-3
390	.279597	.278177	.276768	.275370	.273983	.272605	.271239	.269882	.268536	.267200	-3
400	.265873	.264557	.263250	.261953	.260666	.259388	.258119	.256860	.255609	.254368	-3
410	.253136	.251913	.250699	.249493	.248297	.247108	.245929	.244757	.243594	.242440	-3
420	.241293	.240155	.239024	.237902	.236787	.235681	.234582	.233491	.232407	.231331	-3
430	.230262	.229201	.228147	.227100	.226060	.225028	.224003	.222984	.221973	.220968	-3
440	.219970	.218979	.217995	.217017	.216046	.215081	.214123	.213171	.212226	.211287	-3
450	.210354	.209427	.208506	.207591	.206683	.205780	.204883	.203992	.203107	.202228	-3
460	.201354	.200486	.199623	.198767	.197915	.197069	.196229	.195394	.194564	.193739	-3
470	.192920	.192106	.191297	.190493	.189694	.188900	.188111	.187327	.186548	.185774	-3
480	.185005	.184240	.183480	.182725	.181974	.181228	.180487	.179750	.179018	.178290	-3
490	.177567	.176848	.176133	.175423	.174717	.174015	.173317	.172624	.171935	.171250	-3

TABLE 15B

Blackbody: Relative Fractional Photon Emittance Complement

λT(m.°K)	0	1	2	3	4	5	6	7	8	9	m
500	.170569	.169892	.169219	.168550	.167885	.167224	.166566	.165913	.165264	.164618	-3
510	.163976	.163338	.162704	.162073	.161446	.160822	.160202	.159586	.158973	.158364	-3
520	.157759	.157156	.156558	.155962	.155370	.154782	.154196	.153614	.153036	.152460	-3
530	.151888	.151319	.150753	.150191	.149631	.149075	.148522	.147972	.147424	.146880	-3
540	.146339	.145801	.145266	.144734	.144205	.143678	.143155	.142634	.142116	.141601	-3
550	.141089	.140580	.140073	.139569	.139068	.138569	.138073	.137580	.137090	.136602	-3
560	.136116	.135634	.135153	.134676	.134201	.133728	.133258	.132790	.132325	.131862	-3
570	.131402	.130944	.130489	.130035	.129585	.129136	.128690	.128246	.127805	.127365	-3
580	.126928	.126494	.126061	.125631	.125203	.124777	.124353	.123931	.123512	.123095	-3
590	.122679	.122266	.121855	.121446	.121039	.120635	.120232	.119831	.119432	.119035	-3
600	.118640	.118247	.117856	.117467	.117080	.116695	.116312	.115930	.115551	.115173	-3
610	.114797	.114423	.114051	.113681	.113312	.112946	.112581	.112217	.111856	.111496	-3
620	.111138	.110782	.110427	.110075	.109723	.109374	.109026	.108680	.108335	.107993	-3
630	.107651	.107312	.106974	.106637	.106302	.105969	.105637	.105307	.104979	.104651	-3
640	.104326	.104002	.103679	.103358	.103039	.102721	.102404	.102089	.101775	.101463	-3
650	.101152	.100843	.100535	.100228	.099923	.099619	.099317	.099016	.098716	.098418	-3
660	.981213	.978257	.975314	.972385	.969469	.966566	.963676	.960799	.957935	.955084	-4
670	.952245	.949419	.946605	.943804	.941016	.938240	.935476	.932724	.929984	.927257	-4
680	.924541	.921837	.919146	.916465	.913797	.911140	.908495	.905862	.903239	.900629	-4
690	.898029	.895441	.892864	.890298	.887743	.885199	.882665	.880143	.877632	.875131	-4
700	.872641	.870162	.867693	.865234	.862787	.860349	.857922	.855505	.853098	.850701	-4
710	.848315	.845938	.843572	.841215	.838868	.836531	.834204	.831887	.829579	.827280	-4
720	.824992	.822712	.820443	.818182	.815931	.813689	.811456	.809233	.807019	.804813	-4
730	.802617	.800430	.798252	.796082	.793922	.791770	.789627	.787492	.785367	.783250	-4
740	.781141	.779041	.776949	.774866	.772791	.770725	.768666	.766616	.764574	.762541	-4
750	.760515	.758498	.756488	.754487	.752493	.750507	.748529	.746559	.744597	.742643	-4
760	.740696	.738757	.736825	.734901	.732984	.731075	.729174	.727279	.725393	.723513	-4
770	.721641	.719776	.717918	.716068	.714224	.712388	.710559	.708737	.706922	.705114	-4
780	.703312	.701518	.699730	.697950	.696176	.694409	.692648	.690894	.689147	.687407	-4
790	.685673	.683946	.682225	.680511	.678803	.677101	.675406	.673717	.672035	.670359	-4
800	.668689	.667026	.665368	.663717	.662072	.660433	.658800	.657173	.655552	.653937	-4
810	.652329	.650726	.649129	.647537	.645952	.644373	.642799	.641231	.639669	.638112	-4
820	.636561	.635016	.633476	.631942	.630414	.628891	.627374	.625862	.624355	.622854	-4
830	.621359	.619868	.618383	.616904	.615430	.613961	.612497	.611039	.609585	.608137	-4
840	.606694	.605256	.603824	.602396	.600973	.599556	.598143	.596736	.595333	.593936	-4
850	.592543	.591155	.589772	.588394	.587021	.585652	.584288	.582929	.581575	.580226	-4
860	.578881	.577541	.576205	.574874	.573548	.572226	.570909	.569597	.568289	.566985	-4
870	.565686	.564392	.563101	.561816	.560534	.559257	.557985	.556716	.555452	.554193	-4
880	.552937	.551686	.550439	.549197	.547958	.546724	.545494	.544268	.543046	.541828	-4
890	.540615	.539405	.538200	.536998	.535801	.534607	.533418	.532232	.531051	.529873	-4
900	.528700	.527530	.526364	.525202	.524044	.522889	.521739	.520592	.519449	.518309	-4
910	.517174	.516042	.514914	.513790	.512669	.511552	.510439	.509329	.508223	.507120	-4
920	.506021	.504926	.503834	.502746	.501661	.500580	.499502	.498427	.497357	.496289	-4
930	.495225	.494165	.493108	.492054	.491003	.489956	.488913	.487872	.486835	.485802	-4
940	.484771	.483744	.482720	.481700	.480682	.479668	.478657	.477649	.476645	.475643	-4
950	.474645	.473650	.472658	.471669	.470683	.469700	.468720	.467744	.466770	.465800	-4
960	.464832	.463868	.462906	.461948	.460992	.460040	.459090	.458144	.457200	.456259	-4
970	.455321	.454386	.453454	.452525	.451598	.450674	.449754	.448836	.447921	.447008	-4
980	.446099	.445192	.444288	.443387	.442488	.441592	.440699	.439809	.438921	.438036	-4
990	.437154	.436274	.435397	.434523	.433651	.432782	.431915	.431051	.430190	.429331	-4

TABLE 15B

Blackbody: Relative Fractional Photon Emittance Complement

λT(m.°K)	0	1	2	3	4	5	6	7	8	9	m
1000	.428475	.420053	.411876	.403935	.396223	.388728	.381445	.374364	.367479	.360762	-4
1100	.354266	.347926	.341754	.335744	.329892	.324192	.318638	.313225	.307950	.302806	-4
1200	.297790	.292898	.288126	.283469	.278924	.274488	.270156	.265926	.261795	.257760	-4
1300	.253817	.249963	.246197	.242516	.238916	.235396	.231953	.228585	.225290	.222065	-4
1400	.218910	.215821	.212797	.209836	.206937	.204097	.201315	.198590	.195920	.193303	-4
1500	.190738	.188224	.185760	.183343	.180974	.178650	.176370	.174134	.171940	.169787	-4
1600	.167675	.165601	.163566	.161568	.159607	.157681	.155790	.153932	.152108	.150315	-4
1700	.148555	.146825	.145125	.143454	.141812	.140198	.138612	.137052	.135519	.134011	-4
1800	.132528	.131069	.129635	.128224	.126836	.125470	.124126	.122804	.121503	.120222	-4
1900	.118961	.117721	.116499	.115296	.114112	.112946	.111798	.110667	.109554	.108457	-4
2000	.107376	.106312	.105263	.104230	.103212	.102208	.101219	.100245	.099285	.098338	-4
2100	.974045	.964845	.955774	.946831	.938013	.929317	.920742	.912284	.903943	.895716	-5
2200	.887600	.879595	.871697	.863905	.856217	.848631	.841145	.833758	.826469	.819274	-5
2300	.812173	.805163	.798244	.791414	.784671	.778014	.771441	.764952	.758544	.752215	-5
2400	.745966	.739795	.733699	.727679	.721733	.715859	.710056	.704324	.698661	.693066	-5
2500	.687538	.682075	.676678	.671344	.666073	.660865	.655716	.650628	.645599	.640628	-5
2600	.635714	.630856	.626054	.621307	.616613	.611972	.607384	.602847	.598360	.593924	-5
2700	.589536	.585197	.580906	.576662	.572464	.568312	.564205	.560142	.556123	.552147	-5
2800	.548213	.544322	.540472	.536662	.532892	.529163	.525472	.521819	.518205	.514628	-5
2900	.511088	.507584	.504116	.500684	.497286	.493923	.490594	.487299	.484037	.480807	-5
3000	.477609	.474444	.471309	.468206	.465133	.462091	.459078	.456094	.453140	.450214	-5
3100	.447316	.444446	.441604	.438789	.436001	.433239	.430503	.427793	.425109	.422450	-5
3200	.419816	.417206	.414621	.412059	.409522	.407007	.404516	.402047	.399601	.397178	-5
3300	.394776	.392396	.390037	.387700	.385384	.383088	.380813	.378557	.376322	.374107	-5
3400	.371911	.369735	.367577	.365438	.363318	.361217	.359133	.357068	.355020	.352990	-5
3500	.350977	.348981	.347003	.345041	.343095	.341166	.339254	.337357	.335476	.333611	-5
3600	.331762	.329927	.328108	.326304	.324515	.322740	.320980	.319235	.317503	.315786	-5
3700	.314082	.312392	.310716	.309053	.307404	.305768	.304145	.302534	.300937	.299352	-5
3800	.297779	.296219	.294671	.293135	.291612	.290100	.288599	.287111	.285634	.284168	-5
3900	.282713	.281270	.279838	.278416	.277006	.275606	.274216	.272837	.271469	.270111	-5
4000	.268763	.267425	.266097	.264779	.263470	.262171	.260882	.259603	.258332	.257071	-5
4100	.255820	.254577	.253343	.252119	.250903	.249696	.248498	.247308	.246127	.244954	-5
4200	.243790	.242634	.241486	.240346	.239214	.238090	.236974	.235866	.234766	.233674	-5
4300	.232589	.231511	.230441	.229379	.228323	.227275	.226235	.225201	.224174	.223155	-5
4400	.222142	.221136	.220137	.219145	.218160	.217181	.216209	.215243	.214284	.213331	-5
4500	.212384	.211444	.210510	.209582	.208660	.207744	.206835	.205931	.205033	.204141	-5
4600	.203255	.202375	.201500	.200631	.199768	.198910	.198057	.197210	.196369	.195533	-5
4700	.194702	.193877	.193057	.192241	.191432	.190627	.189827	.189032	.188243	.187458	-5
4800	.186678	.185903	.185133	.184367	.183607	.182851	.182099	.181353	.180611	.179873	-5
4900	.179140	.178411	.177687	.176967	.176252	.175541	.174834	.174132	.173433	.172739	-5

TABLE 15B

Blackbody: Relative Fractional Photon Emittance Complement

λT(m.°K)	0	1	2	3	4	5	6	7	8	9	m
5000	.172049	.171364	.170682	.170004	.169331	.168661	.167995	.167334	.166676	.166022	-5
5100	.165372	.164725	.164083	.163444	.162809	.162178	.161550	.160926	.160305	.159689	-5
5200	.159075	.158465	.157859	.157256	.156657	.156061	.155468	.154879	.154293	.153711	-5
5300	.153132	.152556	.151983	.151413	.150847	.150284	.149724	.149167	.148613	.148063	-5
5400	.147515	.146971	.146429	.145890	.145355	.144822	.144292	.143765	.143241	.142720	-5
5500	.142202	.141687	.141174	.140664	.140157	.139652	.139151	.138652	.138155	.137662	-5
5600	.137171	.136682	.136197	.135713	.135233	.134755	.134279	.133806	.133336	.132868	-5
5700	.132402	.131939	.131478	.131020	.130564	.130110	.129659	.129210	.128764	.128320	-5
5800	.127878	.127438	.127001	.126566	.126133	.125702	.125273	.124847	.124423	.124001	-5
5900	.123581	.123164	.122748	.122335	.121923	.121514	.121107	.120701	.120298	.119897	-5
6000	.119498	.119101	.118705	.118312	.117921	.117532	.117144	.116759	.116375	.115993	-5
6100	.115614	.115236	.114859	.114485	.114113	.113742	.113373	.113006	.112641	.112277	-5
6200	.111916	.111556	.111197	.110841	.110486	.110133	.109781	.109432	.109083	.108737	-5
6300	.108392	.108049	.107708	.107368	.107029	.106693	.106357	.106024	.105692	.105362	-5
6400	.105033	.104705	.104380	.104055	.103732	.103411	.103091	.102773	.102456	.102141	-5
6500	.101827	.101514	.101203	.100894	.100586	.100279	.099973	.099669	.099367	.099066	-5
6600	.987657	.984672	.981701	.978743	.975798	.972867	.969948	.967043	.964151	.961272	-6
6700	.958405	.955552	.952711	.949883	.947067	.944264	.941474	.938696	.935930	.933176	-6
6800	.930434	.927705	.924987	.922281	.919587	.916905	.914235	.911576	.908929	.906294	-6
6900	.903670	.901057	.898456	.895865	.893286	.890718	.888162	.885616	.883081	.880557	-6
7000	.878044	.875541	.873049	.870568	.868098	.865638	.863188	.860749	.858320	.855901	-6
7100	.853492	.851094	.848706	.846328	.843960	.841601	.839253	.836914	.834585	.832266	-6
7200	.829957	.827657	.825367	.823086	.820814	.818552	.816300	.814056	.811822	.809597	-6
7300	.807381	.805175	.802977	.800788	.798608	.796438	.794275	.792122	.789978	.787842	-6
7400	.785715	.783596	.781486	.779385	.777292	.775207	.773131	.771063	.769003	.766952	-6
7500	.764909	.762874	.760847	.758828	.756817	.754814	.752819	.750832	.748853	.746882	-6
7600	.744918	.742962	.741014	.739074	.737141	.735215	.733298	.731387	.729485	.727589	-6
7700	.725701	.723820	.721947	.720081	.718222	.716370	.714526	.712688	.710858	.709035	-6
7800	.707218	.705409	.703607	.701811	.700022	.698241	.696466	.694697	.692936	.691181	-6
7900	.689433	.687691	.685956	.684228	.682506	.680791	.679082	.677379	.675683	.673993	-6
8000	.672310	.670633	.668962	.667297	.665639	.663987	.662340	.660700	.659067	.657439	-6
8100	.655817	.654201	.652591	.650987	.649389	.647797	.646211	.644630	.643056	.641487	-6
8200	.639924	.638366	.636814	.635268	.633728	.632193	.630663	.629139	.627621	.626108	-6
8300	.624601	.623099	.621603	.620111	.618626	.617145	.615670	.614200	.612736	.611276	-6
8400	.609822	.608373	.606929	.605491	.604057	.602629	.601205	.599787	.598374	.596965	-6
8500	.595562	.594163	.592770	.591381	.589997	.588619	.587244	.585875	.584511	.583151	-6
8600	.581796	.580446	.579100	.577759	.576423	.575091	.573764	.572442	.571124	.569811	-6
8700	.568502	.567198	.565898	.564602	.563311	.562025	.560743	.559465	.558192	.556923	-6
8800	.555658	.554398	.553142	.551890	.550642	.549399	.548160	.546925	.545694	.544467	-6
8900	.543245	.542027	.540812	.539602	.538396	.537194	.535996	.534802	.533611	.532425	-6
9000	.531243	.530065	.528891	.527720	.526553	.525391	.524232	.523077	.521926	.520778	-6
9100	.519635	.518495	.517359	.516226	.515098	.513973	.512851	.511734	.510620	.509509	-6
9200	.508403	.507299	.506200	.505104	.504011	.502923	.501837	.500755	.499677	.498602	-6
9300	.497531	.496463	.495398	.494337	.493279	.492225	.491174	.490127	.489082	.488042	-6
9400	.487004	.485970	.484939	.483911	.482887	.481865	.480847	.479833	.478821	.477813	-6
9500	.476808	.475806	.474807	.473811	.472819	.471829	.470843	.469860	.468880	.467903	-6
9600	.466928	.465957	.464989	.464024	.463063	.462104	.461148	.460195	.459244	.458297	-6
9700	.457353	.456412	.455473	.454538	.453605	.452675	.451749	.450824	.449903	.448985	-6
9800	.448069	.447156	.446246	.445339	.444435	.443533	.442634	.441738	.440844	.439953	-6
9900	.439065	.438180	.437297	.436417	.435539	.434665	.433792	.432923	.432056	.431192	-6

TABLE 16

Blackbody: Photon Emittance (photons/m^2.s)

T(°K)	0	10	20	30	40	50	60	70	80	90	m
100	0.15204	0.20237	0.26273	0.33404	0.41720	0.51314	0.62276	0.74698	0.88671	1.04285	22
200	1.21633	1.40806	1.61894	1.84989	2.10182	2.37565	2.67228	2.99264	3.33762	3.70814	22
300	4.10512	4.52947	4.98210	5.46392	5.97584	6.51878	7.09365	7.70136	8.34283	9.01896	22
400	0.97307	1.04789	1.12645	1.20884	1.29515	1.38548	1.47991	1.57854	1.68146	1.78875	23
500	1.90052	2.01685	2.13783	2.26355	2.39411	2.52959	2.67009	2.81570	2.96651	3.12262	23
600	3.28410	3.45106	3.62358	3.80176	3.98568	4.17544	4.37114	4.57285	4.78068	4.99470	23
700	5.21503	5.44174	5.67492	5.91468	6.16109	6.41426	6.67426	6.94120	7.21517	7.49624	23
800	0.77845	0.80801	0.83831	0.86935	0.90116	0.93373	0.96707	1.00120	1.03612	1.07185	24
900	1.10838	1.14574	1.18393	1.22296	1.26283	1.30357	1.34517	1.38764	1.43100	1.47526	24
1000	1.52042	1.56649	1.61348	1.66140	1.71026	1.76007	1.81084	1.86258	1.91529	1.96898	24
1100	2.02367	2.07937	2.13608	2.19380	2.25256	2.31236	2.37321	2.43512	2.49809	2.56214	24
1200	2.62728	2.69351	2.76084	2.82929	2.89886	2.96956	3.04140	3.11439	3.18854	3.26386	24
1300	3.34035	3.41803	3.49691	3.57699	3.65828	3.74079	3.82454	3.90953	3.99576	4.08326	24
1400	4.17202	4.26206	4.35339	4.44601	4.53994	4.63518	4.73174	4.82964	4.92887	5.02946	24
1500	5.13140	5.23472	5.33941	5.44549	5.55296	5.66184	5.77213	5.88385	5.99700	6.11158	24
1600	6.22762	6.34512	6.46409	6.58454	6.70647	6.82990	6.95483	7.08128	7.20925	7.33876	24
1700	7.46980	7.60240	7.73656	7.87228	8.00959	8.14848	8.28897	8.43106	8.57477	8.72010	24
1800	0.88671	0.90157	0.91659	0.93179	0.94714	0.96267	0.97837	0.99423	1.01027	1.02647	25
1900	1.04285	1.05941	1.07613	1.09304	1.11011	1.12737	1.14480	1.16241	1.18021	1.19818	25
2000	1.21633	1.23467	1.25319	1.27189	1.29078	1.30986	1.32912	1.34857	1.36821	1.38804	25
2100	1.40806	1.42827	1.44867	1.46927	1.49006	1.51105	1.53223	1.55361	1.57519	1.59696	25
2200	1.61894	1.64112	1.66349	1.68608	1.70886	1.73185	1.75504	1.77844	1.80205	1.82587	25
2300	1.84989	1.87412	1.89857	1.92323	1.94809	1.97318	1.99847	2.02399	2.04971	2.07566	25
2400	2.10182	2.12821	2.15481	2.18163	2.20868	2.23594	2.26343	2.29115	2.31909	2.34726	25
2500	2.37565	2.40427	2.43312	2.46220	2.49152	2.52106	2.55084	2.58084	2.61109	2.64157	25
2600	2.67228	2.70324	2.73443	2.76586	2.79753	2.82944	2.86159	2.89398	2.92662	2.95951	25
2700	2.99264	3.02601	3.05963	3.09350	3.12762	3.16199	3.19661	3.23148	3.26661	3.30198	25
2800	3.33762	3.37351	3.40965	3.44605	3.48271	3.51963	3.55681	3.59425	3.63195	3.66992	25
2900	3.70814	3.74664	3.78539	3.82442	3.86371	3.90327	3.94310	3.98320	4.02357	4.06421	25
3000	4.10512	4.14631	4.18777	4.22951	4.27153	4.31382	4.35639	4.39924	4.44237	4.48578	25
3100	4.52947	4.57345	4.61771	4.66225	4.70708	4.75219	4.79760	4.84329	4.88927	4.93554	25
3200	4.98210	5.02895	5.07610	5.12354	5.17127	5.21930	5.26763	5.31625	5.36518	5.41440	25
3300	5.46392	5.51374	5.56387	5.61429	5.66503	5.71606	5.76740	5.81905	5.87101	5.92327	25
3400	5.97584	6.02873	6.08192	6.13543	6.18925	6.24338	6.29783	6.35259	6.40767	6.46307	25
3500	6.51878	6.57482	6.63117	6.68785	6.74485	6.80217	6.85982	6.91779	6.97608	7.03470	25
3600	7.09365	7.15293	7.21254	7.27248	7.33275	7.39335	7.45428	7.51555	7.57715	7.63909	25
3700	7.70136	7.76398	7.82693	7.89022	7.95385	8.01782	8.08213	8.14679	8.21179	8.27714	25
3800	8.34283	8.40887	8.47525	8.54198	8.60907	8.67650	8.74429	8.81242	8.88091	8.94976	25
3900	9.01896	9.08851	9.15842	9.22869	9.29932	9.37031	9.44165	9.51336	9.58543	9.65787	25
4000	0.97307	0.98038	0.98774	0.99512	1.00255	1.01001	1.01751	1.02505	1.03263	1.04024	26
4100	1.04789	1.05557	1.06330	1.07106	1.07886	1.08669	1.09457	1.10248	1.11043	1.11842	26
4200	1.12645	1.13451	1.14261	1.15076	1.15894	1.16716	1.17542	1.18371	1.19205	1.20042	26
4300	1.20884	1.21729	1.22578	1.23432	1.24289	1.25150	1.26015	1.26884	1.27757	1.28634	26
4400	1.29515	1.30400	1.31289	1.32182	1.33080	1.33981	1.34886	1.35795	1.36709	1.37626	26
4500	1.38548	1.39474	1.40403	1.41337	1.42275	1.43218	1.44164	1.45115	1.46069	1.47028	26
4600	1.47991	1.48958	1.49930	1.50906	1.51886	1.52870	1.53858	1.54851	1.55848	1.56849	26
4700	1.57854	1.58864	1.59878	1.60896	1.61919	1.62946	1.63977	1.65013	1.66053	1.67097	26
4800	1.68146	1.69199	1.70256	1.71318	1.72385	1.73455	1.74530	1.75610	1.76694	1.77783	26
4900	1.78875	1.79973	1.81075	1.82181	1.83292	1.84407	1.85527	1.86652	1.87781	1.88914	26

TABLE 16

Blackbody: Photon Emittance (photons/m^2.s)

T(°K)	0	10	20	30	40	50	60	70	80	90	m
5000	1.90052	1.91195	1.92342	1.93494	1.94650	1.95811	1.96976	1.98146	1.99321	2.00501	26
5100	2.01685	2.02873	2.04067	2.05265	2.06468	2.07675	2.08887	2.10104	2.11325	2.12552	26
5200	2.13783	2.15018	2.16259	2.17504	2.18754	2.20009	2.21269	2.22533	2.23802	2.25076	26
5300	2.26355	2.27639	2.28927	2.30221	2.31519	2.32822	2.34130	2.35443	2.36761	2.38083	26
5400	2.39411	2.40743	2.42081	2.43423	2.44771	2.46123	2.47480	2.48842	2.50210	2.51582	26
5500	2.52959	2.54342	2.55729	2.57121	2.58519	2.59921	2.61329	2.62741	2.64159	2.65582	26
5600	2.67009	2.68442	2.69880	2.71324	2.72772	2.74225	2.75684	2.77148	2.78617	2.80091	26
5700	2.81570	2.83055	2.84545	2.86040	2.87540	2.89045	2.90556	2.92072	2.93593	2.95120	26
5800	2.96651	2.98188	2.99731	3.01279	3.02831	3.04390	3.05953	3.07522	3.09097	3.10676	26
5900	3.12262	3.13852	3.15448	3.17049	3.18656	3.20268	3.21885	3.23508	3.25137	3.26771	26
6000	3.28410	3.30055	3.31705	3.33361	3.35022	3.36689	3.38361	3.40039	3.41722	3.43411	26
6100	3.45106	3.46806	3.48511	3.50222	3.51939	3.53662	3.55390	3.57123	3.58862	3.60607	26
6200	3.62358	3.64114	3.65876	3.67643	3.69416	3.71195	3.72980	3.74770	3.76566	3.78368	26
6300	3.80176	3.81989	3.83808	3.85632	3.87463	3.89299	3.91141	3.92989	3.94843	3.96703	26
6400	3.98568	4.00439	4.02316	4.04199	4.06088	4.07983	4.09883	4.11790	4.13702	4.15620	26
6500	4.17544	4.19474	4.21410	4.23352	4.25300	4.27254	4.29214	4.31180	4.33152	4.35130	26
6600	4.37114	4.39103	4.41099	4.43101	4.45109	4.47123	4.49144	4.51170	4.53202	4.55240	26
6700	4.57285	4.59336	4.61392	4.63455	4.65524	4.67599	4.69681	4.71768	4.73862	4.75961	26
6800	4.78067	4.80180	4.82298	4.84423	4.86554	4.88691	4.90834	4.92984	4.95140	4.97302	26
6900	4.99470	5.01645	5.03826	5.06014	5.08207	5.10407	5.12614	5.14826	5.17045	5.19271	26
7000	5.21503	5.23741	5.25986	5.28237	5.30494	5.32758	5.35028	5.37305	5.39588	5.41878	26
7100	5.44174	5.46476	5.48785	5.51101	5.53423	5.55751	5.58087	5.60428	5.62776	5.65131	26
7200	5.67492	5.69860	5.72235	5.74616	5.77003	5.79397	5.81798	5.84206	5.86620	5.89040	26
7300	5.91468	5.93902	5.96342	5.98790	6.01244	6.03705	6.06172	6.08646	6.11127	6.13615	26
7400	6.16109	6.18610	6.21118	6.23633	6.26154	6.28682	6.31217	6.33759	6.36308	6.38863	26
7500	6.41426	6.43995	6.46571	6.49154	6.51743	6.54340	6.56943	6.59554	6.62171	6.64795	26
7600	6.67426	6.70064	6.72709	6.75361	6.78020	6.80686	6.83359	6.86039	6.88725	6.91419	26
7700	6.94120	6.96828	6.99543	7.02265	7.04994	7.07730	7.10473	7.13223	7.15981	7.18745	26
7800	7.21517	7.24295	7.27081	7.29874	7.32674	7.35481	7.38295	7.41117	7.43946	7.46781	26
7900	7.49624	7.52475	7.55332	7.58197	7.61069	7.63948	7.66835	7.69728	7.72629	7.75538	26
8000	7.78453	7.81376	7.84306	7.87244	7.90188	7.93140	7.96100	7.99067	8.02041	8.05023	26
8100	8.08011	8.11008	8.14012	8.17023	8.20041	8.23067	8.26101	8.29141	8.32190	8.35246	26
8200	8.38309	8.41380	8.44458	8.47543	8.50637	8.53737	8.56846	8.59961	8.63085	8.66216	26
8300	8.69354	8.72500	8.75654	8.78815	8.81984	8.85160	8.88344	8.91536	8.94735	8.97942	26
8400	9.01157	9.04379	9.07609	9.10847	9.14092	9.17345	9.20605	9.23874	9.27150	9.30434	26
8500	9.33726	9.37025	9.40332	9.43647	9.46970	9.50300	9.53638	9.56985	9.60338	9.63700	26
8600	9.67070	9.70447	9.73833	9.77226	9.80627	9.84036	9.87452	9.90877	9.94310	9.97750	26
8700	1.00120	1.00465	1.00812	1.01159	1.01507	1.01856	1.02206	1.02556	1.02907	1.03259	27
8800	1.03612	1.03966	1.04320	1.04675	1.05031	1.05388	1.05746	1.06104	1.06464	1.06824	27
8900	1.07185	1.07546	1.07909	1.08272	1.08636	1.09001	1.09367	1.09734	1.10101	1.10469	27
9000	1.10838	1.11208	1.11579	1.11950	1.12323	1.12696	1.13070	1.13445	1.13820	1.14197	27
9100	1.14574	1.14952	1.15331	1.15711	1.16092	1.16473	1.16855	1.17239	1.17623	1.18007	27
9200	1.18393	1.18779	1.19167	1.19555	1.19944	1.20334	1.20725	1.21116	1.21508	1.21902	27
9300	1.22296	1.22691	1.23086	1.23483	1.23881	1.24279	1.24678	1.25078	1.25479	1.25881	27
9400	1.26283	1.26687	1.27091	1.27496	1.27902	1.28309	1.28717	1.29126	1.29535	1.29945	27
9500	1.30357	1.30769	1.31182	1.31596	1.32010	1.32426	1.32842	1.33260	1.33678	1.34097	27
9600	1.34517	1.34937	1.35359	1.35782	1.36205	1.36629	1.37055	1.37481	1.37908	1.38336	27
9700	1.38764	1.39194	1.39624	1.40056	1.40488	1.40921	1.41355	1.41790	1.42226	1.42663	27
9800	1.43100	1.43539	1.43978	1.44419	1.44860	1.45302	1.45745	1.46189	1.46634	1.47079	27
9900	1.47526	1.47973	1.48422	1.48871	1.49321	1.49772	1.50224	1.50677	1.51131	1.51586	27

TABLE 16

Blackbody: Photon Emittance (photons/m^2.s)

T(°K)	0	100	200	300	400	500	600	700	800	900	m
10000	1.52042	1.56649	1.61348	1.66140	1.71026	1.76007	1.81084	1.86258	1.91529	1.96898	27
11000	2.02367	2.07937	2.13608	2.19380	2.25256	2.31236	2.37321	2.43512	2.49809	2.56214	27
12000	2.62728	2.69351	2.76084	2.82929	2.89886	2.96956	3.04140	3.11439	3.18854	3.26386	27
13000	3.34035	3.41803	3.49691	3.57699	3.65828	3.74079	3.82454	3.90953	3.99576	4.08326	27
14000	4.17202	4.26206	4.35339	4.44601	4.53994	4.63518	4.73174	4.82964	4.92887	5.02946	27
15000	5.13140	5.23472	5.33941	5.44549	5.55296	5.66184	5.77213	5.88385	5.99700	6.11158	27
16000	6.22762	6.34512	6.46409	6.58454	6.70647	6.82990	6.95483	7.08128	7.20925	7.33876	27
17000	7.46980	7.60240	7.73656	7.87228	8.00959	8.14848	8.28897	8.43106	8.57477	8.72010	27
18000	0.88671	0.90157	0.91659	0.93179	0.94714	0.96267	0.97837	0.99423	1.01027	1.02647	28
19000	1.04285	1.05941	1.07613	1.09304	1.11011	1.12737	1.14480	1.16241	1.18021	1.19818	28
20000	1.21633	1.23467	1.25319	1.27189	1.29078	1.30986	1.32912	1.34857	1.36821	1.38804	28
21000	1.40806	1.42827	1.44867	1.46927	1.49006	1.51105	1.53223	1.55361	1.57519	1.59696	28
22000	1.61894	1.64112	1.66349	1.68608	1.70886	1.73185	1.75504	1.77844	1.80205	1.82587	28
23000	1.84989	1.87412	1.89857	1.92323	1.94809	1.97318	1.99847	2.02399	2.04971	2.07566	28
24000	2.10182	2.12821	2.15481	2.18163	2.20868	2.23594	2.26343	2.29115	2.31909	2.34726	28
25000	2.37565	2.40427	2.43312	2.46220	2.49152	2.52106	2.55083	2.58084	2.61109	2.64157	28
26000	2.67228	2.70324	2.73443	2.76586	2.79753	2.82944	2.86159	2.89398	2.92662	2.95951	28
27000	2.99264	3.02601	3.05963	3.09350	3.12762	3.16199	3.19661	3.23148	3.26661	3.30198	28
28000	3.33762	3.37351	3.40965	3.44605	3.48271	3.51963	3.55681	3.59425	3.63195	3.66991	28
29000	3.70814	3.74664	3.78539	3.82442	3.86371	3.90327	3.94310	3.98320	4.02357	4.06421	28
30000	4.10512	4.14631	4.18777	4.22951	4.27153	4.31382	4.35639	4.39924	4.44237	4.48578	28
31000	4.52947	4.57345	4.61771	4.66225	4.70708	4.75219	4.79760	4.84329	4.88927	4.93554	28
32000	4.98210	5.02895	5.07610	5.12354	5.17127	5.21930	5.26763	5.31625	5.36518	5.41440	28
33000	5.46392	5.51374	5.56387	5.61429	5.66503	5.71606	5.76740	5.81905	5.87101	5.92327	28
34000	5.97584	6.02873	6.08192	6.13543	6.18925	6.24338	6.29783	6.35259	6.40767	6.46307	28
35000	6.51878	6.57482	6.63117	6.68785	6.74485	6.80217	6.85982	6.91779	6.97608	7.03470	28
36000	7.09365	7.15293	7.21254	7.27248	7.33275	7.39335	7.45428	7.51555	7.57715	7.63909	28
37000	7.70136	7.76398	7.82693	7.89022	7.95385	8.01782	8.08213	8.14679	8.21179	8.27714	28
38000	8.34283	8.40887	8.47525	8.54198	8.60907	8.67650	8.74429	8.81242	8.88091	8.94976	28
39000	9.01896	9.08851	9.15842	9.22869	9.29932	9.37031	9.44165	9.51336	9.58543	9.65787	28
40000	0.97307	0.98038	0.98774	0.99512	1.00255	1.01001	1.01751	1.02505	1.03263	1.04024	29
41000	1.04789	1.05557	1.06330	1.07106	1.07886	1.08669	1.09457	1.10248	1.11043	1.11842	29
42000	1.12645	1.13451	1.14261	1.15076	1.15894	1.16716	1.17542	1.18371	1.19205	1.20042	29
43000	1.20884	1.21729	1.22578	1.23432	1.24289	1.25150	1.26015	1.26884	1.27757	1.28634	29
44000	1.29515	1.30400	1.31289	1.32182	1.33080	1.33981	1.34886	1.35795	1.36709	1.37626	29
45000	1.38548	1.39474	1.40403	1.41337	1.42275	1.43218	1.44164	1.45115	1.46069	1.47028	29
46000	1.47991	1.48958	1.49930	1.50906	1.51886	1.52870	1.53858	1.54851	1.55848	1.56849	29
47000	1.57854	1.58864	1.59878	1.60896	1.61919	1.62946	1.63977	1.65013	1.66053	1.67097	29
48000	1.68146	1.69199	1.70256	1.71318	1.72385	1.73455	1.74530	1.75610	1.76694	1.77783	29
49000	1.78875	1.79973	1.81075	1.82181	1.83292	1.84407	1.85527	1.86652	1.87780	1.88914	29

TABLE 16

Blackbody: Photon Emittance (photons/m^2.s)

T(°K)	0	10	20	30	40	50	60	70	80	90	m
50000	1.90052	1.91195	1.92342	1.93494	1.94650	1.95811	1.96976	1.98146	1.99321	2.00501	29
51000	2.01685	2.02873	2.04067	2.05265	2.06468	2.07675	2.08887	2.10104	2.11325	2.12552	29
52000	2.13783	2.15018	2.16259	2.17504	2.18754	2.20009	2.21269	2.22533	2.23802	2.25076	29
53000	2.26355	2.27639	2.28927	2.30221	2.31519	2.32822	2.34130	2.35443	2.36761	2.38083	29
54000	2.39411	2.40743	2.42081	2.43423	2.44771	2.46123	2.47480	2.48842	2.50210	2.51582	29
55000	2.52959	2.54342	2.55729	2.57121	2.58519	2.59921	2.61329	2.62741	2.64159	2.65582	29
56000	2.67009	2.68442	2.69880	2.71324	2.72772	2.74225	2.75684	2.77148	2.78617	2.80091	29
57000	2.81570	2.83055	2.84545	2.86040	2.87540	2.89045	2.90556	2.92072	2.93593	2.95120	29
58000	2.96651	2.98188	2.99731	3.01278	3.02831	3.04390	3.05953	3.07522	3.09097	3.10676	29
59000	3.12262	3.13852	3.15448	3.17049	3.18656	3.20268	3.21885	3.23508	3.25137	3.26771	29
60000	3.28410	3.30055	3.31705	3.33361	3.35022	3.36689	3.38361	3.40039	3.41722	3.43411	29
61000	3.45106	3.46806	3.48511	3.50222	3.51939	3.53662	3.55390	3.57123	3.58862	3.60607	29
62000	3.62358	3.64114	3.65876	3.67643	3.69416	3.71195	3.72980	3.74770	3.76566	3.78368	29
63000	3.80175	3.81989	3.83808	3.85632	3.87463	3.89299	3.91141	3.92989	3.94843	3.96703	29
64000	3.98568	4.00439	4.02316	4.04199	4.06088	4.07983	4.09883	4.11790	4.13702	4.15620	29
65000	4.17544	4.19474	4.21410	4.23352	4.25300	4.27254	4.29214	4.31180	4.33152	4.35130	29
66000	4.37114	4.39103	4.41099	4.43101	4.45109	4.47123	4.49144	4.51170	4.53202	4.55240	29
67000	4.57285	4.59336	4.61392	4.63455	4.65524	4.67599	4.69681	4.71768	4.73862	4.75961	29
68000	4.78067	4.80180	4.82298	4.84423	4.86554	4.88691	4.90834	4.92984	4.95140	4.97302	29
69000	4.99470	5.01645	5.03826	5.06014	5.08207	5.10407	5.12614	5.14826	5.17045	5.19271	29
70000	5.21503	5.23741	5.25986	5.28237	5.30494	5.32758	5.35028	5.37305	5.39588	5.41878	29
71000	5.44174	5.46476	5.48785	5.51101	5.53423	5.55751	5.58087	5.60428	5.62776	5.65131	29
72000	5.67492	5.69860	5.72235	5.74616	5.77003	5.79397	5.81798	5.84206	5.86620	5.89040	29
73000	5.91468	5.93902	5.96342	5.98790	6.01244	6.03705	6.06172	6.08646	6.11127	6.13615	29
74000	6.16109	6.18610	6.21118	6.23633	6.26154	6.28682	6.31217	6.33759	6.36308	6.38863	29
75000	6.41426	6.43995	6.46571	6.49153	6.51743	6.54340	6.56943	6.59554	6.62171	6.64795	29
76000	6.67426	6.70064	6.72709	6.75361	6.78020	6.80686	6.83359	6.86039	6.88725	6.91419	29
77000	6.94120	6.96828	6.99543	7.02265	7.04994	7.07730	7.10473	7.13223	7.15981	7.18745	29
78000	7.21517	7.24295	7.27081	7.29874	7.32674	7.35481	7.38295	7.41117	7.43945	7.46781	29
79000	7.49624	7.52475	7.55332	7.58197	7.61069	7.63948	7.66835	7.69728	7.72629	7.75537	29
80000	7.78453	7.81376	7.84306	7.87244	7.90188	7.93140	7.96100	7.99067	8.02041	8.05023	29
81000	8.08011	8.11008	8.14012	8.17023	8.20041	8.23067	8.26101	8.29141	8.32190	8.35246	29
82000	8.38309	8.41380	8.44458	8.47543	8.50637	8.53737	8.56846	8.59961	8.63085	8.66216	29
83000	8.69354	8.72500	8.75654	8.78815	8.81984	8.85160	8.88344	8.91536	8.94735	8.97942	29
84000	9.01157	9.04379	9.07609	9.10846	9.14092	9.17345	9.20605	9.23874	9.27150	9.30434	29
85000	9.33726	9.37025	9.40332	9.43647	9.46970	9.50300	9.53638	9.56985	9.60338	9.63700	29
86000	9.67070	9.70447	9.73832	9.77226	9.80627	9.84036	9.87452	9.90877	9.94310	9.97750	29
87000	1.00120	1.00465	1.00812	1.01159	1.01507	1.01856	1.02206	1.02556	1.02907	1.03259	30
88000	1.03612	1.03966	1.04320	1.04675	1.05031	1.05388	1.05746	1.06104	1.06464	1.06824	30
89000	1.07185	1.07546	1.07909	1.08272	1.08636	1.09001	1.09367	1.09734	1.10101	1.10469	30
90000	1.10838	1.11208	1.11579	1.11950	1.12323	1.12696	1.13070	1.13445	1.13820	1.14197	30
91000	1.14574	1.14952	1.15331	1.15711	1.16092	1.16473	1.16855	1.17239	1.17623	1.18007	30
92000	1.18393	1.18779	1.19167	1.19555	1.19944	1.20334	1.20725	1.21116	1.21508	1.21902	30
93000	1.22296	1.22691	1.23086	1.23483	1.23881	1.24279	1.24678	1.25078	1.25479	1.25881	30
94000	1.26283	1.26687	1.27091	1.27496	1.27902	1.28309	1.28717	1.29126	1.29535	1.29945	30
95000	1.30357	1.30769	1.31182	1.31596	1.32010	1.32426	1.32842	1.33260	1.33678	1.34097	30
96000	1.34517	1.34937	1.35359	1.35782	1.36205	1.36629	1.37055	1.37481	1.37908	1.38336	30
97000	1.38764	1.39194	1.39624	1.40056	1.40488	1.40921	1.41355	1.41790	1.42226	1.42663	30
98000	1.43100	1.43539	1.43978	1.44419	1.44860	1.45302	1.45745	1.46189	1.46634	1.47079	30
99000	1.47526	1.47973	1.48422	1.48871	1.49321	1.49772	1.50224	1.50677	1.51131	1.51586	30

E. Blackbody Radiation Tables - Luminous

Table 17—Spectral Luminance

Table 18—Fractional Luminance

TABLE 17

Blackbody Spectral Luminance, Photopic ($cd/m^2.\mu$) | **Scotopic ($sc.cd/m^2.\mu$)**

$\lambda(\mu)$	Photopic 500° K.	600° K.	700° K.	800° K.	900° K.	Scotopic 500° K.	600° K.	700° K.	800° K.	900° K.
0.36	0.1012E-26	0.6183E-21	0.8392E-17	0.1055E-13	0.2717E-11					
0.37	0.2422E-25	0.1032E-19	0.1083E-15	0.1123E-12	0.2488E-10					
0.38	0.5167E-24	0.1564E-18	0.1287E-14	0.1111E-11	0.2137E-09	0.2004E-22	0.6071E-17	0.4993E-13	0.4313E-10	0.8291E-08
0.39	0.9732E-23	0.2132E-17	0.1392E-13	0.1010E-10	0.1698E-08	0.4600E-21	0.1008E-15	0.6580E-12	0.4778E-09	0.8027E-07
0.40	0.1790E-21	0.2884E-16	0.1511E-12	0.9310E-10	0.1376E-07	0.1078E-19	0.1737E-14	0.9107E-11	0.5609E-08	0.8290E-06
0.41	0.2794E-20	0.3361E-15	0.1429E-11	0.7528E-09	0.9850E-07	0.2065E-18	0.2484E-13	0.1056E-09	0.5564E-07	0.7280E-05
0.42	0.4355E-19	0.3965E-14	0.1382E-10	0.6269E-08	0.7304E-06	0.2701E-17	0.2458E-12	0.8570E-09	0.3888E-06	0.4529E-04
0.43	0.5525E-18	0.3856E-13	0.1111E-09	0.4375E-07	0.4563E-05	0.2443E-16	0.1705E-11	0.4917E-08	0.1935E-05	0.2018E-03
0.44	0.4469E-17	0.2420E-12	0.5824E-09	0.2000E-06	0.1877E-04	0.1636E-15	0.8867E-11	0.2133E-07	0.7328E-05	0.6877E-03
0.45	0.2822E-16	0.1200E-11	0.2428E-08	0.7327E-06	0.6216E-04	0.8677E-15	0.3689E-10	0.7466E-07	0.2252E-04	0.1911E-02
0.46	0.1603E-15	0.5407E-11	0.9273E-08	0.2471E-05	0.1903E-03	0.3891E-14	0.1312E-09	0.2250E-06	0.5998E-04	0.4620E-02
0.47	0.8263E-15	0.2232E-10	0.3267E-07	0.7731E-05	0.5429E-03	0.1575E-13	0.4256E-09	0.6229E-06	0.1474E-03	0.1035E-01
0.48	0.4068E-14	0.8886E-10	0.1117E-06	0.2359E-04	0.1516E-02	0.5959E-13	0.1301E-08	0.1636E-05	0.3456E-03	0.2221E-01
0.49	0.1866E-13	0.3325E-09	0.3614E-06	0.6842E-04	0.4039E-02	0.2083E-12	0.3711E-08	0.4034E-05	0.7638E-03	0.4509E-01
0.50	0.8480E-13	0.1242E-08	0.1173E-05	0.2001E-03	0.1088E-01	0.6618E-12	0.9692E-08	0.9161E-05	0.1561E-02	0.8498E-01
0.51	0.3697E-12	0.4486E-08	0.3707E-05	0.5714E-03	0.2875E-01	0.1880E-11	0.2282E-07	0.1886E-04	0.2907E-02	0.1462E 00
0.52	0.1401E-11	0.1419E-07	0.1030E-04	0.1442E-02	0.6729E-01	0.4740E-11	0.4800E-07	0.3486E-04	0.4877E-02	0.2275E 00
0.53	0.4395E-11	0.3740E-07	0.2398E-04	0.3057E-02	0.1326E 00	0.1061E-10	0.9035E-07	0.5794E-04	0.7385E-02	0.3204E 00
0.54	0.1210E-10	0.8714E-07	0.4958E-04	0.5776E-02	0.2337E 00	0.2117E-10	0.1523E-06	0.8670E-04	0.1010E-01	0.4088E 00
0.55	0.3035E-10	0.1859E-06	0.9425E-04	0.1007E-01	0.3810E 00	0.3766E-10	0.2306E-06	0.1169E-03	0.1249E-01	0.4728E 00
0.56	0.7061E-10	0.3700E-06	0.1678E-03	0.1650E-01	0.5851E 00	0.5991E-10	0.3140E-06	0.1424E-03	0.1400E-01	0.4965E 00
0.57	0.1523E-09	0.6869E-06	0.2799E-03	0.2538E-01	0.8456E 00	0.8529E-10	0.3846E-06	0.1567E-03	0.1421E-01	0.4734E 00
0.58	0.3047E-09	0.1188E-05	0.4366E-03	0.3664E-01	0.1148E 01	0.1090E-09	0.4252E-06	0.1562E-03	0.1310E-01	0.4109E 00
0.59	0.5644E-09	0.1913E-05	0.6360E-03	0.4951E-01	0.1464E 01	0.1253E-09	0.4250E-06	0.1412E-03	0.1099E-01	0.3252E 00
0.60	0.9751E-09	0.2887E-05	0.8711E-03	0.6306E-01	0.1762E 01	0.1315E-09	0.3894E-06	0.1175E-03	0.8507E-02	0.2378E 00
0.61	0.1570E-08	0.4080E-05	0.1121E-02	0.7565E-01	0.2002E 01	0.1277E-09	0.3317E-06	0.9116E-04	0.6152E-02	0.1628E 00
0.62	0.2347E-08	0.5371E-05	0.1348E-02	0.8500E-01	0.2133E 01	0.1166E-09	0.2669E-06	0.6699E-04	0.4224E-02	0.1060E 00
0.63	0.3148E-08	0.6372E-05	0.1464E-02	0.8648E-01	0.2063E 01	0.1017E-09	0.2059E-06	0.4733E-04	0.2794E-02	0.6666E-01
0.64	0.3923E-08	0.7049E-05	0.1488E-02	0.8246E-01	0.1872E 01	0.8618E-10	0.1548E-06	0.3269E-04	0.1811E-02	0.4111E-01
0.65	0.4434E-08	0.7099E-05	0.1380E-02	0.7189E-01	0.1555E 01	0.7205E-10	0.1153E-06	0.2243E-04	0.1168E-02	0.2527E-01
0.66	0.4580E-08	0.6557E-05	0.1177E-02	0.5775E-01	0.1192E 01	0.6032E-10	0.8637E-07	0.1550E-04	0.7606E-03	0.1570E-01
0.67	0.4272E-08	0.5488E-05	0.9119E-03	0.4220E-01	0.8330E 00	0.5073E-10	0.6517E-07	0.1082E-04	0.5012E-03	0.9893E-02
0.68	0.3963E-08	0.4583E-05	0.7063E-03	0.3089E-01	0.5837E 00	0.4283E-10	0.4952E-07	0.7633E-05	0.3339E-03	0.6308E-02
0.69	0.3285E-08	0.3430E-05	0.4914E-03	0.2035E-01	0.3684E 00	0.3630E-10	0.3790E-07	0.5430E-05	0.2248E-03	0.4071E-02
0.70	0.2772E-08	0.2620E-05	0.3497E-03	0.1373E-01	0.2385E 00	0.3088E-10	0.2919E-07	0.3896E-05	0.1529E-03	0.2657E-02
0.71	0.2348E-08	0.2015E-05	0.2511E-03	0.9363E-02	0.1562E 00	0.2636E-10	0.2263E-07	0.2819E-05	0.1051E-03	0.1754E-02
0.72	0.1925E-08	0.1504E-05	0.1752E-03	0.6215E-02	0.9972E-01	0.2258E-10	0.1764E-07	0.2055E-05	0.7290E-04	0.1169E-02
0.73	0.1543E-08	0.1100E-05	0.1201E-03	0.4056E-02	0.6266E-01	0.1939E-10	0.1383E-07	0.1510E-05	0.5099E-04	0.7877E-03
0.74	0.1176E-08	0.7680E-06	0.7868E-04	0.2533E-02	0.3771E-01	0.1672E-10	0.1091E-07	0.1118E-05	0.3600E-04	0.5359E-03
0.75	0.8900E-09	0.5327E-06	0.5131E-04	0.1577E-02	0.2265E-01	0.1446E-10	0.8659E-08	0.8339E-06	0.2563E-04	0.3681E-03
0.76	0.6899E-09	0.3797E-06	0.3443E-04	0.1012E-02	0.1403E-01	0.1254E-10	0.6904E-08	0.6261E-06	0.1840E-04	0.2551E-03
0.77	0.5284E-09	0.2679E-06	0.2291E-04	0.6445E-03	0.8636E-02	0.1091E-10	0.5533E-08	0.4732E-06	0.1331E-04	0.1783E-03
0.78	0.3996E-09	0.1870E-06	0.1511E-04	0.4073E-03	0.5279E-02	0.9516E-11	0.4454E-08	0.3599E-06	0.9698E-05	0.1257E-03
0.79	0.2979E-09	0.1290E-06	0.9861E-05	0.2548E-03	0.3198E-02					
0.80	0.2188E-09	0.8782E-07	0.6358E-05	0.1577E-03	0.1918E-02					
0.81	0.1590E-09	0.5928E-07	0.4070E-05	0.9709E-04	0.1144E-02					
0.82	0.1143E-09	0.3966E-07	0.2586E-05	0.5936E-04	0.6790E-03					
0.83	0.8149E-10	0.2633E-07	0.1633E-05	0.3609E-04	0.4009E-03					

$L_{v\lambda}(\lambda,T)$ $L_{v'\lambda}(\lambda,T)$

TABLE 17

Blackbody Spectral Luminance, Photopic (cd/m^2.μ)

λ(μ)	1000°K.	1100°K.	1200°K.	1300°K.	1400°K.	1500°K.	1600°K.	1700°K.	1800°K.	1900°K.
0.36	0.2304E-09	0.8721E-08	0.1801E-06	0.2334E-05	0.2098E-04	0.1407E-03	0.7440E-03	0.3234E-02	0.1193E-01	0.3841E-01
0.37	0.1872E-08	0.6422E-07	0.1222E-05	0.1477E-04	0.1251E-03	0.7975E-03	0.4031E-02	0.1683E-01	0.6000E-01	0.1870E 00
0.38	0.1435E-07	0.4485E-06	0.7897E-05	0.8945E-04	0.7162E-03	0.4346E-02	0.2105E-01	0.8468E-01	0.2918E 00	0.8830E 00
0.39	0.1023E-06	0.2929E-05	0.4792E-04	0.5100E-03	0.3872E-02	0.2243E-01	0.1043E 00	0.4050E 00	0.1352E 01	0.3977E 01
0.40	0.7487E-06	0.1970E-04	0.3005E-03	0.3015E-02	0.2175E-01	0.1206E 00	0.5399E 00	0.2026E 01	0.6564E 01	0.1879E 02
0.41	0.4862E-05	0.1181E-03	0.1686E-02	0.1598E-01	0.1099E 00	0.5847E 00	0.2523E 01	0.9168E 01	0.2886E 02	0.8053E 02
0.42	0.3285E-04	0.7398E-03	0.9913E-02	0.8910E-01	0.5852E 00	0.2991E 01	0.1246E 02	0.4392E 02	0.1345E 03	0.3663E 03
0.43	0.1879E-03	0.3935E-02	0.4964E-01	0.4240E 00	0.2665E 01	0.1311E 02	0.5287E 02	0.1809E 03	0.5400E 03	0.1436E 04
0.44	0.7104E-03	0.1388E-01	0.1653E 00	0.1345E 01	0.8110E 01	0.3848E 02	0.1503E 03	0.5002E 03	0.1456E 04	0.3788E 04
0.45	0.2169E-02	0.3969E-01	0.4473E 00	0.3473E 01	0.2012E 02	0.9224E 02	0.3495E 03	0.1132E 04	0.3219E 04	0.8200E 04
0.46	0.6150E-02	0.1056E 00	0.1129E 01	0.8387E 01	0.4677E 02	0.2074E 03	0.7635E 03	0.2411E 04	0.6701E 04	0.1672E 05
0.47	0.1629E-01	0.2634E 00	0.2678E 01	0.1905E 02	0.1024E 03	0.4401E 03	0.1576E 04	0.4856E 04	0.1320E 05	0.3232E 05
0.48	0.4239E-01	0.6468E 00	0.6265E 01	0.4280E 02	0.2221E 03	0.9260E 03	0.3228E 04	0.9719E 04	0.2588E 05	0.6218E 05
0.49	0.1055E 00	0.1522E 01	0.1408E 02	0.9248E 02	0.4642E 03	0.1879E 04	0.6387E 04	0.1880E 05	0.4908E 05	0.1158E 06
0.50	0.2664E 00	0.3644E 01	0.3224E 02	0.2039E 03	0.9913E 03	0.3902E 04	0.1294E 05	0.3728E 05	0.9547E 05	0.2214E 06
0.51	0.6606E 00	0.8586E 01	0.7278E 02	0.4440E 03	0.2092E 04	0.8017E 04	0.2597E 05	0.7328E 05	0.1842E 06	0.4203E 06
0.52	0.1455E 01	0.1801E 02	0.1465E 03	0.8633E 03	0.3948E 04	0.1474E 05	0.4670E 05	0.1291E 06	0.3190E 06	0.7164E 06
0.53	0.2708E 01	0.3195E 02	0.2498E 03	0.1423E 04	0.6327E 04	0.2305E 05	0.7143E 05	0.1938E 06	0.4706E 06	0.1040E 07
0.54	0.4513E 01	0.5087E 02	0.3829E 03	0.2112E 04	0.9134E 04	0.3248E 05	0.9858E 05	0.2625E 06	0.6271E 06	0.1366E 07
0.55	0.6971E 01	0.7519E 02	0.5455E 03	0.2918E 04	0.1228E 05	0.4269E 05	0.1269E 06	0.3322E 06	0.7811E 06	0.1678E 07
0.56	0.1016E 02	0.1050E 03	0.7358E 03	0.3820E 04	0.1567E 05	0.5327E 05	0.1553E 06	0.3996E 06	0.9253E 06	0.1961E 07
0.57	0.1397E 02	0.1386E 03	0.9382E 03	0.4731E 04	0.1893E 05	0.6300E 05	0.1803E 06	0.4562E 06	0.1040E 07	0.2177E 07
0.58	0.1808E 02	0.1724E 03	0.1129E 04	0.5540E 04	0.2165E 05	0.7055E 05	0.1983E 06	0.4937E 06	0.1110E 07	0.2293E 07
0.59	0.2199E 02	0.2019E 03	0.1281E 04	0.6115E 04	0.2335E 05	0.7459E 05	0.2060E 06	0.5050E 06	0.1120E 07	0.2286E 07
0.60	0.2531E 02	0.2239E 03	0.1377E 04	0.6407E 04	0.2392E 05	0.7495E 05	0.2035E 06	0.4916E 06	0.1076E 07	0.2170E 07
0.61	0.2752E 02	0.2349E 03	0.1402E 04	0.6362E 04	0.2325E 05	0.7149E 05	0.1910E 06	0.4546E 06	0.9827E 06	0.1958E 07
0.62	0.2811E 02	0.2318E 03	0.1345E 04	0.5953E 04	0.2130E 05	0.6433E 05	0.1691E 06	0.3971E 06	0.8477E 06	0.1670E 07
0.63	0.2609E 02	0.2080E 03	0.1173E 04	0.5074E 04	0.1779E 05	0.5280E 05	0.1367E 06	0.3166E 06	0.6679E 06	0.1302E 07
0.64	0.2275E 02	0.1756E 03	0.9646E 03	0.4075E 04	0.1401E 05	0.4088E 05	0.1043E 06	0.2384E 06	0.4970E 06	0.9591E 06
0.65	0.1819E 02	0.1361E 03	0.7281E 03	0.3009E 04	0.1015E 05	0.2913E 05	0.7327E 05	0.1653E 06	0.3408E 06	0.6510E 06
0.66	0.1344E 02	0.9753E 02	0.5086E 03	0.2057E 04	0.6815E 04	0.1924E 05	0.4773E 05	0.1063E 06	0.2169E 06	0.4102E 06
0.67	0.9055E 01	0.6379E 02	0.3245E 03	0.1285E 04	0.4183E 04	0.1163E 05	0.2846E 05	0.6268E 05	0.1264E 06	0.2369E 06
0.68	0.6126E 01	0.4193E 02	0.2083E 03	0.8086E 03	0.2586E 04	0.7083E 04	0.1710E 05	0.3723E 05	0.7434E 05	0.1380E 06
0.69	0.3737E 01	0.2487E 02	0.1207E 03	0.4596E 03	0.1445E 04	0.3901E 04	0.9301E 04	0.2002E 05	0.3957E 05	0.7281E 05
0.70	0.2340E 01	0.1516E 02	0.7196E 02	0.2687E 03	0.8313E 03	0.2212E 04	0.5209E 04	0.1109E 05	0.2171E 05	0.3960E 05
0.71	0.1484E 01	0.9369E 01	0.4349E 02	0.1594E 03	0.4854E 03	0.1274E 04	0.2964E 04	0.6244E 04	0.1210E 05	0.2190E 05
0.72	0.9185E 00	0.5650E 01	0.2567E 02	0.9243E 02	0.2771E 03	0.7177E 03	0.1650E 04	0.3440E 04	0.6610E 04	0.1185E 05
0.73	0.5598E 00	0.3359E 01	0.1495E 02	0.5289E 02	0.1562E 03	0.3993E 03	0.9077E 03	0.1873E 04	0.3567E 04	0.6348E 04
0.74	0.3271E 00	0.1916E 01	0.8357E 01	0.2906E 02	0.8459E 02	0.2135E 03	0.4800E 03	0.9811E 03	0.1852E 04	0.3270E 04
0.75	0.1909E 00	0.1092E 01	0.4670E 01	0.1597E 02	0.4584E 02	0.1142E 03	0.2541E 03	0.5145E 03	0.9631E 03	0.1687E 04
0.76	0.1149E 00	0.6428E 00	0.2697E 01	0.9078E 01	0.2568E 02	0.6328E 02	0.1392E 03	0.2793E 03	0.5185E 03	0.9020E 03
0.77	0.6886E-01	0.3764E 00	0.1550E 01	0.5137E 01	0.1434E 02	0.3491E 02	0.7606E 02	0.1511E 03	0.2784E 03	0.4808E 03
0.78	0.4099E-01	0.2192E 00	0.8869E 00	0.2893E 01	0.7972E 01	0.1918E 02	0.4138E 02	0.8154E 02	0.1490E 03	0.2555E 03
0.79	0.2419E-01	0.1267E 00	0.5035E 00	0.1618E 01	0.4401E 01	0.1047E 02	0.2238E 02	0.4371E 02	0.7927E 02	0.1350E 03
0.80	0.1415E-01	0.7258E-01	0.2835E 00	0.8979E 00	0.2412E 01	0.5680E 01	0.1201E 02	0.2327E 02	0.4190E 02	0.7089E 02
0.81	0.8237E-02	0.4140E-01	0.1590E 00	0.4966E 00	0.1317E 01	0.3070E 01	0.6436E 01	0.1236E 02	0.2209E 02	0.3715E 02
0.82	0.4770E-02	0.2351E-01	0.8884E-01	0.2735E 00	0.7174E 00	0.1654E 01	0.3436E 01	0.6551E 01	0.1162E 02	0.1941E 02
0.83	0.2751E-02	0.1330E-01	0.4946E-01	0.1502E 00	0.3895E 00	0.8892E 00	0.1831E 01	0.3463E 01	0.6102E 01	0.1013E 02

TABLE 17

Blackbody Spectral Luminance, Scotopic (sc.cd/m^2.μ)

$\lambda(\mu)$	1000°K.	1100°K.	1200°K.	1300°K.	1400°K.	1500°K.	1600°K.	1700°K.	1800°K.	1900°K.
0.38	0.5568E-06	0.1740E-04	0.3064E-03	0.3470E-02	0.2779E-01	0.1686E 00	0.8167E 00	0.3285E 01	0.1132E 02	0.3426E 02
0.39	0.4839E-05	0.1384E-03	0.2265E-02	0.2410E-01	0.1830E 00	0.1060E 01	0.4932E 01	0.1914E 02	0.6392E 02	0.1880E 03
0.40	0.4511E-04	0.1186E-02	0.1810E-01	0.1816E 00	0.1310E 01	0.7268E 01	0.3253E 02	0.1220E 03	0.3955E 03	0.1132E 04
0.41	0.3593E-03	0.8730E-02	0.1246E 00	0.1181E 01	0.8127E 01	0.4321E 02	0.1865E 03	0.6776E 03	0.2133E 04	0.5952E 04
0.42	0.2037E-02	0.4588E-01	0.6147E 00	0.5526E 01	0.3629E 02	0.1854E 03	0.7730E 03	0.2723E 04	0.8344E 04	0.2272E 05
0.43	0.8310E-02	0.1740E 00	0.2195E 01	0.1875E 02	0.1178E 03	0.5800E 03	0.2338E 04	0.8001E 04	0.2388E 05	0.6353E 05
0.44	0.2602E-01	0.5086E 00	0.6056E 01	0.4927E 02	0.2970E 03	0.1409E 04	0.5506E 04	0.1832E 05	0.5334E 05	0.1387E 06
0.45	0.6670E-01	0.1220E 01	0.1375E 02	0.1067E 03	0.6187E 03	0.2836E 04	0.1074E 05	0.3481E 05	0.9898E 05	0.2521E 06
0.46	0.1492E 00	0.2563E 01	0.2741E 02	0.2035E 03	0.1135E 04	0.5034E 04	0.1853E 05	0.5852E 05	0.1626E 06	0.4059E 06
0.47	0.3106E 00	0.5022E 01	0.5106E 02	0.3633E 03	0.1953E 04	0.8392E 04	0.3005E 05	0.9260E 05	0.2518E 06	0.6163E 06
0.48	0.6210E 00	0.9474E 01	0.9178E 02	0.6269E 03	0.3254E 04	0.1356E 05	0.4729E 05	0.1423E 06	0.3791E 06	0.9109E 06
0.49	0.1177E 01	0.1699E 02	0.1571E 03	0.1032E 04	0.5182E 04	0.2097E 05	0.7130E 05	0.2098E 06	0.5479E 06	0.1292E 07
0.50	0.2079E 01	0.2844E 02	0.2516E 03	0.1591E 04	0.7736E 04	0.3045E 05	0.1010E 06	0.2909E 06	0.7450E 06	0.1728E 07
0.51	0.3361E 01	0.4368E 02	0.3702E 03	0.2258E 04	0.1064E 05	0.4078E 05	0.1321E 06	0.3728E 06	0.9372E 06	0.2138E 07
0.52	0.4924E 01	0.6091E 02	0.4955E 03	0.2919E 04	0.1335E 05	0.4986E 05	0.1579E 06	0.4368E 06	0.1078E 07	0.2422E 07
0.53	0.6543E 01	0.7719E 02	0.6036E 03	0.3439E 04	0.1528E 05	0.5568E 05	0.1725E 06	0.4681E 06	0.1136E 07	0.2514E 07
0.54	0.7892E 01	0.8895E 02	0.6695E 03	0.3694E 04	0.1597E 05	0.5680E 05	0.1723E 06	0.4591E 06	0.1096E 07	0.2390E 07
0.55	0.8650E 01	0.9329E 02	0.6769E 03	0.3620E 04	0.1524E 05	0.5297E 05	0.1575E 06	0.4122E 06	0.9691E 06	0.2082E 07
0.56	0.8624E 01	0.8915E 02	0.6243E 03	0.3241E 04	0.1329E 05	0.4520E 05	0.1318E 06	0.3390E 06	0.7851E 06	0.1664E 07
0.57	0.7822E 01	0.7761E 02	0.5253E 03	0.2649E 04	0.1060E 05	0.3527E 05	0.1009E 06	0.2554E 06	0.5828E 06	0.1219E 07
0.58	0.6469E 01	0.6169E 02	0.4040E 03	0.1981E 04	0.7744E 04	0.2523E 05	0.7094E 05	0.1766E 06	0.3972E 06	0.8205E 06
0.59	0.4886E 01	0.4485E 02	0.2845E 03	0.1358E 04	0.5187E 04	0.1656E 05	0.4576E 05	0.1121E 06	0.2488E 06	0.5077E 06
0.60	0.3414E 01	0.3020E 02	0.1858E 03	0.8642E 03	0.3227E 04	0.1011E 05	0.2746E 05	0.6631E 05	0.1451E 06	0.2927E 06
0.61	0.2238E 01	0.1910E 02	0.1140E 03	0.5174E 03	0.1890E 04	0.5813E 04	0.1553E 05	0.3697E 05	0.7991E 05	0.1592E 06
0.62	0.1397E 01	0.1152E 02	0.6684E 02	0.2958E 03	0.1058E 04	0.3197E 04	0.8408E 04	0.1973E 05	0.4213E 05	0.8303E 05
0.63	0.8432E 00	0.6723E 01	0.3793E 02	0.1639E 03	0.5751E 03	0.1706E 04	0.4419E 04	0.1023E 05	0.2158E 05	0.4208E 05
0.64	0.4998E 00	0.3858E 01	0.2118E 02	0.8952E 02	0.3078E 03	0.8980E 03	0.2291E 04	0.5236E 04	0.1091E 05	0.2106E 05
0.65	0.2957E 00	0.2212E 01	0.1183E 02	0.4890E 02	0.1650E 03	0.4734E 03	0.1190E 04	0.2686E 04	0.5538E 04	0.1058E 05
0.66	0.1770E 00	0.1284E 01	0.6698E 01	0.2709E 02	0.8976E 02	0.2534E 03	0.6286E 03	0.1401E 04	0.2856E 04	0.5403E 04
0.67	0.1075E 00	0.7575E 00	0.3854E 01	0.1526E 02	0.4968E 02	0.1381E 03	0.3380E 03	0.7443E 03	0.1501E 04	0.2813E 04
0.68	0.6620E-01	0.4531E 00	0.2251E 01	0.8739E 01	0.2794E 02	0.7654E 02	0.1848E 03	0.4023E 03	0.8034E 03	0.1491E 04
0.69	0.4129E-01	0.2749E 00	0.1334E 01	0.5078E 01	0.1597E 02	0.4310E 02	0.1027E 03	0.2212E 03	0.4372E 03	0.8045E 03
0.70	0.2608E-01	0.1689E 00	0.8018E 00	0.2994E 01	0.9263E 01	0.2465E 02	0.5804E 02	0.1235E 03	0.2419E 03	0.4412E 03
0.71	0.1666E-01	0.1051E 00	0.4883E 00	0.1790E 01	0.5450E 01	0.1430E 02	0.3328E 02	0.7011E 02	0.1359E 03	0.2458E 03
0.72	0.1077E-01	0.6627E-01	0.3011E 00	0.1084E 01	0.3250E 01	0.8418E 01	0.1935E 02	0.4035E 02	0.7754E 02	0.1390E 03
0.73	0.7038E-02	0.4223E-01	0.1879E 00	0.6649E 00	0.1963E 01	0.5020E 01	0.1141E 02	0.2355E 02	0.4485E 02	0.7981E 02
0.74	0.4648E-02	0.2722E-01	0.1187E 00	0.4129E 00	0.1201E 01	0.3033E 01	0.6820E 01	0.1394E 02	0.2631E 02	0.4646E 02
0.75	0.3102E-02	0.1774E-01	0.7591E-01	0.2596E 00	0.7450E 00	0.1857E 01	0.4131E 01	0.8363E 01	0.1565E 02	0.2743E 02
0.76	0.2090E-02	0.1168E-01	0.4905E-01	0.1650E 00	0.4671E 00	0.1150E 01	0.2532E 01	0.5079E 01	0.9429E 01	0.1640E 02
0.77	0.1422E-02	0.7775E-02	0.3202E-01	0.1060E 00	0.2961E 00	0.7211E 00	0.1570E 01	0.3122E 01	0.5750E 01	0.9930E 01
0.78	0.9760E-03	0.5220E-02	0.2111E-01	0.6889E-01	0.1898E 00	0.4568E 00	0.9853E 00	0.1941E 01	0.3547E 01	0.6084E 01

TABLE 17

Blackbody Spectral Luminance, Photopic (cd/m^2.μ)

$\lambda(\mu)$	2000°K.	2100°K.	2200°K.	2300°K.	2400°K.	2500°K.	2600°K.	2700°K.	2800°K.	2900°K.
0.36	0.1099E 00	0.2847E 00	0.6764E 00	0.1490E 01	0.3073E 01	0.5983E 01	0.1106E 02	0.1955E 02	0.3317E 02	0.5427E 02
0.37	0.5205E 00	0.1313E 01	0.3048E 01	0.6574E 01	0.1329E 02	0.2542E 02	0.4624E 02	0.8046E 02	0.1345E 03	0.2172E 03
0.38	0.2391E 01	0.5891E 01	0.1337E 02	0.2825E 02	0.5610E 02	0.1054E 03	0.1888E 03	0.3238E 03	0.5343E 03	0.8517E 03
0.39	0.1050E 02	0.2527E 02	0.5617E 02	0.1164E 03	0.2272E 03	0.4202E 03	0.7412E 03	0.1253E 04	0.2042E 04	0.3216E 04
0.40	0.4842E 02	0.1140E 03	0.2483E 03	0.5056E 03	0.9701E 03	0.1766E 04	0.3072E 04	0.5129E 04	0.8254E 04	0.1285E 05
0.41	0.2027E 03	0.4676E 03	0.9995E 03	0.1999E 04	0.3776E 04	0.6777E 04	0.1162E 05	0.1917E 05	0.3049E 05	0.4698E 05
0.42	0.9024E 03	0.2040E 04	0.4282E 04	0.8427E 04	0.1567E 05	0.2774E 05	0.4699E 05	0.7656E 05	0.1204E 06	0.1836E 06
0.43	0.3465E 04	0.7686E 04	0.1585E 05	0.3072E 05	0.5632E 05	0.9837E 05	0.1646E 06	0.2651E 06	0.4127E 06	0.6232E 06
0.44	0.8957E 04	0.1951E 05	0.3960E 05	0.7557E 05	0.1366E 06	0.2356E 06	0.3897E 06	0.6210E 06	0.9571E 06	0.1431E 07
0.45	0.1902E 05	0.4072E 05	0.8136E 05	0.1530E 06	0.2731E 06	0.4654E 06	0.7611E 06	0.1200E 07	0.1832E 07	0.2716E 07
0.46	0.3809E 05	0.8021E 05	0.1578E 06	0.2929E 06	0.5162E 06	0.8693E 06	0.1406E 07	0.2196E 07	0.3321E 07	0.4882E 07
0.47	0.7234E 05	0.1499E 06	0.2908E 06	0.5327E 06	0.9275E 06	0.1544E 07	0.2474E 07	0.3826E 07	0.5737E 07	0.8364E 07
0.48	0.1368E 06	0.2793E 06	0.5345E 06	0.9666E 06	0.1663E 07	0.2741E 07	0.4348E 07	0.6664E 07	0.9907E 07	0.1433E 08
0.49	0.2508E 06	0.5046E 06	0.9528E 06	0.1702E 07	0.2897E 07	0.4727E 07	0.7426E 07	0.1128E 08	0.1663E 08	0.2388E 08
0.50	0.4722E 06	0.9369E 06	0.1746E 07	0.3084E 07	0.5195E 07	0.8392E 07	0.1306E 08	0.1968E 08	0.2880E 08	0.4105E 08
0.51	0.8832E 06	0.1728E 07	0.3184E 07	0.5560E 07	0.9269E 07	0.1483E 08	0.2289E 08	0.3422E 08	0.4970E 08	0.7035E 08
0.52	0.1483E 07	0.2867E 07	0.5219E 07	0.9017E 07	0.1488E 08	0.2360E 08	0.3613E 08	0.5359E 08	0.7727E 08	0.1086E 09
0.53	0.2126E 07	0.4058E 07	0.7303E 07	0.1248E 08	0.2042E 08	0.3210E 08	0.4875E 08	0.7177E 08	0.1027E 09	0.1435E 09
0.54	0.2755E 07	0.5197E 07	0.9252E 07	0.1566E 08	0.2538E 08	0.3957E 08	0.5962E 08	0.8715E 08	0.1239E 09	0.1721E 09
0.55	0.3341E 07	0.6228E 07	0.1097E 08	0.1840E 08	0.2955E 08	0.4570E 08	0.6835E 08	0.9922E 08	0.1402E 09	0.1935E 09
0.56	0.3856E 07	0.7109E 07	0.1239E 08	0.2060E 08	0.3281E 08	0.5035E 08	0.7476E 08	0.1078E 09	0.1514E 09	0.2078E 09
0.57	0.4231E 07	0.7717E 07	0.1332E 08	0.2194E 08	0.3467E 08	0.5280E 08	0.7787E 08	0.1115E 09	0.1558E 09	0.2126E 09
0.58	0.4406E 07	0.7954E 07	0.1360E 08	0.2221E 08	0.3482E 08	0.5265E 08	0.7712E 08	0.1098E 09	0.1524E 09	0.2069E 09
0.59	0.434[illegible] 07	0.7762E 07	0.1315E 08	0.2130E 08	0.3314E 08	0.4976E 08	0.7242E 08	0.1025E 09	0.1415E 09	0.1911E 09
0.60	0.4078E 07	0.7219E 07	0.1213E 08	0.1948E 08	0.3008E 08	0.4487E 08	0.6489E 08	0.9132E 08	0.1254E 09	0.1685E 09
0.61	0.3643E 07	0.6389E 07	0.1064E 08	0.1696E 08	0.2601E 08	0.3854E 08	0.5540E 08	0.7753E 08	0.1059E 09	0.1416E 09
0.62	0.3077E 07	0.5347E 07	0.8837E 07	0.1397E 08	0.2128E 08	0.3133E 08	0.4478E 08	0.6233E 08	0.8473E 08	0.1127E 09
0.63	0.2375E 07	0.4091E 07	0.6708E 07	0.1053E 08	0.1593E 08	0.2331E 08	0.3313E 08	0.4587E 08	0.6205E 08	0.8222E 08
0.64	0.1733E 07	0.2960E 07	0.4815E 07	0.7509E 07	0.1128E 08	0.1641E 08	0.2319E 08	0.3195E 08	0.4302E 08	0.5675E 08
0.65	0.1165E 07	0.1974E 07	0.3188E 07	0.4938E 07	0.7375E 07	0.1066E 08	0.1499E 08	0.2055E 08	0.2754E 08	0.3618E 08
0.66	0.7281E 06	0.1223E 07	0.1961E 07	0.3018E 07	0.4479E 07	0.6442E 07	0.9010E 07	0.1229E 08	0.1640E 08	0.2145E 08
0.67	0.4169E 06	0.6952E 06	0.1106E 07	0.1691E 07	0.2496E 07	0.3570E 07	0.4969E 07	0.6747E 07	0.8965E 07	0.1168E 08
0.68	0.2408E 06	0.3986E 06	0.6301E 06	0.9574E 06	0.1404E 07	0.1998E 07	0.2768E 07	0.3742E 07	0.4951E 07	0.6426E 07
0.69	0.1260E 06	0.2071E 06	0.3252E 06	0.4911E 06	0.7165E 06	0.1014E 07	0.1398E 07	0.1882E 07	0.2480E 07	0.3207E 07
0.70	0.6802E 05	0.1109E 06	0.1731E 06	0.2599E 06	0.3772E 06	0.5313E 06	0.7290E 06	0.9772E 06	0.1282E 07	0.1652E 07
0.71	0.3733E 05	0.6048E 05	0.9378E 05	0.1399E 06	0.2020E 06	0.2833E 06	0.3870E 06	0.5165E 06	0.6755E 06	0.8671E 06
0.72	0.2006E 05	0.3228E 05	0.4976E 05	0.7386E 05	0.1061E 06	0.1480E 06	0.2013E 06	0.2677E 06	0.3487E 06	0.4461E 06
0.73	0.1066E 05	0.1705E 05	0.2612E 05	0.3857E 05	0.5512E 05	0.7657E 05	0.1037E 06	0.1373E 06	0.1782E 06	0.2273E 06
0.74	0.5455E 04	0.8666E 04	0.1320E 05	0.1938E 05	0.2757E 05	0.3813E 05	0.5144E 05	0.6787E 05	0.8779E 05	0.1115E 06
0.75	0.2796E 04	0.4415E 04	0.6688E 04	0.9772E 04	0.1383E 05	0.1905E 05	0.2559E 05	0.3364E 05	0.4337E 05	0.5494E 05
0.76	0.1484E 04	0.2330E 04	0.3510E 04	0.5103E 04	0.7192E 04	0.9862E 04	0.1319E 05	0.1728E 05	0.2221E 05	0.2805E 05
0.77	0.7862E 03	0.1226E 04	0.1838E 04	0.2660E 04	0.3732E 04	0.5096E 04	0.6795E 04	0.8870E 04	0.1136E 05	0.1430E 05
0.78	0.4152E 03	0.6442E 03	0.9604E 03	0.1383E 04	0.1932E 04	0.2627E 04	0.3490E 04	0.4541E 04	0.5797E 04	0.7278E 04
0.79	0.2180E 03	0.3364E 03	0.4991E 03	0.7154E 03	0.9952E 03	0.1348E 04	0.1784E 04	0.2314E 04	0.2945E 04	0.3687E 04
0.80	0.1138E 03	0.1746E 03	0.2578E 03	0.3678E 03	0.5096E 03	0.6879E 03	0.9074E 03	0.1172E 04	0.1488E 04	0.1857E 04
0.81	0.5929E 02	0.9051E 02	0.1329E 03	0.1889E 03	0.2606E 03	0.3505E 03	0.4608E 03	0.5936E 03	0.7511E 03	0.9352E 03
0.82	0.3081E 02	0.4679E 02	0.6842E 02	0.9680E 02	0.1330E 03	0.1782E 03	0.2335E 03	0.3000E 03	0.3785E 03	0.4700E 03
0.83	0.1598E 02	0.2416E 02	0.3516E 02	0.4954E 02	0.6783E 02	0.9057E 02	0.1182E 03	0.1514E 03	0.1906E 03	0.2360E 03

TABLE 17

Blackbody Spectral Luminance, Scotopic (sc.cd/m^2.μ)

$\lambda(\mu)$	2000°K.	2100°K.	2200°K.	2300°K.	2400°K.	2500°K.	2600°K.	2700°K.	2800°K.	2900°K.
0.38	0.9279E 02	0.2285E 03	0.5187E 03	0.1096E 04	0.2176E 04	0.4091E 04	0.7326E 04	0.1256E 05	0.2073E 05	0.3304E 05
0.39	0.4963E 03	0.1194E 04	0.2655E 04	0.5504E 04	0.1073E 05	0.1986E 05	0.3503E 05	0.5925E 05	0.9653E 05	0.1520E 06
0.40	0.2917E 04	0.6870E 04	0.1496E 05	0.3046E 05	0.5845E 05	0.1064E 06	0.1851E 06	0.3090E 06	0.4973E 06	0.7745E 06
0.41	0.1498E 05	0.3456E 05	0.7387E 05	0.1478E 06	0.2791E 06	0.5009E 06	0.8595E 06	0.1416E 07	0.2254E 07	0.3472E 07
0.42	0.5596E 05	0.1265E 06	0.2655E 06	0.5226E 06	0.9721E 06	0.1720E 07	0.2914E 07	0.4747E 07	0.7469E 07	0.1138E 08
0.43	0.1532E 06	0.3399E 06	0.7013E 06	0.1358E 07	0.2490E 07	0.4350E 07	0.7279E 07	0.1172E 08	0.1825E 08	0.2756E 08
0.44	0.3281E 06	0.7147E 06	0.1450E 07	0.2768E 07	0.5005E 07	0.8632E 07	0.1427E 08	0.2274E 08	0.3505E 08	0.5244E 08
0.45	0.5848E 06	0.1252E 07	0.2501E 07	0.4705E 07	0.8397E 07	0.1430E 08	0.2340E 08	0.3689E 08	0.5632E 08	0.8350E 08
0.46	0.9245E 06	0.1947E 07	0.3831E 07	0.7109E 07	0.1252E 08	0.2110E 08	0.3414E 08	0.5331E 08	0.8063E 08	0.1185E 09
0.47	0.1379E 07	0.2859E 07	0.5546E 07	0.1015E 08	0.1768E 08	0.2945E 08	0.4717E 08	0.7296E 08	0.1093E 09	0.1594E 09
0.48	0.2004E 07	0.4092E 07	0.7830E 07	0.1415E 08	0.2437E 08	0.4016E 08	0.6369E 08	0.9762E 08	0.1451E 09	0.2099E 09
0.49	0.2800E 07	0.5633E 07	0.1063E 08	0.1900E 08	0.3234E 08	0.5277E 08	0.8290E 08	0.1259E 09	0.1857E 09	0.2666E 09
0.50	0.3685E 07	0.7312E 07	0.1363E 08	0.2407E 08	0.4054E 08	0.6549E 08	0.1019E 09	0.1536E 09	0.2248E 09	0.3204E 09
0.51	0.4493E 07	0.8795E 07	0.1619E 08	0.2828E 08	0.4715E 08	0.7546E 08	0.1164E 09	0.1740E 09	0.2528E 09	0.3579E 09
0.52	0.5018E 07	0.9698E 07	0.1765E 08	0.3049E 08	0.5034E 08	0.7984E 08	0.1222E 09	0.1812E 09	0.2613E 09	0.3674E 09
0.53	0.5136E 07	0.9803E 07	0.1764E 08	0.3017E 08	0.4933E 08	0.7756E 08	0.1177E 09	0.1733E 09	0.2482E 09	0.3468E 09
0.54	0.4819E 07	0.9088E 07	0.1617E 08	0.2739E 08	0.4438E 08	0.6920E 08	0.1042E 09	0.1523E 09	0.2167E 09	0.3010E 09
0.55	0.4145E 07	0.7727E 07	0.1361E 08	0.2282E 08	0.3667E 08	0.5671E 08	0.8481E 08	0.1231E 09	0.1740E 09	0.2401E 09
0.56	0.3272E 07	0.6032E 07	0.1052E 08	0.1748E 08	0.2784E 08	0.4272E 08	0.6343E 08	0.9147E 08	0.1285E 09	0.1763E 09
0.57	0.2369E 07	0.4321E 07	0.7462E 07	0.1228E 08	0.1941E 08	0.2956E 08	0.4360E 08	0.6247E 08	0.8723E 08	0.1190E 09
0.58	0.1576E 07	0.2845E 07	0.4867E 07	0.7947E 07	0.1245E 08	0.1883E 08	0.2758E 08	0.3928E 08	0.5453E 08	0.7403E 08
0.59	0.9646E 06	0.1724E 07	0.2922E 07	0.4732E 07	0.7361E 07	0.1105E 08	0.1608E 08	0.2276E 08	0.3143E 08	0.4245E 08
0.60	0.5501E 06	0.9738E 06	0.1636E 07	0.2628E 07	0.4058E 07	0.6053E 07	0.8754E 07	0.1231E 08	0.1691E 08	0.2273E 08
0.61	0.2962E 06	0.5195E 06	0.8656E 06	0.1379E 07	0.2115E 07	0.3134E 07	0.4505E 07	0.6304E 07	0.8613E 07	0.1151E 08
0.62	0.1529E 06	0.2657E 06	0.4391E 06	0.6947E 06	0.1057E 07	0.1557E 07	0.2225E 07	0.3097E 07	0.4210E 07	0.5604E 07
0.63	0.7676E 05	0.1322E 06	0.2167E 06	0.3404E 06	0.5148E 06	0.7533E 06	0.1070E 07	0.1482E 07	0.2005E 07	0.2656E 07
0.64	0.3806E 05	0.6501E 05	0.1057E 06	0.1649E 06	0.2478E 06	0.3605E 06	0.5095E 06	0.7019E 06	0.9450E 06	0.1246E 07
0.65	0.1894E 05	0.3209E 05	0.5181E 05	0.8025E 05	0.1198E 06	0.1733E 06	0.2436E 06	0.3340E 06	0.4476E 06	0.5880E 06
0.66	0.9590E 04	0.1611E 05	0.2583E 05	0.3974E 05	0.5900E 05	0.8485E 05	0.1186E 06	0.1619E 06	0.2160E 06	0.2826E 06
0.67	0.4951E 04	0.8256E 04	0.1314E 05	0.2008E 05	0.2964E 05	0.4240E 05	0.5900E 05	0.8013E 05	0.1064E 06	0.1387E 06
0.68	0.2602E 04	0.4307E 04	0.6810E 04	0.1034E 05	0.1518E 05	0.2160E 05	0.2991E 05	0.4043E 05	0.5350E 05	0.6944E 05
0.69	0.1392E 04	0.2288E 04	0.3593E 04	0.5426E 04	0.7917E 04	0.1120E 05	0.1545E 05	0.2079E 05	0.2740E 05	0.3543E 05
0.70	0.7578E 03	0.1236E 04	0.1929E 04	0.2896E 04	0.4202E 04	0.5920E 04	0.8123E 04	0.1088E 05	0.1429E 05	0.1841E 05
0.71	0.4191E 03	0.6790E 03	0.1052E 04	0.1571E 04	0.2268E 04	0.3180E 04	0.4344E 04	0.5799E 04	0.7584E 04	0.9735E 04
0.72	0.2353E 03	0.3787E 03	0.5837E 03	0.8664E 03	0.1244E 04	0.1736E 04	0.2361E 04	0.3140E 04	0.4090E 04	0.5233E 04
0.73	0.1340E 03	0.2143E 03	0.3284E 03	0.4849E 03	0.6930E 03	0.9626E 03	0.1303E 04	0.1726E 04	0.2241E 04	0.2857E 04
0.74	0.7750E 02	0.1231E 03	0.1875E 03	0.2754E 03	0.3918E 03	0.5418E 03	0.7309E 03	0.9643E 03	0.1247E 04	0.1585E 04
0.75	0.4544E 02	0.7176E 02	0.1087E 03	0.1588E 03	0.2248E 03	0.3096E 03	0.4159E 03	0.5468E 03	0.7049E 03	0.8930E 03
0.76	0.2699E 02	0.4237E 02	0.6383E 02	0.9280E 02	0.1307E 03	0.1793E 03	0.2400E 03	0.3143E 03	0.4039E 03	0.5101E 03
0.77	0.1623E 02	0.2533E 02	0.3797E 02	0.5494E 02	0.7708E 02	0.1052E 03	0.1403E 03	0.1831E 03	0.2346E 03	0.2954E 03
0.78	0.9886E 01	0.1533E 02	0.2286E 02	0.3292E 02	0.4600E 02	0.6256E 02	0.8311E 02	0.1081E 03	0.1380E 03	0.1733E 03

TABLE 17

Blackbody Spectral Luminance, Photopic (cd/m^2.μ)

$\lambda(\mu)$	3000° K.	3100° K.	3200° K.	3300° K.	3400° K.	3500° K.	3600° K.	3700° K.	3800° K.	3900° K.
0.36	0.8592E 02	0.1320E 03	0.1975E 03	0.2884E 03	0.4119E 03	0.5763E 03	0.7914E 03	0.1068E 04	0.1419E 04	0.1859E 04
0.37	0.3397E 03	0.5160E 03	0.7637E 03	0.1103E 04	0.1560E 04	0.2164E 04	0.2946E 04	0.3945E 04	0.5203E 04	0.6764E 04
0.38	0.1316E 04	0.1977E 04	0.2896E 04	0.4146E 04	0.5810E 04	0.7986E 04	0.1078E 05	0.1433E 05	0.1876E 05	0.2422E 05
0.39	0.4915E 04	0.7309E 04	0.1060E 05	0.1503E 05	0.2088E 05	0.2848E 05	0.3816E 05	0.5035E 05	0.6545E 05	0.8396E 05
0.40	0.1943E 05	0.2861E 05	0.4112E 05	0.5781E 05	0.7966E 05	0.1077E 06	0.1433E 06	0.1878E 06	0.2426E 06	0.3092E 06
0.41	0.7032E 05	0.1025E 06	0.1460E 06	0.2036E 06	0.2784E 06	0.3740E 06	0.4941E 06	0.6430E 06	0.8254E 06	0.1045E 07
0.42	0.2722E 06	0.3935E 06	0.5558E 06	0.7689E 06	0.1043E 07	0.1391E 07	0.1826E 07	0.2362E 07	0.3013E 07	0.3797E 07
0.43	0.9155E 06	0.1311E 07	0.1838E 07	0.2523E 07	0.3400E 07	0.4504E 07	0.5874E 07	0.7552E 07	0.9582E 07	0.1201E 08
0.44	0.2084E 07	0.2963E 07	0.4120E 07	0.5616E 07	0.7516E 07	0.9894E 07	0.1282E 08	0.1639E 08	0.2068E 08	0.2579E 08
0.45	0.3922E 07	0.5531E 07	0.7635E 07	0.1033E 08	0.1374E 08	0.1798E 08	0.2317E 08	0.2946E 08	0.3698E 08	0.4589E 08
0.46	0.6995E 07	0.9792E 07	0.1342E 08	0.1804E 08	0.2385E 08	0.3102E 08	0.3976E 08	0.5029E 08	0.6283E 08	0.7760E 08
0.47	0.1189E 08	0.1652E 08	0.2250E 08	0.3007E 08	0.3950E 08	0.5109E 08	0.6515E 08	0.8199E 08	0.1019E 09	0.1253E 09
0.48	0.2022E 08	0.2792E 08	0.3777E 08	0.5017E 08	0.6553E 08	0.8431E 08	0.1069E 09	0.1339E 09	0.1658E 09	0.2029E 09
0.49	0.3347E 08	0.4590E 08	0.6172E 08	0.8151E 08	0.1059E 09	0.1355E 09	0.1711E 09	0.2133E 09	0.2629E 09	0.3205E 09
0.50	0.5715E 08	0.7788E 08	0.1040E 09	0.1367E 09	0.1766E 09	0.2250E 09	0.2827E 09	0.3510E 09	0.4307E 09	0.5231E 09
0.51	0.9730E 08	0.1317E 09	0.1751E 09	0.2287E 09	0.2942E 09	0.3729E 09	0.4665E 09	0.5766E 09	0.7049E 09	0.8528E 09
0.52	0.1493E 09	0.2010E 09	0.2658E 09	0.3454E 09	0.4420E 09	0.5578E 09	0.6949E 09	0.8554E 09	0.1041E 10	0.1255E 10
0.53	0.1961E 09	0.2626E 09	0.3453E 09	0.4466E 09	0.5689E 09	0.7148E 09	0.8867E 09	0.1087E 10	0.1319E 10	0.1584E 10
0.54	0.2338E 09	0.3114E 09	0.4073E 09	0.5243E 09	0.6649E 09	0.8318E 09	0.1027E 10	0.1255E 10	0.1517E 10	0.1817E 10
0.55	0.2614E 09	0.3464E 09	0.4510E 09	0.5778E 09	0.7296E 09	0.9091E 09	0.1119E 10	0.1362E 10	0.1640E 10	0.1958E 10
0.56	0.2792E 09	0.3681E 09	0.4769E 09	0.6084E 09	0.7650E 09	0.9495E 09	0.1164E 10	0.1412E 10	0.1696E 10	0.2017E 10
0.57	0.2841E 09	0.3728E 09	0.4809E 09	0.6108E 09	0.7650E 09	0.9459E 09	0.1155E 10	0.1397E 10	0.1672E 10	0.1983E 10
0.58	0.2752E 09	0.3594E 09	0.4616E 09	0.5839E 09	0.7284E 09	0.8974E 09	0.1092E 10	0.1316E 10	0.1571E 10	0.1858E 10
0.59	0.2529E 09	0.3288E 09	0.4205E 09	0.5298E 09	0.6586E 09	0.8085E 09	0.9813E 09	0.1178E 10	0.1402E 10	0.1653E 10
0.60	0.2220E 09	0.2873E 09	0.3659E 09	0.4593E 09	0.5688E 09	0.6960E 09	0.8421E 09	0.1008E 10	0.1196E 10	0.1406E 10
0.61	0.1857E 09	0.2394E 09	0.3037E 09	0.3797E 09	0.4687E 09	0.5716E 09	0.6894E 09	0.8232E 09	0.9739E 09	0.1142E 10
0.62	0.1472E 09	0.1890E 09	0.2388E 09	0.2976E 09	0.3661E 09	0.4450E 09	0.5352E 09	0.6372E 09	0.7518E 09	0.8796E 09
0.63	0.1069E 09	0.1366E 09	0.1721E 09	0.2137E 09	0.2619E 09	0.3175E 09	0.3807E 09	0.4520E 09	0.5320E 09	0.6208E 09
0.64	0.7350E 08	0.9361E 08	0.1174E 09	0.1453E 09	0.1776E 09	0.2146E 09	0.2566E 09	0.3039E 09	0.3567E 09	0.4154E 09
0.65	0.4667E 08	0.5922E 08	0.7405E 08	0.9133E 08	0.1112E 09	0.1340E 09	0.1598E 09	0.1888E 09	0.2211E 09	0.2569E 09
0.66	0.2757E 08	0.3486E 08	0.4343E 08	0.5341E 08	0.6488E 08	0.7795E 08	0.9271E 08	0.1092E 09	0.1276E 09	0.1479E 09
0.67	0.1495E 08	0.1884E 08	0.2340E 08	0.2868E 08	0.3474E 08	0.4163E 08	0.4939E 08	0.5806E 08	0.6767E 08	0.7827E 08
0.68	0.8196E 07	0.1029E 08	0.1274E 08	0.1557E 08	0.1881E 08	0.2248E 08	0.2660E 08	0.3120E 08	0.3629E 08	0.4188E 08
0.69	0.4076E 07	0.5102E 07	0.6297E 07	0.7674E 07	0.9245E 07	0.1102E 08	0.1301E 08	0.1522E 08	0.1766E 08	0.2034E 08
0.70	0.2093E 07	0.2611E 07	0.3214E 07	0.3906E 07	0.4693E 07	0.5580E 07	0.6572E 07	0.7673E 07	0.8887E 07	0.1021E 08
0.71	0.1094E 07	0.1361E 07	0.1671E 07	0.2025E 07	0.2427E 07	0.2879E 07	0.3383E 07	0.3941E 07	0.4555E 07	0.5227E 07
0.72	0.5615E 06	0.6963E 06	0.8520E 06	0.1029E 07	0.1231E 07	0.1457E 07	0.1708E 07	0.1986E 07	0.2291E 07	0.2624E 07
0.73	0.2852E 06	0.3526E 06	0.4303E 06	0.5188E 06	0.6188E 06	0.7306E 06	0.8549E 06	0.9919E 06	0.1142E 07	0.1305E 07
0.74	0.1395E 06	0.1720E 06	0.2094E 06	0.2518E 06	0.2996E 06	0.3530E 06	0.4122E 06	0.4774E 06	0.5486E 06	0.6260E 06
0.75	0.6852E 05	0.8425E 05	0.1022E 06	0.1227E 06	0.1456E 06	0.1712E 06	0.1995E 06	0.2306E 06	0.2646E 06	0.3014E 06
0.76	0.3488E 05	0.4278E 05	0.5180E 05	0.6200E 05	0.7344E 05	0.8616E 05	0.1002E 06	0.1156E 06	0.1323E 06	0.1505E 06
0.77	0.1773E 05	0.2169E 05	0.2620E 05	0.3129E 05	0.3698E 05	0.4330E 05	0.5027E 05	0.5789E 05	0.6617E 05	0.7514E 05
0.78	0.9001E 04	0.1098E 05	0.1323E 05	0.1576E 05	0.1859E 05	0.2173E 05	0.2517E 05	0.2894E 05	0.3303E 05	0.3745E 05
0.79	0.4548E 04	0.5534E 04	0.6653E 04	0.7911E 04	0.9312E 04	0.1086E 05	0.1256E 05	0.1441E 05	0.1642E 05	0.1859E 05
0.80	0.2285E 04	0.2774E 04	0.3328E 04	0.3949E 04	0.4639E 04	0.5400E 04	0.6234E 04	0.7142E 04	0.8126E 04	0.9185E 04
0.81	0.1147E 04	0.1390E 04	0.1663E 04	0.1969E 04	0.2309E 04	0.2683E 04	0.3092E 04	0.3537E 04	0.4018E 04	0.4535E 04
0.82	0.5754E 03	0.6953E 03	0.8304E 03	0.9812E 03	0.1148E 04	0.1331E 04	0.1532E 04	0.1750E 04	0.1985E 04	0.2237E 04
0.83	0.2882E 03	0.3475E 03	0.4142E 03	0.4885E 03	0.5706E 03	0.6607E 03	0.7590E 03	0.8655E 03	0.9802E 03	0.1103E 04

TABLE 17

Blackbody Spectral Luminance, Scotopic (sc.cd/m^2.μ)

$\lambda(\mu)$	3000°K.	3100°K.	3200°K.	3300°K.	3400°K.	3500°K.	3600°K.	3700°K.	3800°K.	3900°K.
0.38	0.5106E 05	0.7672E 05	0.1123E 06	0.1608E 06	0.2254E 06	0.3098E 06	0.4184E 06	0.5560E 06	0.7279E 06	0.9398E 06
0.39	0.2323E 06	0.3454E 06	0.5011E 06	0.7106E 06	0.9873E 06	0.1346E 07	0.1804E 07	0.2379E 07	0.3094E 07	0.3968E 07
0.40	0.1171E 07	0.1724E 07	0.2477E 07	0.3483E 07	0.4799E 07	0.6493E 07	0.8639E 07	0.1131E 08	0.1461E 08	0.1863E 08
0.41	0.5197E 07	0.7580E 07	0.1079E 08	0.1505E 08	0.2058E 08	0.2764E 08	0.3652E 08	0.4752E 08	0.6100E 08	0.7730E 08
0.42	0.1688E 08	0.2440E 08	0.3447E 08	0.4768E 08	0.6471E 08	0.8629E 08	0.1132E 09	0.1464E 09	0.1869E 09	0.2355E 09
0.43	0.4048E 08	0.5802E 08	0.8129E 08	0.1116E 09	0.1503E 09	0.1992E 09	0.2598E 09	0.3340E 09	0.4237E 09	0.5311E 09
0.44	0.7636E 08	0.1085E 09	0.1509E 09	0.2057E 09	0.2753E 09	0.3624E 09	0.4698E 09	0.6005E 09	0.7578E 09	0.9449E 09
0.45	0.1205E 09	0.1700E 09	0.2347E 09	0.3177E 09	0.4225E 09	0.5528E 09	0.7125E 09	0.9058E 09	0.1137E 10	0.1411E 10
0.46	0.1698E 09	0.2376E 09	0.3257E 09	0.4381E 09	0.5789E 09	0.7530E 09	0.9652E 09	0.1220E 10	0.1525E 10	0.1883E 10
0.47	0.2267E 09	0.3151E 09	0.4290E 09	0.5733E 09	0.7532E 09	0.9742E 09	0.1242E 10	0.1563E 10	0.1943E 10	0.2389E 10
0.48	0.2963E 09	0.4089E 09	0.5532E 09	0.7349E 09	0.9600E 09	0.1235E 10	0.1566E 10	0.1962E 10	0.2428E 10	0.2973E 10
0.49	0.3737E 09	0.5124E 09	0.6890E 09	0.9099E 09	0.1182E 10	0.1513E 10	0.1910E 10	0.2381E 10	0.2935E 10	0.3578E 10
0.50	0.4460E 09	0.6078E 09	0.8123E 09	0.1066E 10	0.1378E 10	0.1756E 10	0.2206E 10	0.2739E 10	0.3361E 10	0.4082E 10
0.51	0.4949E 09	0.6704E 09	0.8910E 09	0.1163E 10	0.1496E 10	0.1897E 10	0.2373E 10	0.2933E 10	0.3586E 10	0.4338E 10
0.52	0.5050E 09	0.6801E 09	0.8989E 09	0.1168E 10	0.1495E 10	0.1886E 10	0.2350E 10	0.2893E 10	0.3522E 10	0.4246E 10
0.53	0.4739E 09	0.6345E 09	0.8343E 09	0.1079E 10	0.1374E 10	0.1726E 10	0.2142E 10	0.2626E 10	0.3186E 10	0.3828E 10
0.54	0.4088E 09	0.5445E 09	0.7123E 09	0.9168E 09	0.1162E 10	0.1454E 10	0.1797E 10	0.2195E 10	0.2654E 10	0.3177E 10
0.55	0.3244E 09	0.4298E 09	0.5596E 09	0.7169E 09	0.9053E 09	0.1128E 10	0.1388E 10	0.1690E 10	0.2036E 10	0.2429E 10
0.56	0.2369E 09	0.3123E 09	0.4047E 09	0.5162E 09	0.6491E 09	0.8056E 09	0.9880E 09	0.1198E 10	0.1439E 10	0.1711E 10
0.57	0.1591E 09	0.2087E 09	0.2692E 09	0.3420E 09	0.4283E 09	0.5296E 09	0.6472E 09	0.7824E 09	0.9365E 09	0.1110E 10
0.58	0.9846E 08	0.1285E 09	0.1651E 09	0.2088E 09	0.2605E 09	0.3210E 09	0.3909E 09	0.4710E 09	0.5621E 09	0.6647E 09
0.59	0.5618E 08	0.7304E 08	0.9340E 08	0.1176E 09	0.1462E 09	0.1795E 09	0.2179E 09	0.2618E 09	0.3114E 09	0.3673E 09
0.60	0.2994E 08	0.3876E 08	0.4936E 08	0.6196E 08	0.7673E 08	0.9388E 08	0.1135E 09	0.1360E 09	0.1613E 09	0.1897E 09
0.61	0.1510E 08	0.1946E 08	0.2469E 08	0.3088E 08	0.3811E 08	0.4648E 08	0.5606E 08	0.6694E 08	0.7919E 08	0.9289E 08
0.62	0.7318E 07	0.9393E 07	0.1187E 08	0.1479E 08	0.1819E 08	0.2211E 08	0.2659E 08	0.3166E 08	0.3736E 08	0.4371E 08
0.63	0.3454E 07	0.4417E 07	0.5561E 07	0.6905E 07	0.8466E 07	0.1025E 08	0.1230E 08	0.1460E 08	0.1719E 08	0.2006E 08
0.64	0.1614E 07	0.2056E 07	0.2579E 07	0.3192E 07	0.3901E 07	0.4714E 07	0.5636E 07	0.6675E 07	0.7836E 07	0.9124E 07
0.65	0.7585E 06	0.9624E 06	0.1203E 07	0.1484E 07	0.1808E 07	0.2178E 07	0.2598E 07	0.3069E 07	0.3594E 07	0.4175E 07
0.66	0.3631E 06	0.4591E 06	0.5721E 06	0.7034E 06	0.8546E 06	0.1026E 07	0.1221E 07	0.1438E 07	0.1681E 07	0.1948E 07
0.67	0.1775E 06	0.2237E 06	0.2779E 06	0.3406E 06	0.4126E 06	0.4944E 06	0.5865E 06	0.6894E 06	0.8036E 06	0.9295E 06
0.68	0.8858E 05	0.1112E 06	0.1377E 06	0.1683E 06	0.2033E 06	0.2429E 06	0.2875E 06	0.3372E 06	0.3921E 06	0.4526E 06
0.69	0.4504E 05	0.5637E 05	0.6958E 05	0.8480E 05	0.1021E 06	0.1217E 06	0.1437E 06	0.1682E 06	0.1952E 06	0.2248E 06
0.70	0.2332E 05	0.2910E 05	0.3581E 05	0.4352E 05	0.5229E 05	0.6218E 05	0.7323E 05	0.8550E 05	0.9902E 05	0.1138E 06
0.71	0.1229E 05	0.1528E 05	0.1876E 05	0.2273E 05	0.2725E 05	0.3232E 05	0.3798E 05	0.4425E 05	0.5114E 05	0.5868E 05
0.72	0.6586E 04	0.8168E 04	0.9994E 04	0.1208E 05	0.1444E 05	0.1709E 05	0.2004E 05	0.2330E 05	0.2687E 05	0.3078E 05
0.73	0.3585E 04	0.4433E 04	0.5409E 04	0.6522E 04	0.7779E 04	0.9185E 04	0.1074E 05	0.1247E 05	0.1435E 05	0.1641E 05
0.74	0.1983E 04	0.2445E 04	0.2975E 04	0.3578E 04	0.4258E 04	0.5017E 04	0.5857E 04	0.6783E 04	0.7795E 04	0.8895E 04
0.75	0.1113E 04	0.1369E 04	0.1662E 04	0.1994E 04	0.2367E 04	0.2783E 04	0.3243E 04	0.3749E 04	0.4300E 04	0.4899E 04
0.76	0.6343E 03	0.7779E 03	0.9419E 03	0.1127E 04	0.1335E 04	0.1566E 04	0.1822E 04	0.2102E 04	0.2407E 04	0.2737E 04
0.77	0.3663E 03	0.4480E 03	0.5412E 03	0.6463E 03	0.7639E 03	0.8944E 03	0.1038E 04	0.1195E 04	0.1366E 04	0.1551E 04
0.78	0.2143E 03	0.2614E 03	0.3150E 03	0.3754E 03	0.4428E 03	0.5174E 03	0.5995E 03	0.6891E 03	0.7865E 03	0.8917E 03

$L_{v\lambda}(\lambda,T)$

TABLE 17

Blackbody Spectral Luminance, Photopic (cd/m^2·μ)

λ(μ)	4000°K.	4100°K.	4200°K.	4300°K.	4400°K.	4500°K.	4600°K.	4700°K.	4800°K.	4900°K.
0.36	0.2401E 04	0.3064E 04	0.3865E 04	0.4823E 04	0.5957E 04	0.7290E 04	0.8843E 04	0.1063E 05	0.1270E 05	0.1505E 05
0.37	0.8679E 04	0.1100E 05	0.1378E 05	0.1710E 05	0.2100E 05	0.2556E 05	0.3084E 05	0.3692E 05	0.4387E 05	0.5177E 05
0.38	0.3087E 05	0.3890E 05	0.4846E 05	0.5977E 05	0.7301E 05	0.8841E 05	0.1061E 06	0.1264E 06	0.1496E 06	0.1757E 06
0.39	0.1063E 06	0.1332E 06	0.1650E 06	0.2024E 06	0.2460E 06	0.2964E 06	0.3542E 06	0.4202E 06	0.4949E 06	0.5790E 06
0.40	0.3894E 06	0.4850E 06	0.5977E 06	0.7294E 06	0.8822E 06	0.1058E 07	0.1258E 07	0.1486E 07	0.1744E 07	0.2032E 07
0.41	0.1309E 07	0.1622E 07	0.1989E 07	0.2416E 07	0.2908E 07	0.3472E 07	0.4114E 07	0.4840E 07	0.5655E 07	0.6566E 07
0.42	0.4730E 07	0.5829E 07	0.7113E 07	0.8599E 07	0.1030E 08	0.1225E 08	0.1446E 08	0.1694E 08	0.1972E 08	0.2282E 08
0.43	0.1488E 08	0.1825E 08	0.2217E 08	0.2668E 08	0.3184E 08	0.3771E 08	0.4433E 08	0.5176E 08	0.6005E 08	0.6924E 08
0.44	0.3181E 08	0.3884E 08	0.4696E 08	0.5629E 08	0.6691E 08	0.7894E 08	0.9246E 08	0.1075E 09	0.1243E 09	0.1429E 09
0.45	0.5634E 08	0.6847E 08	0.8245E 08	0.9843E 08	0.1165E 09	0.1370E 09	0.1599E 09	0.1854E 09	0.2137E 09	0.2448E 09
0.46	0.9483E 08	0.1147E 09	0.1376E 09	0.1636E 09	0.1931E 09	0.2262E 09	0.2631E 09	0.3041E 09	0.3494E 09	0.3992E 09
0.47	0.1525E 09	0.1838E 09	0.2196E 09	0.2602E 09	0.3059E 09	0.3572E 09	0.4142E 09	0.4773E 09	0.5468E 09	0.6229E 09
0.48	0.2460E 09	0.2953E 09	0.3516E 09	0.4151E 09	0.4864E 09	0.5661E 09	0.6544E 09	0.7519E 09	0.8590E 09	0.9760E 09
0.49	0.3870E 09	0.4629E 09	0.5490E 09	0.6461E 09	0.7547E 09	0.8755E 09	0.1009E 10	0.1156E 10	0.1317E 10	0.1493E 10
0.50	0.6292E 09	0.7499E 09	0.8865E 09	0.1039E 10	0.1210E 10	0.1400E 10	0.1609E 10	0.1839E 10	0.2090E 10	0.2363E 10
0.51	0.1022E 10	0.1214E 10	0.1430E 10	0.1672E 10	0.1942E 10	0.2240E 10	0.2567E 10	0.2926E 10	0.3317E 10	0.3741E 10
0.52	0.1499E 10	0.1775E 10	0.2085E 10	0.2431E 10	0.2814E 10	0.3237E 10	0.3701E 10	0.4208E 10	0.4759E 10	0.5355E 10
0.53	0.1886E 10	0.2226E 10	0.2606E 10	0.3030E 10	0.3499E 10	0.4014E 10	0.4578E 10	0.5192E 10	0.5859E 10	0.6579E 10
0.54	0.2156E 10	0.2537E 10	0.2962E 10	0.3434E 10	0.3954E 10	0.4526E 10	0.5149E 10	0.5827E 10	0.6560E 10	0.7351E 10
0.55	0.2316E 10	0.2717E 10	0.3164E 10	0.3658E 10	0.4202E 10	0.4797E 10	0.5446E 10	0.6149E 10	0.6908E 10	0.7725E 10
0.56	0.2379E 10	0.2783E 10	0.3232E 10	0.3727E 10	0.4271E 10	0.4865E 10	0.5510E 10	0.6209E 10	0.6961E 10	0.7769E 10
0.57	0.2332E 10	0.2721E 10	0.3152E 10	0.3626E 10	0.4145E 10	0.4711E 10	0.5325E 10	0.5987E 10	0.6700E 10	0.7463E 10
0.58	0.2179E 10	0.2535E 10	0.2930E 10	0.3362E 10	0.3835E 10	0.4349E 10	0.4905E 10	0.5505E 10	0.6148E 10	0.6837E 10
0.59	0.1934E 10	0.2245E 10	0.2587E 10	0.2963E 10	0.3372E 10	0.3816E 10	0.4295E 10	0.4811E 10	0.5364E 10	0.5954E 10
0.60	0.1641E 10	0.1900E 10	0.2185E 10	0.2496E 10	0.2835E 10	0.3202E 10	0.3598E 10	0.4022E 10	0.4476E 10	0.4961E 10
0.61	0.1329E 10	0.1535E 10	0.1761E 10	0.2008E 10	0.2276E 10	0.2566E 10	0.2877E 10	0.3211E 10	0.3568E 10	0.3947E 10
0.62	0.1021E 10	0.1176E 10	0.1347E 10	0.1532E 10	0.1734E 10	0.1950E 10	0.2183E 10	0.2433E 10	0.2698E 10	0.2981E 10
0.63	0.7191E 09	0.8269E 09	0.9447E 09	0.1072E 10	0.1211E 10	0.1360E 10	0.1519E 10	0.1690E 10	0.1872E 10	0.2064E 10
0.64	0.4800E 09	0.5508E 09	0.6280E 09	0.7117E 09	0.8020E 09	0.8991E 09	0.1003E 10	0.1113E 10	0.1231E 10	0.1356E 10
0.65	0.2962E 09	0.3392E 09	0.3860E 09	0.4366E 09	0.4911E 09	0.5497E 09	0.6122E 09	0.6788E 09	0.7495E 09	0.8243E 09
0.66	0.1702E 09	0.1945E 09	0.2209E 09	0.2494E 09	0.2801E 09	0.3130E 09	0.3480E 09	0.3853E 09	0.4248E 09	0.4666E 09
0.67	0.8987E 08	0.1025E 09	0.1162E 09	0.1309E 09	0.1468E 09	0.1638E 09	0.1818E 09	0.2010E 09	0.2214E 09	0.2428E 09
0.68	0.4799E 08	0.5464E 08	0.6183E 08	0.6958E 08	0.7788E 08	0.8674E 08	0.9617E 08	0.1061E 09	0.1167E 09	0.1278E 09
0.69	0.2327E 08	0.2645E 08	0.2987E 08	0.3356E 08	0.3751E 08	0.4171E 08	0.4618E 08	0.5091E 08	0.5591E 08	0.6117E 08
0.70	0.1166E 08	0.1323E 08	0.1492E 08	0.1673E 08	0.1867E 08	0.2073E 08	0.2292E 08	0.2524E 08	0.2768E 08	0.3025E 08
0.71	0.5956E 07	0.6745E 07	0.7595E 07	0.8505E 07	0.9476E 07	0.1050E 08	0.1160E 08	0.1275E 08	0.1397E 08	0.1525E 08
0.72	0.2985E 07	0.3375E 07	0.3794E 07	0.4242E 07	0.4719E 07	0.5227E 07	0.5763E 07	0.6330E 07	0.6925E 07	0.7550E 07
0.73	0.1482E 07	0.1673E 07	0.1878E 07	0.2097E 07	0.2330E 07	0.2576E 07	0.2838E 07	0.3113E 07	0.3402E 07	0.3704E 07
0.74	0.7098E 06	0.7999E 06	0.8965E 06	0.9995E 06	0.1109E 07	0.1225E 07	0.1347E 07	0.1476E 07	0.1611E 07	0.1753E 07
0.75	0.3412E 06	0.3839E 06	0.4296E 06	0.4783E 06	0.5300E 06	0.5847E 06	0.6424E 06	0.7031E 06	0.7666E 06	0.8331E 06
0.76	0.1701E 06	0.1911E 06	0.2136E 06	0.2375E 06	0.2629E 06	0.2896E 06	0.3178E 06	0.3475E 06	0.3785E 06	0.4109E 06
0.77	0.8479E 05	0.9514E 05	0.1061E 06	0.1179E 06	0.1303E 06	0.1434E 06	0.1572E 06	0.1716E 06	0.1868E 06	0.2025E 06
0.78	0.4220E 05	0.4728E 05	0.5269E 05	0.5843E 05	0.6451E 05	0.7092E 05	0.7765E 05	0.8470E 05	0.9208E 05	0.9977E 05
0.79	0.2091E 05	0.2340E 05	0.2605E 05	0.2885E 05	0.3181E 05	0.3494E 05	0.3821E 05	0.4164E 05	0.4522E 05	0.4896E 05
0.80	0.1032E 05	0.1153E 05	0.1281E 05	0.1418E 05	0.1562E 05	0.1713E 05	0.1872E 05	0.2038E 05	0.2211E 05	0.2391E 05
0.81	0.5089E 04	0.5679E 04	0.6305E 04	0.6967E 04	0.7666E 04	0.8399E 04	0.9168E 04	0.9971E 04	0.1080E 05	0.1168E 05
0.82	0.2507E 04	0.2794E 04	0.3098E 04	0.3420E 04	0.3758E 04	0.4114E 04	0.4486E 04	0.4875E 04	0.5279E 04	0.5700E 04
0.83	0.1234E 04	0.1374E 04	0.1522E 04	0.1678E 04	0.1842E 04	0.2015E 04	0.2195E 04	0.2383E 04	0.2578E 04	0.2782E 04

TABLE 17

Blackbody Spectral Luminance, Scotopic (sc.cd/m^2.μ)

$\lambda(\mu)$	4000° K.	4100° K.	4200° K.	4300° K.	4400° K.	4500° K.	4600° K.	4700° K.	4800° K.	4900° K.
0.38	0.1198E 07	0.1509E 07	0.1880E 07	0.2319E 07	0.2833E 07	0.3430E 07	0.4118E 07	0.4907E 07	0.5804E 07	0.6818E 07
0.39	0.5027E 07	0.6295E 07	0.7800E 07	0.9568E 07	0.1162E 08	0.1401E 08	0.1674E 08	0.1986E 08	0.2339E 08	0.2736E 08
0.40	0.2346E 08	0.2922E 08	0.3601E 08	0.4394E 08	0.5315E 08	0.6374E 08	0.7584E 08	0.8958E 08	0.1050E 09	0.1224E 09
0.41	0.9681E 08	0.1199E 09	0.1470E 09	0.1785E 09	0.2149E 09	0.2566E 09	0.3041E 09	0.3577E 09	0.4179E 09	0.4853E 09
0.42	0.2933E 09	0.3615E 09	0.4411E 09	0.5332E 09	0.6391E 09	0.7599E 09	0.8968E 09	0.1050E 10	0.1223E 10	0.1415E 10
0.43	0.6582E 09	0.8072E 09	0.9804E 09	0.1180E 10	0.1408E 10	0.1668E 10	0.1960E 10	0.2289E 10	0.2655E 10	0.3062E 10
0.44	0.1165E 10	0.1422E 10	0.1720E 10	0.2061E 10	0.2451E 10	0.2891E 10	0.3386E 10	0.3940E 10	0.4555E 10	0.5236E 10
0.45	0.1732E 10	0.2105E 10	0.2534E 10	0.3026E 10	0.3583E 10	0.4212E 10	0.4916E 10	0.5701E 10	0.6570E 10	0.7528E 10
0.46	0.2301E 10	0.2785E 10	0.3341E 10	0.3973E 10	0.4688E 10	0.5491E 10	0.6387E 10	0.7383E 10	0.8483E 10	0.9691E 10
0.47	0.2908E 10	0.3505E 10	0.4187E 10	0.4962E 10	0.5834E 10	0.6810E 10	0.7897E 10	0.9101E 10	0.1042E 11	0.1187E 11
0.48	0.3603E 10	0.4327E 10	0.5150E 10	0.6081E 10	0.7126E 10	0.8292E 10	0.9586E 10	0.1101E 11	0.1258E 11	0.1429E 11
0.49	0.4320E 10	0.5167E 10	0.6129E 10	0.7212E 10	0.8425E 10	0.9774E 10	0.1126E 11	0.1290E 11	0.1470E 11	0.1666E 11
0.50	0.4910E 10	0.5852E 10	0.6918E 10	0.8115E 10	0.9449E 10	0.1093E 11	0.1256E 11	0.1435E 11	0.1631E 11	0.1844E 11
0.51	0.5199E 10	0.6176E 10	0.7277E 10	0.8509E 10	0.9879E 10	0.1139E 11	0.1306E 11	0.1488E 11	0.1687E 11	0.1903E 11
0.52	0.5071E 10	0.6004E 10	0.7052E 10	0.8222E 10	0.9519E 10	0.1095E 11	0.1252E 11	0.1423E 11	0.1609E 11	0.1811E 11
0.53	0.4556E 10	0.5377E 10	0.6297E 10	0.7320E 10	0.8452E 10	0.9697E 10	0.1106E 11	0.1254E 11	0.1415E 11	0.1589E 11
0.54	0.3770E 10	0.4436E 10	0.5180E 10	0.6005E 10	0.6915E 10	0.7914E 10	0.9004E 10	0.1019E 11	0.1147E 11	0.1285E 11
0.55	0.2873E 10	0.3371E 10	0.3925E 10	0.4539E 10	0.5213E 10	0.5952E 10	0.6757E 10	0.7629E 10	0.8571E 10	0.9585E 10
0.56	0.2018E 10	0.2361E 10	0.2742E 10	0.3163E 10	0.3624E 10	0.4128E 10	0.4675E 10	0.5268E 10	0.5907E 10	0.6592E 10
0.57	0.1306E 10	0.1523E 10	0.1765E 10	0.2030E 10	0.2321E 10	0.2638E 10	0.2981E 10	0.3352E 10	0.3751E 10	0.4179E 10
0.58	0.7795E 09	0.9071E 09	0.1048E 10	0.1202E 10	0.1371E 10	0.1555E 10	0.1754E 10	0.1969E 10	0.2199E 10	0.2445E 10
0.59	0.4295E 09	0.4986E 09	0.5747E 09	0.6581E 09	0.7490E 09	0.8476E 09	0.9541E 09	0.1068E 10	0.1191E 10	0.1322E 10
0.60	0.2213E 09	0.2563E 09	0.2947E 09	0.3368E 09	0.3825E 09	0.4320E 09	0.4853E 09	0.5426E 09	0.6039E 09	0.6692E 09
0.61	0.1080E 09	0.1248E 09	0.1432E 09	0.1633E 09	0.1851E 09	0.2086E 09	0.2340E 09	0.2611E 09	0.2901E 09	0.3210E 09
0.62	0.5074E 08	0.5848E 08	0.6695E 08	0.7618E 08	0.8617E 08	0.9695E 08	0.1085E 09	0.1209E 09	0.1341E 09	0.1481E 09
0.63	0.2323E 08	0.2672E 08	0.3052E 08	0.3466E 08	0.3913E 08	0.4395E 08	0.4911E 08	0.5462E 08	0.6049E 08	0.6672E 08
0.64	0.1054E 08	0.1209E 08	0.1379E 08	0.1563E 08	0.1761E 08	0.1974E 08	0.2203E 08	0.2446E 08	0.2705E 08	0.2979E 08
0.65	0.4814E 07	0.5512E 07	0.6272E 07	0.7095E 07	0.7982E 07	0.8933E 07	0.9949E 07	0.1103E 08	0.1218E 08	0.1339E 08
0.66	0.2241E 07	0.2562E 07	0.2909E 07	0.3285E 07	0.3689E 07	0.4122E 07	0.4584E 07	0.5075E 07	0.5595E 07	0.6145E 07
0.67	0.1067E 07	0.1217E 07	0.1380E 07	0.1555E 07	0.1743E 07	0.1945E 07	0.2160E 07	0.2388E 07	0.2629E 07	0.2883E 07
0.68	0.5187E 06	0.5905E 06	0.6682E 06	0.7519E 06	0.8416E 06	0.9374E 06	0.1039E 07	0.1147E 07	0.1261E 07	0.1381E 07
0.69	0.2571E 06	0.2922E 06	0.3301E 06	0.3708E 06	0.4144E 06	0.4609E 06	0.5103E 06	0.5626E 06	0.6178E 06	0.6759E 06
0.70	0.1299E 06	0.1474E 06	0.1662E 06	0.1864E 06	0.2080E 06	0.2310E 06	0.2554E 06	0.2812E 06	0.3084E 06	0.3370E 06
0.71	0.6687E 05	0.7573E 05	0.8527E 05	0.9548E 05	0.1063E 06	0.1179E 06	0.1302E 06	0.1432E 06	0.1569E 06	0.1712E 06
0.72	0.3501E 05	0.3958E 05	0.4450E 05	0.4976E 05	0.5536E 05	0.6131E 05	0.6760E 05	0.7425E 05	0.8123E 05	0.8856E 05
0.73	0.1863E 05	0.2103E 05	0.2361E 05	0.2636E 05	0.2929E 05	0.3239E 05	0.3567E 05	0.3913E 05	0.4276E 05	0.4657E 05
0.74	0.1008E 05	0.1136E 05	0.1273E 05	0.1420E 05	0.1575E 05	0.1740E 05	0.1914E 05	0.2097E 05	0.2289E 05	0.2490E 05
0.75	0.5545E 04	0.6240E 04	0.6983E 04	0.7775E 04	0.8615E 04	0.9504E 04	0.1044E 05	0.1142E 05	0.1246E 05	0.1354E 05
0.76	0.3094E 04	0.3476E 04	0.3885E 04	0.4319E 04	0.4780E 04	0.5267E 04	0.5780E 04	0.6318E 04	0.6882E 04	0.7472E 04
0.77	0.1751E 04	0.1964E 04	0.2192E 04	0.2434E 04	0.2691E 04	0.2962E 04	0.3246E 04	0.3545E 04	0.3857E 04	0.4184E 04
0.78	0.1004E 04	0.1125E 04	0.1254E 04	0.1391E 04	0.1536E 04	0.1688E 04	0.1848E 04	0.2016E 04	0.2192E 04	0.2375E 04

TABLE 17

Blackbody Spectral Luminance, Photopic (cd/m^2.μ)

λ(μ)	5000°K.	5100°K.	5200°K.	5300°K.	5400°K.	5500°K.	5600°K.	5700°K.	5800°K.	5900°K.
0.36	0.1772E 05	0.2073E 05	0.2410E 05	0.2787E 05	0.3204E 05	0.3666E 05	0.4175E 05	0.4732E 05	0.5341E 05	0.6004E 05
0.37	0.6067E 05	0.7067E 05	0.8184E 05	0.9425E 05	0.1079E 06	0.1231E 06	0.1396E 06	0.1578E 06	0.1775E 06	0.1989E 06
0.38	0.2051E 06	0.2379E 06	0.2745E 06	0.3150E 06	0.3596E 06	0.4085E 06	0.4620E 06	0.5203E 06	0.5835E 06	0.6519E 06
0.39	0.6731E 06	0.7780E 06	0.8942E 06	0.1022E 07	0.1163E 07	0.1317E 07	0.1485E 07	0.1667E 07	0.1864E 07	0.2077E 07
0.40	0.2354E 07	0.2711E 07	0.3105E 07	0.3538E 07	0.4013E 07	0.4530E 07	0.5092E 07	0.5701E 07	0.6358E 07	0.7064E 07
0.41	0.7578E 07	0.8697E 07	0.9929E 07	0.1127E 08	0.1275E 08	0.1435E 08	0.1609E 08	0.1796E 08	0.1998E 08	0.2214E 08
0.42	0.2625E 08	0.3003E 08	0.3417E 08	0.3870E 08	0.4363E 08	0.4898E 08	0.5476E 08	0.6097E 08	0.6765E 08	0.7480E 08
0.43	0.7938E 08	0.9053E 08	0.1027E 09	0.1160E 09	0.1304E 09	0.1460E 09	0.1628E 09	0.1808E 09	0.2001E 09	0.2208E 09
0.44	0.1633E 09	0.1857E 09	0.2102E 09	0.2367E 09	0.2654E 09	0.2964E 09	0.3297E 09	0.3654E 09	0.4035E 09	0.4441E 09
0.45	0.2790E 09	0.3164E 09	0.3570E 09	0.4010E 09	0.4485E 09	0.4997E 09	0.5545E 09	0.6132E 09	0.6757E 09	0.7422E 09
0.46	0.4537E 09	0.5131E 09	0.5774E 09	0.6470E 09	0.7220E 09	0.8025E 09	0.8886E 09	0.9804E 09	0.1078E 10	0.1181E 10
0.47	0.7060E 09	0.7963E 09	0.8940E 09	0.9994E 09	0.1112E 10	0.1233E 10	0.1363E 10	0.1501E 10	0.1647E 10	0.1802E 10
0.48	0.1103E 10	0.1241E 10	0.1390E 10	0.1550E 10	0.1722E 10	0.1906E 10	0.2102E 10	0.2310E 10	0.2530E 10	0.2763E 10
0.49	0.1683E 10	0.1889E 10	0.2111E 10	0.2350E 10	0.2605E 10	0.2877E 10	0.3166E 10	0.3473E 10	0.3798E 10	0.4140E 10
0.50	0.2658E 10	0.2977E 10	0.3320E 10	0.3687E 10	0.4079E 10	0.4496E 10	0.4939E 10	0.5408E 10	0.5903E 10	0.6425E 10
0.51	0.4199E 10	0.4693E 10	0.5222E 10	0.5787E 10	0.6390E 10	0.7031E 10	0.7710E 10	0.8427E 10	0.9184E 10	0.9980E 10
0.52	0.5998E 10	0.6689E 10	0.7428E 10	0.8217E 10	0.9056E 10	0.9946E 10	0.1088E 11	0.1188E 11	0.1292E 11	0.1402E 11
0.53	0.7353E 10	0.8183E 10	0.9070E 10	0.1001E 11	0.1101E 11	0.1208E 11	0.1320E 11	0.1438E 11	0.1562E 11	0.1692E 11
0.54	0.8199E 10	0.9107E 10	0.1007E 11	0.1110E 11	0.1219E 11	0.1335E 11	0.1456E 11	0.1584E 11	0.1719E 11	0.1859E 11
0.55	0.8600E 10	0.9535E 10	0.1053E 11	0.1158E 11	0.1270E 11	0.1388E 11	0.1512E 11	0.1643E 11	0.1779E 11	0.1923E 11
0.56	0.8633E 10	0.9555E 10	0.1053E 11	0.1157E 11	0.1266E 11	0.1382E 11	0.1503E 11	0.1631E 11	0.1764E 11	0.1904E 11
0.57	0.8279E 10	0.9146E 10	0.1006E 11	0.1104E 11	0.1206E 11	0.1315E 11	0.1428E 11	0.1547E 11	0.1671E 11	0.1801E 11
0.58	0.7570E 10	0.8350E 10	0.9176E 10	0.1004E 11	0.1096E 11	0.1193E 11	0.1294E 11	0.1400E 11	0.1511E 11	0.1626E 11
0.59	0.6582E 10	0.7248E 10	0.7953E 10	0.8696E 10	0.9478E 10	0.1029E 11	0.1115E 11	0.1205E 11	0.1299E 11	0.1397E 11
0.60	0.5475E 10	0.6020E 10	0.6595E 10	0.7201E 10	0.7838E 10	0.8506E 10	0.9204E 10	0.9933E 10	0.1069E 11	0.1148E 11
0.61	0.4350E 10	0.4775E 10	0.5224E 10	0.5697E 10	0.6193E 10	0.6712E 10	0.7254E 10	0.7819E 10	0.8408E 10	0.9019E 10
0.62	0.3280E 10	0.3596E 10	0.3928E 10	0.4278E 10	0.4644E 10	0.5027E 10	0.5427E 10	0.5843E 10	0.6276E 10	0.6726E 10
0.63	0.2268E 10	0.2483E 10	0.2710E 10	0.2947E 10	0.3195E 10	0.3455E 10	0.3725E 10	0.4007E 10	0.4299E 10	0.4602E 10
0.64	0.1488E 10	0.1627E 10	0.1773E 10	0.1926E 10	0.2085E 10	0.2252E 10	0.2426E 10	0.2607E 10	0.2794E 10	0.2988E 10
0.65	0.9032E 09	0.9862E 09	0.1073E 10	0.1164E 10	0.1259E 10	0.1358E 10	0.1462E 10	0.1569E 10	0.1680E 10	0.1795E 10
0.66	0.5106E 09	0.5568E 09	0.6052E 09	0.6559E 09	0.7087E 09	0.7637E 09	0.8209E 09	0.8802E 09	0.9417E 09	0.1005E 10
0.67	0.2653E 09	0.2890E 09	0.3138E 09	0.3397E 09	0.3667E 09	0.3947E 09	0.4239E 09	0.4541E 09	0.4853E 09	0.5176E 09
0.68	0.1395E 09	0.1518E 09	0.1646E 09	0.1780E 09	0.1920E 09	0.2065E 09	0.2215E 09	0.2370E 09	0.2531E 09	0.2697E 09
0.69	0.6669E 08	0.7247E 08	0.7851E 08	0.8480E 08	0.9135E 08	0.9814E 08	0.1051E 09	0.1124E 09	0.1200E 09	0.1277E 09
0.70	0.3294E 08	0.3575E 08	0.3869E 08	0.4175E 08	0.4493E 08	0.4823E 08	0.5165E 08	0.5518E 08	0.5882E 08	0.6258E 08
0.71	0.1659E 08	0.1799E 08	0.1944E 08	0.2096E 08	0.2254E 08	0.2417E 08	0.2586E 08	0.2760E 08	0.2940E 08	0.3126E 08
0.72	0.8204E 07	0.8886E 07	0.9597E 07	0.1033E 08	0.1110E 08	0.1189E 08	0.1271E 08	0.1356E 08	0.1443E 08	0.1533E 08
0.73	0.4021E 07	0.4351E 07	0.4694E 07	0.5051E 07	0.5421E 07	0.5803E 07	0.6199E 07	0.6606E 07	0.7026E 07	0.7458E 07
0.74	0.1900E 07	0.2054E 07	0.2215E 07	0.2381E 07	0.2553E 07	0.2731E 07	0.2914E 07	0.3104E 07	0.3299E 07	0.3499E 07
0.75	0.9025E 06	0.9747E 06	0.1049E 07	0.1127E 07	0.1207E 07	0.1291E 07	0.1376E 07	0.1465E 07	0.1556E 07	0.1649E 07
0.76	0.4447E 06	0.4798E 06	0.5163E 06	0.5540E 06	0.5931E 06	0.6334E 06	0.6750E 06	0.7178E 06	0.7618E 06	0.8069E 06
0.77	0.2190E 06	0.2361E 06	0.2538E 06	0.2722E 06	0.2911E 06	0.3107E 06	0.3308E 06	0.3516E 06	0.3729E 06	0.3947E 06
0.78	0.1077E 06	0.1160E 06	0.1246E 06	0.1336E 06	0.1427E 06	0.1522E 06	0.1620E 06	0.1720E 06	0.1823E 06	0.1929E 06
0.79	0.5283E 05	0.5686E 05	0.6103E 05	0.6534E 05	0.6978E 05	0.7436E 05	0.7907E 05	0.8392E 05	0.8888E 05	0.9398E 05
0.80	0.2578E 05	0.2773E 05	0.2973E 05	0.3181E 05	0.3395E 05	0.3615E 05	0.3842E 05	0.4074E 05	0.4313E 05	0.4557E 05
0.81	0.1258E 05	0.1352E 05	0.1449E 05	0.1548E 05	0.1652E 05	0.1758E 05	0.1866E 05	0.1978E 05	0.2093E 05	0.2210E 05
0.82	0.6136E 04	0.6588E 04	0.7055E 04	0.7536E 04	0.8032E 04	0.8542E 04	0.9065E 04	0.9602E 04	0.1015E 05	0.1071E 05
0.83	0.2992E 04	0.3210E 04	0.3435E 04	0.3667E 04	0.3905E 04	0.4150E 04	0.4402E 04	0.4660E 04	0.4924E 04	0.5195E 04

TABLE 17

Blackbody Spectral Luminance, Scotopic (sc.cd/m^2.μ)

$\lambda(\mu)$	5000°K.	5100°K.	5200°K.	5300°K.	5400°K.	5500°K.	5600°K.	5700°K.	5800°K.	5900°K.
0.38	0.7958E 07	0.9233E 07	0.1065E 08	0.1222E 08	0.1395E 08	0.1585E 08	0.1792E 08	0.2018E 08	0.2264E 08	0.2529E 08
0.39	0.3181E 08	0.3677E 08	0.4226E 08	0.4832E 08	0.5498E 08	0.6226E 08	0.7020E 08	0.7881E 08	0.8813E 08	0.9818E 08
0.40	0.1418E 09	0.1633E 09	0.1870E 09	0.2132E 09	0.2417E 09	0.2729E 09	0.3068E 09	0.3435E 09	0.3830E 09	0.4256E 09
0.41	0.5601E 09	0.6428E 09	0.7338E 09	0.8336E 09	0.9425E 09	0.1061E 10	0.1189E 10	0.1327E 10	0.1476E 10	0.1636E 10
0.42	0.1628E 10	0.1862E 10	0.2119E 10	0.2400E 10	0.2706E 10	0.3037E 10	0.3395E 10	0.3781E 10	0.4195E 10	0.4638E 10
0.43	0.3510E 10	0.4003E 10	0.4543E 10	0.5130E 10	0.5768E 10	0.6457E 10	0.7200E 10	0.7998E 10	0.8853E 10	0.9765E 10
0.44	0.5984E 10	0.6804E 10	0.7699E 10	0.8671E 10	0.9723E 10	0.1085E 11	0.1207E 11	0.1338E 11	0.1478E 11	0.1626E 11
0.45	0.8579E 10	0.9727E 10	0.1097E 11	0.1233E 11	0.1379E 11	0.1536E 11	0.1705E 11	0.1885E 11	0.2077E 11	0.2281E 11
0.46	0.1101E 11	0.1245E 11	0.1401E 11	0.1570E 11	0.1752E 11	0.1947E 11	0.2156E 11	0.2379E 11	0.2617E 11	0.2868E 11
0.47	0.1346E 11	0.1518E 11	0.1704E 11	0.1905E 11	0.2121E 11	0.2352E 11	0.2599E 11	0.2862E 11	0.3141E 11	0.3437E 11
0.48	0.1616E 11	0.1818E 11	0.2036E 11	0.2271E 11	0.2523E 11	0.2792E 11	0.3079E 11	0.3384E 11	0.3707E 11	0.4048E 11
0.49	0.1879E 11	0.2109E 11	0.2357E 11	0.2623E 11	0.2908E 11	0.3212E 11	0.3535E 11	0.3877E 11	0.4240E 11	0.4622E 11
0.50	0.2074E 11	0.2323E 11	0.2591E 11	0.2877E 11	0.3183E 11	0.3508E 11	0.3854E 11	0.4220E 11	0.4607E 11	0.5014E 11
0.51	0.2136E 11	0.2387E 11	0.2656E 11	0.2944E 11	0.3251E 11	0.3577E 11	0.3922E 11	0.4287E 11	0.4672E 11	0.5077E 11
0.52	0.2028E 11	0.2262E 11	0.2512E 11	0.2779E 11	0.3062E 11	0.3363E 11	0.3682E 11	0.4018E 11	0.4372E 11	0.4743E 11
0.53	0.1776E 11	0.1976E 11	0.2191E 11	0.2419E 11	0.2661E 11	0.2918E 11	0.3189E 11	0.3474E 11	0.3774E 11	0.4089E 11
0.54	0.1433E 11	0.1592E 11	0.1761E 11	0.1941E 11	0.2132E 11	0.2334E 11	0.2547E 11	0.2771E 11	0.3006E 11	0.3252E 11
0.55	0.1067E 11	0.1183E 11	0.1306E 11	0.1437E 11	0.1576E 11	0.1722E 11	0.1876E 11	0.2038E 11	0.2208E 11	0.2386E 11
0.56	0.7325E 10	0.8107E 10	0.8938E 10	0.9818E 10	0.1074E 11	0.1172E 11	0.1275E 11	0.1384E 11	0.1497E 11	0.1615E 11
0.57	0.4635E 10	0.5121E 10	0.5636E 10	0.6182E 10	0.6757E 10	0.7363E 10	0.7998E 10	0.8665E 10	0.9361E 10	0.1008E 11
0.58	0.2708E 10	0.2986E 10	0.3282E 10	0.3594E 10	0.3923E 10	0.4268E 10	0.4631E 10	0.5010E 10	0.5406E 10	0.5818E 10
0.59	0.1461E 10	0.1609E 10	0.1766E 10	0.1931E 10	0.2105E 10	0.2287E 10	0.2478E 10	0.2678E 10	0.2886E 10	0.3102E 10
0.60	0.7386E 09	0.8120E 09	0.8897E 09	0.9714E 09	0.1057E 10	0.1147E 10	0.1241E 10	0.1340E 10	0.1442E 10	0.1549E 10
0.61	0.3537E 09	0.3883E 09	0.4248E 09	0.4632E 09	0.5036E 09	0.5458E 09	0.5899E 09	0.6358E 09	0.6837E 09	0.7334E 09
0.62	0.1630E 09	0.1787E 09	0.1952E 09	0.2126E 09	0.2308E 09	0.2498E 09	0.2697E 09	0.2904E 09	0.3119E 09	0.3342E 09
0.63	0.7331E 08	0.8026E 08	0.8757E 08	0.9523E 08	0.1032E 09	0.1116E 09	0.1203E 09	0.1294E 09	0.1389E 09	0.1487E 09
0.64	0.3269E 08	0.3574E 08	0.3894E 08	0.4230E 08	0.4581E 08	0.4948E 08	0.5329E 08	0.5726E 08	0.6137E 08	0.6564E 08
0.65	0.1467E 08	0.1602E 08	0.1744E 08	0.1892E 08	0.2047E 08	0.2208E 08	0.2376E 08	0.2550E 08	0.2730E 08	0.2917E 08
0.66	0.6725E 07	0.7333E 07	0.7971E 07	0.8638E 07	0.9334E 07	0.1005E 08	0.1081E 08	0.1159E 08	0.1240E 08	0.1323E 08
0.67	0.3151E 07	0.3432E 07	0.3727E 07	0.4034E 07	0.4354E 07	0.4688E 07	0.5034E 07	0.5392E 07	0.5763E 07	0.6147E 07
0.68	0.1508E 07	0.1641E 07	0.1779E 07	0.1924E 07	0.2075E 07	0.2231E 07	0.2394E 07	0.2562E 07	0.2736E 07	0.2915E 07
0.69	0.7369E 06	0.8007E 06	0.8674E 06	0.9370E 06	0.1009E 07	0.1084E 07	0.1162E 07	0.1242E 07	0.1326E 07	0.1411E 07
0.70	0.3670E 06	0.3984E 06	0.4311E 06	0.4652E 06	0.5006E 06	0.5374E 06	0.5755E 06	0.6148E 06	0.6554E 06	0.6973E 06
0.71	0.1862E 06	0.2019E 06	0.2183E 06	0.2353E 06	0.2530E 06	0.2714E 06	0.2903E 06	0.3099E 06	0.3301E 06	0.3509E 06
0.72	0.9623E 05	0.1042E 06	0.1125E 06	0.1212E 06	0.1302E 06	0.1395E 06	0.1491E 06	0.1590E 06	0.1693E 06	0.1798E 06
0.73	0.5055E 05	0.5470E 05	0.5902E 05	0.6350E 05	0.6815E 05	0.7296E 05	0.7793E 05	0.8305E 05	0.8833E 05	0.9376E 05
0.74	0.2700E 05	0.2919E 05	0.3147E 05	0.3383E 05	0.3627E 05	0.3880E 05	0.4141E 05	0.4410E 05	0.4687E 05	0.4972E 05
0.75	0.1466E 05	0.1584E 05	0.1706E 05	0.1832E 05	0.1963E 05	0.2098E 05	0.2237E 05	0.2381E 05	0.2529E 05	0.2680E 05
0.76	0.8086E 04	0.8725E 04	0.9388E 04	0.1007E 05	0.1078E 05	0.1151E 05	0.1227E 05	0.1305E 05	0.1385E 05	0.1467E 05
0.77	0.4523E 04	0.4876E 04	0.5242E 04	0.5621E 04	0.6013E 04	0.6417E 04	0.6833E 04	0.7261E 04	0.7701E 04	0.8152E 04
0.78	0.2566E 04	0.2763E 04	0.2968E 04	0.3180E 04	0.3399E 04	0.3625E 04	0.3858E 04	0.4096E 04	0.4342E 04	0.4593E 04

TABLE 17

Blackbody Spectral Luminance, Photopic ($cd/m^2 \cdot \mu$)

λ(μ)	6000°K.	6100°K.	6200°K.	6300°K.	6400°K.	6500°K.	6600°K.	6700°K.	6800°K.	6900°K.
0.36	0.6722E 05	0.7499E 05	0.8336E 05	0.9236E 05	0.1020E 06	0.1123E 06	0.1233E 06	0.1350E 06	0.1474E 06	0.1605E 06
0.37	0.2220E 06	0.2469E 06	0.2737E 06	0.3025E 06	0.3332E 06	0.3659E 06	0.4007E 06	0.4376E 06	0.4768E 06	0.5181E 06
0.38	0.7256E 06	0.8049E 06	0.8898E 06	0.9806E 06	0.1077E 07	0.1180E 07	0.1289E 07	0.1405E 07	0.1527E 07	0.1656E 07
0.39	0.2305E 07	0.2551E 07	0.2813E 07	0.3092E 07	0.3389E 07	0.3705E 07	0.4039E 07	0.4392E 07	0.4764E 07	0.5155E 07
0.40	0.7821E 07	0.8631E 07	0.9495E 07	0.1041E 08	0.1138E 08	0.1242E 08	0.1351E 08	0.1466E 08	0.1587E 08	0.1714E 08
0.41	0.2445E 08	0.2692E 08	0.2955E 08	0.3234E 08	0.3529E 08	0.3841E 08	0.4171E 08	0.4517E 08	0.4881E 08	0.5262E 08
0.42	0.8242E 08	0.9054E 08	0.9916E 08	0.1082E 09	0.1179E 09	0.1281E 09	0.1388E 09	0.1500E 09	0.1618E 09	0.1742E 09
0.43	0.2427E 09	0.2661E 09	0.2908E 09	0.3170E 09	0.3445E 09	0.3736E 09	0.4041E 09	0.4360E 09	0.4695E 09	0.5045E 09
0.44	0.4873E 09	0.5331E 09	0.5815E 09	0.6325E 09	0.6863E 09	0.7428E 09	0.8020E 09	0.8641E 09	0.9289E 09	0.9966E 09
0.45	0.8127E 09	0.8873E 09	0.9660E 09	0.1048E 10	0.1136E 10	0.1227E 10	0.1323E 10	0.1423E 10	0.1527E 10	0.1636E 10
0.46	0.1291E 10	0.1407E 10	0.1529E 10	0.1658E 10	0.1792E 10	0.1934E 10	0.2081E 10	0.2235E 10	0.2396E 10	0.2563E 10
0.47	0.1966E 10	0.2139E 10	0.2320E 10	0.2511E 10	0.2711E 10	0.2920E 10	0.3138E 10	0.3365E 10	0.3602E 10	0.3848E 10
0.48	0.3009E 10	0.3268E 10	0.3540E 10	0.3824E 10	0.4122E 10	0.4434E 10	0.4758E 10	0.5095E 10	0.5446E 10	0.5810E 10
0.49	0.4501E 10	0.4880E 10	0.5278E 10	0.5694E 10	0.6129E 10	0.6582E 10	0.7054E 10	0.7544E 10	0.8053E 10	0.8581E 10
0.50	0.6974E 10	0.7549E 10	0.8152E 10	0.8782E 10	0.9439E 10	0.1012E 11	0.1083E 11	0.1157E 11	0.1233E 11	0.1313E 11
0.51	0.1081E 11	0.1169E 11	0.1260E 11	0.1356E 11	0.1455E 11	0.1559E 11	0.1666E 11	0.1778E 11	0.1893E 11	0.2012E 11
0.52	0.1517E 11	0.1638E 11	0.1764E 11	0.1895E 11	0.2031E 11	0.2173E 11	0.2320E 11	0.2472E 11	0.2630E 11	0.2793E 11
0.53	0.1829E 11	0.1971E 11	0.2120E 11	0.2275E 11	0.2435E 11	0.2602E 11	0.2775E 11	0.2954E 11	0.3139E 11	0.3330E 11
0.54	0.2007E 11	0.2160E 11	0.2320E 11	0.2486E 11	0.2659E 11	0.2838E 11	0.3023E 11	0.3214E 11	0.3412E 11	0.3615E 11
0.55	0.2072E 11	0.2228E 11	0.2390E 11	0.2558E 11	0.2732E 11	0.2913E 11	0.3100E 11	0.3292E 11	0.3491E 11	0.3696E 11
0.56	0.2049E 11	0.2200E 11	0.2357E 11	0.2520E 11	0.2689E 11	0.2864E 11	0.3045E 11	0.3231E 11	0.3423E 11	0.3620E 11
0.57	0.1936E 11	0.2077E 11	0.2223E 11	0.2374E 11	0.2530E 11	0.2692E 11	0.2859E 11	0.3031E 11	0.3208E 11	0.3390E 11
0.58	0.1746E 11	0.1871E 11	0.2000E 11	0.2134E 11	0.2272E 11	0.2415E 11	0.2562E 11	0.2714E 11	0.2870E 11	0.3030E 11
0.59	0.1498E 11	0.1603E 11	0.1712E 11	0.1825E 11	0.1941E 11	0.2061E 11	0.2185E 11	0.2312E 11	0.2443E 11	0.2577E 11
0.60	0.1230E 11	0.1315E 11	0.1403E 11	0.1494E 11	0.1587E 11	0.1684E 11	0.1783E 11	0.1886E 11	0.1991E 11	0.2099E 11
0.61	0.9653E 10	0.1030E 11	0.1098E 11	0.1168E 11	0.1241E 11	0.1315E 11	0.1391E 11	0.1470E 11	0.1551E 11	0.1633E 11
0.62	0.7191E 10	0.7672E 10	0.8170E 10	0.8683E 10	0.9211E 10	0.9755E 10	0.1031E 11	0.1088E 11	0.1147E 11	0.1207E 11
0.63	0.4916E 10	0.5240E 10	0.5575E 10	0.5920E 10	0.6275E 10	0.6640E 10	0.7014E 10	0.7399E 10	0.7793E 10	0.8197E 10
0.64	0.3189E 10	0.3396E 10	0.3610E 10	0.3830E 10	0.4056E 10	0.4289E 10	0.4527E 10	0.4772E 10	0.5023E 10	0.5279E 10
0.65	0.1914E 10	0.2036E 10	0.2163E 10	0.2293E 10	0.2426E 10	0.2564E 10	0.2704E 10	0.2848E 10	0.2996E 10	0.3147E 10
0.66	0.1070E 10	0.1138E 10	0.1208E 10	0.1279E 10	0.1353E 10	0.1428E 10	0.1505E 10	0.1585E 10	0.1666E 10	0.1748E 10
0.67	0.5509E 09	0.5852E 09	0.6205E 09	0.6568E 09	0.6940E 09	0.7322E 09	0.7712E 09	0.8113E 09	0.8521E 09	0.8939E 09
0.68	0.2869E 09	0.3045E 09	0.3226E 09	0.3412E 09	0.3603E 09	0.3799E 09	0.3999E 09	0.4204E 09	0.4413E 09	0.4626E 09
0.69	0.1357E 09	0.1440E 09	0.1524E 09	0.1611E 09	0.1700E 09	0.1791E 09	0.1884E 09	0.1979E 09	0.2077E 09	0.2176E 09
0.70	0.6645E 08	0.7042E 08	0.7451E 08	0.7869E 08	0.8298E 08	0.8737E 08	0.9186E 08	0.9644E 08	0.1011E 09	0.1058E 09
0.71	0.3317E 08	0.3512E 08	0.3713E 08	0.3919E 08	0.4130E 08	0.4346E 08	0.4567E 08	0.4792E 08	0.5021E 08	0.5255E 08
0.72	0.1625E 08	0.1720E 08	0.1817E 08	0.1917E 08	0.2019E 08	0.2123E 08	0.2230E 08	0.2338E 08	0.2449E 08	0.2562E 08
0.73	0.7902E 07	0.8357E 07	0.8823E 07	0.9301E 07	0.9789E 07	0.1028E 08	0.1079E 08	0.1131E 08	0.1184E 08	0.1238E 08
0.74	0.3705E 07	0.3915E 07	0.4131E 07	0.4352E 07	0.4578E 07	0.4809E 07	0.5044E 07	0.5284E 07	0.5529E 07	0.5777E 07
0.75	0.1745E 07	0.1843E 07	0.1943E 07	0.2046E 07	0.2151E 07	0.2258E 07	0.2367E 07	0.2479E 07	0.2592E 07	0.2707E 07
0.76	0.8533E 06	0.9007E 06	0.9492E 06	0.9988E 06	0.1049E 07	0.1101E 07	0.1153E 07	0.1207E 07	0.1262E 07	0.1317E 07
0.77	0.4171E 06	0.4400E 06	0.4635E 06	0.4874E 06	0.5118E 06	0.5368E 06	0.5622E 06	0.5880E 06	0.6143E 06	0.6411E 06
0.78	0.2037E 06	0.2148E 06	0.2261E 06	0.2376E 06	0.2494E 06	0.2615E 06	0.2737E 06	0.2861E 06	0.2988E 06	0.3117E 06
0.79	0.9919E 05	0.1045E 06	0.1099E 06	0.1155E 06	0.1211E 06	0.1269E 06	0.1328E 06	0.1388E 06	0.1449E 06	0.1510E 06
0.80	0.4807E 05	0.5063E 05	0.5324E 05	0.5590E 05	0.5861E 05	0.6138E 05	0.6419E 05	0.6705E 05	0.6996E 05	0.7291E 05
0.81	0.2330E 05	0.2453E 05	0.2578E 05	0.2705E 05	0.2835E 05	0.2968E 05	0.3102E 05	0.3239E 05	0.3378E 05	0.3519E 05
0.82	0.1129E 05	0.1187E 05	0.1247E 05	0.1308E 05	0.1371E 05	0.1434E 05	0.1498E 05	0.1564E 05	0.1630E 05	0.1698E 05
0.83	0.5471E 04	0.5752E 04	0.6040E 04	0.6333E 04	0.6631E 04	0.6934E 04	0.7242E 04	0.7555E 04	0.7873E 04	0.8196E 04

TABLE 17

Blackbody Spectral Luminance, Scotopic (sc.cd/m^2.μ)

$\lambda(\mu)$	6000°K.	6100°K.	6200°K.	6300°K.	6400°K.	6500°K.	6600°K.	6700°K.	6800°K.	6900°K.
0.38	0.2815E 08	0.3122E 08	0.3452E 08	0.3804E 08	0.4180E 08	0.4579E 08	0.5003E 08	0.5452E 08	0.5927E 08	0.6427E 08
0.39	0.1089E 09	0.1205E 09	0.1329E 09	0.1461E 09	0.1602E 09	0.1751E 09	0.1909E 09	0.2076E 09	0.2251E 09	0.2437E 09
0.40	0.4712E 09	0.5200E 09	0.5721E 09	0.6274E 09	0.6862E 09	0.7484E 09	0.8142E 09	0.8835E 09	0.9564E 09	0.1033E 10
0.41	0.1807E 10	0.1990E 10	0.2184E 10	0.2390E 10	0.2608E 10	0.2839E 10	0.3082E 10	0.3338E 10	0.3607E 10	0.3889E 10
0.42	0.5111E 10	0.5615E 10	0.6149E 10	0.6715E 10	0.7314E 10	0.7945E 10	0.8609E 10	0.9306E 10	0.1003E 11	0.1080E 11
0.43	0.1073E 11	0.1176E 11	0.1286E 11	0.1401E 11	0.1523E 11	0.1652E 11	0.1787E 11	0.1928E 11	0.2076E 11	0.2231E 11
0.44	0.1785E 11	0.1952E 11	0.2129E 11	0.2317E 11	0.2513E 11	0.2720E 11	0.2937E 11	0.3165E 11	0.3402E 11	0.3650E 11
0.45	0.2498E 11	0.2727E 11	0.2970E 11	0.3224E 11	0.3492E 11	0.3774E 11	0.4068E 11	0.4376E 11	0.4697E 11	0.5031E 11
0.46	0.3135E 11	0.3416E 11	0.3713E 11	0.4024E 11	0.4352E 11	0.4694E 11	0.5052E 11	0.5426E 11	0.5816E 11	0.6221E 11
0.47	0.3749E 11	0.4078E 11	0.4425E 11	0.4788E 11	0.5169E 11	0.5568E 11	0.5983E 11	0.6417E 11	0.6868E 11	0.7337E 11
0.48	0.4408E 11	0.4787E 11	0.5185E 11	0.5603E 11	0.6039E 11	0.6495E 11	0.6970E 11	0.7464E 11	0.7978E 11	0.8512E 11
0.49	0.5025E 11	0.5448E 11	0.5892E 11	0.6356E 11	0.6841E 11	0.7347E 11	0.7874E 11	0.8422E 11	0.8990E 11	0.9579E 11
0.50	0.5442E 11	0.5891E 11	0.6362E 11	0.6853E 11	0.7366E 11	0.7900E 11	0.8455E 11	0.9031E 11	0.9629E 11	0.1024E 12
0.51	0.5502E 11	0.5947E 11	0.6413E 11	0.6899E 11	0.7405E 11	0.7931E 11	0.8478E 11	0.9045E 11	0.9632E 11	0.1023E 12
0.52	0.5133E 11	0.5541E 11	0.5966E 11	0.6409E 11	0.6871E 11	0.7350E 11	0.7847E 11	0.8363E 11	0.8895E 11	0.9446E 11
0.53	0.4419E 11	0.4763E 11	0.5122E 11	0.5496E 11	0.5884E 11	0.6287E 11	0.6705E 11	0.7137E 11	0.7583E 11	0.8044E 11
0.54	0.3509E 11	0.3778E 11	0.4057E 11	0.4348E 11	0.4650E 11	0.4962E 11	0.5286E 11	0.5621E 11	0.5966E 11	0.6322E 11
0.55	0.2571E 11	0.2764E 11	0.2965E 11	0.3174E 11	0.3390E 11	0.3614E 11	0.3846E 11	0.4085E 11	0.4332E 11	0.4586E 11
0.56	0.1738E 11	0.1867E 11	0.2000E 11	0.2138E 11	0.2282E 11	0.2430E 11	0.2583E 11	0.2741E 11	0.2904E 11	0.3071E 11
0.57	0.1084E 11	0.1163E 11	0.1244E 11	0.1329E 11	0.1417E 11	0.1507E 11	0.1601E 11	0.1697E 11	0.1796E 11	0.1898E 11
0.58	0.6248E 10	0.6693E 10	0.7156E 10	0.7634E 10	0.8129E 10	0.8640E 10	0.9166E 10	0.9709E 10	0.1026E 11	0.1083E 11
0.59	0.3327E 10	0.3561E 10	0.3803E 10	0.4053E 10	0.4312E 10	0.4578E 10	0.4853E 10	0.5136E 10	0.5426E 10	0.5724E 10
0.60	0.1659E 10	0.1774E 10	0.1892E 10	0.2015E 10	0.2141E 10	0.2272E 10	0.2406E 10	0.2544E 10	0.2686E 10	0.2831E 10
0.61	0.7849E 09	0.8383E 09	0.8935E 09	0.9504E 09	0.1009E 10	0.1069E 10	0.1131E 10	0.1195E 10	0.1261E 10	0.1328E 10
0.62	0.3573E 09	0.3813E 09	0.4060E 09	0.4315E 09	0.4577E 09	0.4848E 09	0.5125E 09	0.5410E 09	0.5703E 09	0.6002E 09
0.63	0.1588E 09	0.1693E 09	0.1801E 09	0.1913E 09	0.2027E 09	0.2145E 09	0.2266E 09	0.2391E 09	0.2518E 09	0.2648E 09
0.64	0.7004E 08	0.7459E 08	0.7929E 08	0.8412E 08	0.8909E 08	0.9420E 08	0.9945E 08	0.1048E 09	0.1103E 09	0.1159E 09
0.65	0.3110E 08	0.3310E 08	0.3515E 08	0.3726E 08	0.3943E 08	0.4166E 08	0.4395E 08	0.4629E 08	0.4869E 08	0.5114E 08
0.66	0.1410E 08	0.1499E 08	0.1591E 08	0.1685E 08	0.1782E 08	0.1881E 08	0.1983E 08	0.2087E 08	0.2194E 08	0.2303E 08
0.67	0.6542E 07	0.6949E 07	0.7369E 07	0.7799E 07	0.8241E 07	0.8695E 07	0.9159E 07	0.9634E 07	0.1012E 08	0.1061E 08
0.68	0.3100E 07	0.3291E 07	0.3486E 07	0.3688E 07	0.3894E 07	0.4105E 07	0.4321E 07	0.4543E 07	0.4769E 07	0.5000E 07
0.69	0.1500E 07	0.1591E 07	0.1684E 07	0.1780E 07	0.1878E 07	0.1979E 07	0.2082E 07	0.2187E 07	0.2295E 07	0.2404E 07
0.70	0.7404E 06	0.7847E 06	0.8301E 06	0.8768E 06	0.9246E 06	0.9735E 06	0.1023E 07	0.1074E 07	0.1126E 07	0.1179E 07
0.71	0.3724E 06	0.3944E 06	0.4169E 06	0.4400E 06	0.4637E 06	0.4880E 06	0.5127E 06	0.5380E 06	0.5638E 06	0.5900E 06
0.72	0.1907E 06	0.2018E 06	0.2132E 06	0.2249E 06	0.2368E 06	0.2490E 06	0.2615E 06	0.2743E 06	0.2872E 06	0.3005E 06
0.73	0.9934E 05	0.1050E 06	0.1109E 06	0.1169E 06	0.1230E 06	0.1293E 06	0.1357E 06	0.1422E 06	0.1489E 06	0.1557E 06
0.74	0.5264E 05	0.5563E 05	0.5870E 05	0.6184E 05	0.6505E 05	0.6833E 05	0.7167E 05	0.7508E 05	0.7856E 05	0.8209E 05
0.75	0.2836E 05	0.2995E 05	0.3159E 05	0.3326E 05	0.3496E 05	0.3670E 05	0.3848E 05	0.4029E 05	0.4213E 05	0.4401E 05
0.76	0.1551E 05	0.1637E 05	0.1726E 05	0.1816E 05	0.1908E 05	0.2002E 05	0.2098E 05	0.2195E 05	0.2294E 05	0.2395E 05
0.77	0.8615E 04	0.9088E 04	0.9572E 04	0.1006E 05	0.1057E 05	0.1108E 05	0.1161E 05	0.1214E 05	0.1268E 05	0.1324E 05
0.78	0.4851E 04	0.5114E 04	0.5384E 04	0.5659E 04	0.5940E 04	0.6226E 04	0.6517E 04	0.6814E 04	0.7115E 04	0.7422E 04

TABLE 17

Blackbody Spectral Luminance, Photopic (cd/m^2.μ)

$\lambda(\mu)$	7000°K.	7100°K.	7200°K.	7300°K.	7400°K.	7500°K.	7600°K.	7700°K.	7800°K.	7900°K.
0.36	0.1744E 06	0.1891E 06	0.2045E 06	0.2207E 06	0.2378E 06	0.2556E 06	0.2743E 06	0.2938E 06	0.3141E 06	0.3353E 06
0.37	0.5617E 06	0.6076E 06	0.6559E 06	0.7065E 06	0.7595E 06	0.8150E 06	0.8729E 06	0.9332E 06	0.9961E 06	0.1061E 07
0.38	0.1792E 07	0.1934E 07	0.2084E 07	0.2240E 07	0.2404E 07	0.2575E 07	0.2753E 07	0.2939E 07	0.3131E 07	0.3331E 07
0.39	0.5567E 07	0.5998E 07	0.6450E 07	0.6922E 07	0.7414E 07	0.7928E 07	0.8462E 07	0.9017E 07	0.9594E 07	0.1019E 08
0.40	0.1848E 08	0.1987E 08	0.2133E 08	0.2285E 08	0.2444E 08	0.2609E 08	0.2780E 08	0.2958E 08	0.3143E 08	0.3334E 08
0.41	0.5661E 08	0.6079E 08	0.6514E 08	0.6967E 08	0.7439E 08	0.7929E 08	0.8438E 08	0.8965E 08	0.9510E 08	0.1007E 09
0.42	0.1871E 09	0.2005E 09	0.2146E 09	0.2291E 09	0.2443E 09	0.2600E 09	0.2763E 09	0.2932E 09	0.3106E 09	0.3286E 09
0.43	0.5410E 09	0.5790E 09	0.6186E 09	0.6597E 09	0.7023E 09	0.7464E 09	0.7921E 09	0.8393E 09	0.8881E 09	0.9384E 09
0.44	0.1067E 10	0.1140E 10	0.1216E 10	0.1295E 10	0.1377E 10	0.1461E 10	0.1549E 10	0.1639E 10	0.1733E 10	0.1828E 10
0.45	0.1749E 10	0.1867E 10	0.1989E 10	0.2115E 10	0.2246E 10	0.2381E 10	0.2520E 10	0.2664E 10	0.2812E 10	0.2965E 10
0.46	0.2736E 10	0.2916E 10	0.3103E 10	0.3296E 10	0.3495E 10	0.3701E 10	0.3913E 10	0.4132E 10	0.4357E 10	0.4588E 10
0.47	0.4103E 10	0.4367E 10	0.4640E 10	0.4923E 10	0.5214E 10	0.5515E 10	0.5825E 10	0.6144E 10	0.6472E 10	0.6808E 10
0.48	0.6188E 10	0.6578E 10	0.6982E 10	0.7398E 10	0.7828E 10	0.8270E 10	0.8726E 10	0.9194E 10	0.9675E 10	0.1016E 11
0.49	0.9127E 10	0.9691E 10	0.1027E 11	0.1087E 11	0.1149E 11	0.1213E 11	0.1278E 11	0.1346E 11	0.1415E 11	0.1485E 11
0.50	0.1395E 11	0.1479E 11	0.1567E 11	0.1657E 11	0.1749E 11	0.1844E 11	0.1942E 11	0.2042E 11	0.2145E 11	0.2250E 11
0.51	0.2136E 11	0.2263E 11	0.2394E 11	0.2529E 11	0.2667E 11	0.2810E 11	0.2956E 11	0.3106E 11	0.3259E 11	0.3416E 11
0.52	0.2960E 11	0.3133E 11	0.3312E 11	0.3495E 11	0.3683E 11	0.3876E 11	0.4074E 11	0.4277E 11	0.4484E 11	0.4697E 11
0.53	0.3526E 11	0.3729E 11	0.3937E 11	0.4151E 11	0.4370E 11	0.4595E 11	0.4826E 11	0.5062E 11	0.5304E 11	0.5550E 11
0.54	0.3825E 11	0.4041E 11	0.4262E 11	0.4490E 11	0.4723E 11	0.4963E 11	0.5207E 11	0.5458E 11	0.5714E 11	0.5975E 11
0.55	0.3907E 11	0.4123E 11	0.4346E 11	0.4574E 11	0.4808E 11	0.5047E 11	0.5292E 11	0.5542E 11	0.5798E 11	0.6059E 11
0.56	0.3823E 11	0.4031E 11	0.4245E 11	0.4465E 11	0.4689E 11	0.4919E 11	0.5153E 11	0.5393E 11	0.5638E 11	0.5888E 11
0.57	0.3577E 11	0.3769E 11	0.3965E 11	0.4167E 11	0.4373E 11	0.4584E 11	0.4799E 11	0.5019E 11	0.5243E 11	0.5472E 11
0.58	0.3194E 11	0.3363E 11	0.3536E 11	0.3713E 11	0.3893E 11	0.4078E 11	0.4267E 11	0.4459E 11	0.4656E 11	0.4855E 11
0.59	0.2715E 11	0.2856E 11	0.3000E 11	0.3148E 11	0.3299E 11	0.3453E 11	0.3610E 11	0.3771E 11	0.3934E 11	0.4101E 11
0.60	0.2209E 11	0.2322E 11	0.2438E 11	0.2556E 11	0.2677E 11	0.2800E 11	0.2926E 11	0.3054E 11	0.3184E 11	0.3317E 11
0.61	0.1718E 11	0.1805E 11	0.1893E 11	0.1984E 11	0.2076E 11	0.2170E 11	0.2266E 11	0.2364E 11	0.2464E 11	0.2565E 11
0.62	0.1269E 11	0.1332E 11	0.1397E 11	0.1462E 11	0.1530E 11	0.1598E 11	0.1668E 11	0.1739E 11	0.1811E 11	0.1884E 11
0.63	0.8609E 10	0.9031E 10	0.9462E 10	0.9901E 10	0.1034E 11	0.1080E 11	0.1127E 11	0.1174E 11	0.1222E 11	0.1271E 11
0.64	0.5541E 10	0.5809E 10	0.6082E 10	0.6361E 10	0.6644E 10	0.6934E 10	0.7228E 10	0.7527E 10	0.7831E 10	0.8140E 10
0.65	0.3301E 10	0.3458E 10	0.3619E 10	0.3782E 10	0.3949E 10	0.4118E 10	0.4291E 10	0.4466E 10	0.4644E 10	0.4825E 10
0.66	0.1833E 10	0.1919E 10	0.2007E 10	0.2096E 10	0.2188E 10	0.2280E 10	0.2375E 10	0.2470E 10	0.2568E 10	0.2666E 10
0.67	0.9365E 09	0.9800E 09	0.1024E 10	0.1069E 10	0.1115E 10	0.1162E 10	0.1209E 10	0.1257E 10	0.1306E 10	0.1356E 10
0.68	0.4844E 09	0.5066E 09	0.5292E 09	0.5522E 09	0.5756E 09	0.5994E 09	0.6236E 09	0.6481E 09	0.6730E 09	0.6982E 09
0.69	0.2277E 09	0.2380E 09	0.2485E 09	0.2592E 09	0.2700E 09	0.2810E 09	0.2922E 09	0.3036E 09	0.3151E 09	0.3268E 09
0.70	0.1107E 09	0.1157E 09	0.1207E 09	0.1258E 09	0.1310E 09	0.1363E 09	0.1417E 09	0.1471E 09	0.1526E 09	0.1582E 09
0.71	0.5494E 08	0.5736E 08	0.5983E 08	0.6234E 08	0.6489E 08	0.6748E 08	0.7010E 08	0.7276E 08	0.7546E 08	0.7819E 08
0.72	0.2676E 08	0.2793E 08	0.2912E 08	0.3033E 08	0.3155E 08	0.3280E 08	0.3406E 08	0.3534E 08	0.3663E 08	0.3794E 08
0.73	0.1293E 08	0.1349E 08	0.1406E 08	0.1463E 08	0.1522E 08	0.1581E 08	0.1641E 08	0.1702E 08	0.1764E 08	0.1826E 08
0.74	0.6030E 07	0.6288E 07	0.6549E 07	0.6814E 07	0.7084E 07	0.7357E 07	0.7633E 07	0.7914E 07	0.8198E 07	0.8485E 07
0.75	0.2825E 07	0.2944E 07	0.3065E 07	0.3188E 07	0.3312E 07	0.3439E 07	0.3567E 07	0.3696E 07	0.3827E 07	0.3960E 07
0.76	0.1374E 07	0.1431E 07	0.1489E 07	0.1548E 07	0.1608E 07	0.1669E 07	0.1730E 07	0.1792E 07	0.1855E 07	0.1919E 07
0.77	0.6682E 06	0.6958E 06	0.7238E 06	0.7522E 06	0.7810E 06	0.8102E 06	0.8397E 06	0.8696E 06	0.8998E 06	0.9304E 06
0.78	0.3248E 06	0.3380E 06	0.3515E 06	0.3651E 06	0.3790E 06	0.3930E 06	0.4071E 06	0.4215E 06	0.4360E 06	0.4507E 06
0.79	0.1573E 06	0.1637E 06	0.1701E 06	0.1767E 06	0.1833E 06	0.1900E 06	0.1968E 06	0.2036E 06	0.2106E 06	0.2176E 06
0.80	0.7590E 05	0.7894E 05	0.8202E 05	0.8514E 05	0.8831E 05	0.9150E 05	0.9474E 05	0.9802E 05	0.1013E 06	0.1046E 06
0.81	0.3662E 05	0.3807E 05	0.3955E 05	0.4104E 05	0.4255E 05	0.4407E 05	0.4562E 05	0.4718E 05	0.4876E 05	0.5035E 05
0.82	0.1766E 05	0.1835E 05	0.1906E 05	0.1977E 05	0.2049E 05	0.2122E 05	0.2195E 05	0.2270E 05	0.2345E 05	0.2421E 05
0.83	0.8523E 04	0.8854E 04	0.9190E 04	0.9529E 04	0.9873E 04	0.1022E 05	0.1057E 05	0.1092E 05	0.1128E 05	0.1164E 05

TABLE 17

Blackbody Spectral Luminance, Scotopic (sc.cd/m^2.μ)

λ(μ)	7000°K.	7100°K.	7200°K.	7300°K.	7400°K.	7500°K.	7600°K.	7700°K.	7800°K.	7900°K.
0.38	0.6953E 08	0.7506E 08	0.8086E 08	0.8694E 08	0.9329E 08	0.9992E 08	0.1068E 09	0.1140E 09	0.1215E 09	0.1292E 09
0.39	0.2631E 09	0.2835E 09	0.3048E 09	0.3271E 09	0.3504E 09	0.3747E 09	0.3999E 09	0.4262E 09	0.4534E 09	0.4817E 09
0.40	0.1113E 10	0.1197E 10	0.1285E 10	0.1377E 10	0.1472E 10	0.1572E 10	0.1675E 10	0.1782E 10	0.1893E 10	0.2008E 10
0.41	0.4184E 10	0.4493E 10	0.4814E 10	0.5149E 10	0.5498E 10	0.5860E 10	0.6236E 10	0.6626E 10	0.7029E 10	0.7446E 10
0.42	0.1160E 11	0.1243E 11	0.1330E 11	0.1421E 11	0.1515E 11	0.1612E 11	0.1713E 11	0.1818E 11	0.1926E 11	0.2037E 11
0.43	0.2392E 11	0.2560E 11	0.2735E 11	0.2917E 11	0.3106E 11	0.3301E 11	0.3503E 11	0.3712E 11	0.3927E 11	0.4150E 11
0.44	0.3908E 11	0.4176E 11	0.4455E 11	0.4745E 11	0.5044E 11	0.5354E 11	0.5675E 11	0.6006E 11	0.6347E 11	0.6699E 11
0.45	0.5379E 11	0.5740E 11	0.6115E 11	0.6504E 11	0.6906E 11	0.7321E 11	0.7750E 11	0.8192E 11	0.8648E 11	0.9117E 11
0.46	0.6642E 11	0.7079E 11	0.7532E 11	0.8000E 11	0.8484E 11	0.8983E 11	0.9499E 11	0.1002E 12	0.1057E 12	0.1113E 12
0.47	0.7823E 11	0.8327E 11	0.8848E 11	0.9387E 11	0.9943E 11	0.1051E 12	0.1110E 12	0.1171E 12	0.1234E 12	0.1298E 12
0.48	0.9064E 11	0.9636E 11	0.1022E 12	0.1083E 12	0.1146E 12	0.1211E 12	0.1278E 12	0.1346E 12	0.1417E 12	0.1489E 12
0.49	0.1018E 12	0.1081E 12	0.1146E 12	0.1214E 12	0.1283E 12	0.1354E 12	0.1427E 12	0.1502E 12	0.1579E 12	0.1658E 12
0.50	0.1088E 12	0.1154E 12	0.1222E 12	0.1293E 12	0.1365E 12	0.1439E 12	0.1515E 12	0.1594E 12	0.1674E 12	0.1756E 12
0.51	0.1086E 12	0.1151E 12	0.1217E 12	0.1286E 12	0.1357E 12	0.1429E 12	0.1503E 12	0.1580E 12	0.1658E 12	0.1738E 12
0.52	0.1001E 12	0.1059E 12	0.1120E 12	0.1182E 12	0.1245E 12	0.1311E 12	0.1378E 12	0.1446E 12	0.1516E 12	0.1588E 12
0.53	0.8519E 11	0.9008E 11	0.9511E 11	0.1002E 12	0.1055E 12	0.1110E 12	0.1165E 12	0.1222E 12	0.1281E 12	0.1340E 12
0.54	0.6689E 11	0.7066E 11	0.7454E 11	0.7852E 11	0.8260E 11	0.8678E 11	0.9106E 11	0.9544E 11	0.9992E 11	0.1044E 12
0.55	0.4847E 11	0.5116E 11	0.5392E 11	0.5675E 11	0.5965E 11	0.6262E 11	0.6566E 11	0.6877E 11	0.7194E 11	0.7518E 11
0.56	0.3244E 11	0.3421E 11	0.3602E 11	0.3788E 11	0.3978E 11	0.4173E 11	0.4373E 11	0.4576E 11	0.4784E 11	0.4996E 11
0.57	0.2003E 11	0.2110E 11	0.2220E 11	0.2333E 11	0.2448E 11	0.2566E 11	0.2687E 11	0.2810E 11	0.2936E 11	0.3064E 11
0.58	0.1142E 11	0.1203E 11	0.1264E 11	0.1328E 11	0.1392E 11	0.1458E 11	0.1526E 11	0.1595E 11	0.1665E 11	0.1736E 11
0.59	0.6030E 10	0.6343E 10	0.6664E 10	0.6992E 10	0.7328E 10	0.7670E 10	0.8019E 10	0.8376E 10	0.8739E 10	0.9108E 10
0.60	0.2980E 10	0.3132E 10	0.3288E 10	0.3448E 10	0.3611E 10	0.3777E 10	0.3947E 10	0.4119E 10	0.4295E 10	0.4474E 10
0.61	0.1397E 10	0.1467E 10	0.1539E 10	0.1613E 10	0.1688E 10	0.1765E 10	0.1843E 10	0.1922E 10	0.2003E 10	0.2086E 10
0.62	0.6309E 09	0.6622E 09	0.6943E 09	0.7270E 09	0.7603E 09	0.7944E 09	0.8290E 09	0.8643E 09	0.9002E 09	0.9366E 09
0.63	0.2782E 09	0.2918E 09	0.3057E 09	0.3199E 09	0.3344E 09	0.3491E 09	0.3642E 09	0.3794E 09	0.3950E 09	0.4108E 09
0.64	0.1217E 09	0.1275E 09	0.1335E 09	0.1397E 09	0.1459E 09	0.1523E 09	0.1587E 09	0.1653E 09	0.1720E 09	0.1788E 09
0.65	0.5364E 08	0.5620E 08	0.5881E 08	0.6147E 08	0.6417E 08	0.6693E 08	0.6973E 08	0.7258E 08	0.7547E 08	0.7841E 08
0.66	0.2414E 08	0.2528E 08	0.2643E 08	0.2761E 08	0.2881E 08	0.3003E 08	0.3128E 08	0.3254E 08	0.3382E 08	0.3512E 08
0.67	0.1112E 08	0.1163E 08	0.1216E 08	0.1270E 08	0.1324E 08	0.1379E 08	0.1436E 08	0.1493E 08	0.1551E 08	0.1610E 08
0.68	0.5235E 07	0.5475E 07	0.5719E 07	0.5968E 07	0.6221E 07	0.6478E 07	0.6739E 07	0.7004E 07	0.7273E 07	0.7546E 07
0.69	0.2516E 07	0.2630E 07	0.2746E 07	0.2864E 07	0.2984E 07	0.3105E 07	0.3229E 07	0.3355E 07	0.3482E 07	0.3611E 07
0.70	0.1234E 07	0.1289E 07	0.1345E 07	0.1402E 07	0.1460E 07	0.1519E 07	0.1579E 07	0.1639E 07	0.1701E 07	0.1763E 07
0.71	0.6168E 06	0.6440E 06	0.6718E 06	0.6999E 06	0.7285E 06	0.7576E 06	0.7870E 06	0.8169E 06	0.8472E 06	0.8779E 06
0.72	0.3140E 06	0.3277E 06	0.3416E 06	0.3557E 06	0.3701E 06	0.3847E 06	0.3995E 06	0.4145E 06	0.4297E 06	0.4451E 06
0.73	0.1626E 06	0.1696E 06	0.1767E 06	0.1840E 06	0.1913E 06	0.1988E 06	0.2063E 06	0.2140E 06	0.2217E 06	0.2296E 06
0.74	0.8569E 05	0.8934E 05	0.9306E 05	0.9683E 05	0.1006E 06	0.1045E 06	0.1084E 06	0.1124E 06	0.1164E 06	0.1205E 06
0.75	0.4591E 05	0.4785E 05	0.4982E 05	0.5181E 05	0.5384E 05	0.5589E 05	0.5797E 05	0.6008E 05	0.6221E 05	0.6437E 05
0.76	0.2498E 05	0.2602E 05	0.2708E 05	0.2815E 05	0.2924E 05	0.3035E 05	0.3146E 05	0.3260E 05	0.3374E 05	0.3490E 05
0.77	0.1380E 05	0.1437E 05	0.1494E 05	0.1553E 05	0.1613E 05	0.1673E 05	0.1734E 05	0.1795E 05	0.1858E 05	0.1921E 05
0.78	0.7733E 04	0.8049E 04	0.8369E 04	0.8694E 04	0.9023E 04	0.9357E 04	0.9694E 04	0.1003E 05	0.1038E 05	0.1073E 05

TABLE 17

Blackbody Spectral Luminance, Photopic (cd/m^2.μ)

λ(μ)	8000°K.	8100°K.	8200°K.	8300°K.	8400°K.	8500°K.	8600°K.	8700°K.	8800°K.	8900°K.
0.36	0.3573E 06	0.3802E 06	0.4040E 06	0.4287E 06	0.4542E 06	0.4806E 06	0.5078E 06	0.5360E 06	0.5650E 06	0.5949E 06
0.37	0.1129E 07	0.1199E 07	0.1272E 07	0.1348E 07	0.1426E 07	0.1507E 07	0.1590E 07	0.1676E 07	0.1764E 07	0.1855E 07
0.38	0.3539E 07	0.3754E 07	0.3976E 07	0.4206E 07	0.4444E 07	0.4689E 07	0.4941E 07	0.5201E 07	0.5468E 07	0.5743E 07
0.39	0.1081E 08	0.1145E 08	0.1211E 08	0.1279E 08	0.1349E 08	0.1422E 08	0.1497E 08	0.1573E 08	0.1652E 08	0.1733E 08
0.40	0.3531E 08	0.3735E 08	0.3946E 08	0.4163E 08	0.4386E 08	0.4616E 08	0.4853E 08	0.5096E 08	0.5345E 08	0.5601E 08
0.41	0.1065E 09	0.1125E 09	0.1187E 09	0.1251E 09	0.1317E 09	0.1384E 09	0.1453E 09	0.1524E 09	0.1597E 09	0.1672E 09
0.42	0.3471E 09	0.3663E 09	0.3860E 09	0.4062E 09	0.4270E 09	0.4484E 09	0.4703E 09	0.4928E 09	0.5158E 09	0.5394E 09
0.43	0.9902E 09	0.1043E 10	0.1098E 10	0.1154E 10	0.1212E 10	0.1271E 10	0.1332E 10	0.1395E 10	0.1458E 10	0.1524E 10
0.44	0.1927E 10	0.2029E 10	0.2133E 10	0.2240E 10	0.2350E 10	0.2463E 10	0.2578E 10	0.2696E 10	0.2816E 10	0.2940E 10
0.45	0.3122E 10	0.3283E 10	0.3448E 10	0.3617E 10	0.3791E 10	0.3969E 10	0.4151E 10	0.4337E 10	0.4527E 10	0.4721E 10
0.46	0.4825E 10	0.5069E 10	0.5319E 10	0.5575E 10	0.5837E 10	0.6106E 10	0.6380E 10	0.6660E 10	0.6946E 10	0.7237E 10
0.47	0.7154E 10	0.7508E 10	0.7871E 10	0.8242E 10	0.8622E 10	0.9011E 10	0.9407E 10	0.9812E 10	0.1022E 11	0.1064E 11
0.48	0.1067E 11	0.1119E 11	0.1172E 11	0.1226E 11	0.1282E 11	0.1338E 11	0.1396E 11	0.1455E 11	0.1515E 11	0.1576E 11
0.49	0.1558E 11	0.1632E 11	0.1708E 11	0.1786E 11	0.1865E 11	0.1946E 11	0.2028E 11	0.2112E 11	0.2198E 11	0.2285E 11
0.50	0.2358E 11	0.2469E 11	0.2581E 11	0.2696E 11	0.2814E 11	0.2934E 11	0.3056E 11	0.3180E 11	0.3307E 11	0.3435E 11
0.51	0.3577E 11	0.3741E 11	0.3909E 11	0.4080E 11	0.4255E 11	0.4433E 11	0.4614E 11	0.4798E 11	0.4986E 11	0.5177E 11
0.52	0.4914E 11	0.5136E 11	0.5362E 11	0.5593E 11	0.5828E 11	0.6067E 11	0.6311E 11	0.6559E 11	0.6811E 11	0.7068E 11
0.53	0.5802E 11	0.6060E 11	0.6322E 11	0.6590E 11	0.6862E 11	0.7139E 11	0.7421E 11	0.7708E 11	0.8000E 11	0.8296E 11
0.54	0.6242E 11	0.6514E 11	0.6792E 11	0.7074E 11	0.7362E 11	0.7655E 11	0.7952E 11	0.8255E 11	0.8562E 11	0.8874E 11
0.55	0.6325E 11	0.6597E 11	0.6873E 11	0.7154E 11	0.7441E 11	0.7732E 11	0.8028E 11	0.8328E 11	0.8633E 11	0.8943E 11
0.56	0.6142E 11	0.6402E 11	0.6666E 11	0.6934E 11	0.7208E 11	0.7485E 11	0.7767E 11	0.8053E 11	0.8344E 11	0.8639E 11
0.57	0.5705E 11	0.5942E 11	0.6183E 11	0.6429E 11	0.6678E 11	0.6931E 11	0.7189E 11	0.7450E 11	0.7714E 11	0.7983E 11
0.58	0.5059E 11	0.5266E 11	0.5477E 11	0.5691E 11	0.5908E 11	0.6129E 11	0.6353E 11	0.6581E 11	0.6811E 11	0.7045E 11
0.59	0.4270E 11	0.4442E 11	0.4618E 11	0.4795E 11	0.4976E 11	0.5159E 11	0.5345E 11	0.5534E 11	0.5725E 11	0.5919E 11
0.60	0.3452E 11	0.3589E 11	0.3729E 11	0.3870E 11	0.4014E 11	0.4160E 11	0.4308E 11	0.4457E 11	0.4609E 11	0.4763E 11
0.61	0.2668E 11	0.2773E 11	0.2879E 11	0.2987E 11	0.3096E 11	0.3207E 11	0.3319E 11	0.3433E 11	0.3549E 11	0.3665E 11
0.62	0.1959E 11	0.2035E 11	0.2112E 11	0.2190E 11	0.2269E 11	0.2349E 11	0.2430E 11	0.2512E 11	0.2596E 11	0.2680E 11
0.63	0.1320E 11	0.1371E 11	0.1422E 11	0.1474E 11	0.1526E 11	0.1580E 11	0.1634E 11	0.1688E 11	0.1743E 11	0.1799E 11
0.64	0.8454E 10	0.8772E 10	0.9095E 10	0.9422E 10	0.9754E 10	0.1009E 11	0.1043E 11	0.1077E 11	0.1112E 11	0.1147E 11
0.65	0.5009E 10	0.5195E 10	0.5384E 10	0.5575E 10	0.5769E 10	0.5965E 10	0.6163E 10	0.6364E 10	0.6567E 10	0.6773E 10
0.66	0.2767E 10	0.2868E 10	0.2971E 10	0.3075E 10	0.3181E 10	0.3288E 10	0.3396E 10	0.3505E 10	0.3616E 10	0.3727E 10
0.67	0.1406E 10	0.1457E 10	0.1509E 10	0.1561E 10	0.1614E 10	0.1667E 10	0.1722E 10	0.1776E 10	0.1832E 10	0.1888E 10
0.68	0.7238E 09	0.7498E 09	0.7760E 09	0.8026E 09	0.8295E 09	0.8567E 09	0.8842E 09	0.9120E 09	0.9401E 09	0.9685E 09
0.69	0.3386E 09	0.3506E 09	0.3628E 09	0.3751E 09	0.3875E 09	0.4000E 09	0.4127E 09	0.4256E 09	0.4385E 09	0.4516E 09
0.70	0.1639E 09	0.1696E 09	0.1754E 09	0.1813E 09	0.1873E 09	0.1933E 09	0.1993E 09	0.2055E 09	0.2116E 09	0.2179E 09
0.71	0.8096E 08	0.8376E 08	0.8660E 08	0.8946E 08	0.9236E 08	0.9529E 08	0.9825E 08	0.1012E 09	0.1042E 09	0.1073E 09
0.72	0.3927E 08	0.4062E 08	0.4198E 08	0.4335E 08	0.4474E 08	0.4614E 08	0.4756E 08	0.4899E 08	0.5044E 08	0.5190E 08
0.73	0.1890E 08	0.1954E 08	0.2018E 08	0.2084E 08	0.2150E 08	0.2216E 08	0.2284E 08	0.2352E 08	0.2421E 08	0.2490E 08
0.74	0.8776E 07	0.9069E 07	0.9367E 07	0.9667E 07	0.9970E 07	0.1027E 08	0.1058E 08	0.1089E 08	0.1121E 08	0.1153E 08
0.75	0.4094E 07	0.4230E 07	0.4367E 07	0.4506E 07	0.4646E 07	0.4787E 07	0.4930E 07	0.5074E 07	0.5219E 07	0.5365E 07
0.76	0.1984E 07	0.2049E 07	0.2114E 07	0.2181E 07	0.2248E 07	0.2316E 07	0.2384E 07	0.2453E 07	0.2522E 07	0.2592E 07
0.77	0.9613E 06	0.9925E 06	0.1024E 07	0.1055E 07	0.1088E 07	0.1120E 07	0.1153E 07	0.1186E 07	0.1219E 07	0.1253E 07
0.78	0.4655E 06	0.4805E 06	0.4956E 06	0.5109E 06	0.5263E 06	0.5418E 06	0.5575E 06	0.5733E 06	0.5893E 06	0.6053E 06
0.79	0.2247E 06	0.2319E 06	0.2391E 06	0.2464E 06	0.2538E 06	0.2612E 06	0.2687E 06	0.2762E 06	0.2838E 06	0.2915E 06
0.80	0.1080E 06	0.1114E 06	0.1149E 06	0.1183E 06	0.1218E 06	0.1254E 06	0.1289E 06	0.1325E 06	0.1362E 06	0.1398E 06
0.81	0.5196E 05	0.5359E 05	0.5523E 05	0.5688E 05	0.5855E 05	0.6023E 05	0.6193E 05	0.6364E 05	0.6536E 05	0.6710E 05
0.82	0.2498E 05	0.2575E 05	0.2653E 05	0.2732E 05	0.2812E 05	0.2892E 05	0.2973E 05	0.3054E 05	0.3136E 05	0.3219E 05
0.83	0.1201E 05	0.1238E 05	0.1275E 05	0.1313E 05	0.1351E 05	0.1389E 05	0.1427E 05	0.1466E 05	0.1505E 05	0.1544E 05

TABLE 17

Blackbody Spectral Luminance, Scotopic (sc.cd/m^2.μ)

$\lambda(\mu)$	8000°K.	8100°K.	8200°K.	8300°K.	8400°K.	8500°K.	8600°K.	8700°K.	8800°K.	8900°K.
0.38	0.1373E 09	0.1456E 09	0.1542E 09	0.1632E 09	0.1724E 09	0.1819E 09	0.1917E 09	0.2017E 09	0.2121E 09	0.2228E 09
0.39	0.5109E 09	0.5412E 09	0.5724E 09	0.6047E 09	0.6380E 09	0.6723E 09	0.7075E 09	0.7438E 09	0.7811E 09	0.8194E 09
0.40	0.2127E 10	0.2250E 10	0.2377E 10	0.2508E 10	0.2642E 10	0.2781E 10	0.2924E 10	0.3070E 10	0.3220E 10	0.3374E 10
0.41	0.7876E 10	0.8321E 10	0.8778E 10	0.9250E 10	0.9735E 10	0.1023E 11	0.1074E 11	0.1127E 11	0.1180E 11	0.1236E 11
0.42	0.2153E 11	0.2271E 11	0.2393E 11	0.2519E 11	0.2648E 11	0.2780E 11	0.2916E 11	0.3056E 11	0.3199E 11	0.3345E 11
0.43	0.4379E 11	0.4615E 11	0.4857E 11	0.5107E 11	0.5362E 11	0.5625E 11	0.5894E 11	0.6169E 11	0.6451E 11	0.6740E 11
0.44	0.7061E 11	0.7433E 11	0.7815E 11	0.8207E 11	0.8609E 11	0.9022E 11	0.9444E 11	0.9876E 11	0.1031E 12	0.1076E 12
0.45	0.9599E 11	0.1009E 12	0.1060E 12	0.1112E 12	0.1165E 12	0.1220E 12	0.1276E 12	0.1333E 12	0.1391E 12	0.1451E 12
0.46	0.1171E 12	0.1230E 12	0.1291E 12	0.1353E 12	0.1417E 12	0.1482E 12	0.1548E 12	0.1616E 12	0.1686E 12	0.1756E 12
0.47	0.1364E 12	0.1431E 12	0.1500E 12	0.1571E 12	0.1644E 12	0.1718E 12	0.1793E 12	0.1870E 12	0.1949E 12	0.2029E 12
0.48	0.1563E 12	0.1639E 12	0.1717E 12	0.1796E 12	0.1877E 12	0.1960E 12	0.2045E 12	0.2131E 12	0.2219E 12	0.2309E 12
0.49	0.1739E 12	0.1822E 12	0.1907E 12	0.1993E 12	0.2082E 12	0.2172E 12	0.2264E 12	0.2358E 12	0.2454E 12	0.2551E 12
0.50	0.1840E 12	0.1926E 12	0.2014E 12	0.2104E 12	0.2196E 12	0.2289E 12	0.2384E 12	0.2482E 12	0.2580E 12	0.2681E 12
0.51	0.1819E 12	0.1903E 12	0.1988E 12	0.2075E 12	0.2164E 12	0.2255E 12	0.2347E 12	0.2441E 12	0.2536E 12	0.2633E 12
0.52	0.1662E 12	0.1737E 12	0.1813E 12	0.1891E 12	0.1971E 12	0.2052E 12	0.2134E 12	0.2218E 12	0.2303E 12	0.2390E 12
0.53	0.1401E 12	0.1463E 12	0.1527E 12	0.1591E 12	0.1657E 12	0.1724E 12	0.1792E 12	0.1862E 12	0.1932E 12	0.2004E 12
0.54	0.1091E 12	0.1139E 12	0.1187E 12	0.1237E 12	0.1287E 12	0.1338E 12	0.1390E 12	0.1443E 12	0.1497E 12	0.1551E 12
0.55	0.7848E 11	0.8185E 11	0.8528E 11	0.8877E 11	0.9232E 11	0.9593E 11	0.9960E 11	0.1033E 12	0.1071E 12	0.1109E 12
0.56	0.5212E 11	0.5432E 11	0.5656E 11	0.5884E 11	0.6115E 11	0.6351E 11	0.6590E 11	0.6833E 11	0.7080E 11	0.7330E 11
0.57	0.3194E 11	0.3327E 11	0.3462E 11	0.3599E 11	0.3739E 11	0.3881E 11	0.4025E 11	0.4171E 11	0.4319E 11	0.4470E 11
0.58	0.1809E 11	0.1883E 11	0.1959E 11	0.2035E 11	0.2113E 11	0.2192E 11	0.2272E 11	0.2354E 11	0.2436E 11	0.2520E 11
0.59	0.9485E 10	0.9867E 10	0.1025E 11	0.1065E 11	0.1105E 11	0.1146E 11	0.1187E 11	0.1229E 11	0.1271E 11	0.1314E 11
0.60	0.4657E 10	0.4842E 10	0.5030E 10	0.5221E 10	0.5415E 10	0.5611E 10	0.5811E 10	0.6013E 10	0.6218E 10	0.6425E 10
0.61	0.2169E 10	0.2254E 10	0.2341E 10	0.2429E 10	0.2518E 10	0.2608E 10	0.2699E 10	0.2792E 10	0.2886E 10	0.2980E 10
0.62	0.9737E 09	0.1011E 10	0.1049E 10	0.1088E 10	0.1127E 10	0.1167E 10	0.1207E 10	0.1248E 10	0.1290E 10	0.1332E 10
0.63	0.4268E 09	0.4431E 09	0.4596E 09	0.4764E 09	0.4934E 09	0.5106E 09	0.5280E 09	0.5456E 09	0.5635E 09	0.5816E 09
0.64	0.1856E 09	0.1926E 09	0.1997E 09	0.2069E 09	0.2142E 09	0.2216E 09	0.2290E 09	0.2366E 09	0.2442E 09	0.2520E 09
0.65	0.8140E 08	0.8442E 08	0.8749E 08	0.9060E 08	0.9375E 08	0.9693E 08	0.1001E 09	0.1034E 09	0.1067E 09	0.1100E 09
0.66	0.3644E 08	0.3778E 08	0.3913E 08	0.4051E 08	0.4190E 08	0.4330E 08	0.4473E 08	0.4617E 08	0.4762E 08	0.4909E 08
0.67	0.1670E 08	0.1730E 08	0.1792E 08	0.1854E 08	0.1917E 08	0.1980E 08	0.2045E 08	0.2110E 08	0.2175E 08	0.2242E 08
0.68	0.7822E 07	0.8102E 07	0.8386E 07	0.8674E 07	0.8964E 07	0.9258E 07	0.9556E 07	0.9856E 07	0.1016E 08	0.1046E 08
0.69	0.3742E 07	0.3874E 07	0.4008E 07	0.4144E 07	0.4281E 07	0.4420E 07	0.4561E 07	0.4702E 07	0.4846E 07	0.4990E 07
0.70	0.1826E 07	0.1890E 07	0.1955E 07	0.2020E 07	0.2086E 07	0.2153E 07	0.2221E 07	0.2289E 07	0.2358E 07	0.2428E 07
0.71	0.9090E 06	0.9404E 06	0.9722E 06	0.1004E 07	0.1037E 07	0.1069E 07	0.1103E 07	0.1136E 07	0.1170E 07	0.1204E 07
0.72	0.4607E 06	0.4764E 06	0.4924E 06	0.5085E 06	0.5248E 06	0.5413E 06	0.5579E 06	0.5747E 06	0.5916E 06	0.6087E 06
0.73	0.2376E 06	0.2456E 06	0.2537E 06	0.2620E 06	0.2703E 06	0.2787E 06	0.2871E 06	0.2957E 06	0.3043E 06	0.3130E 06
0.74	0.1246E 06	0.1288E 06	0.1330E 06	0.1373E 06	0.1416E 06	0.1460E 06	0.1504E 06	0.1548E 06	0.1593E 06	0.1638E 06
0.75	0.6655E 05	0.6876E 05	0.7099E 05	0.7324E 05	0.7551E 05	0.7781E 05	0.8013E 05	0.8247E 05	0.8483E 05	0.8721E 05
0.76	0.3607E 05	0.3725E 05	0.3845E 05	0.3966E 05	0.4088E 05	0.4211E 05	0.4335E 05	0.4461E 05	0.4587E 05	0.4714E 05
0.77	0.1985E 05	0.2049E 05	0.2115E 05	0.2180E 05	0.2247E 05	0.2314E 05	0.2381E 05	0.2450E 05	0.2518E 05	0.2588E 05
0.78	0.1108E 05	0.1144E 05	0.1180E 05	0.1216E 05	0.1253E 05	0.1290E 05	0.1327E 05	0.1365E 05	0.1403E 05	0.1441E 05

TABLE 17

Blackbody Spectral Luminance, Photopic (cd/m^2.μ)

λ(μ)	9000°K.	9100°K.	9200°K.	9300°K.	9400°K.	9500°K.	9600°K.	9700°K.	9800°K.	9900°K.
0.36	0.6257E 06	0.6574E 06	0.6900E 06	0.7234E 06	0.7578E 06	0.7930E 06	0.8291E 06	0.8660E 06	0.9039E 06	0.9426E 06
0.37	0.1949E 07	0.2045E 07	0.2143E 07	0.2245E 07	0.2348E 07	0.2455E 07	0.2564E 07	0.2675E 07	0.2789E 07	0.2905E 07
0.38	0.6025E 07	0.6315E 07	0.6612E 07	0.6916E 07	0.7228E 07	0.7547E 07	0.7873E 07	0.8206E 07	0.8547E 07	0.8895E 07
0.39	0.1816E 08	0.1901E 08	0.1989E 08	0.2078E 08	0.2169E 08	0.2263E 08	0.2358E 08	0.2456E 08	0.2555E 08	0.2657E 08
0.40	0.5863E 08	0.6131E 08	0.6406E 08	0.6687E 08	0.6974E 08	0.7267E 08	0.7567E 08	0.7872E 08	0.8184E 08	0.8501E 08
0.41	0.1748E 09	0.1827E 09	0.1907E 09	0.1988E 09	0.2072E 09	0.2157E 09	0.2244E 09	0.2332E 09	0.2423E 09	0.2515E 09
0.42	0.5635E 09	0.5881E 09	0.6133E 09	0.6390E 09	0.6652E 09	0.6920E 09	0.7192E 09	0.7470E 09	0.7753E 09	0.8040E 09
0.43	0.1590E 10	0.1658E 10	0.1728E 10	0.1799E 10	0.1871E 10	0.1944E 10	0.2019E 10	0.2096E 10	0.2173E 10	0.2252E 10
0.44	0.3066E 10	0.3194E 10	0.3325E 10	0.3458E 10	0.3594E 10	0.3733E 10	0.3874E 10	0.4017E 10	0.4163E 10	0.4311E 10
0.45	0.4919E 10	0.5121E 10	0.5326E 10	0.5536E 10	0.5749E 10	0.5966E 10	0.6187E 10	0.6411E 10	0.6640E 10	0.6871E 10
0.46	0.7535E 10	0.7838E 10	0.8147E 10	0.8461E 10	0.8781E 10	0.9106E 10	0.9436E 10	0.9772E 10	0.1011E 11	0.1045E 11
0.47	0.1107E 11	0.1151E 11	0.1195E 11	0.1240E 11	0.1286E 11	0.1333E 11	0.1381E 11	0.1429E 11	0.1478E 11	0.1528E 11
0.48	0.1639E 11	0.1702E 11	0.1767E 11	0.1832E 11	0.1899E 11	0.1967E 11	0.2035E 11	0.2105E 11	0.2176E 11	0.2248E 11
0.49	0.2374E 11	0.2464E 11	0.2556E 11	0.2649E 11	0.2744E 11	0.2840E 11	0.2937E 11	0.3036E 11	0.3136E 11	0.3238E 11
0.50	0.3567E 11	0.3700E 11	0.3835E 11	0.3972E 11	0.4112E 11	0.4253E 11	0.4397E 11	0.4542E 11	0.4690E 11	0.4839E 11
0.51	0.5371E 11	0.5568E 11	0.5768E 11	0.5971E 11	0.6178E 11	0.6387E 11	0.6599E 11	0.6813E 11	0.7031E 11	0.7251E 11
0.52	0.7328E 11	0.7593E 11	0.7861E 11	0.8133E 11	0.8410E 11	0.8689E 11	0.8973E 11	0.9260E 11	0.9551E 11	0.9845E 11
0.53	0.8597E 11	0.8902E 11	0.9212E 11	0.9526E 11	0.9844E 11	0.1016E 12	0.1049E 12	0.1082E 12	0.1115E 12	0.1149E 12
0.54	0.9191E 11	0.9512E 11	0.9837E 11	0.1016E 12	0.1050E 12	0.1084E 12	0.1118E 12	0.1153E 12	0.1188E 12	0.1223E 12
0.55	0.9257E 11	0.9575E 11	0.9898E 11	0.1022E 12	0.1055E 12	0.1089E 12	0.1123E 12	0.1157E 12	0.1192E 12	0.1227E 12
0.56	0.8938E 11	0.9240E 11	0.9547E 11	0.9858E 11	0.1017E 12	0.1049E 12	0.1081E 12	0.1113E 12	0.1146E 12	0.1180E 12
0.57	0.8255E 11	0.8530E 11	0.8810E 11	0.9092E 11	0.9378E 11	0.9667E 11	0.9959E 11	0.1025E 12	0.1055E 12	0.1085E 12
0.58	0.7281E 11	0.7521E 11	0.7764E 11	0.8009E 11	0.8257E 11	0.8508E 11	0.8762E 11	0.9019E 11	0.9278E 11	0.9539E 11
0.59	0.6115E 11	0.6313E 11	0.6514E 11	0.6717E 11	0.6922E 11	0.7130E 11	0.7340E 11	0.7551E 11	0.7765E 11	0.7982E 11
0.60	0.4919E 11	0.5076E 11	0.5235E 11	0.5396E 11	0.5559E 11	0.5724E 11	0.5890E 11	0.6058E 11	0.6227E 11	0.6398E 11
0.61	0.3783E 11	0.3903E 11	0.4024E 11	0.4146E 11	0.4269E 11	0.4394E 11	0.4520E 11	0.4647E 11	0.4776E 11	0.4905E 11
0.62	0.2765E 11	0.2852E 11	0.2939E 11	0.3027E 11	0.3116E 11	0.3206E 11	0.3296E 11	0.3388E 11	0.3480E 11	0.3574E 11
0.63	0.1856E 11	0.1913E 11	0.1971E 11	0.2029E 11	0.2088E 11	0.2148E 11	0.2208E 11	0.2268E 11	0.2329E 11	0.2391E 11
0.64	0.1183E 11	0.1219E 11	0.1255E 11	0.1292E 11	0.1329E 11	0.1366E 11	0.1404E 11	0.1442E 11	0.1480E 11	0.1519E 11
0.65	0.6980E 10	0.7190E 10	0.7402E 10	0.7615E 10	0.7831E 10	0.8049E 10	0.8268E 10	0.8490E 10	0.8713E 10	0.8939E 10
0.66	0.3840E 10	0.3954E 10	0.4069E 10	0.4186E 10	0.4303E 10	0.4421E 10	0.4540E 10	0.4661E 10	0.4782E 10	0.4904E 10
0.67	0.1944E 10	0.2001E 10	0.2059E 10	0.2117E 10	0.2176E 10	0.2235E 10	0.2294E 10	0.2354E 10	0.2415E 10	0.2476E 10
0.68	0.9972E 09	0.1026E 10	0.1055E 10	0.1084E 10	0.1114E 10	0.1144E 10	0.1174E 10	0.1205E 10	0.1235E 10	0.1266E 10
0.69	0.4649E 09	0.4782E 09	0.4916E 09	0.5052E 09	0.5189E 09	0.5327E 09	0.5466E 09	0.5606E 09	0.5747E 09	0.5890E 09
0.70	0.2242E 09	0.2306E 09	0.2370E 09	0.2435E 09	0.2500E 09	0.2566E 09	0.2632E 09	0.2699E 09	0.2766E 09	0.2834E 09
0.71	0.1103E 09	0.1134E 09	0.1166E 09	0.1197E 09	0.1229E 09	0.1261E 09	0.1293E 09	0.1326E 09	0.1358E 09	0.1391E 09
0.72	0.5337E 08	0.5485E 08	0.5635E 08	0.5785E 08	0.5937E 08	0.6090E 08	0.6244E 08	0.6400E 08	0.6556E 08	0.6713E 08
0.73	0.2560E 08	0.2630E 08	0.2701E 08	0.2773E 08	0.2845E 08	0.2917E 08	0.2990E 08	0.3064E 08	0.3138E 08	0.3213E 08
0.74	0.1185E 08	0.1217E 08	0.1249E 08	0.1282E 08	0.1315E 08	0.1349E 08	0.1382E 08	0.1416E 08	0.1450E 08	0.1484E 08
0.75	0.5513E 07	0.5662E 07	0.5811E 07	0.5962E 07	0.6115E 07	0.6268E 07	0.6422E 07	0.6577E 07	0.6733E 07	0.6890E 07
0.76	0.2663E 07	0.2734E 07	0.2806E 07	0.2878E 07	0.2951E 07	0.3024E 07	0.3098E 07	0.3172E 07	0.3247E 07	0.3322E 07
0.77	0.1287E 07	0.1321E 07	0.1355E 07	0.1390E 07	0.1424E 07	0.1459E 07	0.1495E 07	0.1530E 07	0.1566E 07	0.1602E 07
0.78	0.6215E 06	0.6378E 06	0.6543E 06	0.6708E 06	0.6874E 06	0.7042E 06	0.7211E 06	0.7380E 06	0.7551E 06	0.7722E 06
0.79	0.2992E 06	0.3070E 06	0.3148E 06	0.3227E 06	0.3307E 06	0.3387E 06	0.3467E 06	0.3548E 06	0.3629E 06	0.3711E 06
0.80	0.1435E 06	0.1472E 06	0.1509E 06	0.1546E 06	0.1584E 06	0.1622E 06	0.1660E 06	0.1699E 06	0.1737E 06	0.1776E 06
0.81	0.6885E 05	0.7061E 05	0.7238E 05	0.7416E 05	0.7595E 05	0.7776E 05	0.7957E 05	0.8140E 05	0.8323E 05	0.8508E 05
0.82	0.3302E 05	0.3385E 05	0.3470E 05	0.3554E 05	0.3640E 05	0.3725E 05	0.3812E 05	0.3898E 05	0.3985E 05	0.4073E 05
0.83	0.1584E 05	0.1624E 05	0.1664E 05	0.1704E 05	0.1745E 05	0.1785E 05	0.1826E 05	0.1868E 05	0.1909E 05	0.1951E 05

TABLE 17

Blackbody Spectral Luminance, Scotopic (sc.cd/m^2.μ)

$\lambda(\mu)$	9000°K.	9100°K.	9200°K.	9300°K.	9400°K.	9500°K.	9600°K.	9700°K.	9800°K.	9900°K.
0.38	0.2337E 09	0.2450E 09	0.2565E 09	0.2683E 09	0.2804E 09	0.2928E 09	0.3054E 09	0.3184E 09	0.3316E 09	0.3451E 09
0.39	0.8587E 09	0.8989E 09	0.9402E 09	0.9824E 09	0.1025E 10	0.1069E 10	0.1114E 10	0.1161E 10	0.1208E 10	0.1256E 10
0.40	0.3532E 10	0.3694E 10	0.3859E 10	0.4029E 10	0.4202E 10	0.4378E 10	0.4559E 10	0.4743E 10	0.4930E 10	0.5122E 10
0.41	0.1292E 11	0.1350E 11	0.1409E 11	0.1469E 11	0.1531E 11	0.1594E 11	0.1658E 11	0.1724E 11	0.1790E 11	0.1858E 11
0.42	0.3494E 11	0.3647E 11	0.3803E 11	0.3963E 11	0.4125E 11	0.4291E 11	0.4460E 11	0.4632E 11	0.4808E 11	0.4986E 11
0.43	0.7034E 11	0.7335E 11	0.7642E 11	0.7956E 11	0.8275E 11	0.8601E 11	0.8933E 11	0.9270E 11	0.9614E 11	0.9963E 11
0.44	0.1123E 12	0.1170E 12	0.1218E 12	0.1266E 12	0.1316E 12	0.1367E 12	0.1419E 12	0.1471E 12	0.1525E 12	0.1579E 12
0.45	0.1512E 12	0.1574E 12	0.1637E 12	0.1702E 12	0.1767E 12	0.1834E 12	0.1902E 12	0.1971E 12	0.2041E 12	0.2112E 12
0.46	0.1829E 12	0.1902E 12	0.1977E 12	0.2053E 12	0.2131E 12	0.2210E 12	0.2290E 12	0.2371E 12	0.2454E 12	0.2538E 12
0.47	0.2111E 12	0.2195E 12	0.2279E 12	0.2366E 12	0.2453E 12	0.2543E 12	0.2633E 12	0.2725E 12	0.2818E 12	0.2913E 12
0.48	0.2401E 12	0.2493E 12	0.2588E 12	0.2684E 12	0.2782E 12	0.2881E 12	0.2982E 12	0.3084E 12	0.3188E 12	0.3293E 12
0.49	0.2650E 12	0.2751E 12	0.2853E 12	0.2957E 12	0.3063E 12	0.3170E 12	0.3279E 12	0.3389E 12	0.3501E 12	0.3615E 12
0.50	0.2783E 12	0.2887E 12	0.2993E 12	0.3100E 12	0.3209E 12	0.3319E 12	0.3431E 12	0.3545E 12	0.3660E 12	0.3776E 12
0.51	0.2732E 12	0.2832E 12	0.2934E 12	0.3038E 12	0.3142E 12	0.3249E 12	0.3357E 12	0.3466E 12	0.3576E 12	0.3688E 12
0.52	0.2478E 12	0.2568E 12	0.2658E 12	0.2750E 12	0.2844E 12	0.2938E 12	0.3034E 12	0.3132E 12	0.3230E 12	0.3329E 12
0.53	0.2076E 12	0.2150E 12	0.2225E 12	0.2301E 12	0.2378E 12	0.2455E 12	0.2534E 12	0.2614E 12	0.2695E 12	0.2777E 12
0.54	0.1607E 12	0.1663E 12	0.1720E 12	0.1777E 12	0.1836E 12	0.1895E 12	0.1955E 12	0.2016E 12	0.2077E 12	0.2139E 12
0.55	0.1148E 12	0.1188E 12	0.1228E 12	0.1268E 12	0.1309E 12	0.1351E 12	0.1393E 12	0.1436E 12	0.1479E 12	0.1522E 12
0.56	0.7583E 11	0.7840E 11	0.8101E 11	0.8364E 11	0.8631E 11	0.8901E 11	0.9174E 11	0.9450E 11	0.9729E 11	0.1001E 12
0.57	0.4622E 11	0.4776E 11	0.4932E 11	0.5091E 11	0.5251E 11	0.5413E 11	0.5576E 11	0.5742E 11	0.5909E 11	0.6078E 11
0.58	0.2604E 11	0.2690E 11	0.2777E 11	0.2864E 11	0.2953E 11	0.3043E 11	0.3134E 11	0.3226E 11	0.3318E 11	0.3412E 11
0.59	0.1358E 11	0.1402E 11	0.1446E 11	0.1491E 11	0.1537E 11	0.1583E 11	0.1630E 11	0.1677E 11	0.1724E 11	0.1772E 11
0.60	0.6635E 10	0.6847E 10	0.7062E 10	0.7279E 10	0.7499E 10	0.7721E 10	0.7945E 10	0.8171E 10	0.8400E 10	0.8631E 10
0.61	0.3077E 10	0.3174E 10	0.3272E 10	0.3371E 10	0.3472E 10	0.3573E 10	0.3676E 10	0.3779E 10	0.3883E 10	0.3989E 10
0.62	0.1374E 10	0.1417E 10	0.1460E 10	0.1504E 10	0.1548E 10	0.1593E 10	0.1638E 10	0.1683E 10	0.1729E 10	0.1776E 10
0.63	0.5998E 09	0.6183E 09	0.6369E 09	0.6558E 09	0.6748E 09	0.6940E 09	0.7134E 09	0.7330E 09	0.7528E 09	0.7727E 09
0.64	0.2598E 09	0.2677E 09	0.2757E 09	0.2837E 09	0.2919E 09	0.3001E 09	0.3084E 09	0.3167E 09	0.3252E 09	0.3337E 09
0.65	0.1134E 09	0.1168E 09	0.1202E 09	0.1237E 09	0.1272E 09	0.1308E 09	0.1343E 09	0.1379E 09	0.1416E 09	0.1452E 09
0.66	0.5058E 08	0.5208E 08	0.5360E 08	0.5513E 08	0.5667E 08	0.5823E 08	0.5980E 08	0.6138E 08	0.6298E 08	0.6459E 08
0.67	0.2309E 08	0.2377E 08	0.2445E 08	0.2514E 08	0.2584E 08	0.2654E 08	0.2725E 08	0.2796E 08	0.2868E 08	0.2940E 08
0.68	0.1077E 08	0.1108E 08	0.1140E 08	0.1172E 08	0.1204E 08	0.1236E 08	0.1269E 08	0.1302E 08	0.1335E 08	0.1368E 08
0.69	0.5136E 07	0.5284E 07	0.5432E 07	0.5582E 07	0.5734E 07	0.5886E 07	0.6040E 07	0.6195E 07	0.6350E 07	0.6508E 07
0.70	0.2498E 07	0.2569E 07	0.2641E 07	0.2713E 07	0.2785E 07	0.2859E 07	0.2933E 07	0.3007E 07	0.3082E 07	0.3157E 07
0.71	0.1239E 07	0.1274E 07	0.1309E 07	0.1344E 07	0.1380E 07	0.1416E 07	0.1452E 07	0.1488E 07	0.1525E 07	0.1562E 07
0.72	0.6260E 06	0.6434E 06	0.6609E 06	0.6786E 06	0.6964E 06	0.7144E 06	0.7325E 06	0.7507E 06	0.7690E 06	0.7875E 06
0.73	0.3218E 06	0.3307E 06	0.3396E 06	0.3486E 06	0.3576E 06	0.3668E 06	0.3760E 06	0.3852E 06	0.3945E 06	0.4039E 06
0.74	0.1683E 06	0.1729E 06	0.1776E 06	0.1822E 06	0.1869E 06	0.1916E 06	0.1964E 06	0.2012E 06	0.2060E 06	0.2108E 06
0.75	0.8960E 05	0.9202E 05	0.9446E 05	0.9691E 05	0.9938E 05	0.1018E 06	0.1043E 06	0.1069E 06	0.1094E 06	0.1119E 06
0.76	0.4843E 05	0.4972E 05	0.5103E 05	0.5234E 05	0.5366E 05	0.5499E 05	0.5633E 05	0.5768E 05	0.5904E 05	0.6040E 05
0.77	0.2658E 05	0.2728E 05	0.2799E 05	0.2870E 05	0.2942E 05	0.3014E 05	0.3087E 05	0.3160E 05	0.3234E 05	0.3308E 05
0.78	0.1479E 05	0.1518E 05	0.1557E 05	0.1597E 05	0.1636E 05	0.1676E 05	0.1716E 05	0.1757E 05	0.1797E 05	0.1838E 05

TABLE 17

Blackbody Spectral Luminance, Photopic (cd/m^2.μ)

$\lambda(\mu)$	10000°K.	11000°K.	12000°K.	13000°K.	14000°K.	15000°K.	16000°K.	17000°K.	18000°K.	19000°K.
0.36	0.9821E 06	0.1424E 07	0.1946E 07	0.2542E 07	0.3204E 07	0.3926E 07	0.4702E 07	0.5524E 07	0.6389E 07	0.7291E 07
0.37	0.3024E 07	0.4345E 07	0.5894E 07	0.7651E 07	0.9595E 07	0.1170E 08	0.1396E 08	0.1635E 08	0.1885E 08	0.2146E 08
0.38	0.9250E 07	0.1317E 08	0.1774E 08	0.2290E 08	0.2858E 08	0.3471E 08	0.4126E 08	0.4817E 08	0.5540E 08	0.6291E 08
0.39	0.2760E 08	0.3900E 08	0.5219E 08	0.6698E 08	0.8320E 08	0.1006E 09	0.1192E 09	0.1388E 09	0.1592E 09	0.1804E 09
0.40	0.8825E 08	0.1237E 09	0.1645E 09	0.2100E 09	0.2597E 09	0.3131E 09	0.3697E 09	0.4292E 09	0.4911E 09	0.5552E 09
0.41	0.2608E 09	0.3631E 09	0.4799E 09	0.6097E 09	0.7509E 09	0.9020E 09	0.1061E 10	0.1229E 10	0.1403E 10	0.1583E 10
0.42	0.8333E 09	0.1151E 10	0.1514E 10	0.1914E 10	0.2348E 10	0.2812E 10	0.3301E 10	0.3812E 10	0.4343E 10	0.4891E 10
0.43	0.2333E 10	0.3204E 10	0.4189E 10	0.5274E 10	0.6445E 10	0.7693E 10	0.9006E 10	0.1037E 11	0.1179E 11	0.1326E 11
0.44	0.4462E 10	0.6090E 10	0.7922E 10	0.9932E 10	0.1209E 11	0.1439E 11	0.1680E 11	0.1932E 11	0.2192E 11	0.2460E 11
0.45	0.7106E 10	0.9642E 10	0.1248E 11	0.1558E 11	0.1892E 11	0.2245E 11	0.2615E 11	0.3000E 11	0.3398E 11	0.3807E 11
0.46	0.1081E 11	0.1458E 11	0.1879E 11	0.2338E 11	0.2829E 11	0.3348E 11	0.3891E 11	0.4454E 11	0.5036E 11	0.5634E 11
0.47	0.1578E 11	0.2118E 11	0.2717E 11	0.3368E 11	0.4064E 11	0.4797E 11	0.5562E 11	0.6356E 11	0.7174E 11	0.8014E 11
0.48	0.2320E 11	0.3098E 11	0.3959E 11	0.4891E 11	0.5883E 11	0.6928E 11	0.8017E 11	0.9144E 11	0.1030E 12	0.1149E 12
0.49	0.3341E 11	0.4440E 11	0.5650E 11	0.6957E 11	0.8346E 11	0.9805E 11	0.1132E 12	0.1289E 12	0.1450E 12	0.1616E 12
0.50	0.4991E 11	0.6601E 11	0.8369E 11	0.1027E 12	0.1229E 12	0.1440E 12	0.1660E 12	0.1887E 12	0.2121E 12	0.2359E 12
0.51	0.7474E 11	0.9841E 11	0.1243E 12	0.1521E 12	0.1815E 12	0.2124E 12	0.2443E 12	0.2773E 12	0.3112E 12	0.3458E 12
0.52	0.1014E 12	0.1329E 12	0.1674E 12	0.2043E 12	0.2432E 12	0.2839E 12	0.3261E 12	0.3696E 12	0.4141E 12	0.4596E 12
0.53	0.1183E 12	0.1545E 12	0.1939E 12	0.2360E 12	0.2804E 12	0.3267E 12	0.3746E 12	0.4239E 12	0.4744E 12	0.5259E 12
0.54	0.1259E 12	0.1638E 12	0.2049E 12	0.2487E 12	0.2948E 12	0.3428E 12	0.3924E 12	0.4435E 12	0.4957E 12	0.5490E 12
0.55	0.1262E 12	0.1636E 12	0.2040E 12	0.2470E 12	0.2922E 12	0.3392E 12	0.3877E 12	0.4375E 12	0.4885E 12	0.5404E 12
0.56	0.1213E 12	0.1567E 12	0.1948E 12	0.2353E 12	0.2778E 12	0.3219E 12	0.3674E 12	0.4141E 12	0.4618E 12	0.5105E 12
0.57	0.1116E 12	0.1436E 12	0.1780E 12	0.2146E 12	0.2528E 12	0.2925E 12	0.3333E 12	0.3753E 12	0.4181E 12	0.4616E 12
0.58	0.9803E 11	0.1257E 12	0.1555E 12	0.1869E 12	0.2198E 12	0.2539E 12	0.2890E 12	0.3250E 12	0.3617E 12	0.3990E 12
0.59	0.8200E 11	0.1048E 12	0.1293E 12	0.1551E 12	0.1821E 12	0.2100E 12	0.2387E 12	0.2681E 12	0.2981E 12	0.3286E 12
0.60	0.6571E 11	0.8375E 11	0.1030E 12	0.1233E 12	0.1446E 12	0.1665E 12	0.1890E 12	0.2121E 12	0.2355E 12	0.2594E 12
0.61	0.5036E 11	0.6400E 11	0.7856E 11	0.9388E 11	0.1098E 12	0.1263E 12	0.1432E 12	0.1605E 12	0.1781E 12	0.1960E 12
0.62	0.3668E 11	0.4648E 11	0.5692E 11	0.6790E 11	0.7930E 11	0.9108E 11	0.1031E 12	0.1154E 12	0.1280E 12	0.1408E 12
0.63	0.2453E 11	0.3100E 11	0.3789E 11	0.4511E 11	0.5260E 11	0.6033E 11	0.6826E 11	0.7635E 11	0.8458E 11	0.9293E 11
0.64	0.1558E 11	0.1964E 11	0.2395E 11	0.2846E 11	0.3314E 11	0.3797E 11	0.4291E 11	0.4795E 11	0.5307E 11	0.5827E 11
0.65	0.9165E 10	0.1152E 11	0.1402E 11	0.1663E 11	0.1934E 11	0.2213E 11	0.2498E 11	0.2789E 11	0.3085E 11	0.3385E 11
0.66	0.5027E 10	0.6305E 10	0.7657E 10	0.9070E 10	0.1053E 11	0.1203E 11	0.1357E 11	0.1514E 11	0.1673E 11	0.1834E 11
0.67	0.2537E 10	0.3175E 10	0.3849E 10	0.4551E 10	0.5278E 10	0.6025E 10	0.6789E 10	0.7566E 10	0.8356E 10	0.9156E 10
0.68	0.1297E 10	0.1620E 10	0.1959E 10	0.2314E 10	0.2680E 10	0.3056E 10	0.3440E 10	0.3831E 10	0.4227E 10	0.4629E 10
0.69	0.6033E 09	0.7515E 09	0.9075E 09	0.1070E 10	0.1237E 10	0.1409E 10	0.1585E 10	0.1764E 10	0.1945E 10	0.2129E 10
0.70	0.2902E 09	0.3607E 09	0.4349E 09	0.5120E 09	0.5915E 09	0.6731E 09	0.7563E 09	0.8409E 09	0.9267E 09	0.1013E 10
0.71	0.1424E 09	0.1767E 09	0.2127E 09	0.2500E 09	0.2885E 09	0.3280E 09	0.3682E 09	0.4091E 09	0.4506E 09	0.4925E 09
0.72	0.6872E 08	0.8507E 08	0.1022E 09	0.1200E 09	0.1383E 09	0.1570E 09	0.1762E 09	0.1956E 09	0.2153E 09	0.2352E 09
0.73	0.3288E 08	0.4062E 08	0.4873E 08	0.5715E 08	0.6580E 08	0.7465E 08	0.8368E 08	0.9284E 08	0.1021E 09	0.1115E 09
0.74	0.1518E 08	0.1872E 08	0.2243E 08	0.2626E 08	0.3021E 08	0.3425E 08	0.3835E 08	0.4253E 08	0.4675E 08	0.5102E 08
0.75	0.7048E 07	0.8676E 07	0.1037E 08	0.1213E 08	0.1394E 08	0.1579E 08	0.1767E 08	0.1958E 08	0.2151E 08	0.2347E 08
0.76	0.3397E 07	0.4174E 07	0.4986E 07	0.5825E 07	0.6686E 07	0.7566E 07	0.8461E 07	0.9369E 07	0.1028E 08	0.1121E 08
0.77	0.1638E 07	0.2009E 07	0.2396E 07	0.2796E 07	0.3207E 07	0.3626E 07	0.4052E 07	0.4484E 07	0.4921E 07	0.5362E 07
0.78	0.7895E 06	0.9668E 06	0.1151E 07	0.1342E 07	0.1537E 07	0.1737E 07	0.1940E 07	0.2145E 07	0.2353E 07	0.2563E 07
0.79	0.3793E 06	0.4638E 06	0.5516E 06	0.6423E 06	0.7352E 06	0.8299E 06	0.9262E 06	0.1023E 07	0.1122E 07	0.1222E 07
0.80	0.1815E 06	0.2216E 06	0.2632E 06	0.3062E 06	0.3501E 06	0.3950E 06	0.4405E 06	0.4866E 06	0.5333E 06	0.5803E 06
0.81	0.8693E 05	0.1059E 06	0.1257E 06	0.1460E 06	0.1668E 06	0.1881E 06	0.2096E 06	0.2315E 06	0.2535E 06	0.2758E 06
0.82	0.4161E 05	0.5064E 05	0.6001E 05	0.6966E 05	0.7953E 05	0.8958E 05	0.9978E 05	0.1101E 06	0.1205E 06	0.1310E 06
0.83	0.1993E 05	0.2422E 05	0.2866E 05	0.3324E 05	0.3792E 05	0.4268E 05	0.4751E 05	0.5240E 05	0.5735E 05	0.6233E 05

TABLE 17

Blackbody Spectral Luminance, Scotopic (sc.cd/$m^2 \cdot \mu$)

$\lambda(\mu)$	10000°K.	11000°K.	12000°K.	13000°K.	14000°K.	15000°K.	16000°K.	17000°K.	18000°K.	19000°K.
0.38	0.3588E 09	0.5112E 09	0.6886E 09	0.8886E 09	0.1108E 10	0.1347E 10	0.1601E 10	0.1869E 10	0.2149E 10	0.2440E 10
0.39	0.1304E 10	0.1843E 10	0.2467E 10	0.3166E 10	0.3932E 10	0.4758E 10	0.5636E 10	0.6561E 10	0.7526E 10	0.8527E 10
0.40	0.5317E 10	0.7454E 10	0.9912E 10	0.1265E 11	0.1565E 11	0.1886E 11	0.2227E 11	0.2585E 11	0.2959E 11	0.3345E 11
0.41	0.1928E 11	0.2683E 11	0.3547E 11	0.4506E 11	0.5550E 11	0.6667E 11	0.7848E 11	0.9086E 11	0.1037E 12	0.1170E 12
0.42	0.5167E 11	0.7143E 11	0.9390E 11	0.1187E 12	0.1456E 12	0.1744E 12	0.2047E 12	0.2364E 12	0.2693E 12	0.3033E 12
0.43	0.1031E 12	0.1417E 12	0.1852E 12	0.2332E 12	0.2850E 12	0.3402E 12	0.3983E 12	0.4589E 12	0.5217E 12	0.5865E 12
0.44	0.1634E 12	0.2230E 12	0.2901E 12	0.3638E 12	0.4430E 12	0.5272E 12	0.6156E 12	0.7077E 12	0.8031E 12	0.9012E 12
0.45	0.2184E 12	0.2964E 12	0.3837E 12	0.4792E 12	0.5817E 12	0.6902E 12	0.8040E 12	0.9224E 12	0.1044E 13	0.1170E 13
0.46	0.2623E 12	0.3540E 12	0.4562E 12	0.5675E 12	0.6867E 12	0.8127E 12	0.9444E 12	0.1081E 13	0.1222E 13	0.1367E 13
0.47	0.3009E 12	0.4038E 12	0.5182E 12	0.6423E 12	0.7748E 12	0.9146E 12	0.1060E 13	0.1211E 13	0.1367E 13	0.1528E 13
0.48	0.3399E 12	0.4539E 12	0.5800E 12	0.7165E 12	0.8619E 12	0.1014E 13	0.1174E 13	0.1339E 13	0.1509E 13	0.1683E 13
0.49	0.3730E 12	0.4956E 12	0.6308E 12	0.7767E 12	0.9317E 12	0.1094E 13	0.1264E 13	0.1439E 13	0.1619E 13	0.1804E 13
0.50	0.3894E 12	0.5151E 12	0.6531E 12	0.8017E 12	0.9592E 12	0.1124E 13	0.1295E 13	0.1473E 13	0.1655E 13	0.1841E 13
0.51	0.3802E 12	0.5006E 12	0.6325E 12	0.7740E 12	0.9238E 12	0.1080E 13	0.1243E 13	0.1411E 13	0.1583E 13	0.1759E 13
0.52	0.3430E 12	0.4498E 12	0.5663E 12	0.6910E 12	0.8227E 12	0.9604E 12	0.1103E 13	0.1250E 13	0.1400E 13	0.1554E 13
0.53	0.2859E 12	0.3734E 12	0.4686E 12	0.5703E 12	0.6774E 12	0.7892E 12	0.9050E 12	0.1024E 13	0.1146E 13	0.1270E 13
0.54	0.2202E 12	0.2864E 12	0.3583E 12	0.4349E 12	0.5155E 12	0.5995E 12	0.6863E 12	0.7755E 12	0.8669E 12	0.9600E 12
0.55	0.1566E 12	0.2030E 12	0.2531E 12	0.3065E 12	0.3625E 12	0.4208E 12	0.4810E 12	0.5429E 12	0.6061E 12	0.6705E 12
0.56	0.1029E 12	0.1329E 12	0.1653E 12	0.1997E 12	0.2357E 12	0.2731E 12	0.3117E 12	0.3514E 12	0.3919E 12	0.4331E 12
0.57	0.6249E 11	0.8041E 11	0.9971E 11	0.1201E 12	0.1415E 12	0.1637E 12	0.1866E 12	0.2101E 12	0.2341E 12	0.2584E 12
0.58	0.3506E 11	0.4497E 11	0.5562E 11	0.6688E 11	0.7865E 11	0.9084E 11	0.1034E 12	0.1162E 12	0.1293E 12	0.1427E 12
0.59	0.1821E 11	0.2328E 11	0.2872E 11	0.3446E 11	0.4045E 11	0.4665E 11	0.5303E 11	0.5956E 11	0.6622E 11	0.7299E 11
0.60	0.8863E 10	0.1129E 11	0.1390E 11	0.1664E 11	0.1950E 11	0.2246E 11	0.2550E 11	0.2861E 11	0.3178E 11	0.3499E 11
0.61	0.4095E 10	0.5204E 10	0.6388E 10	0.7634E 10	0.8931E 10	0.1027E 11	0.1164E 11	0.1305E 11	0.1448E 11	0.1593E 11
0.62	0.1822E 10	0.2310E 10	0.2829E 10	0.3374E 10	0.3941E 10	0.4526E 10	0.5126E 10	0.5739E 10	0.6363E 10	0.6997E 10
0.63	0.7928E 09	0.1001E 10	0.1224E 10	0.1457E 10	0.1699E 10	0.1949E 10	0.2205E 10	0.2467E 10	0.2733E 10	0.3003E 10
0.64	0.3422E 09	0.4314E 09	0.5260E 09	0.6252E 09	0.7280E 09	0.8340E 09	0.9425E 09	0.1053E 10	0.1165E 10	0.1280E 10
0.65	0.1489E 09	0.1872E 09	0.2278E 09	0.2703E 09	0.3143E 09	0.3597E 09	0.4060E 09	0.4533E 09	0.5014E 09	0.5501E 09
0.66	0.6621E 08	0.8305E 08	0.1008E 09	0.1194E 09	0.1387E 09	0.1585E 09	0.1787E 09	0.1994E 09	0.2204E 09	0.2416E 09
0.67	0.3013E 08	0.3771E 08	0.4571E 08	0.5405E 08	0.6268E 08	0.7155E 08	0.8062E 08	0.8985E 08	0.9923E 08	0.1087E 09
0.68	0.1402E 08	0.1750E 08	0.2118E 08	0.2500E 08	0.2896E 08	0.3302E 08	0.3717E 08	0.4140E 08	0.4568E 08	0.5003E 08
0.69	0.6666E 07	0.8303E 07	0.1002E 08	0.1182E 08	0.1367E 08	0.1557E 08	0.1751E 08	0.1949E 08	0.2149E 08	0.2352E 08
0.70	0.3233E 07	0.4019E 07	0.4845E 07	0.5705E 07	0.6591E 07	0.7499E 07	0.8427E 07	0.9369E 07	0.1032E 08	0.1129E 08
0.71	0.1599E 07	0.1984E 07	0.2388E 07	0.2807E 07	0.3239E 07	0.3682E 07	0.4134E 07	0.4593E 07	0.5059E 07	0.5530E 07
0.72	0.8060E 06	0.9978E 06	0.1198E 07	0.1407E 07	0.1622E 07	0.1842E 07	0.2067E 07	0.2295E 07	0.2526E 07	0.2759E 07
0.73	0.4134E 06	0.5107E 06	0.6127E 06	0.7184E 06	0.8272E 06	0.9385E 06	0.1052E 07	0.1167E 07	0.1283E 07	0.1401E 07
0.74	0.2157E 06	0.2660E 06	0.3187E 06	0.3732E 06	0.4293E 06	0.4866E 06	0.5450E 06	0.6043E 06	0.6643E 06	0.7249E 06
0.75	0.1145E 06	0.1410E 06	0.1686E 06	0.1972E 06	0.2266E 06	0.2567E 06	0.2873E 06	0.3183E 06	0.3497E 06	0.3815E 06
0.76	0.6178E 05	0.7591E 05	0.9066E 05	0.1059E 06	0.1215E 06	0.1375E 06	0.1538E 06	0.1703E 06	0.1870E 06	0.2039E 06
0.77	0.3383E 05	0.4150E 05	0.4949E 05	0.5775E 05	0.6623E 05	0.7488E 05	0.8368E 05	0.9261E 05	0.1016E 06	0.1107E 06
0.78	0.1879E 05	0.2301E 05	0.2741E 05	0.3195E 05	0.3661E 05	0.4136E 05	0.4619E 05	0.5108E 05	0.5603E 05	0.6103E 05

TABLE 18

Blackbody Fractional Luminance, Photopic (cd/m^2)

λ(μ)	500° K.	600° K.	700° K.	800° K.	900° K.
0.36	0.0000E 00	0.0000E 00	0.0000E 00	0.0000E 00	0.0000E 00
0.37	0.5668E-27	0.1886E-21	0.1666E-17	0.1521E-14	0.3059E-12
0.38	0.1187E-25	0.2903E-20	0.2064E-16	0.1606E-13	0.2856E-11
0.39	0.2041E-24	0.3672E-19	0.2099E-15	0.1388E-12	0.2176E-10
0.40	0.3582E-23	0.4786E-18	0.2211E-14	0.1246E-11	0.1724E-09
Luminance	0.3582E-23	0.4786E-18	0.2211E-14	0.1246E-11	0.1724E-09
0.41	0.4861E-22	0.4802E-17	0.1780E-13	0.8469E-11	0.1025E-08
0.42	0.7684E-21	0.5918E-16	0.1841E-12	0.7700E-10	0.8439E-08
0.43	0.1022E-19	0.6140E-15	0.1600E-11	0.5862E-09	0.5800E-07
0.44	0.9103E-19	0.4345E-14	0.9628E-11	0.3126E-08	0.2817E-06
0.45	0.6047E-18	0.2316E-13	0.4393E-10	0.1271E-07	0.1048E-05
0.46	0.3453E-17	0.1065E-12	0.1735E-09	0.4483E-07	0.3388E-05
0.47	0.1778E-16	0.4451E-12	0.6248E-09	0.1445E-06	0.1002E-04
0.48	0.8593E-16	0.1755E-11	0.2131E-08	0.4422E-06	0.2821E-04
0.49	0.3901E-15	0.6553E-11	0.6920E-08	0.1293E-05	0.7611E-04
0.50	0.1715E-14	0.2381E-10	0.2195E-07	0.3706E-05	0.2014E-03
Luminance	0.1715E-14	0.2381E-10	0.2195E-07	0.3706E-05	0.2014E-03
0.51	0.5662E-14	0.6143E-10	0.4694E-07	0.6827E-05	0.3285E-03
0.52	0.2750E-13	0.2604E-09	0.1811E-06	0.2459E-04	0.1123E-02
0.53	0.9832E-13	0.8060E-09	0.5075E-06	0.6410E-04	0.2771E-02
0.54	0.2959E-12	0.2100E-08	0.1196E-05	0.1405E-03	0.5751E-02
0.55	0.7925E-12	0.4880E-08	0.2520E-05	0.2755E-03	0.1068E-01
0.56	0.1945E-11	0.1042E-07	0.4889E-05	0.4983E-03	0.1831E-01
0.57	0.4436E-11	0.2077E-07	0.8870E-05	0.8443E-03	0.2946E-01
0.58	0.9436E-11	0.3880E-07	0.1514E-04	0.1349E-02	0.4481E-01
0.59	0.1875E-10	0.6811E-07	0.2438E-04	0.2041E-02	0.6467E-01
0.60	0.3490E-10	0.1126E-06	0.3715E-04	0.2932E-02	0.8888E-01
Luminance	0.3491E-10	0.1126E-06	0.3718E-04	0.2936E-02	0.8908E-01
0.61	0.2620E-10	0.6354E-07	0.1663E-04	0.1083E-02	0.2789E-01
0.62	0.6582E-10	0.1484E-06	0.3697E-04	0.2322E-02	0.5819E-01
0.63	0.1205E-09	0.2523E-06	0.5984E-04	0.3629E-02	0.8859E-01
0.64	0.1894E-09	0.3689E-06	0.8345E-04	0.4896E-02	0.1166E 00
0.65	0.2695E-09	0.4900E-06	0.1060E-03	0.6039E-02	0.1408E 00
0.66	0.3542E-09	0.6048E-06	0.1258E-03	0.6985E-02	0.1599E 00
0.67	0.4357E-09	0.7041E-06	0.1418E-03	0.7701E-02	0.1737E 00
0.68	0.5094E-09	0.7852E-06	0.1538E-03	0.8215E-02	0.1832E 00
0.69	0.5737E-09	0.8491E-06	0.1627E-03	0.8572E-02	0.1896E 00
0.70	0.6257E-09	0.8960E-06	0.1687E-03	0.8804E-02	0.1935E 00
Luminance	0.6606E-09	0.1008E-05	0.2059E-03	0.1174E-01	0.2826E 00
0.71	0.4368E-10	0.3581E-07	0.4319E-05	0.1572E-03	0.2575E-02
0.72	0.7960E-10	0.6266E-07	0.7351E-05	0.2622E-03	0.4230E-02
0.73	0.1083E-09	0.8226E-07	0.9427E-05	0.3308E-03	0.5271E-02
0.74	0.1304E-09	0.9613E-07	0.1080E-04	0.3742E-03	0.5907E-02
0.75	0.1469E-09	0.1056E-06	0.1169E-04	0.4010E-03	0.6285E-02
0.76	0.1594E-09	0.1122E-06	0.1227E-04	0.4178E-03	0.6514E-02
0.77	0.1689E-09	0.1168E-06	0.1266E-04	0.4284E-03	0.6655E-02
0.78	0.1761E-09	0.1201E-06	0.1291E-04	0.4351E-03	0.6740E-02
0.79	0.1814E-09	0.1223E-06	0.1308E-04	0.4393E-03	0.6792E-02
0.80	0.1853E-09	0.1238E-06	0.1319E-04	0.4419E-03	0.6823E-02
Luminance	0.8460E-09	0.1132E-05	0.2191E-03	0.1218E-01	0.2895E 00
0.81	0.2814E-11	0.1013E-08	0.6793E-07	0.1591E-05	0.1849E-04
0.82	0.4829E-11	0.1689E-08	0.1109E-06	0.2561E-05	0.2944E-04
0.83	0.6257E-11	0.2136E-08	0.1380E-06	0.3149E-05	0.3588E-04
Luminance	0.8522E-09	0.1134E-05	0.2192E-03	0.1218E-01	0.2895E 00

xEn = x 10^n. E.g. 1.234E 05 = 1.234 x 10^5

$\int_o^\lambda L_{v\lambda} d\lambda$

Scotopic ($sc.cd/m^2$)

λ(μ)	500° K.	600° K.	700° K.	800° K.	900° K.
0.38	0.0000E 00	0.0000E 00	0.0000E 00	0.0000E 00	0.0000E 00
0.39	0.8724E-23	0.1529E-17	0.8546E-14	0.5532E-11	0.8509E-09
0.40	0.1974E-21	0.2612E-16	0.1196E-12	0.6686E-10	0.9182E-08
Luminance	0.1974E-21	0.2612E-16	0.1196E-12	0.6686E-10	0.9182E-08
0.41	0.3570E-20	0.3521E-15	0.1303E-11	0.6193E-09	0.7489E-07
0.42	0.5212E-19	0.4032E-14	0.1259E-10	0.5277E-08	0.5796E-06
0.43	0.5085E-18	0.3093E-13	0.8152E-10	0.3014E-07	0.3005E-05
0.44	0.3653E-17	0.1765E-12	0.3956E-09	0.1296E-06	0.1178E-04
0.45	0.2056E-16	0.7978E-12	0.1531E-08	0.4474E-06	0.3723E-04
0.46	0.9617E-16	0.3020E-11	0.4994E-08	0.1307E-05	0.1000E-03
0.47	0.3956E-15	0.1012E-10	0.1448E-07	0.3409E-05	0.2401E-03
0.48	0.1498E-14	0.3141E-10	0.3904E-07	0.8274E-05	0.5379E-03
0.49	0.5290E-14	0.9145E-10	0.9915E-07	0.1898E-04	0.1141E-02
0.50	0.1725E-13	0.2481E-09	0.2361E-06	0.4102E-04	0.2288E-02
Luminance	0.1725E-13	0.2481E-09	0.2361E-06	0.4102E-04	0.2288E-02
0.51	0.3408E-13	0.3720E-09	0.2854E-06	0.4166E-04	0.2009E-02
0.52	0.1207E-12	0.1165E-08	0.8229E-06	0.1130E-03	0.5211E-02
0.53	0.3170E-12	0.2684E-08	0.1734E-05	0.2236E-03	0.9833E-02
0.54	0.7137E-12	0.5294E-08	0.3128E-05	0.3786E-03	0.1588E-01
0.55	0.1430E-11	0.9324E-08	0.5053E-05	0.5754E-03	0.2308E-01
0.56	0.2591E-11	0.1493E-07	0.7457E-05	0.8020E-03	0.3086E-01
0.57	0.4277E-11	0.2197E-07	0.1017E-04	0.1038E-02	0.3851E-01
0.58	0.6477E-11	0.2995E-07	0.1295E-04	0.1263E-02	0.4536E-01
0.59	0.9063E-11	0.3812E-07	0.1554E-04	0.1458E-02	0.5095E-01
0.60	0.1182E-10	0.4577E-07	0.1775E-04	0.1612E-02	0.5514E-01
Luminance	0.1184E-10	0.4602E-07	0.1798E-04	0.1653E-02	0.5743E-01
0.61	0.2712E-11	0.6611E-08	0.1737E-05	0.1134E-03	0.2928E-02
0.62	0.5202E-11	0.1197E-07	0.3026E-05	0.1922E-03	0.4858E-02
0.63	0.7371E-11	0.1611E-07	0.3940E-05	0.2445E-03	0.6077E-02
0.64	0.9193E-11	0.1920E-07	0.4568E-05	0.2783E-03	0.6827E-02
0.65	0.1069E-10	0.2148E-07	0.4994E-05	0.2999E-03	0.7283E-02
0.66	0.1192E-10	0.2315E-07	0.5283E-05	0.3136E-03	0.7562E-02
0.67	0.1293E-10	0.2438E-07	0.5480E-05	0.3225E-03	0.7734E-02
0.68	0.1377E-10	0.2530E-07	0.5617E-05	0.3284E-03	0.7842E-02
0.69	0.1446E-10	0.2599E-07	0.5713E-05	0.3323E-03	0.7911E-02
0.70	0.1504E-10	0.2652E-07	0.5781E-05	0.3348E-03	0.7955E-02
Luminance	0.2688E-10	0.7254E-07	0.2376E-04	0.1988E-02	0.6539E-01
0.71	0.4875E-12	0.3996E-09	0.4820E-07	0.1754E-05	0.2873E-04
0.72	0.8981E-12	0.7064E-09	0.8283E-07	0.2954E-05	0.4764E-04
0.73	0.1245E-11	0.9434E-09	0.1079E-06	0.3782E-05	0.6021E-04
0.74	0.1539E-11	0.1127E-08	0.1262E-06	0.4359E-05	0.6865E-04
0.75	0.1790E-11	0.1271E-08	0.1397E-06	0.4765E-05	0.7438E-04
0.76	0.2005E-11	0.1385E-08	0.1497E-06	0.5052E-05	0.7830E-04
0.77	0.2189E-11	0.1474E-08	0.1571E-06	0.5258E-05	0.8102E-04
0.78	0.2347E-11	0.1546E-08	0.1627E-06	0.5406E-05	0.8291E-04
Luminance	0.2923E-10	0.7409E-07	0.2393E-04	0.1993E-02	0.6547E-01

$\int_o^\lambda L_{v'\lambda} d\lambda$

$\int_0^{\lambda} L_{v\lambda} d\lambda$

TABLE 18

Blackbody Fractional Luminance, Photopic (cd/m²)

λ(μ)	1000°K.	1100°K.	1200°K.	1300°K.	1400°K.	1500°K.	1600°K.	1700°K.	1800°K.	1900°K.
0.36	0.0000E 00	0.0000E 00	0.0000E 00	0.0000E 00	0.0000E 00	0.0000E 00	0.0000E 00	0.0000E 00	0.0000E 00	0.0000E 00
0.37	0.2131E-10	0.6870E-09	0.1241E-07	0.1438E-06	0.1174E-05	0.7247E-05	0.3563E-04	0.1452E-03	0.5067E-03	0.1549E-02
0.38	0.1806E-09	0.5380E-08	0.9114E-07	0.9998E-06	0.7794E-05	0.4622E-04	0.2195E-03	0.8684E-03	0.2948E-02	0.8806E-02
0.39	0.1244E-08	0.3417E-07	0.5411E-06	0.5607E-05	0.4163E-04	0.2368E-03	0.1084E-02	0.4153E-02	0.1370E-01	0.3990E-01
0.40	0.8926E-08	0.2258E-06	0.3339E-05	0.3266E-04	0.2308E-03	0.1258E-02	0.5549E-02	0.2056E-01	0.6590E-01	0.1868E 00
Luminance	0.8926E-08	0.2258E-06	0.3339E-05	0.3266E-04	0.2308E-03	0.1258E-02	0.5549E-02	0.2056E-01	0.6590E-01	0.1868E 00
0.41	0.4757E-07	0.1099E-05	0.1505E-04	0.1379E-03	0.9207E-03	0.4772E-02	0.2014E-01	0.7176E-01	0.2220E 00	0.6100E 00
0.42	0.3621E-06	0.7852E-05	0.1020E-03	0.8944E-03	0.5751E-02	0.2886E-01	0.1184E 00	0.4117E 00	0.1246E 01	0.3358E 01
0.43	0.2294E-05	0.4655E-04	0.5726E-03	0.4791E-02	0.2961E-01	0.1436E 00	0.5720E 00	0.1937E 01	0.5729E 01	0.1511E 02
0.44	0.1034E-04	0.1977E-03	0.2313E-02	0.1855E-01	0.1106E 00	0.5202E 00	0.2016E 01	0.6668E 01	0.1931E 02	0.5001E 02
0.45	0.3589E-04	0.6477E-03	0.7229E-02	0.5573E-01	0.3212E 00	0.1467E 01	0.5545E 01	0.1793E 02	0.5091E 02	0.1295E 03
0.46	0.1082E-03	0.1846E-02	0.1967E-01	0.1458E 00	0.8132E 00	0.3609E 01	0.1330E 02	0.4209E 02	0.1172E 03	0.2932E 03
0.47	0.2992E-03	0.4830E-02	0.4916E-01	0.3508E 00	0.1893E 01	0.8169E 01	0.2938E 02	0.9099E 02	0.2486E 03	0.6115E 03
0.48	0.7877E-03	0.1204E-01	0.1171E 00	0.8050E 00	0.4206E 01	0.1764E 02	0.6196E 02	0.1878E 03	0.5035E 03	0.1217E 04
0.49	0.1991E-02	0.2888E-01	0.2689E 00	0.1780E 01	0.9014E 01	0.3680E 02	0.1261E 03	0.3744E 03	0.9857E 03	0.2344E 04
0.50	0.4946E-02	0.6812E-01	0.6076E 00	0.3878E 01	0.1902E 02	0.7562E 02	0.2532E 03	0.7362E 03	0.1902E 04	0.4451E 04
Luminance	0.4946E-02	0.6812E-01	0.6076E 00	0.3878E 01	0.1903E 02	0.7563E 02	0.2532E 03	0.7362E 03	0.1902E 04	0.4451E 04
0.51	0.7287E-02	0.9203E-01	0.7617E 00	0.4555E 01	0.2110E 02	0.7968E 02	0.2548E 03	0.7109E 03	0.1769E 04	0.4001E 04
0.52	0.2392E-01	0.2922E 00	0.2354E 01	0.1376E 02	0.6254E 02	0.2323E 03	0.7323E 03	0.2017E 04	0.4966E 04	0.1112E 05
0.53	0.5652E-01	0.6672E 00	0.5225E 01	0.2983E 02	0.1328E 03	0.4851E 03	0.1507E 04	0.4098E 04	0.9974E 04	0.2210E 05
0.54	0.1123E 00	0.1281E 01	0.9758E 01	0.5442E 02	0.2376E 03	0.8533E 03	0.2612E 04	0.7014E 04	0.1688E 05	0.3704E 05
0.55	0.2000E 00	0.2205E 01	0.1633E 02	0.8903E 02	0.3813E 03	0.1346E 04	0.4065E 04	0.1078E 05	0.2566E 05	0.5578E 05
0.56	0.3290E 00	0.3509E 01	0.2530E 02	0.1348E 03	0.5670E 03	0.1970E 04	0.5867E 04	0.1537E 05	0.3622E 05	0.7801E 05
0.57	0.5086E 00	0.5254E 01	0.3690E 02	0.1925E 03	0.7949E 03	0.2720E 04	0.7994E 04	0.2070E 05	0.4830E 05	0.1031E 06
0.58	0.7443E 00	0.7457E 01	0.5109E 02	0.2611E 03	0.1059E 04	0.3575E 04	0.1037E 05	0.2658E 05	0.6143E 05	0.1300E 06
0.59	0.1035E 01	0.1008E 02	0.6746E 02	0.3382E 03	0.1351E 04	0.4495E 04	0.1289E 05	0.3271E 05	0.7494E 05	0.1574E 06
0.60	0.1375E 01	0.1303E 02	0.8533E 02	0.4203E 03	0.1654E 04	0.5436E 04	0.1542E 05	0.3879E 05	0.8815E 05	0.1839E 06
Luminance	0.1380E 01	0.1309E 02	0.8594E 02	0.4242E 03	0.1673E 04	0.5511E 04	0.1568E 05	0.3952E 05	0.9005E 05	0.1883E 06
0.61	0.3751E 00	0.3145E 01	0.1850E 02	0.8290E 02	0.2997E 03	0.9133E 03	0.2420E 04	0.5721E 04	0.1229E 05	0.2435E 05
0.62	0.7661E 00	0.6315E 01	0.3663E 02	0.1622E 03	0.5807E 03	0.1754E 04	0.4615E 04	0.1083E 05	0.2314E 05	0.4564E 05
0.63	0.1143E 01	0.9272E 01	0.5309E 02	0.2325E 03	0.8250E 03	0.2473E 04	0.6463E 04	0.1508E 05	0.3206E 05	0.6294E 05
0.64	0.1477E 01	0.1181E 02	0.6689E 02	0.2902E 03	0.1021E 04	0.3041E 04	0.7904E 04	0.1836E 05	0.3884E 05	0.7597E 05
0.65	0.1755E 01	0.1386E 02	0.7769E 02	0.3344E 03	0.1169E 04	0.3462E 04	0.8955E 04	0.2072E 05	0.4368E 05	0.8516E 05
0.66	0.1967E 01	0.1537E 02	0.8549E 02	0.3656E 03	0.1271E 04	0.3750E 04	0.9664E 04	0.2229E 05	0.4687E 05	0.9116E 05
0.67	0.2115E 01	0.1640E 02	0.9067E 02	0.3859E 03	0.1337E 04	0.3931E 04	0.1010E 05	0.2325E 05	0.4880E 05	0.9477E 05
0.68	0.2213E 01	0.1706E 02	0.9393E 02	0.3984E 03	0.1377E 04	0.4039E 04	0.1036E 05	0.2381E 05	0.4992E 05	0.9684E 05
0.69	0.2276E 01	0.1748E 02	0.9593E 02	0.4060E 03	0.1400E 04	0.4102E 04	0.1051E 05	0.2413E 05	0.5055E 05	0.9800E 05
0.70	0.2314E 01	0.1772E 02	0.9708E 02	0.4102E 03	0.1413E 04	0.4136E 04	0.1059E 05	0.2430E 05	0.5088E 05	0.9860E 05
Luminance	0.3695E 01	0.3082E 02	0.1830E 03	0.8345E 03	0.3086E 04	0.9648E 04	0.2627E 05	0.6383E 05	0.1409E 06	0.2869E 06
0.71	0.2411E-01	0.1503E 00	0.6912E 00	0.2512E 01	0.7595E 01	0.1981E 02	0.4584E 02	0.9611E 02	0.1856E 03	0.3344E 03
0.72	0.3914E-01	0.2417E 00	0.1102E 01	0.3981E 01	0.1196E 02	0.3107E 02	0.7160E 02	0.1495E 03	0.2878E 03	0.5171E 03
0.73	0.4831E-01	0.2961E 00	0.1342E 01	0.4823E 01	0.1443E 02	0.3734E 02	0.8579E 02	0.1787E 03	0.3431E 03	0.6152E 03
0.74	0.5376E-01	0.3277E 00	0.1478E 01	0.5294E 01	0.1580E 02	0.4076E 02	0.9343E 02	0.1942E 03	0.3724E 03	0.6667E 03
0.75	0.5690E-01	0.3455E 00	0.1554E 01	0.5550E 01	0.1652E 02	0.4257E 02	0.9744E 02	0.2023E 03	0.3874E 03	0.6930E 03
0.76	0.5875E-01	0.3557E 00	0.1596E 01	0.5692E 01	0.1693E 02	0.4355E 02	0.9959E 02	0.2066E 03	0.3954E 03	0.7068E 03
0.77	0.5986E-01	0.3617E 00	0.1621E 01	0.5773E 01	0.1715E 02	0.4409E 02	0.1007E 03	0.2089E 03	0.3997E 03	0.7141E 03
0.78	0.6052E-01	0.3652E 00	0.1635E 01	0.5818E 01	0.1727E 02	0.4439E 02	0.1014E 03	0.2102E 03	0.4019E 03	0.7180E 03
0.79	0.6091E-01	0.3672E 00	0.1643E 01	0.5843E 01	0.1734E 02	0.4455E 02	0.1017E 03	0.2109E 03	0.4032E 03	0.7201E 03
0.80	0.6113E-01	0.3684E 00	0.1647E 01	0.5857E 01	0.1738E 02	0.4464E 02	0.1019E 03	0.2112E 03	0.4038E 03	0.7212E 03
Luminance	0.3756E 01	0.3119E 02	0.1846E 03	0.8403E 03	0.3104E 04	0.9693E 04	0.2637E 05	0.6404E 05	0.1413E 06	0.2877E 06
0.81	0.1316E-03	0.6558E-03	0.2499E-02	0.7756E-02	0.2047E-01	0.4747E-01	0.9910E-01	0.1897E 00	0.3379E 00	0.5665E 00
0.82	0.2077E-03	0.1027E-02	0.3893E-02	0.1202E-01	0.3159E-01	0.7301E-01	0.1519E 00	0.2901E 00	0.5155E 00	0.8622E 00
0.83	0.2514E-03	0.1237E-02	0.4668E-02	0.1436E-01	0.3763E-01	0.8672E-01	0.1800E 00	0.3431E 00	0.6086E 00	0.1016E 01
Luminance	0.3756E 01	0.3119E 02	0.1846E 03	0.8403E 03	0.3104E 04	0.9693E 04	0.2637E 05	0.6404E 05	0.1413E 06	0.2877E 06

xEn - x 10^n. E.g. 1.234E 05 = 1.234 x 10^5

TABLE 18

Blackbody Fractional Luminance, Scotopic (sc.cd/m^2)

λ(μ)	1000°K.	1100°K.	1200°K.	1300°K.	1400°K.	1500°K.	1600°K.	1700°K.	1800°K.	1900°K.
0.38	0.0000E 00	0.0000E 00	0.0000E 00	0.0000E 00	0.0000E 00	0.0000E 00	0.0000E 00	0.0000E 00	0.0000E 00	0.0000E 00
0.39	0.4784E-07	0.1293E-05	0.2019E-04	0.2066E-03	0.1516E-02	0.8537E-02	0.3872E-01	0.1470E 00	0.4813E 00	0.1391E 01
0.40	0.4719E-06	0.1186E-04	0.1743E-03	0.1696E-02	0.1192E-01	0.6468E-01	0.2840E 00	0.1048E 01	0.3346E 01	0.9456E 01
Luminance	0.4719E-06	0.1186E-04	0.1743E-03	0.1696E-02	0.1192E-01	0.6468E-01	0.2840E 00	0.1048E 01	0.3346E 01	0.9456E 01
0.41	0.3472E-05	0.8019E-04	0.1097E-02	0.1004E-01	0.6706E-01	0.3474E 00	0.1465E 01	0.5221E 01	0.1615E 02	0.4436E 02
0.42	0.2491E-04	0.5409E-03	0.7038E-02	0.6174E-01	0.3973E 00	0.1995E 01	0.8193E 01	0.2849E 02	0.8631E 02	0.2326E 03
0.43	0.1196E-03	0.2441E-02	0.3018E-01	0.2536E 00	0.1573E 01	0.7657E 01	0.3058E 02	0.1038E 03	0.3079E 03	0.8143E 03
0.44	0.4361E-03	0.8386E-02	0.9869E-01	0.7958E 00	0.4766E 01	0.2250E 02	0.8757E 02	0.2905E 03	0.8440E 03	0.2192E 04
0.45	0.1284E-02	0.2335E-01	0.2622E 00	0.2033E 01	0.1178E 02	0.5407E 02	0.2052E 03	0.6663E 03	0.1898E 04	0.4848E 04
0.46	0.3227E-02	0.5558E-01	0.5971E 00	0.4461E 01	0.2504E 02	0.1118E 03	0.4146E 03	0.1318E 04	0.3690E 04	0.9271E 04
0.47	0.7266E-02	0.1187E 00	0.1222E 01	0.8809E 01	0.4797E 02	0.2086E 03	0.7563E 03	0.2358E 04	0.6485E 04	0.1604E 05
0.48	0.1527E-01	0.2371E 00	0.2339E 01	0.1627E 02	0.8602E 02	0.3647E 03	0.1292E 04	0.3953E 04	0.1068E 05	0.2604E 05
0.49	0.3046E-01	0.4499E 00	0.4259E 01	0.2862E 02	0.1469E 03	0.6075E 03	0.2107E 04	0.6322E 04	0.1680E 05	0.4034E 05
0.50	0.5756E-01	0.8103E 00	0.7373E 01	0.4793E 02	0.2392E 03	0.9655E 03	0.3279E 04	0.9661E 04	0.2527E 05	0.5981E 05
Luminance	0.5756E-01	0.8103E 00	0.7373E 01	0.4793E 02	0.2392E 03	0.9655E 03	0.3279E 04	0.9662E 04	0.2527E 05	0.5982E 05
0.51	0.4467E-01	0.5651E 00	0.4684E 01	0.2804E 02	0.1300E 03	0.4915E 03	0.1573E 04	0.4392E 04	0.1093E 05	0.2475E 05
0.52	0.1117E 00	0.1374E 01	0.1113E 02	0.6535E 02	0.2981E 03	0.1110E 04	0.3512E 04	0.9702E 04	0.2393E 05	0.5371E 05
0.53	0.2033E 00	0.2429E 01	0.1921E 02	0.1106E 03	0.4964E 03	0.1824E 04	0.5700E 04	0.1558E 05	0.3809E 05	0.8478E 05
0.54	0.3169E 00	0.3680E 01	0.2845E 02	0.1608E 03	0.7103E 03	0.2576E 04	0.7959E 04	0.2154E 05	0.5221E 05	0.1153E 06
0.55	0.4451E 00	0.5032E 01	0.3808E 02	0.2115E 03	0.9213E 03	0.3301E 04	0.1009E 05	0.2708E 05	0.6513E 05	0.1429E 06
0.56	0.5768E 00	0.6366E 01	0.4726E 02	0.2585E 03	0.1111E 04	0.3941E 04	0.1194E 05	0.3179E 05	0.7598E 05	0.1657E 06
0.57	0.7002E 00	0.7566E 01	0.5525E 02	0.2982E 03	0.1268E 04	0.4459E 04	0.1341E 05	0.3548E 05	0.8434E 05	0.1831E 06
0.58	0.8055E 00	0.8552E 01	0.6161E 02	0.3290E 03	0.1387E 04	0.4843E 04	0.1448E 05	0.3812E 05	0.9025E 05	0.1952E 06
0.59	0.8877E 00	0.9293E 01	0.6624E 02	0.3509E 03	0.1470E 04	0.5104E 04	0.1519E 05	0.3987E 05	0.9408E 05	0.2030E 06
0.60	0.9467E 00	0.9805E 01	0.6935E 02	0.3651E 03	0.1523E 04	0.5268E 04	0.1564E 05	0.4093E 05	0.9639E 05	0.2076E 06
Luminance	0.1004E 01	0.1061E 02	0.7673E 02	0.4131E 03	0.1762E 04	0.6233E 04	0.1892E 05	0.5059E 05	0.1216E 06	0.2675E 06
0.61	0.3945E-01	0.3312E 00	0.1951E 01	0.8750E 01	0.3167E 02	0.9655E 02	0.2561E 03	0.6056E 03	0.1301E 04	0.2581E 04
0.62	0.6439E-01	0.5337E 00	0.3111E 01	0.1382E 02	0.4967E 02	0.1504E 03	0.3969E 03	0.9340E 03	0.1998E 04	0.3948E 04
0.63	0.7954E-01	0.6527E 00	0.3773E 01	0.1666E 02	0.5952E 02	0.1794E 03	0.4715E 03	0.1105E 04	0.2359E 04	0.4648E 04
0.64	0.8849E-01	0.7208E 00	0.4143E 01	0.1820E 02	0.6479E 02	0.1947E 03	0.5101E 03	0.1193E 04	0.2541E 04	0.4998E 04
0.65	0.9373E-01	0.7595E 00	0.4347E 01	0.1904E 02	0.6759E 02	0.2027E 03	0.5301E 03	0.1238E 04	0.2633E 04	0.5172E 04
0.66	0.9682E-01	0.7815E 00	0.4461E 01	0.1950E 02	0.6909E 02	0.2069E 03	0.5404E 03	0.1261E 04	0.2679E 04	0.5260E 04
0.67	0.9866E-01	0.7944E 00	0.4525E 01	0.1975E 02	0.6990E 02	0.2091E 03	0.5459E 03	0.1273E 04	0.2703E 04	0.5305E 04
0.68	0.9978E-01	0.8019E 00	0.4563E 01	0.1989E 02	0.7036E 02	0.2104E 03	0.5489E 03	0.1279E 04	0.2716E 04	0.5328E 04
0.69	0.1004E 00	0.8064E 00	0.4584E 01	0.1997E 02	0.7061E 02	0.2110E 03	0.5505E 03	0.1283E 04	0.2723E 04	0.5341E 04
0.70	0.1008E 00	0.8091E 00	0.4597E 01	0.2002E 02	0.7076E 02	0.2114E 03	0.5514E 03	0.1285E 04	0.2727E 04	0.5348E 04
Luminance	0.1105E 01	0.1142E 02	0.8132E 02	0.4331E 03	0.1833E 04	0.6445E 04	0.1947E 05	0.5188E 05	0.1244E 06	0.2728E 06
0.71	0.2691E-03	0.1678E-02	0.7712E-02	0.2803E-01	0.8475E-01	0.2210E 00	0.5115E 00	0.1072E 01	0.2070E 01	0.3731E 01
0.72	0.4407E-03	0.2721E-02	0.1240E-01	0.4480E-01	0.1346E 00	0.3496E 00	0.8055E 00	0.1682E 01	0.3238E 01	0.5817E 01
0.73	0.5515E-03	0.3378E-02	0.1530E-01	0.5497E-01	0.1644E 00	0.4253E 00	0.9768E 00	0.2034E 01	0.3905E 01	0.7001E 01
0.74	0.6237E-03	0.3796E-02	0.1711E-01	0.6120E-01	0.1825E 00	0.4706E 00	0.1078E 01	0.2240E 01	0.4293E 01	0.7683E 01
0.75	0.6713E-03	0.4066E-02	0.1825E-01	0.6508E-01	0.1935E 00	0.4980E 00	0.1138E 01	0.2362E 01	0.4521E 01	0.8081E 01
0.76	0.7031E-03	0.4242E-02	0.1898E-01	0.6752E-01	0.2004E 00	0.5148E 00	0.1175E 01	0.2436E 01	0.4657E 01	0.8316E 01
0.77	0.7244E-03	0.4357E-02	0.1945E-01	0.6907E-01	0.2047E 00	0.5252E 00	0.1198E 01	0.2481E 01	0.4739E 01	0.8458E 01
0.78	0.7389E-03	0.4434E-02	0.1976E-01	0.7007E-01	0.2074E 00	0.5318E 00	0.1212E 01	0.2508E 01	0.4789E 01	0.8544E 01
Luminance	0.1105E 01	0.1142E 02	0.8134E 02	0.4332E 03	0.1833E 04	0.6445E 04	0.1947E 05	0.5188E 05	0.1244E 06	0.2728E 06

xEn = x 10^n. E.g. 1.234E 05 = 1.234 x 10^5

$\int_o^\lambda L_{v',\lambda}\, d\lambda$

$\int_0^\lambda L_{v\lambda} d\lambda$

TABLE 18

Blackbody Fractional Luminance, Photopic (cd/m²)

λ(μ)	2000°K.	2100°K.	2200°K.	2300°K.	2400°K.	2500°K.	2600°K.	2700°K.	2800°K.	2900°K.
0.36	0.0000E 00	0.0000E 00	0.0000E 00	0.0000E 00	0.0000E 00	0.0000E 00	0.0000E 00	0.0000E 00	0.0000E 00	0.0000E 00
0.37	0.4239E-02	0.1053E-01	0.2410E-01	0.5131E-01	0.1025E 00	0.1940E 00	0.3495E 00	0.6026E 00	0.9994E 00	0.1600E 01
0.38	0.2357E-01	0.5747E-01	0.1292E 00	0.2707E 00	0.5335E 00	0.9959E 00	0.1771E 01	0.3020E 01	0.4957E 01	0.7863E 01
0.39	0.1044E 00	0.2494E 00	0.5503E 00	0.1133E 01	0.2200E 01	0.4048E 01	0.7108E 01	0.1197E 02	0.1943E 02	0.3049E 02
0.40	0.4776E 00	0.1116E 01	0.2417E 01	0.4892E 01	0.9339E 01	0.1693E 02	0.2932E 02	0.4876E 02	0.7821E 02	0.1214E 03
Luminance	0.4776E 00	0.1116E 01	0.2417E 01	0.4892E 01	0.9339E 01	0.1693E 02	0.2932E 02	0.4876E 02	0.7821E 02	0.1214E 03
0.41	0.1514E 01	0.3450E 01	0.7290E 01	0.1443E 02	0.2700E 02	0.4804E 02	0.8178E 02	0.1338E 03	0.2114E 03	0.3236E 03
0.42	0.8195E 01	0.1837E 02	0.3827E 02	0.7481E 02	0.1382E 03	0.2434E 03	0.4101E 03	0.6650E 03	0.1041E 04	0.1581E 04
0.43	0.3621E 02	0.7983E 02	0.1638E 03	0.3157E 03	0.5763E 03	0.1002E 04	0.1671E 04	0.2683E 04	0.4163E 04	0.6269E 04
0.44	0.1177E 03	0.2557E 03	0.5175E 03	0.9851E 03	0.1777E 04	0.3059E 04	0.5050E 04	0.8035E 04	0.1236E 05	0.1847E 05
0.45	0.3004E 03	0.6430E 03	0.1284E 04	0.2417E 04	0.4314E 04	0.7354E 04	0.1203E 05	0.1898E 05	0.2898E 05	0.4300E 05
0.46	0.6695E 03	0.1413E 04	0.2788E 04	0.5186E 04	0.9162E 04	0.1546E 05	0.2508E 05	0.3925E 05	0.5950E 05	0.8764E 05
0.47	0.1375E 04	0.2863E 04	0.5579E 04	0.1026E 05	0.1794E 05	0.3000E 05	0.4824E 05	0.7488E 05	0.1126E 06	0.1648E 06
0.48	0.2697E 04	0.5540E 04	0.1066E 05	0.1939E 05	0.3356E 05	0.5560E 05	0.8861E 05	0.1364E 06	0.2038E 06	0.2961E 06
0.49	0.5116E 04	0.1036E 05	0.1971E 05	0.3546E 05	0.6075E 05	0.9972E 05	0.1575E 06	0.2407E 06	0.3569E 06	0.5150E 06
0.50	0.9571E 04	0.1914E 05	0.3595E 05	0.6396E 05	0.1084E 06	0.1764E 06	0.2764E 06	0.4190E 06	0.6167E 06	0.8839E 06
Luminance	0.9571E 04	0.1914E 05	0.3596E 05	0.6397E 05	0.1084E 06	0.1764E 06	0.2764E 06	0.4190E 06	0.6167E 06	0.8840E 06
0.51	0.8339E 04	0.1620E 05	0.2965E 05	0.5147E 05	0.8534E 05	0.1358E 06	0.2087E 06	0.3106E 06	0.4494E 06	0.6337E 06
0.52	0.2297E 05	0.4429E 05	0.8044E 05	0.1387E 06	0.2286E 06	0.3620E 06	0.5534E 06	0.8198E 06	0.1180E 07	0.1658E 07
0.53	0.4526E 05	0.8655E 05	0.1560E 06	0.2673E 06	0.4379E 06	0.6896E 06	0.1048E 07	0.1546E 07	0.2217E 07	0.3102E 07
0.54	0.7517E 05	0.1426E 06	0.2553E 06	0.4345E 06	0.7076E 06	0.1108E 07	0.1677E 07	0.2461E 07	0.3514E 07	0.4897E 07
0.55	0.1122E 06	0.2113E 06	0.3756E 06	0.6354E 06	0.1028E 07	0.1603E 07	0.2414E 07	0.3527E 07	0.5017E 07	0.6965E 07
0.56	0.1556E 06	0.2909E 06	0.5138E 06	0.8639E 06	0.1391E 07	0.2157E 07	0.3233E 07	0.4705E 07	0.6666E 07	0.9222E 07
0.57	0.2041E 06	0.3789E 06	0.6650E 06	0.1111E 07	0.1781E 07	0.2748E 07	0.4103E 07	0.5947E 07	0.8396E 07	0.1157E 08
0.58	0.2556E 06	0.4713E 06	0.8224E 06	0.1367E 07	0.2180E 07	0.3350E 07	0.4982E 07	0.7195E 07	0.1012E 08	0.1391E 08
0.59	0.3074E 06	0.5634E 06	0.9778E 06	0.1618E 07	0.2569E 07	0.3932E 07	0.5825E 07	0.8385E 07	0.1176E 08	0.1612E 08
0.60	0.3569E 06	0.6506E 06	0.1123E 07	0.1851E 07	0.2928E 07	0.4465E 07	0.6595E 07	0.9465E 07	0.1324E 08	0.1810E 08
Luminance	0.3664E 06	0.6697E 06	0.1159E 07	0.1915E 07	0.3036E 07	0.4642E 07	0.6871E 07	0.9884E 07	0.1385E 08	0.1898E 08
0.61	0.4508E 05	0.7870E 05	0.1305E 06	0.2073E 06	0.3168E 06	0.4679E 06	0.6707E 06	0.9361E 06	0.1275E 07	0.1702E 07
0.62	0.8409E 05	0.1461E 06	0.2416E 06	0.3824E 06	0.5825E 06	0.8580E 06	0.1226E 07	0.1707E 07	0.2322E 07	0.3091E 07
0.63	0.1155E 06	0.2000E 06	0.3296E 06	0.5202E 06	0.7902E 06	0.1161E 07	0.1656E 07	0.2301E 07	0.3122E 07	0.4150E 07
0.64	0.1389E 06	0.2399E 06	0.3943E 06	0.6206E 06	0.9407E 06	0.1379E 07	0.1963E 07	0.2723E 07	0.3691E 07	0.4898E 07
0.65	0.1553E 06	0.2676E 06	0.4388E 06	0.6894E 06	0.1043E 07	0.1527E 07	0.2170E 07	0.3007E 07	0.4069E 07	0.5394E 07
0.66	0.1659E 06	0.2853E 06	0.4672E 06	0.7329E 06	0.1107E 07	0.1619E 07	0.2299E 07	0.3182E 07	0.4303E 07	0.5700E 07
0.67	0.1722E 06	0.2958E 06	0.4839E 06	0.7583E 06	0.1144E 07	0.1672E 07	0.2374E 07	0.3283E 07	0.4437E 07	0.5873E 07
0.68	0.1758E 06	0.3018E 06	0.4932E 06	0.7725E 06	0.1165E 07	0.1702E 07	0.2414E 07	0.3338E 07	0.4509E 07	0.5967E 07
0.69	0.1778E 06	0.3050E 06	0.4983E 06	0.7802E 06	0.1176E 07	0.1718E 07	0.2436E 07	0.3367E 07	0.4547E 07	0.6017E 07
0.70	0.1789E 06	0.3067E 06	0.5009E 06	0.7841E 06	0.1182E 07	0.1726E 07	0.2447E 07	0.3382E 07	0.4567E 07	0.6041E 07
Luminance	0.5453E 06	0.9765E 06	0.1660E 07	0.2699E 07	0.4219E 07	0.6368E 07	0.9319E 07	0.1326E 08	0.1842E 08	0.2503E 08
0.71	0.5680E 03	0.9175E 03	0.1418E 04	0.2112E 04	0.3042E 04	0.4256E 04	0.5802E 04	0.7731E 04	0.1009E 05	0.1293E 05
0.72	0.8763E 03	0.1412E 04	0.2179E 04	0.3238E 04	0.4655E 04	0.6502E 04	0.8851E 04	0.1177E 05	0.1535E 05	0.1966E 05
0.73	0.1040E 04	0.1674E 04	0.2579E 04	0.3827E 04	0.5496E 04	0.7667E 04	0.1042E 05	0.1386E 05	0.1805E 05	0.2309E 05
0.74	0.1126E 04	0.1809E 04	0.2785E 04	0.4129E 04	0.5924E 04	0.8259E 04	0.1122E 05	0.1490E 05	0.1941E 05	0.2481E 05
0.75	0.1169E 04	0.1878E 04	0.2888E 04	0.4280E 04	0.6137E 04	0.8551E 04	0.1161E 05	0.1542E 05	0.2007E 05	0.2565E 05
0.76	0.1192E 04	0.1913E 04	0.2941E 04	0.4357E 04	0.6246E 04	0.8700E 04	0.1181E 05	0.1568E 05	0.2040E 05	0.2607E 05
0.77	0.1204E 04	0.1932E 04	0.2969E 04	0.4397E 04	0.6302E 04	0.8777E 04	0.1191E 05	0.1581E 05	0.2057E 05	0.2628E 05
0.78	0.1210E 04	0.1941E 04	0.2984E 04	0.4418E 04	0.6331E 04	0.8816E 04	0.1196E 05	0.1588E 05	0.2066E 05	0.2639E 05
0.79	0.1213E 04	0.1946E 04	0.2991E 04	0.4429E 04	0.6346E 04	0.8836E 04	0.1199E 05	0.1592E 05	0.2070E 05	0.2645E 05
0.80	0.1215E 04	0.1949E 04	0.2995E 04	0.4434E 04	0.6354E 04	0.8847E 04	0.1200E 05	0.1593E 05	0.2072E 05	0.2648E 05
Luminance	0.5465E 06	0.9784E 06	0.1663E 07	0.2704E 07	0.4225E 07	0.6377E 07	0.9331E 07	0.1328E 08	0.1844E 08	0.2505E 08
0.81	0.9018E 00	0.1373E 01	0.2013E 01	0.2854E 01	0.3932E 01	0.5279E 01	0.6930E 01	0.8916E 01	0.1126E 02	0.1401E 02
0.82	0.1370E 01	0.2082E 01	0.3048E 01	0.4316E 01	0.5937E 01	0.7962E 01	0.1044E 02	0.1341E 02	0.1694E 02	0.2104E 02
0.83	0.1612E 01	0.2448E 01	0.3579E 01	0.5063E 01	0.6959E 01	0.9324E 01	0.1221E 02	0.1569E 02	0.1979E 02	0.2458E 02
Luminance	0.5465E 06	0.9784E 06	0.1663E 07	0.2704E 07	0.4225E 07	0.6377E 07	0.9331E 07	0.1328E 08	0.1844E 08	0.2505E 08

xEn = x 10^n. E.g. 1.234E 05 = 1.234 x 10^5

TABLE 18

Blackbody Fractional Luminance, Scotopic (sc.cd/m²)

λ(μ)	2000°K.	2100°K.	2200°K.	2300°K.	2400°K.	2500°K.	2600°K.	2700°K.	2800°K.	2900°K.
0.38	0.0000E 00	0.0000E 00	0.0000E 00	0.0000E 00	0.0000E 00	0.0000E 00	0.0000E 00	0.0000E 00	0.0000E 00	0.0000E 00
0.39	0.3615E 01	0.8579E 01	0.1882E 02	0.3856E 02	0.7444E 02	0.1363E 03	0.2383E 03	0.3997E 03	0.6460E 03	0.1010E 04
0.40	0.2408E 02	0.5613E 02	0.1211E 03	0.2445E 03	0.4656E 03	0.8420E 03	0.1454E 04	0.2414E 04	0.3864E 04	0.5987E 04
Luminance	0.2408E 02	0.5613E 02	0.1211E 03	0.2445E 03	0.4656E 03	0.8420E 03	0.1454E 04	0.2414E 04	0.3864E 04	0.5987E 04
0.41	0.1101E 03	0.2507E 03	0.5298E 03	0.1048E 04	0.1961E 04	0.3490E 04	0.5939E 04	0.9717E 04	0.1535E 05	0.2349E 05
0.42	0.5679E 03	0.1273E 04	0.2654E 04	0.5190E 04	0.9596E 04	0.1689E 05	0.2847E 05	0.4617E 05	0.7234E 05	0.1098E 06
0.43	0.1954E 04	0.4316E 04	0.8871E 04	0.1712E 05	0.3130E 05	0.5453E 05	0.9102E 05	0.1462E 06	0.2272E 06	0.3425E 06
0.44	0.5177E 04	0.1126E 05	0.2285E 05	0.4358E 05	0.7879E 05	0.1358E 06	0.2246E 06	0.3580E 06	0.5517E 06	0.8255E 06
0.45	0.1127E 05	0.2420E 05	0.4847E 05	0.9142E 05	0.1635E 06	0.2793E 06	0.4580E 06	0.7239E 06	0.1107E 07	0.1645E 07
0.46	0.2125E 05	0.4503E 05	0.8915E 05	0.1663E 06	0.2947E 06	0.4990E 06	0.8114E 06	0.1272E 07	0.1933E 07	0.2854E 07
0.47	0.3626E 05	0.7590E 05	0.1485E 06	0.2744E 06	0.4818E 06	0.8087E 06	0.1304E 07	0.2032E 07	0.3067E 07	0.4500E 07
0.48	0.5808E 05	0.1200E 06	0.2325E 06	0.4252E 06	0.7398E 06	0.1231E 07	0.1972E 07	0.3050E 07	0.4574E 07	0.6671E 07
0.49	0.8880E 05	0.1814E 06	0.3475E 06	0.6295E 06	0.1085E 07	0.1792E 07	0.2849E 07	0.4377E 07	0.6522E 07	0.9458E 07
0.50	0.1300E 06	0.2625E 06	0.4978E 06	0.8933E 06	0.1527E 07	0.2503E 07	0.3950E 07	0.6029E 07	0.8930E 07	0.1287E 08
Luminance	0.1300E 06	0.2626E 06	0.4979E 06	0.8936E 06	0.1527E 07	0.2504E 07	0.3951E 07	0.6031E 07	0.8934E 07	0.1288E 08
0.51	0.5161E 05	0.1003E 06	0.1836E 06	0.3189E 06	0.5289E 06	0.8425E 06	0.1294E 07	0.1927E 07	0.2788E 07	0.3933E 07
0.52	0.1111E 06	0.2147E 06	0.3905E 06	0.6745E 06	0.1113E 07	0.1764E 07	0.2700E 07	0.4004E 07	0.5773E 07	0.8116E 07
0.53	0.1742E 06	0.3342E 06	0.6046E 06	0.1038E 07	0.1705E 07	0.2692E 07	0.4104E 07	0.6063E 07	0.8711E 07	0.1220E 08
0.54	0.2354E 06	0.4490E 06	0.8078E 06	0.1381E 07	0.2258E 07	0.3550E 07	0.5392E 07	0.7939E 07	0.1137E 08	0.1589E 08
0.55	0.2900E 06	0.5502E 06	0.9852E 06	0.1677E 07	0.2731E 07	0.4280E 07	0.6479E 07	0.9511E 07	0.1358E 08	0.1894E 08
0.56	0.3346E 06	0.6321E 06	0.1127E 07	0.1912E 07	0.3104E 07	0.4850E 07	0.7322E 07	0.1072E 08	0.1528E 08	0.2126E 08
0.57	0.3682E 06	0.6930E 06	0.1232E 07	0.2084E 07	0.3375E 07	0.5260E 07	0.7926E 07	0.1158E 08	0.1648E 08	0.2289E 08
0.58	0.3914E 06	0.7347E 06	0.1303E 07	0.2199E 07	0.3555E 07	0.5532E 07	0.8322E 07	0.1214E 08	0.1726E 08	0.2395E 08
0.59	0.4061E 06	0.7608E 06	0.1347E 07	0.2270E 07	0.3666E 07	0.5697E 07	0.8562E 07	0.1248E 08	0.1773E 08	0.2458E 08
0.60	0.4147E 06	0.7761E 06	0.1372E 07	0.2311E 07	0.3729E 07	0.5791E 07	0.8697E 07	0.1267E 08	0.1799E 08	0.2493E 08
Luminance	0.5448E 06	0.1038E 07	0.1870E 07	0.3205E 07	0.5256E 07	0.8295E 07	0.1264E 08	0.1870E 08	0.2692E 08	0.3781E 08
0.61	0.4779E 04	0.8346E 04	0.1385E 05	0.2200E 05	0.3363E 05	0.4968E 05	0.7123E 05	0.9944E 05	0.1355E 06	0.1808E 06
0.62	0.7287E 04	0.1268E 05	0.2100E 05	0.3327E 05	0.5073E 05	0.7480E 05	0.1070E 06	0.1491E 06	0.2029E 06	0.2704E 06
0.63	0.8557E 04	0.1486E 05	0.2456E 05	0.3884E 05	0.5914E 05	0.8706E 05	0.1244E 06	0.1731E 06	0.2354E 06	0.3133E 06
0.64	0.9187E 04	0.1593E 05	0.2629E 05	0.4154E 05	0.6319E 05	0.9294E 05	0.1327E 06	0.1845E 06	0.2507E 06	0.3334E 06
0.65	0.9498E 04	0.1646E 05	0.2714E 05	0.4285E 05	0.6513E 05	0.9574E 05	0.1366E 06	0.1899E 06	0.2578E 06	0.3428E 06
0.66	0.9653E 04	0.1672E 05	0.2755E 05	0.4349E 05	0.6607E 05	0.9709E 05	0.1385E 06	0.1925E 06	0.2613E 06	0.3473E 06
0.67	0.9732E 04	0.1685E 05	0.2776E 05	0.4380E 05	0.6654E 05	0.9776E 05	0.1394E 06	0.1937E 06	0.2629E 06	0.3495E 06
0.68	0.9773E 04	0.1692E 05	0.2787E 05	0.4396E 05	0.6678E 05	0.9809E 05	0.1399E 06	0.1943E 06	0.2638E 06	0.3505E 06
0.69	0.9794E 04	0.1695E 05	0.2792E 05	0.4405E 05	0.6690E 05	0.9827E 05	0.1401E 06	0.1947E 06	0.2642E 06	0.3511E 06
0.70	0.9806E 04	0.1697E 05	0.2795E 05	0.4409E 05	0.6696E 05	0.9836E 05	0.1402E 06	0.1948E 06	0.2644E 06	0.3513E 06
Luminance	0.5546E 06	0.1055E 07	0.1898E 07	0.3249E 07	0.5323E 07	0.8393E 07	0.1278E 08	0.1890E 08	0.2719E 08	0.3816E 08
0.71	0.6338E 01	0.1023E 02	0.1583E 02	0.2356E 02	0.3394E 02	0.4748E 02	0.6474E 02	0.8626E 02	0.1126E 03	0.1443E 03
0.72	0.9856E 01	0.1588E 02	0.2450E 02	0.3641E 02	0.5235E 02	0.7312E 02	0.9954E 02	0.1324E 03	0.1726E 03	0.2210E 03
0.73	0.1183E 02	0.1904E 02	0.2933E 02	0.4352E 02	0.6249E 02	0.8718E 02	0.1185E 03	0.1575E 03	0.2052E 03	0.2625E 03
0.74	0.1297E 02	0.2083E 02	0.3206E 02	0.4752E 02	0.6817E 02	0.9501E 02	0.1290E 03	0.1714E 03	0.2231E 03	0.2852E 03
0.75	0.1363E 02	0.2187E 02	0.3362E 02	0.4980E 02	0.7139E 02	0.9944E 02	0.1350E 03	0.1792E 03	0.2332E 03	0.2979E 03
0.76	0.1401E 02	0.2247E 02	0.3453E 02	0.5112E 02	0.7325E 02	0.1019E 03	0.1384E 03	0.1836E 03	0.2389E 03	0.3051E 03
0.77	0.1424E 02	0.2283E 02	0.3507E 02	0.5190E 02	0.7433E 02	0.1034E 03	0.1403E 03	0.1862E 03	0.2421E 03	0.3092E 03
0.78	0.1438E 02	0.2305E 02	0.3539E 02	0.5236E 02	0.7497E 02	0.1043E 03	0.1415E 03	0.1877E 03	0.2441E 03	0.3116E 03
Luminance	0.5546E 06	0.1055E 07	0.1898E 07	0.3249E 07	0.5324E 07	0.8393E 07	0.1279E 08	0.1890E 08	0.2719E 08	0.3816E 08

xEn = x 10^n. E.g. 1.234E 05 = 1.234 x 10^5

$\int_o^\lambda L_{v'\lambda} d\lambda$

$\int_0^\lambda L_{v\lambda} d\lambda$

TABLE 18

Blackbody Fractional Luminance, Photopic (cd/m²)

λ(μ)	3000°K.	3100°K.	3200°K.	3300°K.	3400°K.	3500°K.	3600°K.	3700°K.	3800°K.	3900°K.
0.36	0.0000E 00	0.0000E 00	0.0000E 00	0.0000E 00	0.0000E 00	0.0000E 00	0.0000E 00	0.0000E 00	0.0000E 00	0.0000E 00
0.37	0.2484E 01	0.3748E 01	0.5512E 01	0.7918E 01	0.1113E 02	0.1535E 02	0.2080E 02	0.2772E 02	0.3639E 02	0.4711E 02
0.38	0.1209E 02	0.1809E 02	0.2639E 02	0.3763E 02	0.5256E 02	0.7201E 02	0.9696E 02	0.1284E 03	0.1677E 03	0.2159E 03
0.39	0.4645E 02	0.6887E 02	0.9962E 02	0.1409E 03	0.1953E 03	0.2657E 03	0.3554E 03	0.4679E 03	0.6072E 03	0.7776E 03
0.40	0.1830E 03	0.2688E 03	0.3853E 03	0.5405E 03	0.7433E 03	0.1003E 04	0.1333E 04	0.1743E 04	0.2248E 04	0.2861E 04
Luminance	0.1830E 03	0.2688E 03	0.3853E 03	0.5405E 03	0.7433E 03	0.1003E 04	0.1333E 04	0.1743E 04	0.2248E 04	0.2861E 04
0.41	0.4815E 03	0.6984E 03	0.9896E 03	0.1373E 04	0.1868E 04	0.2498E 04	0.3287E 04	0.4262E 04	0.5450E 04	0.6883E 04
0.42	0.2336E 04	0.3364E 04	0.4737E 04	0.6532E 04	0.8840E 04	0.1175E 05	0.1539E 05	0.1985E 05	0.2528E 05	0.3179E 05
0.43	0.9186E 04	0.1313E 05	0.1835E 05	0.2515E 05	0.3382E 05	0.4472E 05	0.5823E 05	0.7475E 05	0.9469E 05	0.1185E 06
0.44	0.2687E 05	0.3816E 05	0.5301E 05	0.7220E 05	0.9656E 05	0.1270E 06	0.1645E 06	0.2102E 06	0.2651E 06	0.3304E 06
0.45	0.6213E 05	0.8768E 05	0.1210E 06	0.1640E 06	0.2182E 06	0.2856E 06	0.3684E 06	0.4686E 06	0.5886E 06	0.7308E 06
0.46	0.1258E 06	0.1764E 06	0.2423E 06	0.3264E 06	0.4322E 06	0.5631E 06	0.7230E 06	0.9159E 06	0.1145E 07	0.1417E 07
0.47	0.2350E 06	0.3277E 06	0.4476E 06	0.5998E 06	0.7902E 06	0.1024E 07	0.1310E 07	0.1652E 07	0.2059E 07	0.2538E 07
0.48	0.4196E 06	0.5816E 06	0.7898E 06	0.1053E 07	0.1380E 07	0.1781E 07	0.2267E 07	0.2849E 07	0.3537E 07	0.4343E 07
0.49	0.7254E 06	0.9994E 06	0.1349E 07	0.1790E 07	0.2335E 07	0.3000E 07	0.3803E 07	0.4758E 07	0.5884E 07	0.7199E 07
0.50	0.1236E 07	0.1694E 07	0.2275E 07	0.3002E 07	0.3897E 07	0.4985E 07	0.6290E 07	0.7839E 07	0.9656E 07	0.1177E 08
Luminance	0.1237E 07	0.1694E 07	0.2275E 07	0.3002E 07	0.3898E 07	0.4986E 07	0.6291E 07	0.7840E 07	0.9659E 07	0.1177E 08
0.51	0.8735E 06	0.1179E 07	0.1562E 07	0.2035E 07	0.2610E 07	0.3300E 07	0.4119E 07	0.5080E 07	0.6196E 07	0.7482E 07
0.52	0.2277E 07	0.3064E 07	0.4046E 07	0.5255E 07	0.6720E 07	0.8474E 07	0.1055E 08	0.1297E 08	0.1579E 08	0.1903E 08
0.53	0.4243E 07	0.5689E 07	0.7488E 07	0.9695E 07	0.1236E 08	0.1554E 08	0.1930E 08	0.2369E 08	0.2877E 08	0.3458E 08
0.54	0.6675E 07	0.8919E 07	0.1170E 08	0.1510E 08	0.1921E 08	0.2409E 08	0.2984E 08	0.3654E 08	0.4428E 08	0.5312E 08
0.55	0.9460E 07	0.1259E 08	0.1648E 08	0.2121E 08	0.2690E 08	0.3366E 08	0.4160E 08	0.5083E 08	0.6146E 08	0.7359E 08
0.56	0.1248E 08	0.1657E 08	0.2162E 08	0.2776E 08	0.3512E 08	0.4384E 08	0.5406E 08	0.6591E 08	0.7954E 08	0.9506E 08
0.57	0.1562E 08	0.2068E 08	0.2691E 08	0.3446E 08	0.4349E 08	0.5417E 08	0.6666E 08	0.8113E 08	0.9772E 08	0.1166E 09
0.58	0.1872E 08	0.2472E 08	0.3208E 08	0.4099E 08	0.5163E 08	0.6418E 08	0.7884E 08	0.9577E 08	0.1151E 09	0.1372E 09
0.59	0.2164E 08	0.2850E 08	0.3691E 08	0.4706E 08	0.5916E 08	0.7341E 08	0.9002E 08	0.1091E 09	0.1311E 09	0.1559E 09
0.60	0.2424E 08	0.3187E 08	0.4119E 08	0.5242E 08	0.6579E 08	0.8151E 08	0.9980E 08	0.1208E 09	0.1449E 09	0.1722E 09
Luminance	0.2548E 08	0.3356E 08	0.4347E 08	0.5543E 08	0.6969E 08	0.8649E 08	0.1060E 09	0.1287E 09	0.1546E 09	0.1840E 09
0.61	0.2227E 07	0.2865E 07	0.3627E 07	0.4528E 07	0.5579E 07	0.6793E 07	0.8181E 07	0.9755E 07	0.1152E 08	0.1350E 08
0.62	0.4038E 07	0.5185E 07	0.6554E 07	0.8169E 07	0.1005E 08	0.1222E 08	0.1469E 08	0.1750E 08	0.2065E 08	0.2417E 08
0.63	0.5412E 07	0.6938E 07	0.8757E 07	0.1090E 08	0.1339E 08	0.1626E 08	0.1954E 08	0.2325E 08	0.2741E 08	0.3204E 08
0.64	0.6379E 07	0.8167E 07	0.1029E 08	0.1280E 08	0.1571E 08	0.1906E 08	0.2288E 08	0.2720E 08	0.3205E 08	0.3744E 08
0.65	0.7018E 07	0.8977E 07	0.1130E 08	0.1404E 08	0.1723E 08	0.2089E 08	0.2506E 08	0.2976E 08	0.3504E 08	0.4091E 08
0.66	0.7410E 07	0.9471E 07	0.1192E 08	0.1480E 08	0.1814E 08	0.2199E 08	0.2636E 08	0.3130E 08	0.3684E 08	0.4299E 08
0.67	0.7631E 07	0.9750E 07	0.1226E 08	0.1522E 08	0.1865E 08	0.2260E 08	0.2709E 08	0.3215E 08	0.3783E 08	0.4414E 08
0.68	0.7751E 07	0.9900E 07	0.1245E 08	0.1545E 08	0.1893E 08	0.2292E 08	0.2747E 08	0.3260E 08	0.3835E 08	0.4474E 08
0.69	0.7814E 07	0.9979E 07	0.1255E 08	0.1557E 08	0.1907E 08	0.2309E 08	0.2767E 08	0.3284E 08	0.3862E 08	0.4505E 08
0.70	0.7845E 07	0.1001E 08	0.1259E 08	0.1562E 08	0.1914E 08	0.2318E 08	0.2777E 08	0.3295E 08	0.3875E 08	0.4520E 08
Luminance	0.3332E 08	0.4358E 08	0.5607E 08	0.7105E 08	0.8883E 08	0.1096E 09	0.1338E 09	0.1616E 09	0.1933E 09	0.2292E 09
0.71	0.1630E 05	0.2025E 05	0.2482E 05	0.3005E 05	0.3597E 05	0.4263E 05	0.5004E 05	0.5825E 05	0.6726E 05	0.7711E 05
0.72	0.2475E 05	0.3072E 05	0.3761E 05	0.4549E 05	0.5441E 05	0.6443E 05	0.7558E 05	0.8791E 05	0.1014E 06	0.1162E 06
0.73	0.2906E 05	0.3604E 05	0.4409E 05	0.5329E 05	0.6371E 05	0.7540E 05	0.8840E 05	0.1027E 06	0.1185E 06	0.1357E 06
0.74	0.3121E 05	0.3868E 05	0.4731E 05	0.5716E 05	0.6830E 05	0.8080E 05	0.9471E 05	0.1100E 06	0.1269E 06	0.1453E 06
0.75	0.3225E 05	0.3996E 05	0.4886E 05	0.5902E 05	0.7051E 05	0.8340E 05	0.9773E 05	0.1135E 06	0.1309E 06	0.1498E 06
0.76	0.3277E 05	0.4060E 05	0.4963E 05	0.5995E 05	0.7161E 05	0.8468E 05	0.9922E 05	0.1152E 06	0.1328E 06	0.1521E 06
0.77	0.3304E 05	0.4093E 05	0.5002E 05	0.6041E 05	0.7216E 05	0.8532E 05	0.9996E 05	0.1161E 06	0.1338E 06	0.1532E 06
0.78	0.3317E 05	0.4109E 05	0.5022E 05	0.6065E 05	0.7243E 05	0.8565E 05	0.1003E 06	0.1165E 06	0.1343E 06	0.1537E 06
0.79	0.3324E 05	0.4117E 05	0.5032E 05	0.6076E 05	0.7257E 05	0.8581E 05	0.1005E 06	0.1167E 06	0.1346E 06	0.1540E 06
0.80	0.3328E 05	0.4121E 05	0.5037E 05	0.6082E 05	0.7264E 05	0.8589E 05	0.1006E 06	0.1168E 06	0.1347E 06	0.1541E 06
Luminance	0.3336E 08	0.4362E 08	0.5612E 08	0.7112E 08	0.8890E 08	0.1097E 09	0.1339E 09	0.1617E 09	0.1935E 09	0.2293E 09
0.81	0.1717E 02	0.2078E 02	0.2484E 02	0.2939E 02	0.3443E 02	0.3997E 02	0.4604E 02	0.5262E 02	0.5973E 02	0.6738E 02
0.82	0.2578E 02	0.3116E 02	0.3724E 02	0.4402E 02	0.5154E 02	0.5981E 02	0.6884E 02	0.7864E 02	0.8923E 02	0.1006E 03
0.83	0.3008E 02	0.3635E 02	0.4342E 02	0.5130E 02	0.6004E 02	0.6964E 02	0.8013E 02	0.9151E 02	0.1037E 03	0.1169E 03
Luminance	0.3336E 08	0.4362E 08	0.5612E 08	0.7112E 08	0.8890E 08	0.1097E 09	0.1339E 09	0.1617E 09	0.1935E 09	0.2293E 09

xEn - x 10^n. E.g. 1.234E 05 = 1.234 x 10^5

TABLE 18

Blackbody Fractional Luminance, Scotopic (sc.cd/m^2)

$\lambda(\mu)$	3000°K.	3100°K.	3200°K.	3300°K.	3400°K.	3500°K.	3600°K.	3700°K.	3800°K.	3900°K.
0.38	0.0000E 00	0.0000E 00	0.0000E 00	0.0000E 00	0.0000E 00	0.0000E 00	0.0000E 00	0.0000E 00	0.0000E 00	0.0000E 00
0.39	0.1533E 04	0.2265E 04	0.3267E 04	0.4608E 04	0.6369E 04	0.8641E 04	0.1152E 05	0.1514E 05	0.1960E 05	0.2504E 05
0.40	0.9009E 04	0.1320E 05	0.1890E 05	0.2646E 05	0.3634E 05	0.4900E 05	0.6498E 05	0.8488E 05	0.1093E 06	0.1389E 06
Luminance	0.9009E 04	0.1320E 05	0.1890E 05	0.2646E 05	0.3634E 05	0.4900E 05	0.6498E 05	0.8488E 05	0.1093E 06	0.1389E 06
0.41	0.3495E 05	0.5069E 05	0.7182E 05	0.9964E 05	0.1355E 06	0.1813E 06	0.2385E 06	0.3092E 06	0.3954E 06	0.4993E 06
0.42	0.1623E 06	0.2338E 06	0.3292E 06	0.4540E 06	0.6145E 06	0.8174E 06	0.1070E 07	0.1381E 07	0.1758E 07	0.2211E 07
0.43	0.5023E 06	0.7187E 06	0.1005E 07	0.1378E 07	0.1855E 07	0.2455E 07	0.3199E 07	0.4109E 07	0.5208E 07	0.6522E 07
0.44	0.1202E 07	0.1709E 07	0.2377E 07	0.3241E 07	0.4339E 07	0.5713E 07	0.7409E 07	0.9474E 07	0.1195E 08	0.1491E 08
0.45	0.2381E 07	0.3365E 07	0.4654E 07	0.6312E 07	0.8409E 07	0.1102E 08	0.1423E 08	0.1812E 08	0.2278E 08	0.2831E 08
0.46	0.4106E 07	0.5771E 07	0.7940E 07	0.1071E 08	0.1421E 08	0.1854E 08	0.2384E 08	0.3024E 08	0.3789E 08	0.4693E 08
0.47	0.6437E 07	0.8998E 07	0.1231E 08	0.1654E 08	0.2184E 08	0.2839E 08	0.3637E 08	0.4597E 08	0.5740E 08	0.7086E 08
0.48	0.9489E 07	0.1319E 08	0.1797E 08	0.2404E 08	0.3161E 08	0.4092E 08	0.5222E 08	0.6577E 08	0.8185E 08	0.1007E 09
0.49	0.1338E 08	0.1851E 08	0.2510E 08	0.3343E 08	0.4377E 08	0.5644E 08	0.7177E 08	0.9010E 08	0.1117E 09	0.1371E 09
0.50	0.1812E 08	0.2495E 08	0.3369E 08	0.4467E 08	0.5826E 08	0.7486E 08	0.9486E 08	0.1186E 09	0.1467E 09	0.1795E 09
Luminance	0.1813E 08	0.2497E 08	0.3371E 08	0.4470E 08	0.5830E 08	0.7491E 08	0.9492E 08	0.1187E 09	0.1468E 09	0.1797E 09
0.51	0.5423E 07	0.7323E 07	0.9704E 07	0.1264E 08	0.1621E 08	0.2051E 08	0.2560E 08	0.3157E 08	0.3852E 08	0.4651E 08
0.52	0.1115E 08	0.1501E 08	0.1984E 08	0.2579E 08	0.3300E 08	0.4164E 08	0.5188E 08	0.6386E 08	0.7775E 08	0.9373E 08
0.53	0.1672E 08	0.2246E 08	0.2961E 08	0.3838E 08	0.4901E 08	0.6171E 08	0.7672E 08	0.9426E 08	0.1145E 09	0.1378E 09
0.54	0.2171E 08	0.2908E 08	0.3825E 08	0.4949E 08	0.6306E 08	0.7926E 08	0.9836E 08	0.1206E 09	0.1464E 09	0.1759E 09
0.55	0.2582E 08	0.3451E 08	0.4531E 08	0.5851E 08	0.7443E 08	0.9339E 08	0.1157E 09	0.1417E 09	0.1718E 09	0.2061E 09
0.56	0.2894E 08	0.3861E 08	0.5061E 08	0.6525E 08	0.8289E 08	0.1038E 09	0.1285E 09	0.1573E 09	0.1904E 09	0.2283E 09
0.57	0.3112E 08	0.4146E 08	0.5428E 08	0.6990E 08	0.8871E 08	0.1110E 09	0.1373E 09	0.1678E 09	0.2030E 09	0.2432E 09
0.58	0.3252E 08	0.4329E 08	0.5662E 08	0.7286E 08	0.9239E 08	0.1155E 09	0.1428E 09	0.1744E 09	0.2109E 09	0.2526E 09
0.59	0.3335E 08	0.4437E 08	0.5799E 08	0.7459E 08	0.9454E 08	0.1182E 09	0.1460E 09	0.1783E 09	0.2155E 09	0.2579E 09
0.60	0.3381E 08	0.4496E 08	0.5874E 08	0.7553E 08	0.9570E 08	0.1196E 09	0.1477E 09	0.1803E 09	0.2179E 09	0.2608E 09
Luminance	0.5194E 08	0.6993E 08	0.9246E 08	0.1202E 09	0.1540E 09	0.1945E 09	0.2426E 09	0.2991E 09	0.3648E 09	0.4405E 09
0.61	0.2367E 06	0.3045E 06	0.3856E 06	0.4815E 06	0.5933E 06	0.7225E 06	0.8703E 06	0.1037E 07	0.1226E 07	0.1436E 07
0.62	0.3534E 06	0.4541E 06	0.5744E 06	0.7162E 06	0.8817E 06	0.1072E 07	0.1290E 07	0.1537E 07	0.1815E 07	0.2125E 07
0.63	0.4091E 06	0.5251E 06	0.6636E 06	0.8269E 06	0.1017E 07	0.1236E 07	0.1487E 07	0.1770E 07	0.2089E 07	0.2444E 07
0.64	0.4351E 06	0.5582E 06	0.7051E 06	0.8782E 06	0.1079E 07	0.1312E 07	0.1577E 07	0.1877E 07	0.2214E 07	0.2590E 07
0.65	0.4473E 06	0.5736E 06	0.7243E 06	0.9018E 06	0.1108E 07	0.1346E 07	0.1618E 07	0.1926E 07	0.2271E 07	0.2656E 07
0.66	0.4530E 06	0.5808E 06	0.7333E 06	0.9129E 06	0.1122E 07	0.1362E 07	0.1637E 07	0.1948E 07	0.2297E 07	0.2686E 07
0.67	0.4558E 06	0.5843E 06	0.7376E 06	0.9182E 06	0.1128E 07	0.1370E 07	0.1646E 07	0.1959E 07	0.2310E 07	0.2700E 07
0.68	0.4571E 06	0.5860E 06	0.7398E 06	0.9208E 06	0.1131E 07	0.1374E 07	0.1651E 07	0.1964E 07	0.2316E 07	0.2707E 07
0.69	0.4578E 06	0.5869E 06	0.7408E 06	0.9220E 06	0.1133E 07	0.1376E 07	0.1653E 07	0.1966E 07	0.2318E 07	0.2711E 07
0.70	0.4582E 06	0.5873E 06	0.7413E 06	0.9227E 06	0.1133E 07	0.1376E 07	0.1654E 07	0.1968E 07	0.2320E 07	0.2712E 07
Luminance	0.5240E 08	0.7052E 08	0.9320E 08	0.1211E 09	0.1551E 09	0.1959E 09	0.2443E 09	0.3011E 09	0.3671E 09	0.4432E 09
0.71	0.1819E 03	0.2260E 03	0.2769E 03	0.3353E 03	0.4014E 03	0.4756E 03	0.5583E 03	0.6499E 03	0.7504E 03	0.8603E 03
0.72	0.2783E 03	0.3454E 03	0.4229E 03	0.5115E 03	0.6118E 03	0.7244E 03	0.8498E 03	0.9884E 03	0.1140E 04	0.1306E 04
0.73	0.3303E 03	0.4095E 03	0.5011E 03	0.6056E 03	0.7240E 03	0.8567E 03	0.1004E 04	0.1167E 04	0.1346E 04	0.1542E 04
0.74	0.3587E 03	0.4446E 03	0.5436E 03	0.6568E 03	0.7847E 03	0.9282E 03	0.1087E 04	0.1264E 04	0.1457E 04	0.1668E 04
0.75	0.3745E 03	0.4640E 03	0.5672E 03	0.6850E 03	0.8182E 03	0.9675E 03	0.1133E 04	0.1317E 04	0.1518E 04	0.1737E 04
0.76	0.3835E 03	0.4749E 03	0.5804E 03	0.7008E 03	0.8369E 03	0.9894E 03	0.1159E 04	0.1346E 04	0.1551E 04	0.1775E 04
0.77	0.3886E 03	0.4811E 03	0.5879E 03	0.7097E 03	0.8475E 03	0.1001E 04	0.1173E 04	0.1362E 04	0.1570E 04	0.1797E 04
0.78	0.3915E 03	0.4847E 03	0.5922E 03	0.7149E 03	0.8536E 03	0.1009E 04	0.1181E 04	0.1372E 04	0.1581E 04	0.1809E 04
Luminance	0.5240E 08	0.7052E 08	0.9320E 08	0.1211E 09	0.1551E 09	0.1959E 09	0.2443E 09	0.3011E 09	0.3671E 09	0.4432E 09

xEn = x 10^n. E.g. 1.234E 05 = 1.234 x 10^5

$\int_0^\lambda L_{v',\lambda} d\lambda$

$\int_o^\lambda L_{v\lambda} d\lambda$

TABLE 18

Blackbody Fractional Luminance, Photopic (cd/m²)

λ(μ)	4000°K.	4100°K.	4200°K.	4300°K.	4400°K.	4500°K.	4600°K.	4700°K.	4800°K.	4900°K.
0.36	0.0000E 00	0.0000E 00	0.0000E 00	0.0000E 00	0.0000E 00	0.0000E 00	0.0000E 00	0.0000E 00	0.0000E 00	0.0000E 00
0.37	0.6020E 02	0.7602E 02	0.9493E 02	0.1173E 03	0.1436E 03	0.1742E 03	0.2096E 03	0.2502E 03	0.2964E 03	0.3488E 03
0.38	0.2746E 03	0.3451E 03	0.4290E 03	0.5280E 03	0.6438E 03	0.7780E 03	0.9325E 03	0.1109E 04	0.1309E 04	0.1536E 04
0.39	0.9835E 03	0.1229E 04	0.1521E 04	0.1864E 04	0.2262E 04	0.2723E 04	0.3250E 04	0.3851E 04	0.4531E 04	0.5296E 04
0.40	0.3599E 04	0.4476E 04	0.5509E 04	0.6716E 04	0.8114E 04	0.9722E 04	0.1155E 05	0.1363E 05	0.1598E 05	0.1861E 05
Luminance	0.3599E 04	0.4476E 04	0.5509E 04	0.6716E 04	0.8114E 04	0.9722E 04	0.1155E 05	0.1363E 05	0.1598E 05	0.1861E 05
0.41	0.8591E 04	0.1060E 05	0.1296E 05	0.1570E 05	0.1885E 05	0.2245E 05	0.2654E 05	0.3115E 05	0.3631E 05	0.4207E 05
0.42	0.3951E 05	0.4860E 05	0.5920E 05	0.7144E 05	0.8549E 05	0.1014E 06	0.1195E 06	0.1399E 06	0.1626E 06	0.1879E 06
0.43	0.1466E 06	0.1796E 06	0.2179E 06	0.2620E 06	0.3124E 06	0.3696E 06	0.4341E 06	0.5063E 06	0.5869E 06	0.6761E 06
0.44	0.4073E 06	0.4971E 06	0.6008E 06	0.7199E 06	0.8556E 06	0.1009E 07	0.1181E 07	0.1374E 07	0.1588E 07	0.1825E 07
0.45	0.8976E 06	0.1091E 07	0.1315E 07	0.1570E 07	0.1861E 07	0.2188E 07	0.2555E 07	0.2964E 07	0.3418E 07	0.3918E 07
0.46	0.1734E 07	0.2102E 07	0.2524E 07	0.3006E 07	0.3551E 07	0.4165E 07	0.4851E 07	0.5613E 07	0.6456E 07	0.7384E 07
0.47	0.3095E 07	0.3739E 07	0.4476E 07	0.5314E 07	0.6260E 07	0.7321E 07	0.8505E 07	0.9817E 07	0.1126E 08	0.1285E 08
0.48	0.5278E 07	0.6354E 07	0.7583E 07	0.8976E 07	0.1054E 08	0.1229E 08	0.1424E 08	0.1640E 08	0.1878E 08	0.2138E 08
0.49	0.8719E 07	0.1046E 08	0.1244E 08	0.1468E 08	0.1720E 08	0.2001E 08	0.2312E 08	0.2656E 08	0.3033E 08	0.3446E 08
0.50	0.1420E 08	0.1699E 08	0.2014E 08	0.2370E 08	0.2769E 08	0.3212E 08	0.3702E 08	0.4242E 08	0.4833E 08	0.5478E 08
Luminance	0.1420E 08	0.1699E 08	0.2015E 08	0.2371E 08	0.2769E 08	0.3213E 08	0.3703E 08	0.4243E 08	0.4835E 08	0.5480E 08
0.51	0.8949E 07	0.1061E 08	0.1248E 08	0.1457E 08	0.1689E 08	0.1945E 08	0.2226E 08	0.2534E 08	0.2869E 08	0.3232E 08
0.52	0.2271E 08	0.2688E 08	0.3156E 08	0.3678E 08	0.4256E 08	0.4895E 08	0.5594E 08	0.6358E 08	0.7188E 08	0.8086E 08
0.53	0.4120E 08	0.4866E 08	0.5703E 08	0.6635E 08	0.7666E 08	0.8801E 08	0.1004E 09	0.1139E 09	0.1286E 09	0.1445E 09
0.54	0.6316E 08	0.7446E 08	0.8711E 08	0.1011E 09	0.1167E 09	0.1337E 09	0.1524E 09	0.1727E 09	0.1948E 09	0.2185E 09
0.55	0.8734E 08	0.1027E 09	0.1200E 09	0.1392E 09	0.1603E 09	0.1835E 09	0.2088E 09	0.2364E 09	0.2662E 09	0.2983E 09
0.56	0.1126E 09	0.1323E 09	0.1543E 09	0.1786E 09	0.2054E 09	0.2349E 09	0.2669E 09	0.3018E 09	0.3395E 09	0.3800E 09
0.57	0.1379E 09	0.1617E 09	0.1883E 09	0.2178E 09	0.2502E 09	0.2856E 09	0.3243E 09	0.3662E 09	0.4114E 09	0.4601E 09
0.58	0.1620E 09	0.1898E 09	0.2207E 09	0.2549E 09	0.2924E 09	0.3335E 09	0.3782E 09	0.4267E 09	0.4789E 09	0.5351E 09
0.59	0.1839E 09	0.2152E 09	0.2499E 09	0.2883E 09	0.3305E 09	0.3765E 09	0.4266E 09	0.4808E 09	0.5392E 09	0.6020E 09
0.60	0.2029E 09	0.2371E 09	0.2751E 09	0.3171E 09	0.3631E 09	0.4133E 09	0.4679E 09	0.5269E 09	0.5905E 09	0.6588E 09
Luminance	0.2171E 09	0.2541E 09	0.2953E 09	0.3408E 09	0.3908E 09	0.4454E 09	0.5049E 09	0.5694E 09	0.6389E 09	0.7136E 09
0.61	0.1569E 08	0.1810E 08	0.2075E 08	0.2363E 08	0.2676E 08	0.3014E 08	0.3377E 08	0.3765E 08	0.4180E 08	0.4621E 08
0.62	0.2806E 08	0.3235E 08	0.3704E 08	0.4215E 08	0.4769E 08	0.5366E 08	0.6008E 08	0.6694E 08	0.7426E 08	0.8204E 08
0.63	0.3717E 08	0.4281E 08	0.4898E 08	0.5570E 08	0.6297E 08	0.7081E 08	0.7922E 08	0.8822E 08	0.9782E 08	0.1080E 09
0.64	0.4340E 08	0.4995E 08	0.5712E 08	0.6491E 08	0.7334E 08	0.8242E 08	0.9217E 08	0.1025E 09	0.1136E 09	0.1254E 09
0.65	0.4741E 08	0.5454E 08	0.6233E 08	0.7080E 08	0.7996E 08	0.8983E 08	0.1004E 09	0.1117E 09	0.1237E 09	0.1365E 09
0.66	0.4979E 08	0.5726E 08	0.6542E 08	0.7429E 08	0.8387E 08	0.9420E 08	0.1052E 09	0.1171E 09	0.1296E 09	0.1430E 09
0.67	0.5111E 08	0.5876E 08	0.6712E 08	0.7620E 08	0.8602E 08	0.9659E 08	0.1079E 09	0.1200E 09	0.1329E 09	0.1465E 09
0.68	0.5180E 08	0.5955E 08	0.6801E 08	0.7720E 08	0.8714E 08	0.9783E 08	0.1093E 09	0.1215E 09	0.1345E 09	0.1484E 09
0.69	0.5216E 08	0.5995E 08	0.6847E 08	0.7771E 08	0.8771E 08	0.9847E 08	0.1100E 09	0.1223E 09	0.1354E 09	0.1493E 09
0.70	0.5233E 08	0.6015E 08	0.6869E 08	0.7796E 08	0.8798E 08	0.9877E 08	0.1103E 09	0.1226E 09	0.1358E 09	0.1497E 09
Luminance	0.2694E 09	0.3143E 09	0.3640E 09	0.4188E 09	0.4788E 09	0.5442E 09	0.6153E 09	0.6921E 09	0.7747E 09	0.8634E 09
0.71	0.8780E 05	0.9936E 05	0.1117E 06	0.1251E 06	0.1392E 06	0.1543E 06	0.1703E 06	0.1872E 06	0.2049E 06	0.2235E 06
0.72	0.1322E 06	0.1496E 06	0.1682E 06	0.1881E 06	0.2094E 06	0.2319E 06	0.2558E 06	0.2811E 06	0.3076E 06	0.3354E 06
0.73	0.1544E 06	0.1746E 06	0.1962E 06	0.2194E 06	0.2441E 06	0.2703E 06	0.2981E 06	0.3274E 06	0.3582E 06	0.3905E 06
0.74	0.1652E 06	0.1867E 06	0.2099E 06	0.2346E 06	0.2610E 06	0.2890E 06	0.3186E 06	0.3498E 06	0.3826E 06	0.4171E 06
0.75	0.1704E 06	0.1925E 06	0.2163E 06	0.2418E 06	0.2690E 06	0.2978E 06	0.3282E 06	0.3604E 06	0.3942E 06	0.4296E 06
0.76	0.1729E 06	0.1954E 06	0.2195E 06	0.2453E 06	0.2728E 06	0.3020E 06	0.3329E 06	0.3655E 06	0.3997E 06	0.4357E 06
0.77	0.1741E 06	0.1968E 06	0.2211E 06	0.2471E 06	0.2748E 06	0.3042E 06	0.3352E 06	0.3680E 06	0.4025E 06	0.4387E 06
0.78	0.1748E 06	0.1975E 06	0.2219E 06	0.2479E 06	0.2757E 06	0.3052E 06	0.3364E 06	0.3693E 06	0.4039E 06	0.4401E 06
0.79	0.1751E 06	0.1978E 06	0.2222E 06	0.2484E 06	0.2762E 06	0.3057E 06	0.3370E 06	0.3699E 06	0.4045E 06	0.4408E 06
0.80	0.1752E 06	0.1980E 06	0.2224E 06	0.2486E 06	0.2764E 06	0.3060E 06	0.3372E 06	0.3702E 06	0.4049E 06	0.4412E 06
Luminance	0.2696E 09	0.3145E 09	0.3642E 09	0.4190E 09	0.4791E 09	0.5445E 09	0.6156E 09	0.6924E 09	0.7751E 09	0.8638E 09
0.81	0.7556E 02	0.8427E 02	0.9351E 02	0.1032E 03	0.1135E 03	0.1243E 03	0.1357E 03	0.1475E 03	0.1598E 03	0.1727E 03
0.82	0.1127E 03	0.1257E 03	0.1394E 03	0.1539E 03	0.1692E 03	0.1853E 03	0.2021E 03	0.2196E 03	0.2379E 03	0.2569E 03
0.83	0.1310E 03	0.1460E 03	0.1620E 03	0.1788E 03	0.1965E 03	0.2151E 03	0.2345E 03	0.2549E 03	0.2760E 03	0.2980E 03
Luminance	0.2696E 09	0.3145E 09	0.3642E 09	0.4190E 09	0.4791E 09	0.5445E 09	0.6156E 09	0.6924E 09	0.7751E 09	0.8638E 09

xEn = x 10^n. E.g. 1.234E 05 = 1.234 x 10^5

TABLE 18

Blackbody Fractional Luminance, Scotopic (sc.cd/m^2)

$\lambda(\mu)$	4000° K.	4100° K.	4200° K.	4300° K.	4400° K.	4500° K.	4600° K.	4700° K.	4800° K.	4900° K.
0.38	0.0000E 00	0.0000E 00	0.0000E 00	0.0000E 00	0.0000E 00	0.0000E 00	0.0000E 00	0.0000E 00	0.0000E 00	0.0000E 00
0.39	0.3160E 05	0.3944E 05	0.4870E 05	0.5955E 05	0.7215E 05	0.8668E 05	0.1033E 06	0.1222E 06	0.1435E 06	0.1675E 06
0.40	0.1745E 06	0.2168E 06	0.2666E 06	0.3247E 06	0.3919E 06	0.4691E 06	0.5571E 06	0.6568E 06	0.7691E 06	0.8948E 06
Luminance	0.1745E 06	0.2168E 06	0.2666E 06	0.3247E 06	0.3919E 06	0.4691E 06	0.5571E 06	0.6568E 06	0.7691E 06	0.8948E 06
0.41	0.6232E 06	0.7694E 06	0.9406E 06	0.1139E 07	0.1367E 07	0.1628E 07	0.1924E 07	0.2258E 07	0.2633E 07	0.3050E 07
0.42	0.2748E 07	0.3381E 07	0.4118E 07	0.4971E 07	0.5949E 07	0.7062E 07	0.8322E 07	0.9739E 07	0.1132E 08	0.1308E 08
0.43	0.8077E 07	0.9898E 07	0.1201E 08	0.1445E 08	0.1723E 08	0.2040E 08	0.2397E 08	0.2797E 08	0.3243E 08	0.3737E 08
0.44	0.1840E 08	0.2247E 08	0.2718E 08	0.3259E 08	0.3875E 08	0.4573E 08	0.5358E 08	0.6236E 08	0.7212E 08	0.8292E 08
0.45	0.3481E 08	0.4237E 08	0.5108E 08	0.6107E 08	0.7241E 08	0.8522E 08	0.9959E 08	0.1156E 09	0.1333E 09	0.1530E 09
0.46	0.5751E 08	0.6978E 08	0.8390E 08	0.1000E 09	0.1182E 09	0.1388E 09	0.1618E 09	0.1874E 09	0.2158E 09	0.2470E 09
0.47	0.8656E 08	0.1047E 09	0.1255E 09	0.1492E 09	0.1761E 09	0.2062E 09	0.2399E 09	0.2772E 09	0.3185E 09	0.3639E 09
0.48	0.1227E 09	0.1480E 09	0.1770E 09	0.2099E 09	0.2470E 09	0.2886E 09	0.3350E 09	0.3863E 09	0.4430E 09	0.5051E 09
0.49	0.1665E 09	0.2003E 09	0.2389E 09	0.2827E 09	0.3319E 09	0.3870E 09	0.4482E 09	0.5159E 09	0.5903E 09	0.6719E 09
0.50	0.2174E 09	0.2610E 09	0.3105E 09	0.3665E 09	0.4293E 09	0.4995E 09	0.5773E 09	0.6632E 09	0.7576E 09	0.8607E 09
Luminance	0.2176E 09	0.2612E 09	0.3107E 09	0.3668E 09	0.4297E 09	0.4999E 09	0.5779E 09	0.6639E 09	0.7583E 09	0.8616E 09
0.51	0.5564E 08	0.6599E 08	0.7763E 08	0.9063E 08	0.1050E 09	0.1210E 09	0.1385E 09	0.1577E 09	0.1785E 09	0.2011E 09
0.52	0.1119E 09	0.1325E 09	0.1556E 09	0.1814E 09	0.2101E 09	0.2416E 09	0.2763E 09	0.3141E 09	0.3552E 09	0.3997E 09
0.53	0.1644E 09	0.1943E 09	0.2279E 09	0.2654E 09	0.3069E 09	0.3525E 09	0.4026E 09	0.4573E 09	0.5166E 09	0.5807E 09
0.54	0.2095E 09	0.2473E 09	0.2897E 09	0.3369E 09	0.3891E 09	0.4466E 09	0.5095E 09	0.5780E 09	0.6524E 09	0.7328E 09
0.55	0.2452E 09	0.2891E 09	0.3384E 09	0.3931E 09	0.4536E 09	0.5201E 09	0.5929E 09	0.6721E 09	0.7580E 09	0.8507E 09
0.56	0.2712E 09	0.3196E 09	0.3737E 09	0.4338E 09	0.5002E 09	0.5731E 09	0.6528E 09	0.7396E 09	0.8336E 09	0.9350E 09
0.57	0.2888E 09	0.3401E 09	0.3974E 09	0.4610E 09	0.5313E 09	0.6084E 09	0.6927E 09	0.7844E 09	0.8836E 09	0.9907E 09
0.58	0.2997E 09	0.3528E 09	0.4120E 09	0.4778E 09	0.5504E 09	0.6301E 09	0.7171E 09	0.8117E 09	0.9142E 09	0.1024E 10
0.59	0.3060E 09	0.3600E 09	0.4204E 09	0.4873E 09	0.5612E 09	0.6423E 09	0.7309E 09	0.8272E 09	0.9314E 09	0.1043E 10
0.60	0.3093E 09	0.3639E 09	0.4248E 09	0.4924E 09	0.5670E 09	0.6488E 09	0.7382E 09	0.8353E 09	0.9404E 09	0.1053E 10
Luminance	0.5270E 09	0.6251E 09	0.7356E 09	0.8592E 09	0.9967E 09	0.1148E 10	0.1316E 10	0.1499E 10	0.1698E 10	0.1915E 10
0.61	0.1670E 07	0.1927E 07	0.2209E 07	0.2516E 07	0.2849E 07	0.3209E 07	0.3595E 07	0.4010E 07	0.4452E 07	0.4922E 07
0.62	0.2468E 07	0.2846E 07	0.3260E 07	0.3711E 07	0.4200E 07	0.4727E 07	0.5294E 07	0.5901E 07	0.6548E 07	0.7236E 07
0.63	0.2838E 07	0.3271E 07	0.3745E 07	0.4261E 07	0.4820E 07	0.5423E 07	0.6071E 07	0.6764E 07	0.7503E 07	0.8289E 07
0.64	0.3006E 07	0.3463E 07	0.3964E 07	0.4509E 07	0.5100E 07	0.5736E 07	0.6420E 07	0.7152E 07	0.7932E 07	0.8760E 07
0.65	0.3082E 07	0.3550E 07	0.4063E 07	0.4621E 07	0.5225E 07	0.5877E 07	0.6577E 07	0.7325E 07	0.8123E 07	0.8971E 07
0.66	0.3117E 07	0.3590E 07	0.4108E 07	0.4672E 07	0.5283E 07	0.5941E 07	0.6648E 07	0.7404E 07	0.8210E 07	0.9066E 07
0.67	0.3133E 07	0.3609E 07	0.4130E 07	0.4696E 07	0.5309E 07	0.5971E 07	0.6681E 07	0.7441E 07	0.8250E 07	0.9110E 07
0.68	0.3141E 07	0.3618E 07	0.4140E 07	0.4707E 07	0.5322E 07	0.5985E 07	0.6697E 07	0.7458E 07	0.8269E 07	0.9131E 07
0.69	0.3145E 07	0.3622E 07	0.4145E 07	0.4713E 07	0.5328E 07	0.5992E 07	0.6704E 07	0.7466E 07	0.8278E 07	0.9141E 07
0.70	0.3147E 07	0.3624E 07	0.4147E 07	0.4716E 07	0.5331E 07	0.5995E 07	0.6708E 07	0.7471E 07	0.8283E 07	0.9146E 07
Luminance	0.5301E 09	0.6287E 09	0.7397E 09	0.8640E 09	0.1002E 10	0.1154E 10	0.1322E 10	0.1506E 10	0.1707E 10	0.1924E 10
0.71	0.9796E 03	0.1108E 04	0.1247E 04	0.1395E 04	0.1554E 04	0.1722E 04	0.1900E 04	0.2088E 04	0.2286E 04	0.2494E 04
0.72	0.1487E 04	0.1681E 04	0.1891E 04	0.2115E 04	0.2354E 04	0.2608E 04	0.2876E 04	0.3160E 04	0.3458E 04	0.3771E 04
0.73	0.1754E 04	0.1983E 04	0.2229E 04	0.2492E 04	0.2773E 04	0.3071E 04	0.3386E 04	0.3718E 04	0.4068E 04	0.4435E 04
0.74	0.1897E 04	0.2144E 04	0.2410E 04	0.2694E 04	0.2996E 04	0.3317E 04	0.3657E 04	0.4015E 04	0.4392E 04	0.4787E 04
0.75	0.1975E 04	0.2232E 04	0.2508E 04	0.2803E 04	0.3117E 04	0.3450E 04	0.3803E 04	0.4175E 04	0.4566E 04	0.4976E 04
0.76	0.2018E 04	0.2280E 04	0.2562E 04	0.2863E 04	0.3183E 04	0.3523E 04	0.3883E 04	0.4263E 04	0.4662E 04	0.5080E 04
0.77	0.2042E 04	0.2307E 04	0.2592E 04	0.2896E 04	0.3220E 04	0.3564E 04	0.3928E 04	0.4312E 04	0.4715E 04	0.5137E 04
0.78	0.2056E 04	0.2323E 04	0.2609E 04	0.2915E 04	0.3241E 04	0.3587E 04	0.3953E 04	0.4339E 04	0.4745E 04	0.5170E 04
Luminance	0.5301E 09	0.6287E 09	0.7397E 09	0.8640E 09	0.1002E 10	0.1154E 10	0.1322E 10	0.1506E 10	0.1707E 10	0.1924E 10

xEn = x 10^n. E.g. 1.234E 05 = 1.234 x 10^5

$\int_o^\lambda L_{v'\lambda}\, d\lambda$

TABLE 18

$\int_o^\lambda L_{v\lambda} d\lambda$

Blackbody Fractional Luminance, Photopic (cd/m^2)

λ(μ)	5000°K.	5100°K.	5200°K.	5300°K.	5400°K.	5500°K.	5600°K.	5700°K.	5800°K.	5900°K.
0.36	0.0000E 00	0.0000E 00	0.0000E 00	0.0000E 00	0.0000E 00	0.0000E 00	0.0000E 00	0.0000E 00	0.0000E 00	0.0000E 00
0.37	0.4078E 03	0.4739E 03	0.5475E 03	0.6291E 03	0.7191E 03	0.8181E 03	0.9264E 03	0.1044E 04	0.1172E 04	0.1311E 04
0.38	0.1790E 04	0.2074E 04	0.2389E 04	0.2737E 04	0.3121E 04	0.3541E 04	0.4001E 04	0.4500E 04	0.5042E 04	0.5626E 04
0.39	0.6152E 04	0.7104E 04	0.8158E 04	0.9320E 04	0.1059E 05	0.1199E 05	0.1350E 05	0.1515E 05	0.1693E 05	0.1885E 05
0.40	0.2153E 05	0.2478E 05	0.2837E 05	0.3231E 05	0.3662E 05	0.4131E 05	0.4641E 05	0.5193E 05	0.5788E 05	0.6428E 05
Luminance	0.2153E 05	0.2478E 05	0.2837E 05	0.3231E 05	0.3662E 05	0.4131E 05	0.4641E 05	0.5193E 05	0.5788E 05	0.6428E 05
0.41	0.4845E 05	0.5550E 05	0.6324E 05	0.7170E 05	0.8093E 05	0.9094E 05	0.1017E 06	0.1134E 06	0.1259E 06	0.1394E 06
0.42	0.2159E 06	0.2467E 06	0.2804E 06	0.3173E 06	0.3573E 06	0.4007E 06	0.4475E 06	0.4979E 06	0.5519E 06	0.6096E 06
0.43	0.7746E 06	0.8827E 06	0.1000E 07	0.1129E 07	0.1269E 07	0.1419E 07	0.1582E 07	0.1756E 07	0.1943E 07	0.2142E 07
0.44	0.2085E 07	0.2371E 07	0.2682E 07	0.3020E 07	0.3386E 07	0.3781E 07	0.4205E 07	0.4660E 07	0.5145E 07	0.5663E 07
0.45	0.4467E 07	0.5068E 07	0.5721E 07	0.6429E 07	0.7194E 07	0.8017E 07	0.8900E 07	0.9845E 07	0.1085E 08	0.1192E 08
0.46	0.8400E 07	0.9508E 07	0.1071E 08	0.1201E 08	0.1341E 08	0.1492E 08	0.1654E 08	0.1826E 08	0.2010E 08	0.2205E 08
0.47	0.1459E 08	0.1648E 08	0.1853E 08	0.2074E 08	0.2312E 08	0.2567E 08	0.2841E 08	0.3132E 08	0.3441E 08	0.3770E 08
0.48	0.2421E 08	0.2729E 08	0.3063E 08	0.3422E 08	0.3808E 08	0.4221E 08	0.4662E 08	0.5131E 08	0.5629E 08	0.6157E 08
0.49	0.3895E 08	0.4381E 08	0.4906E 08	0.5471E 08	0.6077E 08	0.6724E 08	0.7414E 08	0.8147E 08	0.8924E 08	0.9745E 08
0.50	0.6178E 08	0.6935E 08	0.7751E 08	0.8627E 08	0.9564E 08	0.1056E 09	0.1162E 09	0.1275E 09	0.1395E 09	0.1521E 09
Luminance	0.6180E 08	0.6937E 08	0.7754E 08	0.8630E 08	0.9568E 08	0.1056E 09	0.1163E 09	0.1276E 09	0.1395E 09	0.1521E 09
0.51	0.3623E 08	0.4044E 08	0.4495E 08	0.4977E 08	0.5490E 08	0.6034E 08	0.6611E 08	0.7219E 08	0.7861E 08	0.8535E 08
0.52	0.9055E 08	0.1009E 09	0.1120E 09	0.1239E 09	0.1365E 09	0.1499E 09	0.1641E 09	0.1790E 09	0.1947E 09	0.2112E 09
0.53	0.1616E 09	0.1800E 09	0.1996E 09	0.2205E 09	0.2427E 09	0.2662E 09	0.2911E 09	0.3173E 09	0.3448E 09	0.3737E 09
0.54	0.2441E 09	0.2715E 09	0.3008E 09	0.3319E 09	0.3649E 09	0.3999E 09	0.4368E 09	0.4757E 09	0.5166E 09	0.5594E 09
0.55	0.3328E 09	0.3698E 09	0.4092E 09	0.4511E 09	0.4955E 09	0.5425E 09	0.5921E 09	0.6443E 09	0.6990E 09	0.7564E 09
0.56	0.4235E 09	0.4701E 09	0.5197E 09	0.5724E 09	0.6282E 09	0.6872E 09	0.7493E 09	0.8147E 09	0.8833E 09	0.9551E 09
0.57	0.5123E 09	0.5681E 09	0.6274E 09	0.6905E 09	0.7572E 09	0.8276E 09	0.9018E 09	0.9798E 09	0.1061E 10	0.1147E 10
0.58	0.5953E 09	0.6595E 09	0.7278E 09	0.8003E 09	0.8770E 09	0.9579E 09	0.1043E 10	0.1132E 10	0.1226E 10	0.1324E 10
0.59	0.6691E 09	0.7408E 09	0.8169E 09	0.8977E 09	0.9830E 09	0.1073E 10	0.1167E 10	0.1267E 10	0.1371E 10	0.1480E 10
0.60	0.7318E 09	0.8096E 09	0.8923E 09	0.9799E 09	0.1072E 10	0.1170E 10	0.1272E 10	0.1380E 10	0.1493E 10	0.1610E 10
Luminance	0.7936E 09	0.8790E 09	0.9698E 09	0.1066E 10	0.1168E 10	0.1275E 10	0.1389E 10	0.1507E 10	0.1632E 10	0.1762E 10
0.61	0.5089E 08	0.5583E 08	0.6104E 08	0.6651E 08	0.7225E 08	0.7826E 08	0.8454E 08	0.9108E 08	0.9788E 08	0.1049E 09
0.62	0.9029E 08	0.9899E 08	0.1081E 09	0.1178E 09	0.1279E 09	0.1384E 09	0.1494E 09	0.1609E 09	0.1729E 09	0.1853E 09
0.63	0.1187E 09	0.1301E 09	0.1421E 09	0.1547E 09	0.1679E 09	0.1817E 09	0.1961E 09	0.2111E 09	0.2266E 09	0.2428E 09
0.64	0.1379E 09	0.1511E 09	0.1649E 09	0.1795E 09	0.1947E 09	0.2106E 09	0.2272E 09	0.2445E 09	0.2625E 09	0.2811E 09
0.65	0.1500E 09	0.1643E 09	0.1793E 09	0.1951E 09	0.2116E 09	0.2288E 09	0.2468E 09	0.2655E 09	0.2849E 09	0.3051E 09
0.66	0.1571E 09	0.1720E 09	0.1877E 09	0.2042E 09	0.2214E 09	0.2394E 09	0.2582E 09	0.2777E 09	0.2980E 09	0.3190E 09
0.67	0.1610E 09	0.1762E 09	0.1923E 09	0.2091E 09	0.2267E 09	0.2451E 09	0.2643E 09	0.2843E 09	0.3050E 09	0.3265E 09
0.68	0.1630E 09	0.1784E 09	0.1946E 09	0.2117E 09	0.2295E 09	0.2481E 09	0.2675E 09	0.2877E 09	0.3086E 09	0.3303E 09
0.69	0.1640E 09	0.1795E 09	0.1958E 09	0.2129E 09	0.2309E 09	0.2496E 09	0.2691E 09	0.2894E 09	0.3104E 09	0.3323E 09
0.70	0.1645E 09	0.1800E 09	0.1964E 09	0.2136E 09	0.2315E 09	0.2503E 09	0.2698E 09	0.2902E 09	0.3113E 09	0.3332E 09
Luminance	0.9581E 09	0.1059E 10	0.1166E 10	0.1279E 10	0.1399E 10	0.1526E 10	0.1658E 10	0.1798E 10	0.1943E 10	0.2096E 10
0.71	0.2431E 06	0.2634E 06	0.2846E 06	0.3067E 06	0.3296E 06	0.3534E 06	0.3779E 06	0.4033E 06	0.4294E 06	0.4563E 06
0.72	0.3645E 06	0.3950E 06	0.4266E 06	0.4596E 06	0.4938E 06	0.5292E 06	0.5658E 06	0.6036E 06	0.6425E 06	0.6826E 06
0.73	0.4243E 06	0.4596E 06	0.4964E 06	0.5346E 06	0.5742E 06	0.6153E 06	0.6577E 06	0.7015E 06	0.7467E 06	0.7931E 06
0.74	0.4531E 06	0.4907E 06	0.5299E 06	0.5706E 06	0.6128E 06	0.6566E 06	0.7018E 06	0.7484E 06	0.7965E 06	0.8460E 06
0.75	0.4667E 06	0.5054E 06	0.5457E 06	0.5875E 06	0.6310E 06	0.6759E 06	0.7224E 06	0.7704E 06	0.8198E 06	0.8707E 06
0.76	0.4732E 06	0.5124E 06	0.5533E 06	0.5957E 06	0.6397E 06	0.6852E 06	0.7323E 06	0.7809E 06	0.8310E 06	0.8825E 06
0.77	0.4765E 06	0.5159E 06	0.5570E 06	0.5997E 06	0.6440E 06	0.6898E 06	0.7372E 06	0.7861E 06	0.8365E 06	0.8884E 06
0.78	0.4780E 06	0.5176E 06	0.5588E 06	0.6017E 06	0.6461E 06	0.6921E 06	0.7396E 06	0.7886E 06	0.8392E 06	0.8912E 06
0.79	0.4788E 06	0.5185E 06	0.5597E 06	0.6026E 06	0.6471E 06	0.6932E 06	0.7408E 06	0.7899E 06	0.8405E 06	0.8926E 06
0.80	0.4792E 06	0.5189E 06	0.5602E 06	0.6031E 06	0.6476E 06	0.6937E 06	0.7413E 06	0.7905E 06	0.8411E 06	0.8932E 06
Luminance	0.9586E 09	0.1059E 10	0.1166E 10	0.1280E 10	0.1400E 10	0.1526E 10	0.1659E 10	0.1798E 10	0.1944E 10	0.2097E 10
0.81	0.1860E 03	0.1997E 03	0.2140E 03	0.2287E 03	0.2438E 03	0.2594E 03	0.2754E 03	0.2918E 03	0.3086E 03	0.3258E 03
0.82	0.2766E 03	0.2970E 03	0.3181E 03	0.3399E 03	0.3623E 03	0.3854E 03	0.4091E 03	0.4334E 03	0.4582E 03	0.4837E 03
0.83	0.3208E 03	0.3444E 03	0.3688E 03	0.3940E 03	0.4199E 03	0.4466E 03	0.4740E 03	0.5020E 03	0.5308E 03	0.5603E 03
Luminance	0.9586E 09	0.1059E 10	0.1166E 10	0.1280E 10	0.1400E 10	0.1526E 10	0.1659E 10	0.1798E 10	0.1944E 10	0.2097E 10

xEn = x 10^n. E.g. 1.234E 05 = 1.234 x 10^5

TABLE 18

Blackbody Fractional Luminance, Scotopic (sc.cd/m^2)

$\lambda(\mu)$	5000°K.	5100°K.	5200°K.	5300°K.	5400°K.	5500°K.	5600°K.	5700°K.	5800°K.	5900°K.
0.38	0.0000E 00	0.0000E 00	0.0000E 00	0.0000E 00	0.0000E 00	0.0000E 00	0.0000E 00	0.0000E 00	0.0000E 00	0.0000E 00
0.39	0.1943E 06	0.2241E 06	0.2570E 06	0.2932E 06	0.3330E 06	0.3763E 06	0.4235E 06	0.4746E 06	0.5299E 06	0.5893E 06
0.40	0.1034E 07	0.1189E 07	0.1361E 07	0.1548E 07	0.1754E 07	0.1977E 07	0.2220E 07	0.2482E 07	0.2765E 07	0.3069E 07
Luminance	0.1034E 07	0.1189E 07	0.1361E 07	0.1548E 07	0.1754E 07	0.1977E 07	0.2220E 07	0.2482E 07	0.2765E 07	0.3069E 07
0.41	0.3513E 07	0.4023E 07	0.4584E 07	0.5198E 07	0.5867E 07	0.6592E 07	0.7377E 07	0.8223E 07	0.9132E 07	0.1010E 08
0.42	0.1503E 08	0.1717E 08	0.1952E 08	0.2209E 08	0.2488E 08	0.2790E 08	0.3116E 08	0.3467E 08	0.3843E 08	0.4245E 08
0.43	0.4283E 08	0.4883E 08	0.5538E 08	0.6252E 08	0.7027E 08	0.7864E 08	0.8766E 08	0.9734E 08	0.1077E 09	0.1187E 09
0.44	0.9480E 08	0.1078E 09	0.1220E 09	0.1374E 09	0.1542E 09	0.1722E 09	0.1916E 09	0.2124E 09	0.2346E 09	0.2583E 09
0.45	0.1745E 09	0.1981E 09	0.2237E 09	0.2516E 09	0.2817E 09	0.3141E 09	0.3489E 09	0.3861E 09	0.4258E 09	0.4681E 09
0.46	0.2812E 09	0.3186E 09	0.3592E 09	0.4032E 09	0.4506E 09	0.5016E 09	0.5563E 09	0.6147E 09	0.6769E 09	0.7431E 09
0.47	0.4135E 09	0.4676E 09	0.5262E 09	0.5897E 09	0.6580E 09	0.7314E 09	0.8099E 09	0.8936E 09	0.9828E 09	0.1077E 10
0.48	0.5730E 09	0.6468E 09	0.7267E 09	0.8130E 09	0.9058E 09	0.1005E 10	0.1111E 10	0.1224E 10	0.1345E 10	0.1473E 10
0.49	0.7608E 09	0.8574E 09	0.9618E 09	0.1074E 10	0.1195E 10	0.1324E 10	0.1462E 10	0.1609E 10	0.1765E 10	0.1930E 10
0.50	0.9730E 09	0.1094E 10	0.1226E 10	0.1367E 10	0.1519E 10	0.1681E 10	0.1854E 10	0.2038E 10	0.2233E 10	0.2439E 10
Luminance	0.9740E 09	0.1095E 10	0.1227E 10	0.1369E 10	0.1521E 10	0.1683E 10	0.1856E 10	0.2040E 10	0.2235E 10	0.2442E 10
0.51	0.2255E 09	0.2517E 09	0.2798E 09	0.3098E 09	0.3418E 09	0.3757E 09	0.4116E 09	0.4495E 09	0.4895E 09	0.5315E 09
0.52	0.4477E 09	0.4993E 09	0.5545E 09	0.6134E 09	0.6760E 09	0.7425E 09	0.8128E 09	0.8870E 09	0.9651E 09	0.1047E 10
0.53	0.6498E 09	0.7239E 09	0.8032E 09	0.8878E 09	0.9777E 09	0.1072E 10	0.1173E 10	0.1279E 10	0.1391E 10	0.1508E 10
0.54	0.8193E 09	0.9120E 09	0.1011E 10	0.1116E 10	0.1228E 10	0.1347E 10	0.1473E 10	0.1605E 10	0.1744E 10	0.1890E 10
0.55	0.9504E 09	0.1057E 10	0.1171E 10	0.1292E 10	0.1421E 10	0.1558E 10	0.1702E 10	0.1854E 10	0.2014E 10	0.2181E 10
0.56	0.1044E 10	0.1160E 10	0.1285E 10	0.1418E 10	0.1558E 10	0.1707E 10	0.1865E 10	0.2030E 10	0.2204E 10	0.2387E 10
0.57	0.1105E 10	0.1229E 10	0.1360E 10	0.1500E 10	0.1648E 10	0.1805E 10	0.1971E 10	0.2145E 10	0.2328E 10	0.2520E 10
0.58	0.1143E 10	0.1270E 10	0.1405E 10	0.1550E 10	0.1702E 10	0.1864E 10	0.2035E 10	0.2214E 10	0.2403E 10	0.2600E 10
0.59	0.1164E 10	0.1293E 10	0.1431E 10	0.1577E 10	0.1733E 10	0.1897E 10	0.2070E 10	0.2253E 10	0.2444E 10	0.2645E 10
0.60	0.1175E 10	0.1305E 10	0.1444E 10	0.1592E 10	0.1748E 10	0.1914E 10	0.2089E 10	0.2273E 10	0.2466E 10	0.2668E 10
Luminance	0.2149E 10	0.2401E 10	0.2672E 10	0.2961E 10	0.3270E 10	0.3598E 10	0.3945E 10	0.4313E 10	0.4701E 10	0.5110E 10
0.61	0.5420E 07	0.5947E 07	0.6502E 07	0.7085E 07	0.7697E 07	0.8338E 07	0.9007E 07	0.9704E 07	0.1042E 08	0.1118E 08
0.62	0.7964E 07	0.8734E 07	0.9545E 07	0.1039E 08	0.1129E 08	0.1222E 08	0.1320E 08	0.1422E 08	0.1527E 08	0.1637E 08
0.63	0.9121E 07	0.1000E 08	0.1092E 08	0.1189E 08	0.1291E 08	0.1398E 08	0.1509E 08	0.1625E 08	0.1746E 08	0.1871E 08
0.64	0.9638E 07	0.1056E 08	0.1154E 08	0.1256E 08	0.1364E 08	0.1476E 08	0.1593E 08	0.1715E 08	0.1842E 08	0.1974E 08
0.65	0.9868E 07	0.1081E 08	0.1181E 08	0.1286E 08	0.1396E 08	0.1510E 08	0.1630E 08	0.1755E 08	0.1885E 08	0.2019E 08
0.66	0.9972E 07	0.1092E 08	0.1193E 08	0.1299E 08	0.1410E 08	0.1526E 08	0.1647E 08	0.1773E 08	0.1904E 08	0.2040E 08
0.67	0.1002E 08	0.1098E 08	0.1199E 08	0.1305E 08	0.1417E 08	0.1533E 08	0.1655E 08	0.1781E 08	0.1913E 08	0.2049E 08
0.68	0.1004E 08	0.1100E 08	0.1202E 08	0.1308E 08	0.1420E 08	0.1537E 08	0.1658E 08	0.1785E 08	0.1917E 08	0.2054E 08
0.69	0.1005E 08	0.1101E 08	0.1203E 08	0.1310E 08	0.1421E 08	0.1538E 08	0.1660E 08	0.1787E 08	0.1919E 08	0.2056E 08
0.70	0.1006E 08	0.1102E 08	0.1204E 08	0.1310E 08	0.1422E 08	0.1539E 08	0.1661E 08	0.1788E 08	0.1920E 08	0.2057E 08
Luminance	0.2159E 10	0.2412E 10	0.2684E 10	0.2974E 10	0.3284E 10	0.3613E 10	0.3962E 10	0.4331E 10	0.4721E 10	0.5131E 10
0.71	0.2712E 04	0.2939E 04	0.3176E 04	0.3422E 04	0.3678E 04	0.3943E 04	0.4217E 04	0.4499E 04	0.4791E 04	0.5091E 04
0.72	0.4098E 04	0.4440E 04	0.4796E 04	0.5166E 04	0.5550E 04	0.5948E 04	0.6360E 04	0.6785E 04	0.7223E 04	0.7674E 04
0.73	0.4819E 04	0.5220E 04	0.5637E 04	0.6071E 04	0.6521E 04	0.6987E 04	0.7469E 04	0.7966E 04	0.8479E 04	0.9006E 04
0.74	0.5200E 04	0.5632E 04	0.6081E 04	0.6548E 04	0.7032E 04	0.7534E 04	0.8052E 04	0.8587E 04	0.9138E 04	0.9706E 04
0.75	0.5405E 04	0.5853E 04	0.6319E 04	0.6804E 04	0.7306E 04	0.7827E 04	0.8364E 04	0.8919E 04	0.9491E 04	0.1008E 05
0.76	0.5517E 04	0.5974E 04	0.6449E 04	0.6943E 04	0.7456E 04	0.7986E 04	0.8534E 04	0.9100E 04	0.9682E 04	0.1028E 05
0.77	0.5580E 04	0.6041E 04	0.6521E 04	0.7020E 04	0.7538E 04	0.8074E 04	0.8628E 04	0.9199E 04	0.9788E 04	0.1039E 05
0.78	0.5615E 04	0.6079E 04	0.6562E 04	0.7064E 04	0.7584E 04	0.8123E 04	0.8680E 04	0.9255E 04	0.9847E 04	0.1045E 05
Luminance	0.2159E 10	0.2412E 10	0.2684E 10	0.2974E 10	0.3284E 10	0.3613E 10	0.3962E 10	0.4331E 10	0.4721E 10	0.5131E 10

xEn = x 10^n. E.g. 1.234E 05 = 1.234 x 10^5

$\int_o^\lambda L_{v'\lambda} d\lambda$

$\int_0^\lambda L_{v\lambda}\, d\lambda$

TABLE 18

Blackbody Fractional Luminance, Photopic (cd/m^2)

λ(μ)	6000°K.	6100°K.	6200°K.	6300°K.	6400°K.	6500°K.	6600°K.	6700°K.	6800°K.	6900°K.
0.36	0.0000E 00	0.0000E 00	0.0000E 00	0.0000E 00	0.0000E 00	0.0000E 00	0.0000E 00	0.0000E 00	0.0000E 00	0.0000E 00
0.37	0.1461E 04	0.1623E 04	0.1796E 04	0.1981E 04	0.2179E 04	0.2390E 04	0.2613E 04	0.2850E 04	0.3101E 04	0.3365E 04
0.38	0.6256E 04	0.6933E 04	0.7657E 04	0.8431E 04	0.9255E 04	0.1013E 05	0.1106E 05	0.1204E 05	0.1308E 05	0.1417E 05
0.39	0.2092E 05	0.2313E 05	0.2549E 05	0.2801E 05	0.3069E 05	0.3352E 05	0.3653E 05	0.3970E 05	0.4305E 05	0.4657E 05
0.40	0.7114E 05	0.7847E 05	0.8628E 05	0.9460E 05	0.1034E 06	0.1127E 06	0.1226E 06	0.1329E 06	0.1439E 06	0.1554E 06
Luminance	0.7114E 05	0.7847E 05	0.8628E 05	0.9460E 05	0.1034E 06	0.1127E 06	0.1226E 06	0.1329E 06	0.1439E 06	0.1554E 06
0.41	0.1537E 06	0.1690E 06	0.1853E 06	0.2025E 06	0.2207E 06	0.2400E 06	0.2602E 06	0.2815E 06	0.3039E 06	0.3273E 06
0.42	0.6712E 06	0.7367E 06	0.8063E 06	0.8798E 06	0.9575E 06	0.1039E 07	0.1125E 07	0.1216E 07	0.1310E 07	0.1409E 07
0.43	0.2354E 07	0.2579E 07	0.2817E 07	0.3069E 07	0.3335E 07	0.3614E 07	0.3908E 07	0.4216E 07	0.4537E 07	0.4874E 07
0.44	0.6212E 07	0.6795E 07	0.7411E 07	0.8061E 07	0.8746E 07	0.9465E 07	0.1021E 08	0.1100E 08	0.1183E 08	0.1269E 08
0.45	0.1306E 08	0.1426E 08	0.1553E 08	0.1687E 08	0.1828E 08	0.1976E 08	0.2130E 08	0.2292E 08	0.2461E 08	0.2637E 08
0.46	0.2412E 08	0.2630E 08	0.2861E 08	0.3103E 08	0.3358E 08	0.3624E 08	0.3903E 08	0.4195E 08	0.4499E 08	0.4815E 08
0.47	0.4117E 08	0.4483E 08	0.4869E 08	0.5275E 08	0.5700E 08	0.6145E 08	0.6610E 08	0.7096E 08	0.7601E 08	0.8127E 08
0.48	0.6714E 08	0.7301E 08	0.7919E 08	0.8567E 08	0.9246E 08	0.9956E 08	0.1069E 09	0.1146E 09	0.1227E 09	0.1310E 09
0.49	0.1061E 09	0.1152E 09	0.1248E 09	0.1348E 09	0.1453E 09	0.1563E 09	0.1677E 09	0.1796E 09	0.1920E 09	0.2048E 09
0.50	0.1654E 09	0.1793E 09	0.1940E 09	0.2093E 09	0.2253E 09	0.2420E 09	0.2594E 09	0.2775E 09	0.2963E 09	0.3158E 09
Luminance	0.1654E 09	0.1794E 09	0.1941E 09	0.2094E 09	0.2254E 09	0.2421E 09	0.2595E 09	0.2776E 09	0.2964E 09	0.3159E 09
0.51	0.9242E 08	0.9982E 08	0.1075E 09	0.1156E 09	0.1240E 09	0.1327E 09	0.1418E 09	0.1511E 09	0.1608E 09	0.1709E 09
0.52	0.2286E 09	0.2467E 09	0.2656E 09	0.2852E 09	0.3057E 09	0.3270E 09	0.3491E 09	0.3719E 09	0.3956E 09	0.4200E 09
0.53	0.4040E 09	0.4357E 09	0.4687E 09	0.5030E 09	0.5388E 09	0.5759E 09	0.6143E 09	0.6541E 09	0.6952E 09	0.7377E 09
0.54	0.6043E 09	0.6511E 09	0.6999E 09	0.7507E 09	0.8034E 09	0.8581E 09	0.9148E 09	0.9735E 09	0.1034E 10	0.1096E 10
0.55	0.8164E 09	0.8790E 09	0.9442E 09	0.1012E 10	0.1082E 10	0.1155E 10	0.1231E 10	0.1309E 10	0.1389E 10	0.1473E 10
0.56	0.1030E 10	0.1108E 10	0.1189E 10	0.1274E 10	0.1362E 10	0.1453E 10	0.1547E 10	0.1644E 10	0.1745E 10	0.1848E 10
0.57	0.1236E 10	0.1329E 10	0.1426E 10	0.1526E 10	0.1631E 10	0.1739E 10	0.1851E 10	0.1966E 10	0.2085E 10	0.2208E 10
0.58	0.1426E 10	0.1533E 10	0.1643E 10	0.1758E 10	0.1878E 10	0.2001E 10	0.2129E 10	0.2261E 10	0.2396E 10	0.2536E 10
0.59	0.1593E 10	0.1711E 10	0.1834E 10	0.1962E 10	0.2094E 10	0.2230E 10	0.2372E 10	0.2517E 10	0.2668E 10	0.2822E 10
0.60	0.1733E 10	0.1861E 10	0.1993E 10	0.2131E 10	0.2274E 10	0.2421E 10	0.2574E 10	0.2731E 10	0.2893E 10	0.3060E 10
Luminance	0.1898E 10	0.2040E 10	0.2188E 10	0.2341E 10	0.2499E 10	0.2664E 10	0.2834E 10	0.3009E 10	0.3190E 10	0.3376E 10
0.61	0.1122E 09	0.1198E 09	0.1276E 09	0.1357E 09	0.1440E 09	0.1526E 09	0.1614E 09	0.1705E 09	0.1797E 09	0.1893E 09
0.62	0.1981E 09	0.2114E 09	0.2251E 09	0.2393E 09	0.2539E 09	0.2689E 09	0.2843E 09	0.3001E 09	0.3164E 09	0.3330E 09
0.63	0.2595E 09	0.2768E 09	0.2947E 09	0.3131E 09	0.3321E 09	0.3517E 09	0.3717E 09	0.3923E 09	0.4134E 09	0.4351E 09
0.64	0.3004E 09	0.3203E 09	0.3409E 09	0.3621E 09	0.3840E 09	0.4065E 09	0.4296E 09	0.4533E 09	0.4776E 09	0.5025E 09
0.65	0.3259E 09	0.3475E 09	0.3698E 09	0.3927E 09	0.4164E 09	0.4407E 09	0.4656E 09	0.4912E 09	0.5175E 09	0.5444E 09
0.66	0.3408E 09	0.3633E 09	0.3865E 09	0.4104E 09	0.4351E 09	0.4604E 09	0.4864E 09	0.5131E 09	0.5405E 09	0.5685E 09
0.67	0.3487E 09	0.3717E 09	0.3955E 09	0.4199E 09	0.4451E 09	0.4710E 09	0.4976E 09	0.5249E 09	0.5528E 09	0.5814E 09
0.68	0.3528E 09	0.3761E 09	0.4001E 09	0.4248E 09	0.4502E 09	0.4764E 09	0.5033E 09	0.5308E 09	0.5591E 09	0.5880E 09
0.69	0.3549E 09	0.3783E 09	0.4024E 09	0.4272E 09	0.4528E 09	0.4791E 09	0.5061E 09	0.5338E 09	0.5622E 09	0.5913E 09
0.70	0.3559E 09	0.3793E 09	0.4035E 09	0.4284E 09	0.4540E 09	0.4804E 09	0.5075E 09	0.5352E 09	0.5637E 09	0.5928E 09
Luminance	0.2254E 10	0.2420E 10	0.2591E 10	0.2769E 10	0.2953E 10	0.3144E 10	0.3341E 10	0.3544E 10	0.3754E 10	0.3969E 10
0.71	0.4840E 06	0.5125E 06	0.5416E 06	0.5715E 06	0.6021E 06	0.6334E 06	0.6653E 06	0.6979E 06	0.7312E 06	0.7651E 06
0.72	0.7239E 06	0.7662E 06	0.8096E 06	0.8541E 06	0.8997E 06	0.9462E 06	0.9938E 06	0.1042E 07	0.1091E 07	0.1142E 07
0.73	0.8409E 06	0.8899E 06	0.9402E 06	0.9917E 06	0.1044E 07	0.1098E 07	0.1153E 07	0.1209E 07	0.1266E 07	0.1325E 07
0.74	0.8969E 06	0.9491E 06	0.1002E 07	0.1057E 07	0.1113E 07	0.1170E 07	0.1229E 07	0.1289E 07	0.1350E 07	0.1412E 07
0.75	0.9230E 06	0.9766E 06	0.1031E 07	0.1088E 07	0.1145E 07	0.1204E 07	0.1264E 07	0.1326E 07	0.1388E 07	0.1452E 07
0.76	0.9355E 06	0.9899E 06	0.1045E 07	0.1102E 07	0.1161E 07	0.1220E 07	0.1281E 07	0.1344E 07	0.1407E 07	0.1472E 07
0.77	0.9416E 06	0.9963E 06	0.1052E 07	0.1109E 07	0.1168E 07	0.1228E 07	0.1290E 07	0.1352E 07	0.1416E 07	0.1481E 07
0.78	0.9446E 06	0.9995E 06	0.1055E 07	0.1113E 07	0.1172E 07	0.1232E 07	0.1294E 07	0.1356E 07	0.1420E 07	0.1486E 07
0.79	0.9461E 06	0.1001E 07	0.1057E 07	0.1115E 07	0.1174E 07	0.1234E 07	0.1296E 07	0.1358E 07	0.1422E 07	0.1488E 07
0.80	0.9468E 06	0.1001E 07	0.1058E 07	0.1115E 07	0.1174E 07	0.1235E 07	0.1296E 07	0.1359E 07	0.1424E 07	0.1489E 07
Luminance	0.2255E 10	0.2421E 10	0.2592E 10	0.2770E 10	0.2955E 10	0.3145E 10	0.3342E 10	0.3546E 10	0.3755E 10	0.3970E 10
0.81	0.3434E 03	0.3614E 03	0.3797E 03	0.3984E 03	0.4175E 03	0.4369E 03	0.4566E 03	0.4766E 03	0.4970E 03	0.5177E 03
0.82	0.5098E 03	0.5363E 03	0.5635E 03	0.5911E 03	0.6193E 03	0.6480E 03	0.6772E 03	0.7068E 03	0.7369E 03	0.7674E 03
0.83	0.5903E 03	0.6211E 03	0.6524E 03	0.6844E 03	0.7169E 03	0.7500E 03	0.7837E 03	0.8179E 03	0.8527E 03	0.8879E 03
Luminance	0.2255E 10	0.2421E 10	0.2592E 10	0.2770E 10	0.2955E 10	0.3145E 10	0.3342E 10	0.3546E 10	0.3755E 10	0.3970E 10

xEn = x 10^n. E.g. 1.234E 05 = 1.234 x 10^5

TABLE 18

Blackbody Fractional Luminance, Scotopic (sc.cd/m^2)

$\lambda(\mu)$	6000°K.	6100°K.	6200°K.	6300°K.	6400°K.	6500°K.	6600°K.	6700°K.	6800°K.	6900°K.
0.38	0.0000E 00	0.0000E 00	0.0000E 00	0.0000E 00	0.0000E 00	0.0000E 00	0.0000E 00	0.0000E 00	0.0000E 00	0.0000E 00
0.39	0.6531E 06	0.7215E 06	0.7944E 06	0.8720E 06	0.9545E 06	0.1041E 07	0.1134E 07	0.1231E 07	0.1334E 07	0.1442E 07
0.40	0.3395E 07	0.3742E 07	0.4113E 07	0.4507E 07	0.4924E 07	0.5366E 07	0.5832E 07	0.6324E 07	0.6840E 07	0.7382E 07
Luminance	0.3395E 07	0.3742E 07	0.4113E 07	0.4507E 07	0.4924E 07	0.5366E 07	0.5832E 07	0.6324E 07	0.6840E 07	0.7382E 07
0.41	0.1114E 08	0.1225E 08	0.1343E 08	0.1467E 08	0.1600E 08	0.1739E 08	0.1886E 08	0.2040E 08	0.2202E 08	0.2371E 08
0.42	0.4674E 08	0.5131E 08	0.5615E 08	0.6128E 08	0.6669E 08	0.7240E 08	0.7840E 08	0.8470E 08	0.9130E 08	0.9821E 08
0.43	0.1305E 09	0.1430E 09	0.1563E 09	0.1703E 09	0.1851E 09	0.2006E 09	0.2170E 09	0.2341E 09	0.2520E 09	0.2708E 09
0.44	0.2835E 09	0.3102E 09	0.3384E 09	0.3682E 09	0.3996E 09	0.4326E 09	0.4672E 09	0.5035E 09	0.5414E 09	0.5809E 09
0.45	0.5129E 09	0.5604E 09	0.6106E 09	0.6635E 09	0.7191E 09	0.7776E 09	0.8388E 09	0.9028E 09	0.9696E 09	0.1039E 10
0.46	0.8132E 09	0.8874E 09	0.9656E 09	0.1047E 10	0.1134E 10	0.1225E 10	0.1320E 10	0.1419E 10	0.1522E 10	0.1630E 10
0.47	0.1177E 10	0.1283E 10	0.1394E 10	0.1512E 10	0.1635E 10	0.1764E 10	0.1898E 10	0.2039E 10	0.2186E 10	0.2338E 10
0.48	0.1608E 10	0.1750E 10	0.1900E 10	0.2057E 10	0.2222E 10	0.2395E 10	0.2576E 10	0.2764E 10	0.2960E 10	0.3164E 10
0.49	0.2105E 10	0.2289E 10	0.2482E 10	0.2685E 10	0.2897E 10	0.3119E 10	0.3351E 10	0.3593E 10	0.3844E 10	0.4105E 10
0.50	0.2656E 10	0.2885E 10	0.3125E 10	0.3377E 10	0.3641E 10	0.3917E 10	0.4204E 10	0.4503E 10	0.4814E 10	0.5137E 10
Luminance	0.2659E 10	0.2889E 10	0.3129E 10	0.3382E 10	0.3646E 10	0.3922E 10	0.4210E 10	0.4510E 10	0.4821E 10	0.5145E 10
0.51	0.5756E 09	0.6217E 09	0.6699E 09	0.7201E 09	0.7725E 09	0.8269E 09	0.8833E 09	0.9418E 09	0.1002E 10	0.1065E 10
0.52	0.1133E 10	0.1223E 10	0.1317E 10	0.1414E 10	0.1516E 10	0.1622E 10	0.1732E 10	0.1846E 10	0.1963E 10	0.2085E 10
0.53	0.1631E 10	0.1760E 10	0.1894E 10	0.2033E 10	0.2179E 10	0.2329E 10	0.2486E 10	0.2648E 10	0.2815E 10	0.2988E 10
0.54	0.2043E 10	0.2203E 10	0.2369E 10	0.2542E 10	0.2723E 10	0.2910E 10	0.3104E 10	0.3304E 10	0.3512E 10	0.3726E 10
0.55	0.2357E 10	0.2540E 10	0.2730E 10	0.2929E 10	0.3135E 10	0.3350E 10	0.3571E 10	0.3801E 10	0.4038E 10	0.4283E 10
0.56	0.2577E 10	0.2777E 10	0.2984E 10	0.3200E 10	0.3425E 10	0.3657E 10	0.3898E 10	0.4148E 10	0.4405E 10	0.4671E 10
0.57	0.2721E 10	0.2930E 10	0.3149E 10	0.3376E 10	0.3612E 10	0.3856E 10	0.4110E 10	0.4371E 10	0.4642E 10	0.4921E 10
0.58	0.2807E 10	0.3023E 10	0.3247E 10	0.3481E 10	0.3723E 10	0.3975E 10	0.4236E 10	0.4505E 10	0.4783E 10	0.5070E 10
0.59	0.2855E 10	0.3074E 10	0.3302E 10	0.3539E 10	0.3785E 10	0.4041E 10	0.4305E 10	0.4578E 10	0.4861E 10	0.5152E 10
0.60	0.2879E 10	0.3100E 10	0.3330E 10	0.3569E 10	0.3817E 10	0.4074E 10	0.4341E 10	0.4616E 10	0.4900E 10	0.5193E 10
Luminance	0.5539E 10	0.5989E 10	0.6460E 10	0.6951E 10	0.7464E 10	0.7997E 10	0.8551E 10	0.9126E 10	0.9722E 10	0.1033E 11
0.61	0.1196E 08	0.1277E 08	0.1360E 08	0.1446E 08	0.1535E 08	0.1626E 08	0.1720E 08	0.1817E 08	0.1916E 08	0.2017E 08
0.62	0.1751E 08	0.1868E 08	0.1990E 08	0.2115E 08	0.2245E 08	0.2378E 08	0.2514E 08	0.2655E 08	0.2799E 08	0.2946E 08
0.63	0.2000E 08	0.2134E 08	0.2273E 08	0.2415E 08	0.2563E 08	0.2714E 08	0.2870E 08	0.3029E 08	0.3193E 08	0.3361E 08
0.64	0.2110E 08	0.2251E 08	0.2397E 08	0.2548E 08	0.2702E 08	0.2862E 08	0.3026E 08	0.3194E 08	0.3366E 08	0.3543E 08
0.65	0.2159E 08	0.2303E 08	0.2452E 08	0.2606E 08	0.2764E 08	0.2927E 08	0.3094E 08	0.3266E 08	0.3442E 08	0.3622E 08
0.66	0.2181E 08	0.2326E 08	0.2477E 08	0.2632E 08	0.2791E 08	0.2956E 08	0.3125E 08	0.3298E 08	0.3476E 08	0.3658E 08
0.67	0.2191E 08	0.2337E 08	0.2488E 08	0.2644E 08	0.2804E 08	0.2969E 08	0.3139E 08	0.3313E 08	0.3491E 08	0.3674E 08
0.68	0.2195E 08	0.2342E 08	0.2493E 08	0.2649E 08	0.2810E 08	0.2975E 08	0.3145E 08	0.3319E 08	0.3498E 08	0.3681E 08
0.69	0.2198E 08	0.2344E 08	0.2496E 08	0.2652E 08	0.2813E 08	0.2978E 08	0.3148E 08	0.3323E 08	0.3502E 08	0.3685E 08
0.70	0.2199E 08	0.2345E 08	0.2497E 08	0.2653E 08	0.2814E 08	0.2980E 08	0.3150E 08	0.3324E 08	0.3503E 08	0.3687E 08
Luminance	0.5561E 10	0.6013E 10	0.6485E 10	0.6978E 10	0.7492E 10	0.8027E 10	0.8583E 10	0.9159E 10	0.9757E 10	0.1037E 11
0.71	0.5400E 04	0.5717E 04	0.6043E 04	0.6376E 04	0.6717E 04	0.7066E 04	0.7423E 04	0.7787E 04	0.8158E 04	0.8536E 04
0.72	0.8137E 04	0.8613E 04	0.9101E 04	0.9601E 04	0.1011E 05	0.1063E 05	0.1117E 05	0.1171E 05	0.1227E 05	0.1283E 05
0.73	0.9548E 04	0.1010E 05	0.1067E 05	0.1126E 05	0.1185E 05	0.1247E 05	0.1309E 05	0.1373E 05	0.1438E 05	0.1504E 05
0.74	0.1028E 05	0.1088E 05	0.1150E 05	0.1213E 05	0.1277E 05	0.1343E 05	0.1410E 05	0.1478E 05	0.1548E 05	0.1619E 05
0.75	0.1068E 05	0.1130E 05	0.1194E 05	0.1259E 05	0.1326E 05	0.1394E 05	0.1463E 05	0.1534E 05	0.1607E 05	0.1681E 05
0.76	0.1089E 05	0.1153E 05	0.1218E 05	0.1284E 05	0.1352E 05	0.1421E 05	0.1492E 05	0.1565E 05	0.1638E 05	0.1714E 05
0.77	0.1101E 05	0.1165E 05	0.1231E 05	0.1298E 05	0.1366E 05	0.1437E 05	0.1508E 05	0.1581E 05	0.1656E 05	0.1732E 05
0.78	0.1108E 05	0.1172E 05	0.1238E 05	0.1305E 05	0.1374E 05	0.1445E 05	0.1517E 05	0.1590E 05	0.1665E 05	0.1742E 05
Luminance	0.5561E 10	0.6013E 10	0.6485E 10	0.6978E 10	0.7492E 10	0.8027E 10	0.8583E 10	0.9159E 10	0.9757E 10	0.1037E 11

xEn - x 10^n. E.g. 1.234E 05 = 1.234 x 10^5

$\int_o^\lambda L_{v'\lambda} d\lambda$

$\int_0^{\lambda} L_{v\lambda}\, d\lambda$

TABLE 18

Blackbody Fractional Luminance, Photopic (cd/m^2)

λ(μ)	7000°K.	7100°K.	7200°K.	7300°K.	7400°K.	7500°K.	7600°K.	7700°K.	7800°K.	7900°K.
0.36	0.0000E 00	0.0000E 00	0.0000E 00	0.0000E 00	0.0000E 00	0.0000E 00	0.0000E 00	0.0000E 00	0.0000E 00	0.0000E 00
0.37	0.3644E 04	0.3937E 04	0.4245E 04	0.4567E 04	0.4904E 04	0.5256E 04	0.5624E 04	0.6006E 04	0.6404E 04	0.6818E 04
0.38	0.1532E 05	0.1653E 05	0.1779E 05	0.1912E 05	0.2050E 05	0.2194E 05	0.2345E 05	0.2501E 05	0.2664E 05	0.2832E 05
0.39	0.5026E 05	0.5414E 05	0.5819E 05	0.6242E 05	0.6684E 05	0.7145E 05	0.7624E 05	0.8121E 05	0.8638E 05	0.9173E 05
0.40	0.1674E 06	0.1800E 06	0.1931E 06	0.2069E 06	0.2212E 06	0.2360E 06	0.2515E 06	0.2675E 06	0.2841E 06	0.3013E 06
Luminance	0.1674E 06	0.1800E 06	0.1931E 06	0.2069E 06	0.2212E 06	0.2360E 06	0.2515E 06	0.2675E 06	0.2841E 06	0.3013E 06
0.41	0.3517E 06	0.3773E 06	0.4039E 06	0.4316E 06	0.4604E 06	0.4903E 06	0.5213E 06	0.5534E 06	0.5866E 06	0.6209E 06
0.42	0.1513E 07	0.1621E 07	0.1733E 07	0.1850E 07	0.1971E 07	0.2097E 07	0.2227E 07	0.2362E 07	0.2501E 07	0.2645E 07
0.43	0.5225E 07	0.5590E 07	0.5970E 07	0.6364E 07	0.6773E 07	0.7197E 07	0.7635E 07	0.8088E 07	0.8555E 07	0.9037E 07
0.44	0.1359E 08	0.1452E 08	0.1549E 08	0.1649E 08	0.1754E 08	0.1861E 08	0.1973E 08	0.2088E 08	0.2206E 08	0.2328E 08
0.45	0.2820E 08	0.3011E 08	0.3208E 08	0.3413E 08	0.3625E 08	0.3844E 08	0.4070E 08	0.4303E 08	0.4543E 08	0.4791E 08
0.46	0.5144E 08	0.5486E 08	0.5839E 08	0.6206E 08	0.6585E 08	0.6976E 08	0.7380E 08	0.7796E 08	0.8224E 08	0.8665E 08
0.47	0.8673E 08	0.9239E 08	0.9825E 08	0.1043E 09	0.1105E 09	0.1170E 09	0.1237E 09	0.1305E 09	0.1376E 09	0.1448E 09
0.48	0.1397E 09	0.1486E 09	0.1579E 09	0.1675E 09	0.1774E 09	0.1876E 09	0.1981E 09	0.2089E 09	0.2201E 09	0.2315E 09
0.49	0.2181E 09	0.2319E 09	0.2461E 09	0.2608E 09	0.2760E 09	0.2916E 09	0.3077E 09	0.3242E 09	0.3412E 09	0.3586E 09
0.50	0.3359E 09	0.3568E 09	0.3783E 09	0.4005E 09	0.4234E 09	0.4469E 09	0.4711E 09	0.4960E 09	0.5215E 09	0.5477E 09
Luminance	0.3361E 09	0.3569E 09	0.3785E 09	0.4007E 09	0.4236E 09	0.4471E 09	0.4714E 09	0.4963E 09	0.5218E 09	0.5480E 09
0.51	0.1813E 09	0.1919E 09	0.2029E 09	0.2143E 09	0.2259E 09	0.2378E 09	0.2501E 09	0.2626E 09	0.2755E 09	0.2887E 09
0.52	0.4452E 09	0.4712E 09	0.4979E 09	0.5254E 09	0.5536E 09	0.5825E 09	0.6122E 09	0.6427E 09	0.6738E 09	0.7056E 09
0.53	0.7814E 09	0.8265E 09	0.8729E 09	0.9206E 09	0.9695E 09	0.1019E 10	0.1071E 10	0.1123E 10	0.1177E 10	0.1232E 10
0.54	0.1161E 10	0.1227E 10	0.1295E 10	0.1365E 10	0.1437E 10	0.1511E 10	0.1586E 10	0.1663E 10	0.1742E 10	0.1823E 10
0.55	0.1558E 10	0.1646E 10	0.1737E 10	0.1830E 10	0.1926E 10	0.2023E 10	0.2124E 10	0.2226E 10	0.2331E 10	0.2438E 10
0.56	0.1955E 10	0.2065E 10	0.2177E 10	0.2293E 10	0.2411E 10	0.2533E 10	0.2657E 10	0.2784E 10	0.2914E 10	0.3046E 10
0.57	0.2334E 10	0.2464E 10	0.2597E 10	0.2734E 10	0.2874E 10	0.3018E 10	0.3164E 10	0.3315E 10	0.3468E 10	0.3624E 10
0.58	0.2680E 10	0.2828E 10	0.2980E 10	0.3136E 10	0.3295E 10	0.3459E 10	0.3626E 10	0.3797E 10	0.3971E 10	0.4149E 10
0.59	0.2981E 10	0.3145E 10	0.3313E 10	0.3485E 10	0.3661E 10	0.3841E 10	0.4026E 10	0.4214E 10	0.4406E 10	0.4603E 10
0.60	0.3232E 10	0.3408E 10	0.3589E 10	0.3774E 10	0.3964E 10	0.4158E 10	0.4356E 10	0.4559E 10	0.4766E 10	0.4977E 10
Luminance	0.3568E 10	0.3765E 10	0.3967E 10	0.4175E 10	0.4387E 10	0.4605E 10	0.4828E 10	0.5055E 10	0.5288E 10	0.5525E 10
0.61	0.1990E 09	0.2090E 09	0.2191E 09	0.2295E 09	0.2402E 09	0.2510E 09	0.2620E 09	0.2732E 09	0.2847E 09	0.2963E 09
0.62	0.3500E 09	0.3675E 09	0.3853E 09	0.4034E 09	0.4220E 09	0.4409E 09	0.4601E 09	0.4797E 09	0.4997E 09	0.5199E 09
0.63	0.4572E 09	0.4798E 09	0.5030E 09	0.5266E 09	0.5506E 09	0.5752E 09	0.6002E 09	0.6256E 09	0.6515E 09	0.6778E 09
0.64	0.5279E 09	0.5540E 09	0.5805E 09	0.6077E 09	0.6353E 09	0.6635E 09	0.6923E 09	0.7215E 09	0.7512E 09	0.7814E 09
0.65	0.5719E 09	0.6000E 09	0.6287E 09	0.6580E 09	0.6878E 09	0.7183E 09	0.7493E 09	0.7808E 09	0.8129E 09	0.8455E 09
0.66	0.5972E 09	0.6265E 09	0.6564E 09	0.6869E 09	0.7180E 09	0.7498E 09	0.7820E 09	0.8149E 09	0.8483E 09	0.8823E 09
0.67	0.6107E 09	0.6406E 09	0.6712E 09	0.7023E 09	0.7341E 09	0.7665E 09	0.7995E 09	0.8330E 09	0.8671E 09	0.9018E 09
0.68	0.6176E 09	0.6478E 09	0.6787E 09	0.7102E 09	0.7423E 09	0.7750E 09	0.8083E 09	0.8422E 09	0.8767E 09	0.9117E 09
0.69	0.6210E 09	0.6514E 09	0.6824E 09	0.7141E 09	0.7464E 09	0.7792E 09	0.8127E 09	0.8468E 09	0.8814E 09	0.9166E 09
0.70	0.6226E 09	0.6531E 09	0.6842E 09	0.7159E 09	0.7483E 09	0.7812E 09	0.8148E 09	0.8489E 09	0.8837E 09	0.9189E 09
Luminance	0.4190E 10	0.4418E 10	0.4651E 10	0.4891E 10	0.5136E 10	0.5386E 10	0.5643E 10	0.5904E 10	0.6172E 10	0.6444E 10
0.71	0.7996E 06	0.8347E 06	0.8705E 06	0.9068E 06	0.9436E 06	0.9811E 06	0.1019E 07	0.1057E 07	0.1096E 07	0.1136E 07
0.72	0.1193E 07	0.1245E 07	0.1298E 07	0.1352E 07	0.1407E 07	0.1463E 07	0.1519E 07	0.1576E 07	0.1634E 07	0.1693E 07
0.73	0.1384E 07	0.1445E 07	0.1506E 07	0.1568E 07	0.1632E 07	0.1696E 07	0.1761E 07	0.1827E 07	0.1895E 07	0.1962E 07
0.74	0.1475E 07	0.1539E 07	0.1605E 07	0.1671E 07	0.1738E 07	0.1807E 07	0.1876E 07	0.1947E 07	0.2018E 07	0.2090E 07
0.75	0.1517E 07	0.1583E 07	0.1650E 07	0.1719E 07	0.1788E 07	0.1858E 07	0.1929E 07	0.2002E 07	0.2075E 07	0.2149E 07
0.76	0.1537E 07	0.1604E 07	0.1672E 07	0.1741E 07	0.1811E 07	0.1883E 07	0.1955E 07	0.2028E 07	0.2102E 07	0.2177E 07
0.77	0.1547E 07	0.1614E 07	0.1683E 07	0.1752E 07	0.1823E 07	0.1894E 07	0.1967E 07	0.2041E 07	0.2115E 07	0.2191E 07
0.78	0.1552E 07	0.1619E 07	0.1688E 07	0.1758E 07	0.1828E 07	0.1900E 07	0.1973E 07	0.2047E 07	0.2122E 07	0.2197E 07
0.79	0.1554E 07	0.1622E 07	0.1690E 07	0.1760E 07	0.1831E 07	0.1903E 07	0.1976E 07	0.2050E 07	0.2125E 07	0.2201E 07
0.80	0.1555E 07	0.1623E 07	0.1692E 07	0.1762E 07	0.1832E 07	0.1904E 07	0.1977E 07	0.2051E 07	0.2126E 07	0.2202E 07
Luminance	0.4192E 10	0.4420E 10	0.4653E 10	0.4892E 10	0.5137E 10	0.5388E 10	0.5644E 10	0.5906E 10	0.6174E 10	0.6447E 10
0.81	0.5386E 03	0.5599E 03	0.5814E 03	0.6033E 03	0.6253E 03	0.6477E 03	0.6703E 03	0.6931E 03	0.7162E 03	0.7396E 03
0.82	0.7984E 03	0.8298E 03	0.8617E 03	0.8939E 03	0.9265E 03	0.9595E 03	0.9929E 03	0.1026E 04	0.1060E 04	0.1095E 04
0.83	0.9237E 03	0.9600E 03	0.9967E 03	0.1033E 04	0.1071E 04	0.1109E 04	0.1148E 04	0.1187E 04	0.1226E 04	0.1266E 04
Luminance	0.4192E 10	0.4420E 10	0.4653E 10	0.4892E 10	0.5137E 10	0.5388E 10	0.5644E 10	0.5906E 10	0.6174E 10	0.6447E 10

xEn = x 10^n. E.g. 1.234E 05 = 1.234 x 10^5

TABLE 18

Blackbody Fractional Luminance, Scotopic (sc.cd/m²)

λ(μ)	7000°K.	7100°K.	7200°K.	7300°K.	7400°K.	7500°K.	7600°K.	7700°K.	7800°K.	7900°K.
0.38	0.0000E 00	0.0000E 00	0.0000E 00	0.0000E 00	0.0000E 00	0.0000E 00	0.0000E 00	0.0000E 00	0.0000E 00	0.0000E 00
0.39	0.1555E 07	0.1674E 07	0.1798E 07	0.1928E 07	0.2063E 07	0.2203E 07	0.2350E 07	0.2502E 07	0.2659E 07	0.2822E 07
0.40	0.7951E 07	0.8545E 07	0.9166E 07	0.9813E 07	0.1048E 08	0.1118E 08	0.1191E 08	0.1267E 08	0.1345E 08	0.1426E 08
Luminance	0.7951E 07	0.8545E 07	0.9166E 07	0.9813E 07	0.1048E 08	0.1118E 08	0.1191E 08	0.1267E 08	0.1345E 08	0.1426E 08
0.41	0.2549E 08	0.2734E 08	0.2927E 08	0.3127E 08	0.3336E 08	0.3553E 08	0.3777E 08	0.4010E 08	0.4250E 08	0.4499E 08
0.42	0.1054E 09	0.1129E 09	0.1207E 09	0.1289E 09	0.1373E 09	0.1461E 09	0.1552E 09	0.1646E 09	0.1743E 09	0.1843E 09
0.43	0.2903E 09	0.3106E 09	0.3318E 09	0.3538E 09	0.3766E 09	0.4002E 09	0.4246E 09	0.4499E 09	0.4760E 09	0.5029E 09
0.44	0.6221E 09	0.6650E 09	0.7095E 09	0.7557E 09	0.8036E 09	0.8532E 09	0.9044E 09	0.9573E 09	0.1011E 10	0.1068E 10
0.45	0.1111E 10	0.1187E 10	0.1265E 10	0.1346E 10	0.1430E 10	0.1517E 10	0.1607E 10	0.1699E 10	0.1795E 10	0.1893E 10
0.46	0.1742E 10	0.1859E 10	0.1980E 10	0.2105E 10	0.2234E 10	0.2368E 10	0.2506E 10	0.2648E 10	0.2794E 10	0.2945E 10
0.47	0.2497E 10	0.2661E 10	0.2832E 10	0.3008E 10	0.3190E 10	0.3379E 10	0.3573E 10	0.3773E 10	0.3979E 10	0.4190E 10
0.48	0.3375E 10	0.3595E 10	0.3822E 10	0.4057E 10	0.4299E 10	0.4549E 10	0.4807E 10	0.5073E 10	0.5346E 10	0.5627E 10
0.49	0.4376E 10	0.4657E 10	0.4947E 10	0.5247E 10	0.5557E 10	0.5877E 10	0.6205E 10	0.6544E 10	0.6892E 10	0.7249E 10
0.50	0.5472E 10	0.5818E 10	0.6177E 10	0.6546E 10	0.6928E 10	0.7321E 10	0.7726E 10	0.8142E 10	0.8570E 10	0.9009E 10
Luminance	0.5480E 10	0.5827E 10	0.6186E 10	0.6556E 10	0.6939E 10	0.7332E 10	0.7738E 10	0.8155E 10	0.8583E 10	0.9023E 10
0.51	0.1129E 10	0.1196E 10	0.1264E 10	0.1335E 10	0.1407E 10	0.1482E 10	0.1558E 10	0.1637E 10	0.1717E 10	0.1799E 10
0.52	0.2210E 10	0.2339E 10	0.2473E 10	0.2609E 10	0.2750E 10	0.2894E 10	0.3042E 10	0.3193E 10	0.3349E 10	0.3507E 10
0.53	0.3166E 10	0.3350E 10	0.3539E 10	0.3733E 10	0.3932E 10	0.4137E 10	0.4347E 10	0.4561E 10	0.4781E 10	0.5006E 10
0.54	0.3946E 10	0.4174E 10	0.4407E 10	0.4648E 10	0.4894E 10	0.5147E 10	0.5406E 10	0.5672E 10	0.5943E 10	0.6221E 10
0.55	0.4535E 10	0.4794E 10	0.5061E 10	0.5336E 10	0.5617E 10	0.5906E 10	0.6201E 10	0.6504E 10	0.6814E 10	0.7130E 10
0.56	0.4945E 10	0.5226E 10	0.5516E 10	0.5814E 10	0.6119E 10	0.6432E 10	0.6753E 10	0.7081E 10	0.7416E 10	0.7759E 10
0.57	0.5208E 10	0.5504E 10	0.5808E 10	0.6121E 10	0.6441E 10	0.6770E 10	0.7106E 10	0.7450E 10	0.7802E 10	0.8162E 10
0.58	0.5365E 10	0.5669E 10	0.5982E 10	0.6303E 10	0.6632E 10	0.6969E 10	0.7315E 10	0.7668E 10	0.8030E 10	0.8399E 10
0.59	0.5451E 10	0.5760E 10	0.6077E 10	0.6402E 10	0.6736E 10	0.7079E 10	0.7429E 10	0.7788E 10	0.8154E 10	0.8529E 10
0.60	0.5495E 10	0.5806E 10	0.6125E 10	0.6453E 10	0.6790E 10	0.7134E 10	0.7487E 10	0.7849E 10	0.8218E 10	0.8595E 10
Luminance	0.1097E 11	0.1163E 11	0.1231E 11	0.1301E 11	0.1372E 11	0.1446E 11	0.1522E 11	0.1600E 11	0.1680E 11	0.1761E 11
0.61	0.2121E 08	0.2228E 08	0.2336E 08	0.2447E 08	0.2560E 08	0.2676E 08	0.2793E 08	0.2913E 08	0.3035E 08	0.3159E 08
0.62	0.3097E 08	0.3252E 08	0.3410E 08	0.3571E 08	0.3736E 08	0.3903E 08	0.4074E 08	0.4248E 08	0.4425E 08	0.4605E 08
0.63	0.3533E 08	0.3708E 08	0.3888E 08	0.4071E 08	0.4258E 08	0.4449E 08	0.4643E 08	0.4841E 08	0.5042E 08	0.5246E 08
0.64	0.3724E 08	0.3908E 08	0.4097E 08	0.4290E 08	0.4487E 08	0.4687E 08	0.4891E 08	0.5099E 08	0.5311E 08	0.5526E 08
0.65	0.3807E 08	0.3996E 08	0.4189E 08	0.4386E 08	0.4587E 08	0.4791E 08	0.5000E 08	0.5212E 08	0.5428E 08	0.5648E 08
0.66	0.3844E 08	0.4035E 08	0.4229E 08	0.4428E 08	0.4631E 08	0.4837E 08	0.5048E 08	0.5262E 08	0.5480E 08	0.5702E 08
0.67	0.3861E 08	0.4052E 08	0.4248E 08	0.4447E 08	0.4651E 08	0.4858E 08	0.5070E 08	0.5285E 08	0.5504E 08	0.5726E 08
0.68	0.3869E 08	0.4061E 08	0.4256E 08	0.4456E 08	0.4660E 08	0.4868E 08	0.5080E 08	0.5295E 08	0.5514E 08	0.5737E 08
0.69	0.3873E 08	0.4065E 08	0.4261E 08	0.4461E 08	0.4665E 08	0.4873E 08	0.5085E 08	0.5300E 08	0.5520E 08	0.5743E 08
0.70	0.3874E 08	0.4066E 08	0.4262E 08	0.4463E 08	0.4667E 08	0.4875E 08	0.5087E 08	0.5303E 08	0.5522E 08	0.5745E 08
Luminance	0.1101E 11	0.1167E 11	0.1235E 11	0.1305E 11	0.1377E 11	0.1451E 11	0.1527E 11	0.1605E 11	0.1685E 11	0.1767E 11
0.71	0.8921E 04	0.9313E 04	0.9712E 04	0.1011E 05	0.1052E 05	0.1094E 05	0.1136E 05	0.1179E 05	0.1223E 05	0.1267E 05
0.72	0.1341E 05	0.1400E 05	0.1459E 05	0.1520E 05	0.1582E 05	0.1644E 05	0.1708E 05	0.1772E 05	0.1837E 05	0.1903E 05
0.73	0.1572E 05	0.1640E 05	0.1710E 05	0.1781E 05	0.1853E 05	0.1926E 05	0.2000E 05	0.2075E 05	0.2151E 05	0.2228E 05
0.74	0.1692E 05	0.1766E 05	0.1840E 05	0.1917E 05	0.1994E 05	0.2072E 05	0.2152E 05	0.2232E 05	0.2314E 05	0.2397E 05
0.75	0.1756E 05	0.1832E 05	0.1910E 05	0.1989E 05	0.2069E 05	0.2150E 05	0.2232E 05	0.2316E 05	0.2400E 05	0.2486E 05
0.76	0.1790E 05	0.1868E 05	0.1947E 05	0.2027E 05	0.2109E 05	0.2192E 05	0.2276E 05	0.2361E 05	0.2447E 05	0.2534E 05
0.77	0.1809E 05	0.1887E 05	0.1967E 05	0.2048E 05	0.2131E 05	0.2214E 05	0.2299E 05	0.2385E 05	0.2472E 05	0.2560E 05
0.78	0.1819E 05	0.1898E 05	0.1979E 05	0.2060E 05	0.2143E 05	0.2227E 05	0.2312E 05	0.2399E 05	0.2486E 05	0.2575E 05
Luminance	0.1101E 11	0.1167E 11	0.1235E 11	0.1305E 11	0.1377E 11	0.1451E 11	0.1527E 11	0.1605E 11	0.1685E 11	0.1767E 11

xEn = x 10^n. E.g. 1.234E 05 = 1.234 x 10^5

$\int_0^\lambda L_{v'\lambda} d\lambda$

TABLE 18

$\int_0^\lambda L_{v\lambda} d\lambda$

Blackbody Fractional Luminance, Photopic (cd/m²)

λ(μ)	8000°K.	8100°K.	8200°K.	8300°K.	8400°K.	8500°K.	8600°K.	8700°K.	8800°K.	8900°K.
0.36	0.0000E 00	0.0000E 00	0.0000E 00	0.0000E 00	0.0000E 00	0.0000E 00	0.0000E 00	0.0000E 00	0.0000E 00	0.0000E 00
0.37	0.7247E 04	0.7692E 04	0.8152E 04	0.8629E 04	0.9121E 04	0.9628E 04	0.1015E 05	0.1069E 05	0.1124E 05	0.1181E 05
0.38	0.3007E 05	0.3188E 05	0.3375E 05	0.3569E 05	0.3768E 05	0.3974E 05	0.4186E 05	0.4404E 05	0.4629E 05	0.4859E 05
0.39	0.9727E 05	0.1030E 06	0.1089E 06	0.1150E 06	0.1213E 06	0.1278E 06	0.1344E 06	0.1413E 06	0.1483E 06	0.1556E 06
0.40	0.3191E 06	0.3374E 06	0.3564E 06	0.3759E 06	0.3960E 06	0.4167E 06	0.4379E 06	0.4598E 06	0.4822E 06	0.5052E 06
Luminance	0.3191E 06	0.3374E 06	0.3564E 06	0.3759E 06	0.3960E 06	0.4167E 06	0.4379E 06	0.4598E 06	0.4822E 06	0.5052E 06
0.41	0.6563E 06	0.6928E 06	0.7304E 06	0.7691E 06	0.8089E 06	0.8497E 06	0.8916E 06	0.9346E 06	0.9787E 06	0.1023E 07
0.42	0.2793E 07	0.2946E 07	0.3103E 07	0.3264E 07	0.3430E 07	0.3601E 07	0.3775E 07	0.3954E 07	0.4138E 07	0.4325E 07
0.43	0.9534E 07	0.1004E 08	0.1057E 08	0.1111E 08	0.1166E 08	0.1223E 08	0.1281E 08	0.1341E 08	0.1402E 08	0.1464E 08
0.44	0.2454E 08	0.2583E 08	0.2716E 08	0.2852E 08	0.2992E 08	0.3135E 08	0.3282E 08	0.3432E 08	0.3585E 08	0.3742E 08
0.45	0.5045E 08	0.5306E 08	0.5575E 08	0.5850E 08	0.6132E 08	0.6421E 08	0.6716E 08	0.7018E 08	0.7327E 08	0.7643E 08
0.46	0.9118E 08	0.9582E 08	0.1005E 09	0.1054E 09	0.1104E 09	0.1156E 09	0.1208E 09	0.1262E 09	0.1316E 09	0.1372E 09
0.47	0.1523E 09	0.1599E 09	0.1678E 09	0.1758E 09	0.1840E 09	0.1924E 09	0.2010E 09	0.2098E 09	0.2187E 09	0.2279E 09
0.48	0.2432E 09	0.2552E 09	0.2675E 09	0.2801E 09	0.2930E 09	0.3062E 09	0.3196E 09	0.3333E 09	0.3473E 09	0.3616E 09
0.49	0.3764E 09	0.3947E 09	0.4135E 09	0.4326E 09	0.4522E 09	0.4722E 09	0.4926E 09	0.5134E 09	0.5347E 09	0.5563E 09
0.50	0.5745E 09	0.6020E 09	0.6301E 09	0.6588E 09	0.6882E 09	0.7181E 09	0.7487E 09	0.7798E 09	0.8115E 09	0.8439E 09
Luminance	0.5749E 09	0.6023E 09	0.6305E 09	0.6592E 09	0.6885E 09	0.7185E 09	0.7491E 09	0.7803E 09	0.8120E 09	0.8444E 09
0.51	0.3021E 09	0.3158E 09	0.3299E 09	0.3442E 09	0.3588E 09	0.3736E 09	0.3888E 09	0.4042E 09	0.4198E 09	0.4358E 09
0.52	0.7382E 09	0.7714E 09	0.8053E 09	0.8399E 09	0.8751E 09	0.9110E 09	0.9476E 09	0.9847E 09	0.1022E 10	0.1060E 10
0.53	0.1289E 10	0.1346E 10	0.1405E 10	0.1464E 10	0.1525E 10	0.1587E 10	0.1650E 10	0.1714E 10	0.1780E 10	0.1846E 10
0.54	0.1905E 10	0.1989E 10	0.2075E 10	0.2163E 10	0.2252E 10	0.2342E 10	0.2434E 10	0.2528E 10	0.2623E 10	0.2720E 10
0.55	0.2547E 10	0.2658E 10	0.2772E 10	0.2888E 10	0.3005E 10	0.3125E 10	0.3247E 10	0.3371E 10	0.3497E 10	0.3625E 10
0.56	0.3182E 10	0.3320E 10	0.3460E 10	0.3604E 10	0.3749E 10	0.3898E 10	0.4048E 10	0.4202E 10	0.4357E 10	0.4515E 10
0.57	0.3784E 10	0.3947E 10	0.4113E 10	0.4282E 10	0.4454E 10	0.4629E 10	0.4806E 10	0.4987E 10	0.5170E 10	0.5356E 10
0.58	0.4331E 10	0.4516E 10	0.4704E 10	0.4896E 10	0.5091E 10	0.5290E 10	0.5491E 10	0.5696E 10	0.5904E 10	0.6115E 10
0.59	0.4803E 10	0.5007E 10	0.5214E 10	0.5426E 10	0.5641E 10	0.5859E 10	0.6081E 10	0.6307E 10	0.6536E 10	0.6768E 10
0.60	0.5193E 10	0.5412E 10	0.5635E 10	0.5862E 10	0.6094E 10	0.6328E 10	0.6567E 10	0.6809E 10	0.7055E 10	0.7305E 10
Luminance	0.5767E 10	0.6014E 10	0.6266E 10	0.6522E 10	0.6782E 10	0.7047E 10	0.7316E 10	0.7590E 10	0.7867E 10	0.8149E 10
0.61	0.3081E 09	0.3201E 09	0.3323E 09	0.3446E 09	0.3572E 09	0.3699E 09	0.3828E 09	0.3958E 09	0.4090E 09	0.4224E 09
0.62	0.5405E 09	0.5615E 09	0.5827E 09	0.6043E 09	0.6261E 09	0.6483E 09	0.6707E 09	0.6935E 09	0.7165E 09	0.7398E 09
0.63	0.7045E 09	0.7317E 09	0.7592E 09	0.7872E 09	0.8155E 09	0.8442E 09	0.8733E 09	0.9028E 09	0.9326E 09	0.9628E 09
0.64	0.8121E 09	0.8433E 09	0.8749E 09	0.9070E 09	0.9396E 09	0.9725E 09	0.1005E 10	0.1039E 10	0.1074E 10	0.1108E 10
0.65	0.8786E 09	0.9123E 09	0.9464E 09	0.9810E 09	0.1016E 10	0.1051E 10	0.1087E 10	0.1124E 10	0.1161E 10	0.1198E 10
0.66	0.9168E 09	0.9518E 09	0.9873E 09	0.1023E 10	0.1059E 10	0.1096E 10	0.1134E 10	0.1172E 10	0.1210E 10	0.1249E 10
0.67	0.9370E 09	0.9728E 09	0.1009E 10	0.1045E 10	0.1083E 10	0.1120E 10	0.1159E 10	0.1197E 10	0.1237E 10	0.1276E 10
0.68	0.9473E 09	0.9834E 09	0.1020E 10	0.1057E 10	0.1094E 10	0.1133E 10	0.1171E 10	0.1210E 10	0.1250E 10	0.1290E 10
0.69	0.9524E 09	0.9887E 09	0.1025E 10	0.1062E 10	0.1100E 10	0.1139E 10	0.1177E 10	0.1217E 10	0.1257E 10	0.1297E 10
0.70	0.9548E 09	0.9912E 09	0.1028E 10	0.1065E 10	0.1103E 10	0.1141E 10	0.1180E 10	0.1220E 10	0.1260E 10	0.1300E 10
Luminance	0.6722E 10	0.7006E 10	0.7294E 10	0.7587E 10	0.7886E 10	0.8189E 10	0.8497E 10	0.8810E 10	0.9128E 10	0.9450E 10
0.71	0.1176E 07	0.1216E 07	0.1257E 07	0.1298E 07	0.1340E 07	0.1383E 07	0.1425E 07	0.1469E 07	0.1512E 07	0.1556E 07
0.72	0.1752E 07	0.1813E 07	0.1873E 07	0.1935E 07	0.1997E 07	0.2060E 07	0.2123E 07	0.2187E 07	0.2252E 07	0.2317E 07
0.73	0.2031E 07	0.2101E 07	0.2171E 07	0.2242E 07	0.2314E 07	0.2387E 07	0.2460E 07	0.2534E 07	0.2609E 07	0.2684E 07
0.74	0.2163E 07	0.2237E 07	0.2312E 07	0.2387E 07	0.2464E 07	0.2541E 07	0.2619E 07	0.2698E 07	0.2777E 07	0.2857E 07
0.75	0.2224E 07	0.2300E 07	0.2377E 07	0.2455E 07	0.2533E 07	0.2612E 07	0.2692E 07	0.2773E 07	0.2855E 07	0.2937E 07
0.76	0.2253E 07	0.2330E 07	0.2408E 07	0.2486E 07	0.2566E 07	0.2646E 07	0.2727E 07	0.2809E 07	0.2892E 07	0.2975E 07
0.77	0.2267E 07	0.2345E 07	0.2423E 07	0.2502E 07	0.2582E 07	0.2663E 07	0.2744E 07	0.2826E 07	0.2910E 07	0.2993E 07
0.78	0.2274E 07	0.2352E 07	0.2430E 07	0.2509E 07	0.2590E 07	0.2671E 07	0.2752E 07	0.2835E 07	0.2918E 07	0.3002E 07
0.79	0.2277E 07	0.2355E 07	0.2434E 07	0.2513E 07	0.2593E 07	0.2674E 07	0.2756E 07	0.2839E 07	0.2922E 07	0.3007E 07
0.80	0.2279E 07	0.2357E 07	0.2435E 07	0.2515E 07	0.2595E 07	0.2676E 07	0.2758E 07	0.2841E 07	0.2924E 07	0.3009E 07
Luminance	0.6725E 10	0.7008E 10	0.7296E 10	0.7590E 10	0.7888E 10	0.8192E 10	0.8500E 10	0.8813E 10	0.9130E 10	0.9453E 10
0.81	0.7631E 03	0.7869E 03	0.8109E 03	0.8351E 03	0.8595E 03	0.8841E 03	0.9089E 03	0.9339E 03	0.9591E 03	0.9845E 03
0.82	0.1130E 04	0.1165E 04	0.1200E 04	0.1236E 04	0.1272E 04	0.1308E 04	0.1345E 04	0.1382E 04	0.1419E 04	0.1456E 04
0.83	0.1306E 04	0.1346E 04	0.1387E 04	0.1429E 04	0.1470E 04	0.1512E 04	0.1554E 04	0.1597E 04	0.1640E 04	0.1683E 04
Luminance	0.6725E 10	0.7008E 10	0.7296E 10	0.7590E 10	0.7888E 10	0.8192E 10	0.8500E 10	0.8813E 10	0.9130E 10	0.9453E 10

$xEn = x\ 10^n$. E.g. $1.234E\ 05 = 1.234 \times 10^5$

TABLE 18

Blackbody Fractional Luminance, Scotopic (sc.cd/m^2)

λ(μ)	8000°K.	8100°K.	8200°K.	8300°K.	8400°K.	8500°K.	8600°K.	8700°K.	8800°K.	8900°K.
0.38	0.0000E 00	0.0000E 00	0.0000E 00	0.0000E 00	0.0000E 00	0.0000E 00	0.0000E 00	0.0000E 00	0.0000E 00	0.0000E 00
0.39	0.2991E 07	0.3165E 07	0.3345E 07	0.3531E 07	0.3723E 07	0.3920E 07	0.4122E 07	0.4330E 07	0.4544E 07	0.4764E 07
0.40	0.1509E 08	0.1596E 08	0.1685E 08	0.1776E 08	0.1871E 08	0.1968E 08	0.2068E 08	0.2170E 08	0.2275E 08	0.2383E 08
Luminance	0.1509E 08	0.1596E 08	0.1685E 08	0.1776E 08	0.1871E 08	0.1968E 08	0.2068E 08	0.2170E 08	0.2275E 08	0.2383E 08
0.41	0.4755E 08	0.5019E 08	0.5292E 08	0.5572E 08	0.5860E 08	0.6156E 08	0.6459E 08	0.6771E 08	0.7090E 08	0.7417E 08
0.42	0.1946E 09	0.2053E 09	0.2162E 09	0.2275E 09	0.2390E 09	0.2509E 09	0.2631E 09	0.2756E 09	0.2883E 09	0.3014E 09
0.43	0.5306E 09	0.5591E 09	0.5884E 09	0.6185E 09	0.6494E 09	0.6811E 09	0.7136E 09	0.7469E 09	0.7809E 09	0.8158E 09
0.44	0.1125E 10	0.1185E 10	0.1246E 10	0.1309E 10	0.1373E 10	0.1439E 10	0.1507E 10	0.1576E 10	0.1647E 10	0.1719E 10
0.45	0.1994E 10	0.2098E 10	0.2204E 10	0.2313E 10	0.2426E 10	0.2540E 10	0.2658E 10	0.2778E 10	0.2901E 10	0.3026E 10
0.46	0.3100E 10	0.3259E 10	0.3422E 10	0.3590E 10	0.3761E 10	0.3937E 10	0.4116E 10	0.4300E 10	0.4487E 10	0.4678E 10
0.47	0.4408E 10	0.4631E 10	0.4860E 10	0.5094E 10	0.5334E 10	0.5580E 10	0.5831E 10	0.6088E 10	0.6350E 10	0.6617E 10
0.48	0.5915E 10	0.6211E 10	0.6514E 10	0.6824E 10	0.7142E 10	0.7466E 10	0.7798E 10	0.8137E 10	0.8483E 10	0.8835E 10
0.49	0.7616E 10	0.7991E 10	0.8376E 10	0.8770E 10	0.9174E 10	0.9586E 10	0.1000E 11	0.1043E 11	0.1087E 11	0.1132E 11
0.50	0.9459E 10	0.9920E 10	0.1039E 11	0.1087E 11	0.1136E 11	0.1187E 11	0.1238E 11	0.1291E 11	0.1345E 11	0.1399E 11
Luminance	0.9474E 10	0.9936E 10	0.1040E 11	0.1089E 11	0.1138E 11	0.1189E 11	0.1240E 11	0.1293E 11	0.1347E 11	0.1402E 11
0.51	0.1883E 10	0.1968E 10	0.2056E 10	0.2145E 10	0.2236E 10	0.2329E 10	0.2423E 10	0.2519E 10	0.2617E 10	0.2716E 10
0.52	0.3669E 10	0.3835E 10	0.4004E 10	0.4176E 10	0.4352E 10	0.4531E 10	0.4713E 10	0.4898E 10	0.5087E 10	0.5278E 10
0.53	0.5236E 10	0.5470E 10	0.5709E 10	0.5953E 10	0.6202E 10	0.6455E 10	0.6712E 10	0.6974E 10	0.7241E 10	0.7511E 10
0.54	0.6504E 10	0.6793E 10	0.7088E 10	0.7389E 10	0.7696E 10	0.8008E 10	0.8325E 10	0.8648E 10	0.8976E 10	0.9310E 10
0.55	0.7453E 10	0.7783E 10	0.8119E 10	0.8462E 10	0.8811E 10	0.9166E 10	0.9528E 10	0.9895E 10	0.1026E 11	0.1064E 11
0.56	0.8109E 10	0.8467E 10	0.8831E 10	0.9202E 10	0.9580E 10	0.9965E 10	0.1035E 11	0.1075E 11	0.1115E 11	0.1156E 11
0.57	0.8529E 10	0.8903E 10	0.9285E 10	0.9674E 10	0.1007E 11	0.1047E 11	0.1088E 11	0.1130E 11	0.1172E 11	0.1215E 11
0.58	0.8776E 10	0.9161E 10	0.9553E 10	0.9952E 10	0.1035E 11	0.1077E 11	0.1119E 11	0.1162E 11	0.1205E 11	0.1249E 11
0.59	0.8911E 10	0.9301E 10	0.9699E 10	0.1010E 11	0.1051E 11	0.1093E 11	0.1136E 11	0.1179E 11	0.1223E 11	0.1268E 11
0.60	0.8980E 10	0.9373E 10	0.9773E 10	0.1018E 11	0.1059E 11	0.1101E 11	0.1144E 11	0.1188E 11	0.1232E 11	0.1277E 11
Luminance	0.1845E 11	0.1930E 11	0.2018E 11	0.2107E 11	0.2198E 11	0.2291E 11	0.2385E 11	0.2482E 11	0.2580E 11	0.2680E 11
0.61	0.3285E 08	0.3413E 08	0.3543E 08	0.3675E 08	0.3809E 08	0.3944E 08	0.4082E 08	0.4221E 08	0.4362E 08	0.4505E 08
0.62	0.4788E 08	0.4974E 08	0.5162E 08	0.5354E 08	0.5548E 08	0.5744E 08	0.5944E 08	0.6146E 08	0.6350E 08	0.6557E 08
0.63	0.5454E 08	0.5665E 08	0.5879E 08	0.6097E 08	0.6317E 08	0.6540E 08	0.6767E 08	0.6996E 08	0.7228E 08	0.7463E 08
0.64	0.5744E 08	0.5966E 08	0.6192E 08	0.6420E 08	0.6652E 08	0.6887E 08	0.7125E 08	0.7366E 08	0.7610E 08	0.7856E 08
0.65	0.5871E 08	0.6097E 08	0.6327E 08	0.6561E 08	0.6797E 08	0.7037E 08	0.7280E 08	0.7526E 08	0.7775E 08	0.8027E 08
0.66	0.5927E 08	0.6155E 08	0.6387E 08	0.6623E 08	0.6862E 08	0.7104E 08	0.7349E 08	0.7597E 08	0.7848E 08	0.8102E 08
0.67	0.5952E 08	0.6182E 08	0.6415E 08	0.6651E 08	0.6891E 08	0.7133E 08	0.7380E 08	0.7629E 08	0.7881E 08	0.8136E 08
0.68	0.5964E 08	0.6194E 08	0.6427E 08	0.6664E 08	0.6904E 08	0.7147E 08	0.7394E 08	0.7643E 08	0.7896E 08	0.8152E 08
0.69	0.5969E 08	0.6199E 08	0.6433E 08	0.6670E 08	0.6910E 08	0.7154E 08	0.7401E 08	0.7650E 08	0.7903E 08	0.8159E 08
0.70	0.5972E 08	0.6202E 08	0.6436E 08	0.6673E 08	0.6913E 08	0.7157E 08	0.7404E 08	0.7654E 08	0.7907E 08	0.8163E 08
Luminance	0.1851E 11	0.1937E 11	0.2024E 11	0.2114E 11	0.2205E 11	0.2298E 11	0.2393E 11	0.2489E 11	0.2588E 11	0.2688E 11
0.71	0.1312E 05	0.1357E 05	0.1403E 05	0.1449E 05	0.1495E 05	0.1543E 05	0.1590E 05	0.1639E 05	0.1687E 05	0.1736E 05
0.72	0.1970E 05	0.2037E 05	0.2106E 05	0.2175E 05	0.2245E 05	0.2315E 05	0.2387E 05	0.2459E 05	0.2531E 05	0.2605E 05
0.73	0.2306E 05	0.2385E 05	0.2465E 05	0.2545E 05	0.2627E 05	0.2709E 05	0.2793E 05	0.2877E 05	0.2961E 05	0.3047E 05
0.74	0.2481E 05	0.2565E 05	0.2651E 05	0.2738E 05	0.2825E 05	0.2914E 05	0.3003E 05	0.3093E 05	0.3184E 05	0.3276E 05
0.75	0.2573E 05	0.2661E 05	0.2750E 05	0.2839E 05	0.2930E 05	0.3022E 05	0.3114E 05	0.3208E 05	0.3302E 05	0.3397E 05
0.76	0.2623E 05	0.2712E 05	0.2802E 05	0.2894E 05	0.2986E 05	0.3080E 05	0.3174E 05	0.3269E 05	0.3365E 05	0.3462E 05
0.77	0.2650E 05	0.2740E 05	0.2831E 05	0.2924E 05	0.3017E 05	0.3111E 05	0.3206E 05	0.3302E 05	0.3399E 05	0.3497E 05
0.78	0.2665E 05	0.2755E 05	0.2847E 05	0.2940E 05	0.3034E 05	0.3129E 05	0.3224E 05	0.3321E 05	0.3418E 05	0.3517E 05
Luminance	0.1851E 11	0.1937E 11	0.2024E 11	0.2114E 11	0.2205E 11	0.2298E 11	0.2393E 11	0.2489E 11	0.2588E 11	0.2688E 11

xEn - x 10^n. E.g. 1.234E 05 = 1.234 x 10^5

$\int_0^\lambda L_{v',\lambda} d\lambda$

$\int_0^\lambda L_{v\lambda}\, d\lambda$

TABLE 18

Blackbody Fractional Luminance, Photopic (cd/m^2)

λ(μ)	9000°K.	9100°K.	9200°K.	9300°K.	9400°K.	9500°K.	9600°K.	9700°K.	9800°K.	9900°K.
0.36	0.0000E 00	0.0000E 00	0.0000E 00	0.0000E 00	0.0000E 00	0.0000E 00	0.0000E 00	0.0000E 00	0.0000E 00	0.0000E 00
0.37	0.1240E 05	0.1300E 05	0.1362E 05	0.1425E 05	0.1490E 05	0.1557E 05	0.1625E 05	0.1694E 05	0.1765E 05	0.1838E 05
0.38	0.5096E 05	0.5339E 05	0.5588E 05	0.5843E 05	0.6104E 05	0.6371E 05	0.6644E 05	0.6923E 05	0.7208E 05	0.7499E 05
0.39	0.1630E 06	0.1706E 06	0.1784E 06	0.1864E 06	0.1945E 06	0.2029E 06	0.2114E 06	0.2201E 06	0.2290E 06	0.2380E 06
0.40	0.5287E 06	0.5528E 06	0.5775E 06	0.6027E 06	0.6285E 06	0.6549E 06	0.6818E 06	0.7092E 06	0.7372E 06	0.7657E 06
Luminance	0.5287E 06	0.5528E 06	0.5775E 06	0.6027E 06	0.6285E 06	0.6549E 06	0.6818E 06	0.7092E 06	0.7372E 06	0.7657E 06
0.41	0.1070E 07	0.1117E 07	0.1165E 07	0.1214E 07	0.1265E 07	0.1316E 07	0.1368E 07	0.1421E 07	0.1476E 07	0.1531E 07
0.42	0.4517E 07	0.4713E 07	0.4913E 07	0.5117E 07	0.5326E 07	0.5538E 07	0.5754E 07	0.5975E 07	0.6199E 07	0.6428E 07
0.43	0.1528E 08	0.1593E 08	0.1659E 08	0.1727E 08	0.1796E 08	0.1866E 08	0.1938E 08	0.2011E 08	0.2085E 08	0.2160E 08
0.44	0.3902E 08	0.4066E 08	0.4232E 08	0.4402E 08	0.4575E 08	0.4751E 08	0.4931E 08	0.5113E 08	0.5299E 08	0.5487E 08
0.45	0.7964E 08	0.8293E 08	0.8627E 08	0.8968E 08	0.9315E 08	0.9669E 08	0.1002E 09	0.1039E 09	0.1076E 09	0.1114E 09
0.46	0.1429E 09	0.1487E 09	0.1546E 09	0.1606E 09	0.1667E 09	0.1730E 09	0.1793E 09	0.1857E 09	0.1923E 09	0.1989E 09
0.47	0.2372E 09	0.2467E 09	0.2563E 09	0.2661E 09	0.2761E 09	0.2863E 09	0.2966E 09	0.3071E 09	0.3178E 09	0.3286E 09
0.48	0.3762E 09	0.3910E 09	0.4060E 09	0.4214E 09	0.4370E 09	0.4528E 09	0.4689E 09	0.4852E 09	0.5018E 09	0.5186E 09
0.49	0.5783E 09	0.6008E 09	0.6236E 09	0.6468E 09	0.6703E 09	0.6943E 09	0.7186E 09	0.7432E 09	0.7683E 09	0.7936E 09
0.50	0.8768E 09	0.9102E 09	0.9442E 09	0.9788E 09	0.1013E 10	0.1049E 10	0.1085E 10	0.1122E 10	0.1159E 10	0.1197E 10
Luminance	0.8773E 09	0.9108E 09	0.9448E 09	0.9794E 09	0.1014E 10	0.1050E 10	0.1086E 10	0.1123E 10	0.1160E 10	0.1198E 10
0.51	0.4519E 09	0.4684E 09	0.4851E 09	0.5020E 09	0.5192E 09	0.5366E 09	0.5542E 09	0.5721E 09	0.5902E 09	0.6085E 09
0.52	0.1100E 10	0.1139E 10	0.1179E 10	0.1220E 10	0.1261E 10	0.1303E 10	0.1346E 10	0.1389E 10	0.1432E 10	0.1476E 10
0.53	0.1913E 10	0.1981E 10	0.2050E 10	0.2121E 10	0.2192E 10	0.2264E 10	0.2337E 10	0.2411E 10	0.2486E 10	0.2562E 10
0.54	0.2818E 10	0.2918E 10	0.3019E 10	0.3121E 10	0.3225E 10	0.3330E 10	0.3437E 10	0.3545E 10	0.3654E 10	0.3764E 10
0.55	0.3754E 10	0.3886E 10	0.4019E 10	0.4155E 10	0.4292E 10	0.4430E 10	0.4571E 10	0.4713E 10	0.4857E 10	0.5002E 10
0.56	0.4676E 10	0.4838E 10	0.5003E 10	0.5170E 10	0.5339E 10	0.5510E 10	0.5684E 10	0.5859E 10	0.6037E 10	0.6216E 10
0.57	0.5545E 10	0.5736E 10	0.5930E 10	0.6127E 10	0.6326E 10	0.6527E 10	0.6731E 10	0.6938E 10	0.7146E 10	0.7357E 10
0.58	0.6329E 10	0.6546E 10	0.6766E 10	0.6989E 10	0.7214E 10	0.7443E 10	0.7674E 10	0.7907E 10	0.8144E 10	0.8383E 10
0.59	0.7004E 10	0.7242E 10	0.7484E 10	0.7729E 10	0.7977E 10	0.8228E 10	0.8482E 10	0.8739E 10	0.8999E 10	0.9262E 10
0.60	0.7558E 10	0.7814E 10	0.8074E 10	0.8337E 10	0.8603E 10	0.8873E 10	0.9145E 10	0.9421E 10	0.9700E 10	0.9981E 10
Luminance	0.8435E 10	0.8725E 10	0.9019E 10	0.9316E 10	0.9618E 10	0.9923E 10	0.1023E 11	0.1054E 11	0.1086E 11	0.1118E 11
0.61	0.4359E 09	0.4496E 09	0.4635E 09	0.4774E 09	0.4916E 09	0.5058E 09	0.5202E 09	0.5348E 09	0.5495E 09	0.5643E 09
0.62	0.7634E 09	0.7872E 09	0.8113E 09	0.8356E 09	0.8602E 09	0.8850E 09	0.9101E 09	0.9354E 09	0.9609E 09	0.9867E 09
0.63	0.9933E 09	0.1024E 10	0.1055E 10	0.1086E 10	0.1118E 10	0.1150E 10	0.1183E 10	0.1216E 10	0.1249E 10	0.1282E 10
0.64	0.1143E 10	0.1179E 10	0.1214E 10	0.1251E 10	0.1287E 10	0.1324E 10	0.1361E 10	0.1399E 10	0.1437E 10	0.1475E 10
0.65	0.1236E 10	0.1274E 10	0.1312E 10	0.1351E 10	0.1391E 10	0.1430E 10	0.1471E 10	0.1511E 10	0.1552E 10	0.1593E 10
0.66	0.1289E 10	0.1328E 10	0.1368E 10	0.1409E 10	0.1450E 10	0.1491E 10	0.1533E 10	0.1575E 10	0.1618E 10	0.1661E 10
0.67	0.1316E 10	0.1357E 10	0.1398E 10	0.1439E 10	0.1481E 10	0.1523E 10	0.1566E 10	0.1609E 10	0.1652E 10	0.1696E 10
0.68	0.1331E 10	0.1372E 10	0.1413E 10	0.1455E 10	0.1497E 10	0.1540E 10	0.1583E 10	0.1626E 10	0.1670E 10	0.1714E 10
0.69	0.1338E 10	0.1379E 10	0.1420E 10	0.1462E 10	0.1505E 10	0.1548E 10	0.1591E 10	0.1634E 10	0.1678E 10	0.1723E 10
0.70	0.1341E 10	0.1382E 10	0.1424E 10	0.1466E 10	0.1508E 10	0.1551E 10	0.1595E 10	0.1638E 10	0.1682E 10	0.1727E 10
Luminance	0.9776E 10	0.1010E 11	0.1044E 11	0.1078E 11	0.1112E 11	0.1147E 11	0.1182E 11	0.1218E 11	0.1254E 11	0.1290E 11
0.71	0.1601E 07	0.1645E 07	0.1690E 07	0.1736E 07	0.1782E 07	0.1828E 07	0.1874E 07	0.1921E 07	0.1969E 07	0.2016E 07
0.72	0.2383E 07	0.2450E 07	0.2517E 07	0.2584E 07	0.2652E 07	0.2721E 07	0.2790E 07	0.2859E 07	0.2929E 07	0.3000E 07
0.73	0.2760E 07	0.2837E 07	0.2914E 07	0.2992E 07	0.3071E 07	0.3150E 07	0.3230E 07	0.3310E 07	0.3391E 07	0.3472E 07
0.74	0.2938E 07	0.3020E 07	0.3102E 07	0.3185E 07	0.3268E 07	0.3353E 07	0.3437E 07	0.3523E 07	0.3609E 07	0.3695E 07
0.75	0.3020E 07	0.3104E 07	0.3188E 07	0.3274E 07	0.3359E 07	0.3446E 07	0.3533E 07	0.3620E 07	0.3709E 07	0.3798E 07
0.76	0.3059E 07	0.3144E 07	0.3230E 07	0.3316E 07	0.3403E 07	0.3490E 07	0.3578E 07	0.3667E 07	0.3756E 07	0.3846E 07
0.77	0.3078E 07	0.3163E 07	0.3249E 07	0.3336E 07	0.3423E 07	0.3511E 07	0.3600E 07	0.3689E 07	0.3779E 07	0.3869E 07
0.78	0.3087E 07	0.3173E 07	0.3259E 07	0.3346E 07	0.3433E 07	0.3522E 07	0.3610E 07	0.3700E 07	0.3790E 07	0.3881E 07
0.79	0.3091E 07	0.3177E 07	0.3263E 07	0.3351E 07	0.3438E 07	0.3527E 07	0.3616E 07	0.3705E 07	0.3795E 07	0.3886E 07
0.80	0.3094E 07	0.3179E 07	0.3266E 07	0.3353E 07	0.3441E 07	0.3529E 07	0.3618E 07	0.3708E 07	0.3798E 07	0.3889E 07
Luminance	0.9779E 10	0.1011E 11	0.1044E 11	0.1078E 11	0.1113E 11	0.1147E 11	0.1183E 11	0.1218E 11	0.1254E 11	0.1291E 11
0.81	0.1010E 04	0.1035E 04	0.1061E 04	0.1087E 04	0.1113E 04	0.1140E 04	0.1166E 04	0.1193E 04	0.1220E 04	0.1247E 04
0.82	0.1494E 04	0.1532E 04	0.1570E 04	0.1609E 04	0.1647E 04	0.1686E 04	0.1725E 04	0.1764E 04	0.1804E 04	0.1844E 04
0.83	0.1726E 04	0.1770E 04	0.1814E 04	0.1858E 04	0.1903E 04	0.1948E 04	0.1993E 04	0.2038E 04	0.2084E 04	0.2130E 04
Luminance	0.9779E 10	0.1011E 11	0.1044E 11	0.1078E 11	0.1113E 11	0.1147E 11	0.1183E 11	0.1218E 11	0.1254E 11	0.1291E 11

xEn = x 10^n. E.g. 1.234E 05 = 1.234 x 10^5

TABLE 18

Blackbody Fractional Luminance, Scotopic (sc.cd/m²)

λ(μ)	9000°K.	9100°K.	9200°K.	9300°K.	9400°K.	9500°K.	9600°K.	9700°K.	9800°K.	9900°K.
0.38	0.0000E 00	0.0000E 00	0.0000E 00	0.0000E 00	0.0000E 00	0.0000E 00	0.0000E 00	0.0000E 00	0.0000E 00	0.0000E 00
0.39	0.4989E 07	0.5219E 07	0.5455E 07	0.5697E 07	0.5944E 07	0.6196E 07	0.6453E 07	0.6716E 07	0.6985E 07	0.7258E 07
0.40	0.2494E 08	0.2607E 08	0.2722E 08	0.2841E 08	0.2961E 08	0.3085E 08	0.3211E 08	0.3339E 08	0.3470E 08	0.3603E 08
Luminance	0.2494E 08	0.2607E 08	0.2722E 08	0.2841E 08	0.2961E 08	0.3085E 08	0.3211E 08	0.3339E 08	0.3470E 08	0.3603E 08
0.41	0.7751E 08	0.8093E 08	0.8442E 08	0.8799E 08	0.9163E 08	0.9535E 08	0.9914E 08	0.1030E 09	0.1069E 09	0.1109E 09
0.42	0.3148E 09	0.3284E 09	0.3424E 09	0.3566E 09	0.3712E 09	0.3860E 09	0.4011E 09	0.4164E 09	0.4321E 09	0.4480E 09
0.43	0.8514E 09	0.8877E 09	0.9248E 09	0.9626E 09	0.1001E 10	0.1040E 10	0.1080E 10	0.1121E 10	0.1162E 10	0.1204E 10
0.44	0.1793E 10	0.1868E 10	0.1945E 10	0.2024E 10	0.2103E 10	0.2185E 10	0.2267E 10	0.2352E 10	0.2437E 10	0.2524E 10
0.45	0.3154E 10	0.3285E 10	0.3418E 10	0.3554E 10	0.3692E 10	0.3833E 10	0.3976E 10	0.4121E 10	0.4269E 10	0.4420E 10
0.46	0.4874E 10	0.5073E 10	0.5276E 10	0.5482E 10	0.5693E 10	0.5907E 10	0.6124E 10	0.6345E 10	0.6570E 10	0.6798E 10
0.47	0.6890E 10	0.7168E 10	0.7451E 10	0.7739E 10	0.8032E 10	0.8330E 10	0.8633E 10	0.8941E 10	0.9254E 10	0.9571E 10
0.48	0.9195E 10	0.9561E 10	0.9934E 10	0.1031E 11	0.1069E 11	0.1109E 11	0.1149E 11	0.1189E 11	0.1230E 11	0.1272E 11
0.49	0.1177E 11	0.1223E 11	0.1271E 11	0.1319E 11	0.1367E 11	0.1417E 11	0.1467E 11	0.1518E 11	0.1570E 11	0.1623E 11
0.50	0.1455E 11	0.1511E 11	0.1569E 11	0.1628E 11	0.1687E 11	0.1748E 11	0.1809E 11	0.1871E 11	0.1935E 11	0.1999E 11
Luminance	0.1457E 11	0.1514E 11	0.1572E 11	0.1630E 11	0.1690E 11	0.1751E 11	0.1812E 11	0.1875E 11	0.1938E 11	0.2002E 11
0.51	0.2817E 10	0.2920E 10	0.3024E 10	0.3130E 10	0.3237E 10	0.3345E 10	0.3455E 10	0.3567E 10	0.3680E 10	0.3794E 10
0.52	0.5473E 10	0.5670E 10	0.5871E 10	0.6074E 10	0.6281E 10	0.6490E 10	0.6702E 10	0.6916E 10	0.7133E 10	0.7353E 10
0.53	0.7786E 10	0.8065E 10	0.8348E 10	0.8636E 10	0.8927E 10	0.9222E 10	0.9521E 10	0.9823E 10	0.1013E 11	0.1044E 11
0.54	0.9648E 10	0.9992E 10	0.1034E 11	0.1069E 11	0.1105E 11	0.1141E 11	0.1178E 11	0.1215E 11	0.1253E 11	0.1291E 11
0.55	0.1103E 11	0.1142E 11	0.1182E 11	0.1222E 11	0.1263E 11	0.1304E 11	0.1346E 11	0.1388E 11	0.1431E 11	0.1474E 11
0.56	0.1198E 11	0.1240E 11	0.1283E 11	0.1327E 11	0.1371E 11	0.1416E 11	0.1461E 11	0.1506E 11	0.1553E 11	0.1600E 11
0.57	0.1259E 11	0.1303E 11	0.1348E 11	0.1393E 11	0.1440E 11	0.1486E 11	0.1534E 11	0.1582E 11	0.1630E 11	0.1679E 11
0.58	0.1294E 11	0.1340E 11	0.1386E 11	0.1432E 11	0.1480E 11	0.1528E 11	0.1576E 11	0.1626E 11	0.1675E 11	0.1726E 11
0.59	0.1313E 11	0.1360E 11	0.1406E 11	0.1454E 11	0.1502E 11	0.1550E 11	0.1599E 11	0.1649E 11	0.1700E 11	0.1751E 11
0.60	0.1323E 11	0.1370E 11	0.1417E 11	0.1464E 11	0.1513E 11	0.1562E 11	0.1611E 11	0.1661E 11	0.1712E 11	0.1763E 11
Luminance	0.2781E 11	0.2884E 11	0.2989E 11	0.3095E 11	0.3203E 11	0.3313E 11	0.3424E 11	0.3537E 11	0.3651E 11	0.3766E 11
0.61	0.4649E 08	0.4795E 08	0.4943E 08	0.5092E 08	0.5243E 08	0.5395E 08	0.5549E 08	0.5704E 08	0.5861E 08	0.6019E 08
0.62	0.6766E 08	0.6978E 08	0.7192E 08	0.7408E 08	0.7626E 08	0.7847E 08	0.8070E 08	0.8294E 08	0.8521E 08	0.8750E 08
0.63	0.7700E 08	0.7941E 08	0.8183E 08	0.8429E 08	0.8677E 08	0.8927E 08	0.9180E 08	0.9435E 08	0.9692E 08	0.9952E 08
0.64	0.8106E 08	0.8359E 08	0.8614E 08	0.8872E 08	0.9132E 08	0.9395E 08	0.9661E 08	0.9929E 08	0.1020E 09	0.1047E 09
0.65	0.8282E 08	0.8540E 08	0.8800E 08	0.9064E 08	0.9330E 08	0.9598E 08	0.9869E 08	0.1014E 09	0.1041E 09	0.1069E 09
0.66	0.8360E 08	0.8620E 08	0.8882E 08	0.9148E 08	0.9416E 08	0.9687E 08	0.9961E 08	0.1023E 09	0.1051E 09	0.1079E 09
0.67	0.8394E 08	0.8655E 08	0.8919E 08	0.9186E 08	0.9455E 08	0.9727E 08	0.1000E 09	0.1027E 09	0.1055E 09	0.1084E 09
0.68	0.8411E 08	0.8672E 08	0.8936E 08	0.9204E 08	0.9473E 08	0.9746E 08	0.1002E 09	0.1029E 09	0.1057E 09	0.1086E 09
0.69	0.8418E 08	0.8680E 08	0.8944E 08	0.9212E 08	0.9482E 08	0.9754E 08	0.1003E 09	0.1030E 09	0.1058E 09	0.1087E 09
0.70	0.8422E 08	0.8684E 08	0.8948E 08	0.9216E 08	0.9486E 08	0.9759E 08	0.1003E 09	0.1031E 09	0.1059E 09	0.1087E 09
Luminance	0.2789E 11	0.2893E 11	0.2998E 11	0.3105E 11	0.3213E 11	0.3323E 11	0.3434E 11	0.3547E 11	0.3661E 11	0.3777E 11
0.71	0.1786E 05	0.1836E 05	0.1886E 05	0.1937E 05	0.1988E 05	0.2039E 05	0.2091E 05	0.2144E 05	0.2196E 05	0.2249E 05
0.72	0.2679E 05	0.2753E 05	0.2829E 05	0.2904E 05	0.2981E 05	0.3058E 05	0.3136E 05	0.3214E 05	0.3292E 05	0.3372E 05
0.73	0.3133E 05	0.3220E 05	0.3308E 05	0.3397E 05	0.3486E 05	0.3576E 05	0.3666E 05	0.3757E 05	0.3849E 05	0.3942E 05
0.74	0.3369E 05	0.3462E 05	0.3557E 05	0.3652E 05	0.3747E 05	0.3844E 05	0.3941E 05	0.4039E 05	0.4137E 05	0.4236E 05
0.75	0.3493E 05	0.3590E 05	0.3688E 05	0.3786E 05	0.3885E 05	0.3985E 05	0.4086E 05	0.4187E 05	0.4289E 05	0.4391E 05
0.76	0.3560E 05	0.3658E 05	0.3758E 05	0.3858E 05	0.3959E 05	0.4060E 05	0.4163E 05	0.4266E 05	0.4370E 05	0.4474E 05
0.77	0.3596E 05	0.3695E 05	0.3796E 05	0.3897E 05	0.3999E 05	0.4101E 05	0.4205E 05	0.4309E 05	0.4414E 05	0.4519E 05
0.78	0.3616E 05	0.3716E 05	0.3817E 05	0.3918E 05	0.4021E 05	0.4124E 05	0.4228E 05	0.4333E 05	0.4438E 05	0.4544E 05
Luminance	0.2789E 11	0.2893E 11	0.2998E 11	0.3105E 11	0.3213E 11	0.3323E 11	0.3434E 11	0.3547E 11	0.3661E 11	0.3777E 11

xEn = x 10^n. E.g. 1.234E 05 = 1.234 x 10^5

$\int_o^\lambda L_{v'\lambda} d\lambda$

$\int_0^\lambda L_{v\lambda} d\lambda$

TABLE 18

Blackbody Fractional Luminance, Photopic (cd/m²)

λ(μ)	10000°K.	11000°K.	12000°K.	13000°K.	14000°K.	15000°K.	16000°K.	17000°K.	18000°K.	19000°K.
0.36	0.0000E 00	0.0000E 00	0.0000E 00	0.0000E 00	0.0000E 00	0.0000E 00	0.0000E 00	0.0000E 00	0.0000E 00	0.0000E 00
0.37	0.1912E 05	0.2733E 05	0.3692E 05	0.4775E 05	0.5970E 05	0.7265E 05	0.8647E 05	0.1010E 06	0.1163E 06	0.1322E 06
0.38	0.7795E 05	0.1107E 06	0.1487E 06	0.1916E 06	0.2387E 06	0.2896E 06	0.3438E 06	0.4010E 06	0.4607E 06	0.5228E 06
0.39	0.2473E 06	0.3488E 06	0.4662E 06	0.5978E 06	0.7419E 06	0.8970E 06	0.1062E 07	0.1235E 07	0.1416E 07	0.1604E 07
0.40	0.7947E 06	0.1113E 07	0.1478E 07	0.1886E 07	0.2331E 07	0.2809E 07	0.3315E 07	0.3847E 07	0.4401E 07	0.4974E 07
Luminance	0.7947E 06	0.1113E 07	0.1478E 07	0.1886E 07	0.2331E 07	0.2809E 07	0.3315E 07	0.3847E 07	0.4401E 07	0.4974E 07
0.41	0.1587E 07	0.2200E 07	0.2899E 07	0.3672E 07	0.4512E 07	0.5410E 07	0.6357E 07	0.7349E 07	0.8379E 07	0.9444E 07
0.42	0.6660E 07	0.9185E 07	0.1204E 08	0.1521E 08	0.1863E 08	0.2228E 08	0.2613E 08	0.3016E 08	0.3433E 08	0.3864E 08
0.43	0.2237E 08	0.3068E 08	0.4007E 08	0.5041E 08	0.6156E 08	0.7343E 08	0.8592E 08	0.9895E 08	0.1124E 09	0.1263E 09
0.44	0.5679E 08	0.7750E 08	0.1008E 09	0.1263E 09	0.1539E 09	0.1831E 09	0.2138E 09	0.2458E 09	0.2789E 09	0.3130E 09
0.45	0.1152E 09	0.1565E 09	0.2029E 09	0.2536E 09	0.3081E 09	0.3658E 09	0.4264E 09	0.4894E 09	0.5545E 09	0.6215E 09
0.46	0.2057E 09	0.2782E 09	0.3593E 09	0.4478E 09	0.5427E 09	0.6431E 09	0.7482E 09	0.8575E 09	0.9703E 09	0.1086E 10
0.47	0.3396E 09	0.4574E 09	0.5887E 09	0.7317E 09	0.8846E 09	0.1046E 10	0.1215E 10	0.1390E 10	0.1571E 10	0.1757E 10
0.48	0.5357E 09	0.7186E 09	0.9218E 09	0.1142E 10	0.1377E 10	0.1626E 10	0.1885E 10	0.2154E 10	0.2431E 10	0.2716E 10
0.49	0.8193E 09	0.1094E 10	0.1399E 10	0.1729E 10	0.2080E 10	0.2451E 10	0.2837E 10	0.3237E 10	0.3648E 10	0.4071E 10
0.50	0.1235E 10	0.1643E 10	0.2094E 10	0.2581E 10	0.3098E 10	0.3642E 10	0.4209E 10	0.4795E 10	0.5398E 10	0.6016E 10
Luminance	0.1236E 10	0.1644E 10	0.2095E 10	0.2582E 10	0.3100E 10	0.3645E 10	0.4212E 10	0.4799E 10	0.5402E 10	0.6021E 10
0.51	0.6271E 09	0.8238E 09	0.1038E 10	0.1269E 10	0.1512E 10	0.1767E 10	0.2031E 10	0.2304E 10	0.2583E 10	0.2868E 10
0.52	0.1521E 10	0.1994E 10	0.2509E 10	0.3061E 10	0.3643E 10	0.4252E 10	0.4883E 10	0.5533E 10	0.6199E 10	0.6879E 10
0.53	0.2638E 10	0.3450E 10	0.4333E 10	0.5278E 10	0.6274E 10	0.7313E 10	0.8390E 10	0.9499E 10	0.1063E 11	0.1179E 11
0.54	0.3876E 10	0.5056E 10	0.6340E 10	0.7710E 10	0.9154E 10	0.1065E 11	0.1221E 11	0.1382E 11	0.1546E 11	0.1713E 11
0.55	0.5150E 10	0.6704E 10	0.8392E 10	0.1019E 11	0.1208E 11	0.1405E 11	0.1610E 11	0.1820E 11	0.2035E 11	0.2254E 11
0.56	0.6398E 10	0.8313E 10	0.1038E 11	0.1260E 11	0.1492E 11	0.1735E 11	0.1985E 11	0.2242E 11	0.2506E 11	0.2774E 11
0.57	0.7571E 10	0.9819E 10	0.1225E 11	0.1484E 11	0.1757E 11	0.2040E 11	0.2333E 11	0.2634E 11	0.2942E 11	0.3256E 11
0.58	0.8624E 10	0.1116E 11	0.1392E 11	0.1684E 11	0.1992E 11	0.2312E 11	0.2642E 11	0.2981E 11	0.3328E 11	0.3681E 11
0.59	0.9527E 10	0.1232E 11	0.1534E 11	0.1855E 11	0.2191E 11	0.2542E 11	0.2903E 11	0.3274E 11	0.3654E 11	0.4040E 11
0.60	0.1026E 11	0.1326E 11	0.1649E 11	0.1993E 11	0.2353E 11	0.2728E 11	0.3114E 11	0.3511E 11	0.3917E 11	0.4330E 11
Luminance	0.1150E 11	0.1490E 11	0.1859E 11	0.2251E 11	0.2663E 11	0.3092E 11	0.3536E 11	0.3991E 11	0.4457E 11	0.4932E 11
0.61	0.5792E 09	0.7351E 09	0.9013E 09	0.1076E 10	0.1258E 10	0.1445E 10	0.1638E 10	0.1835E 10	0.2035E 10	0.2239E 10
0.62	0.1012E 10	0.1283E 10	0.1572E 10	0.1875E 10	0.2191E 10	0.2517E 10	0.2851E 10	0.3192E 10	0.3540E 10	0.3892E 10
0.63	0.1316E 10	0.1666E 10	0.2040E 10	0.2432E 10	0.2839E 10	0.3260E 10	0.3691E 10	0.4132E 10	0.4580E 10	0.5035E 10
0.64	0.1513E 10	0.1915E 10	0.2343E 10	0.2792E 10	0.3258E 10	0.3739E 10	0.4233E 10	0.4737E 10	0.5249E 10	0.5770E 10
0.65	0.1635E 10	0.2068E 10	0.2528E 10	0.3011E 10	0.3513E 10	0.4031E 10	0.4562E 10	0.5104E 10	0.5656E 10	0.6215E 10
0.66	0.1704E 10	0.2154E 10	0.2633E 10	0.3136E 10	0.3658E 10	0.4196E 10	0.4748E 10	0.5311E 10	0.5884E 10	0.6466E 10
0.67	0.1740E 10	0.2200E 10	0.2688E 10	0.3201E 10	0.3733E 10	0.4282E 10	0.4845E 10	0.5419E 10	0.6004E 10	0.6597E 10
0.68	0.1758E 10	0.2222E 10	0.2716E 10	0.3233E 10	0.3771E 10	0.4325E 10	0.4893E 10	0.5473E 10	0.6063E 10	0.6662E 10
0.69	0.1768E 10	0.2234E 10	0.2729E 10	0.3249E 10	0.3789E 10	0.4346E 10	0.4917E 10	0.5500E 10	0.6092E 10	0.6694E 10
0.70	0.1772E 10	0.2239E 10	0.2736E 10	0.3257E 10	0.3798E 10	0.4356E 10	0.4928E 10	0.5512E 10	0.6106E 10	0.6708E 10
Luminance	0.1327E 11	0.1714E 11	0.2132E 11	0.2577E 11	0.3043E 11	0.3528E 11	0.4029E 11	0.4542E 11	0.5068E 11	0.5603E 11
0.71	0.2064E 07	0.2558E 07	0.3076E 07	0.3614E 07	0.4169E 07	0.4737E 07	0.5316E 07	0.5905E 07	0.6501E 07	0.7105E 07
0.72	0.3071E 07	0.3803E 07	0.4571E 07	0.5369E 07	0.6190E 07	0.7031E 07	0.7889E 07	0.8760E 07	0.9644E 07	0.1053E 08
0.73	0.3554E 07	0.4400E 07	0.5287E 07	0.6208E 07	0.7156E 07	0.8126E 07	0.9116E 07	0.1012E 08	0.1114E 08	0.1217E 08
0.74	0.3782E 07	0.4681E 07	0.5623E 07	0.6601E 07	0.7608E 07	0.8639E 07	0.9689E 07	0.1075E 08	0.1183E 08	0.1293E 08
0.75	0.3887E 07	0.4810E 07	0.5777E 07	0.6781E 07	0.7815E 07	0.8873E 07	0.9952E 07	0.1104E 08	0.1215E 08	0.1328E 08
0.76	0.3937E 07	0.4871E 07	0.5850E 07	0.6866E 07	0.7912E 07	0.8983E 07	0.1007E 08	0.1118E 08	0.1230E 08	0.1344E 08
0.77	0.3960E 07	0.4900E 07	0.5885E 07	0.6907E 07	0.7959E 07	0.9036E 07	0.1013E 08	0.1124E 08	0.1237E 08	0.1352E 08
0.78	0.3972E 07	0.4914E 07	0.5902E 07	0.6926E 07	0.7981E 07	0.9061E 07	0.1016E 08	0.1128E 08	0.1241E 08	0.1355E 08
0.79	0.3978E 07	0.4921E 07	0.5910E 07	0.6936E 07	0.7992E 07	0.9073E 07	0.1017E 08	0.1129E 08	0.1243E 08	0.1357E 08
0.80	0.3980E 07	0.4924E 07	0.5914E 07	0.6940E 07	0.7997E 07	0.9079E 07	0.1018E 08	0.1130E 08	0.1243E 08	0.1358E 08
Luminance	0.1327E 11	0.1715E 11	0.2133E 11	0.2578E 11	0.3044E 11	0.3529E 11	0.4030E 11	0.4544E 11	0.5069E 11	0.5604E 11
0.81	0.1274E 04	0.1552E 04	0.1840E 04	0.2137E 04	0.2441E 04	0.2751E 04	0.3065E 04	0.3383E 04	0.3705E 04	0.4029E 04
0.82	0.1884E 04	0.2293E 04	0.2719E 04	0.3156E 04	0.3604E 04	0.4060E 04	0.4524E 04	0.4993E 04	0.5466E 04	0.5944E 04
0.83	0.2176E 04	0.2648E 04	0.3138E 04	0.3643E 04	0.4159E 04	0.4685E 04	0.5218E 04	0.5759E 04	0.6304E 04	0.6855E 04
Luminance	0.1327E 11	0.1715E 11	0.2133E 11	0.2578E 11	0.3044E 11	0.3529E 11	0.4030E 11	0.4544E 11	0.5069E 11	0.5604E 11

$xEn = x\ 10^n$. E.g. $1.234E.05 = 1.234 \times 10^5$

TABLE 18

Blackbody Fractional Luminance, Scotopic (sc.cd/m^2)

λ(μ)	10000°K.	11000°K.	12000°K.	13000°K.	14000°K.	15000°K.	16000°K.	17000°K.	18000°K.	19000°K.
0.38	0.0000E 00	0.0000E 00	0.0000E 00	0.0000E 00	0.0000E 00	0.0000E 00	0.0000E 00	0.0000E 00	0.0000E 00	0.0000E 00
0.39	0.7537E 07	0.1059E 08	0.1412E 08	0.1807E 08	0.2238E 08	0.2702E 08	0.3194E 08	0.3712E 08	0.4252E 08	0.4811E 08
0.40	0.3739E 08	0.5227E 08	0.6933E 08	0.8834E 08	0.1090E 09	0.1312E 09	0.1548E 09	0.1795E 09	0.2052E 09	0.2318E 09
Luminance	0.3739E 08	0.5227E 08	0.6933E 08	0.8834E 08	0.1090E 09	0.1312E 09	0.1548E 09	0.1795E 09	0.2052E 09	0.2318E 09
0.41	0.1150E 09	0.1593E 09	0.2099E 09	0.2659E 09	0.3267E 09	0.3917E 09	0.4603E 09	0.5321E 09	0.6067E 09	0.6838E 09
0.42	0.4642E 09	0.6402E 09	0.8400E 09	0.1060E 10	0.1299E 10	0.1554E 10	0.1822E 10	0.2103E 10	0.2394E 10	0.2695E 10
0.43	0.1247E 10	0.1712E 10	0.2237E 10	0.2816E 10	0.3441E 10	0.4105E 10	0.4805E 10	0.5535E 10	0.6292E 10	0.7072E 10
0.44	0.2613E 10	0.3570E 10	0.4647E 10	0.5831E 10	0.7106E 10	0.8460E 10	0.9883E 10	0.1136E 11	0.1290E 11	0.1448E 11
0.45	0.4572E 10	0.6220E 10	0.8071E 10	0.1009E 11	0.1227E 11	0.1458E 11	0.1700E 11	0.1953E 11	0.2214E 11	0.2482E 11
0.46	0.7030E 10	0.9527E 10	0.1232E 11	0.1537E 11	0.1865E 11	0.2212E 11	0.2575E 11	0.2953E 11	0.3344E 11	0.3746E 11
0.47	0.9894E 10	0.1336E 11	0.1722E 11	0.2144E 11	0.2596E 11	0.3074E 11	0.3574E 11	0.4093E 11	0.4630E 11	0.5181E 11
0.48	0.1314E 11	0.1769E 11	0.2275E 11	0.2826E 11	0.3415E 11	0.4036E 11	0.4687E 11	0.5362E 11	0.6058E 11	0.6772E 11
0.49	0.1676E 11	0.2249E 11	0.2885E 11	0.3575E 11	0.4313E 11	0.5090E 11	0.5902E 11	0.6744E 11	0.7612E 11	0.8503E 11
0.50	0.2064E 11	0.2760E 11	0.3532E 11	0.4368E 11	0.5260E 11	0.6200E 11	0.7180E 11	0.8196E 11	0.9242E 11	0.1031E 12
Luminance	0.2068E 11	0.2765E 11	0.3539E 11	0.4377E 11	0.5271E 11	0.6213E 11	0.7195E 11	0.8213E 11	0.9262E 11	0.1033E 12
0.51	0.3910E 10	0.5137E 10	0.6479E 10	0.7917E 10	0.9437E 10	0.1102E 11	0.1267E 11	0.1437E 11	0.1612E 11	0.1790E 11
0.52	0.7576E 10	0.9934E 10	0.1250E 11	0.1526E 11	0.1817E 11	0.2121E 11	0.2437E 11	0.2761E 11	0.3094E 11	0.3435E 11
0.53	0.1075E 11	0.1407E 11	0.1769E 11	0.2157E 11	0.2565E 11	0.2992E 11	0.3434E 11	0.3890E 11	0.4357E 11	0.4833E 11
0.54	0.1329E 11	0.1738E 11	0.2182E 11	0.2657E 11	0.3158E 11	0.3681E 11	0.4222E 11	0.4779E 11	0.5350E 11	0.5933E 11
0.55	0.1518E 11	0.1982E 11	0.2486E 11	0.3024E 11	0.3592E 11	0.4184E 11	0.4797E 11	0.5428E 11	0.6074E 11	0.6733E 11
0.56	0.1647E 11	0.2148E 11	0.2692E 11	0.3274E 11	0.3886E 11	0.4524E 11	0.5185E 11	0.5865E 11	0.6561E 11	0.7272E 11
0.57	0.1729E 11	0.2253E 11	0.2822E 11	0.3430E 11	0.4070E 11	0.4737E 11	0.5428E 11	0.6138E 11	0.6865E 11	0.7607E 11
0.58	0.1776E 11	0.2314E 11	0.2898E 11	0.3521E 11	0.4177E 11	0.4860E 11	0.5568E 11	0.6295E 11	0.7040E 11	0.7800E 11
0.59	0.1802E 11	0.2347E 11	0.2939E 11	0.3570E 11	0.4234E 11	0.4926E 11	0.5643E 11	0.6379E 11	0.7133E 11	0.7903E 11
0.60	0.1815E 11	0.2363E 11	0.2959E 11	0.3594E 11	0.4262E 11	0.4959E 11	0.5680E 11	0.6421E 11	0.7180E 11	0.7954E 11
Luminance	0.3883E 11	0.5129E 11	0.6498E 11	0.7971E 11	0.9534E 11	0.1117E 12	0.1287E 12	0.1463E 12	0.1644E 12	0.1829E 12
0.61	0.6178E 08	0.7842E 08	0.9616E 08	0.1148E 09	0.1342E 09	0.1542E 09	0.1748E 09	0.1958E 09	0.2172E 09	0.2390E 09
0.62	0.8981E 08	0.1139E 09	0.1395E 09	0.1665E 09	0.1946E 09	0.2235E 09	0.2533E 09	0.2836E 09	0.3145E 09	0.3459E 09
0.63	0.1021E 09	0.1294E 09	0.1585E 09	0.1891E 09	0.2209E 09	0.2537E 09	0.2874E 09	0.3218E 09	0.3568E 09	0.3924E 09
0.64	0.1074E 09	0.1361E 09	0.1667E 09	0.1988E 09	0.2322E 09	0.2667E 09	0.3020E 09	0.3382E 09	0.3749E 09	0.4122E 09
0.65	0.1097E 09	0.1390E 09	0.1702E 09	0.2030E 09	0.2371E 09	0.2722E 09	0.3083E 09	0.3451E 09	0.3826E 09	0.4207E 09
0.66	0.1108E 09	0.1403E 09	0.1718E 09	0.2048E 09	0.2392E 09	0.2747E 09	0.3110E 09	0.3482E 09	0.3860E 09	0.4244E 09
0.67	0.1112E 09	0.1409E 09	0.1725E 09	0.2056E 09	0.2401E 09	0.2757E 09	0.3122E 09	0.3495E 09	0.3875E 09	0.4260E 09
0.68	0.1114E 09	0.1411E 09	0.1728E 09	0.2060E 09	0.2406E 09	0.2762E 09	0.3128E 09	0.3501E 09	0.3882E 09	0.4268E 09
0.69	0.1115E 09	0.1413E 09	0.1729E 09	0.2062E 09	0.2408E 09	0.2764E 09	0.3130E 09	0.3504E 09	0.3885E 09	0.4271E 09
0.70	0.1116E 09	0.1413E 09	0.1730E 09	0.2063E 09	0.2409E 09	0.2766E 09	0.3132E 09	0.3506E 09	0.3886E 09	0.4273E 09
Luminance	0.3895E 11	0.5143E 11	0.6515E 11	0.7992E 11	0.9558E 11	0.1120E 12	0.1290E 12	0.1467E 12	0.1648E 12	0.1833E 12
0.71	0.2303E 05	0.2854E 05	0.3432E 05	0.4033E 05	0.4651E 05	0.5285E 05	0.5931E 05	0.6588E 05	0.7253E 05	0.7927E 05
0.72	0.3451E 05	0.4274E 05	0.5138E 05	0.6034E 05	0.6957E 05	0.7903E 05	0.8867E 05	0.9846E 05	0.1083E 06	0.1184E 06
0.73	0.4035E 05	0.4994E 05	0.6001E 05	0.7046E 05	0.8122E 05	0.9223E 05	0.1034E 06	0.1148E 06	0.1264E 06	0.1381E 06
0.74	0.4336E 05	0.5366E 05	0.6446E 05	0.7566E 05	0.8720E 05	0.9901E 05	0.1110E 06	0.1232E 06	0.1356E 06	0.1482E 06
0.75	0.4495E 05	0.5561E 05	0.6679E 05	0.7839E 05	0.9033E 05	0.1025E 06	0.1150E 06	0.1276E 06	0.1405E 06	0.1534E 06
0.76	0.4580E 05	0.5665E 05	0.6803E 05	0.7984E 05	0.9200E 05	0.1044E 06	0.1171E 06	0.1300E 06	0.1430E 06	0.1562E 06
0.77	0.4626E 05	0.5721E 05	0.6870E 05	0.8062E 05	0.9289E 05	0.1054E 06	0.1182E 06	0.1312E 06	0.1444E 06	0.1577E 06
0.78	0.4651E 05	0.5753E 05	0.6907E 05	0.8105E 05	0.9339E 05	0.1060E 06	0.1188E 06	0.1319E 06	0.1451E 06	0.1585E 06
Luminance	0.3895E 11	0.5143E 11	0.6515E 11	0.7992E 11	0.9558E 11	0.1120E 12	0.1290E 12	0.1467E 12	0.1648E 12	0.1833E 12

xEn = x 10^n. E.g. 1.234E 05 = 1.234 x 10^5

$\int_o^\lambda L_{v'\lambda}\, d\lambda$

F. Functions for Perfect Lens, Circular Aperture, Monochromatic Radiation

Table 19—Point Spread Function

Table 20—Encircled Energy Complement

Table 21—Line Spread Function

Table 22—Modulation Transfer Function

TABLE 19

Point Spread Function

r_r	.000	.001	.002	.003	.004	.005	.006	.007	.008	.009	m	Δ_2
0.00	.000000	.999998	.999990	.999978	.999961	.999938	.999911	.999879	.999842	.999800	0	−4
0.01	.999753	.999701	.999645	.999583	.999516	.999445	.999368	.999287	.999201	.999110	0	−4
0.02	.999013	.998912	.998806	.998695	.998580	.998459	.998333	.998203	.998067	.997927	0	−4
0.03	.997781	.997631	.997476	.997316	.997151	.996981	.996806	.996627	.996442	.996253	0	−4
0.04	.996059	.995859	.995655	.995446	.995233	.995014	.994790	.994562	.994329	.994090	0	−4
0.05	.993847	.993599	.993347	.993089	.992827	.992559	.992287	.992010	.991728	.991442	0	−4
0.06	.991150	.990854	.990553	.990247	.989936	.989620	.989300	.988975	.988645	.988310	0	−4
0.07	.987970	.987626	.987277	.986923	.986564	.986201	.985833	.985460	.985082	.984699	0	−4
0.08	.984312	.983920	.983523	.983122	.982716	.982305	.981889	.981469	.981044	.980614	0	−4
0.09	.980180	.979741	.979297	.978848	.978395	.977937	.977475	.977008	.976536	.976059	0	−4
0.10	.975578	.975092	.974602	.974107	.973607	.973103	.972594	.972081	.971563	.971040	0	−4
0.11	.970513	.969981	.969445	.968904	.968359	.967809	.967254	.966695	.966132	.965564	0	−4
0.12	.964991	.964414	.963832	.963246	.962656	.962061	.961461	.960857	.960249	.959636	0	−3
0.13	.959018	.958397	.957770	.957140	.956505	.955865	.955222	.954573	.953921	.953264	0	−3
0.14	.952602	.951937	.951267	.950592	.949914	.949231	.948543	.947852	.947156	.946456	0	−3
0.15	.945751	.945042	.944329	.943612	.942891	.942165	.941435	.940701	.939962	.939220	0	−3
0.16	.938473	.937722	.936967	.936207	.935444	.934676	.933904	.933128	.932348	.931564	0	−3
0.17	.930776	.929984	.929187	.928387	.927582	.926773	.925961	.925144	.924323	.923499	0	−3
0.18	.922670	.921837	.921001	.920160	.919315	.918467	.917614	.916758	.915897	.915033	0	−3
0.19	.914165	.913293	.912417	.911537	.910653	.909766	.908874	.907979	.907080	.906177	0	−3
0.20	.905271	.904360	.903446	.902528	.901606	.900681	.899752	.898819	.897882	.896942	0	−3
0.21	.895998	.895050	.894099	.893144	.892185	.891223	.890257	.889287	.888314	.887338	0	−3
0.22	.886357	.885374	.884386	.883395	.882401	.881403	.880402	.879397	.878388	.877376	0	−2
0.23	.876361	.875342	.874320	.873295	.872266	.871233	.870197	.869158	.868116	.867070	0	−2
0.24	.866021	.864968	.863912	.862853	.861791	.860725	.859656	.858584	.857508	.856430	0	−2
0.25	.855348	.854263	.853175	.852083	.850989	.849891	.848790	.847686	.846579	.845469	0	−2
0.26	.844356	.843239	.842120	.840998	.839872	.838744	.837612	.836478	.835340	.834200	0	−2
0.27	.833057	.831910	.830761	.829609	.828454	.827296	.826135	.824972	.823805	.822636	0	−2
0.28	.821464	.820289	.819111	.817931	.816748	.815562	.814373	.813181	.811987	.810790	0	−2
0.29	.809591	.808389	.807184	.805976	.804766	.803553	.802338	.801120	.799900	.798677	0	−2
0.30	.797451	.796223	.794992	.793759	.792524	.791286	.790045	.788802	.787557	.786309	0	−1
0.31	.785058	.783806	.782551	.781293	.780034	.778772	.777507	.776241	.774972	.773701	0	−1
0.32	.772427	.771151	.769873	.768593	.767311	.766026	.764739	.763451	.762159	.760866	0	−1
0.33	.759571	.758274	.756974	.755673	.754369	.753063	.751756	.750446	.749134	.747820	0	−1
0.34	.746505	.745187	.743868	.742546	.741223	.739897	.738570	.737241	.735910	.734577	0	−1
0.35	.733243	.731907	.730568	.729228	.727887	.726543	.725198	.723851	.722502	.721152	0	−1
0.36	.719800	.718446	.717091	.715734	.714375	.713015	.711653	.710290	.708925	.707559	0	−1
0.37	.706191	.704821	.703450	.702078	.700704	.699328	.697952	.696573	.695194	.693812	0	−0
0.38	.692430	.691046	.689661	.688274	.686887	.685497	.684107	.682715	.681322	.679928	0	−0
0.39	.678533	.677136	.675738	.674339	.672939	.671537	.670135	.668731	.667326	.665920	0	−0
0.40	.664513	.663105	.661696	.660286	.658875	.657463	.656049	.654635	.653220	.651804	0	−0
0.41	.650387	.648969	.647550	.646130	.644710	.643288	.641866	.640443	.639019	.637594	0	−0
0.42	.636169	.634742	.633315	.631887	.630459	.629029	.627599	.626169	.624737	.623305	0	−0
0.43	.621873	.620439	.619006	.617571	.616136	.614700	.613264	.611827	.610390	.608952	0	0
0.44	.607514	.606075	.604636	.603197	.601756	.600316	.598875	.597434	.595992	.594550	0	0
0.45	.593108	.591665	.590222	.588778	.587335	.585891	.584447	.583002	.581557	.580112	0	0
0.46	.578667	.577222	.575777	.574331	.572885	.571439	.569993	.568547	.567101	.565654	0	0
0.47	.564208	.562761	.561315	.559868	.558422	.556975	.555529	.554082	.552636	.551190	0	0
0.48	.549743	.548297	.546851	.545405	.543959	.542514	.541068	.539623	.538178	.536733	0	0
0.49	.535288	.533843	.532399	.530955	.529511	.528068	.526624	.525182	.523739	.522297	0	0

TABLE 19

Point Spread Function

r_r	.000	.001	.002	.003	.004	.005	.006	.007	.008	.009	m	Δ_2
0.50	.520855	.519413	.517972	.516532	.515091	.513651	.512212	.510773	.509334	.507896	0	0
0.51	.506459	.505021	.503585	.502149	.500713	.499278	.497844	.496410	.494977	.493544	0	1
0.52	.492112	.490681	.489250	.487820	.486390	.484961	.483533	.482106	.480679	.479253	0	1
0.53	.477828	.476404	.474980	.473557	.472135	.470714	.469293	.467874	.466455	.465037	0	1
0.54	.463620	.462204	.460789	.459374	.457961	.456548	.455137	.453726	.452317	.450908	0	1
0.55	.449500	.448094	.446688	.445283	.443880	.442477	.441076	.439675	.438276	.436878	0	1
0.56	.435481	.434085	.432690	.431297	.429904	.428513	.427123	.425734	.424346	.422959	0	1
0.57	.421574	.420190	.418807	.417426	.416045	.414666	.413289	.411912	.410537	.409164	0	1
0.58	.407791	.406420	.405050	.403682	.402315	.400950	.399586	.398223	.396862	.395502	0	1
0.59	.394143	.392787	.391431	.390077	.388725	.387374	.386024	.384676	.383330	.381985	0	2
0.60	.380642	.379300	.377960	.376621	.375285	.373949	.372615	.371283	.369953	.368624	0	2
0.61	.367297	.365971	.364648	.363325	.362005	.360686	.359369	.358054	.356740	.355429	0	2
0.62	.354119	.352810	.351504	.350199	.348896	.347595	.346296	.344998	.343703	.342409	0	2
0.63	.341117	.339827	.338538	.337252	.335968	.334685	.333404	.332126	.330849	.329574	0	2
0.64	.328301	.327030	.325761	.324493	.323228	.321965	.320704	.319445	.318188	.316932	0	2
0.65	.315679	.314428	.313179	.311932	.310687	.309444	.308203	.306965	.305728	.304493	0	2
0.66	.303261	.302030	.300802	.299576	.298352	.297130	.295911	.294693	.293478	.292265	0	2
0.67	.291054	.289845	.288638	.287434	.286231	.285031	.283834	.282638	.281445	.280254	0	2
0.68	.279065	.277878	.276694	.275512	.274332	.273155	.271980	.270807	.269636	.268468	0	2
0.69	.267302	.266138	.264977	.263818	.262662	.261507	.260356	.259206	.258059	.256914	0	2
0.70	.255772	.254632	.253494	.252359	.251226	.250096	.248968	.247842	.246719	.245599	0	2
0.71	.244480	.243364	.242251	.241140	.240032	.238926	.237822	.236721	.235623	.234527	0	2
0.72	.233433	.232342	.231254	.230168	.229084	.228003	.226925	.225849	.224775	.223704	0	3
0.73	.222636	.221570	.220507	.219446	.218388	.217333	.216280	.215229	.214182	.213136	0	3
0.74	.212094	.211054	.210016	.208981	.207949	.206919	.205892	.204868	.203846	.202827	0	3
0.75	.201810	.200797	.199785	.198777	.197771	.196767	.195767	.194768	.193773	.192780	0	3
0.76	.191790	.190803	.189818	.188836	.187857	.186880	.185906	.184934	.183966	.183000	0	3
0.77	.182037	.181076	.180118	.179163	.178210	.177261	.176314	.175369	.174428	.173489	0	3
0.78	.172552	.171619	.170688	.169760	.168835	.167912	.166992	.166075	.165161	.164249	0	3
0.79	.163340	.162434	.161531	.160630	.159732	.158837	.157945	.157055	.156168	.155284	0	3
0.80	.154403	.153524	.152648	.151775	.150905	.150037	.149173	.148311	.147451	.146595	0	3
0.81	.145741	.144890	.144042	.143197	.142354	.141514	.140677	.139843	.139011	.138182	0	3
0.82	.137357	.136533	.135713	.134895	.134081	.133268	.132459	.131653	.130849	.130048	0	3
0.83	.129250	.128455	.127662	.126872	.126085	.125301	.124520	.123741	.122965	.122192	0	3
0.84	.121422	.120654	.119890	.119128	.118369	.117612	.116859	.116108	.115360	.114614	0	3
0.85	.113872	.113132	.112395	.111661	.110930	.110201	.109476	.108753	.108032	.107315	0	3
0.86	.106600	.105888	.105179	.104473	.103769	.103068	.102370	.101675	.100982	.100292	0	3
0.87	.099605	.098921	.098239	.097561	.096885	.096211	.095541	.094873	.094208	.093546	0	3
0.88	.092886	.092229	.091575	.090924	.090275	.089630	.088986	.088346	.087708	.087073	0	3
0.89	.086441	.085812	.085185	.084561	.083940	.083321	.082705	.082092	.081481	.080874	0	3
0.90	.080269	.079666	.079066	.078469	.077875	.077284	.076695	.076108	.075525	.074944	0	3
0.91	.074366	.073790	.073217	.072647	.072079	.071515	.070952	.070393	.069836	.069281	0	3
0.92	.068730	.068181	.067634	.067091	.066550	.066011	.065475	.064942	.064412	.063884	0	3
0.93	.063358	.062835	.062315	.061798	.061283	.060770	.060260	.059753	.059249	.058747	0	3
0.94	.058247	.057750	.057256	.056764	.056275	.055788	.055304	.054823	.054344	.053867	0	3
0.95	.053393	.052922	.052453	.051987	.051523	.051061	.050603	.050146	.049692	.049241	0	2
0.96	.048792	.048346	.047902	.047461	.047022	.046585	.046152	.045720	.045291	.044864	0	2
0.97	.044440	.044019	.043599	.043183	.042768	.042356	.041947	.041539	.041135	.040732	0	2
0.98	.040332	.039935	.039540	.039147	.038757	.038369	.037983	.037600	.037219	.036840	0	2
0.99	.036464	.036090	.035719	.035350	.034983	.034618	.034256	.033896	.033539	.033183	0	2

TABLE 19

Point Spread Function

r_r	.00	.01	.02	.03	.04	.05	.06	.07	.08	.09	m
1.00	.328305	.294259	.262452	.232826	.205322	.179882	.156444	.134944	.115319	.097504	-1
1.10	.081432	.067037	.054253	.043010	.033241	.024878	.017853	.012097	.007543	.004123	-1
1.20	.001770	.000418	.000000	.000454	.001713	.003717	.006402	.009709	.013579	.017954	-1
1.30	.022778	.027997	.033557	.039408	.045500	.051785	.058219	.064756	.071355	.077977	-1
1.40	.084583	.091138	.097607	.103959	.110164	.116195	.122025	.127631	.132991	.138085	-1
1.50	.142896	.147409	.151608	.155482	.159021	.162216	.165061	.167550	.169681	.171451	-1
1.60	.172860	.173909	.174601	.174940	.174931	.174579	.173894	.172883	.171555	.169922	-1
1.70	.167995	.165785	.163306	.160571	.157594	.154389	.150973	.147359	.143564	.139603	-1
1.80	.135493	.131249	.126889	.122427	.117881	.113266	.108597	.103891	.099163	.094428	-1
1.90	.089700	.084994	.080323	.075701	.071139	.066652	.062250	.057943	.053744	.049660	-1
2.00	.457023	.418782	.381958	.346620	.312833	.280652	.250128	.221303	.194212	.168885	-2
2.10	.145342	.123600	.103666	.085543	.069225	.054704	.041962	.030977	.021722	.014164	-2
2.20	.008264	.003981	.001267	.000071	.000338	.002008	.005020	.009309	.014806	.021441	-2
2.30	.029142	.037835	.047443	.057890	.069099	.080990	.093485	.106506	.119974	.133812	-2
2.40	.147941	.162287	.176775	.191331	.205884	.220365	.234706	.248844	.262714	.276259	-2
2.50	.289420	.302145	.314383	.326086	.337210	.347714	.357562	.366719	.375156	.382845	-2
2.60	.389764	.395894	.401217	.405723	.409402	.412250	.414263	.415444	.415798	.415331	-2
2.70	.414056	.411985	.409137	.405529	.401184	.396126	.390383	.383983	.376956	.369336	-2
2.80	.361156	.352453	.343264	.333627	.323581	.313166	.302423	.291394	.280120	.268643	-2
2.90	.257005	.245247	.233411	.221537	.209665	.197837	.186089	.174459	.162985	.151701	-2
3.00	.140642	.129839	.119325	.109127	.099274	.089791	.080703	.072031	.063796	.056016	-2
3.10	.048707	.041883	.035556	.029736	.024431	.019647	.015387	.011654	.008447	.005764	-2
3.20	.003601	.001953	.000811	.000166	.000007	.000321	.001094	.002311	.003955	.006006	-2
3.30	.008446	.011254	.014408	.017887	.021666	.025723	.030031	.034568	.039306	.044220	-2
3.40	.049285	.054475	.059763	.065125	.070533	.075964	.081392	.086792	.092142	.097417	-2
3.50	.102595	.107655	.112576	.117337	.121921	.126308	.130483	.134429	.138132	.141579	-2
3.60	.144758	.147657	.150268	.152583	.154594	.156295	.157683	.158755	.159509	.159944	-2
3.70	.160061	.159863	.159353	.158535	.157415	.156000	.154297	.152315	.150064	.147555	-2
3.80	.144800	.141809	.138597	.135178	.131565	.127773	.123818	.119716	.115483	.111134	-2
3.90	.106688	.102159	.097566	.092926	.088254	.083567	.078883	.074217	.069586	.065004	-2
4.00	.604882	.560519	.517098	.474756	.433624	.393826	.355479	.318695	.283576	.250217	-3
4.10	.218705	.189120	.161532	.136003	.112585	.091325	.072256	.055406	.040794	.028427	-3
4.20	.018308	.010427	.004768	.001307	.000012	.000841	.003748	.008676	.015564	.024343	-3
4.30	.034938	.047269	.061250	.076789	.093790	.112153	.131774	.152547	.174361	.197103	-3
4.40	.220660	.244915	.269751	.295051	.320697	.346573	.372561	.398546	.424416	.450058	-3
4.50	.475363	.500226	.524543	.548215	.571147	.593247	.614429	.634612	.653717	.671674	-3
4.60	.688416	.703884	.718022	.730782	.742121	.752004	.760399	.767283	.772637	.776452	-3
4.70	.778721	.779445	.778631	.776292	.772448	.767122	.760344	.752150	.742580	.731679	-3
4.80	.719498	.706091	.691517	.675838	.659121	.641434	.622850	.603444	.583294	.562478	-3
4.90	.541079	.519178	.496858	.474205	.451302	.428234	.405085	.381938	.358877	.335982	-3

TABLE 19

Point Spread Function

r_r	.00	.01	.02	.03	.04	.05	.06	.07	.08	.09	m
5.00	.313333	.291009	.269084	.247633	.226726	.206430	.186811	.167930	.149845	.132610	
5.10	.116275	.100887	.086486	.073112	.060798	.049572	.039459	.030479	.022647	.015974	−3
5.20	.010465	.006123	.002944	.000921	.000043	.000293	.001651	.004092	.007590	.012111	−3
5.30	.017621	.024080	.031447	.039676	.048720	.058528	.069047	.080222	.091997	.104313	−3
5.40	.117110	.130327	.143901	.157771	.171873	.186143	.200519	.214938	.229338	.243656	−3
5.50	.257832	.271808	.285525	.298927	.311960	.324572	.336712	.348335	.359393	.369845	−3
5.60	.379653	.388778	.397189	.404854	.411747	.417844	.423125	.427574	.431176	.433923	−3
5.70	.435807	.436827	.436981	.436275	.434715	.432312	.429081	.425036	.420200	.414594	−3
5.80	.408243	.401178	.393427	.385024	.376005	.366408	.356271	.345635	.334544	.323042	−3
5.90	.311173	.298983	.286521	.273833	.260968	.247974	.234899	.221792	.208699	.195670	−3
6.00	.182750	.169985	.157420	.145099	.133064	.121354	.110011	.099070	.088567	.078536	−3
6.10	.069007	.060010	.051570	.043713	.036460	.029830	.023840	.018503	.013831	.009833	−3
6.20	.006513	.003876	.001922	.000649	.000051	.000123	.000853	.002229	.004238	.006861	−3
6.30	.010080	.013873	.018217	.023086	.028454	.034292	.040569	.047254	.054314	.061715	−3
6.40	.069421	.077397	.085606	.094010	.102573	.111256	.120021	.128831	.137647	.146433	−3
6.50	.155151	.163765	.172240	.180541	.188635	.196488	.204070	.211350	.218302	.224896	−3
6.60	.231109	.236917	.242298	.247232	.251702	.255692	.259188	.262178	.264653	.266605	−3
6.70	.268028	.268920	.269279	.269107	.268405	.267181	.265439	.263190	.260445	.257216	−3
6.80	.253519	.249368	.244783	.239784	.234390	.228625	.222511	.216075	.209341	.202338	−3
6.90	.195091	.187630	.179984	.172182	.164254	.156231	.148142	.140017	.131887	.123782	−3
7.00	.115731	.107763	.099908	.092192	.084642	.077286	.070149	.063254	.056624	.050282	−3
7.10	.044248	.038541	.033179	.028178	.023554	.019318	.015483	.012058	.009052	.006470	−3
7.20	.004319	.002601	.001317	.000467	.000048	.000058	.000491	.001340	.002596	.004250	−3
7.30	.006290	.008703	.011475	.014590	.018032	.021783	.025824	.030135	.034695	.039483	−3
7.40	.044476	.049652	.054988	.060459	.066041	.071710	.077441	.083210	.088992	.094764	−3
7.50	.100500	.106177	.111772	.117262	.122624	.127838	.132882	.137736	.142382	.146801	−3
7.60	.150976	.154892	.158533	.161886	.164939	.167681	.170102	.172193	.173947	.175359	−3
7.70	.176425	.177141	.177507	.177522	.177187	.176505	.175480	.174118	.172424	.170408	−3
7.80	.168077	.165443	.162516	.159310	.155837	.152111	.148149	.143966	.139580	.135006	−3
7.90	.130265	.125374	.120352	.115219	.109995	.104700	.099353	.093976	.088587	.083208	−3
8.00	.077858	.072556	.067323	.062176	.057134	.052216	.047437	.042816	.038367	.034106	−3
8.10	.030048	.026204	.022588	.019212	.016084	.013216	.010615	.008288	.006241	.004480	−3
8.20	.003007	.001826	.000939	.000344	.000042	.000031	.000306	.000865	.001702	.002811	−3
8.30	.004183	.005812	.007687	.009799	.012136	.014687	.017439	.020378	.023492	.026766	−3
8.40	.030184	.033731	.037391	.041149	.044987	.048890	.052840	.056820	.060815	.064806	−3
8.50	.068778	.072715	.076599	.080415	.084148	.087783	.091305	.094700	.097955	.101058	−3
8.60	.103995	.106757	.109332	.111711	.113884	.115845	.117585	.119099	.120381	.121427	−3
8.70	.122234	.122800	.123122	.123201	.123037	.122631	.121987	.121106	.119994	.118656	−3
8.80	.117098	.115326	.113348	.111172	.108808	.106265	.103554	.100686	.097672	.094524	−3
8.90	.091256	.087878	.084406	.080852	.077231	.073555	.069840	.066098	.062346	.058595	−3
9.00	.548617	.511581	.474985	.438961	.403641	.369150	.335612	.303145	.271862	.241872	−4
9.10	.213276	.186171	.160646	.136785	.114664	.094350	.075905	.059382	.044827	.032278	−4
9.20	.021763	.013304	.006915	.002600	.000357	.000173	.002029	.005900	.011749	.019534	−4
9.30	.029206	.040709	.053978	.068944	.085531	.103657	.123234	.144169	.166367	.189724	−4
9.40	.214136	.239493	.265684	.292594	.320108	.348106	.376470	.405080	.433815	.462556	−4
9.50	.491185	.519582	.547632	.575222	.602239	.628575	.654126	.678789	.702468	.725069	−4
9.60	.746506	.766695	.785559	.803027	.819032	.833516	.846426	.857714	.867340	.875273	−4
9.70	.881485	.885958	.888678	.889641	.888849	.886309	.882037	.876055	.868392	.859083	−4
9.80	.848170	.835699	.821724	.806305	.789504	.771392	.752043	.731534	.709949	.687374	−4
9.90	.663898	.639614	.614618	.589007	.562881	.536340	.509486	.482421	.455250	.428073	−4

TABLE 19

Point Spread Function

r_r	.0	.1	.2	.3	.4	.5	.6	.7	.8	.9	m
10.00	.400993	.156800	.016248	.021186	.157370	.362912	.553852	.656450	.633908	.497967	-4
11.00	.301921	.118625	.012448	.015851	.119018	.275684	.422178	.501938	.486145	.383032	-4
12.00	.232974	.091901	.009745	.012166	.092180	.214313	.329145	.392351	.380960	.300913	-4
13.00	.183519	.072637	.007770	.009539	.072841	.169887	.261557	.312479	.304058	.240687	-4
14.00	.147127	.058402	.006295	.007617	.058555	.136938	.211277	.252896	.246537	.195518	-4
15.00	.119755	.047656	.005170	.006178	.047772	.111988	.173101	.207548	.202657	.160981	-4
16.00	.098773	.039393	.004298	.005080	.039483	.092748	.143596	.172425	.168601	.134121	-4
17.00	.082420	.032935	.003612	.004227	.033005	.077675	.120432	.144801	.141769	.112920	-4
18.00	.069486	.027814	.003064	.003555	.027871	.065699	.101995	.122776	.120341	.095961	-4
19.00	.059123	.023703	.002621	.003018	.023748	.056064	.087137	.105000	.103022	.082236	-4
20.00	.507221	.203629	.022600	.025844	.204002	.482242	.750300	.904974	.888739	.710078	-5
21.00	.438405	.176224	.019622	.022296	.176530	.417806	.650661	.785469	.772021	.617343	-5
22.00	.381494	.153523	.017144	.019370	.153779	.364358	.567912	.686113	.674878	.540077	-5
23.00	.334024	.134559	.015067	.016934	.134774	.319651	.498620	.602832	.593373	.475186	-5
24.00	.294113	.118595	.013312	.014889	.118777	.281968	.440158	.532501	.524481	.420288	-5
25.00	.260316	.105060	.011820	.013161	.105213	.249983	.390489	.472699	.465854	.373531	-5
26.00	.231504	.093508	.010542	.011691	.093640	.222659	.348021	.421527	.415650	.333460	-5
27.00	.206793	.083589	.009442	.010431	.083703	.199176	.311494	.377483	.372407	.298922	-5
28.00	.185476	.075025	.008490	.009346	.075124	.178883	.279906	.339367	.334961	.268993	-5
29.00	.166992	.067592	.007661	.008407	.067678	.161255	.252448	.306214	.302372	.242930	-5
30.00	.150885	.061110	.006937	.007589	.061185	.145869	.228467	.277243	.273877	.220129	-5
31.00	.136784	.055430	.006302	.006874	.055496	.132380	.207430	.251814	.248853	.200093	-5
32.00	.124387	.050433	.005742	.006246	.050491	.120505	.188898	.229403	.226787	.182418	-5
33.00	.113444	.046019	.005246	.005692	.046070	.110008	.172510	.209574	.207254	.166764	-5
34.00	.103748	.042106	.004805	.005202	.042151	.100696	.157964	.191966	.189902	.152852	-5
35.00	.095126	.038624	.004413	.004766	.038664	.092406	.145008	.176276	.174434	.140444	-5
36.00	.087433	.035515	.004062	.004378	.035551	.085001	.133432	.162251	.160601	.129345	-5
37.00	.080548	.032732	.003748	.004031	.032764	.078367	.123055	.149675	.148193	.119384	-5
38.00	.074368	.030231	.003465	.003720	.030261	.072406	.113727	.138366	.137031	.110421	-5
39.00	.068804	.027980	.003210	.003439	.028006	.067034	.105319	.128168	.126962	.102333	-5
40.00	.063781	.025946	.002979	.003187	.025970	.062181	.097719	.118948	.117857	.095016	-5
41.00	.059236	.024105	.002770	.002958	.024126	.057785	.090834	.110592	.109601	.088380	-5
42.00	.551122	.224336	.025798	.027508	.224529	.537939	.845799	*******	*******	.823483	-6
43.00	.513625	.209134	.024067	.025626	.209312	.501622	.788877	.960875	.952661	.768529	-6
44.00	.479454	.195276	.022488	.023911	.195439	.468501	.736947	.897801	.890297	.718359	-6
45.00	.448252	.182616	.021045	.022345	.182762	.438232	.689477	.840128	.833260	.672463	-6
46.00	.419696	.171026	.019722	.020914	.171162	.410518	.646000	.787292	.780992	.630394	-6
47.00	.393517	.160397	.018508	.019602	.160521	.385090	.606100	.738794	.733006	.591762	-6
48.00	.369469	.150630	.017391	.018398	.150746	.361721	.569421	.694200	.688872	.556223	-6
49.00	.347343	.141641	.016363	.017290	.141747	.340204	.535644	.653124	.648212	.523475	-6

TABLE 19

Point Spread Function

r_r	.0	.1	.2	.3	.4	.5	.6	.7	.8	.9	m
50.00	.326948	.133354	.015414	.016269	.133451	.320361	.504484	.615225	.610689	.493248	
51.00	.308119	.125699	.014537	.015327	.125790	.302032	.475697	.580203	.576007	.465303	-6
52.00	.290709	.118622	.013725	.014457	.118704	.285074	.449057	.547789	.543903	.439431	-6
53.00	.274585	.112063	.012973	.013651	.112141	.269363	.424372	.517746	.514140	.415440	-6
54.00	.259633	.105980	.012274	.012904	.106053	.254785	.401463	.489860	.486511	.393166	-6
55.00	.245747	.100331	.011625	.012210	.100397	.241240	.380172	.463941	.460826	.372457	-6
56.00	.232833	.095074	.011021	.011566	.095137	.228640	.360362	.439819	.436917	.353175	-6
57.00	.220810	.090180	.010458	.010966	.090238	.216900	.341904	.417340	.414635	.335203	-6
58.00	.209599	.085615	.009933	.010407	.085670	.205952	.324686	.396368	.393842	.318429	-6
59.00	.199135	.081353	.009442	.009885	.081404	.195728	.308604	.376778	.374416	.302756	-6
60.00	.189357	.077370	.008983	.009397	.077417	.186169	.293568	.358457	.356248	.288095	-6
61.00	.180207	.073642	.008553	.008941	.073687	.177224	.279493	.341306	.339236	.274365	-6
62.00	.171639	.070150	.008151	.008514	.070191	.168841	.266303	.325231	.323290	.261495	-6
63.00	.163604	.066875	.007773	.008114	.066915	.160980	.253930	.310150	.308328	.249416	-6
64.00	.156063	.063802	.007418	.007738	.063838	.153600	.242312	.295988	.294276	.238071	-6
65.00	.148980	.060914	.007085	.007385	.060947	.146662	.231391	.282674	.281064	.227403	-6
66.00	.142317	.058197	.006771	.007054	.058229	.140137	.221118	.270148	.268632	.217364	-6
67.00	.136047	.055640	.006475	.006742	.055670	.133993	.211444	.258350	.256922	.207906	-6
68.00	.130140	.053230	.006197	.006448	.053258	.128203	.202326	.247230	.245883	.198990	-6
69.00	.124569	.050957	.005934	.006171	.050984	.122743	.193724	.236739	.235468	.190575	-6
70.00	.119313	.048812	.005686	.005909	.048837	.117588	.185604	.226833	.225632	.182629	-6
71.00	.114347	.046786	.005451	.005663	.046810	.112718	.177931	.217472	.216337	.175119	-6
72.00	.109654	.044870	.005229	.005429	.044893	.108112	.170675	.208619	.207545	.168014	-6
73.00	.105214	.043058	.005019	.005209	.043079	.103755	.163809	.200241	.199224	.161289	-6
74.00	.101010	.041341	.004820	.005000	.041362	.099628	.157306	.192305	.191341	.154919	-6
75.00	.097027	.039715	.004632	.004802	.039734	.095717	.151143	.184783	.183869	.148879	-6
76.00	.093252	.038173	.004453	.004614	.038191	.092009	.145297	.177648	.176781	.143149	-6
77.00	.089669	.036710	.004283	.004436	.036727	.088490	.139749	.170876	.170052	.137709	-6
78.00	.086268	.035320	.004122	.004268	.035337	.085147	.134480	.164443	.163661	.132542	-6
79.00	.083036	.034000	.003969	.004107	.034016	.081971	.129472	.158330	.157586	.127630	-6
80.00	.079964	.032745	.003823	.003955	.032760	.078951	.124711	.152516	.151808	.122958	-6
81.00	.077042	.031551	.003685	.003810	.031565	.076078	.120179	.146983	.146309	.118511	-6
82.00	.074260	.030414	.003553	.003672	.030428	.073342	.115865	.141714	.141073	.114275	-6
83.00	.071611	.029331	.003427	.003540	.029344	.070736	.111754	.136695	.136083	.110240	-6
84.00	.069086	.028300	.003307	.003415	.028311	.068252	.107836	.131909	.131326	.106392	-6
85.00	.066679	.027316	.003193	.003296	.027326	.065883	.104099	.127344	.126788	.102721	-6
86.00	.064381	.026376	.003083	.003182	.026387	.063622	.100533	.122988	.122457	.099217	-6
87.00	.062189	.025480	.002979	.003073	.025490	.061464	.097128	.118828	.118321	.095871	-6
88.00	.060094	.024623	.002879	.002969	.024633	.059402	.093875	.114854	.114369	.092673	-6
89.00	.058094	.023805	.002784	.002870	.023814	.057431	.090765	.111055	.110591	.089617	-6
90.00	.056180	.023023	.002693	.002775	.023032	.055546	.087791	.107421	.106978	.086693	-6
91.00	.054350	.022274	.002606	.002685	.022283	.053744	.084946	.103945	.103520	.083894	-6
92.00	.052598	.021557	.002523	.002598	.021566	.052018	.082223	.100617	.100210	.081215	-6
93.00	.050921	.020871	.002443	.002515	.020879	.050365	.079614	.097429	.097039	.078649	-6
94.00	.049315	.020214	.002366	.002435	.020221	.048782	.077115	.094374	.094001	.076190	-6
95.00	.047775	.019584	.002293	.002359	.019591	.047264	.074719	.091446	.091089	.073833	-6
96.00	.046298	.018980	.002222	.002286	.018987	.045809	.072422	.088638	.088295	.071571	-6
97.00	.044882	.018401	.002155	.002216	.018407	.044413	.070217	.085944	.085614	.069401	-6
98.00	.043524	.017844	.002090	.002148	.017851	.043073	.068102	.083358	.083041	.067318	-6
99.00	.042219	.017311	.002028	.002084	.017317	.041786	.066070	.080874	.080570	.065317	-6

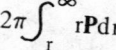

$2\pi \int_r^\infty r\mathbf{P}dr$

TABLE 20

Encircled Energy Complement

r_r	.000	.001	.002	.003	.004	.005	.006	.007	.008	.009	m	Δ_2
0.00	1.00	.999998	.999990	.999978	.999961	.999938	.999911	.999879	.999842	.999800	0	-4
0.01	.999753	.999701	.999645	.999583	.999516	.999445	.999369	.999287	.999201	.999110	0	-4
0.02	.999014	.998912	.998806	.998696	.998580	.998459	.998333	.998203	.998067	.997927	0	-4
0.03	.997782	.997632	.997477	.997317	.997152	.996982	.996807	.996628	.996443	.996254	0	-4
0.04	.996060	.995861	.995657	.995448	.995234	.995016	.994793	.994564	.994331	.994093	0	-4
0.05	.993850	.993603	.993350	.993093	.992831	.992564	.992292	.992015	.991734	.991448	0	-4
0.06	.991157	.990861	.990560	.990255	.989944	.989629	.989310	.988985	.988656	.988321	0	-4
0.07	.987983	.987639	.987291	.986937	.986579	.986217	.985849	.985477	.985101	.984719	0	-4
0.08	.984333	.983942	.983546	.983146	.982741	.982331	.981917	.981498	.981074	.980646	0	-4
0.09	.980213	.979775	.979333	.978886	.978434	.977978	.977517	.977052	.976582	.976107	0	-4
0.10	.975628	.975145	.974656	.974163	.973666	.973164	.972658	.972147	.971631	.971111	0	-4
0.11	.970586	.970057	.969524	.968986	.968443	.967896	.967345	.966789	.966228	.965664	0	-3
0.12	.965094	.964521	.963943	.963360	.962773	.962182	.961586	.960986	.960382	.959773	0	-3
0.13	.959160	.958543	.957921	.957295	.956665	.956030	.955391	.954748	.954101	.953449	0	-3
0.14	.952793	.952133	.951468	.950799	.950126	.949449	.948768	.948082	.947393	.946699	0	-3
0.15	.946001	.945299	.944593	.943882	.943168	.942449	.941726	.941000	.940269	.939534	0	-3
0.16	.938795	.938052	.937305	.936554	.935799	.935040	.934277	.933510	.932739	.931964	0	-3
0.17	.931185	.930402	.929615	.928825	.928030	.927232	.926429	.925623	.924813	.923999	0	-3
0.18	.923182	.922360	.921535	.920706	.919873	.919036	.918196	.917352	.916504	.915652	0	-3
0.19	.914797	.913938	.913076	.912209	.911339	.910466	.909588	.908707	.907823	.906935	0	-3
0.20	.906043	.905148	.904249	.903347	.902441	.901531	.900619	.899702	.898782	.897859	0	-2
0.21	.896932	.896002	.895068	.894131	.893191	.892247	.891299	.890349	.889395	.888438	0	-2
0.22	.887477	.886513	.885546	.884575	.883601	.882624	.881644	.880661	.879674	.878684	0	-2
0.23	.877691	.876695	.875695	.874693	.873687	.872678	.871666	.870651	.869633	.868612	0	-2
0.24	.867588	.866561	.865531	.864498	.863461	.862422	.861380	.860335	.859287	.858236	0	-2
0.25	.857182	.856126	.855066	.854004	.852938	.851870	.850799	.849726	.848649	.847570	0	-2
0.26	.846488	.845403	.844316	.843226	.842133	.841037	.839939	.838838	.837735	.836629	0	-2
0.27	.835520	.834409	.833295	.832179	.831060	.829938	.828815	.827688	.826559	.825428	0	-1
0.28	.824294	.823158	.822019	.820878	.819734	.818589	.817440	.816290	.815137	.813982	0	-1
0.29	.812824	.811665	.810503	.809338	.808172	.807003	.805832	.804659	.803484	.802307	0	-1
0.30	.801127	.799946	.798762	.797576	.796388	.795198	.794006	.792812	.791616	.790418	0	-1
0.31	.789218	.788016	.786812	.785606	.784399	.783189	.781978	.780764	.779549	.778332	0	-1
0.32	.777113	.775892	.774670	.773446	.772220	.770992	.769763	.768532	.767299	.766064	0	-1
0.33	.764828	.763590	.762351	.761110	.759867	.758623	.757378	.756130	.754882	.753631	0	-0
0.34	.752379	.751126	.749872	.748615	.747358	.746099	.744838	.743577	.742314	.741049	0	-0
0.35	.739783	.738516	.737248	.735978	.734707	.733435	.732162	.730887	.729611	.728334	0	-0
0.36	.727056	.725777	.724496	.723215	.721932	.720648	.719363	.718078	.716791	.715503	0	-0
0.37	.714214	.712924	.711633	.710341	.709048	.707755	.706460	.705165	.703868	.702571	0	-0
0.38	.701273	.699974	.698675	.697374	.696073	.694771	.693468	.692165	.690860	.689556	0	-0
0.39	.688250	.686944	.685637	.684329	.683021	.681713	.680403	.679094	.677783	.676472	0	-0
0.40	.675161	.673849	.672536	.671224	.669910	.668596	.667282	.665968	.664653	.663337	0	0
0.41	.662021	.660705	.659389	.658072	.656755	.655438	.654120	.652803	.651485	.650166	0	0
0.42	.648848	.647529	.646211	.644892	.643572	.642253	.640934	.639615	.638295	.636976	0	0
0.43	.635656	.634336	.633017	.631697	.630378	.629058	.627738	.626419	.625100	.623780	0	0
0.44	.622461	.621142	.619823	.618505	.617186	.615868	.614549	.613231	.611914	.610596	0	0
0.45	.609279	.607962	.606645	.605329	.604013	.602697	.601382	.600067	.598752	.597438	0	0
0.46	.596124	.594811	.593498	.592186	.590874	.589562	.588251	.586941	.585631	.584321	0	1
0.47	.583012	.581704	.580396	.579089	.577783	.576477	.575171	.573867	.572563	.571260	0	1
0.48	.569957	.568655	.567354	.566054	.564754	.563456	.562157	.560860	.559564	.558268	0	1
0.49	.556973	.555679	.554386	.553094	.551803	.550513	.549223	.547935	.546647	.545360	0	1

TABLE 20

Encircled Energy Complement

r_r	.000	.001	.002	.003	.004	.005	.006	.007	.008	.009	m	Δ_2
0.50	.544075	.542790	.541506	.540224	.538942	.537662	.536382	.535104	.533826	.532550	0	1
0.51	.531275	.530001	.528728	.527456	.526185	.524916	.523647	.522380	.521115	.519850	0	1
0.52	.518586	.517324	.516063	.514804	.513545	.512288	.511032	.509778	.508525	.507273	0	1
0.53	.506023	.504774	.503526	.502280	.501035	.499791	.498549	.497309	.496070	.494832	0	1
0.54	.493596	.492361	.491128	.489896	.488666	.487437	.486210	.484985	.483761	.482538	0	2
0.55	.481318	.480098	.478881	.477665	.476450	.475238	.474027	.472817	.471610	.470404	0	2
0.56	.469199	.467997	.466796	.465597	.464399	.463204	.462010	.460818	.459627	.458439	0	2
0.57	.457252	.456067	.454884	.453703	.452524	.451346	.450170	.448996	.447825	.446655	0	2
0.58	.445486	.444320	.443156	.441993	.440833	.439675	.438518	.437364	.436211	.435060	0	2
0.59	.433912	.432765	.431621	.430478	.429338	.428199	.427063	.425929	.424796	.423666	0	2
0.60	.422538	.421412	.420288	.419167	.418047	.416929	.415814	.414701	.413590	.412481	0	2
0.61	.411374	.410270	.409167	.408067	.406969	.405873	.404780	.403688	.402599	.401513	0	2
0.62	.400428	.399346	.398266	.397188	.396112	.395039	.393968	.392899	.391833	.390769	0	2
0.63	.389708	.388648	.387591	.386536	.385484	.384434	.383387	.382341	.381299	.380258	0	2
0.64	.379220	.378184	.377151	.376120	.375092	.374066	.373042	.372021	.371002	.369986	0	2
0.65	.368972	.367961	.366952	.365946	.364942	.363940	.362941	.361945	.360951	.359959	0	2
0.66	.358970	.357984	.357000	.356019	.355040	.354063	.353090	.352118	.351150	.350184	0	3
0.67	.349220	.348259	.347300	.346345	.345391	.344440	.343492	.342547	.341604	.340663	0	3
0.68	.339726	.338791	.337858	.336928	.336001	.335076	.334154	.333235	.332318	.331404	0	3
0.69	.330492	.329583	.328677	.327774	.326873	.325975	.325079	.324186	.323296	.322408	0	3
0.70	.321524	.320641	.319762	.318885	.318011	.317140	.316271	.315405	.314541	.313681	0	3
0.71	.312823	.311968	.311115	.310265	.309418	.308574	.307732	.306893	.306057	.305223	0	3
0.72	.304393	.303565	.302739	.301917	.301097	.300280	.299465	.298654	.297845	.297039	0	3
0.73	.296235	.295435	.294637	.293842	.293049	.292260	.291473	.290689	.289907	.289128	0	3
0.74	.288353	.287579	.286809	.286041	.285277	.284514	.283755	.282999	.282245	.281494	0	3
0.75	.280745	.280000	.279257	.278517	.277780	.277045	.276314	.275585	.274858	.274135	0	3
0.76	.273414	.272696	.271981	.271269	.270559	.269852	.269148	.268447	.267748	.267052	0	3
0.77	.266359	.265669	.264981	.264297	.263615	.262935	.262259	.261585	.260914	.260246	0	3
0.78	.259580	.258917	.258257	.257600	.256945	.256293	.255644	.254998	.254354	.253713	0	3
0.79	.253075	.252440	.251807	.251177	.250550	.249925	.249304	.248685	.248068	.247455	0	3
0.80	.246844	.246236	.245630	.245027	.244427	.243830	.243235	.242643	.242054	.241467	0	3
0.81	.240883	.240302	.239724	.239148	.238575	.238004	.237436	.236871	.236309	.235749	0	3
0.82	.235192	.234637	.234085	.233536	.232990	.232446	.231905	.231366	.230830	.230297	0	3
0.83	.229766	.229238	.228712	.228190	.227669	.227152	.226637	.226124	.225615	.225107	0	3
0.84	.224603	.224101	.223601	.223104	.222610	.222118	.221629	.221143	.220659	.220177	0	3
0.85	.219698	.219222	.218748	.218277	.217808	.217342	.216878	.216417	.215958	.215502	0	2
0.86	.215048	.214597	.214149	.213702	.213259	.212818	.212379	.211943	.211509	.211077	0	2
0.87	.210649	.210222	.209798	.209377	.208958	.208541	.208127	.207715	.207306	.206899	0	2
0.88	.206494	.206092	.205692	.205295	.204900	.204507	.204117	.203729	.203343	.202960	0	2
0.89	.202579	.202201	.201825	.201451	.201079	.200710	.200343	.199979	.199617	.199257	0	2
0.90	.198899	.198544	.198191	.197840	.197491	.197145	.196801	.196459	.196120	.195782	0	2
0.91	.195447	.195115	.194784	.194456	.194129	.193805	.193484	.193164	.192846	.192531	0	2
0.92	.192218	.191907	.191598	.191292	.190987	.190685	.190384	.190086	.189790	.189496	0	2
0.93	.189205	.188915	.188627	.188342	.188058	.187777	.187497	.187220	.186945	.186671	0	2
0.94	.186400	.186131	.185864	.185599	.185336	.185074	.184815	.184558	.184303	.184050	0	2
0.95	.183798	.183549	.183302	.183056	.182813	.182571	.182331	.182094	.181858	.181624	0	2
0.96	.181392	.181161	.180933	.180707	.180482	.180259	.180038	.179819	.179602	.179387	0	2
0.97	.179173	.178961	.178751	.178543	.178336	.178132	.177929	.177728	.177528	.177331	0	2
0.98	.177135	.176940	.176748	.176557	.176368	.176181	.175995	.175811	.175629	.175448	0	2
0.99	.175269	.175092	.174916	.174742	.174570	.174399	.174230	.174062	.173896	.173732	0	2

$$2\pi \int_r^{\infty} r\mathbf{P}dr$$

TABLE 20

Encircled Energy Complement

r_r	.00	.01	.02	.03	.04	.05	.06	.07	.08	.09	m	Δ_2
1.00	.173569	.172026	.170633	.169382	.168264	.167272	.166397	.165632	.164969	.164400	0	86
1.10	.163918	.163514	.163181	.162912	.162699	.162535	.162414	.162329	.162272	.162239	0	17
1.20	.162222	.162216	.162215	.162214	.162208	.162192	.162161	.162111	.162038	.161938	0	−29
1.30	.161808	.161645	.161445	.161207	.160927	.160604	.160237	.159822	.159361	.158850	0	−48
1.40	.158291	.157681	.157022	.156314	.155555	.154748	.153893	.152990	.152041	.151048	0	−42
1.50	.150011	.148933	.147815	.146659	.145468	.144243	.142987	.141703	.140392	.139057	0	−20
1.60	.137702	.136329	.134940	.133538	.132126	.130707	.129284	.127859	.126436	.125016	0	6
1.70	.123602	.122198	.120805	.119426	.118064	.116721	.115398	.114099	.112825	.111577	0	29
1.80	.110359	.109171	.108015	.106892	.105804	.104752	.103736	.102758	.101819	.100919	0	39
1.90	.100058	.099237	.098456	.097715	.097014	.096352	.095731	.095148	.094604	.094098	0	37
2.00	.093628	.093195	.092797	.092433	.092102	.091803	.091534	.091294	.091081	.090895	0	24
2.10	.090732	.090593	.090474	.090375	.090294	.090229	.090177	.090139	.090110	.090091	0	7
2.20	.090079	.090073	.090070	.090069	.090069	.090068	.090064	.090057	.090043	.090023	0	−7
2.30	.089994	.089956	.089908	.089847	.089774	.089687	.089586	.089469	.089337	.089187	0	−16
2.40	.089021	.088837	.088635	.088414	.088176	.087919	.087643	.087349	.087036	.086706	0	−17
2.50	.086358	.085992	.085609	.085210	.084795	.084365	.083920	.083462	.082990	.082507	0	−10
2.60	.082012	.081507	.080993	.080470	.079940	.079404	.078862	.078316	.077768	.077217	0	−0
2.70	.076665	.076114	.075564	.075016	.074471	.073931	.073397	.072868	.072347	.071834	0	9
2.80	.071330	.070836	.070353	.069881	.069422	.068975	.068541	.068121	.067716	.067325	0	15
2.90	.066949	.066589	.066245	.065917	.065605	.065308	.065029	.064765	.064517	.064285	0	16
3.00	.064069	.063869	.063684	.063513	.063357	.063215	.063087	.062971	.062868	.062777	0	11
3.10	.062697	.062628	.062568	.062518	.062476	.062442	.062415	.062394	.062378	.062367	0	4
3.20	.062360	.062355	.062353	.062353	.062353	.062352	.062351	.062349	.062344	.062336	0	−3
3.30	.062324	.062308	.062287	.062260	.062228	.062189	.062143	.062089	.062028	.061958	0	−8
3.40	.061880	.061792	.061696	.061591	.061476	.061351	.061217	.061073	.060920	.060757	0	−8
3.50	.060584	.060402	.060211	.060011	.059803	.059585	.059360	.059127	.058887	.058639	0	−6
3.60	.058385	.058125	.057859	.057588	.057313	.057033	.056750	.056464	.056175	.055885	0	−0
3.70	.055593	.055300	.055008	.054715	.054424	.054135	.053847	.053562	.053280	.053002	0	4
3.80	.052729	.052460	.052196	.051937	.051685	.051439	.051199	.050967	.050742	.050525	0	8
3.90	.050316	.050114	.049921	.049737	.049561	.049394	.049235	.049085	.048944	.048812	0	9
4.00	.048688	.048573	.048466	.048368	.048278	.048195	.048120	.048052	.047992	.047938	0	6
4.10	.047891	.047850	.047814	.047784	.047758	.047738	.047721	.047708	.047698	.047691	0	2
4.20	.047686	.047683	.047682	.047681	.047681	.047681	.047680	.047679	.047677	.047672	0	−1
4.30	.047666	.047657	.047646	.047631	.047613	.047591	.047565	.047534	.047499	.047459	0	−4
4.40	.047413	.047363	.047307	.047245	.047178	.047104	.047025	.046940	.046850	.046753	0	−5
4.50	.046650	.046542	.046428	.046308	.046182	.046052	.045916	.045775	.045630	.045480	0	−3
4.60	.045326	.045168	.045006	.044840	.044672	.044501	.044327	.044151	.043973	.043794	0	−0
4.70	.043614	.043433	.043252	.043070	.042889	.042709	.042530	.042352	.042176	.042002	0	2
4.80	.041830	.041661	.041495	.041332	.041173	.041017	.040866	.040719	.040576	.040438	0	5
4.90	.040305	.040176	.040053	.039935	.039822	.039715	.039613	.039517	.039426	.039340	0	5

TABLE 20

Encircled Energy Complement

r_r	.00	.01	.02	.03	.04	.05	.06	.07	.08	.09	m	Δ_2
5.00	.039260	.039186	.039116	.039052	.038993	.038940	.038891	.038846	.038806	.038771	0	4
5.10	.038740	.038712	.038689	.038669	.038652	.038638	.038626	.038617	.038611	.038606	0	2
5.20	.038602	.038600	.038599	.038599	.038599	.038599	.038598	.038598	.038596	.038594	0	−0
5.30	.038590	.038584	.038577	.038568	.038556	.038542	.038525	.038505	.038483	.038456	0	−2
5.40	.038427	.038394	.038357	.038317	.038273	.038225	.038173	.038117	.038057	.037993	0	−3
5.50	.037925	.037853	.037777	.037697	.037614	.037527	.037436	.037342	.037245	.037144	0	−2
5.60	.037041	.036934	.036825	.036714	.036600	.036485	.036368	.036249	.036128	.036007	0	−0
5.70	.035885	.035762	.035639	.035515	.035392	.035269	.035147	.035025	.034905	.034786	0	2
5.80	.034668	.034552	.034438	.034326	.034216	.034109	.034005	.033903	.033805	.033709	0	3
5.90	.033617	.033528	.033443	.033361	.033282	.033208	.033137	.033070	.033006	.032946	0	4
6.00	.032890	.032838	.032790	.032745	.032703	.032665	.032631	.032599	.032571	.032546	0	3
6.10	.032524	.032505	.032488	.032473	.032461	.032451	.032443	.032437	.032432	.032428	0	1
6.20	.032426	.032424	.032423	.032423	.032423	.032423	.032423	.032422	.032421	.032420	0	−0
6.30	.032417	.032413	.032408	.032402	.032394	.032384	.032372	.032358	.032342	.032324	0	−1
6.40	.032303	.032280	.032254	.032226	.032195	.032161	.032124	.032084	.032042	.031996	0	−2
6.50	.031948	.031897	.031843	.031786	.031726	.031664	.031599	.031532	.031462	.031390	0	−1
6.60	.031316	.031240	.031162	.031082	.031000	.030917	.030832	.030746	.030660	.030572	0	−0
6.70	.030484	.030395	.030306	.030216	.030127	.030038	.029949	.029861	.029773	.029686	0	1
6.80	.029601	.029516	.029433	.029352	.029272	.029193	.029117	.029043	.028971	.028901	0	2
6.90	.028833	.028768	.028705	.028645	.028587	.028533	.028480	.028431	.028384	.028340	0	3
7.00	.028299	.028260	.028224	.028191	.028160	.028132	.028106	.028083	.028062	.028043	0	2
7.10	.028027	.028012	.028000	.027989	.027980	.027972	.027966	.027961	.027958	.027955	0	1
7.20	.027953	.027952	.027951	.027951	.027951	.027951	.027951	.027950	.027950	.027948	0	−0
7.30	.027946	.027944	.027940	.027935	.027930	.027922	.027914	.027904	.027892	.027878	0	−1
7.40	.027863	.027846	.027827	.027805	.027782	.027757	.027730	.027700	.027668	.027634	0	−1
7.50	.027598	.027560	.027519	.027477	.027432	.027386	.027337	.027287	.027234	.027180	0	−1
7.60	.027124	.027067	.027008	.026948	.026886	.026823	.026760	.026695	.026629	.026563	0	−0
7.70	.026496	.026429	.026361	.026294	.026226	.026158	.026091	.026024	.025958	.025892	0	1
7.80	.025827	.025762	.025699	.025637	.025576	.025516	.025458	.025402	.025346	.025293	0	2
7.90	.025241	.025191	.025143	.025097	.025053	.025011	.024971	.024933	.024897	.024863	0	2
8.00	.024832	.024802	.024774	.024749	.024725	.024703	.024683	.024665	.024649	.024635	0	2
8.10	.024622	.024611	.024601	.024593	.024586	.024580	.024575	.024571	.024568	.024566	0	1
8.20	.024565	.024564	.024563	.024563	.024563	.024563	.024563	.024562	.024562	.024561	0	−0
8.30	.024560	.024558	.024555	.024551	.024547	.024541	.024535	.024527	.024518	.024507	0	−0
8.40	.024496	.024482	.024467	.024451	.024433	.024414	.024392	.024370	.024345	.024319	0	−1
8.50	.024291	.024261	.024230	.024197	.024162	.024126	.024088	.024049	.024008	.023966	0	−0
8.60	.023922	.023877	.023831	.023784	.023736	.023687	.023637	.023587	.023536	.023484	0	0
8.70	.023431	.023379	.023326	.023273	.023220	.023167	.023114	.023061	.023009	.022957	0	1
8.80	.022906	.022856	.022806	.022757	.022709	.022662	.022616	.022572	.022528	.022486	0	1
8.90	.022445	.022406	.022368	.022332	.022297	.022263	.022232	.022202	.022173	.022146	0	2
9.00	.022121	.022098	.022076	.022055	.022037	.022019	.022004	.021989	.021977	.021965	0	1
9.10	.021955	.021946	.021938	.021931	.021926	.021921	.021917	.021914	.021912	.021910	0	1
9.20	.021909	.021908	.021908	.021907	.021907	.021907	.021907	.021907	.021907	.021906	0	0
9.30	.021905	.021903	.021901	.021898	.021895	.021890	.021885	.021879	.021872	.021864	0	−0
9.40	.021854	.021844	.021832	.021819	.021805	.021789	.021772	.021754	.021734	.021713	0	−0
9.50	.021691	.021667	.021642	.021616	.021588	.021559	.021529	.021497	.021465	.021431	0	−0
9.60	.021396	.021360	.021324	.021286	.021247	.021208	.021168	.021127	.021086	.021044	0	0
9.70	.021002	.020960	.020917	.020875	.020832	.020789	.020747	.020704	.020662	.020621	0	0
9.80	.020579	.020539	.020498	.020459	.020420	.020382	.020345	.020309	.020274	.020240	0	1
9.90	.020207	.020175	.020144	.020115	.020087	.020060	.020034	.020010	.019987	.019965	0	1

$$2\pi \int_r^\infty r\mathbf{P}dr$$

$$2\pi \int_r^\infty r\mathbf{P}\,dr$$

TABLE 20

Encircled Energy Complement

r_r	.0	.1	.2	.3	.4	.5	.6	.7	.8	.9	m	Δ_2
10.00	.199444	.198089	.197713	.197680	.197270	.195939	.193528	.190296	.186817	.183746	-1	900
11.00	.181575	.180451	.180137	.180111	.179771	.178666	.176656	.173958	.171046	.168470	-1	750
12.00	.166644	.165697	.165432	.165409	.165123	.164191	.162491	.160203	.157730	.155538	-1	636
13.00	.153982	.153173	.152946	.152927	.152683	.151885	.150428	.148464	.146338	.144451	-1	545
14.00	.143109	.142410	.142212	.142196	.141986	.141296	.140033	.138329	.136481	.134839	-1	473
15.00	.133669	.133059	.132886	.132872	.132689	.132086	.130982	.129489	.127868	.126426	-1	414
16.00	.125398	.124861	.124708	.124696	.124535	.124004	.123029	.121711	.120278	.119002	-1	365
17.00	.118091	.117614	.117479	.117468	.117325	.116853	.115987	.114814	.113538	.112401	-1	325
18.00	.111588	.111163	.111041	.111031	.110904	.110483	.109708	.108658	.107514	.106494	-1	291
19.00	.105764	.105382	.105273	.105264	.105149	.104771	.104074	.103127	.102097	.101177	-1	262
20.00	.100518	.100173	.100074	.100066	.099963	.099620	.098990	.098133	.097199	.096365	-1	237
21.00	.095768	.095454	.095364	.095357	.095263	.094952	.094379	.093600	.092750	.091991	-1	215
22.00	.091446	.091160	.091078	.091072	.090986	.090703	.090179	.089467	.088691	.087996	-1	197
23.00	.087498	.087236	.087161	.087155	.087077	.086817	.086337	.085684	.084972	.084334	-1	180
24.00	.083876	.083636	.083566	.083561	.083489	.083250	.082809	.082208	.081552	.080964	-1	166
25.00	.080543	.080321	.080257	.080252	.080186	.079965	.079558	.079003	.078397	.077854	-1	153
26.00	.077464	.077258	.077199	.077194	.077133	.076929	.076552	.076038	.075477	.074973	-1	142
27.00	.074612	.074421	.074366	.074362	.074305	.074116	.073766	.073288	.072766	.072298	-1	132
28.00	.071962	.071785	.071734	.071730	.071677	.071501	.071175	.070730	.070244	.069808	-1	123
29.00	.069494	.069329	.069281	.069277	.069228	.069064	.068760	.068345	.067890	.067483	-1	114
30.00	.067190	.067035	.066991	.066987	.066941	.066788	.066503	.066115	.065690	.065308	-1	107
31.00	.065034	.064889	.064847	.064844	.064801	.064657	.064390	.064026	.063627	.063269	-1	100
32.00	.063012	.062876	.062836	.062833	.062793	.062658	.062407	.062065	.061690	.061354	-1	94
33.00	.061111	.060983	.060946	.060943	.060906	.060778	.060542	.060220	.059867	.059551	-1	89
34.00	.059322	.059202	.059167	.059164	.059128	.059008	.058786	.058482	.058149	.057850	-1	84
35.00	.057635	.057521	.057488	.057486	.057452	.057339	.057129	.056842	.056527	.056245	-1	79
36.00	.056041	.055934	.055903	.055900	.055868	.055761	.055562	.055291	.054993	.054726	-1	75
37.00	.054533	.054431	.054402	.054399	.054369	.054268	.054080	.053822	.053540	.053287	-1	71
38.00	.053104	.053007	.052979	.052977	.052949	.052852	.052674	.052430	.052162	.051921	-1	67
39.00	.051748	.051656	.051630	.051627	.051600	.051509	.051339	.051107	.050853	.050624	-1	64
40.00	.050459	.050372	.050347	.050345	.050319	.050232	.050071	.049850	.049608	.049390	-1	61
41.00	.049233	.049150	.049126	.049124	.049100	.049017	.048863	.048653	.048423	.048215	-1	58
42.00	.048066	.047987	.047964	.047962	.047938	.047859	.047713	.047513	.047293	.047095	-1	55
43.00	.046952	.046876	.046855	.046853	.046830	.046755	.046615	.046424	.046214	.046025	-1	53
44.00	.045889	.045817	.045796	.045794	.045773	.045701	.045567	.045384	.045184	.045003	-1	50
45.00	.044873	.044804	.044784	.044782	.044762	.044693	.044565	.044390	.044198	.044025	-1	48
46.00	.043901	.043834	.043815	.043814	.043794	.043728	.043606	.043439	.043255	.043089	-1	46
47.00	.042970	.042906	.042888	.042886	.042868	.042805	.042687	.042527	.042351	.042192	-1	44
48.00	.042077	.042017	.041999	.041998	.041980	.041919	.041807	.041653	.041484	.041331	-1	42
49.00	.041221	.041163	.041146	.041145	.041128	.041070	.040962	.040814	.040651	.040505	-1	41

TABLE 20

Encircled Energy Complement

r_r	.0	.1	.2	.3	.4	.5	.6	.7	.8	.9	m	Δ_2
50.00	.040400	.040344	.040327	.040326	.040310	.040254	.040150	.040008	.039852	.039711		
51.00	.039610	.039556	.039541	.039539	.039523	.039470	.039370	.039233	.039083	.038948	-1	39
52.00	.038851	.038799	.038784	.038783	.038767	.038716	.038620	.038488	.038344	.038214	-1	38
53.00	.038120	.038070	.038055	.038054	.038040	.037990	.037897	.037771	.037632	.037506	-1	36
54.00	.037416	.037368	.037354	.037353	.037339	.037291	.037202	.037080	.036946	.036825	-1	35
55.00	.036738	.036691	.036678	.036677	.036663	.036617	.036531	.036413	.036284	.036168	-1	34
56.00	.036083	.036039	.036026	.036025	.036012	.035967	.035884	.035771	.035646	.035533	-1	32
57.00	.035452	.035409	.035397	.035396	.035383	.035340	.035260	.035150	.035030	.034921	-1	31
58.00	.034843	.034801	.034789	.034788	.034776	.034734	.034657	.034551	.034434	.034329	-1	30
59.00	.034254	.034213	.034202	.034201	.034189	.034149	.034074	.033972	.033859	.033758	-1	29
60.00	.033684	.033645	.033634	.033633	.033622	.033583	.033511	.033412	.033303	.033204	-1	28
61.00	.033133	.033096	.033085	.033084	.033073	.033035	.032965	.032870	.032764	.032669	-1	27
62.00	.032600	.032564	.032553	.032553	.032542	.032505	.032438	.032345	.032243	.032151	-1	26
63.00	.032084	.032049	.032039	.032038	.032027	.031992	.031927	.031837	.031738	.031649	-1	26
64.00	.031584	.031550	.031540	.031539	.031529	.031495	.031431	.031344	.031248	.031162	-1	25
65.00	.031100	.031066	.031057	.031056	.031046	.031013	.030951	.030867	.030774	.030690	-1	24
66.00	.030629	.030597	.030588	.030587	.030578	.030546	.030486	.030404	.030314	.030232	-1	23
67.00	.030173	.030142	.030133	.030132	.030123	.030092	.030034	.029954	.029867	.029788	-1	23
68.00	.029731	.029700	.029692	.029691	.029682	.029652	.029595	.029518	.029433	.029356	-1	22
69.00	.029301	.029271	.029263	.029262	.029253	.029224	.029169	.029094	.029012	.028937	-1	21
70.00	.028883	.028855	.028846	.028846	.028837	.028809	.028755	.028682	.028602	.028530	-1	21
71.00	.028477	.028449	.028441	.028441	.028432	.028405	.028353	.028282	.028204	.028134	-1	20
72.00	.028083	.028056	.028048	.028047	.028039	.028012	.027962	.027893	.027817	.027748	-1	19
73.00	.027699	.027672	.027665	.027664	.027656	.027630	.027581	.027514	.027440	.027373	-1	19
74.00	.027325	.027300	.027292	.027292	.027284	.027258	.027211	.027145	.027074	.027009	-1	18
75.00	.026962	.026937	.026929	.026929	.026922	.026897	.026850	.026787	.026717	.026653	-1	18
76.00	.026608	.026583	.026576	.026576	.026569	.026544	.026499	.026437	.026369	.026307	-1	17
77.00	.026263	.026239	.026232	.026232	.026225	.026201	.026157	.026097	.026030	.025970	-1	17
78.00	.025927	.025904	.025897	.025897	.025890	.025867	.025824	.025765	.025700	.025642	-1	17
79.00	.025599	.025577	.025570	.025570	.025563	.025541	.025499	.025441	.025378	.025321	-1	16
80.00	.025280	.025258	.025252	.025251	.025245	.025223	.025182	.025126	.025064	.025009	-1	16
81.00	.024968	.024947	.024941	.024940	.024934	.024913	.024873	.024818	.024758	.024704	-1	15
82.00	.024665	.024644	.024638	.024637	.024631	.024610	.024571	.024518	.024459	.024406	-1	15
83.00	.024368	.024348	.024342	.024341	.024335	.024315	.024277	.024225	.024168	.024116	-1	15
84.00	.024078	.024059	.024053	.024052	.024046	.024027	.023990	.023939	.023883	.023832	-1	14
85.00	.023796	.023776	.023771	.023770	.023764	.023745	.023709	.023659	.023605	.023555	-1	14
86.00	.023519	.023501	.023495	.023495	.023489	.023470	.023435	.023386	.023333	.023285	-1	14
87.00	.023250	.023231	.023226	.023225	.023220	.023201	.023167	.023119	.023067	.023020	-1	13
88.00	.022986	.022968	.022963	.022962	.022957	.022939	.022905	.022859	.022808	.022761	-1	13
89.00	.022728	.022710	.022705	.022705	.022700	.022682	.022649	.022604	.022554	.022509	-1	13
90.00	.022476	.022459	.022454	.022453	.022448	.022431	.022399	.022354	.022305	.022261	-1	12
91.00	.022229	.022213	.022208	.022207	.022202	.022185	.022154	.022110	.022063	.022019	-1	12
92.00	.021988	.021972	.021967	.021966	.021962	.021945	.021914	.021872	.021825	.021783	-1	12
93.00	.021752	.021736	.021731	.021731	.021726	.021710	.021680	.021638	.021592	.021551	-1	12
94.00	.021521	.021505	.021501	.021500	.021496	.021480	.021450	.021410	.021365	.021324	-1	11
95.00	.021295	.021280	.021275	.021275	.021270	.021254	.021225	.021186	.021142	.021102	-1	11
96.00	.021074	.021058	.021054	.021054	.021049	.021034	.021005	.020966	.020924	.020885	-1	11
97.00	.020857	.020842	.020837	.020837	.020833	.020818	.020790	.020752	.020710	.020672	-1	11
98.00	.020644	.020630	.020625	.020625	.020621	.020606	.020579	.020541	.020500	.020463	-1	10
99.00	.020436	.020422	.020418	.020417	.020413	.020399	.020372	.020335	.020295	.020258	-1	10

$$2\pi\int_r^\infty rPdr$$

TABLE 21

Line Spread Function

x_r	.000	.001	.002	.003	.004	.005	.006	.007	.008	.009	m	Δ_2
0.00	*******	.999997	.999989	.999976	.999958	.999934	.999905	.999871	.999832	.999787	0	-4
0.01	.999737	.999682	.999621	.999555	.999484	.999408	.999326	.999240	.999148	.999050	-0	-4
0.02	.998948	.998840	.998727	.998609	.998485	.998356	.998222	.998083	.997938	.997789	-0	-4
0.03	.997634	.997473	.997308	.997137	.996961	.996780	.996594	.996402	.996206	.996004	-0	-4
0.04	.995797	.995584	.995367	.995144	.994916	.994683	.994444	.994201	.993952	.993698	-0	-4
0.05	.993439	.993174	.992905	.992630	.992351	.992066	.991776	.991480	.991180	.990874	-0	-4
0.06	.990564	.990248	.989927	.989601	.989269	.988933	.988592	.988245	.987893	.987537	-0	-4
0.07	.987175	.986808	.986436	.986059	.985676	.985289	.984897	.984499	.984097	.983690	-0	-4
0.08	.983277	.982859	.982437	.982009	.981577	.981139	.980696	.980248	.979796	.979338	-0	-4
0.09	.978875	.978408	.977935	.977458	.976975	.976488	.975995	.975498	.974995	.974488	-0	-4
0.10	.973976	.973459	.972937	.972410	.971878	.971342	.970800	.970254	.969703	.969146	-0	-4
0.11	.968585	.968020	.967449	.966874	.966293	.965708	.965118	.964524	.963924	.963320	-0	-4
0.12	.962711	.962097	.961478	.960855	.960227	.959594	.958957	.958315	.957668	.957016	-0	-4
0.13	.956360	.955699	.955033	.954363	.953688	.953009	.952324	.951636	.950942	.950244	-0	-4
0.14	.949541	.948834	.948122	.947406	.946685	.945960	.945230	.944495	.943756	.943012	-0	-4
0.15	.942264	.941512	.940755	.939993	.939227	.938457	.937682	.936903	.936119	.935331	-0	-3
0.16	.934538	.933741	.932940	.932134	.931324	.930510	.929691	.928868	.928041	.927209	-0	-3
0.17	.926373	.925533	.924689	.923840	.922987	.922130	.921268	.920403	.919533	.918659	-0	-3
0.18	.917781	.916898	.916012	.915121	.914226	.913327	.912424	.911517	.910606	.909690	-0	-3
0.19	.908771	.907848	.906920	.905989	.905053	.904114	.903170	.902223	.901272	.900316	-0	-3
0.20	.899357	.898394	.897427	.896456	.895481	.894502	.893519	.892533	.891542	.890548	-0	-3
0.21	.889550	.888548	.887543	.886533	.885520	.884504	.883483	.882459	.881431	.880399	-0	-3
0.22	.879364	.878325	.877282	.876236	.875186	.874132	.873075	.872014	.870950	.869882	-0	-3
0.23	.868811	.867736	.866657	.865575	.864490	.863401	.862309	.861213	.860113	.859011	-0	-2
0.24	.857905	.856795	.855682	.854566	.853447	.852324	.851198	.850068	.848935	.847799	-0	-2
0.25	.846660	.845517	.844372	.843222	.842070	.840915	.839756	.838595	.837430	.836262	-0	-2
0.26	.835091	.833916	.832739	.831559	.830375	.829189	.827999	.826807	.825611	.824413	-0	-2
0.27	.823211	.822007	.820800	.819589	.818376	.817160	.815941	.814720	.813495	.812268	-0	-2
0.28	.811037	.809804	.808569	.807330	.806089	.804845	.803598	.802349	.801096	.799842	-0	-2
0.29	.798584	.797324	.796061	.794796	.793528	.792258	.790985	.789709	.788431	.787150	-0	-2
0.30	.785867	.784581	.783293	.782003	.780710	.779414	.778117	.776816	.775514	.774209	-0	-1
0.31	.772902	.771592	.770280	.768966	.767650	.766331	.765010	.763687	.762362	.761034	-0	-1
0.32	.759704	.758373	.757039	.755702	.754364	.753024	.751682	.750337	.748991	.747642	-0	-1
0.33	.746291	.744939	.743584	.742228	.740869	.739509	.738147	.736783	.735417	.734049	-0	-1
0.34	.732679	.731307	.729934	.728559	.727182	.725803	.724422	.723040	.721656	.720271	-0	-1
0.35	.718883	.717494	.716104	.714711	.713317	.711922	.710525	.709126	.707726	.706324	-0	-1
0.36	.704921	.703516	.702110	.700702	.699293	.697883	.696471	.695057	.693643	.692227	-0	-0
0.37	.690809	.689390	.687970	.686549	.685126	.683702	.682277	.680850	.679423	.677994	-0	-0
0.38	.676564	.675133	.673700	.672267	.670832	.669396	.667960	.666522	.665083	.663643	-0	-0
0.39	.662202	.660760	.659317	.657873	.656428	.654983	.653536	.652088	.650640	.649190	-0	-0
0.40	.647740	.646289	.644837	.643385	.641931	.640477	.639022	.637566	.636110	.634653	-0	-0
0.41	.633195	.631736	.630277	.628818	.627357	.625896	.624435	.622973	.621510	.620047	-0	-0
0.42	.618583	.617119	.615654	.614189	.612723	.611257	.609790	.608323	.606856	.605388	-0	0
0.43	.603920	.602452	.600983	.599514	.598045	.596575	.595105	.593635	.592165	.590694	-0	0
0.44	.589223	.587752	.586281	.584810	.583338	.581867	.580395	.578923	.577452	.575980	-0	0
0.45	.574508	.573036	.571564	.570092	.568620	.567149	.565677	.564205	.562733	.561262	-0	0
0.46	.559791	.558319	.556848	.555377	.553906	.552436	.550966	.549495	.548025	.546556	-0	0
0.47	.545086	.543617	.542149	.540680	.539212	.537744	.536277	.534810	.533343	.531877	-0	1
0.48	.530411	.528946	.527481	.526017	.524553	.523089	.521626	.520164	.518702	.517241	-0	1
0.49	.515780	.514320	.512861	.511402	.509943	.508486	.507029	.505573	.504117	.502662	-0	

TABLE 21

Line Spread Function

x_r	.000	.001	.002	.003	.004	.005	.006	.007	.008	.009	m	Δ_2
0.50	.501208	.499755	.498302	.496850	.495399	.493949	.492499	.491051	.489603	.488156	-0	1
0.51	.486710	.485264	.483820	.482377	.480934	.479492	.478052	.476612	.475173	.473736	-0	1
0.52	.472299	.470863	.469429	.467995	.466562	.465131	.463701	.462271	.460843	.459416	-0	1
0.53	.457990	.456565	.455142	.453719	.452298	.450878	.449459	.448042	.446626	.445211	-0	1
0.54	.443797	.442384	.440973	.439563	.438155	.436748	.435342	.433937	.432534	.431132	-0	1
0.55	.429732	.428333	.426936	.425540	.424145	.422752	.421360	.419970	.418581	.417194	-0	2
0.56	.415809	.414425	.413042	.411661	.410282	.408904	.407528	.406153	.404780	.403409	-0	2
0.57	.402039	.400671	.399305	.397940	.396577	.395216	.393856	.392498	.391142	.389788	-0	2
0.58	.388435	.387084	.385735	.384388	.383042	.381699	.380357	.379017	.377679	.376343	-0	2
0.59	.375008	.373676	.372345	.371016	.369690	.368365	.367042	.365721	.364402	.363085	-0	2
0.60	.361770	.360457	.359145	.357836	.356529	.355224	.353921	.352620	.351321	.350024	-0	2
0.61	.348730	.347437	.346146	.344858	.343572	.342287	.341005	.339725	.338448	.337172	-0	2
0.62	.335898	.334627	.333358	.332091	.330827	.329564	.328304	.327046	.325790	.324537	-0	2
0.63	.323286	.322037	.320790	.319546	.318304	.317064	.315826	.314591	.313358	.312128	-0	2
0.64	.310900	.309674	.308451	.307230	.306011	.304795	.303581	.302370	.301161	.299954	-0	2
0.65	.298750	.297548	.296349	.295152	.293958	.292766	.291577	.290390	.289205	.288023	-0	3
0.66	.286844	.285667	.284493	.283321	.282152	.280985	.279821	.278659	.277500	.276343	-0	3
0.67	.275189	.274038	.272889	.271743	.270599	.269458	.268320	.267184	.266051	.264920	-0	3
0.68	.263793	.262667	.261545	.260425	.259308	.258193	.257081	.255972	.254865	.253762	-0	3
0.69	.252660	.251562	.250466	.249373	.248283	.247195	.246111	.245028	.243949	.242873	-0	3
0.70	.241799	.240728	.239659	.238594	.237531	.236471	.235414	.234359	.233307	.232259	-0	3
0.71	.231213	.230169	.229129	.228091	.227056	.226024	.224995	.223969	.222945	.221925	-0	3
0.72	.220907	.219892	.218880	.217870	.216864	.215860	.214860	.213862	.212867	.211875	-0	3
0.73	.210886	.209899	.208916	.207935	.206958	.205983	.205011	.204042	.203076	.202113	-0	3
0.74	.201153	.200195	.199241	.198289	.197341	.196395	.195452	.194513	.193576	.192642	-0	3
0.75	.191711	.190783	.189858	.188936	.188016	.187100	.186187	.185276	.184369	.183465	-0	3
0.76	.182563	.181664	.180769	.179876	.178987	.178100	.177216	.176335	.175458	.174583	-0	3
0.77	.173711	.172842	.171976	.171113	.170254	.169397	.168543	.167692	.166844	.165999	-0	3
0.78	.165157	.164318	.163481	.162648	.161818	.160991	.160167	.159346	.158528	.157713	-0	3
0.79	.156901	.156091	.155285	.154482	.153682	.152885	.152090	.151299	.150511	.149726	-0	3
0.80	.148943	.148164	.147388	.146614	.145844	.145077	.144313	.143551	.142793	.142037	-0	3
0.81	.141285	.140536	.139789	.139046	.138305	.137568	.136833	.136102	.135373	.134648	-0	3
0.82	.133925	.133205	.132489	.131775	.131064	.130357	.129652	.128950	.128251	.127555	-0	3
0.83	.126862	.126172	.125485	.124801	.124120	.123442	.122767	.122095	.121425	.120759	-0	3
0.84	.120096	.119435	.118778	.118123	.117471	.116823	.116177	.115534	.114894	.114257	-0	3
0.85	.113623	.112992	.112363	.111738	.111116	.110496	.109879	.109266	.108655	.108047	-0	3
0.86	.107442	.106840	.106240	.105644	.105050	.104460	.103872	.103287	.102705	.102126	-0	3
0.87	.101549	.100976	.100405	.099838	.099273	.098711	.098151	.097595	.097041	.096491	-0	3
0.88	.959426	.953975	.948552	.943157	.937791	.932452	.927141	.921859	.916604	.911377	-1	28
0.89	.906178	.901006	.895863	.890747	.885659	.880598	.875566	.870560	.865582	.860632	-1	27
0.90	.855709	.850813	.845945	.841104	.836290	.831503	.826743	.822011	.817305	.812627	-1	27
0.91	.807975	.803351	.798753	.794182	.789638	.785120	.780629	.776165	.771727	.767315	-1	26
0.92	.762930	.758572	.754240	.749934	.745654	.741401	.737173	.732972	.728796	.724647	-1	26
0.93	.720523	.716426	.712354	.708308	.704287	.700292	.696323	.692379	.688461	.684568	-1	25
0.94	.680700	.676858	.673041	.669249	.665482	.661740	.658023	.654331	.650664	.647022	-1	25
0.95	.643404	.639811	.636243	.632699	.629180	.625685	.622215	.618768	.615347	.611949	-1	24
0.96	.608575	.605226	.601900	.598599	.595321	.592067	.588837	.585630	.582447	.579288	-1	23
0.97	.576152	.573039	.569950	.566884	.563841	.560822	.557825	.554852	.551901	.548974	-1	23
0.98	.546069	.543186	.540327	.537490	.534676	.531884	.529114	.526367	.523642	.520940	-1	22
0.99	.518259	.515600	.512964	.510349	.507756	.505185	.502636	.500108	.497602	.495117	-1	21

TABLE 21

Line Spread Function

x_r	.00	.01	.02	.03	.04	.05	.06	.07	.08	.09	m
1.00	.492654	.469182	.447771	.428347	.410834	.395156	.381236	.368995	.358356	.349238	-1
1.10	.341562	.335251	.330223	.326402	.323708	.322065	.321396	.321624	.322676	.324479	-1
1.20	.326960	.330049	.333677	.337777	.342285	.347137	.352272	.357631	.363156	.368795	-1
1.30	.374494	.380204	.385877	.391469	.396937	.402241	.407345	.412213	.416813	.421116	-1
1.40	.425095	.428726	.431986	.434857	.437321	.439364	.440973	.442140	.442856	.443117	-1
1.50	.442918	.442260	.441143	.439570	.437546	.435077	.432172	.428840	.425094	.420944	-1
1.60	.416407	.411496	.406229	.400622	.394695	.388467	.381957	.375187	.368177	.360950	-1
1.70	.353527	.345930	.338183	.330308	.322327	.314263	.306139	.297977	.289798	.281624	-1
1.80	.273476	.265375	.257340	.249392	.241548	.233827	.226246	.218820	.211567	.204499	-1
1.90	.197631	.190975	.184543	.178346	.172393	.166693	.161253	.156081	.151181	.146558	-1
2.00	.142215	.138155	.134380	.130889	.127682	.124758	.122115	.119748	.117654	.115828	-1
2.10	.114263	.112955	.111894	.111074	.110486	.110121	.109968	.110019	.110262	.110686	-1
2.20	.111280	.112032	.112931	.113963	.115118	.116382	.117744	.119189	.120707	.122285	-1
2.30	.123909	.125569	.127252	.128947	.130642	.132325	.133987	.135616	.137202	.138737	-1
2.40	.140211	.141614	.142939	.144179	.145325	.146372	.147313	.148143	.148857	.149451	-1
2.50	.149921	.150264	.150477	.150558	.150506	.150321	.150001	.149548	.148962	.148245	-1
2.60	.147398	.146424	.145326	.144107	.142770	.141321	.139764	.138103	.136344	.134492	-1
2.70	.132554	.130535	.128442	.126282	.124060	.121784	.119462	.117099	.114703	.112282	-1
2.80	.109842	.107390	.104934	.102481	.100037	.097609	.095204	.092828	.090488	.088189	-1
2.90	.859371	.837382	.815973	.795193	.775089	.755705	.737081	.719254	.702256	.686118	-2
3.00	.670867	.656524	.643109	.630638	.619121	.608568	.598982	.590365	.582713	.576021	-2
3.10	.570280	.565475	.561592	.558611	.556510	.555264	.554846	.555225	.556368	.558242	-2
3.20	.560808	.564027	.567860	.572264	.577196	.582609	.588460	.594701	.601285	.608164	-2
3.30	.615290	.622616	.630093	.637673	.645310	.652956	.660567	.668097	.675503	.682741	-2
3.40	.689772	.696556	.703054	.709231	.715052	.720486	.725503	.730073	.734173	.737778	-2
3.50	.740867	.743422	.745427	.746868	.747734	.748016	.747708	.746806	.745310	.743219	-2
3.60	.740539	.737274	.733432	.729025	.724065	.718566	.712545	.706021	.699013	.691544	-2
3.70	.683637	.675318	.666612	.657548	.648153	.638459	.628495	.618292	.607883	.597300	-2
3.80	.586576	.575743	.564835	.553884	.542924	.531986	.521103	.510306	.499626	.489093	-2
3.90	.478736	.468584	.458663	.449000	.439619	.430544	.421796	.413396	.405362	.397714	-2
4.00	.390465	.383630	.377222	.371251	.365726	.360653	.356039	.351887	.348198	.344972	-2
4.10	.342209	.339904	.338052	.336646	.335678	.335139	.335017	.335298	.335970	.337016	-2
4.20	.338419	.340163	.342228	.344594	.347242	.350148	.353291	.356649	.360198	.363914	-2
4.30	.367773	.371751	.375824	.379967	.384156	.388367	.392575	.396757	.400890	.404951	-2
4.40	.408918	.412770	.416485	.420044	.423428	.426619	.429599	.432353	.434866	.437124	-2
4.50	.439114	.440825	.442246	.443370	.444188	.444695	.444885	.444754	.444301	.443525	-2
4.60	.442426	.441005	.439266	.437213	.434851	.432187	.429229	.425985	.422465	.418680	-2
4.70	.414643	.410365	.405861	.401145	.396231	.391136	.385876	.380467	.374928	.369275	-2
4.80	.363526	.357700	.351815	.345889	.339940	.333987	.328048	.322141	.316283	.310492	-2
4.90	.304785	.299178	.293687	.288327	.283114	.278060	.273180	.268486	.263989	.259701	-2

TABLE 21

Line Spread Function

x_r	.00	.01	.02	.03	.04	.05	.06	.07	.08	.09	m
5.00	.255631	.251788	.248180	.244815	.241698	.238835	.236229	.233884	.231802	.229984	-2
5.10	.228429	.227138	.226107	.225334	.224815	.224546	.224519	.224730	.225170	.225831	-2
5.20	.226705	.227781	.229049	.230498	.232117	.233893	.235815	.237868	.240039	.242315	-2
5.30	.244682	.247126	.249632	.252186	.254774	.257381	.259993	.262596	.265175	.267718	-2
5.40	.270210	.272639	.274991	.277255	.279419	.281471	.283401	.285198	.286854	.288359	-2
5.50	.289705	.290885	.291892	.292720	.293364	.293819	.294082	.294151	.294023	.293697	-2
5.60	.293172	.292449	.291530	.290416	.289110	.287616	.285937	.284079	.282047	.279847	-2
5.70	.277487	.274974	.272315	.269520	.266597	.263555	.260405	.257157	.253821	.250407	-2
5.80	.246928	.243393	.239815	.236205	.232573	.228933	.225294	.221668	.218067	.214501	-2
5.90	.210981	.207518	.204122	.200803	.197570	.194432	.191398	.188477	.185675	.183001	-2
6.00	.180461	.178061	.175806	.173701	.171751	.169959	.168329	.166863	.165562	.164427	-2
6.10	.163459	.162659	.162024	.161554	.161246	.161098	.161106	.161267	.161576	.162028	-2
6.20	.162618	.163340	.164187	.165153	.166230	.167411	.168689	.170053	.171498	.173012	-2
6.30	.174588	.176217	.177889	.179596	.181326	.183073	.184825	.186575	.188312	.190028	-2
6.40	.191713	.193360	.194959	.196503	.197984	.199394	.200726	.201972	.203128	.204186	-2
6.50	.205141	.205987	.206721	.207338	.207834	.208207	.208453	.208570	.208557	.208413	-2
6.60	.208137	.207729	.207190	.206521	.205723	.204799	.203751	.202581	.201294	.199893	-2
6.70	.198383	.196769	.195055	.193247	.191351	.189374	.187321	.185199	.183015	.180776	-2
6.80	.178490	.176164	.173806	.171422	.169021	.166611	.164199	.161793	.159400	.157028	-2
6.90	.154683	.152375	.150108	.147891	.145730	.143630	.141598	.139641	.137762	.135967	-2
7.00	.134262	.132649	.131134	.129719	.128408	.127204	.126108	.125123	.124250	.123490	-2
7.10	.122844	.122311	.121891	.121583	.121387	.121300	.121321	.121446	.121674	.122001	-2
7.20	.122422	.122936	.123536	.124219	.124980	.125814	.126715	.127677	.128696	.129765	-2
7.30	.130878	.132028	.133210	.134417	.135643	.136881	.138124	.139367	.140603	.141825	-2
7.40	.143028	.144206	.145351	.146460	.147525	.148543	.149507	.150413	.151256	.152031	-2
7.50	.152736	.153366	.153917	.154387	.154774	.155073	.155285	.155406	.155436	.155374	-2
7.60	.155219	.154971	.154631	.154199	.153675	.153062	.152361	.151574	.150702	.149750	-2
7.70	.148720	.147614	.146438	.145193	.143885	.142518	.141096	.139624	.138106	.136548	-2
7.80	.134955	.133331	.131683	.130016	.128335	.126645	.124952	.123262	.121580	.119911	-2
7.90	.118260	.116633	.115035	.113470	.111943	.110459	.109023	.107638	.106308	.105038	-2
8.00	.103829	.102687	.101613	.100610	.099681	.098828	.098052	.097355	.096737	.096201	-2
8.10	.957451	.953710	.950781	.948659	.947338	.946805	.947047	.948047	.949786	.952242	-3
8.20	.955390	.959201	.963647	.968695	.974312	.980460	.987102	.994197	.001706	.009585	-3
8.30	.101779	.102628	.103500	.104392	.105297	.106213	.107133	.108054	.108971	.109878	-2
8.40	.110772	.111649	.112503	.113330	.114127	.114890	.115614	.116297	.116934	.117522	-2
8.50	.118059	.118542	.118968	.119334	.119640	.119882	.120060	.120172	.120217	.120194	-2
8.60	.120103	.119944	.119717	.119422	.119060	.118631	.118138	.117580	.116961	.116281	-2
8.70	.115543	.114749	.113902	.113004	.112059	.111069	.110038	.108969	.107865	.106731	-2
8.80	.105570	.104386	.103182	.101964	.100734	.099497	.098257	.097017	.095783	.094558	-2
8.90	.933448	.921487	.909730	.898214	.886974	.876044	.865458	.855247	.845441	.836067	-3
9.00	.827152	.818720	.810794	.803394	.796537	.790240	.784517	.779377	.774832	.770887	-3
9.10	.767546	.764811	.762682	.761156	.760229	.759891	.760134	.760946	.762313	.764218	-3
9.20	.766644	.769570	.772974	.776834	.781123	.785815	.790881	.796293	.802020	.808029	-3
9.30	.814288	.820764	.827423	.834230	.841151	.848149	.855190	.862239	.869260	.876219	-3
9.40	.883081	.889813	.896383	.902756	.908904	.914794	.920400	.925692	.930644	.935233	-3
9.50	.939435	.943228	.946592	.949510	.951966	.953945	.955435	.956427	.956912	.956883	-3
9.60	.956338	.955273	.953690	.951589	.948977	.945858	.942242	.938137	.933557	.928515	-3
9.70	.923026	.917109	.910782	.904065	.896980	.889552	.881803	.873761	.865451	.856901	-3
9.80	.848141	.839198	.830103	.820885	.811577	.802208	.792809	.783411	.774045	.764741	-3
9.90	.755530	.746441	.737503	.728744	.720193	.711874	.703814	.696038	.688567	.681425	-3

TABLE 21

Line Spread Function

x_r	.0	.1	.2	.3	.4	.5	.6	.7	.8	.9	m
10.00	.674631	.629252	.628994	.666370	.720401	.765052	.779148	.754023	.696155	.624059	-3
11.00	.560843	.525385	.525491	.555467	.598833	.634929	.646796	.627342	.581561	.524164	-3
12.00	.473679	.445372	.445682	.470169	.505610	.535288	.545373	.529977	.493042	.446483	-3
13.00	.405423	.382413	.382835	.403151	.432563	.457317	.465964	.453553	.423257	.384884	-3
14.00	.350967	.331974	.332452	.349533	.374265	.395171	.402645	.392482	.367276	.335214	-3
15.00	.306821	.290933	.291435	.305963	.326997	.344847	.351356	.342920	.321688	.294579	-3
16.00	.270533	.257089	.257593	.270075	.288145	.303530	.309239	.302152	.284073	.260914	-3
17.00	.240341	.228847	.229343	.240162	.255823	.269197	.274236	.268218	.252678	.232710	-3
18.00	.214948	.205034	.205514	.214966	.228647	.240360	.244833	.239677	.226205	.208846	-3
19.00	.193388	.184767	.185227	.193545	.205579	.215907	.219900	.215444	.203677	.188475	-3
20.00	.174925	.167375	.167813	.175178	.185832	.194995	.198576	.194696	.184348	.170948	-3
21.00	.158993	.152336	.152752	.159312	.168798	.176972	.180198	.176798	.167641	.155758	-3
22.00	.145147	.139245	.139638	.145511	.154001	.161331	.164249	.161250	.153102	.142507	-3
23.00	.133039	.127777	.128149	.133432	.141067	.147670	.150320	.147661	.140373	.130879	-3
24.00	.122390	.117675	.118026	.122799	.129696	.135669	.138085	.135714	.129165	.120619	-3
25.00	.112972	.108729	.109061	.113391	.119646	.125070	.127279	.125155	.119245	.111520	-3
26.00	.104604	.100770	.101083	.105025	.110719	.115663	.117690	.115779	.110424	.103414	-3
27.00	.971352	.936563	.939520	.975543	.027553	.072765	.091409	.074154	.025454	.961613	-4
28.00	.904403	.872728	.875522	.908541	.956201	.997678	.014877	.999234	.954794	.896459	-4
29.00	.844160	.815226	.817867	.848223	.892028	.930189	.946095	.931864	.891182	.837713	-4
30.00	.789756	.763244	.765742	.793728	.834102	.869307	.884052	.871064	.833712	.784561	-4
31.00	.740460	.716096	.718461	.744329	.781637	.814199	.827899	.816009	.781619	.736314	-4
32.00	.695649	.673199	.675440	.699408	.733968	.764157	.776914	.765998	.734253	.692386	-4
33.00	.654796	.634057	.636182	.658441	.690529	.718581	.730483	.720435	.691059	.652278	-4
34.00	.617447	.598243	.600259	.620976	.650833	.676955	.688080	.678808	.651562	.615557	-4
35.00	.583211	.565389	.567304	.586625	.614463	.638835	.649253	.640677	.615352	.581855	-4
36.00	.551753	.535177	.536998	.555051	.581057	.603840	.613612	.605662	.582074	.550847	-4
37.00	.522778	.507331	.509063	.525963	.550301	.571638	.580819	.573433	.551421	.522255	-4
38.00	.496032	.481609	.483259	.499106	.521923	.541939	.550578	.543703	.523123	.495833	-4
39.00	.471292	.457801	.459373	.474258	.495684	.514490	.522632	.516221	.496947	.471368	-4
40.00	.448361	.435721	.437221	.451223	.471375	.489072	.496755	.490765	.472684	.448671	-4
41.00	.427068	.415206	.416637	.429829	.448809	.465487	.472747	.467142	.450153	.427575	-4
42.00	.407259	.396110	.397477	.409923	.427826	.443565	.450434	.445179	.429194	.407934	-4
43.00	.388800	.378305	.379612	.391370	.408280	.423152	.429659	.424726	.409663	.389616	-4
44.00	.371571	.361678	.362929	.374050	.390042	.404114	.410286	.405648	.391434	.372505	-4
45.00	.355464	.346127	.347324	.357857	.373000	.386330	.392190	.387823	.374393	.356498	-4
46.00	.340385	.331560	.332707	.342695	.357050	.369692	.375262	.371145	.358440	.341501	-4
47.00	.326247	.317897	.318996	.328477	.342101	.354105	.359403	.355517	.343484	.327432	-4
48.00	.312973	.305063	.306118	.315127	.328071	.339480	.344527	.340853	.329444	.314214	-4
49.00	.300495	.292993	.294006	.302576	.314887	.325742	.330553	.327077	.316246	.301782	-4

TABLE 21

Line Spread Function

x_r	.0	.1	.2	.3	.4	.5	.6	.7	.8	.9	m
50.00	.288750	.281627	.282600	.290761	.302481	.312819	.317410	.314117	.303825	.290073	-4
51.00	.277682	.270912	.271847	.279625	.290794	.300649	.305033	.301910	.292120	.279033	-4
52.00	.267240	.260799	.261698	.269118	.279772	.289175	.293364	.290400	.281079	.268612	-4
53.00	.257376	.251243	.252108	.259193	.269364	.278343	.282351	.279533	.270651	.258764	-4
54.00	.248050	.242204	.243037	.249808	.259526	.268108	.271945	.269265	.260792	.249448	-4
55.00	.239223	.233646	.234448	.240924	.250217	.258426	.262102	.259550	.251461	.240626	-4
56.00	.230859	.225534	.226308	.232506	.241399	.249259	.252783	.250351	.242622	.232265	-4
57.00	.222928	.217839	.218585	.224523	.233040	.240569	.243951	.241631	.234240	.224332	-4
58.00	.215398	.210532	.211252	.216944	.225107	.232326	.235572	.233358	.226285	.216799	-4
59.00	.208245	.203587	.204283	.209743	.217573	.224498	.227617	.225501	.218727	.209639	-4
60.00	.201443	.196981	.197653	.202895	.210410	.217059	.220057	.218034	.211542	.202828	-4
61.00	.194969	.190693	.191342	.196377	.203595	.209983	.212867	.210931	.204704	.196344	-4
62.00	.188803	.184702	.185329	.190169	.197106	.203246	.206022	.204169	.198193	.190166	-4
63.00	.182926	.178989	.179596	.184251	.190922	.196828	.199502	.197726	.191987	.184275	-4
64.00	.177319	.173538	.174125	.178605	.185024	.190709	.193286	.191583	.186067	.178654	-4
65.00	.171967	.168333	.168901	.173215	.179396	.184870	.187355	.185721	.180417	.173286	-4
66.00	.166853	.163359	.163909	.168066	.174020	.179295	.181692	.180123	.175020	.168157	-4
67.00	.161965	.158603	.159136	.163143	.168883	.173968	.176281	.174774	.169862	.163252	-4
68.00	.157289	.154052	.154569	.158434	.163969	.168875	.171108	.169659	.164927	.158559	-4
69.00	.152813	.149694	.150196	.153926	.159267	.164001	.166159	.164766	.160205	.154066	-4
70.00	.148525	.145520	.146006	.149607	.154764	.159336	.161421	.160080	.155682	.149760	-4
71.00	.144416	.141518	.141990	.145469	.150449	.154866	.156883	.155591	.151348	.145633	-4
72.00	.140475	.137679	.138137	.141499	.146312	.150581	.152532	.151288	.147193	.141674	-4
73.00	.136694	.133994	.134440	.137691	.142344	.146471	.148360	.147161	.143206	.137875	-4
74.00	.133063	.130456	.130889	.134034	.138534	.142528	.144357	.143200	.139378	.134226	-4
75.00	.129576	.127057	.127477	.130521	.134876	.138741	.140512	.139396	.135702	.130721	-4
76.00	.126224	.123789	.124197	.127144	.131360	.135102	.136819	.135742	.132169	.127351	-4
77.00	.123001	.120646	.121043	.123897	.127981	.131605	.133270	.132229	.128773	.124109	-4
78.00	.119899	.117621	.118007	.120773	.124729	.128242	.129856	.128850	.125505	.120990	-4
79.00	.116914	.114709	.115085	.117766	.121601	.125006	.126572	.125599	.122360	.117987	-4
80.00	.114039	.111904	.112270	.114870	.118588	.121890	.123410	.122470	.119332	.115095	-4
81.00	.111269	.109201	.109557	.112079	.115686	.118889	.120365	.119456	.116414	.112307	-4
82.00	.108599	.106594	.106941	.109389	.112889	.115998	.117432	.116551	.113603	.109620	-4
83.00	.106024	.104081	.104419	.106795	.110192	.113211	.114604	.113751	.110892	.107028	-4
84.00	.103540	.101655	.101984	.104292	.107591	.110523	.111876	.111050	.108276	.104527	-4
85.00	.101142	.099313	.099634	.101876	.105081	.107929	.109245	.108444	.105752	.102113	-4
86.00	.988264	.970516	.973644	.995433	.026574	.054253	.067053	.059291	.033152	.997811	-5
87.00	.965897	.948665	.951716	.972898	.003168	.030077	.042529	.035001	.009615	.975285	-5
88.00	.944282	.927546	.930523	.951121	.980554	.006721	.018839	.011536	.986873	.953513	-5
89.00	.923386	.907126	.910030	.930067	.958695	.984150	.995946	.988859	.964890	.932462	-5
90.00	.903176	.887375	.890208	.909705	.937559	.962329	.973814	.966934	.943632	.912101	-5
91.00	.883624	.868262	.871029	.890005	.917114	.941224	.952411	.945729	.923069	.892400	-5
92.00	.864701	.849762	.852463	.870940	.897330	.920805	.931703	.925212	.903170	.873330	-5
93.00	.846380	.831848	.834486	.852480	.878180	.901042	.911663	.905355	.883906	.854866	-5
94.00	.828636	.814496	.817073	.834602	.859636	.881908	.892261	.886129	.865253	.836981	-5
95.00	.811445	.797683	.800200	.817281	.841673	.863376	.873471	.867508	.847183	.819652	-5
96.00	.794785	.781385	.783846	.800495	.824267	.845421	.855266	.849467	.829672	.802855	-5
97.00	.778632	.765583	.767988	.784220	.807395	.828020	.837624	.831982	.812698	.786570	-5
98.00	.762969	.750257	.752608	.768438	.791036	.811149	.820521	.815030	.796240	.770776	-5
99.00	.747773	.735387	.737686	.753127	.775169	.794789	.803936	.798590	.780276	.755452	-5

TABLE 22

Modulation Transfer Function, Perfect Lens, Monochromatic

ν_r	.0000	.0001	.0002	.0003	.0004	.0005	.0006	.0007	.0008	.0009
0.000	1.00	.9998727	.9997454	.9996180	.9994907	.9993634	.9992361	.9991087	.9989814	.9988541
0.001	.9987268	.9985994	.9984721	.9983448	.9982175	.9980901	.9979628	.9978355	.9977082	.9975809
0.002	.9974535	.9973262	.9971989	.9970716	.9969442	.9968169	.9966896	.9965623	.9964349	.9963076
0.003	.9961803	.9960530	.9959256	.9957983	.9956710	.9955437	.9954164	.9952890	.9951617	.9950344
0.004	.9949071	.9947797	.9946524	.9945251	.9943978	.9942705	.9941431	.9940158	.9938885	.9937612
0.005	.9936338	.9935065	.9933792	.9932519	.9931245	.9929972	.9928699	.9927426	.9926153	.9924879
0.006	.9923606	.9922333	.9921060	.9919787	.9918513	.9917240	.9915967	.9914694	.9913420	.9912147
0.007	.9910874	.9909601	.9908328	.9907054	.9905781	.9904508	.9903235	.9901962	.9900688	.9899415
0.008	.9898142	.9896869	.9895596	.9894322	.9893049	.9891776	.9890503	.9889230	.9887956	.9886683
0.009	.9885410	.9884137	.9882864	.9881590	.9880317	.9879044	.9877771	.9876498	.9875225	.9873951
0.010	.9872678	.9871405	.9870132	.9868859	.9867586	.9866312	.9865039	.9863766	.9862493	.9861220
0.011	.9859947	.9858673	.9857400	.9856127	.9854854	.9853581	.9852308	.9851035	.9849761	.9848488
0.012	.9847215	.9845942	.9844669	.9843396	.9842122	.9840849	.9839576	.9838303	.9837030	.9835757
0.013	.9834484	.9833210	.9831937	.9830664	.9829391	.9828118	.9826845	.9825572	.9824299	.9823025
0.014	.9821752	.9820479	.9819206	.9817933	.9816660	.9815387	.9814114	.9812841	.9811567	.9810294
0.015	.9809021	.9807748	.9806475	.9805202	.9803929	.9802656	.9801383	.9800110	.9798837	.9797563
0.016	.9796290	.9795017	.9793744	.9792471	.9791198	.9789925	.9788652	.9787379	.9786106	.9784833
0.017	.9783560	.9782287	.9781014	.9779741	.9778467	.9777194	.9775921	.9774648	.9773375	.9772102
0.018	.9770829	.9769556	.9768283	.9767010	.9765737	.9764464	.9763191	.9761918	.9760645	.9759372
0.019	.9758099	.9756826	.9755553	.9754280	.9753007	.9751734	.9750461	.9749188	.9747915	.9746642
0.020	.9745369	.9744096	.9742823	.9741550	.9740277	.9739004	.9737731	.9736458	.9735185	.9733912
0.021	.9732639	.9731366	.9730093	.9728821	.9727548	.9726275	.9725002	.9723729	.9722456	.9721183
0.022	.9719910	.9718637	.9717364	.9716091	.9714818	.9713545	.9712272	.9711000	.9709727	.9708454
0.023	.9707181	.9705908	.9704635	.9703362	.9702089	.9700816	.9699543	.9698271	.9696998	.9695725
0.024	.9694452	.9693179	.9691906	.9690633	.9689360	.9688088	.9686815	.9685542	.9684269	.9682996
0.025	.9681723	.9680450	.9679178	.9677905	.9676632	.9675359	.9674086	.9672813	.9671541	.9670268
0.026	.9668995	.9667722	.9666450	.9665177	.9663904	.9662631	.9661358	.9660086	.9658813	.9657540
0.027	.9656267	.9654994	.9653722	.9652449	.9651176	.9649903	.9648631	.9647358	.9646085	.9644812
0.028	.9643540	.9642267	.9640994	.9639721	.9638449	.9637176	.9635903	.9634631	.9633358	.9632085
0.029	.9630812	.9629540	.9628267	.9626994	.9625722	.9624449	.9623176	.9621904	.9620631	.9619358
0.030	.9618085	.9616813	.9615540	.9614268	.9612995	.9611722	.9610450	.9609177	.9607904	.9606632
0.031	.9605359	.9604086	.9602814	.9601541	.9600269	.9598996	.9597723	.9596451	.9595178	.9593906
0.032	.9592633	.9591360	.9590088	.9588815	.9587543	.9586270	.9584997	.9583725	.9582452	.9581180
0.033	.9579907	.9578635	.9577362	.9576090	.9574817	.9573545	.9572272	.9571000	.9569727	.9568455
0.034	.9567182	.9565910	.9564637	.9563365	.9562092	.9560820	.9559547	.9558275	.9557002	.9555730
0.035	.9554457	.9553185	.9551912	.9550640	.9549367	.9548095	.9546823	.9545550	.9544278	.9543005
0.036	.9541733	.9540460	.9539188	.9537916	.9536643	.9535371	.9534098	.9532826	.9531554	.9530281
0.037	.9529009	.9527737	.9526464	.9525192	.9523920	.9522647	.9521375	.9520103	.9518830	.9517558
0.038	.9516286	.9515013	.9513741	.9512469	.9511196	.9509924	.9508652	.9507379	.9506107	.9504835
0.039	.9503563	.9502290	.9501018	.9499746	.9498473	.9497201	.9495929	.9494657	.9493385	.9492112
0.040	.9490840	.9489568	.9488296	.9487023	.9485751	.9484479	.9483207	.9481935	.9480663	.9479390
0.041	.9478118	.9476846	.9475574	.9474302	.9473030	.9471757	.9470485	.9469213	.9467941	.9466669
0.042	.9465397	.9464125	.9462852	.9461580	.9460308	.9459036	.9457764	.9456492	.9455220	.9453948
0.043	.9452676	.9451404	.9450132	.9448860	.9447588	.9446316	.9445044	.9443772	.9442500	.9441227
0.044	.9439955	.9438684	.9437412	.9436139	.9434867	.9433596	.9432324	.9431052	.9429780	.9428508
0.045	.9427236	.9425964	.9424692	.9423420	.9422148	.9420876	.9419604	.9418332	.9417060	.9415788
0.046	.9414517	.9413245	.9411973	.9410701	.9409429	.9408157	.9406885	.9405613	.9404342	.9403070
0.047	.9401798	.9400526	.9399254	.9397982	.9396711	.9395439	.9394167	.9392895	.9391623	.9390352
0.048	.9389080	.9387808	.9386536	.9385265	.9383993	.9382721	.9381449	.9380178	.9378906	.9377634
0.049	.9376362	.9375091	.9373819	.9372547	.9371276	.9370004	.9368732	.9367461	.9366189	.9364917

TABLE 22

Modulation Transfer Function, Perfect Lens, Monochromatic

ν_r	.0000	.0001	.0002	.0003	.0004	.0005	.0006	.0007	.0008	.0009
0.050	.9363646	.9362374	.9361102	.9359831	.9358559	.9357288	.9356016	.9354744	.9353473	.9352201
0.051	.9350929	.9349658	.9348386	.9347115	.9345843	.9344572	.9343300	.9342029	.9340757	.9339486
0.052	.9338214	.9336942	.9335671	.9334399	.9333128	.9331857	.9330585	.9329313	.9328042	.9326771
0.053	.9325499	.9324228	.9322956	.9321685	.9320413	.9319142	.9317871	.9316599	.9315328	.9314056
0.054	.9312785	.9311514	.9310242	.9308971	.9307700	.9306428	.9305157	.9303886	.9302614	.9301343
0.055	.9300072	.9298800	.9297529	.9296258	.9294986	.9293715	.9292444	.9291172	.9289901	.9288630
0.056	.9287359	.9286088	.9284816	.9283545	.9282274	.9281003	.9279731	.9278460	.9277189	.9275918
0.057	.9274647	.9273375	.9272104	.9270833	.9269562	.9268291	.9267020	.9265749	.9264478	.9263206
0.058	.9261935	.9260664	.9259393	.9258122	.9256851	.9255580	.9254309	.9253038	.9251767	.9250496
0.059	.9249225	.9247954	.9246683	.9245412	.9244141	.9242870	.9241599	.9240328	.9239057	.9237786
0.060	.9236515	.9235244	.9233973	.9232702	.9231431	.9230160	.9228889	.9227619	.9226348	.9225077
0.061	.9223806	.9222535	.9221264	.9219993	.9218723	.9217452	.9216181	.9214910	.9213639	.9212368
0.062	.9211098	.9209827	.9208556	.9207285	.9206014	.9204744	.9203473	.9202202	.9200932	.9199661
0.063	.9198390	.9197119	.9195849	.9194578	.9193307	.9192037	.9190766	.9189495	.9188225	.9186954
0.064	.9185683	.9184413	.9183142	.9181872	.9180601	.9179330	.9178060	.9176789	.9175519	.9174248
0.065	.9172978	.9171707	.9170436	.9169166	.9167895	.9166625	.9165354	.9164084	.9162813	.9161543
0.066	.9160272	.9159002	.9157731	.9156461	.9155191	.9153920	.9152650	.9151379	.9150109	.9148839
0.067	.9147568	.9146298	.9145027	.9143757	.9142487	.9141216	.9139946	.9138676	.9137405	.9136135
0.068	.9134865	.9133595	.9132324	.9131054	.9129784	.9128514	.9127243	.9125973	.9124703	.9123433
0.069	.9122162	.9120892	.9119622	.9118352	.9117082	.9115812	.9114541	.9113271	.9112001	.9110731
0.070	.9109461	.9108191	.9106921	.9105650	.9104380	.9103110	.9101840	.9100570	.9099300	.9098030
0.071	.9096760	.9095490	.9094220	.9092950	.9091680	.9090410	.9089140	.9087870	.9086600	.9085330
0.072	.9084060	.9082790	.9081520	.9080251	.9078981	.9077711	.9076441	.9075171	.9073901	.9072631
0.073	.9071361	.9070091	.9068822	.9067552	.9066282	.9065012	.9063742	.9062473	.9061203	.9059933
0.074	.9058663	.9057394	.9056124	.9054854	.9053584	.9052315	.9051045	.9049775	.9048506	.9047236
0.075	.9045966	.9044697	.9043427	.9042157	.9040888	.9039618	.9038349	.9037079	.9035809	.9034540
0.076	.9033270	.9032001	.9030731	.9029462	.9028192	.9026923	.9025653	.9024384	.9023114	.9021845
0.077	.9020575	.9019306	.9018036	.9016767	.9015497	.9014228	.9012959	.9011689	.9010420	.9009150
0.078	.9007881	.9006612	.9005342	.9004073	.9002804	.9001534	.9000265	.8998996	.8997727	.8996457
0.079	.8995188	.8993919	.8992649	.8991380	.8990111	.8988842	.8987573	.8986304	.8985034	.8983765
0.080	.8982496	.8981227	.8979958	.8978689	.8977419	.8976150	.8974881	.8973612	.8972343	.8971074
0.081	.8969805	.8968536	.8967267	.8965998	.8964729	.8963460	.8962191	.8960922	.8959653	.8958384
0.082	.8957115	.8955846	.8954577	.8953308	.8952039	.8950770	.8949501	.8948232	.8946964	.8945695
0.083	.8944426	.8943157	.8941888	.8940619	.8939351	.8938082	.8936813	.8935544	.8934275	.8933007
0.084	.8931738	.8930469	.8929200	.8927932	.8926663	.8925394	.8924126	.8922857	.8921588	.8920320
0.085	.8919051	.8917782	.8916514	.8915245	.8913977	.8912708	.8911439	.8910171	.8908902	.8907634
0.086	.8906365	.8905097	.8903828	.8902560	.8901291	.8900023	.8898754	.8897486	.8896217	.8894949
0.087	.8893681	.8892412	.8891144	.8889875	.8888607	.8887339	.8886070	.8884802	.8883534	.8882265
0.088	.8880997	.8879729	.8878461	.8877192	.8875924	.8874656	.8873387	.8872119	.8870851	.8869583
0.089	.8868315	.8867046	.8865778	.8864510	.8863242	.8861974	.8860706	.8859438	.8858170	.8856901
0.090	.8855633	.8854365	.8853097	.8851829	.8850561	.8849293	.8848025	.8846757	.8845489	.8844221
0.091	.8842953	.8841685	.8840417	.8839149	.8837881	.8836614	.8835346	.8834078	.8832810	.8831542
0.092	.8830274	.8829006	.8827739	.8826471	.8825203	.8823935	.8822667	.8821400	.8820132	.8818864
0.093	.8817596	.8816329	.8815061	.8813793	.8812526	.8811258	.8809990	.8808723	.8807455	.8806187
0.094	.8804920	.8803652	.8802385	.8801117	.8799849	.8798582	.8797314	.8796047	.8794779	.8793512
0.095	.8792244	.8790977	.8789709	.8788442	.8787175	.8785907	.8784640	.8783372	.8782105	.8780838
0.096	.8779570	.8778303	.8777035	.8775768	.8774501	.8773234	.8771966	.8770699	.8769432	.8768164
0.097	.8766897	.8765630	.8764363	.8763096	.8761828	.8760561	.8759294	.8758027	.8756760	.8755492
0.098	.8754225	.8752958	.8751691	.8750424	.8749157	.8747890	.8746623	.8745356	.8744089	.8742822
0.099	.8741555	.8740288	.8739021	.8737754	.8736487	.8735220	.8733953	.8732686	.8731420	.8730153

TABLE 22

Modulation Transfer Function, Perfect Lens, Monochromatic

ν_r	.0000	.0001	.0002	.0003	.0004	.0005	.0006	.0007	.0008	.0009
0.100	.8728886	.8727619	.8726352	.8725085	.8723818	.8722552	.8721285	.8720018	.8718751	.8717485
0.101	.8716218	.8714951	.8713684	.8712418	.8711151	.8709884	.8708618	.8707351	.8706084	.8704818
0.102	.8703551	.8702285	.8701018	.8699751	.8698485	.8697218	.8695952	.8694685	.8693419	.8692152
0.103	.8690886	.8689619	.8688353	.8687086	.8685820	.8684554	.8683287	.8682021	.8680755	.8679488
0.104	.8678222	.8676955	.8675689	.8674423	.8673157	.8671890	.8670624	.8669358	.8668092	.8666825
0.105	.8665559	.8664293	.8663027	.8661761	.8660494	.8659228	.8657962	.8656696	.8655430	.8654164
0.106	.8652898	.8651632	.8650366	.8649100	.8647834	.8646568	.8645302	.8644036	.8642770	.8641504
0.107	.8640238	.8638972	.8637706	.8636440	.8635174	.8633908	.8632643	.8631377	.8630111	.8628845
0.108	.8627579	.8626313	.8625048	.8623782	.8622516	.8621250	.8619985	.8618719	.8617453	.8616188
0.109	.8614922	.8613656	.8612391	.8611125	.8609859	.8608594	.8607328	.8606063	.8604797	.8603532
0.110	.8602266	.8601001	.8599735	.8598470	.8597204	.8595939	.8594673	.8593408	.8592143	.8590877
0.111	.8589612	.8588346	.8587081	.8585816	.8584550	.8583285	.8582020	.8580754	.8579489	.8578224
0.112	.8576959	.8575693	.8574428	.8573163	.8571898	.8570633	.8569368	.8568102	.8566837	.8565572
0.113	.8564307	.8563042	.8561777	.8560512	.8559247	.8557982	.8556717	.8555452	.8554187	.8552922
0.114	.8551657	.8550392	.8549127	.8547862	.8546597	.8545333	.8544068	.8542803	.8541538	.8540273
0.115	.8539008	.8537744	.8536479	.8535214	.8533949	.8532685	.8531420	.8530155	.8528891	.8527626
0.116	.8526361	.8525097	.8523832	.8522567	.8521303	.8520038	.8518774	.8517509	.8516245	.8514980
0.117	.8513715	.8512451	.8511187	.8509922	.8508658	.8507393	.8506129	.8504864	.8503600	.8502336
0.118	.8501071	.8499807	.8498543	.8497278	.8496014	.8494750	.8493486	.8492221	.8490957	.8489693
0.119	.8488429	.8487164	.8485900	.8484636	.8483372	.8482108	.8480844	.8479580	.8478316	.8477051
0.120	.8475787	.8474523	.8473260	.8471995	.8470731	.8469467	.8468203	.8466940	.8465676	.8464412
0.121	.8463148	.8461884	.8460620	.8459356	.8458093	.8456829	.8455565	.8454301	.8453037	.8451774
0.122	.8450510	.8449246	.8447982	.8446719	.8445455	.8444191	.8442928	.8441664	.8440400	.8439137
0.123	.8437873	.8436610	.8435346	.8434083	.8432819	.8431556	.8430292	.8429029	.8427765	.8426502
0.124	.8425238	.8423975	.8422712	.8421448	.8420185	.8418922	.8417658	.8416395	.8415132	.8413868
0.125	.8412605	.8411342	.8410079	.8408815	.8407552	.8406289	.8405026	.8403763	.8402499	.8401236
0.126	.8399973	.8398710	.8397447	.8396184	.8394921	.8393658	.8392395	.8391132	.8389869	.8388606
0.127	.8387343	.8386080	.8384817	.8383555	.8382292	.8381029	.8379766	.8378503	.8377240	.8375978
0.128	.8374715	.8373452	.8372189	.8370926	.8369664	.8368401	.8367138	.8365876	.8364613	.8363351
0.129	.8362088	.8360825	.8359563	.8358300	.8357038	.8355775	.8354513	.8353250	.8351988	.8350725
0.130	.8349463	.8348200	.8346938	.8345675	.8344413	.8343151	.8341888	.8340626	.8339364	.8338101
0.131	.8336839	.8335577	.8334315	.8333053	.8331790	.8330528	.8329266	.8328004	.8326742	.8325479
0.132	.8324217	.8322955	.8321693	.8320431	.8319169	.8317907	.8316645	.8315383	.8314121	.8312859
0.133	.8311597	.8310335	.8309073	.8307812	.8306550	.8305288	.8304026	.8302764	.8301502	.8300241
0.134	.8298979	.8297717	.8296455	.8295194	.8293932	.8292670	.8291409	.8290147	.8288885	.8287624
0.135	.8286362	.8285101	.8283839	.8282577	.8281316	.8280054	.8278793	.8277531	.8276270	.8275008
0.136	.8273747	.8272486	.8271224	.8269963	.8268702	.8267440	.8266179	.8264918	.8263656	.8262395
0.137	.8261134	.8259873	.8258611	.8257350	.8256089	.8254828	.8253567	.8252306	.8251045	.8249784
0.138	.8248523	.8247261	.8246000	.8244739	.8243478	.8242217	.8240956	.8239696	.8238435	.8237174
0.139	.8235913	.8234652	.8233391	.8232130	.8230869	.8229609	.8228348	.8227087	.8225826	.8224566
0.140	.8223305	.8222044	.8220783	.8219523	.8218262	.8217002	.8215741	.8214480	.8213220	.8211959
0.141	.8210699	.8209438	.8208178	.8206917	.8205657	.8204396	.8203136	.8201876	.8200615	.8199355
0.142	.8198094	.8196834	.8195574	.8194314	.8193053	.8191793	.8190533	.8189273	.8188012	.8186752
0.143	.8185492	.8184232	.8182972	.8181712	.8180452	.8179191	.8177932	.8176671	.8175411	.8174151
0.144	.8172891	.8171631	.8170372	.8169112	.8167852	.8166592	.8165332	.8164072	.8162812	.8161552
0.145	.8160293	.8159033	.8157773	.8156513	.8155254	.8153994	.8152734	.8151475	.8150215	.8148955
0.146	.8147696	.8146436	.8145177	.8143917	.8142658	.8141398	.8140138	.8138879	.8137620	.8136360
0.147	.8135101	.8133841	.8132582	.8131323	.8130063	.8128804	.8127545	.8126285	.8125026	.8123767
0.148	.8122508	.8121248	.8119989	.8118730	.8117471	.8116212	.8114953	.8113694	.8112435	.8111175
0.149	.8109916	.8108657	.8107398	.8106139	.8104880	.8103622	.8102363	.8101104	.8099845	.8098586

TABLE 22

Modulation Transfer Function, Perfect Lens, Monochromatic

ν_r	.0000	.0001	.0002	.0003	.0004	.0005	.0006	.0007	.0008	.0009
0.150	.8097327	.8096068	.8094809	.8093551	.8092292	.8091033	.8089774	.8088516	.8087257	.8085998
0.151	.8084740	.8083481	.8082222	.8080964	.8079705	.8078447	.8077188	.8075930	.8074671	.8073413
0.152	.8072154	.8070896	.8069637	.8068379	.8067121	.8065862	.8064604	.8063346	.8062087	.8060829
0.153	.8059571	.8058313	.8057054	.8055796	.8054538	.8053280	.8052022	.8050764	.8049505	.8048247
0.154	.8046989	.8045731	.8044473	.8043215	.8041957	.8040699	.8039441	.8038183	.8036926	.8035668
0.155	.8034410	.8033152	.8031894	.8030636	.8029379	.8028121	.8026863	.8025605	.8024348	.8023090
0.156	.8021832	.8020575	.8019317	.8018059	.8016802	.8015544	.8014287	.8013029	.8011772	.8010514
0.157	.8009257	.8007999	.8006742	.8005484	.8004227	.8002970	.8001712	.8000455	.7999198	.7997941
0.158	.7996683	.7995426	.7994169	.7992912	.7991655	.7990397	.7989140	.7987883	.7986626	.7985369
0.159	.7984112	.7982855	.7981598	.7980341	.7979084	.7977827	.7976570	.7975313	.7974056	.7972799
0.160	.7971542	.7970286	.7969029	.7967772	.7966515	.7965259	.7964002	.7962745	.7961488	.7960232
0.161	.7958975	.7957718	.7956462	.7955205	.7953949	.7952692	.7951436	.7950179	.7948923	.7947666
0.162	.7946410	.7945153	.7943897	.7942641	.7941384	.7940128	.7938872	.7937616	.7936359	.7935103
0.163	.7933847	.7932591	.7931334	.7930078	.7928822	.7927566	.7926310	.7925054	.7923798	.7922542
0.164	.7921286	.7920030	.7918774	.7917518	.7916262	.7915006	.7913750	.7912494	.7911238	.7909983
0.165	.7908727	.7907471	.7906215	.7904959	.7903704	.7902448	.7901192	.7899937	.7898681	.7897426
0.166	.7896170	.7894914	.7893659	.7892403	.7891148	.7889892	.7888637	.7887381	.7886126	.7884871
0.167	.7883615	.7882360	.7881104	.7879849	.7878594	.7877339	.7876083	.7874828	.7873573	.7872318
0.168	.7871063	.7869808	.7868553	.7867297	.7866042	.7864787	.7863532	.7862277	.7861022	.7859767
0.169	.7858512	.7857257	.7856003	.7854748	.7853493	.7852238	.7850983	.7849728	.7848474	.7847219
0.170	.7845964	.7844709	.7843455	.7842200	.7840946	.7839691	.7838436	.7837182	.7835927	.7834673
0.171	.7833418	.7832164	.7830909	.7829655	.7828400	.7827146	.7825892	.7824637	.7823383	.7822129
0.172	.7820874	.7819620	.7818366	.7817112	.7815858	.7814604	.7813349	.7812095	.7810841	.7809587
0.173	.7808333	.7807079	.7805825	.7804571	.7803317	.7802063	.7800809	.7799555	.7798301	.7797047
0.174	.7795794	.7794540	.7793286	.7792032	.7790779	.7789525	.7788271	.7787018	.7785764	.7784510
0.175	.7783257	.7782003	.7780749	.7779496	.7778242	.7776989	.7775736	.7774482	.7773229	.7771975
0.176	.7770722	.7769468	.7768215	.7766962	.7765709	.7764455	.7763202	.7761949	.7760696	.7759442
0.177	.7758189	.7756936	.7755683	.7754430	.7753177	.7751924	.7750671	.7749418	.7748165	.7746912
0.178	.7745659	.7744406	.7743153	.7741901	.7740648	.7739395	.7738142	.7736889	.7735637	.7734384
0.179	.7733131	.7731879	.7730626	.7729373	.7728121	.7726868	.7725616	.7724363	.7723111	.7721858
0.180	.7720606	.7719353	.7718101	.7716848	.7715596	.7714344	.7713091	.7711839	.7710587	.7709335
0.181	.7708082	.7706830	.7705578	.7704326	.7703074	.7701821	.7700569	.7699317	.7698065	.7696813
0.182	.7695561	.7694309	.7693058	.7691806	.7690554	.7689302	.7688050	.7686798	.7685546	.7684295
0.183	.7683043	.7681791	.7680539	.7679288	.7678036	.7676784	.7675533	.7674281	.7673030	.7671778
0.184	.7670527	.7669275	.7668024	.7666772	.7665521	.7664269	.7663018	.7661767	.7660515	.7659264
0.185	.7658013	.7656762	.7655510	.7654259	.7653008	.7651757	.7650506	.7649255	.7648003	.7646752
0.186	.7645501	.7644250	.7642999	.7641748	.7640498	.7639247	.7637996	.7636745	.7635494	.7634243
0.187	.7632992	.7631742	.7630491	.7629240	.7627990	.7626739	.7625488	.7624238	.7622987	.7621736
0.188	.7620486	.7619235	.7617985	.7616734	.7615484	.7614233	.7612983	.7611733	.7610482	.7609232
0.189	.7607982	.7606731	.7605481	.7604231	.7602981	.7601730	.7600480	.7599230	.7597980	.7596730
0.190	.7595480	.7594230	.7592980	.7591730	.7590480	.7589230	.7587980	.7586730	.7585480	.7584231
0.191	.7582981	.7581731	.7580481	.7579232	.7577982	.7576732	.7575482	.7574233	.7572983	.7571734
0.192	.7570484	.7569235	.7567985	.7566735	.7565486	.7564237	.7562987	.7561738	.7560489	.7559239
0.193	.7557990	.7556740	.7555491	.7554242	.7552993	.7551744	.7550494	.7549245	.7547996	.7546747
0.194	.7545498	.7544249	.7543000	.7541751	.7540502	.7539253	.7538004	.7536755	.7535506	.7534258
0.195	.7533009	.7531760	.7530511	.7529262	.7528014	.7526765	.7525516	.7524268	.7523019	.7521771
0.196	.7520522	.7519274	.7518025	.7516776	.7515528	.7514280	.7513031	.7511783	.7510534	.7509286
0.197	.7508038	.7506790	.7505541	.7504293	.7503045	.7501797	.7500549	.7499301	.7498052	.7496804
0.198	.7495556	.7494308	.7493060	.7491812	.7490564	.7489316	.7488069	.7486821	.7485573	.7484325
0.199	.7483077	.7481830	.7480582	.7479334	.7478086	.7476839	.7475591	.7474344	.7473096	.7471848

TABLE 22

Modulation Transfer Function, Perfect Lens, Monochromatic

ν_r	.0000	.0001	.0002	.0003	.0004	.0005	.0006	.0007	.0008	.0009
0.200	.7470601	.7469353	.7468106	.7466858	.7465611	.7464364	.7463116	.7461869	.7460622	.7459374
0.201	.7458127	.7456880	.7455633	.7454385	.7453138	.7451891	.7450644	.7449397	.7448150	.7446903
0.202	.7445656	.7444409	.7443162	.7441915	.7440668	.7439421	.7438174	.7436927	.7435681	.7434434
0.203	.7433187	.7431940	.7430694	.7429447	.7428200	.7426954	.7425707	.7424461	.7423214	.7421968
0.204	.7420721	.7419475	.7418228	.7416982	.7415735	.7414489	.7413243	.7411997	.7410750	.7409504
0.205	.7408258	.7407012	.7405765	.7404519	.7403273	.7402027	.7400781	.7399535	.7398289	.7397043
0.206	.7395797	.7394551	.7393305	.7392059	.7390814	.7389568	.7388322	.7387076	.7385831	.7384585
0.207	.7383339	.7382094	.7380848	.7379602	.7378357	.7377111	.7375866	.7374620	.7373375	.7372129
0.208	.7370884	.7369639	.7368393	.7367148	.7365903	.7364657	.7363412	.7362167	.7360922	.7359677
0.209	.7358431	.7357186	.7355941	.7354696	.7353451	.7352206	.7350961	.7349716	.7348471	.7347226
0.210	.7345981	.7344737	.7343492	.7342247	.7341002	.7339758	.7338513	.7337268	.7336024	.7334779
0.211	.7333534	.7332290	.7331045	.7329801	.7328556	.7327312	.7326067	.7324823	.7323579	.7322334
0.212	.7321090	.7319846	.7318601	.7317357	.7316113	.7314869	.7313625	.7312381	.7311137	.7309892
0.213	.7308648	.7307404	.7306160	.7304917	.7303673	.7302429	.7301185	.7299941	.7298697	.7297453
0.214	.7296210	.7294966	.7293722	.7292478	.7291235	.7289991	.7288748	.7287504	.7286261	.7285017
0.215	.7283774	.7282530	.7281287	.7280043	.7278800	.7277557	.7276313	.7275070	.7273827	.7272584
0.216	.7271340	.7270097	.7268854	.7267611	.7266368	.7265125	.7263882	.7262639	.7261396	.7260153
0.217	.7258910	.7257667	.7256424	.7255181	.7253939	.7252696	.7251453	.7250210	.7248968	.7247725
0.218	.7246482	.7245240	.7243997	.7242755	.7241512	.7240270	.7239027	.7237785	.7236542	.7235300
0.219	.7234058	.7232815	.7231573	.7230331	.7229089	.7227846	.7226604	.7225362	.7224120	.7222878
0.220	.7221636	.7220394	.7219152	.7217910	.7216668	.7215426	.7214184	.7212942	.7211700	.7210459
0.221	.7209217	.7207975	.7206733	.7205492	.7204250	.7203008	.7201767	.7200525	.7199284	.7198042
0.222	.7196801	.7195559	.7194318	.7193076	.7191835	.7190594	.7189352	.7188111	.7186870	.7185629
0.223	.7184387	.7183146	.7181905	.7180664	.7179423	.7178182	.7176941	.7175700	.7174459	.7173218
0.224	.7171977	.7170736	.7169495	.7168254	.7167014	.7165773	.7164532	.7163292	.7162051	.7160810
0.225	.7159570	.7158329	.7157089	.7155848	.7154607	.7153367	.7152127	.7150886	.7149646	.7148406
0.226	.7147165	.7145925	.7144685	.7143444	.7142204	.7140964	.7139724	.7138484	.7137244	.7136004
0.227	.7134764	.7133524	.7132284	.7131044	.7129804	.7128564	.7127324	.7126084	.7124845	.7123605
0.228	.7122365	.7121125	.7119886	.7118646	.7117407	.7116167	.7114928	.7113688	.7112449	.7111209
0.229	.7109970	.7108730	.7107491	.7106252	.7105012	.7103773	.7102534	.7101295	.7100055	.7098816
0.230	.7097577	.7096338	.7095099	.7093860	.7092621	.7091382	.7090143	.7088904	.7087665	.7086426
0.231	.7085188	.7083949	.7082710	.7081471	.7080233	.7078994	.7077755	.7076517	.7075278	.7074040
0.232	.7072801	.7071563	.7070324	.7069086	.7067847	.7066609	.7065371	.7064132	.7062894	.7061656
0.233	.7060418	.7059179	.7057941	.7056703	.7055465	.7054227	.7052989	.7051751	.7050513	.7049275
0.234	.7048037	.7046799	.7045561	.7044324	.7043086	.7041848	.7040610	.7039373	.7038135	.7036897
0.235	.7035660	.7034422	.7033185	.7031947	.7030710	.7029472	.7028235	.7026998	.7025760	.7024523
0.236	.7023285	.7022048	.7020811	.7019574	.7018337	.7017099	.7015862	.7014625	.7013388	.7012151
0.237	.7010914	.7009677	.7008440	.7007203	.7005967	.7004730	.7003493	.7002256	.7001020	.6999783
0.238	.6998546	.6997310	.6996073	.6994836	.6993600	.6992363	.6991127	.6989890	.6988654	.6987418
0.239	.6986181	.6984945	.6983709	.6982472	.6981236	.6980000	.6978764	.6977528	.6976291	.6975055
0.240	.6973819	.6972583	.6971347	.6970111	.6968876	.6967640	.6966404	.6965168	.6963932	.6962696
0.241	.6961461	.6960225	.6958989	.6957754	.6956518	.6955283	.6954047	.6952811	.6951576	.6950341
0.242	.6949105	.6947870	.6946635	.6945399	.6944164	.6942929	.6941693	.6940458	.6939223	.6937988
0.243	.6936753	.6935518	.6934283	.6933048	.6931813	.6930578	.6929343	.6928108	.6926873	.6925639
0.244	.6924404	.6923169	.6921934	.6920700	.6919465	.6918230	.6916996	.6915761	.6914527	.6913292
0.245	.6912058	.6910823	.6909589	.6908355	.6907120	.6905886	.6904652	.6903417	.6902183	.6900949
0.246	.6899715	.6898481	.6897247	.6896013	.6894779	.6893545	.6892311	.6891077	.6889843	.6888609
0.247	.6887375	.6886142	.6884908	.6883674	.6882441	.6881207	.6879973	.6878740	.6877506	.6876273
0.248	.6875039	.6873806	.6872572	.6871339	.6870106	.6868872	.6867639	.6866406	.6865173	.6863939
0.249	.6862706	.6861473	.6860240	.6859007	.6857774	.6856541	.6855308	.6854075	.6852842	.6851609

TABLE 22

Modulation Transfer Function, Perfect Lens, Monochromatic

ν_r	.0000	.0001	.0002	.0003	.0004	.0005	.0006	.0007	.0008	.0009
0.250	.6850376	.6849144	.6847911	.6846678	.6845446	.6844213	.6842980	.6841748	.6840515	.6839283
0.251	.6838050	.6836818	.6835585	.6834353	.6833120	.6831888	.6830656	.6829424	.6828191	.6826959
0.252	.6825727	.6824495	.6823263	.6822031	.6820799	.6819567	.6818335	.6817103	.6815871	.6814639
0.253	.6813407	.6812175	.6810943	.6809712	.6808480	.6807248	.6806017	.6804785	.6803554	.6802322
0.254	.6801091	.6799859	.6798628	.6797396	.6796165	.6794934	.6793702	.6792471	.6791240	.6790009
0.255	.6788777	.6787546	.6786315	.6785084	.6783853	.6782622	.6781391	.6780160	.6778929	.6777698
0.256	.6776468	.6775237	.6774006	.6772775	.6771545	.6770314	.6769083	.6767853	.6766622	.6765392
0.257	.6764161	.6762931	.6761700	.6760470	.6759240	.6758009	.6756779	.6755549	.6754318	.6753088
0.258	.6751858	.6750628	.6749398	.6748168	.6746938	.6745708	.6744478	.6743248	.6742018	.6740788
0.259	.6739559	.6738329	.6737099	.6735869	.6734640	.6733410	.6732180	.6730951	.6729721	.6728492
0.260	.6727262	.6726033	.6724803	.6723574	.6722345	.6721115	.6719886	.6718657	.6717428	.6716199
0.261	.6714970	.6713740	.6712511	.6711282	.6710053	.6708824	.6707595	.6706367	.6705138	.6703909
0.262	.6702680	.6701451	.6700223	.6698994	.6697765	.6696537	.6695308	.6694080	.6692851	.6691623
0.263	.6690394	.6689166	.6687937	.6686709	.6685481	.6684253	.6683024	.6681796	.6680568	.6679340
0.264	.6678112	.6676884	.6675656	.6674428	.6673200	.6671972	.6670744	.6669516	.6668288	.6667061
0.265	.6665833	.6664605	.6663378	.6662150	.6660922	.6659695	.6658467	.6657240	.6656012	.6654785
0.266	.6653557	.6652330	.6651103	.6649875	.6648648	.6647421	.6646194	.6644967	.6643740	.6642513
0.267	.6641285	.6640059	.6638832	.6637605	.6636378	.6635151	.6633924	.6632697	.6631471	.6630244
0.268	.6629017	.6627791	.6626564	.6625337	.6624111	.6622884	.6621658	.6620431	.6619205	.6617979
0.269	.6616752	.6615526	.6614300	.6613074	.6611847	.6610621	.6609395	.6608169	.6606943	.6605717
0.270	.6604491	.6603265	.6602039	.6600813	.6599587	.6598362	.6597136	.6595910	.6594684	.6593459
0.271	.6592233	.6591008	.6589782	.6588557	.6587331	.6586106	.6584880	.6583655	.6582430	.6581204
0.272	.6579979	.6578754	.6577529	.6576304	.6575078	.6573853	.6572628	.6571403	.6570178	.6568953
0.273	.6567729	.6566504	.6565279	.6564054	.6562829	.6561605	.6560380	.6559155	.6557931	.6556706
0.274	.6555482	.6554257	.6553033	.6551808	.6550584	.6549359	.6548135	.6546911	.6545687	.6544462
0.275	.6543238	.6542014	.6540790	.6539566	.6538342	.6537118	.6535894	.6534670	.6533446	.6532222
0.276	.6530999	.6529775	.6528551	.6527327	.6526104	.6524880	.6523657	.6522433	.6521210	.6519986
0.277	.6518763	.6517539	.6516316	.6515093	.6513869	.6512646	.6511423	.6510200	.6508977	.6507753
0.278	.6506530	.6505307	.6504084	.6502861	.6501638	.6500416	.6499193	.6497970	.6496747	.6495524
0.279	.6494302	.6493079	.6491856	.6490634	.6489411	.6488189	.6486966	.6485744	.6484521	.6483299
0.280	.6482077	.6480854	.6479632	.6478410	.6477188	.6475966	.6474744	.6473521	.6472299	.6471077
0.281	.6469855	.6468634	.6467412	.6466190	.6464968	.6463746	.6462524	.6461303	.6460081	.6458860
0.282	.6457638	.6456416	.6455195	.6453973	.6452752	.6451531	.6450309	.6449088	.6447867	.6446645
0.283	.6445424	.6444203	.6442982	.6441761	.6440540	.6439319	.6438098	.6436877	.6435656	.6434435
0.284	.6433214	.6431993	.6430773	.6429552	.6428331	.6427111	.6425890	.6424669	.6423449	.6422228
0.285	.6421008	.6419787	.6418567	.6417347	.6416126	.6414906	.6413686	.6412466	.6411246	.6410026
0.286	.6408805	.6407585	.6406365	.6405145	.6403925	.6402706	.6401486	.6400266	.6399046	.6397827
0.287	.6396607	.6395387	.6394167	.6392948	.6391728	.6390509	.6389289	.6388070	.6386851	.6385631
0.288	.6384412	.6383193	.6381973	.6380754	.6379535	.6378316	.6377097	.6375878	.6374659	.6373440
0.289	.6372221	.6371002	.6369783	.6368564	.6367346	.6366127	.6364908	.6363690	.6362471	.6361252
0.290	.6360034	.6358815	.6357597	.6356378	.6355160	.6353942	.6352723	.6351505	.6350287	.6349069
0.291	.6347851	.6346632	.6345414	.6344196	.6342978	.6341760	.6340542	.6339324	.6338107	.6336889
0.292	.6335671	.6334453	.6333236	.6332018	.6330800	.6329583	.6328365	.6327148	.6325930	.6324713
0.293	.6323496	.6322278	.6321061	.6319844	.6318626	.6317409	.6316192	.6314975	.6313758	.6312541
0.294	.6311324	.6310107	.6308890	.6307673	.6306456	.6305240	.6304023	.6302806	.6301589	.6300373
0.295	.6299156	.6297940	.6296723	.6295507	.6294290	.6293074	.6291857	.6290641	.6289425	.6288209
0.296	.6286992	.6285776	.6284560	.6283344	.6282128	.6280912	.6279696	.6278480	.6277264	.6276048
0.297	.6274832	.6273617	.6272401	.6271185	.6269970	.6268754	.6267539	.6266323	.6265107	.6263892
0.298	.6262677	.6261461	.6260246	.6259031	.6257815	.6256600	.6255385	.6254170	.6252955	.6251740
0.299	.6250525	.6249310	.6248095	.6246880	.6245665	.6244450	.6243235	.6242021	.6240806	.6239591

TABLE 22

Modulation Transfer Function, Perfect Lens, Monochromatic

ν_r	.0000	.0001	.0002	.0003	.0004	.0005	.0006	.0007	.0008	.0009
0.300	.6238377	.6237162	.6235948	.6234733	.6233519	.6232304	.6231090	.6229876	.6228661	.6227447
0.301	.6226233	.6225019	.6223805	.6222590	.6221376	.6220162	.6218948	.6217735	.6216521	.6215307
0.302	.6214093	.6212879	.6211665	.6210452	.6209238	.6208024	.6206811	.6205597	.6204384	.6203170
0.303	.6201957	.6200744	.6199530	.6198317	.6197104	.6195891	.6194677	.6193464	.6192251	.6191038
0.304	.6189825	.6188612	.6187399	.6186186	.6184974	.6183761	.6182548	.6181335	.6180123	.6178910
0.305	.6177697	.6176485	.6175272	.6174060	.6172847	.6171635	.6170423	.6169210	.6167998	.6166786
0.306	.6165574	.6164362	.6163149	.6161937	.6160725	.6159514	.6158301	.6157090	.6155878	.6154666
0.307	.6153454	.6152242	.6151031	.6149819	.6148607	.6147396	.6146184	.6144973	.6143761	.6142550
0.308	.6141339	.6140127	.6138916	.6137705	.6136494	.6135283	.6134071	.6132860	.6131649	.6130438
0.309	.6129227	.6128016	.6126806	.6125595	.6124384	.6123173	.6121962	.6120752	.6119541	.6118331
0.310	.6117120	.6115910	.6114699	.6113489	.6112278	.6111068	.6109858	.6108647	.6107437	.6106227
0.311	.6105017	.6103807	.6102597	.6101387	.6100177	.6098967	.6097757	.6096547	.6095338	.6094128
0.312	.6092918	.6091708	.6090499	.6089289	.6088080	.6086870	.6085661	.6084451	.6083242	.6082033
0.313	.6080823	.6079614	.6078405	.6077196	.6075987	.6074778	.6073569	.6072360	.6071151	.6069942
0.314	.6068733	.6067524	.6066315	.6065107	.6063898	.6062689	.6061480	.6060272	.6059063	.6057855
0.315	.6056646	.6055438	.6054230	.6053021	.6051813	.6050605	.6049397	.6048189	.6046980	.6045772
0.316	.6044564	.6043356	.6042148	.6040941	.6039733	.6038525	.6037317	.6036109	.6034902	.6033694
0.317	.6032487	.6031279	.6030071	.6028864	.6027657	.6026449	.6025242	.6024035	.6022827	.6021620
0.318	.6020413	.6019206	.6017999	.6016792	.6015585	.6014378	.6013171	.6011964	.6010757	.6009550
0.319	.6008344	.6007137	.6005930	.6004724	.6003517	.6002311	.6001104	.5999898	.5998691	.5997485
0.320	.5996279	.5995072	.5993866	.5992660	.5991454	.5990248	.5989042	.5987836	.5986630	.5985424
0.321	.5984218	.5983012	.5981806	.5980600	.5979395	.5978189	.5976983	.5975778	.5974572	.5973367
0.322	.5972161	.5970956	.5969751	.5968545	.5967340	.5966135	.5964930	.5963724	.5962519	.5961314
0.323	.5960109	.5958904	.5957699	.5956495	.5955290	.5954085	.5952880	.5951675	.5950471	.5949266
0.324	.5948062	.5946857	.5945652	.5944448	.5943244	.5942039	.5940835	.5939631	.5938427	.5937222
0.325	.5936018	.5934814	.5933610	.5932406	.5931202	.5929998	.5928794	.5927590	.5926387	.5925183
0.326	.5923979	.5922775	.5921572	.5920368	.5919165	.5917961	.5916758	.5915555	.5914351	.5913148
0.327	.5911945	.5910741	.5909538	.5908335	.5907132	.5905929	.5904726	.5903523	.5902320	.5901117
0.328	.5899914	.5898712	.5897509	.5896306	.5895103	.5893901	.5892698	.5891496	.5890293	.5889091
0.329	.5887888	.5886686	.5885484	.5884282	.5883079	.5881877	.5880675	.5879473	.5878271	.5877069
0.330	.5875867	.5874665	.5873463	.5872262	.5871060	.5869858	.5868656	.5867455	.5866253	.5865052
0.331	.5863850	.5862649	.5861447	.5860246	.5859045	.5857843	.5856642	.5855441	.5854240	.5853039
0.332	.5851838	.5850637	.5849436	.5848235	.5847034	.5845833	.5844632	.5843432	.5842231	.5841030
0.333	.5839830	.5838629	.5837429	.5836228	.5835028	.5833828	.5832627	.5831427	.5830227	.5829026
0.334	.5827826	.5826626	.5825426	.5824226	.5823026	.5821826	.5820626	.5819427	.5818227	.5817027
0.335	.5815827	.5814628	.5813428	.5812229	.5811029	.5809830	.5808630	.5807431	.5806231	.5805032
0.336	.5803833	.5802634	.5801435	.5800235	.5799036	.5797837	.5796639	.5795440	.5794241	.5793042
0.337	.5791843	.5790644	.5789446	.5788247	.5787048	.5785850	.5784651	.5783453	.5782254	.5781056
0.338	.5779858	.5778660	.5777461	.5776263	.5775065	.5773867	.5772669	.5771471	.5770273	.5769075
0.339	.5767877	.5766679	.5765481	.5764284	.5763086	.5761888	.5760691	.5759493	.5758296	.5757098
0.340	.5755901	.5754704	.5753506	.5752309	.5751112	.5749914	.5748717	.5747520	.5746323	.5745126
0.341	.5743929	.5742732	.5741536	.5740339	.5739142	.5737945	.5736749	.5735552	.5734355	.5733159
0.342	.5731962	.5730766	.5729569	.5728373	.5727177	.5725981	.5724784	.5723588	.5722392	.5721196
0.343	.5720000	.5718804	.5717608	.5716412	.5715216	.5714021	.5712825	.5711629	.5710433	.5709238
0.344	.5708042	.5706847	.5705651	.5704456	.5703261	.5702065	.5700870	.5699675	.5698480	.5697284
0.345	.5696089	.5694894	.5693699	.5692504	.5691309	.5690115	.5688920	.5687725	.5686530	.5685336
0.346	.5684141	.5682947	.5681752	.5680557	.5679363	.5678169	.5676974	.5675780	.5674586	.5673392
0.347	.5672197	.5671003	.5669809	.5668615	.5667421	.5666227	.5665033	.5663840	.5662646	.5661452
0.348	.5660258	.5659065	.5657871	.5656678	.5655484	.5654291	.5653097	.5651904	.5650711	.5649518
0.349	.5648324	.5647131	.5645938	.5644745	.5643552	.5642359	.5641166	.5639973	.5638780	.5637588

TABLE 22

Modulation Transfer Function, Perfect Lens, Monochromatic

ν_r	.0000	.0001	.0002	.0003	.0004	.0005	.0006	.0007	.0008	.0009
0.350	.5636395	.5635202	.5634010	.5632817	.5631624	.5630432	.5629240	.5628047	.5626855	.5625663
0.351	.5624470	.5623278	.5622086	.5620894	.5619702	.5618510	.5617318	.5616126	.5614934	.5613742
0.352	.5612550	.5611359	.5610167	.5608975	.5607784	.5606592	.5605401	.5604209	.5603018	.5601826
0.353	.5600635	.5599444	.5598253	.5597062	.5595870	.5594679	.5593488	.5592297	.5591106	.5589916
0.354	.5588725	.5587534	.5586343	.5585153	.5583962	.5582772	.5581581	.5580390	.5579200	.5578010
0.355	.5576819	.5575629	.5574439	.5573249	.5572058	.5570868	.5569678	.5568488	.5567298	.5566108
0.356	.5564919	.5563729	.5562539	.5561349	.5560160	.5558970	.5557781	.5556591	.5555402	.5554212
0.357	.5553023	.5551834	.5550644	.5549455	.5548266	.5547077	.5545888	.5544699	.5543510	.5542321
0.358	.5541132	.5539943	.5538754	.5537566	.5536377	.5535188	.5534000	.5532811	.5531623	.5530434
0.359	.5529246	.5528057	.5526869	.5525681	.5524493	.5523305	.5522116	.5520929	.5519741	.5518553
0.360	.5517365	.5516177	.5514989	.5513801	.5512614	.5511426	.5510238	.5509051	.5507863	.5506676
0.361	.5505488	.5504301	.5503114	.5501926	.5500739	.5499552	.5498365	.5497178	.5495991	.5494804
0.362	.5493617	.5492430	.5491243	.5490057	.5488870	.5487683	.5486497	.5485310	.5484124	.5482937
0.363	.5481751	.5480564	.5479378	.5478192	.5477005	.5475819	.5474633	.5473447	.5472261	.5471075
0.364	.5469889	.5468703	.5467518	.5466332	.5465146	.5463960	.5462775	.5461589	.5460404	.5459218
0.365	.5458033	.5456848	.5455662	.5454477	.5453292	.5452106	.5450921	.5449736	.5448551	.5447366
0.366	.5446181	.5444997	.5443812	.5442627	.5441442	.5440257	.5439073	.5437888	.5436704	.5435519
0.367	.5434335	.5433151	.5431966	.5430782	.5429598	.5428414	.5427229	.5426045	.5424861	.5423677
0.368	.5422493	.5421310	.5420126	.5418942	.5417758	.5416575	.5415391	.5414208	.5413024	.5411841
0.369	.5410657	.5409474	.5408290	.5407107	.5405924	.5404741	.5403558	.5402375	.5401192	.5400009
0.370	.5398826	.5397643	.5396460	.5395277	.5394095	.5392912	.5391729	.5390547	.5389364	.5388182
0.371	.5386999	.5385817	.5384635	.5383453	.5382270	.5381088	.5379906	.5378724	.5377542	.5376360
0.372	.5375178	.5373997	.5372815	.5371633	.5370451	.5369270	.5368088	.5366907	.5365725	.5364544
0.373	.5363362	.5362181	.5361000	.5359818	.5358637	.5357456	.5356275	.5355094	.5353913	.5352732
0.374	.5351551	.5350370	.5349190	.5348009	.5346828	.5345648	.5344467	.5343287	.5342106	.5340926
0.375	.5339745	.5338565	.5337385	.5336205	.5335025	.5333844	.5332664	.5331484	.5330304	.5329125
0.376	.5327945	.5326765	.5325585	.5324406	.5323226	.5322046	.5320867	.5319687	.5318508	.5317329
0.377	.5316149	.5314970	.5313791	.5312612	.5311432	.5310254	.5309074	.5307896	.5306717	.5305538
0.378	.5304359	.5303180	.5302002	.5300823	.5299644	.5298466	.5297287	.5296109	.5294930	.5293752
0.379	.5292574	.5291396	.5290217	.5289039	.5287861	.5286683	.5285505	.5284327	.5283150	.5281972
0.380	.5280794	.5279616	.5278439	.5277261	.5276083	.5274906	.5273729	.5272551	.5271374	.5270196
0.381	.5269019	.5267842	.5266665	.5265488	.5264311	.5263134	.5261957	.5260780	.5259603	.5258427
0.382	.5257250	.5256073	.5254897	.5253720	.5252544	.5251367	.5250191	.5249014	.5247838	.5246662
0.383	.5245486	.5244310	.5243134	.5241957	.5240782	.5239606	.5238430	.5237254	.5236078	.5234902
0.384	.5233727	.5232551	.5231376	.5230200	.5229025	.5227849	.5226674	.5225499	.5224323	.5223148
0.385	.5221973	.5220798	.5219623	.5218448	.5217273	.5216098	.5214924	.5213749	.5212574	.5211400
0.386	.5210225	.5209050	.5207876	.5206702	.5205527	.5204353	.5203178	.5202004	.5200830	.5199656
0.387	.5198482	.5197308	.5196134	.5194960	.5193786	.5192612	.5191439	.5190265	.5189091	.5187918
0.388	.5186744	.5185571	.5184397	.5183224	.5182051	.5180878	.5179704	.5178531	.5177358	.5176185
0.389	.5175012	.5173839	.5172666	.5171494	.5170321	.5169148	.5167975	.5166803	.5165630	.5164458
0.390	.5163285	.5162113	.5160940	.5159768	.5158596	.5157424	.5156252	.5155080	.5153908	.5152736
0.391	.5151564	.5150392	.5149220	.5148048	.5146877	.5145705	.5144533	.5143362	.5142191	.5141019
0.392	.5139848	.5138676	.5137505	.5136334	.5135163	.5133992	.5132821	.5131650	.5130479	.5129308
0.393	.5128137	.5126966	.5125796	.5124625	.5123454	.5122284	.5121113	.5119943	.5118772	.5117602
0.394	.5116432	.5115262	.5114091	.5112921	.5111751	.5110581	.5109411	.5108242	.5107072	.5105902
0.395	.5104732	.5103562	.5102393	.5101223	.5100054	.5098884	.5097715	.5096546	.5095376	.5094207
0.396	.5093038	.5091869	.5090700	.5089531	.5088362	.5087193	.5086024	.5084855	.5083686	.5082518
0.397	.5081349	.5080180	.5079012	.5077844	.5076675	.5075507	.5074338	.5073170	.5072002	.5070834
0.398	.5069666	.5068498	.5067330	.5066162	.5064994	.5063826	.5062658	.5061491	.5060323	.5059156
0.399	.5057988	.5056821	.5055653	.5054486	.5053318	.5052151	.5050984	.5049817	.5048650	.5047483

TABLE 22

Modulation Transfer Function, Perfect Lens, Monochromatic

ν_r	.0000	.0001	.0002	.0003	.0004	.0005	.0006	.0007	.0008	.0009
0.400	.5046316	.5045149	.5043982	.5042815	.5041648	.5040482	.5039315	.5038149	.5036982	.5035816
0.401	.5034649	.5033483	.5032316	.5031150	.5029984	.5028818	.5027652	.5026486	.5025320	.5024154
0.402	.5022988	.5021822	.5020657	.5019491	.5018325	.5017160	.5015994	.5014829	.5013663	.5012498
0.403	.5011333	.5010167	.5009002	.5007837	.5006672	.5005507	.5004342	.5003177	.5002012	.5000848
0.404	.4999683	.4998518	.4997353	.4996189	.4995024	.4993860	.4992695	.4991531	.4990367	.4989203
0.405	.4988038	.4986874	.4985710	.4984546	.4983382	.4982218	.4981055	.4979891	.4978727	.4977564
0.406	.4976400	.4975236	.4974073	.4972909	.4971746	.4970583	.4969419	.4968256	.4967093	.4965930
0.407	.4964767	.4963604	.4962441	.4961278	.4960115	.4958953	.4957790	.4956627	.4955465	.4954302
0.408	.4953140	.4951977	.4950815	.4949653	.4948490	.4947328	.4946166	.4945004	.4943842	.4942680
0.409	.4941518	.4940356	.4939194	.4938033	.4936871	.4935709	.4934548	.4933386	.4932225	.4931063
0.410	.4929902	.4928741	.4927580	.4926418	.4925257	.4924096	.4922935	.4921774	.4920614	.4919453
0.411	.4918292	.4917131	.4915971	.4914810	.4913649	.4912489	.4911329	.4910168	.4909008	.4907848
0.412	.4906687	.4905527	.4904367	.4903207	.4902047	.4900887	.4899728	.4898568	.4897408	.4896248
0.413	.4895089	.4893929	.4892770	.4891610	.4890451	.4889292	.4888132	.4886973	.4885814	.4884655
0.414	.4883496	.4882337	.4881178	.4880019	.4878860	.4877702	.4876543	.4875384	.4874226	.4873067
0.415	.4871909	.4870750	.4869592	.4868434	.4867276	.4866117	.4864959	.4863801	.4862643	.4861485
0.416	.4860328	.4859170	.4858012	.4856854	.4855697	.4854539	.4853382	.4852224	.4851067	.4849909
0.417	.4848752	.4847595	.4846438	.4845281	.4844123	.4842966	.4841810	.4840653	.4839496	.4838339
0.418	.4837182	.4836026	.4834869	.4833713	.4832556	.4831400	.4830243	.4829087	.4827931	.4826775
0.419	.4825619	.4824463	.4823307	.4822151	.4820995	.4819839	.4818683	.4817527	.4816372	.4815216
0.420	.4814061	.4812905	.4811750	.4810595	.4809439	.4808284	.4807129	.4805974	.4804819	.4803664
0.421	.4802509	.4801354	.4800199	.4799044	.4797890	.4796735	.4795580	.4794426	.4793271	.4792117
0.422	.4790963	.4789808	.4788654	.4787500	.4786346	.4785192	.4784038	.4782884	.4781730	.4780576
0.423	.4779422	.4778269	.4777115	.4775962	.4774808	.4773655	.4772501	.4771348	.4770195	.4769041
0.424	.4767888	.4766735	.4765582	.4764429	.4763276	.4762123	.4760971	.4759818	.4758665	.4757513
0.425	.4756360	.4755207	.4754055	.4752903	.4751750	.4750598	.4749446	.4748294	.4747142	.4745990
0.426	.4744838	.4743686	.4742534	.4741382	.4740230	.4739079	.4737927	.4736776	.4735624	.4734473
0.427	.4733321	.4732170	.4731019	.4729868	.4728717	.4727565	.4726414	.4725264	.4724113	.4722962
0.428	.4721811	.4720660	.4719510	.4718359	.4717209	.4716058	.4714908	.4713758	.4712607	.4711457
0.429	.4710307	.4709157	.4708007	.4706857	.4705707	.4704557	.4703407	.4702258	.4701108	.4699958
0.430	.4698809	.4697659	.4696510	.4695360	.4694211	.4693062	.4691913	.4690764	.4689615	.4688465
0.431	.4687316	.4686168	.4685019	.4683870	.4682721	.4681573	.4680424	.4679276	.4678127	.4676979
0.432	.4675830	.4674682	.4673534	.4672386	.4671238	.4670090	.4668942	.4667794	.4666646	.4665498
0.433	.4664350	.4663203	.4662055	.4660908	.4659760	.4658613	.4657465	.4656318	.4655171	.4654024
0.434	.4652877	.4651730	.4650583	.4649436	.4648289	.4647142	.4645995	.4644849	.4643702	.4642555
0.435	.4641409	.4640262	.4639116	.4637970	.4636824	.4635677	.4634531	.4633385	.4632239	.4631093
0.436	.4629947	.4628802	.4627656	.4626510	.4625364	.4624219	.4623073	.4621928	.4620783	.4619637
0.437	.4618492	.4617347	.4616202	.4615057	.4613912	.4612767	.4611622	.4610477	.4609332	.4608187
0.438	.4607043	.4605898	.4604754	.4603609	.4602465	.4601321	.4600176	.4599032	.4597888	.4596744
0.439	.4595600	.4594456	.4593312	.4592168	.4591024	.4589881	.4588737	.4587593	.4586450	.4585306
0.440	.4584163	.4583020	.4581876	.4580733	.4579590	.4578447	.4577304	.4576161	.4575018	.4573875
0.441	.4572732	.4571590	.4570447	.4569305	.4568162	.4567020	.4565877	.4564735	.4563593	.4562450
0.442	.4561308	.4560166	.4559024	.4557882	.4556740	.4555598	.4554457	.4553315	.4552173	.4551032
0.443	.4549890	.4548749	.4547607	.4546466	.4545325	.4544184	.4543042	.4541901	.4540760	.4539619
0.444	.4538478	.4537338	.4536197	.4535056	.4533916	.4532775	.4531634	.4530494	.4529354	.4528213
0.445	.4527073	.4525933	.4524793	.4523653	.4522513	.4521373	.4520233	.4519093	.4517953	.4516814
0.446	.4515674	.4514534	.4513395	.4512255	.4511116	.4509977	.4508838	.4507698	.4506559	.4505420
0.447	.4504281	.4503142	.4502003	.4500865	.4499726	.4498587	.4497449	.4496310	.4495172	.4494033
0.448	.4492895	.4491757	.4490618	.4489480	.4488342	.4487204	.4486066	.4484928	.4483790	.4482653
0.449	.4481515	.4480377	.4479240	.4478102	.4476965	.4475827	.4474690	.4473553	.4472415	.4471278

TABLE 22

Modulation Transfer Function, Perfect Lens, Monochromatic

ν_r	.0000	.0001	.0002	.0003	.0004	.0005	.0006	.0007	.0008	.0009
0.450	.4470141	.4469004	.4467867	.4466730	.4465594	.4464457	.4463320	.4462184	.4461047	.4459911
0.451	.4458774	.4457638	.4456501	.4455365	.4454229	.4453093	.4451957	.4450821	.4449685	.4448549
0.452	.4447413	.4446278	.4445142	.4444006	.4442871	.4441735	.4440600	.4439465	.4438329	.4437194
0.453	.4436059	.4434924	.4433789	.4432654	.4431519	.4430384	.4429250	.4428115	.4426980	.4425846
0.454	.4424711	.4423577	.4422442	.4421308	.4420174	.4419040	.4417906	.4416772	.4415638	.4414504
0.455	.4413370	.4412236	.4411102	.4409969	.4408835	.4407702	.4406568	.4405435	.4404301	.4403168
0.456	.4402035	.4400902	.4399769	.4398636	.4397503	.4396370	.4395237	.4394105	.4392972	.4391839
0.457	.4390707	.4389574	.4388442	.4387310	.4386177	.4385045	.4383913	.4382781	.4381649	.4380517
0.458	.4379385	.4378253	.4377121	.4375990	.4374858	.4373727	.4372595	.4371464	.4370332	.4369201
0.459	.4368070	.4366939	.4365808	.4364676	.4363545	.4362415	.4361284	.4360153	.4359022	.4357892
0.460	.4356761	.4355631	.4354500	.4353370	.4352239	.4351109	.4349979	.4348849	.4347719	.4346589
0.461	.4345459	.4344329	.4343199	.4342070	.4340940	.4339811	.4338681	.4337552	.4336422	.4335293
0.462	.4334164	.4333034	.4331905	.4330776	.4329647	.4328518	.4327390	.4326261	.4325132	.4324003
0.463	.4322875	.4321746	.4320618	.4319490	.4318361	.4317233	.4316105	.4314977	.4313849	.4312721
0.464	.4311593	.4310465	.4309337	.4308209	.4307082	.4305954	.4304827	.4303699	.4302572	.4301445
0.465	.4300317	.4299190	.4298063	.4296936	.4295809	.4294682	.4293555	.4292428	.4291302	.4290175
0.466	.4289048	.4287922	.4286796	.4285669	.4284543	.4283417	.4282290	.4281164	.4280038	.4278912
0.467	.4277786	.4276660	.4275535	.4274409	.4273283	.4272158	.4271032	.4269907	.4268782	.4267656
0.468	.4266531	.4265406	.4264281	.4263156	.4262031	.4260906	.4259781	.4258656	.4257532	.4256407
0.469	.4255282	.4254158	.4253034	.4251909	.4250785	.4249661	.4248537	.4247412	.4246288	.4245164
0.470	.4244041	.4242917	.4241793	.4240669	.4239546	.4238422	.4237299	.4236175	.4235052	.4233929
0.471	.4232806	.4231682	.4230559	.4229436	.4228313	.4227191	.4226068	.4224945	.4223822	.4222700
0.472	.4221577	.4220455	.4219332	.4218210	.4217088	.4215966	.4214844	.4213722	.4212600	.4211478
0.473	.4210356	.4209234	.4208112	.4206991	.4205869	.4204748	.4203626	.4202505	.4201384	.4200262
0.474	.4199141	.4198020	.4196899	.4195778	.4194657	.4193537	.4192416	.4191295	.4190174	.4189054
0.475	.4187934	.4186813	.4185693	.4184572	.4183452	.4182332	.4181212	.4180092	.4178972	.4177852
0.476	.4176733	.4175613	.4174493	.4173374	.4172254	.4171135	.4170015	.4168896	.4167777	.4166658
0.477	.4165539	.4164420	.4163301	.4162182	.4161063	.4159944	.4158826	.4157707	.4156588	.4155470
0.478	.4154351	.4153233	.4152115	.4150997	.4149879	.4148761	.4147642	.4146525	.4145407	.4144289
0.479	.4143171	.4142054	.4140936	.4139819	.4138701	.4137584	.4136467	.4135349	.4134232	.4133115
0.480	.4131998	.4130881	.4129764	.4128647	.4127531	.4126414	.4125298	.4124181	.4123065	.4121948
0.481	.4120832	.4119716	.4118599	.4117483	.4116367	.4115251	.4114136	.4113020	.4111904	.4110788
0.482	.4109673	.4108557	.4107442	.4106326	.4105211	.4104096	.4102980	.4101865	.4100750	.4099635
0.483	.4098520	.4097405	.4096291	.4095176	.4094061	.4092947	.4091832	.4090718	.4089604	.4088489
0.484	.4087375	.4086261	.4085147	.4084033	.4082919	.4081805	.4080691	.4079578	.4078464	.4077351
0.485	.4076237	.4075124	.4074010	.4072897	.4071784	.4070671	.4069557	.4068444	.4067332	.4066219
0.486	.4065106	.4063993	.4062881	.4061768	.4060655	.4059543	.4058431	.4057318	.4056206	.4055094
0.487	.4053982	.4052870	.4051758	.4050646	.4049534	.4048422	.4047311	.4046199	.4045088	.4043976
0.488	.4042865	.4041754	.4040642	.4039531	.4038420	.4037309	.4036198	.4035087	.4033976	.4032866
0.489	.4031755	.4030644	.4029534	.4028423	.4027313	.4026203	.4025093	.4023982	.4022872	.4021762
0.490	.4020652	.4019542	.4018433	.4017323	.4016213	.4015104	.4013994	.4012885	.4011775	.4010666
0.491	.4009557	.4008448	.4007339	.4006230	.4005121	.4004012	.4002903	.4001794	.4000686	.3999577
0.492	.3998468	.3997360	.3996252	.3995143	.3994035	.3992927	.3991819	.3990711	.3989603	.3988495
0.493	.3987387	.3986280	.3985172	.3984064	.3982957	.3981849	.3980742	.3979635	.3978528	.3977420
0.494	.3976313	.3975206	.3974099	.3972993	.3971886	.3970779	.3969672	.3968566	.3967459	.3966353
0.495	.3965247	.3964140	.3963034	.3961928	.3960822	.3959716	.3958610	.3957504	.3956399	.3955293
0.496	.3954187	.3953082	.3951976	.3950871	.3949765	.3948660	.3947555	.3946450	.3945345	.3944240
0.497	.3943135	.3942030	.3940925	.3939821	.3938716	.3937612	.3936507	.3935403	.3934298	.3933194
0.498	.3932090	.3930986	.3929882	.3928778	.3927674	.3926570	.3925467	.3924363	.3923259	.3922156
0.499	.3921053	.3919949	.3918846	.3917743	.3916639	.3915536	.3914433	.3913331	.3912228	.3911125

TABLE 22

Modulation Transfer Function, Perfect Lens, Monochromatic

ν_r	.0000	.0001	.0002	.0003	.0004	.0005	.0006	.0007	.0008	.0009
0.500	.3910022	.3908920	.3907817	.3906715	.3905612	.3904510	.3903408	.3902305	.3901203	.3900101
0.501	.3898999	.3897898	.3896796	.3895694	.3894592	.3893491	.3892389	.3891288	.3890186	.3889085
0.502	.3887984	.3886883	.3885782	.3884681	.3883580	.3882479	.3881378	.3880277	.3879177	.3878076
0.503	.3876976	.3875875	.3874775	.3873675	.3872574	.3871474	.3870374	.3869274	.3868174	.3867075
0.504	.3865975	.3864875	.3863776	.3862676	.3861577	.3860477	.3859378	.3858279	.3857180	.3856081
0.505	.3854982	.3853883	.3852784	.3851685	.3850586	.3849488	.3848389	.3847291	.3846192	.3845094
0.506	.3843996	.3842898	.3841800	.3840702	.3839604	.3838506	.3837408	.3836310	.3835213	.3834115
0.507	.3833017	.3831920	.3830823	.3829725	.3828628	.3827531	.3826434	.3825337	.3824240	.3823143
0.508	.3822047	.3820950	.3819853	.3818757	.3817660	.3816564	.3815468	.3814371	.3813275	.3812179
0.509	.3811083	.3809987	.3808891	.3807796	.3806700	.3805604	.3804509	.3803413	.3802318	.3801223
0.510	.3800127	.3799032	.3797937	.3796842	.3795747	.3794652	.3793558	.3792463	.3791368	.3790274
0.511	.3789179	.3788085	.3786990	.3785896	.3784802	.3783708	.3782614	.3781520	.3780426	.3779332
0.512	.3778238	.3777145	.3776051	.3774958	.3773864	.3772771	.3771678	.3770584	.3769491	.3768398
0.513	.3767305	.3766212	.3765120	.3764027	.3762934	.3761842	.3760749	.3759657	.3758564	.3757472
0.514	.3756380	.3755288	.3754195	.3753104	.3752012	.3750920	.3749828	.3748736	.3747645	.3746553
0.515	.3745462	.3744370	.3743279	.3742188	.3741097	.3740006	.3738915	.3737824	.3736733	.3735642
0.516	.3734552	.3733461	.3732370	.3731280	.3730190	.3729099	.3728009	.3726919	.3725829	.3724739
0.517	.3723649	.3722559	.3721469	.3720380	.3719290	.3718200	.3717111	.3716022	.3714932	.3713843
0.518	.3712754	.3711665	.3710576	.3709487	.3708398	.3707309	.3706221	.3705132	.3704044	.3702955
0.519	.3701867	.3700779	.3699690	.3698602	.3697514	.3696426	.3695338	.3694250	.3693163	.3692075
0.520	.3690987	.3689900	.3688813	.3687725	.3686638	.3685551	.3684464	.3683376	.3682289	.3681202
0.521	.3680116	.3679029	.3677942	.3676856	.3675769	.3674683	.3673596	.3672510	.3671424	.3670338
0.522	.3669252	.3668166	.3667080	.3665994	.3664908	.3663823	.3662737	.3661652	.3660566	.3659481
0.523	.3658396	.3657310	.3656225	.3655140	.3654055	.3652971	.3651886	.3650801	.3649716	.3648632
0.524	.3647547	.3646463	.3645379	.3644294	.3643210	.3642126	.3641042	.3639958	.3638874	.3637791
0.525	.3636707	.3635623	.3634540	.3633456	.3632373	.3631290	.3630206	.3629123	.3628040	.3626957
0.526	.3625874	.3624791	.3623709	.3622626	.3621543	.3620461	.3619378	.3618296	.3617214	.3616131
0.527	.3615049	.3613967	.3612885	.3611803	.3610722	.3609640	.3608558	.3607477	.3606395	.3605314
0.528	.3604232	.3603151	.3602070	.3600989	.3599908	.3598827	.3597746	.3596665	.3595585	.3594504
0.529	.3593424	.3592343	.3591263	.3590182	.3589102	.3588022	.3586942	.3585862	.3584782	.3583702
0.530	.3582623	.3581543	.3580463	.3579384	.3578304	.3577225	.3576146	.3575067	.3573987	.3572908
0.531	.3571829	.3570751	.3569672	.3568593	.3567514	.3566436	.3565358	.3564279	.3563201	.3562123
0.532	.3561044	.3559966	.3558888	.3557810	.3556733	.3555655	.3554577	.3553500	.3552422	.3551345
0.533	.3550267	.3549190	.3548113	.3547036	.3545959	.3544882	.3543805	.3542728	.3541651	.3540575
0.534	.3539498	.3538422	.3537345	.3536269	.3535193	.3534117	.3533041	.3531965	.3530889	.3529813
0.535	.3528737	.3527662	.3526586	.3525511	.3524435	.3523360	.3522284	.3521209	.3520134	.3519059
0.536	.3517984	.3516909	.3515835	.3514760	.3513685	.3512611	.3511536	.3510462	.3509388	.3508313
0.537	.3507239	.3506165	.3505091	.3504017	.3502944	.3501870	.3500796	.3499723	.3498649	.3497576
0.538	.3496503	.3495429	.3494356	.3493283	.3492210	.3491137	.3490065	.3488992	.3487919	.3486847
0.539	.3485774	.3484702	.3483629	.3482557	.3481485	.3480413	.3479341	.3478269	.3477197	.3476125
0.540	.3475053	.3473982	.3472910	.3471839	.3470768	.3469696	.3468625	.3467554	.3466483	.3465412
0.541	.3464341	.3463270	.3462200	.3461129	.3460059	.3458988	.3457918	.3456847	.3455777	.3454707
0.542	.3453637	.3452567	.3451497	.3450427	.3449358	.3448288	.3447219	.3446149	.3445080	.3444010
0.543	.3442941	.3441872	.3440803	.3439734	.3438665	.3437596	.3436528	.3435459	.3434390	.3433322
0.544	.3432253	.3431185	.3430117	.3429049	.3427981	.3426913	.3425845	.3424777	.3423709	.3422642
0.545	.3421574	.3420507	.3419439	.3418372	.3417305	.3416237	.3415170	.3414103	.3413036	.3411970
0.546	.3410903	.3409836	.3408770	.3407703	.3406637	.3405570	.3404504	.3403438	.3402372	.3401306
0.547	.3400240	.3399174	.3398108	.3397043	.3395977	.3394912	.3393846	.3392781	.3391716	.3390651
0.548	.3389585	.3388520	.3387456	.3386391	.3385326	.3384261	.3383197	.3382132	.3381068	.3380004
0.549	.3378939	.3377875	.3376811	.3375747	.3374683	.3373619	.3372556	.3371492	.3370428	.3369365

TABLE 22

Modulation Transfer Function, Perfect Lens, Monochromatic

ν_r	.0000	.0001	.0002	.0003	.0004	.0005	.0006	.0007	.0008	.0009
0.550	.3368301	.3367238	.3366175	.3365112	.3364049	.3362986	.3361923	.3360860	.3359797	.3358735
0.551	.3357672	.3356609	.3355547	.3354485	.3353422	.3352360	.3351298	.3350236	.3349174	.3348113
0.552	.3347051	.3345989	.3344928	.3343866	.3342805	.3341744	.3340682	.3339621	.3338560	.3337499
0.553	.3336438	.3335378	.3334317	.3333256	.3332196	.3331135	.3330075	.3329015	.3327954	.3326894
0.554	.3325834	.3324774	.3323714	.3322655	.3321595	.3320535	.3319476	.3318416	.3317357	.3316298
0.555	.3315238	.3314179	.3313120	.3312061	.3311003	.3309944	.3308885	.3307827	.3306768	.3305710
0.556	.3304651	.3303593	.3302535	.3301477	.3300419	.3299361	.3298303	.3297245	.3296188	.3295130
0.557	.3294073	.3293015	.3291958	.3290901	.3289843	.3288786	.3287729	.3286673	.3285616	.3284559
0.558	.3283502	.3282446	.3281389	.3280333	.3279277	.3278221	.3277164	.3276108	.3275052	.3273997
0.559	.3272941	.3271885	.3270830	.3269774	.3268719	.3267663	.3266608	.3265553	.3264498	.3263443
0.560	.3262388	.3261333	.3260278	.3259224	.3258169	.3257115	.3256060	.3255006	.3253952	.3252898
0.561	.3251843	.3250790	.3249736	.3248682	.3247628	.3246575	.3245521	.3244468	.3243414	.3242361
0.562	.3241308	.3240255	.3239202	.3238149	.3237096	.3236043	.3234990	.3233938	.3232885	.3231833
0.563	.3230781	.3229728	.3228676	.3227624	.3226572	.3225520	.3224468	.3223417	.3222365	.3221314
0.564	.3220262	.3219211	.3218160	.3217108	.3216057	.3215006	.3213955	.3212904	.3211854	.3210803
0.565	.3209752	.3208702	.3207652	.3206601	.3205551	.3204501	.3203451	.3202401	.3201351	.3200301
0.566	.3199251	.3198202	.3197152	.3196103	.3195053	.3194004	.3192955	.3191906	.3190857	.3189808
0.567	.3188759	.3187710	.3186662	.3185613	.3184565	.3183516	.3182468	.3181420	.3180372	.3179324
0.568	.3178276	.3177228	.3176180	.3175132	.3174085	.3173037	.3171990	.3170942	.3169895	.3168848
0.569	.3167801	.3166754	.3165707	.3164660	.3163613	.3162567	.3161520	.3160474	.3159427	.3158381
0.570	.3157335	.3156289	.3155243	.3154197	.3153151	.3152105	.3151060	.3150014	.3148969	.3147923
0.571	.3146878	.3145833	.3144788	.3143742	.3142698	.3141653	.3140608	.3139563	.3138519	.3137474
0.572	.3136430	.3135385	.3134341	.3133297	.3132253	.3131209	.3130165	.3129121	.3128077	.3127034
0.573	.3125990	.3124947	.3123903	.3122860	.3121817	.3120774	.3119731	.3118688	.3117645	.3116603
0.574	.3115560	.3114517	.3113475	.3112432	.3111390	.3110348	.3109306	.3108264	.3107222	.3106180
0.575	.3105138	.3104097	.3103055	.3102014	.3100972	.3099931	.3098890	.3097849	.3096808	.3095767
0.576	.3094726	.3093685	.3092644	.3091604	.3090563	.3089523	.3088483	.3087442	.3086402	.3085362
0.577	.3084322	.3083282	.3082242	.3081203	.3080163	.3079124	.3078084	.3077045	.3076006	.3074967
0.578	.3073928	.3072889	.3071850	.3070811	.3069772	.3068734	.3067695	.3066657	.3065618	.3064580
0.579	.3063542	.3062504	.3061466	.3060428	.3059390	.3058353	.3057315	.3056277	.3055240	.3054203
0.580	.3053165	.3052128	.3051091	.3050054	.3049017	.3047981	.3046944	.3045907	.3044871	.3043834
0.581	.3042798	.3041762	.3040725	.3039689	.3038653	.3037618	.3036582	.3035546	.3034510	.3033475
0.582	.3032440	.3031404	.3030369	.3029334	.3028299	.3027264	.3026229	.3025194	.3024159	.3023125
0.583	.3022090	.3021056	.3020021	.3018987	.3017953	.3016919	.3015885	.3014851	.3013817	.3012784
0.584	.3011750	.3010717	.3009683	.3008650	.3007617	.3006583	.3005550	.3004517	.3003485	.3002452
0.585	.3001419	.3000386	.2999354	.2998322	.2997289	.2996257	.2995225	.2994193	.2993161	.2992129
0.586	.2991097	.2990066	.2989034	.2988003	.2986971	.2985940	.2984909	.2983878	.2982846	.2981816
0.587	.2980785	.2979754	.2978723	.2977693	.2976662	.2975632	.2974602	.2973571	.2972541	.2971511
0.588	.2970481	.2969452	.2968422	.2967392	.2966363	.2965333	.2964304	.2963274	.2962245	.2961216
0.589	.2960187	.2959158	.2958130	.2957101	.2956072	.2955044	.2954015	.2952987	.2951959	.2950931
0.590	.2949902	.2948874	.2947847	.2946819	.2945791	.2944763	.2943736	.2942709	.2941681	.2940654
0.591	.2939627	.2938600	.2937573	.2936546	.2935519	.2934493	.2933466	.2932440	.2931413	.2930387
0.592	.2929361	.2928335	.2927309	.2926283	.2925257	.2924231	.2923205	.2922180	.2921154	.2920129
0.593	.2919104	.2918079	.2917054	.2916029	.2915004	.2913979	.2912954	.2911930	.2910905	.2909881
0.594	.2908856	.2907832	.2906808	.2905784	.2904760	.2903736	.2902712	.2901689	.2900665	.2899642
0.595	.2898618	.2897595	.2896572	.2895549	.2894526	.2893503	.2892480	.2891457	.2890435	.2889412
0.596	.2888390	.2887367	.2886345	.2885323	.2884301	.2883279	.2882257	.2881235	.2880214	.2879192
0.597	.2878170	.2877149	.2876128	.2875107	.2874085	.2873065	.2872044	.2871023	.2870002	.2868981
0.598	.2867961	.2866940	.2865920	.2864900	.2863880	.2862860	.2861840	.2860820	.2859800	.2858780
0.599	.2857761	.2856741	.2855722	.2854702	.2853683	.2852664	.2851645	.2850626	.2849607	.2848589

TABLE 22

Modulation Transfer Function, Perfect Lens, Monochromatic

ν_r	.0000	.0001	.0002	.0003	.0004	.0005	.0006	.0007	.0008	.0009
0.600	.2847570	.2846551	.2845533	.2844515	.2843496	.2842478	.2841460	.2840442	.2839424	.2838407
0.601	.2837389	.2836371	.2835354	.2834336	.2833319	.2832302	.2831285	.2830268	.2829251	.2828234
0.602	.2827217	.2826201	.2825184	.2824168	.2823151	.2822135	.2821119	.2820103	.2819087	.2818071
0.603	.2817055	.2816040	.2815024	.2814008	.2812993	.2811978	.2810963	.2809948	.2808933	.2807918
0.604	.2806903	.2805888	.2804874	.2803859	.2802845	.2801830	.2800816	.2799802	.2798788	.2797774
0.605	.2796760	.2795746	.2794733	.2793719	.2792706	.2791693	.2790679	.2789666	.2788653	.2787640
0.606	.2786627	.2785614	.2784602	.2783589	.2782577	.2781564	.2780552	.2779540	.2778528	.2777516
0.607	.2776504	.2775492	.2774480	.2773469	.2772457	.2771446	.2770435	.2769423	.2768412	.2767401
0.608	.2766390	.2765379	.2764369	.2763358	.2762348	.2761337	.2760327	.2759317	.2758306	.2757296
0.609	.2756286	.2755277	.2754267	.2753257	.2752248	.2751238	.2750229	.2749220	.2748210	.2747201
0.610	.2746192	.2745183	.2744175	.2743166	.2742157	.2741149	.2740141	.2739132	.2738124	.2737116
0.611	.2736108	.2735100	.2734092	.2733085	.2732077	.2731070	.2730062	.2729055	.2728048	.2727041
0.612	.2726034	.2725027	.2724020	.2723013	.2722007	.2721000	.2719994	.2718987	.2717981	.2716975
0.613	.2715969	.2714963	.2713957	.2712952	.2711946	.2710941	.2709935	.2708930	.2707925	.2706919
0.614	.2705914	.2704909	.2703905	.2702900	.2701895	.2700891	.2699886	.2698882	.2697878	.2696874
0.615	.2695870	.2694866	.2693862	.2692858	.2691854	.2690851	.2689847	.2688844	.2687841	.2686838
0.616	.2685835	.2684832	.2683829	.2682826	.2681824	.2680821	.2679819	.2678816	.2677814	.2676812
0.617	.2675810	.2674808	.2673806	.2672804	.2671803	.2670801	.2669800	.2668798	.2667797	.2666796
0.618	.2665795	.2664794	.2663793	.2662792	.2661792	.2660791	.2659791	.2658790	.2657790	.2656790
0.619	.2655790	.2654790	.2653790	.2652790	.2651791	.2650791	.2649792	.2648792	.2647793	.2646794
0.620	.2645795	.2644796	.2643797	.2642799	.2641800	.2640801	.2639803	.2638805	.2637806	.2636808
0.621	.2635810	.2634812	.2633814	.2632817	.2631819	.2630822	.2629824	.2628827	.2627830	.2626833
0.622	.2625835	.2624839	.2623842	.2622845	.2621848	.2620852	.2619856	.2618859	.2617863	.2616867
0.623	.2615871	.2614875	.2613879	.2612883	.2611888	.2610892	.2609897	.2608902	.2607906	.2606911
0.624	.2605916	.2604921	.2603927	.2602932	.2601937	.2600943	.2599949	.2598954	.2597960	.2596966
0.625	.2595972	.2594978	.2593984	.2592991	.2591997	.2591004	.2590010	.2589017	.2588024	.2587031
0.626	.2586038	.2585045	.2584052	.2583060	.2582067	.2581075	.2580082	.2579090	.2578098	.2577106
0.627	.2576114	.2575122	.2574130	.2573139	.2572147	.2571156	.2570165	.2569173	.2568182	.2567191
0.628	.2566200	.2565210	.2564219	.2563228	.2562238	.2561247	.2560257	.2559267	.2558277	.2557287
0.629	.2556297	.2555307	.2554317	.2553328	.2552338	.2551349	.2550360	.2549371	.2548382	.2547393
0.630	.2546404	.2545415	.2544427	.2543438	.2542450	.2541461	.2540473	.2539485	.2538497	.2537509
0.631	.2536521	.2535533	.2534546	.2533558	.2532571	.2531584	.2530597	.2529609	.2528622	.2527636
0.632	.2526649	.2525662	.2524676	.2523689	.2522703	.2521717	.2520730	.2519744	.2518758	.2517773
0.633	.2516787	.2515801	.2514816	.2513830	.2512845	.2511860	.2510875	.2509890	.2508905	.2507920
0.634	.2506935	.2505951	.2504966	.2503982	.2502997	.2502013	.2501029	.2500045	.2499061	.2498078
0.635	.2497094	.2496110	.2495127	.2494144	.2493160	.2492177	.2491194	.2490211	.2489229	.2488246
0.636	.2487263	.2486281	.2485298	.2484316	.2483334	.2482352	.2481370	.2480388	.2479406	.2478425
0.637	.2477443	.2476462	.2475480	.2474499	.2473518	.2472537	.2471556	.2470575	.2469595	.2468614
0.638	.2467633	.2466653	.2465673	.2464693	.2463713	.2462733	.2461753	.2460773	.2459793	.2458814
0.639	.2457834	.2456855	.2455876	.2454897	.2453918	.2452939	.2451960	.2450981	.2450003	.2449024
0.640	.2448046	.2447068	.2446089	.2445111	.2444133	.2443155	.2442178	.2441200	.2440223	.2439245
0.641	.2438268	.2437291	.2436314	.2435337	.2434360	.2433383	.2432406	.2431430	.2430453	.2429477
0.642	.2428500	.2427524	.2426548	.2425572	.2424597	.2423621	.2422645	.2421670	.2420694	.2419719
0.643	.2418744	.2417769	.2416794	.2415819	.2414844	.2413870	.2412895	.2411921	.2410946	.2409972
0.644	.2408998	.2408024	.2407050	.2406076	.2405102	.2404129	.2403155	.2402182	.2401209	.2400236
0.645	.2399263	.2398290	.2397317	.2396344	.2395372	.2394399	.2393427	.2392454	.2391482	.2390510
0.646	.2389538	.2388566	.2387595	.2386623	.2385651	.2384680	.2383709	.2382737	.2381766	.2380795
0.647	.2379824	.2378854	.2377883	.2376912	.2375942	.2374972	.2374001	.2373031	.2372061	.2371091
0.648	.2370122	.2369152	.2368182	.2367213	.2366243	.2365274	.2364305	.2363336	.2362367	.2361398
0.649	.2360429	.2359461	.2358492	.2357524	.2356556	.2355588	.2354619	.2353651	.2352684	.2351716

TABLE 22

Modulation Transfer Function, Perfect Lens, Monochromatic

ν_r	.0000	.0001	.0002	.0003	.0004	.0005	.0006	.0007	.0008	.0009
0.650	.2350748	.2349781	.2348813	.2347846	.2346879	.2345912	.2344945	.2343978	.2343011	.2342044
0.651	.2341078	.2340111	.2339145	.2338179	.2337213	.2336247	.2335281	.2334315	.2333349	.2332384
0.652	.2331418	.2330453	.2329488	.2328523	.2327558	.2326593	.2325628	.2324663	.2323699	.2322734
0.653	.2321770	.2320806	.2319842	.2318878	.2317914	.2316950	.2315986	.2315023	.2314059	.2313096
0.654	.2312133	.2311169	.2310206	.2309243	.2308281	.2307318	.2306355	.2305393	.2304430	.2303468
0.655	.2302506	.2301544	.2300582	.2299620	.2298658	.2297697	.2296735	.2295774	.2294813	.2293852
0.656	.2292891	.2291930	.2290969	.2290008	.2289047	.2288087	.2287127	.2286166	.2285206	.2284246
0.657	.2283286	.2282326	.2281367	.2280407	.2279448	.2278488	.2277529	.2276570	.2275611	.2274652
0.658	.2273693	.2272734	.2271776	.2270817	.2269859	.2268901	.2267942	.2266984	.2266026	.2265069
0.659	.2264111	.2263153	.2262196	.2261238	.2260281	.2259324	.2258367	.2257410	.2256453	.2255496
0.660	.2254540	.2253583	.2252627	.2251671	.2250715	.2249759	.2248803	.2247847	.2246891	.2245936
0.661	.2244980	.2244025	.2243069	.2242114	.2241159	.2240204	.2239249	.2238295	.2237340	.2236386
0.662	.2235431	.2234477	.2233523	.2232569	.2231615	.2230661	.2229708	.2228754	.2227801	.2226847
0.663	.2225894	.2224941	.2223988	.2223035	.2222082	.2221130	.2220177	.2219225	.2218272	.2217320
0.664	.2216368	.2215416	.2214464	.2213512	.2212561	.2211609	.2210658	.2209706	.2208755	.2207804
0.665	.2206853	.2205902	.2204952	.2204001	.2203050	.2202100	.2201150	.2200200	.2199250	.2198300
0.666	.2197350	.2196400	.2195450	.2194501	.2193552	.2192602	.2191653	.2190704	.2189755	.2188806
0.667	.2187858	.2186909	.2185961	.2185012	.2184064	.2183116	.2182168	.2181220	.2180272	.2179325
0.668	.2178377	.2177430	.2176482	.2175535	.2174588	.2173641	.2172694	.2171747	.2170801	.2169854
0.669	.2168908	.2167961	.2167015	.2166069	.2165123	.2164178	.2163232	.2162286	.2161341	.2160395
0.670	.2159450	.2158505	.2157560	.2156615	.2155670	.2154726	.2153781	.2152836	.2151892	.2150948
0.671	.2150004	.2149060	.2148116	.2147172	.2146228	.2145285	.2144342	.2143398	.2142455	.2141512
0.672	.2140569	.2139626	.2138683	.2137741	.2136798	.2135856	.2134914	.2133971	.2133029	.2132088
0.673	.2131146	.2130204	.2129263	.2128321	.2127380	.2126439	.2125497	.2124556	.2123616	.2122675
0.674	.2121734	.2120794	.2119853	.2118913	.2117973	.2117033	.2116093	.2115153	.2114213	.2113274
0.675	.2112334	.2111395	.2110456	.2109516	.2108577	.2107639	.2106700	.2105761	.2104823	.2103884
0.676	.2102946	.2102008	.2101069	.2100132	.2099194	.2098256	.2097318	.2096381	.2095443	.2094506
0.677	.2093569	.2092632	.2091695	.2090758	.2089822	.2088885	.2087949	.2087012	.2086076	.2085140
0.678	.2084204	.2083268	.2082333	.2081397	.2080461	.2079526	.2078591	.2077656	.2076721	.2075786
0.679	.2074851	.2073916	.2072982	.2072047	.2071113	.2070179	.2069245	.2068311	.2067377	.2066443
0.680	.2065509	.2064576	.2063643	.2062709	.2061776	.2060843	.2059910	.2058977	.2058045	.2057112
0.681	.2056180	.2055248	.2054315	.2053383	.2052451	.2051520	.2050588	.2049656	.2048725	.2047793
0.682	.2046862	.2045931	.2045000	.2044069	.2043138	.2042208	.2041277	.2040347	.2039416	.2038486
0.683	.2037556	.2036626	.2035696	.2034767	.2033837	.2032908	.2031978	.2031049	.2030120	.2029191
0.684	.2028262	.2027333	.2026405	.2025476	.2024548	.2023620	.2022691	.2021763	.2020835	.2019908
0.685	.2018980	.2018053	.2017125	.2016198	.2015271	.2014344	.2013417	.2012490	.2011563	.2010636
0.686	.2009710	.2008784	.2007857	.2006931	.2006005	.2005079	.2004154	.2003228	.2002302	.2001377
0.687	.2000452	.1999527	.1998602	.1997677	.1996752	.1995827	.1994903	.1993978	.1993054	.1992130
0.688	.1991206	.1990282	.1989358	.1988434	.1987511	.1986587	.1985664	.1984741	.1983818	.1982895
0.689	.1981972	.1981049	.1980126	.1979204	.1978282	.1977359	.1976437	.1975515	.1974593	.1973672
0.690	.1972750	.1971828	.1970907	.1969986	.1969064	.1968144	.1967223	.1966302	.1965381	.1964461
0.691	.1963540	.1962620	.1961700	.1960780	.1959860	.1958940	.1958020	.1957101	.1956181	.1955262
0.692	.1954343	.1953424	.1952505	.1951586	.1950667	.1949748	.1948830	.1947911	.1946993	.1946075
0.693	.1945157	.1944239	.1943322	.1942404	.1941486	.1940569	.1939652	.1938735	.1937818	.1936901
0.694	.1935984	.1935067	.1934151	.1933234	.1932318	.1931402	.1930486	.1929570	.1928654	.1927739
0.695	.1926823	.1925908	.1924992	.1924077	.1923162	.1922247	.1921333	.1920418	.1919503	.1918589
0.696	.1917675	.1916760	.1915846	.1914932	.1914019	.1913105	.1912191	.1911278	.1910365	.1909451
0.697	.1908538	.1907625	.1906713	.1905800	.1904887	.1903975	.1903063	.1902150	.1901238	.1900326
0.698	.1899415	.1898503	.1897591	.1896680	.1895768	.1894857	.1893946	.1893035	.1892124	.1891214
0.699	.1890303	.1889393	.1888482	.1887572	.1886662	.1885752	.1884842	.1883933	.1883023	.1882113

TABLE 22

Modulation Transfer Function, Perfect Lens, Monochromatic

ν_r	.0000	.0001	.0002	.0003	.0004	.0005	.0006	.0007	.0008	.0009
0.700	.1881204	.1880295	.1879386	.1878477	.1877568	.1876659	.1875751	.1874842	.1873934	.1873026
0.701	.1872118	.1871210	.1870302	.1869394	.1868487	.1867579	.1866672	.1865765	.1864857	.1863951
0.702	.1863044	.1862137	.1861230	.1860324	.1859418	.1858511	.1857605	.1856699	.1855793	.1854888
0.703	.1853982	.1853077	.1852171	.1851266	.1850361	.1849456	.1848551	.1847647	.1846742	.1845838
0.704	.1844933	.1844029	.1843125	.1842221	.1841317	.1840414	.1839510	.1838607	.1837703	.1836800
0.705	.1835897	.1834994	.1834091	.1833189	.1832286	.1831384	.1830481	.1829579	.1828677	.1827775
0.706	.1826873	.1825972	.1825070	.1824169	.1823268	.1822367	.1821465	.1820565	.1819664	.1818763
0.707	.1817863	.1816962	.1816062	.1815162	.1814262	.1813362	.1812462	.1811563	.1810663	.1809764
0.708	.1808864	.1807965	.1807066	.1806168	.1805269	.1804370	.1803472	.1802573	.1801675	.1800777
0.709	.1799879	.1798981	.1798084	.1797186	.1796289	.1795391	.1794494	.1793597	.1792700	.1791803
0.710	.1790906	.1790010	.1789114	.1788217	.1787321	.1786425	.1785529	.1784633	.1783738	.1782842
0.711	.1781947	.1781052	.1780156	.1779261	.1778366	.1777472	.1776577	.1775683	.1774788	.1773894
0.712	.1773000	.1772106	.1771212	.1770318	.1769425	.1768531	.1767638	.1766745	.1765852	.1764959
0.713	.1764066	.1763173	.1762281	.1761388	.1760496	.1759604	.1758712	.1757820	.1756928	.1756036
0.714	.1755145	.1754254	.1753362	.1752471	.1751580	.1750689	.1749799	.1748908	.1748018	.1747127
0.715	.1746237	.1745347	.1744457	.1743567	.1742677	.1741788	.1740898	.1740009	.1739120	.1738231
0.716	.1737342	.1736453	.1735564	.1734676	.1733788	.1732899	.1732011	.1731123	.1730235	.1729348
0.717	.1728460	.1727572	.1726685	.1725798	.1724911	.1724024	.1723137	.1722250	.1721364	.1720477
0.718	.1719591	.1718705	.1717819	.1716933	.1716047	.1715162	.1714276	.1713391	.1712505	.1711620
0.719	.1710735	.1709851	.1708966	.1708081	.1707197	.1706313	.1705428	.1704544	.1703660	.1702777
0.720	.1701893	.1701009	.1700126	.1699243	.1698359	.1697476	.1696594	.1695711	.1694828	.1693946
0.721	.1693063	.1692181	.1691299	.1690417	.1689535	.1688654	.1687772	.1686891	.1686010	.1685128
0.722	.1684247	.1683367	.1682486	.1681605	.1680725	.1679844	.1678964	.1678084	.1677204	.1676324
0.723	.1675445	.1674565	.1673686	.1672806	.1671927	.1671048	.1670169	.1669291	.1668412	.1667533
0.724	.1666655	.1665777	.1664899	.1664021	.1663143	.1662265	.1661388	.1660510	.1659633	.1658756
0.725	.1657879	.1657002	.1656125	.1655249	.1654372	.1653496	.1652620	.1651744	.1650868	.1649992
0.726	.1649116	.1648241	.1647365	.1646490	.1645615	.1644740	.1643865	.1642990	.1642116	.1641241
0.727	.1640367	.1639493	.1638619	.1637745	.1636871	.1635997	.1635124	.1634250	.1633377	.1632504
0.728	.1631631	.1630758	.1629886	.1629013	.1628141	.1627268	.1626396	.1625524	.1624652	.1623781
0.729	.1622909	.1622037	.1621166	.1620295	.1619424	.1618553	.1617682	.1616811	.1615941	.1615071
0.730	.1614200	.1613330	.1612460	.1611590	.1610720	.1609851	.1608982	.1608112	.1607243	.1606374
0.731	.1605505	.1604636	.1603768	.1602899	.1602031	.1601163	.1600295	.1599427	.1598559	.1597691
0.732	.1596824	.1595956	.1595089	.1594222	.1593355	.1592488	.1591621	.1590755	.1589888	.1589022
0.733	.1588156	.1587290	.1586424	.1585558	.1584693	.1583827	.1582962	.1582097	.1581232	.1580367
0.734	.1579502	.1578637	.1577773	.1576908	.1576044	.1575180	.1574316	.1573452	.1572588	.1571725
0.735	.1570861	.1569998	.1569135	.1568272	.1567409	.1566547	.1565684	.1564821	.1563959	.1563097
0.736	.1562235	.1561373	.1560511	.1559650	.1558788	.1557927	.1557066	.1556205	.1555344	.1554483
0.737	.1553622	.1552762	.1551901	.1551041	.1550181	.1549321	.1548461	.1547602	.1546742	.1545883
0.738	.1545023	.1544164	.1543305	.1542447	.1541588	.1540729	.1539871	.1539013	.1538154	.1537296
0.739	.1536439	.1535581	.1534723	.1533866	.1533009	.1532151	.1531294	.1530438	.1529581	.1528724
0.740	.1527868	.1527011	.1526155	.1525299	.1524443	.1523588	.1522732	.1521876	.1521021	.1520166
0.741	.1519311	.1518456	.1517601	.1516747	.1515892	.1515038	.1514183	.1513329	.1512475	.1511622
0.742	.1510768	.1509915	.1509061	.1508208	.1507355	.1506502	.1505649	.1504796	.1503944	.1503092
0.743	.1502239	.1501387	.1500535	.1499683	.1498832	.1497980	.1497129	.1496278	.1495426	.1494575
0.744	.1493725	.1492874	.1492023	.1491173	.1490323	.1489473	.1488623	.1487773	.1486923	.1486074
0.745	.1485224	.1484375	.1483526	.1482677	.1481828	.1480979	.1480131	.1479282	.1478434	.1477586
0.746	.1476738	.1475890	.1475042	.1474195	.1473347	.1472500	.1471653	.1470806	.1469959	.1469113
0.747	.1468266	.1467420	.1466573	.1465727	.1464881	.1464035	.1463190	.1462344	.1461499	.1460654
0.748	.1459808	.1458963	.1458119	.1457274	.1456429	.1455585	.1454741	.1453897	.1453053	.1452209
0.749	.1451365	.1450522	.1449678	.1448835	.1447992	.1447149	.1446306	.1445463	.1444621	.1443778

TABLE 22

Modulation Transfer Function, Perfect Lens, Monochromatic

ν_r	.0000	.0001	.0002	.0003	.0004	.0005	.0006	.0007	.0008	.0009
0.750	.1442936	.1442094	.1441252	.1440410	.1439569	.1438727	.1437886	.1437044	.1436203	.1435363
0.751	.1434522	.1433681	.1432841	.1432000	.1431160	.1430320	.1429480	.1428640	.1427801	.1426961
0.752	.1426122	.1425283	.1424444	.1423605	.1422766	.1421927	.1421089	.1420250	.1419412	.1418574
0.753	.1417736	.1416899	.1416061	.1415223	.1414386	.1413549	.1412712	.1411875	.1411038	.1410202
0.754	.1409365	.1408529	.1407693	.1406857	.1406021	.1405185	.1404350	.1403514	.1402679	.1401844
0.755	.1401009	.1400174	.1399340	.1398505	.1397671	.1396837	.1396002	.1395168	.1394335	.1393501
0.756	.1392667	.1391834	.1391001	.1390168	.1389335	.1388502	.1387670	.1386837	.1386005	.1385173
0.757	.1384341	.1383509	.1382677	.1381845	.1381014	.1380183	.1379352	.1378521	.1377690	.1376859
0.758	.1376028	.1375198	.1374368	.1373538	.1372708	.1371878	.1371048	.1370219	.1369389	.1368560
0.759	.1367731	.1366902	.1366073	.1365245	.1364416	.1363588	.1362760	.1361932	.1361104	.1360276
0.760	.1359449	.1358621	.1357794	.1356967	.1356140	.1355313	.1354486	.1353660	.1352833	.1352007
0.761	.1351181	.1350355	.1349529	.1348704	.1347878	.1347053	.1346228	.1345403	.1344578	.1343753
0.762	.1342928	.1342104	.1341280	.1340455	.1339631	.1338808	.1337984	.1337160	.1336337	.1335514
0.763	.1334691	.1333868	.1333045	.1332222	.1331400	.1330577	.1329755	.1328933	.1328111	.1327290
0.764	.1326468	.1325646	.1324825	.1324004	.1323183	.1322362	.1321542	.1320721	.1319901	.1319080
0.765	.1318260	.1317440	.1316621	.1315801	.1314981	.1314162	.1313343	.1312524	.1311705	.1310886
0.766	.1310068	.1309249	.1308431	.1307613	.1306795	.1305977	.1305160	.1304342	.1303525	.1302708
0.767	.1301891	.1301074	.1300257	.1299440	.1298624	.1297808	.1296991	.1296175	.1295360	.1294544
0.768	.1293728	.1292913	.1292098	.1291283	.1290468	.1289653	.1288838	.1288024	.1287210	.1286396
0.769	.1285582	.1284768	.1283954	.1283140	.1282327	.1281514	.1280701	.1279888	.1279075	.1278263
0.770	.1277450	.1276638	.1275826	.1275014	.1274202	.1273390	.1272579	.1271767	.1270956	.1270145
0.771	.1269334	.1268523	.1267713	.1266902	.1266092	.1265282	.1264472	.1263662	.1262852	.1262043
0.772	.1261233	.1260424	.1259615	.1258806	.1257997	.1257189	.1256380	.1255572	.1254764	.1253956
0.773	.1253148	.1252340	.1251533	.1250725	.1249918	.1249111	.1248304	.1247498	.1246691	.1245885
0.774	.1245078	.1244272	.1243466	.1242660	.1241855	.1241049	.1240244	.1239439	.1238634	.1237829
0.775	.1237024	.1236220	.1235415	.1234611	.1233807	.1233003	.1232199	.1231395	.1230592	.1229789
0.776	.1228985	.1228182	.1227380	.1226577	.1225774	.1224972	.1224170	.1223368	.1222566	.1221764
0.777	.1220963	.1220161	.1219360	.1218559	.1217758	.1216957	.1216156	.1215356	.1214556	.1213755
0.778	.1212955	.1212156	.1211356	.1210556	.1209757	.1208958	.1208159	.1207360	.1206561	.1205762
0.779	.1204964	.1204166	.1203368	.1202570	.1201772	.1200974	.1200177	.1199379	.1198582	.1197785
0.780	.1196988	.1196192	.1195395	.1194599	.1193803	.1193007	.1192211	.1191415	.1190619	.1189824
0.781	.1189029	.1188234	.1187439	.1186644	.1185849	.1185055	.1184260	.1183466	.1182672	.1181879
0.782	.1181085	.1180291	.1179498	.1178705	.1177912	.1177119	.1176326	.1175534	.1174741	.1173949
0.783	.1173157	.1172365	.1171573	.1170782	.1169990	.1169199	.1168408	.1167617	.1166826	.1166036
0.784	.1165245	.1164455	.1163665	.1162875	.1162085	.1161295	.1160506	.1159717	.1158927	.1158138
0.785	.1157350	.1156561	.1155772	.1154984	.1154196	.1153408	.1152620	.1151832	.1151044	.1150257
0.786	.1149470	.1148683	.1147896	.1147109	.1146323	.1145536	.1144750	.1143964	.1143178	.1142392
0.787	.1141606	.1140821	.1140036	.1139251	.1138466	.1137681	.1136896	.1136112	.1135327	.1134543
0.788	.1133759	.1132976	.1132192	.1131408	.1130625	.1129842	.1129059	.1128276	.1127493	.1126711
0.789	.1125928	.1125146	.1124364	.1123582	.1122801	.1122019	.1121238	.1120457	.1119676	.1118895
0.790	.1118114	.1117333	.1116553	.1115773	.1114993	.1114213	.1113433	.1112654	.1111874	.1111095
0.791	.1110316	.1109537	.1108758	.1107980	.1107201	.1106423	.1105645	.1104867	.1104089	.1103312
0.792	.1102534	.1101757	.1100980	.1100203	.1099426	.1098650	.1097873	.1097097	.1096321	.1095545
0.793	.1094769	.1093994	.1093218	.1092443	.1091668	.1090893	.1090118	.1089343	.1088569	.1087795
0.794	.1087021	.1086247	.1085473	.1084699	.1083926	.1083153	.1082379	.1081606	.1080834	.1080061
0.795	.1079289	.1078516	.1077744	.1076972	.1076201	.1075429	.1074658	.1073886	.1073115	.1072344
0.796	.1071573	.1070803	.1070032	.1069262	.1068492	.1067722	.1066952	.1066183	.1065413	.1064644
0.797	.1063875	.1063106	.1062337	.1061569	.1060800	.1060032	.1059264	.1058496	.1057728	.1056961
0.798	.1056193	.1055426	.1054659	.1053892	.1053125	.1052359	.1051592	.1050826	.1050060	.1049294
0.799	.1048528	.1047763	.1046997	.1046232	.1045467	.1044702	.1043938	.1043173	.1042409	.1041644

TABLE 22

Modulation Transfer Function, Perfect Lens, Monochromatic

ν_r	.0000	.0001	.0002	.0003	.0004	.0005	.0006	.0007	.0008	.0009
0.800	.1040880	.1040117	.1039353	.1038589	.1037826	.1037063	.1036300	.1035537	.1034774	.1034012
0.801	.1033249	.1032487	.1031725	.1030964	.1030202	.1029440	.1028679	.1027918	.1027157	.1026396
0.802	.1025635	.1024875	.1024115	.1023355	.1022595	.1021835	.1021075	.1020316	.1019557	.1018798
0.803	.1018039	.1017280	.1016522	.1015763	.1015005	.1014247	.1013489	.1012731	.1011974	.1011216
0.804	.1010459	.1009702	.1008945	.1008189	.1007432	.1006676	.1005920	.1005164	.1004408	.1003652
0.805	.1002897	.1002141	.1001386	.1000631	.0999876	.0999122	.0998367	.0997613	.0996859	.0996105
0.806	.0995352	.0994598	.0993845	.0993091	.0992338	.0991586	.0990833	.0990080	.0989328	.0988576
0.807	.0987824	.0987072	.0986320	.0985569	.0984817	.0984066	.0983315	.0982565	.0981814	.0981064
0.808	.0980313	.0979563	.0978813	.0978064	.0977314	.0976565	.0975815	.0975066	.0974317	.0973569
0.809	.0972820	.0972072	.0971324	.0970576	.0969828	.0969080	.0968333	.0967586	.0966839	.0966092
0.810	.0965345	.0964598	.0963852	.0963106	.0962360	.0961614	.0960868	.0960123	.0959377	.0958632
0.811	.0957887	.0957142	.0956398	.0955653	.0954909	.0954165	.0953421	.0952677	.0951933	.0951190
0.812	.0950447	.0949704	.0948961	.0948218	.0947476	.0946733	.0945991	.0945249	.0944507	.0943766
0.813	.0943024	.0942283	.0941542	.0940801	.0940060	.0939320	.0938579	.0937839	.0937099	.0936359
0.814	.0935620	.0934880	.0934141	.0933402	.0932663	.0931924	.0931185	.0930447	.0929709	.0928971
0.815	.0928233	.0927495	.0926758	.0926020	.0925283	.0924546	.0923809	.0923073	.0922336	.0921600
0.816	.0920864	.0920128	.0919392	.0918657	.0917921	.0917186	.0916451	.0915716	.0914981	.0914247
0.817	.0913513	.0912779	.0912045	.0911311	.0910577	.0909844	.0909111	.0908378	.0907645	.0906912
0.818	.0906180	.0905448	.0904715	.0903983	.0903252	.0902520	.0901789	.0901058	.0900326	.0899596
0.819	.0898865	.0898135	.0897404	.0896674	.0895944	.0895214	.0894485	.0893755	.0893026	.0892297
0.820	.0891568	.0890840	.0890111	.0889383	.0888655	.0887927	.0887199	.0886471	.0885744	.0885017
0.821	.0884290	.0883563	.0882836	.0882110	.0881384	.0880658	.0879932	.0879206	.0878480	.0877755
0.822	.0877030	.0876305	.0875580	.0874855	.0874131	.0873407	.0872683	.0871959	.0871235	.0870511
0.823	.0869788	.0869065	.0868342	.0867619	.0866896	.0866174	.0865452	.0864730	.0864008	.0863286
0.824	.0862565	.0861843	.0861122	.0860401	.0859681	.0858960	.0858240	.0857519	.0856799	.0856080
0.825	.0855360	.0854641	.0853921	.0853202	.0852483	.0851765	.0851046	.0850328	.0849610	.0848892
0.826	.0848174	.0847456	.0846739	.0846022	.0845305	.0844588	.0843871	.0843155	.0842438	.0841722
0.827	.0841006	.0840291	.0839575	.0838860	.0838144	.0837430	.0836715	.0836000	.0835286	.0834572
0.828	.0833857	.0833144	.0832430	.0831716	.0831003	.0830290	.0829577	.0828864	.0828152	.0827440
0.829	.0826727	.0826016	.0825304	.0824592	.0823881	.0823170	.0822459	.0821748	.0821037	.0820327
0.830	.0819616	.0818906	.0818196	.0817487	.0816777	.0816068	.0815359	.0814650	.0813941	.0813233
0.831	.0812524	.0811816	.0811108	.0810400	.0809693	.0808985	.0808278	.0807571	.0806864	.0806157
0.832	.0805451	.0804745	.0804039	.0803333	.0802627	.0801922	.0801216	.0800511	.0799806	.0799102
0.833	.0798397	.0797693	.0796988	.0796284	.0795581	.0794877	.0794174	.0793470	.0792767	.0792065
0.834	.0791362	.0790660	.0789957	.0789255	.0788553	.0787852	.0787150	.0786449	.0785748	.0785047
0.835	.0784346	.0783646	.0782946	.0782245	.0781546	.0780846	.0780146	.0779447	.0778748	.0778049
0.836	.0777350	.0776652	.0775953	.0775255	.0774557	.0773859	.0773162	.0772464	.0771767	.0771070
0.837	.0770373	.0769677	.0768980	.0768284	.0767588	.0766892	.0766196	.0765501	.0764806	.0764111
0.838	.0763416	.0762721	.0762026	.0761332	.0760638	.0759944	.0759251	.0758557	.0757864	.0757171
0.839	.0756478	.0755785	.0755093	.0754400	.0753708	.0753016	.0752324	.0751633	.0750942	.0750250
0.840	.0749560	.0748869	.0748178	.0747488	.0746798	.0746108	.0745418	.0744728	.0744039	.0743350
0.841	.0742661	.0741972	.0741284	.0740595	.0739907	.0739219	.0738531	.0737844	.0737156	.0736469
0.842	.0735782	.0735095	.0734409	.0733722	.0733036	.0732350	.0731664	.0730979	.0730293	.0729608
0.843	.0728923	.0728239	.0727554	.0726870	.0726185	.0725501	.0724818	.0724134	.0723451	.0722767
0.844	.0722084	.0721402	.0720719	.0720037	.0719354	.0718672	.0717991	.0717309	.0716628	.0715946
0.845	.0715265	.0714585	.0713904	.0713224	.0712544	.0711864	.0711184	.0710504	.0709825	.0709146
0.846	.0708467	.0707788	.0707109	.0706431	.0705753	.0705075	.0704397	.0703720	.0703042	.0702365
0.847	.0701688	.0701011	.0700335	.0699658	.0698982	.0698306	.0697631	.0696955	.0696280	.0695605
0.848	.0694930	.0694255	.0693581	.0692906	.0692232	.0691558	.0690885	.0690211	.0689538	.0688865
0.849	.0688192	.0687519	.0686847	.0686174	.0685502	.0684831	.0684159	.0683487	.0682816	.0682145

TABLE 22

Modulation Transfer Function, Perfect Lens, Monochromatic

ν_r	.0000	.0001	.0002	.0003	.0004	.0005	.0006	.0007	.0008	.0009
0.850	.0681474	.0680804	.0680133	.0679463	.0678793	.0678123	.0677454	.0676784	.0676115	.0675446
0.851	.0674777	.0674109	.0673441	.0672772	.0672104	.0671437	.0670769	.0670102	.0669435	.0668768
0.852	.0668101	.0667435	.0666768	.0666102	.0665436	.0664771	.0664105	.0663440	.0662775	.0662110
0.853	.0661446	.0660781	.0660117	.0659453	.0658789	.0658126	.0657462	.0656799	.0656136	.0655473
0.854	.0654811	.0654149	.0653486	.0652825	.0652163	.0651501	.0650840	.0650179	.0649518	.0648858
0.855	.0648197	.0647537	.0646877	.0646217	.0645557	.0644898	.0644239	.0643580	.0642921	.0642263
0.856	.0641604	.0640946	.0640288	.0639630	.0638973	.0638316	.0637659	.0637002	.0636345	.0635689
0.857	.0635032	.0634376	.0633721	.0633065	.0632410	.0631754	.0631099	.0630445	.0629790	.0629136
0.858	.0628482	.0627828	.0627174	.0626521	.0625867	.0625214	.0624562	.0623909	.0623257	.0622604
0.859	.0621952	.0621301	.0620649	.0619998	.0619347	.0618696	.0618045	.0617395	.0616744	.0616094
0.860	.0615444	.0614795	.0614145	.0613496	.0612847	.0612198	.0611550	.0610902	.0610253	.0609606
0.861	.0608958	.0608310	.0607663	.0607016	.0606369	.0605723	.0605076	.0604430	.0603784	.0603139
0.862	.0602493	.0601848	.0601203	.0600558	.0599913	.0599269	.0598624	.0597980	.0597337	.0596693
0.863	.0596050	.0595407	.0594764	.0594121	.0593478	.0592836	.0592194	.0591552	.0590911	.0590269
0.864	.0589628	.0588987	.0588346	.0587706	.0587066	.0586426	.0585786	.0585146	.0584507	.0583867
0.865	.0583228	.0582590	.0581951	.0581313	.0580675	.0580037	.0579399	.0578762	.0578124	.0577487
0.866	.0576851	.0576214	.0575578	.0574942	.0574306	.0573670	.0573035	.0572399	.0571764	.0571129
0.867	.0570495	.0569861	.0569226	.0568593	.0567959	.0567325	.0566692	.0566059	.0565426	.0564794
0.868	.0564161	.0563529	.0562897	.0562266	.0561634	.0561003	.0560372	.0559741	.0559110	.0558480
0.869	.0557850	.0557220	.0556590	.0555961	.0555332	.0554703	.0554074	.0553445	.0552817	.0552189
0.870	.0551561	.0550933	.0550306	.0549679	.0549052	.0548425	.0547799	.0547172	.0546546	.0545920
0.871	.0545295	.0544669	.0544044	.0543419	.0542794	.0542170	.0541546	.0540921	.0540298	.0539674
0.872	.0539051	.0538428	.0537805	.0537182	.0536559	.0535937	.0535315	.0534693	.0534072	.0533451
0.873	.0532829	.0532209	.0531588	.0530968	.0530347	.0529727	.0529108	.0528488	.0527869	.0527250
0.874	.0526631	.0526012	.0525394	.0524776	.0524158	.0523540	.0522923	.0522306	.0521689	.0521072
0.875	.0520455	.0519839	.0519223	.0518607	.0517992	.0517376	.0516761	.0516146	.0515532	.0514917
0.876	.0514303	.0513689	.0513075	.0512462	.0511848	.0511235	.0510623	.0510010	.0509398	.0508785
0.877	.0508174	.0507562	.0506951	.0506339	.0505728	.0505118	.0504507	.0503897	.0503287	.0502677
0.878	.0502067	.0501458	.0500849	.0500240	.0499632	.0499023	.0498415	.0497807	.0497199	.0496592
0.879	.0495985	.0495378	.0494771	.0494164	.0493558	.0492952	.0492346	.0491741	.0491135	.0490530
0.880	.0489925	.0489321	.0488716	.0488112	.0487508	.0486905	.0486301	.0485698	.0485095	.0484492
0.881	.0483890	.0483287	.0482685	.0482083	.0481482	.0480881	.0480280	.0479679	.0479078	.0478478
0.882	.0477878	.0477278	.0476678	.0476079	.0475479	.0474880	.0474282	.0473683	.0473085	.0472487
0.883	.0471889	.0471292	.0470695	.0470098	.0469501	.0468904	.0468308	.0467712	.0467116	.0466520
0.884	.0465925	.0465330	.0464735	.0464141	.0463546	.0462952	.0462358	.0461764	.0461171	.0460578
0.885	.0459985	.0459392	.0458800	.0458208	.0457616	.0457024	.0456432	.0455841	.0455250	.0454660
0.886	.0454069	.0453479	.0452889	.0452299	.0451709	.0451120	.0450531	.0449942	.0449354	.0448765
0.887	.0448177	.0447590	.0447002	.0446415	.0445828	.0445241	.0444654	.0444068	.0443482	.0442896
0.888	.0442310	.0441725	.0441140	.0440555	.0439970	.0439386	.0438802	.0438218	.0437634	.0437051
0.889	.0436468	.0435885	.0435302	.0434720	.0434137	.0433556	.0432974	.0432392	.0431811	.0431230
0.890	.0430650	.0430069	.0429489	.0428909	.0428329	.0427750	.0427171	.0426592	.0426013	.0425435
0.891	.0424857	.0424279	.0423701	.0423124	.0422546	.0421970	.0421393	.0420816	.0420240	.0419664
0.892	.0419089	.0418513	.0417938	.0417363	.0416788	.0416214	.0415640	.0415066	.0414492	.0413919
0.893	.0413346	.0412773	.0412200	.0411628	.0411056	.0410484	.0409912	.0409341	.0408770	.0408199
0.894	.0407628	.0407058	.0406488	.0405918	.0405348	.0404779	.0404210	.0403641	.0403072	.0402504
0.895	.0401936	.0401368	.0400800	.0400233	.0399666	.0399099	.0398533	.0397966	.0397400	.0396835
0.896	.0396269	.0395704	.0395139	.0394574	.0394010	.0393445	.0392881	.0392318	.0391754	.0391191
0.897	.0390628	.0390065	.0389503	.0388941	.0388379	.0387817	.0387256	.0386695	.0386134	.0385573
0.898	.0385013	.0384453	.0383893	.0383333	.0382774	.0382215	.0381656	.0381098	.0380540	.0379982
0.899	.0379424	.0378866	.0378309	.0377752	.0377195	.0376639	.0376083	.0375527	.0374971	.0374416

TABLE 22

Modulation Transfer Function, Perfect Lens, Monochromatic

ν_r	.0000	.0001	.0002	.0003	.0004	.0005	.0006	.0007	.0008	.0009
0.900	.0373861	.0373306	.0372751	.0372197	.0371643	.0371089	.0370536	.0369982	.0369429	.0368877
0.901	.0368324	.0367772	.0367220	.0366668	.0366117	.0365566	.0365015	.0364464	.0363914	.0363364
0.902	.0362814	.0362264	.0361715	.0361166	.0360617	.0360069	.0359520	.0358972	.0358425	.0357877
0.903	.0357330	.0356783	.0356236	.0355690	.0355144	.0354598	.0354053	.0353507	.0352962	.0352418
0.904	.0351873	.0351329	.0350785	.0350241	.0349698	.0349155	.0348612	.0348069	.0347527	.0346985
0.905	.0346443	.0345902	.0345360	.0344819	.0344279	.0343738	.0343198	.0342658	.0342118	.0341579
0.906	.0341040	.0340501	.0339963	.0339424	.0338887	.0338349	.0337811	.0337274	.0336737	.0336201
0.907	.0335664	.0335128	.0334593	.0334057	.0333522	.0332987	.0332452	.0331918	.0331384	.0330850
0.908	.0330316	.0329783	.0329250	.0328717	.0328184	.0327652	.0327120	.0326589	.0326057	.0325526
0.909	.0324995	.0324465	.0323935	.0323405	.0322875	.0322346	.0321816	.0321288	.0320759	.0320231
0.910	.0319703	.0319175	.0318647	.0318120	.0317593	.0317067	.0316540	.0316014	.0315488	.0314963
0.911	.0314438	.0313913	.0313388	.0312864	.0312339	.0311816	.0311292	.0310769	.0310246	.0309723
0.912	.0309201	.0308679	.0308157	.0307635	.0307114	.0306593	.0306072	.0305552	.0305032	.0304512
0.913	.0303992	.0303473	.0302954	.0302435	.0301917	.0301399	.0300881	.0300363	.0299846	.0299329
0.914	.0298812	.0298296	.0297780	.0297264	.0296748	.0296233	.0295718	.0295203	.0294689	.0294175
0.915	.0293661	.0293147	.0292634	.0292121	.0291608	.0291096	.0290584	.0290072	.0289561	.0289049
0.916	.0288538	.0288028	.0287517	.0287007	.0286498	.0285988	.0285479	.0284970	.0284461	.0283953
0.917	.0283445	.0282937	.0282430	.0281923	.0281416	.0280909	.0280403	.0279897	.0279391	.0278886
0.918	.0278381	.0277876	.0277372	.0276867	.0276363	.0275860	.0275357	.0274854	.0274351	.0273848
0.919	.0273346	.0272844	.0272343	.0271842	.0271341	.0270840	.0270340	.0269840	.0269340	.0268840
0.920	.0268341	.0267842	.0267344	.0266846	.0266348	.0265850	.0265353	.0264856	.0264359	.0263862
0.921	.0263366	.0262870	.0262375	.0261880	.0261385	.0260890	.0260396	.0259902	.0259408	.0258914
0.922	.0258421	.0257928	.0257436	.0256944	.0256452	.0255960	.0255469	.0254978	.0254487	.0253997
0.923	.0253507	.0253017	.0252527	.0252038	.0251549	.0251061	.0250572	.0250085	.0249597	.0249110
0.924	.0248622	.0248136	.0247649	.0247163	.0246677	.0246192	.0245707	.0245222	.0244737	.0244253
0.925	.0243769	.0243285	.0242802	.0242319	.0241836	.0241354	.0240872	.0240390	.0239909	.0239428
0.926	.0238947	.0238466	.0237986	.0237506	.0237027	.0236547	.0236068	.0235590	.0235111	.0234633
0.927	.0234156	.0233678	.0233201	.0232724	.0232248	.0231772	.0231296	.0230821	.0230345	.0229870
0.928	.0229396	.0228922	.0228448	.0227974	.0227501	.0227028	.0226555	.0226083	.0225611	.0225139
0.929	.0224668	.0224197	.0223726	.0223256	.0222786	.0222316	.0221847	.0221378	.0220909	.0220440
0.930	.0219972	.0219504	.0219037	.0218570	.0218103	.0217636	.0217170	.0216704	.0216238	.0215773
0.931	.0215308	.0214844	.0214379	.0213916	.0213452	.0212989	.0212526	.0212063	.0211601	.0211139
0.932	.0210677	.0210216	.0209755	.0209294	.0208834	.0208374	.0207914	.0207455	.0206996	.0206537
0.933	.0206078	.0205620	.0205163	.0204705	.0204248	.0203792	.0203335	.0202879	.0202423	.0201968
0.934	.0201513	.0201058	.0200604	.0200150	.0199696	.0199243	.0198790	.0198337	.0197884	.0197432
0.935	.0196981	.0196529	.0196078	.0195628	.0195177	.0194727	.0194277	.0193828	.0193379	.0192930
0.936	.0192482	.0192034	.0191586	.0191139	.0190692	.0190245	.0189799	.0189353	.0188907	.0188462
0.937	.0188017	.0187573	.0187128	.0186684	.0186241	.0185798	.0185355	.0184912	.0184470	.0184028
0.938	.0183586	.0183145	.0182704	.0182264	.0181824	.0181384	.0180945	.0180505	.0180067	.0179628
0.939	.0179190	.0178753	.0178315	.0177878	.0177441	.0177005	.0176569	.0176134	.0175698	.0175263
0.940	.0174829	.0174395	.0173961	.0173527	.0173094	.0172661	.0172229	.0171797	.0171365	.0170934
0.941	.0170502	.0170072	.0169641	.0169211	.0168782	.0168352	.0167924	.0167495	.0167067	.0166639
0.942	.0166211	.0165784	.0165358	.0164931	.0164505	.0164079	.0163654	.0163229	.0162804	.0162380
0.943	.0161956	.0161533	.0161110	.0160687	.0160264	.0159842	.0159420	.0158999	.0158578	.0158157
0.944	.0157737	.0157317	.0156898	.0156478	.0156060	.0155641	.0155223	.0154805	.0154388	.0153971
0.945	.0153554	.0153138	.0152722	.0152307	.0151892	.0151477	.0151062	.0150648	.0150235	.0149821
0.946	.0149408	.0148996	.0148584	.0148172	.0147760	.0147349	.0146939	.0146528	.0146118	.0145709
0.947	.0145300	.0144891	.0144482	.0144074	.0143667	.0143259	.0142852	.0142446	.0142040	.0141634
0.948	.0141228	.0140823	.0140419	.0140014	.0139610	.0139207	.0138804	.0138401	.0137999	.0137597
0.949	.0137195	.0136794	.0136393	.0135993	.0135592	.0135193	.0134793	.0134395	.0133996	.0133598

TABLE 22

Modulation Transfer Function, Perfect Lens, Monochromatic

ν_r	.0000	.0001	.0002	.0003	.0004	.0005	.0006	.0007	.0008	.0009
0.950	.0133200	.0132803	.0132406	.0132009	.0131613	.0131217	.0130822	.0130427	.0130032	.0129638
0.951	.0129244	.0128850	.0128457	.0128065	.0127672	.0127280	.0126889	.0126498	.0126107	.0125717
0.952	.0125327	.0124937	.0124548	.0124159	.0123771	.0123383	.0122996	.0122608	.0122222	.0121835
0.953	.0121449	.0121064	.0120679	.0120294	.0119909	.0119526	.0119142	.0118759	.0118376	.0117994
0.954	.0117612	.0117230	.0116849	.0116469	.0116088	.0115708	.0115329	.0114950	.0114571	.0114193
0.955	.0113815	.0113438	.0113060	.0112684	.0112308	.0111932	.0111556	.0111181	.0110807	.0110433
0.956	.0110059	.0109686	.0109313	.0108940	.0108568	.0108197	.0107825	.0107455	.0107084	.0106714
0.957	.0106345	.0105975	.0105607	.0105238	.0104871	.0104503	.0104136	.0103769	.0103403	.0103037
0.958	.0102672	.0102307	.0101943	.0101579	.0101215	.0100852	.0100489	.0100127	.0099765	.0099403
0.959	.0099042	.0098682	.0098321	.0097962	.0097602	.0097243	.0096885	.0096527	.0096169	.0095812
0.960	.0095455	.0095099	.0094743	.0094388	.0094033	.0093678	.0093324	.0092971	.0092617	.0092265
0.961	.0091912	.0091560	.0091209	.0090858	.0090507	.0090157	.0089808	.0089458	.0089110	.0088761
0.962	.0088413	.0088066	.0087719	.0087372	.0087026	.0086681	.0086336	.0085991	.0085646	.0085303
0.963	.0084959	.0084616	.0084274	.0083932	.0083590	.0083249	.0082909	.0082569	.0082229	.0081890
0.964	.0081551	.0081213	.0080875	.0080537	.0080200	.0079864	.0079528	.0079192	.0078857	.0078523
0.965	.0078188	.0077855	.0077522	.0077189	.0076857	.0076525	.0076193	.0075863	.0075532	.0075202
0.966	.0074873	.0074544	.0074215	.0073887	.0073560	.0073233	.0072906	.0072580	.0072255	.0071930
0.967	.0071605	.0071281	.0070957	.0070634	.0070311	.0069989	.0069667	.0069346	.0069025	.0068705
0.968	.0068385	.0068066	.0067747	.0067429	.0067111	.0066794	.0066477	.0066161	.0065845	.0065530
0.969	.0065215	.0064901	.0064587	.0064273	.0063961	.0063648	.0063336	.0063025	.0062714	.0062404
0.970	.0062094	.0061785	.0061476	.0061168	.0060860	.0060553	.0060246	.0059940	.0059634	.0059329
0.971	.0059025	.0058720	.0058417	.0058114	.0057811	.0057509	.0057207	.0056906	.0056606	.0056306
0.972	.0056006	.0055708	.0055409	.0055111	.0054814	.0054517	.0054221	.0053925	.0053630	.0053335
0.973	.0053041	.0052748	.0052454	.0052162	.0051870	.0051579	.0051288	.0050997	.0050707	.0050418
0.974	.0050129	.0049841	.0049554	.0049267	.0048980	.0048694	.0048409	.0048124	.0047839	.0047556
0.975	.0047272	.0046990	.0046708	.0046426	.0046145	.0045865	.0045585	.0045306	.0045027	.0044749
0.976	.0044471	.0044194	.0043918	.0043642	.0043367	.0043092	.0042818	.0042545	.0042272	.0041999
0.977	.0041727	.0041456	.0041186	.0040916	.0040646	.0040377	.0040109	.0039841	.0039574	.0039308
0.978	.0039042	.0038776	.0038512	.0038248	.0037984	.0037721	.0037459	.0037197	.0036936	.0036676
0.979	.0036416	.0036157	.0035898	.0035640	.0035382	.0035126	.0034870	.0034614	.0034359	.0034105
0.980	.0033851	.0033598	.0033345	.0033094	.0032843	.0032592	.0032342	.0032093	.0031844	.0031596
0.981	.0031349	.0031102	.0030856	.0030611	.0030366	.0030122	.0029878	.0029636	.0029394	.0029152
0.982	.0028911	.0028671	.0028432	.0028193	.0027955	.0027717	.0027480	.0027244	.0027009	.0026774
0.983	.0026540	.0026306	.0026074	.0025842	.0025610	.0025379	.0025149	.0024920	.0024692	.0024464
0.984	.0024236	.0024010	.0023784	.0023559	.0023335	.0023111	.0022888	.0022666	.0022444	.0022224
0.985	.0022003	.0021784	.0021566	.0021348	.0021130	.0020914	.0020698	.0020483	.0020269	.0020056
0.986	.0019843	.0019631	.0019420	.0019210	.0019000	.0018791	.0018583	.0018376	.0018169	.0017963
0.987	.0017758	.0017554	.0017351	.0017148	.0016946	.0016745	.0016545	.0016345	.0016146	.0015948
0.988	.0015751	.0015555	.0015360	.0015165	.0014971	.0014778	.0014586	.0014395	.0014205	.0014015
0.989	.0013826	.0013638	.0013451	.0013265	.0013080	.0012895	.0012712	.0012529	.0012347	.0012166
0.990	.0011986	.0011807	.0011629	.0011451	.0011275	.0011099	.0010925	.0010751	.0010578	.0010406
0.991	.0010235	.0010066	.0009896	.0009728	.0009561	.0009395	.0009230	.0009066	.0008903	.0008740
0.992	.0008579	.0008419	.0008260	.0008101	.0007944	.0007788	.0007633	.0007479	.0007326	.0007174
0.993	.0007023	.0006873	.0006724	.0006577	.0006430	.0006285	.0006140	.0005997	.0005855	.0005714
0.994	.0005574	.0005435	.0005298	.0005161	.0005026	.0004892	.0004759	.0004628	.0004498	.0004369
0.995	.0004241	.0004114	.0003989	.0003865	.0003743	.0003621	.0003501	.0003382	.0003265	.0003149
0.996	.0003035	.0002922	.0002810	.0002700	.0002591	.0002484	.0002378	.0002274	.0002172	.0002071
0.997	.0001971	.0001874	.0001778	.0001683	.0001591	.0001500	.0001411	.0001324	.0001238	.0001155
0.998	.0001073	.0000994	.0000916	.0000841	.0000768	.0000697	.0000629	.0000562	.0000499	.0000438
0.999	.0000379	.0000324	.0000271	.0000222	.0000176	.0000134	.0000096	.0000062	.0000034	.0000011

G. Additional MTF's, Perfect Lens, Circular Aperture

Table 23—Defocused Lens, Monochromatic Radiation

Table 24—Integrals of MTF and $(MTF)^2$ for Table 23

Table 25—Focused Lens, Polychromatic Radiation

TABLE 23

MTF, Defocused Lens, Monochromatic – Δ = 0.1 Rayleigh Units [Δ ≈ d/2λF²]

ν_r	.000	.001	.002	.003	.004	.005	.006	.007	.008	.009	Δ_2
.00	1.000000	.998727	.997453	.996180	.994906	.993633	.992359	.991085	.989811	.988537	0
.01	.987263	.985989	.984715	.983440	.982166	.980892	.979617	.978343	.977068	.975793	0
.02	.974518	.973244	.971969	.970694	.969419	.968144	.966869	.965594	.964319	.963044	0
.03	.961768	.960493	.959218	.957943	.956667	.955392	.954117	.952841	.951566	.950291	0
.04	.949015	.947740	.946464	.945189	.943914	.942638	.941363	.940087	.938812	.937536	0
.05	.936261	.934986	.933710	.932435	.931159	.929884	.928609	.927333	.926058	.924783	0
.06	.923508	.922232	.920957	.919682	.918407	.917132	.915857	.914582	.913307	.912032	0
.07	.910757	.909482	.908208	.906933	.905658	.904384	.903109	.901835	.900560	.899286	0
.08	.898012	.896738	.895463	.894189	.892915	.891642	.890368	.889094	.887820	.886547	0
.09	.885273	.884000	.882727	.881453	.880180	.878907	.877634	.876362	.875089	.873816	0
.10	.872544	.871271	.869999	.868727	.867455	.866183	.864911	.863639	.862368	.861096	0
.11	.859825	.858554	.857283	.856012	.854741	.853470	.852200	.850929	.849659	.848389	0
.12	.847119	.845849	.844579	.843310	.842040	.840771	.839502	.838233	.836964	.835696	0
.13	.834427	.833159	.831891	.830623	.829355	.828087	.826820	.825553	.824286	.823019	0
.14	.821752	.820485	.819219	.817953	.816687	.815421	.814155	.812890	.811624	.810359	0
.15	.809094	.807830	.806565	.805301	.804037	.802773	.801509	.800246	.798983	.797720	0
.16	.796457	.795194	.793932	.792670	.791408	.790146	.788884	.787623	.786362	.785101	0
.17	.783841	.782580	.781320	.780060	.778801	.777541	.776282	.775023	.773764	.772506	0
.18	.771248	.769990	.768732	.767475	.766217	.764960	.763704	.762447	.761191	.759935	0
.19	.758680	.757424	.756169	.754914	.753660	.752405	.751151	.749898	.748644	.747391	0
.20	.746138	.744885	.743633	.742381	.741129	.739877	.738626	.737375	.736125	.734874	0
.21	.733624	.732375	.731125	.729876	.728627	.727379	.726130	.724882	.723635	.722387	0
.22	.721140	.719894	.718647	.717401	.716155	.714910	.713665	.712420	.711176	.709932	0
.23	.708688	.707444	.706201	.704958	.703716	.702474	.701232	.699990	.698749	.697508	0
.24	.696268	.695028	.693788	.692549	.691310	.690071	.688832	.687594	.686357	.685119	0
.25	.683883	.682646	.681410	.680174	.678938	.677703	.676468	.675234	.674000	.672766	0
.26	.671533	.670300	.669068	.667835	.666604	.665372	.664141	.662911	.661680	.660451	0
.27	.659221	.657992	.656763	.655535	.654307	.653080	.651852	.650626	.649399	.648174	0
.28	.646948	.645723	.644498	.643274	.642050	.640827	.639604	.638381	.637159	.635937	0
.29	.634715	.633495	.632274	.631054	.629834	.628615	.627396	.626178	.624960	.623742	0
.30	.622525	.621308	.620092	.618876	.617661	.616446	.615231	.614017	.612804	.611591	0
.31	.610378	.609166	.607954	.606742	.605532	.604321	.603111	.601902	.600693	.599484	0
.32	.598276	.597068	.595861	.594654	.593448	.592242	.591037	.589832	.588628	.587424	0
.33	.586220	.585017	.583815	.582613	.581411	.580210	.579010	.577810	.576610	.575411	0
.34	.574213	.573015	.571817	.570620	.569423	.568227	.567032	.565837	.564642	.563448	0
.35	.562254	.561061	.559869	.558677	.557485	.556294	.555104	.553914	.552725	.551536	1
.36	.550347	.549159	.547972	.546785	.545599	.544413	.543228	.542043	.540859	.539676	1
.37	.538493	.537310	.536128	.534947	.533766	.532585	.531405	.530226	.529048	.527869	1
.38	.526692	.525515	.524338	.523162	.521987	.520812	.519638	.518464	.517291	.516118	1
.39	.514946	.513775	.512604	.511434	.510264	.509095	.507927	.506759	.505591	.504424	1
.40	.503258	.502093	.500927	.499763	.498599	.497436	.496273	.495111	.493950	.492789	1
.41	.491628	.490469	.489310	.488151	.486993	.485836	.484679	.483523	.482368	.481213	1
.42	.480059	.478905	.477752	.476600	.475448	.474297	.473147	.471997	.470847	.469699	1
.43	.468551	.467403	.466257	.465111	.463965	.462820	.461676	.460533	.459390	.458248	1
.44	.457106	.455965	.454825	.453685	.452546	.451408	.450270	.449133	.447997	.446861	1
.45	.445726	.444592	.443458	.442325	.441193	.440061	.438930	.437800	.436670	.435541	1
.46	.434413	.433285	.432158	.431032	.429907	.428782	.427658	.426534	.425411	.424289	1
.47	.423168	.422047	.420927	.419808	.418689	.417571	.416454	.415338	.414222	.413107	1
.48	.411992	.410879	.409766	.408654	.407542	.406431	.405321	.404212	.403103	.401996	1
.49	.400889	.399782	.398676	.397572	.396467	.395364	.394261	.393159	.392058	.390958	1

TABLE 23

MTF, Defocused Lens, Monochromatic – Δ = 0.1 Rayleigh Units [$\Delta \approx d/2\lambda F^2$]

ν_r	.000	.001	.002	.003	.004	.005	.006	.007	.008	.009	Δ_2
.50	.389858	.388759	.387661	.386563	.385467	.384371	.383276	.382181	.381087	.379995	1
.51	.378902	.377811	.376721	.375631	.374542	.373453	.372366	.371279	.370193	.369108	1
.52	.368024	.366940	.365858	.364776	.363694	.362614	.361534	.360456	.359378	.358300	1
.53	.357224	.356148	.355074	.354000	.352927	.351854	.350783	.349712	.348642	.347573	1
.54	.346505	.345437	.344371	.343305	.342240	.341176	.340113	.339050	.337989	.336928	1
.55	.335868	.334809	.333751	.332694	.331637	.330581	.329527	.328473	.327420	.326367	1
.56	.325316	.324266	.323216	.322167	.321119	.320072	.319026	.317981	.316937	.315893	1
.57	.314851	.313809	.312768	.311728	.310689	.309651	.308614	.307577	.306542	.305508	1
.58	.304474	.303441	.302409	.301379	.300349	.299320	.298291	.297264	.296238	.295213	1
.59	.294188	.293165	.292142	.291120	.290100	.289080	.288061	.287043	.286026	.285010	1
.60	.283995	.282981	.281968	.280956	.279945	.278935	.277925	.276917	.275910	.274904	1
.61	.273898	.272894	.271890	.270888	.269886	.268886	.267887	.266888	.265891	.264894	1
.62	.263899	.262904	.261911	.260918	.259927	.258936	.257947	.256959	.255971	.254985	1
.63	.253999	.253015	.252032	.251050	.250068	.249088	.248109	.247131	.246154	.245178	1
.64	.244203	.243229	.242256	.241285	.240314	.239344	.238376	.237408	.236442	.235476	1
.65	.234512	.233549	.232587	.231626	.230666	.229707	.228749	.227793	.226837	.225883	1
.66	.224929	.223977	.223026	.222076	.221127	.220179	.219233	.218287	.217343	.216400	1
.67	.215458	.214517	.213577	.212638	.211701	.210764	.209829	.208895	.207962	.207030	1
.68	.206100	.205170	.204242	.203315	.202389	.201464	.200541	.199619	.198697	.197778	1
.69	.196859	.195941	.195025	.194110	.193196	.192283	.191372	.190462	.189553	.188645	1
.70	.187738	.186833	.185929	.185026	.184124	.183224	.182325	.181427	.180530	.179635	1
.71	.178741	.177848	.176957	.176067	.175178	.174290	.173404	.172519	.171635	.170752	1
.72	.169871	.168991	.168113	.167235	.166359	.165485	.164612	.163740	.162869	.162000	1
.73	.161132	.160265	.159400	.158536	.157673	.156812	.155952	.155094	.154237	.153381	1
.74	.152527	.151674	.150823	.149972	.149124	.148276	.147430	.146586	.145743	.144901	1
.75	.144061	.143222	.142385	.141549	.140714	.139881	.139049	.138219	.137390	.136563	1
.76	.135737	.134913	.134090	.133269	.132449	.131631	.130814	.129998	.129185	.128372	2
.77	.127561	.126752	.125944	.125138	.124333	.123530	.122728	.121928	.121130	.120333	2
.78	.119537	.118744	.117951	.117161	.116372	.115584	.114798	.114014	.113231	.112450	2
.79	.111671	.110893	.110116	.109342	.108569	.107798	.107028	.106260	.105494	.104729	2
.80	.103966	.103205	.102445	.101687	.100931	.100177	.099424	.098673	.097923	.097176	2
.81	.096430	.095686	.094943	.094203	.093464	.092727	.091992	.091258	.090526	.089796	2
.82	.089068	.088342	.087617	.086895	.086174	.085455	.084737	.084022	.083309	.082597	2
.83	.081887	.081179	.080473	.079769	.079067	.078367	.077668	.076972	.076277	.075585	2
.84	.074894	.074206	.073519	.072834	.072151	.071471	.070792	.070115	.069440	.068768	2
.85	.068097	.067428	.066762	.066097	.065435	.064774	.064116	.063460	.062806	.062154	2
.86	.061504	.060856	.060210	.059567	.058926	.058287	.057650	.057015	.056382	.055752	2
.87	.055124	.054498	.053875	.053253	.052634	.052017	.051403	.050790	.050181	.049573	2
.88	.048968	.048365	.047764	.047166	.046570	.045977	.045386	.044797	.044211	.043627	2
.89	.043046	.042467	.041891	.041317	.040746	.040177	.039611	.039048	.038487	.037928	3
.90	.037372	.036819	.036268	.035721	.035175	.034633	.034093	.033556	.033021	.032489	3
.91	.031961	.031434	.030911	.030391	.029873	.029358	.028846	.028337	.027831	.027328	3
.92	.026828	.026330	.025836	.025345	.024857	.024372	.023890	.023411	.022935	.022462	3
.93	.021993	.021527	.021064	.020604	.020148	.019695	.019245	.018799	.018356	.017916	3
.94	.017480	.017048	.016619	.016194	.015772	.015354	.014939	.014528	.014121	.013718	4
.95	.013319	.012923	.012531	.012144	.011760	.011381	.011005	.010634	.010266	.009904	4
.96	.009545	.009191	.008841	.008495	.008155	.007818	.007487	.007160	.006838	.006521	5
.97	.006209	.005902	.005600	.005304	.005013	.004727	.004447	.004173	.003904	.003642	6
.98	.003385	.003135	.002891	.002654	.002424	.002200	.001984	.001776	.001575	.001383	8
.99	.001199	.001024	.000858	.000702	.000557	.000435	.000107	.000070	.000038	.000013	7

TABLE 23

MTF, Defocused Lens, Monochromatic – Δ= 0.2 Rayleigh Units [Δ ≈ d/2λF²]

ν_r	.000	.001	.002	.003	.004	.005	.006	.007	.008	.009	Δ_2
.00	1.000000	.998727	.997453	.996179	.994904	.993629	.992354	.991078	.989802	.988525	0
.01	.987249	.985972	.984694	.983416	.982138	.980860	.979581	.978302	.977023	.975743	0
.02	.974463	.973183	.971902	.970622	.969340	.968059	.966777	.965496	.964213	.962931	0
.03	.961648	.960365	.959082	.957799	.956515	.955231	.953947	.952663	.951379	.950094	0
.04	.948809	.947524	.946239	.944953	.943668	.942382	.941096	.939810	.938523	.937237	0
.05	.935950	.934663	.933376	.932089	.930802	.929515	.928227	.926939	.925652	.924364	0
.06	.923076	.921788	.920499	.919211	.917923	.916634	.915346	.914057	.912768	.911479	0
.07	.910191	.908902	.907613	.906323	.905034	.903745	.902456	.901167	.899877	.898588	0
.08	.897299	.896009	.894720	.893430	.892141	.890851	.889562	.888272	.886983	.885693	0
.09	.884404	.883114	.881825	.880535	.879246	.877956	.876667	.875378	.874088	.872799	0
.10	.871510	.870221	.868932	.867643	.866354	.865065	.863776	.862487	.861198	.859910	0
.11	.858621	.857333	.856044	.854756	.853468	.852179	.850891	.849603	.848316	.847028	0
.12	.845740	.844453	.843166	.841878	.840591	.839304	.838017	.836731	.835444	.834158	0
.13	.832871	.831585	.830299	.829014	.827728	.826442	.825157	.823872	.822587	.821302	0
.14	.820018	.818733	.817449	.816165	.814881	.813597	.812314	.811030	.809747	.808464	0
.15	.807182	.805899	.804617	.803335	.802053	.800772	.799490	.798209	.796928	.795648	0
.16	.794367	.793087	.791807	.790527	.789248	.787969	.786690	.785411	.784133	.782854	0
.17	.781576	.780299	.779022	.777744	.776468	.775191	.773915	.772639	.771363	.770088	0
.18	.768813	.767538	.766263	.764989	.763715	.762442	.761168	.759895	.758623	.757350	0
.19	.756078	.754807	.753535	.752264	.750993	.749723	.748453	.747183	.745914	.744645	0
.20	.743376	.742108	.740839	.739572	.738304	.737037	.735771	.734505	.733239	.731973	0
.21	.730708	.729443	.728179	.726915	.725651	.724388	.723125	.721862	.720600	.719338	0
.22	.718077	.716816	.715555	.714295	.713035	.711776	.710516	.709258	.708000	.706742	0
.23	.705484	.704227	.702971	.701715	.700459	.699204	.697949	.696694	.695440	.694186	0
.24	.692933	.691680	.690428	.689176	.687925	.686674	.685423	.684173	.682923	.681674	0
.25	.680425	.679177	.677929	.676681	.675434	.674188	.672942	.671696	.670451	.669206	0
.26	.667962	.666718	.665475	.664232	.662990	.661748	.660507	.659266	.658026	.656786	0
.27	.655546	.654307	.653069	.651831	.650593	.649357	.648120	.646884	.645649	.644414	0
.28	.643179	.641945	.640712	.639479	.638246	.637014	.635783	.634552	.633322	.632092	1
.29	.630863	.629634	.628405	.627178	.625950	.624724	.623498	.622272	.621047	.619822	1
.30	.618598	.617375	.616152	.614929	.613707	.612486	.611265	.610045	.608825	.607606	1
.31	.606388	.605169	.603952	.602735	.601519	.600303	.599088	.597873	.596659	.595445	1
.32	.594232	.593020	.591808	.590597	.589386	.588176	.586966	.585757	.584549	.583341	1
.33	.582134	.580927	.579721	.578516	.577311	.576106	.574903	.573700	.572497	.571295	1
.34	.570094	.568893	.567693	.566493	.565294	.564096	.562898	.561701	.560505	.559309	1
.35	.558113	.556919	.555725	.554531	.553338	.552146	.550954	.549763	.548573	.547383	1
.36	.546194	.545005	.543818	.542630	.541444	.540258	.539072	.537888	.536703	.535520	1
.37	.534337	.533155	.531973	.530792	.529612	.528432	.527253	.526075	.524897	.523720	1
.38	.522544	.521368	.520193	.519018	.517845	.516671	.515499	.514327	.513156	.511985	1
.39	.510815	.509646	.508478	.507310	.506143	.504976	.503810	.502645	.501480	.500317	1
.40	.499153	.497991	.496829	.495668	.494507	.493347	.492188	.491030	.489872	.488715	1
.41	.487559	.486403	.485248	.484093	.482940	.481787	.480635	.479483	.478332	.477182	1
.42	.476032	.474884	.473736	.472588	.471442	.470296	.469150	.468006	.466862	.465719	1
.43	.464576	.463434	.462293	.461153	.460014	.458875	.457736	.456599	.455462	.454326	1
.44	.453191	.452056	.450923	.449789	.448657	.447525	.446394	.445264	.444135	.443006	1
.45	.441878	.440751	.439624	.438498	.437373	.436249	.435125	.434003	.432880	.431759	1
.46	.430638	.429519	.428399	.427281	.426163	.425047	.423930	.422815	.421701	.420587	1
.47	.419474	.418361	.417250	.416139	.415029	.413919	.412811	.411703	.410596	.409490	1
.48	.408385	.407280	.406176	.405073	.403970	.402869	.401768	.400668	.399569	.398470	1
.49	.397373	.396276	.395180	.394084	.392990	.391896	.390803	.389711	.388619	.387529	1

TABLE 23

MTF, Defocused Lens, Monochromatic – $\Delta = 0.2$ Rayleigh Units $[\Delta \approx d/2\lambda F^2]$

ν_r	.000	.001	.002	.003	.004	.005	.006	.007	.008	.009	Δ_2
.50	.386439	.385350	.384262	.383174	.382088	.381002	.379917	.378833	.377750	.376667	1
.51	.375585	.374504	.373424	.372345	.371266	.370189	.369112	.368036	.366960	.365886	1
.52	.364812	.363740	.362668	.361596	.360526	.359457	.358388	.357320	.356253	.355187	1
.53	.354122	.353057	.351994	.350931	.349869	.348808	.347748	.346688	.345630	.344572	1
.54	.343515	.342459	.341404	.340350	.339296	.338244	.337192	.336141	.335091	.334042	1
.55	.332994	.331946	.330900	.329854	.328809	.327765	.326722	.325680	.324639	.323598	1
.56	.322559	.321520	.320482	.319446	.318410	.317374	.316340	.315307	.314275	.313243	1
.57	.312212	.311183	.310154	.309126	.308099	.307073	.306047	.305023	.304000	.302977	1
.58	.301956	.300935	.299915	.298897	.297879	.296862	.295846	.294831	.293816	.292803	1
.59	.291791	.290779	.289769	.288759	.287751	.286743	.285736	.284731	.283726	.282722	1
.60	.281719	.280717	.279716	.278716	.277717	.276719	.275722	.274726	.273730	.272736	1
.61	.271743	.270750	.269759	.268769	.267779	.266791	.265803	.264817	.263832	.262847	1
.62	.261864	.260881	.259900	.258919	.257940	.256961	.255984	.255007	.254032	.253057	1
.63	.252084	.251111	.250140	.249169	.248200	.247231	.246264	.245298	.244332	.243368	1
.64	.242405	.241443	.240481	.239521	.238562	.237604	.236647	.235691	.234736	.233783	1
.65	.232830	.231878	.230927	.229978	.229029	.228082	.227135	.226190	.225246	.224303	1
.66	.223361	.222420	.221480	.220541	.219603	.218666	.217731	.216796	.215863	.214931	1
.67	.214000	.213070	.212141	.211213	.210286	.209361	.208436	.207513	.206591	.205670	1
.68	.204750	.203831	.202914	.201997	.201082	.200168	.199255	.198343	.197432	.196522	1
.69	.195614	.194707	.193801	.192896	.191992	.191090	.190188	.189288	.188389	.187491	1
.70	.186595	.185699	.184805	.183912	.183020	.182130	.181241	.180352	.179466	.178580	1
.71	.177695	.176812	.175930	.175049	.174170	.173292	.172415	.171539	.170664	.169791	1
.72	.168919	.168048	.167179	.166311	.165444	.164578	.163714	.162851	.161989	.161128	1
.73	.160269	.159411	.158555	.157699	.156845	.155993	.155141	.154291	.153443	.152595	1
.74	.151749	.150905	.150061	.149219	.148379	.147540	.146702	.145865	.145030	.144196	1
.75	.143364	.142533	.141703	.140875	.140048	.139223	.138398	.137576	.136755	.135935	1
.76	.135116	.134299	.133484	.132670	.131857	.131046	.130236	.129428	.128621	.127815	1
.77	.127012	.126209	.125408	.124609	.123811	.123014	.122219	.121425	.120633	.119843	2
.78	.119054	.118267	.117481	.116696	.115913	.115132	.114352	.113574	.112797	.112022	2
.79	.111249	.110477	.109706	.108938	.108170	.107405	.106641	.105879	.105118	.104359	2
.80	.103601	.102845	.102091	.101338	.100588	.099838	.099091	.098345	.097600	.096858	2
.81	.096117	.095378	.094640	.093904	.093170	.092438	.091707	.090979	.090251	.089526	2
.82	.088802	.088080	.087360	.086642	.085925	.085211	.084498	.083787	.083077	.082370	2
.83	.081664	.080960	.080258	.079558	.078860	.078163	.077469	.076776	.076085	.075396	2
.84	.074709	.074024	.073341	.072660	.071980	.071303	.070628	.069954	.069283	.068613	2
.85	.067946	.067280	.066617	.065955	.065296	.064638	.063983	.063330	.062678	.062029	2
.86	.061382	.060737	.060094	.059453	.058815	.058178	.057544	.056911	.056281	.055653	2
.87	.055028	.054404	.053783	.053164	.052547	.051932	.051320	.050710	.050102	.049496	2
.88	.048893	.048292	.047694	.047097	.046503	.045912	.045323	.044736	.044151	.043570	2
.89	.042990	.042413	.041838	.041266	.040696	.040129	.039564	.039002	.038443	.037886	3
.90	.037331	.036779	.036230	.035683	.035139	.034598	.034059	.033523	.032990	.032459	3
.91	.031931	.031406	.030884	.030364	.029848	.029334	.028823	.028315	.027809	.027307	3
.92	.026808	.026311	.025818	.025327	.024840	.024356	.023874	.023396	.022921	.022449	3
.93	.021980	.021515	.021052	.020593	.020137	.019685	.019236	.018790	.018347	.017908	3
.94	.017473	.017041	.016612	.016187	.015766	.015348	.014934	.014523	.014117	.013714	4
.95	.013315	.012919	.012528	.012141	.011757	.011378	.011002	.010631	.010264	.009901	4
.96	.009543	.009189	.008839	.008494	.008153	.007817	.007486	.007159	.006837	.006520	5
.97	.006208	.005902	.005600	.005303	.005012	.004727	.004447	.004172	.003904	.003641	6
.98	.003385	.003135	.002891	.002654	.002424	.002200	.001984	.001776	.001575	.001383	8
.99	.001199	.001024	.000858	.000702	.000557	.000435	.000107	.000070	.000038	.000013	7

TABLE 23

MTF, Defocused Lens, Monochromatic – $\Delta = 0.3$ Rayleigh Units [$\Delta \approx d/2\lambda F^2$]

ν_r	.000	.001	.002	.003	.004	.005	.006	.007	.008	.009	Δ_2
.00	1.000000	.998726	.997452	.996176	.994900	.993623	.992345	.991066	.989787	.988506	0
.01	.987225	.985943	.984660	.983377	.982092	.980807	.979521	.978235	.976948	.975660	0
.02	.974371	.973082	.971792	.970501	.969210	.967918	.966625	.965332	.964038	.962743	0
.03	.961448	.960152	.958856	.957559	.956262	.954964	.953665	.952366	.951066	.949766	0
.04	.948466	.947164	.945863	.944560	.943258	.941955	.940651	.939347	.938042	.936738	0
.05	.935432	.934126	.932820	.931514	.930207	.928899	.927591	.926283	.924975	.923666	0
.06	.922357	.921047	.919737	.918427	.917116	.915805	.914494	.913183	.911871	.910559	0
.07	.909247	.907934	.906621	.905308	.903995	.902681	.901368	.900054	.898740	.897425	0
.08	.896111	.894796	.893481	.892166	.890850	.889535	.888219	.886904	.885588	.884272	0
.09	.882956	.881639	.880323	.879007	.877690	.876373	.875057	.873740	.872423	.871106	0
.10	.869789	.868472	.867155	.865837	.864520	.863203	.861886	.860569	.859251	.857934	0
.11	.856617	.855300	.853982	.852665	.851348	.850031	.848714	.847397	.846080	.844763	0
.12	.843446	.842129	.840813	.839496	.838180	.836863	.835547	.834231	.832915	.831599	0
.13	.830283	.828967	.827652	.826336	.825021	.823706	.822391	.821076	.819761	.818447	0
.14	.817132	.815818	.814504	.813191	.811877	.810564	.809251	.807938	.806625	.805313	0
.15	.804001	.802689	.801377	.800065	.798754	.797443	.796133	.794822	.793512	.792202	0
.16	.790892	.789583	.788274	.786965	.785657	.784349	.783041	.781733	.780426	.779119	0
.17	.777813	.776506	.775201	.773895	.772590	.771285	.769980	.768676	.767372	.766069	0
.18	.764766	.763463	.762161	.760859	.759557	.758256	.756955	.755655	.754355	.753055	0
.19	.751756	.750458	.749159	.747861	.746564	.745267	.743970	.742674	.741378	.740083	0
.20	.738788	.737494	.736200	.734906	.733613	.732321	.731028	.729737	.728446	.727155	0
.21	.725865	.724575	.723286	.721997	.720709	.719421	.718134	.716847	.715561	.714275	1
.22	.712990	.711705	.710421	.709138	.707855	.706572	.705290	.704009	.702728	.701447	1
.23	.700167	.698888	.697609	.696331	.695053	.693776	.692500	.691224	.689948	.688674	1
.24	.687399	.686126	.684853	.683580	.682308	.681037	.679766	.678496	.677227	.675958	1
.25	.674689	.673422	.672154	.670888	.669622	.668357	.667092	.665828	.664565	.663302	1
.26	.662040	.660778	.659517	.658257	.656997	.655738	.654480	.653222	.651965	.650708	1
.27	.649453	.648198	.646943	.645689	.644436	.643183	.641932	.640680	.639430	.638180	1
.28	.636931	.635682	.634435	.633187	.631941	.630695	.629450	.628206	.626962	.625719	1
.29	.624476	.623235	.621994	.620753	.619514	.618275	.617037	.615799	.614562	.613326	1
.30	.612091	.610856	.609622	.608389	.607157	.605925	.604694	.603463	.602234	.601005	1
.31	.599776	.598549	.597322	.596096	.594871	.593646	.592422	.591199	.589977	.588755	1
.32	.587534	.586314	.585095	.583876	.582658	.581441	.580225	.579009	.577794	.576580	1
.33	.575366	.574154	.572942	.571731	.570520	.569311	.568102	.566894	.565686	.564480	1
.34	.563274	.562069	.560865	.559661	.558458	.557256	.556055	.554855	.553655	.552456	1
.35	.551258	.550061	.548864	.547668	.546473	.545279	.544086	.542893	.541701	.540510	1
.36	.539320	.538130	.536941	.535753	.534566	.533380	.532194	.531010	.529826	.528642	1
.37	.527460	.526279	.525098	.523918	.522739	.521560	.520383	.519206	.518030	.516855	1
.38	.515680	.514507	.513334	.512162	.510991	.509820	.508651	.507482	.506314	.505147	1
.39	.503981	.502815	.501651	.500487	.499324	.498162	.497000	.495840	.494680	.493521	1
.40	.492363	.491206	.490049	.488893	.487739	.486585	.485431	.484279	.483128	.481977	1
.41	.480827	.479678	.478530	.477382	.476236	.475090	.473945	.472801	.471658	.470515	1
.42	.469374	.468233	.467093	.465954	.464815	.463678	.462541	.461406	.460271	.459137	1
.43	.458003	.456871	.455739	.454609	.453479	.452350	.451222	.450094	.448968	.447842	1
.44	.446717	.445593	.444470	.443348	.442226	.441106	.439986	.438867	.437749	.436631	1
.45	.435515	.434400	.433285	.432171	.431058	.429946	.428835	.427724	.426615	.425506	1
.46	.424398	.423291	.422185	.421079	.419975	.418871	.417769	.416667	.415566	.414465	1
.47	.413366	.412268	.411170	.410073	.408977	.407882	.406788	.405695	.404602	.403511	1
.48	.402420	.401330	.400241	.399153	.398066	.396979	.395894	.394809	.393725	.392642	1
.49	.391560	.390479	.389398	.388319	.387240	.386163	.385086	.384010	.382935	.381860	1

TABLE 23

MTF, Defocused Lens, Monochromatic – Δ = 0.3 Rayleigh Units [$\Delta \approx d/2\lambda F^2$]

ν_r	.000	.001	.002	.003	.004	.005	.006	.007	.008	.009	Δ_2
.50	.380787	.379714	.378643	.377572	.376502	.375433	.374365	.373298	.372231	.371166	1
.51	.370101	.369037	.367974	.366912	.365851	.364791	.363732	.362673	.361615	.360559	1
.52	.359503	.358448	.357394	.356341	.355288	.354237	.353186	.352137	.351088	.350040	1
.53	.348993	.347947	.346902	.345857	.344814	.343771	.342730	.341689	.340649	.339610	1
.54	.338572	.337535	.336499	.335463	.334429	.333395	.332362	.331331	.330300	.329270	1
.55	.328241	.327212	.326185	.325159	.324133	.323109	.322085	.321062	.320040	.319019	1
.56	.317999	.316980	.315962	.314945	.313928	.312913	.311898	.310885	.309872	.308860	1
.57	.307849	.306839	.305830	.304822	.303815	.302809	.301803	.300799	.299795	.298793	1
.58	.297791	.296790	.295790	.294791	.293793	.292796	.291800	.290805	.289811	.288818	1
.59	.287825	.286834	.285843	.284854	.283865	.282878	.281891	.280905	.279920	.278936	1
.60	.277954	.276972	.275991	.275010	.274031	.273053	.272076	.271100	.270124	.269150	1
.61	.268177	.267204	.266233	.265262	.264293	.263324	.262357	.261390	.260424	.259460	1
.62	.258496	.257533	.256571	.255611	.254651	.253692	.252734	.251777	.250821	.249867	1
.63	.248913	.247960	.247008	.246057	.245107	.244158	.243210	.242263	.241317	.240372	1
.64	.239428	.238485	.237543	.236602	.235663	.234724	.233786	.232849	.231913	.230978	1
.65	.230044	.229112	.228180	.227249	.226319	.225391	.224463	.223536	.222611	.221686	1
.66	.220763	.219840	.218919	.217999	.217079	.216161	.215244	.214328	.213412	.212498	1
.67	.211585	.210673	.209763	.208853	.207944	.207036	.206130	.205224	.204320	.203416	1
.68	.202514	.201613	.200713	.199814	.198916	.198019	.197123	.196229	.195335	.194443	1
.69	.193552	.192661	.191772	.190884	.189997	.189112	.188227	.187344	.186461	.185580	1
.70	.184700	.183821	.182943	.182066	.181191	.180317	.179443	.178571	.177700	.176831	1
.71	.175962	.175094	.174228	.173363	.172499	.171636	.170775	.169914	.169055	.168197	1
.72	.167340	.166485	.165630	.164777	.163925	.163074	.162225	.161376	.160529	.159683	1
.73	.158839	.157995	.157153	.156312	.155472	.154634	.153796	.152960	.152125	.151292	1
.74	.150460	.149629	.148799	.147971	.147143	.146317	.145493	.144670	.143848	.143027	1
.75	.142207	.141389	.140573	.139757	.138943	.138130	.137318	.136508	.135699	.134892	1
.76	.134086	.133281	.132477	.131675	.130875	.130075	.129277	.128480	.127685	.126891	1
.77	.126099	.125308	.124518	.123730	.122943	.122157	.121373	.120591	.119809	.119030	1
.78	.118251	.117474	.116699	.115925	.115152	.114381	.113612	.112844	.112077	.111312	1
.79	.110548	.109786	.109025	.108266	.107509	.106752	.105998	.105245	.104493	.103743	2
.80	.102995	.102248	.101503	.100759	.100017	.099276	.098537	.097800	.097064	.096329	2
.81	.095597	.094866	.094136	.093409	.092683	.091958	.091235	.090514	.089795	.089077	2
.82	.088361	.087646	.086933	.086222	.085513	.084805	.084099	.083395	.082693	.081992	2
.83	.081293	.080596	.079900	.079207	.078515	.077825	.077136	.076450	.075765	.075082	2
.84	.074401	.073722	.073045	.072369	.071696	.071024	.070354	.069686	.069020	.068356	2
.85	.067694	.067034	.066375	.065719	.065064	.064412	.063761	.063113	.062466	.061822	2
.86	.061179	.060539	.059900	.059264	.058630	.057998	.057367	.056739	.056113	.055489	2
.87	.054868	.054248	.053630	.053015	.052402	.051791	.051182	.050576	.049971	.049369	2
.88	.048769	.048172	.047576	.046983	.046392	.045804	.045218	.044634	.044052	.043473	2
.89	.042896	.042322	.041750	.041180	.040613	.040049	.039486	.038927	.038369	.037815	3
.90	.037262	.036713	.036166	.035621	.035079	.034540	.034003	.033469	.032938	.032409	3
.91	.031883	.031359	.030839	.030321	.029806	.029294	.028784	.028278	.027774	.027273	3
.92	.026775	.026280	.025788	.025298	.024812	.024329	.023849	.023372	.022898	.022427	3
.93	.021959	.021495	.021033	.020575	.020120	.019668	.019220	.018775	.018333	.017895	3
.94	.017460	.017029	.016601	.016177	.015756	.015339	.014925	.014515	.014109	.013706	4
.95	.013308	.012913	.012522	.012135	.011752	.011373	.010998	.010627	.010260	.009898	4
.96	.009540	.009186	.008837	.008492	.008151	.007815	.007484	.007158	.006836	.006519	5
.97	.006207	.005901	.005599	.005303	.005012	.004726	.004446	.004172	.003903	.003641	6
.98	.003385	.003134	.002891	.002654	.002423	.002200	.001984	.001776	.001575	.001383	8
.99	.001199	.001024	.000858	.000702	.000557	.000435	.000107	.000070	.000038	.000013	7

TABLE 23

MTF, Defocused Lens, Monochromatic – $\Delta = 0.4$ Rayleigh Units $[\Delta \approx d/2\lambda F^2]$

ν_r	.000	.001	.002	.003	.004	.005	.006	.007	.008	.009	Δ_2
.00	1.000000	.998726	.997450	.996173	.994895	.993614	.992333	.991050	.989765	.988479	0
.01	.987192	.985903	.984612	.983321	.982028	.980733	.979438	.978141	.976842	.975543	0
.02	.974242	.972940	.971637	.970332	.969026	.967720	.966412	.965102	.963792	.962481	0
.03	.961168	.959854	.958540	.957224	.955907	.954589	.953270	.951950	.950629	.949308	0
.04	.947985	.946661	.945336	.944011	.942684	.941357	.940029	.938700	.937370	.936039	0
.05	.934707	.933375	.932042	.930708	.929373	.928038	.926702	.925365	.924027	.922689	0
.06	.921350	.920011	.918670	.917329	.915988	.914646	.913303	.911960	.910616	.909271	0
.07	.907926	.906581	.905235	.903888	.902541	.901194	.899846	.898497	.897148	.895799	0
.08	.894449	.893099	.891749	.890398	.889046	.887695	.886343	.884990	.883637	.882284	0
.09	.880931	.879577	.878223	.876869	.875515	.874160	.872805	.871450	.870094	.868739	0
.10	.867383	.866027	.864671	.863314	.861958	.860601	.859245	.857888	.856531	.855174	0
.11	.853816	.852459	.851102	.849744	.848387	.847029	.845672	.844314	.842957	.841599	0
.12	.840241	.838884	.837526	.836169	.834811	.833454	.832096	.830739	.829382	.828025	0
.13	.826668	.825311	.823954	.822597	.821241	.819884	.818528	.817172	.815816	.814461	0
.14	.813105	.811750	.810394	.809039	.807685	.806330	.804976	.803622	.802268	.800915	0
.15	.799561	.798208	.796856	.795503	.794151	.792799	.791448	.790097	.788746	.787395	0
.16	.786045	.784695	.783346	.781997	.780648	.779300	.777952	.776604	.775257	.773910	0
.17	.772564	.771218	.769872	.768527	.767183	.765838	.764495	.763151	.761809	.760466	0
.18	.759124	.757783	.756442	.755102	.753762	.752423	.751084	.749745	.748408	.747070	1
.19	.745734	.744397	.743062	.741727	.740392	.739058	.737725	.736392	.735060	.733728	1
.20	.732397	.731067	.729737	.728408	.727079	.725751	.724424	.723097	.721771	.720446	1
.21	.719121	.717797	.716473	.715151	.713828	.712507	.711186	.709866	.708547	.707228	1
.22	.705910	.704593	.703276	.701960	.700645	.699331	.698017	.696704	.695391	.694080	1
.23	.692769	.691459	.690150	.688841	.687533	.686226	.684920	.683614	.682310	.681006	1
.24	.679703	.678400	.677098	.675798	.674498	.673198	.671900	.670602	.669305	.668009	1
.25	.666714	.665420	.664126	.662834	.661542	.660251	.658960	.657671	.656382	.655095	1
.26	.653808	.652522	.651237	.649952	.648669	.647386	.646104	.644824	.643544	.642265	1
.27	.640986	.639709	.638432	.637157	.635882	.634608	.633335	.632063	.630792	.629522	1
.28	.628253	.626984	.625717	.624450	.623185	.621920	.620656	.619393	.618131	.616870	1
.29	.615610	.614350	.613092	.611835	.610578	.609323	.608068	.606815	.605562	.604310	1
.30	.603059	.601809	.600560	.599313	.598065	.596819	.595574	.594330	.593087	.591845	1
.31	.590604	.589363	.588124	.586885	.585648	.584412	.583176	.581942	.580708	.579476	1
.32	.578244	.577013	.575784	.574555	.573328	.572101	.570875	.569651	.568427	.567204	1
.33	.565982	.564762	.563542	.562323	.561105	.559889	.558673	.557458	.556244	.555032	1
.34	.553820	.552609	.551399	.550190	.548983	.547776	.546570	.545365	.544161	.542959	1
.35	.541757	.540556	.539356	.538158	.536960	.535763	.534568	.533373	.532179	.530987	1
.36	.529795	.528604	.527415	.526226	.525038	.523852	.522666	.521482	.520298	.519116	1
.37	.517934	.516754	.515574	.514396	.513218	.512042	.510867	.509692	.508519	.507346	1
.38	.506175	.505005	.503835	.502667	.501500	.500334	.499169	.498004	.496841	.495679	1
.39	.494518	.493358	.492199	.491041	.489884	.488728	.487573	.486419	.485266	.484114	1
.40	.482963	.481813	.480664	.479516	.478369	.477224	.476079	.474935	.473792	.472650	1
.41	.471510	.470370	.469231	.468094	.466957	.465821	.464687	.463553	.462421	.461289	1
.42	.460158	.459029	.457900	.456773	.455646	.454521	.453397	.452273	.451151	.450029	1
.43	.448909	.447789	.446671	.445554	.444437	.443322	.442208	.441094	.439982	.438871	1
.44	.437760	.436651	.435543	.434436	.433329	.432224	.431120	.430017	.428914	.427813	1
.45	.426713	.425614	.424515	.423418	.422322	.421227	.420133	.419039	.417947	.416856	1
.46	.415766	.414676	.413588	.412501	.411415	.410330	.409245	.408162	.407080	.405999	1
.47	.404918	.403839	.402761	.401684	.400607	.399532	.398458	.397384	.396312	.395241	1
.48	.394170	.393101	.392033	.390965	.389899	.388833	.387769	.386705	.385643	.384581	1
.49	.383521	.382461	.381403	.380345	.379289	.378233	.377178	.376125	.375072	.374020	1

TABLE 23

MTF, Defocused Lens, Monochromatic – Δ = 0.4 Rayleigh Units [$\Delta \approx d/2\lambda F^2$]

ν_r	.000	.001	.002	.003	.004	.005	.006	.007	.008	.009	Δ_2
.50	.372970	.371920	.370871	.369823	.368777	.367731	.366686	.365642	.364599	.363557	1
.51	.362516	.361476	.360437	.359399	.358362	.357326	.356290	.355256	.354223	.353191	1
.52	.352159	.351129	.350099	.349071	.348044	.347017	.345991	.344967	.343943	.342921	1
.53	.341899	.340878	.339858	.338839	.337822	.336805	.335789	.334774	.333760	.332746	1
.54	.331734	.330723	.329713	.328704	.327695	.326688	.325681	.324676	.323671	.322668	1
.55	.321665	.320664	.319663	.318663	.317664	.316666	.315669	.314673	.313678	.312684	1
.56	.311691	.310699	.309708	.308717	.307728	.306740	.305752	.304766	.303780	.302796	1
.57	.301812	.300829	.299847	.298866	.297887	.296908	.295930	.294953	.293976	.293001	1
.58	.292027	.291054	.290081	.289110	.288140	.287170	.286201	.285234	.284267	.283301	1
.59	.282337	.281373	.280410	.279448	.278487	.277527	.276568	.275609	.274652	.273696	1
.60	.272740	.271786	.270833	.269880	.268928	.267978	.267028	.266079	.265132	.264185	1
.61	.263239	.262294	.261350	.260407	.259465	.258523	.257583	.256644	.255706	.254768	1
.62	.253832	.252896	.251962	.251028	.250096	.249164	.248233	.247303	.246375	.245447	1
.63	.244520	.243594	.242669	.241745	.240822	.239900	.238979	.238058	.237139	.236221	1
.64	.235304	.234387	.233472	.232558	.231644	.230732	.229820	.228910	.228000	.227091	1
.65	.226184	.225277	.224371	.223467	.222563	.221660	.220759	.219858	.218958	.218059	1
.66	.217161	.216264	.215369	.214474	.213580	.212687	.211795	.210904	.210014	.209125	1
.67	.208237	.207350	.206464	.205579	.204695	.203812	.202930	.202049	.201170	.200291	1
.68	.199413	.198536	.197660	.196785	.195911	.195038	.194167	.193296	.192426	.191557	1
.69	.190690	.189823	.188957	.188093	.187229	.186367	.185505	.184645	.183785	.182927	1
.70	.182070	.181213	.180358	.179504	.178651	.177799	.176948	.176098	.175249	.174402	1
.71	.173555	.172709	.171865	.171021	.170179	.169338	.168498	.167658	.166820	.165984	1
.72	.165148	.164313	.163479	.162647	.161816	.160985	.160156	.159328	.158501	.157675	1
.73	.156851	.156027	.155205	.154384	.153563	.152744	.151927	.151110	.150294	.149480	1
.74	.148667	.147855	.147044	.146234	.145426	.144618	.143812	.143007	.142203	.141401	1
.75	.140599	.139799	.139000	.138202	.137406	.136611	.135816	.135023	.134232	.133441	1
.76	.132652	.131864	.131077	.130292	.129508	.128725	.127943	.127163	.126383	.125605	1
.77	.124829	.124053	.123279	.122507	.121735	.120965	.120196	.119429	.118662	.117898	1
.78	.117134	.116372	.115611	.114851	.114093	.113336	.112581	.111827	.111074	.110323	1
.79	.109573	.108824	.108077	.107331	.106587	.105844	.105102	.104362	.103623	.102886	1
.80	.102150	.101416	.100683	.099951	.099221	.098493	.097765	.097040	.096316	.095593	1
.81	.094872	.094152	.093434	.092718	.092003	.091289	.090577	.089867	.089158	.088451	2
.82	.087745	.087041	.086338	.085637	.084938	.084240	.083544	.082849	.082157	.081465	2
.83	.080776	.080088	.079401	.078717	.078034	.077353	.076673	.075995	.075319	.074645	2
.84	.073972	.073301	.072632	.071965	.071299	.070635	.069973	.069313	.068654	.067998	2
.85	.067343	.066690	.066039	.065389	.064742	.064096	.063452	.062811	.062171	.061533	2
.86	.060897	.060262	.059630	.059000	.058372	.057745	.057121	.056499	.055878	.055260	2
.87	.054644	.054030	.053418	.052807	.052199	.051594	.050990	.050388	.049789	.049191	2
.88	.048596	.048003	.047412	.046823	.046237	.045653	.045071	.044491	.043914	.043338	2
.89	.042765	.042195	.041627	.041061	.040497	.039936	.039377	.038821	.038267	.037715	2
.90	.037166	.036620	.036076	.035534	.034995	.034459	.033925	.033393	.032864	.032338	3
.91	.031815	.031294	.030776	.030260	.029747	.029237	.028730	.028225	.027724	.027225	3
.92	.026729	.026236	.025745	.025258	.024773	.024292	.023813	.023338	.022865	.022396	3
.93	.021930	.021466	.021006	.020550	.020096	.019645	.019198	.018754	.018314	.017876	3
.94	.017443	.017012	.016585	.016162	.015742	.015325	.014913	.014503	.014098	.013696	4
.95	.013298	.012904	.012514	.012127	.011745	.011366	.010992	.010621	.010255	.009893	4
.96	.009535	.009182	.008833	.008488	.008148	.007812	.007481	.007155	.006834	.006517	5
.97	.006206	.005899	.005598	.005301	.005011	.004725	.004445	.004171	.003903	.003640	6
.98	.003384	.003134	.002890	.002653	.002423	.002200	.001984	.001776	.001575	.001383	8
.99	.001199	.001023	.000858	.000702	.000557	.000435	.000107	.000070	.000038	.000013	7

TABLE 23

MTF, Defocused Lens, Monochromatic – Δ = 0.5 Rayleigh Units [$\Delta \approx d/2\lambda F^2$]

ν_r	.000	.001	.002	.003	.004	.005	.006	.007	.008	.009	Δ_2
.00	1.000000	.998726	.997449	.996169	.994888	.993604	.992317	.991028	.989737	.988444	-1
.01	.987149	.985851	.984551	.983249	.981945	.980638	.979330	.978020	.976707	.975393	-1
.02	.974076	.972758	.971437	.970115	.968791	.967465	.966137	.964807	.963476	.962143	-1
.03	.960808	.959471	.958133	.956792	.955451	.954107	.952762	.951416	.950068	.948718	-1
.04	.947367	.946014	.944660	.943304	.941947	.940589	.939229	.937868	.936505	.935141	0
.05	.933776	.932410	.931042	.929673	.928303	.926931	.925559	.924185	.922810	.921434	0
.06	.920057	.918679	.917300	.915920	.914539	.913156	.911773	.910389	.909004	.907618	0
.07	.906231	.904843	.903454	.902065	.900675	.899284	.897892	.896499	.895106	.893711	0
.08	.892316	.890921	.889525	.888128	.886730	.885332	.883933	.882534	.881134	.879734	0
.09	.878332	.876931	.875529	.874126	.872723	.871320	.869916	.868512	.867107	.865702	0
.10	.864296	.862891	.861484	.860078	.858671	.857264	.855857	.854449	.853041	.851633	0
.11	.850225	.848816	.847407	.845999	.844590	.843180	.841771	.840362	.838952	.837543	0
.12	.836133	.834723	.833313	.831904	.830494	.829084	.827674	.826264	.824855	.823445	0
.13	.822036	.820626	.819217	.817807	.816398	.814989	.813580	.812171	.810763	.809355	0
.14	.807946	.806538	.805131	.803723	.802316	.800909	.799502	.798096	.796689	.795284	0
.15	.793878	.792473	.791068	.789663	.788259	.786855	.785452	.784049	.782646	.781244	0
.16	.779842	.778441	.777040	.775640	.774240	.772840	.771441	.770043	.768645	.767247	1
.17	.765851	.764454	.763058	.761663	.760268	.758874	.757481	.756088	.754695	.753304	1
.18	.751913	.750522	.749132	.747743	.746355	.744967	.743580	.742193	.740807	.739422	1
.19	.738038	.736654	.735271	.733889	.732508	.731127	.729747	.728368	.726990	.725612	1
.20	.724235	.722859	.721484	.720110	.718736	.717364	.715992	.714621	.713250	.711881	1
.21	.710513	.709145	.707778	.706413	.705048	.703684	.702320	.700958	.699597	.698236	1
.22	.696877	.695519	.694161	.692804	.691449	.690094	.688740	.687387	.686036	.684685	1
.23	.683335	.681986	.680638	.679292	.677946	.676601	.675257	.673914	.672573	.671232	1
.24	.669892	.668554	.667216	.665880	.664544	.663210	.661877	.660545	.659214	.657884	1
.25	.656555	.655227	.653900	.652574	.651250	.649926	.648604	.647283	.645963	.644644	1
.26	.643326	.642010	.640694	.639380	.638066	.636754	.635443	.634134	.632825	.631518	1
.27	.630211	.628906	.627602	.626299	.624998	.623697	.622398	.621100	.619803	.618508	1
.28	.617213	.615920	.614628	.613337	.612048	.610759	.609472	.608186	.606901	.605618	1
.29	.604336	.603054	.601775	.600496	.599219	.597943	.596668	.595394	.594122	.592850	1
.30	.591580	.590312	.589044	.587778	.586513	.585250	.583987	.582726	.581466	.580208	1
.31	.578950	.577694	.576439	.575186	.573933	.572682	.571433	.570184	.568937	.567691	1
.32	.566447	.565203	.563961	.562720	.561481	.560243	.559006	.557770	.556536	.555302	1
.33	.554071	.552840	.551611	.550383	.549156	.547931	.546707	.545484	.544263	.543042	1
.34	.541824	.540606	.539390	.538175	.536961	.535749	.534537	.533328	.532119	.530912	1
.35	.529706	.528501	.527298	.526096	.524895	.523696	.522498	.521301	.520105	.518911	1
.36	.517718	.516526	.515336	.514147	.512959	.511772	.510587	.509403	.508221	.507039	1
.37	.505859	.504681	.503503	.502327	.501152	.499979	.498807	.497636	.496466	.495297	1
.38	.494130	.492965	.491800	.490637	.489475	.488314	.487155	.485997	.484840	.483684	1
.39	.482530	.481377	.480225	.479075	.477926	.476778	.475631	.474486	.473342	.472199	1
.40	.471058	.469917	.468779	.467641	.466504	.465369	.464235	.463103	.461971	.460841	1
.41	.459712	.458585	.457458	.456333	.455210	.454087	.452966	.451846	.450727	.449609	1
.42	.448493	.447378	.446264	.445151	.444040	.442930	.441821	.440713	.439607	.438502	1
.43	.437398	.436295	.435194	.434094	.432994	.431897	.430800	.429705	.428611	.427518	1
.44	.426426	.425335	.424246	.423158	.422071	.420986	.419901	.418818	.417736	.416655	1
.45	.415575	.414497	.413420	.412344	.411269	.410195	.409123	.408051	.406981	.405912	1
.46	.404845	.403778	.402713	.401648	.400585	.399523	.398463	.397403	.396345	.395288	1
.47	.394232	.393177	.392123	.391070	.390019	.388969	.387920	.386872	.385825	.384779	1
.48	.383735	.382691	.381649	.380608	.379568	.378529	.377492	.376455	.375420	.374385	1
.49	.373352	.372320	.371289	.370259	.369231	.368203	.367176	.366151	.365127	.364104	1

TABLE 23

MTF, Defocused Lens, Monochromatic – Δ = 0.5 Rayleigh Units [$\Delta \approx d/2\lambda F^2$]

ν_r	.000	.001	.002	.003	.004	.005	.006	.007	.008	.009	Δ_2
.50	.363082	.362061	.361041	.360022	.359004	.357988	.356972	.355958	.354945	.353932	1
.51	.352921	.351911	.350902	.349895	.348888	.347882	.346877	.345874	.344871	.343870	1
.52	.342869	.341870	.340872	.339875	.338879	.337884	.336890	.335897	.334905	.333914	1
.53	.332924	.331935	.330948	.329961	.328975	.327991	.327007	.326025	.325043	.324063	1
.54	.323083	.322105	.321127	.320151	.319176	.318201	.317228	.316256	.315284	.314314	1
.55	.313345	.312377	.311409	.310443	.309478	.308514	.307551	.306588	.305627	.304667	1
.56	.303708	.302750	.301792	.300836	.299881	.298927	.297973	.297021	.296070	.295119	1
.57	.294170	.293222	.292274	.291328	.290382	.289438	.288494	.287552	.286610	.285669	1
.58	.284730	.283791	.282853	.281917	.280981	.280046	.279112	.278179	.277247	.276316	1
.59	.275386	.274457	.273529	.272602	.271675	.270750	.269826	.268902	.267980	.267058	1
.60	.266137	.265218	.264299	.263381	.262464	.261548	.260633	.259719	.258806	.257894	1
.61	.256983	.256072	.255163	.254254	.253347	.252440	.251534	.250630	.249726	.248823	1
.62	.247921	.247020	.246120	.245220	.244322	.243425	.242528	.241632	.240738	.239844	1
.63	.238951	.238059	.237168	.236278	.235389	.234501	.233613	.232727	.231841	.230957	1
.64	.230073	.229190	.228309	.227428	.226548	.225668	.224790	.223913	.223036	.222161	1
.65	.221286	.220413	.219540	.218668	.217797	.216927	.216058	.215190	.214323	.213456	1
.66	.212591	.211726	.210863	.210000	.209138	.208277	.207417	.206558	.205700	.204843	1
.67	.203986	.203131	.202276	.201423	.200570	.199719	.198868	.198018	.197169	.196321	1
.68	.195474	.194627	.193782	.192938	.192094	.191252	.190410	.189570	.188730	.187891	1
.69	.187053	.186216	.185380	.184545	.183711	.182878	.182046	.181215	.180384	.179555	1
.70	.178726	.177899	.177072	.176247	.175422	.174598	.173775	.172954	.172133	.171313	1
.71	.170494	.169676	.168859	.168043	.167228	.166414	.165601	.164789	.163977	.163167	1
.72	.162358	.161550	.160743	.159936	.159131	.158327	.157524	.156721	.155920	.155120	1
.73	.154321	.153522	.152725	.151929	.151134	.150339	.149546	.148754	.147963	.147173	1
.74	.146384	.145596	.144809	.144023	.143238	.142454	.141671	.140890	.140109	.139329	1
.75	.138551	.137773	.136997	.136221	.135447	.134674	.133902	.133131	.132361	.131592	1
.76	.130825	.130058	.129292	.128528	.127765	.127003	.126242	.125482	.124723	.123965	1
.77	.123209	.122454	.121700	.120947	.120195	.119444	.118695	.117946	.117199	.116453	1
.78	.115708	.114965	.114222	.113481	.112741	.112003	.111265	.110529	.109794	.109060	1
.79	.108327	.107596	.106866	.106137	.105410	.104683	.103958	.103234	.102512	.101791	1
.80	.101071	.100353	.099635	.098919	.098205	.097492	.096780	.096069	.095360	.094652	1
.81	.093946	.093241	.092537	.091834	.091134	.090434	.089736	.089039	.088344	.087650	1
.82	.086958	.086267	.085577	.084889	.084202	.083517	.082834	.082151	.081471	.080792	2
.83	.080114	.079438	.078763	.078090	.077419	.076749	.076080	.075414	.074748	.074085	2
.84	.073423	.072762	.072104	.071447	.070791	.070137	.069485	.068834	.068186	.067538	2
.85	.066893	.066249	.065607	.064967	.064328	.063692	.063057	.062423	.061792	.061162	2
.86	.060534	.059908	.059284	.058662	.058041	.057422	.056805	.056191	.055578	.054966	2
.87	.054357	.053750	.053145	.052541	.051940	.051341	.050743	.050148	.049554	.048963	2
.88	.048374	.047787	.047202	.046619	.046038	.045459	.044882	.044308	.043736	.043166	2
.89	.042598	.042032	.041469	.040907	.040348	.039792	.039237	.038685	.038136	.037588	2
.90	.037043	.036501	.035960	.035423	.034887	.034354	.033824	.033296	.032771	.032248	3
.91	.031728	.031210	.030695	.030182	.029672	.029165	.028661	.028159	.027660	.027163	3
.92	.026670	.026179	.025691	.025206	.024724	.024244	.023768	.023294	.022824	.022356	3
.93	.021892	.021430	.020972	.020517	.020065	.019616	.019170	.018728	.018288	.017853	3
.94	.017420	.016991	.016565	.016143	.015724	.015309	.014897	.014489	.014084	.013683	4
.95	.013286	.012893	.012503	.012117	.011736	.011358	.010984	.010614	.010248	.009887	4
.96	.009530	.009177	.008828	.008484	.008144	.007809	.007478	.007152	.006831	.006515	5
.97	.006203	.005897	.005596	.005300	.005009	.004724	.004444	.004170	.003902	.003640	6
.98	.003384	.003134	.002890	.002653	.002423	.002200	.001984	.001775	.001575	.001382	8
.99	.001198	.001023	.000858	.000702	.000557	.000435	.000107	.000070	.000038	.000013	7

TABLE 23

MTF, Defocused Lens, Monochromatic – Δ = 1.0 Rayleigh Units $[\Delta \approx d/2\lambda F^2]$

ν_r	.000	.001	.002	.003	.004	.005	.006	.007	.008	.009	Δ_2
.00	1.000000	.998722	.997434	.996136	.994829	.993513	.992187	.990851	.989507	.988153	-8
.01	.986791	.985420	.984040	.982651	.981254	.979848	.978434	.977011	.975580	.974142	-7
.02	.972695	.971240	.969778	.968308	.966830	.965345	.963852	.962352	.960845	.959331	-6
.03	.957809	.956281	.954746	.953204	.951655	.950100	.948538	.946970	.945395	.943815	-5
.04	.942228	.940635	.939036	.937431	.935820	.934204	.932582	.930954	.929321	.927683	-4
.05	.926040	.924391	.922737	.921078	.919414	.917745	.916072	.914393	.912710	.911023	-4
.06	.909331	.907635	.905934	.904229	.902520	.900807	.899090	.897369	.895644	.893916	-3
.07	.892184	.890448	.888708	.886966	.885219	.883470	.881717	.879961	.878202	.876440	-2
.08	.874675	.872907	.871136	.869363	.867587	.865808	.864027	.862243	.860457	.858668	-1
.09	.856878	.855085	.853290	.851492	.849693	.847892	.846089	.844285	.842478	.840670	-1
.10	.838860	.837049	.835236	.833422	.831606	.829789	.827971	.826152	.824331	.822510	0
.11	.820687	.818864	.817039	.815214	.813388	.811561	.809734	.807906	.806077	.804248	0
.12	.802418	.800588	.798758	.796927	.795096	.793265	.791434	.789602	.787771	.785939	0
.13	.784108	.782277	.780446	.778615	.776784	.774954	.773124	.771294	.769465	.767637	0
.14	.765809	.763981	.762154	.760328	.758502	.756678	.754854	.753030	.751208	.749387	1
.15	.747566	.745747	.743929	.742111	.740295	.738480	.736666	.734854	.733043	.731233	1
.16	.729424	.727617	.725811	.724007	.722204	.720403	.718603	.716805	.715009	.713214	2
.17	.711421	.709630	.707840	.706053	.704267	.702483	.700701	.698921	.697142	.695366	2
.18	.693592	.691820	.690050	.688282	.686516	.684753	.682992	.681233	.679476	.677721	2
.19	.675969	.674219	.672471	.670726	.668984	.667243	.665506	.663770	.662037	.660307	3
.20	.658579	.656854	.655132	.653412	.651695	.649980	.648268	.646559	.644853	.643149	3
.21	.641449	.639751	.638055	.636363	.634674	.632987	.631303	.629622	.627945	.626270	3
.22	.624598	.622929	.621263	.619600	.617940	.616283	.614630	.612979	.611331	.609687	3
.23	.608046	.606408	.604773	.603141	.601512	.599887	.598265	.596646	.595030	.593417	3
.24	.591808	.590202	.588600	.587000	.585404	.583812	.582222	.580636	.579054	.577475	3
.25	.575899	.574326	.572757	.571191	.569629	.568070	.566515	.564963	.563414	.561869	3
.26	.560328	.558790	.557255	.555724	.554196	.552672	.551151	.549634	.548121	.546611	4
.27	.545104	.543601	.542102	.540606	.539113	.537624	.536139	.534657	.533179	.531705	4
.28	.530234	.528766	.527303	.525842	.524386	.522933	.521483	.520037	.518595	.517156	4
.29	.515721	.514290	.512862	.511438	.510017	.508600	.507187	.505777	.504371	.502969	4
.30	.501570	.500174	.498783	.497395	.496010	.494629	.493252	.491878	.490508	.489142	4
.31	.487779	.486420	.485064	.483712	.482364	.481019	.479678	.478341	.477007	.475676	4
.32	.474350	.473026	.471707	.470391	.469078	.467769	.466464	.465163	.463864	.462570	4
.33	.461279	.459991	.458707	.457427	.456150	.454877	.453607	.452341	.451078	.449819	4
.34	.448563	.447311	.446063	.444818	.443576	.442338	.441103	.439872	.438644	.437420	3
.35	.436199	.434982	.433768	.432558	.431351	.430147	.428947	.427750	.426557	.425367	3
.36	.424181	.422997	.421818	.420641	.419468	.418299	.417133	.415970	.414810	.413654	3
.37	.412501	.411352	.410205	.409062	.407923	.406786	.405653	.404523	.403397	.402274	3
.38	.401154	.400037	.398923	.397813	.396706	.395602	.394501	.393404	.392309	.391218	3
.39	.390130	.389045	.387963	.386885	.385809	.384737	.383668	.382602	.381539	.380479	3
.40	.379422	.378368	.377317	.376269	.375225	.374183	.373144	.372109	.371076	.370046	3
.41	.369020	.367996	.366975	.365957	.364942	.363930	.362921	.361915	.360912	.359911	3
.42	.358914	.357919	.356927	.355939	.354952	.353969	.352989	.352011	.351036	.350064	3
.43	.349095	.348128	.347164	.346203	.345245	.344289	.343336	.342386	.341438	.340493	3
.44	.339551	.338612	.337675	.336740	.335809	.334879	.333953	.333029	.332108	.331189	3
.45	.330273	.329359	.328448	.327539	.326633	.325730	.324828	.323930	.323034	.322140	2
.46	.321249	.320360	.319473	.318589	.317708	.316828	.315951	.315077	.314205	.313335	2
.47	.312467	.311602	.310739	.309879	.309020	.308164	.307311	.306459	.305610	.304763	2
.48	.303918	.303075	.302235	.301396	.300560	.299726	.298895	.298065	.297237	.296412	2
.49	.295589	.294767	.293948	.293131	.292316	.291503	.290692	.289883	.289076	.288271	2

TABLE 23

MTF, Defocused Lens, Monochromatic – Δ = 1.0 Rayleigh Units $[\Delta \approx d/2\lambda F^2]$

ν_r	.000	.001	.002	.003	.004	.005	.006	.007	.008	.009	Δ_2
.50	.287468	.286667	.285868	.285071	.284276	.283483	.282692	.281902	.281115	.280329	2
.51	.279545	.278764	.277984	.277205	.276429	.275654	.274882	.274111	.273342	.272574	2
.52	.271809	.271045	.270283	.269522	.268764	.268007	.267251	.266498	.265746	.264995	2
.53	.264247	.263500	.262754	.262010	.261268	.260528	.259789	.259051	.258315	.257581	2
.54	.256848	.256117	.255387	.254659	.253932	.253207	.252483	.251761	.251040	.250320	1
.55	.249602	.248886	.248170	.247457	.246744	.246033	.245323	.244615	.243908	.243202	1
.56	.242498	.241795	.241093	.240393	.239694	.238996	.238299	.237603	.236909	.236216	1
.57	.235525	.234834	.234145	.233457	.232770	.232084	.231399	.230716	.230033	.229352	1
.58	.228672	.227993	.227315	.226638	.225962	.225287	.224614	.223941	.223269	.222599	1
.59	.221929	.221261	.220593	.219926	.219261	.218596	.217932	.217270	.216608	.215947	1
.60	.215287	.214628	.213970	.213312	.212656	.212000	.211346	.210692	.210039	.209387	1
.61	.208736	.208085	.207435	.206786	.206138	.205491	.204845	.204199	.203554	.202909	1
.62	.202266	.201623	.200981	.200340	.199699	.199059	.198420	.197781	.197143	.196506	1
.63	.195869	.195233	.194598	.193963	.193329	.192695	.192062	.191430	.190798	.190167	1
.64	.189537	.188907	.188277	.187648	.187020	.186392	.185765	.185138	.184512	.183886	1
.65	.183261	.182636	.182012	.181388	.180765	.180142	.179519	.178897	.178276	.177655	0
.66	.177034	.176414	.175794	.175175	.174556	.173937	.173319	.172701	.172083	.171466	0
.67	.170850	.170233	.169617	.169002	.168386	.167771	.167156	.166542	.165928	.165314	0
.68	.164701	.164088	.163475	.162863	.162250	.161638	.161027	.160415	.159804	.159193	0
.69	.158583	.157972	.157362	.156752	.156142	.155533	.154924	.154315	.153706	.153098	0
.70	.152489	.151881	.151273	.150665	.150058	.149451	.148843	.148236	.147630	.147023	0
.71	.146417	.145810	.145204	.144598	.143993	.143387	.142782	.142176	.141571	.140966	0
.72	.140361	.139757	.139152	.138548	.137943	.137339	.136735	.136131	.135527	.134924	0
.73	.134320	.133717	.133113	.132510	.131907	.131304	.130702	.130099	.129496	.128894	0
.74	.128291	.127689	.127087	.126485	.125883	.125281	.124680	.124078	.123477	.122875	0
.75	.122274	.121673	.121072	.120471	.119870	.119269	.118669	.118068	.117468	.116868	0
.76	.116267	.115667	.115068	.114468	.113868	.113269	.112669	.112070	.111471	.110872	0
.77	.110273	.109674	.109076	.108477	.107879	.107281	.106683	.106085	.105487	.104890	0
.78	.104292	.103695	.103098	.102501	.101905	.101308	.100712	.100116	.099520	.098924	0
.79	.098329	.097733	.097138	.096543	.095949	.095354	.094760	.094166	.093573	.092979	0
.80	.092386	.091793	.091200	.090608	.090016	.089424	.088833	.088242	.087651	.087060	0
.81	.086470	.085880	.085290	.084701	.084112	.083524	.082936	.082348	.081761	.081174	0
.82	.080587	.080001	.079415	.078830	.078245	.077660	.077077	.076493	.075910	.075328	0
.83	.074746	.074164	.073583	.073003	.072423	.071843	.071264	.070686	.070109	.069531	1
.84	.068955	.068379	.067804	.067229	.066656	.066082	.065510	.064938	.064367	.063796	1
.85	.063227	.062658	.062090	.061522	.060956	.060390	.059825	.059261	.058697	.058135	1
.86	.057573	.057013	.056453	.055894	.055336	.054779	.054223	.053668	.053114	.052561	1
.87	.052009	.051458	.050909	.050360	.049812	.049266	.048721	.048176	.047633	.047092	1
.88	.046551	.046012	.045474	.044937	.044402	.043867	.043335	.042803	.042273	.041744	1
.89	.041217	.040691	.040167	.039644	.039123	.038603	.038085	.037568	.037053	.036540	2
.90	.036028	.035518	.035010	.034503	.033998	.033495	.032994	.032494	.031997	.031501	2
.91	.031007	.030515	.030025	.029537	.029052	.028568	.028086	.027606	.027129	.026654	2
.92	.026180	.025710	.025241	.024775	.024311	.023849	.023390	.022933	.022479	.022027	3
.93	.021577	.021131	.020687	.020245	.019806	.019370	.018937	.018506	.018079	.017654	3
.94	.017232	.016813	.016397	.015985	.015575	.015168	.014765	.014365	.013968	.013575	3
.95	.013184	.012798	.012415	.012035	.011659	.011287	.010918	.010553	.010192	.009835	4
.96	.009482	.009133	.008788	.008447	.008111	.007779	.007451	.007128	.006809	.006495	5
.97	.006186	.005881	.005582	.005288	.004998	.004715	.004436	.004163	.003896	.003635	6
.98	.003379	.003130	.002887	.002651	.002421	.002198	.001983	.001775	.001574	.001382	8
.99	.001198	.001023	.000858	.000702	.000557	.000435	.000107	.000070	.000038	.000013	7

TABLE 23

MTF, Defocused Lens, Monochromatic – Δ = 1.5 Rayleigh Units [$\Delta \approx d/2\lambda F^2$]

ν_r	.000	.001	.002	.003	.004	.005	.006	.007	.008	.009	Δ_2
.00	1.000000	.998716	.997409	.996081	.994732	.993361	.991969	.990556	.989123	.987669	-19
.01	.986195	.984701	.983188	.981655	.980102	.978531	.976941	.975332	.973705	.972059	-17
.02	.970396	.968715	.967017	.965301	.963568	.961819	.960052	.958270	.956471	.954656	-15
.03	.952826	.950980	.949119	.947243	.945351	.943446	.941525	.939591	.937642	.935680	-13
.04	.933704	.931714	.929712	.927696	.925668	.923627	.921573	.919508	.917431	.915341	-11
.05	.913241	.911128	.909005	.906871	.904726	.902570	.900404	.898228	.896042	.893846	-9
.06	.891640	.889425	.887201	.884967	.882725	.880474	.878215	.875947	.873671	.871387	-7
.07	.869096	.866797	.864490	.862176	.859856	.857528	.855193	.852853	.850505	.848152	-5
.08	.845792	.843427	.841056	.838679	.836297	.833910	.831518	.829120	.826719	.824312	-4
.09	.821901	.819486	.817067	.814644	.812217	.809787	.807353	.804915	.802475	.800031	-2
.10	.797585	.795136	.792684	.790230	.787773	.785314	.782853	.780390	.777926	.775460	-1
.11	.772992	.770523	.768053	.765581	.763109	.760635	.758161	.755687	.753212	.750736	0
.12	.748260	.745785	.743309	.740833	.738357	.735882	.733408	.730933	.728460	.725987	1
.13	.723516	.721045	.718575	.716107	.713640	.711174	.708710	.706248	.703787	.701329	2
.14	.698872	.696417	.693964	.691514	.689066	.686620	.684177	.681736	.679298	.676863	3
.15	.674431	.672002	.669576	.667152	.664733	.662316	.659903	.657493	.655087	.652684	4
.16	.650285	.647890	.645498	.643111	.640728	.638348	.635973	.633602	.631235	.628872	4
.17	.626514	.624161	.621812	.619467	.617127	.614792	.612462	.610136	.607816	.605500	5
.18	.603189	.600884	.598583	.596288	.593998	.591713	.589434	.587160	.584892	.582629	5
.19	.580371	.578119	.575873	.573632	.571398	.569168	.566945	.564728	.562516	.560311	6
.20	.558111	.555918	.553730	.551549	.549374	.547205	.545042	.542885	.540735	.538591	6
.21	.536453	.534322	.532197	.530078	.527967	.525861	.523762	.521670	.519584	.517505	7
.22	.515432	.513366	.511307	.509254	.507209	.505169	.503137	.501112	.499093	.497081	7
.23	.495076	.493078	.491087	.489102	.487125	.485155	.483191	.481235	.479285	.477342	7
.24	.475407	.473478	.471557	.469642	.467735	.465835	.463942	.462055	.460176	.458304	7
.25	.456440	.454582	.452731	.450888	.449052	.447222	.445400	.443586	.441778	.439977	7
.26	.438184	.436398	.434619	.432847	.431082	.429325	.427575	.425831	.424096	.422367	7
.27	.420645	.418931	.417223	.415523	.413830	.412145	.410466	.408795	.407130	.405473	7
.28	.403823	.402180	.400545	.398916	.397295	.395680	.394073	.392473	.390880	.389294	7
.29	.387715	.386143	.384578	.383020	.381470	.379926	.378389	.376860	.375337	.373822	7
.30	.372313	.370811	.369316	.367829	.366348	.364874	.363407	.361947	.360493	.359047	7
.31	.357607	.356175	.354749	.353330	.351917	.350512	.349113	.347721	.346336	.344957	7
.32	.343585	.342220	.340862	.339510	.338165	.336826	.335494	.334169	.332850	.331538	7
.33	.330232	.328933	.327640	.326354	.325074	.323801	.322534	.321274	.320020	.318772	6
.34	.317531	.316296	.315067	.313845	.312629	.311419	.310215	.309018	.307827	.306642	6
.35	.305463	.304290	.303124	.301963	.300809	.299660	.298518	.297382	.296251	.295127	6
.36	.294009	.292896	.291790	.290689	.289594	.288505	.287422	.286345	.285273	.284207	6
.37	.283147	.282093	.281044	.280001	.278964	.277932	.276906	.275885	.274870	.273861	6
.38	.272857	.271858	.270865	.269878	.268896	.267919	.266948	.265982	.265021	.264066	5
.39	.263116	.262171	.261231	.260297	.259368	.258444	.257525	.256611	.255702	.254799	5
.40	.253900	.253007	.252118	.251235	.250356	.249483	.248614	.247750	.246891	.246037	5
.41	.245188	.244344	.243504	.242669	.241839	.241013	.240193	.239377	.238565	.237758	5
.42	.236956	.236158	.235365	.234576	.233792	.233013	.232237	.231467	.230700	.229938	4
.43	.229181	.228427	.227678	.226934	.226193	.225457	.224725	.223997	.223274	.222554	4
.44	.221839	.221128	.220420	.219717	.219018	.218323	.217632	.216945	.216262	.215583	4
.45	.214908	.214236	.213569	.212905	.212245	.211589	.210937	.210288	.209643	.209002	4
.46	.208365	.207731	.207101	.206474	.205851	.205231	.204616	.204003	.203394	.202789	3
.47	.202187	.201588	.200993	.200401	.199813	.199228	.198646	.198068	.197493	.196921	3
.48	.196352	.195787	.195225	.194666	.194110	.193557	.193007	.192461	.191917	.191377	3
.49	.190839	.190305	.189773	.189245	.188719	.188196	.187677	.187160	.186646	.186134	3

TABLE 23

MTF, Defocused Lens, Monochromatic – Δ = 1.5 Rayleigh Units [$\Delta \approx d/2\lambda F^2$]

ν_r	.000	.001	.002	.003	.004	.005	.006	.007	.008	.009	Δ_2
.50	.185626	.185120	.184617	.184117	.183620	.183125	.182633	.182144	.181657	.181173	3
.51	.180691	.180212	.179736	.179262	.178791	.178322	.177856	.177392	.176930	.176471	2
.52	.176015	.175561	.175109	.174659	.174212	.173767	.173324	.172884	.172446	.172010	2
.53	.171576	.171144	.170715	.170288	.169863	.169439	.169018	.168600	.168183	.167768	2
.54	.167355	.166944	.166535	.166128	.165723	.165320	.164919	.164519	.164122	.163726	2
.55	.163332	.162940	.162550	.162161	.161774	.161389	.161006	.160624	.160244	.159866	2
.56	.159489	.159114	.158740	.158368	.157997	.157628	.157261	.156895	.156530	.156167	1
.57	.155806	.155446	.155087	.154729	.154373	.154019	.153665	.153313	.152962	.152613	1
.58	.152265	.151918	.151572	.151228	.150884	.150542	.150201	.149861	.149522	.149185	1
.59	.148848	.148513	.148178	.147845	.147512	.147181	.146850	.146521	.146192	.145865	1
.60	.145538	.145212	.144887	.144563	.144240	.143917	.143596	.143275	.142955	.142636	1
.61	.142317	.141999	.141682	.141366	.141050	.140735	.140421	.140107	.139794	.139481	1
.62	.139169	.138858	.138547	.138236	.137927	.137617	.137308	.137000	.136692	.136384	0
.63	.136077	.135771	.135464	.135158	.134853	.134548	.134243	.133938	.133634	.133330	0
.64	.133026	.132723	.132419	.132116	.131813	.131511	.131208	.130906	.130604	.130302	0
.65	.130000	.129698	.129396	.129094	.128793	.128491	.128190	.127888	.127586	.127285	0
.66	.126983	.126682	.126380	.126078	.125776	.125474	.125172	.124870	.124567	.124265	0
.67	.123962	.123659	.123356	.123053	.122750	.122446	.122142	.121838	.121533	.121228	0
.68	.120923	.120618	.120312	.120006	.119699	.119392	.119085	.118777	.118469	.118161	0
.69	.117852	.117543	.117233	.116923	.116612	.116301	.115989	.115677	.115364	.115051	0
.70	.114737	.114423	.114108	.113792	.113476	.113159	.112842	.112524	.112205	.111886	0
.71	.111566	.111246	.110924	.110602	.110280	.109956	.109632	.109307	.108982	.108655	0
.72	.108328	.108001	.107672	.107343	.107012	.106681	.106349	.106017	.105683	.105349	0
.73	.105014	.104678	.104341	.104003	.103664	.103325	.102984	.102643	.102301	.101958	0
.74	.101614	.101268	.100922	.100575	.100228	.099879	.099529	.099178	.098826	.098473	0
.75	.098120	.097765	.097409	.097052	.096694	.096336	.095976	.095615	.095253	.094890	0
.76	.094526	.094161	.093795	.093427	.093059	.092690	.092319	.091948	.091575	.091202	0
.77	.090827	.090451	.090074	.089696	.089317	.088937	.088556	.088173	.087790	.087405	0
.78	.087020	.086633	.086245	.085856	.085466	.085075	.084682	.084289	.083894	.083499	0
.79	.083102	.082704	.082305	.081905	.081504	.081102	.080698	.080294	.079888	.079482	0
.80	.079074	.078665	.078255	.077844	.077432	.077019	.076605	.076190	.075773	.075356	0
.81	.074938	.074518	.074098	.073676	.073254	.072830	.072406	.071980	.071553	.071126	0
.82	.070697	.070268	.069837	.069406	.068973	.068540	.068106	.067671	.067234	.066797	0
.83	.066359	.065921	.065481	.065040	.064599	.064157	.063714	.063270	.062825	.062379	0
.84	.061933	.061486	.061038	.060590	.060140	.059690	.059240	.058788	.058336	.057883	0
.85	.057430	.056976	.056521	.056066	.055610	.055154	.054697	.054240	.053782	.053324	0
.86	.052865	.052405	.051946	.051485	.051025	.050564	.050103	.049641	.049179	.048717	0
.87	.048255	.047792	.047329	.046866	.046403	.045939	.045476	.045012	.044548	.044084	0
.88	.043620	.043157	.042693	.042229	.041765	.041302	.040838	.040375	.039912	.039449	0
.89	.038986	.038524	.038062	.037600	.037139	.036678	.036217	.035757	.035297	.034838	1
.90	.034380	.033922	.033464	.033008	.032552	.032096	.031642	.031188	.030735	.030283	1
.91	.029832	.029382	.028933	.028485	.028037	.027591	.027146	.026703	.026260	.025819	1
.92	.025379	.024940	.024503	.024067	.023633	.023200	.022769	.022339	.021912	.021485	2
.93	.021061	.020638	.020217	.019798	.019381	.018966	.018553	.018142	.017733	.017327	2
.94	.016922	.016520	.016121	.015724	.015329	.014937	.014547	.014160	.013776	.013395	3
.95	.013016	.012641	.012268	.011899	.011532	.011169	.010809	.010452	.010099	.009749	3
.96	.009403	.009060	.008722	.008387	.008056	.007728	.007405	.007087	.006772	.006462	4
.97	.006156	.005855	.005559	.005267	.004980	.004699	.004422	.004151	.003886	.003626	6
.98	.003372	.003124	.002882	.002646	.002418	.002195	.001980	.001773	.001573	.001381	8
.99	.001197	.001023	.000857	.000702	.000557	.000435	.000107	.000070	.000038	.000013	7

TABLE 23

MTF, Defocused Lens, Monochromatic – $\Delta = 2.0$ Rayleigh Units $[\Delta \approx d/2\lambda F^2]$

ν_r	.000	.001	.002	.003	.004	.005	.006	.007	.008	.009	Δ_2
.00	1.000000	.998707	.997375	.996004	.994596	.993149	.991665	.990143	.988585	.986991	-35
.01	.985361	.983696	.981996	.980261	.978492	.976689	.974853	.972984	.971082	.969149	-31
.02	.967183	.965187	.963159	.961102	.959014	.956896	.954749	.952574	.950370	.948138	-27
.03	.945879	.943592	.941279	.938939	.936573	.934182	.931766	.929325	.926859	.924370	-23
.04	.921857	.919321	.916762	.914181	.911578	.908954	.906308	.903641	.900954	.898247	-19
.05	.895521	.892775	.890010	.887227	.884426	.881607	.878771	.875918	.873048	.870162	-15
.06	.867260	.864342	.861409	.858462	.855500	.852524	.849534	.846531	.843515	.840487	-12
.07	.837446	.834392	.831328	.828252	.825165	.822067	.818959	.815841	.812713	.809576	-8
.08	.806431	.803276	.800113	.796942	.793763	.790576	.787383	.784183	.780976	.777763	-5
.09	.774544	.771319	.768089	.764854	.761614	.758369	.755121	.751868	.748612	.745352	-2
.10	.742089	.738823	.735555	.732284	.729012	.725737	.722461	.719184	.715905	.712626	0
.11	.709346	.706065	.702785	.699505	.696225	.692945	.689667	.686389	.683113	.679838	1
.12	.676565	.673294	.670025	.666758	.663494	.660233	.656974	.653719	.650467	.647218	4
.13	.643973	.640733	.637496	.634263	.631035	.627812	.624593	.621379	.618171	.614968	5
.14	.611770	.608578	.605392	.602211	.599037	.595869	.592708	.589553	.586405	.583263	7
.15	.580129	.577002	.573882	.570769	.567664	.564567	.561477	.558396	.555322	.552257	8
.16	.549200	.546151	.543111	.540079	.537057	.534043	.531038	.528042	.525055	.522077	9
.17	.519109	.516151	.513201	.510262	.507332	.504412	.501502	.498602	.495712	.492832	10
.18	.489962	.487103	.484254	.481415	.478587	.475770	.472963	.470167	.467382	.464607	11
.19	.461843	.459091	.456349	.453619	.450899	.448191	.445494	.442808	.440134	.437471	11
.20	.434819	.432179	.429550	.426933	.424327	.421733	.419151	.416580	.414021	.411473	12
.21	.408938	.406414	.403902	.401402	.398914	.396437	.393973	.391520	.389080	.386651	12
.22	.384234	.381830	.379437	.377056	.374687	.372331	.369986	.367654	.365333	.363025	12
.23	.360728	.358444	.356172	.353912	.351664	.349428	.347204	.344992	.342792	.340604	12
.24	.338428	.336265	.334113	.331973	.329846	.327730	.325627	.323535	.321456	.319388	12
.25	.317332	.315288	.313256	.311237	.309228	.307232	.305248	.303275	.301314	.299365	12
.26	.297428	.295503	.293589	.291687	.289797	.287918	.286051	.284195	.282351	.280519	12
.27	.278698	.276888	.275090	.273304	.271528	.269765	.268012	.266271	.264541	.262822	11
.28	.261115	.259419	.257733	.256059	.254396	.252744	.251103	.249473	.247854	.246246	11
.29	.244649	.243062	.241487	.239922	.238367	.236824	.235291	.233768	.232257	.230755	11
.30	.229264	.227784	.226314	.224854	.223405	.221966	.220537	.219118	.217710	.216312	10
.31	.214923	.213545	.212177	.210818	.209470	.208131	.206802	.205483	.204174	.202874	10
.32	.201584	.200303	.199033	.197771	.196519	.195277	.194043	.192820	.191605	.190400	9
.33	.189204	.188017	.186839	.185670	.184510	.183360	.182218	.181085	.179961	.178845	9
.34	.177739	.176641	.175552	.174471	.173399	.172336	.171281	.170234	.169196	.168166	8
.35	.167144	.166131	.165126	.164129	.163141	.162160	.161187	.160223	.159266	.158317	8
.36	.157376	.156443	.155518	.154600	.153690	.152788	.151893	.151006	.150127	.149255	7
.37	.148390	.147533	.146683	.145840	.145004	.144176	.143355	.142541	.141734	.140935	7
.38	.140142	.139356	.138577	.137805	.137040	.136281	.135529	.134784	.134046	.133314	7
.39	.132589	.131871	.131159	.130453	.129754	.129061	.128374	.127694	.127020	.126352	6
.40	.125691	.125035	.124386	.123743	.123106	.122474	.121849	.121230	.120616	.120008	6
.41	.119406	.118810	.118220	.117635	.117056	.116482	.115914	.115352	.114795	.114243	5
.42	.113697	.113157	.112621	.112091	.111566	.111047	.110533	.110023	.109519	.109020	5
.43	.108526	.108038	.107554	.107075	.106601	.106132	.105667	.105208	.104753	.104303	5
.44	.103858	.103418	.102982	.102551	.102124	.101702	.101284	.100871	.100462	.100058	4
.45	.099658	.099263	.098872	.098485	.098102	.097724	.097350	.096980	.096614	.096253	4
.46	.095895	.095541	.095192	.094846	.094505	.094167	.093833	.093503	.093177	.092855	4
.47	.092537	.092222	.091911	.091604	.091301	.091001	.090704	.090412	.090123	.089837	4
.48	.089555	.089276	.089001	.088729	.088461	.088196	.087935	.087676	.087421	.087170	3
.49	.086921	.086676	.086434	.086195	.085959	.085726	.085497	.085270	.085047	.084826	3

TABLE 23

MTF, Defocused Lens, Monochromatic – Δ = 2.0 Rayleigh Units [$\Delta \approx d/2\lambda F^2$]

ν_r	.000	.001	.002	.003	.004	.005	.006	.007	.008	.009	Δ_2
.50	.084609	.084394	.084183	.083974	.083768	.083565	.083365	.083168	.082974	.082782	3
.51	.082593	.082407	.082223	.082042	.081864	.081689	.081516	.081345	.081177	.081012	3
.52	.080849	.080689	.080531	.080376	.080223	.080072	.079924	.079779	.079635	.079494	2
.53	.079355	.079218	.079084	.078952	.078822	.078694	.078569	.078445	.078324	.078205	2
.54	.078088	.077973	.077860	.077749	.077640	.077533	.077428	.077324	.077223	.077124	2
.55	.077026	.076931	.076837	.076745	.076655	.076567	.076480	.076395	.076312	.076230	2
.56	.076151	.076072	.075996	.075921	.075848	.075776	.075706	.075637	.075570	.075504	1
.57	.075440	.075377	.075316	.075256	.075198	.075141	.075085	.075031	.074978	.074926	1
.58	.074876	.074827	.074779	.074732	.074687	.074643	.074600	.074558	.074517	.074477	1
.59	.074439	.074401	.074365	.074330	.074296	.074262	.074230	.074199	.074168	.074139	1
.60	.074111	.074083	.074056	.074031	.074006	.073981	.073958	.073936	.073914	.073893	1
.61	.073873	.073853	.073834	.073816	.073799	.073782	.073766	.073750	.073735	.073721	1
.62	.073707	.073693	.073681	.073668	.073657	.073645	.073634	.073624	.073614	.073604	0
.63	.073595	.073586	.073578	.073569	.073561	.073554	.073546	.073539	.073532	.073526	0
.64	.073519	.073513	.073507	.073501	.073495	.073489	.073484	.073478	.073472	.073467	0
.65	.073461	.073456	.073450	.073445	.073439	.073434	.073428	.073422	.073416	.073410	0
.66	.073404	.073397	.073390	.073383	.073376	.073369	.073361	.073353	.073345	.073336	0
.67	.073327	.073318	.073308	.073298	.073288	.073277	.073265	.073253	.073241	.073228	0
.68	.073215	.073201	.073186	.073171	.073155	.073139	.073122	.073104	.073086	.073067	0
.69	.073048	.073027	.073006	.072984	.072961	.072938	.072913	.072888	.072862	.072835	0
.70	.072808	.072779	.072749	.072719	.072687	.072655	.072621	.072587	.072551	.072515	0
.71	.072477	.072438	.072398	.072358	.072315	.072272	.072228	.072182	.072135	.072087	0
.72	.072038	.071987	.071936	.071883	.071828	.071772	.071715	.071657	.071597	.071536	0
.73	.071473	.071409	.071344	.071277	.071208	.071138	.071067	.070994	.070919	.070843	-1
.74	.070766	.070686	.070605	.070523	.070439	.070353	.070266	.070177	.070086	.069993	-1
.75	.069899	.069803	.069705	.069606	.069504	.069401	.069296	.069190	.069081	.068971	-1
.76	.068858	.068744	.068628	.068510	.068390	.068268	.068144	.068018	.067891	.067761	-1
.77	.067629	.067495	.067359	.067222	.067082	.066940	.066796	.066650	.066501	.066351	-1
.78	.066199	.066044	.065888	.065729	.065568	.065405	.065240	.065072	.064902	.064731	-1
.79	.064557	.064380	.064202	.064021	.063838	.063653	.063466	.063276	.063084	.062890	-1
.80	.062693	.062495	.062294	.062090	.061885	.061677	.061467	.061254	.061039	.060822	-1
.81	.060603	.060381	.060157	.059931	.059702	.059471	.059238	.059002	.058764	.058524	-1
.82	.058281	.058036	.057789	.057539	.057287	.057033	.056777	.056518	.056257	.055993	-1
.83	.055728	.055460	.055189	.054917	.054642	.054365	.054086	.053804	.053520	.053234	-1
.84	.052946	.052655	.052362	.052067	.051770	.051471	.051169	.050866	.050560	.050252	-1
.85	.049942	.049629	.049315	.048999	.048680	.048360	.048037	.047713	.047386	.047058	-1
.86	.046727	.046395	.046060	.045724	.045386	.045046	.044704	.044360	.044015	.043667	-1
.87	.043318	.042967	.042615	.042260	.041905	.041547	.041188	.040827	.040465	.040101	-1
.88	.039736	.039369	.039000	.038631	.038260	.037887	.037513	.037138	.036762	.036385	0
.89	.036006	.035626	.035245	.034863	.034480	.034096	.033710	.033324	.032938	.032550	0
.90	.032161	.031772	.031382	.030991	.030599	.030207	.029815	.029422	.029028	.028634	0
.91	.028240	.027845	.027450	.027055	.026659	.026264	.025868	.025473	.025077	.024682	0
.92	.024286	.023891	.023496	.023101	.022707	.022313	.021920	.021527	.021135	.020743	1
.93	.020352	.019962	.019572	.019184	.018796	.018410	.018025	.017640	.017257	.016876	1
.94	.016495	.016117	.015739	.015363	.014989	.014617	.014246	.013878	.013511	.013146	2
.95	.012784	.012423	.012065	.011709	.011356	.011005	.010657	.010312	.009969	.009630	3
.96	.009293	.008960	.008629	.008302	.007979	.007659	.007342	.007029	.006721	.006416	4
.97	.006115	.005818	.005526	.005238	.004955	.004677	.004403	.004135	.003872	.003614	5
.98	.003362	.003115	.002875	.002641	.002413	.002192	.001977	.001771	.001571	.001380	8
.99	.001196	.001022	.000857	.000702	.000557	.000435	.000107	.000070	.000038	.000013	7

TABLE 23

MTF, Defocused Lens, Monochromatic – Δ = 2.5 Rayleigh Units [Δ ≈ d/2λF²]

ν_T	.000	.001	.002	.003	.004	.005	.006	.007	.008	.009	Δ_2
.00	1.000000	.998696	.997331	.995906	.994420	.992876	.991273	.989612	.987895	.986120	-56
.01	.984290	.982405	.980465	.978471	.976424	.974325	.972173	.969971	.967718	.965415	-49
.02	.963063	.960663	.958215	.955720	.953179	.950592	.947960	.945284	.942564	.939801	-42
.03	.936996	.934149	.931262	.928334	.925367	.922361	.919317	.916235	.913117	.909963	-35
.04	.906773	.903548	.900290	.896998	.893673	.890317	.886929	.883510	.880061	.876583	-28
.05	.873076	.869541	.865978	.862389	.858774	.855133	.851467	.847777	.844064	.840327	-22
.06	.836568	.832787	.828986	.825163	.821321	.817460	.813580	.809681	.805766	.801833	-16
.07	.797884	.793919	.789939	.785945	.781936	.777914	.773879	.769832	.765773	.761702	-10
.08	.757621	.753530	.749429	.745319	.741200	.737073	.732938	.728796	.724648	.720494	-5
.09	.716334	.712168	.707998	.703824	.699646	.695465	.691281	.687095	.682907	.678717	0
.10	.674527	.670335	.666144	.661953	.657762	.653572	.649384	.645198	.641014	.636833	3
.11	.632655	.628480	.624309	.620142	.615980	.611823	.607670	.603524	.599383	.595249	6
.12	.591121	.587000	.582886	.578780	.574682	.570592	.566510	.562437	.558373	.554319	9
.13	.550274	.546239	.542214	.538200	.534196	.530203	.526222	.522252	.518294	.514347	12
.14	.510413	.506491	.502582	.498685	.494802	.490932	.487075	.483232	.479403	.475588	14
.15	.471787	.468001	.464229	.460472	.456730	.453003	.449291	.445594	.441914	.438249	16
.16	.434599	.430966	.427349	.423748	.420164	.416596	.413045	.409511	.405993	.402492	17
.17	.399009	.395543	.392094	.388662	.385248	.381852	.378473	.375112	.371769	.368443	18
.18	.365136	.361847	.358575	.355322	.352087	.348871	.345673	.342493	.339331	.336188	18
.19	.333064	.329958	.326871	.323802	.320752	.317721	.314708	.311714	.308739	.305783	19
.20	.302845	.299926	.297026	.294145	.291283	.288439	.285614	.282808	.280021	.277253	19
.21	.274503	.271772	.269060	.266367	.263692	.261036	.258399	.255781	.253181	.250599	19
.22	.248037	.245493	.242967	.240460	.237972	.235501	.233050	.230616	.228201	.225804	18
.23	.223426	.221065	.218723	.216398	.214092	.211804	.209534	.207281	.205046	.202830	18
.24	.200630	.198449	.196285	.194138	.192009	.189898	.187804	.185727	.183667	.181624	17
.25	.179599	.177590	.175598	.173624	.171666	.169724	.167800	.165892	.164000	.162125	16
.26	.160266	.158424	.156597	.154787	.152993	.151215	.149453	.147706	.145975	.144260	16
.27	.142561	.140877	.139208	.137555	.135917	.134294	.132686	.131093	.129515	.127952	15
.28	.126403	.124869	.123350	.121845	.120355	.118879	.117417	.115970	.114536	.113116	14
.29	.111711	.110319	.108940	.107576	.106225	.104887	.103563	.102252	.100954	.099669	13
.30	.098397	.097139	.095893	.094660	.093439	.092231	.091036	.089853	.088682	.087524	12
.31	.086377	.085243	.084121	.083011	.081912	.080825	.079750	.078687	.077634	.076594	11
.32	.075564	.074546	.073539	.072543	.071558	.070584	.069621	.068668	.067726	.066795	11
.33	.065874	.064963	.064063	.063173	.062293	.061423	.060564	.059714	.058874	.058044	10
.34	.057223	.056412	.055611	.054819	.054036	.053263	.052499	.051744	.050998	.050261	9
.35	.049533	.048814	.048104	.047403	.046710	.046025	.045349	.044682	.044022	.043372	8
.36	.042729	.042094	.041468	.040849	.040239	.039636	.039041	.038454	.037874	.037302	8
.37	.036737	.036180	.035631	.035088	.034553	.034025	.033504	.032991	.032484	.031984	7
.38	.031491	.031005	.030526	.030053	.029587	.029128	.028675	.028228	.027788	.027354	6
.39	.026927	.026506	.026091	.025682	.025279	.024882	.024491	.024106	.023726	.023353	6
.40	.022985	.022623	.022267	.021916	.021570	.021230	.020896	.020567	.020243	.019925	5
.41	.019611	.019303	.019000	.018702	.018409	.018121	.017838	.017560	.017287	.017018	5
.42	.016754	.016495	.016241	.015991	.015746	.015505	.015269	.015038	.014810	.014587	4
.43	.014369	.014154	.013944	.013738	.013536	.013339	.013145	.012956	.012770	.012589	4
.44	.012411	.012237	.012067	.011901	.011739	.011581	.011426	.011275	.011127	.010983	4
.45	.010843	.010706	.010573	.010443	.010317	.010194	.010074	.009958	.009845	.009736	3
.46	.009629	.009526	.009426	.009330	.009236	.009146	.009058	.008974	.008892	.008814	3
.47	.008739	.008666	.008597	.008530	.008466	.008405	.008347	.008292	.008239	.008189	3
.48	.008142	.008097	.008056	.008016	.007980	.007946	.007914	.007885	.007859	.007835	2
.49	.007814	.007795	.007778	.007764	.007752	.007742	.007735	.007731	.007728	.007728	2

TABLE 23

MTF, Defocused Lens, Monochromatic – Δ = 2.5 Rayleigh Units [Δ ≈ d/2λF²]

ν_r	.000	.001	.002	.003	.004	.005	.006	.007	.008	.009	Δ_2
.50	.007730	.007734	.007741	.007749	.007760	.007773	.007789	.007806	.007825	.007847	2
.51	.007870	.007896	.007924	.007953	.007985	.008019	.008054	.008092	.008131	.008173	2
.52	.008216	.008261	.008308	.008357	.008408	.008461	.008515	.008571	.008629	.008689	2
.53	.008750	.008813	.008878	.008945	.009013	.009083	.009154	.009227	.009302	.009379	2
.54	.009457	.009536	.009617	.009700	.009784	.009870	.009958	.010046	.010137	.010229	1
.55	.010322	.010417	.010513	.010610	.010710	.010810	.010912	.011015	.011120	.011226	1
.56	.011333	.011441	.011551	.011663	.011775	.011889	.012004	.012121	.012238	.012357	1
.57	.012477	.012599	.012721	.012845	.012970	.013096	.013223	.013352	.013481	.013612	1
.58	.013744	.013876	.014010	.014146	.014282	.014419	.014557	.014697	.014837	.014978	1
.59	.015121	.015264	.015408	.015554	.015700	.015847	.015996	.016145	.016295	.016446	1
.60	.016598	.016751	.016904	.017059	.017214	.017370	.017527	.017685	.017844	.018003	1
.61	.018164	.018325	.018487	.018649	.018813	.018977	.019141	.019307	.019473	.019640	1
.62	.019808	.019976	.020145	.020314	.020484	.020655	.020826	.020998	.021171	.021344	1
.63	.021517	.021692	.021866	.022042	.022217	.022393	.022570	.022747	.022925	.023103	0
.64	.023281	.023460	.023640	.023819	.023999	.024180	.024360	.024541	.024723	.024905	0
.65	.025087	.025269	.025451	.025634	.025817	.026000	.026184	.026367	.026551	.026735	0
.66	.026919	.027103	.027287	.027472	.027656	.027841	.028025	.028210	.028395	.028579	0
.67	.028764	.028948	.029133	.029317	.029502	.029686	.029870	.030054	.030238	.030422	0
.68	.030605	.030788	.030971	.031154	.031337	.031519	.031701	.031883	.032064	.032245	0
.69	.032426	.032606	.032786	.032965	.033144	.033323	.033501	.033678	.033855	.034032	0
.70	.034208	.034383	.034558	.034732	.034905	.035078	.035250	.035421	.035592	.035762	0
.71	.035931	.036099	.036266	.036433	.036599	.036764	.036928	.037091	.037253	.037414	0
.72	.037574	.037733	.037892	.038049	.038205	.038359	.038513	.038666	.038817	.038967	0
.73	.039116	.039264	.039411	.039556	.039700	.039842	.039984	.040123	.040262	.040399	0
.74	.040534	.040668	.040801	.040932	.041061	.041189	.041316	.041440	.041563	.041684	−1
.75	.041804	.041922	.042038	.042153	.042265	.042376	.042485	.042592	.042697	.042800	−1
.76	.042902	.043001	.043098	.043194	.043287	.043378	.043467	.043554	.043639	.043722	−1
.77	.043802	.043881	.043957	.044031	.044102	.044171	.044238	.044303	.044365	.044425	−1
.78	.044482	.044537	.044589	.044639	.044687	.044732	.044774	.044814	.044851	.044886	−2
.79	.044917	.044947	.044973	.044997	.045018	.045036	.045052	.045065	.045075	.045082	−2
.80	.045086	.045087	.045086	.045081	.045074	.045064	.045051	.045034	.045015	.044993	−2
.81	.044967	.044939	.044908	.044873	.044835	.044795	.044751	.044704	.044653	.044600	−2
.82	.044544	.044484	.044421	.044355	.044285	.044212	.044136	.044057	.043975	.043889	−2
.83	.043800	.043707	.043612	.043513	.043410	.043304	.043195	.043083	.042967	.042848	−2
.84	.042725	.042600	.042470	.042338	.042202	.042062	.041920	.041773	.041624	.041471	−2
.85	.041315	.041155	.040992	.040826	.040656	.040483	.040307	.040127	.039944	.039757	−2
.86	.039568	.039375	.039178	.038979	.038776	.038569	.038360	.038147	.037932	.037713	−2
.87	.037490	.037265	.037036	.036805	.036570	.036332	.036091	.035847	.035600	.035350	−2
.88	.035097	.034841	.034582	.034320	.034055	.033788	.033517	.033244	.032968	.032690	−2
.89	.032408	.032125	.031838	.031549	.031257	.030963	.030667	.030368	.030066	.029762	−1
.90	.029456	.029148	.028838	.028525	.028210	.027893	.027575	.027254	.026931	.026607	−1
.91	.026280	.025952	.025622	.025291	.024958	.024623	.024287	.023950	.023611	.023271	0
.92	.022930	.022587	.022244	.021899	.021554	.021208	.020861	.020513	.020164	.019815	0
.93	.019465	.019115	.018765	.018414	.018063	.017712	.017361	.017010	.016659	.016308	0
.94	.015958	.015607	.015258	.014909	.014560	.014213	.013866	.013520	.013175	.012831	1
.95	.012489	.012147	.011808	.011469	.011133	.010798	.010465	.010134	.009805	.009478	2
.96	.009153	.008831	.008512	.008195	.007881	.007570	.007261	.006957	.006655	.006357	3
.97	.006062	.005771	.005484	.005201	.004923	.004648	.004379	.004114	.003853	.003598	5
.98	.003349	.003104	.002866	.002633	.002407	.002187	.001974	.001768	.001569	.001378	8
.99	.001195	.001021	.000856	.000701	.000557	.000434	.000107	.000070	.000038	.000013	7

TABLE 23

MTF, Defocused Lens, Monochromatic – Δ = 3.0 Rayleigh Units [$\Delta \approx d/2\lambda F^2$]

ν_r	.000	.001	.002	.003	.004	.005	.006	.007	.008	.009	Δ_2
.00	1.000000	.998682	.997277	.995785	.994206	.992543	.990795	.988964	.987051	.985056	-80
.01	.982982	.980828	.978596	.976286	.973901	.971440	.968905	.966297	.963616	.960865	-70
.02	.958044	.955154	.952195	.949171	.946080	.942924	.939705	.936424	.933081	.929677	-59
.03	.926215	.922694	.919115	.915481	.911792	.908048	.904252	.900404	.896506	.892557	-49
.04	.888560	.884516	.880425	.876289	.872108	.867884	.863618	.859311	.854964	.850577	-38
.05	.846153	.841691	.837194	.832661	.828095	.823495	.818864	.814201	.809509	.804788	-28
.06	.800039	.795263	.790460	.785633	.780782	.775907	.771010	.766092	.761154	.756196	-18
.07	.751220	.746226	.741215	.736188	.731147	.726091	.721022	.715941	.710848	.705744	-10
.08	.700631	.695508	.690378	.685239	.680094	.674944	.669788	.664628	.659464	.654297	-2
.09	.649128	.643958	.638787	.633617	.628447	.623278	.618112	.612948	.607788	.602631	4
.10	.597480	.592334	.587194	.582060	.576934	.571815	.566705	.561604	.556512	.551431	10
.11	.546360	.541300	.536251	.531215	.526192	.521182	.516185	.511203	.506235	.501282	15
.12	.496345	.491424	.486519	.481631	.476760	.471907	.467072	.462256	.457458	.452679	19
.13	.447920	.443181	.438462	.433763	.429086	.424429	.419794	.415181	.410590	.406022	22
.14	.401476	.396953	.392453	.387977	.383524	.379095	.374691	.370311	.365955	.361624	25
.15	.357319	.353038	.348783	.344553	.340349	.336171	.332019	.327893	.323794	.319721	26
.16	.315674	.311654	.307662	.303696	.299757	.295845	.291961	.288104	.284274	.280472	27
.17	.276697	.272950	.269230	.265539	.261875	.258239	.254631	.251050	.247498	.243973	28
.18	.240477	.237008	.233568	.230155	.226770	.223413	.220085	.216784	.213511	.210266	28
.19	.207048	.203859	.200697	.197563	.194457	.191378	.188327	.185304	.182308	.179340	27
.20	.176398	.173484	.170598	.167738	.164906	.162100	.159321	.156570	.153844	.151146	27
.21	.148474	.145828	.143209	.140616	.138049	.135508	.132994	.130504	.128041	.125603	26
.22	.123190	.120803	.118441	.116104	.113792	.111505	.109243	.107005	.104791	.102602	24
.23	.100437	.098296	.096179	.094086	.092016	.089970	.087947	.085947	.083970	.082016	23
.24	.080085	.078176	.076290	.074426	.072585	.070765	.068967	.067191	.065436	.063703	21
.25	.061991	.060300	.058630	.056980	.055351	.053743	.052155	.050587	.049039	.047511	20
.26	.046003	.044514	.043044	.041594	.040162	.038750	.037356	.035981	.034624	.033285	18
.27	.031965	.030662	.029378	.028110	.026861	.025628	.024413	.023215	.022033	.020868	17
.28	.019720	.018588	.017472	.016373	.015289	.014221	.013169	.012132	.011110	.010104	15
.29	.009112	.008136	.007174	.006227	.005294	.004375	.003471	.002580	.001703	.000840	14
.30	-.000009	-.000845	-.001668	-.002478	-.003274	-.004058	-.004829	-.005588	-.006334	-.007068	12
.31	-.007790	-.008499	-.009197	-.009883	-.010558	-.011221	-.011872	-.012512	-.013142	-.013760	11
.32	-.014367	-.014964	-.015550	-.016126	-.016691	-.017246	-.017791	-.018326	-.018851	-.019366	10
.33	-.019872	-.020368	-.020855	-.021333	-.021801	-.022260	-.022711	-.023152	-.023585	-.024009	9
.34	-.024425	-.024832	-.025231	-.025622	-.026004	-.026379	-.026746	-.027105	-.027456	-.027800	8
.35	-.028136	-.028465	-.028787	-.029101	-.029409	-.029709	-.030003	-.030290	-.030569	-.030843	7
.36	-.031110	-.031370	-.031624	-.031872	-.032113	-.032349	-.032578	-.032802	-.033019	-.033231	6
.37	-.033437	-.033638	-.033833	-.034023	-.034207	-.034386	-.034560	-.034728	-.034892	-.035050	5
.38	-.035204	-.035352	-.035496	-.035635	-.035770	-.035900	-.036025	-.036146	-.036263	-.036375	4
.39	-.036483	-.036587	-.036687	-.036783	-.036874	-.036962	-.037046	-.037126	-.037202	-.037275	4
.40	-.037344	-.037409	-.037471	-.037529	-.037584	-.037635	-.037683	-.037728	-.037770	-.037808	3
.41	-.037843	-.037876	-.037905	-.037931	-.037954	-.037975	-.037992	-.038007	-.038019	-.038028	3
.42	-.038035	-.038039	-.038040	-.038039	-.038035	-.038029	-.038021	-.038010	-.037996	-.037981	2
.43	-.037963	-.037943	-.037920	-.037896	-.037869	-.037840	-.037810	-.037777	-.037742	-.037705	2
.44	-.037666	-.037626	-.037583	-.037539	-.037493	-.037445	-.037395	-.037343	-.037290	-.037235	2
.45	-.037179	-.037121	-.037061	-.037000	-.036937	-.036872	-.036806	-.036739	-.036670	-.036600	1
.46	-.036528	-.036455	-.036381	-.036305	-.036228	-.036149	-.036070	-.035989	-.035906	-.035823	1
.47	-.035738	-.035652	-.035565	-.035477	-.035387	-.035297	-.035205	-.035112	-.035019	-.034924	1
.48	-.034828	-.034731	-.034633	-.034534	-.034434	-.034332	-.034230	-.034127	-.034023	-.033919	1
.49	-.033813	-.033706	-.033598	-.033490	-.033380	-.033270	-.033159	-.033047	-.032934	-.032820	1

TABLE 23

MTF, Defocused Lens, Monochromatic – Δ = 3.0 Rayleigh Units $[\Delta \approx d/2\lambda F^2]$

ν_r	.000	.001	.002	.003	.004	.005	.006	.007	.008	.009	Δ_2
.50	-.032705	-.032590	-.032474	-.032357	-.032239	-.032120	-.032001	-.031880	-.031759	-.031637	1
.51	-.031515	-.031392	-.031268	-.031143	-.031017	-.030891	-.030764	-.030636	-.030508	-.030378	1
.52	-.030248	-.030118	-.029986	-.029854	-.029722	-.029588	-.029454	-.029319	-.029183	-.029047	1
.53	-.028910	-.028773	-.028634	-.028495	-.028356	-.028215	-.028074	-.027933	-.027790	-.027647	1
.54	-.027503	-.027359	-.027214	-.027068	-.026922	-.026775	-.026627	-.026478	-.026329	-.026180	1
.55	-.026029	-.025878	-.025726	-.025574	-.025420	-.025267	-.025112	-.024957	-.024801	-.024644	1
.56	-.024487	-.024329	-.024171	-.024011	-.023851	-.023691	-.023529	-.023367	-.023204	-.023041	1
.57	-.022877	-.022712	-.022546	-.022380	-.022213	-.022045	-.021877	-.021708	-.021538	-.021367	1
.58	-.021196	-.021024	-.020851	-.020678	-.020503	-.020328	-.020153	-.019976	-.019799	-.019621	1
.59	-.019443	-.019263	-.019083	-.018902	-.018721	-.018538	-.018355	-.018171	-.017986	-.017801	1
.60	-.017615	-.017428	-.017240	-.017051	-.016862	-.016672	-.016481	-.016290	-.016097	-.015904	1
.61	-.015710	-.015515	-.015320	-.015123	-.014926	-.014728	-.014530	-.014330	-.014130	-.013929	1
.62	-.013727	-.013524	-.013320	-.013116	-.012911	-.012705	-.012498	-.012291	-.012083	-.011873	1
.63	-.011664	-.011453	-.011241	-.011029	-.010816	-.010602	-.010387	-.010172	-.009956	-.009739	1
.64	-.009521	-.009302	-.009083	-.008862	-.008641	-.008420	-.008197	-.007974	-.007750	-.007525	1
.65	-.007299	-.007073	-.006845	-.006617	-.006389	-.006159	-.005929	-.005698	-.005467	-.005234	1
.66	-.005001	-.004767	-.004533	-.004298	-.004062	-.003825	-.003588	-.003350	-.003111	-.002872	1
.67	-.002632	-.002391	-.002150	-.001908	-.001665	-.001422	-.001178	-.000934	-.000689	-.000443	1
.68	-.000197	.000050	.000297	.000545	.000794	.001042	.001292	.001542	.001792	.002043	0
.69	.002294	.002546	.002799	.003051	.003304	.003558	.003812	.004066	.004321	.004576	0
.70	.004831	.005086	.005342	.005599	.005855	.006112	.006369	.006626	.006883	.007141	0
.71	.007398	.007656	.007914	.008172	.008430	.008688	.008947	.009205	.009463	.009721	0
.72	.009980	.010238	.010496	.010754	.011012	.011269	.011527	.011784	.012041	.012298	0
.73	.012555	.012811	.013067	.013323	.013578	.013833	.014087	.014341	.014595	.014848	0
.74	.015101	.015353	.015604	.015855	.016105	.016354	.016603	.016851	.017099	.017345	0
.75	.017591	.017836	.018080	.018323	.018565	.018806	.019047	.019286	.019524	.019761	0
.76	.019997	.020231	.020465	.020697	.020928	.021158	.021386	.021613	.021839	.022063	0
.77	.022286	.022507	.022727	.022945	.023162	.023377	.023590	.023801	.024011	.024219	-1
.78	.024425	.024629	.024832	.025032	.025231	.025427	.025622	.025814	.026004	.026192	-1
.79	.026378	.026562	.026743	.026922	.027099	.027273	.027445	.027615	.027782	.027946	-2
.80	.028108	.028267	.028424	.028578	.028729	.028878	.029023	.029166	.029306	.029444	-2
.81	.029578	.029709	.029837	.029963	.030085	.030204	.030320	.030432	.030542	.030648	-2
.82	.030751	.030851	.030947	.031040	.031130	.031216	.031298	.031378	.031453	.031525	-3
.83	.031594	.031658	.031720	.031777	.031831	.031881	.031927	.031969	.032008	.032043	-3
.84	.032074	.032101	.032124	.032143	.032158	.032169	.032177	.032180	.032179	.032174	-3
.85	.032165	.032152	.032134	.032113	.032087	.032058	.032024	.031986	.031943	.031897	-3
.86	.031846	.031791	.031732	.031669	.031601	.031529	.031453	.031372	.031287	.031198	-3
.87	.031105	.031008	.030906	.030800	.030689	.030575	.030456	.030333	.030205	.030074	-3
.88	.029938	.029798	.029654	.029505	.029353	.029196	.029035	.028871	.028702	.028529	-3
.89	.028352	.028171	.027986	.027797	.027604	.027407	.027206	.027002	.026794	.026582	-3
.90	.026366	.026147	.025924	.025697	.025467	.025233	.024996	.024756	.024512	.024265	-2
.91	.024014	.023761	.023504	.023244	.022981	.022715	.022447	.022175	.021901	.021624	-2
.92	.021344	.021062	.020777	.020490	.020200	.019909	.019615	.019319	.019020	.018720	-1
.93	.018418	.018115	.017809	.017502	.017194	.016884	.016573	.016260	.015947	.015632	0
.94	.015317	.015000	.014683	.014366	.014048	.013729	.013410	.013091	.012772	.012453	0
.95	.012135	.011816	.011498	.011181	.010864	.010548	.010233	.009919	.009606	.009294	1
.96	.008984	.008676	.008369	.008065	.007762	.007462	.007164	.006868	.006575	.006285	3
.97	.005998	.005714	.005434	.005157	.004883	.004614	.004349	.004088	.003831	.003579	5
.98	.003333	.003091	.002855	.002624	.002399	.002181	.001969	.001764	.001566	.001376	8
.99	.001194	.001020	.000856	.000701	.000557	.000434	.000107	.000070	.000038	.000013	7

TABLE 23

MTF, Defocused Lens, Monochromatic – Δ = 3.5 Rayleigh Units [$\Delta \approx d/2\lambda F^2$]

ν_r	.000	.001	.002	.003	.004	.005	.006	.007	.008	.009	Δ_2
.00	1.000000	.998667	.997213	.995642	.993953	.992149	.990230	.988198	.986054	.983800	-109
.01	.981437	.978966	.976390	.973708	.970924	.968038	.965051	.961966	.958784	.955506	-95
.02	.952134	.948669	.945114	.941468	.937735	.933916	.930012	.926025	.921956	.917807	-79
.03	.913580	.909276	.904897	.900444	.895919	.891325	.886661	.881930	.877134	.872273	-63
.04	.867351	.862367	.857325	.852225	.847069	.841858	.836595	.831281	.825916	.820504	-47
.05	.815045	.809542	.803994	.798405	.792775	.787107	.781400	.775658	.769882	.764072	-32
.06	.758231	.752360	.746460	.740533	.734580	.728602	.722602	.716580	.710537	.704476	-18
.07	.698397	.692301	.686191	.680067	.673930	.667783	.661625	.655459	.649285	.643105	-5
.08	.636920	.630730	.624539	.618345	.612151	.605958	.599766	.593577	.587392	.581212	5
.09	.575038	.568870	.562711	.556561	.550420	.544290	.538172	.532067	.525975	.519898	14
.10	.513836	.507790	.501761	.495750	.489757	.483784	.477831	.471899	.465988	.460100	22
.11	.454235	.448394	.442577	.436785	.431019	.425279	.419566	.413880	.408223	.402594	29
.12	.396994	.391424	.385884	.380375	.374897	.369451	.364037	.358656	.353307	.347992	33
.13	.342711	.337464	.332251	.327074	.321931	.316824	.311754	.306719	.301721	.296759	37
.14	.291835	.286947	.282098	.277286	.272512	.267776	.263078	.258419	.253799	.249218	39
.15	.244675	.240172	.235708	.231283	.226898	.222552	.218246	.213980	.209753	.205566	40
.16	.201420	.197313	.193246	.189219	.185231	.181284	.177377	.173510	.169682	.165895	40
.17	.162147	.158439	.154771	.151143	.147554	.144004	.140494	.137023	.133592	.130199	39
.18	.126846	.123531	.120255	.117018	.113819	.110659	.107537	.104452	.101406	.098398	38
.19	.095427	.092493	.089597	.086737	.083915	.081129	.078379	.075666	.072989	.070347	36
.20	.067742	.065171	.062636	.060136	.057671	.055240	.052844	.050481	.048153	.045858	34
.21	.043596	.041368	.039172	.037009	.034879	.032780	.030714	.028679	.026675	.024703	31
.22	.022761	.020851	.018970	.017120	.015299	.013508	.011746	.010014	.008310	.006634	29
.23	.004987	.003368	.001777	.000213	-.001324	-.002834	-.004317	-.005774	-.007205	-.008610	26
.24	-.009989	-.011344	-.012673	-.013977	-.015257	-.016512	-.017744	-.018952	-.020137	-.021298	23
.25	-.022436	-.023552	-.024646	-.025717	-.026766	-.027794	-.028801	-.029786	-.030751	-.031695	21
.26	-.032618	-.033522	-.034406	-.035270	-.036115	-.036941	-.037748	-.038537	-.039307	-.040059	18
.27	-.040793	-.041510	-.042209	-.042891	-.043557	-.044205	-.044837	-.045453	-.046053	-.046638	16
.28	-.047206	-.047760	-.048298	-.048822	-.049331	-.049825	-.050306	-.050772	-.051224	-.051663	14
.29	-.052089	-.052502	-.052901	-.053288	-.053662	-.054024	-.054374	-.054712	-.055038	-.055353	12
.30	-.055656	-.055948	-.056229	-.056499	-.056759	-.057008	-.057247	-.057476	-.057695	-.057904	10
.31	-.058103	-.058294	-.058475	-.058647	-.058810	-.058964	-.059110	-.059247	-.059376	-.059497	8
.32	-.059610	-.059715	-.059813	-.059903	-.059986	-.060061	-.060130	-.060191	-.060246	-.060294	7
.33	-.060336	-.060371	-.060400	-.060423	-.060440	-.060451	-.060456	-.060456	-.060450	-.060439	5
.34	-.060423	-.060401	-.060375	-.060344	-.060307	-.060267	-.060221	-.060172	-.060117	-.060059	4
.35	-.059997	-.059930	-.059859	-.059785	-.059707	-.059625	-.059540	-.059452	-.059360	-.059264	3
.36	-.059166	-.059064	-.058960	-.058852	-.058742	-.058629	-.058513	-.058394	-.058274	-.058150	2
.37	-.058025	-.057897	-.057766	-.057634	-.057500	-.057363	-.057225	-.057085	-.056943	-.056799	2
.38	-.056653	-.056506	-.056358	-.056208	-.056056	-.055903	-.055749	-.055594	-.055437	-.055279	1
.39	-.055120	-.054960	-.054799	-.054638	-.054475	-.054311	-.054147	-.053982	-.053816	-.053649	1
.40	-.053482	-.053314	-.053146	-.052977	-.052808	-.052638	-.052468	-.052298	-.052127	-.051956	0
.41	-.051785	-.051613	-.051442	-.051270	-.051098	-.050926	-.050754	-.050582	-.050410	-.050238	0
.42	-.050067	-.049895	-.049723	-.049552	-.049380	-.049209	-.049038	-.048868	-.048697	-.048527	0
.43	-.048358	-.048188	-.048019	-.047850	-.047682	-.047514	-.047346	-.047179	-.047013	-.046847	0
.44	-.046681	-.046516	-.046351	-.046187	-.046023	-.045860	-.045698	-.045536	-.045375	-.045214	0
.45	-.045054	-.044894	-.044735	-.044577	-.044420	-.044263	-.044107	-.043951	-.043796	-.043642	0
.46	-.043488	-.043336	-.043184	-.043032	-.042882	-.042732	-.042582	-.042434	-.042286	-.042139	0
.47	-.041993	-.041847	-.041702	-.041558	-.041415	-.041272	-.041130	-.040989	-.040849	-.040709	0
.48	-.040571	-.040432	-.040295	-.040158	-.040023	-.039887	-.039753	-.039619	-.039486	-.039354	0
.49	-.039223	-.039092	-.038962	-.038833	-.038704	-.038576	-.038449	-.038323	-.038197	-.038072	0

TABLE 23

MTF, Defocused Lens, Monochromatic – Δ = 3.5 Rayleigh Units [$\Delta \approx d/2\lambda F^2$]

ν_r	.000	.001	.002	.003	.004	.005	.006	.007	.008	.009	Δ_2
.50	-.037947	-.037824	-.037701	-.037578	-.037456	-.037335	-.037215	-.037095	-.036976	-.036858	0
.51	-.036740	-.036623	-.036506	-.036390	-.036275	-.036160	-.036045	-.035932	-.035819	-.035706	0
.52	-.035594	-.035483	-.035372	-.035261	-.035151	-.035042	-.034933	-.034824	-.034717	-.034609	0
.53	-.034502	-.034395	-.034289	-.034183	-.034078	-.033973	-.033868	-.033764	-.033660	-.033557	0
.54	-.033454	-.033351	-.033249	-.033146	-.033045	-.032943	-.032842	-.032741	-.032640	-.032539	0
.55	-.032439	-.032339	-.032239	-.032139	-.032040	-.031940	-.031841	-.031742	-.031643	-.031544	0
.56	-.031446	-.031347	-.031248	-.031150	-.031051	-.030953	-.030854	-.030756	-.030658	-.030559	0
.57	-.030461	-.030362	-.030264	-.030165	-.030066	-.029968	-.029869	-.029769	-.029670	-.029571	0
.58	-.029471	-.029372	-.029272	-.029172	-.029071	-.028971	-.028870	-.028768	-.028667	-.028565	0
.59	-.028463	-.028361	-.028258	-.028155	-.028052	-.027948	-.027844	-.027739	-.027634	-.027528	0
.60	-.027422	-.027316	-.027209	-.027101	-.026993	-.026885	-.026776	-.026666	-.026556	-.026445	1
.61	-.026334	-.026222	-.026109	-.025996	-.025882	-.025767	-.025652	-.025536	-.025419	-.025302	1
.62	-.025184	-.025065	-.024945	-.024825	-.024703	-.024581	-.024458	-.024334	-.024210	-.024084	1
.63	-.023958	-.023831	-.023702	-.023573	-.023443	-.023312	-.023180	-.023047	-.022913	-.022779	1
.64	-.022643	-.022506	-.022368	-.022229	-.022089	-.021948	-.021805	-.021662	-.021518	-.021372	1
.65	-.021225	-.021078	-.020929	-.020779	-.020627	-.020475	-.020321	-.020167	-.020011	-.019853	1
.66	-.019695	-.019535	-.019374	-.019212	-.019049	-.018884	-.018718	-.018551	-.018382	-.018212	1
.67	-.018041	-.017869	-.017695	-.017520	-.017343	-.017166	-.016987	-.016806	-.016624	-.016441	1
.68	-.016257	-.016071	-.015884	-.015695	-.015505	-.015314	-.015121	-.014927	-.014731	-.014534	1
.69	-.014336	-.014137	-.013936	-.013733	-.013529	-.013324	-.013118	-.012910	-.012700	-.012490	1
.70	-.012277	-.012064	-.011849	-.011633	-.011415	-.011196	-.010976	-.010755	-.010532	-.010307	1
.71	-.010082	-.009855	-.009626	-.009397	-.009166	-.008934	-.008700	-.008466	-.008230	-.007992	1
.72	-.007754	-.007514	-.007273	-.007031	-.006788	-.006543	-.006298	-.006051	-.005803	-.005554	1
.73	-.005304	-.005053	-.004800	-.004547	-.004293	-.004037	-.003781	-.003524	-.003266	-.003006	1
.74	-.002746	-.002485	-.002223	-.001961	-.001697	-.001433	-.001168	-.000902	-.000636	-.000368	1
.75	-.000101	.000168	.000437	.000707	.000977	.001248	.001519	.001790	.002063	.002335	0
.76	.002608	.002881	.003154	.003428	.003702	.003976	.004250	.004525	.004799	.005074	0
.77	.005348	.005623	.005897	.006171	.006445	.006719	.006993	.007266	.007539	.007812	0
.78	.008084	.008356	.008627	.008898	.009168	.009437	.009706	.009974	.010241	.010508	0
.79	.010773	.011038	.011302	.011564	.011826	.012086	.012346	.012604	.012860	.013116	0
.80	.013370	.013622	.013874	.014123	.014371	.014617	.014862	.015105	.015346	.015585	-1
.81	.015823	.016058	.016291	.016523	.016752	.016979	.017203	.017426	.017646	.017863	-1
.82	.018078	.018291	.018501	.018708	.018913	.019115	.019314	.019510	.019704	.019894	-2
.83	.020081	.020266	.020447	.020625	.020799	.020971	.021139	.021304	.021465	.021623	-3
.84	.021777	.021927	.022074	.022218	.022357	.022493	.022624	.022752	.022876	.022996	-3
.85	.023112	.023224	.023331	.023435	.023534	.023629	.023720	.023806	.023888	.023966	-3
.86	.024039	.024107	.024171	.024230	.024285	.024335	.024381	.024422	.024458	.024489	-4
.87	.024515	.024537	.024554	.024566	.024573	.024575	.024572	.024564	.024552	.024534	-4
.88	.024511	.024484	.024451	.024413	.024371	.024323	.024270	.024212	.024149	.024081	-4
.89	.024008	.023930	.023847	.023759	.023666	.023568	.023465	.023357	.023244	.023126	-4
.90	.023003	.022875	.022743	.022606	.022463	.022316	.022165	.022008	.021847	.021682	-4
.91	.021511	.021337	.021157	.020974	.020786	.020593	.020397	.020196	.019991	.019782	-3
.92	.019568	.019351	.019130	.018906	.018677	.018445	.018209	.017970	.017727	.017481	-2
.93	.017232	.016980	.016724	.016466	.016205	.015941	.015674	.015405	.015133	.014859	-1
.94	.014583	.014305	.014024	.013742	.013458	.013173	.012886	.012597	.012308	.012017	0
.95	.011725	.011433	.011140	.010846	.010552	.010258	.009963	.009669	.009375	.009081	0
.96	.008788	.008495	.008204	.007913	.007624	.007336	.007049	.006765	.006482	.006201	2
.97	.005923	.005647	.005374	.005104	.004837	.004574	.004314	.004057	.003805	.003557	4
.98	.003314	.003075	.002841	.002613	.002391	.002174	.001963	.001760	.001563	.001374	7
.99	.001192	.001019	.000855	.000700	.000556	.000434	.000107	.000070	.000038	.000013	7

TABLE 23

MTF, Defocused Lens, Monochromatic – Δ = 4.0 Rayleigh Units [Δ ≈ d/2λF²]

ν_r	.000	.001	.002	.003	.004	.005	.006	.007	.008	.009	Δ_2
.00	1.000000	.998648	.997140	.995477	.993661	.991694	.989578	.987314	.984905	.982352	-143
.01	.979657	.976822	.973849	.970740	.967497	.964122	.960618	.956986	.953228	.949347	-122
.02	.945345	.941224	.936986	.932633	.928168	.923593	.918910	.914121	.909229	.904236	-100
.03	.899144	.893955	.888673	.883298	.877834	.872282	.866646	.860526	.855127	.849249	-77
.04	.843295	.837268	.831170	.825003	.818769	.812471	.806110	.799690	.793212	.786679	-54
.05	.780093	.773455	.766769	.760036	.753259	.746439	.739579	.732681	.725747	.718779	-33
.06	.711780	.704750	.697693	.690610	.683503	.676375	.669226	.662060	.654877	.647681	-13
.07	.640472	.633252	.626024	.618789	.611548	.604305	.597059	.589814	.582570	.575330	3
.08	.568094	.560865	.553643	.546432	.539231	.532043	.524869	.517710	.510568	.503444	18
.09	.496340	.489256	.482195	.475157	.468144	.461156	.454195	.447263	.440360	.433487	30
.10	.426646	.419837	.413062	.406321	.399616	.392947	.386316	.379723	.373169	.366655	40
.11	.360182	.353750	.347361	.341014	.334712	.328454	.322241	.316074	.309953	.303880	47
.12	.297853	.291875	.285946	.280066	.274235	.268455	.262725	.257045	.251418	.245841	52
.13	.240317	.234846	.229426	.224060	.218747	.213488	.208282	.203130	.198032	.192989	54
.14	.188000	.183065	.178186	.173361	.168590	.163875	.159215	.154610	.150060	.145564	55
.15	.141124	.136739	.132409	.128134	.123914	.119748	.115638	.111581	.107580	.103632	54
.16	.099739	.095900	.092114	.088383	.084704	.081079	.077508	.073989	.070522	.067108	52
.17	.063747	.060437	.057178	.053971	.050815	.047710	.044655	.041651	.038696	.035791	50
.18	.032935	.030127	.027369	.024658	.021995	.019380	.016812	.014290	.011815	.009386	46
.19	.007002	.004664	.002370	.000121	-.002084	-.004246	-.006364	-.008440	-.010473	-.012464	42
.20	-.014414	-.016322	-.018190	-.020017	-.021805	-.023553	-.025262	-.026933	-.028565	-.030160	38
.21	-.031717	-.033238	-.034722	-.036170	-.037582	-.038959	-.040302	-.041610	-.042885	-.044125	33
.22	-.045333	-.046509	-.047652	-.048764	-.049844	-.050893	-.051912	-.052901	-.053860	-.054791	29
.23	-.055692	-.056565	-.057410	-.058228	-.059018	-.059782	-.060520	-.061232	-.061918	-.062579	25
.24	-.063215	-.063828	-.064416	-.064981	-.065522	-.066041	-.066538	-.067012	-.067465	-.067897	21
.25	-.068308	-.068699	-.069069	-.069420	-.069751	-.070063	-.070356	-.070632	-.070889	-.071128	18
.26	-.071350	-.071555	-.071744	-.071916	-.072072	-.072213	-.072338	-.072448	-.072544	-.072625	14
.27	-.072692	-.072745	-.072784	-.072811	-.072824	-.072825	-.072814	-.072790	-.072755	-.072708	12
.28	-.072650	-.072581	-.072501	-.072411	-.072310	-.072200	-.072080	-.071950	-.071812	-.071664	9
.29	-.071508	-.071343	-.071170	-.070989	-.070800	-.070603	-.070400	-.070189	-.069971	-.069746	7
.30	-.069515	-.069278	-.069034	-.068785	-.068529	-.068269	-.068002	-.067731	-.067455	-.067174	5
.31	-.066888	-.066598	-.066303	-.066005	-.065702	-.065396	-.065086	-.064772	-.064455	-.064135	3
.32	-.063812	-.063486	-.063157	-.062825	-.062491	-.062155	-.061816	-.061476	-.061133	-.060789	2
.33	-.060442	-.060094	-.059745	-.059394	-.059042	-.058689	-.058335	-.057980	-.057624	-.057267	1
.34	-.056909	-.056551	-.056193	-.055834	-.055475	-.055115	-.054756	-.054396	-.054037	-.053678	0
.35	-.053319	-.052960	-.052601	-.052243	-.051886	-.051529	-.051173	-.050817	-.050462	-.050108	0
.36	-.049755	-.049403	-.049052	-.048702	-.048353	-.048005	-.047659	-.047314	-.046970	-.046627	0
.37	-.046286	-.045946	-.045608	-.045272	-.044937	-.044603	-.044271	-.043941	-.043613	-.043286	-1
.38	-.042962	-.042639	-.042318	-.041999	-.041681	-.041366	-.041053	-.040742	-.040432	-.040125	-1
.39	-.039820	-.039517	-.039216	-.038917	-.038621	-.038326	-.038034	-.037744	-.037456	-.037170	-1
.40	-.036887	-.036606	-.036327	-.036051	-.035776	-.035505	-.035235	-.034968	-.034703	-.034440	-1
.41	-.034180	-.033922	-.033667	-.033414	-.033163	-.032915	-.032669	-.032425	-.032184	-.031945	-1
.42	-.031709	-.031475	-.031243	-.031014	-.030787	-.030563	-.030341	-.030121	-.029904	-.029689	-1
.43	-.029477	-.029267	-.029059	-.028854	-.028651	-.028450	-.028252	-.028056	-.027863	-.027672	-1
.44	-.027483	-.027296	-.027112	-.026930	-.026751	-.026574	-.026399	-.026226	-.026056	-.025888	-1
.45	-.025722	-.025558	-.025397	-.025238	-.025081	-.024926	-.024774	-.024624	-.024476	-.024330	-1
.46	-.024186	-.024044	-.023905	-.023768	-.023632	-.023499	-.023368	-.023239	-.023113	-.022988	-1
.47	-.022865	-.022744	-.022626	-.022509	-.022394	-.022282	-.022171	-.022062	-.021956	-.021851	-1
.48	-.021748	-.021647	-.021548	-.021450	-.021355	-.021262	-.021170	-.021080	-.020992	-.020906	-1
.49	-.020821	-.020739	-.020658	-.020579	-.020501	-.020426	-.020352	-.020279	-.020209	-.020140	-1

TABLE 23

MTF, Defocused Lens, Monochromatic – Δ = 4.0 Rayleigh Units [$\Delta \approx d/2\lambda F^2$]

ν_r	.000	.001	.002	.003	.004	.005	.006	.007	.008	.009	Δ_2
.50	-.020072	-.020007	-.019943	-.019880	-.019819	-.019760	-.019702	-.019646	-.019592	-.019538	-1
.51	-.019487	-.019437	-.019388	-.019341	-.019295	-.019251	-.019208	-.019166	-.019126	-.019087	0
.52	-.019050	-.019014	-.018979	-.018946	-.018913	-.018883	-.018853	-.018825	-.018797	-.018771	0
.53	-.018747	-.018723	-.018701	-.018680	-.018660	-.018641	-.018623	-.018606	-.018590	-.018576	0
.54	-.018562	-.018550	-.018538	-.018528	-.018518	-.018510	-.018502	-.018495	-.018490	-.018485	0
.55	-.018481	-.018478	-.018475	-.018474	-.018473	-.018474	-.018475	-.018476	-.018479	-.018482	0
.56	-.018486	-.018491	-.018496	-.018502	-.018509	-.018516	-.018524	-.018533	-.018542	-.018552	0
.57	-.018562	-.018573	-.018584	-.018596	-.018608	-.018621	-.018634	-.018648	-.018662	-.018676	0
.58	-.018691	-.018706	-.018722	-.018738	-.018754	-.018770	-.018787	-.018804	-.018821	-.018839	0
.59	-.018856	-.018874	-.018892	-.018910	-.018928	-.018946	-.018964	-.018983	-.019001	-.019020	0
.60	-.019038	-.019056	-.019075	-.019093	-.019111	-.019130	-.019148	-.019165	-.019183	-.019201	0
.61	-.019218	-.019235	-.019252	-.019269	-.019285	-.019301	-.019317	-.019332	-.019347	-.019362	0
.62	-.019376	-.019390	-.019403	-.019416	-.019428	-.019440	-.019452	-.019462	-.019473	-.019482	1
.63	-.019491	-.019500	-.019507	-.019515	-.019521	-.019526	-.019531	-.019535	-.019539	-.019541	1
.64	-.019543	-.019544	-.019544	-.019543	-.019541	-.019538	-.019534	-.019529	-.019524	-.019517	1
.65	-.019509	-.019500	-.019490	-.019479	-.019466	-.019453	-.019438	-.019422	-.019405	-.019387	1
.66	-.019367	-.019346	-.019324	-.019301	-.019276	-.019249	-.019222	-.019193	-.019162	-.019130	1
.67	-.019097	-.019062	-.019025	-.018987	-.018947	-.018906	-.018863	-.018819	-.018773	-.018725	2
.68	-.018675	-.018624	-.018571	-.018517	-.018460	-.018402	-.018342	-.018280	-.018216	-.018151	2
.69	-.018083	-.018014	-.017943	-.017870	-.017795	-.017718	-.017639	-.017558	-.017475	-.017390	2
.70	-.017303	-.017214	-.017122	-.017029	-.016934	-.016836	-.016737	-.016635	-.016531	-.016425	2
.71	-.016317	-.016207	-.016094	-.015980	-.015863	-.015743	-.015622	-.015498	-.015373	-.015244	2
.72	-.015114	-.014982	-.014847	-.014709	-.014570	-.014428	-.014284	-.014138	-.013989	-.013838	2
.73	-.013685	-.013530	-.013372	-.013212	-.013050	-.012885	-.012718	-.012549	-.012377	-.012204	2
.74	-.012028	-.011849	-.011669	-.011486	-.011301	-.011113	-.010924	-.010732	-.010538	-.010342	2
.75	-.010144	-.009943	-.009740	-.009536	-.009329	-.009120	-.008909	-.008695	-.008480	-.008263	2
.76	-.008044	-.007822	-.007599	-.007374	-.007147	-.006918	-.006687	-.006454	-.006220	-.005983	2
.77	-.005745	-.005505	-.005264	-.005021	-.004776	-.004530	-.004282	-.004032	-.003781	-.003529	1
.78	-.003275	-.003020	-.002764	-.002506	-.002247	-.001987	-.001725	-.001463	-.001199	-.000935	1
.79	-.000669	-.000403	-.000136	.000133	.000401	.000671	.000941	.001212	.001484	.001756	0
.80	.002028	.002301	.002574	.002847	.003120	.003394	.003668	.003942	.004215	.004489	0
.81	.004762	.005035	.005308	.005581	.005853	.006124	.006395	.006665	.006935	.007203	0
.82	.007471	.007738	.008004	.008269	.008532	.008795	.009056	.009315	.009574	.009830	-1
.83	.010085	.010339	.010590	.010840	.011088	.011333	.011577	.011819	.012058	.012295	-1
.84	.012529	.012762	.012991	.013218	.013442	.013664	.013882	.014098	.014310	.014520	-2
.85	.014726	.014929	.015128	.015325	.015517	.015707	.015892	.016074	.016252	.016426	-3
.86	.016596	.016763	.016925	.017083	.017236	.017386	.017531	.017672	.017808	.017940	-4
.87	.018067	.018189	.018307	.018420	.018528	.018631	.018730	.018823	.018911	.018994	-4
.88	.019072	.019145	.019212	.019274	.019331	.019382	.019428	.019469	.019504	.019533	-5
.89	.019557	.019576	.019589	.019596	.019597	.019593	.019583	.019568	.019547	.019520	-5
.90	.019487	.019449	.019405	.019355	.019299	.019238	.019171	.019098	.019020	.018936	-5
.91	.018846	.018751	.018650	.018544	.018432	.018315	.018192	.018063	.017930	.017791	-4
.92	.017647	.017497	.017343	.017183	.017018	.016849	.016674	.016495	.016311	.016122	-4
.93	.015929	.015731	.015529	.015323	.015112	.014898	.014679	.014456	.014230	.014000	-3
.94	.013767	.013530	.013289	.013046	.012800	.012550	.012298	.012043	.011786	.011527	-1
.95	.011265	.011001	.010736	.010469	.010200	.009930	.009658	.009386	.009113	.008839	0
.96	.008564	.008290	.008015	.007740	.007466	.007192	.006919	.006647	.006375	.006106	1
.97	.005837	.005571	.005306	.005044	.004784	.004527	.004273	.004022	.003775	.003532	4
.98	.003292	.003057	.002826	.002601	.002381	.002166	.001957	.001755	.001559	.001371	7
.99	.001190	.001018	.000854	.000700	.000556	.000434	.000107	.000070	.000038	.000013	7

TABLE 23

MTF, Defocused Lens, Monochromatic – Δ = 4.5 Rayleigh Units [Δ ≈ d/2λF²]

ν_T	.000	.001	.002	.003	.004	.005	.006	.007	.008	.009	Δ_2
.00	1.000000	.998627	.997057	.995290	.993331	.991180	.988840	.986314	.983604	.980712	-180
.01	.977642	.974395	.970974	.967383	.963623	.959698	.955610	.951362	.946958	.942399	-153
.02	.937690	.932833	.927831	.922687	.917404	.911985	.906434	.900753	.894946	.889016	-122
.03	.882966	.876799	.870518	.864127	.857629	.851027	.844323	.837523	.830627	.823641	-90
.04	.816566	.809407	.802165	.794846	.787451	.779983	.772447	.764844	.757179	.749454	-59
.05	.741672	.733836	.725949	.718015	.710036	.702014	.693954	.685858	.677729	.669569	-29
.06	.661382	.653170	.644936	.636683	.628412	.620128	.611832	.603527	.595216	.586901	-3
.07	.578585	.570270	.561957	.553651	.545353	.537065	.528789	.520528	.512284	.504059	19
.08	.495855	.487674	.479517	.471388	.463288	.455218	.447181	.439178	.431211	.423281	38
.09	.415391	.407542	.399735	.391973	.384255	.376585	.368963	.361390	.353869	.346400	52
.10	.338984	.331623	.324318	.317069	.309879	.302747	.295676	.288665	.281717	.274831	63
.11	.268008	.261250	.254557	.247930	.241370	.234876	.228451	.222093	.215805	.209586	69
.12	.203437	.197359	.191351	.185415	.179549	.173756	.168035	.162386	.156810	.151307	73
.13	.145876	.140519	.135235	.130025	.124887	.119823	.114833	.109916	.105072	.100301	73
.14	.095604	.090980	.086429	.081950	.077544	.073211	.068950	.064760	.060643	.056597	71
.15	.052621	.048717	.044883	.041120	.037426	.033801	.030246	.026759	.023340	.019989	68
.16	.016705	.013488	.010338	.007253	.004233	.001278	-.001612	-.004439	-.007202	-.009903	63
.17	-.012542	-.015119	-.017635	-.020091	-.022487	-.024824	-.027102	-.029322	-.031485	-.033592	57
.18	-.035642	-.037637	-.039576	-.041462	-.043294	-.045074	-.046801	-.048476	-.050101	-.051675	50
.19	-.053200	-.054675	-.056103	-.057483	-.058816	-.060102	-.061343	-.062539	-.063691	-.064799	44
.20	-.065864	-.066887	-.067868	-.068808	-.069707	-.070567	-.071388	-.072171	-.072916	-.073624	37
.21	-.074295	-.074931	-.075531	-.076097	-.076629	-.077127	-.077593	-.078027	-.078430	-.078801	31
.22	-.079142	-.079454	-.079737	-.079991	-.080218	-.080417	-.080589	-.080735	-.080856	-.080952	25
.23	-.081023	-.081070	-.081094	-.081095	-.081074	-.081031	-.080967	-.080882	-.080777	-.080652	20
.24	-.080508	-.080346	-.080165	-.079966	-.079750	-.079517	-.079268	-.079003	-.078722	-.078427	15
.25	-.078117	-.077793	-.077455	-.077104	-.076740	-.076363	-.075975	-.075574	-.075163	-.074740	11
.26	-.074307	-.073864	-.073412	-.072949	-.072478	-.071998	-.071510	-.071013	-.070509	-.069997	7
.27	-.069479	-.068953	-.068422	-.067884	-.067340	-.066791	-.066236	-.065676	-.065112	-.064543	4
.28	-.063970	-.063393	-.062813	-.062229	-.061642	-.061052	-.060459	-.059863	-.059266	-.058666	2
.29	-.058065	-.057462	-.056857	-.056252	-.055645	-.055037	-.054429	-.053821	-.053212	-.052603	0
.30	-.051994	-.051385	-.050776	-.050168	-.049561	-.048955	-.048349	-.047745	-.047142	-.046540	0
.31	-.045940	-.045341	-.044744	-.044149	-.043556	-.042966	-.042377	-.041790	-.041206	-.040625	-1
.32	-.040046	-.039470	-.038896	-.038326	-.037758	-.037194	-.036632	-.036074	-.035519	-.034967	-2
.33	-.034418	-.033873	-.033332	-.032794	-.032260	-.031729	-.031202	-.030679	-.030159	-.029644	-3
.34	-.029132	-.028624	-.028121	-.027621	-.027125	-.026634	-.026146	-.025663	-.025183	-.024708	-3
.35	-.024238	-.023771	-.023309	-.022850	-.022397	-.021947	-.021502	-.021061	-.020624	-.020192	-3
.36	-.019764	-.019340	-.018921	-.018506	-.018096	-.017690	-.017288	-.016890	-.016497	-.016109	-3
.37	-.015724	-.015344	-.014968	-.014597	-.014230	-.013867	-.013509	-.013155	-.012805	-.012460	-3
.38	-.012118	-.011781	-.011448	-.011120	-.010795	-.010475	-.010159	-.009847	-.009540	-.009236	-3
.39	-.008936	-.008641	-.008350	-.008062	-.007779	-.007500	-.007224	-.006953	-.006686	-.006422	-3
.40	-.006162	-.005907	-.005655	-.005407	-.005162	-.004922	-.004685	-.004452	-.004222	-.003997	-3
.41	-.003775	-.003556	-.003341	-.003130	-.002922	-.002718	-.002518	-.002320	-.002127	-.001936	-2
.42	-.001749	-.001566	-.001385	-.001208	-.001035	-.000864	-.000697	-.000533	-.000373	-.000215	-2
.43	-.000061	.000091	.000239	.000384	.000526	.000666	.000802	.000935	.001065	.001193	-2
.44	.001317	.001439	.001558	.001674	.001787	.001898	.002006	.002111	.002213	.002313	-2
.45	.002410	.002504	.002596	.002686	.002773	.002857	.002939	.003018	.003095	.003169	-1
.46	.003241	.003311	.003378	.003443	.003506	.003567	.003625	.003681	.003734	.003786	-1
.47	.003835	.003882	.003927	.003970	.004010	.004049	.004085	.004120	.004152	.004183	-1
.48	.004211	.004238	.004262	.004285	.004305	.004324	.004341	.004356	.004369	.004380	-1
.49	.004390	.004397	.004403	.004407	.004410	.004410	.004409	.004406	.004402	.004396	-1

TABLE 23

MTF, Defocused Lens, Monochromatic – $\Delta = 4.5$ Rayleigh Units $[\Delta \approx d/2\lambda F^2]$

ν_r	.000	.001	.002	.003	.004	.005	.006	.007	.008	.009	Δ_2
.50	.004388	.004378	.004367	.004355	.004340	.004324	.004307	.004288	.004267	.004245	-1
.51	.004222	.004196	.004170	.004142	.004112	.004081	.004049	.004015	.003980	.003943	0
.52	.003905	.003865	.003825	.003783	.003739	.003694	.003648	.003601	.003552	.003502	0
.53	.003451	.003398	.003345	.003290	.003233	.003176	.003117	.003058	.002997	.002934	0
.54	.002871	.002807	.002741	.002674	.002607	.002538	.002468	.002396	.002324	.002251	0
.55	.002177	.002101	.002025	.001948	.001869	.001790	.001709	.001628	.001546	.001462	0
.56	.001378	.001293	.001207	.001120	.001032	.000943	.000853	.000763	.000671	.000579	0
.57	.000486	.000392	.000297	.000201	.000105	.000008	-.000090	-.000189	-.000289	-.000389	0
.58	-.000490	-.000592	-.000694	-.000797	-.000901	-.001005	-.001110	-.001216	-.001323	-.001430	0
.59	-.001537	-.001646	-.001755	-.001864	-.001974	-.002084	-.002196	-.002307	-.002419	-.002532	0
.60	-.002645	-.002758	-.002873	-.002987	-.003102	-.003217	-.003333	-.003449	-.003565	-.003682	0
.61	-.003799	-.003917	-.004034	-.004152	-.004271	-.004389	-.004508	-.004627	-.004746	-.004865	0
.62	-.004985	-.005105	-.005224	-.005344	-.005464	-.005585	-.005705	-.005825	-.005945	-.006066	0
.63	-.006186	-.006306	-.006426	-.006546	-.006666	-.006786	-.006906	-.007025	-.007145	-.007264	0
.64	-.007383	-.007501	-.007620	-.007738	-.007856	-.007973	-.008090	-.008207	-.008323	-.008439	0
.65	-.008555	-.008670	-.008784	-.008898	-.009011	-.009124	-.009236	-.009348	-.009459	-.009569	1
.66	-.009678	-.009787	-.009895	-.010002	-.010109	-.010214	-.010319	-.010423	-.010525	-.010627	1
.67	-.010728	-.010828	-.010927	-.011025	-.011121	-.011217	-.011311	-.011404	-.011496	-.011587	1
.68	-.011676	-.011765	-.011851	-.011937	-.012021	-.012103	-.012185	-.012264	-.012342	-.012419	2
.69	-.012494	-.012567	-.012639	-.012709	-.012777	-.012844	-.012909	-.012972	-.013033	-.013093	2
.70	-.013150	-.013206	-.013259	-.013311	-.013360	-.013408	-.013453	-.013497	-.013538	-.013577	2
.71	-.013614	-.013648	-.013681	-.013711	-.013739	-.013764	-.013787	-.013808	-.013826	-.013841	2
.72	-.013855	-.013865	-.013874	-.013879	-.013882	-.013882	-.013880	-.013875	-.013867	-.013857	3
.73	-.013844	-.013828	-.013809	-.013788	-.013763	-.013736	-.013706	-.013673	-.013637	-.013598	3
.74	-.013556	-.013511	-.013463	-.013412	-.013358	-.013300	-.013240	-.013177	-.013111	-.013041	3
.75	-.012968	-.012893	-.012814	-.012732	-.012646	-.012558	-.012466	-.012371	-.012273	-.012172	3
.76	-.012067	-.011960	-.011849	-.011735	-.011617	-.011497	-.011373	-.011246	-.011115	-.010982	3
.77	-.010845	-.010706	-.010563	-.010416	-.010267	-.010114	-.009959	-.009800	-.009638	-.009473	3
.78	-.009305	-.009134	-.008960	-.008783	-.008603	-.008420	-.008234	-.008045	-.007853	-.007659	3
.79	-.007461	-.007261	-.007058	-.006852	-.006644	-.006433	-.006220	-.006004	-.005785	-.005564	2
.80	-.005341	-.005115	-.004887	-.004656	-.004424	-.004189	-.003952	-.003713	-.003472	-.003229	2
.81	-.002984	-.002738	-.002490	-.002240	-.001988	-.001735	-.001480	-.001224	-.000967	-.000708	1
.82	-.000448	-.000187	.000074	.000337	.000601	.000866	.001131	.001397	.001663	.001930	0
.83	.002197	.002465	.002733	.003001	.003268	.003536	.003804	.004071	.004338	.004604	0
.84	.004870	.005135	.005400	.005663	.005926	.006188	.006448	.006707	.006965	.007221	0
.85	.007476	.007729	.007980	.008229	.008477	.008722	.008965	.009205	.009444	.009679	-2
.86	.009912	.010143	.010370	.010595	.010816	.011035	.011250	.011461	.011669	.011874	-3
.87	.012075	.012272	.012465	.012655	.012840	.013021	.013198	.013370	.013538	.013702	-4
.88	.013860	.014014	.014164	.014308	.014447	.014582	.014711	.014835	.014953	.015066	-4
.89	.015174	.015277	.015373	.015464	.015550	.015629	.015703	.015771	.015832	.015888	-5
.90	.015938	.015982	.016020	.016051	.016076	.016095	.016108	.016115	.016115	.016109	-5
.91	.016097	.016078	.016053	.016022	.015984	.015940	.015889	.015833	.015770	.015700	-5
.92	.015625	.015543	.015455	.015361	.015261	.015155	.015042	.014924	.014800	.014670	-5
.93	.014535	.014393	.014246	.014094	.013936	.013773	.013604	.013430	.013252	.013068	-4
.94	.012880	.012687	.012489	.012287	.012080	.011869	.011655	.011436	.011214	.010988	-3
.95	.010759	.010526	.010291	.010052	.009811	.009567	.009321	.009072	.008822	.008570	-1
.96	.008316	.008061	.007805	.007548	.007290	.007032	.006773	.006514	.006256	.005998	0
.97	.005741	.005485	.005230	.004976	.004725	.004475	.004228	.003983	.003741	.003503	3
.98	.003268	.003036	.002809	.002587	.002369	.002157	.001950	.001749	.001555	.001368	7
.99	.001188	.001016	.000853	.000699	.000556	.000434	.000107	.000070	.000038	.000013	7

TABLE 23

MTF, Defocused Lens, Monochromatic – Δ = 5.0 Rayleigh Units [$\Delta \approx d/2\lambda F^2$]

ν_r	.000	.001	.002	.003	.004	.005	.006	.007	.008	.009	Δ_2
.00	1.000000	.998604	.996963	.995082	.992961	.990605	.988015	.985196	.982151	.978882	-222
.01	.975393	.971687	.967768	.963640	.959305	.954768	.950033	.945103	.939982	.934674	-186
.02	.929183	.923513	.917669	.911653	.905471	.899126	.892623	.885966	.879160	.872207	-145
.03	.865113	.857882	.850519	.843027	.835411	.827676	.819825	.811862	.803794	.795622	-101
.04	.787353	.778989	.770536	.761998	.753379	.744682	.735914	.727076	.718174	.709212	-59
.05	.700193	.691123	.682004	.672841	.663638	.654398	.645126	.635825	.626499	.617151	-20
.06	.607787	.598408	.589018	.579622	.570222	.560822	.551425	.542034	.532654	.523286	13
.07	.513934	.504602	.495292	.486006	.476749	.467522	.458329	.449172	.440054	.430977	41
.08	.421944	.412957	.404019	.395133	.386299	.377521	.368801	.360140	.351541	.343005	64
.09	.334535	.326132	.317799	.309536	.301345	.293228	.285187	.277223	.269337	.261530	80
.10	.253804	.246161	.238600	.231124	.223733	.216429	.209211	.202082	.195041	.188091	90
.11	.181230	.174461	.167783	.161198	.154705	.148305	.141999	.135786	.129668	.123644	94
.12	.117715	.111881	.106141	.100497	.094948	.089494	.084135	.078871	.073702	.068627	95
.13	.063648	.058762	.053971	.049273	.044669	.040158	.035740	.031414	.027180	.023037	91
.14	.018986	.015024	.011153	.007371	.003677	.000071	-.003447	-.006878	-.010224	-.013484	85
.15	-.016659	-.019751	-.022760	-.025686	-.028531	-.031295	-.033979	-.036585	-.039112	-.041562	77
.16	-.043935	-.046233	-.048457	-.050606	-.052683	-.054688	-.056622	-.058486	-.060282	-.062008	68
.17	-.063668	-.065262	-.066791	-.068255	-.069656	-.070995	-.072272	-.073490	-.074648	-.075747	58
.18	-.076790	-.077776	-.078706	-.079582	-.080405	-.081175	-.081894	-.082563	-.083182	-.083752	49
.19	-.084275	-.084751	-.085181	-.085567	-.085909	-.086208	-.086465	-.086681	-.086857	-.086994	39
.20	-.087092	-.087153	-.087178	-.087167	-.087122	-.087042	-.086930	-.086785	-.086609	-.086403	30
.21	-.086167	-.085902	-.085609	-.085289	-.084943	-.084570	-.084173	-.083752	-.083308	-.082840	23
.22	-.082351	-.081841	-.081310	-.080760	-.080190	-.079602	-.078996	-.078373	-.077734	-.077079	16
.23	-.076409	-.075724	-.075025	-.074313	-.073588	-.072851	-.072102	-.071343	-.070572	-.069792	10
.24	-.069003	-.068204	-.067397	-.066582	-.065760	-.064931	-.064095	-.063253	-.062406	-.061553	5
.25	-.060696	-.059834	-.058969	-.058100	-.057228	-.056353	-.055476	-.054596	-.053716	-.052834	1
.26	-.051951	-.051067	-.050183	-.049299	-.048416	-.047533	-.046651	-.045770	-.044890	-.044013	-1
.27	-.043137	-.042263	-.041392	-.040524	-.039659	-.038796	-.037938	-.037082	-.036231	-.035383	-3
.28	-.034539	-.033700	-.032866	-.032035	-.031210	-.030390	-.029575	-.028765	-.027960	-.027161	-5
.29	-.026368	-.025581	-.024799	-.024023	-.023254	-.022490	-.021733	-.020982	-.020238	-.019500	-6
.30	-.018769	-.018044	-.017327	-.016616	-.015912	-.015215	-.014525	-.013841	-.013165	-.012496	-6
.31	-.011834	-.011180	-.010532	-.009892	-.009259	-.008633	-.008015	-.007403	-.006799	-.006203	-6
.32	-.005613	-.005031	-.004457	-.003889	-.003329	-.002776	-.002231	-.001692	-.001161	-.000637	-6
.33	-.000120	.000389	.000891	.001387	.001875	.002356	.002830	.003297	.003757	.004210	-6
.34	.004656	.005095	.005528	.005954	.006373	.006785	.007190	.007589	.007982	.008368	-6
.35	.008747	.009120	.009487	.009847	.010201	.010549	.010890	.011226	.011555	.011879	-5
.36	.012196	.012508	.012813	.013113	.013407	.013696	.013979	.014256	.014528	.014794	-4
.37	.015055	.015311	.015561	.015806	.016046	.016281	.016511	.016736	.016956	.017171	-4
.38	.017381	.017587	.017787	.017983	.018175	.018362	.018544	.018723	.018896	.019066	-3
.39	.019231	.019392	.019549	.019702	.019850	.019995	.020136	.020273	.020406	.020536	-3
.40	.020662	.020784	.020902	.021017	.021128	.021236	.021341	.021442	.021540	.021635	-2
.41	.021726	.021815	.021900	.021982	.022061	.022138	.022211	.022281	.022349	.022414	-2
.42	.022476	.022535	.022592	.022646	.022697	.022746	.022793	.022837	.022878	.022917	-1
.43	.022954	.022989	.023021	.023051	.023079	.023105	.023129	.023150	.023170	.023187	-1
.44	.023203	.023216	.023228	.023238	.023246	.023252	.023256	.023259	.023260	.023259	-1
.45	.023256	.023252	.023246	.023239	.023230	.023219	.023207	.023194	.023179	.023162	0
.46	.023145	.023125	.023105	.023083	.023060	.023035	.023009	.022982	.022954	.022924	0
.47	.022893	.022861	.022828	.022794	.022758	.022722	.022684	.022646	.022606	.022565	0
.48	.022523	.022480	.022436	.022391	.022345	.022299	.022251	.022202	.022152	.022102	0
.49	.022050	.021998	.021944	.021890	.021835	.021779	.021723	.021665	.021607	.021548	0

TABLE 23

MTF, Defocused Lens, Monochromatic – Δ = 5.0 Rayleigh Units [$\Delta \approx d/2\lambda F^2$]

ν_r	.000	.001	.002	.003	.004	.005	.006	.007	.008	.009	Δ_2
.50	.021488	.021427	.021365	.021303	.021240	.021176	.021111	.021045	.020979	.020912	0
.51	.020844	.020776	.020707	.020637	.020566	.020495	.020423	.020350	.020276	.020202	0
.52	.020127	.020052	.019975	.019898	.019820	.019742	.019663	.019583	.019503	.019421	0
.53	.019340	.019257	.019174	.019090	.019005	.018920	.018834	.018747	.018660	.018572	0
.54	.018483	.018394	.018304	.018213	.018121	.018029	.017936	.017843	.017749	.017654	0
.55	.017558	.017462	.017365	.017267	.017169	.017069	.016970	.016869	.016768	.016666	0
.56	.016563	.016460	.016355	.016251	.016145	.016039	.015932	.015824	.015715	.015606	0
.57	.015496	.015385	.015274	.015161	.015048	.014935	.014820	.014705	.014589	.014472	0
.58	.014354	.014236	.014117	.013997	.013876	.013755	.013633	.013510	.013386	.013261	0
.59	.013136	.013010	.012883	.012755	.012627	.012498	.012367	.012237	.012105	.011973	0
.60	.011839	.011705	.011570	.011435	.011299	.011161	.011023	.010885	.010745	.010605	0
.61	.010464	.010322	.010179	.010036	.009892	.009747	.009601	.009455	.009307	.009159	0
.62	.009011	.008861	.008711	.008560	.008408	.008256	.008103	.007949	.007795	.007639	0
.63	.007483	.007327	.007170	.007012	.006853	.006694	.006534	.006374	.006213	.006051	0
.64	.005889	.005726	.005562	.005398	.005234	.005069	.004903	.004737	.004570	.004403	0
.65	.004236	.004068	.003899	.003731	.003561	.003392	.003222	.003052	.002881	.002710	0
.66	.002539	.002367	.002195	.002023	.001851	.001679	.001506	.001334	.001161	.000988	0
.67	.000815	.000642	.000469	.000296	.000123	-.000050	-.000222	-.000395	-.000568	-.000740	0
.68	-.000912	-.001084	-.001256	-.001427	-.001598	-.001769	-.001939	-.002109	-.002279	-.002448	0
.69	-.002616	-.002784	-.002951	-.003118	-.003284	-.003449	-.003614	-.003778	-.003941	-.004103	1
.70	-.004264	-.004425	-.004584	-.004742	-.004900	-.005056	-.005211	-.005365	-.005518	-.005670	1
.71	-.005820	-.005969	-.006116	-.006263	-.006407	-.006550	-.006692	-.006832	-.006971	-.007107	2
.72	-.007243	-.007376	-.007507	-.007637	-.007765	-.007890	-.008014	-.008136	-.008256	-.008373	2
.73	-.008489	-.008602	-.008713	-.008821	-.008927	-.009031	-.009133	-.009231	-.009328	-.009421	3
.74	-.009513	-.009601	-.009687	-.009770	-.009850	-.009927	-.010002	-.010073	-.010141	-.010207	3
.75	-.010269	-.010329	-.010385	-.010438	-.010487	-.010534	-.010577	-.010617	-.010653	-.010686	3
.76	-.010716	-.010742	-.010764	-.010783	-.010799	-.010810	-.010818	-.010823	-.010823	-.010820	4
.77	-.010813	-.010803	-.010788	-.010770	-.010748	-.010722	-.010692	-.010658	-.010620	-.010578	4
.78	-.010532	-.010482	-.010428	-.010370	-.010308	-.010242	-.010172	-.010098	-.010019	-.009937	4
.79	-.009851	-.009760	-.009666	-.009567	-.009464	-.009358	-.009247	-.009132	-.009013	-.008891	4
.80	-.008764	-.008633	-.008498	-.008360	-.008217	-.008071	-.007920	-.007766	-.007608	-.007447	4
.81	-.007281	-.007112	-.006940	-.006763	-.006583	-.006400	-.006213	-.006023	-.005830	-.005633	3
.82	-.005433	-.005230	-.005023	-.004814	-.004602	-.004386	-.004168	-.003947	-.003724	-.003498	3
.83	-.003269	-.003038	-.002804	-.002568	-.002330	-.002090	-.001848	-.001604	-.001358	-.001111	2
.84	-.000861	-.000611	-.000359	-.000105	.000149	.000405	.000662	.000919	.001178	.001437	1
.85	.001696	.001956	.002216	.002477	.002737	.002997	.003257	.003517	.003776	.004035	0
.86	.004293	.004550	.004806	.005061	.005314	.005567	.005817	.006066	.006313	.006559	-1
.87	.006802	.007043	.007282	.007518	.007751	.007982	.008210	.008435	.008656	.008875	-2
.88	.009090	.009301	.009509	.009712	.009912	.010108	.010300	.010487	.010670	.010848	-4
.89	.011021	.011190	.011354	.011512	.011666	.011814	.011957	.012095	.012226	.012352	-5
.90	.012473	.012587	.012696	.012798	.012895	.012985	.013069	.013146	.013217	.013282	-5
.91	.013340	.013391	.013436	.013474	.013506	.013530	.013548	.013559	.013563	.013560	-6
.92	.013550	.013534	.013510	.013480	.013442	.013398	.013347	.013289	.013225	.013153	-6
.93	.013075	.012990	.012898	.012800	.012695	.012584	.012467	.012343	.012213	.012077	-5
.94	.011935	.011787	.011634	.011474	.011309	.011139	.010963	.010783	.010597	.010407	-4
.95	.010212	.010012	.009808	.009600	.009388	.009172	.008953	.008730	.008505	.008276	-2
.96	.008045	.007811	.007575	.007337	.007097	.006855	.006613	.006369	.006125	.005880	0
.97	.005635	.005390	.005146	.004902	.004659	.004417	.004177	.003939	.003704	.003471	3
.98	.003240	.003014	.002790	.002571	.002356	.002147	.001942	.001743	.001550	.001364	7
.99	.001185	.001014	.000852	.000698	.000555	.000434	.000107	.000070	.000038	.000013	7

TABLE 23

MTF, Defocused Lens, Monochromatic – Δ = 6.0 Rayleigh Units [Δ ≈ d/2λF²]

ν_r	.000	.001	.002	.003	.004	.005	.006	.007	.008	.009	Δ_2
.00	1.000000	.998550	.996748	.994598	.992106	.989274	.986108	.982612	.978792	.974652	-318
.01	.970198	.965436	.960371	.955009	.949356	.943418	.937201	.930713	.923959	.916945	-258
.02	.909680	.902169	.894419	.886438	.878232	.869808	.861175	.852338	.843305	.834084	-188
.03	.824682	.815106	.805363	.795461	.785408	.775210	.764875	.754410	.743823	.733121	-114
.04	.722312	.711401	.700397	.689307	.678138	.666897	.655590	.644225	.632809	.621347	-44
.05	.609848	.598317	.586761	.575186	.563598	.552005	.540411	.528823	.517247	.505689	17
.06	.494153	.482647	.471174	.459741	.448353	.437014	.425730	.414505	.403344	.392252	69
.07	.381234	.370292	.359432	.348658	.337974	.327382	.316888	.306494	.296204	.286022	107
.08	.275949	.265990	.256147	.246422	.236820	.227341	.217988	.208764	.199670	.190709	133
.09	.181883	.173192	.164640	.156226	.147953	.139823	.131835	.123991	.116292	.108739	146
.10	.101333	.094074	.086962	.079998	.073183	.066516	.059998	.053629	.047408	.041335	148
.11	.035411	.029635	.024006	.018524	.013188	.007998	.002953	-.001947	-.006705	-.011320	142
.12	-.015793	-.020126	-.024320	-.028375	-.032294	-.036077	-.039725	-.043241	-.046625	-.049880	130
.13	-.053005	-.056004	-.058877	-.061626	-.064253	-.066760	-.069148	-.071419	-.073575	-.075617	113
.14	-.077548	-.079369	-.081082	-.082688	-.084191	-.085592	-.086892	-.088094	-.089199	-.090210	95
.15	-.091128	-.091955	-.092694	-.093346	-.093914	-.094398	-.094802	-.095127	-.095375	-.095548	75
.16	-.095647	-.095676	-.095636	-.095528	-.095355	-.095118	-.094820	-.094462	-.094046	-.093574	56
.17	-.093047	-.092468	-.091839	-.091160	-.090434	-.089663	-.088848	-.087991	-.087093	-.086157	39
.18	-.085183	-.084174	-.083130	-.082055	-.080948	-.079812	-.078647	-.077457	-.076241	-.075001	24
.19	-.073739	-.072456	-.071154	-.069833	-.068495	-.067141	-.065772	-.064390	-.062996	-.061590	11
.20	-.060175	-.058751	-.057319	-.055880	-.054435	-.052985	-.051532	-.050076	-.048617	-.047158	1
.21	-.045699	-.044239	-.042782	-.041327	-.039874	-.038425	-.036981	-.035542	-.034109	-.032682	-5
.22	-.031262	-.029850	-.028446	-.027051	-.025665	-.024290	-.022924	-.021570	-.020226	-.018895	-11
.23	-.017576	-.016269	-.014975	-.013695	-.012428	-.011175	-.009937	-.008713	-.007504	-.006310	-14
.24	-.005131	-.003968	-.002821	-.001690	-.000575	.000524	.001606	.002672	.003721	.004753	-16
.25	.005768	.006766	.007746	.008710	.009656	.010586	.011497	.012392	.013269	.014129	-16
.26	.014971	.015797	.016604	.017395	.018169	.018925	.019664	.020387	.021092	.021780	-16
.27	.022452	.023107	.023746	.024368	.024973	.025563	.026136	.026694	.027236	.027762	-15
.28	.028272	.028767	.029247	.029712	.030162	.030597	.031017	.031423	.031815	.032193	-13
.29	.032557	.032907	.033243	.033566	.033876	.034173	.034458	.034729	.034989	.035236	-11
.30	.035470	.035693	.035905	.036105	.036293	.036471	.036638	.036794	.036939	.037074	-9
.31	.037199	.037315	.037420	.037516	.037602	.037680	.037748	.037807	.037858	.037901	-7
.32	.037935	.037961	.037980	.037991	.037994	.037990	.037978	.037960	.037935	.037903	-6
.33	.037865	.037820	.037770	.037713	.037651	.037583	.037509	.037430	.037346	.037257	-4
.34	.037163	.037064	.036961	.036854	.036742	.036626	.036505	.036381	.036254	.036123	-3
.35	.035988	.035850	.035708	.035564	.035417	.035267	.035114	.034958	.034800	.034640	-1
.36	.034477	.034312	.034146	.033977	.033806	.033634	.033460	.033285	.033108	.032929	0
.37	.032750	.032569	.032387	.032204	.032021	.031836	.031651	.031465	.031278	.031091	0
.38	.030903	.030715	.030527	.030339	.030150	.029961	.029772	.029583	.029395	.029206	0
.39	.029018	.028830	.028642	.028454	.028267	.028081	.027894	.027709	.027524	.027340	1
.40	.027156	.026973	.026791	.026610	.026429	.026249	.026071	.025893	.025716	.025540	1
.41	.025366	.025192	.025020	.024848	.024678	.024509	.024341	.024175	.024009	.023845	1
.42	.023683	.023521	.023361	.023202	.023045	.022889	.022735	.022581	.022430	.022280	1
.43	.022131	.021984	.021838	.021694	.021551	.021410	.021270	.021132	.020995	.020860	2
.44	.020727	.020595	.020464	.020335	.020208	.020083	.019959	.019836	.019715	.019596	2
.45	.019478	.019362	.019248	.019135	.019024	.018914	.018806	.018699	.018595	.018491	2
.46	.018389	.018289	.018191	.018094	.017998	.017905	.017812	.017722	.017632	.017545	2
.47	.017459	.017374	.017291	.017210	.017130	.017051	.016975	.016899	.016825	.016753	1
.48	.016682	.016612	.016544	.016478	.016413	.016349	.016287	.016226	.016166	.016108	1
.49	.016051	.015996	.015942	.015890	.015838	.015789	.015740	.015693	.015647	.015602	1

TABLE 23

MTF, Defocused Lens, Monochromatic – Δ = 6.0 Rayleigh Units [$\Delta \approx d/2\lambda F^2$]

ν_r	.000	.001	.002	.003	.004	.005	.006	.007	.008	.009	Δ_2
.50	.015559	.015517	.015476	.015436	.015398	.015361	.015325	.015290	.015257	.015225	1
.51	.015194	.015164	.015135	.015107	.015081	.015055	.015031	.015007	.014985	.014964	1
.52	.014944	.014925	.014907	.014890	.014874	.014859	.014844	.014831	.014819	.014808	1
.53	.014797	.014788	.014779	.014771	.014764	.014758	.014753	.014748	.014745	.014742	1
.54	.014740	.014738	.014737	.014738	.014738	.014740	.014742	.014744	.014748	.014752	1
.55	.014756	.014762	.014767	.014774	.014781	.014788	.014796	.014804	.014813	.014822	0
.56	.014832	.014842	.014853	.014863	.014875	.014886	.014898	.014911	.014923	.014936	0
.57	.014949	.014962	.014976	.014990	.015004	.015018	.015032	.015046	.015060	.015075	0
.58	.015090	.015104	.015119	.015133	.015148	.015163	.015177	.015191	.015206	.015220	0
.59	.015234	.015248	.015261	.015275	.015288	.015301	.015314	.015326	.015338	.015350	0
.60	.015361	.015372	.015383	.015393	.015402	.015411	.015420	.015428	.015436	.015443	0
.61	.015449	.015455	.015460	.015464	.015468	.015471	.015473	.015475	.015475	.015475	0
.62	.015474	.015472	.015470	.015466	.015461	.015456	.015449	.015442	.015433	.015424	0
.63	.015413	.015401	.015388	.015374	.015359	.015342	.015324	.015305	.015285	.015264	0
.64	.015241	.015216	.015191	.015164	.015135	.015105	.015074	.015041	.015007	.014971	−1
.65	.014934	.014895	.014854	.014812	.014768	.014722	.014675	.014626	.014576	.014524	−1
.66	.014469	.014414	.014356	.014296	.014235	.014172	.014107	.014040	.013971	.013901	−1
.67	.013828	.013753	.013677	.013598	.013518	.013436	.013351	.013265	.013176	.013086	−1
.68	.012993	.012899	.012802	.012703	.012603	.012500	.012395	.012288	.012179	.012068	−1
.69	.011954	.011839	.011722	.011602	.011480	.011357	.011231	.011103	.010973	.010841	−1
.70	.010707	.010571	.010433	.010293	.010151	.010007	.009861	.009713	.009563	.009411	−1
.71	.009258	.009102	.008944	.008785	.008624	.008461	.008296	.008130	.007961	.007792	−1
.72	.007620	.007447	.007272	.007096	.006918	.006739	.006558	.006376	.006193	.006008	0
.73	.005822	.005635	.005446	.005257	.005066	.004874	.004682	.004488	.004294	.004099	0
.74	.003903	.003706	.003509	.003311	.003113	.002914	.002715	.002515	.002315	.002116	0
.75	.001916	.001716	.001516	.001316	.001116	.000917	.000718	.000519	.000321	.000123	1
.76	−.000074	−.000270	−.000465	−.000659	−.000853	−.001045	−.001236	−.001426	−.001615	−.001802	1
.77	−.001988	−.002172	−.002354	−.002535	−.002714	−.002890	−.003065	−.003238	−.003408	−.003576	2
.78	−.003742	−.003905	−.004065	−.004223	−.004378	−.004530	−.004679	−.004825	−.004968	−.005108	3
.79	−.005244	−.005377	−.005507	−.005633	−.005755	−.005874	−.005988	−.006099	−.006206	−.006308	4
.80	−.006407	−.006501	−.006591	−.006676	−.006757	−.006834	−.006906	−.006973	−.007036	−.007093	5
.81	−.007146	−.007194	−.007237	−.007275	−.007308	−.007335	−.007358	−.007375	−.007387	−.007393	5
.82	−.007395	−.007391	−.007381	−.007366	−.007346	−.007320	−.007288	−.007251	−.007209	−.007161	6
.83	−.007108	−.007048	−.006984	−.006914	−.006838	−.006757	−.006670	−.006578	−.006481	−.006378	5
.84	−.006270	−.006156	−.006037	−.005913	−.005784	−.005650	−.005510	−.005366	−.005217	−.005063	5
.85	−.004904	−.004740	−.004572	−.004399	−.004222	−.004041	−.003855	−.003666	−.003472	−.003275	4
.86	−.003073	−.002868	−.002660	−.002448	−.002234	−.002016	−.001795	−.001571	−.001345	−.001116	2
.87	−.000885	−.000652	−.000417	−.000180	.000059	.000299	.000540	.000782	.001026	.001270	1
.88	.001514	.001759	.002004	.002249	.002493	.002737	.002981	.003224	.003465	.003705	0
.89	.003944	.004181	.004417	.004650	.004880	.005109	.005334	.005557	.005777	.005993	−2
.90	.006206	.006414	.006620	.006820	.007017	.007209	.007396	.007579	.007756	.007929	−4
.91	.008095	.008257	.008412	.008562	.008705	.008842	.008973	.009097	.009215	.009326	−6
.92	.009430	.009527	.009616	.009699	.009774	.009842	.009902	.009955	.010000	.010037	−7
.93	.010066	.010088	.010102	.010107	.010105	.010095	.010077	.010052	.010018	.009976	−7
.94	.009927	.009870	.009805	.009732	.009652	.009564	.009469	.009367	.009257	.009141	−6
.95	.009017	.008887	.008750	.008607	.008457	.008302	.008140	.007973	.007800	.007622	−4
.96	.007439	.007252	.007060	.006863	.006663	.006459	.006251	.006041	.005828	.005612	−1
.97	.005394	.005175	.004954	.004732	.004509	.004286	.004063	.003840	.003618	.003397	1
.98	.003178	.002961	.002747	.002535	.002327	.002123	.001924	.001729	.001539	.001356	6
.99	.001180	.001010	.000849	.000697	.000554	.000433	.000107	.000069	.000038	.000013	7

TABLE 23

MTF, Defocused Lens, Monochromatic – Δ = 7.0 Rayleigh Units [Δ ≈ d/2λF²]

ν_r	.000	.001	.002	.003	.004	.005	.006	.007	.008	.009	Δ_2
.00	1.000000	.998486	.996493	.994028	.991095	.987703	.983856	.979563	.974831	.969668	-430
.01	.964082	.958082	.951677	.944875	.937685	.930119	.922185	.913894	.905256	.896281	-335
.02	.886981	.877365	.867446	.857234	.846740	.835977	.824954	.813685	.802179	.790450	-223
.03	.778509	.766367	.754036	.741528	.728855	.716027	.703057	.689956	.676736	.663408	-107
.04	.649983	.636472	.622887	.609238	.595537	.581793	.568018	.554221	.540413	.526604	0
.05	.512804	.499023	.485269	.471553	.457882	.444267	.430715	.417235	.403835	.390523	88
.06	.377307	.364195	.351193	.338308	.325548	.312918	.300425	.288076	.275875	.263829	154
.07	.251942	.240220	.228667	.217288	.206087	.195069	.184236	.173593	.163143	.152889	196
.08	.142834	.132980	.123330	.113886	.104650	.095624	.086808	.078205	.069816	.061640	214
.09	.053680	.045935	.038406	.031093	.023996	.017114	.010447	.003994	-.002245	-.008271	213
.10	-.014086	-.019690	-.025085	-.030273	-.035255	-.040033	-.044608	-.048984	-.053162	-.057144	196
.11	-.060933	-.064531	-.067941	-.071166	-.074207	-.077069	-.079753	-.082263	-.084602	-.086772	168
.12	-.088778	-.090622	-.092308	-.093838	-.095217	-.096447	-.097532	-.098475	-.099281	-.099951	135
.13	-.100490	-.100901	-.101188	-.101355	-.101404	-.101339	-.101164	-.100883	-.100498	-.100014	100
.14	-.099433	-.098759	-.097996	-.097147	-.096215	-.095204	-.094117	-.092958	-.091728	-.090433	66
.15	-.089074	-.087656	-.086180	-.084651	-.083071	-.081443	-.079770	-.078055	-.076300	-.074509	37
.16	-.072684	-.070828	-.068943	-.067032	-.065097	-.063142	-.061167	-.059176	-.057171	-.055154	12
.17	-.053127	-.051092	-.049051	-.047007	-.044961	-.042914	-.040870	-.038829	-.036794	-.034766	-6
.18	-.032746	-.030736	-.028737	-.026752	-.024780	-.022824	-.020886	-.018965	-.017063	-.015181	-19
.19	-.013321	-.011484	-.009669	-.007879	-.006113	-.004374	-.002660	-.000975	.000683	.002312	-28
.20	.003913	.005483	.007023	.008533	.010012	.011459	.012874	.014258	.015609	.016927	-32
.21	.018213	.019466	.020686	.021872	.023026	.024146	.025232	.026286	.027306	.028293	-32
.22	.029247	.030168	.031056	.031911	.032734	.033525	.034283	.035010	.035705	.036368	-30
.23	.037001	.037603	.038175	.038716	.039228	.039710	.040163	.040588	.040984	.041353	-27
.24	.041694	.042009	.042296	.042558	.042794	.043004	.043190	.043352	.043489	.043604	-22
.25	.043695	.043764	.043811	.043837	.043841	.043825	.043789	.043734	.043659	.043566	-18
.26	.043454	.043325	.043179	.043016	.042837	.042643	.042433	.042208	.041969	.041716	-13
.27	.041449	.041170	.040878	.040574	.040259	.039932	.039595	.039247	.038889	.038522	-9
.28	.038145	.037760	.037367	.036966	.036557	.036141	.035718	.035289	.034854	.034413	-5
.29	.033967	.033515	.033059	.032599	.032135	.031667	.031195	.030721	.030243	.029763	-2
.30	.029281	.028797	.028311	.027824	.027335	.026846	.026355	.025865	.025374	.024883	0
.31	.024392	.023902	.023412	.022923	.022435	.021949	.021463	.020979	.020497	.020017	2
.32	.019539	.019063	.018589	.018118	.017649	.017183	.016720	.016260	.015803	.015349	3
.33	.014899	.014452	.014008	.013568	.013132	.012700	.012271	.011846	.011426	.011009	4
.34	.010596	.010188	.009784	.009384	.008988	.008597	.008210	.007828	.007450	.007077	5
.35	.006708	.006344	.005984	.005629	.005279	.004933	.004592	.004256	.003924	.003597	5
.36	.003274	.002956	.002643	.002335	.002031	.001732	.001437	.001147	.000862	.000581	5
.37	.000305	.000034	-.000233	-.000496	-.000754	-.001007	-.001257	-.001501	-.001742	-.001978	4
.38	-.002209	-.002437	-.002660	-.002879	-.003094	-.003304	-.003510	-.003713	-.003911	-.004105	4
.39	-.004296	-.004482	-.004665	-.004843	-.005018	-.005189	-.005356	-.005520	-.005680	-.005836	4
.40	-.005989	-.006138	-.006284	-.006426	-.006565	-.006701	-.006833	-.006962	-.007088	-.007210	3
.41	-.007330	-.007446	-.007559	-.007669	-.007776	-.007881	-.007982	-.008081	-.008176	-.008269	3
.42	-.008359	-.008447	-.008532	-.008614	-.008694	-.008771	-.008845	-.008918	-.008987	-.009055	2
.43	-.009120	-.009183	-.009243	-.009301	-.009357	-.009411	-.009463	-.009513	-.009560	-.009606	2
.44	-.009649	-.009691	-.009731	-.009769	-.009805	-.009839	-.009871	-.009902	-.009931	-.009958	2
.45	-.009983	-.010007	-.010029	-.010050	-.010069	-.010086	-.010102	-.010117	-.010130	-.010142	1
.46	-.010152	-.010161	-.010168	-.010174	-.010179	-.010182	-.010185	-.010186	-.010185	-.010184	1
.47	-.010181	-.010177	-.010172	-.010166	-.010159	-.010151	-.010141	-.010131	-.010119	-.010106	1
.48	-.010093	-.010078	-.010063	-.010046	-.010028	-.010010	-.009991	-.009970	-.009949	-.009927	1
.49	-.009904	-.009880	-.009855	-.009829	-.009803	-.009775	-.009747	-.009718	-.009688	-.009658	1

TABLE 23

MTF, Defocused Lens, Monochromatic – Δ = 7.0 Rayleigh Units [$\Delta \approx d/2\lambda F^2$]

ν_r	.000	.001	.002	.003	.004	.005	.006	.007	.008	.009	Δ:
.50	-.009626	-.009594	-.009561	-.009528	-.009493	-.009458	-.009422	-.009385	-.009348	-.009309	1
.51	-.009270	-.009231	-.009190	-.009149	-.009107	-.009064	-.009021	-.008977	-.008932	-.008887	1
.52	-.008841	-.008794	-.008746	-.008698	-.008649	-.008600	-.008549	-.008498	-.008447	-.008394	1
.53	-.008341	-.008287	-.008233	-.008178	-.008122	-.008065	-.008008	-.007950	-.007891	-.007832	1
.54	-.007772	-.007711	-.007649	-.007587	-.007524	-.007460	-.007396	-.007331	-.007265	-.007198	1
.55	-.007131	-.007063	-.006994	-.006925	-.006855	-.006784	-.006712	-.006640	-.006566	-.006493	1
.56	-.006418	-.006342	-.006266	-.006189	-.006111	-.006033	-.005954	-.005874	-.005793	-.005711	1
.57	-.005629	-.005546	-.005462	-.005377	-.005291	-.005205	-.005118	-.005030	-.004941	-.004852	1
.58	-.004761	-.004670	-.004578	-.004486	-.004392	-.004298	-.004203	-.004107	-.004010	-.003913	1
.59	-.003814	-.003715	-.003615	-.003515	-.003413	-.003311	-.003208	-.003104	-.002999	-.002894	1
.60	-.002788	-.002681	-.002573	-.002464	-.002355	-.002245	-.002135	-.002023	-.001911	-.001798	1
.61	-.001684	-.001570	-.001455	-.001339	-.001223	-.001106	-.000988	-.000869	-.000750	-.000631	1
.62	-.000510	-.000389	-.000268	-.000146	-.000023	.000100	.000224	.000348	.000473	.000598	0
.63	.000723	.000850	.000976	.001103	.001230	.001358	.001486	.001615	.001743	.001872	0
.64	.002001	.002131	.002261	.002390	.002520	.002651	.002781	.002911	.003041	.003172	0
.65	.003302	.003432	.003563	.003693	.003823	.003953	.004082	.004212	.004341	.004469	0
.66	.004598	.004726	.004854	.004981	.005107	.005233	.005359	.005484	.005608	.005732	0
.67	.005855	.005977	.006098	.006218	.006338	.006456	.006573	.006690	.006805	.006919	0
.68	.007032	.007144	.007254	.007363	.007471	.007577	.007682	.007785	.007887	.007987	-1
.69	.008085	.008182	.008277	.008370	.008461	.008550	.008637	.008722	.008805	.008886	-1
.70	.008965	.009041	.009116	.009187	.009257	.009324	.009388	.009450	.009510	.009566	-2
.71	.009621	.009672	.009720	.009766	.009809	.009849	.009886	.009920	.009951	.009978	-2
.72	.010003	.010025	.010043	.010058	.010070	.010078	.010083	.010085	.010083	.010078	-2
.73	.010069	.010057	.010041	.010021	.009998	.009972	.009942	.009908	.009870	.009829	-3
.74	.009784	.009735	.009683	.009627	.009567	.009503	.009436	.009365	.009290	.009211	-3
.75	.009129	.009043	.008953	.008860	.008762	.008662	.008557	.008449	.008337	.008222	-3
.76	.008104	.007981	.007856	.007727	.007594	.007458	.007319	.007177	.007032	.006883	-2
.77	.006732	.006577	.006420	.006260	.006097	.005931	.005763	.005592	.005418	.005243	-1
.78	.005065	.004885	.004702	.004518	.004332	.004144	.003954	.003763	.003571	.003377	0
.79	.003182	.002985	.002788	.002590	.002391	.002192	.001992	.001792	.001591	.001391	0
.80	.001190	.000990	.000790	.000591	.000392	.000194	-.000003	-.000199	-.000394	-.000587	1
.81	-.000779	-.000969	-.001157	-.001343	-.001527	-.001709	-.001888	-.002065	-.002239	-.002410	3
.82	-.002578	-.002742	-.002903	-.003061	-.003215	-.003365	-.003512	-.003654	-.003792	-.003925	4
.83	-.004054	-.004178	-.004298	-.004412	-.004522	-.004626	-.004725	-.004819	-.004907	-.004989	6
.84	-.005066	-.005136	-.005201	-.005260	-.005313	-.005359	-.005399	-.005433	-.005461	-.005481	6
.85	-.005496	-.005504	-.005505	-.005499	-.005487	-.005468	-.005442	-.005410	-.005371	-.005325	7
.86	-.005272	-.005212	-.005146	-.005073	-.004994	-.004908	-.004815	-.004716	-.004611	-.004499	6
.87	-.004381	-.004256	-.004126	-.003990	-.003848	-.003700	-.003547	-.003388	-.003223	-.003054	5
.88	-.002880	-.002701	-.002517	-.002329	-.002136	-.001939	-.001739	-.001534	-.001327	-.001116	3
.89	-.000902	-.000685	-.000466	-.000244	-.000020	.000205	.000432	.000661	.000890	.001120	1
.90	.001351	.001582	.001813	.002043	.002273	.002502	.002730	.002956	.003180	.003403	-1
.91	.003623	.003840	.004055	.004266	.004474	.004678	.004878	.005074	.005266	.005452	-4
.92	.005633	.005809	.005980	.006144	.006303	.006455	.006601	.006740	.006872	.006997	-6
.93	.007114	.007224	.007327	.007421	.007508	.007586	.007656	.007718	.007771	.007816	-8
.94	.007852	.007879	.007898	.007908	.007909	.007901	.007885	.007859	.007825	.007782	-8
.95	.007731	.007671	.007602	.007525	.007440	.007347	.007246	.007137	.007020	.006896	-6
.96	.006765	.006627	.006482	.006331	.006174	.006011	.005842	.005669	.005490	.005307	-4
.97	.005119	.004928	.004734	.004536	.004336	.004134	.003930	.003724	.003518	.003312	0
.98	.003106	.002900	.002696	.002494	.002293	.002096	.001902	.001712	.001527	.001347	5
.99	.001173	.001006	.000846	.000695	.000553	.000432	.000107	.000069	.000038	.000013	7

TABLE 23

MTF, Defocused Lens, Monochromatic – Δ = 8.0 Rayleigh Units [Δ ≈ d/2λF²]

ν_r	.000	.001	.002	.003	.004	.005	.006	.007	.008	.009	Δ_2
.00	1.000000	.998412	.996199	.993369	.989930	.985892	.981262	.976053	.970275	.963939	-557
.01	.957058	.949644	.941710	.933270	.924339	.914929	.905058	.894739	.883988	.872822	-415
.02	.861256	.849307	.836992	.824327	.811331	.798019	.784409	.770519	.756365	.741967	-244
.03	.727340	.712503	.697473	.682267	.666903	.651398	.635768	.620032	.604204	.588303	-73
.04	.572344	.556343	.540316	.524279	.508247	.492234	.476256	.460327	.444461	.428671	76
.05	.412970	.397372	.381890	.366534	.351317	.336251	.321345	.306611	.292059	.277698	191
.06	.263537	.249586	.235852	.222344	.209069	.196033	.183245	.170709	.158432	.146419	264
.07	.134674	.123203	.112010	.101097	.090469	.080128	.070076	.060316	.050850	.041678	295
.08	.032802	.024223	.015940	.007954	.000264	-.007131	-.014232	-.021039	-.027555	-.033781	290
.09	-.039720	-.045374	-.050746	-.055838	-.060654	-.065198	-.069473	-.073482	-.077230	-.080720	257
.10	-.083958	-.086948	-.089694	-.092201	-.094474	-.096518	-.098339	-.099941	-.101330	-.102511	208
.11	-.103490	-.104273	-.104864	-.105271	-.105498	-.105551	-.105436	-.105158	-.104725	-.104140	151
.12	-.103411	-.102543	-.101541	-.100412	-.099161	-.097794	-.096316	-.094734	-.093051	-.091275	94
.13	-.089410	-.087462	-.085436	-.083337	-.081170	-.078940	-.076652	-.074312	-.071923	-.069490	44
.14	-.067019	-.064513	-.061976	-.059413	-.056829	-.054226	-.051609	-.048982	-.046349	-.043712	3
.15	-.041077	-.038444	-.035819	-.033204	-.030603	-.028017	-.025450	-.022905	-.020383	-.017889	-25
.16	-.015423	-.012988	-.010586	-.008219	-.005889	-.003599	-.001348	.000860	.003025	.005144	-44
.17	.007218	.009244	.011221	.013149	.015027	.016853	.018628	.020349	.022018	.023633	-53
.18	.025193	.026699	.028151	.029547	.030889	.032175	.033407	.034583	.035704	.036771	-54
.19	.037783	.038741	.039646	.040497	.041295	.042040	.042733	.043376	.043967	.044508	-49
.20	.045000	.045443	.045838	.046186	.046488	.046744	.046956	.047123	.047248	.047331	-41
.21	.047372	.047374	.047336	.047260	.047147	.046998	.046813	.046595	.046343	.046058	-31
.22	.045743	.045398	.045023	.044620	.044190	.043734	.043253	.042747	.042219	.041669	-21
.23	.041097	.040505	.039894	.039265	.038618	.037955	.037276	.036583	.035877	.035157	-12
.24	.034426	.033683	.032930	.032168	.031397	.030619	.029833	.029041	.028244	.027442	-4
.25	.026635	.025826	.025013	.024199	.023383	.022566	.021749	.020933	.020117	.019303	2
.26	.018491	.017681	.016875	.016072	.015273	.014479	.013690	.012906	.012127	.011355	6
.27	.010589	.009830	.009078	.008334	.007598	.006869	.006149	.005438	.004735	.004042	9
.28	.003358	.002683	.002019	.001364	.000719	.000084	-.000540	-.001154	-.001757	-.002349	11
.29	-.002931	-.003501	-.004061	-.004610	-.005147	-.005674	-.006189	-.006693	-.007186	-.007667	11
.30	-.008138	-.008597	-.009045	-.009481	-.009907	-.010322	-.010725	-.011117	-.011499	-.011869	11
.31	-.012229	-.012578	-.012916	-.013244	-.013561	-.013868	-.014165	-.014451	-.014727	-.014993	10
.32	-.015249	-.015496	-.015733	-.015960	-.016178	-.016387	-.016586	-.016777	-.016958	-.017131	9
.33	-.017295	-.017451	-.017599	-.017738	-.017869	-.017993	-.018108	-.018216	-.018316	-.018410	7
.34	-.018495	-.018574	-.018646	-.018712	-.018770	-.018823	-.018868	-.018908	-.018942	-.018970	6
.35	-.018992	-.019008	-.019019	-.019025	-.019025	-.019021	-.019011	-.018997	-.018978	-.018955	5
.36	-.018927	-.018895	-.018859	-.018819	-.018775	-.018728	-.018676	-.018621	-.018563	-.018502	3
.37	-.018437	-.018370	-.018299	-.018226	-.018150	-.018072	-.017991	-.017907	-.017822	-.017734	2
.38	-.017644	-.017552	-.017458	-.017363	-.017266	-.017167	-.017067	-.016965	-.016862	-.016758	1
.39	-.016652	-.016545	-.016438	-.016329	-.016220	-.016110	-.015999	-.015887	-.015775	-.015662	1
.40	-.015549	-.015436	-.015322	-.015208	-.015093	-.014979	-.014864	-.014750	-.014635	-.014521	0
.41	-.014406	-.014292	-.014178	-.014064	-.013950	-.013837	-.013724	-.013612	-.013500	-.013388	0
.42	-.013277	-.013167	-.013057	-.012948	-.012840	-.012732	-.012625	-.012518	-.012413	-.012308	0
.43	-.012204	-.012101	-.011999	-.011897	-.011797	-.011698	-.011599	-.011502	-.011405	-.011310	0
.44	-.011215	-.011122	-.011029	-.010938	-.010848	-.010759	-.010671	-.010584	-.010498	-.010413	0
.45	-.010330	-.010248	-.010166	-.010086	-.010008	-.009930	-.009854	-.009778	-.009704	-.009631	0
.46	-.009560	-.009489	-.009420	-.009352	-.009285	-.009220	-.009155	-.009092	-.009030	-.008969	0
.47	-.008910	-.008852	-.008794	-.008739	-.008684	-.008630	-.008578	-.008527	-.008477	-.008428	0
.48	-.008381	-.008334	-.008289	-.008245	-.008202	-.008161	-.008120	-.008081	-.008042	-.008005	0
.49	-.007969	-.007934	-.007901	-.007868	-.007836	-.007806	-.007776	-.007748	-.007721	-.007694	0

TABLE 23

MTF, Defocused Lens, Monochromatic – $\Delta = 8.0$ Rayleigh Units $[\Delta \approx d/2\lambda F^2]$

ν_r	.000	.001	.002	.003	.004	.005	.006	.007	.008	.009	Δ_2
.50	-.007669	-.007645	-.007622	-.007600	-.007579	-.007559	-.007539	-.007521	-.007504	-.007488	0
.51	-.007473	-.007458	-.007445	-.007432	-.007421	-.007410	-.007400	-.007391	-.007383	-.007376	0
.52	-.007369	-.007364	-.007359	-.007355	-.007351	-.007349	-.007347	-.007346	-.007346	-.007346	0
.53	-.007348	-.007349	-.007352	-.007355	-.007359	-.007363	-.007368	-.007374	-.007380	-.007387	0
.54	-.007395	-.007403	-.007411	-.007420	-.007429	-.007439	-.007450	-.007461	-.007472	-.007484	0
.55	-.007496	-.007508	-.007521	-.007534	-.007548	-.007562	-.007576	-.007590	-.007605	-.007619	0
.56	-.007635	-.007650	-.007665	-.007681	-.007697	-.007713	-.007729	-.007745	-.007761	-.007777	0
.57	-.007793	-.007809	-.007825	-.007842	-.007858	-.007874	-.007889	-.007905	-.007921	-.007936	0
.58	-.007951	-.007966	-.007981	-.007996	-.008010	-.008024	-.008037	-.008050	-.008063	-.008076	0
.59	-.008088	-.008099	-.008110	-.008121	-.008131	-.008140	-.008149	-.008157	-.008165	-.008172	1
.60	-.008178	-.008184	-.008189	-.008193	-.008196	-.008198	-.008200	-.008201	-.008201	-.008200	1
.61	-.008198	-.008195	-.008191	-.008186	-.008180	-.008173	-.008165	-.008156	-.008146	-.008134	1
.62	-.008121	-.008107	-.008092	-.008076	-.008058	-.008039	-.008018	-.007996	-.007973	-.007948	1
.63	-.007922	-.007895	-.007866	-.007835	-.007803	-.007769	-.007734	-.007697	-.007658	-.007618	2
.64	-.007576	-.007533	-.007487	-.007440	-.007392	-.007341	-.007289	-.007235	-.007179	-.007121	2
.65	-.007062	-.007000	-.006937	-.006872	-.006804	-.006735	-.006665	-.006592	-.006517	-.006440	2
.66	-.006361	-.006281	-.006198	-.006113	-.006027	-.005938	-.005848	-.005755	-.005661	-.005564	2
.67	-.005466	-.005365	-.005263	-.005159	-.005052	-.004944	-.004834	-.004722	-.004608	-.004492	2
.68	-.004375	-.004255	-.004134	-.004011	-.003886	-.003759	-.003631	-.003500	-.003368	-.003235	2
.69	-.003100	-.002963	-.002825	-.002685	-.002544	-.002401	-.002257	-.002112	-.001965	-.001817	1
.70	-.001668	-.001517	-.001366	-.001213	-.001059	-.000905	-.000749	-.000593	-.000436	-.000278	1
.71	-.000120	.000039	.000199	.000359	.000519	.000680	.000841	.001002	.001163	.001324	0
.72	.001485	.001646	.001806	.001967	.002126	.002286	.002445	.002603	.002760	.002916	0
.73	.003072	.003226	.003379	.003532	.003682	.003832	.003979	.004125	.004270	.004412	-1
.74	.004553	.004692	.004828	.004963	.005095	.005224	.005351	.005476	.005597	.005716	-2
.75	.005832	.005945	.006055	.006162	.006265	.006365	.006462	.006555	.006644	.006729	-3
.76	.006811	.006889	.006962	.007032	.007097	.007159	.007215	.007268	.007316	.007359	-4
.77	.007398	.007432	.007462	.007487	.007506	.007521	.007531	.007536	.007536	.007531	-4
.78	.007521	.007506	.007486	.007461	.007430	.007394	.007353	.007307	.007256	.007199	-4
.79	.007138	.007071	.006999	.006922	.006840	.006753	.006662	.006565	.006463	.006357	-4
.80	.006245	.006130	.006009	.005884	.005755	.005621	.005484	.005342	.005196	.005046	-3
.81	.004892	.004735	.004575	.004411	.004243	.004073	.003900	.003724	.003545	.003364	-1
.82	.003181	.002996	.002809	.002620	.002429	.002238	.002045	.001851	.001656	.001461	0
.83	.001266	.001071	.000876	.000681	.000487	.000293	.000101	-.000091	-.000280	-.000468	2
.84	-.000654	-.000838	-.001020	-.001198	-.001374	-.001547	-.001717	-.001883	-.002045	-.002204	4
.85	-.002358	-.002508	-.002653	-.002793	-.002928	-.003058	-.003183	-.003302	-.003415	-.003523	6
.86	-.003624	-.003718	-.003807	-.003889	-.003963	-.004032	-.004093	-.004147	-.004193	-.004232	7
.87	-.004264	-.004289	-.004305	-.004314	-.004316	-.004309	-.004295	-.004273	-.004243	-.004205	8
.88	-.004160	-.004107	-.004046	-.003977	-.003901	-.003817	-.003725	-.003627	-.003521	-.003408	7
.89	-.003287	-.003160	-.003027	-.002887	-.002740	-.002587	-.002429	-.002265	-.002095	-.001920	5
.90	-.001740	-.001555	-.001366	-.001172	-.000975	-.000774	-.000570	-.000363	-.000153	.000059	2
.91	.000273	.000489	.000706	.000923	.001142	.001360	.001579	.001797	.002013	.002229	0
.92	.002443	.002655	.002864	.003071	.003275	.003475	.003671	.003862	.004050	.004232	-4
.93	.004409	.004580	.004746	.004905	.005057	.005203	.005342	.005473	.005596	.005711	-7
.94	.005819	.005917	.006008	.006089	.006162	.006225	.006279	.006324	.006360	.006385	-9
.95	.006402	.006408	.006405	.006393	.006371	.006339	.006297	.006247	.006187	.006118	-8
.96	.006040	.005953	.005857	.005753	.005641	.005521	.005394	.005259	.005117	.004969	-6
.97	.004814	.004654	.004488	.004317	.004142	.003963	.003780	.003594	.003406	.003215	-1
.98	.003024	.002831	.002638	.002446	.002254	.002065	.001877	.001693	.001512	.001336	4
.99	.001165	.001000	.000842	.000693	.000552	.000432	.000106	.000069	.000038	.000013	7

TABLE 23

MTF, Defocused Lens, Monochromatic – Δ = 9.0 Rayleigh Units [Δ ≈ d/2λF²]

ν_r	.000	.001	.002	.003	.004	.005	.006	.007	.008	.009	Δ_2
.00	1.000000	.998328	.995866	.992624	.988611	.983842	.978328	.972086	.965128	.957474	-697
.01	.949138	.940140	.930499	.920233	.909365	.897913	.885901	.873350	.860284	.846724	-492
.02	.832696	.818223	.803328	.788037	.772375	.756366	.740034	.723406	.706507	.689360	-246
.03	.671992	.654427	.636690	.618805	.600797	.582689	.564505	.546269	.528003	.509730	-6
.04	.491472	.473250	.455087	.437001	.419014	.401145	.383412	.365835	.348430	.331215	190
.05	.314206	.297418	.280867	.264567	.248532	.232774	.217305	.202138	.187283	.172750	322
.06	.158549	.144687	.131174	.118015	.105219	.092790	.080734	.069055	.057758	.046844	384
.07	.036318	.026181	.016434	.007077	-.001888	-.010463	-.018648	-.026445	-.033856	-.040883	384
.08	-.047530	-.053800	-.059696	-.065224	-.070389	-.075194	-.079646	-.083750	-.087513	-.090942	335
.09	-.094042	-.096822	-.099288	-.101448	-.103310	-.104881	-.106170	-.107186	-.107936	-.108430	257
.10	-.108675	-.108681	-.108456	-.108010	-.107350	-.106488	-.105430	-.104186	-.102766	-.101178	168
.11	-.099430	-.097533	-.095493	-.093321	-.091025	-.088613	-.086093	-.083474	-.080764	-.077971	83
.12	-.075103	-.072167	-.069171	-.066122	-.063027	-.059894	-.056730	-.053540	-.050331	-.047110	12
.13	-.043883	-.040655	-.037432	-.034219	-.031023	-.027847	-.024696	-.021576	-.018491	-.015444	-38
.14	-.012440	-.009482	-.006575	-.003721	-.000923	.001814	.004490	.007100	.009643	.012117	-69
.15	.014519	.016848	.019101	.021279	.023378	.025399	.027339	.029199	.030978	.032674	-81
.16	.034288	.035820	.037269	.038636	.039920	.041122	.042243	.043282	.044240	.045119	-79
.17	.045919	.046640	.047285	.047853	.048347	.048767	.049114	.049391	.049598	.049737	-67
.18	.049809	.049817	.049761	.049644	.049467	.049232	.048940	.048594	.048195	.047746	-50
.19	.047247	.046701	.046110	.045475	.044799	.044083	.043329	.042540	.041716	.040860	-31
.20	.039974	.039060	.038118	.037152	.036162	.035151	.034121	.033072	.032007	.030927	-14
.21	.029834	.028730	.027616	.026493	.025363	.024228	.023088	.021946	.020802	.019658	0
.22	.018515	.017375	.016238	.015105	.013979	.012859	.011747	.010644	.009550	.008468	11
.23	.007396	.006337	.005291	.004259	.003242	.002240	.001254	.000284	-.000669	-.001604	18
.24	-.002521	-.003419	-.004299	-.005159	-.005999	-.006819	-.007619	-.008399	-.009157	-.009895	21
.25	-.010611	-.011306	-.011979	-.012631	-.013261	-.013869	-.014456	-.015021	-.015564	-.016085	22
.26	-.016585	-.017063	-.017520	-.017956	-.018370	-.018763	-.019135	-.019487	-.019818	-.020129	20
.27	-.020420	-.020691	-.020943	-.021175	-.021389	-.021583	-.021759	-.021917	-.022058	-.022180	18
.28	-.022286	-.022374	-.022447	-.022503	-.022543	-.022567	-.022577	-.022572	-.022552	-.022518	14
.29	-.022471	-.022410	-.022336	-.022250	-.022152	-.022041	-.021919	-.021786	-.021642	-.021488	10
.30	-.021323	-.021149	-.020966	-.020773	-.020572	-.020362	-.020145	-.019920	-.019687	-.019448	7
.31	-.019202	-.018949	-.018691	-.018426	-.018157	-.017882	-.017602	-.017318	-.017030	-.016737	4
.32	-.016441	-.016142	-.015839	-.015534	-.015225	-.014915	-.014602	-.014288	-.013971	-.013653	1
.33	-.013334	-.013014	-.012693	-.012372	-.012050	-.011727	-.011405	-.011083	-.010761	-.010440	0
.34	-.010119	-.009799	-.009480	-.009162	-.008845	-.008529	-.008215	-.007903	-.007592	-.007284	-1
.35	-.006977	-.006672	-.006369	-.006069	-.005771	-.005475	-.005182	-.004892	-.004604	-.004319	-2
.36	-.004037	-.003758	-.003482	-.003209	-.002938	-.002671	-.002407	-.002146	-.001889	-.001635	-2
.37	-.001384	-.001136	-.000892	-.000651	-.000413	-.000179	.000051	.000279	.000502	.000723	-2
.38	.000939	.001153	.001363	.001569	.001772	.001971	.002167	.002360	.002549	.002734	-2
.39	.002917	.003095	.003271	.003443	.003611	.003777	.003939	.004097	.004253	.004405	-2
.40	.004554	.004700	.004842	.004982	.005118	.005251	.005382	.005509	.005633	.005754	-2
.41	.005873	.005988	.006100	.006210	.006317	.006421	.006523	.006621	.006718	.006811	-2
.42	.006902	.006990	.007076	.007159	.007240	.007319	.007395	.007468	.007540	.007609	-1
.43	.007676	.007741	.007803	.007864	.007922	.007978	.008032	.008085	.008135	.008183	-1
.44	.008230	.008274	.008317	.008358	.008397	.008434	.008470	.008504	.008536	.008567	-1
.45	.008596	.008623	.008649	.008674	.008697	.008718	.008738	.008757	.008774	.008790	0
.46	.008805	.008818	.008830	.008841	.008850	.008858	.008865	.008871	.008876	.008880	0
.47	.008882	.008884	.008884	.008883	.008881	.008879	.008875	.008870	.008864	.008858	0
.48	.008850	.008841	.008832	.008822	.008810	.008798	.008785	.008771	.008757	.008741	0
.49	.008725	.008708	.008690	.008671	.008652	.008632	.008611	.008589	.008567	.008543	0

TABLE 23

MTF, Defocused Lens, Monochromatic – $\Delta = 9.0$ Rayleigh Units $[\Delta \approx d/2\lambda F^2]$

ν_T	.000	.001	.002	.003	.004	.005	.006	.007	.008	.009	Δ_2
.50	.008520	.008495	.008470	.008444	.008417	.008390	.008362	.008333	.008303	.008273	0
.51	.008242	.008211	.008179	.008146	.008113	.008079	.008044	.008009	.007973	.007936	0
.52	.007899	.007861	.007822	.007783	.007743	.007702	.007661	.007619	.007577	.007534	0
.53	.007490	.007446	.007400	.007355	.007308	.007261	.007214	.007165	.007116	.007066	0
.54	.007016	.006965	.006913	.006861	.006808	.006754	.006700	.006644	.006589	.006532	0
.55	.006475	.006417	.006358	.006299	.006239	.006178	.006117	.006054	.005991	.005928	0
.56	.005863	.005798	.005732	.005666	.005598	.005530	.005461	.005392	.005321	.005250	0
.57	.005178	.005106	.005032	.004958	.004883	.004808	.004731	.004654	.004576	.004497	0
.58	.004418	.004338	.004257	.004175	.004092	.004009	.003925	.003840	.003755	.003669	0
.59	.003582	.003494	.003406	.003317	.003227	.003136	.003045	.002953	.002860	.002767	0
.60	.002673	.002579	.002483	.002387	.002291	.002194	.002096	.001998	.001899	.001799	0
.61	.001699	.001599	.001498	.001396	.001294	.001192	.001089	.000986	.000882	.000778	0
.62	.000674	.000569	.000464	.000358	.000253	.000147	.000041	-.000066	-.000172	-.000279	0
.63	-.000385	-.000492	-.000599	-.000705	-.000812	-.000919	-.001025	-.001132	-.001238	-.001344	0
.64	-.001450	-.001555	-.001660	-.001765	-.001870	-.001974	-.002077	-.002180	-.002283	-.002384	1
.65	-.002486	-.002586	-.002686	-.002785	-.002883	-.002980	-.003076	-.003171	-.003265	-.003359	1
.66	-.003451	-.003541	-.003631	-.003719	-.003806	-.003891	-.003975	-.004058	-.004139	-.004218	2
.67	-.004296	-.004372	-.004446	-.004518	-.004589	-.004657	-.004724	-.004788	-.004851	-.004911	2
.68	-.004969	-.005025	-.005078	-.005129	-.005177	-.005224	-.005267	-.005308	-.005346	-.005382	3
.69	-.005415	-.005445	-.005472	-.005497	-.005518	-.005537	-.005552	-.005564	-.005574	-.005580	3
.70	-.005583	-.005583	-.005579	-.005573	-.005563	-.005549	-.005532	-.005512	-.005489	-.005462	3
.71	-.005431	-.005397	-.005360	-.005319	-.005274	-.005226	-.005175	-.005120	-.005061	-.004999	4
.72	-.004933	-.004864	-.004791	-.004715	-.004635	-.004552	-.004465	-.004375	-.004281	-.004184	3
.73	-.004084	-.003980	-.003873	-.003763	-.003650	-.003534	-.003415	-.003292	-.003167	-.003039	3
.74	-.002908	-.002774	-.002638	-.002499	-.002358	-.002214	-.002068	-.001919	-.001769	-.001617	2
.75	-.001462	-.001306	-.001148	-.000989	-.000828	-.000666	-.000503	-.000338	-.000173	-.000007	1
.76	.000160	.000328	.000496	.000664	.000832	.001000	.001168	.001336	.001503	.001669	0
.77	.001835	.001999	.002163	.002325	.002486	.002645	.002802	.002958	.003111	.003262	-1
.78	.003410	.003556	.003699	.003839	.003976	.004110	.004240	.004366	.004489	.004608	-3
.79	.004723	.004834	.004940	.005042	.005139	.005232	.005319	.005401	.005479	.005551	-4
.80	.005617	.005678	.005734	.005783	.005827	.005865	.005897	.005923	.005943	.005957	-5
.81	.005964	.005965	.005960	.005948	.005931	.005906	.005876	.005839	.005795	.005745	-5
.82	.005689	.005627	.005558	.005483	.005402	.005316	.005223	.005124	.005019	.004909	-5
.83	.004793	.004672	.004546	.004414	.004278	.004137	.003991	.003841	.003686	.003528	-3
.84	.003366	.003200	.003031	.002859	.002684	.002506	.002326	.002144	.001960	.001775	-1
.85	.001588	.001400	.001212	.001023	.000834	.000646	.000458	.000271	.000085	-.000100	1
.86	-.000283	-.000463	-.000641	-.000817	-.000989	-.001159	-.001324	-.001486	-.001643	-.001795	5
.87	-.001943	-.002086	-.002223	-.002355	-.002481	-.002600	-.002713	-.002820	-.002919	-.003011	7
.88	-.003096	-.003174	-.003243	-.003305	-.003359	-.003405	-.003442	-.003471	-.003491	-.003503	9
.89	-.003506	-.003501	-.003486	-.003463	-.003431	-.003391	-.003341	-.003283	-.003217	-.003141	8
.90	-.003058	-.002966	-.002867	-.002759	-.002643	-.002520	-.002390	-.002252	-.002108	-.001957	7
.91	-.001800	-.001636	-.001467	-.001293	-.001114	-.000930	-.000742	-.000550	-.000354	-.000156	3
.92	.000046	.000249	.000454	.000661	.000868	.001076	.001284	.001491	.001697	.001902	0
.93	.002105	.002305	.002503	.002697	.002887	.003074	.003255	.003431	.003602	.003767	-5
.94	.003926	.004077	.004222	.004359	.004488	.004609	.004721	.004825	.004919	.005005	-8
.95	.005080	.005146	.005202	.005248	.005284	.005310	.005325	.005330	.005324	.005308	-9
.96	.005282	.005245	.005199	.005142	.005075	.004999	.004914	.004819	.004716	.004604	-8
.97	.004484	.004356	.004221	.004078	.003930	.003775	.003615	.003450	.003281	.003108	-3
.98	.002932	.002754	.002574	.002393	.002211	.002030	.001850	.001671	.001496	.001324	4
.99	.001156	.000994	.000838	.000690	.000550	.000431	.000106	.000069	.000038	.000013	7

TABLE 23

MTF, Defocused Lens, Monochromatic – Δ = 10.0 Rayleigh Units [Δ ≈ d/2λF²]

ν_r	.000	.001	.002	.003	.004	.005	.006	.007	.008	.009	Δ_2
.00	1.000000	.998235	.995494	.991791	.987138	.981554	.975056	.967664	.959398	.950281	-850
.01	.940338	.929593	.918073	.905806	.892819	.879143	.864808	.849845	.834287	.818165	-562
.02	.801513	.784364	.766753	.748713	.730280	.711487	.692369	.672960	.653296	.633411	-221
.03	.613337	.593110	.572763	.552327	.531837	.511323	.490817	.470349	.449949	.429647	97
.04	.409469	.389444	.369598	.349957	.330545	.311385	.292501	.273913	.255642	.237708	337
.05	.220129	.202921	.186102	.169686	.153686	.138116	.122988	.108311	.094096	.080350	470
.06	.067081	.054295	.041997	.030191	.018881	.008069	-.002245	-.012060	-.021376	-.030195	497
.07	-.038520	-.046354	-.053701	-.060566	-.066953	-.072870	-.078324	-.083321	-.087870	-.091980	439
.08	-.095660	-.098919	-.101768	-.104217	-.106278	-.107962	-.109279	-.110244	-.110867	-.111160	329
.09	-.111138	-.110812	-.110195	-.109301	-.108143	-.106733	-.105085	-.103212	-.101127	-.098844	199
.10	-.096374	-.093732	-.090930	-.087980	-.084895	-.081687	-.078369	-.074951	-.071446	-.067864	76
.11	-.064218	-.060517	-.056773	-.052995	-.049193	-.045377	-.041556	-.037739	-.033934	-.030150	-20
.12	-.026395	-.022676	-.018999	-.015373	-.011804	-.008297	-.004859	-.001494	.001792	.004995	-82
.13	.008109	.011132	.014059	.016888	.019614	.022236	.024751	.027157	.029453	.031635	-111
.14	.033705	.035659	.037499	.039223	.040831	.042324	.043701	.044964	.046112	.047148	-112
.15	.048071	.048884	.049588	.050184	.050675	.051062	.051347	.051533	.051622	.051617	-93
.16	.051519	.051333	.051060	.050703	.050265	.049750	.049160	.048497	.047766	.046970	-65
.17	.046111	.045193	.044218	.043190	.042113	.040989	.039821	.038613	.037367	.036087	-33
.18	.034776	.033436	.032071	.030683	.029276	.027852	.026414	.024964	.023505	.022041	-5
.19	.020572	.019102	.017633	.016167	.014706	.013253	.011810	.010378	.008959	.007555	15
.20	.006169	.004801	.003453	.002127	.000824	-.000455	-.001709	-.002935	-.004134	-.005303	29
.21	-.006443	-.007552	-.008630	-.009675	-.010687	-.011666	-.012611	-.013521	-.014396	-.015236	35
.22	-.016041	-.016810	-.017543	-.018240	-.018901	-.019526	-.020115	-.020668	-.021186	-.021668	35
.23	-.022114	-.022526	-.022904	-.023247	-.023556	-.023831	-.024074	-.024284	-.024462	-.024609	31
.24	-.024725	-.024811	-.024867	-.024894	-.024893	-.024864	-.024808	-.024726	-.024619	-.024486	25
.25	-.024329	-.024150	-.023947	-.023723	-.023477	-.023211	-.022926	-.022622	-.022300	-.021961	17
.26	-.021605	-.021233	-.020846	-.020446	-.020031	-.019604	-.019165	-.018715	-.018254	-.017783	10
.27	-.017303	-.016814	-.016318	-.015814	-.015304	-.014788	-.014267	-.013741	-.013211	-.012677	3
.28	-.012141	-.011602	-.011062	-.010520	-.009978	-.009435	-.008892	-.008350	-.007809	-.007270	-1
.29	-.006733	-.006198	-.005666	-.005137	-.004612	-.004090	-.003573	-.003060	-.002552	-.002050	-4
.30	-.001552	-.001061	-.000575	-.000096	.000378	.000844	.001304	.001757	.002202	.002640	-6
.31	.003071	.003495	.003910	.004318	.004717	.005109	.005493	.005868	.006235	.006594	-7
.32	.006944	.007287	.007620	.007945	.008262	.008571	.008870	.009162	.009445	.009720	-7
.33	.009986	.010244	.010494	.010735	.010968	.011194	.011411	.011620	.011821	.012015	-7
.34	.012200	.012378	.012549	.012712	.012868	.013016	.013157	.013292	.013419	.013539	-6
.35	.013653	.013760	.013860	.013954	.014042	.014124	.014200	.014269	.014333	.014391	-5
.36	.014444	.014491	.014533	.014570	.014602	.014629	.014651	.014668	.014681	.014689	-3
.37	.014693	.014693	.014688	.014680	.014668	.014652	.014632	.014609	.014583	.014553	-2
.38	.014520	.014484	.014445	.014404	.014359	.014312	.014263	.014211	.014156	.014100	-1
.39	.014041	.013980	.013917	.013853	.013786	.013718	.013649	.013578	.013505	.013431	0
.40	.013356	.013280	.013202	.013124	.013044	.012964	.012883	.012801	.012719	.012636	0
.41	.012552	.012468	.012383	.012298	.012213	.012127	.012041	.011956	.011870	.011784	0
.42	.011698	.011612	.011526	.011440	.011354	.011269	.011184	.011099	.011015	.010931	0
.43	.010847	.010764	.010681	.010599	.010517	.010436	.010355	.010275	.010196	.010118	1
.44	.010040	.009963	.009886	.009810	.009736	.009661	.009588	.009516	.009444	.009373	1
.45	.009303	.009235	.009166	.009099	.009033	.008968	.008904	.008840	.008778	.008716	1
.46	.008656	.008597	.008538	.008481	.008424	.008369	.008315	.008261	.008209	.008158	1
.47	.008108	.008058	.008010	.007963	.007917	.007872	.007828	.007785	.007743	.007702	1
.48	.007662	.007623	.007585	.007548	.007513	.007478	.007444	.007411	.007379	.007348	1
.49	.007319	.007290	.007262	.007235	.007209	.007184	.007160	.007137	.007115	.007094	1

TABLE 23

MTF, Defocused Lens, Monochromatic – Δ = 10.0 Rayleigh Units [Δ ≈ $d/2\lambda F^2$]

ν_r	.000	.001	.002	.003	.004	.005	.006	.007	.008	.009	Δ_2
.50	.007073	.007054	.007035	.007018	.007001	.006985	.006971	.006956	.006943	.006931	1
.51	.006919	.006909	.006899	.006890	.006881	.006874	.006867	.006861	.006856	.006851	1
.52	.006847	.006844	.006842	.006840	.006839	.006839	.006839	.006840	.006842	.006844	1
.53	.006847	.006850	.006854	.006858	.006863	.006869	.006875	.006881	.006888	.006896	0
.54	.006904	.006912	.006921	.006930	.006940	.006950	.006960	.006971	.006981	.006993	0
.55	.007004	.007016	.007028	.007040	.007053	.007065	.007078	.007091	.007104	.007117	0
.56	.007130	.007143	.007157	.007170	.007184	.007197	.007210	.007223	.007237	.007250	0
.57	.007263	.007275	.007288	.007300	.007312	.007324	.007336	.007347	.007358	.007369	0
.58	.007379	.007389	.007399	.007408	.007416	.007424	.007432	.007439	.007445	.007451	0
.59	.007456	.007461	.007465	.007468	.007470	.007472	.007473	.007473	.007472	.007470	0
.60	.007468	.007464	.007460	.007454	.007448	.007440	.007431	.007422	.007411	.007399	0
.61	.007386	.007372	.007356	.007339	.007321	.007302	.007281	.007259	.007235	.007211	0
.62	.007184	.007156	.007127	.007096	.007064	.007030	.006995	.006958	.006919	.006879	-1
.63	.006837	.006794	.006748	.006701	.006653	.006602	.006550	.006496	.006440	.006383	-1
.64	.006324	.006263	.006200	.006135	.006068	.006000	.005929	.005857	.005783	.005707	-1
.65	.005629	.005550	.005468	.005385	.005300	.005213	.005124	.005033	.004941	.004846	-1
.66	.004750	.004652	.004553	.004451	.004348	.004244	.004137	.004029	.003919	.003808	-1
.67	.003695	.003581	.003465	.003348	.003230	.003110	.002988	.002866	.002742	.002617	0
.68	.002491	.002364	.002236	.002107	.001977	.001846	.001714	.001582	.001449	.001316	0
.69	.001182	.001047	.000912	.000777	.000642	.000507	.000372	.000236	.000101	-.000034	0
.70	-.000168	-.000302	-.000436	-.000569	-.000701	-.000833	-.000964	-.001093	-.001222	-.001349	1
.71	-.001475	-.001600	-.001723	-.001845	-.001965	-.002083	-.002199	-.002313	-.002425	-.002535	2
.72	-.002642	-.002747	-.002850	-.002950	-.003047	-.003141	-.003232	-.003320	-.003405	-.003486	3
.73	-.003565	-.003639	-.003711	-.003778	-.003842	-.003902	-.003958	-.004010	-.004058	-.004101	4
.74	-.004141	-.004176	-.004207	-.004233	-.004255	-.004272	-.004285	-.004292	-.004296	-.004294	5
.75	-.004288	-.004276	-.004260	-.004239	-.004213	-.004182	-.004146	-.004106	-.004060	-.004009	5
.76	-.003954	-.003893	-.003828	-.003758	-.003683	-.003603	-.003519	-.003430	-.003337	-.003239	4
.77	-.003136	-.003029	-.002918	-.002803	-.002684	-.002561	-.002434	-.002304	-.002170	-.002032	3
.78	-.001891	-.001747	-.001601	-.001451	-.001299	-.001144	-.000987	-.000828	-.000667	-.000505	2
.79	-.000341	-.000175	-.000009	.000158	.000326	.000494	.000663	.000831	.000999	.001167	0
.80	.001334	.001499	.001664	.001827	.001988	.002148	.002305	.002459	.002611	.002760	-2
.81	.002906	.003048	.003187	.003322	.003453	.003579	.003700	.003817	.003929	.004036	-4
.82	.004137	.004233	.004322	.004406	.004484	.004555	.004620	.004678	.004730	.004774	-6
.83	.004812	.004843	.004866	.004882	.004891	.004893	.004887	.004874	.004853	.004825	-6
.84	.004789	.004746	.004695	.004638	.004573	.004500	.004421	.004335	.004242	.004143	-6
.85	.004036	.003924	.003805	.003681	.003551	.003415	.003274	.003128	.002977	.002822	-3
.86	.002663	.002500	.002334	.002164	.001991	.001816	.001639	.001460	.001280	.001099	0
.87	.000917	.000735	.000553	.000371	.000191	.000012	-.000165	-.000340	-.000513	-.000682	3
.88	-.000848	-.001010	-.001168	-.001321	-.001469	-.001612	-.001750	-.001881	-.002006	-.002124	7
.89	-.002235	-.002338	-.002434	-.002522	-.002602	-.002674	-.002737	-.002791	-.002836	-.002872	9
.90	-.002898	-.002916	-.002923	-.002921	-.002910	-.002889	-.002858	-.002818	-.002768	-.002709	9
.91	-.002640	-.002562	-.002475	-.002380	-.002275	-.002162	-.002041	-.001913	-.001776	-.001633	7
.92	-.001482	-.001325	-.001162	-.000994	-.000820	-.000641	-.000458	-.000272	-.000082	.000111	3
.93	.000306	.000503	.000700	.000899	.001097	.001295	.001492	.001687	.001879	.002069	-2
.94	.002255	.002438	.002616	.002788	.002956	.003117	.003271	.003418	.003558	.003690	-7
.95	.003813	.003928	.004033	.004129	.004215	.004290	.004356	.004411	.004455	.004488	-10
.96	.004511	.004522	.004522	.004511	.004489	.004456	.004412	.004358	.004293	.004218	-9
.97	.004133	.004038	.003935	.003822	.003702	.003573	.003437	.003295	.003146	.002992	-4
.98	.002833	.002670	.002503	.002334	.002163	.001991	.001819	.001647	.001478	.001310	3
.99	.001147	.000987	.000834	.000687	.000549	.000430	.000106	.000069	.000038	.000013	7

TABLE 23

MTF, Defocused Lens, Monochromatic – Δ= 15.0 Rayleigh Units [$\Delta \approx d/2\lambda F^2$]

ν_r	.000	.001	.002	.003	.004	.005	.006	.007	.008	.009	Δ_2
.00	1.000000	.997621	.993049	.986321	.977485	.966593	.953709	.938905	.922258	.903853	-1757
.01	.883782	.862141	.839033	.814563	.788843	.761986	.734108	.705327	.675762	.645534	-663
.02	.614762	.583567	.552066	.520378	.488617	.456895	.425321	.394002	.363037	.332526	453
.03	.302560	.273227	.244610	.216784	.189821	.163786	.138739	.114731	.091810	.070017	1127
.04	.049385	.029942	.011712	-.005291	-.021056	-.035578	-.048859	-.060905	-.071729	-.081345	1207
.05	-.089777	-.097048	-.103189	-.108234	-.112219	-.115184	-.117173	-.118231	-.118405	-.117746	833
.06	-.116306	-.114136	-.111290	-.107823	-.103790	-.099246	-.094245	-.088841	-.083089	-.077042	296
.07	-.070751	-.064266	-.057637	-.050910	-.044132	-.037346	-.030593	-.023912	-.017340	-.010912	-143
.08	-.004660	.001388	.007203	.012761	.018042	.023025	.027693	.032033	.036033	.039683	-349
.09	.042977	.045909	.048478	.050682	.052524	.054007	.055137	.055920	.056365	.056483	-327
.10	.056284	.055782	.054990	.053923	.052597	.051028	.049233	.047231	.045038	.042673	-171
.11	.040155	.037502	.034732	.031865	.028918	.025909	.022855	.019775	.016684	.013598	5
.12	.010533	.007504	.004525	.001608	-.001233	-.003987	-.006643	-.009191	-.011623	-.013930	125
.13	-.016105	-.018141	-.020033	-.021777	-.023368	-.024805	-.026084	-.027204	-.028166	-.028969	159
.14	-.029614	-.030104	-.030440	-.030625	-.030664	-.030559	-.030317	-.029941	-.029437	-.028812	122
.15	-.028071	-.027220	-.026268	-.025220	-.024083	-.022866	-.021575	-.020218	-.018803	-.017337	51
.16	-.015827	-.014281	-.012706	-.011110	-.009500	-.007881	-.006262	-.004648	-.003047	-.001463	-17
.17	.000098	.001629	.003126	.004584	.005998	.007364	.008678	.009937	.011137	.012275	-61
.18	.013349	.014356	.015295	.016163	.016959	.017683	.018333	.018909	.019411	.019840	-73
.19	.020194	.020475	.020684	.020822	.020891	.020891	.020824	.020692	.020498	.020243	-60
.20	.019930	.019561	.019139	.018666	.018146	.017580	.016972	.016325	.015642	.014926	-32
.21	.014179	.013405	.012607	.011787	.010949	.010096	.009230	.008355	.007473	.006586	-3
.22	.005698	.004811	.003928	.003051	.002182	.001323	.000477	-.000354	-.001169	-.001965	18
.23	-.002742	-.003497	-.004228	-.004935	-.005617	-.006271	-.006897	-.007495	-.008063	-.008600	30
.24	-.009106	-.009581	-.010023	-.010433	-.010811	-.011156	-.011468	-.011747	-.011993	-.012208	32
.25	-.012390	-.012540	-.012658	-.012746	-.012803	-.012831	-.012829	-.012799	-.012741	-.012656	27
.26	-.012545	-.012409	-.012248	-.012064	-.011857	-.011629	-.011380	-.011112	-.010825	-.010521	17
.27	-.010200	-.009864	-.009514	-.009150	-.008774	-.008386	-.007988	-.007581	-.007166	-.006744	7
.28	-.006315	-.005880	-.005442	-.004999	-.004554	-.004108	-.003660	-.003212	-.002765	-.002319	0
.29	-.001876	-.001435	-.000998	-.000565	-.000137	.000285	.000702	.001112	.001515	.001911	-7
.30	.002298	.002678	.003048	.003409	.003761	.004103	.004435	.004757	.005069	.005369	-10
.31	.005659	.005937	.006205	.006461	.006706	.006939	.007161	.007371	.007570	.007758	-10
.32	.007934	.008098	.008252	.008394	.008525	.008645	.008754	.008853	.008940	.009018	-9
.33	.009085	.009142	.009189	.009227	.009255	.009274	.009283	.009284	.009277	.009261	-7
.34	.009237	.009205	.009165	.009118	.009064	.009003	.008936	.008862	.008782	.008696	-5
.35	.008604	.008507	.008405	.008299	.008187	.008071	.007951	.007827	.007699	.007568	-2
.36	.007434	.007296	.007156	.007013	.006868	.006720	.006571	.006420	.006267	.006113	0
.37	.005957	.005801	.005643	.005485	.005326	.005167	.005008	.004848	.004689	.004529	0
.38	.004370	.004211	.004053	.003896	.003739	.003583	.003428	.003274	.003121	.002969	1
.39	.002819	.002670	.002522	.002376	.002231	.002089	.001947	.001808	.001670	.001535	2
.40	.001401	.001269	.001139	.001011	.000884	.000760	.000638	.000519	.000401	.000285	2
.41	.000172	.000060	-.000049	-.000156	-.000261	-.000364	-.000465	-.000563	-.000660	-.000754	2
.42	-.000846	-.000936	-.001024	-.001110	-.001193	-.001275	-.001355	-.001432	-.001508	-.001581	2
.43	-.001653	-.001723	-.001790	-.001856	-.001920	-.001982	-.002042	-.002101	-.002157	-.002212	2
.44	-.002265	-.002317	-.002367	-.002415	-.002461	-.002506	-.002549	-.002591	-.002631	-.002670	2
.45	-.002707	-.002743	-.002777	-.002810	-.002841	-.002872	-.002901	-.002928	-.002954	-.002980	1
.46	-.003003	-.003026	-.003048	-.003068	-.003087	-.003105	-.003122	-.003138	-.003153	-.003167	1
.47	-.003180	-.003192	-.003203	-.003213	-.003222	-.003230	-.003238	-.003244	-.003250	-.003255	1
.48	-.003259	-.003262	-.003264	-.003266	-.003267	-.003267	-.003266	-.003265	-.003263	-.003260	1
.49	-.003256	-.003252	-.003247	-.003242	-.003236	-.003229	-.003222	-.003214	-.003205	-.003196	1

TABLE 23

MTF, Defocused Lens, Monochromatic – Δ= 15.0 Rayleigh Units [$\Delta \approx d/2\lambda F^2$]

ν_r	.000	.001	.002	.003	.004	.005	.006	.007	.008	.009	Δ_2
.50	-.003186	-.003176	-.003164	-.003153	-.003141	-.003128	-.003114	-.003100	-.003086	-.003071	1
.51	-.003055	-.003039	-.003022	-.003005	-.002987	-.002968	-.002949	-.002929	-.002909	-.002888	1
.52	-.002867	-.002845	-.002823	-.002800	-.002776	-.002752	-.002727	-.002702	-.002676	-.002649	1
.53	-.002622	-.002594	-.002566	-.002537	-.002508	-.002477	-.002447	-.002415	-.002383	-.002351	1
.54	-.002317	-.002284	-.002249	-.002214	-.002178	-.002142	-.002104	-.002067	-.002028	-.001989	1
.55	-.001949	-.001909	-.001868	-.001826	-.001784	-.001740	-.001697	-.001652	-.001607	-.001561	1
.56	-.001514	-.001467	-.001419	-.001370	-.001321	-.001271	-.001220	-.001169	-.001117	-.001064	1
.57	-.001010	-.000956	-.000901	-.000846	-.000790	-.000733	-.000676	-.000617	-.000559	-.000500	1
.58	-.000440	-.000379	-.000318	-.000256	-.000194	-.000132	-.000068	-.000005	.000060	.000124	0
.59	.000189	.000255	.000321	.000387	.000454	.000521	.000588	.000656	.000724	.000792	0
.60	.000860	.000929	.000997	.001066	.001135	.001203	.001272	.001341	.001409	.001478	0
.61	.001546	.001614	.001682	.001749	.001816	.001883	.001949	.002015	.002080	.002144	0
.62	.002208	.002271	.002333	.002395	.002455	.002515	.002574	.002631	.002688	.002743	0
.63	.002797	.002850	.002901	.002951	.003000	.003046	.003092	.003135	.003177	.003217	-1
.64	.003255	.003292	.003326	.003358	.003388	.003416	.003442	.003465	.003486	.003505	-1
.65	.003521	.003535	.003546	.003555	.003560	.003564	.003564	.003562	.003556	.003548	-2
.66	.003537	.003524	.003507	.003487	.003464	.003438	.003409	.003377	.003342	.003304	-2
.67	.003262	.003218	.003171	.003120	.003067	.003010	.002951	.002888	.002823	.002754	-2
.68	.002683	.002609	.002532	.002453	.002370	.002286	.002198	.002109	.002017	.001922	-1
.69	.001826	.001728	.001627	.001525	.001421	.001316	.001209	.001100	.000991	.000880	0
.70	.000769	.000657	.000544	.000431	.000317	.000203	.000090	-.000024	-.000137	-.000249	1
.71	-.000361	-.000471	-.000581	-.000689	-.000796	-.000901	-.001004	-.001105	-.001204	-.001300	2
.72	-.001394	-.001484	-.001572	-.001657	-.001738	-.001815	-.001889	-.001959	-.002025	-.002087	4
.73	-.002144	-.002197	-.002245	-.002288	-.002326	-.002359	-.002387	-.002410	-.002427	-.002439	5
.74	-.002445	-.002446	-.002442	-.002431	-.002415	-.002393	-.002366	-.002332	-.002293	-.002249	6
.75	-.002199	-.002143	-.002082	-.002016	-.001944	-.001867	-.001786	-.001699	-.001608	-.001512	4
.76	-.001412	-.001309	-.001201	-.001089	-.000975	-.000857	-.000736	-.000613	-.000487	-.000359	2
.77	-.000230	-.000100	.000032	.000164	.000296	.000429	.000561	.000692	.000822	.000951	0
.78	.001078	.001202	.001324	.001443	.001559	.001671	.001779	.001883	.001982	.002077	-4
.79	.002165	.002249	.002326	.002397	.002462	.002520	.002571	.002615	.002652	.002682	-6
.80	.002703	.002718	.002724	.002723	.002713	.002696	.002671	.002638	.002597	.002549	-7
.81	.002493	.002429	.002358	.002280	.002196	.002104	.002007	.001903	.001794	.001680	-4
.82	.001560	.001437	.001309	.001177	.001042	.000905	.000766	.000624	.000482	.000339	0
.83	.000196	.000054	-.000087	-.000227	-.000364	-.000499	-.000630	-.000758	-.000881	-.000999	5
.84	-.001112	-.001219	-.001319	-.001413	-.001499	-.001578	-.001648	-.001711	-.001764	-.001809	9
.85	-.001844	-.001870	-.001886	-.001892	-.001889	-.001876	-.001853	-.001821	-.001778	-.001726	10
.86	-.001665	-.001595	-.001515	-.001428	-.001332	-.001228	-.001117	-.001000	-.000876	-.000746	6
.87	-.000611	-.000472	-.000329	-.000182	-.000034	.000117	.000268	.000420	.000571	.000721	0
.88	.000869	.001014	.001155	.001293	.001425	.001551	.001671	.001784	.001889	.001986	-7
.89	.002074	.002153	.002222	.002280	.002328	.002365	.002391	.002406	.002409	.002401	-11
.90	.002381	.002350	.002308	.002254	.002190	.002115	.002030	.001935	.001832	.001720	-8
.91	.001600	.001472	.001339	.001199	.001055	.000907	.000756	.000603	.000449	.000294	0
.92	.000141	-.000011	-.000161	-.000306	-.000447	-.000583	-.000712	-.000833	-.000947	-.001051	9
.93	-.001146	-.001230	-.001304	-.001366	-.001416	-.001453	-.001478	-.001490	-.001489	-.001475	13
.94	-.001448	-.001407	-.001354	-.001289	-.001211	-.001122	-.001021	-.000910	-.000790	-.000661	9
.95	-.000523	-.000379	-.000228	-.000072	.000087	.000250	.000414	.000579	.000743	.000906	-1
.96	.001065	.001221	.001371	.001514	.001650	.001778	.001896	.002004	.002101	.002186	-11
.97	.002259	.002319	.002365	.002397	.002415	.002420	.002410	.002386	.002349	.002298	-12
.98	.002234	.002159	.002072	.001974	.001866	.001750	.001627	.001497	.001362	.001224	-2
.99	.001084	.000944	.000805	.000669	.000538	.000425	.000104	.000068	.000038	.000013	6

TABLE 23

MTF, Defocused Lens, Monochromatic – Δ = 20.0 Rayleigh Units [Δ ≈ d/2λF²]

ν_r	.000	.001	.002	.003	.004	.005	.006	.007	.008	.009	Δ_2
.00	1.000000	.996761	.989632	.978699	.964075	.945902	.924347	.899600	.871871	.841392	-2750
.01	.808408	.773181	.735983	.697094	.656801	.615393	.573163	.530398	.487383	.444397	28
.02	.401707	.359573	.318240	.277938	.238883	.201271	.165281	.131074	.098787	.068541	2040
.03	.040433	.014541	-.009080	-.030394	-.049384	-.066053	-.080424	-.092536	-.102443	-.110218	2133
.04	-.115944	-.119720	-.121652	-.121860	-.120470	-.117615	-.113432	-.108064	-.101655	-.094349	896
.05	-.086291	-.077624	-.068487	-.059016	-.049340	-.039585	-.029868	-.020297	-.010976	-.001996	-341
.06	.006558	.014614	.022106	.028980	.035190	.040702	.045490	.049536	.052834	.055383	-747
.07	.057192	.058277	.058662	.058375	.057450	.055928	.053853	.051272	.048237	.044799	-401
.08	.041015	.036938	.032626	.028135	.023518	.018831	.014125	.009450	.004854	.000380	123
.09	-.003930	-.008038	-.011910	-.015517	-.018832	-.021832	-.024500	-.026822	-.028788	-.030391	362
.10	-.031630	-.032505	-.033022	-.033189	-.033016	-.032517	-.031710	-.030612	-.029245	-.027631	247
.11	-.025794	-.023759	-.021551	-.019198	-.016727	-.014163	-.011534	-.008866	-.006185	-.003516	-11
.12	-.000882	.001695	.004192	.006591	.008872	.011021	.013022	.014861	.016529	.018017	-180
.13	.019316	.020422	.021331	.022041	.022553	.022868	.022989	.022921	.022670	.022244	-174
.14	.021651	.020902	.020005	.018974	.017820	.016556	.015194	.013749	.012234	.010663	-55
.15	.009049	.007406	.005748	.004087	.002436	.000807	-.000788	-.002339	-.003835	-.005266	64
.16	-.006625	-.007902	-.009092	-.010188	-.011184	-.012077	-.012862	-.013538	-.014102	-.014553	113
.17	-.014892	-.015119	-.015235	-.015243	-.015145	-.014944	-.014646	-.014254	-.013773	-.013209	83
.18	-.012567	-.011855	-.011077	-.010242	-.009355	-.008424	-.007455	-.006456	-.005434	-.004395	16
.19	-.003346	-.002295	-.001246	-.000208	.000815	.001817	.002792	.003735	.004642	.005508	-40
.20	.006329	.007101	.007822	.008489	.009098	.009649	.010139	.010567	.010933	.011236	-62
.21	.011476	.011653	.011767	.011820	.011813	.011747	.011624	.011446	.011215	.010934	-49
.22	.010605	.010231	.009815	.009361	.008870	.008347	.007795	.007217	.006616	.005996	-18
.23	.005360	.004712	.004055	.003392	.002726	.002060	.001398	.000742	.000095	-.000540	12
.24	-.001161	-.001766	-.002352	-.002918	-.003460	-.003979	-.004471	-.004936	-.005372	-.005778	30
.25	-.006153	-.006497	-.006809	-.007088	-.007335	-.007548	-.007728	-.007875	-.007989	-.008070	33
.26	-.008119	-.008137	-.008124	-.008082	-.008010	-.007910	-.007782	-.007629	-.007451	-.007249	24
.27	-.007025	-.006779	-.006514	-.006231	-.005930	-.005614	-.005284	-.004941	-.004587	-.004223	10
.28	-.003851	-.003472	-.003086	-.002697	-.002304	-.001910	-.001515	-.001121	-.000729	-.000340	-2
.29	.000045	.000425	.000799	.001166	.001526	.001876	.002217	.002548	.002867	.003176	-10
.30	.003472	.003756	.004027	.004285	.004529	.004758	.004974	.005175	.005362	.005534	-14
.31	.005691	.005833	.005961	.006074	.006173	.006257	.006326	.006382	.006424	.006452	-13
.32	.006467	.006469	.006458	.006435	.006400	.006353	.006295	.006226	.006147	.006058	-9
.33	.005960	.005852	.005736	.005612	.005480	.005341	.005195	.005043	.004885	.004721	-4
.34	.004552	.004379	.004202	.004022	.003838	.003651	.003461	.003270	.003077	.002883	0
.35	.002688	.002492	.002296	.002100	.001904	.001710	.001516	.001323	.001132	.000943	2
.36	.000755	.000570	.000388	.000208	.000030	-.000144	-.000315	-.000483	-.000647	-.000808	4
.37	-.000965	-.001119	-.001268	-.001414	-.001555	-.001693	-.001826	-.001955	-.002080	-.002201	4
.38	-.002318	-.002430	-.002538	-.002641	-.002741	-.002836	-.002927	-.003014	-.003097	-.003175	4
.39	-.003250	-.003321	-.003387	-.003450	-.003509	-.003564	-.003615	-.003663	-.003708	-.003748	3
.40	-.003786	-.003820	-.003851	-.003878	-.003903	-.003925	-.003943	-.003959	-.003973	-.003983	3
.41	-.003991	-.003997	-.004000	-.004001	-.004000	-.003996	-.003991	-.003984	-.003974	-.003963	2
.42	-.003951	-.003936	-.003921	-.003903	-.003885	-.003865	-.003844	-.003821	-.003798	-.003773	1
.43	-.003748	-.003722	-.003695	-.003667	-.003638	-.003609	-.003579	-.003549	-.003519	-.003487	0
.44	-.003456	-.003424	-.003392	-.003360	-.003328	-.003295	-.003262	-.003230	-.003197	-.003165	0
.45	-.003132	-.003100	-.003068	-.003036	-.003004	-.002972	-.002941	-.002910	-.002879	-.002849	0
.46	-.002819	-.002789	-.002760	-.002731	-.002703	-.002675	-.002647	-.002620	-.002594	-.002568	0
.47	-.002543	-.002518	-.002494	-.002470	-.002447	-.002425	-.002403	-.002381	-.002361	-.002341	0
.48	-.002321	-.002303	-.002284	-.002267	-.002250	-.002234	-.002218	-.002203	-.002189	-.002175	0
.49	-.002162	-.002150	-.002138	-.002127	-.002116	-.002106	-.002097	-.002088	-.002080	-.002073	0

TABLE 23

MTF, Defocused Lens, Monochromatic – $\Delta = 20.0$ Rayleigh Units [$\Delta \approx d/2\lambda F^2$]

ν_r	.000	.001	.002	.003	.004	.005	.006	.007	.008	.009	Δ_2
.50	-.002066	-.002060	-.002054	-.002049	-.002045	-.002041	-.002038	-.002035	-.002033	-.002032	0
.51	-.002031	-.002030	-.002030	-.002031	-.002032	-.002033	-.002035	-.002038	-.002041	-.002044	0
.52	-.002048	-.002052	-.002057	-.002062	-.002068	-.002073	-.002080	-.002086	-.002093	-.002100	0
.53	-.002108	-.002115	-.002124	-.002132	-.002140	-.002149	-.002158	-.002167	-.002176	-.002186	0
.54	-.002195	-.002205	-.002215	-.002224	-.002234	-.002244	-.002254	-.002264	-.002273	-.002283	0
.55	-.002293	-.002302	-.002311	-.002320	-.002329	-.002338	-.002346	-.002354	-.002362	-.002370	0
.56	-.002377	-.002383	-.002390	-.002395	-.002401	-.002406	-.002410	-.002413	-.002417	-.002419	1
.57	-.002421	-.002422	-.002422	-.002422	-.002420	-.002418	-.002415	-.002412	-.002407	-.002401	1
.58	-.002394	-.002387	-.002378	-.002368	-.002357	-.002345	-.002332	-.002318	-.002302	-.002285	1
.59	-.002267	-.002247	-.002227	-.002204	-.002181	-.002156	-.002130	-.002102	-.002073	-.002042	1
.60	-.002010	-.001977	-.001941	-.001905	-.001867	-.001827	-.001786	-.001743	-.001699	-.001653	2
.61	-.001606	-.001557	-.001506	-.001454	-.001401	-.001346	-.001290	-.001232	-.001173	-.001113	1
.62	-.001051	-.000988	-.000923	-.000858	-.000791	-.000723	-.000654	-.000584	-.000513	-.000441	1
.63	-.000368	-.000295	-.000221	-.000146	-.000070	.000006	.000082	.000158	.000235	.000312	0
.64	.000389	.000466	.000542	.000619	.000695	.000770	.000845	.000919	.000992	.001064	0
.65	.001135	.001205	.001274	.001341	.001406	.001470	.001532	.001592	.001650	.001705	-1
.66	.001759	.001809	.001857	.001903	.001946	.001985	.002022	.002055	.002086	.002112	-2
.67	.002136	.002156	.002172	.002184	.002193	.002198	.002199	.002196	.002189	.002178	-3
.68	.002163	.002143	.002120	.002093	.002061	.002026	.001986	.001942	.001895	.001843	-3
.69	.001788	.001729	.001666	.001599	.001530	.001457	.001381	.001301	.001219	.001135	-2
.70	.001047	.000958	.000866	.000773	.000678	.000581	.000484	.000385	.000286	.000186	0
.71	.000086	-.000013	-.000112	-.000211	-.000308	-.000405	-.000499	-.000592	-.000682	-.000770	2
.72	-.000855	-.000937	-.001016	-.001091	-.001162	-.001229	-.001292	-.001350	-.001403	-.001451	5
.73	-.001494	-.001531	-.001562	-.001588	-.001608	-.001621	-.001629	-.001630	-.001624	-.001613	6
.74	-.001595	-.001570	-.001539	-.001502	-.001459	-.001410	-.001354	-.001293	-.001226	-.001154	5
.75	-.001077	-.000995	-.000908	-.000817	-.000722	-.000624	-.000522	-.000417	-.000310	-.000201	2
.76	-.000091	.000021	.000133	.000245	.000357	.000468	.000578	.000686	.000791	.000894	-2
.77	.000993	.001089	.001180	.001266	.001347	.001423	.001493	.001556	.001612	.001661	-6
.78	.001703	.001738	.001764	.001783	.001793	.001795	.001789	.001774	.001751	.001720	-7
.79	.001681	.001633	.001578	.001515	.001445	.001368	.001284	.001194	.001099	.000998	-4
.80	.000893	.000784	.000671	.000555	.000437	.000318	.000198	.000078	-.000042	-.000161	1
.81	-.000277	-.000390	-.000501	-.000606	-.000707	-.000803	-.000892	-.000975	-.001050	-.001118	8
.82	-.001177	-.001228	-.001269	-.001301	-.001323	-.001335	-.001338	-.001330	-.001312	-.001283	10
.83	-.001245	-.001198	-.001140	-.001074	-.000999	-.000916	-.000825	-.000727	-.000622	-.000512	6
.84	-.000397	-.000278	-.000156	-.000031	.000095	.000222	.000348	.000473	.000596	.000716	-2
.85	.000831	.000941	.001045	.001142	.001232	.001314	.001386	.001448	.001501	.001543	-10
.86	.001573	.001592	.001600	.001596	.001580	.001553	.001514	.001464	.001403	.001332	-9
.87	.001251	.001161	.001062	.000956	.000844	.000725	.000602	.000476	.000347	.000216	0
.88	.000086	-.000044	-.000171	-.000294	-.000413	-.000527	-.000633	-.000732	-.000821	-.000901	10
.89	-.000971	-.001029	-.001075	-.001109	-.001130	-.001138	-.001133	-.001114	-.001083	-.001039	13
.90	-.000982	-.000913	-.000833	-.000743	-.000643	-.000534	-.000417	-.000294	-.000166	-.000034	4
.91	.000100	.000235	.000371	.000504	.000634	.000760	.000879	.000992	.001095	.001189	-9
.92	.001273	.001344	.001403	.001449	.001480	.001498	.001501	.001490	.001464	.001424	-13
.93	.001370	.001303	.001223	.001133	.001031	.000920	.000802	.000677	.000546	.000412	-2
.94	.000277	.000141	.000007	-.000125	-.000251	-.000371	-.000483	-.000585	-.000677	-.000756	12
.95	-.000823	-.000875	-.000913	-.000935	-.000942	-.000932	-.000907	-.000866	-.000810	-.000739	15
.96	-.000655	-.000558	-.000450	-.000331	-.000204	-.000070	.000069	.000211	.000355	.000499	0
.97	.000640	.000777	.000908	.001030	.001143	.001246	.001335	.001411	.001472	.001517	-15
.98	.001547	.001560	.001556	.001536	.001499	.001448	.001382	.001303	.001212	.001111	-9
.99	.001002	.000886	.000766	.000645	.000524	.000417	.000101	.000067	.000037	.000013	6

TABLE 23

MTF, Defocused Lens, Monochromatic – Δ = 25.0 Rayleigh Units [$\Delta \approx d/2\lambda F^2$]

ν_T	.000	.001	.002	.003	.004	.005	.006	.007	.008	.009	Δ_2
.00	1.000000	.995656	.985250	.968955	.947014	.919733	.887479	.850670	.809772	.765287	-3586
.01	.717748	.667711	.615746	.562428	.508331	.454018	.400034	.346903	.295114	.245125	1799
.02	.197349	.152157	.109868	.070755	.035036	.002876	-.025613	-.050370	-.071384	-.088695	3704
.03	-.102387	-.112584	-.119451	-.123186	-.124017	-.122193	-.117988	-.111688	-.103587	-.093987	1499
.04	-.083190	-.071491	-.059181	-.046536	-.033817	-.021269	-.009112	.002452	.013249	.023130	-915
.05	.031972	.039677	.046177	.051426	.055404	.058117	.059589	.059867	.059014	.057112	-1049
.06	.054253	.050542	.046092	.041024	.035461	.029527	.023349	.017047	.010740	.004538	106
.07	-.001455	-.007144	-.012446	-.017287	-.021604	-.025349	-.028484	-.030982	-.032831	-.034028	652
.08	-.034583	-.034513	-.033847	-.032623	-.030885	-.028683	-.026072	-.023114	-.019870	-.016406	221
.09	-.012786	-.009075	-.005336	-.001631	.001982	.005451	.008725	.011760	.014518	.016966	-309
.10	.019078	.020834	.022218	.023225	.023851	.024101	.023985	.023517	.022716	.021607	-308
.11	.020215	.018572	.016710	.014662	.012467	.010159	.007776	.005354	.002930	.000537	31
.12	-.001792	-.004026	-.006138	-.008101	-.009894	-.011497	-.012894	-.014074	-.015026	-.015745	233
.13	-.016229	-.016479	-.016497	-.016293	-.015873	-.015252	-.014442	-.013460	-.012325	-.011054	135
.14	-.009668	-.008188	-.006636	-.005032	-.003399	-.001758	-.000129	.001469	.003016	.004496	-67
.15	.005892	.007190	.008377	.009443	.010377	.011174	.011826	.012332	.012688	.012895	-148
.16	.012954	.012870	.012646	.012289	.011806	.011207	.010500	.009697	.008809	.007847	-72
.17	.006824	.005753	.004646	.003515	.002374	.001233	.000106	-.000996	-.002064	-.003086	45
.18	-.004055	-.004961	-.005797	-.006556	-.007233	-.007824	-.008324	-.008732	-.009046	-.009264	95
.19	-.009388	-.009419	-.009358	-.009209	-.008975	-.008661	-.008271	-.007811	-.007286	-.006704	58
.20	-.006070	-.005392	-.004676	-.003931	-.003163	-.002379	-.001586	-.000792	-.000003	.000773	-11
.21	.001532	.002267	.002973	.003644	.004276	.004864	.005406	.005897	.006335	.006719	-54
.22	.007045	.007314	.007524	.007675	.007767	.007802	.007779	.007702	.007571	.007389	-50
.23	.007158	.006882	.006563	.006204	.005809	.005382	.004927	.004447	.003945	.003427	-16
.24	.002896	.002356	.001810	.001263	.000718	.000178	-.000353	-.000872	-.001375	-.001861	18
.25	-.002326	-.002768	-.003185	-.003575	-.003936	-.004267	-.004566	-.004832	-.005065	-.005264	34
.26	-.005429	-.005559	-.005655	-.005716	-.005744	-.005739	-.005702	-.005633	-.005534	-.005407	29
.27	-.005252	-.005071	-.004865	-.004637	-.004388	-.004119	-.003833	-.003532	-.003217	-.002890	12
.28	-.002553	-.002209	-.001858	-.001502	-.001145	-.000786	-.000429	-.000074	.000276	.000621	-5
.29	.000959	.001289	.001609	.001918	.002215	.002499	.002769	.003024	.003264	.003489	-15
.30	.003696	.003887	.004060	.004216	.004354	.004474	.004576	.004660	.004727	.004776	-17
.31	.004808	.004822	.004820	.004802	.004768	.004719	.004656	.004578	.004487	.004383	-12
.32	.004266	.004139	.004001	.003853	.003695	.003529	.003356	.003175	.002988	.002796	-4
.33	.002599	.002398	.002193	.001986	.001777	.001567	.001356	.001145	.000934	.000725	1
.34	.000517	.000312	.000109	-.000091	-.000287	-.000479	-.000666	-.000849	-.001027	-.001199	5
.35	-.001366	-.001527	-.001681	-.001829	-.001971	-.002106	-.002235	-.002356	-.002470	-.002578	7
.36	-.002678	-.002772	-.002858	-.002937	-.003009	-.003075	-.003133	-.003185	-.003229	-.003268	7
.37	-.003300	-.003325	-.003344	-.003358	-.003365	-.003367	-.003363	-.003354	-.003340	-.003321	5
.38	-.003297	-.003268	-.003235	-.003198	-.003157	-.003112	-.003063	-.003011	-.002956	-.002897	3
.39	-.002836	-.002773	-.002707	-.002638	-.002568	-.002496	-.002422	-.002346	-.002269	-.002191	1
.40	-.002112	-.002031	-.001950	-.001869	-.001787	-.001704	-.001622	-.001539	-.001456	-.001373	0
.41	-.001291	-.001209	-.001127	-.001046	-.000965	-.000885	-.000806	-.000727	-.000650	-.000573	0
.42	-.000498	-.000423	-.000350	-.000277	-.000206	-.000136	-.000067	-.000000	.000066	.000131	0
.43	.000194	.000256	.000317	.000376	.000434	.000490	.000545	.000599	.000651	.000702	0
.44	.000751	.000799	.000846	.000891	.000935	.000977	.001018	.001058	.001097	.001134	0
.45	.001170	.001204	.001237	.001269	.001300	.001330	.001358	.001386	.001412	.001437	0
.46	.001461	.001484	.001506	.001527	.001546	.001565	.001583	.001600	.001616	.001631	0
.47	.001646	.001659	.001671	.001683	.001694	.001704	.001714	.001722	.001730	.001738	0
.48	.001744	.001750	.001755	.001760	.001764	.001767	.001769	.001772	.001773	.001774	0
.49	.001774	.001774	.001773	.001772	.001770	.001768	.001765	.001762	.001758	.001753	0

TABLE 23

MTF, Defocused Lens, Monochromatic – Δ=25.0 Rayleigh Units [Δ ≈ d/2λF²]

ν_r	.000	.001	.002	.003	.004	.005	.006	.007	.008	.009	Δ_2
.50	.001749	.001743	.001738	.001731	.001725	.001717	.001710	.001702	.001693	.001684	0
.51	.001675	.001665	.001654	.001643	.001632	.001620	.001608	.001595	.001582	.001568	0
.52	.001554	.001539	.001524	.001508	.001492	.001475	.001458	.001441	.001422	.001404	0
.53	.001384	.001365	.001344	.001324	.001302	.001280	.001258	.001235	.001211	.001187	0
.54	.001162	.001137	.001111	.001084	.001057	.001029	.001001	.000972	.000942	.000912	0
.55	.000882	.000850	.000818	.000786	.000753	.000719	.000685	.000650	.000614	.000578	0
.56	.000542	.000505	.000467	.000429	.000390	.000351	.000311	.000271	.000231	.000190	0
.57	.000148	.000106	.000064	.000022	-.000021	-.000064	-.000108	-.000151	-.000195	-.000239	0
.58	-.000283	-.000328	-.000372	-.000416	-.000460	-.000504	-.000548	-.000592	-.000636	-.000679	0
.59	-.000722	-.000765	-.000807	-.000849	-.000890	-.000930	-.000970	-.001009	-.001047	-.001085	1
.60	-.001121	-.001156	-.001191	-.001224	-.001256	-.001287	-.001316	-.001344	-.001371	-.001396	2
.61	-.001419	-.001440	-.001460	-.001478	-.001494	-.001508	-.001520	-.001530	-.001538	-.001544	2
.62	-.001547	-.001548	-.001547	-.001543	-.001537	-.001529	-.001518	-.001504	-.001488	-.001469	3
.63	-.001447	-.001423	-.001396	-.001367	-.001335	-.001300	-.001263	-.001223	-.001180	-.001135	3
.64	-.001088	-.001038	-.000986	-.000932	-.000875	-.000816	-.000756	-.000693	-.000628	-.000562	2
.65	-.000494	-.000425	-.000354	-.000282	-.000210	-.000136	-.000061	.000014	.000089	.000164	0
.66	.000240	.000315	.000390	.000464	.000538	.000610	.000681	.000751	.000819	.000885	-1
.67	.000949	.001011	.001070	.001127	.001180	.001230	.001278	.001321	.001361	.001397	-3
.68	.001429	.001457	.001480	.001499	.001513	.001523	.001528	.001528	.001523	.001514	-4
.69	.001499	.001479	.001455	.001425	.001390	.001351	.001307	.001259	.001206	.001148	-3
.70	.001087	.001022	.000953	.000880	.000805	.000726	.000645	.000561	.000476	.000388	-1
.71	.000300	.000211	.000121	.000031	-.000059	-.000148	-.000236	-.000322	-.000407	-.000489	2
.72	-.000568	-.000644	-.000717	-.000786	-.000850	-.000909	-.000964	-.001013	-.001057	-.001095	6
.73	-.001126	-.001151	-.001170	-.001181	-.001186	-.001184	-.001175	-.001159	-.001136	-.001106	7
.74	-.001069	-.001025	-.000975	-.000919	-.000857	-.000789	-.000716	-.000638	-.000556	-.000469	4
.75	-.000379	-.000286	-.000191	-.000094	.000005	.000104	.000203	.000301	.000398	.000493	-1
.76	.000586	.000675	.000761	.000842	.000918	.000989	.001053	.001111	.001162	.001205	-6
.77	.001240	.001267	.001286	.001296	.001298	.001290	.001274	.001249	.001215	.001173	-7
.78	.001123	.001065	.000999	.000927	.000848	.000763	.000673	.000578	.000480	.000378	-2
.79	.000274	.000169	.000064	-.000041	-.000145	-.000247	-.000345	-.000440	-.000530	-.000614	6
.80	-.000692	-.000762	-.000825	-.000880	-.000925	-.000961	-.000988	-.001004	-.001010	-.001005	10
.81	-.000990	-.000965	-.000929	-.000883	-.000828	-.000764	-.000691	-.000610	-.000523	-.000429	6
.82	-.000329	-.000226	-.000119	-.000009	.000101	.000212	.000321	.000428	.000532	.000631	-4
.83	.000725	.000812	.000891	.000962	.001024	.001075	.001116	.001146	.001164	.001170	-11
.84	.001164	.001147	.001117	.001076	.001024	.000961	.000888	.000806	.000717	.000619	-6
.85	.000516	.000409	.000297	.000184	.000070	-.000044	-.000155	-.000263	-.000366	-.000463	6
.86	-.000553	-.000634	-.000705	-.000766	-.000815	-.000851	-.000875	-.000886	-.000884	-.000868	13
.87	-.000839	-.000797	-.000743	-.000676	-.000599	-.000513	-.000417	-.000314	-.000205	-.000091	5
.88	.000025	.000143	.000261	.000376	.000488	.000594	.000694	.000785	.000866	.000936	-10
.89	.000994	.001039	.001071	.001088	.001090	.001078	.001051	.001010	.000955	.000887	-12
.90	.000808	.000718	.000619	.000512	.000399	.000282	.000163	.000044	-.000074	-.000189	4
.91	-.000298	-.000400	-.000493	-.000575	-.000645	-.000701	-.000744	-.000771	-.000783	-.000779	16
.92	-.000759	-.000723	-.000672	-.000607	-.000528	-.000438	-.000338	-.000229	-.000113	.000008	5
.93	.000131	.000254	.000375	.000492	.000603	.000705	.000797	.000877	.000943	.000994	-14
.94	.001030	.001049	.001051	.001036	.001004	.000955	.000892	.000814	.000724	.000623	-10
.95	.000513	.000396	.000275	.000153	.000030	-.000089	-.000202	-.000307	-.000402	-.000485	12
.96	-.000554	-.000607	-.000644	-.000662	-.000663	-.000645	-.000609	-.000556	-.000486	-.000401	15
.97	-.000303	-.000194	-.000076	.000049	.000177	.000307	.000435	.000559	.000675	.000782	-9
.98	.000877	.000958	.001024	.001072	.001102	.001114	.001106	.001081	.001037	.000978	-15
.99	.000903	.000816	.000720	.000615	.000507	.000409	.000097	.000066	.000037	.000013	5

TABLE 23

MTF, Defocused Lens, Monochromatic – $\Delta = 30.0$ Rayleigh Units [$\Delta \approx d/2\lambda F^2$]

ν_r	.000	.001	.002	.003	.004	.005	.006	.007	.008	.009	Δ_2
.00	1.000000	.994307	.979912	.957135	.926434	.888401	.843741	.793256	.737831	.678408	-3996
.01	.615971	.551521	.486061	.420570	.355988	.293200	.233018	.176171	.123289	.074902	4495
.02	.031428	-.006825	-.039664	-.067006	-.088876	-.105398	-.116791	-.123353	-.125457	-.123535	4026
.03	-.118066	-.109562	-.098558	-.085597	-.071219	-.055949	-.040286	-.024696	-.009601	.004624	-869
.04	.017658	.029236	.039154	.047263	.053477	.057762	.060139	.060676	.059484	.056713	-1579
.05	.052541	.047173	.040828	.033739	.026142	.018269	.010348	.002592	-.004804	-.011666	535
.06	-.017842	-.023210	-.027674	-.031165	-.033647	-.035108	-.035563	-.035052	-.033639	-.031404	822
.07	-.028445	-.024873	-.020807	-.016373	-.011700	-.006916	-.002144	.002498	.006901	.010967	-335
.08	.014614	.017769	.020378	.022401	.023815	.024611	.024796	.024391	.023430	.021957	-510
.09	.020028	.017706	.015060	.012163	.009093	.005926	.002737	-.000401	-.003417	-.006250	184
.10	-.008844	-.011148	-.013123	-.014738	-.015969	-.016805	-.017240	-.017281	-.016939	-.016235	361
.11	-.015198	-.013861	-.012261	-.010443	-.008451	-.006333	-.004137	-.001910	.000300	.002449	-60
.12	.004496	.006402	.008133	.009660	.010960	.012014	.012809	.013337	.013597	.013591	-264
.13	.013328	.012821	.012087	.011146	.010023	.008745	.007340	.005837	.004267	.002662	-35
.14	.001051	-.000535	-.002070	-.003527	-.004881	-.006113	-.007203	-.008137	-.008903	-.009492	177
.15	-.009899	-.010122	-.010163	-.010025	-.009717	-.009248	-.008631	-.007879	-.007010	-.006041	100
.16	-.004990	-.003876	-.002720	-.001541	-.000359	.000807	.001940	.003022	.004038	.004973	-79
.17	.005816	.006556	.007184	.007693	.008080	.008341	.008476	.008486	.008373	.008144	-116
.18	.007804	.007361	.006824	.006203	.005509	.004754	.003949	.003108	.002242	.001365	-11
.19	.000488	-.000376	-.001217	-.002023	-.002784	-.003493	-.004139	-.004718	-.005223	-.005648	79
.20	-.005992	-.006251	-.006424	-.006511	-.006514	-.006433	-.006273	-.006037	-.005731	-.005358	66
.21	-.004926	-.004441	-.003910	-.003341	-.002740	-.002117	-.001478	-.000831	-.000184	.000455	-6
.22	.001081	.001686	.002263	.002808	.003316	.003780	.004199	.004567	.004883	.005145	-54
.23	.005350	.005498	.005590	.005625	.005604	.005529	.005402	.005225	.005002	.004736	-42
.24	.004429	.004087	.003713	.003311	.002887	.002444	.001987	.001520	.001049	.000577	0
.25	.000108	-.000352	-.000801	-.001235	-.001649	-.002042	-.002411	-.002751	-.003063	-.003343	31
.26	-.003590	-.003803	-.003980	-.004122	-.004228	-.004298	-.004331	-.004330	-.004295	-.004226	33
.27	-.004125	-.003994	-.003835	-.003648	-.003437	-.003204	-.002949	-.002677	-.002389	-.002088	13
.28	-.001777	-.001456	-.001130	-.000800	-.000470	-.000140	.000187	.000509	.000823	.001128	-8
.29	.001422	.001704	.001972	.002224	.002460	.002678	.002878	.003059	.003220	.003360	-19
.30	.003480	.003580	.003658	.003715	.003752	.003769	.003765	.003743	.003701	.003642	-17
.31	.003565	.003471	.003362	.003238	.003101	.002951	.002790	.002618	.002437	.002247	-7
.32	.002051	.001848	.001641	.001430	.001216	.001000	.000784	.000568	.000353	.000141	2
.33	-.000068	-.000274	-.000475	-.000670	-.000860	-.001043	-.001218	-.001386	-.001546	-.001697	9
.34	-.001840	-.001973	-.002096	-.002210	-.002314	-.002408	-.002493	-.002567	-.002631	-.002685	10
.35	-.002729	-.002763	-.002788	-.002803	-.002809	-.002806	-.002795	-.002775	-.002747	-.002711	8
.36	-.002668	-.002618	-.002561	-.002497	-.002428	-.002353	-.002272	-.002187	-.002098	-.002004	4
.37	-.001907	-.001806	-.001702	-.001596	-.001487	-.001377	-.001265	-.001152	-.001038	-.000923	1
.38	-.000808	-.000693	-.000578	-.000464	-.000350	-.000237	-.000126	-.000016	.000092	.000199	-1
.39	.000304	.000406	.000507	.000605	.000700	.000793	.000883	.000971	.001055	.001137	-2
.40	.001216	.001291	.001364	.001434	.001501	.001564	.001625	.001682	.001737	.001788	-2
.41	.001836	.001882	.001925	.001964	.002001	.002035	.002067	.002096	.002122	.002145	-2
.42	.002167	.002186	.002202	.002216	.002229	.002239	.002247	.002253	.002257	.002260	-1
.43	.002260	.002259	.002257	.002253	.002248	.002241	.002233	.002224	.002214	.002203	0
.44	.002191	.002178	.002164	.002149	.002134	.002118	.002101	.002084	.002066	.002048	0
.45	.002030	.002011	.001992	.001972	.001953	.001933	.001913	.001893	.001873	.001854	0
.46	.001834	.001814	.001794	.001775	.001755	.001736	.001717	.001699	.001680	.001662	0
.47	.001644	.001627	.001609	.001593	.001576	.001560	.001545	.001529	.001515	.001500	0
.48	.001486	.001473	.001460	.001448	.001436	.001424	.001413	.001402	.001392	.001383	1
.49	.001374	.001365	.001357	.001350	.001343	.001336	.001330	.001325	.001320	.001315	0

TABLE 23

MTF, Defocused Lens, Monochromatic – Δ = 30.0 Rayleigh Units [$\Delta \approx d/2\lambda F^2$]

ν_r	.000	.001	.002	.003	.004	.005	.006	.007	.008	.009	Δ_2
.50	.001311	.001308	.001305	.001302	.001300	.001298	.001297	.001296	.001296	.001296	0
.51	.001297	.001298	.001299	.001301	.001303	.001306	.001309	.001312	.001316	.001320	0
.52	.001324	.001328	.001333	.001339	.001344	.001350	.001356	.001362	.001368	.001375	0
.53	.001381	.001388	.001395	.001402	.001409	.001416	.001424	.001431	.001438	.001445	0
.54	.001452	.001459	.001466	.001473	.001480	.001486	.001493	.001499	.001504	.001510	0
.55	.001515	.001520	.001524	.001528	.001531	.001534	.001537	.001539	.001540	.001541	0
.56	.001541	.001540	.001539	.001537	.001534	.001530	.001526	.001520	.001514	.001506	0
.57	.001498	.001489	.001478	.001467	.001454	.001440	.001425	.001409	.001392	.001373	0
.58	.001354	.001333	.001310	.001287	.001262	.001235	.001208	.001179	.001148	.001117	0
.59	.001084	.001049	.001014	.000977	.000938	.000899	.000858	.000816	.000773	.000728	0
.60	.000683	.000636	.000588	.000540	.000490	.000440	.000389	.000337	.000284	.000231	0
.61	.000177	.000123	.000069	.000014	-.000040	-.000095	-.000150	-.000204	-.000258	-.000312	0
.62	-.000365	-.000417	-.000468	-.000519	-.000568	-.000616	-.000663	-.000709	-.000752	-.000794	2
.63	-.000834	-.000872	-.000908	-.000941	-.000972	-.001000	-.001026	-.001049	-.001069	-.001086	3
.64	-.001100	-.001110	-.001118	-.001122	-.001122	-.001119	-.001113	-.001102	-.001089	-.001071	4
.65	-.001050	-.001026	-.000997	-.000965	-.000930	-.000891	-.000849	-.000804	-.000755	-.000704	3
.66	-.000649	-.000592	-.000533	-.000471	-.000406	-.000340	-.000273	-.000203	-.000133	-.000062	1
.67	.000010	.000083	.000155	.000227	.000299	.000369	.000439	.000507	.000573	.000637	-1
.68	.000699	.000757	.000813	.000865	.000913	.000958	.000998	.001034	.001065	.001091	-4
.69	.001111	.001127	.001137	.001142	.001141	.001134	.001121	.001103	.001079	.001049	-5
.70	.001014	.000974	.000928	.000878	.000822	.000762	.000698	.000630	.000559	.000485	-2
.71	.000408	.000329	.000248	.000166	.000084	.000001	-.000081	-.000162	-.000242	-.000320	2
.72	-.000395	-.000467	-.000535	-.000599	-.000658	-.000712	-.000760	-.000803	-.000839	-.000868	7
.73	-.000891	-.000906	-.000913	-.000914	-.000906	-.000891	-.000869	-.000839	-.000802	-.000757	7
.74	-.000706	-.000649	-.000586	-.000517	-.000443	-.000365	-.000283	-.000198	-.000111	-.000022	2
.75	.000067	.000157	.000246	.000333	.000418	.000500	.000578	.000651	.000719	.000781	-5
.76	.000836	.000883	.000923	.000954	.000977	.000991	.000996	.000991	.000977	.000954	-8
.77	.000922	.000881	.000832	.000775	.000710	.000639	.000561	.000479	.000391	.000301	-2
.78	.000208	.000113	.000018	-.000076	-.000169	-.000259	-.000345	-.000426	-.000501	-.000569	7
.79	-.000630	-.000682	-.000726	-.000759	-.000783	-.000796	-.000798	-.000789	-.000770	-.000739	11
.80	-.000699	-.000649	-.000589	-.000521	-.000445	-.000363	-.000274	-.000181	-.000085	.000014	2
.81	.000113	.000212	.000309	.000402	.000491	.000574	.000650	.000718	.000777	.000826	-9
.82	.000863	.000890	.000904	.000907	.000897	.000874	.000840	.000795	.000738	.000672	-9
.83	.000596	.000512	.000422	.000326	.000227	.000125	.000023	-.000079	-.000178	-.000273	4
.84	-.000362	-.000443	-.000516	-.000579	-.000632	-.000672	-.000699	-.000713	-.000714	-.000701	14
.85	-.000675	-.000635	-.000583	-.000519	-.000445	-.000362	-.000270	-.000173	-.000071	.000034	3
.86	.000140	.000244	.000346	.000442	.000532	.000613	.000684	.000744	.000791	.000824	-13
.87	.000843	.000847	.000837	.000811	.000771	.000718	.000652	.000575	.000488	.000393	-7
.88	.000292	.000186	.000079	-.000029	-.000133	-.000233	-.000327	-.000411	-.000485	-.000546	12
.89	-.000594	-.000627	-.000645	-.000647	-.000632	-.000602	-.000556	-.000496	-.000423	-.000338	11
.90	-.000244	-.000143	-.000036	.000074	.000184	.000292	.000395	.000491	.000578	.000653	-10
.91	.000715	.000762	.000793	.000807	.000805	.000784	.000747	.000695	.000627	.000547	-12
.92	.000455	.000355	.000248	.000138	.000026	-.000083	-.000187	-.000284	-.000371	-.000445	12
.93	-.000505	-.000549	-.000576	-.000584	-.000575	-.000547	-.000501	-.000439	-.000363	-.000273	13
.94	-.000174	-.000066	.000046	.000161	.000273	.000382	.000482	.000573	.000650	.000712	-14
.95	.000758	.000784	.000792	.000781	.000750	.000701	.000636	.000555	.000462	.000359	-9
.96	.000249	.000136	.000024	-.000086	-.000188	-.000281	-.000360	-.000424	-.000470	-.000497	19
.97	-.000504	-.000490	-.000456	-.000403	-.000332	-.000245	-.000145	-.000036	.000081	.000200	3
.98	.000319	.000433	.000538	.000633	.000713	.000776	.000820	.000844	.000848	.000830	-20
.99	.000793	.000738	.000666	.000582	.000488	.000400	.000093	.000064	.000037	.000013	4

TABLE 23

MTF, Defocused Lens, Monochromatic – Δ 35.0 Rayleigh Units [Δ ≈ d/2λF²]

ν_r	.000	.001	.002	.003	.004	.005	.006	.007	.008	.009	Δ_2
.00	1.000000	.992714	.973629	.943288	.902495	.852283	.793881	.728679	.658182	.583967	-3717
.01	.507639	.430788	.354941	.281533	.211864	.147073	.088118	.035755	-.009471	-.047227	7470
.02	-.077392	-.100045	-.115453	-.124051	-.126418	-.123252	-.115341	-.103535	-.088716	-.071772	2126
.03	-.053566	-.034918	-.016579	.000784	.016602	.030416	.041879	.050758	.056935	.060399	-2712
.04	.061238	.059628	.055819	.050120	.042886	.034502	.025362	.015861	.006380	-.002732	370
.05	-.011160	-.018636	-.024947	-.029933	-.033493	-.035584	-.036218	-.035459	-.033415	-.030233	1137
.06	-.026093	-.021197	-.015762	-.010010	-.004162	.001568	.006986	.011917	.016213	.019756	-752
.07	.022461	.024274	.025177	.025181	.024329	.022689	.020354	.017433	.014052	.010343	-326
.08	.006444	.002493	-.001377	-.005044	-.008395	-.011336	-.013787	-.015690	-.017008	-.017720	605
.09	-.017830	-.017357	-.016340	-.014833	-.012902	-.010623	-.008081	-.005363	-.002561	.000236	-3
.10	.002945	.005485	.007786	.009789	.011443	.012712	.013572	.014013	.014035	.013653	-403
.11	.012892	.011787	.010382	.008725	.006874	.004888	.002825	.000748	-.001287	-.003223	98
.12	-.005011	-.006606	-.007971	-.009076	-.009900	-.010429	-.010660	-.010594	-.010243	-.009625	267
.13	-.008765	-.007692	-.006441	-.005050	-.003559	-.002009	-.000440	.001106	.002591	.003980	-95
.14	.005242	.006348	.007277	.008010	.008535	.008845	.008940	.008823	.008502	.007992	-189
.15	.007309	.006475	.005513	.004448	.003309	.002123	.000919	-.000276	-.001435	-.002533	61
.16	-.003547	-.004457	-.005247	-.005901	-.006410	-.006766	-.006965	-.007009	-.006899	-.006643	147
.17	-.006249	-.005730	-.005100	-.004374	-.003570	-.002706	-.001802	-.000876	.000051	.000962	-15
.18	.001840	.002667	.003428	.004112	.004705	.005201	.005590	.005869	.006035	.006089	-112
.19	.006030	.005865	.005598	.005236	.004790	.004268	.003682	.003044	.002366	.001662	-26
.20	.000944	.000225	-.000482	-.001167	-.001817	-.002422	-.002974	-.003465	-.003888	-.004238	73
.21	-.004511	-.004704	-.004816	-.004848	-.004800	-.004676	-.004479	-.004215	-.003888	-.003506	56
.22	-.003075	-.002604	-.002099	-.001571	-.001026	-.000474	.000078	.000621	.001147	.001651	-22
.23	.002124	.002561	.002957	.003308	.003609	.003857	.004051	.004189	.004271	.004296	-55
.24	.004267	.004184	.004049	.003867	.003640	.003372	.003068	.002731	.002368	.001983	-21
.25	.001582	.001169	.000750	.000330	-.000085	-.000492	-.000886	-.001263	-.001617	-.001947	25
.26	-.002249	-.002520	-.002758	-.002961	-.003128	-.003258	-.003350	-.003404	-.003421	-.003401	37
.27	-.003346	-.003256	-.003134	-.002981	-.002800	-.002593	-.002364	-.002114	-.001846	-.001564	14
.28	-.001271	-.000970	-.000663	-.000354	-.000045	.000260	.000559	.000849	.001129	.001395	-12
.29	.001647	.001881	.002097	.002293	.002467	.002620	.002750	.002857	.002940	.003000	-23
.30	.003036	.003049	.003039	.003007	.002954	.002881	.002787	.002676	.002548	.002404	-15
.31	.002247	.002076	.001894	.001703	.001504	.001298	.001087	.000873	.000657	.000440	0
.32	.000225	.000012	-.000198	-.000403	-.000602	-.000793	-.000977	-.001152	-.001317	-.001471	10
.33	-.001615	-.001747	-.001867	-.001974	-.002069	-.002151	-.002220	-.002276	-.002319	-.002349	13
.34	-.002367	-.002372	-.002365	-.002347	-.002317	-.002276	-.002225	-.002164	-.002094	-.002016	9
.35	-.001929	-.001835	-.001734	-.001627	-.001514	-.001397	-.001275	-.001150	-.001021	-.000891	2
.36	-.000759	-.000625	-.000491	-.000357	-.000224	-.000091	.000040	.000169	.000296	.000420	-2
.37	.000541	.000659	.000773	.000883	.000988	.001089	.001186	.001278	.001364	.001446	-4
.38	.001522	.001593	.001659	.001719	.001774	.001824	.001868	.001907	.001941	.001970	-4
.39	.001993	.002012	.002026	.002035	.002040	.002040	.002036	.002028	.002017	.002001	-3
.40	.001982	.001959	.001933	.001904	.001872	.001838	.001801	.001761	.001720	.001676	-1
.41	.001631	.001584	.001535	.001485	.001434	.001382	.001329	.001275	.001221	.001166	0
.42	.001111	.001055	.000999	.000944	.000888	.000832	.000777	.000722	.000668	.000614	0
.43	.000561	.000508	.000456	.000404	.000354	.000304	.000255	.000208	.000161	.000115	1
.44	.000070	.000026	-.000017	-.000059	-.000099	-.000139	-.000177	-.000215	-.000251	-.000286	1
.45	-.000320	-.000353	-.000385	-.000416	-.000445	-.000474	-.000501	-.000528	-.000553	-.000578	1
.46	-.000601	-.000624	-.000645	-.000666	-.000685	-.000704	-.000722	-.000739	-.000755	-.000770	1
.47	-.000784	-.000798	-.000811	-.000823	-.000834	-.000845	-.000855	-.000864	-.000873	-.000881	1
.48	-.000888	-.000895	-.000901	-.000907	-.000912	-.000916	-.000920	-.000923	-.000926	-.000928	1
.49	-.000930	-.000931	-.000932	-.000933	-.000933	-.000932	-.000931	-.000930	-.000928	-.000926	0

TABLE 23

MTF, Defocused Lens, Monochromatic – $\Delta = 35.0$ Rayleigh Units $[\Delta \approx d/2\lambda F^2]$

ν_r	.000	.001	.002	.003	.004	.005	.006	.007	.008	.009	Δ_2
.50	-.000923	-.000920	-.000917	-.000913	-.000908	-.000904	-.000899	-.000893	-.000887	-.000881	0
.51	-.000874	-.000867	-.000860	-.000852	-.000843	-.000834	-.000825	-.000816	-.000806	-.000795	0
.52	-.000784	-.000773	-.000761	-.000749	-.000736	-.000723	-.000709	-.000695	-.000681	-.000666	0
.53	-.000650	-.000634	-.000617	-.000600	-.000583	-.000564	-.000546	-.000527	-.000507	-.000487	1
.54	-.000466	-.000444	-.000423	-.000400	-.000377	-.000354	-.000330	-.000305	-.000280	-.000255	1
.55	-.000228	-.000202	-.000175	-.000147	-.000119	-.000090	-.000061	-.000032	-.000002	.000028	0
.56	.000059	.000090	.000121	.000153	.000185	.000217	.000249	.000281	.000314	.000347	0
.57	.000379	.000412	.000445	.000477	.000510	.000542	.000574	.000606	.000637	.000668	0
.58	.000698	.000728	.000758	.000786	.000814	.000841	.000867	.000892	.000916	.000939	0
.59	.000961	.000981	.001000	.001018	.001034	.001049	.001062	.001073	.001083	.001091	-1
.60	.001097	.001100	.001102	.001102	.001100	.001095	.001088	.001079	.001068	.001054	-1
.61	.001038	.001020	.000999	.000976	.000951	.000923	.000893	.000861	.000826	.000789	-1
.62	.000750	.000709	.000666	.000622	.000575	.000527	.000477	.000425	.000372	.000319	0
.63	.000264	.000208	.000152	.000095	.000037	-.000020	-.000077	-.000134	-.000190	-.000246	1
.64	-.000301	-.000354	-.000406	-.000456	-.000505	-.000551	-.000595	-.000636	-.000675	-.000711	3
.65	-.000743	-.000772	-.000798	-.000819	-.000837	-.000851	-.000861	-.000866	-.000867	-.000864	4
.66	-.000856	-.000843	-.000826	-.000805	-.000779	-.000749	-.000715	-.000676	-.000633	-.000587	4
.67	-.000537	-.000484	-.000427	-.000368	-.000306	-.000242	-.000176	-.000109	-.000041	.000028	1
.68	.000097	.000166	.000235	.000302	.000368	.000432	.000493	.000551	.000606	.000658	-3
.69	.000705	.000748	.000786	.000819	.000846	.000868	.000883	.000893	.000896	.000892	-5
.70	.000883	.000867	.000844	.000816	.000781	.000740	.000694	.000642	.000586	.000525	-3
.71	.000460	.000392	.000320	.000246	.000171	.000094	.000017	-.000060	-.000135	-.000209	2
.72	-.000281	-.000350	-.000414	-.000475	-.000530	-.000580	-.000623	-.000660	-.000690	-.000713	7
.73	-.000727	-.000734	-.000732	-.000723	-.000705	-.000679	-.000646	-.000605	-.000557	-.000502	7
.74	-.000441	-.000374	-.000303	-.000227	-.000149	-.000068	.000014	.000097	.000179	.000260	0
.75	.000338	.000413	.000484	.000549	.000608	.000660	.000705	.000741	.000769	.000787	-8
.76	.000796	.000795	.000784	.000764	.000734	.000695	.000648	.000592	.000529	.000459	-6
.77	.000384	.000304	.000220	.000134	.000047	-.000040	-.000125	-.000208	-.000287	-.000361	5
.78	-.000429	-.000489	-.000541	-.000584	-.000617	-.000640	-.000652	-.000652	-.000641	-.000620	11
.79	-.000587	-.000544	-.000491	-.000429	-.000359	-.000283	-.000200	-.000114	-.000024	.000066	1
.80	.000157	.000246	.000331	.000412	.000486	.000552	.000610	.000657	.000694	.000718	-11
.81	.000731	.000731	.000718	.000693	.000656	.000608	.000549	.000480	.000404	.000320	-6
.82	.000232	.000139	.000046	-.000048	-.000139	-.000226	-.000308	-.000382	-.000446	-.000501	10
.83	-.000544	-.000574	-.000590	-.000593	-.000583	-.000558	-.000520	-.000469	-.000407	-.000335	10
.84	-.000254	-.000166	-.000073	.000022	.000119	.000214	.000305	.000391	.000469	.000537	-9
.85	.000594	.000639	.000669	.000685	.000686	.000672	.000642	.000599	.000542	.000474	-11
.86	.000395	.000308	.000215	.000118	.000019	-.000078	-.000172	-.000260	-.000339	-.000408	11
.87	-.000464	-.000507	-.000534	-.000546	-.000541	-.000520	-.000483	-.000431	-.000366	-.000289	12
.88	-.000203	-.000109	-.000010	.000091	.000191	.000287	.000377	.000458	.000527	.000583	-12
.89	.000623	.000648	.000655	.000645	.000617	.000573	.000514	.000441	.000357	.000264	-8
.90	.000165	.000063	-.000038	-.000137	-.000228	-.000311	-.000381	-.000437	-.000477	-.000499	18
.91	-.000503	-.000488	-.000456	-.000406	-.000340	-.000261	-.000171	-.000072	.000030	.000135	1
.92	.000237	.000333	.000421	.000496	.000557	.000601	.000627	.000633	.000620	.000588	-18
.93	.000538	.000471	.000390	.000298	.000198	.000094	-.000011	-.000112	-.000206	-.000289	11
.94	-.000359	-.000412	-.000446	-.000461	-.000455	-.000428	-.000382	-.000318	-.000238	-.000146	13
.95	-.000045	.000061	.000168	.000272	.000369	.000455	.000526	.000580	.000615	.000628	-20
.96	.000620	.000590	.000541	.000473	.000390	.000296	.000193	.000087	-.000019	-.000118	5
.97	-.000209	-.000285	-.000344	-.000384	-.000402	-.000397	-.000370	-.000321	-.000253	-.000168	17
.98	-.000070	.000037	.000149	.000260	.000366	.000462	.000545	.000609	.000654	.000676	-21
.99	.000675	.000652	.000608	.000545	.000468	.000390	.000087	.000062	.000036	.000013	3

TABLE 23

MTF, Defocused Lens, Monochromatic – Δ = 40.0 Rayleigh Units [Δ ≈ d/2λF²]

ν_r	.000	.001	.002	.003	.004	.005	.006	.007	.008	.009	Δ_2
.00	1.000000	.990878	.966412	.927477	.875382	.811808	.738745	.658405	.573143	.485366	-2514
.01	.397449	.311649	.230036	.154424	.086325	.026909	-.023016	-.063012	-.093004	-.113256	9739
.02	-.124343	-.127101	-.122583	-.111999	-.096659	-.077917	-.057112	-.035519	-.014306	.005510	-1397
.03	.023082	.037758	.049089	.056830	.060927	.061509	.058859	.053391	.045617	.036116	-1726
.04	.025502	.014391	.003374	-.007010	-.016294	-.024100	-.030149	-.034267	-.036386	-.036538	1968
.05	-.034844	-.031508	-.026798	-.021026	-.014538	-.007687	-.000824	.005723	.011662	.016747	-854
.06	.020785	.023645	.025257	.025614	.024765	.022814	.019906	.016226	.011978	.007387	-344
.07	.002676	-.001937	-.006247	-.010078	-.013280	-.015741	-.017386	-.018181	-.018129	-.017270	807
.08	-.015678	-.013453	-.010720	-.007618	-.004294	-.000901	.002415	.005517	.008284	.010613	-437
.09	.012425	.013665	.014305	.014342	.013798	.012718	.011165	.009220	.006976	.004533	-197
.10	.001997	-.000530	-.002948	-.005166	-.007106	-.008701	-.009904	-.010680	-.011017	-.010916	438
.11	-.010396	-.009491	-.008250	-.006729	-.004994	-.003117	-.001172	.000768	.002632	.004356	-140
.12	.005881	.007159	.008153	.008836	.009195	.009226	.008939	.008353	.007500	.006415	-230
.13	.005143	.003734	.002239	.000711	-.000797	-.002237	-.003562	-.004732	-.005713	-.006479	215
.14	-.007012	-.007300	-.007343	-.007145	-.006721	-.006089	-.005277	-.004314	-.003235	-.002076	80
.15	-.000875	.000329	.001501	.002605	.003610	.004489	.005219	.005783	.006169	.006370	-184
.16	.006385	.006220	.005883	.005389	.004757	.004007	.003164	.002253	.001303	.000339	-12
.17	-.000610	-.001521	-.002368	-.003132	-.003793	-.004337	-.004753	-.005032	-.005172	-.005172	140
.18	-.005036	-.004771	-.004387	-.003896	-.003315	-.002659	-.001946	-.001197	-.000430	.000336	0
.19	.001081	.001790	.002444	.003032	.003539	.003956	.004276	.004494	.004607	.004615	-104
.20	.004521	.004328	.004045	.003679	.003241	.002742	.002194	.001610	.001005	.000391	-8
.21	-.000219	-.000811	-.001374	-.001897	-.002369	-.002783	-.003131	-.003407	-.003609	-.003734	77
.22	-.003780	-.003750	-.003646	-.003470	-.003230	-.002930	-.002577	-.002181	-.001748	-.001289	27
.23	-.000812	-.000327	.000158	.000634	.001092	.001525	.001926	.002288	.002606	.002875	-48
.24	.003092	.003255	.003362	.003413	.003407	.003347	.003235	.003074	.002867	.002619	-40
.25	.002335	.002021	.001681	.001322	.000949	.000569	.000188	-.000189	-.000555	-.000907	15
.26	-.001239	-.001547	-.001827	-.002075	-.002289	-.002466	-.002606	-.002707	-.002768	-.002789	39
.27	-.002772	-.002718	-.002627	-.002503	-.002347	-.002162	-.001952	-.001719	-.001468	-.001201	15
.28	-.000922	-.000636	-.000346	-.000054	.000234	.000516	.000789	.001049	.001294	.001522	-16
.29	.001730	.001917	.002081	.002220	.002335	.002423	.002485	.002521	.002531	.002516	-24
.30	.002476	.002413	.002327	.002220	.002094	.001950	.001791	.001617	.001431	.001236	-9
.31	.001033	.000824	.000611	.000397	.000183	-.000029	-.000237	-.000439	-.000634	-.000821	9
.32	-.000997	-.001162	-.001315	-.001454	-.001579	-.001690	-.001785	-.001865	-.001930	-.001978	16
.33	-.002011	-.002028	-.002030	-.002016	-.001989	-.001948	-.001893	-.001826	-.001748	-.001658	11
.34	-.001559	-.001451	-.001334	-.001211	-.001082	-.000947	-.000809	-.000667	-.000523	-.000378	1
.35	-.000232	-.000088	.000056	.000197	.000336	.000471	.000601	.000727	.000848	.000962	-5
.36	.001071	.001173	.001267	.001355	.001435	.001507	.001572	.001628	.001677	.001718	-7
.37	.001751	.001776	.001794	.001804	.001807	.001803	.001792	.001775	.001751	.001721	-5
.38	.001685	.001644	.001599	.001548	.001493	.001434	.001372	.001306	.001237	.001165	-2
.39	.001091	.001016	.000938	.000860	.000780	.000700	.000619	.000538	.000457	.000376	0
.40	.000296	.000217	.000138	.000061	-.000015	-.000089	-.000162	-.000232	-.000301	-.000368	2
.41	-.000433	-.000496	-.000556	-.000614	-.000670	-.000723	-.000774	-.000823	-.000869	-.000912	2
.42	-.000953	-.000992	-.001028	-.001062	-.001093	-.001122	-.001149	-.001173	-.001195	-.001215	2
.43	-.001233	-.001249	-.001263	-.001276	-.001286	-.001295	-.001301	-.001307	-.001311	-.001313	1
.44	-.001314	-.001313	-.001312	-.001309	-.001305	-.001300	-.001294	-.001287	-.001279	-.001270	1
.45	-.001261	-.001251	-.001241	-.001229	-.001218	-.001206	-.001193	-.001181	-.001168	-.001154	0
.46	-.001141	-.001127	-.001113	-.001100	-.001086	-.001072	-.001058	-.001044	-.001031	-.001017	0
.47	-.001004	-.000991	-.000978	-.000965	-.000952	-.000940	-.000928	-.000916	-.000905	-.000894	0
.48	-.000883	-.000873	-.000863	-.000853	-.000844	-.000835	-.000827	-.000819	-.000811	-.000804	0
.49	-.000797	-.000791	-.000785	-.000779	-.000774	-.000769	-.000765	-.000761	-.000758	-.000755	0

TABLE 23

MTF, Defocused Lens, Monochromatic – Δ = 40.0 Rayleigh Units [$\Delta \approx d/2\lambda F^2$]

ν_r	.000	.001	.002	.003	.004	.005	.006	.007	.008	.009	Δ_2
.50	-.000752	-.000750	-.000749	-.000747	-.000746	-.000746	-.000746	-.000746	-.000747	-.000748	0
.51	-.000749	-.000751	-.000753	-.000755	-.000758	-.000761	-.000765	-.000768	-.000772	-.000776	0
.52	-.000781	-.000785	-.000790	-.000795	-.000800	-.000806	-.000811	-.000817	-.000823	-.000829	0
.53	-.000835	-.000841	-.000847	-.000853	-.000859	-.000865	-.000870	-.000876	-.000882	-.000887	0
.54	-.000893	-.000898	-.000902	-.000907	-.000911	-.000915	-.000919	-.000922	-.000924	-.000926	0
.55	-.000928	-.000929	-.000930	-.000930	-.000929	-.000928	-.000926	-.000923	-.000919	-.000915	1
.56	-.000909	-.000903	-.000896	-.000888	-.000879	-.000869	-.000857	-.000845	-.000832	-.000817	1
.57	-.000802	-.000785	-.000767	-.000748	-.000728	-.000706	-.000683	-.000659	-.000634	-.000608	1
.58	-.000580	-.000551	-.000521	-.000490	-.000458	-.000424	-.000390	-.000354	-.000318	-.000280	1
.59	-.000242	-.000203	-.000163	-.000123	-.000081	-.000040	.000002	.000045	.000087	.000130	0
.60	.000172	.000215	.000257	.000299	.000341	.000382	.000422	.000461	.000499	.000536	0
.61	.000572	.000606	.000639	.000670	.000699	.000726	.000751	.000773	.000793	.000811	-2
.62	.000826	.000838	.000847	.000853	.000856	.000856	.000853	.000847	.000837	.000824	-2
.63	.000807	.000788	.000765	.000738	.000709	.000676	.000641	.000602	.000561	.000518	-2
.64	.000471	.000423	.000372	.000320	.000266	.000211	.000155	.000098	.000041	-.000017	0
.65	-.000074	-.000131	-.000187	-.000242	-.000295	-.000346	-.000395	-.000441	-.000485	-.000525	3
.66	-.000562	-.000595	-.000624	-.000648	-.000668	-.000683	-.000693	-.000699	-.000698	-.000693	5
.67	-.000682	-.000667	-.000645	-.000619	-.000588	-.000552	-.000511	-.000466	-.000417	-.000364	4
.68	-.000307	-.000248	-.000187	-.000123	-.000058	.000008	.000074	.000140	.000206	.000269	0
.69	.000331	.000391	.000447	.000499	.000547	.000591	.000629	.000661	.000688	.000708	-5
.70	.000721	.000728	.000727	.000720	.000705	.000684	.000655	.000620	.000579	.000532	-5
.71	.000479	.000422	.000360	.000294	.000226	.000155	.000083	.000011	-.000062	-.000133	1
.72	-.000202	-.000268	-.000330	-.000388	-.000440	-.000487	-.000526	-.000559	-.000583	-.000600	8
.73	-.000608	-.000607	-.000598	-.000579	-.000553	-.000518	-.000476	-.000426	-.000369	-.000307	6
.74	-.000240	-.000169	-.000094	-.000018	.000060	.000137	.000212	.000285	.000355	.000419	-4
.75	.000478	.000529	.000573	.000608	.000635	.000651	.000658	.000654	.000640	.000616	-9
.76	.000583	.000540	.000488	.000429	.000363	.000292	.000215	.000136	.000055	-.000026	0
.77	-.000106	-.000183	-.000256	-.000323	-.000384	-.000437	-.000480	-.000514	-.000537	-.000548	11
.78	-.000548	-.000536	-.000513	-.000479	-.000434	-.000380	-.000317	-.000246	-.000170	-.000089	4
.79	-.000006	.000079	.000163	.000244	.000320	.000391	.000454	.000508	.000551	.000583	-10
.80	.000603	.000610	.000604	.000586	.000554	.000511	.000457	.000393	.000321	.000242	-6
.81	.000158	.000072	-.000016	-.000101	-.000183	-.000260	-.000329	-.000388	-.000437	-.000473	12
.82	-.000496	-.000505	-.000500	-.000481	-.000448	-.000402	-.000344	-.000276	-.000199	-.000116	7
.83	-.000029	.000061	.000150	.000236	.000317	.000390	.000453	.000504	.000543	.000567	-13
.84	.000575	.000569	.000547	.000510	.000459	.000397	.000323	.000241	.000154	.000063	-2
.85	-.000029	-.000118	-.000202	-.000278	-.000344	-.000398	-.000438	-.000462	-.000470	-.000461	17
.86	-.000436	-.000395	-.000340	-.000272	-.000194	-.000107	-.000015	.000078	.000171	.000259	-3
.87	.000340	.000411	.000469	.000512	.000540	.000549	.000541	.000516	.000474	.000416	-14
.88	.000345	.000264	.000174	.000081	-.000015	-.000107	-.000194	-.000272	-.000337	-.000388	15
.89	-.000422	-.000438	-.000435	-.000414	-.000375	-.000318	-.000248	-.000166	-.000075	.000021	5
.90	.000118	.000212	.000300	.000377	.000441	.000489	.000520	.000530	.000521	.000492	-19
.91	.000445	.000381	.000304	.000216	.000120	.000022	-.000074	-.000165	-.000246	-.000314	13
.92	-.000365	-.000398	-.000411	-.000402	-.000373	-.000325	-.000259	-.000179	-.000088	.000010	7
.93	.000110	.000207	.000298	.000377	.000441	.000487	.000513	.000517	.000500	.000461	-20
.94	.000403	.000328	.000241	.000145	.000045	-.000054	-.000147	-.000229	-.000297	-.000346	18
.95	-.000374	-.000379	-.000362	-.000323	-.000263	-.000186	-.000096	.000003	.000106	.000207	-1
.96	.000300	.000382	.000447	.000492	.000515	.000515	.000490	.000444	.000377	.000295	-15
.97	.000201	.000100	-.000001	-.000097	-.000183	-.000253	-.000303	-.000331	-.000334	-.000313	25
.98	-.000267	-.000200	-.000116	-.000018	.000087	.000194	.000296	.000388	.000464	.000521	-19
.99	.000554	.000563	.000546	.000507	.000447	.000381	.000082	.000060	.000036	.000013	2

TABLE 23

MTF, Defocused Lens, Monochromatic – Δ = 45.0 Rayleigh Units [Δ ≈ d/2λF²]

ν_r	.000	.001	.002	.003	.004	.005	.006	.007	.008	.009	Δ_2
.00	1.000000	.988801	.958276	.909770	.845302	.767457	.679255	.584002	.485130	.386043	-214
.01	.289963	.199800	.118036	.046638	-.012996	-.060066	-.094364	-.116241	-.126550	-.126565	10294
.02	-.117889	-.102347	-.081881	-.058443	-.033899	-.009944	.011972	.030687	.045366	.055510	-4535
.03	.060947	.061810	.058498	.051627	.041970	.030400	.017826	.005138	-.006846	-.017424	1407
.04	-.026037	-.032296	-.035989	-.037077	-.035686	-.032084	-.026655	-.019865	-.012231	-.004284	313
.05	.003464	.010548	.016574	.021240	.024345	.025794	.025599	.023868	.020794	.016638	-1081
.06	.011709	.006342	.000881	-.004348	-.009050	-.012978	-.015941	-.017819	-.018556	-.018169	1125
.07	-.016734	-.014386	-.011303	-.007694	-.003789	.000179	.003985	.007423	.010318	.012536	-677
.08	.013983	.014614	.014430	.013475	.011834	.009624	.006988	.004083	.001076	-.001872	60
.09	-.004607	-.006994	-.008922	-.010310	-.011107	-.011295	-.010887	-.009927	-.008485	-.006652	391
.10	-.004535	-.002249	.000084	.002348	.004433	.006243	.007699	.008743	.009339	.009474	-460
.11	.009157	.008419	.007308	.005891	.004244	.002452	.000602	-.001217	-.002923	-.004442	187
.12	-.005710	-.006679	-.007313	-.007596	-.007526	-.007117	-.006398	-.005411	-.004207	-.002845	158
.13	-.001390	.000094	.001542	.002892	.004092	.005094	.005863	.006373	.006610	.006574	-273
.14	.006274	.005729	.004969	.004031	.002957	.001795	.000592	-.000601	-.001740	-.002780	98
.15	-.003684	-.004422	-.004969	-.005310	-.005437	-.005353	-.005065	-.004591	-.003952	-.003178	136
.16	-.002300	-.001354	-.000375	.000601	.001538	.002404	.003170	.003812	.004310	.004652	-156
.17	.004829	.004839	.004687	.004381	.003938	.003373	.002711	.001975	.001191	.000387	-20
.18	-.000411	-.001177	-.001888	-.002522	-.003060	-.003489	-.003797	-.003978	-.004030	-.003955	128
.19	-.003757	-.003446	-.003034	-.002536	-.001970	-.001353	-.000705	-.000046	.000604	.001228	-26
.20	.001807	.002327	.002774	.003137	.003408	.003582	.003656	.003631	.003510	.003299	-90
.21	.003005	.002638	.002210	.001733	.001221	.000689	.000150	-.000382	-.000893	-.001371	33
.22	-.001804	-.002184	-.002501	-.002750	-.002926	-.003027	-.003052	-.003002	-.002880	-.002691	67
.23	-.002441	-.002137	-.001788	-.001402	-.000989	-.000559	-.000122	.000312	.000734	.001134	-20
.24	.001505	.001839	.002130	.002372	.002563	.002698	.002778	.002800	.002767	.002679	-53
.25	.002541	.002356	.002129	.001865	.001571	.001252	.000916	.000569	.000217	-.000131	3
.26	-.000471	-.000795	-.001099	-.001378	-.001626	-.001841	-.002019	-.002158	-.002257	-.002314	41
.27	-.002331	-.002307	-.002244	-.002144	-.002009	-.001843	-.001649	-.001430	-.001192	-.000937	16
.28	-.000671	-.000397	-.000121	.000154	.000423	.000683	.000929	.001160	.001371	.001560	-21
.29	.001725	.001864	.001976	.002060	.002115	.002142	.002141	.002112	.002056	.001976	-24
.30	.001872	.001746	.001601	.001439	.001263	.001074	.000875	.000670	.000460	.000249	-1
.31	.000039	-.000168	-.000370	-.000563	-.000747	-.000920	-.001079	-.001224	-.001353	-.001465	17
.32	-.001560	-.001637	-.001695	-.001735	-.001757	-.001760	-.001746	-.001715	-.001667	-.001604	15
.33	-.001526	-.001435	-.001332	-.001218	-.001094	-.000963	-.000824	-.000680	-.000532	-.000381	3
.34	-.000229	-.000077	.000074	.000222	.000367	.000507	.000642	.000770	.000891	.001003	-7
.35	.001107	.001202	.001288	.001363	.001428	.001483	.001527	.001561	.001585	.001598	-9
.36	.001601	.001595	.001579	.001554	.001520	.001478	.001429	.001372	.001309	.001240	-5
.37	.001165	.001085	.001001	.000914	.000823	.000730	.000635	.000538	.000441	.000343	0
.38	.000246	.000149	.000053	-.000041	-.000133	-.000224	-.000311	-.000396	-.000477	-.000555	3
.39	-.000629	-.000700	-.000766	-.000828	-.000887	-.000940	-.000990	-.001035	-.001075	-.001111	4
.40	-.001143	-.001170	-.001193	-.001212	-.001227	-.001237	-.001244	-.001247	-.001246	-.001242	3
.41	-.001235	-.001224	-.001210	-.001193	-.001174	-.001152	-.001128	-.001102	-.001073	-.001043	2
.42	-.001010	-.000977	-.000942	-.000905	-.000868	-.000830	-.000790	-.000751	-.000710	-.000669	0
.43	-.000628	-.000587	-.000545	-.000504	-.000462	-.000421	-.000380	-.000339	-.000299	-.000259	0
.44	-.000220	-.000182	-.000144	-.000106	-.000070	-.000034	.000001	.000035	.000068	.000101	0
.45	.000132	.000163	.000192	.000221	.000249	.000276	.000302	.000327	.000351	.000374	0
.46	.000396	.000418	.000438	.000458	.000476	.000494	.000511	.000528	.000543	.000558	0
.47	.000571	.000585	.000597	.000609	.000620	.000630	.000640	.000649	.000657	.000665	0
.48	.000672	.000679	.000685	.000690	.000695	.000700	.000704	.000707	.000710	.000713	0
.49	.000715	.000717	.000718	.000719	.000719	.000719	.000719	.000718	.000717	.000716	0

TABLE 23

MTF, Defocused Lens, Monochromatic – $\Delta = 45.0$ Rayleigh Units $[\Delta \approx d/2\lambda F^2]$

ν_r	.000	.001	.002	.003	.004	.005	.006	.007	.008	.009	Δ_2
.50	.000714	.000712	.000709	.000706	.000703	.000699	.000695	.000691	.000686	.000681	0
.51	.000675	.000669	.000663	.000657	.000650	.000642	.000634	.000626	.000618	.000609	0
.52	.000599	.000590	.000580	.000569	.000558	.000546	.000534	.000522	.000509	.000496	0
.53	.000482	.000468	.000453	.000438	.000422	.000406	.000389	.000372	.000355	.000336	0
.54	.000318	.000298	.000279	.000259	.000238	.000217	.000195	.000173	.000151	.000128	0
.55	.000105	.000081	.000057	.000032	.000008	-.000017	-.000043	-.000068	-.000094	-.000120	0
.56	-.000146	-.000172	-.000199	-.000225	-.000251	-.000277	-.000303	-.000329	-.000355	-.000380	0
.57	-.000405	-.000429	-.000453	-.000476	-.000499	-.000521	-.000542	-.000562	-.000582	-.000600	1
.58	-.000617	-.000633	-.000648	-.000661	-.000673	-.000683	-.000692	-.000699	-.000705	-.000709	2
.59	-.000711	-.000711	-.000709	-.000705	-.000699	-.000691	-.000680	-.000668	-.000653	-.000637	2
.60	-.000618	-.000597	-.000574	-.000548	-.000521	-.000492	-.000460	-.000427	-.000392	-.000355	2
.61	-.000316	-.000276	-.000235	-.000192	-.000148	-.000104	-.000058	-.000012	.000034	.000080	0
.62	.000127	.000173	.000219	.000264	.000308	.000351	.000392	.000432	.000470	.000505	-1
.63	.000538	.000569	.000597	.000622	.000643	.000661	.000675	.000686	.000693	.000696	-3
.64	.000694	.000689	.000679	.000666	.000648	.000626	.000600	.000570	.000536	.000498	-3
.65	.000458	.000414	.000367	.000318	.000266	.000213	.000158	.000102	.000045	-.000012	0
.66	-.000069	-.000124	-.000179	-.000232	-.000283	-.000331	-.000377	-.000418	-.000456	-.000490	4
.67	-.000518	-.000542	-.000560	-.000573	-.000580	-.000581	-.000577	-.000566	-.000549	-.000526	6
.68	-.000498	-.000464	-.000424	-.000380	-.000332	-.000279	-.000223	-.000164	-.000103	-.000040	2
.69	.000023	.000087	.000150	.000212	.000272	.000329	.000382	.000432	.000476	.000515	-5
.70	.000547	.000574	.000593	.000605	.000609	.000606	.000595	.000576	.000550	.000517	-6
.71	.000477	.000431	.000379	.000322	.000261	.000196	.000129	.000060	-.000009	-.000077	1
.72	-.000144	-.000208	-.000269	-.000324	-.000374	-.000418	-.000455	-.000483	-.000503	-.000514	9
.73	-.000516	-.000509	-.000493	-.000468	-.000433	-.000391	-.000342	-.000285	-.000223	-.000157	5
.74	-.000086	-.000014	.000059	.000132	.000203	.000270	.000333	.000391	.000441	.000483	-7
.75	.000517	.000541	.000554	.000557	.000549	.000531	.000502	.000464	.000416	.000360	-7
.76	.000298	.000229	.000156	.000081	.000004	-.000072	-.000146	-.000215	-.000279	-.000336	7
.77	-.000385	-.000423	-.000451	-.000468	-.000473	-.000465	-.000446	-.000415	-.000373	-.000321	10
.78	-.000261	-.000193	-.000119	-.000041	.000038	.000117	.000194	.000267	.000334	.000393	-7
.79	.000442	.000480	.000506	.000520	.000519	.000506	.000479	.000440	.000390	.000329	-9
.80	.000260	.000185	.000105	.000023	-.000058	-.000137	-.000211	-.000278	-.000335	-.000381	11
.81	-.000414	-.000434	-.000439	-.000430	-.000406	-.000369	-.000318	-.000257	-.000186	-.000108	7
.82	-.000025	.000059	.000143	.000222	.000296	.000361	.000415	.000456	.000483	.000494	-14
.83	.000489	.000469	.000433	.000384	.000322	.000250	.000170	.000086	-.000001	-.000086	1
.84	-.000166	-.000239	-.000302	-.000353	-.000389	-.000409	-.000412	-.000398	-.000367	-.000321	16
.85	-.000261	-.000190	-.000109	-.000023	.000065	.000152	.000234	.000309	.000372	.000422	-13
.86	.000455	.000472	.000470	.000450	.000413	.000361	.000294	.000216	.000131	.000042	-3
.87	-.000048	-.000133	-.000212	-.000279	-.000332	-.000369	-.000387	-.000387	-.000367	-.000329	18
.88	-.000274	-.000205	-.000124	-.000036	.000055	.000146	.000231	.000307	.000371	.000419	-15
.89	.000448	.000457	.000447	.000416	.000367	.000301	.000223	.000136	.000044	-.000049	0
.90	-.000136	-.000215	-.000280	-.000330	-.000359	-.000368	-.000356	-.000323	-.000271	-.000202	17
.91	-.000120	-.000029	.000065	.000158	.000245	.000320	.000381	.000423	.000444	.000442	-21
.92	.000419	.000374	.000311	.000233	.000144	.000050	-.000045	-.000134	-.000213	-.000277	15
.93	-.000322	-.000345	-.000345	-.000322	-.000277	-.000212	-.000132	-.000041	.000056	.000152	0
.94	.000241	.000319	.000381	.000421	.000439	.000432	.000401	.000349	.000277	.000191	-13
.95	.000097	-.000001	-.000094	-.000178	-.000247	-.000296	-.000322	-.000323	-.000298	-.000250	24
.96	-.000181	-.000096	-.000001	.000099	.000196	.000284	.000356	.000409	.000437	.000440	-25
.97	.000417	.000369	.000300	.000215	.000119	.000021	-.000074	-.000158	-.000225	-.000270	22
.98	-.000290	-.000282	-.000248	-.000189	-.000109	-.000015	.000088	.000191	.000287	.000370	-13
.99	.000432	.000471	.000482	.000466	.000425	.000373	.000075	.000058	.000035	.000013	1

TABLE 23

MTF, Defocused Lens, Monochromatic – Δ = 50.0 Rayleigh Units [$\Delta \approx d/2\lambda F^2$]

ν_T	.000	.001	.002	.003	.004	.005	.006	.007	.008	.009	Δ_2
.00	1.000000	.986482	.949236	.890244	.812485	.719748	.616393	.507095	.396568	.289309	3269
.01	.189363	.100126	.024193	-.036723	-.081834	-.111243	-.125873	-.127342	-.117808	-.099794	8480
.02	-.076001	-.049129	-.021711	.004027	.026277	.043715	.055541	.061468	.061690	.056811	-5101
.03	.047757	.035676	.021830	.007491	-.006157	-.018093	-.027526	-.033928	-.037050	-.036920	3253
.04	-.033811	-.028203	-.020730	-.012117	-.003120	.005535	.013202	.019358	.023631	.025822	-2082
.05	.025900	.023996	.020380	.015434	.009613	.003407	-.002697	-.008250	-.012871	-.016272	1220
.06	-.018270	-.018798	-.017896	-.015709	-.012465	-.008453	-.004003	.000543	.004855	.008637	-529
.07	.011647	.013710	.014727	.014676	.013610	.011652	.008977	.005802	.002368	-.001080	-13
.08	-.004308	-.007105	-.009304	-.010781	-.011470	-.011357	-.010485	-.008941	-.006855	-.004385	384
.09	-.001708	.000996	.003551	.005798	.007607	.008879	.009555	.009613	.009075	.007995	-540
.10	.006462	.004585	.002493	.000322	-.001794	-.003728	-.005370	-.006633	-.007453	-.007799	475
.11	-.007666	-.007078	-.006085	-.004759	-.003189	-.001472	.000287	.001986	.003532	.004842	-235
.12	.005849	.006507	.006791	.006697	.006242	.005462	.004410	.003151	.001761	.000317	-52
.13	-.001101	-.002420	-.003572	-.004502	-.005168	-.005544	-.005619	-.005397	-.004900	-.004161	242
.14	-.003224	-.002142	-.000973	.000220	.001380	.002448	.003377	.004123	.004657	.004959	-231
.15	.005020	.004845	.004448	.003855	.003098	.002217	.001257	.000264	-.000716	-.001639	58
.16	-.002463	-.003156	-.003688	-.004042	-.004206	-.004178	-.003965	-.003580	-.003045	-.002388	123
.17	-.001638	-.000831	-.000002	.000813	.001582	.002272	.002858	.003319	.003638	.003807	-149
.18	.003823	.003689	.003414	.003012	.002503	.001909	.001255	.000568	-.000125	-.000797	21
.19	-.001424	-.001983	-.002456	-.002826	-.003082	-.003219	-.003234	-.003130	-.002913	-.002594	102
.20	-.002186	-.001707	-.001175	-.000609	-.000030	.000541	.001087	.001588	.002029	.002397	-72
.21	.002681	.002874	.002972	.002974	.002883	.002703	.002442	.002110	.001720	.001285	-44
.22	.000820	.000340	-.000141	-.000608	-.001047	-.001445	-.001793	-.002080	-.002300	-.002448	72
.23	-.002522	-.002521	-.002447	-.002304	-.002098	-.001835	-.001525	-.001177	-.000801	-.000408	17
.24	-.000009	.000385	.000765	.001119	.001440	.001721	.001955	.002137	.002264	.002334	-56
.25	.002346	.002303	.002205	.002058	.001864	.001631	.001364	.001071	.000759	.000436	-10
.26	.000109	-.000214	-.000527	-.000822	-.001093	-.001334	-.001543	-.001713	-.001844	-.001933	42
.27	-.001979	-.001982	-.001944	-.001865	-.001749	-.001599	-.001418	-.001212	-.000984	-.000739	16
.28	-.000483	-.000221	.000042	.000301	.000552	.000790	.001011	.001211	.001387	.001537	-25
.29	.001659	.001751	.001813	.001844	.001844	.001814	.001756	.001670	.001560	.001427	-22
.30	.001274	.001104	.000920	.000725	.000523	.000317	.000109	-.000096	-.000296	-.000488	8
.31	-.000670	-.000839	-.000993	-.001131	-.001250	-.001350	-.001430	-.001489	-.001527	-.001544	21
.32	-.001541	-.001517	-.001474	-.001413	-.001335	-.001241	-.001134	-.001014	-.000883	-.000743	9
.33	-.000597	-.000445	-.000291	-.000135	.000021	.000175	.000325	.000469	.000607	.000738	-7
.34	.000859	.000970	.001070	.001159	.001235	.001299	.001350	.001388	.001413	.001425	-12
.35	.001425	.001412	.001388	.001352	.001306	.001250	.001184	.001111	.001029	.000942	-6
.36	.000848	.000750	.000647	.000542	.000435	.000326	.000217	.000109	.000001	-.000104	2
.37	-.000207	-.000306	-.000402	-.000494	-.000580	-.000662	-.000738	-.000808	-.000871	-.000929	6
.38	-.000980	-.001024	-.001062	-.001093	-.001118	-.001136	-.001147	-.001152	-.001152	-.001145	6
.39	-.001133	-.001115	-.001093	-.001065	-.001033	-.000997	-.000957	-.000913	-.000866	-.000816	3
.40	-.000764	-.000709	-.000653	-.000594	-.000535	-.000474	-.000413	-.000351	-.000289	-.000227	0
.41	-.000166	-.000104	-.000044	.000016	.000074	.000132	.000188	.000242	.000295	.000347	-1
.42	.000396	.000444	.000489	.000533	.000574	.000614	.000651	.000686	.000720	.000751	-1
.43	.000780	.000806	.000831	.000854	.000875	.000894	.000911	.000926	.000939	.000951	-1
.44	.000961	.000969	.000976	.000982	.000986	.000989	.000990	.000990	.000989	.000988	0
.45	.000985	.000981	.000976	.000971	.000965	.000958	.000950	.000942	.000934	.000925	0
.46	.000916	.000906	.000896	.000886	.000875	.000865	.000854	.000844	.000833	.000822	0
.47	.000811	.000801	.000790	.000780	.000769	.000759	.000749	.000739	.000730	.000721	0
.48	.000712	.000703	.000694	.000686	.000678	.000671	.000664	.000657	.000651	.000645	0
.49	.000639	.000634	.000629	.000624	.000620	.000616	.000612	.000609	.000607	.000604	0

TABLE 23

MTF, Defocused Lens, Monochromatic – Δ = 50.0 Rayleigh Units [$\Delta \approx d/2\lambda F^2$]

ν_r	.000	.001	.002	.003	.004	.005	.006	.007	.008	.009	Δ_2
.50	.000602	.000601	.000600	.000599	.000598	.000598	.000599	.000599	.000600	.000601	0
.51	.000603	.000605	.000607	.000609	.000612	.000615	.000618	.000622	.000626	.000630	0
.52	.000634	.000638	.000642	.000647	.000652	.000657	.000662	.000667	.000672	.000677	0
.53	.000682	.000687	.000692	.000696	.000701	.000706	.000710	.000715	.000719	.000722	0
.54	.000726	.000729	.000732	.000734	.000736	.000738	.000739	.000739	.000739	.000738	0
.55	.000737	.000735	.000732	.000728	.000724	.000719	.000713	.000706	.000698	.000689	0
.56	.000679	.000668	.000656	.000644	.000630	.000614	.000598	.000581	.000562	.000543	0
.57	.000522	.000500	.000477	.000453	.000428	.000401	.000374	.000346	.000317	.000286	0
.58	.000255	.000223	.000191	.000157	.000124	.000089	.000055	.000019	-.000016	-.000051	0
.59	-.000087	-.000122	-.000157	-.000191	-.000225	-.000259	-.000291	-.000323	-.000353	-.000382	1
.60	-.000410	-.000436	-.000461	-.000483	-.000504	-.000522	-.000538	-.000552	-.000563	-.000572	3
.61	-.000577	-.000580	-.000580	-.000577	-.000571	-.000562	-.000550	-.000534	-.000516	-.000494	3
.62	-.000470	-.000443	-.000412	-.000380	-.000344	-.000307	-.000267	-.000225	-.000181	-.000136	1
.63	-.000090	-.000042	.000005	.000054	.000102	.000149	.000196	.000242	.000286	.000329	-1
.64	.000369	.000407	.000442	.000474	.000502	.000526	.000547	.000563	.000574	.000581	-4
.65	.000583	.000580	.000572	.000559	.000541	.000518	.000491	.000459	.000423	.000383	-3
.66	.000339	.000292	.000242	.000190	.000136	.000080	.000024	-.000032	-.000087	-.000142	1
.67	-.000194	-.000244	-.000291	-.000335	-.000374	-.000409	-.000438	-.000462	-.000480	-.000491	6
.68	-.000496	-.000495	-.000486	-.000471	-.000449	-.000422	-.000387	-.000348	-.000303	-.000253	5
.69	-.000200	-.000143	-.000084	-.000023	.000039	.000101	.000161	.000220	.000275	.000327	-3
.70	.000375	.000416	.000452	.000481	.000502	.000516	.000521	.000519	.000508	.000488	-7
.71	.000461	.000427	.000385	.000336	.000283	.000224	.000162	.000097	.000031	-.000035	0
.72	-.000100	-.000163	-.000222	-.000276	-.000325	-.000367	-.000400	-.000426	-.000442	-.000448	10
.73	-.000445	-.000432	-.000410	-.000378	-.000337	-.000289	-.000234	-.000173	-.000107	-.000039	3
.74	.000031	.000101	.000169	.000234	.000294	.000347	.000394	.000431	.000459	.000476	-9
.75	.000483	.000478	.000462	.000435	.000397	.000351	.000296	.000234	.000166	.000095	-3
.76	.000022	-.000051	-.000121	-.000188	-.000249	-.000303	-.000347	-.000381	-.000404	-.000414	12
.77	-.000412	-.000398	-.000371	-.000332	-.000283	-.000225	-.000160	-.000089	-.000014	.000062	1
.78	.000137	.000208	.000274	.000332	.000381	.000418	.000443	.000454	.000452	.000435	-13
.79	.000405	.000363	.000310	.000247	.000176	.000101	.000023	-.000055	-.000130	-.000200	5
.80	-.000261	-.000313	-.000352	-.000378	-.000389	-.000385	-.000366	-.000332	-.000286	-.000227	12
.81	-.000159	-.000084	-.000005	.000076	.000154	.000228	.000294	.000350	.000392	.000420	-13
.82	.000433	.000429	.000409	.000373	.000324	.000262	.000190	.000111	.000029	-.000053	0
.83	-.000131	-.000203	-.000265	-.000314	-.000348	-.000366	-.000367	-.000350	-.000316	-.000266	15
.84	-.000204	-.000130	-.000050	.000034	.000118	.000197	.000269	.000329	.000376	.000405	-16
.85	.000417	.000410	.000384	.000342	.000284	.000213	.000133	.000049	-.000037	-.000120	3
.86	-.000195	-.000258	-.000307	-.000338	-.000351	-.000343	-.000316	-.000271	-.000210	-.000137	13
.87	-.000054	.000033	.000120	.000202	.000275	.000334	.000377	.000400	.000403	.000384	-20
.88	.000346	.000290	.000220	.000138	.000050	-.000038	-.000123	-.000199	-.000261	-.000306	17
.89	-.000331	-.000334	-.000315	-.000275	-.000217	-.000144	-.000059	.000030	.000120	.000205	-4
.90	.000279	.000337	.000376	.000393	.000388	.000360	.000310	.000243	.000163	.000074	-7
.91	-.000018	-.000106	-.000185	-.000249	-.000295	-.000318	-.000317	-.000292	-.000245	-.000179	19
.92	-.000098	-.000008	.000085	.000175	.000255	.000320	.000364	.000386	.000382	.000354	-24
.93	.000303	.000232	.000148	.000055	-.000038	-.000126	-.000202	-.000260	-.000295	-.000305	25
.94	-.000290	-.000249	-.000186	-.000105	-.000014	.000082	.000175	.000257	.000322	.000366	-21
.95	.000384	.000375	.000340	.000281	.000203	.000112	.000016	-.000078	-.000161	-.000227	17
.96	-.000271	-.000288	-.000277	-.000239	-.000177	-.000096	-.000002	.000097	.000191	.000274	-11
.97	.000338	.000377	.000388	.000371	.000326	.000257	.000171	.000075	-.000022	-.000111	8
.98	-.000185	-.000236	-.000259	-.000253	-.000217	-.000154	-.000071	.000027	.000130	.000228	-3
.99	.000313	.000378	.000416	.000424	.000403	.000365	.000069	.000055	.000034	.000013	0

TABLE 23

MTF, Defocused Lens, Monochromatic – Δ = 60.0 Rayleigh Units [Δ ≈ d/2λF²]

ν_r	.000	.001	.002	.003	.004	.005	.006	.007	.008	.009	Δ_2
.00	1.000000	.981126	.928519	.846080	.739652	.616494	.484650	.352282	.227020	.115391	13633
.01	.022381	-.048848	-.097062	-.122884	-.128571	-.117659	-.094533	-.063958	-.030617	.001291	-1431
.02	.028378	.048271	.059715	.062541	.057548	.046291	.030827	.013433	-.003663	-.018518	2241
.03	-.029652	-.036162	-.037759	-.034733	-.027869	-.018307	-.007383	.003532	.013195	.020608	-2249
.04	.025110	.026422	.024642	.020203	.013801	.006295	-.001400	-.008415	-.014019	-.017689	1935
.05	-.019148	-.018381	-.015614	-.011274	-.005931	-.000223	.005212	.009806	.013116	.014863	-1563
.06	.014948	.013455	.010627	.006833	.002521	-.001834	-.005778	-.008925	-.010992	-.011822	1238
.07	-.011391	-.009804	-.007278	-.004113	-.000656	.002734	.005723	.008033	.009467	.009923	-977
.08	.009399	.007990	.005870	.003274	.000471	-.002262	-.004669	-.006534	-.007705	-.008100	776
.09	-.007713	-.006611	-.004925	-.002834	-.000544	.001727	.003773	.005418	.006530	.007027	-613
.10	.006889	.006151	.004900	.003265	.001402	-.000518	-.002327	-.003873	-.005033	-.005721	472
.11	-.005895	-.005558	-.004756	-.003572	-.002117	-.000522	.001078	.002551	.003781	.004676	-334
.12	.005174	.005248	.004907	.004189	.003165	.001924	.000570	-.000787	-.002043	-.003103	195
.13	-.003892	-.004359	-.004478	-.004250	-.003703	-.002888	-.001873	-.000739	.000426	.001537	-54
.14	.002514	.003290	.003816	.004061	.004015	.003690	.003118	.002344	.001430	.000441	-73
.15	-.000551	-.001480	-.002282	-.002908	-.003321	-.003498	-.003434	-.003140	-.002643	-.001979	166
.16	-.001197	-.000351	.000503	.001311	.002022	.002593	.002994	.003202	.003211	.003024	-194
.17	.002659	.002143	.001510	.000803	.000064	-.000660	-.001328	-.001901	-.002349	-.002648	148
.18	-.002786	-.002758	-.002571	-.002239	-.001784	-.001235	-.000625	.000010	.000636	.001217	-43
.19	.001724	.002132	.002420	.002576	.002596	.002481	.002242	.001893	.001455	.000954	-63
.20	.000415	-.000133	-.000662	-.001146	-.001563	-.001893	-.002122	-.002242	-.002250	-.002148	110
.21	-.001944	-.001649	-.001280	-.000855	-.000397	.000074	.000535	.000966	.001348	.001665	-64
.22	.001905	.002058	.002121	.002092	.001976	.001778	.001510	.001184	.000816	.000422	-25
.23	.000018	-.000378	-.000750	-.001084	-.001368	-.001591	-.001747	-.001829	-.001838	-.001774	73
.24	-.001641	-.001447	-.001200	-.000911	-.000591	-.000252	.000091	.000427	.000743	.001030	-29
.25	.001276	.001475	.001621	.001710	.001740	.001711	.001626	.001489	.001306	.001084	-38
.26	.000831	.000556	.000269	-.000021	-.000305	-.000574	-.000820	-.001036	-.001215	-.001354	41
.27	-.001450	-.001499	-.001503	-.001461	-.001376	-.001253	-.001094	-.000906	-.000695	-.000466	17
.28	-.000227	.000015	.000254	.000484	.000698	.000892	.001060	.001199	.001306	.001380	-33
.29	.001418	.001422	.001391	.001328	.001234	.001112	.000967	.000803	.000622	.000431	-10
.30	.000233	.000034	-.000162	-.000351	-.000529	-.000692	-.000837	-.000962	-.001064	-.001142	24
.31	-.001194	-.001221	-.001222	-.001198	-.001150	-.001080	-.000990	-.000881	-.000757	-.000620	13
.32	-.000472	-.000318	-.000160	-.000001	.000157	.000309	.000455	.000591	.000716	.000828	-12
.33	.000924	.001005	.001069	.001115	.001143	.001154	.001147	.001123	.001083	.001028	-14
.34	.000960	.000878	.000786	.000685	.000576	.000461	.000341	.000220	.000098	-.000024	1
.35	-.000143	-.000258	-.000367	-.000470	-.000566	-.000653	-.000730	-.000798	-.000855	-.000901	11
.36	-.000936	-.000959	-.000972	-.000973	-.000964	-.000945	-.000916	-.000878	-.000831	-.000777	8
.37	-.000716	-.000648	-.000576	-.000498	-.000418	-.000334	-.000249	-.000163	-.000076	.000010	0
.38	.000095	.000178	.000258	.000335	.000409	.000478	.000543	.000603	.000658	.000708	-4
.39	.000752	.000790	.000822	.000849	.000870	.000885	.000894	.000898	.000897	.000890	-4
.40	.000879	.000863	.000842	.000818	.000789	.000757	.000722	.000684	.000644	.000601	-1
.41	.000556	.000510	.000462	.000413	.000364	.000314	.000263	.000213	.000163	.000113	0
.42	.000064	.000016	-.000032	-.000078	-.000123	-.000167	-.000209	-.000249	-.000288	-.000325	2
.43	-.000361	-.000394	-.000426	-.000456	-.000484	-.000510	-.000534	-.000557	-.000577	-.000596	2
.44	-.000613	-.000628	-.000642	-.000654	-.000664	-.000673	-.000681	-.000687	-.000692	-.000695	1
.45	-.000697	-.000699	-.000699	-.000698	-.000696	-.000694	-.000690	-.000686	-.000682	-.000676	1
.46	-.000670	-.000664	-.000657	-.000650	-.000642	-.000634	-.000626	-.000618	-.000609	-.000601	0
.47	-.000592	-.000583	-.000575	-.000566	-.000557	-.000549	-.000541	-.000532	-.000524	-.000516	0
.48	-.000509	-.000501	-.000494	-.000487	-.000480	-.000474	-.000468	-.000462	-.000456	-.000451	0
.49	-.000446	-.000442	-.000437	-.000434	-.000430	-.000427	-.000424	-.000422	-.000419	-.000418	0

TABLE 23

MTF, Defocused Lens, Monochromatic – Δ = 60.0 Rayleigh Units $[\Delta \approx d/2\lambda F^2]$

ν_r	.000	.001	.002	.003	.004	.005	.006	.007	.008	.009	Δ_2
.50	-.000416	-.000415	-.000414	-.000414	-.000414	-.000414	-.000415	-.000416	-.000417	-.000418	0
.51	-.000420	-.000422	-.000424	-.000427	-.000430	-.000433	-.000436	-.000440	-.000443	-.000447	0
.52	-.000451	-.000455	-.000459	-.000464	-.000468	-.000473	-.000477	-.000482	-.000486	-.000491	0
.53	-.000495	-.000499	-.000504	-.000508	-.000512	-.000515	-.000519	-.000522	-.000524	-.000527	0
.54	-.000529	-.000531	-.000532	-.000532	-.000533	-.000532	-.000531	-.000529	-.000527	-.000524	1
.55	-.000520	-.000515	-.000510	-.000503	-.000496	-.000488	-.000479	-.000468	-.000457	-.000445	1
.56	-.000432	-.000417	-.000402	-.000386	-.000368	-.000349	-.000330	-.000309	-.000287	-.000264	1
.57	-.000241	-.000216	-.000190	-.000164	-.000137	-.000109	-.000080	-.000051	-.000021	.000009	0
.58	.000039	.000069	.000100	.000130	.000161	.000190	.000220	.000248	.000276	.000303	0
.59	.000329	.000354	.000377	.000399	.000419	.000437	.000453	.000467	.000478	.000487	-1
.60	.000494	.000498	.000499	.000497	.000493	.000485	.000475	.000462	.000445	.000426	-2
.61	.000404	.000379	.000352	.000322	.000290	.000256	.000219	.000181	.000142	.000101	0
.62	.000060	.000018	-.000024	-.000066	-.000107	-.000147	-.000186	-.000223	-.000259	-.000291	2
.63	-.000321	-.000348	-.000372	-.000391	-.000407	-.000418	-.000425	-.000427	-.000424	-.000417	5
.64	-.000405	-.000388	-.000367	-.000341	-.000310	-.000276	-.000239	-.000198	-.000154	-.000108	2
.65	-.000060	-.000012	.000038	.000087	.000135	.000182	.000227	.000269	.000308	.000342	-3
.66	.000372	.000397	.000417	.000430	.000438	.000439	.000433	.000421	.000403	.000378	-5
.67	.000348	.000312	.000271	.000226	.000177	.000126	.000072	.000018	-.000037	-.000091	1
.68	-.000143	-.000192	-.000237	-.000278	-.000313	-.000341	-.000363	-.000378	-.000384	-.000383	8
.69	-.000373	-.000355	-.000330	-.000297	-.000258	-.000213	-.000162	-.000108	-.000051	.000008	2
.70	.000067	.000125	.000181	.000233	.000280	.000321	.000355	.000380	.000397	.000405	-8
.71	.000403	.000392	.000371	.000341	.000303	.000257	.000205	.000148	.000087	.000025	-1
.72	-.000038	-.000099	-.000157	-.000210	-.000257	-.000296	-.000327	-.000347	-.000357	-.000355	11
.73	-.000343	-.000319	-.000286	-.000243	-.000192	-.000135	-.000073	-.000007	.000059	.000123	0
.74	.000185	.000241	.000290	.000330	.000359	.000377	.000383	.000376	.000357	.000326	-11
.75	.000285	.000233	.000175	.000110	.000042	-.000027	-.000094	-.000157	-.000214	-.000262	9
.76	-.000299	-.000324	-.000337	-.000335	-.000319	-.000291	-.000249	-.000198	-.000137	-.000070	6
.77	.000001	.000072	.000141	.000205	.000261	.000307	.000341	.000361	.000366	.000356	-14
.78	.000330	.000291	.000240	.000179	.000111	.000038	-.000036	-.000107	-.000173	-.000229	9
.79	-.000274	-.000305	-.000320	-.000320	-.000302	-.000269	-.000222	-.000163	-.000095	-.000021	6
.80	.000055	.000129	.000198	.000257	.000304	.000336	.000352	.000350	.000330	.000294	-16
.81	.000242	.000179	.000107	.000030	-.000048	-.000122	-.000188	-.000243	-.000282	-.000305	17
.82	-.000309	-.000294	-.000260	-.000211	-.000147	-.000074	.000005	.000084	.000160	.000227	-8
.83	.000282	.000321	.000340	.000340	.000320	.000281	.000225	.000156	.000079	-.000003	-3
.84	-.000083	-.000156	-.000218	-.000265	-.000292	-.000299	-.000284	-.000249	-.000195	-.000127	15
.85	-.000048	.000035	.000117	.000193	.000256	.000303	.000329	.000334	.000316	.000277	-20
.86	.000218	.000146	.000064	-.000021	-.000103	-.000176	-.000234	-.000273	-.000290	-.000282	24
.87	-.000252	-.000201	-.000132	-.000051	.000035	.000121	.000198	.000261	.000306	.000327	-22
.88	.000324	.000296	.000245	.000176	.000094	.000007	-.000080	-.000158	-.000221	-.000263	21
.89	-.000281	-.000273	-.000240	-.000184	-.000110	-.000024	.000066	.000152	.000227	.000283	-17
.90	.000316	.000323	.000302	.000257	.000189	.000107	.000016	-.000074	-.000154	-.000219	17
.91	-.000260	-.000274	-.000261	-.000219	-.000155	-.000073	.000018	.000110	.000194	.000261	-16
.92	.000305	.000321	.000307	.000265	.000199	.000116	.000023	-.000070	-.000153	-.000217	18
.93	-.000257	-.000267	-.000246	-.000196	-.000124	-.000035	.000061	.000153	.000231	.000288	-21
.94	.000316	.000313	.000279	.000217	.000135	.000040	-.000056	-.000142	-.000209	-.000249	27
.95	-.000258	-.000234	-.000179	-.000100	-.000007	.000092	.000183	.000257	.000305	.000320	-31
.96	.000301	.000251	.000174	.000081	-.000018	-.000111	-.000186	-.000234	-.000249	-.000228	35
.97	-.000175	-.000096	.000001	.000102	.000196	.000270	.000315	.000325	.000300	.000241	-32
.98	.000157	.000059	-.000041	-.000129	-.000194	-.000227	-.000223	-.000182	-.000109	-.000014	23
.99	.000092	.000193	.000278	.000335	.000357	.000351	.000054	.000049	.000033	.000013	-3

TABLE 23

MTF, Defocused Lens, Monochromatic – Δ = 70.0 Rayleigh Units [Δ ≈ d/2λF²]

ν_r	.000	.001	.002	.003	.004	.005	.006	.007	.008	.009	Δ_2
.00	1.000000	.974821	.904422	.795748	.659062	.506720	.351743	.206394	.080919	-.017398	27158
.01	-.084763	-.120950	-.128958	-.114310	-.084121	-.046073	-.007420	.025825	.049472	.061417	*****
.02	.061613	.051769	.034852	.014487	-.005665	-.022448	-.033619	-.038068	-.035851	-.028047	5588
.03	-.016477	-.003352	.009107	.019007	.025036	.026603	.023861	.017609	.009117	-.000116	-740
.04	-.008603	-.015097	-.018755	-.019231	-.016690	-.011737	-.005297	.001552	.007750	.012409	-1538
.05	.014933	.015083	.012989	.009101	.004103	-.001203	-.006023	-.009681	-.011713	-.011920	1826
.06	-.010373	-.007393	-.003484	.000745	.004678	.007773	.009638	.010074	.009093	.006901	-1210
.07	.003862	.000435	-.002890	-.005663	-.007533	-.008289	-.007880	-.006414	-.004135	-.001384	473
.08	.001454	.003999	.005930	.007021	.007163	.006379	.004809	.002686	.000305	-.002020	56
.09	-.003996	-.005389	-.006046	-.005914	-.005040	-.003562	-.001684	.000348	.002278	.003877	-331
.10	.004962	.005422	.005225	.004417	.003117	.001496	-.000245	-.001897	-.003271	-.004217	427
.11	-.004642	-.004514	-.003867	-.002792	-.001425	.000070	.001521	.002769	.003681	.004168	-424
.12	.004190	.003760	.002939	.001828	.000555	-.000737	-.001912	-.002848	-.003454	-.003677	383
.13	-.003506	-.002970	-.002137	-.001099	.000030	.001133	.002099	.002836	.003277	.003386	-330
.14	.003162	.002638	.001873	.000948	-.000042	-.001001	-.001837	-.002475	-.002861	-.002964	282
.15	-.002783	-.002343	-.001693	-.000897	-.000033	.000819	.001583	.002191	.002593	.002761	-234
.16	.002684	.002377	.001873	.001220	.000478	-.000285	-.001004	-.001620	-.002083	-.002359	188
.17	-.002429	-.002293	-.001968	-.001485	-.000888	-.000228	.000441	.001064	.001593	.001989	-132
.18	.002224	.002284	.002169	.001891	.001478	.000962	.000386	-.000205	-.000767	-.001258	70
.19	-.001646	-.001902	-.002013	-.001974	-.001791	-.001482	-.001070	-.000588	-.000072	.000443	-1
.20	.000920	.001327	.001639	.001837	.001909	.001855	.001681	.001401	.001037	.000613	-58
.21	.000160	-.000294	-.000720	-.001090	-.001383	-.001583	-.001679	-.001669	-.001554	-.001345	94
.22	-.001057	-.000709	-.000322	.000078	.000470	.000829	.001136	.001374	.001532	.001602	-87
.23	.001583	.001478	.001294	.001044	.000742	.000407	.000058	-.000288	-.000612	-.000896	39
.24	-.001128	-.001296	-.001394	-.001418	-.001368	-.001249	-.001069	-.000837	-.000566	-.000272	25
.25	.000033	.000333	.000613	.000861	.001067	.001221	.001318	.001354	.001330	.001248	-57
.26	.001113	.000932	.000714	.000470	.000210	-.000054	-.000310	-.000548	-.000758	-.000934	36
.27	-.001067	-.001154	-.001192	-.001181	-.001123	-.001021	-.000880	-.000706	-.000507	-.000290	17
.28	-.000066	.000159	.000375	.000576	.000753	.000902	.001017	.001096	.001136	.001138	-38
.29	.001101	.001029	.000924	.000790	.000634	.000459	.000274	.000083	-.000107	-.000290	7
.30	-.000461	-.000614	-.000746	-.000852	-.000931	-.000981	-.001001	-.000991	-.000952	-.000886	27
.31	-.000795	-.000684	-.000554	-.000411	-.000259	-.000101	.000058	.000214	.000362	.000499	-10
.32	.000623	.000729	.000817	.000884	.000930	.000953	.000953	.000933	.000891	.000830	-18
.33	.000753	.000659	.000554	.000438	.000315	.000188	.000059	-.000069	-.000193	-.000311	6
.34	-.000420	-.000520	-.000607	-.000681	-.000741	-.000785	-.000814	-.000828	-.000825	-.000808	15
.35	-.000777	-.000732	-.000675	-.000608	-.000531	-.000446	-.000354	-.000259	-.000160	-.000060	1
.36	.000040	.000138	.000233	.000323	.000407	.000485	.000554	.000616	.000668	.000711	-9
.37	.000744	.000767	.000780	.000783	.000777	.000762	.000738	.000706	.000666	.000619	-6
.38	.000567	.000509	.000447	.000381	.000313	.000242	.000170	.000098	.000027	-.000044	1
.39	-.000113	-.000179	-.000243	-.000303	-.000359	-.000411	-.000459	-.000502	-.000539	-.000572	5
.40	-.000600	-.000622	-.000639	-.000651	-.000658	-.000660	-.000657	-.000649	-.000638	-.000622	4
.41	-.000602	-.000579	-.000553	-.000524	-.000492	-.000457	-.000421	-.000383	-.000344	-.000303	1
.42	-.000262	-.000220	-.000177	-.000135	-.000093	-.000051	-.000009	.000031	.000071	.000110	0
.43	.000147	.000184	.000219	.000252	.000284	.000315	.000343	.000370	.000396	.000419	-1
.44	.000441	.000462	.000480	.000497	.000512	.000526	.000538	.000549	.000558	.000566	0
.45	.000572	.000577	.000581	.000584	.000586	.000587	.000587	.000586	.000584	.000581	0
.46	.000578	.000574	.000569	.000564	.000559	.000553	.000547	.000540	.000533	.000526	0
.47	.000519	.000512	.000505	.000497	.000490	.000482	.000475	.000468	.000461	.000454	0
.48	.000447	.000440	.000434	.000427	.000421	.000415	.000410	.000405	.000400	.000395	0
.49	.000390	.000386	.000383	.000379	.000376	.000373	.000371	.000368	.000367	.000365	0

TABLE 23

MTF, Defocused Lens, Monochromatic – Δ = 70.0 Rayleigh Units [$\Delta \approx d/2\lambda F^2$]

ν_r	.000	.001	.002	.003	.004	.005	.006	.007	.008	.009	Δ_2
.50	.000364	.000363	.000363	.000362	.000362	.000363	.000363	.000364	.000366	.000367	0
.51	.000369	.000371	.000373	.000376	.000379	.000382	.000385	.000388	.000392	.000395	0
.52	.000399	.000403	.000407	.000411	.000415	.000419	.000424	.000428	.000432	.000436	0
.53	.000440	.000443	.000447	.000450	.000453	.000456	.000458	.000460	.000462	.000463	0
.54	.000464	.000464	.000464	.000463	.000461	.000459	.000456	.000453	.000448	.000443	0
.55	.000437	.000430	.000422	.000413	.000403	.000392	.000380	.000367	.000354	.000339	0
.56	.000323	.000305	.000287	.000268	.000248	.000227	.000205	.000182	.000159	.000134	0
.57	.000109	.000083	.000057	.000031	.000004	-.000023	-.000051	-.000078	-.000105	-.000131	0
.58	-.000158	-.000183	-.000208	-.000232	-.000254	-.000276	-.000296	-.000314	-.000331	-.000345	2
.59	-.000358	-.000368	-.000376	-.000381	-.000384	-.000384	-.000382	-.000376	-.000368	-.000357	3
.60	-.000343	-.000326	-.000307	-.000284	-.000259	-.000232	-.000203	-.000171	-.000138	-.000103	2
.61	-.000067	-.000030	.000008	.000046	.000083	.000121	.000157	.000192	.000225	.000256	-1
.62	.000285	.000311	.000333	.000352	.000367	.000378	.000385	.000387	.000384	.000377	-4
.63	.000365	.000348	.000327	.000302	.000273	.000240	.000203	.000164	.000122	.000079	-1
.64	.000035	-.000010	-.000055	-.000099	-.000141	-.000181	-.000217	-.000251	-.000279	-.000303	5
.65	-.000322	-.000335	-.000342	-.000342	-.000336	-.000324	-.000306	-.000281	-.000251	-.000215	5
.66	-.000175	-.000131	-.000085	-.000036	.000015	.000065	.000114	.000161	.000205	.000245	-3
.67	.000281	.000310	.000332	.000348	.000356	.000355	.000347	.000331	.000307	.000275	-6
.68	.000238	.000194	.000146	.000094	.000041	-.000014	-.000068	-.000120	-.000169	-.000212	5
.69	-.000250	-.000280	-.000303	-.000316	-.000320	-.000314	-.000299	-.000274	-.000241	-.000200	8
.70	-.000152	-.000099	-.000043	.000015	.000074	.000130	.000183	.000230	.000271	.000302	-7
.71	.000324	.000336	.000336	.000326	.000304	.000272	.000231	.000182	.000127	.000067	-3
.72	.000005	-.000056	-.000115	-.000169	-.000217	-.000255	-.000283	-.000300	-.000304	-.000295	13
.73	-.000274	-.000242	-.000198	-.000146	-.000087	-.000024	.000041	.000105	.000165	.000218	-5
.74	.000263	.000296	.000317	.000324	.000318	.000297	.000263	.000217	.000162	.000100	-6
.75	.000033	-.000034	-.000100	-.000160	-.000212	-.000252	-.000279	-.000292	-.000289	-.000271	15
.76	-.000238	-.000191	-.000134	-.000070	-.000000	.000070	.000137	.000198	.000248	.000286	-12
.77	.000309	.000315	.000304	.000276	.000234	.000178	.000113	.000042	-.000031	-.000101	3
.78	-.000164	-.000217	-.000256	-.000279	-.000283	-.000269	-.000238	-.000191	-.000130	-.000061	9
.79	.000014	.000088	.000158	.000219	.000266	.000296	.000307	.000299	.000271	.000226	-17
.80	.000166	.000096	.000019	-.000058	-.000130	-.000191	-.000239	-.000268	-.000277	-.000264	21
.81	-.000232	-.000181	-.000115	-.000040	.000039	.000117	.000187	.000243	.000282	.000301	-20
.82	.000297	.000270	.000224	.000160	.000085	.000004	-.000076	-.000149	-.000209	-.000250	18
.83	-.000270	-.000266	-.000238	-.000189	-.000122	-.000044	.000040	.000121	.000193	.000250	-15
.84	.000286	.000297	.000284	.000247	.000188	.000114	.000030	-.000055	-.000134	-.000199	14
.85	-.000244	-.000265	-.000260	-.000228	-.000173	-.000101	-.000016	.000071	.000153	.000222	-12
.86	.000270	.000293	.000288	.000256	.000199	.000124	.000037	-.000052	-.000134	-.000201	15
.87	-.000245	-.000262	-.000250	-.000210	-.000145	-.000064	.000027	.000116	.000194	.000253	-18
.88	.000286	.000289	.000262	.000207	.000131	.000041	-.000051	-.000136	-.000204	-.000246	25
.89	-.000258	-.000238	-.000189	-.000114	-.000025	.000070	.000158	.000230	.000276	.000290	-30
.90	.000272	.000222	.000146	.000055	-.000041	-.000130	-.000201	-.000244	-.000255	-.000230	35
.91	-.000174	-.000093	.000002	.000100	.000187	.000251	.000286	.000285	.000248	.000181	-30
.92	.000091	-.000008	-.000104	-.000183	-.000235	-.000252	-.000232	-.000177	-.000094	.000004	16
.93	.000105	.000194	.000258	.000288	.000280	.000234	.000157	.000061	-.000042	-.000136	9
.94	-.000207	-.000245	-.000243	-.000201	-.000127	-.000030	.000075	.000172	.000246	.000286	-33
.95	.000285	.000244	.000168	.000069	-.000037	-.000134	-.000206	-.000242	-.000236	-.000188	42
.96	-.000105	-.000002	.000106	.000201	.000267	.000294	.000277	.000218	.000127	.000020	-16
.97	-.000087	-.000174	-.000227	-.000237	-.000201	-.000126	-.000023	.000089	.000191	.000266	-26
.98	.000300	.000287	.000230	.000139	.000029	-.000080	-.000167	-.000218	-.000221	-.000177	48
.99	-.000092	.000019	.000135	.000237	.000305	.000338	.000037	.000042	.000031	.000013	-6

TABLE 23

MTF, Defocused Lens, Monochromatic – Δ = 80.0 Rayleigh Units $[\Delta \approx d/2\lambda F^2]$

ν_r	.000	.001	.002	.003	.004	.005	.006	.007	.008	.009	Δ_2
.00	1.000000	.967580	.877132	.740103	.573078	.395226	.225500	.080015	-.029949	-.099064	40848
.01	-.128072	-.122955	-.093387	-.050775	-.006266	.030968	.054894	.063107	.056639	.039202	*****
.02	.016070	-.007159	-.025686	-.036328	-.037916	-.031259	-.018731	-.003608	.010694	.021313	-3683
.03	.026448	.025603	.019544	.010017	-.000706	-.010328	-.016988	-.019593	-.017959	-.012767	3558
.04	-.005351	.002623	.009515	.014031	.015446	.013706	.009379	.003496	-.002681	-.007924	934
.05	-.011272	-.012199	-.010680	-.007162	-.002444	.002497	.006706	.009425	.010225	.009066	-1959
.06	.006274	.002462	-.001604	-.005156	-.007565	-.008448	-.007722	-.005602	-.002545	.000849	337
.07	.003952	.006220	.007287	.007020	.005530	.003136	.000301	-.002460	-.004669	-.005967	911
.08	-.006170	-.005291	-.003528	-.001220	.001219	.003372	.004887	.005537	.005249	.004109	-851
.09	.002339	.000256	-.001789	-.003462	-.004508	-.004782	-.004268	-.003081	-.001433	.000397	182
.10	.002113	.003451	.004215	.004308	.003738	.002617	.001137	-.000465	-.001941	-.003072	344
.11	-.003702	-.003752	-.003236	-.002249	-.000952	.000455	.001762	.002783	.003380	.003481	-495
.12	.003087	.002272	.001162	-.000075	-.001264	-.002239	-.002873	-.003088	-.002869	-.002258	391
.13	-.001351	-.000279	.000809	.001766	.002470	.002835	.002823	.002447	.001766	.000878	-206
.14	-.000097	-.001032	-.001809	-.002334	-.002548	-.002433	-.002014	-.001350	-.000531	.000339	51
.15	.001152	.001811	.002242	.002398	.002269	.001878	.001278	.000545	-.000232	-.000961	47
.16	-.001560	-.001964	-.002133	-.002053	-.001740	-.001235	-.000600	.000091	.000762	.001338	-93
.17	.001761	.001989	.002001	.001803	.001421	.000900	.000298	-.000321	-.000890	-.001354	106
.18	-.001668	-.001803	-.001751	-.001520	-.001140	-.000652	-.000105	.000444	.000942	.001342	-97
.19	.001608	.001719	.001667	.001461	.001125	.000692	.000206	-.000287	-.000743	-.001121	78
.20	-.001389	-.001525	-.001521	-.001381	-.001119	-.000763	-.000345	.000098	.000527	.000906	-49
.21	.001203	.001397	.001474	.001429	.001270	.001012	.000679	.000299	-.000097	-.000476	16
.22	-.000810	-.001072	-.001246	-.001319	-.001288	-.001158	-.000941	-.000656	-.000325	.000026	20
.23	.000371	.000685	.000946	.001137	.001245	.001265	.001198	.001051	.000834	.000566	-51
.24	.000265	-.000048	-.000351	-.000624	-.000851	-.001017	-.001113	-.001134	-.001081	-.000960	69
.25	-.000778	-.000550	-.000290	-.000014	.000259	.000513	.000734	.000909	.001029	.001088	-60
.26	.001084	.001019	.000898	.000730	.000524	.000293	.000050	-.000190	-.000415	-.000614	27
.27	-.000776	-.000893	-.000961	-.000976	-.000941	-.000856	-.000729	-.000566	-.000377	-.000171	17
.28	.000042	.000250	.000444	.000614	.000754	.000858	.000921	.000943	.000921	.000860	-39
.29	.000762	.000632	.000478	.000306	.000124	-.000060	-.000237	-.000402	-.000546	-.000666	25
.30	-.000756	-.000814	-.000838	-.000828	-.000785	-.000713	-.000613	-.000491	-.000352	-.000202	12
.31	-.000045	.000111	.000262	.000402	.000526	.000631	.000713	.000771	.000802	.000807	-26
.32	.000786	.000740	.000671	.000583	.000478	.000361	.000234	.000103	-.000029	-.000157	3
.33	-.000279	-.000390	-.000488	-.000571	-.000635	-.000681	-.000707	-.000714	-.000700	-.000668	19
.34	-.000619	-.000554	-.000476	-.000386	-.000288	-.000184	-.000077	.000031	.000137	.000239	-3
.35	.000334	.000420	.000496	.000561	.000612	.000650	.000675	.000685	.000681	.000663	-12
.36	.000633	.000592	.000539	.000478	.000409	.000333	.000253	.000169	.000085	-.000000	0
.37	-.000083	-.000164	-.000240	-.000310	-.000375	-.000432	-.000481	-.000522	-.000554	-.000577	9
.38	-.000591	-.000596	-.000592	-.000580	-.000559	-.000532	-.000497	-.000457	-.000411	-.000360	5
.39	-.000306	-.000248	-.000189	-.000128	-.000066	-.000005	.000056	.000115	.000173	.000227	-2
.40	.000279	.000327	.000371	.000410	.000446	.000476	.000502	.000524	.000540	.000552	-4
.41	.000558	.000561	.000559	.000552	.000542	.000528	.000510	.000490	.000466	.000440	-2
.42	.000411	.000380	.000348	.000314	.000279	.000243	.000207	.000170	.000133	.000096	0
.43	.000059	.000023	-.000012	-.000046	-.000080	-.000112	-.000143	-.000173	-.000201	-.000228	1
.44	-.000254	-.000277	-.000300	-.000320	-.000339	-.000357	-.000373	-.000387	-.000400	-.000412	1
.45	-.000422	-.000430	-.000438	-.000444	-.000449	-.000452	-.000455	-.000457	-.000457	-.000457	1
.46	-.000456	-.000454	-.000452	-.000449	-.000445	-.000441	-.000436	-.000431	-.000426	-.000420	0
.47	-.000415	-.000408	-.000402	-.000396	-.000389	-.000383	-.000376	-.000370	-.000364	-.000357	0
.48	-.000351	-.000345	-.000339	-.000334	-.000328	-.000323	-.000318	-.000313	-.000308	-.000304	0
.49	-.000300	-.000296	-.000293	-.000290	-.000287	-.000285	-.000282	-.000280	-.000279	-.000278	0

TABLE 23

MTF, Defocused Lens, Monochromatic – Δ = 80.0 Rayleigh Units [$\Delta \approx d/2\lambda F^2$]

ν_r	.000	.001	.002	.003	.004	.005	.006	.007	.008	.009	Δ_2
.50	-.000277	-.000276	-.000276	-.000276	-.000276	-.000277	-.000277	-.000279	-.000280	-.000282	0
.51	-.000284	-.000286	-.000288	-.000291	-.000294	-.000297	-.000300	-.000303	-.000307	-.000311	0
.52	-.000314	-.000318	-.000322	-.000326	-.000330	-.000334	-.000338	-.000342	-.000345	-.000349	0
.53	-.000352	-.000356	-.000359	-.000361	-.000364	-.000366	-.000367	-.000369	-.000369	-.000370	1
.54	-.000369	-.000369	-.000367	-.000365	-.000362	-.000358	-.000354	-.000348	-.000342	-.000335	1
.55	-.000327	-.000318	-.000308	-.000297	-.000285	-.000272	-.000257	-.000242	-.000226	-.000209	1
.56	-.000191	-.000172	-.000152	-.000131	-.000109	-.000087	-.000063	-.000040	-.000016	.000009	0
.57	.000034	.000059	.000084	.000109	.000133	.000157	.000181	.000204	.000225	.000246	0
.58	.000266	.000284	.000300	.000314	.000327	.000337	.000345	.000351	.000354	.000355	-2
.59	.000352	.000347	.000339	.000329	.000315	.000299	.000279	.000258	.000234	.000207	-1
.60	.000178	.000148	.000116	.000082	.000048	.000013	-.000022	-.000058	-.000092	-.000126	1
.61	-.000158	-.000188	-.000216	-.000241	-.000263	-.000282	-.000297	-.000308	-.000314	-.000316	5
.62	-.000313	-.000306	-.000294	-.000277	-.000256	-.000231	-.000201	-.000169	-.000133	-.000094	3
.63	-.000054	-.000012	.000030	.000072	.000113	.000152	.000189	.000223	.000253	.000279	-4
.64	.000299	.000314	.000322	.000325	.000321	.000310	.000293	.000270	.000241	.000207	-4
.65	.000169	.000126	.000081	.000034	-.000013	-.000061	-.000106	-.000149	-.000189	-.000223	5
.66	-.000252	-.000274	-.000288	-.000295	-.000294	-.000284	-.000267	-.000241	-.000209	-.000170	6
.67	-.000126	-.000077	-.000026	.000026	.000078	.000128	.000174	.000216	.000251	.000279	-7
.68	.000298	.000308	.000308	.000298	.000278	.000249	.000212	.000168	.000118	.000064	-3
.69	.000008	-.000048	-.000102	-.000153	-.000197	-.000233	-.000261	-.000278	-.000283	-.000277	12
.70	-.000260	-.000232	-.000194	-.000147	-.000094	-.000036	.000024	.000083	.000140	.000191	-4
.71	.000234	.000267	.000289	.000299	.000295	.000278	.000249	.000208	.000158	.000101	-6
.72	.000039	-.000024	-.000086	-.000143	-.000193	-.000233	-.000260	-.000274	-.000273	-.000257	15
.73	-.000227	-.000184	-.000131	-.000070	-.000004	.000062	.000126	.000183	.000231	.000266	-11
.74	.000287	.000292	.000280	.000252	.000210	.000156	.000093	.000025	-.000044	-.000110	3
.75	-.000169	-.000217	-.000250	-.000267	-.000266	-.000248	-.000212	-.000161	-.000100	-.000030	7
.76	.000042	.000112	.000176	.000228	.000265	.000285	.000285	.000265	.000228	.000174	-15
.77	.000109	.000036	-.000039	-.000110	-.000173	-.000222	-.000253	-.000265	-.000255	-.000225	21
.78	-.000176	-.000112	-.000039	.000039	.000114	.000181	.000235	.000270	.000284	.000275	-22
.79	.000244	.000193	.000126	.000050	-.000030	-.000107	-.000174	-.000224	-.000254	-.000261	24
.80	-.000243	-.000202	-.000142	-.000068	.000014	.000095	.000169	.000228	.000267	.000281	-23
.81	.000270	.000233	.000175	.000101	.000017	-.000067	-.000144	-.000205	-.000245	-.000259	26
.82	-.000246	-.000206	-.000144	-.000065	.000021	.000107	.000183	.000240	.000273	.000278	-27
.83	.000254	.000204	.000132	.000046	-.000043	-.000127	-.000195	-.000241	-.000257	-.000244	31
.84	-.000200	-.000133	-.000048	.000044	.000132	.000206	.000257	.000279	.000268	.000227	-30
.85	.000159	.000072	-.000022	-.000112	-.000187	-.000237	-.000256	-.000241	-.000194	-.000121	26
.86	-.000030	.000066	.000156	.000226	.000269	.000278	.000251	.000192	.000109	.000012	-12
.87	-.000085	-.000169	-.000228	-.000254	-.000244	-.000197	-.000121	-.000026	.000074	.000166	-8
.88	.000236	.000274	.000274	.000236	.000165	.000071	-.000032	-.000128	-.000204	-.000247	32
.89	-.000251	-.000216	-.000145	-.000050	.000055	.000153	.000230	.000273	.000275	.000236	-41
.90	.000161	.000062	-.000045	-.000143	-.000216	-.000252	-.000244	-.000195	-.000110	-.000006	21
.91	.000103	.000196	.000259	.000281	.000257	.000191	.000095	-.000015	-.000121	-.000203	23
.92	-.000248	-.000247	-.000199	-.000114	-.000006	.000107	.000203	.000265	.000282	.000250	-48
.93	.000175	.000070	-.000045	-.000149	-.000223	-.000253	-.000232	-.000164	-.000062	.000055	15
.94	.000165	.000246	.000283	.000268	.000204	.000102	-.000016	-.000128	-.000212	-.000251	46
.95	-.000235	-.000169	-.000065	.000056	.000169	.000252	.000287	.000266	.000194	.000085	-36
.96	-.000038	-.000150	-.000226	-.000250	-.000217	-.000133	-.000016	.000109	.000215	.000279	-41
.97	.000287	.000237	.000139	.000015	-.000108	-.000202	-.000245	-.000229	-.000155	-.000040	41
.98	.000090	.000205	.000280	.000297	.000251	.000154	.000027	-.000099	-.000195	-.000237	54
.99	-.000215	-.000133	-.000010	.000125	.000241	.000321	.000020	.000035	.000029	.000013	-9

TABLE 23

MTF, Defocused Lens, Monochromatic – $\Delta = 90.0$ Rayleigh Units $[\Delta \approx d/2\lambda F^2]$

ν_r	.000	.001	.002	.003	.004	.005	.006	.007	.008	.009	Δ_2
.00	1.000000	.959417	.846858	.680085	.484159	.286669	.112734	-.019157	-.099727	-.129480	50818
.01	-.117334	-.077778	-.027284	.019216	.051030	.063155	.056299	.035661	.008929	-.015983	1820
.02	-.032885	-.038562	-.033126	-.019507	-.002334	.013455	.023899	.026891	.022489	.012643	-5442
.03	.000489	-.010581	-.017805	-.019686	-.016241	-.008845	.000251	.008566	.014041	.015515	-4000
.04	.012936	.007276	.000203	-.006398	-.010904	-.012341	-.010570	-.006254	-.000637	.004807	-172
.05	.008750	.010325	.009296	.006077	.001582	-.003027	-.006634	-.008435	-.008105	-.005844	1931
.06	-.002296	.001631	.004992	.007033	.007346	.005950	.003255	-.000051	-.003180	-.005427	882
.07	-.006325	-.005735	-.003863	-.001190	.001648	.004013	.005401	.005550	.004483	.002488	-927
.08	.000039	-.002311	-.004061	-.004860	-.004577	-.003318	-.001391	.000770	.002700	.004006	-624
.09	.004439	.003946	.002663	.000886	-.001004	-.002617	-.003637	-.003881	-.003329	-.002121	656
.10	-.000522	.001137	.002528	.003387	.003566	.003054	.001975	.000556	-.000921	-.002170	226
.11	-.002965	-.003170	-.002767	-.001853	-.000613	.000711	.001874	.002667	.002959	.002712	-537
.12	.001989	.000935	-.000252	-.001357	-.002187	-.002605	-.002550	-.002048	-.001199	-.000159	190
.13	.000886	.001760	.002318	.002477	.002221	.001606	.000744	-.000214	-.001110	-.001798	208
.14	-.002172	-.002182	-.001837	-.001202	-.000385	.000477	.001249	.001812	.002085	.002033	-323
.15	.001674	.001071	.000323	-.000454	-.001141	-.001639	-.001878	-.001831	-.001511	-.000974	218
.16	-.000303	.000401	.001037	.001514	.001769	.001773	.001532	.001085	.000499	-.000139	-53
.17	-.000742	-.001229	-.001536	-.001627	-.001495	-.001163	-.000681	-.000116	.000456	.000960	-66
.18	.001334	.001532	.001534	.001344	.000991	.000524	.000003	-.000506	-.000940	-.001249	126
.19	-.001398	-.001372	-.001181	-.000850	-.000422	.000051	.000511	.000905	.001191	.001338	-138
.20	.001331	.001177	.000896	.000522	.000099	-.000325	-.000704	-.000998	-.001177	-.001224	131
.21	-.001138	-.000931	-.000628	-.000263	.000125	.000496	.000811	.001041	.001164	.001170	-116
.22	.001062	.000853	.000565	.000229	-.000122	-.000453	-.000734	-.000939	-.001051	-.001062	101
.23	-.000973	-.000794	-.000544	-.000247	.000068	.000374	.000644	.000854	.000987	.001035	-85
.24	.000995	.000871	.000678	.000432	.000156	-.000128	-.000395	-.000625	-.000799	-.000905	68
.25	-.000936	-.000892	-.000777	-.000603	-.000383	-.000137	.000118	.000360	.000574	.000742	-44
.26	.000855	.000906	.000891	.000814	.000681	.000504	.000295	.000070	-.000155	-.000365	15
.27	-.000547	-.000688	-.000780	-.000819	-.000803	-.000734	-.000618	-.000464	-.000283	-.000085	16
.28	.000116	.000308	.000479	.000619	.000722	.000781	.000795	.000763	.000689	.000577	-36
.29	.000436	.000273	.000098	-.000079	-.000248	-.000401	-.000529	-.000627	-.000690	-.000716	37
.30	-.000704	-.000656	-.000575	-.000466	-.000335	-.000189	-.000035	.000119	.000265	.000398	-13
.31	.000511	.000600	.000661	.000692	.000692	.000663	.000606	.000524	.000422	.000304	-15
.32	.000176	.000043	-.000089	-.000216	-.000331	-.000431	-.000513	-.000573	-.000610	-.000623	24
.33	-.000613	-.000579	-.000525	-.000452	-.000364	-.000264	-.000156	-.000043	.000069	.000178	-3
.34	.000280	.000372	.000451	.000515	.000562	.000592	.000603	.000597	.000573	.000533	-15
.35	.000479	.000412	.000335	.000250	.000160	.000067	-.000027	-.000117	-.000203	-.000282	7
.36	-.000353	-.000414	-.000463	-.000500	-.000524	-.000535	-.000533	-.000519	-.000494	-.000457	11
.37	-.000411	-.000356	-.000295	-.000228	-.000157	-.000084	-.000010	.000064	.000135	.000202	-2
.38	.000265	.000323	.000374	.000418	.000453	.000481	.000501	.000512	.000515	.000509	-7
.39	.000496	.000476	.000449	.000416	.000377	.000334	.000287	.000237	.000185	.000132	0
.40	.000077	.000023	-.000030	-.000083	-.000133	-.000180	-.000225	-.000266	-.000303	-.000337	4
.41	-.000366	-.000390	-.000411	-.000426	-.000438	-.000444	-.000447	-.000445	-.000439	-.000430	4
.42	-.000417	-.000401	-.000382	-.000360	-.000335	-.000309	-.000281	-.000251	-.000221	-.000189	1
.43	-.000156	-.000124	-.000091	-.000058	-.000025	.000007	.000038	.000069	.000099	.000127	0
.44	.000154	.000181	.000205	.000229	.000251	.000271	.000290	.000307	.000323	.000338	0
.45	.000351	.000362	.000372	.000381	.000388	.000395	.000400	.000404	.000407	.000408	0
.46	.000409	.000410	.000409	.000407	.000405	.000402	.000399	.000395	.000391	.000387	0
.47	.000382	.000376	.000371	.000365	.000360	.000354	.000348	.000342	.000336	.000330	0
.48	.000325	.000319	.000314	.000308	.000303	.000298	.000293	.000289	.000284	.000280	0
.49	.000277	.000273	.000270	.000267	.000264	.000262	.000260	.000258	.000257	.000256	0

TABLE 23

MTF, Defocused Lens, Monochromatic – Δ = 90.0 Rayleigh Units [$\Delta \approx d/2\lambda F^2$]

ν_r	.000	.001	.002	.003	.004	.005	.006	.007	.008	.009	Δ_2
.50	.000255	.000254	.000254	.000254	.000255	.000255	.000256	.000258	.000259	.000261	0
.51	.000263	.000265	.000268	.000270	.000273	.000276	.000280	.000283	.000287	.000290	0
.52	.000294	.000298	.000302	.000306	.000309	.000313	.000317	.000321	.000324	.000327	0
.53	.000331	.000333	.000336	.000338	.000340	.000341	.000342	.000343	.000343	.000342	0
.54	.000341	.000339	.000336	.000332	.000328	.000323	.000317	.000310	.000302	.000293	0
.55	.000283	.000272	.000260	.000247	.000233	.000217	.000201	.000184	.000166	.000147	0
.56	.000127	.000106	.000085	.000063	.000040	.000017	-.000007	-.000030	-.000054	-.000077	0
.57	-.000101	-.000124	-.000146	-.000168	-.000188	-.000208	-.000226	-.000242	-.000257	-.000270	2
.58	-.000281	-.000289	-.000295	-.000299	-.000300	-.000298	-.000293	-.000286	-.000276	-.000262	3
.59	-.000246	-.000227	-.000206	-.000182	-.000155	-.000127	-.000097	-.000065	-.000033	.000001	1
.60	.000035	.000068	.000101	.000133	.000164	.000193	.000219	.000243	.000263	.000280	-3
.61	.000292	.000301	.000304	.000303	.000298	.000287	.000272	.000252	.000227	.000199	-3
.62	.000167	.000132	.000095	.000055	.000015	-.000026	-.000067	-.000106	-.000143	-.000177	3
.63	-.000208	-.000233	-.000254	-.000269	-.000278	-.000280	-.000276	-.000265	-.000247	-.000223	6
.64	-.000194	-.000159	-.000119	-.000077	-.000031	.000015	.000062	.000107	.000150	.000189	-3
.65	.000223	.000251	.000272	.000286	.000291	.000288	.000276	.000256	.000228	.000193	-6
.66	.000151	.000105	.000056	.000004	-.000047	-.000097	-.000143	-.000184	-.000219	-.000246	8
.67	-.000263	-.000271	-.000268	-.000255	-.000231	-.000199	-.000158	-.000111	-.000058	-.000003	3
.68	.000053	.000108	.000158	.000202	.000238	.000265	.000280	.000283	.000275	.000254	-11
.69	.000221	.000179	.000129	.000073	.000014	-.000046	-.000104	-.000156	-.000200	-.000235	10
.70	-.000257	-.000265	-.000260	-.000241	-.000208	-.000164	-.000111	-.000051	.000013	.000076	0
.71	.000136	.000189	.000232	.000262	.000277	.000277	.000262	.000231	.000186	.000131	-10
.72	.000068	.000001	-.000066	-.000128	-.000182	-.000224	-.000252	-.000262	-.000255	-.000230	17
.73	-.000190	-.000136	-.000072	-.000002	.000068	.000135	.000193	.000238	.000267	.000277	-18
.74	.000268	.000239	.000194	.000134	.000065	-.000009	-.000081	-.000148	-.000202	-.000240	16
.75	-.000259	-.000256	-.000232	-.000189	-.000130	-.000059	.000017	.000094	.000163	.000219	-12
.76	.000258	.000275	.000269	.000241	.000191	.000126	.000050	-.000031	-.000108	-.000174	10
.77	-.000224	-.000253	-.000258	-.000237	-.000194	-.000131	-.000054	.000029	.000110	.000182	-9
.78	.000236	.000268	.000274	.000253	.000208	.000141	.000061	-.000025	-.000108	-.000179	12
.79	-.000230	-.000256	-.000254	-.000223	-.000167	-.000092	-.000005	.000083	.000164	.000227	-16
.80	.000265	.000275	.000254	.000204	.000131	.000044	-.000048	-.000133	-.000202	-.000245	25
.81	-.000259	-.000239	-.000190	-.000115	-.000025	.000069	.000156	.000225	.000266	.000275	-32
.82	.000249	.000192	.000111	.000016	-.000080	-.000165	-.000227	-.000257	-.000251	-.000210	35
.83	-.000139	-.000047	.000053	.000147	.000222	.000267	.000275	.000246	.000182	.000093	-24
.84	-.000008	-.000108	-.000190	-.000244	-.000260	-.000236	-.000175	-.000086	.000018	.000121	0
.85	.000206	.000262	.000278	.000252	.000188	.000095	-.000012	-.000116	-.000200	-.000250	34
.86	-.000259	-.000223	-.000150	-.000049	.000061	.000163	.000239	.000277	.000269	.000217	-44
.87	.000129	.000020	-.000091	-.000186	-.000246	-.000261	-.000228	-.000153	-.000049	.000067	11
.88	.000172	.000248	.000281	.000264	.000199	.000099	-.000017	-.000130	-.000216	-.000260	43
.89	-.000252	-.000194	-.000097	.000022	.000138	.000230	.000279	.000274	.000218	.000120	-41
.90	-.000001	-.000120	-.000212	-.000260	-.000253	-.000191	-.000088	.000036	.000155	.000244	-29
.91	.000285	.000267	.000194	.000083	-.000045	-.000161	-.000240	-.000265	-.000229	-.000141	53
.92	-.000019	.000111	.000218	.000279	.000281	.000221	.000114	-.000018	-.000142	-.000232	35
.93	-.000266	-.000235	-.000147	-.000021	.000113	.000224	.000284	.000280	.000211	.000095	-47
.94	-.000042	-.000167	-.000247	-.000265	-.000214	-.000106	.000031	.000164	.000260	.000294	-61
.95	.000257	.000158	.000022	-.000116	-.000222	-.000267	-.000239	-.000145	-.000009	.000134	7
.96	.000245	.000296	.000272	.000179	.000042	-.000102	-.000216	-.000266	-.000239	-.000143	70
.97	-.000002	.000145	.000256	.000302	.000267	.000163	.000017	-.000128	-.000232	-.000265	72
.98	-.000216	-.000098	.000054	.000197	.000290	.000304	.000236	.000105	-.000050	-.000183	22
.99	-.000253	-.000240	-.000145	.000004	.000162	.000293	.000003	.000027	.000027	.000012	-13

TABLE 23

MTF, Defocused Lens, Monochromatic – Δ = 100.0 Rayleigh Units [$\Delta \approx d/2\lambda F^2$]

ν_r	.000	.001	.002	.003	.004	.005	.006	.007	.008	.009	Δ_2
.00	1.000000	.950347	.813833	.616687	.394774	.185300	.018753	-.086943	-.128779	-.117257	53358
.01	-.071945	-.015446	.032502	.059319	.061356	.043057	.014095	-.014419	-.033482	-.038543	14002
.02	-.030155	-.013007	.006142	.020854	.026965	.023586	.012947	-.000687	-.012603	-.019168	5351
.03	-.018845	-.012443	-.002608	.007198	.013863	.015563	.012201	.005274	-.002740	-.009242	1513
.04	-.012323	-.011294	-.006793	-.000474	.005577	.009519	.010294	.007884	.003241	-.002078	-675
.05	-.006421	-.008545	-.007953	-.004993	-.000692	.003591	.006593	.007503	.006175	.003125	-1722
.06	-.000663	-.004052	-.006083	-.006247	-.004595	-.001695	.001566	.004245	.005617	.005361	-1626
.07	.003629	.000975	-.001822	-.003984	-.004945	-.004500	-.002833	-.000454	.001964	.003768	-613
.08	.004502	.004016	.002493	.000378	-.001746	-.003324	-.003966	-.003544	-.002211	-.000347	531
.09	.001549	.002994	.003637	.003347	.002234	.000607	-.001108	-.002484	-.003193	-.003086	816
.10	-.002220	-.000832	.000721	.002059	.002870	.002977	.002381	.001247	-.000140	-.001446	81
.11	-.002369	-.002709	-.002407	-.001553	-.000363	.000883	.001901	.002468	.002473	.001932	-544
.12	.000984	-.000150	-.001215	-.001980	-.002288	-.002088	-.001438	-.000492	.000540	.001437	-135
.13	.002014	.002159	.001855	.001179	.000281	-.000647	-.001415	-.001873	-.001936	-.001604	395
.14	-.000954	-.000126	.000713	.001396	.001794	.001839	.001531	.000940	.000186	-.000581	-13
.15	-.001218	-.001606	-.001681	-.001438	-.000930	-.000257	.000451	.001066	.001478	.001620	-270
.16	.001472	.001069	.000489	-.000161	-.000766	-.001222	-.001453	-.001427	-.001154	-.000687	194
.17	-.000111	.000476	.000974	.001305	.001416	.001297	.000971	.000498	-.000043	-.000564	20
.18	-.000982	-.001234	-.001284	-.001130	-.000801	-.000352	.000145	.000614	.000983	.001200	-151
.19	.001237	.001091	.000790	.000381	-.000073	-.000506	-.000856	-.001075	-.001135	-.001031	164
.20	-.000783	-.000427	-.000017	.000391	.000740	.000985	.001095	.001059	.000885	.000599	-111
.21	.000241	-.000140	-.000494	-.000776	-.000953	-.001004	-.000925	-.000731	-.000447	-.000111	52
.22	.000235	.000548	.000792	.000938	.000972	.000893	.000712	.000452	.000146	-.000169	-9
.23	-.000459	-.000691	-.000839	-.000890	-.000840	-.000697	-.000479	-.000211	.000077	.000354	-10
.24	.000590	.000762	.000853	.000856	.000773	.000613	.000395	.000143	-.000119	-.000363	17
.25	-.000567	-.000711	-.000783	-.000778	-.000698	-.000552	-.000355	-.000128	.000109	.000333	-11
.26	.000526	.000669	.000752	.000768	.000718	.000608	.000448	.000253	.000041	-.000170	1
.27	-.000363	-.000522	-.000635	-.000694	-.000695	-.000640	-.000535	-.000388	-.000212	-.000022	15
.28	.000168	.000344	.000492	.000602	.000668	.000684	.000652	.000575	.000459	.000313	-29
.29	.000149	-.000023	-.000190	-.000340	-.000465	-.000557	-.000610	-.000622	-.000593	-.000526	38
.30	-.000426	-.000300	-.000157	-.000005	.000145	.000285	.000407	.000504	.000571	.000604	-32
.31	.000603	.000569	.000504	.000413	.000301	.000175	.000042	-.000091	-.000216	-.000328	14
.32	-.000421	-.000491	-.000535	-.000551	-.000539	-.000501	-.000439	-.000357	-.000258	-.000149	11
.33	-.000033	.000082	.000193	.000294	.000381	.000451	.000501	.000530	.000537	.000522	-21
.34	.000486	.000432	.000362	.000279	.000187	.000090	-.000009	-.000106	-.000198	-.000280	9
.35	-.000351	-.000408	-.000450	-.000476	-.000484	-.000476	-.000452	-.000414	-.000362	-.000300	11
.36	-.000228	-.000151	-.000069	.000013	.000094	.000172	.000244	.000308	.000363	.000407	-9
.37	.000440	.000461	.000470	.000467	.000453	.000427	.000392	.000348	.000297	.000240	-5
.38	.000178	.000113	.000047	-.000018	-.000082	-.000144	-.000201	-.000253	-.000299	-.000338	7
.39	-.000370	-.000394	-.000410	-.000418	-.000419	-.000411	-.000397	-.000376	-.000349	-.000317	5
.40	-.000280	-.000238	-.000194	-.000148	-.000100	-.000051	-.000002	.000047	.000093	.000138	-1
.41	.000181	.000221	.000257	.000289	.000318	.000343	.000363	.000379	.000391	.000398	-3
.42	.000402	.000401	.000396	.000388	.000377	.000362	.000344	.000324	.000302	.000277	-1
.43	.000251	.000224	.000195	.000166	.000136	.000106	.000076	.000046	.000016	-.000013	1
.44	-.000041	-.000068	-.000094	-.000120	-.000144	-.000166	-.000188	-.000207	-.000226	-.000243	1
.45	-.000258	-.000273	-.000285	-.000297	-.000306	-.000315	-.000322	-.000328	-.000333	-.000337	1
.46	-.000340	-.000342	-.000343	-.000343	-.000342	-.000341	-.000339	-.000336	-.000333	-.000329	0
.47	-.000325	-.000321	-.000316	-.000311	-.000306	-.000301	-.000295	-.000290	-.000284	-.000279	0
.48	-.000274	-.000268	-.000263	-.000258	-.000253	-.000248	-.000244	-.000240	-.000235	-.000232	0
.49	-.000228	-.000225	-.000222	-.000219	-.000216	-.000214	-.000212	-.000211	-.000210	-.000209	0

TABLE 23

MTF, Defocused Lens, Monochromatic – Δ= 100.0 Rayleigh Units [$\Delta \approx d/2\lambda F^2$]

ν_r	.000	.001	.002	.003	.004	.005	.006	.007	.008	.009	Δ_2
.50	-.000208	-.000208	-.000208	-.000208	-.000208	-.000209	-.000210	-.000212	-.000213	-.000215	0
.51	-.000218	-.000220	-.000223	-.000226	-.000229	-.000232	-.000235	-.000239	-.000243	-.000246	0
.52	-.000250	-.000254	-.000258	-.000262	-.000266	-.000270	-.000273	-.000277	-.000280	-.000284	0
.53	-.000286	-.000289	-.000291	-.000293	-.000294	-.000295	-.000296	-.000295	-.000295	-.000293	1
.54	-.000291	-.000288	-.000284	-.000279	-.000273	-.000266	-.000259	-.000250	-.000240	-.000230	1
.55	-.000218	-.000205	-.000191	-.000175	-.000159	-.000142	-.000124	-.000105	-.000085	-.000064	1
.56	-.000043	-.000021	.000002	.000024	.000047	.000070	.000093	.000116	.000138	.000159	0
.57	.000180	.000200	.000218	.000234	.000249	.000262	.000273	.000282	.000288	.000292	-2
.58	.000292	.000290	.000285	.000277	.000266	.000252	.000235	.000216	.000193	.000168	-1
.59	.000141	.000112	.000081	.000049	.000016	-.000017	-.000050	-.000083	-.000115	-.000145	2
.60	-.000173	-.000198	-.000220	-.000239	-.000254	-.000264	-.000270	-.000271	-.000267	-.000258	5
.61	-.000244	-.000225	-.000202	-.000174	-.000142	-.000107	-.000070	-.000030	.000010	.000051	0
.62	.000091	.000130	.000165	.000198	.000226	.000249	.000266	.000277	.000280	.000277	-6
.63	.000267	.000250	.000226	.000196	.000160	.000120	.000077	.000031	-.000016	-.000063	0
.64	-.000107	-.000149	-.000186	-.000217	-.000241	-.000257	-.000264	-.000262	-.000252	-.000232	9
.65	-.000204	-.000168	-.000126	-.000078	-.000028	.000025	.000077	.000126	.000171	.000210	-5
.66	.000241	.000263	.000274	.000274	.000264	.000242	.000210	.000169	.000121	.000067	-4
.67	.000010	-.000047	-.000102	-.000152	-.000195	-.000229	-.000251	-.000261	-.000258	-.000242	13
.68	-.000213	-.000173	-.000123	-.000066	-.000006	.000056	.000115	.000169	.000214	.000248	-10
.69	.000268	.000274	.000264	.000239	.000201	.000150	.000091	.000027	-.000039	-.000103	2
.70	-.000160	-.000206	-.000240	-.000258	-.000259	-.000242	-.000209	-.000162	-.000103	-.000036	8
.71	.000034	.000103	.000165	.000216	.000252	.000271	.000271	.000251	.000213	.000160	-15
.72	.000095	.000022	-.000051	-.000121	-.000181	-.000227	-.000254	-.000260	-.000245	-.000209	21
.73	-.000156	-.000088	-.000013	.000065	.000138	.000200	.000245	.000270	.000272	.000250	-23
.74	.000206	.000144	.000069	-.000012	-.000092	-.000162	-.000218	-.000252	-.000261	-.000245	26
.75	-.000204	-.000143	-.000066	.000018	.000102	.000176	.000233	.000267	.000274	.000254	-27
.76	.000207	.000139	.000056	-.000032	-.000117	-.000188	-.000238	-.000262	-.000255	-.000219	29
.77	-.000157	-.000077	.000014	.000104	.000183	.000242	.000273	.000272	.000240	.000179	-27
.78	.000096	.000003	-.000091	-.000172	-.000232	-.000262	-.000258	-.000220	-.000152	-.000064	21
.79	.000034	.000129	.000208	.000260	.000279	.000260	.000208	.000127	.000029	-.000072	-3
.80	-.000162	-.000229	-.000263	-.000259	-.000216	-.000141	-.000044	.000060	.000157	.000233	-21
.81	.000275	.000276	.000237	.000163	.000065	-.000043	-.000144	-.000221	-.000263	-.000262	42
.82	-.000219	-.000139	-.000036	.000075	.000175	.000248	.000282	.000270	.000214	.000123	-34
.83	.000012	-.000100	-.000194	-.000254	-.000269	-.000236	-.000161	-.000055	.000062	.000170	-8
.84	.000248	.000284	.000270	.000207	.000108	-.000010	-.000126	-.000217	-.000267	-.000265	51
.85	-.000211	-.000116	.000004	.000124	.000223	.000279	.000282	.000232	.000136	.000014	-25
.86	-.000109	-.000210	-.000266	-.000268	-.000212	-.000112	.000014	.000139	.000237	.000287	-47
.87	.000278	.000211	.000100	-.000031	-.000154	-.000242	-.000276	-.000248	-.000164	-.000040	38
.88	.000094	.000209	.000280	.000289	.000234	.000127	-.000008	-.000139	-.000237	-.000278	57
.89	-.000252	-.000164	-.000035	.000104	.000221	.000287	.000286	.000217	.000097	-.000046	-21
.90	-.000176	-.000260	-.000278	-.000224	-.000112	.000032	.000169	.000267	.000298	.000255	-74
.91	.000148	.000004	-.000139	-.000244	-.000282	-.000243	-.000137	.000009	.000155	.000262	-38
.92	.000301	.000260	.000150	.000001	-.000146	-.000251	-.000284	-.000234	-.000116	.000039	36
.93	.000185	.000282	.000301	.000236	.000105	-.000054	-.000195	-.000276	-.000273	-.000186	84
.94	-.000041	.000120	.000249	.000306	.000274	.000161	.000002	-.000155	-.000261	-.000284	84
.95	-.000215	-.000075	.000092	.000234	.000306	.000285	.000176	.000014	-.000149	-.000261	52
.96	-.000284	-.000211	-.000065	.000107	.000249	.000313	.000277	.000152	-.000019	-.000180	10
.97	-.000276	-.000274	-.000173	-.000008	.000165	.000287	.000315	.000238	.000084	-.000095	-23
.98	-.000236	-.000287	-.000231	-.000085	.000097	.000252	.000323	.000284	.000149	-.000033	-46
.99	-.000196	-.000278	-.000249	-.000118	.000068	.000251	-.000014	.000019	.000024	.000012	-17

TABLE 24

Integrals of MTF and $(MTF)^2$

Δ	$\int T d\nu_r$	$\int T^2 d\nu_r$	Δ	$\int T d\nu_r$	$\int T^2 d\nu_r$	Δ	$\int T d\nu_r$	$\int T^2 d\nu_r$
0	.424413	.272421						
0.1	.423773	.271770	5	.068711	.052847	50	.006534	.005393
0.2	.421866	.269834	6	.062762	.044304	60	.005310	.004495
0.3	.418706	.266650	7	.048898	.038012	70	.004636	.003854
0.4	.414322	.262284	8	.039128	.033340	80	.003982	.003373
0.5	.408754	.256823	9	.037243	.029668	90	.003592	.002999
1.0	.365230	.217336	10	.035503	.026748	100	.003184	.002699
1.5	.303131	.170479	15	.022042	.017888			
2.0	.234804	.130808	20	.015878	.013440			
2.5	.172130	.103720	25	.013037	.010763			
3.0	.123454	.086659	30	.011058	.008976			
3.5	.091964	.074942	35	.009264	.007697			
4.0	.075870	.065810	40	.007962	.006737			
4.5	.070103	.058534	45	.007178	.005990			

TABLE 25

MTF, Perfect Lens, Polychromatic Blackbody at 3000° K

υ_s	Normal observer		Photoelectric detectors							Photographic emulsions		
c/mm	Photopic	Scotopic	S-10	S-11	S-12	S-20	CdSe	Si diode	Vidicon	Blue sensitive	Ortho chromatic	Pan chromatic
50	.9637	.9672	.9650	.9676	.9689	.9632	.9513	.9470	.9607	.9692	.9669	.9648
100	.9273	.9345	.9301	.9352	.9378	.9264	.9027	.8942	.9214	.9384	.9338	.9297
150	.8911	.9018	.8952	.9028	.9067	.8897	.8542	.8415	.8822	.9076	.9007	.8946
200	.8549	.8691	.8603	.8705	.8757	.8530	.8059	.7891	.8431	.8769	.8677	.8596
250	.8189	.8366	.8256	.8383	.8448	.8165	.7580	.7372	.8042	.8462	.8348	.8248
300	.7830	.8042	.7911	.8062	.8139	.7802	.7104	.6858	.7655	.8157	.8020	.7900
350	.7473	.7719	.7567	.7742	.7832	.7441	.6633	.6350	.7271	.7853	.7694	.7554
400	.7118	.7397	.7225	.7424	.7526	.7082	.6167	.5850	.6889	.7550	.7369	.7211
450	.6765	.7077	.6885	.7108	.7222	.6726	.5708	.5358	.6511	.7248	.7046	.6869
500	.6416	.6760	.6548	.6794	.6919	.6373	.5255	.4876	.6136	.6949	.6725	.6531
550	.6069	.6445	.6214	.6482	.6618	.6024	.4811	.4406	.5766	.6651	.6407	.6195
600	.5726	.6132	.5884	.6173	.6320	.5679	.4376	.3949	.5401	.6356	.6091	.5862
650	.5387	.5822	.5557	.5866	.6024	.5337	.3951	.3507	.5040	.6063	.5779	.5533
700	.5052	.5515	.5234	.5562	.5730	.5001	.3537	.3081	.4686	.5772	.5469	.5208
750	.4722	.5211	.4915	.5262	.5440	.4670	.3136	.2674	.4338	.5485	.5163	.4887
800	.4397	.4911	.4601	.4966	.5152	.4345	.2749	.2289	.3997	.5200	.4861	.4571
850	.4077	.4615	.4292	.4673	.4867	.4025	.2378	.1928	.3664	.4919	.4562	.4260
900	.3763	.4323	.3988	.4384	.4587	.3713	.2024	.1599	.3339	.4642	.4269	.3954
950	.3455	.4035	.3691	.4100	.4310	.3408	.1690	.1308	.3024	.4368	.3980	.3654
1000	.3154	.3753	.3400	.3821	.4037	.3111	.1380	.1061	.2719	.4099	.3696	.3361
1050	.2861	.3475	.3116	.3547	.3768	.2822	.1097	.0856	.2424	.3834	.3417	.3074
1100	.2576	.3204	.2840	.3279	.3504	.2544	.0846	.0688	.2142	.3573	.3144	.2795
1150	.2299	.2938	.2572	.3017	.3245	.2275	.0626	.0552	.1875	.3318	.2878	.2524
1200	.2032	.2678	.2313	.2761	.2991	.2018	.0440	.0441	.1624	.3068	.2619	.2261
1250	.1775	.2425	.2064	.2512	.2743	.1775	.0290	.0351	.1391	.2824	.2366	.2008
1300	.1528	.2180	.1825	.2271	.2501	.1547	.0181	.0279	.1178	.2587	.2122	.1765
1350	.1295	.1942	.1599	.2038	.2266	.1337	.0114	.0221	.0987	.2356	.1886	.1533
1400	.1074	.1713	.1388	.1813	.2037	.1145	.0082	.0174	.0817	.2132	.1659	.1314
1450	.0869	.1493	.1192	.1599	.1816	.0974	.0062	.0137	.0668	.1916	.1442	.1108
1500	.0680	.1283	.1014	.1395	.1602	.0822	.0047	.0107	.0540	.1708	.1235	.0918
1550	.0513	.1084	.0853	.1203	.1398	.0690	.0037	.0084	.0432	.1509	.1041	.0747
1600	.0369	.0898	.0711	.1024	.1202	.0575	.0029	.0065	.0341	.1321	.0860	.0600
1650	.0252	.0724	.0587	.0861	.1017	.0476	.0023	.0050	.0266	.1143	.0694	.0479
1700	.0162	.0567	.0480	.0715	.0843	.0392	.0018	.0039	.0205	.0979	.0547	.0383
1750	.0098	.0428	.0389	.0587	.0682	.0320	.0015	.0030	.0156	.0830	.0424	.0308
1800	.0055	.0310	.0313	.0476	.0535	.0260	.0012	.0023	.0118	.0700	.0328	.0248
1850	.0029	.0215	.0249	.0383	.0404	.0210	.0009	.0018	.0089	.0590	.0255	.0201
1900	.0015	.0143	.0197	.0305	.0289	.0168	.0007	.0013	.0066	.0500	.0199	.0164
1950	.0007	.0091	.0155	.0241	.0195	.0134	.0006	.0010	.0049	.0425	.0155	.0134
2000	.0003	.0055	.0120	.0188	.0129	.0105	.0005	.0008	.0036	.0362	.0120	.0108
2050	.0002	.0032	.0093	.0146	.0092	.0082	.0004	.0006	.0026	.0308	.0092	.0087
2100	.0001	.0018	.0071	.0112	.0065	.0063	.0003	.0004	.0019	.0260	.0069	.0069
2150	.0000	.0009	.0053	.0085	.0046	.0048	.0002	.0003	.0014	.0218	.0052	.0054
2200	.0000	.0004	.0040	.0064	.0032	.0036	.0002	.0002	.0010	.0180	.0038	.0042
2250	.0000	.0002	.0029	.0047	.0022	.0027	.0001	.0002	.0007	.0147	.0028	.0032
2300	.0000	.0001	.0021	.0034	.0015	.0020	.0001	.0001	.0004	.0118	.0020	.0024
2350	.0000	.0000	.0015	.0025	.0010	.0014	.0001	.0001	.0003	.0092	.0014	.0018
2400	.0000	.0000	.0010	.0017	.0007	.0010	.0000	.0001	.0002	.0071	.0010	.0013
2450	.0000	.0000	.0007	.0012	.0004	.0007	.0000	.0000	.0001	.0052	.0006	.0009
2500	.0000	.0000	.0005	.0008	.0002	.0004	.0000	.0000	.0001	.0037	.0004	.0006
2550	.0000	.0000	.0003	.0005	.0001	.0003	.0000	.0000	.0000	.0025	.0003	.0004
2600	.0000	.0000	.0002	.0003	.0001	.0002	.0000	.0000	.0000	.0016	.0002	.0002
2650	.0000	.0000	.0001	.0001	.0000	.0001	.0000	.0000	.0000	.0009	.0001	.0001
2700	.0000	.0000	.0000	.0001	.0000	.0000	.0000	.0000	.0000	.0005	.0000	.0001
2750	.0000	.0000	.0000	.0000	.0000	.0000	.0000	.0000	.0000	.0002	.0000	.0000
2800	.0000	.0000	.0000	.0000	.0000	.0000	.0000	.0000	.0000	.0000	.0000	.0000

TABLE 25

MTF, Perfect Lens, Polychromatic Balckbodv at 5000° K

	Normal observer		Photoelectric detectors								Photographic emulsions		
υ_8 c/mm	Photopic	Scotopic	S-10	S-11	S-12	S-20	CdSe	Si diode	Vidieon	Blue sensitive	Ortho chromatic	Pan chromatic
50	.9643	.9679	.9681	.9698	.9696	.9672	.9537	.9536	.9641	.9721	.9686	.9677
100	.9286	.9358	.9362	.9396	.9393	.9345	.9075	.9072	.9283	.9442	.9372	.9353
150	.8930	.9037	.9043	.9094	.9090	.9018	.8614	.8610	.8925	.9163	.9058	.9031
200	.8575	.8717	.8726	.8793	.8787	.8692	.8156	.8150	.8568	.8885	.8746	.8709
250	.8221	.8398	.8409	.8492	.8485	.8366	.7699	.7693	.8213	.8607	.8433	.8388
300	.7868	.8080	.8093	.8192	.8184	.8042	.7246	.7240	.7858	.8330	.8122	.8068
350	.7517	.7763	.7778	.7894	.7884	.7720	.6798	.6792	.7506	.8054	.7813	.7749
400	.7168	.7448	.7465	.7597	.7586	.7399	.6354	.6348	.7156	.7779	.7504	.7432
450	.6822	.7134	.7154	.7301	.7288	.7080	.5915	.5911	.6809	.7505	.7197	.7117
500	.6478	.6823	.6845	.7007	.6993	.6763	.5483	.5482	.6465	.7233	.6892	.6804
550	.6137	.6513	.6538	.6715	.6699	.6449	.5057	.5060	.6124	.6962	.6589	.6493
600	.5799	.6206	.6234	.6424	.6407	.6138	.4640	.4648	.5786	.6692	.6289	.6185
650	.5465	.5901	.5932	.6137	.6118	.5830	.4232	.4245	.5453	.6424	.5991	.5880
700	.5135	.5600	.5633	.5851	.5831	.5525	.3833	.3855	.5124	.6159	.5696	.5577
750	.4809	.5301	.5338	.5569	.5546	.5224	.3445	.3478	.4800	.5895	.5403	.5279
800	.4489	.5006	.5046	.5289	.5264	.4926	.3070	.3115	.4481	.5634	.5114	.4984
850	.4173	.4714	.4758	.5012	.4986	.4634	.2708	.2770	.4168	.5376	.4829	.4693
900	.3863	.4427	.4475	.4739	.4710	.4346	.2361	.2444	.3861	.5120	.4547	.4406
950	.3558	.4143	.4195	.4469	.4439	.4063	.2030	.2144	.3561	.4867	.4269	.4124
1000	.3261	.3864	.3921	.4204	.4171	.3786	.1719	.1871	.3268	.4617	.3996	.3847
1050	.2970	.3590	.3652	.3942	.3907	.3514	.1431	.1628	.2984	.4370	.3728	.3576
1100	.2687	.3322	.3389	.3685	.3647	.3249	.1169	.1412	.2709	.4127	.3464	.3310
1150	.2412	.3058	.3131	.3433	.3392	.2992	.0933	.1221	.2443	.3887	.3206	.3051
1200	.2145	.2801	.2880	.3186	.3142	.2741	.0728	.1053	.2189	.3652	.2954	.2798
1250	.1888	.2550	.2637	.2945	.2897	.2500	.0556	.0905	.1947	.3421	.2707	.2553
1300	.1642	.2306	.2400	.2709	.2657	.2268	.0422	.0775	.1719	.3194	.2468	.2315
1350	.1406	.2069	.2173	.2480	.2424	.2047	.0329	.0662	.1505	.2973	.2235	.2086
1400	.1183	.1840	.1955	.2258	.2197	.1838	.0273	.0564	.1308	.2756	.2010	.1867
1450	.0974	.1620	.1748	.2043	.1977	.1641	.0231	.0478	.1126	.2545	.1794	.1657
1500	.0781	.1408	.1552	.1836	.1764	.1458	.0196	.0404	.0962	.2340	.1586	.1460
1550	.0605	.1207	.1369	.1638	.1559	.1288	.0168	.0340	.0814	.2141	.1389	.1276
1600	.0451	.1017	.1199	.1449	.1362	.1132	.0144	.0285	.0683	.1950	.1202	.1109
1650	.0322	.0839	.1043	.1272	.1175	.0989	.0124	.0238	.0568	.1766	.1028	.0960
1700	.0218	.0675	.0900	.1107	.0997	.0859	.0106	.0198	.0468	.1591	.0868	.0831
1750	.0139	.0527	.0772	.0956	.0831	.0742	.0091	.0164	.0383	.1425	.0727	.0718
1800	.0084	.0398	.0657	.0818	.0678	.0636	.0077	.0135	.0310	.1271	.0606	.0619
1850	.0048	.0289	.0554	.0695	.0538	.0542	.0066	.0111	.0250	.1131	.0505	.0533
1900	.0026	.0202	.0464	.0585	.0414	.0458	.0055	.0090	.0200	.1003	.0420	.0458
1950	.0014	.0136	.0386	.0488	.0309	.0384	.0046	.0073	.0158	.0887	.0347	.0392
2000	.0007	.0087	.0318	.0404	.0229	.0318	.0039	.0058	.0124	.0781	.0284	.0332
2050	.0004	.0054	.0259	.0331	.0175	.0262	.0032	.0046	.0096	.0683	.0230	.0279
2100	.0002	.0032	.0209	.0268	.0134	.0213	.0026	.0037	.0074	.0593	.0184	.0232
2150	.0001	.0017	.0167	.0215	.0101	.0171	.0021	.0028	.0056	.0509	.0146	.0190
2200	.0000	.0009	.0131	.0170	.0075	.0135	.0017	.0022	.0041	.0432	.0114	.0154
2250	.0000	.0004	.0102	.0132	.0055	.0105	.0013	.0016	.0030	.0361	.0087	.0122
2300	.0000	.0002	.0077	.0102	.0040	.0081	.0010	.0012	.0021	.0297	.0066	.0096
2350	.0000	.0001	.0058	.0076	.0028	.0060	.0008	.0009	.0015	.0239	.0049	.0073
2400	.0000	.0000	.0042	.0056	.0020	.0044	.0006	.0006	.0010	.0188	.0036	.0054
2450	.0000	.0000	.0030	.0040	.0013	.0031	.0004	.0004	.0006	.0143	.0025	.0039
2500	.0000	.0000	.0020	.0027	.0008	.0021	.0003	.0003	.0004	.0105	.0017	.0027
2550	.0000	.0000	.0013	.0018	.0005	.0014	.0002	.0002	.0002	.0073	.0012	.0018
2600	.0000	.0000	.0008	.0011	.0003	.0008	.0001	.0001	.0001	.0048	.0007	.0011
2650	.0000	.0000	.0004	.0006	.0002	.0005	.0001	.0001	.0000	.0029	.0004	.0006
2700	.0000	.0000	.0002	.0003	.0001	.0002	.0000	.0000	.0000	.0015	.0002	.0003
2750	.0000	.0000	.0001	.0001	.0000	.0001	.0000	.0000	.0000	.0006	.0001	.0001
2800	.0000	.0000	.0000	.0000	.0000	.0000	.0000	.0000	.0000	.0002	.0000	.0000

TABLE 25

MTF, Perfect Lens, Polychromatic CIE Illuminant A

ν_s c/mm	Normal observer: Photopic	Normal observer: Scotopic	Photoelectric detectors: S-10	S-11	S-12	S-20	CdSe	Si diode	Vidieon	Photographic emulsions: Blue sensitive	Ortho chromatic	Pan chromatic
50	.9636	.9672	.9647	.9673	.9688	.9627	.9531	.9555	.9606	.9689	.9667	.9646
100	.9272	.9343	.9294	.9347	.9376	.9255	.9062	.9111	.9212	.9378	.9334	.9292
150	.8908	.9015	.8941	.9021	.9065	.8883	.8594	.8668	.8819	.9067	.9002	.8939
200	.8546	.8688	.8590	.8696	.8754	.8513	.8129	.8227	.8427	.8758	.8671	.8586
250	.8185	.8362	.8239	.8371	.8444	.8143	.7666	.7788	.8037	.8448	.8340	.8235
300	.7825	.8037	.7890	.8048	.8135	.7776	.7206	.7352	.7649	.8140	.8011	.7885
350	.7467	.7713	.7543	.7726	.7827	.7410	.6750	.6920	.7263	.7833	.7683	.7537
400	.7112	.7391	.7198	.7406	.7520	.7048	.6299	.6491	.6880	.7528	.7357	.7191
450	.6759	.7071	.6855	.7087	.7215	.6687	.5854	.6068	.6501	.7224	.7032	.6848
500	.6408	.6752	.6515	.6771	.6912	.6331	.5415	.5650	.6125	.6921	.6710	.6507
550	.6061	.6436	.6178	.6457	.6610	.5977	.4983	.5238	.5754	.6621	.6390	.6169
600	.5717	.6123	.5845	.6145	.6311	.5628	.4558	.4833	.5388	.6323	.6073	.5834
650	.5378	.5812	.5515	.5837	.6014	.5284	.4143	.4436	.5026	.6028	.5759	.5503
700	.5042	.5504	.5189	.5531	.5720	.4944	.3737	.4047	.4671	.5735	.5448	.5176
750	.4711	.5200	.4868	.5229	.5429	.4610	.3342	.3667	.4322	.5445	.5141	.4853
800	.4385	.4900	.4551	.4931	.5141	.4281	.2958	.3298	.3980	.5159	.4837	.4535
850	.4065	.4603	.4240	.4636	.4856	.3959	.2588	.2941	.3646	.4875	.4538	.4222
900	.3751	.4310	.3935	.4346	.4574	.3644	.2232	.2596	.3320	.4596	.4243	.3915
950	.3443	.4022	.3635	.4060	.4297	.3337	.1893	.2264	.3003	.4320	.3953	.3613
1000	.3141	.3739	.3343	.3780	.4023	.3038	.1571	.1949	.2696	.4049	.3668	.3318
1050	.2848	.3461	.3057	.3505	.3754	.2748	.1268	.1651	.2401	.3782	.3388	.3031
1100	.2562	.3189	.2780	.3235	.3490	.2468	.0989	.1372	.2117	.3520	.3115	.2750
1150	.2285	.2923	.2511	.2972	.3230	.2199	.0736	.1116	.1847	.3263	.2848	.2478
1200	.2018	.2663	.2251	.2715	.2976	.1942	.0514	.0889	.1593	.3012	.2588	.2215
1250	.1761	.2410	.2002	.2466	.2728	.1699	.0334	.0701	.1358	.2767	.2335	.1961
1300	.1515	.2165	.1763	.2224	.2486	.1472	.0204	.0550	.1144	.2528	.2090	.1718
1350	.1281	.1927	.1538	.1991	.2250	.1263	.0124	.0431	.0951	.2297	.1854	.1486
1400	.1061	.1698	.1328	.1766	.2021	.1075	.0088	.0336	.0782	.2072	.1627	.1267
1450	.0856	.1478	.1134	.1552	.1800	.0907	.0065	.0262	.0635	.1856	.1410	.1062
1500	.0669	.1268	.0958	.1349	.1586	.0760	.0049	.0203	.0510	.1648	.1203	.0873
1550	.0502	.1070	.0801	.1158	.1382	.0633	.0038	.0157	.0404	.1449	.1010	.0703
1600	.0359	.0883	.0663	.0981	.1187	.0523	.0029	.0121	.0317	.1261	.0829	.0558
1650	.0244	.0711	.0543	.0819	.1002	.0430	.0023	.0093	.0245	.1085	.0665	.0440
1700	.0156	.0555	.0441	.0676	.0828	.0351	.0018	.0071	.0187	.0922	.0519	.0348
1750	.0094	.0417	.0355	.0551	.0668	.0285	.0014	.0054	.0141	.0775	.0398	.0277
1800	.0052	.0301	.0283	.0444	.0522	.0230	.0011	.0041	.0106	.0647	.0304	.0221
1850	.0027	.0207	.0224	.0354	.0391	.0184	.0009	.0031	.0079	.0541	.0234	.0177
1900	.0013	.0136	.0176	.0280	.0278	.0146	.0007	.0023	.0058	.0455	.0181	.0143
1950	.0007	.0086	.0137	.0219	.0185	.0115	.0006	.0018	.0043	.0385	.0141	.0116
2000	.0003	.0052	.0105	.0170	.0121	.0090	.0004	.0013	.0031	.0326	.0108	.0094
2050	.0002	.0030	.0081	.0131	.0085	.0070	.0003	.0010	.0023	.0276	.0082	.0075
2100	.0001	.0016	.0061	.0100	.0060	.0054	.0003	.0007	.0016	.0233	.0061	.0059
2150	.0000	.0008	.0046	.0075	.0042	.0040	.0002	.0005	.0011	.0194	.0045	.0046
2200	.0000	.0004	.0034	.0056	.0029	.0030	.0001	.0004	.0008	.0160	.0033	.0036
2250	.0000	.0002	.0025	.0041	.0020	.0022	.0001	.0003	.0005	.0131	.0024	.0027
2300	.0000	.0001	.0018	.0030	.0013	.0016	.0001	.0002	.0004	.0105	.0017	.0020
2350	.0000	.0000	.0013	.0021	.0009	.0011	.0001	.0001	.0002	.0082	.0012	.0015
2400	.0000	.0000	.0009	.0015	.0006	.0008	.0000	.0001	.0001	.0063	.0008	.0010
2450	.0000	.0000	.0006	.0010	.0004	.0005	.0000	.0001	.0001	.0046	.0006	.0007
2500	.0000	.0000	.0004	.0007	.0002	.0003	.0000	.0000	.0001	.0033	.0004	.0005
2550	.0000	.0000	.0002	.0004	.0001	.0002	.0000	.0000	.0000	.0022	.0002	.0003
2600	.0000	.0000	.0001	.0002	.0001	.0001	.0000	.0000	.0000	.0014	.0001	.0002
2650	.0000	.0000	.0001	.0001	.0000	.0001	.0000	.0000	.0000	.0008	.0001	.0001
2700	.0000	.0000	.0000	.0001	.0000	.0000	.0000	.0000	.0000	.0004	.0000	.0000
2750	.0000	.0000	.0000	.0000	.0000	.0000	.0000	.0000	.0000	.0002	.0000	.0000
2800	.0000	.0000	.0000	.0000	.0000	.0000	.0000	.0000	.0000	.0000	.0000	.0000

TABLE 25

MTF, Perfect Lens, Polychromatic DIE Illuminant B

	Normal observer		Photoelectric detectors							Photographic emulsions		
υ_8 c/mm	Photopic	Scotopic	S-10	S-11	S-12	S-20	CdSe	Si diode	Vidicon	Blue sensitive	Ortho chromatic	Pan chromatic
50	.9643	.9679	.9673	.9690	.9694	.9663	.9547	.9592	.9638	.9705	.9680	.9667
100	.9285	.9358	.9347	.9381	.9387	.9327	.9096	.9185	.9276	.9410	.9361	.9335
150	.8929	.9037	.9021	.9072	.9081	.8990	.8645	.8779	.8915	.9116	.9042	.9003
200	.8573	.8718	.8695	.8763	.8776	.8655	.8196	.8373	.8555	.8822	.8723	.8671
250	.8219	.8399	.8371	.8455	.8471	.8321	.7749	.7970	.8197	.8529	.8406	.8341
300	.7865	.8081	.8048	.8149	.8168	.7988	.7306	.7570	.7839	.8236	.8089	.8012
350	.7514	.7764	.7726	.7843	.7865	.7656	.6866	.7172	.7484	.7945	.7774	.7684
400	.7165	.7449	.7405	.7539	.7564	.7327	.6431	.6777	.7131	.7655	.7460	.7358
450	.6818	.7135	.7087	.7236	.7264	.6999	.6000	.6386	.6781	.7366	.7148	.7034
500	.6474	.6823	.6771	.6935	.6966	.6674	.5576	.5999	.6434	.7079	.6838	.6712
550	.6132	.6514	.6457	.6636	.6669	.6352	.5157	.5618	.6089	.6793	.6530	.6393
600	.5794	.6207	.6146	.6340	.6375	.6033	.4746	.5242	.5749	.6510	.6224	.6077
650	.5460	.5902	.5838	.6045	.6083	.5717	.4343	.4872	.5413	.6228	.5921	.5763
700	.5130	.5601	.5533	.5754	.5793	.5405	.3949	.4508	.5081	.5949	.5621	.5453
750	.4804	.5302	.5231	.5465	.5507	.5096	.3564	.4153	.4754	.5673	.5324	.5147
800	.4483	.5007	.4934	.5179	.5223	.4792	.3191	.3805	.4433	.5399	.5031	.4845
850	.4167	.4716	.4640	.4897	.4942	.4493	.2829	.3466	.4118	.5128	.4741	.4547
900	.3856	.4428	.4351	.4618	.4664	.4199	.2480	.3137	.3808	.4860	.4455	.4253
950	.3552	.4145	.4067	.4344	.4391	.3910	.2146	.2819	.3506	.4595	.4174	.3965
1000	.3254	.3866	.3788	.4073	.4121	.3627	.1828	.2513	.3212	.4334	.3897	.3682
1050	.2963	.3592	.3514	.3807	.3855	.3351	.1528	.2220	.2925	.4077	.3625	.3405
1100	.2679	.3323	.3247	.3546	.3593	.3082	.1248	.1942	.2648	.3825	.3358	.3134
1150	.2404	.3060	.2986	.3289	.3337	.2820	.0992	.1681	.2380	.3576	.3097	.2870
1200	.2138	.2803	.2732	.3039	.3085	.2567	.0764	.1441	.2124	.3332	.2842	.2613
1250	.1881	.2552	.2485	.2794	.2839	.2323	.0572	.1227	.1880	.3094	.2593	.2364
1300	.1634	.2308	.2247	.2556	.2598	.2089	.0424	.1042	.1651	.2861	.2351	.2124
1350	.1399	.2071	.2018	.2324	.2364	.1867	.0321	.0880	.1436	.2633	.2117	.1893
1400	.1176	.1842	.1800	.2100	.2136	.1658	.0260	.0740	.1239	.2412	.1891	.1672
1450	.0967	.1621	.1593	.1883	.1915	.1463	.0215	.0620	.1059	.2197	.1673	.1462
1500	.0774	.1410	.1398	.1676	.1701	.1283	.0179	.0517	.0896	.1989	.1466	.1265
1550	.0600	.1209	.1217	.1478	.1496	.1118	.0149	.0429	.0751	.1789	.1268	.1083
1600	.0446	.1018	.1051	.1290	.1299	.0967	.0125	.0354	.0623	.1598	.1082	.0919
1650	.0317	.0840	.0898	.1115	.1112	.0831	.0105	.0291	.0512	.1416	.0909	.0776
1700	.0214	.0676	.0761	.0953	.0935	.0708	.0088	.0237	.0415	.1244	.0752	.0654
1750	.0137	.0528	.0638	.0805	.0770	.0598	.0073	.0192	.0333	.1085	.0614	.0549
1800	.0083	.0399	.0530	.0673	.0617	.0501	.0061	.0154	.0265	.0941	.0499	.0460
1850	.0047	.0290	.0435	.0556	.0479	.0415	.0050	.0123	.0208	.0813	.0404	.0385
1900	.0026	.0203	.0354	.0454	.0357	.0340	.0041	.0097	.0162	.0701	.0325	.0320
1950	.0014	.0136	.0283	.0365	.0255	.0275	.0033	.0076	.0124	.0603	.0260	.0265
2000	.0007	.0087	.0223	.0289	.0178	.0219	.0026	.0058	.0093	.0516	.0204	.0217
2050	.0004	.0053	.0174	.0226	.0129	.0172	.0021	.0044	.0069	.0440	.0158	.0175
2100	.0002	.0030	.0133	.0174	.0093	.0133	.0016	.0033	.0050	.0372	.0120	.0139
2150	.0001	.0016	.0100	.0132	.0066	.0101	.0012	.0024	.0036	.0312	.0089	.0109
2200	.0000	.0008	.0074	.0098	.0046	.0076	.0009	.0017	.0025	.0258	.0065	.0084
2250	.0000	.0003	.0054	.0072	.0031	.0055	.0007	.0012	.0017	.0210	.0047	.0063
2300	.0000	.0001	.0039	.0052	.0021	.0040	.0005	.0008	.0011	.0168	.0033	.0047
2350	.0000	.0000	.0027	.0037	.0014	.0028	.0004	.0006	.0007	.0132	.0023	.0034
2400	.0000	.0000	.0019	.0026	.0009	.0020	.0003	.0004	.0004	.0102	.0016	.0024
2450	.0000	.0000	.0013	.0018	.0005	.0013	.0002	.0002	.0003	.0076	.0011	.0017
2500	.0000	.0000	.0008	.0012	.0003	.0009	.0001	.0002	.0001	.0055	.0007	.0011
2550	.0000	.0000	.0005	.0007	.0002	.0006	.0001	.0001	.0001	.0038	.0005	.0008
2600	.0000	.0000	.0003	.0004	.0001	.0003	.0000	.0001	.0000	.0025	.0003	.0005
2650	.0000	.0000	.0002	.0002	.0001	.0002	.0000	.0000	.0000	.0015	.0002	.0003
2700	.0000	.0000	.0001	.0001	.0000	.0001	.0000	.0000	.0000	.0008	.0001	.0001
2750	.0000	.0000	.0000	.0000	.0000	.0000	.0000	.0000	.0000	.0003	.0000	.0000
2800	.0000	.0000	.0000	.0000	.0000	.0000	.0000	.0000	.0000	.0001	.0000	.0000

TABLE 25

MTF, Perfect Lens, Polychromatic P-11 Phosphor

ν_g c/mm	Normal observer: Photopic	Normal observer: Scotopic	Photoelectric detectors: S-10	S-11	S-12	S-20	CdSe	Si diode	Vidicon	Photographic emulsions: Blue sensitive	Ortho chromatic	Pan chromatic
50	.9675	.9692	.9701	.9702	.9696	.9703	.9700	.9697	.9696	.9701	.9699	.9701
100	.9350	.9385	.9403	.9405	.9391	.9405	.9400	.9395	.9393	.9401	.9398	.9403
150	.9025	.9078	.9105	.9108	.9087	.9109	.9101	.9093	.9090	.9102	.9098	.9105
200	.8701	.8771	.8807	.8811	.8784	.8812	.8802	.8791	.8788	.8804	.8798	.8807
250	.8378	.8465	.8510	.8515	.8481	.8516	.8504	.8490	.8486	.8506	.8498	.8510
300	.8056	.8160	.8214	.8220	.8180	.8221	.8207	.8190	.8185	.8210	.8200	.8214
350	.7735	.7856	.7919	.7926	.7879	.7928	.7911	.7891	.7885	.7914	.7902	.7919
400	.7416	.7554	.7625	.7633	.7579	.7635	.7616	.7593	.7587	.7619	.7606	.7625
450	.7098	.7253	.7333	.7341	.7281	.7343	.7322	.7297	.7290	.7326	.7311	.7333
500	.6783	.6953	.7042	.7051	.6985	.7054	.7030	.7002	.6994	.7034	.7018	.7042
550	.6470	.6656	.6753	.6763	.6690	.6765	.6739	.6709	.6700	.6744	.6727	.6753
600	.6159	.6360	.6465	.6476	.6398	.6479	.6451	.6418	.6409	.6456	.6437	.6465
650	.5851	.6067	.6180	.6192	.6107	.6195	.6165	.6129	.6119	.6170	.6150	.6180
700	.5546	.5776	.5897	.5910	.5819	.5913	.5881	.5843	.5832	.5887	.5865	.5897
750	.5244	.5488	.5617	.5630	.5534	.5633	.5599	.5559	.5548	.5605	.5582	.5617
800	.4945	.5203	.5339	.5353	.5251	.5356	.5320	.5278	.5266	.5327	.5303	.5339
850	.4651	.4921	.5064	.5078	.4972	.5082	.5044	.5000	.4987	.5052	.5026	.5064
900	.4360	.4643	.4792	.4807	.4696	.4811	.4772	.4725	.4712	.4779	.4753	.4792
950	.4074	.4368	.4523	.4539	.4423	.4544	.4502	.4454	.4440	.4510	.4482	.4524
1000	.3792	.4097	.4258	.4275	.4154	.4280	.4237	.4186	.4172	.4245	.4216	.4259
1050	.3516	.3830	.3997	.4014	.3889	.4019	.3975	.3922	.3908	.3984	.3954	.3998
1100	.3245	.3568	.3740	.3758	.3628	.3763	.3717	.3663	.3648	.3726	.3695	.3741
1150	.2979	.3310	.3487	.3505	.3372	.3511	.3464	.3408	.3393	.3473	.3441	.3489
1200	.2720	.3057	.3239	.3258	.3121	.3263	.3215	.3158	.3143	.3225	.3192	.3241
1250	.2467	.2810	.2996	.3015	.2875	.3020	.2971	.2913	.2897	.2982	.2948	.2998
1300	.2222	.2569	.2758	.2777	.2634	.2782	.2733	.2674	.2658	.2744	.2709	.2760
1350	.1984	.2334	.2525	.2544	.2400	.2550	.2500	.2440	.2424	.2512	.2476	.2528
1400	.1754	.2105	.2298	.2318	.2171	.2324	.2273	.2213	.2197	.2285	.2250	.2302
1450	.1533	.1883	.2078	.2098	.1950	.2104	.2053	.1993	.1976	.2065	.2029	.2082
1500	.1321	.1669	.1865	.1884	.1735	.1890	.1840	.1779	.1762	.1852	.1816	.1869
1550	.1119	.1463	.1658	.1678	.1528	.1684	.1634	.1573	.1557	.1647	.1611	.1664
1600	.0929	.1265	.1460	.1479	.1330	.1485	.1436	.1376	.1359	.1449	.1413	.1466
1650	.0752	.1077	.1270	.1288	.1140	.1295	.1246	.1188	.1171	.1260	.1224	.1277
1700	.0588	.0900	.1089	.1107	.0960	.1113	.1066	.1009	.0993	.1081	.1046	.1097
1750	.0441	.0734	.0918	.0935	.0791	.0942	.0897	.0842	.0826	.0912	.0878	.0928
1800	.0315	.0581	.0758	.0774	.0633	.0782	.0740	.0687	.0672	.0755	.0722	.0771
1850	.0211	.0443	.0612	.0626	.0489	.0634	.0595	.0547	.0532	.0612	.0581	.0628
1900	.0133	.0323	.0480	.0493	.0360	.0500	.0466	.0422	.0409	.0485	.0455	.0500
1950	.0079	.0223	.0364	.0375	.0250	.0383	.0353	.0315	.0303	.0375	.0346	.0387
2000	.0044	.0144	.0266	.0274	.0165	.0282	.0258	.0225	.0216	.0283	.0253	.0290
2050	.0023	.0087	.0185	.0192	.0109	.0198	.0180	.0154	.0147	.0207	.0176	.0209
2100	.0011	.0048	.0122	.0127	.0069	.0133	.0119	.0100	.0094	.0147	.0116	.0143
2150	.0005	.0023	.0076	.0079	.0040	.0083	.0074	.0060	.0057	.0100	.0071	.0093
2200	.0002	.0010	.0044	.0046	.0022	.0049	.0044	.0034	.0032	.0064	.0041	.0057
2250	.0001	.0004	.0024	.0024	.0011	.0026	.0024	.0018	.0016	.0039	.0022	.0032
2300	.0000	.0001	.0012	.0012	.0005	.0013	.0012	.0009	.0008	.0022	.0011	.0017
2350	.0000	.0000	.0005	.0005	.0002	.0006	.0005	.0004	.0003	.0011	.0005	.0008
2400	.0000	.0000	.0002	.0002	.0001	.0002	.0002	.0001	.0001	.0005	.0002	.0003
2450	.0000	.0000	.0001	.0001	.0000	.0001	.0001	.0000	.0000	.0002	.0001	.0001
2500	.0000	.0000	.0000	.0000	.0000	.0000	.0000	.0000	.0000	.0000	.0000	.0000
2550	.0000	.0000	.0000	.0000	.0000	.0000	.0000	.0000	.0000	.0000	.0000	.0000
2600	.0000	.0000	.0000	.0000	.0000	.0000	.0000	.0000	.0000	.0000	.0000	.0000
2650	.0000	.0000	.0000	.0000	.0000	.0000	.0000	.0000	.0000	.0000	.0000	.0000
2700	.0000	.0000	.0000	.0000	.0000	.0000	.0000	.0000	.0000	.0000	.0000	.0000
2750	.0000	.0000	.0000	.0000	.0000	.0000	.0000	.0000	.0000	.0000	.0000	.0000
2800	.0000	.0000	.0000	.0000	.0000	.0000	.0000	.0000	.0000	.0000	.0000	.0000

TABLE 25

MTF, Perfect Lens, Polychromatic P-16 Phosphor

υ_R	Normal observer		Photoelectric detectors							Photographic emulsions		
c/mm	Photopic	Scotopic	S-10	S-11	S-12	S-20	CdSe	Si diode	Vidicon	Blue sensitive	Ortho chromatic	Pan chromatic
50	.9713	.9722	.9749	.9749	.9744	.9749	.9750	.9746	.9744	.9757	.9748	.9751
100	.9427	.9443	.9497	.9498	.9488	.9498	.9500	.9493	.9488	.9514	.9495	.9502
150	.9141	.9165	.9246	.9248	.9232	.9248	.9251	.9240	.9232	.9272	.9243	.9254
200	.8855	.8888	.8996	.8998	.8976	.8997	.9002	.8987	.8977	.9029	.8992	.9005
250	.8570	.8611	.8745	.8748	.8721	.8747	.8753	.8734	.8722	.8787	.8740	.8757
300	.8285	.8334	.8495	.8499	.8467	.8498	.8505	.8483	.8467	.8546	.8490	.8510
350	.8002	.8059	.8246	.8250	.8213	.8249	.8257	.8231	.8214	.8305	.8240	.8263
400	.7720	.7784	.7998	.8002	.7960	.8001	.8010	.7980	.7961	.8064	.7990	.8017
450	.7438	.7510	.7750	.7755	.7707	.7754	.7763	.7731	.7708	.7825	.7741	.7771
500	.7158	.7238	.7503	.7508	.7456	.7507	.7518	.7482	.7457	.7586	.7494	.7527
550	.6880	.6967	.7257	.7263	.7206	.7262	.7274	.7234	.7207	.7348	.7247	.7283
600	.6603	.6698	.7012	.7019	.6956	.7017	.7030	.6987	.6958	.7111	.7001	.7041
650	.6328	.6430	.6769	.6776	.6709	.6774	.6788	.6742	.6710	.6875	.6757	.6799
700	.6055	.6164	.6527	.6534	.6462	.6533	.6547	.6498	.6464	.6641	.6514	.6559
750	.5785	.5900	.6286	.6294	.6217	.6292	.6308	.6255	.6219	.6407	.6272	.6321
800	.5516	.5638	.6047	.6055	.5974	.6053	.6070	.6014	.5976	.6175	.6032	.6084
850	.5250	.5379	.5809	.5818	.5733	.5816	.5834	.5775	.5734	.5945	.5794	.5848
900	.4987	.5122	.5574	.5583	.5493	.5581	.5599	.5537	.5495	.5716	.5558	.5614
950	.4727	.4867	.5340	.5349	.5256	.5348	.5367	.5302	.5257	.5489	.5323	.5382
1000	.4469	.4616	.5108	.5118	.5020	.5116	.5136	.5069	.5022	.5264	.5091	.5152
1050	.4216	.4367	.4878	.4889	.4787	.4887	.4908	.4837	.4789	.5040	.4860	.4925
1100	.3965	.4122	.4651	.4662	.4557	.4660	.4682	.4609	.4559	.4819	.4632	.4699
1150	.3719	.3880	.4426	.4437	.4329	.4435	.4458	.4382	.4331	.4600	.4407	.4476
1200	.3476	.3641	.4204	.4215	.4103	.4213	.4236	.4159	.4105	.4383	.4184	.4255
1250	.3237	.3406	.3984	.3996	.3881	.3994	.4018	.3938	.3883	.4168	.3964	.4037
1300	.3003	.3176	.3768	.3780	.3662	.3777	.3802	.3720	.3664	.3957	.3747	.3821
1350	.2774	.2949	.3554	.3566	.3446	.3564	.3589	.3505	.3447	.3747	.3532	.3609
1400	.2550	.2727	.3343	.3356	.3233	.3354	.3379	.3293	.3235	.3541	.3322	.3400
1450	.2331	.2510	.3136	.3149	.3024	.3146	.3173	.3085	.3025	.3337	.3114	.3193
1500	.2118	.2298	.2932	.2946	.2818	.2943	.2970	.2881	.2820	.3137	.2910	.2991
1550	.1910	.2092	.2733	.2746	.2617	.2743	.2770	.2680	.2619	.2940	.2710	.2791
1600	.1710	.1891	.2537	.2550	.2420	.2547	.2575	.2484	.2421	.2746	.2514	.2596
1650	.1516	.1696	.2345	.2358	.2227	.2355	.2383	.2291	.2228	.2556	.2322	.2405
1700	.1329	.1508	.2157	.2171	.2039	.2168	.2196	.2104	.2040	.2370	.2134	.2217
1750	.1151	.1326	.1974	.1988	.1857	.1985	.2013	.1921	.1857	.2188	.1951	.2035
1800	.0981	.1153	.1796	.1810	.1679	.1807	.1835	.1743	.1679	.2010	.1773	.1857
1850	.0821	.0987	.1623	.1637	.1507	.1634	.1662	.1570	.1507	.1837	.1601	.1683
1900	.0671	.0830	.1456	.1470	.1341	.1467	.1495	.1403	.1341	.1668	.1433	.1516
1950	.0533	.0683	.1295	.1308	.1182	.1305	.1333	.1243	.1181	.1505	.1272	.1353
2000	.0409	.0546	.1139	.1153	.1029	.1150	.1177	.1088	.1027	.1346	.1118	.1197
2050	.0302	.0422	.0991	.1004	.0885	.1001	.1028	.0941	.0882	.1194	.0970	.1047
2100	.0213	.0313	.0850	.0863	.0748	.0860	.0886	.0802	.0744	.1047	.0830	.0904
2150	.0141	.0219	.0717	.0730	.0621	.0726	.0752	.0671	.0616	.0907	.0698	.0769
2200	.0086	.0143	.0593	.0605	.0504	.0602	.0626	.0550	.0497	.0774	.0576	.0642
2250	.0047	.0084	.0479	.0490	.0397	.0487	.0510	.0439	.0390	.0649	.0463	.0524
2300	.0023	.0044	.0376	.0386	.0303	.0383	.0405	.0340	.0295	.0532	.0362	.0416
2350	.0010	.0019	.0285	.0294	.0222	.0291	.0311	.0253	.0213	.0425	.0274	.0320
2400	.0004	.0007	.0208	.0216	.0156	.0213	.0230	.0181	.0147	.0328	.0199	.0237
2450	.0001	.0003	.0144	.0151	.0103	.0148	.0163	.0123	.0095	.0244	.0139	.0168
2500	.0000	.0001	.0094	.0099	.0064	.0097	.0108	.0078	.0056	.0171	.0092	.0112
2550	.0000	.0000	.0058	.0062	.0037	.0060	.0068	.0046	.0031	.0114	.0057	.0070
2600	.0000	.0000	.0031	.0034	.0019	.0033	.0038	.0024	.0014	.0068	.0032	.0040
2650	.0000	.0000	.0016	.0017	.0009	.0017	.0020	.0012	.0006	.0037	.0017	.0020
2700	.0000	.0000	.0006	.0007	.0004	.0007	.0008	.0005	.0002	.0017	.0007	.0009
2750	.0000	.0000	.0002	.0002	.0001	.0002	.0002	.0001	.0000	.0005	.0002	.0002
2800	.0000	.0000	.0000	.0000	.0000	.0000	.0000	.0000	.0000	.0001	.0000	.0000

H. Miscellaneous Tables

Table 26—Spectral Efficiency of Combinations of Light Sources and Detectors

Table 27—Radiator and Detector Characteristics for Tables 25 and 26

Table 28—Spectral Luminous Efficiency, Standard Observer

Table 29—Tristimulus Values, Equal-Energy Spectrum

Table 30—Photometric Units, Conversion Factors

Table 31—Useful Formulae

$\int c d\lambda / \int M_\lambda d\lambda$

TABLE 26A

Spectral Efficiency, Source – Detector Combinations

		Normal observer		Photoelectric detectors							Photographic emulsions			Luminous efficacies	
		Photopic	Scotopic	S-10	S-11	S-12	S-20	CdSe	Si diode	Vidicon	Blue Sensitive	Ortho chromatic	Pan chromatic	Photopic (lm/W)	Scotopic (sc. lm/W)
Blackbody	3000° K	.0303	.0180	.0414	.0269	.0055	.0483	.0481	.2794	.0704	0.016	0.049	0.045	20.6	31.4
	5000° K	.1192	.1028	.2275	.1847	.0389	.2366	.0871	.4918	.2594	0.142	0.291	0.235	81.0	179.6
Standard illuminants$_a$	A	.1777	.1008	.2380	.1497	.0302	.2833	.2563	.8254	.4157	0.085	0.273	0.259	120.8	176.0
	B	.2565	.2196	.4493	.3526	.0799	.4682	.1697	.7379	.5479	0.219	0.580	0.454	174.4	383.5
Phosphors (JEDEC)	P–11	.2010	.6033	.9504	.9159	.2701	.8783	.0401	.5371	.7306	0.429	1.243	0.688	136.7	1053.4
	P–16	.0032	.0424	.8593	.8858	.0945	.9057	.0400	.3292	.3615	1.341	0.941	0.969	2.2	74.1
REFERENCE WAVELENGTH (μ)		.5550	.5070	.4500	.4400	.5020	.4200	.7400	.9400	.5500	.4000	.4000	.4000		

TABLE 26A

Additional System Efficiencies***

	System efficiency											
	Photocathodes								Normal observer		Luminous efficacy	
	S1	S4	S9	S10	S11	S17	S20	S25	Photopic	Scotopic	lm/W	sc. lm/W
Phosphors												
P1	0.278	0.498	0.843	0.807	0.687	0.892	0.700	0.853	0.768	0.743	522	1297
P4	0.310	0.549	0.725	0.767	0.661	0.734	0.724	0.861	0.402	0.452	273	789
P7	0.312	0.611	0.748	0.805	0.709	0.773	0.771	0.882	0.411	0.388	279	677
P11	0.217	0.816	0.930	0.949	0.914	0.954	0.877	0.953	0.201	0.601	137	1049
P15	0.385	0.701	0.815	0.855	0.787	0.871	0.802	0.904	0.376	0.495	256	864
P16	0.830	0.970	0.698	0.853	0.880	0.855	0.902	0.922	0.003	0.042	2	73
P20	0.395	0.284	0.560	0.612	0.427	0.563	0.583	0.782	0.707	0.354	481	618
P22B	0.217	0.893	0.923	0.974	0.960	0.948	0.927	0.979	0.808	0.477	549	833
P22G	0.278	0.495	0.844	0.807	0.686	0.896	0.699	0.855	0.784	0.747	533	1304
P22R	0.632	0.036	0.084	0.264	0.055	0.077	0.368	0.623	0.225	0.008	153	14
P24	0.279	0.545	0.807	0.806	0.696	0.827	0.725	0.869	0.540	0.621	367	1084
P31	0.276	0.533	0.826	0.811	0.698	0.853	0.722	0.868	0.626	0.651	426	1137
NaI	0.534	0.923	0.784	0.885	0.889	0.889	0.900	0.933	0.046	0.224	31	391
Sun*												
In space	0.535	0.308	0.344	0.388	0.328	0.380	0.406	0.547	0.179	0.172	122	300
Sea level**	0.536	0.236	0.302	0.348	0.277	0.315	0.360	0.513	0.197	0.175	134	306
Day sky	0.537	0.520	0.508	0.556	0.508	0.589	0.581	0.700	0.170	0.218	116	381
Blackbodies*												
6000°K	0.533	0.308	0.333	0.376	0.320	0.375	0.397	0.521	0.167	0.159	114	278
3000°K	0.512	0.053	0.078	0.102	0.067	0.080	0.120	0.232	0.075	0.044	51	77
2870°K	0.504	0.044	0.067	0.090	0.057	0.069	0.106	0.216	0.067	0.038	46	66
2854°K	0.500	0.042	0.066	0.088	0.055	0.068	0.103	0.211	0.065	0.037	44	65
2810°K	0.493	0.039	0.060	0.081	0.051	0.062	0.097	0.150	0.061	0.034	41	59
2042°K	0.401	0.008	0.014	0.023	0.011	0.014	0.033	0.090	0.018	0.007	12	12

*Calculated only for .300–1.200 μm wavelength interval.
**+2 air masses, *cf.* [28].
***Except for the last two columns, these data are from Eberhardt[26] with permission.

$\int c d\lambda / \int M_\lambda d\lambda$

TABLE 26B

Spectral Efficiency, Blackbody-S Curves and Blackbody-Vision Combinations

S1 Peak at 0.800μ Eff. at 2854°K = 0.1856

Temp	0	100	200	300	400	500	600	700	800	900
0	0.000E 00	0.000E 00	0.328E-23	0.772E-15	0.113E-10	0.358E-08	0.165E-06	0.252E-05	0.192E-04	0.920E-04
1000	0.317E-03	0.863E-03	0.196E-02	0.388E-02	0.689E-02	0.112E-01	0.170E-01	0.244E-01	0.334E-01	0.439E-01
2000	0.0558	0.0689	0.0830	0.0979	0.1134	0.1292	0.1452	0.1613	0.1771	0.1928
3000	0.2081	0.2229	0.2372	0.2510	0.2643	0.2769	0.2890	0.3005	0.3114	0.3218
4000	0.3316	0.3409	0.3497	0.3581	0.3659	0.3734	0.3804	0.3870	0.3933	0.3992
5000	0.4048	0.4100	0.4149	0.4195	0.4238	0.4278	0.4316	0.4351	0.4383	0.4413
6000	0.4441	0.4466	0.4489	0.4509	0.4528	0.4544	0.4559	0.4571	0.4582	0.4590
7000	0.4597	0.4602	0.4605	0.4607	0.4607	0.4606	0.4603	0.4598	0.4592	0.4585
8000	0.4577	0.4567	0.4556	0.4544	0.4530	0.4516	0.4500	0.4484	0.4467	0.4449
9000	0.4430	0.4410	0.4389	0.4368	0.4346	0.4323	0.4300	0.4276	0.4251	0.4226

S3 Peak at 0.420μ Eff. at 2854°K = 0.0580

Temp	0	100	200	300	400	500	600	700	800	900
0	0.000E 00	0.000E 00	0.289E-31	0.515E-20	0.184E-14	0.360E-11	0.531E-09	0.180E-07	0.246E-06	0.184E-05
1000	0.905E-05	0.328E-04	0.952E-04	0.232E-03	0.495E-03	0.948E-03	0.166E-02	0.273E-02	0.422E-02	0.620E-02
2000	0.876E-02	0.119E-01	0.157E-01	0.203E-01	0.255E-01	0.315E-01	0.381E-01	0.455E-01	0.535E-01	0.620E-01
3000	0.0712	0.0808	0.0910	0.1015	0.1123	0.1234	0.1347	0.1461	0.1576	0.1692
4000	0.1807	0.1922	0.2035	0.2147	0.2257	0.2365	0.2470	0.2571	0.2670	0.2766
5000	0.2857	0.2946	0.3030	0.3110	0.3187	0.3259	0.3328	0.3392	0.3453	0.3510
6000	0.3562	0.3611	0.3656	0.3698	0.3736	0.3770	0.3801	0.3829	0.3853	0.3875
7000	0.3893	0.3909	0.3921	0.3932	0.3939	0.3945	0.3948	0.3948	0.3947	0.3944
8000	0.3938	0.3931	0.3923	0.3912	0.3900	0.3887	0.3872	0.3856	0.3839	0.3821
9000	0.3802	0.3782	0.3760	0.3738	0.3716	0.3692	0.3668	0.3643	0.3618	0.3592

S4 Peak at 0.400μ Eff. at 2854°K = 0.0157

Temp	0	100	200	300	400	500	600	700	800	900
0	0.000E 00	0.000E 00	0.000E 00	0.297E-27	0.447E-20	0.911E-16	0.692E-13	0.803E-11	0.287E-09	0.468E-08
1000	0.438E-07	0.273E-06	0.125E-05	0.455E-05	0.137E-04	0.355E-04	0.813E-04	0.168E-03	0.320E-03	0.567E-03
2000	0.947E-03	0.149E-02	0.227E-02	0.330E-02	0.465E-02	0.635E-02	0.844E-02	0.109E-01	0.139E-01	0.173E-01
3000	0.0213	0.0257	0.0306	0.0360	0.0419	0.0482	0.0550	0.0621	0.0696	0.0774
4000	0.0855	0.0938	0.1024	0.1111	0.1199	0.1288	0.1378	0.1467	0.1557	0.1646
5000	0.1734	0.1822	0.1908	0.1992	0.2075	0.2155	0.2234	0.2310	0.2384	0.2456
6000	0.2525	0.2591	0.2655	0.2715	0.2774	0.2830	0.2882	0.2932	0.2980	0.3024
7000	0.3066	0.3105	0.3142	0.3176	0.3207	0.3236	0.3263	0.3287	0.3309	0.3329
8000	0.3347	0.3363	0.3376	0.3388	0.3398	0.3406	0.3413	0.3417	0.3421	0.3422
9000	0.3422	0.3421	0.3418	0.3415	0.3410	0.3403	0.3396	0.3388	0.3378	0.3368

S4 Filtered Peak at 0.400μ Eff. at 2854°K = 0.0155

Temp	0	100	200	300	400	500	600	700	800	900
0	0.000E 00	0.000E 00	0.000E 00	0.297E-27	0.447E-20	0.911E-16	0.692E-13	0.803E-11	0.287E-09	0.468E-08
1000	0.438E-07	0.273E-06	0.125E-05	0.455E-05	0.137E-04	0.355E-04	0.813E-04	0.168E-03	0.320E-03	0.567E-03
2000	0.945E-03	0.149E-02	0.226E-02	0.329E-02	0.463E-02	0.631E-02	0.838E-02	0.108E-01	0.137E-01	0.171E-01
3000	0.0210	0.0253	0.0301	0.0353	0.0409	0.0470	0.0534	0.0601	0.0672	0.0745
4000	0.0820	0.0897	0.0976	0.1056	0.1136	0.1216	0.1297	0.1377	0.1456	0.1534
5000	0.1611	0.1687	0.1761	0.1833	0.1903	0.1971	0.2036	0.2099	0.2160	0.2218
6000	0.2273	0.2326	0.2376	0.2424	0.2469	0.2511	0.2551	0.2588	0.2622	0.2654
7000	0.2684	0.2711	0.2736	0.2759	0.2779	0.2797	0.2814	0.2828	0.2840	0.2851
8000	0.2859	0.2866	0.2872	0.2876	0.2878	0.2879	0.2878	0.2876	0.2873	0.2869
9000	0.2863	0.2857	0.2849	0.2841	0.2832	0.2821	0.2810	0.2798	0.2786	0.2773

TABLE 26B

Spectral Efficiency, Blackbody-S Curves and Blackbody-Vision Combinations

S5 Peak at 0.340μ Eff. at 2854°K = 0.0130

Temp	0	100	200	300	400	500	600	700	800	900
0	0.000E 00	0.000E 00	0.000E 00	0.136E-27	0.280E-20	0.670E-16	0.555E-13	0.676E-11	0.248E-09	0.409E-08
1000	0.384E-07	0.239E-06	0.109E-05	0.395E-05	0.118E-04	0.303E-04	0.692E-04	0.142E-03	0.269E-03	0.476E-03
2000	0.791E-03	0.124E-02	0.188E-02	0.274E-02	0.385E-02	0.526E-02	0.699E-02	0.909E-02	0.115E-01	0.144E-01
3000	0.0177	0.0215	0.0256	0.0303	0.0353	0.0408	0.0467	0.0529	0.0595	0.0665
4000	0.0738	0.0813	0.0891	0.0971	0.1053	0.1137	0.1221	0.1307	0.1393	0.1480
5000	0.1566	0.1653	0.1739	0.1824	0.1908	0.1991	0.2073	0.2153	0.2231	0.2308
6000	0.2383	0.2456	0.2526	0.2595	0.2661	0.2725	0.2786	0.2845	0.2902	0.2956
7000	0.3008	0.3057	0.3104	0.3148	0.3190	0.3229	0.3266	0.3301	0.3334	0.3364
8000	0.3392	0.3418	0.3442	0.3464	0.3484	0.3502	0.3518	0.3532	0.3545	0.3555
9000	0.3565	0.3572	0.3578	0.3583	0.3586	0.3587	0.3588	0.3587	0.3585	0.3581

S5 Filtered Peak at 0.340μ Eff. at 2854°K = 0.0128

Temp	0	100	200	300	400	500	600	700	800	900
0	0.000E 00	0.000E 00	0.000E 00	0.136E-27	0.280E-20	0.670E-16	0.555E-13	0.676E-11	0.248E-09	0.409E-08
1000	0.384E-07	0.239E-06	0.109E-05	0.395E-05	0.118E-04	0.303E-04	0.691E-04	0.142E-03	0.269E-03	0.475E-03
2000	0.789E-03	0.124E-02	0.187E-02	0.272E-02	0.382E-02	0.520E-02	0.690E-02	0.894E-02	0.113E-01	0.141E-01
3000	0.0172	0.0208	0.0247	0.0290	0.0337	0.0387	0.0440	0.0496	0.0555	0.0616
4000	0.0679	0.0743	0.0809	0.0876	0.0943	0.1011	0.1079	0.1147	0.1214	0.1281
5000	0.1346	0.1411	0.1474	0.1536	0.1596	0.1654	0.1711	0.1765	0.1818	0.1868
6000	0.1917	0.1963	0.2007	0.2049	0.2088	0.2126	0.2161	0.2194	0.2225	0.2253
7000	0.2280	0.2305	0.2328	0.2348	0.2367	0.2384	0.2400	0.2413	0.2425	0.2435
8000	0.2444	0.2452	0.2457	0.2462	0.2465	0.2467	0.2468	0.2467	0.2466	0.2463
9000	0.2460	0.2455	0.2450	0.2443	0.2436	0.2429	0.2420	0.2411	0.2401	0.2390

S6 Peak at 0.275μ Eff. at 2854°K = 0.490E-03

Temp	0	100	200	300	400	500	600	700	800	900
0	0.000E 00	0.000E 00	0.000E 00	0.000E 00	0.906E-32	0.561E-25	0.182E-20	0.294E-17	0.734E-15	0.528E-13
1000	0.159E-11	0.256E-10	0.258E-09	0.180E-08	0.954E-08	0.401E-07	0.140E-06	0.424E-06	0.113E-05	0.270E-05
2000	0.593E-05	0.120E-04	0.229E-04	0.411E-04	0.702E-04	0.114E-03	0.180E-03	0.273E-03	0.402E-03	0.576E-03
3000	0.804E-03	0.109E-02	0.146E-02	0.192E-02	0.247E-02	0.314E-02	0.393E-02	0.486E-02	0.592E-02	0.715E-02
4000	0.853E-02	0.100E-01	0.118E-01	0.137E-01	0.158E-01	0.180E-01	0.205E-01	0.232E-01	0.260E-01	0.290E-01
5000	0.0322	0.0356	0.0391	0.0428	0.0466	0.0506	0.0548	0.0590	0.0634	0.0678
6000	0.0724	0.0770	0.0818	0.0865	0.0914	0.0963	0.1012	0.1061	0.1111	0.1160
7000	0.1210	0.1259	0.1308	0.1357	0.1406	0.1454	0.1501	0.1548	0.1595	0.1640
8000	0.1685	0.1730	0.1773	0.1816	0.1857	0.1898	0.1938	0.1977	0.2015	0.2051
9000	0.2087	0.2122	0.2156	0.2189	0.2220	0.2251	0.2280	0.2309	0.2336	0.2363

S6 Filtered Peak at 0.275μ Eff. at 2854°K = 0.295E-03

Temp	0	100	200	300	400	500	600	700	800	900
0	0.000E 00	0.000E 00	0.000E 00	0.000E 00	0.906E-32	0.561E-25	0.182E-20	0.294E-17	0.733E-15	0.527E-13
1000	0.158E-11	0.254E-10	0.254E-09	0.176E-08	0.916E-08	0.379E-07	0.130E-06	0.384E-06	0.996E-06	0.232E-05
2000	0.493E-05	0.968E-05	0.177E-04	0.307E-04	0.506E-04	0.796E-04	0.120E-03	0.175E-03	0.248E-03	0.340E-03
3000	0.457E-03	0.598E-03	0.769E-03	0.969E-03	0.120E-02	0.146E-02	0.176E-02	0.209E-02	0.246E-02	0.286E-02
4000	0.329E-02	0.375E-02	0.425E-02	0.476E-02	0.531E-02	0.587E-02	0.646E-02	0.706E-02	0.768E-02	0.831E-02
5000	0.895E-02	0.960E-02	0.102E-01	0.109E-01	0.115E-01	0.122E-01	0.128E-01	0.134E-01	0.141E-01	0.147E-01
6000	0.0153	0.0159	0.0165	0.0171	0.0176	0.0181	0.0187	0.0192	0.0197	0.0201
7000	0.0206	0.0210	0.0214	0.0218	0.0221	0.0225	0.0228	0.0231	0.0234	0.0237
8000	0.0240	0.0242	0.0244	0.0246	0.0248	0.0250	0.0251	0.0253	0.0254	0.0255
9000	0.0256	0.0257	0.0258	0.0258	0.0259	0.0259	0.0259	0.0260	0.0260	0.0260

TABLE 26B

Spectral Efficiency, Blackbody-S Curves and Blackbody-Vision Combinations

S7 Peak at 0.585μ Eff. at 2854°K = 0.1638

Temp	0	100	200	300	400	500	600	700	800	900
0	0.000E 00	0.000E 00	0.452E-27	0.313E-17	0.249E-12	0.206E-09	0.173E-07	0.398E-06	0.405E-05	0.240E-04
1000	0.982E-04	0.305E-03	0.774E-03	0.168E-02	0.324E-02	0.566E-02	0.916E-02	0.139E-01	0.200E-01	0.276E-01
2000	0.0366	0.0472	0.0591	0.0724	0.0868	0.1024	0.1188	0.1361	0.1540	0.1723
3000	0.1911	0.2101	0.2292	0.2483	0.2674	0.2863	0.3050	0.3234	0.3414	0.3591
4000	0.3763	0.3930	0.4092	0.4249	0.4400	0.4546	0.4686	0.4820	0.4948	0.5071
5000	0.5187	0.5298	0.5403	0.5502	0.5595	0.5683	0.5765	0.5842	0.5914	0.5980
6000	0.6042	0.6098	0.6150	0.6197	0.6240	0.6278	0.6312	0.6342	0.6368	0.6391
7000	0.6409	0.6424	0.6436	0.6444	0.6450	0.6452	0.6452	0.6448	0.6443	0.6434
8000	0.6424	0.6411	0.6395	0.6378	0.6359	0.6338	0.6315	0.6291	0.6265	0.6238
9000	0.6209	0.6179	0.6147	0.6115	0.6081	0.6047	0.6011	0.5975	0.5938	0.5900

S8 Peak at 0.366μ Eff. at 2854°K = 0.0217

Temp	0	100	200	300	400	500	600	700	800	900
0	0.000E 00	0.000E 00	0.243E-35	0.206E-23	0.327E-17	0.167E-13	0.488E-11	0.279E-09	0.578E-08	0.609E-07
1000	0.400E-06	0.186E-05	0.673E-05	0.199E-04	0.504E-04	0.112E-03	0.227E-03	0.423E-03	0.734E-03	0.119E-02
2000	0.186E-02	0.276E-02	0.396E-02	0.550E-02	0.741E-02	0.974E-02	0.125E-01	0.157E-01	0.194E-01	0.237E-01
3000	0.0284	0.0336	0.0393	0.0455	0.0522	0.0592	0.0667	0.0745	0.0826	0.0910
4000	0.0997	0.1086	0.1176	0.1267	0.1360	0.1453	0.1546	0.1639	0.1731	0.1823
5000	0.1913	0.2003	0.2090	0.2176	0.2260	0.2342	0.2421	0.2498	0.2573	0.2645
6000	0.2714	0.2780	0.2844	0.2905	0.2963	0.3018	0.3070	0.3120	0.3167	0.3210
7000	0.3252	0.3290	0.3326	0.3359	0.3390	0.3418	0.3443	0.3467	0.3488	0.3507
8000	0.3523	0.3538	0.3550	0.3561	0.3570	0.3576	0.3582	0.3585	0.3587	0.3587
9000	0.3586	0.3584	0.3580	0.3575	0.3568	0.3561	0.3552	0.3542	0.3532	0.3520

S9 Peak at 0.480μ Eff. at 2854°K = 0.0239

Temp	0	100	200	300	400	500	600	700	800	900
0	0.000E 00	0.000E 00	0.000E 00	0.245E-28	0.145E-20	0.632E-16	0.751E-13	0.114E-10	0.481E-09	0.665E-08
1000	0.856E-07	0.549E-06	0.255E-05	0.925E-05	0.276E-04	0.705E-04	0.158E-03	0.322E-03	0.602E-03	0.104E-02
2000	0.170E-02	0.264E-02	0.392E-02	0.559E-02	0.771E-02	0.103E-01	0.134E-01	0.171E-01	0.214E-01	0.262E-01
3000	0.0316	0.0376	0.0440	0.0510	0.0584	0.0662	0.0744	0.0829	0.0917	0.1007
4000	0.1099	0.1192	0.1285	0.1379	0.1473	0.1566	0.1658	0.1749	0.1839	0.1927
5000	0.2012	0.2096	0.2177	0.2255	0.2331	0.2403	0.2473	0.2540	0.2604	0.2664
6000	0.2722	0.2776	0.2827	0.2876	0.2921	0.2963	0.3002	0.3039	0.3072	0.3103
7000	0.3131	0.3157	0.3180	0.3200	0.3218	0.3234	0.3248	0.3259	0.3269	0.3276
8000	0.3282	0.3286	0.3288	0.3288	0.3287	0.3285	0.3280	0.3275	0.3268	0.3261
9000	0.3252	0.3241	0.3230	0.3218	0.3205	0.3191	0.3177	0.3161	0.3145	0.3128

S10 Peak at 0.450μ Eff. at 2854°K = 0.0322

Temp	0	100	200	300	400	500	600	700	800	900
0	0.000E 00	0.000E 00	0.000E 00	0.404E-24	0.138E-17	0.108E-13	0.414E-11	0.284E-09	0.670E-08	0.775E-07
1000	0.546E-06	0.267E-05	0.100E-04	0.305E-04	0.787E-04	0.178E-03	0.363E-03	0.677E-03	0.117E-02	0.191E-02
2000	0.296E-02	0.438E-02	0.624E-02	0.859E-02	0.114E-01	0.149E-01	0.190E-01	0.237E-01	0.290E-01	0.350E-01
3000	0.0415	0.0487	0.0564	0.0645	0.0732	0.0823	0.0917	0.1014	0.1113	0.1215
4000	0.1318	0.1422	0.1526	0.1630	0.1734	0.1837	0.1938	0.2038	0.2135	0.2231
5000	0.2324	0.2414	0.2501	0.2585	0.2667	0.2744	0.2819	0.2890	0.2957	0.3021
6000	0.3082	0.3139	0.3193	0.3243	0.3290	0.3334	0.3374	0.3411	0.3445	0.3476
7000	0.3504	0.3529	0.3552	0.3572	0.3589	0.3604	0.3616	0.3626	0.3634	0.3640
8000	0.3643	0.3645	0.3645	0.3643	0.3640	0.3635	0.3628	0.3620	0.3611	0.3600
9000	0.3588	0.3575	0.3561	0.3546	0.3530	0.3513	0.3496	0.3477	0.3458	0.3438

TABLE 26B

Spectral Efficiency, Blackbody-S Curves and Blackbody-Vision Combinations

S11 Peak at 0.440μ Eff. at 2854° K = 0.0202

Temp	0	100	200	300	400	500	600	700	800	900
0	0.000E 00	0.000F 00	0.000E 00	0.325E-26	0.223E-19	0.277E-15	0.157E-12	0.154E-10	0.509E-09	0.793E-08
1000	0.723E-07	0.442E-06	0.200E-05	0.715E-05	0.212E-04	0.541E-04	0.122E-03	0.249E-03	0.469E-03	0.819E-03
2000	0.134E-02	0.210E-02	0.315E-02	0.453E-02	0.630E-02	0.850E-02	0.111E-01	0.143E-01	0.180E-01	0.222E-01
3000	0.0270	0.0323	0.0381	0.0445	0.0512	0.0585	0.0661	0.0740	0.0823	0.0908
4000	0.0995	0.1084	0.1175	0.1266	0.1357	0.1448	0.1539	0.1629	0.1718	0.1806
5000	0.1892	0.1976	0.2058	0.2138	0.2215	0.2290	0.2362	0.2431	0.2497	0.2560
6000	0.2621	0.2678	0.2732	0.2784	0.2832	0.2877	0.2920	0.2960	0.2996	0.3030
7000	0.3062	0.3090	0.3116	0.3140	0.3161	0.3180	0.3196	0.3211	0.3223	0.3233
8000	0.3241	0.3248	0.3252	0.3255	0.3257	0.3256	0.3254	0.3251	0.3247	0.3241
9000	0.3234	0.3225	0.3216	0.3206	0.3194	0.3182	0.3169	0.3155	0.3140	0.3125

S12 Peak at 0.502μ Eff. at 2854° K = 0.407E-02

Temp	0	100	200	300	400	500	600	700	800	900
0	0.000E 00	0.000E 00	0.000E 00	0.194E-31	0.372E-23	0.317E-18	0.605E-15	0.136E-12	0.824E-11	0.204E-09
1000	0.270E-08	0.222E-07	0.128E-06	0.559E-06	0.195E-05	0.570E-05	0.144E-04	0.323E-04	0.657E-04	0.122E-03
2000	0.213E-03	0.350E-03	0.546E-03	0.812E-03	0.116E-02	0.160E-02	0.215E-02	0.282E-02	0.360E-02	0.450E-02
3000	0.552E-02	0.666E-02	0.791E-02	0.927E-02	0.107E-01	0.122E-01	0.139E-01	0.156E-01	0.173E-01	0.191E-01
4000	0.0210	0.0228	0.0247	0.0266	0.0285	0.0304	0.0322	0.0340	0.0358	0.0375
5000	0.0392	0.0408	0.0424	0.0440	0.0454	0.0468	0.0481	0.0494	0.0506	0.0517
6000	0.0528	0.0538	0.0547	0.0556	0.0564	0.0572	0.0579	0.0585	0.0590	0.0596
7000	0.0600	0.0604	0.0608	0.0611	0.0613	0.0616	0.0617	0.0619	0.0619	0.0620
8000	0.0620	0.0620	0.0620	0.0619	0.0618	0.0616	0.0615	0.0613	0.0611	0.0609
9000	0.0606	0.0603	0.0601	0.0597	0.0594	0.0591	0.0588	0.0584	0.0580	0.0576

S13 Peak at 0.440μ Eff. at 2854° K = 0.0205

Temp	0	100	200	300	400	500	600	700	800	900
0	0.000E 00	0.000E 00	0.000E 00	0.209E-27	0.403E-20	0.945E-16	0.793E-13	0.994E-11	0.377E-09	0.641E-08
1000	0.618E-07	0.392E-06	0.182E-05	0.665E-05	0.200E-04	0.517E-04	0.117E-03	0.242E-03	0.458E-03	0.805E-03
2000	0.133E-02	0.208E-02	0.313E-02	0.452E-02	0.630E-02	0.853E-02	0.112E-01	0.144E-01	0.182E-01	0.226E-01
3000	0.0275	0.0330	0.0390	0.0456	0.0527	0.0603	0.0683	0.0767	0.0856	0.0947
4000	0.1041	0.1138	0.1236	0.1336	0.1437	0.1539	0.1641	0.1742	0.1844	0.1944
5000	0.2043	0.2141	0.2237	0.2331	0.2423	0.2513	0.2600	0.2684	0.2766	0.2845
6000	0.2921	0.2994	0.3064	0.3131	0.3195	0.3256	0.3314	0.3369	0.3421	0.3469
7000	0.3515	0.3558	0.3598	0.3635	0.3669	0.3701	0.3730	0.3756	0.3780	0.3802
8000	0.3821	0.3838	0.3853	0.3865	0.3876	0.3885	0.3891	0.3896	0.3900	0.3901
9000	0.3901	0.3899	0.3896	0.3892	0.3886	0.3879	0.3870	0.3861	0.3850	0.3838

S14 Peak at 1.500μ Eff. at 2854° K = 0.4656

Temp	0	100	200	300	400	500	600	700	800	900
0	0.000E 00	0.953E-28	0.903E-13	0.837E-08	0.240E-05	0.670E-04	0.579E-03	0.257E-02	0.759E-02	0.170E-01
1000	0.0317	0.0516	0.0762	0.1043	0.1349	0.1669	0.1991	0.2308	0.2614	0.2904
2000	0.3175	0.3426	0.3655	0.3862	0.4049	0.4215	0.4362	0.4490	0.4602	0.4698
3000	0.4779	0.4848	0.4904	0.4949	0.4984	0.5010	0.5028	0.5038	0.5041	0.5039
4000	0.5030	0.5017	0.4999	0.4977	0.4951	0.4923	0.4891	0.4856	0.4819	0.4781
5000	0.4740	0.4697	0.4653	0.4608	0.4562	0.4514	0.4466	0.4417	0.4368	0.4318
6000	0.4267	0.4217	0.4166	0.4115	0.4064	0.4012	0.3961	0.3910	0.3860	0.3809
7000	0.3758	0.3708	0.3658	0.3609	0.3560	0.3511	0.3463	0.3415	0.3367	0.3320
8000	0.3274	0.3228	0.3182	0.3137	0.3093	0.3049	0.3006	0.2963	0.2920	0.2879
9000	0.2837	0.2797	0.2757	0.2717	0.2678	0.2640	0.2602	0.2565	0.2528	0.2491

$\int sM_\lambda d\lambda / \int M_\lambda d\lambda$

TABLE 26B

Spectral Efficiency, Blackbody-S Curves and Blackbody-Vision Combinations

S15 Peak at 0.580μ Eff. at 2854°K = 0.0464

Temp	0	100	200	300	400	500	600	700	800	900
0	0.000E 00	0.000E 00	0.800E-34	0.536E-22	0.461E-16	0.162E-12	0.375E-10	0.183E-08	0.342E-07	0.333E-06
1000	0.206E-05	0.911E-05	0.312E-04	0.883E-04	0.213E-03	0.455E-03	0.876E-03	0.155E-02	0.256E-02	0.398E-02
2000	0.590E-02	0.836E-02	0.114E-01	0.151E-01	0.194E-01	0.243E-01	0.299E-01	0.360E-01	0.427E-01	0.498E-01
3000	0.0573	0.0651	0.0732	0.0815	0.0899	0.0983	0.1068	0.1152	0.1235	0.1317
4000	0.1397	0.1475	0.1550	0.1622	0.1692	0.1758	0.1821	0.1881	0.1938	0.1991
5000	0.2041	0.2088	0.2131	0.2171	0.2207	0.2241	0.2271	0.2299	0.2323	0.2345
6000	0.2364	0.2381	0.2395	0.2407	0.2417	0.2424	0.2430	0.2433	0.2435	0.2435
7000	0.2434	0.2431	0.2426	0.2420	0.2413	0.2405	0.2396	0.2385	0.2374	0.2362
8000	0.2349	0.2335	0.2320	0.2305	0.2289	0.2273	0.2256	0.2239	0.2221	0.2203
9000	0.2185	0.2166	0.2147	0.2128	0.2109	0.2089	0.2069	0.2050	0.2030	0.2010

S16 Peak at 0.730μ Eff. at 2854°K = 0.1067

Temp	0	100	200	300	400	500	600	700	800	900
0	0.000E 00	0.000E 00	0.153E-25	0.198E-16	0.688E-12	0.358E-09	0.228E-07	0.440E-06	0.399E-05	0.219E-04
1000	0.850E-04	0.254E-03	0.628E-03	0.133E-02	0.252E-02	0.434E-02	0.693E-02	0.104E-01	0.147E-01	0.201E-01
2000	0.0264	0.0336	0.0416	0.0503	0.0597	0.0695	0.0798	0.0902	0.1009	0.1116
3000	0.1223	0.1328	0.1432	0.1534	0.1633	0.1728	0.1820	0.1908	0.1992	0.2072
4000	0.2148	0.2219	0.2287	0.2350	0.2408	0.2463	0.2514	0.2561	0.2604	0.2643
5000	0.2679	0.2711	0.2740	0.2766	0.2788	0.2808	0.2825	0.2840	0.2852	0.2861
6000	0.2868	0.2873	0.2876	0.2877	0.2877	0.2874	0.2870	0.2865	0.2858	0.2849
7000	0.2840	0.2829	0.2817	0.2804	0.2790	0.2775	0.2760	0.2743	0.2726	0.2708
8000	0.2690	0.2671	0.2652	0.2632	0.2611	0.2591	0.2570	0.2548	0.2527	0.2505
9000	0.2483	0.2460	0.2438	0.2415	0.2393	0.2370	0.2347	0.2324	0.2302	0.2279

S17 Peak at 0.490μ Eff. at 2854°K = 0.0245

Temp	0	100	200	300	400	500	600	700	800	900
0	0.000E 00	0.000E 00	0.000E 00	0.214E-26	0.253E-19	0.416E-15	0.263E-12	0.264E-10	0.848E-09	0.127E-07
1000	0.111E-06	0.660E-06	0.290E-05	0.101E-04	0.295E-04	0.742E-04	0.165E-03	0.333E-03	0.618E-03	0.106E-02
2000	0.174E-02	0.269E-02	0.399E-02	0.570E-02	0.786E-02	0.105E-01	0.137E-01	0.175E-01	0.219E-01	0.269E-01
3000	0.0325	0.0387	0.0455	0.0529	0.0607	0.0691	0.0779	0.0870	0.0965	0.1063
4000	0.1164	0.1266	0.1370	0.1475	0.1580	0.1686	0.1791	0.1896	0.2000	0.2102
5000	0.2203	0.2301	0.2398	0.2492	0.2584	0.2673	0.2760	0.2843	0.2923	0.3001
6000	0.3075	0.3146	0.3213	0.3278	0.3339	0.3397	0.3451	0.3503	0.3551	0.3596
7000	0.3639	0.3678	0.3714	0.3748	0.3778	0.3806	0.3832	0.3854	0.3875	0.3893
8000	0.3908	0.3922	0.3933	0.3942	0.3949	0.3954	0.3958	0.3959	0.3959	0.3958
9000	0.3954	0.3950	0.3944	0.3936	0.3927	0.3917	0.3906	0.3894	0.3881	0.3867

S18 Peak at 0.450μ Eff. at 2854°K = 0.295

Temp	0	100	200	300	400	500	600	700	800	900
0	0.000E 00	0.000E 00	0.000E 00	0.109E-22	0.156E-16	0.702E-13	0.181E-10	0.931E-09	0.174E-07	0.168E-06
1000	0.102E-05	0.446E-05	0.151E-04	0.423E-04	0.102E-03	0.218E-03	0.423E-03	0.758E-03	0.127E-02	0.201E-02
2000	0.304E-02	0.441E-02	0.616E-02	0.836E-02	0.110E-01	0.141E-01	0.178E-01	0.220E-01	0.268E-01	0.320E-01
3000	0.0377	0.0439	0.0505	0.0575	0.0648	0.0724	0.0802	0.0882	0.0964	0.1046
4000	0.1129	0.1212	0.1295	0.1376	0.1457	0.1535	0.1613	0.1688	0.1761	0.1831
5000	0.1899	0.1964	0.2026	0.2085	0.2142	0.2195	0.2245	0.2293	0.2337	0.2378
6000	0.2416	0.2452	0.2484	0.2514	0.2541	0.2565	0.2587	0.2607	0.2623	0.2638
7000	0.2651	0.2661	0.2669	0.2676	0.2680	0.2683	0.2684	0.2684	0.2682	0.2678
8000	0.2673	0.2667	0.2660	0.2652	0.2642	0.2632	0.2620	0.2608	0.2595	0.2581
9000	0.2566	0.2551	0.2535	0.2519	0.2502	0.2484	0.2466	0.2448	0.2430	0.2411

TABLE 26B

Spectral Efficiency, Blackbody-S Curves and Blackbody-Vision Combinations

S19 Peak at 0.330μ Eff. at 2854°K = 0.0103

Temp	0	100	200	300	400	500	600	700	800	900
0	0.000E 00	0.000E 00	0.000E 00	0.162E-25	0.999E-19	0.110E-14	0.524E-12	0.415E-10	0.107E-08	0.133E-07
1000	0.990E-07	0.506E-06	0.196E-05	0.614E-05	0.163E-04	0.380E-04	0.796E-04	0.152E-03	0.272E-03	0.457E-03
2000	0.730E-03	0.111E-02	0.163E-02	0.232E-02	0.320E-02	0.430E-02	0.565E-02	0.727E-02	0.919E-02	0.114E-01
3000	0.0139	0.0169	0.0201	0.0237	0.0277	0.0321	0.0368	0.0419	0.0474	0.0531
4000	0.0592	0.0657	0.0724	0.0794	0.0866	0.0942	0.1019	0.1098	0.1180	0.1262
5000	0.1347	0.1432	0.1519	0.1607	0.1695	0.1784	0.1873	0.1963	0.2053	0.2142
6000	0.2232	0.2321	0.2409	0.2497	0.2584	0.2671	0.2757	0.2841	0.2925	0.3008
7000	0.3089	0.3169	0.3248	0.3325	0.3401	0.3475	0.3548	0.3620	0.3690	0.3758
8000	0.3825	0.3890	0.3954	0.4015	0.4075	0.4134	0.4191	0.4246	0.4299	0.4351
9000	0.4401	0.4450	0.4496	0.4542	0.4585	0.4627	0.4667	0.4706	0.4743	0.4779

S19 Filtered Peak at 0.330μ Eff. at 2854°K = 0.0100

Temp	0	100	200	300	400	500	600	700	800	900
0	0.000E 00	0.000E 00	0.000E 00	0.162E-25	0.999E-19	0.110E-14	0.524E-12	0.415E-10	0.107E-08	0.133E-07
1000	0.990E-07	0.506E-06	0.196E-05	0.614E-05	0.163E-04	0.380E-04	0.795E-04	0.152E-03	0.272E-03	0.457E-03
2000	0.728E-03	0.110E-02	0.162E-02	0.230E-02	0.317E-02	0.424E-02	0.555E-02	0.711E-02	0.895E-02	0.110E-01
3000	0.0134	0.0161	0.0191	0.0223	0.0259	0.0297	0.0337	0.0380	0.0425	0.0472
4000	0.0520	0.0570	0.0621	0.0673	0.0726	0.0779	0.0832	0.0885	0.0938	0.0991
5000	0.1043	0.1094	0.1145	0.1194	0.1242	0.1289	0.1335	0.1379	0.1422	0.1463
6000	0.1503	0.1540	0.1577	0.1611	0.1644	0.1675	0.1705	0.1732	0.1758	0.1783
7000	0.1806	0.1827	0.1846	0.1864	0.1881	0.1896	0.1910	0.1922	0.1933	0.1943
8000	0.1951	0.1958	0.1964	0.1969	0.1973	0.1976	0.1978	0.1979	0.1979	0.1978
9000	0.1976	0.1974	0.1970	0.1966	0.1962	0.1956	0.1950	0.1944	0.1937	0.1929

S20 Peak at 0.420μ Eff. at 2854°K = 0.0383

Temp	0	100	200	300	400	500	600	700	800	900
0	0.000E 00	0.000E 00	0.297E-34	0.231E-22	0.280E-16	0.118E-12	0.299E-10	0.152E-08	0.285E-07	0.274E-06
1000	0.166E-05	0.719E-05	0.241E-04	0.669E-04	0.159E-03	0.336E-03	0.642E-03	0.113E-02	0.187E-02	0.292E-02
2000	0.434E-02	0.620E-02	0.855E-02	0.114E-01	0.149E-01	0.189E-01	0.236E-01	0.290E-01	0.349E-01	0.414E-01
3000	0.0485	0.0562	0.0644	0.0730	0.0821	0.0916	0.1014	0.1114	0.1217	0.1322
4000	0.1427	0.1534	0.1641	0.1748	0.1854	0.1959	0.2063	0.2165	0.2266	0.2364
5000	0.2460	0.2553	0.2644	0.2732	0.2816	0.2898	0.2976	0.3050	0.3122	0.3190
6000	0.3254	0.3316	0.3373	0.3428	0.3479	0.3526	0.3571	0.3612	0.3650	0.3686
7000	0.3718	0.3747	0.3773	0.3797	0.3818	0.3837	0.3853	0.3866	0.3878	0.3887
8000	0.3894	0.3899	0.3902	0.3903	0.3902	0.3900	0.3896	0.3890	0.3883	0.3875
9000	0.3865	0.3854	0.3842	0.3829	0.3814	0.3799	0.3782	0.3765	0.3747	0.3728

S21 Peak at 0.440μ Eff. at 2854°K = 0.0209

Temp	0	100	200	300	400	500	600	700	800	900
0	0.000E 00	0.000E 00	0.000E 00	0.633E-27	0.931E-20	0.182E-15	0.133E-12	0.150E-10	0.523E-09	0.834E-08
1000	0.766E-07	0.469E-06	0.211E-05	0.755E-05	0.223E-04	0.568E-04	0.128E-03	0.261E-03	0.489E-03	0.853E-03
2000	0.140E-02	0.218E-02	0.326E-02	0.469E-02	0.652E-02	0.880E-02	0.115E-01	0.148E-01	0.186E-01	0.230E-01
3000	0.0280	0.0335	0.0395	0.0461	0.0532	0.0608	0.0688	0.0773	0.0860	0.0951
4000	0.1045	0.1141	0.1238	0.1337	0.1437	0.1538	0.1639	0.1739	0.1839	0.1938
5000	0.2036	0.2132	0.2227	0.2319	0.2410	0.2498	0.2584	0.2667	0.2747	0.2825
6000	0.2900	0.2972	0.3040	0.3106	0.3169	0.3229	0.3285	0.3339	0.3390	0.3437
7000	0.3482	0.3524	0.3563	0.3599	0.3633	0.3664	0.3692	0.3718	0.3741	0.3762
8000	0.3781	0.3797	0.3811	0.3824	0.3834	0.3842	0.3848	0.3853	0.3856	0.3857
9000	0.3857	0.3855	0.3852	0.3847	0.3841	0.3834	0.3825	0.3816	0.3805	0.3794

TABLE 26B

Spectral Efficiency, Blackbody-S Curves and Blackbody-Vision Combinations

S23 Peak at 0.238μ Eff. at 2854° K = 0.128E-04

Temp	0	100	200	300	400	500	600	700	800	900
0	0.000E 00	0.000E 00	0.000E 00	0.000E 00	0.000E 00	0.000E 00	0.952E-29	0.242E-24	0.467E-21	0.163E-18
1000	0.173E-16	0.782E-15	0.186E-13	0.272E-12	0.272E-11	0.200E-10	0.115E-09	0.543E-09	0.215E-08	0.743E-08
2000	0.226E-07	0.622E-07	0.155E-06	0.360E-06	0.777E-06	0.157E-05	0.302E-05	0.552E-05	0.965E-05	0.162E-04
3000	0.262E-04	0.411E-04	0.626E-04	0.928E-04	0.134E-03	0.189E-03	0.262E-03	0.356E-03	0.476E-03	0.625E-03
4000	0.809E-03	0.103E-02	0.130E-02	0.161E-02	0.199E-02	0.242E-02	0.292E-02	0.349E-02	0.414E-02	0.486E-02
5000	0.567E-02	0.657E-02	0.756E-02	0.865E-02	0.983E-02	0.111E-01	0.124E-01	0.139E-01	0.155E-01	0.172E-01
6000	0.0190	0.0209	0.0229	0.0249	0.0271	0.0294	0.0317	0.0342	0.0367	0.0393
7000	0.0420	0.0447	0.0476	0.0504	0.0534	0.0564	0.0594	0.0625	0.0656	0.0688
8000	0.0719	0.0752	0.0784	0.0816	0.0849	0.0882	0.0914	0.0947	0.0979	0.1012
9000	0.1044	0.1077	0.1109	0.1141	0.1172	0.1203	0.1234	0.1265	0.1295	0.1325

S25 Peak at 0.420μ Eff. at 2854° K = 0.0602

Temp	0	100	200	300	400	500	600	700	800	900
0	0.000E 00	0.000E 00	0.745E-30	0.356E-19	0.690E-14	0.955E-11	0.113E-08	0.330E-07	0.405E-06	0.279E-05
1000	0.128E-04	0.445E-04	0.123E-03	0.291E-03	0.605E-03	0.113E-02	0.194E-02	0.313E-02	0.476E-02	0.691E-02
2000	0.964E-02	0.130E-01	0.170E-01	0.217E-01	0.271E-01	0.332E-01	0.400E-01	0.475E-01	0.556E-01	0.643E-01
3000	0.0736	0.0834	0.0937	0.1044	0.1155	0.1269	0.1386	0.1505	0.1625	0.1747
4000	0.1869	0.1992	0.2115	0.2237	0.2358	0.2477	0.2595	0.2712	0.2826	0.2937
5000	0.3046	0.3152	0.3255	0.3355	0.3452	0.3545	0.3635	0.3721	0.3804	0.3883
6000	0.3959	0.4031	0.4099	0.4164	0.4225	0.4283	0.4337	0.4387	0.4435	0.4479
7000	0.4519	0.4557	0.4591	0.4623	0.4651	0.4677	0.4699	0.4719	0.4737	0.4752
8000	0.4765	0.4775	0.4783	0.4788	0.4792	0.4794	0.4794	0.4791	0.4788	0.4782
9000	0.4775	0.4766	0.4756	0.4744	0.4731	0.4717	0.4702	0.4685	0.4668	0.4649

Visual Spectral Efficiency (Photopic), Blackbody Radiation
Peak at 0.555μ Eff. at 2854° K = 0.239

Temp	0	100	200	300	400	500	600	700	800	900
0	0.000E 00	0.000E 00	0.000E 00	0.107E-25	0.419E-19	0.609E-15	0.421E-12	0.467E-10	0.160E-08	0.247E-07
1000	0.217E-06	0.127E-05	0.544E-05	0.183E-04	0.514E-04	0.123E-03	0.264E-03	0.510E-03	0.907E-03	0.150E-02
2000	0.235E-02	0.349E-02	0.497E-02	0.681E-02	0.904E-02	0.116E-01	0.146E-01	0.180E-01	0.218E-01	0.258E-01
3000	0.0302	0.0348	0.0396	0.0445	0.0496	0.0547	0.0599	0.0650	0.0701	0.0751
4000	0.0800	0.0848	0.0894	0.0938	0.0981	0.1021	0.1059	0.1096	0.1130	0.1161
5000	0.1191	0.1218	0.1243	0.1266	0.1287	0.1305	0.1322	0.1337	0.1350	0.1361
6000	0.1371	0.1378	0.1385	0.1390	0.1393	0.1395	0.1396	0.1396	0.1395	0.1392
7000	0.1389	0.1385	0.1380	0.1374	0.1367	0.1360	0.1352	0.1344	0.1335	0.1325
8000	0.1316	0.1305	0.1295	0.1284	0.1273	0.1261	0.1250	0.1238	0.1226	0.1214
9000	0.1201	0.1189	0.1176	0.1164	0.1151	0.1138	0.1126	0.1113	0.1100	0.1087

Visual Spectral Efficiency (Scotopic), Blackbody Radiation
Peak at 0.507μ Eff. at 2854° K = 0.0136

Temp	0	100	200	300	400	500	600	700	800	900
0	0.000E 00	0.000E 00	0.000E 00	0.397E-28	0.261E-21	0.733E-17	0.962E-14	0.179E-11	0.929E-10	0.200E-08
1000	0.230E-07	0.168E-06	0.873E-06	0.346E-05	0.111E-04	0.303E-04	0.721E-04	0.153E-03	0.296E-03	0.531E-03
2000	0.890E-03	0.140E-02	0.212E-02	0.307E-02	0.428E-02	0.578E-02	0.758E-02	0.971E-02	0.121E-01	0.149E-01
3000	0.0179	0.0213	0.0249	0.0288	0.0329	0.0371	0.0415	0.0461	0.0507	0.0553
4000	0.0600	0.0647	0.0693	0.0739	0.0784	0.0828	0.0871	0.0913	0.0953	0.0991
5000	0.1028	0.1063	0.1097	0.1128	0.1158	0.1186	0.1213	0.1237	0.1260	0.1281
6000	0.1300	0.1317	0.1333	0.1347	0.1360	0.1371	0.1381	0.1389	0.1397	0.1402
7000	0.1407	0.1411	0.1413	0.1414	0.1415	0.1414	0.1413	0.1411	0.1408	0.1404
8000	0.1399	0.1394	0.1389	0.1382	0.1376	0.1368	0.1361	0.1352	0.1344	0.1335
9000	0.1326	0.1316	0.1307	0.1297	0.1286	0.1276	0.1265	0.1254	0.1243	0.1232

TABLE 26B

Spectral Efficiency, Blackbody-S Curves and Blackbody-Vision Combinations

Luminous Efficacy (Lumens/watt)

	0	100	200	300	400	500	600	700	800	900
1000	0.2081E-03	0.1180E-02	0.4934E-02	0.1630E-01	0.4477E-01	0.1060E 00	0.2230E 00	0.4249E 00	0.7460E 00	0.1223E 01
2000	0.1892E 01	0.2787E 01	0.3935E 01	0.5354E 01	0.7057E 01	0.9046E 01	0.1131E 02	0.1384E 02	0.1662E 02	0.1963E 02
3000	0.2282E 02	0.2617E 02	0.2965E 02	0.3323E 02	0.3686E 02	0.4053E 02	0.4419E 02	0.4783E 02	0.5142E 02	0.5493E 02
4000	0.5836E 02	0.6167E 02	0.6486E 02	0.6791E 02	0.7082E 02	0.7358E 02	0.7619E 02	0.7863E 02	0.8091E 02	0.8303E 02
5000	0.8499E 02	0.8678E 02	0.8843E 02	0.8992E 02	0.9125E 02	0.9245E 02	0.9351E 02	0.9443E 02	0.9522E 02	0.9589E 02
6000	0.9644E 02	0.9688E 02	0.9722E 02	0.9745E 02	0.9759E 02	0.9765E 02	0.9761E 02	0.9750E 02	0.9732E 02	0.9707E 02
7000	0.9675E 02	0.9638E 02	0.9594E 02	0.9546E 02	0.9493E 02	0.9436E 02	0.9375E 02	0.9310E 02	0.9242E 02	0.9171E 02
8000	0.9097E 02	0.9021E 02	0.8942E 02	0.8862E 02	0.8779E 02	0.8695E 02	0.8610E 02	0.8524E 02	0.8436E 02	0.8348E 02
9000	0.8259E 02	0.8170E 02	0.8080E 02	0.7989E 02	0.7899E 02	0.7808E 02	0.7718E 02	0.7627E 02	0.7537E 02	0.7447E 02

Luminous Efficacy (Lumens/watt)

	0	100	200	300	400	500	600	700	800	900
1000	0.6127E-04	0.4325E-03	0.2173E-02	0.8405E-02	0.2644E-01	0.7055E-01	0.1646E 00	0.3442E 00	0.6566E 00	0.1160E 01
2000	0.1920E 01	0.3008E 01	0.4491E 01	0.6434E 01	0.8891E 01	0.1190E 02	0.1550E 02	0.1971E 02	0.2451E 02	0.2990E 02
3000	0.3584E 02	0.4231E 02	0.4925E 02	0.5661E 02	0.6432E 02	0.7234E 02	0.8060E 02	0.8902E 02	0.9756E 02	0.1061E 03
4000	0.1147E 03	0.1232E 03	0.1317E 03	0.1400E 03	0.1481E 03	0.1560E 03	0.1637E 03	0.1710E 03	0.1781E 03	0.1849E 03
5000	0.1914E 03	0.1976E 03	0.2034E 03	0.2088E 03	0.2140E 03	0.2188E 03	0.2232E 03	0.2273E 03	0.2311E 03	0.2346E 03
6000	0.2377E 03	0.2406E 03	0.2431E 03	0.2454E 03	0.2474E 03	0.2491E 03	0.2506E 03	0.2518E 03	0.2528E 03	0.2536E 03
7000	0.2542E 03	0.2545E 03	0.2547E 03	0.2547E 03	0.2545E 03	0.2542E 03	0.2537E 03	0.2531E 03	0.2523E 03	0.2514E 03
8000	0.2504E 03	0.2493E 03	0.2481E 03	0.2468E 03	0.2454E 03	0.2439E 03	0.2424E 03	0.2408E 03	0.2391E 03	0.2374E 03
9000	0.2356E 03	0.2337E 03	0.2319E 03	0.2299E 03	0.2280E 03	0.2260E 03	0.2240E 03	0.2220E 03	0.2199E 03	0.2179E 03

TABLE 27A

Radiator and Detector Characteristics

Wavelength (μ)	Radiation sources: Blackbody 3000°K	Blackbody 5000°K	Standard illuminant A	Standard illuminant B	Phosphor P-11	Phosphor P-16	Detectors: Normal observer Photopic	Normal observer Scotopic	Photoelectric detectors S-10	S-11	S-12	S-20
0.35	.0255	.4764	.0051	.0883	.0000	.2000	.0000	.0000	.5540	.6210	.0310	.6800
0.36	.0324	.5200	.0064	.0964	.0000	.5400	.0000	.0000	.6780	.7800	.0420	.7600
0.37	.0405	.5628	.0081	.1043	.0000	.8800	.0000	.0000	.7720	.8380	.0580	.8250
0.38	.0499	.6044	.0098	.1120	.0000	.9999	.0000	.0006	.8380	.8700	.0740	.8800
0.39	.0605	.6446	.0121	.1565	.0300	.9700	.0001	.0022	.8720	.8870	.0900	.9270
0.40	.0725	.6831	.0147	.2065	.0900	.8600	.0004	.0093	.9100	.9050	.1060	.9600
0.41	.0859	.7197	.0177	.2605	.2000	.6800	.0012	.0348	.9400	.9290	.1220	.9900
0.42	.1006	.7542	.0210	.3160	.3430	.4600	.0040	.0966	.9650	.9690	.1380	*****
0.43	.1166	.7864	.0247	.3655	.5900	.2600	.0116	.1998	.9840	.9910	.1540	.9950
0.44	.1340	.8163	.0287	.4040	.7600	.1350	.0230	.3281	.9960	*****	.1700	.9750
0.45	.1525	.8439	.0331	.4270	.9500	.0680	.0380	.4550	*****	.9940	.1860	.9520
0.46	.1723	.8690	.0378	.4415	.9999	.0350	.0600	.5672	.9970	.9780	.2040	.9270
0.47	.1932	.8918	.0429	.4600	.9300	.0180	.0910	.6756	.9850	.9580	.2240	.8950
0.48	.2150	.9122	.0482	.4760	.8400	.0120	.1390	.7931	.9670	.9310	.2650	.8520
0.49	.2379	.9302	.0539	.4825	.7200	.0070	.2080	.9043	.9450	.9020	.5100	.8200
0.50	.2615	.9459	.0599	.4710	.6000	.0020	.3230	.9818	.9200	.8640	.9000	.7910
0.51	.2859	.9595	.0661	.4535	.4500	.0005	.5030	.9966	.8920	.8170	.4000	.7620
0.52	.3108	.9709	.0725	.4475	.3500	.0000	.7100	.9352	.8590	.7680	.0580	.7330
0.53	.3363	.9803	.0791	.4610	.2350	.0000	.8620	.8110	.8210	.7050	.0360	.7040
0.54	.3622	.9877	.0860	.4845	.1550	.0000	.9540	.6497	.7790	.6400	.0280	.6750
0.55	.3884	.9933	.0929	.5050	.0830	.0000	.9950	.4808	.7330	.5710	.0220	.6450
0.56	.4147	.9971	.1000	.5140	.0600	.0000	.9950	.3288	.6830	.4960	.0180	.6170
0.57	.4412	.9993	.1072	.5130	.0350	.0000	.9520	.2076	.6300	.4200	.0140	.5880
0.58	.4676	*****	.1144	.5050	.0170	.0000	.8700	.1212	.5750	.3460	.0110	.5590
0.59	.4939	.9992	.1217	.4960	.0000	.0000	.7570	.0655	.5190	.2680	.0080	.5300
0.60	.5200	.9971	.1290	.4900	.0000	.0000	.6310	.0332	.4630	.1970	.0050	.5010
0.61	.5458	.9938	.1363	.4925	.0000	.0000	.5030	.0159	.4080	.1290	.0000	.4720
0.62	.5712	.9893	.1436	.4985	.0000	.0000	.3810	.0074	.3560	.0790	.0000	.4430
0.63	.5962	.9838	.1508	.5050	.0000	.0000	.2650	.0033	.3090	.0470	.0000	.4140
0.64	.6207	.9774	.1580	.5110	.0000	.0000	.1750	.0015	.2670	.0290	.0000	.3850
0.65	.6446	.9700	.1650	.5195	.0000	.0000	.1070	.0007	.2290	.0180	.0000	.3560
0.66	.6680	.9619	.1720	.5250	.0000	.0000	.0610	.0003	.1950	.0120	.0000	.3270
0.67	.6906	.9530	.1788	.5245	.0000	.0000	.0320	.0001	.1630	.0080	.0000	.2980
0.68	.7125	.9435	.1854	.5195	.0000	.0000	.0170	.0001	.1330	.0060	.0000	.2690
0.69	.7337	.9335	.1919	.5080	.0000	.0000	.0082	.0000	.1060	.0050	.0000	.2400
0.70	.7542	.9229	.1983	.4955	.0000	.0000	.0041	.0000	.0830	.0040	.0000	.2110
0.71	.7738	.9119	.2044	.4810	.0000	.0000	.0021	.0000	.0640	.0030	.0000	.1820
0.72	.7926	.9004	.2104	.4645	.0000	.0000	.0010	.0000	.0480	.0020	.0000	.1530
0.73	.8105	.8886	.2161	.4470	.0000	.0000	.0005	.0000	.0350	.0010	.0000	.1260
0.74	.8276	.8766	.2217	.4345	.0000	.0000	.0002	.0000	.0250	.0000	.0000	.1030
0.75	.8439	.8643	.2270	.4260	.0000	.0000	.0001	.0000	.0170	.0000	.0000	.0850
0.76	.8593	.8517	.2321	.4235	.0000	.0000	.0001	.0000	.0100	.0000	.0000	.0680
0.77	.8738	.8390	.2370	.4270	.0000	.0000	.0000	.0000	.0000	.0000	.0000	.0520
0.78	.8874	.8262	.2417	.4305	.0000	.0000	.0000	.0000	.0000	.0000	.0000	.0400
0.79	.9002	.8133	.2450	.4300	.0000	.0000	.0000	.0000	.0000	.0000	.0000	.0290
0.80	.9122	.8003	.2482	.4300	.0000	.0000	.0000	.0000	.0000	.0000	.0000	.0190
0.81	.9233	.7872	.2511	.4300	.0000	.0000	.0000	.0000	.0000	.0000	.0000	.0110
0.82	.9335	.7742	.2540	.4300	.0000	.0000	.0000	.0000	.0000	.0000	.0000	.0050
0.83	.9430	.7611	.2569	.4300	.0000	.0000	.0000	.0000	.0000	.0000	.0000	.0000
0.84	.9516	.7481	.0000	.0000	.0000	.0000	.0000	.0000	.0000	.0000	.0000	.0000
0.85	.9595	.7352	.0000	.0000	.0000	.0000	.0000	.0000	.0000	.0000	.0000	.0000
0.86	.9666	.7223	.0000	.0000	.0000	.0000	.0000	.0000	.0000	.0000	.0000	.0000
0.87	.9729	.7094	.0000	.0000	.0000	.0000	.0000	.0000	.0000	.0000	.0000	.0000
0.88	.9786	.6967	.0000	.0000	.0000	.0000	.0000	.0000	.0000	.0000	.0000	.0000
0.89	.9835	.6841	.0000	.0000	.0000	.0000	.0000	.0000	.0000	.0000	.0000	.0000
0.90	.9877	.6716	.0000	.0000	.0000	.0000	.0000	.0000	.0000	.0000	.0000	.0000

* THIS VALUE IS UNITY

TABLE 27A

Radiator and Detector Characteristics

Wavelength (μ)	Detectors: Photoelectric detectors: CdSe	Si diode	Vidicon	Photographic emulsions: Blue sensitive	Ortho chromatic	Pan chromatic
0.35	.0400	.1000	.0000	3.5500	0.7940	1.0000
0.36	.0400	.2000	.0952	2.5100	0.7940	1.0000
0.37	.0400	.2400	.2190	1.9040	0.7940	1.0000
0.38	.0400	.3000	.3048	1.5850	0.7940	1.0000
0.39	.0400	.3200	.3714	1.2590	0.8320	1.0000
0.40	.0400	.3700	.4286	1.0000	1.0000	1.0000
0.41	.0400	.4000	.4762	0.7240	1.1480	0.9552
0.42	.0400	.4200	.5429	0.5760	1.2020	0.9123
0.43	.0400	.4400	.5810	0.4790	1.2590	0.8512
0.44	.0400	.4800	.6286	0.3890	1.2590	0.7580
0.45	.0400	.5000	.6476	0.3460	1.2590	0.6920
0.46	.0400	.5200	.6857	0.3160	1.2590	0.6310
0.47	.0400	.5300	.7238	0.3160	1.2300	0.6030
0.48	.0400	.5500	.7619	0.3524	1.1750	0.5620
0.49	.0400	.5900	.8190	0.4080	1.0470	0.5500
0.50	.0400	.6000	.8571	0.5020	1.0000	0.5500
0.51	.0400	.6200	.8857	0.5380	1.1750	0.6030
0.52	.0400	.6400	.9238	0.5760	1.4800	0.7240
0.53	.0400	.6500	.9429	0.5890	1.6590	0.8912
0.54	.0400	.6800	.9810	0.5760	1.8100	0.9552
0.55	.0400	.7000	*****	0.5020	1.9500	0.9773
0.56	.0400	.7200	.9619	0.3160	2.0900	0.9552
0.57	.0500	.7300	.9238	0.1260	1.7770	0.9552
0.58	.0500	.7400	.8762	0.0450	0.8710	0.9335
0.59	.0600	.7600	.8286	0.0159	0.2000	1.0000
0.60	.0600	.8000	.7714	0.0048	0.0415	1.0960
0.61	.0700	.8000	.7143	0.0014	0.0086	1.0000
0.62	.0700	.8000	.6667	0.0004	0.0018	0.6920
0.63	.0800	.8100	.6190	0.0001	0.0004	0.4370
0.64	.0900	.8200	.5810	0.0000	0.0001	0.2690
0.65	.1000	.8300	.5524	0.0000	0.0000	0.1520
0.66	.1100	.8300	.5333	0.0000	0.0000	0.0415
0.67	.1200	.8400	.4857	0.0000	0.0000	0.0113
0.68	.1300	.8500	.4476	0.0000	0.0000	0.0031
0.69	.1400	.8600	.3905	0.0000	0.0000	0.0008
0.70	.1500	.8700	.3524	0.0000	0.0000	0.0002
0.71	.1700	.8800	.3238	0.0000	0.0000	0.0000
0.72	.5000	.8900	.2952	0.0000	0.0000	0.0000
0.73	.8600	.9000	.2762	0.0000	0.0000	0.0000
0.74	*****	.9000	.2476	0.0000	0.0000	0.0000
0.75	.8700	.9000	.2095	0.0000	0.0000	0.0000
0.76	.5500	.9000	.1619	0.0000	0.0000	0.0000
0.77	.4000	.9000	.1429	0.0000	0.0000	0.0000
0.78	.3400	.9000	.1238	0.0000	0.0000	0.0000
0.79	.3000	.9100	.1048	0.0000	0.0000	0.0000
0.80	.2500	.9100	.0952	0.0000	0.0000	0.0000
0.81	.2000	.9100	.0857	0.0000	0.0000	0.0000
0.82	.1700	.9100	.0762	0.0000	0.0000	0.0000
0.83	.1500	.9200	.0667	0.0000	0.0000	0.0000
0.84	.1300	.9200	.0571	0.0000	0.0000	0.0000
0.85	.1100	.9300	.0476	0.0000	0.0000	0.0000
0.86	.1000	.9300	.0381	0.0000	0.0000	0.0000
0.87	.0900	.9300	.0381	0.0000	0.0000	0.0000
0.88	.0800	.9400	.0000	0.0000	0.0000	0.0000
0.89	.0700	.9500	.0000	0.0000	0.0000	0.0000
0.90	.0700	.9600	.0000	0.0000	0.0000	0.0000

Wavelength (μ)	Blackbody: 3000° K	5000° K	Photoelectric detectors: CdSe	Si diode
0.91	.9913	.6592	.0600	.9700
0.92	.9942	.6470	.0600	.9800
0.93	.9965	.6349	.0600	.9900
0.94	.9982	.6230	.0600	*****
0.95	.9993	.6112	.0600	.9900
0.96	.9999	.5996	.0600	.9800
0.97	*****	.5882	.0500	.9500
0.98	.9995	.5769	.0500	.9100
0.99	.9986	.5658	.0500	.8800
1.00	.9971	.5549	.0500	.8200
1.01	.9953	.5442	.0000	.7900
1.02	.9930	.5336	.0000	.7100
1.03	.9903	.5232	.0000	.6300
1.04	.9873	.5130	.0000	.6000
1.05	.9838	.5030	.0000	.5400
1.06	.9801	.4932	.0000	.5000
1.07	.9760	.4836	.0000	.4300
1.08	.9716	.4741	.0000	.3900
1.09	.9669	.4648	.0000	.3300
1.10	.9619	.4557	.0000	.2500
1.11	.9567	.4468	.0000	.2100
1.12	.9512	.4380	.0000	.1600
1.13	.9455	.4294	.0000	.1200
1.14	.9396	.4210	.0000	.0900
1.15	.9335	.4127	.0000	.0700
1.16	.9272	.4047	.0000	.0520
1.17	.9207	.3967	.0000	.0400
1.18	.9141	.3890	.0000	.0000
1.19	.9073	.3814	.0000	.0000
1.20	.9004	.3740	.0000	.0000
1.21	.8934	.3667	.0000	.0000
1.22	.8863	.3595	.0000	.0000
1.23	.8790	.3526	.0000	.0000
1.24	.8717	.3457	.0000	.0000
1.25	.8643	.3390	.0000	.0000
1.26	.8568	.3324	.0000	.0000
1.27	.8492	.3261	.0000	.0000
1.28	.8416	.3198	.0000	.0000
1.29	.8339	.3136	.0000	.0000
1.30	.8262	.3076	.0000	.0000

TABLE 27B

S-Curve Sensitivities, Relative

LAMDA	S1	S3	S4	S5	S6	S7	S8	S9	S10	S11	S12	S13
0.30	0.909	0.000	0.048	0.900	0.900	0.430	0.048	0.000	0.000	0.000	0.009	0.565
0.31	0.270	0.000	0.260	0.940	0.810	0.430	0.170	0.075	0.000	0.008	0.010	0.600
0.32	0.520	0.000	0.546	0.970	0.700	0.870	0.568	0.270	0.100	0.059	0.012	0.645
0.33	0.810	0.000	0.685	0.990	0.570	1.140	0.815	0.375	0.250	0.178	0.016	0.690
0.34	1.105	0.000	0.782	1.000	0.460	1.070	0.929	0.440	0.400	0.393	0.023	0.740
0.35	1.350	0.565	0.858	0.990	0.370	0.990	0.974	0.515	0.554	0.621	0.031	0.790
0.36	1.480	0.700	0.915	0.975	0.300	0.940	0.994	0.560	0.678	0.780	0.042	0.835
0.37	1.350	0.860	0.955	0.950	0.240	0.900	0.998	0.620	0.772	0.838	0.058	0.880
0.38	1.100	0.930	0.981	0.925	0.190	0.860	0.992	0.650	0.838	0.870	0.074	0.915
0.39	0.820	0.965	0.995	0.900	0.150	0.840	0.982	0.695	0.872	0.887	0.090	0.940
0.40	0.561	0.985	1.000	0.870	0.115	0.840	0.964	0.725	0.910	0.905	0.106	0.965
0.41	0.360	0.997	0.997	0.840	0.075	0.860	0.945	0.765	0.940	0.929	0.122	0.980
0.42	0.235	1.000	0.986	0.810	0.050	0.880	0.923	0.810	0.965	0.969	0.138	0.990
0.43	0.190	0.998	0.968	0.780	0.025	0.900	0.900	0.875	0.984	0.991	0.154	0.995
0.44	0.177	0.993	0.946	0.750	0.010	0.910	0.875	0.905	0.996	1.000	0.170	1.000
0.45	0.171	0.982	0.920	0.715	0.000	0.930	0.848	0.945	1.000	0.994	0.186	0.992
0.46	0.170	0.970	0.889	0.680	0.000	0.945	0.818	0.970	0.997	0.978	0.204	0.982
0.47	0.174	0.950	0.853	0.645	0.000	0.955	0.785	0.990	0.985	0.958	0.224	0.970
0.48	0.182	0.930	0.810	0.610	0.000	0.960	0.748	1.000	0.967	0.931	0.265	0.952
0.49	0.194	0.900	0.761	0.575	0.000	0.965	0.708	0.990	0.945	0.902	0.510	0.930
0.50	0.209	0.870	0.706	0.535	0.000	0.975	0.665	0.980	0.920	0.864	0.900	0.893
0.51	0.227	0.840	0.645	0.495	0.000	0.980	0.620	0.960	0.892	0.817	0.400	0.852
0.52	0.248	0.810	0.578	0.450	0.000	0.985	0.573	0.920	0.859	0.768	0.058	0.810
0.53	0.271	0.780	0.506	0.405	0.000	0.985	0.527	0.870	0.821	0.705	0.036	0.750
0.54	0.295	0.750	0.431	0.360	0.000	0.990	0.482	0.810	0.779	0.640	0.028	0.680
0.55	0.321	0.720	0.360	0.310	0.000	0.995	0.438	0.740	0.733	0.571	0.022	0.600
0.56	0.349	0.690	0.294	0.260	0.000	0.995	0.396	0.670	0.683	0.496	0.018	0.500
0.57	0.378	0.665	0.235	0.210	0.000	0.995	0.357	0.600	0.630	0.420	0.014	0.410
0.58	0.408	0.640	0.184	0.165	0.000	1.000	0.321	0.510	0.575	0.346	0.011	0.325
0.59	0.439	0.615	0.140	0.130	0.000	1.000	0.287	0.420	0.519	0.268	0.008	0.245
0.60	0.471	0.590	0.104	0.100	0.000	1.000	0.254	0.330	0.463	0.197	0.005	0.170
0.61	0.504	0.565	0.076	0.070	0.000	1.000	0.223	0.220	0.408	0.129	0.000	0.105
0.62	0.539	0.540	0.054	0.050	0.000	1.000	0.194	0.140	0.356	0.079	0.000	0.068
0.63	0.575	0.515	0.038	0.035	0.000	1.000	0.168	0.080	0.309	0.047	0.000	0.043
0.64	0.611	0.492	0.026	0.025	0.000	1.000	0.144	0.035	0.267	0.029	0.000	0.028
0.65	0.647	0.469	0.017	0.015	0.000	1.000	0.123	0.010	0.229	0.018	0.000	0.020
0.66	0.683	0.446	0.011	0.010	0.000	0.995	0.105	0.000	0.195	0.012	0.000	0.012
0.67	0.718	0.424	0.007	0.006	0.000	0.995	0.089	0.000	0.163	0.008	0.000	0.008
0.68	0.753	0.402	0.004	0.003	0.000	0.990	0.075	0.000	0.133	0.006	0.000	0.005
0.69	0.787	0.380	0.002	0.001	0.000	0.985	0.064	0.000	0.106	0.005	0.000	0.002
0.70	0.820	0.360	0.001	0.000	0.000	0.980	0.053	0.000	0.083	0.004	0.000	0.000
0.71	0.851	0.340	0.000	0.000	0.000	0.980	0.044	0.000	0.064	0.003	0.000	0.000
0.72	0.881	0.320	0.000	0.000	0.000	0.970	0.036	0.000	0.048	0.002	0.000	0.000
0.73	0.909	0.301	0.000	0.000	0.000	0.965	0.029	0.000	0.035	0.001	0.000	0.000
0.74	0.933	0.282	0.000	0.000	0.000	0.960	0.023	0.000	0.025	0.000	0.000	0.000
0.75	0.954	0.264	0.000	0.000	0.000	0.945	0.018	0.000	0.017	0.000	0.000	0.000
0.76	0.971	0.246	0.000	0.000	0.000	0.930	0.012	0.000	0.010	0.000	0.000	0.000
0.77	0.984	0.228	0.000	0.000	0.000	0.920	0.008	0.000	0.000	0.000	0.000	0.000
0.78	0.993	0.210	0.000	0.000	0.000	0.900	0.005	0.000	0.000	0.000	0.000	0.000
0.79	0.998	0.192	0.000	0.000	0.000	0.875	0.003	0.000	0.000	0.000	0.000	0.000
0.80	1.000	0.175	0.000	0.000	0.000	0.845	0.002	0.000	0.000	0.000	0.000	0.000
0.81	0.998	0.158	0.000	0.000	0.000	0.805	0.001	0.000	0.000	0.000	0.000	0.000
0.82	0.993	0.140	0.000	0.000	0.000	0.770	0.000	0.000	0.000	0.000	0.000	0.000
0.83	0.983	0.122	0.000	0.000	0.000	0.730	0.000	0.000	0.000	0.000	0.000	0.000
0.84	0.969	0.105	0.000	0.000	0.000	0.695	0.000	0.000	0.000	0.000	0.000	0.000
0.85	0.951	0.088	0.000	0.000	0.000	0.650	0.000	0.000	0.000	0.000	0.000	0.000
0.86	0.929	0.070	0.000	0.000	0.000	0.610	0.000	0.000	0.000	0.000	0.000	0.000
0.87	0.903	0.052	0.000	0.000	0.000	0.565	0.000	0.000	0.000	0.000	0.000	0.000
0.88	0.874	0.035	0.000	0.000	0.000	0.520	0.000	0.000	0.000	0.000	0.000	0.000
0.89	0.841	0.020	0.000	0.000	0.000	0.480	0.000	0.000	0.000	0.000	0.000	0.000

LAMDA	S14
0.30	0.010
0.35	0.110
0.40	0.210
0.45	0.290
0.50	0.335
0.55	0.370
0.60	0.405
0.65	0.438
0.70	0.472
0.75	0.506
0.80	0.540
0.85	0.574
0.90	0.608
0.95	0.642
1.00	0.676
1.05	0.710
1.10	0.744
1.15	0.778
1.20	0.812
1.25	0.846
1.30	0.880
1.35	0.914
1.40	0.948
1.45	0.980
1.50	1.000
1.55	0.980
1.60	0.935
1.65	0.860
1.70	0.650
1.75	0.350
1.80	0.200
1.85	0.120
1.90	0.070
1.95	0.030
2.00	0.007

TABLE 27B

S-Curve Sensitivities, Relative

LAMDA	S15	S16	S17	S18	S19	S20	S21	S23	S25
0.30	0.000	0.029	0.290	0.000	0.899	0.140	0.555	0.045	0.475
0.31	0.040	0.086	0.540	0.000	0.940	0.250	0.610	0.020	0.550
0.32	0.070	0.151	0.670	0.000	0.980	0.350	0.662	0.015	0.660
0.33	0.100	0.211	0.730	0.000	1.000	0.480	0.705	0.000	0.740
0.34	0.115	0.238	0.770	0.000	0.982	0.590	0.750	0.000	0.805
0.35	0.130	0.254	0.795	0.000	0.950	0.680	0.788	0.000	0.885
0.36	0.140	0.269	0.815	0.000	0.913	0.760	0.820	0.000	0.925
0.37	0.150	0.274	0.830	0.000	0.874	0.825	0.850	0.000	0.955
0.38	0.160	0.280	0.845	0.000	0.835	0.880	0.870	0.000	0.973
0.39	0.170	0.288	0.850	0.760	0.796	0.927	0.885	0.000	0.983
0.40	0.180	0.293	0.860	0.329	0.757	0.960	0.898	0.000	0.992
0.41	0.190	0.298	0.872	0.886	0.713	0.990	0.914	0.000	0.998
0.42	0.200	0.303	0.887	0.930	0.670	1.000	0.940	0.000	1.000
0.43	0.220	0.308	0.905	0.962	0.627	0.995	0.988	0.000	0.997
0.44	0.240	0.313	0.927	0.986	0.584	0.975	1.000	0.000	0.990
0.45	0.265	0.318	0.950	1.000	0.541	0.952	0.999	0.000	0.982
0.46	0.295	0.324	0.975	0.990	0.498	0.927	0.989	0.000	0.962
0.47	0.335	0.330	0.988	0.968	0.455	0.895	0.965	0.000	0.927
0.48	0.380	0.335	0.995	0.932	0.412	0.852	0.933	0.000	0.900
0.49	0.440	0.340	1.000	0.887	0.369	0.820	0.900	0.000	0.855
0.50	0.500	0.346	0.995	0.836	0.327	0.791	0.865	0.000	0.825
0.51	0.570	0.352	0.990	0.780	0.286	0.762	0.830	0.000	0.795
0.52	0.645	0.358	0.975	0.722	0.248	0.733	0.795	0.000	0.765
0.53	0.720	0.365	0.930	0.671	0.210	0.704	0.760	0.000	0.743
0.54	0.800	0.371	0.886	0.620	0.183	0.675	0.698	0.000	0.717
0.55	0.870	0.378	0.820	0.570	0.163	0.645	0.632	0.000	0.687
0.56	0.935	0.385	0.750	0.522	0.145	0.617	0.565	0.000	0.660
0.57	0.980	0.392	0.600	0.480	0.130	0.588	0.465	0.000	0.635
0.58	1.000	0.399	0.450	0.437	0.115	0.559	0.323	0.000	0.612
0.59	0.980	0.407	0.300	0.398	0.105	0.530	0.201	0.000	0.587
0.60	0.945	0.416	0.190	0.360	0.096	0.501	0.160	0.000	0.567
0.61	0.890	0.425	0.130	0.326	0.087	0.472	0.120	0.000	0.545
0.62	0.830	0.435	0.100	0.295	0.078	0.443	0.094	0.000	0.525
0.63	0.760	0.445	0.075	0.266	0.069	0.414	0.070	0.000	0.505
0.64	0.690	0.455	0.062	0.240	0.060	0.385	0.047	0.000	0.483
0.65	0.620	0.466	0.050	0.217	0.051	0.356	0.031	0.000	0.463
0.66	0.530	0.483	0.039	0.193	0.042	0.327	0.022	0.000	0.443
0.67	0.440	0.504	0.030	0.171	0.034	0.298	0.015	0.000	0.423
0.68	0.370	0.539	0.022	0.150	0.028	0.269	0.009	0.000	0.403
0.69	0.300	0.594	0.015	0.134	0.023	0.240	0.005	0.000	0.380
0.70	0.240	0.670	0.008	0.118	0.018	0.211	0.002	0.000	0.362
0.71	0.200	0.800	0.002	0.100	0.013	0.182	0.000	0.000	0.347
0.72	0.160	0.930	0.000	0.087	0.008	0.153	0.000	0.000	0.328
0.73	0.130	1.000	0.000	0.076	0.003	0.126	0.000	0.000	0.313
0.74	0.100	0.960	0.000	0.064	0.001	0.103	0.000	0.000	0.295
0.75	0.075	0.870	0.000	0.053	0.000	0.085	0.000	0.000	0.277
0.76	0.060	0.800	0.000	0.045	0.000	0.068	0.000	0.000	0.262
0.77	0.048	0.744	0.000	0.036	0.000	0.052	0.000	0.000	0.243
0.78	0.037	0.695	0.000	0.029	0.000	0.040	0.000	0.000	0.230
0.79	0.027	0.653	0.000	0.021	0.000	0.029	0.000	0.000	0.210
0.80	0.022	0.612	0.000	0.017	0.000	0.019	0.000	0.000	0.195
0.81	0.017	0.572	0.000	0.000	0.000	0.011	0.000	0.000	0.177
0.82	0.012	0.533	0.000	0.000	0.000	0.005	0.000	0.000	0.164
0.83	0.007	0.496	0.000	0.000	0.000	0.000	0.000	0.000	0.146
0.84	0.003	0.460	0.000	0.000	0.000	0.000	0.000	0.000	0.130
0.85	0.000	0.426	0.000	0.000	0.000	0.000	0.000	0.000	0.114
0.86	0.000	0.394	0.000	0.000	0.000	0.000	0.000	0.000	0.098
0.87	0.000	0.362	0.000	0.000	0.000	0.000	0.000	0.000	0.082
0.88	0.000	0.332	0.000	0.000	0.000	0.000	0.000	0.000	0.070
0.89	0.000	0.303	0.000	0.000	0.000	0.000	0.000	0.000	0.058

Ultraviolet

LAMDA	S6	S7	S8	S17	S19	S23	S25
0.15	0.000	0.000	0.000	0.000	0.006	0.000	0.000
0.16	0.000	0.000	0.000	0.000	0.020	0.000	0.000
0.17	0.000	0.000	0.000	0.000	0.090	0.000	0.000
0.18	0.000	0.000	0.000	0.000	0.420	0.200	0.000
0.19	0.000	0.000	0.000	0.000	0.675	0.360	0.000
0.20	0.000	0.000	0.000	0.000	0.750	0.530	0.000
0.21	0.100	0.000	0.000	0.000	0.800	0.685	0.000
0.22	0.270	0.000	0.000	0.000	0.820	0.865	0.000
0.23	0.460	0.000	0.000	0.000	0.797	0.975	0.002
0.24	0.640	0.000	0.000	0.000	0.739	0.990	0.011
0.25	0.810	0.015	0.000	0.000	0.700	0.935	0.024
0.26	0.930	0.040	0.000	0.000	0.730	0.685	0.052
0.27	0.990	0.080	0.000	0.000	0.771	0.420	0.150
0.28	0.990	0.150	0.000	0.010	0.813	0.190	0.250
0.29	0.950	0.435	0.005	0.100	0.856	0.080	0.360

Infrared

LAMDA	S1	S7	S16	S25
0.90	0.803	0.435	0.275	0.044
0.91	0.761	0.385	0.250	0.032
0.92	0.716	0.335	0.225	0.022
0.93	0.670	0.300	0.200	0.012
0.94	0.622	0.250	0.175	0.004
0.95	0.573	0.200	0.152	0.001
0.96	0.523	0.150	0.132	0.000
0.97	0.473	0.100	0.113	0.000
0.98	0.423	0.060	0.097	0.000
0.99	0.375	0.040	0.081	0.000
1.00	0.329	0.025	0.068	0.000
1.01	0.285	0.015	0.054	0.000
1.02	0.245	0.010	0.042	0.000
1.03	0.208	0.005	0.032	0.000
1.04	0.175	0.000	0.024	0.000
1.05	0.148	0.000	0.018	0.000
1.06	0.124	0.000	0.014	0.000
1.07	0.104	0.000	0.010	0.000
1.08	0.088	0.000	0.007	0.000
1.09	0.074	0.000	0.005	0.000
1.10	0.063	0.000	0.003	0.000
1.11	0.053	0.000	0.000	0.000
1.12	0.044	0.000	0.000	0.000
1.13	0.036	0.000	0.000	0.000
1.14	0.029	0.000	0.000	0.000
1.15	0.023	0.000	0.000	0.000
1.16	0.017	0.000	0.000	0.000
1.17	0.012	0.000	0.000	0.000
1.18	0.008	0.000	0.000	0.000
1.19	0.005	0.000	0.000	0.000
1.20	0.002	0.000	0.000	0.000

TABLE 28A

Spectral Luminous Efficiency, Photopic

λ(μ)	0	1	2	3	4	5	6	7	8	9	EXP
0.36	0.3917	0.4394	0.4930	0.5532	0.6208	0.6965	0.7813	0.8767	0.9840	1.1040	-5
0.37	0.1239	0.1389	0.1556	0.1744	0.1958	0.2202	0.2484	0.2804	0.3153	0.3522	-4
0.38	0.3900	0.4283	0.4691	0.5159	0.5718	0.6400	0.7234	0.8221	0.9351	1.0610	-4
0.39	0.1200	0.1350	0.1515	0.1702	0.1918	0.2170	0.2469	0.2812	0.3185	0.3573	-3
0.40	0.3960	0.4337	0.4730	0.5179	0.5722	0.6400	0.7246	0.8255	0.9412	1.0700	-3
0.41	0.1210	0.1362	0.1531	0.1720	0.1935	0.2180	0.2455	0.2764	0.3118	0.3526	-2
0.42	0.4000	0.4546	0.5159	0.5829	0.6546	0.7300	0.8087	0.8909	0.9768	1.0660	-2
0.43	0.1160	0.1257	0.1358	0.1463	0.1572	0.1684	0.1801	0.1921	0.2045	0.2172	-1
0.44	0.2300	0.2429	0.2561	0.2696	0.2835	0.2980	0.3131	0.3288	0.3452	0.3623	-1
0.45	0.3800	0.3985	0.4177	0.4377	0.4584	0.4800	0.5024	0.5257	0.5498	0.5746	-1
0.46	0.6000	0.6260	0.6528	0.6804	0.7091	0.7390	0.7702	0.8027	0.8367	0.8723	-1
0.47	0.9098	0.9492	0.9905	1.0340	1.0790	1.1260	1.1750	1.2270	1.2800	1.3350	-1
0.48	0.1390	0.1447	0.1505	0.1565	0.1627	0.1693	0.1762	0.1836	0.1913	0.1994	0
0.49	0.2080	0.2171	0.2267	0.2369	0.2475	0.2586	0.2702	0.2823	0.2951	0.3086	0
0.50	0.3230	0.3384	0.3547	0.3717	0.3893	0.4073	0.4256	0.4443	0.4634	0.4829	0
0.51	0.5030	0.5236	0.5445	0.5657	0.5870	0.6082	0.6293	0.6503	0.6709	0.6908	0
0.52	0.7100	0.7282	0.7455	0.7620	0.7778	0.7932	0.8081	0.8225	0.8363	0.8495	0
0.53	0.8620	0.8738	0.8850	0.8955	0.9054	0.9149	0.9237	0.9321	0.9399	0.9472	0
0.54	0.9540	0.9603	0.9660	0.9713	0.9760	0.9803	0.9841	0.9875	0.9903	0.9928	0
0.55	0.9950	0.9967	0.9981	0.9991	0.9997	1.0000	0.9999	0.9993	0.9983	0.9969	0
0.56	0.9950	0.9926	0.9897	0.9864	0.9827	0.9786	0.9741	0.9692	0.9639	0.9581	0
0.57	0.9520	0.9455	0.9385	0.9312	0.9235	0.9154	0.9070	0.8983	0.8892	0.8798	0
0.58	0.8700	0.8599	0.8494	0.8386	0.8276	0.8163	0.8048	0.7931	0.7812	0.7692	0
0.59	0.7570	0.7448	0.7324	0.7200	0.7075	0.6949	0.6822	0.6695	0.6567	0.6438	0
0.60	0.6310	0.6182	0.6053	0.5925	0.5796	0.5668	0.5540	0.5411	0.5284	0.5156	0
0.61	0.5030	0.4905	0.4780	0.4657	0.4534	0.4412	0.4291	0.4170	0.4050	0.3930	0
0.62	0.3810	0.3689	0.3568	0.3448	0.3328	0.3210	0.3093	0.2979	0.2866	0.2756	0
0.63	0.2650	0.2548	0.2449	0.2353	0.2261	0.2170	0.2082	0.1995	0.1912	0.1830	0
0.64	0.1750	0.1672	0.1596	0.1523	0.1451	0.1382	0.1315	0.1250	0.1188	0.1128	0
0.65	1.0700	1.0150	0.9619	0.9112	0.8626	0.8160	0.7712	0.7283	0.6871	0.6477	-1
0.66	0.6100	0.5740	0.5396	0.5067	0.4755	0.4458	0.4176	0.3908	0.3656	0.3420	-1
0.67	0.3200	0.2996	0.2808	0.2633	0.2471	0.2320	0.2180	0.2050	0.1928	0.1812	-1
0.68	1.7000	1.5900	1.4840	1.3810	1.2830	1.1920	1.1070	1.0270	0.9533	0.8846	-2
0.69	0.8210	0.7624	0.7085	0.6591	0.6138	0.5723	0.5343	0.4996	0.4676	0.4380	-2
0.70	0.4102	0.3838	0.3589	0.3354	0.3134	0.2929	0.2738	0.2560	0.2393	0.2237	-2
0.71	0.2091	0.1954	0.1825	0.1704	0.1590	0.1484	0.1384	0.1291	0.1204	0.1123	-2
0.72	1.0470	0.9766	0.9111	0.8501	0.7932	0.7400	0.6901	0.6433	0.5995	0.5585	-3
0.73	0.5200	0.4839	0.4501	0.4183	0.3887	0.3611	0.3354	0.3114	0.2892	0.2685	-3
0.74	0.2492	0.2313	0.2147	0.1993	0.1850	0.1719	0.1598	0.1486	0.1383	0.1288	-3
0.75	1.2000	1.1190	1.0430	0.9734	0.9085	0.8480	0.7915	0.7386	0.6892	0.6430	-4
0.76	0.6000	0.5598	0.5223	0.4872	0.4545	0.4240	0.3956	0.3692	0.3445	0.3215	-4
0.77	0.3000	0.2799	0.2611	0.2436	0.2272	0.2120	0.1978	0.1845	0.1722	0.1606	-4
0.78	1.4990	1.3990	1.3050	1.2180	1.1360	1.0600	0.9886	0.9217	0.8592	0.8009	-5
0.79	0.7466	0.6960	0.6488	0.6049	0.5639	0.5258	0.4902	0.4570	0.4260	0.3972	-5
0.80	0.3703	0.3452	0.3218	0.3000	0.2797	0.2608	0.2431	0.2267	0.2113	0.1970	-5
0.81	1.8370	1.7120	1.5960	1.4880	1.3870	1.2930	1.2060	1.1240	1.0480	0.9771	-6
0.82	0.9109	0.8493	0.7917	0.7381	0.6881	0.6415	0.5981	0.5576	0.5198	0.4846	-6
0.83	0.4518										-6

TABLE 28B

Spectral Luminous Efficiency, Scotopic

λ(μ)	0	1	2	3	4	5	6	7	8	9	EXP
0.38	0.5893	0.6647	0.7521	0.8537	0.9716	1.1080	1.2680	1.4530	1.6680	1.9180	-3
0.39	0.2209	0.2547	0.2939	0.3394	0.3921	0.4531	0.5236	0.6049	0.6984	0.8059	-2
0.40	0.9292	1.0700	1.2310	1.4130	1.6190	1.8520	2.1130	2.4050	2.7300	3.0890	-2
0.41	0.3483	0.3916	0.4386	0.4897	0.5448	0.6041	0.6677	0.7357	0.8080	0.8849	-1
0.42	0.9661	1.0510	1.1410	1.2350	1.3340	1.4350	1.5410	1.6510	1.7640	1.8790	-1
0.43	0.1998	0.2119	0.2243	0.2369	0.2496	0.2625	0.2755	0.2886	0.3017	0.3149	0
0.44	0.3281	0.3412	0.3543	0.3673	0.3803	0.3931	0.4058	0.4183	0.4307	0.4429	0
0.45	0.4550	0.4669	0.4786	0.4902	0.5015	0.5129	0.5240	0.5349	0.5458	0.5565	0
0.46	0.5672	0.5778	0.5884	0.5991	0.6097	0.6204	0.6312	0.6422	0.6533	0.6644	0
0.47	0.6756	0.6871	0.6986	0.7102	0.7219	0.7337	0.7454	0.7574	0.7693	0.7811	0
0.48	0.7930	0.8048	0.8166	0.8281	0.8397	0.8509	0.8620	0.8730	0.8837	0.8941	0
0.49	0.9043	0.9139	0.9234	0.9324	0.9410	0.9491	0.9568	0.9638	0.9703	0.9763	0
0.50	0.9817	0.9865	0.9904	0.9938	0.9966	0.9984	0.9995	1.0000	0.9995	0.9984	0
0.51	0.9966	0.9936	0.9901	0.9858	0.9808	0.9750	0.9685	0.9612	0.9532	0.9445	0
0.52	0.9352	0.9253	0.9147	0.9036	0.8919	0.8796	0.8668	0.8535	0.8397	0.8257	0
0.53	0.8110	0.7960	0.7807	0.7652	0.7492	0.7332	0.7166	0.7002	0.6834	0.6667	0
0.54	0.6497	0.6327	0.6156	0.5985	0.5814	0.5644	0.5475	0.5306	0.5139	0.4973	0
0.55	0.4808	0.4645	0.4484	0.4325	0.4170	0.4015	0.3864	0.3715	0.3569	0.3427	0
0.56	0.3288	0.3151	0.3018	0.2888	0.2762	0.2639	0.2519	0.2403	0.2291	0.2182	0
0.57	0.2076	0.1975	0.1876	0.1782	0.1690	0.1602	0.1517	0.1436	0.1358	0.1284	0
0.58	1.2120	1.1430	1.0780	1.0150	0.9557	0.8989	0.8449	0.7934	0.7447	0.6986	-1
0.59	0.6548	0.6133	0.5741	0.5372	0.5022	0.4694	0.4383	0.4091	0.3816	0.3558	-1
0.60	0.3315	0.3087	0.2874	0.2674	0.2487	0.2312	0.2147	0.1994	0.1851	0.1718	-1
0.61	1.5930	1.4770	1.3690	1.2690	1.1750	1.0880	1.0070	0.9322	0.8624	0.7974	-2
0.62	0.7374	0.6817	0.6301	0.5822	0.5379	0.4969	0.4590	0.4238	0.3913	0.3613	-2
0.63	0.3335	0.3079	0.2842	0.2623	0.2421	0.2235	0.2062	0.1903	0.1757	0.1621	-2
0.64	1.4970	1.3820	1.2760	1.1780	1.0880	1.0050	0.9281	0.8574	0.7925	0.7325	-3
0.65	0.6772	0.6262	0.5792	0.5358	0.4958	0.4590	0.4249	0.3935	0.3645	0.3377	-3
0.66	0.3129	0.2901	0.2689	0.2493	0.2313	0.2146	0.1991	0.1848	0.1716	0.1593	-3
0.67	1.4800	1.3750	1.2770	1.1870	1.1040	1.0260	0.9543	0.8876	0.8258	0.7686	-4
0.68	0.7155	0.6660	0.6203	0.5777	0.5381	0.5014	0.4673	0.4356	0.4061	0.3787	-4
0.69	0.3533	0.3295	0.3075	0.2870	0.2679	0.2501	0.2336	0.2182	0.2038	0.1905	-4
0.70	1.7800	1.6640	1.5560	1.4540	1.3600	1.2730	1.1910	1.1140	1.0430	0.9763	-5
0.71	0.9143	0.8562	0.8020	0.7513	0.7040	0.6598	0.6184	0.5798	0.5438	0.5099	-5
0.72	0.4783	0.4487	0.4211	0.3951	0.3709	0.3482	0.3270	0.3070	0.2884	0.2710	-5
0.73	0.2546	0.2393	0.2250	0.2115	0.1989	0.1870	0.1759	0.1655	0.1557	0.1466	-5
0.74	1.3790	1.2990	1.2230	1.1510	1.0840	1.0220	0.9625	0.9070	0.8549	0.8057	-6
0.75	0.7596	0.7163	0.6755	0.6371	0.6010	0.5670	0.5351	0.5050	0.4767	0.4500	-6
0.76	0.4249	0.4012	0.3790	0.3580	0.3382	0.3196	0.3021	0.2855	0.2699	0.2552	-6
0.77	0.2413	0.2282	0.2159	0.2042	0.1932	0.1829	0.1731	0.1638	0.1551	0.1468	-6
0.78	0.1390										-6

TABLE 29.

Tristimulus Values (C.I.E. 1931, Standard Observer)

Wave-length μ	Tristimulus Values of Equal-Energy Spectrum $\bar{x}_\lambda$	$\bar{y}_\lambda$	$\bar{z}_\lambda$
.380	0.0014	0.0000	0.0065
.385	.0022	0.0001	0.0105
.390	.0042	0.0001	0.0201
395	.0076	0.0002	0.0362
.400	.0143	0.0004	0.0679
.405	.0232	0.0006	0.1102
.410	.0435	0.0012	0.2074
.415	.0776	0.0022	0.3713
.420	.1344	0.0040	0.6456
.425	.2148	0.0073	1.0391
.430	.2839	0.0116	1.3856
.435	.3285	0.0168	1.6230
.440	.3483	0.0230	1.7471
.445	.3481	0.0298	1.7826
.450	.3362	0.0380	1.7721
.455	.3187	0.0480	1.7441
.460	.2908	0.0600	1.6692
.465	.2511	0.0739	1.5281
.470	.1954	0.0910	1.2876
.475	.1421	0.1126	1.0419
.480	.0956	0.1390	0.8130

Wave-length, μ	Tristimulus Values of Equal-Energy Spectrum $\bar{x}_\lambda$	$\bar{y}_\lambda$	$\bar{z}_\lambda$
.485	.0580	0.1693	0.6162
.490	.0320	0.2080	0.4652
.495	.0147	0.2586	0.3533
.500	.0049	0.3230	0.2720
.505	.0024	0.4073	0.2123
.510	.0093	0.5030	0.1582
.515	.0291	0.6082	0.1117
.520	.0633	0.7100	0.0782
.525	.1096	0.7932	0.0573
.530	.1655	0.8620	0.0422
.535	.2257	0.9149	0.0298
.540	.2904	0.9540	0.0203
.545	.3597	0.9803	0.0134
.550	.4334	0.9950	0.0087
.555	.5121	1.0002	0.0057
.560	.5945	0.9950	0.0039
.565	.6784	.09786	0.0027
.570	.7621	0.9520	0.0021
.575	.8425	0.9154	0.0018
.580	.9163	0.8700	0.0017

Wave-length μ	Tristimulus Values of Equal-Energy Spectrum $\bar{x}_\lambda$	$\bar{y}_\lambda$	$\bar{z}_\lambda$
.580	0.9163	0.8700	0.0017
.585	0.9786	.8163	.0014
.590	1.0263	.7570	.0011
.595	1.0567	.6949	.0010
.600	1.0622	.6310	.0008
.605	1.0456	.5668	.0006
.610	1.0026	.5030	.0003
.615	0.9384	.4412	.0002
.620	0.8544	.3810	.0002
.625	0.7514	.3210	.0001
.630	0.6424	.2650	.0000
.635	0.5419	.2170	.0000
.640	0.4479	.1750	.0000
.645	0.3608	.1382	.0000
.650	0.2835	.1070	.0000
.655	0.2187	.0816	.0000
.660	0.1649	.0610	.0000
.665	0.1212	.0446	.0000
.670	0.0874	.0320	.0000
.675	0.0636	.0232	.0000
.680	0.0468	.0170	.0000

Wave-length, μ	Tristimulus Values of Equal-Energy Spectrum $\bar{x}_\lambda$	$\bar{y}_\lambda$	$\bar{z}_\lambda$
.685	0.0329	.0119	.0000
.690	0.0227	.0082	.0000
.695	0.0158	.0057	.0000
.700	0.0114	.0041	.0000
.705	0.0081	.0029	.0000
.710	0.0058	.0021	.0000
.715	0.0041	.0015	.0000
.720	0.0029	.0010	.0000
.725	0.0020	.0007	.0000
.730	0.0014	.0005	.0000
.735	0.0010	.0004	.0000
.740	0.0007	.0003	.0000
.745	0.0005	.0002	.0000
.750	0.0003	.0001	.0000
.755	0.0002	.0001	.0000
.760	0.0002	.0001	.0000
.765	0.0001	.0000	.0000
.770	0.0001	.0000	.0000
.775	0.0000	.0000	.0000
.780	0.0000	.0000	.0000
Totals	21.3713	21.3714	21.3715

TABLE 30.

Photometric Units, Conversion Factors, Etc.

TABLE 30A.

Luminance Conversion Factors

	Candela/m²	Stilbs	Candela/in²	Candela/ft²	Apostilbs	Lamberts	Milli-Lamberts	Foot-Lamberts
1 Candela/m² (nit) =	1	10^{-4}	0.6452×10^{-3}	0.0929	3.142	0.3142×10^{-3}	0.3142	0.2919
1 Stilb (cd/cm²) =	10^4	1	6.452	929	31,420	3.142	3142	2919
1 Candle/in² =	1550	0.155	1	144	4869	0.4869	486.9	452.4
1 Candle/ft² =	10.76	1.076×10^{-3}	6.944×10^{-3}	1	33.82	3.382×10^{-3}	3.382	3.142
1 Apostilb =	0.3183	31.83×10^{-6}	0.2054×10^{-3}	0.02957	1	10^{-4}	0.1	0.0929
1 Lambert =	3183	0.3183	2.054	295.7	10^4	1	1000	929
1 Milli-Lambert =	3.183	0.3183×10^{-3}	2.054×10^{-3}	0.2957	10	10^{-3}	1	0.929
1 Foot-Lambert =	3.426	0.3426×10^{-3}	2.210×10^{-3}	0.3183	10.76	1.076×10^{-3}	1.076	1

TABLE 30B.

Illumination Conversion Factors

	Foot-candles	Luxes	Phots	Milli-phots
1 Ft. candle =	1	10.76	1.076×10^{-3}	1.076
1 Lux (lm/m²) =	0.0929	1	10^{-4}	0.1
1 Phot (lm/cm²) =	929	10^4	1	1000
1 Milliphot =	0.929	10	10^{-3}	1

For a Plane, Lambert-law Radiator:

Emittance = $\pi \times L_o$, L_o is normal Luminance (30.1)

Intensity (at angle θ from normal) = $I_o \cos\theta$, I_o is normal intensity (30.2)

Temperature-conversion Formulae

Let T_C, T_K, T_F, T_R represent temperature expressed in degrees Celsius (or centigrade), Kelvin, Fahrenheit, and Rankine, respectively. Then

$$T_C = T_K - 273.16 = \frac{5}{9}(T_F - 32) \quad (30.3)$$

$$T_K = T_C + 273.16 = \frac{5}{9}(T_F + 459.69) \quad (30.4)$$

$$T_F = \frac{9}{5}T_C + 32 = \frac{9}{5}T_K \overset{*}{-} 459.69 \quad (30.5)$$

$$T_R = T_F + 459.69 \quad (30.6)$$

TABLE 31. Useful Formulae

Formulae for Lens Imaging Parameters[4h]

$$f = \frac{ab}{a+b} = \frac{cm}{(m+1)^2} = \frac{am}{m+1} = \frac{b}{m+1} \qquad (31.1)$$

$$a = \frac{bf}{b-f} = \frac{c}{m+1} = \frac{f(m+1)}{m} = b/m = \tfrac{1}{2}(c \pm \beta) \qquad (31.2)$$

$$b = \frac{af}{a-f} = \frac{cm}{m+1} = f(m+1) = am = \tfrac{1}{2}(c \mp \beta) \qquad (31.3)$$

$$m = b/a = \frac{f}{a-f} = \frac{b}{f} - 1 = \frac{c - 2f \pm \beta}{2f} \qquad (31.4)$$

where

- f is the effective focal length of the lens (positive if converging lens)
- a, b are object and image conjugates, resp. (positive for real object and image, respectively)
- m is the magnification: image size / object size
- $c =$ $a + b$ is object to image distance (with negligible lens thickness)
- $\beta =$ $\sqrt{c^2 - 4fc}$

Formulae for Perfect-lens Imaging in Air[4i]

Diameter of central lobe of diffraction pattern

$$1.22\,\lambda F \approx \tfrac{2}{3} F\mu \text{ in visible light} \qquad (31.5)$$

Cut-off frequency of optical transfer function

$$(\lambda F)^{-1} \approx 1{,}800/F \text{ cycles/mm in visible light} \qquad (31.6)$$

Tolerance on defocusing (Rayleigh's ¼-limit)

$$2\lambda F^2 \approx F^2 \mu \text{ in visible light} \qquad (31.7)$$

Illumination in image plane[4j]

$$\frac{\pi L}{4F^2} \qquad (31.8)$$

where λ is the wavelength

- $F =$ $(2\ \text{n.a.})^{-1}$ is the "effective" image-plane F/number (n.a. = numerical aperture)
- L is the object-plane luminance

Limits of Useful Magnification[4i]

Microscope: 1000 x n.a. (31.9)

Telescope: Radius (in mm) of the objective lens aperture (31.10)

Blackbody Radiation, Summary of Formulae

The spectral radiant emittance is

$$M_{e\lambda} = \frac{c_1}{\lambda^5 (e^{c_2/\lambda T} - 1)} \text{ (Planck's law).} \qquad (31.11)$$

The total radiant emittance is

$$M_e(T) = \sigma T^4 \text{ (Stefan-Boltzmann's law).} \qquad (31.12)$$

The peak spectral radiant emittance is

$$M_{e\lambda max}(T) = c_p T_5 \qquad (31.13)$$

and occurs at

$$\lambda_0 = \frac{c_2}{c^* T} \text{ (Wien's displacement law).} \qquad (31.14)$$

The spectral photon emittance is

$$N_\lambda(T) = 2\pi c/\lambda^4 (e^{c_2/\lambda T} - 1). \qquad (31.15)$$

The total photon emittance is

$$N(T) = c_N T^3. \qquad (31.16)$$

The accepted values for the constants are

$$c_1 = 3.7415 \times 10^8\ W\mu^4/m^2$$

$$c_2 = 14{,}387.9\mu°K$$

$$\sigma = 5.6697 \times 10^{-8}\ W/m^2\ (°K)^4$$

$c_p = 1.2865 \times 10^{-11}$ W/m$^2\mu$(°K)^5

$\frac{c_2}{c^*} = 2897.8\mu$°K.

$2\pi c = 1.883652 \times 10^8$ m/s

$c_N = 2.53043\ 10^{15}$ photons/m^2 · s · (°K)^3

Fourier Transform, Correlation, and Convolution

For a treatment of the theory of the Fourier transform, the reader is referred to any of a number of good texts, e.g., Reference 29; for a compilation of transforms, to Reference 30. Here we present only some fundamental theorems, results connecting correlation and convolution, and Fourier transforms of special interest in applied optics.

We define the following operations for any square-integrable functions, $f(x)$ & $g(x)$.

Fourier transform:

$$F(y) = \mathcal{F}[f(x)] = \int_{-\infty}^{\infty} f(x) \exp(i2\pi xy)\, dx \quad (31.17)$$

and

$$\mathcal{F}^{-1}[F(y)] = \int_{-\infty}^{\infty} F(y) \exp(-i2\pi xy)\, dy \quad (31.18)$$

Correlation:

$$f \odot g = \int_{-\infty}^{\infty} f(u)g(x+u)\, du \quad (31.19)$$

Convolution:

$$f \circledast g = \int_{-\infty}^{\infty} f(u)g(x-u)\, du \quad (31.20)$$

Inverse of a function:

$$\hat{f}(x) = f(-x), \quad (31.21)$$

and denote the complex conjugate by an asterisk (*).

Significance of the inverse transform:

$$\mathcal{F}^{-1}\Big\{\mathcal{F}[f(x)]\Big\} = f(x). \quad (31.22)$$

At a discontinuity this yields: $\frac{1}{2}[f(x-) + f(x+)]$.

Convolution Theorem

$$\mathcal{F}[f \circledast g] = FG. \quad (31.23)$$

Parseval's Theorem

$$\int_{-\infty}^{\infty} F(y)\, G^*(y)\, dy = \int_{-\infty}^{\infty} f(x)\, g^*(x)\, dx \quad (31.24)$$

and, specifically,

$$\int_{-\infty}^{\infty} |F(y)|^2\, dy = \int_{-\infty}^{\infty} |f(x)|^2\, dx. \quad (31.24a)$$

Wiener-Khintchine Theorem

$$\mathcal{F}[f \odot g^*] = FG^* \quad (31.25)$$

and, specifically,

$$\mathcal{F}[f \odot f^*] = |F|^2. \quad (31.25a)$$

Fourier Transform Modifications

Complex conjugate

$$F^* \equiv \mathcal{F}^*[f] = \mathcal{F}[\hat{f}^*] = \mathcal{F}[\hat{f^*}] = \mathcal{F}^{-1}[f^*]. \quad (31.26)$$

Inverse

$$\hat{F} \equiv \hat{\mathcal{F}}[f] = \mathcal{F}^{-1}[f] = \mathcal{F}[\hat{f}] = \mathcal{F}^*[f^*] \quad (31.27)$$

and, hence,

$$\mathcal{F}\Big\{\mathcal{F}[f]\Big\} = \hat{f} \text{ or } F = \mathcal{F}^{-1}[\hat{f}]. \quad (31.27a)$$

Convolution and Correlation Relationships

Commutativity

$$f \circledast g = g \circledast f \qquad (31.28)$$

$$f \odot g = g \,\hat{\odot}\, f \qquad (31.29)$$

Convolution vs. correlation

$$f \circledast \hat{g} = f \odot g \qquad (31.30)$$

$$f \odot \hat{g} = f \circledast g \qquad (31.31)$$

Symmetry and Realness

Symmetrical functions

$$[f \equiv \hat{f}] : F \equiv \hat{F} \qquad (31.32)$$

Real functions

$$[f \equiv f^*] : F^* \equiv \hat{F} \qquad (31.33)$$

Function symmetrical and real

$$[f = \hat{f} = f^*] : F = \hat{F} = F^* \qquad (31.34)$$

Fourier Transform Pairs

The following functions are useful to represent modulation transfer functions. The corresponding line spread functions are given in the second column and the area of application in the last column.

MTF	Line Spread Function	Application
$1, \nu < \nu_c$; $0, \nu \geqslant \nu$	$\dfrac{\sin 2\pi\nu_c x}{\pi x}$	Rectangular aperture, coherent illumination
$1 - \nu/\nu_c, \nu < \nu_c$; $0, \quad , \nu \geqslant \nu_c$	$\nu_c\left(\dfrac{\sin \pi\nu_c x}{\pi\nu_c x}\right)^2$	Rectangular aperture, incoherent illumination
$\dfrac{2}{\pi}\left[\cos^{-1}\dfrac{\nu}{\nu_c} - \dfrac{\nu}{\nu_c}\sqrt{1-\left(\dfrac{\nu}{\nu_c}\right)^2}\right]$	$\dfrac{3\pi}{2}\dfrac{\mathbf{H}_1(2\pi\nu_c x)}{(2\pi\nu_c x)^2}$	Circular aperture, incoherent illumination
$\dfrac{b^2}{b^2+\nu^2} = \dfrac{a^2}{a^2+(2\pi\nu)^2}$	$\pi b e^{-2\pi b\lvert x\rvert} = \dfrac{a}{2}e^{-a\lvert x\rvert}$	Photogr. emulsions
$e^{-a\nu} = e^{-2\pi b\nu}$	$\dfrac{2a}{a^2+(2\pi x)^2} = \dfrac{b/\pi}{b^2+x^2}$	Photogr. emulsions
$e^{-s\nu^2} = e^{-(2\pi\nu\sigma)^2}$	$\sqrt{\dfrac{\pi}{s}}\,e^{-\pi^2x^2/s} = \dfrac{e^{-x^2/2\sigma^2}}{\sqrt{2\pi\sigma^2}}$	Cathode ray tubes

Here a, b, s, σ are constants,

$\nu_c = w/f\lambda = (F\lambda)^{-1}$ is the cut-off frequency,
w is the aperture diameter or width,
f is the image distance,
λ is the radiation wavelength in image space, and
F is the "effective" F/number.